GENETICS
From Genes to Genomes

Third Edition

Leland H. Hartwell
Fred Hutchinson Cancer Research Center

Leroy Hood
The Institute for Systems Biology

Michael L. Goldberg
Cornell University

Ann E. Reynolds
Fred Hutchinson Cancer Research Center

Lee M. Silver
Princeton University

Ruth C. Veres

Boston Burr Ridge, IL Dubuque, IA New York San Francisco St. Louis
Bangkok Bogotá Caracas Kuala Lumpur Lisbon London Madrid Mexico City
Milan Montreal New Delhi Santiago Seoul Singapore Sydney Taipei Toronto

The McGraw·Hill Companies

GENETICS: FROM GENES TO GENOMES, THIRD EDITION

Published by McGraw-Hill, a business unit of The McGraw-Hill Companies, Inc., 1221 Avenue of the Americas, New York, NY 10020.

This book is printed on acid-free paper.

1 2 3 4 5 6 7 8 9 0 DOW/DOW 0 9 8 7 6

ISBN 978–0–07–110215–5
MHID 0–07–110215–9

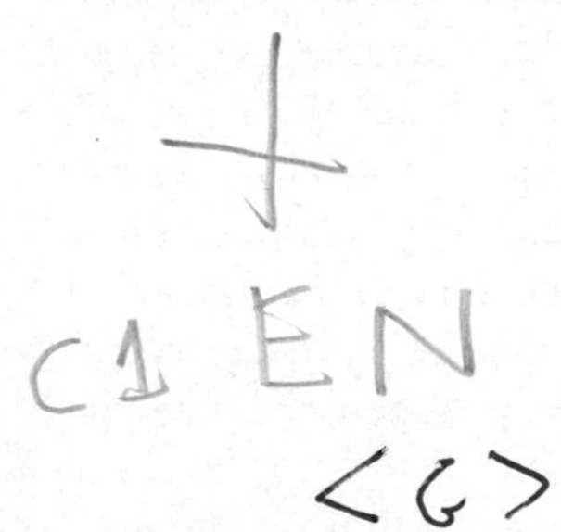

www.mhhe.com

About the Authors

Dr. Leland Hartwell is President and Director of Seattle's Fred Hutchinson Cancer Research Center and Professor of Genome Sciences at the University of Washington.

Dr. Hartwell's primary research contributions were in identifying genes that control cell division in yeast including those necessary for the division process as well as those necessary for the fidelity of genome reproduction. Subsequently many of these same genes have been found to control cell division in humans and often to be the site of alteration in cancer cells.

Dr. Hartwell is a member of the National Academy of Sciences and has received the Albert Lasker Basic Medical Research Award, the Gairdner Foundation International Award, the Alfred P. Sloan Award in Cancer Research, and the 2001 Nobel Prize in Physiology or Medicine.

Dr. Lee Hood received an M.D. from the Johns Hopkins Medical School and a Ph.D. in biochemistry from the California Institute of Technology. His research interests include immunology, cancer biology, development, and the development of biological instrumentation (for example, the protein sequencer and the automated fluorescent DNA sequencer). His early research played a key role in unraveling the mysteries of antibody diversity. More recently he has pioneered systems approaches to biology and medicine.

Dr. Hood has taught molecular evolution, immunology, molecular biology, genomics and biochemistry and has co-authored textbooks in biochemistry, molecular biology, and immunology, as well as *The Code of Codes*—a monograph about the Human Genome Project. He was one of the first advocates for the Human Genome Project and directed one of the federal genome centers that sequenced the human genome. Dr. Hood is currently the president (and co-founder) of the cross-disciplinary Institute for Systems Biology in Seattle, Washington.

Dr. Hood has received a variety of awards, including the Albert Lasker Award for Medical Research (1987), the Distinguished Service Award from the National Association of Teachers (1998) and the Lemelson/MIT Award for Invention (2003). He is the 2002 recipient of the Kyoto Prize in Advanced Biotechnology—an award recognizing his pioneering work in developing the protein and DNA synthesizers and sequencers that provide the technical foundation of modern biology. He is deeply involved in K-12 science education. His hobbies include running, mountain climbing, and reading.

Dr. Michael Goldberg is a professor at Cornell University, where he teaches introductory genetics. He was an undergraduate at Yale University and received his Ph.D. in biochemistry from Stanford University. Dr. Goldberg performed postdoctoral research at the Biozentrum of the University of Basel (Switzerland) and at Harvard University, and he received an NIH Fogarty Senior International Fellowship for study at Imperial College (England) and at the University of Rome (Italy). His current research uses the tools of *Drosophila* genetics to investigate the mechanisms that ensure proper chromosome segregation during mitosis and meiosis.

Dr. Ann Reynolds is an educator and author. She began teaching genetics and biology in 1990, and her research has included studies of gene regulation in *E. coli,* chromosome structure and DNA replication in yeast, and chloroplast gene expression in marine algae. She is a graduate of Mount Holyoke College and received her Ph.D. from Tufts University. Dr. Reynolds was a postdoctoral fellow in the Harvard University Department of Molecular Biology and Genome Sciences at the University of Washington. She was also an author and producer of the laserdisc and CD-ROM *Genetics: Fundamentals to Frontiers.*

Dr. Lee M. Silver received B.A. and M.S. degrees in physics from the University of Pennsylvania, and a Ph.D. in biophysics from Harvard University. He obtained further training at New York's Memorial Sloan-Kettering Cancer Center, Cold Spring Harbor Laboratory, and the Pasteur Institute in Paris, France. Since 1984, he has been a professor at Princeton University, currently in the Department of Molecular Biology and the Woodrow Wilson School of Public and International Affairs. He also has joint appointments in the Program in Science, Technology, and Environmental Policy, the Center for Health and Wellbeing, the Office of Population Research, and the Princeton Environmental Institute, all at Princeton University.

Dr. Silver has published over 200 articles in the fields of mammalian genetics, evolution, reproduction, embryology, computer modeling, and behavioral science, and other scholarly papers on topics at the interfaces among biotechnology, law, politics, and religion. He has been elected to the governing boards of the Genetics Society of America and the International Mammalian Genome Society, and was a member of the New Jersey Bioethics Commission Task Force formed to recommend reproductive policy for the New Jersey State Legislature. Silver has been elected a lifetime fellow of the American Association for the Advancement of Science (AAAS) and he received a prestigious MERIT Award for outstanding research in genetics from the National Institutes of Health.

Dr. Silver's other books include *Remaking Eden: How Genetic Engineering and Cloning will Transform the American Family,* published in 16 languages, *Mouse Genetics,* and *Challenging Nature: The Clash of Science and Spirituality at the New Frontiers of Life.* He has also written popular articles for *The New York Times, Washington Post, Time Magazine,* and *Newsweek International.* Further information about Dr. Silver is available at www.leemsilver.net.

Ruth C. Veres is a science writer and editor with 35 years of experience in textbook publishing. She received her B.A. from Swarthmore College, obtained M.A. degrees from Columbia University and Tufts University, and taught writing and languages at the University of California at Berkeley.

In addition to developing and editing more than 30 texts in the fields of political science, economics, psychology, nutrition, chemistry, and biology, Veres has coauthored a book on the immune system and an introductory biology text. She is currently working on a book with Dr. Lee Hood that looks at biological information and the emergence of systems biology.

Contributors

Genetics research tends to proceed down highly specialized paths. A number of experts in specific areas generously provided information in their areas of expertise. We thank them for their contributions to this edition of our text.

Ian Duncan, *Washington University, St. Louis*
Sylvia Fromherz, *University of Colorado at Boulder*
Gail E. Gasparich, *Towson University*
Bernadette Holdener, *State University of New York, Stony Brook*
Nancy M. Hollingsworth, *State University of New York, Stony Brook*
Kari Beth Krieger, *University of Wisconsin, Green Bay*
Debra Nero, *Cornell University*
Eric Richards, *Washington University, St. Louis*
Kenneth Shull, *Appalachian State University*

Brief Contents

Contents

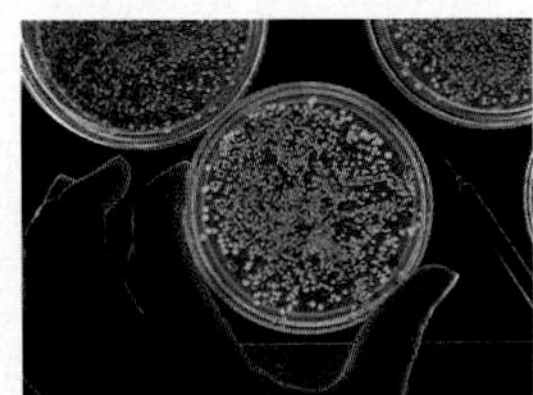

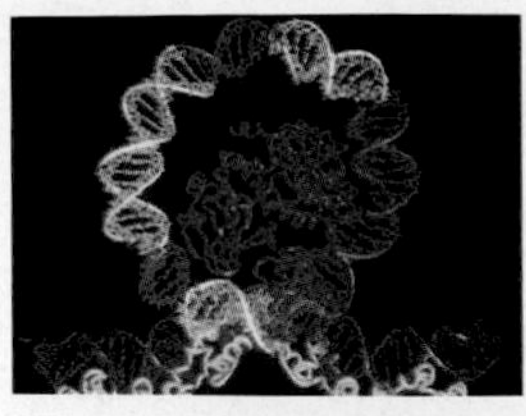

PART V
How Genes Are Regulated 609

PART VI
How Genes Change 757

Preface

A Note from the Authors

The science of genetics is less than 150 years old, but its accomplishments within that short time have been astonishing. Gregor Mendel first described genes as abstract units of inheritance in 1865; his work was ignored and then "rediscovered" in 1900. Thomas Hunt Morgan and his students provided experimental verification of the idea that genes reside within chromosomes during the years 1910-1920. By 1944, Oswald Avery and his coworkers had established that genes are made of DNA. James Watson and Francis Crick published their pathbreaking structure of DNA in 1953. Remarkably, less than 50 years later (in 2001), an international consortium of investigators deciphered the sequence of the 3 billion nucleotides in the human genome. Twentieth-century genetics made it possible to identify individual genes and to understand a great deal about their functions.

Today, scientists are able to access the enormous amounts of genetic data generated by the sequencing of many organisms' genomes. Analysis of these data will result in a deeper understanding of the complex molecular interactions within and among vast networks of genes, proteins, and other molecules that help bring organisms to life. Finding new methods and tools for analyzing these data will be a significant part of genetics in the twenty-first century.

Our third edition of *Genetics: From Genes to Genomes* emphasizes both the core concepts of genetics and the cutting-edge discoveries, modern tools, and analytic methods that will keep the science of genetics moving forward.

Our Focus—An Integrated Approach

Genetics: From Genes to Genomes represents a new approach to an undergraduate course in genetics. It reflects the way we, the authors, currently view the molecular basis of life. We integrate:

- **Formal genetics:** the rules by which genes are transmitted.
- **Molecular genetics:** the structure of DNA and how it directs the structure of proteins.
- **Genomics and systems biology:** the new technologies that allow a comprehensive analysis of the entire gene set and its expression in an organism.
- **Human genetics:** how genes contribute to health and disease.
- **The unity of life-forms:** the synthesis of information from many different organisms into coherent models that explain many biological systems.
- **Molecular evolution:** the molecular mechanisms by which biological systems and whole organisms have evolved and diverged.

The strength of this integrated approach is that students who complete the book will have a strong command of genetics as it is practiced today by academic and corporate researchers. These scientists are rapidly changing our understanding of living organisms, including ourselves; increasing our ability to prevent, diagnose, and treat disease and to engineer new life-forms for food and medical uses; and, ultimately, creating the ability to replace or correct detrimental genes.

The Genetic Way of Thinking

To encourage a genetic way of thinking, we begin the book with a presentation of Mendelian principles and the chromosomal basis of inheritance. From the outset, however, the integration of Mendelian genetics with fundamental molecular mechanisms is central to our approach. Chapter 1 presents the foundation of this integration. In Chapter 2, we tie Mendel's studies of pea-shape inheritance to the action of an enzyme that determines whether a pea is round or wrinkled. In the same chapter, we point to the relatedness of patterns of heredity in all organisms by using Mendelian principles to look at heredity in humans. Starting in Chapter 6, we focus on the physical characteristics of DNA, the implications and uses of mutations, and how the double helix structure of DNA encodes, copies, and transmits biological information. Beginning in Chapter 9 we look at modern genetic techniques, including such biotechnology tools as gene cloning, hybridization, PCR, and microarrays, exploring how researchers use them to reveal the modular construction and genetic relatedness of genomes. We then show how the complete genome sequences of humans and model organisms provide insights into the architecture and evolution of genomes; how modular genomic construction has contributed to the relatively rapid evolution of life and helped generate the enormous diversity of life-forms we see around us.

Genetic portrait chapters on the website (www.mhhe.com/hartwell3) contain detailed discussions of model organisms, which clarify that their use in the study of human biology is possible only because of the genetic relatedness of all organisms. Throughout our book, we present the scientific reasoning of some of the ingenious researchers who have carried out genetic analysis, from Mendel, to Watson and Crick, to the collaborators on the Human Genome Project.

Student-Friendly Features

We have taken great pains to help the student make the leap to a deeper understanding of genetics. Numerous features of this book were developed with that goal in mind.

- ***One Voice*** The role of our science writer, Ruth Veres, is to create one voice for our author team. With more than 30 years' experience in life science textbook publishing, Ms. Veres is uniquely suited to this task. By working closely with everyone on the team, she has created the friendly, engaging reading style that helps students master the concepts throughout this book. This team approach provides the student with the focus and continuity required to make the book successful in the classroom.
- ***Visualizing Genetics*** The highly specialized art program developed for this book integrates photos and line art in a manner that provides the most engaging visual presentation of genetics available. Our Feature Figure illustrations break down complex processes into step-by-step illustrations that lead to greater student understanding. All illustrations are rendered with a consistent color theme—for example, all presentations of phosphate groups are the same color, as are all presentations of mRNA.
- ***Problem Solving*** Developing strong problem-solving skills is vital for every genetics student. The authors have carefully created problem sets at the end of each chapter that allow students to improve upon their problem-solving ability.
- **Social and Ethical Issues** questions require critical thinking analysis of the scientific issues that impact our society.
- **Solved Problems** provide insight into the step-by-step process of problem solving.
- **Review Problems** offer a variety of levels of questions that develop excellent problem-solving skills.
- **Accessibility** Our intention is to bring cutting-edge content to the student level. A number of more complex illustrations are revised and segmented to help the student follow the process. Legends have been streamlined to highlight only the most important ideas, and throughout the book, topics have been revised to focus on the most critical information.

New to the Third Edition

- The **End-of-Chapter Problem Sets Have Been Extensively Revised** and include over 100 new problems. The problems are now organized by chapter section and in order of increasing difficulty within each section for ease of use by instructors and students. Each chapter contains a variety of problem types including: *Social & Ethical Issues* which prompt the student to apply problem-solving skills to real-world situations that scientific breakthroughs have forced us to face as a society; *Solved Problems* which cover topical material with complete answers to aid the student in understanding the problem solving process; and *Problems & Questions* that allow students to develop their own problem-solving skills. Answers to selected problems are in the back of the book.
- **New Chapter: Chapter 12 *Systems Biology and Proteomics*** provides a framework for thinking about what a biological system is and describes tools for analyzing the genes and proteins of a system, as well as computational tools for integrating and modeling this information to begin to explain a system's emergent properties.
- **Content Updates** throughout make this the most current and modern book available. Every chapter reflects the updated information generated by the breakthroughs of the past few years. For example,
  - Chapter 18, *Gene Regulation in Eukaryotes,* discusses the latest on RNAi technology.
  - Chapter 22, *Evolution at the Molecular Level,* includes information on network evolution and comparative genome evolution.
- **"Tools of Genetics" boxed essays are new** to this edition. They explain various techniques geneticists use to look at DNA, genes, other aspects of the genome, and proteins, with examples of interesting applications in biology and medicine.
- An **"On Our Website" Feature,** located at the end of each chapter, directs students and teachers to additional, more detailed information on specialized topics not found in the textbook. This information is in the form of new content, references, or links to other websites.
- **Interactive Web Exercises** offer students an interactive way to analyze genetic data on the Web and complete exercises that test their understanding of the data.
- A **New Design** is more user friendly and emphasizes the pedagogical structure and features of the presentation.

A Word About the Portraits of Model Organisms

Five **Genetic Portraits** are included on the book-specific website at www.mhhe.com/hartwell3 as easy-to-download PDF files. The Genetic Portraits are also available as a printed supplement upon request. Each Genetic Portrait profiles a different model organism whose study has contributed to genetic research. The five selected were the ones chosen as the focus of the Human Genome Project. They are:

Saccharomyces cerevisiae: Genetic Portrait of Yeast
Arabidopsis thaliana: Genetic Portrait of a Model Plant
Caenorhabditis elegans: Genetic Portrait of a Simple Multicellular Organism
Drosophila melanogaster: Genetic Portrait of the Fruit Fly
Mus musculus: Genetic Portrait of the House Mouse

We anticipate that instructors will choose to cover one or two portraits during the semester. Students may then use the specifics of the selected model organism to build an understanding of the principles and applications discussed in the book. The unique genetic manipulations and properties of each of the models make them important for addressing different biological questions using genetic analysis. In the portraits, we explain how biologists learned that the evolutionary relatedness of all organisms permits the extrapolation from a model to the analysis of other living forms. The portraits should thus help students understand how insights from one model organism can suggest general principles applicable to other organisms, including humans.

Guided Tour

Students and instructors can become acquainted with the key features of this book by browsing through the Guided Tour starting on the next page. These pages constitute a visual exposition of the book's pedagogy and art program.

Guided Tour

Integrating Genetic Concepts

Genetics: From Genes to Genomes takes an integrated approach in its presentation of genetics, thereby giving students a strong command of genetics as it is practiced today by academic and corporate researchers. Principles are related throughout the text in examples, essays, case histories, and Connections sections to make sure students fully understand the relationships between topics.

Fast Forward Essays

This feature is one of the methods used to integrate the Mendelian principles presented early in the book with the molecular principles that will follow.

Tools of Genetics Essays

Current readings explain various techniques and tools used by geneticists, including examples of applications in biology and medicine.

Genetics and Society Essays

Dramatic essays explore the social and ethical issues created by the multiple applications of modern genetic research.

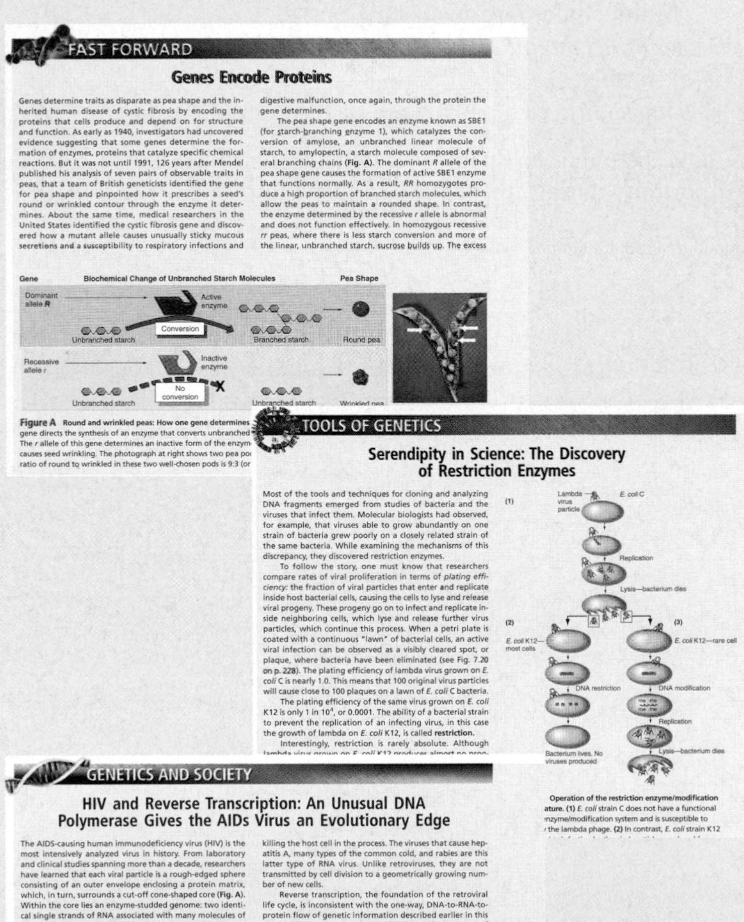

FAST FORWARD

Genes Encode Proteins

Genes determine traits as disparate as pea shape and the inherited human disease of cystic fibrosis by encoding the proteins that cells produce and depend on for structure and function. As early as 1940, investigators had uncovered evidence suggesting that some genes determine the formation of enzymes, proteins that catalyze specific chemical reactions. But it was not until 1991, 126 years after Mendel published his analysis of seven pairs of observable traits in peas, that a team of British geneticists identified the gene for pea shape and pinpointed how it prescribes a seed's round or wrinkled contour through the enzyme it determines. About the same time, medical researchers in the United States identified the cystic fibrosis gene and discovered how a mutant allele causes unusually sticky mucous secretions and a susceptibility to respiratory infections and digestive malfunction, once again, through the protein the gene determines.

The pea shape gene encodes an enzyme known as SBE1 (for starch-branching enzyme 1), which catalyzes the conversion of amylose, an unbranched linear molecule of starch, to amylopectin, a starch molecule composed of several branching chains (**Fig. A**). The dominant *R* allele of the pea shape gene causes the formation of active SBE1 enzyme that functions normally. As a result, *RR* homozygotes produce a high proportion of branched starch molecules, which allow the peas to maintain a rounded shape. In contrast, the enzyme determined by the recessive *r* allele is abnormal and does not function effectively. In homozygous recessive *rr* peas, where there is less starch conversion and more of the linear, unbranched starch, sucrose builds up. The excess

Figure A **Round and wrinkled peas: How one gene determines** gene directs the synthesis of an enzyme that converts unbranched The *r* allele of this gene determines an inactive form of the enzym causes seed wrinkling. The photograph at right shows two pea po ratio of round to wrinkled in these two well-chosen pods is 9:3 (or

TOOLS OF GENETICS

Serendipity in Science: The Discovery of Restriction Enzymes

Most of the tools and techniques for cloning and analyzing DNA fragments emerged from studies of bacteria and the viruses that infect them. Molecular biologists had observed, for example, that viruses able to grow abundantly on one strain of bacteria grew poorly on a closely related strain of the same bacteria. While examining the mechanisms of this discrepancy, they discovered restriction enzymes.

To follow the story, one must know that researchers compare rates of viral proliferation in terms of *plating efficiency*: the fraction of viral particles that enter and replicate inside host bacterial cells, causing the cells to lyse and release viral progeny. These progeny go on to infect and replicate inside neighboring cells, which lyse and release further virus particles, which continue this process. When a petri plate is coated with a continuous "lawn" of bacterial cells, an active viral infection can be observed as a visibly cleared spot, or plaque, where bacteria have been eliminated (see Fig. 7.20 on p. 228). The plating efficiency of lambda virus grown on *E. coli* C is nearly 1.0. This means that 100 original virus particles will cause close to 100 plaques on a lawn of *E. coli* C bacteria.

The plating efficiency of the same virus grown on *E. coli* K12 is only 1 in 10^4, or 0.0001. The ability of a bacterial strain to prevent the replication of an infecting virus, in this case the growth of lambda on *E. coli* K12, is called **restriction.**

Interestingly, restriction is rarely absolute. Although

Operation of the restriction enzyme/modification ature. **(1)** *E. coli* strain C does not have a functional nzyme/modification system and is susceptible to the lambda phage. **(2)** In contrast, *E. coli* strain K12

GENETICS AND SOCIETY

HIV and Reverse Transcription: An Unusual DNA Polymerase Gives the AIDs Virus an Evolutionary Edge

The AIDS-causing human immunodeficiency virus (HIV) is the most intensively analyzed virus in history. From laboratory and clinical studies spanning more than a decade, researchers have learned that each viral particle is a rough-edged sphere consisting of an outer envelope enclosing a protein matrix, which, in turn, surrounds a cut-off cone-shaped core (**Fig. A**). Within the core lies an enzyme-studded genome: two identical single strands of RNA associated with many molecules of an unusual DNA polymerase known as **reverse transcriptase.**

During infection, the AIDS virus binds to and injects its cone-shaped core into cells of the human immune system (**Fig. B**). It next uses reverse transcriptase to copy its RNA genome into double-stranded DNA molecules in the cytoplasm of the host cell. The double helixes then travel to the nucleus where another enzyme inserts them into a host chromosome. Once integrated into a host-cell chromosome, the viral genome can do one of two things. It can commandeer the host cell's protein synthesis machinery to make hundreds of new viral particles that bud off from the parent cell, taking with them part of the cell membrane and sometimes resulting in the host cell's death. Alternately, it can lie latent inside the host chromosome, which then copies and transmits the viral genome to two new cells with each cell division.

The events of this life cycle make HIV a **retrovirus:** an RNA virus that after infecting a host cell copies its own single strands of RNA into double helixes of DNA, which a viral enzyme then integrates into a host chromosome. RNA viruses that are not retroviruses simply infect a host cell and then use the cellular machinery to make more of themselves, often killing the host cell in the process. The viruses that cause hepatitis A, many types of the common cold, and rabies are this latter type of RNA virus. Unlike retroviruses, they are not transmitted by cell division to a geometrically growing number of new cells.

Reverse transcription, the foundation of the retroviral life cycle, is inconsistent with the one-way, DNA-to-RNA-to-protein flow of genetic information described earlier in this chapter. Because it was so unexpected, the phenomenon of reverse transcription encountered great resistance in the scientific community when first reported by Howard Temin of the University of Wisconsin and David Baltimore, then of MIT. It is now an established fact. In the normal order of molecular genetics, DNA polymerases construct DNA polymers from a DNA template, replicating a cell's double-helical genome for transmission to daughter cells during the cell cycle; and RNA polymerases construct RNA polymers from a DNA template, copying the information of genes into an RNA transcript that is then translated into protein. Reverse transcriptase, by comparison, is a remarkable DNA polymerase that can construct a DNA polymer from either an RNA or a DNA template.

Figure A Structure of the AIDS virus

Figure B Life cycle of the AIDS virus

Comprehensive Examples

Comprehensive Examples are extensive case histories or research synopses that, through text and art, summarize the main points in the preceding section or chapter and show how they relate to each other.

18.4 Sex Determination in *Drosophila:* A Comprehensive Example of Gene Regulation

Male and female *Drosophila* exhibit many sex-specific differences in morphology, biochemistry, behavior, and function of the germ line (**Fig. 18.20**). By examining the phenotypes of flies with different chromosomal constitutions, researchers confirmed that the ratio of X to autosomal chromosomes (X:A) helps determine sex, fertility, and viability (**Table 18.2**). They then carried out genetic experiments that showed that the X:A ratio influences sex through three independent pathways: One determines whether the flies look and act like males or females; another determines whether germ cells devel
perm; and a third produces dosage compen
oubling the rate of transcription of X-lir

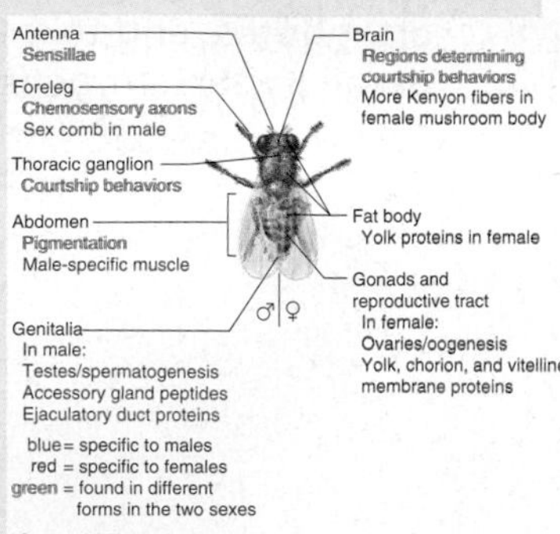

Figure 18.20 **Sex-specific traits in *Drosophila*.** Objects or traits shown in *blue* are specific to males. Objects or traits shown in *red* are specific to females. Objects or traits shown in *green* are found in different forms in the two sexes.

TABLE 18.2 How Chromosomal Constitution Affects Phenotype in *Drosophila*

Sex Chromosomes	X:A	Sex Phenotype
Autosomal Diploids		
XO	0.5	Male (sterile)
XY	0.5	Male
XX	1.0	Female
XXY	1.0	Female
Autosomal Triploids		
XXX	1.0	Female
XYY	0.33	Male
XXY	0.66	Intersex

Connections

Each chapter closes with a Connections section that serves as a bridge between the topics in the just-completed chapter and those in the upcoming chapter or chapters.

Connections

The existence of numerous controls in each of several cell-cycle pathways suggests that evolution has erected many barriers in multicellular animals to the uncontrolled reproduction of "selfish" cells. At the same time, the hundreds of genes contributing to normal cell-cycle regulation provide hundreds of targets for cancer-producing mutations.

Variations on the theme of cell-cycle regulation play a key role in the development of eukaryotic organisms. During the development of multicellular organisms, cells must not only control their cell cycles, they must also adopt different fates and differentiate into different tissues. In *Drosophila,* for example, after fertilization, nuclear division occurs without cell division for the first 13 cycles; during these cycles, the nuclei go through many rapid S and M phases without any intervening G_1 or G_2 (**Fig. 19.25**). In cycles 10–13, the synthesis and degradation of cyclinB regulates mitosis. Sometime during cycles 14–16, a G_2 phase appears, and distinct patches of cells with different-length cycles become evident within the embryo. The differences in cycle time between the different cell types is the result of variable G_2 phases. Late in G_2, *CDC25* activates cyclin-dependent kinases to control the timing of mitosis. Many tissues stop dividing at cycle 16, but a few continue. In the still-dividing cells, a G_1 phase appears. Some of these cells will arrest in G_1 during larval growth, only to start dividing again in response to signals relayed during metamorphosis.

In Chapter 20, we present the basic principles of development and describe how biologists have used genetic analysis in various model organisms to examine development at the cellular and molecular levels.

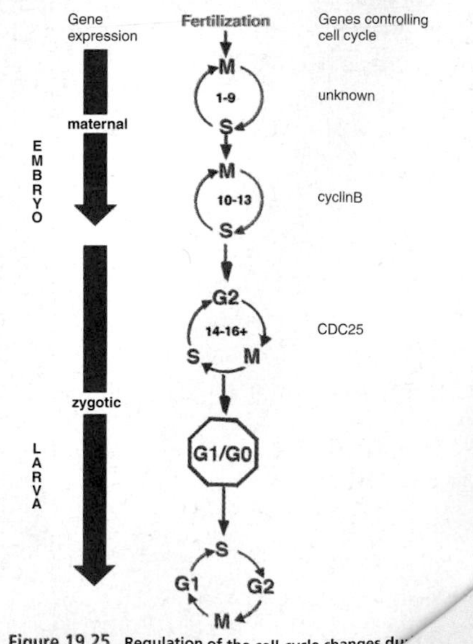

Figure 19.25 **Regulation of the cell-cycle changes du *Drosophila* development.** Each step of development regulators that act as barriers to uncontrolled repr "selfish" cells. Some of these regulators, such as *CDC25,* are known; others are not.

Visualizing Genetics

Full-color illustrations and photographs bring the printed word to life. These visual reinforcements support and further clarify the topics discussed throughout the text.

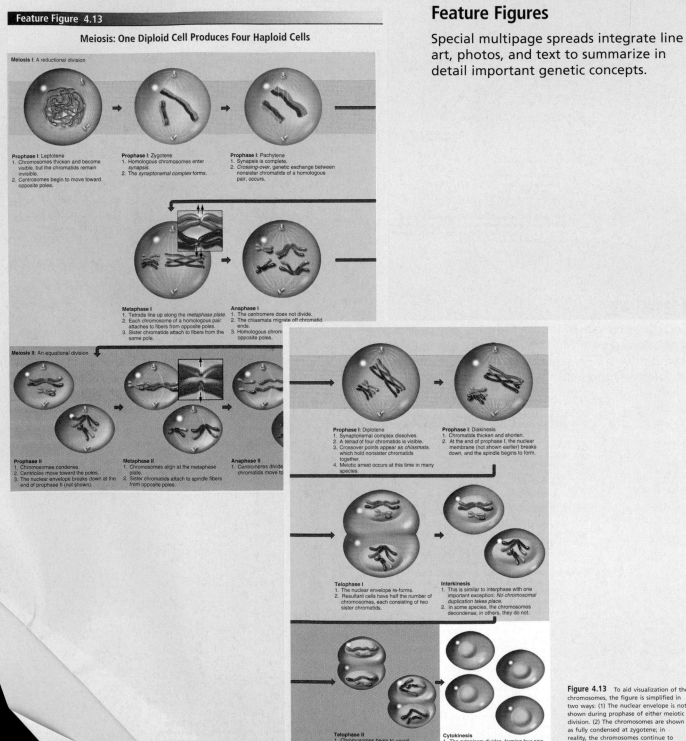

Feature Figures

Special multipage spreads integrate line art, photos, and text to summarize in detail important genetic concepts.

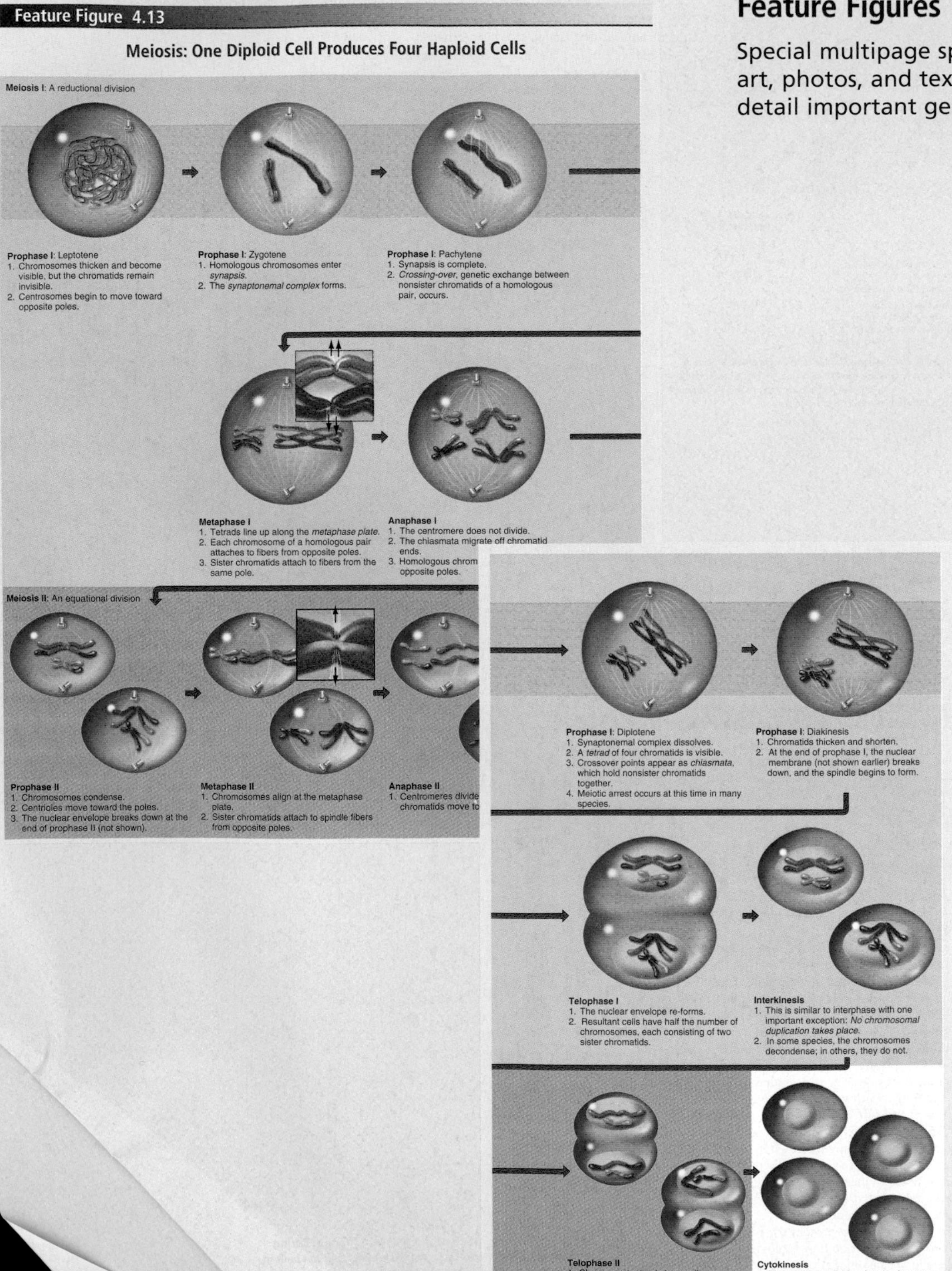

Feature Figure 4.13
Meiosis: One Diploid Cell Produces Four Haploid Cells
Meiosis I: A reductional division
Prophase I: Leptotene
1. Chromosomes thicken and become visible, but the chromatids remain invisible.
2. Centrosomes begin to move toward opposite poles.
Prophase I: Zygotene
1. Homologous chromosomes enter synapsis.
2. The synaptonemal complex forms.
Prophase I: Pachytene
1. Synapsis is complete.
2. Crossing-over, genetic exchange between nonsister chromatids of a homologous pair, occurs.
Metaphase I
1. Tetrads line up along the metaphase plate.
2. Each chromosome of a homologous pair attaches to fibers from opposite poles.
3. Sister chromatids attach to fibers from the same pole.
Anaphase I
1. The centromere does not divide.
2. The chiasmata migrate off chromatid ends.
Meiosis II: An equational division
Prophase II
1. Chromosomes condense.
2. Centrioles move toward the poles.
3. The nuclear envelope breaks down at the end of prophase II (not shown).
Metaphase II
1. Chromosomes align at the metaphase plate.
2. Sister chromatids attach to spindle fibers from opposite poles.
Anaphase II
Prophase I: Diplotene
1. Synaptonemal complex dissolves.
2. A tetrad of four chromatids is visible.
3. Crossover points appear as chiasmata, which hold nonsister chromatids together.
4. Meiotic arrest occurs at this time in many species.
Prophase I: Diakinesis
1. Chromatids thicken and shorten.
2. At the end of prophase I, the nuclear membrane (not shown earlier) breaks down, and the spindle begins to form.
Telophase I
1. The nuclear envelope re-forms.
2. Resultant cells have half the number of chromosomes, each consisting of two sister chromatids.
Interkinesis
1. This is similar to interphase with one important exception: No chromosomal duplication takes place.
2. In some species, the chromosomes decondense; in others, they do not.
Telophase II
1. Chromosomes begin to uncoil.
2. Nuclear envelopes and nucleoli (not shown) re-form.
Cytokinesis
1. The cytoplasm divides, forming four new haploid cells.

Figure 4.13 To aid visualization of the chromosomes, the figure is simplified in two ways: (1) The nuclear envelope is not shown during prophase of either meiotic division. (2) The chromosomes are shown as fully condensed at zygotene; in reality, the chromosomes continue to condense throughout prophase such that full condensation does not occur until diakinesis.

Process Figures

Step-by-step descriptions allow the student to walk through a compact summary of important details.

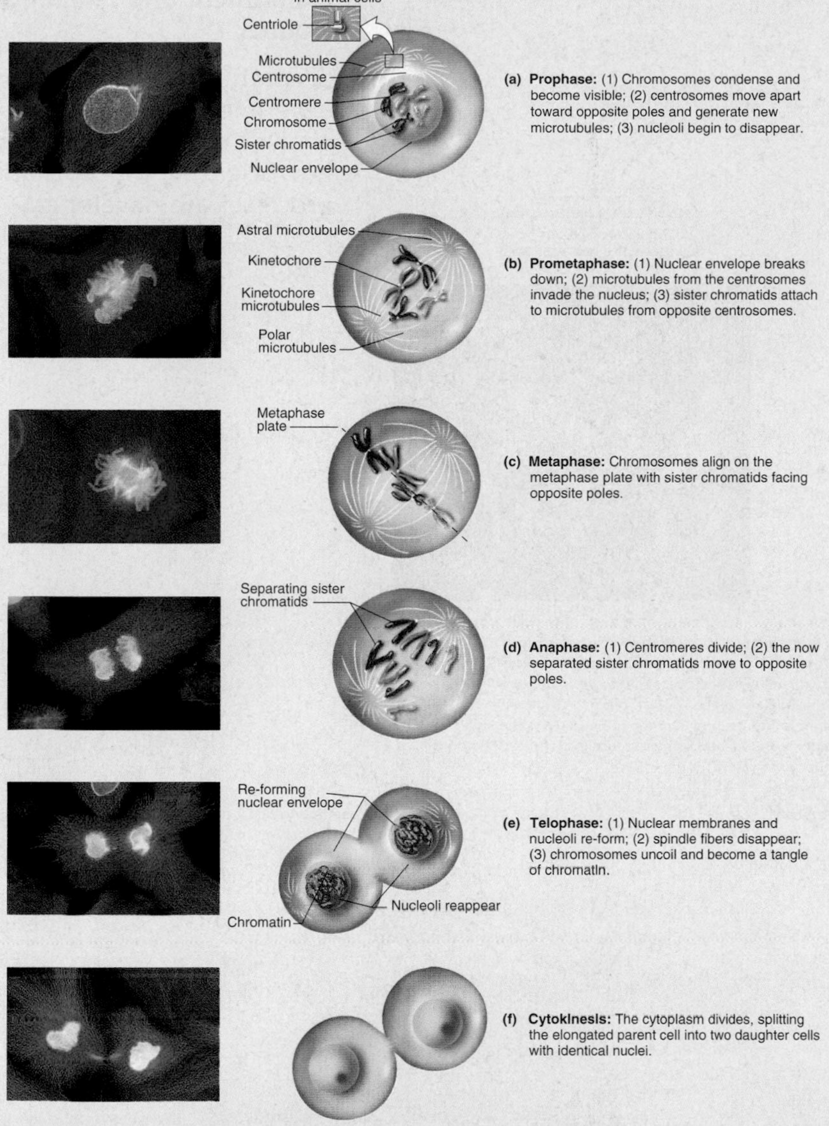

In animal cells
Centriole
Microtubules
Centrosome
Centromere
Chromosome
Sister chromatids
Nuclear envelope
(a) **Prophase:** (1) Chromosomes condense and become visible; (2) centrosomes move apart toward opposite poles and generate new microtubules; (3) nucleoli begin to disappear.
Astral microtubules
Kinetochore
Kinetochore microtubules
Polar microtubules
(b) **Prometaphase:** (1) Nuclear envelope breaks down; (2) microtubules from the centrosomes invade the nucleus; (3) sister chromatids attach to microtubules from opposite centrosomes.
Metaphase plate
(c) **Metaphase:** Chromosomes align on the metaphase plate with sister chromatids facing opposite poles.
Separating sister chromatids
(d) **Anaphase:** (1) Centromeres divide; (2) the now separated sister chromatids move to opposite poles.
Re-forming nuclear envelope
Chromatin
Nucleoli reappear
(e) **Telophase:** (1) Nuclear membranes and nucleoli re-form; (2) spindle fibers disappear; (3) chromosomes uncoil and become a tangle of chromatin.
(f) **Cytokinesis:** The cytoplasm divides, splitting the elongated parent cell into two daughter cells with identical nuclei.

Figure 4.8 Mitosis maintains the chromosome number of the parent cell nucleus in the two daughter nuclei. In the photomicrographs of newt lung cells, chromosomes are stained *blue* and microtubules appear either *green* or *yellow*.

Micrographs

Stunning micrographs bring the genetics world to life.

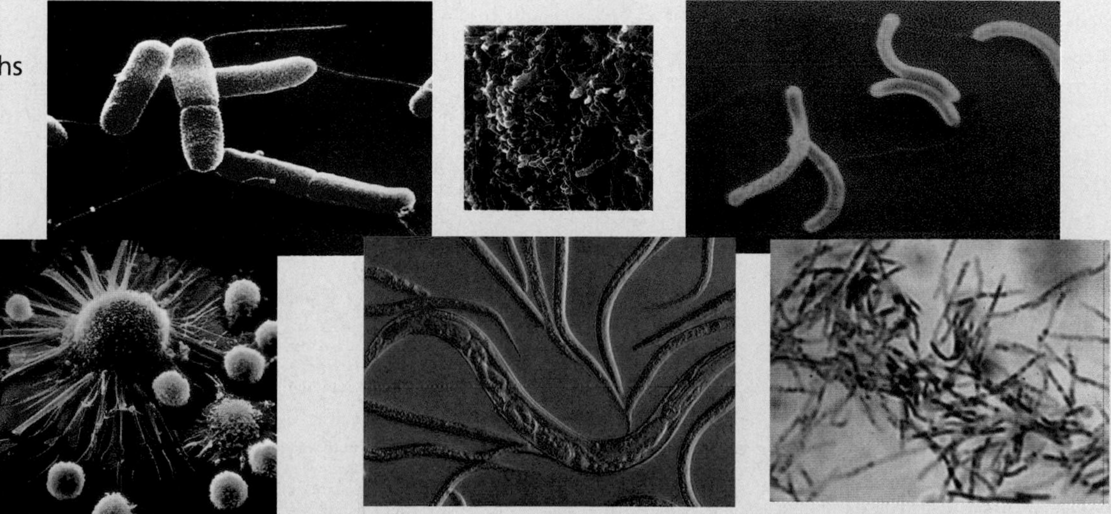

(a)

1. Drop cells onto a glass slide.

2. Gently denature DNA by treating briefly with DNase.

Fluorescent probes

Fluorescent dye

3. Add hybridization probes labeled with fluorescent dye and wash away unhybridized probe.

Fluorescence microscope

Eyepiece

Barrier filter 2 (further blockage of stray UV rays)

UV source

Mirror to UV light; transparent to visible light

Barrier filter 1 (blocks dangerous short UV rays, allows needed long UV rays to pass through)

Objective lens

Object

4. Expose to ultraviolet (UV) light. Take picture of fluorescent chromosomes.

(b)

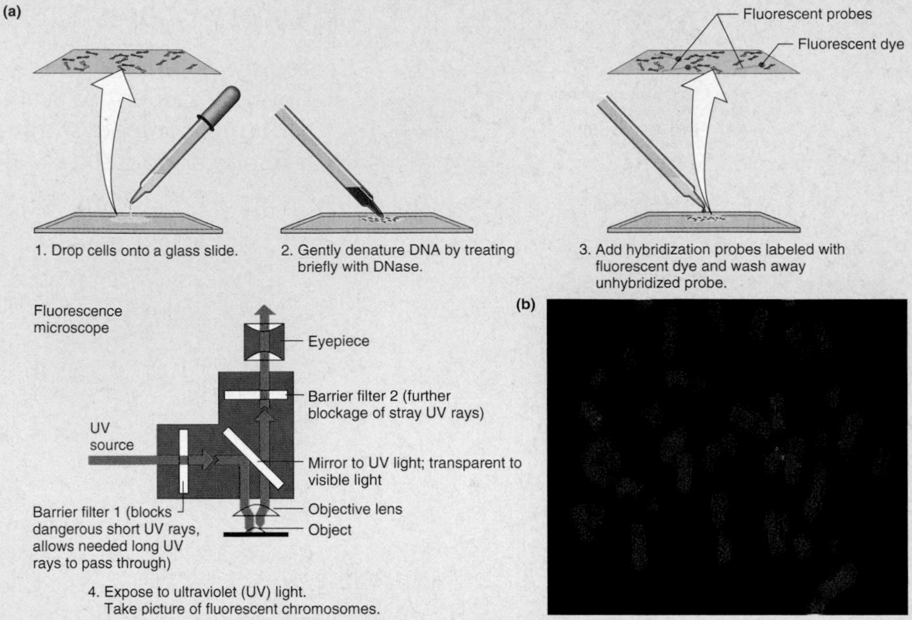

Figure 10.8 The FISH protocol. (a) The technique. (1) First, drop cells arrested in the metaphase stage of the cell cycle onto a micrc scope slide. The force of the droplet hitting the slide causes the cells to burst open with the chromosomes spread apart. (2) Next, fix the chromosomes and gently denature the DNA within them such that the overall chromosomal structure is maintained even though each DNA double helix opens up at numerous points. (3) Label a DNA probe with a fluorescent dye, add it to the slide, incubate the probe with the slide long enough for hybridization to occur, and wash away unhybridized probe. (4) Now place the slide under a special micrc scope that focuses ultraviolet (UV) light on the chromosomes. The UV light causes the bound probe to fluoresce in the visible range of the spectrum. You can view the fluorescence through the eyepiece and photograph it. **(b)** A fluorescence micrograph. Photograph of a baby hamster kidney cell subjected to FISH analysis. It shows the microtubular structure.

Experiment and Technique Figures

Illustrations of performed experiments and genetic analysis techniques highlight how scientific concepts and processes are developed.

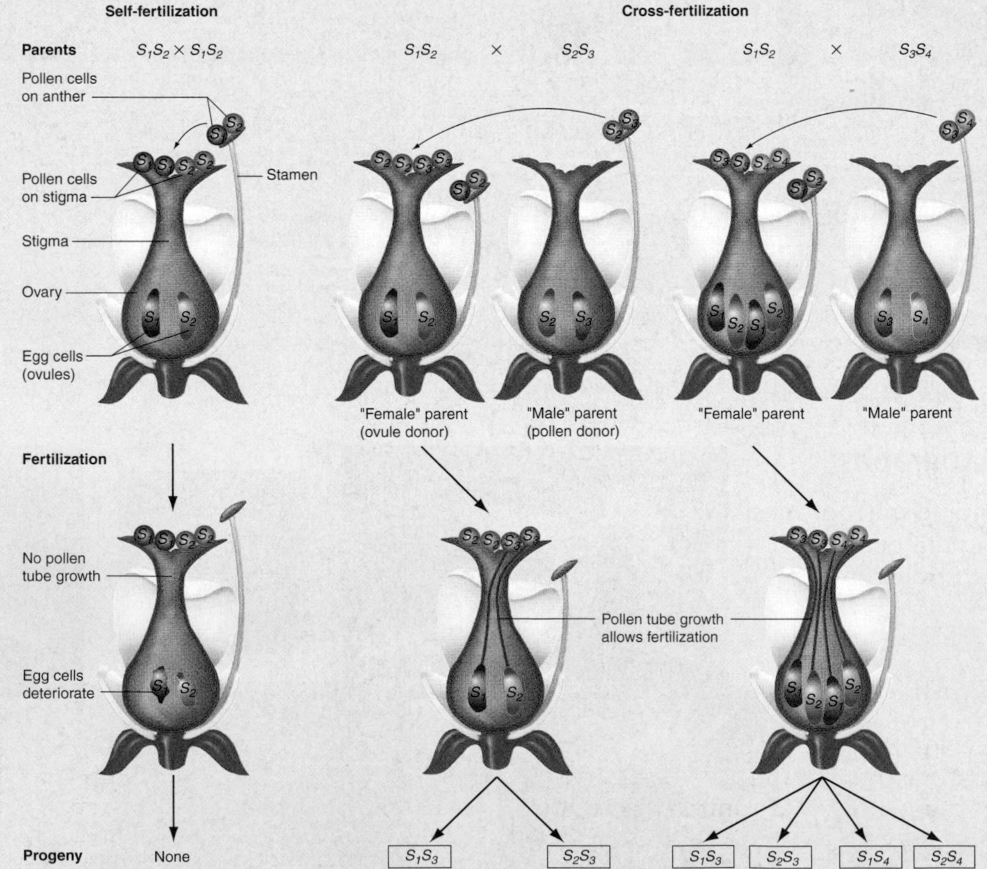

Figure 3.8 Plant incompatibility systems promote outbreeding and allele proliferation. A pollen grain carrying an allele of the self-incompatibility gene that is identical to either of the two alleles carried by a potential female parent is unable to grow a pollen tube; as a result, fertilization cannot take place. Because all the pollen grains produced by any one plant have one of the two alleles carried by the female reproductive parts of the same plant, self-fertilization is impossible.

Comparative Figures

Comparison illustrations lay out the basic differences of often confusing principles.

Solving Genetics Problems

The best way for students to assess and increase their understanding of genetics is to practice through problems. Found at the end of each chapter, problem sets assist students in evaluating their grasp of key concepts and allow them to apply what they have learned to real-life issues.

Review Problems

Problems are organized by chapter section and in order of increasing difficulty to help students develop strong problem-solving skills. The answers to select problems can be found in the back of this text.

Problems

Vocabulary

1. The following is a list of mutational changes. For each of the specific mutations described, indicate which of the terms in the right-hand column applies, either as a description of the mutation or as a possible cause. More than one term from the right column can apply to each statement in the left column.

1. an A–T base pair in the wild-type gene is changed to a G–C pair	a. transition
2. an A–T base pair is changed to a T–A pair	b. base substitution
3. the sequence AAGCTTATCG is changed to AAGCTATCG	c. transversion
4. the sequence AAGCTTATCG is changed to AAGCTTTATCG	d. inversion
5. the sequence AACGTTATCG is changed to AATGTTATCG	e. translocation
6. the sequence AACGTCACACACACATCG is changed to AACGTCACATCG	f. deletion
7. the gene map in a given chromosome arm is changed from *bog-rad-fox1-fox2-try-duf* (where *fox1* and *fox2* are highly homologous, recently diverged genes) to *bog-rad-fox1-fox3-fox2-try-duf* (where *fox3* is a new gene with one end similar to *fox1* and the other similar to *fox2*)	g. insertion
8. the gene map in a chromosome is changed from *bog-rad-fox1-fox2-try-duf* to *bog-rad-fox2-fox1-try-duf*	h. deamination
9. the gene map in a given chromosome is changed from *bog-rad-fox1-fox2-try-duf* to *bog-rad-fox1-mel-qui-txu-sqm*	i. X-ray irradiation
	j. intercalator
	k. unequal crossing-over

Section 7.1

2. The DNA sequence of a gene from three independently isolated mutants is given here. Using this information, what is the sequence of the wild-type gene in this region?

```
mutant 1    ACCGTAATCGACTGGTAAACTTTGCGCG
mutant 2    ACCGTAGTCGACCGGTAAACTTTGCGCG
mutant 3    ACCGTAGTCGACTGGTTAACTTTGCGCG
```

3. Over a period of several years, a large hospital kept track of the number of births of babies displaying the trait achondroplasia. Achondroplasia is a very rare autosomal dominant condition resulting in dwarfism with abnormal body proportions. After 120,000 births, it was noted that there had been 27 babies born with achondroplasia. One physician was interested in determining how [illegible] new mutation [illegible] in his area w[illegible] families of th[illegible] the dwarf bab[illegible]

4. Among mammals, measurements of the rate of generation of autosomal recessive mutations have been made almost exclusively in mice, while many measurements of the rate of generation of dominant mutations have been made both in mice and in humans. Why do you think there has been this difference?

5. In a genetics lab, Kim and Maria infected a sample from an *E. coli* culture with a particular virulent bacteriophage. They noticed that most of the cells were lysed, but a few survived. The survival rate in their sample was about 1×10^{-4}. Kim was sure the bacteriophage induced the resistance in the cells, while Maria thought that resistant mutants probably already existed in the sample of cells they used. Earlier, for a different experiment, they had spread a dilute suspension of *E. coli* onto solid medium in a large petri dish, and, after seeing that about 10^5 colonies were growing up, they had replica-plated that plate onto three other plates. Kim and Maria decided to use these plates to test their theories. They pipette a suspension of the bacteriophage onto each of the three replica plates. What should they see if Kim is right? What should they see if Maria is right?

6. Suppose you wanted to study genes controlling the structure of bacterial cell surfaces. You decide to start by isolating bacterial mutants that are resistant to infection by a bacteriophage that binds to the cell surface. The selection procedure is simple: Spread cells from a culture of sensitive bacteria on a petri plate, expose them to a high concentration of phages, and pick the bacterial colonies that grow. To set up the selection you could (1) spread cells from a single liquid culture of sensitive bacteria on many different plates and pick every resistant colony *or* (2) start many different cultures, each grown from a single colony of sensitive bacteria, spread one plate from each culture, and then pick a single mutant from each plate. Which method would ensure that you are isolating many independent mutations?

7. A wild-type male *Drosophila* was exposed to a large dose of X-rays and was then mated to an unirradiated female, one of whose X chromosomes carried both a dominant mutation for the trait *Bar* eyes and several inversions. Many F_1 females from this mating were recovered who had the *Bar*, multiply inverted X chromo-[illegible]

Solved Problems

Solved problems offer step-by-step guidance needed to understand the problem-solving process.

Solved Problems

I. Mutations can often be reverted to wild type by treatment with mutagens. The type of mutagen that will reverse a mutation gives us information about the nature of the original mutation. The mutagen EMS almost exclusively causes transitions; proflavin is an intercalating agent that causes insertion or deletion of a base; ultraviolet (UV) light causes single-base substitutions. Cultures of several *E. coli* met^- mutants were treated with three mutagens separately and spread onto a plate lacking methionine to look for revertants. (In the chart, − indicates that no colonies grew, and + indicates that some met^+ revertant colonies grew.)

	Mutagen treatment		
Mutant number	EMS	Proflavin	UV light
1	+	−	+
2	−	+	−
3	−	−	−
4	−	−	+

a. Given the results, what can you say about the nature of the original mutation in each of the strains?

b. Experimental controls are designed to eliminate possible explanations for the results, thereby ensuring that data are interpretable. In the experiment described, we scored the presence or absence of colonies. How do we know if colonies that appear on plates are mutagen-induced revertants? What else could they be? What control would enable us to be confident of our revertant analysis?

Answer

To answer this question, you need to understand the concepts of mutation and reversion.

a. Mutation 1 is reverted by the mutagen that causes transitions, *so mutation 1 must have been a transition*. Consistent with this conclusion is the fact the UV light can also revert the mutation and the intercalating agent proflavin does not cause reversion. *Mutation 2 is reverted by proflavin and therefore must be either an insertion or a deletion of a base.* The other two mutagens do not revert mutation 2. Mutation 3 is not reverted by any of these mutagenic agents. It is therefore not a single-base substitution, a single-base insertion, or a single-base deletion. *Mutation 3 could be a deletion of several bases or an inversion.* Mutation 4 is reverted by UV light, so it is a single-base change, but it is not a transition, since EMS did not revert the mutation. *Mutation 4 must be a transversion.*

Social and Ethical Issues

These challenging problems stir discussion and debate. The issues are presented within the context of real-life case studies and require the student to consider not only scientific issues but legal and ethical issues as well.

Social and Ethical Issues

1. Chemicals that are mutagenic are identified by the Ames test, which measures the level of mutagenesis in bacteria. The susceptibility of humans to mutagenic chemicals may vary depending on the genetic makeup of the individual. The dose that affects one person may be different from that which affects another. However, there are few, if any, reliable tests that determine a person's level of susceptibility. If this is true, is it a good idea to translate the results of the Ames test of mutability in bacteria to a prediction of carcinogenicity in humans? Often, reports of Ames test results on a chemical make newspaper headlines. Is this a useful and honest way to report findings that could affect human health, or do people need to consider other variables to make an informed decision?

2. Mr. and Mrs. Aswari have a child with fragile X syndrome (see the Genetics and Society box on p. 216–217). They want to have a second child but are considering egg donation because genetic screening has indicated that Mrs. Aswari carries a premutation allele with 120 CGG repeats. If you were the Aswari's genetic counselor, what would you tell them about their risk of having a second child with fragile X syndrome? What are the ethical issues related to genetic screening when (1) a result indicates no risk, (2) a result indicates that the phenotype being screened for will be exhibited, and (3) an intermediary result does not clearly fall into either category?

Media and Supplements

For the Instructor

ARIS Presentation Center

Build instructional materials where-ever, when-ever, and how-ever you want! ARIS Presentation Center is an online digital library containing assets such as photos, artwork, animations, PowerPoints, and other media types that can be used to create customized lectures, visually enhanced tests and quizzes, compelling course websites, or attractive printed support materials.

Nothing could be easier! Accessed from the instructor side of your textbook's ARIS website, Presentation Center's dynamic search engine allows you to explore by discipline, course, textbook chapter, asset type, or keyword. Simply browse, select, and download the files you need to build engaging course materials. All assets are copyright McGraw-Hill Higher Education but can be used by instructors for classroom purposes.

Instructor's Testing and Resource CD-ROM

This cross-platform CD features a computerized test bank that uses testing software to quickly create customized exams. The user-friendly program allows instructors to search for questions by topic, format, or difficulty level; edit existing questions or add new ones; and scramble questions for multiple versions of the same test.

McGraw-Hill ARIS—Assessment, Review, and Instruction System

McGraw-Hill's ARIS is a complete, online electronic homework and course management system, designed for greater ease of use than any other system available. Created specifically for *Genetics: From Genes to Genomes,* third edition, instructors can create and share course materials and assignments with colleagues with a few clicks of the mouse. For instructors, personal response system questions, all PowerPoint lectures, and assignable content are directly tied to text-specific materials in *Genetics: From Genes to Genomes*. Instructors can also edit questions, import their own content, and create announcements and due dates for assignments. Also included on the ARIS site is the *Instructors's Manual and Integration Guide*. This manual provides a guide to integrating all the available resources for *Genetics: From Genes to Genomes* into your course presentations.

ARIS has automatic grading and reporting of easy-to-assign homework, quizzing, and testing. All student activity within McGraw-Hill's ARIS is automatically recorded and available to the instructor through a fully integrated grade book that can be downloaded to Excel.

For students, there are multiple-choice quizzes, animations with quizzing, web interactive exercises, and even more materials that may be used for self-study or in combination with assigned materials.

Go to aris.mhhe.com to learn more and register!

PageOut
McGraw-Hill's exclusive tool for creating your own website for your genetics course. It requires no knowledge of coding and is hosted by McGraw-Hill.

Course Management Systems
ARIS content compatible with online course management systems like WebCT and Blackboard makes putting together your course website easy. Contact your local McGraw-Hill sales representative for details.

Transparencies
150 four-color illustrations from the book will be available to adopters.

For the Student

Solutions Manual/Study Guide
ISBN 978-0-07-299587-9
MHID 0-07-299587-4
Extensively revised by Dr. Debra Nero of Cornell University, this manual presents the solutions to the end-of-chapter problems and questions along with the step-by-step logic of each solution. The manual also includes a synopsis, the objectives, and problem-solving tips for each chapter. Key figures and tables from the textbook are referenced throughout to guide student study.

McGraw-Hill's ARIS
(Assessment Review and Instruction System)
Makes homework meaningful—and manageable—for instructors and students.

Explore this dynamic site for a variety of study tools.

- **Self-quizzes** test your understanding of key concepts.
- **Flash cards** ease learning of new vocabulary.
- **Animations** bring key genetic concepts to life and are followed by a **quiz** to test your understanding.
- **On Our Website** content includes downloadable PDF files of new material, and easy access to articles and websites related to the concepts of individual chapters.
- **Interactive Web Exercises** guide your exploration of key genetic databases. These exercises include an assignment so you can apply what you've discovered in the databases to the concepts covered in the textbook chapter.

Go to aris.mhhe.com to learn more or go directly to this book's ARIS site at www.mhhe.com/hartwell3

***Genetics: From Genes to Genomes* CD-ROM,** developed with the content of this book, covers the most challenging concepts in the introductory genetics course. The CD presents animations of basic genetic processes, interactive exercises, and simulations involving fundamental principles. Additional quizzing options allow students to self-test and identify those areas needing additional study. Glossary definitions can be reached via hot links. A correlation guide linking book topics to the related CD material is included on the CD.

Acknowledgments

The creation of a project of this scope is never solely the work of the authors. We are grateful to our colleagues around the world who took the time to review this manuscript and make suggestions for its improvement. Their willingness to share their experiences and expertise was a tremendous help to us.

Third Editewers

Dr. Michael Abler, *University of Wisconsin, La Crosse*
Amelia J. Ahern-Rindell, *University of Portland*
Robert Angus, *University of Alabama, Birmingham*
Arthur R. Ayers, *Albertson College of Idaho*
Vernon W. Bauer, *Francis Marion University*
Robert E. Braun, *University of Washington School of Medicine*
Kirk Cammarata, *Texas A&M University, Corpus Christi*
James J. Campanella, Ph.D., *Montclair State University*
J. Aaron Cassill, *University of Texas, San Antonio*
Kerry L. Cheesman, Ph.D., *Capital University*
Richard W. Cheney, Jr., *Christopher Newport University*
Yury O. Chernoff, *Georgia Institute of Technology*
Ruth Chesnut, *Eastern Illinois University*
Michael J. Christoffers, *North Dakota State University*
Thomas W. Cline, *University of California, Berkeley*
Bruce Cochrane, *University of South Florida, Tampa*
Bernard P. Duncker, *University of Waterloo*
Christine Dupont, *University of Waterloo*
Bert Ely, *University of South Carolina*
William F. Ettinger, *Gonzaga University*
Ann P. Evancoe, *Hudson Valley Community College*
Rebecca V. Ferrell, *Metropolitan State College of Denver*
Victor Fet, *Marshall University*
David Foltz, *Louisiana State University*
Wayne C. Forrester, *Indiana University*
Robert G. Fowler, *San Jose State University*
Sylvia Fromherz, *University of Colorado at Boulder*
Julia Frugoli, *Clemson University*
Anne M. Galbraith, *University of Wisconsin, La Crosse*
Gail E. Gasparich, *Towson University*
Dr. Nabarun Ghosh, Ph.D., *West Texas A&M University*
Susan Godfrey, *University of Pittsburgh*
Michael A. Goldman, Ph.D., *San Francisco State University*
Elliott Goldstein, *Arizona State University*
Deborah J. Good, *Virginia Polytechnic Institute and State University*
Nels H. Granholm, Ph.D., *South Dakota State University*
Robert Gregerson, *Lyon College*
Martha Hamblin, *Cornell University*
Pamela L. Hanratty, *Indiana University*
Pamela K. Hanson, *Birmingham-Southern College*
Stephen C. Hedman, *University of Minnesota*
Peter W. Hoffman, Ph.D., *College of Notre Dame of Maryland*
Bruce Hofkin, *University of New Mexico*
Nancy M. Hollingsworth, *State University of New York, Stony Brook*
Laura L. Mays Hoopes, *Pomona College*
Kamal M. Ibrahim, *Southern Illinois University*
Bob Ivarie, *University of Georgia*
Bradley Jett, Ph.D., *Oklahoma Baptist University*
Gregg Jongeward, *University of the Pacific*
Todd Kelson, *Brigham Young University, Idaho*
Stephen T. Kilpatrick, *University of Pittsburgh, Johnstown*
Deborah A. Kimbrell, *University of California, Davis*
Bruce Kohorn, *Bowdoin College*
Sidney Kushner, *University of Georgia*
John C. Larkin, *Louisiana State University*
Howard Laten, *Loyola University, Chicago*
Elena Levine Keeling, *California Polytechnical State University*
Roger Lightner, *University of Arkansas, Fort Smith*
Paul S. Lovett, *University of Maryland*
Hiten Madhani, *University of California, San Francisco*
James Makowski, Ph.D., *Messiah College*
Alfred R. Martin, Ph.D., *Benedictine University*
Debra McDonough, *University of New England*
Kim S. McKim, *Rutgers University*
Scott D. Michaels, *Indiana University*
Robert Moss, Ph.D., *Wofford College*
Mary Rengo Murnik, *Ferris State University*
Stuart J. Newfeld, *Arizona State University*
John C. Osterman, *University of Nebraska, Lincoln*
Dr. David K. Peyton, *Morehead State University*
Gregory J. Podgorski, *Utah State University*
James V. Price, *Utah Valley State College*
Rongsun Pu, *Kean University*
David H. Reed, *University of Mississippi*
Jennifer L. Regan, *University of Southern Mississippi*
David L. Remington, *University of North Carolina, Greensboro*
Inder Saxena, *University of Texas, Austin*
Malcolm P. Schug, *University of North Carolina, Greensboro*
Jeff Sekelsky, *University of North Carolina, Chapel Hill*
Monica M. Skinner, Ph.D., *Oregon State University*
Chris Somerville, *Carnegie Institution and Stanford University*
James H. Thomas, *University of Washington*
Doug Thrower, *University of California, Santa Barbara*
Jonathan E. Visick, *North Central College*
Alan S. Waldman, *University of South Carolina*
Dr. Sarah Ward, *Colorado State University*
Ted Weinert, *University of Arizona*
David R. Wessner, *Davidson College*
Matthew M. White, *Ohio University*
Robert Wiggers, *Stephen F. Austin State University*
David Wofford, *University of Florida, Gainesville*
Yang Yen, *South Dakota State University*
Jianzhi Zhang, *University of Michigan*

Second-Edition Reviewers

Lawrence R. Aaronson, *Utica College*
Ruth Ballard, *California State University, Sacramento*
Mary Bedell, *University of Georgia*
Michelle Bell, *Xavier University*
Michael Benedik, *University of Houston*
Susan Bergeson, *University of Texas*
David Carroll, *Florida Institute of Technology*
Helen Chamberlin, *Ohio State University*
Ruth Chesnut, *Eastern Illinois University*
Bruce Cochrane, *University of South Florida*
Claire Cronmiller, *University of Virginia*
Mike Dalbey, *University of California, Santa Cruz*
David Durica, *University of Oklahoma*
David Duvernell, *Southern Illinois University*
Sarah Elgin, *Washington University*
Johnny El-Rady, *University of South Florida*
Victor Fet, *Marshall University*
Janice Fisher, *University of Texas, Austin*
David Foltz, *Louisiana State University*
Jim Ford, *Stanford University*
David Fromson, *California State University, Fullerton*
Anne Galbraith, *University of Wisconsin, La Crosse*
Peter Gergen, *SUNY Stony Brook*
Elliot Goldstein, *Arizona State University*
James Haber, *Brandeis University*
Ralph Hillman, *Temple University*
Nancy Hollingsworth, *SUNY Stony Brook*
Jackie Peltier Horn, *Houston Baptist University*
Shelley Jansky, *University of Wisconsin, Stevens Point*
Eric N. Jellen, *Brigham Young University*
Cheryl Jorcyk, *Boise State University*
Kathleen Karrer, *Marquette University*
Sean Kimbro, *Clark Atlanta University*
Rebecca Kohn, *Ursinus College*
Stanley Maloy, *University of Illinois*
Steve McCommas, *Southern Illinois University, Edwardsville*
Elliot M. Meyerowitz, *California Institute of Technology*
Leilani Miller, *Santa Clara University*
Roderick M. Morgan, *Grand Valley State University*
Nick Norton, *Southeastern Louisiana University*
Valerie Oke, *University of Pittsburgh*
Greg Podgorski, *Utah State University*
Jim Prince, *California State University, Fresno*
Inder Saxena, *University of Texas*
Daniel Schoen, *McGill University*
Mark Seeger, *Ohio State University*
Malcolm Shields, *University of West Florida*
David Smith, *University of Delaware*
Harold Vaessin, *Ohio State University*
Dan Wells, *University of Houston*
Andrew Wood, *Southern Illinois University*

First-Edition Reviewers

Ken Belanger, *University of North Carolina, Chapel Hill*
John Belote, *Syracuse University*
Anna Berkovitz, *Purdue University*
John Botsford, *New Mexico State University*
Michael Breindl, *San Diego State University*
Bruce Chase, *University of Nebraska, Omaha*
Lee Chatfield, *University of Central Lancashire*
Alan Christensen, *University of Nebraska*
Bruce Cochrane, *University of South Florida*
James Curran, *Wake Forest University*
Rowland Davis, *University of California, Irvine*
Paul Demchick, *Barton College*
Stephen D'Surney, *University of Mississippi*
Rick Duhrkopf, *Baylor University*
Susan Dutcher, *University of Colorado*
DuWayne Englert, *Southern Illinois University*
Bentley Fane, *University of Arkansas*
Victoria Finnerty, *Emory University*
David Foltz, *Louisiana State University*
David Futch, *San Diego State University*
Ann Gerber, *University of North Dakota*
Richard Gethmann, *University of Maryland, Baltimore County*
Mike Goldman, *San Francisco State University*
Elliott Goldstein, *Arizona State University*
Nels Granholm, *South Dakota State University*
Charles Green, *Rowan College of New Jersey*
Poonam Gulati, *University of Houston*
Stephen Hedman, *University of Minnesota*
Ralph Hillman, *Temple University*
Christine Holler-Dinsmore, *Fort Peck Community College*
Martin Hollingsworth, *Tallahassee Community College*
Nancy Hollingsworth, *State University of New York, Stony Brook*
Andrew Hoyt, *Johns Hopkins University*
Lynne Hunter, *University of Pittsburgh*
Robert Ivarie, *University of Georgia*
R. C. Jackson, *Texas Technological University*
Duane Johnson, *Colorado State University*
Chris Kaiser, *Massachusetts Institute of Technology*
Kenneth J. Kemphues, *Cornell University*
Susan Kracher, *Purdue University*
Alan Koetz, *Illinois State University*
Andrew Lambertsson, *University of Oslo*
Don Lee, *University of Nebraska*
John Locke, *University of Alberta*
Larry Loeb, *University of Washington*
Robertson McClung, *Dartmouth College*
Peter Meacock, *University of Leicester*
John Merriam, *University of California, Los Angeles*
Beth Montelone, *Kansas State University*

Patricia Moore, *Transylvania University*
Gail Patt, *Boston University*
Michael Perlin, *University of Louisville*
Richard Richardson, *University of Texas, Austin*
Mary Rykowski, *University of Arizona*
Mark Sanders, *University of California, Davis*
Randall Scholl, *Ohio State University*
David Sheppard, *University of Delaware*
Anthea Stavroulakis, *Kingsborough Community College*
John Sternick, *Mansfield University*
David Sullivan, *Syracuse University*
William Thwaites, *San Diego State University*
Akfi Uzman, *University of Houston*
Peter Webster, *University of Massachusetts*
Dean Whited, *North Dakota State University*
John Williamson, *Davidson College*
John Zamora, *Middle Tennessee State University*
Stephan Zweifel, *Carleton College*

A special thank you to Nancy Hollingsworth and Martha Hamblin for their extensive feedback and to Malcolm Schug, Ronald Strohmeyer, and Michael Windelspecht for their work on the ancillary materials that accompany this book.

We would also like to thank the highly skilled publishing professionals at McGraw-Hill who guided the development and production of the third edition of *Genetics: From Genes to Genomes*: Patrick Reidy for his sponsorship and support; Rose Koos for her organizational skills and tireless work to tie up all the loose ends; and Joyce Berendes and her production team for their careful attention to detail and ability to move the schedule along.

Genetics: The Study of Biological Information

Chapter 1

Genetics, the science of heredity, is at its core the study of biological information. All living organisms—from single-celled bacteria and protozoa to multicellular plants and animals—must store, replicate, transmit to the next generation, and use vast quantities of information to develop, reproduce, and survive in their environments (**Fig. 1.1**). Geneticists examine how organisms pass biological information on to their progeny and how they use it during their lifetime.

This book introduces the field of genetics as it exists in the first decade of the twenty-first century. Seven overarching themes recur throughout our presentation:

Information can be stored in many ways including the patterns of letters and words in books and the sequence of nucleotides in DNA molecules.

- The biological information fundamental to life is encoded in the DNA molecule.
- Biological function emerges primarily from protein molecules.
- Complex biological systems emerge from the functioning of regulatory networks that specify the behavior of genes and proteins.
- All living forms are descended from a common ancestor and therefore closely related at the molecular level.
- The modular construction of genomes has allowed the rapid evolution of biological complexity.
- Genetic techniques permit the dissection of biological complexity.
- Our focus is on human genetics.

In the remainder of this chapter, we introduce these themes. It will help to keep them in mind as you delve into the details of genetics.

1.1 The Biological Information Fundamental to Life Is Encoded in the DNA Molecule

The process of evolution has taken close to 4 billion years to generate the amazingly efficient mechanisms for storing, replicating, expressing, and diversifying biological information seen in organisms now inhabiting the earth. The linear DNA molecule stores biological information in units known as nucleotides. Within each DNA molecule, the sequence of the four letters of the DNA alphabet—G, C, A, and T—specify which proteins an organism will make as well as when and where protein synthesis will occur. The letters refer to the bases—guanine, cytosine, adenine, and thymine—that are components of the nucleotide building blocks of DNA. The DNA molecule itself is a double strand of nucleotides carrying complementary G–C or A–T base pairs (**Fig. 1.2**). These complementary base pairs can bind together through hydrogen bonds. The molecular complementarity of double-stranded DNA is its most important property and the key to understanding how DNA functions.

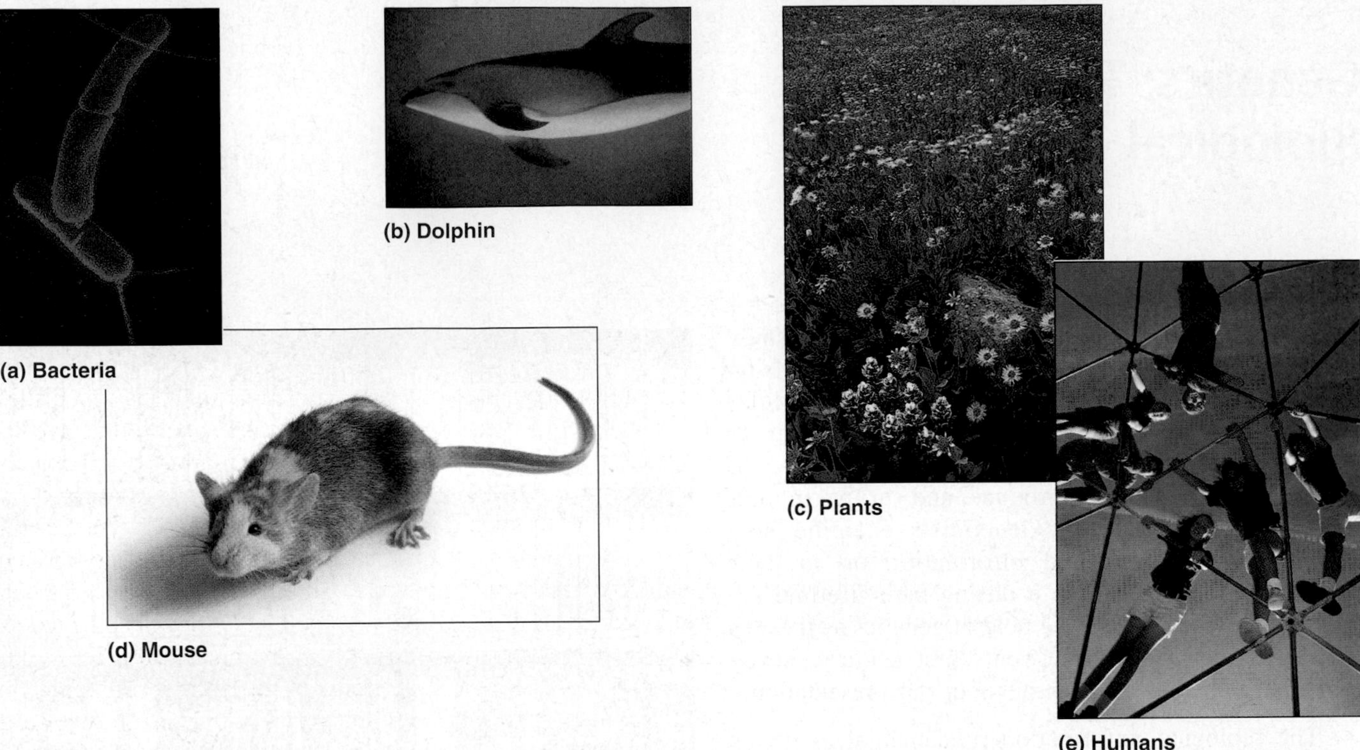

Figure 1.1 The biological information in DNA generates an enormous diversity of living organisms.

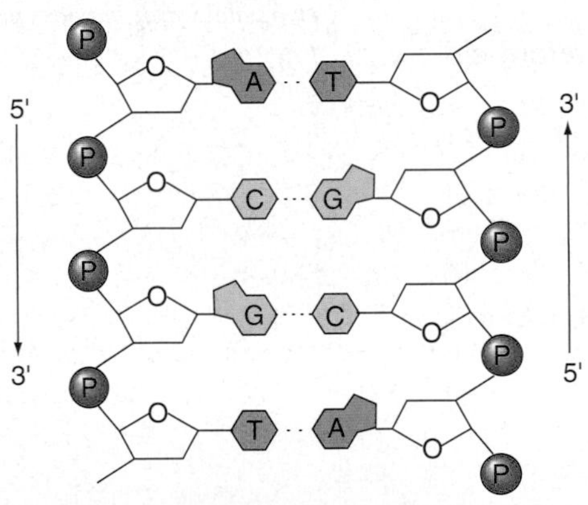

Figure 1.2 Complementary base pairs are a key feature of the DNA molecule. A single strand of DNA is composed of nucleotide subunits each consisting of a deoxyribose sugar (depicted here as a *white pentagon*), a phosphate (depicted as a *yellow circle*), and one of four nitrogenous bases—adenine, thymine, cytosine, or guanine (designated as *lavender* or *green* A's, T's, C's, or G's). The chemical structure of the bases enables A to associate tightly with T, and C to associate tightly with G through hydrogen bonding. As a result, A and T form one kind of complementary base pair, while C and G form another kind of complementary base pair. The association through base pairing of two complementary DNA strands produces a DNA double helix. The *arrows* labeled 5′ to 3′ show that the two strands of the double helix have opposite orientations relative to chemically distinct 5′ and 3′ ends.

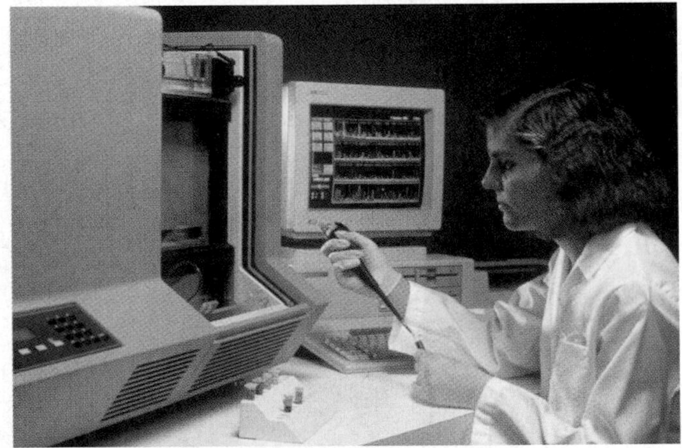

Figure 1.3 An automated DNA sequencer. This instrument can sequence about 1,000,000 base pairs a day.

Although the DNA molecule is three-dimensional, most of its information is one-dimensional and digital. The information is one-dimensional because it is encoded as a specific sequence of letters along the length of the molecule. It is digital because each unit of information—one of the four letters of the DNA alphabet—is discrete. Because genetic information is digital, it can be stored as readily in a computer memory as in a DNA molecule. Indeed, the combined power of DNA sequencers (**Fig. 1.3**), computers, and DNA synthesizers makes it possible to interpret, store, replicate, and transmit genetic information electronically from one place to another anywhere on the planet. Such electronic wizardry works something like this: A DNA sequencer reads the base sequence of a DNA molecule. The sequence information is stored in a computer. The computer transmits the information via satellite from New York to a receiver in Hong Kong or Paris. There, the information is fed into a DNA synthesizer, which makes an exact replica of a portion of the originally sequenced DNA molecule.

The DNA regions that encode proteins are called *genes.* Just as the limited number of letters in a written alphabet places no restrictions on the stories one can tell, so too the limited number of letters in the genetic code alphabet places no restrictions on the kinds of proteins and thus the kinds of organisms genetic information can define. The basic genetic language is virtually the same for all organisms, whether single-cell bacteria or multicellular humans. The differences are in the content and amount of information, and in when and where that information is *expressed,* that is, converted to protein.

Within the cells of an organism, DNA molecules carrying the genes are assembled into *chromosomes:* organelles that package and manage the storage, duplication, expression, and evolution of DNA (**Fig. 1.4**). The entire collection of chromosomes in each cell of an organism is its *genome.* Human cells, for example, contain 24 distinct kinds of chromosomes carrying approximately 3×10^9 base pairs and roughly 20,000–30,000 genes. The amount of information that can be encoded in this size genome is equivalent to 6 million pages of text containing 250 words per page, with each letter corresponding to one *base pair,* or pair of nucleotides. It may seem incredible that it takes only 3 billion base pairs of genetic information to develop a human being, from its basic body plan to the initiation of consciousness. To appreciate the long journey from a finite amount of genetic information easily storable on a computer disk to the production of a human being, it is necessary to examine proteins, the molecules that determine how complex systems of cells, tissues, and organisms function.

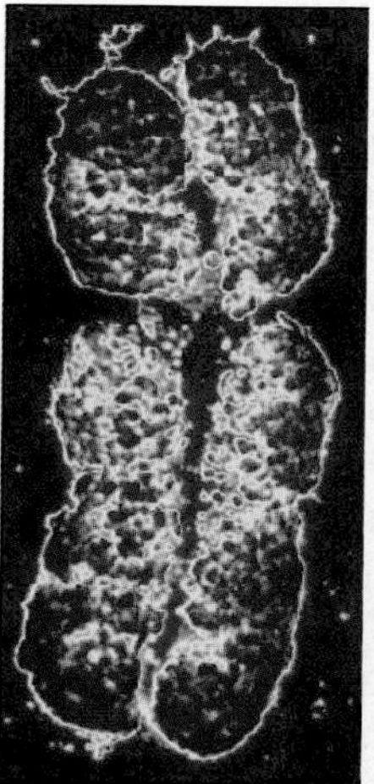

Figure 1.4 One of 24 different types of human chromosomes. Each chromosome contains thousands of genes.

1.2 Biological Function Emerges Primarily from Protein Molecules

Although there is no single characteristic that distinguishes living organisms from inanimate matter, you would have little trouble deciding which entities in a group of 20 objects are alive. Over time, these living organisms, governed by the laws of physics and chemistry as well as a genetic program, would be able to reproduce themselves. Most of the organisms would also have an elaborate and complicated structure. Consider the fly. It lays eggs, which hatch into larvae, which metamorphose at the appropriate time into adult flies. Yet another characteristic of life is the ability to move. Animals swim, fly, walk, or run, while plants grow toward or away from light. Still another characteristic of living organisms is the capacity to adapt selectively to the environment, whether it be a robin choosing materials to build a nest or a vine weaving its way up a fence. Finally, a key characteristic of living organisms is the ability to use sources of energy and matter to grow, that is, the ability to convert foreign material into their own body parts. The chemical and physical reactions that carry out these conversions are known as *metabolism.*

Most properties of living organisms ultimately arise from the class of molecules known as *proteins*—large polymers composed of hundreds to thousands of amino-acid subunits strung together in long chains; each chain folds into a specific three-dimensional conformation dictated by the sequence of its amino acids (**Fig. 1.5**). There are 20 different amino acids. The information in the DNA of genes dictates, via a genetic code, the order of amino acids in a protein molecule.

You can think of proteins as constructed from a set of 20 different kinds of snap beads distinguished by color and shape; if you were to arrange the beads in any order, make strings of a thousand beads each, and then fold or twist the chains into shapes dictated by the order of their beads, you would be able to make a nearly infinite number of different three-dimensional shapes. The astonishing diversity of three-dimensional protein structure generates the extraordinary diversity of protein function that is the basis of each organism's complex and adaptive behavior. The structure and shape of the hemoglobin protein, for example, allow it to transport oxygen in the bloodstream and release it to the tissues. The proteins myosin and actin can slide together to allow muscle contraction. Chymotrypsin and elastase are enzymes that help break down other proteins. Most of the

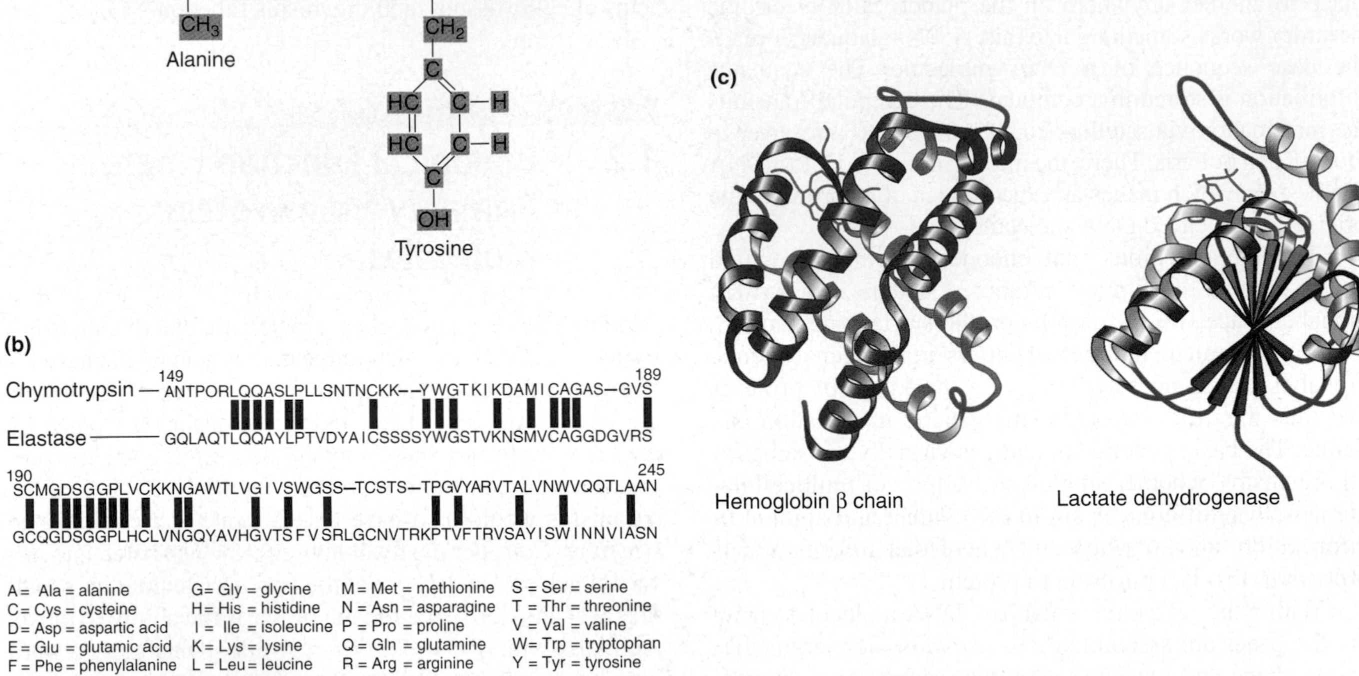

Figure 1.5 Proteins are polymers of amino acids that fold in three dimensions. The specific sequence of amino acids in a chain determines the precise three-dimensional shape of the protein. **(a)** Chemical formulas for two amino acids: alanine with a relatively simple CH_3 side chain and tyrosine with a more complex aromatic side chain. All amino acids have a basic amino group (–NH) at one end and an acidic carboxyl group (–COOH) at the other. The specific side chain (here the simple $-CH_3$ or the more complex $-CH_2$ plus aromatic ring structure) determines the amino acid's chemical properties. **(b)** A comparison of equivalent segments in the chains of two digestive proteins, chymotrypsin and elastase. The *red lines* connect sites in the two sequences that carry identical amino acids; the two chains differ at all the other sites shown. Thus, even though these two proteins are evolutionarily related to each other, they differ at enough amino acids that their structures and functions are not identical. **(c)** Schematic drawings of the hemoglobin β chain (*green*) and lactate dehydrogenase (*purple*) show the different three-dimensional shapes determined by different amino-acid sequences. The β chain is part of the complex hemoglobin molecule, which binds and delivers oxygen to body tissues. Lactate dehydrogenase is an enzyme that catalyzes energy conversions in microorganisms such as yeast and in the muscle cells of animals.

properties associated with life emerge from the constellation of protein molecules that an organism synthesizes according to instructions contained in its DNA.

1.3 Complex Systems Arise from DNA-Protein and Protein-Protein Interactions

In addition to DNA and protein, a third level of biological information encompasses dynamic interactions among DNA, protein, and other types of molecules as well as interactions among cells and tissues. These complex interactive networks represent **biological systems** that function both within individual cells and among groups of cells within an organism. Here we use *biological system* to mean any complex network of interacting molecules or groups of cells that function in a coordinated manner through dynamic signaling. There are several layers of biological systems. The human pancreas, for example, is an isolated biological system that operates within the larger biological system of the human body and mind. A whole community of animals, such as a colony of ants that functions in a highly coordinated manner, is also a biological system.

The information that defines any biological system is four-dimensional because it is constantly changing over the three dimensions of space and the one dimension of time. One of the most complex examples of this level of biological information (other than an entire human being) is the human brain with its 10^{11} (100,000,000,000) neurons connected through perhaps 10^{18} (1,000,000,000,000,000,000) junctions known as synapses. From this enormous biological network,

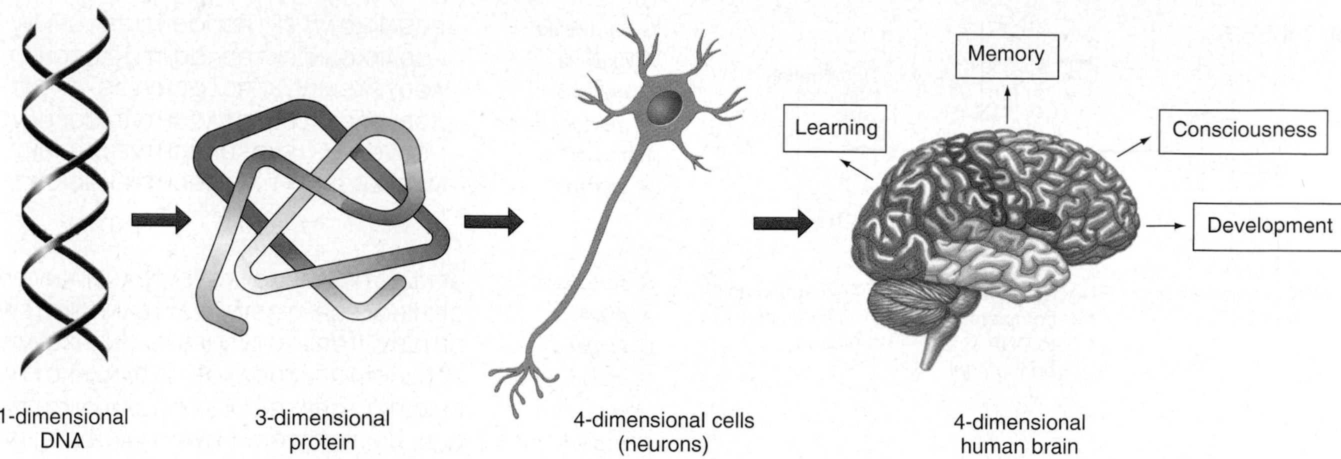

Figure 1.6 **Diagram of the conversion of biological information from a one- to a three- and finally a four-dimensional state.**

based ultimately on the information in DNA and protein, arises properties such as memory, consciousness, and the ability to learn (**Fig. 1.6**).

1.4 All Living Things Are Closely Related at the Molecular Level

The evolution of biological information is a fascinating story spanning the 4 billion years of earth's history. Many biologists think that RNA was the first information-processing molecule to appear. Very similar to DNA, RNA molecules are also composed of four subunits: the bases G, C, A, and U (for uracil, which replaces the T of DNA). Like DNA, RNA has the capacity to store, replicate, mutate, and express information; like proteins, RNA can fold in three dimensions to produce molecules capable of catalyzing the chemistry of life. RNA molecules, however, are intrinsically unstable. Thus, it is probable that the more stable DNA took over the linear information storage and replication functions of RNA, while proteins, with their far greater capacity for diversity, preempted the functions derived from RNA's three-dimensional folding. The information contained in the sequence of DNA nucleotides then came to specify the sequence of amino acids in the proteins. With this division of labor, RNA became an intermediary in converting the information in DNA into the sequence of amino acids in protein. The DNA letters G, C, A, and T are informationally equivalent to the RNA letters G, C, A, and U (**Fig. 1.7a**). The separation that placed information storage in DNA and biological function in proteins was so successful that all organisms alive today descend from the first organisms that happened upon this molecular specialization.

The evidence for the common origin of all living forms is present in their DNA sequences. All living organisms use essentially the same genetic code in which various triplet groupings of the 4 letters of the DNA and RNA alphabets encode the 20 letters of the amino-acid alphabet. Via the code, the order of bases in any organism's DNA specifies the amino-acid sequence of its proteins (**Fig. 1.7b**).

The relatedness of all living organisms is also evident from comparisons of genes with similar functions in very different organisms. For example, there is striking similarity between the genes for many proteins in bacteria, yeast, plants, worms, flies, mice, and humans (**Fig. 1.8**). Moreover, it is often possible to place a gene from one organism into the genome of a very different organism and see it function normally in the new environment. Human genes that help regulate cell division, for example, can replace related genes in yeast and enable the yeast cells to function normally.

One of the most striking examples of relatedness at this level of biological information was uncovered in studies of eye development. Both insects and vertebrates (including humans) have eyes, but they are of very different types (**Fig. 1.9**). Biologists had long assumed that the evolution of eyes occurred independently in the lineages leading to present-day insects and present-day vertebrates. Indeed, in many evolution textbooks, eyes are used as an example of *convergent evolution,* that is, of evolution in which structurally unrelated but functionally analogous organs emerge in different species as a result of natural selection. Studies of a gene called the *Pax6* gene have turned this view upside down. *Pax6* is one of nine genes encoding proteins with a "paired box" structure that enables them to bind to DNA and regulate the expression of other genes.

Mutations in the *Pax6* gene lead to a failure of eye development in both people (with a condition known as anirida) and mice, and molecular studies have suggested that *Pax6* might play a central role in the initiation of eye development in all vertebrates. Remarkably, when the human *Pax6* gene is expressed in cells along the surface of the fruit fly body, it induces numerous little eyes to develop there. This result demonstrates that there was a single origin of the eye in an ancestor common to flies and people

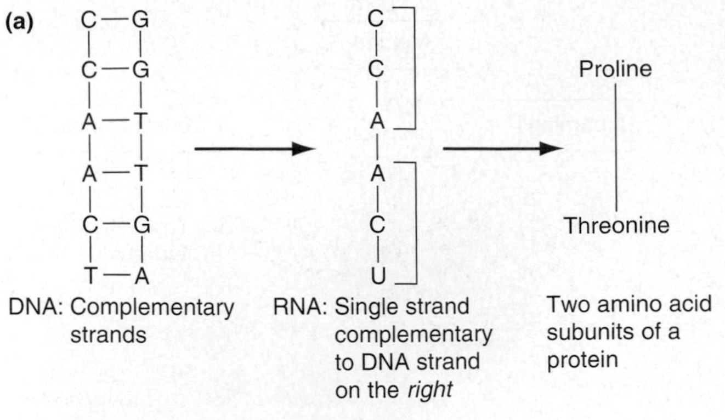

(b)

First letter	Second letter: U	C	A	G	Third letter
U	UUU Phe	UCU Ser	UAU Tyr	UGU Cys	U
U	UUC Phe	UCC Ser	UAC Tyr	UGC Cys	C
U	UUA Leu	UCA Ser	UAA Stop	UGA Stop	A
U	UUG Leu	UCG Ser	UAG Stop	UGG Trp	G
C	CUU Leu	CCU Pro	CAU His	CGU Arg	U
C	CUC Leu	CCC Pro	CAC His	CGC Arg	C
C	CUA Leu	CCA Pro	CAA Gln	CGA Arg	A
C	CUG Leu	CCG Pro	CAG Gln	CGG Arg	G
A	AUU Ile	ACU Thr	AAU Asn	AGU Ser	U
A	AUC Ile	ACC Thr	AAC Asn	AGC Ser	C
A	AUA Ile	ACA Thr	AAA Lys	AGA Arg	A
A	AUG Met	ACG Thr	AAG Lys	AGG Arg	G
G	GUU Val	GCU Ala	GAU Asp	GGU Gly	U
G	GUC Val	GCC Ala	GAC Asp	GGC Gly	C
G	GUA Val	GCA Ala	GAA Glu	GGA Gly	A
G	GUG Val	GCG Ala	GAG Glu	GGG Gly	G

Figure 1.7 RNA is an intermediary in the conversion of DNA information into protein via the genetic code. (a) The linear bases of DNA are copied through molecular complementarity into the linear bases of RNA. The bases of RNA are read three at a time, that is, as triplets, to encode the amino-acid subunits of proteins. **(b)** The genetic code dictionary specifies the relationship between RNA triplets and the amino-acid subunits of proteins. (Note that this table uses three-letter abbreviations of the 20 amino acids, whose names are spelled out in Fig. 1.5.)

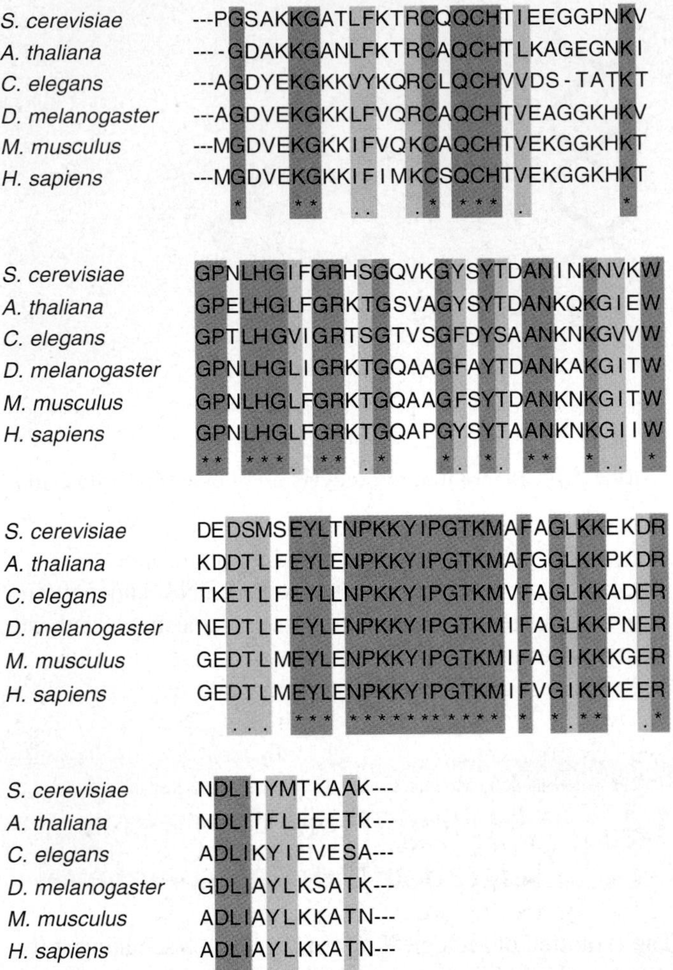

Figure 1.8 Comparisons of gene products in different species provide evidence for the relatedness of living organisms. This chart shows the amino-acid sequence for equivalent portions of the cytochrome C protein in six species: *Saccharomyces cerevisiae* (yeast), *Arabidopsis thaliana* (a weedlike flowering plant), *Caenorhabditis elegans* (a nematode), *Drosophila melanogaster* (the fruit fly), *Mus musculus* (the house mouse), and *Homo sapiens* (humans). Consult Fig. 1.5 for the key to amino-acid names. Cytochrome C functions during cellular respiration and photosynthesis. The most abundant and most stable of the cytochromes, its form and function have been conserved throughout evolution. As the chart shows, there are many sequence similarities among the six types of organisms.

and that, after 600 million years of divergent evolution, both vertebrates and insects still share the same main control switch for initiating eye development. Indeed, the eyes of all multicellular organisms, not just those of vertebrates and insects, may have a common evolutionary origin. Studies of many other genes have shown that the entire developmental program of flies and people uses many of the same genes. These genes have duplicated and evolved divergent functions, but they still retain their ancestral relationship markings.

The utility of the finding of relatedness and unity at all levels of biological information cannot be overstated. It means that in many cases, the experimental manipulation of organisms known as *model organisms* can shed light on complex networks in humans. Model organisms for genetic analysis are amenable to breeding experiments and direct manipulation of their genomes. If genes similar to human genes function in simple model organisms such as fruit flies or bacteria scientists can determine gene function and regulation in these experimentally manipulable organisms and

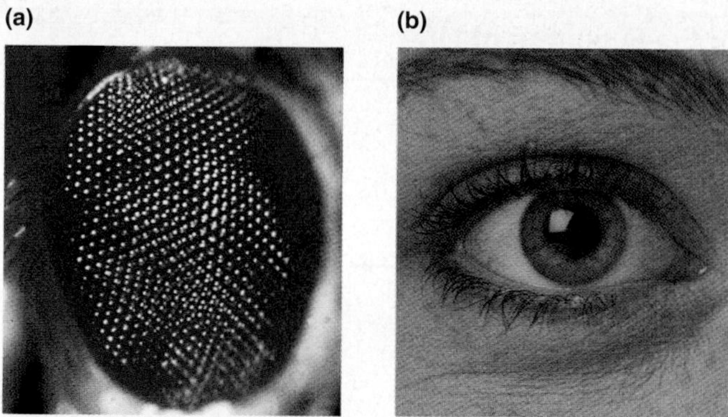

Figure 1.9 The eyes of insects and humans have a common ancestor. **(a)** A fly eye and **(b)** human eye.

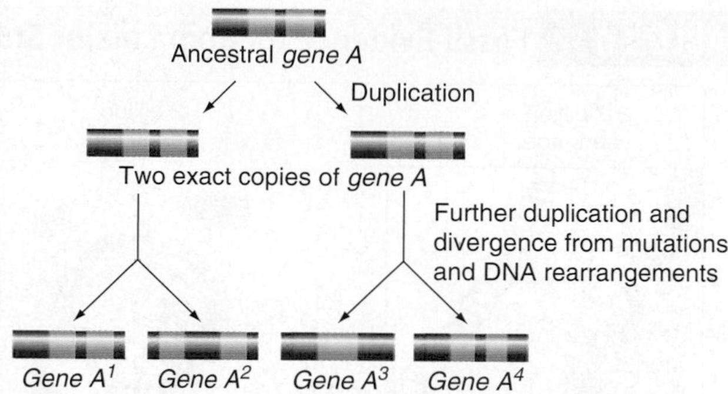

Figure 1.10 How genes arise by duplication and divergence. Duplications of ancestral *gene A* followed by mutations and DNA rearrangements generate a family of related genes. The *dark blue* and *red* bands indicate the different exons of the genes while the *light blue* bands represent introns.

bring these insights to an understanding of the human organism. The same is true of the shared informational pathways such as DNA replication and protein synthesis. You can visit our website at www.mhhe.com/hartwell3 for detailed genetic portraits of five key model organisms: the yeast *S. cerevisiae,* the simple plant known as *A. thaliana,* the roundworm *C. elegans,* the fruit fly *D. melanogaster,* and the house mouse *M. musculus.*

The close relatedness of all living organisms at the molecular level has great significance for an understanding of biology. It makes it possible to combine bits and pieces learned from different organisms into a global understanding of molecular and cellular biology that is valid for all organisms. And even though controlled experimentation with humans is usually impossible, the relatedness of all organisms allows us to learn about human biology from mice, flies, worms, peas, yeast, and other organisms that are accessible to experimentation.

1.5 The Modular Construction of Genomes Has Allowed the Rapid Evolution of Complexity

We have seen that roughly 20,000–30,000 genes direct human growth and development. How did such complexity arise? Recent technical advances have enabled researchers to complete structural analyses of the entire genome of more than 250 organisms. The information obtained reveals that families of genes have arisen by duplication of a primordial gene; after duplication, mutations and rearrangements may cause the two copies to diverge from each other (**Fig. 1.10**). In both mice and humans, for example, five different hemoglobin genes produce five different hemoglobin molecules at successive stages of development, with each protein functioning in a slightly different way to fulfill different needs for oxygen transport. The set of five hemoglobin genes arose from a single primordial gene by several duplications followed by slight divergences in structure.

Duplication followed by divergence underlies the evolution of new genes with new functions. This principle appears to have been built into the genome structure of all eukaryotic organisms. The protein-coding region of most genes is subdivided into as many as 10 or more small pieces (called *exons*), separated by DNA that does not code for protein (called *introns*) as shown in Fig. 1.10. This modular construction facilitates the rearrangement of different modules from different genes to create new combinations during evolution. It is likely that this process of modular reassortment facilitated the rapid diversification of living forms about 570 million years ago (see Fig. 1.10).

The tremendous advantage of the duplication and divergence of existing pieces of genetic information is evident in the history of life's evolution (**Table 1.1**). *Prokaryotic* cells such as bacteria, which do not have a membrane-bounded nucleus, evolved about 3.7 billion years ago; *eukaryotic* cells such as algae, which have a membrane-bounded nucleus, emerged around 2 billion years ago; and multicellular eukaryotic organisms appeared 600–700 million years ago. Then, at about 570 million years ago, within the relatively short evolutionary time of roughly 20–50 million years known as the Cambrian explosion, the multicellular life-forms diverged into a bewildering array of organisms, including primitive vertebrates.

A fascinating question is, since it took eukaryotic cells almost 2 billion years to evolve from prokaryotic cells and multicellular organisms three-quarters of a billion years to evolve from single-celled eukaryotes, how could the multicellular forms achieve such enormous diversity in only 20–50 million years? The answer lies, in part, in the hierarchic organization of the information encoded in chromosomes. Exons are arranged into genes; genes duplicate and diverge to generate multigene families; and multigene families sometimes rapidly expand to gene superfamilies containing hundreds of related genes. In both mouse and human adults, for example, the immune system is encoded

TABLE 1.1 Fossil Evidence for Some Major Stages in the Evolution of Life

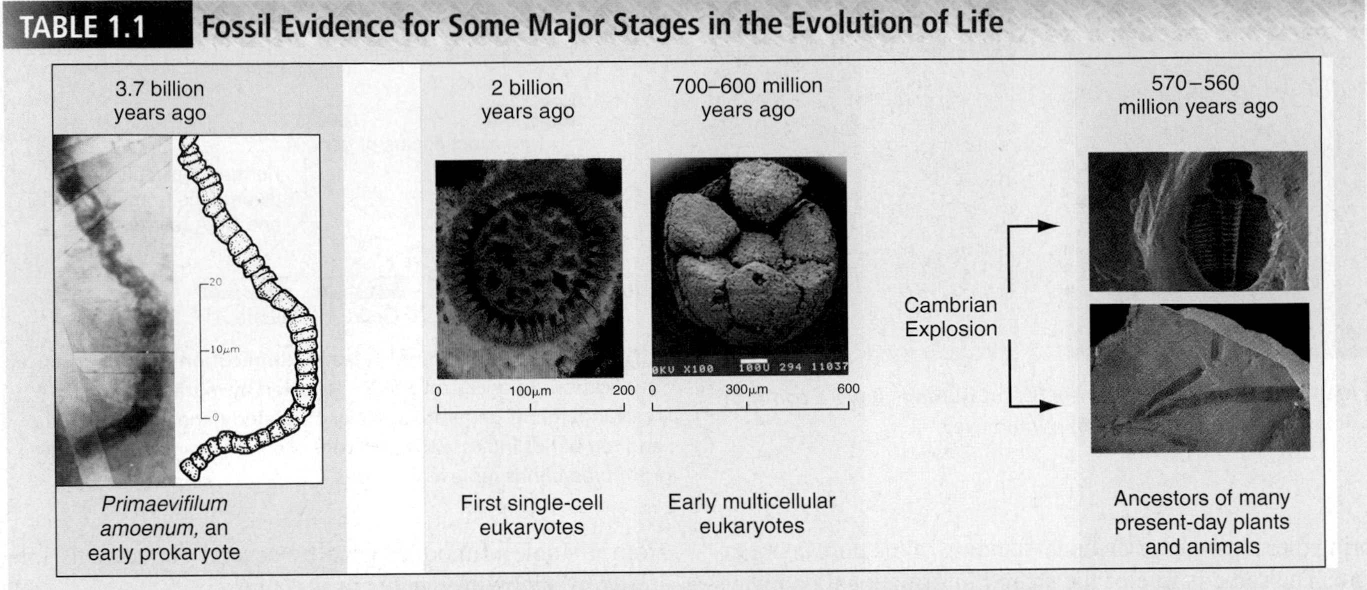

by a gene superfamily composed of hundreds of closely related but slightly divergent genes. With the emergence of each successively larger informational unit, evolution gains the ability to duplicate increasingly complex informational cartridges through single genetic events.

Figure 1.11 Two-winged and four-winged flies. Geneticists converted a contemporary normal two-winged fly to a four-winged insect resembling the fly's evolutionary antecedent. They accomplished this by mutating a key element in the fly's regulatory network. Note the club-shaped halteres behind the wings of the fly at the top.

Probably even more important for the evolution of complexity is the rapid change of regulatory networks that specify how genes behave (that is, when, where, and to what degree they are expressed) during development. For example, the two-winged fly evolved from a four-winged ancestor not because of changes in gene-encoded structural proteins, but rather because of a rewiring of the regulatory network, which converted one pair of wings into two balancing organs known as haltere (**Fig. 1.11**).

1.6 Genetic Techniques Permit the Dissection of Complexity

The complexity of living systems has developed over 4 billion years from the continuous amplification and refinement of genetic information. The simplest bacterial cells contain about 1000 genes that interact in complex networks. Yeast cells, the simplest eukaryotic cells, contain about 6,000 genes. Nematodes (roundworms) and fruit flies, the simplest multicellular organisms, contain roughly 14,000– 19,000 genes; humans may have as many as 30,000 genes. The Human Genome Project, in addition to completing the sequencing of the entire human genome, has sequenced the genomes of *E. coli,* yeast, the nematode, the fruit fly, and the mouse (**Fig. 1.12**). Each of these organisms has provided valuable insights into biology in general and human biology in particular.

With genetic techniques, researchers can dissect the complexity of a genome piece by piece, although the task is daunting. The logic used in genetic dissection is quite simple: inactivate a gene in a model organism and observe the consequences. For example, loss of a gene for visual pigment produces fruit flies with white eyes instead of eyes

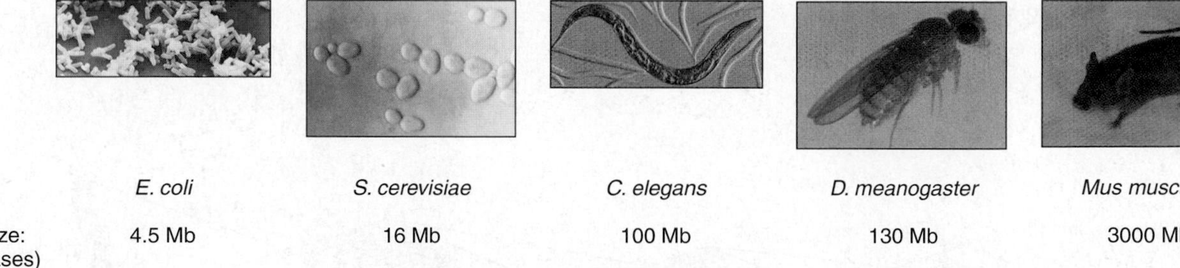

Organism	*E. coli*	*S. cerevisiae*	*C. elegans*	*D. meanogaster*	*Mus musculus*
Genome size: (in megabases)	4.5 Mb	16 Mb	100 Mb	130 Mb	3000 Mb
Number of genes	4500	6200	19,200	13,900	20,000–30,000

Figure 1.12 Five model organisms whose genomes were sequenced as part of the Human Genome Project. The chart indicates genome size in millions of base pairs, or megabases (Mb). It also shows the approximate number of genes for each organism.

of the normal red color. One can thus conclude that the protein product of this gene plays a key role in the development of eye pigmentation. From their study of model organisms, researchers are amassing a detailed picture of the complexity of living systems.

However, even though the power of genetic techniques is astonishing, the complexity of biological systems is difficult to comprehend. We have seen that the human organism carries 20,000–30,000 genes and that each human being arises from the networks of interactions created by these genes and the proteins they encode. Knowing everything there is to know about each of these genes and proteins would not, however, reveal how a human results from a particular ensemble of genes and proteins. For example, the human nervous system is a network of 10^{11} neurons with perhaps 10^{18} connections. The complexity of the system is far too great to be encoded by a simple correspondence between genes and neurons or genes and connections. Moreover, the remarkable properties of the system, such as learning, memory, and personality, do not arise solely from the genes and proteins; network interactions and the environment also play a role. The goal of understanding higher-order processes that arise from interacting networks of genes, proteins, cells, and organs is one of the most challenging aspects of modern biology. Genetics provides useful tools for tackling this challenge, but the concepts and information needed to achieve this understanding are as yet unknown.

The new global tools of genomics—such as high-throughput DNA sequencers, genotypers, and large-scale DNA arrays (also called DNA chips)—have the capacity to analyze thousands of genes rapidly and accurately. These global tools are not specific to a particular system or organism; rather, they can be used to study the genes of all living things.

The DNA chip is a powerful example of a global genomic tool. Individual chips are subdivided into arrays of microscopic blocks that each contain a unique string of DNA units (**Fig. 1.13a**). When a chip is exposed to a complex mixture of fluorescently labeled nucleic acid—such as DNA or RNA from any cell type or sample—the unique string in each microscopic block can bind to and detect a specific complementary sequence. This type of binding is known as *hybridization* (**Fig. 1.13b**). A computer-driven microscope can then analyze the bound sequences of the hundreds of thousands of blocks on the chip, and special software can enter this information into a database (**Fig. 1.13c**).

The potential of DNA chips is enormous for both research and clinical purposes. Already chips with over 400,000 different detectors can provide simultaneous information on the presence or absence of 400,000 discrete DNA or RNA sequences in a complex sample. And they can do it within hours. Here is one example. Now that the sequence of all human genes is known, unique stretches of DNA representing each of the 20,000–30,000 human genes can be placed on a chip and used to determine the complete set of genes copied into RNA in any human cell type at any stage of development or differentiation. Computer-driven comparisons can be used to contrast the genes expressed (that is copied to RNA for translation to protein) in different cell types, for example, in neurons and muscle cells, making it possible to determine which genes of the human genome contribute to the construction of various cell types. Scientists have already created catalogues of the genes expressed in different cell types and have discovered that some genes, called "housekeeping genes," are expressed in nearly all cell types, while other genes are expressed only in certain specialized cells. This knowledge of the relation between particular genes and particular cell types is helping us understand how the cellular specialization necessary for the construction of all human organs arises.

In medicine, clinical researchers have used DNA chip technology to identify genes whose expression increases or decreases when tumor cells are treated with an experimental cancer drug (Fig. 1.13b-c). Changes in the patterns of gene expression may provide clues to the mechanisms by which the drug might inhibit tumor growth. In a related but slightly different application of the same idea, researchers can assess the inherent differences between breast cancers that respond well to a particular drug therapy and those that do not (that is, that recur despite treatment). Knowledge of these patients since microarray analysis of their tumors can predict with considerable accuracy whether a specific drug will be effective against their particular type of cancer.

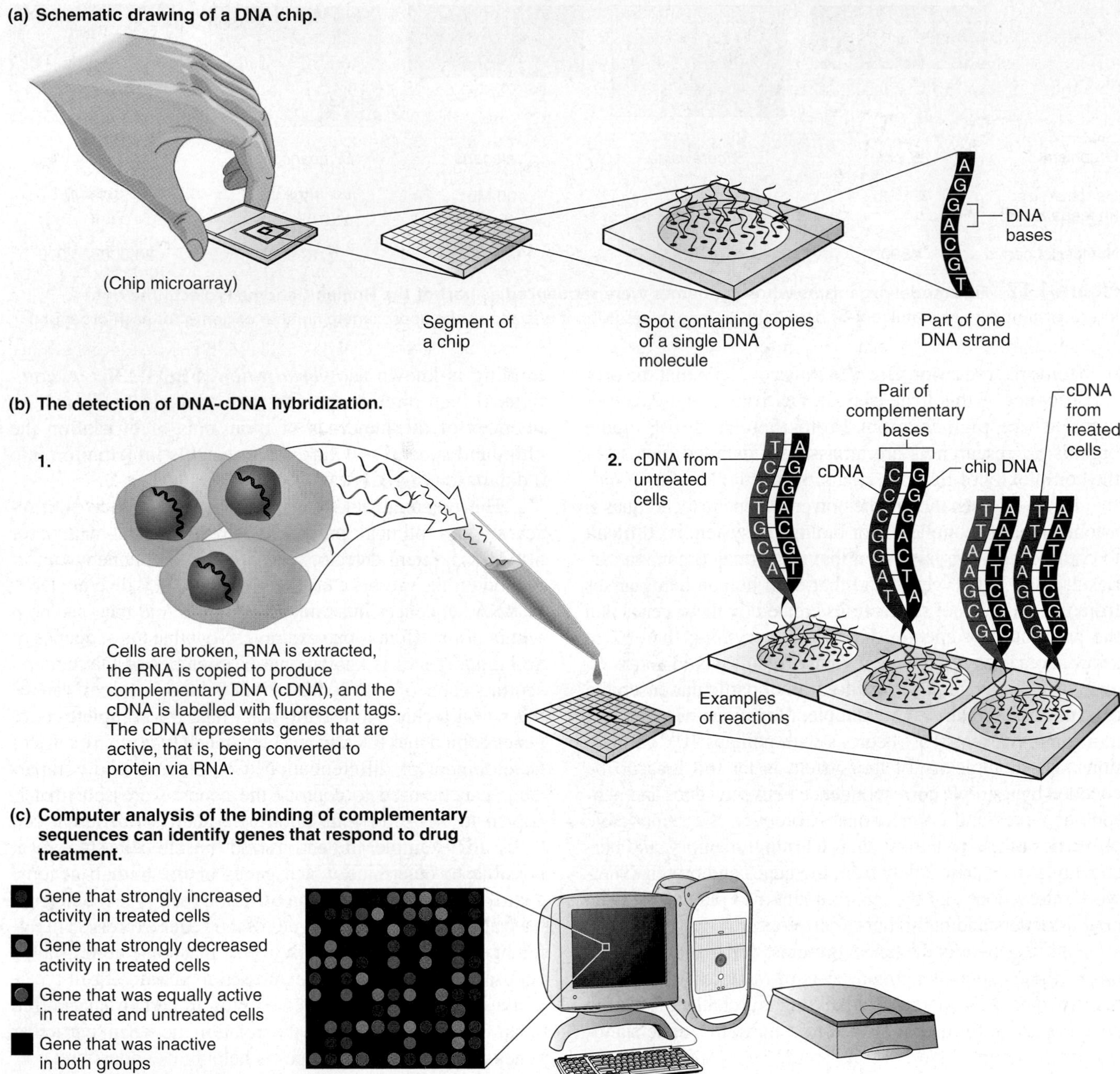
(a) Schematic drawing of a DNA chip.
(Chip microarray)
Segment of a chip
Spot containing copies of a single DNA molecule
DNA bases
Part of one DNA strand
(b) The detection of DNA-cDNA hybridization.
1.
Cells are broken, RNA is extracted, the RNA is copied to produce complementary DNA (cDNA), and the cDNA is labelled with fluorescent tags. The cDNA represents genes that are active, that is, being converted to protein via RNA.
2. cDNA from untreated cells
Pair of complementary bases
cDNA
chip DNA
cDNA from treated cells
Examples of reactions
(c) Computer analysis of the binding of complementary sequences can identify genes that respond to drug treatment.
Gene that strongly increased activity in treated cells
Gene that strongly decreased activity in treated cells
Gene that was equally active in treated and untreated cells
Gene that was inactive in both groups

Figure 1.13 One use of a DNA chip. **(a)** Schematic drawing of the components of a DNA chip. **(b)** 1. Preparing complementary DNA, or cDNA, with a fluorescent tag from the RNA of a group of cells. 2. The hybridization of chip DNA to fluorescent cDNA from untreated and drug-treated cells. **(c)** Computerized analysis of chip hybridizations makes it possible to compare gene activity in any two types of cells.

1.7 Our Focus Is on Human Genetics

In the mid-1990s, a majority of scientists who responded to a survey conducted by *Science* magazine rated genetics as the most important field of science for the next decade. One reason is that the powerful tools of genetics open up the possibility of understanding biology, including human biology, from the molecular level up to the level of the whole organism. In combination with an appreciation of the relatedness of all living organisms, the potential of genetic analysis heralds an era that promises to help reveal more about who we are.

The Human Genome Project, by changing the way we view biology and genetics, has led to a significant paradigm change: the systems approach to biology and medicine. The systems approach seeks to study the relationships of all the elements in a biological system as it undergoes genetic perturbation or biological activation (see Chapter 12). This is a fundamental change from the study of complex systems one gene or protein at a time.

How Human Genetics Is Leading Us Toward Predictive and Preventive Medicine

Over the next 25 years, geneticists will identify hundreds of genes with variations that predispose people to many types of disease: cardiovascular, cancerous, immunological, mental, metabolic. Some mutations will always cause disease; others will only predispose to disease. For example, a change in a specific single DNA base (that is, a change in one DNA unit) in the *β-globin* gene will nearly always cause sickle-cell anemia, a painful, life-threatening condition that leads to severe anemia. By contrast, a mutation in the *breast cancer 1 (BRCA1)* gene has only a 70% chance of causing breast cancer in a woman carrying one copy of the mutation; this conditional state arises because the *BRCA1* gene interacts with environmental factors that affect the probability of activating the cancerous condition and because various forms of other genes modify expression of the *BRCA1* gene. Defining and analyzing the multiple factors contributing to genetic predispositions may be an important element in understanding and designing therapies for some diseases. Physicians may be able to use DNA diagnostics—a collection of techniques for characterizing genes—to analyze an individual's DNA for genes that predispose to some diseases. With this genetic profile, they may be able to write out a probabilistic health history for some medical conditions. Many people will benefit from genetically based diagnoses and forecasts. This will move us into the era of predictive medicine.

As scientists come to understand the complex systems in which disease genes operate, they may be able to design therapeutic drugs to block and/or reverse the effects of mutant genes. If taken before the onset of disease, such drugs could prevent occurrence or minimize symptoms of the gene-based disease. This will usher in the era of preventive medicine. Although the discussion here has focused on genetic conditions rather than infectious diseases, it is possible that ongoing analyses of microbial and human genomes will lead to procedures for controlling the virulence of some pathogens.

The New Scope of Human Genetics and the New Potential of Predictive and Preventive Medicine Intensify the Need to Confront Many Social Issues

Genetics began as a separate biological discipline dedicated to determining the rules governing the frequency of appearance of alternative traits in siblings and other related individuals. At the beginning of the twenty-first century, it has become the central focus and tool in the study of complex biological systems created by the interactions of molecular entities. Although biological information is similar to other types of information from a strictly technical point of view, it is as different as can be in its meaning and impact on individual human beings and human society as a whole. The difference lies in the personal nature of the unique genetic profile carried by each person from birth. Within this basic level of biological information are complex life codes that provide greater or lower susceptibility or resistance to many diseases, as well as greater or lesser potential for the expression of many physiologic, physical, and neurological attributes that distinguish people from each other. Until now, almost all this information has remained hidden away. But if research continues at its present pace, in less than a decade it will become possible to read a large part of a person's genetic profile, and with this information will come the power to make some limited predictions about future possibilities and risks.

As we will see in many of the Genetics and Society boxes throughout this book, society can use genetic information not only to help people but also to restrict their lives (for example, by denying insurance or employment). We believe that just as our society respects an individual's right to privacy in other realms, it should also respect the privacy of an individual's genetic profile and work against all types of discrimination.

Another issue raised by the potential for detailed genetic profiles is the interpretation or misinterpretation of that information. Without accurate interpretation, the information becomes useless at best and harmful at worst. Proper interpretation of genetic information requires some understanding of statistical concepts such as risk and probability. To help people understand these concepts, widespread education in this area will be essential. Since the media play an enormous role in the lives of most people, public education could begin with media reports on new genetic findings that are well reasoned and accurate. It will also be essential to bring kindergarten through high-school education up-to-date so that children can learn the concepts and implications of modern human biology as a science of information.

Yet another pressing issue concerns the regulation and control of the new technology. With the sequencing of the entire human genome, government funds appropriated for the sequencing project can be redirected toward analyzing genetic variation among humans as well as various aspects of genome structure and organization. The question of whether the government should establish guidelines for the use of genetic and genomic information, reflecting society's social and ethical values, remains in open debate.

To many people, the most frightening potential of the new genetics is the development of technology that can alter or add to the genes present within the *germ line* (reproductive cell precursors) of human embryos. This technology (referred to as "transgenic technology" in scientific discourse and "genetic engineering" in public discussions) has become routine in hundreds of laboratories working with various animals other than humans.

Some people caution that developing the power to alter our own genomes is a step we should not take, arguing that if genetic information and technology are misused (as they certainly have been in the past), the consequences could be horrific. Attempts to use genetic information for social purposes were prevalent in the early twentieth century, leading to enforced sterilization of individuals thought to be inferior, to laws that prohibited interracial marriage, and to laws prohibiting immigration of certain ethnic groups. The scientific basis of these actions has been thoroughly discredited. Others agree that we must not repeat the mistakes of the past, but warn that if the new technologies could help children and adults lead healthier, happier lives, we need to think very carefully about whether the reasons for objecting outright to their use are valid. Most agree that the biological revolution we are living through will have a greater impact on human society than any technological revolution of the past and that education and public debate are the key to preparing for the consequences of this revolution.

The focus on human genetics in this book looks forward into the new era of biology and genetic analysis. As we gain increasingly sophisticated knowledge about the human genetic makeup, it will not only become possible to cure human diseases that now resist therapy; it will also become possible to have an impact on our own evolution (through, for example, germ-line alterations). We have seen that these new possibilities raise serious moral and ethical issues that will demand wisdom and humility. It is in the hope of educating young people for the moral and ethical challenges awaiting the next generation that we write this book.

Connections

Genetics, the study of biological information, is also the study of the DNA and RNA molecules that store, replicate, transmit, and evolve information for the construction of proteins. With their extraordinary diversity of structure and function, proteins generate the complex and adaptive behaviors of all living organisms. At the molecular level, all living things are closely related. As a result, observations of model organisms as different as yeast and mice can provide insights into general biological principles as well as human biology.

Remarkably, more than 75 years before the discovery of DNA, Gregor Mendel, an Augustinian monk working in what is now Brno in the Czech Republic, delineated the basic laws of gene transmission with no knowledge of the molecular basis of heredity. He accomplished this by following simple traits, such as flower or seed color, that come in two discrete forms, such as white and purple or yellow and green, through several generations. He used the pea plant (*Pisum sativum*) as his experimental organism and set up carefully controlled matings between plants that differed in one or a few traits. We now know that his findings apply to all sexually reproducing organisms. Chapter 2 describes Mendel's studies and insights, which became the foundation of the field of genetics.

On Our Website

www.mhhe.com/hartwell3
Chapter 1

Annotated Suggested Readings and Links to Other Websites

- Additional information about DNA
- Conversion of DNA to RNA to protein
- More about systems biology and predictive/preventive medicine

Mendel's Breakthrough: Patterns, Particles, and Principles of Heredity

Chapter 2

A quick glance at an extended family portrait is likely to reveal children who resemble one parent or the other or who look like a combination of the two, with perhaps wavy hair from the father, a broad nose from the mother, and a skin color in between the two parents' (**Fig. 2.1**). Some children, however, look unlike any of the assembled relatives and more like a throwback to a great, great grandparent. What causes the similarities and differences of appearance and the skipping of generations?

Although Mendel's laws can predict the probability that an individual will have a particular genetic makeup, the chance meeting of particular male and female gametes determines an individual's actual genetic fate.

The answers lie in our **genes,** the basic units of biological information, and in **heredity,** the way genes transmit biochemical, anatomical, and behavioral traits from parents to offspring. Each of us starts out as a single fertilized egg cell that develops, by division and differentiation, into a mature adult made up of 10^{14} (a hundred trillion) specialized cells, including muscle cells capable of contraction, brain cells structured for rapid communication, red blood cells tailored for transporting oxygen, and hair cells that carry pigment for black, brown, blond, or flaming red hair. By current estimates, only about 25,000 genes control this amazing developmental process. Passed from parents to offspring through egg and sperm, these genes underlie the formation of every heritable trait. Such traits are as diverse as the shape of your hairline, the tendency to bald as you age, the timbre of your voice, the way you clasp your hands, even your susceptibility to heart disease and certain cancers. And they all run in families in predictable patterns that impose some possibilities and exclude others.

Genetics, the science of heredity, pursues a precise explanation of the biological structures and mechanisms that determine inheritance. Geneticists seek to identify genes, to learn how they determine particular traits, and to understand how genes work together to create a person, a plant, or a protozoan. In some instances, the relationship between gene and trait is remarkably simple. A change in a single gene, for example, results in sickle-cell anemia by causing construction of a defective hemoglobin molecule, the oxygen-carrying protein in red blood cells; when oxygen is in short supply, red blood cells carrying the abnormal hemoglobin become sickle-shaped and clog small blood vessels. In other instances, the correlations between genes and traits are bewilderingly complex. An example is the genetic basis of facial features, in which many genes determine a large number of molecules that interact in a variety of ways to generate the combination we recognize as a friend's face.

Gregor Mendel (1822–1884; **Fig. 2.2**), a stocky, bespectacled Augustinian monk and expert plant breeder, discovered the basic principles of genetics in the mid–nineteenth century. He published his findings in 1866, just seven years after

Figure 2.1 A family portrait. The extended family shown here includes members of four generations.

Darwin's *On the Origin of Species* appeared in print. Mendel lived and worked in Brunn, Austria (now Brno in the Czech Republic), a nineteenth-century center of learning in the sciences and humanities, situated in the rich agricultural valley of the province of Moravia. Here he examined the inheritance of such clear-cut alternative traits in pea plants as purple versus white flowers or yellow versus green seeds. In so doing, he discovered why some of these traits disappeared in one generation and then reappeared in another. By rigorously analyzing the patterns of transmission through generations, he inferred genetic laws that allowed him to make verifiable predictions about which traits would appear, disappear, and then reappear in which generations. Simple and straightforward, Mendel's laws are based on the hypothesis that observable traits such as seed color are determined by independent units of inheritance not visible to the naked eye. We now call these units *genes.* The concept of the gene continues to change as research deepens and refines our understanding of genetic phenomena. Today, a gene is recognized as a region of DNA that encodes a specific protein or a particular type of RNA. In the beginning, however, it was an abstraction—an imagined particle with no physical features, whose function was to control a visible trait—that Mendel proposed as the explanation for the results of his plant breeding experiments.

Figure 2.2 Gregor Mendel. Photographed around 1862 holding one of his experimental plants. His work formed the basis for our understanding and continued exploration of the science of genetics.

We begin our study of genetics with a detailed look at what Mendel's laws are and how they were discovered. In subsequent chapters, we discuss logical extensions to these laws and describe how Mendel's successors grounded the abstract concept of hereditary units (genes) in an actual biological molecule (DNA). Today, geneticists integrate the analytical tools of Mendel with modern molecular techniques as they continue to probe the nature of the hereditary material and examine exactly how it is passed from parent to offspring, how it acts in organisms to produce the visible and invisible traits that define individuals, and how it evolves over time.

Four general themes emerge from our detailed discussion of Mendel's work. The first is that variation, as expressed in alternative forms of a trait (a high-pitched voice or a low one, a green pea or a yellow one), is widespread in nature. This genetic diversity provides the raw material for the continuously evolving variety of life we see around us. Second, observable variation is essential for following genes. If all the traits of all offspring resembled their parents', Mendel would have had no basis for discerning and analyzing patterns of transmission. Third, variation is not distributed solely by chance; rather, it is inherited according to genetic laws that explain why like begets both like and unlike. Dogs beget other dogs; pea plants beget pea plants; and people, people. But there are hundreds of breeds of dogs, and even within a breed—Labrador retrievers, for instance—two black dogs could have a litter of black, brown, and golden puppies (**Fig. 2.3**). Mendel's insights help explain why this is so. Fourth, the laws Mendel discovered about heredity apply equally well to all sexually reproducing organisms, from protozoans to peas to people.

Figure 2.3 Like begets like and unlike. A Labrador retriever with her litter of pups.

Our presentation of Mendelian genetics examines

- The background: The historical puzzle of inheritance and how Mendel's innovative experimental approach helped resolve it.
- The work itself: Genetic analysis according to Mendel, including a discussion of Mendel's seminal experiments and analytical tools.
- The medical significance: Two comprehensive examples of Mendelian inheritance in humans.

2.1 Background: The Historical Puzzle of Inheritance

There are several steps to understanding genetic phenomena: the careful observation over time of groups of organisms, such as human families, herds of cattle, or fields of corn or tomatoes; the rigorous analysis of systematically recorded information gleaned from these observations; and the development of a theoretical framework that can explain the origin of these phenomena and their relationships. In the mid–nineteenth century, Gregor Mendel became the first person to combine the three approaches and reveal the true basis of heredity. For many thousands of years before that, the elementary selective breeding of domesticated plants and animals, with no guarantee of what a particular mating would produce, was the only genetic practice.

Artificial Selection Was the First Applied Genetic Technique

A rudimentary use of genetics was the driving force behind a key transition in human civilization, allowing hunters and gatherers to settle in villages and survive as shepherds and farmers. Even before recorded history, people practiced applied genetics as they domesticated plants and animals for their own uses. From a large litter of semitamed wolves, for example, they sent the savage and the misbehaving to the stew pot while sparing the alert sentries and friendly companions for longer life and eventual mating. As a result of this **artificial selection**—purposeful control over mating by choice of parents for the next generation—the domestic dog *(Canis domesticus)* slowly arose from ancestral wolves *(Canis lupus)*. The oldest bones identified indisputably as dog (and not wolf) are a skull excavated from a 20,000-year-old Alaskan settlement. Many millennia of evolution guided by artificial selection have produced massive Great Danes and minuscule Chihuahuas as well as hundreds of other modern breeds of dog. The amazing range of size, shape, and behavior bears witness to the enormous amount of genetic variation in ancient canines and the degree of differentiation that artificial selection can produce. By 10,000 years ago, people had used this same kind of genetic manipulation to develop economically valuable herds of reindeer, sheep, goats, pigs, and cattle that produced life-sustaining meat, hides, and wools.

Farmers also carried out artificial selection of plants, storing seed from the healthiest and tastiest individuals for the next planting, eventually producing strains that grew better, produced more, and were easier to cultivate and harvest. In this way, scrawny weedlike plants gradually, with human guidance, turned into rice, wheat, barley, lentils, and dates in Asia; corn, squash, tomatoes, potatoes, and peppers in North and South America; yams, peanuts, and gourds in Africa. Later, plant breeders recognized male and female organs in plants and carried out artificial pollination. An Assyrian frieze carved in the ninth century B.C., pictured in **Fig. 2.4,** is the oldest known visual record of this kind of genetic experiment. It depicts priests brushing the flowers of female date palms with selected male pollen. By this method of artificial selection, early practical geneticists produced several hundred varieties of dates, each differing in specific observable qualities, such as the fruit's size, color, or taste. A 1929 botanical survey of three oases in Egypt turned up 400 varieties of date-bearing palms, twentieth-century evidence of the natural and artificially generated variation among these trees.

Figure 2.4 The earliest known record of applied genetics. In this 2800-year-old Assyrian relief from the Northwest Palace of Assurnasirpal II (883–859 B.C.), priests wearing bird masks artificially pollinate flowers of female date palms.

The Puzzle of Passing on Desirable Traits

In 1822, the year of Mendel's birth, what people in Moravia understood about the basic principles of heredity was not much different from what the people of ancient Assyria had understood. By the nineteenth century, plant and animal breeders had created many strains in which offspring often carried a prized parental trait. Using such strains, they could produce plants or animals with desired characteristics for food and fiber, but they could not always predict why a valued trait would sometimes disappear and then reappear in only some offspring. For example, selective breeding practices had resulted in valuable flocks of merino sheep producing large quantities of soft, fine wool, but at the 1837 annual meeting of the Moravian Sheep Breeders Society, one breeder's dilemma epitomized the state of the art. He possessed an outstanding ram that would be priceless "if its advantages are inherited by its offspring," but "if they are not inherited, then it is worth no more than the cost of wool, meat, and skin." Which would it be? According to the meeting's recorded minutes, current breeding practices offered no definite answers. In his concluding remarks at this sheep-breeders meeting, the Abbot Cyril Napp pointed to a possible way out. He proposed that breeders could improve their ability to predict what traits would appear in the offspring by finding the answers to three basic questions: What is inherited? How is it inherited? What is the role of chance in heredity?

This is where matters stood in 1843 when 21-year-old Gregor Mendel entered the monastery in Brno, presided over by the same Abbot Napp. Although Mendel was a monk trained in theology, he was not a rank amateur in science, discovering new and exciting things in a vacuum. Moravia was a center of learning and scientific activity, providing access to the latest techniques and literature; Mendel, for instance, was able to acquire a copy of Darwin's *On the Origin of Species* shortly after it was translated into German in 1863. As a youth, Mendel had been a student of such outstanding caliber that he was allowed to enter the monastery to pursue his studies, while a brother-in-law took over management of the family farm. Abbot Napp, recognizing Mendel's intellectual abilities, sent him to the University of Vienna—all expenses paid—where he prescribed his own course of study based on his particular interests. His choices were an unusual mix: physics (with the renowned Christian Doppler, discoverer of the "Doppler effect"), mathematics, chemistry, botany, paleontology, and plant physiology. The cross-pollination of ideas from several disciplines would play a significant role in Mendel's discoveries. One year after he returned to Brno, he began his series of seminal genetic experiments. **Figure 2.5** shows where Mendel worked and the microscope he used.

A New Experimental Approach

Before Mendel, many misconceptions clouded people's thinking about heredity. Two of the prevailing errors were particularly misleading. The first was that one parent contributes most to an offspring's inherited features; Nicolaas Hartsoeker, one of the earliest microscopists, contended in 1694 that it was the male, by way of a fully formed homunculus inside the sperm (**Fig. 2.6**). Another deceptive notion was the concept of *blended inheritance,* the idea that parental traits become mixed and forever changed in the offspring, as when blue and yellow pigment merge to green on a painter's palette. The theory of blending may have grown out of a natural tendency for

(a)

(b)

Figure 2.5 Mendel's garden and microscope. (a) Gregor Mendel's garden was part of his monastery's property in Brno. **(b)** Mendel used this microscope to examine plant reproductive organs and to pursue his interests in natural history.

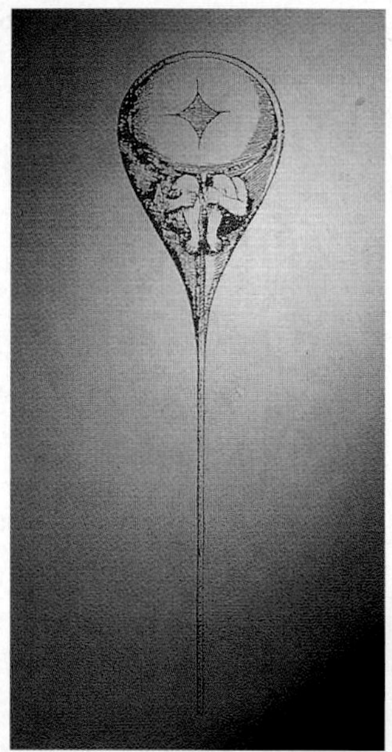

Figure 2.6 The homunculus: A misconception. Well into the nineteenth century, many prominent microscopists believed they saw a fully formed, miniature fetus crouched within the head of a sperm.

parents to see a combination of their own traits in their offspring. While blending could account for children who look like a combination of their parents, it could not explain obvious differences between biological brothers and sisters nor the persistence of variation within extended families.

The experiments Mendel devised would lay these myths to rest by providing precise, verifiable answers to the three questions Abbot Napp had raised almost 15 years earlier: What is inherited? How is it inherited? What is the role of chance in heredity? A key component of Mendel's breakthrough was the way he set up his experiments.

What did Mendel do differently from those who preceded him? First, he chose the garden pea *(Pisum sativum)* as his experimental organism (**Fig. 2.7a** and **b**). Peas grew well in Brno and, with male and female organs in the same flower, they were normally self-fertilizing. In **self-fertilization** (or *selfing*), both egg and pollen come from the same plant. The particular anatomy of pea flowers, however, makes it easy to prevent self-fertilization and instead to **cross-fertilize** (or *cross*) two individuals by brushing pollen from one plant onto a female organ of another plant, as illustrated in Fig. 2.7c. Peas offered yet another advantage. For each successive generation, Mendel could obtain large numbers of individuals within a relatively short growing season. By comparison, if he had worked with sheep, each mating would have generated only a few offspring and the time between generations would have been several years.

Second, Mendel examined the inheritance of clear-cut alternative forms of particular traits—purple versus white flowers, yellow versus green peas. Using such "either-or" traits, he could distinguish and trace unambiguously the transmission of one or the other observed characteristic, because there were no intermediate forms. (The opposite of these so-called *discrete traits* are *continuous traits,* such as height and skin color in humans. Continuous traits show many intermediate forms.)

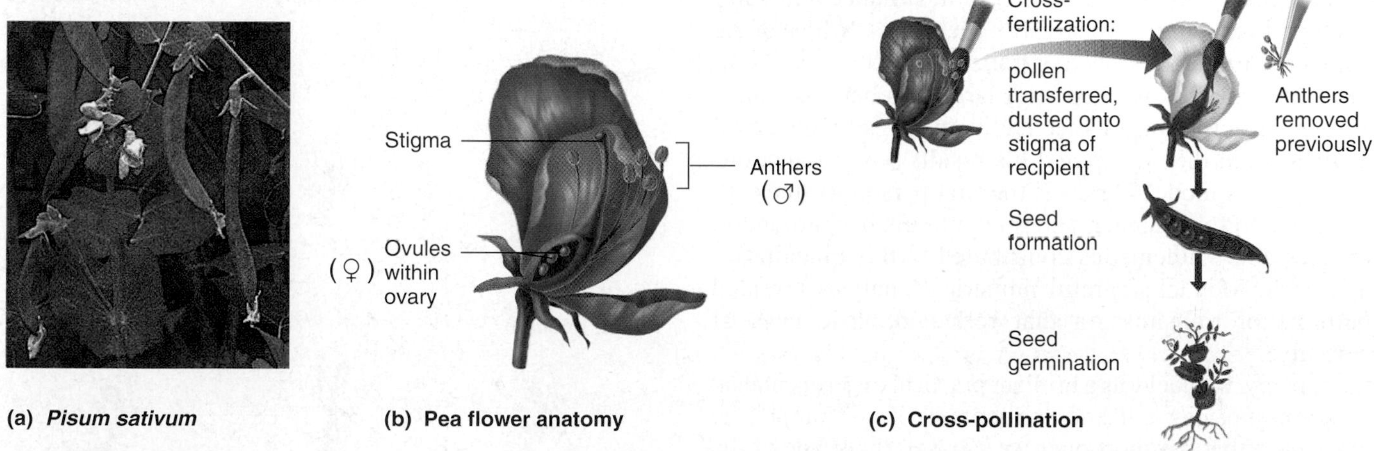

Figure 2.7 Mendel's experimental organism: The garden pea. **(a)** Pea plants with white flowers. **(b)** Pollen is produced in the anthers. Mature pollen lands on the stigma, which is connected to the ovary (which becomes the pea pod). After landing, the pollen grows a tube that extends through the stigma to one of the ovules (immature seeds), allowing fertilization to take place. **(c)** To prevent self-fertilization, breeders remove the anthers from the female parents (here, the white flower) before the plant produces mature pollen. They then transfer pollen with a paintbrush from the anthers of the male parent (here, the purple flower) to the stigma of the female parent. Each fertilized ovule becomes an individual pea (mature seed) that can grow into a new pea plant. All of the peas produced from one flower are encased in the same pea pod, but these peas form from different pollen grains and ovules.

Third, Mendel collected and perpetuated lines of peas that bred true. Matings within such **pure-breeding lines** produce offspring carrying specific parental traits that remain constant from generation to generation. Mendel observed his pure-breeding lines for up to eight generations. Plants with white flowers always produced offspring with white flowers; plants with purple flowers produced only offspring with purple flowers. Mendel called constant but mutually exclusive, alternative traits, such as purple versus white flowers or yellow versus green seeds, "antagonistic pairs" and settled on seven such pairs for his study (**Fig. 2.8**). In his experiments, he not only perpetuated pure-breeding stocks for each member of a pair, but he also cross-fertilized pairs of plants to produce **hybrids,** offspring of genetically dissimilar parents, for each pair of antagonistic traits. Figure 2.8 shows the appearance of the hybrids he studied.

Fourth, Mendel, the expert plant breeder, carefully controlled his matings, going to great lengths to ensure that the progeny he observed really resulted from the specific fertilizations he intended. Thus he painstakingly prevented the intrusion of any foreign pollen and assured self- or cross-pollination as the experiment demanded. Not only did this allow him to carry out controlled breedings of selected traits, he could also make **reciprocal crosses.** In such crosses, he reversed the traits of the male and female parents, thus controlling whether a particular trait was transmitted via the egg cell within the ovule or via a sperm cell within the pollen. For example, he could use pollen from a purple flower to fertilize the eggs of a white flower and also use pollen from a white flower to fertilize the eggs of a purple flower. Because the progeny of these reciprocal crosses were similar, he disproved the idea that one parent contributes more to the next generation, demonstrating instead that they contribute equally to inheritance. "It is immaterial to the form of the hybrid," he wrote, "which of the parental types was the seed or pollen plant."

Fifth, Mendel worked with large numbers of plants, counted all offspring, subjected his findings to numerical analysis, and then compared his results with predictions based on his models. He was the first person to study inheritance in this manner, and no doubt his background in physics and mathematics contributed to this quantitative approach. Mendel's careful numerical analysis revealed patterns of transmission that reflected basic laws of heredity.

Finally, Mendel was a brilliant practical experimentalist. When comparing tall and short plants, for example, he made sure that the short ones were out of the shade of the tall ones so their growth would not be stunted. Eventually he focused on certain traits of the pea seeds themselves, such as their color or shape, rather than on traits of the plants arising from the seeds. In this way, he could observe many more individuals from the limited space of the monastery garden, and he could evaluate the results of a cross in a single growing season.

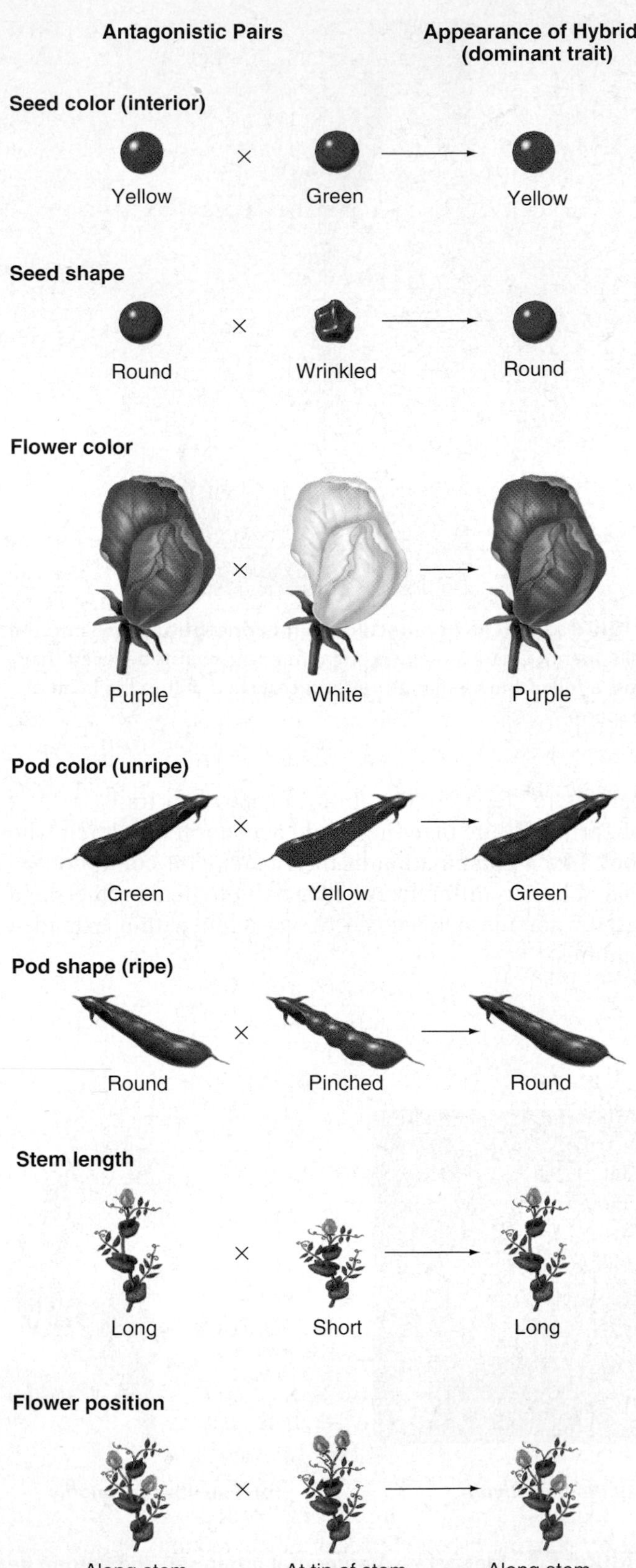

Figure 2.8 The mating of parents with antagonistic traits produces hybrids. Note that each of the hybrids for the seven antagonistic traits studied by Mendel resembles only one of the parents. The parental trait that shows up in the hybrid is known as the "dominant" trait.

In short, Mendel purposely set up a simplified "black-and-white" experimental system and then figured out how it worked. He did not look at the vast number of variables that determine the development of a prize ram nor at the origin of differences between species. Rather, he looked at discrete traits that came in two mutually exclusive forms and asked questions that could be answered by observation and computation.

2.2 Genetic Analysis According to Mendel

In early 1865 at the age of 43, Gregor Mendel presented a paper entitled "Experiments on Plant Hybrids" before the Natural Science Society of Brno. Despite its modest heading, it was a scientific paper of uncommon clarity and simplicity that summarized a decade of original observations and experiments. In it Mendel describes in detail the transmission of visible characteristics in pea plants, defines unseen but logically deduced units (genes) that determine why these visible traits appear, and analyzes the function of these discrete determinants in simple mathematical terms to reveal previously unsuspected principles of heredity.

Published the following year, the paper would eventually become the cornerstone of modern genetics. Its stated purpose was to see whether there is a "generally applicable law governing the formation and development of hybrids." Let us examine its insights.

Monohybrid Crosses Reveal Units of Inheritance and the Law of Segregation

Once he had isolated pure-breeding lines for several sets of characteristics, Mendel carried out a series of matings between individuals that differed in only one trait, such as seed color or stem length. In each cross, one parent carries one form of the trait, and the other parent carries an alternative form of the same trait. **Figure 2.9** illustrates one such mating. Early in the spring of 1854, for example, Mendel planted pure-breeding green peas and pure-breeding yellow peas and allowed them to grow into the **parental (P) generation.** Later that spring when the plants had flowered, he dusted the female stigma of "green-pea" plant flowers with pollen from "yellow-pea" plants. He also performed the reciprocal cross, dusting "yellow-pea" plant stigmas with "green-pea" pollen. In the fall, when he collected and separately analyzed the progeny peas of these reciprocal crosses, he found that in both cases, the peas were all yellow. These yellow peas, progeny of the P generation, were the beginning of what we now call the **first filial (F_1)** generation. To learn whether the green trait had disappeared entirely or remained intact but hidden in these F_1 yellow peas, Mendel planted them and allowed the F_1 plants that grew from them to self-fertilize. Such experiments involving hybrids for a single trait are often called **monohybrid crosses.** He then harvested and counted the peas of the resulting **second filial (F_2)** generation, progeny of the F_1 generation. Among the progeny of one series of F_1 self-fertilizations, there were 6022 yellow and 2001 green F_2 peas, an almost perfect ratio of 3 yellow : 1 green. F_1 plants derived from the reciprocal of the original cross produced a similar ratio of yellow to green F_2 progeny.

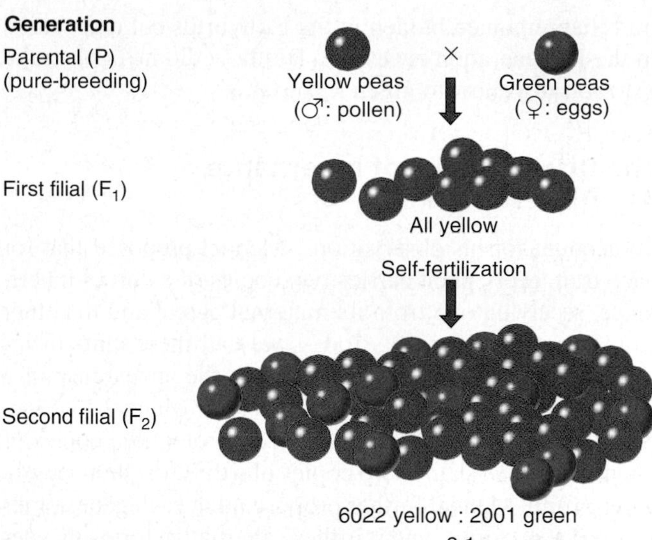

Figure 2.9 Analyzing a monohybrid cross. Cross-pollination of pure-breeding parental plants produces F_1 hybrids, all of which resemble one of the parents. Self-pollination of F_1 plants gives rise to an F_2 generation with a 3:1 ratio of individuals resembling the two original parental types. For simplicity, we do not show the plants that produce the peas or that grow from the planted peas.

Reappearance of the Recessive Trait Disproves Blending

These results were irrefutable evidence that blending had not occurred. If it had, the information necessary to make green peas would have been irretrievably lost in the F_1 hybrids. Instead, the information remained intact and was able to direct the formation of 2001 green peas actually harvested from the second filial generation. These green peas were indistinguishable from their green grandparents. Mendel concluded that there must be two types of yellow peas: those that breed true like the yellow peas of the P generation, and those that can yield some green offspring like the yellow F_1 hybrids. This second type somehow contains latent information for green peas. He called the trait that appeared in all the F_1 hybrids—in this case, yellow seeds—**dominant** (see Fig. 2.8) and the "antagonistic" green-pea trait that remained hidden in the F_1 hybrids but reappeared in the F_2 generation **recessive.** But how did he explain the 3:1 ratio of yellow to green F_2 peas?

The Discrete Units of Inheritance Are Alleles of Genes

To account for his observations, Mendel proposed that for each trait, every plant carries two copies of a unit of inheritance, receiving one from its maternal parent and the other from the paternal parent. Today, we call these units of inheritance *genes.* Each unit determines the appearance of a specific characteristic. The pea plants in Mendel's collection had two copies of a gene for seed color, two copies of another for seed shape, two copies of a third for stem length, and so forth. Mendel further proposed that each gene comes in alternative forms, and it is these alternative forms that determined the contrasting characteristics he was studying. Today we call the alternative forms of a single gene **alleles.** The gene for pea color, for example, has yellow and green alleles; the gene for pea shape has round and wrinkled alleles. (The Fast Forward box "Genes Encode Proteins," pp. 22–23, describes the biochemical and molecular mechanisms by which different alleles determine different traits.) In Mendel's monohybrid crosses, one allele of each gene was dominant, the other recessive. In the P generation, one parent carried two dominant alleles for the trait under consideration; the other parent, two recessive alleles. The hybrids of the F_1 generation carried one dominant and one recessive allele for the trait. Individuals having two different alleles for a single trait are **monohybrids.**

The Law of Segregation Explains How Genes Are Transmitted

If a plant has two copies of every gene, how does it pass only one copy of each to its progeny? And how then do the offspring end up with two copies of these same genes, one from each parent? Mendel drew on his background in plant physiology and answered these questions in terms of the two biological mechanisms behind reproduction: gamete formation and the random union of gametes at fertilization. **Gametes** are the specialized cells—eggs within the ovules of the female parent and sperm cells within the pollen grains—that carry genes between generations. He imagined that during the formation of pollen and eggs, the two copies of each gene in the parent separate (or *segregate*) so that each gamete receives only one allele for each trait (**Fig. 2.10a**). Thus, each egg and each pollen grain receives only one allele for pea color (either yellow or green). At fertilization, pollen with one or the other allele unites at random with an egg carrying one or the other allele, restoring the two copies of the gene for each trait in the fertilized egg, or **zygote** (Fig. 2.10b). If the pollen carries yellow and the egg green, the result will be a hybrid yellow pea like the F_1 monohybrids that resulted when pure-breeding parents

(a) The two alleles for each trait separate during gamete formation.

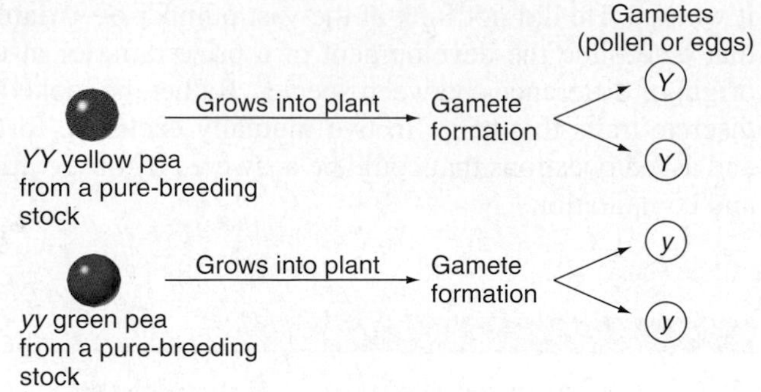

(b) Two gametes, one from each parent, unite at random at fertilization.

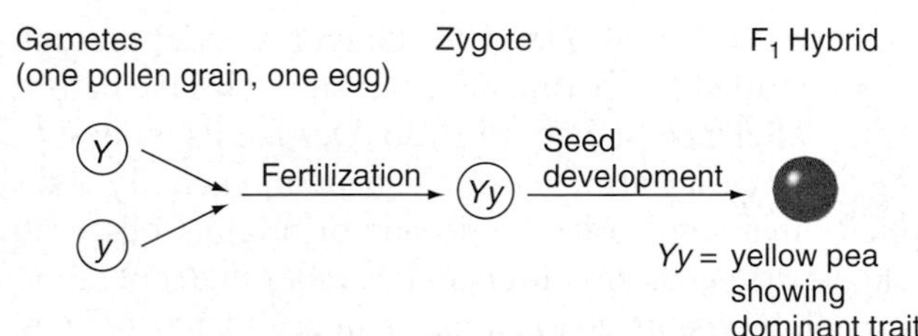

Figure 2.10 The law of segregation. (a) The two identical alleles of pure-breeding plants separate (segregate) during gamete formation. As a result, each pollen grain or egg carries only one of each pair of parental alleles. **(b)** Cross-pollination and fertilization between pure-breeding parents with antagonistic traits result in F_1 hybrid zygotes with two different alleles, one from each type of parent, for each trait. For the seed color gene, a *Yy* hybrid zygote will develop into a yellow pea.

of opposite types mated. If the yellow pollen unites with a yellow egg, the result will be a yellow pea that grows into a pure-breeding plant like those of the P generation that produced only yellow peas. And finally, if pollen carrying the allele for green peas fertilizes a green-carrying egg, the progeny will be a pure-breeding green pea.

Mendel's **law of segregation** encapsulates this general principle of heredity: *The two alleles for each trait separate (segregate) during gamete formation, and then unite at random, one from each parent, at fertilization.* Throughout this book, the term **segregation** refers to such *equal segregation* in which one allele, and only one allele, of each gene goes to each gamete. Note that the law of segregation makes a clear distinction between organisms, whose cells have two copies of each gene, and gametes, which bear only a single allele of each gene.

Figure 2.11 shows a simple way of visualizing the results of the segregation and random union of alleles during gamete formation and fertilization. Mendel invented a system of symbols that allowed him to analyze all his crosses in the same way. He designated dominant alleles with a

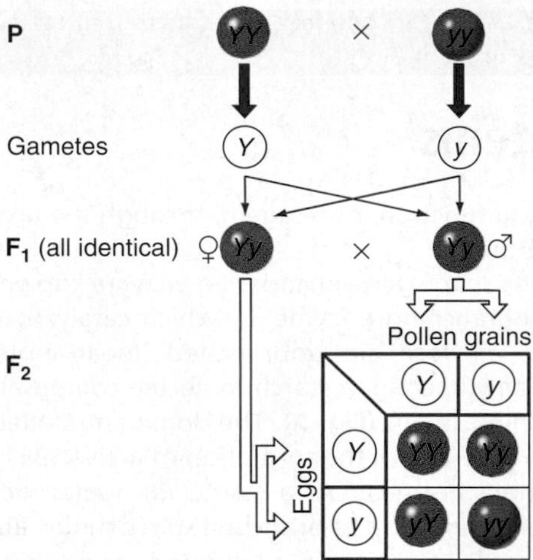

Figure 2.11 The Punnett square: Visual summary of a cross. This Punnett square illustrates the combinations that can arise when an F_1 hybrid undergoes gamete formation and self-fertilization. The F_2 generation should have a 3:1 ratio of yellow to green peas.

capital *A, B,* or *C* and recessive ones with a lowercase *a, b,* or *c.* Modern geneticists have adopted this convention for naming genes in peas and many other organisms, but they often add some reference to the trait in question—a *Y* for yellow or an *R* for round. Throughout this book, we present gene symbols in italics. In Fig. 2.11, we denote the dominant yellow allele by a capital *Y* and the recessive green allele by a lower case *y.* The pure-breeding plants of the parental generation are either *YY* (yellow peas) or *yy* (green peas). The *YY* parent can produce only *Y* gametes, the *yy* parent only *y* gametes. You can see from the diagram why every cross between *YY* and *yy* produces exactly the same result—a *Yy* hybrid—no matter which parent (male or female) donates which particular allele.

Next, to visualize what happens when the *Yy* hybrids self-fertilize, we set up a Punnett square (named after British mathematician Reginald Punnett, who introduced it in 1906; Fig. 2.11). The square provides a simple and convenient method for tracking the kinds of gametes produced as well as all the possible combinations that might occur at fertilization. As the Punnett square shows, each hybrid produces two kinds of gametes, *Y* and *y,* in a ratio of 1:1. Thus, half the pollen and half the eggs carry *Y,* the other half *y.* At fertilization, 1/4 of the progeny will be *YY,* 1/4 *Yy,* 1/4 *yY,* and 1/4 *yy.* Since the gametic source of an allele (egg or pollen) for the traits Mendel studied had no influence on the allele's effect, *Yy* and *yY* are equivalent. This means that 1/2 of the progeny are yellow *Yy* hybrids, 1/4 *YY* true-breeding yellows, and 1/4 true-breeding *yy* greens. The diagram illustrates how the segregation of alleles during gamete formation and the random union of egg and pollen at fertilization can produce the 3:1 ratio of yellow to green that Mendel observed in the F_2 generation.

Mendel's Results Reflect Basic Rules of Probability

Though you may not have realized it, the Punnett square illustrates two simple rules of probability that are central to the analysis of genetic crosses. These rules predict the likelihood that an event will occur. The **product rule** states that the probability of two or more *independent events* occurring together is the *product* of the probabilities that each event will occur by itself.

With independent events

Probability of event 1 *and* event 2 =

Probability of event 1 × probability of event 2

Consecutive coin tosses are obviously independent events; a heads in one toss neither increases nor decreases the probability of a heads in the next toss. If you toss two coins at the same time, the results are also independent events. A heads for one coin neither increases nor decreases the probability of a heads for the other coin. Thus, the probability of a given combination is the product of their independent probabilities. For example, the probability that both coins will turn up heads is

$$1/2 \times 1/2 = 1/4$$

Similarly, the formation of egg and pollen are independent events; in a hybrid plant, the probability is 1/2 that a given gamete will carry *Y* and 1/2 that it will carry *y.* Since fertilization happens at random, the probability that a particular combination of maternal and paternal alleles will occur simultaneously in the same zygote is the product of the independent probabilities of these alleles being packaged in egg and sperm. Thus, to find the chance of a *Y* egg (formed as the result of one event) uniting with a *Y* sperm (the result of an independent event), you simply multiply 1/2 × 1/2 to get 1/4. This is the same fraction of *YY* progeny seen in the Punnett square of Fig. 2.11, which demonstrates that the Punnett square is simply another way of depicting the product rule.

While we can describe the moment of random fertilization as the simultaneous occurrence of two independent events, we can also say that two different fertilization events are mutually exclusive. For instance, if *Y* combines with *Y,* it cannot also combine with *y* in the same zygote. A second rule of probability, the **sum rule,** states that the probability of either of two such *mutually exclusive events* occurring is the *sum* of their individual probabilities.

With mutually exclusive events

Probability of event 1 *or* event 2 =

Probability of event 1 + probability of event 2

To find the likelihood that an offspring of a *Yy* hybrid self-fertilization will be a hybrid like the parents, you add 1/4 (the probability of maternal *Y* uniting with paternal *y*) and 1/4 (the probability of the mutually exclusive event

Genes Encode Proteins

Genes determine traits as disparate as pea shape and the inherited human disease of cystic fibrosis by encoding the proteins that cells produce and depend on for structure and function. As early as 1940, investigators had uncovered evidence suggesting that some genes determine the formation of enzymes, proteins that catalyze specific chemical reactions. But it was not until 1991, 126 years after Mendel published his analysis of seven pairs of observable traits in peas, that a team of British geneticists identified the gene for pea shape and pinpointed how it prescribes a seed's round or wrinkled contour through the enzyme it determines. About the same time, medical researchers in the United States identified the cystic fibrosis gene and discovered how a mutant allele causes unusually sticky mucous secretions and a susceptibility to respiratory infections and digestive malfunction, once again, through the protein the gene determines.

The pea shape gene encodes an enzyme known as SBE1 (for starch-branching enzyme 1), which catalyzes the conversion of amylose, an unbranched linear molecule of starch, to amylopectin, a starch molecule composed of several branching chains (**Fig. A**). The dominant *R* allele of the pea shape gene causes the formation of active SBE1 enzyme that functions normally. As a result, *RR* homozygotes produce a high proportion of branched starch molecules, which allow the peas to maintain a rounded shape. In contrast, the enzyme determined by the recessive *r* allele is abnormal and does not function effectively. In homozygous recessive *rr* peas, where there is less starch conversion and more of the linear, unbranched starch, sucrose builds up. The excess

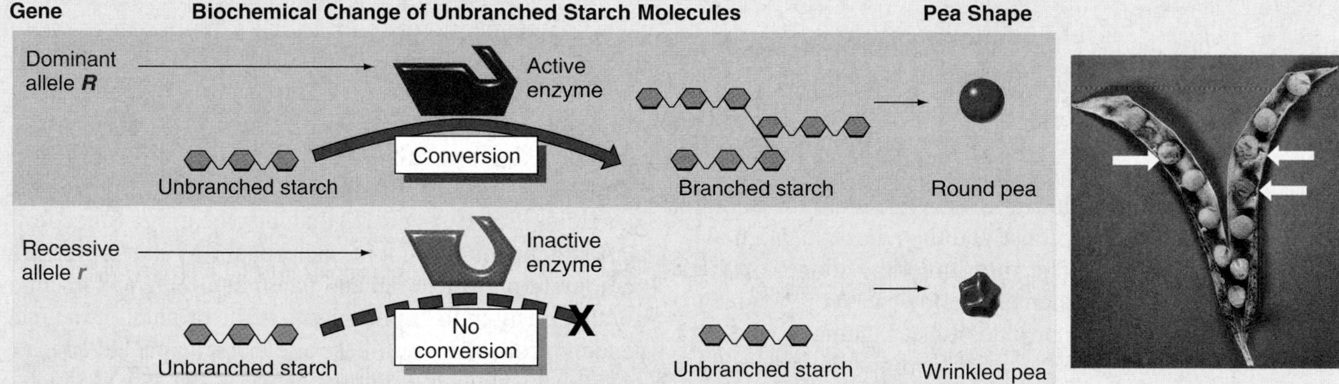

Figure A Round and wrinkled peas: How one gene determines an enzyme that affects pea shape. The *R* allele of the pea shape gene directs the synthesis of an enzyme that converts unbranched starch to branched starch, indirectly leading to round pea shape. The *r* allele of this gene determines an inactive form of the enzyme, leading to a buildup of linear, unbranched starch that ultimately causes seed wrinkling. The photograph at right shows two pea pods, each of which contains wrinkled (arrows) and round peas; the ratio of round to wrinkled in these two well-chosen pods is 9:3 (or 3:1).

where paternal *Y* unites with maternal *y*) to get 1/2, again the same result as in the Punnett square. In another use of the rule of the sum, you could predict the ratio of yellow to green F_2 progeny. The fraction of F_2 peas that will be yellow is the sum of 1/4 (the event producing *YY*) plus 1/4 (the mutually exclusive event generating *Yy*) plus 1/4 (the third mutually exclusive event producing *yY*) to get 3/4. The remaining 1/4 of the F_2 progeny will be green. So the yellow-to-green ratio is 3/4 to 1/4, or more simply, 3:1.

Further Crosses Confirm Ratios Predicted by the Law of Segregation

Although Mendel's law of segregation explains the data from his pea crosses, he performed additional experiments to rule out other possibilities. In the rigorous check of his hypotheses illustrated in **Fig. 2.12,** he allowed self-fertilization of all the plants in the F_2 generation and counted the types of F_3 progeny. What kind of plants would you anticipate from a planting of green peas of the F_2 generation, if the law of segregation does, in fact, apply? Mendel found that the plants that developed from F_2 green peas all produced only F_3 green peas, and when the resulting F_3 plants self-fertilized, the next generation also produced green peas (not shown). This is what we (and Mendel) would expect of pure-breeding lines carrying two copies of the recessive allele. The yellow peas were a different story. When Mendel allowed 518 F_2 plants that developed from yellow peas to self-fertilize, he observed that 166, roughly 1/3 of the total, were pure-breeding yellow

sucrose modifies osmotic pressure, causing water to enter the young seeds. As the seeds mature, they lose water, shrink, and wrinkle. The single dominant allele in *Rr* heterozygotes apparently produces enough of the normal enzyme to prevent wrinkling. Although the biochemistry may seem complex, the genetic process is straightforward. A specific gene determines a specific enzyme whose activity affects pea shape. Knowledge of how SBE1 works to determine pea shape provides a physical reality for Mendel's concept of hereditary units.

The human disease of cystic fibrosis (CF) was first described in 1938, but doctors and scientists did not understand the biochemical mechanism that produced the serious respiratory and digestive malfunctions associated with the disease. As a result, treatments could do little more than relieve some of the symptoms, and most CF sufferers died before the age of 30. In 1989, molecular geneticists identified and cloned the cystic fibrosis gene (see Chapters 9 and 10 for specific techniques and procedures). Once they had located the gene, they could find the protein it encodes and clarify how a defect in just one protein can cause a cascade of biochemically determined symptoms. They could also work toward a cure.

The cystic fibrosis gene determines a protein that forges a channel through the cell membrane (**Fig. B**). Known as the cystic fibrosis transmembrane conductance regulator (CFTR), the protein regulates the flow of chloride ions into and out of the cell. The normal allele of this gene produces a CFTR protein that correctly regulates the back-and-forth exchange of ions, which, in turn, determines the cell's osmotic pressure and the flow of water through the cell membrane. In people with cystic fibrosis, however, the cells carry two recessive alleles that produce only an abnormal form of the CFTR protein. The abnormal protein cannot be inserted into the cell membranes, so patients lack functional CFTR chloride channels. Since chloride ions do not flow in and out of the cells as required, the control of osmotic pressure breaks down, the cells retain water, and a thick, dehydrated mucus builds up outside the cells. In cells lining the airways and the ducts of secretory organs such as the pancreas, this single biochemical defect in an ion channel produces clogging and blockages that result in respiratory and digestive malfunction.

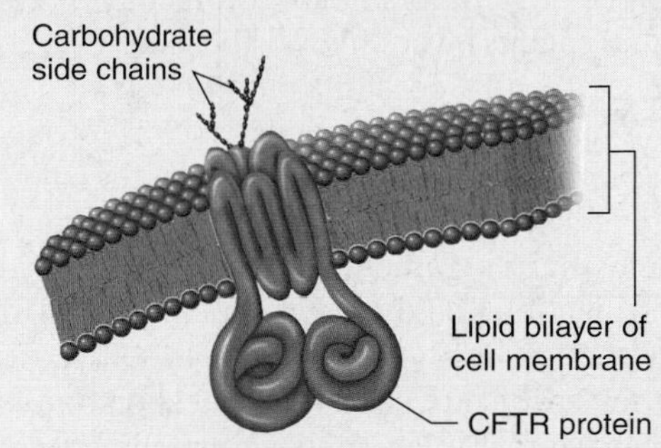

Figure B The cystic fibrosis gene encodes a cell membrane protein. A model of the normal CFTR protein that regulates the passage of chloride ions through the cell membrane. A small change in the gene that codes for CFTR results in an altered protein that prevents proper flow of chloride ions, leading to the varied symptoms of cystic fibrosis.

Cloning of the cystic fibrosis gene brought not only a protein-based explanation of disease symptoms but also the promise of a cure. In the early 1990s, medical researchers successfully used the cloned cystic fibrosis gene to relieve disease symptoms in mice. They did this by placing the normal allele of the gene into respiratory tissue, which could then produce a functional CFTR protein. Such encouraging results in these small mammals suggested that in the not-too-distant future, gene therapy might bestow relatively normal health on people suffering from this once life-threatening genetic disorder. Unfortunately, human trials of CFTR gene therapy have not yet achieved clear success.

Although Mendel defined genes as independent units of heredity that reside in cells, he did not have the biochemical information even to speculate on precisely what they are and how they work. Chapters 6–8 describe the chemical composition of genes and clarify how they carry information for making proteins.

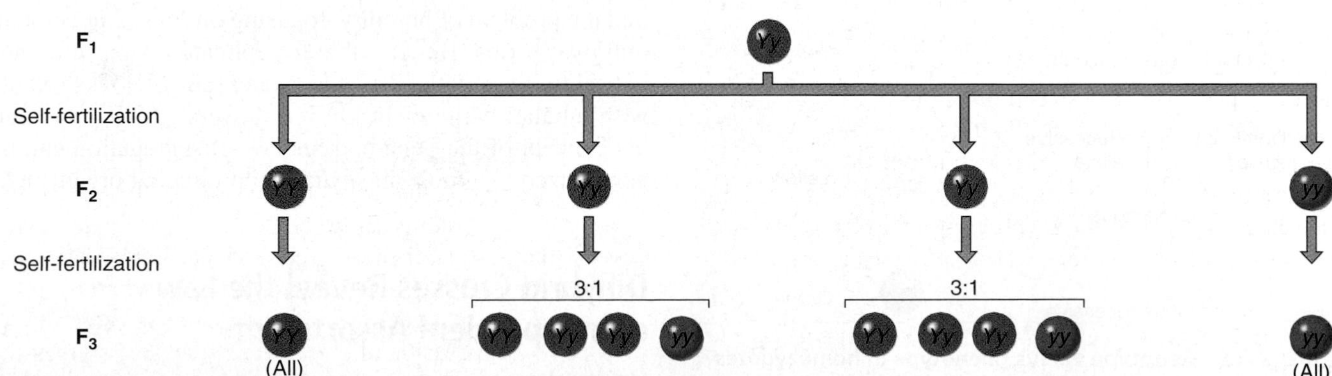

Figure 2.12 Yellow F_2 peas are of two types: Pure breeding and hybrid. The distribution of a pair of contrasting alleles (*Y* and *y*) after two generations of self-fertilization. The homozygous individuals of each generation breed true, whereas the hybrids do not.

through several generations, but the other 352 (2/3 of the total yellow F_2 plants) were hybrids because they gave rise to yellow and green F_3 peas in a ratio of 3:1.

It took Mendel years to conduct such rigorous experiments on seven pairs of pea traits, but in the end, he was able to conclude that the segregation of dominant and recessive alleles during gamete formation and their random union at fertilization could indeed explain the 3:1 ratios he observed whenever he allowed hybrids to self-fertilize. His results, however, raised yet another question, one of some importance to future plant and animal breeders. Plants showing a dominant trait, such as yellow peas, can be either pure-breeding (*YY*) or hybrid (*Yy*). How can you distinguish one from the other? For self-fertilizing plants, the answer is simply to observe the appearance of the next generation. But how would you distinguish pure-breeding from hybrid individuals showing the dominant trait in species that do not self-fertilize?

Testcrosses Establish Genotype

Before describing Mendel's answer, we need to define a few more terms. An observable characteristic, such as yellow or green pea seeds, is a **phenotype,** while the actual pair of alleles present in an individual is its **genotype.** A *YY* or a *yy* genotype is called **homozygous,** because the two copies of the gene that determine the particular trait in question are the same. In contrast, a genotype with two different alleles for a trait is **heterozygous;** in other words, it is a hybrid for that trait (**Fig. 2.13**). An individual with a homozygous genotype is a **homozygote;** one with a heterozygous genotype is a **heterozygote.** Note that the phenotype of a heterozygote (that is, of a hybrid) defines which allele is dominant: because *Yy* peas are yellow, the yellow allele *Y* is dominant to the *y* allele for green. If you know the genotype and the dominance relation of the alleles, you can accurately predict the phenotype. The reverse is not true,

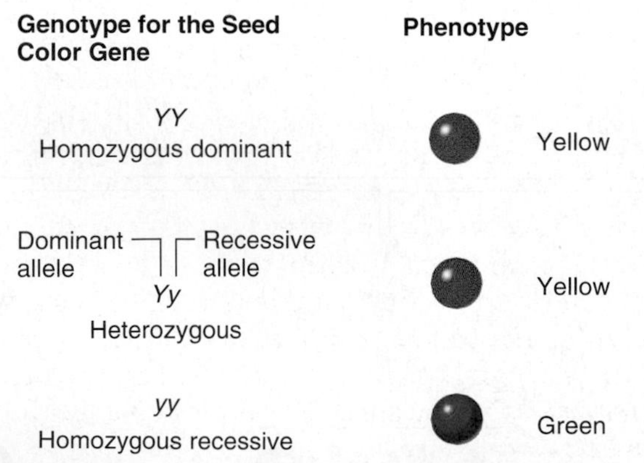

Figure 2.13 Genotype versus phenotype in homozygotes and heterozygotes. The relationship between genotype and phenotype with a pair of contrasting alleles where one allele (*Y*) shows complete dominance over the other (*y*).

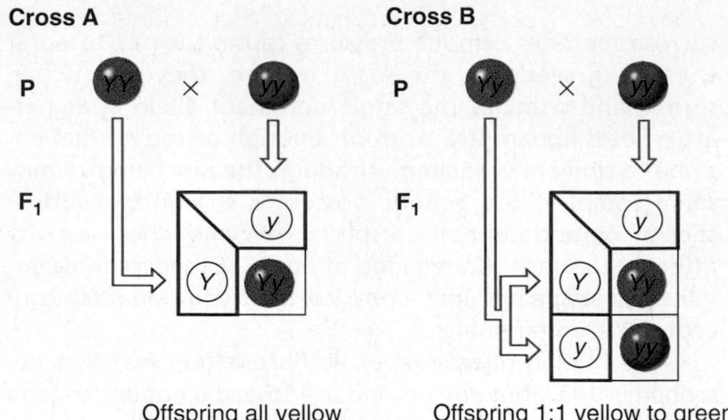

Figure 2.14 How a testcross reveals genotype. An individual of unknown genotype, but dominant phenotype, is crossed with a homozygous recessive. If the unknown genotype is homozygous, all progeny will exhibit the dominant phenotype, as shown in *cross A.* If the unknown genotype is heterozygous, half the progeny will exhibit the dominant trait, half the recessive trait, as shown in *cross B.*

however, because some phenotypes can derive from more than one genotype. For example, the phenotype of yellow peas can result from either the *YY* or the *Yy* genotype.

With these distinctions in mind, we can look at the method Mendel devised for deciphering the unknown genotype, we'll call it *Y*–, responsible for a dominant phenotype; the dash represents the unknown second allele, either *Y* or *y*. This method, called the **testcross,** is a mating in which an individual showing the dominant phenotype, for instance, a *Y*– plant grown from a yellow pea, is crossed with an individual expressing the recessive phenotype, in this case a *yy* plant grown from a green pea. As the Punnett squares in **Fig. 2.14** illustrate, if the dominant phenotype in question derives from a homozygous *YY* genotype, all the offspring of the testcross will show the dominant yellow phenotype. But if the dominant parent of unknown genotype is a heterozygous hybrid (*Yy*), 1/2 of the progeny will be yellow peas, the other half green. In this way, the testcross establishes the genotype behind a dominant phenotype, resolving any uncertainty.

As we mentioned earlier, Mendel deliberately simplified the problem of heredity, focusing on traits that come in only two forms. He was able to replicate his basic monohybrid findings with corn, beans, and four-o'clocks (plants with tubular, white or bright red flowers). As it turns out, his concept of the gene and his law of segregation can be generalized to almost all sexually reproducing organisms.

Dihybrid Crosses Reveal the Law of Independent Assortment

Having determined from monohybrid crosses that genes are inherited according to the law of segregation. Mendel turned his attention to the simultaneous inheritance of two or more apparently unrelated traits in peas. He asked how two pairs of alleles would segregate in a **dihybrid** individual, that is, in a plant that is heterozygous for two genes at the same time. To construct such a dihybrid, he mated true-breeding plants grown from yellow round peas (*YY RR*) with true-breeding plants grown from green wrinkled peas (*yy rr*). From this cross he obtained a dihybrid F_1 generation (*Yy Rr*) showing only the two dominant phenotypes, yellow and round (**Fig. 2.15**). He then allowed these F_1 dihybrids to self-fertilize to produce the F_2 generation. Mendel could not predict the outcome of this mating. Would all the F_2 progeny be **parental types** that looked like either the original yellow round parent or the green wrinkled parent? Or would some new combinations of phenotypes occur that were not seen in the parental lines, such as yellow wrinkled or green round peas? New phenotypic combinations like these are called **recombinant types.** When Mendel counted the F_2 generation of one experiment, he found 315 yellow round peas, 101 yellow wrinkled, 108 green round, and 32 green wrinkled. There were, in fact, yellow wrinkled and green round recombinant phenotypes, providing evidence that some shuffling of alleles had taken place.

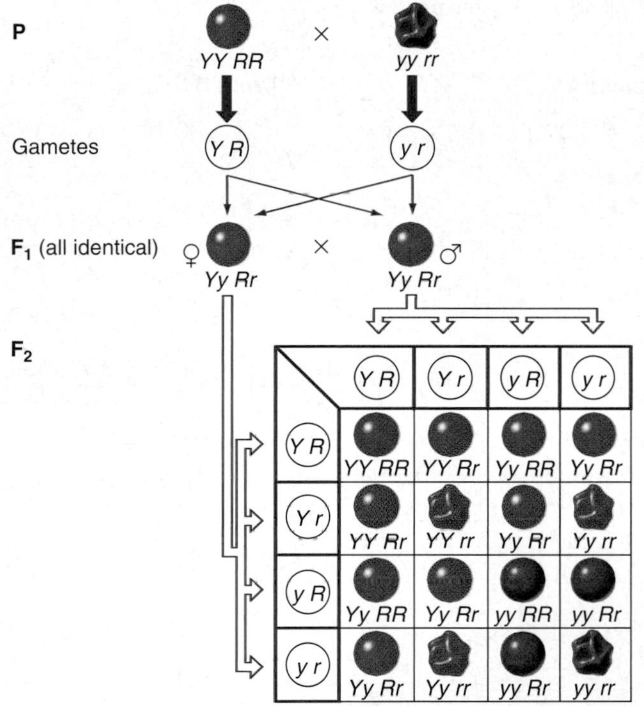

Type	Genotype	Phenotype	Number	Phenotypic Ratio
Parental	*Y– R–*	yellow round	315	9/16
Recombinant	*yy R–*	green round	108	3/16
Recombinant	*Y– rr*	yellow wrinkled	101	3/16
Parental	*yy rr*	green wrinkled	32	1/16

Ratio of yellow (dominant) to green (recessive) = 12:4 or 3:1

Ratio of round (dominant) to wrinkled (recessive) = 12:4 or 3:1

Figure 2.15 A dihybrid cross produces parental types and recombinant types. In this dihybrid cross, pure-breeding parents (P) produce a genetically uniform generation of F_1 dihybrids. Self-pollination or cross-pollination of the F_1 plants yields the characteristic F_2 phenotypic ratio of 9:3:3:1.

From the observed ratios, Mendel inferred the biological mechanism of that shuffling—the **independent assortment** of gene pairs during gamete formation. Because the genes for pea color and for pea shape assort independently, the allele for pea shape in a *Y* carrying gamete could with equal likelihood be either *R* or *r*. Thus, the presence of a particular allele of one gene, say, the dominant *Y* for pea color, provides no information whatsoever about the allele of the second gene. Each dihybrid of the F_1 generation can therefore make four kinds of gametes: *Y R, Y r, y R,* and *y r.* In a large number of gametes, the four kinds will appear in an almost perfect ratio of 1:1:1:1, or put another way, roughly 1/4 of the eggs and 1/4 of the pollen will contain each of the four possible combinations of alleles. That "the different kinds of germinal cells [eggs or pollen] of a hybrid are produced on the average in equal numbers" was yet another one of Mendel's incisive insights.

At fertilization then, in a mating of dihybrids, 4 different kinds of eggs can combine with any one of 4 different kinds of pollen, producing a total of 16 possible zygotes. Once again, a Punnett square is a convenient way to visualize the process. If you look at the square in Fig. 2.15, you will see that some of the 16 potential allelic combinations are identical. In fact, there are only nine different genotypes—*YY RR, YY Rr, Yy RR, Yy Rr, yy RR, yy Rr, YY rr, Yy rr,* and *yy rr*—because the source of the alleles (egg or pollen) does not make any difference. If you look at the combinations of traits determined by the nine genotypes, you will see only four phenotypes—yellow round, yellow wrinkled, green round, and green wrinkled—in a ratio of 9:3:3:1. If, however, you look at just pea color or just pea shape, you can see that each trait is inherited in the 3:1 ratio predicted by Mendel's law of segregation. In the Punnett square, there are 12 yellows for every 4 greens and 12 rounds for every 4 wrinkleds. In other words, the ratio of each dominant trait (yellow or round) to its antagonistic recessive trait (green or wrinkled) is 12:4, or 3:1. This means that the inheritance of the gene for pea color is unaffected by the inheritance of the gene for pea shape, and vice versa.

The preceding analysis became the basis of Mendel's second general genetic principle, the **law of independent assortment:** *During gamete formation, different pairs of alleles segregate independently of each other* (**Fig. 2.16**). The independence of their segregation and the subsequent random union of gametes at fertilization determine the phenotypes observed. Using the product rule for assessing the probability

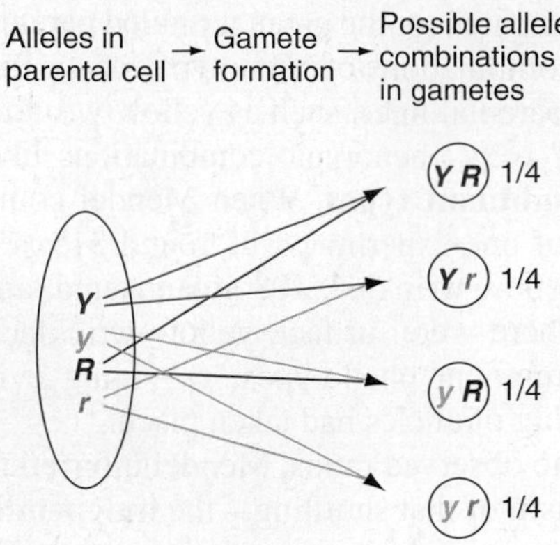

Figure 2.16 The law of independent assortment. In a dihybrid cross, each pair of alleles assorts independently during gamete formation. In the gametes, *Y* is equally likely to be found with *R* or *r* (that is, *Y R* = *Y r*); the same is true for *y* (that is, *y R* = *y r*). As a result, all four possible types of gametes (*Y R*, *Y r*, *y R*, and *y r*) are produced in equal frequency among a large population of gametes.

of independent events, you can see mathematically how the 9:3:3:1 phenotypic ratio observed in a dihybrid cross derives from two separate 3:1 phenotypic ratios. If the two sets of alleles assort independently, the yellow-to-green ratio in the F_2 generation will be 3/4 : 1/4, and likewise, the round-to-wrinkled ratio will be 3/4 : 1/4. To find the probability that two independent events such as yellow and round will occur simultaneously in the same plant, you multiply as follows:

Probability of yellow round = $3/4 \times 3/4 = 9/16$

Probability of yellow wrinkled = $3/4 \times 1/4 = 3/16$

Probability of green round = $1/4 \times 3/4 = 3/16$

Probability of green wrinkled = $1/4 \times 1/4 = 1/16$

Thus, in a population of F_2 plants, there will be a 9:3:3:1 phenotypic ratio of yellow round to yellow wrinkled to green round to green wrinkled.

A convenient way to keep track of the probabilities of each potential outcome in a genetic cross is to construct a **branched-line diagram** (**Fig. 2.17**), which shows all the possibilities for each gene in a sequence of columns. In Fig. 2.17, the first column shows the two possible pea color phenotypes; and the second column demonstrates that each pea color can occur with either of two pea shapes. Again, the 9:3:3:1 ratio of phenotypes is apparent.

An understanding of dihybrid crosses has many applications. Suppose, for example, that you work for a wholesale nursery, and your assignment is to grow pure-breeding plants guaranteed to produce yellow round peas. How would you proceed? One answer would be to plant the peas produced from a dihybrid cross that have the desired yellow round phenotype. Only one out of nine of such progeny—those grown from peas with a *YY RR* genotype—will be appropriate for your uses. To find these plants, you could subject each yellow round candidate to a testcross for genotype with a green wrin-

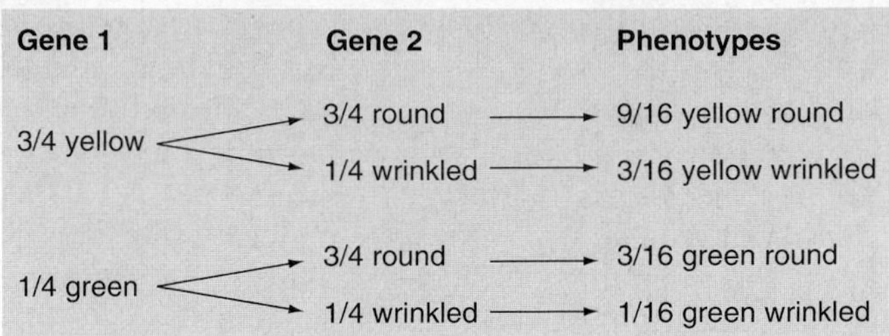

Figure 2.17 Following crosses with branched-line diagrams. A branched-line diagram, which uses a series of columns to track every gene in a cross, provides an organized overview of all possible outcomes. This branched-line diagram of a dihybrid cross generates the same phenotypic ratios as the Punnett square in Fig. 2.15, showing that the two methods are equivalent ways of looking at the same process.

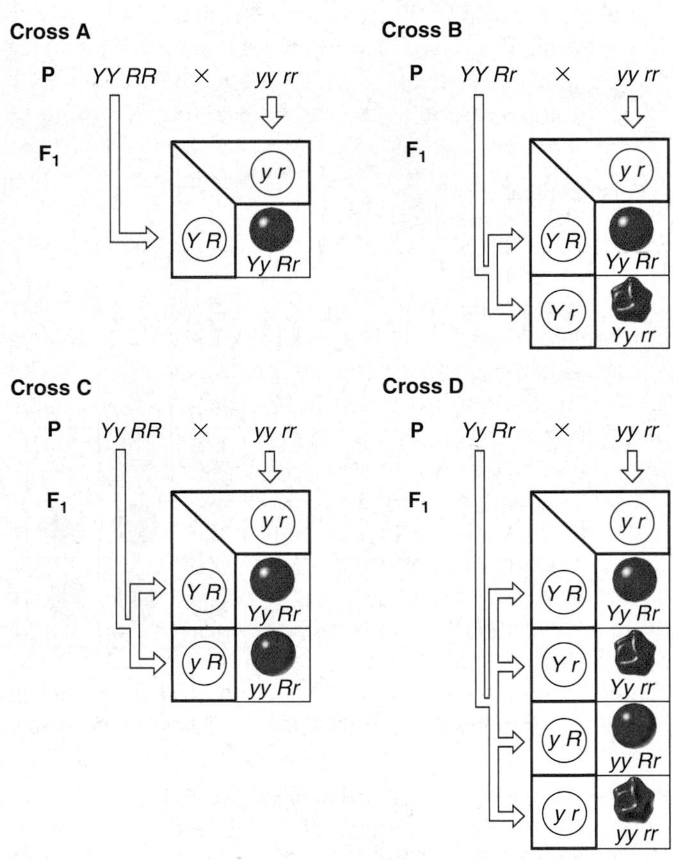

Figure 2.18 Testcrosses on dihybrids. Testcrosses involving two pairs of independently assorting alleles yield different, predictable results depending on the tested individual's genotype for the two genes in question.

kled (*yy rr*) plant, as illustrated in **Fig. 2.18.** If the testcross yields all yellow round offspring (testcross A), you can sell your test plant, because you know it is homozygous for both pea color and pea shape. If your testcross yields 1/2 yellow round and 1/2 yellow wrinkled (testcross B), or 1/2 yellow round and 1/2 green round (testcross C), you know that the candidate plant in question is genetically homozygous for one trait and heterozygous for the other and must therefore be discarded. Finally, if the testcross yields 1/4 yellow round, 1/4 yellow wrinkled, 1/4 green round, and 1/4 green wrinkled (testcross D), you know that the plant is a heterozygote for both the pea color and the pea shape genes.

Using Mendel's Laws: Probabilities and Predictions

Mendel performed several sets of dihybrid crosses and also carried out **multihybrid crosses:** matings between the F_1 progeny of true-breeding parents that differed in three or more traits. In all of these experiments, he observed numbers and ratios very close to what he expected on the basis of his two general biological principles: the alleles of a gene segregate during the formation of egg or pollen, and the alleles of different genes assort independently of each other. Mendel's laws of inheritance, in conjunction with the mathematical rules of probability, provide geneticists with powerful tools for predicting and interpreting the results of genetic crosses. But as with all tools, they have their limitations. We examine here both the power and the limitations of Mendelian analysis.

First, the power: Using simple Mendelian analysis, it is possible to make accurate predictions about the offspring of extremely complex crosses. Suppose you want to predict the occurrence of one specific genotype in a cross involving several independently assorting genes. For example, if hybrids that are heterozygous for four traits are allowed to self-fertilize—*Aa Bb Cc Dd* × *Aa Bb Cc Dd*—what proportion of their progeny will have the genotype *AA bb Cc Dd*? You could set up a Punnett square to answer the question. Since for each trait there are two different alleles, the number of different eggs or sperm is found by raising 2 to the power of the number of differing traits (2^n, where n is the number of traits). By this calculation, each hybrid parent in this cross would make 16 different kinds of gametes: 4 different traits implies $2^4 = 16$ kinds of gametes. The Punnett square depicting such a cross would thus contain 256 boxes (16 × 16). This may be fine if you live in a monastery with a bit of time on your hands, but not if you're taking a 1-hour exam. It would be much simpler to analyze the problem by breaking down the multihybrid cross into four independently assorting monohybrid crosses. Remember that the genotypic ratios of each monohybrid cross are 1 homozygote for the dominant allele, to 2 heterozygotes, to 1 homozygote for the recessive allele = $\underline{1}/4 : \underline{2}/4 : \underline{1}/4$. Thus, you can find the probability of *AA bb Cc Dd* by multiplying the probability of each independent event: *AA* (1/4 of the progeny produced by *Aa* × *Aa*); *bb* (1/4); *Cc* (2/4); *Dd* (2/4):

$$1/4 \times 1/4 \times 2/4 \times 2/4 = 4/256 = 1/64$$

The Punnett square approach would provide the same answer, but it would require much more time.

If instead of a specific genotype, you want to predict the probability of a certain phenotype, you can again use the product rule as long as you know the phenotypic ratios produced by each pair of alleles in the cross. For example, if in the multihybrid cross of *Aa Bb Cc Dd* × *Aa Bb Cc Dd,* you want to know how many offspring will show the dominant A trait (genotype *AA* or *Aa* = 1/4 + 2/4, or 3/4), the recessive b trait (genotype *bb* = 1/4), the dominant C trait (genotype *CC* or *Cc* = 3/4), and the dominant D trait (genotype *DD* or *Dd* = 3/4), you simply multiply

$$3/4 \times 1/4 \times 3/4 \times 3/4 = 27/256$$

In this way, the rules of probability make it possible to predict the outcome of very complex crosses.

You can see from these examples that particular problems in genetics are amenable to particular modes of analysis. As a rule of thumb, Punnett squares are excellent for visualizing simple crosses involving a few genes, but they become unwieldy in the dissection of more complicated matings. Direct calculations of probabilities, such as those in the two preceding problems, are useful when you want to know the chances of one or a few outcomes of complex crosses. If, however, you want to know all the outcomes of a multihybrid cross, a branched-line diagram is the best way to go as it will keep track of the possibilities in an organized fashion.

Now, the limitations of Mendelian analysis: Like Mendel, if you were to breed pea plants or corn or any other organism, you would most likely observe some deviation from the ratios you expected in each generation. What can account for such variation? One element is chance, as witnessed in the common coin toss experiment. With each throw, the probability of the coin coming up heads is equal to the likelihood it will come up tails. But if you toss a coin 10 times, you may get 30% (3) heads and 70% (7) tails, or vice versa. If you toss it 100 times, you are more likely to get a result closer to the expected 50% heads and 50% tails. The larger the number of trials, the lower the probability that chance significantly skews the data. This is one reason Mendel worked with large numbers of pea plants. Mendel's laws, in fact, have great predictive power for populations of organisms, but they do not tell us what will happen in any one individual. With a garden full of self-fertilizing monohybrid pea plants, for example, you can expect that 3/4 of the F_2 progeny will show the dominant phenotype and 1/4 the recessive, but you cannot predict the phenotype of any particular F_2 plant. In Chapter 5, we discuss mathematical methods for assessing whether the chance variation observed in a sample of individuals within a population is compatible with a genetic hypothesis.

Why Mendel's Work Was Unappreciated Before 1900

Mendel's insights into the workings of heredity were a breakthrough of monumental proportions. By counting and

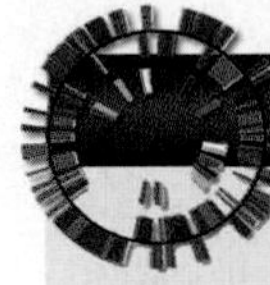

TOOLS OF GENETICS

Genetic Engineering Can Make Plants into Living Chemical Factories

For millenia, farmers used selective breeding to obtain crop plants or domestic animals with desired phenotypic characteristics, such as hardiness, improved yields, or better taste. Then, beginning in the early twentieth century, breeders were able to apply Mendel's laws to the inheritance of many traits and make probability-based predictions about the outcomes of crosses. Understanding Mendel's laws increased the efficiency and success of their breeding programs. However, even with the application of these basic rules of genetics, plant and animal breeders cannot always achieve their goals. Except in the most highly inbred lines, the phenotypes produced by most crosses result from complex interactions involving many genes whose cumulative effects are difficult to predict. Geneticists are also limited by the availability of various alleles. Most mutations generating new alleles of genes occur extremely rarely. As a result, breeders cannot detect many alleles that might potentially be useful.

Beginning in the 1980s, a revolution in genetics took place that made it possible to overcome these limitations. Scientists developed techniques that allowed them to study and then manipulate DNA, the molecule of which genes are made. You will learn about these methods later in this book. For now, it is important to understand simply that these new tools of genetic engineering allow researchers to remove a specific gene from an organism, change the gene in virtually any way they desire, and even move a gene from one organism to an individual of a different species. There are two major advantages of genetic engineering as compared to any type of selective breeding program. First, genetic engineering is extremely efficient in that breeders can specifically target a gene they think might have an interesting effect on phenotype. Second, researchers can now use their imaginations to make new alleles of genes (or even new genes!) that could otherwise never be found. In both of these respects, this kind of planned manipulation frees genetics from a reliance on rare natural mutations that occur only randomly due to chance events.

One of the most exciting potential applications of these new tools is the genetic engineering of plants to convert them into factories that inexpensively make useful biomolecules such as vitamins or vaccines. Consider, for example, potato plants containing a foreign gene (a *transgene*) from the hepatitis B virus that specifies a protein found on the viral surface. If the potatoes could use this gene to make a large amount of the viral protein, then people who ate these potatoes might develop an immune response to that protein. The immune response would protect them from infection by hepatitis B; in other words, such potatoes would act as an "edible vaccine" against the virus. This approach is of particular interest in developing countries, where the costs of manufacturing and delivering traditional vaccines are prohibitive.

analyzing data from thousands of pea plant crosses, he inferred the existence of genes—independent units that determine the observable patterns of inheritance for particular traits. His work explained the reappearance of "hidden" traits, disproved the idea of blended inheritance, and showed that mother and father make an equal genetic contribution to the next generation. The model of heredity that he formulated was so specific that he could test predictions based on it by observation and experiment. In the process of developing his model. Mendel deduced the two basic principles of gene transmission: the segregation of each gene's alleles during gamete formation followed by a random union at fertilization, and the independent assortment of the alleles for two or more different genes. Finally, he tied the transmission of traits to the function of particular cells—the gametes—and also recognized that knowledge of the general laws of heredity were a prerequisite for understanding "the evolutionary history of organic forms [organisms]."

With the exception of Abbot Napp, none of Mendel's contemporaries appreciated the importance of his research. Mendel did not teach at a prestigious university and was not well known outside Brno. Even in Brno, members of the Natural Science Society were disappointed when he presented "Experiments on Plant Hybrids" to them. They wanted to view and discuss intriguing mutants and lovely flowers, so they did not appreciate his numerical analyses. Mendel, it seems, was far ahead of his time. Sadly, despite written requests from Mendel that others try to replicate his studies, no one repeated his experiments. Several citations of his paper between 1866 and 1900 referred to his expertise as a plant breeder but made no mention of his laws. Moreover, at the time Mendel presented his work, no one had yet seen the structures within cells, the *chromosomes,* that actually segregate and assort independently. That would happen only in the next few decades (as described in Chapter 4). If scientists had been able to see these structures, they might have more readily accepted Mendel's

Edible vaccines can be grown in a field rather than made in a laboratory; they do not require refrigeration; and they can be administered orally instead of being injected by medical personnel. As of 2005, the basic idea of an edible vaccine appeared to be feasible: Volunteers eating such genetically engineered potatoes, in fact, have mounted an immune response against hepatitis B. However, various technical difficulties have to date prevented the widespread application of this concept. For example, the immune response in different people who ingested edible vaccines has been quite variable. In addition, cooking the potatoes destroys the vaccine, and few volunteers have been eager to eat sizeable helpings of raw potatoes.

Although the use of transgenic plants to produce edible vaccines is clearly some years in the future, plants genetically engineered in other ways have already had a huge economic impact. Crop plants such as corn and cotton have been genetically engineered to express the gene for a protein called Bt. This protein, made naturally by the bacterium *Bacillus thuringiensis,* is lethal to insects that ingest it but not to other animals. If an insect pest such as a corn borer eats part of a corn plant making the Bt protein, the corn borer will die. In this sense, the engineered corn manufactures its own insecticide, reducing the need for costly chemical pesticides that damage the environment. This approach has already shown itself to be very successful: Approximately one-third of all corn currently grown in the United States contains *Bt* transgenes. In a second example of genetic engineering in crop plants, researchers enhanced the nutritional content of rice to create so-called "golden rice." This engineered plant gets its name from its color, which is the result of beta-carotene production that is enabled by a transgene derived from daffodils and another transgene from a soil bacterium. Humans require beta-carotene as a precursor for the production of vitamin A. The World Health Organization estimates that vitamin A deficiency is a major cause of blindness in developing countries. The ability to provide close to the recommended daily allowance of beta-carotene in countries where an estimate 2 billion people suffer from vitamin A deficiency would have a substantial impact on global health and the global cost of health care.

Despite its promise, many people are uncomfortable with the concept of genetically modified (GM) crops. Some critics, for example, have raised concerns about this technology's potential negative effects on human health, agricultural communities (particularly in developing countries), and the environment. Researchers who are developing GM crops respond that prior to the advent of genetic engineering, plant breeders altered crops in astonishing ways simply by mating various plants together, and that the occasional exchange of genetic information between different species has occurred naturally throughout evolution. These researchers thus argue that the potential risks associated with any crop produced by genetic engineering are no different than those associated with time-honored methods of plant breeding or even the simple propagation of plants in nature. In the Genetics and Society box entitled "The Use of Recombinant DNA Technology to Produce Pest-Resistant Crops" on p. 320 of Chapter 9, we describe a way to evaluate GM crops such as Bt corn. This method balances potential benefits against dangers that are calculated relative to risks associated with traditional agricultural products long accepted by society.

ideas, because the chromosomes are actual physical structures that behave exactly as Mendel predicted.

In an ironic twist of history, Mendel revealed a heritable source of variation—the occurrence, at fertilization, of new combinations of independently assorting alleles, which may give rise to recombinant phenotypes—but Charles Darwin (1809–1882), who was unfamiliar with Mendel's work, was plagued in his later years by critics who found his explanations for the persistence of variation in organisms insufficient. Darwin considered such variation a cornerstone of his theory of evolution, maintaining that natural selection would favor particular variants in a given population in a given environment. If the selected combinations of variant traits were passed on to subsequent generations, this transmission of variation would propel evolution. He could not, however, say how that transmission might occur. Had he been aware of Mendel's laws, he might not have been backed into such an uncomfortable corner. Nearly a century after Mendel published his findings, historians found an uncut copy of Mendel's paper in Darwin's study; Darwin had received but never read it.

For 34 years, Mendel's laws lay dormant—untested, unconfirmed, and unapplied. Then in 1900, 16 years after Mendel's death, Carl Correns, Hugo de Vries, and Erich von Tschermak independently rediscovered and acknowledged his work (**Fig. 2.19**). The scientific community had finally caught up with Mendel. Within a decade, investigators had coined many of the modern terms we have been using: phenotype, genotype, homozygote, heterozygote, gene, and genetics, the label given to the twentieth-century science of heredity. Mendel's paper provided the new discipline's foundation. His principles and analytic techniques endure today, guiding geneticists and evolutionary biologists in their studies of genetic variation. The Tools of Genetics box starting on p. 28 explains how modern-day "genetic engineers" apply Mendel's laws to help them artificially manipulate genes and genomes in new ways not achieved by natural evolution on earth.

(a) Gregor Mendel

(b) Carl Correns

(c) Hugo de Vries

(d) Erich von Tschermak

Figure 2.19 The science of genetics begins with the rediscovery of Mendel. Working independently near the beginning of the twentieth century, Correns, de Vries, and von Tschermak each came to the same conclusions as those Mendel summarized in his laws.

2.3 Mendelian Inheritance in Humans: Two Comprehensive Examples

Although many human traits clearly run in families, most do not show a simple Mendelian pattern of inheritance. Suppose, for example, that you have brown eyes, but both your parents' eyes are blue. Since blue is normally considered recessive to brown, does this mean that you are adopted or that your father isn't really your father? Not necessarily, because eye color is influenced by more than one gene.

Like eye color, most common and obvious human phenotypes arise from the interaction of many genes. In contrast, single-gene traits in people usually involve an abnormality that is disabling or life-threatening. Examples are the progressive mental retardation and other neurological damage of Huntington disease and the clogged lungs and potential respiratory failure of cystic fibrosis. A defective allele of a single gene gives rise to Huntington disease; defective alleles of a different gene are responsible for cystic fibrosis. There were roughly 3500 such single-gene traits known in humans in 2005, and the number continues to grow as new studies confirm the genetic basis of more traits. **Table 2.1** lists some of the most common single-gene traits in humans.

Determining a genetic defect's pattern of transmission is not always an easy task since people make slippery genetic subjects. Their generation time is long, and the families they produce are relatively small, which makes statistical analysis difficult. They do not base their choice of mates on purely genetic considerations. There are thus no pure-breeding lines and no controlled matings. And there is rarely a true F_2 generation (like the one in which Mendel observed the 3:1 ratios from which he derived his rules) because brothers and sisters almost never mate. Geneticists circumvent these difficulties by working with a large number of families or with several generations of a very large family. This allows them to study the large numbers of genetically related individuals needed to establish the inheritance patterns of specific traits. A family history, known as a **pedigree,** is an orderly diagram of a family's relevant genetic features, extending back to at least both sets of grandparents and preferably through as many more generations as possible. From systematic pedigree analysis in the light of Mendel's laws, geneticists can tell if a trait is determined by alternative alleles of a single gene and whether a single-gene trait is dominant or recessive. Because Mendel's principles are so simple and straightforward, a little logic can go a long way in explaining how traits are inherited in humans.

Figure 2.20 shows how to interpret a family pedigree diagram. Squares (□) represent males, circles (○) are females, diamonds (◇) indicate that the sex is unspecified; family members affected by the trait in question are indicated by a filled-in symbol (for example, ■). A single horizontal line connecting a male and a female (□—○) represents a mating, a double connecting line (□═○) designates a **consanguineous** mating, that is, a mating between relatives, and a horizontal line above a series of symbols (○□○) indicates the children of the same parents (a *sibship*) arranged and numbered from left to right in order of their birth. Roman numerals to the left or right of the diagram indicate the generations.

To reach a conclusion about the mode of inheritance of a family trait, human geneticists must use a pedigree that supplies sufficient information. For example, they could not determine whether the allele causing the disease depicted at the bottom of Fig. 2.20 is dominant or recessive solely on the basis of the simple pedigree shown. The data are consistent with both possibilities. If the trait is dominant, then the father and the affected son are heterozygotes, while the mother and the unaffected son are homozygotes for the recessive normal allele. If

TABLE 2.1 Some of the Most Common Single-Gene Traits in Humans

Disease	Effect	Incidence of Disease
Caused by a Recessive Allele		
Thalassemia (chromosome 16 or 11)	Reduced amounts of hemoglobin; anemia, bone and spleen enlargement	1/10 in parts of Italy
Sickle-cell anemia (chromosome 11)	Abnormal hemoglobin; sickle-shaped red cells, anemia, blocked circulation; increased resistance to malaria	1/625 African-Americans
Cystic fibrosis (chromosome 7)	Defective cell membrane protein; excessive mucous production; digestive and respiratory failure	1/2000 Caucasians
Tay-Sachs disease (chromosome 15)	Missing enzyme; buildup of fatty deposit in brain; buildup destroys mental development	1/3000 Eastern European Jews
Phenylketonuria (PKU) (chromosome 12)	Missing enzyme; mental deficiency	1/10,000 Caucasians
Caused by a Dominant Allele		
Hypercholesterolemia (chromosome 19)	Missing protein that removes cholesterol from the blood; heart attack by age 50	1/122 French Canadians
Huntington disease (chromosome 4)	Progressive mental and neurological damage; neurologic disorders by ages 40–70	1/25,000 Caucasians

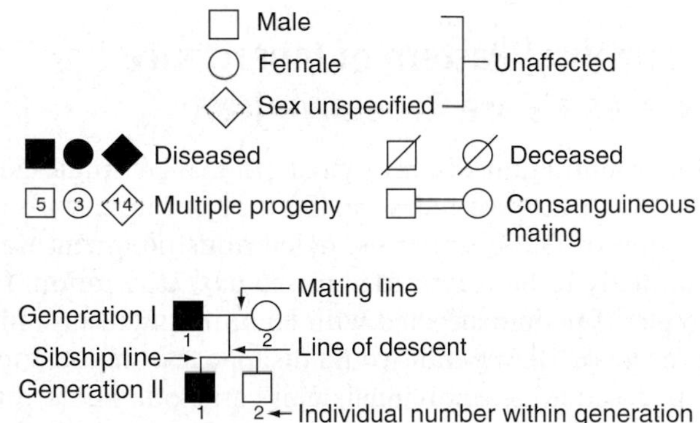

Figure 2.20 Symbols used in pedigree analysis. In the simple pedigree at the bottom, I.1 is the father, I.2 is the mother, and II.1 and II.2 are their sons. The father and the first son are both affected by the disease trait.

instead the trait is recessive, the father and affected son are homozygotes for the recessive disease-causing allele, while the mother and the unaffected son are heterozygotes.

Several kinds of additional information could help resolve this uncertainty. Human geneticists would particularly want to know the frequency at which the trait in question is found in the population from which the family came. If the trait is rare in the population, then the allele giving rise to the trait should also be rare, and the most likely hypothesis would require that the fewest genetically unrelated people carry the allele. Only the father in Fig. 2.20 would need to have a dominant disease-causing allele, but both parents would need to carry a recessive disease-causing allele (the father two copies and the mother one). However, even the information that the trait is rare does not allow us to draw the firm conclusion that it is inherited in a dominant fashion. The pedigree in the figure is so limited that we cannot be sure the two parents are themselves unrelated. As we discuss later in more detail, related parents might have both received a rare recessive allele from their common ancestor. This example illustrates why human geneticists try to collect family histories that cover several generations.

We now look at more extensive pedigrees for the dominant trait of Huntington disease and for the recessive condition of cystic fibrosis. The patterns by which these traits appear in the pedigrees provide important clues that can indicate modes of inheritance and allow geneticists to assign genotypes to family members.

A Vertical Pattern of Inheritance Indicates a Rare Dominant Trait

Huntington disease is named for George Huntington, the New York physician who first described its course. This illness usually shows up in middle age and slowly destroys its victims both mentally and physically. Symptoms include intellectual deterioration, severe depression, and jerky,

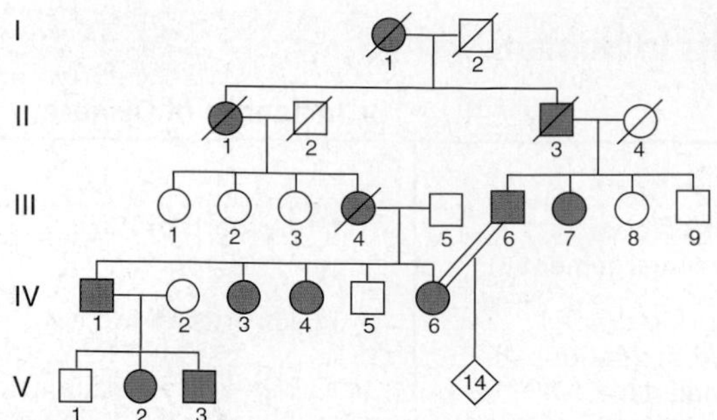

Figure 2.21 Huntington disease: A rare dominant trait. All individuals represented by filled-in symbols are heterozygotes (except I-1, who could have been homozygous for the dominant *HD* disease allele); all individuals represented by open symbols are homozygotes for the recessive HD^+ normal allele. Among the 14 children of the consanguineous mating, DNA testing shows that some are *HD HD*, some are *HD* HD^+, and some are HD^+ HD^+. Human geneticists often employ the diamond designation to mask information about sex or genetic status so as to preserve patient confidentiality.

irregular movements, all caused by the progressive death of nerve cells. If one parent develops the symptoms, his or her children have a 50% probability of suffering from the disease, provided they live to adulthood. Because symptoms are not present at birth and manifest themselves only later in life, Huntington disease is known as a **late-onset** genetic condition.

How would you proceed in assigning genotypes to the individuals in the Huntington disease pedigree depicted in **Fig. 2.21**? First, you would need to find out if the disease-producing allele is dominant or recessive. Several clues suggest that Huntington disease is transmitted by a dominant allele of a single gene. Everyone who develops the disease has at least one parent who shows the trait, and in several generations, approximately half of the offspring are affected. The pattern of affected individuals is thus vertical: If you trace back through the ancestors of any affected individual, you would see at least one affected person in each generation, giving a continuous line of family members with the disease. When a disease is rare in the population as a whole, a vertical pattern is strong evidence that a dominant allele causes the trait; the alternative would require that many unrelated people carry a rare recessive allele. (A recessive trait that is extremely common might also show up in every generation; we examine this possibility in Problem 34 on p. 43.)

In tracking a dominant allele through a pedigree, you can view every mating between an affected and an unaffected partner as analogous to a testcross. If some of the offspring do not have Huntington's, you know the parent showing the trait is a heterozygote. You can check your genotype assignments against the answers in the caption to Fig. 2.21.

No effective treatment yet exists for Huntington disease, and because of its late onset, there was until the 1980s no way for children of a Huntington's parent to know before middle age—usually until well after their own childbearing years—whether they carried the Huntington disease allele (*HD*). Children of Huntington's parents have a 50% probability of inheriting *HD* and a 25% probability of passing the defective allele on to one of their children. In the mid-1980s, with new knowledge of the gene, molecular geneticists developed a DNA test that determines whether an individual carries the *HD* allele. Because of the lack of effective treatment for the disease, some young adults whose parents died of Huntington's prefer not to be tested so that they will not prematurely learn their own fate. However, other at-risk individuals employ the test for the *HD* allele to guide their decisions about having children. If someone whose parent had Huntington disease does not have *HD*, he or she has no chance of developing the disease or of transmitting it to offspring. If the test shows the presence of *HD*, the at-risk person and his or her partner might chose to conceive a child, obtain a prenatal diagnosis of the fetus, and then, depending on their beliefs, elect an abortion if the fetus is affected. The Genetics and Society box "Developing Guidelines for Genetic Screening" on pp. 34–35 discusses significant social and ethical issues raised by information obtained from family pedigrees and molecular tests.

A Horizontal Pattern of Inheritance Indicates a Rare Recessive Trait

Unlike Huntington disease, most confirmed single-gene traits in humans are recessive. This is because, with the exception of late-onset traits, deleterious dominant traits are unlikely to be transmitted to the next generation. For example, if people affected with Huntington disease died by the age of 10, the trait would disappear from the population. In contrast, individuals can carry one allele for a recessive trait without ever being affected by any symptoms. **Figure 2.22** shows three pedigrees for cystic fibrosis (CF), the most commonly inherited recessive disease among Caucasian children in the United States. A double dose of the recessive *CF* allele causes a fatal disorder in which the lungs, pancreas, and other organs become clogged with a thick, viscous mucus that can interfere with breathing and digestion. One in every 2000 white Americans is born with cystic fibrosis, and only 10% of them survive into their 30s.

There are two salient features of the CF pedigrees. First, the family pattern of people showing the trait is often horizontal: The parents, grandparents, and great-grandparents of children born with CF do not themselves manifest the disease, while several brothers and sisters in a single generation may. A horizontal pedigree pattern is a strong indication that the trait is recessive. The unaffected parents are heterozygous **carriers:** they bear a

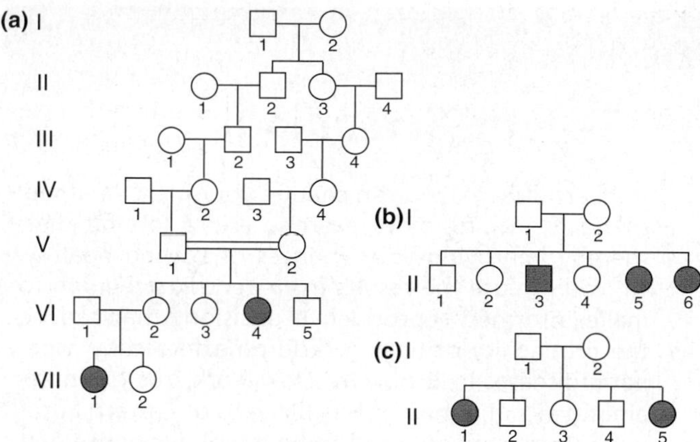

Figure 2.22 Cystic fibrosis: A recessive condition. In (a), the two affected individuals (VI-4 and VII-1) are $CF\,CF$; that is, homozygotes for the recessive disease allele. Their unaffected parents must be carriers, so V-1, V-2, VI-1, and VI-2 must all be $CF\,CF^{+}$ (where CF^{+} is the dominant normal allele). It is probable that II-2, II-3, III-2, III-4, IV-2, and IV-4 are also carriers of the same *CF* allele inherited from a common ancestor in the first generation. We cannot determine which of these founders (I-1 or I-2) was a carrier, so we must designate their genotypes as $CF^{+}-$. Because the *CF* allele is relatively rare, it is likely that II-1, II-4, III-1, III-3, IV-1, and IV-3 are $CF^{+}CF^{+}$ homozygotes. The genotype of the remaining unaffected people (VI-3, VI-5, and VII-2) is uncertain ($CF^{+}-$). **(b** and **c)** Two families in which carrier parents produce multiple children with cystic fibrosis demonstrate horizontal patterns of inheritance. Without further information, the unaffected children in each pedigree must be regarded as having a $CF^{+}-$ genotype.

TABLE 2.2 How to Recognize Dominant and Recessive Traits in Pedigrees

Dominant Traits

1. Affected children always have at least one affected parent.
2. As a result, dominant traits show a *vertical pattern* of inheritance: the trait shows up in every generation.
3. Two affected parents can produce unaffected children, if both parents are heterozygotes.

Recessive Traits

1. Affected individuals can be the children of two unaffected carriers, particularly as a result of consanguineous matings.
2. All the children of two affected parents should be affected.
3. *Rare* recessive traits show a *horizontal pattern* of inheritance: the trait first appears among several members of one generation and is not seen in earlier generations.
4. Recessive traits may show a vertical pattern of inheritance if the trait is extremely common in the population.

dominant normal allele that masks the effects of the recessive abnormal one. An estimated 12 million Americans are carriers of the recessive *CF* allele. **Table 2.2** summarizes some of the clues found in pedigrees that can help you decide whether a trait is caused by a dominant or a recessive allele.

The second salient feature of the CF pedigrees is that many of the couples who produce afflicted children are blood relatives; that is, their mating is consanguineous (as indicated by the double line). In Fig. 2.22a, the consanguineous mating in generation V is between third cousins. Of course, children with cystic fibrosis can also have unrelated carrier parents, but because relatives share genes, their offspring have a much greater than average chance of receiving two copies of a rare allele. Whether or not they are related, carrier parents are both heterozygotes. Thus among their offspring, the proportion of unaffected to affected children is expected to be 3:1. To look at it another way, the chances are that one out of four children of two heterozygous carriers will be homozygous CF sufferers. You can gauge your understanding of this inheritance pattern by assigning a genotype to each person in Fig. 2.22 and then checking your answers against the caption. Note that for several individuals, such as the generation I individuals in part a of the figure, it is impossible to assign a full genotype. We know that one of these people must be the carrier who supplied the original *CF* allele, but we do not know if it was the male or the female. As with an ambiguous dominant phenotype in peas, the unknown second allele is indicated by a dash.

In Fig. 2.22a, a mating between the unrelated carriers VI-1 and VI-2 produced a child with cystic fibrosis. How likely is such a marriage between unrelated carriers for a recessive genetic condition? The answer depends on the gene in question and the particular population into which a person is born. As Table 2.1 on p. 31 shows, the incidence of genetic diseases (and thus the frequency of their carriers) varies markedly among populations. Such variation reflects the distinct genetic histories of different groups. The area of genetics that analyzes differences among groups of individuals is called *population genetics,* a subject we cover in detail in Chapter 21. Notice that in Fig. 2.22a, several unrelated, unaffected people, such as II-1 and II-4, married into the family under consideration. Although it is highly probable that these individuals are homozygotes for the normal allele of the gene ($CF^{+}CF^{+}$), there is a small chance (whose magnitude depends on the population) that any one of them could be a carrier of the disease.

Genetic researchers identified the cystic fibrosis gene in 1989, but they are still in the process of developing a gene therapy that would ameliorate the disease's debilitating symptoms (review the Fast Forward box "Genes Encode Proteins" on pp. 22–23).

GENETICS AND SOCIETY

Developing Guidelines for Genetic Screening

In the early 1970s, the United States launched a national screening program for carriers of sickle-cell anemia, a recessive genetic disease that afflicts roughly 1 in 600 African-Americans. The disease is caused by a particular allele, called $Hb\beta^S$, of the β-globin gene; the dominant normal allele is $Hb\beta^A$. The protein determined by the β-globin gene is one component of the oxygen-carrying hemoglobin molecule. $Hb\beta^S\ Hb\beta^S$ homozygotes have a decrease in oxygen supply, tire easily, and often develop heart failure from stress on the circulatory system.

The national screening program for sickle-cell anemia was based on a simple test of hemoglobin mobility: normal and "sickling" hemoglobins move at different rates in a gel. People who participated in the screening program could use the test results to make informed reproductive decisions. A healthy man, for example, who learned he was a carrier (that is, that he was a $Hb\beta^S\ Hb\beta^A$ heterozygote), would not have to worry about having an affected child if his mate were a noncarrier. If, however, they were both carriers, they could choose either not to conceive or to conceive in spite of the 25% risk of bearing an afflicted child. In the 1980s, newly developed techniques allowing direct prenatal detection of the fetal genotype provided additional options. Depending on their beliefs, a couple could decide to continue a pregnancy only if the fetus were not a homozygote for the $Hb\beta^S$ allele, or knowing that their child would have sickle-cell anemia, they could learn how to deal with the symptoms of the condition.

The original sickle-cell screening program, based on detection of the abnormal hemoglobin protein, was unfortunately not an unqualified success, largely because of insufficient educational follow-through. Many who learned they were carriers mistakenly thought they had the disease. Moreover, because employers and insurance companies obtained access to the information without receiving sufficient instruction as to its meaning, some $Hb\beta^S\ Hb\beta^A$ heterozygotes were denied jobs or health insurance for no acceptable reason. Problems of public relations and education thus made a reliable screening test into a source of dissent and alienation.

Today, with the ability to look directly at the genotype of individuals born or unborn, it is becoming feasible to screen families at risk not only for sickle-cell anemia but for a growing number of other genetic disorders as well. The need to establish guidelines for genetic screening thus becomes more and more pressing. Several related questions reveal the complexity of the issue.

1. *Why carry out genetic screening at all?* There are two basic reasons. The first is to obtain information that will benefit individuals. For example, if you learn at an early age that you have a genetic predisposition to heart disease, you can change your lifestyle if necessary to include more exercise and a low-fat diet, thereby improving your chances of staying healthy. Or, you can use the results from genetic screening to make informed reproductive decisions that reduce the probability of having children affected by a genetic disease. In Brooklyn, New York, for example, among a community of Hasidic Jews of Eastern European descent, there used to be a high incidence of a fatal neurodegenerative syndrome known as Tay-Sachs disease. In this traditional, Old World community, marriages are arranged by rabbis or matchmakers who, by encouraging testing for the abnormal allele, helped eradicate the disease. With confidential access to test results, a rabbi could counsel against marriages between two carriers. The second reason for genetic screening, which often conflicts with the first, is to benefit groups within society. Insurance companies and employers, for example, would like to be able to find out who is at risk for various genetic conditions.

2. *When is a test accurate and comprehensive enough to be used as the basis for screening?* The accuracy of standard genetic tests for cystic fibrosis is more than 90%. Because it is not 100%, a few people who test negative may actually be carriers. In contrast, the tests for Huntington disease and the sickle-cell trait pick up close to 100% of those who carry the abnormal allele. In addition to the problem of false negatives, all genetic tests occasionally produce a false positive in which an individual is diagnosed with the disease even though he or she is actually unaffected. What all this means is that some people might decide to have children or not to have children on the basis of inaccurate information. While the accuracy of the tests continues to improve, the question remains: Do the benefits outweigh both the costs and the risks of inaccuracies? Answers to this question depend not only on the severity of the disease, the availability of treatments, and the accuracy of the particular test, but more importantly on who is making the decision. Benefits and risks have very different meanings to individuals, physicians, employers, and the government.

3. *Once an accurate test becomes available at reasonable cost, should screening be required or optional?* This is partly a societal decision because the public treasury bears a large part of the cost of caring for the

sufferers of genetic diseases. But it is also a personal decision. For most inherited diseases, there is as yet no cure. Since the psychological burden of anticipating a fatal late-onset disease for which there is no treatment can be devastating, some people might decide not to be tested. Other reasons people may not want to be tested include religious beliefs and concerns about confidentiality. On the other hand, timely information about the presence of an abnormal gene that causes a condition for which there is a therapy can save lives. An example is the genetic test for hemochromatosis, an inherited disorder affecting 1 of every 200 people in the United States. The disease deposits iron in the heart, liver, and other organs of the body, and by the fifth or sixth decade of life, those organs break down. The simple, ancient practice of bloodletting, if begun early in life, prevents the complications of hemochromatosis. Timely information may also affect childbearing decisions and thereby reduce the incidence of a disease in the population.

4. *If a screening program is established, who should be tested?* The answer depends on what the test is trying to accomplish as well as on the fact that genetic screening is expensive. Ultimately, the cost of a procedure must be weighed against the usefulness of the data it provides. For the 1 in 200 people inheriting the hemochromatosis mutation, widespread testing could reduce health-care costs of affected individuals later in life. Because the disease is relatively common, health-care savings for society as a whole would be significant. By contrast, to reduce the risk of a child being born with a rarer inherited disease, a program might try to target groups with the highest incidence. In the United States, only one-tenth as many African-Americans as Caucasians are affected by cystic fibrosis, and Asians almost never have the disease. Should all racial groups be tested or only Caucasians? Because of the expense, DNA testing for cystic fibrosis and other relatively rare genetic diseases has not yet been carried out on large populations. Rather it has been reserved for couples or individuals whose family history puts them at risk for a severely debilitating disease.

5. *Should private employers and insurance companies be allowed to test their clients and employees?* Some employers advocate genetic screening to reduce the incidence of occupational disease, arguing that they can use data from genetic tests to make sure employees are not assigned to environments that might cause them harm. People with sickle-cell disease, for example, may be at increased risk for a life-threatening episode of severe sickling if exposed to carbon monoxide or trace amounts of cyanide. Critics of this position say that screening violates workers' rights, including the right to privacy, and increases racial and ethnic discrimination in the workplace. Many critics also oppose informing insurance companies of the results of genetic screening, as these companies may deny coverage to people with inherited medical problems or just the possibility of developing such problems. According to many medical ethicists, discrimination of this sort will grow unless countries pass laws, similar to one enacted in France, that ensure genetic information is confidential, to be given out only at the discretion of the tested individual. In 2000, President Clinton issued an executive order banning genetic discrimination by federal agencies in the hiring or promotion of any employee. By 2005, 41 states had enacted legislation preventing genetic discrimination in health insurance and barring genetic discrimination in the workplace.

   A recent high-profile case illustrates some of these issues. The Chicago Bulls, before signing a contract with the basketball player Eddy Curry, wanted him to take a DNA test to find out if he had a genetic predisposition for hypertrophic cardiomyopathy (a potentially fatal condition). The Bulls requested this test because Curry had suffered from episodes of heart arrythmia. Curry refused, citing privacy issues and stating that the test would not be in his or his family's best interest. After a battery of health exams—but not the DNA test—Curry was deemed fit to play, but he was traded to another team and eventually signed a six-year, $56 million contract with the New York Knicks.

6. *Finally, how should people be educated about the meaning of test results?* In one small-community screening program, people identified as carriers of the recessive, life-threatening blood disorder known as β-thalassemia were ostracized and as a result, ended up marrying one another. This only made medical matters worse as it greatly increased the chances that their children would be born with two copies of the defective allele and thus the disease. By contrast, in Ferrara, Italy, where there used to be 30 new cases of β-thalassemia every year, extensive screening was so successfully combined with intensive education that the 1980s passed with no more than a few new cases of the disease.

Given all of these considerations, what kind of guidelines would you like to see established to ensure that genetic screening reaches the right people at the right time and that information gained from such screening is used for the right purposes?

Connections

Mendel answered the three basic questions about heredity as follows: To "What is inherited?" he replied, "alleles of genes." To "How is it inherited?" he responded, "according to the principles of segregation and independent assortment." And to "What is the role of chance in heredity?" he said, "for each individual, inheritance is determined by chance, but within a population, this chance operates in a context of strictly defined probabilities." Knowing all this, we can understand why a trait may disappear in the F_1 hybrid generation and reappear in the F_2 generation or why with individuals that are pure-breeding, like begets like, while with hybrids, like can beget both like and unlike.

Within a decade of the 1900 rediscovery of Mendel's work, numerous breeding studies had shown that Mendel's laws hold true not only for seven pairs of antagonistic characteristics in peas, but for an enormous diversity of traits in a wide variety of sexually reproducing plant and animal species, including four-o'clock flowers, beans, corn, wheat, fruit flies, chickens, mice, horses, and humans. Some of these same breeding studies, however, raised a challenge to the new genetics. For certain traits in certain species, the studies uncovered unanticipated phenotypic ratios, or the results included F_1 and F_2 progeny with novel phenotypes that resembled neither pure-breeding parent's. Mendel himself observed such "puzzling phenomena" when he replicated his pea experiments with beans. A mating between bean plants with white flowers and those with crimson flowers produced a whole range of colors in the offspring, from purple to pale violet to white.

These phenomena could not be explained by Mendel's hypothesis that for each gene, two alternative alleles, one completely dominant, the other recessive, determine a single trait. Mendel, in fact, correctly explained the unexpected outcomes he observed by proposing that two genes (instead of one) determine flower color in beans. In humans, we now know that most common traits, including skin color, eye color, and height, are determined by interactions between two or more genes. We also know that within a given population, there may be more than two alleles for some of those genes. Chapter 3 shows how the genetic analysis of such complex traits, that is, traits produced by complex interactions between genes and between genes and the environment, extended rather than contradicted Mendel's laws of inheritance. Each trait is still determined by the alleles of genes that behave during gamete formation and fertilization exactly as Mendel envisioned. What is new is the relationship between genotype and phenotype. It is not always as straightforward as in the analysis of pea color or of human genetic diseases such as cystic fibrosis.

Essential Concepts

1. Discrete units called *genes* control the appearance of inherited traits.
2. Genes come in alternative forms called *alleles* that are responsible for the expression of different forms of a trait.
3. Body cells of sexually reproducing organisms carry two copies of each gene. When the two copies of a gene are the same allele, the individual is *homozygous* for that gene. When the two copies of a gene are different alleles, the individual is *heterozygous* for that gene.
4. The *genotype* is a description of the allelic combination of the two copies of a gene present in an individual. The *phenotype* is the observable form of the trait that the individual expresses.
5. A cross between two parental lines (P) that are pure-breeding for alternative alleles of a gene will produce a *first filial* (F_1) *generation* of hybrids that are heterozygous. The phenotype expressed by these hybrids is determined by the *dominant* allele of the pair, and this phenotype is the same as that expressed by individuals homozygous for the dominant allele. The phenotype associated with the *recessive* allele will reappear only in the F_2 generation in individuals homozygous for this allele. In crosses between F_1 heterozygotes, the dominant and recessive phenotypes will appear in the F_2 generation in a ratio of 3:1.
6. The two copies of each gene segregate during the formation of gametes. As a result, each egg and each sperm or pollen grain contains only one copy, and thus, only one allele, of each gene. Male and female gametes unite at random at fertilization. Mendel described this process as the *law of segregation.*
7. The segregation of alleles of any one gene is independent of the segregation of the alleles of other genes. Mendel described this process as the *law of independent assortment.* According to this law, crosses between *Aa Bb* F_1 dihybrids will generate F_2 progeny with a phenotypic ratio of 9 (*A– B–*) : 3 (*A– bb*) : 3 (*aa B–*) : 1 (*aa bb*).

On Our Website

www.mhhe.com/hartwell3
Chapter 2

Annotated Suggested Readings and Links to Other Websites

- More about Mendel and the early history of genetics
- More on the practice of human genetics
- An online database of human genetic diseases (OMIM)

Specialized Topics

- The binomial expansion: application of an advanced statistical method to genetics
- Conditional probabilities (Bayesian analysis): application of another advanced statistical method to genetic analysis

Social and Ethical Issues

1. A key tool for research on inherited conditions in humans is the family pedigree. To reach conclusions about the genetic basis of a trait, human geneticists need to collect data from as many family members as possible. One way to obtain this data is to ask people affected by a genetic condition about their relatives. However, gaining information in this way may create ethical problems, because these relatives become research subjects without their knowledge or approval. Suppose, for example, that an interviewee mentions that his aunt has had precancerous colon polyps removed. The aunt might consider the dissemination of this information embarrassing, and if the information were to find its way into public databases, she could be subject to discrimination by her employer or insurance company. In 1999, the federal Office for Human Research Protections (OHRP) received a loss-of-privacy complaint for exactly this reason and temporarily shut down a genetic research program as a result. On the other hand, genetic research using human pedigrees might lead to important new therapies and preventive measures.

   One way to deal with genetic privacy concerns is to require that researchers obtain written informed consent from all family members before a pedigree is published. However, many scientists regard such a requirement as too restrictive: much effort would be required to obtain this consent, and fewer people would enroll in genetic studies if they knew all their relatives had to be contacted. In your opinion, should the right to genetic privacy take precedence over research into human genetic conditions that could benefit a large number of people in the future? If informed consent were not obtained, are there ways to ensure that a person's genetic information will remain private?

2. Louise, a single mother with four children, has a job in an aeronautics factory where she is exposed to the element beryllium. Some individuals are susceptible to the damaging effects of this element, developing a respiratory illness called chronic beryllium disease. Workers in the factory were recently tested for susceptibility, and Louise tested positive. She needs the job and wants to continue working at the factory until she finds another job. The company wants her to quit immediately. Should the individual or the company have the right to decide whether a susceptible person continues to work in a potentially damaging environment? If Louise insisted on working in the factory, should she still have the right to sue the company for compensation if she later developed chronic beryllium disease?

Solved Problems

Solving Genetics Problems

The best way to evaluate and increase your understanding of the material in the chapter is to apply your knowledge in solving genetics problems. Genetics word problems are like puzzles. Take them in slowly—don't be overwhelmed by the whole problem. Identify useful facts given in the problem, and use the facts to deduce additional information. Use genetic principles and logic to work toward the solutions. The more problems you do, the easier they become. In doing problems, you will not only solidify your understanding of genetic concepts, but you will also develop basic analytical skills that are applicable in many disciplines.

Solving genetics problems requires more than simply plugging numbers into formulas. Each problem is unique and requires thoughtful evaluation of the information given and the question being asked. The following are general guidelines you can follow in approaching these word problems:

a. Read through the problem once to get some sense of the concepts involved.
b. Go back through the problem, noting all the information supplied to you. For example, genotypes or phenotypes of offspring or parents may be given to you or implied in the problem. Represent the known information in a symbolic format—assign symbols for alleles; use these symbols to indicate genotypes; make a diagram of the crosses including genotypes and phenotypes given or implied. Be sure that you do not assign different letters of the alphabet to two alleles of the same gene, as this can cause confusion. Also, be careful to discriminate clearly between the upper and lowercases of letters, such as *C*(*c*) or *S*(*s*).
c. Now, reassess the question and work toward the solution using the information given. Make sure you answer the question being asked!
d. When you finish the problem, check to see that the answer makes sense. You can often check solutions by working backwards; that is, see if you can reconstruct the data from your answer.
e. After you have completed a question and checked your answer, spend a minute to think about which major concepts were involved in the solution. This is a critical step for improving your understanding of genetics.

For each chapter, the logic involved in solving two or three types of problems is described in detail.

I. In cats, white patches are caused by the dominant allele *P*, while *pp* individuals are solid-colored. Short hair is caused by a dominant allele *S*, while *ss* cats have long hair. A long-haired cat with patches whose mother was solid-colored and short-haired mates with a short-haired, solid-colored cat whose mother was long-haired and solid-colored. What kinds of kittens can arise from this mating, and in what proportions?

Answer

The solution to this problem requires an understanding of dominance/recessiveness, gamete formation, and the independent assortment of alleles of two genes in a cross.

First make a representation of the known information:

Mothers:	solid, short-haired		solid, long-haired
Cross:	cat 1		cat 2
	patches, long-haired	×	solid, short-haired

What genotypes can you assign? Any cat showing a recessive phenotype must be homozygous for the recessive allele. Therefore the long-haired cats are *ss;* solid cats are *pp*. Cat 1 is long-haired, so it must be homozygous for the recessive allele (*ss*). This cat has the dominant phenotype of patches and could be either *PP* or *Pp*, but since the mother was *pp* and could only contribute a *p* allele in her gametes, the cat must be *Pp*. Cat 1's full genotype is *Pp ss*. Similarly, cat 2 is solid-colored, so it must be homozygous for the recessive allele (*pp*). Because this cat is short-haired, it could have either the *SS* or *Ss* genotype. Its mother was long-haired (*ss*) and could only contribute an *s* allele in her gamete, so cat 2 must be heterozygous *Ss*. The full genotype is *pp Ss*.

The cross is therefore between a *Pp ss* (cat 1) and a *pp Ss* (cat 2). To determine the types of kittens, first establish the types of gametes that can be produced by each cat and then set up a Punnett square to determine the genotypes of the offspring. Cat 1 (*Pp ss*) produces *Ps* and *ps* gametes in equal proportions. Cat 2 (*pp Ss*) produces *pS* and *ps* gametes in equal proportions. *Four types of kittens can result from this mating with equal probability:* Pp Ss *(patches, short-haired),* Pp ss *(patches, long-haired),* pp Ss *(solid, short-haired), and* pp ss *(solid, long-haired).*

		Cat 1	
		P s	*p s*
Cat 2	*p S*	*Pp Ss*	*pp Ss*
	p s	*Pp ss*	*pp ss*

You could also work through this problem using the product rule of probability instead of a Punnett square. The principles are the same: gametes produced in equal amounts by either parent are combined at random.

Cat 1 gamete		Cat 2 gamete		Progeny	
1/2 *P s*	×	1/2 *p S*	→	1/4 *Pp Ss*	patches, short-haired
1/2 *P s*	×	1/2 *p s*	→	1/4 *Pp ss*	patches, long-haired
1/2 *p s*	×	1/2 *p S*	→	1/4 *pp Ss*	solid-colored, short-haired
1/2 *p s*	×	1/2 *p s*	→	1/4 *pp ss*	solid-colored, long-haired

II. In tomatoes, red fruit is dominant to yellow fruit, and purple stems are dominant to green stems. The progeny from one mating consisted of 305 red fruit, purple stem plants; 328 red fruit, green stem plants; 110 yellow fruit, purple stem plants; and 97 yellow fruit, green stem plants. What were the genotypes of the parents in this cross?

Answer

This problem requires an understanding of independent assortment in a dihybrid cross as well as the ratios predicted from monohybrid crosses.

Designate the alleles:

R = red, r = yellow

P = purple stems, p = green stems

In genetics problems, the ratios of offspring can indicate the genotype of parents. You will usually need to total the number of progeny and approximate the ratio of offspring in each of the different classes. For this problem, in which the inheritance of two traits is given, consider each trait independently. For red fruit, there are 305 + 328 = 633 red-fruited plants out of a total of 840 plants. This value (633/840) is close to 3/4. About 1/4 of the plants have yellow fruit (110 + 97 = 207/840). From Mendel's work, you know that a 3:1 phenotypic ratio results from crosses between plants that are hybrid (heterozygous) for one gene. Therefore, the genotype for fruit color of each parent must have been *Rr*.

For stem color, 305 + 110 or 415/840 plants had purple stems. About half had purple stems, and the other half (328 + 97) had green stems. A 1:1 phenotypic ratio occurs when a heterozygote is mated to a homozygous recessive (as in a testcross). The parents' genotypes must have been *Pp* and *pp* for stem color.

The complete genotype of the parent plants in this cross was Rr Pp × Rr pp.

III. Tay-Sachs is a recessive lethal disease in which there is neurological deterioration early in life. This disease is rare in the population overall but is found at relatively high frequency in Ashkenazi Jews from Central Europe. A woman whose maternal uncle had the disease is trying to determine the probability that she and her husband could have an affected child. Her father does not come from a high-risk population. Her husband's sister died of the disease at an early age.

a. Draw the pedigree of the individuals described. Include the genotypes where possible.
b. Determine the probability that the couple's first child will be affected.

Answer

This problem requires an understanding of dominance/ recessiveness and probability. Designate the alleles:

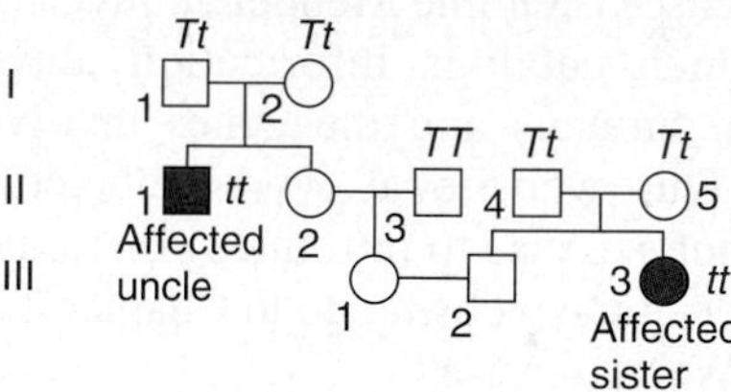

The genotypes of the two affected individuals, the woman's uncle (II-1) and the husband's sister (III-3) are *tt*. Because the uncle was affected, his parents must have been heterozygous. There was a 1/4 chance that these parents had a homozygous recessive (affected) child, a 2/4 chance that they had a heterozygous child (carrier), and a 1/4 chance they had a homozygous dominant (unaffected) child. However, you have been told that the woman's mother (II-2) is unaffected, so the mother could only have had a heterozygous or a homozygous dominant genotype. Consider the probability that these two genotypes will occur. If you were looking at a Punnett square, there would be only three combinations of alleles possible for the normal mother. Two of these are heterozygous combinations and one is homozygous dominant. There is a 2/3 chance (2 out of the 3 possible cases) that the mother was a carrier. The father was not from a high-risk population, so we can assume that he is homozygous dominant. There is a 2/3 chance that the wife's mother was heterozygous and if so, a 1/2 chance that the wife inherited a recessive allele from her mother. Because both conditions are necessary for inheritance of a recessive allele, the individual probabilities are multiplied, and the probability that the wife (III-1) is heterozygous is 2/3 × 1/2.

The husband (III-2) has a sister who died from the disease; therefore, his parents must have been heterozygous. The probability that he is a carrier is 2/3 (using the same rationale as for II-2). The probability that the man and woman are both carriers is 2/3 × 1/2 × 2/3. Since there is a 1/4 probability that a particular child of two carriers will be affected, *the overall probability that the first child of this couple (III-1 and III-2) will be affected is 2/3 × 1/2 × 2/3 × 1/4 = 4/72, or 1/18.*

Problems

Interactive Web Exercise

The National Center for Biotechnology Information (NCBI) at the National Institutes of Health maintains several databases that are a treasure trove for geneticists. One of these databases is Online Mendelian Inheritance in Man (OMIM), which catalogs information about inherited conditions in humans and the genes involved in these syndromes. Our website at www.mhhe.com/hartwell3 contains a brief exercise to introduce you to the use of this database; once at the website, go to Chapter 2 and click on "Interactive Web Exercise".

Vocabulary

1. For each of the terms in the left column, choose the best matching phrase in the right column.

a. phenotype	1. having two identical alleles of a given gene
b. alleles	2. the allele expressed in the phenotype of the heterozygote
c. independent assortment	3. alternate forms of a gene
d. gametes	4. observable characteristic
e. gene	5. a cross between individuals both heterozygous for two genes
f. segregation	6. alleles of one gene separate into gametes randomly with respect to alleles of other genes
g. heterozygote	7. reproductive cells containing only one copy of each gene
h. dominant	8. the allele that does not contribute to the phenotype of the heterozygote
i. F_1	9. the cross of an individual of ambiguous genotype with a homozygous recessive individual
j. testcross	10. an individual with two different alleles of a gene
k. genotype	11. the heritable entity that determines a characteristic
l. recessive	12. the alleles an individual has
m. dihybrid cross	13. the separation of the two alleles of a gene into different gametes
n. homozygote	14. offspring of the P generation

Section 2.1

2. During the millennia in which selective breeding was practiced, why did breeders fail to uncover the principle that traits are governed by discrete units of inheritance (that is, by genes)?

3. Describe the characteristics of the garden pea that made it a good organism for Mendel's analysis of the basic principles of inheritance. Evaluate how easy or difficult it would be to make a similar study of inheritance in humans by considering the same attributes you described for the pea.

Section 2.2

4. An albino corn snake is crossed with a normal-colored corn snake. The offspring are all normal-colored. When these first generation progeny snakes are crossed among themselves, they produce 32 normal-colored snakes and 10 albino snakes.

a. Which of these phenotypes is controlled by the dominant allele?

b. In these snakes, albino color is determined by a recessive allele *a,* and normal pigmentation is determined by *A* allele. A normal-colored female snake is involved in a testcross. This cross produces 10 normal-colored and 11 albino offspring. What are the genotypes of the parents and the offspring?

5. Two short-haired cats mate and produce six short-haired and two long-haired kittens. What does this information suggest about how hair length is inherited?

6. Piebald spotting is a condition found in humans in which there are patches of skin that lack pigmentation. The condition results from the inability of pigment-producing cells to migrate properly during development. Two adults with piebald spotting have one child who has this trait and a second child with normal skin pigmentation.

a. Is the piebald spotting trait dominant or recessive? What information led you to this answer?

b. What are the genotypes of the parents?

7. As a *Drosophila* research geneticist, you keep stocks of flies of specific genotypes. You have a fly that has normal wings (dominant phenotype). Flies with short wings are homozygous for a recessive allele of the wing-length gene. You need to know if this fly with normal wings is pure-breeding or heterozygous for the wing-length trait. What cross would you do to determine the genotype, and what results would you expect for each possible genotype?

8. A mutant cucumber plant has flowers that fail to open when mature. Crosses can be done with this plant by manually opening and pollinating the flowers with pollen from another plant. When closed × open crosses were done, all the F_1 progeny were open. The F_2 plants were 145 open and 59 closed. A cross of closed × F_1 gave 81 open and 77 closed. How is the closed trait inherited? What evidence led you to your conclusion?

9. In a particular population of mice, certain individuals display a phenotype called "short tail," which is inherited as a dominant trait. Some individuals display a recessive trait called "dilute," which affects coat color. Which of these traits would be easier to eliminate from the population by selective breeding? Why?

10. In humans, a dimple in the chin is a dominant characteristic.

a. A man who does not have a chin dimple has children with a woman with a chin dimple whose mother lacked the dimple. What proportion of their children would be expected to have a chin dimple?

b. A man with a chin dimple and a woman who lacks the dimple produce a child who lacks a dimple. What is the man's genotype?

c. A man with a chin dimple and a nondimpled woman produce eight children, all having the chin dimple. Can you be certain of the man's genotype? Why or why not? What genotype is more likely, and why?

11. Among native Americans, two types of earwax (cerumen) are seen, dry and sticky. A geneticist studied the inheritance of this trait by observing the types of offspring produced by different kinds of matings. He observed the following numbers:

Parents	Number of mating pairs	Offspring Sticky	Offspring Dry
Sticky × sticky	10	32	6
Sticky × dry	8	21	9
Dry × dry	12	0	42

a. How is earwax type inherited?

b. Why are there no 3:1 or 1:1 ratios in the data shown in the chart?

12. Imagine you have just purchased a black stallion of unknown genotype. You mate him to a red mare, and she delivers twin foals, one red and one black. Can you tell from these results how color is inherited, assuming that alternative alleles of a single gene are involved? What crosses could you do to work this out?

13. If you roll a die (singular of dice), what is the probability you will roll: (a) a 6? (b) an even number? (c) a number divisible by 3? (d) If you roll a pair of dice, what is the probability that you will roll two 6s? (e) an even number on one and an odd number on the other? (f) matching numbers? (g) two numbers both over 4?

14. In a standard deck of playing cards, there are four suits (red suits = hearts and diamonds, black suits = spades and clubs). Each suit has thirteen cards: Ace (A), 2, 3, 4, 5, 6, 7, 8, 9, 10, and the face cards Jack (J), Queen (Q), and King (K). In a single draw, what is the probability that you will draw a face card? A red card? A red face card?

15. How many genetically different eggs could be formed by women with the following genotypes?

a. *Aa bb CC DD*

b. *AA Bb Cc dd*

c. *Aa Bb cc Dd*

d. *Aa Bb Cc Dd*

16. What is the probability of producing a child that will phenotypically resemble either one of the two parents in the following four crosses? How many phenotypically different kinds of progeny could potentially result from each of the four crosses?

a. *Aa Bb Cc Dd* × *aa bb cc dd*

b. *aa bb cc dd* × *AA BB CC DD*

c. *Aa Bb Cc Dd* × *Aa Bb Cc Dd*

d. *aa bb cc dd* × *aa bb cc dd*

17. A mouse sperm of genotype *a B C D E* fertilizes an egg of genotype *a b c D e*. What are all the possibilities for the genotypes of (a) the zygote and (b) a sperm or egg of the baby mouse that develops from this fertilization?

18. Galactosemia is a recessive human disease that is treatable by restricting lactose and glucose in the diet. Susan Smithers and her husband are both heterozygous for the galactosemia gene.

a. Susan is pregnant with twins. If she has fraternal (nonidentical) twins, what is the probability both of the twins will be girls who have galactosemia?

b. If the twins are identical, what is the probability that both will be girls and have galactosemia?

For parts c–g, assume that none of the children is a twin.

c. If Susan and her husband have four children, what is the probability that none of the four will have galactosemia?

d. If the couple has four children, what is the probability that at least one child will have galactosemia?

e. If the couple has four children, what is the probability that the first two will have galactosemia and the second two will not?

f. If the couple has three children, what is the probability that two of the children will have galactosemia and one will not, regardless of order?

g. If the couple has four children with galactosemia, what is the probability that their next child will have galactosemia?

19. Albinism is a condition in which pigmentation is lacking. In humans, the result is white hair, nonpigmented skin, and pink eyes. The trait in humans is caused by a recessive allele. Two normal parents have an albino child. What are the parents' genotypes? What is the probability that the next child will be albino?

20. A cross between two pea plants, both of which grew from yellow round seeds, gave the following numbers of seeds: 156 yellow round and 54 yellow wrinkled. What are the genotypes of the parent plants? (Yellow and round are dominant traits.)

21. A third-grader decided to breed guinea pigs for her school science project. She went to a pet store and bought a male with smooth black fur and a female with rough white fur. She wanted to study the inheritance of those features and was sorry to see that the first litter of eight contained only rough black animals. To her disappointment, the second litter from those same parents contained seven rough black animals. Soon the first litter had begun to produce F_2 offspring, and they showed a variety of coat types. Before long, the child had 125 F_2 guinea pigs. Eight of them had smooth white coats, 25 had smooth black coats, 23 were rough and white, and 69 were rough and black.

a. How are the coat color and texture characteristics inherited? What evidence supports your conclusions?

b. What phenotypes and proportions of offspring should the girl expect if she mates one of the smooth white F_2 females to an F_1 male?

22. The self-fertilization of an F_1 pea plant produced from a parent plant homozygous for yellow and wrinkled seeds and a parent homozygous for green and round seeds resulted in a pod containing seven F_2 peas. (Yellow and round are dominant.) What is the probability that all seven peas in the pod are yellow and round?

23. The achoo syndrome (sneezing in response to bright light) and trembling chin (triggered by anxiety) are both dominant traits in humans.

a. What is the probability that the first child of parents who are heterozygous for both the achoo gene and trembling chin will have achoo syndrome but lack the trembling chin?

b. What is the probability that the first child will not have achoo syndrome or trembling chin?

24. A pea plant from a pure-breeding strain that is tall, has green pods, and has purple flowers that are terminal is crossed to a plant from a pure-breeding strain that is dwarf, has yellow pods, and has white flowers that are axial. The F_1 plants are all tall and have purple axial flowers as well as green pods.

a. What phenotypes do you expect to see in the F_2?

b. What phenotypes and ratios would you predict in the progeny from crossing an F_1 plant to the dwarf parent?

25. The following chart shows the results of different matings between jimsonweed plants that had either purple or white flowers and spiny or smooth pods. Determine the dominant allele for the two traits and indicate the genotypes of the parents for each of the crosses.

Parents	Offspring phenotypes			
	Purple Spiny	White Spiny	Purple Smooth	White Smooth
a. purple spiny × purple spiny	94	32	28	11
b. purple spiny × purple smooth	40	0	38	0
c. purple spiny × white spiny	34	30	0	0
d. purple spiny × white spiny	89	92	31	27
e. purple smooth × purple smooth	0	0	36	11
f. white spiny × white spiny	0	45	0	16

26. A pea plant heterozygous for plant height, pod shape, and flower color was selfed. The progeny consisted of 272 tall, inflated pods, purple flowers; 92 tall, inflated, white flowers; 88 tall, flat pods, purple; 93 dwarf, inflated, purple; 35 tall, flat, white; 31 dwarf, inflated, white; 29 dwarf, flat, purple; 11 dwarf, flat, white. Which alleles are dominant in this cross?

27. In the fruit fly *Drosophila melanogaster*, the following genes and mutations are known:

Wingsize: recessive allele for tiny wings *t;* dominant allele for normal wings *T.*

Eye shape: recessive allele for narrow eyes *n;* dominant allele for normal (oval) eyes *N.*

For each of the following crosses, give the genotypes of each of the parents.

Cross	Male Wings	Male Eyes		Female Wings	Female Eyes	Offspring
1	tiny	oval	×	tiny	oval	78 tiny wings, oval eyes 24 tiny wings, narrow eyes
2	normal	narrow	×	tiny	oval	45 normal wings, oval eyes 40 normal wings, narrow eyes 38 tiny wings, oval eyes 44 tiny wings, narrow eyes
3	normal	narrow	×	normal	oval	35 normal wings, oval eyes 29 normal wings, narrow eyes 10 tiny wings, oval eyes 11 tiny wings, narrow eyes
4	normal	narrow	×	normal	oval	62 normal wings, oval eyes 19 tiny wings, oval eyes

28. Based on the information you discovered in Problem 27 above, answer the following:

a. A female fruit fly with genotype *Tt nn* is mated to a male of genotype *Tt Nn*. What is the probability that any one of their offspring will have normal phenotypes for both characters?

b. What phenotypes would you expect among the offspring of this cross? If you obtained 200 progeny, how many of each phenotypic class would you expect?

Section 2.3

29. For each of the following human pedigrees, indicate whether the inheritance pattern is recessive or dominant. What feature(s) of the pedigree did you use to determine the inheritance? Give the genotypes of affected individuals and of individuals who carry the disease allele.

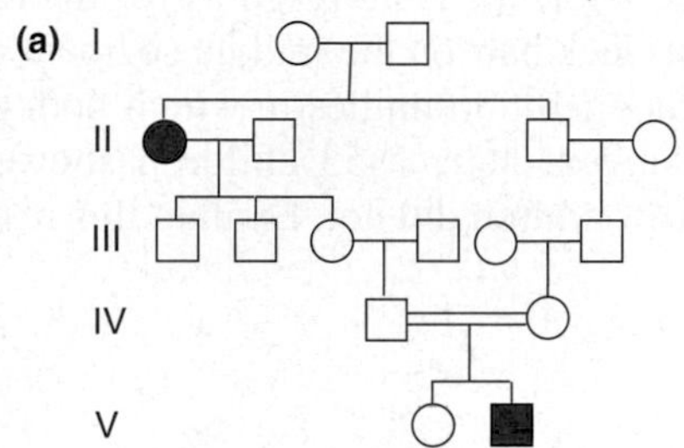

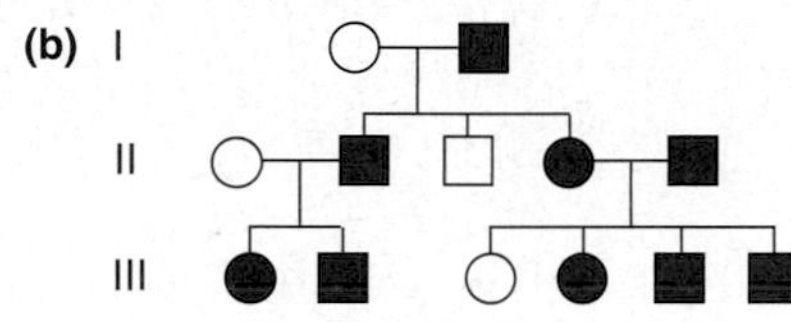

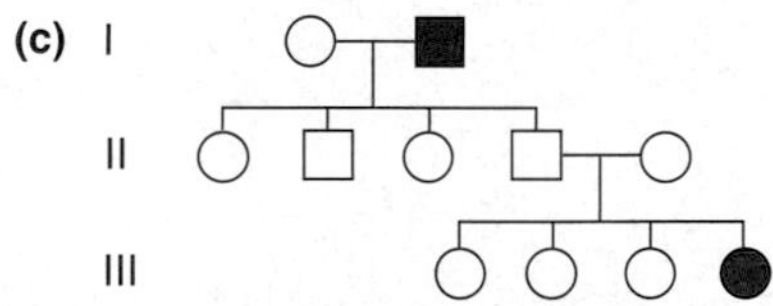

30. Consider the pedigree that follows for cutis laxa, a connective tissue disorder in which the skin hangs in loose folds.

a. Assuming complete penetrance and that the trait is rare, what is the apparent mode of inheritance?

b. What is the probability that individual II-2 is a carrier?

c. What is the probability that individual II-3 is a carrier?

d. What is the probability that individual III-1 is affected by the disease?

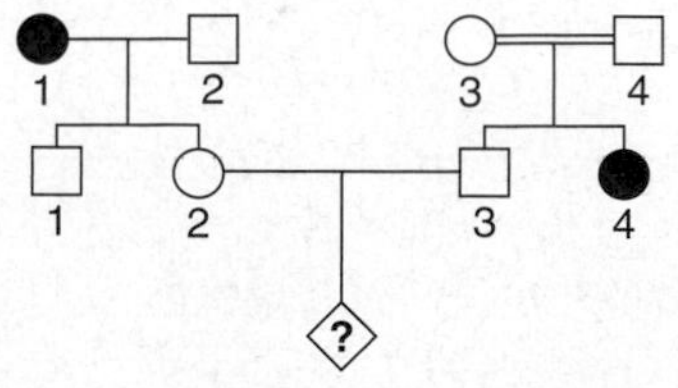

31. A young couple went to see a genetic counselor because each had a sibling affected with cystic fibrosis. (Cystic fibrosis is a recessive disease, and neither member of the couple nor any of their four parents is affected.)

a. What is the probability that the female of this couple is a carrier?

b. What are the chances that their child will be affected with cystic fibrosis?

c. What is the probability that their child will be a carrier of the cystic fibrosis mutation?

32. Huntington disease is a rare fatal, degenerative neurological disease in which individuals start to show symptoms, on average, in their 40s. It is caused by a dominant allele. Joe, a man in his 20s, just learned that his father has Huntington disease.

a. What is the probability that Joe will also develop the disease?

b. Joe and his new wife have been eager to start a family. What is the probability that their first child will eventually develop the disease?

33. Is the disease shown in the following pedigree dominant or recessive? Why? Based on this limited pedigree, do you think the disease allele is rare or common in the population? Why?

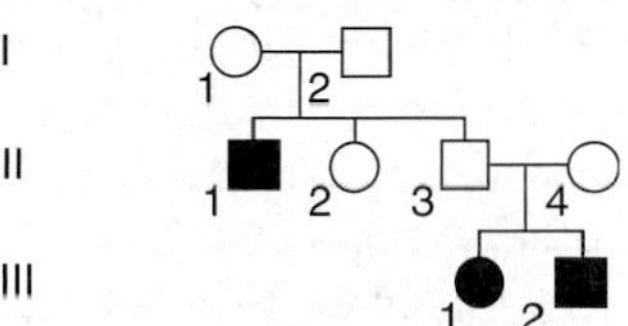

34. Figure 2.21 on p. 32 shows the inheritance of Huntington disease in a family from a small village near Lake Maracaibo in Venezuela. The village was founded by a small number of immigrants, and generations of their descendents have remained concentrated in this isolated location. The allele for Huntington disease has remained unusually prevalent there.

a. Why could you not conclude definitively that the disease is the result of a dominant or a recessive allele solely by looking at this pedigree?

b. Is there any information you could glean from the family's history that might imply the disease is due to a dominant rather than a recessive allele?

35. The common grandfather of two first cousins has hereditary hemochromatosis, a recessive condition causing an abnormal buildup of iron in the body. Neither of the cousins has the disease nor do any of their relatives.

a. If the first cousins mated with each other and had a child, what is the chance that the child would have hemochromatosis? Assume that the unrelated, unaffected parents of the cousins are not carriers.

b. How would your calculation change if you knew that 1 out of every 10 unaffected people in the population (including the unrelated parents of these cousins) was a carrier for hemochromatosis?

36. People with nail-patella syndrome have poorly developed or absent kneecaps and nails. Individuals with alkaptonuria have arthritis as well as urine that darkens when exposed to air. Both nail-patella syndrome and alkaptonuria are rare phenotypes. In the following pedigree, vertical red lines indicate individuals with nail-patella syndrome, while horizontal green lines denote individuals with alkaptonuria.

a. What are the most likely modes of inheritance of nail-patella syndrome and alkaptonuria? What genotypes can you ascribe to each of the individuals in the pedigree for both of these phenotypes?

b. In a mating between IV-2 and IV-5, what is the chance that the child produced would have both nail-patella syndrome and alkaptonuria? Nail-patella syndrome alone? Alkaptonuria alone? Neither defect?

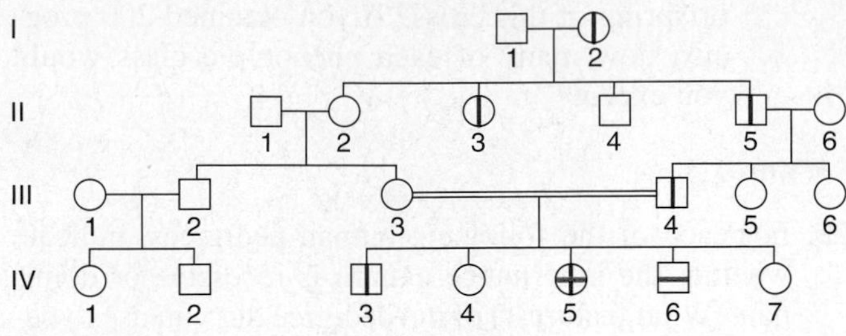

37. Midphalangeal hair (hair on top of the middle segment of the fingers) is a common phenotype caused by a dominant allele *M*. Homozygotes for the recessive allele (*mm*) lack hair on the middle segment of their fingers. Among 1000 families in which both parents had midphalangeal hair, 1853 children showed the trait while 209 children did not. Explain this result.

Extensions to Mendel: Complexities in Relating Genotype to Phenotype

Chapter 3

Unlike the pea traits that Mendel examined, most human characteristics—height, hair color, skin color, voice quality, artistic or athletic ability, to name a few—do not fall neatly into just two opposing phenotypic categories. And because people cannot be simply divided into two contrasting camps for each trait (brown hair versus blond, high voice versus low), they seem to defy Mendelian analysis. The same can be said of traits expressed by many of the world's food crops; the size, shape, succulence, and nutrient content of the plants' edible parts vary over a wide range of values. Lentils (*Lens culinaris*) provide a graphic illustration of this variation. A legume like peas and beans, lentils are grown in many parts of the world as a rich source of both protein and carbohydrate. The mature plants set fruit in the form of diminutive pods that contain two small seeds, which can be ground into meal or used in soups, salads, and stews. These seeds come in an intriguing array of colors and patterns (**Fig. 3.1**), and since particular couplings of color and pattern are preferred in the cuisines of different cultures, there is considerable commercial interest in producing specific seed coat combinations. But crosses between pure-breeding lines of lentils result in some startling surprises in relation to Mendel's laws. A cross between pure-breeding tan and pure-breeding gray parents, for example, yields an all brown F_1 generation; and when these novel hybrids are left to self-pollinate, the F_2 plants produce not only tan, gray, and brown lentils, but also green.

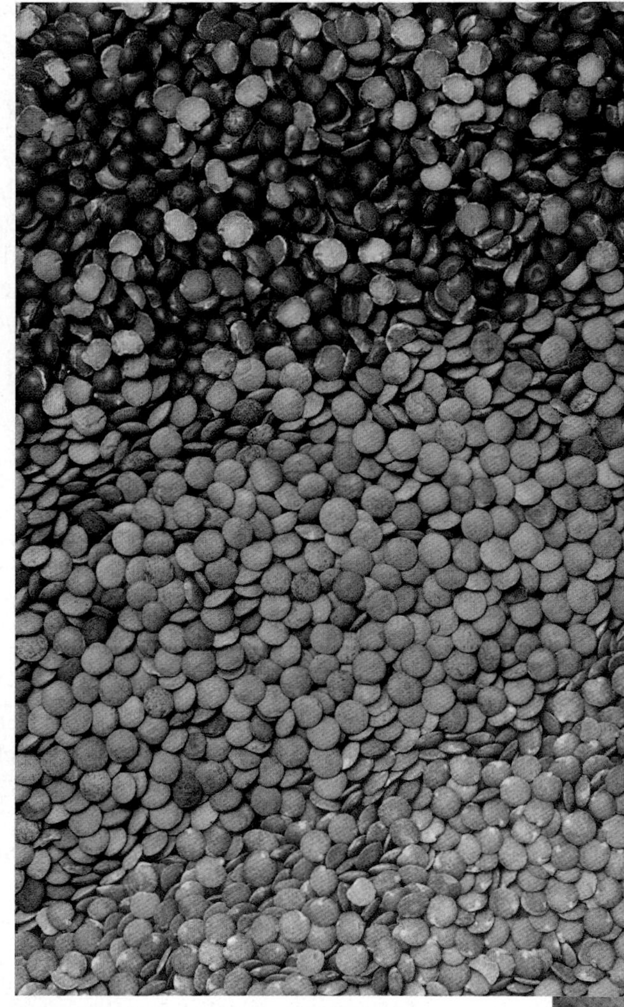

In this array of red, green, and brown lentils, some of the seeds have speckled patterns, while others are clear.

Beginning with the first decade of the twentieth century, geneticists subjected all kinds of puzzling variation in plants and animals to controlled breeding tests, using Mendel's 3:1 phenotypic ratio as a guideline in interpreting the results. If the traits under analysis behaved as predicted by Mendel's laws, it meant they were determined by a single gene with alternative alleles that show clear-cut dominance and recessiveness. Many traits, however, did not behave in this way. For some, there was no definitive dominance and recessiveness, or there were more than two alternative alleles in a particular cross. Other traits turned out to be **multifactorial,** that is, determined by two or more factors, including multiple genes interacting with each other or one or more genes interacting with the environment. The seed coat color of lentils is a multifactorial trait. Because such traits arise from an intricate network of interactions, they do not necessarily generate straightforward Mendelian phenotypic ratios. Nonetheless, simple extensions of Mendel's hypotheses that clarify the relationship between genotype and phenotype have allowed geneticists to explain all of the observed deviations without challenging Mendel's basic laws.

As we look at the genetic factors that make it possible for plants and animals to generate the enormous amount of phenotypic variation seen in the natural world, we encounter the following general theme: In their attempt to sort things out, geneticists usually limit the number of variables under investigation at any one time.

Figure 3.1 Some phenotypic variation poses a challenge to Mendelian analysis. As we will see in this chapter, combinations of alleles at two genes determine differences between the colors of lentil seed coats, while combinations of the multiple alleles of a third gene produce variations in patterns of lentil speckling.

Mendel, for example, compared and mated pure-breeding, inbred strains of peas that differed from each other by one or a few traits. By establishing a genetic background that was uniform, or homogenized, for all other traits, he was able to visualize the actions of single genes in isolation. Similarly, twentieth-century geneticists used inbred populations of fruit flies, mice, and other experimental organisms to study specific traits. Of course, geneticists cannot approach people in this way. Human populations are far from inbred, and if a phenotype arises from many genes that are not held constant, it may be difficult to see how a particular gene affects a particular trait. For this reason, until recently the genetic basis of much human variation remained a mystery. The advent of molecular biology in the 1970s, however, provided new tools that geneticists now use to unravel the genetics of complex human traits.

We divide our presentation of extensions to Mendelian analysis into two broad categories:

- *Single-gene inheritance* in which
  a. pairs of alleles show deviations from complete dominance and recessiveness,
  b. different forms of the gene are not limited to two alleles, and
  c. one gene may determine more than one trait.
- *Multifactorial inheritance* in which the phenotype arises from
  a. the interaction of more than one gene, and
  b. the interaction of genes with the environment.

3.1 Extensions to Mendel for Single-Gene Inheritance

William Bateson, an early interpreter and defender of Mendel, who coined the terms "genetics," "allelomorph" (later shortened to "allele"), "homozygote," and "heterozygote," entreated the audience at a 1908 lecture: "Treasure your exceptions! . . . Keep them always uncovered and in sight. Exceptions are like the rough brickwork of a growing building which tells that there is more to come and shows where the next construction is to be." Consistent exceptions to simple Mendelian ratios revealed unexpected patterns of single-gene inheritance. By distilling the significance of these patterns, Bateson and other early geneticists extended the scope of Mendelian analysis and obtained a deeper understanding of the relationship between genotype and phenotype. We now look at the major extensions to Mendelian analysis elucidated over the last century.

Dominance Is Not Always Complete

A consistent working definition of dominance and recessiveness depends on the F_1 hybrids that arise from a mating between two pure-breeding lines. If a hybrid is identical to one parent for the trait under consideration, the allele carried by that parent is deemed dominant to the allele carried by the parent whose trait is not expressed in the hybrid. If, for example, a mating between a pure-breeding white line and a pure-breeding blue line produces F_1 hybrids that are white, the white allele of the gene for color is dominant to the blue allele. If the F_1 hybrids are blue, the blue allele is dominant to the white one (**Fig. 3.2**).

Mendel described and relied on complete dominance in sorting out his ratios and laws, but it is not the only kind of dominance he observed. Figure 3.2 diagrams two situations in which neither allele of a gene is completely dominant. As the figure shows, crosses between true-breeding strains can produce hybrids with phenotypes that differ from both parents. We now explain how these phenotypes arise.

In Incomplete Dominance, the F_1 Hybrid Resembles Neither Pure-Breeding Parent

A cross between pure late-blooming and pure early-blooming pea plants results in an F_1 generation that blooms in between the two extremes. This is just one of many examples of **incomplete dominance,** in which the hybrid does not resemble either pure-breeding parent. F_1 hybrids that differ from both parents often express a phenotype that is intermediate between those of the pure-breeding parents. Thus, with incomplete dominance, neither parental allele is

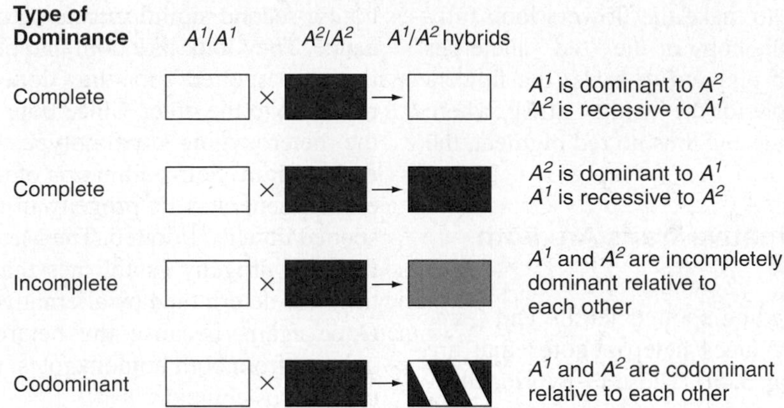

Figure 3.2 Different dominance relationships. The phenotype of the heterozygote defines the dominance relationship between two alleles of the same gene (here, A^1 and A^2). Dominance is complete when the hybrid resembles one of the two pure-breeding parents. Dominance is incomplete when the hybrid resembles neither parent; its novel phenotype is usually intermediate between the parental phenotypes. Codominance occurs when the hybrid shows the traits from both pure-breeding parents.

(a) *Antirrhinum majus* (snapdragons)

(b) A Punnett square for incomplete dominance

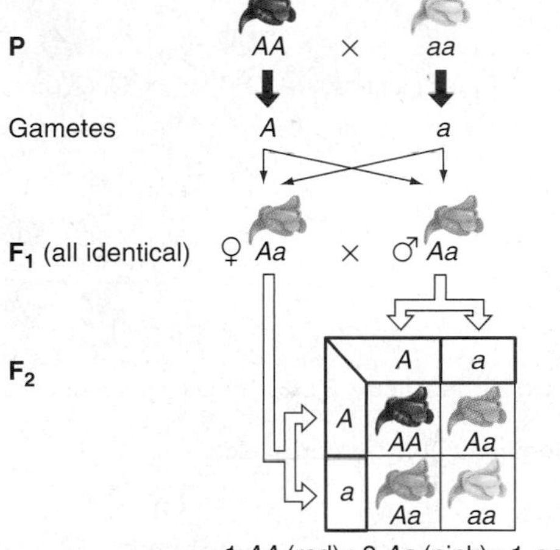

Figure 3.3 Pink flowers are the result of incomplete dominance. (a) Color differences in these snapdragons reflect the activity of one pair of alleles. **(b)** The F_1 hybrids from a cross of pure-breeding red and white strains of snapdragons have pink blossoms, resembling neither parent. Flower colors in the F_2 appear in the ratio of 1 red : 2 pink : 1 white. This ratio signifies that the alleles of a single gene determine these three colors.

dominant or recessive to the other; both contribute to the F_1 phenotype. Mendel observed plants that bloomed midway between two extremes when he cultivated various types of pure-breeding peas for his hybridization studies, but he did not pursue the implications. Blooming time was not one of the seven characteristics he chose to analyze in detail, almost certainly because in peas, the time of bloom was not as clear-cut as seed shape or flower color.

In many plant species, flower color serves as a striking example of incomplete dominance. With the tubular flowers of four-o'clocks or the floret clusters of snapdragons, for instance, a cross between pure-breeding red-flowered parents and pure-breeding white yields hybrids with pink blossoms, as if a painter had mixed red and white pigments to get pink (**Fig. 3.3a**). If allowed to self-pollinate, the F_1 pink-blooming plants produce F_2 progeny bearing red, pink, and white flowers in a ratio of 1:2:1 (Fig. 3.3b). This is the familiar *genotypic* ratio of an ordinary single-gene F_1 self-cross. What is new is that because the heterozygotes look unlike either homozygote, the *phenotypic* ratios are an exact reflection of the genotypic ratios. The biochemical explanation for this type of incomplete dominance is that each allele of the gene under analysis specifies an alternative form of a protein molecule with an enzymatic role in pigment production. If the "white" allele does not give rise to a functional enzyme, no pigment appears. Thus, in snapdragons and four-o'clocks, two "red" alleles per cell produce a double dose of a red-producing enzyme, which generates enough pigment to make the flowers look fully red. In the heterozygote, one copy of the "red" allele per cell results in only enough pigment to make the flowers look pink. In the homozygote for the "white" allele, where there is no functional enzyme and thus no red pigment, the flowers appear white.

In Codominance, Alternative Traits Are Both Visible in the F_1 Hybrid

A cross between pure-breeding spotted lentils and pure-breeding dotted lentils produces heterozygotes that are both spotted and dotted (**Fig. 3.4a**). These F_1 hybrids illustrate a second significant departure from complete dominance. They look like both parents, which means that neither the "spotted" nor the "dotted" allele is dominant or recessive to the other. Since both traits show up equally in the heterozygote's phenotype, the alleles are termed **codominant.** Self-pollination of the spotted/dotted F_1 generation generates F_2 progeny in the ratio of 1 spotted : 2 spotted/dotted : 1 dotted. The Mendelian 1:2:1 ratio among these F_2 progeny establishes that the spotted and dotted traits are determined by alternative alleles of a single gene. Once again, because the heterozygotes can be distinguished from both homozygotes, the phenotypic and genotypic ratios coincide.

In humans, some of the complex membrane-anchored molecules that distinguish different types of red blood cells exhibit codominance. For example, one gene (I) with alleles I^A and I^B controls the presence of a sugar polymer that protrudes from the red blood cell membrane. The alternative alleles each encode a slightly different form of an enzyme that causes production of a slightly different form of the complex sugar. In heterozygous individuals, the red blood cells carry both the I^A-determined and the I^B-determined sugars on their surface, whereas the cells of homozygous individuals display the products of either I^A or I^B alone (Fig. 3.4b). As this example illustrates, alleles are often codominant for phenotypes expressed at the molecular level.

Figure 3.2 on p. 47 summarizes the differences between complete dominance, incomplete dominance, and codominance for phenotypes reflected in color variations. With complete dominance, F_1 progeny look like one of the true-breeding parents. With incomplete dominance, hybrids resemble neither of the parents and thus display neither pure-breeding trait. With codominance, the phenotypes of both pure-breeding lines show up simultaneously in the F_1 hybrid. Determinations of dominance relationships depend on what phenotype appears in the F_1 generation. Complete dominance, as we saw in Chapter 2, results in a 3:1 ratio of phenotypes in the F_2, while both incomplete dominance and codominance yield 1:2:1 F_2 ratios.

(a) Codominant lentil coat patterns

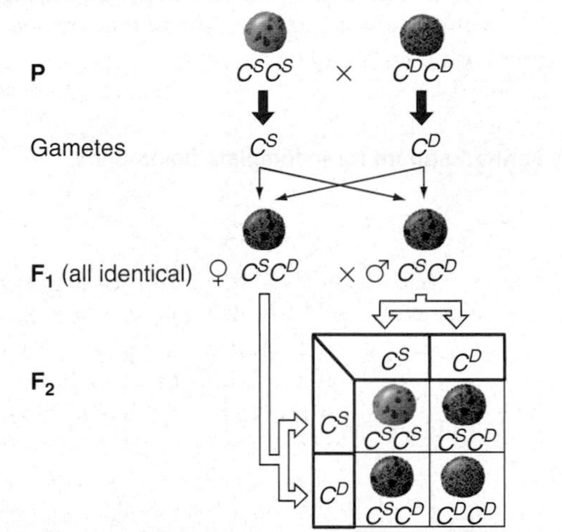

1 C^SC^S (spotted) : 2 C^SC^D (spotted/dotted) : 1 C^DC^D (dotted)

(b) Codominant blood group alleles

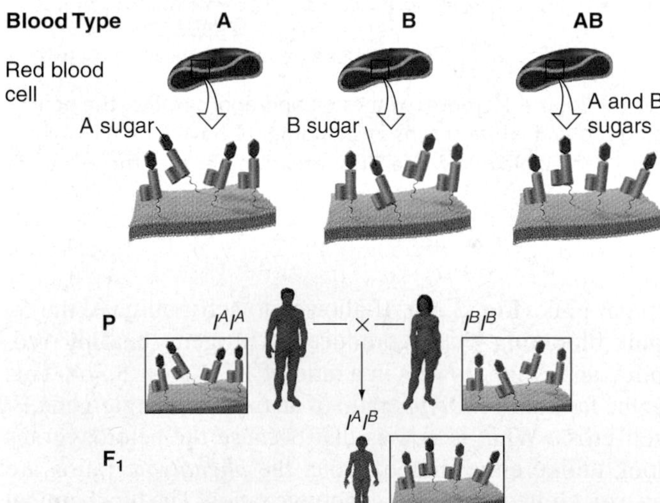

Figure 3.4 In codominance, F_1 hybrids display the traits of both parents. **(a)** A cross between pure-breeding spotted lentils and pure-breeding dotted lentils produces heterozygotes that are both spotted and dotted. Because each genotype has its own corresponding phenotype, the F_2 ratio is 1:2:1. **(b)** The I^A and I^B blood group alleles are codominant because the red blood cells of an I^AI^B heterozygote have both kinds of sugars at their surface.

Variations on Complete Dominance Do Not Negate Mendel's Law of Segregation

The dominance relations of a gene's alleles do not affect the alleles' transmission. Whether two alternative alleles of a single gene show complete dominance, incomplete dominance, or codominance depends on the kinds of proteins determined by the alleles and the biochemical function of those proteins in the cell. These same phenotypic dominance relations, however, have no bearing on the segregation of the alleles during gamete formation. As Mendel proposed, cells still carry two copies of each gene, and these copies—a pair of either similar or dissimilar alleles—segregate during gamete formation. Fertilization then restores two alleles to each cell without reference to whether the alleles are the same or different. Variations in dominance relations thus do not detract from Mendel's laws of segregation. Rather, they reflect differences in the way gene products control the production of phenotypes, adding a level of complexity to the task of sorting out the visible results of gene transmission and inferring genotype from phenotype.

A Gene May Have More Than Two Alleles

Mendel analyzed "either-or" traits controlled by genes with two alternative alleles, but for many traits, there are more than two alternatives. Human blood types provide an example. If a person with blood type A mates with a person with blood type B, it is possible in some cases for the couple to have a child that is neither A nor B, but a third blood type called O. The reason? The gene for the ABO blood types has three alleles: I^A, I^B, and i (**Fig. 3.5a**). Allele I^A gives rise to blood type A by specifying an enzyme that adds sugar A, I^B results in blood type B by specifying an enzyme that adds sugar B; i does not produce a functional sugar-adding enzyme. Alleles I^A and I^B are both dominant to i, and blood type O is therefore a result of homozygosity for allele i.

Note in Fig. 3.5a that the A phenotype can arise from two genotypes, I^AI^A or I^Ai. The same is true for the B blood type, which can be produced by I^BI^B or I^Bi. But a combination of the two alleles I^AI^B generates blood type AB.

We can draw several conclusions from these observations. First, as already stated, a given gene may have more than two alleles, or **multiple alleles;** in our example, the series of alleles is denoted I^A, I^B, i.

Second, although the ABO blood group gene has three alleles, each person carries only two of the alternatives— I^AI^A, I^BI^B, I^AI^B, I^Ai, I^Bi, or ii. There are thus six possible ABO genotypes. Because each individual carries no more than two alleles for each gene, no matter how many alleles there are in a series, Mendel's law of segregation remains intact, since in a sexually reproducing organism, the two alleles of a gene separate during gamete formation.

Third, an allele is not inherently dominant or recessive; its dominance or recessiveness is always relative to a second allele. In other words, dominance relations are unique to a pair of alleles. In our example, I^A is completely dominant to i, but it is codominant with I^B. Given these dominance relations, the six genotypes possible with I^A, I^B, and i generate four different phenotypes: blood groups A, B, AB, and O. With this background, you can understand how a type A and a type B parent could produce a type O child: The parents must be I^Ai and I^Bi heterozygotes, and the child receives an i allele from each parent.

An understanding of the genetics of the ABO system has had profound medical and legal repercussions. Matching ABO blood types is a prerequisite of successful blood transfusions, because people make antibodies to foreign blood cell molecules. A person whose cells carry only A molecules, for example, produces anti-B antibodies; B people manufacture anti-A antibodies; AB individuals make neither type of antibody; and O individuals produce both anti-A and anti-B antibodies (Fig. 3.5b). These antibodies cause coagulation of cells displaying the foreign molecules (Fig. 3.5c). As a result, people with blood type O have historically been known as universal donors because their red blood cells carry no surface molecules that will stimulate an antibody attack. In contrast, people with blood type AB are considered universal recipients, because they make neither anti-A nor anti-B antibodies, which, if present, would target the surface molecules of incoming blood cells.

Information about ABO blood types can also be used as legal evidence in court, to exclude the possibility of paternity or criminal guilt. In a paternity suit, for example, if the mother is type A and her child is type B, logic dictates that the I^B allele must have come from the father, whose

(a)

Genotypes	Corresponding Phenotypes: Type(s) of Molecule on Cell
I^AI^A I^Ai	A
I^BI^B I^Bi	B
I^AI^B	AB
ii	O

(b)

Blood Type	Antibodies in Serum
A	Antibodies against B
B	Antibodies against A
AB	No antibodies against A or B
O	Antibodies against A and B

(c)

Blood Type of Recipient	Donor Blood Type (Red Cells) A	B	AB	O
A	+	−	−	+
B	−	+	−	+
AB	+	+	+	+
O	−	−	−	+

Figure 3.5 ABO blood types are determined by three alleles of one gene. **(a)** Six genotypes produce the four blood group phenotypes. **(b)** Blood serum contains antibodies against foreign red blood cell molecules. **(c)** If a recipient's serum has antibodies against the sugars on a donor's red blood cells, the blood types of recipient and donor are incompatible and coagulation of red blood cells will occur during transfusions. In this table, a plus (+) indicates compatibility, and a minus (−) indicates incompatibility. Although in some cases antibodies in the donor's blood can react with the recipient's red blood cells (for example, a type O donor and an AB recipient), this is almost never a problem because the small volume of transfused antibody is diluted in the recipient's bloodstream.

genotype may be I^AI^B, I^BI^B, or I^Bi. In 1944, the actress Joan Barry (phenotype A) sued Charlie Chaplin (phenotype O) for support of a child (phenotype B) whom she claimed he fathered. The scientific evidence contradicting her claim (the father had to carry an I^B allele and Chaplin was apparently *ii*) was admissible in court but did not convince the jury, and Chaplin had to pay. Today, the molecular genotyping of DNA provides a powerful tool to help establish paternity, guilt, or innocence, but juries still often find it difficult to evaluate such evidence.

Lentils offer another example of multiple alleles. A gene for seed coat pattern has five alleles: spotted, dotted, clear (pattern absent), and two types of marbled. Reciprocal crosses between pairs of pure-breeding lines of all patterns (marbled-1 × marbled-2, marbled-1 × spotted, marbled-2 × spotted, and so forth) have clarified the dominance relations of all possible pairs of the alleles to reveal a **dominance series** in which alleles are listed in order from most dominant to most recessive. For example, crosses of marbled-1 with marbled-2, or of marbled-1 with spotted or dotted or clear, produce the marbled-1 phenotype in the F_1 generation and a ratio of three marbled-1 to one of any of the other phenotypes in the F_2. This indicates that the marbled-1 allele is completely dominant to each of the other four alleles. Analogous crosses with the remaining four phenotypes reveal the dominance series shown in **Fig. 3.6.** Recall that dominance relations are meaningful only when comparing two alleles; an allele, such as marbled-2, can be recessive to a second allele (marbled-1) but dominant to a third and fourth (dotted and clear). The fact that all tested pairings of lentil seed coat pattern alleles yielded a 3:1 ratio in the F_2 generation (except for spotted × dotted, which yielded the 1:2:1 phenotypic ratio reflective of codominance) indicates that these lentil seed coat patterns are determined by different alleles of the same gene.

In some multiple allelic series, each allele is codominant with every other allele, and every distinct genotype therefore produces a distinct phenotype. This happens particularly with traits detectable only at the molecular level. An extreme example is the group of three major genes that encode a class of cell surface molecules in humans and other mammals known as **histocompatibility antigens.** Carried by all of the body's cells except the red blood cells and sperm, histocompatibility antigens play a critical role in stimulating a proper immune response that destroys intruders (viral or bacterial, for example) while leaving the body's own tissues intact. Because each of the three major histocompatibility genes (called *HLA-A, HLA-B,* and *HLA-C* in humans) has between 20 and 100 alleles, the number of possible allelic combinations creates a powerful potential for the phenotypic variation of cell surface molecules. Other than identical (that is, *monozygotic*) twins, no two people are likely to carry the same array of cell surface molecules.

Parental Generation: Parental seed coat pattern in cross Parent 1 × Parent 2	F_1 Generation: F_1 phenotype	F_2 Generation: Total F_2 frequencies and phenotypes			Apparent phenotypic ratio
marbled-1 × clear	marbled-1	798	296		3:1
marbled-2 × clear	marbled-2	123	46		3:1
spotted × clear	spotted	283	107		3:1
dotted × clear	dotted	1,706	522		3:1
marbled-1 × marbled-2	marbled-1	272	72		3:1
marbled-1 × spotted	marbled-1	499	147		3:1
marbled-1 × dotted	marbled-1	1,597	549		3:1
marbled-2 × dotted	marbled-2	182	70		3:1
spotted × dotted	spotted/dotted	168	339	157	1:2:1

Dominance series: marbled-1 > marbled-2 > spotted = dotted > clear

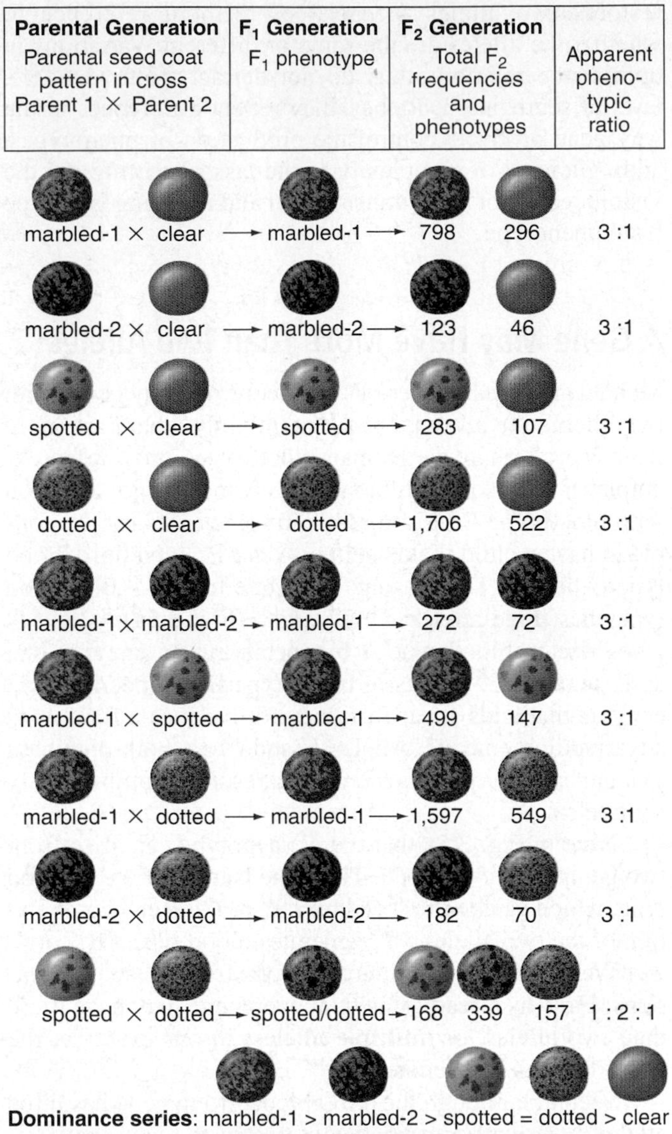

Figure 3.6 How to establish the dominance relations between multiple alleles. Pure-breeding lentils with different seed coat patterns are crossed in pairs, and the F_1 progeny are self-fertilized to produce an F_2 generation. The 3:1 or 1:2:1 F_2 monohybrid ratios from all of these crosses indicate that different alleles of a single gene determine all the traits. The phenotypes of the F_1 hybrids establish the dominance relationships between the alleles, as shown at the *bottom.* Note that spotted and dotted alleles are codominant to each other, but each is recessive to the marbled alleles and dominant to clear.

Mutations Are the Source of New Alleles

How do the multiple alleles of an allelic series arise? The answer is that chance alterations of the genetic material, known as **mutations,** arise spontaneously in nature. Once they occur in gamete-producing cells, they are faithfully inherited. Mutations that have phenotypic consequences can be counted, and such counting reveals that they occur at low frequency. The frequency of gametes carrying a mutation in a particular gene varies anywhere from 1 in 10,000 to 1 in 1,000,000. This range exists because different genes have different mutation rates.

Mutations make it possible to follow gene transmission. If, for example, a mutation specifies an alteration in an enzyme that normally produces yellow so that it now makes green, the new phenotype (green) will make it possible to recognize the new mutant allele. In fact, it takes at least two alleles, that is, some form of variation, to "see" the transmission of a gene. Thus, in segregation studies, geneticists can analyze only genes with variants; they have no way of following a gene that comes in only one form. If all peas were yellow, Mendel would not have been able to decipher the transmission patterns of the gene for the seed color trait. We discuss mutations in greater detail in Chapter 7.

The continued existence of a new allele within a population—that is, within a group of interbreeding organisms of the same species that inhabit the same area—depends on many factors, particularly, its contribution to the reproductive success of the organism. Alleles that confer a survival advantage allow individuals carrying them to produce more offspring; these alleles will tend to increase in the population. Other alleles are deleterious and are lost from the population. Still other alleles are neutral; more than 99.9% of these are eventually lost from a large population because they provide no selective advantage, but occasionally, purely by chance, one may become established.

Since each organism carries two copies of every gene, you can calculate the number of copies of a gene in a given population by multiplying the number of individuals by 2. Each allele of the gene accounts for a percentage of the total number of gene copies, and that percentage is known as the **allele frequency.** An allele whose frequency is greater than 1% is by definition a **wild-type allele,** often designated by a superscript plus sign ($^+$). An allele with a frequency of less than 1% is considered a **mutant allele.** (A mutation is a newly arisen mutant allele.) In mice, for example, one of the main genes determining coat color is the *agouti* gene. The wild-type allele (*A*) produces fur with each hair having yellow and black bands that blend together from a distance to give the appearance of dark gray, or agouti. Researchers have identified in the laboratory 14 distinguishable mutant alleles for the *agouti* gene. One of these (a^t) is recessive to the wild type and gives rise to a black coat on the back and a yellow coat on the belly; another (*a*) is also recessive to *A* and produces a pure black coat (**Fig. 3.7**). In nature, wild-type agoutis (*AA*) survive to reproduce, while very few black-backed or pure black mutants (a^ta^t or *aa*) do so because their dark coat makes it hard for them to evade the eyes of predators. As a result, *A* is present at a frequency of much more than 99% and is thus the only wild-type allele in mice for the *agouti* gene. A gene with only one wild-type allele is **monomorphic.**

(a) *Mus musculus* (house mouse) coat colors

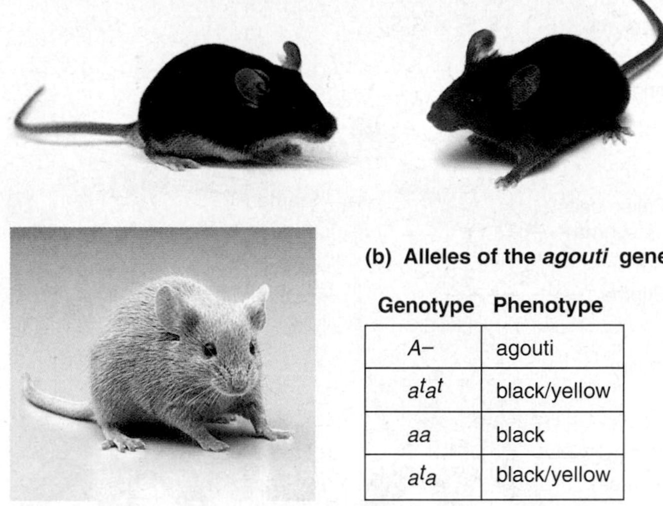

(b) Alleles of the *agouti* gene

Genotype	Phenotype
$A-$	agouti
a^ta^t	black/yellow
aa	black
a^ta	black/yellow

(c) Evidence for a dominance series

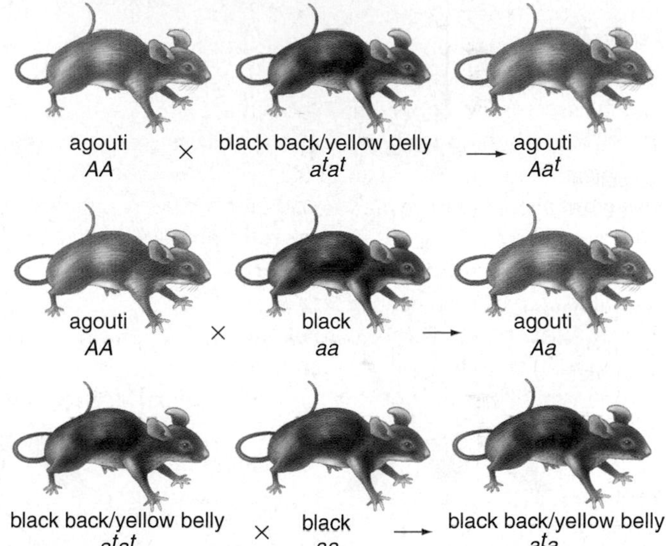

Dominance series: $A > a^t > a$

Figure 3.7 The mouse *agouti* gene: One wild-type allele, many mutant alleles. (a) Black-backed, yellow-bellied (*top left*); black (*top right*); and agouti (*bottom*) mice. **(b)** Genotypes and corresponding phenotypes for alleles of the *agouti* gene. **(c)** Crosses between pure-breeding lines make it possible to arrange these three alleles in a dominance series. Interbreeding of the F_1 hybrids (not shown) yields 3:1 phenotypic ratios of F_2 progeny, indicating that *A*, a^t, and *a* are in fact alleles of the same gene.

In contrast, some genes have more than one wild-type allele, which makes them **polymorphic.** For example, in the ABO blood type system, all three alleles—I^A, I^B, and *i*—have a frequency greater than 1%. They are thus all wild-type alleles, and the gene is polymorphic.

A rather unusual mechanism leading to the proliferation of many different alleles occurs in the mating systems

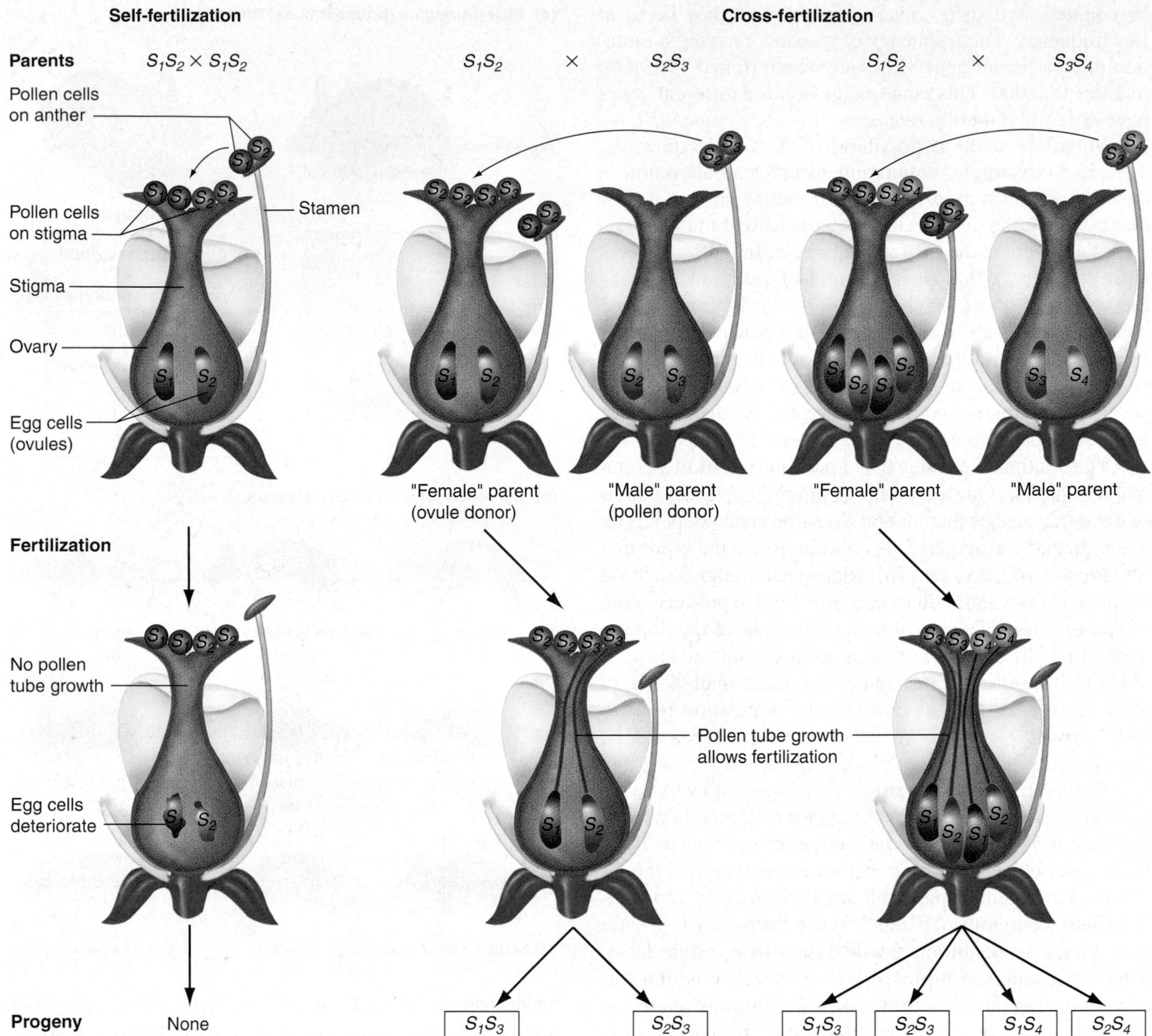

Figure 3.8 Plant incompatibility systems promote outbreeding and allele proliferation. A pollen grain carrying an allele of the self-incompatibility gene that is identical to either of the two alleles carried by a potential female parent is unable to grow a pollen tube; as a result, fertilization cannot take place. Because all the pollen grains produced by any one plant have one of the two alleles carried by the female reproductive parts of the same plant, self-fertilization is impossible.

of wild species of tomatoes and petunias. Evolution of an "incompatibility" gene whose alleles determine acceptance or rejection of pollen has allowed these plants to prevent self-fertilization and promote outbreeding. In this form of incompatibility, a plant cannot accept pollen carrying an allele identical to either of its own incompatibility alleles. If, for example, pollen carrying allele S_1 of the incompatibility gene lands on the stigma of a plant that also carries S_1 as one of its incompatibility alleles, a pollen tube will not grow (**Fig. 3.8**). Every plant is thus heterozygous for the incompatibility gene, since the pollen grain and female reproductive organs needed to form the plant cannot share alleles. Plants carrying rare alleles (that have arisen relatively recently by mutation and are not present in many other plants) will be able to send pollen to and receive pollen from most of the other plants in their population. In some species with this type of mating system, geneticists have detected as many as 92 alleles for the incompatibility gene. Because this incompatibility mechanism encourages the proliferation of new mutants, there are numerous wild-type alleles as well as many mutants. This is an extreme case of multiple alleles, not seen with most genes.

One Gene May Contribute to Several Visible Characteristics

Mendel derived his laws from studies in which one gene determined one trait; but, always the careful observer, he himself noted possible departures. In listing the traits selected for his pea experiments, he remarked that specific seed coat colors are always associated with specific flower colors.

The phenomenon of a single gene determining a number of distinct and seemingly unrelated characteristics is known as **pleiotropy.** Since geneticists now know that each gene determines a specific protein and that each protein can have a cascade of effects on an organism, we can understand how pleiotropy arises. Among the aboriginal Maori people of New Zealand, for example, many of the men develop respiratory problems and are also sterile. Researchers have found that the fault lies with the recessive allele of a single gene. The gene's normal dominant allele specifies a protein necessary for the action of cilia and flagella, both of which are hairlike structures extending from the surfaces of some cells. In men who are homozygous for the recessive allele, cilia that normally clear the airways fail to work effectively, and flagella that normally propel sperm fail to do their job. Thus, one gene determines a protein that indirectly affects both respiratory function and reproduction.

Some Alleles May Cause Lethality

A significant variation of pleiotropy occurs in alleles that not only produce a visible phenotype but also affect viability. Mendel assumed that all genotypes are equally viable such that representatives of each genotype have an equal rate of survival. If this were not true and a large percentage of, say, homozygotes for a particular allele died before germination or birth, you would not be able to count them after birth, and this would alter the 1:2:1 genotypic ratios and the 3:1 phenotypic ratios predicted for the F_2 generation.

Consider the inheritance of coat color in mice. As mentioned earlier, wild-type agouti (*AA*) animals have black and yellow striped hairs that appear dark gray to the eye. One of the 14 mutant alleles of the *agouti* gene gives rise to mice with a much lighter, almost yellow color. When inbred *AA* mice are mated to yellow mice, one always observes a 1:1 ratio of the two coat colors among the offspring (**Fig. 3.9a**). From this result, we can draw three conclusions: (1) All yellow mice must carry the *agouti* allele even though they do not express it; (2) yellow is therefore dominant to agouti; and (3) all yellow mice are heterozygotes. (Note again that dominance and recessiveness are defined in the context of each pair of alleles. Even though, as previously mentioned, agouti (*A*) is dominant to the a^t and *a* mutations for black coat color, it can still be recessive to the yellow coat color allele.) If we designate the allele for yellow as A^y, the yellow mice in the preceding cross are A^yA heterozygotes, and the agoutis, *AA* homozygotes. So far, no surprises. But a mating of yellow to yellow produces a skewed phenotypic ratio of two yellow mice to one agouti (Fig. 3.9b). Among these progeny, matings between agouti mice show that the agoutis are all pure-breeding and therefore *AA* homozygotes as expected. There are, however, no pure-breeding yellow mice among the progeny. When the yellow mice are mated to each other, they unfailingly produce 2/3 yellow and 1/3 agouti offspring, a ratio of 2:1, so they must therefore be heterozygotes. In short, one can never obtain pure-breeding yellow mice.

How can we explain this phenomenon? The Punnett square in Fig. 3.9b suggests an answer. Two copies of the A^y allele prove fatal to the animal carrying them, whereas one copy of the allele produces a yellow coat. This means that the A^y allele affects two different traits: It is dominant to *A* in the determination of coat color, but it is recessive to

(a) All yellow mice are heterozygotes.

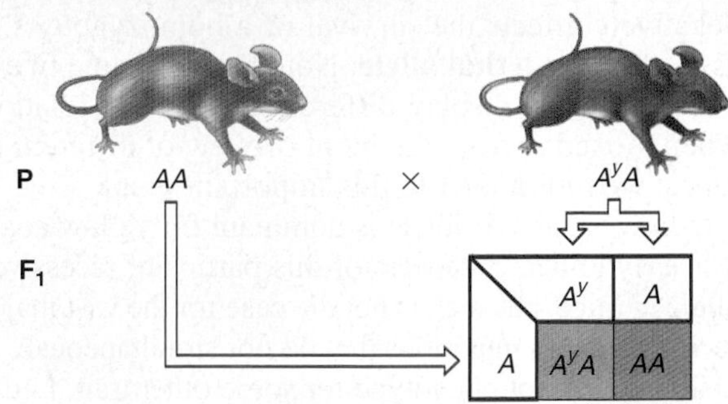

(b) Two copies of A^y cause lethality.

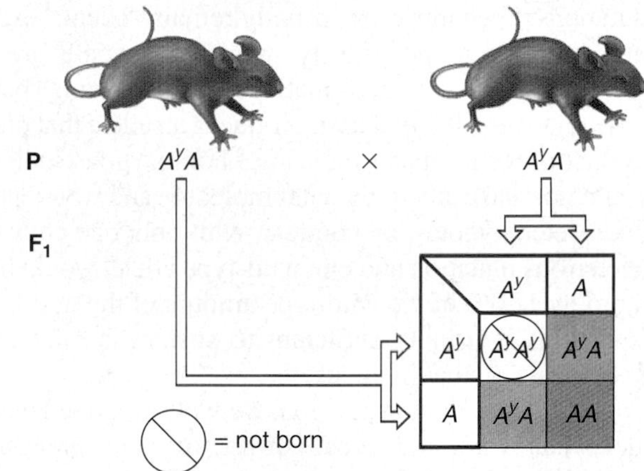

Figure 3.9 **A^y: A recessive lethal allele that also produces a dominant coat color phenotype.** **(a)** A cross between inbred agouti mice and yellow mice yields a 1:1 ratio of yellow to agouti progeny. The yellow mice are therefore A^yA heterozygotes, and for the trait of coat color, A^y (for yellow) is dominant to *A* (for agouti). **(b)** Yellow mice do not breed true. In a yellow × yellow cross, the 2:1 ratio of yellow to agouti progeny indicates that the A^y allele is a recessive lethal.

A in the production of lethality. An allele, such as A^y, that negatively affects the survival of a homozygote is known as a **recessive lethal allele.** Note that the same two alleles (A^y and A) can display different dominance relationships when looked at from the point of view of different phenotypes; we return later to this important point.

Because the A^y allele is dominant for yellow coat color, it is easy to detect carriers of this particular recessive lethal allele in mice, but such is not the case for the vast majority of recessive lethal mutations that do not simultaneously show a visible dominant phenotype for some other trait. Lethal mutations can arise in many different genes, and as a result, most animals, including humans, carry some recessive lethal mutations. Such mutations usually remain "silent," except in rare cases of homozygosity, which in people are often caused by consanguineous matings (that is, matings between close relatives). If a mutation produces an allele that prevents production of a crucial molecule, homozygous individuals would not make any of the vital molecule and would not survive. Heterozygotes, by contrast, with only one copy of the deleterious mutation and one wild-type allele, would be able to produce 50% of the wild-type amount of the normal molecule; this is usually sufficient to sustain normal cellular processes such that life goes on.

In the preceding discussion, we have described recessive alleles that result in the death of homozygotes prenatally, *in utero*. With some mutations, however, homozygotes may survive beyond birth and die later from the deleterious consequences of the genetic defect. An example is seen in human infants with Tay-Sachs disease. The seemingly normal newborns remain healthy for five to six months but then develop blindness, paralysis, mental retardation, and other symptoms of a deteriorating nervous system; the disease usually proves fatal by the age of six. Tay-Sachs disease results from the absence of an active lysosomal enzyme called hexosaminidase A, whose lack leads to the accumulation of a toxic waste product inside nerve cells. The approximate incidence of Tay-Sachs among live births is 1/35,000 worldwide, but it is 1/3000 among Jewish people of Eastern European descent. Reliable tests that detect carriers, in combination with genetic counseling and educational programs, have all but eliminated the disease in the United States.

Recessive alleles causing prenatal or early childhood lethality can only be passed on to subsequent generations by heterozygous carriers, because affected homozygotes die before they can mate. However, for late-onset diseases causing death in adults, homozygous patients can pass on the lethal allele before they become debilitated. An example is provided by the degenerative disease Friedreich ataxia: Some homozygotes first display symptoms of ataxia (loss of muscle coordination) at age 30–35 and die about five years later from heart failure. Dominant alleles causing late-onset lethality can also be transmitted to subsequent generations; Figure 2.21 on p. 32 illustrates this for the inheritance of Huntington disease. By contrast, if the lethality caused by a dominant allele occurs instead during fetal development or early childhood, the allele will not be passed on, so all dominant early lethal mutant alleles must be new mutations.

Table 3.1 summarizes Mendel's basic assumptions about dominance, the number and viability of one gene's alleles, and the effects of each gene on phenotype, and then compares these assumptions with the extensions contributed by his twentieth-century successors. Through carefully controlled monohybrid crosses, these later geneticists analyzed the transmission patterns of the alleles of single genes, challenging and then confirming the law of segregation.

A Comprehensive Example: Sickle-Cell Disease Illustrates Many Extensions to Mendel's Analysis of Single-Gene Inheritance

Sickle-cell disease is the result of a faulty hemoglobin molecule. Hemoglobin is composed of two types of polypeptide

TABLE 3.1 For Traits Determined by One Gene: Extensions to Mendel's Analysis Explain Alterations of the 3:1 Monohybrid Ratio

What Mendel Described	Extension	Extension's Effect on Heterozygous Phenotype	Extension's Effect on Ratios Resulting from an $F_1 \times F_1$ Cross
Complete dominance	Incomplete dominance Codominance	Unlike either homozygote	Phenotypes coincide with genotypes in a ratio of 1:2:1
Two alleles	Multiple alleles	Multiplicity of phenotypes	A series of 3:1 ratios
All alleles are equally viable	Recessive lethal alleles	No effect	2:1 instead of 3:1
One gene determines one trait	Pleiotropy: one gene influences several traits	Several traits affected in different ways, depending on dominance relations	Different ratios, depending on dominance relations for each affected trait

chains, alpha (α) globin and beta (β) globin, each specified by a different gene: $Hb\alpha$ for α globin and $Hb\beta$ for β globin. Normal red blood cells are packed full of millions upon millions of hemoglobin molecules, each of which picks up oxygen in the lungs and transports it to all the body's tissues.

Multiple Alleles

The β-globin gene has a normal wild-type allele ($Hb\beta^A$) that gives rise to fully functional β-globin, as well as close to 400 mutant alleles that have been identified so far. Some of these mutant alleles result in the production of hemoglobin that carries oxygen only inefficiently. Other mutant alleles prevent the production of β-globin, causing a hemolytic (blood-destroying) disease called *β-thalassemia*. Here, we discuss the most common mutant allele of the β-globin gene, $Hb\beta^S$, which specifies an abnormal polypeptide that causes sickling of red blood cells (**Fig. 3.10a**).

Pleiotropy

The $Hb\beta^S$ allele of the β-globin gene affects more than one trait (Fig. 3.10b). Hemoglobin molecules in the red blood cells of homozygous $Hb\beta^S Hb\beta^S$ individuals behave aberrantly after releasing their oxygen. Instead of remaining soluble in the cytoplasm, they aggregate to form long fibers that deform the red blood cell from a normal biconcave disk to a sickle shape (see Fig. 3.10a). The deformed cells clog the small blood vessels, reducing oxygen flow to the tissues and giving rise to muscle cramps, shortness of breath, and fatigue. The sickled cells are also very fragile and easily broken. Consumption of fragmented cells by phagocytic white blood cells leads to a low red blood cell count, a condition called anemia. On the positive side, $Hb\beta^S Hb\beta^S$ homozygotes are resistant to malaria, because although the organism that causes the disease, *Plasmodium falciparum,* can multiply with impunity in normal red blood cells, it cannot do so in cells that sickle. Infection by *P. falciparum* causes sickle-shaped cells to break down before the malaria organism has a chance to multiply.

Recessive Lethality

People who are homozygous for the recessive $Hb\beta^S$ allele often develop heart failure because of stress on the circulatory system. Many sickle-cell sufferers die in childhood, adolescence, or early adulthood.

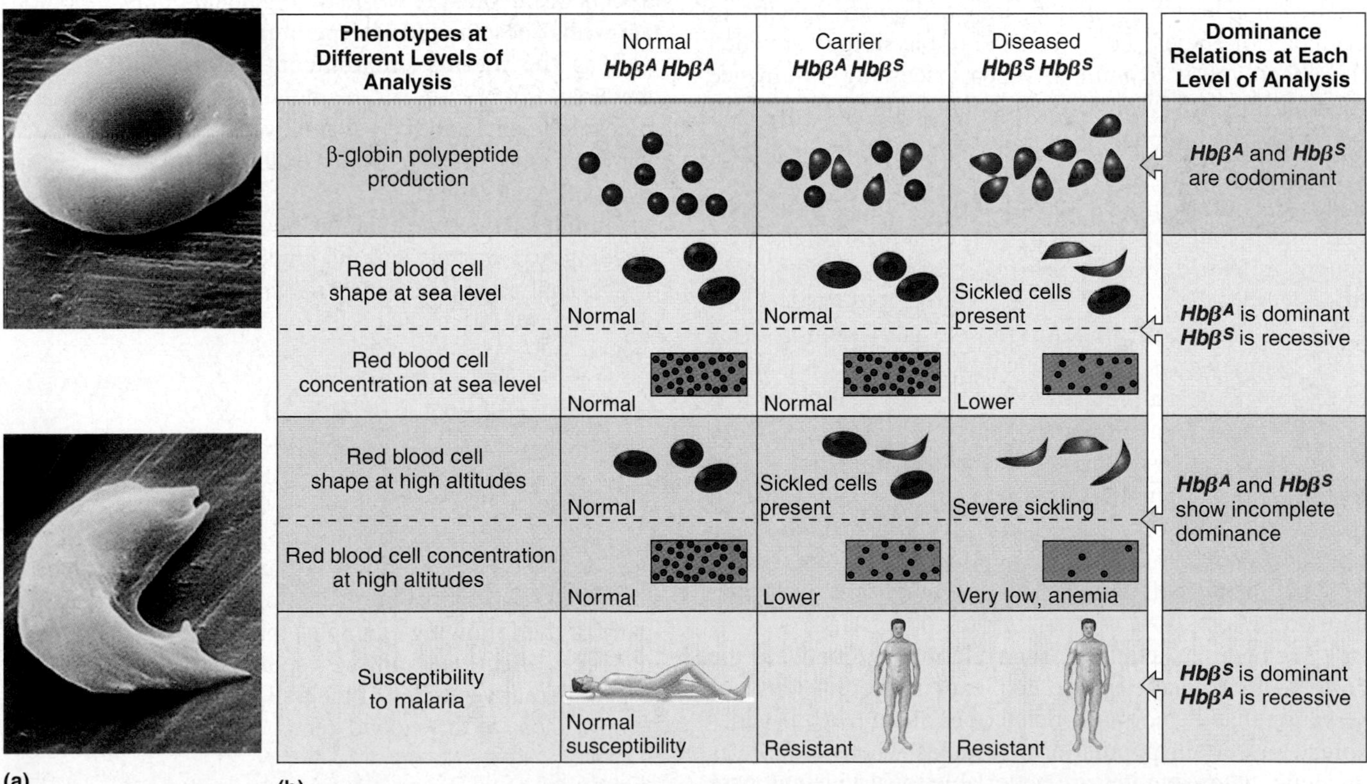

Figure 3.10 Pleiotropy of sickle-cell anemia: Dominance relations vary with the phenotype under consideration. (a) A normal red blood cell (*top*) is easy to distinguish from the sickled cell in the scanning electron micrograph at the *bottom.* **(b)** Different levels of analysis identify various phenotypes. Dominance relationships between the $Hb\beta^S$ and $Hb\beta^A$ alleles of the $Hb\beta$ gene vary with the phenotype and sometimes even change with the environment.

Different Dominance Relations

Comparisons of heterozygous carriers of the sickle-cell allele—individuals whose cells contain one $Hb\beta^A$ and one $Hb\beta^S$ allele—with homozygous $Hb\beta^A Hb\beta^A$ (normal) and homozygous $Hb\beta^S Hb\beta^S$ (diseased) individuals make it possible to distinguish different dominance relations for different phenotypic aspects of sickle-cell anemia (Fig. 3.10b). At the molecular level—the production of β globin—both alleles are expressed such that $Hb\beta^A$ and $Hb\beta^S$ are *codominant.* At the cellular level, in their effect on red blood cell shape, the $Hb\beta^A$ and $Hb\beta^S$ alleles show *incomplete dominance.* Although under normal oxygen conditions, the great majority of a heterozygote's red blood cells have the normal biconcave shape, when oxygen levels drop, sickling occurs in some cells. All $Hb\beta^A Hb\beta^S$ cells, however, are resistant to malaria because like the $Hb\beta^S Hb\beta^S$ cells described previously, they break down before the malarial organism has a chance to reproduce. Thus for the trait of resistance to malaria, the $Hb\beta^S$ allele is *dominant* to the $Hb\beta^A$ allele. But luckily for the heterozygote, for the phenotypes of anemia or death, $Hb\beta^S$ is *recessive* to $Hb\beta^A$. A corollary of this observation is that in its effect on general health under normal environmental conditions and its effect on red blood cell count, the $Hb\beta^A$ allele is *dominant* to $Hb\beta^S$. Thus, for the β-globin gene, as for other genes, dominance and recessiveness are not an inherent quality of alleles in isolation; rather, they are specific to each pair of alleles and to the level of physiology at which the phenotype is examined. When discussing dominance relationships, it is therefore essential to define the particular phenotype under analysis.

In the 1940s, the incomplete dominance of the $Hb\beta^A$ and $Hb\beta^S$ alleles in the expression of red blood cell shape had significant repercussions for certain soldiers who fought in World War II. Aboard transport planes flying troops across the Pacific, several heterozygous carriers suffered sickling crises similar to those usually seen in $Hb\beta^S Hb\beta^S$ homozygotes. The explanation is as follows. The heterozygous red blood cells of a carrier produce both normal and abnormal hemoglobin molecules. At sea level, these molecules together deliver sufficient oxygen, although less than the normal amount, to the body's tissues, but with a decrease in the amount of oxygen available at the high-flying altitudes, the hemoglobin picks up less oxygen, the rate of red blood cell sickling increases, and symptoms of the disease occur.

The complicated dominance relationships between the $Hb\beta^S$ and $Hb\beta^A$ alleles also help explain the puzzling observation that the normally deleterious allele $Hb\beta^S$ is widespread in certain populations. In areas where malaria is endemic, heterozygotes are better able to survive and pass on their genes than are either type of homozygote. $Hb\beta^S Hb\beta^S$ individuals often die of sickle-cell disease, while those with the genotype $Hb\beta^A Hb\beta^A$ often die of malaria. Heterozygotes, however, are relatively immune to both conditions, so high frequencies of both alleles persist in tropical environments where malaria is found. We explore this phenomenon in more quantitative detail in Chapter 21 on population genetics.

New therapies have improved the medical condition of many $Hb\beta^S Hb\beta^S$ individuals, but these treatments have significant shortcomings; as a result, sickle-cell disease remains a major health problem. The Fast Forward box "Gene Therapy for Sickle-Cell Disease in Mice" on p. 57 describes recent success in using genetic engineering to counteract red blood cell sickling in mice whose genomes carry human $Hb\beta^S$ alleles. Researchers hope that similar types of "gene therapies" will one day lead to a cure for sickle-cell disease in humans.

3.2 Extensions to Mendel for Multifactorial Inheritance

Although some traits are indeed determined by allelic variations of a single gene, the vast majority of common traits in all organisms are *multifactorial,* arising from the action of two or more genes, or from interactions between genes and the environment. In genetics, the term *environment* has an unusually broad meaning that encompasses all aspects of the outside world an organism comes into contact with. These include temperature, diet, and exercise as well as the uterine environment before birth (minute chemical differences, for example, can make a difference in development) and the psychological environment afterward (the amount of stress, for example, can affect various levels of phenotype).

In this section, we examine how geneticists again used breeding experiments and the guidelines of Mendelian ratios to analyze the complex network of interactions that give rise to multifactorial traits.

Two Genes Can Interact to Determine One Trait

Two genes can interact in several ways to determine a single trait, such as the color of a flower, a seed coat, a chicken's feathers, or a dog's fur, and each type of interaction produces its own signature of phenotypic ratios. In many of the following examples that show how two genes interact to affect one trait, we use big *A* and little *a* to represent alternative alleles of the first gene and big *B* and little *b* for those of the second gene.

Novel Phenotypes Can Emerge from the Combined Action of the Alleles of Two Genes

In the chapter opening, we described a mating of tan and gray lentils that produced a uniformly brown F_1 generation

Gene Therapy for Sickle-Cell Disease in Mice

The most widespread inherited blood disorder in the United States is sickle-cell disease, which affects approximately 80,000 Americans. It is caused, we have seen, by homozygosity for the $Hb\beta^S$ allele of the gene specifying the β-globin constituent of hemoglobin. Because heterozygotes for this allele are partially protected from malaria, $Hb\beta^S$ is fairly common in people of African, Indian, Mediterranean, and Middle Eastern descent; 1 in 13 African-Americans is a carrier of the sickle-cell allele.

When red blood cells become sickle-shaped, they become sticky and block the flow of blood through blood vessels, causing pain and organ damage. Damage to the spleen interferes with the immune system, making patients highly susceptible to bacterial infections. And because the sickled red blood cells are rapidly destroyed by the body, the blood loses much of its ability to transport oxygen, and the patient becomes anemic. Before the 1980s, most people with sickle-cell disease died during childhood. However, advances in medical care have improved the outlook for many of these patients so that about half of them now live beyond the age of 50.

The main therapies in use today include aggressive treatment with antibiotics to ward off infection, treatment with the drug hydroxyurea (which stimulates the production of other kinds of hemoglobin), and bone marrow transplantation (which replaces the patient's red-blood-cell-forming hematopoietic stem cells with those of a healthy donor). Unfortunately, these treatments are not ideal. Hydroxyurea has toxic side effects, and bone marrow transplantation, besides being inherently risky, can be carried out successfully only with a donor whose tissues are perfectly matched with the patient's so that the transplanted bone marrow will not be rejected. As a result, medical researchers are exploring an alternative: the possibility of developing gene therapy for sickle-cell disease in humans.

Hopes that this alternative approach might eventually provide a cure for sickle-cell disease were raised in 2001 when a research team from Harvard Medical School announced the successful use of gene therapy to treat mice that had been genetically engineered to have sickling red blood cells. These transgenic mice (called SAD mice) express an allelic form of the human $Hb\beta$ gene, closely related to $Hb\beta^S$.

The research team began by removing the bone marrow from the SAD model mice and isolating the hematopoietic stem cells from the marrow. They next used genetic engineering to add an antisickling transgene to these stem cells. The transgene was a synthetically mutated allele of the human $Hb\beta$ gene; it encoded a special β-globin protein designed to prevent sickling in red blood cells that also contain $Hb\beta^S$. When the genetically modified stem cells were transplanted back into the SAD mice, healthy, nonsickling red blood cells were produced. The new genetically modified transgene thus counteracted the effects of the Hb^S allele and prevented sickling, as predicted.

The idea of human gene therapy based on the success of the Harvard research team is very promising, particularly because doctors could, in theory, add a transgene to hematopoietic stem cells derived from the sickle-cell patient so that there would be no threat of tissue rejection when these engineered stem cells are transplanted back into the patient. However, before this potential for human gene therapy can become a reality, researchers must overcome several potential problems. First, the method is not guaranteed to work in humans: Though SAD mice do have red blood cells that sickle, these mice do not exhibit all aspects of sickle-cell disease in humans. Another difficulty is how to make sure the therapeutic gene gets into enough target cells to make a difference. The Harvard group resolved this issue in mice by using a modified version of the HIV virus causing AIDS (**a**cquired **i**mmune **d**eficiency **s**yndrome) to transport the genetically engineered antisickling transgene into the stem cells. It remains to be proven that cells treated with a modified HIV virus will be safe when reintroduced into the human body. Finally, successful gene therapy of this type requires that the patient's bone marrow be populated exclusively by hematopoietic stem cells able to give rise to red blood cells that cannot sickle; all the stem cells without the transgene must be removed. The Harvard researchers did this in mice by destroying the bone marrow in the SAD mice with large doses of X-rays before putting the transgene-containing stem cells back into the mice. However, such a treatment in humans would be extremely toxic. Despite these potential complications, the successful application of gene therapy to a mouse model for sickle-cell disease suggests an exciting pathway for future clinical research aimed at a cure for this debilitating disease.

and then an F_2 generation containing lentils with brown, tan, gray, and green seed coats. An understanding of how this can happen emerges from experimental results demonstrating that the ratio of the four F_2 colors is 9 brown: 3 tan: 3 gray: 1 green (**Fig. 3.11a**). Recall from Chapter 2 that this is the same ratio Mendel observed in his analysis of the F_2 generations from dihybrid crosses following two independently assorting genes. In Mendel's studies, each of the four classes consisted of plants that expressed a combination of two unrelated traits. With lentils, however, we are looking at a single trait—seed coat color. The simplest explanation for the parallel ratios is that a combination of genotypes at two independently assorting genes interacts to produce the phenotype of seed coat color in lentils.

Results obtained from self-crosses with the various types of F_2 lentil plants support this explanation. Self-crosses of

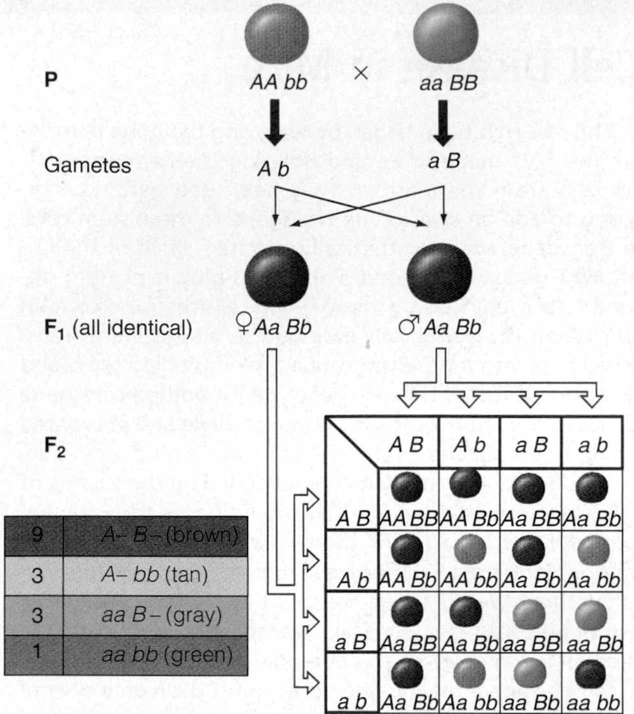

(b) Self-pollination of the F_2 to produce an F_3

Phenotypes of F_2 Individual	Observed F_3 Phenotypes	Expected Proportion of F_2 Population*
Green	Green	1/16
Tan	Tan	1/16
Tan	Tan, green	2/16
Gray	Gray, green	2/16
Gray	Gray	1/16
Brown	Brown	1/16
Brown	Brown, tan	2/16
Brown	Brown, gray	2/16
Brown	Brown, gray, tan, green	4/16

*This 1:1:2:2:1:1:2:2:4 F_2 genotypic ratio corresponds to a 9 brown:3 tan:3 gray:1 green F_2 phenotypic ratio.

(c) Sorting out the dominance relations by select crosses

Seed Coat Color of Parents	F_2 Phenotypes and Frequencies	Ratio
Tan × green	231 tan, 85 green	3:1
Gray × green	2586 gray, 867 green	3:1
Brown × gray	964 brown, 312 gray	3:1
Brown × tan	255 brown, 76 tan	3:1
Brown × green	57 brown, 18 gray, 13 tan, 4 green	9:3:3:1

Figure 3.11 How two genes interact to produce new colors in lentils. **(a)** In a cross of pure-breeding tan and gray lentils, all the F_1 hybrids are brown, but four different phenotypes appear among the F_2 progeny. The 9:3:3:1 ratio of F_2 phenotypes suggests that seed coat color is determined by two independently assorting genes in a dihybrid cross. **(b)** Expected results of selfing individual F_2 plants of the indicated phenotypes to produce an F_3 generation, if seed coat color is determined by the interaction of two genes. The third column shows the proportion of the F_2 population that would be expected to produce the observed F_3 phenotypes. **(c)** Other two-generation crosses involving pure-breeding parental lines also support the two-gene hypothesis. In this table, the F_1 hybrid generation has been omitted.

F_2 green individuals show that they are pure-breeding, producing an F_3 generation that is entirely green. Tans generate either all tan or tan plus green; grays similarly produce all gray or gray plus green; and browns either breed true, or instead they assort into two colors (either tan plus brown or gray plus brown), or into all four colors (Fig. 3.11b). The two-gene hypothesis explains why there is

- only one green genotype: pure-breeding *aa bb,* but
- two types of tans: pure-breeding *AA bb* as well as tan- and green-producing *Aa bb,* and
- two types of grays: pure-breeding *aa BB* and gray- and green-producing *aa Bb,* yet
- four types of browns: true-breeding *AA BB,* brown- and tan-producing *AA Bb,* brown- and gray-producing *Aa BB,* and *Aa Bb* dihybrids that give rise to plants producing lentils of all four colors.

In short, for the two genes that determine seed coat color, both dominant alleles must be present to yield brown (*A– B–*); the dominant allele of one gene produces tan (*A– bb*); the dominant allele of the other specifies gray (*aa B–*); and the complete absence of dominant alleles (that is, the double recessive) yields green (*aa bb*). Thus, the four color phenotypes arise from four **genotypic classes,** with each class defined in terms of the presence or absence of the dominant alleles of two genes: (1) both present (*A– B–*), (2) one present (*A– bb*), (3) the other present (*aa B–*), and (4) neither present (*aa bb*). Note that the *A–* notation means that the second allele of this gene can be either *A* or *a,* while *B–* denotes a second allele of either *B* or *b.* Note also that only with a two-gene system in which the dominance and recessiveness of alleles at both genes is complete can the nine different genotypes of the F_2 generation be categorized into the four genotypic classes described. With incomplete dominance or codominance, the F_2 genotypes could not be grouped together in this simple way, as they would give rise to more than four phenotypes.

Further crosses between plants carrying lentils of different colors confirmed the two-gene hypothesis (review Fig. 3.11c). Thus, the 9 : 3 : 3 : 1 phenotypic ratio of brown to tan to gray to green in an F_2 descended from pure-breeding tan and pure-breeding gray lentils tells us not only that two genes assorting independently interact to produce the seed coat color, but also that each genotypic class (*A– B–*, *A– bb*, *aa B–*, and *aa bb*) determines a particular phenotype.

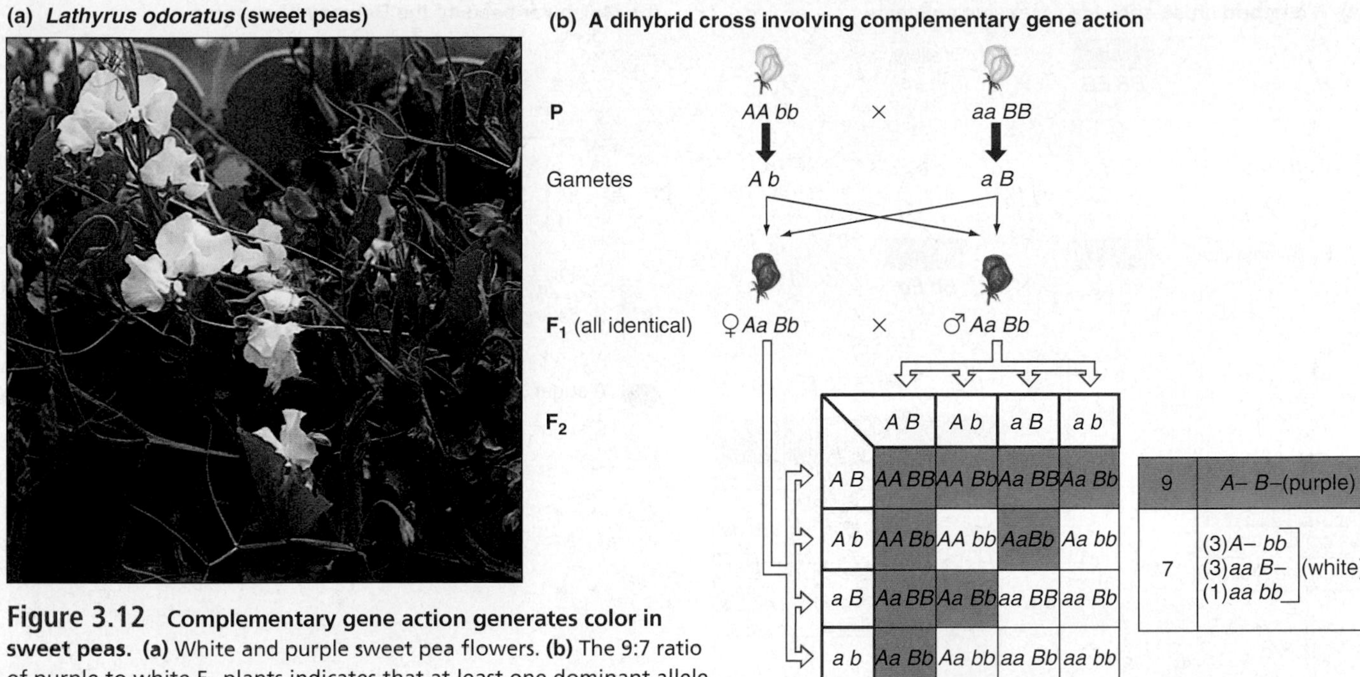

Figure 3.12 **Complementary gene action generates color in sweet peas. (a)** White and purple sweet pea flowers. **(b)** The 9:7 ratio of purple to white F_2 plants indicates that at least one dominant allele for each gene is necessary for the development of purple color.

This is not always the case. In some two-gene interactions, the four F_2 genotypic classes produce fewer than four observable phenotypes, because some of the phenotypes include two or more genotypic classes. For example, in the first decade of the twentieth century, William Bateson conducted a cross between two lines of pure-breeding white-flowered sweet peas (**Fig. 3.12**). Quite unexpectedly, all of the F_1 progeny were purple. Self-pollination of these novel hybrids produced a ratio of 9 purple : 7 white in the F_2 generation. The explanation? Two genes work in tandem to produce purple sweet-pea flowers, and a dominant allele of both genes must be present to produce that color. A simple biochemical explanation for this type of **complementary gene action** is shown in **Fig. 3.13.** Because it takes two enzymes catalyzing two separate biochemical reactions to change a colorless precursor into a colorful pigment, only the *A– B–*genotypic class, which produces active forms of both required enzymes, can generate colored flowers. The other three genotypic classes (*A– bb, aa B–,* and *aa bb*) become grouped together with respect to phenotype since they do not specify functional forms of one or the other requisite enzyme and thus give rise to no color, which is the same as white (review the Fast Forward box "Genes Encode Proteins" in Chapter 2). It is easy to see how the "7" part of the 9:7 ratio encompasses the 3:3:1 of the 9:3:3:1 ratio of two genes in action. The 9:7 ratio is the phenotypic signature of this type of complementary gene interaction in which the dominant alleles of two genes acting together (*A– B–*) produce color or some other trait, while the other three genotypic classes (*A– bb, aa B–,* and *aa bb*) do not (see Fig. 3.12b).

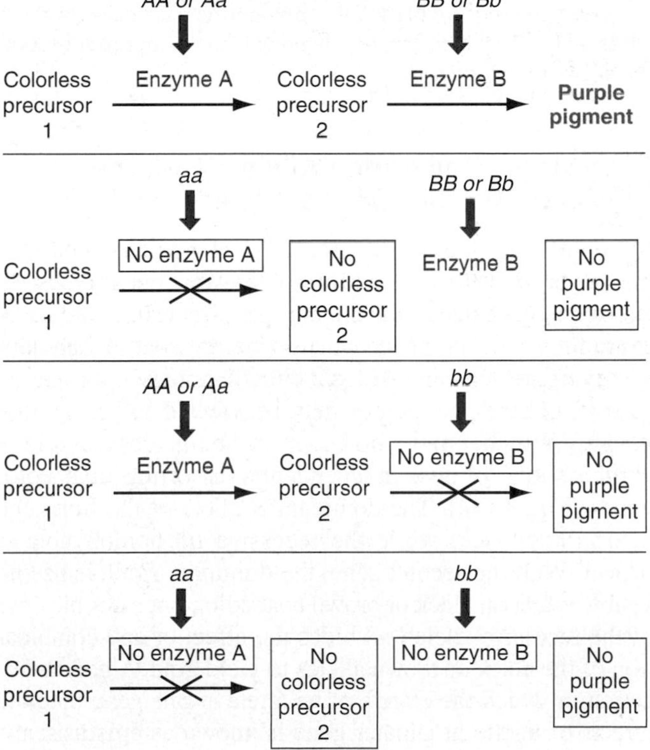

Figure 3.13 **A possible biochemical explanation for complementary gene action in the generation of sweet pea color.** Enzymes specified by the dominant alleles of two genes are both necessary for completion of a biochemical pathway for pigment production. The recessive alleles of both genes specify inactive enzymes. In *aa* homozygotes, no intermediate precursor 2 is created, so even if enzyme B is available, it cannot create the pathway's end product of purple pigment.

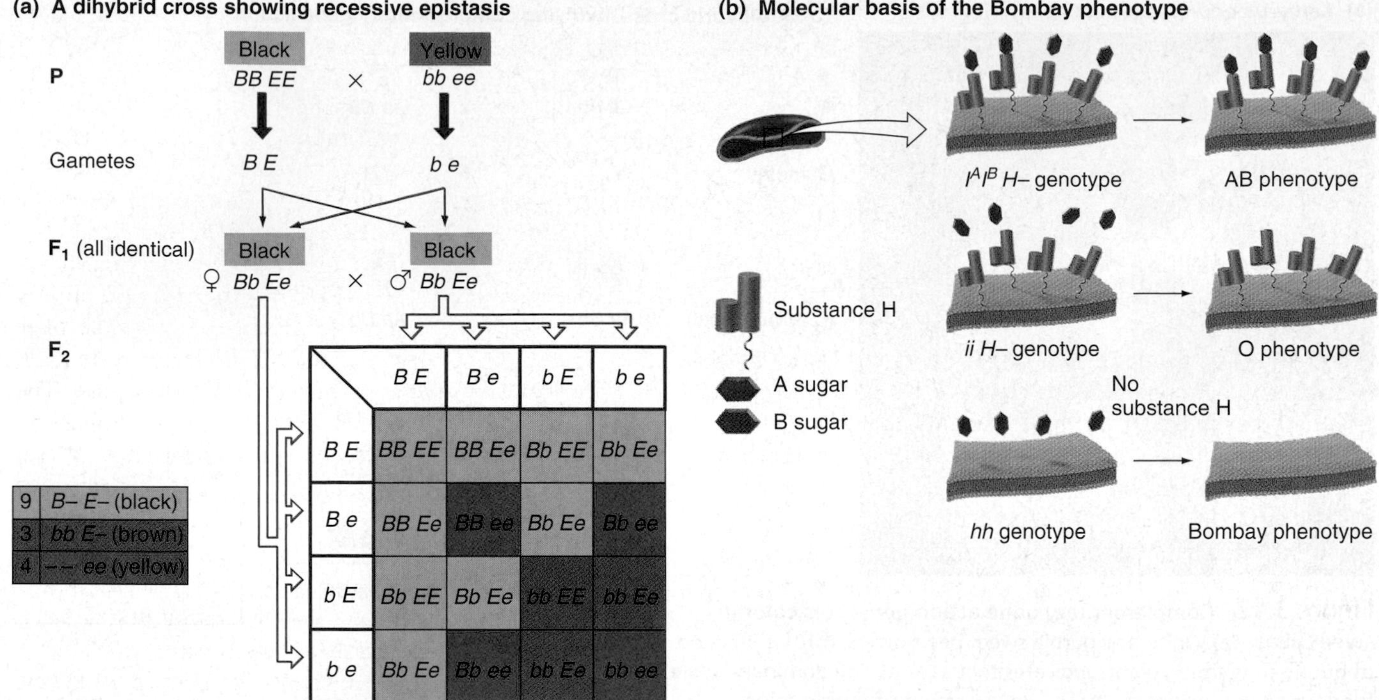

Figure 3.14 **Recessive epistasis: Coat color in Labrador retrievers and a rare human blood type.** **(a)** Golden Labrador retrievers are homozygous for the recessive *e* allele, which masks the effects of the *B* or *b* alleles of a second coat color gene. In *E–* dogs, a *B–* genotype produces black and a *bb* genotype produces brown. **(b)** Homozygosity for the *h* Bombay allele is epistatic to the *I* gene determining ABO blood types. *hh* individuals fail to produce substance H, which is needed for the addition of A or B sugars at the surface of red blood cells.

In Epistasis, One Gene's Alleles Mask the Effects of Another Gene's Alleles

In some gene interactions, the four Mendelian genotypic classes produce fewer than four observable phenotypes because one gene masks the phenotypic effects of another. An example is seen in the sleek, short-haired coat of Labrador retrievers, which can be black, chocolate brown, or golden yellow. (These phenotypes may be viewed in Fig. 2.3 on p. 14.) Which color shows up depends on the allelic combinations of two independently assorting coat color genes (**Fig. 3.14a**). The dominant *B* allele of the first gene determines black, while the recessive *bb* homozygote is brown. With the second gene, the dominant *E* allele has no visible effect on black or brown coat color, but a double dose of the recessive allele (*ee*) hides the effect of any combination of the black or brown alleles to yield gold. A gene interaction in which the effects of an allele at one gene hide the effects of alleles at another gene is known as **epistasis;** the allele that is doing the masking (in this case, the *e* allele of the *E* gene) is **epistatic** to the gene that is being masked (the *hypostatic gene*). In our example, where homozygosity for a recessive *e* allele of the second gene is required to hide the effects of another gene, the masking phenomenon is called **recessive epistasis** (because the allele causing the epistasis is recessive), and the recessive *ee* homozygote is considered epistatic to any allelic combination at the first gene.

Let's look at the phenomenon in greater detail. Crosses between pure-breeding black retrievers (*BB EE*) and one type of pure-breeding golden retriever (*bb ee*) create an F_1 generation of dihybrid black retrievers (*Bb Ee*). Crosses between these F_1 dihybrids produce an F_2 generation with nine black dogs (*B– E–*) for every three brown (*bb E–*) and four gold (– – *ee*) (Fig. 3.14a). Note that there are only three phenotypic classes because the two genotypic classes without a dominant *E* allele—the three *B– ee* and the one *bb ee*—combine to produce golden phenotypes. The telltale ratio of recessive epistasis in the F_2 generation is thus 9 : 3 : 4, with the 4 representing a combination of 3 (*B– ee*) + 1 (*bb ee*). Because the *ee* genotype completely masks the influence of the other gene for coat color, you cannot tell by looking at a golden Labrador what its genotype is for the black or brown (*B* or *b*) gene.

An understanding of recessive epistasis made it possible to resolve an intriguing puzzle in human genetics. In rare instances, two parents who appear to have blood type O, and thus genotype *ii,* produce a child who is either blood type A (genotype I^Ai) or blood type B (genotype I^Bi). This phenomenon occurs because an extremely rare trait, called the Bombay phenotype after its discovery in Bombay, India, superficially resembles blood type O. As Fig. 3.14b shows, the Bombay phenotype actually arises from homozygosity for a mutant recessive allele (*hh*) of a second gene that masks the effects of any ABO alleles that might be present.

Here's how it works at the molecular level. In the construction of the red blood cell surface molecules that determine blood type, type A individuals make an enzyme that adds polysaccharide A onto a base consisting of a sugar polymer known as substance H; type B individuals make an altered form of the enzyme that adds polysaccharide B onto the base; and type O individuals make neither A-adding nor B-adding enzyme and thus have an exposed substance H in the membranes of their red blood cells. All people of A, B, or O phenotype carry at least one dominant wild-type *H* allele for the second gene and thus produce some substance H. In contrast, the rare Bombay-phenotype individuals, with genotype *hh* for the second gene, do not make substance H at all, so even if they make an enzyme that would add A or B to this polysaccharide base, they have nothing to add it onto; as a result, they appear to be type O. For this reason, homozygosity for the recessive *h* allele of the H-substance gene masks the effects of the *ABO* gene, making the *hh* genotype epistatic to any combination of I^A, I^B, and *i* alleles.

A person who carries I^A, I^B, or both I^A and I^B but is also an *hh* homozygote for the H-substance gene may appear to be type O, but he or she will be able to pass along an I^A or I^B allele in sperm or egg. The offspring receiving, let's say, an I^A allele for the ABO gene and a recessive *h* allele for the H-substance gene from its mother plus an *i* allele and a dominant *H* allele from its father would have blood type A (genotype I^Ai, *Hh*), even though neither of its parents is phenotype A or AB.

Epistasis can also be caused by a dominant allele. In summer squash, two genes influence the color of the fruit (**Fig. 3.15a**). With one gene, the dominant allele (*A–*) determines yellow, while homozygotes for the recessive allele (*aa*) are green. A second gene's dominant allele (*B–*) produces white, while *bb* fruit may be either yellow or green, depending on the genotype of the first gene. In the interaction between these two genes, the presence of *B* hides the effects of either *A–* or *aa,* producing white fruit, and *B–* is thus epistatic to any genotype of the *Aa* gene. The recessive *b* allele has no effect on fruit color determined by the *Aa* gene. Epistasis in which the dominant allele of one gene hides the effects of another gene is called **dominant epistasis.** In a cross between white F_1 dihybrids (*Aa Bb*), the F_2 phenotypic ratio is 12 white : 3 yellow : 1 green (Fig. 3.15a). The "12" includes two genotypic classes: 9 *A– B–* and 3 *aa B–*. Another way of looking at this same phenomenon is that dominant epistasis restores the 3:1 ratio for the dominant epistatic phenotype (12 white) versus all other phenotypes (4 green plus yellow).

A variation of this ratio is seen in the feather color of certain chickens (Fig. 3.15b). White leghorns have a

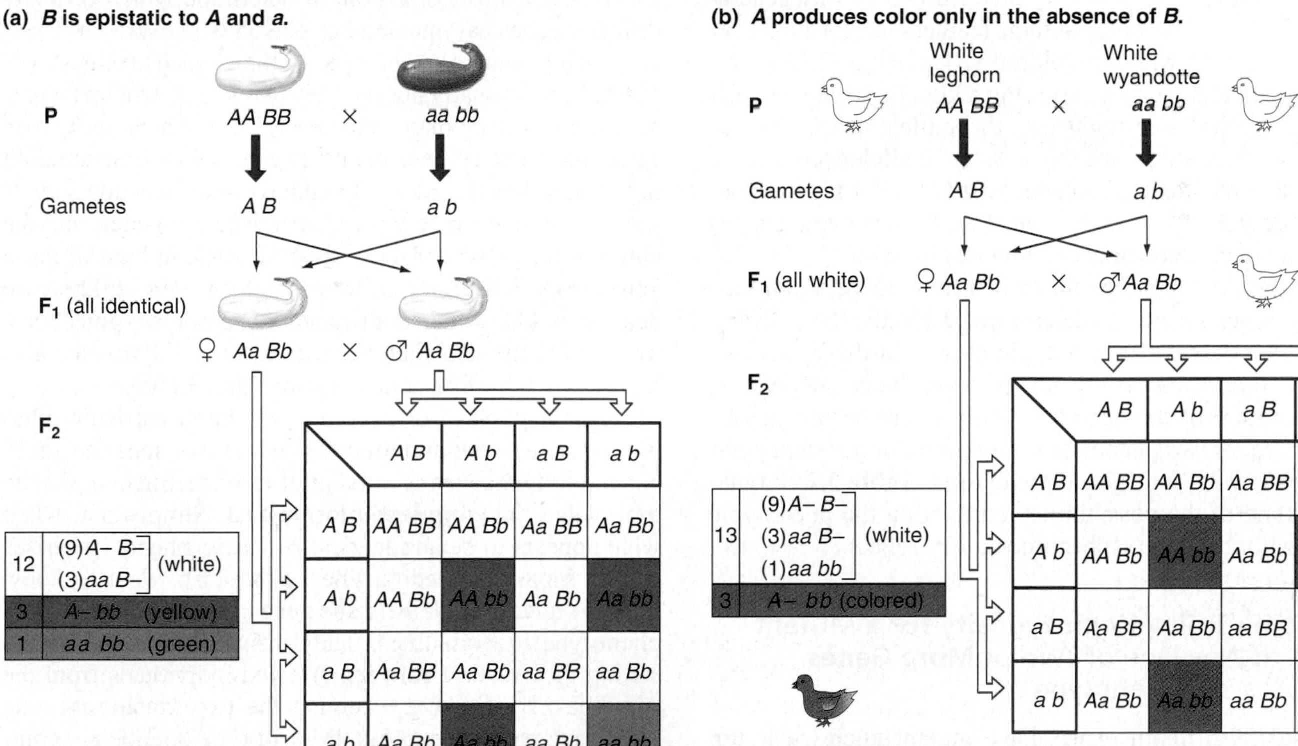

Figure 3.15 **Dominant epistasis produces telltale phenotypic ratios of 12:3:1 or 13:3.** **(a)** In summer squash, the dominant *B* allele causes white color and is sufficient to mask the effects of any combination of *A* and *a* alleles. As a result, yellow (*A–*) or green (*aa*) color is expressed only in *bb* individuals. **(b)** In the F_2 generation resulting from a dihybrid cross between white leghorn and white wyandotte chickens, the ratio of white birds to birds with color is 13:3. This is because at least one copy of *A* and the absence of *B* is needed to produce color.

TABLE 3.2 Summary of Discussed Gene Interactions

Gene Interaction	F_2 Genotypic Ratios from an F_1 Dihybrid Cross — Example	*A– B–*	*A– bb*	*aa B–*	*aa bb*	F_2 Phenotypic Ratio
None: Four distinct F_2 phenotypes	Lentil: seed coat color (see Fig. 3.11)		3	3	1	9:3:3:1
Complementary: One dominant allele of each of two genes is necessary to produce phenotype	Sweet pea: flower color (see Fig. 3.12b)		3	3	1	9:7
Recessive epistasis: Homozygous recessive of one gene masks both alleles of another gene	Retriever: coat color (see Fig. 3.14a)		3	3	1	9:3:4
Dominant epistasis I: Dominant allele of one gene hides effects of both alleles of another gene	Summer squash: color (see Fig. 3.15a)		3		1	12:3:1
Dominant epistasis II: Dominant allele of one gene hides effects of dominant allele of another gene	Chicken: feather color (see Fig. 3.15b)		3			13:3

doubly dominant *AA BB* genotype for feather color; white wyandottes are homozygous recessive for both genes (*aa bb*). A cross between these two pure-breeding white strains produces an all-white dihybrid (*Aa Bb*) F_1 generation, but birds with color in their feathers appear in the F_2, and the ratio of white to colored is 13:3 (Fig. 3.15b). We can explain this ratio by assuming a kind of dominant epistasis in which *B* is epistatic to *A;* the *A* allele (in the absence of *B*) produces color; and the *a, B,* and *b* alleles produce no color. The interaction is characterized by a 13:3 ratio because the 9 *A– B–*, 3 *aa B–*, and 1 *aa bb* genotypic classes combine to produce only one phenotype: white.

So far we have seen that when two independently assorting genes interact to determine a trait, the 9:3:3:1 ratio of the four Mendelian genotypic classes in the F_2 generation can produce a variety of phenotypic ratios, depending on the nature of the gene interactions. The result may be four, three, or two phenotypes, composed of different combinations of the four genotypic classes. **Table 3.2** summarizes some of the possibilities, correlating the phenotypic ratios with the genetic phenomena they reflect.

For Some Traits, Homozygosity for a Mutant Allele at Any One of Two or More Genes Produces the Phenotype

Close to 50 different genes have mutant alleles that can cause deafness in humans. This is because it takes many genes to generate the developmental pathway that brings about hearing, and a loss of function in any part of the pathway, for instance, in one small bone of the middle ear, can result in deafness. In other words, it takes a dominant wild-type allele at each of these 50 genes to produce normal hearing. Thus, deafness is a **heterogeneous trait:** A mutation at any one of a number of genes can give rise to the same phenotype.

It is not always possible to determine which of many different genes has mutated in a person who expresses a heterogeneous mutant phenotype. In the case of deafness, for example, it is usually not possible to discover whether a particular nonhearing man and a particular nonhearing woman carry mutations at the same gene, unless they have children together. If they have only children who can hear, the parents most likely carry mutations at two different genes, and the children carry one normal, wild-type allele at both of those genes (**Fig. 3.16a**). By contrast, if all of their children are deaf, it is likely that both parents are homozygous for a mutation in the same gene, and all of their children are also homozygous for this same mutation (Fig. 3.16b).

This method of discovering whether a particular phenotype arises from mutations in the same or separate genes is a naturally occurring version of an experimental genetic tool called the **complementation test.** Simply put, when what appears to be an identical *recessive* phenotype arises in two separate breeding lines, geneticists want to know whether mutations at the same gene are responsible for the phenotype in both lines. They answer this question by setting up a mating between affected individuals from the two lines. If offspring receiving the two mutations—one from each parent—express the wild-type phenotype, complementation has occurred. The observation of complementation means that the original mutations affected two different genes, and for both genes, the normal allele from one parent can provide what the mutant allele of the same gene from the other parent cannot. Figure 3.16a illustrates one example of this phenomenon in humans. By contrast, if

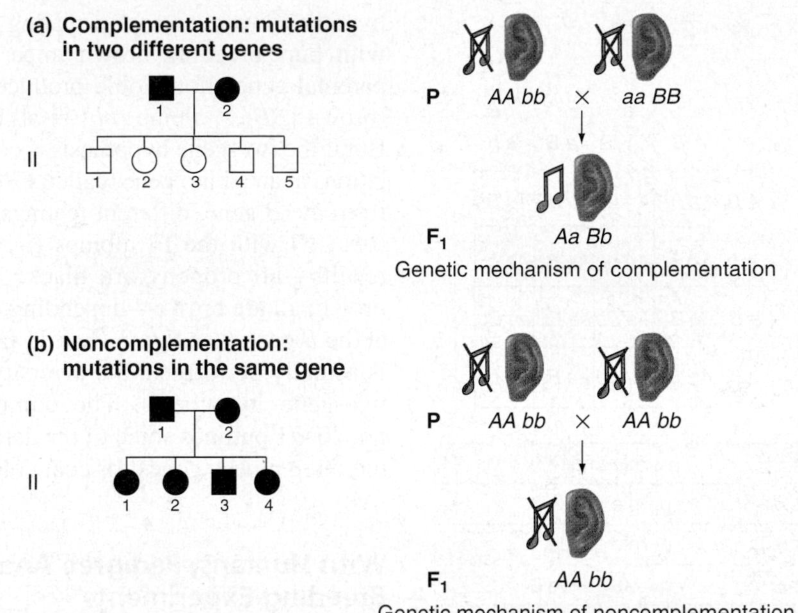

Figure 3.16 Genetic heterogeneity in humans: Mutations in many genes can cause deafness. Hearing requires the coordinated function of many complex structures within the ear. **(a)** Two deaf parents can have hearing offspring. This situation is an example of genetic complementation; it occurs if the nonhearing parents are homozygous for recessive mutations in different genes. **(b)** Two deaf parents may produce all deaf children. This happens when complementation does not occur because both parents carry mutations in the same gene.

offspring receiving two recessive mutant alleles—again, one from each parent—express the mutant phenotype, complementation does not occur because the two mutations independently alter the same gene (review Fig. 3.16b). Thus, the occurrence of complementation reveals genetic heterogeneity. Note that complementation tests cannot be used if either of the mutations is dominant to the wild type. Chapter 7 includes an in-depth discussion of complementation tests and their uses.

In sum, there are several genetic variations on the theme of multifactorial traits: (1) genes can interact to generate novel phenotypes, (2) the dominant alleles of two interacting genes can both be necessary for the production of a particular phenotype, (3) one gene's alleles can mask the effects of another's, and (4) mutant alleles at one of two or more different genes can result in the same phenotype. In examining each of these categories, for the sake of simplicity, we have looked at examples in which one allele of each gene in a pair showed complete dominance over the other. But for any type of gene interaction, the alleles of one or both genes may exhibit incomplete dominance or codominance, and these possibilities increase the potential for phenotypic diversity. For example, **Fig. 3.17** shows how incomplete dominance at both genes in a dihybrid cross generates additional phenotypic variation.

Although the possibilities for variation are manifold, none of the observed departures from Mendelian phenotypic ratios contradicts Mendel's genetic laws of segregation and independent assortment. The alleles of each gene still segregate as he proposed. Interactions between the alleles of many genes simply make it harder to unravel the complex relation of genotype to phenotype.

Breeding Studies Help Decide How a Trait Is Inherited

How do you know whether a particular trait is caused by the alleles of one gene or by two genes interacting in one of a number of possible ways? Breeding tests can usually resolve the issue. Phenotypic ratios diagnostic of a particular mode of inheritance (for instance, the 9:7 or 13:3 ratios indicating that two genes are interacting) can provide the first clues and suggest hypotheses. Further breeding studies can then show which hypothesis is correct. We have seen, for example, that yellow coat color in mice is determined by a dominant allele of the *agouti* gene, which also acts as a recessive lethal. We now look at two other mouse genes for coat color. Since we have already designated alleles of the *agouti* gene as *Aa,* we use *Bb* and *Cc* to designate the alleles of these additional genes.

A mating of one strain of pure-breeding white albino mice with pure-breeding brown results in black hybrids; and a cross between the black F_1 hybrids produces 90 black, 30 brown, and 40 albino offspring. What is the genetic constitution of these phenotypes? We could assume that we are seeing the 9 : 3 : 4 ratio of recessive epistasis and hypothesize that two genes, one epistatic to the other, interact to produce

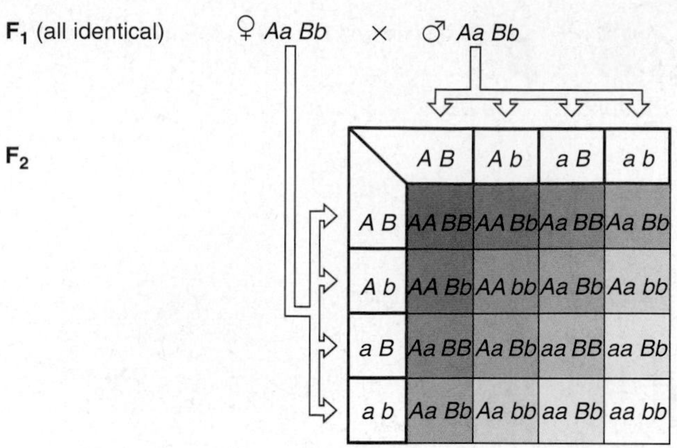

1	*AA BB*	purple shade 9
2	*AA Bb*	purple shade 8
2	*Aa BB*	purple shade 7
1	*AA bb*	purple shade 6
4	*Aa Bb*	purple shade 5
1	*aa BB*	purple shade 4
2	*Aa bb*	purple shade 3
2	*aa Bb*	purple shade 2
1	*aa bb*	purple shade 1 (white)

Figure 3.17 With incomplete dominance, the interaction of two genes can produce nine different phenotypes for a single trait. In this example, two genes produce purple pigments. Alleles *A* and *a* of the first gene exhibit incomplete dominance, as do alleles *B* and *b* of the second gene. Because the two alleles of each gene can combine to generate three different phenotypes, double heterozygotes have the potential for producing progeny of nine (3 × 3) different colors in a ratio of 1:2:2:1:4:1:2:2:1.

the three mouse phenotypes (**Fig. 3.18a**). But how do we know if this hypothesis is correct? We might also explain the data—160 progeny in a ratio of 90:30:40—by the activity of one gene (Fig. 3.18b). According to this one-gene hypothesis, albinos would be homozygotes for one allele (*bb*), brown mice would be homozygotes for a second allele (*BB*), and black mice would be heterozygotes (*Bb*) that have their own "intermediate" phenotype because *B* shows incomplete dominance over *b.* Under this system, a mating of black (*Bb*) to black (*Bb*) would be expected to produce 1 *BB* brown : 2 *Bb* black : 1 *bb* albino, or 40 brown : 80 black : 40 albino. Is it possible that the 30 brown, 90 black, and 40 albino mice actually counted were obtained from the inheritance of a single gene? Intuitively, the answer is yes: the ratios 40:80:40 and 30:90:40 do not seem that different. We know that if we flip a coin 100 times, it doesn't always come up 50 heads : 50 tails; sometimes it's 60:40 just by chance. So, how can we decide between the two-gene versus the one-gene model?

The answer is that we can use other types of crosses to verify or refute the hypotheses. For instance, if the one-gene hypothesis were correct, a mating of pure white F_2 albinos with pure-breeding brown mice similar to those of the parental generation would produce all black heterozygotes (brown [*BB*] × albino [*bb*] = all black [*Bb*]) (Fig. 3.18b). But if the two-gene hypothesis is correct, with recessive mutations at an albino gene (called *C*) epistatic to all expression from the *B* gene, different matings of pure-breeding brown (*bb CC*) with the F_2 albinos (– – *cc*) will give different results—all progeny are black; half are black and half brown; all are brown—depending on the albino's genotype at the *B* gene (see Fig. 3.18a). In fact, when the experiment is actually performed, the diversity of results confirms the two-gene hypothesis. The comprehensive example on pp. 70–71 outlines some of the details of the interactions of the three mouse genes for coat color.

With Humans, Pedigree Analysis Replaces Breeding Experiments

Breeding experiments cannot be applied to humans, for obvious reasons. But a careful examination of as many family pedigrees as possible can help elucidate the genetic basis of a particular condition. In a form of albinism known as ocular-cutaneous albinism (OCA), for example, people with the inherited condition have little or no pigment in their skin, hair, and eyes (**Fig. 3.19a**). The horizontal inheritance pattern seen in Fig. 3.19b suggests that OCA is determined by the recessive allele of one gene, with albino family members being homozygotes for that allele. But a 1952 paper on albinism reported a family in which two albino parents produced three normally pigmented children (Fig. 3.19c). How would you explain this phenomenon? The answer is that albinism is another example of heterogeneity: Mutant alleles at any one of several different genes can cause the condition. The reported mating was, in effect, an inadvertent complementation test, which showed that one parent was homozygous for an OCA-causing mutation in gene *A,* while the other parent was homozygous for an OCA-causing mutation in a different gene, *B* (compare with Fig. 3.16 on p. 63).

The Same Genotype Does Not Always Produce the Same Phenotype

In our discussion of gene interactions so far, we have looked at examples in which a genotype reliably fashions a particular phenotype. But this is not always what happens. Sometimes a genotype is not expressed at all; that is, even though the genotype is present, the expected phenotype does not appear. Other times, the trait caused by a genotype is expressed to varying degrees or in a variety of ways in different individuals. Factors that alter the phenotypic expression of genotype include modifier genes, the environment (in the broadest sense, as defined earlier), and chance.

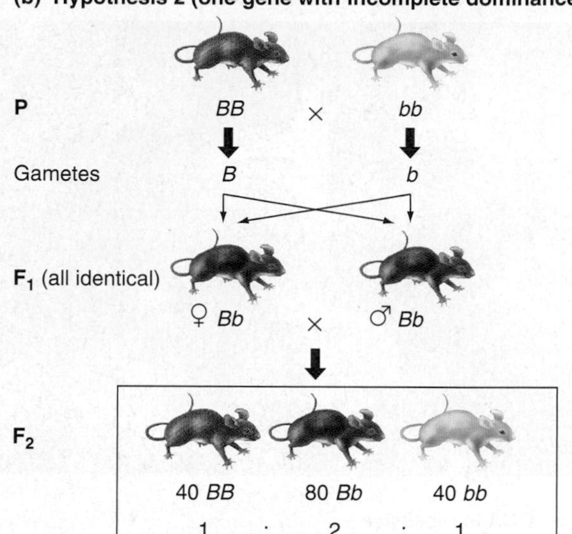

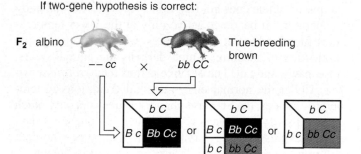

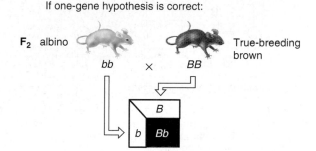

Figure 3.18 Specific breeding tests can help decide between hypotheses. Either of two hypotheses could explain the results of a cross tracking coat color in mice. **(a)** In one hypothesis, two genes interact with recessive epistasis to produce a 9:3:4 ratio. **(b)** In the other hypothesis, a single gene with incomplete dominance between the alleles generates the observed results. One way to decide between these models is to cross each of several albino F_2 mice with true-breeding brown mice. The two-gene model predicts several different outcomes depending on the – – *cc* albino's genotype at the *B* gene. The one-gene model predicts that all progeny of the crosses will be black.

Phenotype Often Depends on Penetrance and Expressivity

Retinoblastoma, the most malignant form of eye cancer, arises from a dominant mutation of one gene, but only 75% of people who carry the mutant allele develop the disease. Geneticists use the term **penetrance** to describe how many members of a population with a particular genotype show the expected phenotype. Penetrance can be *complete* (100%), as in the traits that Mendel studied, or *incomplete,* as in retinoblastoma (see the Genetics and Society box "Disease Prevention Versus the Right to Privacy" on pp. 68–69 for another example of incomplete penetrance). For retinoblastoma, the penetrance is 75%. In some people who show the trait, only one eye is affected, while in other individuals with the phenotype, both eyes are diseased. **Expressivity** refers to the degree or intensity with which a particular genotype is expressed in a phenotype. Expressivity can be *variable,* as in retinoblastoma (one or both eyes affected), or *unvarying,* as in pea color. As we will see, the incomplete penetrance and variable expressivity of retinoblastoma are the result of chance, but in other cases, it is modifier genes and/or the environment that causes such variations in the appearance of phenotype.

Modifier Genes Produce Secondary Effects on Phenotype

Not all genes that influence the appearance of a trait contribute equally to the phenotype. Major genes have a large influence, while **modifier genes** have a more subtle, secondary effect. Modifier genes alter the phenotypes produced by the alleles of other genes. There is no formal distinction between major and modifier genes. Rather, there is a continuum between the two, and the cutoff is arbitrary.

Modifier genes influence the length of a mouse's tail. The mutant *T* allele of the tail-length gene causes a

(a) Ocular-cutaneous albinism (OCA)

(b) OCA is recessive

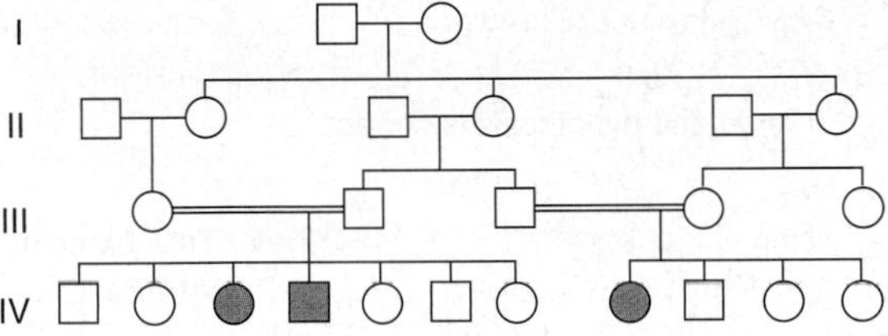

(c) Complementation for albinism

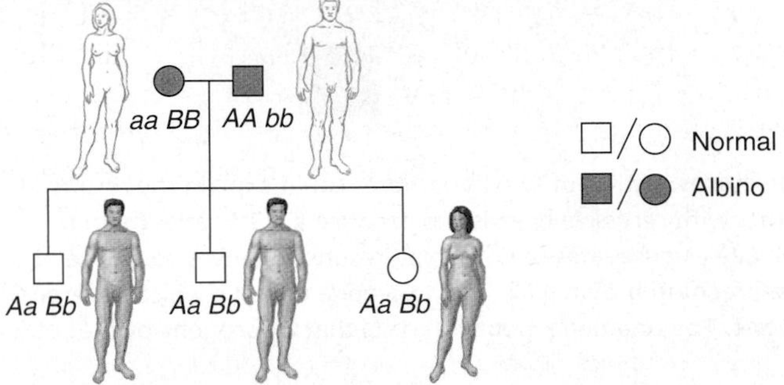

Figure 3.19 Family pedigrees help unravel the genetic basis of ocular-cutaneous albinism (OCA). (a) An albino Nigerian girl and her sister celebrating the conclusion of the All Africa games. **(b)** A pedigree following the inheritance of OCA in an inbred family indicates that the trait is recessive. **(c)** A family in which two albino parents have nonalbino children demonstrates that homozygosity for a recessive allele of either of two genes can cause OCA.

shortening of the normally long wild-type tail. But not all mice carrying the *T* mutation have the same length tail. A comparison of several inbred lines points to modifier genes as the cause of this variable expressivity. In one inbred line, mice carrying the *T* mutation have tails that are approximately 75% as long as normal tails; in another inbred line, the tails are 50% normal length; and in a third line, the tails are only 10% as long as wild-type tails. Because all members of each inbred line grow the same length tail, no matter what the environment (for example, diet, cage temperature, or bedding), geneticists conclude it is genes and not the environment or chance that determines the length of a mutant mouse's tail. Different inbred lines most likely carry different alleles of the modifier genes that determine exactly how short the tail will be when the *T* mutation is present. Researchers have not yet identified the modifier genes or ascertained what they do, but the fact that all members of one inbred line differ in exactly the same way from all members of the second and third lines reinforces the interpretation that differences in *T*-produced tail length are caused by modifier genes.

The Environment Can Affect the Phenotypic Expression of a Genotype

Temperature is one element of the environment that can have a visible effect on phenotype. For example, temperature influences the unique coat color pattern of Siamese cats (**Fig. 3.20**). These domestic felines are homozygous for one of the multiple alleles of a gene that encodes an enzyme catalyzing the production of the dark pigment melanin. The form of the enzyme generated by the variant "*Siamese*" allele does not function at the cat's normal body temperature. It becomes active only at the lower temperatures found in the cat's extremities, where it promotes the production of melanin, which darkens the animal's ears, nose, paws, and tail. The enzyme is thus *temperature sensitive.* Under the normal environmental conditions in temperate climates, the Siamese phenotype does not vary much in expressivity from one cat to another, but one can imagine the expression of a very different phenotype—no dark extremities—in equatorial deserts, where the ambient temperature is at or above normal body temperature.

Temperature can also affect survivability. In one type of experimentally bred fruit fly *(Drosophila melanogaster),* some individuals develop and multiply normally at temperatures between 18°C and 29°C; but if the thermometer climbs beyond that cutoff for a short time, they become reversibly paralyzed, and if the temperature remains high for more than a few hours, they die. These insects carry a temperature-sensitive allele of the *shibire* gene, which encodes a protein essential for nerve cell transmission. This type of allele is known as a **conditional lethal** because it is lethal only under certain conditions. The range of temperatures under which the insects remain viable is the **permissive** range; the lethal temperatures above that are **restrictive.** Thus, at one temperature, the allele gives rise to a phenotype that is indistinguishable from the wild type, while at another temperature, the same allele generates a mutant phenotype (in this case, lethality). Flies with the wild-type *shibire* allele are viable even at the higher temperatures. The fact that some mutations are lethal only under certain conditions clearly illustrates that the environment can affect the penetrance of a phenotype. Since flies with temperature-sensitive *shibire* mutations can survive only under carefully controlled environmental conditions, they do not survive in nature, where the mutation is more likely to be penetrant at some point in the life cycle.

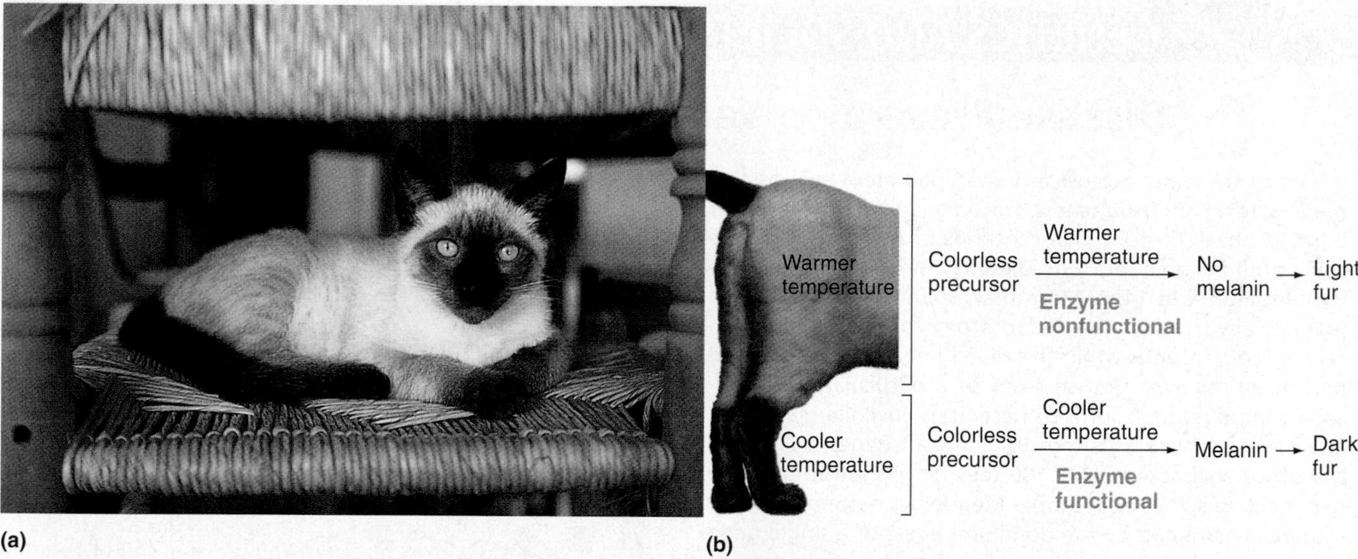

Figure 3.20 In Siamese cats, temperature affects coat color. (a) A Siamese cat. (b) Melanin is produced only in the cooler extremities. This is because Siamese cats are homozygous for a mutation that renders an enzyme involved in melanin synthesis temperature sensitive. The mutant enzyme is active at lower temperatures but inactive at higher temperatures.

Even in genetically normal individuals, exposure to chemicals or other environmental agents can have phenotypic consequences that are similar to those caused by mutant alleles of specific genes. A change in phenotype arising from such environmental agents that mimics the effects of a mutation in a gene is known as a **phenocopy.** Phenocopies are not heritable because they do not arise from a change in a gene. In humans, ingestion of the sedative thalidomide by pregnant women in the early 1960s produced a phenocopy of a rare dominant trait called *phocomelia.* By disrupting limb development in otherwise normal fetuses, the drug mimicked the effect of the phocomelia-causing mutation. When this became evident, thalidomide was withdrawn from the market.

Some types of environmental change may have a positive effect on an organism's survivability, as in the following example, where a straightforward application of medical science artificially reduces the penetrance of a mutant phenotype. Children born with the recessive trait known as phenylketonuria, or PKU, will develop a range of neurological problems, including convulsive seizures and mental retardation, unless they are put on a special diet. Homozygosity for the mutant PKU allele eliminates the activity of a gene encoding the enzyme phenylalanine hydroxylase. This enzyme normally converts the amino acid phenylalanine to the amino acid tyrosine. Absence of the enzyme causes a buildup of phenylalanine, and this buildup results in neurological problems. Today, there is a reliable blood test for detecting the condition in newborns. Once a baby with PKU is identified, the doctor or other health-care practitioner prescribes a protective diet that excludes phenylalanine; the diet must also provide enough calories to prevent the infant's body from breaking down its own proteins and thereby releasing the damaging amino acid from within. Such dietary therapy—a simple change in the environment—although somewhat unpleasant because it is highly limiting, now enables many PKU infants to develop into healthy adults.

Finally, two of the top killer diseases in the United States—cardiovascular disease and lung cancer—also illustrate how the environment can alter phenotype by influencing both expressivity and penetrance. People may inherit a propensity to heart disease, but the environmental factors of diet and exercise contribute to the occurrence (penetrance) and seriousness (expressivity) of their condition. Similarly, some people are born genetically prone to lung cancer, but whether or not they get it (penetrance) is strongly determined by whether or not they choose to smoke.

Thus, various aspects of an organism's environment, including temperature, diet, and exercise, interact with its genotype to generate the functional phenotype, the ultimate combination of traits that determines what a plant or animal looks like and how it behaves.

Chance Can Affect Penetrance and Expressivity

The retinoblastoma example described earlier concerns people who are born with the disease mutation in one allele of the retinoblastoma gene in every cell. Whether or not they show the phenotype and whether they show it in one or both eyes depend on additional genetic events that occur randomly to alter the second allele of the gene in specific body cells. Examples of random events that can trigger the onset of the disease include exposure to cosmic rays that alter the genetic material in retinal cells or mistakes made by retinal cell machinery during cell division. Events of this type provide the second "hit"—a mutation in the second copy of the retinoblastoma gene—necessary

GENETICS AND SOCIETY

Disease Prevention Versus the Right to Privacy

In one of the most extensive human pedigrees ever assembled, a team of researchers traced a familial pattern of blindness back through five centuries of related individuals to its origin in a couple who died in a small town in northwestern France in 1495. More than 30,000 French men and women alive today descended from that one fifteenth-century couple, and within this direct lineage reside close to half of all reported French cases of a particular blindness-producing disease, a form of hereditary juvenile glaucoma. The massive genealogic tree for the trait (when posted on the office wall, it was over 100 feet long) showed that the genetic defect follows a simple Mendelian pattern of transmission determined by the dominant allele of a single gene (**Fig. A**). The pedigree also showed that the dominant genetic defect displays incomplete penetrance: Not all people receiving the dominant allele from one or the other of their parents become blind; these sighted carriers may unknowingly pass the blindness-causing dominant allele to their children. Thus, with hereditary juvenile glaucoma, a parent, grandparent, or great-grandparent may possess the defective dominant allele but not show up as an affected patient in the family pedigree.

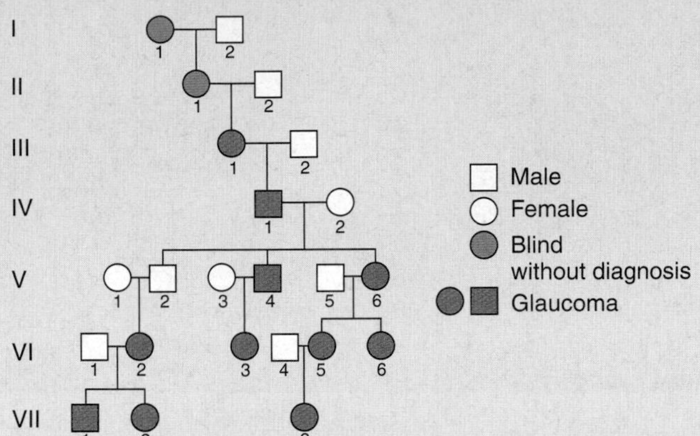

Figure A A pedigree showing the transmission of juvenile glaucoma. A small part of the genealogic tree: The vertical transmission pattern over seven generations shows that a dominant allele of a single gene causes juvenile glaucoma. The lack of glaucoma in V-2 followed by its reappearance in VI-2 reveals that the trait is incompletely penetrant. As a result, sighted heterozygotes may unknowingly pass the condition on to their children.

Unfortunately, people do not know they have the disease until their vision starts to deteriorate, and by that time, their optic fibers have sustained irreversible damage, and blindness is all but inevitable. Surprisingly, the existence of medical therapies that make it possible to arrest the nerve

to turn a normal retinal cell into a cancerous one. The phenotype of retinoblastoma thus results from a specific heritable mutation in a specific gene, but the incomplete penetrance and variable expressivity of the disease depend on chance genetic events that affect the other allele in certain cells.

By contributing to incomplete penetrance and variable expressivity, modifier genes, the environment, and chance give rise to phenotypic variation. Unlike dominant epistasis or recessive lethality, however, the probability of penetrance and the level of expressivity cannot be derived from the original Mendelian principles of segregation and independent assortment; they are determined empirically by observation and counting.

Even Continuous Variation Can Be Explained by Extensions to Mendelian Analysis

In Mendel's experiments, height in pea plants was determined by two segregating alleles of one gene (in the wild, it is determined by many genes, but in Mendel's inbred populations, the alleles of all these other genes were invariant). The phenotypes that resulted from these alternative alleles were clear-cut, either short or tall, and pea plant height was therefore known as a **discontinuous trait.** In contrast, because people do not produce inbred populations, height in humans is determined by segregating alleles of many different genes whose interaction with each other and the environment produces phenotypes showing continuous variation; height in humans is thus an example of a **continuous trait.** Within most outbred human populations, individual heights vary over a range of values that when charted on a graph produce a bell curve (**Fig. 3.21a** on p. 70). In fact, many human traits, including height, weight, and skin color, show continuous variation, rather than the clear-cut alternatives analyzed by Mendel. They also appear to blend and "unblend." Think for a moment of skin color. Children of marriages between people of African and Northern European descent, for example, often seem to be a blend of their parents' skin color. Matings of these F_1 individuals produce offspring with a wide range of skin pigmentation; a few may be as light as the original Northern European parent, a few as dark as the original African parent, but most will fall in a range between the two (Fig. 3.21b). For these reasons, early human geneticists were slow to accept Mendelian analysis. Because they were working with outbred populations, they found very few examples of "either-or" Mendelian traits in normal, healthy people.

deterioration of hereditary juvenile glaucoma and prevent blindness created a quandary in the late 1980s. Because treatment, to be effective, has to begin before symptoms of impending blindness show up, information in the pedigree could have helped doctors pinpoint people who are at risk, even if neither of their parents is blind. The researchers who compiled the massive family history therefore wanted to give physicians the names of at-risk individuals living in their area; with such information, the doctors could keep a close watch on certain patients and when necessary, recommend drugs or surgery to prevent blindness. However, a long-standing French law protecting personal privacy forbids public circulation of the names in genetic pedigrees. The French government agency interpreting this law maintained that if the actual names in the glaucoma pedigree were made public, potential carriers of the disease might suffer discrimination in hiring or insurance.

France thus faced a serious ethical dilemma: On the one hand, giving out names could save perhaps thousands of people from blindness; on the other hand, laws designed to protect personal privacy and forestall discrimination precluded the dissemination of specific names. The solution adopted by the French government at the time was a massive educational program to alert the general public to the problem so that concerned families could seek medical advice. This approach addressed the legal issues but was only partially helpful in dealing with the medical problem. The government's educational program helped identify some teenagers who were beginning to suffer from glaucoma, but many affected individuals escaped detection.

By 1997, molecular geneticists had identified the gene whose dominant mutant allele causes juvenile glaucoma. This gene specifies a protein called myocilin whose normal function in the eye is at present unknown. The mutant allele encodes a form of myocilin that misfolds and accumulates abnormally in the trabecular meshwork, a group of tiny canals in the eye through which fluid drains into the bloodstream. The accumulation of misfolded myocilin clogs the meshwork, thereby decreasing the outflow of excess vitreous humor in the eye. The resulting increased pressure within the eye (glaucoma) eventually damages the optic nerve, and this damage leads to blindness.

Knowledge of the specific disease-causing mutations in the *myocilin* gene has more recently resulted in the development of diagnostic tests based on the direct analysis of genotype. (We describe methods for direct genotype analysis in Chapters 9 and 11.) These DNA-based tests can not only identify individuals at risk, but they can also improve disease management, because detection of the mutant allele before the optic nerve is permanently damaged allows for timely treatment. Interestingly, if such tests become sufficiently inexpensive in the future, they could resolve France's ethical dilemma. With a low enough test cost, doctors could routinely administer the tests to all newborns and immediately identify nearly all potentially affected children; private information in a pedigree would thus not be needed. Physicians could then work with the parents to provide preventative therapies early in life.

By 1930, however, studies of corn and tobacco conclusively demonstrated that it is possible to provide a Mendelian explanation of continuous variation by simply increasing the number of genes contributing to a phenotype. The more genes, the more phenotypic classes, and the more classes, the more the variation appears continuous. As a hypothetical example, consider a series of genes (*A*, *B*, *C*, . . .) all affecting the height of pole beans. For each gene, there are two alleles, a *"0"* allele that contributes nothing to height and a *"1"* allele that increases the height of a plant by one unit. All alleles exhibit incomplete dominance relative to alternative alleles at the same gene. The phenotypes determined by all these genes are additive. What would be the result of a two-generation cross between pure-breeding plants carrying only *0* alleles at each height gene and pure-breeding plants carrying only *1* alleles at each height gene? If only one gene were responsible for height, and if environmental effects could be discounted, the F_2 population would be distributed among three classes: homozygous A^0A^0 plants with 0 height (they lie prostrate on the ground); heterozygous A^0A^1 plants with a height of 1; and homozygous A^1A^1 plants with a height of 2 (**Fig. 3.22a** on p. 371). This distribution of heights over three phenotypic classes does not make a continuous curve. But for two genes, there will be five phenotypic classes in the F_2 generation (Fig. 3.22b); for three genes, seven classes (Fig. 3.22c); and for four genes, nine classes (not shown). The distributions produced by three and four genes begin to approach continuous variation, and if we add a small contribution from environmental variation, a smooth curve will appear. After all, we would expect bean plants to grow better in good soil, with ample sunlight and water. The environmental component effectively converts the stepped bar graph to a continuous curve by producing some variation in expressivity within each genotypic class. Moreover, additional variation might arise from more than two alleles at some genes (Fig. 3.22d), unequal contribution to the phenotype by the various genes involved (review Fig. 3.17 on p. 64), interactions with modifier genes, and chance. Thus, from what we now know about the relation between genotype and phenotype, it is possible to see how just a handful of genes that behave according to known Mendelian principles can easily generate continuous variation. In fact, studies suggest that most of the variation in skin color between people of Northern European and African descent (see Figure 3.21b) can be explained by the action of three to four genes.

(a)

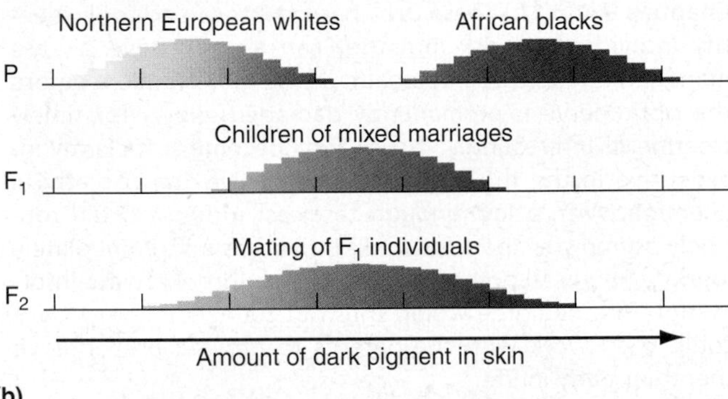

(b)

Figure 3.21 Continuous traits in humans. (a) Women runners at the start of a 5th Avenue mile race in New York City demonstrate that height is a trait showing continuous variation. **(b)** The skin color of most F_1 offspring is usually between the parental extremes, while the next (F_2) generation exhibits a broader distribution of continuous variation.

Continuous traits (also called **quantitative traits**) vary quantitatively over a range of values and can usually be measured: the length of a tobacco flower in millimeters, the amount of milk produced by a cow per day in liters, or the height of a person in meters. They are usually **polygenic**—controlled by multiple genes—and show the additive effects of a large number of alleles, which creates an enormous potential for variation within a population. Polygenic traits are, by definition, multifactorial, but not all multifactorial traits are polygenic (some, for instance, are determined by the way the alleles of one gene interact with each other and the environment). We discuss the analysis and distribution of multifactorial traits in Chapter 21 on population genetics.

The Mouse's Coat Color: A Comprehensive Example of Multiple Alleles and Multifactorial Traits

Most field mice are a dark gray (agouti), but mice bred for specific mutations in the laboratory can be gray, tan, yellow, brown, black, or various combinations thereof. Here we look at the alleles of three of the genes that make such variation possible. Our review underscores how the various allelic interactions of just a handful of genes can produce an astonishing diversity of phenotypes.

Gene 1—Agouti or Other Color Patterns

The *agouti* gene determines the distribution of color on each hair and has multiple alleles. The wild-type allele A specifies the bands of yellow and black that give the agouti appearance; A^y gets rid of the black and thus produces solid yellow; a gets rid of the yellow and thus produces solid black; and a^t specifies black on the back and yellow on the stomach. The dominance series for this set of *agouti* gene alleles is $A^y > A > a^t > a$. However, although A^y is dominant to all other alleles for coat color, it is recessive to all the others for lethality: A^yA^y homozygotes die before birth, while A^yA, A^ya^t, or A^ya heterozygotes survive.

Gene 2—Black or Brown

A second gene specifies whether the dark color of each hair is black or brown. This gene has two alleles: B is dominant and designates black; b is recessive and generates brown. Since the A^y allele at the *agouti* gene completely eliminates the dark band of each hair, it acts in a dominant epistatic manner to the B gene. With all other *agouti* alleles, however, it is possible to distinguish the effects of the two different B alleles on phenotype. The $A– B–$ genotype gives rise to the wild-type agouti having black with yellow hairs. The $A– bb$ genotype generates a color referred to as cinnamon (with hairs having stripes of brown and yellow); $aa\ bb$ is all brown; and $a^ta^t\ bb$ is brown on the back and yellow on the stomach. A cross between two F_1 hybrid animals of genotype $A^ya\ Bb$ would yield an F_2 generation with yellow (A^ya – –), black ($aa\ B–$), and brown ($aa\ bb$) animals in a ratio of 8:3:1. This ratio reflects the dominant epistasis of A^y and the loss of a class of four (A^yA^y – –) due to prenatal lethality.

Gene 3—Albino or Pigmented

Like other mammals, mice have a third gene influencing coat color. A recessive allele (c) abolishes the function of the enzyme that leads to the formation of dark pigment melanin, making this allele epistatic to all other coat color genes. As a result, cc homozygotes are pure white, while $C–$ mice are agouti, black, brown, yellow, or yellow and black (or other colors and patterns), depending on what alleles they carry at the A and B genes, as well as at some 50 other genes known to play a role in determining the coat color of mice. Adding to the complex color potential are other alleles that geneticists have uncovered for the albino gene; these cause only a partial inactivation of the melanin-producing enzyme and thus have a partial epistatic effect on phenotype.

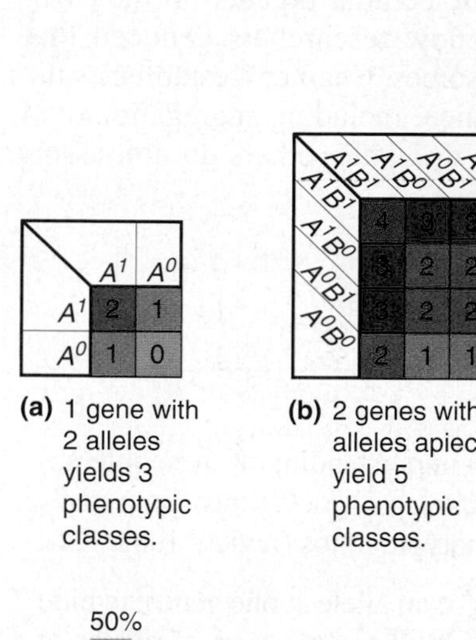

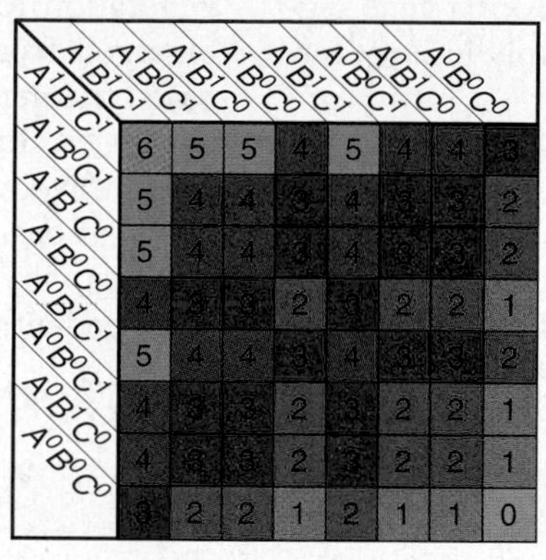

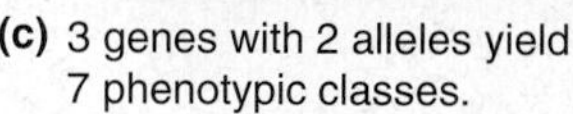

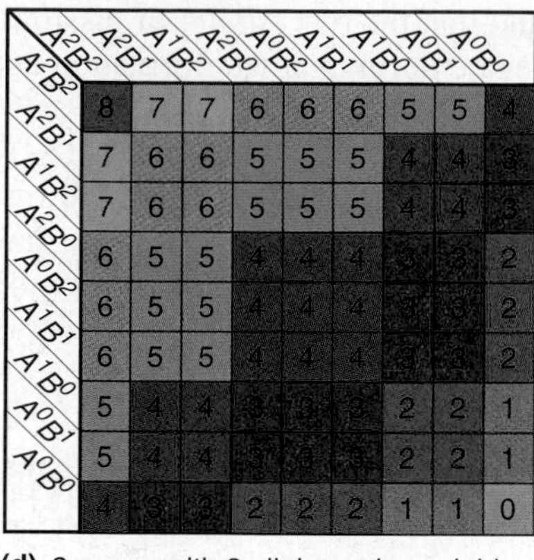

(a) 1 gene with 2 alleles yields 3 phenotypic classes.

(b) 2 genes with 2 alleles apiece yield 5 phenotypic classes.

(c) 3 genes with 2 alleles yield 7 phenotypic classes.

(d) 2 genes with 3 alleles apiece yield 9 phenotypic classes.

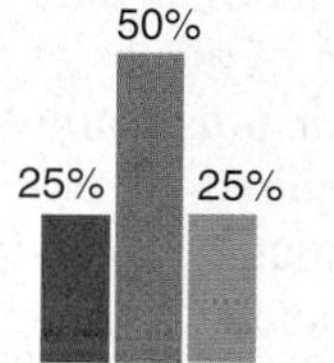

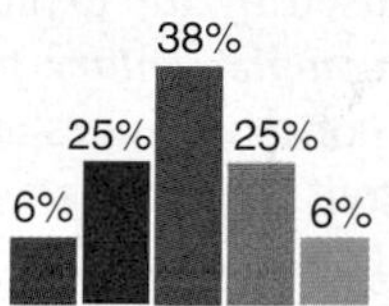

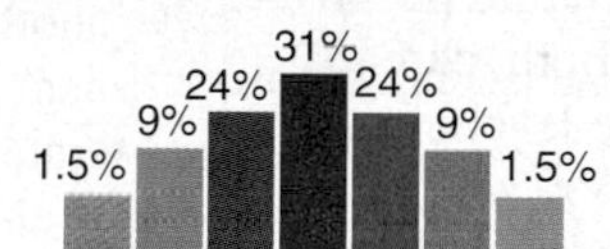

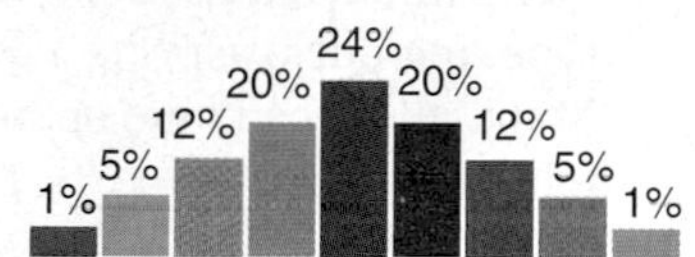

Figure 3.22 A Mendelian explanation of continuous variation. The more genes or alleles, the more possible phenotypic classes, and the greater the similarity to continuous variation. In these examples, several pairs of incompletely dominant alleles have additive effects. Percentages denote frequencies of each genotype expressed as fractions of the total population.

This comprehensive example of coat color in mice gives some idea of the potential for variation from just a few genes, some with multiple alleles. Amazingly, this is just the tip of the iceberg. When you realize that both mice and humans carry roughly 25,000 genes, the number of interactions that connect various alleles of these genes in the expression of phenotype is in the millions, if not the billions. The potential for variation and diversity among individuals is staggering indeed.

Connections

Part of Mendel's genius was to look at the genetic basis of variation through a very narrow window, focusing his first glimpse of the mechanisms of inheritance on such fundamental phenomena that the import of his insights remains undiminished to this day. Mendel worked on just a handful of traits in inbred populations of one species, and for each trait, one gene with one completely dominant and one recessive allele determined two distinguishable, or discontinuous, phenotypes. Both the dominant and recessive alleles showed complete penetrance and negligible differences of expressivity. He knew from other examples that phenotype does not always reflect genotype, but he probably did not foresee how many complexities could cloud the correlation between the two.

In the first few decades of the twentieth century, many questioned the general applicability of Mendelian analysis, for it seemed to shed little light on the complex inheritance patterns of most plant and animal traits or on the mechanisms producing continuous variation. Simple embellishments, however, clarified the genetic basis of continuous variation and provided explanations for other apparent exceptions to Mendelian analysis. These embellishments included the ideas that dominance need not be complete, that one gene can have multiple alleles, that one gene can determine more than one trait, that several genes can contribute to the same trait, and that the expression of genes can be affected in a variety of ways by other genes, the environment, and chance. Each embellishment extends the range of Mendelian analysis and deepens our understanding of the genetic basis of variation. And no matter how broad the view, Mendel's basic conclusions, embodied in his first law—that genes are the discrete units of heredity and that the two similar or dissimilar copies of a gene segregate during gamete formation and randomly unite at fertilization—remain valid.

But what about Mendel's second law that genes assort independently? As it turns out, its application is not as universal as that of the law of segregation. Many genes do assort independently, but some do not; rather, they appear to be linked and transmitted together from generation to generation. An understanding of this fact emerged from studies that located Mendel's hereditary units, the genes, in specific cellular organelles, the chromosomes. In describing how researchers deduced that genes travel with chromosomes, Chapter 4 establishes the physical basis of inheritance, including segregation, and clarifies why some genes do, while others do not, assort independently.

Essential Concepts

1. The F_1 phenotype generated by each pair of alleles defines the dominance relationship between these alleles. One allele is not always completely dominant or completely recessive to another. With *incomplete dominance,* the F_1 hybrid phenotype resembles neither parent. With *codominance,* the F_1 hybrid phenotype includes aspects derived from both parents. Many allele pairs are codominant at the level of protein production.
2. One gene can contribute to multiple traits (*pleiotropy*); for such a gene, the dominance relation between any two alleles can vary according to the particular phenotype under consideration.
3. A single gene may have any number of alleles, each of which can cause the appearance of different phenotypes. New alleles arise by mutation. Alleles whose frequency is greater than 1% in a population are considered *wild types;* those whose frequency is below 1% are considered *mutants.* When two or more wild-type alleles exist for a gene, the gene is considered *polymorphic;* a gene with only one wild-type allele is *monomorphic.*
4. Two or more genes may interact in several ways to affect the production of a single trait. It is often possible to arrive at an understanding of these interactions by observing characteristic deviations from traditional Mendelian phenotypic ratios (review Table 3.2).
5. In *epistasis,* the action of an allele at one gene can hide traits normally caused by the expression of alleles at another gene. In *complementary gene action,* dominant alleles of two or more genes are required to generate a particular trait. In *heterogeneity,* mutant alleles at any one of two or more genes are sufficient to elicit a phenotype. The *complementation test* can reveal whether a particular phenotype seen in two individuals arises from mutations in the same or separate genes.
6. In many cases, the route from genotype to phenotype can be modified by the environment, chance, or other genes. A phenotype shows *incomplete penetrance* when it is expressed in fewer than 100% of individuals with the same genotype. A phenotype shows *variable expressivity* when it is expressed at a quantitatively different level in different individuals with the same genotype.
7. A *continuous trait* can have any value of expression between two extremes. Most traits of this type are *polygenic,* that is, determined by the interactions of multiple genes.

On Our Website

www.mhhe.com/hartwell3
Chapter 3

Annotated Suggested Readings and Links to Other Websites

- Additional historical examples of complications in Mendelian analysis
- Recently discovered interesting genetic systems

Specialized Topics

- Use of chi-square analysis to test the likelihood that the experimental outcomes of a cross can be explained by a particular hypothesis for the mode of inheritance (This is a different use of chi-square analysis than the one we present in Chapter 5, where we introduce the technique as a way to determine whether two genes are linked to each other.)

Social and Ethical Issues

1. John and Nancy's daughter has a blood test performed by their family doctor. The doctor knows that both John and Nancy have the AB blood type. But when results come back on their daughter's blood, the doctor discovers that her blood type is O. Neither parent is heterozygous for the Bombay allele. The only conclusion the doctor can reach from this result is that babies must have been switched at the hospital. The doctor is now in an awkward position. John and Nancy and their healthy daughter make a very happy family. If he explains to them the implication of his findings, their lives will be turned upside down. If not, he will be withholding personal information that one could argue rightly belongs to them. What would you do if you were the doctor?
2. A man who has committed several violent rapes is on trial, and his lawyer argues that the man cannot help himself because his behavior is the result of his genetic makeup. The man's father and brother also have a history of committing violent crimes. In defense of this position, the lawyer cites studies reported in 1993 on a small Dutch family indicating that some people inherit a propensity toward violent behavior. Affected individuals in this family have a mutation in the monoamine oxidase gene involved in the breakdown of neurotransmitters. Do you think the sentencing of the man on trial should depend in part on this type of scientific study? Our judicial system is based on the assumption that people act with free will and are therefore responsible for their actions. Are people with known genetic alterations responsible for their actions? Should we make exceptions in the case of genetic predispositions? What role should society play in helping people whose behaviors may be influenced by the alleles they carry?
3. A teenage girl was diagnosed with chronic myelogenous leukemia, for which the treatment is a bone marrow transplant. Successful tissue transplantation depends on matches with the donor's alleles of the polymorphic HLA genes (described on page 50). A matching donor could not be found, and her parents decided to have another child, hoping that the younger sibling could serve as a donor of bone marrow for the older child. Is it ethical to conceive a child for the purpose of tissue or organ donation to a sibling? Consider situations in which the donor's life will be compromised by the donation as well as cases in which the donor's life will not be compromised.

Solved Problems

I. Imagine you purchased an albino mouse (genotype *cc*) in a pet store. The *c* allele is epistatic to other coat color genes. How would you go about determining the genotype of this mouse at the brown locus? (In pigmented mice, *BB* and *Bb* are black, *bb* is brown.)

Answer

This problem requires an understanding of gene interactions, specifically epistasis. You have been placed in the role of experimenter and need to design crosses that will answer the question. To determine the alleles of the B gene present, you need to eliminate the blocking action of the cc genotype. Because only the recessive c allele is epistatic, when a C allele is present, no epistasis will occur. To introduce a C allele during the mating, the test mouse you mate to your albino can have the genotype CC or Cc. (If the mouse is Cc, half of the progeny will be albino and will not contribute useful information, but the nonalbinos from this cross would be informative.) What alleles of the B gene should the test mouse carry? To make this decision, work through the expected results using each of the possible genotypes.

Test mouse genotype		Albino mouse	Expected progeny
BB	×	BB	all black
	×	*Bb*	all black
	×	*bb*	all black
Bb	×	BB	all black
	×	*Bb*	3/4 black, 1/4 brown
	×	*bb*	1/2 black, 1/2 brown
bb	×	BB	all black
	×	*Bb*	1/2 black, 1/2 brown
	×	*bb*	all brown

From these hypothetical crosses, you can see that a test mouse with either the *Bb* or *bb* genotype would yield distinct outcomes for each of the three possible albino mouse genotypes. However, a *bb* test mouse would be more useful and less ambiguous. First, it is easier to identify a mouse with the *bb* genotype because a brown mouse must have this homozygous recessive genotype. Second, the results are completely different for each of the three possible genotypes when you use the *bb* test mouse. (In contrast, a *Bb* test mouse would yield both black and brown progeny whether the albino mouse was *Bb* or *bb;* the only distinguishing feature is the ratio.) *To determine the full genotype of the albino mouse, you should cross it to a brown mouse (which could be* CC bb *or* Cc bb*).*

II. In a particular kind of wildflower, the wild-type flower color is deep purple, and the plants are true-breeding. In one true-breeding mutant stock, the flowers have a reduced pigmentation, resulting in a lavender color. In a different true-breeding mutant stock, the flowers have no pigmentation and are thus white. When a lavender-flowered plant from the first mutant stock was crossed to a white-flowered plant from the second mutant stock, all the F_1 plants had purple flowers. The F_1 plants were then allowed to self-fertilize to produce an F_2 generation. The 277 F_2 plants were 157 purple: 71 white : 49 lavender. Explain how flower color is inherited. Is this trait controlled by the alleles of a single gene? What kinds of progeny would be produced if lavender F_2 plants were allowed to self-fertilize?

Answer

Are there any modes of single-gene inheritance compatible with the data? The observations that the F_1 plants look different from either of their parents and that the F_2 generation is composed of plants with three different phenotypes exclude complete dominance. The ratio of the three phenotypes in the F_2 plants has some resemblance to the 1:2:1 ratio expected from codominance or incomplete dominance, but the results would then imply that purple plants must be heterozygotes. This conflicts with the information provided that purple plants are true-breeding.

Consider now the possibility that two genes are involved. From a cross between plants heterozygous for two genes (*W* and *P*), the F_2 generation would contain a 9:3:3:1 ratio of the genotypes *W– P–*, *W– pp*, *ww P–*, and *ww pp* (where the dash indicates that the allele could be either a dominant or a recessive form). Are there any combinations of the 9:3:3:1 ratio that would be close to that seen in the F_2 generation in this example? The numbers seem close to a 9:4:3 ratio. What hypothesis would support combining two of the classes (3 + 1)? If *w* is epistatic to the *P* gene, then the *ww P–* and *ww pp* genotypic classes would have the same white phenotype. With this explanation, 1/3 of the F_2 lavender plants would be *WW pp*, and the remaining 2/3 would be *Ww pp*. Upon self-fertilization, *WW pp* plants would produce only lavender (*WW pp*) progeny, while *Ww pp* plants would produce a 3:1 ratio of lavender (*W– pp*) and white (*ww pp*) progeny.

III. Huntington disease (HD) is a rare dominant condition in humans that results in a slow but inexorable deterioration of the nervous system. HD shows what might be called "age-dependent penetrance," which is to say that the probability that a person with the HD genotype will express the phenotype varies with age. Assume that 50% of those inheriting the *HD* allele will express the symptoms by age 40. Susan is a 35-year-old woman whose father has HD. She currently shows no symptoms. What is the probability that Susan will show symptoms in five years?

Answer

This problem involves probability and penetrance. Two conditions are necessary for Susan to show symptoms of the disease. There is a 1/2 (50%) chance that she inherited the mutant allele from her father and a 1/2 (50%) chance that she will express the phenotype by age 40. Since these are independent events, *the probability is the product of the individual probabilities, or 1/4.*

Problems

Interactive Web Exercise

PubMed is a database maintained by the National Center for Biotechnology Information (NCBI) that provides synopses of, and in many cases direct access to, published biomedical journal articles. This database is invaluable to genetics researchers, as well as all biologists and physicians. Our website at www.mhhe.com/hartwell3 contains a brief exercise introducing you to the resources at PubMed; once at the website, go to Chapter 3 and click on "Interactive Web Exercise".

Vocabulary

1. For each of the terms in the left column, choose the best matching phrase in the right column.

a. epistasis	1. one gene affecting more than one phenotype
b. modifier gene	2. the alleles of one gene mask the effects of alleles of another gene
c. conditional lethal	3. both parental phenotypes are expressed in the F_1 hybrids
d. permissive condition	4. a heritable change in a gene
e. reduced penetrance	5. cell surface molecules that are involved in the immune system and are highly variable
f. multifactorial trait	6. genes whose alleles alter phenotypes produced by the action of other genes
g. incomplete dominance	7. less than 100% of the individuals possessing a particular genotype express it in their phenotype
h. codominance	8. environmental conditions that allow conditional lethals to live

i. histocompatibility antigens	9. a trait produced by the interaction of alleles of at least two genes or from interactions between gene and environment
j. mutation	10. individuals with the same genotype have related phenotypes that vary in intensity
k. pleiotropy	11. a genotype that is lethal in some situations (for example, high temperature) but viable in others
l. variable expressivity	12. the heterozygote resembles neither homozygote

Section 3.1

2. In four-o'clocks, the allele for red flowers is incompletely dominant over the allele for white flowers, so heterozygotes have pink flowers. What ratios of flower colors would you expect among the offspring of the following crosses: (a) pink × pink, (b) white × pink, (c) red × red, (d) red × pink, (e) white × white, and (f) red × white? If you specifically wanted to produce pink flowers, which of these crosses would be most efficient?

3. A cross between two plants that both have yellow flowers produces 80 offspring plants, of which 38 have yellow flowers, 22 have red flowers, and 20 have white flowers. If one assumes that this variation in color is due to inheritance at a single locus, what is the genotype associated with each flower color, and how can you describe the inheritance of flower color?

4. In *Drosophila melanogaster*, very dark (ebony) body color is determined by the *e* allele. The e^+ allele produces the normal wild-type honey-colored body. In heterozygotes for the two alleles, a dark marking called the trident can be seen on the thorax, but otherwise the body is honey-colored. The e^+ allele is thus considered to be incompletely dominant to the *e* allele.

a. When female e^+e^+ flies are crossed to male e^+e flies, what is the probability that progeny will have the dark trident marking?

b. Animals with the trident marking mate among themselves. Of 300 progeny, how many would be expected to have a trident, how many ebony bodies, and how many honey-colored bodies?

5. A wild legume with white flowers and long pods is crossed to one with purple flowers and short pods. The F_1 offspring are allowed to self-fertilize, and the F_2 generation has 301 long purple, 99 short purple, 612 long pink, 195 short pink, 295 long white, and 98 short white. How are these traits being inherited?

6. In radishes, color and shape are each controlled by a single locus with two incompletely dominant alleles. Color may be red (*RR*), purple (*Rr*), or white (*rr*) and shape can be long (*LL*), oval (*Ll*), or round (*ll*). What phenotypic classes and proportions would you expect among the offspring of a cross between two plants heterozygous at both loci?

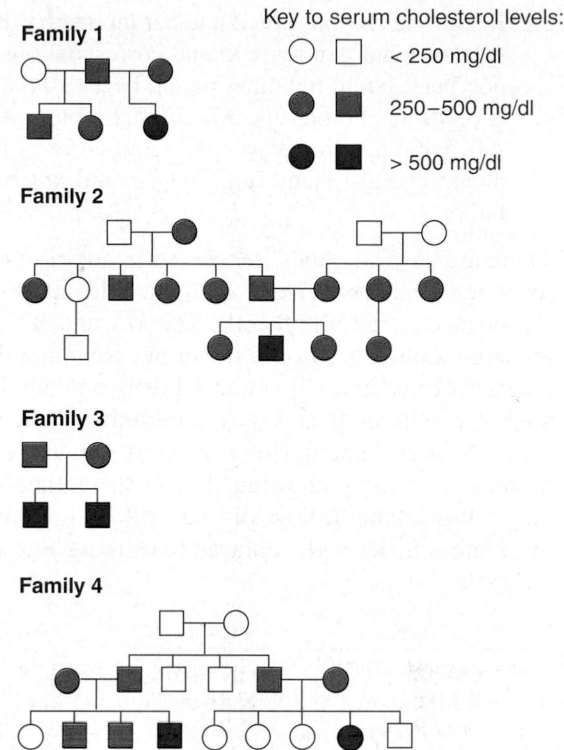

7. Familial hypercholesterolemia (FH) is an inherited trait in humans that results in higher than normal serum cholesterol levels (measured in milligrams of cholesterol per deciliter of blood [mg/dl]). People with serum cholesterol levels that are roughly twice normal have a 25 times higher frequency of heart attacks than unaffected individuals. People with serum cholesterol levels three or more times higher than normal have severely blocked arteries and almost always die before they reach the age of 20. The pedigrees above show the occurrence of FH in four Japanese families:

a. What is the most likely mode of inheritance of FH based on this data? Are there any individuals in any of these pedigrees who do not fit your hypothesis? What special conditions might account for such individuals?

b. Why do individuals in the same phenotypic class (unfilled, yellow, or orange symbols) show such variation in their levels of serum cholesterol?

8. Describe briefly

a. The genotype of a person who has sickle-cell anemia.

b. The genotype of a person with a normal phenotype who has a child with sickle-cell anemia.

c. The *total* number of different alleles of the β-globin gene that could be carried by five children with the same mother and father.

9. Assuming no involvement of the Bombay phenotype:

a. If a girl has blood type O, what could be the genotypes and corresponding phenotypes of her parents?

b. If a girl has blood type B and her mother has blood type A, what genotype(s) and corresponding phenotype(s) could the other parent have?
c. If a girl has blood type AB and her mother is also AB, what are the genotype(s) and corresponding phenotype(s) of any male who could *not* be her father?

10. There are several genes in humans in addition to the *ABO* gene that give rise to recognizable antigens on the surface of red blood cells. The *MN* and *Rh* genes are two examples. The *Rh* locus can contain either a positive or negative allele, with positive being dominant to negative. *M* and *N* are codominant alleles of the *MN* gene. The following chart shows several mothers and their children. For each mother–child pair, choose the father of the child from among the males in the right column, assuming one child per male.

	Mother	Child	Males
a.	O M Rh pos	B MN Rh neg	O M Rh neg
b.	B MN Rh neg	O N Rh neg	A M Rh pos
c.	O M Rh pos	A M Rh neg	O MN Rh pos
d.	AB N Rh neg	B MN Rh neg	B MN Rh pos

11. Alleles of the gene that determines seed coat patterns in lentils can be organized in a dominance series: marbled > spotted = dotted (codominant alleles) > clear. A lentil plant homozygous for the marbled seed coat pattern allele was crossed to one homozygous for the spotted pattern allele. In another cross, a homozygous dotted lentil plant was crossed to one homozygous for clear. An F_1 plant from the first cross was then mated to an F_1 plant from the second cross.
a. What phenotypes in what proportions are expected from this mating between the two F_1 types?
b. What are the expected phenotypes of the F_1 plants from the two original parental crosses?

12. In clover plants, the pattern on the leaves is determined by a single gene with multiple alleles that are related in a dominance series. Seven different alleles of this gene are known; an allele that determines the absence of a pattern is recessive to the other six alleles, each of which produces a different pattern. All heterozygous combinations of alleles show complete dominance.
a. How many different kinds of leaf patterns (including the absence of a pattern) are possible in a population of clover plants in which all seven alleles are represented?
b. What is the largest number of different genotypes that could be associated with any one phenotype? Is there any phenotype that could be represented by only a single genotype?
c. In a particular field, you find that the large majority of clover plants lack a pattern on their leaves, even though you can identify a few plants representative of all possible pattern types. Explain this finding.

13. In a population of rabbits, you find three different coat color phenotypes: chinchilla (C), himalaya (H), and albino (A). To understand the inheritance of coat colors, you cross individual rabbits with each other and note the results in the following table.

Cross number	Parental phenotypes	Phenotypes of progeny
1	H × H	3/4 H:1/4 A
2	H × A	1/2 H:1/2 A
3	C × C	3/4 C:1/4 H
4	C × H	all C
5	C × C	3/4 C:1/4 A
6	H × A	all H
7	C × A	1/2 C:1/2 A
8	A × A	all A
9	C × H	1/2 C:1/2 H
10	C × H	1/2 C:1/4 H:1/4 A

a. What can you conclude about the inheritance of coat color in this population of rabbits?
b. Ascribe genotypes to the parents in each of the 10 crosses.
c. What kinds of progeny would you expect, and in what proportions, if you crossed the chinchilla parents in crosses #9 and #10?

14. Some plant species have an incompatibility system different from that shown in Fig. 3.8. In this alternate kind of incompatibility, a mating cannot produce viable seeds if the male parent shares an incompatibility allele with the female parent. (Just as with the kind of incompatibility system shown in Fig. 3.8, this system ensures that all plants are heterozygous for the incompatibility gene.) Five plants were isolated from a wild population of a species with this alternate type of incompatiblity. The results of matings between each pair of plants are given here (– means no seeds were produced; + means seeds were produced). How many different alleles of the incompatibility gene are present in this group of five plants? What are the genotypes of the plants?

	1	2	3	4	5
1	–	–	–	+	–
2		–	+	+	+
3			–	+	–
4				–	–
5					–

15. Fruit flies with one allele for curly wings (*Cy*) and one allele for normal wings (Cy^+) have curly wings. When two curly-winged flies were crossed, 203 curly-winged and 98 normal-winged flies were obtained. In fact, all crosses between curly-winged flies produce nearly the same curly : normal ratio among the progeny.

a. What is the approximate phenotypic ratio in these offspring?
b. Suggest an explanation for these data.
c. If a curly-winged fly was mated to a normal-winged fly, how many flies of each type would you expect among 180 total offspring?

16. Spherocytosis is an inherited blood disease in which the erythrocytes (red blood cells) are spherical instead of biconcave. This condition is inherited in a dominant fashion, with Sph^- dominant to Sph^+. In people with spherocytosis, the spleen "reads" the spherical red blood cells as defective, and it removes them from the bloodstream, leading to anemia. The spleen in different people removes the spherical erythrocytes with different efficiencies. Some people with spherical erythrocytes suffer severe anemia and some mild anemia, yet others have spleens that function so poorly there are no symptoms of anemia at all. When 2400 people with the genotype Sph^- Sph^+ were examined, it was found that 2250 had anemia of varying severity, but 150 had no symptoms.
a. Does this description of people with spherocytosis represent incomplete penetrance, variable expressivity, or both? Explain your answer. Can you derive any values from the numerical data to measure penetrance or expressivity?
b. Suggest a treatment for spherocytosis and describe how the incomplete penetrance and/or variable expressivity of the condition might affect this treatment.

17. In a species of tropical fish, a colorful orange and black variety called montezuma occurs. When two montezumas, are crossed, 2/3 of the progeny are montezuma, and 1/3 are the wild-type, dark grayish green color. Montezuma is a single-gene trait, and montezuma fish are never true-breeding.
a. Explain the inheritance pattern seen here and show how your explanation accounts for the phenotypic ratios given.
b. In this same species, the morphology of the dorsal fin is altered from normal to ruffled by homozygosity for a recessive allele designated *f*. What progeny would you expect to obtain, and in what proportions, from the cross of a montezuma fish homozygous for normal fins to a green, ruffled fish?
c. What phenotypic ratios of progeny would be expected from the crossing of two of the montezuma progeny from part *b*?

18. You have come into contact with two unrelated patients who express what you think is a rare phenotype—a dark spot on the bottom of the foot. According to a medical source, this phenotype is seen in 1 in every 100,000 people in the population. The two patients give their family histories to you, and you generate the pedigrees shown at the top of the next column.

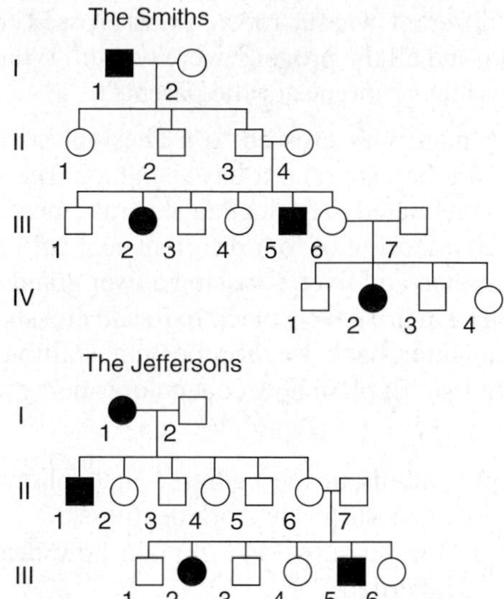

a. Given that this trait is rare, do you think the inheritance is dominant or recessive? Are there any special conditions that appear to apply to the inheritance?
b. Which nonexpressing members of these families must carry the mutant allele?
c. If this trait is instead quite common in the population, what alternative explanation would you propose for the inheritance?
d. Based on this new explanation (part c), which nonexpressing members of these families must have the genotype normally causing the trait?

19. Polycystic kidney disease is a dominant trait that causes the growth of numerous cysts in the kidneys. The condition eventually leads to kidney failure. A child with polycystic kidney disease is born to a couple, neither of whom shows the disease. What possibilities might explain this outcome?

Section 3.2

20. A rooster with a particular comb morphology called walnut was crossed to a hen with a type of comb morphology known as single. The F_1 progeny all had walnut combs. When F_1 males and females were crossed to each other, 93 walnut and 11 single combs were seen among the F_2 progeny, but there were also 29 birds with a new kind of comb called rose and 32 birds with another new comb type called pea.
a. Explain how comb morphology is inherited.
b. What progeny would result from crossing a homozygous rose-combed hen with a homozygous pea-combed rooster? What phenotypes and ratios would be seen in the F_2 progeny?
c. A particular walnut rooster was crossed to a pea hen, and the progeny consisted of 12 walnut, 11 pea, 3 rose, and 4 single chickens. What are the likely genotypes of the parents?

d. A different walnut rooster was crossed to a rose hen, and all the progeny were walnut. What are the possible genotypes of the parents?

21. A black mare was crossed to a chestnut stallion and produced a bay son and a bay daughter. The two offspring were mated to each other several times, and they produced offspring of four different coat colors: black, bay, chestnut, and liver. Crossing a liver grandson back to the black mare gave a black foal, and crossing a liver granddaughter back to the chestnut stallion gave a chestnut foal. Explain how coat color is being inherited in these horses.

22. Filled-in symbols in the pedigree that follows designate individuals suffering from deafness.

a. Study the pedigree and explain how deafness is being inherited.

b. What is the genotype of the individuals in generation V? Why are they not affected?

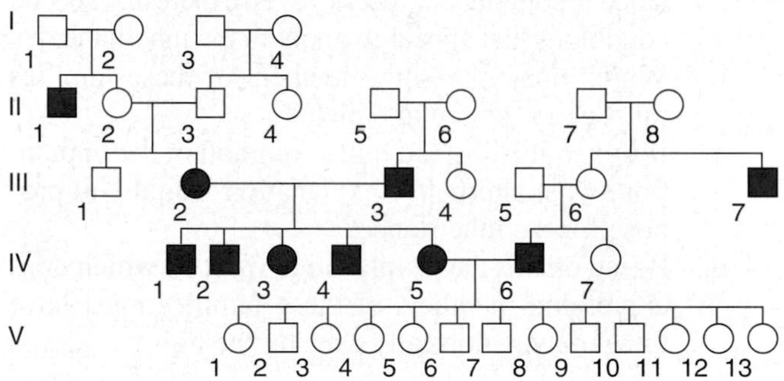

23. You do a cross between two true-breeding strains of zucchini. One has green fruit and the other has yellow fruit. The F_1 plants are all green, but when these are crossed, the F_2 plants consist of 9 green : 7 yellow.

a. Explain this result. What were the genotypes of the two parental strains?

b. Indicate the phenotypes, with frequencies, of the progeny of a testcross of the F_1 plants.

24. Two true-breeding white strains of the plant *Illegitimati noncarborundum* were mated, and the F_1 progeny were all white. When the F_1 plants were allowed to self-fertilize, 126 white-flowered and 33 purple-flowered F_2 plants grew.

a. How could you describe inheritance of flower color? Describe how specific alleles influence each other and therefore affect phenotype.

b. A white F_2 plant is allowed to self-fertilize. Of the progeny, 3/4 are white-flowered, and 1/4 are purple-flowered. What is the genotype of the white F_2 plant?

c. A purple F_2 plant is allowed to self-fertilize. Of the progeny, 3/4 are purple-flowered, and 1/4 are white-flowered. What is the genotype of the purple F_2 plant?

d. Two white F_2 plants are crossed with each other. Of the progeny, 1/2 are purple-flowered, and 1/2 are white-flowered. What are the genotypes of the two white F_2 plants?

25. Explain the difference between epistasis and dominance. How many loci are involved in each case?

26. As you will learn in later chapters, duplication of genes is an important evolutionary mechanism. As a result, many cases are known in which a species has two or more nearly identical genes.

a. Suppose there are two genes, *A* and *B*, that specify production of the same enzyme. An abnormal phenotype results only if an individual does not make any of that enzyme. What ratio of normal versus abnormal progeny would result from a mating between two parents of genotype *Aa Bb*, where *A* and *B* represent alleles that specify production of the enzyme, while *a* and *b* are alleles that do not?

b. Suppose now that there are three genes specifying production of this enzyme, and again that a single functional allele is sufficient for a wild-type phenotype. What ratio of normal versus abnormal progeny would result from a mating between two triply heterozygous parents?

27. "Secretors" (genotypes *SS* and *Ss*) secrete their A and B blood group antigens into their saliva and other body fluids, while "nonsecretors" (*ss*) do not. What would be the apparent phenotypic blood group proportions among the offspring of an I^AI^B *Ss* woman and an I^AI^A *Ss* man if typing was done using saliva?

28. Normally, wild violets have yellow petals with dark brown markings and erect stems. Imagine you discover a plant with white petals, no markings, and prostrate stems. What experiment could you perform to determine whether the non-wild-type phenotypes are due to several different mutant genes or to the pleiotropic effects of alleles at a single locus? Explain how your experiment would settle the question.

29. The following table shows the responses of blood samples from the individuals in the pedigree to anti-A and anti-B sera. A "+" in the anti-A row indicates that the red blood cells of that individual were clumped by anti-A serum and therefore the individual made A antigens, and a "−" indicates no clumping. The same notation is used to describe the test for the anti-B antigens.

a. Deduce the blood type of each individual from the data in the table.

b. Assign genotypes for the blood groups as accurately as you can from this data, explaining the pattern of inheritance shown in the pedigree. Assume that all genetic relationships are as presented in the pedigree (that is, there are no cases of false paternity).

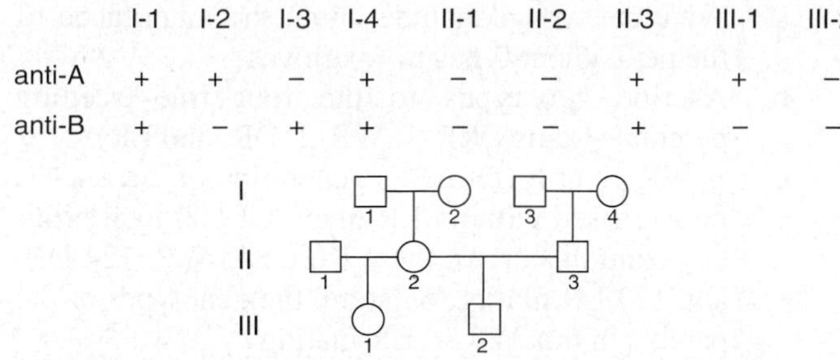

I-1 I-2 I-3 I-4 II-1 II-2 II-3 III-1 III-2
anti-A + + − + − − + + −
anti-B + − + + − − + − −
I
1 2 3 4
II
1 2 3
III
1 2

30. Three different pure-breeding strains of corn that produce ears with white kernels were crossed to each other. In each case, the F_1 plants were all red, while both red and white kernels were observed in the F_2 generation in a 9:7 ratio. These results are tabulated here.

	F_1	F_2
white-1 × white-2	red	9 red : 7 white
white-1 × white-3	red	9 red : 7 white
white-2 × white-3	red	9 red : 7 white

a. How many genes are involved in determining kernel color in these three strains?
b. Define your symbols and show the genotypes for the pure-breeding strains white-1, white-2, and white-3.
c. Diagram the cross between white-1 and white-2, showing the genotypes and phenotypes of the F_1 and F_2 progeny. Explain the observed 9:7 ratio.

31. In mice, the A^y allele of the *agouti* gene is a recessive lethal allele, but it is dominant for yellow coat color. What phenotypes and ratios of offspring would you expect from the cross of a mouse heterozygous at the agouti locus (genotype A^yA) and also at the albino locus (*Cc*) to an albino mouse (*cc*) heterozygous at the agouti locus (A^yA)?

32. A student whose hobby was fishing pulled a very unusual carp out of Cayuga Lake: It had no scales on its body. She decided to investigate whether this strange nude phenotype had a genetic basis. She therefore obtained some inbred carp that were pure-breeding for the wild-type scale phenotype (body covered with scales in a regular pattern) and crossed them with her nude fish. To her surprise, the F_1 progeny consisted of wild-type fish and fish with a single linear row of scales on each side in a 1:1 ratio.

a. Can a single gene with two alleles account for this result? Why or why not?
b. To follow up on the first cross, the student allowed the linear fish from the F_1 generation to mate with each other. The progeny of this cross consisted of fish with four phenotypes: linear, wild type, nude, and scattered (the latter had a few scales scattered irregularly on the body). The ratio of these phenotypes was 6:3:2:1, respectively. How many genes appear to be involved in determining these phenotypes?
c. In parallel, the student allowed the phenotypically wild-type fish from the F_1 generation to mate with each other and observed, among their progeny, wild-type and scattered carp in a ratio of 3:1. How many genes with how many alleles appear to determine the difference between wild-type and scattered carp?
d. The student confirmed the conclusions of part c by crossing those scattered carp with her pure-breeding wild-type stock. Diagram the genotypes and phenotypes of the parental, F_1, and F_2 generations for this cross and indicate the ratios observed.
e. The student attempted to generate a true-breeding nude stock of fish by inbreeding. However, she found that this was impossible. Every time she crossed two nude fish, she found nude and scattered fish in the progeny, in a 2:1 ratio. (The scattered fish from these crosses bred true.) Diagram the phenotypes and genotypes of this gene in a nude × nude cross and explain the altered Mendelian ratio.
f. The student now felt she could explain all of her results. Diagram the genotypes in the linear × linear cross performed by the student (in part b). Show the genotypes of the four phenotypes observed among the progeny and explain the 6:3:2:1 ratio.

33. You picked up two mice (one female and one male) that had escaped from experimental cages in the animal facility. One mouse is yellow in color, and the other is brown agouti. You know that this mouse colony has animals with different alleles at only three coat color genes: the agouti or nonagouti or yellow alleles of the *A* gene, the black or brown allele of the *B* gene, and the albino or nonalbino alleles of the *C* gene. However, you don't know which alleles of these genes are actually present in each of the animals that you've captured. To determine the genotypes, you breed them together. The first litter has only three pups. One is albino, one is brown (nonagouti), and the third is black agouti.

a. What alleles of the *A, B,* and *C* genes are present in the two mice you caught?
b. After raising several litters from these two parents, you have many offspring. How many different coat color phenotypes (in total) do you expect to see expressed in the population of offspring? What are the phenotypes and corresponding genotypes?

34. Figure 3.17 on p. 64 and Fig. 3.22b on p. 71 both show traits that are determined by two genes, each of which has two incompletely dominant alleles. But in Fig. 3.17 the gene interaction produces nine different phenotypes, while the situation depicted in Fig. 3.22b shows only five possible phenotypic classes. How can you explain this difference in the amount of phenotypic variation?

35. Three genes in fruit flies affect a particular trait, and one dominant allele of *each* gene is necessary to get a wild-type phenotype.

a. What phenotypic ratios would you predict among the progeny if you crossed triply heterozygous flies?

b. You cross a particular wild-type male in succession with three tester strains. In the cross with one tester strain (*AA bb cc*), only 1/4 of the progeny are wild type. In the crosses involving the other two tester strains (*aa BB cc* and *aa bb CC*), half of the progeny are wild type. What is the genotype of the wild-type male?

36. The garden flower *Salpiglossis sinuata* ("painted tongue") comes in many different colors. Several crosses are made between true-breeding parental strains to produce F_1 plants, which are in turn self-fertilized to produce F_2 progeny.

Parents	F_1 phenotypes	F_2 phenotypes
red × blue	all red	102 red, 33 blue
lavender × blue	all lavender	149 lavender, 51 blue
lavender × red	all bronze	84 bronze, 43 red, 41 lavender
red × yellow	all red	133 red, 58 yellow, 43 blue
yellow × blue	all lavender	183 lavender, 81 yellow, 59 blue

a. State a hypothesis explaining the inheritance of flower color in painted tongues.

b. Assign genotypes to the parents, F_1 progeny, and F_2 progeny for all five crosses.

c. In a cross between true-breeding yellow and true-breeding lavender plants, all of the F_1 progeny are bronze. If you used these F_1 plants to produce an F_2 generation, what phenotypes in what ratios would you expect? Are there any genotypes that might produce a phenotype that you cannot predict from earlier experiments, and if so, how might this alter the phenotypic ratios among the F_2 progeny?

37. In foxgloves, there are three different petal phenotypes: white with red spots (WR), dark red (DR), and light red (LR). There are actually two different kinds of true-breeding WR strains (WR-1 and WR-2) that can be distinguished by two-generation intercrosses with true-breeding DR and LR strains:

Cross	Parental	Phenotypes: F_1	F_2: WR	F_2: LR	F_2: DR
1	WR-1 × LR	all WR	480	39	119
2	WR-1 × DR	all WR	99	32	0
3	DR × LR	all DR	0	43	132
4	WR-2 × LR	all WR	193	64	0
5	WR-2 × DR	all WR	286	24	74

a. What can you conclude about the inheritance of the petal phenotypes in foxgloves?

b. Ascribe genotypes to the four true-breeding parental strains (WR-1, WR-2, DR, and LR).

c. A WR plant from the F_2 generation of cross #1 is now crossed with an LR plant. Of 500 total progeny from this cross, there were 253 WR, 124 DR, and 123 LR plants. What are the genotypes of the parents in this WR × LR mating?

38. In a culture of fruit flies, matings between any two flies with hairy wings (wings abnormally containing additional small hairs along their edges) always produce both hairy-winged and normal-winged flies in a 2:1 ratio. You now take hairy-winged flies from this culture and cross them with four types of normal-winged flies; the results for each cross are shown in the following table. Assuming that there are only two possible alleles of the hairy-winged gene (one for hairy wings and one for normal wings), what can you say about the genotypes of the four types of normal-winged flies?

Type of normal-winged flies	Progeny obtained from cross with hairy-winged flies: Fraction with normal wings	Fraction with hairy wings
1	1/2	1/2
2	1	0
3	3/4	1/4
4	2/3	1/3

39. A married man and woman, both of whom are deaf, carry some recessive mutant alleles in three different "hearing genes": *d1* is recessive to *D1*, *d2* is recessive to *D2*, and *d3* is recessive to *D3*. Homozygosity for a mutant allele at any one of these three genes causes deafness. In addition, homozygosity for any two of the three genes together in the same genome will cause prenatal lethality (and spontaneous abortion) with a penetrance of 25%. Furthermore, homozygosity for the mutant alleles of all three genes will cause prenatal lethality with a penetrance of 75%. If the genotypes of the mother and father are as indicated here, what is the likelihood that a live-born child will be deaf?

Mother: *D1 d1, D2 d2, d3 d3*

Father: *d1 d1, D2 d2, D3 d3*

The Chromosome Theory of Inheritance

Chapter 4

Down syndrome was the first human genetic disorder attributable not to a gene mutation but to an abnormal number of chromosomes. In the spherical, membrane-bounded nuclei of plant and animal cells prepared for viewing under the microscope, chromosomes appear as brightly colored, threadlike bodies. The nuclei of normal human cells carry 23 pairs of chromosomes for a total of 46. There are noticeable differences in size and shape among the 23 pairs, but within each pair, the two chromosomes appear to match exactly. (The only exceptions are the male's sex chromosomes, designated X and Y, which constitute an unmatched pair.) Children born with Down syndrome have 47 chromosomes in each cell nucleus because they carry three, instead of the normal pair, of a very small chromosome referred to as number 21. The aberrant genotype, known as trisomy 21, gives rise to an abnormal phenotype, including a wide skull that is flatter than normal at the back, an unusually large tongue, learning disabilities caused by the abnormal development of the hippocampus and other parts of the brain, and a propensity to respiratory infections as well as heart disorders, rapid aging, and leukemia (**Fig. 4.1**).

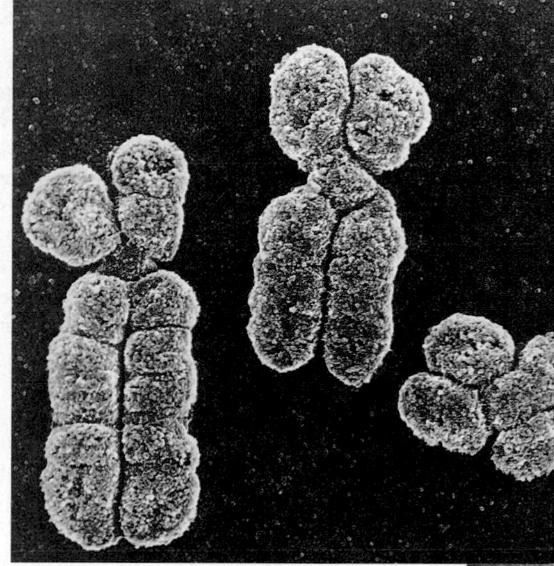

Each of these three human chromosomes carries hundreds to thousands of genes.

How can one extra copy of a chromosome that is itself of normal size and shape cause such wide-ranging phenotypic effects? The answer has two parts. First and foremost, chromosomes are the cellular organelles responsible for transmitting genetic information. That is, each chromosome carries a number of genes that interact with each other and those of other chromosomes to determine the development, behavior, and appearance of an individual. (Note that we use the term "organelle" in its most general sense to designate a large, complex structure of characteristic shape that performs essential functions within the cell, rather than in its more specific sense of an intracellular, membrane-bounded entity.) In this chapter, we describe how geneticists concluded that chromosomes are the carriers of genes, an idea that became known as the **chromosome theory of inheritance.** The second part of the answer is that proper development depends not just on what type of genetic material is present but also on how much of it there is. Thus the mechanisms governing gene transmission during cell division must vigilantly maintain each cell's chromosome number.

Proof that genes are located on chromosomes comes from both breeding experiments and the microscopic examination of cells. As we analyze the evidence, you will see that the behavior of chromosomes during one type of nuclear division called *meiosis* accounts for the segregation and independent assortment of genes proposed by Mendel. Meiosis figures prominently in the process by which most sexually reproducing organisms generate the gametes—eggs or sperm—that at fertilization unite to form the first cell of the next generation. This first cell is the fertilized egg, or *zygote.* Once the union of egg and sperm creates the zygote, a second kind of nuclear division, known as *mitosis,* occurs during the millions of cell divisions that propel development from a single cell to a complex multicellular organism. Mitosis is a conservative process that provides each of the many cells in an individual with the same number and types of chromosomes.

The precise chromosome-parceling mechanisms of meiosis and mitosis are so crucial to the normal functioning of an organism that when the machinery does

Figure 4.1 Down syndrome: One extra chromosome 21 has widespread phenotypic consequences. Trisomy 21 usually causes changes in the physical characteristics of the head and body, as well as in the potential for learning. Many children with Down syndrome, such as the fifth grader in the striped shirt at the center of the photograph, nonetheless participate fully in regular activities.

not function properly, errors in chromosome distribution can have dire repercussions on the individual's health and survival. Down syndrome, for example, is the result of a failure of chromosome segregation during meiosis. The meiotic error gives rise to an egg or sperm carrying an extra chromosome 21, which if incorporated in the zygote at fertilization, is passed on via mitosis to every cell of the developing embryo. Trisomy—three copies of a chromosome instead of two—can occur with other chromosomes, but in nearly all such cases, the condition is prenatally lethal and results in a miscarriage.

Two themes emerge in our discussion of meiosis and mitosis. First, it was direct microscopic observations of chromosomes during gamete formation that led early twentieth-century investigators to recognize that *chromosome movements parallel the behavior of Mendel's genes, so chromosomes are likely to carry the genetic material.* This chromosome theory of inheritance was proposed in 1902 and confirmed in the following 15 years through elegant experiments performed mainly on the fruit fly *Drosophila melanogaster.* Second, the chromosome theory transformed the concept of a gene from an abstract particle to a physical reality—part of a chromosome that could be seen and manipulated.

Our discussion of the connection between Mendel's laws of heredity and the behavior of chromosomes during cell division examines

- Observations and experiments that placed the hereditary material in the cell nucleus, specifically on the chromosomes.
- Mitosis, which distributes chromosomes equally to each of two daughter cells, ensuring that every cell in an organism carries the same set of chromosomes.
- Meiosis, which distributes one member of each chromosome pair to the cells that will become the gametes.
- Gametogenesis, a process including both meiotic and mitotic divisions, by which specialized germ cells multiply and differentiate into gametes.
- Validation of the chromosome theory of inheritance.

4.1 Chromosomes Contain the Genetic Material

One of the first questions asked at the birth of an infant—Is it a boy or a girl?—acknowledges that male and female are mutually exclusive characteristics like the yellow versus green of Mendel's peas. What's more, among humans and most other sexually reproducing species, there is a roughly 1:1 ratio of the two genders. In any randomly chosen human population, for example, approximately 50% are women and 50% men. Both males and females produce cells specialized for reproduction—sperm or eggs—that serve as a physical link to the next generation. In bridging the gap between generations, these gametes must each contribute half of the genetic material for making a normal, healthy son or daughter. Whatever part of the gamete carries this material, its structure and function must be able to account for the either-or aspect of sex determination as well as the generally observed 1:1 ratio of men to women. As we will see, these two features of sex determination were among the earliest clues to the cellular basis of heredity.

Evidence That Genes Reside in the Nucleus

The nature of the link between sex and reproduction remained a mystery until Anton van Leeuwenhoek, one of the earliest and most astute of microscopists, discovered in 1667 that semen contains spermatozoa (literally "sperm animals"). He imagined that these microscopic creatures might enter the egg and somehow achieve fertilization, but it was not possible to confirm this hypothesis for another 200 years. Then, during a 20-year period starting in 1854 (about the same time Gregor Mendel was beginning his pea experiments), microscopists studying fertilization in frogs and sea urchins observed the union of male and female gametes and recorded the details of the process in a series of drawings. These drawings, as well as later micrographs (photographs shot through a microscope), clearly show that egg and sperm nuclei are the only elements contributed equally by maternal and paternal gametes. This observation implies that something in the nucleus contains the hereditary material. In humans, the nuclei of the gametes are less than 2 millionth of a meter in diameter. It is indeed remarkable that the genetic link between generations is packaged within such an exceedingly small space.

Evidence That Genes Reside in Chromosomes

Further investigations, some dependent on technical innovations in microscopy, suggested that yet smaller, discrete structures within the nucleus are the repository of genetic information. In the 1880s, for example, a newly discovered combination of organic and inorganic dyes revealed the existence of the long, brightly staining, threadlike bodies within the nucleus that we call **chromosomes** (literally "colored bodies"). It was now possible to follow the movement of chromosomes during different kinds of cell division.

In embryonic cells, the chromosomal threads split lengthwise in two just before cell division, and each of the two newly forming daughter cells receives one-half of every split thread. The kind of nuclear division followed by cell division that results in two daughter cells containing the same number and type of chromosomes as the original parent cell is called **mitosis** (from the Greek *mitos* meaning "thread" and *-osis* meaning "formation" or "increase"). In contrast to mitosis, in the cells that give rise to male and female gametes, a key event is the segregation of the chromosomes composing each pair such that the resulting gametes receive only one chromosome from each chromosome pair. The gametes thus have half the number of chromosomes as other cells within the same organism. The kind of nuclear division that generates egg or sperm cells containing half the number of chromosomes found in other cells within the same organism is called **meiosis** (from the Greek word for "diminution").

At Fertilization, Haploid Gametes Unite to Produce Diploid Zygotes

In the first decade of the twentieth century, cytologists—scientists who use the microscope to study cell structure—showed that the chromosomes in a fertilized egg actually consist of two matching sets, one contributed by the maternal gamete, the other by the paternal gamete. The corresponding maternal and paternal chromosomes appear alike in size and shape, forming pairs (with one exception—the *sex chromosomes*—which we discuss in a later section.) While the zygote carries two complete chromosome sets, each gamete contains one-half the number of chromosomes in the zygote, or one complete set. Gametes and other cells that carry only a single set of chromosomes are called **haploid** (from the Greek word for "single"). Zygotes and other cells carrying two matching sets are **diploid** (from the Greek word for "double"). The number of chromosomes in a normal haploid cell is designated by the shorthand symbol ***n;*** the number of chromosomes in a normal diploid cell is then ***2n.*** **Figure 4.2** shows diploid cells as well as the haploid gametes that arise from them in *Drosophila* where $2n = 8$ and $n = 4$. In humans, $2n = 46$; $n = 23$. You can see how the halving of chromosome number during meiosis and gamete

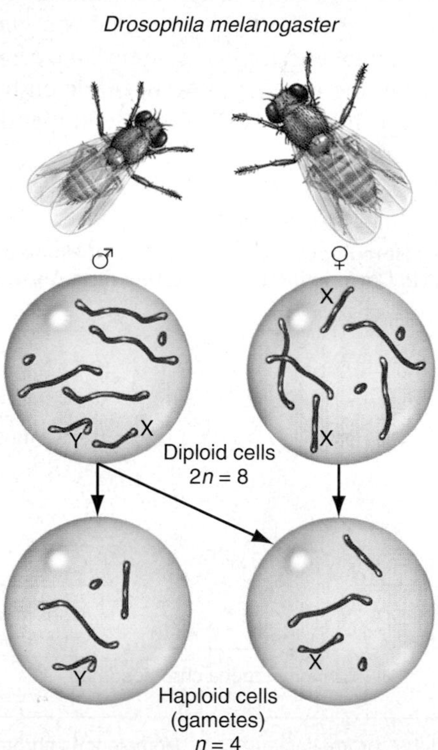

Figure 4.2 Diploid versus haploid: 2*n* versus *n*. Most body cells are diploid: They carry a maternal and paternal copy of each chromosome. Meiosis generates haploid gametes with only one copy of each chromosome. In *Drosophila,* diploid cells have eight chromosomes ($2n = 8$), while gametes have four chromosomes ($n = 4$). The fusion of gametes at fertilization yields a diploid zygote (*not shown*). Note that the chromosomes in this diagram are pictured before their replication. The X and Y chromosomes determine the sex of the individual.

formation, followed by the union of two gametes' chromosomes at fertilization, allows a constant $2n$ number of chromosomes to be maintained from generation to generation in all individuals of a species (except, of course, for those carrying rare abnormalities such as trisomy 21). The chromosomes of every pair must segregate from each other during meiosis so that the haploid gametes will each have a complete set of chromosomes. After fertilization forms the zygote, the process of mitosis then ensures that all the cells of the developing individual have identical diploid chromosome sets.

The Number and Shape of Chromosomes Vary from Species to Species

Scientists analyze the chromosomal makeup of a cell when the chromosomes are most visible—at a specific moment in the cell cycle of growth and division, just before the nucleus divides. At this point, known as *metaphase* (described in detail later), individual chromosomes have duplicated and condensed from thin threads into compact rodlike structures. Each chromosome now consists of two identical halves known as **sister chromatids** attached to each other at a specific location called the **centromere** (**Fig. 4.3**). In **metacentric** chromosomes, the centromere is more or less in the middle; in **acrocentric** chromosomes, the centromere is very close to one end. Modern high-resolution microscopy has failed to find any chromosomes in which the centromere is exactly at one end. As a result, the sister chromatids of all chromosomes actually have two "arms" separated by a centromere, even if one of the arms is very short.

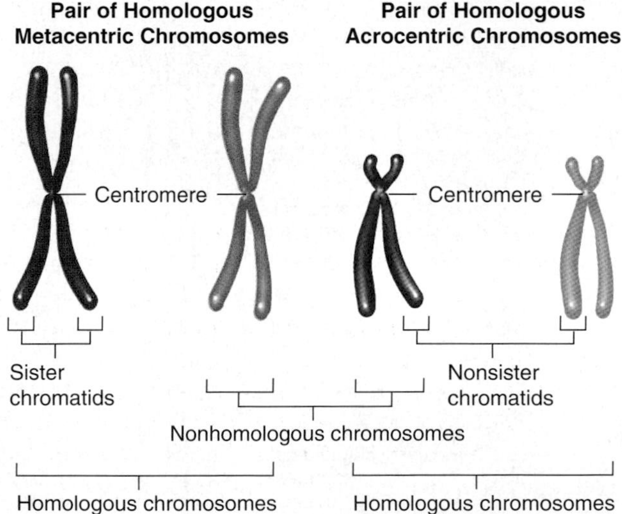

Figure 4.3 Metaphase chromosomes can be classified by centromere position. Before cell division, each chromosome replicates into two sister chromatids connected at a centromere. In highly condensed metaphase chromosomes, the centromere can appear near the middle (making it a metacentric chromosome), very near an end (forming an acrocentric chromosome), or anywhere in between. In a diploid cell, the chromosomes come in homologous pairs, with one chromosome in the pair from the mother and the other from the father.

Cells in metaphase can be fixed and stained with one of several dyes that highlight the chromosomes and accentuate the centromeres. In each kind of chromosome, under the right conditions, the dyes also produce characteristic banding patterns of lighter and darker regions, making it possible to distinguish matching chromosome pairs. Chromosomes that match in size, shape, and banding are called **homologous chromosomes,** or **homologs.** The two homologs of each pair contain the same set of genes, although for some of those genes, they may carry different alleles. The differences between alleles are too small to show up in the microscope.

Figure 4.3 introduces a system of notation employed throughout the remainder of this book, using color to indicate degrees of relatedness between chromosomes. Thus, sister chromatids—the two daughter molecules of a duplicated chromosome that are joined at a single centromere—which are identical at the completion of replication, appear in the same shade of the same color. Homologous chromosomes, which carry the same genes but may vary in the identity of particular alleles, are pictured in different shades (light or dark) of the same color. *Nonhomologous chromosomes,* which carry completely unrelated sets of genetic information, appear in different colors.

To study the chromosomes of a single organism, geneticists cut images of the stained chromosomes from a micrograph and, by convention, arrange them in homologous pairs of decreasing size to produce a **karyotype.** Karyotype assembly can now be speeded and automated by computerized image analysis. **Figure 4.4** shows the karyotype of a human male, with 46 chromosomes arranged in 22 matching pairs of chromosomes and one nonmatching pair. The 44 chromosomes in matching pairs are known as **autosomes.** The two unmatched chromosomes in this male karyotype are called sex chromosomes, because they determine the sex of the individual. (We discuss sex chromosomes in more detail in subsequent sections.)

As you can see in Fig. 4.4, autosome number 21 is one of the smallest chromosomes. Interestingly, modern methods of DNA analysis can reveal differences between the maternally and paternally derived chromosomes of a homologous pair, and medical researchers can use this information to trace the origin of an extra chromosome, such as the extra 21 that causes Down syndrome. Data collected in studies of this type indicate that in 80% of cases, the third chromosome 21 comes from the egg; in 20%, from the sperm. The Genetics and Society box on p. 87 describes how physicians use karyotype analysis and a technique called *amniocentesis* to diagnose Down syndrome prenatally, roughly three months after a fetus is conceived.

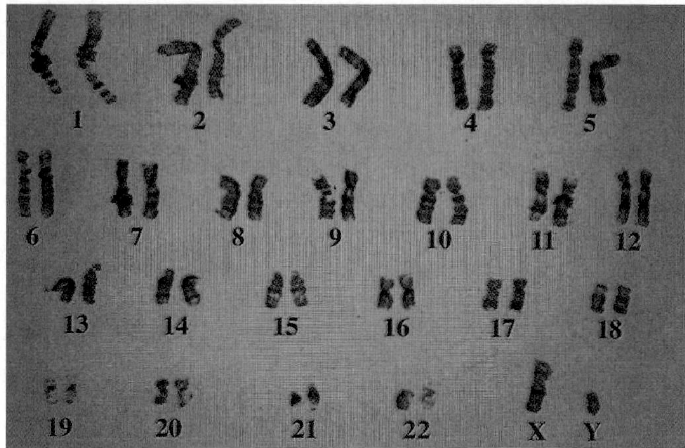

Figure 4.4 Karyotype of a human male. Photos of metaphase human chromosomes are paired and arranged in order of decreasing size. In a normal human male karyotype, there are 22 pairs of autosomes, as well as an X and a Y ($2n$ = 46). Homologous chromosomes share the same characteristic pattern of dark and light bands.

Figure 4.5 The great lubber grasshopper. In this mating pair, the smaller male is astride the female.

Through thousands of karyotypes on normal individuals, cytologists have verified that the cells of each species carry a distinctive diploid number of chromosomes. Among three species of fruit flies, for example, *Drosophila melanogaster* carries 8 (4 pairs of) chromosomes, *Drosophila obscura* carries 10 (5 pairs), and *Drosophila virilis,* 12 (6 pairs). Differences in the size, shape, and number of chromosomes reflect differences in the assembled genetic material that determines what each species looks like and how it functions. Mendel's peas contain 14 (7 pairs of) chromosomes in each diploid cell, macaroni wheat has 28 (14 pairs), giant sequoia trees 22 (11 pairs), goldfish 94 (47 pairs), dogs 78 (39 pairs), and people 46 (23 pairs). As these figures show, the number of chromosomes does not always correlate with the size or complexity of the organism.

In Many Species, One Chromosome Pair Determines an Individual's Sex

Walter S. Sutton, a young American graduate student at Columbia University in the first decade of the twentieth century, was one of the earliest cytologists to realize that particular chromosomes carry the information for determining sex. In one study, he obtained cells from the testes of the great lubber grasshopper (*Brachystola magna;* **Fig. 4.5**) and followed them through the meiotic divisions that produce sperm. He observed that prior to meiosis, precursor cells within the testes of a great lubber grasshopper contain a total of 24 chromosomes. Of these, 22 are found in 11 matched pairs and are thus autosomes. The remaining 2 chromosomes are unmatched. He called the larger of these the X chromosome and the smaller the Y chromosome. After meiosis, the sperm produced within these testes are of two equally prevalent types: one-half have a set of 11 autosomes plus an X chromosome, while the other half have a set of 11 autosomes plus a Y. By comparison, all of the eggs produced by females of the species carry an 11-plus-X set of chromosomes like the set found in the first class of sperm. When a sperm with an X chromosome fertilizes an egg, an XX female grasshopper results; when a Y-containing sperm fuses with an egg, an XY male develops. Sutton concluded that the X and Y chromosomes determine sex.

Several researchers studying other organisms soon verified that in many sexually reproducing species, two distinct chromosomes—known as the **sex chromosomes**—provide the basis of sex determination. One sex carries two copies of the same chromosome (a matching pair), while the other sex has one of each type of sex chromosome (an unmatched pair). The cells of normal human females, for example, contain 23 pairs of chromosomes. The two chromosomes of each pair, including the sex-determining X chromosomes, appear to be identical in size and shape. In males, however, there is one unmatched pair of chromosomes: the larger of these is the X; the smaller, the Y (Fig. 4.4 and **Fig. 4.6a**). Apart from this difference in sex chromosomes, the two sexes are not distinguishable at any other pair of chromosomes. Thus, geneticists can designate women as XX and men as XY and represent sexual reproduction as a simple cross between XX × XY.

If sex is an inherited trait determined by a pair of sex chromosomes that separate to different cells during gamete formation, such a cross could account for both the mutual exclusion of genders and the near 1:1 ratio of males to females, which are among the most striking features of sex determination (Fig. 4.6b). And if chromosomes carry information defining the two contrasting sex phenotypes, it is not extravagant to infer that chromosomes also carry genetic information specifying other characteristics as well.

(a)

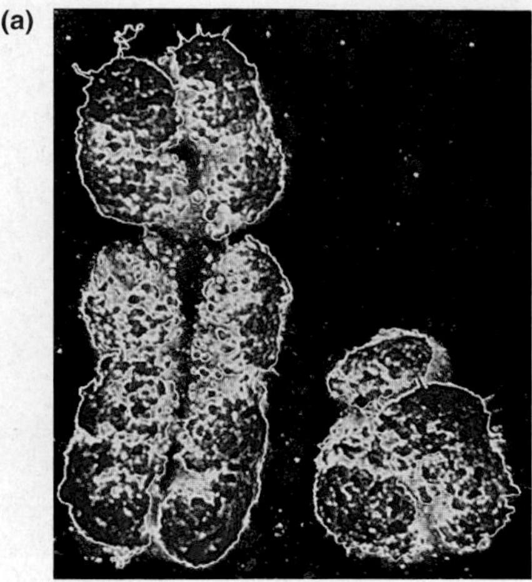

(b)

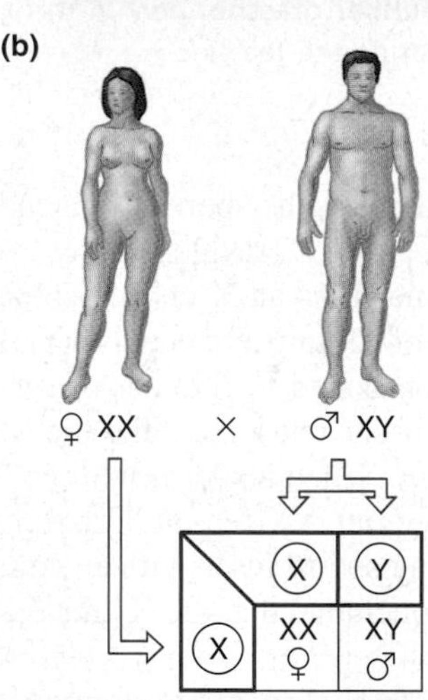

Figure 4.6 How the X and Y chromosomes determine sex in humans. (a) This colorized micrograph shows the human X chromosome on the *left* and the human Y on the *right.* **(b)** Children can receive only an X chromosome from their mother, but they can inherit either an X or a Y from their father.

There Is Variation Between Species in How Chromosomes Determine an Individual's Sex

As we have just seen, humans and other mammals have a pair of sex chromosomes that are identical in the XX female but different in the XY male. Several studies have shown that in humans, it is the presence or absence of the Y that actually makes the difference; that is, any person carrying a Y chromosome will look like a male. For example, rare humans with two X and one Y chromosomes (XXY) are males displaying certain abnormalities collectively called *Klinefelter syndrome.* Klinefelter males are typically tall, thin, and sterile, and they may be mentally retarded. That these individuals are males shows that two X chromosomes are insufficient for female development in the presence of a Y. In contrast, humans carrying an X and no second sex chromosome (XO) are females with *Turner syndrome.* Turner females are usually sterile, lack secondary sexual characteristics such as pubic hair, are of short stature, and have folds of skin between their necks and shoulders (webbed necks). Even though these individuals have only one X chromosome, they develop as females because they have no Y chromosome.

Other species show variations on this XX = female/XY = male chromosomal strategy of sex determination. In fruit flies, for example, although normal females are XX and normal males XY (see Fig. 4.2), it is ultimately the ratio of X chromosomes to autosomes (and not the presence or absence of the Y) that determines sex. In female *Drosophila,* the ratio is 1:1 (there are two X chromosomes and two copies of each autosome); in males, the ratio is 1:2 (there is one X chromosome but two copies of each autosome). Curiously, a rarely observed abnormal intermediate ratio of 2:3 produces intersex flies that display both male and female characteristics. Although the Y chromosome in *Drosophila* does not determine whether a fly looks like a male, it is necessary for male fertility; XO flies are thus sterile males. **Table 4.1** compares how humans and *Drosophila* respond to unusual complements of sex

TABLE 4.1 Sex Determination in Fruit Flies and Humans

	Complement of Sex Chromosomes						
	XXX	XX	XXY	XO	XY	XYY	OY
Drosophila	Dies	Normal female	Normal female	Sterile male	Normal male	Normal male	Dies
Humans	Nearly normal female	Normal female	Klinefelter male (sterile); tall, thin	Turner female (sterile); webbed neck	Normal male	Normal or nearly normal male	Dies

Humans can tolerate extra X chromosomes better than *Drosophila* (compare the fate of XXX individuals). Complete absence of an X chromosome is lethal to both fruit flies and humans because this chromosome carries essential genes not found on other chromosomes. Additional Y chromosomes have little effect in either species.

Using Amniocentesis for the Prenatal Diagnosis of Down Syndrome

Mendel and the geneticists who followed in his path observed phenotypes in related individuals and used this information to infer genotypes. In the latter part of the twentieth century, geneticists developed methods that allowed them to reverse this strategy: Through observations of chromosomes and the DNA in genes, they were able to define an individual's genotype directly and use this information to predict aspects of the individual's phenotype, even before these traits manifested themselves. Doctors have used this basic strategy to diagnose, before birth, whether or not a baby will be born with a genetic condition.

The first prerequisite for prenatal diagnosis is to obtain fetal cells whose DNA and chromosomes can be analyzed for genotype, and the most frequently used method for acquiring these cells is **amniocentesis (Fig. A)**. To carry out this procedure, a doctor inserts a needle through a pregnant woman's abdominal wall into the amniotic sac in which the fetus is growing; this procedure is performed about 16 weeks after the woman's last menstrual period. By using ultrasound imaging to define the location of the needle, the physician can minimize the chance of injuring the fetus. The doctor then withdraws some of the amniotic fluid, in which the fetus is suspended, back through the needle into a syringe. The fluid contains living cells called *amniocytes* that were shed by the fetus. When placed in a culture medium, these fetal cells undergo several rounds of mitosis and increase in number. Once enough fetal cells are available, clinicians look at the chromosomes and genes in those cells. In later chapters, we describe techniques that allow the direct examination of the DNA constituting particular disease genes.

Amniocentesis also provides fetal cells for a less technically sophisticated purpose: the diagnosis of Down syndrome through the analysis of chromosomes by karyotyping. Because the risk of Down syndrome increases rapidly with the age of the mother, more than half the pregnant women in North America who are over the age of 35 currently undergo amniocentesis. The cultured fetal cells are treated with a drug that disrupts the mitotic spindle. This disruption arrests the cells in a metaphase-like state when they are most condensed and their banding patterns are easy to observe (see Fig. 4.4 on p. 85). Although the goal of this karyotyping is to learn whether the fetus is trisomic for chromosome 21, many other abnormalities in chromosome number or shape may show up when the karyotype is examined.

The availability of amniocentesis and other techniques of prenatal diagnosis is intimately entwined with the personal and societal issue of abortion. The large majority of amniocentesis procedures are performed with the understanding that a fetus whose genotype indicates a genetic disorder such as Down syndrome will be aborted. Some prospective parents who are opposed to abortion still elect to undergo amniocentesis so that they can better prepare for an affected child, but this is rare. The ethical and political aspects of the abortion debate influence many of the practical questions underlying prenatal diagnosis. For example, parents must decide which genetic conditions would be sufficiently severe that they would be willing to abort the fetus. They must also assess the small but measurable risk that amniocentesis might harm the fetus. The normal risk of miscarriage at 16 weeks of gestation is about 2–3%; amniocentesis increases that risk by about 0.5% (about 1 in 200 procedures). From the economic point of view, society must decide who should pay for prenatal diagnosis procedures. In the United States, almost all private insurance companies and most state Medicare programs cover at least some of the approximately $1500 cost of amniocentesis.

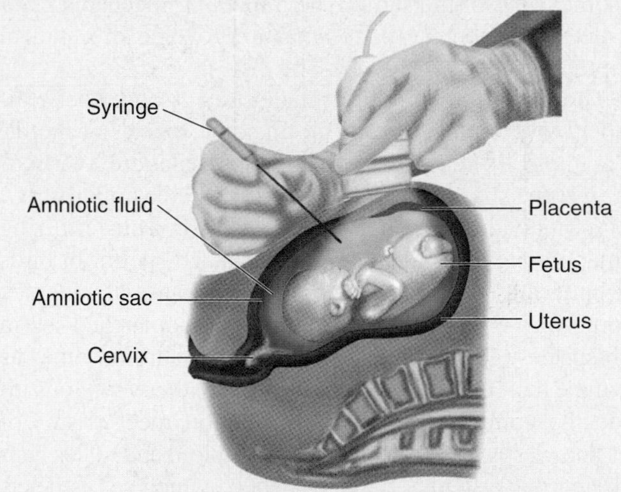

Figure A Obtaining fetal cells by amniocentesis. A physician guides insertion of the needle into the amniotic sac using ultrasound imaging and extracts amniotic fluid containing fetal cells into the syringe.

In current practice, the risks and costs of prenatal testing generally restrict amniocentesis to women over age 35 or to mothers whose fetuses are at high risk for a testable genetic condition because of family history. The personal and societal equations determining the frequency of prenatal testing may, however, need to be overhauled in the not-to-distant future because of technological advances that will simplify the procedures and thereby minimize the costs and risks. In 2005, it was reported that a specialized examination of the fetus by ultrasound in conjunction with tests for abnormal levels of two proteins in the mother's blood serum can detect about 85% of Down syndrome fetuses, with a 5% rate of false positives. When these techniques become more widely adopted, karyotype analysis of cells obtained by amniocentesis (which is close to 100% accurate) would most often be used as a last resort to verify a diagnosis. Clinicians may also soon be able to take advantage of new methods currently under evaluation to purify the very small number of fetal cells that find their way into the mother's bloodstream during pregnancy. Collecting these cells from the mother's blood would be much less invasive and expensive than amniocentesis and would pose no risk to the fetus, yet karyotype analysis would be just as accurate.

chromosomes. Differences between the two species arise in part because the genes they carry on their sex chromosomes are not identical and in part because the strategies they use to deal with the presence of additional sex chromosomes are not the same. The molecular mechanisms of sex determination in *Drosophila* are covered in detail in Chapter 18.

The XX = female / XY = male strategy of sex determination is by no means universal. In some species of moths, for example, the females are XX, but the males are XO. In *C. elegans* (one species of nematode), males are similarly XO, but XX individuals are not females; they are self-fertilizing hermaphrodites that produce both eggs and sperm. In birds and butterflies, it is the males that have the matching sex chromosomes, while the females have an unmatched set; in such species, geneticists represent the sex chromosomes as ZZ in the male and ZW in the female. Yet other variations include the complicated sex-determination mechanisms of bees and wasps, in which females are diploid and males haploid, and the systems of certain fish, in which sex is determined by changes in the environment, such as fluctuations in temperature. **Table 4.2** summarizes some of the astonishing variety in the ways that different species have solved the problem of assigning gender to individuals.

TABLE 4.2 Mechanisms of Sex Determination

	♀	♂
Humans and *Drosophila*	XX	XY
Moths and *C. elegans*	XX (hermaphrodites in *C. elegans*)	XO
Birds and Butterflies	ZW	ZZ
Bees and Wasps	Diploid	Haploid
Lizards and Alligators	Cool temperature	Warm temperature
Tortoises and Turtles	Warm temperature	Cool temperature
Anemone Fish	Older adults	Young adults

In the species highlighted in *blue*, sex is determined by sex chromosomes. Note that the sex of the offspring is determined by the type of gamete received from the parent of the **heterogametic sex** (that is, the gender producing two different types of gametes: XY, XO, or ZW in the table). The species highlighted in *yellow* have identical chromosomes in the two sexes, and sex is determined instead by environmental or other factors. Anenome fish (bottom row) undergo a sex change from male to female as they age; in other species (not shown), animals are females as young adults and change to males as they mature.

Despite these differences, it soon became clear that chromosomes can carry the genetic information specifying sexual identity and probably many other characteristics as well. Sutton and other early adherents of the chromosome theory realized that the perpetuation of life itself therefore depends on the proper distribution of chromosomes during cell division. In the next sections, as we examine step by step the microscopic maneuvers by which cells maintain their crucial chromosomal constancy, you will see that the behavior of chromosomes during mitosis and meiosis is exactly that expected of organelles carrying genes.

4.2 Mitosis Ensures That Every Cell in an Organism Carries the Same Chromosomes

The fertilized human egg is a single diploid cell that preserves its genetic identity unchanged through more than 100 generations of cells as it divides again and again to produce a full-term infant ready to be born. As the newborn infant develops into a toddler, a teenager, and an adult, yet more cell divisions fuel continued growth and maturation. Mitosis, the nuclear division that apportions chromosomes in equal fashion to two daughter cells, is the cellular mechanism that preserves genetic information through all these generations of cells. In this section, we take a close look at how the nuclear division of mitosis fits into the overall scheme of cell growth and division.

If you were to peer through a microscope and follow the history of one cell through time, you would see that for much of your observation, the chromosomes resemble a mass of extremely fine tangled string—called **chromatin**—surrounded by the **nuclear envelope.** Each convoluted thread of chromatin is composed mainly of DNA (which carries the genetic information) and protein (which serves as a scaffold for packaging and managing that information, as described in Chapter 13). You would also be able to distinguish one or two darker areas of chromatin called **nucleoli** (singular, *nucleolus,* literally "small nucleus"); nucleoli play a key role in the manufacture of ribosomes, organelles that function in protein synthesis (see Chapter 13 for details). Although the chromatin-laden nucleus appears quiescent during this period between cell divisions, it houses a great deal of invisible activity necessary for the growth and survival of the cell. One particularly important part of this activity is the accurate duplication of all the chromosomal material.

With continued vigilance, you would observe during a very short period in the life history of the cell a dramatic change in the nuclear landscape: The chromatin condenses into discrete threads and then each chromosome compacts even further into the twin rods clamped together at the centromere that can be identified in karyotype analysis (review Fig. 4.3 on p. 84). Each rod in a duo is called a **chromatid;** as described earlier, it is an exact duplicate of the other sister chromatid to which it is connected. Continued observation would reveal the doubled chromosomes beginning to jostle around and then lining up at the midplane of the cell. At this point, the sister chromatids comprising each chromosome separate to opposite poles of the now elongating cell, where they become identical sets of chromosomes. Each of the two identical sets eventually ends up enclosed in a separate nucleus in a separate cell. The two cells, known as *daughter cells,* are thus genetically identical.

The repeating pattern of cell growth (an increase in size) followed by division (the splitting of one cell into two) is called the **cell cycle** (**Fig. 4.7**). Only a small part of the cell cycle is spent in division; the period between divisions is called **interphase.**

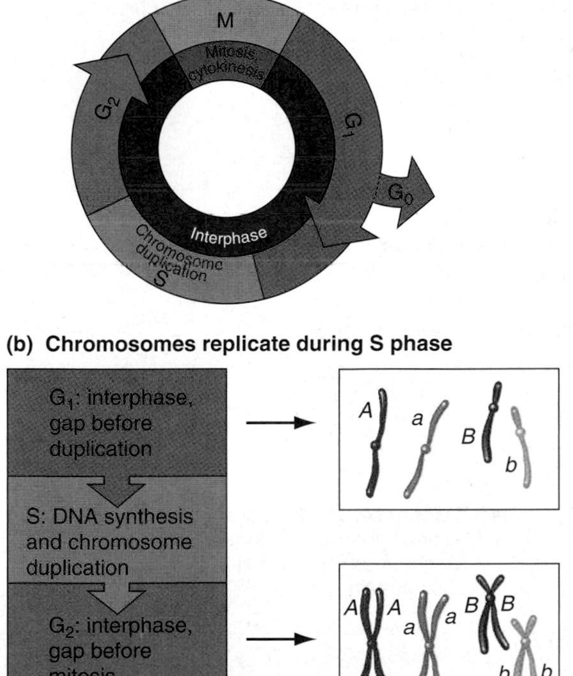

Figure 4.7 The cell cycle: An alternation between interphase and mitosis. (a) Chromosomes replicate to form sister chromatids during synthesis (S phase); the sister chromatids segregate to daughter cells during mitosis (M phase). The gaps between the S and M phases are called the G_1 and G_2 phases. It is during the gaps that most cell growth takes place. **(b)** Interphase consists of the G_1, S, and G_2 phases together. In multicellular organisms, some terminally differentiated cells stop dividing and arrest in a so-called G_0 stage, as shown in part (a).

During Interphase, Cells Grow and Replicate Their Chromosomes

Interphase consists of three parts: gap 1 (**G_1**), synthesis (**S**), and gap 2 (**G_2**) (Fig. 4.7). G_1 lasts from the birth of a new cell to the onset of chromosome replication; for the genetic material, it is a period when the chromosomes are neither duplicating nor dividing. During this time, the cell achieves most of its growth by using the information from its genes to make and assemble the materials it needs to function normally. G_1 varies in length more than any other phase of the cell cycle. In rapidly dividing cells of the human embryo, for example, G_1 is as short as a few hours. In contrast, mature brain cells become arrested in a resting form of G_1 known as **G_0** and do not normally divide again during a person's lifetime.

Synthesis (S) is the time when the cell duplicates its genetic material by synthesizing DNA. During duplication, each chromosome doubles to produce identical sister chromatids that will become visible when the chromosomes condense at the beginning of mitosis. The two sister chromatids remain joined to each other at the centromere. Note that this joined structure is considered a single chromosome as long as the connection between sister chromatids is maintained. The replication of chromosomes during S phase is critical; it is because of this exact copying of the genetic information that both daughter cells can receive identical sets of chromosomes at the conclusion of mitosis.

Gap 2 (G_2) is the interval between chromosome duplication and the beginning of mitosis. During this time, the cell may grow (usually less than during G_1); it also synthesizes proteins that are essential to the subsequent steps of mitosis itself.

In addition, during interphase an array of fine microtubules crucial for many interphase processes becomes visible outside the nucleus. The microtubules radiate out into the cytoplasm from a single organizing center known as the **centrosome,** usually located near the nuclear envelope. In animal cells, the discernible core of each centrosome is a pair of small, darkly staining bodies called **centrioles** (**Fig. 4.8a**); the microtubule-organizing center of plants does not contain centrioles. During the S and G_2 stages of interphase, the centrosomes replicate, producing two centrosomes that remain in extremely close proximity.

During Mitosis (M Phase), Sister Chromatids Separate and Are Apportioned to Different Daughter Nuclei

Although the rigorously choreographed events of nuclear and cellular division occur as a dynamic and continuous process, for the sake of clarity, we take the traditional approach of analyzing the process in separate stages marked

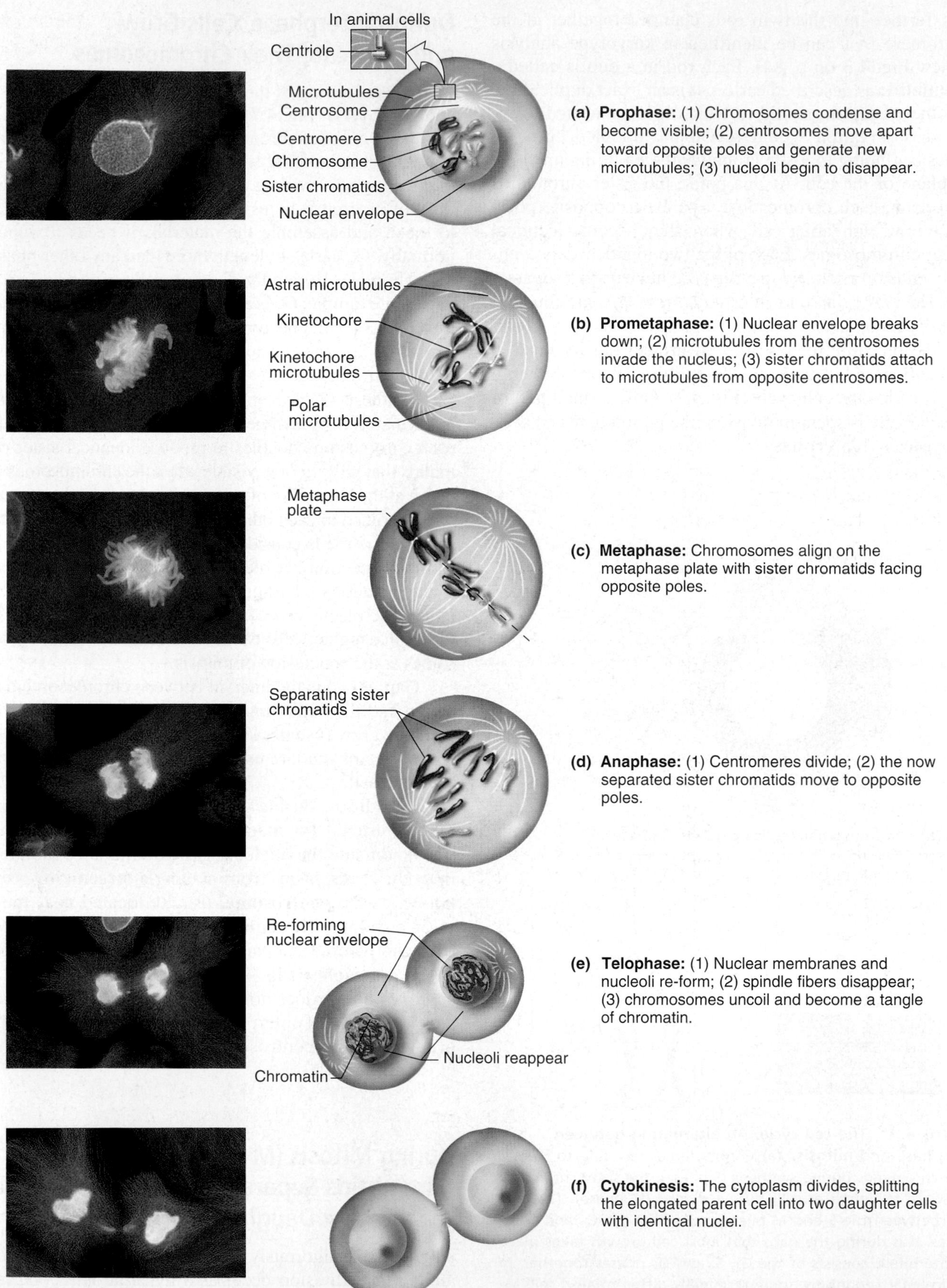
In animal cells
Centriole
Microtubules
Centrosome
Centromere
Chromosome
Sister chromatids
Nuclear envelope
(a) Prophase: (1) Chromosomes condense and become visible; (2) centrosomes move apart toward opposite poles and generate new microtubules; (3) nucleoli begin to disappear.
Astral microtubules
Kinetochore
Kinetochore microtubules
Polar microtubules
(b) Prometaphase: (1) Nuclear envelope breaks down; (2) microtubules from the centrosomes invade the nucleus; (3) sister chromatids attach to microtubules from opposite centrosomes.
Metaphase plate
(c) Metaphase: Chromosomes align on the metaphase plate with sister chromatids facing opposite poles.
Separating sister chromatids
(d) Anaphase: (1) Centromeres divide; (2) the now separated sister chromatids move to opposite poles.
Re-forming nuclear envelope
Nucleoli reappear
Chromatin
(e) Telophase: (1) Nuclear membranes and nucleoli re-form; (2) spindle fibers disappear; (3) chromosomes uncoil and become a tangle of chromatin.
(f) Cytokinesis: The cytoplasm divides, splitting the elongated parent cell into two daughter cells with identical nuclei.

Figure 4.8 Mitosis maintains the chromosome number of the parent cell nucleus in the two daughter nuclei. In the photomicrographs of newt lung cells, chromosomes are stained *blue* and microtubules appear either *green* or *yellow.*

by visible cytological events. The artist's sketches in Fig. 4.8 illustrate these stages in the nematode *Ascaris,* whose diploid cells contain only four chromosomes (two pairs of homologous chromosomes).

Prophase: Chromosomes Condense (Fig. 4.8a)

During the whole of interphase, the cell nucleus remains intact, and the chromosomes are indistinguishable aggregates of chromatin. At **prophase** (from the Greek *pro-* meaning "before"), the gradual emergence, or **condensation,** of individual chromosomes from the undifferentiated mass of chromatin marks the beginning of mitosis. Each emerging chromosome has already been duplicated during interphase and thus consists of sister chromatids attached at the centromere. At this stage in *Ascaris* cells, there are therefore four chromosomes with a total of eight chromatids. The progressive appearance of an array of individual chromosomes is a truly impressive event: Interphase DNA molecules as long as 3–4 cm condense into discrete chromosomes whose length is measured in microns (millionths of a meter). This is equivalent to compacting a 200 m length of thin string (as long as two football fields) into a cylinder 8 mm long and 1 mm wide. Chromosomes condense only during mitosis (and meiosis—see below), when the duplicated chromosomes convert from a metabolically active state to a condition suitable for subsequent transport to daughter cells.

Another visible change in chromatin also takes place during prophase: The darkly staining nucleoli begin to break down and disappear. As a result, the manufacture of ribosomes ceases, providing one indication that general cellular metabolism shuts down so that the cell can focus its energy on chromosome movements and cellular division.

In addition to the nuclear events of chromosome condensation and nucleolar breakdown, several important processes that characterize prophase occur outside the nucleus in the cytoplasm. The centrosomes, which replicated during interphase, now move apart and become clearly distinguishable as two separate entities in the light microscope. At the same time, the interphase scaffolding of long, stable microtubules disappears and is replaced by a set of microtubules that rapidly grow from and contract back toward their centrosomal organizing centers. The centrosomes continue to move apart, migrating around the nuclear envelope toward opposite ends of the nucleus, apparently propelled by forces exerted between interdigitated microtubules extending from both centrosomes.

Prometaphase: The Spindle Forms (Fig. 4.8b)

Prometaphase ("before middle stage") begins with the breakdown of the nuclear envelope, which allows microtubules extending from the two centrosomes to invade the nucleus. Chromosomes attach to these microtubules through the **kinetochore,** a structure in the centromere region of each chromatid that is specialized for conveyance. Each kinetochore contains proteins that act as molecular motors, enabling the chromosome to slide along the microtubule. When the kinetochore of a chromosome originally contacts a microtubule at prometaphase, the kinetochore-based motor moves the chromosome toward the centrosome from which that microtubule radiates. Microtubules growing from the two centrosomes randomly capture chromosomes by the kinetochore of one of the two sister chromatids. As a result, it is sometimes possible to observe groups of chromosomes congregating in the vicinity of each centrosome. In this early part of prometaphase, for each chromosome, one chromatid's kinetochore is attached to a microtubule, but the sister chromatid's kinetochore remains unattached.

During prometaphase, three different types of microtubule fibers together form the **mitotic spindle;** all of these microtubules originate from the centrosomes, which function as the two "poles" of the spindle apparatus. Microtubules that extend between a centrosome and the kinetochore of a chromatid are called **kinetochore microtubules,** or **centromeric fibers.** Microtubules from each centrosome that are directed toward the middle of the cell are **polar microtubules;** polar microtubules originating in opposite centrosomes interdigitate near the cell's equator. Finally, there are short, unstable **astral microtubules** that extend out from the centrosome toward the cell's periphery.

Near the end of prometaphase, the kinetochore of each chromosome's previously unattached sister chromatid now associates with microtubules extending from the opposite centrosome. This event orients each chromosome such that one sister chromatid faces one pole of the cell, the other, the opposite pole. Experimental manipulation has shown that if both kinetochores become attached to microtubules from the same pole, the configuration is unstable; one of the kinetochores will repeatedly detach from the spindle until it associates with microtubules from the other pole. The attachment of sister chromatids to opposite spindle poles is the only stable arrangement.

Metaphase: Chromosomes Align at the Cell's Equator (Fig. 4.8c)

During **metaphase** ("middle stage"), the connection of sister chromatids to opposite spindle poles sets in motion a series of jostling movements that cause the chromosomes to move toward an imaginary equator halfway between the two poles. The imaginary midline is called the **metaphase plate.** When the chromosomes are aligned along it, the forces pulling and pushing them toward or away from each pole are in a balanced equilibrium. As a result, any movement away from the metaphase plate is rapidly compensated by tension that restores the chromosome to its position equidistant between the poles.

Anaphase: Sister Chromatids Move to Opposite Spindle Poles (Fig. 4.8d)

The nearly simultaneous severing of the centromeric connection between the sister chromatids of all chromosomes indicates that **anaphase** (from the Greek *ana-* meaning "up" as in "up toward the poles") is underway. The separation of sister chromatids allows each chromatid to be pulled toward the spindle pole to which it is connected by its kinetochore microtubules; as the chromatid moves toward the pole, its kinetochore microtubules shorten. Since the arms of the chromatids lag behind the kinetochores, metacentric chromatids have a characteristic V shape during anaphase. The connection of sister chromatids to microtubules emanating from opposite spindle poles means that the genetic information migrating toward one pole is exactly the same as its counterpart moving toward the opposite pole.

Telophase: Identical Sets of Chromosomes Are Enclosed in Two Nuclei (Fig. 4.8e)

The final transformation of chromosomes and the nucleus during mitosis happens at **telophase** (from the Greek *telo-* meaning "end"). Telophase is like a rewind of prophase. The spindle fibers begin to disperse; a nuclear envelope forms around the group of chromatids at each pole; and one or more nucleoli reappears. The former chromatids now function as independent chromosomes, which decondense (uncoil) and dissolve into a tangled mass of chromatin. Mitosis, the division of one nucleus into two identical nuclei, is over.

Cytokinesis: The Cytoplasm Divides (Fig. 4.8f)

In the final stage of cell division, the daughter nuclei emerging at the end of telophase are packaged into two separate daughter cells. This final stage of division is called **cytokinesis** (literally "cell movement"). During cytokinesis, the elongated parent cell separates into two smaller independent daughter cells with identical nuclei. Cytokinesis usually begins during anaphase, but it is not completed until after telophase. The mechanism by which cells accomplish cytokinesis differs in animals and plants. In animal cells, cytoplasmic division depends on a **contractile ring** that pinches the cell into two approximately equal halves, similar to the way the pulling of a string closes the opening of a bag of marbles (**Fig. 4.9a**). Intriguingly, some types of molecules that form the contractile ring also participate in the mechanism responsible for muscle contraction. In plants, whose cells are surrounded by a rigid cell wall, a membrane-enclosed disk, known as the **cell plate,** forms inside the cell near the equator and then grows rapidly outward, thereby dividing the cell in two (Fig. 4.9b).

During cytokinesis, a large number of important organelles and other cellular components, including ribosomes, mitochondria, membranous structures such as Golgi bodies, and (in plants) chloroplasts, must be parcelled out to the emerging daughter cells. The mechanism accomplishing this task does not appear to predetermine which organelle is destined for which daughter cell. Instead, because most cells contain many copies of these cytoplasmic structures, each new cell is bound to receive at least a few representatives of each component. This original complement of structures is enough to sustain the cell until synthetic activity can repopulate the cytoplasm with organelles.

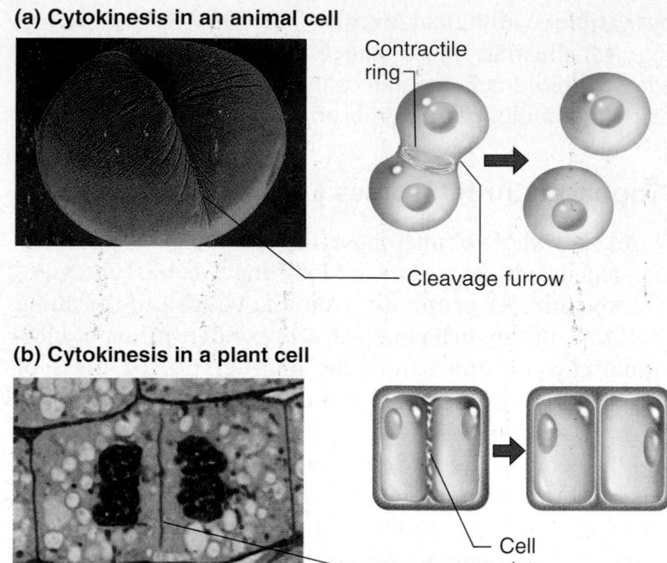

Figure 4.9 Cytokinesis: The cytoplasm divides, producing two daughter cells. **(a)** In this dividing frog zygote, the contractile ring at the cell's periphery has contracted to form a cleavage furrow visible from the exterior that will eventually pinch the cell in two. **(b)** In this dividing onion root cell, a cell plate that began forming near the equator of the cell expands to the periphery, separating the two daughter cells.

Sometimes cytoplasmic division does not immediately follow nuclear division, and the result is a cell containing more than one nucleus. An animal cell with two or more nuclei is known as a **syncytium.** The early embryos of fruit flies are multinucleated syncytia (**Fig. 4.10**), as are the precursors of spermatozoa in humans and many other animals. A multinucleate plant tissue is called a **coenocyte;** coconut milk is a nutrient-rich food composed of coenocytes.

In summary, through mitosis plus cytokinesis, a parental cell generates two daughter cells that have the same number and kind of chromosomes as each other and as the original parental cell.

Regulatory Checkpoints Ensure Correct Chromosome Separation During Mitosis

The cell cycle is a complex sequence of precisely coordinated events. In higher organisms, a cell's "decision" to

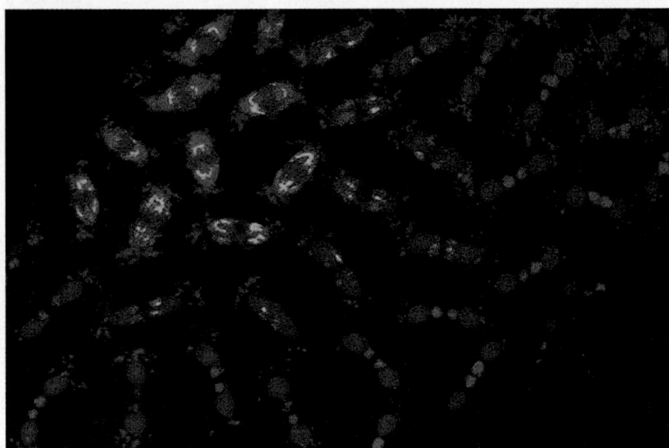

Figure 4.10 If cytokinesis does not follow mitosis, one cell may contain many nuclei. In fertilized *Drosophila* eggs, 13 rounds of mitosis take place without being followed by cytokinesis. The result is a single-celled syncytial embryo that contains several thousand nuclei. The photograph shows part of an embryo in which the nuclei are all dividing; chromosomes are in *red,* and spindle fibers are in *green.* Nuclei at the *upper left* are in metaphase, while nuclei toward the *bottom right* are progressively later in anaphase. Membranes eventually grow around these nuclei, dividing the embryo into cells.

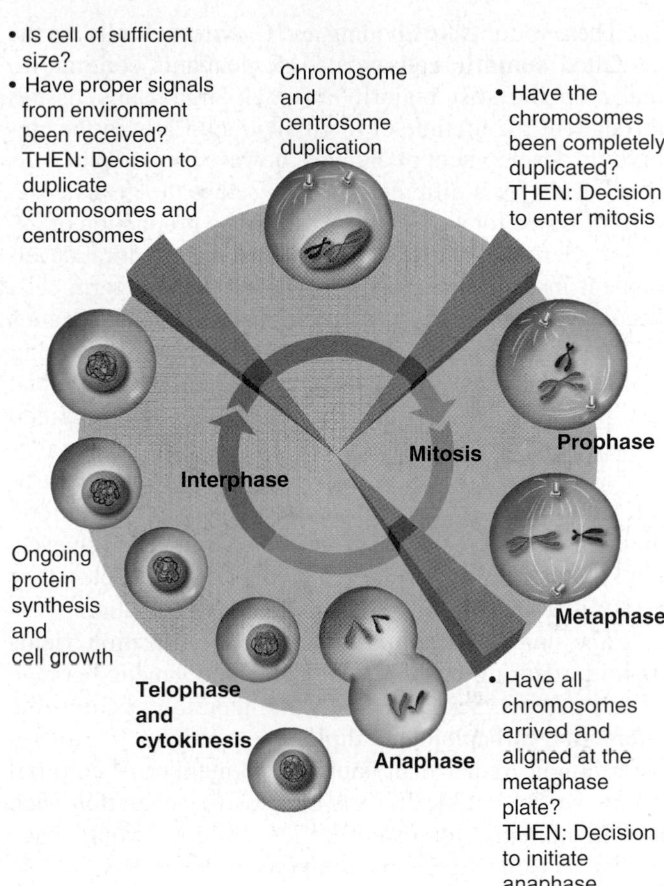

Figure 4.11 Checkpoints help regulate the cell cycle. Cellular checkpoints (*red wedges*) ensure that important events in the cell cycle occur in the proper sequence. At each checkpoint, the cell evaluates one or more prior events before deciding whether to proceed to the next step of the cell cycle. (For simplicity, we show here only two chromosomes per cell.)

divide depends on both intrinsic factors, such as conditions within the cell that register a sufficient size for division, and signals from the environment, such as hormonal cues or contacts with neighboring cells that encourage or restrain division. Once a cell has initiated events leading to division, usually during the G_1 period of interphase, everything else follows like clockwork. A number of **checkpoints**—moments at which the cell evaluates the results of previous steps—allow the sequential coordination of cell-cycle events. Consequently, under normal circumstances, the chromosomes replicate before they condense, and the doubled chromosomes separate to opposite poles only after correct metaphase alignment of sister chromatids ensures equal distribution to the daughter nuclei (**Fig. 4.11**). In one illustration of the molecular basis of checkpoints, even a single kinetochore that has not attached to spindle fibers generates a molecular signal that prevents the sister chromatids of all chromosomes from separating at their centromeres. This signal makes the beginning of anaphase dependent on the prior proper alignment of all the chromosomes at metaphase. As a result of the multiple checkpoints operating during the cell cycle, each daughter cell reliably receives the right number of chromosomes.

Breakdown of the mitotic machinery can produce division mistakes that have crucial consequences for the cell. Improper chromosome segregation, for example, can cause serious malfunction or even the death of daughter cells. As the Fast Forward box "How Gene Mutations Cause Errors in Mitosis" on p. 95 explains, one source of improper segregation is gene mutations that disrupt mitotic structures, such as the spindle, kinetochores, or centrosomes. Other problems occur in cells where the normal restraints on cell division, such as checkpoints, have broken down. Such cells may divide uncontrollably, leading to a tumor. We present the details of cell-cycle regulation, checkpoint controls, and cancer formation in Chapter 19.

4.3 Meiosis Produces Haploid Germ Cells, the Gametes

During the many rounds of cell division within an embryo, most cells either grow and divide via the mitotic cell cycle just described, or they stop growing and become arrested in

G_0. These mitotically dividing and G_0-arrested cells are the so-called **somatic cells** whose descendants continue to make up the vast majority of each organism's tissues throughout the lifetime of the individual. Early in the embryonic development of animals, however, a group of cells is set aside for a different fate. These are the **germ cells:** cells destined for a specialized role in the production of gametes. Germ cells arise later in plants, during floral development instead of during embryogenesis. The germ cells become incorporated in the reproductive organs—ovaries and testes in animals; ovaries and anthers in flowering plants—where they ultimately undergo meiosis, the special two-part cell division that produces gametes (eggs and sperm or pollen) containing half the number of chromosomes as other body cells. Normal somatic cells have a finite lifetime; they undergo a certain number of mitotic divisions and then die. In contrast, the germ cells achieve a kind of immortality because, in giving rise to gametes, they transmit an individual's genes to the next generation.

The union of haploid gametes at fertilization yields diploid offspring that carry the combined genetic heritage of two parents. Sexual reproduction therefore requires the alternation of haploid and diploid generations. If gametes were diploid rather than haploid, the number of chromosomes would double in each successive generation such that in humans, for example, the children would have 92 chromosomes per cell, the grandchildren 184, and so on. Meiosis prevents this lethal, exponential accumulation of chromosomes. We now turn to a detailed examination of meiosis, focusing on how the first round of this two-part nuclear division gives rise to genetic variation as it halves a germ cell's chromosome number from two sets to one.

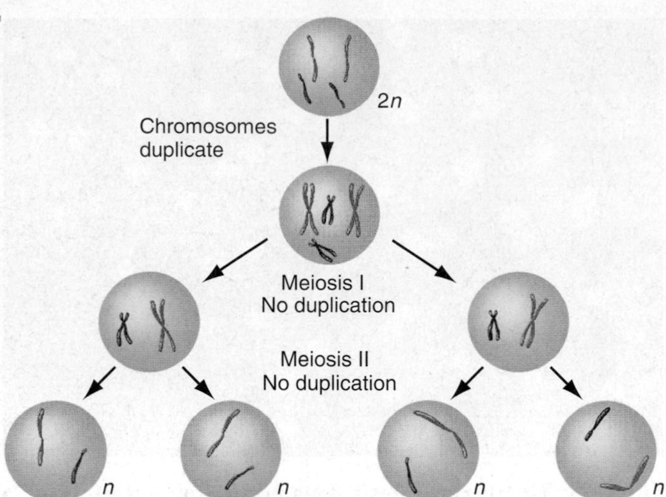

Figure 4.12 An overview of meiosis: The chromosomes replicate once, while the nuclei divide twice. In this figure, all four chromatids of each chromosome pair are shown in the same shade of the same color. Note that the chromosomes duplicate before meiosis I, but they do not duplicate between meiosis I and meiosis II.

Meiosis Consists of One Round of Chromosome Replication But Two Rounds of Nuclear Division

Unlike mitosis, meiosis consists of two successive nuclear divisions, logically named **division I of meiosis** and **division II of meiosis,** or simply **meiosis I** and **meiosis II.** With each round, the cell passes through a prophase, metaphase, anaphase, and telophase followed by cytokinesis. In meiosis I, the parent nucleus divides to form two daughter nuclei; in meiosis II, each of the two daughter nuclei divides, resulting in four nuclei (**Fig. 4.12**). These four nuclei—the final products of meiosis—become partitioned in four separate daughter cells because cytokinesis occurs after both rounds of division. The chromosomes duplicate at the start of meiosis I, but they do not duplicate in meiosis II, which explains why the gametes generated by meiosis contain half the number of chromosomes found in other body cells. A close look at each round of meiotic division reveals the mechanisms by which each gamete comes to receive one full haploid set of chromosomes.

During Meiosis I, Homologous Chromosomes Pair, Exchange Parts, and Then Segregate from Each Other

The events of meiosis I are unique among nuclear divisions (**Fig. 4.13,** meiosis I pp 96-97). The process begins with the replication of chromosomes, after which each one consists of two sister chromatids. A key to understanding meiosis I is the observation that the centromeres joining these chromatids remain intact throughout the entire division, rather than splitting as in mitosis. As the division proceeds, homologous chromosomes align across the cellular equator to form a coupling that ensures proper chromosome segregation to separate nuclei. Moreover, during the time homologous chromosomes face each other across the equator, the maternal and paternal chromosomes of each homologous pair exchange parts, creating new combinations of alleles at different genes along the chromosomes. Afterward, the two homologous chromosomes, each consisting of two sister chromatids connected at a single, unsplit centromere, are pulled to opposite poles of the spindle. As a result, it is homologous chromosomes (rather than sister chromatids as in mitosis) that segregate into different daughter cells at the conclusion of the first meiotic division. With this overview in mind, let us take a closer look at the specific events of meiosis I, bearing in mind that we analyze a dynamic, flowing sequence of cellular events by breaking it down somewhat arbitrarily into the easily pictured, traditional phases.

How Gene Mutations Cause Errors in Mitosis

During each cell cycle, the chromosomes participate in a tightly patterned choreography that leads to duplication before condensation, midplane alignment before sister chromatid separation, and nuclear division before cytokinesis. Because of these sequential steps, synchronized in both time and space, the chromosomes convey a complete set of genes to each of two newly forming daughter cells. Not surprisingly, some of the very genes they carry encode proteins that shepherd them through the dance.

A variety of proteins, some assembled into structures such as centrosomes and microtubule fibers, make up the molecular machinery that helps coordinate the orderly progression of events in mitosis. Because a particular gene specifies each protein, we might predict that mutant alleles generating defects in particular proteins could disrupt the dance. Cells homozygous for a mutant allele might be unable to complete chromosome duplication or mitosis or cytokinesis because a required component of the molecular machinery is missing or unable to function. Experiments on organisms as disparate as yeast and fruit flies have borne out this prediction. Here we describe the effects of a mutation in one of the many *Drosophila* genes critical for proper chromosome segregation.

Although most mistakes in mitosis are eventually lethal to a multicellular organism, some mutant cells may manage to divide early in development, and when prepared for viewing under the microscope, these cells actually allow us to see the effects of defective mitosis. To understand these effects, we first present part of a normal mitosis as a basis for comparison. **Figure A** (*left panel*) shows the eight condensed metaphase chromosomes of a wild-type male fruit fly (*Drosophila melanogaster*): two pairs of large metacentric autosomes with the centromere in the center, a pair of dot-like autosomes that are so small it is not possible to see the centromere region, an acrocentric X chromosome with the centromere very close to one end, and a metacentric Y chromosome. Because most of the Y chromosome consists of a special form of chromatin known as heterochromatin, the two Y sister chromatids remain so tightly connected that they often appear as one.

Figure B (*left panel*) shows the results of aberrant mitosis in an animal homozygous for a mutation in a gene called *zw10* that encodes a component of the chromosomal kinetochores. The mutation disrupted mitotic chromosome segregation during early development, producing cells with the wrong number of chromosomes. The problem in chromosome segregation probably occurred during anaphase of the previous cell division. Figure A (*right panel*) shows a normal anaphase separation leading to the wild-type chromosome complement. Figure B (*right panel*) portrays an aberrant anaphase separation in a mutant animal that could lead to an abnormal chromosome complement similar to that depicted in the left panel of the same figure; you can see that many more chromatids are migrating to one spindle pole than to the other.

The smooth unfolding of each cell cycle depends on a diverse array of proteins. Some form the structures that transport chromosomes (like the kinetochores in this example) or that cleave cells in two; others help control when and where things happen. Particular genes specify each of the proteins active in mitosis and cytokinesis, and each protein makes a contribution to the coordinated events of the cell cycle. As a result, a mutation in any of a number of genes can disrupt the meticulously choreographed mitotic mechanism that, when functioning normally, creates generation upon generation of chromosomally identical diploid daughter cells.

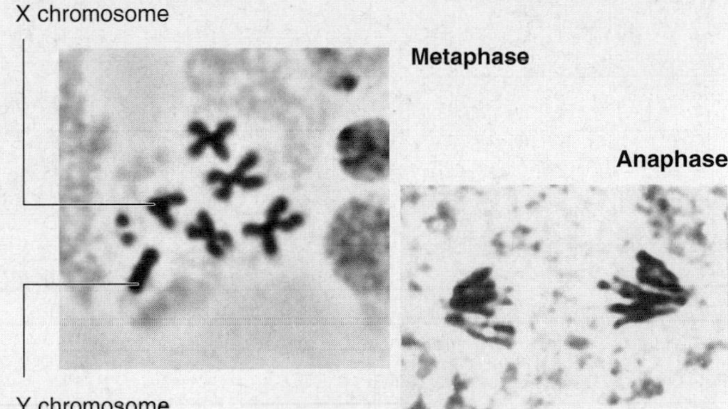

Figure A Metaphase and anaphase chromosomes in a wild-type male fruit fly.

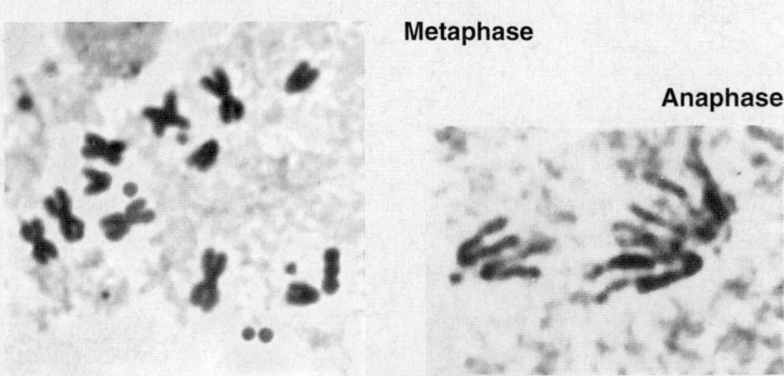

Figure B Metaphase and anaphase chromosomes in a mutant fly. These cells are from a *Drosophila* male hemizygous for an X-linked mutation called *zw10.* The mutant metaphase cell (*left*) contains extra chromosomes as compared with the wild-type metaphase cell in Fig. A. In the mutant anaphase cell (*right*), more chromatids are moving toward one spindle pole than toward the other.

Feature Figure 4.13

Meiosis: One Diploid Cell Produces Four Haploid Cells

Meiosis I: A reductional division

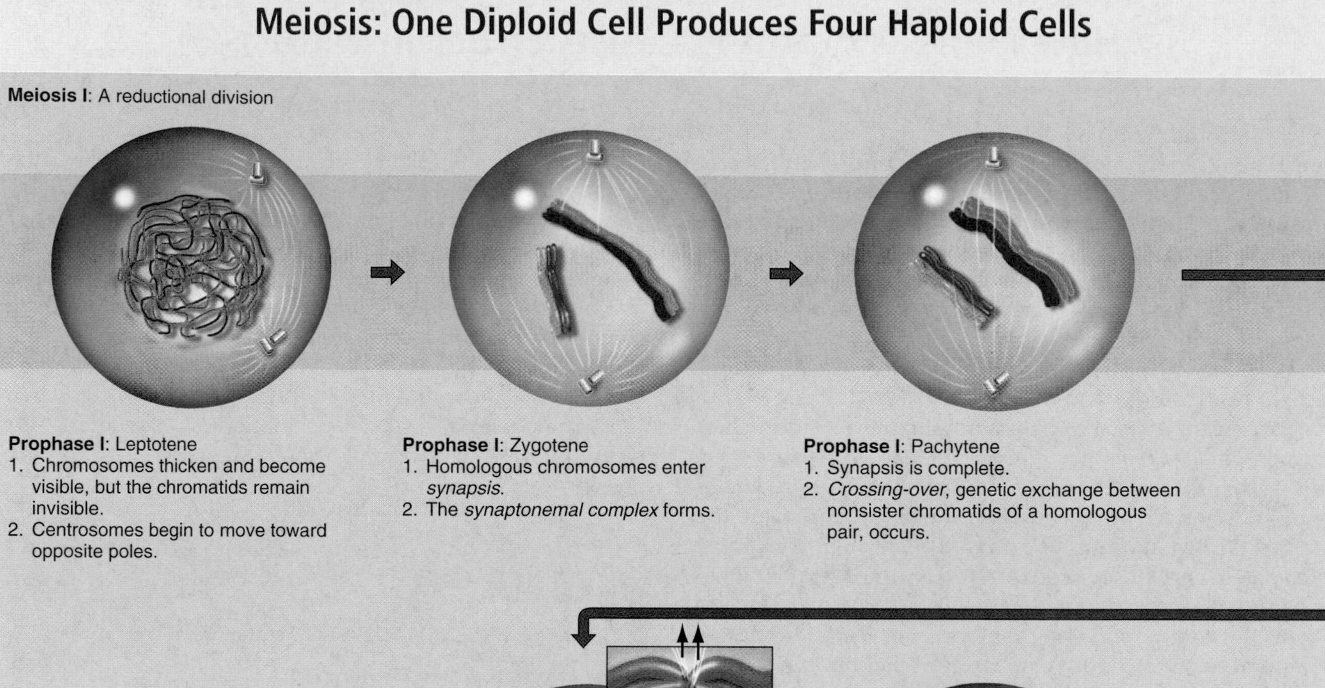

Prophase I: Leptotene
1. Chromosomes thicken and become visible, but the chromatids remain invisible.
2. Centrosomes begin to move toward opposite poles.

Prophase I: Zygotene
1. Homologous chromosomes enter *synapsis*.
2. The *synaptonemal complex* forms.

Prophase I: Pachytene
1. Synapsis is complete.
2. *Crossing-over*, genetic exchange between nonsister chromatids of a homologous pair, occurs.

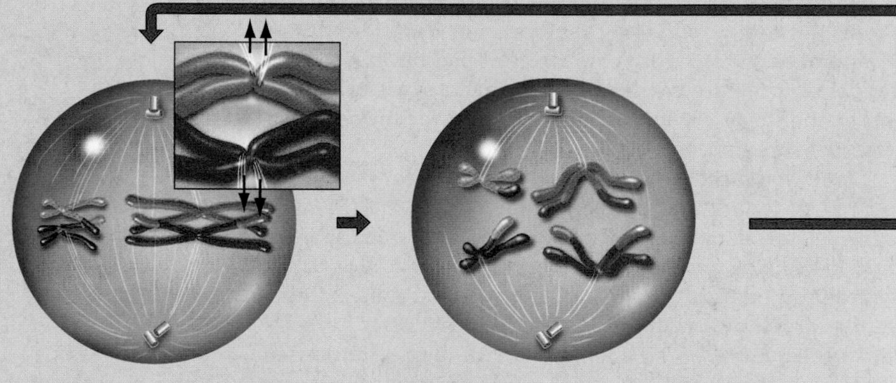

Metaphase I
1. Tetrads line up along the *metaphase plate*.
2. Each chromosome of a homologous pair attaches to fibers from opposite poles.
3. Sister chromatids attach to fibers from the same pole.

Anaphase I
1. The centromere does not divide.
2. The chiasmata migrate off chromatid ends.
3. Homologous chromosomes move to opposite poles.

Meiosis II: An equational division

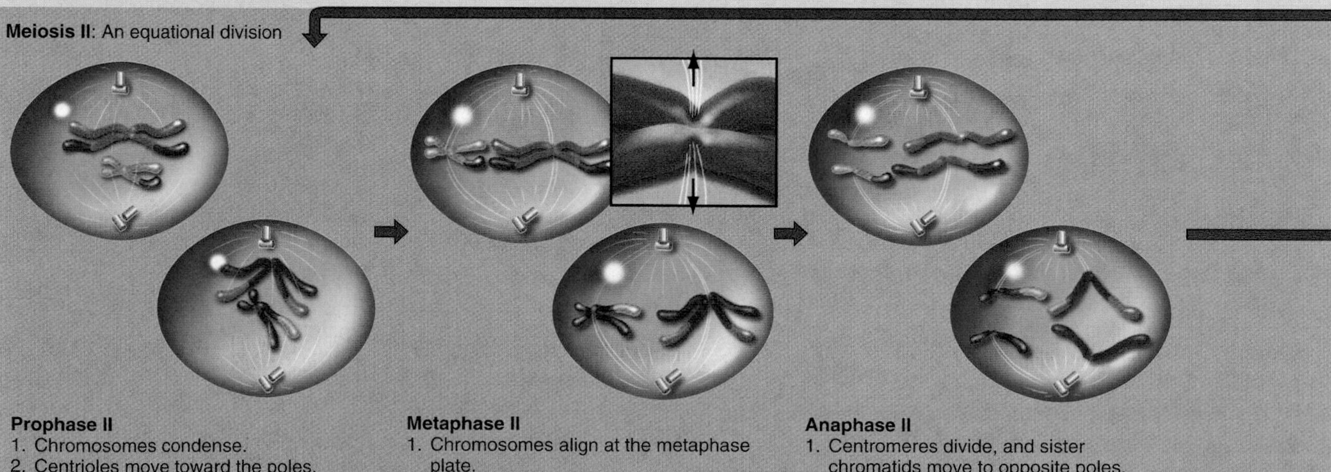

Prophase II
1. Chromosomes condense.
2. Centrioles move toward the poles.
3. The nuclear envelope breaks down at the end of prophase II (not shown).

Metaphase II
1. Chromosomes align at the metaphase plate.
2. Sister chromatids attach to spindle fibers from opposite poles.

Anaphase II
1. Centromeres divide, and sister chromatids move to opposite poles.

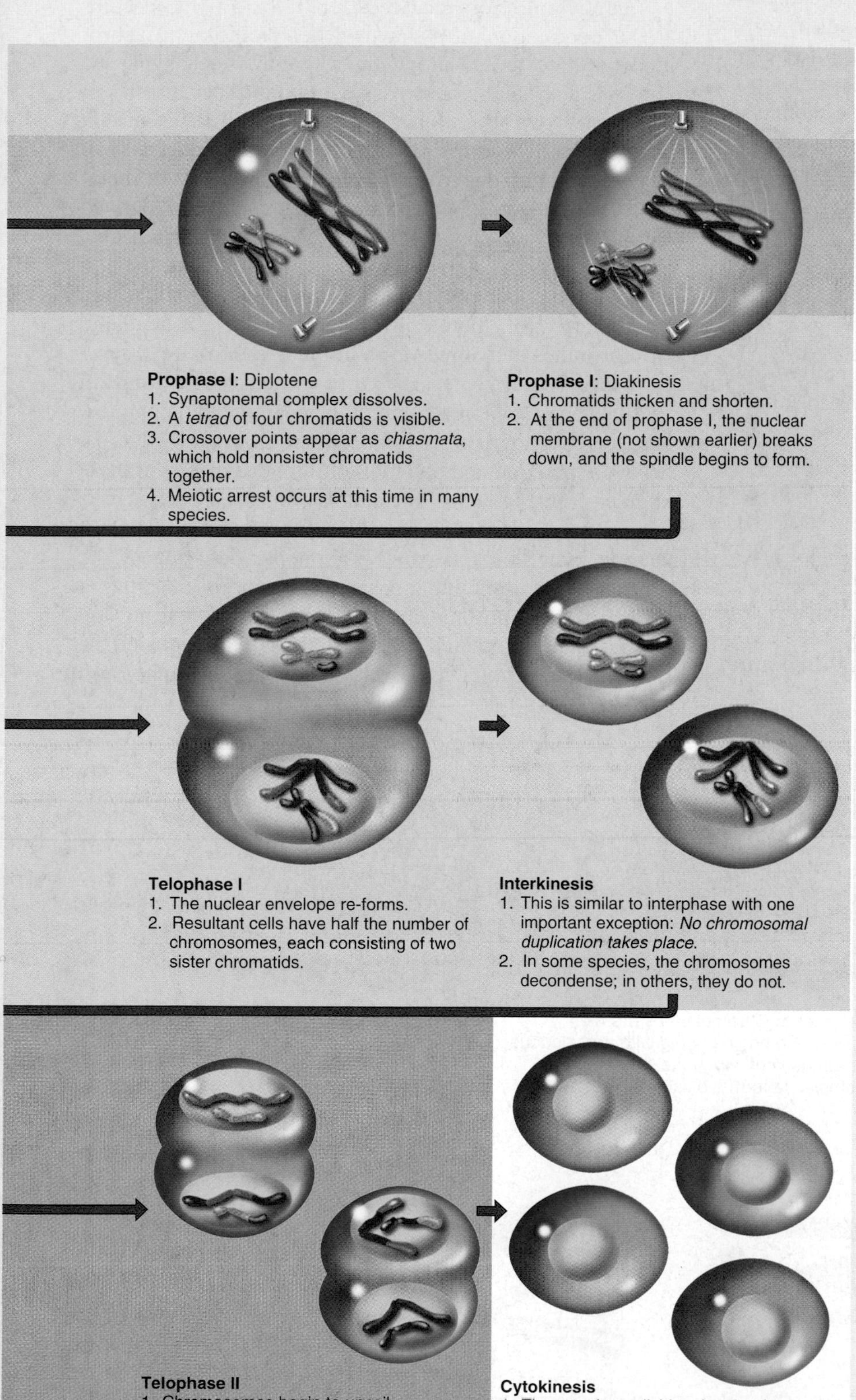

Prophase I: Diplotene
1. Synaptonemal complex dissolves.
2. A *tetrad* of four chromatids is visible.
3. Crossover points appear as *chiasmata*, which hold nonsister chromatids together.
4. Meiotic arrest occurs at this time in many species.
Prophase I: Diakinesis
1. Chromatids thicken and shorten.
2. At the end of prophase I, the nuclear membrane (not shown earlier) breaks down, and the spindle begins to form.
Telophase I
1. The nuclear envelope re-forms.
2. Resultant cells have half the number of chromosomes, each consisting of two sister chromatids.
Interkinesis
1. This is similar to interphase with one important exception: *No chromosomal duplication takes place.*
2. In some species, the chromosomes decondense; in others, they do not.
Telophase II
1. Chromosomes begin to uncoil.
2. Nuclear envelopes and nucleoli (not shown) re-form.
Cytokinesis
1. The cytoplasm divides, forming four new haploid cells.

Figure 4.13 To aid visualization of the chromosomes, the figure is simplified in two ways: (1) The nuclear envelope is not shown during prophase of either meiotic division. (2) The chromosomes are shown as fully condensed at zygotene; in reality, the chromosomes continue to condense throughout prophase such that full condensation does not occur until diakinesis.

Prophase I: During This Longest, Most Complex Phase of Meiosis, Crossing-Over Occurs

Among the critical events of **prophase I** are the condensation of chromatin, the pairing of homologous chromosomes, and the reciprocal exchange of genetic information between these paired homologs (Fig. 4.13, meiosis I). These complicated events can take many days, months, or even years to complete. For example, in the female germ cells of several species, including our own, meiosis is suspended at prophase I until ovulation. This may sound surprising, but it will become clear when we discuss the details of egg formation later in this chapter.

Leptotene (from the Greek for "thin" and "delicate") is the first definable substage of prophase I, the time when the long, thin chromosomes begin to thicken (see **Fig. 4.14a** for a more detailed view). Each chromosome has already duplicated prior to prophase I (as in mitosis) and thus consists of two sister chromatids affixed at a centromere. At this point, however, these sister chromatids are so tightly bound together that they are not yet visible as separate entities.

Zygotene (from the Greek for "conjugation") begins as each chromosome seeks out its homologous partner and the matching chromosomes become zipped together in a process known as **synapsis.** The "zipper" itself is an elaborate protein structure called the **synaptonemal complex** that aligns the homologs with remarkable precision, juxtaposing the corresponding genetic regions of the chromosome pair (Fig. 4.14b).

Pachytene (from the Greek for "thick" or "fat") begins at the completion of synapsis when homologous chromosomes are united along their length. Each synapsed chromosome pair is known as a **bivalent** (because it encompasses two chromosomes), or a **tetrad** (because it contains four chromatids, which, as meiosis proceeds, will be parcelled out, one to each of the four products of meiosis). On one side of the bivalent is a maternally derived chromosome, on the other side a paternally derived one. Because X and Y chromosomes are not identical, they do not synapse completely; there is, however, a small region of similarity (or "homology") between the X and the Y chromosomes that allows for a limited amount of pairing.

During pachytene, structures called **recombination nodules** begin to appear along the synaptonemal complex, and an exchange of parts between nonsister (that is, between maternal and paternal) chromatids occurs at these nodules (see Fig. 4.14c for details). Such an exchange is known as **crossing-over;** it results in the **recombination** of genetic material. You can envision the outcome of the process by imagining that the two chromatids constituting a maternal chromosome are dark green pieces of string and the paternal chromatids are two light green pieces of the same length. If you cut one dark green and one light green

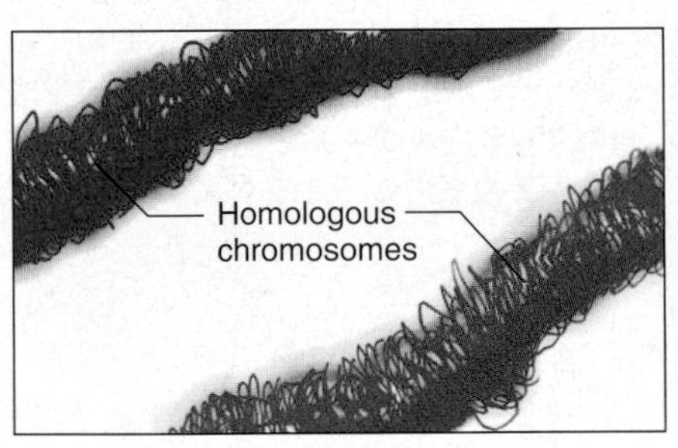

(a) Leptotene: Threadlike chromosomes begin to condense and thicken, becoming visible as discrete structures. Although the chromosomes have duplicated, the sister chromatids of each chromosome are not yet visible in the microscope.

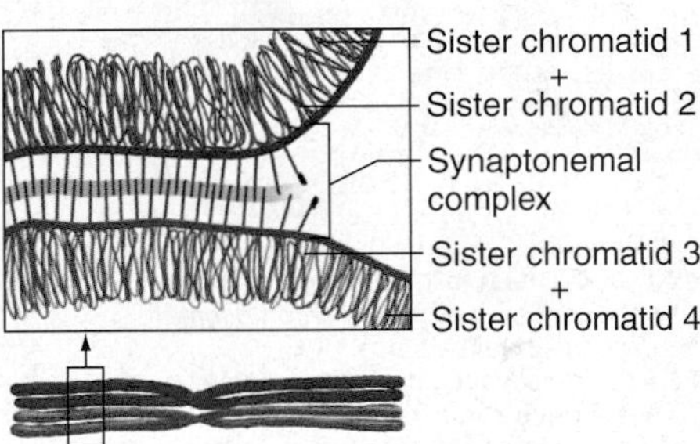

(b) Zygotene: Chromosomes are clearly visible and begin pairing with homologous chromosomes along the synaptonemal complex to form a bivalent, or tetrad.

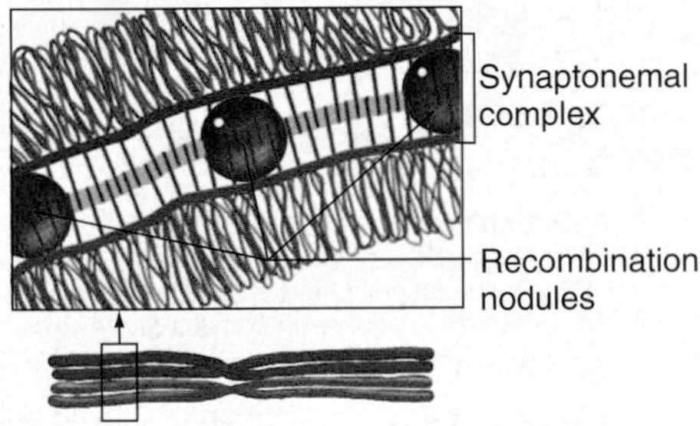

(c) Pachytene: Full synapsis of homologues. Recombination nodules appear along the synaptonemal complex.

(d) Diplotene: Bivalent appears to pull apart slightly but remains connected at crossover sites, called chiasmata.

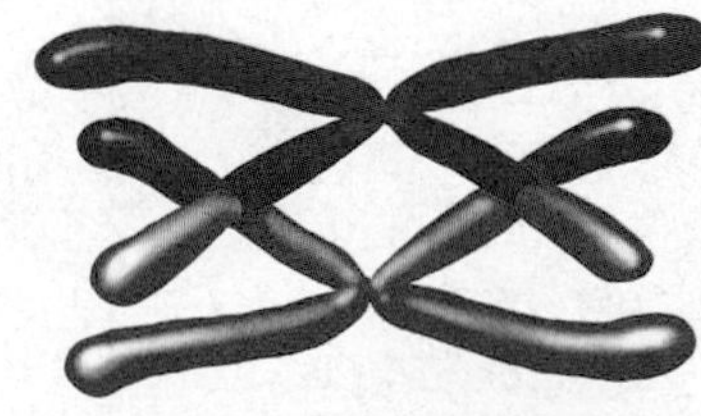

(e) Diakinesis: Further condensation of chromatids. Nonsister chromatids that have exchanged parts by crossing-over remain closely associated at chiasmata.

Figure 4.14 Prophase I of meiosis at very high magnification.

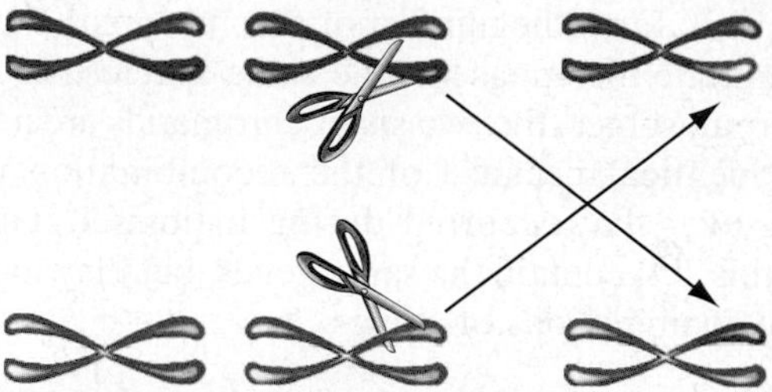

Figure 4.15 **How crossing-over produces recombined chromosomes.** Crossing-over and genetic recombination occur after the chromosomes have replicated into sister chromatids. Crossing-over involves cutting, at the same position, one chromatid from each chromatid pair. The subsequent reshuffling and splicing of the cut regions produces two recombinant chromatids of the original size.

string at the same position and retie dark to light and light to dark, you still have two pieces of string of equal length, but they are recombined mosaics of the originals (**Fig. 4.15**). As a result of crossing-over, each homologous chromosome in a pair may no longer be of purely maternal or purely paternal origin; each one, however, is still a complete entity, with no gaps in its genetic information.

Diplotene (from the Greek for "twofold" or "double") is signaled by the gradual dissolution of the synaptonemal zipper complex and a slight separation of regions of the homologous chromosomes (see Fig. 4.14d). The aligned homologous chromosomes of each bivalent nonetheless remain very tightly merged at intervals along their length called **chiasmata** (singular, *chiasma*), which represent the sites where crossing-over occurred.

Diakinesis (from the Greek for "double movement") is accompanied by further condensation of the chromatids. Because of this chromatid thickening and shortening, it can now clearly be seen that each tetrad consists of four separate chromatids, or viewed in another way, that the two homologous chromosomes of a bivalent are each composed of two sister chromatids held together at a centromere (see Fig. 4.14e). Nonsister chromatids that have undergone crossing-over remain closely associated at chiasmata. The end of diakinesis is analogous to the prometaphase of mitosis: The nuclear envelope breaks down, and the microtubules of the spindle apparatus begin to form.

When Do Chromosome Pairing and Recombination Actually Take Place? We have described the events of prophase I as they appear in the light microscope. But recent genetic experiments using the unicellular yeast *Saccharomyces cerevisiae* indicate that at least in this organism, chromosomes somehow begin to pair during and possibly even prior to leptotene, well before synaptonemal complexes become visible during zygotene, and that recombination starts during leptotene or zygotene, before recombination nodules become visible during pachytene. Surprisingly, these experiments also show that recombination can occur even if a synaptonemal complex that tightly juxtaposes the homologous chromosomes does not form. Many investigators are now trying to understand how visible structures such as the synaptonemal complex, recombination nodules, and chiasmata contribute to chromosome pairing and the initiation and completion of genetic recombination.

Metaphase I: Homologous Chromosomes Attach to Spindle Fibers Growing from Opposite Poles

There is an essential difference between the spindle formed during meiosis I and that formed during mitosis. As we have seen, during mitosis, each sister chromatid has a kinetochore that becomes attached to microtubules emanating from opposite spindle poles. In contrast, during meiosis I, the kinetochores of sister chromatids fuse so that each chromosome contains only a single functional kinetochore; there are thus no oppositely directed forces that could later pull the sister chromatids apart during the anaphase of meiosis I. Instead, during **metaphase I** (Fig 4.13, meiosis I), it is the kinetochores of homologous chromosomes that attach to microtubules from opposite spindle poles. As a result, in chromosomes aligned at the metaphase plate, the kinetochores of maternally and paternally derived chromosomes face opposite spindle poles, positioning the homologous chromosomes to which they are connected to move in opposite directions. Because the alignment and hookup of each bivalent is independent of every other bivalent's, the chromosomes facing each pole are a random mix of maternal and paternal origin.

Anaphase I: Homologous Chromosomes Move to Opposite Poles of the Spindle

At the onset of **anaphase I,** the chiasmata joining homologous chromosomes dissolve, which allows the maternal and paternal homologs to begin to move toward opposite spindle poles (see Fig. 4.13, meiosis I). Note that in the first meiotic division, the centromeres do not divide as they do in mitosis. Thus, from each homologous pair, one chromosome consisting of two sister chromatids joined at their centromere segregates to each spindle pole.

Recombination through crossing-over plays an important role in the proper segregation of homologous chromosomes during the first meiotic division. This is because chiasmata, in holding homologous chromosomes together, ensure that their kinetochores remain attached to opposite spindle poles throughout metaphase. When recombination does not occur within a bivalent, mistakes in hookup and conveyance may cause homologous chromosomes to move to the same pole instead of segregating to opposite poles. In some organisms, however, proper segregation of nonrecombinant chromosomes nonetheless occurs through other pairing processes. Investigators do not yet completely understand the nature of these processes and are currently evaluating several models to explain them.

Telophase I: Nuclear Envelopes Reform

The telophase of the first meiotic division, or **telophase I,** takes place when nuclear membranes begin to form around the chromosomes that have moved to the poles. Each of the incipient daughter nuclei contains one-half the number of chromosomes in the original parent nucleus, but each chromosome consists of two sister chromatids joined at the centromere (see Fig. 4.13, meiosis I). Because the number of chromosomes is reduced to one-half the normal diploid number, meiosis I is often called a ***reductional division.***

In most species, cytokinesis follows telophase I, with daughter nuclei becoming enclosed in separate daughter cells. A short interphase then ensues. During this time, the chromosomes usually decondense, in which case they must recondense during the prophase of the subsequent second meiotic division. In some cases, however, the chromosomes simply stay condensed. Most importantly, there is no S phase during the interphase between meiosis I and meiosis II; that is, the chromosomes do not replicate during meiotic interphase. The relatively brief interphase between meiosis I and meiosis II is known as **interkinesis.**

During Meiosis II, Sister Chromatids Separate to Produce Haploid Gametes

The second meiotic division (meiosis II) proceeds in a fashion very similar to that of mitosis, but since the number of chromosomes in each dividing nucleus has already been reduced by half, the resulting daughter cells are haploid. The same process occurs in each of the two daughter cells generated by meiosis I, producing four haploid cells at the end of this second meiotic round (see Fig. 4.13, meiosis II).

Prophase II: The Chromosomes Condense as in the Prophase of Mitosis

If the chromosomes decondensed during the preceding interphase, they recondense during **prophase II.** At the end of prophase II, the nuclear envelope breaks down, and the spindle apparatus re-forms.

Metaphase II: Chromosomes Align at the Metaphase Plate

The kinetochores of sister chromatids attach to microtubule fibers emanating from opposite poles of the spindle apparatus, just as in mitotic metaphase. There are nonetheless two significant features of **metaphase II** that distinguish it from mitosis. First, the number of chromosomes is one-half that in mitotic metaphase of the same species. Second, in most chromosomes, the two sister chromatids are no longer strictly identical because of the recombination through crossing-over that occurred during meiosis I. The sister chromatids still contain the same genes, but they may carry different combinations of alleles.

Anaphase II: Sister Chromatids Separate to Opposite Spindle Poles

Just as in mitosis, severing of the centromeric connection between sister chromatids allows them to move toward opposite spindle poles during **anaphase II.**

Telophase II: Nuclear Membrane Formation, Followed by Cytokinesis, Creates Four Haploid Daughter Cells

Membranes form around each of four daughter nuclei in **telophase II,** and cytokinesis places each nucleus in a separate cell. The result is four haploid gametes. Note that at the end of meiosis II, each daughter cell (that is, each gamete) has the same number of chromosomes as the parental cell present at the beginning of this division. For this reason, meiosis II is termed an ***equational division.***

A Summary of the Significant Events of Meiosis

During meiosis, duplication of chromosomes followed by two nuclear divisions reduces the chromosome number by half (review Fig. 4.12 on p. 94). This reduction in chromosome number is not haphazard because in metaphase I, homologous chromosomes pair up and then separate to different daughter cells. At the end of a normal meiosis I, each daughter cell contains one complete set of chromosomes, with each chromosome consisting of two attached sister chromatids. During meiosis II, these two sister chromatids separate to daughter nuclei.

Segregational errors during either meiotic division can lead to aberrations, such as trisomies, in the next generation. If, for example, the homologs of a chromosome pair do not segregate during meiosis I (a mistake known as **nondisjunction**), they may travel together to the same pole and eventually become part of the same gamete. Such an error in the segregation of homologous chromosomes may at fertilization result in any one of a large variety of possible trisomies. Most autosomal trisomies, as we already mentioned, are lethal *in utero;* one exception is trisomy 21, the genetic basis of Down syndrome. Like trisomy 21, extra sex chromosomes may also be nonlethal but cause a variety of mental and physical abnormalities, such as those seen in Klinefelter syndrome (see Table 4.1 on p. 86).

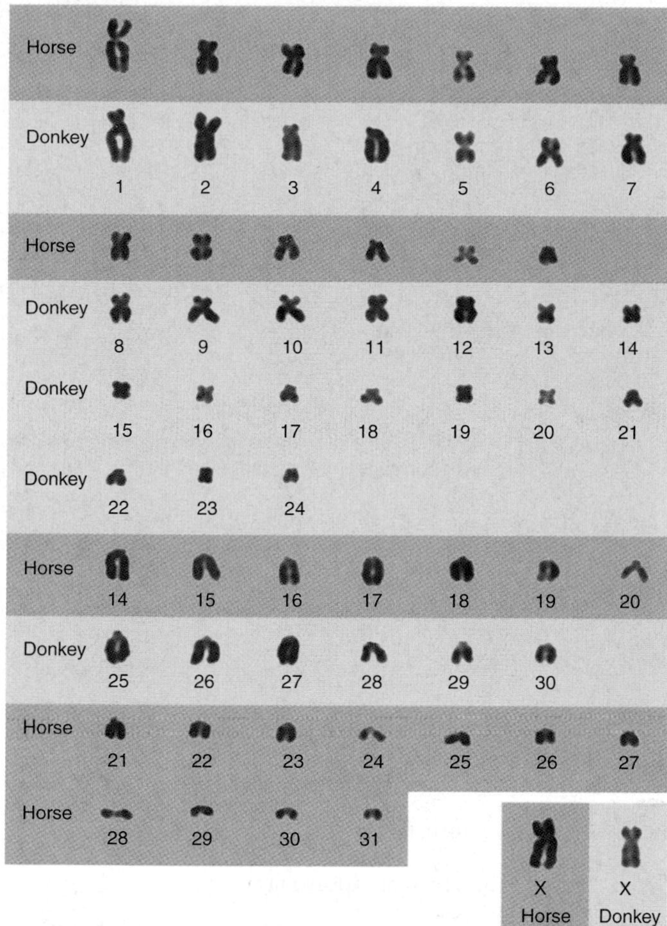

Figure 4.16 Hybrid sterility: When chromosomes cannot pair during meiosis I, they will segregate improperly. The mating of a male donkey (*Equus asinus; green*) and a female horse (*Equus caballus; peach color*) produces a mule with 63 chromosomes. In this karyotype of a female mule, the first 13 donkey and horse chromosomes are homologous and pictured in pairs. Starting at chromosome 14, the donkey and horse chromosomes are numbered and pictured separately because they are too dissimilar to pair with each other during meiosis I.

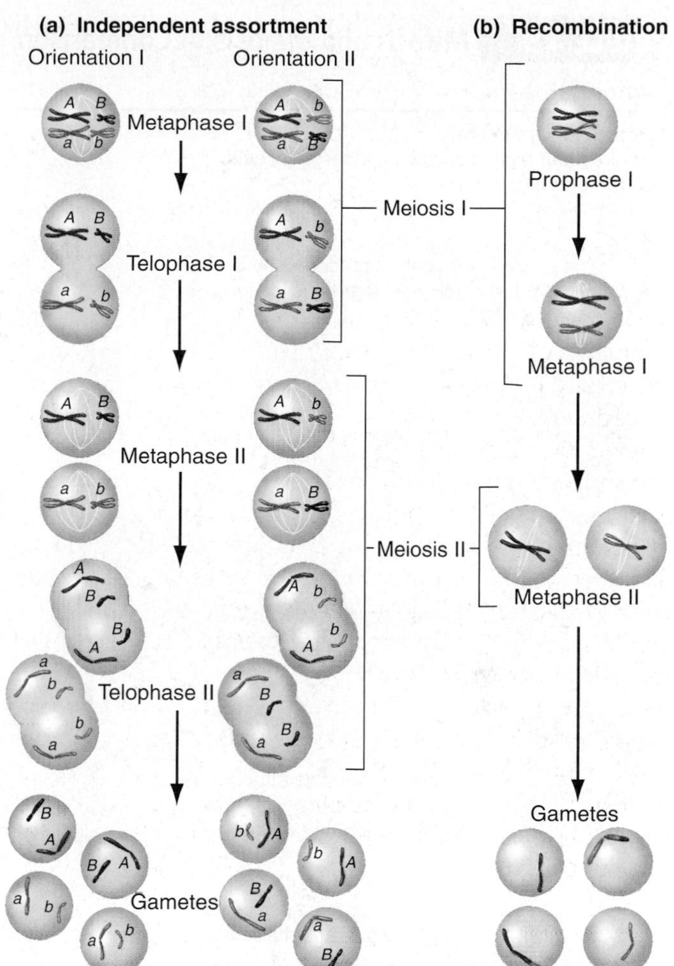

Figure 4.17 How meiosis contributes to genetic diversity. **(a)** The variation resulting from the independent assortment of nonhomologous chromosomes increases with the number of chromosomes in the genome. **(b)** Crossing-over between homologous chromosomes ensures that each gamete produced by an individual is unique.

In contrast to rare mistakes in the segregation of one pair of chromosomes, some hybrid animals carry nonhomologous chromosomes that can never pair up and segregate properly. **Figure 4.16** shows the two dissimilar sets of chromosomes carried by the diploid cells of a mule; the set inherited from the donkey father contains 31 chromosomes, while the set from the horse mother has 32 chromosomes. Viable gametes cannot form in these animals, so mules are sterile.

Meiosis Contributes to Genetic Diversity

The wider the assortment of different gene combinations among members of a species, the greater the chance that at least some individuals will carry combinations of alleles that allow survival in a changing environment. Two aspects of meiosis contribute to genetic diversity in a population. First, since it is purely a matter of chance toward which pole the paternal and maternal homologs of each bivalent migrate during the first meiotic division, different gametes carry a different mix of maternal and paternal chromosomes. **Figure 4.17a** shows how two different patterns of homolog migration produce four different mixes of parental chromosomes in the gametes. The amount of potential variation generated by this random independent assortment increases with the number of chromosomes. In *Ascaris*, for example, where $n = 2$ (the chromosome complement shown in Fig. 4.17a), the random assortment of homologs could produce only 2^2, or 4 types of gametes; in a person, however, where $n = 23$, this same mechanism alone could generate 2^{23}, or more than 8 million genetically

TABLE 4.3 Mitosis and Meiosis: A Comparison

Mitosis

Occurs in somatic cells
Haploid and diploid cells can undergo mitosis
One round of division

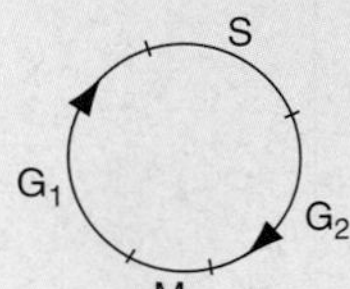

Mitosis is preceded by S phase (chromosome duplication).

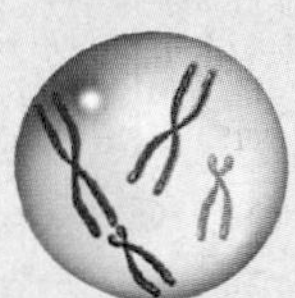

Homologous chromosomes do not pair.

Genetic exchange between homologous chromosomes is very rare.

Sister chromatids attach to spindle fibers from opposite poles during metaphase.

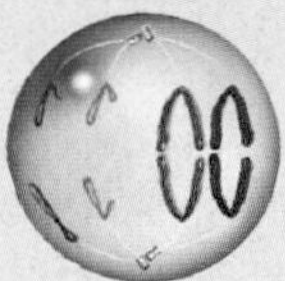

The centromere splits at the beginning of anaphase.

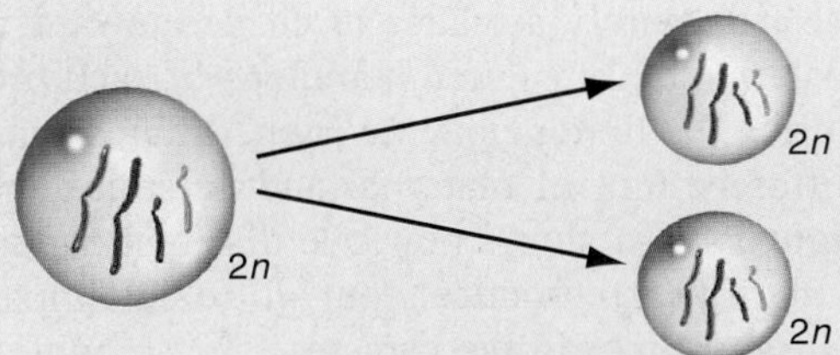

Mitosis produces two new daughter cells, identical to each other and the original cell. Mitosis is thus genetically conservative.

Meiosis

Occurs in germ cells as part of the sexual cycle
Two rounds of division, meiosis I and meiosis II
Only diploid cells undergo meiosis

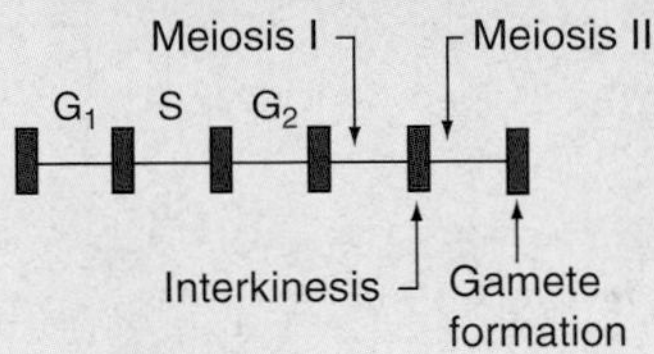

Chromosomes duplicate prior to meiosis I but not before meiosis II.

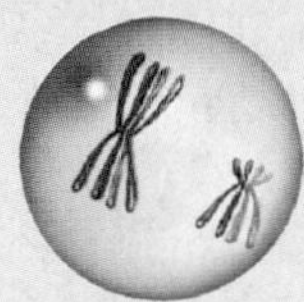

During prophase of meiosis I, homologous chromosomes pair (synapse) along their length.

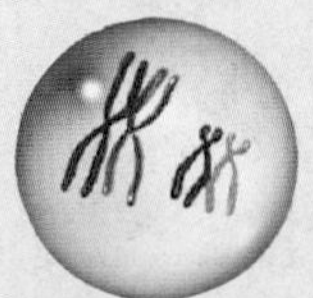

Crossing-over occurs between homologous chromosomes during prophase of meiosis I.

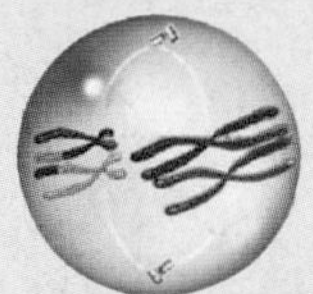

Homologous chromosomes (not sister chromatids) attach to spindle fibers from opposite poles during metaphase I.

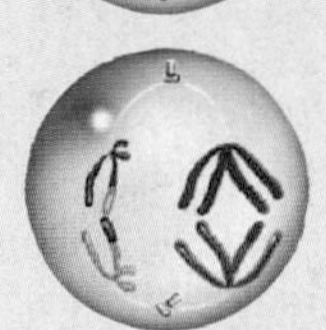

The centromere does not split during meiosis I.

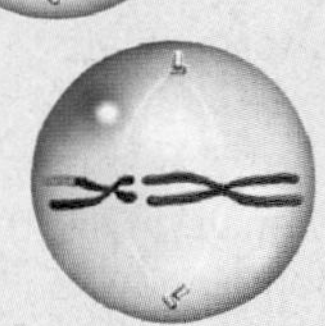

Sister chromatids attach to spindle fibers from opposite poles during metaphase II.

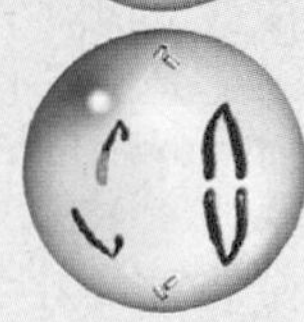

The centromere splits at the beginning of anaphase II.

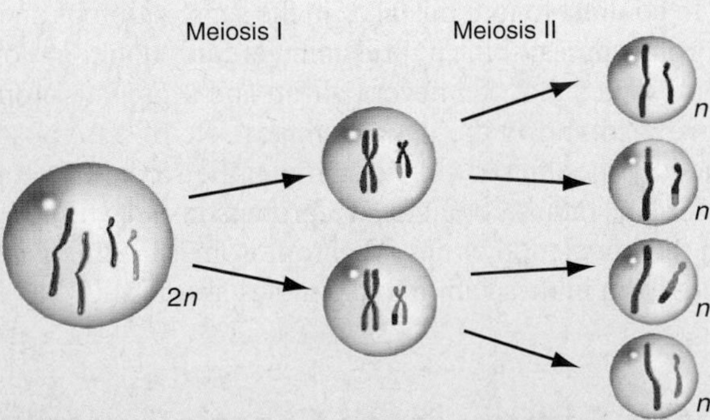

Meiosis produces four haploid cells, one (egg) or all (sperm) of which can become gametes. None of these is identical to each other or to the original cell, because meiosis results in combinatorial change.

different kinds of gametes. A second feature of meiosis—the reshuffling of genetic information through crossing-over during prophase I—ensures an even greater amount of genetic diversity in gametes. Because crossing-over recombines maternally and paternally derived genes, each chromosome in each different gamete produced by a person could consist of different combinations of maternal and paternal information (Fig. 4.17b). Of course, sexual reproduction adds yet another means of generating genetic diversity. At fertilization, any of a vast number of genetically diverse sperm can fertilize an egg with its own distinctive genetic constitution. It is thus not very surprising that, with the exception of identical twins, the 6 billion people in the world are all genetically unique.

Meiosis and Mitosis: A Comparison

Mitosis occurs in all types of eukaryotic cells (that is, cells with a membrane-bounded nucleus) and is a conservative mechanism that preserves the genetic status quo. Mitosis followed by cytokinesis produces growth by increasing the number of cells. It also promotes the continual replacement of roots, stems, and leaves in plants and the regeneration of blood cells, intestinal tissues, and skin in animals. Meiosis, on the other hand, occurs only in sexually reproducing organisms, in just a few specialized germ cells within the reproductive organs that produce haploid gametes. It is not a conservative mechanism; rather, the extensive combinatorial changes arising from meiosis are one source of the genetic variation that fuels evolution. **Table 4.3** illustrates the significant contrasts between the two mechanisms of cell division.

4.4 Gametogenesis Requires Both Mitotic and Meiotic Divisions

In all sexually reproducing animals, the embryonic germ cells (collectively known as the **germ line**) undergo a series of mitotic divisions that yield a collection of specialized diploid cells, which subsequently divide by meiosis to produce haploid cells. As with other biological processes, there are variations on this general pattern. In some species, the haploid cells resulting from meiosis are the gametes themselves, while in other species, those cells must undergo a specific plan of differentiation to fulfill that function. Moreover, in certain organisms, the four haploid products of a single meiosis do not all become gametes. Gamete formation, or **gametogenesis,** thus gives rise to haploid gametes marked not only by the events of meiosis per se but also by cellular events that precede and follow meiosis. Nevertheless, although the specifics of gametogenesis vary widely from species to species as well as from sex to sex within the same species, the general pattern of mitotic divisions followed by a meiotic halving of chromosome number is nearly universal. Here we illustrate gametogenesis with a description of egg and sperm formation in humans. The details of gamete formation in several other organisms appear throughout the book in discussions of specific experimental studies; they also appear in the Genetic Portraits on our website (www.mhhe.com/hartwell3).

Oogenesis in Humans: Asymmetrical Meiotic Divisions Produce One Large Ovum

The end product of egg formation in humans is a large, nutrient-rich ovum whose stored resources can sustain the early embryo. The process, known as **oogenesis** (**Fig. 4.18**), begins when diploid germ cells in the ovary, called **oogonia** (singular, *oogonium*), multiply rapidly by mitosis and produce a large number of **primary oocytes,** which then undergo meiosis. For each primary oocyte, meiosis I results in the formation of two daughter cells that differ in size, so this division is asymmetric. The larger of these cells, the **secondary oocyte,** receives over 95% of the cytoplasm. The other small sister cell is known as the first **polar body.** During meiosis II, the secondary oocyte undergoes another asymmetrical division to produce a large haploid **ovum** and a small, haploid second polar body. The first polar body usually arrests its development and does not undergo the second meiotic division. However, the first polar body does divide in a small proportion of cases, producing two haploid polar bodies. The two (or rarely, three) small polar bodies apparently serve no function and disintegrate, leaving one large haploid ovum as the functional gamete. Thus, only one of the three (or rarely, four) products of a single meiosis serves as a female gamete. A normal human ovum carries 22 autosomes and an X sex chromosome.

Oogenesis begins in the fetus. By six months after conception, the fetal ovaries are fully formed and contain about half a million primary oocytes arrested in the diplotene of meiosis I. These cells, with their homologous chromosomes locked in synapsis, are the only oocytes the female will produce, so a girl is born with all the oocytes she will ever possess. From the onset of puberty, at about age 12, until menopause, some 35–40 years later, most women release one primary oocyte each month (from alternate ovaries), amounting to roughly 480 oocytes released during the reproductive years. The remaining primary oocytes disintegrate during menopause. At ovulation, a released oocyte completes meiosis I and proceeds as far as the metaphase of meiosis II. If the oocyte is then fertilized, that is, penetrated by a sperm nucleus, it quickly completes meiosis II. The nuclei of the sperm and ovum then fuse to form the diploid nucleus of the zygote, and the zygote divides by mitosis to produce a functional embryo. In contrast, unfertilized oocytes exit the body during the menses stage of the menstrual cycle.

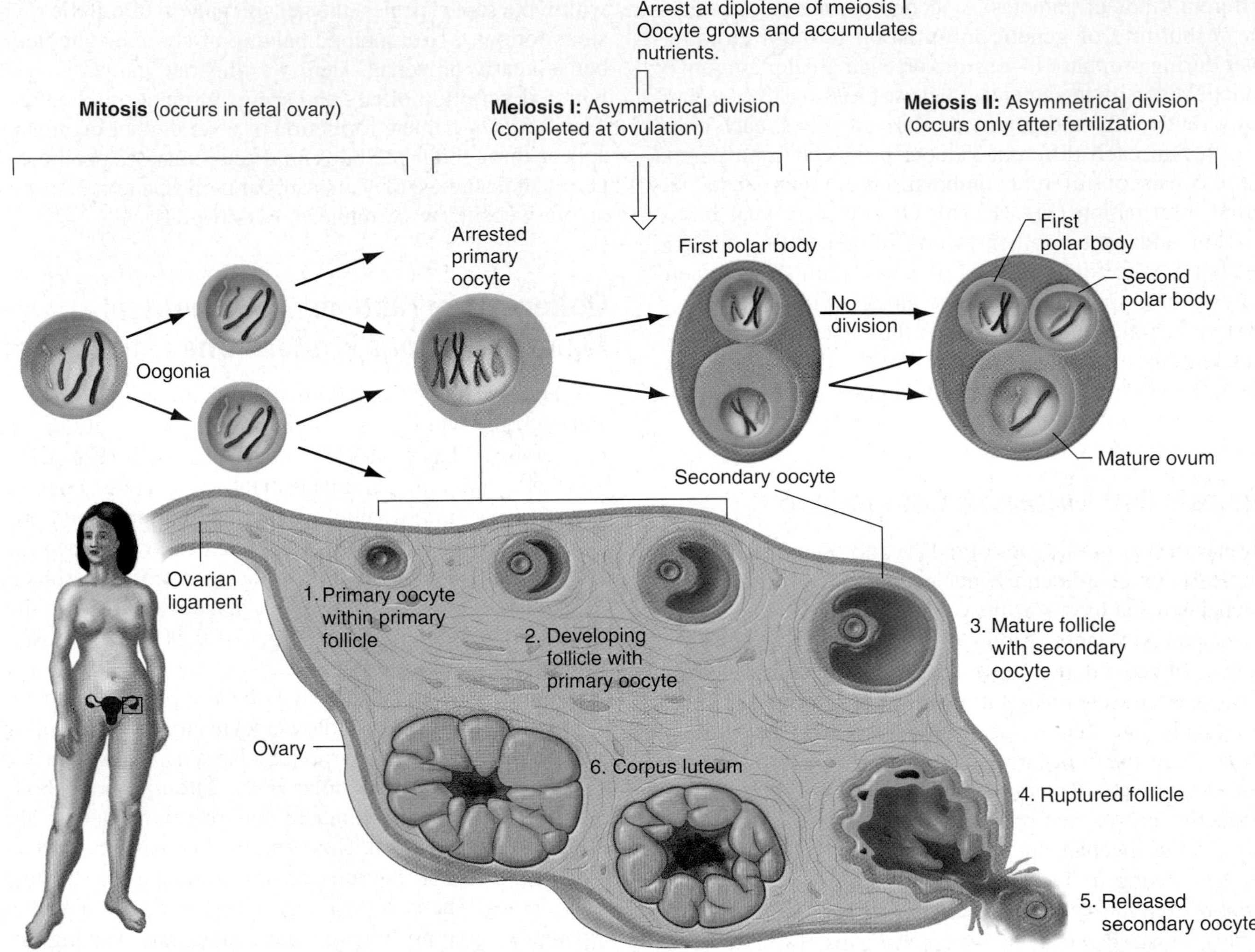

Figure 4.18 **In humans, egg formation begins in the fetal ovaries and arrests during the prophase of meiosis I.** Fetal ovaries contain about 500,000 primary oocytes arrested in the diplotene of meiosis I. If the egg released during a menstrual cycle is fertilized, meiosis is completed. Only one of the three (rarely, four) cells produced by meiosis serves as the functional gamete, or ovum.

Interestingly, the long interval before completion of meiosis in oocytes released by women in their 30s, 40s, and 50s may contribute to an observed correlation between maternal age and meiotic segregational errors, including those that produce trisomies. Women in their mid-20s, for example, run a very small risk of trisomy 21; only 0.05% of children born to women of this age have Down syndrome. During the later childbearing years, however, the risk rapidly rises; at age 35, it is 0.9% of live births, and at age 45, it is 3%. You would not expect this age-related increase in risk if meiosis were completed before the mother's birth.

Spermatogenesis in Humans: Symmetrical Meiotic Divisions Produce Four Sperm

The production of sperm, or **spermatogenesis** (**Fig. 4.19**), begins in the male testes in germ cells known as **spermatogonia.** Mitotic divisions of the spermatogonia produce many diploid cells, the **primary spermatocytes.** Unlike primary oocytes, primary spermatocytes undergo a symmetrical meiosis I, producing two **secondary spermatocytes,** each of which undergoes a symmetrical meiosis II. At the conclusion of meiosis, each original primary spermatocyte thus yields four equivalent haploid **spermatids.** These spermatids then mature by developing a characteristic whiplike tail and by concentrating all their chromosomal material in a head, thereby becoming functional **sperm.** A human sperm, much smaller than the ovum it will fertilize, contains 22 autosomes and *either* an X *or* a Y sex chromosome.

The timing of sperm production differs radically from that of egg formation. The meiotic divisions allowing conversion of primary spermatocytes to spermatids begin only at puberty, but meiosis then continues throughout a man's life. The entire process of spermatogenesis takes about 48–60 days: 16–20 for meiosis I, 16–20 for meiosis II, and 16–20 for the maturation of spermatids into fully functional sperm. Within each testis after puberty, millions of

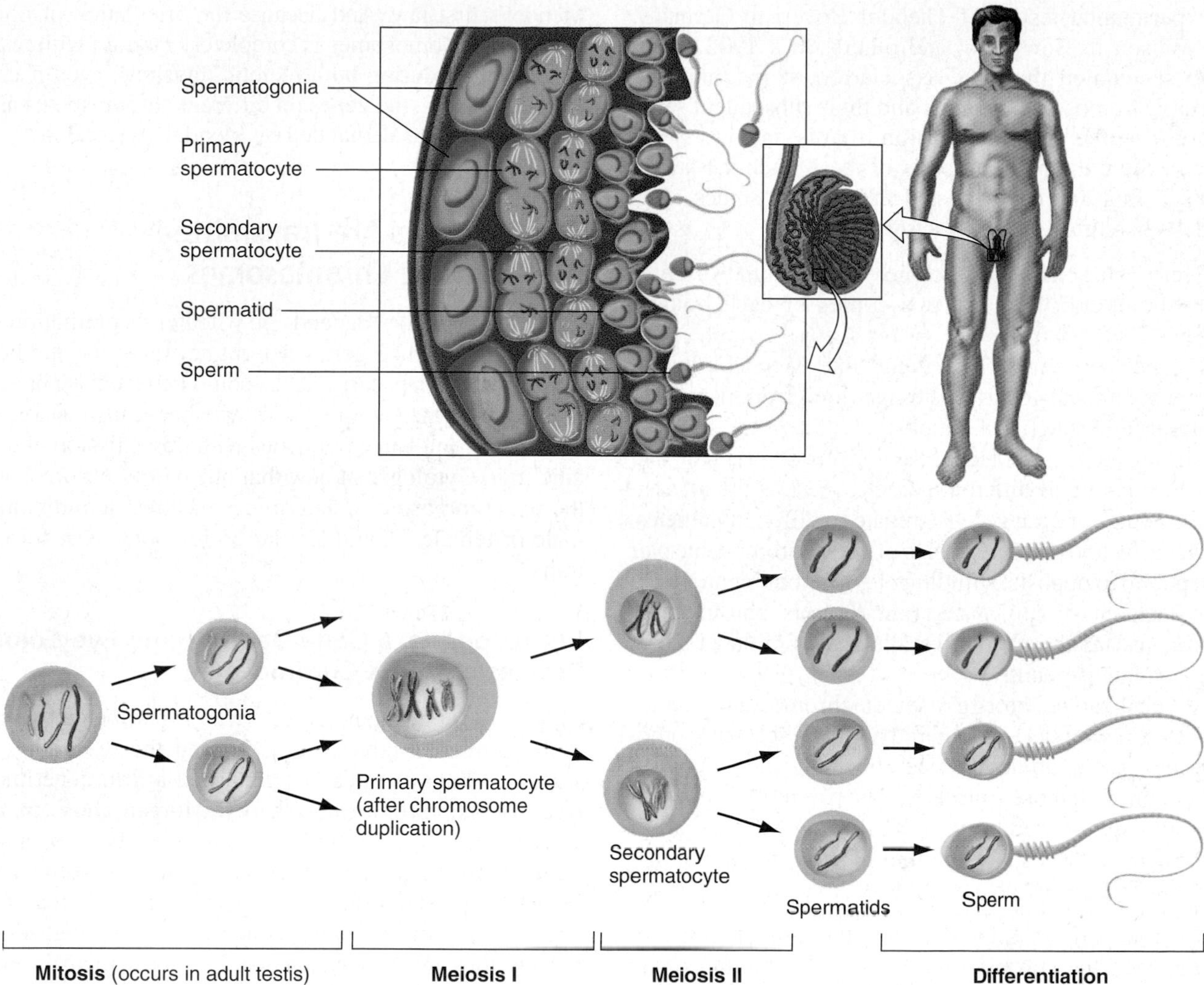

Figure 4.19 Human sperm form continuously in the testes after puberty. Spermatogonia are located near the exterior of seminiferous tubules in a human testis. Once they divide to produce the primary spermatocytes, the subsequent stages of spermatogenesis—meiotic divisions in the spermatocytes and maturation of spermatids into sperm—occur successively closer to the middle of the tubule. This allows for release of the mature sperm into the central lumen of the tubule for ejaculation.

sperm are always in production, and a single ejaculate can contain up to 300 million. Over a lifetime, a man can produce billions of sperm, almost equally divided between those bearing an X and those bearing a Y chromosome.

4.5 Validation of the Chromosome Theory

So far, we have presented two circumstantial lines of evidence in support of the chromosome theory of inheritance. First, the phenotype of sexual identity is associated with the inheritance of particular chromosomes. Second, the events of mitosis, meiosis, and gametogenesis ensure a constant number of chromosomes in the somatic cells of all members of a species over time; one would expect the genetic material to exhibit this kind of stability even in organisms with very different modes of reproduction. Final acceptance of the chromosome theory depended on researchers going beyond the circumstantial evidence to a rigorous demonstration of two key points: (1) that the inheritance of genes corresponds with the inheritance of chromosomes in every detail and (2) that the transmission of particular chromosomes coincides with the transmission of specific traits other than sex determination.

The Chromosome Theory Correlates Mendel's Laws with Chromosome Behavior During Meiosis

Walter Sutton first outlined the chromosome theory of inheritance in 1902–1903, building on the theoretical ideas and experimental results of Theodor Boveri in Germany, E. B. Wilson in New York, and others. In a 1902 paper, Sutton speculated that "the association of paternal and maternal chromosomes in pairs and their subsequent separation during the reducing division [that is, meiosis I] . . . may constitute the physical basis of the Mendelian law of heredity." In 1903, he suggested that chromosomes carry Mendel's hereditary units because

1. Every cell contains two copies of each kind of chromosome, and there are two copies of each kind of gene.
2. The chromosome complement, like Mendel's genes, appears unchanged as it is transmitted from parents to offspring through generations.
3. During meiosis, homologous chromosomes pair and then separate to different gametes, just as the alternative alleles of each gene segregate to different gametes.
4. Maternal and paternal copies of each chromosome pair separate to opposite spindle poles without regard to the assortment of any other homologous chromosome pair, just as the alternative alleles of unrelated genes assort independently.
5. At fertilization, an egg's set of chromosomes unites with a randomly encountered sperm's set of chromosomes, just as alleles obtained from one parent unite at random with those from the other parent.
6. In all cells derived from the fertilized egg, one-half of the chromosomes and one-half of the genes are of maternal origin, the other half of paternal origin.

The two parts of **Table 4.4** show the intimate relationship between the chromosome theory of inheritance and Mendel's laws of segregation and independent assortment. If Mendel's genes for pea shape and pea color are assigned to different (that is, nonhomologous) chromosomes, the behavior of chromosomes can be seen to parallel the behavior of genes. Walter Sutton's observation of these parallels led him to propose that chromosomes and genes are physically connected in some manner. Meiosis ensures that each gamete will contain only a single chromatid of a bivalent and thus only a single allele of any gene on that chromatid (Table 4.4a). The independent behavior of two bivalents during meiosis means that the genes carried on different chromosomes will assort into gametes independently (Table 4.4b).

From a review of Fig. 4.17 (on p. 101), which follows two different chromosome pairs through the process of meiosis, you might wonder whether crossing-over abolishes the clear correspondence between Mendel's laws and the movement of chromosomes. The answer is no. Each chromatid of a homologous chromosome pair contains only one copy of a given gene, and only one chromatid from each pair of homologs is incorporated into each gamete. Because alternative alleles remain on different chromatids even after crossing-over has occurred, alternative alleles still segregate to different gametes as demanded by Mendel's first law. And because the orientation of nonhomologous chromosomes is completely random with respect to each other during both meiotic divisions, even if crossing-over occurs, the genes on different chromosomes assort independently, as demanded by Mendel's second law.

Specific Traits Are Transmitted with Specific Chromosomes

The fate of a theory depends on whether its predictions can be validated. Since genes determine traits, the prediction that chromosomes carry genes could be tested by breeding experiments that would show whether transmission of a specific chromosome coincides with transmission of a specific trait. Cytologists knew that one pair of chromosomes, the sex chromosomes, determines whether an individual is male or female. Would similar correlations exist for other traits?

In *Drosophila,* a Gene Determining Eye Color Resides on the X Chromosome

Thomas Hunt Morgan, an American experimental biologist with training in embryology, headed the research group whose findings eventually established a firm experimental base for the chromosome theory. Morgan chose to work with the fruit fly *Drosophila melanogaster* because it is extremely prolific and has a very short generation time, taking only 12 days to develop from a fertilized egg into a mature adult capable of producing hundreds of offspring. Morgan fed his flies mashed bananas and housed them in empty milk bottles capped with wads of cotton.

In 1910, a white-eyed male appeared among a large group of flies with brick-red eyes. A mutation had apparently altered a gene determining eye color, changing it from the normal wild-type allele specifying red to a new allele that produced white. When Morgan allowed the white-eyed male to mate with its red-eyed sisters, all the flies of the F_1 generation had red eyes; the red allele was clearly dominant to the white (**Fig. 4.20,** cross A).

Establishing a pattern of nomenclature for *Drosophila* geneticists, Morgan named the gene identified by the abnormal white eye color, the *white* gene, for the mutation that revealed its existence. The normal wild-type allele of the *white* gene, abbreviated w^+, is for brick-red eyes, while the counterpart mutant *w* allele results in white eye color. The superscript + signifies the wild type. By writing the gene name and abbreviation in lowercase, Morgan symbolized that the mutant *w* allele is recessive to the wild-type w^+. (If a mutation results in a dominant non-wild-type phenotype, the first letter of the gene name or of its abbreviation is written in uppercase; thus the mutation known as *Bar* eyes is dominant to the wild-type Bar^+ allele. See the *Guidelines for Gene Nomenclature* on p. N–1, directly following Chapter 22.)

TABLE 4.4 How the Chromosome Theory of Inheritance Explains Mendel's Laws

(a) The Law of Segregation

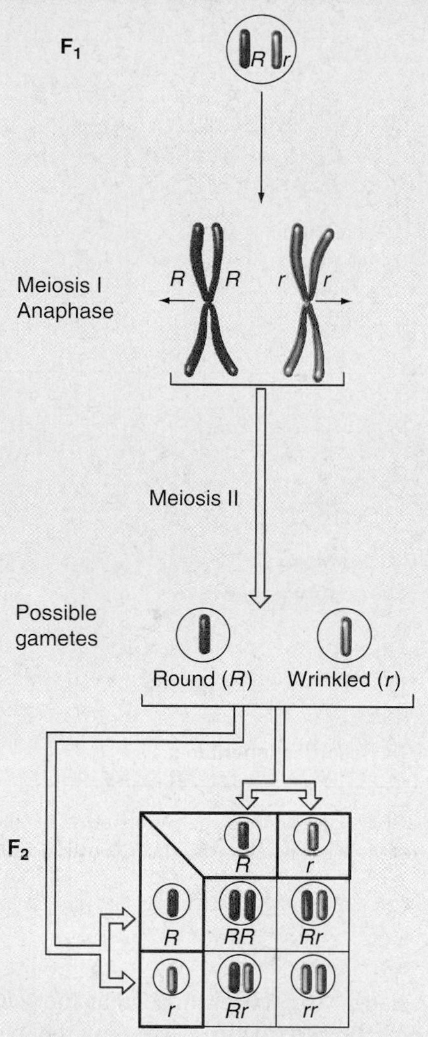

Recall that Mendel's law of segregation was based on the results obtained by the self-fertilization of F_1 hybrid peas for traits such as pea shape. Sutton assumed that in the hybrids, the allele for round-seeded peas (*R*) is found on one chromosome, and the allele for wrinkled peas (*r*) is on the homologous chromosome. The pairing between the two homologous chromosomes during prophase through metaphase of meiosis I makes sure that the homologs will separate to opposite spindle poles during anaphase of meiosis I. Sutton realized this process ensures that at the end of meiosis II, two types of gametes will be produced: half will have *R*, and half will have *r*, but no gametes will have both alleles. Thus, the separation of homologous chromosomes at meiosis I corresponds to the segregation of alleles. As the Punnett square shows, fertilization of 50% *R* and 50% *r* eggs with the same proportion of *R* and *r* pollen leads to Mendel's 3:1 ratio in the F_2 generation.

(b) The Law of Independent Assortment

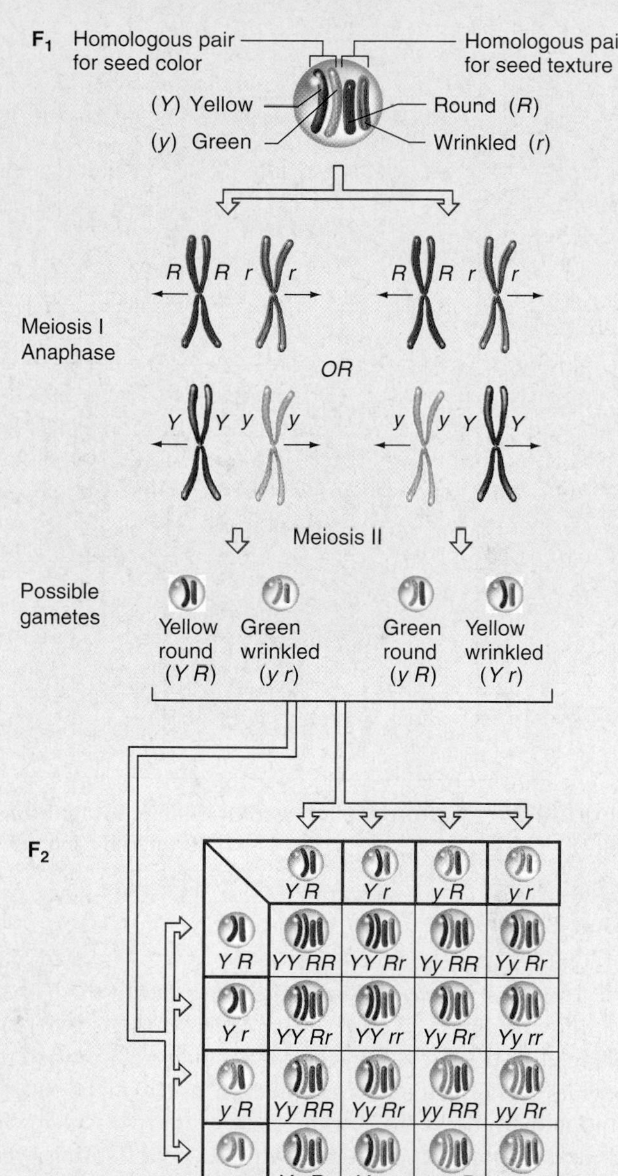

Sutton rationalized Mendel's law of independent assortment as follows. One pair of homologous chromosomes carries the gene for seed texture. A second pair of homologous chromosomes carries the gene for seed color. Each homologous pair aligns at random at the metaphase plate during meiosis I, orienting its two chromosomes toward opposite poles of the cell independently of all other homologous pairs. This means that there are two equally likely configurations for any two chromosome pairs in terms of their migration toward the poles during anaphase of meiosis I. As a result, a dihybrid individual will produce four equally likely types of gametes with regard to the two traits in question. And as the Punnett square affirms, this means that the assortment of traits carried by one pair of chromosomes is independent of the assortment of traits carried by all other homologous pairs. Mendel's 9:3:3:1 ratio reflects this independent assortment.

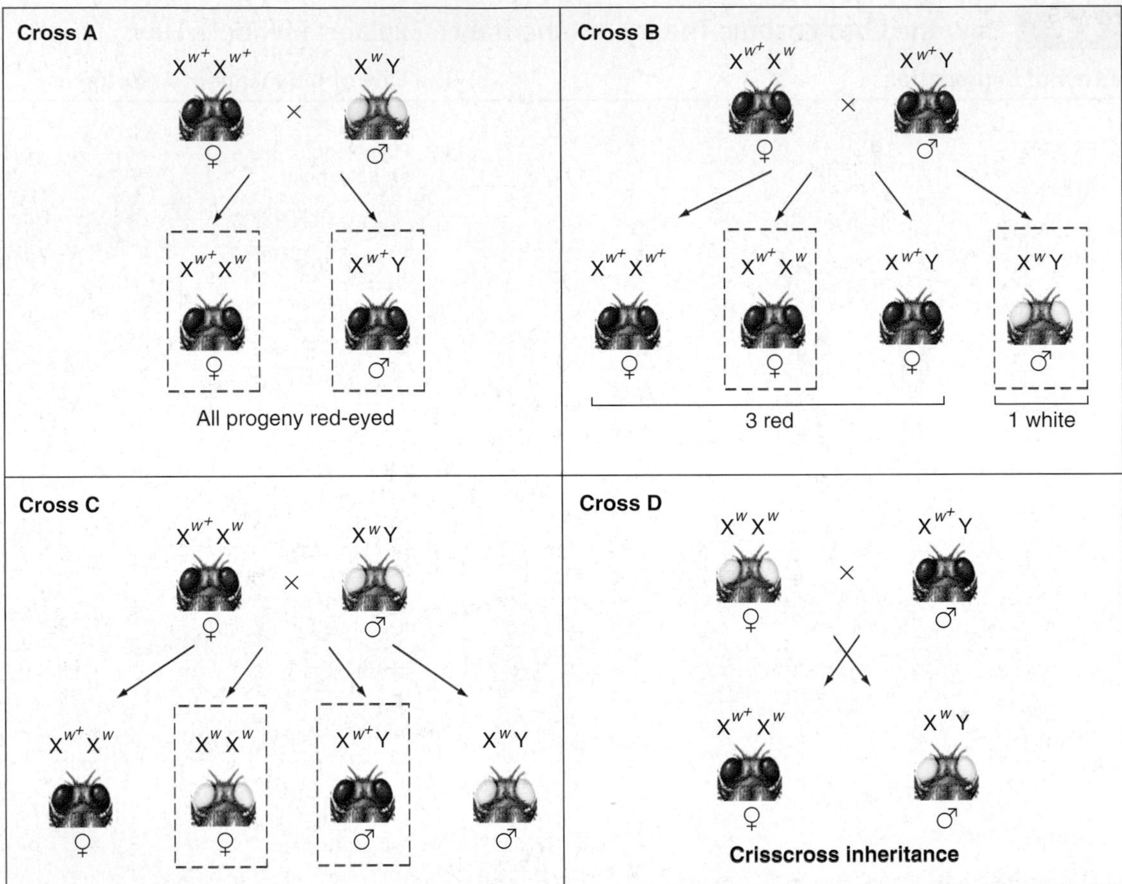

Figure 4.20 A *Drosophila* eye color gene is located on the X chromosome. X-linkage explains the inheritance of alleles of the *white* gene in this series of crosses performed by Thomas Hunt Morgan. The progeny of Crosses A, B, and C outlined with green dotted boxes are those used as the parents in the next cross of the series.

Morgan then crossed the red-eyed males of the F_1 generation with their red-eyed sisters (Fig. 4.20, cross B) and obtained an F_2 generation with the predicted 3:1 ratio of red to white eyes. But there was something askew in the pattern: Among the red-eyed offspring, there were two females for every one male, and all the white-eyed offspring were males. This result was surprisingly different from the equal transmission to both sexes of the Mendelian traits discussed in Chapters 2 and 3. In these fruit flies, the ratio of various phenotypes was not the same in male and female progeny. By mating F_2 red-eyed females with their white-eyed brothers (Fig. 4.20, cross C), Morgan obtained some females with white eyes, which then allowed him to mate a white-eyed female with a red-eyed wild-type male (Fig. 4.20, cross D). The result was exclusively red-eyed daughters and white-eyed sons. The pattern seen in cross D is known as **crisscross inheritance** because the males inherit their eye color from their mothers, while the daughters inherit their eye color from their fathers. Note in Fig. 4.20 that the results of the reciprocal crosses red female × white male (cross A) and white female × red male (cross D) are not identical, again in contrast with Mendel's findings.

From the data, Morgan reasoned that the *white* gene for eye color is **X linked,** that is, carried by the X chromosome. (Note that while symbols for genes and alleles are italicized, symbols for chromosomes are not.) The Y chromosome carries no allele of this gene for eye color. Males, therefore, have only one copy of the gene, which they inherit from their mother along with their only X chromosome; their Y chromosome must come from their father. Thus, males are **hemizygous** for this eye color gene, because their diploid cells have half the number of alleles carried by the female on their two X chromosomes.

If the single *white* gene on the X chromosome of a male is the wild-type w^+ allele, he will have red eyes and a genotype that can be written $X^{w+}Y$. (Here we designate the chromosome [X or Y] together with the allele it carries, to emphasize that certain genes are X linked.) In contrast to an $X^{w+}Y$ male, a hemizygous X^wY male whose lone X carries the *w* allele would have a phenotype of white eyes. Females with two X chromosomes can be one of three genotypes: X^wX^w (white-eyed), X^wX^{w+} (red-eyed because w^+ is dominant to *w*), or $X^{w+}X^{w+}$ (red-eyed).

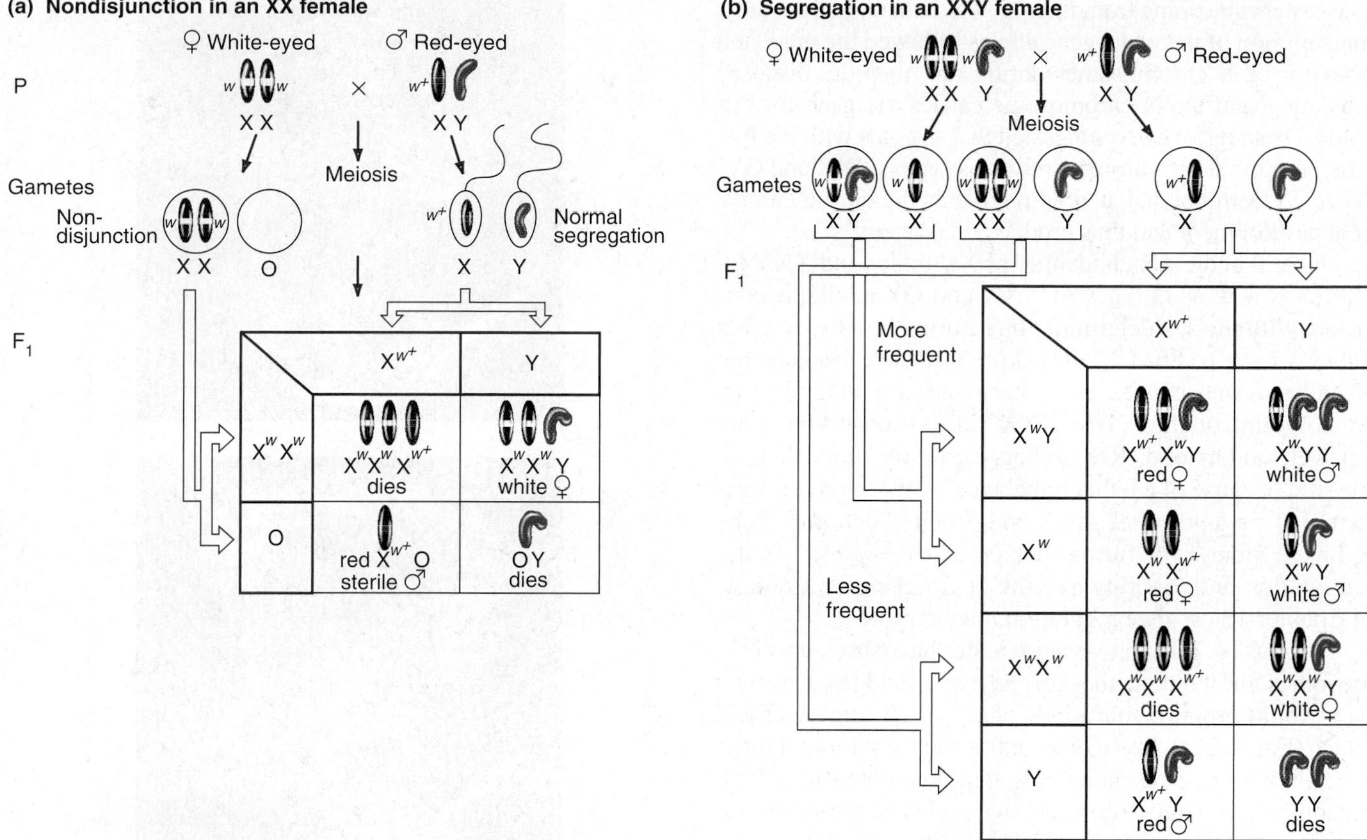

Figure 4.21 **Nondisjunction: Rare mistakes in meiosis help confirm the chromosome theory.** **(a)** Rare events of nondisjunction in an XX female produce XX and O eggs. The results of normal disjunction in the female are not shown. XO males are sterile because the missing Y chromosome is needed for male fertility in *Drosophila*. **(b)** In an XXY female, the three sex chromosomes can pair and segregate in two ways, producing progeny with unusual sex chromosome complements.

As shown in Fig. 4.20, Morgan's assumption that the gene for eye color is X linked explains the results of his breeding experiments. Crisscross inheritance, for example, results because the only X chromosome in sons of a white-eyed mother ($X^w X^w$) must carry the *w* allele, so the sons will be white-eyed. In contrast, because daughters of a red-eyed (X^{w+} Y) father must receive a w^+-bearing X chromosome from their father, they will have red eyes.

Analysis of Rare Mistakes in Meiosis Provided Further Support for the Chromosome Theory

Although Morgan's work strongly supported the hypothesis that the gene for eye color lies on the X chromosome, he himself continued to question the absolute validity of the chromosome theory until Calvin Bridges, one of his top students, found another key piece of evidence. Bridges repeated the cross Morgan had performed between white-eyed females and red-eyed males, but this time he did the experiment on a larger scale. As expected, the progeny of this cross consisted mostly of red-eyed females and white-eyed males. However, about 1 in every 2000 males had red eyes, and about the same small fraction of females had white eyes.

Bridges hypothesized that these exceptions arose through rare events in which the X chromosomes fail to separate during meiosis in females. He called such failures in chromosome segregation *nondisjunction.* As **Fig. 4.21a** shows, nondisjunction would result in some eggs with two X chromosomes and others with none. Fertilization of these chromosomally abnormal eggs could produce four types of zygotes: XXY (with two X chromosomes from the egg and a Y from the sperm), XXX (with two Xs from the egg and one X from the sperm), XO (with the lone sex chromosome from the sperm and no sex chromosome from the egg), and OY (with the only sex chromosome again coming from the sperm). When Bridges examined the sex chromosomes of the rare white-eyed females produced in his large-scale cross, he found that they were indeed XXY individuals who must have received two X chromosomes and with them two *w* alleles from their white-eyed $X^w X^w$ mothers. The exceptional red-eyed males emerging from the cross were XO; their eye color showed that they must have obtained their sole sex chromosome from their X^{w+} Y fathers. In this study, transmission of the *white* gene alleles followed the predicted behavior of X chromosomes during rare meiotic mistakes, indicating that the X chromosome carries the gene for eye color. These results also suggested that zygotes with the two other abnormal sex chromosome karyotypes (XXX and OY) expected from nondisjunction in females die during embryonic development and thus produce no progeny.

Note that the sexual identity of XO males and XXY females, as well as the lack of XXX and OY adults, is consistent with the sex determination information presented in Table 4.1 on p. 86. OY flies do not survive because the X chromosome carries genes necessary for viability that are not found on the Y. Why XXX flies do not survive is not yet well understood. Researchers currently think that the flies die because of a lethal imbalance between proteins determined by autosomal genes and proteins determined by X-linked genes. For further details of *Drosophila* sex determination and viability, see the comprehensive example in Chapter 18 on gene regulation in eukaryotes.

Because XXY white-eyed females have three sex chromosomes rather than the normal two, Bridges reasoned they would produce four kinds of eggs: XY and X, or XX and Y (Fig. 4.21b). You can visualize the formation of these four kinds of eggs by imagining that when the three chromosomes pair and disjoin during meiosis, two chromosomes must go to one pole and one chromosome to the other. With this kind of segregation, there are only two possible results; either one X and the Y go to one pole and the second X to the other (yielding XY and X gametes), or the two Xs go to one pole and the Y to the other (yielding XX and Y gametes). The first of these two scenarios occurs more often because it comes about when the two similar X chromosomes pair with each other, ensuring that they will go to opposite poles during the first meiotic division. The second, less likely possibility happens only if the two X chromosomes fail to pair with each other.

Bridges next predicted that fertilization of these four kinds of eggs by normal sperm would generate an array of sex chromosome karyotypes associated with specific eye color phenotypes in the progeny. Bridges verified all his predictions when he analyzed the eye color and sex chromosomes of a large number of offspring. For instance, he showed cytologically that all of the white-eyed females emerging from the cross in Fig. 4.21b had two X chromosomes and one Y chromosome, while one-half of the white-eyed males had a single X chromosome and two Y chromosomes. Bridges' painstaking observations provided compelling evidence that specific genes do in fact reside on specific chromosomes.

Figure 4.22 Red-green colorblindness is an X-linked recessive trait in humans. How the world looks to a person with either normal color vision (*top*) or a kind of red-green colorblindness known as deuteranopia (*bottom*).

X- and Y-Linked Traits in Humans Are Identified by Pedigree Analysis

A person unable to tell red from green would find it nearly impossible to distinguish the rose, scarlet, and magenta in the flowers of a garden bouquet from the delicately variegated greens in their foliage, or to complete a complex electrical circuit by fastening red-clad metallic wires to red ones and green to green. Such a person has most likely inherited some form of red-green colorblindness, a recessive condition that runs in families and affects mostly males. Among Western Europeans in North America and Europe, 8% of men but only 0.44% of women have this vision defect. **Figure 4.22** suggests to readers with normal color vision what people with red-green colorblindness actually see.

In 1911, E. B. Wilson, a contributor to the chromosome theory of inheritance, combined familiarity with studies of colorblindness and recent knowledge of sex determination by the X and Y chromosomes to make the first assignment of a human gene to a particular chromosome. The gene for red-green colorblindness, he said, lies on the X because the condition usually passes from a maternal grandfather through an unaffected carrier mother to roughly 50% of the grandsons.

(a) X-linked recessive: Hemophilia

I Queen Victoria | Prince Albert — ⊙ Carrier ■ Hemophiliac

II Victoria, Edward VII, Alfred, Louise, Arthur, Beatrice, Helena; Alice | Louis IV; Leopold | Helene

III Alix | Nicholas II

IV Alexis, Rupert

(b) X-linked dominant: Hypophosphatemia

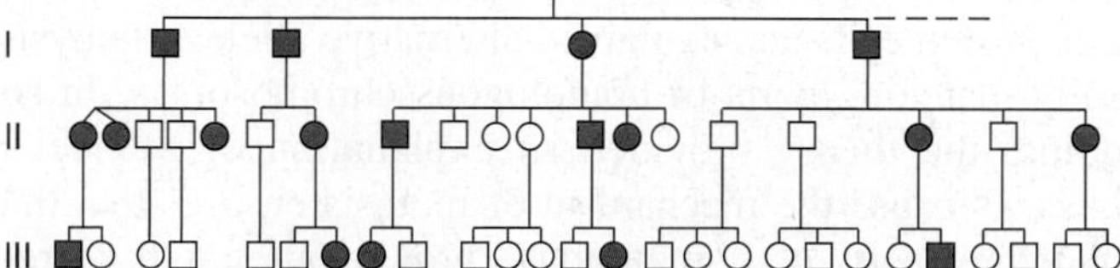

Figure 4.23 X-linked traits may be recessive or dominant. (a) Pedigree showing inheritance of the recessive X-linked trait hemophilia in Queen Victoria's family. (b) Pedigree showing the inheritance of the dominant X-linked trait hypophosphatemia, commonly referred to as vitamin D–resistant rickets.

TABLE 4.5 Pedigree Patterns Suggesting Sex-Linked Inheritance

X-Linked Recessive Trait

1. The trait appears in more males than females since a female must receive two copies of the rare defective allele to display the phenotype, whereas a hemizygous male with only one copy will show it.
2. The mutation will never pass from father to son because sons receive only a Y chromosome from their father.
3. An affected male passes the X-linked mutation to all his daughters, who are thus unaffected carriers. One-half of the sons of these carrier females will inherit the defective allele and thus the trait.
4. The trait often skips a generation as the mutation passes from grandfather through a carrier daughter to grandson.
5. The trait can appear in successive generations when a sister of an affected male is a carrier. If she is, one-half her sons will be affected.
6. With the rare affected female, all her sons will be affected and all her daughters will be carriers.

X-Linked Dominant Trait

1. More females than males show the aberrant trait.
2. The trait is seen in every generation because it is dominant.
3. All the daughters but none of the sons of an affected male will be affected. This criterion is the most useful for distinguishing an X-linked dominant trait from an autosomal dominant trait.
4. One-half the sons and one-half the daughters of an affected female will be affected.

Y-Linked Inheritance

1. The trait is seen only in males.
2. All male descendants of an affected man will exhibit the trait.
3. Not only do females not exhibit the trait, they also cannot transmit it.

Several years after Wilson made this gene assignment, pedigree analysis established that various forms of hemophilia, or "bleeders disease" (in which the blood fails to clot properly), also result from mutations on the X chromosome that give rise to a relatively rare, recessive trait. In this context, rare means "infrequent in the population." The family histories under review, including one following the descendants of Queen Victoria of England (**Fig. 4.23a**), showed that relatively rare X-linked traits appear more often in males than in females and often skip generations. The clues that suggest X-linked recessive inheritance in a pedigree are summarized in **Table 4.5**.

Unlike colorblindness and hemophilia, some—although very few—of the known rare mutations on the X chromosome are dominant to the wild-type allele. With such dominant X-linked mutations, more females than males show the aberrant phenotype. This is because all the daughters of an affected male but none of the sons will have the condition, while one-half the sons and one-half the daughters of an affected female will receive the dominant allele and therefore show the phenotype (see Table 4.5). Vitamin D–resistant rickets, or hypophosphatemia, is an example of an X-linked dominant trait. Figure 4.23b presents the pedigree of a family affected by this disease.

Theoretically, phenotypes caused by mutations on the Y chromosome should also be identifiable by pedigree analysis. Such traits would pass from an affected father to all of his sons, and from them to all future male descendants. Females would neither exhibit nor transmit a Y-linked phenotype (see Table 4.5). However, besides the determination of maleness itself, as well as a contribution to sperm formation and thus male fertility, no clear-cut Y-linked visible traits have turned up. The paucity of known Y-linked traits in humans reflects the fact that the small Y chromosome contains very few genes. Indeed, one would expect the Y chromosome to have only a limited effect on phenotype because normal XX females do perfectly well without it.

Autosomal Genes Can Also Affect Phenotypic Differences Between the Sexes

Not all genes that produce differences in the two sexes reside on the X or Y chromosomes. Some autosomal genes

Figure 4.24 Male pattern baldness, a sex-influenced trait. Apparently, geneticists will not have therapies for male pattern baldness even by the twenty-third century. This trait is thought to be governed mainly by an allele of an autosomal gene that is dominant in men and recessive in women.

govern traits that appear in one sex but not the other, or traits that are expressed differently in the two sexes.

Sex-limited traits affect a structure or process that is found in one sex but not the other. Mutations in genes for sex-limited traits can influence only the phenotype of the sex that expresses those structures or processes. A curious example of a sex-limited trait occurs in *Drosophila* males homozygous for an autosomal recessive mutation known as *stuck,* which affects the ability of mutant males to retract their penis and release the claspers by which they hold on to female genitalia during copulation. The mutant males have difficulty separating from females after mating. In extreme cases, both individuals die, forever caught in their embrace. Because females lack penises and claspers, homozygous *stuck* mutant females can mate normally.

Sex-influenced traits show up in both sexes, but expression of such traits may differ between the two sexes because of hormonal differences. Pattern baldness, a condition in which hair is lost prematurely from the top of the head but not from the sides (**Fig. 4.24**), is a sex-influenced trait in humans. Men heterozygous for the balding allele lose their hair while still in their 20s, whereas heterozygous women do not show any significant hair loss. In contrast, homozygotes in both sexes become bald (though the onset of baldness in homozygous women is usually much later in life than in homozygous men). This sex-influenced trait is thus dominant in men, recessive in women.

The Chromosome Theory Integrates Many Aspects of Gene Behavior

Mendel had assumed that genes are located in cells. The chromosome theory assigned the genes to a specific structure within cells and explained alternative alleles as physically matching parts of homologous chromosomes. In so doing, the theory provided an explanation of Mendel's laws. Because the mechanism of meiosis ensures that the matching parts of homologous chromosomes will segregate to different gametes (except in rare instances of nondisjunction), meiosis in specialized germ cells ensures that alleles segregate from each other as predicted by Mendel's first law. Because each homologous chromosome pair aligns independently of all others at meiosis I, genes carried on different chromosomes will assort independently, as predicted by Mendel's second law. The chromosome theory is also able to explain the creation of new alleles through mutation, a spontaneous change in a particular gene, that is, in a particular part of a chromosome. If a mutation occurs in the germ line, it can be transmitted to subsequent generations. Finally, through mitotic cell division in the embryo and after birth, each cell in a multicellular organism receives the same chromosomes—and thus the same maternal and paternal alleles of each gene—as the zygote received from the egg and sperm at fertilization. This means that, with a few exceptions (to be discussed in later chapters), an individual's genome—the chromosomes and genes he or she carries—remains constant throughout life.

Connections

T. H. Morgan and his students, collectively known as the *Drosophila* group, acknowledged that Mendelian genetics could exist independently of chromosomes. "Why then, we are often asked, do you drag in the chromosomes? Our answer is that since the chromosomes furnish exactly the kind of mechanism that Mendelian laws call for, and since there is an ever-increasing body of information that points clearly to the chromosomes as the bearers of the Mendelian factors, it would be folly to close one's eyes to so patent a relation. Moreover, as biologists, we are interested in heredity not primarily as a mathematical formulation, but rather as a problem concerning the cell, the egg, and the sperm."

The *Drosophila* group went on to find several X-linked mutations in addition to white eyes. One made the body yellow instead of brown, another shortened the wings, yet another made bent instead of straight body bristles. These findings raised several compelling questions. First, if the genes for all of these traits are physically linked together on the X chromosome, does this linkage affect their ability to assort independently, and if so, how?

Second, does each gene have an exact chromosomal address, and if so, does this specific location in any way affect its transmission? In Chapter 5 we describe how the *Drosophila* group and others analyzed the transmission patterns of genes on the same chromosome in terms of known chromosome movements during meiosis, and then used the information obtained to localize genes at specific chromosomal positions.

Essential Concepts

1. *Chromosomes* are cell organelles specialized for the storage and transmission of genetic material. Genes are located on chromosomes and travel with them during cell division and gamete formation.
2. In sexually reproducing organisms, *somatic cells* carry a precise number of homologous pairs of chromosomes, which is characteristic of the species. One chromosome of each pair is of maternal origin; the other, paternal.
3. During the first division of *meiosis,* homologous chromosomes in *germ cells* segregate from each other. As a result, each gamete receives one member of each matching pair, as predicted by Mendel's first law.
4. Also during the first meiotic division, the independent alignment of each pair of homologous chromosomes at the cellular midplane results in the independent assortment of genes carried on different chromosomes, as predicted by Mendel's second law.
5. *Crossing-over* and the independent alignment of homologs during the first meiotic division generate diversity.
6. The second meiotic division generates gametes with a *haploid* number of chromosomes (n).
7. Fertilization—the union of egg and sperm—restores the *diploid* number of chromosomes ($2n$) to the zygote.
8. *Mitosis* underlies the growth and development of the individual. Through mitosis, diploid cells produce identical diploid progeny cells. During mitosis, the sister chromatids of every chromosome separate to each of two daughter cells; before the next cell division, the chromosomes again duplicate to form sister chromatids.
9. The discovery of *sex linkage,* by which specific genes could be assigned to the X chromosome, provided considerable support for the chromosome theory of inheritance. Later, the analysis of rare mistakes in meiotic chromosome segregation (*nondisjunction*) yielded more detailed proof that specific genes are carried on specific chromosomes.

On Our Website

www.mhhe.com/hartwell3
Chapter 4

Annotated Suggested Readings and Links to Other Websites

- More on the history of the chromosome theory of inheritance
- Mechanisms of sex determination in various organisms
- Recent research into the biochemical mechanisms underlying mitosis and meiosis
- Further examples of sex-linked inheritance in humans

Specialized Topics

- Chromosome behavior during mitosis and meiosis

Social and Ethical Issues

1. Gregg and Jennifer are in their late 20s and just had a baby born with *cri du chat* syndrome. The physical and mental manifestations are microcephaly (small head), mental retardation, and a catlike cry. This dominant condition is due to a loss of a portion of one copy of chromosome 5 and is readily detectable by karyotype analysis. The policy of the health organization to which Gregg and Jennifer belong is to do routine amniocentesis on pregnant women over 35 years of age and on patients from a family with a history of a

detectable genetic condition. (See the Genetics and Society box on p. 87 for a description of amniocentesis.) Jennifer fit into neither category and so was not tested. Jennifer and Gregg feel their doctor was remiss in not ordering the amniocentesis, and they want to bring a malpractice suit against the doctor and the health organization to which they belong. Is there just cause to bring this lawsuit? Should people have a say in which tests they receive? Should doctors be required to give all tests possible? What effect could this have on the cost of health insurance?

2. Jack and Jessica are going to have a baby, and because Jessica is 38 years old, she had amniocentesis. The karyotype results indicate that the baby has three copies of chromosome 21 (trisomy 21 or Down syndrome). A genetic counselor and doctor have discussed with the couple the potential emotional and financial concerns in raising a Down syndrome child at home versus sending the child to an institution (many of which are state-supported). Jack and Jessica have decided to continue the pregnancy and raise their child at home. Given their decision to have the child knowing it will be affected, should they be eligible to receive state funds to defray the costs of special educational and therapeutic programs that they may need to raise their child at home? Should society pay for expenses accrued by children with genetic disorders if it is possible to screen for these disorders before birth?

3. In 1965, a study on mentally subnormal males with violent tendencies in a Scottish institution was published; it indicated that 7 of 197, or 3.5%, had the karyotype XYY. These individuals were, on average, 6 inches taller than the XY males in the institution. This percentage (3.5%) of XYY males in this institution seemed abnormally high when compared with that seen in the general population (about 0.1%). A large-scale study was initiated in 1968 in Boston to screen newborns for the XYY karyotype and to follow up on the development of these boys. In 1974, a group of scientists opposed this study, feeling that it was based on preliminary observations and could be damaging to the boys involved. A major concern was that the boys would be unfairly branded as abnormal and violent. In addition, the parents had not been fully informed of the purpose of the testing. What is a fair way to approach the scientific question of the effects of an extra Y chromosome in humans? Can one separate social and cultural impacts from biological impacts of an extra Y chromosome?

Solved Problems

I. In humans, chromosome 16 sometimes has a heavily staining area in the long arm near the centromere. This feature can be seen through the microscope but has no effect on the phenotype of the person carrying it. When such a "blob" exists on a particular copy of chromosome 16, it is a constant feature of that chromosome and is inherited. A couple conceived a child, but the fetus had multiple abnormalities and was miscarried. When the chromosomes of the fetus were studied, it was discovered that it was trisomic for chromosome 16, and that two of the three chromosome 16s had large blobs. Both chromosome 16 homologs in the mother lacked blobs, but the father was heterozygous for blobs. Which parent experienced nondisjunction, and in which meiotic division did it occur?

Answer

This problem requires an understanding of nondisjunction during meiosis. When individual chromosomes contain some distinguishing feature that allows one homolog to be distinguished from another, it is possible to follow the path of the two homologs through meiosis. In this case, since the fetus had two chromosome 16s with the blob, we can conclude that the extra chromosome came from the father (the only parent with a blobbed chromosome). In which meiotic division did the nondisjunction occur? When nondisjunction occurs during meiosis I, homologs fail to segregate to opposite poles. If this occurred in the father, the chromosome with the blob and the normal chromosome 16 would segregate into the same cell (a secondary spermatocyte). After meiosis II, the gametes resulting from this cell would carry both types of chromosomes. If such sperm fertilized a normal egg, the zygote would have two copies of the normal chromosome 16 and one of the chromosome with a blob. On the other hand, if nondisjunction occurred during meiosis II in the father in a secondary spermatocyte containing the blobbed chromosome 16, sperm with two copies of the blob-marked chromosome would be produced. After fertilization with a normal egg, the result would be a zygote of the type seen in this spontaneous abortion. *Therefore, the nondisjunction occurred in meiosis II in the father.*

II. (a) What sex ratio would you expect among the offspring of a cross between a normal male mouse and a female mouse heterozygous for a recessive X-linked lethal gene? (b) What would be the expected sex ratio among the offspring of a cross between a normal hen and a rooster heterozygous for a recessive Z-linked lethal allele?

Answer

This problem deals with sex-linked inheritance and sex determination.

a. Mice have a sex determination system of XX = female and XY = male. A normal male mouse (X^RY) × a heterozygous female mouse (X^RX^r) would result in X^RX^R, X^RX^r, X^RY, and X^rY mice. The X^rY mice would die, so there would be a 2:1 ratio of females to males.

b. The sex determination system in birds is ZZ = male and ZW = female. A normal hen (Z^RW) × a heterozygous rooster (Z^RZ^r) would result in Z^RZ^R, Z^RZ^r, Z^RW, and Z^rW chickens. Because the Z^rW offspring do not live, the ratio of females to males would be 1:2.

III. A woman with normal color vision whose father was color-blind mates with a man with normal color vision.
a. What do you expect to see among their offspring?
b. What would you expect if it was the normal man's father who was color-blind?

Answer

This problem involves sex-linked inheritance.

a. The woman's father has a genotype of $X^{cb}Y$. Because the woman had to inherit an X from her father, she must have an X^{cb} chromosome, but because she has normal color vision, her other X chromosome must be X^{CB}. The man she mates with has normal color vision and therefore has an $X^{CB}Y$ genotype. Their children could with equal probability be $X^{CB}X^{CB}$ (normal female), $X^{CB}X^{cb}$ (carrier female), $X^{CB}Y$ (normal male), or $X^{cb}Y$ (color-blind male).

b. If the man with normal color vision had a color-blind father, the X^{cb} chromosome would not have been passed on to him, since a male does not inherit an X chromosome from his father. The man has the genotype $X^{CB}Y$ and cannot pass on the color-blind allele.

Problems

Vocabulary

1. Choose the best matching phrase in the right column for each of the terms in the left column.

a. meiosis	1. X and Y
b. gametes	2. chromosomes that do not differ between the sexes
c. karyotype	3. one of the two identical halves of a replicated chromosome
d. mitosis	4. microtubule organizing centers at the spindle poles
e. interphase	5. cells in the testes that undergo meiosis
f. syncytium	6. division of the cytoplasm
g. synapsis	7. haploid germ cells that unite at fertilization
h. sex chromosomes	8. an animal cell containing more than one nucleus
i. cytokinesis	9. pairing of homologous chromosomes
j. anaphase	10. one diploid cell gives rise to two diploid cells
k. chromatid	11. the array of chromosomes in a given cell
l. autosomes	12. the part of the cell cycle during which the chromosomes are not visible
m. centromere	13. one diploid cell gives rise to four haploid cells
n. centrosomes	14. cell produced by meiosis that does not become a gamete
o. polar body	15. the time during mitosis when sister chromatids separate
p. spermatocytes	16. connection between sister chromatids

Section 4.1

2. Humans have 46 chromosomes in each somatic cell.
a. How many chromosomes does a child receive from its father?
b. How many autosomes and how many sex chromosomes are present in each somatic cell?
c. How many chromosomes are present in a human ovum?
d. How many sex chromosomes are present in a human ovum?

3. The figure that follows (on p. 116) shows the metaphase chromosomes of a male of a particular species. These chromosomes are prepared as they would be for a karyotype, but they have not yet been ordered in pairs of decreasing size.
a. How many centromeres are shown?
b. How many chromosomes are shown?
c. How many chromatids are shown?
d. How many pairs of homologous chromosomes are shown?
e. How many chromosomes on the figure are metacentric? Acrocentric?
f. What is the likely mode of sex determination in this species? What would you predict to be different about the karyotype of a female in this species?

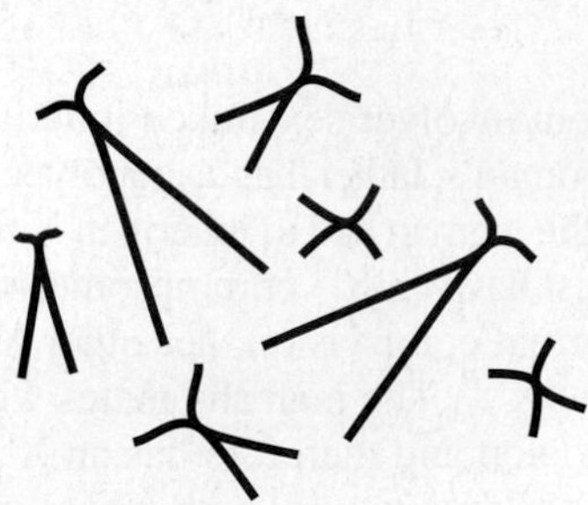

Section 4.2

4. One oak tree cell with 14 chromosomes undergoes mitosis. How many daughter cells are formed, and what is the chromosome number in each cell?

5. Indicate which of the cells illustrated matches each of the following stages of mitosis:
a. anaphase
b. prophase
c. metaphase
d. G_2
e. telophase/cytokinesis

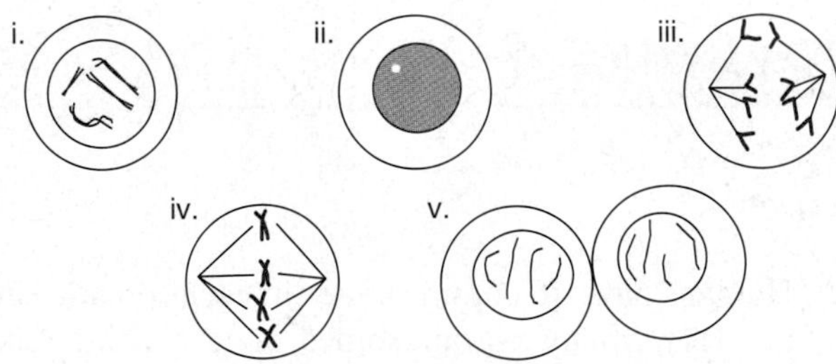

6. a. What are the four major stages of the cell cycle?
b. Which stages are included in interphase?
c. What events distinguish G_1, S, and G_2?

7. Answer the questions that follow for each stage of the cell cycle (G_1, S, G_2, prophase, metaphase, anaphase, telophase). If necessary, use an arrow to indicate a change that occurs during a particular cell cycle stage (for example, 1 → 2 or yes → no).
a. How many chromatids comprise each chromosome during this stage?
b. Is the nucleolus present?
c. Is the mitotic spindle organized?
d. Is the nuclear membrane present?

8. Is there any reason that mitosis could not occur in a cell whose genome is haploid?

Section 4.3

9. One oak tree cell with 14 chromosomes undergoes meiosis. How many cells will result from this process, and what is the chromosome number in each cell?

10. Which type(s) of cell division (mitosis, meiosis I, meiosis II) reduce(s) the chromosome number by half? Which type(s) of cell division can be classified as reductional? Which type(s) of cell division can be classified as equational?

11. Complete the following statements using as many of the following terms as are appropriate: mitosis, meiosis I (first meiotic division), meiosis II (second meiotic division), and none (not mitosis nor meiosis I nor meiosis II).
a. The spindle apparatus is present in cells undergoing _____.
b. Chromosome replication occurs just prior to _____.
c. The cells resulting from _____ in a haploid cell have a ploidy of *n*.
d. The cells resulting from _____ in a diploid cell have a ploidy of *n*.
e. Homologous chromosome pairing regularly occurs during _____.
f. Nonhomologous chromosome pairing regularly occurs during _____.
g. Physical recombination leading to the production of recombinant progeny classes occurs during _____.
h. Centromere division occurs during _____.
i. Nonsister chromatids are found in the same cell during _____.

12. The five cells shown in figures a–e are all from the same individual. For each cell, indicate whether it is in mitosis, meiosis I, or meiosis II. What stage of cell division is represented in each case? What is *n* in this organism?

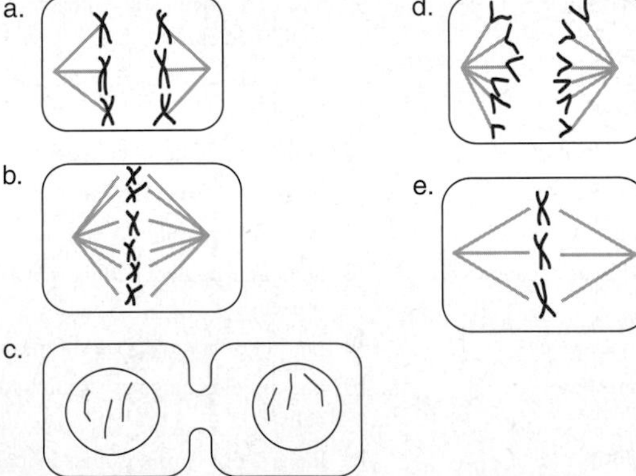

13. One of the first microscopic observations of chromosomes in cell division was published in 1905 by Nettie Stevens. Because it was hard to reproduce photographs at the time, she recorded these observations as *camera lucida* sketches. One such drawing, of a completely normal cell division in the mealworm *Tenebrio molitor,* is shown here. The techniques of the time were relatively unsophisticated by today's standards, and they did not allow her to resolve chromosomal structures that must have been present.

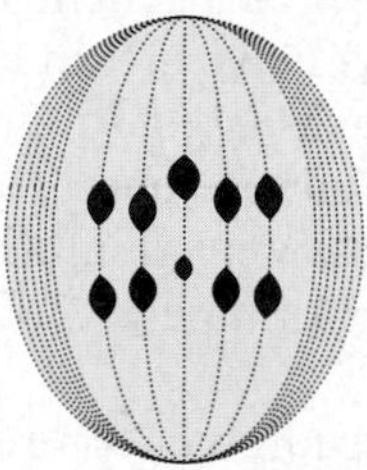

a. Describe in as much detail as possible the kind of cell division and the stage of division depicted in the drawing.
b. What chromosomal structure(s) cannot be resolved in the drawing?
c. How many chromosomes are present in normal *Tenebrio molitor* gametes?

14. A person is simultaneously heterozygous for two autosomal genetic traits. One is a recessive condition for albinism (alleles *A* and *a*); this albinism gene is found near the centromere on the long arm of an acrocentric autosome. The other trait is the dominantly inherited Huntington disease (alleles *HD* and HD^+). The Huntington gene is located near the telomere of one of the arms of a metacentric autosome. Draw all copies of the two relevant chromosomes in this person as they would appear during metaphase of (a) mitosis, (b) meiosis I, and (c) meiosis II. In each figure, label the location on every chromatid of the alleles for these two genes, assuming that no recombination takes place.

15. Assuming (i) that the two chromosomes in an homologous pair carry different alleles of some genes and (ii) that no crossing-over takes place, how many genetically different offspring could any one human couple potentially produce? Which of these two assumptions (i or ii) is more realistic?

16. In the moss *Polytrichum commune*, the haploid chromosome number is 7. A haploid male gamete fuses with a haploid female gamete to form a diploid cell that divides and develops into the multicellular sporophyte. Cells of the sporophyte then undergo meiosis to produce haploid cells called spores. What is the probability that an individual spore will contain a set of chromosomes all of which came from the male gamete? Assume no recombination.

17. Is there any reason that meiosis could not occur in an organism whose genome is always haploid?

18. Sister chromatids are held together through metaphase of mitosis by complexes of *cohesin* proteins that form rubber band-like rings bundling the two sister chromatids. Cohesin rings are found both at centromeres and at many locations scattered along the length of the chromosomes. The rings are destroyed by protease enzymes at the beginning of anaphase, allowing the sister chromatids to separate.
a. Cohesin complexes between sister chromatids are also responsible for keeping homologous chromosomes together until anaphase of meiosis I. With this point in mind, which of the two diagrams that follow (i or ii) properly represents the arrangement of chromatids during prophase through metaphase of meiosis I? Explain.
b. What does your answer to part (a) allow you to infer about the nature of cohesin complexes at the centromere versus those along the chromosome arms? Suggest a molecular hypothesis to explain your inference.

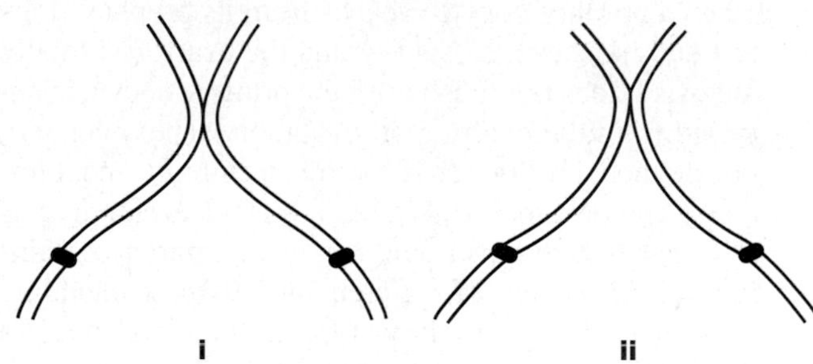

Section 4.4

19. In humans,
a. How many sperm develop from 100 primary spermatocytes?
b. How many sperm develop from 100 secondary spermatocytes?
c. How many sperm develop from 100 spermatids?
d. How many ova develop from 100 primary oocytes?
e. How many ova develop from 100 secondary oocytes?
f. How many ova develop from 100 polar bodies?

20. Somatic cells of chimpanzees contain 48 chromosomes.

How many chromatids and chromosomes are present at (a) anaphase of mitosis, (b) anaphase I of meiosis, (c) anaphase II of meiosis, (d) G_1 prior to mitosis, (e) G_2 prior to mitosis, (f) G_1 prior to meiosis I, and (g) prophase of meiosis I?

How many chromatids or chromosomes are present in (h) an oogonial cell prior to S phase, (i) a spermatid, (j) a primary oocyte arrested prior to ovulation, (k) a secondary oocyte arrested prior to fertilization, (l) a second polar body, and (m) a chimpanzee sperm?

21. In a certain strain of turkeys, unfertilized eggs sometimes develop parthenogenetically to produce diploid offspring. (Females have ZW and males have ZZ sex chromosomes. Assume that WW cells are inviable.) What distribution of sexes would you expect to see among the parthenogenetic offspring according to each of the following models for how parthenogenesis occurs?

a. The eggs develop without ever going through meiosis.
b. The eggs go all the way through meiosis and then duplicate their chromosomes to become diploid.
c. The eggs go through meiosis I, and the chromatids separate to create diploidy.
d. The egg goes all the way through meiosis and then fuses at random with one of its three polar bodies (this assumes the first polar body goes through meiosis II).

22. Female mammals, including women, sometimes develop benign tumors called "ovarian teratomas" or "dermoid cysts" in their ovaries. Such a tumor begins when a primary oocyte escapes from its prophase I arrest and finishes meiosis I within the ovary. (Normally meiosis I does not finish until the primary oocyte is expelled from the ovary upon ovulation.) The secondary oocyte then develops as if it were an embryo, and it implants and develops within the follicle. Development is disorganized, however, and results in a tumor containing a wide variety of differentiated tissues, including teeth, hair, bone, muscle, nerve, and many others. If a dermoid cyst forms in a woman whose genotype is *Aa*, what are the possible genotypes of the cyst?

Section 4.5

23. A system of sex determination known as haplodiploidy is found in honeybees. Females are diploid, and males (drones) are haploid. Male offspring result from the development of unfertilized eggs. Sperm are produced by mitosis in males and fertilize eggs in the females. Ivory eye is a recessive characteristic in honeybees; wild-type eyes are brown.

a. What progeny would result from an ivory-eyed queen and a brown-eyed drone? Give both genotype and phenotype for progeny produced from fertilized and nonfertilized eggs.
b. What would result from crossing a daughter from the mating in part a. with a brown-eyed drone?

24. Imagine you have two pure-breeding lines of canaries, one with yellow feathers and the other with brown feathers. In crosses between these two strains, yellow female × brown male gives only brown sons and daughters, while brown female × yellow male gives only brown sons and yellow daughters. Propose a hypothesis to explain these results.

25. Barred feather pattern is a Z-linked dominant trait in chickens. What offspring would you expect from (a) the cross of a barred hen to a nonbarred rooster? (b) the cross of an F_1 rooster from part (a) to one of his sisters?

26. Each of the four pedigrees shown here represents a human family within which a genetic disease is segregating. Affected individuals are indicated by filled-in symbols. One of the diseases is transmitted as an autosomal recessive condition, one as an X-linked recessive, one as an autosomal dominant, and one as an X-linked dominant. Assume all four traits are rare in the population.

a. Indicate which pedigree represents which mode of inheritance, and explain how you know.
b. For each pedigree, how would you advise the parents of the chance that their child (indicated by the hexagon shape) will have the condition?

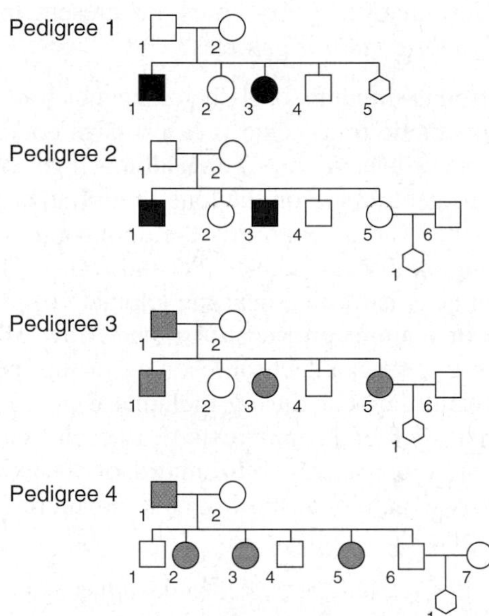

27. In a vial of *Drosophila,* a research student noticed several female flies (but no male flies) with "bag" wings each consisting of a large, liquid-filled blister instead of the usual smooth wing blade. When bag-winged females were crossed with wild-type males, 1/3 of the progeny were bag-winged females, 1/3 were normal-winged females, and 1/3 were normal-winged males. Explain these results.

28. Duchenne muscular dystrophy (DMD) is caused by a relatively rare X-linked recessive allele. It results in progressive muscular wasting and usually leads to death before age 20.

a. What is the probability that the first son of a woman whose brother is affected will be affected?
b. What is the probability that the second son of a woman whose brother is affected will be affected, if her first son was affected?

c. What is the probability that a child of an unaffected man whose brother is affected will be affected?
d. An affected man mates with his unaffected first cousin; there is otherwise no history of DMD in this family. If the mothers of this man and his mate were sisters, what is the probability that the couple's first child will be an affected boy? An affected girl? An unaffected child?
e. If two of the parents of the couple in part (d) were brother and sister, what is the probability that the couple's first child will be an affected boy? An affected girl? An unaffected child?

29. The following is a pedigree of a family in which a rare form of colorblindness is found (filled-in symbols). Indicate as much as you can about the genotypes of all the individuals in the pedigree.

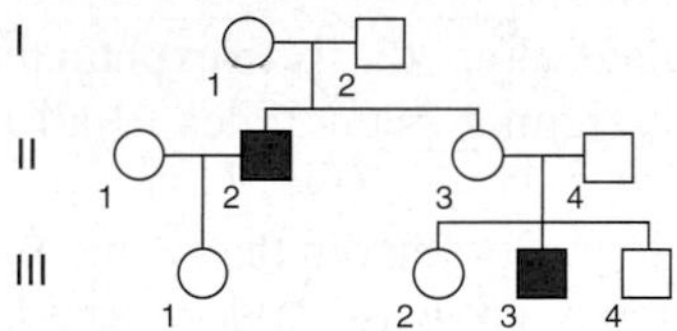

30. In 1995, doctors reported a Chinese family in which retinitis pigmentosa (progressive degeneration of the retina leading to blindness) affected only males. All six sons of affected males were affected, but all of the five daughters of affected males (and all of the children of these daughters) were unaffected.
a. What is the likelihood that this form of retinitis pigmentosa is due to an autosomal mutation showing complete dominance?
b. What other possibilities could explain the inheritance of retinitis pigmentosa in this family? Which of these possibilities do you think is most likely?

31. The pedigree shown here indicates the occurrence of albinism in a group of Hopi Indians, among whom the trait is unusually frequent. Assume that the trait is fully penetrant (all individuals with a genotype that could give rise to albinism will display this condition).
a. Is albinism in this population caused by a recessive or a dominant allele?
b. Is the gene sex-linked or autosomal?

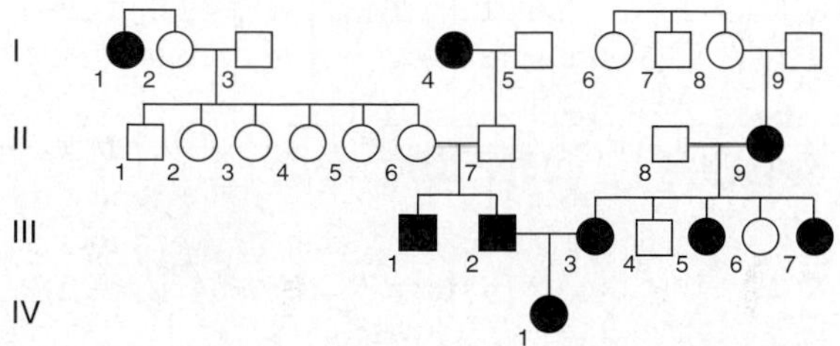

What are the genotypes of the following individuals?
c. individual I-1
d. individual I-8
e. individual I-9
f. individual II-6
g. individual II-8
h. individual III-4

32. When Calvin Bridges observed a large number of offspring from a cross of white-eyed female *Drosophila* to red-eyed males, he observed very rare white-eyed females and red-eyed males among the offspring. He was able to show that these exceptions resulted from nondisjunction, such that the white-eyed females had received two Xs from the egg and a Y from the sperm, while the red-eyed males had received no sex chromosome from the egg and an X from the sperm. What progeny would have arisen from these same kinds of nondisjunctional events if they had occurred in the male parent? What would their eye colors have been?

33. In *Drosophila,* a cross was made between a yellow-bodied male with vestigial (not fully developed) wings and a wild-type female (brown body). The F_1 generation consisted of wild-type males and wild-type females. F_1 males and females were crossed, and the F_2 progeny consisted of 16 yellow-bodied males with vestigial wings, 48 yellow-bodied males with normal wings, 15 males with brown bodies and vestigial wings, 49 wild-type males, 31 brown-bodied females with vestigial wings, and 97 wild-type females. Explain the inheritance of the two genes in question based on these results.

34. Consider the following pedigrees from human families containing a male with Klinefelter syndrome (a set of abnormalities seen in XXY individuals; indicated with shaded boxes). In each, *A* and *B* refer to codominant alleles of the X-linked *G6PD* gene. The *phenotypes* of each individual (A, B, or AB) are shown on the pedigree. Indicate if nondisjunction occurred in the mother or father of the son with Klinefelter syndrome for each of the three examples. Can you tell if the nondisjunction was in the first or second meiotic division?

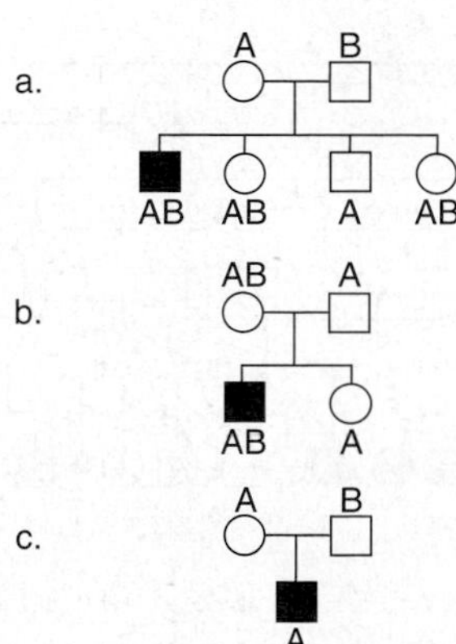

35. The pedigree at the bottom of the page shows five generations of a family that exhibits congenital hypertrichosis, a rare condition in which affected individuals are born with unusually abundant amounts of hair on their faces and upper bodies. The two small black dots in the pedigree indicate miscarriages.

a. What can you conclude about the inheritance of hypertrichosis in this family, assuming complete penetrance of the trait?

b. On what basis can you exclude other modes of inheritance?

c. How many husbands did III-2 and III-9 have?

36. In *Drosophila,* the autosomal recessive *brown* eye color mutation displays interactions with both the X-linked recessive *vermilion* mutation and the autosomal recessive *scarlet* mutation. Flies homozygous for *brown* and simultaneously hemizygous or homozygous for *vermilion* have white eyes. Flies simultaneously homozygous for both the *brown* and *scarlet* mutations also have white eyes. Predict the F_1 and F_2 progeny of crossing the following true-breeding parents:

a. vermilion females × brown males

b. brown females × vermilion males

c. scarlet females × brown males

d. brown females × scarlet males

37. Several different antigens can be detected in blood tests. The following four traits were tested for each individual shown:

ABO type	(I^A and I^B codominant, *i* recessive)
Rh type	(Rh^+ dominant to Rh^-)
MN type	(*M* and *N* codominant)
$Xg^{(a)}$ type	($Xg^{(a+)}$ dominant to $Xg^{(a-)}$)

All of these blood type genes are autosomal, except for $Xg^{(a)}$, which is X linked.

Mother	AB	Rh^-	MN	$Xg^{(a+)}$
Daughter	A	Rh^+	MN	$Xg^{(a-)}$
Alleged father 1	AB	Rh^+	M	$Xg^{(a+)}$
Alleged father 2	A	Rh^-	N	$Xg^{(a-)}$
Alleged father 3	B	Rh^+	N	$Xg^{(a-)}$
Alleged father 4	O	Rh^-	MN	$Xg^{(a-)}$

a. Which, if any, of the alleged fathers could be the real father?

b. Would your answer to part (a) change if the daughter had Turner syndrome (the abnormal phenotype seen in XO individuals)? If so, how?

38. In 1919, Calvin Bridges began studying an X-linked recessive mutation causing eosin-colored eyes in *Drosophila.* Within an otherwise true-breeding culture of eosin-eyed flies, he noticed rare variants that had much lighter cream-colored eyes. By intercrossing these variants, he was able to make a true-breeding cream-eyed stock. Bridges now crossed males from this cream-eyed stock with true-breeding wild-type females. All the F_1 progeny had red (wild-type) eyes. When F_1 flies were intercrossed, the F_2 progeny were 104 females with red eyes, 52 males with red eyes, 44 males with eosin eyes, and 14 males with cream eyes. Assume this represents an 8:4:3:1 ratio.

a. Formulate a hypothesis to explain the F_1 and F_2 results, assigning phenotypes to all possible genotypes.

b. What do you predict in the F_1 and F_2 generations if the parental cross is between true-breeding eosin-eyed males and true-breeding cream-eyed females?

c. What do you predict in the F_1 and F_2 generations if the parental cross is between true-breeding eosin-eyed females and true-breeding cream-eyed males?

39. As we learned in this chapter, the *white* mutation of *Drosophila* studied by Thomas Hunt Morgan is X linked and recessive to wild type. When true-breeding white-eyed males carrying this mutation were crossed with true-breeding purple-eyed females, all the F_1 progeny had wild-type (red) eyes. When the F_1 progeny were intercrossed, the F_2 progeny emerged in the ratio 3/8 wild-type females: 1/4 white-eyed males: 3/16 wild-type males: 1/8 purple-eyed females: 1/16 purple-eyed males.

a. Formulate a hypothesis to explain the inheritance of these eye colors.

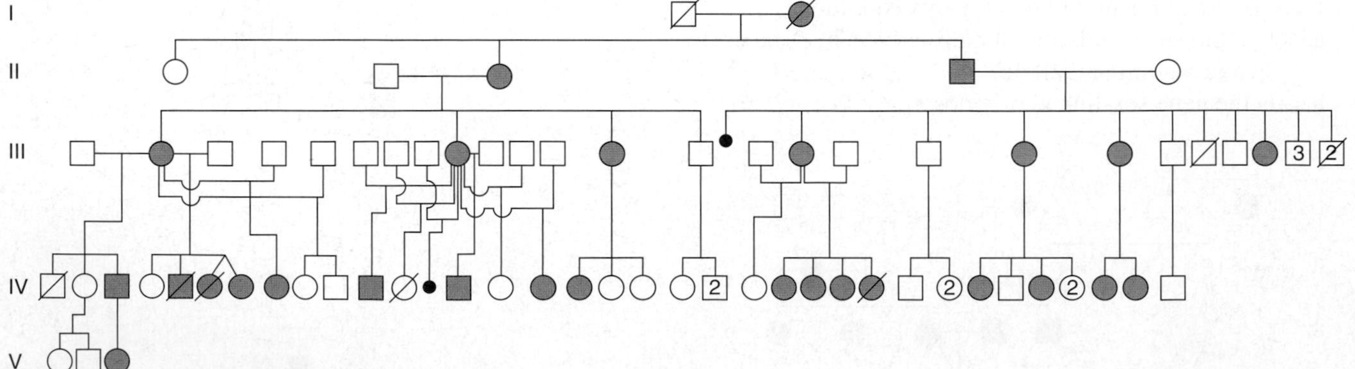

b. Predict the F_1 and F_2 progeny if the parental cross was reversed (that is, if the parental cross was between true-breeding white-eyed females and true-breeding purple-eyed males).

40. The ancestry of a white female tiger bred in a city zoo is depicted in the following pedigree. White tigers are indicated with unshaded symbols. (As you can see, there was considerable inbreeding in this lineage. For example, the white tiger Mohan was mated with his daughter.) In answering the following questions, assume that "white" is determined by allelic differences at a single gene and that the trait is fully penetrant. Explain your answers by citing the relevant information in the pedigree.

a. Could white coat color be caused by a Y-linked allele?

b. Could white coat color be caused by a dominant X-linked allele?

c. Could white coat color be caused by a dominant autosomal allele?

d. Could white coat color be caused by a recessive X-linked allele?

e. Could white coat color be caused by a recessive autosomal allele?

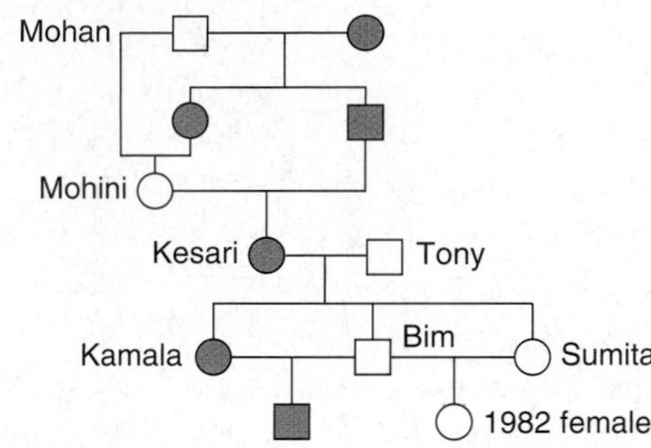

Linkage, Recombination, and the Mapping of Genes on Chromosomes

Chapter 5

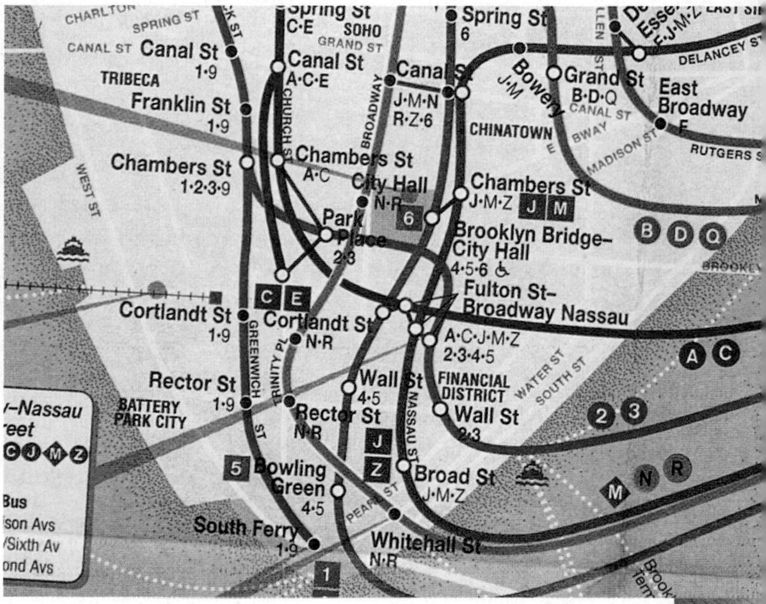

Maps illustrate the spatial relationships of objects, such as the locations of subway stations along subway lines. Genetic maps portray the positions of genes along chromosomes.

In 1928, doctors completed a four-generation pedigree tracing two known X-linked traits: red-green colorblindness and hemophilia A (the more serious X-linked form of "bleeders disease"). The maternal grandfather of the family exhibited both traits, which means that his single X chromosome carried mutant alleles of the two corresponding genes. As expected, neither colorblindness nor hemophilia showed up in his sons and daughters, but two grandsons and one great-grandson inherited both of the X-linked conditions (**Fig. 5.1a**). The fact that none of the descendants of this family's maternal grandfather manifested one of the traits without the other suggests that the mutant alleles did not assort independently during meiosis. Instead they traveled together in the gametes forming one generation and then into the gametes forming the next generation, producing grandsons and great-grandsons with an X chromosome specifying both colorblindness and hemophilia. Genes that travel together more often than not exhibit **genetic linkage.**

In contrast, another pedigree following colorblindness and the slightly different B form of hemophilia, which also arises from a mutation on the X chromosome, revealed a different inheritance pattern. A grandfather with hemophilia B and colorblindness had four grandsons, but only one of them exhibited both conditions. In this family, the genes for colorblindness and hemophilia appeared to assort independently, producing in the male progeny all four possible combinations of the two traits—normal vision and normal blood clotting, colorblindness and hemophilia, colorblindness and normal clotting, and normal vision and hemophilia—in approximately equal frequencies (Fig. 5.1b). Thus, even though the mutant alleles of the two genes were on the same X chromosome in the grandfather, they had to separate to give rise to grandsons III-2 and III-3. This separation of genes on the same chromosome is the result of **recombination,** the occurrence in progeny of new gene combinations not seen in previous generations. (Note that *recombinant progeny* can result in either of two ways: from the recombination of genes on the same chromosome during gamete formation, discussed in this chapter, or from the independent assortment of genes on nonhomologous chromosomes, previously described in Chapter 4.)

As we look at the tools geneticists devised to follow the transmission of genes linked on the same chromosome, you will see that recombination may separate those genes when homologous chromosomes exchange parts during meiosis. The farther apart two genes are, the greater the probability of separation through recombination. Extrapolating from this general rule, you can see that the gene for hemophilia A must be very close to the gene for red-green colorblindness, because, as Fig. 5.1a shows, the two rarely separate. By comparison, the gene for hemophilia B must lie far away from the colorblindness gene, because, as Fig. 5.1b indicates, new combinations of alleles of the two genes occur

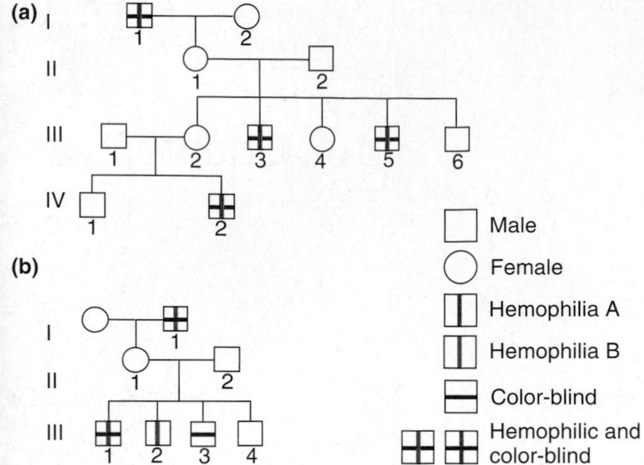

Figure 5.1 Pedigrees indicate that colorblindness and the two forms of hemophilia are X-linked traits. **(a)** Transmission of red-green colorblindness and hemophilia A. Colorblindness and hemophilia A travel together through the pedigree, indicating their genetic linkage. **(b)** Transmission of red-green colorblindness and hemophilia B. Even though the genes for both conditions are X linked, the mutant alleles are inherited together in only one of four grandsons in generation III. These two pedigrees indicate that the gene for colorblindness is close to the hemophilia A gene but far away from the hemophilia B gene on the human X chromosome.

quite often. Geneticists can use data about how often genes separate during transmission to map the genes' relative locations on a chromosome. Such mapping is a key to sorting out and tracking down the components of complex genetic networks; it is also crucial to geneticists' ability to isolate and characterize genes at the molecular level.

Three general themes emerge from our discussion of genetic linkage and recombination. First, maps reflecting the frequency with which linked genes travel together helped validate the chromosome theory of inheritance by confirming that each chromosome carries many genes in a particular order. Second, the twin concepts of linkage and recombination help explain the genetic underpinnings of evolution. Linkage creates a potential for the simultaneous transmission of blocks of genes that function well together, while recombination produces a potential for gene reshuffling that may enhance the chances of survival under changing conditions. Third, chromosome maps led to further refinement in the concept of a gene. For Mendel, the gene was an abstract unit that controlled a trait and assorted independently of other such units. With the chromosome theory, the gene became part of a chromosome. With mapping, the gene acquired a chromosomal address—a precise location that became the focus of more accurate predictions about inheritance.

Our presentation of the transmission patterns of genes on the same chromosome describes

- Linkage and meiotic recombination: Genes linked on the same chromosome usually assort together, instead of independently; such linked genes may nonetheless become separated through recombination during meiosis.
- Mapping: The frequency with which linked genes become separated through meiotic recombination generally reflects the physical distance between them; recombination data thus make it possible to determine the distance between genes along chromosomes.
- Mitotic recombination: Rarely, recombination occurs during mitosis. In multicellular organisms, mitotic recombination can produce genetic mosaicism in which different cells have different genotypes.

5.1 Gene Linkage and Recombination

If people have roughly 20,000–30,000 genes but only 23 pairs of chromosomes, most human chromosomes must carry hundreds, if not thousands, of genes. This is certainly true of the human X chromosome: In 2005, a group of bioinformatics specialists reported that they found 739 protein-encoding genes on this chromosome. This number is likely to grow, at least slightly, as geneticists develop new techniques to analyze the X chromosome's DNA sequence. Moreover, this number does not account for the many genes that do not encode proteins. Recognition that many genes reside on each chromosome raises an important question. If genes on *different* chromosomes assort independently because nonhomologous chromosomes align independently on the spindle during meiosis I, how do genes on the *same* chromosome assort?

Some Genes on the Same Chromosome Assort Together More Often Than Not

We begin our analysis with X-linked *Drosophila* genes because they were the first to be assigned to a specific chromosome and because it is easy to design and follow multigenerational crosses yielding large numbers of *Drosophila* progeny. As we outline various crosses, remember that females carry two alleles for each X-linked gene, while males carry only one. This, of course, is because each female has two X chromosomes, one from each parent, whereas each male has only one, the X inherited from his mother.

We look first at two X-linked genes that determine a fruit fly's eye color and body color. These two genes are said to be **syntenic** because they are located on the same chromosome. The *white* gene was previously introduced in Chapter 4; you will recall that the dominant wild-type allele w^+ specifies red eyes, while the recessive mutant allele w confers white eyes. The alleles of the *yellow* body color gene are y^+ (the dominant wild-type allele for brown bodies) and y (the recessive mutant allele for yellow bodies). To avoid confusion, note that lowercase y and y^+ refer to alleles of the yellow gene, while capital Y refers to the Y chromosome (which does not carry genes for either eye or body color). You should also pay attention to the slash symbol (/), which is used to separate genes found on chromosomes of a pair (either the X and Y chromosomes as in this case, or a pair of X chromosomes or homologous autosomes). Thus $w\,y$/ Y represents the genotype of a male with an X chromosome bearing w and y, as well as a Y chromosome; phenotypically this male has white eyes and a yellow body.

In Dihybrid Crosses, Departures from a 1:1:1:1 Ratio of F$_1$ Gametes Indicate That the Two Genes Are on the Same Chromosome

In a cross between a female with mutant white eyes and a wild-type brown body ($w\,y^+/w\,y^+$) and a male with wild-type red eyes and a mutant yellow body ($w^+\,y$/ Y), the F$_1$ offspring are evenly divided between brown-bodied females with normal red eyes ($w\,y^+/w^+\,y$) and brown-bodied males with mutant white eyes ($w\,y^+$/ Y) (**Fig. 5.2**). Note that the male progeny look like their mother because their phenotype directly reflects the genotype of the single X chromosome they received from her. The same is not true for the F$_1$ females, who received w and y^+ on the X from their mother and $w^+\,y$ on the X from their father. These F$_1$ females are thus dihybrids: With two alleles for each X-linked gene, one derived from each parent, the dominance relations of each pair of alleles determine the female phenotype.

Now comes the significant cross for answering our question about the assortment of genes on the same chromosome. If these two *Drosophila* genes for eye and body color assort independently, as predicted by Mendel's second law, the dihybrid F$_1$ females should make four kinds of gametes, with four different combinations of genes on the X chromosome—$w\,y^+$, $w^+\,y$, $w^+\,y^+$, and $w\,y$. These four types of gametes should occur with equal frequency, that is, in a ratio of 1:1:1:1. If it happens this way, approximately half of the gametes will be of the two **parental types,** carrying either the $w\,y^+$ allele combination seen in the original mother (the female of the P generation) or the w^+y allele combination seen in the original father (the male of the P generation). The remaining half of the gametes will be of two **recombinant types,** in which reshuffling has produced either $w^+\,y^+$ or $w\,y$ allele combinations not seen in the P generation parents of the F$_1$ females.

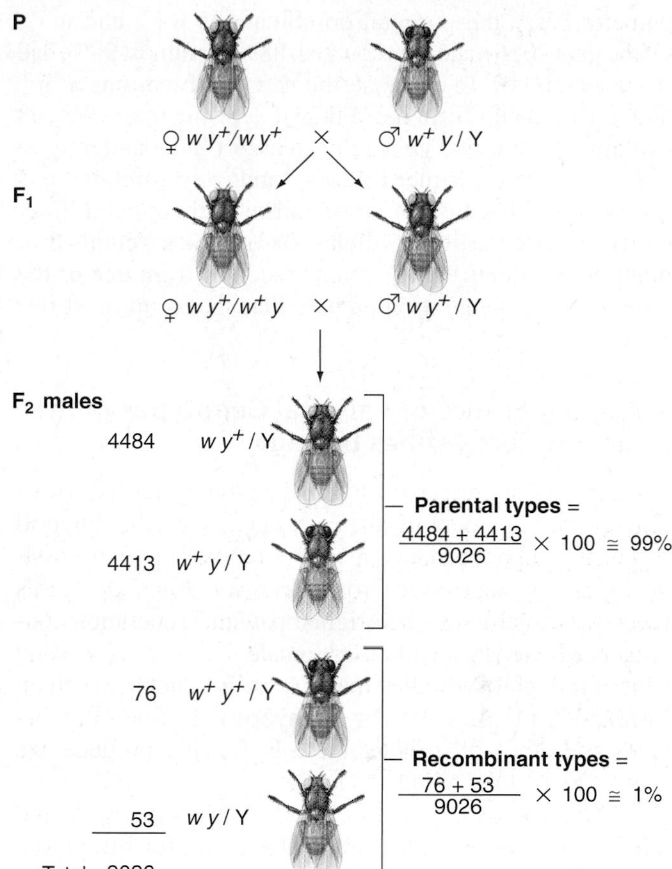

Figure 5.2 When genes are linked, parental combinations outnumber recombinant types. Doubly heterozygous $w\,y^+$ / $w^+\,y$ F$_1$ females produce four types of sons. Sons that look like the father ($w^+\,y$ / Y) or mother ($w\,y^+$ / Y) of the F$_1$ females are parental types. Other sons (w^+y^+ / Y or $w\,y$ / Y) are recombinant types. For these two genes that are very close together on the *Drosophila* X chromosome, there are many more parental than recombinant types among the progeny.

We can see if the 1:1:1:1 ratio of the four kinds of gametes actually materializes by counting the different types of male progeny in the F$_2$ generation, as these sons receive their only X-linked genes from their maternal gamete. The bottom part of Fig. 5.2 depicts the results of a breeding study that produced 9026 F$_2$ males. The relative numbers of the four X-linked gene combinations passed on by the dihybrid F$_1$ females' gametes reflect a significant departure from the 1:1:1:1 ratio expected of independent assortment. By far, the largest numbers of gametes carry the parental combinations $w\,y^+$ and w^+y. Of the total 9026 male flies counted, 8897, or almost 99%, had these genotypes. In contrast, the new combinations w^+y^+ and $w\,y$ made up little more than 1% of the total. We can explain why the two genes fail to assort independently in one of two ways. Either the $w\,y^+$ and $w^+\,y$ combinations are preferred because of some intrinsic chemical affinity between these particular alleles, or it is the parental combination of alleles the F_1 female receives from one or the other of her P generation parents that shows up most frequently.

A Preponderance of Parental Genotypes in the F_2 Generation Defines Linkage

A second set of crosses involving the same genes but with a different arrangement of alleles explains why the dihybrid F_1 females do not produce a 1:1:1:1 ratio of the four possible types of gametes (see Cross Series B in **Fig. 5.3**). In this second set of crosses, the original parental generation consists of red-eyed, brown-bodied females (w^+y^+/w^+y^+) and white-eyed, yellow-bodied males ($w\,y$/Y), and the resultant F_1 females are all $w^+\,y^+/\,w\,y$ dihybrids. To find out what kinds and ratios of gametes these F_1 females produce, we need to look at the telltale F_2 males.

This time, as Cross B in Fig. 5.3 shows, $w^+\,y$/Y and $w\,y^+$/Y are the recombinants that account for little more than 1% of the total, while $w\,y$/Y and $w^+\,y^+$/Y are the parental combinations, which again add up to almost 99%. You can see that there is no preferred association of w^+ and y or of y^+ and w in this cross. Instead, a comparison of the two experiments with these particular X chromosome genes demonstrates that the observed frequencies of the various types of progeny depend on how the arrangement of alleles in the F_1 females originated. We have redrawn Fig 5.2 as Cross Series A in Fig. 5.3 so that you can make this comparison more directly. Note that in both experiments, it is the **parental classes**—the combinations originally present in the P generation—that show up most frequently in the F_2 generation. The reshuffled **recombinant classes** occur less frequently. It is important to appreciate that the designation of "parental" and "recombinant" gametes or progeny of a doubly heterozygous F_1 female is operational, that is, determined by the particular set of alleles she receives from each of her parents.

When genes assort independently, the numbers of parental and recombinant F_2 progeny are equal, because a doubly heterozygous F_1 individual produces an equal number of all four types of gametes. By comparison, two genes are considered **linked** when the number of F_2 progeny with parental genotypes exceeds the number of F_2 progeny with recombinant genotypes. Instead of assorting independently, the genes behave as if they are connected to each other much of the time. The genes for eye and body color that reside on the X chromosome in *Drosophila* are an extreme illustration of the linkage concept. The two genes are so tightly coupled that the parental combinations of alleles—$w^+\,y$ and $w\,y^+$ (in Cross Series A of Fig. 5.3) or $w^+\,y^+$ and $w\,y$ (in Cross Series B)—are reshuffled to form recombinants in only 1 out of every 100 gametes formed. In other words, the two parental allele combinations of these tightly linked genes are inherited together 99 times out of 100.

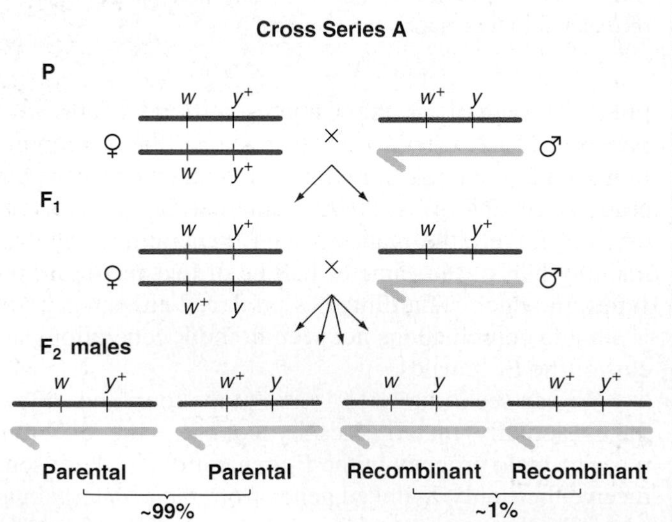

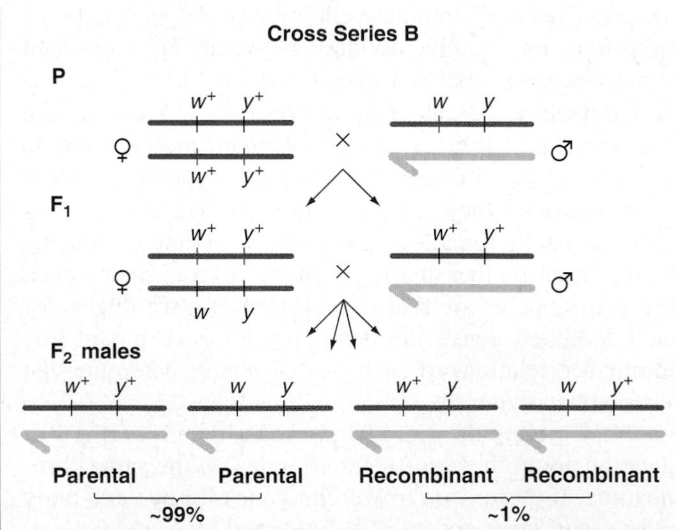

Figure 5.3 Designations of "parental" and "recombinant" relate to past history. Figure 5.2 has been redrawn here as **Cross Series A** for easier comparison with **Cross Series B,** in which the dihybrid F_1 females received a different allelic combination of the *white* and *yellow* genes. Note that the parental and recombinant classes in the two cross series are the opposite of each other. The percentages of recombinant and parental types are nonetheless essentially the same in both experiments, which shows that the frequency of recombination is independent of the particular arrangement of alleles.

Percentages of Parental and Recombinant Classes Vary with the Gene Pair

Linkage is not always this tight. In *Drosophila,* a mutation for miniature wings (*m*) is also found on the X chromosome. A cross of red-eyed females with normal wings ($w^+\ m^+/w^+\ m^+$) and white-eyed males with miniature wings ($w\ m$/Y) yields an F_1 generation containing all red-eyed, normal-winged flies. The genotype of the dihybrid F_1 females is $w^+\ m^+/w\ m$. Of the F_2 males, 67.2% are parental types ($w^+\ m^+$ and $w\ m$), while the remaining 32.8% are recombinants ($w\ m^+$ and $w^+\ m$). This preponderance of parental combinations among the F_2 genotypes reveals that the two genes are linked: The parental combinations of alleles travel together more often than not. But compared to the 99% linkage between the *w* and *y* genes for eye color and body color, the linkage of *w* to *m* is not that tight. The parental combinations for color and wing size are reshuffled in roughly 33 (instead of 1) out of every 100 gametes.

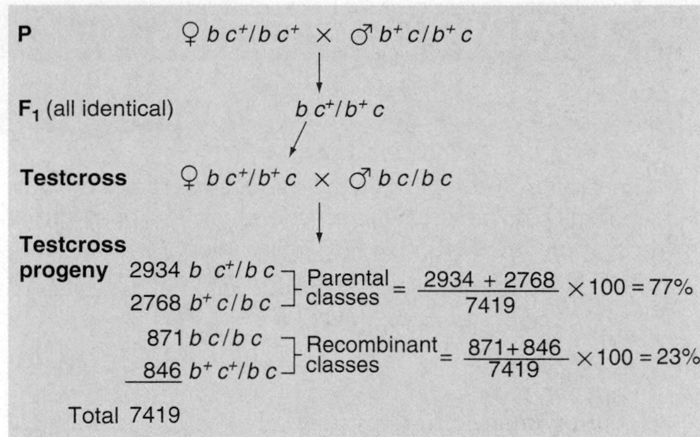

Figure 5.4 **Autosomal genes can also exhibit linkage.** Genes for body color (*b*) and wing shape (*c*) are both autosomal. A testcross shows that the recombination frequency (RF) for this pair of genes is 23%. Because parentals outnumber recombinants, the *b* and *c* genes are genetically linked and must be on the same *Drosophila* autosome.

Autosomal Traits Can Also Exhibit Linkage

Linked autosomal genes are not inherited according to the 9:3:3:1 Mendelian ratio expected for two independently assorting genes. Early twentieth-century geneticists were puzzled by the many experimentally observed departures from this ratio, which they could not explain in terms of the gene interactions discussed in Chapter 3. They found it difficult to interpret these unexpected results because although they knew that individuals receive two copies of each autosomal gene, one from each parent, it was hard to trace which alleles came from which parent. However, by setting up testcrosses in which one parent was homozygous for the recessive alleles of both genes, they were able to analyze the gene combinations received from the gametes of the other, doubly heterozygous parent.

Fruit flies, for example, carry an autosomal gene for body color (in addition to the X-linked *y* gene); the wild type is once again brown, but a recessive mutation in this gene gives rise to black (*b*). A second gene on the same autosome helps determine the shape of a fruit fly's wing, with the wild type having straight edges and a recessive mutation (*c*) producing curves. **Figure 5.4** depicts a cross between black-bodied females with straight wings ($b\ c^+/b\ c^+$) and brown-bodied males with curved wings ($b^+\ c/b^+\ c$). All the F_1 progeny are double heterozygotes ($b\ c^+/b^+\ c$) that are phenotypically wild type. In a testcross of the F_1 females with $b\ c/b\ c$ males, all of the offspring receive the recessive *b* and *c* alleles from their father. The phenotypes of the offspring thus indicate the kinds of gametes received from the mother. For example, a black fly with normal wings would be genotype $b\ c^+/b\ c$; since we know it received the $b\ c$ combination from its father, it must have received $b\ c^+$ from its mother. As Fig. 5.4 shows, roughly 77% of the testcross progeny in one experiment received parental gene combinations (that is, allelic combinations transmitted into the F_1 females by the gametes of each of her parents), while the remaining 23% were recombinants. Because the parental classes outnumbered the recombinant classes, we can conclude that the autosomal genes for black body and curved wings are linked.

To summarize, many pairs of genes on both autosomes and sex chromosomes exhibit linkage: They do not assort independently; instead they are transmitted together more than 50% of the time. (Do not confuse the general concept of linkage with the idea that some genes are X linked, which simply means that they reside on the X chromosome.) For each pair of genes, a particular linkage percentage indicates how often parental combinations travel together. In the preceding examples, the autosomal genes for body color and wing shape were transmitted together 77% of the time, the X-chromosome genes for eye color and wing size were transmitted together 67.2% of the time, and the X-chromosome genes for eye and body color were transmitted together 99% of the time. Linkage is never 100%. No matter how tightly two genes are linked, if you observe enough individuals, you will find some recombinants.

The Chi-Square Test Pinpoints the Probability That Experimental Results Are Evidence for Linkage

How do you know from a particular experiment whether two genes assort independently or are genetically linked? At first glance, this question should pose no problem. Discriminating between the two possibilities involves straightforward calculations based on assumptions well supported by observations. For independently assorting

TOOLS OF GENETICS

The Chi-Square Test

The general protocol for using the chi-square test and evaluating its results can be stated in a series of steps. Two preparatory steps precede the actual chi-square calculation.

1. Use the data obtained from a breeding experiment to answer the following questions:
   a. What is the *total number* of offspring (events) analyzed?
   b. How many different *classes* of offspring (events) are there?
   c. In each class, what is the *number* of offspring (events) *observed?*

2. Calculate how many offspring (events) would be expected for each class if the null hypothesis (here, no linkage) were correct by multiplying the percentage predicted by the null hypothesis (here, 50% parentals and 50% recombinants) by the total number of offspring analyzed in the experiment. You are now ready for the chi-square calculation.

3. To calculate chi square, begin with one class of offspring. Subtract the expected number from the observed number (which gives the deviation from the prediction for this class), square the result, and divide this amount by the expected number. Do this for all classes and sum the individual results (that is, each deviation^2 /expected) together. The final result is the chi-square (χ^2) value. This step is summarized by the equation

$$\chi^2 = \Sigma \frac{(\textit{Number observed} - \textit{Number expected})^2}{\textit{Number expected}}$$

where Σ means "sum of all classes."

4. Compute the **degrees of freedom (df).** The df is a measure of the number of independently varying parameters in the experiment. For the types of experiments we are discussing, the number of degrees of freedom is one less than the number of classes. For example, if the total number of offspring in a testcross fall into four classes and you know the number present in any three, you can easily calculate the number in the fourth. Thus, if N = the number of classes, then the degrees of freedom (df) = $N - 1$. If there are 4 classes, there are only 3 df.

5. Use the chi-square value together with the number of degrees of freedom to determine a ***p* value:** the probability that a deviation from the predicted numbers at least as large as that observed in the experiment will

genes, a dihybrid F_1 female produces four types of gametes in equal numbers, so one-half of the F_2 progeny are of the parental classes and the other half of the recombinant classes. In contrast, for linked genes, the two types of parental classes by definition always outnumber the two types of recombinant classes in the F_2 generation.

The problem is that because real-world genetic transmission is based on chance events, in a particular study even unlinked, independently assorting genes can produce deviations from the 1:1:1:1 ratio, just as in 10 tosses of a coin, you may easily get 6 heads and 4 tails (rather than the predicted 5 and 5). Thus, if a breeding experiment analyzing the transmission of two genes shows a deviation from the equal ratios of parentals and recombinants expected of independent assortment, can we necessarily conclude the two genes are linked? Is it instead possible that the results represent a statistically acceptable chance fluctuation from the mean values expected of unlinked genes that assort independently? Such questions become more pressing in cases where linkage is not all that tight, so that even though the genes are linked, the percentage of recombinant classes approaches 50%.

To answer these kinds of questions, statisticians have devised a quantitative measure that indicates how often an experimentally observed deviation from the predictions of a particular hypothesis will occur solely by chance. This measure of the "goodness of fit" between observed and predicted results is a probability test known as the **chi-square test.** The test is designed to account for the fact that the size of an experimental population (the "sample size") is an important component of statistical significance. To appreciate the role of sample size, it is useful to return to the proverbial coin toss before examining the details of the chi-square test.

We have seen that in 10 tosses of a coin, because of chance, an outcome of 6 heads (60%) and 4 tails (40%) is not unexpected. In contrast, with 1000 tosses of the coin, a result of 600 heads (60%) and 400 tails (40%) would intuitively be highly unlikely. In the first case, a change in the results of one coin toss would alter the expected 5 heads and 5 tails to the observed 6 heads and 4 tails. In the second case, 100 tosses would have to change from tails to heads to generate the stated deviation from the predicted 500 heads and 500 tails. It is reasonable, even likely, that chance events could cause 1 deviation from the predicted number, but not 100. Two important concepts emerge from this simple presentation. First, a comparison of percentages or ratios alone will never allow you to determine whether or not *observed* data are significantly different from *predicted* values. Second, the absolute numbers obtained are important because they reflect the size of the experiment.

occur by chance. Although the p value is arrived at through a numerical analysis, geneticists routinely determine the value by a quick search through a table of critical χ^2 values, such as the one in **Table 5.1.**

6. Evaluate the significance of the p value. You can think of the p value as the probability that the null hypothesis is true and the alternative hypothesis (in this case, that the genes are linked) is wrong. A value greater than 0.05 indicates that in more than 1 in 20 (or more than 5%) of the repetitions of an experiment of the same size, the observed data showing a deviation from the predictions of the null hypothesis could have been obtained by chance even if the null hypothesis were true; the data are therefore *not significant* for rejecting the null hypothesis and showing linkage. Statisticians have arbitrarily selected the 0.05 p value as the boundary between accepting and rejecting the null hypothesis. A p value of less than 0.05 means that you can consider the data showing deviation *significant* and reject the null hypothesis.

TABLE 5.1 Critical Chi-Square Values

Degrees of Freedom	*p* Values						
	Cannot Reject the Null Hypothesis				Null Hypothesis Rejected		
	0.99	0.90	0.50	0.10	0.05	0.01	0.001
	χ^2 Values						
1	—	0.02	.45	2.71	3.84	6.64	10.83
2	0.02	0.21	1.39	4.61	5.99	9.21	13.82
3	0.11	0.58	2.37	6.25	7.81	11.35	16.27
4	0.30	1.06	3.36	7.78	9.49	13.28	18.47
5	0.55	1.61	4.35	9.24	11.07	15.09	20.52

Note: χ^2 values that lie in the reddish-shaded region of this table allow you to reject the null hypothesis with $>$ 95% confidence, and for recombination experiments, to postulate linkage.

The larger the sample size, the closer the observed percentages can be expected to match the values predicted by the experimental hypothesis, *if the hypothesis is correct.* The chi-square test is therefore always calculated with numbers—actual data—and not percentages or proportions.

A critical prerequisite of the chi-square test is the framing of a hypothesis that leads to clear-cut predictions. Although contemporary geneticists use the chi-square test to interpret many kinds of genetic experiments, they use it most often to discover whether data obtained from breeding experiments provide evidence for or against the hypothesis that two genes are linked. But the problem with the general hypothesis that "genes *A* and *B* are linked" is that there is no precise prediction of what to expect in terms of breeding data. This is because, as we have seen, the frequency of recombination varies with each linked gene pair. In contrast, the alternative hypothesis "that genes *A* and *B* are *not* linked" gives rise to a precise prediction: that alleles at different genes will assort independently and produce 50% parental and 50% recombinant progeny. So, whenever a geneticist wants to determine whether two genes are linked, he or she actually tests whether the observed data are consistent with a **null hypothesis** of no linkage. If the chi-square test shows that the observed data differ significantly from those expected with independent assortment, that is, they differ enough not to be reasonably attributable to chance alone, then the researcher can reject the null hypothesis of no linkage and accept the alternative of linkage between the two genes.

The Tools of Genetics box beginning on p. 128 presents the general protocol of the chi-square test. The final result of the calculations is the determination of the numerical probability—the p value—that a particular set of observed experimental results represents a chance deviation from the values predicted by a particular hypothesis. If the probability is high, it is likely that the hypothesis being tested explains the data, and the observed deviation from expected results is considered *insignificant.* If the probability is very low, the observed deviation from expected results becomes *significant.* When this happens, it is unlikely that the hypothesis under consideration explains the data, and the hypothesis can be rejected.

Applying the Chi-Square Test

Figure 5.5 depicts two sets of data obtained from testcross experiments asking whether genes *A* and *B* are linked. We first apply the chi-square analysis to data accumulated in the first experiment. The total number of offspring is 50, of which 31 (that is, 17 + 14) are observed to be parental

Progeny	Experiment 1		Experiment 2	
A B	17		34	
a b	14		28	
A b	8		16	
a B	11		22	
Total	50		100	
Class	**Observed**	**Expected**	**Observed**	**Expected**
Parentals	31	25	62	50
Recombinants	19	25	38	50

Figure 5.5 Applying the chi-square test to see if genes *A* and *B* are linked. The null hypothesis is that the two genes are unlinked. For Experiment 1, $p > 0.05$, so it is not possible to reject the null hypothesis. For Experiment 2, with a data set twice the size, $p < 0.05$. This is below the arbitrary boundary of significance used by most geneticists, which makes it possible to reject the null hypothesis and conclude with greater than 95% confidence that the genes are linked.

types and 19 (8 + 11) recombinant types. Dividing 50 by 2, you get 25, the number of parental or recombinant offspring expected according to the null hypothesis of independent assortment (which predicts that parentals = recombinants). Now, considering first the parental types alone, you square the observed deviation from the expected value, and divide the result by the expected value. After doing the same for the recombinant types, you add the two quotients to obtain the value of chi square.

$$\chi^2 = \frac{(31-25)^2}{25} + \frac{(19-25)^2}{25} = 1.44 + 1.44 = 2.88$$

Since with two classes (parentals and recombinants), the number of degrees of freedom is 1, you scan the chi-square table (see Table 5.1 on p. 129) for $\chi^2 = 2.88$ and df = 1, and you find by extrapolation that the corresponding p value is greater than 0.05 (roughly 0.09). From this, you can conclude that it is not possible to reject the null hypothesis on the basis of this experiment, which means that this data set is not sufficient to demonstrate linkage between *A* and *B*.

If you use the same strategy to calculate a p value for the data observed in the second experiment, where there are a total of 100 offspring and thus an expected number of 50 parentals and 50 recombinants, you get

$$\chi^2 = \frac{(62-50)^2}{50} + \frac{(38-50)^2}{50} = 2.88 + 2.88 = 5.76$$

The number of degrees of freedom (df) remains 1, so Table 5.1 arrives at a p value greater than 0.01 but less than 0.05. In this case, you can consider the difference between the observed and expected values to be significant. As a result, you can reject the null hypothesis of independent assortment and conclude it is likely that genes *A* and *B* are linked.

A certain amount of subjectivity enters into any decision about the significance of a particular p value. As stated in the preceding box, statisticians have arbitrarily selected a p value of 0.05 as the boundary between significance and nonsignificance. Values lower than this indicate there would be less than 5 chances in 100 of obtaining the same results by random sampling if the null hypothesis were true. An extremely low p value thus suggests that the data shows major deviations from values predicted by the hypothesis; these values are significant enough to reject the null hypothesis. The lower the p value, the clearer the case that the data cannot be regarded as chance deviations from the predictions of the hypothesis. For this reason, more conservative scientists often set the boundary of significance at $p = 0.01$ and would thus be less willing to reach a conclusion based on the results of Experiment 2 in Fig. 5.5. In linkage studies, such a low p value means that it is highly unlikely (only 1 chance in 100) that the observed distribution of phenotypes arose from two unlinked genes, and we can conclude, with 99% confidence, that the genes are linked. On the other hand, higher p values (greater than 0.01 or 0.05, depending on the criterion used) do not necessarily mean that two genes are unlinked; it may mean only that the sample size is not large enough to provide an answer. With more data, the p value will normally rise if the null hypothesis of no linkage is correct and fall if there is, in fact, linkage.

Note that in Fig. 5.5 all of the numbers in the second set of data are simply double the numbers in the first set, with the percentages remaining the same. Thus, just by doubling the sample size from 50 to 100 individuals, it was possible to go from no significant difference to a significant difference between the observed and the expected values (using the $p = 0.05$ cutoff). In other words, the larger the sample size, the less the likelihood that a certain percentage deviation from expected results happened simply by chance. Bearing this in mind, you can see that it is not appropriate to use the chi-square test when analyzing very small samples of less than 10, which by their nature are usually insufficient to answer questions concerning linkage. This creates a problem for human geneticists, who cannot "construct" people as they can fruit flies and other experimental organisms nor develop large populations of inbred individuals. To achieve a reasonable sample size for linkage studies in humans, they must pool data from a large number of family pedigrees.

The subjectivity involved in determining the boundary of significance means that the chi-square test *does not* prove linkage or its absence. What it *does* do is provide a quantitative measure of the likelihood that the data from an experiment can be explained by a particular hypothesis. The chi-square analysis is thus a general statistical test for significance; it can be used with many different experimental designs and with hypotheses other than the absence of linkage. As long as it is possible to propose a hypothesis that leads to a predicted set of values for a defined set of data classes, you can readily determine whether or not the observed data are consistent with the hypothesis. If what you are looking for is any one of a range of outcomes that is not the outcome predicted by the hypothesis, that is, if the hypothesis serves as a "straw man," set up only to be knocked down, it is considered a null hypothesis.

When experiments lead to rejection of a null hypothesis, you may need to confirm the alternative. For instance, if you are analyzing the inheritance of two opposing traits on the basis of a null hypothesis that says the two phenotypes result from the segregation of two equally viable alleles of a single gene, you would expect a testcross between an F_1 heterozygote and a recessive homozygote to produce a 1:1 ratio of the two traits in the offspring. If instead, you observe a ratio of 6:4 and the chi-square test produces a p value of 0.009, you can reject the null hypothesis. But you are still left with the question of what the absence of a 1:1 ratio means. There are actually two alternatives: Either the reverse of the null hypothesis is true and the two alleles of the single gene are not equally viable, *or* more than one gene encodes the trait. The chi-square test cannot tell you which possibility is correct, and you would have to study the matter further. The problems at the end of this chapter illustrate several applications of the chi-square test pertinent to genetics.

Recombination Results When Crossing-Over During Meiosis Separates Linked Genes

It is easy to understand how genes that are physically connected on the same chromosome can be transmitted together and thus show genetic linkage. It is not as obvious why all linked genes always show some recombination in a sample population of sufficient size. Do the chromosomes participate in a physical process that gives rise to the reshuffling of linked genes that we call recombination? The answer to this question is of more than passing interest as it provides a basis for gauging relative distances between pairs of genes on a chromosome.

In 1909, the Belgian cytologist Frans Janssens described structures he had observed in the light microscope during prophase of the first meiotic division. He called these structures **chiasmata;** as described in Chapter 4, they seemed to represent regions in which nonsister chromatids of homologous chromosomes cross over each other (review Fig. 4.14 on p. 98). Making inferences from a combination of genetic and cytological data, Thomas Hunt Morgan suggested that the chiasmata observed through the light microscope were sites of chromosome breakage and exchange resulting in genetic recombination.

Reciprocal Exchanges Between Homologous Chromosomes Are the Physical Basis of Recombination

Morgan's idea that the physical breaking and rejoining of chromosomes during meiosis was the basis of genetic recombination seemed reasonable. But although Janssens's chiasmata could be interpreted as signs of the process, before 1930 no one had produced visible evidence that crossing-over between homologous chromosomes actually occurs. The identification of **physical markers,** or cytologically visible abnormalities that make it possible to keep track of specific chromosome parts from one generation to the next, enabled researchers to turn the logical deductions about recombination into facts derived from experimental evidence. In 1931, Harriet Creighton and Barbara McClintock, who studied corn, and Curt Stern, who worked with *Drosophila,* published the results of experiments showing that genetic recombination indeed depends on the reciprocal exchange of parts between maternal and paternal chromosomes. Stern, for example, bred female flies with two different X chromosomes, each containing a distinct physical marker near one of the ends. These same females were also doubly heterozygous for two X-linked **genetic markers**—genes that could serve as points of reference in determining whether particular progeny were the result of recombination.

Figure 5.6 diagrams the chromosomes of these heterozygous females. One X chromosome carried mutations producing carnation eyes (a dark ruby color, abbreviated *car*) that were kidney-shaped (*Bar*); in addition, this chromosome was marked physically by a visible discontinuity, which resulted when the end of the X chromosome was broken off and attached to an autosome. The other X chromosome had wild-type alleles (+) for both the *car* and the *Bar* genes, and its physical marker consisted of part of the Y chromosome that had become connected to the X-chromosome centromere.

Figure 5.6 illustrates how the chromosomes in these *car Bar* / car^+ Bar^+ females were transmitted to male progeny. According to the experimental results, all sons showing a phenotype determined by one or the other parental combination of genes (either *car Bar* or car^+ Bar^+) had an X chromosome that was structurally indistinguishable from one of the original X chromosomes in the mother. In recombinant sons, however, such as those that manifested carnation eye color and normal eye shape (*car* Bar^+/Y), an identifiable exchange of the abnormal features marking the ends of the homologous X chromosomes accompanied the recombination of genes. The evidence thus tied a particular instance of phenotypic recombination to the crossing-over of particular genes located in specifically marked parts of particular chromosomes. This was an elegant demonstration that genetic recombination is associated with the actual reciprocal exchange of segments between homologous chromosomes during meiosis.

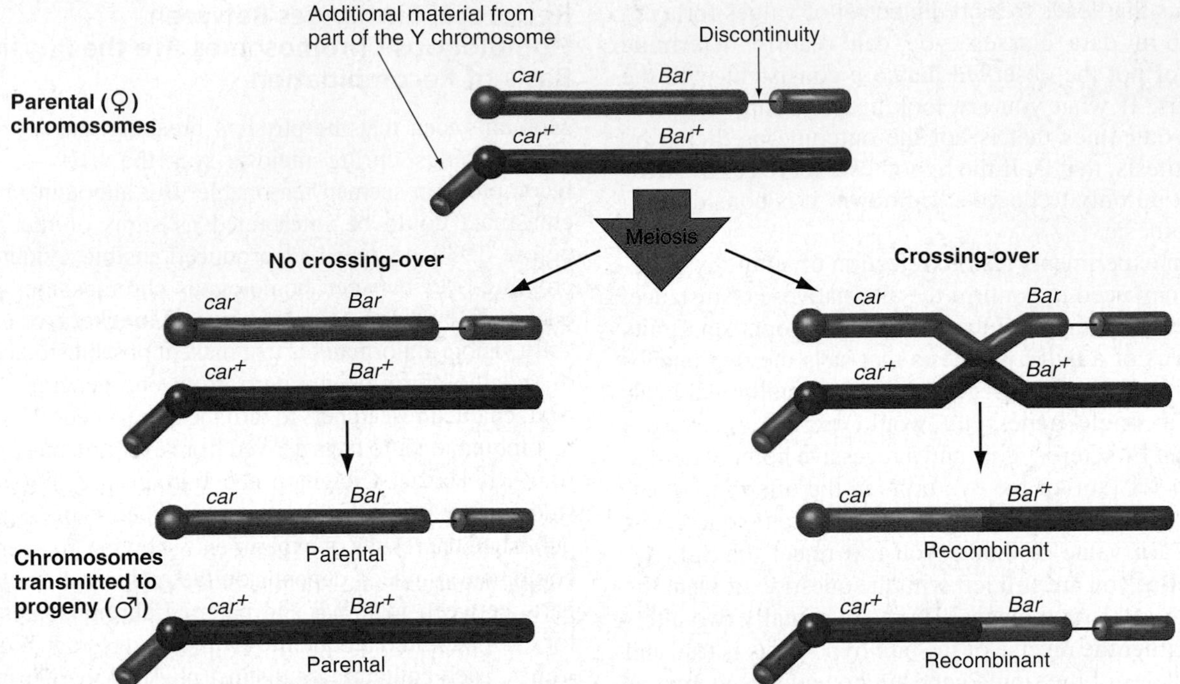

Figure 5.6 Evidence that recombination results from reciprocal exchanges between homologous chromosomes. Genetic recombination between the *car* and *Bar* genes on the *Drosophila* X chromosome is accompanied by the exchange of physical markers observable in the microscope. Note that this depiction of crossing-over is a simplification, as genetic recombination actually occurs after each chromosome has replicated into sister chromatids. Note also that the piece of the X chromosome to the right of the discontinuity is actually attached to an autosome.

Through the Light Microscope: Chiasmata Mark the Sites of Recombination

Figure 5.7 outlines what is currently known about the steps of recombination as they appear in chromosomes viewed through the light microscope. Although this low-resolution view may not represent certain details of recombination with complete accuracy, it nonetheless provides a useful frame of reference. In Fig. 5.7a, the two homologs of each chromosome pair have already replicated, so there are now two pairs of sister chromatids or a total of four chromatids within each bivalent. In Fig. 5.7b, the synaptonemal complex zips together homologous chromosome pairs along their length. The synaptonemal zipper aligns homologous regions of all four chromatids such that allelic DNA sequences are physically near each other (see Fig. 4.14b on p. 98 for a detailed depiction). This proximity facilitates crossing-over between homologous sequences; as we will see in Chapter 6, the biochemical mechanism of recombination requires a close interaction of DNAs on homologous chromosomes that have identical, or nearly identical, nucleotide sequences.

In Fig. 5.7c, the synaptonemal complex begins to disassemble. Although at least some steps of the recombination process occurred while the chromatids were zipped in synapsis, it is only now that the recombination event becomes apparent. As the zipper dissolves, homologous chromosomes remain attached at chiasmata, the actual sites of crossing-over. Visible in the light microscope, chiasmata indicate where chromatid sections have switched from one molecule to another. In Fig. 5.7d, during anaphase I, as the two homologs separate, starting at their centromeres, the ends of the two recombined chromatids pull free of their respective sister chromatids, and the chiasmata shift from their original positions toward a chromosome end, or telomere. This movement of chiasmata is known as **terminalization.** When the chiasmata reach the telomeres, the homologous chromosomes can separate from each other (Fig. 5.7e). Meiosis continues and eventually produces four haploid cells that contain one chromatid—now a chromosome—apiece (Fig. 5.7f). Homologous chromosomes have exchanged parts.

Recombination can also take place apart from meiosis. Indeed, as we explain near the end of this chapter, it sometimes, though rarely, occurs during mitosis. It also occurs with the circular chromosomes of prokaryotic organisms and with cellular organelles such as mitochondria and chloroplasts, which do not undergo meiosis and do not form chiasmata (see Chapters 15 and 16).

Recombination Frequencies for Pairs of Genes Reflect the Distances Between Them

Thomas Hunt Morgan's belief that chiasmata represent sites of physical crossing-over between chromosomes and that such crossing-over may result in recombination, led

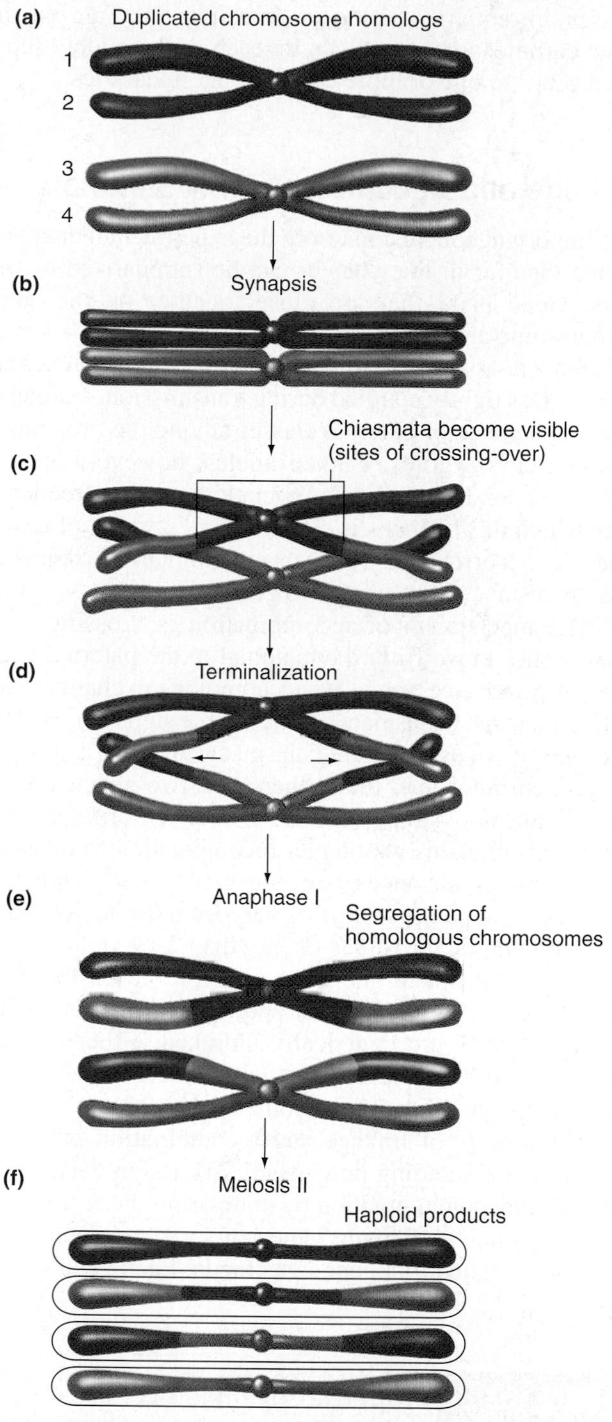

Figure 5.7 Recombination through the light microscope. **(a)** A pair of duplicated homologous chromosomes very early in prophase of meiosis I. **(b)** During leptotene and zygotene of prophase I, the synaptonemal complex helps align corresponding regions of homologous chromosomes, allowing recombination. **(c)** As the synaptonemal complex disassembles during diplotene, homologous chromosomes remain attached at chiasmata. **(d)** and **(e)** The chiasmata terminalize (move toward the chromosome ends), allowing the recombined chromosomes to separate during anaphase and telophase. **(f)** The result of the process is recombinant gametes.

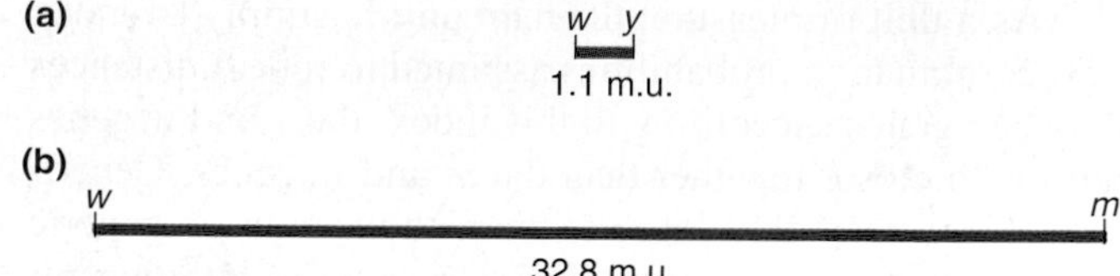

Figure 5.8 Recombination frequencies are the basis of genetic maps. (a) 1.1% of the gametes produced by a female doubly heterozygous for the genes *w* and *y* are recombinant. The recombination frequency (RF) is thus 1.1%, and the genes are approximately 1.1 map units (m.u.) or 1.1 centimorgans (cM) apart. **(b)** The distance between the *w* and *m* genes is longer: 32.8 m.u. (or 32.8 cM).

him to the following logical deduction: Different gene pairs exhibit different linkage rates because genes are arranged in a line along a chromosome. The closer together two genes are on the chromosome, the less their chance of being separated by an event that cuts and recombines the line of genes. To look at it another way, if we assume for the moment that chiasmata can form anywhere along a chromosome with equal likelihood, then the probability of a crossover occurring between two genes increases with the distance separating them. If this is so, the frequency of genetic recombination also must increase with the distance between genes. To illustrate the point, imagine pinning to a wall 10 inches of ribbon with a line of tiny black dots along its length and then repeatedly throwing a dart to see where you will cut the ribbon. You would find that practically every throw of the dart separates a dot at one end of the ribbon from a dot at the other end, while few if any throws separate any two particular dots positioned right next to each other.

Alfred H. Sturtevant, one of Morgan's students, took this idea one step further. He proposed that the percentage of total progeny that were recombinant types, the **recombination frequency (RF),** could be used as a gauge of the physical distance separating any two genes on the same chromosome. Sturtevant arbitrarily defined one RF percentage point as the unit of measure along a chromosome; later, another geneticist named the unit a **centimorgan (cM)** after T. H. Morgan. Mappers often refer to a centimorgan as a **map unit (m.u.).** Although the two terms are interchangeable, researchers prefer one or the other, depending on their experimental organism. *Drosophila* geneticists, for example, use map units while human geneticists use centimorgans. In Sturtevant's system, 1% RF = 1 cM = 1 m.u. A review of the two pairs of X-linked *Drosophila* genes we analyzed earlier shows how his proposal works. Because the X-linked genes for eye color (*w*) and body color (*y*) recombine in 1.1% of F_2 progeny, they are 1.1 m.u. apart (**Fig. 5.8a**). In contrast, the X-linked genes for eye color (*w*) and wing size (*m*) have a recombination frequency of 32.8 and are therefore 32.8 m.u. apart (Fig. 5.8b).

As a unit of measure, the map unit is simply an index of recombination probabilities assumed to reflect distances between genes. According to this index, the *y* and *w* genes are much closer together than the *m* and *w* genes. Geneticists have used this logic to map thousands of genetic markers to the chromosomes of *Drosophila,* building recombination maps step-by-step with closely linked markers. And as we see next, they have learned that markers very far apart on the same chromosome may appear unlinked, even though their recombination distances relative to closely linked intervening markers confirm that they are indeed on the same chromosome.

Experimental Recombination Frequencies Between Two Genes Are Never Greater Than 50%

If the definition of linkage is that the proportion of recombinant classes is less than that of parental classes, a recombination frequency of less than 50% indicates linkage. But what can we conclude about the relative location of genes if there are roughly equal numbers of parental and recombinant progeny? And does it ever happen that recombinants are in the majority?

We already know one situation that can give rise to a recombination frequency of 50%. Genes located on different (that is, nonhomologous) chromosomes will obey Mendel's law of independent assortment because the two chromosomes can line up on the spindle during meiosis I in either of two equally likely configurations (review Fig. 4.17a on p. 101). A dihybrid for these two genes will thus produce all four possible types of gametes (*AB, Ab, aB,* and *ab*) with approximately equal frequency. Importantly, experiments have established that genes located very far apart on the same chromosome also show recombination frequencies of approximately 50%.

Researchers have never observed statistically significant recombination frequencies between two genes greater than 50%, which means that in any cross following two genes, recombinant types are never in the majority. As we explain in more detail later in the chapter, this upper limit of 50% on the recombination frequency between two genes results from two aspects of chromosome behavior during meiosis I. First, multiple crossovers can occur between two genes if they are far apart on the same chromosome, and second, recombination takes place after the chromosomes have replicated into sister chromatids.

For now, simply note that recombination frequencies near 50% suggest either that two genes are on different chromosomes or that they lie far apart on the same chromosome. The only way to tell whether the two genes are syntenic (that is, on the same chromosome) is through a series of matings showing definite linkage with other genes that lie between them. In short, even though crosses between two genes lying very far apart on a chromosome may show no linkage at all (because recombinant and parental classes are equal), you can demonstrate they are on the same chromosome if you can tie each of the widely separated genes to one or more common intermediaries.

Linkage and Recombination: A Summary

The important conclusions from the experimental data presented thus far in this chapter can be summarized as follows. Gene pairs that are close together on the same chromosome are linked and do not follow Mendel's law of independent assortment. Instead, alleles of linked genes are more or less tightly coupled during transmission, leading to a preponderance of parental classes among the progeny of double heterozygotes. Linked alleles, however, can become separated through recombination, and the frequency with which this happens is different for each pair of genes. There is a correlation between recombination frequency and the distance separating two genes.

The mechanism of recombination is crossing-over: Some genes move from the maternal to the paternal chromosome, and vice versa, when homologs exchange parts during meiosis. Chiasmata are the visible signs of crossing-over, and if we assume that chiasmata can occur anywhere along a chromosome, the farther apart two genes are, the greater the opportunity for chiasmata to form between them. This explains why larger recombination frequencies reflect greater distances between genes. Recombination frequencies in pairwise crosses vary from 0% to 50%. Statistically significant values of less than 50% indicate that two genes are linked and must therefore be on the same chromosome. Two genes that show a recombination frequency of 50% are genetically unlinked, either because they are on different chromosomes *or* because they reside far apart on the same chromosome (**Table 5.2**).

Knowledge of linkage and recombination paves the way for understanding how geneticists assign genes a relative chromosomal position by comparing the recombination frequencies of many gene pairs. The chromosomal position assigned to a gene is its **locus.** Each gene's locus

TABLE 5.2 Properties of Linked Versus Unlinked Genes

Linked Genes

Parentals > recombinants (RF < 50%)
Linked genes must be syntenic and sufficiently close together on the same chromosome so that they do not assort independently.

Unlinked Genes

Parentals = recombinants (RF = 50%)
Occurs either when genes are on different chromosomes or when they are sufficiently far apart on the same chromosome.

is the same in all individuals of a species, and the process of determining that locus is known as **mapping.** As you will see, the results of mapping experiments consistently verify the idea that genes are arranged in a line along a chromosome.

5.2 Mapping: Locating Genes Along a Chromosome

Maps are images of the relative positions of objects in space. Whether depicting the floor plan of New York's Metropolitan Museum of Art, the layout of the Roman Forum, or the location of cities served by the railways of Europe, maps turn measurements (how far apart two rooms, ruins, or railway stops are from one another) into patterns of spatial relationships that add a new level of meaning to the original data of distances. Maps that assign genes to locations on particular chromosomes are no exception. By transforming genetic data into spatial arrangements, they sharpen our ability to predict the inheritance patterns of specific traits.

We have seen that recombination frequency (RF) is a measure of the distance separating two genes along a chromosome. We now examine how data from many crosses following two and three genes at a time can be compiled and compared to generate accurate, comprehensive gene/chromosome maps.

Two-Point Crosses: Comparisons Help Establish Relative Gene Positions

In his senior undergraduate thesis, Morgan's student A. H. Sturtevant asked whether data obtained from a large number of two-point crosses (crosses tracing two genes at a time) would support the idea that genes form a definite linear series along a chromosome. He began by looking at X-linked genes in *Drosophila.* **Figure 5.9a** lists his recombination data for several two-point crosses. Recall that the distance between two genes that yields 1% recombinant progeny—an RF of 1%—is 1 m.u. As an example of Sturtevant's reasoning, consider the three genes *w, y,* and *m.* If these genes are arranged in a line (instead of a more complicated branched structure, for example), then one of them must be in the middle, flanked on either side by the other two. The greatest genetic distance should separate the two genes on the outside, and this value should roughly equal the sum of the distances separating the middle gene from each outside gene. The data Sturtevant obtained are consistent with this idea, implying that *w* lies between *y* and *m* (Fig. 5.9b). Note that the left-to-right orientation of this map was selected at random; the map in Fig. 5.9b would be equally correct if it portrayed *y* on the right and *m* on the left.

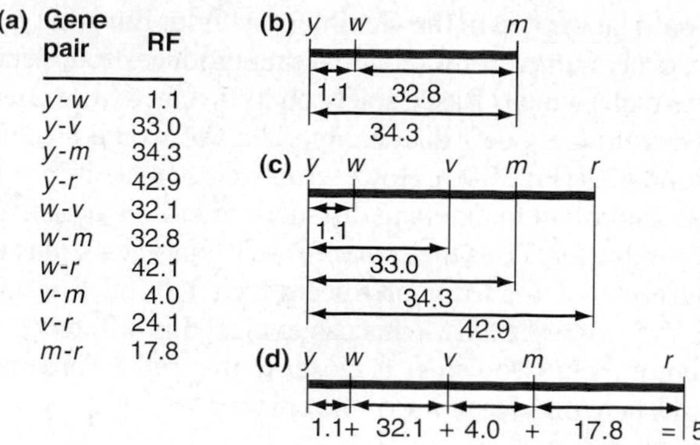

Figure 5.9 Mapping genes by comparisons of two-point crosses. **(a)** Sturtevant's data for the distances between pairs of X-linked genes in *Drosophila.* **(b)** Because the distance between *y* and *m* is greater than the distance between *w* and *m,* the order of genes must be *y-w-m.* **(c)** and **(d)** Maps for five genes on the *Drosophila* X chromosome. The left-to-right orientation is arbitrary. Note that the numerical position of the *r* gene depends on how it is calculated. The best genetic maps are obtained by summing many small intervening distances as in (d).

Recombination Mapping Supports the Idea That Genes Are Arranged in a Line Along a Chromosome

By following exactly the same procedure for each set of three genes, Sturtevant established a self-consistent order for all the genes he investigated on *Drosophila*'s X chromosome (Fig. 5.9c; once again, the left-to-right arrangement is an arbitrary choice). By checking the data for every combination of three genes, you can assure yourself that this ordering makes sense. The fact that the recombination data yield a simple linear map of gene position supports the idea that genes reside in a unique linear order along a chromosome.

Two-Point Crosses Have Their Limitations

Though of great importance, the pairwise mapping of genes has several shortcomings that limit its usefulness. First, in crosses involving only two genes at a time, it may be difficult to determine gene order if some gene pairs lie very close together. For example, in mapping *y, w,* and *m,* 34.3 m.u. separate the outside genes *y* and *m,* while nearly as great a distance (32.8 m.u.) separates the middle *w* from the outside *m* (Fig. 5.9b). Before being able to conclude with any confidence that *y* and *m* are truly farther apart, that is, that the small difference between the values of 34.3 and 32.8 is not the result of sampling error, you would have to examine a very large number of flies and subject the data to a statistical test, such as the chi-square test.

A second problem with Sturtevant's mapping procedure is that the actual distances in his map do not always add up, even approximately. As an example, suppose that the locus of the *y* gene at the far left of the map is regarded as position 0 (Fig. 5.9c). The *w* gene would then lie near position 1, and *m* would be located in the vicinity of 34 m.u. But what about the *r* gene, named for a mutation that produces rudimentary (very small) wings? Based solely on its distance from *y*, as inferred from the $y \leftrightarrow r$ data in Fig. 5.9a, we would place it at position 42.9 (Fig. 5.9c). However, if we calculate its position as the sum of all intervening distances inferred from the data in Fig. 5.9a, that is, as the sum of $y \leftrightarrow w$ plus $w \leftrightarrow v$ plus $v \leftrightarrow m$ plus $m \leftrightarrow r$, the locus of *r* becomes $1.1 + 32.1 + 4.0 + 17.8 = 55.0$ (Fig. 5.9d). What can explain this difference, and which of these two values is closer to the truth? Three-point crosses help provide some of the answers.

Three-Point Crosses: A Faster, More Accurate Way to Map Genes

The simultaneous analysis of three markers makes it possible to obtain enough information to position the three genes in relation to each other from just one set of crosses. To describe this procedure, we look at three genes linked on one of *Drosophila*'s autosomes.

A homozygous female with mutations for vestigial wings (*vg*), black body (*b*), and purple eye color (*pr*) was mated to a wild-type male (**Fig. 5.10a**). All the triply heterozygous F_1 progeny, both male and female, had normal phenotypes for the three characteristics, indicating that the mutations are autosomal recessive. In a testcross of the F_1 females with males having vestigial wings, black body, and purple eyes, the progeny were of eight different phenotypes reflecting eight different genotypes. The order in which the genes in each phenotypic class are listed in Fig. 5.10a is completely arbitrary; instead of *vg b pr*, one could write *b vg pr* or *vg pr b* to indicate the same genotype. This is because at the outset, we do not know the gene order. Deducing it is the goal of the mapping study. Thus, the way the data are tabulated does not necessarily indicate the actual order of genes along the chromosome.

In analyzing the data, we look at two genes at a time (recall that the recombination frequency is always a function of a pair of genes). For the pair *vg* and *b*, the parental combinations are *vg b* and vg^+b^+; the nonparental recombinants are $vg\ b^+$ and vg^+b. To determine whether a particular class of progeny is parental or recombinant for *vg* and *b*, we do not care whether the flies are *pr* or pr^+. Thus, to the nearest tenth of a map unit, the $vg \leftrightarrow b$ distance, calculated as the percentage of recombinants in the total number of progeny, is

$$\frac{252 + 241 + 131 + 118}{4197} \times 100$$

$$= 17.7 \text{ m.u. } (vg \leftrightarrow b \text{ distance})$$

Similarly, since recombinants for the *vg–pr* gene pair are $vg\ pr^+$ and $vg^+\ pr$, the interval between these two genes is

$$\frac{252 + 241 + 13 + 9}{4197} \times 100$$

$$= 12.3 \text{ m.u. } (vg \leftrightarrow pr \text{ distance})$$

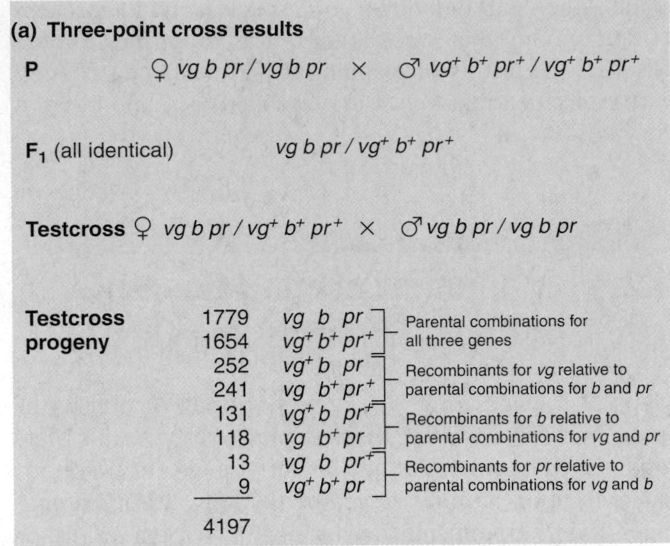

(a) Three-point cross results
P ♀ vg b pr / vg b pr × ♂ vg⁺ b⁺ pr⁺ / vg⁺ b⁺ pr⁺
F1 (all identical) vg b pr / vg⁺ b⁺ pr⁺
Testcross ♀ vg b pr / vg⁺ b⁺ pr⁺ × ♂ vg b pr / vg b pr
Testcross progeny
1779 vg b pr
1654 vg⁺ b⁺ pr⁺
Parental combinations for all three genes
252 vg⁺ b pr
241 vg b⁺ pr⁺
Recombinants for vg relative to parental combinations for b and pr
131 vg⁺ b pr⁺
118 vg b⁺ pr
Recombinants for b relative to parental combinations for vg and pr
13 vg b pr⁺
9 vg⁺ b⁺ pr
Recombinants for pr relative to parental combinations for vg and b
4197

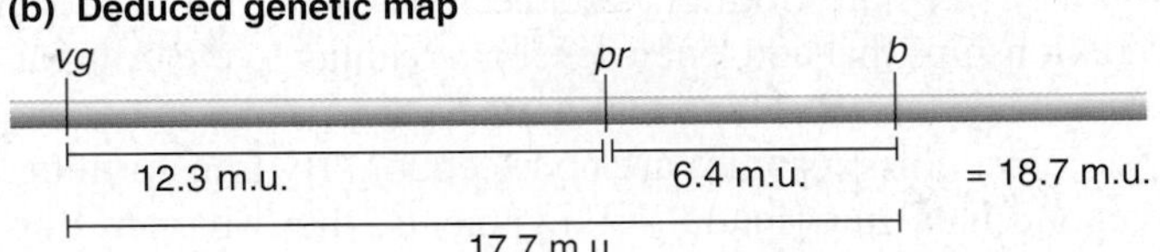

Figure 5.10 Analyzing the results of a three-point cross. **(a)** Results from a three-point testcross of F_1 females simultaneously heterozygous for *vg, b,* and *pr.* **(b)** The gene in the middle must be *pr* because the longest distance is between the other two genes: *vg* and *b*. The most accurate map distances are calculated by summing shorter intervening distances, so 18.7 m.u. is a more accurate estimate of the genetic distance between *vg* and *b* than 17.7 m.u.

while the distance separating the *b–pr* pair is

$$\frac{131 + 118 + 13 + 9}{4197} \times 100$$

$$= 6.4 \text{ m.u. } (b \leftrightarrow pr \text{ distance})$$

These recombination frequencies show that *vg* and *b* are separated by the largest distance (17.7 m.u., as compared with 12.3 and 6.4) and must therefore be the outside genes, flanking *pr* in the middle (Fig. 5.10b). But as with the X-linked *y* and *r* genes analyzed by Sturtevant, the distance separating the outside *vg* and *b* genes (17.7) does not equal the sum of the two intervening distances ($12.3 + 6.4 = 18.7$). In the next section, we learn that the reason for this discrepancy is the rare occurrence of double crossovers.

Three-Point Crosses Allow Correction for Double Crossovers

Figure 5.11 depicts the homologous autosomes of the F_1 females that are heterozygous for the three genes *vg, pr,* and *b*. A close examination of the chromosomes reveals the kinds of crossovers that must have occurred to generate the

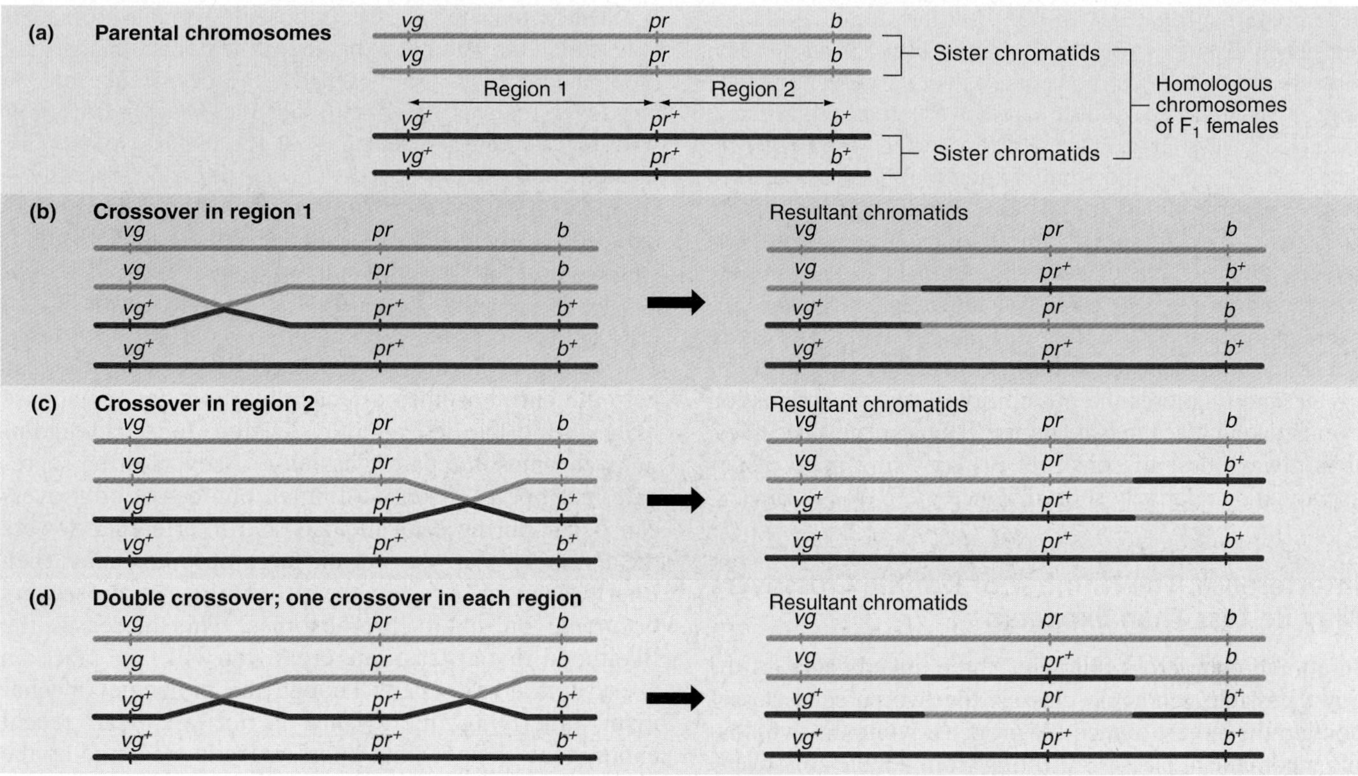

Figure 5.11 **Inferring the location of a crossover event.** Once you establish the order of genes involved in a three-point cross, it is easy to determine which crossover events gave rise to particular recombinant gametes. Note that double crossovers are needed to generate gametes in which the gene in the middle has recombined relative to the parental combinations for the genes at the ends.

classes and numbers of progeny observed. In this and subsequent figures, the chromosomes depicted are in late prophase/early metaphase of meiosis I, when there are four chromatids for each pair of homologous chromosomes. As we have suggested previously and demonstrate more rigorously later, prophase I is the stage at which recombination takes place. Note that we call the space between *vg* and *pr* "region 1" and the space between *pr* and *b* "region 2."

Recall that the progeny from a testcross between triply heterozygous F_1 females and males homozygous for the recessive allele of all three traits fall into eight groups (review Fig. 5.10). Flies in the two largest groups carry the same configurations of genes as did their grandparents of the P generation: *vg b pr* and $vg^+ b^+ pr^+$; they thus represent the parental classes (Fig. 5.11a). The next two groups—$vg^+ b\ pr$ and $vg\ b^+ pr^+$—are composed of recombinants that must be the reciprocal products of a crossover in region 1 between *vg* and *pr* (Fig. 5.11b). Similarly the two groups containing $vg^+\ b\ pr^+$ and $vg\ b^+\ pr$ flies must have resulted from recombination in region 2 between *pr* and *b* (Fig. 5.11c).

But what about the two smallest groups made up of rare $vg\ b\ pr^+$ and $vg^+\ b^+\ pr$ recombinants? What kinds of chromosome exchange could account for them? Most likely, they result from two different crossover events occurring simultaneously, one in region 1, the other in region 2 (Fig. 5.11d). The gametes produced by such double crossovers still have the parental configuration for the outside genes *vg* and *b*, even though not one but two exchanges must have occurred.

Because of the existence of double crossovers, the $vg \leftrightarrow b$ distance of 17.7 m.u. calculated in the previous section does not reflect all of the recombination events producing the gametes that gave rise to the observed progeny. To correct for this oversight, it is necessary to adjust the recombination frequency by adding the double crossovers twice, since each individual in the double crossover groups is the result of two exchanges between *vg* and *b*. The corrected distance is

$$\frac{252 + 241 + 131 + 118 + 13 + 13 + 9 + 9}{4197} \times 100$$

$$= 18.7 \text{ m.u.}$$

This value makes sense because you have accounted for all of the crossovers that occur in region 1 as well as all of the crossovers in region 2. As a result, the corrected value of 18.7 m.u. for the distance between *vg* and *b* is now exactly the same as the sum of the distances between *vg* and *pr* (region 1) and between *pr* and *b* (region 2).

As previously discussed, when Sturtevant originally mapped several X-linked genes in *Drosophila* by two-point crosses, the locus of the rudimentary wings (*r*) gene was ambiguous. A two-point cross involving *y* and *r* gave a recombination frequency of 42.9, but the sum of all the intervening distances was 55.0 (review Fig. 5.9 on p. 135). This discrepancy occurred because the two-point cross (see Fig. 5.9c) ignored double crossovers that might have occurred in the large interval between the *y* and *r* genes. The data summing the smaller intervening distances (see Fig. 5.9d) accounted for at least some of these double crossovers by catching recombinations of gene pairs between *y* and *r*. Moreover, each smaller distance is less likely to encompass a double crossover than a larger distance, so each number for a smaller distance is inherently more accurate. Note that even a three-point cross like the one for *vg, pr,* and *b* ignores the possibility of two recombination events taking place in, say, region 1. For greatest accuracy, it is always best to construct a map using many genes separated by relatively short distances.

Interference: The Number of Double Crossovers May Be Less Than Expected

In a three-point cross following three linked genes, of the eight possible genotypic classes, the two parental classes contain the largest number of progeny, while the two double recombinant classes, resulting from double crossovers, are always the smallest (see Fig. 5.10). We can understand why double-crossover progeny are the rarest by looking at the probability of their occurrence. If an exchange in region 1 of a chromosome does not affect the probability of an exchange in region 2, the probability that both will occur simultaneously is the product of their separate probabilities (recall the product rule in Chapter 2, p. 21). For example, if progeny resulting from recombination in region 1 alone account for 10% of the total progeny (that is, if region 1 is 10 m.u.) and progeny resulting from recombination in region 2 alone account for 20%, the probability of a double crossover (one event in region 1, the second in region 2) is $0.10 \times 0.20 = 0.02$, or 2%. This makes sense because the likelihood of two rare events occurring simultaneously is even less than that of either rare event occurring alone.

As we have seen, in a three-point cross of the *vg pr b* trio of genes, the two classes of progeny arising from double crossovers contain the fewest progeny. The numerical frequencies of observed double crossovers, however, do not coincide with expectations derived from the law of the product. Let's look at the actual numbers. The probability of a single crossover between *vg* and *pr* is 0.123 (corresponding to 12.3 m.u.), and the probability of a single crossover between *pr* and *b* is 0.064 (6.4 m.u.). The product of these probabilities is

$$0.123 \times 0.064 = 0.0079 = 0.79\%$$

But the observed proportion of double crossovers (see Fig. 5.10) was

$$\frac{13 + 9}{4197} \times 100 = 0.52\%$$

The fact that the number of observed double crossovers is less than the number expected if the two exchanges are independent events suggests that the occurrence of one crossover reduces the likelihood that another crossover will occur in an adjacent part of the chromosome. This phenomenon—of crossovers not occurring independently—is called **chromosomal interference.** Interference may exist to ensure that every pair of homologous chromosomes undergoes at least one crossover event. The reasoning behind this hypothesis is straightforward: It is critical that every pair of homologous chromosomes sustain one or more crossover events because such events help the chromosomes orient properly at the metaphase plate during the first meiotic division. Indeed, homologous chromosome pairs without crossovers often segregate improperly. If only a limited number of crossovers can occur during each meiosis and interference lowers the number of crossovers on large chromosomes, then interference also raises the probability of crossovers occurring on small chromosomes. This increases the likelihood that at least one crossover will take place on every homologous pair. Though the molecular mechanism underlying interference is not yet clear, recent experiments suggest that interference is mediated by the synaptonemal complex.

Interference is not uniform and may vary even for different regions of the same chromosome. Investigators can obtain a quantitative measure of the amount of interference in different chromosomal intervals by first calculating a **coefficient of coincidence,** defined as the ratio between the actual frequency of double crossovers observed in an experiment and the number of double crossovers expected on the basis of independent probabilities.

$$\text{Coefficient of coincidence} = \frac{\text{frequency observed}}{\text{frequency expected}}$$

For the three-point cross involving *vg, pr,* and *b,* the coefficient of coincidence is

$$\frac{0.52}{0.79} = 0.66$$

The definition of interference itself is

$$\text{Interference} = 1 - \text{coefficient of coincidence}$$

In this case, it is

$$1 - 0.66 = 0.34$$

To understand the meaning of interference, it is helpful to contrast what happens when there is no interference with what happens when it is complete. If interference is 0, the frequency of observed double crossovers equals expectations, and crossovers in adjacent regions of a chromosome occur independently of each other. If interference is complete, that is, if interference = 1 because the coefficient of coincidence = 0, no double crossovers occur in the experimental progeny because one exchange effectively prevents another. As an example, in a particular three-point cross in mice, the recombination frequency for the pair of genes on the left (region 1) is 20, and for the pair of genes on the right (region 2), it is also 20. Without interference, the expected rate of double crossovers in this chromosomal interval is

$$0.20 \times 0.20 = 0.04, \text{ or } 4\%$$

but when investigators observed 1000 progeny of this cross, they found 0 double recombinants instead of the expected 40.

The Arrangement of Alleles in Double Recombinants Indicates the Relative Order of Three Genes

As we pointed out earlier, in three-point crosses of linked genes, the smallest of the eight possible classes of progeny are the two that contain double recombinants generated by double crossovers. It is possible to use the composition of alleles in these double crossover classes to determine which of the three genes lies in the middle, even without calculating any recombination frequencies. Consider again the progeny of a three-point testcross looking at the *vg, pr,* and *b* genes. The F_1 females are $vg\ pr\ b\ /\ vg^+\ pr^+\ b^+$. As Fig. 5.11d demonstrated, testcross progeny resulting from double crossovers in the trihybrid females of the F_1 generation received gametes from their mothers carrying the allelic combinations $vg\ pr^+\ b$ and $vg^+\ pr\ b^+$. In these individuals, the alleles of the *vg* and *b* genes retain their parental associations ($vg\ b$ and $vg^+\ b^+$), while the *pr* gene has recombined with respect to both the other genes ($pr\ b^+$ and $pr^+\ b$; $vg\ pr^+$ and $vg^+\ pr$). The same is true in all three-point crosses: In those gametes formed by double crossovers, the gene whose alleles have recombined relative to the parental configurations of the other two genes must be the one in the middle.

Comprehensive Example: Using a Three-Point Cross to Reanalyze Sturtevant's Map of the *Drosophila* X Chromosome

The technique of looking at double recombinants to discover which gene has recombined with respect to both other genes allows immediate clarification of gene order even in otherwise difficult cases. Consider the three X-linked genes *y, w,* and *m* that Sturtevant located in his original mapping experiment (see Fig. 5.9 on p. 135). Since the distance between *y* and *m* (34.3 m.u.) appeared slightly larger than the distance separating *w* and *m* (32.8 m.u.), he concluded that *w* was the gene in the middle. But because of the small difference between the two numbers, his conclusion was subject to questions of statistical significance. If, however, we

♀ $w^+\ w\ y^+\ y\ m^+\ m$ × ♂ X / Y

Before data analysis, you do not know the gene order or allele combination on each chromosome.

Male progeny		
2278	$w^+\ y^+\ m$ /Y	Parental class (noncrossover)
2157	$w\ y\ m^+$ /Y	
1203	$w\ y\ m$ /Y	Crossover in region 2 (between *w* and *m*)
1092	$w^+\ y^+\ m^+$ /Y	
49	$w^+\ y\ m$ /Y	Crossover in region 1 (between *y* and *m*)
41	$w\ y^+\ m^+$ /Y	
2	$w^+\ y\ m^+$ /Y	Double crossovers
1	$w\ y^+\ m$ /Y	
6823		

After data analysis, you can conclude that the gene order and allele combinations on the X chromosomes of the F_1 females were $y\ w\ m^+ / y^+\ w^+\ m$.

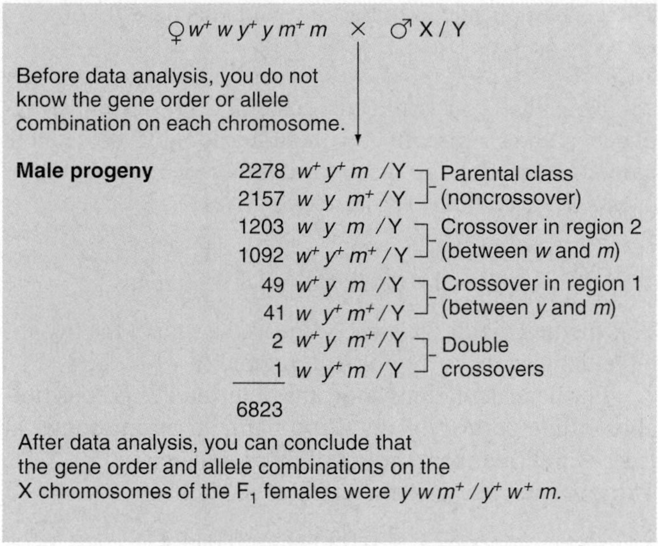

Figure 5.12 How three-point crosses verify Sturtevant's map. The parental classes correspond to the two X chromosomes in the F_1 female. The makeup of the double recombinant classes shows that *w* must be the gene in the middle.

look at a three-point cross following *y, w,* and *m,* these questions disappear.

Figure 5.12 tabulates the classes and numbers of male progeny arising from females heterozygous for the *y, w,* and *m* genes. Since these male progeny receive their only X chromosome from their mothers, their phenotypes directly indicate the gametes produced by the heterozygous females. In each row of the figure's table, the genes appear in an arbitrary order that does not presuppose knowledge of the actual map. As you can see, the two classes of progeny listed at the top of the table outnumber the remaining six classes, which indicates that all three genes are linked to each other. Moreover, these largest groups, which are the parental classes, show that the two X chromosomes of the heterozygous females were $w^+\ y^+\ m$ and $w\ y\ m^+$.

Among the male progeny in Fig. 5.12, the two smallest classes, representing the double crossovers, have X chromosomes carrying $w^+\ y\ m^+$ and $w\ y^+\ m$ combinations, in which the *w* alleles are recombined relative to those of *y* and *m*. The *w* gene must therefore lie between *y* and *m,* verifying Sturtevant's original assessment.

To complete a map based on the *w y m* three-point cross, you can calculate the interval between *y* and *w* (region 1)

$$\frac{49 + 41 + 1 + 2}{6823} \times 100 = 1.3 \text{ m.u.}$$

as well as the interval between *w* and *m* (region 2)

$$\frac{1203 + 1092 + 2 + 1}{6823} \times 100 = 33.7 \text{ m.u.}$$

The genetic distance separating *y* and *m* is the sum of

$$1.3 + 33.7 = 35.0 \text{ m.u.}$$

Note that you could also calculate the distance between *y* and *m* directly by including double crossovers twice, to account for the total number of recombination events detected between these two genes.

$$\text{RF} = (1203 + 1092 + 49 + 41 + 2 + 2 + 1 + 1)/6823 \times 100 = 35.0 \text{ m.u.}$$

This method yields the same value as the sum of the two intervening distances (region 1 + region 2).

Further calculations show that interference is considerable in this portion of the *Drosophila* X chromosome, at least as inferred from the set of data tabulated in Fig. 5.12. The percentage of observed double recombinants was

$$3/6823 = 0.00044, \text{ or } 0.044\%$$

(rounding to the nearest thousandth of a percent), while the percentage of double recombinants expected on the basis of independent probabilities by the law of the product is

$$0.013 \times 0.337 = 0.0044, \text{ or } 0.44\%$$

Thus, the coefficient of coincidence is

$$0.044/0.44 = 0.1$$

and the interference is

$$1 - 0.1 = 0.9$$

How Close Is the Correlation Between a Genetic Map and Physical Reality?

In 1911, when Sturtevant first devised his system of genetic mapping based on recombination frequencies, he acknowledged there was at the time no way to correlate a specific gene position with a physical part of the chromosome. Subsequent experiments, however, have shown that genetic maps do in fact accurately depict the order of genes along chromosomes. Thus, as we will see again and again in the many types of experiments presented throughout the remainder of this book, the *order of genes* revealed by mapping techniques corresponds to the order of those same genes along the DNA molecule of a chromosome.

In contrast, the *actual physical distances between genes*—that is, the amount of DNA separating them—does not always show a direct correspondence to genetic map distances. This is largely because the relationship between recombination frequency and physical distance along a chromosome is not simple. One complicating factor is the existence of double, triple, and even more crossovers. When genes are separated by 1 m.u. or less, double crossovers are not significant because the $0.01 \times 0.01 = 0.0001$ probability of their occurring is so small. But for genes separated by 20, 30, or 40 m.u., the probability of double crossovers skewing the data takes on greater significance. A second confounding factor is the 50% limit on the recombination frequency observable in a cross. This limit reduces the precision of RF as a measure of chromosomal distances. No matter how far apart two genes are on a long chromosome, they will never recombine more than 50% of the time. Yet a third problem is that recombination is not uniform even over the length of a single chromosome: Certain "hotspots" are favored sites of recombination, while other areas—often in the vicinity of centromeres—are "recombination deserts" in which few crossovers ever take place.

Ever since Morgan, Sturtevant, and others began mapping, geneticists have generated mathematical equations called **mapping functions** to compensate for the inaccuracies inherent in relating recombination frequencies to physical distances. These equations generally make large corrections for RF values of widely separated genes, while barely changing the map distances separating genes that lie close together. This reflects the fact that multiple recombination events and the 50% limit on recombination do not confound the calculation of distances between closely linked genes. However, the corrections for large distances are at best imprecise, because mapping functions are based on simplifying assumptions (such as no interference) that rarely apply to the specific situation under consideration. Thus, the best way to create an accurate map is still by summing many smaller intervals, locating widely separated genes through linkage to common intermediaries. Because the most accurate genetic maps are drawn from a large series of genetic crosses, they are subject to continual refinement as more and more newly discovered genes are included.

Rates of recombination may differ from species to species. We know this because recent elucidation of the complete DNA sequences of several organisms' genomes has allowed investigators to compare the actual physical distances between genes (in base pairs of DNA) with genetic map distances. They found that in humans, a map unit corresponds on average to about 1 million base pairs. In yeast, however, where the rate of recombination per length of DNA is much higher than in humans, one map unit is approximately 2500 base pairs. Thus, although map units are useful for estimating distances between the genes of an organism, 1% RF can reflect very different expanses of DNA in different organisms.

Recombination rates sometimes even vary between the two sexes of the same species. *Drosophila* provides an extreme example: No recombination occurs during meiosis in males. If you review the examples already discussed in this chapter, you will discover that they all measure recombination among the progeny of doubly heterozygous

Drosophila females. Problem 17 on p. 161 shows how geneticists can exploit the absence of recombination in *Drosophila* males to establish rapidly that genes far apart on the same chromosome are indeed syntenic.

Multiple-Factor Crosses Help Establish Linkage Groups

Genes chained together by linkage relationships are known collectively as a **linkage group.** When enough genes have been assigned to a particular chromosome, the terms *chromosome* and *linkage group* become synonymous. If you can demonstrate that gene *A* is linked to gene *B, B* to *C, C* to *D,* and *D* to *E,* you can conclude that all of these genes are syntenic. When the genetic map of a genome becomes so dense that it is possible to show that any gene on a chromosome is linked to another gene on the same chromosome, the number of linkage groups equals the number of pairs of homologous chromosomes in the species. Humans have 23 linkage groups, mice have 20, and fruit flies have 4 (**Fig. 5.13**). The total genetic distance along a chromosome, which is obtained by adding many short distances between genes, may be much more than 50 m.u. For example, the two long *Drosophila* autosomes are both slightly more than 100 m.u. in length (Fig. 5.13), while the longest human chromosome is approximately 270 m.u. Recall, however, that even with the longest chromosomes, *pairwise* crosses between genes located at the two ends will not produce more than 50% recombinant progeny.

Linkage mapping has practical applications of great importance. For example, the Fast Forward box "Gene

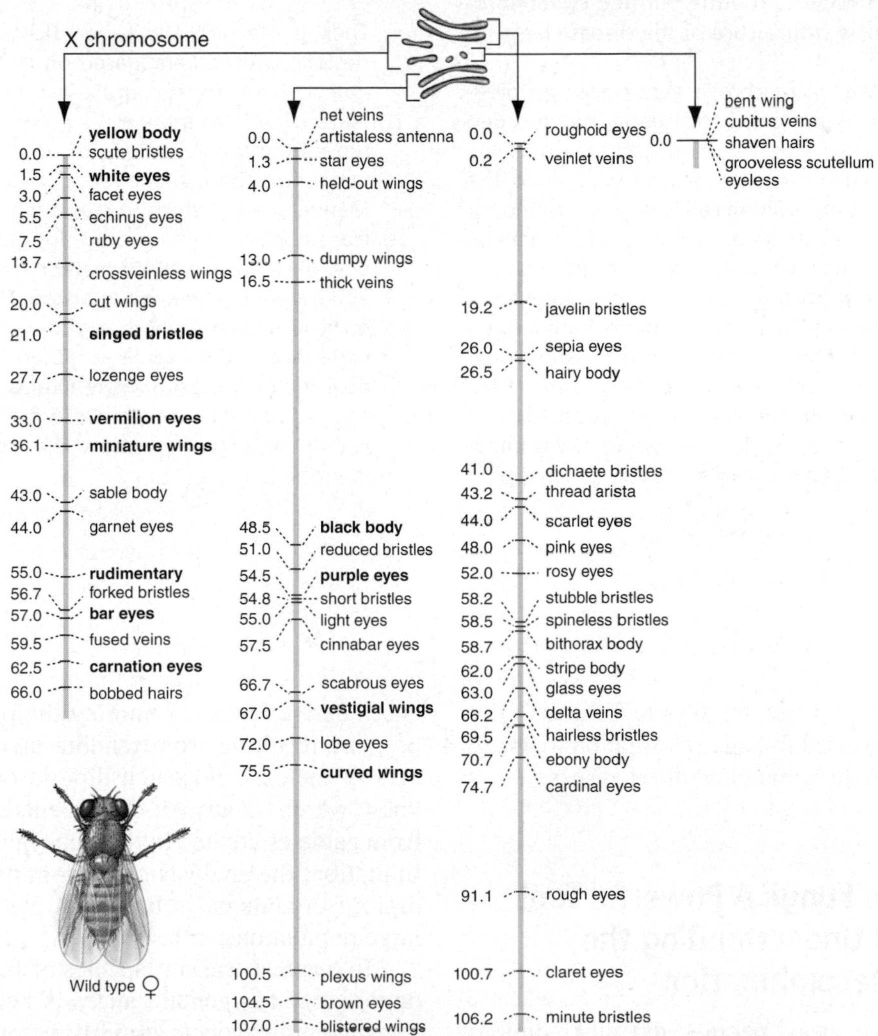

Figure 5.13 ***Drosophila melanogaster*** **has four linkage groups.** A genetic map of the fruit fly, showing the position of many genes influencing visible phenotypes of body morphology, including those used as examples in this chapter (*highlighted in bold*). Because so many *Drosophila* genes have been mapped, each of the four chromosomes can be represented as a single linkage group.

FAST FORWARD

Gene Mapping May Lead to a Cure for Cystic Fibrosis

For 40 years after the symptoms of cystic fibrosis were first described in 1938, there was no molecular clue—no visible chromosomal abnormality transmitted with the disease, no identifiable protein defect carried by affected individuals—suggesting the genetic cause of the disorder. As a result, there was no effective treatment for the 1 in 2000 Caucasian Americans born with the disease, most of whom died before they were 30. In the 1980s, however, geneticists were able to combine recently invented techniques for looking directly at DNA with maps constructed by linkage analysis to pinpoint a precise chromosomal position, or locus, for the cystic fibrosis gene. Knowledge of this locus made it possible to identify and clone the gene, discover the protein it encodes, understand the biochemical mechanisms disrupted by faulty alleles, and develop better therapies for this life-threatening condition. Researchers hope to use the purified cystic fibrosis gene as the basis for a future cure of the disease by gene therapy.

The mappers of the cystic fibrosis gene faced an overwhelming task. They were searching for a gene that encoded an unknown protein, a gene that had not yet even been assigned to a chromosome. It could lie anywhere amid the 23 pairs of chromosomes in a human cell. Imagine looking for a close friend you lost track of years ago, who might now be anywhere in the world. You would first have to find ways to narrow the search to a particular continent (the equivalent of a specific chromosome in the gene mappers' search); then to a country (the long or short arm of the chromosome); next to the state or province, county, city, or town, and street (all increasingly narrow bands of the chromosome); and finally, to a house address (the locus itself). Here, we briefly summarize how researchers applied some of these steps in mapping the cystic fibrosis gene.

- After a review of many family pedigrees containing first-cousin marriages had confirmed that cystic fibrosis is most likely determined by a single gene (the *CF* gene), investigators collected white blood cells from 47 families with two or more affected children, obtaining genetic data from 106 patients, 94 parents, and 44 unaffected siblings.

- They next tried to discover if any other trait is reliably transmitted with the condition. Analyses of the easily obtainable serum enzyme paroxonase showed that its gene (known as the *PON* gene) is indeed linked to the *CF* gene. At first, this knowledge was not that helpful, because the *PON* gene had not yet been assigned to a chromosome.

- Then, in the early 1980s, geneticists developed a large series of DNA markers, based on new techniques that enabled them to recognize variations in the genetic material. A **DNA marker** is a piece of DNA of known size, representing a specific locus, that comes in identifiable variations. These allelic variations segregate according to Mendel's laws, which means it is possible to follow their transmission as you would any gene's. For example, if one allelic form of DNA marker A cosegregates with the abnormal *CF* allele, while another form of marker A assorts with the normal copy of the gene, these identifiable variations in DNA serve as molecular landmarks located near the *CF* gene. Chapters 11 and 12 explain the discovery and use of DNA markers in greater detail; for now, it is only important to know that they exist and can be identified.

Mapping May Lead to a Cure for Cystic Fibrosis" describes how researchers used linkage information to locate the gene for this important human hereditary disease.

Tetrad Analysis in Fungi: A Powerful Tool for Mapping and Understanding the Mechanisms of Recombination

With *Drosophila,* mice, peas, people, and other diploid organisms, each individual represents only one of the four potential gametes generated by each parent in a single meiotic event. Thus, until now, our presentation of linkage, recombination, and mapping has depended on inferences derived from examining the phenotypes of diploid progeny resulting from random unions of random products of meiosis. For such diploid organisms, we do not know which, if any, of the parents' other progeny arose from gametes created in the same meiosis. Because of this limitation, the analysis of random products of meiosis in diploid organisms is based on statistical samplings of large populations.

In contrast, various species of fungi provide a unique opportunity for genetic analysis because they house all four haploid products of each meiosis in a sac called an **ascus** (plural, *asci*). These haploid cells, or **ascospores** (also known as *haplospores*), can germinate and survive as viable haploid individuals that grow and perpetuate themselves by mitosis. The phenotype of such haploid fungi is

By 1986, linkage analyses of hundreds of DNA markers had shown that one marker, known as *D7S15,* is linked with both the *PON* gene and the *CF* gene. Once they had established the fact of linkage, researchers computed recombination frequencies and found that the distance from the DNA marker to the *CF* gene was 15 cM; from the DNA marker to the *PON* gene, 5 cM; and from *PON* to *CF,* 10 cM. After repeating their analyses with an ever-extending group of families, they concluded that the order of the three loci was *D7S15-PON-CF* (**Fig. A**). Since the *CF* gene could lie 15 cM in either of two directions from the DNA marker, the area under investigation was approximately 30 cM. And since the human genome consists of roughly 3000 cM, this step of linkage analysis narrowed the search to 1% of the human genome.

- Next, the DNA marker *D7S15* was localized to the long arm of chromosome 7, which meant that the gene for cystic fibrosis also resides in that same region of chromosome 7. Researchers had now placed the *CF* gene in a certain country on a particular genetic continent.

- Finally, investigators discovered linkage with several other markers on the long arm of chromosome 7. These included a DNA marker known as *J3.11,* a gene for a key molecule of the immune system (the β chain of the T-cell receptor), and an "oncogene" called *met* (a gene which has mutant alleles that contribute to cancer). Two of the markers turned out to be separated from the *CF* gene by a distance of only 1 cM. With all these markers, it became possible to place the *CF* gene in the middle third of chromosome 7's long arm, on band 31 (band 7q31, Fig. A). For families with at least one child who has cystic fibrosis, geneticists using DNA analyses of these closely linked markers could now identify carriers of an abnormal copy of the *CF* gene with substantial confidence.

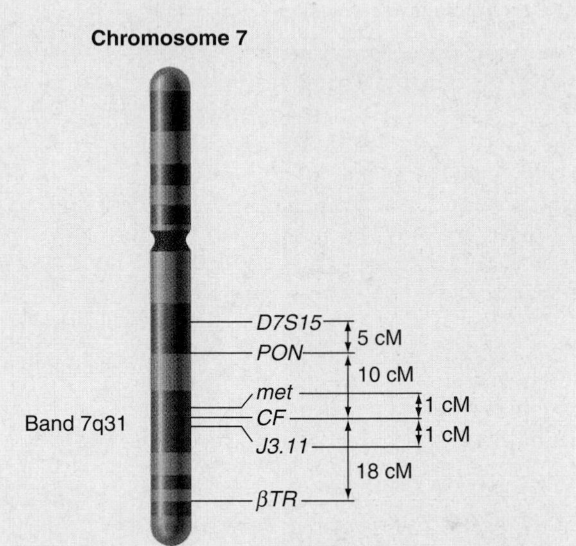

Figure A **How molecular markers helped locate the gene for cystic fibrosis *(CF).***

By 1989, researchers had used this mapping information to identify and clone the *CF* gene on the basis of its location. And by 1992, they had shown it encodes a cell membrane protein that regulates the flow of chloride ions into and out of cells (review the Fast Forward box "Genes Encode Proteins" in Chapter 2). This knowledge has become the basis of new therapies to open up ion flow (older therapies treated the results of flow blockage with only minimal success), as well as gene therapies to introduce normal copies of the *CF* gene into the cells of CF patients. Although only in the early stages of development, such gene therapy holds out hope of an eventual cure for cystic fibrosis.

a direct representation of their genotype, without complications of dominance. **Figure 5.14** illustrates the life cycles of two fungal species that preserve their meiotic products in a sac. One, the normally unicellular baker's yeast (*Saccharomyces cerevisiae*), is sold in supermarkets and contributes to the texture, shape, and flavor of bread; it generates four ascospores with each meiosis. The other, *Neurospora crassa,* is a bread mold that renders the bread on which it grows inedible; it too generates four ascospores with each meiosis, but at the completion of meiosis, each of the four haploid ascospores immediately divides once by mitosis to yield four pairs, for a total of eight haploid cells. The two cells in each pair of *Neurospora* ascospores have the same genotype, because they arose from mitosis.

Haploid cells of both yeast and *Neurospora* normally reproduce vegetatively (that is, asexually) by mitosis. However, sexual reproduction is possible because the haploid cells come in two mating types, and cells of opposite mating types can fuse to form a diploid zygote (Fig. 5.14). In baker's yeast, these diploid cells are stable and can reproduce through successive mitotic cycles. Stress, such as that caused by a scarcity or lack of essential nutrients, induces the diploid cells of yeast to enter meiosis. In bread mold, the diploid zygote instead immediately undergoes meiosis, so the diploid state is only transient.

Mutations in haploid yeast and mold affect many different traits, including the appearance of the cells and their ability to grow under particular conditions. For instance, yeast cells with the *his4* mutation are unable to grow in the

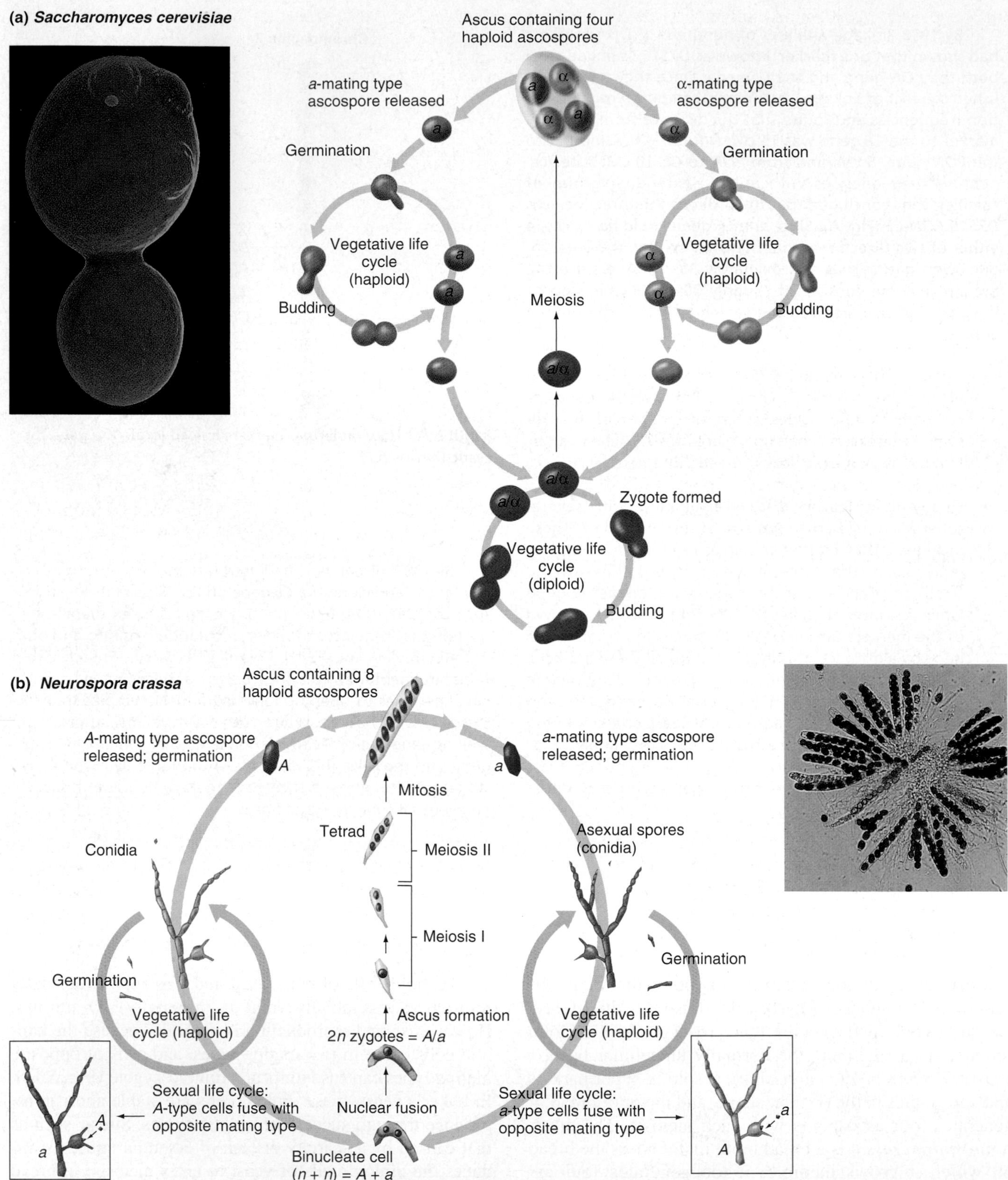

Figure 5.14 The life cycles of the yeast *Saccharomyces cerevisiae* and the bread mold *Neurospora crassa*. Both *S. cerevisiae* and *N. crassa* have two mating types that can fuse to form diploid cells that undergo meiosis. **(a)** Yeast cells can grow vegetatively either as haploids or diploids. The products of meiosis in a diploid cell are four haploid ascospores that are arranged randomly in unordered yeast asci. **(b)** The diploid state in *Neurospora* exists only for a short period. Meiosis in *Neurospora* is followed by mitosis, to give eight haploid ascospores in the ascus. The ordered arrangement of spores in *Neurospora* asci reflects the geometry of the meiotic and mitotic spindles. The photograph showing a budding (mitotically dividing) yeast cell in part (a) is at much higher magnification that the photograph displaying *Neurospora* asci in part (b).

absence of the amino acid histidine, while yeast with the *trp1* mutation cannot grow without an external source of the amino acid tryptophan. Geneticists who specialize in the study of yeast have devised a system of representing genes that is slightly different from the ones for *Drosophila* and mice. They use capital letters (*HIS4*) to designate dominant alleles and lowercase letters (*his4*) to represent recessives. For most of the yeast genes we will discuss, the wild-type alleles are dominant and may be represented by the alternative shorthand "+", while the symbol for the recessive alleles remains the lowercase abbreviation (*his4*). Remember, however, that dominance or recessiveness is relevant only for diploid yeast cells, not for haploid cells that carry only one allele.

After meiosis, the assemblage of four ascospores (or four pairs of ascospores) in a single ascus is called a **tetrad.** (Note that this is a second meaning for the term *tetrad.* In Chapter 4, a tetrad was the four homologous chromatids—two in each chromosome of a bivalent—synapsed during the prophase and metaphase of meiosis I. Here, it is the four products of a single meiosis held together in a sac. Since the four chromatids of a bivalent give rise to the four products of meiosis, the two meanings of tetrad refer to almost the same things.) In yeast, each tetrad is **unordered;** that is, the four meiotic products, known as spores, are arranged at random within the ascus. In *Neurospora crassa,* each tetrad is **ordered,** with the four pairs, or eight haplospores, arranged in a line.

To analyze both unordered and ordered tetrads, researchers can release the spores of each ascus, induce the haploid cells to germinate under appropriate conditions, and then analyze the genetic makeup of the resulting haploid cultures. The data they collect in this way enable them to identify the four products of a single meiosis and compare them with the four products of many other distinct meioses. Ordered tetrads offer another possibility. With the aid of a dissecting microscope, investigators can recover the ascospores in the order in which they occur within the ascus and thereby obtain additional information that is useful for mapping. We look first at the analysis of randomly arranged spores, using the unordered tetrads of yeast as an example. We then describe the additional information that can be gleaned from the microanalysis of ordered tetrads, using *Neurospora* as our model organism.

Tetrads Can Be Characterized by the Number of Parental and Recombinant Spores They Contain

What kinds of tetrads arise when diploid yeast cells heterozygous for two genes on different chromosomes are induced to undergo meiosis? Consider a mating between a haploid strain of yeast of mating type *a,* carrying the *his4* mutation and the wild-type allele of the *TRP1* gene, and a strain of the opposite mating type α that has the genotype *HIS4 trp1*. The resulting *a*/α diploid cells are *his4 / HIS4; trp1 / TRP1,* as shown in **Fig. 5.15a.** (In genetic nomenclature, a semicolon [;] is usually employed to separate genes on nonhomologous chromosomes.) When conditions promote meiosis, the two unlinked genes will assort independently to produce equal frequencies of two different kinds of tetrads. In one kind, all the spores are parental in that the genotype of each spore is the same as one of the parents: *his4 TRP1* or *HIS4 trp1* (Fig. 5.15b). A tetrad that contains four parental class haploid cells is known as a **parental ditype (PD).** Note that *di-*, meaning two, indicates there are two possible parental combinations of alleles; the PD tetrad contains two of each combination. The second kind of tetrad, arising from the equally likely alternative distribution of chromosomes during meiosis, contains four recombinant spores: two *his4 trp1* and two *HIS4 TRP1* (Fig. 5.15c). This kind of tetrad is termed a **nonparental ditype (NPD),** because the two parental classes have recombined to form two reciprocal nonparental combinations of alleles.

A third kind of tetrad also appears when *his4/HIS4; trp1/TRP1* cells undergo meiosis. Called a **tetratype (T)** from the Greek word for "four," it carries four kinds of haploid cells: two different parental class spores (one *his4 TRP1* and one *HIS4 trp1*) and two different recombinants (one *his4 trp1* and one *HIS4 TRP1*). Tetratypes result from a crossover between one of the two genes and the centromere of the chromosome on which it is located (Fig. 5.15d).

Figure 5.15e displays the data from one experiment. Bear in mind that the column headings of PD, NPD, and T refer to tetrads (the group of four cells produced in meiosis) and not to individual haploid cells. Because the spores released from a yeast ascus are not arranged in any particular order, the order in which the spores are listed does not matter. The classification of a tetrad as PD, NPD, or T is based solely on the number of parental and recombinant spores found in the ascus.

When PD = NPD, Two Genes Are Unlinked

A cross following two unlinked genes must give equal numbers of individual parental and recombinant spores. This is simply another way of stating Mendel's second law of independent assortment, which predicts a 50% recombination frequency in such cases. Since T tetrads, regardless of their number, contain two recombinant and two nonrecombinant spores and since all four spores in PD tetrads are parental, the only way 50% of the total progeny spores could be recombinant (as demanded by independent assortment) is if the number of NPDs (with four recombinant spores apiece) is the same as the number of PDs. For this reason, if PD = NPD (as in Fig. 5.15e), the two genes must be unlinked, either because they reside on different chromosomes or because they lie very far apart on the same chromosome.

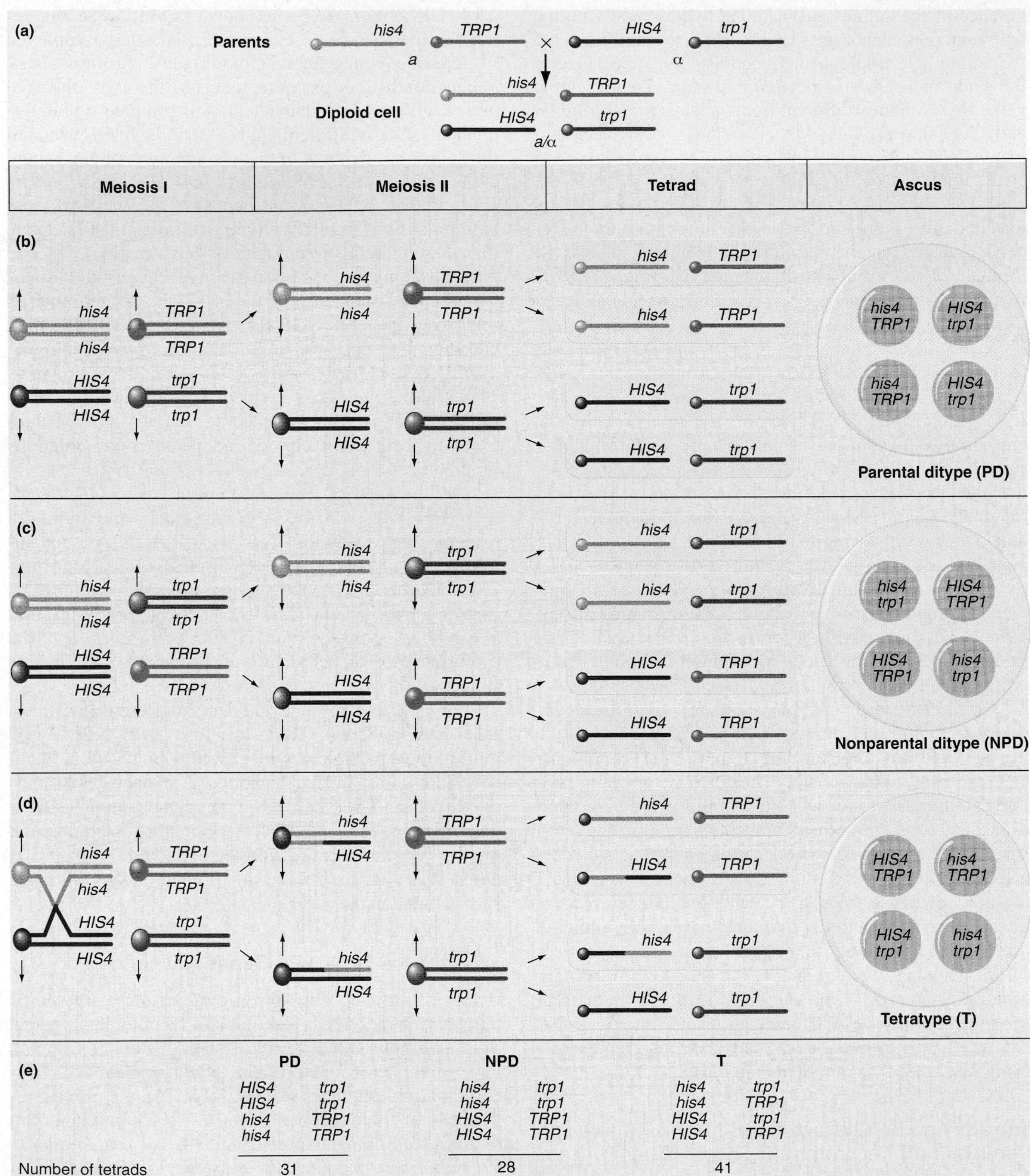

Figure 5.15 How meiosis can generate three kinds of tetrads when two genes are on different chromosomes. **(a)** Parental cross. **(b)** and **(c)** In the absence of recombination, the two equally likely alternative arrangements of two pairs of chromosomes yield either PD or NPD tetrads. T tetrads are made only if either gene recombines with respect to its corresponding centromere, as in **(d)**. Numerical data in **(e)** show that the number of PD tetrads = the number of NPD tetrads when the two genes are unlinked.

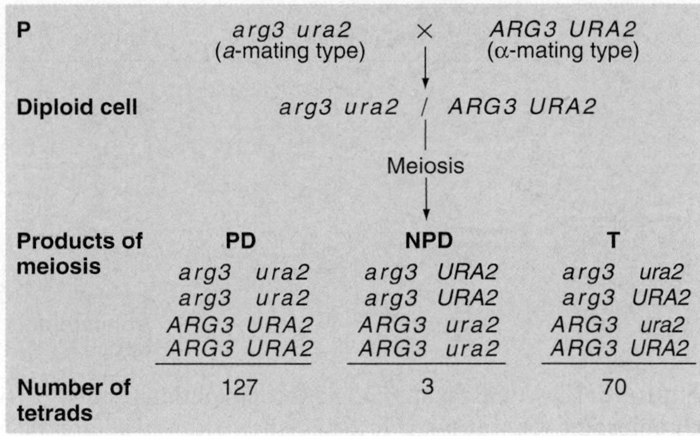

Figure 5.16 **When genes are linked, PDs exceed NPDs.**

When PDs Significantly Outnumber NPDs, Two Genes Are Linked

The genetic definition of linkage is the emergence of more parental types than recombinants among the progeny of a doubly heterozygous parent. In the preceding section, we saw that tetratypes always contribute an equal number of parental and recombinant spores. Thus, with tetrads, linkage exists only when PD >> NPD, that is, when the number of PD tetrads (carrying only parental type spores) substantially exceeds the number of NPD tetrads (containing only recombinants). By analyzing an actual cross involving linked genes, we can see how this follows from the events occurring during meiosis.

A haploid yeast strain containing the *arg3* and *ura2* mutations was mated to a wild-type *ARG3 URA2* haploid strain (**Fig. 5.16**). When the resultant *a*/α diploid was induced to sporulate (that is, undergo meiosis), the 200 tetrads produced had the distribution shown in Fig. 5.16. As you can see, the 127 PD tetrads far outnumber the 3 NPD tetrads, suggesting that the two genes are linked.

Figure 5.17 shows how we can explain the particular kinds of tetrads observed in terms of the various types of crossovers that could occur between the linked genes. If no crossing-over occurs between the two genes, the resulting tetrad must be PD; since none of the four chromatids participates in an exchange, all of the products are of parental configuration (Fig. 5.17a). A single crossover between *ARG3* and *URA2* will generate a tetratype, containing four genetically different spores (Fig. 5.17b). But what about double crossovers? There are actually four different possibilities, depending on which chromatids participate, and each of the four should occur with equal frequency. A double crossover involving only two chromatids (that is, one where both crossovers affect the same two chromatids) produces only parental-type progeny, generating a PD tetrad (Fig. 5.17c). Three-strand double crossovers can occur in the two ways depicted in Fig. 5.17d and e; either way, a tetratype results. Finally, if all four strands take

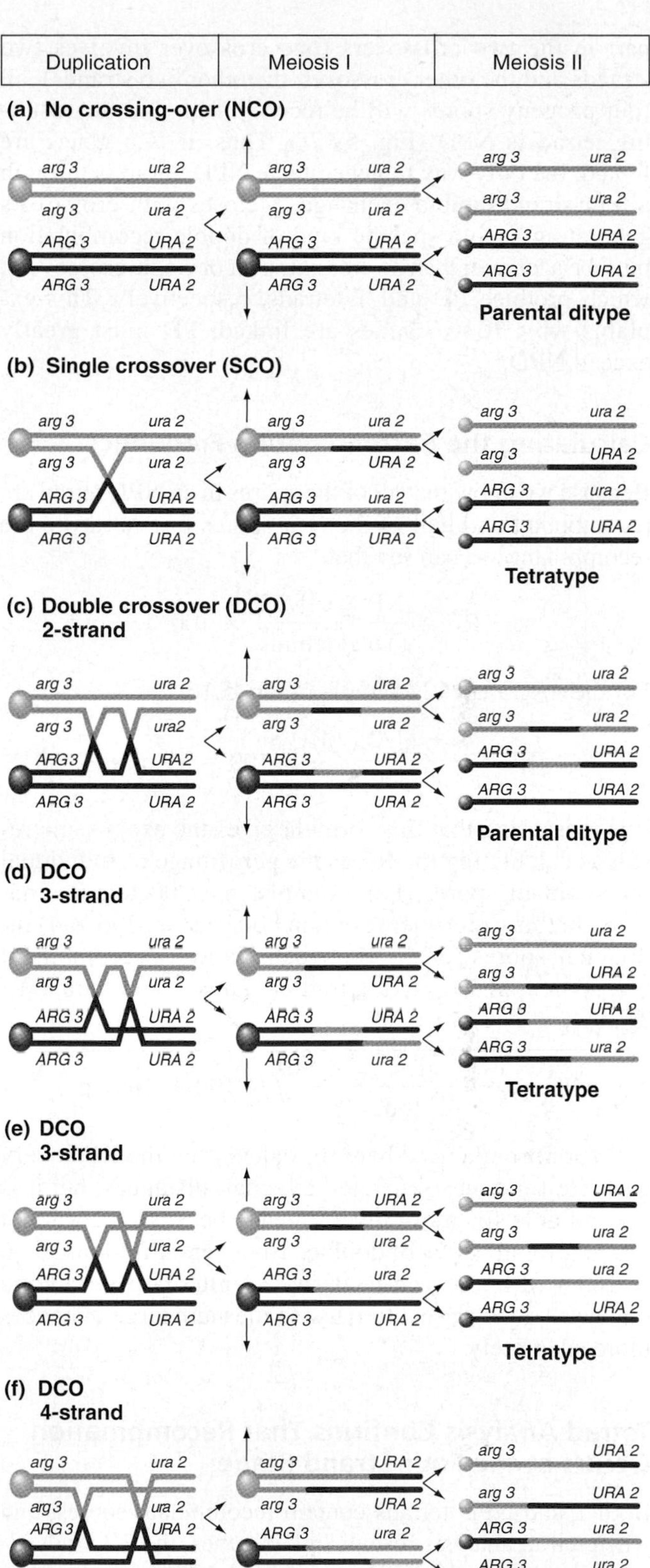

Figure 5.17 **How crossovers between linked genes generate different tetrads.** **(a)** PDs arise when there is no crossing-over. **(b)** Single crossovers between the two genes yield tetratypes. **(c)** to **(f)** Double crossovers between linked genes can generate PD, T, or NPD tetrads, depending on which chromatids participate in the crossovers.

part in the two crossovers (one crossover involves two strands and the other crossover, the other two strands), all four progeny spores will be recombinant, and the resulting tetrad is NPD (Fig. 5.17f). Thus, if two genes are linked, the only way to generate an NPD tetrad is through a four-strand double exchange. Meioses with crossovers generating such a specific kind of double recombination must be a lot rarer than no crossing-over or single crossovers, which produce PD and T tetrads, respectively. This explains why, if two genes are linked, PD must greatly exceed NPD.

Calculating the Recombination Frequency

Because we know that all of the spores in an NPD tetrad are recombinant and half of the four spores in a tetratype are recombinant, we can say that

$$\text{RF} = \frac{\text{NPD} + 1/2\text{T}}{\text{Total tetrads}} \times 100$$

For the *ARG3 URA2* example in Fig. 5.16,

$$\text{RF} = \frac{3 + (1/2)(70)}{200} \times 100 = 19 \text{ m.u.}$$

It is reassuring that this formula gives the exact same result as calculating the RF as the percentage of individual recombinant spores. For example, the 200 tetrads analyzed in this experiment contain 800 (that is, 200 × 4) individual spores; each NPD ascus holds 4 recombinant ascospores, and each T tetrad contains 2 recombinants. Thus,

$$\text{RF} = \frac{(4 \times 3) + (2 \times 70)}{800} \times 100 = 19 \text{ m.u.}$$

The formula used here for calculating the RF is very accurate for genes separated by small distances, but it is less reliable for more distant genes because it does not account for all types of double crossovers. Problem 35 on p. 165 will allow you to derive an alternative equation that yeast geneticists often use to measure large distances more accurately.

Tetrad Analysis Confirms That Recombination Occurs at the Four-Strand Stage

Both T and NPD tetrads contain recombinant spores, and when tetrad analysis reveals linked genes, the T tetrads always outnumber the NPDs, as in the example we have been discussing. This makes sense, because all single and some double crossovers yield tetratypes, while only 1/4 of the rare double crossovers produce NPDs. The very low number of NPDs establishes that recombination occurs after the chromosomes have replicated, when there are four chromatids for each pair of homologs. If recombination took place before chromosome duplication, every single crossover event would yield four recombinant chromatids and generate an NPD tetrad (**Fig. 5.18**). A model assuming that recombination occurs when there are two rather than four chromatids per pair of homologous chromosomes would thus predict more NPD than T tetrads, but experimental observations show just the opposite (see Figs. 5.15e and 5.16).

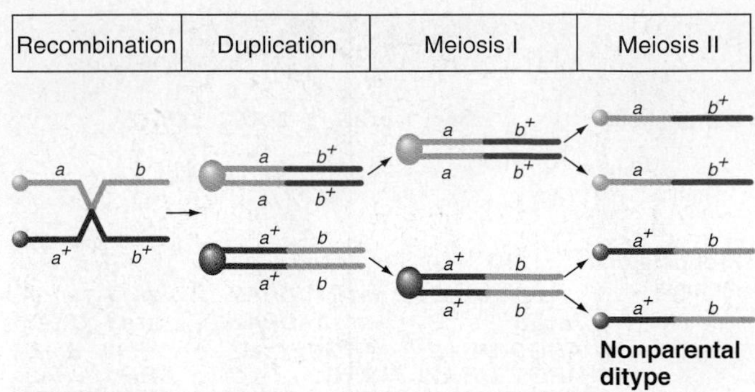

Figure 5.18 A mistaken model: Recombination before chromosome replication. If recombination occurred before the chromosomes duplicated and if two genes were linked, most tetrads containing recombinant spores would be NPDs instead of Ts. Actual results show that the opposite is true.

The fact that recombination takes place after the chromosomes have replicated explains the 50% limit on recombination for genes on the same chromosome. Single crossovers between two genes generate T tetrads containing two out of four spores that are recombinant. Thus, even if one crossover occurred between two such genes in every meiosis, the observed recombination frequency would be 50%. The four kinds of double crossovers depicted in Fig. 5.17 yield either PD tetrads with 0/4 recombinants; T tetrads carrying 2/4 recombinants; differently derived T tetrads, still carrying 2/4 recombinants; and NPD tetrads with 4/4 recombinants. Because these four kinds of double crossovers almost always occur with equal frequency, no more than 50% of the progeny resulting from double (or, in fact, triple or more) crossovers can be recombinant.

Tetrad Analysis Also Demonstrates That Recombination Is Usually Reciprocal

Suppose you are following linked genes *A* and *B* in a cross between *A B* and *a b* strains of yeast. If the recombination that occurs during meiosis is reciprocal, every tetrad with recombinant progeny should contain equal numbers of both classes of recombinants. Observations have in general confirmed this prediction: Every T tetrad carries one *A b* and one *a B* spore, while every NPD tetrad contains two of each type

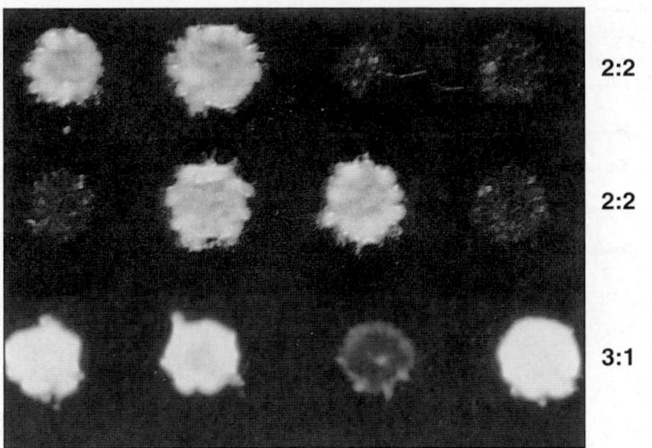

Figure 5.19 In rare tetrads, the two alleles of a gene do not segregate 2:2. Researchers sporulated a *HIS4* / *his4* diploid yeast strain and dissected the four haploid spores from three different tetrads. They then plated these spores on petri plates containing medium without histidine. Each row on the petri plate presents the four spores of a single tetrad. The top two rows show the normal 2:2 segregation of the two alleles of a single gene: two of the spores are *HIS4* and form colonies, whereas the other two spores are *his4* and cannot grow into colonies. The bottom row displays a rare tetrad with an unusual segregation of 3 *HIS4* : 1 *his4*.

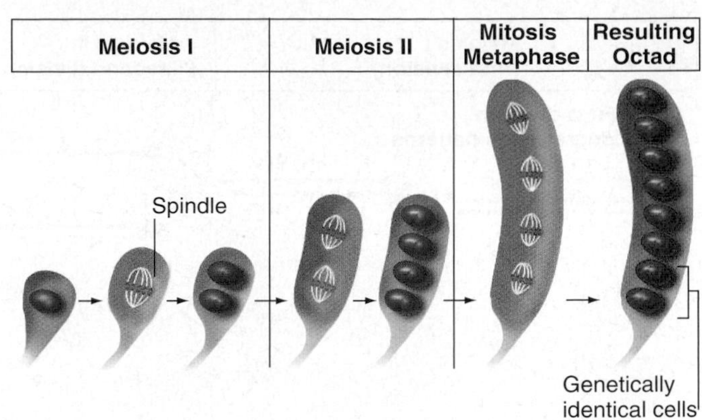

Figure 5.20 How ordered tetrads form. Spindles form parallel to the long axis of the growing *Neurospora* ascus, and the cells cannot slide around each other. The order of ascospores thus reflects meiotic spindle geometry. After meiosis, each haploid cell undergoes mitosis, producing an eight-cell ascus (an octad). The octad consists of four pairs of cells; the two cells of each pair are genetically identical.

of recombinant. We can thus conclude that meiotic recombination is almost always reciprocal, generating two homologous chromosomes that are inverted images of each other.

There are, however, exceptions. Very rarely, a particular cross produces tetrads containing unequal numbers of reciprocal classes, and such tetrads cannot be classified as PD, NPD, or T. In these exceptional tetrads, the two input alleles of one of the genes, instead of segregating at a ratio of 2*A* : 2*a*, produce ratios of 1*A* : 3*a* or 3*A* : 1*a* or even 0*A*: 4*a* or 4*A* : 0*a* (**Fig. 5.19**). In these same tetrads, markers such as *B/b* and *C/c* that flank the *A* or *a* allele on the same chromosome still segregate 2*B* : 2*b* and 2*C* : 2*c*. Moreover, careful phenotypic and genetic tests show that even when alleles do not segregate 2:2, only the original two input alleles occur in the progeny. Thus, recombination, no matter what ratios it generates, does not create new alleles. Geneticists believe that the unusual non-2:2 segregation ratios observed in rare instances result from molecular events at the site of recombination. We discuss these events at the molecular level in Chapter 6. For now, it is simply necessary to know that the unusual ratios exist but are quite rare.

Ordered Tetrads Help Locate Genes in Relation to the Centromere

Analyses of ordered tetrads, such as those produced by the bread mold *Neurospora crassa,* allow you to map the centromere of a chromosome relative to other genetic markers, information that you cannot normally obtain from unordered yeast tetrads. As described earlier, immediately after specialized haploid *Neurospora* cells of different mating types fuse at fertilization, the diploid zygote undergoes meiosis within the confines of a narrow ascus (review Fig. 5.14b on p. 144). At the completion of meiosis, each of the four haploid meiotic products divides once by mitosis, yielding an **octad** of eight haploid ascospores. Dissection of the ascus at this point allows one to determine the phenotype of each of the eight haploid cells.

The cross-sectional diameter of the ascus is so small that cells cannot slip past each other. Moreover, during each division after fertilization, the microtubule fibers of the spindle extend outward from the centrosomes parallel to the long axis of the ascus. These facts have two important repercussions. First, when each of the four products of meiosis divides once by mitosis, the two genetically identical cells that result lie adjacent to each other (**Fig. 5.20**). Because of this feature, starting from either end of the ascus, you can count the octad of ascospores as four cell pairs and analyze it as a tetrad. Second, from the precise positioning of the four ascospore pairs within the ascus, you can infer the arrangement of the four chromatids of each homologous chromosome pair during the two meiotic divisions.

To understand the genetic consequences of the geometry of the ascospores, it is helpful to consider what kinds of tetrads you would expect from the segregation of two alleles of a single gene. (In the following discussion, you will see that *Neurospora* geneticists denote alleles with symbols similar to those used for *Drosophila,* as detailed in the nomenclature guide on p. N-1 of the Appendix.) The mutant *white-spore* allele (*ws*) alters ascospore color from

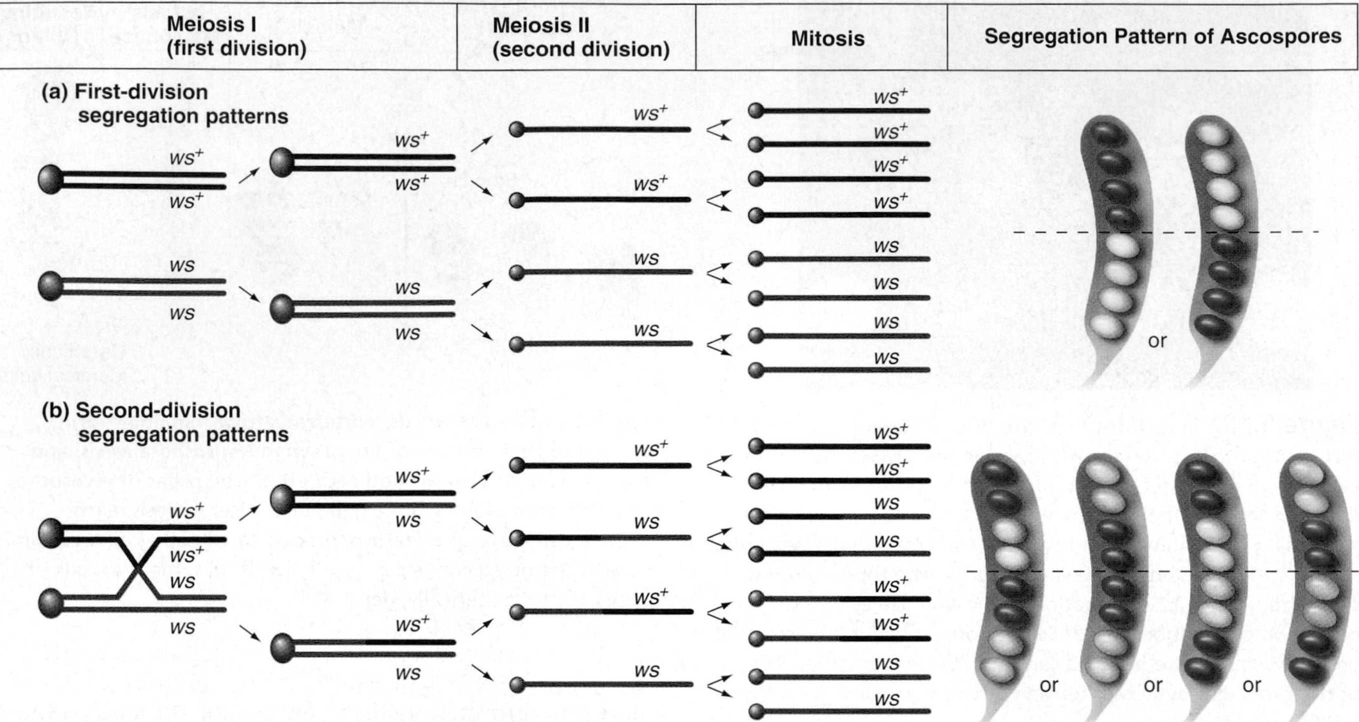

Figure 5.21 **Two segregation patterns in ordered asci.** **(a)** In the absence of a crossover between a gene and its centromere, the two alleles of a gene will separate at the first meiotic division. The result is a first-division segregation pattern in which each allele appears in spores located on only one side of an imaginary line through the middle of the ascus. **(b)** A crossover between a gene and its centromere produces a second-division segregation pattern in which both alleles appear on the same side of the middle line (but in different spores).

wild-type black to white. In the absence of recombination, the two alleles (ws^+ and *ws*) separate from each other at the first meiotic division because the centromeres to which they are attached separate at that stage. The second meiotic division and subsequent mitosis create asci in which the top four ascospores are of one genotype (for instance ws^+) and the bottom four of the other (*ws*). Whether the top four are ws^+ and the bottom four *ws*, or vice versa, depends on the random metaphase I orientation of the homologs that carry the gene relative to the long axis of the developing ascus.

The segregation of two alleles of a single gene at the first meiotic division is thus indicated by an ascus in which an imaginary line drawn between the fourth and the fifth ascospores of the octad cleanly separates haploid products bearing the two alleles. Such an ascus displays a **first-division segregation pattern** (**Fig. 5.21a**).

Suppose now that during meiosis I, a crossover occurs in a heterozygote between the *white-spore* gene and the centromere of the chromosome on which it travels. As Fig. 5.21b illustrates, this can lead to four equally possible ascospore arrangements, each one depending on a particular orientation of the four chromatids during the two meiotic divisions. In all four cases, both ws^+ and *ws* spores are found on both sides of the imaginary line drawn between ascospores 4 and 5, because cells with only one kind of allele do not arise until the end of the second meiotic division. Octads carrying this configuration of spores display a **second-division segregation pattern.**

Since second-division segregation patterns result from meioses in which there has been a crossover between a gene and its centromere, the relative number of asci with this pattern can be used to determine the gene ↔ centromere distance. In an ascus showing second-division segregation, one-half of the ascospores are derived from chromatids that have exchanged parts, while the remaining half arise from chromatids that have not participated in crossovers leading to recombination. To calculate the distance between a gene and its centromere, you therefore simply divide the percentage of second-division segregation octads by 2. Geneticists use information about the location of centromeres to make more accurate genetic maps as well as to study the structure and function of centromeres.

A Numerical Example of Ordered-Tetrad Analysis

In one experiment, a $thr^+ arg^+$ wild-type strain of *Neurospora* was crossed with a *thr arg* double mutant. The *thr* mutants cannot grow in the absence of the amino acid threonine,

(a) A *Neurospora* cross

Tetrad group	A	B	C	D	E	F	G
Segregation pattern	*thr arg* *thr arg* $thr^+ arg^+$ $thr^+ arg^+$	*thr arg* $thr^+ arg$ $thr^+ arg^+$ $thr\ arg^+$	*thr arg* $thr\ arg^+$ $thr^+ arg$ $thr^+ arg^+$	$thr\ arg^+$ $thr^+ arg$ $thr^+ arg^+$ *thr arg*	$thr\ arg^+$ $thr^+ arg$ $thr^+ arg$ $thr\ arg^+$	$thr\ arg^+$ $thr\ arg^+$ $thr^+ arg$ $thr^+ arg$	*thr arg* $thr^+ arg^+$ $thr^+ arg^+$ *thr arg*
Total in group	72	16	11	2	2	1	1

(b) Corresponding genetic map

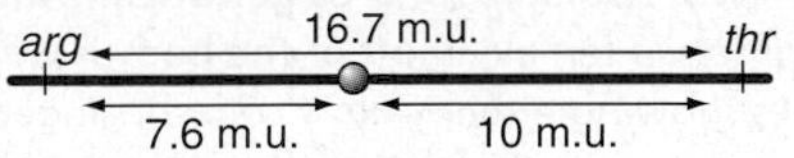

arg — 16.7 m.u. — *thr*

7.6 m.u. — 10 m.u.

Figure 5.22 Genetic mapping by ordered-tetrad analysis: An example. (a) In ordered-tetrad analysis, tetrad classes are defined not only as PD, NPD, or T but also according to whether they show a first- or second-division segregation pattern. Each entry in this table represents a pair of adjacent, identical spores in the actual *Neurospora* octad. **(b)** Genetic map derived from the data in part (a). Ordered-tetrad analysis allows determination of the centromere's position as well as distances between genes.

while *arg* mutants cannot grow without a source of the amino acid arginine; cells carrying the wild-type alleles of both genes can grow in medium that contains neither amino acid. From this cross, 105 octads, considered here as tetrads, were obtained. These tetrads were classified in seven different groups—A, B, C, D, E, F, and G—as shown in **Fig. 5.22a.** For each of the two genes, we can now find the distance between the gene and the centromere of the chromosome on which it is located.

To do this for the *thr* gene, we count the number of tetrads with a second-division segregation pattern for that gene. Drawing an imaginary line through the middle of the tetrads, we see that those in groups B, D, E, and G are the result of second-division segregations for *thr,* while the remainder show first-division patterns. The centromere ↔ *thr* distance is thus

$$\text{Percentage of second-division patterns} = \frac{(1/2)(16 + 2 + 2 + 1)}{105} \times 100 = 10 \text{ m.u.}$$

Similarly, the second-division tetrads for the *arg* gene are in groups C, D, E, and G, so the distance between *arg* and its centromere is

$$\frac{(1/2)(11 + 2 + 2 + 1)}{105} \times 100 = 7.6 \text{ m.u.}$$

To ascertain whether the *thr* and *arg* genes are linked, we need to evaluate the seven tetrad groups in a different way, looking at the combinations of alleles for the two genes to see if the tetrads in that group are PD, NPD, or T. We can then ask whether PD >> NPD. Referring again to Fig. 5.22, we find that groups A and G are PD, because all the ascospores show parental combinations, while groups E and F, with four recombinant spores, are NPD. PD is thus 72 + 1 = 73, while NPD is 1 + 2 = 3. From these data, we can conclude that the two genes are linked.

What is the map distance between *thr* and *arg?* For this calculation, we need to find the numbers of T and NPD tetrads. Tetratypes are found in groups B, C, and D, and we already know that groups E and F carry NPDs. Using the same formula for map distances as the one previously used for yeast,

$$\text{RF} = \frac{\text{NPD} + 1/2\text{T}}{\text{Total tetrads}} \times 100$$

we get

$$\text{RF} = \frac{3 + (1/2)(16 + 11 + 2)}{105} \times 100 = 16.7 \text{ m.u.}$$

Since the distance between *thr* and *arg* is larger than that separating either gene from the centromere, the centromere must lie between *thr* and *arg,* yielding the map in Fig. 5.22b. The distance between the two genes calculated by the formula above (16.7 m.u.) is smaller than the sum of the two gene ↔ centromere distances (10.0 + 7.6 = 17.6 m.u.) because the formula does not account for all of the double crossovers. As always, calculating map positions for more genes with shorter distances between them produces the most accurate picture.

Table 5.3 summarizes some of the procedures for mapping genes in fungi producing ordered and unordered tetrads.

TABLE 5.3 Rules for Tetrad Analysis

For Ordered and Unordered Tetrads

Considering genes two at a time, assign tetrads as PD, NPD, or T.

If PD >> NPD, the two genes are genetically linked.

If PD = NPD, the two genes are genetically independent (unlinked).

The map distance between two genes if they are genetically linked

$$= \frac{\text{NDP} + (1/2)\text{T}}{\text{Total tetrads}} \times 100$$

For Ordered Tetrads Only

The map distance between a gene and its centromere

$$= \frac{(1/2) \times (\text{\# of tetrads showing second-division segregation for this gene})}{\text{Total tetrads}} \times 100$$

5.3 Mitotic Recombination Can Produce Genetic Mosaics

The recombination of genetic material is a critical feature of meiosis. The chiasmata that mark recombination sites keep homologous chromosomes together, allowing them to reach the metaphase plate and subsequently to segregate to opposite spindle poles during meiosis I. Moreover, recombination of genes during meiosis helps ensure the genetic diversity of the gametes. It is thus not surprising that eukaryotic organisms express a variety of enzymes (described in Chapter 6) that specifically initiate meiotic recombination. Recombination can also occur during mitosis. Unlike what happens in

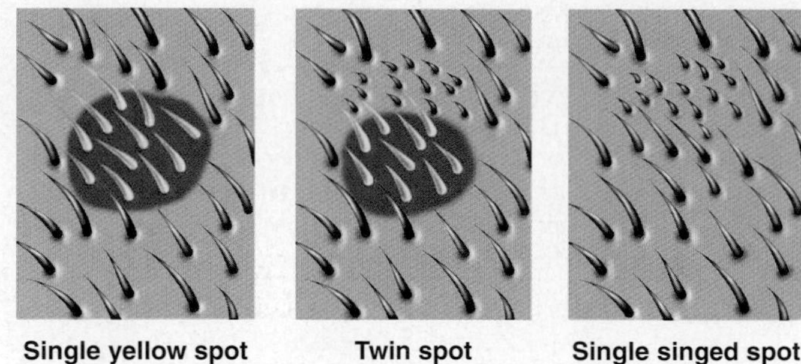

Figure 5.23 Twin spots: A form of genetic mosaicism. In a *y* sn^+ / y^+ *sn Drosophila* female, most of the body is wild type, but aberrant patches showing either yellow color or singed bristles sometimes occur. In some cases, yellow and singed patches are adjacent to each other, a configuration known as *twin spots.*

Transient pairing during mitosis → Mitotic metaphase → Daughter cells

(a) Crossing-over between *sn* and the centromere

sn^+ *y* / *sn* y^+ → sn^+ *y*, *sn* y^+ | sn^+ *y*, *sn* y^+ → Yellow (sn^+ *y*, sn^+ *y*); Singed (*sn* y^+, *sn* y^+) — **Twin spot**

or

sn^+ *y*, *sn* y^+ | *sn* y^+, sn^+ *y* → Wild type (sn^+ *y*, *sn* y^+); Wild type (*sn* y^+, sn^+ *y*) — **Normal tissue**

(b) Crossing-over between *sn* and *y*

sn^+ *y* / *sn* y^+ → sn^+ *y*, sn^+ y^+ | *sn* *y*, *sn* y^+ → Yellow (sn^+ *y*, *sn* *y*) — **Yellow spot**; Wild type (sn^+ y^+, *sn* y^+) — **Normal tissue**

or

sn^+ *y*, sn^+ y^+ | *sn* y^+, *sn* *y* → Wild type (sn^+ *y*, *sn* y^+); Wild type (sn^+ y^+, *sn* *y*) — **Normal tissue**

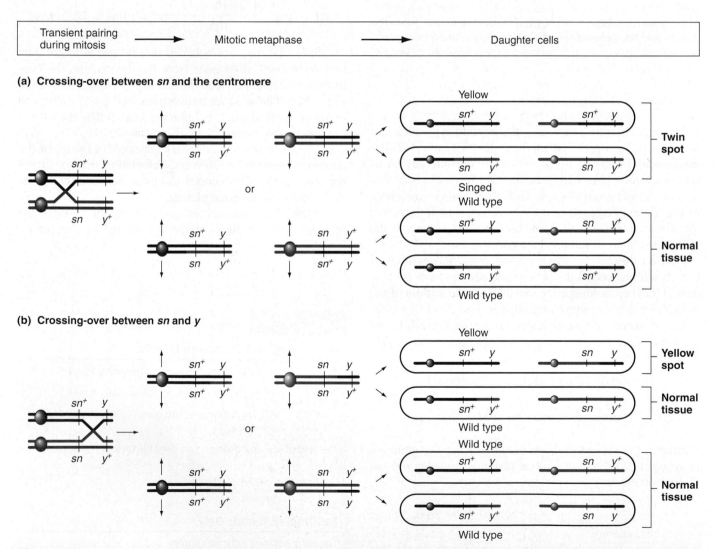

Figure 5.24 Mitotic crossing-over. (a) In a *y* sn^+ / y^+ *sn Drosophila* female, a mitotic crossover between the centromere and *sn* can produce two daughter cells, one homozygous for *y* and the other homozygous for *sn,* that can develop into adjacent aberrant patches (twin spots). This outcome depends on a particular distribution of chromatids at anaphase (*top*). If the chromatids are arranged in the equally likely opposite orientation, only phenotypically normal cells will result (*bottom*). **(b)** Crossovers between *sn* and *y* can generate single yellow patches. In contrast, a single mitotic crossover in these females cannot produce a single singed spot if the *sn* gene is closer to the centromere than the *y* gene. See if you can demonstrate this fact.

meiosis, however, mitotic crossovers are initiated by mistakes in chromosome replication or by chance exposures to radiation that break DNA molecules, rather than by a well-defined cellular program. As a result, mitotic recombination is a rare event, occurring no more frequently than once in a million somatic cell divisions. Nonetheless, the growth of a colony of yeast cells or the development of a complex multicellular organism involves so many cell divisions that geneticists can routinely detect these rare mitotic events.

Curt Stern, an early *Drosophila* geneticist, originally inferred the existence of mitotic recombination from observations of "twin spots" in a few fruit flies. **Twin spots** are adjacent islands of tissue that differ both from each other and from the tissue surrounding them. The distinctive patches arise from homozygous cells with a recessive phenotype growing amid a generally heterozygous cell population displaying the dominant phenotype. In *Drosophila,* the *yellow* (*y*) mutation changes body color from normal brown to yellow, while the *singed bristles* (*sn*) mutation causes body bristles to be short and curled rather than long and straight. Both of these genes are on the X chromosome. In 1936, Stern examined *Drosophila* females of genotype $y\ sn^+ / y^+\ sn$. These heterozygotes were generally wild type in appearance, but Stern noticed that some flies carried patches of yellow body color, others had small areas of singed bristles, and still others displayed twin spots: adjacent patches of yellow cells and cells with singed bristles (**Fig. 5.23**). He assumed that mistakes in the mitotic divisions accompanying fly development could have led to these **mosaic** animals containing tissues of different genotypes. Individual yellow or singed patches could arise from chromosome loss or by mitotic nondisjunction. These errors in mitosis would yield XO cells containing only *y* (but not y^+) or *sn* (but not sn^+) alleles; such cells would show one of the recessive phenotypes.

The twin spots must have a different origin. Stern reasoned that they represented the reciprocal products of mitotic crossing-over between the *sn* gene and the centromere. The mechanism is as follows. During mitosis in a diploid cell, after chromosome duplication, homologous chromosomes occasionally—very occasionally—pair up with each other. While the chromosomes are paired, nonsister chromatids (that is, one chromatid from each of the two homologous chromosomes) can exchange parts by crossing-over. The pairing is transient, and the homologous chromosomes soon resume their independent positions on the mitotic metaphase plate. There, the two chromosomes can line up relative to each other in either of two ways (**Fig. 5.24a**). One of these orientations would yield two daughter cells that remain heterozygous for both genes and thus be indistinguishable from the surrounding wild-type cells. The other orientation, however, will generate two homozygous daughter cells, one $y\ sn^+ / y\ sn^+$, the other $y^+ sn / y^+ sn$. Since the two daughter cells would lie next to each other, subsequent mitotic divisions would produce adjacent patches of *y* and *sn* tissue (that is, twin spots). Note that if crossing-over occurs between *sn* and *y,* single spots of yellow tissue can form, but a reciprocal singed spot cannot be generated in this fashion (Fig. 5.24b).

Diploid yeast cells that are heterozygous for one or more genes exhibit mitotic recombination in the form of **sectors:** portions of a growing colony that have a different genotype than the remainder of the colony. If a diploid yeast cell of genotype *ADE2 / ade2* is placed on a petri plate, its mitotic descendents will grow into a colony. Usually, such colonies will appear white because the dominant wild-type *ADE2* allele specifies that color. However, many colonies will contain red sectors of diploid *ade2 / ade2* cells, which arose as a result of mitotic recombination events between the *ADE2* gene and its centromere (**Fig. 5.25**). (Homozygous *ADE2 / ADE2* cells will also be produced by the same event, but they cannot be distinguished from heterozygotes because both types of cells are white.) The size of the red sectors indicates when mitotic recombination took place. If they are large, it happened early in the growth of the colony, giving the resulting daughter cells a long time to proliferate; if they are small, it happened later.

Mitotic recombination is significant both as an experimental tool and because of the phenotypic consequences of particular mitotic crossovers. Problem 40 on p. 166 at the end of this illustrates how geneticists use mitotic recombination to obtain information for mapping genes relative to each other and to the centromere. Mitotic crossing-over has also been of great value in the study of development because it can generate animals in which different cells have different genotypes (see Chapter 20). Finally, as the Genetics and Society box "Mitotic Recombination and Cancer Formation" explains, mitotic recombination can have major repercussions for human health.

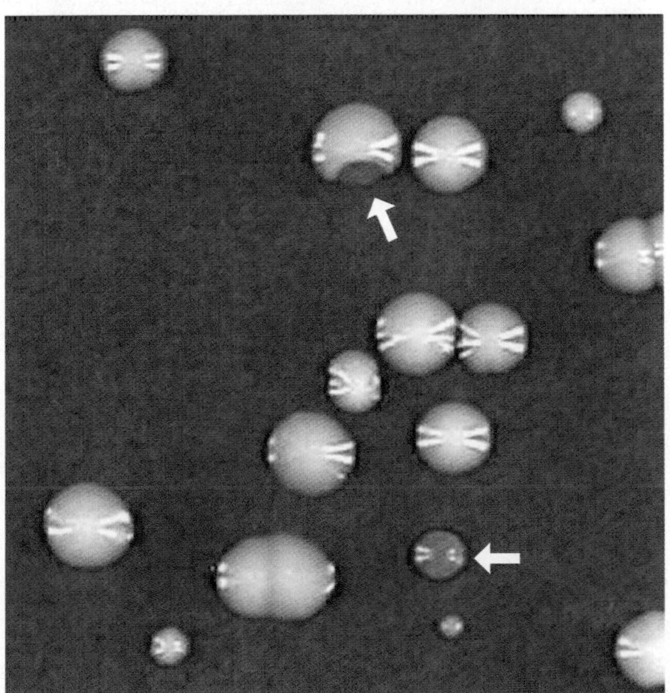

Figure 5.25 Mitotic recombination during the growth of diploid yeast colonies can create sectors. Arrows point to large, red *ade2 / ade2* sectors formed from *ADE2 / ade2* heterozygotes.

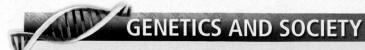

GENETICS AND SOCIETY

Mitotic Recombination and Cancer Formation

In humans, some tumors, such as those found in retinoblastoma, may arise as a result of mitotic recombination. Recall from our discussion of penetrance and expressivity in Chapter 3 that retinoblastoma is the most malignant form of eye cancer. In the United States, a genetic predisposition to the cancer is seen in about 1 in 20,000 live births. The retinoblastoma gene *(RB)* resides on chromosome 13, where the normal wild-type allele *(RB$^+$)* encodes a protein that regulates retinal growth and differentiation. Cells in the eye need at least one copy of the normal wild-type allele to maintain control over cell division. A mutation in *RB* leading to loss of normal gene function is designated *RB$^-$*. If a cell loses both copies of *RB$^+$*, it loses regulatory control, and a tumor results. The normal, wild-type *RB$^+$* allele is thus known as a tumor-suppressor gene.

People with a genetic predisposition to retinoblastoma are born with only one functional copy of the normal *RB$^+$* allele; their second chromosome 13 carries either a nonfunctional *RB$^-$* allele or no *RB* gene at all. If a mutagen (such as radiation) or a mistake in gene replication or segregation destroys or removes the single remaining normal copy of the gene in a retinal cell in either eye, a retinoblastoma tumor will develop at that site. In one study of people with a genetic predisposition to retinoblastoma, cells taken from eye tumors were *RB$^-$* homozygotes, while white blood cells from the same people were *RB$^+$/RB$^-$* heterozygotes. As **Fig. A** shows, mitotic recombination between the *RB* gene and the centromere of the chromosome carrying the gene provides one mechanism by which a cell in an *RB$^+$/RB$^-$* individual could become *RB$^-$/RB$^-$*. Once a homozygous *RB$^-$* cell is generated, it will divide uncontrollably, leading to tumor formation.

Only 40% of retinoblastoma cases follow the preceding scenario. The other 60% occur in people who are born with two normal copies of the *RB* gene. In such people, it takes two mutational events to cause the cancer. The first of these must convert an *RB$^+$* allele to *RB$^-$*, while the second could be a mitotic recombination producing daughter cells that become cancerous because they are homozygous for the newly mutant, nonfunctional allele.

Interestingly, the role of mitotic recombination in the formation of retinoblastoma helps explain the incomplete penetrance and variable expressivity of the disease. People born as *RB$^+$/RB$^-$* heterozygotes may or may not develop the condition (incomplete penetrance). If, as usually happens, they do, they may have it in one or both eyes (variable expressivity). It all depends on whether and in what cells of the body mitotic recombination (or some other "homozygosing" event that affects chromosome 13) occurs.

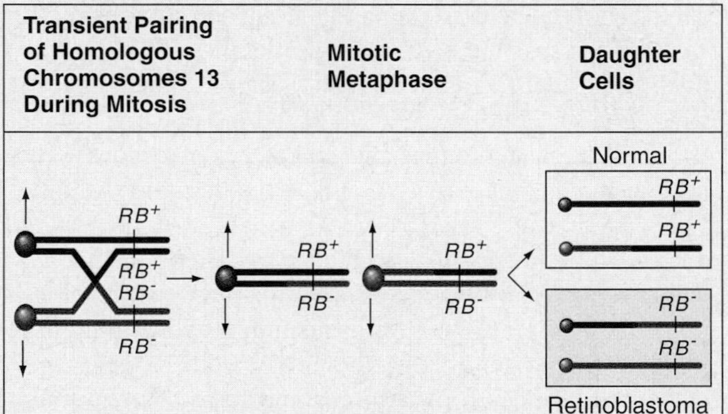

Figure A How mitotic crossing-over can contribute to cancer. Mitotic recombination during retinal growth in an *RB$^-$/RB$^+$* heterozygote may produce an *RB$^-$/RB$^-$* daughter cell that lacks a functional retinoblastoma gene and thus divides out of control. The crossover must occur between the RB gene and its centromere. Only the arrangement of chromatids yielding this result is shown.

Connections

Medical geneticists have used their understanding of linkage, recombination, and mapping to make sense of the pedigrees presented at the beginning of this chapter (see Fig. 5.1 on p. 124). The X-linked gene for red-green colorblindness must lie very close to the gene for hemophilia A because the two are tightly coupled. In fact, an examination of many pedigrees has shown that if a doubly heterozygous woman receives defective alleles of both genes from the same parent, as few as 3% of her sons will have only one of the conditions, as a result of recombination. This means that the genetic distance between the two genes is only 3 m.u. In Fig. 5.1a, the sample size was so small that none of the individuals in the pedigree were recombinant types. On the other hand, even though hemophilia B is also on the X chromosome, it lies far enough away from the red-green colorblindness locus that the two genes recombine relatively freely and may thus appear genetically unlinked, as was the case in Fig. 5.1b. The recombination distance separating the genes for colorblindness and hemophilia B is about 36 m.u. Pedigrees pointing to two different forms of hemophilia, one very closely linked to colorblindness, the other almost not linked at all, provided one of several indications that hemophilia is determined by more than one gene (**Fig. 5.26**).

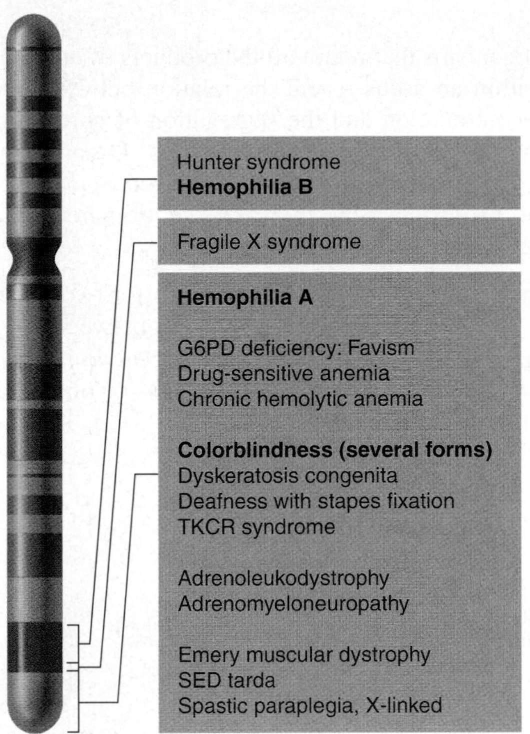

Figure 5.26 **A genetic map of part of the human X chromosome.**

Refining the human chromosome map poses a continuous challenge for medical geneticists. The newfound potential for finding and fitting more and more DNA markers into the map (review the Fast Forward box in this chapter) enormously improves the ability to assign precise locations to genes that cause disease. If several markers flank one such gene and it is possible to test a fetus for those markers, it is also possible to estimate the likelihood that the fetus has received a closely linked disease-causing allele. DNA markers have also been of great importance in the construction of a physical map of the human genome, a process we discuss in Chapters 10 and 11.

The simultaneous occurrence of genetic linkage and recombination is universal among life-forms and presumably confers some advantage to a species beyond allowing geneticists to map its genes. Although definitive conclusions are beyond our reach, speculation about the possible advantages can help us understand why genomes are arranged as they are. Linkage provides the potential for transmitting favorable combinations of genes intact to successive generations, while recombination produces great flexibility in generating new combinations of alleles. Some new combinations may help a species adapt to changing environmental conditions, whereas the inheritance of successfully tested combinations can preserve what has worked in the past. Linkage and recombination thus help chromosomes function as organelles (in the broad sense) that not only transmit but also evolve genetic information.

In Chapters 2–5 of this book, we have examined how genes and chromosomes are transmitted. As important and useful as this knowledge is, it tells us very little about the structure and mode of action of the genetic material. In the next section (Chapters 6–8), we carry our analysis to the level of DNA, the actual molecule of heredity. In Chapter 6, we look at DNA structure and learn how the DNA molecule carries genetic information. In Chapter 7, we describe how geneticists defined the gene as a localized region of DNA containing many nucleotides that together encode the information to make a protein. In Chapter 8, we examine how the cellular machinery interprets the genetic information in genes to produce the multitude of phenotypes that make up an organism.

Essential Concepts

1. Gene pairs that are close together on the same chromosome are genetically linked because they are transmitted together more often than not. The hallmark of linkage is that the number of parental types is greater than the number of recombinant types among the progeny of double heterozygotes.
2. The recombination frequencies of pairs of genes indicate how often two genes are transmitted together. For linked genes, the recombination frequency is less than 50%.
3. Gene pairs that assort independently exhibit a recombination frequency of 50%, since the number of parental types equals the number of recombinants. Genes may assort independently either because they are on different chromosomes or because they are far apart on the same chromosome.
4. Statistical analysis helps determine whether or not two genes assort independently. The probability value (p) calculated by the chi-square test measures the likelihood that a particular set of data supports the null hypothesis of independent assortment, or no linkage. The lower the p value, the less likely is the null hypothesis, and the more likely the linkage. The chi-square test can also be used to determine how well the outcomes of crosses fit other genetic hypotheses (see www.mhhe.com/hartwell3: Chapter 3 for examples).

5. The greater the physical distance between linked genes, the higher the recombination frequency. However, recombination frequencies become more and more inaccurate as the distance between genes increases.

6. Recombination occurs because chromatids of homologous chromosomes exchange parts (that is, cross over) during the prophase of meiosis I, after the chromosomes have replicated.

7. Genetic maps are a visual representation of relative recombination frequencies. The greater the density of genes on the map (and thus the smaller the distance between the genes), the more accurate and useful the map becomes in predicting inheritance.

8. Organisms that retain all the products of one meiosis within an ascus reveal the relation between genetic recombination and the segregation of chromosomes during the two meiotic divisions. Organisms like *Neurospora* that produce ordered octads make it possible to locate a chromosome's centromere on the genetic map.

9. In diploid organisms heterozygous for two alleles of a gene, rare mitotic recombination between the gene and its centromere can produce *genetic mosaics* in which some cells are homozygous for one allele or the other.

On Our Website

www.mhhe.com/hartwell3
Chapter 5

Annotated Suggested Readings and Links to Other Websites

- The early history of genetic mapping
- Construction of a linkage map of the human genome
- New ideas about the significance of chromosomal interference
- Using mitotic recombination to trace cells during development

Specialized Topics

- The derivation and use of mapping functions
- Determining the linkage of human genes using likelihood ratios and LOD scores. A brief introduction to this topic can also be found in the Tools of Genetics box on pp. 416 – 417 of Chapter 11.

Social and Ethical Issues

1. In the initial stages of gene mapping, researchers find DNA markers that are linked to a disease gene (see the Fast Forward box concerning cystic fibrosis on pp. 142–143). It is possible to use such linked DNA markers to test for the presence of a disease-causing allele in a fetus, although these tests are by definition inaccurate. For example, if the distance between the disease gene and the DNA marker is 5 m.u., then 5% of fetuses diagnosed with the disease-causing allele would actually be false positives, while 5% of the fetuses considered free of the disease-causing allele would be false negatives and actually have the disease.

   How accurate must a genetic test be to be used on patients? To what extent does this decision depend upon the severity of the genetic condition? How should one balance the costs and risks of the diagnostic procedure? (For example, there is a very low but apparently real risk of miscarriage when amniocentesis is performed to obtain fetal DNA, as explained in the Genetics and Society box on p. 87 of Chapter 4.) Are most patients intellectually and emotionally prepared to deal with a genetic diagnosis that is less than certain?

2. In 1996, evidence was presented for linkage of a marker to a predisposition to prostate cancer. Using 66 families in which at least three males had prostate cancer, a genetic marker was found that is inherited with high frequency in those men with prostate cancer. Linked to this marker, somewhere in a region of about 10 million base pairs, is a gene that predisposes to prostate cancer. The estimate is that this genetic region is involved in about 3% of the *total* cases of prostate cancer. National newspapers announced these findings with the following headlines:

"Scientists Find Proof That Mutant Gene Can Increase Risk of Prostate Cancer"

"Prostate Cancer Gene Evidence Found"

"Scientists Zero in on Gene Tied to Prostate Cancer"

Do these headlines accurately represent the situation? Could they be misleading in any way? What would you use as a headline for an article on this study? What is the responsibility of the press in presenting new findings? Consider the responsibility of journalists/reporters to their employers (newspapers), their profession, and to the public.

Solved Problems

I. The *Xg* locus on the human X chromosome has two alleles, a^+ and *a*. The a^+ allele causes the presence of the Xg surface antigen on red blood cells, while the recessive *a* allele does not allow antigen to appear. The *Xg* locus is 10 m.u. from the *Sts* locus. The *Sts* allele produces normal activity of the enzyme steroid sulfatase, while the recessive *sts* allele results in the lack of steroid sulfatase activity and the disease ichthyosis (scaly skin). A man with ichthyosis and no Xg antigen has a normal daughter with Xg antigen, who is expecting a child.

a. If the child is a son, what is the probability he will lack antigen and have ichthyosis?

b. What is the probability that a son would have both the antigen and ichthyosis?

c. If the child is a son with ichthyosis, what is the probability he will have Xg antigen?

Answer

a. This problem requires an understanding of how linkage affects the proportions of gametes. First designate the genotype of the individual in which recombination during meiosis affects the transmission of alleles: in this problem, the daughter. The X chromosome she inherited from her father (who had icthyosis and no Xg antigen) must be *sts a*. (No recombination could have separated the genes during meiosis in her father since he has only one X chromosome.) Because the daughter is normal and has the Xg antigen, her other X chromosome (inherited from her mother) must contain the *Sts* and a^+ alleles. Her X chromosomes can be diagrammed as:

sts ——— *a*

Sts ——— a^+

Because the *Sts* and *Xg* loci are 10 m.u. apart on the chromosome, there is a 10% recombination frequency. Ninety percent of the gametes will be parental: *sts a* or *Sts* a^+ (45% of each type) and 10% will be recombinant: *sts* a^+ or *Sts a* (5% of each type). The phenotype of a son directly reflects the genotype of the X chromosome from his mother. *Therefore, the probability that he will lack the Xg antigen and have icthyosis (genotype:* sts a / Y*) is 45/100.*

b. *The probability that he will have the antigen and ichthyosis (genotype:* sts a^+ / Y*) is 5/100.*

c. There are two classes of gametes containing the ichthyosis allele: *sts a* (45%) and *sts* a^+ (5%). If the total number of gametes is 100, then 50 will have the *sts* allele. Of those gametes, 5 (or 10%) will have the a^+ allele. *Therefore there is a 1/10 probability that a son with the* sts *allele will have the Xg antigen.*

II. *Drosophila* females of wild-type appearance but heterozygous for three autosomal genes are mated with males showing three autosomal recessive traits: glassy eyes, coal-colored bodies, and striped thoraxes. One thousand (1000) progeny of this cross are distributed in the following phenotypic classes:

Wild type	27
Striped thorax	11
Coal body	484
Glassy eyes, coal body	8
Glassy eyes, striped thorax	441
Glassy eyes, coal body, striped thorax	29

a. Draw a genetic map based on this data.

b. Show the arrangement of alleles on the two homologous chromosomes in the parent females.

c. Normal-appearing males containing the same chromosomes as the parent females in the preceding cross are mated with females showing glassy eyes, coal-colored bodies, and striped thoraxes. Of 1000 progeny produced, indicate the numbers of the various phenotypic classes you would expect.

Answer

A logical, methodical way to approach a three-point cross is described here.

a. Designate the alleles:

t^+ = wild-type thorax t = striped thorax
g^+ = wild-type eyes g = glassy eyes
c^+ = wild-type body c = coal-colored body

In solving a three-point cross, designate the types of events that gave rise to each group of individuals and the genotypes of the gametes obtained from their mother. (The paternal gametes contain only the recessive alleles of these genes [*t g c*]. They do not change the phenotype and can be ignored.)

Progeny	Number	Type of event	Genotype
1. wild type	27	single crossover	$t^+\ g^+\ c^+$
2. striped thorax	11	single crossover	$t\ g^+\ c^+$
3. coal body	484	parental	$t^+\ g^+\ c$
4. glassy eyes, coal body	8	single crossover	$t^+\ g\ c$
5. glassy eyes, striped thorax	441	parental	$t\ g\ c^+$
6. glassy eyes, coal body, striped thorax	29	single crossover	$t\ g\ c$

Picking out the parental classes is easy. If all the other classes are rare, the two most abundant categories are those gene combinations that have not undergone recombination. Then there should be two sets of two phenotypes that correspond to a single crossover event between the first and second genes, or between the second and third genes. Finally, there should be a pair of classes containing small numbers that result from double crossovers. In this example, there are no flies in the double crossover classes, which would have been in the two missing phenotypic combinations: glassy eyes and coal body, striped thorax.

Look at the most abundant classes to determine which alleles were on each chromosome in the female heterozygous parent. One parental class had the phenotype of coal body (484 flies), so one chromosome in the female must have contained the t^+, g^+, and c alleles. (Notice that we cannot yet say in what order these alleles are located on the chromosome.) The other parental class was glassy eyes and striped thorax, corresponding to a chromosome with the *t, g,* and c^+ alleles.

To determine the order of the genes, compare the $t^+g\ c^+$ double crossover class (not seen in the data) with the most similar parental class ($t\ g\ c^+$). The alleles of g and c retain their parental associations ($g\ c^+$), while the t gene has recombined with respect to both other genes in the double recombinant class. Thus, the t gene is between g and c.

In order to complete the map, calculate the recombination frequencies between the center gene and each of the genes on the ends. For g and t, the nonparental combinations of alleles are in classes 2 and 4, so RF = (11 + 8)/1000 = 19/1000, or 1.9%. For t and c, classes 1 and 6 are nonparental, so RF = (27 + 29)/1000 = 56/1000, or 5.6%.

The genetic map is

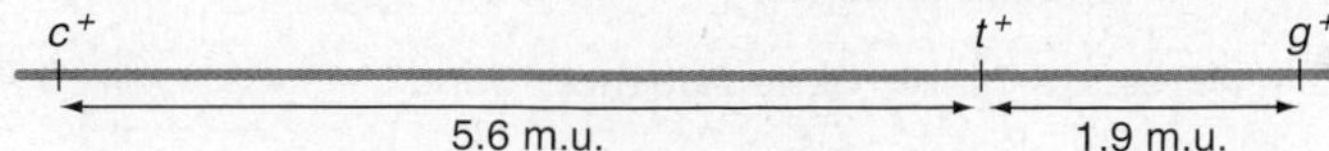

b. The alleles on each chromosome were already determined (c, g^+, t^+ and c^+, g, t). Now that the order of loci has also been determined, the arrangement of the alleles can be indicated.

c t^+ g^+

c^+ t g

c. Males of the same genotype as the starting female ($c\ t^+\ g^+\ /\ c^+\ t\ g$) could produce only two types of gametes: parental types $c\ t^+\ g^+$ and $c^+\ t\ g$ because there is no recombination in male *Drosophila.* The progeny expected from the mating with a homozygous recessive female are thus 500 coal body and 500 glassy eyes, striped thorax flies.

III. The following asci were obtained in *Neurospora* when a wild-type strain ($ad^+\ leu^+$) was crossed to a double mutant strain that cannot grow in the absence of adenine or leucine ($ad^-\ leu^-$). Only one member of each spore pair produced by the final mitosis is shown, since the two cells in a pair have the same genotype. Total asci = 120.

Spore pair	Ascus type				
1–2	$ad^+\ leu^+$	$ad^+\ leu^-$	$ad^+\ leu^+$	$ad^+\ leu^-$	$ad^-\ leu^+$
3–4	$ad^+\ leu^+$	$ad^+\ leu^-$	$ad^+\ leu^-$	$ad^-\ leu^+$	$ad^+\ leu^+$
5–6	$ad^-\ leu^-$	$ad^-\ leu^+$	$ad^-\ leu^+$	$ad^-\ leu^-$	$ad^-\ leu^-$
7–8	$ad^-\ leu^-$	$ad^-\ leu^+$	$ad^-\ leu^-$	$ad^+\ leu^+$	$ad^+\ leu^-$
# of asci	30	30	40	2	18

a. What genetic event causes the alleles of two genes to segregate to different cells at the second meiotic division, and when does this event occur?

b. Provide the best possible map for the two genes and their centromere(s).

Answer

This problem requires an understanding of tetrad analysis and the process (meiosis) that produces the patterns seen in ordered asci.

a. *A crossover between a gene and its centromere causes the segregation of alleles at the second meiotic division. The crossover event occurs during prophase of meiosis I.*

b. Using ordered tetrads you can determine whether two genes are linked, the distance between two genes, and the distance between each gene and its centromere. First designate the five classes of asci shown. The first class is a parental ditype (spores contain the same combinations of alleles as their parents); the second is a nonparental ditype; the last three are tetratypes. Next determine if these genes are linked. The number of PD = number of NPD, so the genes are not linked. When genes are unlinked, the tetratype asci are generated by a crossing-over event between a gene and its centromere. Looking at the *leu* gene, there is a second-division segregation pattern of that gene in the third and fourth asci types. Therefore, the percent of second-division segregation is

$$\frac{40 + 2}{120} \times 100 = 35\%$$

Because only half of the chromatids in the meioses that generated these tetratype asci were involved in the crossover, the map distance between *leu* and its centromere is 35/2, or 17.5 m.u. Asci of the fourth and fifth types show a second-division segregation pattern for the *ad* gene

$$\frac{2 + 18}{120} \times 100 = 16.6\%$$

Dividing 16.6% by 2 gives the recombination frequency and map distance of 8.3 m.u. *The map of these two genes is the following:*

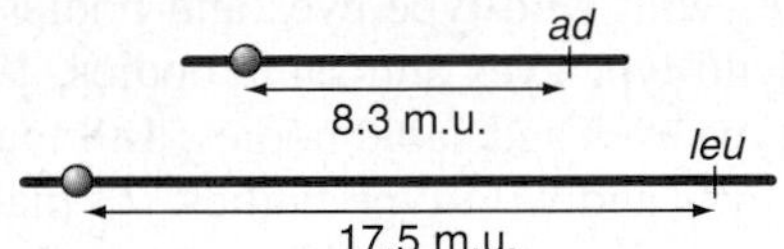

Problems

Vocabulary

1. Choose the phrase from the right column that best fits the term in the left column.

a. recombination	1. a statistical method for testing the fit between observed and expected results
b. linkage	2. an ascus containing spores of four different genotypes
c. chi-square test	3. one crossover along a chromosome makes a second nearby crossover less likely
d. chiasma	4. when two loci recombine in less than 50% of gametes
e. tetratype	5. the relative chromosomal location of a gene
f. locus	6. the ratio of observed double crossovers to expected double crossovers
g. coefficient of coincidence	7. individual composed of cells with different genotypes
h. interference	8. formation of new genetic combinations by exchange of parts between homologs
i. parental ditype	9. when the two alleles of a gene are segregated into different cells at the first meiotic division
j. ascospores	10. an ascus containing only two nonrecombinant kinds of spores
k. first-division segregation	11. structure formed at the spot where crossing-over occurs between homologs
l. mosaic	12. fungal spores contained in a sac

Section 5.1

2. Do the data that Mendel obtained fit his hypotheses? For example, Mendel obtained 315 yellow round, 101 yellow wrinkled, 108 green round, and 32 green wrinkled seeds from the selfing of *Yy Rr* individuals (a total of 556). His hypotheses of segregation and independent assortment predict a 9:3:3:1 ratio in this case. Use the chi-square test to determine whether Mendel's data are significantly different from what he predicted. (The chi-square test did not exist in Mendel's day, so he was not able to test his own data for goodness of fit to his hypotheses.)

3. Two genes control color in corn snakes as follows: *O– B–* snakes are brown, *O– bb* are orange, *oo B–* are black, and *oo bb* are albino. An orange snake was mated to a black snake, and a large number of F_1 progeny were obtained, all of which were brown. When the F_1 snakes were mated to one another, they produced 100 brown offspring, 25 orange, 22 black, and 13 albino.
   a. What are the genotypes of the F_1 snakes?
   b. What proportions of the different colors would have been expected among the F_2 snakes if the two loci assort independently?
   c. Do the observed results differ significantly from what was expected, assuming independent assortment is occurring?
   d. What is the probability that differences this great between observed and expected values would happen by chance?

4. A mouse from a true-breeding population with normal gait was crossed to a mouse displaying an odd gait called "dancing." The F_1 animals all showed normal gait.
   a. If dancing is caused by homozygosity for the recessive allele of a single gene, what proportion of the F_2 mice should be dancers?
   b. If mice must be homozygous for recessive alleles of both of two different genes to have the dancing phenotype, what proportion of the F_2 should be dancers if the two genes are unlinked?
   c. When the F_2 mice were obtained, 42 normal and 8 dancers were seen. Use the chi-square test to determine if these results better fit the one-gene

model from part *a* or the two-gene model from part *b*.

5. In *Drosophila,* males from a true-breeding stock with raspberry-colored eyes were mated to females from a true-breeding stock with sable-colored bodies. In the F_1 generation, all the females had wild-type eye and body color, while all the males had wild-type eye color but sable-colored bodies. When F_1 males and females were mated, the F_2 generation was composed of 216 females with wild-type eyes and bodies, 223 females with wild-type eyes and sable bodies, 191 males with wild-type eyes and sable bodies, 188 males with raspberry eyes and wild-type bodies, 23 males with wild-type eyes and bodies, and 27 males with raspberry eyes and sable bodies. Explain these results by diagramming the crosses, and calculate any relevant map distances.

6. Figure 5.5 on p. 130 applied the chi-square method to test linkage between two genes by asking whether the observed numbers of parental and recombinant classes differed significantly from the expectation of independent assortment that parentals = recombinants. Another possible way to analyze the results from these same experiments is to ask whether the observed frequencies of the four genotypic classes (*A B, a b, A b,* and *a B*) can be explained by a null hypothesis predicting that they should appear in a 1:1:1:1 ratio. In order to consider the relative advantages and disadvantages of analyzing the data in these two different ways answer the following:

a. What is the null hypothesis in each case?
b. Which is a more sensitive test of linkage? (Analyze the data in Fig. 5.5 by the second method.)
c. How would both methods respond to a situation in which one allele of one of the genes causes reduced viability?

7. In mice, the dominant allele *Gs* of the X-linked gene *Greasy* produces shiny fur, while the recessive wild-type Gs^+ allele determines normal fur. The dominant allele *Bhd* of the X-linked *Broadhead* gene causes skeletal abnormalities including broad heads and snouts, while the recessive wild-type Bhd^+ allele yields normal skeletons. Female mice heterozygous for the two alleles of both genes were mated with wild-type males. Among 100 male progeny of this cross, 49 had shiny fur, 48 had skeletal abnormalities, 2 had shiny fur and skeletal abnormalities, and 1 was wild type.

a. Diagram the cross described, and calculate the distance between the two genes.
b. What would have been the results if you had counted 100 female progeny of the cross?

8. *CC DD* and *cc dd* individuals were crossed to each other, and the F_1 generation was backcrossed to the *cc dd* parent; 903 *Cc Dd,* 897 *cc dd,* 98 *Cc dd,* and 102 *cc Dd* offspring resulted.

a. How far apart are the *c* and *d* loci?
b. What progeny and in what frequencies would you expect to result from testcrossing the F_1 generation from a *CC dd* × *cc DD* cross to *cc dd?*

9. If the *a* and *b* loci are 20 m.u. apart in humans and an *A B / a b* woman mates with an *a b / a b* man, what is the probability that their first child will be *A b / a b?*

10. In a particular human family, John and his mother both have brachydactyly (a rare autosomal dominant causing short fingers). John's father has Huntington disease (another rare autosomal dominant). John's wife is phenotypically normal and is pregnant. Two-thirds of people who inherit the Huntington (*HD*) allele show symptoms by age 50, and John is 50 and has no symptoms. Brachydactyly is 90% penetrant.

a. What are the genotypes of John's parents?
b. What are the possible genotypes for John?
c. What is the probability the child will express both brachydactyly and Huntington disease by age 50 if the two genes are unlinked?
d. If these two loci are 20 m.u. apart, how will it change your answer to part *c*?

11. In mice, the autosomal locus coding for the β-globin chain of hemoglobin is 1 m.u. from the albino locus. Assume for the moment that the same is true in humans. The disease sickle-cell anemia is the result of homozygosity for a particular mutation in the β-globin gene.

a. A son is born to an albino man and a woman with sickle-cell anemia. What kinds of gametes will the son form, and in what proportions?
b. A daughter is born to a normal man and a woman who has both albinism and sickle-cell anemia. What kinds of gametes will the daughter form, and in what proportions?
c. If the son in part *a* grows up and marries the daughter in part *b*, what is the probability that a child of theirs will be an albino with sickle-cell anemia?

12. In corn, the allele *A* allows the deposition of anthocyanin (blue) pigment in the kernels (seeds), while *aa* plants have yellow kernels. At a second gene, *W–* produces smooth kernels, while *ww* kernels are wrinkled. A plant with blue smooth kernels was crossed to a plant with yellow wrinkled kernels. The progeny consisted of 1447 blue smooth, 169 blue wrinkled, 186 yellow smooth, and 1510 yellow wrinkled.

a. Are the *a* and *w* loci linked? If so, how far apart are they?
b. What was the genotype of the blue smooth parent? Include the chromosome arrangement of alleles.
c. If a plant grown from a blue wrinkled progeny seed is crossed to a plant grown from a yellow

smooth F_1 seed, what kinds of kernels would be expected, and in what proportions?

13. Albino rabbits (lacking pigment) are homozygous for the recessive *c* allele (*C* allows pigment formation). Rabbits homozygous for the recessive *b* allele make brown pigment, while those with at least one copy of *B* make black pigment. True-breeding brown rabbits were crossed to albinos, which were *BB*. F_1 rabbits, which were all black, were crossed to the double recessive (*bb cc*). The progeny obtained were 34 black, 66 brown, and 100 albino.

a. What phenotypic proportions would have been expected if the *b* and *c* loci were unlinked?

b. How far apart are the two loci?

14. Write the number of *different kinds* of phenotypes, excluding gender, you would see among a large number of progeny from an F_1 mating between individuals of identical genotype that are heterozygous for one or two genes (that is, *Aa* or *Aa Bb*) as indicated. No gene interactions means that the phenotype determined by one gene is not influenced by the genotype of the other gene.

a. One gene; *A* completely dominant to *a*.

b. One gene; *A* and *a* codominant.

c. One gene; *A* incompletely dominant to *a*.

d. Two unlinked genes; no gene interactions; *A* completely dominant to *a*, and *B* completely dominant to *b*.

e. Two genes, 10 m.u. apart; no gene interactions; *A* completely dominant to *a*, and *B* completely dominant to *b*.

f. Two unlinked genes; no gene interactions; *A* and *a* codominant, and *B* incompletely dominant to *b*.

g. Two genes, 10 m.u. apart; *A* completely dominant to *a*, and *B* completely dominant to *b*; and with recessive epistasis between the genes.

h. Two unlinked duplicated genes (that is, *A* and *B* perform the same function); *A* and *B* completely dominant to *a* and *b*, respectively.

i. Two genes, 0 m.u. apart; no gene interactions; *A* completely dominant to *a*, and *B* completely dominant to *b*. (There are two possible answers.)

15. If the *a* and *b* loci are 40 cM apart and an *AA BB* individual and an *aa bb* individual mate:

a. What gametes will the F_1 individuals produce, and in what proportions? What phenotypic classes in what proportions are expected in the F_2 generation (assuming complete dominance for both genes)?

b. If the original cross was *AA bb* × *aa BB*, what gametic proportions would emerge from the F_1? What would be the result in the F_2 generation?

16. A DNA variant has been found linked to a rare autosomal dominant disease in humans and can thus be used as a marker to follow inheritance of the disease allele. In an informative family (in which one parent is heterozygous for both the disease allele and the DNA marker in a known chromosomal arrangement of alleles and his or her mate does not have the same alleles of the DNA variant), the reliability of such a marker as a predictor of the disease in a fetus is related to the map distance between the DNA marker and the gene causing the disease. Imagine that a man affected with the disease (genotype *Dd*) is heterozygous for the V^1 and V^2 forms of the DNA variant, with form V^1 on the same chromosome as the *D* allele and form V^2 on the same chromosome as *d*. His wife is V^3V^3 *dd*, where V^3 is another allele of the DNA marker. Typing of the fetus by amniocentesis reveals that the fetus has the V^2 and V^3 variants of the DNA marker. How likely is it that the fetus has inherited the disease allele *D* if the distance between the *D* locus and the marker locus is (a) 0 m.u., (b) 1 m.u., (c) 5 m.u., (d) 10 m.u., (e) 50 m.u.?

17. In *Drosophila,* the recessive *dp* allele of the *dumpy* gene produces short, curved wings, while the recessive allele *bw* of the *brown* gene causes brown eyes. In a testcross using females heterozygous for both of these genes, the following results were obtained:

wild-type wings, wild-type eyes	178
wild-type wings, brown eyes	185
dumpy wings, wild-type eyes	172
dumpy wings, brown eyes	181

In a testcross using males heterozygous for both of these genes, a different set of results was obtained:

wild-type wings, wild-type eyes	247
dumpy wings, brown eyes	242

a. What can you conclude from the first testcross?

b. What can you conclude from the second testcross?

c. How can you reconcile the data shown in parts *a* and *b*? Can you exploit the difference between these two sets of data to devise a general test for synteny in *Drosophila?*

d. The genetic distance between *dumpy* and *brown* is 91.5 m.u. How could this value be measured?

18. Cinnabar eyes (*cn*) and reduced bristles (*rd*) are autosomal recessive characters in *Drosophila.* A homozygous wild-type female was crossed to a reduced, cinnabar male, and the F_1 males were then crossed to the F_1 females to obtain the F_2. Of the 400 F_2 offspring obtained, 292 were wild type, 9 were cinnabar, 7 were reduced, and 92 were reduced, cinnabar. Explain these results and estimate the distance between the *cn* and *rd* loci.

Section 5.2

19. In *Drosophila,* the recessive allele *mb* of one gene causes missing bristles, the recessive allele *e* of a second gene causes ebony body color, and the recessive allele *k* of a third gene causes kidney-shaped eyes. (Dominant wild-type alleles of all three genes are indicated with a + superscript.) The three different P generation crosses in the table that follows were conducted, and then the resultant F_1 females from each cross were testcrossed to males that were homozygous for the recessive alleles of both genes in question. The phenotypes of the testcross offspring are tabulated as follows. Determine the best genetic map explaining all the data.

Parental cross	Testcross offspring of F_1 females	
$mb^+\ mb^+,\ e^+\ e^+ \times$ $mb\ mb,\ e\ e$	normal bristles, normal body	117
	normal bristles, ebony body	11
	missing bristles, normal body	15
	missing bristles, ebony body	107
$k^+\ k^+,\ e\ e \times k\ k,\ e^+\ e^+$	normal eyes, normal body	11
	normal eyes, ebony body	150
	kidney eyes, normal body	144
	kidney eyes, ebony body	7
$mb^+\ mb^+,\ k^+\ k^+ \times$ $mb\ mb,\ k\ k$	normal bristles, normal eyes	203
	normal bristles, kidney eyes	11
	missing bristles, normal eyes	15
	missing bristles, kidney eyes	193

20. In the tubular flowers of foxgloves, wild-type coloration is red while a mutation called *white* produces white flowers. Another mutation, called *peloria,* causes the flowers at the apex of the stem to be huge. Yet another mutation, called *dwarf,* affects stem length. You cross a white-flowered plant (otherwise phenotypically wild type) to a plant that is dwarf and peloria but has wild-type red flower color. All of the F_1 plants are tall with white, normal-sized flowers. You cross an F_1 plant back to the dwarf and peloria parent, and you see the 543 progeny shown in the chart. (Only mutant traits are noted.)

dwarf, peloria	172
white	162
dwarf, peloria, white	56
wild type	48
dwarf, white	51
peloria	43
dwarf	6
peloria, white	5

a. Which alleles are dominant?
b. What were the genotypes of the parents in the original cross?
c. Draw a map showing the linkage relationships of these three loci.
d. Is there interference? If so, calculate the coefficient of coincidence and the interference value.

21. In *Drosophila,* three autosomal genes have the following map:

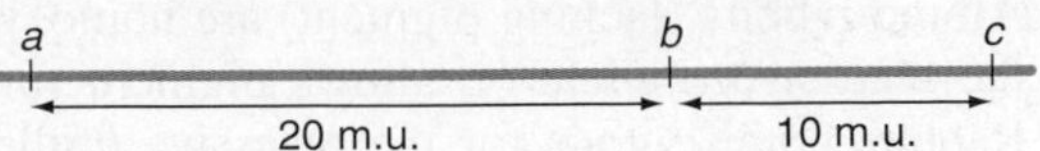

a. Provide the data, in terms of the expected number of flies in the following phenotypic classes, when $a^+\ b^+\ c^+\ /\ a\ b\ c$ females are crossed to $a\ b\ c\ /\ a\ b\ c$ males. Assume 1000 flies were counted and that there is no interference in this region.

a^+	b^+	c^+
a	b	c
a^+	b	c
a	b^+	c^+
a^+	b^+	c
a	b	c^+
a^+	b	c^+
a	b^+	c

b. If the cross were reversed, such that $a^+\ b^+\ c^+\ /\ a\ b\ c$ males are crossed to $a\ b\ c\ /\ a\ b\ c$ females, how many flies would you expect in the same phenotypic classes?

22. A snapdragon with pink petals, black anthers, and long stems was allowed to self-fertilize. From the resulting seeds, 650 adult plants were obtained. The phenotypes of these offspring are listed here.

78	red	long	tan
26	red	short	tan
44	red	long	black
15	red	short	black
39	pink	long	tan
13	pink	short	tan
204	pink	long	black
68	pink	short	black
5	white	long	tan
2	white	short	tan
117	white	long	black
39	white	short	black

a. Using *P* for one allele and *p* for the other, indicate how flower color is inherited.
b. What numbers of red : pink : white would have been expected among these 650 plants?
c. How are anther color and stem length inherited?
d. What was the genotype of the original plant?
e. Do any of the three genes show independent assortment?
f. For any genes that are linked, indicate the arrangements of the alleles on the homologous chromosomes in the original snapdragon, and estimate the distance between the genes.

23. Male *Drosophila* expressing the recessive mutations *sc (scute), ec (echinus), cv (crossveinless),* and *b (black)* were crossed to phenotypically wild-type females, and the 3288 progeny listed were obtained. (Only mutant traits are noted.)

653	black, scute, echinus, crossveinless
670	scute, echinus, crossveinless
675	wild type
655	black
71	black, scute
73	scute
73	black, echinus, crossveinless
74	echinus, crossveinless
87	black, scute, echinus
84	scute, echinus
86	black, crossveinless
83	crossveinless
1	black, scute, crossveinless
1	scute, crossveinless
1	black, echinus
1	echinus

a. Diagram the genotype of the female parent.
b. Map these loci.
c. Is there evidence of interference? Justify your answer with numbers.

24. *Drosophila* females heterozygous for each of three recessive autosome mutations with independent phenotypic effects (thread antennae [*th*], hairy body [*h*], and scarlet eyes [*st*]) were testcrossed to males showing all three mutant phenotypes. The 1000 progeny of this testcross were

thread, hairy, scarlet	432
wild type	429
thread, hairy	37
thread, scarlet	35
hairy	34
scarlet	33

a. Show the arrangement of alleles on the relevant chromosomes in the triply heterozygous females.
b. Draw the best genetic map that explains these data.
c. Calculate any relevant interference values.

25. A true-breeding strain of Virginia tobacco has dominant alleles determining leaf morphology (*M*), leaf color (*C*), and leaf size (*S*). A Carolina strain is homozygous for the recessive alleles of these three genes. These genes are found on the same chromosome as follows:

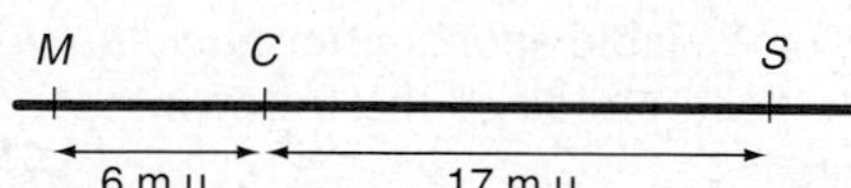

An F_1 hybrid between the two strains is now backcrossed to the Carolina strain. Assuming no interference:

a. What proportion of the backcross progeny will resemble the Virginia strain for all three traits?
b. What proportion of the backcross progeny will resemble the Carolina strain for all three traits?
c. What proportion of the backcross progeny will have the leaf morphology and leaf size of the Virginia strain but the leaf color of the Carolina strain?
d. What proportion of the backcross progeny will have the leaf morphology and leaf color of the Virginia strain but the leaf size of the Carolina strain?

26. a. In *Drosophila,* crosses between F_1 heterozygotes of the form $A\ b\ /\ a\ B$ always yield the same ratio of phenotypes in the F_2 progeny regardless of the distance between the two genes (assuming complete dominance for both autosomal genes). What is this ratio? Would this also be the case if the F_1 heterozygotes were $A\ B\ /\ a\ b$?

b. If you intercrossed F_1 heterozygotes of the form $A\ b\ /\ a\ B$ in mice, the phenotypic ratio among the F_2 progeny would vary with the map distance between the two genes. Is there a simple way to estimate the map distance based on the frequencies of the F_2 phenotypes, assuming rates of recombination are equal in males and females? Could you estimate map distances in the same way if the mouse F_1 heterozygotes were $A\ B\ /\ a\ b$?

27. The following list of four *Drosophila* mutations indicates the symbol for the mutation, the name of the gene, and the mutant phenotype:

Allele symbol	Gene name	Mutant phenotype
dwp	*dwarp*	small body, warped wings
rmp	*rumpled*	deranged bristles
pld	*pallid*	pale wings
rv	*raven*	dark eyes and bodies

You perform the following crosses with the indicated results:

Cross #1: dwarp, rumpled females × pallid, raven males
→ dwarp, rumpled males and wild-type females

Cross #2: pallid, raven females × dwarp, rumpled males
→ pallid, raven males and wild-type females

F_1 females from cross #1 were crossed to males from a true-breeding *dwarp rumpled pallid raven* stock. The 1000 progeny obtained were as follows:

pallid	3
pallid, raven	428
pallid, raven, rumpled	48
pallid, rumpled	23
dwarp, raven	22
dwarp, raven, rumpled	2
dwarp, rumpled	427
dwarp	47

Indicate the best map for these four genes, including all relevant data. Calculate interference values where appropriate.

28. A cross was performed between one haploid strain of yeast with the genotype *a f g* and another haploid strain with the genotype α + + (*a* and α are mating types).

The resulting diploid was sporulated, and a random sample of 101 of the resulting haploid spores was analyzed. The following genotypic frequencies were seen:

α	+	+	31
a	f	g	29
a	f	+	14
α	+	g	13
a	+	g	6
α	f	+	6
a	+	+	1
α	f	g	1

a. Map the loci involved in the cross.
b. Assuming all three genes are on the same chromosome arm, is it possible that a particular ascus could contain an $\alpha\ f\ g$ spore but not an $a + +$ spore? If so, draw a meiosis that could generate such an ascus.

29. *Neurospora* of genotype $a + c$ are crossed with *Neurospora* of genotype $+ b +$. The following tetrads are obtained (note that the genotype of the four spore *pairs* in an ascus are listed, rather than listing all eight spores):

$a + c$	$a\ b\ c$	$+ + c$	$+ b\ c$	$a\ b +$	$a + c$
$a + c$	$a\ b\ c$	$a + c$	$a\ b\ c$	$a\ b +$	$a\ b\ c$
$+ b +$	$+ + +$	$+ b +$	$+ + +$	$+ + c$	$+ + +$
$+ b +$	$+ + +$	$a\ b +$	$a + +$	$+ + c$	$+ b +$
137	141	26	25	2	3

a. In how many cells has meiosis occurred to yield this data?
b. Give the best genetic map to explain these results. Indicate all relevant genetic distances, both between genes and between each gene and its respective centromere.
c. Diagram a meiosis that could give rise to one of the three tetrads in the class at the far right in the list.

30. Two crosses were made in *Neurospora* involving the mating type locus and either the *ad* or *p* genes. In both cases, the mating type locus (*A* or *a*) was one of the loci whose segregation was scored. One cross was $ad\ A \times +\ a$ (cross a), and the other was $p\ A \times +\ a$ (cross b). From cross a, 10 parental ditype, 9 nonparental ditype, and 1 tetratype asci were seen. From cross b, the results were 24 parental ditype, 3 nonparental ditype, and 27 tetratype asci.

a. What are the linkage relationships between the mating type locus and the other two loci?
b. Although these two crosses were performed in *Neurospora,* you cannot use the data given to calculate centromere-to-gene distances for any of these genes. Why not?

31. A cross was performed between a yeast strain that requires methionine and lysine for growth ($met^-\ lys^-$) and another yeast strain, which is $met^+\ lys^+$. One hundred asci were dissected, and colonies were grown from the four spores in each ascus. Cells from these colonies were tested for their ability to grow on petri plates containing either minimal medium (min), min + lysine (lys), min + methionine (met), or min + lys + met. The asci could be divided into two groups based on this analysis:

Group 1: In 89 asci, cells from two of the four spore colonies could grow on all four kinds of media, while the other two spore colonies could grow only on min + lys + met.

Group 2: In 11 asci, cells from one of the four spore colonies could grow on all four kinds of petri plates. Cells from a second one of the four spore colonies could grow only on min + lys plates and on min + lys + met plates. Cells from a third of the four spore colonies could only grow on min + met plates and on min + lys + met. Cells from the remaining colony could only grow on min + lys + met.

a. What are the genotypes of each of the spores within the two types of asci?
b. Are the *lys* and *met* genes linked? If so, what is the map distance between them?
c. If you could extend this analysis to many more asci, you would eventually find some asci with a different pattern. For these asci, describe the phenotypes of the four spores. List these phenotypes as the ability of dissected spores to form colonies on the four kinds of petri plates.

32. The *a, b,* and *c* loci are all on different chromosomes in yeast. When $a +$ yeast were crossed to $+ b$ yeast and the resultant tetrads analyzed, it was found that the number of nonparental ditype tetrads was equal to the number of parental ditypes, but there were no tetratype asci at all. On the other hand, many tetratype asci were seen in the tetrads formed after $a +$ was crossed with $+ c$, and after $b +$ was crossed with $+ c$. Explain these results.

33. Indicate the percentage of tetrads that would have 0, 1, 2, 3, or 4 viable spores after *Saccharomyces cerevisiae* a / α diploids of the following genotypes are sporulated:

a. A true-breeding wild-type strain (with no mutations in any gene essential for viability).
b. A strain heterozygous for a null (completely inactivating) mutation in a single essential gene.

For the remaining parts of this problem, consider crosses between yeast strains of the form $a \times b$, where a and b are both temperature-sensitive mutations in different essential genes. The cross is conducted under permissive (low-temperature) conditions. Indicate the percentage of tetrads that would have 0, 1, 2, 3, or 4 viable spores subsequently measured under restrictive (high-temperature) conditions.

c. a and b are unlinked, and both are 0 m.u. from their respective centromeres.
d. a and b are unlinked; a is 0 m.u. from its centromere, while b is 10 m.u. from its centromere.
e. a and b are 0 m.u. apart.
f. a and b are 10 m.u. apart. Assume all crossovers between a and b are SCOs (single crossovers).
g. In part (f), if a four-strand DCO (double crossover) occurred between a and b, how many of the spores in the resulting tetrad would be viable at high temperature?

34. Two genes are located on the same chromosome as follows:

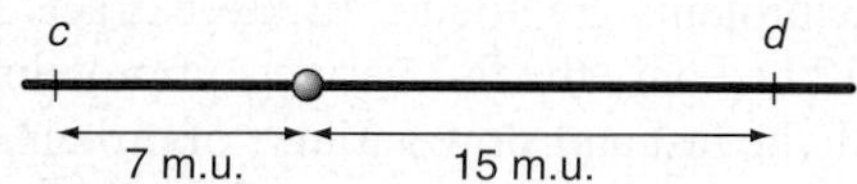

A haploid cross of the form $C\ D \times c\ d$ is made.

a. What proportions of PD, NPD, and T tetrads would you expect if this cross was made between strains of *Saccharomyces cerevisiae* and the interference in this region = 1?
b. If the interference in this region = 0?
c. What kinds of tetrads, and in what proportions, would you expect if this cross was made between strains of *Neurospora crassa* and the interference in this region = 1? (Consider not only whether a tetrad is PD, NPD, or T but also whether the tetrad shows first or second division segregation for each gene.)
d. If the interference in this region = 0?

35. A yeast strain that cannot grow in the absence of the amino acid histidine (*his*$^-$) is mated with a yeast strain that cannot grow in the absence of the amino acid lysine (*lys*$^-$). Among the 400 unordered tetrads resulting from this mating, 233 were PD, 11 were NPD, and 156 were T.

a. What types of spores are in the PD, NPD, and T tetrads?
b. What is the distance in map units between the *his* and *lys* genes?
c. Assuming that none of these tetrads was caused by more than two crossovers between the genes, how can you estimate the number of meioses that generated these 400 tetrads in which zero, one, or two crossovers took place?
d. Based on your answer to part *c*, what is the mean number of crossovers per meiosis in the region between the two genes?
e. The equation RF = 100 × (NPD + 1/2T) / total tetrads accounts for some, but not all, double crossovers between two genes. Which double crossovers are missed? Can you extrapolate from your answer to part *d* to obtain a more accurate equation for calculating map distances between two genes from the results of tetrad analysis?
f. Using your corrected equation from part *e*, what is a more accurate measurement of the distance in map units between the *his* and *lys* genes?

36. A research group has selected three independent *trp*$^-$ haploid strains of *Neurospora,* each of which cannot grow in the absence of the amino acid tryptophan. They first mated these three strains with a wild-type strain of opposite mating type, and then they analyzed the resultant octads. For all three matings, two of the four spore pairs in every octad could grow on minimal medium (that is, in the absence of tryptophan), while the other two spore pairs were unable to grow on this minimal medium.

a. What can you conclude from this result?

In the matings of mutant strains 1 and 2 with wild type, one of the two topmost pairs in some octads had spores that could grow on minimal medium while the other of the two topmost pairs in the same octads had spores that could not grow on minimal medium. In the mating of mutant strain 3 with wild type, either all the spores in the two topmost pairs could grow on minimal medium or all could not grow on minimal medium.

b. What can you conclude from this result?

The researchers next prepared two separate cultures of each mutant strain; one of these cultures was of mating type *A* and the other of mating type *a.* They mated these strains in pairwise fashion, dissected the resultant octads, and determined how many of the individual spores could grow on minimal medium. The results are shown here.

Mating	% of octads with *x* number of spores viable on minimal medium				
	$x = 0$	2	4	6	8
1 × 2	78	22	0	0	0
1 × 3	46	6	48	0	0
2 × 3	42	16	42	0	0

c. For each of the three matings in the table, how many of the 100 octads are PD? NPD? T?
d. Draw a genetic map explaining all of the preceding data. Assume that the sample sizes are sufficiently small that none of the octads are the result of double crossovers.
e. Although this problem describes crosses in *Neurospora,* it does not help in this particular case to present the matings in the table as ordered octads. Why not?
f. Why in this particular problem can you obtain gene–centromere distances from the crosses in the table, even though the data are not presented as ordered octads?

Section 5.3

37. A single yeast cell placed on a solid agar will divide mitotically to produce a colony of about 10^7 cells. A haploid yeast cell that has a mutation in the *ade2* gene will produce a red colony; an $ade2^+$ colony will be white. Some of the colonies formed from diploid yeast cells with a genotype of $ade2^+ / ade2^-$ will contain sectors of red within a white colony.

a. How would you explain these sectors?

b. Although the white colonies are roughly the same size, the red sectors within some of the white colonies vary markedly in size. Why? Do you expect the majority of the red sectors to be relatively large or relatively small?

38. A diploid strain of yeast has a wild-type phenotype but the following genotype:

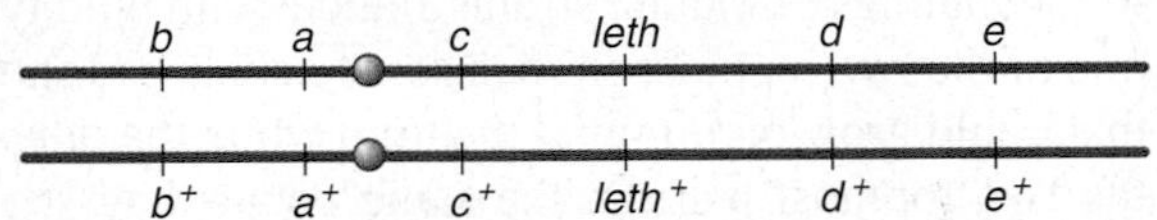

a, b, c, d, and *e* all represent recessive alleles that yield a visible phenotype, and *leth* represents a recessive lethal mutation. All genes are on the same chromosome, and *a* is very tightly linked to its centromere (indicated by a small circle). Which of the following phenotypes could be found in sectors resulting from mitotic recombination in this cell? (1) *a;* (2) *b;* (3) *c;* (4) *d;* (5) *e;* (6) *b e;* (7) *c d;* (8) *c d e;* (9) *d e;* (10) *a b.* Assume that double mitotic crossovers are too rare to be observed.

39. In *Drosophila,* the *yellow* (*y*) gene is near the end of the acrocentric X chromosome, while the *singed* (*sn*) gene is located near the middle of the X chromosome. On the wings of female flies of genotype $y\ sn / y^+\ sn^+$, you can very rarely find patches of yellow tissue within which a small subset of cells also have singed bristles.

a. How can you explain this phenomenon?

b. Would you find similar patches on the wings of females having the genotype $y^+\ sn / y\ sn^+$?

40. Neurofibromas are tumors of the skin that can arise when a skin cell that is originally $NF1^+ / NF1^-$ loses the $NF1^+$ allele. This wild-type allele encodes a functional tumor suppressor protein, while the $NF1^-$ allele encodes a nonfunctional protein.

A patient of genotype $NF1^+ / NF1^-$ has 20 independent tumors in different areas of the skin. Samples are taken of normal, noncancerous cells from this patient, as well as of cells from each of the 20 tumors. Extracts of these samples are analyzed by a technique called gel electrophoresis that can detect variant forms of four different proteins (A, B, C, and D) all encoded by genes that lie on the same autosome as *NF1.* Each protein has a slow (S) and a fast (F) form that are encoded by different alleles (for example, A^S and A^F). In the extract of normal tissue, slow and fast variants of all four proteins are found. In the extracts of the tumors, 12 had only the fast variants of proteins A and D but both the fast and slow variants of proteins B and C; 6 had only the fast variant of protein A but both the fast and slow variants of proteins B, C, and D; and the remaining 2 tumor extracts had only the fast variant of protein A, only the slow variant of protein B, the fast and slow variants of protein C, and only the fast variant of protein D.

a. What kind of genetic event described in this chapter could cause all 20 tumors, assuming that all the tumors are produced by the same mechanism?

b. Draw a genetic map describing these data, assuming that this small sample represents all the types of tumors that could be formed by the same mechanism in this patient. Show which alleles of which genes lie on the two homologous chromosomes. Indicate all relative distances that can be estimated.

c. Another mechanism that can lead to neurofibromas in this patient is a mitotic error producing cells with 45 rather than the normal 46 chromosomes. How can this mechanism cause tumors? How do you know, just from the results described, that none of these 20 tumors is formed by such mitotic errors?

d. Can you think of any other type of error that could produce the results described?

DNA: How the Molecule of Heredity Carries, Replicates, and Recombines Information

Chapter 6

The double-helical structure of DNA provides an explanation for the accurate transmission of genetic information from generation to generation over billions of years.

For nearly 4 billion years, the double-stranded DNA molecule has served as the bearer of genetic information; 3.7 billion years ago the earliest bacterial cells incorporated it into their chromosomes, and about 2 billion years ago, when the eukaryotic precursors of plants, animals, and fungi evolved from these simple cells, their chromosomes also carried a DNA molecule. Since that time, the hardware—the structure of the molecule itself—has not changed. In contrast, evolution has honed and vastly expanded the software—the programs of genetic information that the molecule stores, expresses, and transmits from one generation to the next.

Under special conditions of little or no oxygen, DNA can withstand a wide range of temperature, pressure, and humidity and remain relatively intact for hundreds, thousands, even tens of thousands of years. Molecular sleuths of the 1980s and 1990s have retrieved the evidence: 100-year-old DNA from preserved tissue of the quagga (a partially striped horselike creature that became extinct in the final years of the nineteenth century); 8000-year-old DNA from human skulls found in the swamps of Florida; and 30,000-year-old DNA from a Neanderthal skeleton (**Fig. 6.1**). Surprisingly, this ancient DNA still carries readable sequences—shards of decipherable information that become veritable time machines for the direct viewing of genes in long-vanished organisms and species. Comparisons with equivalent expanses of modern DNA make it possible to identify the precise mutations that have fueled evolution. Analyses of Neanderthal and human DNA, for example, have helped anthropologists settle a long-running debate about the relation of the two species. The genetic evidence shows that Neanderthals and early humans did not intermix to any significant degree and that *Homo sapiens,* upon their arrival in Europe some 40,000 years ago, most likely displaced Neanderthals over a period of time.

Francis Crick, codiscoverer of DNA's double helical structure and a leading theoretician of molecular biology, has written that "almost all aspects of life are engineered at the molecular level, and without understanding molecules, we can only have a very sketchy understanding of life itself." In Chapters 1–5 we examined how Mendel and his successors used data from breeding experiments to deduce the presence and activity of genes, analyze the effect of alternative alleles on phenotype, and tie genes to the movements of chromosomes in time and space. With this approach, it is possible to predict the outcomes of genetic crosses in the absence of detailed knowledge of the molecules and biochemical reactions that underlie the events observed. But without understanding what the molecules that contain genes look like, it is impossible to decipher the exact mechanisms by which genes determine phenotypes, transmit instructions between generations, and evolve new information. For this reason, we shift our perspective in this chapter to an examination of DNA, the molecule that composes the genetic material.

As we extend our analysis to the molecular level, two general themes emerge. First, DNA's genetic functions flow directly from its molecular structure—the way its atoms are arranged in space. Second, most of DNA's genetic functions depend on specialized proteins that interact with the molecule of heredity. One such protein, for example, binds to DNA to begin the process of replication.

In our discussion of the structure and function of DNA, we describe

- How investigators pinpointed DNA as the genetic material.
- The elegant Watson-Crick model of DNA structure.
- How DNA structure provides for the storage of genetic information.
- How DNA structure gives rise to the semiconservative mode of molecular replication.
- How DNA structure promotes the recombination of genetic information.

(a)

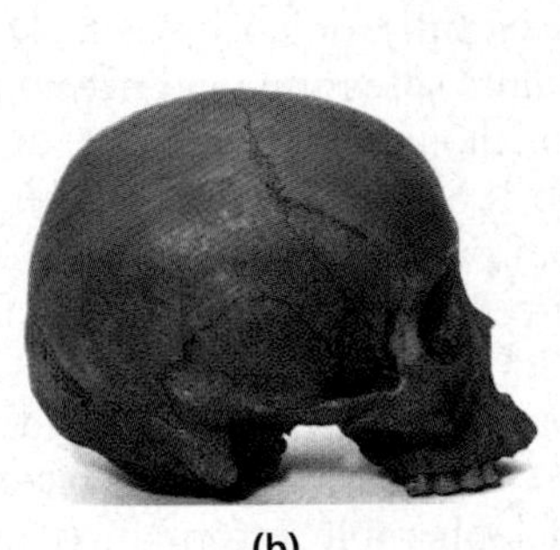

(b)

(c)

Figure 6.1 Ancient DNA still carries information. Molecular biologists have successfully extracted DNA from **(a)** a 100-year-old quagga; **(b)** an 8000-year-old human skull; **(c)** a Neanderthal skeleton excavated in 1856 in Germany's Neander Valley whose DNA was analyzed in the late 1990s. These findings attest to the chemical stability of DNA, the molecule of inheritance.

6.1 Experiments Designate DNA as the Genetic Material

Though accepted today as fact, it was not always obvious that DNA is the genetic material. It took a cohesive pattern of results from experiments performed over more than 50 years to convince the scientific community that DNA is the molecule of heredity. We now present key pieces of the evidence.

Chemical Characterization Localizes DNA in the Chromosomes

In 1869, Friedrich Miescher extracted a weakly acidic, phosphorus-rich material from the nuclei of human white blood cells and named it "nuclein." It was unlike any previously reported chemical compound, and its major component turned out to be DNA, although it also contained some contaminants. The full chemical name of DNA is **deoxyribonucleic acid,** reflecting three characteristics of the substance: one of its constituents is a sugar known as deoxyribose; it is found mainly in cell nuclei; and it is acidic.

After purifying DNA from the nuclein and treating it with various reagents, researchers established that DNA is composed of four different subunits linked in a long chain (**Fig. 6.2**). The four different subunits belong to a class of compounds known as **nucleotides,** the bonds joining one nucleotide to another are covalent **phosphodiester bonds,** and the linked chain of repeating subunits is a type of **polymer.** A procedure first reported in 1923 made it possible to discover where in the cell DNA resides. Named the Feulgen reaction after its designer, the procedure relies on a chemical called the Schiff reagent, which stains DNA red. In a preparation of stained cells, the chromosomes redden, while other areas of the cell remain relatively colorless. The reaction shows that DNA is localized almost exclusively within specific nuclear organelles—the chromosomes.

The finding that DNA is a component of chromosomes does not prove that the molecule has anything to do with

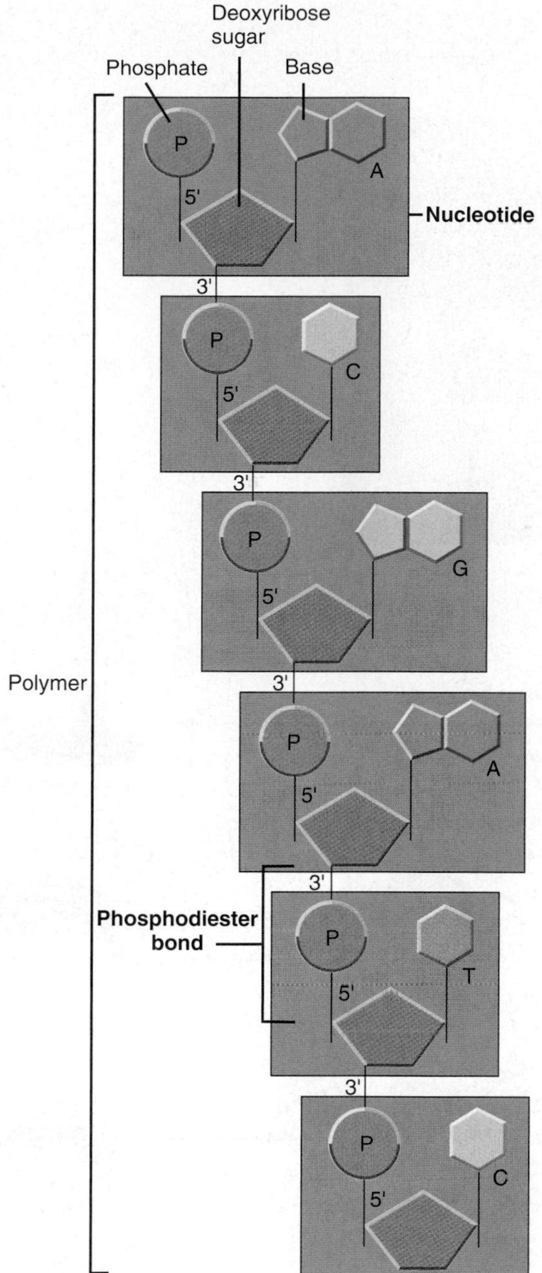

Figure 6.2 The chemical composition of DNA. A single strand of a DNA molecule consists of a chain of nucleotide subunits (*purple boxes*). Each nucleotide is made of the sugar deoxyribose (*tan pentagons*) connected to an inorganic phosphate group (*yellow circles*) and to one of four nitrogenous bases (*purple* or *green polygons*). The phosphodiester bonds that link the nucleotide subunits to each other attach the phosphate group of one nucleotide to the deoxyribose sugar of the preceding nucleotide.

genes. Eukaryotic chromosomes also contain an even greater amount of protein by weight. Because proteins are built of 20 different amino acids whereas DNA carries just four different subunits, many researchers thought proteins had greater potential for diversity and were better suited to serve as the genetic material. These same scientists assumed that even though DNA was an important part of chromosome structure, it was too simple to specify the complexity of genes.

Bacterial Transformation Implicates DNA As the Substance of Genes

Several studies supported the idea that DNA would be the genetic material. The most important of these used single-celled bacteria as their experimental organism. Bacteria carry their genetic material in a single circular chromosome that lies in the nucleoid region of the cell without being enclosed in a nuclear membrane. The single circular chromosome (which makes the cells haploid) and lack of a nuclear membrane (which makes them *prokaryotes,* meaning "before a true nucleus") are two features that distinguish bacteria from the eukaryotic cells of higher organisms. Another feature is their method of cell division. With only one chromosome, bacteria do not undergo meiosis to produce germ cells, and they do not apportion their replicated chromosomes to daughter cells by mitosis; rather, they divide by a process known as binary fission. Even with these acknowledged differences, at least some investigators in the first half of the twentieth century believed that the genetic material of bacteria was similar to that found in eukaryotic organisms.

One prerequisite of genetic studies in bacteria, as with any species, is the detection of alternative forms of a trait within a population. In a 1923 study of *Streptococcus pneumoniae* bacteria grown in laboratory media, Frederick Griffith distinguished two bacterial forms: smooth (S) and rough (R). S is the wild type; a mutation in S gives rise to R. From observation and biochemical analysis, Griffith determined that S forms appear smooth because they synthesize a polysaccharide capsule that surrounds pairs of cells. R forms, which arise spontaneously as mutants of S, are unable to make the capsular polysaccharide, and as a result, their colonies appear to have a rough surface (**Fig. 6.3a**). We now know that the R form lacks an enzyme necessary for synthesis of the capsular polysaccharide. Because the polysaccharide capsule helps protect the bacteria from an animal's immune response, the S bacteria are virulent and kill most laboratory animals exposed to them (Fig. 6.3b.1); by contrast, the R forms fail to cause infection (Fig. 6.3b.2). In humans, the virulent S forms of *S. pneumoniae* can cause pneumonia.

Transformation

In 1928, Griffith published the astonishing finding that genetic information from dead bacterial cells could somehow be transmitted to live cells. He was working with two kinds of bacteria—live R forms and heat-killed S forms. Neither the heat-killed S forms nor the live R forms produced infection when injected into laboratory mice (Fig. 6.3b.2 and 3); but a mixture of the two killed the animals (Fig. 6.3b.4). Bacteria recovered from the blood of the dead animals were living S forms (Fig. 6.3b.4). The ability of a substance to change the genetic characteristics of an organism is known as **transformation.** Something from the heat-killed S bacteria must have transformed the living R bacteria into S.

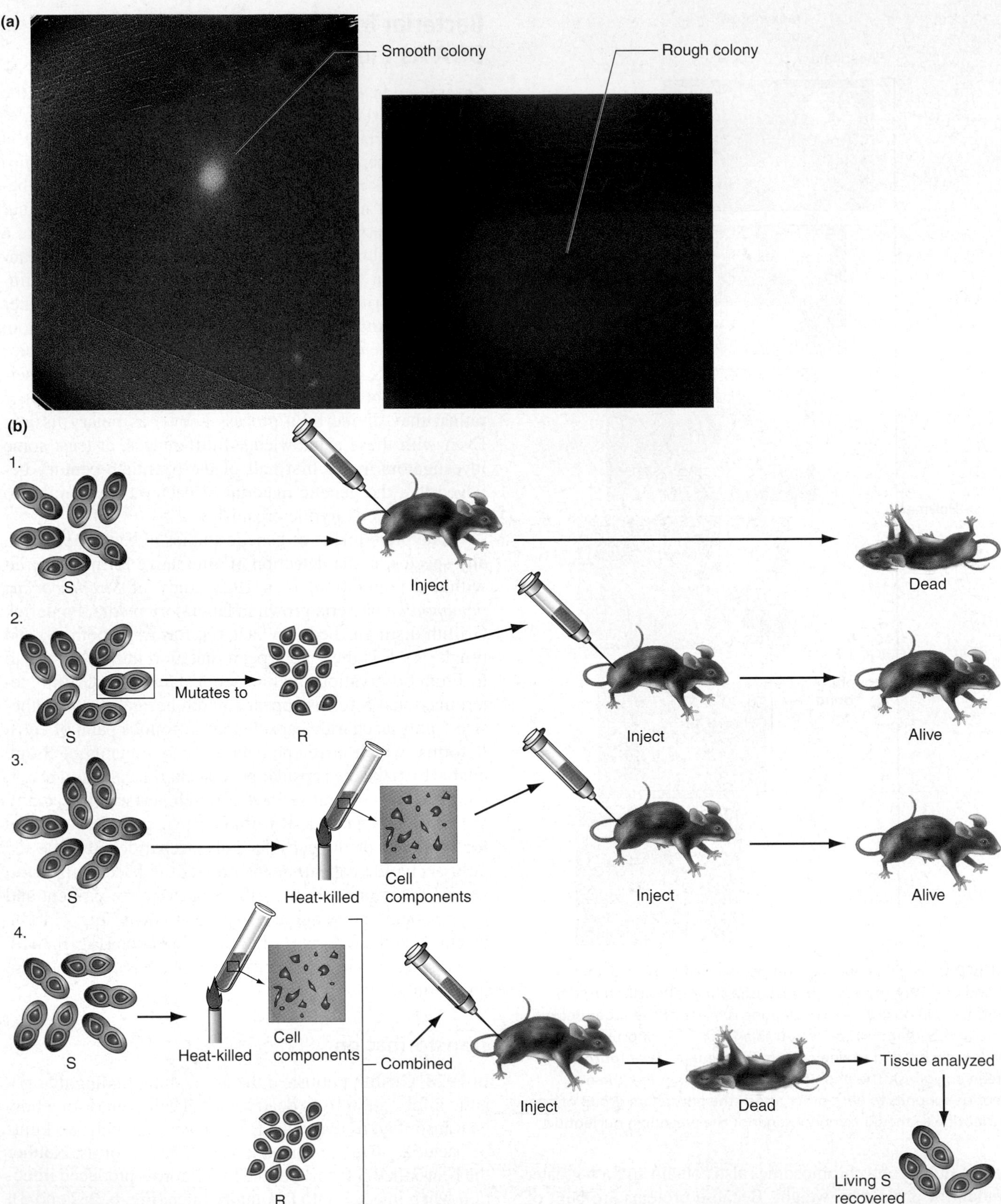

Figure 6.3 Griffith's demonstration of bacterial transformation. **(a)** Smooth (S) and rough (R) colonies of *S. pneumoniae*. **(b)** Griffith's experiment: (1) S bacteria are virulent and can cause lethal infections when injected into mice. (2) Injections of R mutants by themselves do not cause infections that kill mice. (3) Similarly, injections of heat-killed S bacteria do not cause lethal infections. (4) Lethal infection does result, however, from injections of live R bacteria mixed with heat-killed S strains; the blood of the dead host mouse contains living S-type bacteria.

This transformation was permanent and most likely genetic, because all future generations of the bacteria grown in culture were the S form.

Bacterial Transformation Is Caused by DNA

By 1929, two other laboratories had repeated these results, and in 1931, investigators in Oswald T. Avery's laboratory found they could achieve transformation without using any animals at all, simply by growing R-form bacteria in medium in the presence of components from dead S forms (**Fig. 6.4a**). Avery then embarked on a quest that would remain the focus of his work for almost 15 years: "Try to find in that complex mixture, the active principle!" In other words, try to identify the heritable substance in the bacterial extract that induces transformation from harmless R bacteria to pathogenic S bacteria. Avery dubbed the substance he was searching for the "transforming principle" and spent many years trying to purify it sufficiently to be able to identify it unambiguously. He and his coworkers eventually prepared a tangible, active transforming principle. In the final part of their procedure, a long whitish wisp materialized from ice-cold alcohol solution and wound around the glass stirring rod to form a fibrous wad of nearly pure principle (Fig. 6.4b).

Once purified, the transforming principle had to be characterized. In 1944, Avery and two coworkers, Colin MacLeod and Maclyn McCarty, published the cumulative findings of experiments designed to determine the transforming principle's chemical composition (Fig. 6.4c). In these experiments, the purified transforming principle was active at the extraordinarily high dilution of 1 part in 600 million. Although the preparation was almost pure DNA, the investigators nevertheless exposed it to various enzymes to see if some molecule other than DNA could cause transformation. Enzymes that degraded RNA, protein,

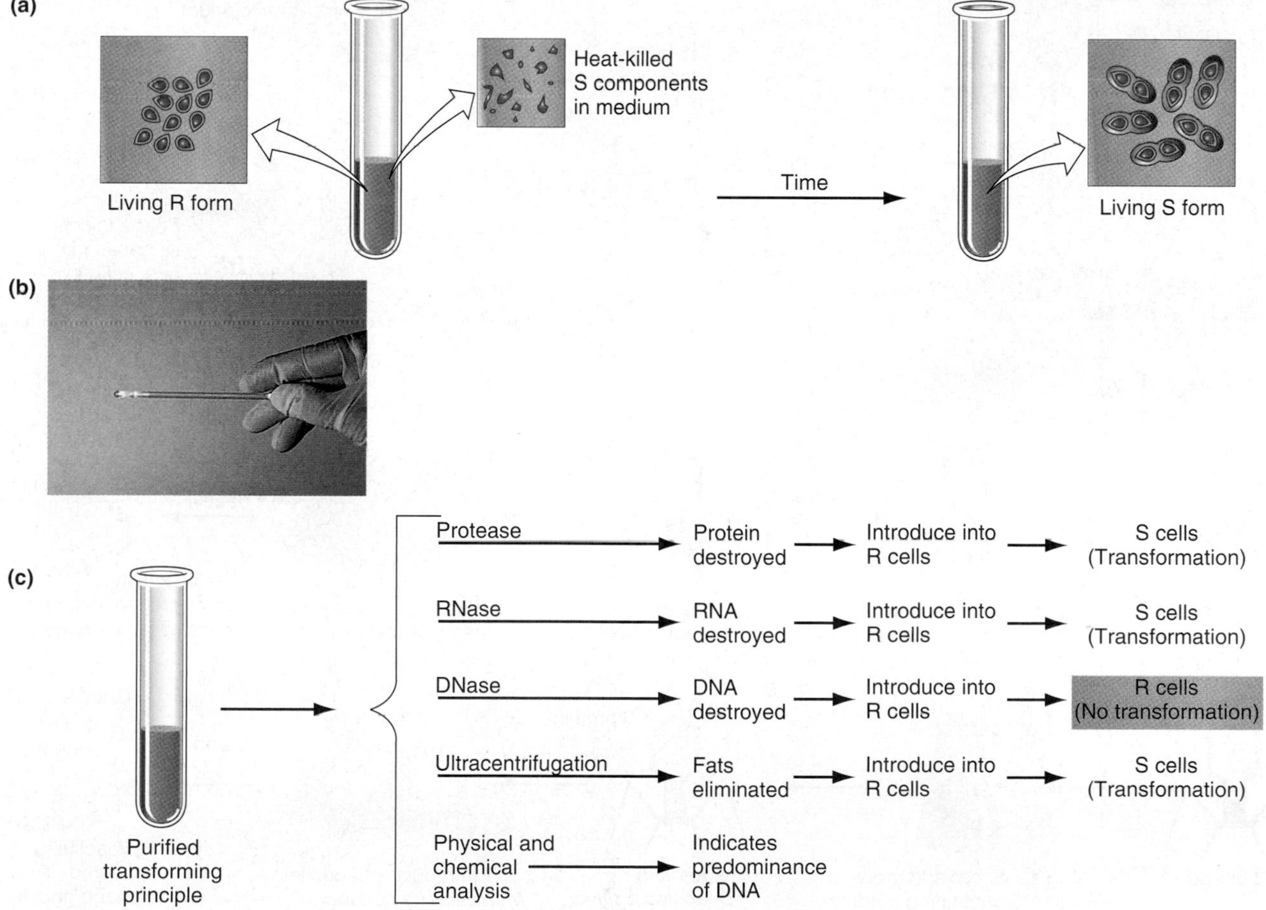

Figure 6.4 The transforming principle is DNA: Experimental confirmation. (a) Bacterial transformation occurs in culture medium containing the remnants of heat-killed S bacteria. This indicates that some "transforming principle" from the heat-killed S bacteria is taken up by the live R bacteria, converting (transforming) them into virulent S strains. **(b)** A wad of purified DNA from heat-killed S-type *S. pneumoniae* wound around a glass stirring rod. Small amounts of this preparation are highly efficient in causing bacterial transformation. **(c)** Chemical fractionation of the transforming principle. Treatment of purified DNA with a DNA-degrading enzyme destroys its ability to cause bacterial transformation, while treatment with enzymes that destroy other macromolecules has no effect on the transforming principle. In contrast to the purified DNA shown in part (b), highly purified preparations of protein, lipids, or carbohydrates do not contain active transforming principle.

or polysaccharide had no effect on the transforming principle, but an enzyme that degrades DNA completely destroyed its activity. The tentative published conclusion was that the transforming principle appeared to be DNA. In a personal letter to his brother, Avery went one step further and confided that the transforming principle "may be a gene."

Despite the paper's abundance of concrete evidence, many within the scientific community still resisted the idea that DNA is the molecule of heredity. They argued that perhaps Avery's results reflected the activity of contaminants; or perhaps genetic transformation was not happening at all, and instead, the transforming principle was somehow triggering a physiological switch in the transformed bacteria. Unconvinced for the moment, these scientists remained attached to the idea that proteins were the prime candidates for the genetic material.

The DNA Molecule Carries the Information Required for the Replication of Bacterial Viruses

Not everyone shared this skepticism. Alfred Hershey and Martha Chase anticipated that they could assess the relative importance of DNA and protein in gene transmission by infecting bacterial cells with viruses called **phages,** short for **bacteriophages** (literally "bacteria eaters"). Viruses are the simplest of organisms. By structure and function, they fall somewhere between living cells capable of reproducing themselves and macromolecules such as proteins. Because viruses depend on their host cell for most of the machinery required for growth and replication, they can be very small indeed and contain very few genes. For many kinds of phage, each particle consists of

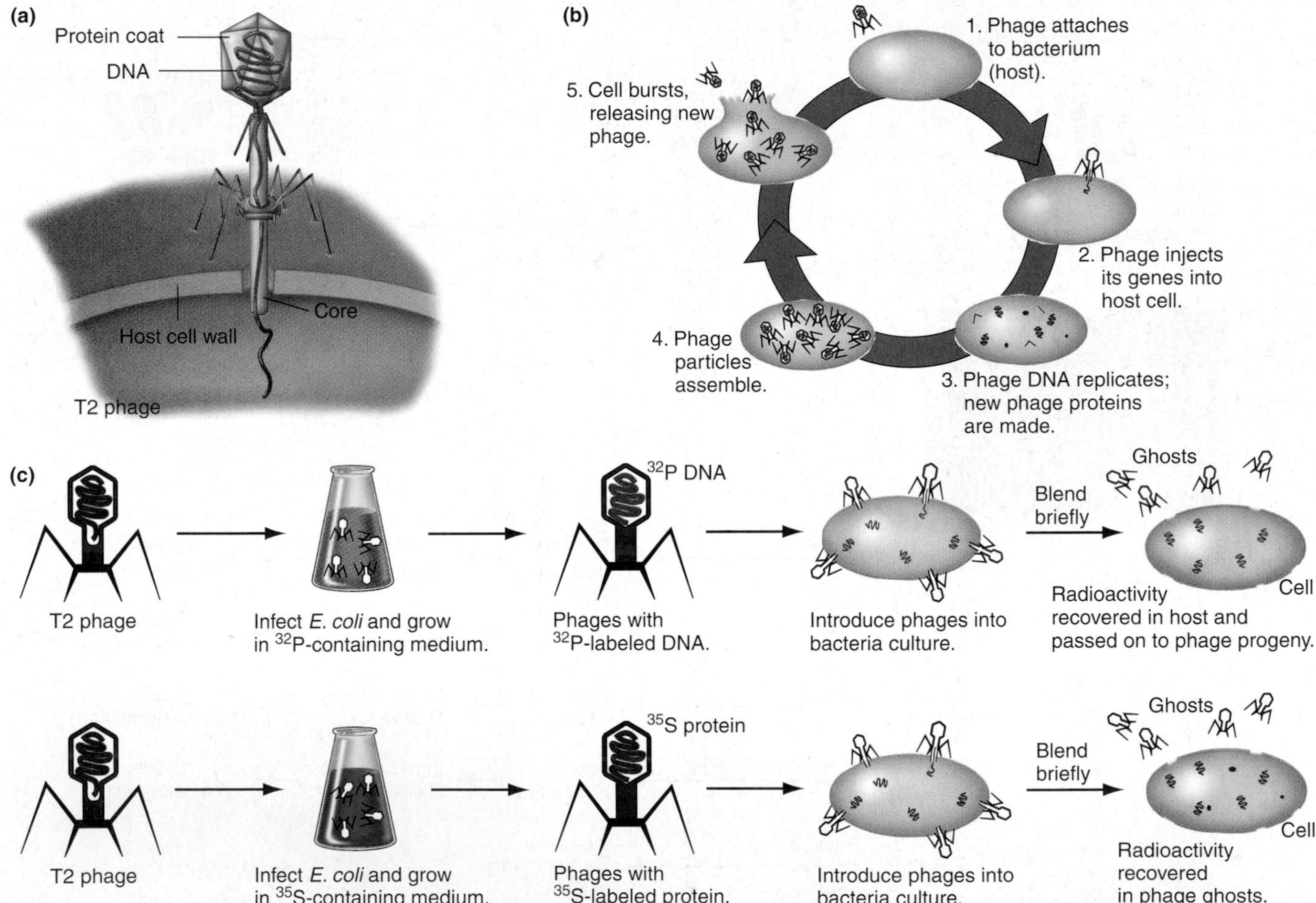

Figure 6.5 Experiments with viruses provide convincing evidence that genes are made of DNA. (a) and **(b)** Bacteriophage T2 structure and life cycle. The phage particle consists of DNA contained within a protein coat. The virus attaches to the bacterial host cell and injects its genes (the DNA) through the bacterial cell wall into the host cell cytoplasm. Inside the host cell, these genes direct the formation of new phage DNA and proteins, which assemble into progeny phages that are released into the environment when the cell bursts. **(c)** The Hershey-Chase Waring blender experiment. T2 bacteriophage particles either with ^{32}P-labeled DNA or with ^{35}S-labeled proteins were used to infect bacterial cells. After a short incubation, Hershey and Chase shook the cultures in a Waring blender and spun the samples in a centrifuge to separate the empty viral ghosts from the heavier infected cells. Most of the ^{35}S-labeled proteins remained with the ghosts, while most of the ^{32}P-labeled DNA was found in the sediment with the T2 gene-containing infected cells.

roughly equal weights of protein and DNA (**Fig. 6.5a**). These phage particles can reproduce themselves only after infecting a bacterial cell. Thirty minutes after infection, the cell bursts and hundreds of newly made phages spill out (Fig. 6.5b). The question is, What substance directs the production of the new phage particles—DNA or protein?

With the invention of the electron microscope in 1939, it became possible to see individual phages, and surprisingly, electron micrographs revealed that the entire phage does not enter the bacterium it infects. Instead, a viral shell—called a *ghost*—remains attached to the outer surface of the bacterial cell wall. Because the empty phage coat remains outside the bacterial cell, one investigator likened phage particles to tiny syringes that bind to the cell surface and inject transforming principle into the host cell.

In their famous Waring blender experiment of 1952, Alfred Hershey and Martha Chase tested the idea that the ghost left on the cell wall is composed of protein, while the injected transforming principle consists of DNA (Fig. 6.5c). A type of phage known as T2 served as their experimental organism. They grew two separate sets of T2 in bacteria maintained in two different culture media, one infused with radioactively labeled phosphorus (^{32}P), the other with radioactively labeled sulfur (^{35}S). Since proteins incorporate sulfur but no phosphorus and DNA contains phosphorus but no sulfur, phages grown on ^{35}S would have radioactively labeled protein while particles grown on ^{32}P would have radioactive DNA. The radioactive tags would enable the investigators to determine the location of each material when the phages infected fresh cultures of bacterial cells.

After exposing one culture of bacteria to ^{32}P-labeled phage and another culture to ^{35}S-labeled phage, Hershey and Chase used a Waring blender to disrupt each culture, effectively separating the viral ghosts from the bacteria harboring the viral genes. Centrifugation of the cultures then separated the heavier infected cells, which ended up in a pellet, from the lighter phage ghosts, which remained suspended in the supernatant. Most of the radioactive ^{32}P (in DNA) went to the pellet, while most of the radioactive ^{35}S (in protein) remained in the supernatant. This confirmed that the extracellular ghosts were indeed mostly protein, while the injected viral material specifying production of more phages was mostly DNA. Bacteria containing the radiolabeled phage DNA behaved just as in a normal phage infection, producing and disgorging hundreds of progeny particles. From these observations, Hershey and Chase concluded that phage genes are made of DNA.

The Hershey-Chase experiment, although less rigorous than the Avery project, had an enormous impact. In the minds of many investigators, it confirmed Avery's results and extended them to viral particles. The spotlight was now on DNA.

6.2 The Watson-Crick Model: DNA Is a Double Helix

Under appropriate conditions, DNA can align in fibers to produce an ordered structure. And just as a crystal chandelier scatters light to produce a distinctive pattern on the wall, DNA fibers scatter X-rays to produce a characteristic diffraction pattern (**Fig. 6.6**). A knowledgeable X-ray crystallographer can interpret DNA's diffraction pattern to elucidate selected aspects of the molecule's three-dimensional structure. When in the spring of 1951 the 23-year-old James Watson learned that the genetic material could project a diffraction pattern, he realized that DNA "must have a regular structure that could be solved in a straightforward fashion."

In this section, we analyze DNA's three-dimensional structure, looking first at significant details of the nucleotide building blocks, then at how those subunits are linked together in a polynucleotide chain, and finally, at how two chains associate to form a double helix.

Nucleotides Are the Basic Building Blocks of DNA

DNA is a long polymer composed of subunits known as *nucleotides.* Each nucleotide consists of a deoxyribose sugar, a phosphate, and a nitrogenous base. Detailed chemical knowledge of these constituents and the way they combine played an important role in Watson and Crick's model building.

Figure 6.7 depicts the chemical composition of deoxyribose, phosphate, and the four nitrogenous bases; how these components come together to form a nucleotide; and how phosphodiester bonds link the nucleotides in a chain. Each individual carbon or nitrogen atom in the central ring structure of a nitrogenous base is assigned a number: from

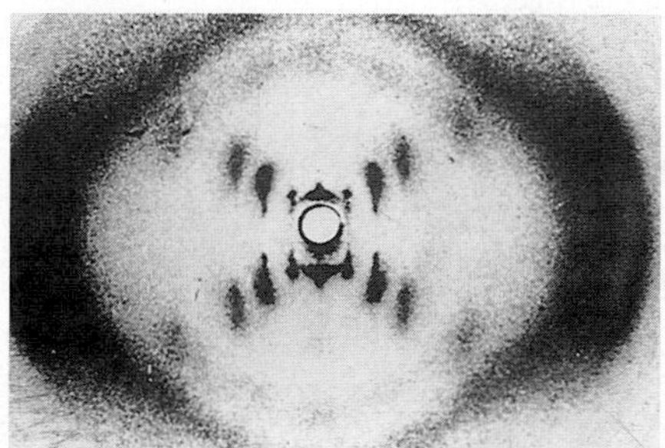

Figure 6.6 X-ray diffraction patterns reflect the helical structure of DNA. Photograph of an X-ray diffraction pattern produced by oriented DNA fibers, taken by Rosalind Franklin and Maurice Wilkins in late 1952. The crosswise pattern of X-ray reflections indicates that DNA is helical.

Feature Figure 6.7

A Detailed Look at DNA's Chemical Constituents

(a) The separate entities

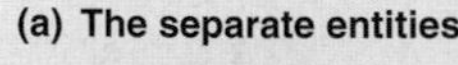

1. Deoxyribose sugar

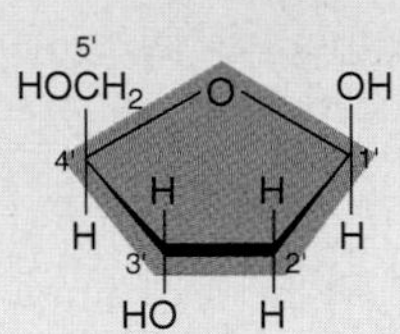

Ribose

2. A phosphate group

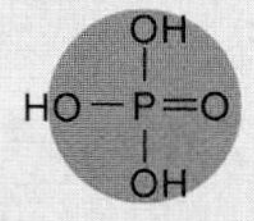

3. Four nitrogenous bases

Purines

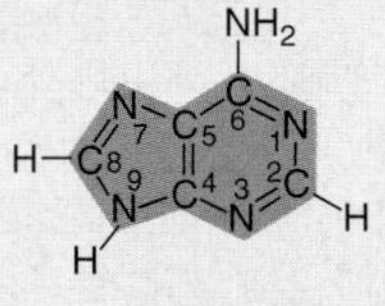

Adenine **(A)**

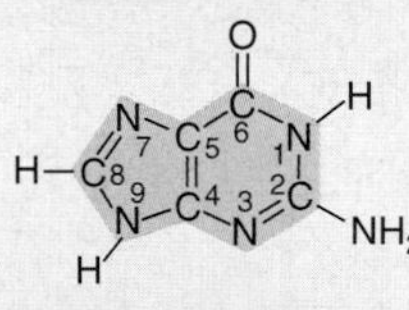

Guanine **(G)**

Pyrimidines

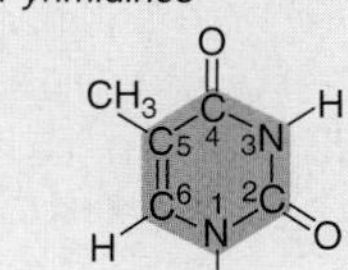

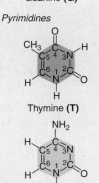

Thymine **(T)**

Cytosine **(C)**

(b) Assembly into a nucleotide

1. Attachment of base to sugar

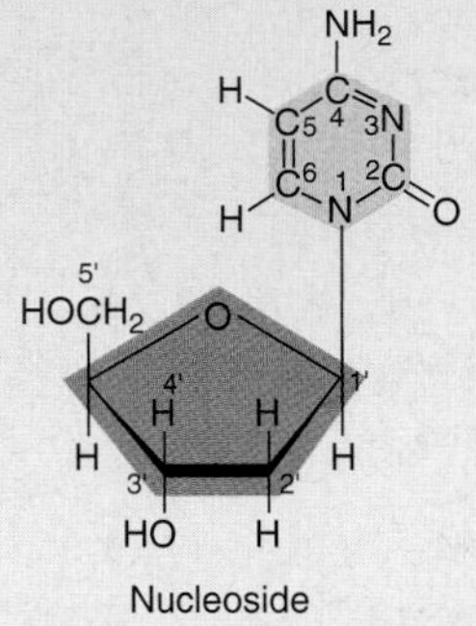

Nucleoside

2. Addition of phosphate

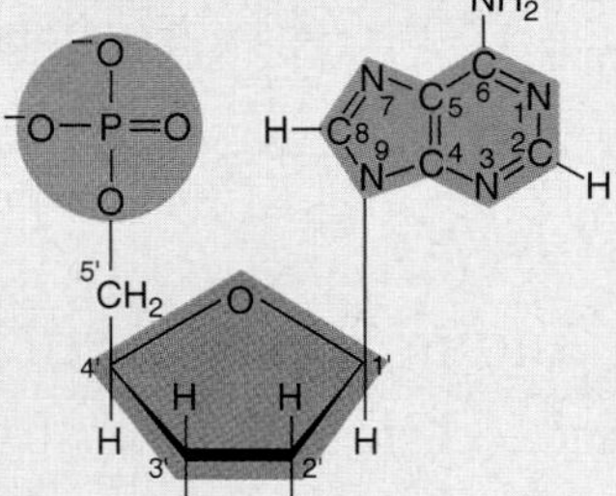

Purine nucleotide

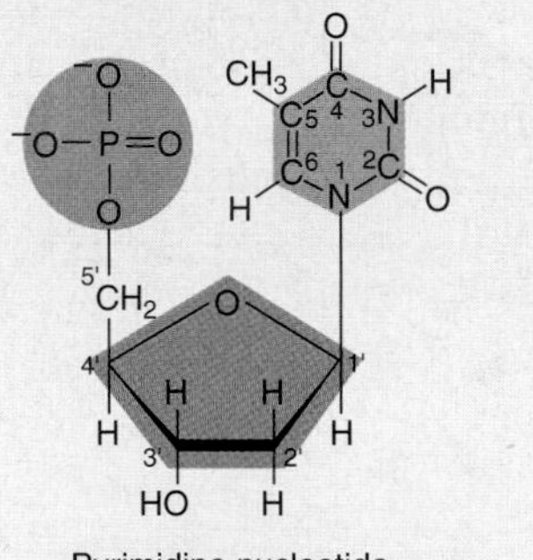

Pyrimidine nucleotide

(c) Nucleotides linked in a directional chain

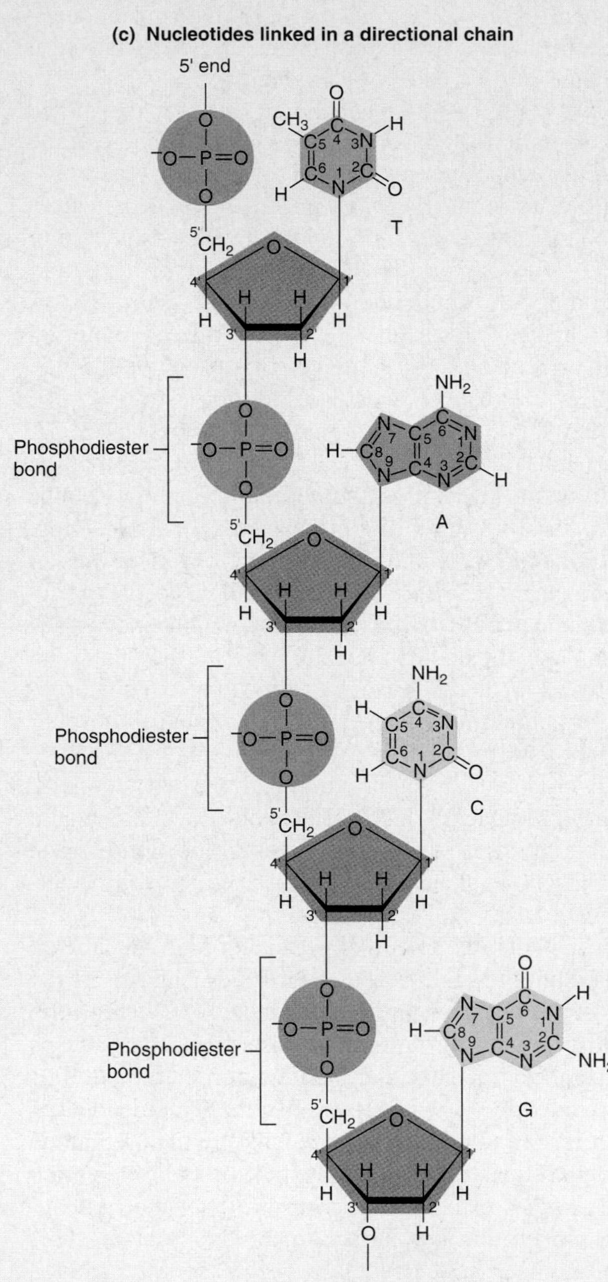

1–9 for purines, and 1–6 for pyrimidines. The carbon atoms of the deoxyribose sugar are distinguished from atoms within the nucleotide base by the use of primed numbers from 1′–5′. Covalent attachment of a base to the 1′ carbon of deoxyribose forms a nucleoside. The addition of a phosphate group to the 5′ carbon forms a complete nucleotide. As Fig. 6.7 shows, a DNA chain composed of many nucleotides has **polarity:** an overall direction. Phosphodiester bonds always form a covalent link between the 3′ carbon of one nucleoside and the 5′ carbon of the following nucleoside. The consistent orientation of the nucleotide building blocks gives a chain overall direction, such that the two ends of a single chain are chemically distinct.

At the 5′ end, the sugar of the terminal nucleotide has a free 5′ carbon atom, free in the sense that it is not linked to another nucleotide. Depending on how the DNA is synthesized or isolated, the 5′ carbon of the nucleotide at the 5′ end may carry either a hydroxyl or a phosphate group. At the other—3′—end of the chain, it is the 3′ carbon of the final nucleotide that is free. Along the chain between the two ends, this 5′-to-3′ polarity is conserved from nucleotide to nucleotide. By convention, a DNA chain is described in terms of its bases, written with the 5′-to-3′ direction going from left to right (unless otherwise noted). The chain depicted in Fig. 6.7c, for instance, would be TACG.

A Directional Base Sequence Can Carry Information

Information can be encoded only in a sequence of symbols whose order varies according to the message to be conveyed. Without this sequence variation, there is no potential for carrying information. Because DNA's sugar-phosphate backbone is chemically identical for every nucleotide in a DNA chain, the only difference between nucleotides is in the identity of the nitrogenous base. Thus, if DNA carries genetic information, that information must consist of variations in the sequence of the A, G, T, and C bases. The information constructed from the 4-letter language of DNA bases is analogous to the information built from the 26-letter alphabet of English or French or Italian. Just as you can combine the 26 letters of the alphabet in different ways to generate the words of a book, so, too, different combinations of the four bases in very long sequences of nucleotides can encode the information for constructing an organism.

The Double Helix Contains Two Antiparallel Chains That Associate by Complementary Base Pairing

Watson and Crick's discovery of the structure of the DNA molecule ranks with Darwin's theory of evolution by natural selection and Mendel's laws of inheritance in its contribution to our understanding of biological phenomena. The Watson-Crick structure, first embodied in a model that superficially resembled the Tinker Toys of preschool children, was based on an interpretation of all the chemical and physical data available at the time. Watson and Crick published their findings in the scientific journal *Nature* in April 1953.

The Physical Data: X-Ray Diffraction Patterns Indicate Molecular Parameters

The diffraction patterns of oriented DNA fibers do not, on their own, contain sufficient information to reveal structure. For instance, the number of diffraction spots, whose intensities and positions constitute the X-ray data (review Fig. 6.6), is considerably less than the number of unknown coordinates of all the atoms in an oriented DNA molecule. Nevertheless, the photographs do reveal a wealth of structural information to the trained eye. Excellent X-ray images produced by Rosalind Franklin and Maurice Wilkins showed that the molecule is spiral-shaped, or helical; the spacing between repeating units along the axis of the helix is 3.4 Å; the helix undergoes one complete turn every 34 Å; the diameter of the molecule is 20 Å. Given this diameter, the molecule must consist of more than one polynucleotide chain.

Key Chemical Data Show That Nucleotides Exhibit Complementarity

If DNA contains more than one chain of nucleotides, what forces hold these chains together? Erwin Chargaff obtained data on the nucleotide composition of DNA from various sources that provided an important clue. Despite large variations between species in the relative amounts of the bases, in every individual, the ratio of A to T is not significantly different from 1:1, and the ratio of G to C is the same (**Table 6.1**). Watson grasped that the roughly 1:1 ratios of A to T and of G to C reflect a significant aspect of the molecule's inherent structure.

To explain Chargaff's ratios in terms of chemical affinities between A and T and between G and C, Watson made cardboard cutouts of the bases in the chemical forms they assume in a normal cellular environment and tried to match these up in various combinations, like pieces in a jigsaw puzzle. He knew that the substituents on purines and pyrimidines play a crucial role in molecular interactions as they can participate in the formation of **hydrogen bonds:** weak electrostatic bonds that result in a partial sharing of hydrogen atoms between reacting groups (**Fig. 6.8**). Some substituents on the nitrogenous bases provide, or "donate," their hydrogen atoms for bonding, while others are able to create, or "accept," bond formation with the donated hydrogens. Watson saw that A and T could be paired together such that two hydrogen bonds formed between them. If G and C were similarly paired, hydrogen bonds could also easily connect the nucleotides carrying these two bases.

TABLE 6.1 Chargaff's Data on Nucleotide Base Composition in the DNA of Various Organisms

	Percentage of Base in DNA				Ratios	
Organism	*A*	*T*	*G*	*C*	*A:T*	*G:C*
Staphylococcus afermentams	12.8	12.9	36.9	37.5	0.99	0.99
Escherichia coli	26.0	23.9	24.9	25.2	1.09	0.99
Yeast	31.3	32.9	18.7	17.1	0.95	1.09
*Caenorhabditis elegans**	31.2	29.1	19.3	20.5	1.07	0.96
*Arabadopsis thaliana**	29.1	29.7	20.5	20.7	0.98	0.99
Drosophila melanogaster	27.3	27.6	22.5	22.5	0.99	1.00
Honeybee	34.4	33.0	16.2	16.4	1.04	0.99
Mus musculus (mouse)	29.2	29.4	21.7	19.7	0.99	1.10
Human (liver)	30.7	31.2	19.3	18.8	0.98	1.03

*Data for *C. elegans* and *A. thaliana* are based on those for close relative organisms.
Note that even though the level of any one nucleotide is different in different organisms, the amount of A always approximately equals the amount of T, and the level of G is always similar to that of C. Moreover, as you can calculate for yourself, the total amount of purines (A plus G) nearly always equals the total amount of pyrimidines (C plus T).

(Watson originally posited two hydrogen bonds between G and C, but there are actually three.) Remarkably, the two pairs—A–T and G–C—had essentially the same shape. This meant that the two pairs could fit in any order between two sugar-phosphate backbones without distorting the structure. It also explained the Chargaff ratios—always equal amounts of A and T and of G and C. Note that both of these base pairs consist of one purine and one pyrimidine. Crick connected the chemical facts with the X-ray data, recognizing that because of the geometry of the base-sugar bonds in nucleotides, the orientation of the bases in Watson's pairing scheme could arise only if the bases were attached to backbones running in opposite directions. **Figure 6.9** illustrates and explains the model Watson and Crick proposed in April 1953: DNA as a double helix.

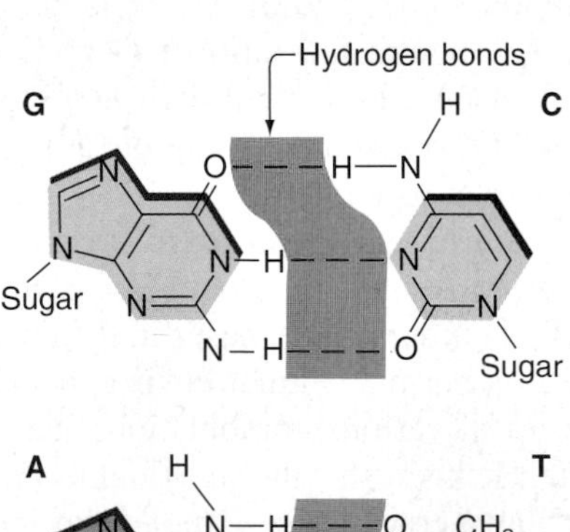

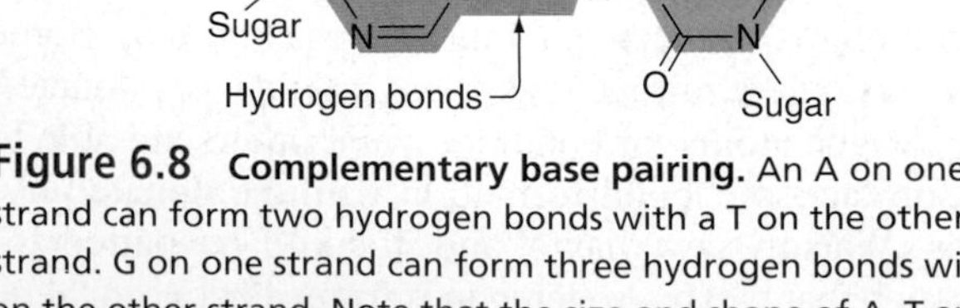

Figure 6.8 Complementary base pairing. An A on one strand can form two hydrogen bonds with a T on the other strand. G on one strand can form three hydrogen bonds with a C on the other strand. Note that the size and shape of A–T and of G–C base pairs are similar, allowing both to fill the same amount of space between the two backbones of the double helix.

The Double Helix May Assume Alternative Forms

Watson and Crick arrived at the double helix model of DNA structure by building models, not by a direct structural determination from the data alone. And even though Watson has written that "a structure this pretty just had to exist," the beauty of the structure is not necessarily evidence of its correctness. At the time of its presentation, the strongest evidence for its correctness was its physical plausibility, its chemical and spatial compatibility with all available data, and its capacity for explaining many biological phenomena.

The Biological Significance of Alternative Overall Configurations Remains Unknown

The majority of naturally occurring DNA molecules have the configuration suggested by Watson and Crick. Such molecules are known as **B-form DNA;** they spiral to the right (**Fig. 6.10a** on p. 179). DNA is, however, more polymorphic than originally assumed. Some of the small double-stranded molecules examined by X-ray crystallography look completely different. One type, for example, contains nucleotide sequences that cause the DNA to assume a **Z form** in which the helix spirals to the left and

Feature Figure 6.9

The Structure of DNA: A Double Helix Composed of Antiparallel Chains Associated by Complementary Base Pairing

(a) In a leap of imagination, Watson and Crick took the known facts about DNA's chemical composition and physical arrangement in space and constructed a wire-frame model that not only united the evidence but also served as a basis for explaining the molecule's function.

(b) In the model (shown on p. 178 at the left), two DNA chains spiral around an axis with the sugar-phosphate backbones on the outside and pairs of bases (one from each chain) meeting in the middle. Although both chains wind around the helix axis in a right-handed sense, chemically one of them runs 5′ to 3′ upward, while the other runs in the opposite direction of 5′ to 3′ downward. In short, the *two chains are antiparallel.* The base pairs are essentially flat and perpendicular to the helix axis, and the planes of the sugars are roughly perpendicular to the base pairs. As the two chains spiral about the helix axis, they wrap around each other once every 10 base pairs, or once every 34 Å. The result is a double helix that looks like a twisted ladder with the two spiraling structural members composed of sugar-phosphate backbones and the rungs consisting of base pairs.

(c) In a space-filling representation of the model (shown on p. 178 at the right), the overall shape is that of a grooved cylinder with a diameter of 20 Å whose axis is the axis of the double helix. The backbones spiral around the axis like threads on a screw, but because there are two backbones, there are two threads, and these two threads are vertically displaced from each other. This displacement of the backbones generates two grooves, one much wider than the other, that also spiral around the helix axis. Biochemists refer to the wider groove as the **major groove** and the narrower one as the **minor groove.**

The *two chains of the double helix are held together by hydrogen bonds between complementary base pairs,* A–T and G–C (see Fig. 6.8). Since the overall shapes of the two base pairs are quite similar, either pair can fit into the structure at each position along the DNA. Moreover, each base pair can be accommodated in the structure in two ways that are the reverse of each other: an A purine may be on strand 1 with its corresponding T pyrimidine on strand 2, or the T pyrimidine may be on strand 1 and the A purine on strand 2. The same is true of G and C base pairs. With complementary base pairing, the sugar-phosphate backbone remains in a relatively regular conformation along the entire length of both DNA chains, and the broad outlines of that conformation are independent of base sequence.

(d) Interestingly, within the double-helical structure, the spatial requirements of the base pairs are satisfied if and only if each pair consists of one small pyrimidine and one large purine, and even then, only for the particular pairing of A–T and G–C. Pyrimidine–pyrimidine pairs are too small for the structure, and purine–purine pairs are too large. In addition, A–C and G–T pairs do not fit well together; that is, they do not easily form hydrogen bonds. Complementary base pairing is thus a logical outgrowth of the molecule's steric requirements. Although any one nucleotide pair forms only two or three hydrogen bonds, the sum of these connections between successive base pairs in a long DNA molecule composed of thousands or millions of nucleotides is one basis of the molecule's great chemical stability.

(a)

Pyrimidine–pyrimidine

Purine–purine

Purine–pyrimidine

20 Å

(d)

(Continued)

Feature Figure 6.9 (*Continued*)

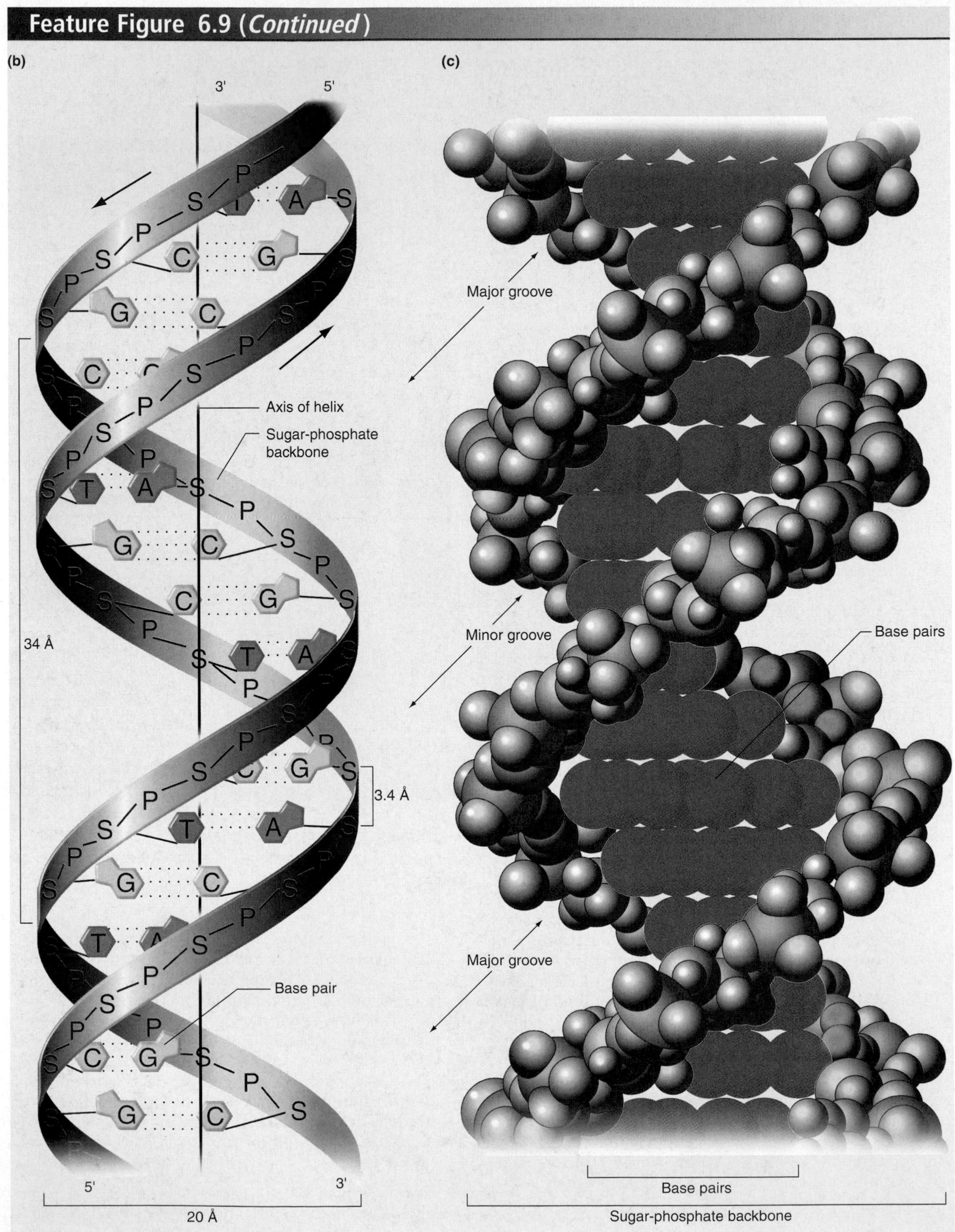

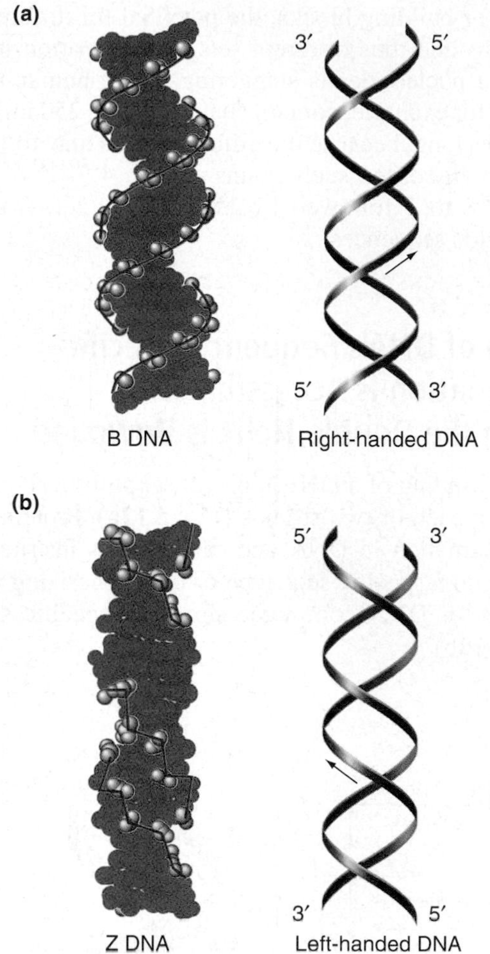

Figure 6.10 **Z DNA is one variant of the double helix.** **(a)** Normal Watson-Crick B-form DNA forms a right-handed helix with a smooth backbone. **(b)** Z-form DNA is left-handed and has an irregular backbone.

the backbone takes on a zigzag shape (Fig. 6.10b). Researchers have observed many kinds of unusual non-B structures *in vitro* (in the test tube, literally "in glass"), and they speculate that some of these might occur at least transiently in living cells. There is some evidence, for instance, that Z DNA might exist in certain chromosomal regions *in vivo* (in the living organism). Whether or not the Z form and other unusual conformations have any biological role remains to be determined.

Some DNA Molecules Are Circular Instead of Linear

The nuclear chromosomes of all eukaryotic organisms are long, linear double helixes, but some smaller chromosomes are circular (**Fig. 6.11a** and **b**). These include the chromosomes of prokaryotic bacteria, the chromosomes of organelles such as the mitochondria and chloroplasts that are found inside eukaryotic cells, and the chromosomes of some viruses, including the papovaviruses that can cause cancers in animals and humans. Such circular chromosomes consist of covalently closed, double-stranded circular DNA molecules. Although neither strand of these circular double helixes has an end, the two strands are still antiparallel in polarity.

Some Viruses Carry Single-Stranded DNA

In some viruses, the genetic material consists of relatively small, single-stranded DNA molecules. Once inside a cell, the single strand serves as a mold for making a second strand, and the resulting double-stranded DNA then governs the production of more virus particles. Examples of viruses carrying single-stranded DNA are bacteriophages ϕX174 and M13, and mammalian parvoviruses, which are associated with fetal death and spontaneous abortion in humans. In both ϕX174 and M13, the single DNA strand is in the form of a covalently closed circle; in the parvoviruses, it is linear (Fig. 6.11c and d).

Alternative B and Z configurations; circularization of the molecule; and single strands that are converted to double helixes before replication and expression—these are minor

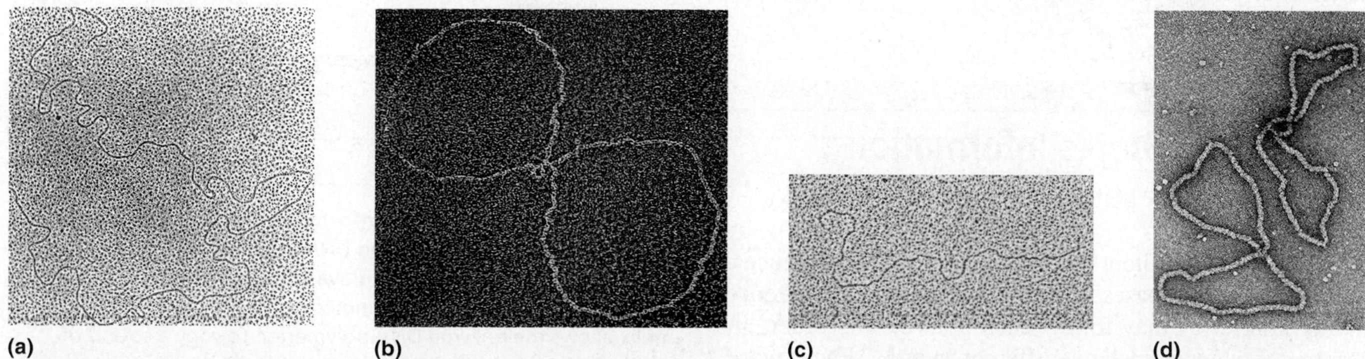

Figure 6.11 **DNA molecules may be linear or circular, double-stranded or single-stranded.** These electron micrographs of naturally occurring DNA molecules show **(a)** a fragment of a long, linear double-stranded human chromosome, **(b)** a circular double-stranded papovavirus chromosome, **(c)** a linear single-stranded parvovirus chromosome, and **(d)** circular single-stranded bacteriophage M13 chromosomes.

variations on the double-helical theme. Despite such experimentally determined departures of detail, the Watson-Crick double helix remains *the* model for thinking about DNA structure. This model describes those features of the molecule that have been preserved through billions of years of evolution.

DNA Structure Is the Foundation of Genetic Function

Without sophisticated techniques for determining base sequence, one cannot distinguish bacterial DNA from human DNA. This is because all DNA molecules have the same general chemical properties and physical structure. Proteins, by comparison, are a much more diverse group of molecules with a much greater complexity of structure and function. In his account of the discovery of the double helix, Crick referred to this difference when he said that "DNA is, at bottom, a much less sophisticated molecule than a highly evolved protein and for this reason reveals its secrets more easily."

There are four basic DNA "secrets," embodied in four questions:

1. How does the molecule carry information?
2. How is that information copied for transmission to future generations?
3. What mechanisms allow the information to change?
4. How does the information govern the expression of phenotype?

The double-helical structure of DNA provides a potential solution to each of these questions, endowing the molecule with the capacity to carry out all the critical functions required of the genetic material.

In the remainder of this chapter, we describe how DNA's structure enables it to carry genetic information, replicate that information with great fidelity, and reorganize the information through recombination. How the information changes through mutation and how the information determines phenotype are the subjects of Chapters 7 and 8.

6.3 DNA Stores Information in the Sequence of Its Bases

The information content of DNA resides in the sequence of its bases. The four bases in each chain are like the letters of an alphabet; they may follow each other in any order, and different sequences spell out different "words." Each "word" has its own meaning, that is, its own effect on phenotype. AGTCAT, for example, means one thing, while CTAGGT means another. Although DNA has only four different letters, or building blocks, the potential for different combinations and thus different sets of information in a long chain of nucleotides is staggering. Some human chromosomes, for example, contain chains that are 250 million nucleotides long; because the different bases may follow each other in any order, such chains contain $4^{250,000,000}$ (which translates to 1 followed by 150,515,000 zeros) potential nucleotide sequences.

Much of DNA's Sequence-Specific Information Is Accessible Only When the Double Helix Is Unwound

The unwinding of a DNA molecule exposes a single file of bases on each of two strands (**Fig. 6.12a**). Proteins "read" the information in DNA and carry out its instructions by binding to a specific sequence or by synthesizing a stretch of RNA or DNA complementary to a specific sequence (Fig. 6.12b).

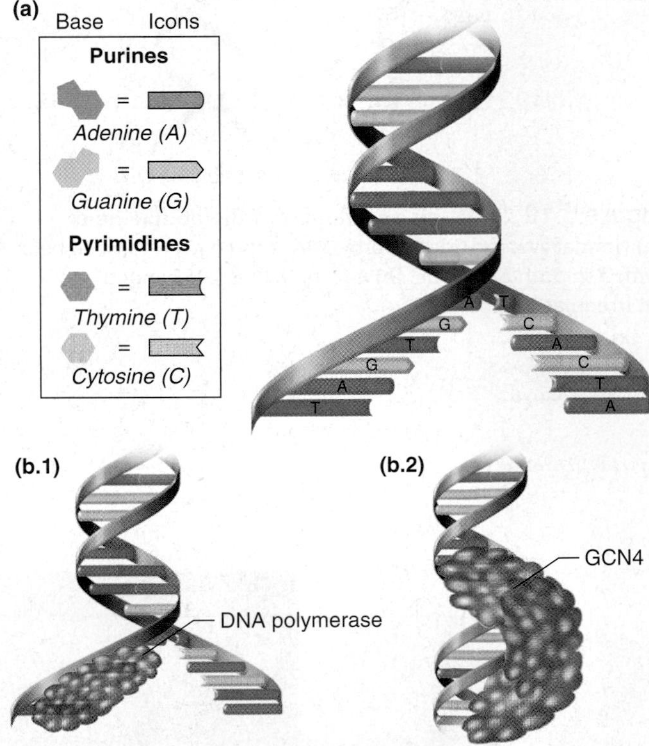

Figure 6.12 DNA stores information in the sequence of its bases. **(a)** A partially unwound DNA double helix. Note that different structural information is available in the double-stranded and unwound regions of the molecule. **(b.1)** Unwinding of the helix allows the enzyme DNA polymerase to copy a strand of DNA. **(b.2)** Some proteins can bind to specific sequences within double-helical DNA. Here, a protein called GCN4 interacts closely with features of a particular sequence of bases that are accessible to the protein in the major groove of the double helix.

Some Genetic Information Is Accessible Even in Intact, Double-Stranded DNA Molecules

This information emerges in part from differences between the four bases that appear in the major and minor grooves and in part from conformational irregularities in the sugar-phosphate backbone. Within the grooves, certain atoms at the periphery of the bases are exposed, and particularly in the major groove, these atoms may provide chemical information. Such information is in the form of spatial patterns of hydrogen bond donors and acceptors as well as in the distinct patterns of particular substituent shapes. Proteins can access this information to "sense" the base sequence in a stretch of DNA without disassembling the double helix (Fig. 6.12b.2). The proteins that help turn genes on and off make use of these subtle conformational differences. Nevertheless, the amount of information exposed in either groove or on the backbone as distinct shapes is relatively limited.

Some proteins can recognize specific base sequences in double-stranded DNA. The Tools of Genetics box "Restriction Enzymes Recognize Specific Base Sequences in DNA" explains how bacteria use enzymatic proteins of this type to stave off viral infection and how geneticists use these same enzymes to cut DNA at particular sites.

A Few Viruses Use RNA As the Repository of Genetic Information

In all cellular forms of life and many viruses, DNA carries the genetic information. Prokaryotes such as *Escherichia coli* bacteria carry their DNA in a double-stranded, covalently closed circular chromosome. Eukaryotic cells package their DNA in two or more double-stranded linear chromosomes. DNA viruses carry it in small molecules that are single- or double-stranded, circular, or linear.

By contrast, some viruses, including those that cause polio and AIDS, use RNA as their genetic material (**Fig. 6.13**).

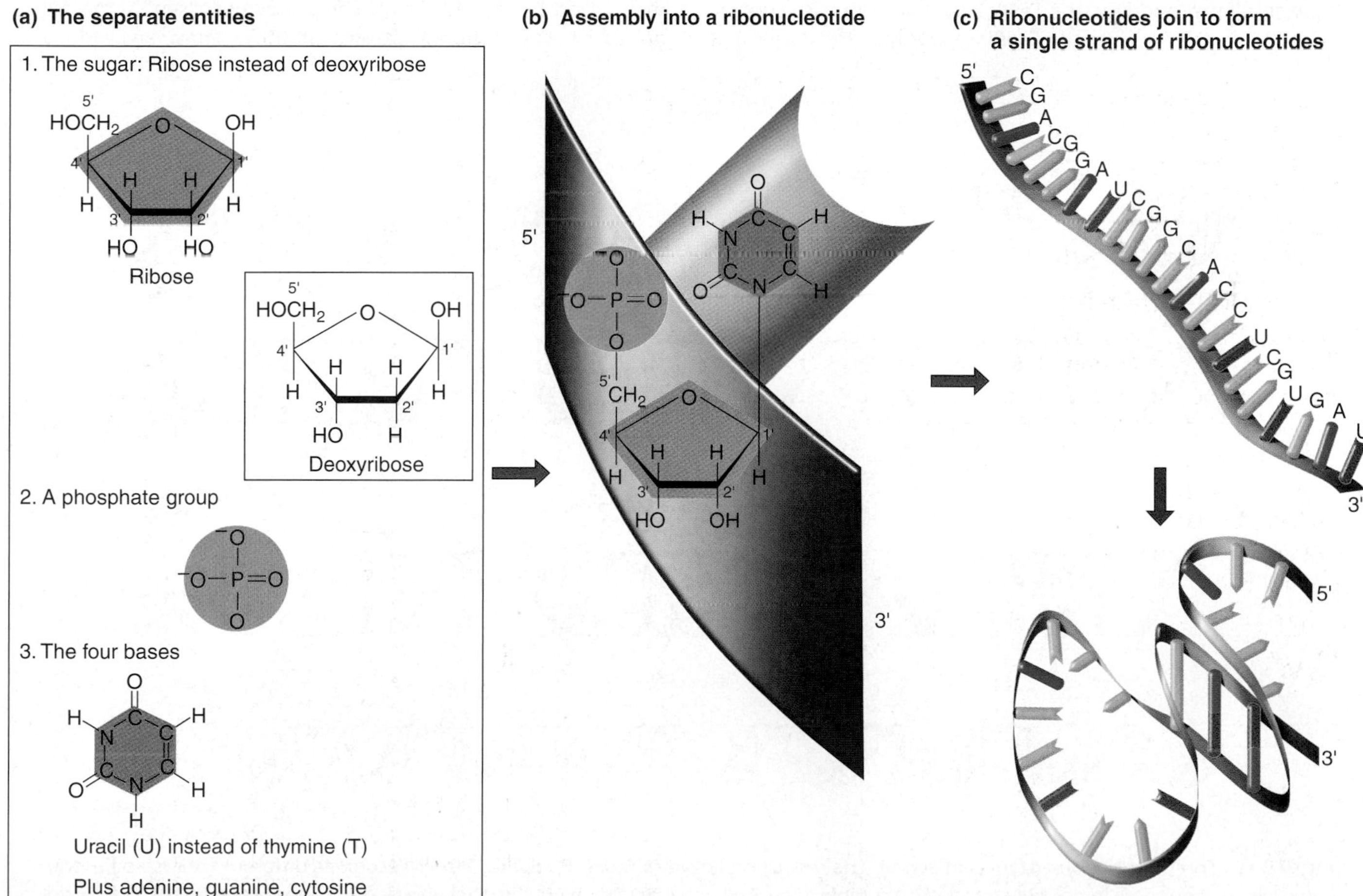

Figure 6.13 RNA: Chemical constituents and complex folding pattern. (a) and **(b)** Each ribonucleotide contains the sugar ribose, an inorganic phosphate group, and a nitrogenous base. RNA contains the pyrimidine uracil (U) instead of the thymine (T) found in DNA. **(c)** Phosphodiester bonds join ribonucleotides into an RNA chain. Most RNA molecules are single-stranded but are sufficiently flexible so that some regions can fold back and form base pairs with other parts of the same molecule.

Restriction Enzymes Recognize Specific Base Sequences in DNA

In many types of bacteria, the unwelcome arrival of viral DNA mobilizes minute molecular weapons known as **restriction enzymes.** Each enzyme has the twofold ability to recognize a specific sequence of four to six base pairs anywhere within a DNA molecule and to sever a covalent bond in the sugar-phosphate backbone at a particular position within or near that sequence on each strand. When a bacterium calls up its reserve of restriction enzymes at the first sign of invasion, the ensuing shredding and dicing of selected stretches of viral DNA incapacitates the virus's genetic material and thereby restricts infection.

Since the early 1970s, geneticists have isolated more than 300 types of restriction enzymes and named them for the bacterial species in which they orginate. *Eco*RI, for instance, comes from *E. coli; Hpa*I is one of several restriction enzymes found in *Haemophilus parainfluenzae; Hind*III derives from *Haemophilus influenzae.* Each enzyme recognizes a different base sequence and cuts the DNA strand at a precise spot in relation to that sequence. *Eco*RI recognizes the sequence 5′...GAATTC...3′ and cleaves between the G and the first A. *Hpa*I recognizes the sequence 5′...GTTAAC...3′ and makes a cut between the second T and the first A. *Hind*III sees the sequence 5′...AAGCTT...3′ and shears between the two As. The DNA of a bacteriophage called lambda (λ) carries the GAATCC sequence recognized by *Eco*RI in five separate places; the enzyme thus cuts the linear lambda DNA at five points, breaking it into six pieces. The DNA of a phage known as ϕX174, however, contains no *Eco*RI recognition sequences. By comparison, the *Hpa*I enzyme can cut λ DNA in 13 places and ϕX174 DNA in 4, while *Hind*III can cut λ DNA in 6 places and ϕX174 in 10.

How, you may wonder, do bacteria keep their arsenal of restriction enzymes from attacking their own DNA? The answer is a group of companion enzymes known as *methylases,* which attach methyl groups ($-CH_3$) to DNA in a precisely defined way. With every cycle of DNA replication, the methylating enzymes recognize the exact same sites that would be recognized by the cell's restriction enzymes, "see" a methyl group on the template strand at those sites, and add a

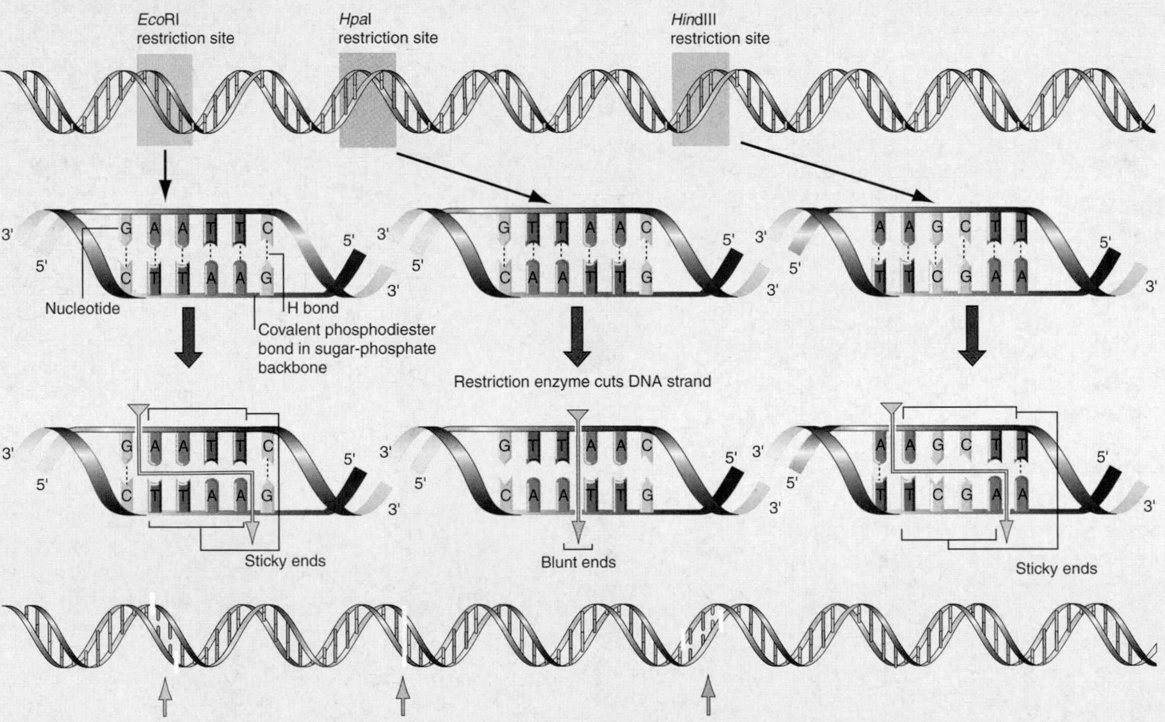

Figure A Three restriction enzymes in action. The restriction enzymes *Eco*RI, *Hpa*I, and *Hind*III recognize different six-base-pair-long symmetrical sequences in double-stranded DNA molecules. Any one enzyme severs the phosphodiester bonds between the same two adjacent nucleotides on each DNA strand. If these bonds are in the middle of the six base recognition site (as with *Hpa*I), the resultant DNA fragments will have blunt ends. If the bonds are offset from the center of the recognition site (as with *Eco*RI and *Hind*III), the products of cleavage will have sticky ends. Note that any sticky end produced by cleavage with a particular restriction enzyme is complementary in sequence to any other sticky end made by the same enzyme.

methyl group at the exact same sites on the newly forming daughter strand. This site-specific methylation of every newly forming strand shields the bacterial DNA from recognition by the cell's own restriction enzymes. The methylases recognize only single-stranded restriction sites that form opposite an already methylated mirror image. Because viral DNA enters the bacterial cell as a double-stranded helix that is not methylated at the restriction enzyme's recognition sites, the methylases do not protect it.

Figure A illustrates the results of three restriction enzymes in action. Note that each recognition sequence in double-stranded DNA is symmetrical; that is, the base sequences on the two strands are identical when each is read in the 5′-to-3′ direction. Thus, each time an enzyme recognizes a short 5′-to-3′ sequence on one strand, it finds the exact same sequence in the 5′-to-3′ direction of the complementary antiparallel strand. In other words, viewed in the 5′-to-3′ direction, the double-stranded recognition sequence is palindromic; like the phrase "TAHITI HAT" or the number 1881, it reads the same backward and forward (although the analogy is not exact because the language of DNA is double-stranded, while English follows only a single line at a time). Note also that the cuts made by restriction enzymes leave two types of DNA ends. *Hpa*I cleaves right in the middle of the six-base sequence, making a smooth cut straight through both chains of the double helix to create what geneticists call "blunt" ends. In contrast, *Eco*RI and *Hin*dIII do not cleave the two strands of the double helix at the same spot in the middle of a recognition sequence, and as a result, the cuts they make are offset in opposite directions on the two strands. When the weak hydrogen bonds between the strands dissociate, these cuts leave short, protruding single-stranded flaps known as **sticky,** or **cohesive, ends.** Like a tiny finger of velcro, each flap can stick to—that is, re-form hydrogen bonds with—a complementary sequence protruding from the end of another piece of DNA.

Even though restriction enzymes evolved in bacteria as a means of protection against viruses, given the appropriate base sequence, they can cut the DNA of any organism. And as **Fig. B** shows, any two DNA molecules with complementary protruding sequences can join together, no matter how unrelated the rest of their base sequences may be. The cutting and splicing together of DNA from two origins creates a **recombinant DNA molecule.**

In the mid-1970s, geneticists took advantage of the activity of restriction enzymes to splice together DNA from any two organisms. In their hands, the enzymes served as precision scissors that, in effect, revolutionized the study of life and gave birth to recombinant DNA technology. Although the sticky ends created by restriction enzymes enable two unrelated DNA molecules to come together by base pairing, another enzyme, known as DNA ligase, is required to stabilize the recombinant molecule. The ligase seals the breaks in the backbones of both strands.

Figure B illustrates one application of recombinant DNA technology: the splicing of the human gene for insulin into a small circle of DNA known as a plasmid, which can replicate inside a bacterial cell. Here is how it works. *Eco*RI is added to solutions of plasmids and human genomic DNA, where it cleaves the DNA molecules. The cleavage converts the circular plasmids to linear DNAs with *Eco*RI sticky ends; it also fragments each copy of the human genomic DNA into hundreds of thousands of pieces, all of which terminate with *Eco*RI sticky ends. When the solutions are then mixed together, the different fragments can adhere to each other in any combination, because of the complementarity of their sticky ends. In one combination, a human genomic fragment containing the insulin gene will re-create a circular DNA molecule

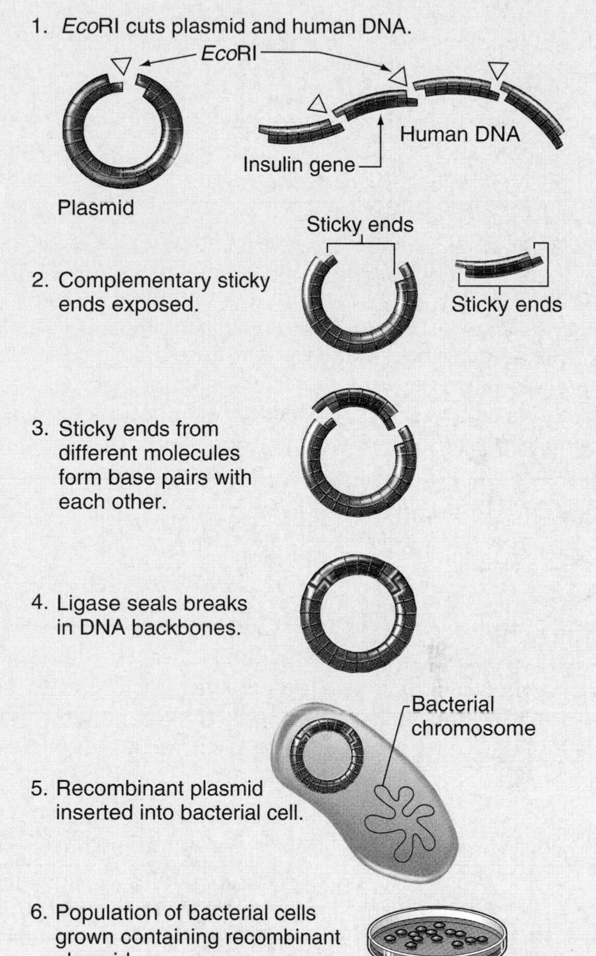
1. EcoRI cuts plasmid and human DNA.
EcoRI
Plasmid
Insulin gene
Human DNA
Sticky ends
2. Complementary sticky ends exposed.
Sticky ends
3. Sticky ends from different molecules form base pairs with each other.
4. Ligase seals breaks in DNA backbones.
Bacterial chromosome
5. Recombinant plasmid inserted into bacterial cell.
6. Population of bacterial cells grown containing recombinant plasmid.

Figure B One use of recombinant DNA technology: Harnessing bacteria to copy the human insulin gene. Geneticists cleave both a small circular chromosome called a plasmid and human chromosomal DNA with the *Eco*RI restriction enzyme. This cleavage converts the plasmid to a linear piece of DNA with *Eco*RI sticky ends, while converting the human DNA into thousands of fragments with *Eco*RI sticky ends; one of these human chromosomal fragments contains the insulin gene. When these pieces of DNA are mixed together, they can adhere to each other in various combinations because of the complementarity of their sticky ends; the enzyme DNA ligase will seal the fragments together by forming the proper phosphodiester bonds. One possible outcome of this process is the creation of a recombinant DNA molecule in which the human insulin gene is spliced into the circular plasmid. *E. coli* cells transformed with this recombinant plasmid can become miniature factories for the synthesis of insulin.

after adhering to the two ends of a linearized plasmid. And just as restriction enzymes operate as scissors, DNA ligase acts as a glue that seals the breaks in the DNA backbone by forming new phosphodiester bonds. Investigators can transform bacteria with the recombinant plasmids containing the insulin gene exactly as Avery transformed bacteria with his "transforming principle." The recombinant DNA molecules will enter some cells. When the bacteria copy their own chromosome in preparation for cell division, they will also make copies of any resident plasmids along with all the genes the plasmids contain. In the illustrated example, the plasmid carrying the gene for insulin also carries sequences that can direct the expression of an inserted gene into protein. As the bacterial culture grows, so does the number of plasmids carrying sequences that direct the expression of the human insulin gene into protein. Eventually, a population of bacteria grows up in which every cell not only contains a copy of the human gene but also makes the insulin encoded by that gene. With this recombinant DNA technology, it became possible to provide diabetic patients with a source of safe and inexpensive medicine to treat the symptoms of their disease.

Techniques for designing and constructing recombinant DNA molecules and for harnessing bacteria to produce large quantities of a particular gene and its protein product hold great promise for medicine, agriculture, and industry. Researchers are currently using the technology to try to develop, among many other projects, a relatively inexpensive nose spray able to kill the viruses that cause the common cold, drugs to help fight cancer and AIDS, corn containing proteins with the nutritional value of beef proteins, and a process for making automobile fuel from discarded corn stalks. These and other recombinant DNA possibilities depend on the use of restriction enzymes to recognize and cut specific base sequences.

There are three major chemical differences between RNA and DNA. First, RNA takes its name from the sugar ribose, which it incorporates instead of the deoxyribose found in DNA (Fig. 6.13a.1 on p. 181). Second, RNA contains the base uracil (U) instead of the base thymine (T); U, like T, base pairs with A (Fig. 6.13a.3). Finally, most RNA molecules are single-stranded and contain far fewer nucleotides than the very long DNA molecules found in nuclear chromosomes. Some completely double-stranded RNA molecules do nonetheless exist. Even within a single-stranded RNA molecule, if folding brings two oppositely oriented regions that carry complementary nucleotide sequences alongside each other, they can form a double-stranded, base-paired stretch within the molecule. This means that, compared to the relatively simple, double-helical shape of a DNA molecule, many RNAs have a complicated structure of short double-stranded segments interspersed with single-stranded loops (Fig. 6.13c).

RNA has the same ability to carry information in the sequence of its bases but is much less stable than DNA. In addition to serving as the genetic material for an array of viruses, RNA fulfills several vital functions in all cells. For example, it participates in gene expression and protein synthesis. It also plays a significant role in DNA replication, which we now describe.

6.4 DNA Replication: Copying Genetic Information for Transmission to the Next Generation

In one of the most famous sentences of the scientific literature, Watson and Crick wrote at the end of their 1953 paper proposing the double helix model: "It has not escaped our notice that the specific pairing we have postulated immediately suggests a possible copying mechanism for the genetic material." This copying, as we saw in Chapter 4, precedes the transmission of chromosomes from one generation to the next via meiosis, and it is also the basis of the chromosome duplication prior to each mitosis that allows all the daughter cells of a developing organism to receive a complete copy of the genetic information.

Complementary Base Pairing Produces Semiconservative Replication: An Overview

In the process of replication postulated by Watson and Crick, the double helix unwinds to expose the bases in each strand of DNA. Each of the two separated strands then acts as a **template,** or molecular mold, for the synthesis of a new second strand (**Fig. 6.14**). The new strand forms as complementary bases align opposite the exposed bases on the parent strand. That is, an A at one position on the original strand signals the addition of a T at the corresponding position on the newly forming strand; a T on the original signifies addition of an A; similarly, G calls for C, and C calls for G. This type of base pairing, which determines the nucleotide sequence of the new strand, is known as **complementary base pairing.** Once the appropriate base has aligned opposite and formed hydrogen bonds with its complement, enzymes join the base's nucleotide to the preceding nucleotide by a phosphodiester bond, eventually linking a whole new line of nucleotides into a continuous strand. This mechanism of complementary base pairing followed by the coupling of successive nucleotides yields two "daughter" double helixes that each contain one of the original DNA strands intact (that is, "conserved") and one completely new strand (**Fig. 6.15a**). For this reason, such a pattern of double helix duplication is called **semiconservative replication:** a copying in which one strand is conserved from the parent molecule and the other is newly synthesized.

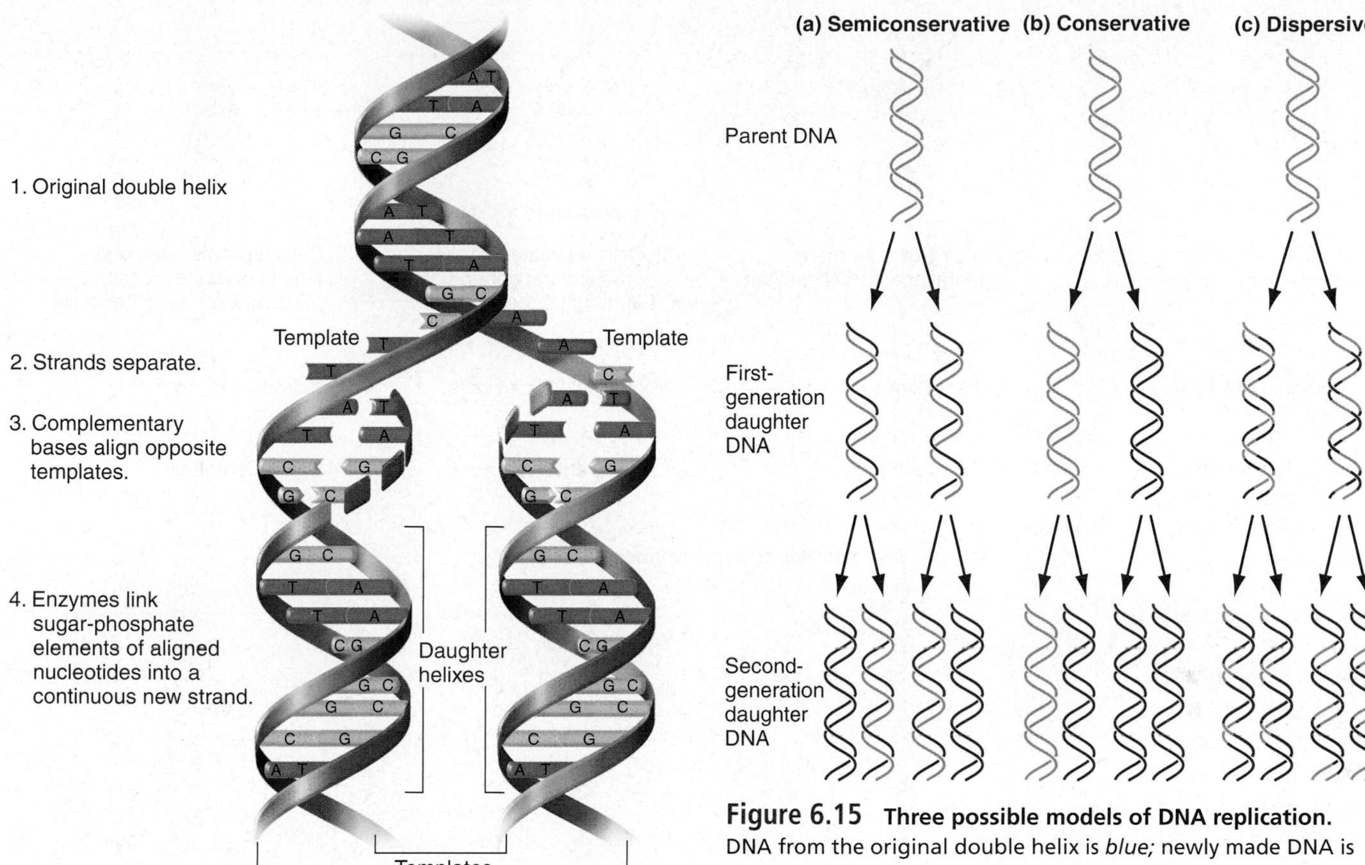

Figure 6.14 The model of DNA replication postulated by Watson and Crick. Unwinding of the double helix allows each of the two strands to serve as a template for the synthesis of a new strand by complementary base pairing. The end result: A single double helix becomes two identical daughter double helixes.

Figure 6.15 Three possible models of DNA replication. DNA from the original double helix is *blue;* newly made DNA is *magenta.* **(a)** Semiconservative replication (the Watson-Crick model). **(b)** Conservative replication: The parental double helix remains intact; both strands of the daughter double helix are newly synthesized. **(c)** Dispersive replication: At completion, both strands of both double helixes contain both original and newly synthesized material.

Watson and Crick's proposal is not the only replication mechanism imaginable. Figure 6.15b and c illustrate two possible alternatives. With *conservative* replication, one of the two "daughter" double helixes would consist entirely of original DNA strands, while the other helix would consist of two newly synthesized strands. With *dispersive* replication, both "daughter" double helixes would carry blocks of original DNA interspersed with blocks of newly synthesized material. These alternatives are less satisfactory than semiconservative replication because they do not immediately suggest a mechanism for copying the information in the sequence of bases, and they do not explain the research data (presented below) as well.

Experimental Proof of Semiconservative Replication

In 1958, Matthew Meselson and Franklin Stahl performed an experiment that confirmed the semiconservative nature of DNA replication (**Fig. 6.16**). The experiment depended on being able to differentiate preexisting "parental" DNA from newly synthesized daughter DNA. To accomplish this, Meselson and Stahl controlled the isotopic composition of the nucleotides incorporated in the newly forming daughter strands as follows. They grew *E. coli* bacteria for many generations on media in which all the nitrogen was the normal isotope ^{14}N; these cultures served as a control. They grew other cultures of *E. coli* for many generations on media in which the only source of nitrogen was the heavy isotope ^{15}N. After several generations of growth on heavy-isotope medium, essentially all the nitrogen atoms in the DNA of these bacterial cells were labeled with (that is, contained) ^{15}N. The cells in some of these cultures were then transferred to new medium in which all the nitrogen was ^{14}N. Any DNA synthesized after the transfer would contain the lighter isotope.

Meselson and Stahl isolated DNA from cells grown in the different nitrogen-isotope cultures and then subjected these DNA samples to *equilibrium density gradient centrifugation,* an analytic technique they had just developed. In a test tube, they dissolved the DNA in a solution of the dense salt cesium chloride (CsCl) and spun

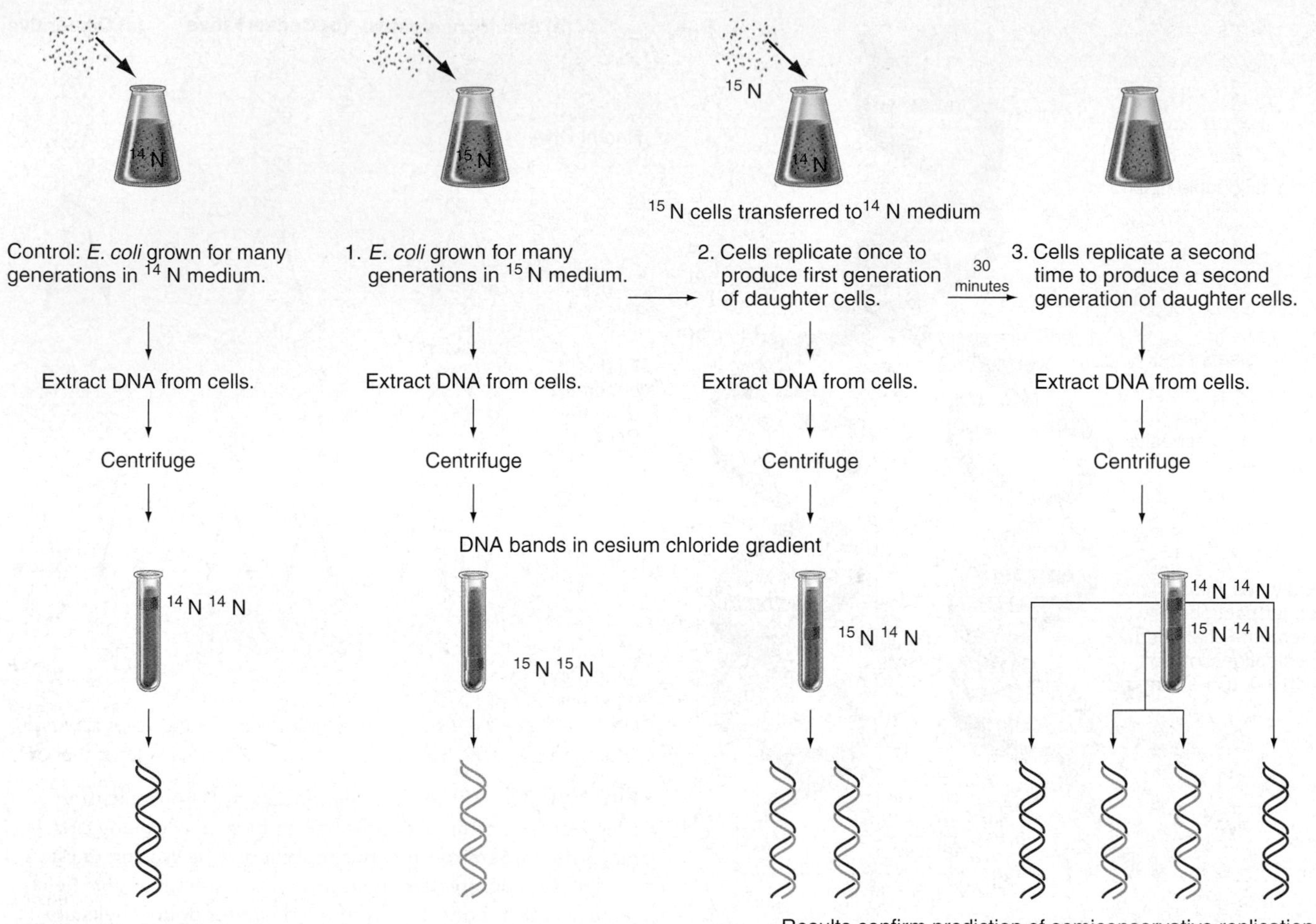

Figure 6.16 How the Meselson-Stahl experiment confirmed semiconservative replication. (1) *E. coli* cells were grown in heavy ^{15}N medium. **(2)** and **(3)** Some of these cells were transferred to ^{14}N medium and allowed to divide either once or twice. When DNA from each of these sets of cells was prepared and centrifuged in a cesium chloride gradient, the density of the extracted DNA conformed to the predictions of the semiconservative mode of replication, as shown at the *bottom* of the figure, where *blue* indicates heavy original DNA and *magenta* depicts light, newly synthesized DNA. The results are inconsistent with the conservative and dispersive models for DNA replication (compare with Fig. 6.15b and c).

these solutions at very high speed (about 50,000 revolutions per minute) in an ultracentrifuge. Over a period of two to three days, the centrifugal force (roughly 250,000 times the force of gravity) causes the formation of a stable gradient of CsCl concentrations, with the highest concentration, and thus highest CsCl density, at the bottom of the tube. The DNA in the tube forms a sharply delineated band at a position where its own density equals that of the CsCl. Since DNA containing ^{15}N is denser than DNA containing ^{14}N, pure ^{15}N DNA will form a band lower, that is, closer to the bottom of the tube, than pure ^{14}N DNA (Fig. 6.16).

As Fig. 6.16 shows, when cells with pure ^{15}N DNA were transferred into ^{14}N medium and allowed to divide once, DNA from the resultant first-generation cells formed a band at a density intermediate between that of pure ^{15}N DNA and that of pure ^{14}N DNA. A logical inference is that the DNA in these cells contains equal amounts of the two isotopes. This finding invalidates the "conservative" model, which predicts the appearance of bands reflecting only pure ^{14}N and pure ^{15}N with no intermediary band. In contrast, DNA extracted from second-generation cells that had undergone a second round of division in the ^{14}N medium produced two observable bands, one at the density corresponding to equal amounts of ^{15}N and ^{14}N, the other at the density of pure ^{14}N. These observations invalidate the "dispersed" model, which predicts a single band between the two bands of the original generation.

Meselson and Stahl's observations are consistent only with semiconservative replication: In the first generation after transfer from the ^{15}N to the ^{14}N medium, one of the two strands in every daughter DNA molecule carries the heavy isotope label; the other, newly synthesized strand carries the lighter ^{14}N isotope. The band at a density intermediate between that of ^{15}N DNA and ^{14}N DNA represents this isotopic hybrid. In the second generation after transfer, half of the DNA molecules have one ^{15}N strand and one ^{14}N strand, while the remaining half carry two ^{14}N strands. The two observable bands—one at the hybrid position, the other at the pure ^{14}N position—reflect this mix.

By confirming the predictions of semiconservative replication, the Meselson-Stahl experiment disproved the conservative and dispersive alternatives. We now know that the semiconservative replication of DNA is nearly universal. To understand material presented in the remainder of this book, it is important to understand precisely how semiconservative replication relates to the structure of chromosomes in eukaryotic cells during the mitotic cell cycle (review Fig. 4.7). Early in interphase, each eukaryotic chromosome contains a single continuous linear double helix of DNA. Later, during the S-phase portion of interphase, the cell replicates the double helix semiconservatively; after this semiconservative replication, each chromosome is composed of two sister chromatids joined at the centromere. Each sister chromatid is a double helix of DNA, with one strand of parental DNA and one strand of newly synthesized DNA. At the conclusion of mitosis, each of the two daughter cells receives one sister chromatid from every chromosome in the cell. This process preserves chromosome number and identity during mitotic cell division because the two sister chromatids are identical in base sequence to each other and to the original parental chromosome.

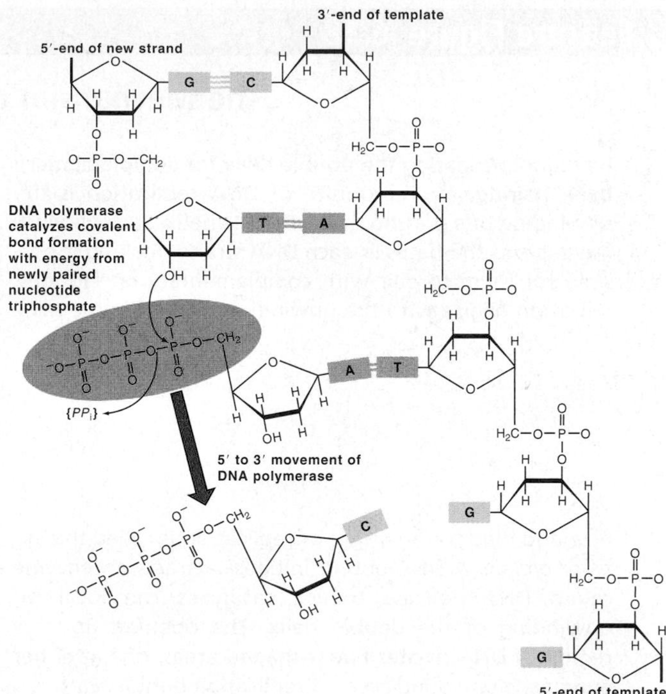

Figure 6.17 DNA synthesis proceeds in a 5′ to 3′ direction. The template strand is shown on the *right* in an antiparallel orientation to the new DNA strand under synthesis on the *left*. In this example, a free molecule of dATP has formed hydrogen bonds with a complementary thymidine base on the template strand. DNA polymerase (*yellow*) catalyzes the cleavage of dATP between the first and second phosphate groups, which releases energy to catalyze the formation of a covalent phosphodiester bond between the terminal 3′-hydroxyl group on the existing DNA strand and the first phosphate of the dATP substrate. Pyrophosphate (*PP_i*) is released as a by-product of the reaction.

The Molecular Mechanism of Replication: Doubling the Double Helix

Watson and Crick's model for semiconservative replication, depicted in Fig. 6.15a, is a simple concept to grasp, but the process through which it occurs is quite complex. Replication does not happen spontaneously any time a mixture of DNA and nucleotides is present. Rather, it occurs at a precise moment in the cell cycle, depends on a network of interacting regulatory elements, requires considerable input of energy, and involves a complex array of the cell's molecular machinery, including a variety of enzymes. The salient details were deduced primarily by the Nobel laureate Arthur Kornberg and members of his laboratory, who purified individual components of the replication machinery from *E. coli.* bacteria. Remarkably, they were eventually able to elicit the reproduction of specific genetic information outside a living cell, in a test tube containing purified enzymes together with DNA template, primer (defined on p. 188), and nucleotide substrates.

Although the biochemistry of DNA replication was elucidated for a single bacterial species, its essential features are conserved—just like the structure of DNA—within all organisms. The energy required to synthesize every DNA molecule found in nature comes from the high-energy phosphate bonds associated with the four deoxynucleotide triphosphate substrates (dATP, dCTP, dGTP, and dTTP; or dNTP as a general moniker) that provide bases for incorporation into the growing DNA strand. As shown in **Fig. 6.17,** this conserved biochemical feature means that DNA synthesis can proceed only from the hydroxyl group present at the 3′ end of an existing polynucleotide. With energy released from severing the triphosphate arm of a dNTP substrate molecule, the DNA polymerase enzyme catalyzes the formation of a new phosophodiester bond. Once this bond is formed, the enzyme proceeds to join up the next nucleotide brought into position by complementary base pairing.

The formation of phosphodiester bonds is just one component of the highly coordinated process by which DNA replication occurs inside a living cell. The entire molecular mechanism, illustrated in **Fig. 6.18,** has two stages: **initiation,** during which proteins open up the double helix and prepare it for complementary base pairing, and **elongation,** during which proteins connect the correct sequence of nucleotides on both newly formed DNA double helixes.

Feature Figure 6.18

The Mechansim of DNA Replication

(a) **Initiation: Preparing the double helix for complementary base pairing.** A prerequisite of DNA replication is the unwinding of a portion of the double helix. Such unwinding exposes the bases in each DNA strand, making it possible for them to pair with complementary nucleotides. Initiation begins with the unwinding of the double helix at a particular short sequence of nucleotides known as the **origin of replication.** Each circular *E. coli* chromosome has a single origin of replication. Several proteins bind to the origin, forming a stable complex in which a small region of DNA is unwound and the two complementary strands are separated. The first of the proteins to recog-

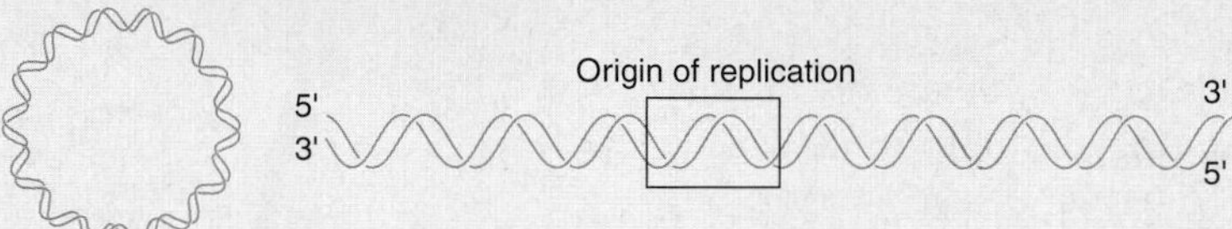

nize and bind to the origin of replication is called the *initiator protein.* A DNA-bound initiator attracts an enzyme called DNA helicase, which catalyzes the localized unwinding of the double helix. The opening up of a region of DNA creates two Y-shaped areas, one at either end of the unwound area, or **replication bubble.** Each Y is called a **replication fork** and consists of the two unwound DNA strands branching out into unpaired (but complementary) single strands. These single strands will serve as **templates**—molecular molds—for fashioning new strands of DNA. The molecule is now ready for replication. (Protein molecules are not drawn to scale.)

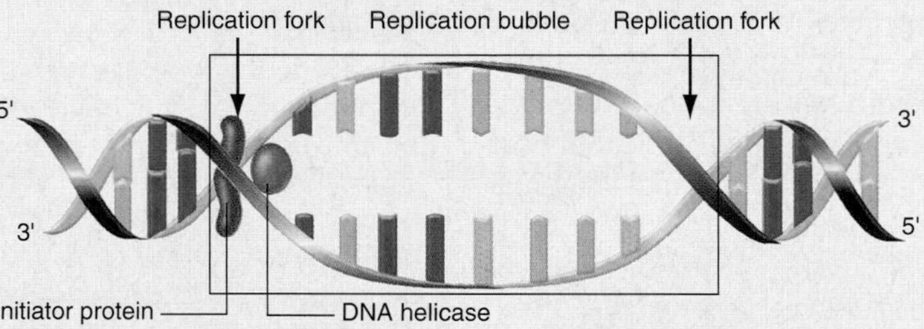

Actual formation of new DNA strands depends on the action of a complex enzyme known as **DNA polymerase III,** which adds nucleotides, one after the other, to the end of a growing DNA strand. DNA polymerase operates according to three strict rules: First, it can copy only DNA that is unwound and maintained in the single-stranded state; second, it adds nucleotides only to the end of an existing chain (that is, it cannot establish the first link in the chain); and third, it functions in only one—the 5′-to-3′—direction. The requirement for a single-stranded template is satisfied by formation of the replication bubble. The requirement for an already existing chain end, however, means that something else must start the about-to-be-constructed chain. That "something else" is RNA. Construction of a very short new strand consisting of a few nucleotides of RNA provides an end to which DNA polymerase can add the nucleotides of DNA. Since this short stretch of RNA prepares the ground for DNA polymerase activity, it is called an RNA **primer.** An enzyme called primase synthesizes the RNA primer at the replication fork, where the primer then base pairs with the single-stranded DNA template. With the double helix unwound and the primer in place, DNA replication can proceed. The third characteristic of DNA polymerase activity—one way only—determines some of the special features of subsequent steps.

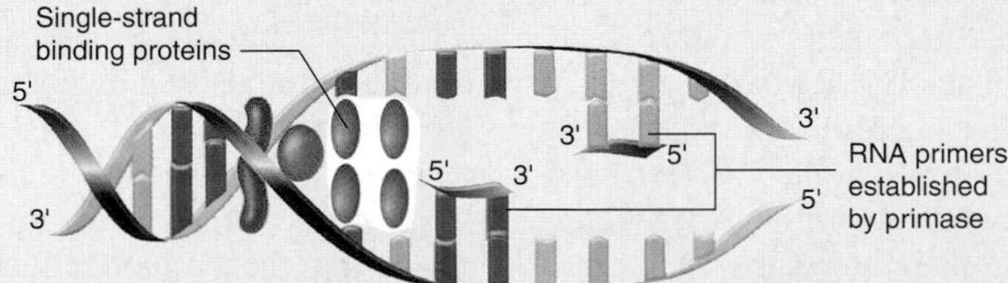

(b) **Elongation: Connecting the correct sequence of nucleotides into a continuous new strand of DNA.** Elongation—the linking together of appropriately aligned nucleotide subunits into a continuous new strand of DNA—is the heart of replication. We have seen that the lineup of bases is determined by complementary base pairing with the template strand. An A on the template strand will form hydrogen bonds only with a T on the forming strand, and a C on the template will pair only with a G. Thus, the order of bases in the template specifies the order of bases in the newly forming strand. Once complementary base pairing has determined the next nucleotide to be added to the newly forming strand, DNA polymerase III catalyzes the actual joining of this properly positioned nucleotide to the preceding nucleotide. The linkage of subunits through the formation of phosphodiester bonds is known as **polymerization.** The DNA polymerase III enzyme first joins the correctly paired nucleotide to the 3′ hydroxyl end of the RNA primer, and then it continues to add the appropriate nucleotides to the 3′ end of the growing chain. As a result, the DNA strand under construction grows in the 5′-to-3′ direction. The fact that the new strand is antiparallel to the template strand means that the DNA polymerase molecule actually moves along that template strand in the 3′-to-5′ direction.

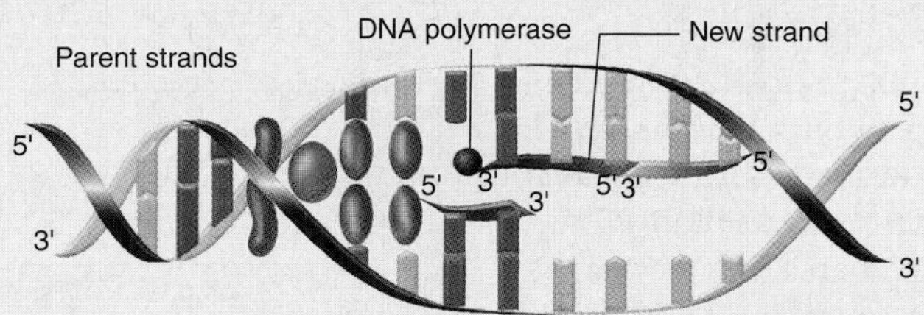

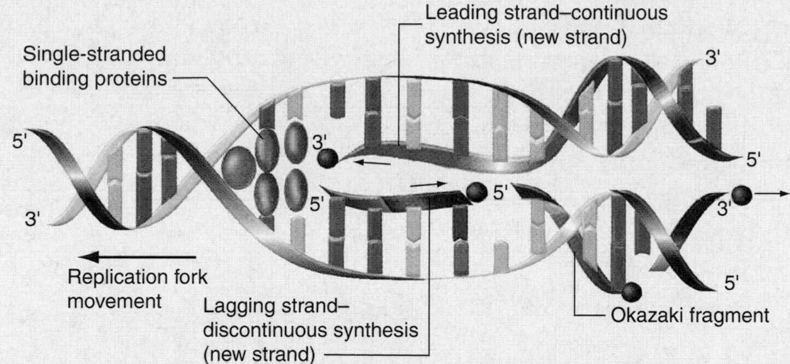

As DNA replication proceeds, helicase progressively unwinds the double helix to expose successive series of nucleotides in their unpaired, single-stranded state. DNA polymerase III can then move in the same direction as the fork to synthesize one of the two new chains under construction at the fork. The enzyme encounters no problems in the polymerization of this chain—called the **leading strand**—because it can add nucleotides continuously to the growing 3′ end as soon as the unraveling fork exposes the corresponding bases on the template strand. The movement of the replication fork, however, presents problems for the synthesis of the second new DNA chain: the **lagging strand.** The polarity of the lagging strand is opposite that of the leading strand, yet as we have seen, DNA polymerase functions only in the 5′-to-3′ direction. This means that in synthesizing the lagging strand, the polymerase must travel in a direction opposite to that of the replication fork. How can the polymerase link the nucleotides of the lagging strand if it cannot approach that strand's recently unwound template at the replication fork?

The answer is that the lagging strand is synthesized *discontinuously* as small fragments of about 1000 bases called **Okazaki fragments** (after two of their discoverers Reiji and Tuneko Okazaki). DNA polymerase III still synthesizes these small fragments in the normal 5′-to-3′ direction, but because the enzyme can add nucleotides only to the 3′ end of an existing strand, each Okazaki fragment is initiated by the synthesis of a short RNA primer. The primase enzyme catalyzes formation of the RNA primer for each upcoming Okazaki fragment as soon as the replication fork has progressed a sufficient distance along the DNA. Polymerase then adds nucleotides to this new primer, creating an Okazaki fragment that extends as far as the 5′ end of the primer of the previously synthesized fragment. Finally, DNA polymerase I and other enzymes replace the RNA primer of the previously made Okazaki fragment with DNA, and an enzyme known as DNA ligase covalently joins successive Okazaki fragments into a continuous strand of DNA. With the completion of both leading and lagging strands, DNA replication is complete.

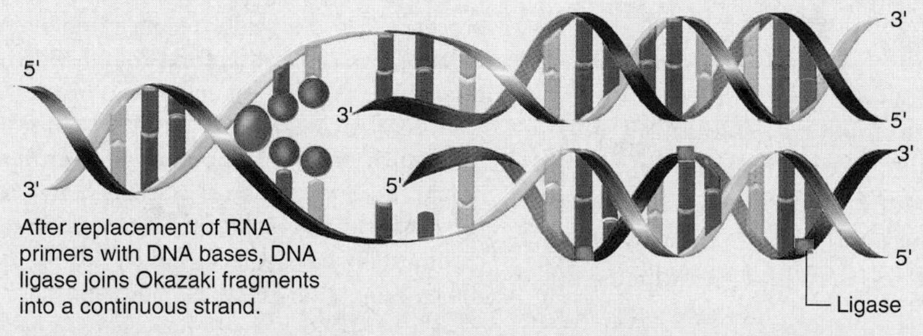

The Mechanics of DNA Replication at the Chromosomal Level

DNA replication, which depends in part on DNA polymerase, is complicated by the way the polymerase enzyme functions. DNA polymerase can lengthen existing DNA chains only by adding nucleotides to the 3′ hydroxy group of the DNA strand, as shown in Fig. 6.17 on p. 187. As a result, when the double helix unwinds during replication, one newly synthesized strand (the *leading strand*) can grow continuously into the opening Y-shaped area, but the other new strand (the *lagging strand*) can come into existence only as a series of smaller Okazaki fragments (named for their discoverers). These fragments must be joined together at a second stage of the process.

As Fig. 6.18 shows, DNA replication depends on the coordinated activity of many different proteins, including two different DNA polymerases called pol I and pol III (*pol* is short for polymerase). Pol III plays the major role in producing the new strands of complementary DNA, while pol I fills in the gaps between newly synthesized Okazaki segments. Other enzymes contribute to the initiation process: DNA helicase unwinds the double helix. A special group of single-stranded binding proteins keep the DNA helix open. An enzyme called primase creates RNA primers to initiate DNA synthesis. The ligase enzyme welds together Okazaki fragments.

It took many years for biochemists and geneticists to discover how the tight collaboration of many proteins drives the intricate mechanism of DNA replication. Today they believe that programmed molecular interactions of this kind underlie most of the biochemical processes that occur in cells. In these processes, a group of proteins, each performing a specialized function, like the workers on an assembly line, cooperate in the manufacture of complex macromolecules.

Recall that the origin of replication has two forks (**Fig. 6.19**). As a result, replication is generally **bidirectional,** with the replication forks moving in opposite directions (Fig. 6.19d). At each fork, polymerase copies both template strands, one in a continuous fashion, the other discontinuously as Okazaki fragments. As replication proceeds, DNA helicase progressively unwinds DNA ahead of each fork, while other proteins separate and stabilize the two single strands. Behind each fork, as elongation replicates chains complementary to both templates, double helixes form spontaneously. In the circular *E. coli* chromosome, there is only one origin of replication. When its two forks, moving in opposite directions, meet at a designated *termination region* about halfway around the circle from the origin of replication, replication is complete (Fig. 6.19d–f).

Not surprisingly, local unwinding of the double helix at a replication fork affects the chromosome as a whole. In *E. coli,* the unwinding of a section of a covalently closed circular chromosome overwinds and distorts the rest of the molecule (Fig. 6.19a and b). Overwinding reduces the

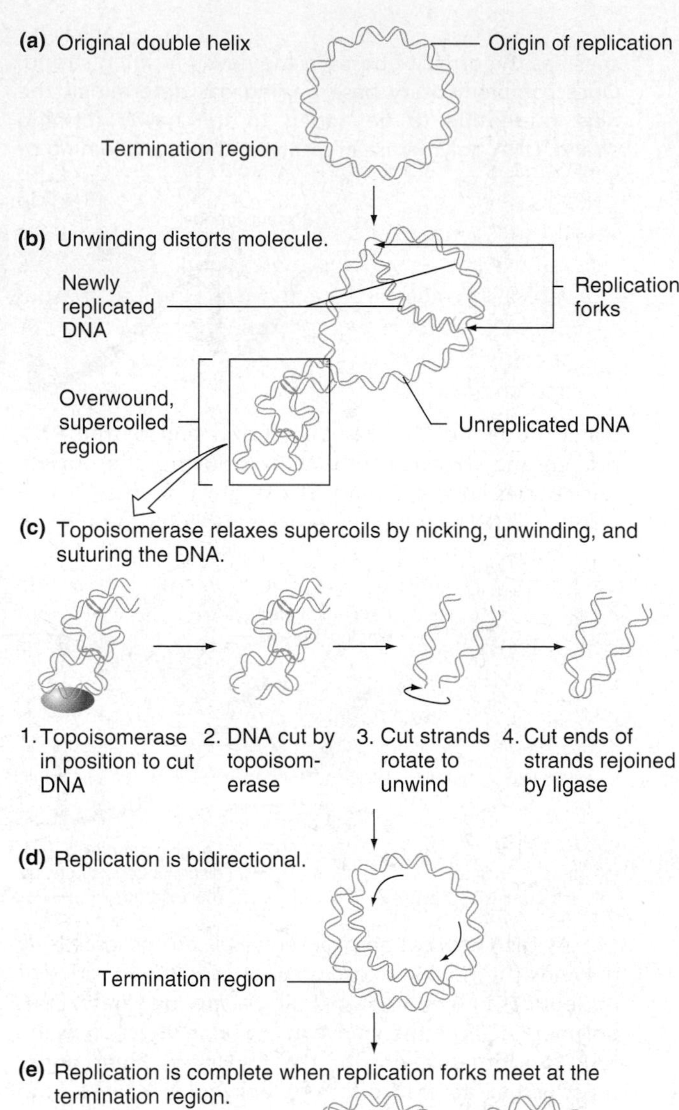
(a) Original double helix
Origin of replication
Termination region
(b) Unwinding distorts molecule.
Newly replicated DNA
Replication forks
Overwound, supercoiled region
Unreplicated DNA
(c) Topoisomerase relaxes supercoils by nicking, unwinding, and suturing the DNA.
1. Topoisomerase in position to cut DNA
2. DNA cut by topoisomerase
3. Cut strands rotate to unwind
4. Cut ends of strands rejoined by ligase
(d) Replication is bidirectional.
Termination region

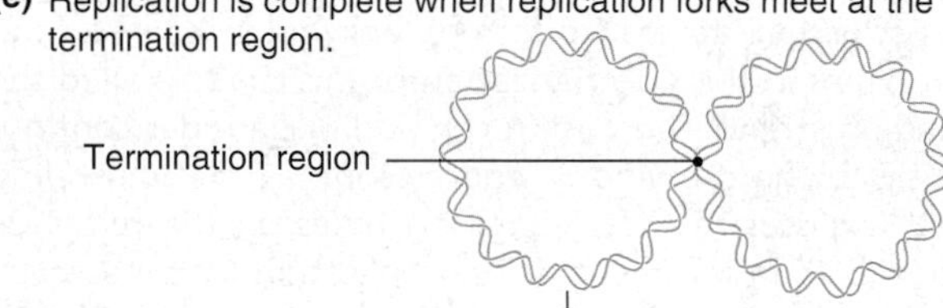
(e) Replication is complete when replication forks meet at the termination region.
Termination region
(f) Topoisomerases separate entwined daughter chromosomes, yielding two daughter molecules.

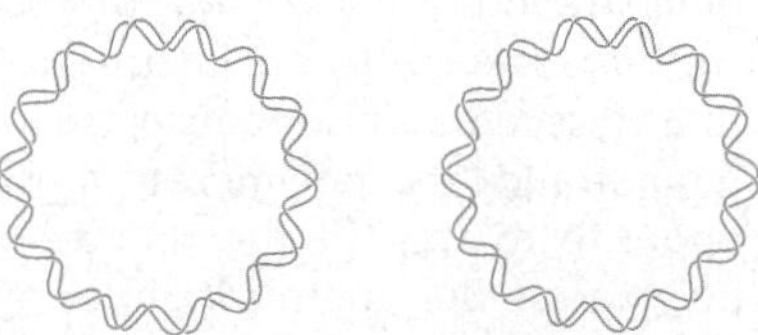

Figure 6.19 The bidirectional replication of a circular bacterial chromosome: An overview. **(a)** and **(b)** Replication proceeds in two directions from a single origin of replication, creating two replication forks that move in opposite directions around the circle. Local unwinding of DNA at the replication forks creates supercoiled twists in the DNA in front of the replication fork. **(c)** The action of topoisomerase enzymes that nick the DNA helps reduce this supercoiling. **(d)** and **(e)** When the two replication forks meet at the replication termination region, the entire chromosome has been copied. **(f)** Topoisomerase enzymes separate the two daughter chromosomes, which are usually entwined.

number of helical turns to less than the one every 10.5 nucleotides characteristic of B-form DNA. The chromosome accommodates the strain of distortion by twisting back upon itself. You can envision the effect by imagining a coiled telephone cord that overwinds and bunches up with use. The additional twisting of the DNA molecule is called **supercoiling.** Movement of the replication fork causes more and more supercoiling. This cumulative supercoiling, if left unchecked, would wind the chromosome up so tight that it would impede the progress of the replication fork. A group of enzymes known as **DNA topoisomerases** help relax the supercoils by nicking (that is, cutting the sugar-phosphate backbone between two adjoining nucleotides in) one or both strands of the DNA (Fig. 6.19c). Just as a telephone cord freed at the receiver end can unwind and restore its normal coiling pattern, the DNA strands, after nicking, can rotate relative to each other and thereby restore the normal coiling density of one helical turn per 10.5 nucleotide pairs. The activity of topoisomerases allows replication to proceed through the entire chromosome by preventing supercoils from accumulating in front of the replication fork. Replication of the original double helix sometimes produces intertwined daughter molecules whose clean separation also depends on topoisomerase activity.

In the much larger, linear chromosomes of eukaryotic cells, bidirectional replication proceeds roughly as just described but from many origins of replication. The multiple origins ensure that copying is completed within the time allotted (that is, within the S period of the cell cycle). Because of the three rules governing DNA polymerase activity (see Fig. 6.18a), replication of the very ends of linear chromosomes is problematic. But eukaryotic chromosomes have evolved specialized termination structures known as **telomeres,** which ensure the maintenance and accurate replication of the two ends of each linear chromosome. (Chapter 13 presents the details of eukaryotic chromosome replication.)

Throughout replication in both circular and linear chromosomes, the disentangling of parent strands and respiraling of daughter helixes occurs continuously as enzymes, including helicases and topoisomerases, wind tight, unwind, and rewind the double helix before and after the passage of polymerase, in part by nicking and rotating the two strands in relation to each other. At completion, replication has produced two identical double-stranded daughter DNA molecules.

Cells Must Ensure the Accuracy .of Their Genetic Information—Before, During, and After Replication

DNA is the sole repository of the vast amount of information required to specify the structure and function of most organisms. In some species, this information may lie in storage for many years and undergo replication many times before it is called on to generate progeny. During this time, the organism must protect the integrity of the information, for even the most minor change—the simple substitution of one base for another—can have disastrous consequences, such as the production of severe genetic disease or even death. Each organism ensures the informational fidelity of its DNA in three important ways.

Redundancy

Since either strand of the double helix can specify the sequence of the other, each double-stranded DNA molecule is informationally redundant. Built into the structure of the molecule itself, this redundancy provides a basis for the repair of errors arising either from chemical alterations sustained during storage or from malfunctions of the replication machinery.

Enzymes That Repair Chemical Damage to DNA

The cell has an array of enzymes devoted to the repair of nearly every imaginable type of chemical damage. We describe how these enzymes carry out their corrections in Chapter 7. The evolution of sophisticated enzymatic mechanisms for detecting and correcting DNA damage underlines the importance of maintaining the fidelity of genetic information.

The Remarkable Precision of the Cellular Replication Machinery

Evolution has perfected the cellular machinery for DNA replication to the point where errors that could creep in during the copying of the information are exceedingly rare. For example, DNA polymerase has acquired the ability to remove unmatched nucleotides from a new strand of DNA; as a result, the enzyme attaches a free nucleotide to a growing leading or lagging strand only if the nucleotide's base is correctly paired with its complementary base on the parent strand. We examine the molecular mechanisms of this type of DNA repair in Chapter 7.

All of these safeguards help ensure that the information content of DNA will be transmitted intact from generation to generation of cells and organisms. However, as we see next, new combinations of existing information arise naturally as a result of recombination.

6.5 Recombination Reshuffles the Information Content of DNA

Mutation, the ultimate source of all new alleles, is a relatively rare phenomenon at any one nucleotide pair on a chromosome. The most important mechanism for generating genomic diversity in sexually reproducing species is the production of new combinations of *already existing alleles*. This type of diversity increases the chances that at least some offspring of a mating pair will be able to survive and reproduce in a changing environment.

New combinations of already existing alleles arise from two different meiotic events: independent assortment, in which each pair of homologous chromosomes segregates free from the influence of other pairs, via random spindle attachment (see Chapter 4); crossing-over, in which two homologous chromosomes exchange parts. Independent assortment can produce gametes carrying new allelic combinations of genes on different chromosomes; but for genes on the same chromosome, independent assortment alone will only conserve the existing combinations of alleles. Crossing-over, however, can generate new allelic combinations of linked genes. The evolution of crossing-over thus compensated for a significant disadvantage of linkage, which would otherwise lock in only existing combinations of alleles on the same chromosome.

Historically, geneticists have used the term "recombination" to indicate the production of new combinations of alleles by any means, including independent assortment. In the remainder of this chapter, we use **recombination** to mean the generation of new allelic combinations—new combinations of nucleotides along a single DNA molecule—through genetic exchange between homologous chromosomes. In our discussion, we refer to the products of recombination as **recombinants:** chromosomes that carry a mix of alleles derived from different homologs.

As we examine recombination at the molecular level, we look first at experiments demonstrating that crossing-over occurs and then at the molecular details of a crossover event.

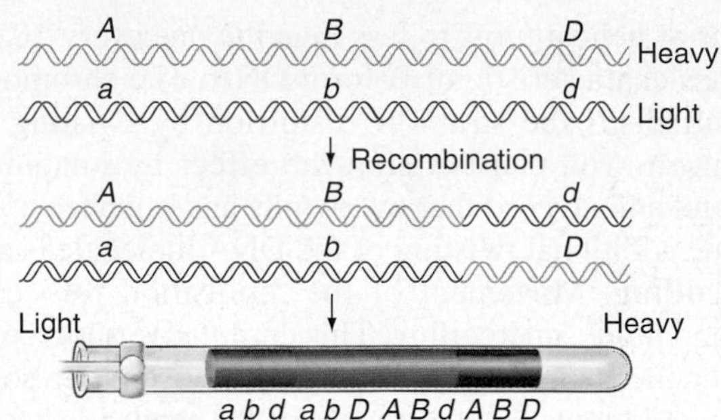

Figure 6.20 DNA molecules break and rejoin during recombination: The experimental evidence. Matthew Meselson and Jean Weigle infected *E. coli* cells with two different genetically marked strains of bacteriophage lambda previously grown in the presence of heavy (^{13}C and ^{15}N) or light (^{12}C and ^{14}N) isotopes of carbon and nitrogen. They then spun the progeny bacteriophages released from the cells on a CsCl density gradient. The position of phages in the gradient reflects the relative density of their DNA. The genetic recombinants had densities intermediate between the heavy and light parents.

During Recombination, DNA Molecules Break and Rejoin

Indirect Physical Evidence of Breaking and Rejoining

When viewed through the light microscope, recombinant chromosomes bearing physical markers appear to result from two homologous chromosomes breaking and exchanging parts as they rejoin (see Figs. 5.6 and 5.7 in Chapter 5). Since the recombined chromosomes, like all other chromosomes, are composed of one long DNA molecule, a logical expectation is that they should show some physical signs of this breakage and rejoining at the molecular level. To evaluate this hypothesis, researchers selected a bacterial virus, **lambda,** as their model organism. Lambda had a distinct experimental advantage for this particular study: It is about half DNA, so the density of the whole virus reflects the density of its DNA.

The experimental approach was similar in principle to the one in which Meselson and Stahl monitored a change in DNA density to follow DNA replication, only in this case, the researchers used the change in DNA density to look at recombination (**Fig. 6.20**). They grew two strains of bacterial viruses that were genetically marked to keep track of recombination, one in medium with a heavy isotope, the other in medium with a light isotope. They then infected the same bacterial cell with the two viruses under conditions that permitted little if any viral replication. With this type of coinfection, recombination could occur between "heavy" and "light" viral DNA molecules. After allowing time for recombination and the repackaging of viral DNA into virus particles, the experimenters isolated the viruses released from the lysed cells and analyzed them on a density gradient. Those viruses that had not participated in recombination formed bands in two distinct positions, one heavy and one light, as expected. Those viruses that had undergone recombination, however, migrated to intermediate densities, which corresponded to the position of the recombination event. If the recombinant derived most of its alleles and hence most of its chromosome from a "heavy" DNA molecule, its density was skewed toward the gradient's heavy region; by comparison, if it derived most of its alleles and chromosome from a "light" DNA molecule, it had a density skewed toward the light region of the gradient. These experimental results demonstrated that recombination at the molecular level results from the breakage and rejoining of DNA molecules.

Extreme Magnification Would Reveal Heteroduplexes That Mark Spots of Recombination

Recall that chiasmata, which are visible in the light microscope, indicate where chromatids from homologous chromosomes have crossed over, or exchanged parts (see Fig. 5.7 on p. 133). A 100,000-fold magnification of the actual site of recombination within a DNA molecule would reveal the breakage, exchange, and rejoining that constitute the molecular mechanism of crossing-over according to the lambda study. Although it is not yet possible to magnify DNA to the point of distinguishing base sequences under the microscope, a variety of molecular and genetic procedures enable geneticists to make deductions equivalent to such a 100,000-fold magnification.

The data obtained provide the following two clues about the mechanism of recombination. First, the products of recombination are almost always in exact register, with not a single base pair lost or gained. Geneticists originally deduced this from observing that recombination usually does not cause mutations; today, we know this to be true from analyses of DNA sequence (which we discuss in Chapter 9). Second, the two strands of a recombinant DNA molecule do not break and rejoin at the same location on the double helix. Instead, the breakpoints on each strand can be offset from each other by hundreds or even thousands of base pairs. The segment of the DNA molecule located between the two breakpoints is called a **heteroduplex region** (from the Greek *hetero* meaning "other" or "different") (**Fig. 6.21**). This name applies not only because one strand of the double

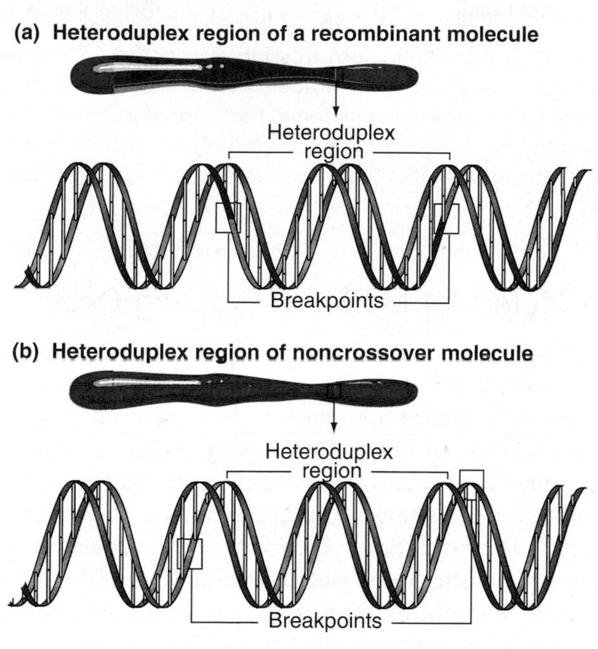

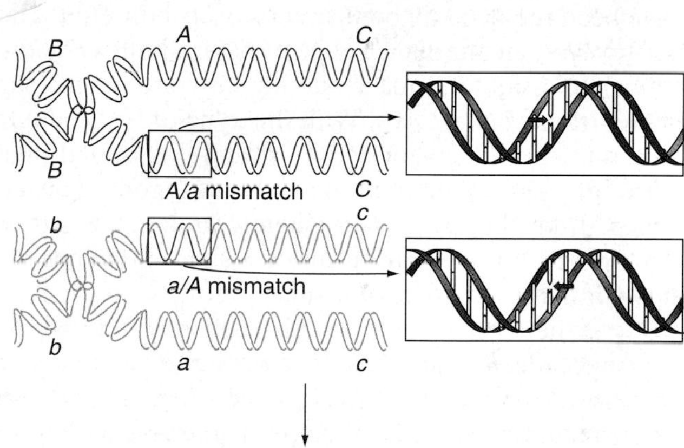

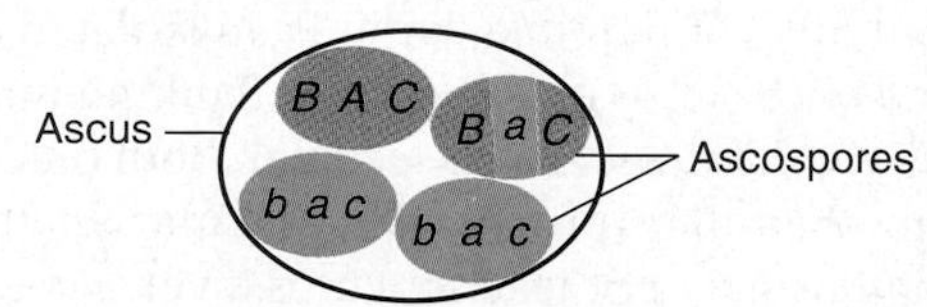

Figure 6.21 Heteroduplex regions occur at sites of genetic exchange. (a) A heteroduplex region separates portions of a chromosome derived from alternative parental homologs after crossing-over. **(b)** A heteroduplex region left behind after an aborted crossover attempt: Sequences from the same parental molecule are found on both sides of the heteroduplex region. The heteroduplexes depicted in (a) and (b) are thought to be two alternative products of the same molecular intermediate (as shown in Fig. 6.23 on pp. 196–199). **(c)** Gene conversion. 1. An aborted crossover during meiosis leaves behind two heteroduplex regions with mismatched bases. 2. DNA repair enzymes eliminate mismatches, converting both heteroduplexes into the *a* allele. 3. The resulting tetrad shows a 1:3 ratio of *A*:*a* alleles.

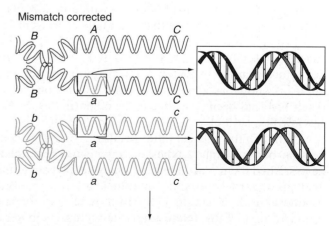

helix in this region is of maternal origin, while the other is paternal, but also because the pairing of maternal and paternal strands may produce mismatches in which bases are not complementary. Such mismatches occur at sites in the sequence that differ between the maternal and paternal homologs involved in the recombination event. The base sequences of homologs differ by 0.1% or less among individuals within any species. Within a heteroduplex, these mismatches prevent proper pairing at the mismatched base pairs, but they do not prevent double helix formation along the neighboring complementary nucleotides. Mismatched heteroduplex molecules do not persist for long. The same DNA repair enzymes that operate to correct mismatches during replication can move in to resolve them during recombination. The outcome of the repair enzymes' work depends on which strand they correct. For example, a repaired G–T mismatch could become either G–C or A–T.

The heteroduplex region of a DNA molecule that has undergone crossing-over has one breakpoint on each strand of the double helix (Fig. 6.21a). Beyond the heteroduplex region, *both* strands of one DNA molecule have been replaced by both strands of its homolog. There is, however, an alternative type of heteroduplex region in which the initiating and resolving cuts are on the same DNA strand (Fig. 6.21b). With this type of heteroduplex, only one short segment of one strand has traded places with one short segment of a homologous nonsister strand. Like the first type of heteroduplex, a short heteroduplex arising from a single-strand exchange may also contain one or a few mismatches.

In either type of heteroduplex, mismatch repair may alter one allele to another. If, for example, the original homologs carried the *A* allele in one segment of two sister chromatids and the *a* allele in the corresponding segment of the other pair of sister chromatids, the *A:a* ratio of alleles in these original homologs would be 2:2. Mismatch repair might change that *A:a* allele ratio from 2:2 to 3:1 (that is, three *A* alleles for every one *a* allele) or 1:3 (one *A* allele for every three *a* alleles; Fig. 6.21c). Any deviation from the expected 2:2 segregation of parental alleles is known as **gene conversion,** because one allele (such as *A*) has been converted to the other (in this case *a*) (review Fig. 5.19 on p. 149).

Although the unusual ratios resulting from gene conversion occur in many types of organisms, geneticists have studied them most intensively in yeast, where tetrad analysis makes it possible to follow all four meiotic products from a single cell (**Fig. 6.22** and review Fig. 5.15 on p. 146). Interestingly, observations in yeast indicate that gene conversion is associated with crossing-over about 50% of the time, but it is an isolated event not associated with a crossover between flanking markers the other 50% of the time. As we see later, both outcomes derive from the same proposed molecular intermediate, which may or may not lead to a crossover.

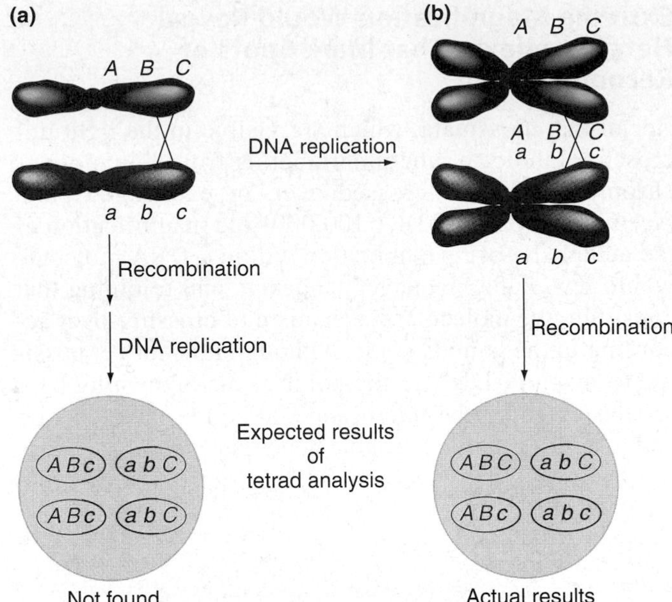

Figure 6.22 When does crossing-over occur? (a) If crossing-over occurs before DNA replication, all four meiotic products would carry a recombinant chromosome. This is not observed. **(b)** If crossing-over occurs after DNA replication, only two meiotic products will carry recombinant chromosomes. This is, in fact, observed.

A Molecular Model of Crossing-Over

A variety of experimental observations provide the framework for a detailed model of crossing-over during meiosis. First, tetrad analysis shows that only two of the four meiotic products from a single cell are affected by any individual recombination event. One member of each pair of sister chromatids remains unchanged (Fig. 6.22). This provides evidence that recombination occurs during meiotic prophase, after completion of DNA replication. The observation that recombination occurs only between homologous regions and is highly accurate, that is, in exact register, suggests an important role for base pairing between complementary strands derived from the two homologs. The observation that crossover sites are often associated with heteroduplex regions further supports the role of base pairing in the recombination process; it also implies that the process is initiated by single-strand exchange between nonsister chromatids. Finally, the observation of heteroduplex regions associated with gene conversion in the absence of crossing-over indicates that not all recombination events lead to crossovers.

The current molecular model for meiotic recombination derives almost entirely from results obtained in experiments on yeast. Researchers have found, however, that the protein Spo11, which plays a crucial role in initiating meiotic recombination in yeast, is also essential for meiotic recombination in other fungi, nematodes, plants, fruit flies, and mammals. This finding suggests that the mechanism of recombination presented in detail in **Fig. 6.23**—and known as the "double-strand-break repair model"—has been conserved throughout the evolution of eukaryotes. In the figure, we focus on the two nonsister chromatids involved in a single recombination event and show the two nonrecombinant chromatids only at the beginning of the process. These two nonrecombinant chromatids, depicted in the outside positions in Fig. 6.23, step 1, remain unchanged throughout recombination.

Only cells undergoing meiosis express the Spo11 protein, which is responsible for a rate of meiotic recombination several orders of magnitude higher than that found in mitotically dividing cells. Meiotic recombination begins when Spo11 makes a double-strand break in one of the four chromatids. In yeast, where meiotic double-strand breaks have been mapped, it is clear that Spo11 has a preference for some genomic sequences over others, resulting in "hot spots" for crossing over.

Unlike meiotic cells, mitotic cells do not usually initiate recombination as part of the normal cell-cycle program; instead, recombination in mitotic cells is a consequence of environmental damage to the DNA. X-rays and ultraviolet light, for example, can cause either double-strand breaks or single-strand nicks. The cell's enzymatic machinery works to repair the damaged DNA site, and recombination is a side effect of this process. Since the physical and chemical insults that damage DNA occur at random along the molecule, they result in a random distribution of nonmeiotic recombination events. Moreover, since such recombination depends on DNA damage, its frequency is low in the absence of damage and increases in proportion to the amount of damage sustained by the cell.

The double-strand-break repair model of meiotic recombination was proposed in 1983, well before the direct observation of any recombination intermediates. Since that time, scientists have seen—at the molecular level—the formation of double-strand breaks, the resection of those breaks to produce 3′ single-strand tails, and intermediate recombination structures in which single strands from two homologs have invaded each other. The double-strand-break repair model for recombination has become established because it explains much of the data so far obtained from genetic and molecular studies as well as the five properties of recombination deduced from breeding experiments:

1. Homologs physically break, exchange parts, and rejoin.
2. Breakage and repair create reciprocal products of recombination.
3. Recombination events can occur anywhere along the DNA molecule.
4. Precision in the exchange—no gain or loss of nucleotide pairs—prevents mutations from occurring during the process.
5. Gene conversion—in which a small segment of information from one homologous chromosome transfers to the other—can give rise to an unequal yield of two different alleles. Fifty percent of gene conversion events are associated with crossing-over between flanking markers, but an equal 50% are not associated with crossover events.

Connections

The Watson-Crick model for the structure of DNA, the single most important biological discovery of the twentieth century, clarified how the genetic material fulfills its primary function of carrying information: Each long, linear or circular molecule carries one of a vast number of potential arrangements of the four nucleotide building blocks (A, T, G, and C). The model also suggested how base complementarity could provide a mechanism for both accurate replication and changes in sequence combinations that arise from recombination events.

Unlike its ability to carry information, DNA's capacities for replication and recombination are not solely properties of the DNA molecule itself. Rather they depend on the cell's complex enzymatic machinery. But even though they rely on the complicated orchestration of many different proteins acting on the DNA, replication and recombination both occur with extremely high fidelity—normally not a single base pair is gained or lost. Occasionally, however, errors do occur, providing the genetic basis of evolution. While most errors are detrimental to the organism, a very small percentage produce dramatic changes in phenotype without killing the individual. For example, although most parts of the X and Y chromosomes are not similar enough to recombine, occasionally an "illegitimate" recombination does occur. Depending on the site of crossing-over, such illegitimate recombination may give rise to an XY individual who is female or an XX individual who is male (**Fig. 6.24** on p. 200). The explanation is as follows. In the first six weeks of development, a human embryo has the potential to become either male or female, but in the critical seventh week, information from a small segment of DNA—the sex-determining region of the Y chromosome containing the *SRY* gene—determines the embryo's sex. An illegitimate recombination between the X and the Y that shifts the *SRY* gene from the Y to the X chromosome creates a Y chromosome lacking *SRY* and an X chromosome with *SRY.*

Feature Figure 6.23

A Model of Recombination at the Molecular Level

Step 1 Double-strand break formation. During meiotic prophase, the meiosis-specific Spo11 protein makes a double-strand break on one of the chromatids by breaking the phosphodiester bonds between adjacent nucleotides on both strands of the DNA.

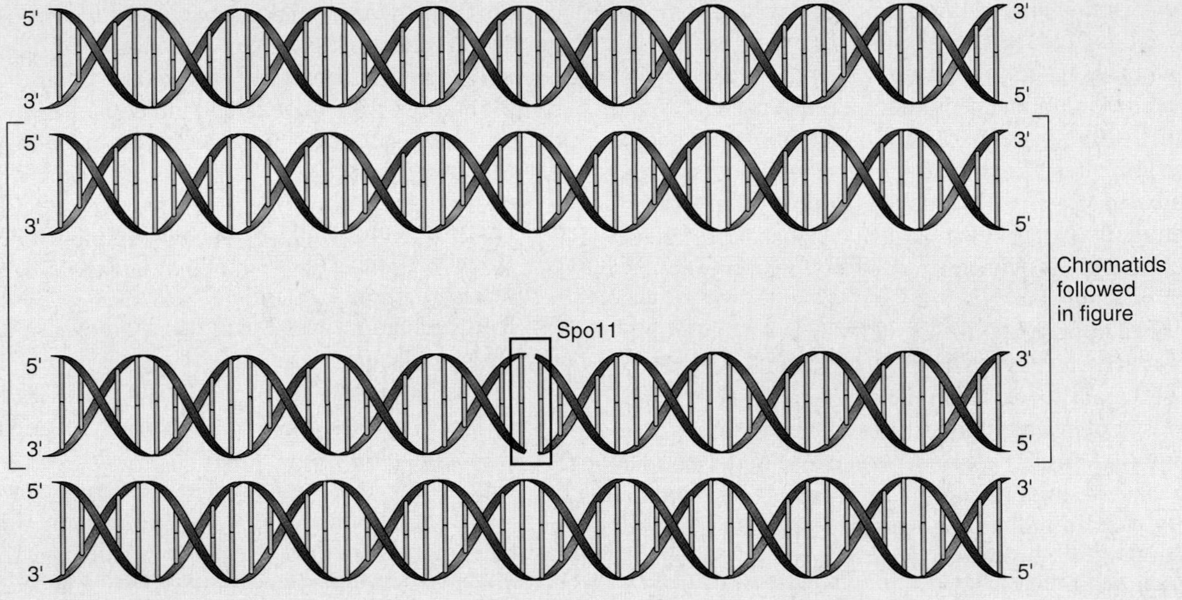

Step 2 Resection. The 5′ ends on each side of the break are degraded to produce two 3′ single-stranded tails.

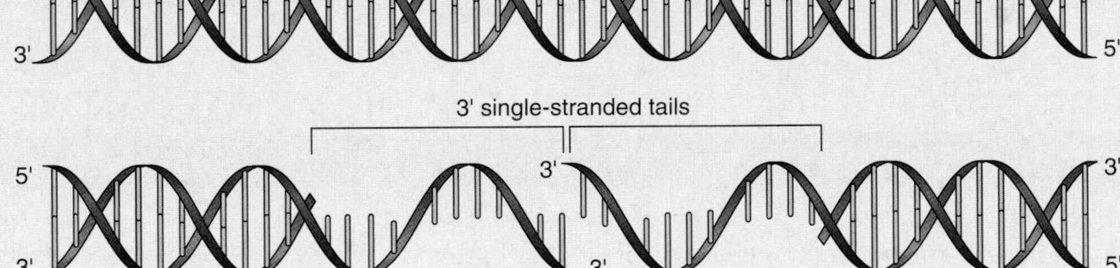

Step 3 First strand invasion (top of p. 197). One single-stranded tail is recognized and bound by an enzyme that also binds to a double helix in the immediate vicinity. In *E. coli*, the enzyme that simultaneously binds to the 3′ tail and the double helix is called *Rec*A (*orange* ovals). It plays a major role in the ensuing steps of the process, although many other enzymes collaborate with it. Their combined efforts open up the *Rec*A-bound double helix, promoting its invasion by the single displaced tail from the other duplex. *Rec*A then moves along the double helix, prying it open in front and releasing it to snap shut behind. With *Rec*A as its guide, the invading strand scans the base sequence it passes in the momentarily unwound stretches of DNA duplex. As soon as it finds a complementary sequence of sufficient length, it becomes immobilized by dozens of hydrogen bonds and forms a stable heteroduplex. Meanwhile, the strand displaced by the invading tail forms a D-loop (for displacement loop), which is stabilized by binding of the single-strand-binding (SSB) protein that played a similar role in DNA replication (see Fig. 6.18 on p.188). D-loops have been observed in electron micrographs of recombining DNA. In eukaryotes, strand invasion is mediated by a *Rec*A-like meiosis-specific protein, Dmcl.

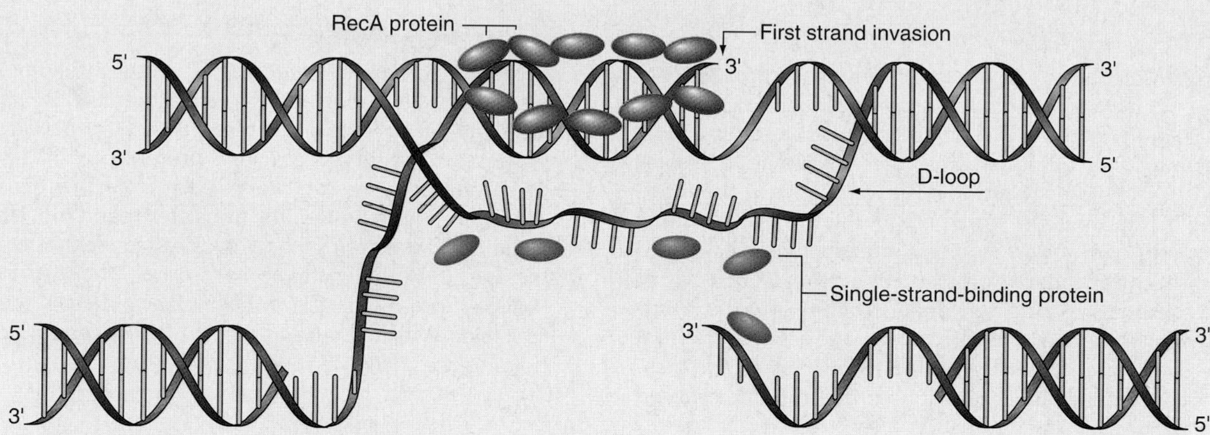

Step 4 Formation of a double Holliday junction. New DNA synthesis (indicated by *dotted string* below) to the invading 3′ tail enlarges the D-loop until the single-stranded bases on the displaced strand can form a complementary base pair with the 3′ tail on the nonsister chromatid. New DNA synthesis from this tail re-creates the DNA duplex on the bottom chromatid. The 5′ end on the right side of the break is then connected to the 3′ end of the invading strand. The resulting X structures are called Holliday junctions after Robin Holliday, the scientist who first proposed them.

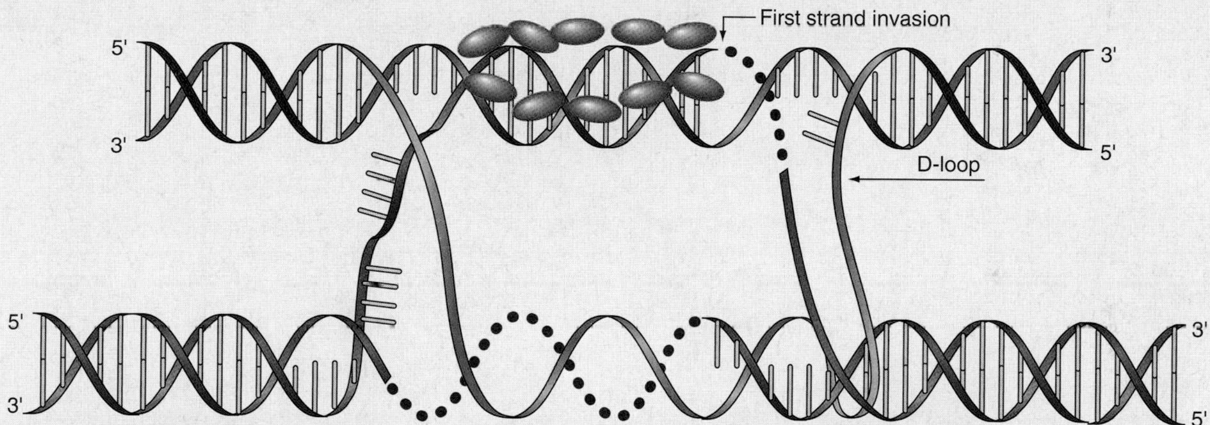

Step 5 Branch migration. The next step, branch migration, results from the tendency of both invading strands to "zip up" by base pairing along the length of their newly formed complementary strands. The DNA double helices unwind in front of this double zippering action, and two newly created heteroduplex molecules rewind behind it. The branches of the two ends of the heteroduplex region (where strands from the two homologous chromosomes cross) move in the direction of the arrows. Branch migration thus lengthens the heteroduplex region of both DNA molecules from tens of base pairs to hundreds or thousands. Because the two invading strands began their scanning from complementary bases at slightly different points on the homologous chromatids, branch migration produces two heteroduplex regions that are somewhat different in length.

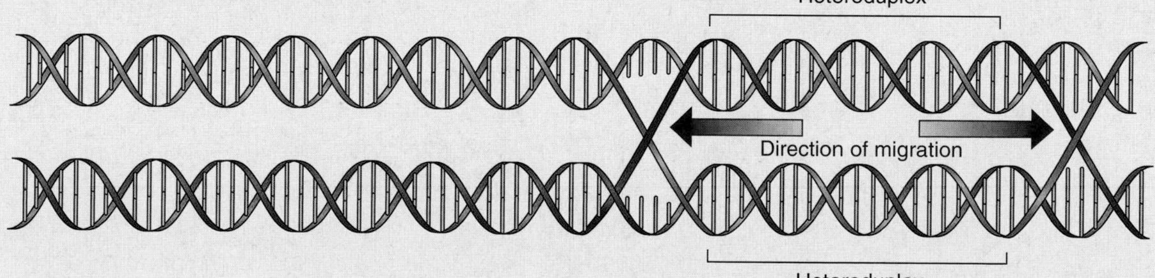

(*Continued*)

Feature Figure 6.23 (*Continued*)

Step 6 The Holliday intermediate. For meiosis to proceed, the two interlocked nonsister chromatids must disengage. There are two equally likely paths to such a resolution of crossing-over. To understand these alternative resolutions, it is helpful to modify the way we view the interlocked intermediate structure. In this figure, we show only one of the two Holliday intermediates associated with each recombination shift. By pushing out each of the four arms of the interlocked structure into the X pattern shown here and then rotating one set of arms from the same original chromatid 180°, we obtain the "isomerized cross-strand exchange configuration" pictured in step 7, commonly referred to as the "Holliday intermediate." It is important to realize that this is simply a different way of looking at the structure for pedagogical purposes. In reality, there is no preferred conformation of chromatid arms relative to each other in this small, localized region. Rather, the arms are free to move about at random, constrained only by the strands that connect the two DNA molecules to each other. The view of the Holliday intermediate, however, clearly reveals that the four single-stranded regions all play an equal role in holding the structure together.

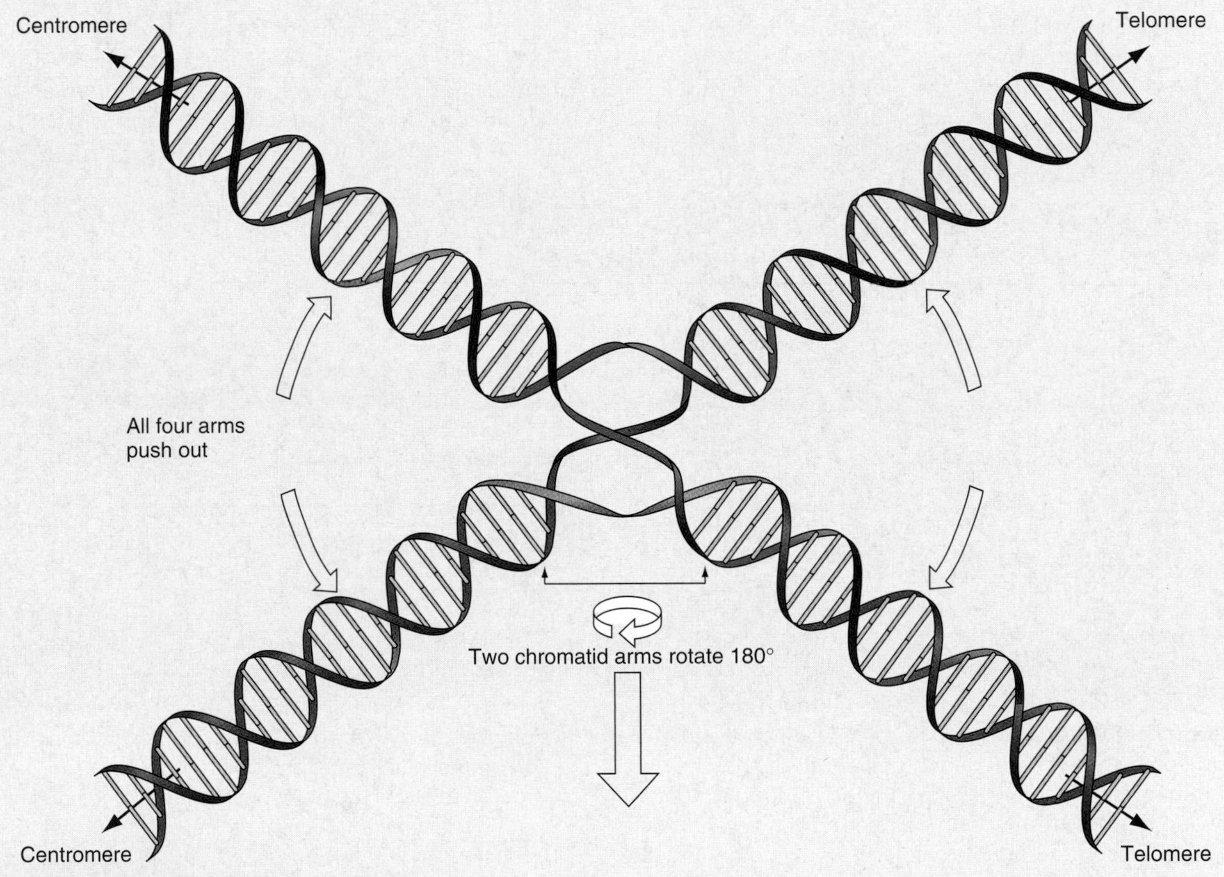

Step 7 Alternative resolutions. If endonucleases make a horizontal cut (as in this illustration) across a Holliday intermediate, the freed centromeric and telomeric strands of both homolog 1 and homolog 2 can become ligated. If the endonucleases make a vertical cut across a strand from homolog 1 and homolog 2, the newly freed strand from the centromeric arm of homolog 1 can now be ligated to the freed strand from the telomeric arm of homolog 2. Likewise, the telomeric strand from homolog 1 can now be ligated to the centromeric strand from homolog 2. This leads to crossing-over between two homologs. However, the resolution of the second Holliday intermediate will determine whether an actual crossing-over event is consummated, as detailed in step 8.

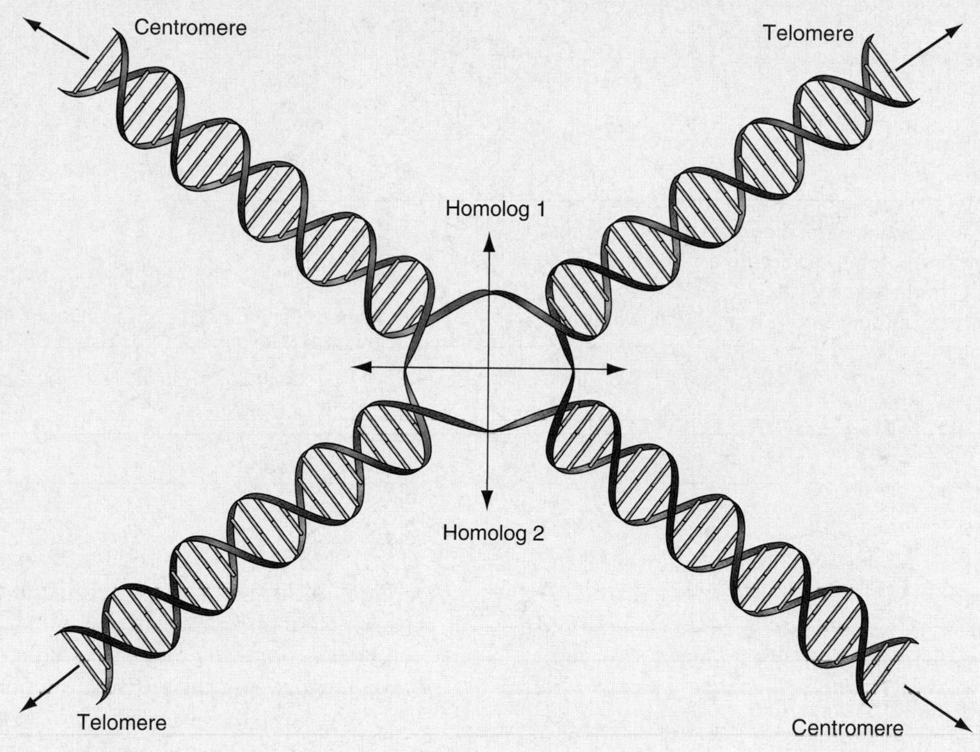

Step 8 Probability of crossover occurring. Because there are two Holliday junctions, both must be resolved. Resolution of both Holliday junctions in the same plane results in a noncrossover chromatid. For a crossover to occur, the two Holliday junctions must be resolved in opposite planes. (Chromatids are shown in initial configuration of step 6.)

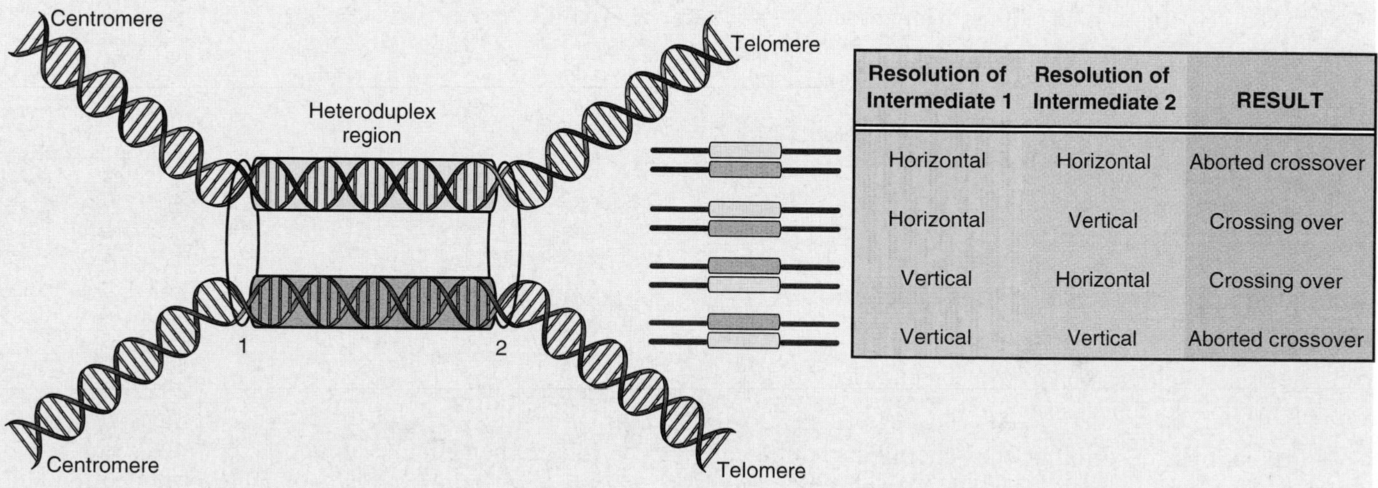

Resolution of Intermediate 1	Resolution of Intermediate 2	RESULT
Horizontal	Horizontal	Aborted crossover
Horizontal	Vertical	Crossing over
Vertical	Horizontal	Crossing over
Vertical	Vertical	Aborted crossover

Fertilization of eggs by sperm with a Y chromosome lacking *SRY* generates XY individuals that develop as females; fertilization of eggs by sperm with an X chromosome containing *SRY* produces XX individuals that develop as males.

How do genes such as *SRY* produce their phenotypic effects? We begin to answer this question in Chapter 7, where we describe how geneticists using mutations as tools of genetic analysis learned that in most cases, a gene is a specific sequence of nucleotides in a discrete region of DNA that encodes a particular protein.

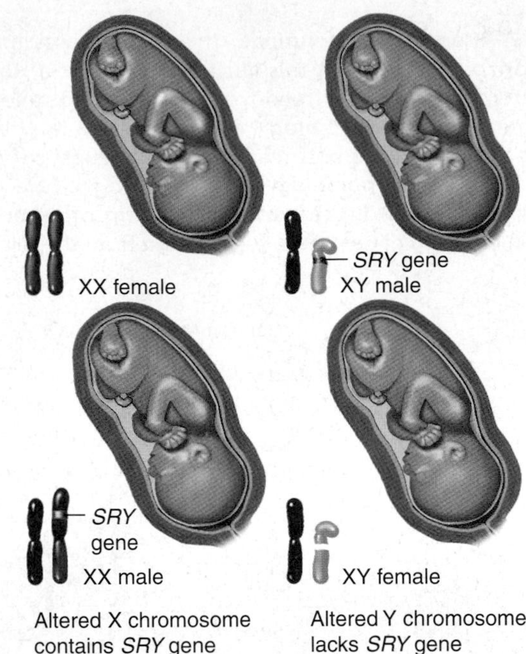

Figure 6.24 Illegitimate recombination may produce an XY female or an XX male. The *SRY* gene, normally located on the Y chromosome, dictates male development; in the absence of *SRY*, an embryo develops as a female. Rare, illegitimate recombination between the X and Y chromosome can create a Y chromosome without *SRY* and an X chromosome with *SRY*. Fertilization with gametes containing these unusual chromosomes will produce XY females or XX males.

Essential Concepts

1. DNA is the nearly universal genetic material. Experiments showing that DNA causes bacterial transformation and is the agent of virus production in phage-infected bacteria demonstrated this fact.
2. According to the Watson-Crick model, proposed in 1953 and confirmed in the succeeding decades, the DNA molecule is a double helix composed of two antiparallel strands of nucleotides; each nucleotide consists of one of four nitrogenous bases (A, T, G, or C), a deoxyribose sugar, and a phosphate. An A on one strand pairs with a T on the other, and a G pairs with a C.
3. DNA carries information in the sequence of its bases, which may follow one another in any order.
4. The DNA molecule reproduces by semiconservative replication. In this type of replication, the two DNA strands separate, and the cellular machinery then synthesizes a complementary strand for each. By producing exact copies of the base sequence information in DNA, semiconservative replication allows life to reproduce itself.
5. Recombination arises from a highly accurate cellular mechanism that includes the base pairing of homologous strands of nonsister chromatids. Recombination generates new combinations of alleles.

On Our Website

www.mhhe.com/hartwell3
Chapter 6

Annotated Suggested Readings and Links to Other Websites

- The original publication by Watson and Crick presenting the double-helical structure of DNA
- Publications describing the chemical nature of the gene and models for DNA replication and recombination.
- More on the recovery and analysis of DNA from extinct organisms.

Specialized Topics

- Three-dimensional, atomic-level models of enzymes operating on DNA to achieve replication and recombination.

Social and Ethical Issues

1. In 1990, researchers reported the recovery and amplification of a small fragment of DNA from a fossilized leaf that was 20 million years old. Subsequent studies suggested that all samples of purported million-year old DNA may actually be the result of unintentional contamination from contemporary organisms. Nevertheless, the recovery of small bits of ancient DNA inspired Michael Crichton to write the best-selling novel *Jurassic Park.* In his story, researchers re-create dinosaurs by isolating and piecing together bits of DNA. The necessity of recovering enough intact long stretches of dinosaur DNA make this a highly improbable scenario, yet the movie portrayed this work with an aura of authenticity. Does a science fiction writer have a responsibility to indicate what is truly possible and where the line between science and fiction is likely to lie? Does this become more of an imperative when a book is made into a movie that reaches a mass audience? Can the reality of what is currently possible be conveyed within the framework of science fiction?

2. Eleanor has just been diagnosed with breast cancer at age 45. An analysis of her DNA shows that a mutant form of the *BRCA1* gene is responsible for the cancer. Mendelian genetics tells us that there is a 50% chance that Eleanor's 18-year-old daughter Erin inherited this mutation from her mother. We also know that the mutation is not completely penetrant—only 70% of women with the mutation develop breast cancer by the age of 60. But a radical bilateral mastectomy, which removes all currently healthy breast tissue, can dramatically decrease or eliminate the possibility of developing breast cancer in the future. Many women feel conflicted about receiving a mastectomy when the possibility also exists that their breast tissue could be healthy for the rest of their lives. Some with a family history of breast cancer choose not to learn whether they actually carry a *BRCA1* mutation. Who should decide whether Erin takes a DNA test that could detect the *BRCA1* mutation? Erin, her parents, or their family physician? Should government regulations mandate that Eleanor and Erin receive specific information about the connection between genotype and phenotype before they make their decision? At what age does a person become competent to make this kind of decision for herself? Should Erin's physician wait until she turns 21 before allowing her to take the test? (Erin's chances of getting cancer before the age of 21 are very low, but if she puts off the decision beyond that age, the probability increases continuously.)

3. The perpetrator of a serious crime has left behind a few hairs, which contain enough DNA to allow detection of specific sequences using a technique called PCR (described in detail in Chapter 9). The police arrested a number of suspects and required them to provide DNA samples so that technicians can determine whether any of their sequences match DNA retrieved from the crime scene. After completed comparisons reveal that the suspects' samples don't match sequences from the crime scene, should the police or other governmental agencies be allowed to store the DNA from these individuals indefinitely, or should they be required to destroy the samples?

4. A decade ago, researchers collected DNA samples from 100 volunteers who gave them permission to use the samples in a specific study aimed at identifying a gene involved in a particular disease. With new discoveries made since the samples were collected, the researchers have become aware that samples from this particular group of 100 people would be very useful in the analysis of another, unrelated disease. Since the volunteers never gave the researchers permission to use their samples in this unexpected manner, the researchers try to contact them to see if they will agree to participate (indirectly) in the new experiment. Unfortunately, 30 of the people cannot be found, but their DNA samples are crucial to the collection of sufficient data for obtaining a significant experimental result. Is it ethical for the researchers to use the DNA from these "lost" people in this particular experiment without permission? Should the answer depend in any way on how important the potential results would be in curing a disease that affects many other people, or is this consideration irrelevant to an ethical decision?

Solved Problems

```
5′ TAAGCGTAACCCGCTAA   CGTATGCGAAC   GGGTCCTATTAACGTGCGTACAC 3′
3′ ATTCGCATTGGGCGATT   GCATACGCTTG   CCCAGGATAATTGCACGCATGTG 5′
```

I. Imagine that the double-stranded DNA molecule shown here was broken at the sites indicated by spaces in the sequence and that before the breaks were repaired, the DNA fragment between the breaks was reversed. What would be the base sequence of the repaired molecule? Explain your reasoning.

Answer

To answer this question, you need to keep in mind the polarity of the DNA strands involved.

The top strand has the polarity left to right of 5′ to 3′. The reversed region must be rejoined with the same polarity. Label the polarity of the strands within the inverting region. To have a 5′-to-3′ polarity maintained on the top strand, the *fragment that is reversed must be flipped over,* so the strand that was formerly on the bottom is now on top.

```
5′             3′5′            3′5′              3′
TAAGCGTAACCCGCTAAGTTCGCATACGGGGTCCTATTAACGTGCGTACAC
ATTCGCATTGGGCGATTCAAGCGTATGCCCCAGGATAATTGCACGCATGTG
3′             3′5′            3′5′              5′
```

II. A new virus has recently been discovered that infects human lymphocytes. The virus can be grown in the laboratory using cultured lymphocytes as host cells. Design an experiment using a radioactive label that would tell you if the virus contains DNA or RNA.

Answer

Use your knowledge of the differences between DNA and RNA to answer this question. RNA contains uracil instead of the thymine found in DNA. *You could set up one culture in which you add radioactive uracil to the media and a second one in which you add radioactive thymine to the culture.* After the viruses have infected cells and produced more new viruses, collect the newly synthesized virus. Determine which culture produced radioactive viruses. If the virus contains RNA, the collected virus grown in media containing radioactive uracil will be radioactive, but the virus grown in radioactive thymine will not be radioactive. If the virus contains DNA, the collected virus from the culture containing radioactive thymine will be radioactive, but the virus from the radioactive uracil culture will not. (You might also consider using radioactively labeled ribose or deoxyribose to differentiate between an RNA- and DNA-containing virus. Technically this does not work as well, because the radioactive sugars are processed by cells before they become incorporated into nucleic acid, thereby obscuring the results.)

III. If you expose a culture of human cells (for example, HeLa cells) to ^{3}H-thymidine during S phase, how would the radioactivity be distributed over a pair of homologous chromosomes at metaphase? Would the radioactivity be in (a) one chromatid of one homolog, (b) both chromatids of one homolog, (c) one chromatid each of both homologs, (d) both chromatids of both homologs, or (e) some other pattern? Choose the correct answer and explain your reasoning.

Answer

This problem requires application of your knowledge of the molecular structure and replication of DNA and how it relates to chromatids and homologs. DNA replication occurs during S phase, so the ^{3}H-thymidine would be incorporated into the new DNA strands. A chromatid is a replicated DNA molecule, and each new DNA molecule contains one new strand of DNA (semiconservative replication). *The radioactivity would be in both chromatids of both homologs (d).*

Problems

Vocabulary

1. For each of the terms in the left column, choose the best matching phrase in the right column.

a. transformation	1. the strand that is synthesized discontinuously during replication
b. bacteriophage	2. the sugar within the nucleotide subunits of DNA
c. pyrimidine	3. a nitrogenous base containing a double ring
d. deoxyribose	4. noncovalent bonds that hold the two strands of the double helix together
e. hydrogen bonds	5. Meselson and Stahl experiment
f. complementary bases	6. Griffith experiment
g. origin	7. structures at ends of eukaryotic chromosomes
h. Okazaki fragments	8. two nitrogenous bases that can pair via hydrogen bonds
i. purine	9. a nitrogenous base containing a single ring
j. topoisomerases	10. a short sequence of bases where unwinding of the double helix for replication begins
k. semiconservative replication	11. a virus that infects bacteria
l. lagging strand	12. short DNA fragments formed by discontinuous replication of one of the strands
m. telomeres	13. enzymes involved in controlling DNA supercoiling

Section 6.1

2. Griffith, in his 1928 experiments, demonstrated that bacterial strains could be genetically transformed. The evidence that DNA was the "transforming principle" responsible for this phenomenon came later. What was the key experiment that Avery, MacCleod, and McCarty performed to prove that DNA was responsible for the genetic change from rough cells into smooth cells?

3. During bacterial transformation, DNA that enters a cell is not an intact chromosome; instead it consists of randomly generated fragments of chromosomal DNA. In a transformation where the donor DNA was from a bacterial strain that was $a^+\ b^+\ c^+$ and the recipient was $a^-b^-c^-$, 55% of the cells that became a^+ were also transformed to c^+. but only 2% of the a^+ cells were b^+. Is gene *b* or *c* closer to gene *a?*

4. Nitrogen and carbon are more abundant in proteins than sulfur. Why did Hershey and Chase use radioactive sulfur instead of nitrogen and carbon to label the protein portion of their bacteriophages in their experiments to determine whether parental protein or parental DNA is necessary for progeny phage production?

Section 6.2

5. Imagine you have three test tubes containing identical solutions of purified, double-stranded human DNA. You expose the DNA in tube 1 to an agent that breaks the sugar-phosphate (phosphodiester) bonds. You expose the DNA in tube 2 to an agent that breaks the bonds that attach the bases to the sugars. You expose the DNA in tube 3 to an agent that breaks the hydrogen bonds. After treatment, how would the structures of the molecules in the three tubes differ?

6. What information about the structure of DNA was obtained from X-ray crystallographic data?

7. If 30% of the bases in human DNA are A, (a) what percentage are C? (b) What percentage are T? (c) What percentage are G?

8. Which of the following statements are true about double-stranded DNA?

a. $A + C = T + G$
b. $A + G = C + T$
c. $A + T = G + C$
d. $A/G = C/T$
e. $A/G = T/C$
f. $(C + A) / (G + T) = 1$

9. A particular virus with DNA as its genetic material has the following proportions of nucleotides: 20% A, 35% T, 25% G, and 20% C. How can you explain this result?

10. When a double-stranded DNA molecule is exposed to high temperature, the two strands separate, and the molecule loses its helical form. We say the DNA has been denatured. (Denaturation also occurs when DNA is exposed to acid or alkaline solutions.)

a. Regions of the DNA that contain many A–T base pairs are the first to become denatured as the temperature of a DNA solution is raised. Thinking about the chemical structure of the DNA molecule, why do you think the A–T-rich regions denature first?
b. If the temperature is lowered, the original DNA strands can reanneal, or renature. In addition to the full double-stranded molecules, some molecules of the type shown here are seen when the molecules are examined under the electron microscope. How can you explain these structures?

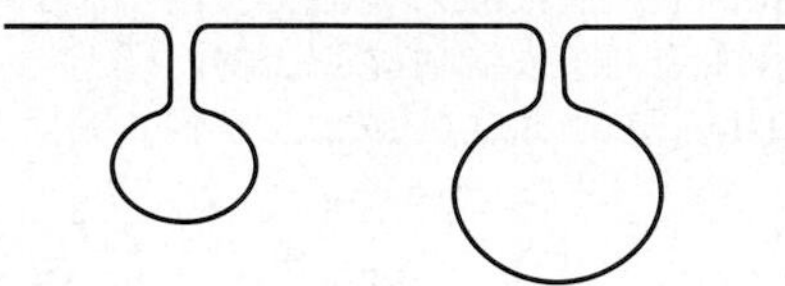

11. A portion of one DNA strand of the human gene responsible for cystic fibrosis is

5′.....ATAGCAGAGCACCATTCTG.....3′

Write the sequence of the corresponding region of the other DNA strand of this gene, noting the polarity. What do the dots before and after the given sequence represent?

Section 6.3

12. The underlying structure of DNA is very simple, consisting of only four possible building blocks.

a. How is it possible for DNA to carry complex genetic information if its structure is so simple?
b. What are these building blocks? Can each block be subdivided into smaller units, and if so, what are they? What kinds of chemical bonds link the building blocks?
c. How does the underlying structure of RNA differ from that of DNA?

13. An RNA virus that infects plant cells is copied into a DNA molecule after it enters the plant cell. What would be the sequence of bases in the first strand of DNA made complementary to the section of viral RNA shown here?

5′ CCCUUGGAACUACAAAGCCGAGAUUAA 3′

14. Bacterial transformation and bacteriophage labeling experiments proved that DNA was the hereditary material in bacteria and in DNA-containing viruses. Some viruses do not contain DNA but have RNA inside the phage particle. An example is the tobacco mosaic virus (TMV) that infects tobacco plants, causing lesions in the leaves. Two different variants of TMV exist that have different forms of a particular protein in the virus particle that can be distinguished. It is possible to reconstitute TMV *in vitro* (in the test tube) by mixing purified proteins and RNA. The reconstituted virus can then be used to infect the host plant cells and produce a new generation of viruses. Design an experiment to show that RNA acts as the hereditary material in TMV.

15. The Tools of Genetics box on pp. 182–184 discusses how restriction enzymes can recognize a short sequence of nucleotides in a long molecule of DNA and can then cut the DNA at that location. In a long DNA molecule with equal proportions of A, C, G, and T in a random sequence, what would be the average spacing (in numbers of nucleotides) between successive occurrences of the sequences recognized by the following restriction enzymes?
   a. *Eco*R1 (5′……GAATTC….. 3′)
   b. *Bam*H1 (5′……GGATCC….. 3′)
   c. *Hae*III (5′……GGCC……… 3′)

Section 6.4

16. In Meselson and Stahl's density shift experiments (diagrammed in Fig. 6.16 on p. 186), describe the results you would expect in each of the following situations:
   a. Conservative replication after two rounds of DNA synthesis on ^{14}N.
   b. Semiconservative replication after three rounds of DNA synthesis on ^{14}N.
   c. Dispersive replication after three rounds of DNA synthesis on ^{14}N.
   d. Conservative replication after three round of DNA synthesis on ^{14}N.

17. When Meselson and Stahl grew *E. coli* in ^{15}N medium for many generations and then transferred to ^{14}N medium for one generation, they found that the bacterial DNA banded at a density intermediate between that of pure ^{15}N DNA and pure ^{14}N DNA following equilibrium density centrifugation. When they allowed the bacteria to replicate one additional time in ^{14}N medium, they observed that half of the DNA remained at the intermediate density, while the other half banded at the density of pure ^{14}N DNA. What would they have seen after an additional generation of growth in ^{14}N medium? After two additional generations?

18. If you expose human tissue culture cells (for example, HeLa cells) to ^{3}H-thymidine just as they enter S phase, then wash this material off the cells and let them go through a second S phase before looking at the chromosomes, how would you expect the ^{3}H to be distributed over a pair of homologous chromosomes? (Ignore the effect recombination could have on this outcome.) Would the radioactivity be in (a) one chromatid of one homolog, (b) both chromatids of one homolog, (c) one chromatid each of both homologs, (d) both chromatids of both homologs, or (e) some other pattern? Choose the correct answer and explain your reasoning. (This problem extends the analysis begun in solved Problem III on p. 202.)

19. Draw a bidirectional replication fork and label the origin of replication, the leading strands, lagging strands, and the 5′ and 3′ ends of all strands shown in your diagram.

20. As Fig. 6.17 on p. 187 shows, DNA polymerase cleaves the high-energy bonds between phosphate groups in nucleotide triphosphates (nucleotides in which three phosphate groups are attached to the 5′-carbon atom of the deoxyribose sugar) and uses this energy to catalyze the formation of a phosphodiester bond when incorporating new nucleotides into the growing chain.
   a. How does this information explain why DNA chains grow during replication in the 5′-to-3′ direction?
   b. The action of the enzyme DNA ligase in joining Okazaki fragments together is shown in Fig. 6.18 on p. 189. Remember that these fragments are connected only after the RNA primers at their ends have been removed. Given this information, infer the type of chemical bond whose formation is catalyzed by DNA ligase and whether or not a source of energy will be required to promote this reaction. Explain why DNA ligase and not DNA polymerase is required to join Okazaki fragments.

21. The bases of one of the strands of DNA in a region where DNA replication begins are shown here. What is the sequence of the primer that is synthesized complementary to the bases in bold? (Indicate the 5′ and 3′ ends of the sequence.)

5′ AGGCCTCGAATTCGTATAGCTTTCAGAAA 3′

22. Replicating structures in DNA can be observed in the electron microscope. Regions being replicated appear as bubbles.
   a. Assuming bidirectional replication, how many origins of replication are active in this DNA molecule?
   b. How many replication forks are present?
   c. Assuming that all replication forks move at the same speed, which origin of replication was activated last?

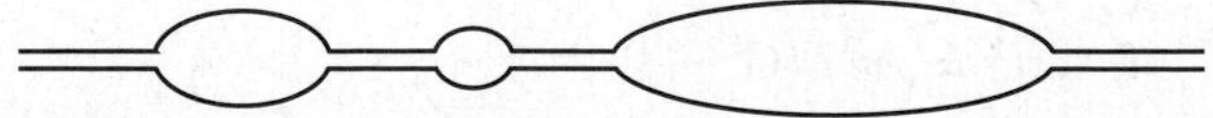

23. Indicate the role of each of the following in DNA replication: (a) topoisomerase, (b) helicase, (c) primase, and (d) ligase.

24. Diagram replication occurring at the end of a double-stranded linear chromosome. Show the leading and lagging strands with their primers. (Indicate the 5′ and 3′ ends of the strands.) What difficulty is encountered in producing copies of both DNA strands at the end of a chromosome?

25. Figure 6.14 on p. 185 depicts Watson and Crick's initial proposal for how the double-helical structure of DNA accounts for DNA replication. Based on our current knowledge, this figure contains a serious error due to oversimplification. Identify the problem with this figure.

26. Researchers have discovered that during replication of the circular DNA chromosome of the animal virus SV40, the two newly completed daughter double helices are intertwined. What would have to happen for the circles to come apart?

27. As we explain in Chapter 9, a DNA synthesizer is a machine that uses automated organic synthesis to create short, single strands of DNA of any given sequence. You have used the machine to create the following three DNA molecules:

(DNA #1) 5′ CTACTACGGATCGGG 3′
(DNA #2) 5′ CCAGTCCCGATCCGT 3′
(DNA #3) 5′ AGTAGCCAGTGGGGAAAAACCCCACTGG 3′

Now you add the DNA molecules either singly or in combination to reaction tubes containing DNA polymerase, dATP, dCTP, dGTP, and dTTP in a buffered solution that allows DNA polymerase to function. For each of the reaction tubes, indicate whether DNA polymerase will synthesize any new DNA molecules, and if so, write the sequence (s) of any such DNAs.

a. DNA #1 plus DNA # 3
b. DNA #2 plus DNA # 3
c. DNA #1 plus DNA # 2
d. DNA #3 only

Section 6.5

28. Bacterial cells were coinfected with two types of bacteriophage lambda: One carried the c^+ allele and the other the *c* allele. After the cells lysed, progeny bacteriophage were collected. When a single such progeny bacteriophage was used to infect a new bacterial cell, it was observed in rare cases that some of the resulting progeny were c^+ and others were *c*. Explain this result.

29. What properties would you expect of an *E. coli* strain that has a mutant allele (null or nonfunctional) of the *recA* gene? Explain.

30. Imagine that you have done a cross between two strains of yeast, one of which has the genotype *A B C* and the other *a b c,* where the letters refer to three rather closely linked genes in the order given. You examine many tetrads resulting from this cross, and you find two that do not contain the expected two *B* and two *b* spores. In tetrad I, the spores are *A B C, A B C, a B c,* and *a b c.* In tetrad II, the spores are *A B C, A b c, a b C,* and *a b c.* How have these unusual tetrads arisen?

31. In yeast, gene conversion occurs equally frequently with recombination of genetic markers flanking the region of gene conversion and without it. Why is this so?

32. From a cross between $e^+ f^+ g^+$ and $e^- f^- g^-$ strains of *Neurospora,* recombination between these linked genes resulted in a few octads containing the following ordered set of spores:

$e^+ f^+ g^+$
$e^+ f^+ g^+$
$e^+ f^- g^+$
$e^+ f^- g^+$
$e^- f^- g^-$
$e^- f^- g^-$
$e^- f^- g^-$
$e^- f^- g^-$

a. Where was recombination initiated?
b. Where did the resolving cut get made?
c. Why do you end up with $2\ f^+ : 6\ f^-$ but $4\ e^+ : 4\ e^-$?

33. DNA fingerprinting, a technique that will be described in Chapter 11, can show whether two different samples of DNA come from the same individual. One form of DNA fingerprinting relies on chromosome regions called microsatellites, which contain many repeats of a short sequence (for example, CACACACA, etc.). The number of repeats is highly variable from individual to individual in a population. Scientists have suggested that this variability could result from recombination. Use the double-strand break model, including strand invasion, to explain how a microsatellite could gain or lose repeats during recombination.

Anatomy and Function of a Gene: Dissection Through Mutation

Chapter 7

A scale played on a piano keyboard and a gene on a chromosome are both a series of simple, linear elements (keys or nucleotide pairs) that produce information. A wrong note or an altered nucleotide pair calls attention to the structure of the musical scale or the gene.

Human chromosome 3 consists of approximately 220 million base pairs and carries 1000–2000 genes. Somewhere on the long arm of the chromosome resides the gene for rhodopsin, a light-sensitive protein active in the rod cells of our retinas. Under the microscope, a lone chromosome 3 released from the nucleus of a human cell looks like a long, undifferentiated looping string of DNA (**Fig. 7.1**). At very high magnification—unfortunately higher than that obtainable with today's microscopes—this genetic twine would appear as a double helix composed of two antiparallel strands of nucleotides held together by the hydrogen bonds of complementary base pairs. But which nucleotides within the scramble make up the rhodopsin gene, and what does the gene look like? Knowledge that DNA is the molecule of heredity does not in itself answer the question "What is a gene?" From Mendel on, geneticists who delineated genes through breeding studies and pedigree analyses considered a gene to be a unit of information that affects phenotype; but the string of DNA in chromosome 3 does not resolve into 1000–2000 observable units. Detailed studies of mutations helped resolve this dilemma.

The rhodopsin gene determines perception of low-intensity light. People who carry the normal, wild-type allele of the gene see well in a dimly lit room and on the road at night. One simple change—a mutation—in the rhodopsin gene, however, diminishes light perception just enough to lead to night blindness; other alterations in the gene cause the destruction of rod cells, resulting in total blindness. Medical researchers have so far identified more than 30 mutations in the rhodopsin gene that affect vision. How can one gene sustain so many different mutations? And how can those mutations have such different phenotypic effects?

In this chapter, we describe how geneticists examined the relationship between mutations, genes, and phenotype as they tried to understand, at the molecular level, what genes are and how they function. The researchers considered **mutations** to be heritable changes in base sequence that affect phenotype. From ingenious experiments engineered to reveal how different mutations can occur in different parts of the same gene and how various mutations in the same gene can have different phenotypic effects, the researchers inferred that physically, a **gene** is usually a specific protein-encoding segment of DNA in a discrete region of a chromosome. (We now know that some genes encode various kinds of RNA that do not get translated into protein.) From these same studies, investigators concluded that a gene is not simply a bead on a string, changeable only as a whole and only in one way, as some had believed. Rather, genes are divisible, and each gene's subunits—the individual nucleotide pairs of DNA—can mutate independently and recombine with each other.

These findings provided fresh insight into gene function, clarifying that the nucleotide sequence of most genes specifies the amino acids that must be strung together to make a protein. The order of amino acids determines each protein's three-dimensional structure—how it folds in space—which, in turn, determines how the protein functions in the cell.

One general theme emerges from our discussion. Knowledge of what genes are and how they work deepens our understanding of Mendelian genetics by providing

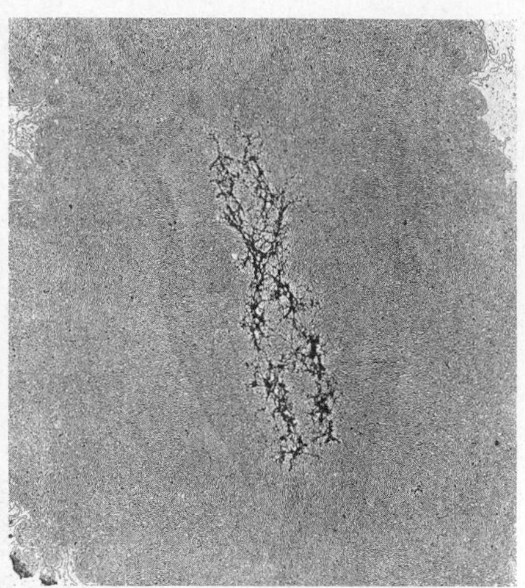

Figure 7.1 The DNA of each human chromosome contains hundreds to thousands of genes. The DNA of this human chromosome has been spread out and magnified 50,000×. Even at this magnification, there are no topological signs that reveal where along the DNA the genes reside. The darker, chromosome-shaped structure in the middle is a scaffold of proteins to which the DNA is attached.

a biochemical explanation for how genotype influences phenotype. One mutation in the rhodopsin gene, for example, causes the substitution of one particular amino acid for another in the construction of the rhodopsin protein, and this specific substitution changes the protein's ability to absorb photons and thus a person's ability to perceive light.

In our presentation of how genes govern the relationship between mutation and phenotype, we examine

- What mutations are; how often they occur; some of the events that cause them; DNA repair systems that minimize their occurrence; and how they affect the survival of an individual and the evolution of species.
- What mutations tell us about gene structure: A gene is a specific sequence of nucleotide pairs in a discrete region of DNA that acts as a unit of function, usually by encoding the instructions for making a particular polypeptide.
- What mutations tell us about gene function: Genes encode proteins by directing the correct assembly of the amino acids that compose each polypeptide; one or more polypeptides compose a protein; mutations that alter a gene's instructions for a sequence of amino acids alter protein structure and function.
- How gene mutations affect light-receiving proteins and vision: a comprehensive example.

7.1 Mutations: Primary Tools of Genetic Analysis

Mutations Are Heritable Changes in Base Sequences That Modify the Information Content of DNA

We saw in Chapter 3 that any allele existing at a frequency of more than 1% in a natural population under study is known as a **wild-type allele.** Many experimental geneticists, however, have a slightly different definition of a "wild-type allele," considering it to be the one allele that dictates the most commonly found phenotype in a population; with this definition, all other alleles—no matter what their frequency in the population—would be mutant alleles. Under either definition, a mutation that changes a wild-type allele of a gene to a different allele is called a **forward mutation.** The resulting novel mutant allele can be either recessive or dominant to the original wild type. Geneticists often diagram forward mutations as $A^+ \rightarrow a$ when the mutation is recessive and as $b^+ \rightarrow B$ when the mutation is dominant. Mutations can also cause a novel mutant allele to revert back to wild type ($a \rightarrow A^+$, or $B \rightarrow b^+$) in a process known as **reverse mutation,** or **reversion.** In this chapter, we designate wild-type alleles, whether recessive or dominant, with a "+".

Mendel originally defined genes by the visible phenotypic effects—yellow or green, round or wrinkled—of their alternative alleles. In fact, the only way he knew that genes existed at all was because alternative alleles for seven particular pea genes had arisen through forward mutations in the genetic material. Close to a century later, knowledge of DNA structure clarified that such mutations are heritable changes in DNA base sequence. DNA thus carries the potential for genetic change in the same place it carries genetic information—the sequence of its bases.

One Way Geneticists Classify Mutations Is by Their Effect on the DNA Molecule

A **substitution** occurs when a base at a certain position in one strand of the DNA molecule is replaced by one of the other three bases (**Fig. 7.2a**); during DNA replication, a base substitution in one strand will cause a new base pair to appear in the daughter molecule generated from that strand. Substitutions can be subdivided into *transitions,* in which one purine (A or G) replaces the other purine, or one pyrimidine (C or T) replaces the other, and *transversions,* in which a purine changes to a pyrimidine, or vice versa.

Other types of mutations produce more complicated rearrangements of DNA sequence. A **deletion** occurs when a block of one or more nucleotide pairs is lost from a DNA molecule; an **insertion** is just the reverse—the addition of one or more nucleotide pairs (Fig. 7.2b and c). Deletions and

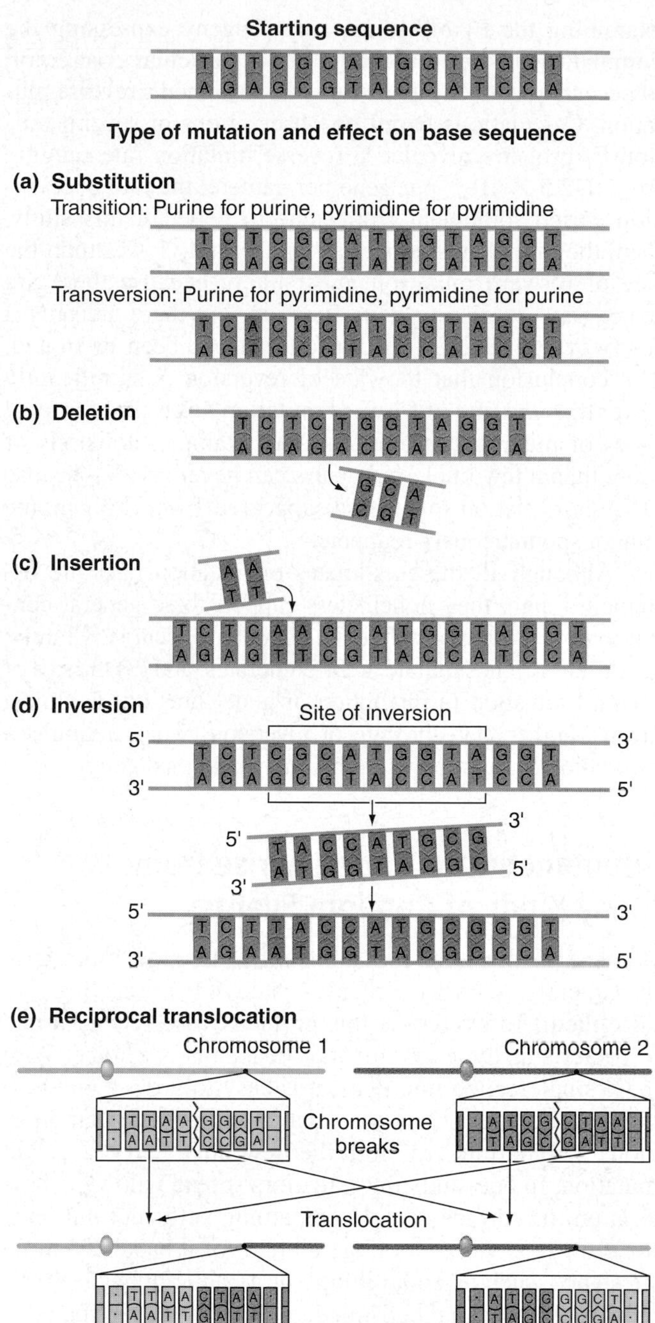

Figure 7.2 Mutations classified by their effect on DNA.

insertions can be as small as a single base pair or as large as megabases (that is, millions of base pairs). Researchers can see the larger changes under the microscope when they observe chromosomes in the context of a karyotype, such as that shown in Fig. 4.4 on p. 85. Even more complex mutations include **inversions,** 180° rotations of a segment of the DNA molecule (Fig. 7.2d), and **reciprocal translocations,** in which parts of two nonhomologous chromosomes change places (Fig. 7.2e). Large-scale DNA rearrangements, including megabase deletions and insertions as well as inversions and translocations, cause major genetic reorganizations that can change either the order of genes along a chromosome or the number of chromosomes in an organism. We discuss these **chromosomal rearrangements,** which affect many genes at a time, in Chapter 14. In this chapter, we focus on mutations that alter only one gene at a time.

Only a small fraction of the mutations in a genome actually alter the nucleotide sequences of genes in a way that affects gene function. By changing one allele to another, these mutations modify the structure or amount of a gene's protein product, and the modification in protein structure or amount influences phenotype. All other mutations either alter genes in a way that does not affect their function or change the DNA between genes. We discuss mutations without observable phenotypic consequences in Chapters 10 and 11; such mutations are very useful for mapping genes and tracking differences between individuals. In the remainder of this chapter, we focus on those mutations that have an impact on gene function and thereby influence phenotype.

Spontaneous Mutations Influencing Phenotypes Occur at a Very Low Rate

Mutations that modify gene function happen so infrequently that geneticists must examine a very large number of individuals from a formerly homogeneous population to detect the new phenotypes that reflect these mutations. In one ongoing study, dedicated investigators have monitored the coat colors of millions of specially bred mice and discovered that on average, a given gene mutates to a recessive allele in roughly 11 out of every 1 million gametes (**Fig. 7.3**). Studies of several other organisms have yielded similar results: an average spontaneous rate of $2–12 \times 10^{-6}$ mutations per gene per gamete.

Looking at the mutation rate from a different perspective, you could ask how many mutations there might be in the genes of an individual. To find out, you would simply multiply the rate of $2–12 \times 10^{-6}$ mutations per gene times 30,000, a generous current estimate of the number of genes in the human genome, to obtain an answer of between 0.06–0.36 mutations per haploid genome. This very rough calculation would mean that, on average, 1 new mutation affecting phenotype could arise in every 4–20 human gametes.

The Mutation Rate Varies from Gene to Gene

Although the average mutation rate per gene is $2–12 \times 10^{-6}$, this number masks considerable variation in the mutation rates for different genes. Experiments with many organisms show that mutation rates range from less than 10^{-9} to more than 10^{-3} per gene per gamete. Variation in the mutation rate of different genes within the same organism reflects differences in gene size (larger genes are larger targets that sustain more mutations) as well as differences in the susceptibility of particular genes to the various mechanisms that cause mutations (described later in this chapter).

(a)

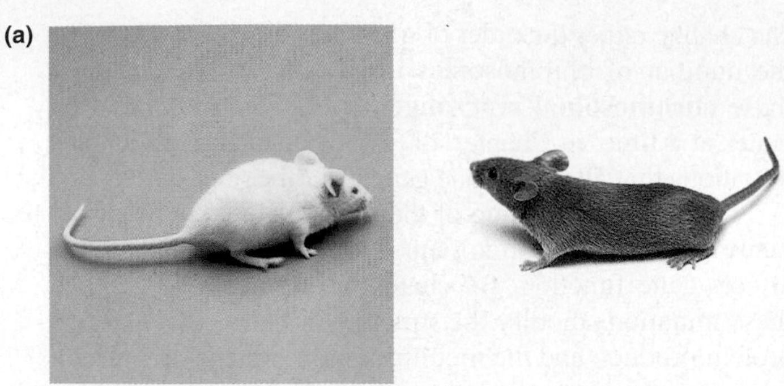

(b)

Locus[a]	Number of gametes tested	Number of mutations	Mutation rate ($\times 10^{-6}$)
a^- *(albino)*	67,395	3	44.5
b^- *(brown)*	919,699	3	3.3
c^- *(nonagouti)*	150,391	5	33.2
d^- *(dilute)*	839,447	10	11.9
ln^- *(leaden)*	243,444	4	16.4
	2,220,376	25	11.2 (average)

[a] Mutation is from wild type to the recessive allele shown.

Figure 7.3 Rates of spontaneous mutation. (a) Mutant mouse coat colors: albino (*left*), brown (*right*). **(b)** Mutation rates from wild type to recessive alleles for five coat color genes. Mice from highly inbred wild-type strains were mated with mice homozygous for recessive coat color alleles. Progeny with mutant coat colors indicated the presence of recessive mutations in gametes produced by the inbred mice.

Estimates of the average mutation rates in bacteria range from 10^{-8} to 10^{-7} mutations per gene per cell division. Although the units here are slightly different than those used for multicellular eukaryotes (because bacteria do not produce gametes), the average rate of mutation in gamete-producing eukaryotes still appears to be considerably higher than that in bacteria. The main reason is that numerous cell divisions take place between the formation of a zygote and meiosis, so mutations that appear in a gamete may have actually occurred many cell generations before the gamete formed. In other words, there are more chances for mutations to accumulate. Some scientists speculate that the diploid genomes of multicellular organisms allow them to tolerate relatively high rates of mutation in their gametes because a zygote would have to receive recessive mutations in the same gene from both gametes for any deleterious effects to occur. In contrast, bacteria would be affected by just a single mutation that disrupted its only copy of the gene.

Forward Mutations Usually Occur More Often Than Reverse Mutations

In the mouse coat color study, when researchers allowed brother and sister mice homozygous for a recessive mutant allele of one of the five mutant coat color genes to mate with each other, they could estimate the rate of reversion by examining the F_1 offspring. Any progeny expressing the dominant wild-type phenotype for a particular coat color, of necessity, carried a gene that had sustained a reverse mutation. Calculations based on observations of several million F_1 progeny revealed a reverse mutation rate ranging from 0–2.5 $\times$ 10^{-6} per gene per gamete; the rate of reversion varied somewhat from gene to gene. In this study, then, the rate of reversion was significantly lower than the rate of forward mutation, most likely because there are many ways to disrupt gene function, but there are only a few ways to restore function once it has been disrupted. The conclusion that the rate of reversion is significantly lower than the rate of forward mutation holds true for most types of mutation. In one extreme example, deletions of more than a few nucleotide pairs can never revert, because DNA information that has disappeared from the genome cannot spontaneously reappear.

Although all these estimates of mutation rates are extremely rough, they nonetheless support three general conclusions: (1) Mutations affecting phenotype occur very rarely; (2) different genes mutate at different rates; and (3) the rate of forward mutation (a disruption of gene function) is almost always higher than the rate of reversion (which requires a restoration of function to a previously disrupted gene).

Spontaneous Mutations Arise from Many Kinds of Random Events

Because spontaneous mutations affecting a gene occur so infrequently, it is very difficult to study the events that produce them. To overcome this problem, researchers turned to bacteria as the experimental organisms of choice. With these single-celled microbes, it is easy to grow many millions of individuals and then search rapidly through enormous populations to find the few that carry a novel mutation. In one study, investigators spread wild-type bacteria on the surface of agar containing sufficient nutrients for growth as well as a large amount of a bacteria-killing substance, such as an antibiotic or a bacteriophage. While most of the bacterial cells died, a few showed resistance to the bactericidal substance and continued to grow and divide. The descendants of a single resistant bacterium, produced by many rounds of binary fission, formed a mound of genetically identical cells called a **colony.**

The few bactericide-resistant colonies that appeared presented a puzzle. Had the cells in the colonies somehow altered their internal biochemistry to produce a life-saving response to the antibiotic or bacteriophage? Or did they carry heritable mutations conferring resistance to the bactericide? And if they did carry mutations, did those mutations arise by chance from random spontaneous events that take place continuously, even in the absence of a bactericidal substance, or did they only arise in response to environmental signals (in this case, the addition of the bactericide) that elicit a compensating change in genetic information?

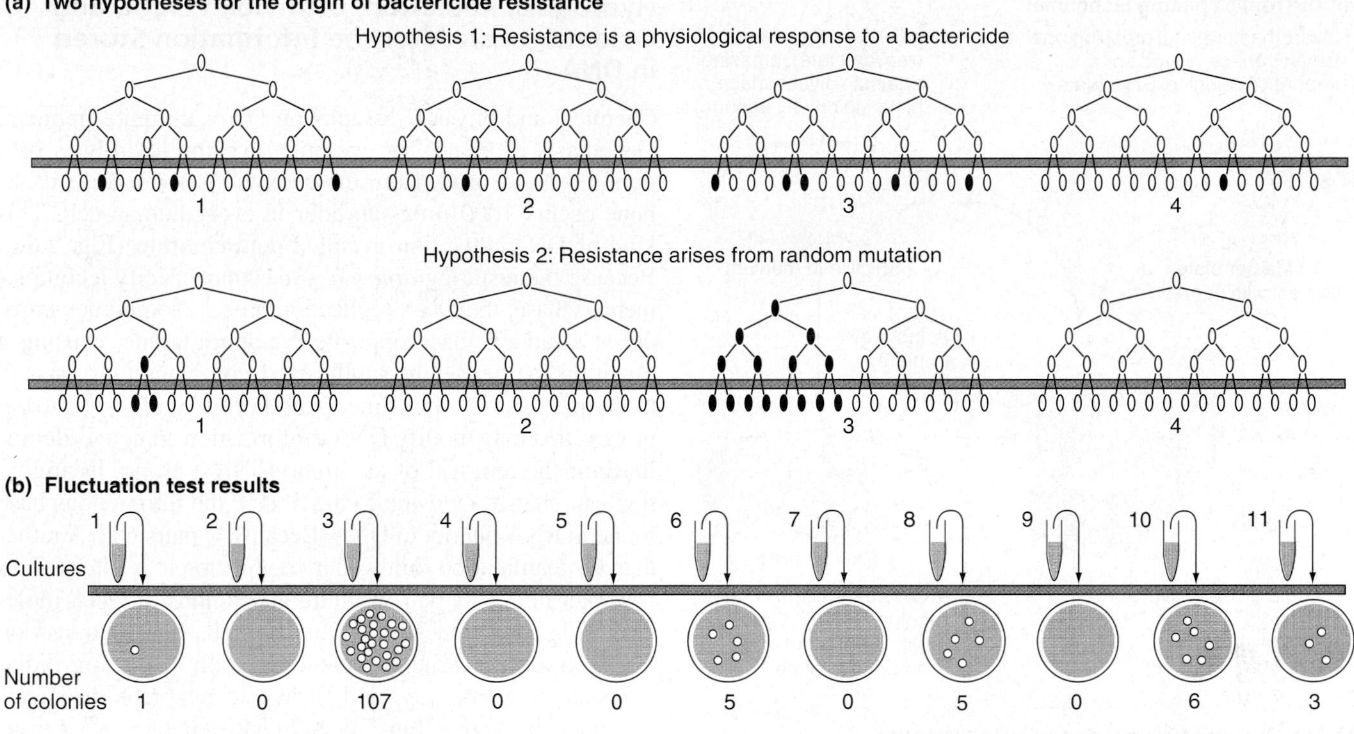

Figure 7.4 The Luria-Delbrück fluctuation experiment. **(a)** Hypothesis 1: If resistance arises only after exposure to a bactericide, all bacterial cultures of equal size should produce roughly the same number of resistant colonies. Hypothesis 2: If random mutations conferring resistance arise before exposure to bactericide, the number of resistant colonies in different cultures should vary (fluctuate) widely. **(b)** Actual results showing large fluctuations suggest that mutations in bacteria occur as spontaneous mistakes independent of exposure to a selective agent.

Mutations Are Chance Occurrences Modifying the Genome at Random

In 1943, Salvador Luria and Max Delbrück devised an experiment to examine the origin of bacterial resistance (**Fig. 7.4**). According to their reasoning, if bacteriophage-resistant colonies arise in direct response to infection by bacteriophages, separate suspensions of bacteria containing equal numbers of cells will generate similar, small numbers of resistant colonies when spread in separate petri plates on nutrient agar suffused with phages. By contrast, if resistance arises from mutations that occur spontaneously even when the phages are not present, different liquid cultures, when spread on separate petri plates, will generate very different numbers of resistant colonies. This is because the mutation conferring resistance can, in theory, arise at any time during the growth of the culture. If it happens early, the cell in which it occurs will produce many mutant progeny prior to petri plating; if it happens later, there will be far fewer mutant progeny when the time for plating arrives. After plating, these numerical differences will show up as fluctuations in the numbers of resistant colonies growing in the different petri plates.

The results of this **fluctuation test** were clear: Most plates supported zero to a few resistant colonies, but a few harbored hundreds of resistant colonies. From this observation of a substantial fluctuation in the number of resistant colonies in different petri plates, Luria and Delbrück concluded that bacterial resistance arises from mutations that exist before exposure to bacteriophage. After exposure, the bactericide in the petri plate becomes a selective agent that kills off nonresistant cells, allowing only the preexisting resistant ones to survive. **Figure 7.5** illustrates how researchers used another technique, known as *replica plating,* to demonstrate even more directly that the mutations conferring bacterial resistance occur before the cells "see" the bactericide that selects for their resistance.

These key experiments showed that bacterial resistance to phages and other bactericides is the result of mutations. And these mutations do not arise in particular genes as a directed response to environmental change; instead, they occur spontaneously as a result of random processes that can happen at any time and hit the genome at any place. Once such random changes occur, however, they usually remain stable. If the resistant mutants of the Luria-Delbrück experiment, for example, were grown for many generations in medium that did not contain bacteriophages, they would nevertheless remain resistant to this bactericidal virus.

We now describe some of the many kinds of random events that cause mutations; afterwards we discuss how cells cope with the damage.

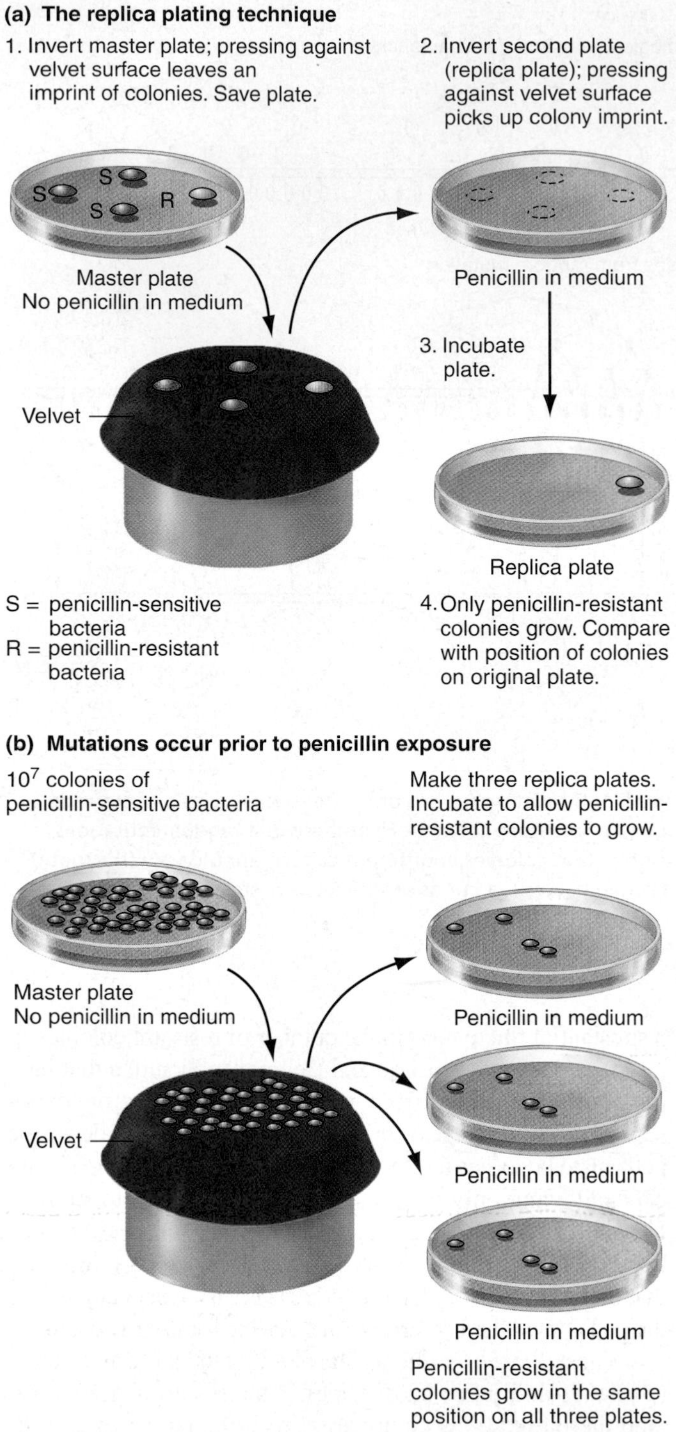

(a) The replica plating technique

1. Invert master plate; pressing against velvet surface leaves an imprint of colonies. Save plate.

2. Invert second plate (replica plate); pressing against velvet surface picks up colony imprint.

Master plate
No penicillin in medium

Penicillin in medium

3. Incubate plate.

Velvet

Replica plate

S = penicillin-sensitive bacteria
R = penicillin-resistant bacteria

4. Only penicillin-resistant colonies grow. Compare with position of colonies on original plate.

(b) Mutations occur prior to penicillin exposure

10^7 colonies of penicillin-sensitive bacteria

Make three replica plates. Incubate to allow penicillin-resistant colonies to grow.

Master plate
No penicillin in medium

Velvet

Penicillin in medium

Penicillin in medium

Penicillin in medium

Penicillin-resistant colonies grow in the same position on all three plates.

Figure 7.5 Replica plating verifies that bacterial resistance is the result of preexisting mutations. (a) Pressing a *master plate* onto a velvet surface transfers some cells from each bacterial colony onto the velvet. Pressing a *replica plate* onto the velvet then transfers some cells from each colony onto the replica plate. Investigators track which colonies on the master plate are able to grow on the replica plate (here, only penicillin-resistant ones). **(b)** A large number of colonies on a master plate without penicillin are sequentially transferred to three replica plates with penicillin. Resistant colonies grow in the same positions on all three replicas, showing that some colonies on the master plate had multiple resistant cells before exposure to the antibiotic.

Hydrolysis, Radiation, Ultraviolet Light, and Oxidation Can Alter the Information Stored in DNA

Chemical and physical assaults on DNA are quite frequent. Geneticists estimate, for example, that the hydrolysis of a purine base, A or G, from the deoxyribose-phosphate backbone occurs 1000 times an hour in every human cell. This kind of DNA alteration is called **depurination (Fig. 7.6a)**. Because the resulting *apurinic site* cannot specify a complementary base, the DNA replication process sometimes introduces a random base opposite the apurinic site, causing a mutation in the newly synthesized complementary strand three-quarters of the time. Another naturally occurring process that may modify DNA's information content is **deamination:** the removal of an amino ($-NH_2$) group. Deamination can change cytosine to uracil (U), the nitrogenous base found in RNA but not in DNA. Because U pairs with A rather than G, deamination followed by replication may alter a C–G base pair to a T–A pair in future generations of DNA molecules (Fig. 7.6b); such a C–G to T–A change is a transition mutation. Other assaults include naturally occurring radiation such as cosmic rays and X-rays, which break the sugar-phosphate backbone (Fig. 7.6c); ultraviolet light, which causes adjacent thymine residues to become chemically linked into thymine–thymine dimers (Fig. 7.6d); and oxidative damage to any of the four bases (Fig. 7.6e). All of these changes alter the information content of the DNA molecule.

Mistakes During DNA Replication Can Also Alter Genetic Information

If the cellular machinery for some reason incorporates the wrong base during replication, for instance, a C opposite an A instead of the expected T, then during the next replication cycle, one of the daughter DNAs will have the normal A–T base pair, while the other will have a mutant G–C. Careful measurements of the fidelity of replication *in vivo,* in both bacteria and human cells, show that such errors are exceedingly rare, occurring less than once in every 10^9 base pairs. That is equivalent to typing this entire book 1000 times while making only one typing error. Considering the complexities of helix unwinding, base pairing, and polymerization, this level of accuracy is amazing. How do cells achieve it?

The replication machinery minimizes errors through successive stages of correction. In the test tube, DNA polymerases replicate DNA with an error rate of about one mistake in every 10^6 bases copied. This rate is about 1000-fold worse than that achieved by the cell. Even so, it is impressively low and is only attained because polymerase molecules provide, along with their polymerization function, a proofreading/editing function in the form of a nuclease that is activated whenever the polymerase makes a mistake. This nuclease portion of the polymerase molecule, called the *3′-to-5′ exonuclease,* recognizes a mispaired base and excises it, allowing the polymerase to copy the nucleotide

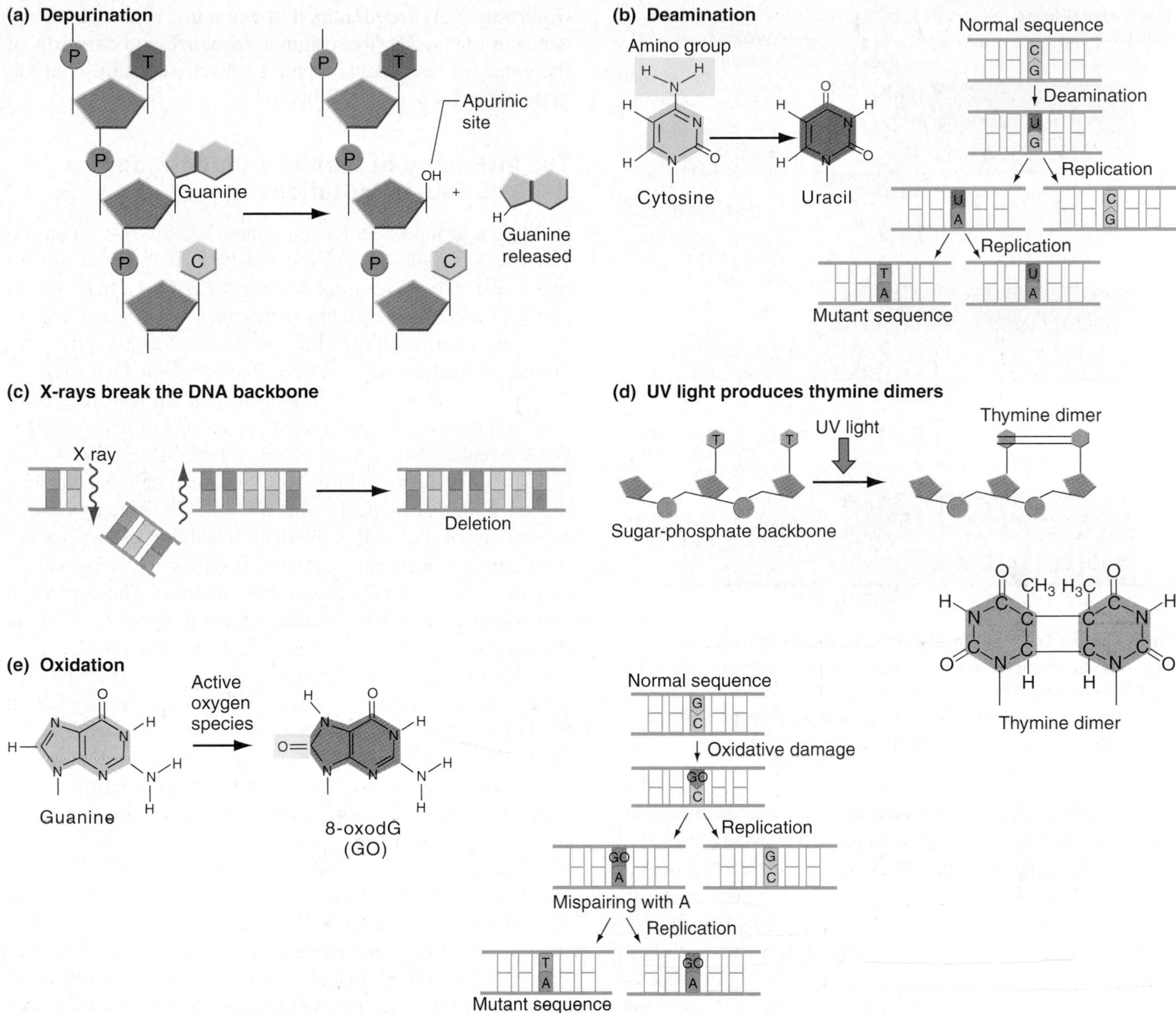

Figure 7.6 How natural processes can change the information stored in DNA. (a) In depurination, the hydrolysis of A or G bases leaves a DNA strand with an unspecified base. **(b)** In deamination, the removal of an amino group from C initiates a process that eventually (after DNA replication) causes a transition. **(c)** X-rays break the sugar-phosphate backbone and thereby split a DNA molecule into smaller pieces, which may be spliced back together improperly. **(d)** Ultraviolet (UV) radiation causes adjacent Ts to form dimers, which can disrupt the readout of genetic information. **(e)** Irradiation causes the formation of *free radicals* (such as oxygen molecules with an unpaired electron) that can damage individual bases. Here, the pairing of the altered base GO with A creates a transversion that changes a G–C base pair to T–A.

correctly on the next try (**Fig. 7.7**). Without its nuclease portion, DNA polymerase would have an error rate of one mistake in every 10^4 bases copied, so its editing function improves the fidelity of replication 100-fold. DNA polymerase *in vivo* is part of a replication system including many other proteins that collectively improve on the error rate another 10-fold, bringing it to within about 100-fold of the fidelity attained by the cell.

The 100-fold higher accuracy of the cell depends on a backup system called *methyl-directed mismatch repair* that notices and corrects residual errors in the newly replicated DNA. We present the details of this repair system later in the chapter when we describe the various ways in which cells attempt to correct mutations once they occur.

Unequal Crossing-Over and Transposons Can Rearrange DNA

Some mutations arise from events other than chemical and physical assaults or replication errors. Significant among the

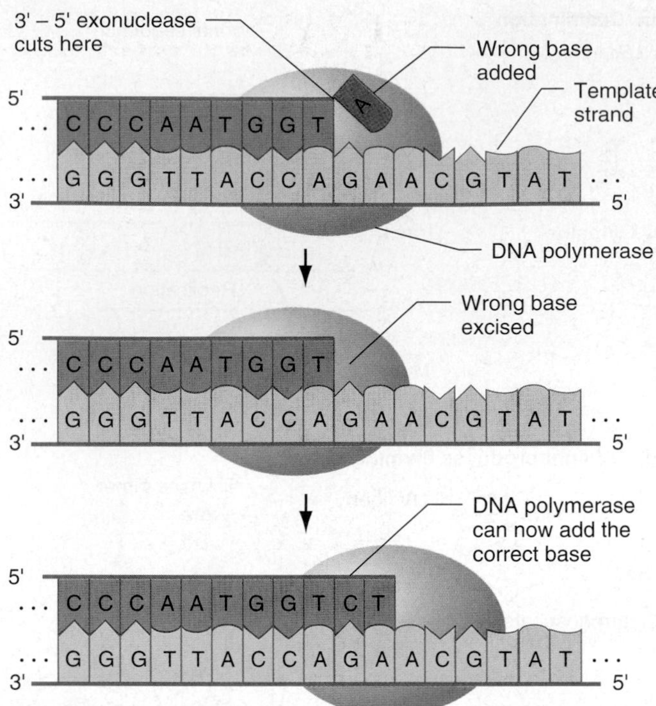

Figure 7.7 DNA polymerase's proofreading function. If DNA polymerase mistakenly adds an incorrect nucleotide at the 3′-end of the strand it is synthesizing, the enzyme's 3′-to-5′ exonuclease activity removes this nucleotide, giving the enzyme a second chance to add the correct nucleotide.

mechanisms giving rise to such mutations is erroneous recombination. For example, in **unequal crossing-over,** two closely related DNA sequences located in different places on two homologous chromosomes can pair with each other during meiosis. If recombination takes place between the mispaired sequences, one homologous chromosome ends up with a duplication (a kind of insertion), while the other homolog sustains a deletion. As **Fig. 7.8a** shows, some forms of red-green colorblindness arise from deletions and duplications in the genes that enable us to perceive red and green wavelengths of light; these reciprocal informational changes are the result of unequal crossing-over.

Another notable mechanism for altering DNA sequence involves the units of DNA known as **transposable elements (TEs).** TEs are DNA segments several hundred to several thousand base pairs long that move (or "transpose" or "jump") from place to place in the genome. If a TE jumps into a gene, it can disrupt the gene's function and cause a mutation. Certain TEs frequently insert themselves into particular genes and not others; this is one reason that mutation rates vary from gene to gene. Although some TEs move by making a new copy that can insert into a different chromosomal location while the initial copy stays put, other TE types actually leave their original position when they move (Fig. 7.8b). Mutations caused by TEs that transpose by this second mechanism are exceptions to the general rule that the rate of reversion is lower than the rate of forward mutation. This is because TE transposition can occur relatively frequently, and when it is accompanied by excision of the TE, the original sequence and function of the gene are restored. Chapter 14 discusses additional genetic consequences of TE behavior.

The Instability of Certain Trinucleotide Repeats Causes Mutations

In 1992, a group of molecular geneticists discovered an unusual and completely unexpected type of mutation in humans: the excessive amplification of a CGG base triplet normally repeated only a few to 50 times in succession. If, for example, a normal allele of a gene carries 5 consecutive repetitions of the base triplet CGG (that is, CGGCGGCGGCGGCGG on one strand), an abnormal allele resulting from mutation could carry 200 repeats in a row. Further investigations revealed that repeats of several trinucleotides—CAG, CTG, and GAA, in addition to CGG—can be unstable such that the number of repeats often increases or decreases in different cells of the same individual. Instability can also occur during the production of gametes, resulting in changes in repeat number from one generation to the next. The expansion and contraction of trinucleotide repeats has now been found not only in humans but in many other species as well.

The rules governing trinucleotide repeat instability appear to be quite complicated, but one general feature is that the larger the number of repeats at a particular location, the higher the probability that expansion and contraction will occur. Usually, tracts with less than 30–50 repetitions of a triplet change in size only infrequently, and the mutations that do occur cause only small variations in the repeat number. Larger tracts involving hundreds of repeats change in size more frequently, and they result in more variation in the number of repetitions. Researchers have not yet determined the precise mechanism of triplet repeat amplification. One possibility is that regions with long trinucleotide repeats form unusual DNA structures that are hard to replicate because they force the copying machinery to slip off, then hop back on, slip off, then hop back on when it gets to this sequence. Such stopping and starting may produce a replication "stutter" that causes synthesis of the same triplet repeat over and over again, expanding the number of copies. This type of error could conversely shrink the size of the trinucleotide repeat tract if, after slipping off, the replication machinery restarts copying at a repeat further down the template sequence. Whatever the mechanism, mutations of long trinucleotide stretches occur quite often, suggesting that the enzymes for excision or mismatch repair are not very efficient at repairing them.

The expansion of trinucleotide repeats is at the root of *fragile X syndrome,* one of the most common forms of human mental retardation, as well as Huntington disease and many other disorders of the nervous system. The Genetics and Society box "Amplified Trinucleotide Repeats May Have Medical Consequences" on pp. 216–217, discusses the fascinating implications of this type of DNA-sequence instability for human genetics and medicine.

Mutagens Induce Mutations

Mutations make genetic analysis possible, but most mutations appear spontaneously at such a low rate that researchers have looked for controlled ways to increase their occurrence. H. J. Muller, an original member of Thomas Hunt Morgan's *Drosophila* group, first showed that exposure to a dose of X-rays higher than the naturally occurring level increases the mutation rate in fruit flies (**Fig. 7.9**). Muller exposed male *Drosophila* to increasingly large

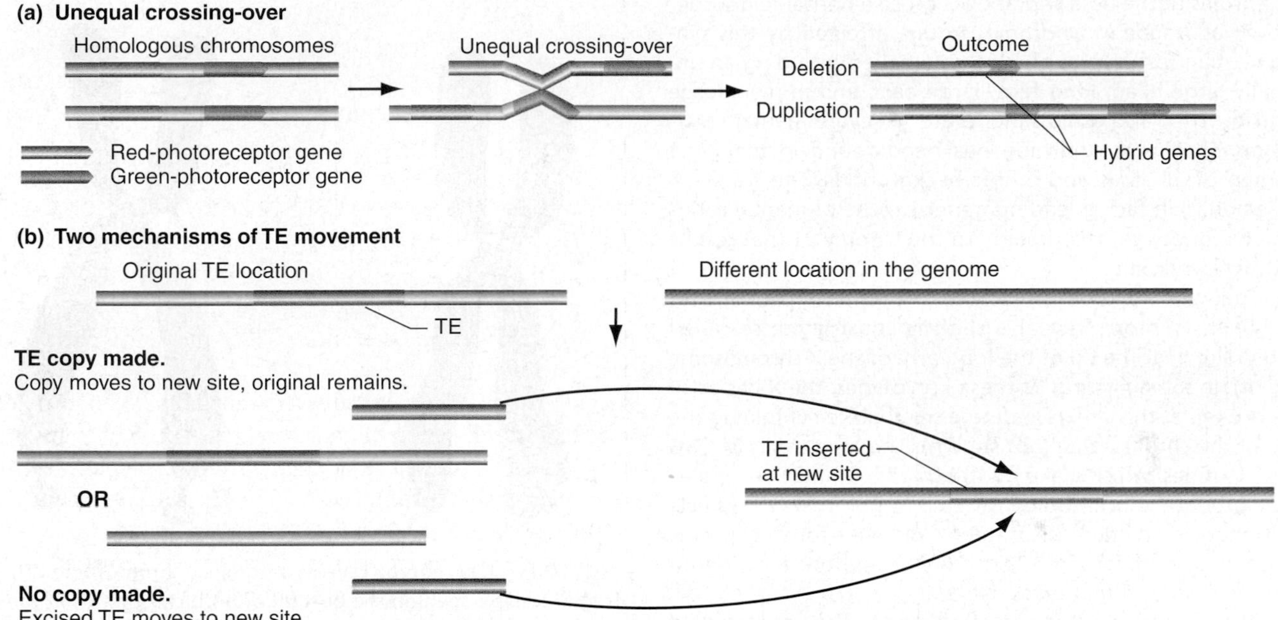

Figure 7.8 How unequal crossing-over and the movement of transposable elements (TEs) change DNA's information content. **(a)** If two nearby regions contain a similar DNA sequence, the two homologous chromosomes may pair out of register during meiosis and produce gametes with either a deletion or a reciprocal duplication of the intervening region. Colorblindness in humans can result from unequal crossing-over between the nearby and highly similar genes for red and green photoreceptors. **(b)** TEs move around the genome. Some TEs copy themselves before moving, while others are excised from their original positions during the transposition process. Insertion of a TE into a gene often has phenotypic consequences.

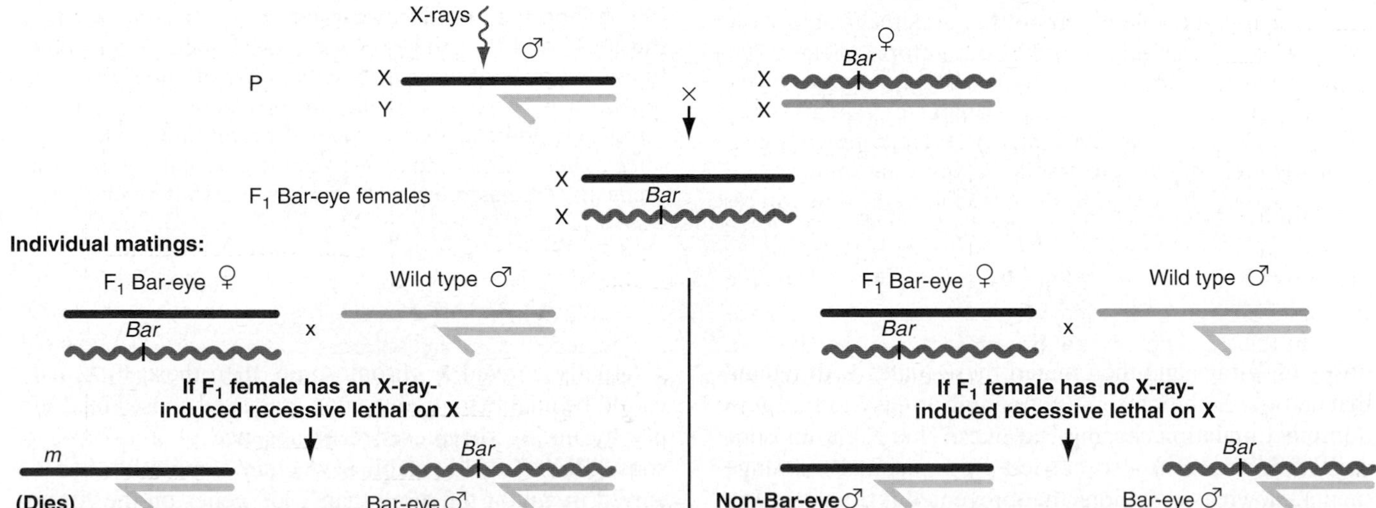

Figure 7.9 Exposure to X-rays increases the mutation rate in *Drosophila*. F_1 females are constructed that have an irradiated paternal X chromosome (*red line*), and a *Bar*-marked "balancer" maternal X chromosome (*wavy blue line*). These chromosomes cannot recombine because the balancer chromosome has multiple inversions (as explained in Chapter 14). Single F_1 females, each representing a single X-ray exposed X chromosome from the father, are then individually mated with wild type males. If the paternal X chromosome in any one F_1 female has an X-ray-induced recessive lethal mutation (*m*), she can produce only Bar-eyed sons (*left*). If the X chromosome has no such mutation, this F_1 female will produce both Bar-eyed and non-Bar-eyed sons (*right*).

GENETICS AND SOCIETY

Amplified Trinucleotide Repeats May Have Medical Consequences

Expansions of the base triplet CGG cause a heritable disorder known as *fragile X syndrome.* Adults affected by this syndrome manifest several physical anomalies, including an unusually large head, long face, large ears, and in men, large testicles. They also exhibit moderate to severe mental retardation. Fragile X syndrome has been found in men and women of all races and ethnic backgrounds. The fragile X mutation is, in fact, a leading genetic cause of mental retardation worldwide, second only to the trisomy 21 that results in Down syndrome.

Specially prepared karyotypes of cells from people with fragile X symptoms reveal a slightly constricted, so-called fragile site near the tip of the long arm of the X chromosome (**Fig. A**). In some pictures of these karyotypes, the X has actually broken at this site, releasing a small piece containing the end of the chromosome (not shown). The long tracts of CGG trinucleotides, which make up the fragile X mutation, apparently produce a localized constricted region that breaks easily. Geneticists named the fragile X disorder for this specific pinpoint of fragility more than 20 years before they identified the mutation that gives rise to it.

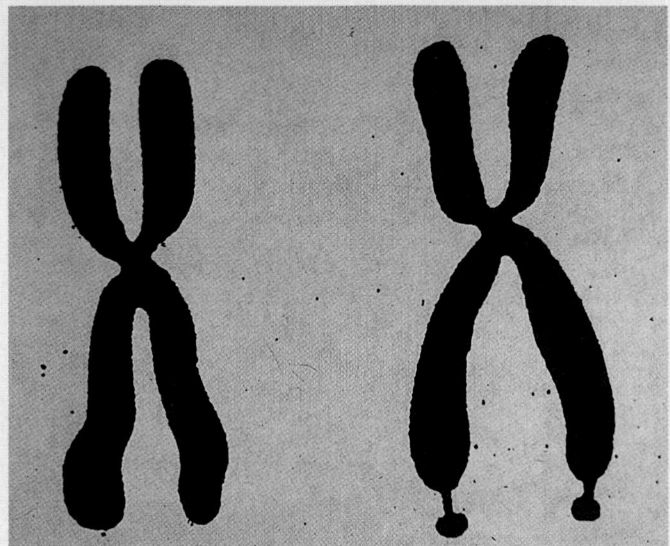

Figure A A karyotype reveals a fragile X chromosome. The fragile X site is seen on the bottom of both chromatids of the X chromosome at the *right*.

The gene in which the fragile X mutation occurs is called *FMR-1* (for fragile-X-associated mental retardation). Near one end of the gene, different people carry a different number of repeats of the sequence CGG, and geneticists now have the molecular tools to quantify these differences. Normal alleles contain 5–54 of these triplet repeats, while the *FMR-1* gene in people with fragile X syndrome contains 200–4000 repeats of the exact same triplet (**Fig. B.1**). The rest of the gene's base sequence is the same in both normal and abnormal alleles.

The triplet repeat mutation that underlies fragile X syndrome has a surprising transmission feature. Alleles with a full-blown mutation are foreshadowed by *premutation alleles* that carry an intermediate number of repeats—more than 50 but fewer than 200 (Fig. B.1). Premutation alleles do not themselves generate fragile X symptoms in most carriers, but they show significant instability and thus forecast the risk of genetic disease in a carrier's progeny. The greater the number of repeats in a premutation allele, the higher the risk of disease in that person's children. For example, if a woman carries a premutation allele with 60 CGG repeats, 17% of her offspring run the risk of exhibiting fragile X syndrome. If she carries a premutation allele with 90 repeats, close to 50% of her offspring will show symptoms. In short, the change that produces a premutation allele increases the likelihood that the *FMR-1* gene will incur more mutations; and the larger the original number of additional CGG repeats, the larger the number of subsequent additions. Interestingly, the expansion of *FMR-1* premutation alleles has some as-yet-unexplained relation to the parental origin of the repeats. Whereas most male carriers transmit their *FMR-1*

doses of X-rays and then mated these males with females that had one X chromosome containing an easy-to-recognize dominant mutation causing Bar eyes. This X chromosome (called a *balancer*) also carried chromosomal rearrangements known as inversions that prevented it from crossing-over with other X chromosomes. (Chapter 14 explains the details of this phenomenon.) Some of the F_1 daughters of this mating were heterozygotes carrying a mutagenized X from their father and a *Bar*-marked X from their mother. If X-rays induced a recessive lethal mutation anywhere on the paternally derived X chromosome, then these F_1 females would be unable to produce non-Bar-eyed sons. Thus, simply by noting the presence or absence of non-Bar-eyed sons, Muller could establish whether a mutation had occurred in any of the more than 1000 genes on the X chromosome that are essential to *Drosophila* viability. He concluded that the greater the X-ray dose, the greater the frequency of recessive lethal mutations.

Any physical or chemical agent that raises the frequency of mutations above the spontaneous rate is called a

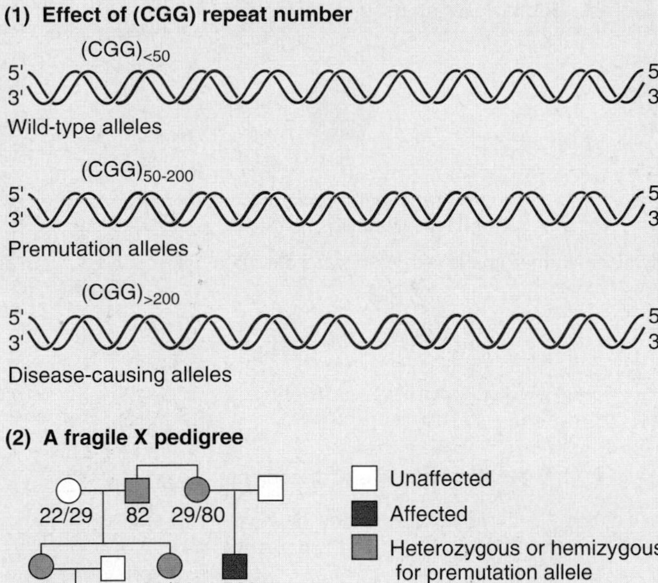

Figure B Amplification of CGG triplet repeats correlates with the fragile X syndrome. **(1)** *FMR-1* genes in unaffected people generally have fewer than 50 CGG repeats. Unstable premutation alleles have between 50 and 200 repeats. Disease-causing alleles have more than 200 CGG repeats, and some have up to 4000. **(2)** A fragile X pedigree showing the number of CGG repeats in different individuals. Fragile X syndrome patients are almost always the progeny of mothers who carried premutation alleles.

allele with only a small change in the number of repeats, many women with premutation alleles bear children with 250–4000 CGG repeats in their *FMR-1* gene (Fig. B.2). One possible explanation is that whatever conditions generate fragile X mutations occur most readily during oogenesis.

This same transmission pattern does not hold for all triplet repeat mutations. In 1993, geneticists found that the mutations causing Huntington disease consist of an expanded CAG trinucleotide repeat within the Huntington chorea gene; but in this case, the amplification occurs most often when the gene is transmitted by the father. The mutations resulting in Huntington disease may thus occur during spermatogenesis. Several other genetic disorders of the nervous system also arise from fluctuating numbers of triplet repeats. Why these repeats are so unstable and prone to expansion (and occasionally shrinkage) is an area of active research.

The CGG trinucleotide repeat expansion underlying fragile X syndrome has interesting implications for genetic counseling. There are thousands of possible alleles of the *FMR-1* gene, ranging from the smallest normal allele isolated to date, with 5 triplet repeats, to the largest abnormal allele so far isolated, with roughly 4000 repeats. This enormous number of alleles gives rise to a slow gradation of normal-to-abnormal phenotypes, with the abnormal phenotypes progressing from mild retardation without other evidence of disease to severe retardation combined with physical anomalies. The relation between genotype and phenotype is clear at both ends of the triplet-repeat spectrum: Individuals whose alleles contain less than 55 repeats are normal, while people with an allele carrying more than 200 repeats are almost always moderately to severely retarded. With an intermediate number of repeats, however, expression of the mental retardation phenotype is highly variable, depending to an unknown degree on chance, the environment, and modifier genes.

This range of variable expressivity leads to an ethical dilemma: Where should medical geneticists draw the line in their assessment of risk? Prospective parents with a history of mental retardation on either side of the family may consult with a counselor to determine their options. The counselor would first test the parents for premutation alleles to establish the possible involvement of fragile X syndrome. He or she would also want to analyze the fetal cells directly by amniocentesis, to determine whether the fetus carries an expanded number of CGG repeats in its *FMR-1* gene. If the results indicate the presence of an allele in the middle range of triplet repeats, the counselor will have to acknowledge that this is a grey area in which prediction can only be imperfect. The prospective parents' difficult decision of whether or not to continue the pregnancy will then rest on the very shaky ground of an inconclusive, overall evaluation of risk.

mutagen. Researchers use many different mutagens to produce mutations for study. With the Watson-Crick model of DNA structure as a guide, they can understand the action of most mutagens at the molecular level. The X-rays used by Muller to induce mutations on the X chromosome, for example, can break the sugar-phosphate backbones of DNA strands, sometimes at the same position on the two strands of the double helix. Multiple double-strand breaks produce DNA fragmentation, and the improper stitching back together of the fragments can cause inversions, deletions, or other rearrangements (see Fig. 7.6c). Another example of a molecular mechanism of mutagenesis involves mutagens known as **base analogs,** which are so similar in chemical structure to the normal nitrogenous bases that the replication machinery can incorporate them into DNA in place of the normal bases (**Fig. 7.10a**). Since a base analog may have pairing properties different from those of the base it replaces, it can cause base substitutions on the complementary strand synthesized in the next round of DNA replication. Other chemical mutagens generate substitutions by

Type of Mutagen	Chemical Action of Mutagen

(a) Replace a base: Base analogs have a chemical structure almost identical to that of a DNA base.

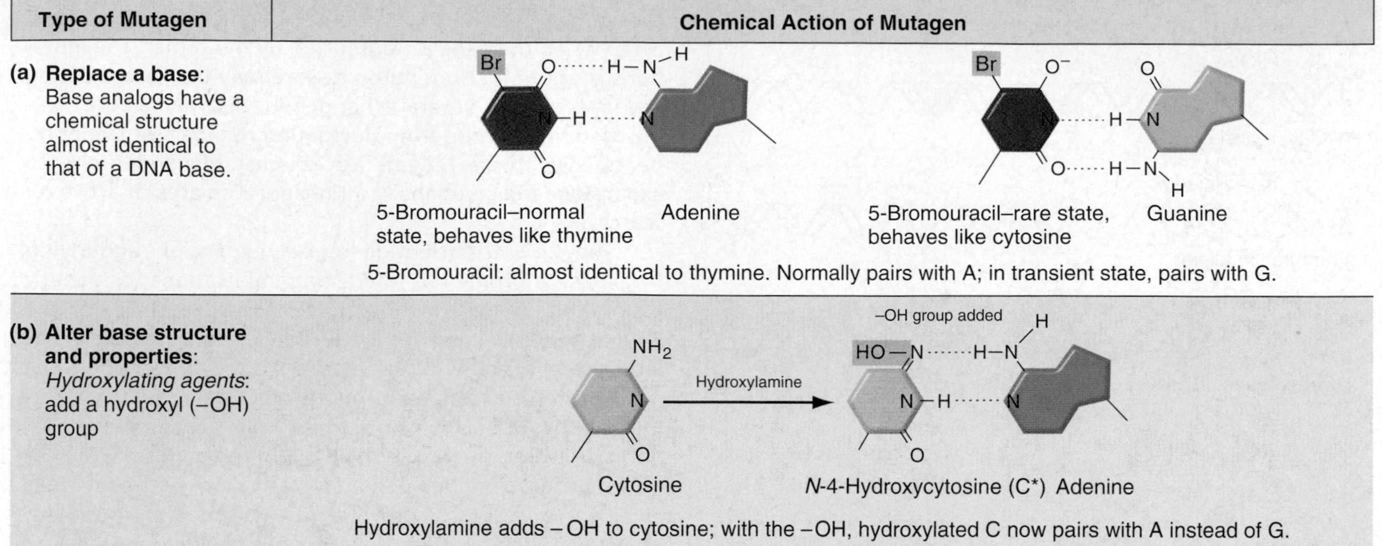

5-Bromouracil: almost identical to thymine. Normally pairs with A; in transient state, pairs with G.

(b) Alter base structure and properties: *Hydroxylating agents*: add a hydroxyl (–OH) group

NH2
N
O
Cytosine
Hydroxylamine
–OH group added
HO—N
H—N
H
N—H
N
O
N-4-Hydroxycytosine (C*) Adenine

Hydroxylamine adds – OH to cytosine; with the – OH, hydroxylated C now pairs with A instead of G.

Alkylating agents: add ethyl ($-CH_2-CH_3$) or methyl ($-CH_3$) groups

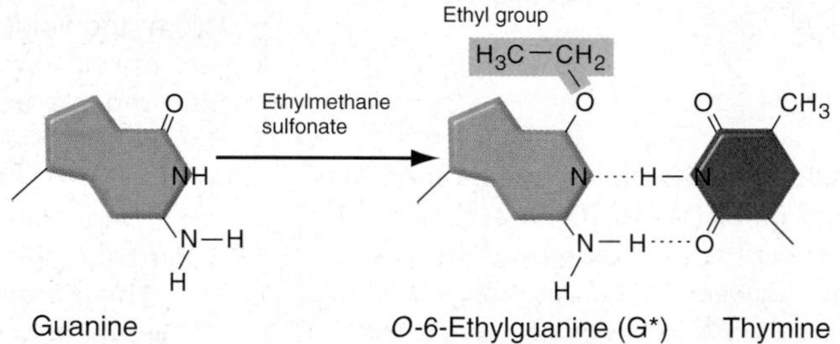

Ethylmethane sulfonate adds an ethyl group to guanine or thymine. Modified G pairs with T above, and modified T pairs with G (not shown).

Deaminating agents: remove amine ($-NH_2$) groups

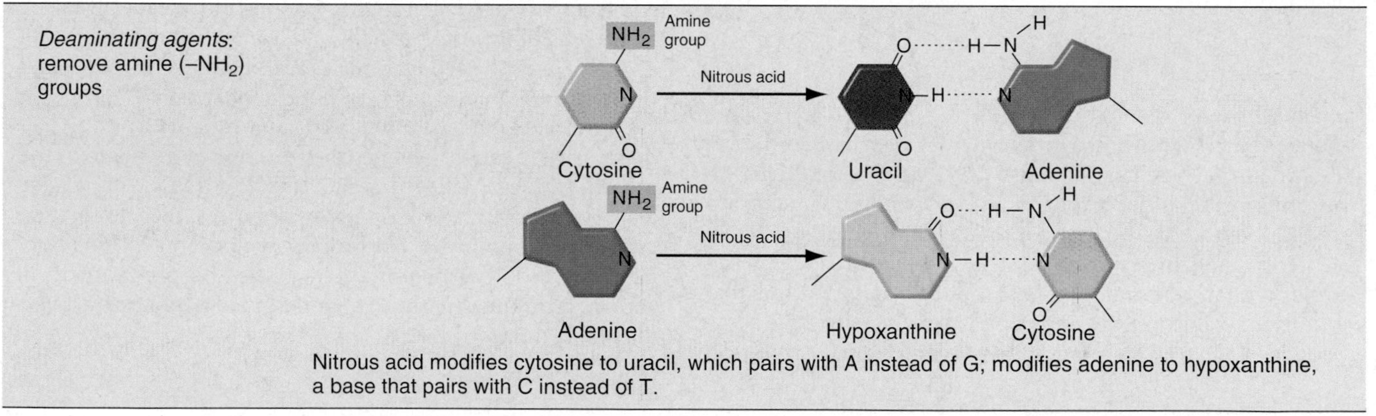

Nitrous acid modifies cytosine to uracil, which pairs with A instead of G; modifies adenine to hypoxanthine, a base that pairs with C instead of T.

(c) Insert between bases: Intercalating agents

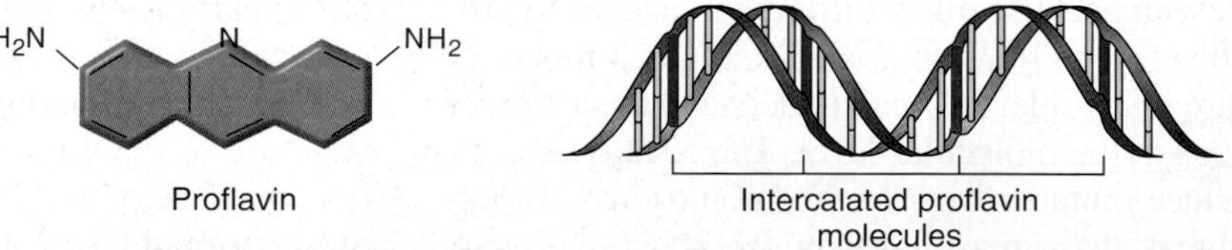

Proflavin intercalates into the double helix. This disrupts DNA metabolism, eventually resulting in deletion or addition of a base pair.

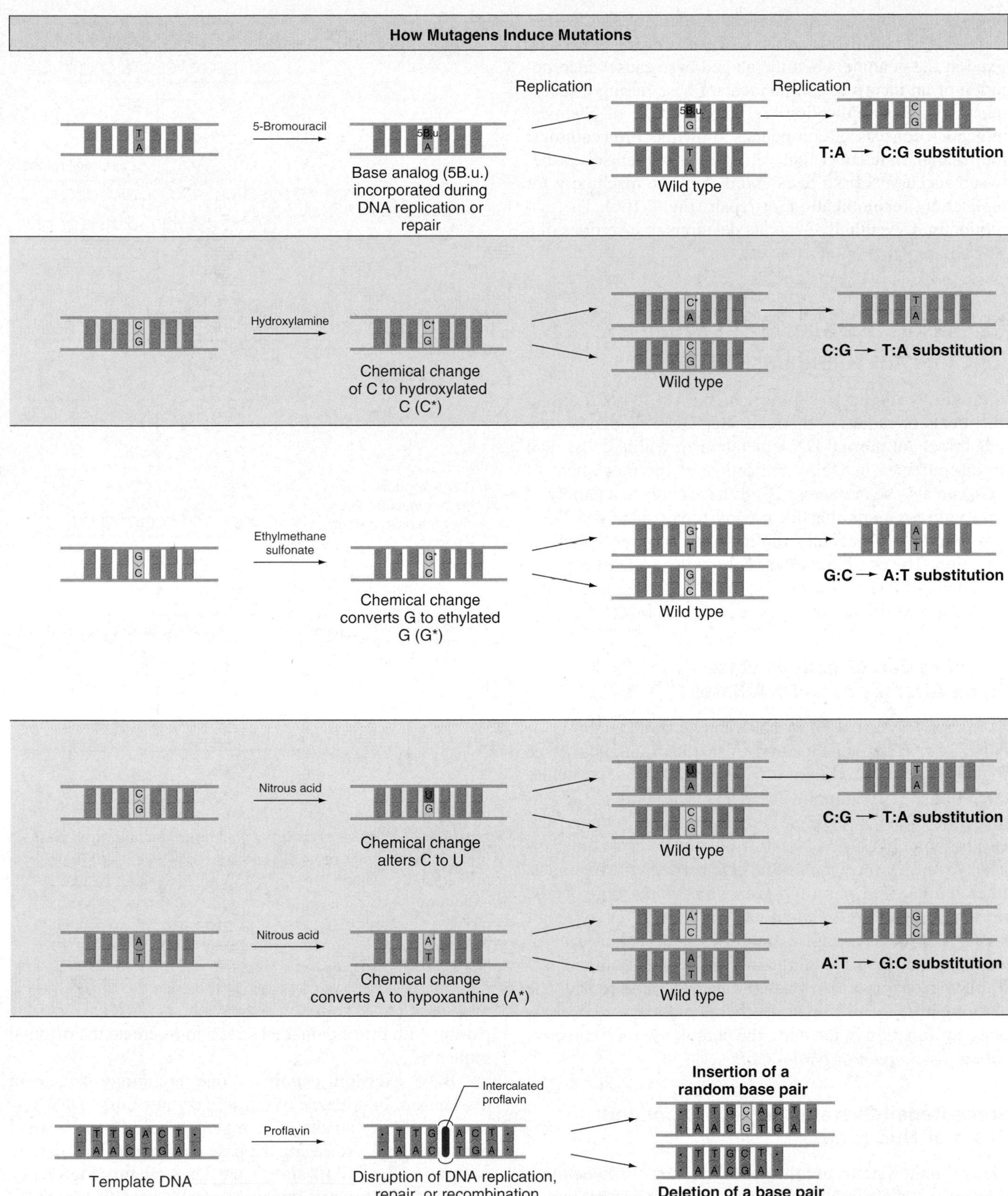

Figure 7.10 How mutagens alter DNA. **(a)** Base analogs incorporated into DNA may pair aberrantly, which allows the addition of incorrect nucleotides to the opposite strand during replication. **(b)** Some mutagens alter the structure of bases such that they pair inappropriately in the next round of replication. **(c)** Intercalating agents are roughly the same size and shape as a base pair of the double helix. Their incorporation into DNA produces insertions or deletions of single base pairs by unknown mechanisms.

directly altering a base's chemical structure and properties (Fig. 7.10b). Again, the effects of these changes become fixed in the genome when the altered base causes incorporation of an incorrect complementary base during a subsequent round of replication. Yet another class of chemical mutagens consists of compounds known as **intercalators:** flat, planar molecules that can sandwich themselves between successive base pairs and disrupt the machinery for replication, recombination, or repair (Fig. 7.10c). The disruption may eventually generate deletions or insertions of a single base pair.

Organisms Use Many DNA Repair Mechanisms to Minimize Mutations

Natural environments expose genomes to many kinds of chemicals or radiation that can alter DNA sequences; the side effects of normal DNA metabolism within cells, such as inaccuracies in DNA replication or the movement of TEs, can also be mutagenic. Cells have evolved a variety of enzymatic systems that locate and repair damaged DNA and thereby dramatically diminish the high potential for mutation. The combination of these repair systems must be extremely efficient, because the rates of spontaneous mutation observed for almost all genes are very low.

Enzymes Can Directly Reverse Some Alterations to DNA Bases

Alkyltransferase enzymes can remove added methyl or ethyl groups from guanine to recreate the original base (see Fig. 7.10b). Other enzymes remedy other base structure alterations. For example, the enzyme *photolyase* recognizes the thymine–thymine dimers produced by exposure to ultraviolet light (review Fig. 7.6d) and reverses the damage by splitting the chemical linkage between the thymines. Interestingly, the photolyase enzyme works only in the presence of visible light. In carrying out its DNA repair tasks, it associates with a small molecule called a *chromophore* that absorbs light in the visible range of the spectrum; the enzyme then uses the energy captured by the chromophore to split thymine–thymine dimers. Because it does not function in the dark, the photolyase mechanism is called *light repair,* or *photorepair.*

Some Repair Systems Remove Damaged Bases or Nucleotides

Many repair systems use the general strategy of *homology-dependent repair* in which they first remove a small region from the DNA strand that contains the altered nucleotide and then use the other strand as a template to resynthesize the removed region. This strategy makes use of one of the great advantages of the double-helical structure: If one strand sustains damage, cells can use complementary base

1. Deaminated DNA with uracil

Uracil released

2. Glycosylase removes uracil, leaving an AP site.

3. AP endonuclease cuts backbone to make a nick at the AP site.

4. DNA exonucleases remove nucleotides near the nick, creating a gap.

5. DNA polymerase synthesizes new DNA to fill in the gap.

6. DNA ligase seals the nick.

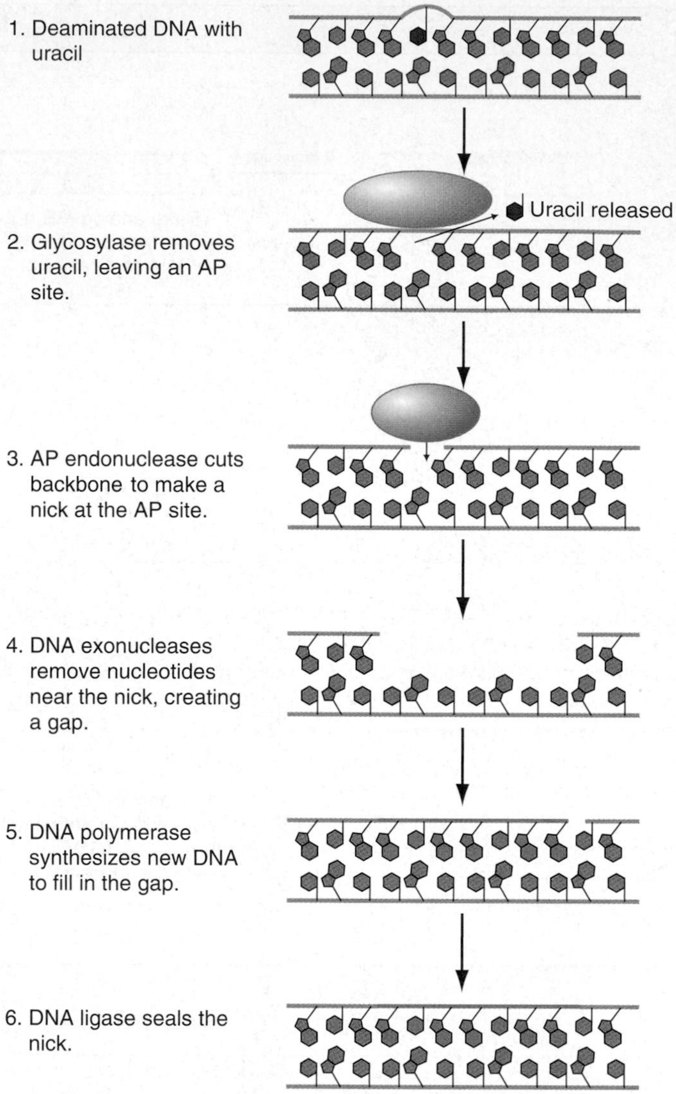

Figure 7.11 Base excision repair removes damaged bases. Glycosylase enzymes remove aberrant bases [like uracil (*red*), formed by the deamination of cytosine], leaving an AP site. AP endonuclease cuts the sugar-phosphate backbone, creating a nick. Exonucleases extend the nick into a gap, which is filled in with the correct information (*green*) by DNA polymerase. DNA ligase reseals the corrected strand.

pairing with the undamaged strand to re-create the original sequence.

Base excision repair is one homology-dependent mechanism. In this type of repair, enzymes called *DNA glycosylases* cleave an altered nitrogenous base from the sugar of its nucleotide, releasing the base and creating an apurinic or apyrimidinic (AP) site in the DNA chain (**Fig. 7.11**). Different glycosylase enzymes cleave specific damaged bases. Base excision repair is particularly important in the removal of uracil from DNA (recall that uracil often results from the natural deamination of cytosine; review Fig. 7.6b). In this repair process, after the enzyme *uracil-DNA glycosylase* has removed uracil from its sugar, leaving an AP site, the

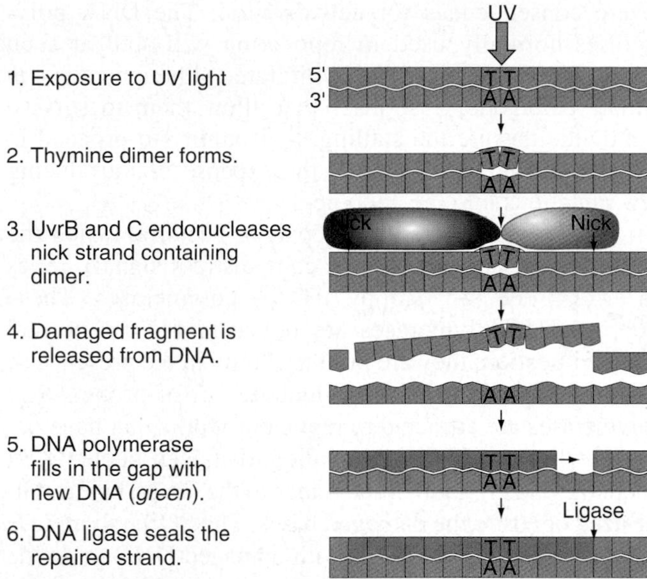
UV
1. Exposure to UV light
5'
3'
T T
A A
3'
5'
2. Thymine dimer forms.
3. UvrB and C endonucleases nick strand containing dimer.
Nick
Nick
4. Damaged fragment is released from DNA.
5. DNA polymerase fills in the gap with new DNA (green).
Ligase
6. DNA ligase seals the repaired strand.

Figure 7.12 Nucleotide excision repair corrects damaged nucleotides. A complex of the UvrA and UvrB proteins (*not shown*) scans DNA for distortions in the double helix caused by DNA damage, such as thymine-thymine dimers. Once a damaged site is found, UvrA dissociates from UvrB, allowing UvrB (*red*) to associate with UvrC (*blue*) at the site of the lesion. These enzymes nick the DNA exactly 4 nucleotides to one side of the damage and 7 nucleotides to the other side, releasing a small fragment of single-stranded DNA and thus leaving a gap in the double helix. DNA polymerases then resynthesize the missing information (*green*), and DNA ligase reseals the now-corrected strand.

enzyme *AP endonuclease* makes a nick in the DNA backbone at the AP site. Other enzymes (known as *DNA exonucleases*) attack the nick and remove nucleotides from its vicinity to create a gap in the previously damaged strand. DNA polymerase fills in the gap by copying the undamaged strand, restoring the original nucleotide in the process. Finally, DNA ligase seals up the backbone of the newly repaired DNA strand.

Nucleotide excision repair (**Fig. 7.12**) removes alterations that base excision cannot repair because there is no DNA glycosylase that recognizes the problem base. Nucleotide excision repair depends on enzyme complexes containing more than one protein molecule. In *E. coli,* these complexes are made of two out of three possible proteins: UvrA, UvrB, and UvrC. One of the complexes (UvrA + UvrB) patrols the DNA for irregularities, detecting lesions that disrupt Watson-Crick base pairing and thus distort the double helix (such as thymine–thymine dimers that have not been corrected by photo repair). A second complex (UvrB + UvrC) cuts the damaged strand in two places that flank the damage. This double-cutting excises a short region of the damaged strand and leaves a gap that will be filled in by DNA polymerase and sealed with DNA ligase.

Methyl-Directed Mismatch Repair Corrects Errors in DNA Replication

DNA polymerase is remarkably accurate in copying DNA, but the DNA replication system still makes about 100 times more mistakes than most cells can tolerate. A backup repair system called **methyl-directed mismatch repair** corrects almost all of these errors (**Fig. 7.13**). Because mismatch repair is active only *after* DNA replication, this system needs to solve a difficult problem. Suppose that a G–C pair were copied to produce two daughter molecules, one of which has the correct G–C base pair, the other an incorrect G–T. The mismatch repair system can easily recognize the incorrectly matched G–T base pair because the improper base pairing distorts the double helix, resulting in abnormal bulges and hollows. But how does the system know whether to correct the pair to a G–C or to an A–T?

Bacteria solve this problem by placing a distinguishing mark on the parental DNA strands at specific places: Everywhere the sequence GATC occurs, the enzyme *adenine methylase* puts a methyl group on the A (Fig. 7.13a). Shortly after replication, the old template strand bears the methyl mark, while the new daughter strand—which contains the wrong nucleotide—is as yet unmarked (Fig. 7.13b). A pair of proteins in *E. coli,* called MutL and MutS, detect and bind to the mismatched nucleotides. MutL and MutS direct another protein, MutH, to nick the newly synthesized strand of DNA at a position across from the nearest methylated GATC; MutH can discriminate the newly synthesized strand because its GATC is *not* methylated (Fig. 7.13c). DNA exonucleases then remove all the nucleotides between the nick and a position just beyond the mismatch, leaving a gap on the new, unmethylated strand (Fig. 7.13d). DNA polymerase can now resynthesize the information using the old, methylated strand as a template, and DNA ligase then seals up the repaired strand. With the completion of replication and repair, enzymes mark the new strand with methyl groups so that its parental origin will be evident in the next round of replication (Fig. 7.13e).

Eukaryotic cells also have a mismatch correction system, but we do not yet know how they distinguish templates from newly replicated strands because, unlike prokaryotes, their GATCs are not tagged with methyl groups, and eukaryotes do not seem to have a protein closely related to MutH.

Error-Prone Repair Systems Are a Last Resort

The repair systems just described are very accurate in repairing DNA damage because they are able to replace damaged nucleotides with a complementary copy of the undamaged strand. However, cells sometimes become exposed to levels or types of mutagens that they cannot handle with these high-fidelity repair systems. Strong doses of UV light, for example, might make more thymine–thymine dimers than the cell can fix. Any unrepaired damage has

(a) Parental strands are marked with methyl groups.

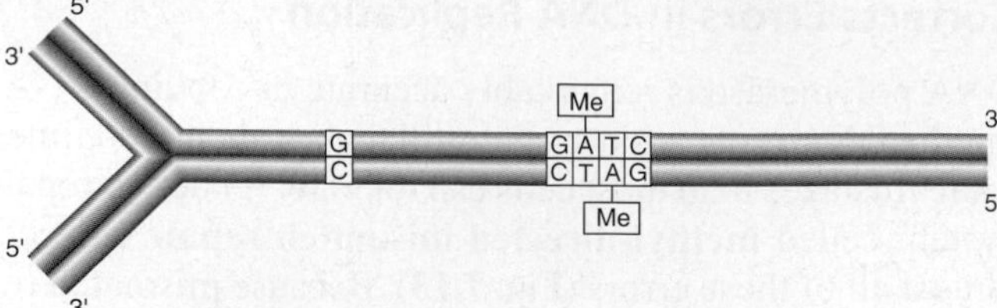

(b) MutS and MutL recognize mismatch in replicated DNA.

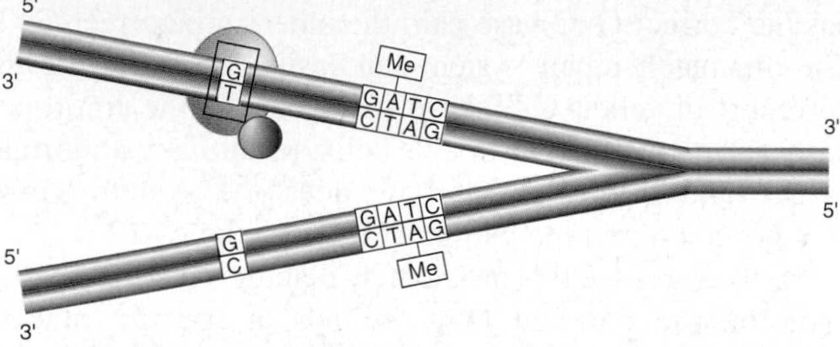

(c) MutL recruits MutH to GATC; MutH makes a nick (*short arrow*) in strand opposite methyl tag.

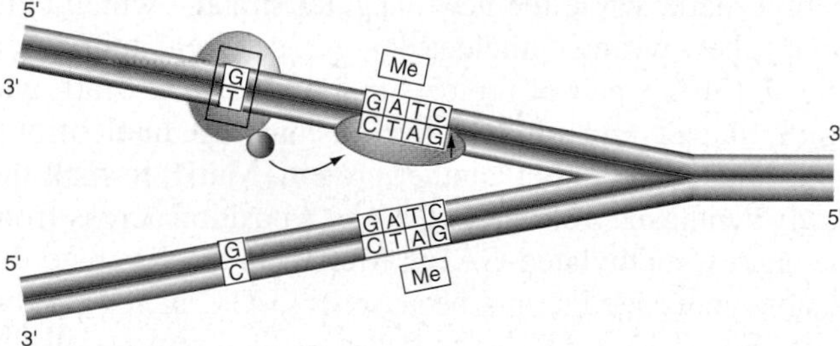

(d) DNA exonucleases (*not shown*) excise DNA from unmethylated new strand.

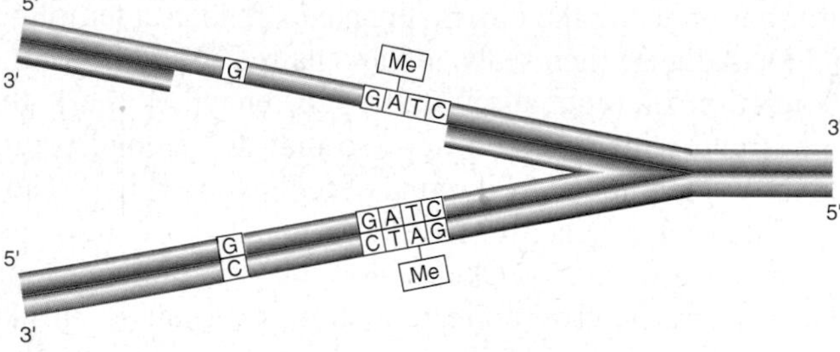

(e) Repair and methylation of newly synthesized DNA strand

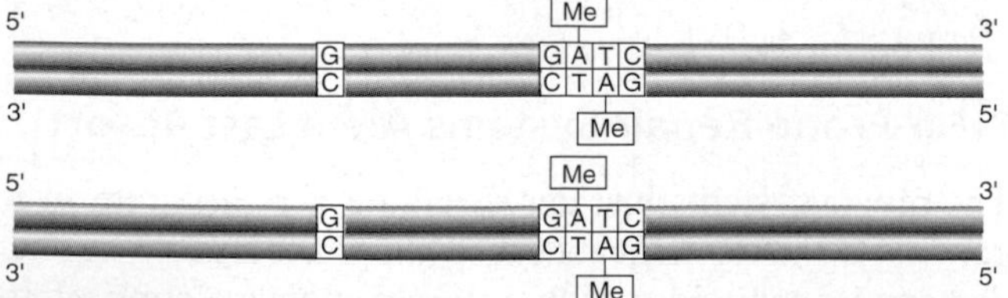

Figure 7.13 Methyl-directed mismatch repair corrects mistakes in replication. Parental strands are *light blue* and newly synthesized strands are *purple*. The MutS protein is *green*, MutL is *dark blue*, and MutH is *yellow*. See text for details.

severe consequences for cell division: The DNA polymerases normally used in replication will stall at such lesions, so the cells cannot proliferate. Although cells can initiate emergency responses that allow them to survive and divide despite the stalling, their ability to proceed in such circumstances comes at the expense of introducing new mutations into the genome.

One type of emergency repair in bacteria, called the **SOS system** (after the Morse code distress signal), relies on error-prone (or "sloppy") DNA polymerases. These sloppy DNA polymerases are not available for normal DNA replication; they are produced only in the presence of DNA damage. The damage-induced, error-prone DNA polymerases are attracted to replication forks that have become stalled at sites of unrepaired, damaged nucleotides. There they add random nucleotides to the strand being synthesized opposite the damaged bases. The SOS polymerase enzymes thus allow the cell with damaged DNA to divide into two daughter cells, but because the sloppy polymerases restore the proper nucleotide only 1/4 of the time, the genomes of these daughter cells carry new mutations. In bacteria, the mutagenic effect of many mutagens either depends on, or is enhanced by, the SOS system.

Another kind of emergency repair system deals with a particularly dangerous kind of DNA lesion: *double-strand breaks,* in which both strands of the double helix are broken at nearby sites (**Fig. 7.14**). Recall from Chapter 6 that

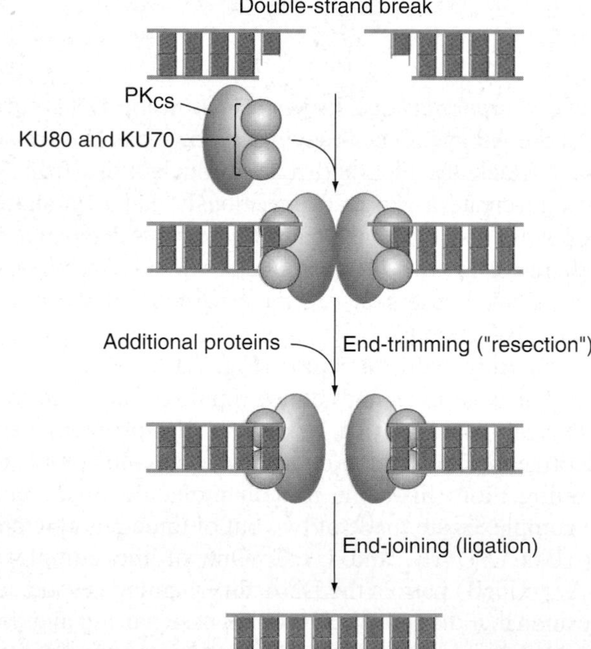

Figure 7.14 Repair of double strand breaks by nonhomologous end-joining. The proteins KU70, KU80, and PK$_{CS}$ bind to DNA ends and bring them together. Other proteins (*not shown*) trim the ends so as to remove any single-stranded regions, and then ligate the two ends together. Note that this mechanism may result in the deletion of nucleotides and is thus potentially mutagenic.

double-strand breaks occur as the first step in meiotic recombination. We do not consider this type of double-strand break here because the mechanism of recombination repairs them with high fidelity and efficiency using complementary base pairing (review Fig. 6.23 on pp. 196–199). However, double-strand breaks can also result from exposure to high-energy radiation such as X-rays (Fig. 7.6c) or highly reactive oxygen species. If left unrepaired, these breaks can lead to a variety of potentially lethal chromosome aberrations, such as large deletions, inversions, or translocations. Cells can restitch the ends formed by double-strand breaks using a mechanism called **nonhomologous end-joining,** which relies on a group of three proteins that bind to the strand ends and bring them close together (Fig. 7.14). After binding, these proteins recruit other proteins that cut back (or resect) any overhanging nucleotides on the ends that do not have a complementary nucleotide to pair with, and then join the two ends together. Because of the resection step, nonhomologous end-joining can result in the loss of DNA and is thus error prone. Evidently, the mutagenic effects of nonhomologous end-joining are less deleterious to the cell than genomic injuries caused by unrepaired double-strand breaks.

Mutations in Genes Encoding DNA Repair Proteins Affect Human Health

Although there are differences of detail between the DNA repair systems of various organisms (such as the presence or absence of methyl tags on GATC to discriminate between old and newly synthesized DNA strands), it is striking that DNA repair mechanisms appear in some form in virtually all organisms. For example, humans have six proteins whose amino acid compositions are about 25% identical with that of the *E. coli* mismatch repair protein MutS. DNA repair systems are thus very old and must have evolved soon after life emerged roughly 3.5 billion years ago. Some scientists believe DNA repair became essential when plants first started to deposit oxygen into the atmosphere, because oxygen favors the formation of free radicals that can damage DNA.

The many known human hereditary diseases associated with the defective processing of DNA damage reveal how crucial DNA repair mechanisms are for survival. In one example, the cells of patients with *Xeroderma pigmentosum* lack the ability to conduct nucleotide excision repair because they are homozygous for mutations in one of seven genes encoding enzymes that normally function in this repair system. These patients cannot efficiently remove thymine–thymine dimers caused by ultraviolet light. Unless they avoid all exposure to sunlight, their skin cells begin to accumulate mutations that eventually lead to skin cancer (**Fig. 7.15**). In another example, researchers have recently learned that hereditary forms of colorectal cancer in humans are associated with mutations in human genes that are closely related to the *E. coli* genes encoding the mismatch-repair proteins MutS and MutL. Chapter 19 discusses the fascinating connections between DNA repair and cancer in more detail.

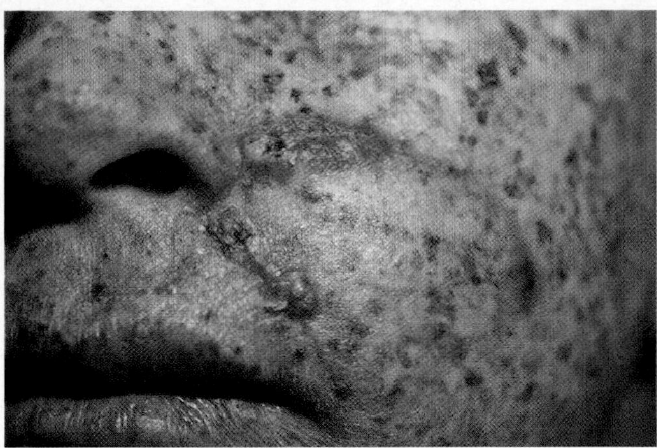

Figure 7.15 Skin lesions in a xeroderma pigmentosum patient. This heritable disease is caused by the lack of a critical enzyme in the nucleotide excision repair system.

Mutations Have Consequences for the Evolution of Species and the Survival of Organisms

"The capacity to blunder slightly is the real marvel of DNA. Without this special attribute, we would still be anaerobic bacteria and there would be no music." In these two sentences, the eminent medical scientist and self-appointed "biology watcher" Lewis Thomas acknowledges that changes in DNA are behind the phenotypic variations that are the raw material on which natural selection has acted for billions of years to drive evolution. The wide-ranging variation in the genetic makeup of the human population—and other populations as well—is, in fact, the result of a balance between the continuous introduction of new mutations, the loss of deleterious mutations because of the selective disadvantage they impose on the individuals that carry them, and the increase in frequency of rare mutations that provide a selective advantage to the individuals carrying them (or that spread through a population by other means; see Chapter 21 for the details of population genetics).

In sexually reproducing multicellular organisms, only germline mutations that can be passed on to the next generation play a role in the evolution of a species. Nevertheless, mutations in somatic cells can still have an impact on the well-being and survival of individuals. Somatic mutations in genes that help regulate the cell cycle may, for example, lead to cancer. The U.S. Food and Drug Administration tries to identify potential cancer-causing agents (known as carcinogens) by using the Ames test to screen

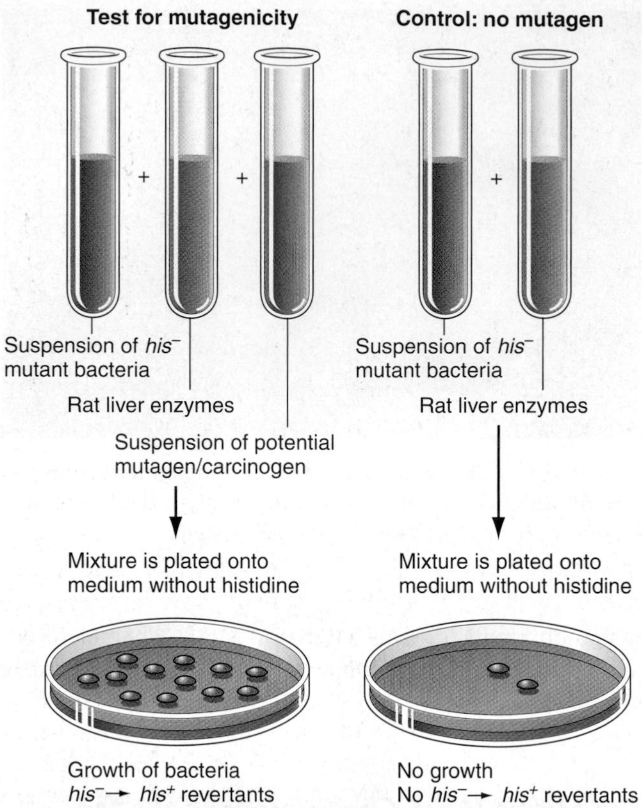

Figure 7.16 The Ames test identifies potential carcinogens through their mutagenicity. A compound to be tested is mixed with cells of a *his*$^-$ strain of *Salmonella typhimurium,* as well as with a solution of rat liver enzymes (which can sometimes convert a harmless compound into a mutagen). Only *his*$^+$ revertants are able to grow on a petri plate without histidine. If this plate (*left*) has more *his*$^+$ revertants than a control plate without histidine (*right*), the compound is considered mutagenic and a potential carcinogen. The rare revertants on the control plate represent the spontaneous rate of mutation.

for chemicals that cause mutations in bacterial cells (**Fig. 7.16**). This test asks whether a particular chemical can induce histidine$^+$ (*his*$^+$) revertants of a special histidine$^-$ (*his*$^-$) mutant strain of the bacterium *Salmonella typhimurium.* The *his*$^+$ revertants can synthesize all the histidine they need from simple compounds in their environment, whereas the original *his*$^-$ mutants cannot make histidine, so they can survive only if histidine is supplied in the environment. The advantage of the test is that only revertants can grow on petri plates that do not contain histidine, so it is possible to examine large numbers of cells from an originally *his*$^-$ culture to find the rare *his*$^+$ revertants induced by the chemical in question. To increase the sensitivity of mutation detection, the *his*$^-$ strain used in the Ames test system contains a second mutation that inactivates the nucleotide excision repair system and thereby prevents the ready repair of mutations, and a third mutation causing defects in the cell wall that allows tested chemicals easier access to the cell interior.

Since most agents that cause mutations in bacteria should also damage the DNA of higher eukaryotic organisms, any mutagen that increases the rate of mutation in bacteria might be expected to cause cancer in people and other mammals. Mammals, however, have complicated metabolic processes capable of inactivating hazardous chemicals. Other biochemical events in mammals can create a mutagenic substance from nonhazardous chemicals. To simulate the action of mammalian metabolism, toxicologists often add a solution of rat liver enzymes to the chemical under analysis by the Ames test (Fig. 7.16). Because this simulation is not perfect, Food and Drug Administration agents ultimately assess whether bacterial mutagens identified by the Ames test can cause cancer in rodents when included in the animals' diets.

In summary, random errors that change the information content of DNA are the ultimate source of variation within and between species. Mutations that help organisms better adapt to changes in their environment will be maintained in populations for many generations. But even though chance mutations provide the raw material of evolution, most changes in genes are deleterious to an organism or its progeny. Cells have evolved several enzymatic systems that either prevent mutations from forming or eliminate those that do. These safeguards enable organisms to keep mutations to a low level that balances their need to evolve with their need to avoid damage to their genomes.

7.2 What Mutations Tell Us About Gene Structure

The science of genetics depends absolutely on mutations because we can track genes in crosses only through the phenotypic effects of their mutant variants. In the 1950s and 1960s, scientists realized they could also use mutations to learn how DNA sequences along a chromosome constitute individual genes. These investigators wanted to collect a large series of mutations in the same gene and analyze how these mutations are arranged with respect to each other. For this approach to be successful, it was necessary to establish that various mutations were, in fact, in the same gene. This was not a trivial exercise, as illustrated by the following situation.

Early *Drosophila* geneticists identified a large number of X-linked recessive mutations affecting the normally red wild-type eye color (**Fig. 7.17**). The first of these to be discovered produced the famous white eyes studied by Morgan's group. Other mutations caused a whole palette of hues to appear in the eyes: darkened shades such as garnet and ruby; bright colors such as vermilion, cherry, and coral; and lighter pigmentations known as apricot, buff, and

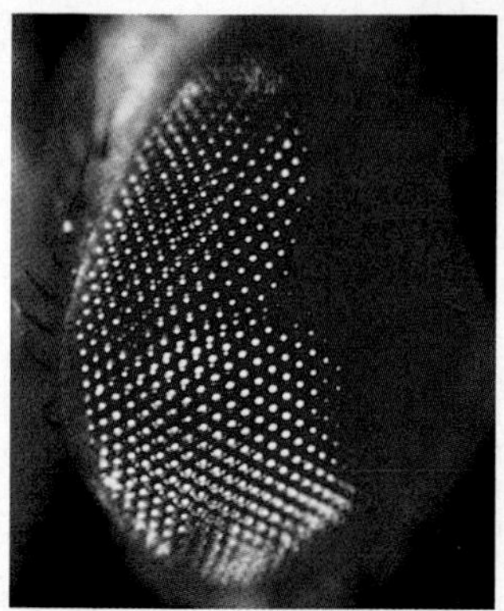
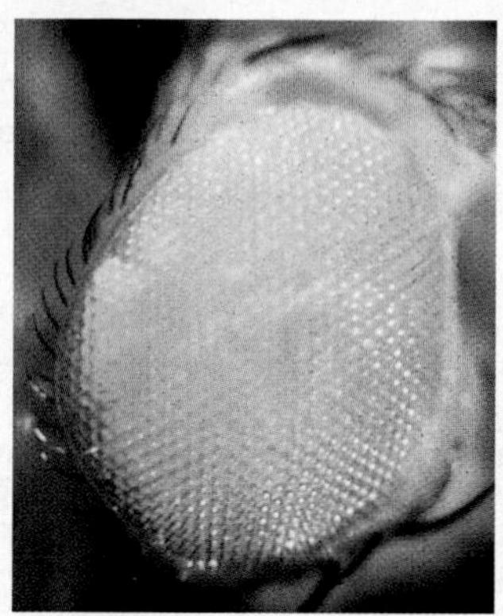
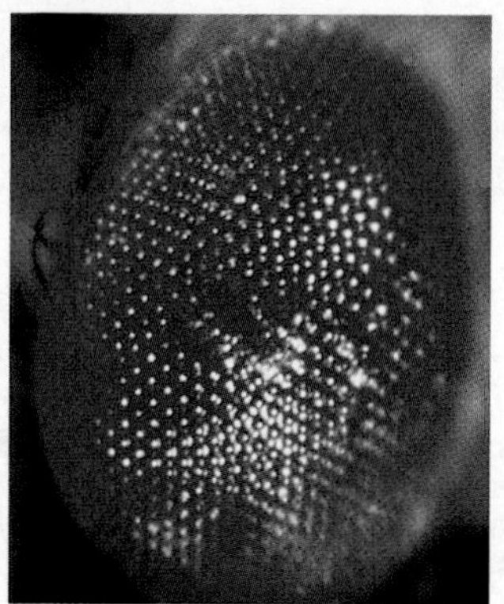
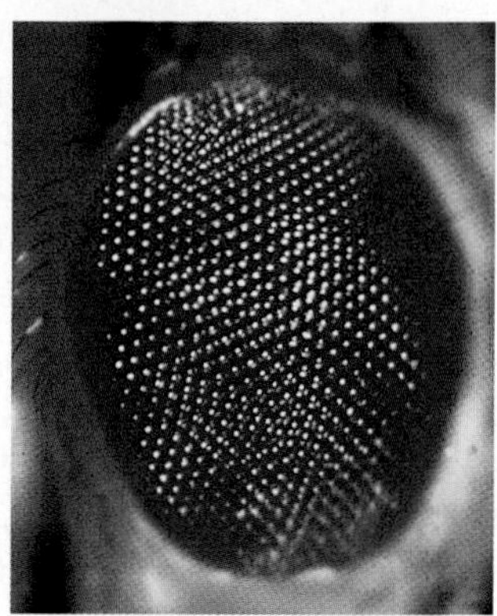

Figure 7.17 ***Drosophila*** **eye color mutations produce a variety of phenotypes.** Flies carrying different X-linked eye color mutations; a wild-type eye is at the *far right.*

carnation. This wide variety of eye color phenotypes posed a puzzle: Were the mutations that caused them multiple alleles of a single gene, or did they affect more than one gene?

Complementation Testing Reveals Whether Two Mutations Are in the Same or Different Genes

Researchers commonly define a gene as a functional unit that directs the appearance of a molecular product that, in turn, contributes to a particular phenotype. They can use this definition to determine whether two mutations are in the same or different genes. If two homologous chromosomes in an individual each carries a mutation recessive to wild type, a normal phenotype will result if the mutations are in different genes. The normal phenotype occurs because almost all recessive mutations disrupt a gene's function (as explained in Chapter 8). The dominant wild-type alleles on each of the two homologs can make up for, or **complement,** the defect in the other chromosome by generating enough of both gene products to yield a normal phenotype (**Fig. 7.18a,** *left*). In contrast, if the recessive mutations on the two homologous chromosomes are in the same gene, no wild-type allele of that gene exists in the individual and neither mutated copy of the gene will be able to perform the normal function. As a result, there will be no complementation and no normal gene product, and a mutant phenotype will appear (Fig. 7.18a, *right*). Ironically, a collection of mutations that do *not* complement each other is known as a **complementation group.** Geneticists often use "complementation group" as a synonym for "gene" because the mutations in a complementation group all affect the same unit of function, thus, the same gene.

A simple test based on the idea of a gene as a unit of function can determine whether or not two mutations are alleles of the same gene. You simply examine the phenotype of a heterozygous individual in which one homolog of a particular chromosome carries one of the recessive mutations and the other homolog carries the other recessive mutation. If the phenotype is wild type, the mutations cannot be in the same gene. This technique for determining whether two homologous chromosomes carry nonallelic mutations of different genes that complement each other or allelic mutations of the same gene that do not complement each other is known as **complementation testing.** For example, because a female *Drosophila* heterozygous for garnet and ruby (*garnet ruby*$^+$ / *garnet*$^+$ *ruby*) has wild-type brick-red eyes, it is possible to conclude that the mutations causing garnet and ruby colors are in different genes.

Complementation testing has, in fact, shown that garnet, ruby, vermilion, and carnation pigmentation are governed by separate genes. But chromosomes carrying mutations yielding white, cherry, coral, apricot, and buff phenotypes fail to complement each other. These mutations therefore make up different alleles of a single gene. *Drosophila* geneticists named this gene the *white,* or *w,* gene after the first mutation observed; they designate the wild-type allele as w^+ and the various mutations as w^1 (the original white-eyed mutation discovered by T. H. Morgan, often simply designated as *w*), w^{cherry}, w^{coral}, $w^{apricot}$, and w^{buff}. As an example, the eyes of a w^1 / $w^{apricot}$ female are a dilute apricot color; because the phenotype of this heterozygote is not wild type, the two mutations are allelic. Figure 7.18b illustrates how researchers collate data from many complementation tests in a **complementation table.** Such a table helps visualize the relationships among a large group of mutants.

In *Drosophila,* mutations in the *w* gene map very close together in the same region of the X chromosome, while mutations in other eye color genes lie elsewhere on the chromosome (Fig. 7.18c). This result suggests that genes are not disjointed entities with parts spread out from one end of a chromosome to another; each gene, in fact, occupies only a relatively small, discrete area of a chromosome. Studies defining genes at the molecular level have shown that most genes consist of 1000–20,000 contiguous base

(a) Complementation testing

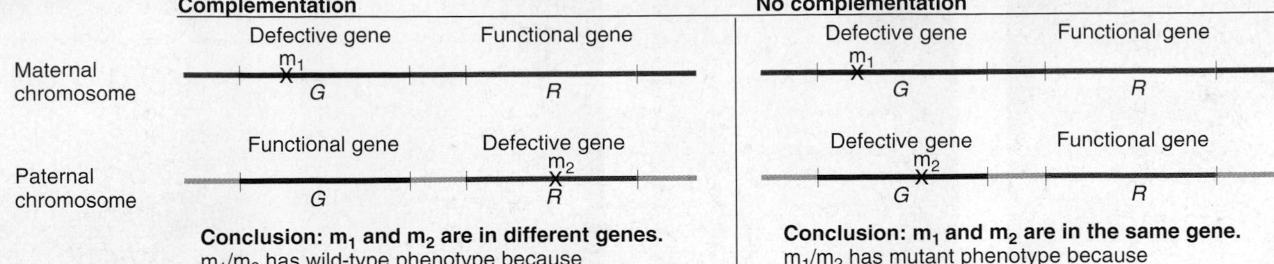

Conclusion: m_1 and m_2 are in different genes. m_1/m_2 has wild-type phenotype because one chromosome supplies gene *G* function, while the other supplies gene *R* function.

Conclusion: m_1 and m_2 are in the same gene. m_1/m_2 has mutant phenotype because organism has no gene *G* function.

(b) A complementation table: X-linked eye color mutations in *Drosophila*

Mutation	white	garnet	ruby	vermilion	cherry	coral	apricot	buff	carnation
white	–	+	+	+	–	–	–	–	+
garnet		–	+	+	+	+	+	+	+
ruby			–	+	+	+	+	+	+
vermilion				–	+	+	+	+	+
cherry					–	–	–	–	+
coral						–	–	–	+
apricot							–	–	+
buff								–	+
carnation									–

(c) Genetic map: X-linked eye color mutations in *Drosophila*

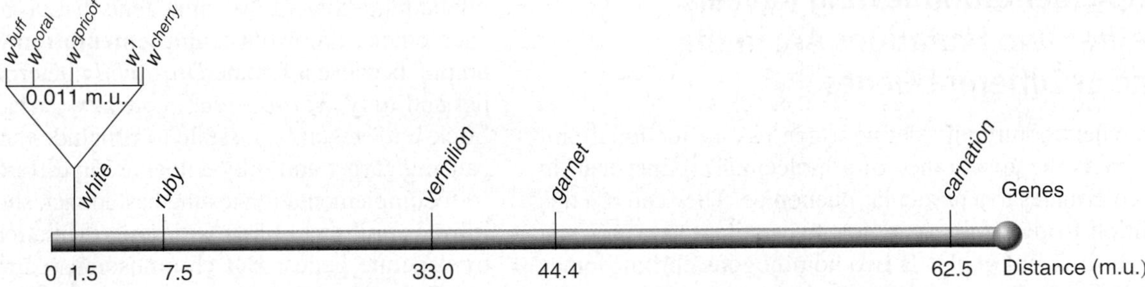

Figure 7.18 Complementation testing of *Drosophila* eye color mutations. (a) A heterozygote has one mutation (m_1) on one chromosome and a different mutation (m_2) on the homologous chromosome. If the mutations are in different genes that contribute to the same trait, the heterozygote will be wild type; the mutations complement each other (*left*). If both mutations affect the same gene, the phenotype will be mutant; the mutations do not complement each other (*right*). Complementation testing makes sense only when both mutations are recessive to wild type. **(b)** This complementation table reveals five complementation groups (five different genes) for eye color. A "+" indicates mutant combinations with wild type eye color; these mutations complement and are thus in different genes. Several mutations fail to complement (–) and are thus alleles of one gene, *white*. **(c)** Recombination mapping shows that mutations in different genes are often far apart, while different mutations in the same gene are very close together.

pairs (bp). In humans, among the shortest genes are the roughly 500-bp-long genes that govern the production of histone proteins, while the longest gene so far identified is the Duchenne muscular dystrophy (*DMD*) gene, which has a length of more than 2 million nucleotide pairs. All known human genes fall somewhere between these extremes. To put these figures in perspective, an average human chromosome is approximately 130 million base pairs in length.

Although complementation testing makes it possible to distinguish mutations in different genes from mutations in the same gene, it does not clarify how the structure of a gene can accommodate different mutations and how these different mutations can alter phenotype in different ways. Does each mutation change the whole gene at a single stroke in a particular way, or does it change only a specific part of a gene, while other mutations alter other parts?

A Gene Is a Linear Sequence of Nucleotide Pairs That Can Mutate Independently and Recombine with Each Other

In the late 1950s, the American geneticist Seymour Benzer used recombination analysis to show that two different mutations that did not complement each other and were therefore known to be in the same gene can in fact change different parts of that gene. He reasoned that if recombination

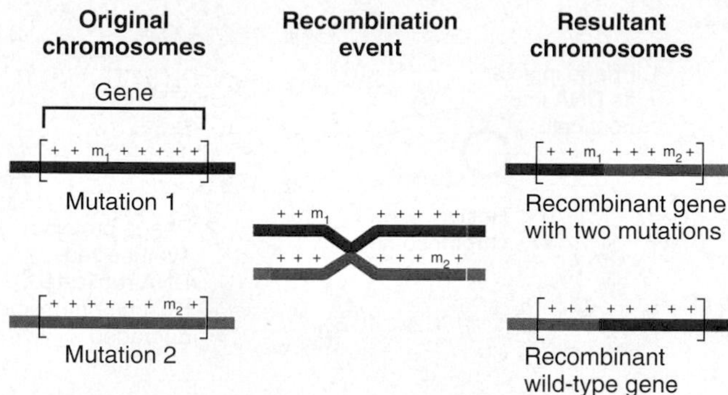

Figure 7.19 How recombination within a gene could generate a wild-type allele. Suppose a gene, indicated by the region between brackets, is composed of many sites that can mutate independently. Recombination between mutations m_1 and m_2 at different sites in the same gene produces a wild-type allele and a reciprocal allele containing both mutations.

can occur not only between genes but within a gene as well, crossovers between homologous chromosomes carrying different mutations known to lie in the same gene could in theory generate a wild-type allele (**Fig. 7.19**). Because mutations affecting a single gene lie very close together on a chromosome, it is necessary to examine a very large number of progeny to see even one crossover event between them. The resolution of the experimental system must thus be extremely high, allowing rapid detection of rare genetic events. For his experimental organism, Benzer chose bacteriophage T4, a DNA virus that infects *Escherichia coli* cells (**Fig. 7.20a.1**). Because each T4 phage that infects a bacterium generates 100–1000 phage progeny in less than an hour, Benzer could easily produce enough rare recombinants for his analysis (Fig. 7.20a.2). Moreover, by exploiting a peculiarity of certain T4 mutations, he devised conditions that allowed only recombinant phages, and not parental phages, to proliferate.

The Experimental System: Analyzing Multiple *rII*⁻ Mutations of Bacteriophage T4

Even though bacteriophages are too small to be seen without the aid of an electron microscope, a simple technique makes it possible to detect their presence with the unaided eye (Fig. 7.20a.3). To do this, researchers mix a population of bacteriophage particles with a much larger number of bacteria and then pour this mixture onto a petri plate, where the cells are immobilized in a nutrient agar. If a single phage infects a single bacterial cell somewhere on this so-called **lawn** of bacteria, the cell produces and releases progeny viral particles that diffuse away to infect adjacent bacteria, which, in turn, produce and release yet more phage progeny. With each release of virus particles, the bacterial host cell dies. Thus, several cycles of phage infection, replication, and release produce a circular cleared area in the plate, called a **plaque,** devoid of living bacterial cells. The rest of the petri plate surface is covered by an opalescent lawn of living bacteria. Most plaques contain from 1 million to 10 million viral progeny of the single bacteriophage that originally infected a cell in that position on the petri plate. Sequential dilution of phage-containing solutions makes it possible to measure the number of phages in a particular plaque and arrive at a countable number of viral particles (Fig. 7.20a.4).

When Benzer first looked for genetic traits associated with bacteriophage T4, he found mutants that, when added to a lawn of *E. coli* B strain bacteria, produced larger plaques with sharper, more clearly rounded edges than those produced by the wild-type bacteriophage (Fig. 7.20b). Because these changes in plaque morphology seemed to result from the abnormally rapid lysis of the host bacteria. Benzer named the mutations giving rise to this phenotype *r* for "rapid lysis." Many *r* mutations map to a region of the T4 chromosome known as the *rII* region; these are called rII^- mutations. An additional property of rII^- mutations makes them ideal for the genetic **fine structure mapping** (the mapping of mutations within a gene) undertaken by Benzer. Wild-type rII^+ bacteriophages form plaques of normal shape and size on cells of both the *E. coli* B strain and a strain known as *E. coli* K(λ). The rII^- mutants, however, have an altered host range: They cannot form plaques with *E. coli* K(λ) cells, although as we have seen, they produce large, unusually distinct plaques with *E. coli* B cells (Fig. 7.20b). The reason that rII^- mutants are unable to infect cells of the K(λ) strain was not clear to Benzer, but their inability to do so enabled him to develop an extremely simple and effective test for rII^+ gene function that he could use to understand gene structure.

The *rII* Region Has Two Genes

Before he could check whether two mutations in the same gene could recombine, Benzer had to be sure he was really looking at two mutations in the same gene. To verify this, he performed customized complementation tests tailored to two significant characteristics of bacteriophage T4: They are haploid (that is, each phage carries a single T4 chromosome), and they can replicate only in a host bacterium. Because T4 phages are haploid, Benzer needed to ensure that two T4 chromosomes entered the same bacterial cell in order to test for complementation between the mutations. In his complementation tests, he simultaneously infected *E. coli* K(λ) cells with two types of T4 chromosomes—one carried one rII^- mutation, the other carried a different rII^- mutation—and then looked for cell lysis (Fig. 7.20c). To ensure that the two kinds of phages would infect almost every bacterial cell, he added many more phages of each type than there were bacteria. If the two rII^- mutations were in different genes, each of the mutant T4 chromosomes would supply one wild-type rII^+ gene function, making up for the lack of that function in the other

Feature Figure 7.20

How Benzer Analyzed the *rII* Genes of Bacteriophage T4

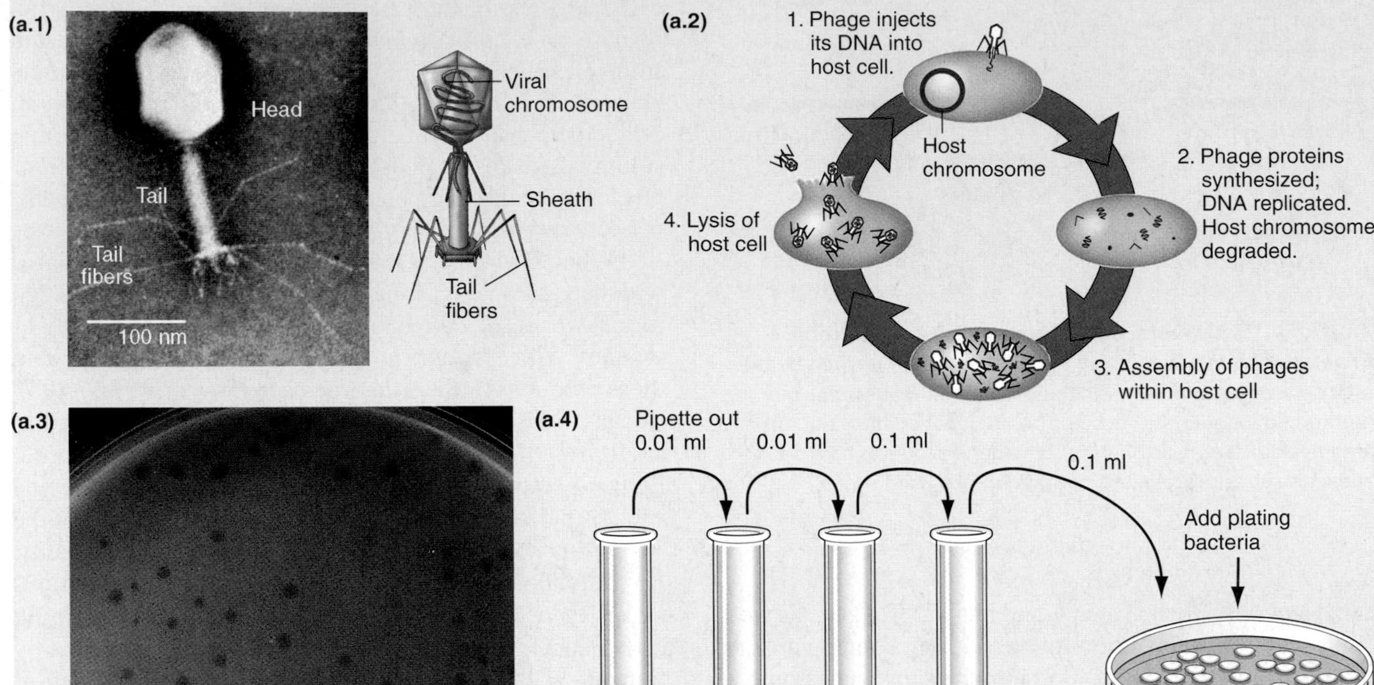

(a) Working with bacteriophage T4

1. Bacteriophage T4 (at a magnification of approximately 100,000×) and in an artist's rendering. The viral chromosome is contained within a protein head. Other proteinaceous parts of the phage particle include the tail fibers, which help the phage attach to host cells, and the sheath, a conduit for injecting the phage chromosome into the host cell.

2. The lytic cycle of bacteriophage T4. A single phage particle infects a host cell; the phage DNA replicates and directs the synthesis of viral protein components using the machinery of the host cell; the new DNA and protein components assemble into new bacteriophage particles. Eventual lysis of the host cell releases up to 1000 progeny bacteriophages into the environment.

3. Clear plaques of bacteriophages in a lawn of bacterial cells. A mixture of bacteriophages and a large number of bacteria are poured onto the agar surface of a petri plate. Uninfected bacterial cells grow, producing an opalescent lawn. A bacterial cell infected by even a single bacteriophage will lyse and release progeny bacteriophages, which can infect adjacent bacteria. Several cycles of infection result in a plaque: a circular cleared area containing millions of bacteriophages genetically identical to the one that originally infected the bacterial cell.

4. Counting bacteriophages by serial dilution. A small sample of a concentrated solution of bacteriophages is transferred to a test tube containing fresh medium, and a small sample of this dilution is transferred to another tube of fresh medium. Successive repeats of this process increase the degree of dilution. A sample of the final dilution, when mixed with bacteria and poured on the agar of a petri plate, yields a countable number of plaques from which it is possible to extrapolate back and calculate the number of bacteriophage particles in the starting solution. The original 1 ml of solution in this illustration contained roughly 2.5×10^7 bacteriophages.

(b) Phenotypic properties of rII^- mutants of bacteriophage T4

1. rII^- mutants, when plated on *E. coli* B cells, produce plaques that are larger and more distinct (with sharper edges) than plaques formed by rII^+ wild-type phage.

2. rII^- mutants are particularly useful for looking at rare recombination events because they have an altered host range. In contrast to rII^+ wild-type phages, rII^- mutants cannot form plaques in lawns of *E. coli* strain K(λ) host bacteria.

(b.1)

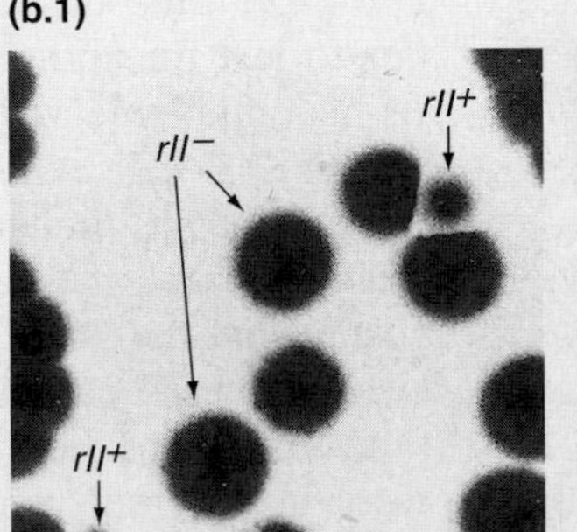

(b.2)

T4 strain	*E. coli* strain B	*E. coli* strain K(λ)
rII^-	Large, distinct	No plaques
rII^+	Small, fuzzy	Small, fuzzy

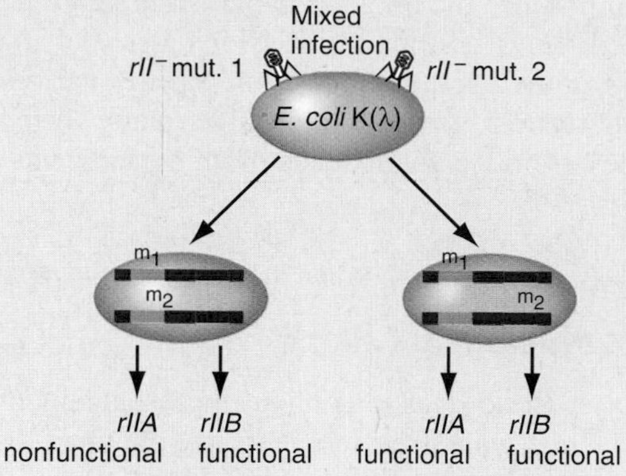

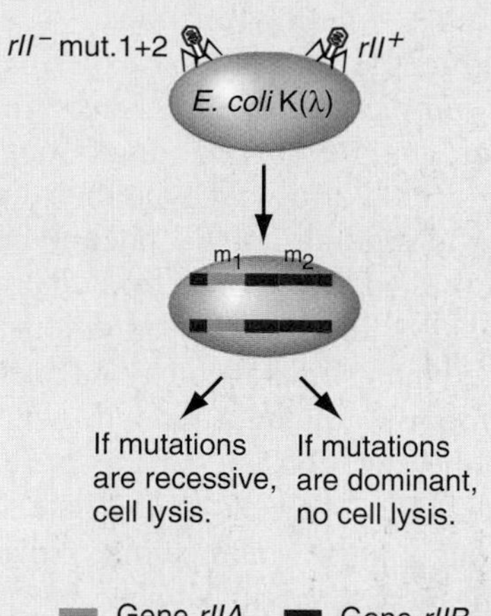

(c) A customized complementation test between *rII*⁻ mutants of bacteriophage T4

1. *E. coli* K(λ) cells are simultaneously infected with an excess of two different *rII*⁻ mutants (m_1 and m_2). Inside the cell, the two mutations will be in *trans;* that is, they lie on different chromosomes. If the two mutations are in the same gene, they will affect the same function and cannot complement each other, so no progeny phages will be produced. If the two mutations are in different genes (*rIIA* and *rIIB*), they will complement each other, leading to progeny phage production and cell lysis.

2. An important control for this complementation test is the simultaneous infection of *E. coli* K(λ) bacteria with a wild-type T4 strain and a T4 strain containing both m_1 and m_2. Inside the infected cells, the two mutations will be in *cis;* that is, they lie on the same chromosome. Release of phage progeny shows that both mutations are recessive to wild type and that there is no interaction between the mutations that prevents the cells from producing progeny phages. Complementation tests are meaningful only if the two mutations tested are both recessive to wild type.

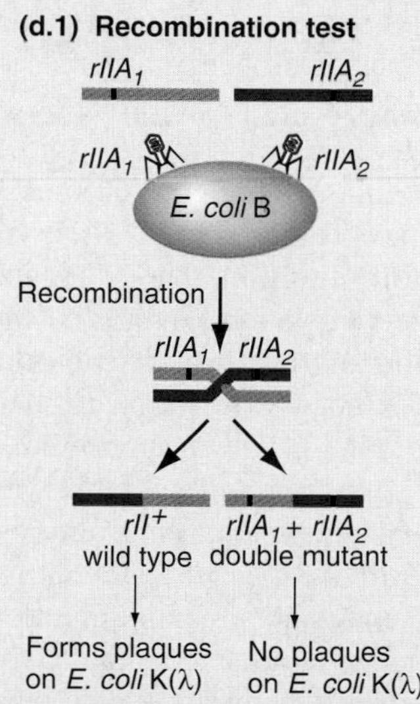

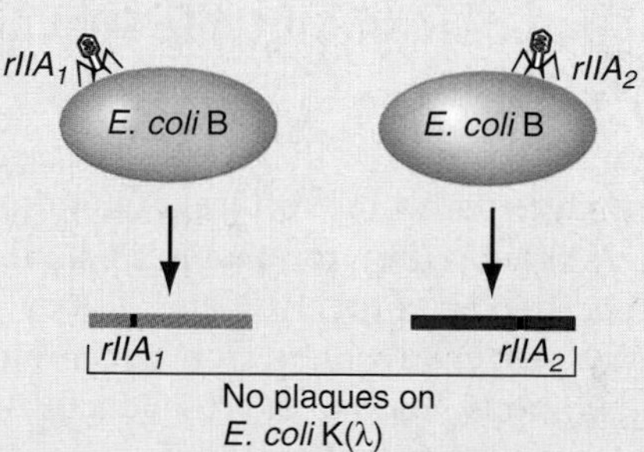

(d) Detecting recombination between two mutations in the same gene

1. *E. coli* B cells are simultaneously infected with a large excess of two different *rIIA*⁻ mutants (*rIIA*$_1$ and *rIIA*$_2$). If no recombination between the two *rIIA*⁻ mutations takes place, progeny phages will carry either of the original mutations and will be phenotypically *rII*⁻. If recombination between the two mutations occurs, one of the products will be an *rII*⁺ recombinant, while the reciprocal product will be a double mutant chromosome containing both *rIIA*$_1$ and *rIIA*$_2$. When the phage progeny subsequently infect *E. coli* K(λ) bacteria, only *rII*⁺ recombinants will be able to form plaques.

2. As a control, *E. coli* B cells are infected with a large amount of only one kind of mutant (*rIIA*$_1$ or *rIIA*$_2$). The only *rII*⁺ phages that can result are revertants of either mutation. This control experiment shows that such revertants are extremely rare and can be ignored among the *rII*⁺ progeny made in the recombination experiment at the *left.* Even if the two *rIIA*⁻ mutations are in adjacent base pairs, the number of *rII*⁺ recombinants obtained is more than 100 times higher than the number of *rII*⁺ revertants the cells infected by a single mutant could produce.

chromosome and resulting in lysis. On the other hand, if the two rII^- mutations were in the same gene, no plaques would appear, because neither mutant chromosome would be able to supply the missing function.

Benzer had to satisfy one final experimental requirement: For the complementation test to be meaningful, he had to make sure that the two rII^- mutations under analysis were each recessive to wild type and did not interact with each other to produce an rII^- phenotype dominant to wild type. He checked these points by a control experiment in which he placed the two rII^- mutations on the same chromosome and then simultaneously infected *E. coli* K(λ) with these double rII^- mutants and with wild-type phages (Fig. 7.20c). If the mutations were recessive and did not interact with each other, the cells would lyse, in which case the complementation test would be interpretable.

The significant distinction between the actual complementation test and the control experiment is in the placement of the two rII^- mutations. In the complementation test, one rII^- mutation is on one chromosome, while the other rII^- mutation is on the other chromosome; two mutations arranged in this way are said to be in the *trans* configuration. In the control experiment, the two mutations are on the same chromosome, in the so-called *cis* configuration. The complete test, including the complementation test and the control experiment, is known as a *cis-trans* test. Benzer called any complementation group identified by the cis-trans test a **cistron,** and some geneticists still use the term "cistron" as a synonym for "gene."

Tests of many different pairs of rII^- mutations showed that they fall into two complementation groups: *rIIA* and *rIIB*. With this knowledge, Benzer could look for two mutations in the same gene and then see if they ever recombine to produce wild-type progeny.

A Gene Is Composed of Mutable Subunits That Can Recombine

When Benzer infected *E. coli* B strain bacteria with a mixture of phages carrying different mutations in the same gene ($rIIA_1$ and $rIIA_2$, for example), he did observe the appearance of rII^+ progeny (Fig. 7.20d). He knew these wild-type progeny resulted from recombination and not reverse mutations because the frequencies of the rII^+ phage particles he observed were much higher than the frequencies of rII^+ revertants seen among progeny produced by infecting B strain bacteria with either mutant alone. On the basis of these observations, he drew three conclusions about gene structure: (1) A gene consists of different parts that can each mutate; (2) recombination between different mutable sites in the same gene can generate a normal, wild-type allele; and (3) a gene performs its normal function only if all of its components are wild type.

From what we now know about the molecular structure of DNA, this all makes perfect sense. The separate mutable sites are different nucleotides within the gene, and recombination can occur between nucleotides within a gene as well as between genes.

But how are the nucleotide pairs that make up a gene arranged—in a continuous row or dispersed in precise patterns around the genome? And do the various mutations that affect gene function alter many different nucleotides or only a small subset within each gene?

A Gene Is a Discrete Linear Set of Nucleotide Pairs

To answer these questions about the arrangement of nucleotides in a gene, Benzer eventually obtained thousands of spontaneous and mutagen-induced rII^- mutations that he mapped with respect to each other. To map the location of a thousand mutants through comparisons of all possible two-point crosses, he would have had to set up a million ($10^3 \times 10^3$) matings. But by taking advantage of deletion mutations, he could obtain the same information with far fewer crosses.

Using Deletions to Map Mutations and Define a Gene

As we have seen, deletions are mutations that remove contiguous nucleotide pairs along a DNA molecule. In crosses between bacteriophages carrying a mutation and bacteriophages carrying deletions of the corresponding region, no wild-type recombinant progeny can arise, because neither chromosome carries the proper information at the location of the mutation. However, if the mutation lies outside the region deleted from the homologous chromosome, wild-type progeny can appear (**Fig. 7.21a**). This is true whether the mutation is a **point mutation,** that is, a mutation of one nucleotide, or is itself a deletion. Crosses between any uncharacterized mutation and a known deletion thus immediately reveal whether or not the mutation resides in the region deleted from the other phage chromosome. This method of mapping with deletions provides a rapid way of finding the general location of a mutation. Using a series of overlapping deletions, Benzer divided the *rII* region into a series of intervals. He could then assign any point mutation to an interval by observing whether or not it recombined to give rII^+ progeny when crossed with the series of deletions (Fig. 7.21b).

Benzer mapped 1612 spontaneous point mutations and several deletions in the *rII* locus of bacteriophage T4 through recombination analysis. He first used recombination to determine the relationship between the deletions. He next found the approximate location of individual point mutations by observing which deletions could recombine with each mutant to yield wild-type progeny. He then performed recombination tests between all point mutations known to lie in the same small region of the chromosome. These results produced a map of the "fine structure" of the region (Fig. 7.21c).

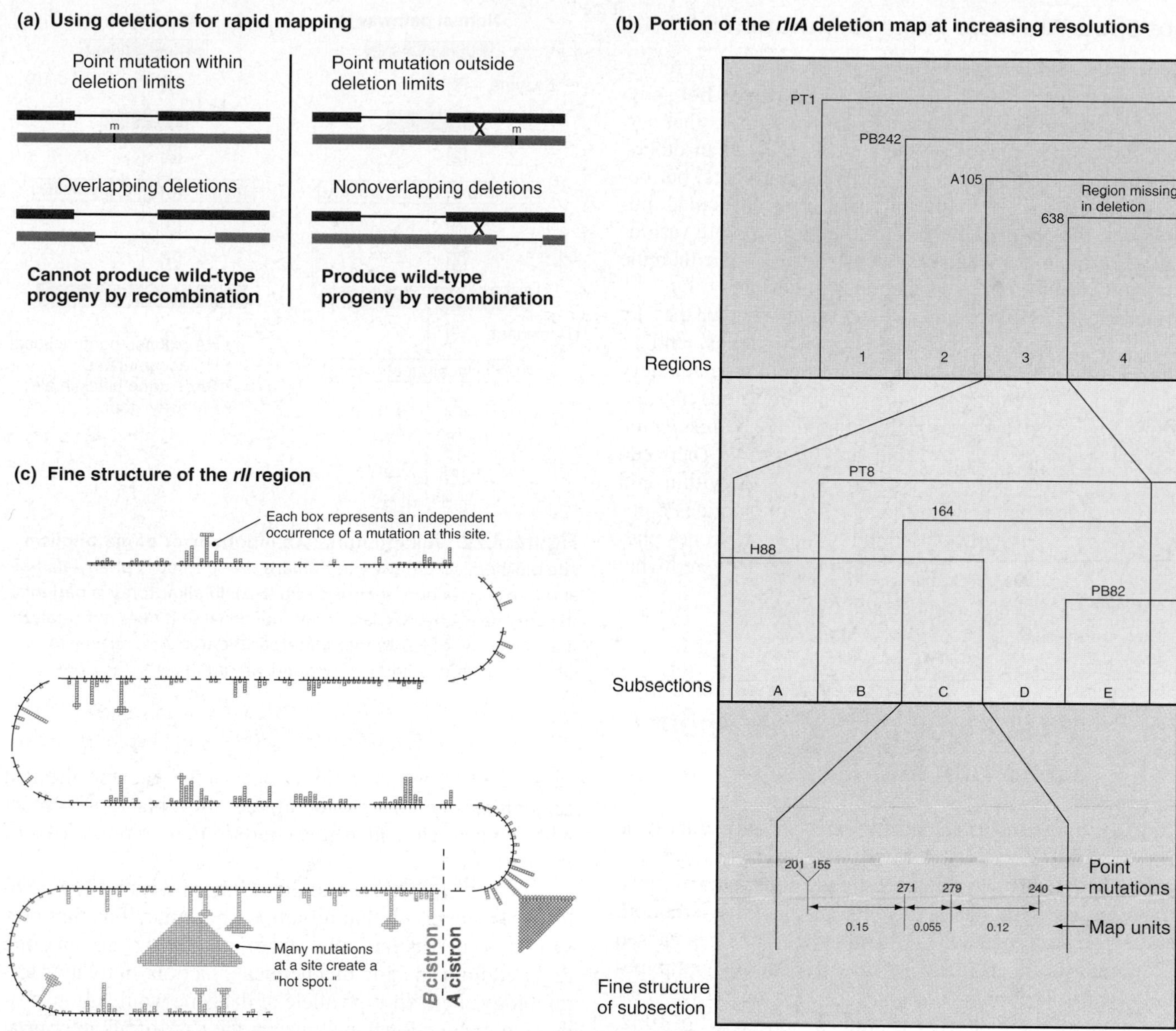

Figure 7.21 **Fine structure mapping of the bacteriophage T4 *rII* genes.** **(a)** A phage cross between a point mutation and a deletion removing the DNA at the position of the mutation cannot yield wild-type recombinants. The same is true if two different deletion mutations overlap each other. **(b)** Large deletions divide the *rII* locus into regions; finer deletions divide each region into subsections. Point mutations, such as 271 (*in red at bottom*), map to region 3 if they do not recombine with deletions PT1, PB242, or A105 but do recombine with deletion 638 (*top*). Point mutations can be mapped to subsections of region 3 using other deletions (*middle*). Recombination tests map point mutations in the same subregion (*bottom*). Point mutations 201 and 155 cannot recombine to yield wild-type recombinants because they affect the same nucleotide pair. **(c)** Benzer's fine structure map. Hot spots are locations with many independent mutations that cannot recombine with each other.

From the observation that the number of mutable sites in the *rII* region is very close to the number of nucleotides estimated to be in this region, Benzer inferred that a mutation can arise from the change of a single nucleotide and that recombination can occur between adjacent nucleotide pairs. From the observation that mutations within the *rII* region form a self-consistent, linear recombination map, he concluded that a gene is composed of a continuous linear sequence of nucleotide pairs within the DNA. And from observations that the positions of mutations in the *rIIA* gene did not overlap those of the *rIIB* gene, he inferred that the nucleotide sequences composing those two genes are separate and distinct. A *gene* is thus a linear set of nucleotide pairs, located within a discrete region of a chromosome, that serves as a unit of function.

Hot Spots Are More Prone to Mutation

Some sites within a gene spontaneously mutate more frequently than others and as a result are known as **hot spots** (Fig. 7.21c). The existence of hot spots suggests that certain nucleotides can be altered more readily than others. Treatment with mutagens also turns up hot spots, but because mutagens have specificities for particular nucleotides, the highly mutable sites that turn up with various mutagens are often at different positions in a gene than the hot spots resulting from spontaneous mutation.

Nucleotides are chemically the same whether they lie within a gene or in the DNA between genes, and as Benzer's experiments show, the molecular machinery responsible for mutation and recombination does not discriminate between those nucleotides that are *intragenic* (within a gene) and those that are *intergenic* (between genes). The main distinction between DNA within and DNA outside a gene is that the array of nucleotides composing a gene has evolved a function that determines phenotype. We now describe how geneticists discovered what that function is.

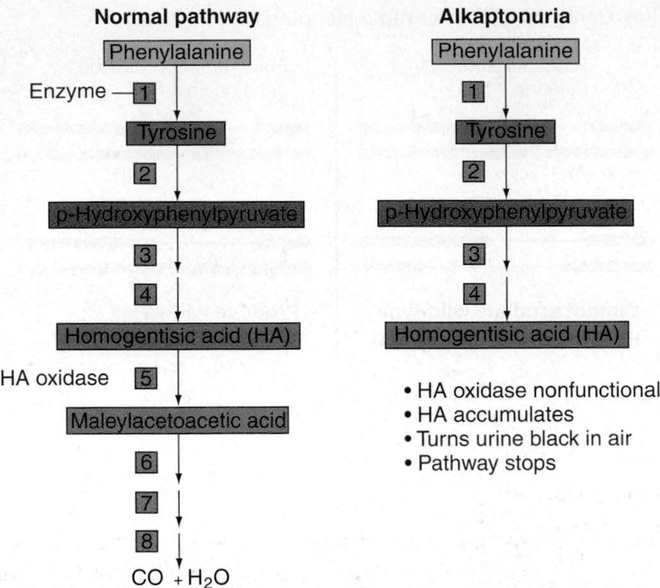

Figure 7.22 Alkaptonuria: An inborn error of metabolism. The biochemical pathway in humans that degrades phenylalanine and tyrosine via homogentisic acid (HA). In alkaptonuria patients, the enzyme HA hydroxylase is not functional so it does not catalyze the conversion of HA to maleylacetoacetic acid. As a result, HA, which oxidizes to a black compound, accumulates in the urine.

7.3 What Mutations Tell Us About Gene Function

Mendel's experiments established that an individual gene can control a visible characteristic, but his laws do not explain how genes actually govern the appearance of traits. Investigators working in the first half of the twentieth century carefully studied the biochemical changes caused by mutations in an effort to understand the genotype-phenotype connection.

In one of the first of these studies, conducted in 1902, the British physician Dr. Archibald Garrod showed that a human genetic disorder known as *alkaptonuria* is determined by the recessive allele of an autosomal gene. Garrod analyzed family pedigrees and performed biochemical analyses on family members with and without the trait. The urine of people with alkaptonuria turns black on exposure to air. Garrod found that a substance known as homogentisic acid, which blackens upon contact with oxygen, accumulates in the urine of alkaptonuria patients. These same people excrete all of the homogentisic acid they ingest, while people not suffering from the condition do not have any homogentisic acid in their urine even after ingesting the substance. From these observations, he concluded that people with alkaptonuria are incapable of metabolizing homogentisic acid to the breakdown products generated by normal individuals (**Fig. 7.22**). Because many biochemical reactions within the cells of organisms are catalyzed by enzymes, Garrod hypothesized that lack of the enzyme that breaks down homogentisic acid is the cause of alkaptonuria; in the absence of this enzyme, the acid accumulates and causes the urine to turn black on contact with oxygen. He called this condition an "inborn error of metabolism."

Garrod studied several other inborn errors of metabolism and suggested that all arose from mutations that prevented a particular gene from producing an enzyme required for a specific biochemical reaction. In today's terminology, the wild-type allele of the gene would allow production of functional enzyme (in the case of alkaptonuria, the enzyme is homogentisic acid oxidase), whereas the mutant allele would not. Because the single wild-type allele in heterozygotes generates sufficient enzyme to prevent accumulation of homogentisic acid and thus alkaptonuria, the mutant allele is recessive.

The One Gene, One Enzyme Hypothesis: A Gene Contains the Information for Producing a Specific Enzyme

George Beadle and Edward Tatum carried out a series of experiments on the bread mold *Neurospora crassa* (whose life cycle was described in Chapter 5) during the 1940s that demonstrated a direct relation between genes and the enzymes that catalyze specific biochemical reactions. Their strategy was simple. They first isolated a number of mutations that disrupted synthesis of the amino acid arginine, a compound needed for *Neurospora* growth. They next hypothesized that different mutations blocked different steps in a particular **biochemical pathway:** the orderly series of reactions that allows *Neurospora* to obtain simple molecules from the environment and convert them step-by-step into successively more complicated molecules culminating in arginine. A study showing that mutations can block discrete biochemical steps in the pathway resulting in arginine would constitute strong evidence that different genes control the appearance of different enzymes, each of which catalyzes a specific reaction in the biochemical pathway.

The Experimental Evidence for "One Gene, One Enzyme"

Figure 7.23a illustrates the experiments Beadle and Tatum performed to test their hypothesis. They first obtained a set of mutagen-induced mutations that prevented *Neurospora* from synthesizing arginine. Cells with any one of these mutations were unable to make arginine and could therefore grow on a minimal medium containing salt and sugar only if it had been supplemented with arginine. A mutant microorganism that can grow on minimal medium only if it has been supplemented with one or more growth factors not required by wild-type strains is known as an **auxotroph.** The cells just mentioned were arginine auxotrophs. (In contrast, a wild-type cell that can synthesize a particular growth factor and thus grow in its absence on minimal medium is a **prototroph** for that factor. In a more general meaning, *prototroph* refers to a wild-type cell that can synthesize all required growth factors and thus grow on minimal medium alone.) Recombination analyses located the auxotrophic arginine-blocking mutations in four distinct regions of the genome, and complementation tests showed that each of the

(a) Isolation of arginine auxotrophs

X-rays
1.
Mutagenized conidia
Wild type
Asci
Fruiting bodies
Crossed with opposite wild type
Ascospores dissected and transferred; one to each culture tube
2. Tubes of complete medium inoculated with single ascospores
Complete medium
Germination, production of conidia
3. Conidia from each culture tested on minimal medium.
Minimal medium
No growth = nutritional mutant
4. Conidia from cultures that fail to grow on minimal medium are tested on minimal medium supplemented with individual amino acids.
Glycine
Leucine
Arginine
Valine
Tyrosine
Proline
Glutamic acid
Asparagine
Serine
Cysteine
Addition of arginine restores growth, reveals arginine auxotroph.

(b) Growth response if nutrient is added to minimal medium

Mutant strain	Supplements: Nothing	Ornithine	Citrulline	Arginino-succinate	Arginine
Wildtype: Arg^+	+	+	+	+	+
$Arg\text{-}E^-$	–	+	+	+	+
$Arg\text{-}F^-$	–	–	+	+	+
$Arg\text{-}G^-$	–	–	–	+	+
$Arg\text{-}H^-$	–	–	–	–	+

(c) Inferred biochemical pathway

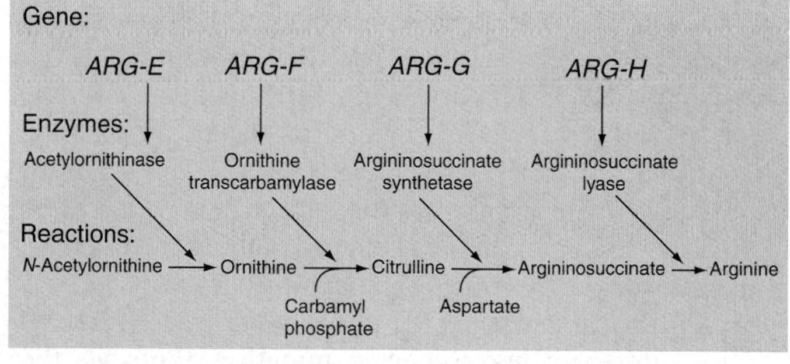

Figure 7.23 Beadle and Tatum's experiment supporting the "one gene, one enzyme" hypothesis. (a) Beadle and Tatum mated an X-ray-mutagenized strain of *Neurospora* with another strain, and they isolated haploid ascospores that grew on complete medium. Cultures that failed to grow on minimal medium were nutritional mutants. Nutritional mutants that could grow on minimal medium plus arginine were arg^- auxotrophs. **(b)** The ability of wild-type and mutant strains to grow on minimal medium supplemented with intermediates in the arginine pathway. **(c)** Each of the four *ARG* genes encodes an enzyme needed to convert one intermediate to the next in the pathway.

four regions correlated with a different complementation group. On the basis of these results, Beadle and Tatum concluded that at least four genes support the biochemical pathway for arginine synthesis. They named the four genes *ARG-E, ARG-F, ARG-G,* and *ARG-H.*

They next asked whether any of the mutant *Neurospora* strains could grow in minimal medium supplemented with any of a number of compounds that are intermediates in the biochemical pathway leading to arginine, instead of with arginine itself. If a mutant grew on medium supplemented in such a way, it would indicate that *Neurospora* is able to convert the intermediate compound into arginine. After finding three such intermediates—ornithine, citrulline, and argininosuccinate—Beadle and Tatum compiled a table describing which arginine auxotrophic mutants were able to grow on minimal medium supplemented with each of the intermediates (Fig. 7.23b).

Interpretation of Results: Genes Encode Enzymes

On the basis of these results, Beadle and Tatum proposed a model of how *Neurospora* cells synthesize arginine (Fig.7.23c). In the linear progression of biochemical reactions by which a cell constructs arginine from the constituents of minimal medium, each intermediate is both the product of one step and the substrate for the next. Each reaction in the precisely ordered sequence is catalyzed by a specific enzyme, and the presence of each enzyme depends on one of the four *ARG* genes. A mutation in one gene blocks the pathway at a particular step because the cell lacks the corresponding enzyme; this lack prevents the cell from growing in the absence of arginine. Supplementing the medium with any intermediate that occurs beyond the blocked reaction will restore growth because the organism has all the remaining enzymes required to convert the intermediate to arginine. Supplementation with an intermediate that occurs before the missing enzyme will not work because the cell is unable to convert the intermediate into arginine.

By inference, because each mutation abolishes the cell's ability to make an enzyme capable of catalyzing a certain reaction, each gene controls the synthesis or activity of an enzyme, or as stated by Beadle and Tatum: one gene, one enzyme. Of course, the gene and the enzyme are not the same thing; rather, the sequence of nucleotides in a gene contains information that somehow encodes the structure of an enzyme molecule.

While the analysis of the arginine pathway studied by Beadle and Tatum was straightforward, studies of biochemical pathways are not always so easy to interpret because some biochemical pathways are not linear progressions of stepwise reactions. For example, a branching pathway occurs if different enzymes act on the same intermediate to convert it into two different end products. If the cell requires both of these end products for growth, a mutation in a gene encoding any of the enzymes required to synthesize the intermediate would make the cell dependent on the addition to minimal medium of both end products. A second possibility is that a cell might employ either of two independent, parallel pathways to synthesize a needed end product. In such a case, a mutation in a gene encoding an enzyme in one of the pathways would be without effect. Only a cell with mutations in genes specifying enzymes in both pathways would display an aberrant phenotype.

Even with nonlinear progressions such as these, careful genetic analysis can reveal the nature of the biochemical pathway on the basis of Beadle and Tatum's insight that genes encode proteins.

Genes Specify the Identity and Order of Amino Acids in Polypeptide Chains

Although the one gene, one enzyme hypothesis was a critical advance in our understanding of how genes influence phenotype, it is an oversimplification. Not all genes govern the construction of enzymes active in biochemical pathways. Enzymes are one class of the molecules known as proteins, and cells contain many different kinds of protein, only some of which behave as enzymes. Among the other types are proteins that provide shape and rigidity to a cell, proteins that transport molecules in and out of cells, proteins that help fold DNA into chromosomes, and proteins that act as hormonal messengers. Genes direct the synthesis of all proteins, enzymes and nonenzymes alike. Moreover, as we see next, genes actually determine the construction of polypeptides, and because some proteins are composed of more than one type of polypeptide, more than one gene determines their construction.

Proteins Are Linear Polymers of Amino Acids Linked by Peptide Bonds

To understand how the information in a gene specifies the production of a particular protein, it is necessary to review the chemical composition of proteins. Proteins are polymers composed of building blocks known as **amino acids.** Cells use mainly 20 different amino acids to synthesize the proteins they need. All of these amino acids have certain basic features, encapsulated by the formula NH_2–CHR–COOH (**Fig. 7.24a**). The –COOH component, also known as *carboxylic acid,* is, as the name implies, acidic; the $-NH_2$ component, also known as an *amino group,* is basic. The R refers to side chains that distinguish each of the 20 amino acids (Fig. 7.24b). An R group can be as simple as a hydrogen atom (in the amino acid glycine) or as complex as a benzene ring (in phenylalanine). Some side chains are relatively neutral and nonreactive, others are acidic, and still others are basic.

During protein synthesis, a cell's protein-building machinery links amino acids by constructing covalent **peptide bonds** that join the –COOH group of one amino acid to the

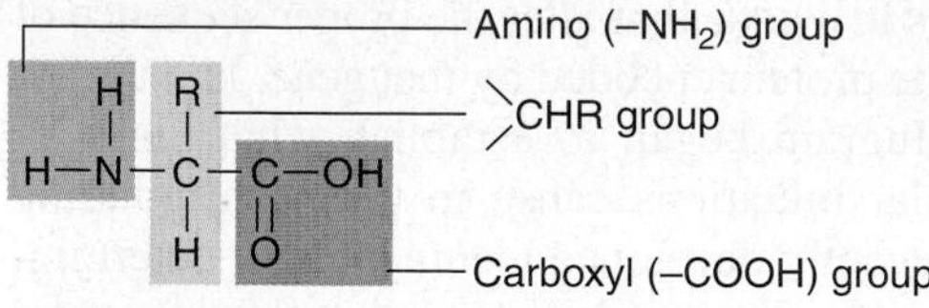

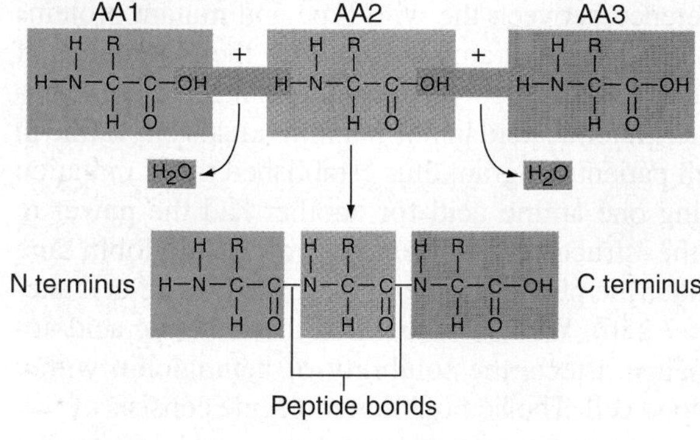

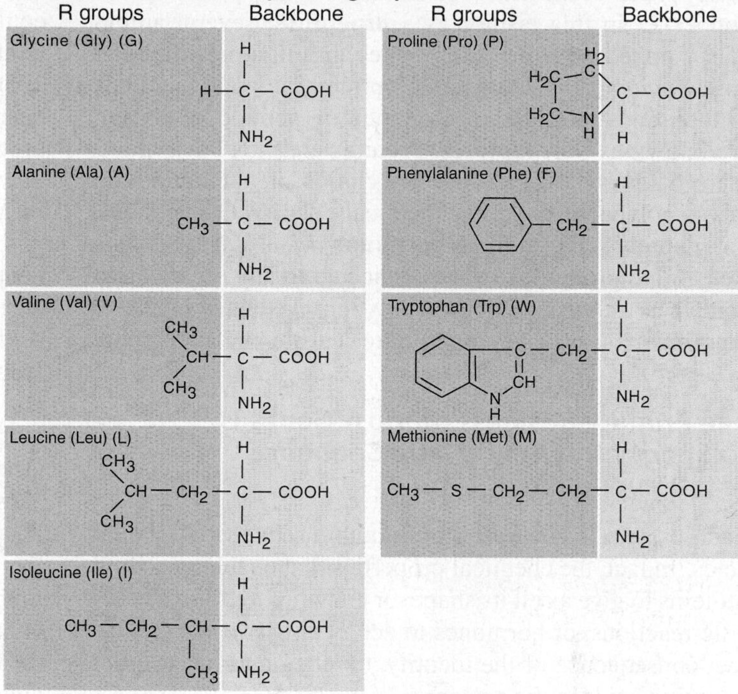

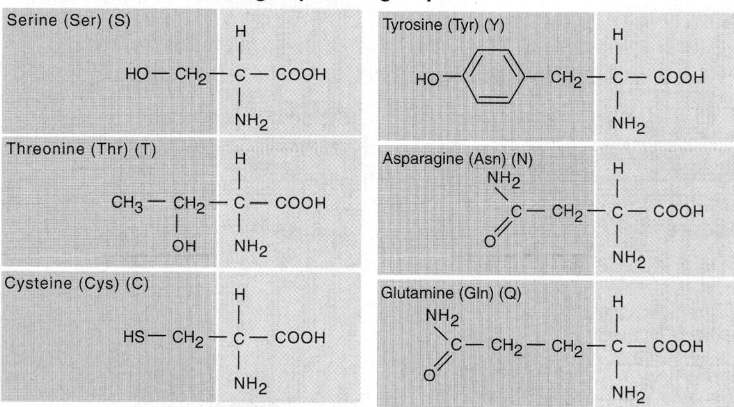

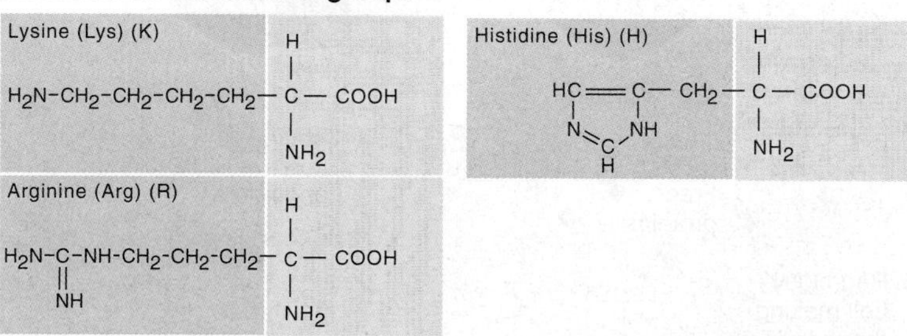

Amino acids with acidic R groups

Aspartic acid (Asp) (D)

Glutamic acid (Glu) (E)

Figure 7.24 Proteins are chains of amino acids linked by peptide bonds. (a) Amino acids contain a basic amino group (–NH_2), an acidic carboxylic acid group (–COOH), and a >CHR moiety, where R stands for one of the 20 different side chains. **(b)** Table of amino acids commonly found in proteins, arranged according to the properties of their R groups. **(c)** One molecule of water is lost when a covalent amide linkage (a peptide bond) is formed between the –COOH of one amino acid and the –NH_2 of the next amino acid. Polypeptides such as the tripeptide shown here have polarity; they extend from an N terminus (with a free amino group) to a C terminus (with a free carboxylic acid group).

$-NH_2$ group of the next (Fig. 7.24c). A pair of amino acids connected in this fashion is a **dipeptide;** several amino acids linked together constitute an **oligopeptide.** The amino acid chains that make up proteins contain hundreds to thousands of amino acids joined by peptide bonds and are known as **polypeptides.** Proteins are thus linear polymers of amino acids. Like the chains of nucleotides in DNA, polypeptides have a chemical polarity. One end of a polypeptide is called the **N terminus** because it contains a free amino group that is not connected to any other amino acid. The other end of the polypeptide chain is the **C terminus,** because it contains a free carboxylic acid group.

The Primary Business of Most Genes Is to Specify the Amino Acid Sequence of a Polypeptide

Each protein is composed of a unique sequence of amino acids. In fact, the chemical properties that enable structural proteins to give a cell its shape, or enzymes to catalyze specific reactions, or hormones to act as messengers are a direct consequence of the identity, number, and linear order of amino acids in the protein.

If genes encode proteins, then at least some mutations could be changes in a gene that alter the proper sequence of amino acids in the protein encoded by that gene. In the mid-1950s, Vernon Ingram began to establish what kinds of changes particular mutations cause in the corresponding protein. Using recently developed techniques for determining the sequence of amino acids in a protein, he compared the amino acid sequence of the normal adult form of hemoglobin (HbA) with that of hemoglobin in the bloodstream of people homozygous for the mutation that causes sickle-cell anemia (HbS). Remarkably, he found only a single amino acid difference between the wild-type and mutant proteins (**Fig. 7.25a**). The sixth amino acid from the N terminus of one of the two types of polypeptide chains making up hemoglobin was glutamic acid in normal individuals but valine in sickle-cell patients. Ingram thus established that a mutation substituting one amino acid for another had the power to change the structure and function of hemoglobin and thereby alter the phenotype from normal to sickle-cell anemia (Fig. 7.25b). We now know that the glutamic acid–to-valine change affects the solubility of hemoglobin within the red blood cell. The hemoglobin molecule consists of two alpha (α) and two beta (β) chains complexed in a tetramer

(a) From mutation to phenotype

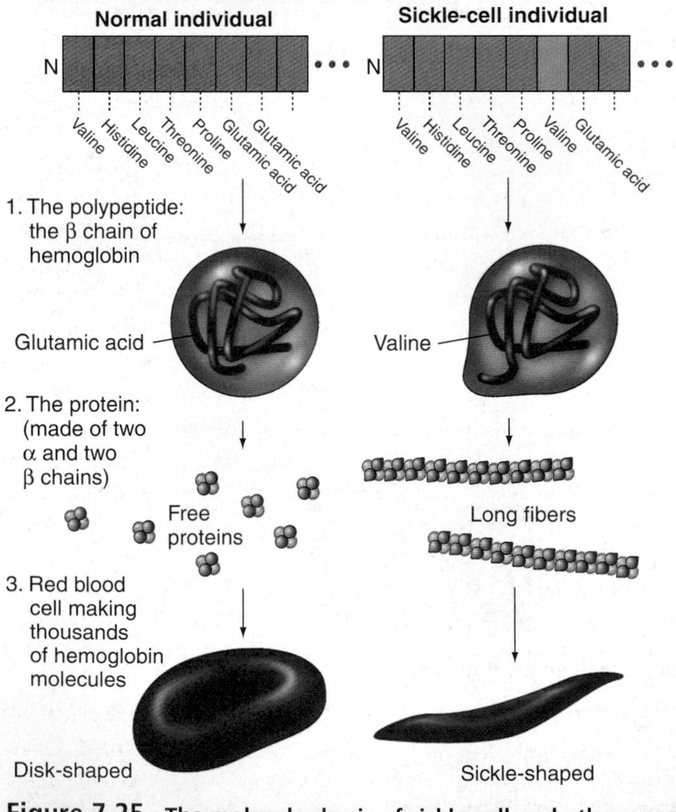

(b) Sickle-cell anemia is pleiotrophic

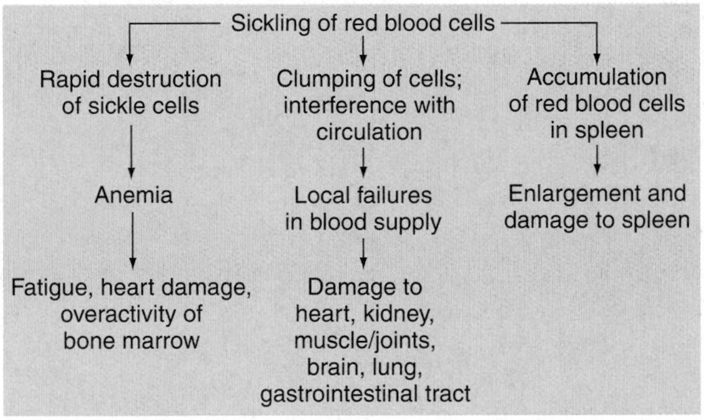

(c) β-chain substitutions/variants

Amino-acid position

	1	2	3 ···	6	7 ···	26 ···	63 ···	67 ···	125 ···	146
Normal (HbA)	Val	His	Leu	Glu	Glu	Glu	His	Val	Glu	His
HbS	Val	His	Leu	Val	Glu	Glu	His	Val	Glu	His
HbC	Val	His	Leu	Lys	Glu	Glu	His	Val	Glu	His
HbG San Jose	Val	His	Leu	Glu	Gly	Glu	His	Val	Glu	His
HbE	Val	His	Leu	Glu	Glu	Lys	His	Val	Glu	His
HbM Saskatoon	Val	His	Leu	Glu	Glu	Glu	Tyr	Val	Glu	His
Hb Zurich	Val	His	Leu	Glu	Glu	Glu	Arg	Val	Glu	His
HbM Milwaukee 1	Val	His	Leu	Glu	Glu	Glu	His	Glu	Glu	His
HbDβ Punjab	Val	His	Leu	Glu	Glu	Glu	His	Val	Gln	His

Figure 7.25 The molecular basis of sickle-cell and other anemias. (a) Substitution of glutamic acid with valine at the sixth amino acid from the N terminus affects the three-dimensional structure of the β chain of hemoglobin. Hemoglobins incorporating the mutant β chain form aggregates that cause red blood cells to sickle. **(b)** Red blood cell sickling has many phenotypic effects. **(c)** Other mutations in the β-chain gene also cause anemias.

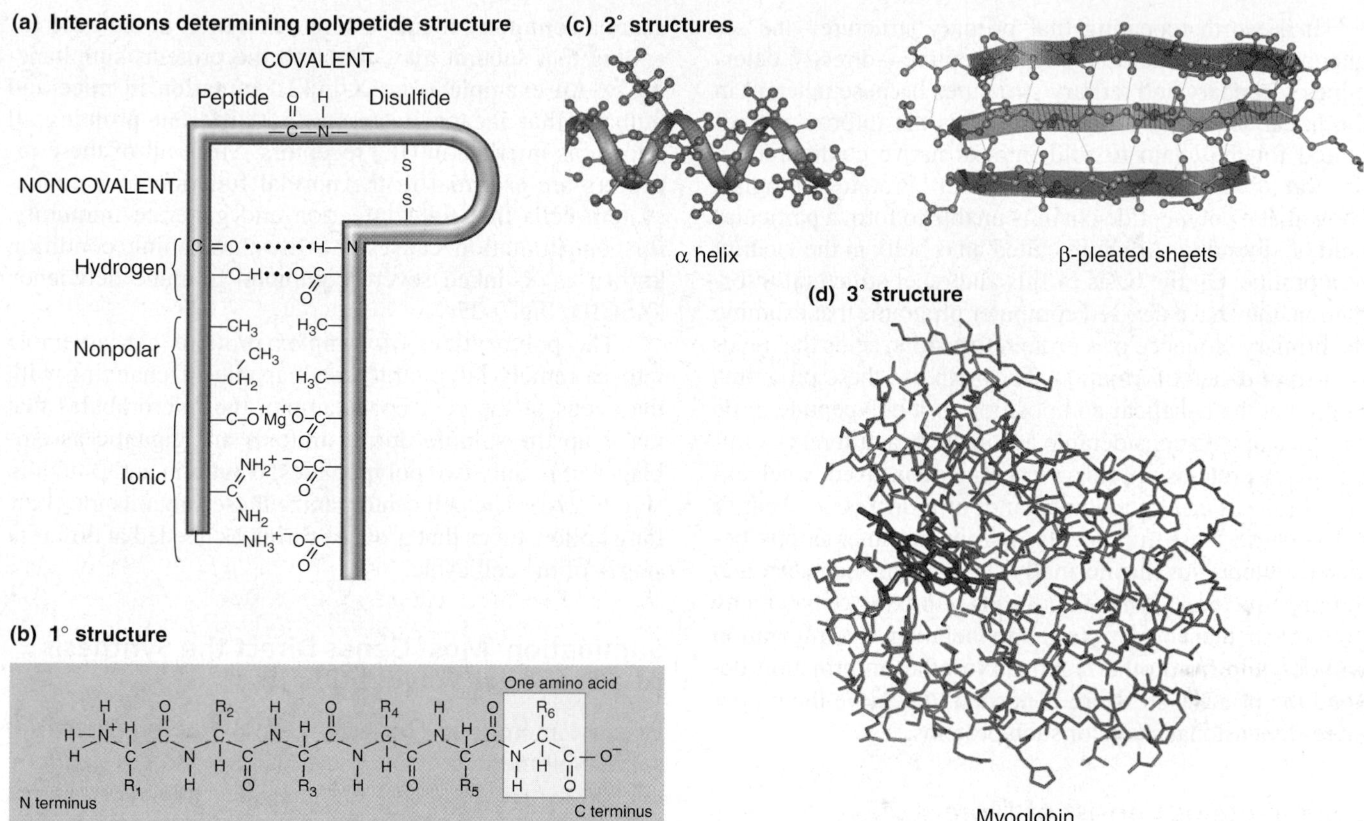

Figure 7.26 Levels of polypeptide structure. (a) Covalent and noncovalent interactions determine the structure of a polypeptide. **(b)** A polypeptide's primary (1°) structure is its amino acid sequence. **(c)** Localized regions form secondary (2°) structures such as α helices and β-pleated sheets. **(d)** The tertiary (3°) structure is the complete three-dimensional arrangement of a polypeptide. In this portrait of myoglobin, the iron-containing heme group, which carries oxygen, is in *red,* while the polypeptide itself is in *green.*

of four chains. At low concentrations of oxygen, the less soluble sickle-cell form of hemoglobin aggregates into long chains that deform the red blood cell (Fig. 7.25a).

Because people suffering from a variety of inherited anemias also have defective hemoglobin molecules, Ingram and other geneticists were able to determine how a large number of different mutations affect the amino acid sequence of hemoglobin (Fig. 7.25c). Most of the altered hemoglobins have a change in only one amino acid. In different patients, the alteration is generally in different amino acids, but occasionally, two independent mutations result in different substitutions for the same amino acid. Geneticists use the term **missense mutation** to describe a genetic alteration that causes the substitution of one amino acid for another.

The Sequence of Amino Acids in Polypeptides Determines a Protein's Three-Dimensional Shape and Thus Its Function

Despite the uniform nature of protein construction—a line of amino acids joined by peptide bonds—each type of polypeptide folds into a unique three-dimensional shape. The linear sequence of amino acids within a polypeptide is its **primary structure.** Each unique primary structure places constraints on how a chain can arrange itself in three-dimensional space. Because the R groups distinguishing the 20 amino acids have dissimilar chemical properties, some amino acids form hydrogen bonds or electrostatic bonds when brought into proximity with other amino acids. Nonpolar amino acids, for example, may become associated with each other by hydrophobic interactions that "hide" them from water, while two cysteine amino acids can form covalent disulfide bridges (–S–S–) through the oxidation of their –SH groups. All of these interactions (**Fig. 7.26a**) help stabilize the polypeptide in a specific three-dimensional conformation. The primary structure (Fig. 7.26b) determines three-dimensional shape by generating localized regions with a characteristic geometry known as **secondary structure** (Fig. 7.26c), as well as the other folds and twists that together with the secondary structure produce the ultimate three-dimensional **tertiary structure** of the entire polypeptide (Fig. 7.26d). Normal tertiary structure—the way a long chain of amino acids naturally folds in three-dimensional space under physiological conditions—is known as a polypeptide's *native configuration.* Various forces, including hydrogen bonds, electrostatic bonds, hydrophobic interactions, and disulfide bridges, help stabilize the native configuration.

It is worth repeating that primary structure—the sequence of amino acids in a polypeptide—directly determines secondary and tertiary structures because inherent in the linear sequence of amino acids is the information required for the chain to fold into its native configuration. Several examples illustrate this fact. Protein biologists know that a polypeptide chain is unable to form a particular kind of secondary structure called an α helix in the vicinity of a proline. On the basis of this and other structural information, they have devised computer programs that examine the primary sequence of a protein for amino acids that favor or disrupt α-helix formation. They can use these programs to predict the α-helical composition of a polypeptide in its native state with considerable accuracy. In a different example, many proteins unfold, or become **denatured,** when exposed to urea and mercaptoethanol or to increasing heat or pH because these treatments disrupt the interactions between amino acids that normally stabilize the secondary and tertiary structures. When conditions return to normal, many proteins spontaneously refold into their native configuration without help from other agents. No other information beyond the primary structure is needed to achieve the proper three-dimensional shape of such proteins.

Some Proteins Consist of More Than One Polypeptide

Certain proteins, such as the rhodopsin that promotes black-and-white vision, consist of one polypeptide. Many others, however, such as the lens crystallin protein, which provides rigidity and transparency to the lenses of our eyes, or the hemoglobin molecule that transports oxygen to and from our tissues, are composed of two or more polypeptide chains that associate in a specific way (**Fig. 7.27a**). The individual polypeptides in an aggregate are known as *subunits,* and the complex of subunits is often referred to as a *multimer.* The three-dimensional configuration of subunits in a multimer is a complex protein's **quaternary structure.** The same forces that stabilize the native form of a polypeptide (that is, hydrogen bonds, electrostatic bonds, hydrophobic interactions, and disulfide bridges) also contribute to the maintenance of quaternary structure. As Fig. 7.27a shows, in some multimers, the two or more interacting subunits are identical polypeptides. These identical chains are encoded by one gene. In other multimers, by contrast, more than one kind of polypeptide makes up the protein (Fig. 7.27b). The different polypeptides in these multimers are encoded by different genes.

Alterations in just one kind of subunit, caused by a mutation in a single gene, can affect the function of a multimer. The adult hemoglobin molecule, for example, consists of two α and two β subunits, with each type of subunit determined by a different gene—one for the α chain and one for the β chain. A mutation in the *Hbβ* gene resulting in an amino acid switch at position 6 in the *β* chain causes sickle-cell anemia. Similarly, if several multimeric proteins share a common subunit, a single mutation in the gene encoding that subunit may affect all the proteins simultaneously. An example is an X-linked mutation in mice and humans that incapacitates several different proteins all known as interleukin (IL) receptors. Since all of these receptors are essential to the normal function of immune-system cells that fight infection and generate immunity, this one mutation causes the life-threatening condition known as X-linked severe combined immune deficiency (XSCID; Fig. 7.27c).

The polypeptides of complex proteins can assemble into extremely large structures capable of changing with the needs of the cell. For example, the microtubules that make up the spindle during mitosis are gigantic assemblages of mainly two polypeptides: α tubulin and β tubulin (Fig. 7.27d). The cell can organize these subunits into very long hollow tubes that grow or shrink as needed at different stages of the cell cycle.

Summation: Most Genes Direct the Synthesis of a Particular Polypeptide

Because more than one gene governs the production of some multimeric proteins and because not all proteins are enzymes, the "one gene, one enzyme" hypothesis is not broad enough to define gene function. A more accurate statement is "one gene, one polypeptide": Each gene governs the construction of a particular polypeptide. As we will see in Chapter 8, even this reformulation does not encompass the function of all genes, as a few genes in all organisms do not determine the construction of proteins; instead, they encode RNAs that are not translated into polypeptides.

Beadle and Tatum's experiments were based on the concept that if each gene encodes a different polypeptide and if each polypeptide plays a specific role in the development, physiology, or behavior of an organism, then a mutation in the gene will block a biological process (like arginine synthesis in *Neurospora*) in a characteristic way. Other scientists soon realized they could use this approach to study virtually any interesting problem in biology. In the Fast Forward box "Using Mutagenesis to Look at Biological Processes" on pp. 240–241, we describe how one biologist found a large group of mutations that disrupted the assembly of bacteriophage T4 particles; by carefully studying the phenotypes caused by these mutations, he inferred the complex pathway that produces an entire bacteriophage.

Knowledge that genes encode proteins by determining the sequence of amino acids in a polypeptide enabled geneticists to analyze how different mutations in a single gene can produce different phenotypes. If each amino acid has a specific effect on the three-dimensional structure of a protein, then changing amino acids at different positions in a polypeptide chain can alter protein function in different ways. For example, most enzymes have an active site that carries out a particular task while other parts of the protein support the shape and position of that site. Mutations that change the

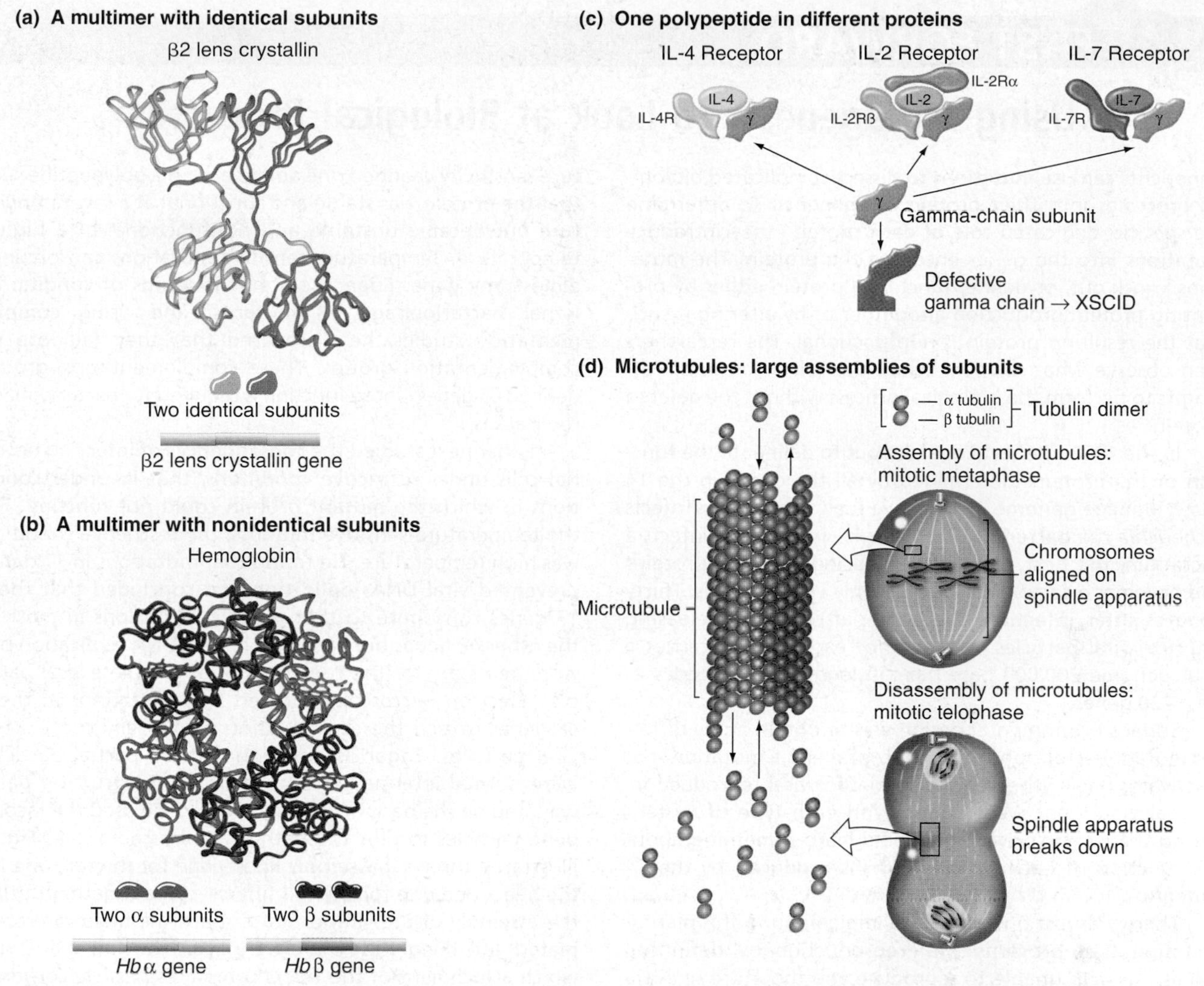

Figure 7.27 Multimeric proteins. (a) β2 lens crystallin contains two copies of one kind of subunit; the subunits are all the product of a single gene. The peptide backbones of the two subunits are shown in different shades of *purple*. **(b)** Hemoglobin is composed of two different kinds of subunits, each encoded by a different gene. **(c)** Three distinct protein receptors for the immune-system molecules called interleukins (ILs; *purple*). All contain a gamma (γ) chain (*yellow*), plus other receptor-specific polypeptides (*green*). A mutant γ chain blocks the function of all three receptors, leading to XSCID, a complete failure of the immune system. **(d)** One α-tubulin and one β-tubulin polypeptide associate to form a tubulin dimer. Many tubulin dimers form a single microtubule. The mitotic spindle is an assembly of many microtubules.

identity of amino acids at the active site may have more serious consequences than those that affect amino acids outside the active site. Some kinds of amino acid substitutions, such as replacement of an amino acid having a basic side chain with an amino acid having an acidic side chain, would be more likely to compromise protein function than substitutions that retain the chemical characteristics of the original amino acid.

Some mutations do not affect the amino acid composition of a protein but still generate an abnormal phenotype. As we see in Chapter 8, such mutations change the amount of normal polypeptide produced by generating altered nucleotide sequences that disrupt the biochemical processes responsible for decoding a gene into a polypeptide.

7.4 How Gene Mutations Affect Light-Receiving Proteins and Vision: A Comprehensive Example

Researchers first described anomalies of color perception in humans close to 200 years ago. Since that time, they have discovered a large number of mutations that modify human vision. By examining the phenotype associated with each mutation and then looking directly at the DNA alterations inherited with the mutation, they have

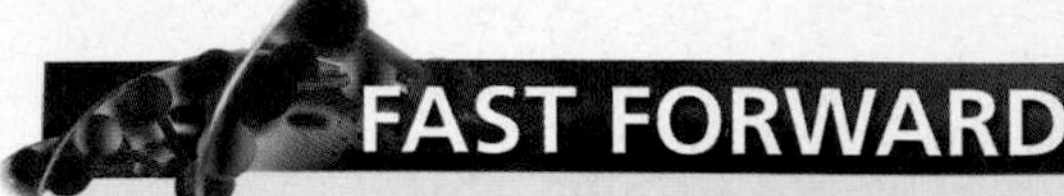

Using Mutagenesis to Look at Biological Processes

Geneticists can use mutations to dissect complicated biological processes into their protein components. To determine the specific, dedicated role of each protein, they introduce mutations into the genes encoding the protein. The mutations knock out, or delete, functional protein either by preventing protein production altogether or by altering it such that the resulting protein is nonfunctional. The researchers then observe what happens when the cell or organism attempts to perform the biological process without the deleted protein.

In the 1960s, Robert Edgar set out to delineate the function of the proteins determined by all the genes in the T4 bacteriophage genome. As we have seen, the T4 virus infects *Escherichia coli* bacteria. After a single viral particle infects a bacterium, the host cell stops producing bacterial proteins and becomes a factory for making only viral proteins. Thirty minutes after infection, the bacterial cell lyses, releasing 100 new viral particles. The head of each particle carries a DNA genome 200,000 base pairs in length that encodes at least 120 genes.

Edgar's experimental design was to obtain many different mutant bacteriophages, each containing a mutation that inactivates one of the genes essential for viral reproduction. By analyzing what went wrong with each type of mutant during the infective cycle, he would learn something about the function of each of the proteins produced by the T4 genome.

There was just one barrier to implementing this plan. A mutation that prevents viral reproduction by definition makes the virus unable to reproduce and therefore unavailable for experimental growth and study. The solution to this dilemma came with the discovery of *conditional lethal mutants:* microbes or other organisms (in this case, viral particles) carrying mutations that are lethal to the organism under one condition but not another. One type of conditional lethal mutant used by Edgar was temperature sensitive; that is, the mutant T4 phage could reproduce at low, but not at high, temperatures. The mutations causing temperature sensitivity changed one amino acid in a polypeptide such that the protein was stable and functional at a low temperature but became unstable and nonfunctional at a higher temperature. Temperature-sensitive mutations can occur in almost any gene. Edgar isolated thousands of conditional lethal bacteriophage T4 mutants, and using complementation studies, he discovered that they fall into 65 complementation groups. These complementation groups defined 65 genes whose function is required for bacteriophage replication.

Edgar next studied the consequences of infecting bacterial cells under *restrictive conditions,* that is, under conditions in which the mutant protein could not function. For the temperature-sensitive mutants, the restrictive condition was high temperature. He found that mutations in 17 genes prevented viral DNA replication and concluded that these 17 genes contribute to that process. Mutations in most of the other 48 genes did not impede viral DNA replication but were necessary for the construction of complete viral particles. Electron microscopy showed that mutations in these 48 genes caused the accumulation of partially constructed viral particles. Edgar assumed that these partial particles were normal intermediates in the virus construction pathway, and on the basis of this assumption, he used the incomplete particles to plot the path of viral assembly. As **Fig. A** illustrates, three subassembly lines—one for the tail, one for the head, and one for the tail fibers—come together during the assembly of the viral product. Once the heads are completed and filled with DNA, they attach to the tails, after which attachment of the fibers completes particle construction. It would have been very difficult to discern this trilateral assembly pathway by any means other than mutagenesis-driven genetic dissection.

Between 1990 and 1995, molecular geneticists determined the complete DNA sequence of the T4 genome, and then using the genetic code dictionary (described in Chapter 8), translated that sequence into coding regions for proteins. In addition to the 65 genes identified by Edgar, another

learned a great deal about the genes influencing human visual perception and the function of the proteins they encode.

Several attributes of human subjects facilitate the experimental analysis of vision. First, people can recognize and describe variations in the way they see, from trivial differences in what the color red looks like, to not seeing any difference between red and green, to not seeing any color at all. Second, the highly developed science of psychophysics provides sensitive, noninvasive tests for accurately defining and comparing phenotypes. One diagnostic test, for example, is based on the fact that people perceive each color as a mixture of three different wavelengths of light—red, green, and blue—and can adjust ratios of red, green, and blue light of different intensities to match an arbitrarily chosen fourth wavelength such as yellow. The mixture of wavelengths does not combine to form the fourth wavelength; it just appears that way to the eye. A person with normal vision, for instance, will select a well-defined proportion of red and green lights to match a particular yellow, but a person who can't tell red from green will permit any proportion of these two color lights to make the same match. Finally, since inherited variations in the visual system rarely affect fecundity or longevity in

55 genes became evident from the sequence. Edgar did not find these genes because they are not essential to viral reproduction under the conditions used in the laboratory. The previously unidentified genes most likely play important roles in the T4 life cycle outside the laboratory, perhaps when the virus infects hosts other than the *E. coli* strain normally used in the laboratory or when the virus grows under different environmental conditions and is competing with other viruses.

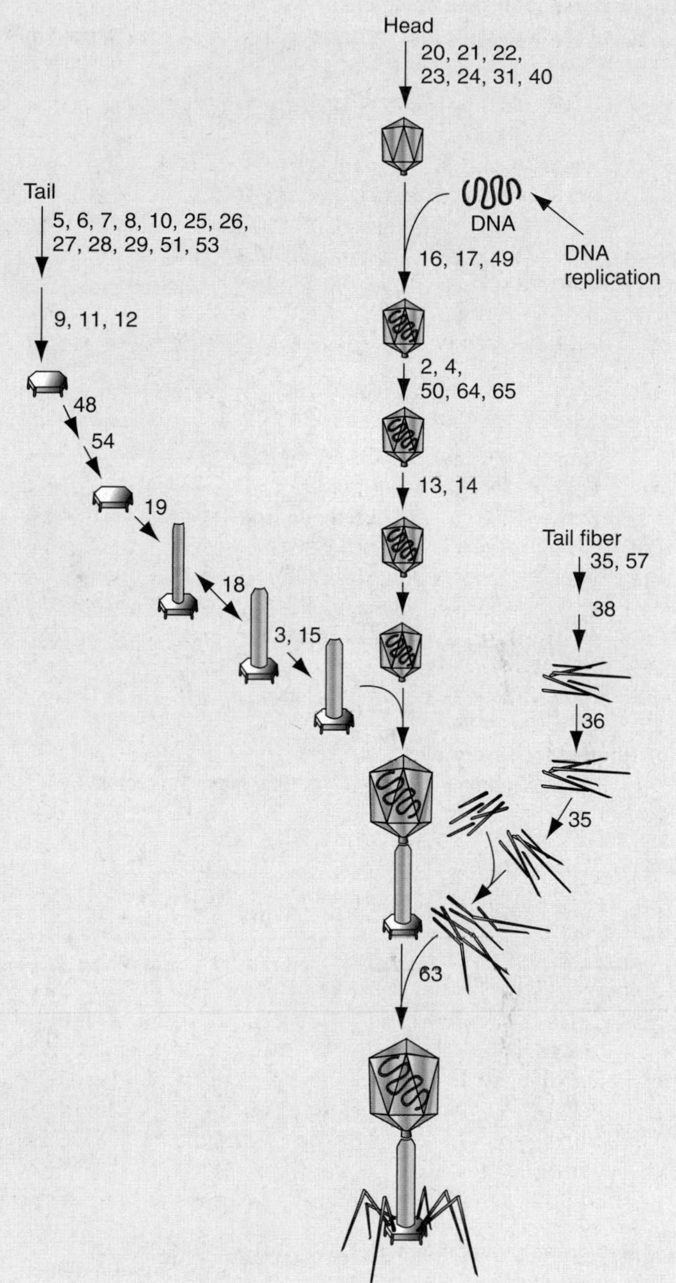

Figure A Steps in the assembly of bacteriophage T4. Robert Edgar determined what kinds of phage structures formed in bacterial cells infected with mutant T4 phage at restrictive temperatures. As an example, a cell infected with a phage carrying a temperature-sensitive mutation in gene 63 filled up with normal-looking phage that lacked tail fibers, and with normal-looking tail fibers. Edgar concluded that gene 63 encodes a protein that allows tail fibers to attach to otherwise completely assembled phage particles.

modern human societies, mutations generating many of the new alleles that change visual perception remain in a population over time.

The Cellular and Molecular Basis of Vision

The Cells

People perceive light through neurons in the retina at the back of the eye (**Fig. 7.28a**). These neurons are of two types: rods and cones. The rods, which make up 95% of all light-receiving neurons, are stimulated by weak light over a range of wavelengths. At higher light intensities, the rods become saturated and no longer send meaningful information to the brain. This is when the cones take over, processing wavelengths of bright light that enable us to see color. The cones come in three forms—one specializes in the reception of red light, a second in the reception of green, and a third in the reception of blue. For each photoreceptor cell, the act of reception consists of absorbing photons from light of a particular wavelength, transducing information about the number and energy of those photons to electrical signals, and transmitting the signals via the optic nerve to the brain.

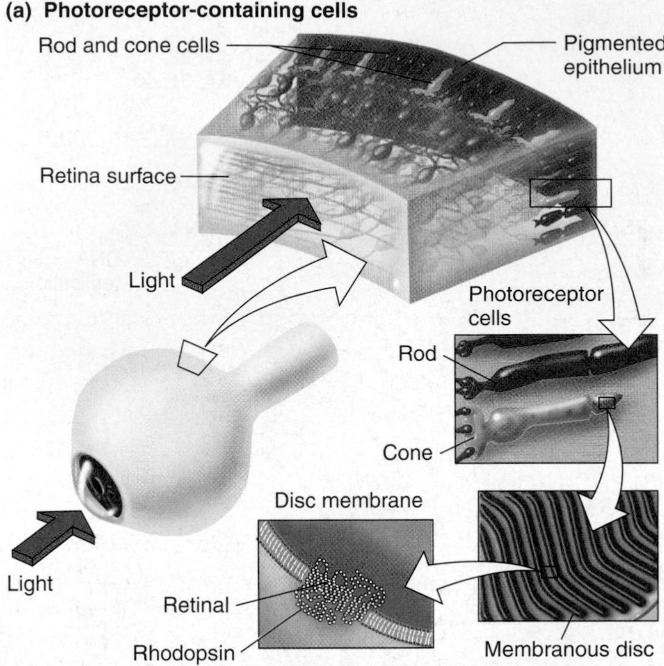

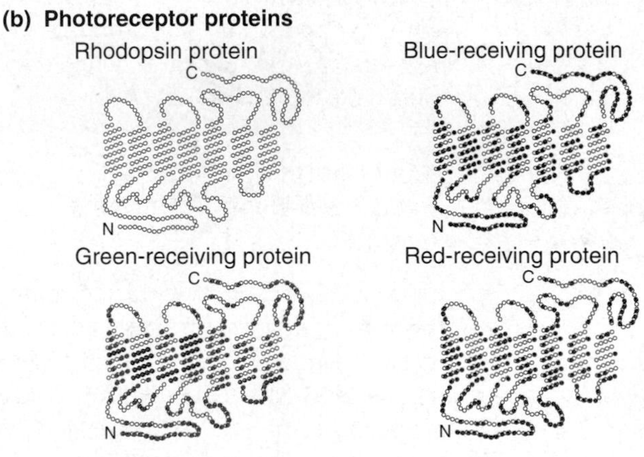

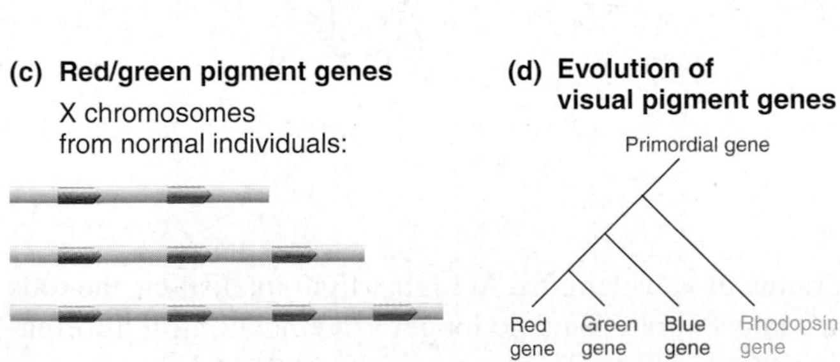

Figure 7.28 The cellular and molecular basis of vision. **(a)** Rod and cone cells in the retina carry membrane-bound photoreceptors. **(b)** The photoreceptor in rod cells is rhodopsin. The blue-, green-, and red-receiving proteins in cone cells are related to rhodopsin. **(c)** One red photoreceptor gene and one to three green photoreceptor genes are clustered on the X chromosome. **(d)** The genes for rhodopsin and the three color receptors probably evolved from a primordial photoreceptor gene through three gene duplication events followed by divergence of the duplicated copies.

Four Genes, Four Polypeptides

The protein that receives photons and triggers the processing of information in rod cells is rhodopsin. It consists of a single polypeptide chain containing 348 amino acids that snakes back and forth across the cell membrane (Fig.7.28b). One lysine within the chain associates with retinal, a carotenoid pigment molecule that actually absorbs photons. The amino acids in the vicinity of the retinal constitute rhodopsin's active site; by positioning the retinal in a particular way, they determine its response to light. Each rod cell contains approximately 100 million molecules of rhodopsin in its specialized membrane. As we have seen, the gene governing the production of rhodopsin is on chromosome 3.

The protein that receives and initiates the processing of photons in the blue cones is a relative of rhodopsin, also consisting of a single polypeptide chain containing 348 amino acids and also encompassing one molecule of retinal. Slightly less than half of the 348 amino acids in the blue-receiving protein are the same as those found in rhodopsin; the rest are different and account for the specialized light-receiving ability of the protein (Fig. 7.28b). The gene for the blue protein is on chromosome 7.

Similarly related to rhodopsin are the red- and green-receiving proteins in the red and green cones. These are also single polypeptides associated with retinal and embedded in the cell membrane, although they are both slightly larger at 364 amino acids in length (Fig. 7.28b). Like the blue protein, the red and green proteins differ from rhodopsin in nearly half of their amino acids; they differ from each other in only four amino acids out of every hundred. Even these small differences, however, are sufficient to differentiate the light sensitivities of the two types of cones and confer on them distinct spectral sensitivities. The genes for the red and green proteins both reside on the X chromosome in a tandem head-to-tail arrangement. Most individuals have one red gene and one to three green genes on their X chromosomes (Fig. 7.28c).

The Rhodopsin Gene Family Evolved by Duplication and Divergence

The similarity in structure and function between the four rhodopsin proteins suggests that the genes encoding these polypeptides arose by duplication of an original photoreceptor gene and then divergence through the accumulation of many mutations. Many of the mutations that promoted the ability to see color must have provided advantages that evolutionary forces selected for over millions of years. The red and green genes are the most similar, differing by less than five nucleotides out of every hundred. This suggests they diverged from each other only in the relatively recent evolutionary past. The less pronounced amino acid similarity of the red or green proteins with the blue protein, and the even lower relatedness between rhodopsin and any color photoreceptor, reflect earlier duplication and divergence events (Fig. 7.28d).

How Mutations in the Rhodopsin Gene Family Influence the Way We See

Many Amino Acid Substitutions in Rhodopsin Result in Partial or Complete Blindness

At least 29 different single nucleotide substitutions in the rhodopsin gene cause an autosomal dominant vision disorder known as *retinitis pigmentosa* that begins with an early loss of rod function followed by a slow progressive degeneration of the peripheral retina. **Figure 7.29a** shows the location of the amino acids affected by these mutations. These amino acid changes result in abnormal rhodopsin proteins that either do not fold properly or, once folded, are unstable. While normal rhodopsin is an essential structural element of rod cell membranes, these nonfunctional mutant proteins are retained in the body of the cell, where they remain unavailable for insertion into the membrane. Rod cells that cannot incorporate enough rhodopsin into their membrane eventually die. Depending on how many rod cells die, partial or complete blindness ensues.

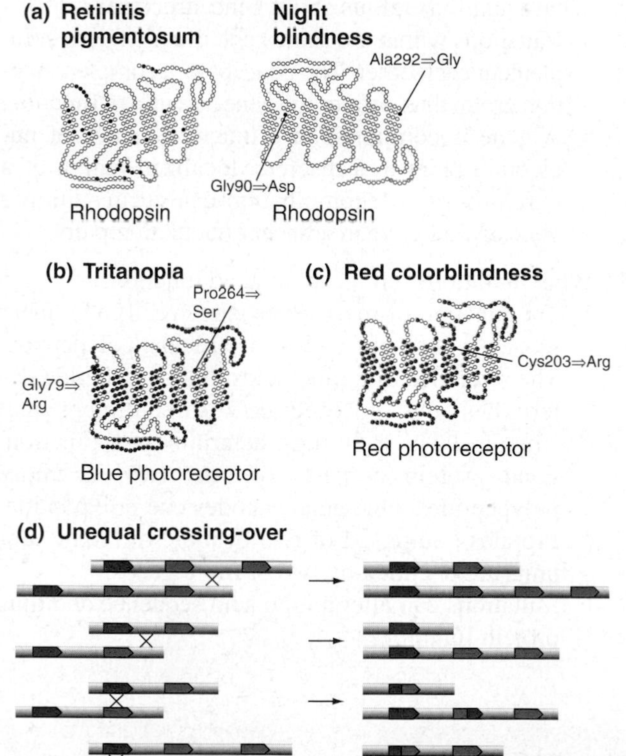

Figure 7.29 How mutations modulate light and color perception. **(a)** Amino-acid substitutions (*black dots*) that disrupt rhodopsin's three-dimensional structure result in retinitis pigmentosum. Other substitutions diminishing rhodopsin's sensitivity to light cause night blindness. **(b)** Substitutions in the blue pigment can produce tritanopia (blue colorblindness). **(c)** Red colorblindness can result from particular mutations that destabilize the red photoreceptor. **(d)** Unequal crossing-over between the red and green genes can alter their arrangement and create hybrid photoreceptor proteins.

Other mutations in the rhodopsin gene cause the far less serious condition of night blindness (Fig. 7.29a). These mutations change the protein's amino acid sequence so that the threshold of stimulation required to trigger the vision cascade increases. With the change, very dim light is no longer enough to initiate vision.

Mutations in the Cone-Cell Pigment Genes Alter Color Vision in Predictable Ways

Vision problems caused by mutations in the cone-cell pigment genes are less severe than vision problems caused by similar defects in the rod cells' rhodopsin genes, most likely because the rods make up 95% of a person's light-receiving neurons, while the cones comprise only about 5%. Some mutations in the blue gene on chromosome 7 cause *tritanopia,* a defect in the ability to discriminate between colors that differ only in the amount of blue light they contain (**Figs.** 7.29b and **7.30**). Mutations in the red gene on the X chromosome can modify or abolish red protein function and as a result, the red cone cells' sensitivity to light. For example, a change at position 203 in the red-receiving protein from cysteine to arginine disrupts one of the disulfide bonds required to support the protein's tertiary structure (see Fig. 7.29c). Without that bond, the protein cannot stably maintain its native configuration, and a person with the mutation has red colorblindness.

Unequal Crossing-Over Between the Red and Green Genes Produces Most Variations in Red-Green Perception

People with normal color vision have a single red gene; some of these normal individuals also have a single adjacent green gene, while others have two or even three green genes. The red and green genes are 96% identical in DNA sequence; the different green genes, 99.9% identical. The proximity and high degree of homology make these genes unusually prone to unequal crossing-over. A variety of

Figure 7.30 How the world looks to a person with tritanopia. Compare with Fig. 4.22 on p. 110.

unequal recombination events produce DNA containing no red gene, no green gene, various combinations of green genes, or hybrid red-green genes (see Fig. 7.29d). These different DNA combinations account for the large majority of the known aberrations in red-green color perception, with the remaining abnormalities stemming from point mutations, as described earlier. Because the accurate perception of red and green depends on the differing ratios of red and green light processed, people with no red or no green gene perceive red and green as the same color (see Fig. 4.22 on p. 110).

In short, we see the way we do in part because four genes direct the production of four polypeptides in the rod and cone cells of our retina. Mutations that alter those polypeptides or their amounts change our perception of light or color.

Connections

Careful studies of mutations showed that genes are linear arrays of mutable elements that direct the assembly of amino acids in a polypeptide. The mutable elements are the nucleotide building blocks of DNA.

Biologists call the parallel between the sequence of nucleotides in a gene and the order of amino acids in a polypeptide **colinearity.** In Chapter 8, we explain how colinearity arises from base pairing, a genetic code, specific enzymes, and macromolecular assemblies like ribosomes that guide the flow of information from DNA through RNA to protein.

Essential Concepts

1. Mutations are alterations in the nucleotide sequence of the DNA molecule that occur by chance and modify the genome at random. Once they occur, mutations can be transmitted from generation to generation when DNA replicates.
   a. Mutations that affect phenotype occur naturally at a very low rate, which varies from gene to gene. Forward mutations usually occur more often than reversions.
   b. The agents of spontaneously occurring mutations include chemical hydrolysis; radiation, such as cosmic rays and ultraviolet light; and mistakes during DNA replication. Mutagens raise the frequency of mutation above the spontaneous rate.
   c. Cells have evolved various enzyme systems that repair DNA and thus minimize mutations.
   d. Mutations are the raw material of evolution. The action of natural selection on heritable mutations is a major agent of evolution.
2. What mutations tell us about gene structure:
   a. Mutations within the same gene usually fail to complement each other. The concept of a complementation group thus defines the gene as a unit of function.
   b. A gene is composed of a linear sequence of nucleotide pairs in a discrete, localized region of a chromosome. Recombination can occur within a gene, even between adjacent nucleotide pairs.
3. What mutations tell us about gene function:
   a. The function of most genes is to specify the linear sequence of amino acids in a particular polypeptide.
   b. The sequence of amino acids in a polypeptide determines the polypeptide's three-dimensional structure, which, in turn, determines its function. Each protein consists of one, two, or more polypeptides. One gene encodes one polypeptide. Proteins composed of two or more different subunits are encoded by two or more genes.
   c. Mutations can alter amino acid sequence and thus protein function.

On Our Website

www.mhhe.com/hartwell3
Chapter 7

Annotated Suggested Readings and Links to Other Websites

- Historical monographs on the nature of mutation, the action of mutagens, DNA repair systems, fine-structure mapping, the "one gene, one polypeptide" hypothesis, and the genetics of human color vision.
- Interesting recent research articles about whether mutations are truly introduced at random, how TEs and trinucleotide repeats affect genomic stability and human health, and examples of the use of genetics to analyze complicated biological processes.

Specialized Topics

- Complications in the interpretation of complementation analysis: a document explaining rare exceptions to the rule that mutations in the same gene are unable to complement each other, as well as other rare cases in which mutations in different genes can fail to complement each other.

Social and Ethical Issues

1. Chemicals that are mutagenic are identified by the Ames test, which measures the level of mutagenesis in bacteria. The susceptibility of humans to mutagenic chemicals may vary depending on the genetic makeup of the individual. The dose that affects one person may be different from that which affects another. However, there are few, if any, reliable tests that determine a person's level of susceptibility. If this is true, is it a good idea to translate the results of the Ames test of mutability in bacteria to a prediction of carcinogenicity in humans? Often, reports of Ames test results on a chemical make newspaper headlines. Is this a useful and honest way to report findings that could affect human health, or do people need to consider other variables to make an informed decision?

2. Mr. and Mrs. Aswari have a child with fragile X syndrome (see the Genetics and Society box on pp. 216–217). They want to have a second child but are considering egg donation because genetic screening has indicated that Mrs. Aswari carries a premutation allele with 120 CGG repeats. If you were the Aswari's genetic counselor, what would you tell them about their risk of having a second child with fragile X syndrome? What are the ethical issues related to genetic screening when (1) a result indicates no risk, (2) a result indicates that the phenotype being screened for will be exhibited, and (3) an intermediary result does not clearly fall into either category?

Solved Problems

I. Mutations can often be reverted to wild type by treatment with mutagens. The type of mutagen that will reverse a mutation gives us information about the nature of the original mutation. The mutagen EMS almost exclusively causes transitions; proflavin is an intercalating agent that causes insertion or deletion of a base; ultraviolet (UV) light causes single-base substitutions. Cultures of several *E. coli met*$^-$ mutants were treated with three mutagens separately and spread onto a plate lacking methionine to look for revertants. (In the chart, − indicates that no colonies grew, and + indicates that some *met*$^+$ revertant colonies grew.)

	Mutagen treatment		
Mutant number	EMS	Proflavin	UV light
1	+	−	+
2	−	+	−
3	−	−	−
4	−	−	+

a. Given the results, what can you say about the nature of the original mutation in each of the strains?

b. Experimental controls are designed to eliminate possible explanations for the results, thereby ensuring that data are interpretable. In the experiment described, we scored the presence or absence of colonies. How do we know if colonies that appear on plates are mutagen-induced revertants? What else could they be? What control would enable us to be confident of our revertant analysis?

Answer

To answer this question, you need to understand the concepts of mutation and reversion.

a. Mutation 1 is reverted by the mutagen that causes transitions, *so mutation 1 must have been a transition.* Consistent with this conclusion is the fact the UV light can also revert the mutation and the intercalating agent proflavin does not cause reversion. *Mutation 2 is reverted by proflavin and therefore must be either an insertion or a deletion of a base.* The other two mutagens do not revert mutation 2. Mutation 3 is not reverted by any of these mutagenic agents. It is therefore not a single-base substitution, a single-base insertion, or a single-base deletion. *Mutation 3 could be a deletion of several bases or an inversion.* Mutation 4 is reverted by UV light, so it is a single-base change, but it is not a transition, since EMS did not revert the mutation. *Mutation 4 must be a transversion.*

b. *The colonies on the plates could arise by spontaneous reversion of the mutation.* Spontaneous reversion should occur with lower frequency than mutagen-induced reversion. The important control here is to *spread each mutant culture without any mutagen treatment onto selective media to assess the level of spontaneous reversion.*

II. Imagine that 10 independently isolated recessive lethal mutations (l^1, l^2, l^3, etc.) map to chromosome 7 in mice. You perform complementation testing by mating all pairwise combinations of heterozygotes bearing these lethal mutations, and you score the absence of complementation by examining pregnant females for dead fetuses. A + in the chart means that the two lethals complemented, and dead embryos were not found. A − indicates that dead embryos were found, at the rate of about one in four conceptions. (The crosses between heterozygous mice would be expected to yield the homozygous recessive showing the lethal phenotype in 1/4 of the embryos.) The lethal mutation in the parental heterozygotes for each cross are listed across the top and down the left side of the chart (that is, l^1 indicates a heterozygote in which one chromosome bears the l^1 mutation and the homologous chromosome is wild type).

	l^1	l^2	l^3	l^4	l^5	l^6	l^7	l^8	l^9	l^{10}
l^1	−	+	+	+	+	−	−	+	+	+
l^2		−	+	+	+	+	+	+	+	−
l^3			−	−	−	+	+	−	−	+
l^4				−	−	+	+	−	−	+
l^5					−	+	+	−	−	+
l^6						−	−	+	+	+
l^7							−	+	+	+
l^8								−	−	+
l^9									−	+
l^{10}										−

How many genes do the 10 lethal mutations represent? What are the complementation groups?

Answer

This problem involves the application of the complementation concept to a set of data. There are two ways to analyze these results. You can focus on the mutations that do complement each other, conclude that they are in different genes, and begin to create a list of mutations in separate genes. Alternatively, you can focus on mutations that do not complement each other and therefore are alleles of the same genes. The latter approach is more efficient when several mutations are involved. For example, l^1 does not complement l^6 and l^7. These three alleles are in one complementation group. l^2 does not complement l^{10}; they are in a second complementation group. l^3 does not complement l^4, l^5, l^8, or l^9, so they form a third complementation group. *There are three complementation groups.* (Note also that for each mutant, the cross between individuals carrying the same alleles resulted in no complementation, because the homozygous recessive lethal was generated.) *The three complementation groups consist of (1)* 1^1, 1^6, 1^7; *(2)* 1^2, 1^{10}; *and (3)* 1^3, 1^4, 1^5, 1^8, 1^9.

III. W, X, and Y are the intermediates (in that order) in a biochemical pathway whose product is Z. Z^- mutants are found in five different complementation groups. *Z1* mutants will grow on Y or Z but not W or X. *Z2* mutants will grow on X, Y, or Z. *Z3* mutants will only grow on Z. *Z4* mutants will grow on Y or Z. Finally, *Z5* mutants will grow on W, X, Y, or Z.

a. Order the five complementation groups in terms of the steps they block.
b. What does this genetic information reveal about the nature of the enzyme that carries out the conversion of X to Y?

Answer

This problem requires that you understand complementation and the connection between genes and enzymes in a biochemical pathway.

a. A biochemical pathway represents an ordered set of reactions that must occur to produce a product. This problem gives the order of intermediates in a pathway for producing product Z. The lack of any enzyme along the way will cause the phenotype of Z^-, but the block can occur at different places along the pathway. If the mutant grows when given an intermediate compound, the enzymatic (and hence gene) defect must be before production of that intermediate compound. The *Z1* mutants that grow on Y or Z (but not on W or X) must have a defect in the enzyme that produces Y. *Z2* mutants have a defect prior to X; *Z3* mutants have a defect prior to Z; *Z4* mutants have a defect prior to Y: *Z5* have a defect prior to W. *The five complementation groups can be placed in order of activity within the biochemical pathway as follows:*

$$\xrightarrow{Z5} W \xrightarrow{Z2} X \xrightarrow{Z1,\,Z4} Y \xrightarrow{Z3} Z$$

b. Mutants *Z1* and *Z4* affect the same step, but because they are in different complementation groups, we know they are in different genes. *Mutations* Z1 *and* Z4 *are probably in genes that encode subunits of a multisubunit enzyme that carries out the conversion of X to Y.* Alternatively, there could be a currently unknown additional intermediate step between X and Y.

Problems

Vocabulary

1. The following is a list of mutational changes. For each of the specific mutations described, indicate which of the terms in the right-hand column applies, either as a description of the mutation or as a possible cause. More than one term from the right column can apply to each statement in the left column.

   1. an A–T base pair in the wild-type gene is changed to a G–C pair
   2. an A–T base pair is changed to a T–A pair
   3. the sequence AAGCTTATCG is changed to AAGCTATCG
   4. the sequence AAGCTTATCG is changed to AAGCTTTATCG
   5. the sequence AACGTTATCG is changed to AATGTTATCG
   6. the sequence AACGTCACACACACATCG is changed to AACGTCACATCG
   7. the gene map in a given chromosome arm is changed from *bog-rad-fox1-fox2-try-duf* (where *fox1* and *fox2* are highly homologous, recently diverged genes) to *bog-rad-fox1-fox3-fox2-try-duf* (where *fox3* is a new gene with one end similar to *fox1* and the other similar to *fox2*)
   8. the gene map in a chromosome is changed from *bog-rad-fox1-fox2-try-duf* to *bog-rad-fox2-fox1-try-duf*
   9. the gene map in a given chromosome is changed from *bog-rad-fox1-fox2-try-duf* to *bog-rad-fox1-mel-qui-txu-sqm*

   a. transition
   b. base substitution
   c. transversion
   d. inversion
   e. translocation
   f. deletion
   g. insertion
   h. deamination
   i. X-ray irradiation
   j. intercalator
   k. unequal crossing-over

Section 7.1

2. The DNA sequence of a gene from three independently isolated mutants is given here. Using this information, what is the sequence of the wild-type gene in this region?

```
mutant 1   ACCGTAATCGACTGGTAAACTTTGCGCG
mutant 2   ACCGTAGTCGACCGGTAAACTTTGCGCG
mutant 3   ACCGTAGTCGACTGGTTAACTTTGCGCG
```

3. Over a period of several years, a large hospital kept track of the number of births of babies displaying the trait achondroplasia. Achondroplasia is a very rare autosomal dominant condition resulting in dwarfism with abnormal body proportions. After 120,000 births, it was noted that there had been 27 babies born with achondroplasia. One physician was interested in determining how many of these dwarf babies result from new mutations and whether the apparent mutation rate in his area was higher than normal. He looked up the families of the 27 dwarf births and discovered that 4 of the dwarf babies had a dwarf parent. What is the apparent mutation rate of the achondroplasia gene in this population? Is it unusually high or low?

4. Among mammals, measurements of the rate of generation of autosomal recessive mutations have been made almost exclusively in mice, while many measurements of the rate of generation of dominant mutations have been made both in mice and in humans. Why do you think there has been this difference?

5. In a genetics lab, Kim and Maria infected a sample from an *E. coli* culture with a particular virulent bacteriophage. They noticed that most of the cells were lysed, but a few survived. The survival rate in their sample was about 1×10^{-4}. Kim was sure the bacteriophage induced the resistance in the cells, while Maria thought that resistant mutants probably already existed in the sample of cells they used. Earlier, for a different experiment, they had spread a dilute suspension of *E. coli* onto solid medium in a large petri dish, and, after seeing that about 10^5 colonies were growing up, they had replica-plated that plate onto three other plates. Kim and Maria decided to use these plates to test their theories. They pipette a suspension of the bacteriophage onto each of the three replica plates. What should they see if Kim is right? What should they see if Maria is right?

6. Suppose you wanted to study genes controlling the structure of bacterial cell surfaces. You decide to start by isolating bacterial mutants that are resistant to infection by a bacteriophage that binds to the cell surface. The selection procedure is simple: Spread cells from a culture of sensitive bacteria on a petri plate, expose them to a high concentration of phages, and pick the bacterial colonies that grow. To set up the selection you could (1) spread cells from a single liquid culture of sensitive bacteria on many different plates and pick every resistant colony *or* (2) start many different cultures, each grown from a single colony of sensitive bacteria, spread one plate from each culture, and then pick a single mutant from each plate. Which method would ensure that you are isolating many independent mutations?

7. A wild-type male *Drosophila* was exposed to a large dose of X-rays and was then mated to an unirradiated female, one of whose X chromosomes carried both a dominant mutation for the trait *Bar* eyes and several inversions. Many F_1 females from this mating were recovered who had the *Bar*, multiply inverted X chromosome from their mother, and an irradiated X chromosome from their fathers. (The inversions ensure that viable offspring of these F_1 females will not have recombinant X chromosomes, as explained in Chapter 14.) After mating to normal males, most F_1 females produced *Bar* and wild-type sons in equal proportions. There were three exceptional F_1 females, however. Female A produced

as many sons as daughters, but half of the sons had *Bar* eyes, and the other half had white eyes. Female B produced half as many sons as daughters, and all of the sons had *Bar* eyes. Female C produced 75% as many sons as daughters. Of these sons, 2/3 had *Bar* eyes, and 1/3 had wild-type eyes. Explain the results obtained with each exceptional F_1 female.

8. A wild-type *Drosophila* female was mated to a wild-type male that had been exposed to X-rays. One of the F_1 females was then mated with a male that had the following recessive markers on the X chromosome: *yellow* body (*y*), *crossveinless* wings (*cv*), *cut* wings (*ct*), *singed* bristles (*sn*), and *miniature* wings (*m*). These markers are known to map in the order *y–cv–ct–sn–m*. The progeny of this second mating were unusual in two respects. First, there were twice as many females as males. Second, while all of the males were wild type in phenotype, 1/2 of the females were wild type, and the other 1/2 exhibited the *ct* and *sn* phenotypes.

a. What did the X-rays do to the irradiated male?
b. Draw the X-chromosome pair present in a progeny female fly produced by the second mating that was phenotypically *ct* and *sn.*
c. If the *ct* and *sn* female fly whose chromosomes were drawn in part (b) was then crossed to a wild-type male, what phenotypic classes would you expect to find among the progeny males?

9. In the experiment shown in Fig. 7.9 on p. 215, H.J. Muller first performed a control in which the P generation males were not exposed to X-rays. He found that 99.7% of the individual F_1 Bar-eyed females produced some male progeny with Bar eyes and some with wild-type (non-Bar) eyes, but 0.3% of these females produced male progeny that were all wild type.

a. If the average spontaneous mutation rate for *Drosophila* genes is 3.5×10^{-6} mutations /gene/gamete, how many genes on the X chromosome can be mutated to produce a recessive lethal allele?
b. As of the year 2005, analysis of the *Drosophila* genome had revealed a total of 2279 genes on the X chromosome. Assuming the X chromosome is typical of the genome, what is the fraction of genes in the fly genome that is essential to survival?
c. Muller now exposed male flies to a specific high dosage of X-rays and found that 12% of F_1 Bar-eyed females produced male progeny that were all wild type. What does this new information say?

10. Figure 7.10 on pp. 218–219 shows examples of base substitutions induced by the mutagens 5-bromouracil, hydroxylamine, ethylmethane sulfonate, and nitrous acid. Which of these mutagens cause transitions, and which cause transversions?

11. So-called *two-way mutagens* can induce both a particular mutation and (when added subsequently to cells whose chromosomes carry this mutation) a reversion of the mutation that restores the original DNA sequence. In contrast, *one-way mutagens* can induce mutations but not exact reversions of these mutations. Based on Fig. 7.10 (pp. 218–219), which of the following mutagens can be classified as one-way and which as two-way?

a. 5-bromouracil
b. hydroxylamine
c. ethylmethane sulfonate
d. nitrous acid
e. proflavin

12. In 1967, J.B. Jenkins treated wild-type male *Drosophila* with the mutagen ethylmethane sulfonate (EMS) and mated them with females homozygous for a recessive mutation called *dumpy* that causes shortened wings. He found some F_1 progeny with two wild-type wings, some with two short wings, and some with one short wing and one wild-type wing. When he mated single F_1 flies with two short wings to *dumpy* homozygotes, he surprisingly found that only about 1/3 of these matings produced any short-winged progeny.

a. Explain these results in light of the mechanism of action of EMS shown in Fig. 7.10 on pp. 218–219.
b. Should the short-winged progeny of the second cross have one or two short wings? Why?

13. Aflatoxin B_1 is a highly mutagenic and carcinogenic compound produced by certain fungi that infect crops such as peanuts. Aflatoxin is a large, bulky molecule that chemically bonds to the base guanine to form the aflatoxin-guanine "adduct" that is pictured below. (In the figure, the aflatoxin is *red,* and the guanine base is *purple.*) This adduct distorts the DNA double helix and blocks replication.

a. What type(s) of DNA repair system is (are) most likely to be involved in repairing the damage caused by exposure of DNA to aflatoxin B_1?
b. Recent evidence suggests that the adduct of guanine and aflatoxin B_1 can attack the bond that connects it to deoxyribose; this liberates the adduced base, forming an apurinic site. How does this new information change your answer to part *a*?

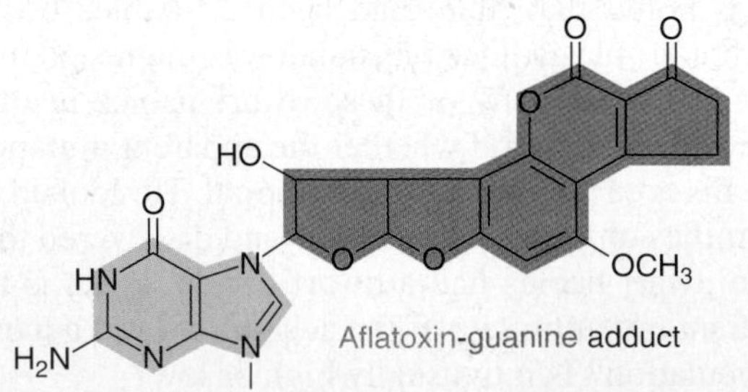

Aflatoxin-guanine adduct

14. When a particular mutagen identified by the Ames test is injected into mice, it causes the appearance of many tumors, showing that this substance is carcinogenic. When cells from these tumors are injected into other mice not exposed to the mutagen, almost all of the new mice develop tumors. However, when mice carrying mutagen-induced tumors are mated to unexposed mice, virtually all of the progeny are tumor free. Why can the tumor be transferred horizontally (by injecting cells) but not vertically (from one generation to the next)?

15. When the *his*$^-$ *Salmonella* strain used in the Ames test is exposed to substance X, no *his*$^+$ revertants are seen. If, however, rat liver supernatant is added to the cells along with substance X, revertants do occur. Is substance X a potential carcinogen for human cells? Explain.

Section 7.2

16. Imagine that you caught a female albino mouse in your kitchen and decided to keep it for a pet. A few months later, while vacationing in Guam, you caught a male albino mouse and decided to take it home for some interesting genetic experiments. You wonder whether the two mice are both albino due to mutations in the same gene. What could you do to find out the answer to this question? Assume that both mutations are recessive.

17. Plant breeders studying genes influencing leaf shape in the plant *Arabidopsis thaliana* identified six independent recessive mutations that resulted in plants that had unusual leaves with serrated rather than smooth edges. The investigators started to perform complementation tests with these mutants, but some of the tests could not be completed because of an accident in the greenhouse. The results of the complementation tests that could be finished are shown in the table that follows.

	1	2	3	4	5	6
1	–	+	–		+	
2		–				–
3			–	–		
4				–		
5					–	+
6						–

a. Exactly what experiment was done to fill in individual boxes in the table with a + or a – ? What does + represent? What does – represent? Why are some boxes in the table filled in *green*?

b. Assuming no complications, what do you expect for the results of the complementation tests that were not performed? That is, complete the table above by placing a + or a – in each of the blank boxes.

c. How many genes are represented among this collection of mutants? Which mutations are in which genes?

18. In humans, albinism is normally inherited in an autosomal recessive fashion. Figure 3.19c on p. 66 shows a pedigree in which two albino parents have several children, none of whom is an albino.

a. Interpret this pedigree in terms of a complementation test.

b. It is very rare to find examples of human pedigrees such as Fig. 3.19c that could be interpreted as a complementation test. This is because most genetic conditions in humans are rare, so it is highly unlikely that unrelated people with the same condition would mate. In the absence of complementation testing, what kinds of experiments could be done to determine whether a particular human disease phenotype can be caused by mutations at more than one gene?

c. Complementation testing requires that the two mutations to be tested both be recessive to wild type. Suppose that two dominant mutations cause similar phenotypes. How could you establish whether these mutations affected the same gene or different genes?

19. a. Seymour Benzer's fine structure analysis of the *rII* region of bacteriophage T4 depended in large part on deletion analysis as shown in Fig. 7.21 on p. 231. But to perform such deletion analysis, Benzer had to know which *rII*$^-$ bacteriophage strains were deletions and which were point mutations. How do you think he was able to distinguish *rII*$^-$ deletions from point mutations?

b. Benzer concluded that recombination can occur between adjacent nucleotide pairs, even within the same gene. How was he able to make this statement? At the time, Benzer had two relevant pieces of information: (i) the total length in μm of the bacteriophage T4 chromosome (measured in the electron microscope) and (ii) many mutations in many bacteriophage T4 genes, including *rIIA* and *rIIB*.

c. Figure 7.21c on p. 231 shows Benzer's fine structure map of point mutations in the *rII* region. A key feature of this map is the existence of "hot spots," which Benzer interpreted as nucleotide pairs that were particularly susceptible to mutation. How could Benzer say that all the independent mutations in a hot spot were due to mutations of the same nucleotide pair?

20. a. You have a test tube containing 5 ml of a solution of bacteriophage, and you would like to estimate the number of bacteriophage in the tube. Assuming the tube actually contains a total of 15 billion bacteriophage, design a serial dilution experiment that would allow you to estimate this number. Ideally, the final plaque-containing plates you count should contain more than 10 and less than 1000 plaques.

b. When you count bacteriophage by the serial dilution method as in part *a,* you are assuming a *plating efficiency* of 100%; that is, the number of plaques on the petri plate exactly represents the number of bacteriophage you mixed with the plating bacteria. Is there any way to test the possibility that only a certain percentage of bacteriophage particles are able to form plaques (so that the plating efficiency would be less than 100%)? Conversely, why is it fair to assume that any plaques are initiated by one rather than multiple bacteriophage particles?

21. You found five T4 *rII*⁻ mutants that will not grow on *E. coli* K(λ). You mixed together all possible combinations of two mutants (as indicated in the following chart), added the mixtures to *E. coli* K(λ), and scored for the ability of the mixtures to grow and make plaques (indicated as a + in the chart).

	1	2	3	4	5
1	−	+	+	−	+
2		−	−	+	−
3			−	+	−
4				−	+
5					−

a. How many genes were identified by this analysis?
b. Which mutants belong to the same complementation groups?

22. The *rosy (ry)* gene of *Drosophila* encodes an enzyme called xanthine dehydrogenase. Flies homozygous for *ry* mutations exhibit a rosy eye color. Heterozygous females were made that had ry^{41} *Sb* on one homolog and *Ly* ry^{564} on the other homolog, where ry^{41} and ry^{564} are two independently isolated alleles of *ry. Ly* (*Lyra* [narrow] wings) and *Sb* (*Stubble* [short] bristles) are dominant markers to the left and right of *ry,* respectively. These females are now mated to males homozygous for ry^{41}. Out of 100,000 progeny, 8 have wild-type eyes, *Lyra* wings, and *Stubble* bristles, while the remainder have rosy eyes.

a. What is the order of these two *ry* mutations relative to the flanking genes *Ly* and *Sb?*
b. What is the genetic distance separating ry^{41} and ry^{564}?

23. Nine *rII*⁻ mutants of bacteriophage T4 were used in pairwise infections of *E. coli* K(λ) hosts. Six of the mutations in these phages are point mutations; the other three are deletions. The ability of the doubly infected cells to produce progeny phages in large numbers is scored in the following chart.

	1	2	3	4	5	6	7	8	9
1	−	−	+	+	−	−	−	+	+
2		−	+	+	−	−	−	+	+
3			−	−	+	−	+	−	−
4				−	+	−	+	−	−
5					−	−	−	+	+
6						−	−	−	−
7							−	+	+
8								−	−
9									−

The same nine mutants were then used in pairwise infections of *E. coli* B hosts. The production of progeny phage that can subsequently lyse *E. coli* K(λ) hosts is now scored. In the table, 0 means the progeny do not produce any plaques on *E. coli* K(λ) cells; − means that only a very few progeny phages produce plaques; and + means that many progeny produce plaques (more than 10 times as many as in the − cases).

	1	2	3	4	5	6	7	8	9
1	−	+	+	+	+	−	−	+	+
2		−	+	+	+	+	−	+	+
3			0	−	+	0	+	+	−
4				−	+	−	+	+	+
5					−	+	−	+	+
6						0	0	−	+
7							0	+	+
8								−	+
9									−

a. Which of the mutants are the three deletions? What criteria did you use to reach your conclusion?
b. If you know that mutation 9 is in the *rIIB* gene, draw the best genetic map possible to explain the data, including the positions of all point mutations and the extent of the three deletions.
c. There should be one uncertainty remaining in your answer to part *b.* How could you resolve this uncertainty?

24. In a haploid yeast strain, eight recessive mutations were found that resulted in a requirement for the amino acid lysine. All the mutations were found to revert at a frequency of about 1×10^{-6}, except mutations 5 and 6, which did not revert. Matings were made between *a* and α cells carrying these mutations. The ability of the resultant diploid strains to grow on minimal medium in the absence of lysine is shown in the following chart (+ means growth and − means no growth.)

	1	2	3	4	5	6	7	8
1	−	+	+	+	+	−	+	−
2	+	−	+	+	+	+	+	+
3	+	+	−	−	−	−	−	+
4	+	+	−	−	−	−	−	+
5	+	+	−	−	−	−	−	+
6	−	+	−	−	−	−	−	−
7	+	+	−	−	−	−	−	+
8	−	+	+	+	+	−	+	−

a. How many complementation groups were revealed by this data? Which point mutations are found within which complementation groups?

The same diploid strains are now induced to undergo sporulation. The vast majority of resultant spores are auxotrophic; that is, they cannot form colonies when plated on minimal medium minus lysine. However, particular diploids can produce rare spores that do form colonies when plated on minimal medium minus lysine (prototrophic spores). The following table shows whether (+) or not (−) any prototrophic spores are formed upon sporulation of the various diploid cells.

	1	2	3	4	5	6	7	8
1	−	+	+	+	+	−	+	+
2	+	−	+	+	+	+	+	+
3	+	+	−	+	−	+	+	+
4	+	+	+	−	−	−	+	+
5	+	+	−	−	−	−	+	+
6	−	+	+	−	−	−	+	+
7	+	+	+	+	+	+	−	+
8	+	+	+	+	+	+	+	−

b. When prototrophic spores occur during sporulation of the diploids just discussed, what ratio of auxotrophic to prototrophic spores would you generally expect to see in any tetrad containing such a prototrophic spore? Explain the ratio you expect.

c. Using the data from all parts of this question, draw the best map of the eight lysine mutations under study. Show the extent of any deletions involved, and indicate the boundaries of the various complementation groups.

Section 7.3

25. The pathway for arginine biosynthesis in *Neurospora crassa* involves several enzymes that produce a series of intermediates.

$$\text{N-acetylornithine} \xrightarrow{argE} \text{ornithine} \xrightarrow{argF} \text{citrulline} \xrightarrow{argG} \text{argininosuccinate} \xrightarrow{argH} \text{arginine}$$

a. If you did a cross between $argE^-$ and $argH^-$ *Neurospora* strains, what would be the distribution of Arg^+ and Arg^- spores within parental ditype and nonparental ditype asci? Give the spore types in the order in which they would appear in the ascus.

b. For each of the spores in your answer to part *a*, what nutrients could you supply in the media to get spore growth?

26. In corn snakes, the wild-type color is brown. One autosomal recessive mutation causes the snake to be orange, and another causes the snake to be black. An orange snake was crossed to a black one, and the F_1 offspring were all brown. Assume that all relevant genes are unlinked.

a. Indicate what phenotypes and ratios you would expect in the F_2 generation of this cross if there is one pigment pathway, with orange and black being different intermediates on the way to brown.

b. Indicate what phenotypes and ratios you would expect in the F_2 generation if orange pigment is a product of one pathway, black pigment is the product of another pathway, and brown is the effect of mixing the two pigments in the skin of the snake.

27. In a certain species of flowering plants with a diploid genome, four enzymes are involved in the generation of flower color. The genes encoding these four enzymes are on different chromosomes. The biochemical pathway involved is as follows; the figure shows that either of two different enzymes is sufficient to convert a blue pigment into a purple pigment.

$$\text{white} \rightarrow \text{green} \rightarrow \text{blue} \genfrac{}{}{0pt}{}{\rightarrow}{\rightarrow} \text{purple}$$

A true-breeding green-flowered plant is mated with a true-breeding blue-flowered plant. All of the plants in the resultant F_1 generation have purple flowers. F_1 plants are allowed to self-fertilize, yielding an F_2 generation. Show genotypes for P, F_1, and F_2 plants, and indicate which genes specify which biochemical steps. Determine the fraction of F_2 plants with the following phenotypes: white flowers, green flowers, blue flowers, and purple flowers. Assume the green-flowered parent is mutant in only a single step of the pathway.

28. The intermediates A, B, C, D, E, and F all occur in the same biochemical pathway. G is the product of the pathway, and mutants 1 through 7 are all G^-, meaning that they cannot produce substance G. The following table shows which intermediates will promote growth in each of the mutants. Arrange the intermediates in order of their occurrence in the pathway, and indicate the step in the pathway at which each mutant strain is blocked. A + in the table indicates that the strain will grow if given that substance, an O means lack of growth.

	Supplements						
Mutant	A	B	C	D	E	F	G
1	+	+	+	+	+	O	+
2	O	O	O	O	O	O	+
3	O	+	+	O	+	O	+
4	O	+	O	O	+	O	+
5	+	+	+	O	+	O	+
6	+	+	+	+	+	+	+
7	O	O	O	O	+	O	+

29. In each of the following cross schemes, two true-breeding plant strains are crossed to make F_1 plants, all of which have purple flowers. The F_1 plants are then self-fertilized to produce F_2 progeny as shown here.

Cross	Parents	F_1	F_2
1	blue × white	all purple	9 purple: 4 white: 3 blue
2	white × white	all purple	9 purple: 7 white
3	red × blue	all purple	9 purple: 3 red: 3 blue: 1 white
4	purple × purple	all purple	15 purple: 1 white

a. For each cross, explain the inheritance of flower color.
b. For each cross, show a possible biochemical pathway that could explain the data.
c. Which of these crosses is compatible with an underlying biochemical pathway involving only a single step that is catalyzed by an enzyme with two dissimilar subunits, both of which are required for enzyme activity?
d. For each of the four crosses, what would you expect in the F_1 and F_2 generations if all relevant genes were tightly linked?

30. The pathways for the biosynthesis of the amino acids glutamine (gln) and proline (pro) involve one or more common intermediates. Auxotrophic yeast mutants numbered 1–7 are isolated that require either glutamine or proline or both amino acids for their growth, as shown in the following table (+ means growth; − no growth). These mutants are also tested for their ability to grow on the intermediates A–E. What is the order of these intermediates in the glutamine and proline pathways, and at which point in the pathway is each mutant blocked?

Mutant	A	B	C	D	E	gln	pro	gln + pro
1	+	−	−	−	+	−	+	+
2	−	−	−	−	−	−	+	+
3	−	−	+	−	−	−	−	+
4	−	−	−	−	−	+	−	+
5	−	−	+	+	−	−	−	+
6	+	−	−	−	−	−	+	+
7	−	+	−	−	−	+	−	+

31. The following noncomplementing *E. coli* mutants were tested for growth on four known precursors of thymine, A–D.

	Precursor/product				
Mutant	A	B	C	D	Thymine
9	+	−	+	−	+
10	−	−	+	−	+
14	+	+	+	−	+
18	+	+	+	+	+
21	−	−	−	−	+

a. Show a simple linear biosynthetic pathway of the four precursors and the end product, thymine. Indicate which step is blocked by each of the five mutations.
b. What precursor would accumulate in the following double mutants: 9 and 10? 10 and 14?

32. In 1952, an article in the *British Medical Journal* reported interesting differences in the behavior of blood plasma obtained from several individuals who suffered from X-linked recessive hemophilia. When mixed together, the cell-free blood plasma from certain combinations of individuals could form clots in the test tube. For example, the following table shows whether (+) or not (−) clots could form in various combinations of plasma from four individuals with hemophilia:

1 and 1	−	2 and 3	+
1 and 2	−	2 and 4	+
1 and 3	+	3 and 3	−
1 and 4	+	3 and 4	−
2 and 2	−	4 and 4	−

What do these data tell you about the inheritance of hemophilia in these individuals? Do these data allow you to exclude any models for the biochemical pathway governing blood clotting?

33. Mutations in an autosomal gene in humans cause a form of hemophilia called von Willebrand disease (vWD). This gene specifies a blood plasma protein cleverly called von Willebrand factor (vWF). vWF stabilizes factor VIII, a blood plasma protein specified by the wild-type hemophilia A gene. Factor VIII is needed to form blood clots. Thus, factor VIII is rapidly destroyed in the absence of vWF.

Which of the following might successfully be employed in the treatment of bleeding episodes in hemophiliac patients? Would the treatments work immediately or only after some delay needed for protein synthesis? Would the treatments have only a short-term or a prolonged effect? Assume that all mutations are null (that is, the mutations result in the complete absence of the protein encoded by the gene) and that the plasma is cell-free.

a. transfusion of plasma from normal blood into a vWD patient
b. transfusion of plasma from a vWD patient into a different vWD patient
c. transfusion of plasma from a hemophilia A patient into a vWD patient
d. transfusion of plasma from normal blood into a hemophilia A patient
e. transfusion of plasma from a vWD patient into a hemophilia A patient
f. transfusion of plasma from a hemophilia A patient into a different hemophilia A patient
g. injection of purified vWF into a vWD patient
h. injection of purified vWF into a hemophilia A patient

i. injection of purified factor VIII into a vWD patient
j. injection of purified factor VIII into a hemophilia A patient

34. Antibodies were made that recognize six proteins that are part of a complex inside the *Caenorhabditis elegans* one-cell embryo. The mother produces proteins that are believed to assemble stepwise into a structure in the egg, beginning at the embryo's inner surface. The antibodies were used to detect the protein location in embryos produced by mutant mothers (who are homozygous recessive for the gene[s] encoding each protein). The *C. elegans* mothers are self-fertilizing hermaphrodites so no wild-type copy of a gene will be introduced during fertilization. In the table, * means the protein was present and at the embryo surface, – means that the protein was not present, and + means that the protein was present but not at the embryo surface. Assume all mutations prevent production of the corresponding protein.

Mutant in gene for	Protein production and location					
protein	A	B	C	D	E	F
A	–	+	*	+	*	+
B	*	–	*	*	*	*
C	*	+	–	+	*	+
D	*	+	*	–	*	+
E	+	+	+	+	–	+
F	*	+	*	*	*	–

Complete the following figure, which shows the construction of the hypothetical protein complex, by writing the letter of the proper protein in each circle. The two proteins marked with arrowheads can assemble into the complex independently of each other, but both are needed for the addition of subsequent proteins to the complex.

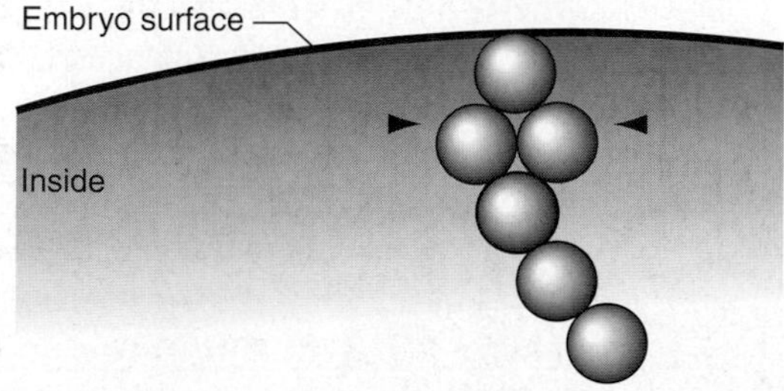

35. Adult hemoglobin is a multimeric protein with four polypeptides, two of which are α globin and two of which are β globin.

a. How many genes are needed to define the structure of the hemoglobin protein?
b. If a person is heterozygous for wild-type alleles and alleles that would yield amino acid substitution variants for both α globin and β globin, how many different kinds of hemoglobin protein would be found in the person's red blood cells and in what proportion? Assume all alleles are expressed at the same level.

36. This problem refers to Fig. A in the Fast Forward box on pp. 240–241. For each part that follows, describe what structures Robert Edgar would have seen in the electron microscope if he examined extracts of *E. coli* cells infected with the indicated temperature-sensitive mutant strains of bacteriophage T4 under restrictive conditions.

a. A strain with a mutation in gene 19
b. A strain with a mutation in gene 16
c. Simultaneous infection with two mutant strains, one in gene 13 and the other in gene 14. The polypeptides produced by genes 13 and 14 associate with each other to form a multimeric protein that governs one step of phage head assembly (see Fig. A on p. 241).
d. A strain whose genome contains mutations in both genes 15 and 35

Section 7.4

37. In addition to the predominant adult hemoglobin, HbA, which contains two α-globin chains and two β-globin chains ($\alpha_2\beta_2$), there is a minor hemoglobin, HbA_2, composed of two α and two δ chains ($\alpha_2\delta_2$). The β- and δ-globin genes are arranged in tandem and are highly homologous. Draw the chromosomes that would result from an event of unequal crossing-over between the β and δ genes.

38. Most mammals, including "New World" primates such as marmosets (a kind of monkey), are *dichromats:* they have only two kinds of rhodopsin-related color receptors. "Old World" primates such as humans and gorillas are *trichromats* with three kinds of color receptors. Primates diverged from other mammals roughly 65 million years ago (Myr), while Old World and New World primates diverged from each other roughly 35 Myr.

a. Using this information, define on Fig. 7.28d (see p. 242) the time span of any events that can be dated.
b. Some New World monkeys have an autosomal color receptor gene and a single X-linked color receptor gene. The X-linked gene has three alleles, each of which encodes a photoreceptor that responds to light of a different wavelength (all three wavelengths are different from that recognized by the autosomal color receptor). How is color vision inherited in these monkeys?
c. About 95% of all light-receiving neurons in humans and other mammals are rod cells containing rhodopsin, a pigment that responds to low-level light of many wavelengths. The remaining 5% of light-receiving neurons are cone cells with pigments that respond to light of specific wavelengths of high intensity. What does this suggest about the lifestyle of the earliest mammals?

Gene Expression: The Flow of Genetic Information from DNA to RNA to Protein

Chapter 8

A dedicated effort to determine the complete nucleotide sequence of the haploid genome in a variety of organisms has been underway since 1990. This massive endeavor has been more successful than many scientists thought possible. By 2001, the sequence of As, Ts, Gs, and Cs in the genomes of more than 20 different species, including the bacterium *Escherichia coli,* the yeast *Saccharomyces cerevisiae,* the fruit fly *Drosophila melanogaster,* the nematode *Caenorhabditis elegans,* the plant *Arabidopsis thaliana,* and humans (*Homo sapiens*), had already been deciphered. With this sequence information in hand, geneticists can consult the *genetic code*—the dictionary equating nucleotide sequence with amino acid sequence—to decide what parts of a genome are likely to be genes. They can also identify genes through matches with nucleotide sequences already known to encode proteins in other organisms. As a result, modern geneticists can use the complete nucleotide sequence of a genome to predict the total number of genes in an organism, and by extension, to discover the number and amino acid sequences of all the polypeptides that determine phenotype. Knowledge of DNA sequence thus opens up powerful new possibilities for understanding an organism's growth and development at the molecular level.

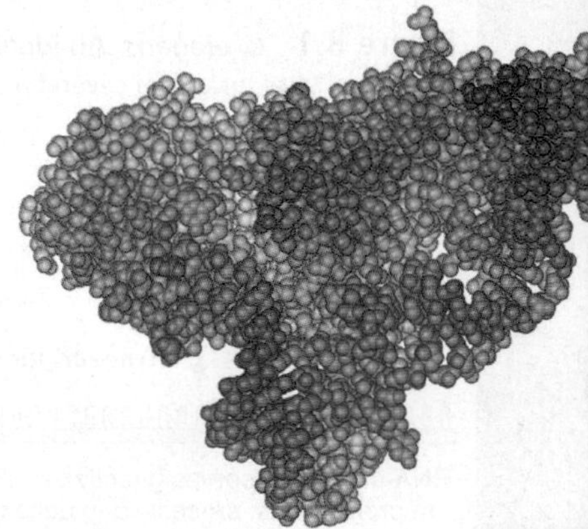

The ability of an aminoacyl-tRNA synthetase (red) to recognize a particular tRNA (blue) and couple it to its corresponding amino acid (not shown) is central to the molecular machinery that converts the language of nucleic acids into the language of proteins.

Studies of the tiny nematode *Caenorhabditis elegans* illustrate the kinds of insights researchers can gain from this DNA-sequence-based approach. *C. elegans* is a roundworm 1 mm in length that lives in soils throughout the world (**Fig. 8.1**). Feeding on bacteria, it grows from fertilized egg to adult—either hermaphrodite or male—in just three days. Each hermaphrodite produces between 250 and 1000 progeny. Because of its small size, short life cycle, and capacity for prolific reproduction, *C. elegans* is an ideal subject for genetic analysis.

The haploid genome of *C. elegans* contains 100 million base pairs distributed among six chromosomes. In 1998, an international team of investigators in the United States and Britain reported the sequencing and preliminary analysis of the entire *C. elegans* genome. Using their knowledge of the concepts explored in this chapter, they found that the nematode genome carries about 20,000 genes. Interestingly, roughly 15% of the genes encode molecules that play some role in **gene expression:** the process by which cells convert DNA sequence information to RNA and then decode the RNA information to the amino acid sequence of a polypeptide (**Fig. 8.2**). The fact that 15% of nematode genes encode components of gene expression suggests the importance of the process to the life of the organism. About 3000 of the genes in the nematode either generate the machinery that enables genes to be interpreted as proteins, or they directly regulate this machinery to determine when, where, and how much of a protein will be made. In this chapter, we describe the cellular mechanisms that carry out gene expression. As intricate as some of the details may appear, the general scheme of gene expression is elegant and straightforward: *Within each cell, genetic information flows from DNA to RNA to protein.* In 1957, Francis Crick proposed that genetic information moves in only one direction and named his concept of a one-way molecular transfer the "Central Dogma" of molecular biology. As Crick explained, "Once 'information' has passed into protein, it cannot get out again."

Inside most cells, as the Central Dogma suggests, genetic information flows from one class of molecule to another in two distinct stages (Fig. 8.2). If you think of genes as

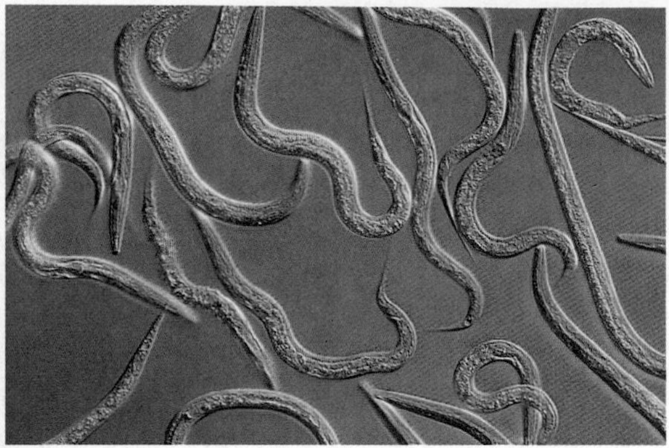

Figure 8.1 ***C. elegans:* An ideal subject for genetic analysis.** Micrograph of several adult worms.

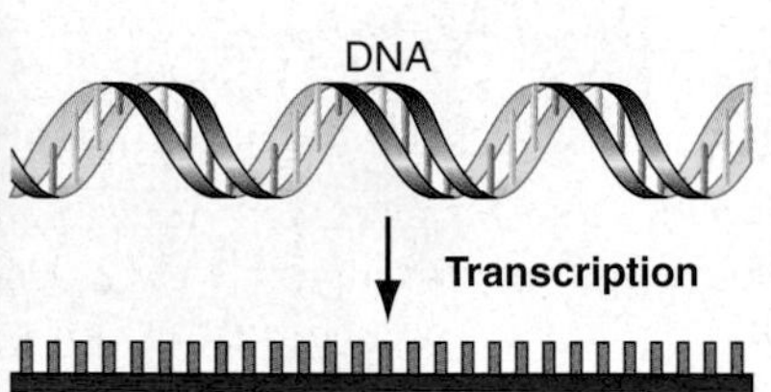

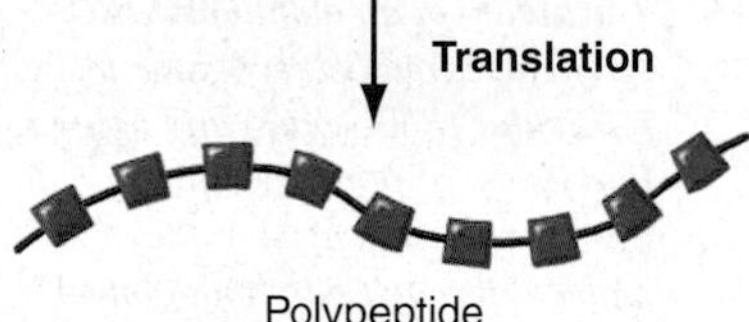

Figure 8.2 Gene expression: The flow of genetic information from DNA via RNA to protein. In transcription, the enzyme RNA polymerase copies DNA to produce an RNA transcript. In translation, the cellular machinery uses instructions in mRNA to synthesize a polypeptide, following the rules of the genetic code.

instructions written in the language of nucleic acids, the cellular machinery first transcribes a set of instructions written in the DNA dialect to the same instructions written in the RNA dialect. The conversion of DNA-encoded information to its RNA-encoded equivalent is known as **transcription.** The product of transcription is a **transcript:** a molecule of **messenger RNA (mRNA)** in prokaryotes, a molecule of RNA that undergoes processing to become an mRNA in eukaryotes. In the second stage of gene expression, the cellular machinery translates the mRNA to its polypeptide equivalent in the language of amino acids. This decoding of nucleotide information to a sequence of amino acids is known as **translation.** It takes place on molecular workbenches called **ribosomes,** which are composed of proteins and ribosomal RNAs (rRNAs), and it depends on the dictionary known as the **genetic code,** which defines each amino acid in terms of specific sequences of three nucleotides. Translation also depends on **transfer RNAs (tRNAs),** small RNA adaptor molecules that place specific amino acids at the correct position in a growing polypeptide chain. The tRNAs can bring amino acids to the right place on the translational machinery because tRNAs and mRNAs have complementary nucleotides that can base pair with each other.

Although Crick's formulation of the Central Dogma was a major milestone in our understanding of biology, it does not explain the behavior of all genes in all organisms. As Crick himself realized, a large subset of genes is transcribed into RNAs that are never translated into proteins. The genes encoding rRNAs and tRNAs belong to this group. Perhaps even more surprising is the finding that certain viruses contain an enzyme that can reverse the DNA-to-RNA flow of information by copying RNA to DNA in a process called **reverse transcription.**

Despite these exceptions, four general themes emerge from our discussion of gene expression. First, the pairing of complementary bases figures prominently in the precise transfer of information from DNA to RNA and from RNA to polypeptide. Second, the polarities of DNA, RNA, and protein molecules help guide the mechanisms of gene expression: the 3′-to-5′ transcription of a template DNA strand yields a polar mRNA that builds from its 5′ to its 3′ end; the 5′-to-3′ translation of this mRNA yields a polar protein running from its amino terminal to its carboxyl terminal. Third, like DNA replication and recombination, gene expression requires an input of energy and the participation of several specific proteins and macromolecular assemblies like ribosomes at different points in the process. Fourth, since the accurate one-way flow of genetic information determines protein structure, mutations that change this information or obstruct its flow can have dramatic effects on phenotype.

As we examine how cells use the sequence information contained in DNA to construct proteins, we present

- The genetic code: How triplets of the four nucleotides unambiguously specify 20 amino acids, making it possible to translate information from a nucleotide chain to a sequence of amino acids.
- Transcription: How RNA polymerase, guided by base pairing, synthesizes a single-stranded mRNA copy of a gene's DNA template.
- Translation: How base pairing between mRNA and tRNAs directs the assembly of a polypeptide on the ribosome.
- Significant differences in gene expression between prokaryotes and eukaryotes.
- A comprehensive example of gene expression in *C. elegans.*
- How mutations affect gene information and expression.

8.1 The Genetic Code: How Precise Groupings of the Four Nucleotides Specify 20 Amino Acids

A code is a system of symbols that equates information in one language with information in another. A useful analogy for the genetic code is the Morse code, which uses dots and dashes to transmit messages over radio or telegraph wires. Various groupings of the dot-dash symbols represent the 26 letters of the English alphabet. Because there are many more letters than the two symbols (dot or dash), groups of one, two, three, or four dots or dashes in various combinations represent individual letters. For example, the symbol for C is dash dot dash dot (– · – ·), the symbol for O is dash dash dash (– – –), D is dash dot dot (– · ·), and E is a single dot (·). Because anywhere from one to four symbols specify each letter, the Morse code requires a symbol for "pause" (in practice, a short interval of time) to signify where one letter ends and the next begins.

In the Genetic Code, a Triplet Codon Represents Each Amino Acid

The language of nucleic acids is written in four nucleotides—A, G, C, and T in the DNA dialect; A, G, C, and U in the RNA dialect—while the language of proteins is written in amino acids. To understand how the sequence of bases in DNA or RNA encodes the order of amino acids in a polypeptide chain, it is essential to know how many distinct amino acids there are. Watson and Crick produced the now accepted list of the 20 amino acids that are genetically encoded by DNA or RNA sequence over lunch one day at a local pub. They created the list by analyzing the amino acid sequence of a variety of naturally occurring polypeptides. Amino acids that are present in only a small number of proteins or in only certain tissues or organisms did not qualify as standard building blocks; Crick and Watson correctly assumed that such amino acids arise when proteins undergo modification after their synthesis. By contrast, amino acids that are present in most, though not necessarily all, proteins made the list. The question then became, How can four nucleotides encode 20 amino acids?

Just as the Morse code conveys information through different groupings of dots and dashes, the four nucleotides encode 20 amino acids through specific groupings of A, G, C, and T or A, G, C, and U. Researchers initially arrived at the number of letters per grouping by deductive reasoning and later confirmed their guess by experiment. They reasoned that if only one nucleotide represented an amino acid, there would be information for only four amino acids: A would encode one amino acid; G, a second amino acid; C, a third; and T, a fourth. If two nucleotides represented each amino acid, there would be $4^2 = 16$ possible combinations of couplets.

Of course, if the code consisted of groups containing one *or* two nucleotides, it would have $4 + 16 = 20$ groups and could account for all the amino acids, but there would be nothing left over to signify the pause required to denote where one group ends and the next begins. Groups of three nucleotides in a row would provide $4^3 = 64$ different triplet combinations, more than enough to code for all the amino acids. If the code consisted of doublets and triplets, a signal denoting pause would once again be necessary. But a triplets-only code would require no symbol for "pause" if the mechanism for counting to three and distinguishing among successive triplets were very reliable.

Although this kind of reasoning—explaining the unknown in terms of the known by looking for the simplest possible explanation—generates a hypothesis, it does not prove it. As it turned out, however, the experiments described later in this chapter did indeed demonstrate that groups of three nucleotides represent all 20 amino acids. Each nucleotide triplet is called a **codon.** Each codon, designated by the bases defining its three nucleotides, specifies one amino acid. For example, GAA is a codon for glutamic acid (Glu), and GUU is a codon for valine (Val). Because the code comes into play only during the translation part of gene expression, that is, during the decoding of messenger RNA to polypeptide, geneticists usually present the code in the RNA dialect of A, G, C, and U, as depicted in **Fig. 8.3.**

First letter	Second letter: U	Second letter: C	Second letter: A	Second letter: G	Third letter
U	UUU Phe	UCU Ser	UAU Tyr	UGU Cys	U
	UUC Phe	UCC Ser	UAC Tyr	UGC Cys	C
	UUA Leu	UCA Ser	UAA Stop	UGA Stop	A
	UUG Leu	UCG Ser	UAG Stop	UGG Trp	G
C	CUU Leu	CCU Pro	CAU His	CGU Arg	U
	CUC Leu	CCC Pro	CAC His	CGC Arg	C
	CUA Leu	CCA Pro	CAA Gln	CGA Arg	A
	CUG Leu	CCG Pro	CAG Gln	CGG Arg	G
A	AUU Ile	ACU Thr	AAU Asn	AGU Ser	U
	AUC Ile	ACC Thr	AAC Asn	AGC Ser	C
	AUA Ile	ACA Thr	AAA Lys	AGA Arg	A
	AUG Met	ACG Thr	AAG Lys	AGG Arg	G
G	GUU Val	GCU Ala	GAU Asp	GGU Gly	U
	GUC Val	GCC Ala	GAC Asp	GGC Gly	C
	GUA Val	GCA Ala	GAA Glu	GGA Gly	A
	GUG Val	GCG Ala	GAG Glu	GGG Gly	G

Figure 8.3 The genetic code: 61 codons represent the 20 amino acids, while 3 codons signify stop. To read the code, find the first letter in the *left column,* the second letter along the *top,* and the third letter in the *right column;* this reading corresponds to the 5′-to-3′ direction along the mRNA.

When speaking of genes, they can substitute T for U to show the same code in the DNA dialect.

If you knew the sequence of nucleotides in a gene or its transcript as well as the sequence of amino acids in the corresponding polypeptide, you could deduce the genetic code without understanding how the cellular machinery uses the code to translate from nucleotides to amino acids. Although techniques for determining both nucleotide and amino acid sequence are available today, this was not true when researchers were trying to crack the genetic code in the 1950s and 1960s. At that time, they could establish a polypeptide's amino acid sequence, but not the nucleotide sequence of DNA or RNA. Because of their inability to read nucleotide sequence, they used an assortment of genetic and biochemical techniques to fathom the code. They began by examining how different mutations in a single gene affected the amino acid sequence of the gene's polypeptide product. In this way, they were able to use the abnormal (specific mutations) to understand the normal (the general relationship between genes and polypeptides).

A Gene's Nucleotide Sequence Is Colinear with the Amino Acid Sequence of the Encoded Polypeptide

We have seen that DNA is a linear molecule with base pairs following one another down the intertwined chains. Proteins, by contrast, have complicated three-dimensional structures. Even so, if unfolded and stretched out from N terminus to C terminus, proteins have a one-dimensional, linear structure—a specific sequence of amino acids. If the information in a gene and its corresponding protein are colinear, the consecutive order of bases in the DNA from the beginning to the end of the gene would stipulate the consecutive order of amino acids from one end to the other of the outstretched protein. Note that this hypothesized relationship implies that both a gene and its protein product have definite polarities with an invariant relation to each other.

Charles Yanofsky, in studying the *E. coli* gene for a subunit of the enzyme tryptophan synthetase during the 1960s, was the first to compare maps of mutations within a gene to the particular amino acid substitutions that resulted. He began by generating a large number of *trp*$^-$ auxotrophic mutants that carried mutations in the *trpA* gene for the tryptophan synthetase subunit. He next made a fine structure recombinational map of these mutations (analogous to Benzer's fine structure map for the *rII* region of bacteriophage T4, discussed in Chapter 7). Yanofsky then purified and determined the amino acid sequence of the mutant tryptophan synthetase subunits. As **Fig. 8.4a** illustrates, his data showed that the order of mutations mapped within the DNA of the gene by recombination was indeed colinear with the positions of the amino acid substitutions occurring in the resulting mutant proteins. In spite of this colinearity in order, distances on the genetic map (measured in map units) do not exactly reflect the number of amino acids between the amino acid substitutions. This is because recombination as seen on this very high resolution map does not occur with an equal probability at every base pair within the gene.

Nonoverlapping Codons Are Set in a Reading Frame

By carefully examining the results of his analysis, Yanofsky, in addition to confirming the existence of colinearity, deduced key features of codons and helped establish many parameters of the genetic code relating nucleotides to amino acids.

A Codon Is Composed of More Than One Nucleotide

Yanofsky observed that different **point mutations** (changes in only one nucleotide pair) may affect the same amino acid. In one example shown in Fig. 8.4a, mutation #23 changed the glycine (Gly) at position 211 of the wild-type polypeptide chain to arginine (Arg), while mutation #46 yielded glutamic acid (Glu) at the same position. In another example, mutation #78 changed the glycine at position 234 to cysteine (Cys), while mutation #58 produced aspartic acid (Asp) at the same position. In both cases, Yanofsky also found that recombination could occur between the two mutations that changed the identity of the same amino acid and such recombination would produce a wild-type tryptophan synthetase gene (Fig. 8.4b). Because the smallest unit of recombination is the base pair, two mutations capable of recombination—in this case, in the same codon because they affect the same amino acid—must be in different (although nearby) nucleotides. Thus, a codon contains more than one nucleotide.

Each Nucleotide Is Part of Only a Single Codon

As Fig. 8.4a illustrates, each of the point mutations in the tryptophan synthetase gene characterized by Yanofsky alters the identity of only a single amino acid. This is also true of the point mutations examined in many other genes, such as the human genes for rhodopsin and hemoglobin (see Chapter 7). Since point mutations change only a single nucleotide pair and most point mutations affect only a single amino acid in a polypeptide, each nucleotide in a gene must influence the identity of only a single amino acid. If, on the contrary, a nucleotide were part of more than one codon, a mutation in that nucleotide would affect more than one amino acid.

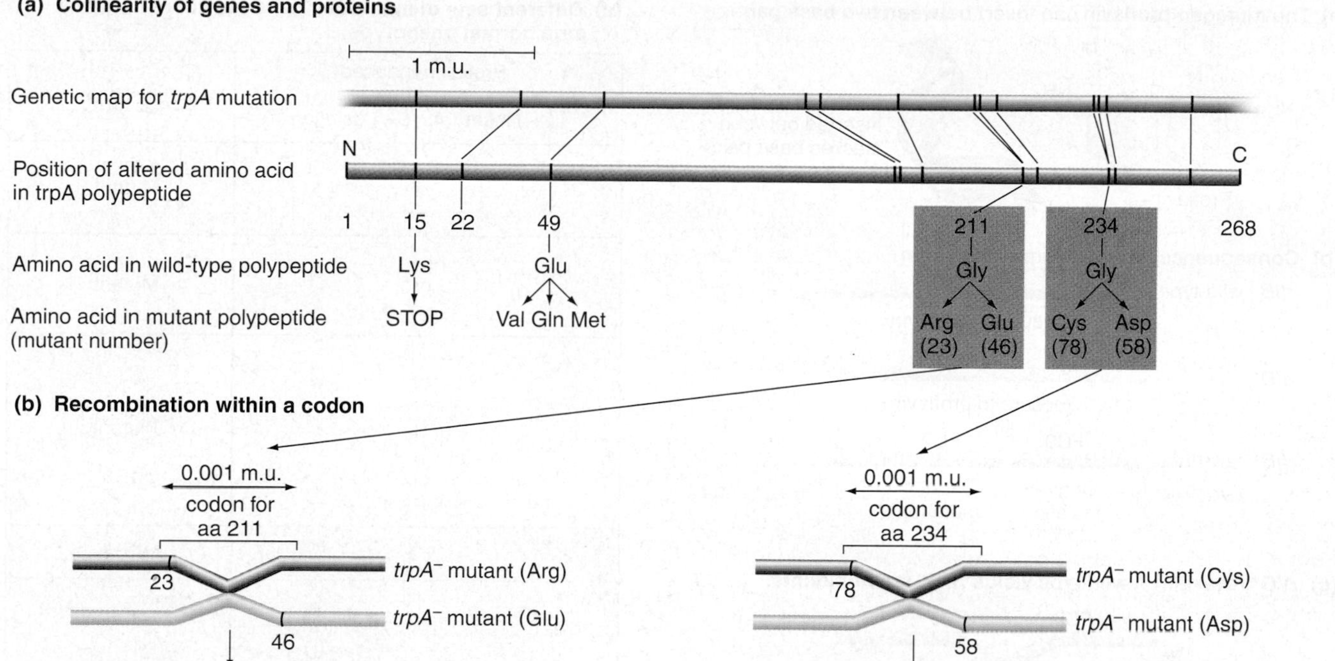

Figure 8.4 Mutations in a gene are colinear with the sequence of amino acids in the encoded polypeptide. (a) The relationship between the genetic map of *E. coli's trpA* gene and the positions of amino acid substitutions in mutant tryptophan synthetase proteins. Each mutation changes only a single amino acid, so each nucleotide is part of only a single codon. **(b)** Codons must include two or more base pairs. When two mutant strains with different amino acids at the same position were crossed, recombination could produce a wild-type allele. Thus, the two point mutations must be in different nucleotide pairs of the same codon.

A Codon Is Composed of Three Nucleotides, and the Starting Point of Each Gene Establishes the Reading Frame

Although the most efficient code that would allow four nucleotides to specify 20 amino acids requires three nucleotides per codon, more complicated scenarios are possible. But in 1955, Francis Crick and Sydney Brenner obtained convincing evidence for the triplet nature of the genetic code in studies of mutations in the bacteriophage T4 *rIIB* gene (**Fig. 8.5**). They induced the mutations with proflavin, an intercalating mutagen that can insert itself between the paired bases stacked in the center of the DNA molecule (Fig. 8.5a). Their assumption was that proflavin would act like other mutagens, causing single-base substitutions. If this were true, it would be possible to generate revertants through treatment with other mutagens that might restore the wild-type DNA sequence. Surprisingly, genes with proflavin-induced mutations did not revert to wild type upon treatment with other mutagens known to cause nucleotide substitutions. Only further exposure to proflavin caused proflavin-induced mutations to revert to wild type (Fig. 8.5b). Crick and Brenner had to explain this observation before they could proceed with their phage experiments. With keen insight, they correctly guessed that proflavin does not cause base substitutions; instead, it causes insertions or deletions. This hypothesis explained why base-substituting mutagens could not revert proflavin-induced mutations; it was also consistent with the structure of proflavin. By intercalating between base pairs, proflavin would distort the double helix and thus interfere with the action of enzymes that function in the repair, replication, or recombination of DNA, eventually causing the deletion or addition of one or more nucleotide pairs to the DNA molecule (review Fig. 7.10c on pp. 218–219).

Crick and Brenner began their experiments with a particular proflavin-induced $rIIB^-$ mutation they called FC0. They next treated this mutant strain with more proflavin to isolate an $rIIB^+$ revertant (Fig. 8.5b). By recombining this revertant with wild-type bacteriophage T4, Crick and Brenner were able to show that the revertant's chromosome actually contained two different $rIIB^-$ mutations (Fig. 8.5c). One was the original FC0 mutation; the other was the newly induced FC7. Either mutation by itself yields a mutant phenotype, but their simultaneous occurrence in the same gene yielded an $rIIB^+$ phenotype. Crick and Brenner reasoned that if the first mutation was the deletion of a single base pair, represented by the symbol (−), then the counteracting mutation must be the insertion of a base pair, represented as (+). The restoration of gene function by one mutation canceling another in the same

(a) The mutagen proflavin can insert between two base pairs.

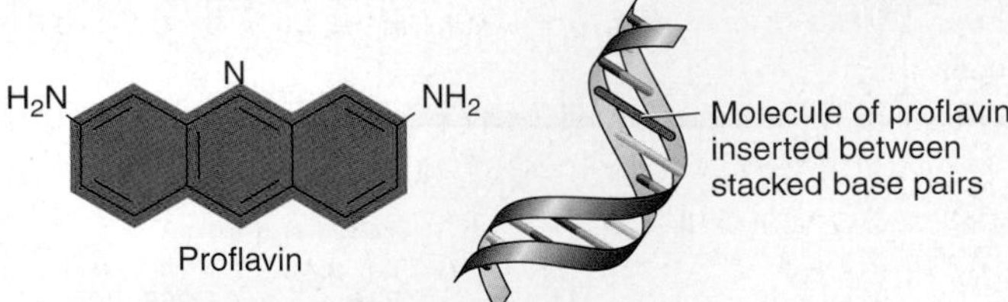

(b) Consequences of exposure to proflavin

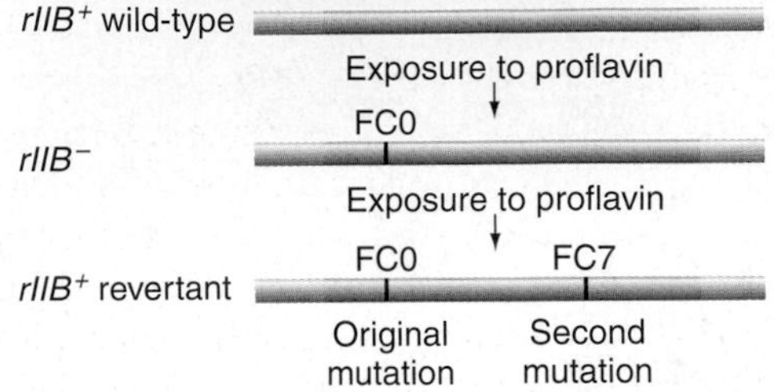

(c) *rIIB*+ revertant X wild type yields *rIIB*− recombinants.

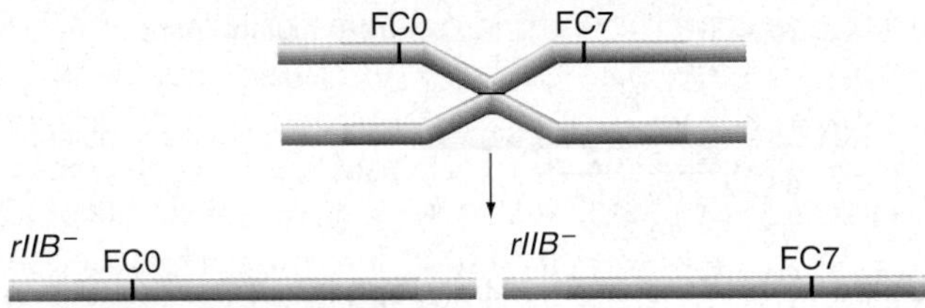

(d) Different sets of mutations generate either a mutant or a normal phenotype.

Proflavin-induced mutations (+) insertion (−) deletion	Phenotype
− or +	Mutant
− − or + +	Mutant
− − − − or − − − − − or + + + + or + + + + +	Mutant
− +	Wild type
− − − or − − − − − − or + + + or + + + + + +	Wild type

Figure 8.5 Studies of frameshift mutations in the bacteriophage T4 *rIIB* gene showed that codons consist of three nucleotides. (a) The mutagen proflavin slips between adjacent base pairs, eventually causing a deletion or insertion. **(b)** Treatment with proflavin produces a mutation at one site (FC0). A second proflavin exposure results in a second mutation (FC7) within the same gene, which suppresses FC0. **(c)** When the revertant is crossed with a wild-type strain, crossing-over separates the two *rIIB*− mutations FC0 and FC7. The reversion to an *rIIB*+ phenotype was thus the result of intragenic suppression. **(d)** Evidence for a triplet code.

gene is known as **intragenic suppression.** On the basis of this reasoning, they went on to establish T4 strains with different numbers of (+) and (−) mutations in the same chromosome. Figure 8.5d tabulates the phenotypes associated with each combination of proflavin-induced mutations.

In analyzing the data, Crick and Brenner assumed that each codon is a trio of nucleotides and for each gene there is a single starting point. This starting point establishes a **reading frame:** the partitioning of groups of three nucleotides such that the sequential interpretation of each succeeding triplet generates the correct order of amino acids in the resulting polypeptide chain. If codons are read in order from a fixed starting point, one mutation will counteract another if the two are equivalent mutations of opposite signs; in such a case, each insertion compensates for each deletion, and this counterbalancing restores the reading frame (**Fig. 8.6a**). The gene would only regain its wild-type activity, however, if the portion of the polypeptide encoded between the two mutations of opposite sign is not required for protein function, because in the double mutant, this region would have an improper amino acid sequence. Similarly, if a gene sustains three or multiples of three changes of the same sign, the encoded polypeptide can still function, because the mutations do not alter the reading frame for the majority of amino acids (Fig. 8.6b). The resulting polypeptide will, however, have one extra or one fewer amino acid than normal (designated by three plus signs (+) or three minus signs (−), respectively), and the region encoded by the part of the gene between the first and the last mutations will not contain the correct amino acids.

By contrast, a single nucleotide inserted into or deleted from a gene alters the reading frame and thereby affects the identity of not only one amino acid but of all other amino acids beyond the point of alteration (Fig. 8.6c). Changes that alter the grouping of nucleotides into codons are called **frameshift mutations:** they shift the reading frame for all codons beyond the point of insertion or deletion, almost always abolishing the function of the polypeptide product.

A review of the evidence tabulated in Fig. 8.5d supports all these points. A single (−) or a single (+) mutation destroyed the function of the *rIIB* gene and produced an *rIIB*− phage. Similarly, any gene with two base changes of the same sign (− − or + +) or with four or five insertions or deletions of the same sign (for example, + + + +) also generated a mutant phenotype. However, genes containing three or multiples of three mutations of the same sign (for example, + + + or − − − − − −), as well as genes containing a (+ −) pair of mutations, generated *rIIB*+ wild-type individuals. In these last examples, intragenic suppression allowed restitution of the reading frame and thereby restored the lost or aberrant genetic function produced by other frameshift mutations in the gene.

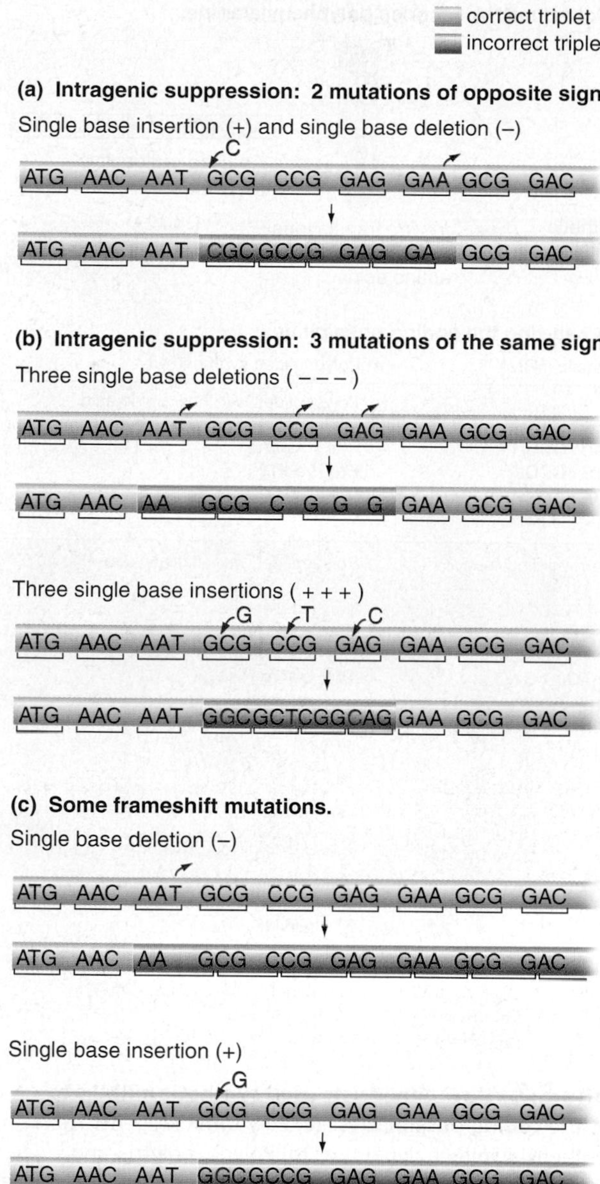

Figure 8.6 Codons consist of three nucleotides read in a defined reading frame. The phenotypic effects of proflavin-induced frameshift mutations depend on whether the reading frame is restored and whether the part of the gene with an altered reading frame specifies an essential or nonessential region of the polypeptide.

Most Amino Acids Are Specified by More Than One Codon

As Fig. 8.6a illustrates, intragenic suppression occurs only if, in the region between two frameshift mutations of opposite sign, a gene still dictates the appearance of amino acids, even if these amino acids are not the same as those appearing in the normal protein. If the frameshifted part of the gene encodes instructions to stop protein synthesis by introducing, for example, a triplet of nucleotides that does not correspond to any amino acid, then wild-type polypeptide production will not continue. This is because polypeptide synthesis would stop before the compensating mutation could reestablish the correct reading frame.

The fact that intragenic suppression occurs as often as it does suggests that the code includes more than one codon for some amino acids. Recall that there are 20 common amino acids but $4^3 = 64$ different combinations of three nucleotides. If each amino acid corresponded to only a single codon, there would be $64 - 20 = 44$ possible triplets not encoding an amino acid. These noncoding triplets would act as "stop" signals and prevent further polypeptide synthesis. If this happened, more than half of all frameshift mutations (44/64) would cause protein synthesis to stop at the first codon after the mutation, and the chances of extending the protein each amino acid farther down the chain would diminish exponentially. As a result, intragenic suppression would rarely occur. However, we have seen that many frameshift mutations of one sign can be offset by mutations of the other sign. The distances between these mutations, estimated by recombination frequencies, are in some cases large enough to code for more than 50 amino acids, which would be possible only if most of the 64 possible triplet codons specified amino acids. Thus, the data of Crick and Brenner provide strong support for the idea that the genetic code is **degenerate:** Two or more nucleotide triplets specify most of the 20 amino acids (see the genetic code in Fig. 8.3 on p. 257).

Cracking the Code: Biochemical Manipulations Revealed Which Codons Represent Which Amino Acids

Although the genetic experiments just described enabled remarkably prescient insights about the nature of the genetic code, they did not make it possible to assign particular codons to their corresponding amino acids. This awaited the discovery of messenger RNA and the development of techniques for synthesizing simple messenger RNA molecules that researchers could use to manufacture simple proteins in the test tube.

The Discovery of Messenger RNAs, Molecules for Transporting Genetic Information

In the 1950s, researchers exposed eukaryotic cells to amino acids tagged with radioactivity and observed that protein synthesis incorporating the radioactive amino acids into polypeptides takes place in the cytoplasm, even though the genes for those polypeptides are sequestered in the cell nucleus. From this discovery, they deduced the existence of an intermediate molecule, made in the nucleus and capable of transporting DNA sequence information to the cytoplasm, where it can direct protein synthesis. RNA was a prime candidate for this intermediary information-carrying molecule. Because of RNA's potential for base pairing with a strand of DNA, one could imagine the cellular machinery copying a strand of DNA into a complementary strand of RNA in a manner analogous to the DNA-to-DNA copying of DNA replication. Subsequent studies in eukaryotes on the incorporation of radioactive uracil (a base found only in RNA) into molecules of RNA showed that although the molecules are synthesized in the nucleus, at least some of them migrate to the cytoplasm. Among those RNA molecules that migrate to the cytoplasm are the messenger RNAs, or mRNAs, depicted in Fig. 8.2 on p. 256. They arise in the nucleus from the transcription of DNA sequence information through base pairing and then move (after processing) to the cytoplasm, where they determine the proper order of amino acids during protein synthesis.

Using Synthetic mRNAs and *In Vitro* Translation to Discover Which Codons Designate Which Amino Acids

Knowledge of mRNA served as the framework for two experimental breakthroughs that led to the deciphering of the genetic code. In the first, biochemists obtained cellular extracts that, with the addition of mRNA, synthesized polypeptides in a test tube. They called these extracts "*in vitro* translational systems." The second breakthrough was the development of techniques enabling the synthesis of artificial mRNAs containing only a few codons of known composition. When added to *in vitro* translational systems, these simple, synthetic mRNAs directed the formation of very simple polypeptides.

In 1961, Marshall Nirenberg and Heinrich Matthaei added a synthetic poly-U (5′... UUUUUUUUUUUU ... 3′) mRNA to a cell-free translational system derived from *E. coli*. With the poly-U mRNA, phenylalanine (Phe) was the only amino acid incorporated into the resulting polypeptide (**Fig. 8.7a**). Because UUU is the only possible triplet in poly-U, UUU must be a codon for phenylalanine. In a similar fashion, Nirenberg and Matthaei showed that CCC encodes proline (Pro), AAA is a codon for lysine (Lys), and GGG encodes glycine (Gly) (Fig. 8.7b).

The chemist Har Gobind Khorana later made mRNAs with repeating dinucleotides, such as poly-UC (5′... UCUCUCUC ... 3′), repeating trinucleotides, such as poly-UUC, and repeating tetranucleotides, such as poly-UAUC, and used them to direct the synthesis of slightly more complex polypeptides. As Fig. 8.7b shows, his results limited the coding possibilities, but some ambiguities remained. For example, poly-UC encodes the polypeptide N ... Ser-Leu-Ser-Leu-Ser-Leu ... C in which serine and leucine alternate with each other. Although the mRNA contains only two different codons (5′ UCU 3′ and 5′ CUC 3′), it is not obvious which corresponds to serine and which to leucine.

Nirenberg and Philip Leder resolved these ambiguities in 1965 with experiments in which they added short, synthetic mRNAs only three nucleotides in length to an *in vitro* translational system containing 1 radioactive amino acid and 19 unlabeled amino acids, all attached to tRNA molecules. They then poured through a filter the mixture of synthetic mRNAs and translational systems containing a tRNA-attached, radioactively labeled amino acid (**Fig. 8.8**). tRNAs carrying an amino acid normally go right through a filter. If, however, a tRNA carrying an amino acid binds to a ribosome, it will stick in the filter, because this larger complex of ribosome, amino-acid-carrying tRNA, and small mRNA cannot pass through the filter. Nirenberg and Leder could thus use this approach to see which small mRNA caused the entrapment of which radioactively labeled amino acid in the filter. For example, they knew from Khorana's earlier work that CUC encoded either serine or leucine. When they added the synthetic triplet CUC to an *in vitro* system where the radioactive amino acid was serine, this tRNA-attached amino

(a) Poly-U mRNA encodes polyphenylalanine.

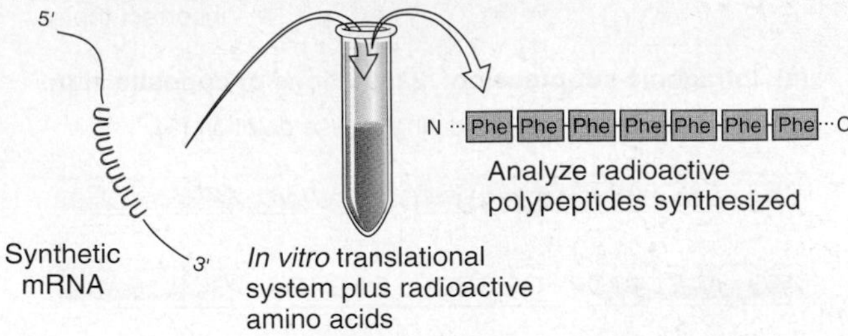

(b) Analyzing the coding possibilities.

Synthetic mRNA	Polypeptides synthesized
	Polypeptides with one amino acid
poly-U UUUU ...	Phe-Phe-Phe ...
poly-C CCCC ...	Pro-Pro-Pro ...
poly-A AAAA ...	Lys-Lys-Lys ...
poly-G GGGG ...	Gly-Gly-Gly ...
Repeating dinucleotides	Polypeptides with alternating amino acids
poly-UC UCUC ...	Ser-Leu-Ser-Leu ...
poly-AG AGAG ...	Arg-Glu-Arg-Glu ...
poly-UG UGUG ...	Cys-Val-Cys-Val ...
poly-AC ACAC ...	Thr-His-Thr-His ...
Repeating trinucleotides	Three polypeptides each with one amino acid
poly-UUC UUCUUCUUC ...	Phe-Phe.... and Ser-Ser.... and Leu-Leu....
poly-AAG AAGAAGAAG ...	Lys-Lys.... and Arg-Arg.... and Glu-Glu....
poly-UUG UUGUUGUUG ...	Leu-Leu.... and Cys-Cys.... and Val-Val....
poly-UAC UACUACUAC ...	Tyr-Tyr.... and Thr-Thr.... and Leu-Leu....
Repeating tetranucleotides	Polypeptides with repeating units of four amino acids
poly-UAUC UAUCUAUC ...	Tyr-Leu-Ser-Ile-Tyr-Leu-Ser-Ile...
poly-UUAC UUACUUAC ...	Leu-Leu-Thr-Tyr-Leu-Leu-Thr-Tyr...
poly-GUAA GUAAGUAA ...	none
poly-GAUA GAUAGAUA ...	none

Figure 8.7 How geneticists used synthetic mRNAs to limit the coding possibilities. **(a)** Poly-U mRNA generates a poly-phenylalanine polypeptide. **(b)** Polydi-, polytri-, and polytetra-nucleotides encode simple polypeptides. Some synthetic mRNAs, such as poly-GUAA, contain stop codons in all three reading frames and thus specify the construction only of short peptides.

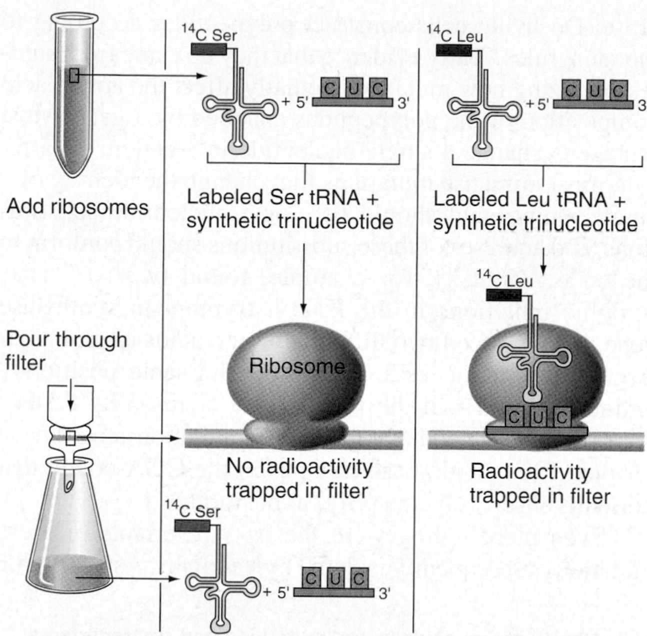

Figure 8.8 Cracking the genetic code with mini-mRNAs. Nirenberg and Leder added trinucleotides of known sequence, in combination with tRNAs charged with a radioactive amino acid, to an *in vitro* extract containing ribosomes. If the trinucleotide specified the radioactive amino acid, the amino acid-bearing tRNA formed a complex with the ribosomes and thus became trapped when the solution was passed through a filter. The experiments shown here indicate that the codon CUC specifies leucine, not serine.

acid passed through the filter, and the filter thus emitted no radiation. But when they added the same triplet to a system where the radioactive amino acid was leucine, the filter lit up with radioactivity, indicating that the radioactively tagged leucine attached to a tRNA had bound to the ribosome-mRNA complex and gotten stuck in the filter. CUC thus encodes leucine, not serine. Nirenberg and Leder used this technique to determine all the codon–amino acid correspondences shown in the genetic code (see Fig. 8.3 on p. 257).

The 5′-to-3′ Direction in mRNA Corresponds to the N-Terminal-to-C-Terminal Direction in the Polypeptide

In studies using synthetic mRNAs, when investigators added the six-nucleotide-long 5′ AAAUUU 3′ to an *in vitro* translational system, the product N-Lys-Phe-C emerged, but no N-Phe-Lys-C appeared. Since AAA is the codon for lysine and UUU is the codon for phenylalanine, this means that the codon closest to the 5′ end of the mRNA encoded the amino acid closest to the N terminus of the corresponding polypeptide. Similarly, the codon nearest the 3′ end of the mRNA encoded the amino acid nearest the C terminus of the resulting polypeptide.

To understand how the polarities of the macromolecules participating in gene expression relate to each other,

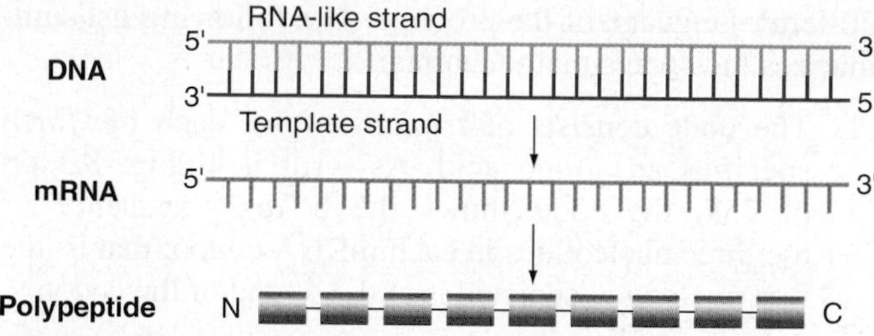

Figure 8.9 Correlation of polarities in DNA, mRNA, and polypeptide. The template strand of DNA is complementary to both the RNA-like DNA strand and the mRNA. The 5′-to-3′ direction in an mRNA corresponds to the N-terminus-to-C-terminus direction in the polypeptide.

remember that although the gene is a segment of a DNA double helix, only one of the two strands serves as a template for the mRNA of any given gene. This strand is known as the **template strand.** The other strand is the **RNA-like strand,** because it has the same polarity and sequence (written in the DNA dialect) as the RNA. Note that some scientists use the terms *sense strand* or *coding strand* as synonyms for the RNA-like strand; in these alternative nomenclatures, the template strand would be the *antisense strand* or the *noncoding strand.* **Figure 8.9** diagrams the respective polarities of a gene's DNA, the mRNA transcript of that DNA, and the resulting polypeptide.

Nonsense Codons Cause Termination of a Polypeptide Chain

Although most of the simple, repetitive RNAs synthesized by Khorana were very long and thus generated very long polypeptides, a few did not. These RNAs had signals that stopped construction of a polypeptide chain. As it turned out, three different triplets—UAA, UAG, and UGA—do not correspond to any of the amino acids. When these codons appear in frame, translation stops. As an example of how investigators established this fact, consider the case of poly-GUAA (review Fig. 8.7b). This mRNA will not generate a long polypeptide because in all possible reading frames, it contains the **stop codon** UAA.

The three stop codons that terminate translation are also known as **nonsense codons.** For historical reasons, researchers often refer to UAA as the *ocher* codon, UAG as the *amber* codon, and UGA as the *opal* codon. The historical basis of this nomenclature is the last name of one of the early investigators—Bernstein—which means "amber" in German; ocher and opal derive from their similarity with amber as semiprecious materials.

The Genetic Code: A Summary

From the experiments just described, researchers learned that the genetic code is a complete, unabridged dictionary equating the 4-letter language of the nucleic acids with the

20-letter language of the proteins. The following list summarizes the code's main features:

1. The code consists of *triplet codons,* each of which specifies an amino acid. As written in Fig. 8.3 on p. 257, the code shows the 5′-to-3′ sequence of the three nucleotides in each mRNA codon; that is, the first nucleotide depicted is at the 5′ end of the codon.

2. The codons are *nonoverlapping.* In the mRNA sequence 5′ GAAGUUGAA 3′, for example, the first three nucleotides (GAA) form one codon; nucleotides 4 through 6 (GUU) form the second; and nucleotides 7 through 9 (GAA), the third. Each nucleotide is part of only one codon.

3. The code includes three *stop,* or *nonsense, codons:* UAA, UAG, and UGA. These codons do not encode an amino acid and thus terminate translation.

4. The code is *degenerate,* which means that in many cases, more than one codon specifies the same amino acid. Despite its degeneracy, the code is unambiguous, because each codon specifies only one amino acid.

5. In reading the transcript of a gene, the cellular machinery scans a single strand of mRNA from a fixed starting point that establishes a *reading frame.* As we see later, the nucleotide triplet AUG, which specifies the amino acid methionine, serves in certain contexts as the **initiation codon,** marking the precise spot in the nucleotide sequence of an mRNA where the code for a particular polypeptide begins. The position of the initiation codon specifies the reading frame for the rest of the sequence by determining how the nucleotides become grouped into triplets. Note that without a set reading frame established by an initiation codon, there would be three possible ways to read a piece of mRNA, depending on whether you begin with the first, second, or third base in the sequence.

6. Moving from the 5′ to the 3′ end of an mRNA, each successive codon is sequentially interpreted into an amino acid, starting at the N terminus and moving toward the C terminus of the resulting polypeptide.

7. Mutations may modify the message encoded in a sequence of nucleotides in three ways. *Frameshift mutations* are nucleotide insertions or deletions that alter the genetic instructions for polypeptide construction by changing the reading frame. *Missense mutations* change a codon for one amino acid to a codon for a different amino acid. *Nonsense mutations* change a codon for an amino acid to a stop codon.

Using Genetics to Verify the Code

The experiments that cracked the genetic code by assigning codons to amino acids were all *in vitro* studies using cell-free extracts and synthetic mRNAs. A logical question thus arose: Do living cells construct polypeptides according to the same rules? Early evidence that they do came from studies analyzing how mutations actually affect the amino acid composition of the polypeptides encoded by a gene. Most mutagens change a single nucleotide in a codon. As a result, most missense mutations that change the identity of a single amino acid should be single-nucleotide substitutions, and analyses of these substitutions should conform to the code. Yanofsky, for example, found two trp^- auxotrophic mutations in the *E. coli* tryptophan synthetase gene that produced two different amino acids (arginine, or Arg, and glutamic acid, or Glu) at the same position—amino acid 211—in the polypeptide chain (**Fig. 8.10a**). According to the code, both of these mutations could have resulted from single-base changes in the GGA codon that normally inserts glycine (Gly) at position 211.

Even more telling were the trp^+ revertants of these mutations subsequently isolated by Yanofsky. As Fig. 8.10a

(a) Altered amino acids in trp^- mutations and trp^+ revertants

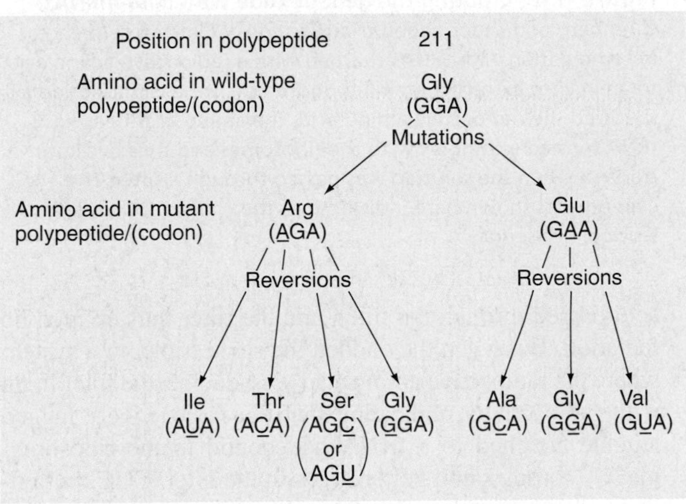

(b) Amino-acid alterations that accompany intragenic suppression

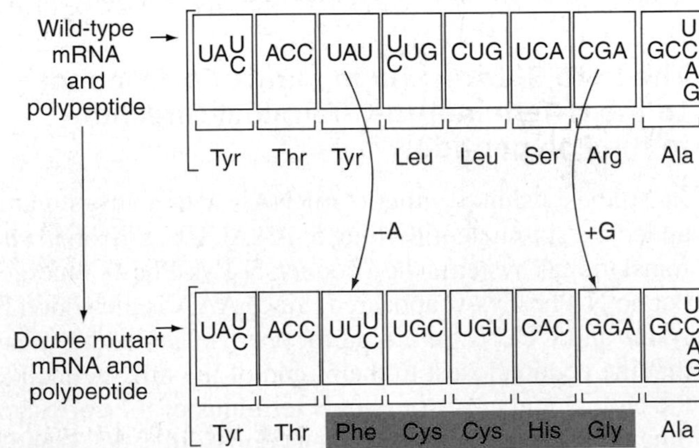

Figure 8.10 Experimental verification of the genetic code. **(a)** Single-base substitutions can explain the amino acid substitutions of trp^- mutations and trp^+ revertants. **(b)** The genetic code predicts the amino acid alterations (*yellow*) that would arise from single-base-pair deletions and suppressing insertions.

illustrates, single-base substitutions could also explain the amino acid changes in these revertants. Note that some of these substitutions restore Gly to position 211 of the polypeptide, while others place amino acids such as Ile, Thr, Ser, Ala, or Val at this site in the tryptophan synthetase molecule. The substitution of these other amino acids for Gly at position 211 in the polypeptide chain is compatible with (that is, largely conserves) the enzyme's function.

Yanofsky obtained better evidence yet that cells use the genetic code *in vivo* by analyzing proflavin-induced frameshift mutations of the tryptophan synthetase gene (Fig. 8.10b). He first treated populations of *E. coli* with proflavin to produce *trp*$^-$ mutants. Subsequent treatment of these mutants with more proflavin generated some *trp*$^+$ revertants among the progeny. The most likely explanation for the revertants was that their tryptophan synthetase gene carried both a single-base-pair deletion and a single-base-pair insertion ($- +$). Upon determining the amino acid sequences of the tryptophan synthetase enzymes made by the revertant strains, Yanofsky found that he could use the genetic code to predict the precise amino acid alterations that had occurred by assuming the revertants had a specific single-base-pair insertion and a specific single-base-pair deletion.

Yanofsky's results helped confirm not only amino acid codon assignments but other parameters of the code as well. His interpretations make sense only if codons do not overlap and are read from a fixed starting point with no pauses or commas separating the adjacent triplets.

In Chapter 9, we describe molecular techniques developed since the 1970s that make it possible to read the exact nucleotide sequence of a piece of DNA. In studies using these techniques, geneticists have determined the DNA sequences of genes encoding proteins of known amino acid sequence. These nucleotide sequences verify that the codons assigned to each amino acid by *in vitro* translational systems incorporating synthetic mRNAs are, in fact, used for translation *in vivo.*

The Genetic Code Is Almost, But Not Quite, Universal

We now know that virtually all cells alive today—from the prokaryotic cells of *E. coli* bacteria through the range of eukaryotic cells in yeast, *C. elegans,* peas, lentils, and people—use the same basic genetic code. One early indication of this uniformity was that a translational system derived from one organism could use the mRNA from another organism to convert genetic information to the encoded protein. Rabbit hemoglobin mRNA, for example, when injected into frog eggs or added to cell-free extracts from wheat germ, directs the synthesis of rabbit hemoglobin proteins. More recently, comparisons of DNA and protein sequences have revealed a perfect correspondence according to the genetic code between codons and amino acids in almost all organisms examined.

The universality of the code is an indication that it evolved very early in the history of life. Once it emerged, it remained constant over billions of years, in part because evolving organisms would have little tolerance for change. A single change in the genetic code could disrupt the production of hundreds or thousands of proteins in a cell—from the DNA polymerase that is essential for replication to the RNA polymerase that is required for gene expression to the tubulin proteins that compose the mitotic spindle—and would therefore be lethal.

Researchers were thus quite amazed to observe a few exceptions to the universality of the code. In some species of the single-celled eukaryotic protozoans known as ciliates, the codons UAA and UAG, which are nonsense codons in most organisms, specify the amino acid glutamine; in other ciliates, UGA, the third stop codon in most organisms, specifies cysteine. These ciliates use the remaining nonsense codons as stop codons. Other systematic changes in the genetic code exist in mitochondria, the semiautonomous, self-reproducing organelles within eukaryotic cells that are the sites of ATP formation. Each mitochondrion has its own chromosomes and its own apparatus for gene expression (which we describe in detail in Chapter 16). In the mitochondria of yeast, CUA specifies threonine instead of leucine. It may be that ciliates and mitochondria tolerated these changes in the genetic code because the changes affected very few proteins. For instance, the nonsense codon UGA might have found only infrequent use in one kind of primitive ciliate, so its switch to a "sense" codon would not have made a tremendous difference in protein production. Similarly, mitochondria might have survived a few changes in the code because they synthesize only a handful of proteins.

The experimental evidence presented so far helped define a nearly universal genetic code. But although cracking the code made it possible to understand the broad outlines of information flow between gene and protein, it did not explain exactly how the cellular machinery accomplishes gene expression. This is our focus as we present the details of transcription and translation.

8.2 Transcription: RNA Polymerase Synthesizes a Single-Stranded RNA Copy of a Gene

Transcription is the process by which the polymerization of ribonucleotides guided by complementary base pairing produces an RNA transcript of a gene. The template for the RNA transcript is one strand of that portion of the DNA double helix that composes the gene.

Details of the Process

Figure 8.11 depicts the basic components of transcription and illustrates key events in the process as it occurs in the bacterium *E. coli.* The following four points are of particular

Feature Figure 8.11

Transcription: How Bacterial Cells Copy the DNA of a Gene into the RNA of a Transcript

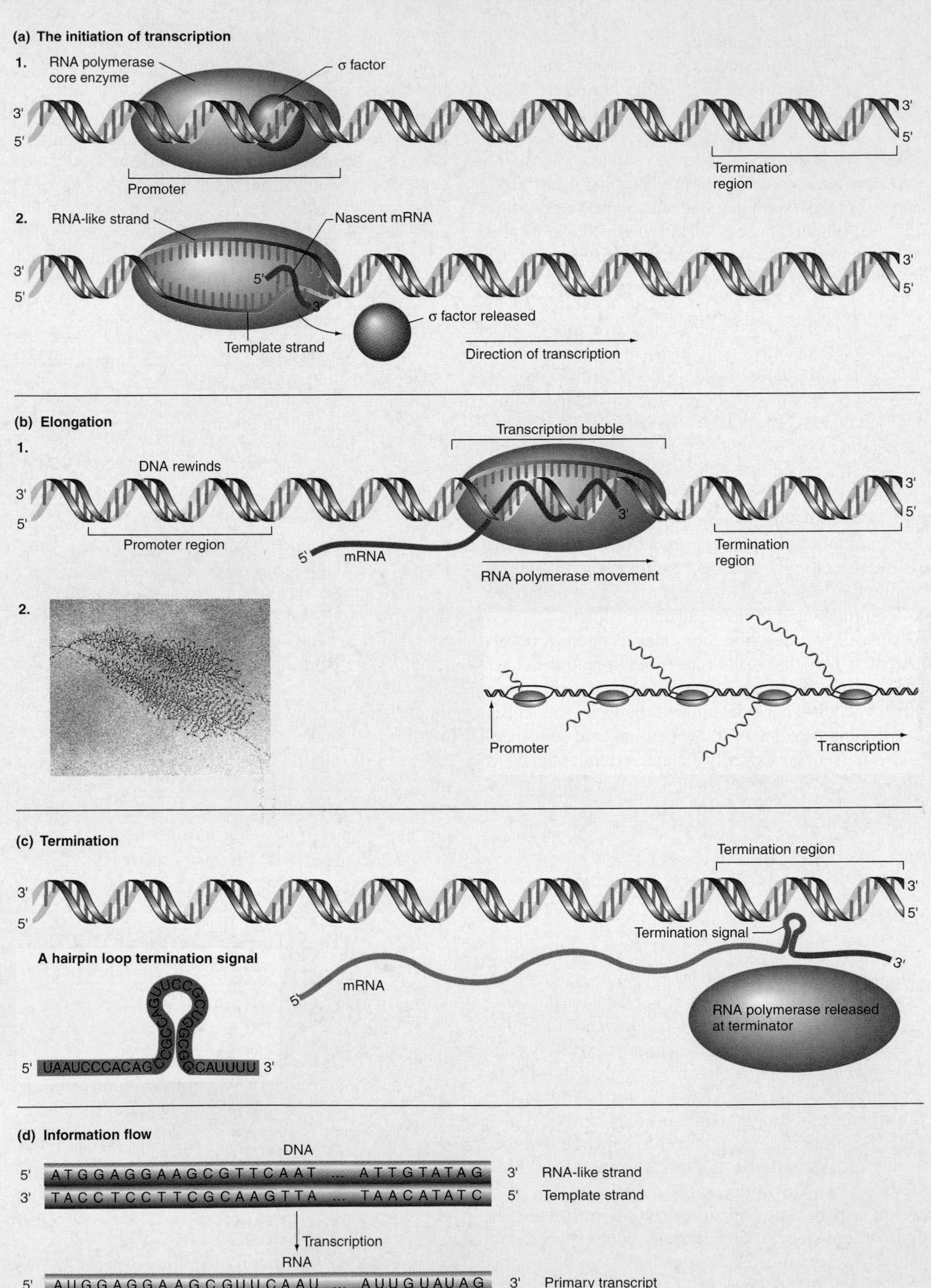

(a) The Initiation of Transcription

1. *RNA polymerase binds to double-stranded DNA at the beginning of the gene to be copied.* RNA polymerase recognizes and binds to **promoters,** specialized DNA sequences near the beginning of a gene where transcription will start. Although the promoters of different genes vary substantially in size and sequence, all promoters contain two characteristic short sequences of 6–10 nucleotide pairs that help bind RNA polymerase. These short sequences are nearly identical in different promoters (Fig. 8.12). In bacteria, the complete RNA polymerase (the *holoenzyme*) consists of a core enzyme, plus a σ (sigma) subunit involved only in initiation. The σ subunit reduces RNA polymerase's general affinity for DNA but simultaneously increases RNA polymerase's affinity for the promoter. As a result, the RNA polymerase holoenzyme can hone in on a promoter and bind tightly to it, forming a so-called *closed promoter complex.*

2. *After binding to the promoter, RNA polymerase unwinds part of the double helix, exposing unpaired bases on the strand that will serve as the template.* The complex formed between the RNA polymerase holoenzyme and an unwound promoter is called an *open promoter complex.* The enzyme identifies the template strand and chooses the two nucleotides with which to initiate copying. Guided by base pairing with these two nucleotides, RNA polymerase aligns the first two ribonucleotides of the new RNA, which will be at the 5′ end of the final RNA product. The end of the gene transcribed into the 5′ end of the mRNA is often called the *5′ end of the gene.* The mRNA grows in the 5′-to-3′ direction because RNA polymerase adds nucleotides to the 3′ end of the growing RNA chain. After aligning the first two ribonucleotides, RNA polymerase catalyzes the formation of a phosphodiester bond between them, resulting in a diribonucleotide. Soon thereafter, the RNA polymerase releases the σ subunit. This release marks the end of initiation.

(b) Elongation: Constructing an RNA Copy of the Gene

1. *When the σ subunit separates from the RNA polymerase, the enzyme loses its enhanced affinity for the promoter sequence and regains its strong generalized affinity for any DNA.* These changes enable the core enzyme to leave the promoter yet remain bound to the gene. The core enzyme now moves along the chromosome, unwinding the double helix to expose the next single-stranded region of the template. The enzyme extends the RNA by linking to the 3′ end of the growing chain a ribonucleotide positioned by complementarity with the template strand. As the enzyme extends the mRNA in the 5′-to-3′ direction, it moves in the antiparallel 3′-to-5′ direction along the DNA template strand.

The region of DNA unwound by RNA polymerase is called the **transcription bubble.** Within the bubble, the nascent RNA chain remains base paired with the DNA template, forming a DNA-RNA hybrid. However, in those parts of the gene behind the bubble that have already been transcribed, the DNA double helix re-forms, displacing the RNA, which hangs out of the transcription complex as a single strand with a free 5′ end.

2. *Once an RNA polymerase has moved off the promoter, other RNA polymerase molecules can move in to initiate transcription.* If the promoter is very strong, that is, if it can rapidly attract RNA polymerase, the gene can undergo transcription by many RNA polymerases simultaneously. Here we show an electron micrograph and an artist's interpretation of simultaneous transcription by several RNA polymerases. As you can see, there is a range of RNAs in production; those in the early steps of elongation are shorter than those in later steps of elongation. The promoter for this gene thus lies very close to where the shortest RNA is emerging from the DNA.

Geneticists often use the direction traveled by RNA polymerase as a reference when discussing various features within a gene. If for example, you started at the 5′ end of a gene at point A and moved along the gene in the same direction as RNA polymerase to point B, you would be traveling in the **downstream** direction. If by contrast, you started at point B and moved in the opposite direction from RNA polymerase to point A, you would be traveling in the **upstream** direction.

(c) Termination: The End of Transcription

RNA sequences that signal the end of transcription are known as **terminators.** There are two types of terminators: *intrinsic terminators,* which cause the RNA polymerase core enzyme to terminate transcription on its own, and *extrinsic terminators,* which require proteins other than RNA polymerase—particularly a polypeptide known as *rho*—to bring about termination. All terminators, whether intrinsic or extrinsic, are specific sequences in the mRNA; they are transcribed from specific DNA regions and they signal the termination of transcription. Terminators often form **hairpin loops** in which nucleotides within the mRNA pair with nearby complementary nucleotides. Upon termination, RNA polymerase and a completed RNA chain are both released from the DNA.

(d) The Product of Transcription Is a Single-Stranded Primary Transcript

The RNA produced by the action of RNA polymerase on a gene is a single strand of nucleotides known as a *primary transcript.* The bases in the primary transcript are complementary to the bases between the initiation and termination sites in the template strand of the gene. The nucleotides carrying these bases include groupings for a start codon, codons specifying all the amino acids in the polypeptide to be built, and a stop codon.

(a) Transcription initiation signals in bacteria

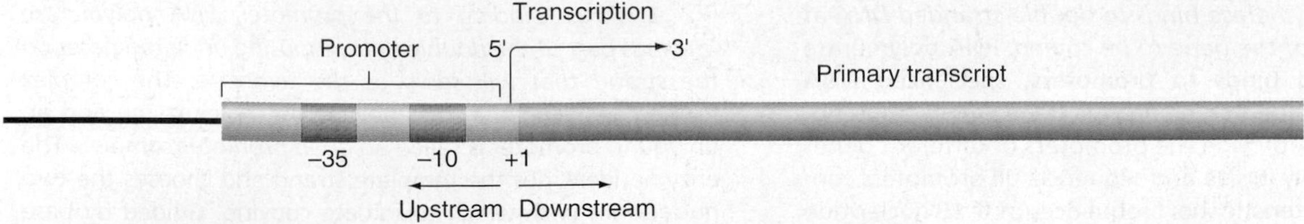

(b) Strong *E. coli* promoters

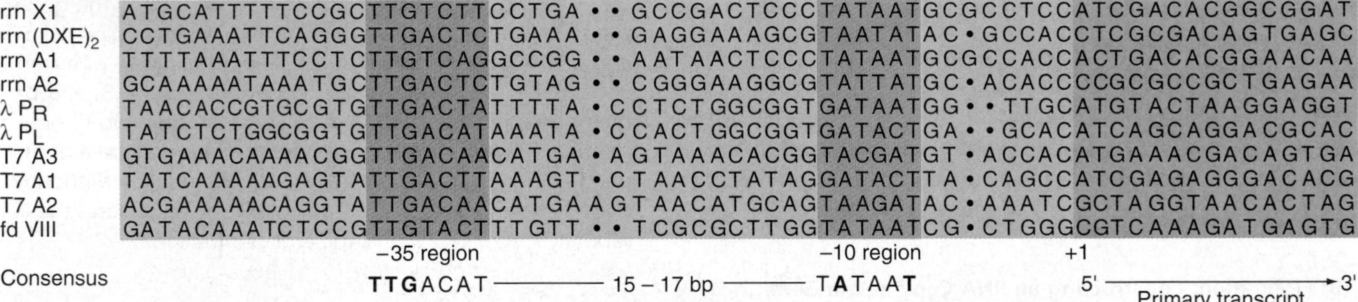

Figure 8.12 The promoters of 10 different bacterial genes. Only the sequence of the RNA-like strand is shown; numbering starts at the first transcribed nucleotide (+1). **(a)** Most promoters are upstream of (that is, before) the start point of transcription. **(b)** All promoters in *E. coli* share two different short stretches of nucleotides (*yellow*) that are essential for recognition of the promoter by RNA polymerase. The most common nucleotides at each position in each stretch constitute a *consensus sequence;* invariant nucleotides within the consensus are in *bold.*

importance: (1) The enzyme **RNA polymerase** catalyzes transcription. (2) DNA sequences near the beginning of genes, called **promoters,** signal RNA polymerase where to begin transcription. As seen in **Fig. 8.12**, most bacterial gene promotors share two short regions in which the promoters have almost identical nucleotide sequences. The nucleotides at these positions have evolved substantial similarity in different promoters because they are the sites at which RNA polymerase makes particularly strong contact with the promoters. (3) RNA polymerase adds nucleotides to the growing RNA polymer in the 5′-to-3′ direction. This polarity is a consequence of the reaction that joins each successive nucleotide to the RNA chain. The chemical mechanism of this reaction is analogous to that underlying the formation of phosphodiester bonds between nucleotides during DNA replication (review Fig. 6.17 on p. 187), with one exception: Transcription uses ribonucleotide triphosphates (ATP, CTP, GTP, and UTP) instead of deoxyribonucleotide triphosphates. Hydrolysis of the high-energy bonds in each ribonucleotide triphosphate provides the energy needed to add that nucleotide to the elongating RNA polymer. (4) Sequences in the RNA products, known as **terminators,** tell RNA polymerase where to stop transcription.

As you examine Fig. 8.11 to see how components of the cellular machinery work together to begin, carry out, and end transcription, bear in mind that a gene consists of two antiparallel strands of DNA, as mentioned earlier. One—the *RNA-like strand*—has the same polarity and sequence (except for T instead of U) as the emerging RNA transcript. The second—the *template strand*—has the opposite polarity and a complementary sequence that enables it to serve as the template for making the RNA transcript. When geneticists refer to the sequence of a gene, they usually mean the sequence of the RNA-like strand.

Although the transcription of all genes in all organisms roughly follows the general scheme shown in Fig. 8.11, there are important variations in detail. For example, the transcription of different genes in bacteria can be initiated by alternative sigma (σ) factors. In eukaryotes, promoters are more complicated than those in bacteria, and there are three different kinds of RNA polymerase that can transcribe different classes of genes. Chapters 17 and 18 describe how prokaryotic and eukaryotic cells can exploit these and other variations to control when, where, and at what level a given gene is expressed. Finally, the Genetics and Society box "HIV and Reverse Transcription: An Unusual DNA Polymerase Gives the AIDS Virus an Evolutionary Edge" on pp. 270–271 describes how the AIDS virus uses an exceptional form of transcription, known as **reverse transcription,** to construct a double strand of DNA from an RNA template.

The result of transcription is a single strand of RNA known as a **primary transcript** (see Fig. 8.11d). In prokaryotic organisms, the RNA produced by transcription is the actual messenger RNA that guides protein synthesis. In eukaryotic organisms, by contrast, most primary transcripts undergo *processing* in the nucleus before they migrate to the cytoplasm to direct protein synthesis. This

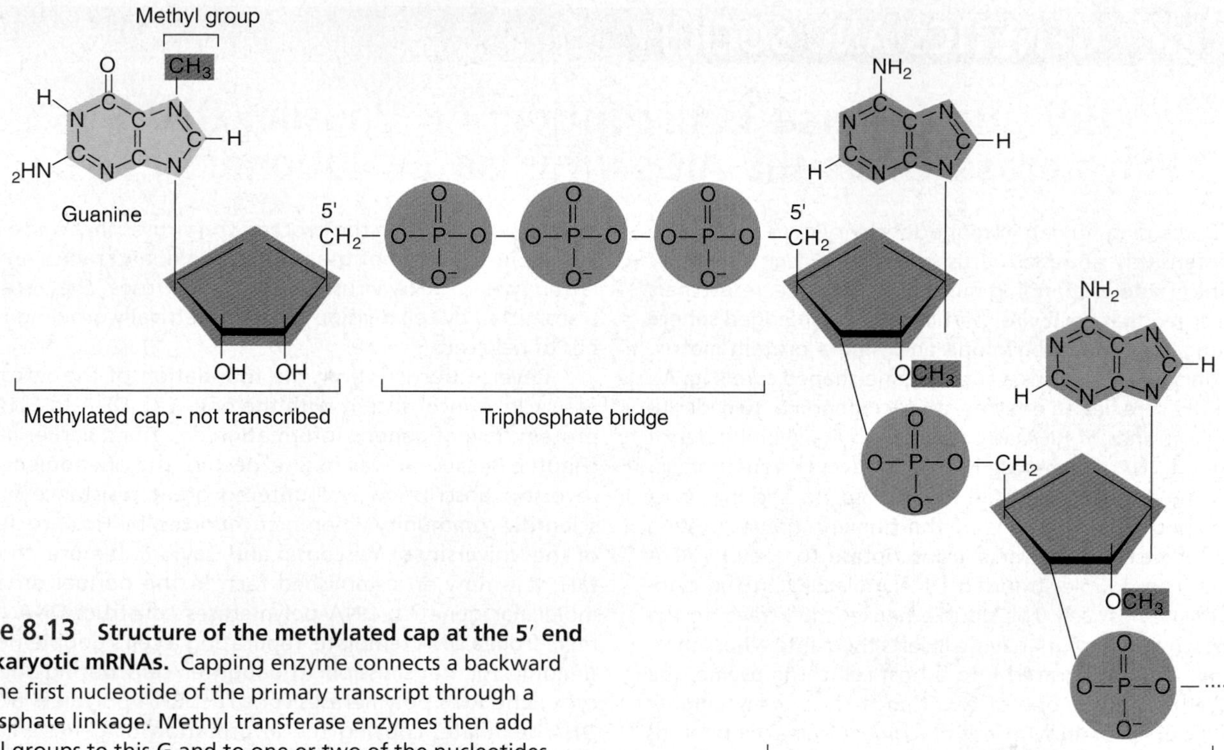

Figure 8.13 Structure of the methylated cap at the 5′ end of eukaryotic mRNAs. Capping enzyme connects a backward G to the first nucleotide of the primary transcript through a triphosphate linkage. Methyl transferase enzymes then add methyl groups to this G and to one or two of the nucleotides first transcribed from the DNA template.

processing has played a fundamental role in the evolution of complex organisms.

In Eukaryotes, RNA Processing After Transcription Produces a Mature mRNA

Some RNA processing in eukaryotes modifies only the 5′ or 3′ ends of the primary transcript, leaving untouched the information content of the rest of the mRNA. Other processing deletes blocks of information from the middle of the primary transcript, so the content of the mature mRNA is related, but not identical, to the complete set of DNA nucleotide pairs in the original gene.

Adding a Methylated Cap at the 5′ End and a Poly-A Tail at the 3′ End

The nucleotide at the 5′ end of a eukaryotic mRNA is a G in reverse orientation from the rest of the molecule; it is connected through a triphosphate linkage to the first nucleotide in the primary transcript. This backward G is not transcribed from the DNA. Instead, a special *capping enzyme* adds it to the primary transcript after polymerization of the transcript's first few nucleotides. Enzymes known as *methyl transferases* then add methyl (–CH_3) groups to the backward G and to one or more of the succeeding nucleotides in the RNA, forming a so-called **methylated cap** (**Fig. 8.13**).

Like the 5′ methylated cap, the 3′ end of most eukaryotic mRNAs is not encoded directly by the gene. In a large majority of eukaryotic mRNAs, the 3′ end consists of 100–200 A's, referred to as a **poly-A tail** (**Fig. 8.14**). Addition of the tail is a two-step process. First, a ribonuclease cleaves the primary

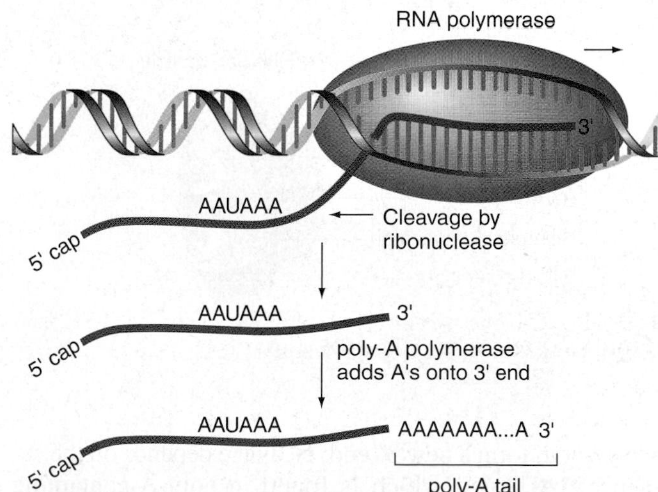

Figure 8.14 How RNA processing adds a tail to the 3′ end of eukaryotic mRNAs. A ribonuclease recognizes AAUAAA in a particular context of the primary transcript and cleaves the transcript 11–30 nucleotides downstream to create a new 3′ end. The enzyme poly-A polymerase then adds 100–200 A's onto this new 3′ end.

GENETICS AND SOCIETY

HIV and Reverse Transcription: An Unusual DNA Polymerase Gives the AIDs Virus an Evolutionary Edge

The AIDS-causing human immunodeficiency virus (HIV) is the most intensively analyzed virus in history. From laboratory and clinical studies spanning more than a decade, researchers have learned that each viral particle is a rough-edged sphere consisting of an outer envelope enclosing a protein matrix, which, in turn, surrounds a cut-off cone-shaped core (**Fig. A**). Within the core lies an enzyme-studded genome: two identical single strands of RNA associated with many molecules of an unusual DNA polymerase known as **reverse transcriptase.**

During infection, the AIDS virus binds to and injects its cone-shaped core into cells of the human immune system (**Fig. B**). It next uses reverse transcriptase to copy its RNA genome into double-stranded DNA molecules in the cytoplasm of the host cell. The double helixes then travel to the nucleus where another enzyme inserts them into a host chromosome. Once integrated into a host-cell chromosome, the viral genome can do one of two things. It can commandeer the host cell's protein synthesis machinery to make hundreds of new viral particles that bud off from the parent cell, taking with them part of the cell membrane and sometimes resulting in the host cell's death. Alternately, it can lie latent inside the host chromosome, which then copies and transmits the viral genome to two new cells with each cell division.

The events of this life cycle make HIV a **retrovirus:** an RNA virus that after infecting a host cell copies its own single strands of RNA into double helixes of DNA, which a viral enzyme then integrates into a host chromosome. RNA viruses that are not retroviruses simply infect a host cell and then use the cellular machinery to make more of themselves, often killing the host cell in the process. The viruses that cause hepatitis A, many types of the common cold, and rabies are this latter type of RNA virus. Unlike retroviruses, they are not transmitted by cell division to a geometrically growing number of new cells.

Reverse transcription, the foundation of the retroviral life cycle, is inconsistent with the one-way, DNA-to-RNA-to-protein flow of genetic information described earlier in this chapter. Because it was so unexpected, the phenomenon of reverse transcription encountered great resistance in the scientific community when first reported by Howard Temin of the University of Wisconsin and David Baltimore, then of MIT. It is now an established fact. In the normal order of molecular genetics, DNA polymerases construct DNA polymers from a DNA template, replicating a cell's double-helical genome for transmission to daughter cells during the cell cycle; and RNA polymerases construct RNA polymers from a DNA template, copying the information of genes into an RNA transcript that is then translated into protein. Reverse transcriptase, by comparison, is a remarkable DNA polymerase that can construct a DNA polymer from either an RNA or a DNA template.

HIV viral particle
Core
Protein matrix
RNA
Reverse transcriptase
Bilipid outer layer

Figure A Structure of the AIDS virus

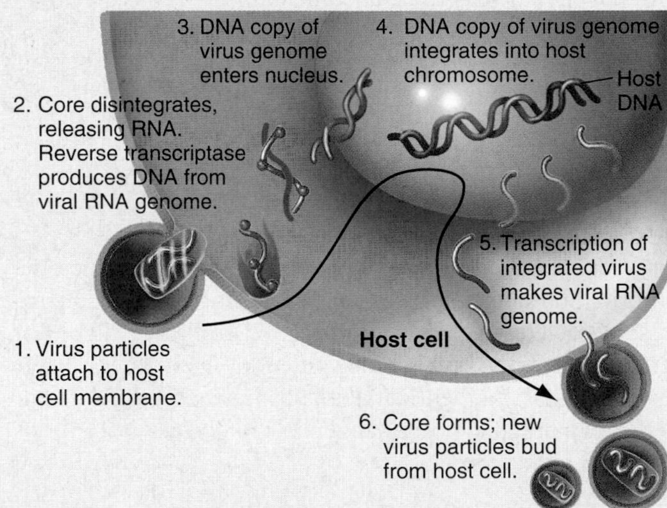
1. Virus particles attach to host cell membrane.
2. Core disintegrates, releasing RNA. Reverse transcriptase produces DNA from viral RNA genome.
3. DNA copy of virus genome enters nucleus.
4. DNA copy of virus genome integrates into host chromosome.
Host DNA
5. Transcription of integrated virus makes viral RNA genome.
Host cell
6. Core forms; new virus particles bud from host cell.

Figure B Life cycle of the AIDS virus

transcript to form a new 3′ end; cleavage depends on the sequence AAUAAA, which is found in poly-A-containing mRNAs 11–30 nucleotides upstream of the position where the tail is added. Next the enzyme *poly-A polymerase* adds A's onto the 3′ end exposed by cleavage.

Unexpectedly, both the methylated cap and the poly-A tail are critical for the efficient translation of the mRNA into protein, even though neither helps specify an amino acid. Recent data indicate that particular *eukaryotic translation initiation factors* bind to the 5′ cap, while *poly-A binding protein* associates with the tail at the 3′ end of the mRNA. The interaction of these proteins shapes the mRNA molecule into a circle. This circularization both enhances the initial steps of translation and stabilizes the mRNA in the cytoplasm by increasing the length of time it can serve as a messenger.

In addition to its comprehensive copying abilities, reverse transcriptase has another feature not seen in most DNA polymerases: inaccuracy. As we saw in Chapter 6, normal DNA polymerases replicate DNA with an error rate of one mistake in every million nucleotides copied. Reverse transcriptase, however, introduces one mutation in every 5000 incorporated nucleotides.

HIV uses this capacity for mutation, in combination with its ability to integrate its genome into the chromosomes of immune-system cells, to gain a tactical advantage over the immune response of its host organism. Cells of the immune system seek to overcome an HIV invasion by multiplying in response to the proliferating viral particles. The numbers are staggering. Each day of infection in every patient, from 100 million to a billion HIV particles are released from infected immune-system cells. As long as the immune system is strong enough to withstand the assault, it may respond by producing as many as 2 billion new cells daily. Many of these new cells produce antibodies targeted against proteins on the surface of the virus. But just when an immune response to a particular HIV protein wipes out those viral particles carrying the targeted protein, virions incorporating new forms of the protein resistant to the current immune response make their appearance. After many years of this complex chase, capture, and destruction by the immune system, the changeable virus outruns the host's immune response and gains the upper hand. Thus, the intrinsic infidelity of HIV's reverse transcriptase, by enhancing the virus's ability to compete in the evolutionary marketplace, increases its threat to human life and health.

The inherent mutability arising from reverse transcriptase's error-prone copying has undermined two therapeutic approaches that have helped control infections other than AIDS: drugs and vaccines. Some of the antiviral drugs approved in the United States for treatment of HIV infection—AZT (zidovudine), ddC (dideoxycytidine), and ddI (dideoxyinosine)—block viral replication by interfering with the action of reverse transcriptase. Each drug is similar to one of the four nucleotides, and when reverse transcriptase incorporates one of the drug molecules rather than a genuine nucleotide into a growing DNA polymer, the enzyme cannot extend the chain any further. However, the drugs are toxic at high doses and thus can be administered only at low doses that do not destroy all viral particles. Because of this limitation and the virus's high rate of mutation, mutant reverse transcriptases soon appear that work even in the presence of the drugs.

Similarly, researchers are having trouble developing safe, effective vaccines. Since HIV infects cells of the immune system and a vaccine works by stimulating immune-system cells to multiply, some of the vaccines tested so far actually increase the activity of the virus; others have only a weak effect on viral replication. Moreover, if it were possible to produce a vaccine that could generate a massive immune response against one, two, or even several HIV proteins at a time, such a vaccine might be effective for only a short while—until enough mutations built up to make the virus resistant.

For these reasons, the AIDS virus will most likely not succumb entirely to drugs or vaccines that target proteins active at various stages of its life cycle, although combinations of these therapeutic tools will remain an important part of the medical arsenal for prolonging an AIDS patient's life. In 1996, for example, medical researchers found that a therapeutic "cocktail" including at least one anti-reverse-transcriptase drug and a relatively new kind of drug known as a protease inhibitor (which blocks the enzymes that cleave a long, inactive polyprotein into shorter, functional viral proteins) could reduce the viral load of some very sick AIDS patients to undetectable levels, thereby relieving their disease symptoms. One year later, however, a clinical study revealed that for slightly more than 50% of patients receiving the drug cocktail in a San Francisco hospital, the treatment lost its effectiveness after six months. For the roughly 50% of patients for whom the cocktail was still effective, the new therapy offers some hope for controlling AIDS unless and until the virus in them evolves resistance to the new treatment.

Thus, after more than a decade of cooperative research and clinical experience, scientists know a great deal about the structure and life cycle of the AIDS virus, but they have been unable to develop drugs and vaccines that stop or even significantly slow its spread for more than a short while. In the mid-1990s, epidemiologists estimated that new infections of the AIDS virus occurred once every 15 seconds. By the year 2005, that rate of infection added up to nearly 40 million HIV-infected people worldwide.

A self-preserving capacity for mutation, perpetuated by reverse transcriptase, is surely one of the main reasons for HIV's success. Ironically, it may also provide a basis for its subjugation. Researchers are studying what happens when the virus increases its mutational load. Their reasoning is as follows: If reverse transcriptase's error rate determines the size and integrity of the viral population in a host organism, greatly accelerated mutagenesis might push the virus beyond the error threshold that allows it to function. In other words, too much mutation might destroy the virus's infectivity, virulence, or capacity to reproduce. If geneticists could figure out how to make this happen, they might be able to give the human immune system the advantage it needs to overrun the virus.

RNA Splicing Removes Introns from the Primary Transcript

Another kind of RNA processing became apparent in the late 1970s, after researchers had developed techniques that enabled them to analyze nucleotide sequences in both DNA and RNA. Using these techniques, which we describe in Chapter 9, they began to compare eukaryotic genes with the mRNAs derived from them. Their expectation was that just as in prokaryotes, the DNA nucleotide sequence of a gene's RNA-like strand would be identical to the RNA nucleotide sequence of the messenger RNA (with the exception of U replacing T in the RNA). Surprisingly, they found that the DNA nucleotide sequences of many eukaryotic genes are much longer than their corresponding mRNAs,

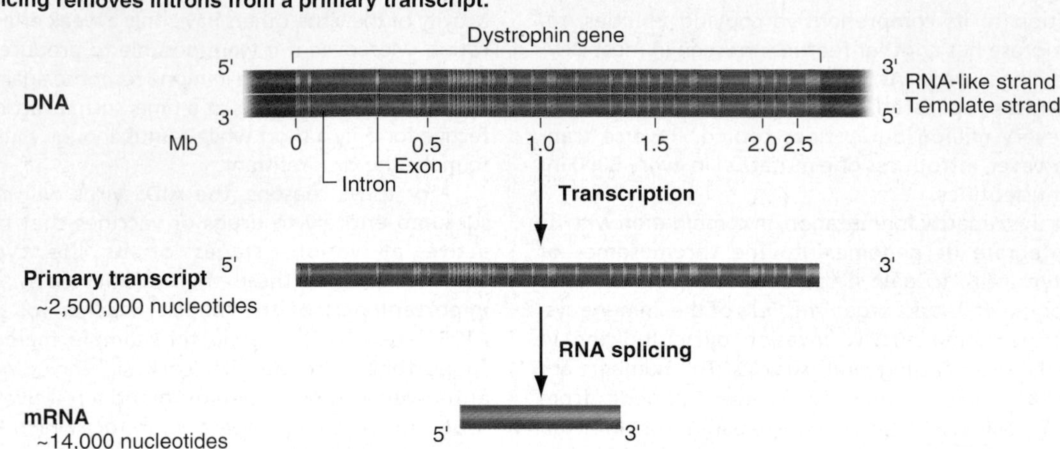

Figure 8.15 **The human dystrophin gene: An extreme example of RNA splicing.** Though the dystrophin gene is 2500 kb (or 2.5 Mb) long, the dystrophin mRNA is only 14 kb long. More than 80 introns are removed from the 2500 kb primary transcript to produce the mature mRNA (which is not drawn to scale).

suggesting that RNA transcripts, in addition to receiving a methylated cap and a poly-A tail, undergo extensive internal processing.

An extreme example of the length difference between primary transcript and mRNA is seen in the human gene for dystrophin (**Fig. 8.15**). Abnormalities in the dystrophin gene underlie the genetic disorder of Duchenne muscular dystrophy (DMD). The dystrophin gene is 2.5 million nucleotides—or 2500 kilobases (kb)—long, whereas the corresponding mRNA is roughly 14,000 nucleotides, or 14kb, in length. Obviously the gene contains DNA sequences that are not present in the mature mRNA. Those regions of the gene that do end up in the mature mRNA are scattered throughout the 2500 kb of DNA.

Sequences found in both a gene's DNA and the mature messenger RNA are called **exons** (for "expressed regions"). The *DMD* gene has more than 80 exons. The sequences found in the DNA of the gene but not in the mature mRNA are known as **introns** (for "intervening regions"). Introns interrupt, or separate, the exon sequences that actually end up in the mature mRNA. Introns can occur anywhere within a gene: between nucleotides that eventually form part of a single codon in the mature mRNA; between nucleotides encoding adjacent codons in the message; or within regions transcribed into parts of the mRNA that are not translated into protein. All of a gene's nucleotides that are transcribed into mRNA codons must be located in exons, but the three nucleotides of a single codon are often present in two different (but successive) exons. Finally, some exons or parts of exons near the beginning and end of a gene may not carry amino-acid-specifying codons. These exons instead encode untranslated regions at the 5′ and 3′ ends of the mRNA that play important roles in regulating the efficiency of mRNA translation into protein.

Exons vary in size from 50 to a few thousand nucleotide pairs (in the *DMD* gene, the mean exon length is 200 bp), while introns range from 50 to over 100,000 nucleotide pairs (in the *DMD* gene, the mean intron length is 35 kb, but one intron is an amazing 400 kb long). The greater size variation seen in introns compared to exons reflects the fact that introns do not appear in mature mRNAs and thus do not have to encode polypeptides. Because of this, there are fewer restraints on the sequence and number of nucleotides within introns. Geneticists know very little about the rules that determine the number, size, and position of introns in different genes. In humans, although a small percentage of genes do not contain introns, the large majority of genes interrupt their coding sequences anywhere from 1–60 times.

How do cells make a mature mRNA from a gene whose coding sequences are interrupted by introns? The answer is that cells first make a primary transcript containing all of a gene's introns and exons, and then they remove the introns from the primary transcript by **RNA splicing,** the process that deletes introns and joins together successive exons to form a mature mRNA consisting only of exons (Fig. 8.15). Because the first and last exons of the primary transcript become the 5′ and 3′ ends of the mRNA, while all intervening introns are spliced out, a gene must have one more exon than introns. To construct the mature mRNA, splicing must be remarkably precise. For example, if an intron lies within a codon, splicing must remove the intron and reconstitute the codon without disrupting the reading frame of the mRNA.

Figure 8.16 illustrates the details of RNA splicing. Three types of short sequences within the primary transcript—**splice donors, splice acceptors,** and **branch sites**—help ensure the specificity of splicing. These sites make it possible to sever the connections between an intron and the exons that precede and follow it, and then to join the formerly separated exons. The mechanism of splicing involves two sequential cuts in the primary transcript. The first cut is at the splice-donor site, at the 5′ end of the intron. After this

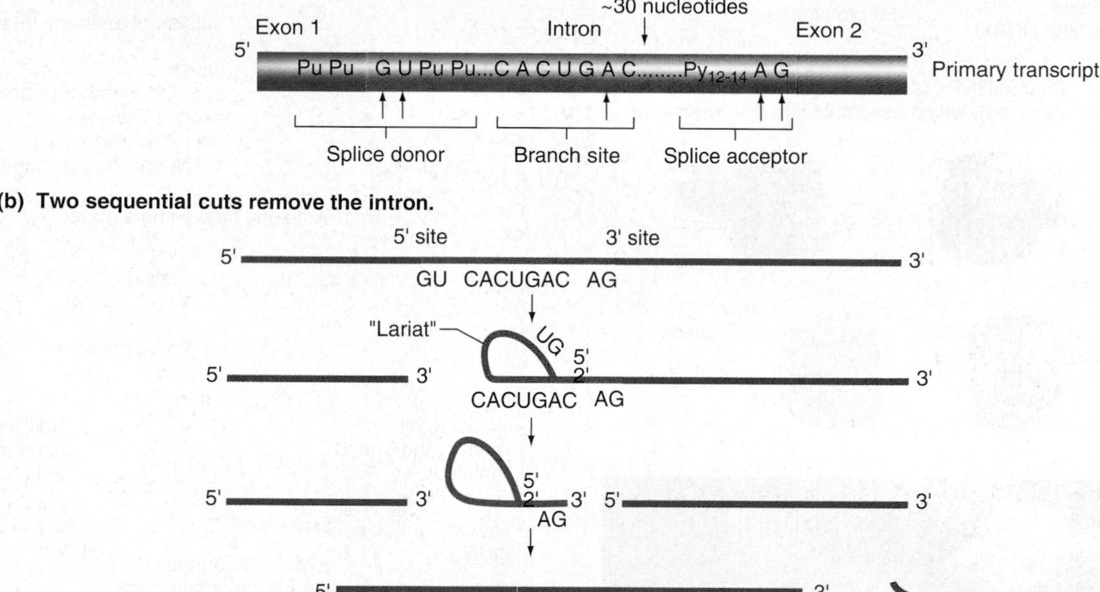

Figure 8.16 How RNA processing splices out introns and joins adjacent exons. **(a)** Three short sequences within the primary transcript determine the specificity of splicing. (1) The splice-donor site occurs where the 3′ end of an exon abuts the 5′ end of an intron. In most splice-donor sites, a GU dinucleotide (*arrows*) that begins the intron is flanked on either side by a few purines (Pu; that is, A or G). (2) The splice-acceptor site is at the 3′ end of the intron where it joins with the next exon. The final nucleotides of the intron are always AG (*arrows*) preceded by 12–14 pyrimidines (Py; that is, C or U). (3) The branch site, which is located within the intron about 30 nucleotides upstream of the splice acceptor, must include an A (*arrow*) and is usually rich in pyrimidines. **(b)** Two sequential cuts, the first at the splice-donor site and the second at the splice-acceptor site, remove the intron (which is subsequently degraded), allowing precise splicing of adjacent exons.

first cut, the new 5′ end of the intron attaches, via a novel 2′–5′ phosphodiester bond, to an A at the branch site located within the intron, forming a so-called *lariat structure*. The second cut is at the splice-acceptor site, at the 3′ end of the intron; this cut removes the intron. The discarded intron is degraded, and the precise splicing of adjacent exons completes the process of intron removal.

Splicing normally requires a complicated intranuclear machine called the **spliceosome,** which ensures that all of the splicing reactions take place in concert (**Fig. 8.17**). The spliceosome consists of four subunits known as small nuclear ribonucleoproteins, or snRNPs (pronounced "snurps"). Each snRNP contains one or two small nuclear RNAs (snRNAs) 100–300 nucleotides long, associated with proteins in a discrete particle. Certain snRNAs can base pair with the splice donor and splice acceptor sequences in the primary transcript, so these snRNAs are particularly important in bringing together the two exons that flank an intron. Given the complexities of spliceosome structure, it is remarkable that a few primary transcripts can splice themselves without the aid of a spliceosome or any additional factor. These rare primary transcripts function as **ribozymes:** RNA molecules that can act as enzymes and catalyze a specific biochemical reaction.

It might seem strange that eukaryotic genes incorporate DNA sequences that are spliced out of the mRNA before translation and thus do not encode amino acids. No one knows exactly why introns exist. One hypothesis proposes that they make it possible to assemble genes from various exon building blocks, which encode modules of protein function. This type of assembly would allow the shuffling of exons to make new genes, a process that appears to have played a key role in the evolution of complex organisms. The exon-as-module proposal is attractive because it is easy to understand the selective advantage of the potential for exon shuffling. Nevertheless, it remains a hypothesis without proof. There is no hard evidence for or against the hypothesis, and introns may have become established through means that scientists have yet to imagine.

Alternative Splicing Produces Different mRNAs from the Same Primary Transcript

Normally, RNA splicing joins together the splice donor and splice acceptor at the opposite ends of an intron, resulting in removal of the intron and fusion of two successive—and now adjacent—exons. For some genes, however, RNA splicing during development is regulated so that at certain

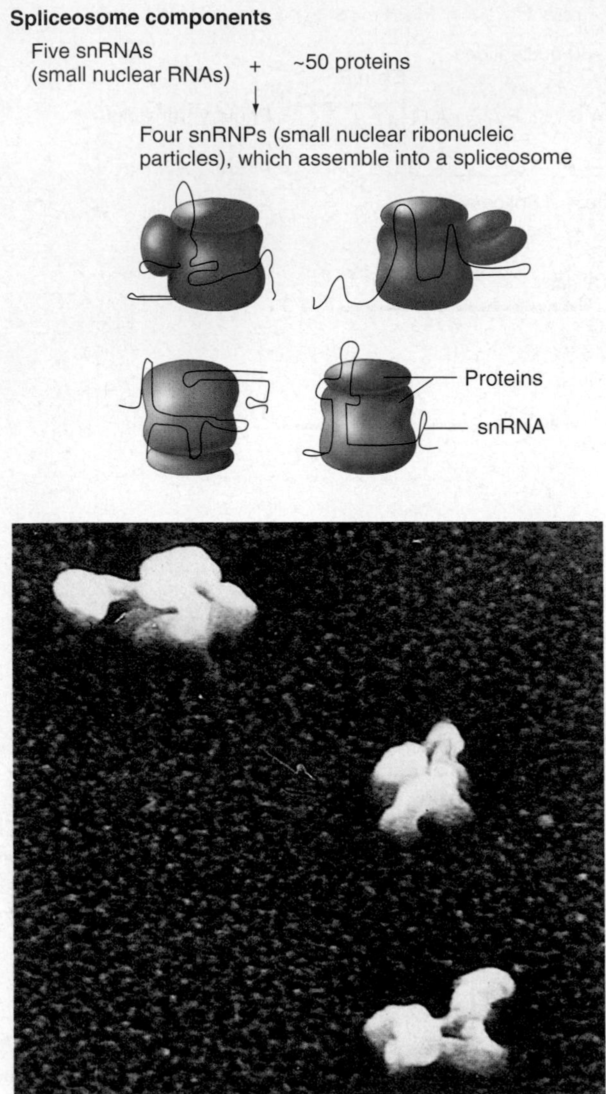

Figure 8.17 Splicing is catalyzed by the spliceosome. (*Top*) The spliceosome is assembled from four snRNP subunits, each of which contains one or two snRNAs and several proteins. (*Bottom*) A view of three spliceosomes in the electron microscope.

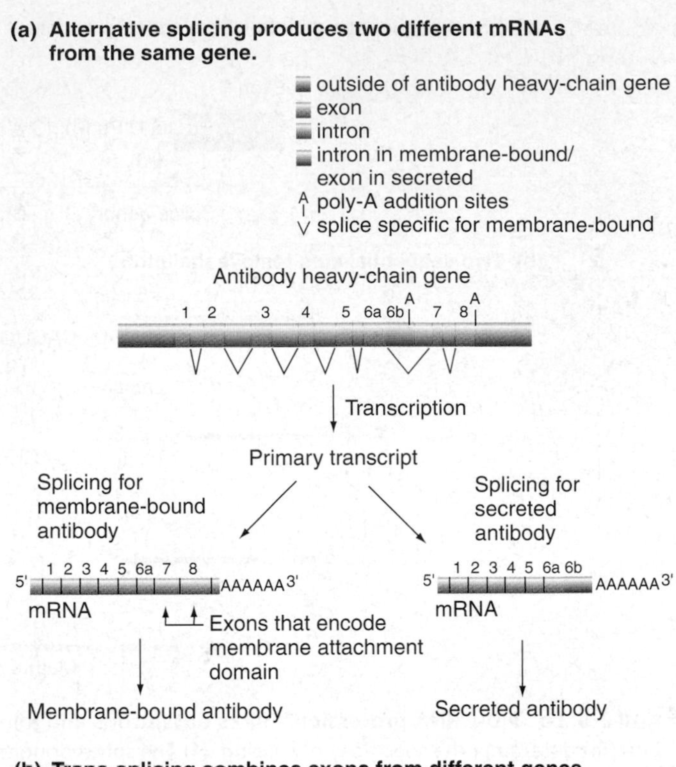

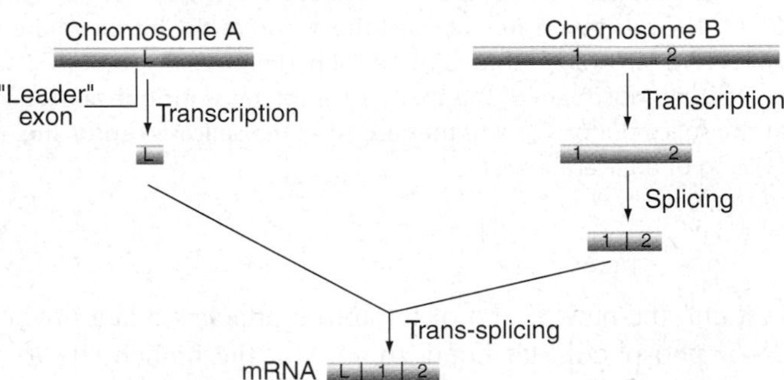

Figure 8.18 Different mRNAs can be produced from the same primary transcript. **(a)** Alternative splicing of the primary transcript for the antibody heavy chain produces mRNAs that encode different kinds of antibody proteins. **(b)** Rare trans-splicing events combine exons from different genes into one mature mRNA.

times or in certain tissues, some splicing signals may be ignored. As an example, splicing may occur between the splice donor site of one intron and the splice acceptor site of a different intron downstream. Such **alternative splicing** produces different mRNA molecules that may encode related proteins with different—though partially overlapping—amino acid sequences. In effect then, alternative splicing can tailor the nucleotide sequence of a primary transcript to produce more than one kind of polypeptide. Alternative splicing largely explains how the 20,000–30,000 genes in the human genome can encode the hundreds of thousands of different proteins estimated to exist in human cells.

In mammals, alternative splicing of the gene encoding the antibody heavy chain determines whether the antibody proteins become embedded in the membrane of the B lymphocyte that makes them or are instead secreted into the blood. The antibody heavy-chain gene has eight exons and seven introns; exon number 6 has a splice-donor site within it. To make the membrane-bound antibody, all exons except for the right-hand part of number 6 are joined to create an mRNA encoding a hydrophobic (water-hating, lipid-loving) C terminus (**Fig. 8.18a**). For the secreted antibody, only the first six exons (including the right part of 6) are spliced together to make an mRNA encoding a heavy chain with a hydrophilic (water-loving) C terminus. The alternate forms of RNA splicing thus allow physiological compartmentalization of the antibody molecule. Membrane-bound antibody molecules serve as receptors for foreign molecules (termed *antigens*); the binding of antigen triggers proliferation and differentiation of the B lymphocyte into an antibody-secreting factory. The secreted antibodies in the blood can interact directly with antigens, including portions of viruses and bacteria, targeting them for destruction.

A rare and unusual strategy of alternative splicing, seen in *C. elegans* and a few other eukaryotes, is **trans-splicing,** in which the spliceosome joins an exon of one gene with an exon of another gene (Fig. 8.18b). Special nucleotide sequences in the RNAs make trans-splicing possible.

In summary, RNA processing may add a methylated cap or a poly-A tail, delete introns from a primary transcript, determine which exons from a primary transcript are joined together to form a mature mRNA, and even join the exons of different genes.

8.3 Translation: Base Pairing Between mRNA and tRNAs Directs Assembly of a Polypeptide on the Ribosome

Virtually all cells—both prokaryotic and eukaryotic—use the same basic genetic code to translate the sequence of nucleotides in a messenger RNA to the sequence of amino acids in the corresponding polypeptide. This process of translation takes place on ribosomes that coordinate the movements of transfer RNAs carrying specific amino acids with the genetic instructions of an mRNA. As we examine the cell's translation machinery, we first describe the structure and function of tRNAs and ribosomes; and we then explain how these components interact during translation.

Transfer RNAs Mediate the Translation of mRNA Codons to Amino Acids

There is no obvious chemical similarity or affinity between the nucleotide triplets of mRNA codons and the amino acids they specify. Rather, *transfer RNAs (tRNAs)* serve as adaptor molecules that mediate the transfer of information from nucleic acid to protein.

tRNAs Carry an Anticodon at One End and an Amino Acid at the Other

tRNAs are short, single-stranded RNA molecules 74–95 nucleotides in length. Several of the nucleotides in tRNAs contain modified bases produced by chemical alterations of the principal A, G, C, and U nucleotides (**Fig. 8.19a**). Each tRNA carries one particular amino acid, and cells must

(a) Some tRNAs contain modified bases.

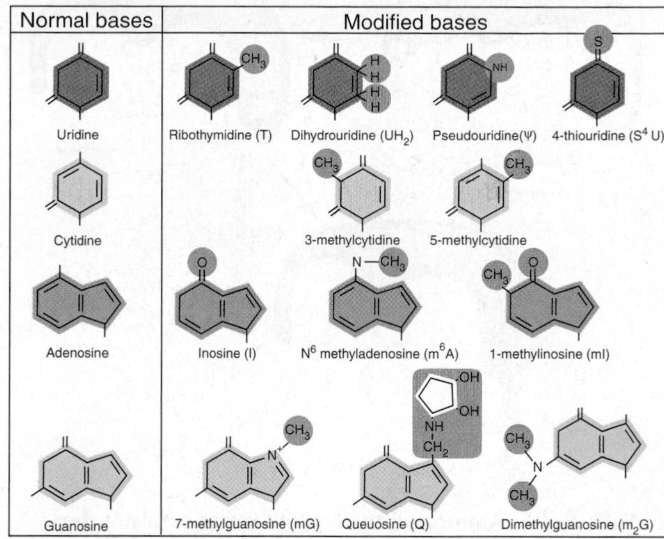

(b) Each tRNA has a primary, secondary, and tertiary structure.

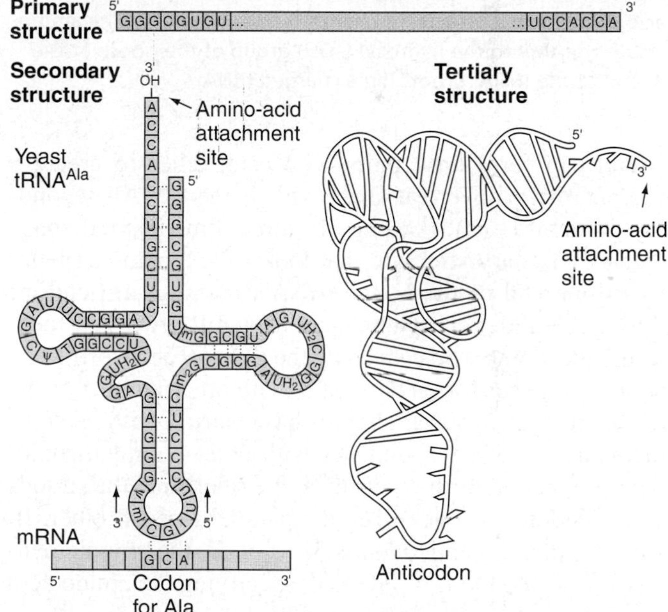

Figure 8.19 tRNAs mediate the transfer of information from nucleic acid to protein. (a) Many tRNAs contain modified bases produced by chemical alterations of A, G, C, and U. **(b)** The primary structures of tRNA molecules fold to form characteristic secondary and tertiary structures. The anticodon and the amino acid attachment site are at opposite ends of the "L" formed by a tRNA.

have at least one tRNA for each of the 20 amino acids specified by the genetic code. The name of a tRNA reflects the amino acid it carries. For example, tRNA^{Gly} carries the amino acid glycine.

As Fig. 8.19b shows, it is possible to consider the structure of a tRNA molecule on three levels. (1) The nucleotide sequence of a tRNA constitutes the primary structure. (2) Short complementary regions within a tRNA's single

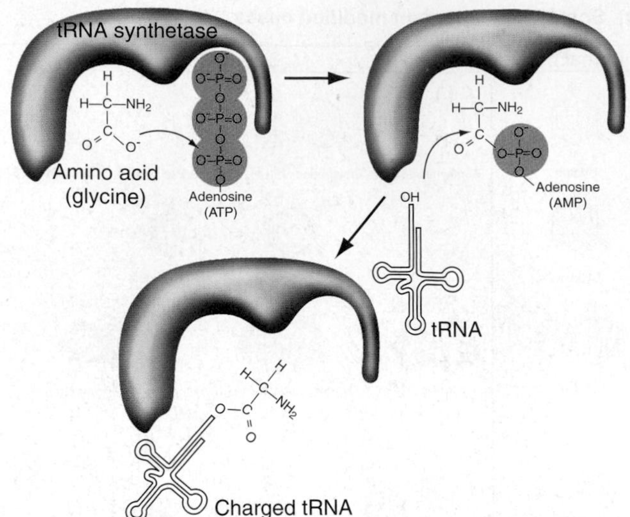

Figure 8.20 Aminoacyl-tRNA synthetases catalyze the attachment of tRNAs to their corresponding amino acids. The aminoacyl tRNA synthetase first uses the energy generated by ATP hydrolysis to activate the amino acid, forming an AMP-amino acid. The enzyme then transfers the carboxyl group of the amino acid from AMP to the hydroxyl (–OH) group of the ribose at the 3′ end of the tRNA, producing a charged tRNA.

strand can form base pairs with each other to create a characteristic cloverleaf shape; this is the tRNA's secondary structure. (3) Folding in three-dimensional space creates a tertiary structure that looks like a compact letter L. At one end of the L, the tRNA carries an **anticodon:** three nucleotides complementary to an mRNA codon specifying the amino acid carried by the tRNA (Fig. 8.19b). The anticodon never forms base pairs with other regions of the tRNA; it is always available for base pairing with its complementary mRNA codon. As with other complementary base sequences, during pairing at the ribosome, the strands of anticodon and codon run antiparallel to each other. If, for example, the anticodon is 3′ CCU 5′, the complementary mRNA codon is 5′ GGA 3′, specifying the amino acid glycine.

At the other end of the L—where the 5′ and 3′ ends of the tRNA strand are found—the tRNA connects to the amino acid specified by the codon. Enzymes known as **aminoacyl-tRNA synthetases** are the catalysts of this attachment (**Fig. 8.20**). These enzymes are extraordinarily specific, recognizing unique features of a particular tRNA—despite its general structural similarities with all other tRNAs—while also recognizing the corresponding amino acid (see the opening figure of this chapter on p. 255). Aminoacyl-tRNA synthetases are, in fact, the only molecules that read the languages of both nucleic acid and protein. They are thus the actual molecular translators. There is at least one aminoacyl-tRNA synthetase for each of the 20 amino acids, and like tRNA, each synthetase functions with only one amino acid. Figure 8.20 shows the two-step process that establishes the covalent bond between an amino acid and the 3′ end of its corresponding tRNA. A tRNA covalently coupled to its amino acid is called a **charged tRNA.** The bond between the amino acid and tRNA contains substantial energy that is later used to drive peptide bond formation.

Base Pairing Between an mRNA Codon and a tRNA Anticodon Directs Amino-Acid Incorporation into a Growing Polypeptide

While attachment of the appropriate amino acid charges a tRNA, the amino acid itself does not play a significant role in determining where it becomes incorporated in a growing polypeptide chain. It is the specific interaction between a tRNA's anticodon and an mRNA's codon that makes that decision. A simple experiment illustrates this point (**Fig. 8.21**). Researchers can subject a charged tRNA to chemical treatments that, without altering the structure of the tRNA, change the amino acid it carries. One treatment replaces the cysteine carried by $tRNA^{Cys}$ with alanine. When investigators then add the $tRNA^{Cys}$ charged with alanine to a cell-free translational system, the system incorporates *alanine* into the growing polypeptide wherever the mRNA contains a cysteine codon complementary to the anticodon of the $tRNA^{Cys}$.

Wobble: Some tRNAs Can Recognize More Than One Codon

Although there is at least one kind of tRNA for each of the 20 amino acids, cells do not necessarily carry tRNAs with anticodons complementary to all of the 61 possible codon triplets in the degenerate genetic code. *E. coli,* for example, makes 79 different tRNAs containing 42 different anticodons. Although several of the 79 tRNAs in this collection obviously have the same anticodon, 61 − 42 = 19 of 61 potential anticodons are not represented. Thus 19 mRNA

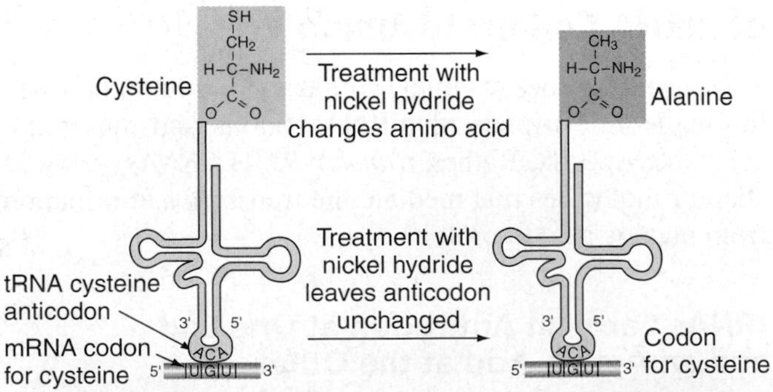

Figure 8.21 Base pairing between an mRNA codon and a tRNA anticodon determines where an amino acid becomes incorporated into a growing polypeptide. A tRNA with an anticodon for cysteine, but carrying the amino acid alanine, adds alanine whenever the mRNA codon for cysteine appears.

codons will not find a complementary anticodon in the *E. coli* collection of tRNAs. Another bacterial species, *Mycoplasma capricolum,* makes 30 different tRNAs carrying 29 different anticodons. In this species, only two tRNAs have the same anticodon, but 61 − 29 = 32 mRNA codons will not find a complementary anticodon in this collection of tRNAs. Although it is understandable that different species assemble their repertoire of tRNAs in different ways, how can an organism construct proper polypeptides if some of the codons in its mRNAs cannot locate tRNAs with complementary anticodons?

The answer is that some tRNAs can recognize more than one codon for the amino acid with which they are charged. That is, the anticodons of these tRNAs can interact with more than one codon for the same amino acid. Although researchers do not fully understand this "promiscuous" base pairing between codons and anticodons, Francis Crick spelled out a few of the rules that govern it. The rules are consistent with the genetic code. By analyzing the genetic code, Crick concluded that the 3′ nucleotide in many codons adds nothing to the specificity of the codon. For example, 5′ GGU 3′, 5′ GGC 3′, 5′ GGA 3′, and 5′ GGG 3′ all encode glycine (review Fig. 8.3 on p. 257). It does not matter whether the 3′ nucleotide is U, C, A, or G as long as the first two letters are GG. The same is true for other amino acids encoded by four different codons, such as valine, where the first two bases must be GU, but the third base can be U, C, A, or G. For amino acids specified by two different codons, the first two bases of the codon are, once again, always the same, while the third base must be either one of the two purines (A or G) or one of the two pyrimidines (U or C). Thus, 5′ CAA 3′ and 5′ CAG 3′ are both codons for glutamine; 5′ CAU 3′ and 5′ CAC 3′ are both codons for histidine. If Pu stands for either purine and Py stands for either pyrimidine, then CAPu represents the codons for glutamine, while CAPy represents the codons for histidine.

In fact, the 5′ nucleotide of a tRNA's anticodon can often pair with more than one kind of nucleotide in the 3′ position of an mRNA's codon. (Recall that after base pairing, the bases in the anticodon run antiparallel to the bases in the codon.) A single tRNA charged with a particular amino acid can thus recognize several or even all of the codons for that amino acid. This flexibility in base pairing between the 3′ nucleotide in the codon and the 5′ nucleotide in the anticodon is known as **wobble** (**Fig. 8.22a**). The combination of normal base pairing at the first two positions of a codon with wobble at the third position clarifies why multiple codons for the same amino acid usually start with the same two letters.

Crick's "wobble rules," shown in Fig. 8.22b, delimit what kind of flexibility in base pairing is consistent with the genetic code. For example, methionine (Met) is specified by a single codon (5′ AUG 3′). As a result, Met-specific tRNAs must have a C at the 5′ end of their anticodons (5′ CAU 3′), because this is the only nucleotide at that position that can base pair only with the G at the 3′ end of the Met codon. By contrast, a single isoleucine-specific tRNA with the modified nucleotide inosine (I) at the 5′ position of the anticodon can recognize all three codons (5′ AUU 3′, 5′ AUC 3′, and 5′ AUA 3′) for isoleucine.

(a)

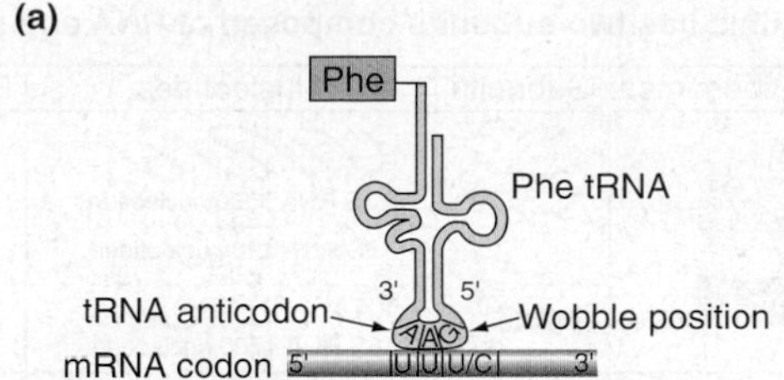

(b)

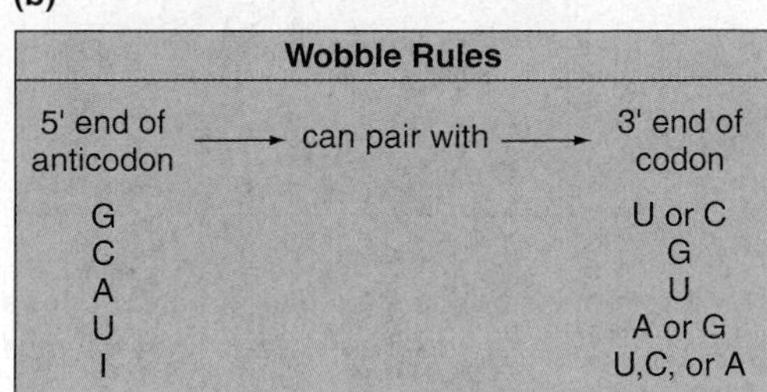

Wobble Rules		
5' end of anticodon	can pair with	3' end of codon
G		U or C
C		G
A		U
U		A or G
I		U,C, or A

Figure 8.22 Wobble: Some tRNAs recognize more than one codon for the amino acid they carry. **(a)** The G at the 5′ end of the anticodon shown here can pair with either U or C at the 3′ end of the codon. **(b)** The chart shows the pairing possibilities for other nucleotides at the 5′ end of an anticodon; I = inosine.

Ribosomes Are the Sites of Polypeptide Synthesis

Ribosomes facilitate polypeptide synthesis in various ways. First, they recognize mRNA features that signal the start of translation. Second, they help ensure accurate interpretation of the genetic code by stabilizing the interactions between tRNAs and mRNAs; without a ribosome, codon-anticodon recognition, mediated by only three base pairs (one of which may wobble), would be extremely weak. Third, they supply the enzymatic activity that links the amino acids in a growing polypeptide chain. Fourth, by moving 5′ to 3′ along an mRNA molecule, they expose the mRNA codons in sequence, ensuring the linear addition of amino acids. Finally, ribosomes help end polypeptide synthesis by dissociating from both the mRNA directing polypeptide construction and the polypeptide product itself.

Ribosomes Are Complex Structures Composed of RNA and Protein

In *E. coli,* ribosomes consist of 3 different **ribosomal RNAs (rRNAs)** and 52 different ribosomal proteins (**Fig. 8.23a**). These components associate to form two

(a) A ribosome has two subunits composed of RNA and protein.

Complete Ribosomes	Subunits	Nucleotides	Proteins
Prokaryotic 70S	50S	23S RNA 3000 nucleotides; 5S RNA 120 nucleotides	31
	30S	16S RNA 1700 nucleotides	21
Eukaryotic 80S	60S	28S RNA 5000 nucleotides; 5.8S RNA 160 nucleotides; 5S RNA 120 nucleotides	~ 45
	40S	18S RNA 2000 nucleotides	~ 33

(b) Different parts of a ribosome have different functions.

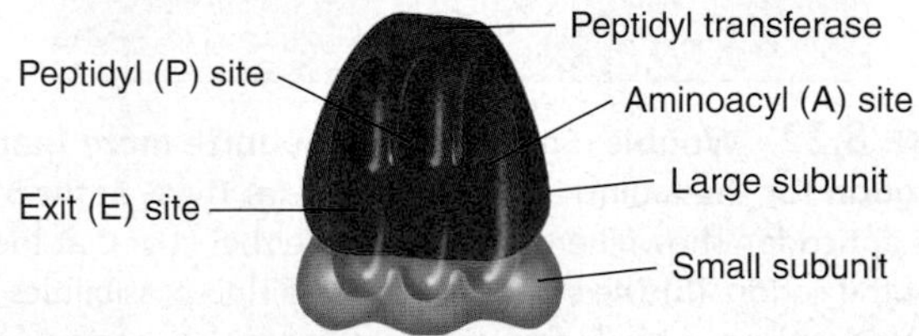

Figure 8.23 The ribosome: Site of polypeptide synthesis. **(a)** A ribosome has two subunits, each composed of rRNA and various proteins. **(b)** The small subunit initially binds to mRNA. The large subunit contributes the enzyme peptidyl transferase, which catalyzes the formation of peptide bonds in the growing polypeptide chain. The two subunits together form the A, P, and E tRNA binding sites.

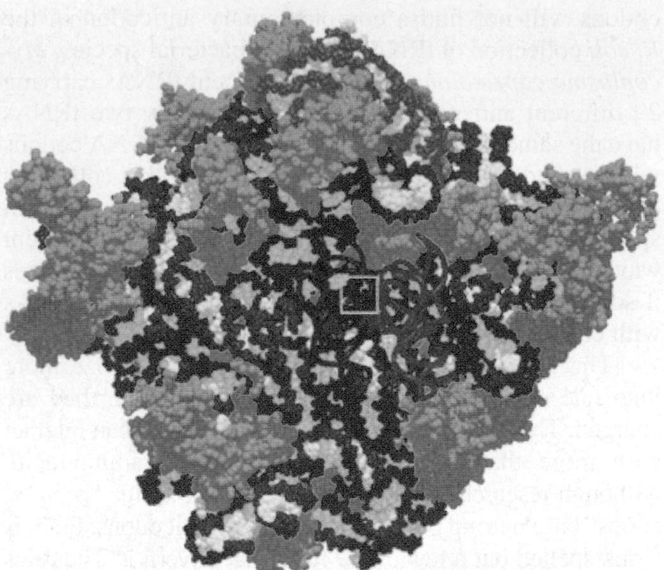

Figure 8.24 The large subunit of a bacterial ribosome. Various ribosomal proteins are *lavender,* 23S rRNA is in *gold* and *white,* and 5S rRNA is *maroon* and *white.* The tRNA in the A site is *green;* the tRNA in the P-site is *red;* no tRNA is shown in the E site. The superimposed box shows the location where new peptide bonds are formed.

different ribosomal subunits called the 30S subunit and the 50S subunit (with S designating a coefficient of sedimentation related to the size and shape of the subunit; the 30S subunit is smaller than the 50S subunit). Before translation begins, the two subunits exist as separate entities in the cytoplasm. Soon after the start of translation, they come together to reconstitute a complete ribosome. Eukaryotic ribosomes have more components than their prokaryotic counterparts, but they still consist of two dissociable subunits.

Different Parts of the Ribosome Have Different Functions

The small 30S subunit is the part of the ribosome that initially binds to mRNA. The larger 50S subunit contributes an enzyme known as **peptidyl transferase,** which catalyzes formation of the peptide bonds joining adjacent amino acids (Fig. 8.23b). Both the small and the large subunits contribute to three distinct tRNA binding areas known as the **aminoacyl** (or **A**) **site,** the **peptidyl** (or **P**) **site,** and the **exit** (or **E**) **site.** Finally, other regions of the ribosome distributed over the two subunits serve as points of contact for some of the additional proteins that play a role in translation.

Using X-ray crystallography and elegant techniques of electron microscopy, researchers have recently gained a remarkably detailed view of the complicated structure of the ribosome. **Fig. 8.24** shows the large subunit of a bacterial ribosome; the small subunit was computationally removed for better visualization of the charged tRNAs occupying the A and P sites. With this illustration, you can see that the rRNAs occupy most of the space in the central part of the ribosome, while the various ribsosomal proteins are studded around the exterior. Surprisingly, no proteins are found close to the region between the two tRNAs where peptide bonds are formed. This finding supports the conclusions of biochemical experiments that peptidyl transferase is actually a function of the 50S subunit's rRNA rather than any protein component of the ribosome; in other words, the rRNA acts as a ribozyme that joins amino acids together.

The Mechanism of Translation

Translation consists of an **initiation** phase that sets the stage for polypeptide synthesis; **elongation,** during which amino acids are added to a growing polypeptide; and a **termination** phase that brings polypeptide synthesis to a halt and enables the ribosome to release a completed chain of amino acids. **Figure 8.25** illustrates the details of the

Feature Figure 8.25

Translation: How Cells Use mRNA and Charged tRNAs to Assemble a Polypeptide on the Ribosome

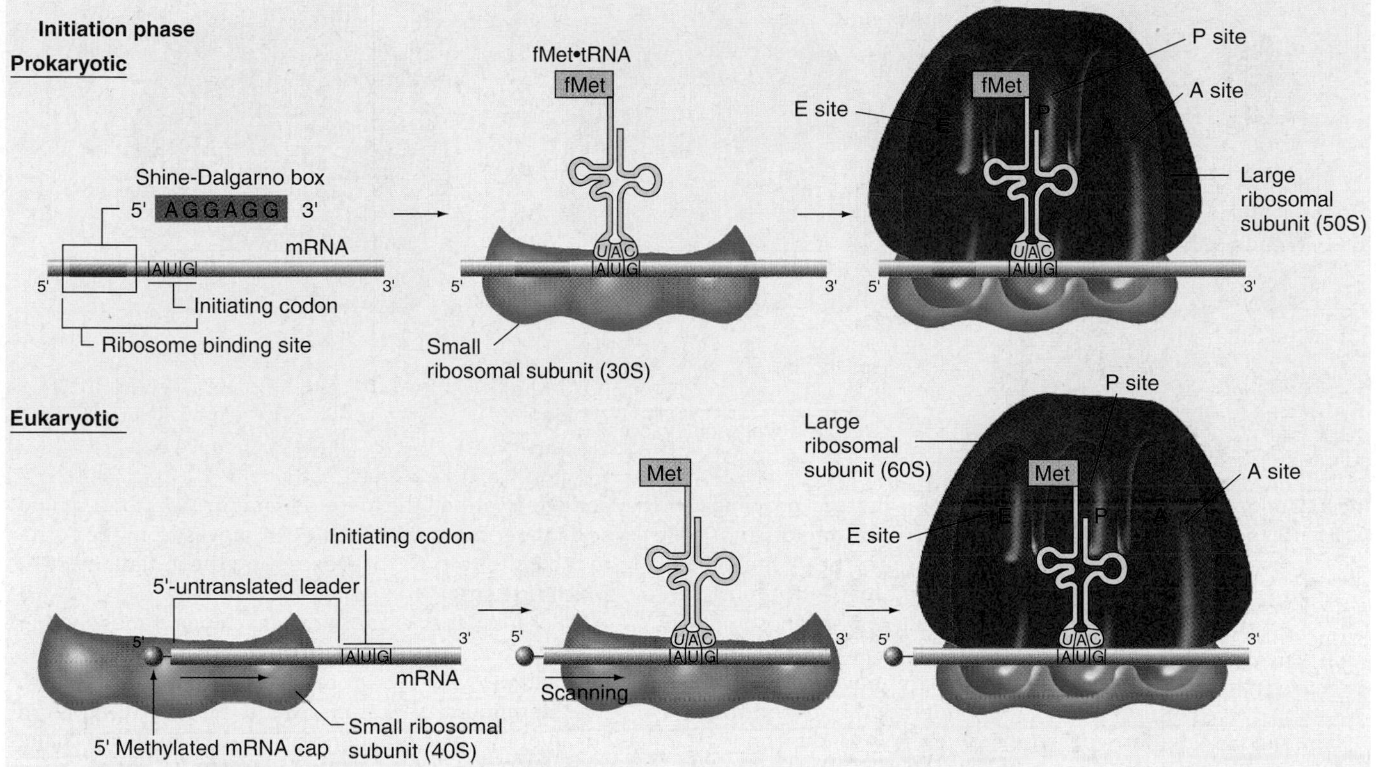

(a) Initiation: Setting the stage for polypeptide synthesis The first three nucleotides of an mRNA do not serve as the first codon to be translated into an amino acid. Instead, a special signal indicates where along the mRNA translation should begin. In prokaryotes, this signal is called the **ribosome binding site,** and it has two important elements. The first is a short sequence of six nucleotides—usually 5′. . . AGGAGG . . . 3′—named the **Shine-Dalgarno box** after its discoverers.

The second element in an mRNA's ribosome binding site is the triplet 5′ AUG 3′, which serves as the initiation codon. A special initiator tRNA, whose 5′ CAU 3′ anticodon is complementary to AUG, recognizes an AUG preceded by the Shine-Dalgarno box of a ribosome binding site. The initiator tRNA carries ***N*-formylmethionine (fMet),** a modified methionine whose amino end is blocked by a formyl group. The specialized fMet tRNA functions only at an initiation site. An AUG codon located within an mRNA's reading frame is recognized by a different tRNA that is charged with an unmodified methionine. This tRNA cannot start translation.

During initiation, the 3′ end of the 16S rRNA in the 30S ribosomal subunit binds to the mRNA's Shine-Dalgarno box (*not shown*), the fMet tRNA binds to the mRNA's initiation codon, and a large 50S ribosomal subunit associates with the small subunit to round out the ribosome. At the end of initiation, the fMet tRNA sits in the P site of the completed ribosome. Proteins known as **initiation factors** (*not shown*) play a transient role in the initiation process.

In eukaryotes, the small ribosomal subunit binds first to the methylated cap at the 5′ end of the mature mRNA. It then migrates to the initiation site—usually the first AUG it encounters as it scans the mRNA in the 5′-to-3′ direction. The initiator tRNA in eukaryotes carries unmodified methionine (Met) instead of fMet.

(*Continued*)

Feature Figure 8.25 (*Continued*)

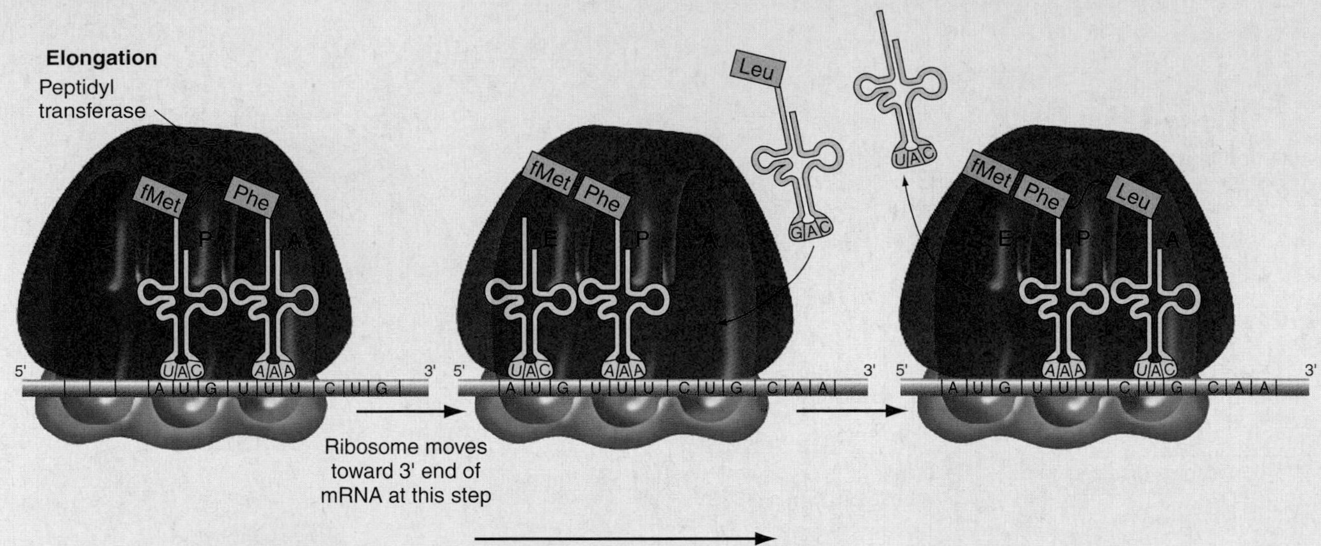

(b) Elongation: The addition of amino acids to a growing polypeptide Proteins known as **elongation factors** (*not shown*) usher the appropriate tRNA into the A site of the ribosome. The anticodon of this charged tRNA must recognize the next codon in the mRNA. The ribosome simultaneously holds the initiating tRNA at its P site and the second tRNA at its A site so that peptidyl transferase can catalyze formation of a peptide bond between the amino acids carried by the two tRNAs. This bond-forming reaction connects the fMet at the P site to the amino acid carried by the tRNA at the A site; it also disconnects fMet from the initiating tRNA. As a result, the tRNA at the A site now carries two amino acids. The N terminus of this dipeptide is fMet; the C terminus is the second amino acid, whose carboxyl group remains covalently linked to its tRNA.

Once formation of the first peptide bond causes the initiating tRNA in the P site to release its amino acid, the ribosome moves, exposing the next mRNA codon. Like the first steps of elongation, the ribosome's movement requires the help of elongation factors and an input of energy. As the ribosome moves, the initiating tRNA, which no longer carries an amino acid, is transferred to the E site, and the other tRNA carrying the dipeptide shifts from the A site to the P site.

The empty A site now receives another tRNA, whose identity is determined by the next codon in the mRNA. The uncharged intiating tRNA is bumped off the E site and leaves the ribosome. Peptidyl transferase then catalyzes formation of a second peptide bond, generating a chain of three amino acids. This tripeptide is connected at its C terminus to the tRNA currently in the A site; the fMet at its N terminus hangs free. With each subsequent round of ribosome movement and peptide bond formation, the peptide chain grows one amino acid longer. Note that each tRNA moves from the A site to the P site to the E site (excepting the initiating tRNA, which first enters the P site).

Because the elongation machinery adds amino acids to the C terminus of the lengthening polypeptide, polypeptide synthesis proceeds from the N terminus to the C terminus. As a result, fMet in prokaryotes (Met in eukaryotes), the first amino acid in the growing chain, will be the N-terminal amino acid of all finished polypeptides prior to protein processing. Moreover, the ribosome must move along the mRNA in the 5′-to-3′ direction so that the polypeptide can grow in the N-to-C direction.

Once a ribosome has moved far enough away from the mRNA's ribosome binding site, that site becomes accessible to other ribosomes. In fact, several ribosomes can work on the same mRNA at one time. A complex of several ribosomes translating from the same mRNA is called a **polyribosome.** This complex allows the simultaneous synthesis of many copies of a polypeptide from a single mRNA.

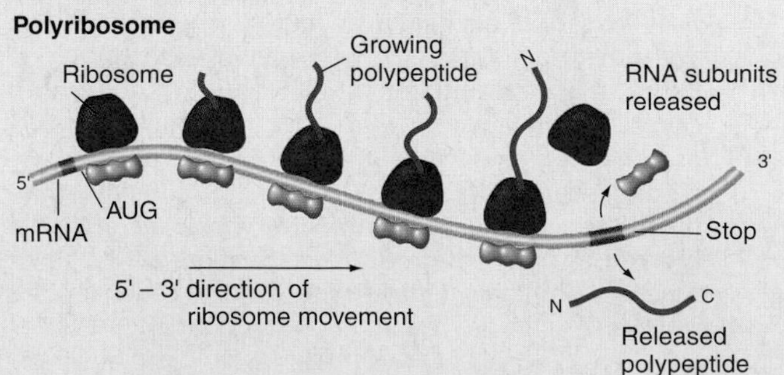

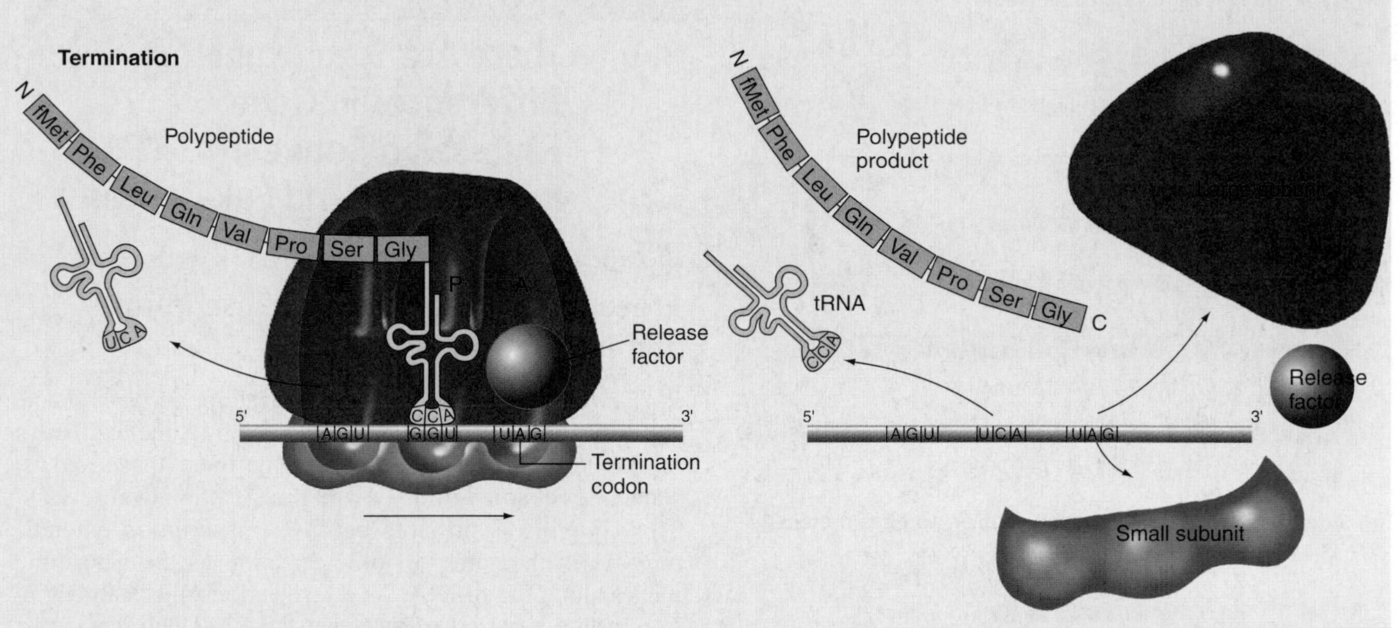

(c) Termination: The ribosome releases the completed polypeptide No normal tRNAs carry anticodons complementary to the three nonsense (stop) codons UAG, UAA, and UGA. Thus, when movement of the ribosome brings a nonsense codon into the ribosome's A site, no tRNAs can bind to that codon through complementary base pairing. Instead, proteins called **release factors** recognize the termination codons and bring polypeptide synthesis to a halt. During termination, three events must occur: The tRNA specifying the C-terminal amino acid releases the completed polypeptide, the same tRNA as well as the mRNA separate from the ribosome, and the ribosome dissociates into its large and small subunits.

process, focusing on translation as it occurs in bacterial cells. As you examine the figure, note the following points about the flow of information during translation. First, the first codon to be translated—the initiation codon—is an AUG set in a special context at the 5′ end of the gene's reading frame (which is *not* directly at the 5′ end of the mRNA). Second, special initiating tRNAs carrying a modified form of methionine called formylmethionine (fMet) recognize the initiation codon. Thus, fMet is the first amino acid incorporated into all bacterial polypeptides and defines their N termini. Third, the ribosome moves along the mRNA in the 5′-to-3′ direction, revealing successive codons in a stepwise fashion. Fourth, at each step of translation, the polypeptide grows by the addition of the next amino acid in the chain to its C terminus. Finally, translation terminates when the ribosome reaches a UAA, UAG, or UGA nonsense codon at the 3′ end of the gene's reading frame. These points explain the biochemical basis of colinearity, that is, the correspondence between the 5′-to-3′ direction in the mRNA and the N-terminus-to-C-terminus direction in the resulting polypeptide.

During elongation, the translation machinery adds about 2–15 amino acids per second to the growing chain. The speed is higher in prokaryotes and lower in eukaryotes. At these rates, construction of an average size 300-amino-acid polypeptide (from an average-length mRNA that is somewhat longer than 1000 nucleotides) could take as little as 20 seconds or as long as 2.5 minutes.

We have deliberately left several details out of Fig. 8.25 so that you can concentrate on key events dictating the flow of information during the translation. In particular, this figure does not depict the important roles played by protein translation factors, which help shepherd mRNAs and tRNAs to their proper locations on the ribosome. Some translation factors also carry GTP to the ribosome, where hydrolysis of the high-energy bonds in the GTP helps power certain molecular movements (such as translocation of the ribosome along the mRNA). The book's website (www.mhhe.com/hartwell3) provides a wealth of information on the details of translation, including links to remarkable animations illustrating each step of the process.

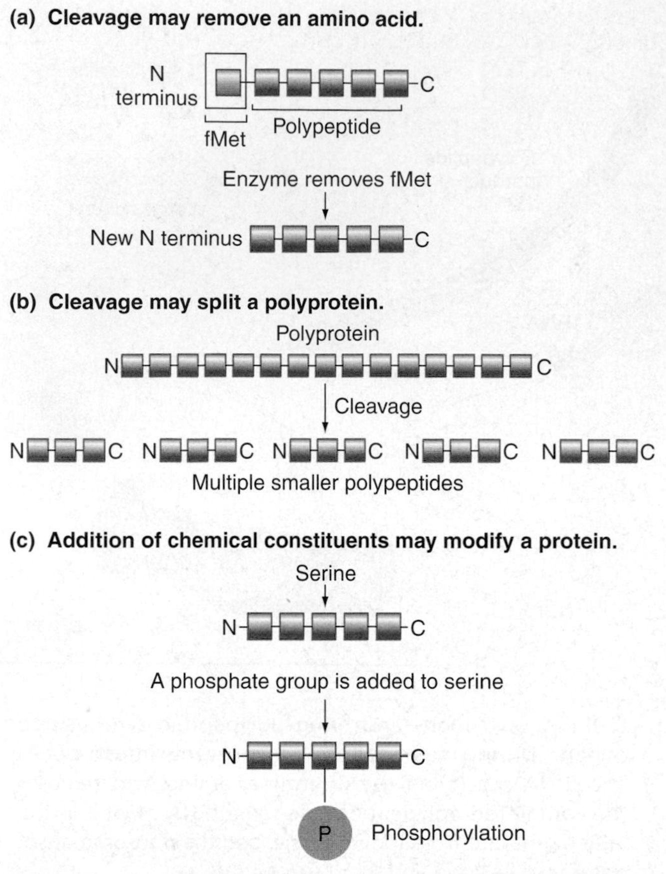

(a) Cleavage may remove an amino acid.

N terminus — fMet — Polypeptide — C

Enzyme removes fMet

New N terminus — C

(b) Cleavage may split a polyprotein.

Polyprotein

N — C

Cleavage

N C N C N C N C N C

Multiple smaller polypeptides

(c) Addition of chemical constituents may modify a protein.

Serine

N — C

A phosphate group is added to serine

N — C

P Phosphorylation

Figure 8.26 Posttranslational processing can modify polypeptide structure. Cleavage may remove an amino acid from the N-terminus end of a polypeptide **(a)** or split a larger *polyprotein* into two or more smaller functional proteins **(b).** **(c)** Chemical reactions may add a phosphate or other functional group to an amino acid in the polypeptide.

Processing After Translation Can Change a Polypeptide's Structure

Protein structure is not irrevocably fixed at the completion of translation. Several different processes may subsequently modify a polypeptide's structure. Cleavage may remove amino acids, such as the N-terminal fMet, from a polypeptide (**Fig. 8.26a**), or it may generate several smaller polypeptides from one larger product of translation (Fig. 8.26b). In the latter case, the larger polypeptide made immediately after translation but before cleavage into smaller polypeptides is often called a **polyprotein.** The addition of chemical constituents, such as phosphate groups, methyl groups, or even carbohydrates, to specific amino acids may also modify a polypeptide after translation (Fig. 8.26c). Such cleavages and additions are known as **posttranslational modifications.** Posttranslational changes to a protein can be very important: For example, the biochemical function of many enzymes directly depends on the addition (or sometimes removal) of phosphate groups.

8.4 There Are Significant Differences in Gene Expression Between Prokaryotes and Eukaryotes

In Eukaryotes, the Nuclear Membrane Prevents the Coupling of Transcription and Translation

In *E. coli* and other prokaryotes, transcription takes place in an open intracellular space undivided by a nuclear membrane; translation occurs in the same open space and is sometimes coupled directly with transcription (**Table 8.1**). This coupling is possible because transcription extends mRNAs in the same 5′-to-3′ direction as the ribosome moves along the mRNA. As a result, ribosomes can begin to translate a partial mRNA that the RNA polymerase is still in the process of transcribing from the DNA. The coupling of transcription and translation has significant consequences for the regulation of gene expression in prokaryotes. For example, in an important regulatory mechanism called *attenuation,* which we describe in Chapter 17, the rate of translation of some mRNAs directly determines the rate at which the corresponding genes are transcribed into these mRNAs. Such coupling cannot occur in eukaryotes because the nuclear envelope physically separates the sites of transcription and RNA processing in the nucleus from the site of translation in the cytoplasm. As a result, translation in eukaryotes can affect the rate at which genes are transcribed only in more indirect ways.

The Initiation of Translation Also Differs Between Prokaryotes and Eukaryotes

In prokaryotes, translation begins at a ribosome binding site on the mRNA, which is defined by a short, characteristic sequence of nucleotides called a *Shine-Dalgarno box* adjacent to an initiating AUG codon (review Fig. 8.25a). There is nothing to prevent an mRNA from having more than one ribosome binding site, and, in fact, many prokaryotic messages are **polycistronic:** They contain the information of several genes (sometimes referred to as *cistrons*), each of which can be translated independently starting at its own ribosome binding site (Table 8.1). In eukaryotes, by contrast, the small ribosomal subunit first binds to the methylated cap at the 5′ end of the mature mRNA and then migrates to the initiation site. This site is almost always the first AUG codon encountered by the ribosomal subunit as it moves along, or scans, the mRNA in the 5′-to-3′ direction (see Fig. 8.25a and Table 8.1). The mRNA region between the 5′ cap and the initiation codon is sometimes referred to

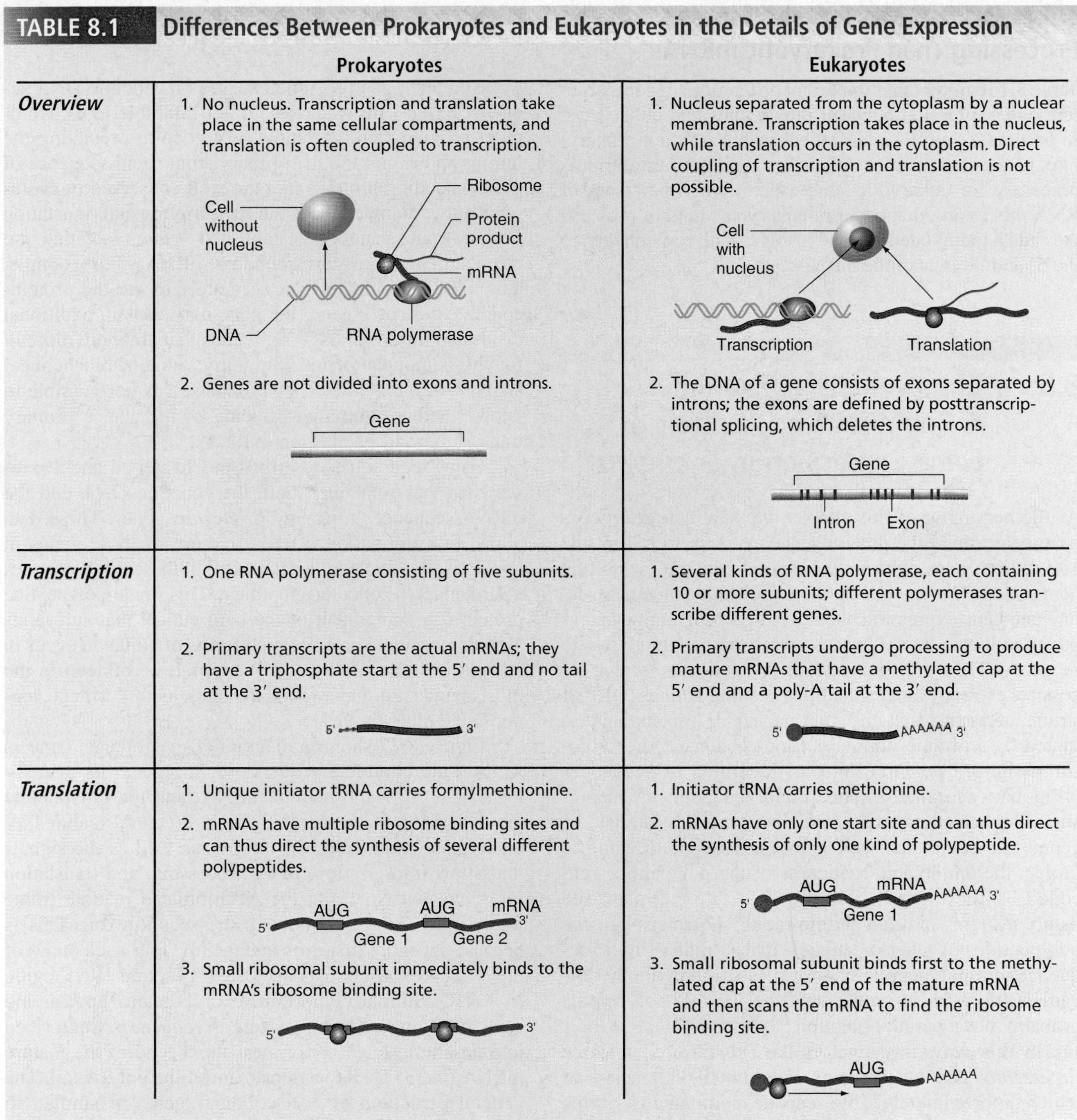

TABLE 8.1 Differences Between Prokaryotes and Eukaryotes in the Details of Gene Expression

	Prokaryotes	Eukaryotes
Overview	1. No nucleus. Transcription and translation take place in the same cellular compartments, and translation is often coupled to transcription.	1. Nucleus separated from the cytoplasm by a nuclear membrane. Transcription takes place in the nucleus, while translation occurs in the cytoplasm. Direct coupling of transcription and translation is not possible.
	2. Genes are not divided into exons and introns.	2. The DNA of a gene consists of exons separated by introns; the exons are defined by posttranscriptional splicing, which deletes the introns.
Transcription	1. One RNA polymerase consisting of five subunits.	1. Several kinds of RNA polymerase, each containing 10 or more subunits; different polymerases transcribe different genes.
	2. Primary transcripts are the actual mRNAs; they have a triphosphate start at the 5′ end and no tail at the 3′ end.	2. Primary transcripts undergo processing to produce mature mRNAs that have a methylated cap at the 5′ end and a poly-A tail at the 3′ end.
Translation	1. Unique initiator tRNA carries formylmethionine.	1. Initiator tRNA carries methionine.
	2. mRNAs have multiple ribosome binding sites and can thus direct the synthesis of several different polypeptides.	2. mRNAs have only one start site and can thus direct the synthesis of only one kind of polypeptide.
	3. Small ribosomal subunit immediately binds to the mRNA's ribosome binding site.	3. Small ribosomal subunit binds first to the methylated cap at the 5′ end of the mature mRNA and then scans the mRNA to find the ribosome binding site.

as either the **5′-untranslated region (5′ UTR)** or the **5′-untranslated leader.** Because of this scanning mechanism, initiation in eukaryotes takes place at only a single site on the mRNA, and each mRNA contains the information for translating only a single kind of polypeptide.

Another translational difference between prokaryotes and eukaryotes is in the composition of the initiating tRNA. In prokaryotes, it carries a modified form of methionine known as formylmethionine, while in eukaryotes, it carries an unmodified methionine (see Table 8.1). This difference means that immediately after translation, prokaryotic polypeptides all have fMet at their N termini, while eukaryotic polypeptides have Met at that same location. Posttranslational cleavage events in both prokaryotes and eukaryotes often create mature proteins that no longer have fMet or Met at their N termini (see Fig. 8.26a).

Eukaryotic mRNAs Require More Processing Than Prokaryotic mRNAs

Table 8.1 reviews other important differences in gene structure and expression between prokaryotes and eukaryotes. In particular, introns interrupt eukaryotic, but not prokaryotic, genes such that the splicing of a primary transcript is necessary for eukaryotic gene expression. Other types of RNA processing that occur in eukaryotes but not prokaryotes add a methylated cap and a poly-A tail, respectively, to the 5′ and 3′ ends of the mRNAs.

8.5 Comprehensive Example: A Computerized Analysis of Gene Expression in *C. elegans*

At the beginning of this chapter, we saw that geneticists have determined the precise sequence of nearly all of the 100 million base pairs in the haploid genome of the tiny nematode *C. elegans*. Using their knowledge of gene structure and gene expression, they have also programmed computers to locate the sequences within the genome likely to be genes. Their programs include instructions to search for possible exons by looking for **open reading frames (ORFs):** strings of amino acid-encoding nucleotide triplets uninterrupted by in-frame nonsense (stop) codons. Other algorithms ignore potential introns, identified as sequences lying between likely splice-donor and splice-acceptor sites. Once the computer has retrieved regions likely to be genes, the researchers ask it to use the genetic code to project the amino acid sequences of the polypeptides encoded by these genes. Finally, they scan computerized databases for similar amino acid sequences in the polypeptides of other organisms. If they find a similar sequence in a polypeptide of known function in another organism, they can conclude that the worm's polypeptide probably has a parallel function.

In this way, investigators have discovered that the *C. elegans* genome contains roughly 20,000 genes, of which approximately 15% encode components of the worm's gene-expression machinery. Many of these gene-expression components are proteins. For example, more than 60 genes encode proteins that function as parts of the ribosome workbench, while more than 300 genes encode **transcription factors:** DNA-binding proteins that regulate transcription. By contrast, a large contingent of expression-related genes produce RNAs that are not translated into protein. There are 659 tRNA genes in the *C. elegans* genome, about 100 rRNA genes, and 72 genes for spliceosomal RNAs. The relatively high numbers of RNA-encoding genes reflect the fact that the genome contains several identical or near-identical copies of these untranslated genes. For example, even though there are 72 spliceosomal RNA genes, there are only 5 different kinds of spliceosomal RNAs.

Computerized predictions based on genomic DNA sequences alone are valuable but not infallible tools. Computer programs are currently very good at predicting the introns and exons and the primary amino acid sequence of genes encoding proteins that are well conserved in evolution. But certain details of the transcription and translation of these genes cannot be established without isolating and characterizing their corresponding mRNAs. For example, although the computer can accurately locate the protein-coding exons of a gene, the gene may contain additional exons and introns at its 5′ or 3′ ends that are more difficult for the computer to find. Similarly, without biochemical analysis of the gene's RNA products, it is not possible to know whether alternative splicing of the gene's primary transcript produces different mRNAs.

Using techniques described in Chapters 9 and 10, researchers have obtained both the genomic DNA and the mRNA sequences for many *C. elegans* genes. These data allow an examination of the structure of these genes in nucleotide-by-nucleotide detail. One of these genes encodes a particular type of collagen protein. This single-polypeptide protein is a component of the hard cuticle that surrounds and protects the worm. Related forms of collagen occur in all multicellular animals. In vertebrates, collagen is the most abundant protein, found in bones, teeth, cartilage, tendons, and other tissues.

Figure 8.27 shows a diagram of the collagen gene as well as the complete sequence of the gene, its primary RNA transcript, the mature mRNA, and the polypeptide product. As you can see, the gene's structural features include three exons and two introns, as well as the signals that allow transcription, RNA processing, and translation into collagen. Note that the ATG-initiated reading frame for the protein begins only in the second exon. This is because the entire first exon and the first four nucleotides of the second exon correspond to the 5′-untranslated region (5′-UTR). Similarly, the third exon contains both amino acid-specifying codons, as well as sequences transcribed into an untranslated region near the 3′ end of the mature mRNA (the 3′-UTR) just upstream of the poly-A tail. The general structure of the collagen gene is similar to the structure of most eukaryotic genes. This is because the basic pattern of gene expression has remained substantially the same throughout evolution, even though the details, such as gene length, exon number, and the spacing or size of the untranslated 5′ and 3′ ends, vary from gene to gene and from organism to organism.

The sequencing of the 100 million nucleotides in the *C. elegans* genome not only led to the identification of 20,000 genes but also helped reveal some uncommon features in the way the worm expresses its genes. In rare instances, worms use trans-splicing to create an mRNA from the primary transcripts of two different genes (review

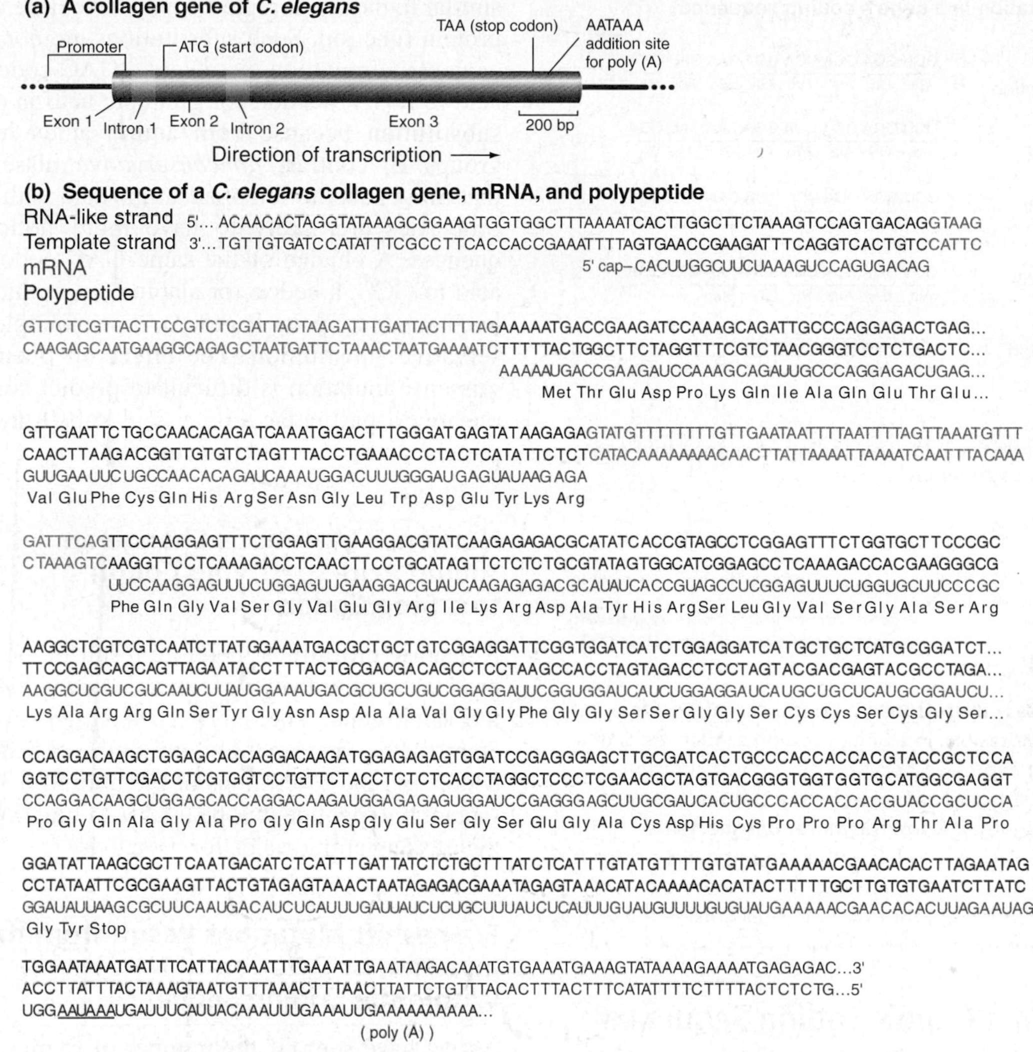

Figure 8.27 Expression of a *C. elegans* gene for collagen. **(a)** Landmarks in the collagen gene. **(b)** Comparison of the sequence of the collagen gene's DNA with the sequence of nucleotides in the mature mRNA (*purple*) pinpoints the start of transcription, the location of exons (*red*) and introns (*green*), and the position of the AAUAAA poly-A addition signal (*underlined in purple*). Translation of the mRNA according to the genetic code determines the amino acids of the protein product.

Fig. 8.18b on p. 274); before observing trans-splicing in the nematode, researchers had seen it mainly in trypanosomes, the single-celled protozoans that cause sleeping sickness. Like bacteria, *C. elegans* transcribes some groups of adjacent genes as one long polycistronic primary transcript; it is one of the very few eukaryotic organisms in which researchers have observed this predominantly prokaryotic phenomenon. Polycistronic transcripts are permissible in *C. elegans* because they are processed by trans-splicing into mature mRNAs for individual genes. The introns that *C. elegans* deletes through RNA splicing are fewer and smaller than the introns found in most other animal genes, and they are unusually rich in A's and T's. Researchers will be able to apply these insights clarifying *C. elegans'* mechanisms of gene expression to studies of the worm's growth and development (see the genetic portrait of *C. elegans* on our website: www.mhhe.com/hartwell3).

8.6 How Mutations Affect Gene Expression and Gene Function

We have seen that the information in DNA is the starting point of gene expression. The cell transcribes that information into mRNA and then translates the mRNA information into protein. Mutations that alter the nucleotide pairs of DNA may modify any of the steps or products of gene expression.

(a) Types of mutation in a gene's coding sequence

Wild-type mRNA	5'	GCU	GGA	GCA	CCA	GGA	CAA	GAU	GGA	3'
Wild-type polypeptide	N	Ala	Gly	Ala	Pro	Gly	Gln	Asp	Gly	C
Silent mutation		GCU	GGA	GCC	CCA	GGA	CAA	GAU	GGA	
		Ala	Gly	Ala	Pro	Gly	Gln	Asp	Gly	
Missense mutation		GCU	GGA	GCA	CCA	AGA	CAA	GAU	GGA	
		Ala	Gly	Ala	Pro	Arg	Gln	Asp	Gly	
Nonsense mutation		GCU	GGA	GCA	CCA	GGA	UAA	GAU	GGA	
		Ala	Gly	Ala	Pro	Gly	Stop			
Frameshift mutation		GCU	GGA	GCC	ACC	AGG	ACA	AGA	UGG	A
		Ala	Gly	Ala	Thr	Arg	Thr	Arg	Trp	

(b) Sites of mutation outside the coding sequence that can disrupt gene expression

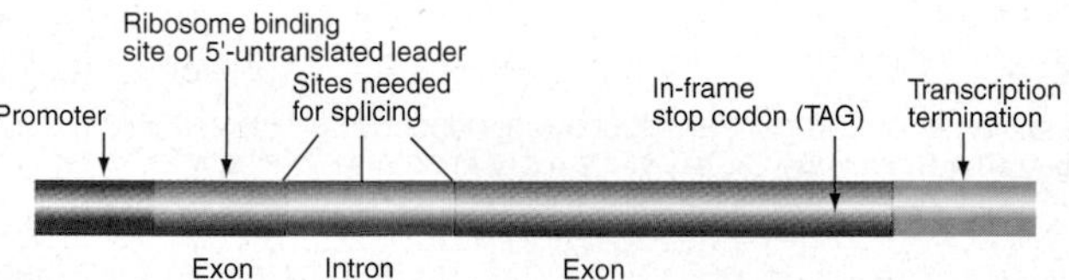

Figure 8.28 How mutations in a gene can affect its expression. (a) Mutations in a gene's coding sequences. *Silent mutations* are not expressed because they do not alter the protein's primary structure. *Missense mutations* replace one amino acid with another. *Nonsense mutations* shorten a polypeptide by replacing a codon with a stop signal. *Frameshift mutations* result in a change in reading frame downstream of the addition or deletion. **(b)** Mutations in sites outside the coding region can also disrupt gene expression.

Mutations in a Gene's Coding Sequence Can Alter the Gene Product

Because of the nature of the genetic code, mutations in a gene's amino acid-encoding exons generate a range of repercussions (**Fig. 8.28a**).

Silent Mutations Do Not Alter the Amino Acid Specified

One consequence of the code's degeneracy is that some mutations, known as **silent mutations,** direct the inclusion of the same amino acid as the unmutated DNA and therefore have no effect on the amino acid composition of the encoded polypeptide or on phenotype. The majority of silent mutations change the third nucleotide of a codon, the position at which most codons for the same amino acid differ. For example, a change from GCA to GCC in a codon would still yield alanine in the protein product.

Missense Mutations Replace One Amino Acid with Another

Missense mutations that cause substitution in the polypeptide of an amino acid with chemical properties similar to the one it replaces may have little or no effect on protein function. Such substitutions are *conservative.* For example, a mutation that alters a GAC codon for aspartic acid to a GAG codon for glutamic acid is a conservative substitution because both amino acids have acidic R groups. By contrast, *nonconservative* missense mutations that cause substitution of an amino acid with very different properties are likely to have more noticeable consequences. A change of the same GAC codon for aspartic acid to GCC, a codon for alanine (an amino acid with an uncharged, nonpolar R group), is an example of a nonconservative substitution. The effect on phenotype of any missense mutation is difficult to predict since it depends on how a particular amino acid substitution changes a protein's structure and function.

Nonsense Mutations Change an Amino Acid-Specifying Codon to a Stop Codon

Nonsense mutations therefore result in the production of proteins smaller than those encoded by wild-type alleles of the same gene. The shorter, *truncated proteins* lack all amino acids between the amino acid encoded by the mutant codon and the C terminus of the normal polypeptide. The mutant polypeptide will be unable to function if it requires the missing amino acids for its activity.

Frameshift Mutations Result from the Insertion or Deletion of Nucleotides Within the Coding Sequence

As we have seen, if the number of extra or missing nucleotides is not divisible by 3, the insertion or deletion will skew the reading frame downstream of the mutation. As a result, **frameshift mutations** cause unrelated amino acids to appear in place of amino acids critical to protein function, destroying or diminishing polypeptide function.

Mutations in a Gene Outside the Coding Sequence Can Also Alter Gene Expression

Mutations that produce a variant phenotype are not restricted to alterations in codons. Since gene expression depends on several signals other than the actual coding sequence, changes in any of these critical signals can disrupt the process (see Fig. 8.28b).

We have seen that promoters and termination signals in the DNA of a gene instruct RNA polymerase where to start and stop transcription. Changes in the sequence of a promoter that make it hard or impossible for RNA polymerase to recognize the site diminish or prevent transcription. Mutations in a termination signal can diminish the amount of mRNA produced and thus the amount of gene product.

In eukaryotes, most primary transcripts have splice-acceptor sites, splice-donor sites, and branch sites that allow splicing to join exons together with precision in the mature mRNA. Changes in a splice-acceptor or -donor site can obstruct splicing, resulting in no mature mRNA and thus no normal polypeptide.

Mature mRNAs have ribosome binding sites and in-frame stop codons indicating where translation should start and stop. Mutations affecting a ribosome binding site would lower the affinity of the mRNA for the small ribosomal subunit; since this affinity helps determine the efficiency of translation, such mutations are likely to diminish the amount of polypeptide product. Mutations in a stop codon would produce longer than normal proteins that might be unstable or nonfunctional.

Most Mutations That Affect Gene Expression Reduce Gene Function

Mutations affect phenotype by changing either the amino acid sequence of a protein or the amount of the protein produced. Any mutation inside or outside a coding region that reduces or abolishes protein activity in one of the many ways previously described is a **loss-of-function mutation.**

Loss-of-Function Alleles Are Usually Recessive

Loss-of-function alleles that completely block the function of a protein are called **null,** or **amorphic, mutations** (**Fig. 8.29**). Such mutations either prevent synthesis of the protein or promote synthesis of a protein incapable of carrying out any function. For example, a deletion of an entire gene would by definition be a null allele. In an A^+/a heterozygote, in which allele a is recessive to wild-type allele A^+, the A^+ allele would generate functional protein, while the null a allele would not. If the amount of protein produced by the single A^+ allele (usually, though not always, half the amount produced in an A^+/A^+ cell) is above the threshold amount sufficient to fulfill the normal biochemical requirements of the cell, the phenotype of the A^+/a heterozygote will be wild type. For the large number of genes that function in this way, A^+/A^+ cells actually make more than twice as much of the protein needed for the normal phenotype.

A **hypomorphic mutation** is a loss-of-function mutation that produces either much less of a protein or a protein with very weak but detectable function (Fig. 8.29). In a B^+/b heterozygote, where b is a hypomorphic allele recessive to wild-type allele B^+, the amount of protein activity will be somewhat greater than half the amount in a B^+/B^+ cell. Usually, this is enough activity to fulfill the normal biochemical requirements of the cell. Most hypomorphic mutations are detectable only in homozygotes and only if the reduction in protein amount or function is sufficient to cause an abnormal phenotype.

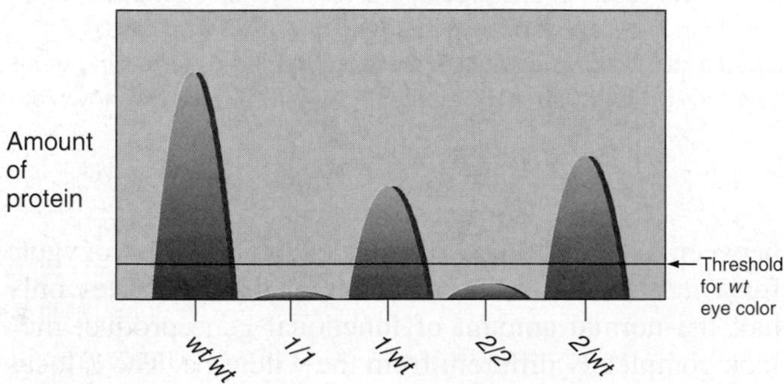

Figure 8.29 Why some mutant alleles are recessive. Researchers subjected fly extracts to rocket immunoelectrophoresis; rocket size reflects the amount of a particular enzyme called xanthine dehydrogenase. Flies need only 10% of the enzyme produced in wild-type strains (*wt/wt*) to have normal eye color. Null allele 1 and hypomorphic allele 2 are recessive to wild type because *1/wt* or *2/wt* heterozygotes have more than the low threshold amount of enzyme needed for normal eye color.

Incomplete Dominance Can Arise When the Phenotype Varies in Proportion to the Amount of Functional Protein

Some combinations of alleles generate phenotypes that vary continuously with the amount of functional gene product. For example, loss-of-function mutations in a single pigment-producing gene can generate a red-to-white spectrum of flower colors, with the white resulting from the absence of an enzyme in a biochemical pathway (**Fig. 8.30**). Consider three alleles of the gene encoding this enzyme: R^+ specifies a high, wild-type amount of the enzyme; r^{50} generates half the normal amount of the same enzyme (or the full amount of an altered form that has half the normal level of activity); and r^0 is a null allele. R^+/r^0

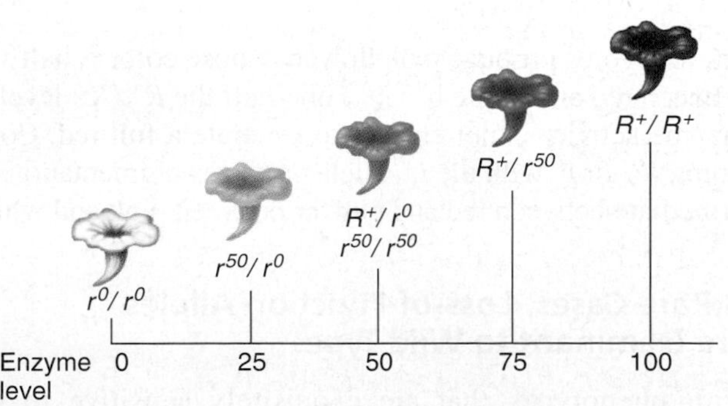

Figure 8.30 When a phenotype varies continuously with levels of protein function, incomplete dominance results.

(a) Haploinsufficiency

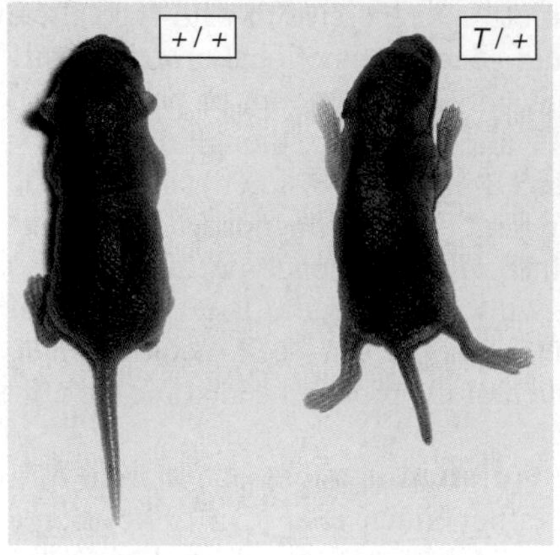

(b) Dominant negative mutations

Functional Enzyme	Nonfunctional Enzyme			
$d^+d^+d^+d^+$	$d^+d^+d^+D$	$d^+d^+D\,d^+$	$d^+D\,d^+d^+$	$D\,d^+d^+d^+$
	$d^+d^+D\,D$	$d^+D\,D\,d^+$	$d^+D\,d^+D$	$D\,d^+D\,d^+$
	$D\,d^+d^+D$	$D\,D\,d^+d^+$	$d^+D\,D\,D$	$D\,d^+D\,D$
	$D\,D\,d^+D$	$D\,D\,D\,d^+$	$D\,D\,D\,D$	

D = dominant mutant subunit
d^+ = wild-type subunit

(c) *Kinky:* A dominant negative mutation

(d) A result of ectopic expression

Figure 8.31 Why some mutant alleles are dominant. (a) Mice heterozygous for a null mutation of the *T* locus (*T*/+) have tails shorter than their wild type litter mates (*t*/+). (b) With proteins composed of four subunits encoded by the same gene, a dominant negative mutant may inactivate 15 out of every 16 multimers. (c) The *Kinky* allele in mice is a dominant negative mutation that causes a kink in the tail. (d) A neomorphic dominant mutation in the fly *Antennapedia* gene causes ectopic expression of a leg-determining gene in structures that normally produce antennae. The photo at *left* shows two legs growing out of the *head*; a normal fly head is shown at *right*.

heterozygotes produce pink flowers whose color is halfway between red and white because one-half the R^+/R^+ level of enzyme activity is not enough to generate a full red. Combining R^+ or r^0 with the r^{50} allele produces pigmentation intermediate between red and pink or between pink and white.

In Rare Cases, Loss-of-Function Alleles Are Dominant to Wild Type

With phenotypes that are exquisitely sensitive to the amount of functional protein produced, even a relatively small change of twofold or less can cause a switch between distinct phenotypes. For example, a heterozygote for a null loss-of-function mutation that generates only half the normal amount of functional gene product may look completely different from the wild type. The *T* locus in mice has just such a mutation, with an easy-to-visualize dominant phenotype (**Fig. 8.31a**). Mice require the wild-type protein product of the *T*-locus gene during embryogenesis for the normal development of the posterior portion of the spinal cord and tail. Embryos heterozygous for a null mutation at the *T* locus produce only half the normal amount of the *T*-determined protein and mature into viable offspring that are normal in all respects except for the absence of the distal two-thirds of their tail. The severely shortened tail reflects the embryo's sensitivity to the level of *T*-gene product available during morphogenesis; half the normal amount of T protein is below the threshold needed for normal development. Geneticists sometimes use the term **haploinsufficiency** to describe situations in which one wild-type allele does not provide enough of a gene product. Only a minority of phenotypes are so sensitive to the amount of a particular protein. Thus, as described earlier, null and hypomorphic alleles usually produce phenotypes that are recessive to wild type.

In another mechanism leading to dominance, some alleles of genes encode subunits of multimers that block the activity of the subunits produced by normal alleles. Such blocking alleles cause a loss of function and are called **dominant negative,** or **antimorphic, alleles.** Consider, for example, a gene encoding a polypeptide that associates with three other identical polypeptides in a four-subunit enzyme. All four subunits are products of the same gene. If a dominant mutant allele *D* directs the synthesis of a polypeptide that can still assemble into aggregates but whose presence in the multimer—even as one subunit out of four—abolishes enzyme function, the chance of a heterozygote producing a multimer composed solely of functional wild-type d^+ subunits is 1 in 16: $(1/2)^4 = 1/16 = 6.25\%$ (Fig. 8.31b). As a result, total enzyme activity in D / d^+ heterozygotes is far less than that seen in wild-type d^+ / d^+ homozygotes. Dominant negative mutations can also affect subunits in multimers composed of more than one type of polypeptide. The *Kinky* allele at the *fused* locus in mice is an example of such a dominant negative mutation (Fig. 8.31c).

Unusual Gain-of-Function Alleles Are Almost Always Dominant

Since there are many ways to interfere with a gene's ability to make sufficient amounts of active protein, the large majority of mutations in most genes are loss-of-function alleles. However, rare mutations that enhance a protein's function or even confer a new activity on a protein produce **gain-of-function alleles.** Because a single such allele by itself can produce sufficient excess protein to alter phenotype, these unusual gain-of-function mutations are almost always dominant to wild-type alleles.

A **hypermorphic mutation** is a gain-of-function mutation that generates either more protein than the wild-type allele or the same amount of a more efficient protein. A hypermorphic mutation in the rhodopsin gene produces a rhodopsin protein that is activated whether or not light is present, resulting in constant, low-level stimulation of rhodopsin in the photoreceptor cells that detect black and white. These cells, known as rod cells, function primarily at night. People with the mutation can still see in bright daylight, but they have congenital night blindness. The blindness probably arises because the constant rhodopsin stimulation prevents adaptation of the rod cells to the very low light intensities present at night.

A very rare class of dominant gain-of-function alleles arises from **neomorphic mutations** that generate a novel phenotype. Some neomorphic mutations produce proteins with a new function, while others cause genes to produce the normal protein but at an inappropriate time or place. A striking example of inappropriate protein production is the *Drosophila* gene *Antennapedia,* active during embryonic and larval stages. Normally, the gene makes its protein product in tissues destined to become legs; the protein ensures that these tissues develop into legs and not, for example, head structures such as antennae. Dominant mutations of the gene cause production of the protein in the head region of the animal, where the *Antennapedia* gene is not normally active. Here, the misplaced protein causes tissues that would normally develop into antennae to develop into legs (Fig. 8.31d). Production of a protein outside of its normal place or time is called **ectopic expression.**

It Is Often Difficult to Predict the Effects of a Mutation

Recall from Fig. 8.28 on p. 286 that mutations can occur within or outside a gene's coding region, and they can affect the amino acid sequence of the encoded protein or any level of gene expression from transcription to RNA processing to translation. Most mutations constitute loss-of-function alleles. This is because many changes in amino acid sequence are likely to disrupt a protein's function, and because most alterations in gene regulatory sites, such as promoters, will make those sites less effective in fulfilling their role in gene expression. Nonetheless, rare mutations at almost any location in a gene can result in a gain of function. Consider, for example, a protein with a region of amino acids near its C terminus that prevents the protein from functioning except under particular conditions. A nonsense mutation that removes the amino acids needed for this negative regulation might be a hypermorphic allele: The protein would work all the time, not just under the proper conditions. In another example, the *Antennapedia* mutation shown in Fig. 8.31d results from an unusual alteration in the gene's promoter that causes *Antennapedia* to be transcribed in the wrong tissues of the animal.

Even when you know how a mutation affects gene function, you cannot always predict whether the mutation will be dominant or recessive to wild type (**Table 8.2** on p. 290). Although most loss-of-function mutations are recessive and almost all gain-of-function mutations are dominant, exceptions to these generalizations exist. The reason is that dominance relations between the wild-type and mutant alleles of genes in diploid organisms depend on how drastically a mutation influences protein production or

TABLE 8.2 Mutations Classified by Their Effects on Protein Function

	Loss-of-Function			Gain-of-Function	
Mutation Type	Hypomorphic (leaky)	Amorphic (null)	Antimorphic (dominant negative)	Hypermorphic	Neomorphic (ectopic expression)
Occurrence	Common	Common	Rare	Rare	Rare
Possible Dominance Relations	Usually recessive to wild type Can be incompletely dominant if phenotype varies continuously with gene product Can be dominant in cases of haploinsufficiency		Usually dominant or incompletely dominant	Usually dominant or incompletely dominant	Usually dominant or incompletely dominant

activity, and how thoroughly phenotype depends on the normal wild-type level of the protein.

Mutations in Genes Encoding the Molecules That Implement Expression May Affect Transcription, mRNA Splicing, or Translation

Gene expression depends on an astonishing number and variety of macromolecules (**Table 8.3**). A separate gene encodes the subunits of each macromolecule. The genes for all the proteins are transcribed and translated the same as any other gene. The genes for all the rRNAs, tRNAs, and snRNAs are transcribed but *not* translated. Many mutations in these genes have a dramatic effect on phenotype.

Mutations Altering Genes Encoding Proteins or RNAs Involved in Gene Expression Are Usually Lethal

This is because such mutations adversely affect the synthesis of all proteins in a cell. Even a 50% reduction in the amount of some of the proteins enumerated in Table 8.3 can have severe repercussions. In *Drosophila,* for example, null mutations in many of the genes encoding the various ribosomal proteins are lethal when homozygous. This same mutation in a heterozygote causes a dominant *Minute* phenotype in which the slow growth of cells delays the fly's development.

Mutations in tRNA Genes Can Suppress Mutations in Protein-Coding Genes

If more than one gene encoded a molecule with the same role in gene expression, a mutation in one of these genes would not be lethal and might even be useful. Bacterial geneticists have found, for example, that mutations in certain tRNA genes can suppress the effect of a nonsense mutation in other genes. The tRNA-gene mutations that have this effect give rise to **nonsense suppressor tRNAs.** Consider, for instance, an otherwise wild-type *E. coli* population with an in-frame UAG nonsense mutation in the tryptophan synthetase gene. All cells in this population make a truncated, nonfunctional form of the tryptophan synthetase enzyme and are thus tryptophan auxotrophs (trp^-) unable to synthesize tryptophan (**Fig. 8.32a**). Subsequent exposure of these auxotrophs to mutagens, however, generates some trp^+ cells that carry two mutations: one is the original tryptophan synthetase nonsense mutation, the second is a mutation in the gene that encodes a tRNA for the amino acid tyrosine. Evidently, the mutation in the tRNA gene suppresses the effect of the nonsense mutation, restoring the function of the tryptophan synthetase gene. As Fig. 8.32b illustrates, the basis of this nonsense suppression is that the $tRNA^{Tyr}$ mutation changes an anticodon that recognizes the codon for tyrosine to an anticodon complementary to the UAG stop codon. The mutant tRNA can therefore insert tyrosine into the polypeptide at the position of the in-frame UAG nonsense mutation, allowing the cell to make at least some full-length enzyme. Similarly, mutations in the anticodons of other tRNA genes can suppress UGA or UAA nonsense mutations.

Cells with a nonsense-suppressing mutation in a tRNA gene can survive only if two conditions coexist with the mutation. First, the cell must have other tRNAs that recognize the same codon as the suppressing tRNA recognized before mutation altered its anticodon. Without such tRNAs, the cell has no way to insert the proper amino acid in response to that codon (in our example, the codon for tyrosine). Second, the suppressing tRNA must have only a weak affinity for the stop codons normally

TABLE 8.3 The Cellular Components of Gene Expression

Function	Cellular Components
*Transcription**	Core RNA polymerase
	Sigma subunit
	Rho factor
Splicing and RNA Processing	snRNAs
	Protein components of spliceosomes
	Additional splicing factors
	Capping enzyme
	Methyl transferases
	Poly-A polymerase
Translation	mRNAs
	tRNAs
	Aminoacyl-tRNA synthetases
	rRNAs
	Protein components of ribosomes
	Translation factors
Protein Processing	Deformylases
	Amino peptidases
	Proteases
	Methylases
	Hydroxylases
	Glycosylases
	Kinases
	Phosphatases

*For simplicity, we list here only proteins from prokaryotic organisms involved in transcription. The cellular components needed for transcription in eukaryotic organisms are more complex; for example, eukaryotes have three different kinds of RNA polymerase, each made of numerous subunits.

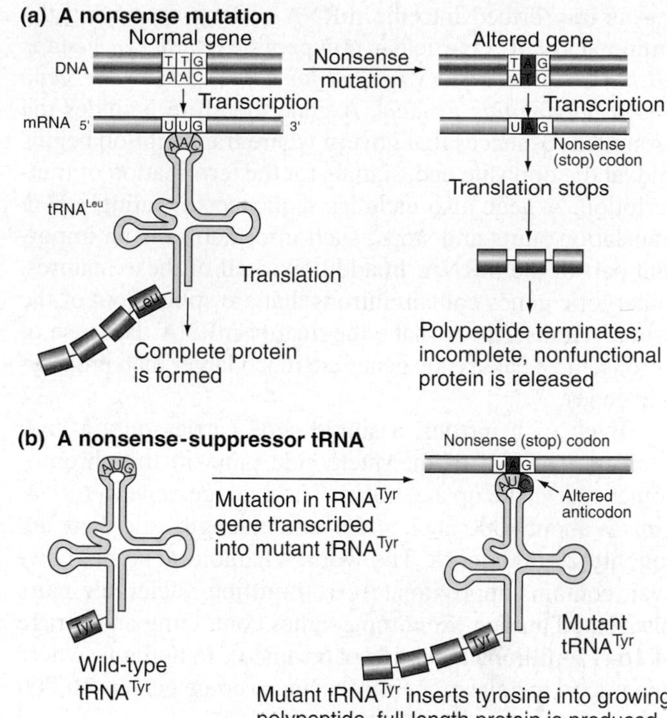

Figure 8.32 Nonsense suppression. (a) A nonsense mutation that generates a stop codon causes production of a truncated, nonfunctional polypeptide. **(b)** A second, nonsense-suppressing mutation in a tRNA gene causes addition of an amino acid in response to the stop codon, allowing production of a full-length polypeptide.

found at the ends of mRNA coding regions. If this were not the case, the suppressing tRNA would wreak havoc in the cell, producing a whole array of aberrant polypeptides that are longer than normal. One way cells guard against this possibility is that for many genes, termination depends on two stop codons in a row. Because a suppressing tRNA's chance of inserting an amino acid at both of these codons is very low, only a small number of extended proteins arise.

Connections

The promoter, a particular region of the DNA double helix, signals the starting point of gene expression. Beginning at this site, RNA polymerase unwinds the DNA and transcribes the simple linear sequence of nucleotides of the gene's template strand into a complementary messenger RNA containing all the information required for construction of a complex polypeptide. The cell then uses the genetic code to translate the mRNA nucleotide sequence into a colinear chain of amino acids whose sequence dictates its three-dimensional folding into a functional protein. Interestingly, the DNA within a cell is much more stable than either the RNA or proteins constructed from it. The half-life of most prokaryotic mRNAs is measured in minutes, while that of eukaryotic mRNAs is measured in hours. The half-life of many proteins is about a day. DNA is so stable it remains intact even after the organism dies.

Our knowledge of gene expression enables us to redefine the concept of a gene. A gene is not simply the DNA that is transcribed into the mRNA codons specifying the amino acids of a particular polypeptide. Rather, *a gene is all the DNA sequences needed for expression of the gene into a polypeptide product.* A gene therefore includes the promoter sequences that govern where transcription begins and, at the opposite end, signals for the termination of transcription. A gene also includes sequences dictating where translation starts and stops; such information is an important part of the mRNA. In addition to all of these features, eukaryotic genes contain introns that are spliced out of the primary transcript to make the mature mRNA. Because of introns, most eukaryotic genes are much larger than prokaryotic genes.

Even with introns, a single gene carries only a very small percentage of the nucleotide pairs in the chromosomes that make up a genome. The average gene in *C. elegans* is about 4000 nucleotide pairs in length, and there are roughly 20,000 genes. The worm's haploid genome, however, contains approximately 100 million nucleotide pairs distributed among six chromosomes containing an average of 16–17 million nucleotide pairs apiece. In humans, where genes tend to have more introns, the average gene is 16,000 nucleotide pairs in length, and there are 20,000–30,000 of them. But the haploid human genome has roughly 3 billion (3,000,000,000) nucleotide pairs distributed among 23 chromosomes containing an average of 130 million nucleotide pairs apiece. How can geneticists sort through these enormous haystacks of genetic information to find the pinpoints of interest: the DNA of a single gene or the small changes within that gene that alter a healthy phenotype to disease?

In Chapters 9–11, we describe how researchers analyze the mass of genetic information in the chromosomes of a genome as they try to discover what parts of the DNA are genes and how those genes influence phenotype. They begin their analysis by breaking the DNA into pieces of manageable size, making many copies of those pieces to obtain enough material for study, and characterizing the pieces down to the level of nucleotide sequence. They then try to reconstruct the DNA sequence of an entire genome by determining the spatial relationship between the many pieces. Finally, they use the knowledge they have obtained to examine the genomic variations that make individuals unique.

Essential Concepts

1. *Gene expression* is the process by which cells convert the DNA sequence of a gene to the RNA sequence of a transcript, and then decode the RNA sequence to the amino acid sequence of a polypeptide.

2. The nearly universal *genetic code* consists of 64 *codons,* each one composed of three nucleotides. Of these codons, 61 specify amino acids, while 3—UAA, UAG, and UGA—are *nonsense* or *stop codons* that do not specify an amino acid.
   a. The code is *degenerate:* More than one codon specifies every amino acid except methionine and tryptophan.
   b. AUG in the context of a ribosome binding site is the *initiation codon;* it establishes a *reading frame* that determines the grouping of nucleotides into *nonoverlapping* triplet codons.

3. *Transcription* is the first stage of gene expression. During transcription, *RNA polymerase* synthesizes a single-stranded *primary transcript* from a DNA template.
   a. RNA polymerase initiates transcription by binding to the *promoter* sequence of the DNA and unwinding the double helix to expose bases for pairing.
   b. RNA polymerase extends the RNA in the 5′-to-3′ direction by catalyzing formation of phosphodiester bonds between successively aligned nucleotides.
   c. *Terminator* sequences in the RNA cause RNA polymerase to dissociate from the DNA.
   d. In prokaryotes, the primary transcript is the *mRNA* that guides polypeptide synthesis.

4. In eukaryotes, *RNA processing* after transcription produces a mature mRNA.
   a. RNA processing adds a methylated cap to the 5′ end and a poly-A tail to the 3′ end of the eukaryotic mRNA.
   b. The *spliceosome* removes *introns* from the primary transcript and precisely splices together the remaining *exons. Alternative splicing* makes it possible to produce different mRNAs from the same primary transcript.

5. *Translation* is the stage of gene expression when the cell synthesizes protein according to instructions in the mRNA.
   a. *tRNAs* carry amino acids to the translation machinery. *Aminoacyl-tRNA synthetases* connect amino acids to their corresponding tRNAs. Each tRNA molecule has an *anticodon* complementary to the mRNA codon specifying the amino acid it carries. Because of *wobble,* some tRNA anticodons recognize more than one mRNA codon.

b. Translation occurs on complex molecular machines called *ribosomes.* Ribosomes have three binding sites for tRNAs—A, P, and E—and also supply the ribozyme known as *peptidyl transferase* that catalyzes formation of a peptide bond between amino acids carried by the tRNAs at the A and P sites.
c. Initiation: To start translation, part of the ribosome binds to a *ribosome binding site* on the mRNA, which includes the AUG initiation codon. Special initiating tRNAs carry the amino acid fMet in prokaryotes or Met in eukaryotes to the ribosomal P site. This amino acid will become the N terminus of the growing polypeptide.
d. Elongation: When the carboxyl group of the amino acid connected to a tRNA at the ribosome's P site becomes attached through a peptide bond to the amino acid carried by the tRNA at the A site, the ribosome travels three nucleotides toward the 3′ end of the mRNA. The 5′-to-3′ direction in the mRNA thus corresponds to the N-terminus-to-C-terminus direction in the polypeptide under construction.
e. Termination: When the ribosome encounters in-frame nonsense (stop) codons, it ends translation by releasing the mRNA and disconnecting the complete polypeptide from the tRNA.

6. *Posttranslational processing* may alter a polypeptide by adding or removing chemical constituents to or from particular amino acids or by cleaving the polypeptide into smaller molecules.

7. Mutations affect gene expression in several ways.
a. Mutations in a gene may modify the message encoded in a sequence of nucleotides. *Silent mutations* usually change the third letter of a codon and have no effect on polypeptide production. *Missense mutations* change the codon for one amino acid to the codon for another amino acid. *Nonsense mutations* change a codon for an amino acid to a stop codon, causing synthesis of a truncated polypeptide. *Frameshift mutations* change the reading frame of a gene, altering the identity of all subsequent amino acids.
b. Mutations outside coding sequences that alter signals required for transcription, mRNA splicing, or translation can modify gene expression by altering the amount, time, or place of protein production.

8. Genotype correlates with gene function and phenotype as follows:
a. *Loss-of-function* mutations reduce or completely block gene expression. Most loss-of-function alleles are recessive to wild-type alleles. However, loss-of-function alleles can have dominant effects, particularly when small perturbations in protein activity disrupt normal phenotypes (as in *haploinsufficiency*).
b. Rare *gain-of-function* mutations cause either increased protein production or synthesis of a protein with enhanced activity. Some gain-of-function alleles confer a novel function on a gene; one example is *ectopic expression* in which the gene product is made in the wrong tissue or at the wrong time in development. Most gain-of-function mutations are dominant.

9. Mutations in genes encoding molecules of the gene-expression machinery are often lethal. Among the exceptions to this rule are mutations in tRNA genes that suppress mutations in polypeptide-encoding genes.

On Our Website

www.mhhe.com/hartwell3
Chapter 8

Annotated Suggested Readings and Links to Other Websites

- Research articles, both historical and recent, describing experiments leading to the elucidation of the genetic code and to our current understanding of the mechanisms responsible for gene expression.
- Animations and high-resolution molecular models illustrating the events occuring during transcription, RNA processing, and translation.
- A database of the *Caenorhabditis elegans* genome.

Specialized Topics

- A comprehensive view of the molecular details of translation, focusing on the roles played by various translation factors in initiation, elongation, and termination.

Social and Ethical Issues

1. An ethnobotanist who works with native peoples of Peru approached a large pharmaceutical corporation, hoping to get some benefits for the native Peruvians as compensation for conveying knowledge on medicinal plants of the rain forest to the company. Scientists in the company found several potential plant-based drugs using information from the indigenous peoples. However, when they tried to purify the drugs from the plants, they found that their yield was low. As an alternative, chemists in the company began to synthesize the drugs. Because the chemists are synthesizing these compounds, the company says it owes neither the native peoples nor the country of Peru anything. Is this a responsible policy for the company? Consider the responsibility the company has to the ethnobotanist, to its stockholders, to the native peoples, and to Peru.

2. In sub-Saharan Africa, where more than 34 million people are infected with HIV, more than 2 million individuals die of AIDS each year. The current treatment for HIV is a "cocktail" of several drugs that costs about $10,000 per year in the United States, an amount that is unaffordable for most Africans. In response, several pharmaceutical companies have relaxed the rules for licensing their patent rights to allow African companies to produce anti-HIV drugs at a decreased cost. Is this type of "corporate philanthropy" an obligation that a pharmaceutical company owes society? If so, how can these companies recoup the $800 million dollars that is, on average, required to develop a new drug? Should the World Health Organization or the governments of more-developed countries share the costs of this philanthropy? Should mechanisms be instituted to prevent the importation of cheaper versions of the drug back into the developed countries? As a general principle, is it fair to index the price of a drug to the ability of a patient to pay?

3. In 1992, the biotechnology company Calgene applied for approval of their rot-resistant tomato, the Flavr Savr® tomato. They had engineered the tomato plant by introducing a copy of the gene encoding a softening enzyme downstream of, and in reverse orientation to, a strong promoter active in the fruit. When the engineered gene is transcribed, an "antisense RNA," complementary in sequence to the softening enzyme mRNA, is produced. The antisense RNA binds to the normal mRNA and blocks translation of the softening enzyme so the tomatoes stay ripe longer. The FDA approved the sale of the tomato without any additional labeling because the DNA that had been introduced into the tomato was tomato DNA. (The FDA's policy on bioengineered foods is to label food as genetically engineered only when a new substance has been introduced that could cause an allergy or when there has been a change in the food's nutritional value.) Several consumer groups were upset by the FDA's action and wanted to see labeling about genetic modification appear on the tomatoes in the stores. Their protest significantly stalled the sale of these tomatoes. Were the consumer groups' actions, which blocked an advance that people might have wanted, warranted? Are these groups important watchdogs of everyone's well-being, or are they hindering progress?

Solved Problems

I. A geneticist examined the amino acid sequence of a particular protein in a variety of *E. coli* mutants. The amino acid in position 40 in the normal enzyme is glycine. The following table shows the substitutions the geneticist found at amino acid position 40 in six mutant forms of the enzyme.

mutant 1	cysteine
mutant 2	valine
mutant 3	serine
mutant 4	aspartic acid
mutant 5	arginine
mutant 6	alanine

Determine the nature of the base substitution that must have occurred in the DNA in each case. Which of these mutants would be capable of recombination with mutant 1 to form a wild-type gene?

Answer

To determine the base substitutions, use the genetic code table (see Fig. 8.3 on p. 257). The original amino acid was glycine, which can be encoded by GGU, GGC, GGA, or GGC. Mutant 1 results in a cysteine at position 40; Cys codons are either UGU or UGC. A change in the base pair in the DNA encoding the first position in the codon (a G–C to T–A transversion) must have occurred, and the original glycine codon must therefore have been either GGU or GGC. Valine (in mutant 2) is encoded by GUN (with N representing any one of the four bases), but assuming that the mutation is a single base change, the Val codon must be either GUU or GUC. The change must have been a G–C to T–A transversion in the DNA for the second position of the codon. To get from glycine to serine (mutant 3) with only one base change, the GGU or GGC would be changed to AGU or AGC, respectively. There was a transition (G–C to A–T) at the first position. Aspartic acid (mutant 4) is encoded by GAU or GAC, so the DNA of mutant 4 is the result of a G–C to A–T transition at position 2. Arginine (mutant 5) is encoded by CGN, so the DNA of mutant 5 must have undergone a G–C to C–G transversion at position 1. Finally, alanine (mutant 6) is encoded by GCN, so the DNA of mutant 6 must have undergone a G–C to C–G transversion at position 2. Mutants 2, 4, and 6 affect a base pair different from that affected by mutant 1, so they could recombine with mutant 1.

In summary, the sequence of nucleotides on the RNA-like strand of the wild-type and mutant genes at this position must be

wild type	5′ G G T/C 3′
mutant 1	5′ T G T/C 3′
mutant 2	5′ G T T/C 3′
mutant 3	5′ A G T/C 3′
mutant 4	5′ G A T/C 3′
mutant 5	5′ C G T/C 3′
mutant 6	5′ G C T/C 3′

II. The double-stranded circular DNA molecule that forms the genome of the SV40 virus can be denatured into single-stranded DNA molecules. Because the base composition of the two strands differs, the strands can be separated on the basis of their density into two strands designated W(atson) and C(rick). When each of the purified preparations of the single strands was mixed with mRNA from cells infected with the virus, hybrids were formed between the RNA and DNA. Closer analysis of these hybridizations showed that RNAs that hybridized with the W preparation were different from RNAs that hybridized with the C preparation. What does this tell you about the transcription templates for the different classes of RNAs?

Answer

An understanding of transcription and the polarity of DNA strands in the double helix are needed to answer this question. *Some genes use one strand of the DNA as a template; others use the opposite strand as a template.* Because of the different polarities of the DNA strands, one set of genes would be transcribed in a clockwise direction on the circular DNA (using say the W strand as the template), and the other set would be transcribed in a counterclockwise direction (with the C strand as template).

III. Geneticists interested in human hemoglobins have found a very large number of mutant forms. Some of these mutant proteins are of normal size, with amino-acid substitutions, while others are short, due to deletions or nonsense mutations. The first extra-long example was named Hb Constant Spring, in which the β globin has several extra amino acids attached at the C-terminal end. What is a plausible explanation for its origin? Is it likely that Hb Constant Spring arose from failure to splice out an intron?

Answer

An understanding of the principles of translation and RNA splicing are needed to answer this question. Because there is an extension on the C-terminal end of the protein, *the mutation probably affected the termination (nonsense) codon rather than affecting splicing of the RNA.* This could have been a base change or a frameshift or a deletion that altered or removed the termination codon. The information in the mRNA beyond the normal stop codon would be translated until another stop codon in the mRNA was reached. A splicing defect could explain Hb Constant Spring only in the more unlikely case that an incorrectly spliced mRNA would encode a protein much longer than normal.

Problems

Interactive Web Exercise

As part of its effort to annotate the human genome, the National Center for Biotechnology Information (NCBI) maintains a database called Sequence View. The files in this database show the structure of genes at the level of base pairs. The Interactive Web Exercise for this chapter at www.mhhe.com/hartwell3 (Chapter 8) provides you with an opportunity to enhance your understanding of gene organization and function by exploring one such file in detail.

Vocabulary

1. For each of the terms in the left column, choose the best matching phrase in the right column.

a. codon	1. removing base sequences corresponding to introns from the primary transcript
b. colinearity	2. UAA, UGA, or UAG
c. reading frame	3. the strand of DNA that has the same base sequence as the primary transcript
d. frameshift mutation	4. a transfer RNA molecule to which the appropriate amino acid has been attached
e. degeneracy of the genetic code	5. a group of three mRNA bases signifying one amino acid
f. nonsense codon	6. most amino acids are not specified by a single codon
g. initiation codon	7. using the information in the nucleotide sequence of a strand of DNA to specify the nucleotide sequence of a strand of RNA
h. template strand	8. the grouping of mRNA bases in threes to be read as codons
i. RNA-like strand	9. AUG in a particular context
j. intron	10. the linear sequence of amino acids in the polypeptide corresponds to the linear sequence of nucleotide pairs in the gene
k. RNA splicing	11. produces different mature mRNAs from the same primary transcript
l. transcription	12. addition or deletion of a number of base pairs other than three into the coding sequence
m. translation	13. a sequence of base pairs within a gene that is not represented by any bases in the mature mRNA
n. alternative splicing	14. the strand of DNA having the base sequence complementary to that of the primary transcript
o. charged tRNA	15. using the information encoded in the nucleotide sequence of an mRNA molecule to specify the amino-acid sequence of a polypeptide molecule
p. reverse transcription	16. copying RNA into DNA

Section 8.1

2. Match the hypothesis from the left column to the observation from the right column that gave rise to it.

a. existence of an intermediate messenger between DNA and protein	1. two mutations affecting the same amino acid can recombine to give wild type
b. the genetic code is nonoverlapping	2. one or two base deletions (or insertions) in a gene disrupt its function; three base deletions (or insertions) are often compatible with function
c. the codon is more than one nucleotide	3. artificial messages containing certain codons produced shorter proteins than messages not containing those codons
d. the genetic code is based on triplets of bases	4. protein synthesis occurs in the cytoplasm, while DNA resides in the nucleus
e. stop codons exist and terminate translation	5. artificial messages with different base sequences gave rise to different proteins in an *in vitro* translation system
f. the amino acid sequence of a protein depends on the base sequence of an mRNA	6. single base substitutions affect only one amino acid in the protein chain

3. How would the artificial mRNA 5′. . GUGUGUGU . . 3′ be read according to each of the following models for the genetic code?

a. two-base, not overlapping
b. two-base, overlapping
c. three-base, not overlapping
d. three-base, overlapping
e. four-base, not overlapping

4. An example of a portion of the T4 *rIIB* gene in which Crick and Brenner had recombined one + and one − mutation is shown here. (The RNA-like strand of the DNA is shown.)

```
wild type   5′ AAA AGT CCA TCA CTT AAT GCC 3′
mutant      5′ AAA GTC CAT CAC TTA ATG GCC 3′
```

a. Where are the + and − mutations in the mutant DNA?
b. What alterations in amino acids occurred in this double mutant, which produces wild-type plaques?
c. How can you explain the fact that amino acids are different in the double mutant compared to the wild-type sequence, yet the phage is wild type?

5. In the *HbS* allele (sickle-cell allele) of the human β-globin gene, the sixth amino acid in the β-globin chain is changed from glutamic acid to valine. In *HbC,* the sixth amino acid in β globin is changed from glutamic acid to lysine. What would be the order of these two mutations within the map of the β-globin gene?

6. The following diagram describes the mRNA sequence of part of the *A* gene and the beginning of the *B* gene of phage ϕX174. In this phage, there are some genes that are read in overlapping reading frames. For example, the code for the *A* gene is used for part of the *B* gene, but the reading frame is displaced by one base. Shown here is the single mRNA with the codons for proteins A and B indicated.

```
aa      5  6  7  8  9  10 11 12 13 14 15 16
A       AlaLysGluTrpAsnAsnSerLeuLysThrLysLeu

mRNA    GCUAAAGAAUGGAACAACUCACUAAAAACCAAGCUG

B                 MetGluGlnLeuThrLysAsnGlnAla
aa                 1  2  3  4  5  6  7  8  9
```

Given the following amino acid changes, indicate the base change that occurred in the mRNA and the consequences for the other protein sequence.

a. Asn at position 10 in protein A is changed to Tyr.
b. Leu at position 12 in protein A is changed to Pro.
c. Gln at position 8 in protein B is changed to Leu.
d. The occurrence of overlapping reading frames is very rare in nature. When it does occur, the extent of the overlap is not very long. Why do you think this is the case?

7. The amino acid sequence of part of a protein has been determined:

N . . . Gly Ala Pro Arg Lys . . . C

A mutation has been induced in the gene encoding this protein using the mutagen proflavin. The resulting mutant protein can be purified and its amino acid sequence determined. The amino acid sequence of the mutant protein is exactly the same as the amino acid sequence of the wild-type protein from the N terminus of the protein to the glycine in the preceding sequence. Starting with this glycine, the sequence of amino acids is changed to the following:

N . . . Gly His Gln Gly Lys . . . C

Using the amino acid sequences, one can determine the sequence of 14 nucleotides from the wild-type gene encoding this protein. What is this sequence?

8. When the artificial mRNA 5′ . . . UCUCUCUC . . . 3′ was added to an *in vitro* protein synthesis system, investigators found that proteins composed of alternating leucine and serine were made. What experiments were done to determine whether leucine was specified by CUC and serine by UCU, or vice versa?

9. Identify all the amino acid-specifying codons where a point mutation (a single base change) could generate a nonsense codon.

10. Translate all the sequences shown in Fig. 8.6 on p. 261, assuming that in each case the RNA-like strand of the gene is depicted.

11. A particular protein has the amino acid sequence

N . . . Ala-Pro-His-Trp-Arg-Lys-Gly-Val-Thr . . . C

within its primary structure. A geneticist studying mutations affecting this protein discovered that several of the mutants produced shortened protein molecules that terminated within this region. In one of them, the His became the terminal amino acid.

a. What DNA single-base change(s) would cause the protein to terminate at the His residue?
b. What other potential sites do you see in the DNA sequence encoding this protein where mutation of a single base pair would cause premature termination of translation?

12. In studying normal and mutant forms of a particular human enzyme, a geneticist came across a particularly interesting mutant form of the enzyme. The normal enzyme is 227 amino acids long, but the mutant form was 312 amino acids long, having that extra 85 amino acids as a block in the middle of the normal sequence. The inserted amino acids do not correspond in any way to the normal protein sequence. What are possible explanations for this phenomenon? How would you distinguish among them?

13. How many possible open reading frames (frames without stop codons) are there that extend through the following sequence?

```
5′... CTTACAGTTTATTGATACGGAGAAGG...3′
3′... GAATGTCAAATAACTATGCCTCTTCC...5′
```

14. a. In Fig. 8.4 on p. 259, the physical map (the number of base pairs) is not exactly equivalent to the genetic map (in map units). Explain this apparent discrepancy.
b. In Fig. 8.4, which region shows the highest rate of recombination, and which the lowest?

15. The sequence of a segment of mRNA, beginning with the initiation codon, is given here, along with the corresponding sequences from several mutant strains.

```
Normal    AUGACACAUCGAGGGGUGGUAAACCCUAAG...
Mutant 1  AUGACACAUCCAGGGGUGGUAAACCCUAAG...
Mutant 2  AUGACACAUCGAGGGUGGUAAACCCUAAG...
Mutant 3  AUGACGCAUCGAGGGGUGGUAAACCCUAAG...
Mutant 4  AUGACACAUCGAGGGGUUGGUAAACCCUAAG...
Mutant 5  AUGACACAUUGAGGGGUGGUAAACCCUAAG...
Mutant 6  AUGACAUUUACCACCCCUCGAUGCCCUAAG...
```

a. Indicate the type of mutation present in each and translate the mutated portion of the sequence into an amino acid sequence in each case.
b. Which of the mutations could be reverted by treatment with EMS (ethylmethane sulfonate; see Fig. 7.10 on pp. 218–219)? With proflavin?

16. You identify a proflavin-generated allele of a gene that produces a 110-amino acid polypeptide rather than the usual 157-amino acid protein. After subjecting this mutant allele to extensive proflavin mutagenesis, you are able to find a number of intragenic suppressors located in the part of the gene between the sequences encoding the N-terminus of the protein and the original mutation but no suppressors located in the region between the original mutation and the sequences encoding the usual C-terminus of the protein. Why do you think this is the case?

Section 8.2

17. Describe the steps in transcription that require complementary base pairing.

18. Chapters 6 and 7 explained that mistakes made by DNA polymerase are corrected either by proofreading mechanisms during DNA replication or by DNA repair systems that operate after replication is complete. The overall rate of errors in DNA replication is about 1×10^{-10}, that is, one error in 10 million base pairs. RNA polymerase also has some proofreading capability, but the overall error rate for transcription is significantly higher (1×10^{-4}, or one error in each 1000 nucleotides). Why can organisms tolerate higher error rates for transcription than for DNA replication?

19. The coding sequence for gene *F* is read from left to right on the following figure. The coding sequence for gene *G* is read from right to left. Which strand of DNA (top or bottom) serves as the template for transcription of each gene?

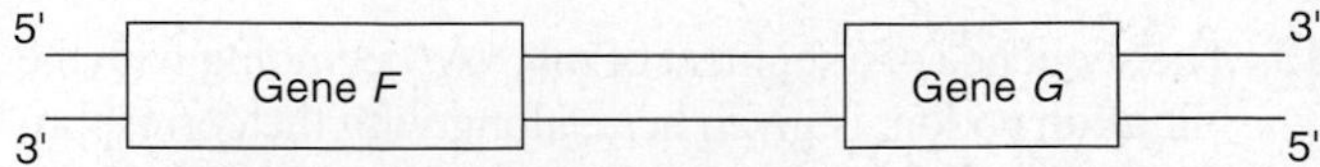

20. If you mixed the mRNA of a human gene with the genomic DNA for the same gene and allowed the RNA and DNA to form a hybrid, what would you be likely to see in the electron microscope? Your figure should include hybridization involving both DNA strands (template and RNA-like) as well as the mRNA.

Section 8.3

21. Describe the steps in translation that require complementary base pairing.

22. Locate as accurately as possible the listed items that are shown on the following figure. Some items are not shown. (a) 5′ end of DNA template strand; (b) 3′ end of mRNA; (c) ribosome; (d) promoter; (e) codon; (f) an amino acid; (g) DNA polymerase; (h) 5′ UTR; (i) centromere; (j) intron; (k) anticodon; (l) N terminus; (m) 5′ end of charged tRNA; (n) RNA polymerase; (o) 3′ end of uncharged tRNA; (p) a nucleotide; (q) mRNA cap; (r) peptide bond; (s) P site; (t) aminoacyl-tRNA synthetase; (u) hydrogen bond; (v) exon; (w) 5′ AUG 3′; (x) potential "wobble" interaction.

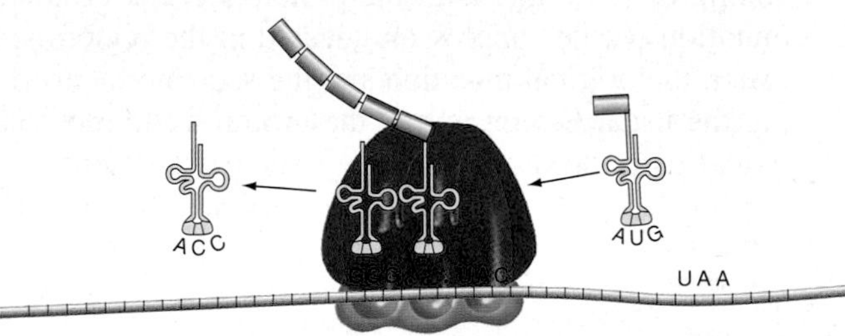

23. Concerning the figure for the previous problem (#22):

a. Which process is being represented?

b. What is the next building block to be added to the growing chain in the figure? To what end of the growing chain will this building block be added? How many building blocks will there be in the chain when it is completed?

c. What other building blocks have a known identity?

d. What details could you add to this figure that would be different in a eukaryotic cell versus a prokaryotic cell?

Section 8.4

24. In prokaryotes, a search for genes in a DNA sequence involves scanning the DNA sequence for long open reading frames (that is, reading frames uninterrupted by stop codons). What problem can you see with this approach in eukaryotes?

25. The yeast gene encoding a protein found in the mitotic spindle was cloned by a laboratory studying mitosis. The gene encodes a protein of 477 amino acids.

a. What is the minimum length in nucleotides of the protein-coding part of this yeast gene?

b. A partial sequence of one DNA strand in an exon containing the middle of the coding region of the yeast gene is given here. What is the sequence of nucleotides of the mRNA in this region of the gene? Show the 5′ and 3′ directionality of your strand.

5′ GTAAGTTAACTTTCGACTAGTCCAGGGT 3′

c. What is the sequence of amino acids in this part of the yeast mitotic spindle protein?

26. The sequence of a complete eukaryotic gene encoding the small protein Met Tyr Arg Gly Ala is shown here. All of the written sequences on the template strand are transcribed into RNA.

5′ CCCCTATGCCCCCCTGGGGGAGGATCAAAACACTTACCTGTACATGGC 3′
3′ GGGGATACGGGGGGACCCCCTCCTAGTTTTGTGAATGGACATGTACCC 5′

a. Which strand is the template strand? Which direction (right to left or left to right) does RNA polymerase move along the template as it transcribes this gene?

b. What is the sequence of the nucleotides in the processed mRNA molecule for this gene? Indicate the 5′ and 3′ polarity of this mRNA.

c. A single base mutation in the gene results in synthesis of the peptide Met Tyr Thr. What is the sequence of nucleotides making up the mRNA produced by this mutant gene?

27. Using recombinant DNA techniques (which will be described in Chapter 9), it is possible to take the DNA of a gene from any source and place it on a chromosome in the nucleus of a yeast cell. When you take the DNA for a human gene and put it into a yeast cell chromosome, the altered yeast cell can make the human protein. But when you remove the DNA for a gene normally present on yeast mitochondrial chromosomes and put it on a yeast chromosome in the nucleus, the yeast cell cannot synthesize the correct protein, even though the gene comes from the same organism. Explain. What would you need to do to ensure that such a yeast cell could make the correct protein?

28. a. The genetic code table shown in Fig. 8.3 on p. 257 applies both to humans and to *E. coli.* Suppose that you have purified a piece of DNA from the human genome containing the entire gene encoding the hormone insulin. You now transform this piece of DNA into *E. coli.* Why can't *E. coli* cells containing the human insulin gene actually make insulin?

b. Pharmaceutical companies have actually been able to obtain *E. coli* cells that make human insulin; such insulin can be purified from the bacterial cells and used to treat diabetic patients. How were the pharmaceutical companies able to create such "bacterial factories" for making insulin?

Section 8.5

29. Arrange the following list of eukaryotic gene elements in the order they would appear in the genome and in the direction traveled by RNA polymerase along the gene. Assume the gene's single intron interrupts the open reading frame. Note that some of these names are abbreviated and thus do not distinguish between elements in DNA versus RNA. For example, "splice-donor site" is an abbreviation for "DNA sequences transcribed into the splice-donor site" because splicing takes place on the gene's RNA transcript, not on the gene itself. Geneticists often use this kind of shorthand for simplicity, even though it is imprecise. (a) splice-donor site; (b) 3′ UTR; (c) promoter; (d) stop codon; (e) nucleotide to which methylated cap is added; (f) initiation codon; (g) transcription terminator; (h) splice-acceptor site; (i) 5′ UTR; (j) poly-A addition site; (k) splice branch site.

30. Concerning the list of eukaryotic gene elements in the previous problem (#29):

a. Which of the element names in the list are abbreviated? (That is, which of these elements actually occur in the gene's primary transcript or mRNA rather than in the gene itself?)

b. Which of the elements in the list are found partly or completely in the first exon of this gene (or the RNA transcribed from this exon)? In the intron? In the second exon?

Section 8.6

31. Do you think each of the following types of mutations would have very severe effects, mild effects, or no effect at all?

a. Nonsense mutations occurring in the sequences encoding amino acids near the N terminus of the protein

b. Nonsense mutations occurring in the sequences encoding amino acids near the C terminus of the protein

c. Frameshift mutations occurring in the sequences encoding amino acids near the N terminus of the protein

d. Frameshift mutations occurring in the sequences encoding amino acids near the C terminus of the protein

e. Silent mutations

f. Conservative missense mutations

g. Nonconservative missense mutations affecting the active site of the protein

h. Nonconservative missense mutations not in the active site of the protein

32. Null mutations are valuable genetic resources because they allow a researcher to determine what happens to an organism in the complete absence of a particular protein. However, it is often not a trivial matter to determine whether a mutation represents the null state of the gene.

a. Geneticists sometimes use the following test for the "nullness" of an allele in a diploid organism: If the abnormal phenotype seen in a homozygote for the allele is identical to that seen in a heterozygote where one chromosome carries the allele in question and the homologous chromosome is known to be completely deleted for the gene, then the allele is null. What is the underlying rationale for this test? What limitations might there be in interpreting such a result?

b. Can you think of other methods to determine whether an allele represents the null state of a particular gene?

33. The following is a list of mutations that have been discovered in a gene that has more than 60 exons and encodes a very large protein of 2532 amino acids. Indicate whether or not each mutation could cause a detectable change in the size or the amount of mRNA and/or a detectable change in the size or the amount of the protein product. (Detectable changes in size or amount must be greater than 1% of normal values.) What kind of change would you predict?

a. Lys576Val (changes amino acid 576 from lysine into valine)

b. Lys576Arg

c. AAG576AAA (changes codon 576 from AAG to AAA)

d. AAG576UAG

e. Met1Arg (there are at least two possible scenarios for this mutation)

f. promoter mutation

g. one base-pair insertion into codon 1841

h. deletion of codon 779

i. IVS18DS, G–A, + 1 (this mutation changes the first nucleotide in the eighteenth intron of the gene, causing exon 18 to be spliced to exon 20, thus skipping exon 19)

j. deletion of the poly-A addition site
k. G-to-A substitution in the 5′ UTR
l. insertion of 1000 base pairs into the sixth intron (this particular insertion does not alter splicing)

34. Considering further the mutations described in the previous problem (#33):
a. Which of the mutations could be null mutations?
b. Which of the mutations would be most likely to result in an allele that is recessive to wild type?
c. Which of the mutations could result in an allele dominant to wild type? What mechanism(s) could explain this dominance?

35. When 1 million cells of a culture of haploid yeast carrying a *met*$^-$ auxotrophic mutation were plated on petri plates lacking methionine (met), five colonies grew. You would expect cells in which the original *met*$^-$ mutation was reversed (by a base change back to the original sequence) would grow on the media lacking methionine, but some of these apparent reversions could be due to a mutation in a different gene that somehow suppresses the original *met*$^-$ mutations. How would you be able to determine if the mutations in your five colonies were due either to a precise reversion of the original *met*$^-$ mutation or to the generation of a suppressor mutation in a gene on another chromosome?

36. a. What are the differences between null, hypomorphic, hypermorphic, dominant negative, and neomorphic mutations?
b. For each of these kinds of mutations, would you predict they would be dominant or recessive to a wild-type allele in producing a mutant phenotype?

37. A mutant *B. adonis* bacterium has a nonsense suppressor tRNA that inserts glutamine (Gln) to match a UAG (but not other nonsense) codons.
a. What is the anticodon of the suppressing tRNA? Indicate the 5′ and 3′ ends.
b. What is the sequence of the template strand of the wild-type tRNA^{Gln}-encoding gene that was altered to produce the suppressor, assuming that only a single-base-pair alteration was involved?
c. What is the *minimum* number of $tRNA^{Gln}$ genes that could be present in a wild-type *B. adonis* cell? Describe the corresponding anticodons.

38. You are studying mutations in a bacterial gene that codes for an enzyme whose amino acid sequence is known. In the wild-type protein, proline is the fifth amino acid from the amino terminal end. In one of your mutants with nonfunctional enzyme, you find a serine at position number 5. You subject this mutant to further mutagenesis and recover three different strains. Strain A has a proline at position number 5 and acts just like wild type. Strain B has tryptophan at position number 5 and also acts like wild type. Strain C has no detectable enzyme function at any temperature, and you can't recover any protein that resembles the enzyme. You mutagenize strain C and recover a strain (C-1) that has enzyme function. The second mutation in C-1 responsible for the recovery of enzyme function does not map at the enzyme locus.
a. What is the nucleotide sequence in both strands of the wild-type gene at this location?
b. Why does strain B have a wild-type phenotype? Why does the original mutant with serine at position 5 lack function?
c. What is the nature of the mutation in strain C?
d. What is the second mutation that arose in C-1?

39. Another class of suppressor mutations, not described in the chapter, are mutations that suppress missense mutations.
a. Why would bacterial strains carrying such missense suppressor mutations generally grow more slowly than strains carrying nonsense suppressor mutations?
b. What other kinds of mutations can you imagine in genes encoding components needed for gene expression that would suppress a missense mutation in a protein-coding gene?

40. Yet another class of suppressor mutations not described in the chapter are mutations in tRNA genes that can suppress frameshift mutations. What would have to be true about a tRNA that could suppress a frameshift mutation involving the insertion of a single base pair?

41. There is at least one nonsense suppressing tRNA known that can suppress more than one type of nonsense codon.
a. What is the anticodon of such a suppressing tRNA?
b. What stop codons would it suppress?
c. What are the amino acids most likely to be carried by this nonsense suppressing tRNA?

42. An investigator was interested in studying UAG nonsense suppressor mutations in bacteria. In one species of bacteria, she was able to select two different mutants of this type, one in the $tRNA^{Tyr}$ gene and the other in the $tRNA^{Gln}$ gene, but in a second species, she was not able to obtain any such nonsense suppressor mutations, even after very extensive effort. What could explain the difference between the two species?

Deconstructing the Genome: DNA at High Resolution

Chapter 9

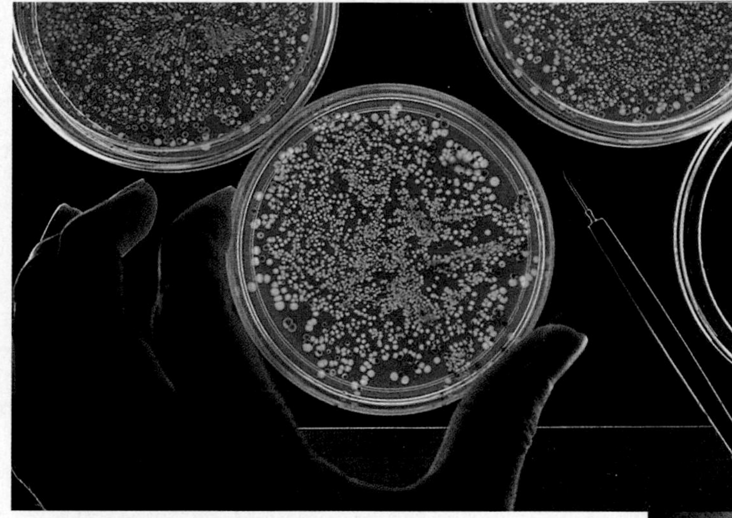

Colonies of bacterial cell clones containing recombinant DNA molecules.

The vivid red color of our blood arises from its life-sustaining ability to carry oxygen. This ability, in turn, derives from billions of red blood cells suspended in proteinaceous solution, each one packed with close to 280 million molecules of the protein pigment known as hemoglobin (**Fig. 9.1a**). The hemoglobin picks up oxygen in the lungs and transports it to tissues throughout the body, where, after its release, the oxygen participates in a multitude of metabolic reactions. A normal adult hemoglobin molecule consists of four polypeptide chains, two alpha (α) and two beta (β) globins, each surrounding an iron-containing small molecular structure known as a heme group (Fig. 9.1b). The iron atom within the heme sustains a reversible interaction with oxygen, binding it firmly enough to hold it on the trip from lungs to body tissue but loosely enough to release it where needed. The intricately folded α and β chains protect the iron-containing hemes from substances in the cell's interior. Each hemoglobin molecule can carry up to four oxygen atoms, one per heme, and it is these oxygenated hemes that impart a scarlet hue to the pigment molecules and thus to the blood cells that carry them.

Interestingly, the genetically determined molecular composition of hemoglobin changes several times during human development, enabling the molecule to adapt its oxygen-transport function to the varying environments of the embryo, fetus, newborn, and adult (Fig. 9.1c). In the first five weeks after conception, the red blood cells carry *embryonic hemoglobin,* which consists of two alphalike zeta (ζ) chains and two betalike epsilon (ϵ) chains. Thereafter, throughout the rest of gestation, the cells contain *fetal hemoglobin,* composed of two bona fide α chains and two β-like gamma (γ) chains. Then, shortly before birth, production of *adult hemoglobin,* composed of two α and two β chains, begins to climb. By the time an infant reaches three months of age, almost all of his or her hemoglobin is of the adult type.

Evolution of the various forms of hemoglobin maximized the delivery of oxygen to an individual's cells at different stages of development. The early embryo, which is not yet associated with a fully functional placenta, has the least access to oxygen in the maternal circulation. The fetus, which is associated with a fully functional placenta, has access to the gases dissolved in the maternal bloodstream, but the amount of oxygen in the maternal circulation is less than that found in the air that enters our lungs after birth. Both embryonic and fetal hemoglobin evolved to bind oxygen more tightly than adult hemoglobin does; they thus facilitate the transfer of maternal oxygen to the embryo or fetus. All the hemoglobins readily release their oxygen to cells, which have an even lower level of oxygen than any source of the gas. After birth, when oxygen is abundantly available in the lungs, adult hemoglobin, with its more relaxed kinetics of oxygen binding, allows for the most efficient pickup and delivery of the vital gas.

What structure and arrangement of hemoglobin genes facilitate the crucial switches in the type of protein produced? And what can go wrong with these genes to generate hemoglobin disorders? Such disorders are the most common genetic diseases in the world and include sickle-cell anemia, which arises from an altered

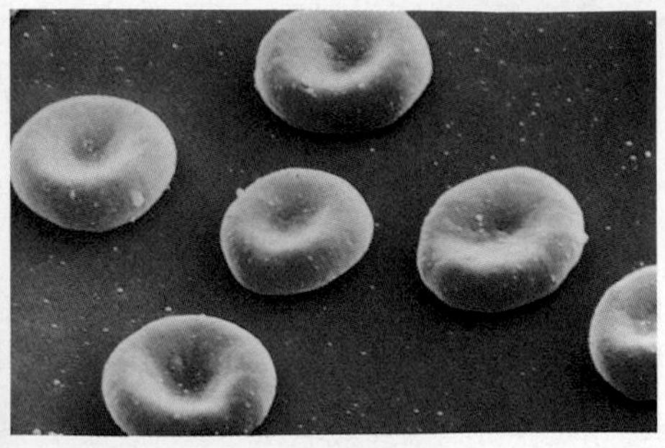

Figure 9.1 Hemoglobin is composed of four polypeptide chains that change during development. (a) Scanning electron micrograph of adult human red blood cells loaded with hemoglobin. **(b)** Adult hemoglobin consists of two α and two β polypeptide chains, each associated with an oxygen-carrying heme group. **(c)** The hemoglobin carried by red blood cells switches during human development from an embryonic form containing two α-like ξ chains and two β-like ϵ chains, to a fetal form containing two α chains and two β-like γ chains, and finally to the adult form containing two α and two β chains. In a small percentage of adult hemoglobin molecules, a β-like δ chain replaces the actual β chain.

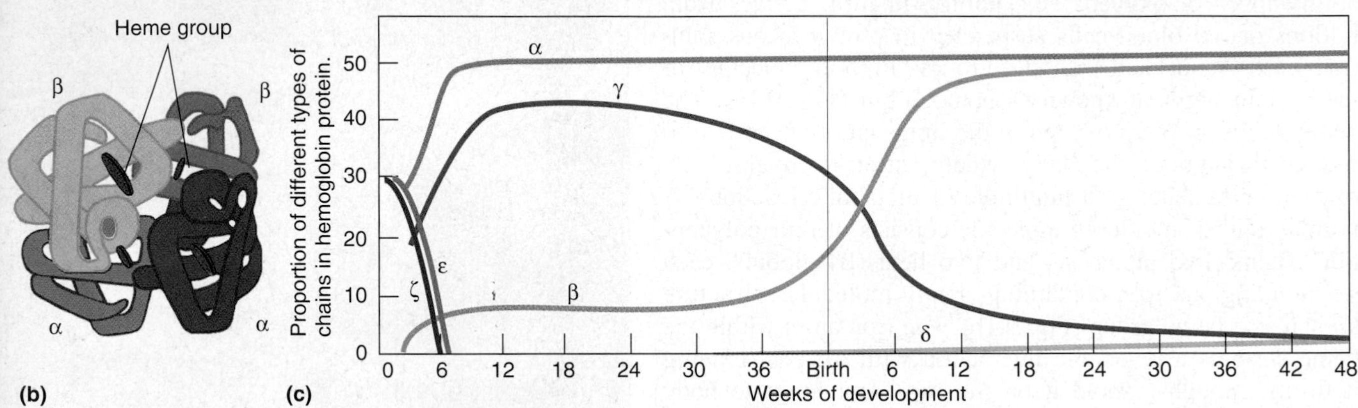

β chain, and thalassemia, which results from decreases in the amount of either α- or β-chain production.

To answer these questions, researchers must have a way of looking at the hemoglobin genes of individuals with normal and abnormal phenotypes. But these genes lie buried in a diploid human genome containing 6 billion base pairs distributed among 46 different strings of DNA (the chromosomes) that range in size from 60 million to 360 million base pairs each. In this chapter, we describe the powerful tools of modern molecular analysis that medical researchers now use to search through these enormously long strings of information for genes that may be only several thousand base pairs in length. These tools take advantage of isolated enzymes and biochemical reactions that occur naturally within the simplest life-forms, bacterial cells. Researchers refer to the whole kit of modern tools and reactions as **recombinant DNA technology** or, simply, **biotechnology.** The most important enzymes in this tool kit are ones that operate on the DNA molecule itself: restriction enzymes that cut DNA, ligases that join two molecules together, and polymerases that synthesize new DNA strands. The most important uncatalyzed biochemical reaction is **hybridization:** the binding together of two DNA strands with complementary nucleotide sequences. Hybridization results from the natural propensity of complementary single-stranded molecules to form stable double helixes. A final tool in the recombinant DNA kit is the natural process of DNA replication, which occurs within growing colonies of cells.

Developed during a technological revolution that began in the mid-1970s, recombinant DNA technology makes it possible to characterize DNA molecules directly, rather than indirectly through the phenotypes they produce. Geneticists use recombinant DNA techniques to gather information unobtainable in any other way or to analyze the results of breeding and cytological studies with greater speed and accuracy than ever before.

Two general themes recur in our discussion. First, the new molecular tools of genetic analysis emerged from a knowledge of DNA structure and function. For example, complementary base pairing between similar or identical stretches of DNA is the basis of hybridization techniques that use labeled DNA molecules as probes to identify related DNA molecules. Second, the speed, sensitivity, and accuracy of the new tools make it possible to answer questions that were impossible to resolve just a decade ago. In one study, for instance, researchers used a protocol for making many copies of a specific DNA segment to trace the activity of the AIDS virus and found that, contrary to prior belief, the virus does not become latent inside all the cells it enters after infecting a person; it remains active in certain cells of the lymph nodes and adenoids.

As we illustrate the power of the modern tools for analyzing a seemingly simple biological system—the hemoglobins—that turned out to be surprisingly complex, we describe how researchers use recombinant DNA technology to carry out five basic operations:

- Cut the enormously long strings of DNA into much smaller fragments with scissorslike *restriction enzymes,* and separate the small fragments according to size through *gel electrophoresis.*
- Isolate, amplify, and purify the fragments through *molecular cloning.*
- Use purified DNA fragments as hybridization *probes* to identify the presence of similar sequences in libraries of clones or in complex mixtures of DNA or RNA molecules.
- Rapidly isolate and amplify previously defined genomic or mRNA sequences from new individuals or cell sources through the *polymerase chain reaction (PCR).*
- Determine the precise *sequence* of bases within isolated DNA fragments.

9.1 Fragmenting Complex Genomes into Bite-Size Pieces for Analysis

Every intact diploid human body cell, including the precursors of red blood cells, carries two nearly identical sets of 3 billion base pairs of information that, when unwound, extend 2 meters in length and contain two copies of 20,000– 30,000 genes. If you could enlarge the cell nucleus to the size of a basketball, the unwound DNA would have the diameter of a fishing line and a length of 200 kilometers. This is much too much material and information to study as a whole. To reduce its complexity, researchers first cut the genome into "bite-size" pieces.

Restriction Enzymes Fragment the Genome at Specific Sites

Researchers use restriction enzymes to cut the DNA released from the nuclei of cells at specific sites. These well-defined cuts generate fragments suitable for manipulation and characterization. A **restriction enzyme** recognizes a specific sequence of bases anywhere within the genome and then severs two covalent bonds (one in each strand) in the sugar-phosphate backbone at particular positions within or near that sequence. The fragments generated by restriction enzymes are referred to as **restriction fragments,** and the act of cutting is often called **digestion.**

Restriction enzymes originate in and can be purified from bacterial cells. The enzymes protect these prokaryotic cells from viral infection by digesting the viral DNA. Bacteria shield their DNA from digestion by their own restriction enzymes through the selective addition of methyl groups ($-CH_3$) to the restriction recognition sites in their DNA. In the test tube, restriction enzymes from bacteria recognize target sequences of 4–8 bp in DNA isolated from any other organism and cut the DNA at or near these sites. **Table 9.1** lists the names, recognition sequences, and microbial origins of just 10 of the more than 100 commonly used restriction enzymes. For the majority of these enzymes, the recognition site contains 4–6 base pairs and exhibits a kind of palindromic symmetry in which the base sequences of each of the two DNA strands are identical when read in the 5′-to-3′ direction. Because of this, base pairs on either side of a central line of symmetry are mirror images of each other. Each enzyme always cuts at the same place relative to its specific recognition sequence, and most enzymes make their cuts in one of two ways: either straight through both DNA strands right at the line of symmetry to produce fragments with **blunt ends,** or displaced equally in opposite directions from the line of symmetry by one or more bases to generate fragments with single-stranded

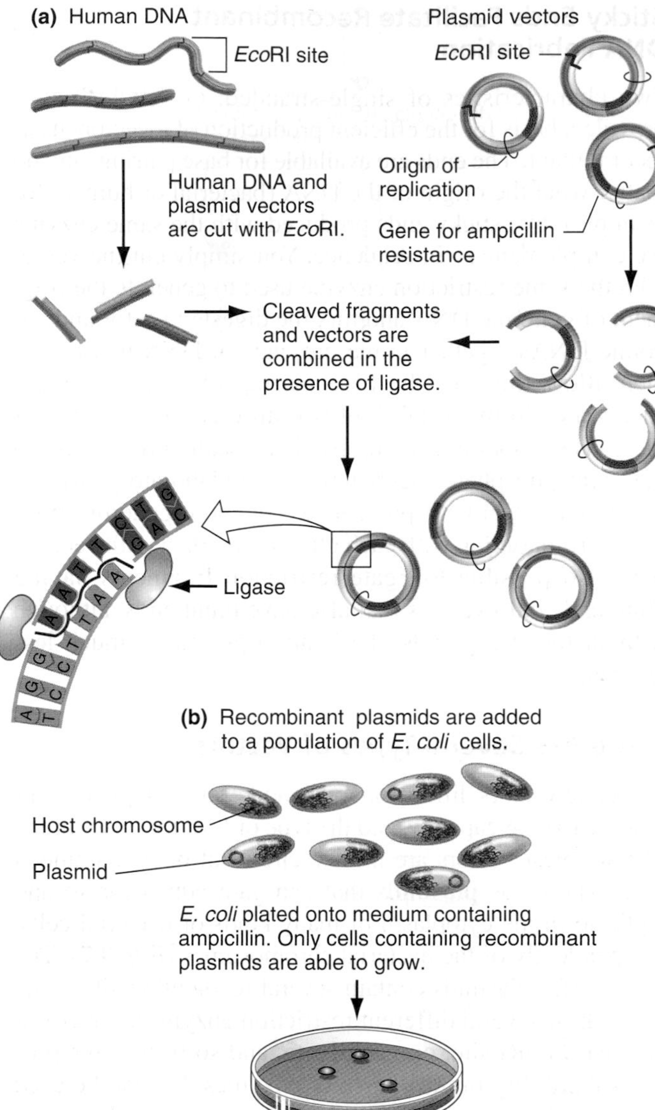
(a) Human DNA
EcoRI site
Plasmid vectors
EcoRI site
Origin of replication
Human DNA and plasmid vectors are cut with EcoRI.
Gene for ampicillin resistance
Cleaved fragments and vectors are combined in the presence of ligase.
Ligase
(b) Recombinant plasmids are added to a population of E. coli cells.
Host chromosome
Plasmid
E. coli plated onto medium containing ampicillin. Only cells containing recombinant plasmids are able to grow.

Figure 9.7 Creating recombinant DNA molecules with plasmid vectors. (a) Human genomic DNA is cut with *Eco*RI to produce a mixture of *Eco*RI restriction fragments. A plasmid vector is also cut with *Eco*RI at its single *Eco*RI recognition site. The cut genomic DNA and the cut plasmid vectors are mixed together in the presence of the enzyme ligase, which sutures the two types of molecules to each other to form circular recombinant DNA molecules. **(b)** *E. coli* cells transformed with recombinant plasmids are recognized by their growth in the presence of ampicillin.

site, allowing the insertion of a foreign DNA fragment, without at the same time splitting the plasmid into many pieces and thereby destroying its continuity and integrity (Fig. 9.7).

A plasmid carrying a foreign insert is known as a *recombinant plasmid.* Each plasmid vector also carries an origin of replication and a gene for resistance to a specific antibiotic. The origin of replication enables it to replicate independently inside a bacterium. The gene for antibiotic resistance confers on the host cell the ability to survive in a medium containing a specific antibiotic; the resistance gene thereby enables experimenters to select for propagation only those bacterial cells that contain a plasmid (**Fig. 9.8**). Antibiotic resistance genes and other vector genes that make it possible to pick out cells harboring a particular DNA molecule are called **selectable markers.** Plasmids fulfill the final requirement for vectors—ease of purification—because they can be purified away from the genomic DNA of the bacterial host by several techniques that take advantage of size and other differences, as described later. Plasmid vector restriction sites useful for cloning are ones that do not interrupt either the vector origin of replication or the coding region of the selectable marker.

Common plasmid vectors are 2–4 kb in length and capable of carrying up to 15 kb of foreign DNA. In some experimental situations, however, it is necessary to isolate DNA fragments that are larger than this limit. A variety of larger-capacity vectors have been developed to accommodate different sizes of DNA fragments. Table 9.2 lists the properties of a few types of vectors and the host cells in which they grow.

Several types of vectors have been constructed from the genomes of naturally occurring viruses such as bacteriophage lambda (λ). Phage λ is a double-stranded DNA virus that infects *E. coli.* As mentioned earlier, each λ chromosome is 48.5 kb long; it can be engineered to receive central inserts of up to 25 kb in length that replace nonessential viral sequences. The remaining viral sequences on both sides of the insert represent vector arms containing genes required to form whole virus particles; these particles infect host cells with extremely high rates of efficiency and then multiply voraciously. Foreign DNA inserted between the two arms of a λ vector will be packaged into the λ virus particles and amplified along with them inside the host cells.

The largest-capacity vectors are *artificial chromosomes:* recombinant DNA molecules formed by combining chromosomal replication and segregation elements with a DNA insert. Such vectors contain the crucial chromosomal elements that allow them to replicate within a particular host. A bacterial artificial chromosome (BAC) can accommodate a DNA insert of 300 kb; a yeast artificial chromosome (YAC) can accommodate a DNA insert of 2000 kb.

The utility of these different cloning vectors becomes apparent when you consider the structure of the β-globin locus. Situated on chromosome human 11, the entire locus, including multiple β-globin-like genes and all the regulatory information for proper expression of these genes, spans 70 kb of DNA. Only a few vectors (including BACs and YACs) can accommodate this much material. By contrast, the collection of exons and introns that constitute a single β-globin gene is only 1.4 kb in length and can thus fit easily into a plasmid. Most other genes are larger than individual β-globin genes, in part because they have greater coding capacity and many more introns; as a result, to be cloned as a single insert, they require a λ, BAC, or occasionally even a YAC vector.

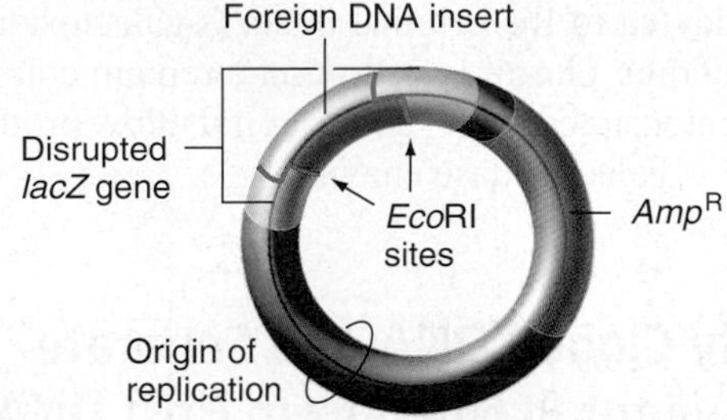

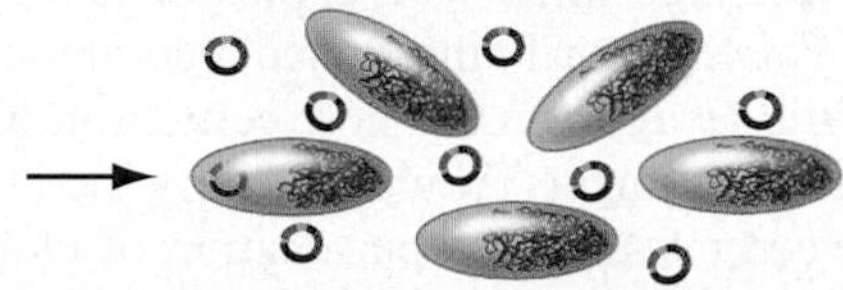

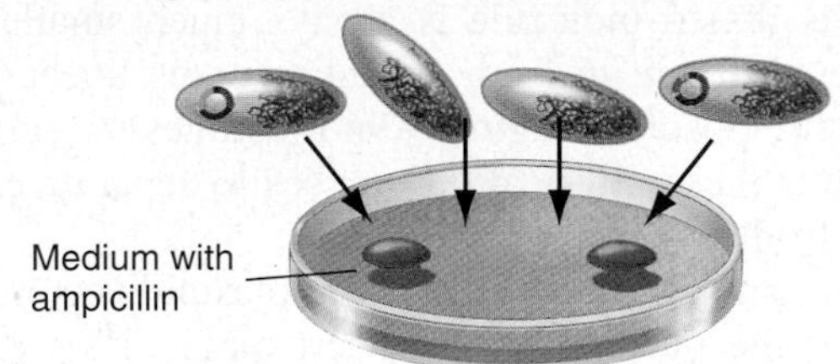

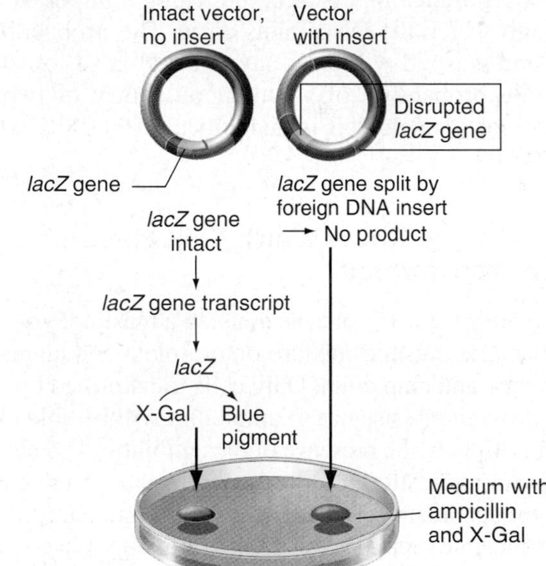

Figure 9.8 How to identify transformed bacterial cells containing plasmids with DNA inserts. **(a)** Plasmid vectors (aqua) are often constructed so that they contain the *E. coli lacZ* gene with a restriction site right in the middle of the gene. If the vector reanneals to itself without inclusion of an insert, the *lacZ* gene will remain uninterrupted; if it accepts an insert, the gene will be interrupted. **(b)** Transformation: When added to a culture of bacteria, plasmids enter about 1 in 1000 cells. **(c)** Only cells transformed by a plasmid carrying a gene for ampicillin resistance will form colonies on petri plates. **(d)** Cells containing vectors that have reannealed to themselves without the inclusion of an insert will express the uninterrupted *lacZ* gene. The polypeptide product of the gene is β-galactosidase. Reaction of this enzyme with a substrate known as X-Gal produces a molecule that turns the cell blue. By contrast, if the plasmid vector contains a fragment of foreign DNA, this insert will interrupt the *lacZ* gene on the plasmid, rendering it unable to produce a functional polypeptide. As a result of this process, called *insertional inactivation,* any cells containing recombinant plasmids will not generate active β-galactosidase and will therefore not turn blue.

Cloning Step 2: Host Cells Take Up and Amplify Vector-Insert Recombinants

Although each type of vector functions in a slightly different way and enters a specific kind of host, the general scheme of entering a host cell and taking advantage of the cellular environment to replicate itself is the same for all. We divide our discussion of this step of cloning into three parts: getting foreign DNA into the host cell; selecting cells that have received a DNA molecule; and distinguishing insert-containing recombinant molecules from vectors without inserts. Figure 9.8 illustrates the three-part process with a plasmid vector containing an origin of replication, the gene for resistance to ampicillin (*amp*^{R}), and the *E. coli lacZ* gene, which encodes the enzyme β-galactosidase. By constructing the vector with a common restriction site like *Eco*RI right in the middle of the *lacZ* gene, researchers can insert foreign DNA into the gene at that location and then use the disruption of *lacZ* gene function to distinguish insert-containing recombinant molecules from vectors without inserts (as described in the caption and later in the text). Many of the plasmid vectors used today incorporate most if not all of the features depicted in Fig. 9.8.

Transformation: Vectors Carry Insert DNA into Cells

Transformation, we saw in Chapter 6, is the process by which a cell or organism takes up a foreign DNA molecule, changing the genetic characteristics of that cell or organism. What we now describe is similar to what Avery and his colleagues did in the transformation experiments that determined DNA was the molecule of heredity (see p. 171 of Chapter 6), but the method outlined here is more efficient.

First, recombinant DNA molecules are added to a suspension of specially prepared *E. coli.* Under conditions favoring entry, such as suspension of the bacterial cells in a cold $CaCl_2$ solution or treatment of the solution with high-voltage electric shock (a technique known as *electroporation*), the plasmids will enter about 1 in 1000 cells (Fig. 9.8b). These protocols increase the permeability of the bacterial cell membrane, in essence, punching temporary holes through which the DNA gains entry. The probability that any one plasmid will enter any one cell is so low (0.001) that the probability of simultaneous entry of two plasmids into a single cell is insignificant ($0.001 \times 0.001 = 0.000001$).

How Do You Know Which Cells Have Been Transformed?

To identify the 0.1% of cells housing a plasmid, you decant the bacteria-plasmid mixture onto a plate containing agar, nutrients, and ampicillin. Only cells transformed by a plasmid providing resistance to ampicillin will be able to grow and multiply in the presence of the antibiotic. The plasmid's origin of replication enables it to replicate in the bacterial cell independently of the bacterial chromosome; in fact, most plasmids replicate so well that a single bacterial cell may end up with hundreds of identical copies of the same plasmid molecule. Each viable plasmid-containing bacterial cell will multiply to produce a colony of tens of millions of genetically identical cells. The colony as a whole is considered a **cellular clone.** Such clones show up on the agar plate as spots about 1 mm in diameter, easily big enough to see yet small enough for many distinct ones to appear on a plate (Fig. 9.8c, see also the chapter opening photo on p. 301). The millions of identical plasmid molecules contained within a colony together make up a **DNA clone.** They can be purified away from other cellular material as described in a following section. The process through which a particular plasmid molecule is created and amplified to millions of copies inside the multiplying cells of a colony derived from a single bacterial progenitor is called **molecular cloning.**

How Do You Know Whether the Plasmids Inside Bacterial Cells Contain an Insert?

If prepared under proper conditions (including, for example, a controlled ratio of fragments to vectors during the ligation procedure), most plasmids will contain an insert. Some plasmids, however, will slip through without one. Figure 9.8d shows how the system we are discussing distinguishes cells with just vectors from cells with vectors containing inserts. The medium on which the transformed, ampicillin-resistant bacteria grow contains, in addition to nutrients and ampicillin, a chemical compound known as X-Gal. This compound serves as a substrate for the reaction catalyzed by the intact β-galactosidase enzyme (the protein encoded by the *lacZ* gene); one product of the reaction is a new, blue-colored chemical. Cells containing vectors without inserts turn blue because they carry the original intact β-galactosidase gene. Cells containing plasmids with inserts remain colorless, because the interrupted *lacZ* gene does not allow production of functional β-galactosidase enzyme.

To Purify Cloned DNA, You Separate Recombinant Plasmid from Host DNA and DNA Insert from Vector

To obtain a test tube filled with copies of just the cloned DNA insert, you first purify the cloned recombinant vector-insert molecule away from the host cell chromosome and other cellular components. You then purify the DNA insert away from vector DNA. The purification of cloned DNA away from the host chromosome depends on the unique physical characteristics that distinguish the two types of DNA. The cloned molecule is always much smaller and sometimes different in form (circular versus linear) from the host chromosome. **Figure 9.9a** illustrates the protocol that exploits these physical differences to separate cloned plasmids from bacterial chromosomes.

To separate genomic inserts from plasmid vectors, you simply cut the purified vector-insert molecules with the same restriction enzyme that you used to make the recombinants in the first place. You then subject the resulting mixture of insert DNA disengaged from vector DNA to gel electrophoresis (Fig. 9.9b). At the completion of electrophoresis, the region of the gel that contains the insert is cut out and the DNA within it purified. Molecular cloning ultimately yields a large amount of purified fragments—billions of identical copies of the original small piece of genomic DNA—that are now ready for study.

Libraries Are Collections of Cloned Fragments

Moving step by step from the DNA of any organism to a single purified DNA fragment is a long and tedious process. Fortunately, scientists do not have to return to step 1 every time they need to purify a new genomic fragment from the same organism. Instead, they can build a **genomic library:** a storable collection of cellular clones that contains copies of every sequence in the whole genome inserted into a suitable vector. Like traditional book libraries, genomic libraries store large amounts of information for retrieval upon request. They make it possible to start a new cloning project at an advanced stage, when the initial cloning step has already been completed and the only difficult task left is to determine which of the many clones in a library contains the DNA sequence of interest. Once the correct cellular or viral clone is identified, it can be amplified to yield a large amount of the desired genomic fragment.

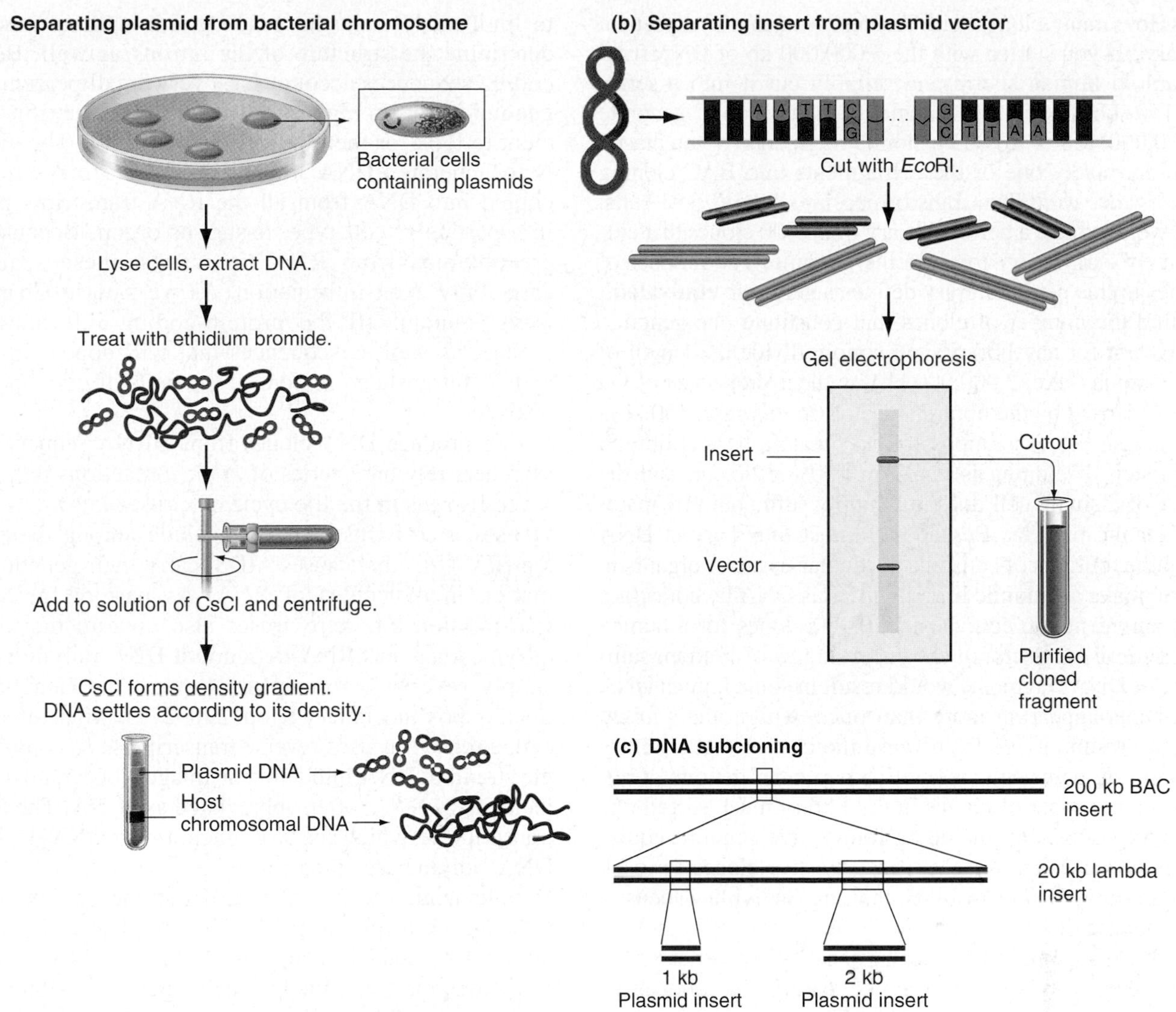

Figure 9.9 **Purifying cloned DNA.** **(a)** Separation of plasmid from the bacterial chromosome. The bacterial chromosome is about 4000 kb long, while the recombinant plasmid is from 4–20 kb in length; in this example, the foreign DNA insert (*green*) is larger than the vector (*aqua*). Cell lysis causes the large bacterial chromosome to fragment into multiple linear pieces, while the small circular plasmid molecules remain intact. Supercoiling of the circular plasmid DNA causes it to bind less ethidium bromide than the bacterial chromosome fragments. The buoyant density of the plasmids is therefore distinguishable from that of the chromosomes when the two are added to a cesium chloride (CsCl) gradient saturated with ethidium bromide. The band containing plasmid is extracted and purified. **(b)** Separation of the genomic DNA insert from vector DNA. The purified plasmid is digested with the same enzyme that was used to construct the recombinant molecule, releasing vector and insert fragments that differ in size. With gel electrophoresis, the two types of fragments can be separated from each other, and the band containing the insert can be cut out and purified. **(c)** Subcloning: From BAC clones to lambda clones to plasmids. The large genomes of higher eukaryotes can be readily divided into ~20,000 BAC clones that each contain ~200 kb of genomic DNA. To study gene-size regions from a particular BAC, it is convenient to subclone 20 kb fragments into a lambda vector. To analyze particular regions of a gene, it is useful to perform another round of subcloning, placing fragments that are 2 kb or smaller into a plasmid vector.

How to Compile a Genomic Library

If you digested the genome of a single cell with a restriction enzyme and ligated every fragment to a vector with 100% efficiency and then transformed all of these recombinant DNA molecules into host cells with 100% efficiency, the resulting set of clones would represent the entire genome in a fragmented form. A hypothetical collection of cellular clones that includes one copy—and one copy only—of every sequence in the entire genome would be a *complete genomic library*.

How many clones will be present in this hypothetical library? If you started with the 3,000,000 kb of DNA from a haploid human sperm and reliably cut it into a series of 150 kb restriction fragments, you would generate 3,000,000/150 = 20,000 genomic fragments. If you placed each and every one of these fragments into BAC cloning vectors that were then transformed into *E. coli* host cells, you would create a perfect library of 20,000 clones that collectively carry every locus in the genome. The number of clones in this perfect library defines a **genomic equivalent.** To find the number of clones that constitute one genomic equivalent for any library, you simply divide the length of the genome (here, 3,000,000 kb) by the average size of the inserts carried by the library's vector (in this case, 150 kb).

In real life, it is impossible to obtain a perfect library. Each step of cloning is far from 100% efficient, and the DNA of a single cell does not supply sufficient raw material for the process. Researchers must thus harvest DNA from the millions of cells in a particular tissue or organism. If you make a genomic library with this DNA by collecting only one genomic equivalent (20,000 clones for a human library in BAC vectors), the uncertainties of random sampling of DNA fragments would result in some human DNA fragments appearing more than once, while others might not be present at all. To increase the chance that a library contains at least one clone with a particular genomic fragment, the number of clones in the library must exceed one genomic equivalent. Including four to five genomic equivalents produces an average of four to five clones for each locus, and a 95% probability that any individual locus is present at least once.

For many types of molecular genetics experiments, it is advantageous to use a vector like a BAC that allows the cloning of large fragments of DNA. This is because the number of clones in a genomic equivalent is much smaller than in libraries constructed with a vector that accepts only smaller fragments, so investigators have to search through fewer clones to find a particular clone of interest. For example, a genomic equivalent of BAC clones with 150 kb inserts contains only 20,000 clones, while a genomic equivalent of plasmid clones with 10 kb inserts would contain 300,000 clones.

But BACs have a disadvantage: It takes a lot more work to characterize the larger inserts they carry. Researchers must often **subclone** smaller restriction fragments from a BAC clone to look more closely at a particular region of genomic DNA (Fig. 9.9c). Sometimes, it is useful to reduce complexity one step at a time, subcloning a BAC DNA fragment into a lambda vector and then subcloning fragments of the lambda clone into a plasmid vector. This process of **subcloning** establishes smaller and smaller DNA clones that are enriched for the sequences of interest.

cDNA Libraries Carry Information from the RNA Transcripts Present in a Particular Tissue

Often, only the information in a gene's coding sequence is of experimental interest, and it would be advantageous to limit analysis to the gene's exons without having to determine the structure of the introns as well. Because coding sequences account for a very small percentage of genomic DNA in higher eukaryotes, however, it is inefficient to look for them in genomic libraries. The solution is to generate **cDNA libraries,** which store sequences copied into DNA from all the RNA transcripts present in a particular cell type, tissue, or organ. Because they are obtained from RNA transcripts, these sequences carry only exon information. As we saw in Chapter 8, exons contain all the protein-coding information of a gene, as well as sequences that will appear in the 5′ and 3′ untranslated regions at the ends of corresponding mRNAs.

To produce DNA clones from mRNA sequences, researchers rely on a series of *in vitro* reactions that mimic several stages in the life cycle of viruses known as **retroviruses.** Retroviruses, which include among their ranks the HIV virus that causes AIDS, carry their genetic information in molecules of RNA. As part of their gene-transmission kit, retroviruses also contain the unusual enzyme known as **RNA-dependent DNA polymerase,** or simply **reverse transcriptase** (review the Genetics and Society box in Chapter 8, pp. 270–271). After infecting a cell, a retrovirus uses reverse transcriptase to copy its single strand of RNA into a mirror-image-like strand of complementary DNA, often abbreviated as **cDNA.** The reverse transcriptase, which can also function as a DNA-dependent DNA polymerase, then makes a second strand of DNA complementary to this first cDNA strand (and equivalent in sequence to the original RNA template). Finally, this double-stranded DNA copy of the retroviral RNA chromosome integrates into the host cell's genome. Although the designation cDNA originally meant a single strand of DNA complementary to an RNA molecule, it now refers to any DNA—single- or double-stranded—derived from an RNA template. In 1975, researchers began to use purified reverse transcriptase to convert mRNA sequences from any tissue in any organism into cDNA sequences for cloning.

Suppose you were interested in studying the structure of a mutant β-globin protein. You have already analyzed hemoglobin obtained from a patient carrying this mutation and found that the alteration affects the amino acid structure of the protein itself and not its regulation, so you now need only look at the sequence of the mutant gene's coding region to understand the primary genetic defect. To establish a library enriched for the mutant gene sequence and lacking all the extraneous information from genomic introns and nontranscribed regions, you would first obtain mRNA from the cytoplasm of the patient's red blood cell precursors (**Fig. 9.10a**). About 80% of the total mRNA in these red blood cells is from the α- and β-hemoglobin genes, so the mRNA preparation contains a much higher proportion of the sequence corresponding to the β-globin gene than do the genomic sequences found in a cell's nuclear DNA.

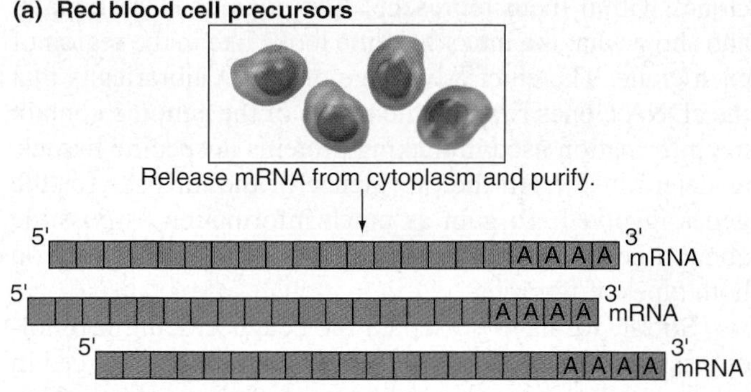

(b) Add oligo(dT) primer. Treat with reverse transcriptase in presence of dATP, dCTP, dGTP, and dTTP.

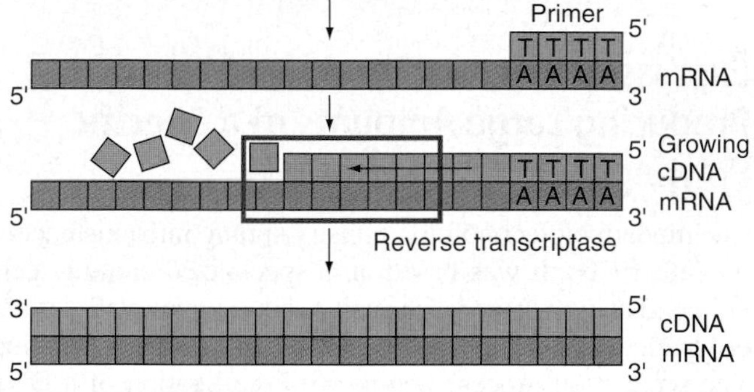

(c) Denature cDNA-mRNA hybrids and digest mRNA with RNase. 3' end of cDNA folds back on itself and acts as primer.

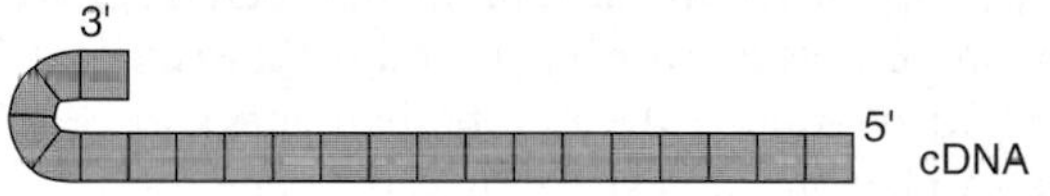

(d) The first cDNA strand acts as a template for synthesis of the second cDNA strand in the presence of the four deoxynucleotides and DNA polymerase

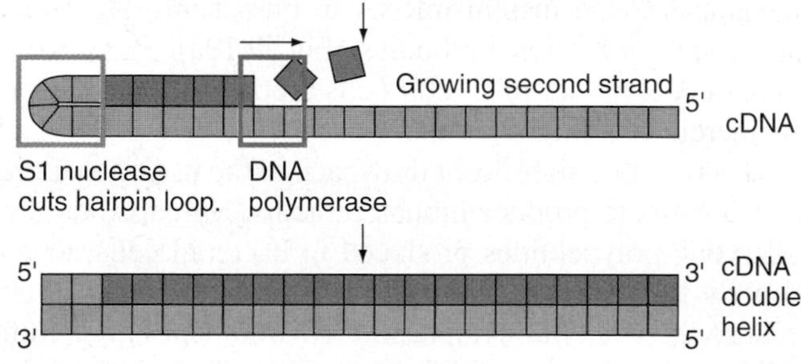

(e) Insert cDNA into vector.

Figure 9.10 Converting RNA transcripts to cDNA. **(a)** Obtain mRNA from red blood cell precursors (which still contain a nucleus). **(b)** Add oligo(dT) primers to initiate reverse transcription of cDNA from the mRNA template with deoxyribonucleotide triphosphates. The result is a hybrid cDNA-mRNA molecule. **(c)** Heat the mixture to separate mRNA and cDNA strands, and then eliminate the mRNA transcript through digestion with RNase. The 3′ end of the cDNA strands will loop around and bind by chance to complementary nucleotides within the same strand, forming the primer necessary to initiate DNA polymerization. **(d)** Add DNA polymerase and the four nucleotides to produce a second cDNA strand complementary to the first. After the reaction is completed, the enzyme S1 nuclease is used to cleave the "hairpin loop" at one end. **(e)** Insert the newly created double-stranded DNA molecule into a vector for cloning.

The addition of reverse transcriptase to the total mRNA preparation—as well as ample amounts of the four deoxyribonucleotide triphosphates, (often abbreviated as "deoxynucleotide triphosphates" or simply "dNPT's") and primers to initiate synthesis—will generate single-stranded cDNA bound to the mRNA template (Fig. 9.10b). The primers used in this reaction would be oligo(dT)—single-stranded fragments of DNA containing about 20 T's in a row—that can bind through hybridization to the poly-A tail at the 3′ end of eukaryotic mRNAs and initiate polymerization of the first cDNA strand. Upon exposure to high temperature, the mRNA-cDNA hybrids separate, or denature, into single strands. The addition of an RNase enzyme that digests the original RNA strands leaves intact single strands of cDNA (Fig. 9.10c). Most of these fold back on themselves at their 3′ end to form transient hairpin loops via base pairing with random complementary nucleotides in nearby sequences in the same strand. These hairpin loops will serve as primers for synthesis of the second DNA strand. Now the addition of DNA polymerase, in the presence of the requisite deoxyribonucleotide triphosphates, initiates the production of a second cDNA strand from the just synthesized single-stranded cDNA template (Fig. 9.10d).

After using restriction enzymes and ligase to insert the double-stranded cDNA into a suitable vector (Fig. 9.10e) and then transforming the vector-insert recombinants into appropriate host cells, you would have a library of double-stranded cDNA fragments, with the cDNA fragment in each individual clone corresponding to an mRNA molecule in the red blood cells that served as your sample. In contrast to a genomic library, this cDNA library does not contain all the DNA sequences in the genome. Instead, it includes only the exons from that part of the genome that the red blood precursors were actively transcribing for translation into protein. Since fully spliced RNA transcripts are on average a few kilobases in length, the cDNA fragments constructed in their image fall within that size range as well. The percentage of clones in the library with cDNA specific for the β-globin protein will reflect the fraction of red blood cell mRNAs that are β-globin messages. For genes expressed infrequently or in very few tissues, you will have to screen many clones of a cDNA library to find the gene of interest. For highly expressed genes, such as the β-globin gene, you will have to screen only a few clones in a red blood cell precursor library.

Genomic Versus cDNA Libraries

Figure 9.11 compares genomic and cDNA libraries. The main advantage of genomic libraries is that the genomic clones within them represent all regions of DNA equally and show what the intact genome looks like in the region of each clone. The chief advantage of cDNA libraries is that the cDNA clones reveal which parts of the genome contain the information used in making proteins in specific tissues, as determined from the prevalence of the mRNAs for the genes involved. To gain as much information as possible about a gene's structure and function, researchers rely on both types of libraries.

So far, we have presented molecular cloning technology as a powerful set of tools used by scientists engaged in basic research to analyze genes and their products. This powerful technology also makes it possible to go beyond analysis to the production of new medicines and other products of commercial value.

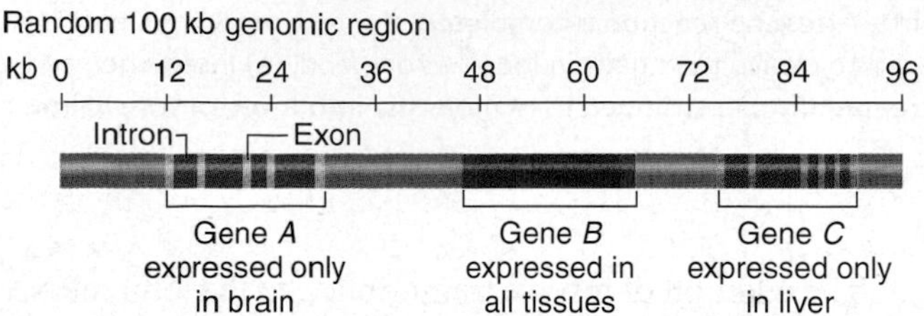

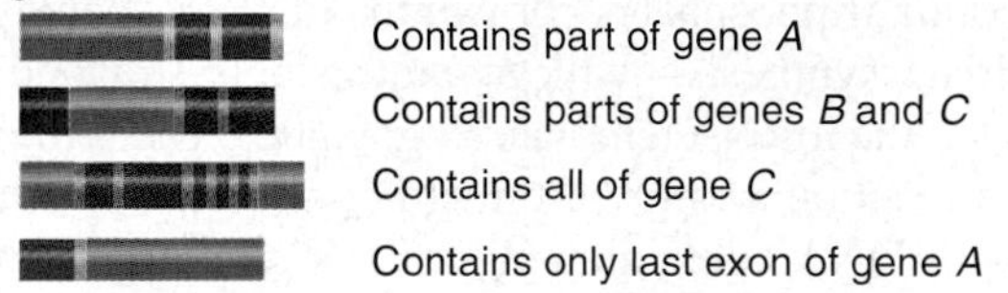

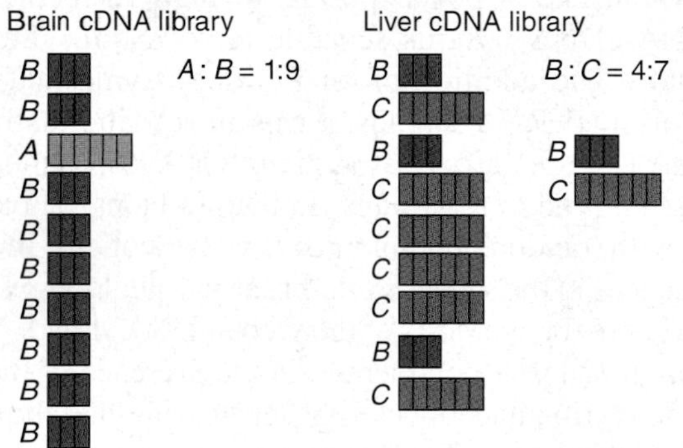

Figure 9.11 A comparison of genomic and cDNA libraries. Every tissue in a multicellular organism can generate the same genomic library, and the DNA fragments in that library collectively carry all the DNA of the genome. Individual genomic fragments may contain no gene, one gene, or more than one gene, or any part of a gene. On average, the clones of a genomic library represent every locus an equal number of times. By contrast, every tissue in a multicellular organism generates a different cDNA library. Clones of a cDNA library represent only the fraction of the genome that is being actively transcribed in that tissue. The frequency with which particular fragments appear in a cDNA library is proportional to the level of the corresponding mRNA in that tissue. Thus, the differential expression of genes within different tissues can be ascertained by the numbers of clones of each type that show up in each library.

Expression Vectors Provide a Means for Producing Large Amounts of a Specific Polypeptide

The modern biotechnology industry sprang into existence in the late 1970s. It was based on a specialized cloning technology that allows the large-scale production of well-defined polypeptides with important medical or commercial value. The production process begins with the ligation of a DNA fragment containing just the coding region of the desired polypeptide to a specialized **expression vector** that contains a *promoter* and other regulatory sequences recognized by the transcriptional apparatus of the host cell (**Fig. 9.12**). The first expression vectors were plasmids that contained an *E. coli* promoter. Among the first human genes to be cloned into a plasmid expression vector was the gene for insulin. In bacterial cells transformed with an expression vector-insert construct, *E. coli* RNA polymerase recognizes the promoter in the vector DNA and proceeds to transcribe the insulin-coding region contained in the insert. The *E. coli* ribosomes then translate the insulin mRNA to human insulin, which can be detected with antibodies (Fig. 9.12a). Purification from a large culture of *E. coli* cells prepares the insulin for commercial use (Fig. 9.12b).

There is one significant drawback to the use of bacterial cells as hosts to produce human proteins. Human and other eukaryotic polypeptides produced in bacterial cells do not properly fold into a functional three-dimensional structure, because the bacterial cytoplasmic environment is not compatible with eukaryotic folding processes. To overcome this problem, scientists developed specialized expression vectors that can both replicate and transcribe insert sequences inside the eukaryotic cells of yeast. Sometimes though, even the yeast cells do not achieve proper polypeptide folding. For these stubborn cases, scientists developed expression vectors that are specific for cultured mammalian cells. There are trade-offs as biotechnologists move from bacterial cells to yeast cells to mammalian cells as factories for producing human proteins. With bacteria, it is possible to obtain large

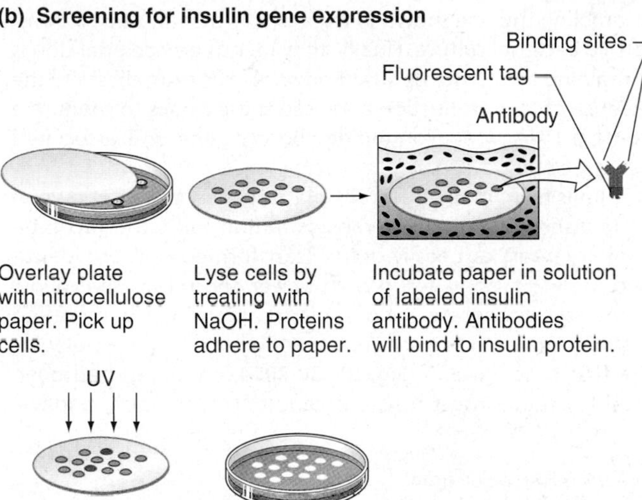
(a) An expression vector allows production of specific polypeptide
Insulin mRNA
E. coli expression vector
Insulin cDNA
+
E. coli promoter
Transform E. coli
Insulin protein
RNA polymerase
Insulin mRNA
Host chromosome
Bacterial cell
(b) Screening for insulin gene expression
Binding sites
Fluorescent tag
Antibody
Overlay plate with nitrocellulose paper. Pick up cells.
Lyse cells by treating with NaOH. Proteins adhere to paper.
Incubate paper in solution of labeled insulin antibody. Antibodies will bind to insulin protein.
UV
Wash filter. Under UV light, identify fluorescent spots. Compare with original plate in order to find bacterial clones containing human insulin gene.

Figure 9.12 An expression vector can be used to produce a desired polypeptide. (a) A cDNA fragment containing just the coding region of insulin is cloned downstream of an *E. coli* promoter in a plasmid vector. The recombinant DNA molecule is transformed into *E. coli* cells, where it is transcribed into insulin mRNA that is translated into insulin protein. **(b)** Screening an expression library with an antibody probe. Clones of an expression library are partially transferred to a nitrocellulose paper disk. The disk is exposed to an NaOH solution that lyses the cells and releases the proteins onto the nitrocellulose paper. The disk is then exposed to a solution containing a fluorescently labeled antibody to insulin. Antibody molecules bind to insulin protein on the disk. Exposure to UV light allows the identification of colonies expressing the gene product (here, insulin) recognized by the antibody.

quantities of proteins most efficiently and cheaply, but in many cases, the proteins are not fully functional. Proteins made in mammalian cells are almost always fully functional, but the process is much more expensive and has a lower yield. Yeast is in the middle on both accounts. Biotechnologists must balance efficiency and functionality in their choice of an appropriate host cell. The Genetics and Society box entitled "The Use of Recombinant DNA Technology to Produce Pest-Resistant Crops" (on pp. 320–321) discusses a successful application of biotechnology to agriculture.

It is possible to produce an entire cDNA library in an expression vector. Such a library is called an *expression library.* Most of the clones in an expression library will be nonproductive, either because the insert contains a coding sequence in the "wrong" direction relative to the promoter or because the cDNA is an incomplete copy of the coding sequence. Nevertheless, if the library is produced in an appropriate manner, a proportion of the clones will produce functional protein. As a result, instead of using a DNA probe to locate a specific cDNA fragment, you can use a fluorescently labeled antibody that binds to the protein product of the gene you wish to clone. To find the gene sequence producing the protein, you simply make a replica of the host cell colonies, lyse these replicas, and expose their contents to the protein-specific antibody labeled with fluorescence. The antibody, by interacting with the protein, will identify the cell colony containing the cDNA sequence responsible for the protein (Fig. 9.12b).

By 2004, the Food and Drug Administration had approved the pharmaceutical use of 187 different proteins produced from recombinant expression vectors, and over 300 million patients had benefited directly from recombinant DNA technology. An additional 370 recombinant medicines and vaccines are currently undergoing clinical studies.

9.3 Hybridization Is Used to Identify Similar DNA Sequences

While you can "screen" an expression library with an antibody that recognizes the protein product of the gene that you want to clone, the DNA inserts in all other types of libraries cannot be detected in such a straightforward fashion. So, once you have collected the hundreds of thousands of human DNA fragments in a genomic or a nonexpressing cDNA library, how do you find the gene you wish to study, the proverbial "needle in the haystack"? For example, how would you go about finding a genomic clone that contained just a portion of the β- globin gene? One way is to take advantage of hybridization— the natural propensity of complementary single-stranded molecules

The Use of Recombinant DNA Technology to Produce Pest-Resistant Crops

The U.S. Department of Agriculture estimates that caterpillar pests such as the European corn borer, corn rootworm, and cotton bollworm are responsible for $1 billion dollars in lost revenue each year in the United States alone. Farmers can spray their crops with pesticides, but the process is costly, labor intensive, not completely effective, and in some cases, harmful to workers in the field.

Although conventional farmers use a variety of laboratory-created pesticides, organic farmers choose to use pesticides that only exist, or are produced, naturally. For the last 30 years, they have taken advantage of the protein-based mechanism evolved by the common soil bacterium *Bacillus thuringiensis kurstaki* (abbreviated *Bt*) to protect itself from being eaten by the same caterpillars that cause so many problems for farmers. About a dozen genes in the microbe code for crystalline (CRY) polypeptides that function as specialized endotoxins. When a caterpillar ingests the bacteria, the CRY proteins bind to specific intestinal membrane sites and disrupt digestion, leading to the insect's rapid death. CRY binding is highly specific for proteins found only in the larvae of moths and butterflies and not in any vertebrate species. Even high dosages of CRY proteins have no toxic or allergenic effects on birds, mammals, reptiles, or amphibians. Organic farmers depend primarily on *Bt* sprays or powder to provide their crops with protection when insect infestations occur.

In the mid-1980s, agricultural molecular biologists realized they could leverage the newly developed tools of recombinant DNA technology to create genetically modified (GM) crops that would be resistant to insect infestation, without the need for pesticide applications. Based on the extensive safety record associated with whole organism *Bt* use and a detailed understanding of the biochemical mechanism of CRY pesticidal action, researchers developed a strategy for creating plants that expressed a *cry* gene within their own cells as follows. They cloned a *cry* family gene named *cry1Ab* into a plasmid vector, cut out the insert with a restriction enzyme, and purified the insert (as illustrated in Fig. 9.9). Next, they ligated a restriction fragment containing a plant gene intron (required to stabilize RNA transcripts) to the 5′ end of the coding region. At the 5′ end of this joined molecule, they added another restriction fragment containing a promoter from a plant virus; at the 3′ end of the construct, they attached a special plant transcription termination signal sequence. They then inserted this four-part *cry*-gene construct into a bacterial plasmid vector resembling the one shown in Fig. 9.8, which was used to transform a bacterial culture. Finally, they identified bacterial clones containing the construct based on antibiotic resistance and the absence of *lacZ* production and used these clones to produce a purified DNA insert containing the *cry* gene and associated genetic elements (**Fig. A.**)

Cells from many different plant species can be grown in petri dishes where DNA transformation with the recombinant *cry* insert can easily occur. Transformed cells are identified, isolated, and then grown under conditions that allow them to regenerate whole plants. Genetically modified plants containing the *cry* gene were grown commercially for the first time in 1996 (**Fig. B**). By 2004, *cry* genes had been used to create insect-resistant canola, cotton, corn, papaya,

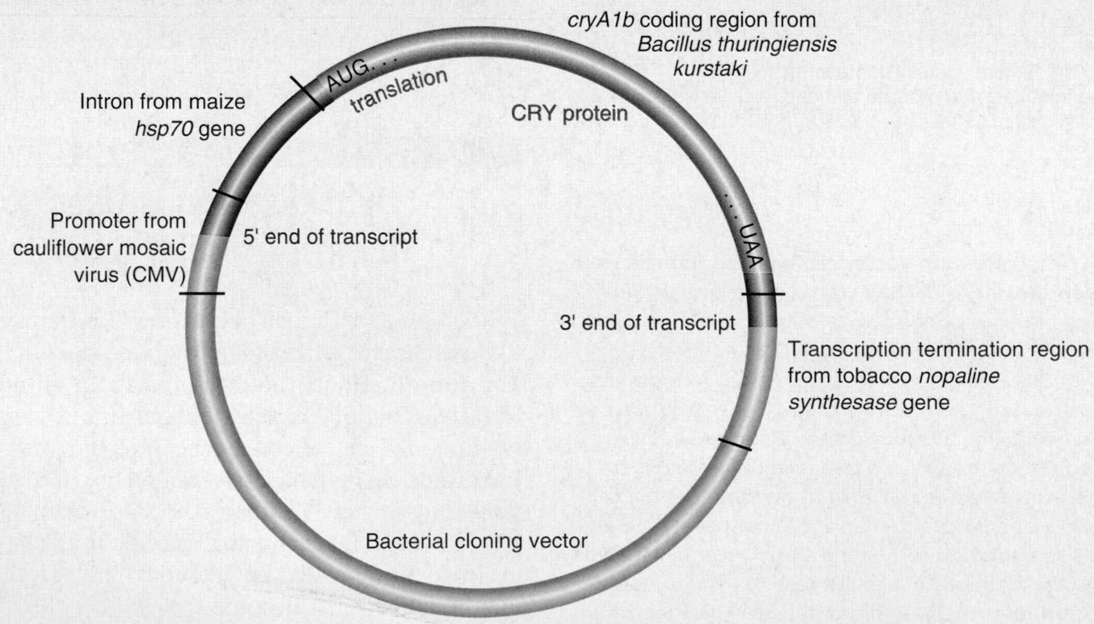

Figure A **DNA construct with recombinant gene that can express CRY protein in plants.**

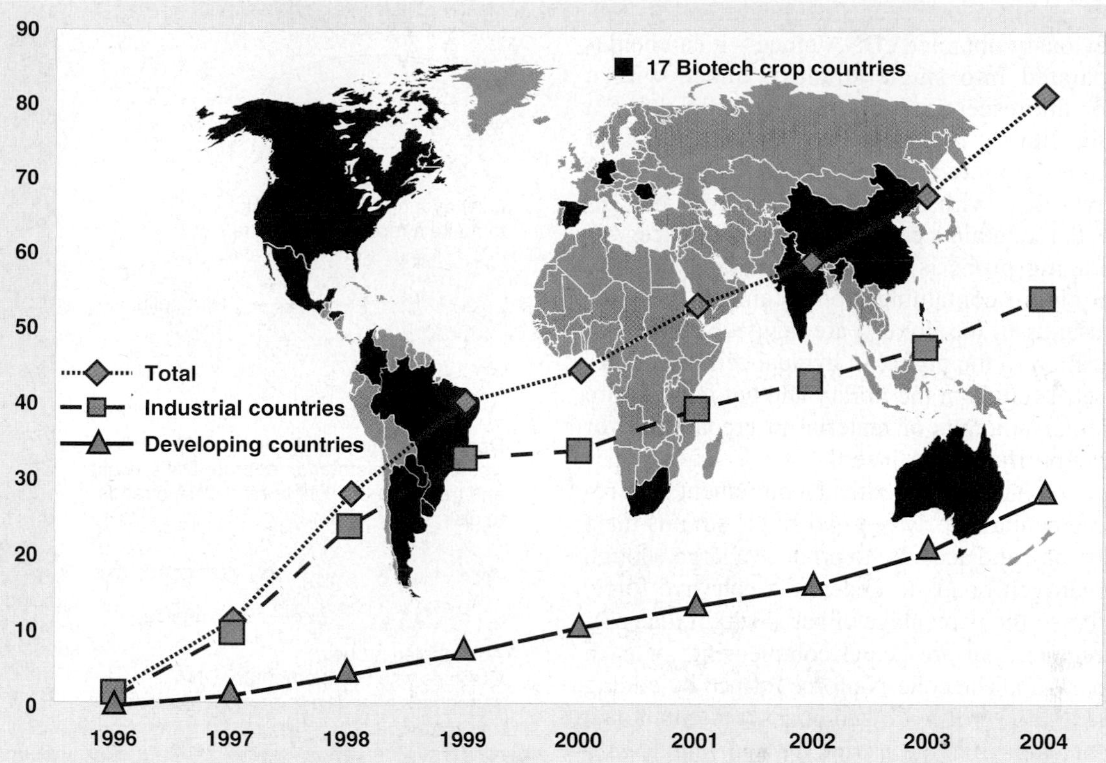

Figure B **Global area of biotech crops (million hectares; 1996–2004).** Increase of 20% 13.3 million hectares, or 32.9 million acres, between 2003 and 2004.

potato, rice, soybean, squash, sugar beet, tomato, and wheat. These and other genetically modified crops were being grown on over 80 million hectares (8 billion acres) of land around the world, in both industrialized and developing countries.

Before genetically modified plants can be grown commercially, they must pass stringent tests for efficacy and safety on a case-by-case basis. The most extensively tested GM food is the corn variety MON810, which carries the *Bt cry1Ab* gene. Standard corn carries about 20,000 genes in its genome. A single *cry1Ab* gene has been added in to the genome of MON810, which means that it represents a 0.00005 fraction of the modified genome. The actual *Bt* protein was detected at a level of three parts per 10 million (0.0000003) in corn and is nonexistent in extracted corn syrup used for soft-drink production. As expected, no difference could be detected between the GM and non-GM variety in amino acids, vitamins, carbohydrates or any other nutritional characteristic. (In contrast, large differences *do* exist between traditional corn varieties bred for different purposes, such as pig feed, corn syrup, or direct human consumption.)

GM foods currently on the American market have not been associated with any kind of negative health effect in any person. This doesn't mean that all future GM plants will be without ill effect and risk free, but risk assessment only makes sense in comparison to the substitute foods that people would eat in the absence of a particular GM product. In some situations, the risk exists for genetically engineered traits to migrate unintentionally into wild plants. Indeed, most scientists take this risk more seriously than alleged health risks. With a scientifically informed regulatory process, the risk of significant eco-harm can be assessed up front and included in the decision to implement, redesign, or reject a particular GM technology on a case-by-case basis. For example, corn poses no risk of transmission to wild plants if it is grown anywhere other than Central America. The wild progenitor of corn is a grass species *teosinte* that is confined to Mexico. Corn pollen cannot naturally fertilize any other wild grass in any other part of the world. However, not every crop will be as easily assessed in terms of potential ecosystem effects.

Over the last decade, the power, precision, and breadth of biotechnology applications have grown even faster than its most ardent early supporters imagined possible. As we saw in the Chapter 2 Tools of Genetics box on pp. 28–29, the centuries-old practice of artificial selection is based on breeding together different species and taking advantage of random genetic combinations and mutations to develop organisms with enhanced value to humans. In contrast, modern biotechnologists begin with a deep scientific understanding of genes and cells and then proceed to a targeted modification of one predetermined biological property at a time. As shown in Fig. B, the global area devoted to planting GM crops continues to increase at a rapid pace in both industrialized and developing countries.

portion of a β-globin sequence is already available—perhaps as a previously obtained cDNA clone—it can be denatured (separated into single strands), linked with a radioactive or fluorescent tag, and then used to probe a whole genome library that is spread out as a series of colonies on one or more petri plates. The tagged DNA probe will hybridize with denatured DNA from the genomic clones that contain a complementary sequence. After nonhybridizing probe is washed away, only the tiny number of β-globin-containing clones (among the hundreds of thousands in the library) are tagged by virtue of their hybridization to the probe. Individual cellular clones can then be retrieved from the library and put into culture to produce larger amounts of material in preparation for recovery of the purified DNA insert.

Hybridization has a single critical requirement: The region of complementarity between two single strands must be sufficiently long and accurate to produce a large enough number of hydrogen bonds to generate a cohesive force. Accuracy refers to the percentage of bases within the complementary regions that are actual complements of each other (C–G or A–T). The cohesive force formed by adding together large numbers of hydrogen bonds will counteract the thermal forces that tend to disrupt the double helix. Individual hydrogen bonds are very weak, and the two (A:T) or three (G:C) hydrogen bonds that form between individual pairs of bases are far too weak to prevent thermal forces from pulling them apart. But if two single strands form hydrogen bonds between 15 or more contiguous base pairs, the combined force is sufficient to counteract the thermal forces that operate even at room temperature. Hybridization can occur between any two single strands of nucleic acid: DNA/DNA, DNA/RNA, or RNA/RNA. In this section, we focus on DNA/DNA hybridization.

We now describe how to screen a library with a DNA probe.

Preparing the Library

Before you can use a DNA probe to screen a library, you must distribute the library's clones on petri plates and place a piece of nitrocellulose paper over each plate, allowing some, but not all, of the cells from each colony to transfer to the nitrocellulose disk replica (**Fig. 9.13**). You now lyse the bacteria on the disk with an alkaline solution, which both releases the DNA from the cells and denatures it into single strands. Baking the disk and treating it with ultraviolet (UV) light will attach the single-stranded DNA molecules from the cells of each colony to the filter at the confined position corresponding to that clone's location in the petri plate. Some of the single-stranded DNA molecules attached to the nitrocellulose disk will be from the recombinant plasmids that replicated in the cells making up the colony. The library's DNA is now ready for screening with DNA probes.

Master plate containing genomic library of mouse clones.

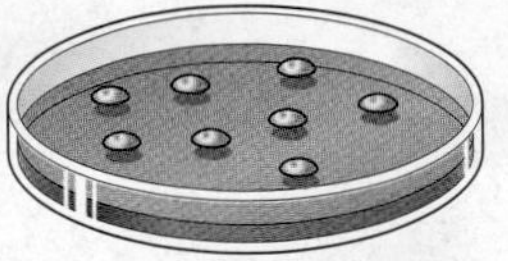

Overlay a nitrocellulose disk to make a replica of the plate.

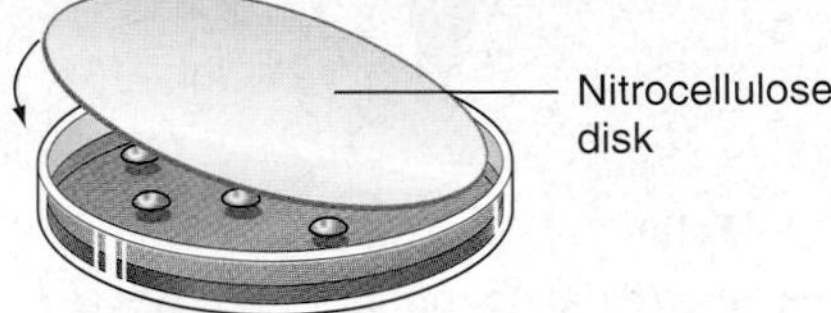

Nitrocellulose disk
Remove disk from plate and use NaOH to lyse cells on plate and denature DNA. Bake and treat with UV light to bind DNA strands to disk.

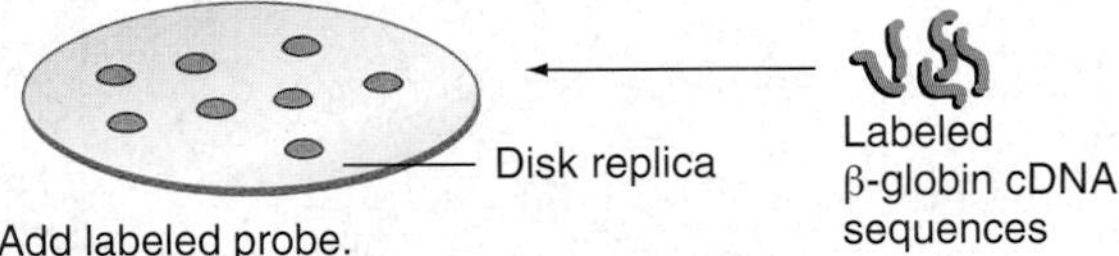

Disk replica
Labeled β-globin cDNA sequences
Add labeled probe.
Colonies with complementary DNA sequences hybridize to probe and restrain it.

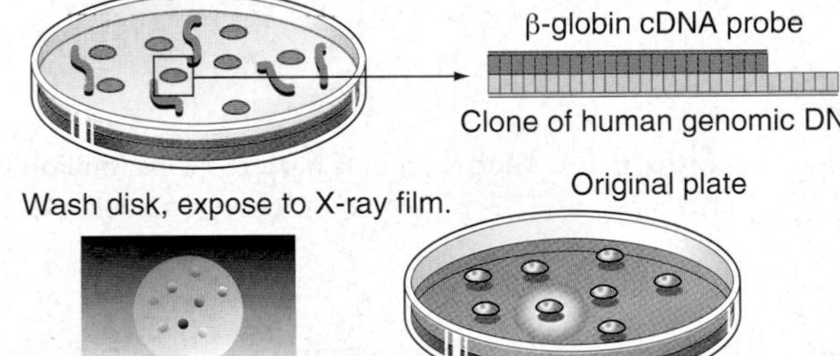

β-globin cDNA probe
Clone of human genomic DNA
Wash disk, expose to X-ray film.
Original plate
Compare with original plate to locate bacterial clone with desired genomic fragment.

Figure 9.13 Screening a library of clones by hybridization to a labeled probe. In the example illustrated here, a human β-globin cDNA clone is used as a probe to identify a clone from a human genomic library containing the β-globin gene. To determine which clone in a library has a DNA fragment of interest, you place a wet nitrocellulose paper disk onto the plate containing cellular clones and then remove it. This procedure allows most cells within the original colonies on the plate to remain in their original positions but transfers a small portion of each colony to the disk. Treating the disk with NaOH causes the cells to lyse and release their DNA; it also causes the DNA to denature. The denatured DNA is then attached to the disk through baking and exposure to UV light. Next, a labeled DNA probe in solution is applied directly to the disk; the solution is left on the disk long enough to promote hybridization between the probe and any complementary strands of DNA that may be present within one or more spots that correspond to one or more of the clones. The nonhybridizing probe is then washed off the disk. If the probe is labeled with a radioactive isotope, a piece of X-ray film is placed up against the disk and exposed to record the location of spots that are radioactive because of hybridization to the probe. This process is referred to as *autoradiography.* By comparing the location of spots on the film with the original bacterial plate, it is possible to identify the clone or clones that contain DNA fragments of interest.

(a) A DNA synthesizer

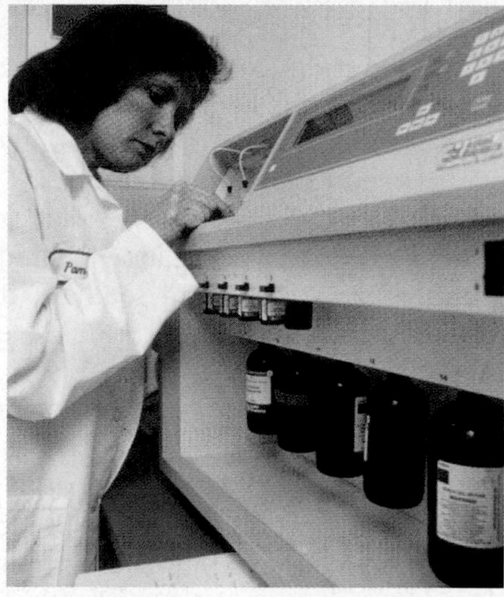

(b) Synthesizing DNA probes based on reverse translation

Protein sequence	Glu	Asp	Met	Trp	Tyr
Degenerate coding sequences	GAA GAG	GAT GAC	ATG	TGG	TAT TAC

Sequences that must be present in the probe

GAAGATATGTGGTAT
GAGGATATGTGGTAT
GAAGACATGTGGTAT
GAGGACATGTGGTAT
GAAGATATGTGGTAC
GAGGATATGTGGTAC
GAAGACATGTGGTAC
GAGGACATGTGGTAC

Figure 9.14 How to make oligonucleotide probes for screening a library. (a) A DNA synthesizer is a machine that automates the addition through chemical reactions of specified nucleotides to the growing DNA chains, known as oligonucleotides. The four small bottles at the left in the middle of the instrument contain solutions of each of the four different deoxynucleotide triphosphates (A, T, C, and G), and the fifth small bottle contains a mixture of the four deoxynucleotide triphosphates for situations where an investigator wants to create an oligonucleotide with a random base in that position. The larger bottles below contain reagents used in the various chemical reactions that the synthesizer carries out. **(b)** Reverse translation. An amino acid sequence can be "reverse translated" into a degenerate DNA sequence, which can be programmed into a DNA synthesizer to create a set of oligonucleotides that must include the one present in the actual genomic DNA.

Constructing the DNA Probes

DNA probes are purified fragments of single-stranded DNA 25 to several thousand nucleotides in length that are subsequently labeled with a radioactive isotope (typically ^{32}P) or a fluorescent dye. DNA probes can be produced from previously cloned fragments of DNA, from purified fragments of DNA amplified by PCR (described in the next section), or from short single strands of chemically synthesized DNA.

In chemical synthesis, an automated DNA synthesizer adds specified nucleotides, one at a time, through chemical reactions, to a growing DNA strand (**Fig. 9.14a**). Modern synthesizers can produce specific sequences up to 100 nucleotides in length. An investigator can instruct the computer controlling the DNA synthesizer to construct a particular sequence of A's, T's, C's, and G's. Within a few hours, the machine produces the desired short DNA chains, which are known as **oligonucleotides.**

To produce labeled probes during chemical synthesis, scientists provide the synthesizer with nucleotides that are prelabeled. To fabricate labeled probes from previously purifed DNA fragments or clones, they use DNA polymerase in combination with prelabeled nucleotides.

With the availability of oligonucleotide synthesis, it is possible to generate probes indirectly from a polypeptide sequence whose corresponding gene coding sequence is unknown, rather than directly from a known DNA sequence. The process by which one turns a polypeptide sequence into a set of corresponding DNA sequences is known as *reverse translation* (Fig. 9.14b). To perform a reverse translation, an investigator first translates the amino acid sequence of a protein into a DNA sequence via the genetic code dictionary. Recall, however, that the genetic code is "degenerate," with more than one codon for many individual amino acids. Without knowing the coding DNA sequence, it is impossible to predict which of several codons is actually present in the genome. To simplify the task, investigators choose peptide sequences containing amino acids encoded by as few potential codons as possible. They must then synthesize a mixture of oligonucleotides containing all possible combinations of codons for each amino acid. This is no problem for an automated DNA synthesizer: An investigator can direct the machine to add in a defined mixture of nucleotides (A and G, for example) at each ambiguous position in the oligonucleotide. With this indirect method of obtaining a DNA probe, researchers can locate and clone genes even if they have only partial coding information based on the proteins the genes encode.

Screening the Library

To find the hemoglobin genes in a genomic library that has been replicated onto nitrocellulose paper disks, you simply expose the library's DNA to a solution containing a β-globin cDNA probe labeled with radioactivity or fluorescence (see Fig. 9.13). After a sufficient period of incubation with radioactively labeled probe, you wash away all probe molecules that have not hybridized to complementary DNA strands stuck to the disk, place a large piece of X-ray film in direct contact with the paper disk to record the radioactive emissions from the hybridizing probe, and then develop the X-ray film to see the dark spots representing the sites of hybridization. An X-ray film that has been processed in this way to identify the locations of hybridizing radioactive probes is called an **autoradiograph,** and the process of creating an autoradiograph is called **autoradiography.** If instead you use fluorescence-labeled probes, you can see the hybridizing clones directly by shining ultraviolet light on the disks. The spots revealed by hybridization correspond to genomic clones containing sequences complementary to the β-globin probe. Knowing the position on the disk that gave rise to these spots, you can return to the original master petri plates and pick out the corresponding colonies containing the cloned inserts of interest. You can then harvest and purify the cloned DNA identified by the globin probes and proceed with further analysis. The DNA clones recovered with this screening procedure would lead to the discovery that the haploid human genome carries one copy of the β-globin gene that has three exons and two introns.

Of great importance to geneticists is the fact that hybridization can occur between single strands that are not completely complementary, including related sequences from different species. In general, two single DNA strands that are longer than 50–100 bp will hybridize so long as the extent of their complementarity is more than 80%, even though mismatches may appear throughout the resulting hybrid molecule. Imperfect hybrids are less stable than perfect ones, but geneticists can exploit this difference in stability to evaluate the similarity between molecules from two different sources. Hybridization, for example, occurs between the mouse and human genes for the cystic fibrosis protein. Researchers can thus use the human genes to identify and isolate the corresponding mouse sequences and then use these sequences to develop a mouse containing a defective cystic fibrosis gene. Such a mouse provides a model for cystic fibrosis in a species that, unlike humans, is amenable to experimental analysis.

Gel Electrophoresis Combined with Hybridization Provides a Tool for Mapping DNA Fragments

We have just seen how researchers use hybridization to screen a library of thousands of clones for particular ones carrying DNA sequences complementary to specific probes. Hybridization with a cloned probe can also provide information about similar DNA regions in a whole-genome sample. The protocol for accomplishing this task combines gel electrophoresis (review discussion on pp. 307–309) with the hybridization of DNA probes to DNA targets immobilized on nitrocellulose paper.

Suppose you had a clone of a gene called *H2K* from the mouse major histocompatibility complex (also known as the MHC). The *H2K* gene plays a critical role in the body's ability to mount an immune response to foreign cells. You want to know whether there are other genes in the mouse genome that are similar to *H2K* and, by extrapolation, also play a role in the immune response. To get an estimate of the number of *H2K*-like genes that exist in the genome, you could turn to a hybridization technique called the **Southern blot,** named for Edward Southern, the British scientist who developed it. **Figure 9.15** illustrates the details of the technique.

Southern blotting can identify individual *H2K*-like DNA sequences within the uncloned expanse of DNA present in a mammalian genome. Cutting the total genomic DNA with *Eco*RI produces about 700,000 different fragments. When you separate these fragments by gel electrophoresis and stain them with ethidium bromide, all you see is a smear, because it is impossible to distinguish 700,000 fragments spread over a distance of some 10 cm (Fig. 9.15). But you can blot the smear of fragments to a nitrocellulose filter paper and probe the resulting Southern blot with a labeled *H2K* clone, which picks out the bands containing the *H2K*-like gene sequences. The result shown in Fig. 9.15 is a pattern of approximately two dozen fragments that constitute a series of related MHC genes within the mouse genome. The Southern blot thus makes it possible to start with a very complex mixture and identify the small number of fragments among hundreds of thousands within a whole genome that are related to your original clone.

Southern blotting can also determine the location of one cloned sequence (such as a 1.4 kb human β-globin cDNA sequence from a plasmid vector) within a larger cloned sequence (such as 30 kb cosmid containing the β-globin locus and surrounding genome). Suppose you want to use these two clones to discover the location of the gene within the cosmid as well as to learn which parts of the full gene are exons and which parts are introns. To answer both of these questions in a very straightforward fashion you would turn to Southern blotting. First you would use gel electrophoresis of restriction-enzyme digested cosmid DNA, and staining with ethidium bromide, to construct a restriction map of the cosmid. Next, you would transfer the restriction fragments from the gel onto a filter paper and probe the filter by hybridization to the labeled β-globin cDNA clone. An autoradiograph will then show the precise cosmid restriction fragments that carry coding regions of the β-globin gene. With very high resolution restriction maps of cosmid subclones and high-resolution gel electrophoresis of small restriction fragments (produced with four-base-cutter restriction enzymes), it would be possible to distinguish restriction fragments containing exons from the other fragments that contain the introns of the β-globin gene or a flanking sequence.

Feature Figure 9.15

Southern Blot Analysis

Genomic DNA was purified from the tissues of seven mice, and each sample was subjected to digestion with the restriction enzymes *Eco*RI. Digested samples were separated by electrophesis in an agarose gel, as illustrated in Fig. 9.5.

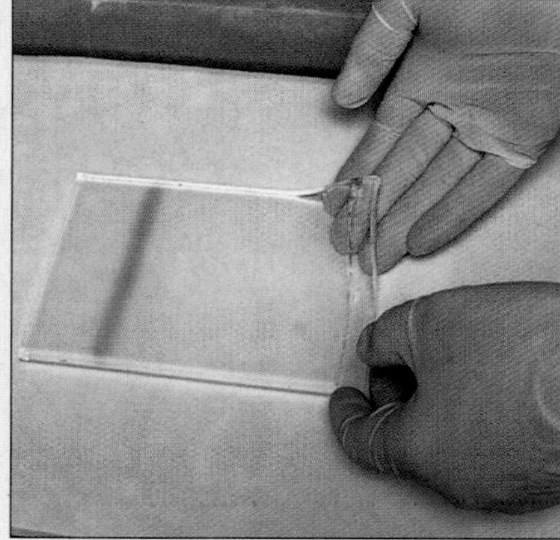

Stain with ethidium bromide to visualize total genomic DNA under UV illumination.

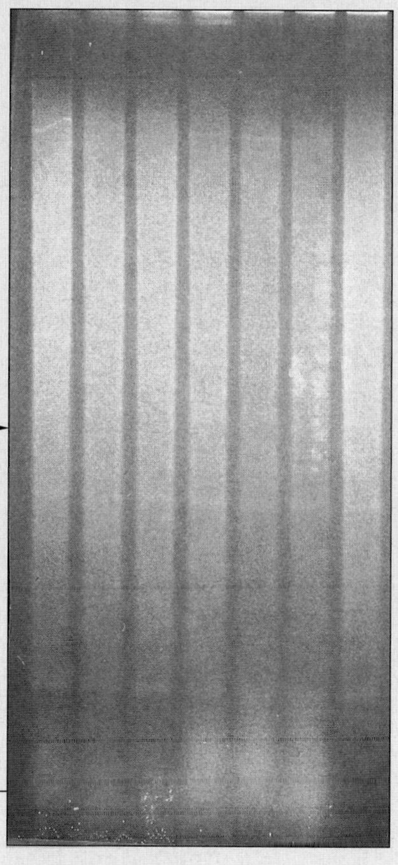

Next you place the gel in a strongly alkaline solution that causes the DNA to denature (that is, to separate into single stands) and then in a neutralizing solution. You now cover the gel containing the separated DNA restriction fragments with a piece of nitrocellulose filter paper. On top of the filter paper, you place a stack of paper towels, and beneath the gel, a sponge saturated with buffer. Within this setup, the dry paper towels act as a blotter, pulling liquid from the buffer-saturated sponge, through the gel, and into the towels themselves. As the liquid is drawn out of the gel, so too is the DNA. But the large DNA molecules do not pass through the filter into the paper towels. Instead, they become trapped in the nitrocellulose at points directly above their locations in the gel, forming a Southern blot, which consists of the nitrocellulose filter containing DNA fragments in a pattern that is a replica of their migration pattern in the gel.

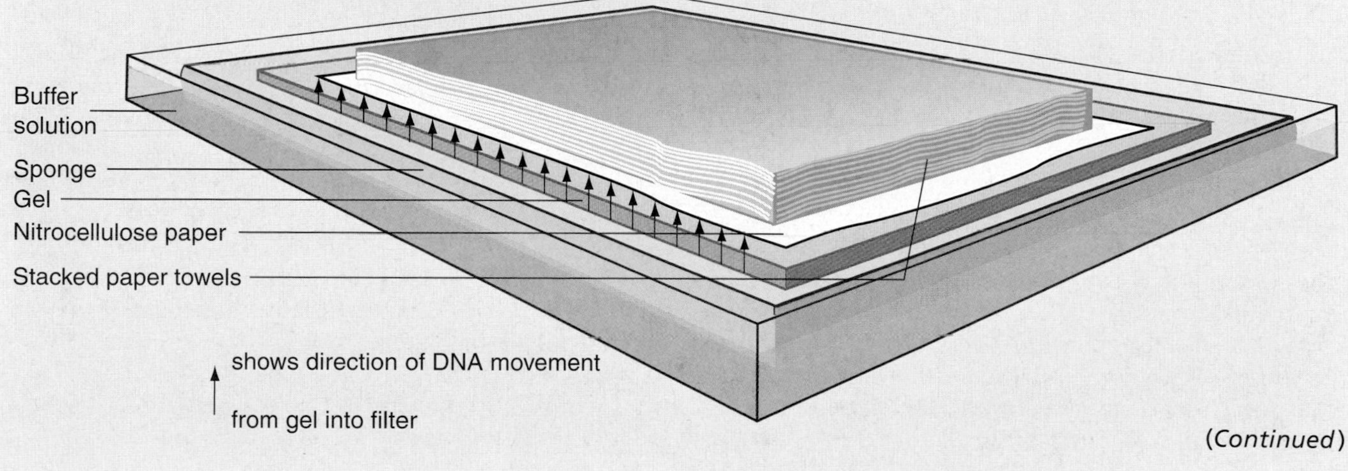

(*Continued*)

Feature Figure 9.15 (*Continued*)

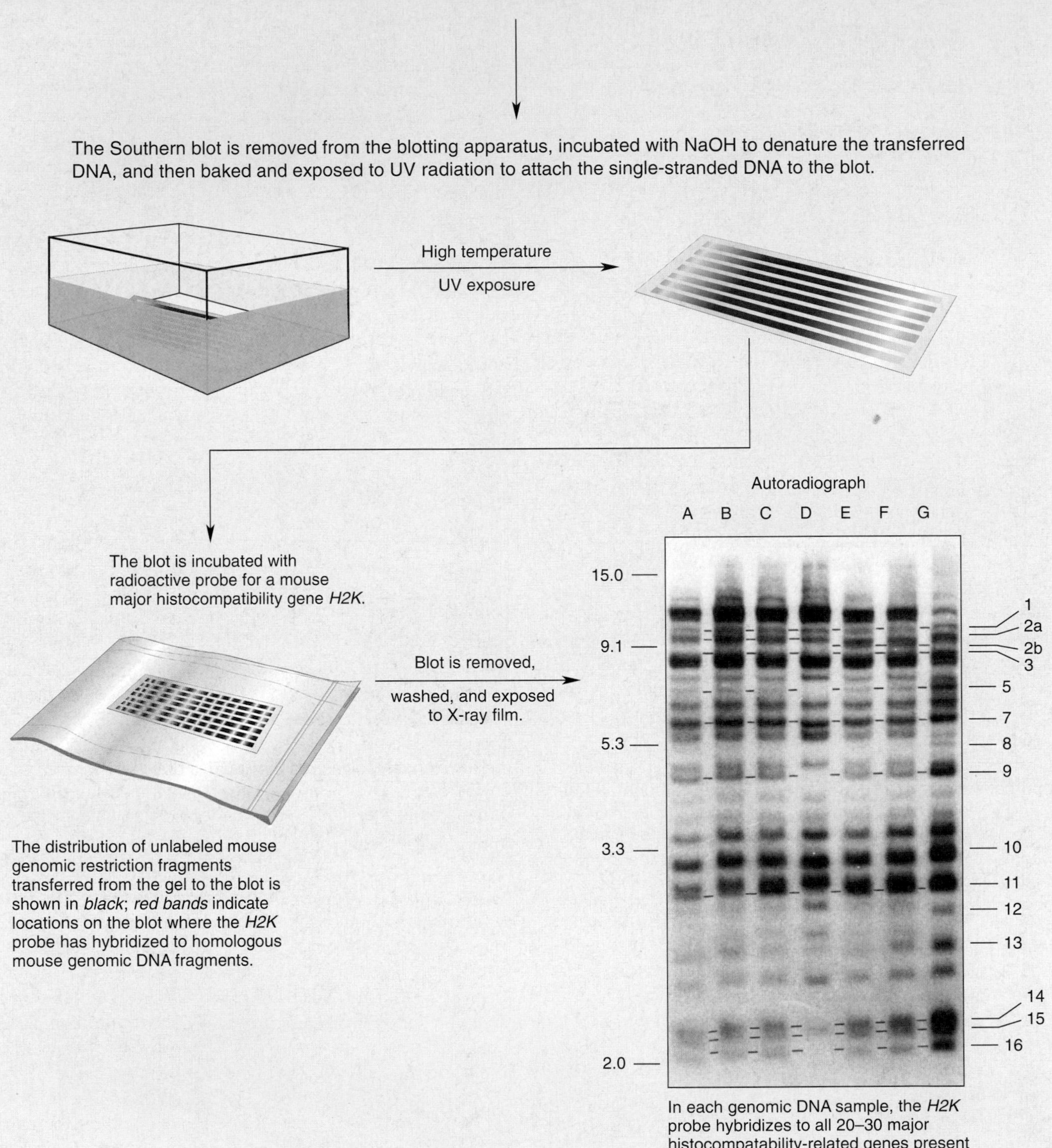

The Southern blot is removed from the blotting apparatus, incubated with NaOH to denature the transferred DNA, and then baked and exposed to UV radiation to attach the single-stranded DNA to the blot.
High temperature
UV exposure
The blot is incubated with radioactive probe for a mouse major histocompatibility gene *H2K*.
Blot is removed, washed, and exposed to X-ray film.
The distribution of unlabeled mouse genomic restriction fragments transferred from the gel to the blot is shown in *black*; *red bands* indicate locations on the blot where the *H2K* probe has hybridized to homologous mouse genomic DNA fragments.
Autoradiograph
A B C D E F G
15.0
9.1
5.3
3.3
2.0
1
2a
2b
3
5
7
8
9
10
11
12
13
14
15
16
In each genomic DNA sample, the *H2K* probe hybridizes to all 20–30 major histocompatability-related genes present within the mouse genome.

9.4 The Polymerase Chain Reaction Provides a Rapid Method for Isolating DNA Fragments

Genes are rare targets in a complex genome: The β-globin gene, for example, spans only about 1400 of the 3,000,000,000 nucleotide pairs in the haploid human genome. Cloning overcomes the problem of studying such rarities by amplifying large amounts of a specific DNA fragment in isolation. But cloning is a tedious, labor-intensive process. Once a sequence is known, or even partially known, molecular biologists now use an alternative method to recover versions of the same sequence from any source material, including related sequences from different species, from different individuals within the same species, and from sequences in RNA transcripts or whole genomes. This alternative method is called the **polymerase chain reaction,** or **PCR.** First developed in 1985, PCR is faster, less expensive, and more flexible in application than cloning. From a complex mixture of DNA—like that present in a person's blood sample—PCR can isolate a purified DNA fragment in just a few hours.

PCR is also extremely efficient. We have seen that in creating a genomic or cDNA library, it is necessary to use a large number of cells from one or more tissues as the source of DNA or mRNA. By contrast, the single copy of a genome present in one sperm cell or the minute amount of severely degraded DNA recovered from the bone marrow of a 30,000-year-old Neanderthal skeleton provides enough material for PCR to make a billion or more copies of a target DNA sequence in an afternoon.

How PCR Achieves the Exponential Accumulation of Target DNA

The polymerase chain reaction is a kind of reiterative loop in which an operation is repeatedly applied to the products of earlier rounds of the same operation. You can liken it to the operation of an imaginary generously paying automatic slot machine. You start the machine by inserting a quarter, at which point the handle cranks, and the machine pays out two quarters; it then reinserts those two coins, cranks, and produces four quarters; reinserts the four, cranks, and spits out eight coins, and so on. By the twenty-second round, this fantasy machine delivers more than 4 million quarters.

The PCR operation brings together and exploits the method of DNA hybridization described earlier in this chapter and the essential features of DNA replication described in Chapter 6. Once a specific genomic region (which may range in size from a few dozen base pairs to 25 kb in length) has been chosen for amplification, an investigator uses knowledge of the sequence (obtained from previous sequencing based on the method described later in this chapter) to synthesize two oligonucleotides that correspond to the two ends of the target region. One oligonucleotide is complementary to one strand of DNA at one end of the region; the other oligonucleotide is complementary to the other strand at the other end of the region. The process of amplification is initiated by the hybridization of these oligonucleotides to denatured DNA molecules within the sample. The oligonucleotides act as primers directing DNA polymerase to create new strands of DNA complementary to both strands between the two primed sections (**Fig. 9.16**).

This initial replication is followed by subsequent rounds in which both the starting DNA and the copies synthesized in previous steps become templates for further replication, resulting in an exponential increase by doubling the number of copies of the replicated region with each step. Figure 9.16 diagrams the steps of the PCR operation, showing how you could use it to obtain many copies of a small portion of the β-globin gene for further study.

PCR Products Can Be Used Just Like Cloned Restriction Fragments

When properly executed, PCR provides all the highly enriched DNA you could want for unambiguous analyses of many types. PCR products can be labeled to produce hybridization probes for detecting related clones within libraries and related sequences within whole genomes spread out on Southern blots. PCR products can be sequenced (as described in the next section) to determine the exact genetic information they contain. Because the products are obtained without cloning, it is possible to amplify and learn the sequence of a specific DNA segment in a very short time. In fact, in checking for a particular hemoglobin mutation, one could start with a blood sample and determine a DNA sequence within two days.

As an analytic tool, PCR has several advantages over cloning. First, it provides the ultimate in sensitivity: The minimum input for the polymerase chain reaction to proceed is a single DNA molecule. Second, as we have seen, it is very fast, requiring no more than a few hours to generate enough amplified DNA for analysis; by comparison, the multiple steps of cloning usually require at least a week to complete. Third, in the sometimes highly competitive research world, it is an agent of democracy: Once the base sequence of the oligonucleotide primers that allow the amplification of a particular target region appears in print, anyone with the relatively small amount of funds needed to synthesize or buy these primers can reproduce the reaction. PCR is nevertheless unsuitable in certain situations. Because the protocol only copies DNA fragments up to 25 kb in length, it cannot amplify larger regions of interest. And because the synthesis of PCR primers depends on sequence information from the vicinity of the target region, the protocol cannot serve as the starting point for the analysis of genes or genomic regions that have not yet been cloned and sequenced.

Feature Figure 9.16

Polymerase Chain Reaction

Suppose you are a physician and wish to understand the molecular details of the β-globin gene mutation that causes the expression of a novel form of anemia in one of your patients. Although the sequence of the normal β-globin gene as well as the sequence of many of its mutant alleles is known, you suspect that your patient may carry a new, as-yet-uncharacterized β-globin allele of unknown sequence. To characterize the potentially novel allele, you could turn to PCR. You begin by preparing a small amount of genomic DNA from skin, blood, semen, or other tissue that is easy to obtain from your patient suffering from the novel anemia. You then synthesize two specific oligonucleotide primers, each a short single-stranded chain of 16–26 nucleotides, whose sequence is dictated by the already known sequence of the wild-type β-globin allele. One of these oligonucleotides (arbitrarily called the "left primer" in the diagram) is equivalent in sequence to a section of DNA along a 5′ strand adjacent to the target region (colored *blue* in the diagram). The second oligonucleotide, the "right primer," is equivalent to a sequence on the opposite adjacent 5′ strand. As you will see, the target DNA amplified by PCR is that stretch of the genome lying between the two primers.

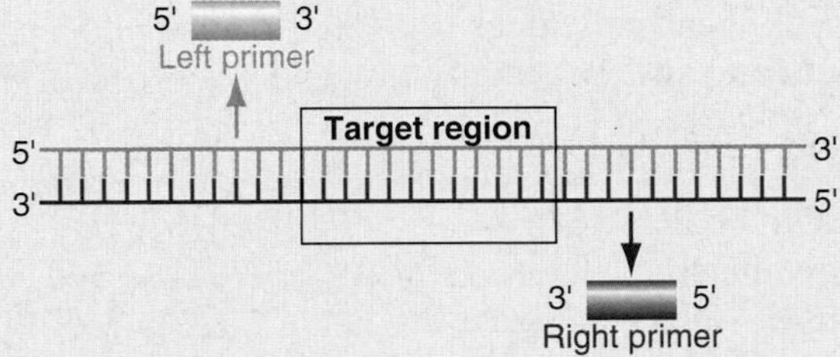

Next you put the patient's genomic DNA in a test tube along with the specially prepared primers, a solution of the four deoxynucleotides, and *Taq* DNA polymerase, a specialized polymerase obtained from *Thermus aquaticus* bacteria living in hot springs. This specialized DNA polymerase remains active at the high temperatures employed during the PCR protocol, which is not true of polymerases from organisms that live under more moderate conditions. Now place the test tube with these components in a machine called a thermal cycler, which repeatedly changes the temperature of incubation according to a preset program with three phases.

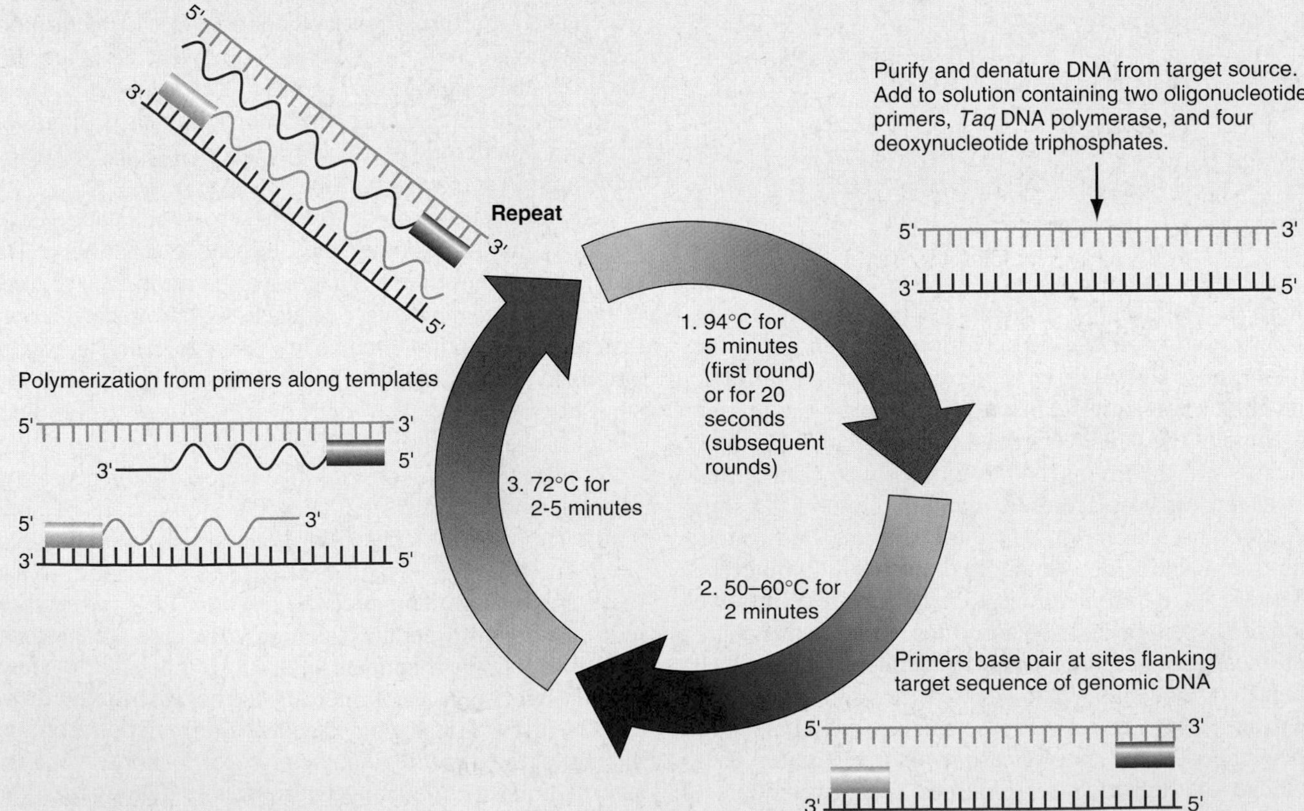

In a typical program, the cycler (1) heats the solution to 94°C for 5 minutes. At this temperature, the target DNA separates into single strands. (2) The temperature is next lowered to 50–60°C for 2 minutes to allow the primers to base pair with complementary sequences in the single-stranded genomic DNA. Specifics of both temperature and timing within these ranges depend on the length and GC:AT ratio of the primer sequences. (3) The thermal cycler then raises the temperature to 72°C, the temperature at which the *Taq* polymerase functions best. Holding the temperature at 72°C for 2–5 minutes (depending on the length of the target sequence) allows DNA polymerization to proceed. At the end of this period, with the completion of DNA synthesis, the first round of PCR is over, and the amount of target DNA has doubled.

To start the next round, the cycler again raises the temperature to 94°C, but this time for only about 20 seconds, to denature the short stretches of DNA consisting of one of the original strands of genomic DNA and a newly synthesized complementary strand initiated by a primer. These short single strands become the templates for the second round of replication because the synthesized primers are able to base pair to them.

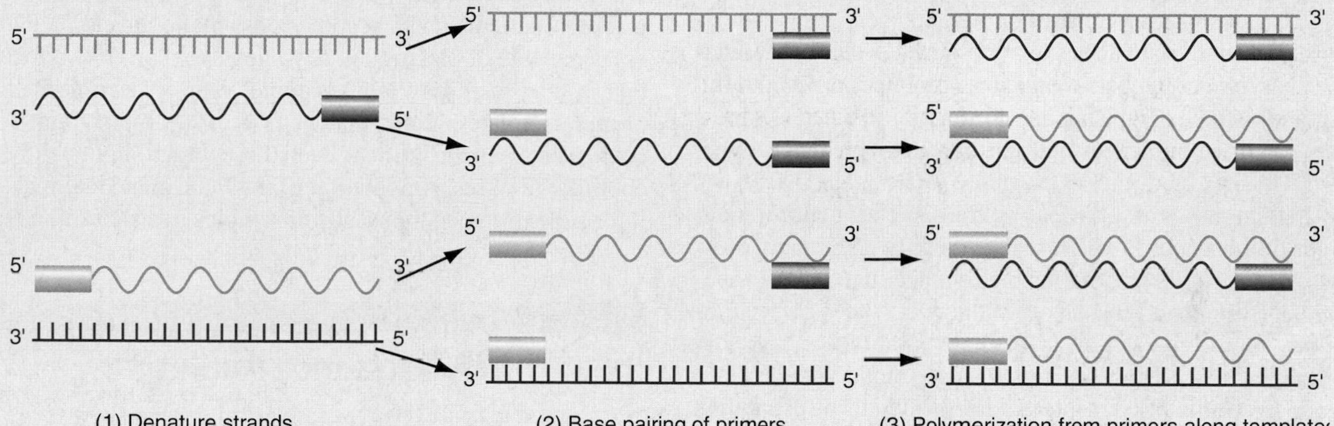

The machine repeats the cycle again and again, generating an exponential increase in the amount of target sequence: 22 repetitions produce over a million copies of the target sequence; 32 repetitions over a billion. The length of the accumulating DNA strands becomes fixed at the length of the DNA between the 5′ ends of the two primers, as shown. This is because, beginning with round 3, the 5′ end of a majority of templates is defined by a primer that has been incorporated in one strand of a PCR product.

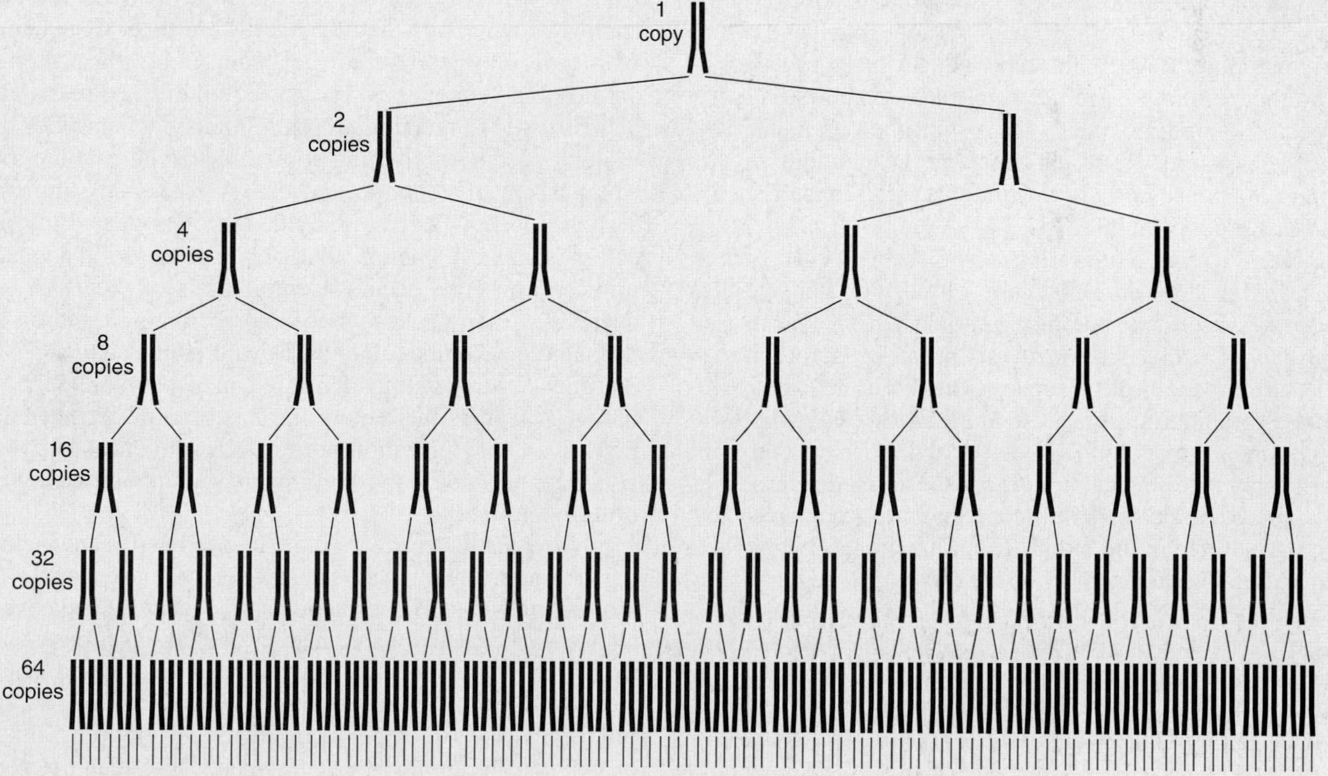

PCR Has Many Uses

PCR is one of the most powerful techniques in molecular biology. Its originator, Kary Mullis, received the 1993 Nobel Prize in chemistry for his 1985 invention of this tool for genetic analysis. By making it possible to amplify a specific sequence directly from complex DNAs (either genomic or cDNA), PCR has made molecular analysis an essential component of genotype detection and gene mapping; we describe its applications in these areas in Chapters 10 and 11. By making it possible to analyze traces of partially degraded DNA from ancient samples (based on related sequences from living organisms) and to compare homologous sequences from a variety of sources, PCR has revolutionized evolutionary studies, enabling researchers to analyze sequences from both living and extinct organisms and to determine the relatedness between these organisms with greater accuracy than ever before (Chapter 22 discusses evolution at the molecular level). By making it possible to analyze the same sequence in hundreds or thousands of individual samples from humans, other animals, or plants without having to make hundreds or thousands of libraries, PCR has facilitated the study of gene diversity at the nucleotide level in populations and greatly simplified the process of monitoring genetic changes in a group over time (see Chapter 21 for the details of population genetics). Finally, PCR has helped bring molecular genetics to many fields outside of traditional genetics. The following example of its use in diagnosing infectious disease provides an inkling of its potential impact on medicine.

AIDS, like other viral diseases, though not inherited, is in one sense a genetic disease, because it is caused by the activity of foreign DNA inside a subgroup of somatic cells. HIV, the virus associated with AIDS, gains entry to a person's body through the bloodstream or lymphatic system, then docks at specific membrane receptors on a few types of white blood cells, fuses with the cell membrane, and releases its RNA chromosome along with several copies of reverse transcriptase into the cell (see the Genetics and Society box on pp. 270–271 in Chapter 8). Once inside the cell, the reverse transcriptase copies the RNA to cDNA. The double-stranded DNA copy of the viral genome then integrates itself into the host genome where, known as a *provirus* or *endogenous retrovirus,* it can lie latent for up to 10 years or become active at any time. When activated, it directs the cellular machinery to make more viral particles.

Standard tests for HIV detect antibodies to the virus, but it may take several months for the antibodies produced by an infected person's immune system to reach levels that are measurable in the blood. Then, in another few months, when ongoing viral activity inside many types of circulating white blood cells subsides, most of the antibodies may disappear from the circulation. This is because once the viral particles have entered the latent state, they are literally in hiding (inside chromosomal DNA) and able to avoid detection by the immune system.

With PCR, it is possible to detect small amounts of virus circulating in the blood or lymph very soon after infection, before antibody production is in full swing. PCR can also detect viral DNA incorporated in the genome of any cell, picking up as little as 1–10 copies of viral DNA per million cells. Thus, with PCR, it becomes possible to confirm and begin treating HIV infection during the critical period before antibodies reach measurable levels. It also becomes possible to follow the progress of each person's HIV infection and tailor therapies accordingly, using a large dose of certain drugs to combat a large amount of viral activity but small doses of perhaps other drugs to prevent a small number of cells from emerging from latency.

In summary, once the base sequence of a particular locus has been determined, it becomes possible to develop a PCR assay capable of picking out any allelic version of that locus that is present in any DNA sample. This is because if the primer templates find a match, PCR will amplify any sequence between the primers, even those with a slightly different order of bases. Automation of the PCR procedure makes it possible to screen thousands of samples in a short time.

9.5 DNA Sequence Analysis

The definitive description of any DNA molecule—from a short fragment in a DNA clone to a whole human chromosome—is the sequence of its bases. The DNA sequence of a gene provides a staggering amount of practical information. Restriction enzyme recognition sites useful for manipulating a gene are immediately visible from the DNA sequence. The codons in open reading frames allow scientists to predict the amino acid sequence of the protein product of the gene. Once the protein's amino acid sequence is known, computer programs can be instructed to model the three-dimensional shape of the protein, thereby predicting something about its function. Comparison of genomic and cDNA sequences immediately shows how a gene is divided into exons and introns and may suggest whether alternative splicing of the gene's primary transcript occurs. Compilations of the DNA sequences of many genes whose expression is turned on by the same environmental cues, as well as the sequences of the genes' surrounding regions, can supply clues about the DNA signals that determine this kind of gene regulation. Even an exploration of the DNA sequences between genes can provide important information about the evolution of genomes.

No matter how large the ultimate goal—up to and including the determination of the sequence of a whole genome—all sequencing projects depend on the protocol described next. It allows the sequence determination of 800–1000 bases at a time from a DNA fragment that has been isolated and amplified by cloning or PCR. For larger sequencing projects, the protocol is repeated on hundreds, thousands, or even millions of individually purified DNA fragments. The final extended sequence is then pieced together like a puzzle according to computer programs that search for overlaps in the sequence information generated by each application of the protocol.

The protocol currently used for determining DNA sequence derives from one of two sequencing technologies developed in the 1970s. One technology, developed by Alan Maxam and Walter Gilbert, is based on the chemical cleavage of DNA molecules at specific nucleotide types. The second technology, developed by Fred Sanger, is based on the enzymatic extension of DNA strands to a defined terminating base. Gilbert and Sanger both won the Nobel Prize for their contribution to DNA sequencing technology. Their techniques have a similar accuracy, which approaches 99.9%. But only the Sanger technique is readily amenable to automation, and it is only with automation that DNA sequencing reaches its full potential.

General Principles of the Procedure

There are two steps to the Sanger method of sequencing, whose object is to reveal the order of base pairs in an isolated DNA molecule. The first step is the generation of a complete series of single-stranded subfragments complementary to a portion of the DNA template under analysis. (Although both strands of a DNA fragment are present in a typical DNA sample, only one is used as a template for sequencing.) Each subfragment differs in length by a single nucleotide from the preceding and succeeding fragments; the graduated set of fragments is known as a *nested array.* A critical feature of the subfragments is that each one is distinguishable according to its terminal 3′ base. Thus, each subfragment has two defining attributes—relative length and one of four possible terminating nucleotides. In the second step of the sequencing process, biologists analyze the mixture of DNA subfragments through polyacrylamide gel electrophoresis, under conditions that allow the separation of DNA molecules differing in length by just a single nucleotide.

The original Sanger sequencing procedure (illustrated in **Fig. 9.17**) begins with the denaturing into single strands of the DNA to be sequenced. The single strands are then mixed in solution with DNA polymerase, the four deoxynucleotide

Feature Figure 9.17

Sanger Sequencing

Begin by mixing the purified, denatured DNA with a labeled oligonucleotide primer that is complementary to a particular site on one strand of the cloned insert. Add DNA polymerase and the four deoxynucleotide triphosphates.

Next divide the mixture into four aliquots and, into each one, add a small amount of a single chain-terminating dideoxyribonucleotide triphosphate abbreviated as "dideoxynucleotide triphosphate" or simply "ddNTP". One aliquot, for example, contains the deoxynucleotides A, T, C, and G spiked with the dideoxynucleotide analog of T. Polymerization from the primer strand continues until, by chance, the dideoxynucleotide is incorporated.

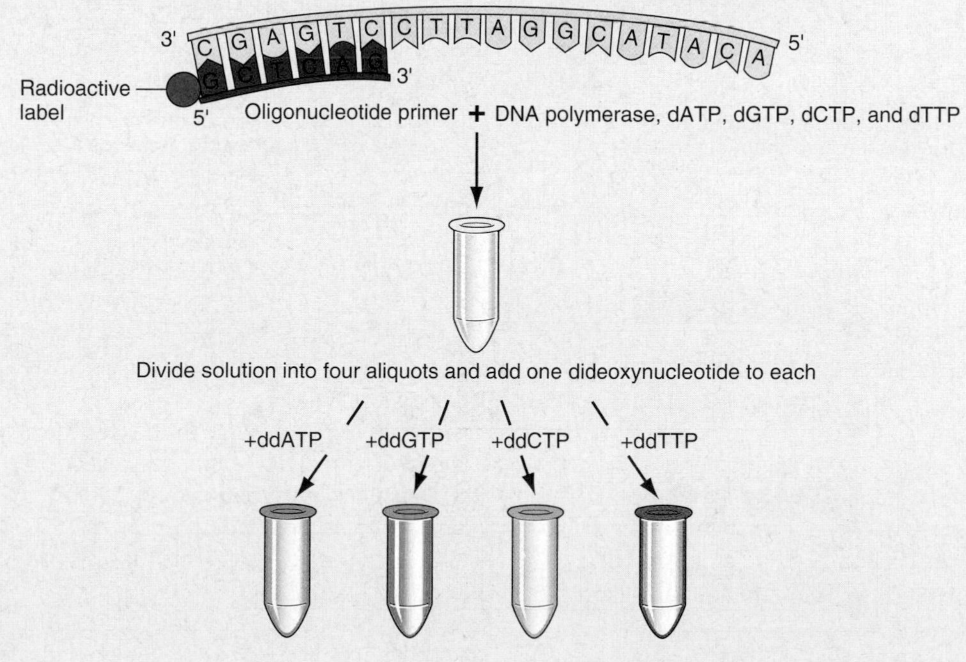

(Continued)

Because a dideoxynucleotide analog has no oxygen at the 3′ position in the sugar, its incorporation prevents the further addition of nucleotides to the strand and thus terminates a growing chain wherever it becomes incorporated in place of an actual deoxynucleotide. The aliquot that has the dideoxy form of thymidine, for example, will generate a population of DNA molecules that terminate at each of the thymidines in the original template strand under analysis.

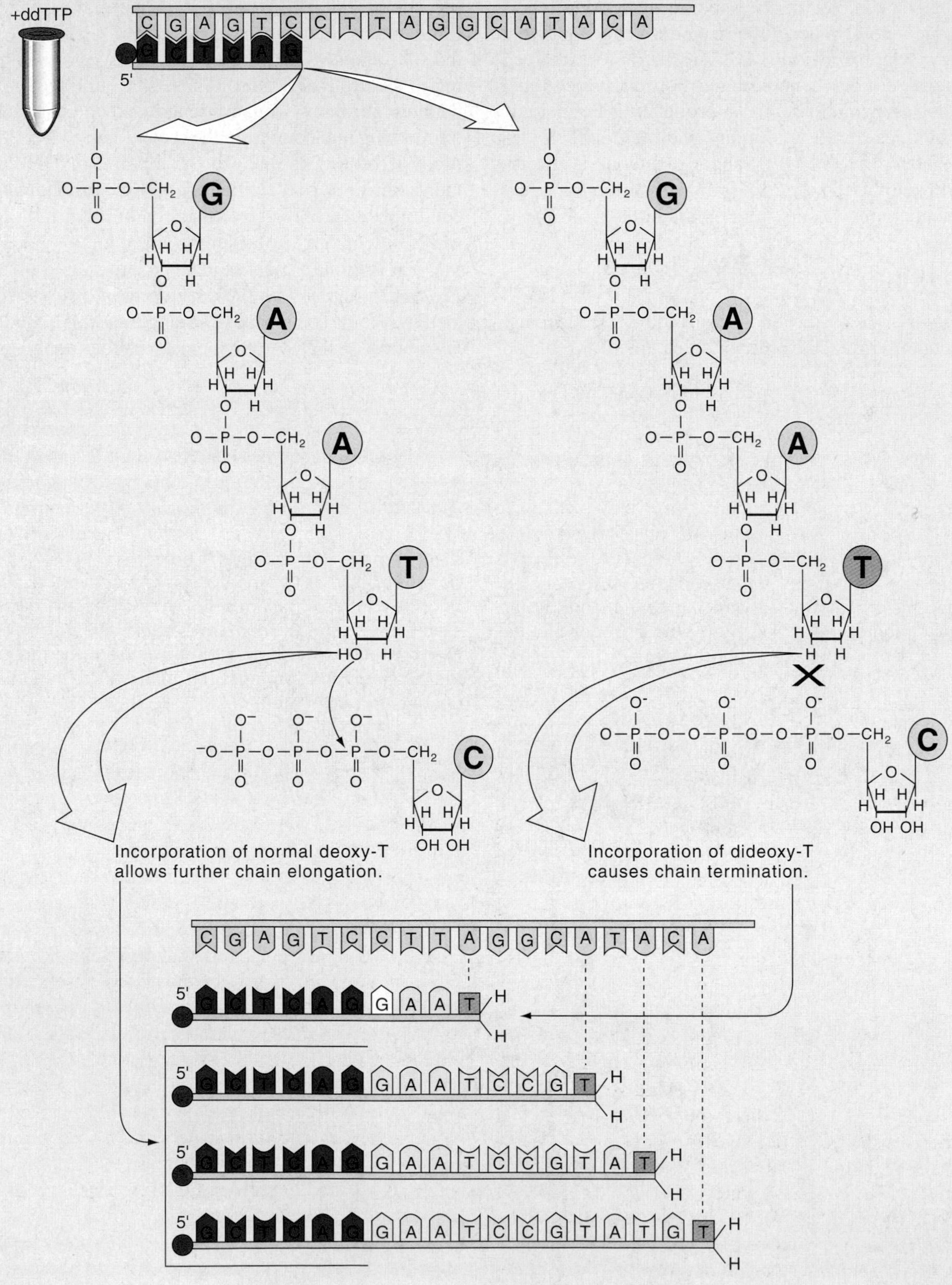

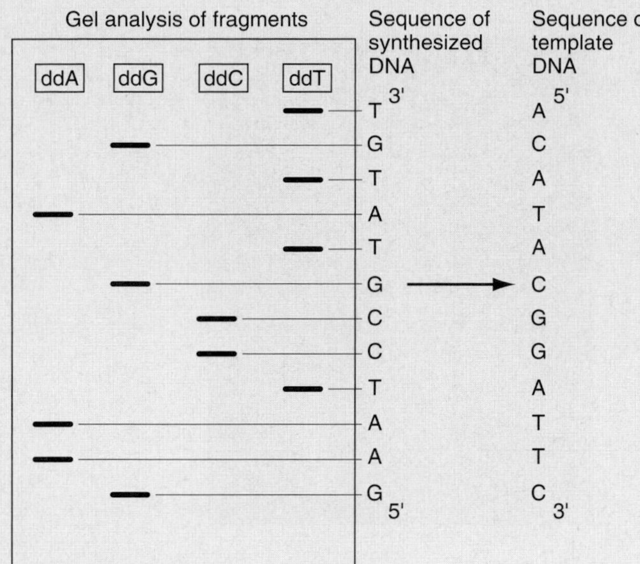

Then use electrophoresis on a polyacrylamide gel to separate the fragments in each of the four aliquots according to size. The resolution of the gel is such that you can distinguish DNA molecules that differ in length by only a single base. The appearance of a DNA fragment of a particular length demonstrates the presence of a particular nucleotide at that position in the strand.

Suppose, for example, that the aliquot polymerized in the presence of dideoxythymidine shows fragments 32, 35, and 39 bases in length. These fragments indicate that thymidine is present at those positions in the strand of nucleotides. In practice, one does not independently determine the exact lengths of each fragment. Instead, one starts at the bottom of the gel, looks at which of the four lanes has a band in it, records that base, then moves up one position and determines which lane has the next band, and so on. In this way, it is possible to read several hundred bases from a single set of reactions.

triphosphates (dATP, dCTP, dGTP, and dTTP), and a radioactively labeled oligonucleotide primer complementary to DNA adjacent to the 3′ end of the template strand under analysis. (Typically, sequencing is performed on purified DNA clones or amplified PCR products. For sequencing DNA clones, the vector-insert recombinant molecule is retained intact, and a primer is used that is complementary to the vector sequence immediately adjacent to the insert DNA. For sequencing PCR products, the PCR primer can also serve as the sequencing primer.) The solution is next divided into four aliquots. To each one, an investigator adds a small amount of a single type of a nucleotide triphosphate lacking the 3′-hydroxyl group that is critical for the formation of the phosophodiester bonds that lead to chain extension (review Fig. 6.7); this nucleotide analogue is called a **dideoxyribonucleotide** (or **dideoxynucleotide**), and it comes in four forms: **ddTTP, ddATP, ddGTP,** or **ddCTP** (abbreviated even further as ddT, ddA, ddG, and ddC).

In each sample reaction tube, the oligonucleotide primer will hybridize at the same location on the template DNA strand. As a primer, it will supply a free 3′ end for DNA chain extension by DNA polymerase. The polymerase will add nucleotides to the growing strand that are complementary to those of the sample's template strand (that is, the actual DNA strand under analysis). The addition of nucleotides will continue until, by chance, a dideoxynucleotide is incorporated instead of a normal nucleotide. The absence of a 3′-hydroxyl group in the dideoxynucleotide prevents the DNA polymerase from forming a phosphodiester bond with any other nucleotide, ending the polymerization for that new strand of DNA.

Next, after allowing enough time for the polymerization of all molecules to reach completion, an investigator releases the templates from the newly synthesized strands by denaturing the DNA at high temperature. Each sample tube now holds a whole collection of single-stranded radioactive DNA chains as well as the nonradioactive single strands of the template DNA. The lengths of the radioactive chains reflect the distance from the 5′ end of the oligonucleotide primer to the position in the sequence at which the specific dideoxynucleotide present in that particular tube was incorporated into the growing chain. (Recall that construction of this chain, which is complementary to the template DNA under analysis, was initiated by a radioactive primer.)

The samples in the four tubes are now electrophoresed in adjacent lanes on a polyacrylamide gel, and the gel is subjected to autoradiography. Because the template strands are not labeled, they do not show up on the autoradiograph. The investigator reads out the sequence of the radioactive strand by starting at the bottom and moving up the autoradiograph, determining which lane carries each subsequent band in the ascending series, as shown in Fig. 9.17. As you ascend the radiograph, each step pictures a chain that is one nucleotide longer than the chain of the step below. Once the sequence of the newly synthesized DNA is known, it is a simple matter to convert this sequence (on paper or by computer) into the complementary sequence of the template strand under analysis.

To automate the DNA sequencing process, molecular geneticists changed the method of labeling the newly formed complementary DNA strands. Instead of placing a

(a) Automated sequencing

1. Generate nested array of fragments; each with a fluorescent label corresponding to the terminating 3' base.

2. Fragments separated by electrophoresis in a single vertical gel lane.

3. As migrating fragments pass through the scanning laser, they fluoresce. A fluorescent detector records the color order of the passing bands. That order is translated into sequence data by a computer.

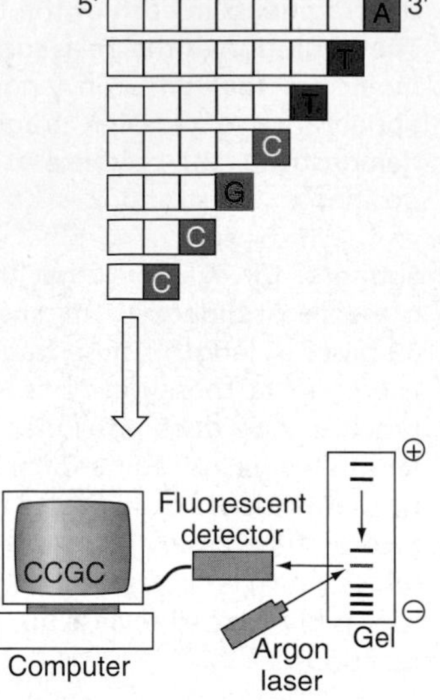

(b) Fluorescent bands in a sequencing gel.

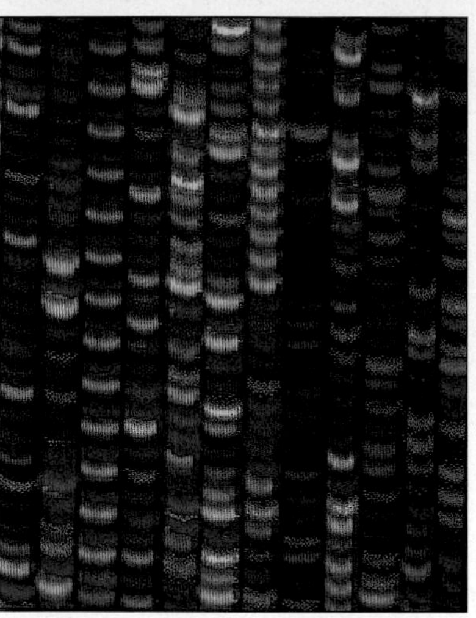

(c) Computer readout

CTNGCTTTGGAGAAAGGCTCCATTGNCAATCAAGACACACAGAGGTGTCCTCTTTTTCCCCTGGTCAGCGNCCAGGTACATNGCACCAAGGCTGCGTAGTGAACTTGNCACCAGNCCATGGAC

(For ease of readability, the yellow color of G nucleotides has been replaced by black.)

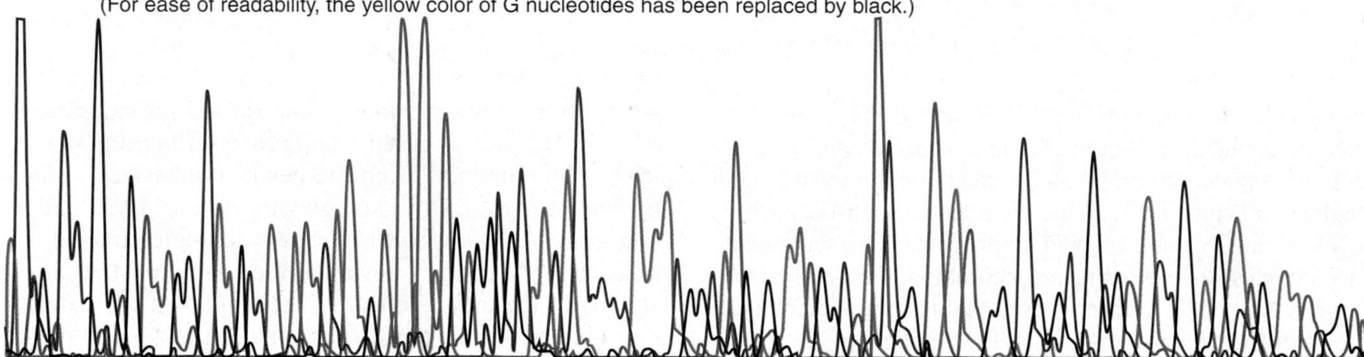

Figure 9.18 Automated sequencing. (a) For automated sequencing, the Sanger protocol is performed with all four fluorescently labeled terminating nucleotides present in a single reaction. At completion of the reaction, DNA fragments terminating at every base in the sequence will be present and color-coded according to the identity of the terminating nucleotide. Separation by gel electrophoresis is next. A laser and accompanying detector are positioned near the bottom of the polyacrylamide gel; as each fragment moves past the laser, the beam determines the color of the fluorescent label on the terminal base and records that color in a computer. **(b)** Output from automated sequencing machine. Each lane displays the sequence obtained with a separate DNA sample and primer. **(c)** The computer reads the sequence of the newly synthesized DNA strand (which is complementary to the template strand) from left to right, which corresponds to the 5'-to-3' direction. The machine records any ambiguity in the base call as an "N"; a technician can resolve most such ambiguities by direct examination of the fluorescence signals.

single radioactive label on the primer oligonucleotide, they labeled each of the four chain-terminating dideoxynucleotides with a different color fluorescent dye. As a result, instead of four separate reactions, as in the original procedure, all four dideoxynucleotides could be combined in a single reaction mixture that could be analyzed in a single lane on a gel (**Fig. 9.18**). A DNA sequencing machine follows the DNA chains of each length in the ascending series through a special detector that can distinguish the different colors associated with each terminating dideoxynucleotide. Thus, in each lane of a gel, it is possible to run a different DNA fragment for complete sequence analysis. Modern DNA sequencing machines can run 100 or more samples simultaneously.

Sequencing Long Regions of DNA

DNA sequences are a form of digital information that is easy to record and analyze with computers. When a project calls for the determination of a long DNA sequence, such as a plasmid containing a 10 kb insert or even a 150 kb BAC clone, multiple rounds of sequencing must be performed to obtain data that the computer can combine into a single final sequence.

There are two basic strategies for sequencing the relatively large DNA inserts of a clone: primer walking and shotgun sequencing.

Primer walking begins with the use of two vector-based primers, flanking the two ends of an insert. Each primer makes it possible to sequence into one strand of the unknown insert for ~800 nucleotides. The information obtained at the end of these sequences can be used to program an automated DNA synthesizer to produce new primers for the next round of sequencing. An investigator repeats this process until he or she has sequenced both strands of the entire insert. Although the two strands are complementary to each other, and thus define each other entirely, it is useful to have two independently obtained sequences to confirm the accuracy of the acquired data.

The primer-walking protocol is a reasonable approach to obtaining a sequence for a relatively small clone of perhaps 5 kb in length. But for a 150 kb BAC clone, under ideal conditions, primer walking would require 300 sequential rounds of DNA sequencing and primer synthesis. This could take a year to complete.

Shotgun sequencing provides a much faster way to determine the DNA sequence of a larger clone (**Fig. 9.19**). To carry out shotgun sequencing, you begin by fragmenting the BAC clone, or other DNA starting material, at random into a large number of smaller pieces. Partial digestion of the DNA with a four-base cutter or through intense shearing with high-frequency, high-intensity sound waves will generate these smaller fragments. You can then clone and sequence them using an oligonucleotide primer complementary to a vector sequence directly adjacent to the unknown insert. You now enter the sequences into a computer programmed to look for sequence overlaps. If the number of sequenced fragments is large enough, it is possible to find overlaps of sequences extending across the entire original clone. Any gaps can be filled in by the primer-walking protocol described above.

The shotgun approach thus relies on redundancy—you have to gather sequence information on three to four times the actual number of base pairs in the original clone to make it work. Even so, it is an efficient method of sequencing. The downside of shotgun sequencing is that only large laboratories equipped with many automated DNA synthesizers able to operate simultaneously can carry it out.

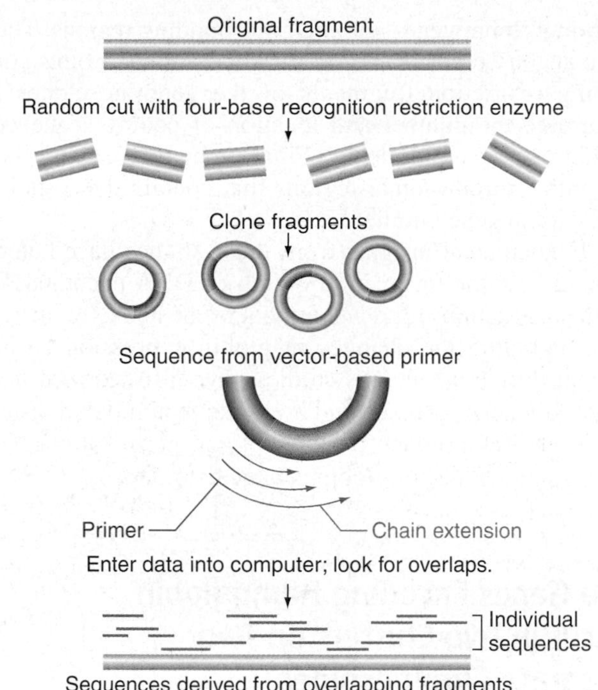

Figure 9.19 Sequencing a long DNA molecule. In shotgun sequencing, the long DNA molecule is broken into a large number of smaller fragments that are cloned and individually sequenced. The individual sequences are aligned by computer to generate a single continuous sequence that spans the entire length of the original molecule.

9.6 Understanding the Genes for Hemoglobin: A Comprehensive Example

Geneticists have used the tools of recombinant DNA technology to analyze the clusters of related genes that make up the α- and β-globin loci. They have isolated mRNA from red blood cell precursors and with the use of reverse transcriptase, produced cDNA libraries. They easily probed these libraries for α-globin and β-globin cDNA clones because 80% of the transcripts within red blood cell precursors code for these polypeptides, and the proportion of α-globin and β-globin cDNA clones in the libraries was equivalently high. They have sequenced individual cDNA clones to identify those that carried coding sequences for either α-globin or β-globin polypeptides. They have used PCR to amplify the globin genes present in many individuals, both normal and diseased, for DNA sequencing analysis.

Geneticists have probed genomic libraries with the α- and β-globin cDNA clones and identified genomic clones carrying globin genes and their surrounding regions. They have also used cDNA clones to probe Southern blots containing restriction fragments of the genomic clones to determine the number and location of coding sequences within each genomic locus. Finally, they have sequenced the entire chromosomal regions that contain the α-globin or β-globin gene families.

Fundamental insights from these studies have helped explain how the linear information of DNA encompasses all the instructions for development of the hemoglobin system, including the changes in globin expression during normal development. The studies have also clarified how the globin genes evolved and how a large number of different mutations produce the phenotypic permutations that give rise to a range of globin-related disorders.

The Genes Encoding Hemoglobin Occur in Two Clusters on Two Separate Chromosomes

The α-globin gene cluster contains three functional genes and spans about 28 kb on chromosome 16 (**Fig. 9.20a**). All the genes in the α-gene cluster point in the same direction; that is, they all use the same strand of DNA as the template for transcription. Moving in the 5′-to-3′ direction along the RNA-like strand, the α or α-like genes appear in the order ζ, α2, α1. The genes in the β-gene cluster, like those in the α-gene cluster, all have the same orientation. The β-globin gene cluster covers 50 kb on chromosome 11 and contains five functional genes in the order ϵ, G_γ, A_γ, δ, and β (Fig. 9.20b). Geneticists refer to the chromosomal region carrying all of the α-globin-like genes as the **α-globin locus** and the region containing the β-globin-like genes as the **β-globin locus.** Note that the term **locus** signifies a location on a chromosome; that location may be as small as a single nucleotide or as large as a cluster of related genes.

The Order of Genes in Each Cluster Correlates with the Sequence of Their Expression During Development

The linear organization of the genes in the α- and β-gene clusters reflects the order in which they are expressed during development. For the α-like chains, that temporal order is ζ during the first five weeks of embryonic life, followed by the two α chains during fetal and adult life. For the β-like chains, the order is ϵ during the first five weeks of embryonic life; then the two γ chains during fetal life; and finally, within a few months of birth, mostly β but also some δ chains (see Fig. 9.1 on p. 302 and Fig. 9.20).

The fact that the order of genes on the chromosomes parallels the order of their expression during development suggests that whatever mechanism turns these genes on and off takes advantage of their relative positions. We now understand what that mechanism is: A *locus control region* (or *LCR*) associated with specialized DNA binding proteins at the 5′ end of each locus works its way down the locus, bending the chromatin back on itself to turn genes on and off in order. We describe this regulatory mechanism in more detail in Chapter 18.

One consequence of a master regulatory element that controls an entire gene complex is seen in a rare medical condition with a surprising prognosis. In some adults, the red blood cell precursors express neither β or δ chains. Although this should be a lethal situation, these adults remain healthy. Cloning and sequence analysis of the β-globin locus from affected adults show that they have a deletion extending across the β and δ genes. Because of this deletion, the master regulatory control can't

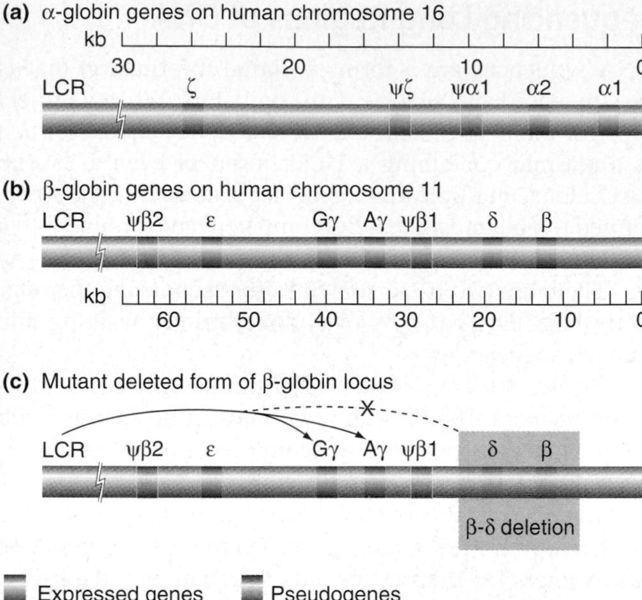

Figure 9.20 The genes for the polypeptide components of human hemoglobin are located in two genomic clusters on two different chromosomes. (a) Schematic representation of the α-globin gene cluster on chromosome 16. The cluster contains three functional genes and two pseudogenes (designated by the symbol ψ); pseudogenes are sequences that look like but do not function as genes. **(b)** Schematic representation of the β-globin gene cluster on chromosome 11; this cluster has five functional genes and two pseudogenes. Upstream from both the α- and β-globin gene clusters lie the locus control regions (LCR), which interact with and activate sequential genes in each complex during human development. **(c)** In this example of a mutant chromosome, the adult β-globin genes β and δ have been deleted; as a result, the LCR cannot switch from activating the fetal genes to activating the adult genes, and the fetal genes remain active in the adult.

switch at birth, as it normally would, from γ-globin production to β- and δ-globin production (Fig. 9.20c). People with this rare condition continue to produce large amounts of fetal γ globin throughout adulthood, and that γ globin is sufficient to maintain a near-normal level of health.

A Variety of Mutations Account for the Diverse Symptoms of Globin-Related Diseases

By comparing DNA sequences from affected individuals with those from normal individuals, researchers have learned that there are two general classes of disorders arising from alterations in the hemoglobin genes. In one class, mutations change the amino acid sequence and thus the three-dimensional structure of the α- or β-globin chain, and these structural changes result in an altered protein whose malfunction causes the destruction of red blood cells. Diseases of this type are known as hemolytic anemias (**Fig. 9.21a**). An example is sickle-cell anemia, caused by an A-to-T substitution in the sixth codon of the β-globin chain. This simple change in DNA sequence alters the sixth amino acid in the chain from glutamic acid to valine, which, in turn, modifies the form and function of the affected hemoglobin molecules. Red blood cells carrying these altered molecules often have abnormal shapes that cause them to block blood vessels or be degraded.

The second major class of hemoglobin-related genetic diseases arises from DNA mutations that reduce or eliminate the production of one of the two globin polypeptides. The disease state resulting from such mutations is known as thalassemia, from the Greek words *thalassa* meaning "sea" and *emia* meaning "blood"; the name arose from the observation that a relatively high rate of this blood disease occurs among people who live near the Mediterranean Sea. Several different types of mutation can cause thalassemia, including those that delete an entire α or β gene, those that alter the sequence in regions that are outside the gene but necessary for its regulation, or those that alter the sequence within the gene such that no protein can be produced. The consequence of these changes in DNA sequence is the total absence or a deficient amount of one or the other of the normal hemoglobin chains. Because there are two α genes (α1 and α2) that see roughly equal expression beginning a few weeks after conception, individuals carrying deletions within the α-globin locus may be missing anywhere from one to four copies of the α-globin genes (Fig. 9.21b). A person lacking only one would be a heterozygote for the deletion of one of two α genes; a fetus missing all four would be a homozygote for deletions of both α genes. The range of mutational possibilities explains the range of phenotypes seen in α-thalassemia. Individuals missing only one of four possible copies of the α genes are normal; those lacking two of the four have a mild anemia, and those without all four die before birth. The fact that the α chains are expressed early in fetal life explains why the α-thalassemias are detrimental *in utero*. By contrast, β-thalassemia major, the disease occurring in people who are homozygotes for most deletions of the single β gene, also usually results in death, but not until soon after birth. These individuals survive that long because the β-like γ globin is expressed in the fetus (review Fig. 9.1 on p. 302).

Comparisons of the altered DNA sequences from affected individuals with wild-type sequences from normal individuals have helped illuminate the sequences necessary for normal hemoglobin expression. In some β-thalassemia patients, for example, disease symptoms arise from the alteration of a few nucleotides adjacent to the 5′ end of the coding region for the β chain. Data of this type have defined sequences that are important for expression of the β-globin locus. One such segment is the TATA box, a sequence found in many eukaryotic promoters (**Fig. 9.22a;** see Chapter 18 for a more detailed discussion). In other thalassemia patients, the entire α-globin locus and adjacent regulatory segments, including the TATA box, are intact, but a mutation has altered the LCR found far to the 5′ side of all the α-like genes. This LCR is necessary for a high level of tissue-specific expression of all α-like genes in red blood cell progenitors (Fig. 9.22b). Mutations in the TATA box or the locus control region, depending on how disruptive they are, produce α- or β-thalassemias of varying severity.

The α- and β-Globin Loci Contain Multiple Genes That Evolved from One Ancestral Gene

From sequencing the DNA of all the human globin genes, researchers can see that the genes form a closely related group, or family, of genes that evolved by duplication and divergence from one ancestral gene (**Fig. 9.23**). All gene sequences evolve continuously by mutation. The two separate products of a duplication, which start out identical, eventually diverge because their sequences evolve independently of each other. The members of a gene family may be grouped together on one chromosome (like the very closely related β-globin genes) or dispersed on different chromosomes (like the less closely related α- and β-globin genes). All the β-like genes are exactly the same length and have two introns at exactly the same positions. All the α-like genes also have two introns at exactly the same positions, but these positions are different from those of the β genes. The sequences of all the β-like genes are more similar to each other than they are to the α-like sequences, and vice versa. These comparisons suggest

(a.1) Major types of structural variants causing hemolytic anemias

Name	Molecular basis of mutation	Change in polypeptide	Pathophysiological effect of mutation	Inheritance
HbS	Single nucleotide substitution	β 6 Glu ↓ Val	Deoxygenated HbS polymerizes → sickle cells → vascular occlusion and hemolysis	Autosomal Recessive
HbC	Single nucleotide substitution	β 6 Glu ↓ Lys	Oxygenated HbC tends to crystallize → less deformable cells → mild hemolysis; the disease in HbS:HbC compounds is like mild sickle-cell anemia	Autosomal Recessive
Hb Hammer-smith	Single nucleotide substitution	β 42 Phe ↓ Ser	An unstable Hb → Hb precipitation → hemolysis; also low O_2 affinity	Autosomal Dominant

(b.1) Clinical results of various α-thalassemia genotypes

Clinical condition	Genotype (ζ α2 α1)	Number of functional α genes	α-chain production
Normal	αα/αα	4	100%
Silent carrier	αα/α–	3	75%
Heterozygous α-thalassemia—mild anemia	α–/α– or αα/– –	2	50%
HbH ($β_4$) disease—moderately severe anemia	α–/– –	1	25%
Homozygous α-thalassemia—lethal	– –/– –	0	0%

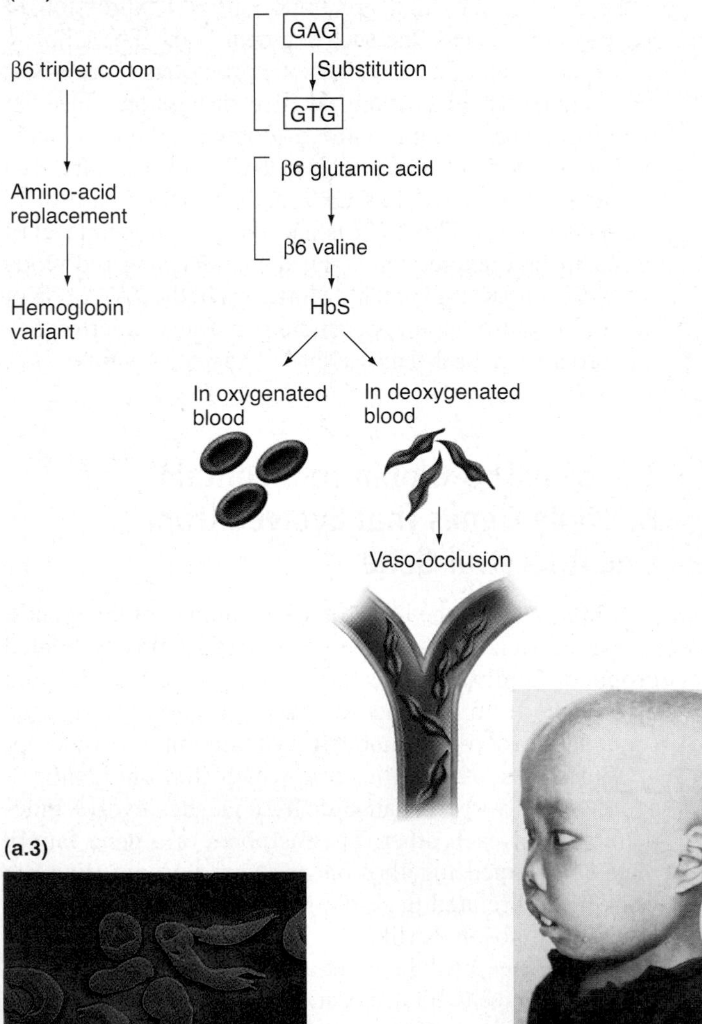

(a.3)

(b.3)

(b.2) A β-thalassemia patient makes only α globin, not β globin.

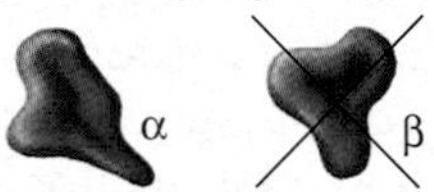

Four α subunits combine to make abnormal hemoglobin.

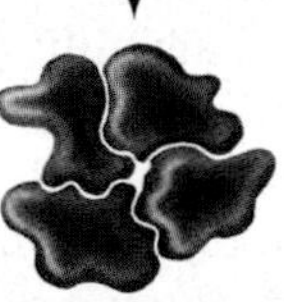

Abnormal hemoglobin molecules clump together, altering shape of red blood cells. Abnormal cells carry reduced amounts of oxygen.

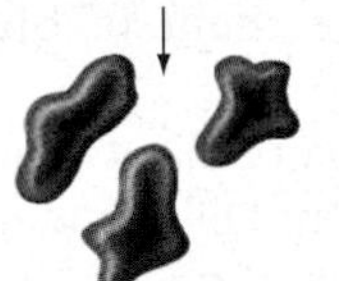

To compensate for reduced oxygen level, medullary cavities of bones enlarge to produce more red blood cells.

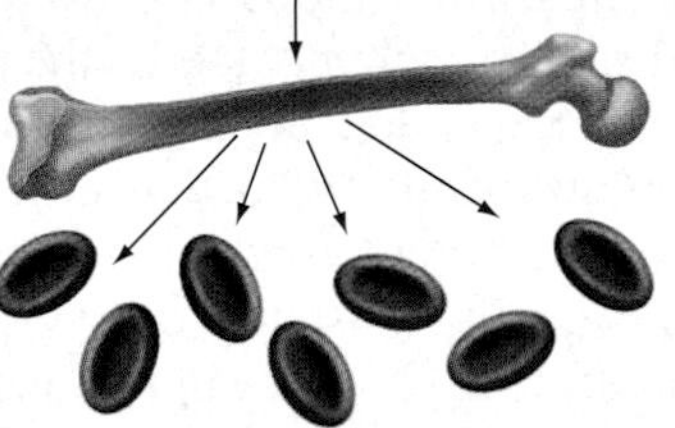

The spleen enlarges to remove excessive number of abnormal red blood cells.

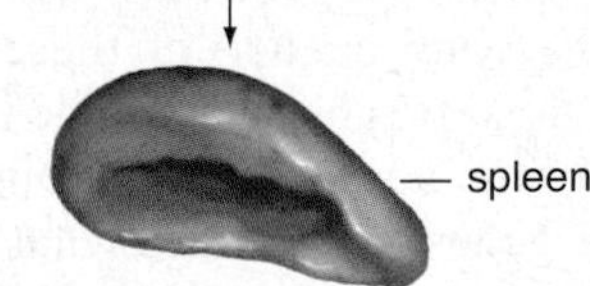

Too few red blood cells in circulation result in anemia.

Figure 9.21 Mutations in the DNA for hemoglobin produce two classes of disease. (a.1) The major types of hemoglobin variants causing hemolytic anemias. **(a.2)** The basis of sickle-cell anemia. **(a.3)** Sickling red blood cells appear as crescents among more rounded nonsickling cells. **(b.1)** Thalassemias associated with deletions in the α-globin polypeptide. **(b.2)** The physiological basis of β-thalassemia major. **(b.3)** Child suffering from β-thalassemia major.

(a) Promoter region of the β-globin gene

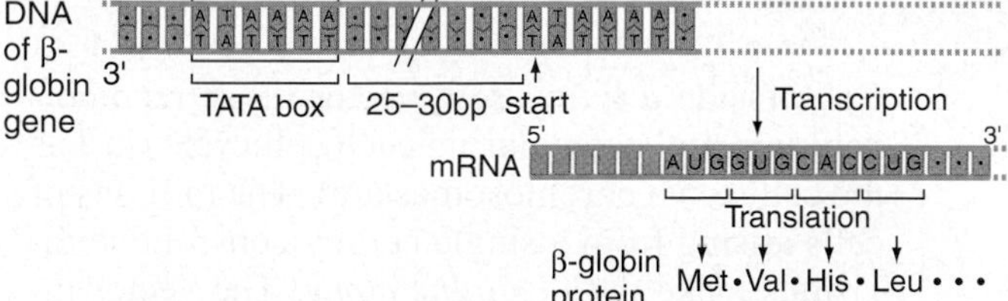

(b) Locus control region of the α-globin gene cluster

Figure 9.22 Regulatory regions affecting globin gene expression. (a) Mutations in the TATA box associated with the β-globin gene can eliminate transcription and cause β-thalassemia. **(b)** A locus control region is present 25–50 kb upstream of the α-globin gene cluster. The function of the LCR is to open up the chromatin domain associated with the complete cluster of α-globin genes. Mutations in the LCR can prevent expression of all the α-globin genes, resulting in severe α-thalassemia.

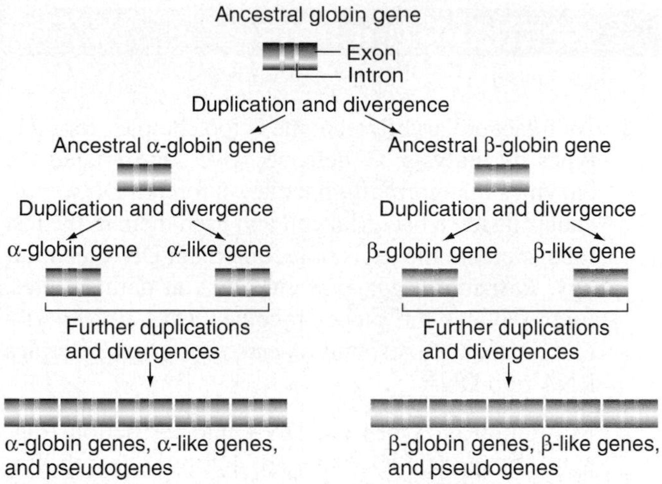

Figure 9.23 Evolution of the globin gene family. Duplication of an ancestral gene followed by divergence of the separate duplication products established the α- and β-globin lineages. Further rounds of duplication and divergence within the separate lineages generated the two sets of genes and pseudogenes of the globin gene family.

that a single ancestral globin gene duplicated, and one copy moved to another chromosome. With time, one of the two gene copies gave rise to the α lineage, the other to the β lineage. Each lineage then underwent further duplications to generate the present array of three α-like and five β-like genes in humans.

Interestingly, the duplications also produced genes that eventually lost the ability to function. Molecular geneticists made this last deduction from data showing two additional α-like sequences within the α locus and one β-like sequence within the β locus that no longer have the capacity for proper expression. The reading frames are interrupted by frameshifts, missense mutations, and nonsense codons, while regions needed to control the expression of the genes have lost key DNA signals. Sequences that look like, but do not function as, a gene are known as **pseudogenes;** they occur throughout all higher eukaryotic genomes.

Connections

The tools of recombinant DNA technology grew out of an understanding of the DNA molecule and its interaction with the enzymes that operate on DNA in normal cells. Geneticists use the tools singly or in combination to look at DNA directly. Through cloning, hybridization, PCR, and sequencing, they have been able to isolate the genes that encode, for example, the hemoglobin proteins; identify sequences near the genes that regulate their expression; determine the complete nucleotide sequence of each gene; and discover the changes in sequence produced by the hundreds of mutations that affect hemoglobin production. The results give a fascinating and detailed picture of how the nucleotides along a DNA molecule determine protein structure and function and how mutations in sequence produce far-ranging and varied effects on human health. The methods of classical genetics that we examined in Chapters 2–5 complement those of recombinant DNA technology to produce an integrated picture of genes and genomes at many levels.

In Chapter 10, we describe how the use of recombinant DNA technology has expanded from the analysis of single genes and gene complexes to the sequencing and examination of whole genomes. Through the automation of sequencing and high-powered computer analysis of the data, scientific teams have determined the DNA sequence of the entire human genome and the genomes of many other organisms as well.

Essential Concepts

1. An intact eukaryotic genome is too complex for most types of analysis. Geneticists have appropriated the enzymes that normally operate on foreign DNA molecules inside a bacterial cell and used them in the test tube to create the tools of recombinant DNA technology. *Restriction enzymes* cut DNA at defined sites, *ligase* splices the pieces together, *DNA polymerase* makes DNA copies, and *reverse transcriptase* copies RNA into DNA.

2. Restriction enzymes cut DNA into restriction fragments at specific sites that are 4–8 bp in length. Enzymes that recognize a 4 bp site digest DNA into fragments that average 256 bp in length; enzymes that recognize 6 bp digest DNA into fragments that average 4 kb in length; enzymes that recognize 8 bp digest DNA into fragments that average 64 kb in length. *Complete digestion* at every site recognized by a particular enzyme can occur under optimal reaction conditions. Under less than optimal conditions (of temperature, enzyme concentration, or duration of reaction), a *partial digestion* occurs in which only a random fraction of restriction sites are digested. Partial digestion enables investigators to obtain a sample of DNA fragments having an average size in-between those obtainable by complete digestion with enzymes that recognize 4, 6, or 8 bp sites. Most restriction enzymes produce DNA molecules with *sticky ends,* but some produce DNA molecules with *blunt ends.*

3. Gel electrophoresis provides a method for separating DNA fragments according to their size. When biologists subject a viral genome, plasmid, or small chromosome to restriction digestion and gel electrophoresis, they can observe the resulting DNA fragments by ethidium bromide staining and determine the size of the fragments by comparing their migration within the gel with the migration of known marker fragments.

4. Cloning DNA fragments:
   a. Restriction fragments and cloning vectors with matching sticky ends can be spliced together to produce *recombinant DNA molecules.* A *cloning vector* is a DNA sequence that can enter a host cell, produce a selectable phenotype, and provide a means of replicating and purifying both itself and any DNA to which it is spliced. Different types of vectors carry different-size inserts, which can serve as the basis for different levels of analysis.
   b. Once inside a living cell, vector-insert recombinants are replicated during each cell cycle (just as the cell's own chromosomes are). The millions of cells arising from a single cell by consecutive divisions make up a *cellular clone.* The vector-insert recombinant molecules inside the cells of a clone, often referred to as *DNA clones,* can be purified by procedures that separate recombinant molecules from host DNA. The insert can then be cut away from the vector by restriction enzymes and purified.
   c. *Genomic libraries* are random collections of vector-insert recombinants containing the DNA fragments of a given species. The most useful libraries carry at least four to five *genomic equivalents. cDNA libraries* carry DNA copies of the RNA transcripts produced in a particular tissue at a particular time. The clones in a cDNA library represent only that part of the genome transcribed and spliced into mRNA in the cells of the specific tissue, organ, or organism.

5. *Hybridization* is the process whereby complementary DNA strands form stable double helixes. Hybridization makes it possible to use previously purified DNA fragments as labeled probes. Biologists use such probes to identify clones containing identical or similar sequences within genomic or cDNA libraries. Hybridization can also be used in combination with gel electrophoresis as part of an analytic technique called *Southern blotting.* Southern blot hybridization allows an investigator to determine the numbers and positions of complementary sequences within isolated DNA fragments or whole genomes of any complexity.

6. The *polymerase chain reaction* (or *PCR*) is a method for the rapid purification and amplification of a single DNA fragment from a complex mixture such as the whole human genome. The DNA fragment to be amplified is defined by a pair of oligonucleotide primers complementary to either end on opposite strands. The PCR procedure operates through a reiterative loop that amplifies the sequence between the primers in an exponential manner. PCR is used in place of cloning to purify DNA fragments whenever sequence information for primers is already available.

7. *Sequencing* provides the ultimate description of a cloned fragment. Automation has increased the speed and scope of sequencing.

On Our Website

www.mhhe.com/hartwell3
Chapter 9

Annotated Suggested Readings and Links to Other Websites

- Foundational articles describing recombinant DNA technology
- Original DNA sequencing articles
- More on the human α- and β-globin loci and their associated diseases
- More on the use of restriction site analysis in the diagnosis of sickle-cell syndrome
- History of the biotechnology industry

Specialized Topics

- Agricultural biotechnology

Social and Ethical Issues

1. Thousands of years ago, people living in the Middle East discovered that when they added a tiny quantity of an extract from the inner lining of the fourth stomach of a freshly killed calf to cow's milk, the milk turned to cheese. In the twentieth century, biochemists isolated and characterized the responsible agent: an enzyme named *chymosin.* The enzyme is encoded by a single gene that is expressed only in animals that chew their cud and only before weaning. Until just two decades ago, cheese makers had to use this animal product as an essential component of the cheese-making process. Then, using the tools of recombinant DNA technology, biotechnologists cloned the cow chymosin gene in a bacterial expression vector, making it possible to produce fully functional chymosin in a culture of cloned bacteria. In 1990, chymosin produced in this way became the first approved product of recombinant DNA technology to enter the U.S. food market. Unlike calf chymosin, recombinant cow chymosin can be standarized; it can also be inexpensively produced, and it doesn't require the killing of any animals. Purified calf chymosin and bacterially produced chymosin are indistinguishable in their structure and enzymatic activities. By 2004, over 90% of the cheeses on supermarket shelves in the United States, Europe, and most other countries around the globe were produced using recombinant chymosin.
   a. Chr. Hansen, the Danish company that produces recombinant chymosin under the trademark of Chy-Max, claims in their promotional literature that it is "nature's own enzyme for clotting milk," and a "natural ingredient for the food industry." What is the definition of the term *natural?* Is it appropriate for a company to use this term to describe a protein made from a mammalian-specific gene inside bacterial cells? Does it matter that the purified chymosin proteins from a calf's stomach and from cloned bacteria are structurally and functionally identical? How important is the *process* relative to the *product* in assessing something's naturalness?
   b. Some people do not want to eat food produced through genetic engineering, and they have called on the FDA to require a "genetically engineered" label on all such products. Food manufacturers argue that since biotech-produced cheese and traditionally produced cheese are indistinguishable, FDA regulations say that they don't have to put special labels on their products to this effect. Both sides are fully aware of survey results showing that 50% or more of Americans would refuse to buy a product carrying a "genetically engineered" label. Do you think that food made with recombinant DNA technology should be labeled as such? What are the benefits and harms of requiring or not requiring such labels?
   c. Acting in response to a request from the organic farmers trade organization, the FDA included in its regulation of organic food labeling the stipulation that no product of recombinant DNA technology was allowable at any point in the food production process. Consequently, cheese produced with the use of Chy-Max cannot be labeled "organic." However, "organic" cheese produced by the traditional method requires the use of an animal ingredient and is, therefore, unsuitable for consumption by vegetarians. This means that dedicated vegetarians who eat only organic food cannot eat any cheese. (Some dairy products that do not require chymosin—such as cream cheese—have cheese in their name but are not true cheeses, according to the accepted definition of that food term.) Do you think organic cheese producers should be required to attach a label

stating "contains animal products" so that vegetarians can be forewarned? What are the benefits and harms of requiring or not requiring such labels?

2. Pigs are a major cause of environmental pollution, as anyone who walks past a pig farm knows. Most of the pollution occurs because the animals are biologically unable to process the organic phosphorus present naturally in feed or grain; as a result, they require mineral phosphorous as an essential growth supplement. Unfortunately, most of the phosphorus in grain passes unabsorbed through the pig gut and into their manure, which is then used by organic farmers to fertilize crops. When it rains, the phosphorus-rich manure runs off into streams and other waterways where it causes eutrophication, algal blooms, depletion of oxygen, dead fish, and greenhouse gases. Natural human gut bacteria contain an enzyme called *phytase* that allows the biological processing of naturally occurring organic phosphorus. And in a reversal of the genetic engineering process described in the preceding Social and Ethical Issue, scientists have inserted a recombinant DNA molecule containing the bacterial *phytase* gene into a pig genome. Animals engineered with genes from other species are called **transgenic,** and the genes themselves are called **transgenes.** The recombinant *phytase* transgene is attached to a special promoter that causes the animal to produce the enzyme only in its salivary secretions. Except for this one difference, transgenic pigs are physiologically indistinguishable from nontransgenic pigs. Transgenic phytase-producing pigs benefit farmers financially because the pig growers no longer have to supplement pig feed with mineral phosphorus. Of greater importance is the fact that the transgenic pigs release 75% less polluting phosphorus in their manure.
   a. Once transgenic-phytase pigs are available commercially, do you think the Environmental Protection Agency should require, or encourage, pig farmers to purchase them in place of traditional animals in order to reduce environmental harm?
   b. As a result of self-formulated regulations, organic pig farmers cannot raise transgenic pigs and cannot use manure from transgenic pigs as fertilizer for crops. Consequently once transgenic-phytase pigs predominate in nonorganic pork production, organic pig farms will become more polluting than the corresponding nonorganic farms. What do you think would be the appropriate course of regulatory action in this situation?

Solved Problems

I. The following map of the plasmid cloning vector pBR322 shows the locations of the ampicillin (amp) and tetracycline (tet) resistance genes as well as two unique restriction enzyme recognition sites, one for *Eco*RI and one for *Bam*HI. You digested this plasmid vector with both *Eco*RI and *Bam*HI enzymes and purified the large *Eco*RI-*Bam*HI vector fragment. You also digested the cellular DNA that you want to insert into the vector with both *Eco*RI and *Bam*HI. After mixing the plasmid vector and the fragments together and ligating, you transformed an ampicillin-sensitive strain of *E. coli* and selected for ampicillin-resistant colonies. If you test all of your selected ampicillin-resistant transformants for tetracycline resistance, what result do you expect, and why?

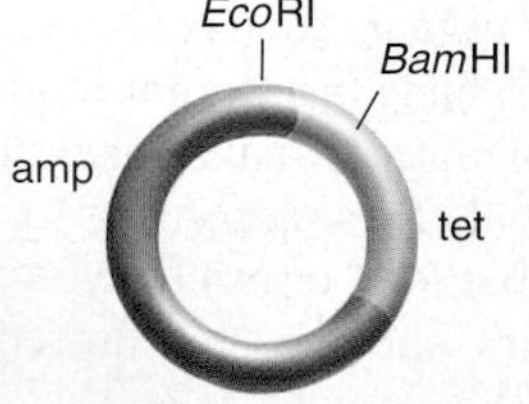

Answer

This problem requires an understanding of vectors and the process of combining DNAs using sticky ends generated by restriction enzymes.

The plasmid must be circular to replicate in *E. coli,* and, in this case, a circular molecule will be formed only if the insert fragment joins with the cut vector DNA. The cut vector will not be able to re-ligate without an inserted fragment because the *Bam*HI and *Eco*RI sticky ends are not complementary and cannot base pair. All ampicillin-resistant colonies therefore contain a *Bam*HI-*Eco*RI fragment ligated to the *Bam*HI-*Eco*RI sites of the vector. Fragments cloned at the *Bam*HI-*Eco*RI site interrupt and therefore inactivate the tetracycline resistance gene. *All ampicillin-resistant clones will be tetracycline sensitive.*

II. The gene for the human peptide hormone somatostatin (encoding nine amino acids) is completely contained on an *Eco*RI (5′ G^AATTC 3′) fragment, which can be cut out of the larger fragment shown at the top of the next page. (The ^ symbol indicates the site where the sugar-phosphate backbone is cut by the restriction enzyme.)

5′ GCCG^AATTCGATCCTATCAACACGAAGTGAAAGTCTTACAACCCATG^AATTCGATTCG 3′
3′ CGGCTTAA^GCTAGGATAGTTGTGCTTCACTTTCAGAATGTTGGGTACTTAA^GCTAAGC 5′

a. What is the amino acid sequence of human somatostatin?
b. Indicate the direction of transcription of this gene.
c. The first step in synthesizing large amounts of human somatostatin for pharmacological treatments involves constructing a so-called fusion gene. In this fusion construct, the N terminus of the protein encoded by the fusion gene consists of the N-terminal half of the *lacZ* gene (encoding β-galactosidase), while the remainder of the product of the fusion gene is human somatostatin. A family of three plasmid vectors for the construction of such a fusion gene has been created. All of these vectors have an ampicillin resistance gene and part of the *lacZ* gene encoding the first 583 amino acids of the β-galactosidase protein. The *Eco*RI fragment (that is, the fragment produced by cutting with *Eco*RI) containing human somatostatin can be inserted into the single *Eco*RI restriction site on the vectors. The sequence of three vectors in the vicinity of the *Eco*RI site is shown here. The numbers refer to amino acids in the β-galactosidase protein with the N-terminal amino acid being number 1. The DNA sequence presented is the same as that of the *lacZ* mRNA (with T's replacing the U's found in RNA). In which of these three vectors must the *Eco*RI fragment containing human somatostatin be inserted to generate a fusion protein with an N-terminal region from β-galactosidase and a C-terminal region from human somatostatin?

582 583
Gly Asn *Eco*RI
pWR590-1 5′ GGCAACCGGGCGAGCTCGAATTCG

582 583
Gly Asn *Eco*RI
pWR590-2 5′ GGCAACCCGGGCGAGCTCGAATTC

582 583
Gly Asn *Eco*RI
pWR590-3 5′ GGCAACGGGGCAGCTCGAATTCGA

Answer

This problem requires an understanding of the sticky ends formed by restriction enzyme digestion and the requirement of appropriate reading frames for the production of proteins.

a. The only complete open reading frame (ATG start codon to a stop codon) is found on the bottom strand (underlined on the following figure). *The amino acid sequence is met-gly-cys-lys-thr-phe-thr-ser-cys.*
b. Based on the amino acid sequence determined for part *a, the gene must be transcribed from right to left.*
c. The cut site for *Eco*RI is after the G at the 5′ end of the *Eco*RI recognition sequence on each strand. For each of the three vectors, the cut will be shifted relative to the reading frame of the *lacZ* gene by one base. The *Eco*RI fragment containing somatostatin can ligate to the vector in two possible orientations, but since we know the sequence on the bottom strand codes for the protein, a fusion protein will be produced only if the fragment is inserted with that coding sequence on the same strand as the vector coding sequences. Consider only this orientation to determine which vector will produce the fusion protein. The *Eco*RI fragment to be inserted into the vector next to the *lacZ* gene has five nucleotides that precede the first codon of the somatostatin gene (see following figure) and therefore requires one more nucleotide to match the reading frame of the vector. For pWR590-1, the cut results in an in-frame end; pWR590-2 has one base extra beyond the reading frame; pWR590-3 has a two-base extension past the reading frame. *The* EcoRI *fragment must be inserted in the vector pWR590-2 to get somatostatin protein produced.*

5′ GCCG^AATTCGATCCTATCAACACGAAGTGAAAGTCTTACAACCCATG^AATTCGATTCG 3′
5′ CGGCTTAA^GCTAGGATAGTTGTGCTTCACTTTCAGAATGTTGGGTACTTAA^GCTAAGC 5′

III. Imagine you have cloned a 14.7 kb piece of DNA, which contains restriction sites as shown here.

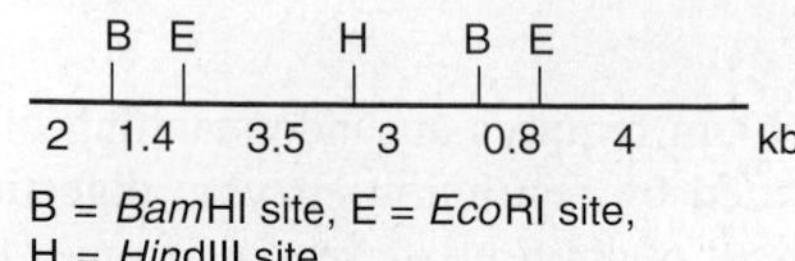

Numbers under the segments represent the sizes of the regions in kilobases (kb). You have labeled the left end of the molecule with ^{32}P.

a. What radioactive bands would you expect to see following electrophoresis if you did a complete digestion with *Bam*HI? *Eco*RI? *Hin*dIII?

b. What radioactive bands would you expect to see following electrophoresis if you did a partial digestion with *Bam*HI? *Eco*RI? *Hin*dIII?

Answer

This problem deals with partial and complete digests and radioactive labeling of fragments.

a. Only the left-most fragment would be seen after complete digestion with any of the three enzymes since only the left end contains radioactivity. *Radioactive bands seen after digestion with* Bam*HI: 2 kb;* Eco*RI: 3.4 kb; and* Hind*III: 6.9 kb.*

b. Partial digestions result in all possible combinations of adjacent fragments but again only those containing the left end of the fragment will be radioactive. For example, partial digestion with *Bam*HI results in the following fragments (asterisks indicate fragments that are radioactive). *Bam*HI: 2*, 7.9, 1.2, 9.9*, 12.7, and 14.7*. *The radioactive fragments are* Bam*HI: 2, 9.9, 14.7 kb;* Eco*RI: 3.4, 10.7, 14.7 kb; and* Hind*III: 6.9, 14.7 kb.*

Problems

Vocabulary

1. Match each of the terms in the left column to the best-fitting phrase from the right column.

a. oligonucleotide	1. a DNA molecule used for transporting, replicating, and purifying a DNA fragment
b. vector	2. a collection of the DNA fragments of a given species, inserted into a vector
c. sticky ends	3. DNA copied from RNA by reverse transcriptase
d. recombinant DNA	4. stable binding of single-stranded DNA molecules to each other
e. reverse translation	5. efficient and rapid technique for amplifying the number of copies of a DNA fragment
f. genomic library	6. computational method for determining the possible sequence of base pairs associated with a particular region of a polypeptide
g. genomic equivalent	7. contains genetic material from two different organisms
h. cDNA	8. the number of DNA fragments that are sufficient in aggregate length to contain the entire genome of a specified organism
i. PCR	9. short single-stranded sequences found at the ends of many restriction fragments
j. hybridization	10. a short DNA fragment that can be synthesized by a machine

Section 9.1

2. Approximately how many restriction fragments would result from the complete digestion of the human genome (3×10^9 bases) with the following restriction enzymes? (The recognition sequence for each enzyme is given in parentheses, where N means any of the four nucleotides.)

a. *Sau*3A (^GATC)
b. *Bam*HI (G^GATCC)
c. *Sfi*I (GGCCNNNN^NGGCC)

3. Why do longer DNA molecules move more slowly than shorter ones during electrophoresis?

4. You have a circular plasmid containing 9 kb of DNA, and you wish to map its *Eco*RI and *Bam*HI sites. When you digest the plasmid with *Eco*RI and run the resulting DNA on a gel, you observe a single band at 9 kb. You get the same result when you digest the DNA with *Bam*HI. When you digest with a mixture of both enzymes, you observe two bands, one 6 kb and the other 3 kb in size. Explain these results. Draw a map of the restriction sites.

5. The linear bacteriophage λ genomic DNA has at each end a single-strand extension of 20 bases. (These are "sticky ends" but are not, in this case, produced by restriction enzyme digestion.) These sticky ends can be ligated to form a circular piece of DNA. In a series of separate tubes, either the linear or circular forms of the DNA are digested to completion with *Eco*RI, *Bam*HI, or a mixture of the two enzymes. The results are shown here.

a. Which of the samples (A or B) represents the circular form of the DNA molecule?

b. What is the total length of the linear form of the DNA molecule?

c. What is the total length of the circular form of the DNA molecule?

d. Draw a restriction map of the linear form of the DNA molecule. Label all restriction enzyme sites as *Eco*RI or *Bam*HI.

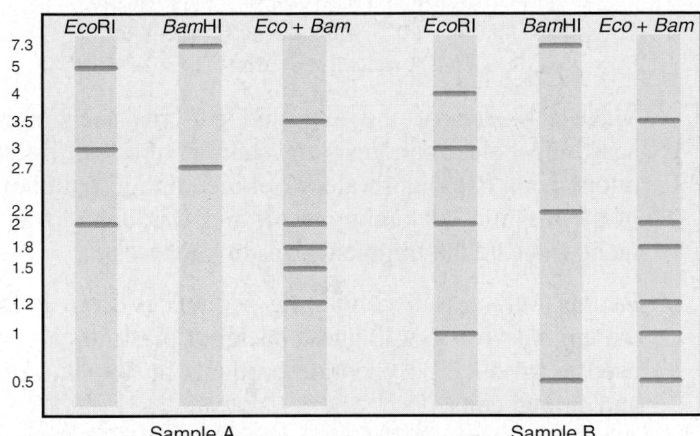

6. The following fragments were found after digestion of a circular plasmid with restriction enzymes as noted. Draw a restriction map of the plasmid.

*Eco*RI: 7.0 kb
*Sal*I: 7.0 kb
*Hin*dIII: 4.0, 2.0, 1.0 kb
*Sal*I + *Hin*dIII: 2.5, 2.0, 1.5, 1.0 kb
*Eco*RI + *Hin*dIII: 4.0, 2.0, 0.6, 0.4 kb
*Eco*RI + *Sal*I: 2.9, 4.1 kb

7. A fragment of human DNA with an *Eco*RI site at one end and a *Hin*dIII site at the other end is labeled with ^{32}P at the *Eco*RI end. This molecule is now partially digested with either restriction enzyme *Hha*I or *Sau*3A, and the partial digests are fractionated on an agarose gel. An autoradiograph of this gel is presented here.

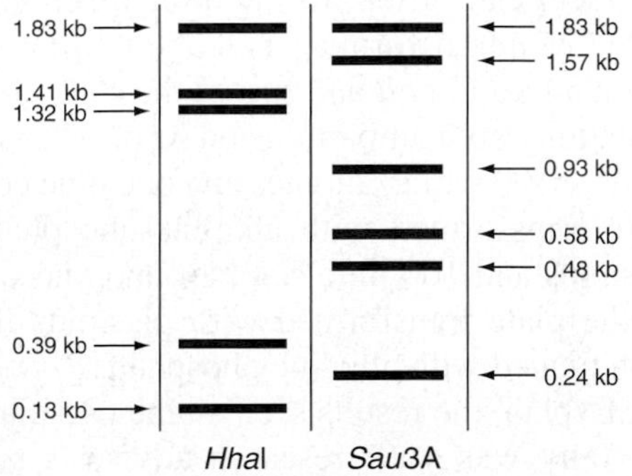

a. How large is the *Eco*RI to *Hin*dIII fragment in base pairs?
b. Draw a map of this fragment showing the locations of the *Hha*I and *Sau*3A recognition sites and the distances of sites from one end.
c. The same fragment, labeled with ^{32}P in the same fashion, is now digested to completion with *Hha*I, and the products are separated on an agarose gel. Sketch the expected pattern for the ethidium bromide stained gel and autoradiography of this gel.

Section 9.2

8. What is an artificial chromosome, and what purpose does it serve for molecular geneticists?

9. What purpose do selectable markers serve in vectors?

10. Why do geneticists studying eukaryotic organisms often construct cDNA libraries, whereas geneticists studying bacteria almost never do? Why would bacterial geneticists have difficulties constructing cDNA libraries even if they wanted to?

11. A plasmid vector pBS281 is cleaved by the enzyme *Bam*HI (G^GATCC), which recognizes only one site in the DNA molecule. Human DNA is partially digested with the enzyme *Mbo*I (^GATC), which recognizes many sites in human DNA. These two digested DNAs are now ligated together. Consider only those molecules in which the pBS281 DNA has been joined with a fragment of human DNA. Answer the following questions concerning the junction between the two different kinds of DNA.

a. What proportion of the junctions between pBS281 and all possible human DNA fragments can be cleaved with *Mbo*I?
b. What proportion of the junctions between pBS281 and all possible human DNA fragments can be cleaved with *Bam*HI?
c. What proportion of the junctions between pBS281 and all possible human DNA fragments can be cleaved with *Xor*II (C^GATCG)?
d. What proportion of the junctions between pBS281 and all possible human DNA fragments can be cleaved with *Eco*RII (Pu Pu A ^ T Py Py)? (Pu and Py stand for purine and pyrimidine, respectively.)
e. What proportion of all possible junctions that can be cleaved with *Bam*HI will result from cases in which the cleavage site in human DNA was not a *Bam*HI site in the human chromosome?

12. Consider four different kinds of human libraries: a genomic library, a brain cDNA library, a liver cDNA library, and a unique genomic sequence (no repetitive DNA) library.

a. Assuming inserts of approximately equal size, which would contain the greatest number of different clones?
b. Would you expect any of these not to overlap the others at all in terms of the sequences it contains? Explain.
c. How do these four libraries differ in terms of the starting material for constructing the clones in the library?

13. As a molecular biologist and horticulturist specializing in snapdragons, you have decided that you need to make a genomic library to characterize the flower color genes of snapdragons.

a. How many genomic equivalents would you like to have represented in your library to be 95% confident of having a clone containing each gene in your library?

b. How do you determine the number of clones that should be isolated and screened to guarantee this number of genomic equivalents?

14. Imagine that you are a molecular geneticist studying a particular gene in which mutations cause a serious human disease. The gene, including its flanking regulatory sequences, spans 200 kb of DNA. The distance from the first to the last coding base is 140 kb, which is divided among 10 exons and 9 introns. The exons contain a total of 9.7 kb, and the introns contain 130.3 kb of DNA. You would like to obtain the following for your work: (a) an intact clone of the whole gene, including flanking sequences; (b) a clone containing the entire coding sequences but no noncoding sequences; and (c) a clone of exon 3, which is the site of the most common disease-causing mutation in this gene. Based on the size of your inserts, what vectors would you use to obtain each of these clones, and why?

15. A 49 bp *Eco*RI fragment containing the somatostatin gene was inserted into the vector pWR590 shown below. The sequence of the inserted fragment is

```
5′ AATTCGATCCTATCAACACGAAGTGAAAGTCTTACAACCCATGAATTCG 3′
3′ GCTAGGATAGTTGTGCTTCACTTTCAGAATGTTGGGTACTTAAGCTTAA 5′
```

Distances between adjacent restriction sites in the pWR590 vector are indicated in the diagram. What are the patterns of restriction digests with *Eco*RI (G^AATTC) or with *Mbo*I (^GATC) before and after cloning the somatostatin gene into the vector? (E = *Eco*RI, M = *Mbo*I)

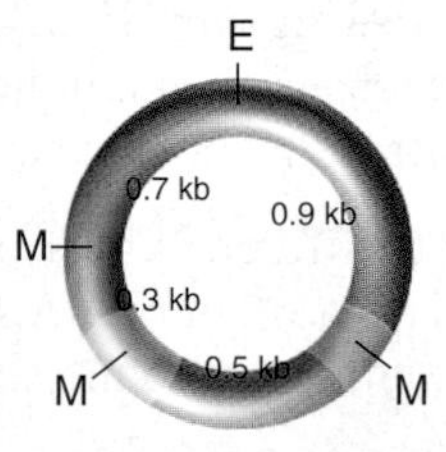

16. The *Notch* gene involved in *Drosophila* development is contained within a restriction fragment of *Drosophila* genomic DNA produced by cleavage with the enzyme *Sal*I. The restriction map of this *Drosophila* fragment for several enzymes (*Sal*I, *Pst*I, and *Xho*I) is shown here; numbers indicate the distances between adjacent restriction sites. This fragment is cloned by sticky-end ligation into the single *Sal*I site of a bacterial plasmid vector that is 5.2 kb long. The plasmid vector has no restriction sites for *Pst*I or *Xho*I enzymes.

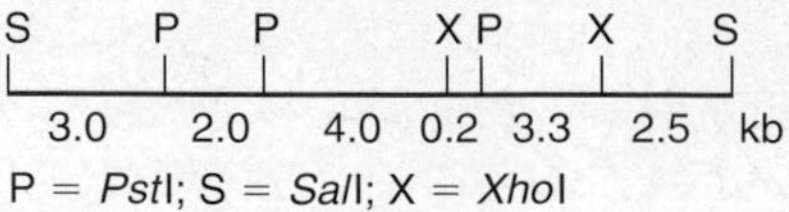

Make a sketch of the expected patterns seen after agarose gel electrophoresis and staining of a *Sal*I digest (alone), of a *Pst*I digest (alone), of a *Xho*I digest (alone), of the plasmid containing vector and *Drosophila* fragment. Indicate the fragment sizes in kilobases.

17. Expression vectors containing sequences from both the animal virus SV40 and a bacterial plasmid can be used to produce polypeptide products in both *E. coli* and animal cells.

a. Give the general characteristics that are necessary for this type of plasmid to be used as a vector in two very different types of cells.

b. What is required for *E. coli* expression of a human protein?

18. Your undergraduate research advisor has assigned you a task: Insert an *Eco*RI-digested fragment of frog DNA into an *E.coli* plasmid that carries a *lacZ* gene with an *Eco*RI site in the middle (see Fig. 9.8 on p. 313). Your advisor suggests that after you digest your plasmid with *Eco*RI, you should treat the plasmid with the enzyme alkaline phosphatase. This enzyme removes phosphate groups that may be located at the 5' ends of DNA strands. You will then add the fragment of frog DNA to the vector and join the two together with the enzyme DNA ligase.

You don't quite follow your advisor's reasoning, so you set up two ligations, one with plasmid that was treated with alkaline phosphatase and the other without such treatment. Otherwise, the ligation mixtures are identical. After the ligation reactions are completed, you transform a small aliquot (portion) of each ligation into *E. coli* and spread the cells on petri plates containing both ampicillin and Xgal. The next day, you observe 100 white colonies and one blue colony on the plate transformed with alkaline-phosphatase-treated plasmids and 100 blue colonies and one white colony on the plate transformed with plasmids that had not been treated with alkaline phosphatase.

a. Explain the results seen on the two plates.

b. Why was your research advisor's suggestion a good one?

c. Why would you normally treat plasmid vectors with alkaline phosphatase but not the DNA fragments you want to add to the vector?

Section 9.3

19. Listed here are steps you would take to clone the *idiosynchratase* gene from the hoot owl, using a probe made from the corresponding gene in bald eagles. Put these into their appropriate order. (There are multiple correct answers.)

a. Mix together BAC vector DNA and hoot owl DNA with ligase.
b. Expose nitrocellulose paper disks to UV radiation and baking.
c. Extract genomic DNA from hoot owl cells.
d. Perform autoradiography.
e. Produce a labeled DNA probe to the *idiosynchratase* gene.
f. Completely digest BAC vector DNA with the restriction enzyme *Hin*dIII.
g. Place nitrocellulose paper disks onto the agar surface to transfer colonies.
h. Incubate DNA probe with nitrocellulose paper disks.
i. Distribute bacteria onto a petri plate containing agar and nutrients and allow growth into colonies.
j. Partially digest hoot owl genomic DNA with the restriction enzyme *Hin*dIII.
k. Transform bacteria.

20. a. Given the following restriction map of a cloned 10 kb piece of DNA, what size fragments would you see after digesting this linear DNA fragment with each of the enzymes or combinations of enzymes listed? (1) *Eco*RI, (2) *Bam*HI, (3) *Eco*RI + *Hin*dIII, (4) *Bam*HI + *Pst*I, and (5) *Eco*RI + *Bam*HI.

b. What fragments in the last three double digests would hybridize on a Southern blot with a probe made from the 4 kb *Bam*HI fragment?

E HH E B P B E

1.5 0.6 1.0 1.2 2.1 1.9 1.7 kb

21. Professor Ozone is hard at work trying to produce huge quantities of the Ozonase enzyme to reduce smog levels in his home town of Los Angeles. The protein is produced naturally in small quantities by an elusive bacterium called Phew. Professor Ozone has already determined a partial amino acid sequence of the protein, and he also has an antibody that recognizes the protein. Now he wants to clone the gene that encodes the protein. Describe two different approaches that he could use to create and identify clones of the gene. Describe how he should go about producing large quantities of the enzyme.

22. You have cloned and characterized a particularly interesting protein-coding gene from the bacterium *Bacillus subtilis,* and you would like to isolate the corresponding, homologous gene from the rare, poorly characterized bacterial species *Beneckea nigripulchritudo* that infects certain shrimp. You decide to make degenerate probes to identify, by hybridization, clones containing this homologous gene. The amount of degeneracy is a potential problem because the more types of different DNA molecules contained in the probe, the worse the signal-to-noise ratio in the hybridization experiment. How can you minimize the degeneracy? Be as specific as possible, mentioning such factors as the length of the probe and the region of DNA you will choose to synthesize by reverse translation.

23. Human genomic DNA was digested with the various restriction enzymes noted in the list below. These digests were subjected to electrophoresis on an agarose gel; the DNA separated in the gel was then stained with ethidium bromide, and a photograph of the fluorescence was taken. The DNA in the gel was transferred to a nitrocellulose filter to make a Southern blot, and this blot was then probed with a radioactive 5 kb-long fragment of cloned human DNA with *Eco*RI sites at both ends. The sizes of the dark bands seen on an X-ray film exposed to the Southern blot for each digest were

*Eco*RI: 5 kb
*Kpn*I: 2.5 kb, 6 kb
*Hin*dIII: 8 kb
*Eco*RI + *Kpn*I: 4 kb, 1 kb
*Eco*RI + *Hin*dIII: 5 kb
*Kpn*I: + *Hin*dIII: 2.5 kb, 4.5 kb

a. Why were these digests separated by electrophoresis on agarose gels rather than polyacrylamide gels?
b. Describe what you would see on the photograph of the ethidium bromide-stained gel.
c. In this problem, the sums of the sizes of all the dark bands seen on the X-ray film of the Southern blot are not the same for all the digests reported. However, in previous problems involving restriction mapping (such as Problems #5, 6, 15, and 16), all the digests of a particular DNA sample produce fragments the sum of whose sizes are the same. Explain this difference.
d. Draw a restriction map that accounts for the results of this Southern blot.
e. Can you orient the restriction map you drew in part (d) to the centromere-to-telomere direction along the human chromosome on which these DNA sequences are located?

Section 9.4

24. Using PCR, you want to amplify a ~1 kb exon of the human autosomal gene encoding the enzyme phenylalanine hydroxylase from the genomic DNA of a patient suffering from the autosomal recessive condition phenylketonuria (PKU).

a. Why might you wish to perform this PCR amplification in the first place, given that the sequence of the human genome has already been determined?

b. Calculate the number of template molecules that are present if you set up a PCR reaction using 1 nanogram (1×10^{-9} grams) of chromosomal DNA as the template. Assume that each haploid genome contains only a single gene for phenylalanine hydroxylase and that the molecular weight of a base pair is 660 grams per mole. The human genome contains 3×10^9 base pairs.

c. Calculate the number of PCR product molecules you would obtain if you perform 25 PCR cycles and the yield from each cycle is exactly twice that of the previous cycle. What would be the mass of these PCR products taken together?

25. Which of the following set(s) of primers could you use to amplify the target DNA sequence below, which is part of the last protein-coding exon of the *CFTR* gene?

```
5′ GGCTAAGATCTGAATTTTCCGAG ... TTGGGCAATAATGTAGCGCCTT 3′
3′ CCGATTCTAGACTTAAAAGGCTC ... AACCCGTTATTACATCGCGGAA 5′
```

a. 5′ GGAAAATTCAGATCTTAG 3′;
5′ TGGGCA ATAATGTAGCGC 3′

b. 5′ GCTAAGATCTGAATTTTC 3′;
3′ ACCCGTTATTACATCGCG 5′

c. 3′ GATTCTAGACTTAAAGGC 5′;
3′ ACCCGTTATTACATCGCG 5′

d. 5′ GCTAAGATCTGAATTTTC 3′;
5′ TGGGCA ATAATGTAGCGC 3′

26. Problem #25 raises several interesting questions about the design of PCR primers.

a. The importance of PCR stems from the fact that it can amplify a single region of DNA from a complex genome. How can you be sure that the two primers you chose as your answer to Problem #25 will amplify only an exon of the CFTR gene from a sample of human genomic DNA?

b. The protocol for PCR shown in Fig. 9.16 on pp. 328–329 states that each of the primers used should be 16–26 nucleotides long. (i) Why do you think the lower limit would be approximately 16? (ii) The upper limit of 26 nucleotides is not absolute. For some applications of PCR, it is possible to use longer primers, but at the risk of introducing potential difficulties. What complications or disadvantages might be associated with longer primers?

c. Suppose that one of the primers you designed in your answer to Problem #25 had a mismatch with a single base in the genomic DNA of a particular individual. Would you be more likely to obtain a PCR product from this genomic DNA if the mismatch were at the 5′ end or at the 3′ end of the primer? Why?

27. You wish to purify large amounts of the part of the CFTR protein that is encoded by the last protein coding exon shown in Problem #25 and that begins with the amino acid sequence

N...Leu Arg Ser Glu Phe Ser Glu...C

and ends with the sequence

N...Trp Ala Ile Met (C terminus)

You will start this process by cloning an appropriate PCR product into the pMore vector, part of whose sequence is shown in the following. The pMore vector makes large amounts of maltose binding protein (MBP) when transformed into *E. coli*. The amino acids shown with the vector sequence correspond to the C-terminal end of MBP. To do the cloning, you will digest both the pMore vector and your PCR product with both the *Eco*R1 (G^AATTC) and *Sal*I (G^TCGAC) restriction enzymes and then ligate the pieces together. The vector has only a single site for each of these enzymes.

```
5′...AGGATTTCAGAATTCGGATCCTCTAGAGTCGACCTGTAGGGCAA...3′
   ArgIleSerGluPheGlySerSerArgValAspLeupMore vector
```

a. Describe the fusion protein that will be made when the PCR product is ligated into the vector. What are the orientations of the parts of MBP and CFTR relative to that of the fusion protein?

b. What advantages might there be for cutting the vector and PCR product with two restriction enzymes instead of one?

c. Design PCR primers that will allow you to construct the desired recombinant DNA molecule. Note (i) that the sequence shown in Problem #25 has neither *Eco*R1 nor *Sal*I sites, (ii) that additional nucleotides can be added to appropriate locations in the PCR primers, and (iii) that restriction enzymes require about 5 nucleotides on either side of the restriction site for the enzymes to work. This problem is extremely difficult, but will help you integrate a great deal of information about gene structure and recombinant DNA technology.

d. MBP can bind to the sugars amylose and maltose. The last 20 amino acids at the C terminus of MBP are not required for this property. It is also possible to synthesize chemically an amylose resin (beads with covalently bound amylose). How would these facts be helpful in allowing you to purify a large amount of a region of the CFTR protein?

Section 9.5

28. Several of the techniques discussed in this chapter, particularly restriction mapping and methods based on DNA hybridization such as Southern blots and screening libraries with oligonucleotide probes, are still often used for studying genes in unusual organisms. However, in the twenty-first century, these techniques are used much more rarely than in the late twentieth century for studying genes in humans or in model organisms such as yeast, *C. elegans, Drosophila,* or mice. What has changed with the millenium, and what new techniques have arisen as replacements?

29. Which of the following processes used in biotechnology relies on specific enzymes? What are those enzymes? What is the basis for any of these processes that are not enzyme based?
a. DNA ligation
b. cleavage of DNA at specific sites
c. cleavage of DNA at nonspecific sites
d. DNA hybridization
e. DNA sequencing
f. cDNA synthesis
g. PCR

30. a. If you are presented with the following sequencing autoradiogram, what can you say about the sequence of the template strand used in these sequencing reactions?
b. If the template for sequencing is the strand that resembles the mRNA, write out the sequence of the mRNA insofar as it can be determined.
c. Is this portion of the genome likely to be within a coding region? Explain your answer.

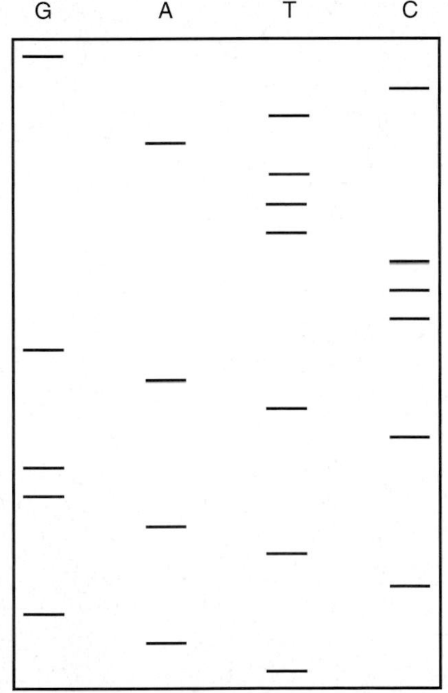

31. You read the following sequence directly from a gel.

5′ TCTAGCCTGAACTAATGC 3′

a. Make a drawing that reproduces the autoradiogram from which this sequence was read.
b. Assuming this sequence is from an exon in the middle of a gene, does this newly synthesized strand or the template strand have the same sequence as the mRNA for the gene (except that T's are present instead of U's)? Justify your answer.
c. Using the genetic code table, give the amino acid sequence of the hexapeptide (six amino acids) translated from the 18-base message. Indicate which is the amino terminal end of the peptide.

32. The following figure portrays a trace derived from the automated sequencing of a certain PCR product produced by the amplification of the genomic DNA from a particular person's cells. The left-to-right orientation of the peaks on the trace corresponds to smaller-to-larger fragments of DNA. The height of the peaks is unimportant. (red = T; green = A; black = G; purple = C)
a. What does the green peak at the left end of the trace signify? Be as precise as possible.
b. Write the sequence of DNA revealed by this trace, indicating the 5′-to-3′ orientation.
c. What do you think is meant by "residue position"? That is, what is located at residue position 1?
d. Explain the apparent anomaly at residue position 370.

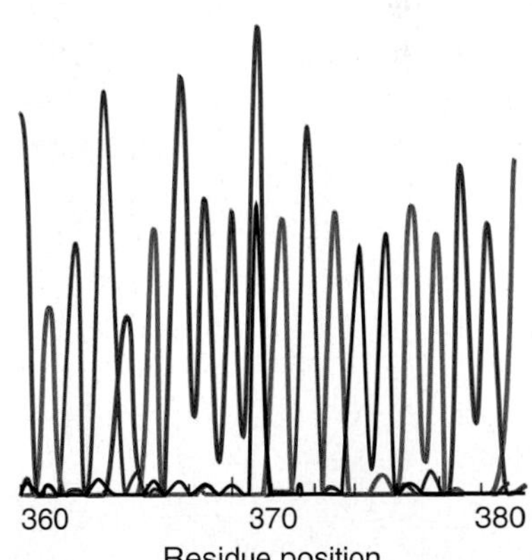

Reconstructing the Genome Through Genetic and Molecular Analysis

Chapter 10

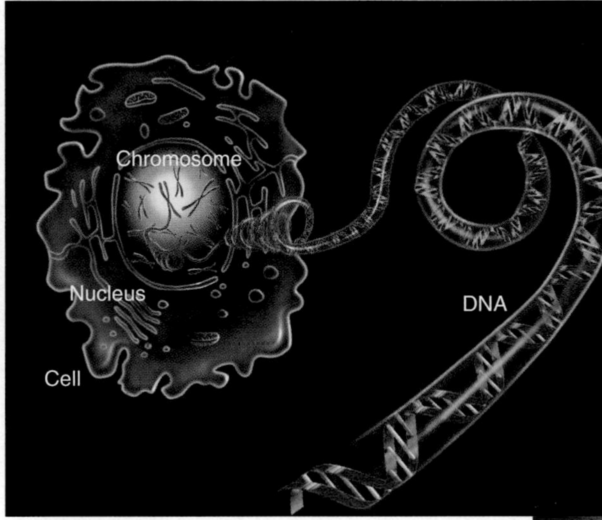

The human genome, present in the nucleus of each cell, contains the instructions for transforming a single fertilized egg cell into an adult with 10^{14} cells. Each genome has ~3 billion letters of the DNA language divided among 24 distinct chromosomes ranging in size from 45 million to 280 million letters. The Human Genome Project undertook the challenging task of deciphering this book of life.

Since the mid-nineteenth century, three advances have radically transformed the field of genetics: Mendel's discovery of fundamental principles in the 1860s, Watson and Crick's elucidation of DNA structure in 1953, and the Human Genome Project from 1990 to the present. In previous chapters, we examined Mendel's ideas and impact, as well as the contributions of Watson and Crick. In this chapter, we discuss the Human Genome Project and the field of *genomics* that it spawned.

The **Human Genome Project** was initiated to sequence and analyze the human genome in conjunction with the genomes of several model organisms. A **genome** is the entire haploid collection of chromosomes present in each cell of an individual organism. The human genome, for example, contains 24 different chromosomes (22 autosomes and the X and Y sex chromosomes). These 24 strings of G's, C's, T's, and A's contain a total of approximately 3 billion nucleotides and range in size from 45 million to 279 million bp. **Genomics,** the study of whole genomes, is a branch of biology dedicated to the development and application of more effective mapping, sequencing, and computational tools. Genomicists use large-scale, or *global,* molecular techniques for linkage analysis, physical mapping, and the sequencing of genomes to generate vast amounts of data, which they then analyze by computer. Sophisticated programs enable genomicists to predict the existence and, in some cases, the general functions of previously undefined genes. Their molecular biologist colleagues can then confirm the computer-generated projections through biochemical studies and genetic engineering in model organisms. **Model organisms,** often used in genomic analysis, have many genetic mechanisms and cellular pathways in common with each other and with humans. Unlike humans, however, they are amenable to classical breeding experiments and direct experimental manipulation of the genome. The fruit fly studied by T. H. Morgan and his group, the nematode worm introduced in Chapter 8, and the common house mouse are three examples of model organisms. For more about model organisms, see Chapter 20 as well as Genetic Portraits A–E on our website: www.mhhe.com/hartwell3.

The history of the Human Genome Project provides a fascinating glimpse into how genomics and the study of model organisms have changed the practice of genetics. In the spring of 1985, the first meeting on the project took place in Santa Cruz, California. The chancellor of the University of California at Santa Cruz had assembled 12 biologists of diverse backgrounds to explore the idea of starting an institute to sequence the human genome. After two days of heated discussion, the 12 biologists concluded that it would, indeed, be possible to develop the technology required to accomplish this then seemingly impossible objective. However, the group was split on whether it would be a good idea scientifically. Two aspects of their discussion were striking. First, the concept of the Human Genome Project introduced the idea of discovery science—a new scientific approach to biology. In **discovery science,** one seeks to identify all the elements of a biological system—for instance, the complete sequences of the 24 chromosomes that contain the 3 billion nucleotides of the human

Figure 10.1 Covers of the February 2001 issues of *Nature* and *Science*. *Nature* published a draft of the human genome sequence obtained by the public effort; *Science* published a draft produced by private efforts.

genome—and place them in a database to enrich the infrastructure of biology. Researchers doing discovery science "discover" the data and make it available for biological analysis. Discovery science stands in contrast to hypothesis-driven approaches to biology in which one asks questions and seeks experimental verification of possible answers. Second, the Human Genome Project required the development of high-throughput automated DNA sequencing technology as well as the computational tools necessary for capturing, storing, and analyzing the vast amounts of cloning, mapping, and sequence data associated with obtaining genome sequences. These technological and computational developments would require integrating the disciplines of engineering, computer science, physics, and mathematics with biology, leading to a cross-disciplinary approach that would eventually transform modern biology and genetics.

Not surprisingly, most biologists initially viewed the Human Genome Project with skepticism, for several reasons. They thought the project would not be very interesting, since many geneticists claimed that only about 2% of the genome represented genes; the remaining 98%, they argued, was "junk." In the mid-1980s, most biologists also believed that the Human Genome Project was not really a scientific endeavor because it was not hypothesis-driven. Many did not understand how the discovery approach to determining the sequence of the human genome would revolutionize the power and potential of genetic and other biological studies. Finally, some viewed the Human Genome Project as big science that would inappropriately compete for funds with the more fruitful and productive small, hypothesis-driven science. Reasonable answers existed for each of these concerns. We now understand, for example, that some of the DNA sequences originally considered to be "junk" may play an important role in sculpting the structure of the genome and that the sequences occasionally evolve into genes or regulatory regions.

In 1988, the National Academy of Sciences appointed a committee (half proponents and half opponents) to consider the scientific merits of the Human Genome Project. After a year of vigorous debate and analysis, the committee unanimously endorsed the project. This was a major turning point in its acceptance by the community of biologists and geneticists. The government-funded Human Genome Project began in 1990 with a projected 15-year time scale and a $3 billion budget for completing the human genome sequence. Through periodic reassessments, the Human Genome Project came to have six distinct objectives:

1. to generate physical, genetic, and sequence maps of the human genome;
2. to sequence the genomes of a variety of model organisms;
3. to develop improved technologies for mapping and sequencing;
4. to develop computational tools for capturing, storing, analyzing, displaying, and distributing map and sequence information;
5. to sequence EST fragments of cDNAs (*expressed sequence tags,* or *ESTs,* are single-sequence runs [~600–1000 bp in length] on the cDNA inserts) and eventually full-length cDNAs encoding the expressed mRNAs in different cell types of humans and mice;
6. to consider the ethical, social, and legal challenges posed by genomic information.

Biology's first large-scale project is close to achieving many of its goals. A rough sequence draft of the human genome was completed in February 2001 (**Fig. 10.1**); in this "draft" the sequence did not yet have an appropriate level of accuracy (an error rate of 1/10,000), and it had some gaps. The final sequence was completed in 2003, two years ahead of the originally proposed 2005 finish date. The early finish was, in part, catalyzed by the 1998 promise of Celera, a private company, to complete a draft of the genome in just three years, employing a novel sequencing strategy. The Celera challenge prompted the federally supported genome effort to move its timetable for sequencing a draft of the human genome ahead by several years. Indeed, this private/public competition benefited biology and genetics through the earlier availability of the human genome sequence.

In addition to the human genome sequence, draft or finished genome sequences existed for eight model organisms by 2002: *Escherichia coli,* representing bacteria; *Saccharomyces cerevisiae,* or the bread yeast, representing single-celled eukaryotes; *Drosophila melanogaster,* the fruit fly, and *Caenorhabditis elegans,* the nematode, together representing multicellular animals of moderate complexity; *Arabidopsis thaliana,* a mustard weed, representing dicotyledonous plants; *Oryza sativa,* or rice, representing monocotyledonous plants; *Fugu rubripes,* the Japanese puffer fish, representing fish and a very compact veretebrate genome; and *Mus musculus,* the house mouse, representing the mammalian class of animals, which includes humans. By mid-2005, more than 50 eukaryotic and 230 prokaryotic genomes had been sequenced. Researchers chose to analyze these models in conjunction with the human genome because the organisms have increasingly complex genomes that are amenable to genetic analysis and that have already been extensively mapped and studied. By comparing the DNA of model organisms to that of humans, researchers seek to uncover genes and other critical DNA elements that are conserved across evolutionary lines. The Human Genome Project catalyzed efforts to sequence these myriad model organism genomes as well as the hundreds of bacterial and archaeal genomes. Such sequencing projects have revolutionized the study of microbiology as well as that of genetics, molecular biology, and plant development.

In other accomplishments, production of a human genomewide genetic map has made it possible to identify more than 1500 genes causing different human diseases. The identification of genes predisposing to specific diseases creates new opportunities for diagnosis and therapy as well as for understanding the biology of the disease. More than 6 million human and 3 million mouse expressed sequence tags (*ESTs* which are portions of cDNA sequences) have been sequenced; and projects are underway to sequence most mouse and human full-length cDNA sequences. These EST and cDNA sequences are playing an important role in defining gene locations, alternative RNA splicing patterns, and exon/intron boundaries in genomic sequences.

The Human Genome Project has also catalyzed the development of high-throughput automated DNA sequencing instruments (for example, a 96-capillary automated DNA sequencer that has a 2000-fold throughput increase over the original sequencing prototype instrument), given rise to a variety of novel strategies for mapping and sequencing genomes, and supported the development of computational tools for dealing with the ensuing explosion of genomic information.

Yet another significant aspect of the project is that it is the first scientific project to include in its statement of purpose consideration of the social, ethical, and legal challenges emerging from its projected results; for example, the issue of genetic privacy. At the suggestion of James Watson, the first director of the Human Genome Project, 3% to 5% of the project's research funds were dedicated to the study of ethical, legal, and social implications of mapping the genome—ELSI is the acronym for this arm of the project. A critical concept is to educate society about the opportunities and challenges of the new genetics so that public and private figures alike can make wise decisions about resource allocations and the rules and regulations governing applications of the new technologies.

Completion of the Human Genome Project is moving biology into a postgenome sequencing era where researchers can use global measurements to analyze complex systems (see Chapter 12). The ultimate goal of genomic sequencing and genetic analysis is to view the genome of any organism as a whole, both to make a connection between every gene and its specific function and to understand how all the genes fit together on the chromosomes. With this knowledge for all genes in the human genome, biologists may one day understand the molecular basis of every genetic disease, as well as the networks of molecular interactions that bring each person to life in a unique way.

As we describe the tools and potential applications of genomics, we encounter a compelling general theme: The social and personal repercussions of the information gained from genomic analysis are generating new areas of biological concern that require close attention as the new knowledge unfolds.

In this chapter, we examine:

- Challenges and strategies of genome analysis, including genome size; features to be analyzed; problems with DNA variations, or polymorphisms; and the development of three types of whole-genome maps—high-density linkage maps, long-range physical maps, and complete sequence maps.
- Major insights emerging from complete and nearly complete genome sequences, including the number and types of genes, the nature of their regulatory sequences, the extent of repeated sequences, genome organization and structure, evolution by lateral gene transfer, and areas of future investigation.
- High throughput tools for analyzing genomes, including DNA sequencers and DNA arrays.

10.1 Analyses of Genomes

The Genomes of Living Organisms Vary Enormously in Size

The genomes of microbes and eukaryotes range from the 700,000 base pairs (700 kb) in a single microbial chromosome to more than 3 billion base pairs (3 gigabase pairs, or 3 Gb) distributed among the 5–96 chromosomes of various mammals to even larger genomes. **Table 10.1** gives the genome sizes of representative microbes, plants, and animals. One amoeba, *Amoeba dubia,* has a genome of more than 600 billion base pairs, and wheat has a genome of 15 billion base pairs. To put these numbers in perspective, the human genome is 200 times larger than the yeast genome and 200 times smaller than the genome of *Amoeba dubia.* Thus, the information content of a genome is not necessarily proportional to its size. The large size of some genomes presents fascinating challenges for their ultimate characterization and analysis.

Genomicists Look at Two Basic Features of Genomes: Sequence and Polymorphisms

Two primary goals of genomic researchers are to determine the sequences of each of the chromosomes in the genome and to identify as many as possible of the **polymorphisms,** or allelic variants, present throughout the genome.

There Are Several Major Challenges to Achieving These Two Objectives

In their quest to sequence and analyze genomic DNA, genomicists must confront various issues.

Questions About Sequencing We've seen that the lengths of chromosomes range from 700 kb to more than 500 million base pairs (500 Mb), and each chromosome is a single continuous double-stranded string of bases. Current technology allows one to sequence, on average, 1000 bp at a time (these 1000 bp are known as a *read*). One question is,

TABLE 10.1 A Comparison of the Developmental Complexity and Genome Features* of Model Organisms

Organism: Type	Organism: Species	Developmental Complexity	Genome Size* (Megabases)	Predicted Genes	# of Genes per Million bp Sequenced	Date Genome Finished
Bacterium	*Escherichia coli*	1-cell prokaryote	4.64	4200	905	1997
Archaebacterium	*Halobacterium sp. NRC-1*	1-cell	2.57	2630 (2411)	1023	2000
Yeast	*Saccharomyces cerevisiae*	1-cell eukaryote	12.07	5800	483	1996
Worm	*Caenorhabditis elegans*	~1000 cells	100	19,099	197	1998
Fly	*Drosophila melanogaster*	~50,000 cells	180	13,601	117	2000
Mustard weed	*Arabidopsis thaliana*	10^{10} cells	125	25,000	221	2000
Pufferfish (draft)	*Fugu rubripes*	10^{12} cells	380	38,000	118	2002
Rice (draft)	*Oryza sativa*	5×10^{10} cells	Ind/jap 466/420	46,000–56,000/ 32,000–50,000	127–155/ 82–128	2002
Mouse	*Mus musculus*	10^{11} cells	3200	25,000	10–13	2005
Human	*Homo sapiens*	10^{14} cells	3200	25,000	18	2003

*Haploid genome size, including heterochromatic DNA.
Note: For the sequenced genomes of model organisms: Gene numbers are taken from the original sequence publications; most numbers have since changed slightly and different sources give different estimates, depending on protocols. For rice, two different strains have been sequenced: *Oryza sativa L. ssp. Japonica* and *Oryza sativa L. ssp. Indica*.

How does one sequence a 500 Mb chromosome 1000 bp at a time? The fact that the DNA sequencing error rate is about 1% per sequence read (that is, per 1000 bp) raises a second question: How accurate should a genome sequence be? It must be accurate enough to readily delineate important chromosomal features, for example, exons and genes. Most geneticists believe that error rates between one error per 1000 bp (1/1000) and one error per 10,000 bp (1/10,000) are acceptable. The Human Genome Project has selected an error rate of less than 1/10,000 for its finished sequence. Genome sequencing faces several challenges in addition to questions of technology and accuracy.

How Does One Distinguish Sequence Errors from Polymorphisms? Higher organisms are diploid; therefore, each genome contains maternal and paternal complements of chromosomes with their attendant polymorphisms. The rate of polymorphisms in humans is about 1 in 500 bp. How does one distinguish these polymorphisms from DNA sequence errors?

Repeated Sequences May Be Hard to Place The genomes of higher organisms are replete with *repeated sequences.* These repetitive sequences range from tens of base pairs, as in simple sequence repeats like $(AT)_n$, to replicated chromosomal blocks that can span hundreds of kilobases (see pp. 396–398 in Chapter 11 as well as pp. 216–217 of the Genetics and Society box in Chapter 7 for more details on DNA repeats). Repetitive sequences may be tandemly arrayed (that is, positioned side by side), or they may be scattered across the genome. All repeated sequences are identical when they arise, but they diverge as they age. They are thus evolutionary signposts that reveal how much time has elapsed since a duplication event occurred. Young repeats, of recent origin, are very similar, if not identical; older repeats, of more ancient origin, may be quite diverged in sequence. It is the young, very similar repeats longer than the 1000 bp reads that constitute a challenge for current approaches to sequencing. If, for example, a 1000 bp read falls in a repeat that has 10,000 copies dispersed around the genome, how do you know which of the repeats you are looking at and where in a particular chromosome it belongs?

Unclonable DNA Cannot Be Sequenced As we saw in Chapter 9, the first step of any initial sequencing is cloning. But, a significant fraction of genomic DNA, often referred to as **heterochromatic DNA,** is not clonable. Heterochromatic DNA may constitute up to 30% or more of the genome, and many believe it contains few genes. How do we know what heterochromatic regions contain if we cannot clone and sequence them?

Given these challenges, how do genome analysts proceed?

A Divide and Conquer Strategy Makes It Possible to Meet Most of the Challenges If one can sequence only 1000 bp at a time, a divide and conquer strategy is the only possible solution for sequencing multimegabase chromosomes. With this approach, the chromosomes are broken into small overlapping pieces, and all pieces are cloned. After the 1000 bp at both ends of the clones are sequenced, the clones, whose ends overlap, are reassembled into their original chromosomal strings. **Figure 10.2** outlines this strategy. The overlapping ends of the cloned DNA inserts are necessary for the final assembly, or joining, of the sequence information (Fig. 10.2b), since there is no way to assemble or determine the order of adjacent fragments with no sequence overlap (Fig. 10.2a). The sequencing of 1000 bp at both ends of each DNA insert provides 2000 bp of sequence per clone analyzed.

To achieve an error rate of 1/10,000, each DNA sequence region must be sequenced multiple times from independently derived clones. Trial and error has shown that the sequencing of every chromosomal region from 10 independent inserts can generate an error rate of less the 1/10,000. This tactic is termed **10-fold sequence coverage.** Ten-fold coverage readily permits the random errors (seen on 1 in 10 of the sequence fragments) to be distinguished from the polymorphisms (seen on 5 in 10 sequence fragments). With 10-fold coverage, the 1000 bp reads are assembled into individual DNA strings computationally by alignment of the overlapping sequences from each insert.

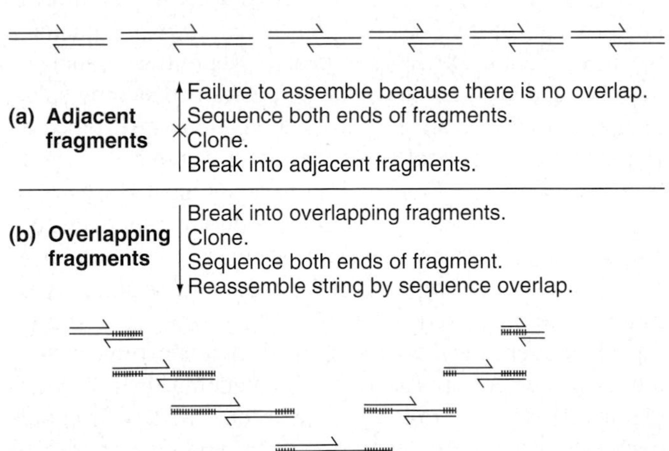

(a) Adjacent fragments
Failure to assemble because there is no overlap.
Sequence both ends of fragments.
Clone.
Break into adjacent fragments.
(b) Overlapping fragments
Break into overlapping fragments.
Clone.
Sequence both ends of fragment.
Reassemble string by sequence overlap.

Figure 10.2 The divide and conquer strategy for DNA sequencing requires overlapping fragments. The *arrows* indicate sequence determined from each end of the clone. The *vertical bars* indicate overlapping sequences between two clones, which make sequence assembly possible. **(a)** The original DNA string is cut into adjacent fragments that cannot be assembled. **(b)** The original DNA string is cut into overlapping fragments that can be assembled.

A critical question in the divide and conquer strategy is whether to generate the overlapping fragments for sequence analyses directly from the genome (a whole-genome approach) or to generate them from large insert clones that researchers have previously ordered in a physical map (a hierarchal approach). We consider this issue when we describe the three types of chromosomal maps—linkage, physical, and sequence—that view the genome at differing levels of resolution. First, however, we provide a brief overview of basic analytic techniques.

Four Relatively Simple Techniques Make Genome Characterization Possible

Cloning, hybridization, PCR amplification, and sequencing (described in detail in Chapter 9), as well as several computational tools, lie at the heart of the various mapping and sequencing strategies for genome analysis.

Review of Techniques for Mapping and Sequencing

Cloning Fragments of genomic DNA ranging in size from 500 bp to more than 1,000,000 bp may be cloned into a variety of vectors (review Table 9.2 on p. 311). The entire collection of cloned DNA fragments for a particular DNA source is termed a **library.** A library may have low coverage of a particular genome with each DNA sequence region on average represented just one time or less, or it may have high coverage with each sequence represented on average 10 times or more.

Hybridization Researchers use hybridization to determine the location of particular DNA sequences in a library of fragments. Recall that hybridization depends on the molecular complementarity of a target DNA sequence with a DNA fragment from the library. A hybridization protocol is carried out after some type of fragment separation, such as the plating out of individual insert colonies on a filter. Through hybridization with specific, defined fragments, the clones containing a particular gene or chromosomal region can be readily identified for mapping or sequencing.

PCR Amplification Using the polymerase chain reaction (PCR), researchers can directly amplify particular regions of the genome up to a millionfold or more. This amplification requires that DNA sequences on either side of the region to be amplified are known so that PCR primers may be selected. Regions ranging from 50 bp to more than 20 kb may be amplified by PCR. A critical point for genomewide analyses is that the pair of PCR primer sequences must be unique (not in a repetitive region). If they are not, the PCR protocol will amplify multiple regions of the genome.

DNA Sequencing An automated DNA sequencer employing the Sanger sequencing method can determine DNA sequences about 1000 bp in length from cloned DNA fragments.

Computational Tools Four types of computational tools are important to genomics: programs for identifying the sequence matches between a particular DNA fragment and a large population of previously sequenced fragments; programs for identifying the overlaps of individual sequences and assembling them into strings of bases; programs for estimating the error rates in different DNA sequences; and programs for identifying genes in chromosomal sequences.

We now describe how researchers use these basic tools in various approaches to the mapping and sequencing of genomes.

Large-Scale Maps Serve As Guides to Whole Genomes

Chromosomal maps show the location of genes, genetic markers, centromeres, telomeres, and other points of interest along the chromosome of an organism. In Chapters 5 and 9, we described techniques for mapping a small number of loci in a relatively small region of a genome. Here we examine how genomicists expand on these techniques to produce genetic, physical, and sequence maps for a whole genome. Geneticists use the terms "linkage" and "genetic" interchangeably when talking about maps made through analyses of recombination frequencies.

The analysis of the human genome focused initially on a genomewide linkage map, then a physical map, and finally a sequence map. These maps were successively integrated into each other. The linkage map was produced by a global genomewide approach; the physical map by a hierarchical divide and conquer approach; and the sequence map by both strategies.

High-Density Linkage Maps: Computerized Analyses of Transmission Data Position Unlimited Numbers of Markers in Relation to Each Other

With the traditional linkage analysis described in Chapter 5, geneticists map only a handful of loci in any one cross because easy-to-follow phenotypes, such as those of color and gross morphology, tend to interfere with each other's expression. In addition, the number of good phenotypic markers is relatively limited (compared, for example, to the number of genes). Moreover, a single phenotype may be encoded by many genes. For example, 34 different genes cause mice to have white spots. Each white-spot gene maps apart from the others, but in a two-point cross following, say, black coat color and white spots, it is impossible to tell which one of the 34 white-spot genes is contributing to the phenotype observed. This problem is particularly acute in human genetics since it is not possible to breed people with combinations of mutations appropriate for linkage analysis. As a result, with loci defined solely by variant phenotypes, the only way to construct an all-encompassing linkage map is through the tedious process of accumulating and pooling information from many two- and three-point crosses.

Geneticists have solved the problems of limited assayable phenotypic traits and traits encoded by multiple genes by using some of the recently developed tools for the direct detection of genotype (described in detail in Chapter 11). With these tools, they can follow the segregation of a very large number of loci in a single experimental plant or animal cross or in a human pedigree, and they use the transmission data to build a high-density linkage map. Such a high-density map shows the relative positions of thousands to millions of closely spaced (that is, densely packed) DNA markers. (Recall from the Fast Forward box on pp.142–143 in Chapter 5 that a DNA marker is a DNA fragment of known size and location that comes in identifiable variants.) Since Chapter 11 covers linkage maps in detail, we now describe only the large-scale maps built from genomewide identification and typing of markers.

The Making of Large-Scale Linkage Maps

The human genome contains millions of DNA polymorphisms scattered across the different chromosomes. Two types of DNA polymorphisms are commonly employed for large-scale linkage mapping: *single nucleotide polymorphisms (SNPs)* in which, for example, the single nucleotide T in one sequence is replaced by a G in the corresponding sequence; and *simple sequence repeats (SSRs),* also called *microsatellites*, in which a set of two to five or so nucleotides is repeated 4–50 or more times (CGTCGTCGTCGT. . .; **Fig. 10.3**). SSRs have a significant tendency to expand or contract during chromosomal replication, and these processes generate length polymorphisms.

SNPs and SSRs may lie within a gene, where they can cause alterations in the resulting gene function ranging from profound to no effect. Both types of DNA variation can also occur in intergenic regions. Hence, these types of polymorphisms may have a profound effect on phenotype or little to no effect at all. Because coding

Single nucleotide polymorphism (SNP)	...GCAA T TCCCGATT... ...GCAA G TCCCGATT...
Simple sequence repeat (SSR)	...GCATTATATATATATC... ...GCATTATAT[ ]C...

Figure 10.3 Two common types of polymorphisms employed for genetic mapping. SNPs occur at rates of 1/500 to 1/1000 nucleotides across the human genome. The SSR illustrated here is a dinucleotide (AT) repeat, whose alleles exhibit the $(AT)_5$ and $(AT)_2$ alternatives. SSRs occur, on average, every 20–40 kb across the genome.

sequences account for only a few percent of eukaryotic genomes, most polymorphisms seem to have little effect on the organism.

Both SNPs and SSRs, which come in identifiable variants, can serve as DNA markers. In making genomewide linkage maps, researchers must be able to rapidly identify and order these DNA markers across the chromosomes. They must also be able to determine the genotypes of large numbers of ordered polymorphisms in populations ranging from hundreds to thousands of individuals.

Genomewide Identification of Genetic Markers

The initial human (and mouse) genetic maps produced by the Human Genome Project employed the highly polymorphic simple sequence repeats (SSRs). As mentioned previously, these tandemly repeated di-, tri-, tetra-, or pentanucleotides readily expand or contract the numbers of their repeats in different individuals, thus generating size polymorphisms. Hybridization probes of simple sequence repeats, for example $(AT)_n$, can identify clone inserts with these same repeat sequences in a genomic library. These clones can then be sequenced to obtain unique 5′ and 3′ adjacent sequences for PCR primers. After the identification of thousands of SSRs, researchers use recombination analyses to map them in families. They then order these genetic markers across the genome to generate global linkage maps.

Recently, the DNA sequence analysis of *orthologous gene* regions and cDNA clones from different individuals has identified millions of potential SNPs. The newly identified SNPs are being verified by suitable recombinational analyses in family pedigrees.

Orthologous genes, defined by their sequence similarities, are genes in two different species that arose from the same gene in the two species' common ancestor. Orthologous genes are part of the more general category of **homologous genes,** which are genes with enough sequence similarity to be evolutionarily related somewhere back in evolutionary history. Orthologous genes are homologous, but homologous genes do not have to be orthologous. They could, for example, be *paralogous.* **Paralogous genes** arise by duplication (sometimes multiple duplications) within the same species, often within the same chromosome; paralogous genes often constitute a multigene family. The differences between paralogous, orthologous, and homologous genes are differences of degree and timing. *Homologous* is a general term that includes all genes with sufficient sequence similarity to have evolved from a common ancestor situated anywhere in evolutionary time. With *orthologous* genes, the common ancestor is the immediate ancestor to the two species. The duplication events that create *paralogous* genes take place in the same species. These three terms have become important to genomicists as they refine their analyses of various genomes. Problem #29 on p. 389 at the end of this chapter will help you explore the meanings of these terms.

Today, genome analysts can determine the order of SNPs and SSRs across the genome by sequence comparisons with the human genome sequence. For example, the current human and mouse linkage maps contain more than 20,000 and 10,000 SSRs, respectively, and more than 4 million and 500,000 SNPs, respectively. With these markers, it is possible to construct very dense maps. The idea is to have a map of ordered genetic markers dense enough to permit the identification of the gene or genes encoding polymorphic phenotypic traits, for example, by linkage analyses in families with the variant gene (or genes). By mid-2005, more than 4 million human SNP markers were being used to create very dense genetic maps.

Genomewide Typing of Genetic Markers

Researchers generally genotype SSRs by examining the polymorphic size differences of allelic PCR products via gel electrophoresis (**Fig. 10.4**). Robots and 384-well PCR machines, together with the automated loading of four-color, 96-capillary DNA sequencing machines, allow thousands of SSRs to be analyzed per day per machine. The typing of SNPs can also be done by the DNA sequence analysis of PCR products, although it is now more often carried out by specialized assays employing oligonucleotide arrays or mass spectrometry, which we discuss later in this chapter and in Chapter 12.

Genomewide linkage maps provide DNA sequences ordered along the chromosomes as well as hybridization

(a) PCR amplification

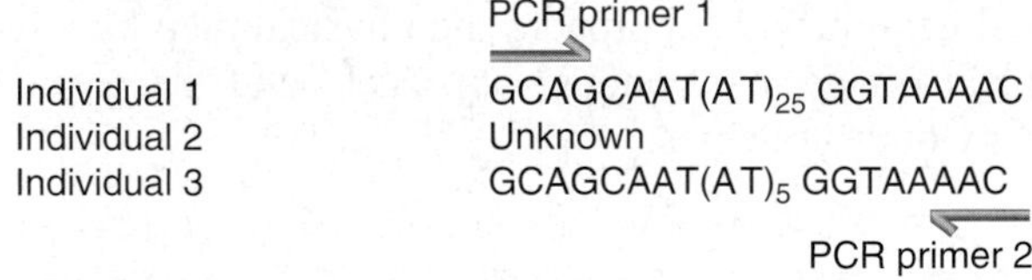

(b) Size separation

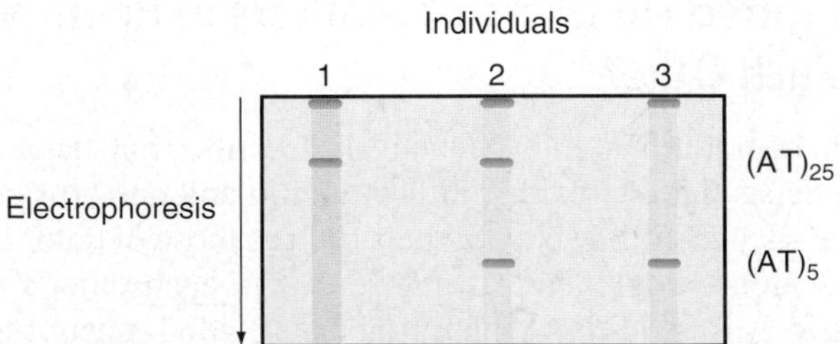

Figure 10.4 The two-stage assay for simple sequence repeats (SSRs). Individuals 1 and 3 are homozygous for the $(AT)_{25}$ and $(AT)_5$ alleles, respectively. Individual 2 has an unknown genotype. In the first stage, PCR primers 1 and 2 amplify the SSRs directly from genomic DNA. In the second stage, electrophoresis separates the alleles by size. Individual 2 turns out to be a heterozygote with both the $(AT)_{25}$ and $(AT)_5$ alleles.

probes. Investigators can use both the ordered sequences and the probes to localize DNA clone inserts and DNA sequences along the individual chromosomes in the making of large-scale, or long-range, physical and sequence maps.

Long-Range Physical Maps: Karyotypes and Genomic Libraries Provide the Basis for Positioning Markers on Chromosomes

A *physical map* is a constellation of overlapping DNA fragments that are ordered and oriented and span each of the chromosomes in a genome. Physical maps, the molecular counterparts of linkage maps, are based on the direct analysis of genomic DNA. Such maps chart the actual number of base pairs (bp), kilobases (kb), or megabases (Mb) that define and separate a locus, or site, from its neighbors in a particular region of a particular chromosome. Linkage and physical maps can be roughly calibrated against one another. For example, in humans, 1 cM $\approx$ 1 Mb, whereas in mice, 1 cM $\approx$ 2 Mb.

As we saw in Fig 9.6 on p. 310, geneticists build relatively short-range physical maps by exposing DNA clones to two or more restriction enzymes and comparing the fragments produced in this way; they then use hybridization probes to ascribe a precise position to the genes and markers included in the mapped clone. The major difference between the physical maps discussed earlier and those developed by genomicists is one of scale. A map of a cosmid clone, for example, would span 30,000 bp of genomic DNA. Here we learn how molecular geneticists map a human chromosome averaging 130,000,000 bp in length (approximately 4300 times longer than a single cosmid).

In developing a genome-wide physical map, the length and stability of the clone inserts are significant when generating the fragments to be used. The inserts should be as long as possible, and they should be very stable (that is, they should not delete or rearrange during vector propagation). The major vector for large clone inserts used to construct physical maps is the bacterial artificial chromosome, or BAC (ranging from 50–300 kb). BAC inserts are stable and simple to purify from their host DNA.

Another consideration in the development of genome-wide physical maps is how to determine the order of the clones across the genome. As noted earlier, the cloned DNA inserts must overlap at their ends. These insert overlaps can be produced by random shearing or partial restriction enzyme digests of genomic DNA before cloning. To cover the entire human genome with BACs containing inserts averaging 200 kb, more than 15,000 BACs are required (3×10^9 bp/genome/ 2×10^5 bp/BAC = 1.5×10^4 BACs/genome).

A Bottom-Up Approach Uses PCR Amplication of Sequence Tagged Sites to Order Clones

The most common approach to producing genomewide physical maps starts with an overlapping library of BAC clones and identifies on those clones unique sequence segments, either from the end sequences of the BACs (which would require sequencing both ends of 30,000 randomly chosen BACs) or from sequenced genetic markers. Next, by synthesizing and placing PCR primers on either side of the identified short unique sequences, one generates **sequence tagged sites (STSs):** short DNA sequences that are readily located and amplified by PCR. A PCR assay using the appropriate primers identifies just a single site in the genome. Because each STS represents a unique sequence in the genome, all clones identified by a particular STS must overlap. Thus, an individual STS can identify all BAC clones that contain a specific site and physically map each of these clones with respect to one another. One can repeat this testing process for as many STSs as are necessary to create a physical map. With a 5- to 10-fold coverage BAC library, one could, in principle, map the whole human genome using 20,000–30,000 STS markers. This is a *bottom-up mapping approach* (**Fig. 10.5**).

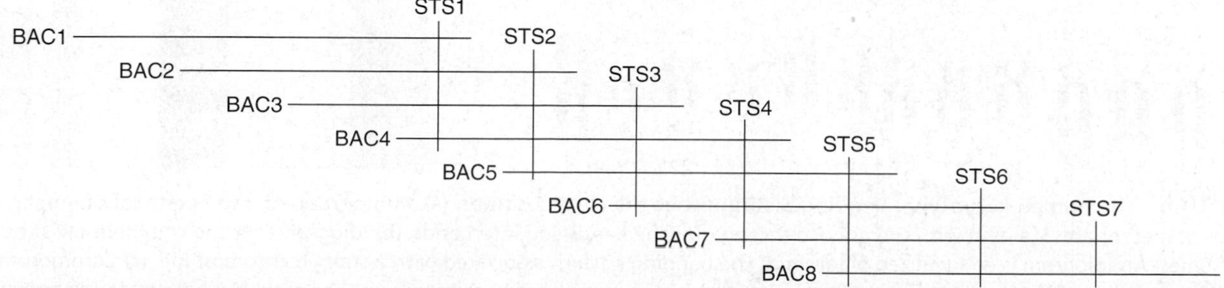

Figure 10.5 A hypothetical physical map generated by the analysis of sequence tagged sites (STSs). Each STS represents a unique segment of the genome that has been amplified by a pair of PCR primers. Here, eight BAC clones are interrogated with seven STS markers. This process can unambiguously determine the physical order of the BAC clones in this example.

The SSR and SNP markers described previously are polymorphic STSs that can also be used to order the clones in a physical map. Ordering by this means permits an integration of the genomewide linkage and physical maps.

In a Top-Down Approach, the FISH Protocol Locates Cloned Loci—Genes or Markers—Directly on the Chromosomes of a Karyotype

We saw in Chapter 4 that the chromosomes of actively dividing cells in metaphase of mitosis, when stained with a Giemsa dye and viewed in the light microscope, show an identifiable series of dark and light regions referred to as bands and interbands (Fig. 4.4 on p. 85 and **Fig. 10.6a**). (The dark regions are A–T-rich and gene-poor, while the light bands are G–C-rich and gene-rich.) The number, intensity, and width of each band and interband are highly reproducible characteristics that a skilled cytogeneticist can use to distinguish each pair of homologs from all other chromosomes in a cell. The visual description of a complete set of chromosomes in one cell of an organism is a *karyotype.* To facilitate the presentation and comparison of data from different karyotypes in the same and diverse species, investigators convert the light and dark bands actually observed under the microscope into black-and-white diagrams of the chromosomes (Fig. 10.6a and b). In these diagrams, the autosomes are numbered in order of descending length. In humans, chromosome 1 is the longest autosome and chromosome 22 is the shortest. The X and Y sex chromosomes are intermediate in size and small, respectively. The shorter arm of each chromosome is designated "p" (for *petit*) and the longer arm is "q" (for queue, the French word for "tail"). Karyotypers number each band and interband, starting at the centromere and moving out along each arm toward the telomere.

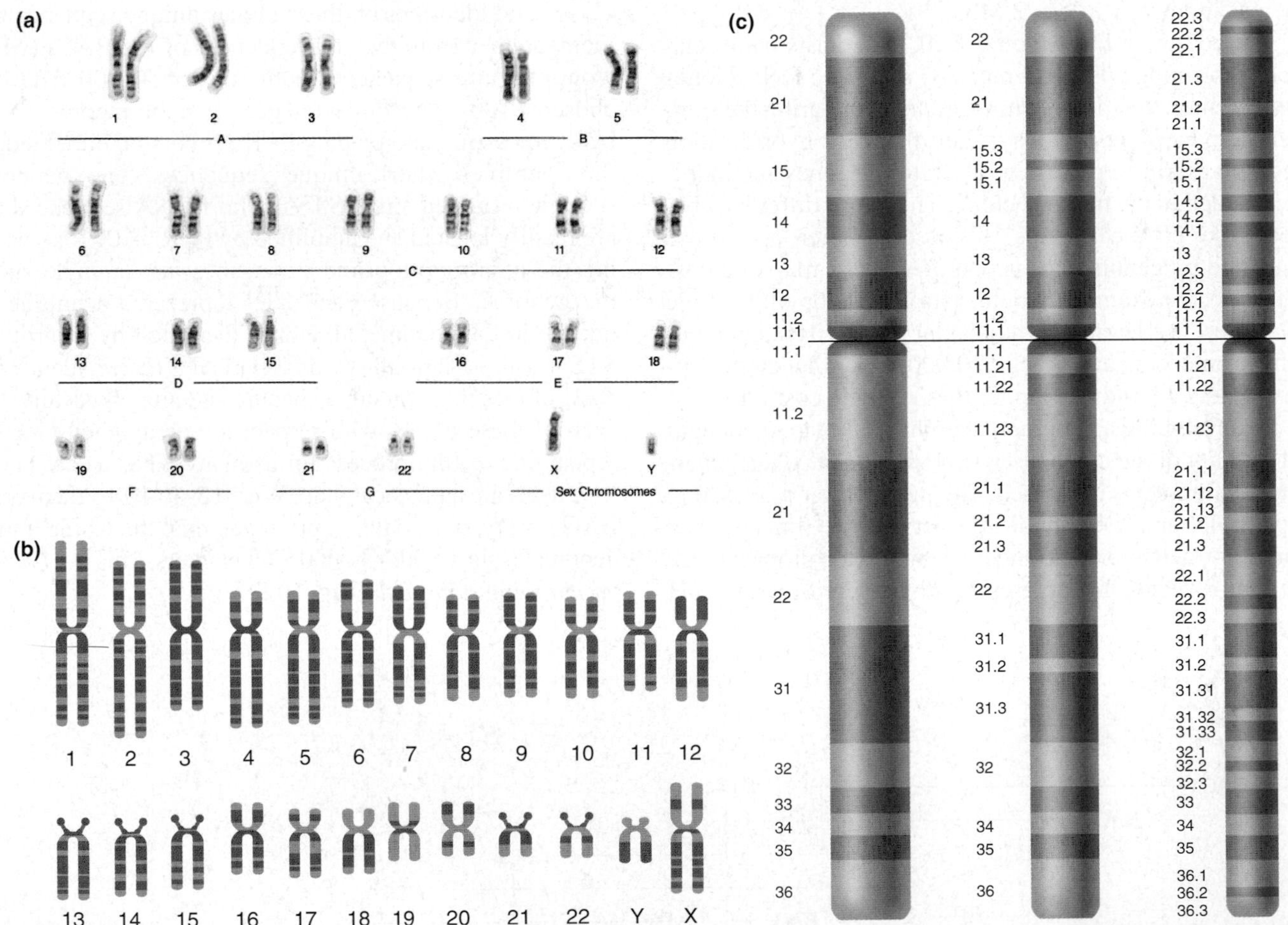

Figure 10.6 The human karyotype: Banding distinguishes the chromosomes. (a) Photograph of a complete set of human chromosomes at metaphase. Staining with Giemsa dye accentuates the bands and interbands. **(b)** Idiograms for the complete set of human chromosomes. An idiogram is an idealized diagram of the banding pattern associated with a stained chromosome. **(c)** Chromosome 7 at three different levels of banding resolution. As staining techniques improve, it becomes possible to resolve what previously appeared as a single band into a series of bands and interbands, producing more and more bands along each chromosome. Thus, at one resolution, 7q31 appears as one band. At a slightly higher resolution, 7q31 becomes two bands (7q31.1 and 7q31.3) flanking an interband (7q31.2); and at an even higher resolution, 7q31.3 itself appears as two bands (7q31.31 and 7q31.33) and an interband (7q31.32).

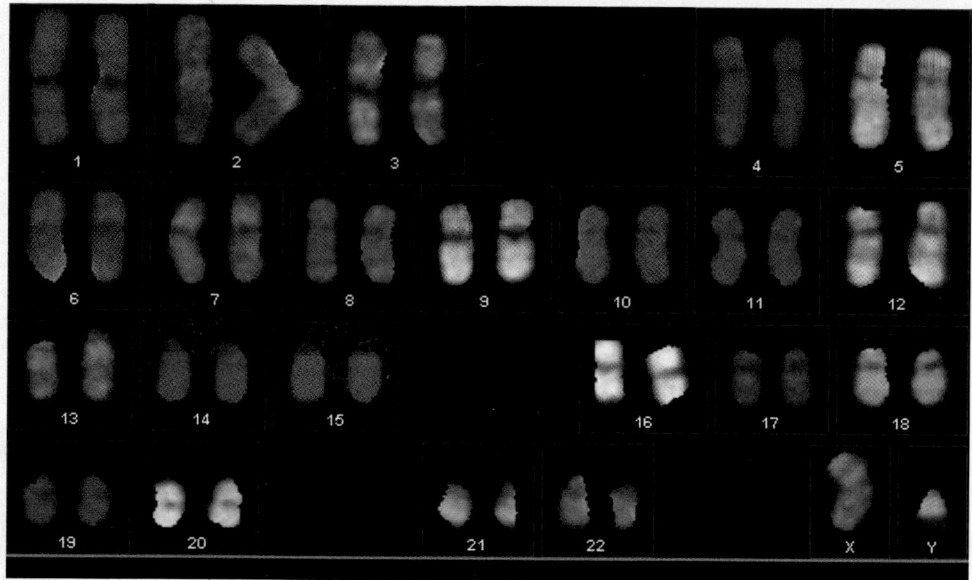

Figure 10.7 **A SKY chromosomal *in situ* hybridization.**

Figure 10.7 shows the elegant, more recent sky (spectral karyotyping) method for constructing karyotypes that makes it easier to identify individual chromosomes. The sky technique is a specialized application of a more general protocol known as **fluorescent *in situ* hybridization,** or **FISH,** which is used to locate regions in the genome that are homologous to a nucleic acid probe. **Figure 10.8** illustrates the steps of the FISH protocol. In spectral karyotyping, each probe for *in situ* hybridization is made from multiple DNAs that originated from positions scattered along the length of individual chromosomes. The probe for chromosome 1 is labeled with a group that fluoresces in one color; the probe for chromosome 2 is tagged with fluorescence of a different color, and so forth, so that each of the 24 human chromosomes in a SKY karyotype can be recognized by its color (Fig. 10.7).

For the purposes of map making, FISH is more commonly used to show the location of a particular DNA sequence within the genome. In a mapping experiment, a single DNA probe is labeled with a fluorescent tag and hybridized with the chromosomes. The site(s) of hybridization appear under the microscope as discrete spots of illumination that indicate the chromosomal location of the DNA in the probe (Fig. 10.8b).

FISH analysis has several advantages over linkage analysis for the mapping of a newly cloned locus to a particular chromosomal site. First, all clones can be mapped by the FISH protocol, whereas only those clones that detect polymorphisms can be mapped by linkage analysis. Second, researchers can perform a FISH analysis on any cloned locus in isolation, whereas linkage mapping requires the analysis of one locus in relation to another. Third, FISH requires only a single sample on a microscope slide, whereas linkage analysis requires genotype information from a large cohort of individuals. For this reason, human geneticists often use the FISH protocol as a rapid way to obtain the general chromosomal position of a newly cloned locus. Finally, the FISH protocol enables direct comparisons between the map positions of cloned loci and the aberrations (such as deletions or translocations) that are visible in the chromosomes of individuals with unusual chromosomes. The main disadvantage of FISH is that its resolving power is much less than that possible with linkage analysis. The metaphase bands and interbands used as guideposts in FISH protocols contain the equivalent of 4–8 Mb or 4–8 cM of chromosomal material. In contrast, we have seen that linkage analysis can resolve map positions down to a fraction of a centimorgan.

Karyotypic analysis thus enables the generation of low-resolution physical maps that locate cloned genes and markers on particular parts of chromosomes. Geneticists use the cloning techniques described in Chapter 9 to expand these maps with a much finer resolution of loci. The ultimate goal of high-resolution physical mapping is the generation of one large contig for each chromosome in a genome. A **contig** (from the word *contiguous*) is a set of two or more overlapping cloned DNA fragments that together cover an uninterrupted stretch of the genome.

Once a combination of approaches to generating whole chromosome contigs has produced an ordered array of overlapping clones for all the chromosomes in the genome, researchers can use subcloning, restriction mapping, and hybridization (as described in Chapter 9) to study the clones of each contig in greater detail. In this way, they can locate and analyze genes and markers of interest. They can then have a computer combine all the information for all the clones into a detailed physical map showing a variety of landmarks across the contigs of all the chromosomes in the genome.

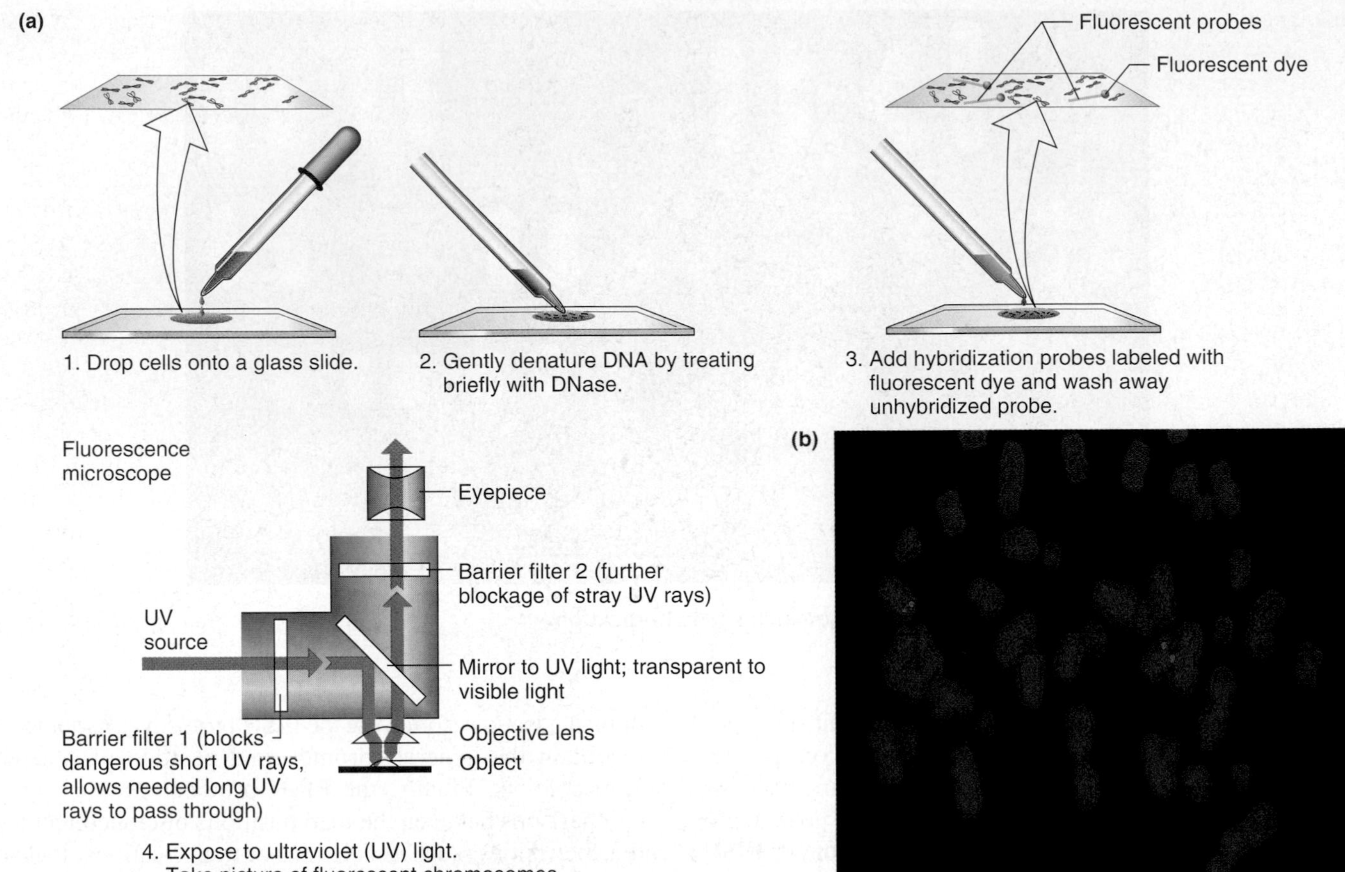
(a)
Fluorescent probes
Fluorescent dye
1. Drop cells onto a glass slide.
2. Gently denature DNA by treating briefly with DNase.
3. Add hybridization probes labeled with fluorescent dye and wash away unhybridized probe.
Fluorescence microscope
Eyepiece
Barrier filter 2 (further blockage of stray UV rays)
UV source
Mirror to UV light; transparent to visible light
Objective lens
Object
Barrier filter 1 (blocks dangerous short UV rays, allows needed long UV rays to pass through)
4. Expose to ultraviolet (UV) light. Take picture of fluorescent chromosomes.
(b)

Figure 10.8 The FISH protocol. (a) The technique. (1) First, drop cells arrested in the metaphase stage of the cell cycle onto a microscope slide. The force of the droplet hitting the slide causes the cells to burst open with the chromosomes spread apart. (2) Next, fix the chromosomes and gently denature the DNA within them such that the overall chromosomal structure is maintained even though each DNA double helix opens up at numerous points. (3) Label a DNA probe with a fluorescent dye, add it to the slide, incubate the probe with the slide long enough for hybridization to occur, and wash away unhybridized probe. (4) Now place the slide under a special microscope that focuses ultraviolet (UV) light on the chromosomes. The UV light causes the bound probe to fluoresce in the visible range of the spectrum. You can view the fluorescence through the eyepiece and photograph it. **(b)** A fluorescence micrograph. Photograph of a baby hamster kidney cell subjected to FISH analysis. The four yellow spots show the locations at which a particular probe hybridizes to the two sister chromatids of two homologous chromosomes.

A Sequence Map Is The Highest-Resolution Genomic Map

Sequence maps show the order of nucleotides in a cloned piece of DNA. The goal of the Human Genome Project is the determination of the complete nucleotide sequence for every chromosome in the genome (of humans and model organisms). Two basic strategies have been employed for sequencing the human genome: the *hierarchical shotgun approach* and the *whole-genome shotgun approach.* The term **shotgun** means that the overlapping insert fragments to be sequenced have been randomly generated either from large insert clones (BACs) or from the whole genome by shearing with sound (sonication) or partial digestion with restriction enzymes.

The Hierarchical Shotgun Sequencing Strategy

The publicly funded effort to obtain a draft sequence of the human genome employed the hierarchical shotgun strategy (**Fig. 10.9**). In this multistep approach, researchers first generate a genomic BAC library, develop a map of overlapping BAC clones across the genome, and then select a set of minimally overlapping BACs across the genome. The final, minimally overlapping set of BACs is called a **minimal tiling path** of BAC clones. To sequence each BAC clone on the minimal tiling path, a researcher randomly shears it into ~2 kb fragments, clones these fragments in plasmids, and then sequences both ends of a sufficient number of plasmid inserts to generate 10-fold coverage across the BAC insert. If the BAC insert is 200 kb and 2000 bp (or 2 kb) are sequenced for each plasmid insert (1000 bp from each end), the sequencing of each BAC would require the sequencing of 200 kb/BAC × 10/2 kb/plasmid = 1000 plasmids/BAC. One advantage of this divide and conquer strategy is that you need to assemble only about 1000 plasmid insert sequences for each BAC. The data from the linkage maps (showing SSRs or SNPs) and the physical maps (showing gene locations and STS maps) are also used to help assemble the sequence maps of the individual chromosomes. The hierarchical shotgun strategy requires a fair amount of front-end work to create the physical map of BAC clones and to generate a plasmid library for each of the 20,000 or so BAC clones to be sequenced.

The Whole-Genome Shotgun Sequencing Strategy

The private company Celera employed the whole-genome shotgun strategy to obtain its draft sequence of the human genome. In this approach, the whole-genome DNA is randomly sheared three times, first to construct a plasmid library with ~2 kb inserts, second to generate a plasmid library with ~10 kb inserts, and third to produce a BAC library of ~200 kb inserts (**Fig. 10.10**). In theory, it should be possible to sequence the 2 kb and 10 kb inserts to attain approximately 6-fold and 3-fold coverage, respectively (3×10^9 [the genome] $\times$ 6/2000 [bp sequenced per insert, assuming as a rough estimate that each sequencing reaction—for each of the two ends of the insert—can sequence 1000 bp.] $= 9 \times 10^6$ plasmid clones; and $3 \times 10^9 \times 3/2000 = 4.5 \times 10^6$ plasmid clones, respectively). One could then sequence the insert ends from the 200 kb BAC library to a 1-fold coverage ($3 \times 10^9 \times 1/2000 = 1.5 \times 10^6$ BACs). The 6-fold plus 3-fold plus 1-fold coverages would amount to a 10-fold coverage. (In the Celera draft, the total coverage was about 6-fold). A genomewide shotgun computer program would then assemble all these sequences into the chromosomal strings.

The whole-genome shotgun strategy has several advantages: It does not require the construction of a physical map. It depends on the construction of only one BAC and two plasmid libraries. It overcomes the problem posed by repeat sequences scattered throughout the genome because paired end sequences from clones of three insert sizes, 2 kb, 10 kb, and 200 kb, make it possible to bridge most lengths of repetitive sequences. It

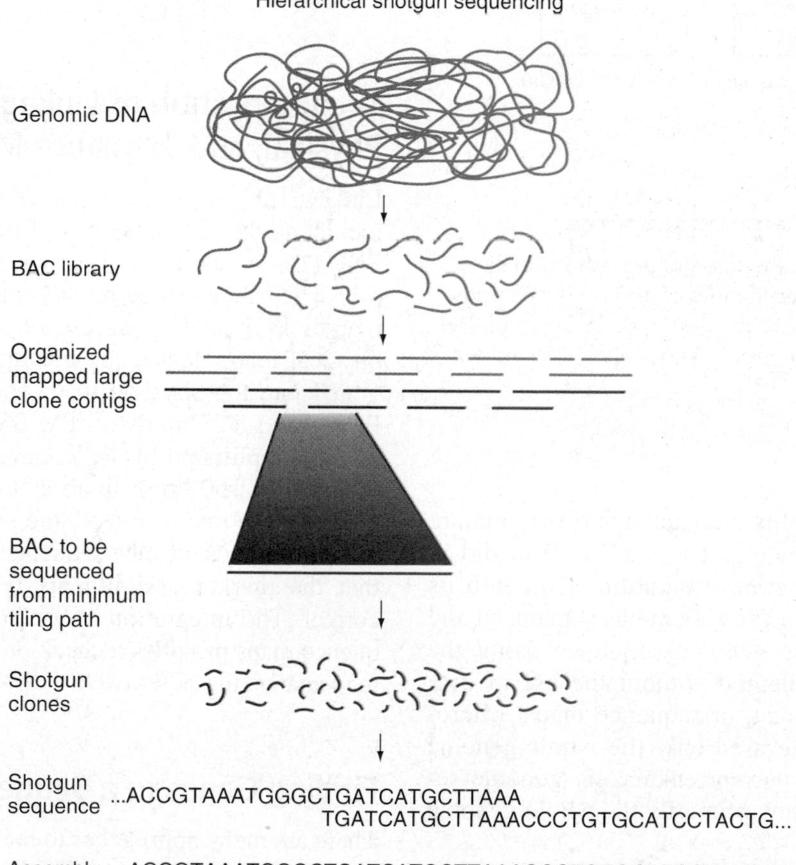

Figure 10.9 Idealized representation of the hierarchical shotgun sequencing strategy. (a) Construct a library by fragmenting the target genome and cloning it into a large fragment cloning vector, in this case, BAC vectors. **(b)** Organize the genomic DNA fragments represented in the library into a physical map, and select and sequence the BAC clones on a minimum tiling path by the random shotgun strategy. The minimum tiling path is indicated by the darker clones. **(c)** Assemble the cloned shotgun sequences to reconstruct the sequence of the genome.

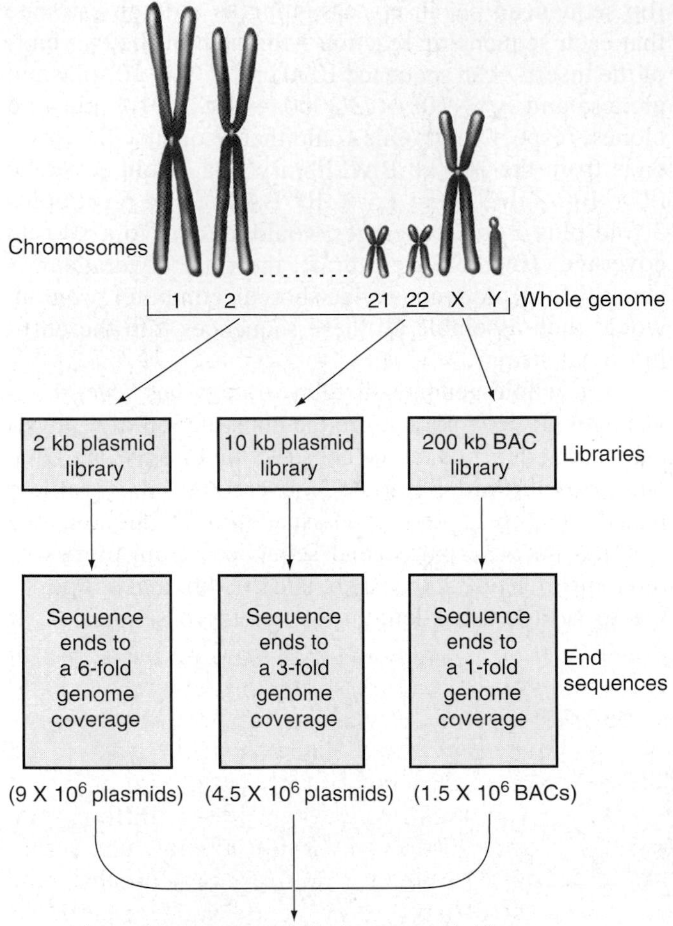

Figure 10.10 Hypothetical whole-genome shotgun sequencing strategy. Only three libraries of differing-sized fragments, here 2 kb, 10 kb, and 200 kb, need to be constructed. The big challenge is to assemble these sequences when they include large numbers of repeats.

relies on only a single highly automated and very mature technology—DNA sequencing. The Celera effort did in the end incorporate data from the public effort into its own data. Since then, however, Celera has sequenced and assembled a draft mouse genome sequence using the whole-genome shotgun method without the use of data from other linkage, physical, or sequence maps. Microbial genome projects have used only the whole-genome shotgun approach. Most current eukaryotic genome sequencing efforts also employ primarily the whole-genome shotgun approach.

Two limitations for all large-scale sequencing efforts are the fact that some genomic sequences cannot be cloned (for example, sequences in the darkly staining, more condensed *heterochromatin* described on pp. 479–480 in Chapter 13), and some sequences rearrange or sustain deletions when cloned (for example, some tandemly arrayed repeats).

The Sequencing of the Human Genome

Most of the sequencing of the draft of the human genome took place in the last year or so of the project. There were three reasons for this. First, DNA sequencing instruments had evolved to run 96 samples (in 96 capillaries) simultaneously and to complete 6 runs in 24 hours. In theory, each instrument could thus sequence about ~0.5 Mb per day (1000 bp/capillary × 96 capillaries × 6 runs/day). Second, an automated factory-like production line generated sufficient DNA fragments to supply the high-throughput sequencers on a daily basis. Finally, several very large sequencing centers emerged with 100–300 sequencing instruments apiece. The theoretical throughput of a 300-instrument center is 150,000,000 bp per day (500,000 × 300). Hence, such a center, in principle, could provide a 10-fold sequence coverage of the human genome in about 200 days (3×10^9 bp in the human genome $\times 10/1.5 \times 10^8$). In practice, the throughput of sequencing machines is considerably lower—perhaps 3-fold or more—because of limitations on providing adequate insert DNA for sequencing, experimental errors, machine failures, and failures in data capture.

The Integration of Linkage, Physical, and Sequence Maps

Linkage, physical, and sequence maps all locate sequence loci along the chromosomes of the genome. The SSR and SNP DNA markers are both amenable to PCR amplification. Thus, they can be readily integrated into the physical map by PCR analysis across all of the insert clones of the physical map. Hence, the linkage map can be fully integrated into the physical map. The physical map employs PCR-based STS markers. The SSR, SNP, and STS markers, all amplifiable by PCR, have unique primer-detected sequences of 50 bp or more that can readily be placed on the sequence map—indeed, the orientation and spacing of the paired PCR primer sequences provides a dual check that the marker assignments for the PCR primers are correct. The integration of the linkage, physical, and sequence maps provides a check on the correct order of each map against the other two.

Finding Genes in a Sequenced Genome

There are many approaches to the identification of genes in chromosomal sequences. First, computational similarity analyses can match EST (expressed sequence tag) or cDNA sequences against their chromosomal counterparts. Recall that such sequences represent transcribed genes. Researchers can use a full-length cDNA sequence to delineate the exons and introns of a gene within the genome sequence. Researchers can also use sufficiently similar EST or cDNA sequences from other animals to identify exons. The similarity programs have associated statistical analyses that determine the significance of various levels of sequence match.

Second, relatively successful computational gene-finding programs are available. These programs analyze the specific features of genes (start and termination sites, intron/exon junctions). They also take advantage of the fact that different animals (and plants and microbes) show individual preferences in the codons used for particular amino acids; these codon preferences can be integrated into the gene-finding programs. The gene-finding programs are 70% to 90% successful in identifying some but not necessarily all the genes or exons in an extended sequence.

Finally, researchers can compare complete genome sequences from appropriately diverged organisms. For the comparison to work, the organisms' genomes should have diverged to the point that the nonfunctional regions of the genome are significantly different while the functional regions remain relatively conserved. Functional chromosomal regions include exons and transcription factor binding sites. In comparisons of the orthologous chromosomal sequences of two organisms, conserved blocks 25 bp to several kilobases in length are automatically candidates for exons or control regions. Comparative genomic analysis is particularly powerful when many of the genes have been defined in one of the two organisms.

Researchers can determine whether hypothetical genes are real genes by scanning across many different cell and tissue types with appropriate PCR primers to identify the genes' expressed mRNA.

The Human Genome Project Has Changed the Practice of Biology, Genetics, and Genomics

The complete human genome sequence, in conjunction with the complete genome sequences of model organisms, has profoundly changed biology, genetics, and genomics. The complete genome sequences have provided a genetic parts list for each organism through the identification of most of its genes as well as the adjacent sequences controlling gene expression.

The genome sequencing achievements to date will expedite future gene-finding and gene-function analyses in several ways. First, the identification of genes in one organism's genome facilitates identification by sequence homology of genes in a second organism. If the second organism's genome sequence is available, its genes can be identified by a computer search. If the second organism's genome is not available and the two genomes are sufficiently similar, PCR primers from the first organism's genome (if sufficiently similar to the second organism's sequences) can be used to identify genes in the second organism's genome.

Second, the comparisons of all the genes within and between organisms has shown that many of the genes are paralogous or orthologous to one another (review definitions on p. 358 of this chapter). Hence, knowledge of how a gene functions in one organism may help us understand its function in a second organism.

Third, the complete genome sequences have revealed that exons often encode discrete **protein domains,** or discrete functional units. Hence, genes with multiple exons encode multiple protein domains analogous to the cars of a train. Each train is composed of many different cars, and each kind of car (engine, flat car, dining car, caboose) has a discrete function. Different trains may carry different combinations of cars and thus execute different functions. Similarly, many genes are composed of distinct exons that encode discrete protein domains. Each gene may express different combinations of domain-encoding regions by the shuffling, addition, or deletion of exons. Through these genetic mechanisms, protein architecture, defined by combinations of domains, may change with evolution. **Figure 10.11,** which shows a series of domains associated with different *transcription factors,* illustrates the idea of different domains and the shifting of domain architecture across the species. Transcription factors are proteins that bind to DNA promoters and other control regions; the DNA binding of transcription factors influences the temporal and spatial patterns of gene expression as well as its amplitude. Differing domains enable transcription factors to bind with different DNA sequences and also to interact uniquely with specific protein cofactors. Just as the makeup of a train's cars provides insights into what the train does, so too the specific gene-encoded domains of a protein are associated with particular functions and provide insights into what the protein does. Biologists may guess at the function of a new protein (or the gene that encodes it) by analogy, after searching for the protein's sequence in a database of all known domains.

Fourth, the genome sequence has estabished ready access to the identification of human polymorphisms, and the genomewide genetic map (using SSRs and SNPs as markers) has made it possible to correlate some of the many DNA polymorphisms and their associated genes with predispositions to disease and interesting aspects of physiology.

Finally, comparisons between sequenced genomes will more readily allow researchers to assemble the sequence fragments of an appropriately related, newly sequenced genome into its chromosomal strings. For example, the mouse and human genomes, which diverged 85 million years ago, exhibit striking similarities in genes as well as their conserved order on the chromosomes. Indeed, the

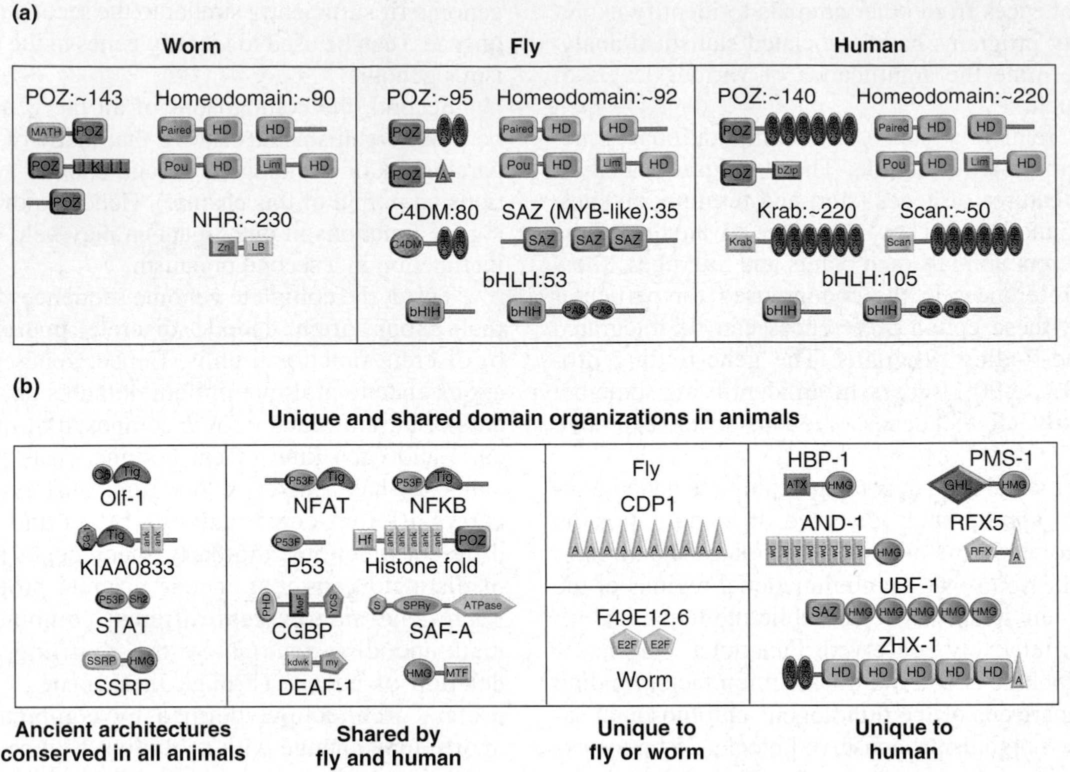

Figure 10.11 **How the domains and architectures of transcription factors have expanded in specific lineages.** **(a)** Specific families of transcription factors have expanded in the worm (nematode), fly (*Drosophila*), and human proteomes. The proteome is the collection of all proteins present in an organism or individual cell type. The diagram shows the approximate numbers of each domain identified in each of the three species. Some domains appear in different animal lineages; others are lineage specific. **(b)** Samples of transcription factor architectures found: in all animals (ancient architectures), in only fruit flies and humans, and uniquely in one lineage. The capital-letter designations indicate something about the different domain structures or functions. For example, HD indicates a homeodomain that can bind to the DNA in a gene's control region and thus help regulate that gene's transcription. (See website at www.mhhe.com/hartwell3 : Chapter 10 for definitions and explanations of abbreviations.)

mouse and human genomes exhibit approximately 180 homologous blocks of chromosomal sequence, ranging in size from 24 kb to 90.5 Mb—for an average of 17.6 Mb (**Fig. 10.12**). Such blocks of linked loci are called **syntenic blocks.** The orders of these blocks, as determined from their chromosomal linkage relationships, are totally different in the two organisms. It is as if one genome had been cut into 180 pieces of varying size and then randomly assembled into the other genome. Within each of the human syntenic blocks is information useful for assembling the corresponding homologous mouse sequences, and vice versa. *Conserved synteny,* in which the same two or more loci are linked in different species, also exists between the human and puffer fish genomes, which diverged more than 400 million years ago. In this case, though, the syntenic blocks are relatively small—averaging about 10 kb in length.

Thus, as these few examples show, successful sequence analysis of the human and model organism genomes has changed the strategies for biology, genomics, and medicine.

10.2 Major Insights from the Human and Model Organism Genome Sequences

In addition to promoting new analytical approaches, the complete sequence of the human genome has provided striking new insights into the architecture, gene organization, structural elements, and evolution of chromosomes. We now describe some of the main lessons and surprises.

There Are Approximately 25,000 Predicted Human Genes

The first surprise to emerge from the human genome sequence was the discovery of just 25,000 genes, a much lower number than expected. A back-of-the-envelope calculation (with no

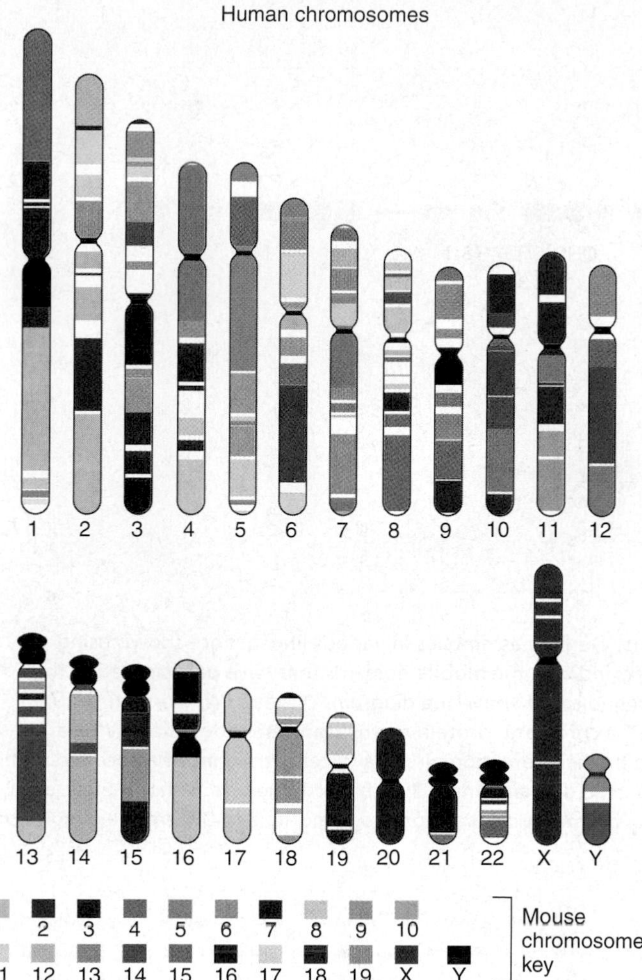

Figure 10.12 Conserved segments or syntenic blocks in the human and mouse genomes. Human chromosomes, with segments containing at least two genes whose order is conserved in the mouse genome; the segments appear as *color blocks.* Each color corresponds to a particular mouse chromosome. Centromeres; subcentromeric heterochromatin of chromosomes 1, 9, and 16; and the repetitive short arms of 13, 14, 15, 21, and 22 are in *black.*

experimental support) done at the initiation of the Human Genome Project had suggested that there might be 100,000 human genes—approximately 1 gene every 30 kb. As we have seen a gene may encode functional RNAs or proteins.

For a variety of reasons, it appears likely that the final gene number will fall around 30,000. First, not all of the genome has been accurately sequenced; hence, some genes have yet to be counted. Second, small genes are difficult to identify. Geneticists estimate that there are hundreds to thousands of additional genes encoding small peptides that are difficult to identify. Third, genomic researchers have discovered many small functional RNAs (about 25 bp long) in a variety of organisms. These small RNAs, which appear to play a regulatory role in controlling the expression levels of complementary mRNA, could number in the thousands. In addition, there are many other types of genes expressed only at the RNA level that are difficult to identify. Finally, some genes are rarely expressed and do not reflect normal codon usage patterns. Such genes would not be detected by ESTs or common gene-finding programs.

Genes Encode Either Noncoding RNAs or Proteins

Noncoding RNA Genes

The noncoding RNAs, probably representing more than 2000 genes, fall into four main categories:

1. transfer RNAs (tRNAs): the adaptors that translate the triplet code of RNA into the amino acid sequence of proteins;
2. ribosomal RNAs (rRNAs): fundamental components of the ribosome;
3. small nucleolar RNAs (snoRNAs): required for rRNA processing and base modification in the nucleolus (an area of the genome that is rich in RNA);
4. small nuclear RNAs (snRNAs): critical components of spliceosomes (the molecular machines that splice out introns from pre-mRNAs in the nucleus).

There are also many additional noncoding RNAs expressed at lower levels.

Protein-Coding Genes Generate the Proteome

From the draft sequence of the human genome, investigators predicted about 25,000 protein-coding genes. The collective translation of the 25,000 predicted genes into proteins is termed the **proteome.** The human proteome contains *homologs*—in this context, proteins related by evolutionary descent—for 61% of the fruit fly proteome, 43% of the worm proteome, and 46% of the yeast proteome. Proteomic analyses have revealed that there are about 1200 gene families containing two or more members, of which 92 (7%) are vertebrate-specific; these vertebrate-specific families encode proteins active in immunity, other types of defense, and the nervous system. Analysts have also shown that there has been a striking increase in the complexity of the proteome from yeast to humans. This increase in complexity has come about in five ways:

1. *More genes.* The human genome has more genes than the genomes of the simpler model organisms, although not nearly as many as one would expect from the increased complexity (see Table 10.1). This means that mechanisms other than the expression of different germline genes must help generate metazoan (multicellular animal) complexity (as discussed in Chapter 22).
2. *The shuffling, increase, or decrease of functional modules.* Many proteins contain discrete functional domains that, as described earlier, can be joined in many

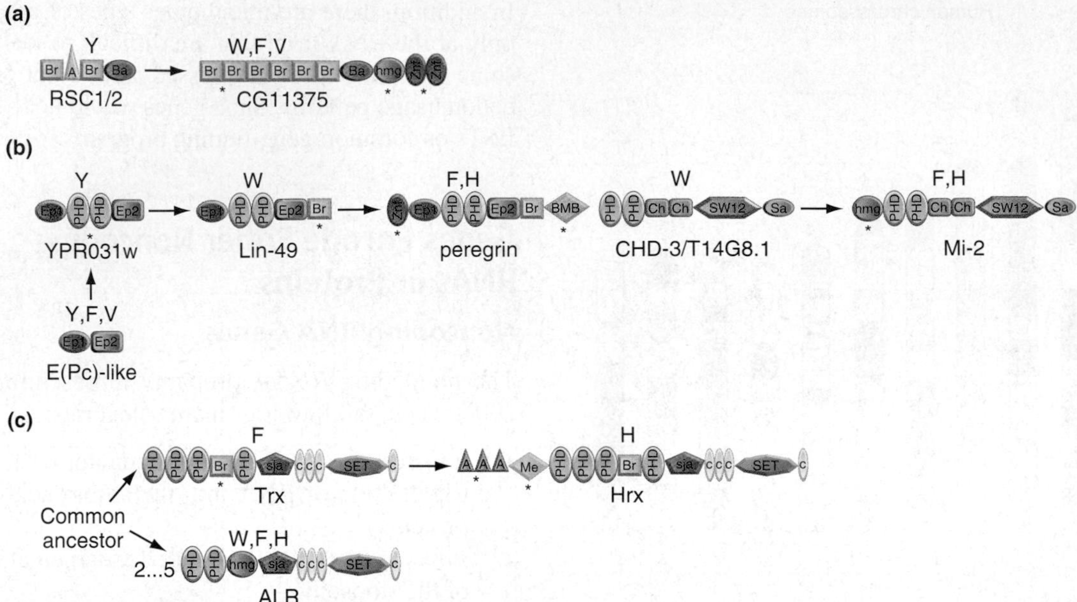

Figure 10.13 Examples of domain accretion in chromatin proteins. Domain assemblies in various lineages are shown using schematic representations of domain architectures (not to scale). *Asterisks* indicate the mobile domains that have participated in the accretion. Species in which a domain architecture has been identified are indicated *above* the diagrams (Y, yeast; W, worm; F, fly; V, vertebrate). Protein names are *below* the diagrams. The domains are SET, a chromatin protein methyltransferase domain; SW12, a superfamily II helicase/ATPase domain; Sa, sant domain; Br, bromo domain; Ch, chromodomain; C, a cysteine triad motif associated with the Msl-2 and SET domains; a, AT hook motif; EP1/EP2, enhancer of polycomb domains 1 and 2; Znf, zinc finger; sja, SET-JOR-associated domain; Me, DNA methylase/Hrx-associated DNA-binding zinc finger; Ba, bromo-associated homology motif. (a–c) Different examples of accretion.

different orders by addition, deletion, or shuffling (**Fig. 10.13**). The number and order of a protein's domains constitute its **domain architecture.** Although the human organism has only a modest increase in the total number of protein domains over the model organisms, it has evolved many new gene arrangements that alter domain architecture, thus creating a far more complex repertoire of proteins. (**Fig. 10.14**).

3. *More paralogs.* Many human gene families have significantly expanded their numbers of paralogs by gene duplication (**Table 10.2**). For example, humans have about 1000 olfactory genes, whereas fish have perhaps 60.
4. *Alternative RNA splicing.* Different pathways of RNA splicing may significantly expand the numbers of proteins. For example, the three human neurexin genes, which encode molecules that guide the appropriate connections of neurons, or nerve cells, may collectively generate more than 2000 splice variants. Humans exhibit significantly more alternative splicing than their model organism counterparts. Key questions include how many of the splice variants encode distinct functional proteins (rather than proteins with the same function) and whether different variants represent different addresses for telling neurons where to go.

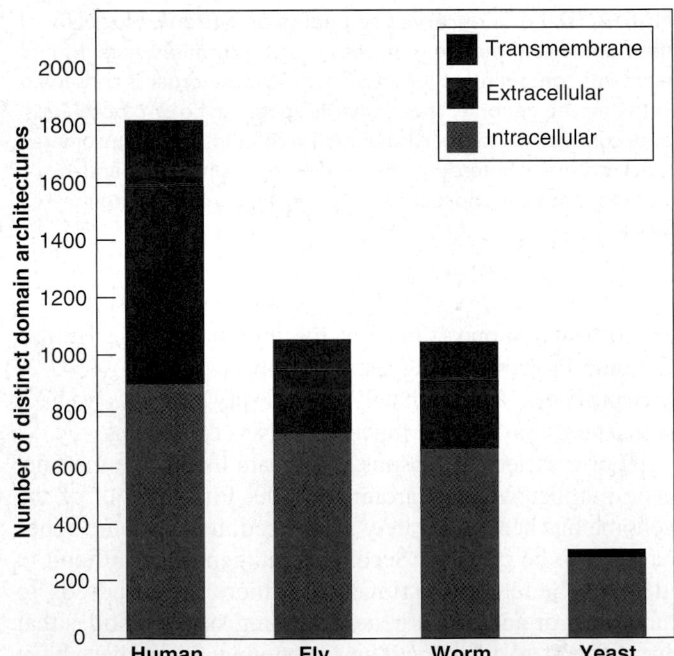

Figure 10.14 Number of distinct domain architectures in four eukaryotic genomes. The number of architectures is split into three cellular environments: intracellular, extracellular, and transmembrane.

TABLE 10.2 The Most Populous Gene Families in the Human Proteome and Other Species

Human		Fly		Worm		Yeast		Mustard Weed		
No. of Genes	Rank	No. of Genes	Rank	No. of Genes	Rank	No. of Genes	Rank	No. of Genes	Rank	
765	(1)	140	(9)	64	(34)	0	(na)	0	(na)	Immunoglobulin domain
706	(2)	357	(1)	151	(10)	48	(7)	115	(20)	C2H2 zinc finger
575	(3)	319	(2)	437	(2)	121	(1)	1049	(1)	Eukaryotic protein kinase
569	(4)	97	(14)	358	(3)	0	(na)	16	(84)	Rhodopsin-like GPCR superfamily
433	(5)	198	(4)	183	(7)	97	(2)	331	(5)	P-loop motif
350	(6)	10	(65)	50	(41)	6	(36)	80	(35)	Reverse transcriptase (RNA-dependent DNA polymerase)
300	(7)	157	(6)	96	(21)	54	(6)	255	(8)	rrm domain
277	(8)	162	(5)	102	(19)	91	(3)	210	(10)	G-protein β WD-40 repeats
276	(9)	105	(13)	107	(17)	19	(23)	120	(18)	Ankyrin repeat
267	(10)	148	(7)	109	(15)	9	(33)	118	(19)	Homeobox domain
252	(11)	77	(22)	71	(31)	27	(17)	27	(73)	PH domain
242	(12)	111	(12)	81	(25)	15	(27)	167	(12)	EF-hand family
222	(13)	81	(20)	113	(14)	0	(na)	17	(83)	EGF-like domain
215	(14)	72	(23)	62	(35)	25	(18)	3	(97)	SH3 domain
210	(15)	114	(11)	126	(12)	85	(12)	379	(4)	RING finger
188	(16)	115	(10)	54	(38)	7	(35)	392	(2)	Leucine-rich repeat
171	(17)	0	(na)	0	(na)	0	(na)	0	(na)	KRAB box
165	(18)	63	(27)	51	(40)	2	(40)	4	(96)	Fibronectin type III domain
162	(19)	70	(24)	66	(33)	2	(40)	15	(85)	PDZ domain
155	(20)	87	(17)	78	(27)	79	(4)	148	(13)	Helicase C-terminal domain

Twenty most populous gene families found in the human proteome compared with equivalent numbers from other species. na, not applicable (used when there are no proteins in an organism in that family).

5. *Chemical modification of proteins.* Human proteins may be modified by more than 400 different chemical reactions, each capable of altering the proteins' functions. Thus, the typical human cell might have perhaps 20,000 different types of mRNAs and perhaps 1 million different proteins. Humans can make more protein modifications than their simple model organism counterparts.

By all of these criteria, the human proteome is considerably more complex than its counterparts in the fly, worm, or yeast.

Repeat Sequences Constitute More Than 50% of the Human Genome

For the purposes of genomic analysis, repeat sequences fall into five classes: (1) transposon-derived repeats scattered across the genome, which constitute more than 45% of the genome; (2) inactive copies of protein-coding genes and small structural RNAs known as *processed pseudogenes* (because they lack introns as well as regulatory sequences); processed pseudogenes arise by retroposition, or the copying of RNA into DNA, followed by the random integration of this cDNA into chromosomes; (3) simple sequence repeats of short multimers two to five nucleotides in length, such as $(AA)_n$, $(AT)_n$, or $(GCC)_n$ (which make up about 3% of the genome); (4) segmental duplications of 10–300 kb copied from one region to another region of the genome (more than 5% of the genome); and (5) blocks of repeated sequences at the centromeres, telomeres, and other chromosomal features.

As mentioned in Chapter 7, *transposons* (also called *transposable elements,* or *TEs*) are units of DNA that move from place to place in the genome. Transposon-derived repeats, which we have seen constitute more than 45% of human DNA, fall into four classes: long interspersed

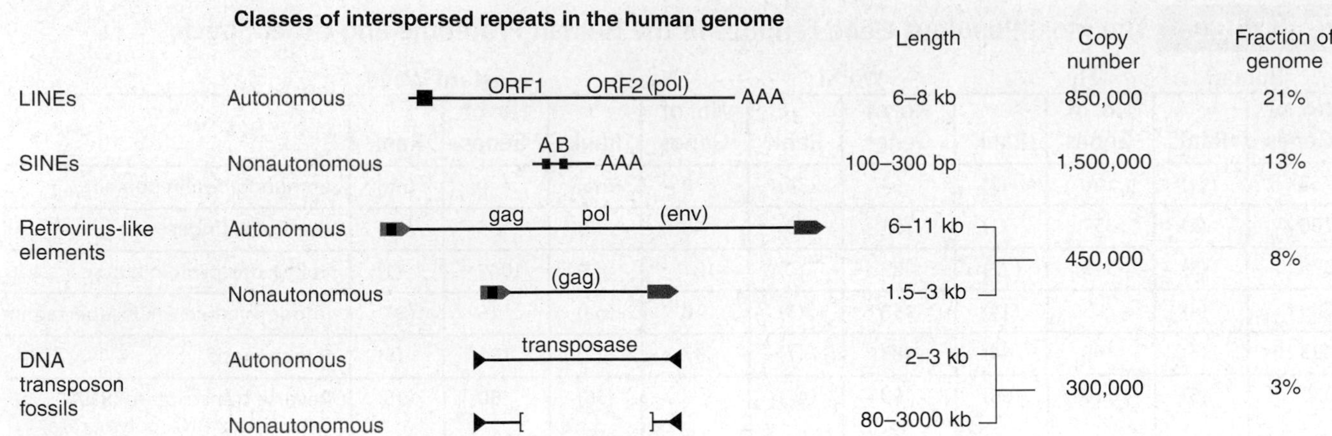

Figure 10.15 **Almost all transposable elements in mammals fall into one of four classes of interspersed repeats (as described in the text).**

elements (LINEs), short interspersed elements (SINEs), long terminal repeat retroposons, and DNA transposons. The first three transpose through RNA intermediates and the fourth transposes directly as DNA. **Figure 10.15** shows the general properties of these repeats. Some transposon-derived repeats have all the machinery required for independent transposition and are termed **autonomous;** others are nonautonomous—they employ the transposition machinery of the autonomous elements. Transposon-derived repeats, which were once considered *junk DNA,* have given rise to at least 47 different human genes. Moreover, they constitute a rich paleontological record of evolutionary events such as mutation and selection because researchers can recognize cohorts of repeats born at the same time and follow their fates in the genomes of different species. Finally, transposon-driven repeats have reshaped the genome by facilitating chromosomal rearrangements and translocations. (See Chapters 14 and 22 for a more detailed discussion of transposons and other types of repeats.)

The Genome Contains Distinct Types of Gene Organization

The organization of genes is strikingly different across the genome. Examples of different types of gene organization include gene families, gene-rich regions, and gene-poor deserts. A fundamental unanswered question is whether these different organizations have biological meaning.

Gene Families

Closely related genes that are members of multigene families abound throughout the genome. Such families may be clustered together on one chromosome or dispersed on several chromosomes. Examples of gene families include the genes that encode histones, hemoglobins, immunoglobins, actins, collagens, and heat-shock proteins. We described the hemoglobin-encoding globin gene family on pp. 337–339 of Chapter 9. Here we take a brief look at the human olfactory multigene family, which has about 1000 members. The olfactory genes, which encode olfactory receptor molecules, evolved in two ways. In the first, one gene underwent multiple duplications to create roughly 20 copies, or a family of approximately 20 paralogs. These family members, which were originally identical, then diverged to paralogs that were quite distinct from one another. Next, a massive duplication event created 30 different families from the original 20-paralog family. In this massive event, the entire family, or portions of it, duplicated and translocated to about 30 sites, or clusters, around the genome. Each of the 30 sites contains all or a portion of the 20 or so paralogs of the original family (**Fig. 10.16**). Interestingly, the equivalent paralogs in each of the 30 clusters, for example, paralog 3 at all 30 sites, are more closely related to each other than to other paralogs, say 2 or 7, in the same cluster. In a second mode of evolution, a single olfactory gene recently duplicated to generate a gene family of mostly similar paralogs. This is the class I family depicted in medium blue in Fig. 10.16.

Perhaps the expansion of olfactory receptor genes reflects an evolutionary response to the selective pressures for a more acute and discriminating sense of smell in some mammals. Clearly, gene families have a variety of mechanisms for generating paralogs. About half of the human olfactory receptor genes are pseudogenes. Why some gene families remain as single clusters and why others are scattered across the genome is a fascinating question. Another is, Why are there so many pseudogenes in some gene families? We discuss the evolution of multigene families in greater detail in Chapter 22.

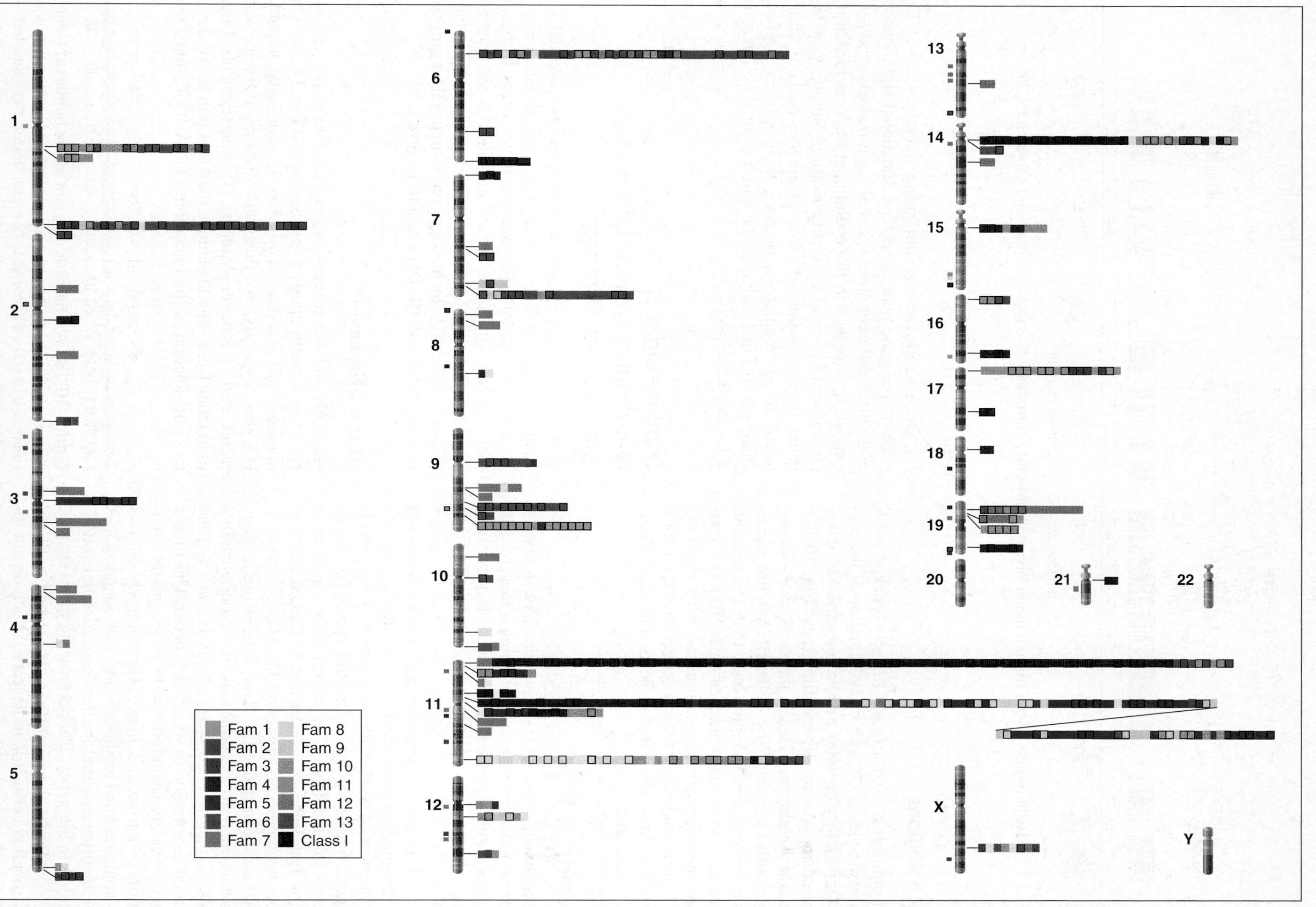

Figure 10.16 Olfactory receptor (OR) gene families. ORs are depicted as *squares,* colored by family (*see key*). All members of the class I family are colored equally. Unclassified ORs are indicated in *light gray. Framed squares* denote intact or functional genes. ORs to the *left* of each chromosome indicate single genes, and those to the *right* are in clusters of two or more. The largest cluster on chromosome 11 is shown split for convenience.

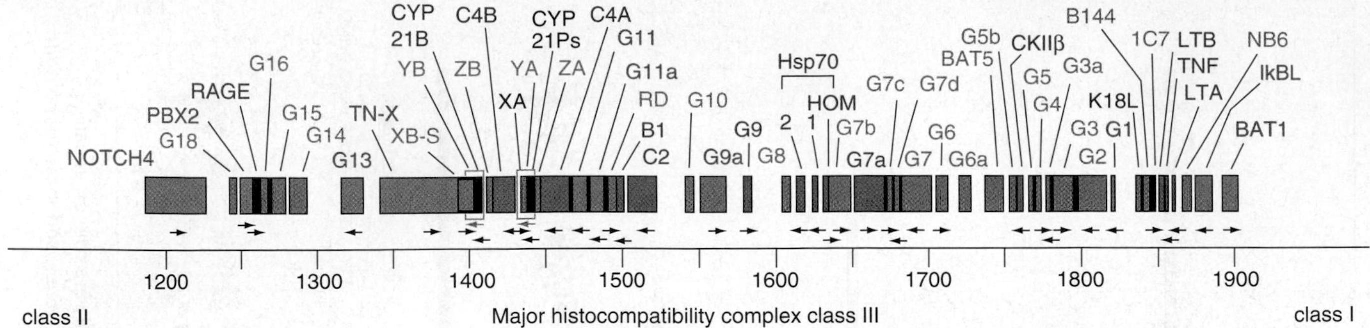

Figure 10.17 **Class III region of the human major histocompatibility complex.** This region contains 60 genes and is 700 kb in length.

Gene-Rich Regions

Some chromosomal regions are densely packed with genes. For example, the 700 kb class III region of the major histocompatibility complex on chromosome 6 contains 60 genes encoding many diverse functions (**Fig. 10.17**). Only one is a pseudogene. This is the most gene-rich region of the human genome; 70% of the DNA in this region is transcribed. Moreover, the region has a high G C content, 54% versus a genomewide average of 45%. High G C content is also seen in other gene-rich areas. Why are these genes so densely packed? Is there a functional explanation or is it a reflection of the chance events that shape chromosonal architecture?

Gene Deserts

The current human genome draft contains 82 gene deserts, which are chromosomal regions containing a megabase or more of DNA with no identifiable genes. Deserts span 144 Mb, or 3% of the genome. The largest desert is 4.1 Mb in length.

One explanation for the existence of gene deserts is that they contain genes that are difficult to identify. An example is the class of genes called *big genes.* A big gene is a single gene whose nuclear transcript spans 500 kb or more of chromosomal DNA. The largest of the big genes is the gene for dystrophin, which spans 2.3 Mb. One hundred twenty-four big genes encompassing 112 Mb of chromosomal DNA have been defined to date. Interestingly, many big genes have modest-sized mRNAs; the exons encoding these RNAs typically encompass about 1% of the total chromosomal gene region in which they occur. This means the exons are widely scattered across large regions composed mainly of introns and, hence, are extremely difficult to find by computational methods alone. Full-length cDNA sequences are often essential for the delineation of exon/intron structures in big genes. But because big genes are synthesized very slowly, their synthesis cannot be completed in rapidly dividing cells. It is in neurons, which do not divide, that a large fraction of the big genes are expressed.

Once again, several intriguing questions arise: Do gene-rich or gene-poor areas have functional significance, or do these different chromosomal organizations represent random fluctuations in the evolutionary events mediating chromosomal evolution and organization? Similarly, do big genes provide some selective advantage, possibly functional and/or regulatory, or does their existence reflect chance events that reshaped the chromosomal architecture at the genomic level?

Combinatorial Strategies May Amplify Genetic Information and Generate Diversity

Combinatorial amplification results from the potential for combining a set of basic elements in many different ways. A simple slot machine, for example, may contain 3 wheels, each carrying 7 different symbols; from its 21 basic elements $(7 + 7 + 7)$, it can generate 343 different combinations $(7^3$, or $7 \times 7 \times 7)$. In biology, combinatorial amplification occurs at both the DNA and RNA levels.

At the DNA Level

Antibody and T-cell receptor genes, for example, are encoded by a multiplicity of gene segments (**Fig. 10.18**). The human T-cell receptor family has 45 functional variable (V) gene segments, two functional diversity (D) gene segments, and 11 functional joining (J) gene segments. In an individual T cell, any D element may first join to any J element by deletion of the intervening DNA. This joined D-J

Figure 10.18 **A schematic diagram of the human β) T-cell receptor gene family.** The variable (antigen recognition) gene is composed of three elements: V, D, and J. During T-cell development, any D may join with any J. Any V may join with any D-J.

element may, in turn, join to any V element—once again by deletion of the intervening DNA—to generate a complete V-D-J gene. This combinatorial process can generate 990 different V-D-J genes (45 × 2 × 11 = 990), although in a given T cell, only one such functional rearrangement occurs. Thus, from 58 gene elements (45 + 2 + 11), a combinatorial joining mechanism can generate 990 V-D-J genes.

T cells, whose receptors are capable of interacting with foreign molecular structures (called antigens), are driven by contact with such antigens to divide and expand their numbers 1000-fold or more. This antigen-triggered expansion by mitosis to a clone of genetically identical cells descended from a common ancestor amplifies the useful combinatorial information and is a key part of every immune response. The particular combinatorial gene arrangements in a few of the original population of T cells produce T-cell receptors more precisely fitted to the antigen. These receptors will then bind the antigen, and such binding will trigger the clonal expansion of the cells that carry the tightly fitting receptors; as a result, the specificity and strength of the immune response will increase.

At the RNA Level

The splicing together of RNA exons in different orders is another way in which combinatorial strategies can increase information and generate diversity. Further diversity results from the initiation of transcription at distinct promoter regions, which create transcripts with different numbers of exons. The three neurexin genes illustrate both of these combinatorial RNA strategies (**Fig. 10.19**). Each neurexin gene contains two promoter regions (producing α and β mRNAs) and five sites at which alternative splicing can occur. Together, these three genes can probably generate more than 2000 alternatively spliced forms of mRNA. A key question is, Do these forms have distinct functions, and if so, how many? Or are they simply the random results of the splicing process? Certainly alternative mRNA splicing will generate distinct functional gene products for some genes, but we have yet to learn the extent to which a single gene may generate functionally distinct alternative mRNA forms.

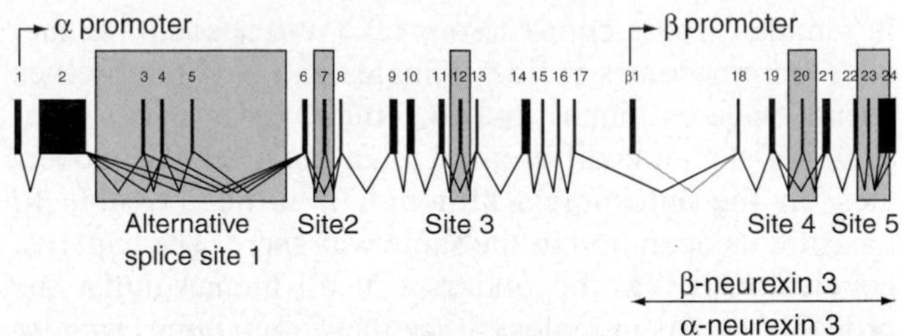

Neurexin Statistics			
Gene	Length in Human	Length in Pufferfish	Number of Potential Alternative Splice Variants in Human
NRXN1	1112 kb	>163 kb	292
NRXN2	117 kb	unknown	194
NRXN3	1692 kb	>181 kb	1764

Figure 10.19 The organization of the three neurexin genes. Each gene has two promoters (α and β) to initiate mRNA synthesis and five sites at which alternative RNA splicing can occur. The *blue rectangles* indicate exons affected by alternative splicing. Numbers at the top of the figure designate exons.

Evolution May Occur by the Lateral Transfer of Genes from One Organism to Another

One of the fundamental dictates of evolutionary theory has been that evolution occurs in a vertical manner; that is, evolutionary changes occur over time as parents pass their rearranged and mutated genomes to their progeny. However, initial analyses of the human genome have suggested that more than 200 human genes may have arisen by direct, or *lateral, transfer* from other organisms such as bacteria. **Lateral gene transfer** (also known as *horizontal transfer*) is the direct transfer of genes from one species into the germ line of another species. Despite some dispute over whether all of the 200 human genes mentioned really represent lateral gene transfers, it appears that lateral gene transfer can and does occur, although we do not yet understand how. Lateral gene transfer is very common in microorganisms. Knowledge of the potential for lateral transfers opens up the possibility that one organism may evolve by using the genes of another organism. This possibility is obviously limited to the exchange of genetic information between organisms that have access to one another's DNA (for example, bacteria in the human intestines).

The possibility of lateral gene transfer suggests that, in principle, the evolutionary potential of one species is enormously enhanced by access to successful evolutionary strategies of other species. It also adds fuel to the debate over genetic engineering, providing a counterpoint to the argument that humans are genetically unique.

Males Appear to Have More Than a Twofold Increased Rate of Mutation in Meiosis over Females

Comparisons of orthologous repeat sequences on the X and Y chromosomes have shown that the rate of mutation in meiosis in males is more than twice as high as the rate in females. These comparisons take into account the fact all Y chromosomes reside in males and X chromosomes spend twice as much time in females as in males. Presumably, the same differential mutation mechanism operates on the autosomes, although it is not possible to measure its operation in the same way there. The tentative conclusion is that the majority of all human mutations probably occurs in males. Since these mutations can generate defects as well as diversity, males may give rise to more defects but also more of the diversity that is the basis of evolution.

The Different Human Races Appear to Have Very Few Uniquely Distinguishing Genes

The human population encompasses a number of different races—Caucasian, Mongoloid, Negroid (or Black), American Indian, and so forth—each of which is distinguishable by certain phenotypic features such as skin color, skin folds at the corners of the eyes, and so on. Genome centers in Europe, Japan, China, and North America have sequenced significant portions of the genome from at least three different races. From these sequences, it appears that the range of polymorphisms within a particular race (for example, Black) can be much greater than the range between any two individuals in different races. Indeed, it is clear that very few genes will turn out to be race-specific. This underscores the basic genetic similarity of all humans.

The Sequences of the Human and Model Organism Genomes Reaffirm That All Living Organisms Evolved from a Common Ancestor

The availability of complete genomic sequences has created an opportunity to analyze large numbers of gene products (mRNAs, proteins) in the context of specific cell types or in the context of the changing patterns of gene expression during the activation of developmental or physiological pathways.

Preliminary analyses support the conclusion of earlier molecular studies that the basic cellular mechanisms of all living organisms (for example, the cell cycle and fundamental features of metabolism) have remarkably similar genetic components. This observation, in turn, supports the idea that life evolved once, and we are all descendants of that fortuitous event. The similarity of basic genetic components also reaffirms that the analysis of appropriate biological systems in model organisms can provide fundamental insights into how the corresponding human systems function.

TABLE 10.3 Sources of Publicly Available Sequence Data and Other Relevant Genomic Information

http://genome. ucsc.edu/ University of Santa Cruz Contains the updated assembly of the human genome sequence.
http://www.ensembl.org EBI/Sanger Centre Allows access to DNA and protein sequences with automatic baseline annotation.
http://www.ncbi.nlm.nih.gov/genome/guide/ NCBI Views of chromosomes and maps and loci with links to other NCBI resources
http://compbio.ornl.gov/channel/ Oak Ridge National Laboratory Provides Java viewers for human genome data.
http://www.ncbi.nlm.nih.gov/Omim/ Online Mendelian Inheritance in Man Contains information about human genes and disease.
http://www.genome.gov/10001618 NHGRI and DOE Contains information, links, and articles on a wide range of social, ethical, and legal issues.

In the Future, Other Features of Chromosomes Will Become Increasingly Important

An understanding of several as yet little understood chromosomal features will become important to study in a global manner as we move into the postgenomic (that is, the post-genome-sequencing) era and develop more effective tools for their characterization. One significant feature is the *chemical modification of bases.* A common modification of DNA is the methylation of C's in the promoter regions of genes. This post-DNA-synthesis event may suppress the readout of individual genes and thereby alter the expression of developmental programs or physiological responses. Other base modifications may also be important.

A second feature is the *interaction of various proteins with the chromosomes.* These interactions can modify the three-dimensional structure of chromosomes and, in so doing, alter their ability to express genes in some regions. For example, the histone proteins catalyze the formation of DNA packaging units known as nucleosomes. Specific DNA-histone interactions thus provide chromosomes with a higher-order structure that allows them to compact; and more compact chromosomal structures are generally incapable of expressing genes (see Chapter 13 for a more thorough discussion of chromosome structure).

A third feature is the *three-dimensional structure of chromosomes in the nucleus.* Such three-dimensional structure may help determine the interactions of specific chromosomal regions with particular regions of the nuclear envelope. The three-dimensional structures of chromosomes are also likely to influence the readouts of the developmental and physiological programs encoded in the genome.

We need to develop far more effective tools for looking at chromosomal chemical modifications, chromosomal interactions with proteins and other macromolecules, and the formation and influence of three-dimensional chromosomal structures.

Table 10.3 lists a variety of websites where you can directly view detailed information on some of the preceding insights into genetics and biology as well as data about other aspects of the Human Genome Project.

10.3 High-Throughput Genomic Platforms Permit the Global Analysis of Genes and Their mRNAs

The Human Genome Project has given us the complete sequences of a variety of genomes. The next step is to carry out global analyses of this basic data to gain a more complete understanding of the information in the genome.

The need for global genomic analysis has driven the development of powerful high-throughput platforms that enable researchers to generate the information necessary to carry out large-scale, automated searches of global databases. In this context, a *platform* denotes all the components needed for an automated acquisition of a set of data. A high-throughput genomic platform might include automated instrumentation to prepare DNA fragments for sequencing, an automated sequencer to carry out the sequencing, and a set of computational tools to capture and store the information obtained from sequencing.

We now describe some of the main instruments developed for large-scale genomic analyses. In our discussion of these tools, we use *genomics* to mean the global analysis of chromosomal and RNA features (such as sequence, genetic markers, mRNAs).

High-Throughput Instruments

Two high-throughput technologies have played a key role in genomic platforms: the DNA sequencer and DNA arrays.

The DNA Sequencer

The automated DNA sequencer employs the dideoxy (Sanger) sequencing strategy described in detail on pp. 331–333 of Chapter 9. The automated format is a three-step process (**Fig. 10.20**), which we briefly summarize. Step 1: DNA polymerase synthesizes a complementary DNA strand from a primer sequence across the unknown DNA insert to be sequenced. The unknown sequences serve as the template strands for the primer-initiated synthesis of complementary strands. The four chain-terminating dideoxy bases are used in separate reactions at specific concentrations such that roughly 0.5% of the growing chains are stopped at each G, C, A, and T across the sequence. The primers for each of the four reactions are labeled with one of four fluorescent dyes. Hence, the A-terminating fragments are labeled with one dye, and the T-, G-, and C-terminating fragments are each labeled with a second, third, and fourth dyes. Step 2: The four different sets of fragments are pooled and then separated by a type of electrophoresis that separates fragments differing by one base. Each of the separated fragments is color-coded with a distinct fluorescent dye, depending on the terminal base in the fragment. Step 3: The specific fluorescence of each fragment identifies its terminal nucleotide. For example, all fragments ending in A will fluorescence red, while those ending in T will fluoresce yellow.

The sequencing instruments described earlier can analyze 96 or, more recently, 384 DNA fragments at a time. Moreover, several large sequencing centers have created sequencing production lines in which the production of DNA fragments for sequencing has been semiautomated, and the capture and quality assessment of the DNA sequence data have been fully automated. The most recent 96-capillary instruments can, in principle, sequence more than 1,500,000 bp in 24 hours.

New DNA sequencing strategies using highly parallel simultaneous analyses (which sequence simultaneously millions of DNA fragments) may permit the rapid and inexpensive sequencing of individual human genomes within the next 10 years.

DNA Arrays

A DNA array is a large set of DNA fragments displayed on a solid support. To make a DNA array, researchers either synthesize DNA fragments encoding portions of individual genes at selected sites on a solid support or "spot" (that is, deposit) presynthesized DNA fragments at precise locations on the solid support (**Fig. 10.21**). They then

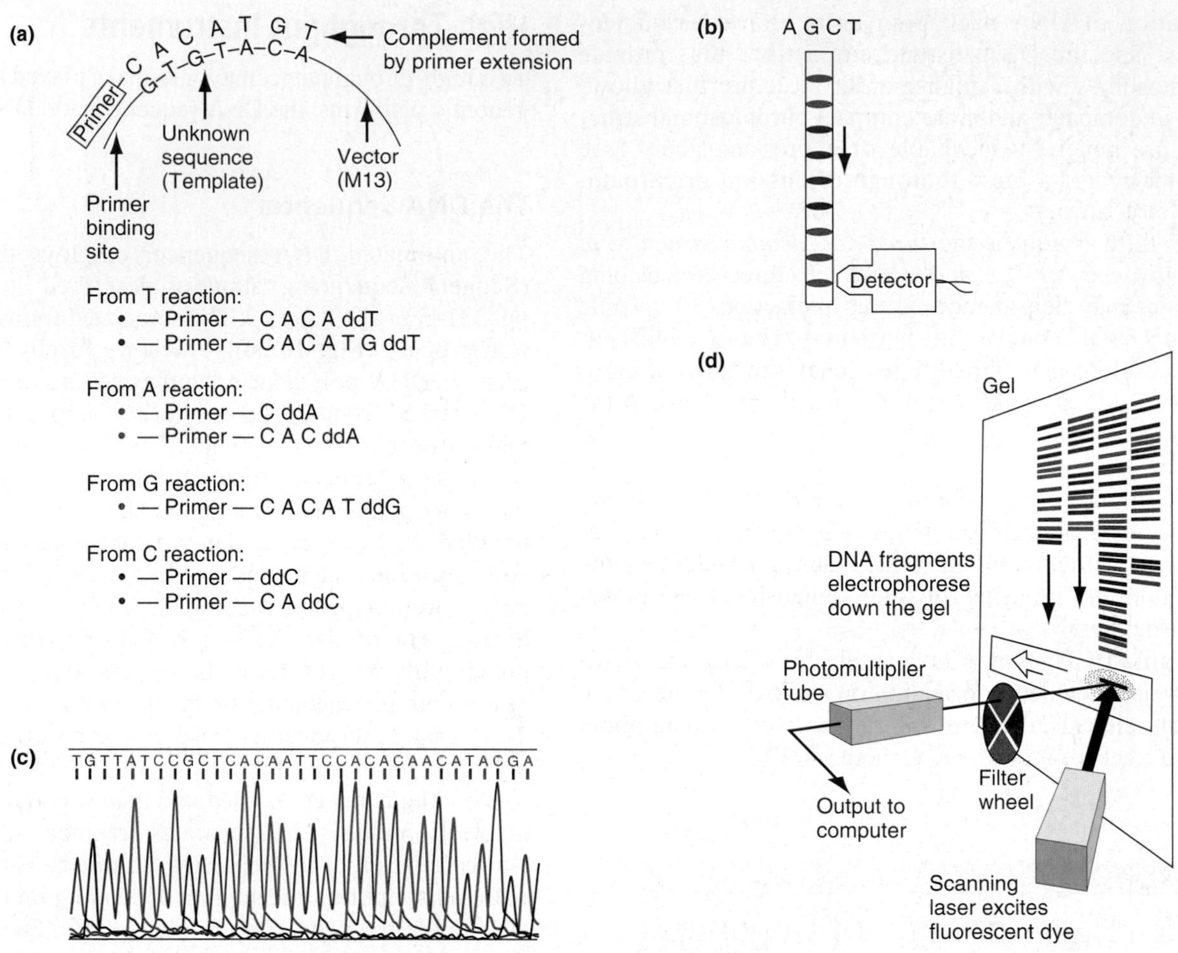

Figure 10.20 **Sanger sequencing scheme. (a)** Each reaction contains the unknown template DNA cloned in a single-strand sequencing vector, a primer oligonucleotide complementary to a region of the vector near the 3′ end of the unknown DNA, a DNA polymerase, all four deoxynucleotide triphosphates (dNTPs), and one dideoxynucleotide triphosphate (ddNTP). The detection label (·) is shown attached to the 5′ end of the primer. **(b)** Schematic depictions of the gel electrophoresis detection system used in DNA sequencing. Real-time detection of bands from four sequencing reactions, each using a different-colored label, run in a single gel track. **(c)** Data from DNA sequencing experiments. Superimposed four-color traces obtained in a sequencing experiment using four-color DNA sequencing on a DNA sequencer. Data shown correspond to the same region of the vector as that in (a). **(d)** Commercial implementation employing fluorescence scanning of slab electrophoresis gel.

hybridize a complex mixture of fluorescent or radio-labeled DNA or RNA fragments (for example, mRNAs or cDNAs) with the array of anchored fragments to analyze in a quantitative manner the levels of expression of individual genes represented on the array. In this manner, they can analyze tens of thousands of gene-expression patterns.

There are three types of DNA arrays.

Macroarrays After the creation of a cDNA library, which is then copied on agarose and placed in microtiter wells, an arrayer robot can pick up DNA from each of 96 or 384 wells and spot it on the nylon membranes forming an array. The robot can individually spot up to 20,000 clones from a cDNA library on the nylon membranes from the microtiter wells. The complex cDNA mixture to be analyzed against the array is radio-labeled, and any hybridization is detected by autoradiography or with a phosphoimager (an instrument for detecting radiolabels). These microtiter plate-based arrays are termed **macroarrays.** Macroarrays, which are relatively inexpensive, are useful for preliminary studies of gene expression in organisms or cell types where only a small fraction of the genes have been sequenced or characterized.

Microarrays PCR amplified gene fragments are spotted on glass slides. A single slide can hold roughly 5000–40,000 PCR products. The mixture of mRNAs or cDNAs to be analyzed is labeled with fluorescent dyes and hybridized. The level of fluorescence in a single DNA spot is proportional to the level of gene expression. These small glass-slide arrays are termed **microarrays.** A researcher can use microarrays

Figure 10.21 An oligonucleotide array. The oligonucleotides are covalently linked to the glass or silicon. Typically, the oligonucleotides range in size from 20–60 bp.

to simultaneously analyze two different cDNA (or RNA) samples taken from one kind of cell in two different states or from two cell types. To distinguish the samples, each is labeled with a different-colored dye (**Fig. 10.22a**). It is possible to measure the levels of the two dyes for each spot on the array to determine patterns of gene expression in the two different cell states or cell types (Fig. 10.22b). The differences in gene expression can be detected only if they vary by a factor of two or more. This technology is relatively inexpensive compared to the third type of array.

Oligonucleotide Arrays This third type of array involves the synthesis or spotting of oligonucleotides at densities of 1 million or more on glass slides. An **oligonucleotide** is a short segment of DNA or RNA that is chemically synthesized to precise specifications. Oligonucleotides are usually 20–60 nucleotides in length. The Affymetrix oligonucleotide array

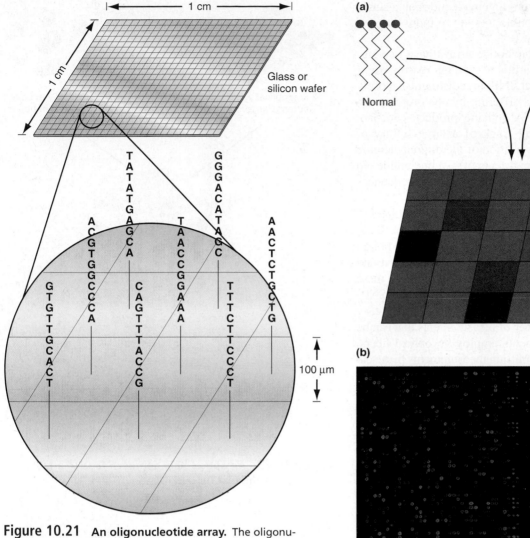

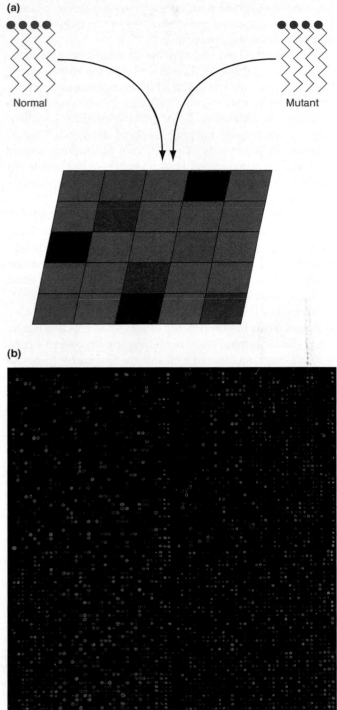

Figure 10.22 Two-color DNA microarrays. (a) Two separate cDNA samples, one from normal yeast, the other from a mutated yeast, are labeled with red and green fluorescent dyes and hybridized to a PCR microarray. When both labeled messages hybridize to a PCR product, a *yellow* color is produced. The *red* hybridizations represent mRNAs (cDNAs) overexpressed in the mutant, and the *green* hybridizations represent mRNAs overexpressed in the normal yeast. Black squares contain a PCR product that fails to hybridize with either cDNA. **(b)** The entire yeast genome of ~6000 genes is represented on this DNA microarray. The array is hybridized with two different populations of labeled mRNA.

can carry up to about 6 million oligonucleotides on a single glass support, making it possible to analyze more than 1.5 million SNPs simultaneously.

The synthesized oligonucleotide arrays have two significant advantages over their macro- and microarray counterparts: First, they can detect SNPs and distinguish closely related gene products (genes differing by one or a few nucleotides) or alternative RNA splicing products. Second, they do not require keeping track of a large library of clones or PCR products. A variation of the oligonucleotide array uses presynthesized fragments 50–70 nucleotides in length for each gene. These 50- to 70-nucleotide sequences can be spotted on glass slides.

One new approach to scaling up the density of oligonucleotide arrays comes from the application of optic fiber technology. It is possible to create 50,000 wells in the end of an optical fiber 1 mm in diameter (**Fig. 10.23**). Fifty thousand beads, each with a different 25-mer, can be placed in these wells and used to interrogate a target sample, either for SNPs or patterns of gene expression. With this approach, it is easy to extend (scale up) the number of measurements that can be made. Since current instruments employ 96 optical fibers, they can carry out close to 5 million measurements per cycle of analysis (50,000, or $5 \times 10^4 \times 96 = 4{,}700{,}000$). These optical arrays can analyze more than 1 million SNPs per day. The objective of all global technologies is high-throughput (the processing of a large amount of information) and low cost per unit of information analyzed.

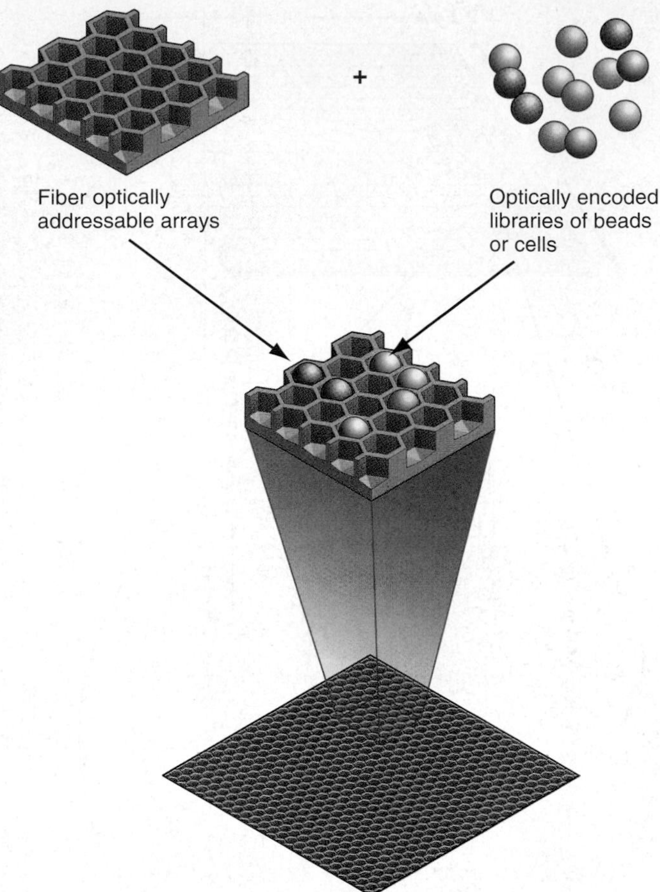

Figure 10.23 **An optical fiber approach to DNA array analyses (see the text for details).**

Global Genomic Strategies and High-Throughput Platforms

The instruments just described are the analytic centers of high-throughput platforms. As mentioned earlier, such platforms include the automated preparation of the samples to be analyzed, the analytic instruments themselves, and automated software for capturing and assessing the quality of the global data sets gathered. We now describe how genomic researchers use the tools in selected platforms.

Analysis of Genomic and RNA Sequences

The platform of choice for genomic and EST or full-length RNA sequencing is the automated DNA sequencer.

Quantitative Analysis of mRNA Levels Researchers employ three types of DNA sequencing to determine the quantities of different mRNAs in a mixture. They have used the automated DNA sequencer, for example, to generate more than 6 million human ESTs from hundreds of different cell types. More recently, they have developed a technique termed **serial analysis of gene expression,** or **SAGE.** With SAGE, they can synthesize small cDNA tags about 15 bp in length from the 3′ ends of mRNA and link these small tags together into 1000 bp DNA that can then be sequenced. In principle, each 1000 bp DNA string contains 15 bp tags from roughly 67 different genes ($15 \times 67 = \sim 1000$). The sequence analyses of thousands of these linked 15 bp tags permit a quantitative estimate of the mRNAs present in the original cell type.

Finally, the very recent massively parallel signature sequence technique (MPSS) appears to be the most powerful approach to quantitatively defining the transcriptomes of individual cell types (**Fig. 10.24**). A **transcriptome** is the population of mRNAs expressed in a single cell or cell type. MPSS is the only technique that readily (and cost effectively) allows the routine identification of most of a cell's rarely expressed mRNAs. In this context, "rarely expressed" means that there are fewer than 10 copies per cell. The MPSS approach could, for example, take a cDNA library of 1 million clones and amplify each clone with a unique PCR tag so that the products from each clone can be attached to a single nylon bead (Fig. 10.24). The million beads are displayed in a flow cell, and then the million different sequences are simultaneously sequenced with fluorescent reporter groups to generate 1 million sequence tags of 17–20 bp. If the genome of the organism whose cell type was used

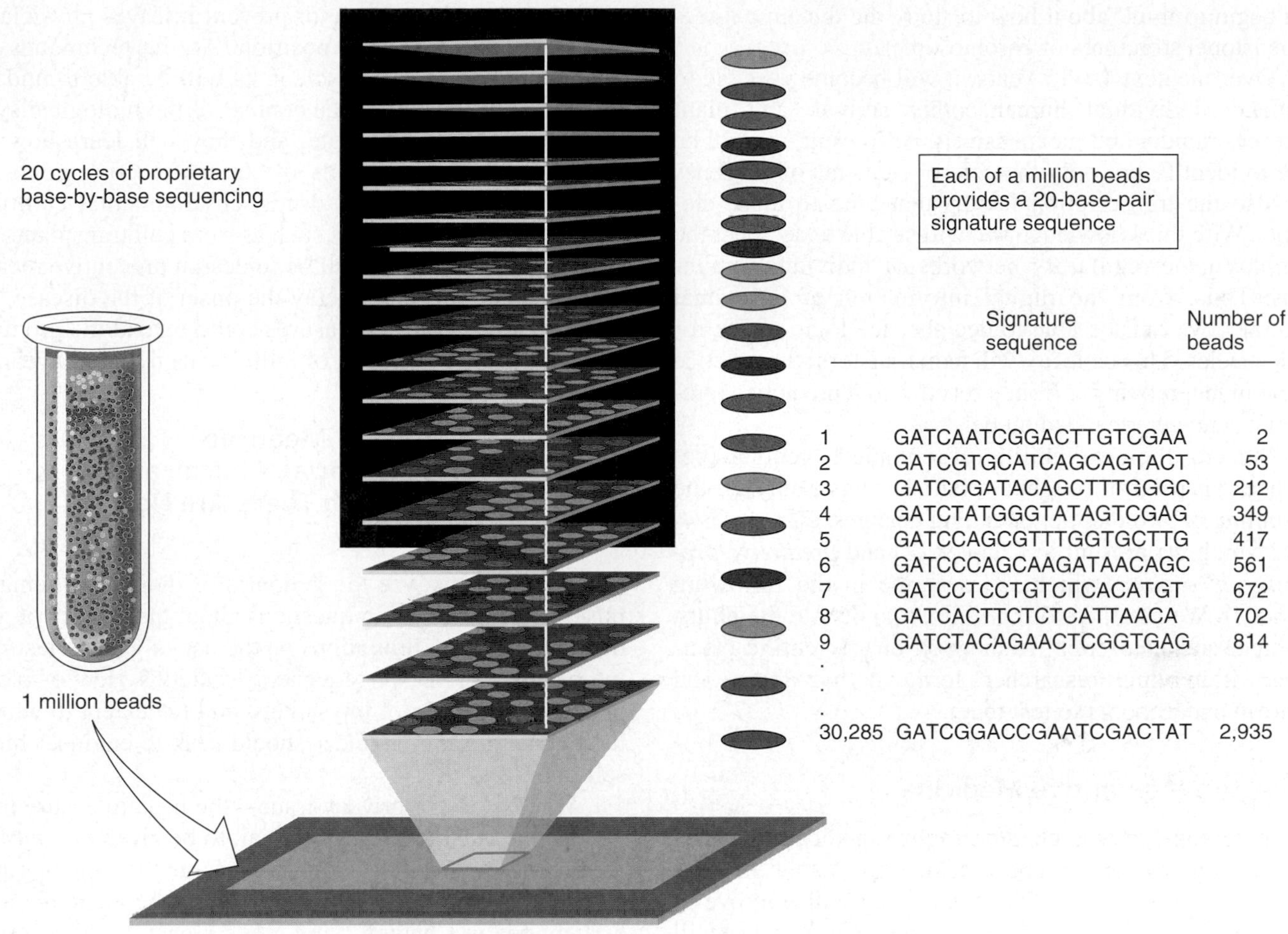

20 cycles of proprietary base-by-base sequencing
Each of a million beads provides a 20-base-pair signature sequence
Signature sequence
Number of beads
1 GATCAATCGGACTTGTCGAA 2
2 GATCGTGCATCAGCAGTACT 53
3 GATCCGATACAGCTTTGGGC 212
4 GATCTATGGGTATAGTCGAG 349
5 GATCCAGCGTTTGGTGCTTG 417
6 GATCCCAGCAAGATAACAGC 561
7 GATCCTCCTGTCTCACATGT 672
8 GATCACTTCTCTCATTAACA 702
9 GATCTACAGAACTCGGTGAG 814
30,285 GATCGGACCGAATCGACTAT 2,935
1 million beads
1 million beads arranged in a monolayer

Figure 10.24 **Lynx Therapeutics sequencing strategy of multiple parallel signature sequencing (MPSS).** The text describes how this recently introduced technique for quantitatively defining the composition of a cell's transcriptome is used.

to make the cDNA library has been sequenced (and, hence, all or most of the genes have been identified), then the MPSS technique can provide a precise quantitative measure of most of the mRNA types in the cell down to those from genes expressed in only a single copy. This sensitivity is significant because many biologically important genes are expressed at very low levels (less than 50 copies per cell). The typical mammalian cell has perhaps 300,000 transcripts representing 20,000 different mRNAs. Ninety percent of the mRNAs are expressed at levels of less than 100 copies per cell.

The DNA array approaches, particularly the microarray and oligonucleotide arrays, also permit a semiquantitative analysis of patterns of gene expression. Two-color microarray analyses have been particularly useful in comparing different tissue or cell types (for example, normal versus cancerous). Most DNA arrays, however, do not accurately detect mRNAs expressed at less than 50 copies per cell.

Analysis of SNPs SNPs can be analyzed by oligonucleotide arrays with the use of molecular hybridization, or they can be analyzed by mass spectrometry described in Chapter 12. There are a variety of different approaches for distinguishing SNPs by mass spectrometry. All involve the analysis of oligonucleotide fragments that reflect the differences in the masses of the single nucleotide polymorphisms. The ability to distinguish SNPs is useful in genomic analysis.

Repercussions of the Human Genome Project and High-Throughput Technology: An Overview

The genome sequences of humans and model organisms have transformed all of biology. Knowledge of these sequences enables us to identify and readily access most human genes, and the ability to do this greatly facilitates our understanding of their functions. We can also use the genome sequences to search for the control elements of each gene, which help regulate the gene's expression (see Chapters 13, 17, and 18). In addition, we can readily do whole genome comparisons within and between species to characterize the fundamental features of evolution; and we can begin to think about how to study the dynamic three-dimensional structures of chromosomes.

Over the next 10–15 years, it will become possible to sequence individual human, other animal, and plant genomes rapidly and inexpensively. As a result, we will be able to identify not only the control elements of all genes but also the transcription factors that bind to those elements. With this knowledge, we will be able to decipher the complex gene regulatory networks of individual organisms. Thus, from the digital information of individual genomes, we will be able to decipher the logic of life for each species. This capacity will help transform the practice of medicine, moving it from a reactive to a predictive, preventive, and personalized mode.

Indeed, the availability of complete genome sequences and high-throughput platforms has catalyzed the beginning of two distinct paradigm changes: *systems biology* (which we present in Chapter 12) and *predictive/preventive medicine* (which we describe in the following sections). We use the word *paradigm* to denote the philosophical and theoretical framework of a scientific discipline within which researchers formulate hypotheses and perform experiments to test them.

Predictive/Preventive Medicine

The potential for predictive/preventive medicine depends on access to whole-genome sequences. Over the next 10 years, technologies for DNA sequencing will improve to the point where an individual human genome sequence will cost less than $1000, and many people will be able to have their genomes sequenced. The inexpensive sequences thus obtained will serve as one basis of predictive medicine because they will provide access to the DNA polymorphisms underlying human variability.

Although most polymorphisms fall outside of genes and thus do little to change human phenotypes, some are responsible for differences in normal physiology (tall or short, thin or fat), and others predispose to disease. With the potential for carrying out dense genomewide genetic analyses, clinical researchers will in the next decade or so identify hundreds of genes that predispose to disease (mostly late-onset diseases such as cancer, cardiovascular disease, immunological diseases, neurological diseases, and metabolic diseases). For example, the breast cancer 1 gene (*BRCA1*), when present in women in a single defective copy, predisposes 70% of them to breast cancer by the time they are 60 years old. Why only 70%? Either there are environmental factors that operate in concert with the defective gene, or there are other modifying, disease-predisposing genes that are present only 70% of the time. In either case, a prediction can be made about the future likelihood of disease for individuals carrying the defective gene. In time, physicians will be able to scan the genomes of the young and provide a probabilistic projection of what the future may hold with regard to a wide variety of diseases. Of course, being able to predict a disease without being able to cure or prevent it leaves physicians in a very uncomfortable position. As the techniques of systems biology mature, scientists will be able to understand defective genes in the context of the biological systems in which they operate, and they will learn how to circumvent the limitations of these defective genes—with, for example, novel drugs, environmental controls (diet), or other approaches such as stem cell transplants or gene therapy. The idea will be to design preventive measures to avoid or greatly delay the onset of the disease. In time, these preventive measures could extend the productive and creative life span of individuals by 10–30 years.

Predictive/Preventive Medicine Raises Challenging Social, Ethical, and Legal Issues for Which There Are No Simple Solutions

The ability to analyze the genomes of individual humans raises a host of pressing questions about the privacy of genetic information, limitations on the use of genetic testing, the patenting of DNA sequences, society's view of older people, the training of physicians, and the extent to which the human genetic engineer should seek to engineer himself or herself.

We discuss the privacy issue—the need to ensure that genetic information is confidential, to be given out only at the discretion of the tested individual—in the Genetics and Society boxes in Chapters 2 and 3. In the Genetics and Society box in Chapter 2 and in the Genetics and Society box "Social and Ethical Issues Surrounding Preimplantation Genetic Diagnosis" in Chapter 11, we outline considerations that make it necessary to establish guidelines for genetic screening; among those considerations are who should be tested and whether testing should be carried out if no therapies or preventive measures exist to help an individual with a specific genetic predisposition. Most geneticists emphasize the need for educating the general public about the benefits and drawbacks of testing. Meanwhile, the U.S. Congress has passed a law ensuring the privacy of medical records. The Genetics and Society box on pp. 381–382 of this chapter discusses the "Patentability of DNA."

With predictive/preventive medicine, many people could remain productive and creative into their nineties. This increase in longevity (which we are already beginning to witness) will require our society to rethink its philosophies of retirement, social security, access to medical care, and related topics.

The physicians trained today will be practicing predictive/preventive medicine in 20 years. How should their training change to prepare them for the new ways of thinking about medicine? And, how can we educate society about the very different nature of predictive/preventive medicine?

Finally, what about the social and ethical concerns raised by genetic engineering, that is, the ability to alter

The Patentability of DNA

Some people argue that genetic information, the naturally occurring raw material of life's evolution, is a common heritage that belongs to everyone. Yet since the mid-1970s, universities and biotechnology companies have sought patents on specific DNA sequences in virtually all types of genomes—plant, human, other animal, bacterial, viral, and plasmid. To understand this apparent paradox, we take a brief look at the history and purpose of U.S. patent law and examine its application to various types of DNA sequences.

Congress passed the first patent act, penned by Thomas Jefferson, in 1790. That same year, the U.S. Patent Office awarded the first patent to Samuel Hopkins of Vermont for the making of potash, a potassium-based compound used to produce fertilizer and soap. One hundred and ninety years later, in 1980, the patent office granted two patents for specific uses of genes. The first was to a microbiologist at General Electric for a genetically engineered oil-eating bacterium that could help clean up spills; the second was to Stanford University and the University of California at San Francisco (UCSF) for the basic gene-splicing and gene-cloning technology invented by researchers at those institutions.

Patent examiners evaluate the patentability of a product or process by three criteria: Is it *new?* Is it *nonobvious?* Is it *useful?* In the DNA arena, the courts have made the following interpretations of patent law. Raw materials of nature, such as wild-type DNA in a living organism, are not novel and thus are not patentable; modified products, such as bacterial DNA altered by a synthetic mutation or human DNA in a mouse genome, are novel and thus eligible for patents by the novelty criterion. DNA-based processes that produce a novel material, such as clones of a gene (consisting of a DNA construct in a vector), are also patentable. Publication in a scientific journal makes an item, such as a DNA sequence, obvious, thus unpatentable per se; but a specific use of a published sequence may be nonobvious and thus patentable. A well-defined use might be a particular test for a particular genetic aberration or a particular process for the manufacture of a particular therapeutic agent.

Since the mid-1980s, patent examiners have used these guidelines to grant DNA-related patents on a case-by-case basis. The cDNA for insulin integrated into a bacterial plasmid has received a patent; this cDNA does not occur naturally in significant quantities, but it is fabricated through a human-devised protocol using reverse transcriptase. By contrast, human cDNA fragments isolated at random from brain cells did not qualify for a patent, in part because the potential use of these fragments could not be defined without identification of the complete cDNAs and the proteins they encode. In another example, researchers at the University of Utah obtained a patent on *BRCA1* sequences to be used as a test for predisposition to breast cancer. These researchers mapped and sequenced the gene and incorporated that knowledge into the design of a socially useful test; thus their application passes muster.

The rationale for granting patents is to encourage innovation—the invention of useful contributions to society—by providing a time-limited monopoly to protect an invention from imitation. This commercial protection is given in exchange for the complete disclosure of the information related to a product or process. In fact, in pursuing a patent, a company protects its interests by making available as much information as possible about the modified product or process. Patents protect use for profit; they do not interfere with research or other noncommercial uses of the information in the patent. Institutions, companies, or individuals holding patents can profit from their inventions themselves or license the right to profit from them to another party. Stanford and UCSF, for example, originally granted many nonexclusive licenses to companies using recombinant DNA technology in the production of various agents for a fee of $10,000 a year. In the first four years of its patent, Stanford received $3 million in licensing fees and patent royalties.

Different countries apply the basic tenets of patent law in different ways. In the United States, the purified form of a gene or protein is patentable because genes and proteins do not exist in nature in purified form. In England, a naturally occurring gene sequence is not patentable no matter what form it is in. And in France, the code on intellectual property declares unpatentable "the human body, its elements and products as well as knowledge of the partial or total structure of a human gene."

There is currently healthy debate over the application of patent criteria originally developed for mechanical and chemical inventions to DNA sequences and various life-forms. The debate focuses on several areas of concern.

Openness Versus Secrecy

How does the patent process affect the open exchange of ideas, which is a prerequisite of basic research? A company putting a great deal of money into the research and development of what it hopes will be a patentable gene therapy may withhold publication of its data or publish only partial results until its patent application is in the pipeline. For example, in the 1980s, before the discovery of the cystic fibrosis gene, one gene diagnostics company published an article about markers for the *CF* gene but, in an effort to protect their work, did not include the fact that those markers were on chromosome 7. As it happens, other groups subsequently found closer markers and then the gene. Still the question remains: Do commercial considerations interfere with the free exchange of ideas and results about genomes? If so, should something be done about this?

Profitability Versus the Social Good

Who has control over genetic data, deciding what applications to pursue with commercial funding? What criteria do they use? Most companies consider potential profitability as the basis for pursuing research and development. Some people wonder whether they should also factor into the

(continued)

equation concerns about serving the poor (who cannot pay much for drugs or therapies) and the relatively few patients suffering from uncommon diseases. For example, about 25 million people in Africa are infected with HIV, the virus that causes AIDS. In the United States and Europe, AIDS can be managed by triple drug therapy—the simultaneous administration of three drugs that attack AIDS with different mechanisms—but this therapy costs more than $12,000 per year. Most infected individuals in Africa cannot afford the treatment. India has started to manufacture generic AIDS drugs costing just a few dollars per day, roughly $1100 per year. India can manufacture these drugs inexpensively because the very costly research to identify promising compounds and develop them into drugs was paid for by European and American pharmaceutical companies. In fact, India is violating drug-company patents when it sells the AIDS drugs inexpensively in Africa. On the other hand, millions of infected people in Africa could never receive appropriate treatment for AIDS without the inexpensive generic drugs manufactured in India. Thus, the challenge to bring up-to-date health care to developing nations poses striking social, ethical, and legal problems.

Development and Funding of Socially Useful Research Applications

Since companies must pour large sums of money into developing DNA-based drugs or therapies, testing them on large numbers of people, and bringing them to market, many researchers—at both universities and commercial enterprises—maintain that the survival of biotechnology companies, particularly small ones, depends on patent protection. Without the possibility of a patent, they argue, companies may not be able to afford to use knowledge of, for example, the *CF* gene and the transmembrane conductance regulator it encodes to develop therapies that may ease and extend the lives of thousands of CF sufferers. Development of drugs or therapies that cost a great deal of money, researchers maintain, simply will not occur without patents. On the other hand, many of the patents providing the basis for development and marketing are on inventions made in university laboratories; and the investigators in some of these laboratories are nonprofit recipients of federal grants that are ultimately funded by taxpayers. Is it proper for companies to use these inventions as the basis for financial gain? Most government agencies allow such patents because they themselves do not have the funds to develop research ideas for the marketplace. Furthermore, nearly all of the major technological advances of the late twentieth century in computers and electronics were based to one degree or another on research funded by federal money. Most people do not question the right of computer companies to profit off extensions of this research.

Thus, the availability of purified DNA, especially specific stretches of the human genome, raises several issues that the framers of patent law could not foresee. Vigorous public debate will help scientists, businesspeople, and lawmakers distill some of the answers. One radical and perhaps impractical proposal to deal with the legitimate concerns on both sides of the public-funding-versus-private-profit issue suggests creation of an international corporation that would hold patents on all human cDNA. This corporation would license the patents for development at auction and reinvest the proceeds from licensing fees in basic research.

specific genes of an organism in specific ways? Many preventative or remedial measures constitute **somatic gene therapy** in which medical practitioners compensate for a faulty gene by inserting a replacement gene into the affected tissue where the gene is expressed. Somatic gene therapy causes biochemical and physiological changes in the genetically modified tissue or tissues that die with the individual. A potential therapy for cystic fibrosis, for example, consists in inserting a wild-type *CFTR* gene into lung cells. This type of genetic engineering is not different in kind from drug therapies aimed at correcting a particular physiological deficiency (for example, the insulin injections aimed at treating diabetes). The alterations resulting from somatic gene therapy affect only the somatic cells of the individual undergoing the therapy and cannot be transmitted to offspring.

Most of the controversy surrounding genetic engineering stems from the potential for **germ-line gene therapy:** modifications of the human germ line. Germ-line gene therapy produces changes in germ cells that are passed on to progeny. Ethical concerns focus on what is and is not appropriate. Should, for example, a couple be able to eliminate a cancer-predisposing gene in their unborn child? Should they be allowed to alter the child's potential for obesity or longevity? Should they have the option of choosing the child's eye color? (See the Genetics and Society box in Chapter 11 for more on this subject.)

Many geneticists and bioethicists support the idea of somatic gene therapy but oppose germ-line therapy. At the same time, they urge serious and open discussion of the issues before perfection of the technology overtakes our ability to control its use.

Connections

One of the triumphs of modern genetics has been the determination of the complete sequence of the human genome and the genomes of many model organisms. With powerful new "shotgun" strategies for sequencing large genomes, genomicists have sequenced more than 250 different microbe, plant, and animal genomes. The genome projects now underway in laboratories around the world will eventually add hundreds of additional genomes to this list.

The Human Genome Project, which supported the development of high-throughput DNA sequencing platforms, has catalyzed the emergence of high-throughput platforms for analyzing mRNA. Based on these tools, analyses of complete genome sequences have provided insights into the architecture and evolution of genomes. The genetics parts lists made available by the genome sequences are transforming the practice of biology and medicine.

In this chapter, we focused on the human genome in a generic sense, as the common framework of genetic information that defines the human species. Indeed, every individual person carries a diploid genome that is 99.9% identical to that carried by every other individual. But the flip side of a 99.9% identity is a 0.1% nonidentity that distinguishes people (other than identical twins) from each other. This 0.1% difference translates into 6 million DNA sequence differences, or polymorphisms, which are responsible for all of the inherited ways in which individuals differ from one another. Of critical interest to medical researchers are the specific DNA polymorphisms that either directly alter or indirectly mark the genes that cause or predispose to disease. In Chapter 11 on the direct detection of individual genotypes, we discuss how researchers identify and use DNA polymorphisms to uncover disease-causing genes and other genes of interest in humans and other species. The discussion includes an examination of complex genetic traits, which arise from interactions among multiple genes and the environment.

Essential Concepts

1. Large-scale maps are the foundation of genomics.
   a. *High-density linkage* or *genetic maps* developed by computer analysis of raw genotyping data show the relative positions of closely spaced DNA markers. High-throughput platforms have been developed for the rapid analysis of large numbers of genetic markers (millions per day). Two types of commonly used genetic markers are single nucleotide polymorphisms (SNPs) and simple sequence repeats (SSRs), or microsatellites.
   b. Long-range *physical maps* chart the features of chromosomes.
      (1) Fluorescent *in situ* hybridization, or *FISH,* locates a cloned locus to a particular band on a particular chromosome.
      (2) DNA hybridization and fingerprinting order the clones of a genomic library into overlapping clusters that span the chromosomes of the species. One approach to assigning chromosomal positions uses the markers of a high-density linkage map and sequence tagged sites (STSs) as hybridization probes to obtain overlapping, large-insert clones that span the chromosomes. Gaps between the clusters of clones (contigs) can be filled in by surveying new large insert libraries with STS markers from either contig end. With a large enough number of clones, it is possible to form contigs that extend across every chromosome.
      (3) Researchers use restriction mapping and hybridization probes to characterize the fine details of each clone in a contig.
      (4) Computers combine the information for individual clones into detailed whole chromosome physical maps.
   c. Long-range sequence maps, compiled from the sequences of subclones, provide a readout of every nucleotide in each chromosome. The subclones are derived either from previously mapped large insert clones (hierarchical shotgun approach) or directly from the genome (whole-genome shotgun approach).
   d. Linkage, physical, and sequence maps can be readily integrated because the markers for the linkage and physical maps (SSRs, SNPs, STSs) use sequence-based PCR primers that generate unique molecular addresses in the genome and thus can be specifically localized on the genome.

2. A variety of major insights have emerged from analyses of human and model organism genomes.
   a. The number of human genes, approximately 25,000 is surprisingly low.
   b. Genes fall into two major classes: noncoding RNA genes, generally representing RNA molecular machines; and protein-coding genes encoding the protein structural components and molecular machines. The collection of proteins in a particular cell is a *proteome,* and the collection of mRNAs, a *transcriptome.* Transcriptomes and proteomes are dynamic and change throughout the time spans of development and physiological responses. Proteins are often composed of one or more domains that encode discrete functions.
   c. Repeat sequences constitute more than 50% of the human genome. Some repeat sequences have evolved to become genes or control sequences. Others may catalyze chromosomal rearrangements and translocations.

d. Evolution can occur by the lateral transfer of genes from one organism to another. Humans, for example, may have as many as 200 genes derived recently from bacteria.
e. The different human races have very few, if any, uniquely distinguishing genes.
f. The sequences of microbes, plants, and animals all employ the same genetic code and show a remarkable similarity among many basic biological systems. This affirms the idea that we all descended from a single common ancestor.

3. The Human Genome Project has catalyzed the development of the DNA sequencer and DNA arrays for genomics. These instruments have been integrated into high-throughput platforms for DNA sequencing and genotyping, as well as for mRNA analysis.
4. The Human Genome Project has spawned paradigm changes in our approaches to biology and medicine. For example, predictive medicine will allow researchers to correlate polymorphisms with disease predisposition and eventually formulate probabilistic health-history predictions for each individual. Preventive medicine will place the defective genes in their biological systems and develop ways to avoid their limitations with, for example, drugs, diet, gene therapy, or stem cell therapy. The potential of predictive/preventive medicine raises social, ethical, and legal questions for which there are no easy answers.

On Our Website

www.mhhe.com/hartwell3
Chapter 10

Annotated Suggested Readings and Links to Other Websites

- Historical papers on the origins of the Human Genome Project
- Historical and contemporary papers on the technologies of the Human Genome Project
- Historical and contemporary papers on the conclusions drawn from the Human Genome Project
- Links to publicly available sequence data and other genomic information
- Historical and contemporary papers on the ethical, social, and legal implications of the Human Genome Project

Specialized Topics

- The first two classic papers on the Human Genome Project

Social and Ethical Issues

1. Several companies were formed in the 1990s with the purpose of sequencing portions of the human genome and using this information to devise new drugs. These companies sought to patent each bit of DNA sequence that they decoded even if they did not know whether it encoded a gene product. Their justification was to ensure that any use of the sequence would yield payback for their research. This strategy failed as shown by the failure of most of these companies. Do you think patents should be granted for each partial gene sequence that has been determined? Do you think patents should be granted for entire gene sequences even if we do not know what the gene does? What criteria do you think are compatible with facilitating private-industry drug development, encouraging research, and protecting intellectual property?
2. Researchers who have similar interests and complementary expertise often collaborate on projects. The Johnston lab has recently started a collaboration with a lab in San Francisco to use a unique tissue culture cell line that the San Francisco lab established. At a scientific meeting, a graduate student from the San Francisco lab heard Marcus Johnston give a seminar talk in which he discussed future work and described this cell line, but he did not mention where it came from. Johnston also did not describe it accurately. The graduate student was very upset with this but then thought perhaps she was being too sensitive and when the work was done and publication occurred, appropriate credit would be given to the San Francisco lab. Should the student speak up at the end of the talk to correct misinformation? Should she approach Johnston later? Should she wait and talk with her advisor first? Should she do nothing and assume the misinformation will be corrected by someone else?

Solved Problems

I. A physical map of overlapping clones (a contig) was available for an area of a human chromosome containing three genes (and part of a fourth gene) that are transcribed within ovarian tumor tissue. The restriction map of the region is shown; tick marks above the line indicate *Bam*H1 restriction sites, while those below the line indicate *Xho*I sites. Sizes of DNA fragments between adjacent restriction sites are given in kilobases. Individual restriction fragments were purified, made radioactive, and used as probes for Northern blots of poly-A-containing RNA derived from ovarian tumors. In a *Northern blot* (described in Fig. 11.20), mRNAs are analyzed just as DNA fragments are in a Southern blot (see Chapter 9). The poly-A-containing mixture of mRNAs from a particular cell type is subjected to electrophoresis—separating the mRNAs according to size (smaller mRNA migrate more rapidly than larger ones). The mRNAs are then transferred from the gel to a nitrocellulose filter and a radio-labeled probe for a particular gene (or gene fragment) is hybridized against the mRNAs on the filter. If size standards are also run, the presence and size of particular mRNAs can be established. The resulting autoradiograms are presented below the restriction map. Using these data, characterize the four genes within the contig in the following ways:

a. What is the length of the mRNA for each of the three complete genes?
b. What is the minimum length of the primary transcript for the largest of these RNAs?
c. What is the minimum number of exons for each gene?
d. What is the minimum number of introns for each gene?

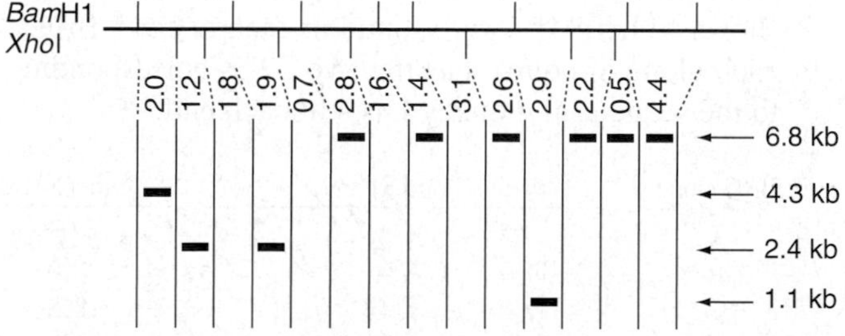

Answer

This problem requires an understanding of primary and processed transcripts and the analysis of RNAs using Northern hybridization.

a. The bands on the Northern blots that hybridize with the probes represent the mRNAs from this region. The three mRNAs corresponding to the genes in this region are *1.1, 2.4,* and *6.8 kb* in length. The 4.3 kb transcript comes from a gene that is only partly contained in the DNA used as a probe since the one DNA fragment that hybridizes is only 2.0 kb in length.

b. The bands on the Northern blot represent processed transcripts (introns have been removed). Primary mRNAs are made by copying from contiguous DNA sequences, including those regions that will be removed by splicing. The minimum length of a primary transcript is based on the sizes of the restriction fragments that hybridize with the RNAs and any intervening fragments. *For the 6.8 kb largest mRNA, the minimum size is 2.8 + 1.6 + 1.4 + 3.1 + 2.6 + 2.9 + 2.2 + 0.5 + 4.4 kb, or 21.5 kb.* We cannot say from these data whether transcription begins before the 2.8 kb fragment or extends beyond the 4.4 kb fragment.

c./d. The minimum number of exons is determined by counting the number of hybridizing fragments or groups of contiguous fragments. This is a minimum number because any of the fragments could contain more than one exon. The nonhybridizing fragments that separate these must contain introns.

Transcript	Minimum number of exons	Minimum number of introns
1.1 kb	one exon	none
2.4 kb	two exons	one
6.8 kb	four exons	three

II. Reverse transcriptase, the enzyme used to synthesize cDNA starting with mRNA as a template, often falls off the template before completely copying the mRNA. When screening for a cDNA clone of a gene, it is therefore not uncommon to isolate partial cDNA clones. What comparison could you make experimentally that would indicate if you had isolated a plasmid clone containing a partial cDNA?

Answer

For this problem, consider the two alternatives: you have the complete cDNA clone or you have a partial cDNA clone. What would be the differences that you could detect experimentally? The cDNA must be as long as its corresponding mRNA to be full length. You need to find out what the full length of the message is to do the comparison. The way to *measure the length of mRNA is to run a Northern gel and hybridize with the cDNA as a probe.* (Another alternative would be to hybridize a fragment from near the 5′ end of the gene to your clone. This would require that you have a fragment that you know is from the 5′ end of the gene.

Problems

Interactive Web Exercises

We highlight here two resources maintained by the National Center for Biotechnology Information (NCBI). The first is Map Viewer, which is a graphical representation of the entire human genome. By navigating through Map Viewer, you will acquire a feeling for the way in which genes are organized on human chromosomes. The second resource is BLAST, a program which enables investigators to find DNAs, RNAs, or proteins homologous to any nucleic acid or protein for which nucleotide or amino acid sequence information is available. To access this material, visit our website at www.mhhe.com/hartwell3, go to Chapter 10, and click on "Interactive Web Exercises".

Vocabulary

1. For each of the terms in the left column, choose the best matching phrase in the right column.

a. sequence tagged sites (STSs)	1. a map showing the order of cloned inserts of DNA
b. whole-genome shotgun sequencing	2. fragments of DNA on a chip.
c. DNA array	3. the presence in the germ line of one organism of a gene from a second organism
d. alternative RNA splicing	4. a genome sequencing strategy that avoids physical mapping
e. lateral gene transfer	5. a polymorphic single-base site
f. physical map	6. the joining together of the exons of a gene in different combinations
g. SNP	7. a map of genetic markers that are separated by less than 1 cm
h. high-density linkage map	8. unique DNA sequences that serve as molecular markers

Section 10.1

2. Would it be possible to construct a complete physical map of the human genome (assuming a 3 Gb size) with 15,000 BAC clones each with a 200 kb insert? Why or why not?

3. Using a simple sequence repeat (SSR) DNA probe, you have isolated several clones from a library. You would like to identify unique sequences adjacent to these repeat DNAs found throughout the genome to establish some SSR-based sequence-tagged sites (STSs).

a. How could you determine if there are unique sequences within the clones you have isolated?

b. The most useful STSs are those associated with SSRs that are polymorphic in the population (that is, have sequence variants). How could you determine if an STS is associated with a polymorphic SSR?

4. To make a set of clones more suitable for the analysis of DNA sequence, a series of cosmid clones was prepared by digesting a BAC clone and subcloning the resulting restriction fragments. The restriction patterns of the inserted fragments are shown next. Arrange these four cosmids into a physical map, showing the order of the clones and the overlap between them.

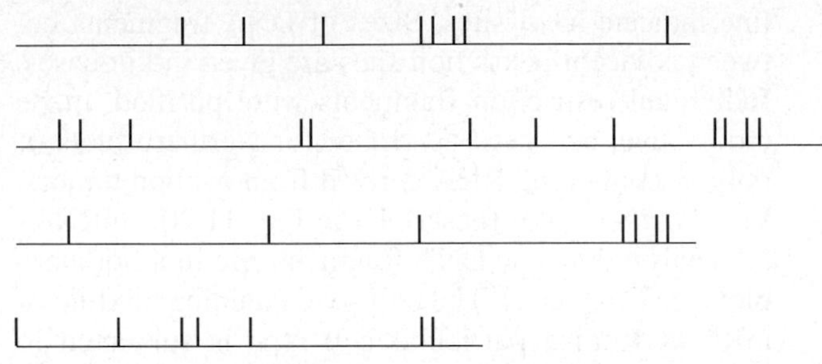

5. In an alternative approach to that described in Problem #4, a cosmid library was probed with an insert from a BAC, and 22 cosmids were obtained. All but two of the cosmids could be arranged as an overlapping group (a contig). This set of cosmids covered the length of the BAC insert. What explanation can you offer for the two cosmids that did not fit into this contig? Why did these hybridize with the BAC probe?

6. During the course of the genome project for the rhesus monkey *Macaca mulatta,* 5 BAC clones (A–E) forming a single contig were obtained. Researchers determined a short (~500 bp) sequence of monkey DNA from each of the two ends of the BAC clones (that is, from where the monkey genomic DNA was joined to the BAC vector). The scientists converted these sequences into sequence tagged sites (STSs) by making PCR primers that could amplify the 500 bp of monkey DNA if it were present in any DNA sample. The table below shows which STSs were found in each of the 5 BACs; each clone of course has the two STSs corresponding to the sequenced monkey DNA at each end.

BAC clone	End STSs	Other STSs
A	1 2	4 5 7 10
B	3 4	2 9
C	5 6	1 8
D	7 8	1 5
E	9 10	2 4

a. Why is it very efficient to determine the sequences of monkey DNA at locations where it is joined to the BAC vector?

b. Diagram a physical map of this region consistent with the data, indicating the relative order of the BAC clones and the location of the STSs.

c. If you wanted to determine the DNA sequence of the entire contig, it would be advantageous to work with the minimal tiling path of the BAC clones. Why? Diagram the minimal tiling path consistent with the data in the table.

d. Estimate the size of the contig in kb.

7. In the course of sequencing a genome, a computer is trying to assemble the following 6 DNA sequences into contigs:

```
5′  CAAATAGCAGCAAATTACAGCAATATGAAG  3′
5′  AAAATGCCCTAAAGGAAATGAGATTTTTAA  3′
5′  TGATCTCTTCATATTGCTGTAATTTGCTGC  3′
5′  GTAGTATCTCCTTTTAAAAATCTCATTTCC  3′
5′  CAATATGAAGAGATCATACAGTCCACTGAA  3′
5′  TCTCATTTCCTTTAGGGCATTTTCAAATTC  3′
```

How many contigs are represented by this set of DNA sequences, and what is the sequence of each contig?

8. Repetitive DNA sequences present a challenge to genome projects. Why is this so? What types of repetitive sequences are most problematic? How can hierarchical shotgun and whole-genome shotgun sequencing strategies deal with this problem?

9. It is often difficult to find genome-unique PCR primers in certain regions of the genome. Offer two explanations.

10. What are two potential difficulties with the whole-genome shotgun sequencing strategy?

11. Figure 10.8b on p. 362 shows a fluorescence *in situ* hybridization (FISH) analysis that reveals the position in the human genome of DNA sequences homologous to a particular probe.

a. Suppose you made two probes, one of which fluoresces in red and the other in green. Estimate how far apart the corresponding sequences on the chromosomes have to be so that you could accurately resolve the order of these sequences along the same chromosome. Assume that the size of the FISH signal is typical for this type of experiment, and that the chromosome containing the homologous sequences is of average size in the human genome.

b. Since the publication of the human genome sequence, human geneticists have done fewer FISH experiments than in the past. Explain why. For what purposes would FISH still be useful today?

12. What are the advantages of the FISH protocol over linkage mapping for the initial characterization of the chromosomal location of a gene?

13. A clear limitation to gene mapping in humans is that family sizes are small, so it is very difficult to collate enough data to get accurate recombination frequencies. A technique that circumvents this problem begins with the purification of DNA from single sperm cells. (Remember that recombination occurs during meiosis. Analysis of a population of sperm provides a large data set for linkage studies.) The DNA from single sperm cells can be used for polymerase chain reaction (PCR) studies. The PCR reactions can be run such that a PCR primer will not hybridize to DNA if there is a single-nucleotide difference between the sequence of the primer and the template DNA to be analyzed. A PCR product will be formed only if both primers hybridize to the template.

A series of three pairs of PCR primers were synthesized according to DNA sequences from human genomic clones. The alpha primers amplify a fragment 100 bp long; the beta primers produce a 200 bp product; the delta primers produce a 400 bp product. Twenty-five sperm obtained from one man were subjected to PCR analysis with these three pairs of primers simultaneously. Refer to the figure at the bottom of this page.

a. Is this man homozygous for any nucleotide polymorphism detected by this method? If so, which one?

b. Determine the genetic map information you can glean from this data.

14. As an example of the computational difficulties in making recombinational maps of DNA polymorphisms, consider three SSR loci that are so polymorphic that any male/female pair would have four recognizable alleles for each of these loci.

a. If the three loci were genetically unlinked, how many different kinds of progeny would you expect, and in what proportion?

b. If two of the three loci were very tightly linked so that it would be very difficult to find recombinants between them, but the third locus was unlinked to both of the other SSR loci, how many different kind of progeny would you expect, and in what proportion?

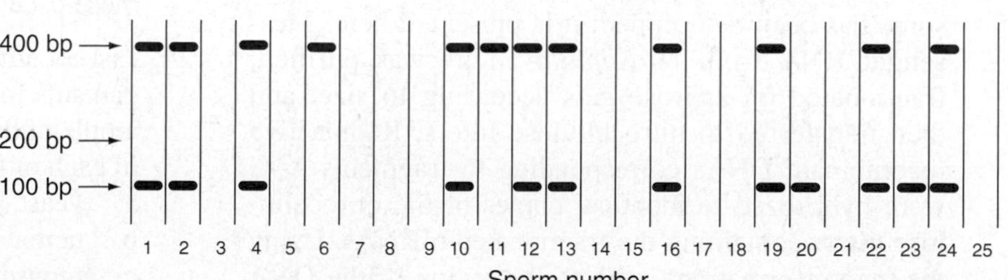

c. Why does the construction of large-scale recombinational maps require computer analysis?
d. Why would it be easier to do this kind of analysis in the *Drosophila* genome than in the human genome?
e. With the publication of the human genome, there is no longer any need to construct large-scale linkage maps to determine the order of DNA polymorphisms along the chromosomes. However, scientists are still searching for human DNA polymorphisms; for example, several pharmaceutical companies joined together to form The SNP Consortium, which reported the identity of more than 1 million SNPs in 2003. Why did these companies invest so much money to search for SNPs if the human genome had already been completed?
f. How many DNA markers evenly spaced along the human genome would you need to obtain a mapping resolution of 1 cM? What would be the average physical spacing of these markers along the genome measured in nucleotides?

15. Large-scale linkage maps require the analysis of many DNA polymorphisms scattered around the genome. Single nucleotide polymorphisms (SNPs), in which a base at a particular location is replaced by one of the three other bases, represent a class of DNA variants that is very useful for constructing such linkage maps. How would you actually find human SNPs that could be employed for this purpose?

Section 10.2

16. Give two different reasons for the much higher ratio of total DNA to DNA that encodes proteins in the human genome compared to bacterial genomes.

17. Using a cDNA library, you isolated two different cDNA clones that hybridized with your probe for a nerve growth factor gene. The beginning and ending sequences of the clones are the same but the middle sequence is different. How can you explain the different cDNAs?

18. What sequence information about a gene is lacking in a cDNA library?

19. A restriction map of part of the *Drosophila* X chromosome has been determined; it is presented here. Messenger RNA from *Drosophila* adults was purified, fractionated on agarose gels according to size, and then transferred to nitrocellulose filters. Radioactive recombinant DNAs corresponding to fragments A–G were hybridized to identical copies of the nitrocellulose filters containing the fractionated mRNAs. Using the map information, indicate the region of the DNA

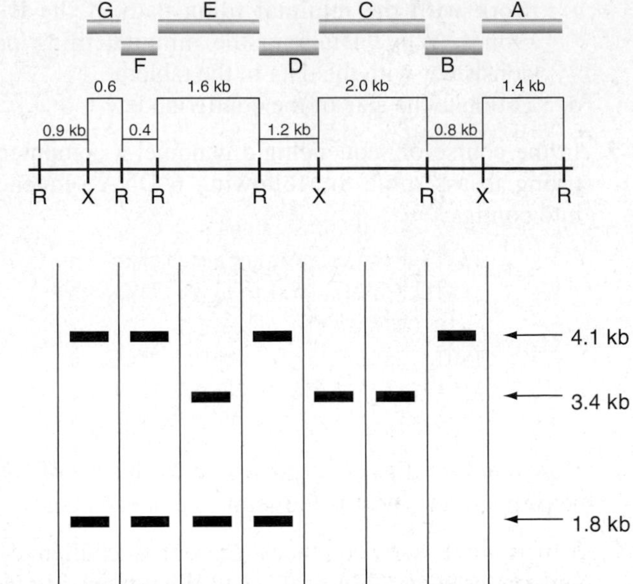

that must be transcribed to form each of three primary transcripts, which are then processed into the three mRNAs observed. Indicate where exons and introns occur in relation to the restriction fragments.

20. The maps of human chromosomes in males and females indicate that the centimorgan distances between genes is different in the two sexes.
a. What could account for this difference in distances?
b. Do you think the physical distances between genes in males and females would differ also? Why or why not?

21. List three independent techniques you could use to identify DNA sequences encoding human genes within a cloned genomic region.

22. Discuss three observations that suggest not all repeat sequences are junk.

23. An interesting phenomenon found in vertebrate DNA is the existence of pseudogenes, nonfunctional copies of a gene found elsewhere in the genome. Pseudogenes appear to be double-stranded DNA copies of mature mRNA inserted into the chromosome. What sequence information would provide a clue(s) that the source of these pseudogenes is cDNA?

24. List an advantage of each of the following model organisms for the analysis of gene function. Give an example of the type of gene you might choose to analyze in each of these model organisms.
a. yeast (*S. cerevisiae*)
b. nematodes (*C. elegans*)
c. mouse (*M. musculus*)

25. With new information from the Human Genome Project, many new genes will be identified for which the function is not known.
   a. What features of the DNA sequence might help you determine the function of a newly identified gene?
   b. What are two other types of analysis that would help you learn more about a new gene?

26. a. If you found a zinc-finger domain (which facilitates DNA binding) in a newly identified gene, what hypothesis would you make about the gene's function?
   b. In another gene, what would a high percentage of similarity throughout the gene with a previously identified gene in the same organism suggest about the origin of the gene?

27. You sequence the genomes of four different organisms and compare their sequences over a short region as shown below.

```
5'  AGGTATATAATTTGCG   3'
5'  CAATATAAAACCCTAC   3'
5'  GCGTATAAAAGAGCTA   3'
5'  TTATATATAAAGAAGT   3'
```

   a. Determine the consensus sequence common to the four regions above.
   b. Why would you want to define the consensus sequence? How would you decide whether the four sequences above were worth comparing to define a consensus?
   c. How could you use this general strategy for defining a consensus sequence to determine which amino acids of a protein are most critical for its function?

28. The human genome has been sequenced, but we still don't have an accurate count of the number of genes. Why not?

29. Chimpanzees have a set of hemoglobin genes very similar to the set in humans (shown in Fig. 9.20 on p. 336). For example, the genomes of both species have $\alpha 1$, $\alpha 2$, β, $G\gamma$, $A\gamma$, δ, ε, and ζ genes.
   a. Of the human and chimpanzee hemoglobin genes, which would be considered homologous? Which paralogous? Which orthologous?
   b. When comparing genomes, geneticists would usually like to know which genes are the most likely to perform similar if not identical functions in different species. This determination can be somewhat complicated in the case of gene families. Would paralogous genes or orthologous genes be more likely to be functionally equivalent? Explain.
   c. Which gene would have the greatest degree of nucleotide homology to the human β gene: the chimpanzee β gene, or the human γ gene? Explain.
   d. Rationalize the pattern of hemoglobin genes in the two species with the existence of duplication and divergence events among the hemoglobin genes depicted in Figure 9.20 on p. 336.

30. A *cladogram* is a branching diagram (a "Tree of Life") used to illustrate evolutionary relationships. The endpoints of the branches represent present-day species or genes or proteins; the branch points tell when these species or molecules diverged from common ancestral forms.
   a. How would a cladogram of gene sequences discriminate whether a particular gene had been laterally transferred (say from bacteria to humans) or vertically inherited over a long period of time from one generation to the next?
   b. If lateral transfer were involved, how might you be able to tell roughly when in the course of evolution the transfer had occurred?
   c. How could you recognize if a gene was somehow lost from the genome of a particular species?

31. From an expression library, you isolated a clone and made an antibody to the protein produced by the cloned gene. When the antibody was used to stain a preparation of a mouse embryo, the protein localized to the brain. Northern analysis showed that there were three mRNA transcripts from the gene present in the brain. Two of these mRNAs were also present in tissues other than the brain. Propose a hypothesis to explain the presence of RNAs in many tissues but protein only in the brain.

32. Complete genome sequences indicate that the human genome has roughly 25,000–30,000 genes, and the worm (nematode) genome has 19,000 genes. Explain how the human genome can encode a creature enormously more complex than the worm with at most only one-third more genes in its genome.

Section 10.3

33. The following factors pose technical challenges for the design of gene therapy experiments. Suggest ways to get around these problems.
   a. large gene size in eukaryotes (up to 100 kb of DNA in the genome)
   b. the need to get genes expressed in specific tissues

34. Consider a microarray experiment similar to the one shown in Fig. 10.22 on p. 377, in which each square represents a PCR-amplified fragment of a different human gene, and the red-labeled probe is cDNA from a human lung tumor while the green-labeled probe is cDNA from normal lung tissue.

a. How would you interpret results in which the fluorescence signal was black? Green? Red? Yellow?

b. If you were searching for an anticancer drug that would inactivate a protein whose activity contributes to cancer, which of the genes represented on the microarray encode proteins you would most likely chose as a potential target for such a drug?

35. A region of the genome from two individuals is amplified by PCR so that the PCR products from one individual are labeled with rhodamine (which fluoresces in red), while the PCR products from the other person are labeled with fluorescein (which fluoresces in green). These PCR products are mixed and hybridized to an oligonucleotide microarray with the following results.

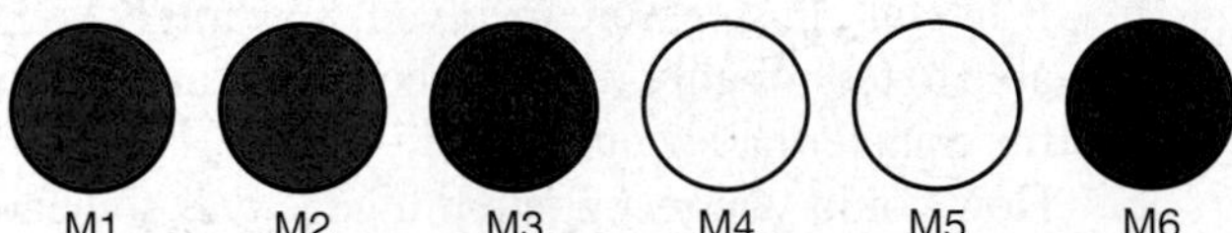

The oligonucleotides on the array are:

```
M1: 5'  ACTTACCGAGAGAACCTGCG  3'
M2: 5'  ACTGACCGAGAGAGCCTGCG  3'
M3: 5'  ACTTACCGAGAGAGCCTGCG  3'
M4: 5'  ACTCACCGAGAGACCCTGCG  3'
M5: 5'  ACTCACCGAGAGATCCTGCG  3'
M6: 5'  ACTGACCGAGAGAACCTGCG  3'
```

a. As accurately as possible, describe the genotypes of the two individuals.

b. Why would you encounter ambiguity in assigning genotypes to these two particular individuals if you sequenced the PCR products directly, rather than by hybridizing them to an oligonucleotide microarray as above?

c. In what way would the oligonucleotide microarray approach be valuable as a diagnostic tool for human genetic diseases?

The Direct Detection of Genotype Distinguishes Individual Genomes

Chapter 11

An eighteenth-century painting depicting the 64 offspring of the ninth Sultan of Yogyakarta and his 13 wives. In this family tree, boys are depicted as fruit, girls as leaves. The two fruits emerging from a single stem on the right *indicate twins. The painting is currently located in the Sultan's Palace inside the central walled-in city (Kraton) of Yogyakarta (on the island of Java in present-day Indonesia).*

A couple whose firstborn suffers from cystic fibrosis learns from the medical diagnosis of that child's symptoms that both parents are carriers of a recessive disease-producing mutation. Together they run a 25% risk of transmitting the life-threatening hereditary condition to a second child. In the early 1990s, one such couple did not want to take that chance and, after genetic counseling, decided to try an experimental protocol: *in vitro* fertilization combined with the direct detection *in vitro* of the embryo's genotype, before its placement in the mother's womb.

The procedure took less than a week. At the start, a team of medical workers, including an obstetrician and a reproductive biologist, obtained 10 eggs from the woman and fertilized them with sperm from the man. Three days later, after the fertilized eggs had undergone several mitotic divisions to generate embryos with 6–10 cells, a research assistant used micropipettes to remove one cell from each embryo (**Fig. 11.1**). Since the natural formation of healthy identical twins results from embryos being split into two groups of cells at this stage, the removal of a single cell would not prevent normal development.

Next, the research assistant used PCR (see fig. 9.16 on pp. 328–329) to amplify from each isolated cell a specific DNA segment of chromosome 7 containing the site of the most common mutation within the gene responsible for cystic fibrosis (the *CFTR* gene). Previous cloning and sequencing of the gene made it possible to obtain primers bracketing this segment. Prior PCR amplification of the parents' DNA had verified that both were carriers of this same mutation.

To *genotype,* that is, detect which alleles were present in the material amplified from each cell, clinicians divided each cell's PCR product into two equal portions, or aliquots. They exposed one aliquot to an allele-specific hybridization probe for the normal *CFTR* allele, the other to an allele-specific probe for the mutant allele. If only the first aliquot from a sample produced a signal, the embryo represented by the sample was a homozygote for the normal allele. If both aliquots were positive for the probe, the embryo was a heterozygote. And if only the second aliquot glowed in response to the probe, the embryo carried two copies of the disease allele.

It took about 8 hours to determine the genotype of each isolated cell and, by inference, of each embryo. On the same day that the embryos were biopsied and genotyped, the doctor, in consultation with the parents, selected for placement in the mother's womb two embryos of known genotype. Of the two embryos they chose, one was a heterozygous carrier, the other a homozygote for the normal allele of the *CFTR* gene. The use of two to three embryos improves the chances of at least one implantation and is a part of most *in vitro* fertilization procedures. The selection of a carrier means that other embryos homozygous for the normal allele were most likely defective in some other way. As it happened, the embryo carrying two normal alleles of the *CFTR* gene implanted into the woman's uterus; the carrier embryo did not.

With this approach, the woman knew that the embryo developing within her womb carried two copies of the normal allele for the *CFTR* gene; as a result, this child, as well as its children, would be free of cystic fibrosis. Nine months later, the mother gave birth to a healthy 7 lb 3 oz baby girl. When evaluated by a pediatrician at her four-week checkup, the infant daughter was found to be completely normal in both physical

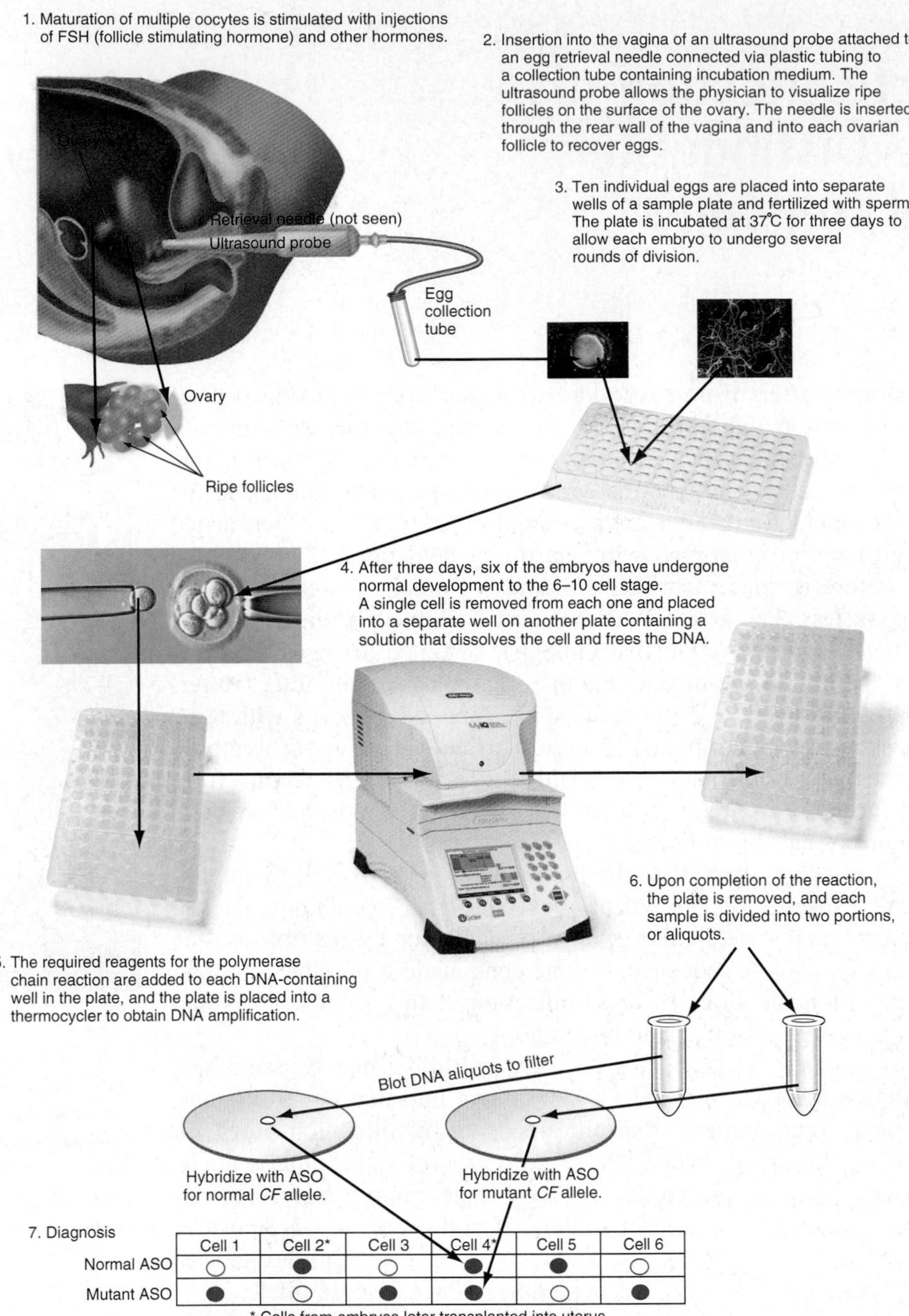

1. Maturation of multiple oocytes is stimulated with injections of FSH (follicle stimulating hormone) and other hormones.
2. Insertion into the vagina of an ultrasound probe attached to an egg retrieval needle connected via plastic tubing to a collection tube containing incubation medium. The ultrasound probe allows the physician to visualize ripe follicles on the surface of the ovary. The needle is inserted through the rear wall of the vagina and into each ovarian follicle to recover eggs.
Ovary
Retrieval needle (not seen)
Ultrasound probe
Egg collection tube
3. Ten individual eggs are placed into separate wells of a sample plate and fertilized with sperm. The plate is incubated at 37°C for three days to allow each embryo to undergo several rounds of division.
Ovary
Ripe follicles
4. After three days, six of the embryos have undergone normal development to the 6–10 cell stage. A single cell is removed from each one and placed into a separate well on another plate containing a solution that dissolves the cell and frees the DNA.
5. The required reagents for the polymerase chain reaction are added to each DNA-containing well in the plate, and the plate is placed into a thermocycler to obtain DNA amplification.
6. Upon completion of the reaction, the plate is removed, and each sample is divided into two portions, or aliquots.
Blot DNA aliquots to filter
Hybridize with ASO for normal *CF* allele.
Hybridize with ASO for mutant *CF* allele.
7. Diagnosis
Cell 1 Cell 2* Cell 3 Cell 4* Cell 5 Cell 6
Normal ASO
Mutant ASO
* Cells from embryos later transplanted into uterus

Figure 11.1 Preimplantation embryo diagnosis. Plucking one cell from an eight-cell embryo for the direct detection of genotype.

and mental development. At the same time, testing at an independent laboratory confirmed that neither of her homologous chromosome 7s carried the *CFTR* mutation.

The couple in this report worked with university doctors and laboratories whose funding and operations are approved by review boards that monitor social and ethical as well as scientific concerns. The Genetics and Society box "Social and Ethical Issues Surrounding Preimplantation Genetic Diagnosis" on pp. 394–395 discusses several of these issues.

In this chapter, we examine how geneticists use an array of molecular tools to detect DNA differences among individuals. Once cloning and sequencing have identified a DNA difference, less cumbersome protocols can use this information for the rapid detection of genotype in many individuals—human, animal, or plant. The techniques are sensitive enough to operate on single hair follicles and even single cells from human embryos.

The ability to detect genotype directly at the DNA level has far-reaching consequences. It provides geneticists with tools for mapping and cloning the genes responsible for human diseases that, like the cystic fibrosis gene, were previously defined only by their effect on phenotype. Once the genes are cloned, DNA genotyping can help predict the probability of future disease—*in vitro, in utero,* or after birth—or reveal the presence of a silent recessive disease allele in a carrier who shows no evidence of disease. The DNA-based technology is also the foundation of DNA fingerprinting, a method for evaluating the most prevalent DNA differences among individuals. DNA fingerprinting can provide legally acceptable evidence for or against a match between a suspect's genome and that of biological material (such as blood, semen, or even a few hairs) left behind at the scene of a crime.

Two general themes emerge from our discussion. First, the ability to distinguish genotypic differences of all kinds extends our concept of a locus and the alleles that define it. Mendel, Morgan, and other geneticists—working before the advent of molecular technology—identified a locus indirectly, inferring its existence from the observation of alternative phenotypes generated by alternative alleles of a gene; in this early era of genetic analysis, loci and genes were considered equivalents. By contrast, modern geneticists, able to look at genotype directly, can pick out genotypic differences in both genes and noncoding DNA regions on the basis of changes in DNA sequence alone. As a result, a **locus** is now a designated location anywhere on a chromosome; it can range in size from a single base pair to more than a megabase pair. A **gene** is simply a coding locus. An **allele** is one of two or more alternative DNA sequences found at a locus—either coding or noncoding—in different homologs of the same chromosome, in a single heterozygous individual, or in different individuals. While different alleles of a gene and its associated regulatory sequences may affect the function of the gene's RNA or polypeptide product with consequent effects on overall phenotype, alleles at noncoding, nonregulatory loci have no effect on phenotype. We will see that researchers can still use DNA variations in noncoding regions as allelic markers to identify, locate, isolate, and follow the transmission of nearby genes.

The second general theme is that techniques for the direct detection of genotype provide access to genetic details about human individuals never before available. This potential for gathering and storing genetic data raises questions about the information so obtained: Who should have access to it? And how should it be used?

Our presentation of strategies for distinguishing genomes describes:

- Four classes of DNA variation that serve as the basis for the direct detection of genotype.
- Genotyping protocols that use hybridization, gel electrophoresis, PCR, microarrays, and sequencing (described in previous chapters) to distinguish alleles of the four types of polymorphic loci. These protocols include DNA fingerprinting, which produces a quick snapshot of a combination of genomic features that can identify an individual.
- Positional cloning: the application of direct genotype detection to linkage analysis, with the ultimate goal of cloning the genes behind any phenotypic differences among people, animals, or plants.
- Ways to look at the genes behind the complex inheritance patterns arising from incomplete penetrance, variable expressivity, and other phenomena that confound straightforward linkage analyses.
- The use of haplotype association studies to map disease loci in humans and other organisms. This novel strategy does not depend on the genetic pattern of allele transmission from parent to child; instead, it relies on chromosomal evolutionary history.

Social and Ethical Issues Surrounding Preimplantation Genetic Diagnosis

- In 1990, preimplantation genetic diagnosis (PGD) was successfully used to screen embryos for regions of the Y chromosome. The goal was to select for female embryos in order to avoid X-linked genetic disorders (including X-linked mental retardation).
- In 1992, clinicians used PGD to select for an embryo that did not carry two copies of the mutant cystic fibrosis allele.
- In 1998, physicians at the Genetics and *In Vitro* Fertilization (IVF) Institute in Fairfax, Virgina, reported the use of Microsort™, a technology to separate sperm cells containing X chromosomes from sperm cells containing Y chromosomes for the purpose of sex selection.
- In 1999, a British medical ethics board approved use of double PGD screen for selection of an embryo that would develop into a child who would not have Fanconi anemia, but would have an HLA genotype match with a diseased sibling. Born in August 2000 and named Adam Nash, this child would become a transplantation donor for the sibling who had Fanconi anemia. With this procedure, the term "savior siblings" was coined.
- In 2004, two London fertility clinics applied for licenses to use PGD to screen embryos for mutations at the *BRCA1* gene that increase the risk of developing breast cancer.

This brief time line highlights major developments in the use of PGD and shows that the uses of this technology have progressed from screening for genetic diseases, to family gender balancing (sex selection), to selecting for individuals who can assist relatives living with a disease, to selecting against alleles that increase the risk of getting a disease.

Nations and groups around the world have responded in different ways to the issues generated by the new reproductive technologies. For example, in 1994, France and Norway passed legislation that limits genetic testing to situations in which the results are medically therapeutic; these laws prohibit the use of genetic testing for sex selection and normal trait enhancement. In contrast, in 2001, the governing board of the American Society for Reproductive Medicine—the major organization representing fertility doctors in the United States—decided that preimplantation sex selection is acceptable when parents already have a child of one sex and want to have another of the opposite sex. And as of 2005, U.S. women could still sell their eggs to the highest bidder in 48 states. Advertisements in some college newspapers offered up to $50,000 for eggs from women who met highly selective criteria of height, appearance, SAT scores, and grades. In all countries of Western Europe, as well as Australia and Canada, egg selling is forbidden.

The range of responses to the issues generated by the new reproductive technologies shows a diversity of approach based in part on national culture and history. It also reflects international apprehension about the potential for misuse and abuse of the new technologies. Here are some of the main concerns.

When Should the Tests Be Used?

The couple in our opening story whose firstborn suffered from cystic fibrosis faced a medical problem. PGD could help them have a second child unaffected by the disease. With no cure at present for CF and no therapy that allows for CF-affected people to look forward to a life of normal health or length, this is an example of medically therapeutic testing. Governmental

11.1 DNA Variation Is Multifaceted and Widespread

Members of the Same Species Show Enormous Sequence Variation in Their Genomes

If molecular geneticists were to clone and sequence the same 250 kb region of genomic DNA encompassing the cystic fibrosis (*CFTR*) gene on chromosome 7 from the two homologs of a healthy person, they would find about one difference in every 1000 bp, or 250 DNA differences in all (**Fig. 11.2**). With so many differences between these allelic sequences, you might suspect that one allele is wild type and the other mutant.

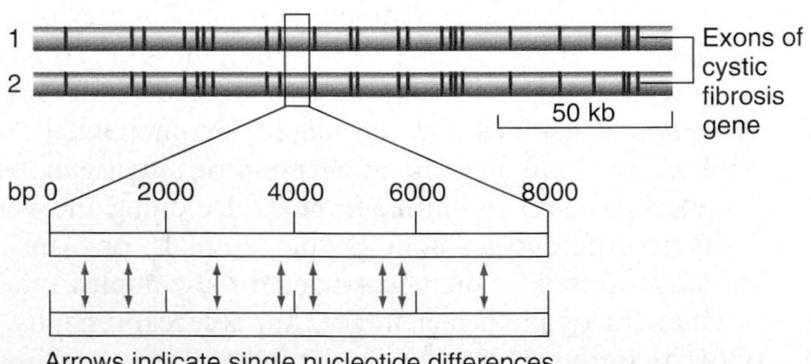

Figure 11.2 Some base-pair differences between DNA cloned from the cystic fibrosis locus of two healthy individuals. The differences highlighted between the two *CFTR* alleles map to a single intron and have no effect on the function of the gene.

and professional committees in most industrialized countries permit PGD for this purpose, although Germany and Japan ban any use of preimplantation genetic testing for any purpose. And even in countries where PGD is permitted to screen for disease, there is opposition from people who hold a religious belief that all human embryos—even those at the earliest stages of development—have a right to life. Others object to PGD because they think people shouldn't have a right to interfere with the natural process of allele segregation.

How Should the Tests Be Carried Out?

The couple screening for CF began by consulting a genetic counselor and then worked with medical practitioners associated with a university laboratory. Most geneticists agree that counseling before a procedure should foster an open discussion of all the issues (including the possibility that, in the case of CF, the tests might give false negatives) and that long-term follow-up should be part of the process. PGD itself, like other forms of genetic testing, should be carried out by highly trained personnel in licensed laboratories. These accredited laboratories operate according to professional standards and have scientific and ethical review boards that monitor all work.

Who Should Have Access to the Technology?

The cost of *in vitro* fertilization and PGD testing averaged $12,000–$15,000 in 2005. Should the government provide tests for people who cannot afford them? How should society decide this issue? (A related discussion of access to medical technology appears in the Genetics and Society box on pp. 34–35 of Chapter 2.)

Should Parents Have the Right to Make Any Genetic Decision?

If, for instance, parents decide to have a child affected by a genetic disease, should they bear all financial responsibility for the child's care, or does society have an obligation to assist with medical treatment? On the other hand, how should physicians handle a request from prospective parents who wish to select against alleles responsible for minor diseases like myopia (nearsightedness) or late-onset diseases like Huntington disease or Alzheimer's disease? What about selection for alleles that provide a child with a relative advantage such as complete resistance to infection by HIV? (Such resistance is inherited naturally by about 10% of individuals in some populations).

Who Should Have Access to Test Results?

Just the parents? The parents and eventually the child? The parents, the child, and certain community institutions, such as schools? Some combination of these plus commercial enterprises, such as insurance companies and places of employment? (We discuss these same questions of privacy in relation to other types of genetic testing in the Genetics and Society boxes in Chapters 2 and 3.)

What Constitutes a Person?

Cultural and religious beliefs, rather than scientific knowledge, are the basis for answers to this question. Some people see PGD as an alternative to abortion that allows a couple to make a decision before pregnancy begins. Others argue that even at the eight-cell stage, a microscopic preimplantation embryo is the equivalent of a human being, and rejection of an embryo is equivalent to killing a human being. The difference between these two positions is in part the result of different religious beliefs about the moral significance of embryonic cells.

Although there are no simple solutions to these complex issues, geneticists around the globe agree on the need for continuous open discussion and tight oversight of the development of new reproductive technologies.

But a geneticist would tell you that none of the observed differences is likely to have any effect on the actual function of the *CFTR* gene; both homologs have wild-type activity. This is because only a small fraction of the total DNA stored in our chromosomes actually encodes or regulates the expression of proteins. Within the 250 kb of the *CFTR* gene, only about 4.5 kb (slightly less than 2%) consists of protein-coding exons or regions that play a role in gene regulation. When mutations occur in coding or regulatory sequences, they are not passed on to future generations, because their effect on gene function is almost always negative. As a consequence, the vast majority of the gene's DNA variation that remains in a population is in noncoding, nonregulatory regions.

Geneticists consider variations at any position in the genome as alternative alleles of a particular chromosomal location, or *locus*. Before the development of recombinant DNA technology, the identification of a locus depended on mutations that gave rise to alternative phenotypes; investigators would conduct breeding experiments and follow these alternative phenotypes by observation. Today, any variation in DNA sequence, once discovered, defines a locus that can be looked at directly, even if it does not affect phenotype.

When two or more alleles exist at a DNA locus, the locus is considered **polymorphic,** and the variations themselves are called **DNA polymorphisms.** When a particular polymorphic DNA locus is useful for mapping studies, disease diagnosis, or any other analytical task, the locus is called a **DNA marker;** like a mile marker on a highway, it marks a particular point in the genome. If that position has no known function, the site is an **anonymous locus.**

Studies of other regions of the human genome indicate that the average frequency of polymorphism is just as high as that found within the cystic fibrosis locus: A sequence difference between homologous chromosomes from the same individual or from two different individuals appears on average once in every 1000 bp. Thus, a comparison of the 3×10^9 bp of DNA in two haploid human genomes (for instance, those in any two sperm or egg cells from different individuals) would exhibit 3 million allelic differences.

In itself, this is an astonishing level of variation. But comparison of a second pair of haploid human genomes would also reveal differences at 3 million loci, and only some of these differences would be the same as those found in the first comparison; others would be at new loci. Indeed, the more human genomes you look at, the more polymorphic loci you will find. The total number of polymorphic human loci may be 100 million or more. This is an enormous reservoir of potential DNA markers.

Geneticists Categorize DNA Polymorphisms in Four Different Classes

For the purposes of genotyping, DNA polymorphisms can be placed into one of four classes based on their impact on DNA sequence, their frequency in the genome, their stability, and their role in genetic analysis (**Table 11.1**). All four types of variation are found throughout the genomes of all multicellular eukaryotic organisms.

Single Nucleotide Polymorphisms (SNPs)

The simplest, most prevalent, and most generally useful class of DNA polymorphisms arises from single base-pair substitutions. Such base-pair changes, which can be triggered by mutagenic chemicals or mistakes in DNA replication, are referred to as **single nucleotide polymorphisms,** or **SNPs.** SNPs account for the vast majority of the total variation that we just discussed. Nearly all SNPs are biallelic (that is, they have only two alleles). The ratio of the two alleles among the chromosomes in a population can range from less than 1:100 to 50:50.

Sequencing the same region of the genome from several individuals can identify large numbers of SNPs. By December 2001, the international SNP Consortium—a nonprofit organization coordinating the efforts of both public and commercial research centers—and private biotechnology companies had identified and mapped over 5 million human SNPs.

Although SNPs in coding sequences can alter the amino-acid sequence of a gene product and have a direct impact on phenotype, the vast majority of SNPs occur at anonymous loci. There is no evolutionary advantage or disadvantage to mutations at these noncoding, nonregulatory loci. For this reason, and because the probability that any particular base in the human genome will mutate is so low (1×10^{-9} per generation), scientists believe that each anonymous SNP represents a single mutational event. This means that if you and a friend share an allele at an anonymous SNP locus, you both got that allele from the same ancestor (who may have lived thousands or even hundreds of thousands of years ago). The fact that every random pair of human beings on the planet shares many unlinked SNP alleles indicates recent common ancestry for all people.

We will see that even though anonymous locus SNPs do not have a direct effect on phenotype, some lie so close to a disease gene or other genes influencing significant phenotypic differences (such as positive or negative responses to a particular medication) that they can serve as *DNA markers:* specific DNA loci with identifiable variations. Medical researchers can use such markers to identify and follow phenotypic differences in groups of people.

Microsatellites

The genomes of humans and other complex organisms are loaded with loci defined by repeated sequences. The repeating unit can be as short as a single base pair or as long as tens of kilobases. The number of repeats can vary

TABLE 11.1 Classes of DNA Polymorphisms

Class	Size of Locus	Number of Alleles	Number of Loci in Population	Rate of Mutation	Use	Method of Detection
SNP	Single base pair	2	100 million	10^{-9}	Linkage and association mapping	PCR followed by ASO hybridization or primer extension
Microsatellite	30–300 bp	2–10	200,000	10^{-3}	Linkage and association mapping	PCR and gel electrophoresis
Multilocus minisatellite	1–20 kb	2–10	30,000	10^{-3}	DNA fingerprinting	Southern blot and hybridization
Indels (deletions and duplications)	1–100 bp	2	N/A	$<10^{-9}$	Linkage and association mapping	PCR and gel electrophoresis

between two and thousands. Here, we discuss the category of repeat sequences known as **microsatellites,** which are DNA elements composed of one-, two-, or three-base sequences repeated in tandem 15–100 times. Microsatellites are also called SSRs (simple sequence repeats; see pp. 357–358 of Chapter 10). Examples are AAAAAAAAAAAAAAA or CACACACACACACACACACACA. In the mammalian genome, the CA-repeat microsatellite occurs on average once in every 30,000 bp. Microsatellites arise spontaneously from random events that initially produce a short repeated sequence with four to five repeat units. Once a short microsatellite mutates into existence, however, it can expand into a longer sequence by the process shown in **Fig. 11.3.**

Unlike SNPs—which are biallelic and do not change after the mutational event that gave rise to them—individual microsatellite loci often mutate into multiple alleles. Research shows that faulty DNA replication is the main mutational mechanism (Fig. 11.3). Because the same short homologous unit (CA, for example) is repeated over and over again, DNA polymerase may develop a stutter during replication; that is, it may slip and make a second copy of the same dinucleotide, or skip over a dinucleotide. Microsatellites are thus highly polymorphic in the number of repeats they carry, with many alleles distinguishable at each microsatellite locus.

New alleles arise at microsatellite loci at an average rate of 10^{-3} per locus per gamete (that is, one in every thousand gametes). This frequency is much greater than the single nucleotide mutation rate of 10^{-9} and results in a large amount of microsatellite variation among unrelated individuals within a population. At the same time, the rate of microsatellite mutation is low enough that changes usually do not occur within a few generations of even a large family; because of this, microsatellites can serve as relatively stable, highly polymorphic DNA markers in linkage studies of human families, other animals, and plants.

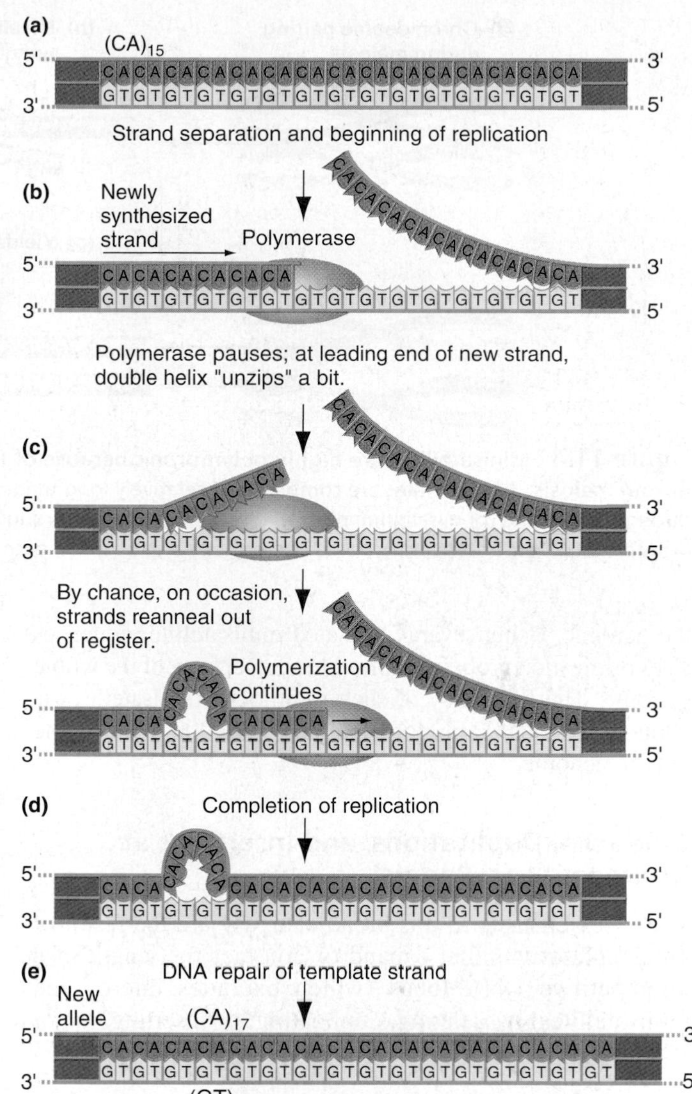

Figure 11.3 Microsatellites are highly polymorphic because of their potential for faulty replication. (a) A microsatellite consisting of 15 tandem repeats of the CA dinucleotide sequence. **(b)** During replication, the strands of the DNA molecule separate and DNA polymerase moves in the 5′-to-3′ direction along each strand. **(c)** Throughout the replication process, pauses can occur if the required nucleotides are, by chance, not in the vicinity of the polymerase. During a pause, the newly synthesized strand may "unzip" or "breathe" apart from the template strand over a length of 10–20 bases. **(d)** When the required nucleotides become available, the newly synthesized strand reanneals to the template and acts as a primer for further replication. After the reannealing in a microsatellite, the new strand may be out of register such that the polymerase begins by adding one or more nucleotides across from a part of the template strand that has already been replicated. **(e)** The resulting DNA molecule will have one or more identical repeats in the newly synthesized strand that were not present in the template strand. DNA repair processes then adjust the template strand to make it the same length as the newly synthesized strand.

Minisatellites

Minisatellites, a second important category of genomic DNA repeats, are larger than microsatellites. The repeating units that compose them are 20–100 bp long, and each unit is repeated up to thousands of times per locus. This gives each minisatellite locus a total length of 0.5–20 kb (**Fig. 11.4**). Like microsatellites, minisatellites arise from random events and are dispersed throughout the genomes of all vertebrates.

Minisatellite loci tend to be highly polymorphic because it is easy for homologous chromosomes with sections of many short repeats to line up out of register along the repeated units during synapsis (Fig. 11.4). Any crossing-over within the misaligned region will be unequal and produce reciprocal homologs that contain either duplications or deletions of one or more of the repeat units. The result is two new alleles—one with fewer, the other with a greater number of units than those on the parental chromosomes.

For geneticists, the most useful minisatellites loci can cross-hybridize to between 5–10 loci distributed throughout

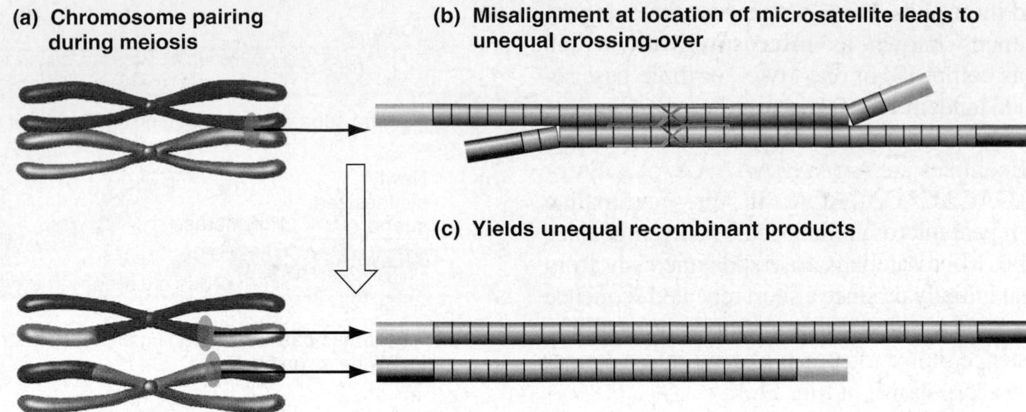

Figure 11.4 Minisatellites are highly polymorphic because of their potential for misalignment and unequal crossing-over during meiosis. Minisatellites are composed of relatively long tandem repeating units of identical sequence. **(a)** Chromosome pairing during meiosis and **(b)** misalignment of microsatellite location produce **(c)** unequal crossing-over and recombinant products that contain different numbers of repeating units than either parental locus; each new recombinant product is a new allele.

the genome. Using several unrelated minisatellite sequences as hybridization probes can provide an overview of the whole genome. The frequency of such multilocus minisatellites is about 1 per 100,000 bp, for a total of ~30,000 in the whole human genome.

Deletions, Duplications, and Insertions at Nonrepeat Loci (Indels)

The DNA changes in this broad category are the result of mutagenic events that expand or contract the length of a nonrepetitive DNA locus (which excludes micro- and minisatellites) by deleting, duplicating, or inserting one or more base pairs. These mutations, commonly called **indels**, can range in size from a one or few base pairs to multiple megabases. Chapter 14 examines the larger scale chromosomal alterations in genomic content.

Small deletions and duplications—from a few base pairs to a kilobase in length—most often arise from unequal crossing-over between nonhomologous sites during meiotic recombination. The material lost as a deletion from one homolog will be gained as a tandem duplication in the other (**Fig. 11.5a**). Small insertions ranging in size from hundreds to thousands of bases can also be caused by transposable elements that integrate at random into the genome (Fig. 11.5b), as described in detail in Chapter 14.

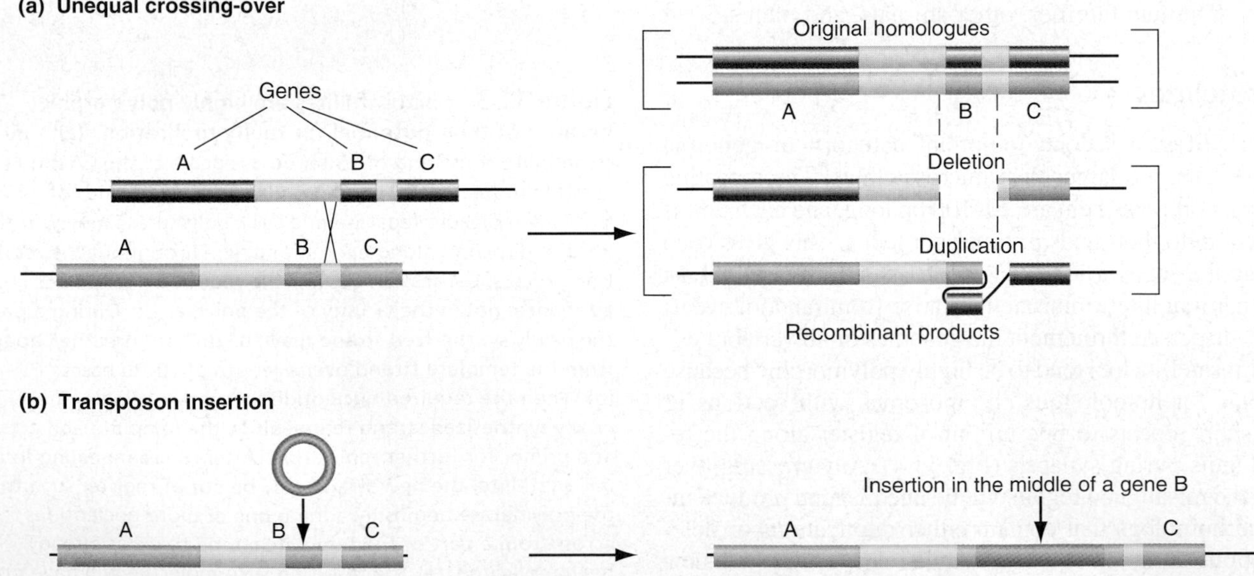

Figure 11.5 Deletion, duplications, and insertions cause variation in genomic content. **(a)** Unequal crossing-over occurs between different loci on homologous chromosomes that have some DNA sequence similarity. One product has a deletion over the region between the crossover sites; the second product has a duplication over the same region. **(b)** A transposon can insert into a chromosome and cause an insertion. If the insertion occurs within a gene's coding region, the gene may no longer function.

Indels of nonrepeated genomic sequences are much less common than SNPs, microsatellites, and minisatellites, but as we see in the next section, they are easy to analyze, and some are associated with particular types of disease mutations.

SNPs, microsatellites, minisatellites, and indels at nonrepeat loci provide the basis for detecting genotypic differences among individuals. Once comparative DNA sequencing or other analytical studies have identified a DNA variation or set of variations, researchers can use faster, less cumbersome tools to follow the transmission of specific variants in families and populations. We now describe the primary techniques for the direct detection of DNA polymorphisms.

11.2 Detecting DNA Genotypes of Different Types of Polymorphisms

One general approach to determining an individual's genotype at a particular polymorphic locus is to extract genomic DNA from the individual, use PCR to amplify the locus, and then sequence the PCR product. This strategy is usually effective, but it does have some drawbacks: It can be difficult to identify heterozygotes, and it is still fairly expensive and time-consuming when done on a large scale. PCR followed by sequencing is thus impractical either in a clinical setting or for genotyping large numbers of loci in large numbers of individuals.

Practical protocols for large-scale DNA genotyping must be inexpensive, rapid, accurate, and easy to perform on hundreds or thousands of samples. Automated DNA genotyping methods that can analyze hundreds of thousands of samples are needed for both large-scale clinical diagnoses and the mapping of complex human traits.

We now describe a variety of genotyping methods suitable for different types of polymorphisms.

There Are Three Approaches to the Direct Detection of SNPs

Southern Blots or PCR Can Detect SNPs That Alter Restriction Sites

A small proportion of SNPs by chance eliminate or create a restriction site recognized by a restriction enzyme. When this happens, researchers can use the restriction enzyme to distinguish between the two alleles. Consider, for example, the SNP shown in **Fig. 11.6.** A single base pair change from A:T in allele 1 to G:C in allele 2 determines the presence or absence of an *Eco*RI restriction site.

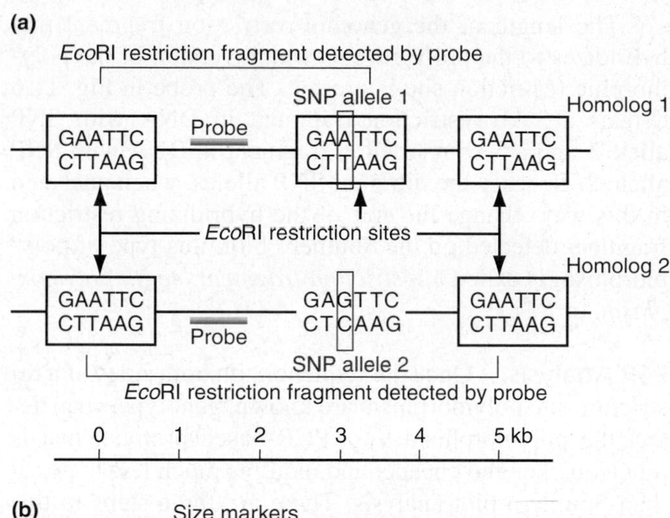

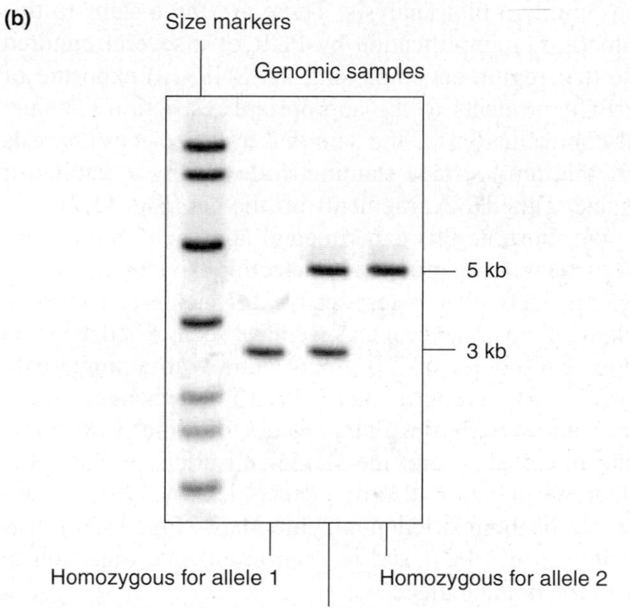

Figure 11.6 SNP-caused restriction fragment length polymorphisms can be detected by Southern blot analysis. (a) A SNP can create a restriction site polymorphism at an *Eco*RI site. The two SNP alleles will produce different-sized restriction fragments that can be detected with a DNA probe from a region between the polymorphic site and an adjacent site. **(b)** Southern blot analysis and hybridization with the probe can distinguish among the three possible genotypes at this SNP locus.

Southern Blot Analysis. Until recently, a Southern blot (see Fig. 9.15 on pp. 325–326) was the tool of choice for detecting a SNP-caused restriction site polymorphism. In the protocol, genomic DNA from the test sample is treated with *Eco*RI, and the resulting restriction fragments are separated by gel electrophoresis and then transferred to a filter paper. The resulting Southern blot is hybridized with a DNA probe obtained from the region between the polymorphic restriction site and an adjacent nonpolymorphic restriction site.

The length of the genomic restriction fragment that hybridizes to the probe reveals which version of the polymorphic restriction site is present. The probe in Fig. 11.6 detects a 3 kb restriction fragment in DNA with SNP allele 1 and a 5 kb restriction fragment in DNA with SNP allele 2. Because the different SNP alleles, when analyzed in this way, change the size of the hybridizing restriction fragment detected on the Southern blot, this type of polymorphism is called a *restriction fragment length polymorphism*, or *RFLP.*

PCR Analysis. Once the sequences on both sides of a restriction site polymorphism are known, genotypers can detect the polymorphism by a PCR-based protocol that is much quicker and cheaper and requires much less material than Southern blot analysis. There are three steps to this protocol: (1) amplification by PCR of a several hundred base-pair region encompassing the SNP; (2) exposure of the PCR products to the appropriate restriction enzyme; and (3) evaluation of the samples by gel electrophoresis and ethidium bromide staining, followed by a reading of the size of the DNA fragments off the gel (**Fig. 11.7**).

We illustrate this experimental approach with a solution to the real-life problem of detecting the mutation at the β-globin locus that is responsible for sickle-cell anemia. Sickle-cell anemia occurs, as we have seen, when a person carries two copies of a β-globin gene with a single-base substitution that replaces an A with a T and changes the encoded amino acid from glutamic acid to valine. The normal allele is called *A*, and the sickle-cell allele *S*. Since the sickle-cell mutation also by chance destroys the recognition site of the restriction enzyme *Mst*II (Fig. 11.7a), it is possible to use PCR and restriction enzyme digestion to detect the mutant allele.

Suppose a carrier couple (both of genotype *AS*) have a child with sickle-cell anemia (genotype *SS*) and want to know the genotype of the fetus they have recently conceived. Through amniocentesis, their doctor recovers fetal cells from the pregnant woman's womb. He or she next subjects the genomic DNA in this sample, as well as in samples from both parents and the first child, to PCR amplification with primers complementary to sequences on either side of the sickle-cell mutation (Fig. 11.7a). The doctor then mixes the restriction enzyme *Mst*II with the PCR products and separates the resulting DNA fragments according to size by gel electrophoresis. It is easy to distinguish DNA from the normal allele—which is digested into two fragments by *Mst*II—from the indigestible DNA of the mutant allele. Figure 11.7b shows the results: The fetus is *AA*, so the younger sibling will neither have sickle-cell anemia nor carry the sickle-cell trait.

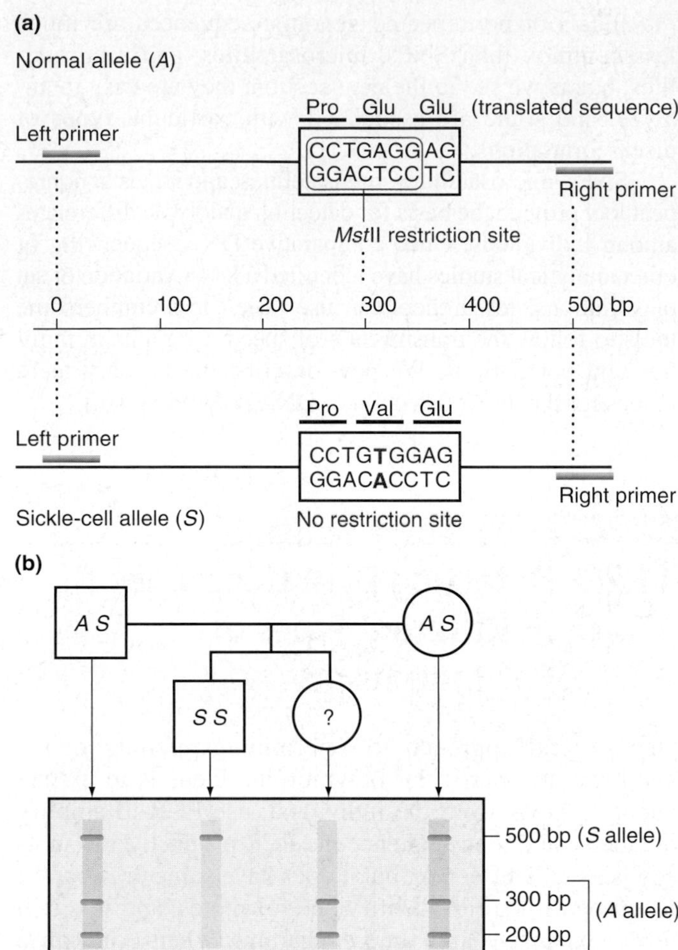

Figure 11.7 Sickle-cell genotypes can be detected rapidly with a PCR-based protocol **(a)** The normal (*A*) and sickle-cell (*S*) alleles at the β-globin locus differ by a single base-pair substitution that changes glutamic acid (Glu) to valine (Val) in the protein product. The base-pair change also eliminates the restriction site *Mst*II. **(b)** PCR amplification of the region containing this SNP, with the primers shown, produces a 500 bp product. Exposure of the normal PCR product to *Mst*II digests this DNA fragment into two smaller fragments of 200 and 300 bp in size; exposure of the mutant PCR product to the restriction enzyme has no effect. Gel electrophoresis and ethidium bromide staining distinguish the three possible genotypes. Results shown are for a hypothetical family in which both parents are carriers, the first child is afflicted with the disease, and the genotype of the fetus is initially unknown. The results of this analysis show that the fetus will be homozygous for the normal *A* allele.

Allele-Specific Oligonucleotide (ASO) Hybridization Can Detect All SNPs

Unfortunately for genotypers, most differences resulting from SNPs do not alter restriction sites. Fortunately, however, there are two general methods for detecting all SNPs. Here we describe a protocol that exploits differences in hybridization between a probe and two target sequences that differ by a single nucleotide.

A SNP is the proverbial needle in a haystack—a single base pair among, for example, the 3 billion in the human genome. The first step in nearly all methods for SNP genotype detection is the amplification by PCR of a short sequence—a few hundred base pairs of DNA—surrounding the SNP. With this step, the SNP becomes one nucleotide in several hundred, rather than one in several billion, and is much easier to detect.

How Allele-Specific-Oligonucleotide Hybridization Detects SNPs. It is only with very short probes—oligonucleotides containing around 20 bases—that single-base changes provide a large enough difference to be readily detected. This is because for very small DNA molecules—those composed of no more than 50 bp—the length of the molecule itself helps determine whether the double helix remains intact or falls apart. The effective length and therefore the strength of the hydrogen-bond forces holding together the double helix of a short-probe/short-target DNA hybrid depends on the longest stretch that does not contain any mismatches. When the two strands do not match exactly, there may not be enough weak hydrogen bonds in a row to hold them together.

If, for example, a 21-base probe hybridizes to a target strand that differs at a single base in the middle of the sequence, the effective length of the resulting double-stranded hybrids is only 10 bp. Since a 10 bp hybrid is significantly less stable than a 21 bp hybrid, one can devise hybridization conditions (by choosing the right temperatures) under which the perfect 21 bp hybrids will remain intact while the imperfect hybrids will not (**Fig. 11.8a**). By comparison, molecules longer than 50 bp can maintain their double helix conformation even with intermittent mismatches. Once a critical number of hydrogen bonds required for double helix stability is achieved, any further increase in the number of hydrogen bonds makes no difference (Fig. 11.8b).

Short oligonucleotides of 18–24 bases that hybridize to only one of the two alleles at a SNP locus under appropriate conditions are known as **allele-specific oligonucleotides,** or **ASOs** (**Fig. 11.9a**). Figure 11.9b–e shows the results of an ASO-based test to distinguish the three possible genotypes at an SNP locus.

The power of the PCR/ASO hybridization protocol is its simplicity. It does not require the running of a gel, and it can be automated if the necessary robotic resources are available. The disadvantage is that the reaction conditions needed for the optimal distinction between two SNP alleles are difficult to predict. Some SNPs are so sensitive to conditions that slight alterations in temperature can lead to false positives or negatives in allele detection.

We now describe a significant clinical application of ASO analysis.

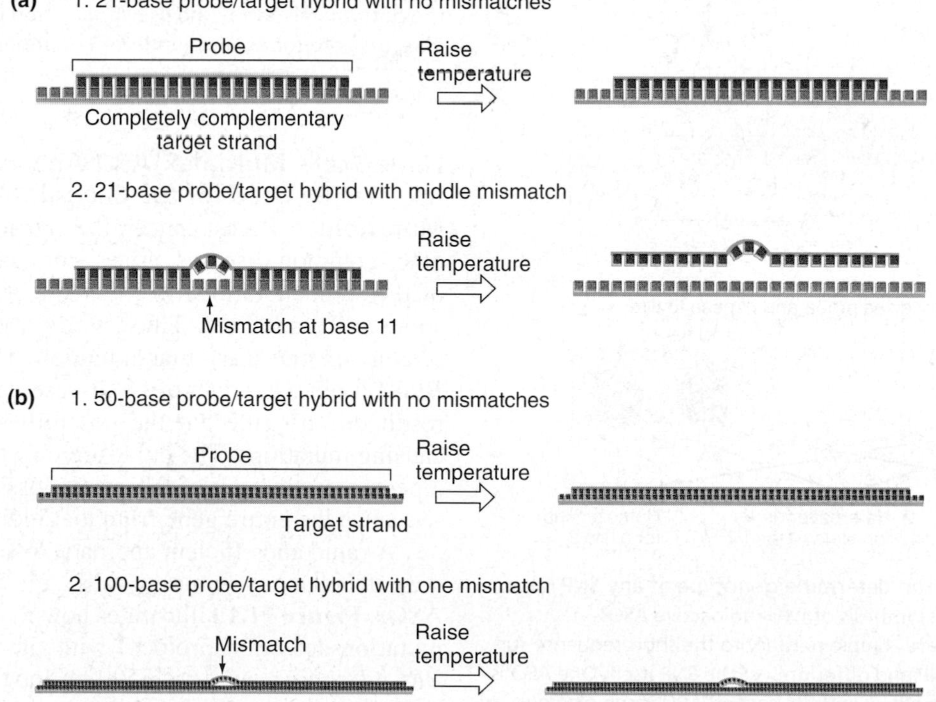

Figure 11.8 Short hybridization probes can distinguish single-base mismatches. **(a)** Researchers allow hybridization between a short 21-base probe and two different target sequences. **(1)** A perfect match between probe and target extends across all 21 bases. When the temperature rises, this hybrid has enough hydrogen bonds to remain intact. **(2)** With a single-base mismatch in the middle of the probe, the effective length of the probe-target hybrid is only 10 bases. When the temperature rises, this hybrid does not have enough hydrogen bonds to remain intact, and it falls apart. **(b)** Researchers allow hybridization to occur with probes of 50 bases and 100 bases. **(1)** A perfect match between a 50-base probe and its target bases achieves a maximum of stability such that any extension in the length of the match has no significant effect on the temperature at which the hybrid falls apart. **(2)** Thus, it is not easy to distinguish a 100 bp hybrid with one mismatched base from a 100 bp hybrid with a perfect match.

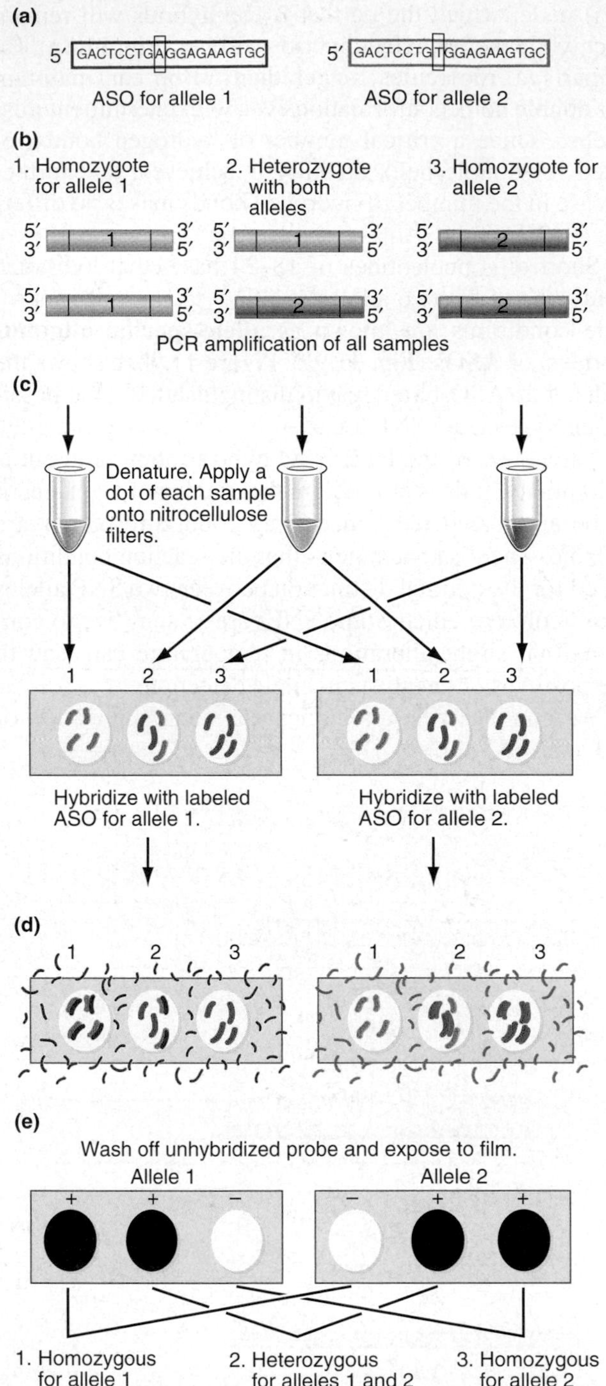

Figure 11.9 ASOs can determine genotype at any SNP locus. **(a)** The first step is the synthesis of two radioactive ASOs—oligonucleotides that are complementary to the short sequence surrounding a defined SNP and differ only at the SNP itself. One ASO is complementary to one allele, and the second ASO is complementary to the other allele at a unique SNP locus. **(b)** Genomic DNA samples can have only one of three genotypes at this SNP locus. **(c)** After amplification of individual samples by PCR, the amplified DNA of each sample is divided into two aliquots, which are denatured and blotted onto different spots on a filter paper. **(d)** One aliquot from each sample is hybridized to the wild-type ASO; the other aliquot is hybridized to the sickle-cell ASO. **(e)** Autoradiography indicates the genotype of each individual.

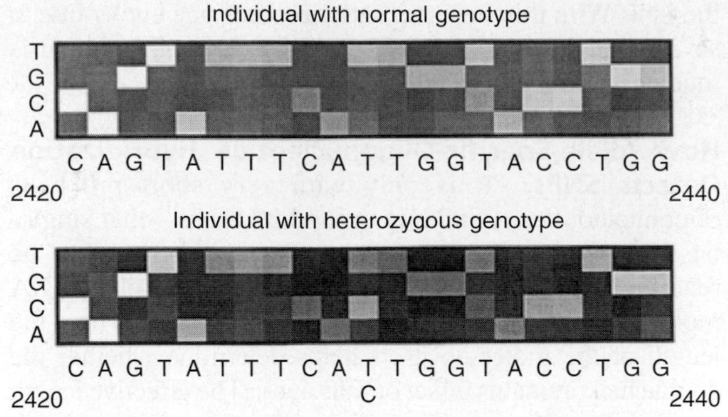

Figure 11.10 The clinical diagnosis of *BRCA1* mutations depends on sequencing with allele-specific oligonucleotides on microarrays. A small section of a microarray with ASOs for each possible base between positions 2420 and 2440 of the *BRCA1* coding region. Each column has four rectangles that separately contain ASOs differing only at the nucleotide position under analysis. *BRCA1* DNA from any one allele can be complementary to only one of the four ASOs in each column, but DNA from individuals heterozygous for a single nucleotide mutation will hybridize to two ASOs in a particular column. In each column which one or two of the four ASOs the lightest signal(s) indicate displays the highest level of hybridization. The readout for each base position is shown below the corresponding column. The *top* microarray has been hybridized to amplified genomic *BRCA1* DNA from an individual with a normal homozygous genotype; the *bottom* microarray shows hybridization to DNA from a heterozygote for a mutation from T–C at position 2430.

Large-Scale Multiplex ASO Analysis with Microarrays Can Be Used for the Clinical Detection of *BRCA1* Mutations. Breast cancer, like cystic fibrosis and many other common diseases, arises from any one of hundreds of different mutations in the same gene—some known, some still undefined. Thus, while a positive result in a screen for one particular mutation, for example, in the *BRCA1* gene, can identify a disease genotype, a negative result doesn't rule out the possibility of other disease-causing mutations. The only sure way to discover whether a person carries any of the many possible mutations is to sequence the entire gene from that individual.

A rapid and efficient approach to sequencing an entire gene from many samples relies on SNP analysis with ASOs. **Figure 11.10** illustrates how to use this whole-gene mutation-detection protocol with the *BRCA1* gene. The *BRCA1* coding region is 5500 bp long. To detect a mutation anywhere in the sequence, for each position, you design four ASOs with identical sequences flanking the position but with a different one of the four bases at the position itself. After repeating this process for all 5500 positions, you have 22,000 different ASOs. You now place each of these on the surface of a microarray.

In our discussion of ASOs, DNA from the genomic samples was attached to a surface and labeled probes were presented in the solution for hybridization. By comparison, in this version of the protocol, it is the probes that are attached to the surface (of the microarray) and the sample that is labeled.

After PCR amplification of the *BRCA1* coding region, the sample is labeled with a fluorescent dye and applied to the microarray under conditions that optimize hybridization only between perfectly complementary sequences. An automated analysis of fluorescence at each ASO now determines which base is at each position in the *BRCA1* coding sequence. Figure 11.10 compares the results from a normal individual carrying two wild-type alleles and a person heterozygous for a single base mutation.

The construction of a microarray chip and determination of the optimal conditions for the analysis of a particular genomic region is time-consuming and expensive. But once the chip is designed and the conditions determined, each use of the technology is relatively inexpensive. This approach is thus most appropriate for detecting mutations at loci like *BRCA1*, which occur often enough in humans to warrant screening high-risk populations for their presence.

Single Nucleotide Primer Extension Can Also Detect SNPs

An alternative to the ASO detection of SNP genotypes is called *DNA-polymerase-assisted single nucleotide primer extension,* or just *primer extension.* It is simpler and more accurate for all loci than the PCR/ASO method.

The primer extension protocol illustrated in **Fig. 11.11** takes advantage of the high fidelity with which DNA polymerase attaches only template-complementary nucleotides to a growing DNA strand during DNA replication. As in all methods of SNP detection, the first step is to amplify by

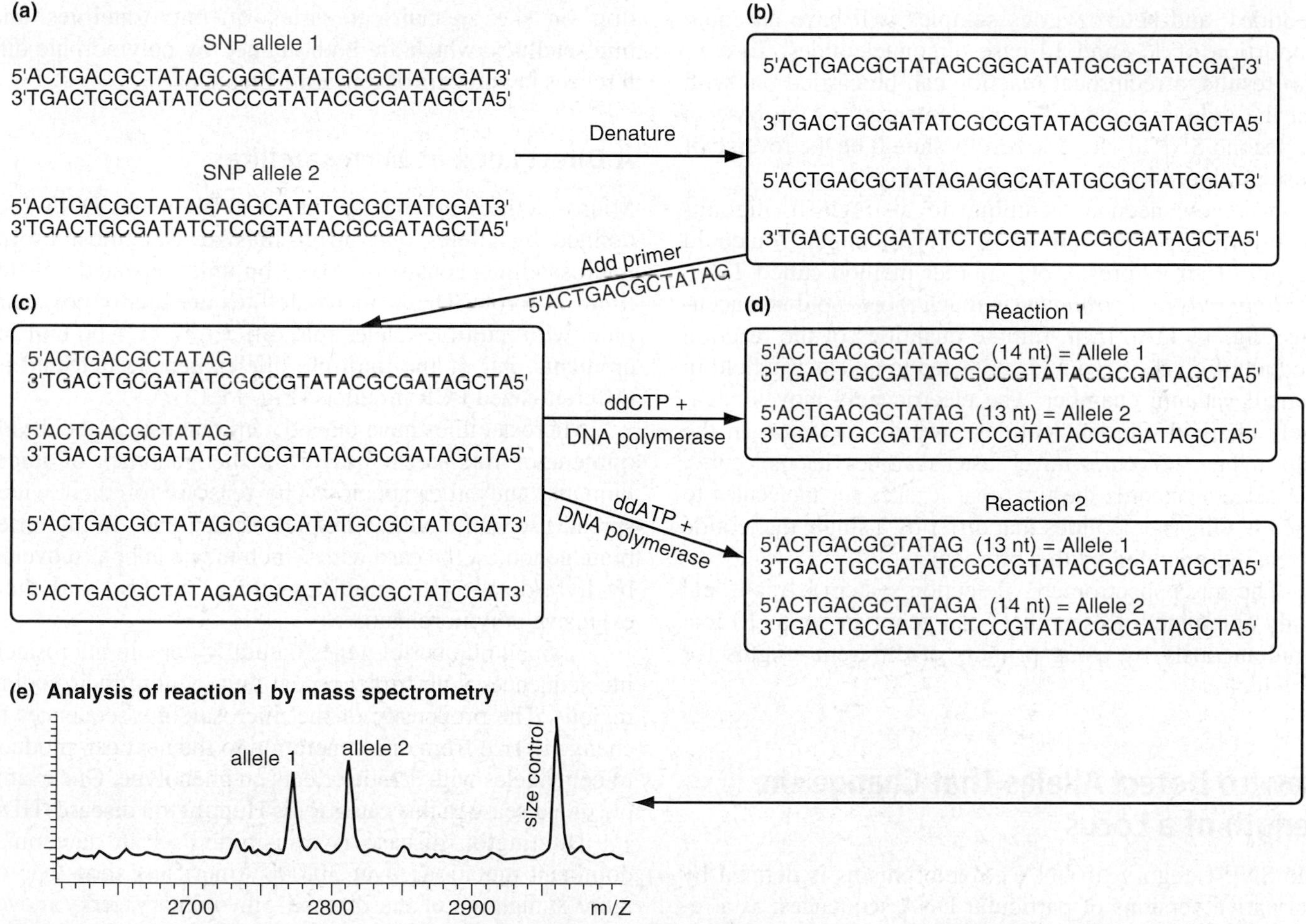

Figure 11.11 SNP detection with DNA polymerase-assisted single nucleotide primer extension. **(a)** The DNA sequence surrounding a SNP locus is amplified by PCR from a heterozygous genomic sample. The two PCR products differ only at the single base pair that defines the SNP locus. **(b)** Denaturation produces four types of single-stranded molecules. **(c)** A 13-nucleotide-long primer complementary to the sequence adjacent to the polymorphic base pair forms hybrids with DNA from both alleles. **(d)** With the addition of DNA polymerase and ddCTP, only primers bound to allele 1 can be extended to a length of 14 nucleotides (nt). In a separate experiment with ddATP, only allele 2 can be extended. **(e)** Mass spectrometry can determine which oligonucleotides (1 and/or 2) are present, and the result indicates the genotype of the genomic sample.

PCR the locus of interest (Fig. 11.11a). The amplified DNA is denatured into single strands and then mixed with an oligonucleotide primer that is complementary to a short sequence of DNA directly adjacent to the polymorphic nucleotide (Fig. 11.11b). The primer will hybridize to the same single strand of DNA from both SNP alleles (since they are identical in the adjacent region; Fig. 11.11c).

DNA polymerase is now added to the reaction along with a dideoxynucleotide type (ddCTP in the figure) that is complementary to one of the two SNP alleles. If the 13-base primer binds to a DNA strand from allele 1 of this SNP, DNA polymerase will extend the primer with a ddCTP, since it is complementary to the G nucleotide at the next position on the template. However, the enzyme cannot extend the primer bound to allele 2 because ddCTP is not the proper nucleotide (Fig. 11.11d).

When the reaction is completed, samples homozygous for allele 1 will contain oligonucleotides that have all been extended to 14 bases; samples homozygous for allele 2 will have only unextended 13-base oligonucleotides; and heterozygous samples will have an equal proportion of 13- and 14-base oligonucleotides. To confirm results, a reciprocal reaction can be carried out with just the dideoxynucleotide complementary to the base in the second SNP allele. The results should be the reverse of those produced in reaction 1 (Fig. 11.11d).

You now need a technique to distinguish oligonucleotides 13 bp long from those 14 bp in length. You could use gel electrophoresis, but another method called *Time-of-Flight mass spectrometry* is much more rapid and accurate (Fig. 11.11e). In it, minute quantities of the reaction products are vaporized and subjected to an electric field in a small vacuum chamber. The electric field moves negatively charged oligonucleotides toward a detector. In the field, lighter molecules travel faster than heavier molecules. The detector records the time that it takes for molecules to reach it. Oligonucleotides that differ by a single nucleotide are easily separated by this protocol.

The mass spectrometry detection system is rapid and easily automated. Researchers can genotype up to 10 loci simultaneously by using primers of different lengths for each locus.

How to Detect Alleles That Change the Length of a Locus

The SNP category of DNA polymorphisms is defined by alternative versions of particular DNA sequences; as a result, all SNP detection methods are based on the ability to detect nucleotide variation. In contrast, the other three categories of DNA polymorphisms—microsatellites, minisatellites, and deletions/duplications/insertions—are all associated with variations in the actual size of the locus. Alleles that differ in size from one another can be directly and easily distinguished by gel electrophoresis.

When the size difference between alleles is less than 1 kb, genotyping can be readily and rapidly accomplished as follows.

You begin by using a pair of primers complementary to sequences on either side of the actual polymorphism to amplify by PCR the locus from an individual's DNA. You then subject the PCR products to gel electrophoresis to separate DNA fragments according to their size. After staining with ethidium bromide, each allele shows up as a specific band. All possible homozygous and heterozygous genotypes can be distinguished in this way (**Fig. 11.12**).

A single researcher can use this protocol to genotype hundreds of samples in a single day, without any specialized equipment (other than the apparatus necessary for PCR and gel electrophoresis). The protocol is also amenable to automation using fluorescently tagged primers and electrophoresis in the same apparatus that is used for automated DNA sequencing (see Fig. 9.18 on p. 334). In Chapter 14, we present an in-depth look at the general category of deletions, duplications, and insertions. Here we focus our attention on the special categories of microsatellites and minisatellites, which are both defined by polymorphic differences in numbers of repeating elements.

A Direct Look at Microsatellites

Microsatellites are by far the most abundant DNA loci defined by alleles that differ in size. The most useful microsatellites consist of 2 or 3 bp units repeated 15–100 times in a row. These microsatellites are highly polymorphic, with multiple alleles that differ in 2 or 3 bp unit increments. All of the multiple alleles can be detected as different-sized PCR products (Fig. 11.12).

Microsatellites have played a crucial role in the development of linkage maps across the genomes of mice, humans, and other species. The reasons for their widespread use include their frequent appearance in all vertebrate genomes, the ease with which they can be discovered by hybridization to simple sequence probes, and their extensive polymorphism.

A small number of genes naturally contain microsatellite sequences with triplet repeat units within their coding regions. The propensity of the microsatellite sequences to change in size from one generation to the next can produce mutant alleles with drastic effects on phenotype. One example of a disease in this category is Huntington disease (HD).

Huntington disease is transmitted as an autosomal dominant mutation. Over 30,000 Americans show one or more symptoms of the disease—involuntary, jerky movements; unsteady gait; mood swings; personality changes; slurred speech; impaired judgment. An additional 150,000 have an affected parent, which gives them a 50:50 chance of expressing the dominant condition themselves as they age. Although symptoms usually show up between the ages of 30 and 50, the first signs of the disease have appeared in people as young as 2 and as old as 83. Some people with a

(a) Determine sequences flanking microsatellites.

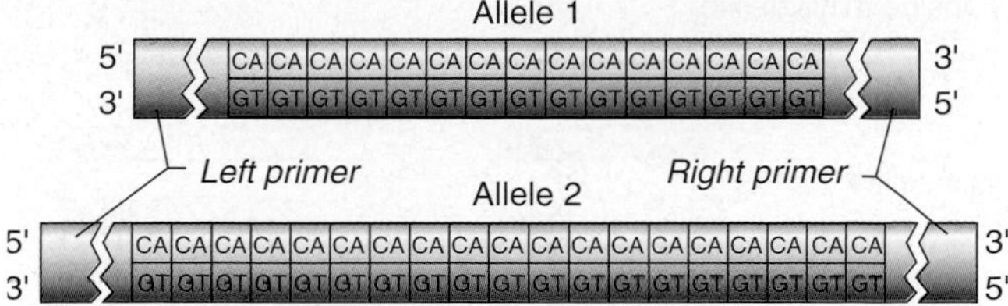

(b) Amplify alleles by PCR.

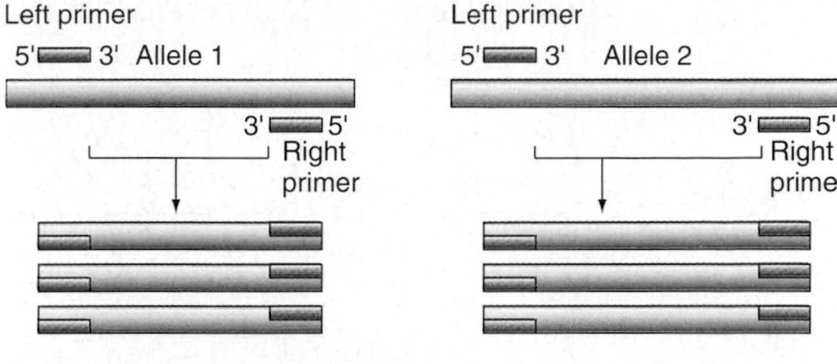

(c) Analyze PCR products by gel electrophoresis and staining.

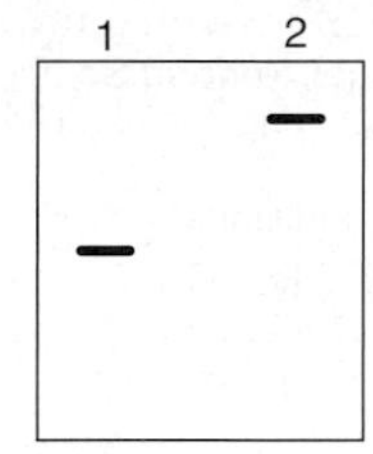

(d) Example of population with three alleles

Allele 1

CA CA CA CA CA CA CA CA CA CA CA CA CA CA CA

Allele 2

CA CA CA CA CA CA CA CA CA CA CA CA CA CA CA CA CA CA CA

Allele 3

CA CA CA CA CA CA CA CA CA CA CA CA CA CA CA CA CA CA CA CA CA

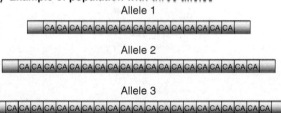

Six diploid genotypes are present in this population.

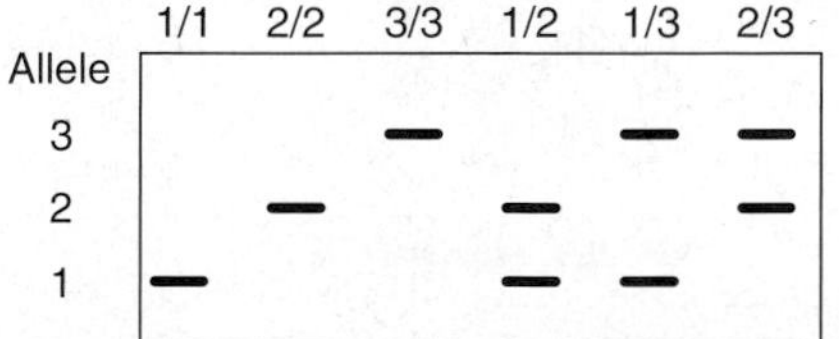

Figure 11.12 Detection of microsatellite polymorphisms by PCR and gel electrophoresis. **(a)** Microsatellite alleles differ in length. Left and right primers are devised based on sequences that flank the microsatellite locus. **(b)** Genomic DNA is amplified by PCR with primers specific for the microsatellite locus. **(c)** Gel electrophoresis and ethidium bromide staining distinguish the alleles from each other. **(d)** Microsatellites are often highly polymorphic with many different alleles present in a population. With just three alleles, there are six possible genotypes. With N alleles, there will be $\frac{N}{2}(N+1)$ genotypes.

family history of HD would like to know their genotype before deciding whether to have a family.

In 1993, after 10 years of intensive research, investigators identified and cloned the *HD* gene. With the gene in hand, they were able to uncover the unusual mutation that causes the disease **(Fig. 11.13a)**. Unlike the vast majority of disease mutations, which result from base-pair changes or the elimination of genetic information, HD is caused by too much genetic information: an expansion of a CAG trinucleotide repeat in the coding sequence, which translates into a string of glutamine amino acids.

It is possible to detect *HD* alleles directly with the same size-based PCR procedure used to detect other microsatellite alleles. The normal allele contains up to 34 repeats while disease-causing alleles carry 42 or more. In general, the greater the number of repeats, the earlier the age of disease onset (Fig. 11.13b). Those who inherit a disease allele invariably get the disease if they live long enough. Thus, although expressivity, which depends on the number of triplet repeats, is variable, penetrance is complete. Several other diseases caused by triplet repeat expansion have been

(a) Basic structure of the *HD* gene's coding region

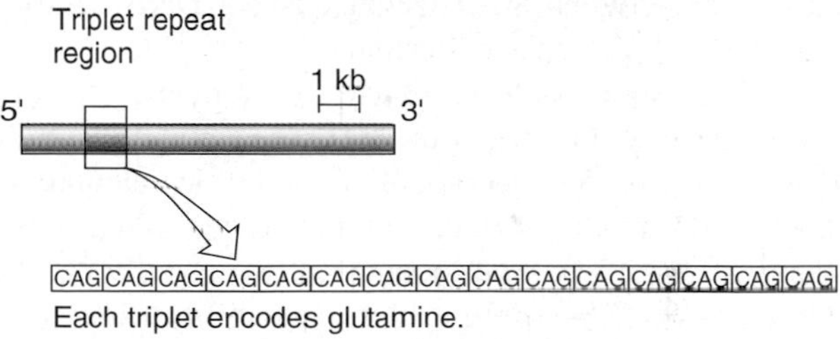

(b) Some alleles at the *HD* locus

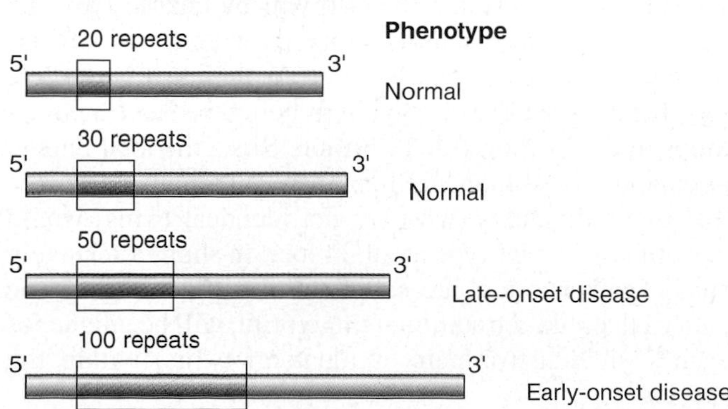

Figure 11.13 Mutations at the Huntington disease locus are caused by expansion of a triplet repeat microsatellite in a coding region. **(a)** Near the 5′ end of the coding region is a repeating triplet sequence that codes for a string of glutamines. **(b)** Different alleles at the HD locus have different numbers of repeating units. Fewer than 34 repeats gives a normal phenotype. As the number of repeats increases beyond 42, the onset of the disease is earlier.

uncovered, including a variety of neurological disorders, and the fragile X syndrome described in the Genetics and Society box on pp. 216–217 in Chapter 7.

A Direct Look at Minisatellites and DNA Fingerprinting

Minisatellites are another type of locus where alleles are distinguished by size, rather than by nucleotide sequence. We have seen that minisatellite alleles vary in length from 0.5–20 kb. Many minisatellite loci are amenable to the PCR analysis described for microsatellites. But the real power of minisatellites lies in the fact that particular minisatellite sequences often occur at a small number of different genomic loci. With restriction enzyme digestion, gel electrophoresis, and Southern blot hybridization using a cross-hybridizing minisatellite probe, researchers can look simultaneously at allelic variation at these multiple unlinked loci (**Fig. 11.14**). (This strategy wouldn't work for microsatellites because their core mono-, di-, or trinucleotide sequences are each present thousands of times in the genome.)

Because minisatellites provide a generalized picture of the whole genome rather than of a single locus and because they are highly polymorphic, they are ideally suited for comparing the overall genetic relationship between two individuals or between an individual and a DNA sample recovered from the scene of a crime.

How many loci would you have to examine to be certain that two DNA samples come from the same individual (or identical twins) and no one else? A simple calculation shows that the probability of two unrelated individuals having identical genotypes at a locus with two equally prevalent alleles is 37.5%—quite a high probability. However, the chance that the same two individuals will be identical at 10 such loci, all unlinked, is only 0.375^{10}, or 0.005%—quite a low probability. The result of 0.005% means there is 1 chance in 20,000 that the two will by chance have the same genotype at 10 unlinked loci. By extension, if you simultaneously detect genotype at 24 unlinked two-allele loci, the chance of two individuals being the same at all 24 drops to 0.375^{24}, or 1 in 17 billion. Since the total human population is less than 8 billion, there is virtually no chance that two individuals (who are not identical twins) would have the same genotype at all 24 loci. In short, a relatively small number of loci are sufficient to produce a genotype pattern that, like a traditional fingerprint, will be unique for each individual (or pair of identical twins) within the species.

DNA Fingerprinting Defined. In 1985, Alec Jeffreys and coworkers made two key findings. Each minisatellite locus is highly polymorphic; and most minisatellites occur at multiple (usually between 2 and 50) sites scattered around the genome. As a result, they realized, minisatellite probes would be perfect reagents for obtaining a **DNA fingerprint:** a pattern produced by the simultaneous detection of genotype at a group of unlinked, highly polymorphic loci.

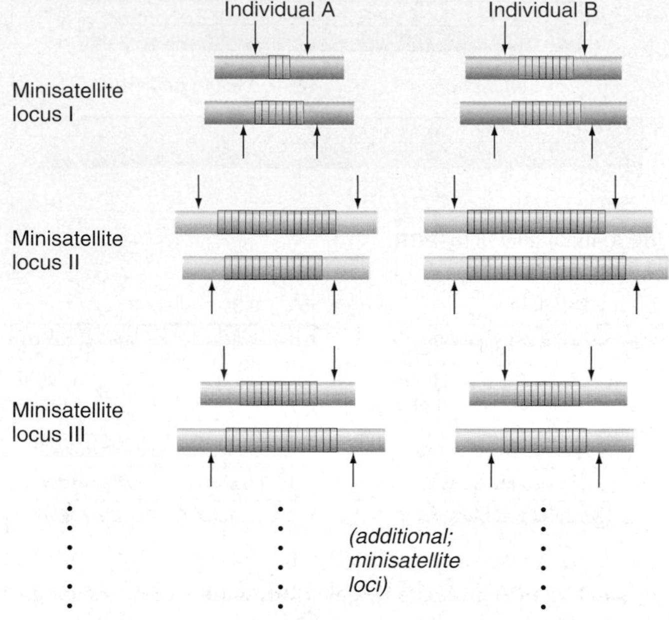
(a) Digest DNA with restriction enzyme that does not cut inside minisatellite.
Individual A
Individual B
Minisatellite locus I
Minisatellite locus II
Minisatellite locus III
(additional; minisatellite loci)
(b) Run DNA samples on a gel. Perform Southern blotting. Hybridize with probe containing minisatellite sequence.

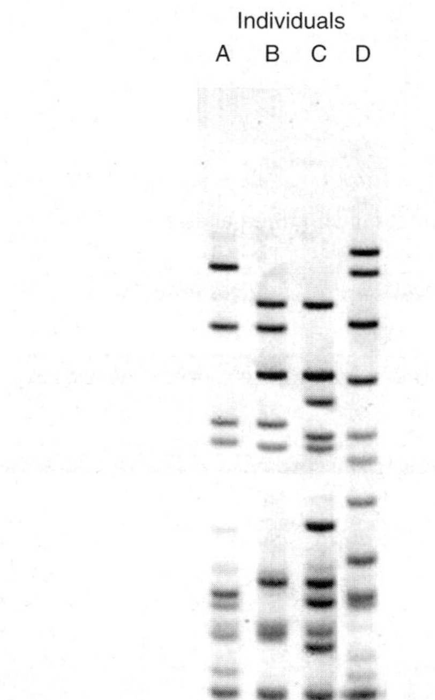
Individuals
A B C D

Figure 11.14 Minisatellite analysis provides a broad comparison of whole genomes. **(a)** Two individuals each carry two alleles at three loci containing the same minisatellite sequence. Additional minisatellite loci are not shown. The *arrows* indicate a restriction site recognized by a particular restriction enzyme. Notice that minisatellite lengths are different both among alleles at a locus and among different loci. **(b)** After restriction enzyme digestion, gel electrophoresis, Southern blotting, and hybridization to the minisatellite probe, researchers can obtain an autoradiograph of the type shown here for four individuals. The darkness (or intensity) of each band depends on the degree of cross-hybridization between the minisatellite probe sequence and the actual sequence within each member of the minisatellite family.

Fingerprinting protocols use restriction enzymes and Southern blotting to detect length polymorphisms at minisatellite loci. The simplest DNA fingerprint reveals which alleles an individual carries for one family of minisatellites. As you saw in Fig. 11.14, to obtain such a fingerprint, you simply digest a sample of genomic DNA with a restriction enzyme that does not cut within the minisatellite itself but does cut within closely flanking sites. You then run the restriction fragments on a gel, perform a Southern blot (as described in Fig. 9.15 on pp. 325–326), and probe the blot with a labeled DNA fragment derived from the minisatellite. The result is the actual DNA fingerprint, which consists of an autoradiograph displaying the pattern of different-sized fragments representing all the loci of that minisatellite in that genome. The most useful minisatellite families have 10–20 members per genome. This range of numbers is small enough to allow the resolution of all the loci as individual bands on an autoradiograph, but large enough to provide true fingerprint information. If one fingerprint is not sufficient to resolve the relationship between two different DNA samples, investigators can always obtain data from two, three, or even more minisatellite families.

Figure 11.15 illustrates an interesting example of the utility of DNA fingerprinting. In 1997, scientists from the Roslin Institute in Scotland announced that they had cloned a sheep by injecting a diploid nucleus from an adult udder cell (grown in culture) into an unfertilized egg whose own genetic material had been removed. Initially, many scientists were skeptical of this result and thought that "Dolly" might actually be the result of a fertilization between some contaminating sperm and the egg. If this were the case, Dolly's genome would be unique. Instead, the results of the fingerprint analysis shown in Fig. 11.15 demonstrated that Dolly's DNA fingerprint was identical to that of the adult udder cell used to clone her. This established beyond a doubt that her genome was indeed a clone of that cell's DNA.

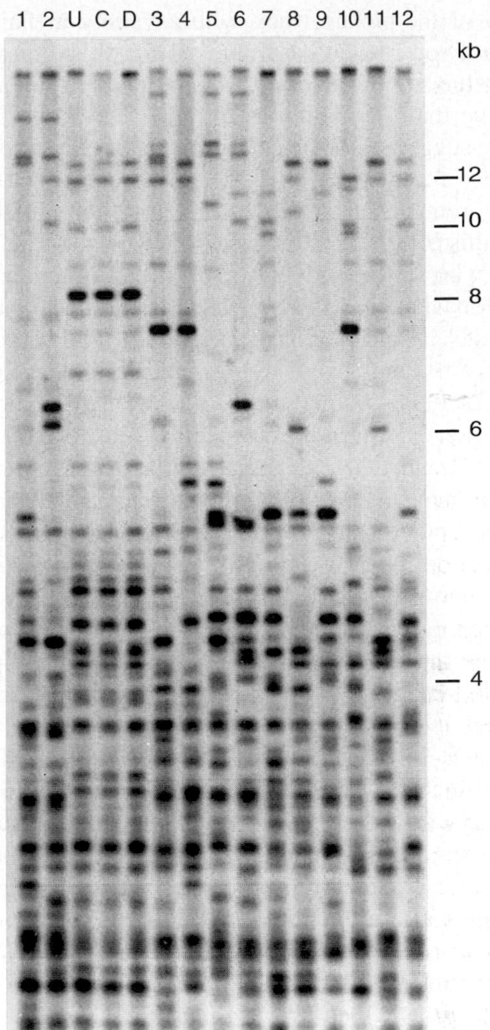

Figure 11.15 DNA fingerprint analysis confirmed that Dolly was cloned from an adult udder cell. Genomic DNA samples were prepared from the donor udder cells (U), from the cell culture prepared from the udder cells (C), from Dolly's blood cells (D), and from control sheep 1–12. These DNAs were digested with the restriction enzyme *Mbo*I and subjected to gel electrophoresis and Southern blot analysis with a minisatellite probe. The DNA fingerprints of the 12 control sheep are all different from each other and from the cells involved in the Dolly experiment. Dolly's DNA fingerprint is identical to the fingerprints of both the udder cells and the derived cell culture. This result provides very strong evidence that Dolly is a clone of the ewe that donated the udder cells.

DNA Fingerprints Can Identify Individuals and Determine Parentage. DNA fingerprints are a powerful tool for forensic analysis. Prosecutors and defense attorneys alike use them to show the likelihood of a suspect's presence at the scene of a crime or to prove the innocence of someone falsely accused. In one case, a man arrested on rape charges in 1981 had blood type alleles that matched those of cells in the semen found on the victim. The crude blood-typing tests available at that time did not have the resolving power to prove that the semen definitely came from the accused. But after the victim picked the accused out of a lineup, the prosecution used the test results as evidence to help obtain a conviction. Eleven years later, a defense lawyer filed an appeal based on the finding that DNA fingerprints of the semen obtained from the victim and from cells of the convicted man were different. With this evidence, the court reversed its decision and in 1993, set the convicted man free. Since that time, over 100 other men unjustly convicted of rape have been released from jail thanks to DNA fingerprint analysis.

Also in 1993, the courts for the first time accepted plant DNA as evidence in a murder trial. The defendant, accused of killing a woman whose body was found abandoned in the Arizona desert, owned a pickup truck. When police searched the back of the truck, they found a few seed pods from a Palo Verde tree. The county sheriff's department asked a molecular geneticist if there was any way to match the pods to an individual tree at the scene of the crime. DNA fingerprinting was used to compare the pods recovered from the bed of the pickup truck with a variety of Palo Verde trees. Only one match was found, and it was to the tree located at the site of the crime. Although this evidence does not prove the defendant was at the scene of the crime, it strongly suggests that the defendant's truck was there.

More recently, DNA fingerprints demonstrated that skeletal remains unearthed at Ekaterinburg in the Ural Mountains of Russia belonged to Czar Nicholas II and his family, who were murdered in 1918 during the Bolshevik revolution. Geneticists established the relationship by comparing DNA from the excavated bones with samples obtained from a number of living relatives of the Romanov family, including Prince Philip, Duke of Edinburgh. This information disproved the claim of Anna Anderson that she was the Grand Duchess Anastasia; in three independent analyses, her DNA (obtained from hair and from biopsy samples removed during an examination for cancer years before her death in 1984) did not match that of members of the Romanov line—living or dead.

When researchers examine groups of polymorphic DNA loci to obtain DNA fingerprints that can help determine the genetic relatedness of DNA samples, the chromosomal location of each locus is unimportant to the analysis. But a knowledge of the chromosomal location of individual DNA markers can be of great use to geneticists who want to determine the chromosomal location of genes responsible for disease phenotypes and other inherited differences among people, animals, and plants. The chromosomal positions of DNA markers linked to genes defined by phenotype alone can serve as a basis for cloning those genes in a process called *positional cloning.*

11.3 Positional Cloning: From DNA Markers to Gene Clones

In Rare Cases, It Is Possible to Move from a Disease Phenotype to the Causative Gene Without Linkage Analysis

Medical researchers learned from the analysis of pedigrees like the one shown in **Fig. 11.16a** that hemophilia A is an X-linked recessive trait governed by a single gene. The main symptom of uncontrollable bleeding in affected individuals prompted geneticists to make an educated guess at the biochemical function of the responsible gene. The function of the wild-type hemophilia A gene, they proposed, is production of a normal clotting factor; mutations that inactivate this factor produce hemophilia A.

Once molecular investigators worked out the details of the blood-clotting cascade (Fig. 11.16b), they could look for clotting factors in normal individuals that were absent in hemophiliacs (Fig. 11.16c). In this way, they identified a protein known as Factor VIII. They next determined the amino acid sequence of Factor VIII and used the genetic code to predict the nucleotide sequence of the corresponding gene. This information allowed them to develop a degenerate oligonucleotide probe that could identify clones of the Factor VIII gene within genomic libraries (Fig. 11.16d). When they sequenced this gene from people suffering from hemophilia A, they found mutations with an absolute correlation to the disease phenotype and thereby verified the gene as the causative agent of the disease.

Of the thousands of genes responsible for known human genetic diseases, only a tiny number can be identified in the manner just outlined. Much more often, even with a clearly defined mutant phenotype, it is hard to make an educated guess about the protein altered by the disease allele.

Cystic fibrosis, for example, is a recessive autosomal genetic condition inherited by 1 child in every 2500 born from two parents of European descent. But because the trait is recessive, the frequency of unaffected carriers in the population is much greater, about 1 in every 36 people. Many carriers come from families where the disease has never appeared, and so the first birth of a child with the disease can come as a complete shock. Children with the disease have a variety of symptoms arising from abnormally viscous secretions in the lungs, pancreas, sweat glands, and several other tissues. Malfunction of the pancreas leads to defective secretion of pancreatic enzymes necessary for normal digestion of food. Hyperviscous secretions in the lungs cause severe infections and breathing difficulties. Even with modern medical treatments that combat some symptoms of the disease, most cystic fibrosis patients die before the age of 30.

Unfortunately, the gross symptoms of cystic fibrosis did not provide insight into the underlying molecular cause of the disease. Hundreds of proteins contribute to the process of cell secretion, and most of these were still unidentified in the 1980s. Without a way to determine which one is defective in cystic fibrosis, investigators had no simple way to work their way from gene function to the amino-acid sequence of a particular protein and from there to the protein's DNA coding sequence.

In Positional Cloning, Linkage Analysis with DNA Markers Helps Identify Disease Genes

The challenge for identifying the defects associated with most human hereditary diseases is that, unlike what happened with hemophilia A, there is no biochemical data to guide researchers to the genes. Without such data, the standard approach is to combine linkage analysis (as described in Chapter 5) with the use of DNA markers described earlier in this chapter to localize the human disease gene to a specific region of chromosomal DNA. Then other techniques can determine which gene—among the small number in this

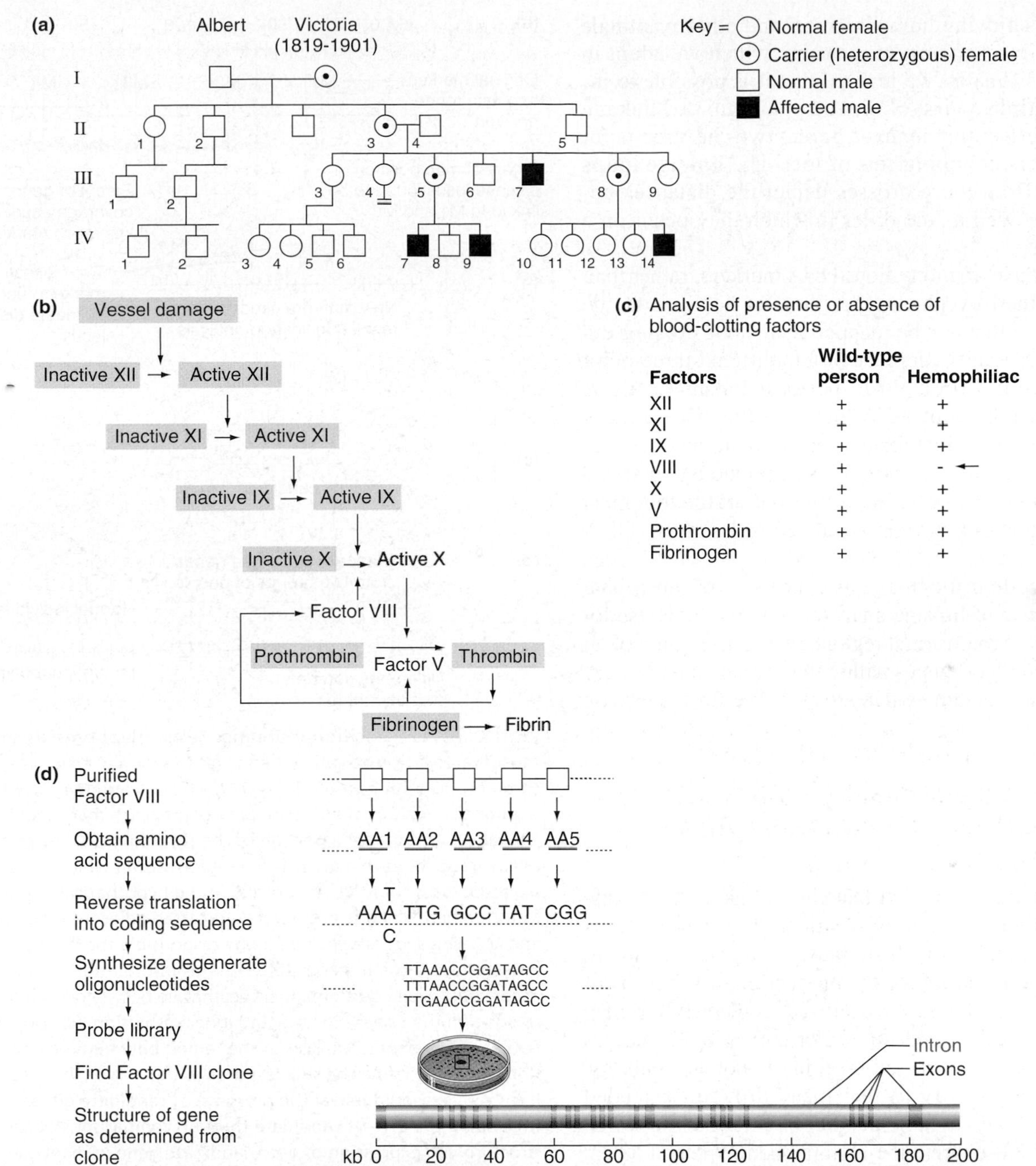

Factors	Wild-type person	Hemophiliac
XII	+	+
XI	+	+
IX	+	+
VIII	+	- ←
X	+	+
V	+	+
Prothrombin	+	+
Fibrinogen	+	+

Figure 11.16 How geneticists identified and cloned the hemophilia A gene. **(a)** A pedigree of the royal family descended from Queen Victoria. This family tree uses the standard pedigree symbols. **(b)** The blood-clotting cascade. Vessel damage induces a cascade of enzymatic events that convert inactive factors to active factors. The cascade results in the transformation of fibrinogen to fibrin and the formation of a clot. **(c)** Blood tests can determine whether an active form of each factor involved in the clotting cascade is present. The results of such analyses show that many hemophiliacs, such as those found in Queen Victoria's pedigree, lack an active Factor VIII in their blood. **(d)** Researchers purified Factor VIII, determined its amino acid sequence, used this information to infer all possible degenerate coding sequences, constructed oligonucleotides for a region with minimal degeneracy, probed a genomic library with these oligonucleotides, and obtained genomic clones of the *Factor VIII* gene. They sequenced these clones to determine the gene's actual coding sequence.

region—contains mutations that correlate with the disease phenotype. This entire protocol is called **positional cloning.**

We saw in Chapter 5 that a simple two-point cross can demonstrate linkage if the two loci under analysis lie close enough together on the same chromosome that the rate of recombination is significantly less than the 50% expected with independent assortment. We also saw that the frequency of recombination between the two loci provides a direct measure of the distance separating them, as recorded in centimorgans (cM), or map units (m.u.).

(Geneticists studying humans, mice, and other mammals use the centimorgan unit of measure, which we adopt in this chapter.) Finally, we learned that it is possible to integrate multiple pairs of linked loci into a "linkage group" by performing many different two- or three-point crosses with overlapping sets of loci. The **linkage maps** constructed from these crosses depict the distances between loci as well as the order in which they occur on a chromosome.

With the use of anonymous DNA markers, rather than markers defined by phenotype, there is no limit to the number of loci that can be mapped in a single cross or extended human family. In place of a traditional three-point cross, it becomes possible to perform linkage analysis combining hundreds of DNA loci with the disease locus of interest. If genetic linkage can be demonstrated between a disease trait and one or more previously mapped DNA markers, then the gene responsible for the trait must lie in the same subchromosomal region as those DNA markers.

We now describe the entire process of positional cloning: the use of linkage analysis to map the disease locus to a small chromosomal region; the determination of all possible "candidate" genes within this region; and the identification of the candidate gene responsible for the disease (**Fig. 11.17**).

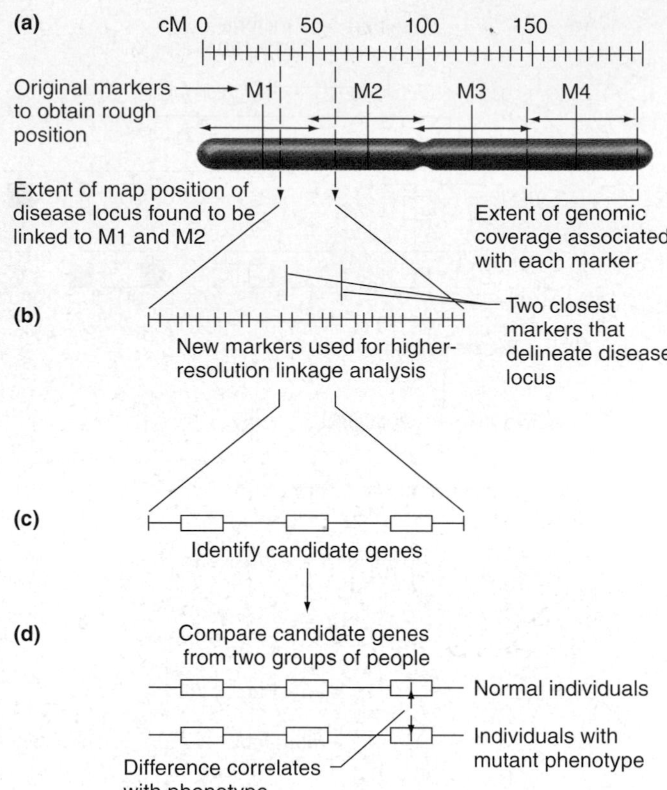

Figure 11.17 Positional cloning: From phenotype to chromosomal location to guilty gene. (a) Diagram of a human chromosome with four markers—M1, M2, M3, and M4—used in the linkage analysis of a disease phenotype. Each marker provides "linkage coverage" of a portion of the chromosome. The region of coverage for each marker is the region within which linkage to a disease locus could be uncovered. In the hypothetical study illustrated here, linkage has been detected to adjacent markers M1 and M2. This suggests that the gene responsible for the disease lies between those markers. **(b)** With this information, an investigator could type additional markers that lie between M1 and M2 to position the disease locus with higher resolution. **(c)** Looking for candidate genes. Analysis of the region between recombination sites that define the smallest area within which the disease locus can lie should reveal the presence of candidate genes. **(d)** Finding the correct candidate through comparisons of the structure and expression of each candidate gene in many diseased and nondiseased individuals. A correlation between a mutant structure or expression for a particular candidate gene and the disease phenotype can provide evidence that a particular gene is responsible for the disease phenotype.

Step 1 of Positional Cloning: Correlating Phenotypic Transmission with One Area of the Genome

Discovery of a DNA marker that shows linkage to the disease locus is the first goal of positional cloning. For traits expressed in plants or small animals, it is a simple matter to set up a single testcross for the production of hundreds of offspring that can be easily analyzed to identify the map position of the trait in question. For human traits such as disease phenotypes, directed breeding is not a possibility. Instead, researchers try to find many different extended families—each with a large number of children—in which some individuals express the mutant phenotype and others do not. The Tools of Genetics box "Using Human Pedigrees and LOD Scores to Calculate the Probability That Two Loci Are Linked" on pp. 416–417 describes the mathematics used to demonstrate linkage between a marker and an abnormal trait in human pedigrees.

The mapping of a human disease locus begins with the genotyping of all members of the disease-carrying families for a series of DNA markers ideally spaced at even intervals of 10 cM on each chromosome. If the distance between markers is greater, the chances of demonstrating linkage between a marker and the disease locus are reduced; if the distance is shorter, investigators would have to do many more genotyping assays without much gain in statistical power.

Since the human genome is approximately 3000 cM long as measured in map distances, you would need approximately 300 markers to map the genome at 10 cM intervals. Because the human genome has been sequenced, you already know the chromosomal positions of these DNA markers. The only unknown map position is that of the disease locus. You can thus view a 301-point linkage analysis as 300 separate two-point crosses, each one a test for linkage between an individual DNA marker and the disease locus. If the number of genotyped individuals is sufficiently large, the disease locus must show linkage to one or more of these markers. Finding linkage to at least one marker of known position will place the disease locus in a particular subregion of a particular chromosome (Fig. 11.17a).

Higher-Resolution Mapping with Additional DNA Markers. The next step is to increase the resolution of the linkage map in the area of the gene. Researchers accomplish this by choosing additional DNA markers that lie closer together in the vicinity of the disease locus. If, for example, the first stage of linkage analysis were to place a disease gene somewhere between markers M1 and M2 on the 175-cM-long chromosome shown in Fig. 11.17, you could query an electronic repository for additional markers between the M1 and M2 positions on the chromosome and then genotype each marker in all members of the pedigree used in the initial linkage analysis (Fig. 11.17b). The goal of this second stage is to identify DNA markers so tightly linked to the disease locus that they show less than 1% recombination in all individuals tested. If you could map the disease locus between two markers separated by 1 cM, the gene would lie within a delineated region of less than 1000 kb, since the human genome contains 3000 Mb of DNA and a total-linkage distance of ~3000 cM. To take advantage of additional markers, it is often necessary to include pedigrees in the study, that is, to increase the probability of observing recombination between markers by looking at the products of a large number of meioses.

In 1984, the Huntington disease (*HD*) locus became the first human disease gene to be successfully mapped by positional cloning. **Figure 11.18** shows the five-generation, 104-member family pedigree used to demonstrate linkage between a previously mapped DNA marker named G8 and the *HD* locus. Preliminary linkage between the *HD* locus and the G8 marker placed the disease gene on human chromosome 4. Further linkage analysis then narrowed down the map position of *HD* to less than 1000 kb.

Step 2 of Positional Cloning: Identifying Candidate Genes

Once the locus of a gene for a particular disease has been narrowed down to a 1000 kb genomic region, researchers catalogue and investigate all the coding sequences within that region. Current estimates of gene number in the human genome center around 25,000. If these genes were spaced evenly throughout the genome, there would be an average density of 25,000 genes/3000 Mb = 8 genes per 1000 kb.

Cataloging All the Genes in the Identified Region. As we saw in Chapter 10, there are several ways to identify coding regions within the human genome. One approach depends on the different forces of natural selection that act upon coding regions versus nonfunctional sequences. Though mutations can occur at any nucleotide in the genome, mutations in genes and their regulatory regions are usually not tolerated. As a result, these sequences evolve much more slowly than nonfunctional sequences. In the 60 million years that mice and humans have been diverging from a common ancestor, nonfunctional sequences in both species have accumulated so many mutations that they no longer show any similarity to each other. In contrast, coding sequences in humans almost always have a conserved homolog in mice, and often, the mouse and human coding sequences will hybridize with each other. Now that both the human and mouse genomes are completely sequenced, cross-hybridization experiments are no longer necessary. Instead, a specialized computer algorithm can be used as a form of what molecular biologists refer to as "*in silico*" *hybridization* (in analogy to *in vitro* and *in vivo* and with a reference to the silicon constituents of computer chips) to identify all DNA regions conserved between the two species (**Fig. 11.19**).

A second independent method of gene identification relies on a different type of computer algorithm that detects exonlike coding regions based on the presence of an open reading frame, a knowledge of codon usage, and the structure of splice sites. A DNA sequence that both looks like a coding region and has been conserved across distant species has a high probability of representing a gene.

A third independent approach for identifying genes is through their appearance in one or more EST clones. As we saw in Chapter 10 millions of random EST clones have been sequenced from cDNA libraries compiled from many different human tissues, and these sequences have been entered in databases. Since cDNAs correspond to RNA transcripts, the presence of a sequence in an EST means that the sequence is likely to be derived from a gene.

Gene Expression Patterns Can Pinpoint Candidate Genes. How can you tell which of the 8 or so protein-coding genes in a region of interest might be the one responsible for a particular trait? The expression pattern of these genes—that is, the level of gene transcription in different tissues at different stages of development—can provide clues. For example, the gene responsible for the neurodegenerative phenotype of Huntington disease must be expressed in the brain, while a gene responsible for breast cancer must be expressed in breast tissue. This criterion is by no means fail-safe, but knowledge of expression patterns often allows a researcher to eliminate most of the genes in the mapped region. The remaining genes are considered "candidates" for the disease locus.

There are two main ways to obtain information about gene-expression patterns. First, in well-studied organisms like humans, the information may already be available in public databases of EST sequences. For example, if a particular gene is represented in many random ESTs from a lung cDNA library but is never found in similar numbers of

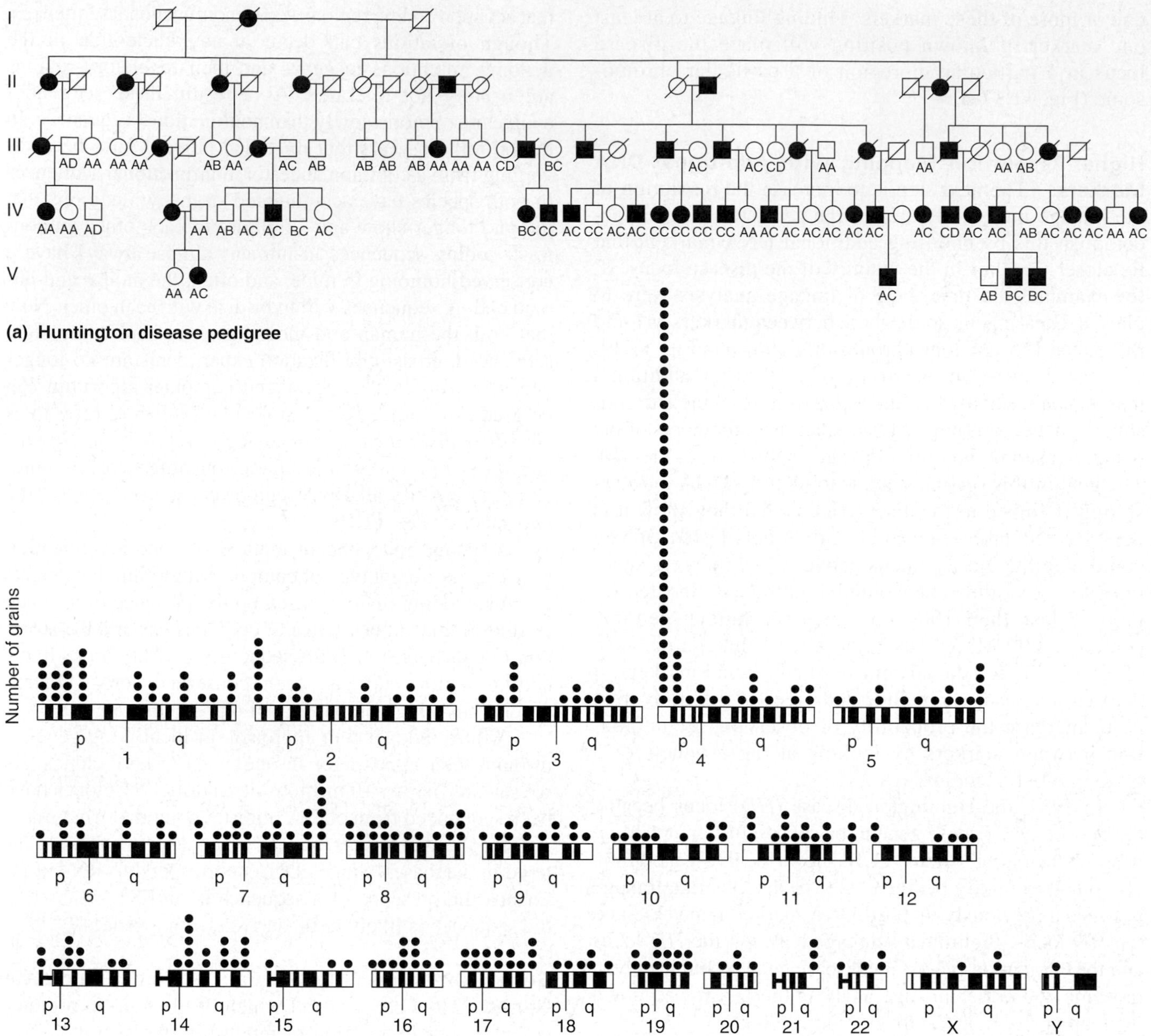

Figure 11.18 Detection of linkage between the DNA marker G8 and a locus responsible for Huntington disease (HD) was the first step in the cloning of the *HD* gene. **(a)** Portion of a large Venezuelan pedigree affected by HD, or Huntington disease. For living members of the pedigree, alleles at the G8 marker locus are indicated (A, B, C, and D). It is easy to see the cotransmission of marker alleles with the mutant and wild-type alleles at the *HD* locus. Pedigree analysis shows that the *HD* locus is within 5 cM of the G8 marker. **(b)** Karyotypic localization of the G8 marker can provide a chromosomal location for the *HD* gene. Researchers carried out a karyotypic localization analysis by a precursor to the FISH technique in which radioactive probe was hybridized to metaphase chromosomes. Investigators counted the number of grains caused by radioactive decay in each chromosomal region. The results of this localization procedure appear in the figure as *dots* above an idiogram of the human karyotype. The tip of the small arm of chromosome 4 has many more grains than any other chromosomal region. This is the location of the G8 marker and, by inference, the *HD* locus. (All the other grains represent background noise.)

random clones from cDNA libraries from other tissues, the gene is probably expressed uniquely in lung tissue.

A second way to determine gene expression patterns is to perform a Northern blot. Recall from Fig. 9.15 on pp. 325–326 that in Southern blotting, researchers separate DNA fragments by electrophoresis, blot the separated fragments to a special filter paper, expose the filter to radioactive probes that hybridize to complementary sequences in the blot, and finally, observe a band on an autoradiograph if the probe has hybridized to a stretch of DNA on the filter. By

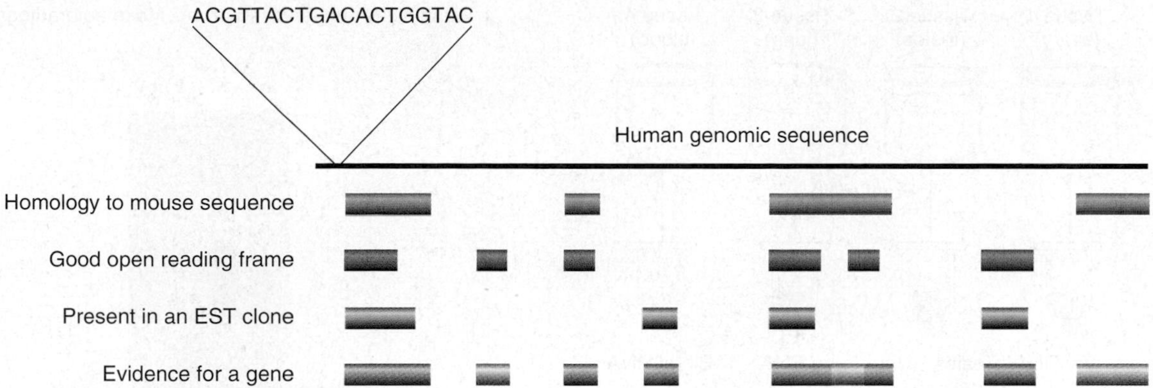

Figure 11.19 Computational analysis of genomic sequence can uncover genes. Genomic sequences can be analyzed computationally in two ways. First, each small region of the human genome can be compared with the whole mouse genome for the presence of homologous sequences. In this hypothetical example, human sequences that exhibit homology to mouse sequences are indicated in *blue*. Second, each region can be analyzed for the presence of an open reading frame containing codons typical of most human genes, bounded by splice-site sequences. Regions that pass this test are shown in *green.* Sequences that pass both tests are considered very likely to represent coding regions. Sequences that pass just one test are possible coding regions. Strong evidence for a transcribed sequence comes when it is found to be present in an EST clone (indicated in *purple.*) *Red* indicates sequences that are definitely or probably transcribed; *orange* shows possible coding regions.

comparison, in Northern blotting, it is the RNA transcripts in the cells of a particular tissue that are separated by gel electrophoresis. The ensuing blotting, probing, and autoradiography are the same as for the Southern blotting (**Fig. 11.20**).

It is possible to infer two types of information from Northern blot data. First, since RNA samples are not enzymatically treated before electrophoresis, the position of bands on the blot provides a direct measure of RNA size; and from the size of an RNA, you can estimate the transcript's coding capacity and thence the size of the protein it encodes. Second, Northern blot analysis of RNA samples from many different tissues enables one to determine which specific tissue a gene is expressed in as well as the relative levels of its expression in all cells where transcription is occurring.

Investigators can use Northern blots to identify the various transcription units in a large cloned genomic region. To accomplish this, they divide the genomic region into many smaller fragments and label each fragment for use as a Northern blot probe. Genomic fragments that do not include coding regions will not light up bands in any RNA sample; fragments from genes expressed in all tissues will light up a band in all RNA samples; and fragments from genes expressed only in specific tissues will light up a band in some RNA samples.

Figure 11.20 shows how Northern blot analysis supported the hypothesis that a particular Y-linked gene called *SRY* is responsible for testes differentiation. When a fragment of this conserved gene is used to probe a Northern blot with RNA samples from several different tissues, a 1.1 kb RNA transcript is found in the testes, but not in the ovaries, lungs, or kidneys. This finding makes the *SRY* gene an excellent candidate for testes-determining factor (TDF). However, there is one more requirement for clinching the case, which we describe in the next section.

Step 3 of Positional Cloning: Finding the Gene Responsible for the Phenotype

At this point in positional cloning, you have a set of candidate genes. They all map to the small region of the genome where the trait locus must lie, and they are transcribed in tissues consistent with your expectations based on the disease or developmental phenotype. How do you determine which one of these candidates is the gene responsible for the trait? Even if you have just one candidate, how do you confirm that it is the correct one and that you have not missed other candidate genes? You must compare groups of individuals with normal and abnormal phenotypes. If the DNA sequence or transcription of the candidate gene is consistently altered in all individuals with the abnormal phenotype, the evidence is very good that your candidate is in fact the gene responsible.

Expression Patterns. Many genetic diseases are caused by mutations that alter the pattern of gene expression in the mutant tissue. Mutations in regulatory regions can prevent transcription, and others at splice junctions can disrupt splicing and lead to aberrant mRNAs and polypeptide synthesis. Other diseases are caused by nonsense mutations in the coding region. These mutations allow the production of properly sized mRNAs but prevent formation of the polypeptide product. Thus, a first step in vetting a candidate gene is to compare both its RNA and polypeptide expression in individuals who express and do not express the trait. RNA expression can be assayed with a Northern blot or by PCR amplification of cDNA with primers specific for the candidate transcript. Polypeptide expression can be assayed with an antibody specific for the wild-type

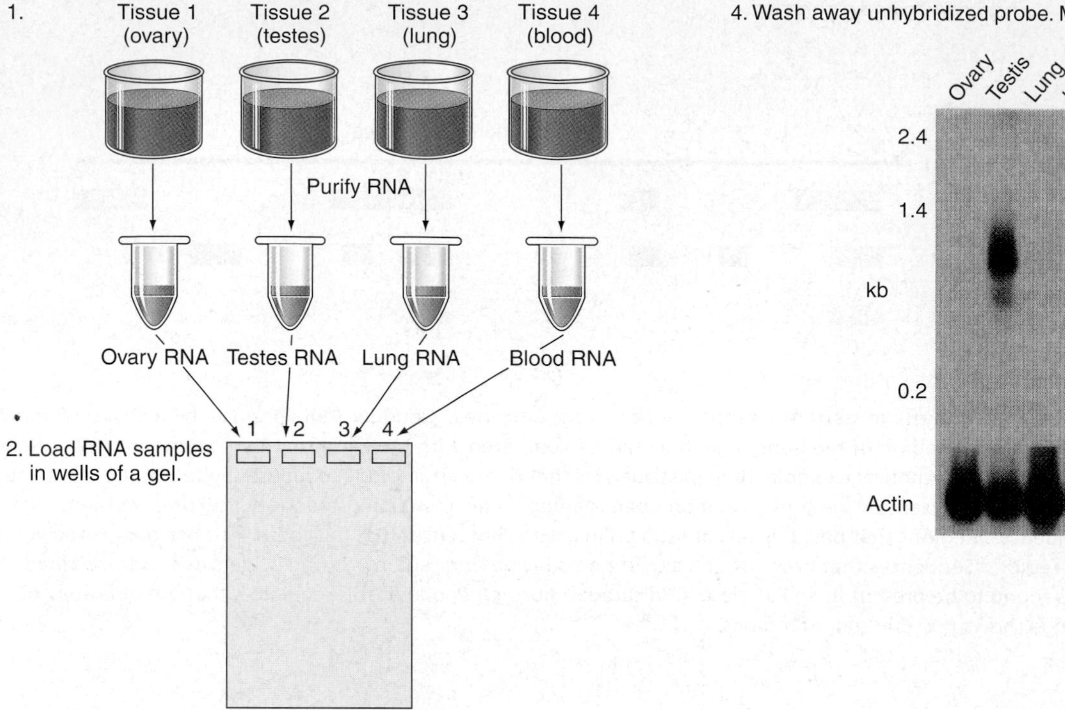

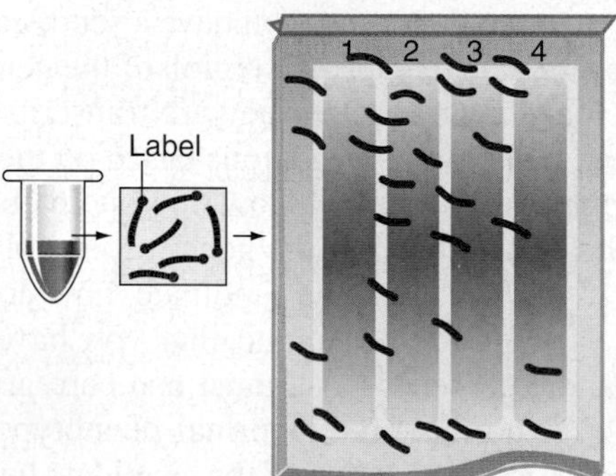

Figure 11.20 Northern blots: Snapshots of gene expression. **(1)** Purify RNA from each tissue to be examined for expression of the gene under investigation; here, you are looking at the *SRY* candidate for the testes-determining factor, so it is important to analyze testicular tissue as well as several control tissues (ovary, lung, and kidney in this experiment). **(2)** Make an agarose gel, load each of the four RNA samples into a different well. Now subject the gel to an electric current that causes the RNA in each sample to migrate along a lane toward the bottom of the gel. The mobility of each RNA transcript in a sample depends on its size: smaller RNAs move faster, while larger RNAs migrate more slowly. When the smallest RNAs reach the bottom of the gel, turn off the current. Staining the RNAs in each lane would produce a smear reflecting the presence of so many RNAs of different sizes that they cannot be resolved from each other. **(3)** Blot the RNA within the gel and fix it to a filter so that each RNA molecule retains its position relative to all the other molecules. Expose the filter to labeled probe and allow the label to hybridize for several hours. The labeled probe in this experiment contains both the *SRY* candidate sequence along with actin DNA. The *actin* gene is transcribed into a small RNA made constitutively by all tissues. **(4)** Wash away unhybridized probe. Place the filter on a film for autoradiography. Develop the film. You will see bands only in those lanes containing a tissue where the gene represented by the probe has been expressed. The presence of the actin band in each lane demonstrates the integrity of all four RNA samples. A second band of ~lkb is found only in the testis lane. The results show that *SRY* is expressed in the testes, but not the ovary, lung, or kidney. This result makes *SRY* a good candidate for the *TDF* locus.

protein. If a candidate gene is not expressed, is underexpressed, is overexpressed, or produces a smaller or larger transcript or polypeptide—or no transcript or polypeptide at all—in all phenotypically abnormal individuals, there is a good chance that it is responsible for the trait.

Sequence Differences. Some mutations, such as the one that causes sickle-cell anemia, produce their effect by substituting one amino acid in the polypeptide for another. The best way to detect subtle missense mutations of this type is by sequencing the coding region of the candidate gene from phenotypically normal and abnormal individuals. If all the phenotypically abnormal people have mutations in the candidate gene that are not present in the control group, this is powerful evidence that the particular candidate and the disease locus are one and the same.

The Ultimate Proof: Transgenic Modification of Phenotype. Incontrovertible evidence that a particular candidate gene is responsible for a particular phenotype must ultimately be obtained in an animal model where it is possible to induce a specific change in the gene and confirm that this change causes the predicted phenotype. The technology used to accomplish this goal is called **transgenic technology.**

The term **transgene** describes any piece of foreign DNA that researchers have inserted into an organism's genome through experimental manipulation of early-stage embryos or germ cells. An individual carrying a transgene is known as a *transgenic* plant or animal. In 1981, researchers developed a general method for transferring foreign DNA (from any organism or synthesized in the laboratory) into the germ line of any mammal. They accomplished the transfer by injecting cloned DNA molecules through very fine pipettes into the haploid egg or sperm nuclei (known as pronuclei) present in newly fertilized eggs (**Fig. 11.21**). They then placed the injected eggs into the uterus of a "foster" mother and allowed them to develop. Up to 50% of the mice born to these foster mothers had integrated the injected DNA into their chromosomes; these transgenic mice later passed the transgene on through their germ cells to their own offspring. Researchers have successfully inserted transgenes into the genomes of cows, pigs, goats, sheep, monkeys, and other mammals.

Transgenic technology was used to confirm the function of the candidate gene *SRY* as the testes-determining factor (*TDF*). Two to five copies of the complete *SRY* gene were injected into each of several fertilized eggs from a mating between two normal parents. The embryos were placed into foster mothers. Two to three weeks after birth, the mice that developed were checked by karyotypic analysis (to determine the combination of sex chromosomes, either XX or XY), by Southern blot analysis (to see whether the transgene had integrated into the genome), and by visual examination (to check for the presence of a male sex organ). The identification of several transgenic XX animals with a penis (Fig. 11.21) proved beyond a doubt that the *SRY* coding region is equivalent to the *TDF* gene.

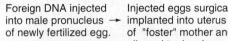

Transgenic analysis

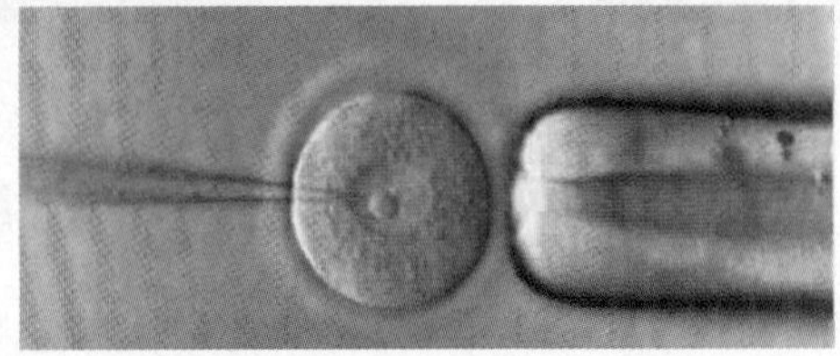

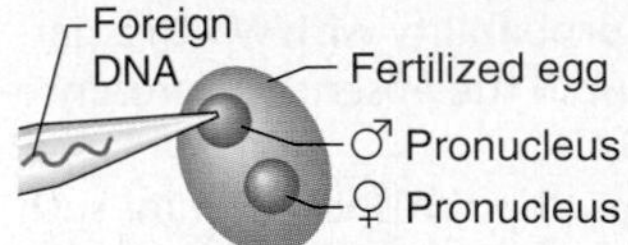
Foreign DNA
Fertilized egg
♂ Pronucleus
♀ Pronucleus
Foreign DNA injected into male pronucleus of newly fertilized egg.

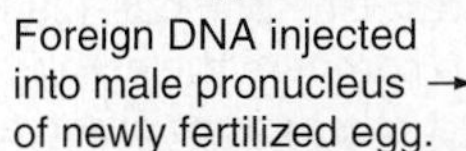
Injected eggs surgically implanted into uterus of "foster" mother and allowed to develop.
Mice are born with foreign DNA in every cell nucleus.

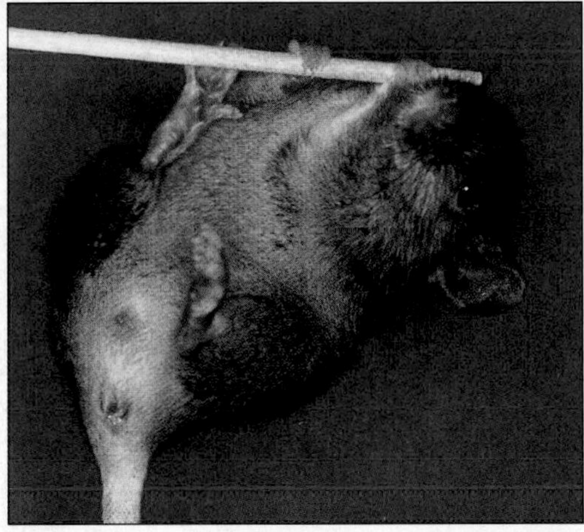

Figure 11.21 Transgenic analysis can prove the equivalence of a candidate gene and a trait locus. To demonstrate that *SRY* is the same as the *TDF* locus, researchers injected a clone containing the *SRY* gene into mouse embryos with two X chromosomes. The transgenic mice that resulted from this protocol had a penis. This observation proves that *SRY* is the equivalent of *TDF*.

Comprehensive Example: Positional Cloning of the Cystic Fibrosis Gene Leads to a Potential Therapy

Figure 11.22 on p. 418 reviews the steps and benefits of positional cloning as medical geneticists used it to identify the cystic fibrosis (*CF*) gene. In 1985, researchers uncovered linkage between the *CF* locus and a marker called *met,* which placed the *CF* gene in the middle of human chromosome 7.

TOOLS OF GENETICS

Using Human Pedigrees and LOD Scores to Calculate the Probability That Two Loci Are Linked

A geneticist cannot obtain a yes or no answer to the question of whether two loci are linked from pedigree data alone. All he or she can do is calculate the probability with which a particular set of data is likely to occur in the absence or presence of linkage between the two loci.

With experimental organisms, the statistical tool for such linkage analysis is the chi-square (χ^2) test (see pp. 128–129 in Chapter 5). This test makes it possible to compare the numbers of recombinant and nonrecombinant chromosomes observed in the offspring of a planned cross and transform the data into linkage probabilities. The (χ^2) test, however, is not applicable to human pedigree analysis. Because human matings are not planned experiments, linkage analysis in humans requires specialized statistical tools. We now describe those tools.

Consider a simple two-generation human pedigree following the unusual phenotype of six fingers on each hand in a family with two parents and eight children (**Fig. A**). The phenotype is caused by a completely penetrant, dominant mutation at a single locus known as *SF* (for *six fingers*). The father expresses the unusual phenotype, while the mother does not. The fact that some of the children do not have six fingers on each hand indicates that the father is a heterozygote for the *SF* mutation with an *SF* / + genotype. The mother, with only five fingers on each hand, is homozygous wild type (+ / +) at the *SF* locus. Like the father, the six-fingered offspring are *SF* / + heterozygotes; like the mother, the normal offspring are + / + homozygotes.

Suppose that you want to evaluate the possibility that the *SF* gene is linked to a polymorphic marker (*M*) with alleles 1 and 2. A single-base-pair difference, detectable by PCR-based protocols, distinguishes the two alleles. If PCR-based analysis demonstrates that the father is heterozygous *M1* / *M2* while the mother is homozygous *M1* / *M1* (as shown in the electrophoresis results depicted below the pedigree in Fig. A), the pedigree would represent a situation analogous to an experimental testcross in which one parent is heterozygous at both loci under analysis (*SF* / +; *M1* / *M2*); the other parent is homozygous at both loci (+ / +; *M1* / *M1*); and the segregation of alleles from each paternal locus to each offspring is distinguishable with absolute confidence.

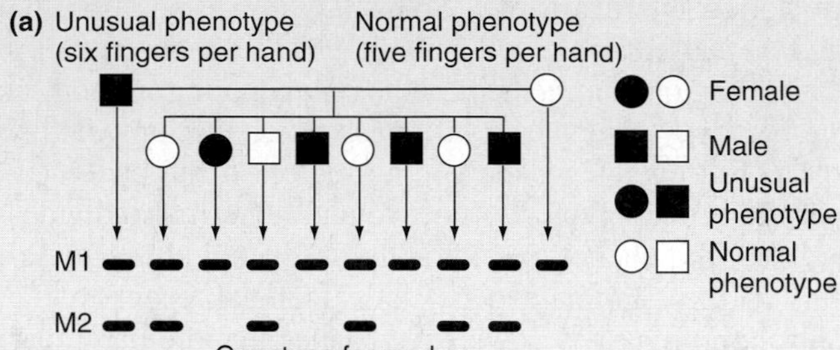

(b)

Odds value (ratios)	LOD score (*Z*)
0.01 (1:100)	-2
0.1 (1:10)	-1
1 (1:1)	0
10 (10:1)	1
100 (100:1)	2
1000 (1000:1)	3

Figure A Using human pedigrees and LOD scores. (a) Two-generation pedigree showing transmission of the hypothetical autosomal dominant phenotype of six fingers and alleles at the *M* marker. **(b)** Relationship between odds values and LOD scores.

If it turns out that the *SF* and the *M* loci are linked, it is possible that the father's mutant *SF* allele is on the same homolog as his *M2* allele, in which case his genotype would be $\frac{SF\leftrightarrow M2}{+\leftrightarrow M1}$, but it is equally possible that his *SF* allele is on the same homolog as his *M1* allele, in which case his genotype would be $\frac{SF\leftrightarrow M1}{+\leftrightarrow M2}$. Each of these two possible allele associations, or *haplotype configurations,* is referred to as a **phase of linkage.**

In experimental testcrosses, the phase of linkage is known at the outset because the haplotypes of the grandparents are known. But since human pedigrees are not "set up,"

Further linkage analysis carried out on a large number of families each having two or more members affected by CF showed that the *met* locus is about 2 cM from the *CF* gene. Next, the investigators identified two additional markers—*XV-2c* and *KM-19*—that in these families displayed no recombination with the disease gene and were thus much closer to it. They then cloned the genomic region between these two markers and identified four candidate genes within it (Fig. 11.22a).

As discussed earlier, children afflicted with cystic fibrosis have a variety of symptoms arising from abnormally viscous secretions in the lungs, pancreas, and sweat glands. Northern blot analyses of the four candidate genes revealed that only one, subsequently called *CFTR* (for cystic fibrosis transmembrane conductance regulator), was expressed in all of these tissues and no others (Fig. 11.22b). When medical geneticists examined the *CFTR* gene in many CF-affected people, they found that in every patient, small deletions or substitutions had destroyed *CFTR* function by either eliminating polypeptide production or disrupting protein structure and function (Fig. 11.22c and d).

the phase of linkage in a pedigree's first-generation parents is usually unknown. This makes it impossible to determine which inherited allelic combinations are recombinant and which are nonrecombinant. How then is it possible to use pedigree data to calculate the probability of linkage?

Geneticists Use Odds of Linkage and LOD Scores to Discover Evidence for or Against Linkage

With just a cursory look at the pedigree in Fig. A, it is clear that in seven out of eight children, the paternal *M2* allele has been inherited with the wild-type phenotype, the paternal *M1* allele with the six-finger phenotype. Intuitively, these results seem more likely if the *M* marker and the *SF* loci are linked than if they are not. (Don't be alarmed by the fact that the wild-type mother is homozygous for the *M1* allele. Since the marker and the disease locus can recombine, different individuals can have different haplotype associations of alleles at linked loci.)

How can you convert this hunch into a numerical estimate of the probability of linkage? The answer is with a statistical approach known as **LOD score analysis.** The "LOD" of this type of analysis is an acronym for "logarithm of the odds"; the odds are the likelihood that linkage exists relative to the likelihood that it does not. Although gamblers speak of odds in terms of ratios, such as 3:2 or 10:1, for mathematical calculations, it is convenient to convert these ratios, through division, to real numerical values. Thus, 3:2 = (3 divided by 2) = 1.5; 10:1 = (10 divided by 1) = 10. If the calculated odds value is greater than 1.0, the likelihood of linkage is greater than the likelihood of no linkage. With an odds value of less than 1.0, the opposite is true.

Using a probability analysis that we will not describe here, it is possible to calculate the odds value for the pedigree in Fig. A as 6.3, a result that supports the hunch of linkage. But, just because the odds favor linkage does not mean that this is the correct interpretation of the data. To be on the safe side, the human genetics community has set an odds value of 1000 (in other words, 1 chance in 1001 of no linkage) as the minimal cutoff for accepting odds calculations as evidence of linkage between two loci.

Because data from a single human pedigree rarely generate a sufficiently high odds value, it helps to obtain data from multiple *SF*-segregating families. But how many families does it take to make the numbers large enough, and how does one combine data from different families? You cannot simply add up the recombinant and nonrecombinant chromosomes (since we don't know with certainty which are which) across all families, as one would do for a testcross because the phase of linkage may be different in different families. Instead, you must use the product rule to combine probabilities derived from many independent families into a single grand probability for or against linkage.

To do this, you convert the odds for each family into a fraction in which the probability of linkage *P*(L) is the numerator and the probability of no linkage *P*(NL) is the denominator. The odds calculated for each family would then be *P*(L) / *P*(NL), and the combined odds calculated for, say, three independent families would be

$$\left[\text{Odds} = \frac{P(\text{L})}{P(\text{NL})} = \frac{P1(\text{L})}{P1(\text{NL})} \times \frac{P2(\text{L})}{P2(\text{NL})} \times \frac{P3(\text{L})}{P3(\text{NL})}\right]$$

If a geneticist obtained an odds value of 6.3 with a single pedigree, he or she would want to figure out how many additional families of similar size would have to be analyzed to reach the 1000 cutoff. If the geneticist assumes that the odds for each family will be ~ 6.3, he or she can multiply 6.3 × 6.3 × 6.3 . . . until crossing the 1000 threshold. But it is more convenient to work with experimental values that can be added—rather than multiplied together—to get a composite value. The mathematical trick for converting multiplication to addition consists in converting odds values into LOD scores through the common logarithmic function based on the following mathematical principle: $\log (A \times B \times C) = (\log A) + (\log B) + (\log C)$.

The table in Fig. A shows the LOD scores that correspond to six key odds values. The cutoff value of 1000 converts to a LOD score of 3.0. The critical feature of LOD scores is that values obtained from multiple families for the same two loci can be summed together to produce a single composite LOD score. As we see in the Fig. A table, the odds value of 6.3, obtained from the simple *SF* pedigree, converts to a LOD score of 0.8. It would take four families with a similar LOD score to push the composite score above the 3.0 cutoff: 0.8 + 0.8 + 0.8 + 0.8 = 3.2.

If the two loci under consideration are not linked, the assumption that additional families will have similar LOD scores would turn out to be wrong. Instead, there would be some negative LOD scores that when added to the initial positive number would give a negative total, supporting a hypothesis of no linkage.

The evidence pointing to the *CFTR* gene as the causative agent of cystic fibrosis is very strong because the disease affects so many people, all of whom show mutations in the gene. If a disease is not as prevalent, it may be difficult to accumulate enough evidence to be as confident that a suspected candidate gene is really the right one. To resolve this dilemma, scientists rely on the fact that most human genes have a homolog in the mouse genome. If the candidate gene is present in the mouse genome, they can create mutations of this gene in mice and see whether the mutations produce the same phenotype as that observed in disease-afflicted people.

We have seen that one method for altering the genomes of experimental animals is the injection of transgenic DNA into the pronucleus of the one-cell embryo. Unfortunately, the DNA injection method does not work to alter existing genes in the genome; it only adds new genes. But with cystic fibrosis and many other human diseases, the phenotype arises from mutations that eliminate gene function. How can scientists eliminate, or "knock out," the function of the homologous gene in mice?

The answer is through another kind of genome-altering technology called **targeted mutagenesis.** This technology

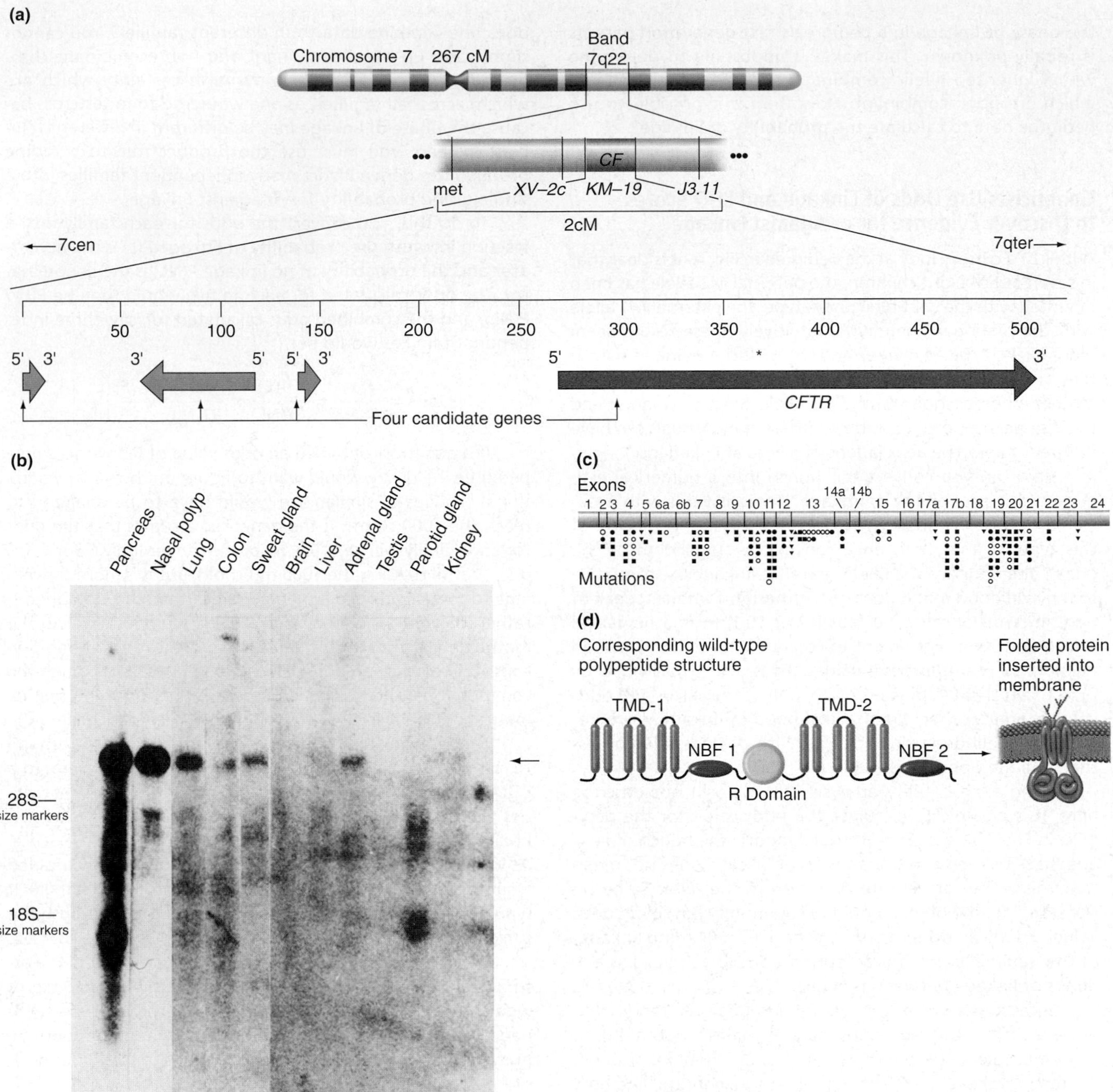

Figure 11.22 Positional cloning of the cystic fibrosis gene: A review. **(a)** Linkage analysis places the *CF* locus on chromosome 7. This idiogram of chromosome 7 shows the positions of markers—*met, XV-2c, KM-19,* and *J3.11*—that define the 2 cM region within which the *CF* locus must lie. Four candidate genes were identified within the region between *XV-2c* and *KM-19*. **(b)** Northern blot analysis reveals that only one of the candidate genes is expressed in the lungs and pancreas, tissues affected by cystic fibrosis (*arrow at right*). This candidate (now known as *CFTR*) is not expressed in the brain, liver, or testes, all tissues that are not affected in diseased individuals. **(c)** Every CF patient has a mutated allele of the *CFTR* gene on both chromosome 7 homologs. The locations of mutations uncovered at different positions within the *CFTR* gene are indicated under the diagram of the gene shown divided into its 24 exons. **(d)** The structure of the wild-type polypeptide schematically divided into its different domains. CFTR is a membrane protein. TMD-1 and TMD-2 are transmembrane domains that anchor the protein in the membrane. The other domains lie on the cytoplasmic side of the properly positioned protein and are involved in intracellular signaling.

is more technically demanding and complex than nuclear injection. Its application for the production of mutations in the mouse *CFTR* gene is described in detail in the genetic portrait of the mouse as an experimental organism on our website (www.mhhe.com/hartwell3). Here we provide a brief summary of the protocol. Essentially, researchers introduce DNA molecules containing a mutant nonfunctional copy of the mouse *CFTR* gene into normal mouse embryonic cells growing in culture. In rare cells (less than one in a million), a molecule containing the mutant sequence undergoes a double recombination event with the homologous wild-type gene within the cell's chromosomes. Sophisticated techniques of selection and screening can identify these rare cells, which contain a mutant *CFTR* gene in place of a wild-type copy. The mutant cells are introduced into normal mouse embryos, where they become incorporated into the germ line of the developing animals. Gametes created from such germ-line cells produce offspring referred to as **knockout mice** because they carry an induced mutation in a targeted gene. When animals heterozygous for the *CFTR* knockout were bred together, they produced progeny homozygous for the *CFTR* mutation. These offspring exhibited the disease symptoms characteristic of cystic fibrosis in people.

Medical scientists now use the ASO-based genotyping protocols to determine *CFTR* genotypes for families concerned about the disease (review Fig. 11.9). These techniques can be applied to adults (to see if they are carriers), to fetuses (to establish what the disease status of the newborn will be), or to preimplantation embryos (to select nonafflicted ones for initiating a pregnancy). Medical researchers also grow cultured cells containing specialized clones of the gene to synthesize the wild-type protein and learn how its absence or inactivation causes disease symptoms. The Fast Forward box on pp. 22–23 in Chapter 2 summarizes our current understanding of the CFTR polypeptide: It is a cell membrane protein that regulates the passage of chloride ions from and into both cells and the extracellular environment of tissues like the lungs.

This understanding of gene function suggests a potential therapy: inserting a copy of the wild-type *CFTR* gene into a small fraction of a CF patient's lung cells. Cells that express the inserted, wild-type gene may be able to produce enough functional protein to eliminate the buildup of fluids in the lungs, which is the leading cause of death from CF. Clinical trials to test the safety and efficacy of such a therapy are in progress in both mice and humans.

11.4 Genetic Dissection of Complex Traits

In humans, only a small fraction of disease traits follow the simple Mendelian pattern of single-gene inheritance seen in cystic fibrosis, Huntington disease, and hemophilia A. Most common characteristics of human appearance, such as height, skin color, the shape of the face, hair type (curly, wavy, or straight), and many essential measures of human physiology important to human health (including blood pressure, cholesterol level, basal metabolism, and susceptibility to infectious diseases) have a more complex pattern of inheritance.

We now review the chief causes of complex inheritance and describe the problems they pose for the linkage mapping and positional cloning protocols described so far for single-gene traits (**Table 11.2**). We then explain how researchers adapt the procedures for analyzing single-gene traits to the more difficult task of identifying, mapping, and characterizing the genes that contribute to complex traits.

TABLE 11.2 Complexities That Alter Traditional Mendelian Ratios

Category of Complexity	Problem in Observed Relationship Between Genotype and Phenotype	Changes Necessitated in Mapping Strategy
Incomplete penetrance	Disease genotype can occur in an individual who does not express the disease phenotype	Eliminate nondiseased individuals from analysis
Phenocopy	Disease phenotype can be expressed by an individual who does not have the disease genotype	Limit studies to families that show evidence for inheritance of the trait
Genetic heterogeneity	In different families, different disease genotypes are responsible for the same disease phenotype	Divide complete set of disease-transmitting families into subgroups (based on various parameters such as average age of onset) and perform linkage analysis separately on each subgroup
Polygenic determination	Mutant alleles at more than one locus influence expression of the disease phenotype in a single individual	Program computer to search for complex patterns of association between the disease trait and multiple DNA markers

How the Causes of Complex Inheritance Patterns Confound Linkage Mapping and Thus Positional Cloning

Incomplete Penetrance: When a Mutant Genotype Does Not Always Cause a Mutant Phenotype

All cancers have a genetic basis; that is, they are the result of mutations in genes that regulate cell proliferation (see Chapter 19 for details). Most of these mutations occur in somatic tissues and are *not* inherited. Cancers arising in this way are termed *sporadic* and account for 90% of all breast cancers. There is a correlation between sporadic breast cancer and several environmental factors, including alcohol consumption. By contrast, about 10% of women with breast cancer have inherited an allele that predisposes them to this condition, as suggested by the observation that their mothers and aunts have a higher than normal incidence of breast cancer as well. Medical investigators used a positional cloning protocol to map and clone the *BRCA1* (BReast CAncer 1) gene, one of several genes that can cause breast cancer. Significantly, only 66% of women who carry a mutant allele at the *BRCA1* locus develop breast cancer by the age of 55. As seen in the first pedigree in **Fig. 11.23,** it is possible for a mother to carry a mutant *BRCA1* allele and remain disease-free, while her daughter becomes afflicted with the disease. Thus, although the mutant *BRCA1* allele predisposes a woman to breast cancer, it does not guarantee that the disease phenotype will occur; that is, it is not completely penetrant. By comparison, a disease such as sickle-cell anemia, in which a mutant genotype always causes a mutant phenotype, is completely penetrant.

The causes of incomplete penetrance vary from trait to trait and from individual to individual. With breast cancer, it seems that chance plays the largest role in determining which predisposed individuals get the disease—through the accumulation of secondary somatic mutations. With heart disease, the individual's environment—especially diet and amount of exercise—plays a large role in determining whether a predisposing genotype results in a mutant phenotype, and if so, at what age.

Incomplete penetrance hampers linkage mapping and positional cloning for one main reason: Individuals who do not express a mutant phenotype may nevertheless carry a mutant genotype. The simplest solution to this problem is to exclude all nondiseased individuals from the analysis. With age-dependent traits like breast cancer and Huntington disease, such exclusion has meant that in disease-carrying families, the majority of children and adults under the age of 40 could not be included in the analysis. As a result, many more families were required for the studies that led to the mapping and cloning of the genes associated with both diseases.

Variable Expressivity: When Expression of a Mutant Trait Differs from Person to Person

Variation in gene expression may be in age of onset, phenotypic severity, or any other measurable parameter. Variable expressivity does not normally interfere with genetic analysis, because geneticists can use any degree of mutant phenotype as evidence for the presence of a mutant allele.

Phenocopy: When the Disease Phenotype Is Not Caused by Any Inherited Predisposing Mutation

The observation that 3% of women who do not carry a mutation at the *BRCA1* locus or have any family history of the disease still develop breast cancer by age 55 suggests that the

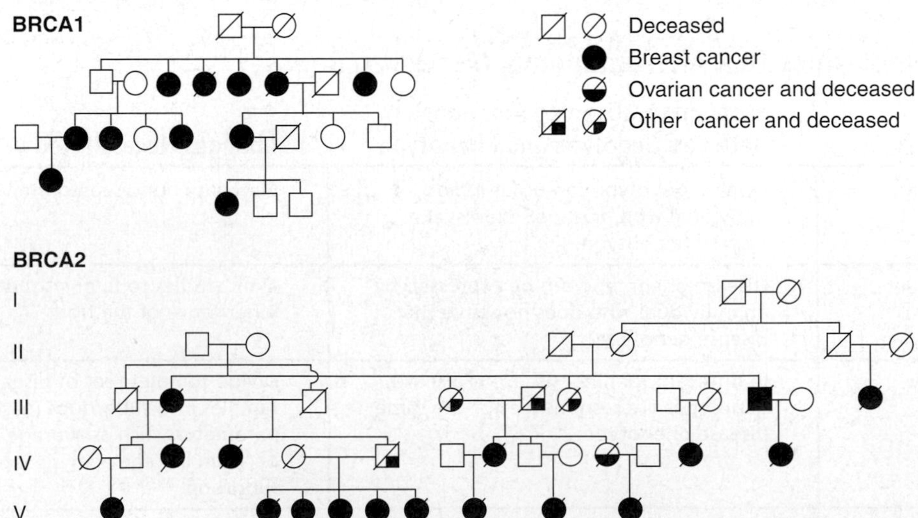

Figure 11.23 Incomplete penetrance and genetic heterogeneity in the inheritance of breast cancer. Both pedigrees in this figure show evidence of the transmission of a dominant mutation with incomplete penetrance that causes breast cancer. Linkage analysis shows that the mutation in the first pedigree resides on chromosome 17, whereas the mutation in the second pedigree is on chromosome 13. The first family is segregating a mutant *BRCA1* allele, while the second family is segregating a mutant *BRCA2* allele.

disease can arise entirely from one or more somatic mutations in the breast cells themselves. This form of the disease is considered a **phenocopy** because it is indistinguishable from the inherited form of the disease yet is not caused by an inherited mutant genotype. The percentage of women who develop phenocopy breast cancer rises to 8% by age 80.

We have seen that researchers focus on families with a history of disease to map predisposing alleles. If a small but significant fraction of women who develop the disease carry wild-type alleles, the correlation between the inheritance of the disease locus (or a locus-linked marker) and expression of the disease will diminish. Phenocopies thus make it more difficult to map disease-causing loci.

Genetic Heterogeneity: When Mutations at More Than One Locus Cause the Same Phenotype

Sometimes it is possible to use sophisticated diagnostic techniques to separate what appears to be a single disease into a set of related diseases caused by mutations in different genes. For example, researchers can distinguish insulin-dependent from insulin-independent diabetes on the basis of their different physiological origins. However, even when the limit of disease subdivision has been reached, what appears to be a homogeneous phenotype may still arise from genetic heterogeneity. The seemingly simple disease of thalassemia is a case in point. Mutations in either the α-globin or the β-globin gene can cause the same phenotype: severe reduction or elimination of the functional hemoglobin molecules produced in red blood cell precursors.

Genetic heterogeneity complicates attempts to map disease-causing loci in the following way. Although individual human families usually segregate only a single mutation responsible for a rare disease, most families do not have enough members to provide sufficient data for determining linkage (see the Tools of Genetics box "Using Human Pedigrees and LOD Scores to Calculate the Probability That Two Loci Are Linked" on pp. 416–417). For this reason, linkage studies in humans almost always combine data from multiple families. But if a disease is heterogeneous, a marker linked to the disease locus in one family may assort independently from a different disease locus in a second family. When data from the two families are combined, the calculated probability of linkage between the marker and the initial disease locus will drop below that obtained with the first family alone.

Genetic heterogeneity is suspected whenever a comprehensive analysis of many families, each with several affected members, fails to map a locus responsible for the disease trait. In such a case, investigators try to divide the complete set of disease-transmitting families into subsets—based on any of several phenotypic parameters—and then combine only the families in each subset for linkage analysis. For example, when researchers combined the data from a large set of breast-cancer-prone families, they found no evidence of linkage to any marker. But when they selected a subset of the families in which the average age of disease onset was less than 47 years, they obtained strong evidence for a disease locus on chromosome 17, named *BRCA1* (Fig. 11.23). It is now clear that mutations in *BRCA1* cause the onset of breast cancer at an earlier age than predisposing alleles at other loci. Classification of families into early-onset and late-onset groups may be helpful with any trait showing age-dependent expression. With traits for which classification by age of onset fails to produce evidence of linkage with a disease locus, classification by other variables, such as severity of the expressed phenotype, may prove helpful.

Once researchers have identified a first locus responsible for a disease, they can use DNA markers at that locus to determine whether it is responsible for the disease in other families with a disease history. A process of elimination may identify a subset of families that must inherit the disease because of a predisposing allele at a different locus or loci. This type of testing and elimination identified a group of families in which *BRCA1* could not be the locus predisposing women to breast cancer. The combined data from these non-*BRCA1* families revealed a second breast cancer locus, on chromosome 13, named *BRCA2* (Fig. 11.23). Mutations at *BRCA1* or *BRCA2* account for many inherited breast cancers, but not all. As of 2005, there was still at least one, and probably more, breast cancer loci yet to be discovered.

Polygenic Inheritance: When Two or More Genes Interact in the Expression of a Phenotype

So far, we have examined ways in which diseases caused by mutations in a single gene are associated with complex patterns of inheritance. But as we saw in Chapter 3, many traits arise from the interaction between two or more genes. Some such polygenic traits are discrete: They either show up or they don't. The occurrence of a heart attack, or myocardial infarction, is a discrete polygenic trait. Other polygenic traits are quantitative: They vary over a continuous range of measurement, from one extreme through the normal range to the opposite extreme. Blood sugar levels, cholesterol levels, and depression are examples of quantitative traits. Loci that influence the expression of such quantitative traits are known as **quantitative trait loci,** or **QTLs.** Although extreme values of QTL expression are considered abnormal, the border between normal and abnormal is arbitrary.

There is virtually an unlimited number of transmission patterns for polygenic traits. A completely penetrant discrete trait may require mutations at multiple loci to cause the abnormal phenotype. With other discrete polygenic traits, penetrance may increase as the number of mutant loci increases. With quantitative polygenic traits, the measured degree of expression (expressivity) may vary with the number of mutant loci present in the individual or with the degree to which different mutations at the same locus alter the level of polypeptide production.

Many other factors can complicate the analysis of polygenic traits. Some members of a set of interacting polygenic loci may make a disproportionately large (or small) contribution to the penetrance or expressivity of the trait. Mutations at some loci in a set may be recessive, while mutations at other loci are dominant or codominant. Some traits may arise from a mixture of polygenic and heterogeneous components. For example, one form of a disease may be caused by mutations at loci A, B, C, and D; a second form, by mutations at B and E; and a third form, by a single mutation at F. The more complex the inheritance pattern of a trait, the more difficult it is to identify the loci involved.

How Geneticists Move from Complex Traits to Sets of Contributing Loci

The mouse is a model mammalian organism in which genetic analysis is a powerful tool for analyzing traits too complex for dissection with human pedigrees. Many of the numerous genetically homogeneous inbred mouse lines express a variety of traits that correspond to similar traits in humans. The analysis of two contrasting genotypes represented by two parental inbred mouse strains is advantageous because it eliminates genetic heterogeneity. Since in a cross between two inbred strains, there are only two alternative alleles at every marker and disease locus, such a cross is the equivalent of a family pedigree documenting several hundred children all born to the same two parents. Another advantage of the mouse is that the possibility of maintaining all animals under identical conditions greatly reduces the impact of environmental differences.

Human geneticists are not the only researchers interested in the dissection of complex traits. Such analysis is also important in agriculture. For example, dairy and crop farmers are interested in milk production in cows and the taste, texture, and shape of fruits. Methods for dissecting complex traits are basically the same in all sexually reproducing experimental organisms. Here we examine the genetics of pulp content in tomatoes, but you could perform analogous experiments to look at complex traits in mice, fruit flies, or nematodes.

The pulp of a tomato is the succulent, edible solid part of the fruit—everything but the water, seeds, and skin. Manufacturers of canned tomatoes, tomato sauce, and tomato paste look for fruit with a high pulp content because it produces the highest product yield per weight of starting material (**Fig. 11.24a**). A strain of tomato plants with a maximum

Figure 11.24 Using linkage analysis to identify loci that contribute to the expression of complex traits, such as pulp content in tomatoes. (a) Tomatoes with an ideal amount of pulp (*right*) and with too little pulp (*left*). **(b)** Researchers use a two-generation cross to map loci that contribute to the amount of pulp in individual tomatoes. The first-generation cross is between parental lines that differ greatly in pulp content. The F_1 plants are all genetically identical, but their gametes contain recombinant chromosomes that assort to different offspring in the F_2 generation. The researchers correlate expression of the trait in the F_2 offspring with transmission of particular chromosomal regions. They then use statistical analysis to identify significant correlations between chromosomal regions and expression of the trait.

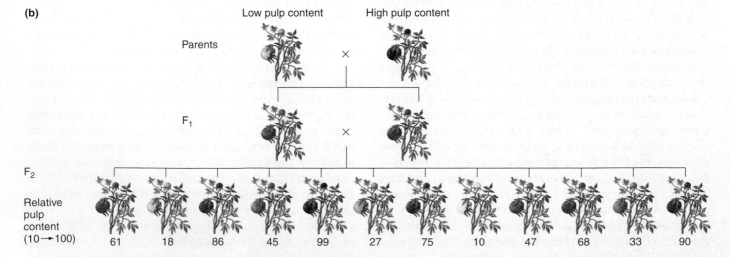

(a)

Too little pulp   Ideal pulp

(b)

Low pulp content   High pulp content

Parents   ×

F_1 ×

F_2

Relative pulp content (10→100)

61 18 86 45 99 27 75 10 47 68 33 90

of pulp content, however, might not have an ideal taste or texture. If you could identify DNA markers closely linked to loci that contribute to pulp content, you could cross an ideal pulp strain with other strains ideal for taste and texture and then follow the markers to identify F_2 plants that had received the ideal pulp alleles and were otherwise genetically the same as parents with the best taste and texture.

The first step in such an analysis is the identification of two inbred strains with extreme, reproducible differences in the expression of the trait of interest.

To find markers linked to loci responsible for high pulp content, you cross plants from the high- and low-pulp strains and then cross the identical F_1 hybrids with each other to generate several hundred F_2 offspring. Because of independent assortment and recombination in the F_1 hybrids, the F_2 generation will have a range of genotype combinations and thus exhibit a range of "pulp" phenotypes (Fig. 11.24b). Some plants will have a pulp content similar to one parent's; others will have a pulp content similar to the other parent's. With most plants, the pulp content will lie somewhere between the two parental extremes. You now determine the genotype at polymorphic DNA markers spaced evenly across the genome, at 20 cM intervals, in each F_2 plant, and analyze the data. In your analysis, you look for correlations between marker genotype and pulp phenotype.

For example, if a marker is closely linked to a semidominant gene with a large influence on pulp content, most of the F_2 plants with high pulp content will be homozygous at this marker for the allele from the high-pulp parental strain, and most of the F_2 plants with low pulp content will be homozygous at this marker for the allele from the low-pulp parental strain. In contrast, markers not linked to a pulp-content locus would show no correlation between marker genotype and plant pulp phenotype.

You could use this basic process to find all tomato loci with a significant effect on pulp content and then perform further analyses to understand the interactions between the various loci. When it is possible to map the genes contributing to every major trait of interest to tomato growers and processors, it will be possible, through further breeding and selection of individuals carrying the appropriate combination of marker alleles, to custom design a tomato plant possessing a combination of the most desirable traits.

Eventually, the techniques of positional cloning described earlier will be used to clone the desirable genes themselves. When this happens, agricultural researchers will be able to use the tools of recombinant DNA technology to improve upon available characteristics and then insert the relevant modified genes directly into the plant genome. By combining traditional genetic analysis, positional cloning, recombinant DNA technology, and transgenic technology, scientists hope to bring forth a new generation of crops that can grow in hostile environments with higher yields and higher nutritional value. If applied wisely, a compendium of genetic tools could help sustain the world's population while avoiding environmental degradation and maximizing the health of the biosphere.

11.5 Haplotype Association Studies for High-Resolution Mapping in Humans

Linkage analysis in humans is not as simple as it is in animals or plants for which researchers design specific crosses to generate large numbers of offspring from well-defined parents. In humans, traditional linkage analysis can be done in retrospect only with families that already exist. And because most human parents have few children, it is extremely difficult to generate high-resolution linkage maps, especially for diseases with complex inheritance patterns.

To overcome these limitations, geneticists developed a novel genetic approach to mapping disease loci that looks at the ancestry of chromosomes in present-day populations. This genetic approach is called *haplotype association analysis* or *linkage disequilibrium (LD) mapping*.

Haplotypes Are Sets of Closely Linked Alleles

A contraction of the phrase "haploid genotype," the term **haplotype** refers to a specific combination of two or more DNA marker alleles situated close together on the same DNA molecule, that is, on the same homolog. As SNPs are the most prominent class of polymorphic DNA markers, they are also the most frequent markers in haplotypes.

Haplotypes are distinguished by a series of closely linked mutations that accumulate one by one over time in the surviving descendants of an original ancestral chromosome (**Fig. 11.25**). The genetic distance over which the multiple loci of a haplotype extend must be short enough that the alleles are likely to remain associated with each other over many generations. In practice, haplotypes can extend in length from 1–100 kb. In humans, a 100 kb genomic distance is roughly equivalent to 0.1 cM, which means that the chance of such a haplotype being severed by a recombination event is 1 in a 1000. Recombination within a 1 kb haplotype will occur at the even lower frequency of 1 in 100,000 generations.

The formation of a SNP at any particular base pair is an extremely rare event (with a probability of $\sim 10^{-9}$ per gamete). The probability of sequentially forming multiple SNPs is the product of the probability of generating each individual SNP. Thus, a particular haplotype containing

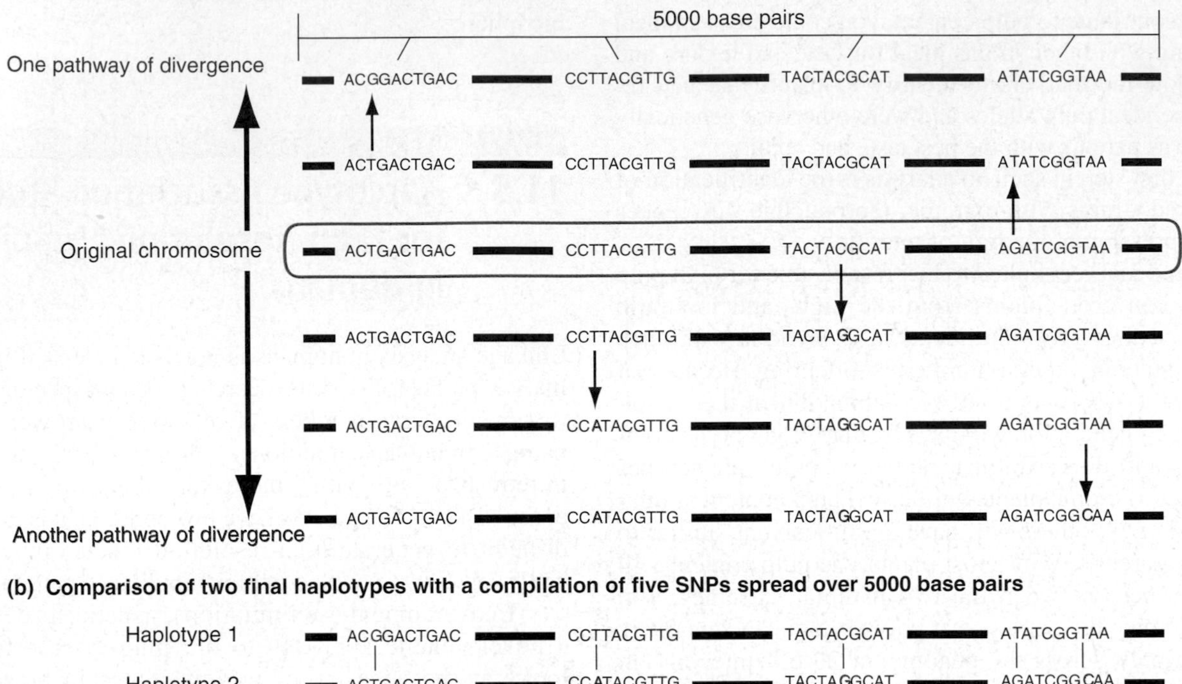

Figure 11.25 Haplotypes are formed by sequential mutations in a small genomic region. (a) Two pathways of divergence from an original 5 kb chromosomal region (*displayed in the middle of the two divergence pathways*). Along one pathway, two mutations accumulate (*blue*); along the second pathway, three mutations accumulate (*red*). **(b)** The surviving descendant chromosomes represent two distinct haplotypes that differ at five separate SNP loci.

four or more SNPs can arise only once in human history (the chance of generating the same haplotype a second time is less than 1 in 10^{36} for four SNPs, which is 10^{26} times greater than the whole human population). Clearly, all individuals who carry a particular haplotype composed of four or more loci must have inherited it from the same ancestor, who may have lived hundreds or even thousands of years ago. Haplotype analysis thus provides a powerful tool for identifying ancient genetic relationships among people who have no recent ancestor in common.

Ancient Disease Loci Are Associated with Particular Haplotypes

Haplotype association analysis is based on two facts. First, a particular disease phenotype in many people from the same population is often the result of a single mutational event that occurred in a long-forgotten common ancestor who lived and died centuries ago. Second, the disease mutation occurs in a haplotype of very closely linked SNPs that is passed down from generation to generation along with the disease mutation itself. While the mutation may be a single-base-change needle in the 10^{-9}-bp-genomic haystack, its associated haplotype can magnify the needle 100,000-fold. That is, instead of looking for a single base change, you can follow a number of nearby, closely associated polymorphisms spread over an area of 100 kb.

To perform haplotype association analysis, you collect DNA samples from a subgroup that expresses the disease of interest, such as schizophrenia, or obesity, and choose an equal number of DNA samples from random, unrelated people who do not exhibit the trait. Now you genotype each individual in both subgroups for haplotypes (groups of SNPs) throughout the entire genome. The vast majority of haplotypes will have no association with the locus responsible for the disease; the frequency of appearance of such haplotypes will be the same in the disease group and the test group. You hope to find a haplotype present at a significantly higher frequency in the disease group than in the control group (**Fig. 11.26**). Alleles at separate loci—in this case, haplotype marker alleles and the disease allele—that are associated with each other at a frequency significantly higher than that expected by chance, are said to be in **linkage disequilibrium.** A finding of linkage disequilibrium not only means that the two (or more) loci are linked to each other, it also means that the disease mutation arose on a chromosome that already carried the SNP alleles in the haplotype.

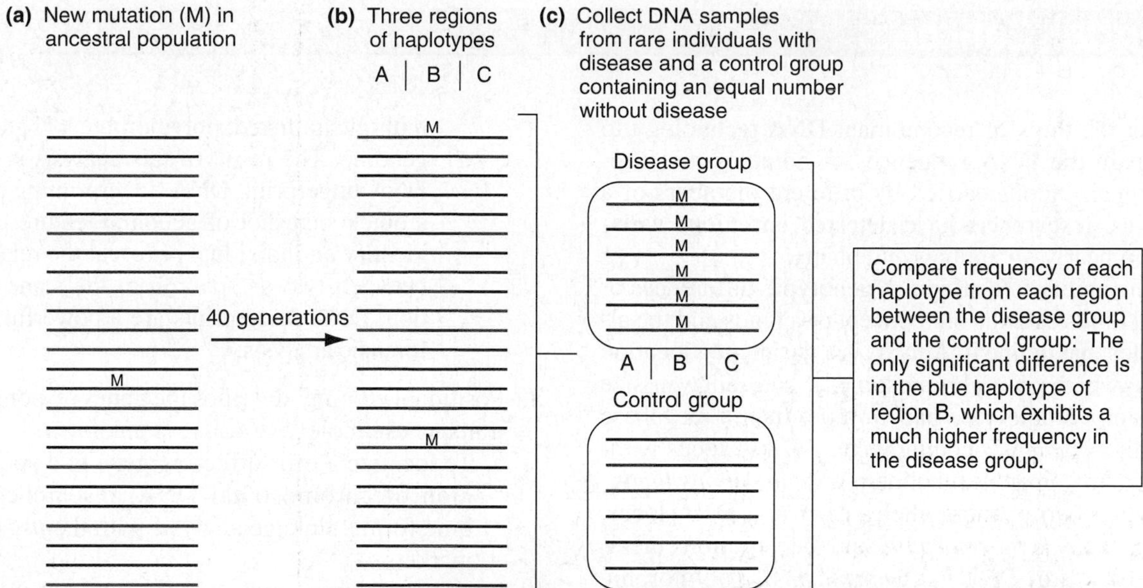

Figure 11.26 Haplotype association allows high-resolution gene mapping. **(a)** Representation of the same subchromosomal region in different individuals within an ancestral population that lived several thousand years ago. Lines of the same color have the same set of alleles because of common descent. The disease mutation (M) occurred on one ancestral chromosome. **(b)** After 40 generations, the original subchromosomal region has broken apart through recombination into three smaller regions (*A on the left, B in the middle, and C on the right*). Each smaller region still occurs in three different haplotypes. **(c)** You now collect DNA samples from a number of unrelated people with the disease and from an equal number of people without the disease and compare the frequency of each haplotype from regions A, B, and C in the disease and control groups. Genotyping of the samples will enable you to compare haplotype frequencies in the two groups. Haplotypes of regions not closely linked to the disease locus will occur with the same frequency in both groups. The only significant difference will be in the one haplotype that encompasses the disease locus (the *blue* haplotype of region B); it has a much higher frequency (here 100%) in the disease group than in the control group.

When genetic mapping by association works, it resolves much more than traditional pedigree analysis. This is because for many inherited diseases, recombination events over the numerous generations between the mutated ancestor and the current population have narrowed to 100 kb or less the DNA markers that remain part of the disease-associated haplotype. Most genomic regions of this size contain fewer than three candidate genes whose contribution to the disease can be evaluated by positional cloning.

Connections

In Chapter 9, we described the new tools of recombinant DNA technology that enable geneticists to look directly at the DNA sequence in individual chromosomes. In Chapter 10, we explained how high-speed automated machines incorporate these tools to generate huge amounts of sequence information; analysis of these data by sophisticated computer algorithms produces whole-genome sequence maps showing the locations of functional units, including the coding sequences of genes. In this chapter, we examined how researchers apply the tools of recombinant DNA technology and genomics to the direct analysis of genotype at the DNA level. Health professionals can use direct genotyping to diagnose hereditary disease; forensic experts can use it to determine the identity and degree of relatedness of DNA samples. Finally, researchers can use direct genotyping to identify and clone the genes responsible for any inherited trait that differs in its expression among individuals within a population. With these new approaches, the human species has become a superb system for genetic analysis.

Will this enormous capacity for obtaining genetic information provide a complete understanding of life? Far from it. Biologists are fully aware that a static list of genes cannot describe life. Rather, life is a dynamic system of molecular interactions and information processing. The study of these dynamic processes at the level of an organism or discrete biological systems is called "systems biology." In Chapter 12, we explore the experimental methods used by systems biologists and the insight into life provided by these approaches.

Essential Concepts

1. Using the tools of recombinant DNA technology to compare the DNA sequences of homologous chromosomal regions carried by different members of a species, researchers have detected enormous variation in nearly all animals and plants.
   a. The ability to distinguish genotypic differences of all kinds extends our concept of a locus and the alleles that define it. A *locus* is a designated location anywhere on a chromosome; it can range in size from a single base pair to more than a megabase pair. A *gene* is a coding locus. A DNA locus without any apparent function is an *anonymous locus.*
   b. When two or more alleles exist at a DNA locus, the locus is *polymorphic,* and the variations themselves are *DNA polymorphisms.* Polymorphic DNA loci that are useful for genetic studies are known as *DNA markers.*
   c. There are four classes of DNA polymorphisms: single nucleotide polymorphisms, or SNPs; microsatellites; minisatellites; and deletions, duplications, and insertions in nonrepeat loci (indels).
2. Once a DNA polymorphism at a particular locus has been uncovered, researchers can use a variety of methods to determine the genotype at that locus in any individual.
   a. There are three ways to detect SNPs.
      (1) A small proportion of SNPs, by chance, eliminate or create a restriction site. PCR provides a rapid, cost-effective method for their detection.
      (2) A general method for detecting SNPs exploits differences in hybridization between a probe and two allelic target sequences in the center of the region of hybridization that differ by a single nucleotide. Probes developed for this purpose are called allele-specific oligonucleotides, or ASOs.
      (3) Primer extension analysis takes advantage of the high fidelity with which DNA polymerase attaches complementary nucleotides to a growing DNA strand during DNA replication.
   b. Polymorphisms in microsatellite and minisatellite loci, and small deletions, duplications, and insertions in nonrepeat loci, change the size of a locus and are easy to detect by PCR.
      (1) Microsatellite variants are detectable in this way as are microsatellite-related disease loci like the one responsible for HD.
      (2) Minisatellite alleles can also be detected by PCR. Simultaneous Southern blot analysis of cross-hybridizing minisatellite loci can identify multiple unlinked, polymorphic loci around the genome. The result of this analysis is called a DNA fingerprint. DNA fingerprinting provides a quick snapshot of genomic features that can identify an individual or reveal the relationship between two samples from the same population. DNA fingerprints are a powerful tool for forensic analysis.
3. Positional cloning identifies the genes responsible for traits whose molecular cause is unknown.
   a. To localize a trait-affecting gene to a specific region of chromosomal DNA, researchers combine formal linkage analysis with the use of DNA markers.
   b. Once they have narrowed the location of the gene to a small enough genomic region, researchers catalogue all possible candidate genes that map to that region, using computational tools that search for coding sequences within genome databases and sequences that are conserved between distantly related species. They can also search for EST sequences in clones derived from cDNA libraries, and they can use Northern blots to identify gene sequences that are transcribed into RNAs in tissues affected by the trait.
   c. To identify the one candidate gene that is responsible for the trait, researchers compare groups of phenotypically normal and abnormal individuals. A finding that the DNA sequence or transcription of the candidate gene is altered in all individuals exhibiting the mutant trait is strong evidence that the candidate gene is responsible for the trait.
   d. Incontrovertible evidence that a candidate gene causes a particular phenotype is best obtained in an experimental animal model where it is possible to induce a specific genetic change and demonstrate the appearance of a predicted phenotype. Two technologies can accomplish this goal: nuclear injection of DNA into embryos to produce transgenic mice, and targeted mutagenesis.
4. Most common, genetically determined trait variation among individuals of a species is inherited in a complex manner.
   a. With incomplete penetrance, a mutant genotype does not always cause a mutant phenotype.
   b. Mutant traits that arise in the absence of a mutant genotype are considered *phenocopies.*
   c. Mutant traits caused by mutations at any one of two or more alternative loci show *genetic heterogeneity.*

d. A phenotype controlled by alleles at multiple loci is a *polygenic* trait.
e. If different combinations of alleles cause quantifiable differences in a trait, the trait is a *quantitative trait,* and the loci involved are *quantitative trait loci,* or *QTLs.*
f. The contemporary method of dissecting complex traits is basically the same in all sexually reproducing experimental organisms.

5. Haplotype association analysis is a novel genetic approach to mapping disease loci that is not dependent on the analysis of genetic transmission of alleles from parent to child.
   a. The term *haplotype* refers to a specific combination of alleles at two or more DNA markers situated close together on the same chromosomal homolog. All individuals who carry a particular haplotype must have inherited it from the same ancestor.
   b. Expression of a specific disease phenotype in many people from the same population is often the result of an ancient mutational event in a haplotype of very closely linked SNPs that has been passed intact from generation to generation.
   c. When alleles at separate loci are associated with each other at a significantly higher frequency than is expected by chance, they are in *linkage disequilibrium.* A finding of linkage disequilibrium is strong evidence that a locus lies in a small, defined chromosomal region.

On Our Website

www.mhhe.com/hartwell3
Chapter 11

Annotated Suggested Readings and Links to Other Websites

- Polymorphism analysis
- Sex determination in mice
- DNA fingerprinting on Dolly
- The SNP consortium
- The HapMap project

Specialized Topics

- Positional cloning of the cystic fibrosis gene

Social and Ethical Issues

1. (a) As a result of the new molecular genotyping and positional cloning tools described in this chapter, it has become possible to identify currently healthy individuals who are predisposed to—but may never express—a host of diseases and other deleterious traits. Insurance companies could use this information as the basis for charging people different rates according to the genotypes they've inherited. Most Americans are opposed to genetic discrimination by insurance companies, and several states have passed laws against the practice. The response from the insurance industry is that they have always used family histories—a weak form of genetic information—to set rates and that laws against the use of genetic information would require them to charge nonpredisposed people more than they would otherwise. Do you think insurance companies should be allowed to set rates according to genetic predisposition or resistance to a relevant trait?

   (b) One form of genetic discrimination is already practiced by all American auto insurance companies who charge hundreds of dollars more per year to insure a person between the ages of 18 and 25 who happens to carry the *SRY* gene. The presence of this particular gene predisposes a young person to engage in reckless driving and costly accidents, although most *SRY*-positive individuals drive safely. Should auto insurance discrimination based on *SRY* be allowed to continue? (*SRY,* of course, is the male-determining gene on the Y chromosome.)

2. Mary is a 25-year-old woman who has a good job with a company where she has room to advance in her career. When she accepted her job at the age of 22, she signed a contract that stipulated that she would tell her supervisor of any change that occurred in her health status. A footnote in the contract stated explicitly that a change in health status was defined not only as a symptomatic change but also as a prognostic change. This footnote was meant to include persons who tested positive for HIV even when they did not yet exhibit the symptoms of AIDS. It seemed clear to Mary that the purpose of this stipulation was to provide the company with grounds for dismissal. Three years after the

beginning of Mary's employment, Mary's 56-year-old mother Elizabeth was diagnosed with breast cancer. Because several other women in the family had breast cancer, there is a high probability that there is an inherited predisposition to breast cancer in the family. Tests indicated that her mother does carry a mutant allele of the *BRCA1* gene, and therefore, Mary has a 50% chance of having a *BRCA1* allele that predisposes her to breast cancer. The counselor encouraged Mary to take the test since there are preventative measures that can be taken if she finds out she is positive. But, if she takes the test and the result shows that she carries the *BRCA1* mutation, she believes that she will have to tell her boss and possibly lose her job. On the other hand, if the test shows that she doesn't have the mutation, a large load will be taken off her mind as she realizes her risk of disease is no more than that of other women without the *BRCA1* mutation. What would you do if you were Mary?

3. Both members of a married couple express the trait of achondroplasia, the dominant inherited form of dwarfism. (The parents are heterozygous for the achondroplasia allele. Homozygous individuals die before age one of severe skeletal abnormalities.) The parents want to have a baby and they have told the genetic counselor at the testing clinic that they will abort a fetus if it *does not* carry a mutant achondroplasia allele. The prospective parents are concerned that they would not be able to discipline a child who became, at a young age, much taller than they are. Who should decide whether it is appropriate for this couple to select a child with achondroplasia—the government, the physicians at the testing clinic, or the couple themselves? According to the 1973 Supreme Court decision in the case of Roe v. Wade, a woman has the right to obtain an abortion for any reason during the first two trimesters of pregnancy. Even if it is legal, do you think it is ethical to abort a fetus simply because it does *not* carry a particular allele that most members of our society consider to be a mutation?

4. Antoine and Naomi's first child has cystic fibrosis. They would like to have a second child but want to make sure the child does not have cystic fibrosis. After *in vitro* fertilization, seven embryos are screened for the presence of the disease. Of the seven, two are determined to have the disease. Prior to preimplantation, Antoine and Naomi asked the doctor to screen the remaining "healthy" five embryos for gender as they want to have only one more child, and they want that child to be a boy to carry on the family name. What would you tell the couple if you were the doctor? Is there a difference, in your opinion, between screening for a genetic disease and gender? If a genetic test for a phenotype exists, should the screening be available (e.g., if Antoine and Naomi asked to select an embryo carrying the genes to express blue eyes and blond hair, should they be able to do so)? What are some potential implications of selecting for alleles that do not have an impact on survival? What types of genetic screening, if any, should be available?

Solved Problems

I. The figure shows the pedigree of a family in which a completely penetrant, autosomal dominant disease is transmitted through two generations, together with a corresponding Southern blot with individual pedigree samples digested with *Eco*RI and probed with a DNA fragment that detects a restriction fragment length polymorphism (RFLP). Do the data suggest the existence of genetic linkage between the RFLP locus and the disease locus? If so, what is the estimated genetic distance between the two loci?

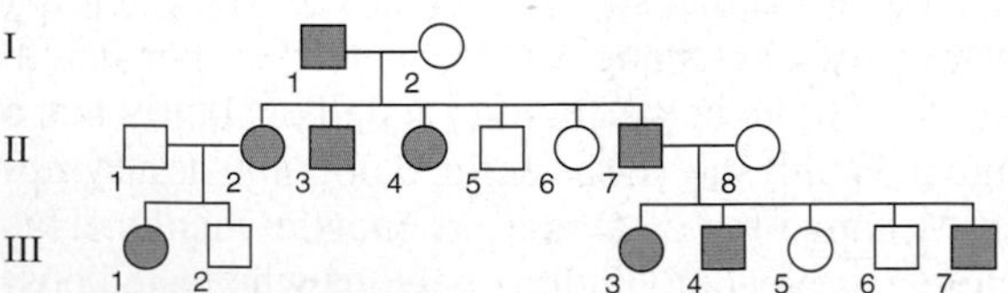

I1 I2 II1 II2 II3 II4 II5 II6 II7 II8 III1 III2 III3 III4 III5 III6 III7

8 kb

7 kb

Answer

To solve this problem, you need to understand how DNA polymorphisms can be followed through a pedigree and how they can be tested for linkage to a locus defined by phenotype alone. First, examine the Southern blot pattern to determine what the forms of the DNA polymorphism are. The two segregating DNA alleles in this pedigree are represented by RFLPs having sizes of 8 kb and 7 kb. Some individuals are heterozygous, carrying both restriction fragments; and some are homozygous, with just the 8 kb fragment or the 7 kb fragment alone.

When two parents have one DNA allele in common, but are different at the second allele, it is possible for a child to inherit the common allele from either or both parents. If the child is homozygous for the allele, then he or she must have received it from both parents. But, if the child has just one copy of the common allele, exclusion analysis can be used to determine which parent had to be the one that transmitted it. For example, children II-5, II-6, and II-7 have a 7 kb allele that could have come only from their father (I-1), because their mother doesn't carry this allele. By exclusion, their second allele—which is 8 kb—must have come from their mother, even though it is present in the genomes of both parents:

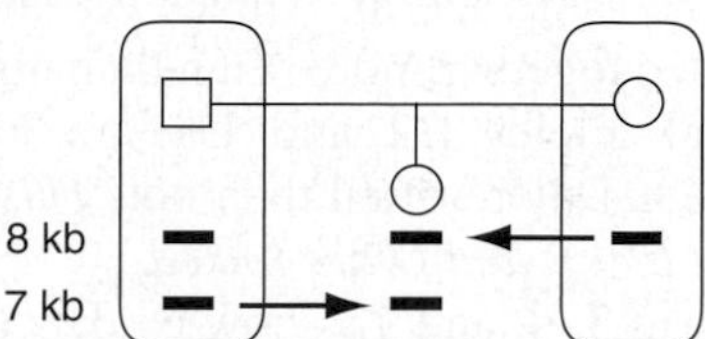

In the second generation set of siblings (II-2 through II-7), inheritance of the paternal 8 kb allele correlates with inheritance of the disease allele in five of six children: II-2, II-3, and II-4 inherit the paternal 8 kb allele along with the disease, and II-5 and II-6 do not receive the paternal 8 kb allele and do not exhibit the disease. The remaining child (II-7) exhibits the disease but inherited the 7 kb allele from his father. There are two possible explanations for this discrepant individual. First, the RFLP locus could be unlinked to the disease locus, and the six out of seven transmission correlation could be a chance event. Second, the loci could indeed be linked with the II-7 child representing a recombination event that brings the disease locus onto the same chromosome as the 7 kb allele.

To distinguish between these possibilities, you need to examine transmission to the third-generation children using the same logic but with different facts. First, in the third-generation family on the left, the diseased parent (II-2) is homozygous for the 8 kb RFLP allele, and thus, no useful data on linkage can be obtained from her children III-1 and III-2. But useful data can be obtained from the third-generation family on the right. If your hypothesis of linkage is correct, it should now be the 7 kb allele that is transmitted in correlation with the disease allele. Indeed, III-3, III-4, and III-7 receive a paternal 7 kb allele together with the disease allele, and III-5 and III-6 do not receive either the 7 kb allele or the disease allele.

When linkage data are combined from the entire pedigree, you find that there are 11 informative offspring. *In 10 of 11 cases, alleles at the disease locus show cotransmission with alleles at the RFLP locus.* This is evidence for linkage between the two loci. Further studies with additional families would be required to confirm this linkage.

II. A clear limitation to gene mapping in humans is that family sizes are small, so it is very difficult to collect enough data to get accurate recombination frequencies. A technique that circumvents this problem begins with the purification of DNA from single sperm cells. (Remember that recombination occurs during meiosis. Analysis of individual sperm in a large population can provide a large data set for linkage studies.) The DNA from single sperm cells can be used for SNP studies. Four pairs of primers were used for PCR amplification of four defined SNP loci from one man's somatic cells and from 21 single sperm that he provided for this research. Each of these primer pairs amplifies a different SNP locus referred to as A, B, C, and D. The four pairs of PCR primers were used simultaneously on each sample of DNA. Each of the amplified DNAs was divided into eight aliquots (identical subsamples), and these aliquots were denatured and spotted onto eight nitrocellulose membrane strips (vertically, as shown in the figure). Each of these strips was then hybridized with a different ASO (allele-specific oligonucleotide). There are two different ASOs for each SNP. For example, ASOs named A1 and A2 detect different alleles at SNP locus A. Black spots indicate that the amplified DNA hybridized to the ASO probe.

a. Based on the results shown, which SNP loci could be X-linked?
b. Which SNP loci could be on the Y chromosome?
c. Which SNP loci must be autosomal and homozygous?
d. Which SNP loci must be autosomal and heterozygous?
e. Do any SNP loci appear to be linked to each other?
f. Ignoring the results from sperm number 21, what is the distance between the two linked SNP loci?
g. How could you map the genomic region defined by SNP locus A?
h. What event could have given rise to sperm number 21?

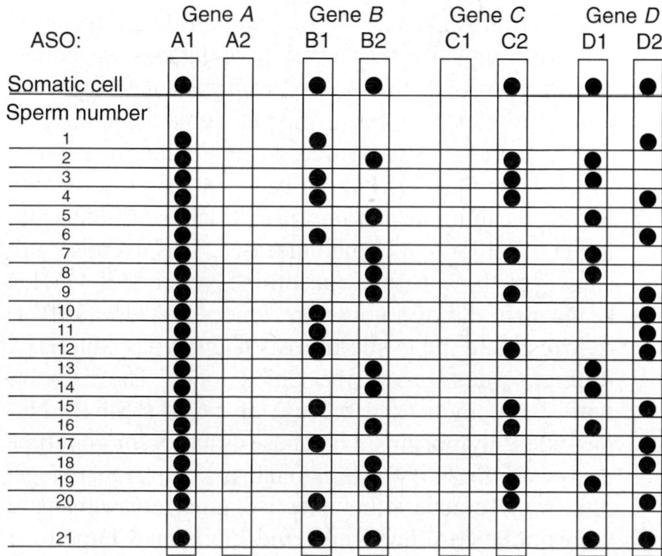

Answer

For this problem, you need to understand how ASOs detect SNP alleles and the advantages and limitations of SNPs as DNA marker loci. An advantage of SNP analysis by PCR is that the technique is so sensitive that the single alleles present within individual sperm cells can be assayed. An ASO result is either positive or negative. If the result is positive (as indicated by a black dot of hybridization in this example), a tested somatic cell sample can be either homozygous or heterozygous for the corresponding ASO allele; a positive ASO result by itself doesn't distinguish between these possibilities. If the result is negative, the sample doesn't contain the ASO allele under analysis, but nothing can be said about the alleles (if any) that are present at the tested locus.

a. Half of the sperm cells will not have an X chromosome and would not be expected to show a positive result with any ASO for any X-linked SNP locus. The other half of the sperm cell will carry the same SNP allele. *Gene C shows this type of pattern.*
b. Similarly, a gene on the Y chromosome would be found in only half the sperm. Again, *gene C is a candidate for a Y-linked gene.*
c. If an individual is homozygous at an autosomal SNP locus, all the sperm from that individual will show hybridization to one ASO for that locus and not any other. *Locus A appears to be homozygous and autosomal.*
d. At a heterozygous SNP locus, one ASO will hybridize to approximately one-half of the sperm samples, and a second ASO will hybridize to those samples that do not hybridize to the first ASO. *SNP loci B and D show this type of pattern.*
e. Alleles at linked loci will segregate together more than 50% of the time and would therefore end up in the same sperm. Alleles B1 and D2 are transmitted together more often than not, and the reciprocal alleles B2 and D1 are also transmitted together more often than not. *This result suggests that loci B and D are linked.*
f. Sperm 3, 9, and 18 show evidence of recombination between alleles at the B and D loci. Three out of 20, or 15%, are recombinant. *The distance between the B and D loci is therefore 15 cM.*
g. Since SNP locus A is homozygous in this individual, it can't be mapped in a linkage analysis. But, SNPs are found at an approximate rate of one in a thousand base pairs. *By sequencing several kilobases of genomic DNA around SNP locus A from this individual, you could identify a nearby SNP locus that is heterozygous and that could be mapped in a linkage analysis.*
h. *Sperm sample number 21 could accidentally have two sperm cells rather than one. It is also possible that a single sperm in this sample has accidently received two copies of the chromosome that carries loci B and D through meiotic nondisjunction.*

Problems

Interactive Web Exercise

The National Center for Biotechnology Information (NCBI) brings together a variety of database resources providing information on the occurrence and impact of mutant alleles that cause mutant human phenotypes. The oldest of these resources is the Online Inheritance in Man (or OMIM), which came into existence long before the human genome project as a printed catalogue of all known human diseases. This exercise will show you how to use the online version of OMIM (www.ncbi.nlm.nih.gov/entrez/query.fcgi?db=OMIM) to investigate the cystic fibrosis phenotypes caused by different mutations at the CFTR locus. The exercise will also demonstrate the use of NCBI's Single Nucleotide Polymorphism database as a tool for genotype analysis (http://www.ncbi.nlm.nih.gov/entrez/query.fcgi?db=Snp). To access this exercise, go to our website at www.mhhe.com/hartwell3 and click on Chapter 11 then Interactive Web Exercise.

Vocabulary

1. Choose the phrase from the right column that best fits the term in the left column.

a. DNA polymorphism	1. DNA element composed of tandemly repeated identical sequences
b. haplotype	2. two different nucleotides appear at the same position in genomic DNA from different individuals
c. RFLP	3. combination of alleles at multiple tightly linked loci that are transmitted together over many generations
d. ASO	4. location on a chromosome
e. SNP	5. a DNA sequence that occurs in two or more variant forms
f. DNA fingerprinting	6. a short oligonucleotide probe that will hybridize to only one allele at a chosen SNP locus
g. minisatellite	7. detection of genotype at a number of unlinked highly polymorphic loci using one probe
h. locus	8. variation in the length of a restriction fragment detected by a particular probe due to nucleotide changes at a restriction site

Section 11.1

2. What advantages do anonymous DNA markers afford for genetic mapping as opposed to traditional allelic markers associated with visible phenotypes? What are the disadvantages of anonymous DNA markers for mapping?

3. Would you characterize the pattern of inheritance of anonymous DNA polymorphisms as recessive, dominant, incompletely dominant, or codominant?

4. Would you be more likely to find SNPs in the protein-coding or in the non-coding DNA of the human genome?

5. Mutations at microsatellite loci occur at a frequency of 1×10^{-3}, which is much higher than the rate of base substitutions at other loci.

a. What is the nature of microsatellite polymorphisms?

b. By what mechanism are these polymorphisms generated?

c. Minisatellites also mutate at a relatively high frequency. Do these mutations occur by the same or a different mechanism?

6. If you were comparing two closely-related but non-identical gene sequences from different individuals of the same species, how would you distinguish whether these sequences represented polymorphisms of a single gene or two different paralogous genes?

Section 11.2

7. Each of the following reagents can be used to detect SNP polymorphisms. Where is the polymorphism located in relation to the probe or primer DNA sequence used in each of these techniques?

a. allele-specific oligonucleotide (ASO)

b. primer-extension oligonucleotide

c. RFLP probe

8. An 18 bp deletion in the *PAX-3* gene causes Waardenburg syndrome (an autosomal dominant condition that is responsible for a small percentage of deafness in humans). What features of this mutation make it amenable to molecular analyses that could not be applied to detection of the mutation in the β-globlin gene responsible for sickle cell anemia?

9. Given the two allelic sequences shown here and the site at which a single-base polymorphism occurs (underlined in the sequence), what sequences would you use as oligonucleotide probes for ASO analysis of genotype by hybridization? (Assume that ASO probes are usually 19 bp in length.)

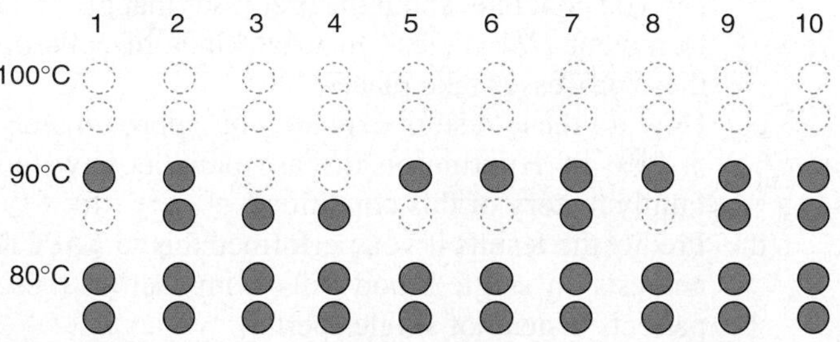


```
allele 1:
5′ GGCATTGCATGCTAACCCTATAAATGCGCTAGGCGTAGTTAGCTGGGAA
   TAAAAAGCT 3′
allele 2:
5′ GGCATTGCATGCTAACCCTATAAATGGGCTAGGCGTAGTTAGCTGGGAA
   TAAAAAGCT 3′
```

10. The ASO technique was used to determine the genotypes of 10 family members with regard to sickle-cell anemia, as shown here. Each pair of dots represents the results of ASO analysis for the DNA from one person. The upper row represents hybridization with the normal oligonucleotide, and the lower row represents the results of hybridization using the mutant oligonucleotide. The three replications of the assay were incubated at 100°C (upper set), 90°C (middle set), and 80°C (lower set).

a. Why do the three replications of the same sample set look different?

b. What are the genotypes of the individuals?

11. Angela and George have one child, and Angela has sickle-cell anemia. They want to have more children but do not want any of them to suffer from this disease. They also do not want to be in a position of having to abort a fetus, so they elect to have *in vitro* fertilization and embryo screening. Briefly list the steps they must take to accomplish this goal.

12. DNA fingerprinting can be used to settle cases concerning paternity. In the DNA fingerprint shown, the mother's DNA sample is in lane 1, the daughter is in lane 2, and two samples of men that could be the father are in lanes in 3 and 4. Can you determine from these data if one of the men must be the father?

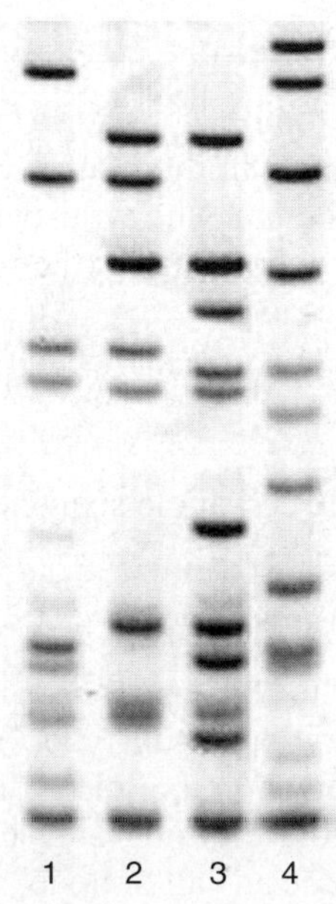

13. The police discovered the body of a woman who had been brutally beaten and raped while working late in her office one evening. They suspected that one of her seven coworkers might be responsible for the crime. To test this possibility, they recovered semen from her vagina and used it to prepare a DNA sample. They also took DNA samples from each of her coworkers. All eight samples were subjected to DNA fingerprinting analysis based on restriction digestion, gel electrophoresis, Southern blotting, and probing with a minisatellite. The results are shown in the photo shown at the top of p. 432.

Is any one of her coworkers likely to be the perpetrator of the crime? Which one? Estimate how likely it is that this particular person is the perpetrator rather than another person unknown to the authorities.

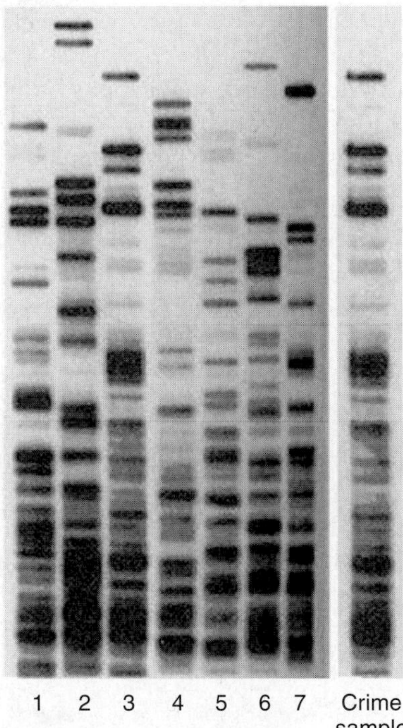

14. Individuals homozygous for a point mutation, changing an A to a T, in the human β-globin gene develop sickle cell anemia. The wild-type gene sequence over this region is shown here (the top strand is the RNA-like coding strand):

```
5′ ATGGTGCACCTGACTCCTGAGGAG 3′
3′ TACCACGTGGACTGAGGACTCCTC 5′
```

and the sickle cell allele sequence is:

```
5′ ATGGTGCACCTGACTCCTGTGGAG 3′
3′ TACCACGTGGACTGAGGACACCTC 5′
```

Design a PCR-based strategy to distinguish the DNA from homozygous wild-type, heterozygous, and homozygous sickle cell allele individuals. Your strategy should exploit the single A to T transversion and should produce an "all or none" response in which amplification will occur with particular sets of primers while no amplification will occur with other sets of primers.

15. The trinucleotide repeat region of the Huntington disease (*HD*) locus in six individuals is amplified by PCR and analyzed by gel electrophoresis as shown in the following figure; the numbers to the right indicate the sizes of the PCR products in bp. Each person whose DNA was analyzed has one affected parent.

a. Which individuals are most likely to be affected by Huntington disease, and in which of these people is the onset of the disease likely to be earliest?
b. Which individuals are least likely to be affected by the disease?
c. Consider the two PCR primers used to amplify the trinucleotide repeat region. If the 5′-end of one of these primers is located 70 nucleotides upstream of the first CAG repeat, what is the maximum distance downstream of the last CAG repeat at which the 5′-end of the other primer could be found?

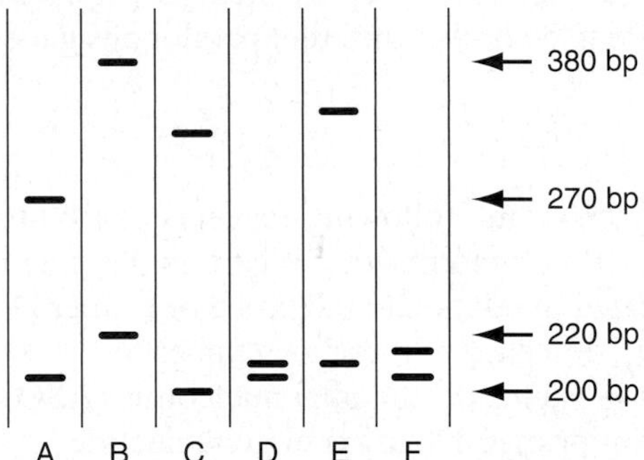

16. Sperm samples were taken from two men just beginning to show the effects of Huntington disease. Individual sperm from these samples were analyzed by PCR for the length of the trinucleotide repeat region in the *HD* gene. In the graphs that follow, the horizontal axes represent the number of CAG repeats in each sperm, and the vertical axes represent the fraction of total sperm of a particular size. The first graph shows the results for a man whose mutant *HD* allele (as measured in somatic cells) contained 62 CAG repeats; the man whose sperm were analyzed in the second graph had a mutant *HD* allele with 48 repeats.

a. What is the approximate CAG repeat number in the HD^+ alleles from both patients?
b. Assuming that these results indicate a trend, what can you conclude about the processes that give rise to mutant *HD* alleles? In what kinds of cells do these processes take place?
c. How do these results explain why approximately 5–10% of Huntington disease patients have no family history of this condition?
d. Predict the results if you performed this same PCR analysis on single blood cells from each of these patients instead of single sperm.

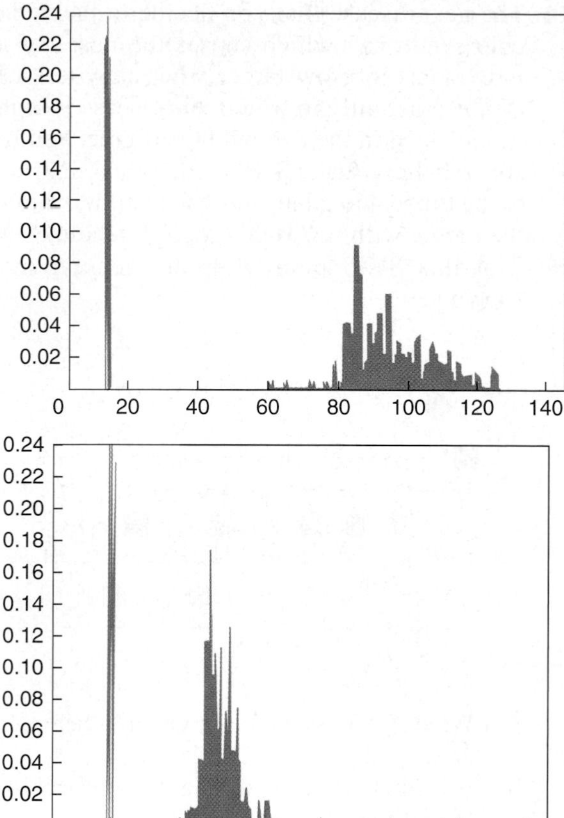

17. A relatively frequent, completely penetrant recessive disease known as the foul mouth syndrome (FM) has been found to be due to a variety of mutations in the *FM* gene, which has recently been cloned. Analysis of Southern blots of human DNA cleaved with the enzyme *Hpa*I and probed with a radioactively labeled fragment of the *FM* gene has revealed that *Hpa*I does not cleave within the gene itself. However, the positions of *Hpa*I sites surrounding the gene vary among individuals, producing at least three RFLPs in the population: the sizes of the RFLP alleles are 13, 10, and 7 kb. Shown here are two small pedigrees of families in which individuals with the disease are shaded in black. Below each pedigree symbol is the corresponding result obtained from the DNA sample in a Southern blot analysis with the FM probe.

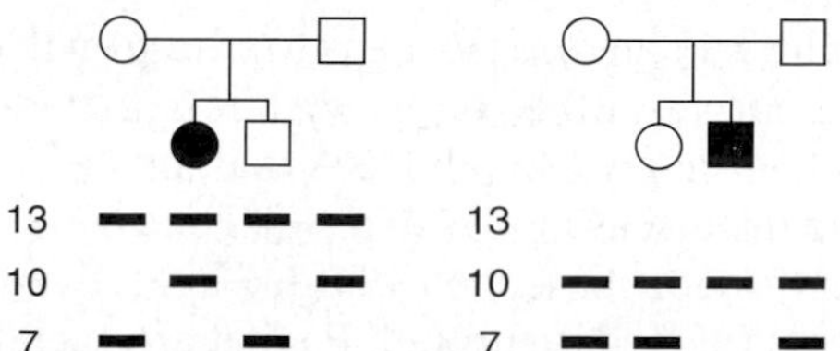

a. Which restriction fragment is associated with the disease mutation in the father shown in the left pedigree?

b. Which restriction fragment is associated with the disease mutation in the mother shown in the right pedigree?

c. If the male child from the left pedigree marries the female child from the right pedigree, what is the probability that their child will be diseased?

d. What is the probability their child will be a carrier?

18. The recessive disease cystic fibrosis displays extensive *allelic heterogeneity:* more than 800 different mutations of the CFTR gene have been shown to be associated with cystic fibrosis worldwide. Approximately 1 in 35 Americans is a carrier for some mutant allele of CFTR. One of these alleles, a deletion of three nucleotides that results in the loss of a single phenylalanine from the encoded protein, accounts for approximately 70% of the mutant alleles in populations of western European descent. With these facts in mind, is it feasible and worthwhile to mount a nationwide screening program for cystic fibrosis, and if so, how should this screening program be conducted?

Section 11.3

19. When a researcher begins to choose DNA markers for linkage analysis of a disease trait in a particular family, what are the first criteria used in this choice?

20. Imagine that you have identified a SNP marker that lies 1 cM away from a locus causing a rare hereditary autosomal dominant disease. You test additional nearby markers and find one that shows no recombination with the disease locus in the one large family that you have used for your linkage analysis. Furthermore, you discover that all afflicted individuals have a G base at this SNP on their mutant chromosomes, while all wild-type chromosomes have a T base at this SNP. You would like to think that you have discovered the disease locus and the causative mutation but realize you need to consider other possibilities.

a. What is another possible interpretation of the results?

b. How would you go about obtaining additional genetic information that could support or eliminate your hypothesis that the base-pair difference is responsible for the disease?

21. Approximately 3% of the population carries a mutant allele at the *CFTR* gene responsible for cystic fibrosis. New disease-causing mutations at this locus arise at a frequency of 1 in 10^4 gametes. A genetic counselor is examining a family in which both parents are known to be carriers for a *CFTR* mutation. Their first child was born with the disease, and the parents have come to the counselor to assess whether the new fetus inside the mother is also diseased, is a carrier, or is completely wild type at the *CF* locus. DNA samples from each family member and the fetus are tested by PCR and gel electrophoresis for a microsatellite marker within one of the *CFTR* gene's introns. The following results are obtained:

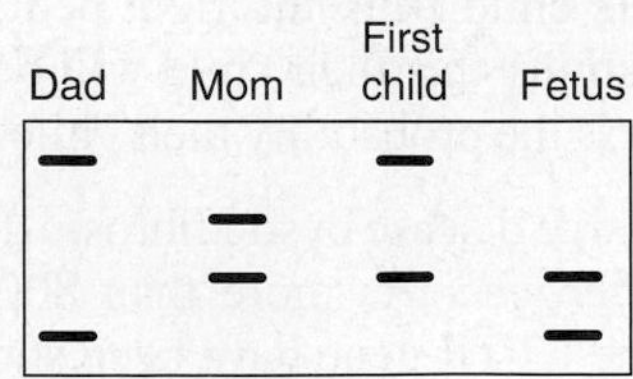

a. What is the probability that the child who will develop from this fetus will exhibit the disease?
b. If this child grows up and gets married, what is the probability that one of her children will be afflicted with the disease?

22. The pedigrees indicated here were obtained with three unrelated families whose members express the same completely penetrant disease caused by a dominant mutation that is linked at a distance of 10 cM from a marker locus with three alleles numbered 1, 2, and 3. The marker alleles present within each live genotype are indicated below the pedigree symbol. The phenotypes of the newly born labeled individuals—A, B, C, and D—are unknown. What is the probability of disease expression in each of these individuals?

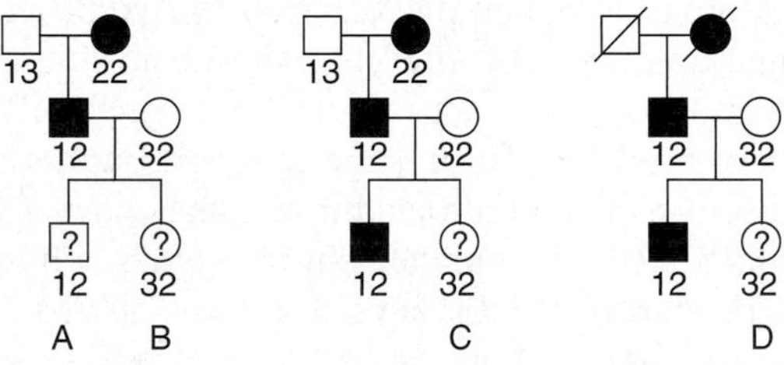

23. One of the difficulties faced by human geneticists is that matings are not performed with a scientific goal in mind, so pedigrees may not always provide desired information. As an example, consider the following matings (W, X, Y, or Z). Which of these matings are informative and which non-informative for testing linkage between anonymous loci A and B? (A1 and A2 are different alleles of locus A, B1 and B2 are different alleles of locus B, etc.) Explain your answer for each mating.

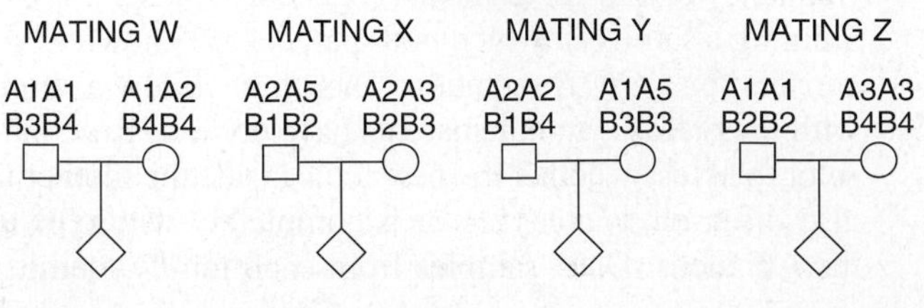

24. The next disease that you decide to tackle is Pinocchio syndrome, which causes the noses of afflicted individuals to grow larger when they tell a lie. You discover a family in which this disease is segregating as indicated in the following pedigree. You have reason to believe that a SNP locus called SNP1 is linked to the Pinocchio gene, and you test each individual in the family with ASOs that recognize allele 1 or allele 2 at this SNP locus with the genotyping results shown here.

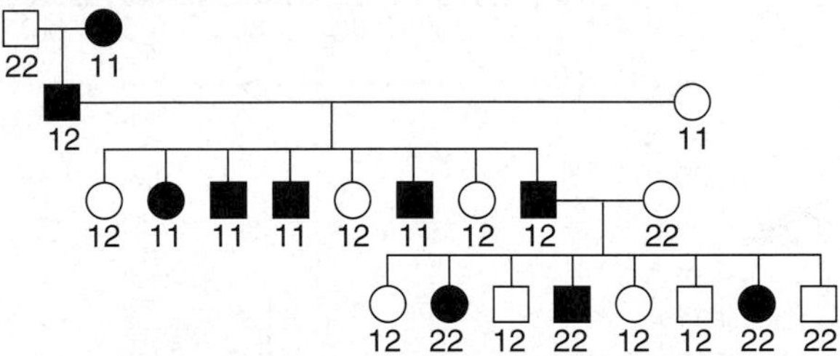

a. What is the most likely genetic basis for Pinocchio syndrome?
b. Is it likely or unlikely that the Pinocchio locus is linked to the SNP1 locus?
c. Sequence analysis shows the SNP1 locus actually resides in the middle of a coding region. How likely is it that this coding region is equivalent to the Pinocchio locus?

25. List three independent conceptual approaches for finding genes within a large cloned and sequenced genomic region.

26. Mice can be genetically engineered to express a hereditary disease that afflicts people. Strains of these mutant mice can provide biomedical scientists with a model for testing possible therapies.

a. What is the best strategy for creating a mouse model for a particular form of hypercholesterolemia that results from a mutant human gene that overexpresses a cholesterol-forming enzyme?
b. What is the best strategy for creating a mouse model for hemophilia?

27. A rare human disease leads to overgrowth of the heart without any other effect on the afflicted individual. Linkage analysis with DNA markers has been used to map the disease locus to a small chromosomal region. This region has been divided into five DNA fragments (named A through E) that are each labeled and used to probe Northern blots containing a single lane of RNA from one of three tissues—liver, heart, and muscle—taken from nondiseased cadavers. The results are shown here.

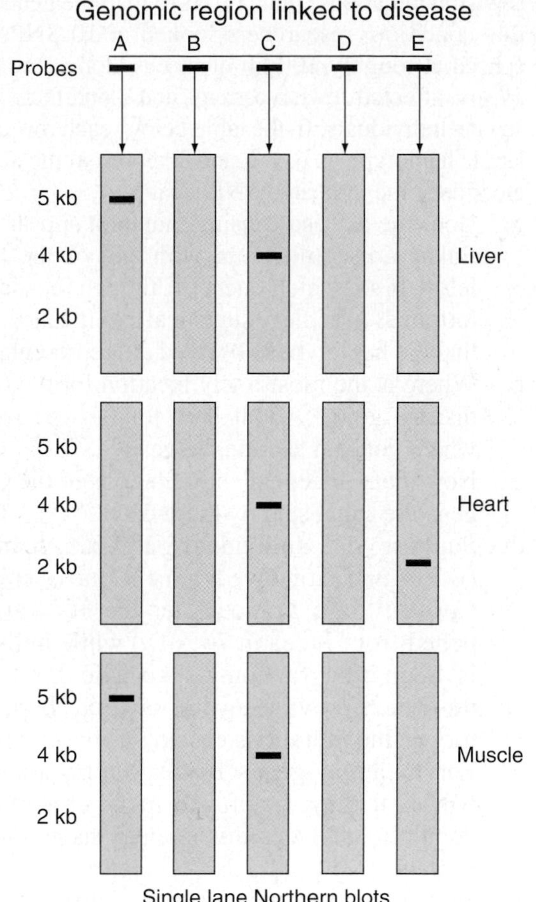

Single lane Northern blots

a. Which of the five DNA fragments is likely to contain a gene?
b. How many genes have been identified in this region by Northern blot analysis?
c. Is it possible that there are more genes in this region than those detected here?
d. Which of the five DNA fragments could possibly contain the gene responsible for this disease?
e. Which fragment is most likely to contain the gene?
f. What is your next step in testing this candidate gene as the causative factor for the disease?

Section 11.4

28. You have decided to study another disease trait that is very rare. You have searched far and wide to come up with an extended family in which a number of nonsibling individuals express the disease. The pedigree is as follows:

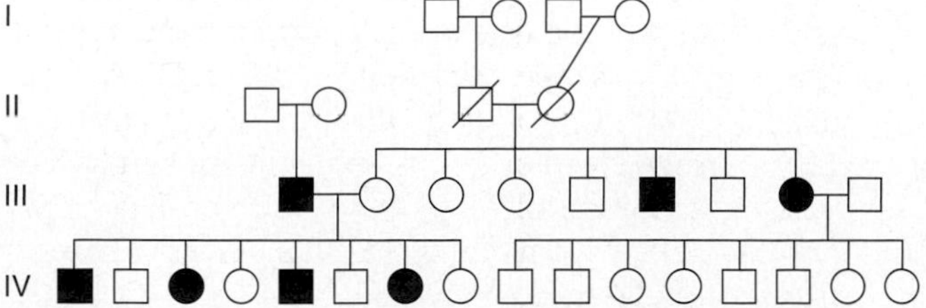

a. What is the most likely genetic basis for the disease?
b. Are there any individuals in the pedigree who *must* carry the disease mutation even though they don't express the disease trait? If so, list those individuals according to their generation number (in Roman numerals) and their number within the generation (counting from left to right across the entire pedigree).

29. You have decided to study another disease trait that is very rare. You have found an extended family in which a number of nonsibling individuals express the disease. One pair of identical twins are indicated with a horizontal line that joins both symbols together and with descent lines that join together at a vertex.

The pedigree is as follows:

a. What is the most likely genetic basis for the disease?

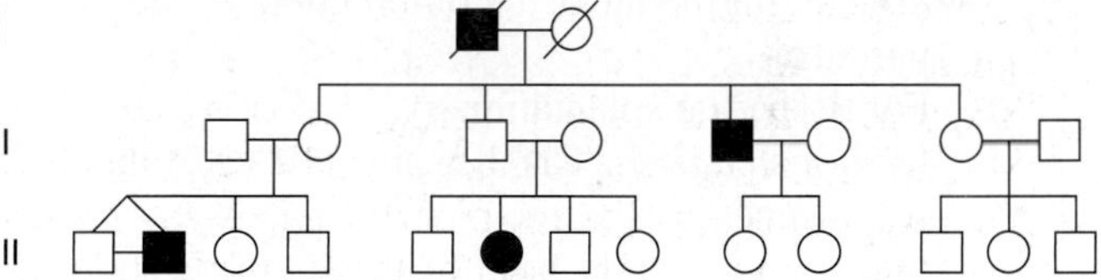

b. Are there any individuals in the pedigree who *must* carry the disease mutation even though they don't express the disease trait? If so, list those individuals according to their generation number (in Roman numerals) and their number within the generation (counting from left to right across the entire pedigree).

30. Among the most prevalent diseases that afflict human beings is heart disease, which can have a severe impact on quality of life as well as result in premature death. While heart disease mostly afflicts those who are older, 1% or 2% of people in their thirties, and even their twenties, suffer from this disease. There are genetic and environmental components to this disease. Use this information to answer the following questions.

a. What strategy might you use to choose families to participate in a linkage study of heart disease-causing genes?
b. Once you have cloned a gene that you believe plays a role in heart disease, how would you confirm this role?

Section 11.5

31. Human chromosome 6 has a region containing several closely-linked genes encoding cell surface proteins called human leukocyte antigens. Three of the genes in this region called *HLA-A, HLA-B,* and *HLA-C* are highly polymorphic: About 25 alleles of *HLA-A,* 50 alleles of *HLA-B,* and 10 alleles of *HLA-C* are known.

a. How many different haplotypes for these three genes are possible in human populations?

b. How many diplotypes (that is, different pairs of haplotypes) are possible?

Now consider the inheritance of HLA alleles in the following family:

	HLA-A		*HLA-C*		*HLA-B*	
Father	A23	A25	C2	C4	B7	B35
Mother	A3	A24	C5	C9	B8	D44
Child #1	A24	A25	C4	C5	B7	B8
Child #2	A3	A23	C2	C9	B35	B44
Child #3	A23	A24	C2	C5	B8	B35
Child #4	A3	A25	C4	C9	B7	B44

c. Diagram the two haplotypes in the father and the two haplotypes in the mother. Because the genes are so closely linked, assume none of these children is the result of recombination events in the parents.

d. For tissue transplantation to succeed, it is best that the donated tissue has the same alleles of the three *HLA* genes as the recipient. What is the chance that the next child born to this family (child #5) would be able to serve as a bone marrow donor to child #1 (his sister) with no danger of rejection due to incompatibility between *HLA-A, HLA-B,* and *HLA-C* antigens?

32. Canavan disease is a recessive, severe neurodegenerative syndrome usually causing death by the age of 18 months. The frequency of Canavan disease is particularly high in Jewish populations. In an effort to map the gene causing this condition, researchers looked at 10 SNPs (1–10) spaced at roughly 100 kb distances along chromosome 17 in 5 affected Jewish patients and 4 unaffected control Jewish individuals. In the table below, each row depicts a single haplotype. G, C, A, and T represent the actual nucleotide at the indicated SNP location.

a. Does the disease-causing mutation appear to be in linkage disequilibrium with any of the SNP alleles? If so, which ones? (Although this data set is too small to achieve statistical significance, assume that the haplotypes observed are representative.)

b. Where is the most likely location for the Canavan disease gene? About how long is the region to which you can ascribe the gene?

c. How many independent mutations of the Canavan gene are suggested by these data?

d. Suppose that individuals 2–9 are Ashkenazic (whose ancestors lived in the Rhine river basin of Germany and France after the Jews were expelled from Judea in 70 A.D.) while individual 1 is Sephardic (a non-Ashkenazic Jew). Would these facts provide any information about the history of the mutations causing Canavan disease?

e. For mapping genes by haplotype association, why is it often helpful to focus on certain subpopulations? Does this strategy have any disadvantages?

f. Human chromosome 17 is an autosome, so each person contains two copies of each region along the chromosome. With this in mind, how could the researchers determine any individual haplotype, such as those shown in the table?

Patient	SNP1	SNP2	SNP3	SNP4	SNP5	SNP6	SNP7	SNP8	SNP9	SNP10
1	G	T	G	T	T	T	C	A	G	T
2	A	T	G	T	T	T	C	A	G	T
3	G	T	G	T	T	T	C	A	G	C
4	A	A	G	T	T	T	C	T	C	C
5	G	A	G	C	C	T	G	A	C	C
Control										
6	A	A	G	T	T	T	C	A	G	T
7	G	T	G	G	C	T	G	A	G	T
8	A	T	C	T	C	G	C	T	C	C
9	G	T	C	G	T	G	G	A	C	T

Systems Biology and Proteomics

Chapter 12

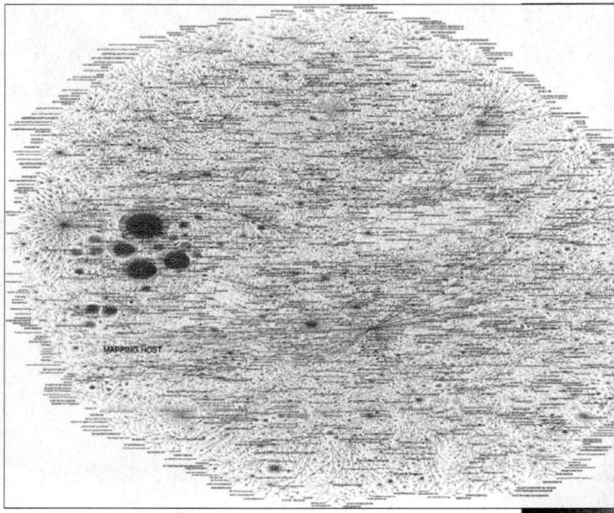

Complex network of major internet providers as of 2005.

Edward Jenner (1749–1823), a dedicated English country doctor, was fascinated by the eighteenth-century rural folk wisdom that milkmaids could not get smallpox. They often got the non-life-threatening cowpox, including blisters on their hands, but they did not get smallpox. Believing there was a connection between these two observations, Jenner hypothesized that the pus in cowpox blisters somehow protects against smallpox.

To test this idea, he took pus from cowpox blisters on the hands of a milkmaid who had regularly milked a cowpox-infected cow and injected some of this pus into a young boy named James Phipps (**Fig. 12.1**). He repeated this process over a number of days in 1796, gradually increasing the amount of pus he put into the boy. James contracted and recovered from cowpox. Jenner then deliberately injected him with material from smallpox lesions. Although the boy fell ill, he fully recovered in a few days and suffered no negative side effects. Jenner had made a brilliant discovery: The human immune system can protect against infection through a specific response to ***vaccination:*** inoculation with a nonvirulent or attenuated infectious agent. We now know that related viruses cause cowpox and smallpox, but the cowpox virus is much less virulent. (*Variolation,* injection with live smallpox virus obtained from a patient with a mild case of the disease, originated in Asia and was practiced in England throughout most of the eighteenth century; while it saved many lives, it was much more dangerous than Jenner's vaccination.) Jenner's insight, which has saved millions of lives in the last 200 years, thus represents one of the earliest recorded instances of safe and effective preventive medicine.

Today, we understand surprisingly little about how to control the immune response to many pathogens. In the two centuries since Jenner's time, only about two dozen safe and effective vaccines have been developed. These vaccines can protect against some but clearly not all infectious diseases. In one example of the difficulty of vaccine development, billions of research dollars have been spent to create an effective vaccine against the AIDS virus, but no AIDS vaccine has yet passed the "safe and effective" test. Indeed, even with the advent of modern medical technology, the current approach to vaccine development is still remarkably similar to Jenner's.

The inability to develop vaccines against any and every infectious disease is ironic because we know more about the molecular components of the vertebrate immune system (especially human and mouse) than perhaps any other eukaryotic biological system. One reason for this discrepancy is that immunologists have studied the components of the immune system one gene and one protein at a time, but they have never studied them all together as the system functions. Consequently, they do not understand the system's two most fundamental emergent properties: **immunity,** the ability to generate immune responses to infections or vaccines; and **tolerance,** the ability to prevent the body from making immune responses to its own proteins. Consider the following analogy. If you could describe all the components of a radio (resistors, transistors, condensers, and so forth), you would still not understand how the radio works because you would not understand how the components connect and interact in interrelated electrical circuits that function together as a unified system.

(a)

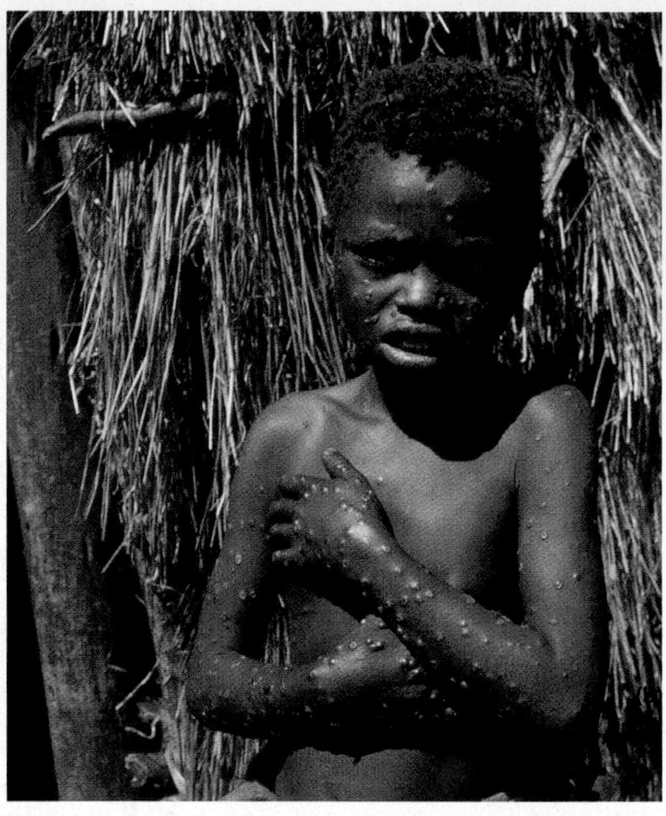

(b)

Figure 12.1 **Edward Jenner tested the first vaccine. (a)** Painting depicting Jenner's vaccination of James Phipps with cowpox pus on May 14, 1796. **(b)** Ethiopian boy scarred by smallpox, 1970. Since 1977, vaccination has eradicated smallpox worldwide.

Practitioners of the newly emerging field of **molecular systems biology** attempt to define all the components of a biological system and understand how they function in conjunction with one another. This pursuit relies on the tools and strategies of both genomics (the global study of genomes) and **proteomics:** the global analysis of most (or ideally, all) of the proteins in a particular cell type or organism. The collection of all the proteins in a cell or organism is known as a **proteome.** It also employs computational and mathematical tools for the modeling and simulation of biological systems. Only when all the gene and protein components of a system are studied as they function together can we begin to understand how the system functions as a whole and how its interacting elements give rise to its emergent properties. When we are able to do this for the immune system, we will acquire the insights to generate vaccines on demand against almost any infectious organism, including perhaps the AIDS virus.

In this chapter, as we examine systems biology and proteomics, we consider

- What is systems biology?
- Why looking at biology as an informational science is central to the practice of systems biology.
- How the high-throughput data platforms emerging from the relatively new field of proteomics make it possible to analyze systemwide proteomic data.
- The practice of systems biology.
- How a systems approach to disease leads to predictive, preventive, and personalized medicine.

12.1 What Is Systems Biology?

Definitions and Background

A **biological system** is a collection of interacting elements that carry out a specific biological task. These elements may be molecules such as proteins, mRNAs, metabolites and other small molecules, and the control elements of genes; or they may be cells such as immune system cells, hormonal network cells, and neuronal network cells that carry out a specific biological task. As mentioned earlier, the elements rarely act independently; rather, they most often function in association with one another. Systems biology seeks to describe the multiple components of a biological system and analyze the complex interactions of these components both within the system and in relation to the components of other systems. The nine proteins in yeast that help convert the sugar galactose to the sugar glucose constitute a system distinct from all other systems encoded by the 6000 genes in yeast; to help regulate the timing and extent of the galactose-to-glucose conversion, these nine elements interact with the proteins of other yeast cell systems. In mammals, the cells of the innate immune system and the adaptive immune system function together to generate immune responses; in carrying out these functions, they interact with cells of the nervous system. Thus, biological systems are defined by their discrete biological functions, but they often interact with other systems.

Some would argue that biologists have been practicing systems approaches to biology for more than 100 years, and this is true. At the turn of the twentieth century, physiologists were interested in homeostasis—certainly a systems problem. So, too, have developmental biologists, neurobiologists, and immunologists been interested in problems that require thinking about systems. What is different about systems biology at the beginning of the twenty-first century is that the Human Genome Project has provided global data-gathering tools and genomic information on a scale never before available. These tools and data are central to the current practice of molecular systems biology.

The following consequences of the Human Genome Project serve as the foundation for the current practice of systems biology:

- The complete genome sequences of humans and model organisms provide genetics-parts lists of all the genes in those organisms, as well as predictions of the proteins those genes encode.
- High-throughput platforms for genomics and proteomics enable the acquisition of global, or comprehensive, data sets of differing types of biological information (all genes, all mRNAs, all proteins, and so forth).
- Powerful computational tools make it possible to acquire, store, analyze, integrate, display, and model biological information.
- Studies of simple model organisms such as *E coli* and yeast allow scientists to compile global data sets from experimental manipulations of less complex biological systems. From their analysis of these data, they can learn how to do systems biology in more complex organisms.
- Comparative genomics allows scientists to begin to determine the logic of life for individual organisms and to discover how that logic has changed in different evolutionary lineages. For example, they can study the energy-generating systems in one organism and compare them to those in other organisms.

How to Think About Systems Biology

There are four fundamental questions one can ask about a biological system:

1. What are the *elements* of the system (proteins, cells, other)?
2. What are the *physical associations* among the elements? It is possible to depict the interactions of a system's elements in a graphic representation of a network where the nodes, or points, represent individual proteins and the edges, or connections, represent physical interactions between the proteins (**Fig. 12.2**). A series of such network graphs can reveal how systems change throughout the development of an organism or during physiological responses to changing environments. Delineating the dynamic behavior of systems is one of the

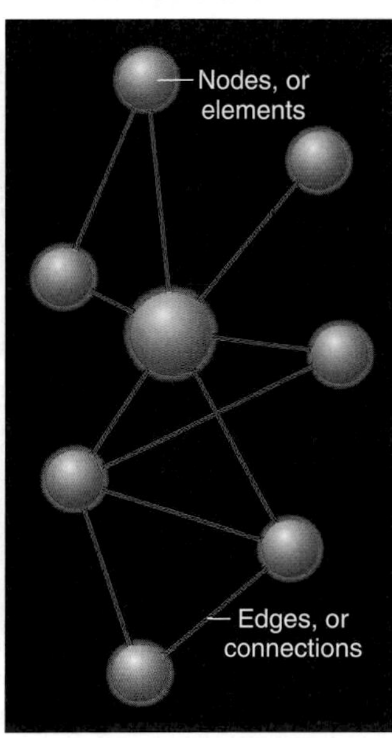

Figure 12.2 A hypothetical biological network represented by a graph. The nodes (*blue circles*) of the graph may represent molecules such as proteins and metabolites or cells (like those of the immune or nervous system). The edges (*lines* connecting some nodes) represent relationships between the elements (for example, protein-protein or protein-DNA interactions). The challenge is how to represent the dynamic changes in relationships that arise in many networks during development or physiological responses.

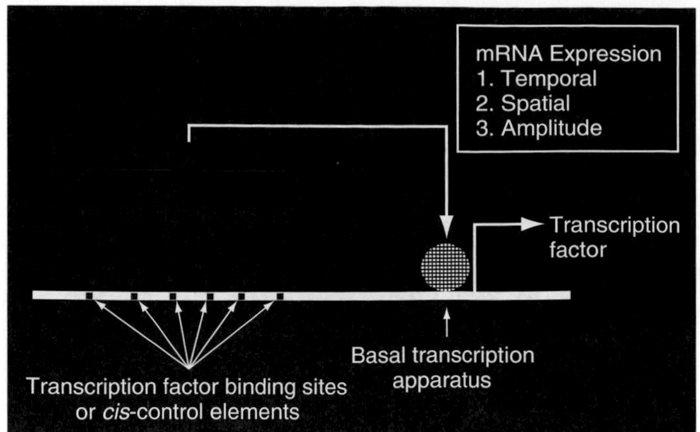

Figure 12.7 Diagram of a gene that encodes a transcription factor interacting with six other transcription factors. The gene is represented by the *white horizontal line* with *small red squares* indicating the *cis*-control sites; the six transcription factors with which the gene interacts are numbered tf1–tf6. The complex of six transcription factors interacts with the basal transcription apparatus; and this interaction regulates when, where, and how much mRNA is synthesized. If the transcription factor encoded by the gene interacting with six other transcription factors controls the expression of yet another transcription factor, as depicted here, then these DNA-protein interactions constitute a linkage in a gene regulatory network.

mediates the synthesis of an RNA transcript, which is then spliced and edited to produce an mRNA. The transcription factors may exhibit positive or negative control on mRNA expression. The interactions of multiple *cis*-control elements and their cognate transcription factors determines where in space, when in time, and to what level the mRNA is expressed.

In interacting with *cis*-control elements, some transcription factors control the expression of two or more genes encoding other transcription factors (Fig. 12.7). Thus, a single transcription factor may affect the behavior of many other transcription factors and thereby create a complex network structure. Gene regulatory networks that encode both subcircuits of transcription factors and the *cis*-control elements with which the factors interact may generate complex feed-forward and feedback regulatory loops (**Fig. 12.8**). In addition, transcription factors in the second layer of a network may control still other transcription factors in a third layer, and so on (Fig. 12.8). Thus gene regulatory networks may maintain, amplify, suppress, or shut off the expression of transcription factor genes.

The complexity of a gene regulatory network is specified in part by the number of layers in the network and the number of genes involved in each layer. The gene regulatory networks delineated by transcription factor-*cis*-control

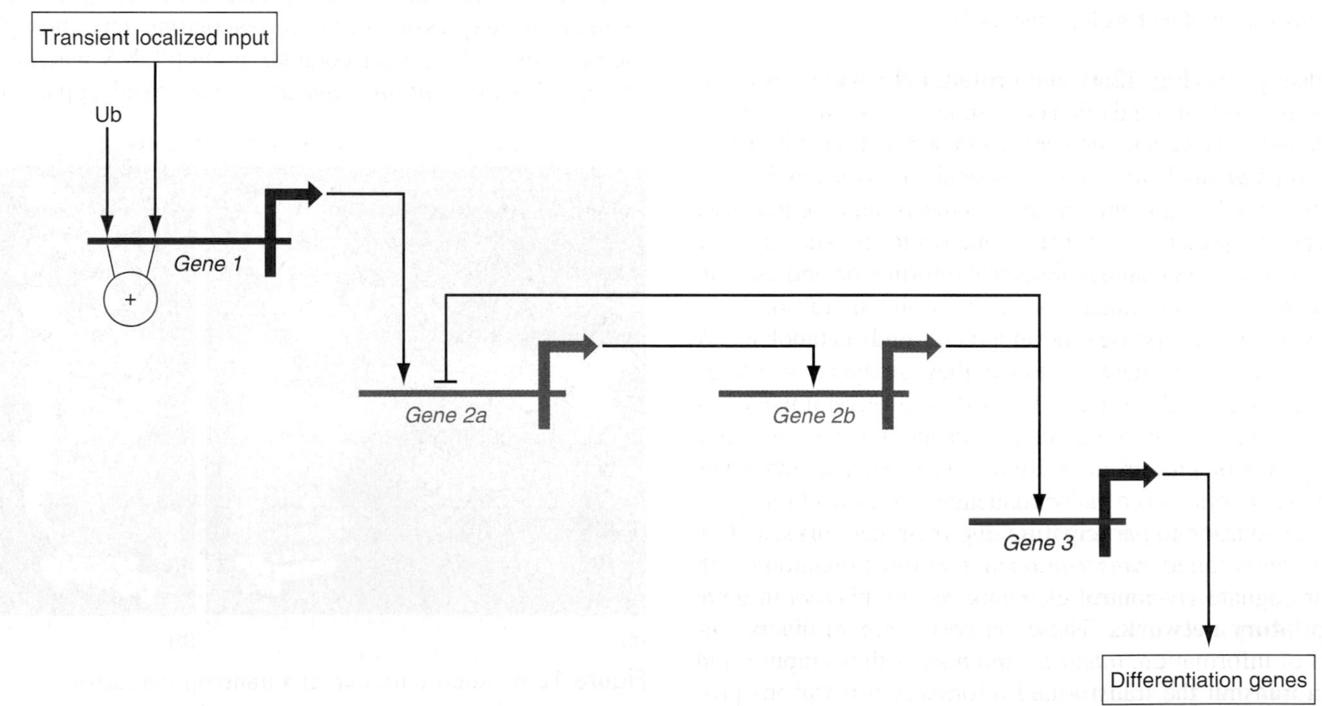

Figure 12.8 Diagram of a gene regulatory network. In this network, transcription factors interact with transcription factor genes. The genes are represented by *horizontal lines* (*red, purple, blue,* and *green*) that fall into three layers. The transcription factor interactions may be positive (indicated by an *arrow* at the interaction site) or negative (indicated by a *bar* at the interaction site—see 2b interacting with 2a). A transcription factor may interact with another transcription factor in a lower layer or it may feed back to another gene in its layer or in an earlier layer. The input signal is represented as affecting the first gene in the layer; and the output (the turning on of genes whose expression leads to cell differentiation) is depicted as coming from the *green* gene on the lowest layer. The complex interactions within the gene regulatory network integrate and modulate the input signals and generate the appropriate output signals.

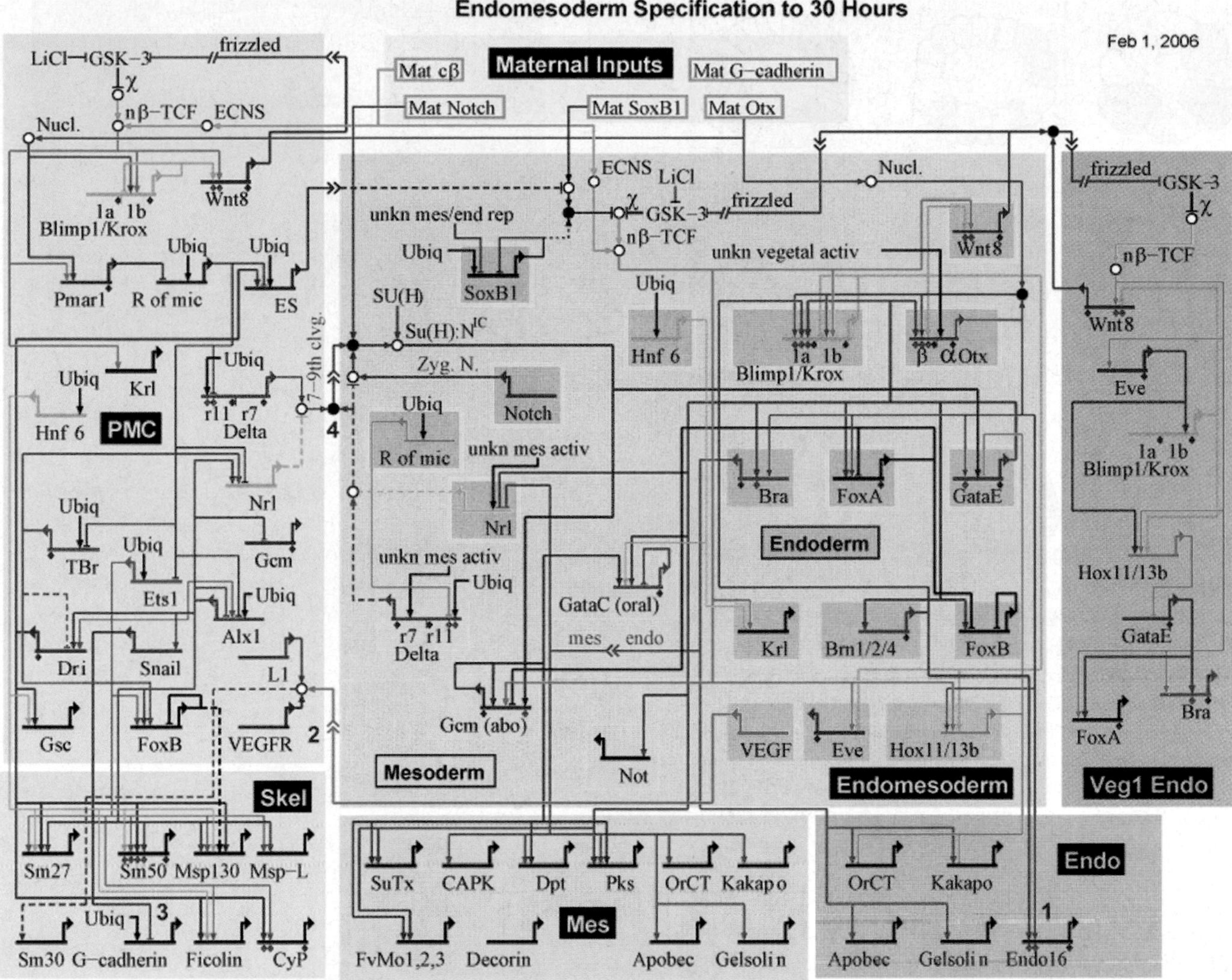

Figure 12.9 **The most complex gene regulatory network delineated to date.** About 50 proteins are represented in this network, and 35 of them are transcription factors. The schematic representation in this figure is similar to the one in Fig. 12.8. The particular network represented here makes it possible to understand many aspects of gut development in sea urchin larvae (depicted in Fig. 12.10).

element interactions integrate and modulate the information received from the environment via various signal transduction pathways and other inputs. They then transmit this information to batteries of genes that carry out metabolism, development, or physiological responses. **Figure 12.9** shows the most complex gene regulatory network analyzed to date: the network that controls endomesodermal, or gut, development in sea urchin larvae (**Fig. 12.10**). The network depicted in Fig. 12.9 accurately predicts the events required for gut development (the blue territory in Fig. 12.10); that is, it explains the emergent properties of this developmental system. Because of its predictive power, it is possible to rationally reengineer this network by modifying the behavior of one or more genes to create predictable new emergent properties (for example, to convert all pigment cells to skeletal cells). Plants with greater nutritional value may be created through the reengineering of appropriate cell types in crop plants.

To summarize: Gene regulatory networks integrate and modulate inputs of biological information and transmit the transformed information to protein networks. The protein networks then use this information to execute specific functions in the organism.

A final point about biological information is that it is hierarchic. It starts with the digital DNA informational core of the genome and progresses through mRNA, protein, molecular machines, networks, cells, networks of cells, and tissues to individual organisms, populations of organisms, and finally ecosystems (**Fig. 12.11**). Environmental information impinges upon and modulates the core digital information at each of these successive steps. To understand biological systems, one must capture and integrate information from

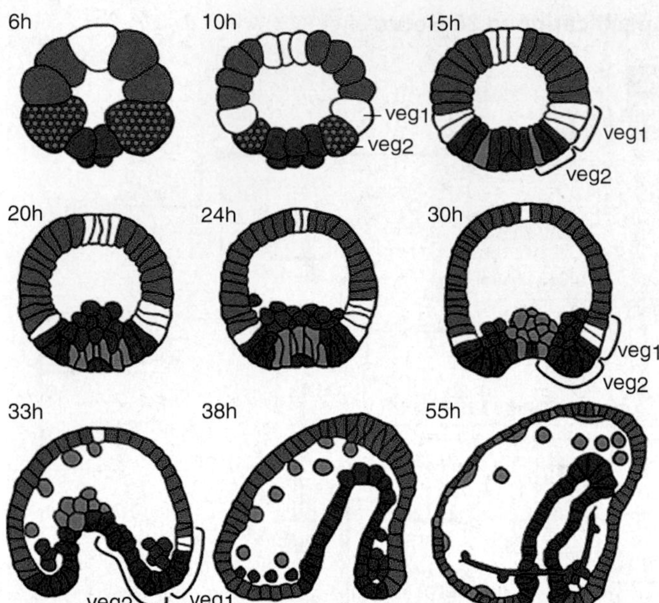

Figure 12.10 Larval development of the sea urchin. Development begins with the fertilized egg and in 72 hours generates a larva with 1800 cells. The drawings illustrate three separate stages of this process: blastula, gastrula, and differentiated larva. The four distinct territories of this development process are indicated by four distinct colors. Endomesodermal, or gut, development is indicated by the *blue*. What is amazing is that the gene regulatory network depicted in Fig. 12.9 can explain most of the process of gut development.

as many different hierarchic levels as possible. Thus, the fundamental informational requirement of systems biology is the acquisition and integration of global data sets from different levels of biological information.

12.3 Global Proteomics Strategies and High-Throughput Platforms Make It Possible to Gather and Analyze Systemwide Protein Data

Global technologies make possible the comprehensive, large-scale study of the components of biological systems. In Chapter 10, we discussed high-throughput platforms for the acquisition of genomic data, including DNA sequence, DNA polymorphisms, and mRNA levels. Here we discuss high-throughput platforms for the acquisition and measurement of proteomic data, including amino acid sequence and the amount, modifications, and compartmentalization of all the proteins in a cell or organism.

The analysis of proteomes is more complex and challenging than the analysis of genomes for two reasons. First,

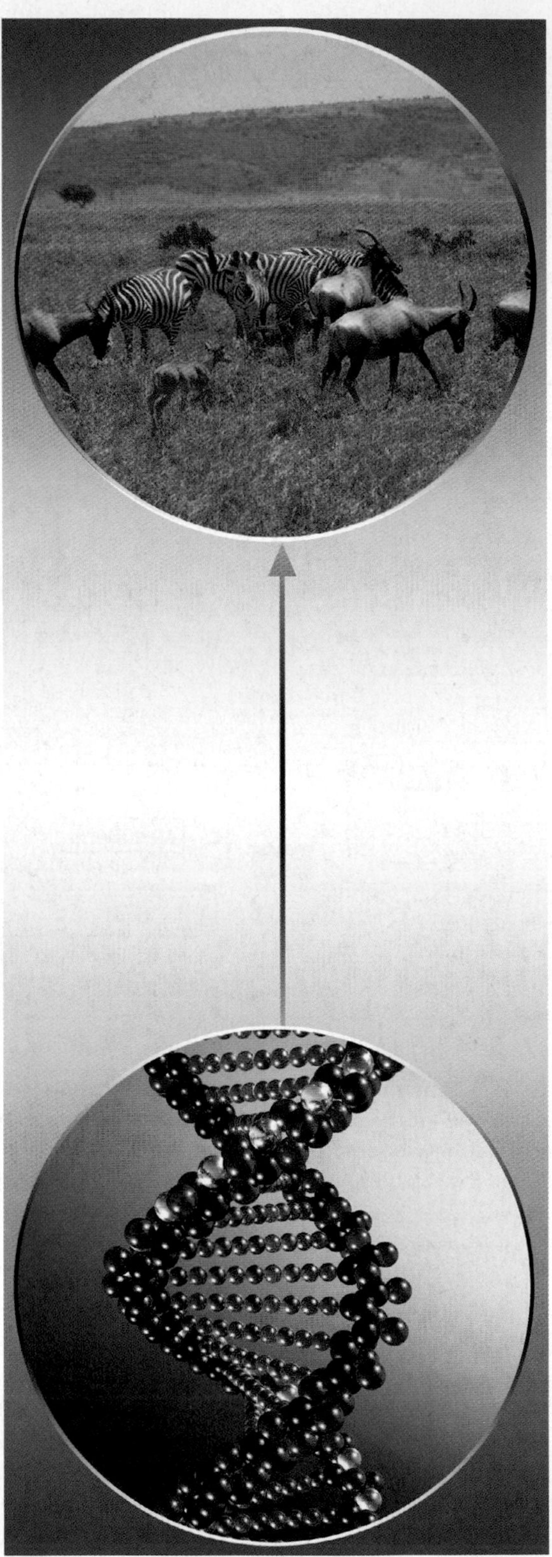

Figure 12.11 Levels of biological information. Starting with the digital DNA core, the successively higher levels of biological information are modified by environmental signals. Thus, during development, organisms employ a combination of nature (digital DNA) and nurture (environmental information). To understand systems, one must capture and integrate as many different types of information as possible.

the range of protein expression in cells is enormous: from one copy to 10^6 copies/cell. Since there is no equivalent of PCR for the amplification of interesting proteins, the tools of protein analysis need to function across concentration differences spanning six orders of magnitude. Second, proteins in complex mixtures have many different features that need to be identified and characterized. These include the gene(s) encoding each protein; chemical modifications; levels of expression in particular cell types, which change across developmental stages or physiological responses; covalent modifications (for example, phosphorylation, glycosylation); interactions with other proteins, macromolecules, or small molecules; compartmentalization within the cell (for example, in the nucleus, cytoplasm, or on the cell surface); state of activation; half-lives; three-dimensional structure; and the relationship between structure and function. In sum, proteins are dynamic molecules with constantly changing concentrations, chemical modifications, interactions, and cellular locations. The ability to analyze variations in all the features of proteins is important to an understanding of how they function in the systems in which they participate.

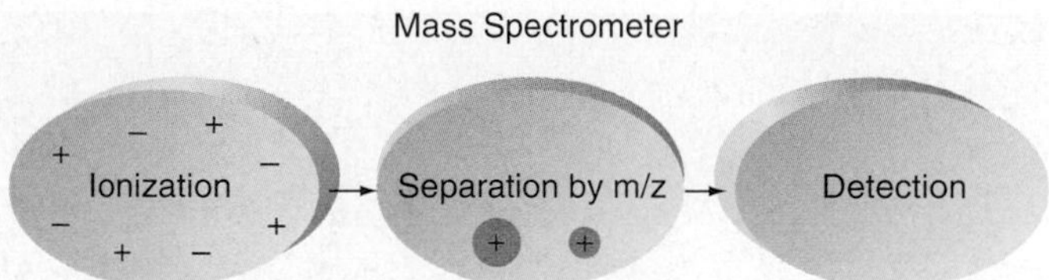

Figure 12.12 Mass spectrometer. The mass spectrometer consists of three components: (1) an ionization source, (2) a mass analyzer, and (3) a detector. The text explains the function of each component.

The Mass Spectrometer

The mass spectrometer, an instrument that measures the masses of a wide variety of molecules, can analyze proteins that can be enriched to a certain level of abundance. The sensitivity of mass spectrometers is continually increasing, so they can be used to analyze more and more proteins from complex mixtures such as cells or blood. For our purposes, interesting molecules include small proteins, peptides, oligonucleotides, lipids, carbohydrates, small RNAs, and other molecules. The mass spectrometer requires that the molecules to be analyzed can be ionized and transferred to a vacuum. The spectrometer determines the masses by measuring the molecules' migration rates in an electric field (small masses migrate more rapidly than large masses). Since mass spectrometry can measure the masses of small molecules more accurately than large ones, proteins or protein mixtures are often converted into smaller peptide fragments by a proteolytic enzyme such as trypsin. Mass spectrometers consist of three components: (1) a source of ionization, (2) a mass analyzer, and (3) a detector (**Fig. 12.12**). The ionization source ionizes, for example, the peptides of a single protein or a mixture of proteins; the mass analyzer separates the ionized fragments according to their mass-to-charge ratios (m/z); and the detector measures the separation times or distances, which researchers can use to determine the fragment masses.

Using different combinations of ionization sources and mass analyzers, commercial mass spectrometers can accurately measure the masses of peptides and small proteins. They can also sequence the peptides by further fragmentation and mass measurement of the smaller derivative peptides. An instrument known as the *ion trap tandem spectrometer* separates different-sized fragments in a first mass spectrometer, selects fragments for further fragmentation by collision with ions, and then measures the masses of the resulting peptide fragments in a second mass spectrometer. In this manner, it can sequence, or determine the order of, amino acids in peptides. The mass spectrometer is thus a powerful instrument for identifying and sequencing peptide fragments from proteins in complex mixtures.

Identifying Proteins in Complex Mixtures

The Human Genome Project and the consequent identification of all the genes in humans (and other species) have revolutionized protein chemistry. They have done this by providing the information to make the mass spectrometer a powerful tool for identifying the protein components of proteomes. For example, from the 25,000 or so gene sequences in the human genome, one can determine the corresponding protein sequences. Once the protein sequences are known, one can computationally determine the mass of each individual protein as well as the mass of the peptides derived from it by enzymatic cleavage (assuming that one knows the amino acid residues at which peptide cleavage occurs). The realization that the proteins of a complex mixture can be proteolytically cleaved and then identified from mass determinations of their individual peptides has given proteomics an enormous boost.

Proteins encoded by genomes where most of the genes have been sequenced and annotated can be identified as follows. First, cells of a particular type are purified. In the next step, proteins are extracted from the cells and digested with trypsin, a protein-splitting enzyme, or *protease,* that cleaves polypeptide chains at arginine and lysine residues. The peptides resulting from treatment with trypsin, known as *tryptic peptides,* are partially fractionated on a *reverse phase column:* a column separating peptides based on their hydrophobicity, that is, their ability to repel water. Successive aliquots from the column are introduced into the mass spectrometer to obtain mass measurements of the partially fractionated tryptic peptides. In an alternative approach, the proteins can be separated by two-dimensional gel electrophoresis (separation by size in one dimension and by

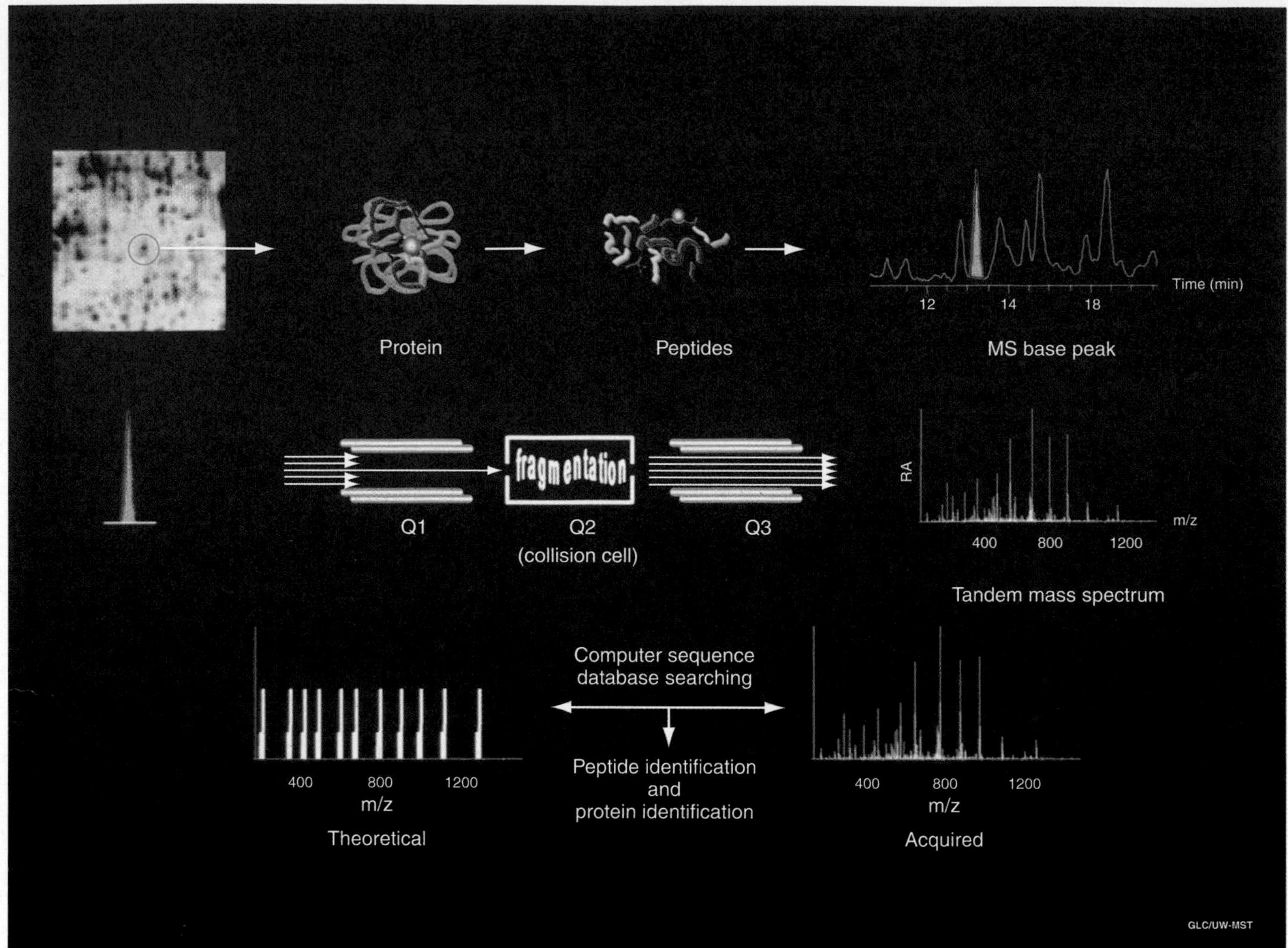

Figure 12.13 A strategy using two-dimensional gels and mass spectrometry to identify proteins in complex mixtures and the genes that encode them. A two-dimensional gel separates proteins by size and charge. Individual spots can be extracted from the gel, digested with trypsin to produce peptides, and analyzed in a tandem mass spectrometer. The first mass spectrometer in this instrument (Q1) separates the peptide fragments (MS base peak); one fragment (*orange*) is then moved to the collision cell (Q2) and fragmented into components, which can be analyzed in the second mass spectrometer (Q3). Researchers can search the mass spectrum obtained in this way against a databases of theoretical fragments to identify peptides and proteins and, by extrapolation, the genes that encode them.

charge in the second), and the protein spots can be cut out from the gel and digested with trypsin. The resulting tryptic peptides are then injected into the mass spectrometer (**Fig. 12.13**). In every yeast cell, for example, there are potentially 6000 gene-encoded proteins and approximately 350,000 tryptic peptides. Most of these peptides have unique masses that are recorded in a database. A computer software program can compare the mass of each peptide experimentally analyzed in a mass spectrometer against the database of theoretical peptide masses for the proteins encoded by the genome. This database has been computationally determined through the identification of all the genes in a specific genome, followed by the determination of genes that encode the proteins from which particular tryptic peptides derive. A single mass spectrometry run can assign hundreds of peptide fragments to the genes that encode them.

Quantifying Changes in Protein Concentration in Different Cell or Tissue States

A technique called *isotope analysis,* which employs prepared reagents known as *isotope-coded affinity tags (ICATs)*, permits researchers to analyze the changing patterns of protein expression in two different cellular states. The ICAT reagent has three components: (1) a biotin tag (biotin is a molecule that binds tightly to avidin molecules, and this binding

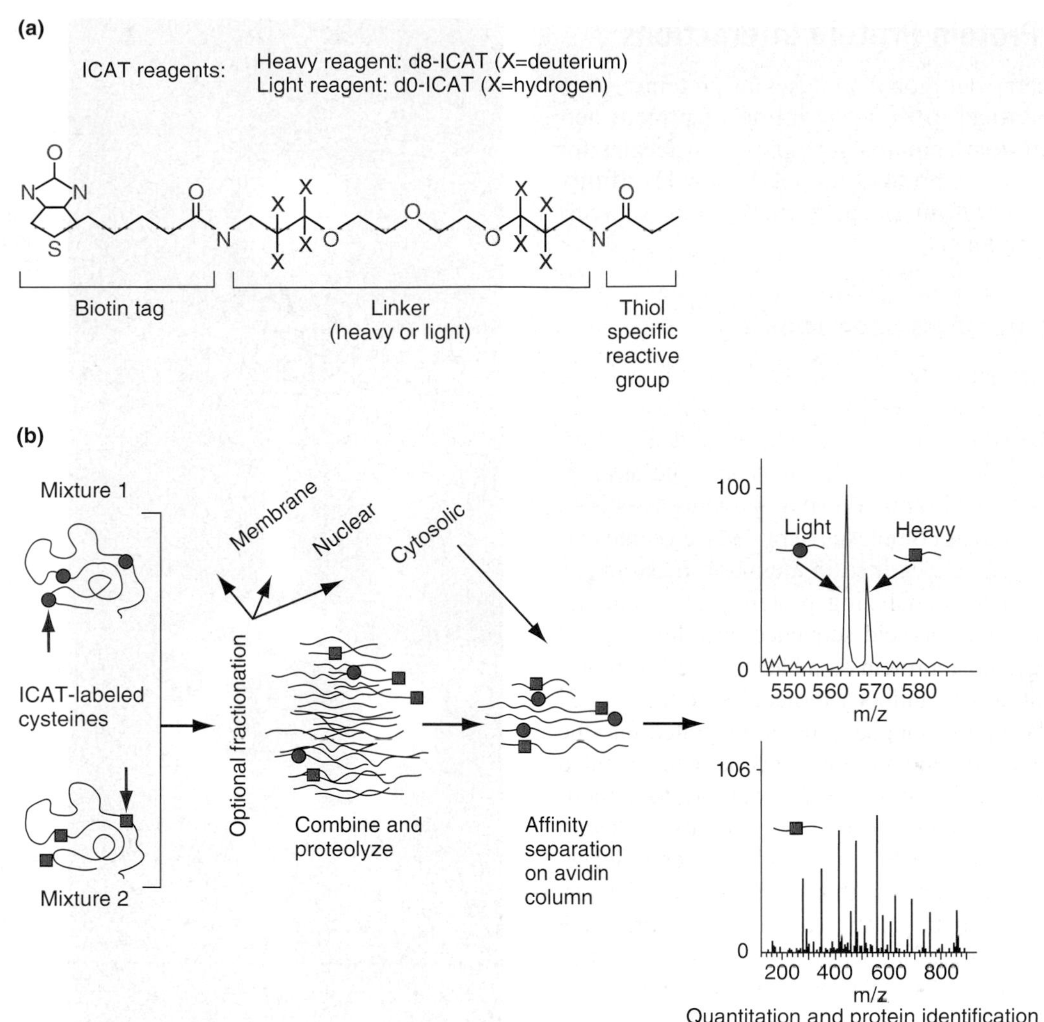

Figure 12.14 **The isotope-coded affinity tag (ICAT) approach to quantifying complex protein mixtures from two different states.** **(a)** The isotope-coded affinity tag reagent (described in the text). **(b)** The strategy for labeling the proteins of two cell types with the light and heavy reagents (also described in the text). The *green circles* indicate covalent linkage of the heavy ICAT reagent (with deuterium) from part (a) to cysteines in proteins from cancer cells. The *yellow squares* indicate covalent linkage of the light ICAT reagent (with hydrogen) from part (a) to cysteines in proteins from normal cells.

provides a means for purifying proteins or peptides of interest); (2) a linker to which eight hydrogens or eight deuteriums can be attached to create light (hydrogen) or heavy (deuterium) chemical isotope forms (differing by 8 mass units); and (3) a chemical group that reacts with the thiol (-SH) group of cysteine amino acids and will thus attach the ICAT reagent to all cysteines in a protein or peptide (**Fig. 12.14a**).

In one example of this isotope analysis technique, normal cells are labeled with the light isotope reagent and cancerous counterparts with the heavy one (**Fig. 12.14b**). After equal quantities of the normal and cancerous cells are mixed together, their proteins are purified and digested with trypsin, and the cysteine-ICAT-labeled peptides are purified by affinity chromatography (that is, poured through an avidin column—a tube filled with avidin; the avidin will bind all biotin-labeled peptides, while all other peptides will flow right through the column). The biotin-labeled peptides are then fractionated on a reverse phase column, and successive aliquots of the biotin-labeled fractionated peptides are analyzed in two ways in the mass spectrometer. First, as shown in Fig. 12.14b, the areas under the curves of the isotope pairs of peptides (hydrogen/deuterium) are determined. These areas are proportional to the concentrations of the peptides (and, by extension, the proteins) expressed in the two cell types. After further fragmentation in the mass spectrometer, the peptide pairs are analyzed to determine their amino acid sequences. With these sequences in hand, it is possible to determine the gene that encodes the protein from the database table of the masses of tryptic fragments. The ICAT technique is thus similar to the two-color microarray studies described in Chapter 10: The ICAT quantifies proteins, the microarrays quantify mRNAs, from two different cellular states. Work is now underway to develop high-throughput platforms for measuring many other features of proteins, such as phosphorylation and activation.

second example, a battery of specific antibodies is arrayed to determine whether the molecules they interact with are present in a given protein mixture. It is also possible to use antibody arrays to compare two different cellular states to see whether the levels of individual proteins change from one state to the other.

Current limitations of protein chips are that they do not allow a global analysis of all the proteins in the proteome being studied (most chips have fewer than a few hundred proteins arrayed) and the individual protein-capture components on the chip after interaction with their cognate proteins do not allow very precise quantification of the recognized proteins from the proteome. Projects are underway to make antibodies against all human proteins. Studies using these antibodies in conjunction with comprehensive protein arrays could quantify the expression patterns of most proteins in the proteomes under investigation.

ChIP/Chip Analyses Identify Protein-DNA interactions

The ability to make global measurements of protein-DNA interactions is critical for the description of gene regulatory networks. The most widely used tool for these measurements is chromatin immunoprecipitation analyzed on a chip—abbreviated as *ChIP/chip analysis.* It employs a combination of genomic and proteomic techniques to measure the interactions of transcription factors (proteins) with their *cis*-control elements (DNA) or the interactions of complex protein machines, such as activator or repressor complexes with their chromosome (DNA) binding sites.

The ChIP/chip technique, also known as *genome-wide localization,* can identify all the genomic sites at which a transcription factor expressed in a particular cell type may bind. The protocol, diagrammed in **Fig. 12.18,** works as follows. Through genetic engineering, a DNA sequence encoding an antibody-binding-site peptide is added to the 5′ or 3′ end of the gene for the transcription factor under investigation. This modified gene is then transformed into the organism's genome; when expressed, the protein product of the gene will have the peptide tag at its N or C terminus. Next, the chromatin is isolated from a cell type of interest, and the transcription factor is cross-linked to its corresponding binding sites by a small reversible cross-linking chemical. The DNA of the chromatin is then fragmented into small pieces (300–500 base pairs in length), and the antibody against the peptide tag is added to the fragmented chromatin. The antibody will bind to the tagged transcription factor and thus isolate the transcription factor as well as the DNA to which it is cross-linked. All the cross-links are then reversed, and both ends of the purified transcription-factor-binding DNA fragments are attached to PCR primer sequences. The specific DNA binding sites are then amplified and labeled with a red fluorescent dye. A control procedure lacking the modified transcription factor gene follows the protocol just described, and the PCR-amplified control DNA fragments are labeled with a green fluorescent dye. DNA or oligonucleotide arrays are prepared that contain the potential DNA sequences of all sites at which *cis*-control elements may be encoded: 5′, 3′, and within the introns of each gene whose *cis*-control regions are to be studied. (Ideally, one would like to study all the genes of the genome at the same time.) The experimental and control samples are mixed, then hybridized against noncoding DNA or oligonucleotide arrays. The ratio of red to green dye is determined for each DNA fragment on the array. High red ratios indicate DNA fragments to which the corresponding transcription factor has bound (the *cis*-control DNA has been enriched in the experimental PCR amplifications). By repeating this process for each of the transcription factors expressed in a cell, one can obtain a global map of the *cis*-control elements to which the transcription factors bind.

Uses of Genomic and Proteomic High-Throughput Platforms

For each interesting organism, researchers would like to use the high-throughput platforms described in Chapter 10 and previously in this chapter to sequence the genome, define quantitatively the transcriptomes and proteomes in each cell type, delineate the nature of proteome interactions in various cell types, and analyze other features of the proteome, such as the location of proteins, their state and level of activation, their half-life, and their three-dimensional structure. For each organism, they would also like to develop strains from different genetic manipulations (for example, separate knockout strains for each gene of the genome). In addition, they would like to obtain many mutations of each gene (by chemical mutagenesis, for example); the global production of mutants is important because one cannot determine the consequences of different levels of mRNA or protein activity from all-or-none knockout mutations. Geneticists would also like to be able to express each protein from an organism in order to study it individually or in combination with other proteins. In yeast, many of these global discovery projects are quite advanced; in other organisms, such programs are just beginning.

High-throughput platforms can also be applied to experimental genetic and environmental manipulations of organisms or cells, for example, to analyze the transcriptome or proteome of knockout strains or chemical mutants, or to compare normal and diseased cells. Yeast researchers, for example, have used knockout strains and global protein interaction data to study a variety of biological systems, including the cell cycle. And medical researchers have carried out microarray analyses of tissues from patients with leukemia, prostate cancer, breast cancer, and melanoma and of the corresponding normal tissues. These analyses have revealed molecular markers that allow phenotypically similar cancers to be separated into distinct groups. This indicates that the cancers probably result from different cellular defects and thus have different prognoses and may require different treatments.

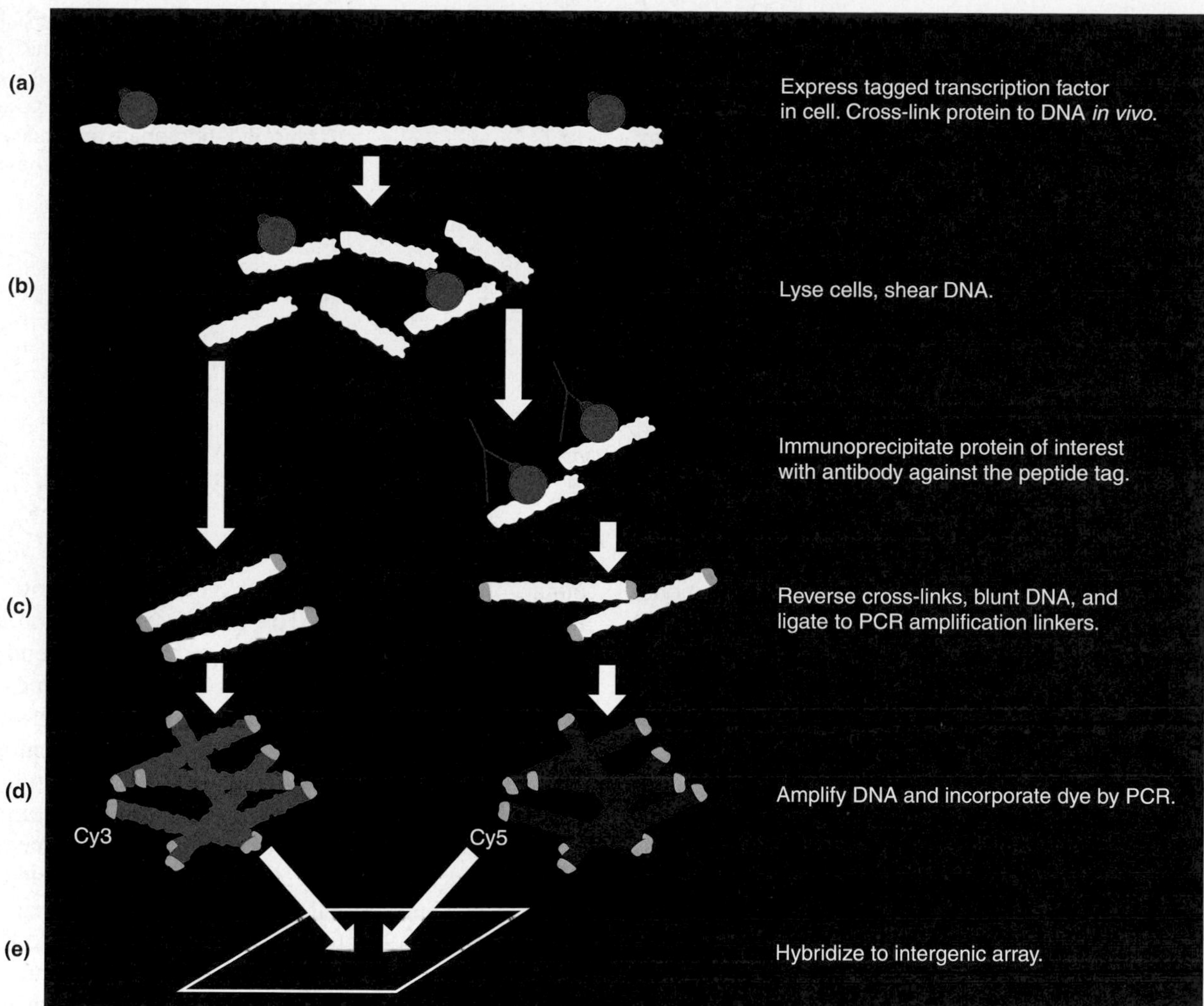

Figure 12.18 Diagram of the CHIP/chip process for genome-wide localization of the DNA binding sites of transcription factors. **(a)** A DNA sequence encoding an antibody binding site is attached to one end of a transcription factor gene. This transcription factor gene is expressed in the appropriate cell type, and the expressed protein (*yellow* with *orange* peptide tag) is allowed to bind to all of its cognate *cis*-control elements. It is then covalently cross-linked to these DNA sites. **(b)** The DNA is sheared into fragments 300–500 bp in length, and these fragments are purified using the specific antibody (*blue*) to the binding site attached to the transcription factor. **(c)** PCR amplification linkers (*aqua*) are now put on both ends of these DNA fragments, and the fragments are amplified incorporating a *red* dye, as shown in the righthand image of **(d)**. In a similar control protocol, investigators leave out the cross-linking and incorporate a *green* dye into the amplified DNA (left images of (c) and (d)). **(e)** The amplified control and experimental DNAs and then mixed and hybridized to a DNA array carrying all the appropriate non-coding (5′ and 3′, as well as intronic) DNAs to which the transcription factor might bind. The color ratios of the individual DNA spots on the chip are analyzed; a high red-to-green spot in a DNA array indicates the presence of a *cis*-control element that binds to the transcription factor under study.

12.4 Putting It All Together: The Practice of Systems Biology

An Algorithm for Using the Systems Approach to Biology

The modern systems approach to biology incorporates the following steps.

1. **Scan the biological literature** for all that is known about the system of interest. A key aspect of this search is access to the entire genome sequence of the organism under consideration, because the discovery of all (or as many as possible) of the genes, RNAs, and proteins in a cell or organism is the foundation of systems analysis. Use the knowledge gained from discovery science and the literature to define the system of interest as best you can, identifying its elements, their relationships, and their contextual changes. Then **develop a preliminary model** about how the system functions. This model may be descriptive (words), graphic (network diagrams), or mathematical, depending on how much information is available.

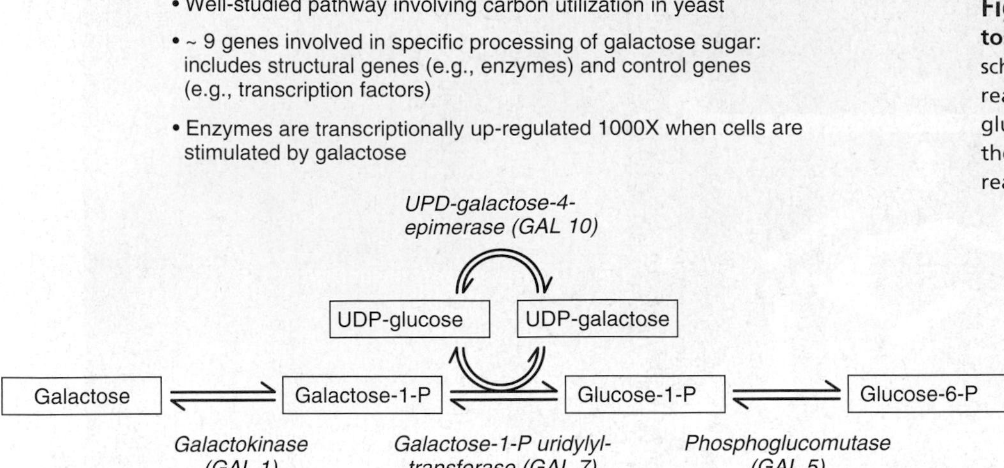
• Well-studied pathway involving carbon utilization in yeast
• ~ 9 genes involved in specific processing of galactose sugar: includes structural genes (e.g., enzymes) and control genes (e.g., transcription factors)
• Enzymes are transcriptionally up-regulated 1000X when cells are stimulated by galactose
UPD-galactose-4-epimerase (GAL 10)
UDP-glucose
UDP-galactose
Galactose
Galactose-1-P
Glucose-1-P
Glucose-6-P
Galactokinase (GAL 1)
Galactose-1-P uridylyl-transferase (GAL 7)
Phosphoglucomutase (GAL 5)

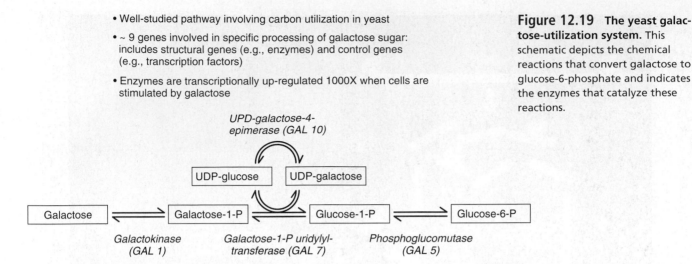

Figure 12.19 The yeast galactose-utilization system. This schematic depicts the chemical reactions that convert galactose to glucose-6-phosphate and indicates the enzymes that catalyze these reactions.

2. **Formulate a hypothesis-driven query** about the model and answer this query through genetic or environmental perturbations of the system. These perturbations may include gene knockouts and changes to the environment such as the addition of nutrients. In conjunction with these perturbations, collect comprehensive data sets of different levels of biological information (DNA sequence, mRNA levels, protein levels, protein-protein or protein-DNA interactions, and so forth).
3. **Integrate** these different types of data either **graphically** or **mathematically** and compare the results against the initially formulated model. Disparities will arise between the new experimental data and predictions based on the original model. Formulate new hypotheses to resolve these discrepancies.
4. **Iterative perturbations:** To test the new hypotheses, design a second round of genetic and environmental perturbations that will generate new global data sets whose integration will make it possible to resolve the discrepancies. Repeat steps 2–4 until model and experimental data are in accord.
5. The **refined final model** should enable biologists to predict the behavior of the system, even with perturbations that have never before been tested; explain the emergent properties of the system; and reengineer the system to produce predictable new emergent properties.

To summarize: Systems biology employs both discovery science and hypothesis-driven science. It requires both the *acquisition of global data sets* from different levels of biological information and the graphic or mathematic *integration of different types of data.*

Example of a Systems Approach to the Galactose Utilization System in Yeast

Researchers have successfully used the general systems approach just outlined to look at how yeast, the simplest eukaryotic cell, turns various genes on and off in the utilization of galactose. The galactose-utilization system in yeast converts the sugar galactose to a second sugar, glucose-6 phosphate, which is a major component in one carbohydrate pathway for energy production. The initial model, illustrated in **Fig. 12.19,** summarizes more than 30 years of galactose-utilization analysis in which researchers looked at one gene or one protein at a time.

The galactose-utilization system has nine elements: four enzymes (encoded by the *GAL 1, GAL 5, GAL 7,* and *GAL 10* genes shown in Fig. 12.19) that catalyze the necessary chemical-modification reactions; a transporter molecule that carries galactose across the yeast membrane and also sets the activity state of the galactose system such that when galactose is absent, the system is shut down; and four transcription factors that turn the system on and off.

To analyze the galactose-utilization system, researchers carried out three types of global experiments. They first constructed nine genetically perturbed yeast strains; in each strain, one of the nine genes encoding the protein elements of the system had been knocked out. The investigators then used global microarrays with all 6000 yeast genes represented to analyze the nine knockout strains, as well as the wild-type strain in both the presence and absence of galactose. These analyses represented 20 perturbations (9 knockout + 1 wild type = 10 yeast strains × 2 system states).

Several interesting observations emerged from these experiments. First, more than eight unexpected gene-expression patterns were recorded, which suggested that the galactose system has additional control features not captured in the original model (depicted in Fig. 12.19). In two instances, researchers formulated new hypotheses, generated new perturbations (double knockouts), and carried out a second round of global microarray analyses, which afforded new insights on the regulation of galactose utilization that could be incorporated into the original model (**Fig. 12.20**). These fundamental new insights helped explain the operation of the galactose system. Second, the

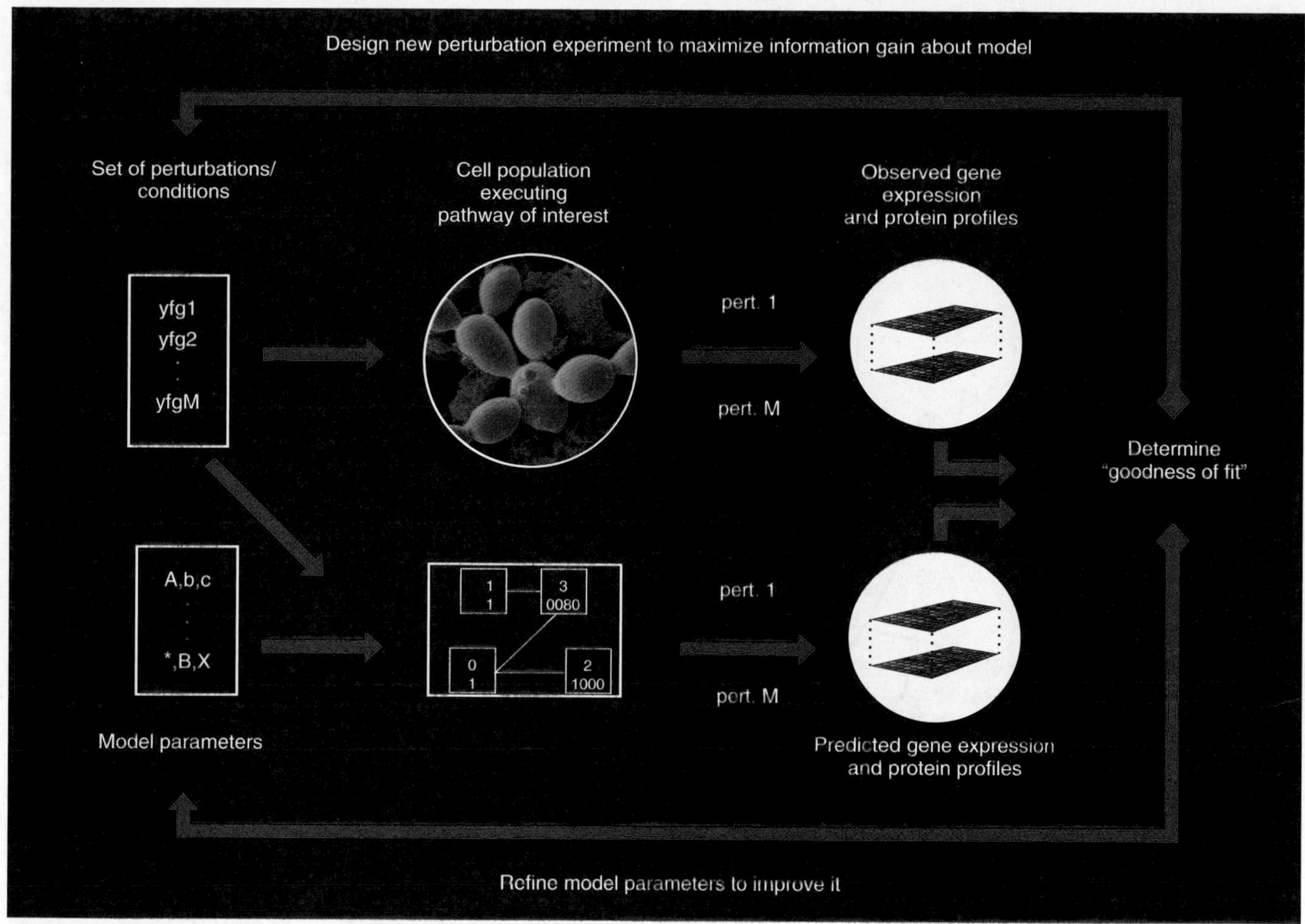

Figure 12.20 Diagram of the iterative nature of the systems approach to biology. To study the galactose-utilization system in yeast using a systems biology approach, researchers constructed a preliminary model based on 30 years of prior genetic, molecular, and cellular studies. They then knocked out (perturbed) the nine genes in the system and globally tested each knockout for mRNA changes, with the system on (galactose present) and off (galactose absent). Eight of the experimental observations failed to match the predictions from the model. For two of these observations, the researchers formulated hypotheses to explain the discrepancies and carried out additional perturbations to discriminate between the possibilities. They then used the new data to refine the model. They repeated (iterated) this process as many times as necessary to bring experiment and theory into accord with one another. In other perturbations, they used the ICAT approach to determine how protein level changed with different perturbations (see Fig. 12.14). The mRNA and protein concentration data were integrated with protein-protein and DNA-protein interaction data to get a more complete picture of the system. These integrative, iterative, and model-refining processes are integral components of systems biology.

expression patterns of 997 genes could be clustered into 16 groups, each exhibiting similar patterns of gene-expression change across the perturbations. Strikingly, some of the genes within each cluster were known to function in different systems in the yeast cell (for example, during the cell cycle, in amino acid synthesis, or in other aspects of carbohydrate metabolism). The investigators hypothesized that each of these systems was directly or indirectly connected to the galactose-utilization system such that perturbations in the galactose system caused changes in the others. A second round of analyses using protein-protein and protein-DNA interactions confirmed this hypothesis.

Databases cataloging more than 15,000 protein-protein interactions and thousands of protein-DNA (transcription factor/promoter region) interactions exist for yeast. Investigators correlated these cataloged interactions with the 997 genes perturbed in the knockout experiments. They then constructed a graph or model of the integrated mRNA-expression changes and the protein-protein and protein-DNA interactions. These data supported the hypothesis that interactions occur between the galactose system and many other metabolic and cellular systems (**Fig. 12.21**). This finding, in turn, pointed to additional global analyses—for example, to define in more detail how the different systems communicate with one another.

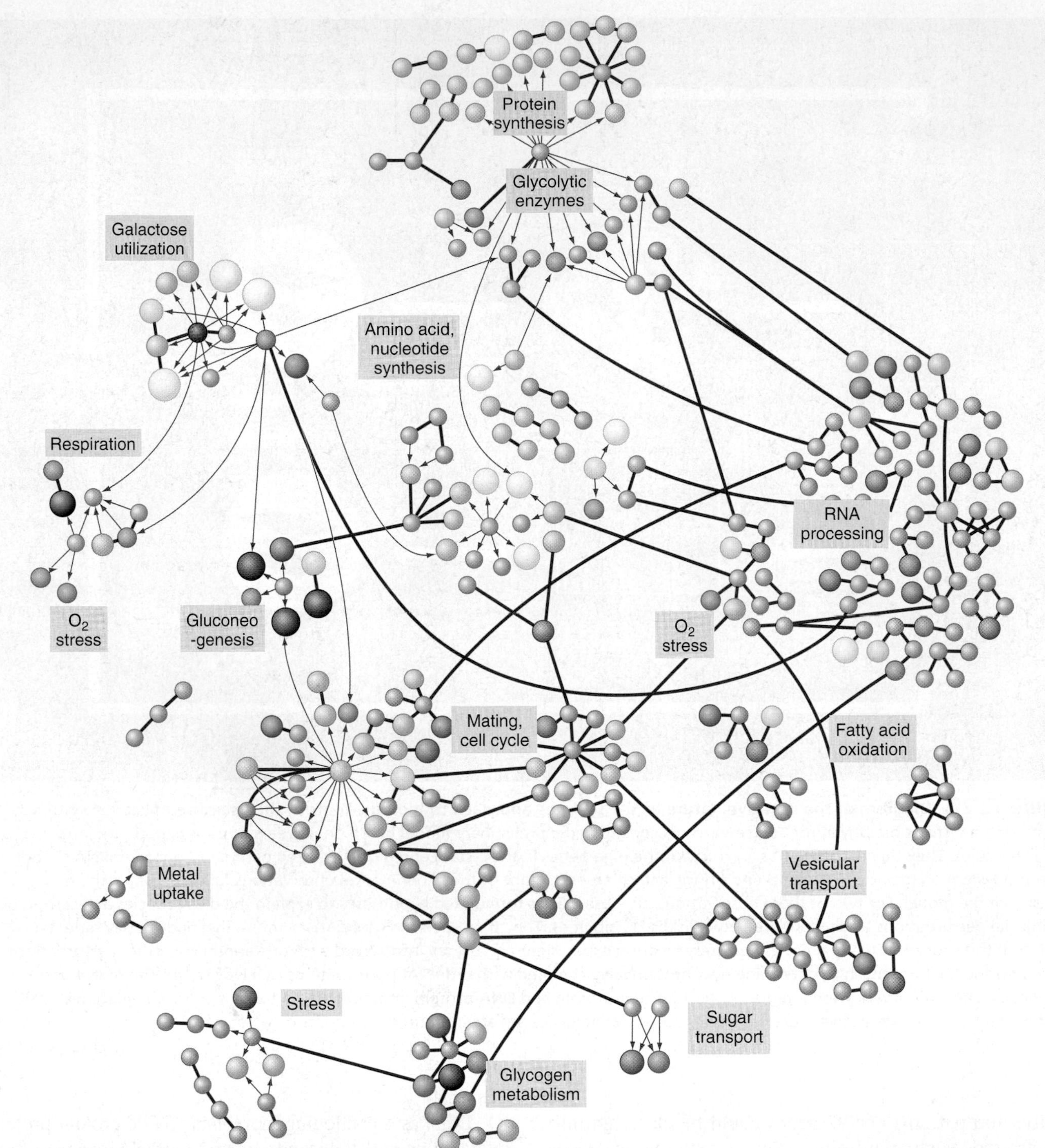

Figure 12.21 How genetic perturbations of the galactose-utilization system in yeast (the knockout of individual genes) affects the network of interactions with other metabolic and functional systems. This network was developed by combining the clusters of mRNAs defined by knockout perturbation experiments and protein-protein and protein-DNA interaction data. The *yellow arrows* indicate protein-DNA interactions (transcription factor activity), and the *blue bars* show protein-protein interactions. The *red circle* indicates where the Gal 4 gene has been knocked out. A *gray scale* depicts levels of mRNA expression: *black* = high levels; *white* = low levels.

Finally, the researchers used the ICAT technology to compare the concentrations of various proteins in two types of wild-type yeast cells: those with the galactose system running and those with the system shut off. They examined about 300 proteins, 30 of which exhibited significant changes in concentration; significantly, 15 of those 30 had no changes in their mRNA concentrations. One conclusion is that for these 15 proteins, the regulation of concentration must operate at a posttranscriptional level.

The galactose-utilization experiments show that in studying biological systems, it is important to analyze many types of data and integrate the various levels of biological information obtained. In the yeast studies, investigators looked at mRNA levels, protein levels, protein-protein interactions, and protein-DNA interactions. The experiments also illustrate how useful global, iterative, hypothesis-driven studies, and model refinement are in the examination of biological systems. The Genetics and Society box on p. 457 examines the challenges faced by professors trying to incorporate systems biology into the university curriculum.

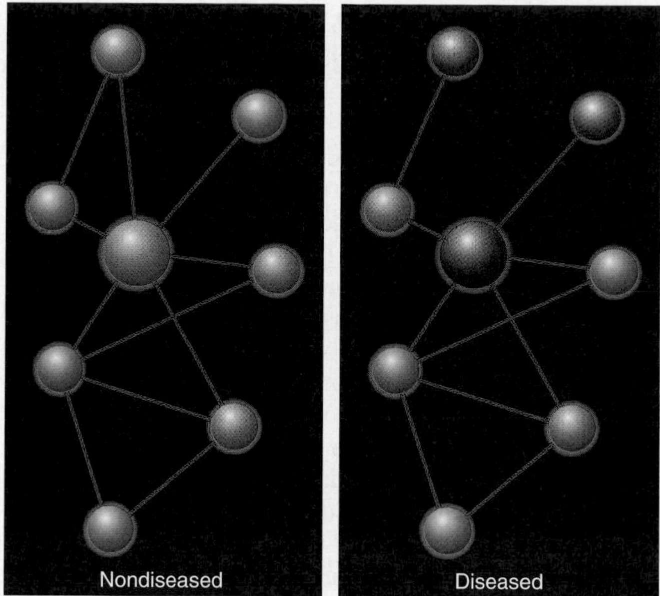

Figure 12.22 Diagram of a hypothetical network in a normal and a diseased cell. As in Fig. 12.2, the *blue circles* represent nodes (proteins), and the *lines* depict the connections (interactions) between nodes. Normal nodes are in *blue*; nodes perturbed by disease to differing extents are in *red, orange,* and *green.*

12.5 A Systems Approach to Disease Leads to Predictive, Preventive, and Personalized Medicine

Overview of the Systems Approach to Disease

Genetic and environmental perturbations that cause cellular networks to alter their patterns of gene expression can lead to disease (**Fig. 12.22**). The disruptions that result in disease may arise from mutated genes, as in various types of cancer, or from infection by microbes, as in AIDS, smallpox, and the flu. The view of disease as perturbations in cellular networks opens the door to new approaches to diagnostics, therapeutics, and, ultimately, prevention. For example, the altered patterns of gene expression in the disease-perturbed networks of a cancerous prostate gland cause changes in the levels and types of proteins expressed by various prostate cells. A significant fraction (roughly 10%) of the proteins with disease-perturbed expression levels are secreted into the blood where they constitute a protein fingerprint that indicates the status of the organ from which they were secreted. In this case, the altered levels of proteins in the blood fingerprint reflect the diseased state of the gland (the fingerprint would be different if the gland were healthy) and even help distinguish the type of disease (for instance, inflammation or cancer). Normal prostate cells secrete a protein called prostate-specific antigen (PSA) into the blood, and this is one component of the blood protein fingerprint for the prostate. In cancer, the levels of PSA increase in the blood, and hence, blood measurements of this protein are routinely used to detect prostate cancer. There are probably 50 additional proteins in the prostate fingerprint, and measuring each of these in the future will give much more accurate diagnoses of prostate cancer. Each major organ or cell type in the body undoubtedly produces a molecular fingerprint. In this way, the blood becomes a window into health and disease.

Some of the proteins in a molecular fingerprint can point to the protein networks that have been perturbed by disease. An understanding of the protein interactions in these networks can lead to new candidates for *drug targets:* proteins whose interactions with complementary drugs will either kill the cell (in the case of cancer) or alter the function of the network back toward normal. One can even imagine the future creation of drugs able to prevent disease by preventing networks from becoming perturbed in the first place. This systems approach will undoubtedly lead to the integration of diagnosis and therapy. It will also lead to a revolution in medicine in which predictive, preventive, and personalized modes will replace the largely reactive current model in which physicians begin treatment only when a person is sick.

This new medicine will require new technologies. Over the next 10 years or so, nanotechnology will revolutionize DNA sequencing, making it possible to rapidly sequence individual human genomes for well under \$1000. More sensitive *molecular imaging* techniques will permit the noninvasive visualization of drug activity and function in model organisms and humans. *Microfluidics* and *nanotechnology* will produce devices that measure, identify, and inexpensively quantify in a fully automated high-throughput platform thousands of proteins from a small drop of blood or assess the information content of individual cells. Let us consider one possible scenario for this revolution in medicine.

Predictive, Preventive, and Personalized Medicine

Prediction

It is possible that within 10 years, physicians will have in place two major approaches to medical prediction. First, all patients will have their genome sequence determined by a nanotechnology device. From this sequence, it will be possible to glean the information for predicting the individual's future health. For instance, a woman might learn that she has a 30% chance of developing cardiovascular disease by age 50; a 40% chance of getting ovarian cancer by age 60; and a 40% chance of developing rheumatoid arthritis by age 65. Second, quantitative measurements of the 1000–2000 proteins in a droplet of blood, also made by a handheld nanotechnology device, will be sent by wireless transmission to a server that will process the data in this molecular fingerprint and send the client/patient and the physician an e-mail stating, for example, "you are fine–do this again in six months" or "see your oncologist." These blood fingerprints will allow the very early detection of disease as well as the stratification (that is, separation) of particular disease types (prostate cancer, for example, is probably four to six diseases rather than one). The fingerprints will also make it possible to follow a patient's response to therapy and to detect adverse drug reactions in their early stages. Because predictive medicine without the ability to treat or prevent is unsatisfactory to most patients, better therapies and prevention will have to emerge along with the predictive tools.

Prevention

The development of efficient nanolaboratories able to measure the protein and mRNA levels as well as protein-protein and protein-DNA interactions in individual cells will give rise to a new kind of preventive medicine that will work hand in hand with the lifestyle measures (eating right and exercising) used today. The new prevention strategies will rely on the delineation of the networks in normal and diseased cells; analyses that clarify their differences; and the identification of key proteins (central nodes in the networks) as potential drug targets. For instance, neuropsychiatrists had known for more than a decade that the protein serotonin (a neurotransmitter) plays a crucial role in the network whose perturbation contributes to clinical depression in humans, but they weren't sure how the serotonin functioned or what went wrong in depression. Studies completed in 2005 show that a protein designated p11 is another key player in that same network. It appears to modulate serotonin activity by influencing the number of serotonin receptors in the membranes of brain neurons that modulate mood. While current drugs, such as Prozac and other selective serotonin reuptake inhibitors, slow the resorption of serotonin from the synapses, future antidepressive drugs that target the p11 node of the network may help the serotonin in the synapses more effectively do its job.

Over the next two decades, the systems approach will not only produce more effective therapeutic agents for treating existing diseases, it will also lead to the development of drugs that can prevent disease by intervening to keep networks from becoming perturbed. Physicians will then be able to counsel patients with greater insight and sophistication. For example, your doctor might explain that although you have a 40% chance of developing ovarian cancer by the age of 60, if you start taking a specific medication when you are 40, you can essentially prevent the disease by reducing that probability to 2%.

Personalization

Because the genome of one person differs from that of another by about 6 million base pairs, we are each susceptible to differing combinations of diseases. Increasingly, medical practitioners will be able to practice personalized medicine by applying the power of predictive and preventive medicine to our individual needs.

Predictive, preventive, and personalized medicine will transform the health-care industry and the practice of medicine. Striking changes based on the systems approach to biology will have a profound effect on all their components, including the research and development of pharmaceutical companies, the research and development of the biotechnology industry, the provision of health insurance, and the curricula of medical schools.

Connections

While all the informational components of a cell or organism play an interactive and integrative role in the execution of various biological functions, proteins are particularly important because they execute the digital information of the genome. Systems biology seeks to understand the emergent properties of a system through high-throughput, global analyses that look at all of the system's elements, all of its interactions, and the biological context in which it functions. Biologists use the systems approach to study the interacting molecules of complex molecular machines as well as the molecules and cells of biological networks.

Chromosomes are one type of complex molecular machine. Composed of many protein molecules in addition to DNA, they package and manage the availability of the information carried by DNA. Their hierarchic organization enables them to go through a cycle of condensation and decondensation with each cell cycle. They also contain specialized regions that govern their duplication and segregation during nuclear division. The activity of large chromosomal regions is controlled not just by local DNA sequence, but by the way the DNA is packaged; some regions, for example, are marked in a way that irreversibly extinguishes all activity, including gene expression. In Chapter 13, we examine how proteins interact with DNA to generate the functional complexity of a chromosome.

GENETICS AND SOCIETY

The Academic Challenges of Systems Biology

Among biologists, systems biology raises some of the same concerns as the Human Genome Project, including the relative merits of big versus small science and whether academic institutions can accommodate the unique organizational needs of systems biology. Big science requires the integration of biology, technology, computation/mathematics, and medicine and, accordingly, requires a cross-disciplinary environment. It depends on teamwork, high-throughput facilities, and a strong computational infrastructure. Small science is practiced by individual academic scientists working most often with up to 10 or so colleagues on a highly focused problem, such as the isolation of one or a few genes or proteins and the identification of their function.

Big and Small Science Are Complementary Approaches

The big-versus-small-science issue has many ramifications for systems biology. Since big science entails large expenses to support a cross-disciplinary environment, high-throughput technical platforms, strong computational infrastructures, and global experiments, the question naturally arises, Will the funding of small science be sacrificed for the financial needs of big science? As far as the Human Genome Project is concerned, one can say that the endeavor probably brought more resources into biology than any other single effort, apart perhaps from the drive to understand and find a cure for cancer. The systems biology approach will empower small science just as its human genome counterpart did. Small science brings the ability to understand the subtle details of biology, while the big science of systems biology's discovery phase generates hypotheses that need to be tested by small science in systems biology's iterative, hypothesis-driven phase. Big science and small science are thus highly complementary approaches to biology and medicine in the twenty-first century.

Can Academia Accommodate the Needs of Systems Biology?

Because the big science of systems biology requires an infrastructure that is distinct from its small science counterpart, it is not clear how well scientists can practice systems biology in academic institutions. Here are several reasons why. First, systems biology requires a cross-disciplinary environment in which biologists, chemists, computer scientists, mathematicians, engineers, and physicists work side by side; interact constantly; and learn to speak the language of each other's discipline. This interdisciplinary team approach enables them to understand and push the biology, but it runs counter to the academic ethos where the goal of tenure requires that young scientists work independently. Moreover, many science departments have strong views about what constitutes good science (often one-gene-at-a-time science) and how students should be educated (often mostly in the relevant discipline). Second, to achieve significant advances, all scientists in a systems biology center must work in groups that focus on a small number of problems. This type of group focus runs counter to the tendency of academics to work independently on problems that interest them. Third, systems biology requires genomic and proteomic high-throughput facilities that are often not available in universities. Fourth, systems biology depends on the rapid establishment of efficient industrial and academic partnerships that bring in outside expertise and leverage the scientific strengths of the systems biology center. Most universities don't easily form such partnerships. Fifth, a systems biology center must be able to raise substantial funds outside the ordinary federal grants and contracts, because creating and maintaining the appropriate infrastructure is expensive. Finally, systems biology centers that use biology to drive the development of new technologies and computational tools will generate enormous amounts of intellectual property. Most academic institutions are not very good at generating patents to effectively protect intellectual property. Nor are they effective in dealing with the dispersion of this intellectual property to existing companies (in the form or intellectual property licenses) or in catalyzing the creation of new start-up companies. An interesting question is whether there are organizational structures within the academic framework that could provide the very different infrastructure that systems biology requires.

Essential Concepts

1. The practice of systems biology requires one to identify the elements of a biological system; measure their changing relationships; measure their relationships to the other systems functioning in the same context (organism or cell); and with this information, attempt to explain the system's emergent properties. The key point about biological systems is that they are dynamic entities that reflect changes in cellular states across evolutionary, developmental, and physiological responses.
2. Biological information consists of the digital information of the genome and environmental signals from outside the genome, which modify the digital genome's output. Gene regulatory networks integrate the inputs of information from signal transduction pathways and transmit information to the batteries of genes that encode the protein networks that carry out metabolism, development, and physiology. To truly understand a system, it is necessary to integrate biological information from many hierarchic levels.
3. Researchers use genetic and environmental perturbations to study biological systems. The idea is to create a preliminary model of the system from preexisting knowledge; perturb the system; gather global genomic, proteomic, genetic and biological assay data sets; integrate the data; compare a visualization of the integrated data against the model; and where discrepancies arise, formulate hypotheses that can be tested by another round of perturbations. This process is repeated until the experimental data and the model are in accord. Biologists can then begin to explain the system's emergent properties, predict its response to never-before-seen perturbations, and even reengineer the system to have predictable, new emergent properties.
4. Proteomics is the global analysis of proteins in the cells or organisms being studied; the collection of these proteins is termed the *proteome*. Proteins have many different features requiring measurement: levels, interactions, localization, modifications, activation, half-lives, three-dimensional structure, and correlation with function. Global techniques for quantifying proteins have been developed using measurements from mass spectrometry and protein chips. Several approaches (yeast two-hybrid and protein pull-down) have been developed for studying the protein-protein interactions that are the fundamental linkages of protein networks. Likewise, ChIP/chip techniques permit the measurement of interactions between transcription factor and *cis*-DNA elements. Such interactions constitute the fundamental links of gene regulatory networks.
5. A study of the galactose-utilization system in yeast illustrates the effectiveness of the systems approach, including genetic and environmental perturbations and the integration of different types of data in a graphical network. The ultimate model resulting from this study demonstrates the interconnectedness of many networks within the yeast cell and the importance of posttranscriptional regulation.
6. A systems approach to disease encompasses the idea that disease arises from perturbed networks. From this simple idea come powerful new approaches to diagnosis, therapy, and prevention. The systems approach to disease is catalyzing a change from the current reactive mode of medicine to a future of predictive, preventive, and personalized medicine.

On Our Website

www.mhhe.com/hartwell3
Chapter 12

Annotated Suggested Readings and Links to Other Websites

- Historical papers on the origins of systems biology
- Historical and contemporary papers on systems approaches to the biology of model organisms
- Historical and contemporary papers on the global and high-throughput technologies and computational tools integral to systems biology
- Contemporary papers on predictive, preventive, and personalized medicine

Specialized Topics

- Classic systems biology analyses

Social and Ethical Issues

1. The cost of medicine is rising faster than the rate of inflation. Some argue that because of this, we should not continue with expensive medical research. Yet the systems approach will in all likelihood decrease the cost of medicine. Here are some of the reasons why. First, predictive and preventive medicine are less expensive than the diagnosis and therapy of reactive medicine. Second, the tools of predictive, preventive, and personalized medicine (for example, for sequencing human genomes and measuring proteins in the blood) will be far less expensive and more effective than the current less precise methods for measuring the parameters of health and disease. Third, systems approaches to the discovery of drug targets will diminish the cost of new drug development. The drug industry now spends about $1 billion per new drug; in the future, companies should be able to spend 20 times less than that. Do you think that the potential future savings related to the application of systems approaches to medicine are worth the current cost of clinical and basic research? In formulating your response, consider the potential for less expensive medical services in developing nations and the possibility of providing insurance to the 45 million uninsured people now living in the United States.

2. The digitization of medicine over the next 10 years may have a far greater impact on society than the digitization of information technology has already had. First, digitized medical records and tests are easy to transfer from one place to another. As a result, they can be made in one place, read by an expert in another, and returned to the patient and care provider in yet another. Second, digitized tools for extracting information from single molecules (such as proteins) and single cells (such as cancerous cells) will make it possible to rapidly and inexpensively compare normal and diseased cells taken by biopsy from the same patient; delineate the disease-perturbed networks; and correlate them with molecular fingerprints in the blood. Accurate diagnosis (including disease stratification), efficient selection of the optimal treatment, and the ability to monitor a patient's progress with a high degree of accuracy will follow. The digitization of information technology and communications has led to big revolutions in the lifestyles of both the developed and developing countries. The digitization of medicine will essentially make it possible to practice predictive, preventive, and personalized medicine inexpensively on a very large scale. Suggest some of the revolutions in health care that will come from the digitization of medicine. How will these changes be similar to or distinct from those which emerged from the digitization of information technology and communications? Compare the advantages of digitization with its potential drawbacks and describe what you consider to be the most effective way to implement the digitization of medicine.

Solved Problems

I. You hypothesize that breast cancer that has been histologically classified as invasive ductal cancer can be further subclassified based on molecular signatures. If you can create such subcategorizations, you may be able to stratify the cancers and develop specific therapies for one or more of the subcategories. You and your collaborators have biopsy tissue from 63 invasive ductal breast cancer patients and clinical data on the course of disease in each of these patients. You plan to use microarray technology to measure gene expression in each of these samples, and then use the results to classify the tissues. When you analyze these samples with microarrays, what other samples should you analyze and why?

Answer

Controls are an important part of any scientific experiment. You need to choose control samples to provide confidence that the subcategories you identify truly reflect a stratification of invasive ductal breast cancer. In particular, you must ensure that the biopsies you receive are neither misdiagnosed nor improperly collected. For example, if some normal breast tissue is mistakenly included as a tumor in your analysis, you may identify normal tissue as a "subcategory of invasive cancer". Thus, you will want to include normal breast tissue as well as other types of breast cancer in your arrays. In addition, your analysis will require comparing gene expression in the cancer samples to a standard reference, so your set of normal controls will serve this important function. Finally, analyzing a comprehensive set of control samples will also help to develop a statistical model for the variability of your measurements and thus increase confidence in your results.

You may also wish to approach other hypotheses with your data. For example, you may want to identify markers specific to breast cancer or breast tissue. Thus, you may wish to include some samples of other normal tissues as well as cancers of these tissues.

II. The following diagram is a small portion of the much larger computerized model of an interaction network in the yeast *Saccharomyces cerevisiae* that is shown as Fig. 12.21 on p. 454. The nodes in the diagram represent either proteins or the genes encoding those proteins. Blue lines connecting nodes indicate protein-protein interactions. Red arrows connecting nodes show protein-DNA interactions, with the protein at the base of the arrow and the DNA sequence at the arrowhead.

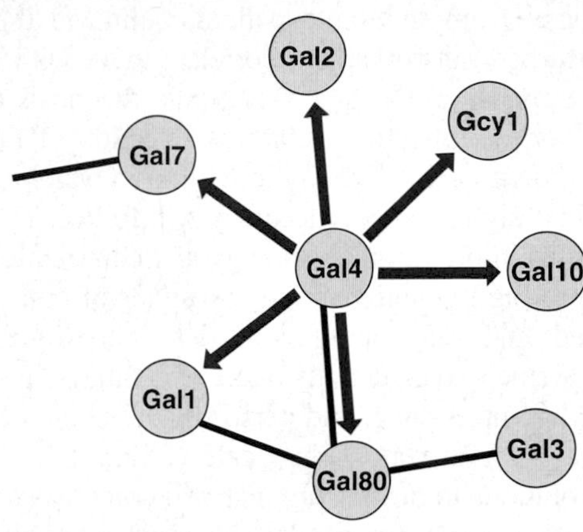

a. What kinds of techniques discussed in this chapter might have been involved in the collection of data for this interaction network?
b. What proteins in the diagram are likely to be in a complex with each other?
c. What protein(s) is (are) likely to act as a transcription factor, and how would this transcription factor operate?
d. The proteins and genes indicated in the diagram enable yeast cells to utilize the sugar galactose as a carbon source. You hypothesize that these genes are regulated at the transcriptional level by the type of carbon source: for example, expression of many of these genes might be increased if the medium contained galactose, but expression might be repressed if the medium contained glucose. What kinds of genome-scale experiments might you do to test such hypotheses?

Answer

a. The data describing protein-protein interactions could have been derived in either of two ways. First, researchers might have used the affinity capture/mass spectrometry technique diagrammed in Fig. 12.15 on p. 448. A second method to uncover protein-protein interactions is the yeast two hybrid approach shown in Fig. 12.16 on p. 449. Databases of the yeast interactome already exist that report the results of systematic tests of each yeast gene fused either to a DNA binding domain or to a DNA activation domain. Protein-DNA interactions could be identified by the ChIP/chip technique shown in Fig. 12.18 on p. 451, if antibodies that recognize the transcription factor(s) among the proteins in the diagram were available.
b. The proteins that could be in a complex must be linked by protein-protein interactions. These include Gal4, Gal80, Gal1, and Gal3.
c. The diagram indicates that the Gal4 protein can bind to DNA sequences in the vicinity of several genes, including the genes encoding Gal1, Gcy1, Gal2, Gal7, Gal10, and Gal80; the DNA sequences are presumably near the promoters of these genes. The binding of Gal4 to these DNA sequences in theory would regulate the transcription of these other genes.
d. One type of experiment to test your hypothesis is a microarray analysis. You would grow yeast cells in the presence of either galactose or glucose, and then compare the levels of the mRNAs for all yeast genes under these two conditions, with particular attention to those genes in the diagram that are potential targets for regulation by a protein complex containing Gal4. Even more interesting would be the microarray analysis of transcription in yeast that had some elements of the galactose utilization system "knocked out" by mutation.

A great deal of information about the small network shown in this diagram has been gleaned from a variety of experiments. Chapter 18 presents a detailed model for the regulation of the genes in yeast involved in the utilization of galactose.

Problems

Interactive Web Exercise

Systems biology depends on the computerized analysis of massive data sets to extract useful information. The interactive web exercise at www.mhhe.com/hartwell3 (click on Chapter 12) provides model data sets and access to computer algorithms that will allow you to explore issues and techniques in systems biology such as the stratification (classification) of tumors by microarray analysis and the use of mass spectrometry to investigate proteomes.

Vocabulary

1. Choose the phrase from the right column that best fits the term in the left column.

a. preventive medicine	1. protein that binds a *cis*-control element to regulate transcription
b. vaccine	2. a property arising from the system as a whole
c. systems biology	3. a set of quantitative measurements that identify the state of a biological system
d. proteome	4. intervention designed to prevent disease
e. immunity	5. technology to manipulate cells, typically with channels on the order of one micron
f. tolerance	6. a protein network describing all the interactions in a cell
g. proteomics	7. defining all the components of a biological system and how they function together
h. genomics	8. technology to make measuments on the scale of nanometers, permitting massive parallelization
i. biological system	9. an antigenic preparation used to produce active immunity to a disease
j. network graph	10. the collection of all the proteins in a cell or organism
k. emergent property	11. the ability to generate immune responses to infections or vaccines
l. *cis*-control elements	12. a network in which the nodes are proteins and the edges are interactions between proteins
m. transcription factor	13. the ability to prevent the body from making immune responses to its own proteins
n. protein networks	14. the global study of genomes
o. interactome	15. the global analysis of the proteins in a particular cell type or organism
p. molecular fingerprint	16. controlling DNA sequences adjacent to a gene
q. microfluidics	17. diagram of a network in which the connections between nodes represent physical interactions between elements
r. nanotechnology	18. a collection of interacting elements that carry out a specific biological task

Section 12.1

2. How has the Human Genome Project enabled systems biology in the 21st century?

3. What are the four fundamental concepts related to defining a biological system?

4. What are the two fundamental components of systems?

Section 12.2

5. If you were to catalog *cis*-control elements, would you be studying protein networks or gene regulatory networks? What other information would be required to gain an understanding of the kind of network you are studying?

Section 12.3

6. You prepare a column in which rabbit antibodies specific to the *Drosophila melanogaster* protein ZW10 are covalently coupled to beads of an inert resin. (See Fig. A on p. 95 of Chapter 4 for a description of the role of the ZW10 protein.) You now prepare large amounts of an extract from early *Drosophila* embryos, and pass this extract over the anti-ZW10 column and over a control column containing antibodies from a rabbit that has not been exposed to any *Drosophila* proteins. You then elute proteins from the columns with buffers

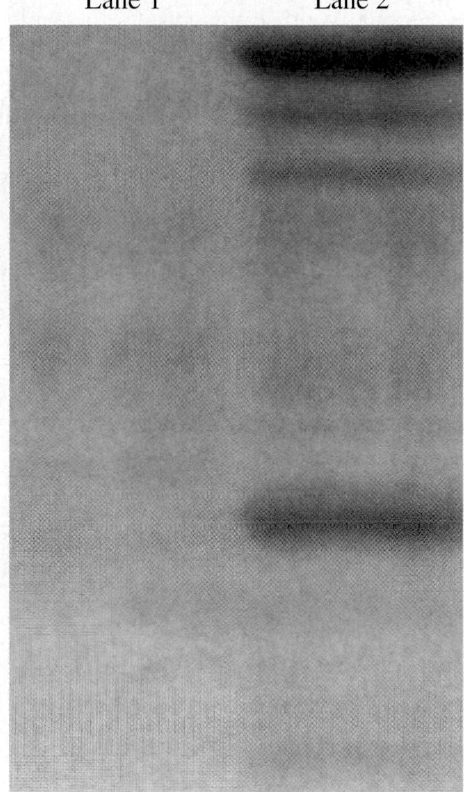

containing high concentrations of salt (NaCl). The high salt disrupts interactions between proteins. Samples from these elutions are now run on a polyacrylamide gel, and the gel is stained with Coomassie Blue, a dye for proteins of all types. Results from this experiment are shown in the following picture; Lane 1 shows the elution from the control column, and Lane 2 shows the elution from the column with anti-ZWIO.

a. What is the purpose of this experiment? What kinds of proteins do you expect to find in the two lanes of the gel?

b. How could you use a mass spectrometer to determine the identity of the protein bands stained with Coomassie Blue? Explain how the mass spectrometer works and how the data you obtain would identify a particular Coomassie Blue-staining band.

c. How could you find the same kind of information if an antibody specific to the ZW10 protein was not available?

7. In an alternative approach to finding proteins that can associate with the ZW10 protein, you prepare a "bait construct" in which the open reading frame of the fly *zw10* gene is fused in frame to sequences encoding the DNA binding domain of the yeast Gal4 protein, as in Fig. 12.16 on p. 449. You transform this bait construct into yeast containing a *HIS3* gene that has been engineered to contain a binding site for the Gal4 protein. That binding site is located just upstream of the *HIS3* gene's promoter.
a. How could you use this system to search for proteins that interact with ZW10? What is the underlying biological basis of the yeast two-hybrid system?
b. What are some of the advantages and disadvantages of this yeast two-hybrid system relative to the antibody columns described in the previous problem (#6) in terms of searching for proteins that interact with ZW10?
c. How could you try to search systematically for all potential protein-protein interactions in *Drosophila* with the yeast two-hybrid system?

8. Two different laboratories carried out yeast two-hybrid studies on the same 6000 yeast proteins and their results are reported in the table below.

Lab	Proteins studied	Numbers of interactions	Discrepancies
A	6000	15,600	7,800
B	6000	15,600	

a. How would you explain these data?
b. Can you suggest experiments to resolve at least some of the discrepancies?

9. Suppose you are conducting an analysis of the entire yeast transcriptome using DNA chips. You examine yeast in two different cellular states, and you determine that about 400 genes increase or decrease their levels of expression in state one as opposed to state two. You now analyze 600 proteins from these two different yeast states using the ICAT proteomics technology. To your amazement, you find that 50 protein levels change when their corresponding mRNA transcripts levels do not. Moreover, you find that 60 protein levels do not change when their corresponding mRNA transcript levels remain the same. Finally, you notice that 30 protein levels changed in opposite directions from that of their mRNA transcript counterparts. What simple explanation can you offer for these observations (apart from error in the measurements)?

10. You hypothesize that the addition of a phosphate group to a particular serine amino acid in a kinase enzyme is important for the ability of this kinase to play a role during mitosis. How could you use two dimensional gel electrophoresis and/or mass spectrometry to determine whether or not the kinase was indeed phosphorylated at this serine during mitosis?

11. In the ChIP/chip technique, the DNA to which the transcription factors are bound is fragmented into 300–500 bp lengths. The typical transcription factor binding site is 6–15 bp in length. How could you increase the resolution of the ChIP/chip technique with regard to the identification of the actual binding site?

12. The figure at the bottom of this page summarizes the results of ChIP/chip analysis for the transcription factor p53, which plays important roles in protecting cells from DNA damage; mutations in the gene encoding p53 may lead to cancer (see Chapter 19 for details).
a. What do you think this figure represents? How did the researchers obtain the data summarized in the figure?
b. What is a likely consensus sequence for p53 binding sites?
c. Suppose that active p53 protein is a dimer of two identical subunits, and that these subunits are asymmetric in three dimensional space (that is, each subunit has a "front" that is different from its "back"). On the basis of the data in the figure, would you conclude that the two subunits in the dimer face each other, or that the front of one dimer is juxtaposed to the back of the other?
d. How might the results summarized in the figure be useful to other scientists?

Section 12.4

13. Answer the following true or false:
a. Systems biology employs both discovery science and hypothesis-driven science.

b. Gene regulatory networks integrate information they receive from signal transduction networks and transmit it to protein networks.
c. A transcription factor binds only a single *cis*-control region in the genome.
d. DNA sequence may be modified by environmental information.
e. The yeast cell contains only about 6000 proteins.
f. The integration of different types of global sets may be carried out with graphical networks.
g. The proteome of the organism is the sum of the proteomes of all cells in all developmental or physiological states.
h. The mass spectrometer currently has the capacity to quantify globally all of the proteins in a given cell type.
i. Protein chips are as global in their measurement capacity as DNA chips.
j. The global localization procedure for identifying transcription factor binding sites also works for other proteins or protein complexes directly bound to DNA or indirectly bound to DNA through other DNA-binding proteins.
k. The galactose utilization system is interconnected to many other cellular systems in yeast.

14. A friend of yours states that in his immunology research he is measuring the levels of 100 cytokines in response to knock out perturbations of interesting genes in his system. He claims that he is doing systems biology. What do you think?

15. Researchers carried out the following experiments on the galactose utilization system in yeast. (1) They grew wild-type yeast in the presence and absence of galactose (that is, with the galactose-utilization system running and shut down, respectively). (2) They used microarrays to quantify all mRNA expression differences between these two states. (3) They used the ICAT approach to quantitate protein expression differences in 300 proteins in these two states. About 30 of the 300 proteins changed expression levels under these conditions. However, for 15 of these 300 proteins, there was no change in their corresponding mRNA levels between the two states. Suggest two explanation for these observations.

Section 12.5

16. Answer the following true or false:
a. A systems approach to disease embodies the concept that diseased cells have some abnormal networks.
b. A protein molecular fingerprint in the blood has the capacity to assess the state (e.g., health or disease) of the cell type from which it was secreted.
c. Systems approaches to disease provide new approaches for the discovery of drug targets.
d. Predictive medicine without the ability to treat the predicted disease raises ethical concerns about whether insurance companies could use this information to modify insurance rates.
e. Predictive, preventive and personalized medicine will require medical education to be greatly modified.

The Eukaryotic Chromosome: An Organelle for Packaging and Managing DNA

Chapter 13

Mouse mitotic metaphase chromosomes stained for karyotype analysis (magnification 600×).

To build an artificial chromosome able to function as part of a cellular genome, what chromosomal components would you need to assemble? Three sets of observations help answer this question. In Chapter 4, you saw that when chromosomes segregate during mitosis or meiosis, some of the cell's newly elongated spindle fibers attach to the **centromere**, the pinched-in portion of the chromosome viewed under the microscope (**Fig. 13.1**). A centromere is therefore essential to proper chromosome segregation. Second, scientists have found that the ends of linear chromosomes remain intact even though linear pieces of DNA, when introduced into a cell, are susceptible to degradation by nucleases. A search for the chromosomal element that prevents degradation revealed that specific repeated DNA sequences and special proteins compose the tips of linear chromosomes. Known as *telomeres,* these protective DNA-protein caps are another critical chromosome component. Finally, you know that DNA must replicate before cell division so that each daughter cell receives a complete copy of the genomic DNA. To ensure that your artificial chromosome replicates properly, you must include at least one **origin of replication.**

In the 1980s, molecular geneticists took these three key chromosomal elements—centromeres, telomeres, and origins of replication—and, using raw materials from the yeast *Saccharomyces cerevisiae,* constructed the first artificial eukaryotic chromosome, known appropriately as a yeast artificial chromosome, or YAC. Single-celled yeast was the organism of choice because it is easy to manipulate, its genetic machinery resembles that found in the cells of higher organisms, and it is so far the only eukaryotic organism whose origins of replication, centromeres, and telomeres have been defined as discrete, small segments of DNA. Yeast cells have 16 chromosomes that range in length from 235,000 to more than 2 million base pairs. Within each chromosome, the centromere, is only about 120 bp long. In contrast, human chromosomes have an average length of 100 million base pairs and a centromere may span 1 million base pairs.

Manipulation of the YAC construction process has produced insights into chromosome function. Plasmids containing only origins of replication but no centromere or telomere replicate but do not segregate properly. Plasmids with origins of replication and a centromere but no telomeres replicate and segregate fairly well if they are circular; if they are linear, they fragment and eventually become lost from the cell. Small DNA molecules carrying all three elements replicate and segregate as linear chromosomes, but unlike natural yeast chromosomes, they do not segregate accurately; instead, they segregate at random. Studies aimed at understanding this phenomenon revealed that the amount of DNA in a chromosome influences that chromosome's function. YACs carrying 11,000 bp show segregation errors in 50% of cell divisions. YACs containing 55,000 bp show segregation errors in 1.5% of cell divisions. With artificial chromosomes extended to more than 100,000 bp, the rate of segregation error falls to 0.3%. Although the longer artificial chromosomes function well, their frequency of segregation error is still 200 times greater than that seen with natural yeast chromosomes of normal size,

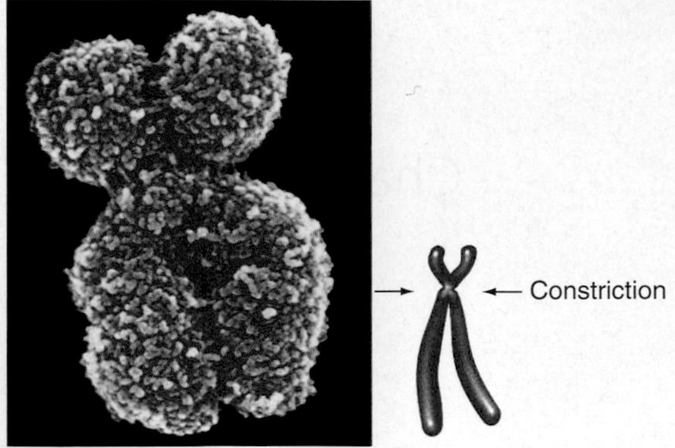

Figure 13.1 A chromosome's centromere looks like a constriction under the microscope. Color-enhanced scanning electron micrograph of a human metaphase chromosome and interpretive drawing show that the centromere corresponds to the major constriction of mitotic chromosomes.

indicating that some subtle aspects of chromosome structure and function remain to be discovered.

The successful construction and analysis of YACs confirms that three noncoding elements are essential to chromosome function: centromeres, telomeres, and origins of replication. Observations of YAC behavior show there is a minimum length required for accurate chromosome segregation. Geneticists can use YACs for further studies of mitosis and meiosis. They can also use them as vectors for cloning large segments of DNA, such as those used in the Human Genome Project. Recall from Table 9.2 on p. 311 that cosmids can accept at most 45,000 bp of foreign DNA, while YACs can accommodate up to 1 million base pairs.

In this chapter, we examine the structure and function of significant features of the eukaryotic chromosome. Our discussion reveals that the chromosome is a dynamic organelle for the packaging, replication, segregation, and expression of the information in a single long molecule of DNA. (As noted earlier in the book, we consider chromosomes "organelles" because they are highly organized cellular structures that have important functions and are transmitted to daughter cells at cell division. This contrasts with the usage of many biologists who consider an entity to be an organelle only if it is surrounded by a membrane.) Each chromosome consists of one DNA molecule combined with a variety of proteins. Flexible DNA-protein interactions condense the chromosome for segregation during mitosis and decondense it for replication or gene expression during interphase. Specific sequences in the chromosomal DNA dictate where spindle attachment occurs for proper segregation; others determine where replication begins.

One general theme emerges from our discussion. Chromosomes have a versatile, modular structure for packaging DNA that supports a remarkable flexibility of form and function.

Our presentation of the chromosome as an organelle for packaging and managing genetic information describes

- The components: one long DNA molecule and many kinds of proteins.
- Chromosome structure: how variable DNA-protein interactions create reversible levels of compaction.
- Specialized elements that ensure the accurate duplication and segregation of chromosomes: telomeres, origins of replication, and centromeres.
- How chromosomal packaging influences the activity of chromosomal elements.

13.1 The Components of Eukaryotic Chromosomes: DNA, Histones, and Nonhistone Proteins

When viewed under the light microscope, chromosomes seem to change shape, character, and position as they pass through the cell cycle. During interphase, they look like tangled masses of spaghetti. By metaphase of mitosis, they appear as a set number of paired bars (the two sister chromatids) facing each other across the cellular midplane. In this section, we describe what chromosomes are made of; then, in succeeding sections, we explain how these chromosomal components associate, dissociate, and reassociate to produce the observed metamorphoses of structure.

Each Chromosome Is Composed of a Single Long Molecule of DNA

Researchers learned from physical analyses that each chromosome within a cell nucleus contains one long linear molecule of DNA. In one early study, they placed chromosomal DNA between two cylinders, stretched the DNA by rotating one of the cylinders, and measured the

TABLE 13.1 Some Molecular Characteristics of the Histone Proteins Found in Calf Thymus*

Histone	Basic/Acidic Amino Acids	Molecular Weight	Total Amino Acid Residues
H1	5.4	23,000	224
H2A	1.4	13,960	129
H2B	1.7	13,774	125
H3	1.8	15,273	135
H4	2.5	11,236	102

*Note the small size of the histones; in addition, their predominantly basic amino acid constitution ensures that they will be positively charged under physiological conditions. The four core histones are in boldface type.

DNA's rate of recoil. Shorter molecules recoil faster than longer ones. When they applied this measure to the DNA in a *Drosophila* chromosome, the length of the DNA molecule was sufficient to account for all the DNA in the chromosome, so the chromosome must contain a single linear molecule.

The complete sequences of many genomes have shown that the single DNA molecule that makes up each chromosome contains genes composed of coding sequences (exons) interspersed with noncoding sequences (introns) as well as different levels of organizational units that affect function. In one type of organizational unit, some genes with similar functions exist in clusters, which may facilitate the coordinated regulation of the related genes. Another organizational feature consists of substantial stretches of noncoding repetitive DNA concentrated in specific chromosomal regions, for example, at centromeres and telomeres. The repetitive sequences at these locations are critical for the function of the chromosomal element.

The Protein Components of Eukaryotic Chromosomes: Histones and Nonhistone Proteins

By itself, DNA does not have the ability to fold up small enough to fit in the cell nucleus. For sufficient compaction, it depends on interactions with two categories of proteins: histones and nonhistone chromosomal proteins. **Chromatin** is the generic term for any complex of DNA and protein found in a cell's nucleus. Chromosomes are the separate pieces of chromatin that behave as a unit during cell division. Chromatin is the same chemical substance that Miescher extracted from the nuclei of white blood cells and named *nuclein* in 1869. Although chromatin is roughly 1/3 DNA, 1/3 histones, and 1/3 nonhistone proteins by weight, it may also contain traces of RNA. Because these RNA bits result mainly from gene transcription and are probably unrelated to chromatin structure, we do not include RNA in our discussion of chromatin components.

Histone Proteins Abound in the Chromatin of All Eukaryotic Cells

Discovered in 1884, **histones** are relatively small proteins with a preponderance of the basic, positively charged amino acids lysine and arginine. The histones' strong positive charge enables them to bind to and neutralize the negatively charged DNA throughout the chromatin. Histones make up half of all chromatin protein by weight and are classified into five types of molecules: H1, H2A, H2B, H3, and H4 (**Table 13.1**). The last four types—H2A, H2B, H3, and H4—form the core of the most rudimentary DNA packaging unit—the **nucleosome**—and are therefore referred to as **core histones.** (We examine the role of these histones in nucleosome structure later.) All five types of histones appear throughout the chromatin of nearly all diploid eukaryotic cells. Histone synthesis occurs during S phase of the cell cycle, when chromosome replication requires more histones for the packaging of newly made DNA. Special regulatory mechanisms tightly correlate DNA and histone synthesis so that both occur at the appropriate time. During protein synthesis, specific modifications occur on amino acids within histones H3 and H4. The acetylation of (addition of acetyl groups to) lysines as well as various methylations are important for the functioning and assembly of chromatin. Variations in the acetylation and methylation patterns in different regions of chromatin result in different functions for those regions.

Interestingly, there is great similarity among particular histone proteins in different organisms. In the H4 proteins of pea plants and calves, for example, all but two of the sequence of 102 amino acids are identical. That histones have changed so little throughout evolution underscores the importance of their contribution to chromatin structure.

Nonhistone Proteins Are a Heterogeneous Group Named by Default

Fully half of the mass of protein in the chromatin of most eukaryotic cells is not composed of histones. Rather, it

(a)

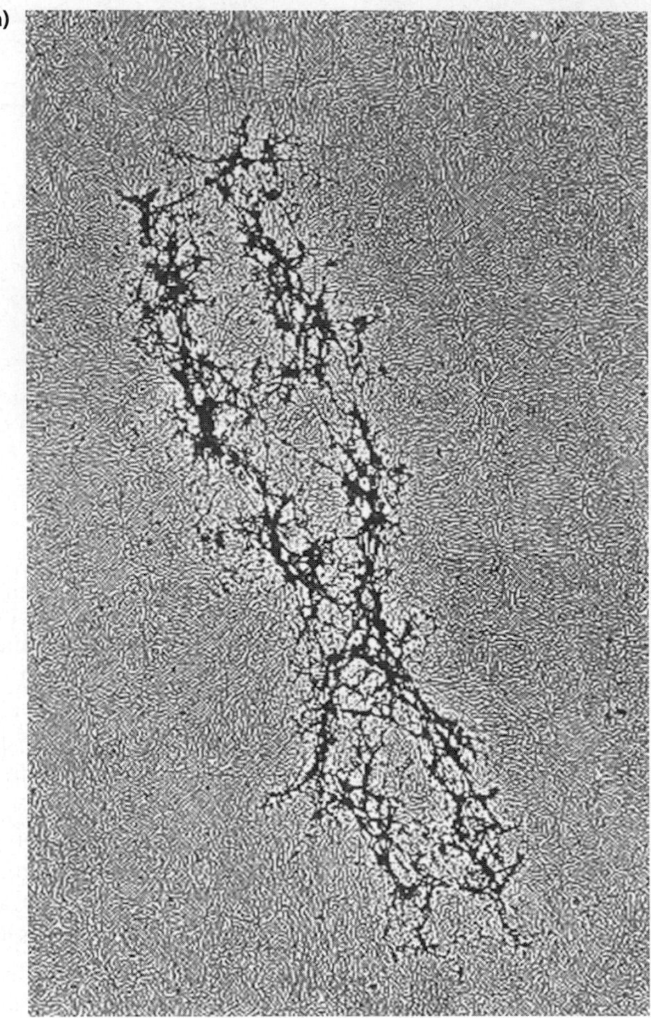

(b)

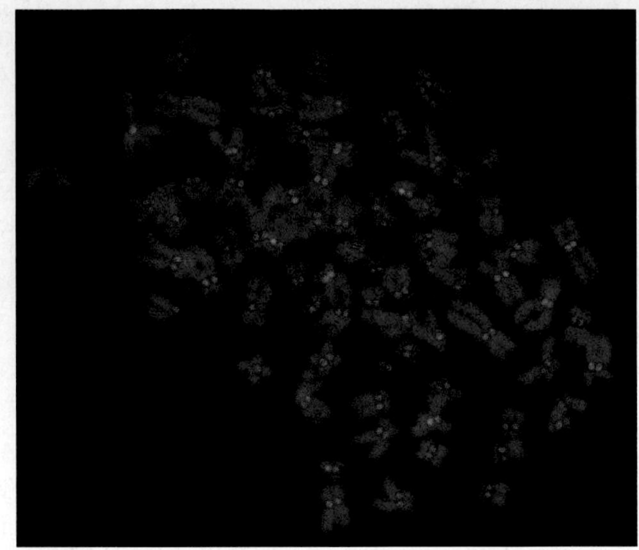

Figure 13.2 Nonhistone proteins have diverse functions. **(a)** Some nonhistone proteins form the chromosome scaffold. When the human chromosome in this picture was gently treated with detergents to remove the histones and some of the nonhistone proteins, a dark scaffold composed of some of the remaining nonhistone proteins, and in the shape of the two sister chromatids, became visible. Loops of DNA freed by the detergent treatment surround the scaffold. **(b)** Some nonhistone proteins power chromosome movements along the spindle during cell division. In this figure, chromosomes are stained in *blue* and a nonhistone protein known as CENP-E is stained in *red.* CENP-E is located at the sister kinetochores of each duplicated chromosome and appears to play a major role in moving separated sister chromatids toward the spindle poles during anaphase.

consists of hundreds or even thousands of different kinds of nonhistone proteins, depending on the organism. The chromatin of a diploid genome contains from 200–2,000,000 molecules of each kind of nonhistone protein. Not surprisingly, this large variety of proteins fulfills many different functions, only a few of which have been defined to date. Some nonhistone proteins play a purely structural role, helping to package DNA into structures distinct from the histone-containing nucleosomes. The proteins that form the structural backbone, or *scaffold,* of the chromosome fall in this category (**Fig. 13.2a**). Others, such as DNA polymerase, are active in replication. Still others are active in chromosome segregation; for example, the motor proteins of kinetochores help move chromosomes along the spindle apparatus and thus expedite the transport of chromosomes from parent to daughter cells during mitosis and meiosis (Fig. 13.2b). By far the largest class of nonhistone proteins foster or regulate transcription and RNA processing during gene expression. Mammals carry 5000–10,000 different proteins of this kind. By interacting with DNA, these proteins influence when, where, and at what rate genes give rise to their protein products.

Unlike the distribution of histones, which are found in similar amounts in all cells of an organism and are dispersed relatively evenly throughout the chromosomes, the distribution of nonhistone proteins is uneven. Nonhistone proteins appear in different amounts and in different proportions in different tissues within the same organism. The difference between universal and sporadic distribution no doubt reflects the functional differences between the two categories of proteins. Whereas histones have the singular task of interacting with DNA to form universal units of compaction, the nonhistone proteins fulfill diverse functions specific to different tissues (for example, tissue-specific transcription factors), to different moments in the life cycle (for example, proteins of the synaptonemal complex), or to different parts of a chromosome (for example, proteins involved in centromere function).

In summary, chromatin is roughly 1/3 DNA, 1/3 histones, and 1/3 nonhistone proteins. We now examine how the components of chromatin associate to package long strands of DNA into chromosomes that can condense or decondense as circumstances require.

13.2 Chromosome Structure: Variable DNA-Protein Interactions Create Reversible Levels of Compaction

Stretched out in a thin, straight thread, the DNA of a single human cell would be 6 feet in length. This is, of course, much longer than the microscopic cell itself, whose dimensions are measured in fractions of millimeters. Several levels of compaction enable the DNA to fit inside the cell (**Table 13.2**). First, the winding of DNA around histones forms small nucleosomes. Next, tight coiling gathers the DNA with nucleosomes together into higher-order structures. Other levels of compaction, which researchers do not yet understand, produce the metaphase chromosomes observable in the microscope.

The Nucleosome: The Fundamental Unit of Chromosomal Packaging Arises from DNA's Association with Histones

The electron micrograph of chromatin in **Fig. 13.3a** shows long, nub-studded fibers bursting from the nucleus of a chick red blood cell. The chromatin fibers resemble beads on a string, with the beads having a diameter of about 100 Å and the string a diameter of about 20 Å (1 Å = 10^{-10} m = 0.1 nm). The 20 Å string is DNA. Figure 13.3b illustrates how DNA wraps around histone cores to form the chromatin fiber's observed beads-on-a-string structure. Each bead is a nucleosome containing roughly 160 bp of DNA wrapped around a core composed of eight histones—two each of H2A, H2B, H3, and H4, arranged as shown in the figure. The 160 bp of DNA wrap twice around this core octamer. An additional 40 bp form **linker DNA,** which connects one nucleosome with the next. Histone H1 lies outside the core, apparently associating with DNA where the DNA enters and leaves the nucleosome. When investigators use specific reagents to remove H1 from the chromatin, some DNA unwinds from each nucleosome, but the nucleosomes do not fall apart; about 140 bp remain wrapped around each core.

One can crystallize the nucleosome cores and subject the crystals to X-ray diffraction analysis. The pictures obtained confirm the nucleosome structure just described and also indicate that the DNA does not coil smoothly around the histone core (Fig. 13.3c). Instead, it bends sharply at some positions and barely at all at others. Because the sharp bending may occur only with some DNA sequences and not others, base sequence helps dictate preferred nucleosome positions along the DNA. It is not yet known exactly what elements of base sequence dictate nucleosome interaction. The bending required for a nucleosome is close to the limit possible for B DNA. The stiffer forms of A DNA and Z DNA cannot bend sufficiently to form nucleosomes. Although DNA sequence and B-form structure help determine where nucleosomes occur, other factors (discussed later in the chapter) also contribute to nucleosome formation.

The spacing and structure of nucleosomes affect genetic function. The nucleosomes of each chromosome are not evenly spaced, but they do have a well-defined arrangement along the chromatin. This arrangement is transmitted with high fidelity from parent to daughter cells. The spacing of nucleosomes along the chromosome is critical because DNA in the regions between nucleosomes is readily available for interactions with proteins that initiate expression, replication, and further compaction. The way in which DNA is wound around a nucleosome also plays a role in determining whether and how certain proteins interact with specific DNA sequences. This is because some DNA sequences in the nucleosome, despite their proximity to the histone core, can still be recognized by nonhistone binding proteins.

TABLE 13.2 Different Levels of Chromosome Compaction

Mechanism	Status	What It Accomplishes
Nucleosome	Confirmed by crystal structure	Condenses naked DNA 7-fold to a 100 Å fiber
Supercoiling	Hypothetical model (although the 300 Å fiber predicted by the model has been seen in the electron microscope)	Causes additional 6-fold compaction, achieving a 40- to 50-fold condensation relative to naked DNA
Radial Loops—Scaffold	Hypothetical model (preliminary experimental support exists for this model)	Through progressive compaction of 300 Å fiber, condenses DNA to rodlike mitotic chromosome that is 10,000 times more compact than naked DNA

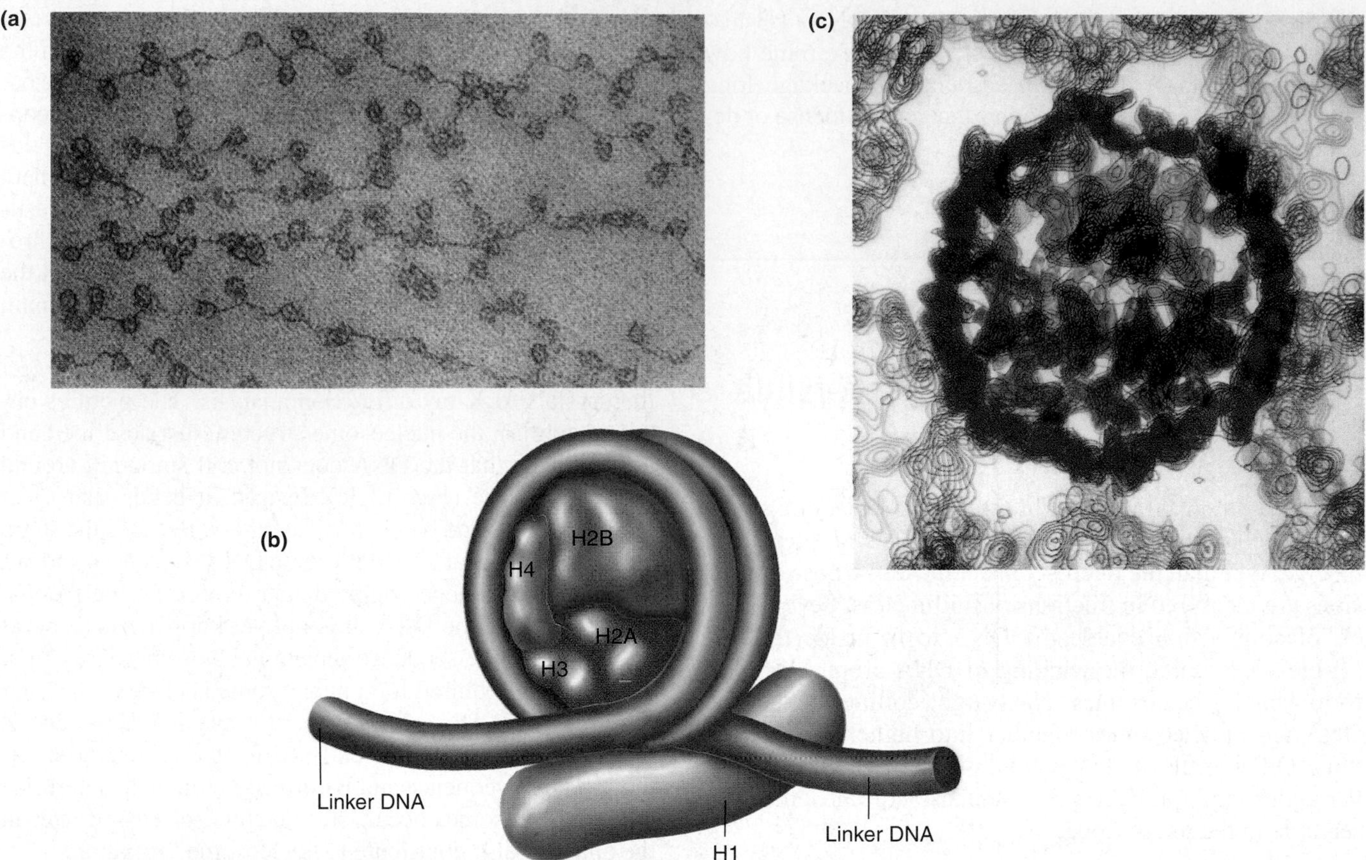

Figure 13.3 Nucleosomes: The basic building blocks of chromatin. **(a)** In the electron microscope, nucleosomes look like beads on a string. **(b)** Nucleosome structure: The DNA in each nucleosome wraps twice around a nucleosome core. An additional molecule of each histone (H2A, H2B, H3, and H4) in the nucleosome core is hidden in this view. Histone H1 associates with the DNA as it enters and leaves the nucleosome. **(c)** The structure of the nucleosome as determined by X-ray crystallography: In this overhead view of the core particle, you can see that the DNA (*orange*) actually bends sharply at several places as it wraps around the core histone octamer (*blue* and *turquoise*).

Packaging into nucleosomes condenses naked DNA about sevenfold. With this condensation, the 2 m of DNA in a diploid human genome shortens to approximately 0.25 m (a little less than a foot) in length. This is still much too long to fit in the nucleus of even the largest cell, and additional compaction is required.

Models of Higher-Level Packaging Seek to Explain the Compaction of Chromosomes in Interphase and Mitotic Cells

The details of chromosomal condensation beyond the nucleosome remain unknown, but researchers have proposed several models to explain the different levels of compaction (see Table 13.2).

Formation of the 300 Å Fiber Through Supercoiling

One model of additional compaction beyond nucleosomal winding proposes that the 100 Å nucleosomal chromatin supercoils into a 300 Å superhelix, achieving a further sixfold chromatin condensation. Support for this model comes in part from electron microscope images of 300 Å fibers that contain about six nucleosomes per turn (**Fig. 13.4a**). Whereas the 100 Å fiber is one nucleosome in width, the 300 Å fiber looks three beads wide. Removal of some H1 from a 300 Å fiber causes it to unwind to 100 Å. Adding back the H1 reinstates the 300 Å fiber. Although electron microscopists can actually see the 300 Å fiber, they do not know its exact structure. Higher levels of compaction are even less well understood.

The Radial Loop-Scaffold Model Seeks to Explain Compaction of the 300 Å Fiber

This model proposes that several nonhistone proteins, including topoisomerase II, bind to chromatin every 60–100 kb and tether the supercoiled, nucleosome-studded 300 Å fiber into structural loops (Fig. 13.4b). Evidence that nonhistone proteins fasten these loops comes from chemical manipulations in which the removal of histones does not cause the chromatin to unfold completely. A complex of proteins

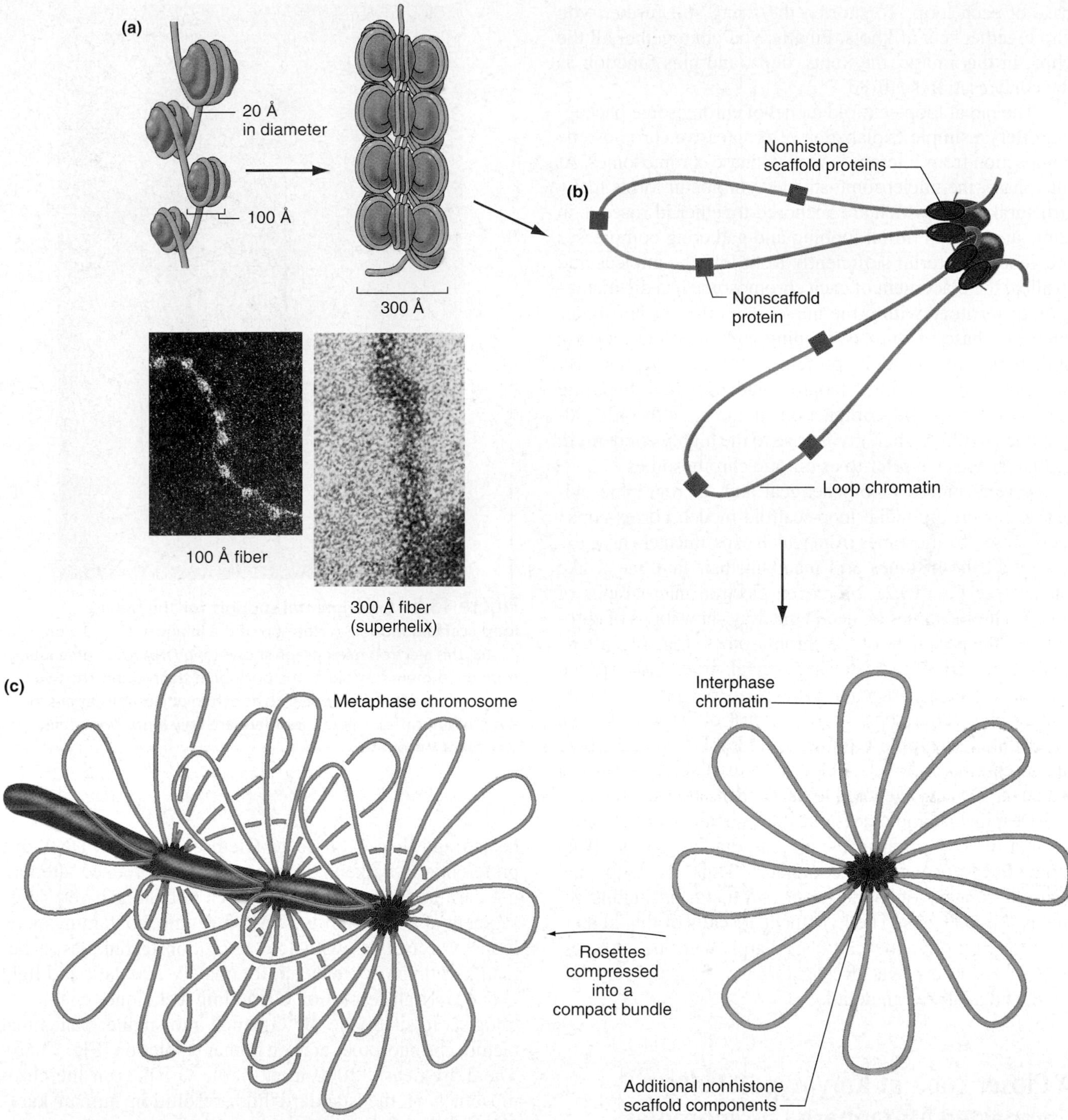

Figure 13.4 Models of higher-level packaging. **(a)** Electron micrographs contrasting the 100 Å fiber (*left*) with the 300 Å fiber (*right*). The line drawings show the probable arrangement of nucleosomes (with *green cores*) in these structures. **(b)** The radial loop-scaffold model for yet higher levels of compaction. According to this model, the 300 Å fiber is first drawn into loops, each including 60–100 kb of DNA (purple), that are tethered at their bases by nonhistone scaffold proteins (*brown* and *orange*) including topoisomerase II. **(c)** Additional nonhistone proteins might gather several loops together into daisylike rosettes and then compress the rosette centers into a compact bundle.

known as **condensins** may act to further condense chromosomes for mitosis. These and other proteins may gather the loops into daisylike rosettes (Fig. 13.4c) and then compress the rosette centers into a compact bundle. A range of nonhistone proteins thus forms the condensation scaffold depicted in Fig. 13.2a. This proposal of looping and gathering is known as the **radial loop-scaffold model** of compaction. To visualize how it achieves condensation, imagine a long piece of string. To shorten it, you knot it at intervals to form loops separated by straight stretches; the knots are at the base of each loop. To shorten the string still further, you clip together sets of knots. Finally, you pin together all the clips. In this image, the knots, clips, and pins function as the condensation scaffold.

The radial loop-scaffold model of chromosome packaging offers a simple explanation of progressive chromosome compaction from interphase to metaphase chromosomes. At interphase, the nucleosome-studded chromatin forms many structural loops, which are anchored together in rosettes in some areas. This initial looping and gathering compresses the genetic material sufficiently to fit into the nucleus and to allow the placement of each chromosome in a distinct region or territory within the nucleus. As the chromosomes enter prophase of mitosis, looping and gathering increase and bundling through protein cross-ties begins. By metaphase, the height of looping, gathering, and bundling achieves a 250-fold compaction of the roughly 40-fold-compacted 300 Å fiber, giving rise to the highly condensed, rodlike shapes we refer to as mitotic chromosomes.

Several pieces of biochemical and micrographic evidence support the radial loop-scaffold model. For example, metaphase chromosomes from which experimenters have extracted all the histones still maintain their familiar X-like shapes (see Fig. 13.2). Moreover, electron micrographs of mitotic chromosomes treated in this way show loops of chromatin at the periphery of the chromosomes (**Fig. 13.5**). And analyses of DNA indicate that special, irregularly spaced AT-rich sequences associate with nonhistone proteins to define the chromatin loops. These stretches of DNA are known as **scaffold-associated regions,** or **SARs.** Found at the base of the chromatin loops, SARs are most likely the sites at which the DNA is anchored to the condensation scaffold.

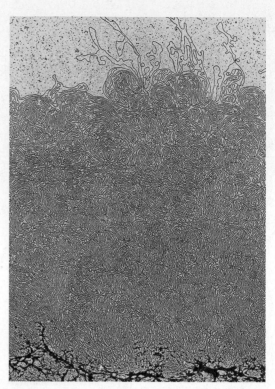

Figure 13.5 Experimental support for the radial loop-scaffold model. A close-up of the image in Fig. 13.2 on p. 468, this electron micrograph shows long DNA loops emanating from the protein scaffold at the *bottom* of the picture. The two ends of each DNA loop appear to attach to adjacent locations in the protein scaffold. Note that there are only loops—not ends—at the *top* of the photo.

Despite bits and pieces of experimental evidence, studies that directly confirm or reject the radial loop-scaffold model have not yet been completed. Thus, the loops and scaffold concept of higher-level chromatin packaging remains a hypothesis. The hypothetical status of this higher-level compaction model contrasts sharply with nucleosomes, which are entities that investigators have isolated, crystallized, and analyzed in detail.

A Closer Look at Karyotypes: Fully Compacted Metaphase Chromosomes Have Unique, Reproducible Banding Patterns That Identify Them

We have seen that different levels of packaging compact the DNA in human metaphase chromosomes 10,000-fold (see Table 13.2). With this amount of compaction, the centromere region and telomeres of each chromosome become visible. We have also seen (in Chapter 4) that various staining techniques reveal a characteristic banding pattern for each metaphase chromosome, establishing a karyotype. In G-banding, for instance, chromosomes are first gently heated and then exposed to Giemsa stain; this DNA dye preferentially darkens certain regions to produce alternating dark and light "G bands." Each G band is a very large segment of DNA from 1–10 Mb in length, containing many loops. On analysis with low-resolution techniques, a human karyotype contains approximately 300 dark and light G bands. High-resolution G-banding techniques enable cytologists to subdivide the G bands into smaller bands that picture chromosomes at even higher resolution (**Fig. 13.6a**). There are nearly 2000 identifiable G bands in the chromosomes of the standard high-resolution human karyotype. G-banding serves as the basis for most karyotyping because G-band preparations are very stable and require only a good light microscope for detection. The karyotypes described in Chapter 4 present a cell's mitotic metaphase chromosomes by size, shape, and banding patterns.

Banding Patterns Are Highly Reproducible, Although No One Knows for Sure What They Represent

Most molecular geneticists think the bands produced by Giesma staining probably do *not* embody differences in

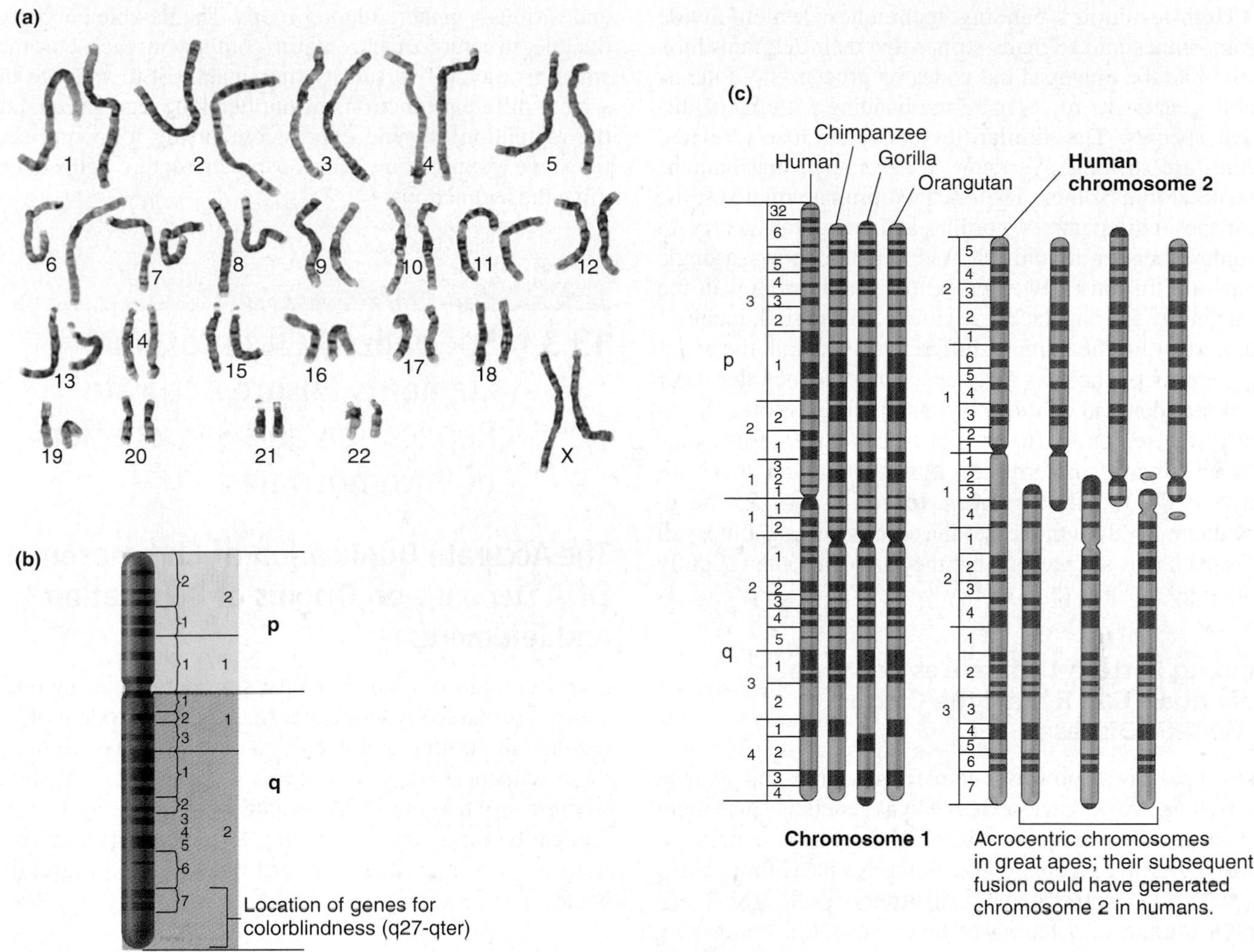

Figure 13.6 Chromosomes can be characterized by their banding patterns. (a) The karyotype of a human female examined by high-resolution G-banding techniques reveals approximately 1000 bands in the haploid complement of chromosomes. (b) Genes for colorblindness in humans have been localized to a small region near the tip of the long arm of the X chromosome. (c) A comparison of banding patterns in the two largest human chromosomes and their counterparts in the great apes.

base composition over long distances but more likely reflect an uneven packaging of loops determined in some way by the spacing and density of short repetitive DNA sequences. Although the detailed biochemical basis of banding is not yet understood, every time a chromosome replicates, whatever underlies its banding pattern is faithfully reproduced. The fact that banding patterns are so highly reproducible from one generation to the next indicates they are an intrinsic property of each chromosome, determined by the DNA sequence itself.

Researchers Can Use Bands to Locate Genes

Geneticists can designate the chromosomal location of a gene by describing its position in relation to the bands on the p (short) or q (long) arm of a particular chromosome. For this purpose, the p and q arms are subdivided into regions, and within each region, the dark and light bands are numbered consecutively. The X-linked genes for colorblindness, for example, reside at q27-qter, which means they are located on the X chromosome's long (q) arm somewhere between the beginning of the seventh band in the second region and the end of the telomere (terminus, or ter; Fig. 13.6b).

Geneticists Can Use Banding Patterns to Analyze Chromosomal Differences Between Species

In all placental mammals, the diploid genome carries roughly 6 billion base pairs, but this amazingly similar amount of DNA is packaged into different numbers of chromosomes with different banding patterns. Deer mice have 4 chromosomes in the nucleus of each body cell; cats have 38; house mice have 40; rabbits 44; humans 46; dogs have 48; cattle have 60; horses have 64; and hippopotamuses have 96.

High-resolution banding techniques, which divide chromosomes into so many stripes that their diagrams look a little like the universal bar codes on grocery-store items, enable geneticists to compare the banding patterns of different species. The similarities between closely related species are striking. We know, for example, that humans have 46 chromosomes. Their nearest primate relatives, the great apes (chimpanzees, gorillas, and orangutans) carry 48 chromosomes per diploid cell. As Fig. 13.6c shows, a single event—the fusion of two acrocentric chromosomes in the great apes to form metacentric chromosome 2 in humans—can account for the numeric difference. Note that the banding patterns of the two great ape chromosomes that later fused are identical to those of human chromosome 2, except in the region of fusion and at the chromosome tips. Human banding patterns are almost identical to chimpanzees' for 13 chromosomes, to gorillas for 9, and to orangutans for 8. In the rest of the chromosomes, almost all the same bands are present, but they are distributed slightly differently.

Banding Pattern Differences Between Individuals Can Reveal the Cause of Genetic Disease

Karyotypes are a powerful tool for research and genetic counseling. As we have seen, medical geneticists use them to diagnose abnormalities related to chromosome number. Down syndrome, for example, is the result of three chromosome 21s. Some genetic disorders result from very slight additions or deletions of genetic material somewhere within the normal number of chromosomes. With high-resolution G-banding techniques, clinicians can uncover some of these discrepancies by comparing the chromosomes of a person suffering from the genetic disorder with those of a standard karyotype.

A striking example of how genetic counselors exploit the phenomenon of exact and highly reproducible chromosomal packaging for clinical purposes is seen in the case history of a young boy identified in the medical literature as BB. With no known family history of genetic disorders, BB suffered from defects in four X-linked traits: Duchenne muscular dystrophy; chronic granulomatous disease, which impairs the infection-fighting ability of white blood cells; the rare McLeod's blood type, which results in bouts of anemia; and retinitis pigmentosa, in which a deterioration of the retina eventually causes blindness. The simultaneous occurrence of four X-linked conditions is so rare that one doctor asked medical geneticists to scrutinize the boy's X chromosome for the cause of the problems. High-resolution banding revealed the answer—deletion of a small white band sandwiched between two larger dark bands, which removed the four genes in question.

In summary, the DNA-protein interactions that define chromatin structure generate several levels of packaging: tightly wound nucleosomes, some form of supercoiling, and variously gathered larger loops. The flexible but reproducible structure of chromatin confers on each chromosome a structural versatility that enables it to adapt to the widely different functions of duplication, segregation, and the regulation of gene expression. In the following sections, we examine how chromosome structure helps determine those functions.

13.3 Specialized Chromosomal Elements Ensure Accurate Replication and Segregation of Chromosomes

The Accurate Duplication of Chromosomal DNA Depends on Origins of Replication and Telomeres

As the chromosomes decondense for copying during replication, special DNA sequences that do not encode proteins regulate the timing and accuracy of the process. Some of these sequences serve as origins of replication that signal where and when the DNA double helix opens up to form replication forks; others function as telomeres that protect the ends of individual chromosomes from progressive decay.

There Are Many Origins of Replication

During replication, the enzyme DNA polymerase assembles a new string of nucleotides according to a DNA template, linking about 50 nucleotides per second in a typical human cell. At this rate and with only one origin of replication, it would take the polymerase about 800 hours, a little more than a month, to copy the 130 million base pairs in an average human chromosome. But the length of the cell cycle in actively dividing human tissues is much shorter, some 24 hours, and S phase (the period of DNA replication) occupies only about a third of this time. Eukaryotic chromosomes meet these time constraints through multiple origins of replication that can function simultaneously (**Fig. 13.7a**). Most mammalian cells carry ~10,000 such origins strategically positioned among the chromosomes. As we saw in Chapter 6, each origin of replication binds proteins that unwind the two strands of the double helix, separating them to produce two mirror-image replication forks. Replication then proceeds in two directions (bidirectionally), going one way at one fork and the opposite way at the other, until the forks run into adjacent forks. As replication opens up a chromosome's DNA, a replication bubble becomes visible in the electron microscope, and with many origins, many bubbles appear. The DNA running

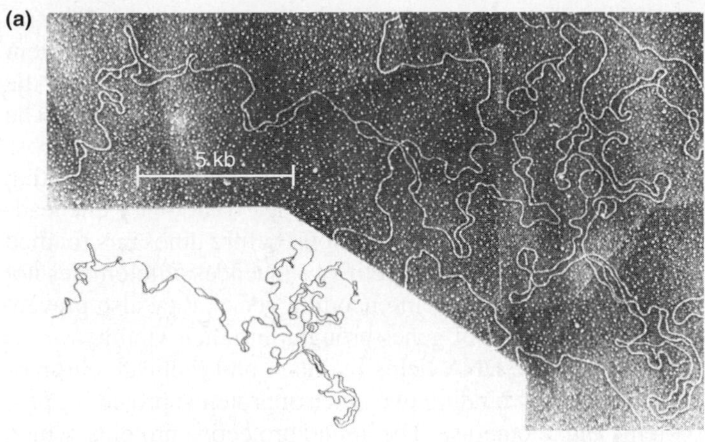

(b)

Consensus region

5' •••CAAATTTCGTCAAAAATGCTAAGAAATAGGTTATTACTTTTATTTAAGTATTGTTTGTGCCTTTTGAAAAGCAAGCATAAAAGATCTAAACATAAAATCTGTAAAATAAC••• 3'
3' •••GTTTAAAGCAGTTTTTACGATTCTTTATCCAATAATGAAAATAAATTCATAACAAACACGGAAAACTTTTCGTTCGTATTTTCTAGATTTGTATTTTAGACATTTTATTG••• 5'

Figure 13.7 Eukaryotic chromosomes have multiple origins of replication. **(a)** Electron micrograph and diagrammatic interpretation, showing a region of replicating DNA from a *Drosophila* embryo. Many origins of replication are active at the same time, creating multiple replicons. **(b)** Structure of the yeast origin of replication ARS1 (the first ARS to be characterized). The *orange-boxed* sequence is the AT-rich consensus region found in all ARS elements. The *blue boxes* are the flanking sequences close to the ARS1 consensus region that promote function.

both ways from one origin of replication to the endpoints, where it merges with DNA from adjoining replication forks, is called a **replication unit,** or **replicon.** As yet unidentified controls tie the number of active origins to the length of S phase. In *Drosophila,* for example, early embryonic cells replicate their DNA in less than 10 minutes. To complete S phase in this short a time, their chromosomes use many more origins of replication than are active later in development when S phase is 6–10 times longer. Thus, all origins of replication are not necessarily active during all the mitotic divisions that create an organism.

The 10,000 origins of replication scattered throughout the chromatin of each mammalian cell nucleus are separated from each other by 30–300 kb of DNA, which suggests that there is at least one origin of replication per loop of DNA. Origins of replication in yeast (known as autonomously replicating sequences, or ARSs) can be isolated by their ability to permit replication of plasmids in yeast cells. ARS's are capable of binding to the enzymes that initiate replication. They consist of an AT-rich region of DNA adjacent to special flanking sequences (Fig. 13.7b). By digesting interphase chromatin with DNase I, an enzyme that fragments the chromatin only at points where the DNA is not protected inside a nucleosome, investigators have determined that origins of replication are accessible regions of DNA devoid of nucleosomes.

Telomeres Preserve the Integrity of Linear Chromosomes

The linear chromosomes of eukaryotic cells terminate at both ends in protective caps called telomeres (**Fig. 13.8**). Composed of DNA associated with proteins, these caps contain no genes but are nevertheless crucial in preserving the structural integrity of each chromosome. Chromosomes unprotected by telomeres fuse end to end, producing entities with two centromeres. During anaphase of mitosis, if the two centromeres are pulled in opposite directions, the DNA between them will rupture, resulting in broken chromosomes that segregate poorly and eventually disappear from the daughter cells.

Cells must thus preserve their telomeres to maintain the normal genetic complement on which their viability depends. But the replication of telomeres is problematic. As we saw in Chapter 6, DNA polymerase, a key component of the replication machinery, functions only in the 5′-to-3′ direction and can add nucleotides only to the 3′ end of an existing chain. With these constraints, the enzyme on its own

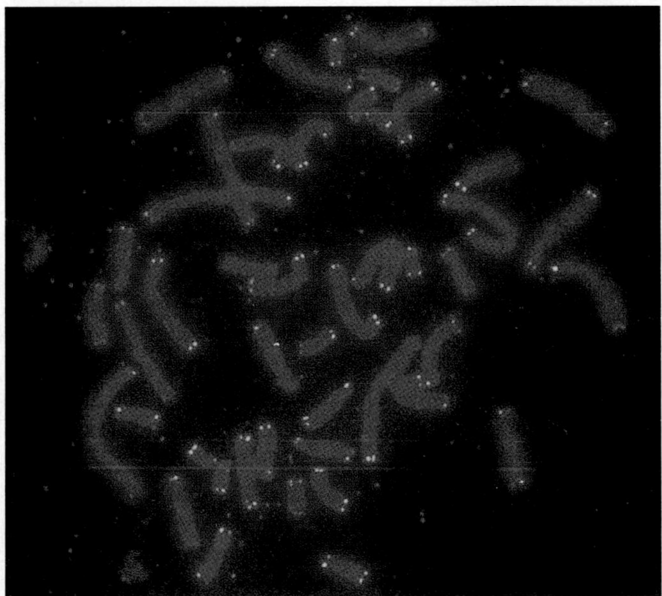

Figure 13.8 Telomeres protect the ends of eukaryotic chromosomes. Human telomeres light up in *yellow* upon *in situ* hybridization with fluorescent probes (FISH) that recognize the base sequence TTAGGG. The telomeres of humans and many other species contain many repeats of this 6 bp motif.

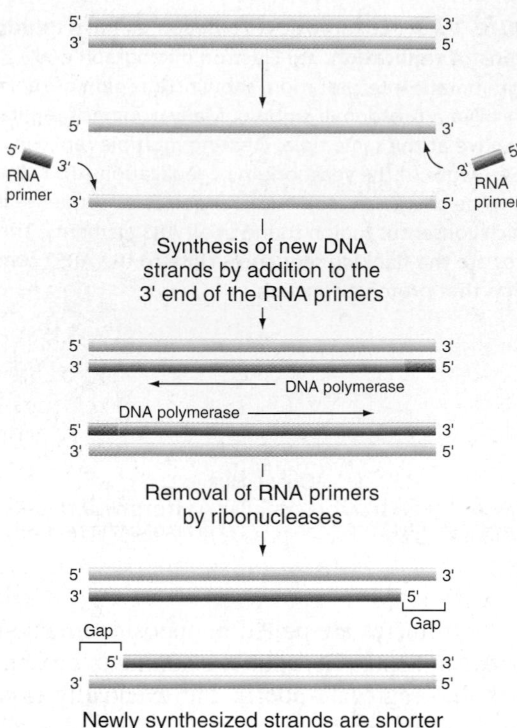

Figure 13.9 DNA polymerase cannot construct the 5′ end of a new DNA strand. Even if an RNA primer at the 5′ end can begin synthesis of a new strand, a gap will remain when ribonucleases eventually remove the primer. The requirement of DNA polymerase for a primer on which to continue polymerization means that the enzyme cannot fill this gap, so newly synthesized strands would always be shorter than parental strands if DNA polymerase were the only player in the production of new ends. In this figure, both parental DNA strands are in *light blue,* the two newly synthesized strands in *dark blue,* and the RNA primers in *red.*

cannot possibly replicate some of the nucleotides at the 5′ ends of the two DNA strands (one of which is in the telomere at one end of the chromosome, the other of which is in the telomere at the other end). In short, DNA polymerase can reconstruct the 3′ end of each newly made DNA strand in a chromosome, but not the 5′ end (**Fig. 13.9**). If left to its own devices, the enzyme would fail to fill in an RNA primer's length of nucleotides at the 5′ end of every new chromosomal strand with each cell cycle. As a result, the chromosomes in successive generations of cells would become shorter and shorter, losing crucial genes as their DNA diminished.

Telomeres and an enzyme called **telomerase** provide a countermeasure to these problematic quirks of DNA polymerase. Telomeres consist of special repetitive DNA sequences. Human telomeres are composed of the base sequence TTAGGG repeated 250–1500 times (see Fig. 13.8). The number of repeats varies with the cell type. Sperm have the longest telomeres. The same exact TTAGGG sequence occurs in the telomeres of all mammals as well as in birds, reptiles, amphibians, bony fish, and many plant species. Some much more distantly related organisms also have repeats in their telomeres but with slightly different sequences. For example, the telomeric repeat in the chromosomes of the ciliate *Tetrahymena* is TTGGGG. The close conservation of these repeated sequences across phyla suggests that they perform a vital function that emerged in the earliest stages of the evolutionary line leading to eukaryotic organisms, long before dinosaurs roamed the earth. The special repetitive sequences of telomeres not only contain no protein-encoding genes, they also prevent the transcription of genes brought into their vicinity.

Telomere DNA helps maintain and replicate chromosome ends by binding two types of proteins: protective proteins and telomerase. The bound protective proteins, which recognize the single-stranded TTAGGG sequences at the very ends of a chromosome, shield these ends from unwanted fusion or degradation, as explained below. When the proteins are dislodged, the telomere attracts telomerase, which can also bind to the single-stranded TTAGGG sequence. The bound enzyme extends the telomere, roughly restoring it to its original length.

Telomerase is an unusual enzyme consisting of protein in association with RNA. Because of this mix, it is called a *ribonucleoprotein.* The RNA portion of the enzyme contains 3' AAUCCC 5' repeats that are complementary to the 5' TTAGGG 3' repeats in telomeres, and they serve as a template for adding new TTAGGG repeats to the end of the telomere (**Fig. 13.10**). In many cells, including the perpetually reproducing cells of yeast and the germ cells of humans, some kind of feedback mechanism appears to maintain the optimal number of repeats at the telomeres. In human somatic cells, low telomerase activity results in the progressive shortening of telomeres as cells divide.

Many studies and observations have shown that telomeres are critical to chromosome function. In addition to preventing a chromosomal shortening during replication that could dismantle vital genes nucleotide by nucleotide, the telomeres maintain the integrity of the chromosomal ends. Broken chromosomes that lack telomeres are recognized as defective by the cellular DNA repair machinery, which often remedies the situation by putting the broken ends back together, restoring the telomeres. Sometimes, however, the unprotected, broken, nontelomeric ends are subject to inappropriate repair resulting in chromosome fusion, or they may attract enzymes that degrade the chromosome entirely. Both fusion and degradation disrupt chromosome number and function. Thus, even though they normally carry no genes, telomeres contain information essential to the duplication, segregation, and stability of chromosomes.

Telomerase Activity Correlates with Cell Proliferation

The activity of telomerase in normal yeast cells ensures the full reconstruction of each chromosome's ends with each DNA replication. In studies where researchers deleted the

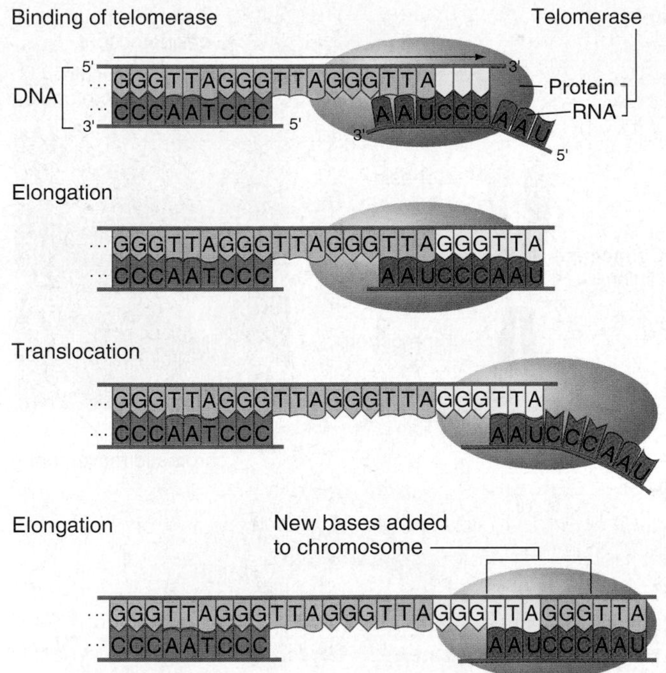

Figure 13.10 **How telomerase extends telomeres.** Telomerase binds to the ends of chromosomes because of complementarity between the 3' AAUCCC 5' repeats of telomerase RNA (*orange*) and the TTAGGG repeats of telomeres. Telomerase RNA 3' AAUCCC 5' repeats serve as templates for adding TTAGGG repeats to the ends of telomeres. After a telomere has acquired a new repeat, the telomerase enzyme moves (translocates) to the newly synthesized end, allowing additional rounds of telomere elongation.

yeast gene for telomerase, the telomeres shortened at the rate of about 3 bp per generation, and after significant loss of telomeric length, the chromosomes began to break, and the yeast cells died. Telomerase activity, it seems, endows normal yeast cells with the potential for immortality; given the proper conditions and continual telomere reconstruction, the cells can reproduce forever.

In humans, the telomerase gene is part of every cell's genome. Germ-line cells, which maintain their chromosomal ends through repeated rounds of DNA replication, express the gene, as do some stem cells; but many normal somatic cells, which have a finite life span, express very little telomerase. In these differentiated somatic cells, the telomeres shorten slightly with each cell division. This shortening helps determine how many times a particular cell is able to divide. In culture, most somatic cells, after dividing for 30 or 40 or 50 generations, show signs of senescence and then die.

Tumor cells are somatic cells gone awry that continue to divide indefinitely. In contrast with normal somatic cells, many human tumor cells that become immortal exhibit high telomerase activity. Cells isolated from human ovarian tumors, for example, express the telomerase enzyme and maintain stable telomeres; cells from normal ovarian tissue do not. Oncologists hypothesize from these and other observations that expression of telomerase in cancerous human cells may keep those cells from losing their telomeres and eventually dying, and thereby perpetuate tumors. Because high telomerase activity is a characteristic of many tumor cells, drug companies are developing cancer treatment drugs that inhibit telomerase activity.

Proteins Are an Essential Part of Reproducing Chromatin Structure

DNA replication is only one step in chromatin duplication. The complex process also includes the synthesis and incorporation of histone and nonhistone proteins to regenerate tissue-specific chromatin structure. Researchers speculate that the process works something like this. Before DNA synthesis can take place, the chromatin fiber must unwind. Next, as DNA replication proceeds, newly formed DNA must associate with histones, either preexisting histones or recently synthesized histones that have just made their way to the nucleus. The synthesis and transport of histones must be tightly coordinated with DNA synthesis since the nascent DNA becomes incorporated into nucleosomes within minutes of its formation. Proteins that mediate assembly of nucleosomes have been identified in several organisms, including yeast and humans. Finally, the nucleosomal DNA must interact in specific ways with a variety of proteins to compact in the same pattern as before. An exception to the exact replication of compaction patterns occurs in differentiating cells. Changes in available nuclear proteins produce slightly different folding patterns that promote the expression of different genes. Studies with mammalian cells have shown that some hormones can induce changes in gene expression if and only if they are present during chromatin replication.

The Segregation of Condensed Chromosomes Depends on Centromeres

When cell nuclei divide at mitosis or meiosis II, the two chromatids of each replicated chromosome must separate from one another and segregate such that each daughter cell receives one and only one chromatid from each chromosome. At meiosis I, homologous chromosomes must pair and segregate such that each daughter cell receives one and only one chromosome from each homologous pair. The centromeres of eukaryotic chromosomes ensure this precise distribution during different kinds of cell division by serving as segregation centers.

Centromeres Appear As Constrictions in the Chromosomes

Centromeric constrictions arise because centromeres are contained within blocks of repetitive, simple noncoding sequences, known as **satellite DNAs,** which have a very different chromatin structure and different higher-order packaging than other chromosomal regions (see Fig. 13.1). There are many different kinds of satellite DNA, each consisting of short sequences 5–300 bp long repeated in tandem thousands or millions of times to form large arrays. The predominant human satellite, "α-satellite," is a noncoding sequence 171 bp in length; it is present in a block of tandem repeats extending over a megabase of DNA in the centromeric region of each chromosome. Various human centromeres also contain sequences unrelated to α-satellite, which give their centromeric regions a complex structure. Although most satellite sequences lie in centromeric regions, some satellites are found outside the centromere on the chromosome arms.

The centromere can occur almost anywhere on a chromosome, except at the very ends (which must, instead, be telomeres). As we have seen, in a metacentric chromosome, the centromere is at or near the middle, while in an acrocentric chromosome, it is near one end.

Centromeres Have Two Functions That Ensure Proper Segregation

First, they are the sites that hold sister chromatids together—up to the beginning of anaphase in mitosis and up to anaphase II in meiosis. There is strong evidence that a multisubunit protein complex called **cohesin** acts as the glue that holds sister chromatids together until segregation takes place (**Fig. 13.11a**). After the chromosomes replicate in S phase of the cell cycle, the cohesin proteins associate in complexes throughout the length of each chromosome. When the cell enters mitosis, cohesin is lost from the arms, but a small amount of the protein remains at the centromere. At anaphase the centromeric cohesin is cleaved, and the sister chromatids separate. Mutations in any of cohesin's subunits result in chromosome segregation errors. If a cell expresses a mutant cohesin that is noncleavable, the number of segregation errors increases. The cohesin proteins have been conserved throughout the evolution of eukaryotes.

Cohesin also plays an important role in sister chromatid behavior during meiosis. Throughout meiosis I, cohesion remains at the centromere to hold sister chromatids together and ensure that they migrate to the same pole. Its behavior during meiosis II is very similar to that in mitosis. Cohesin is cleaved at anaphase II, allowing segregation of sister chromatids to opposite poles.

In addition to holding sister chromatids together, centromeres contribute to proper chromosome segregation through elaboration of a **kinetochore:** a specialized structure composed of DNA and proteins that is the site at which chromosomes attach to the spindle fibers (Fig. 13.11a). Some of the kinetochore proteins are motor proteins that help power chromosome movement during mitosis and meiosis. During mitosis, a kinetochore develops late in prophase on each sister chromatid, at the part of the centromere that faces one or the other cellular pole. By prometaphase, the kinetochores on the two sister chromatids attach to spindle fibers emanating from centrosomes at opposite poles of the cell. Although it is not yet clear what ensures this bipolar attachment, it appears that kinetochores somehow measure the tension arising when sister chromatids that are connected through their centromere are pulled in opposite directions. At the beginning of anaphase, the cohesin complex is split, freeing the sister chromatids to migrate toward opposite poles, with the assistance of the motor proteins in the kinetochore (review Fig. 4.8 on p. 90).

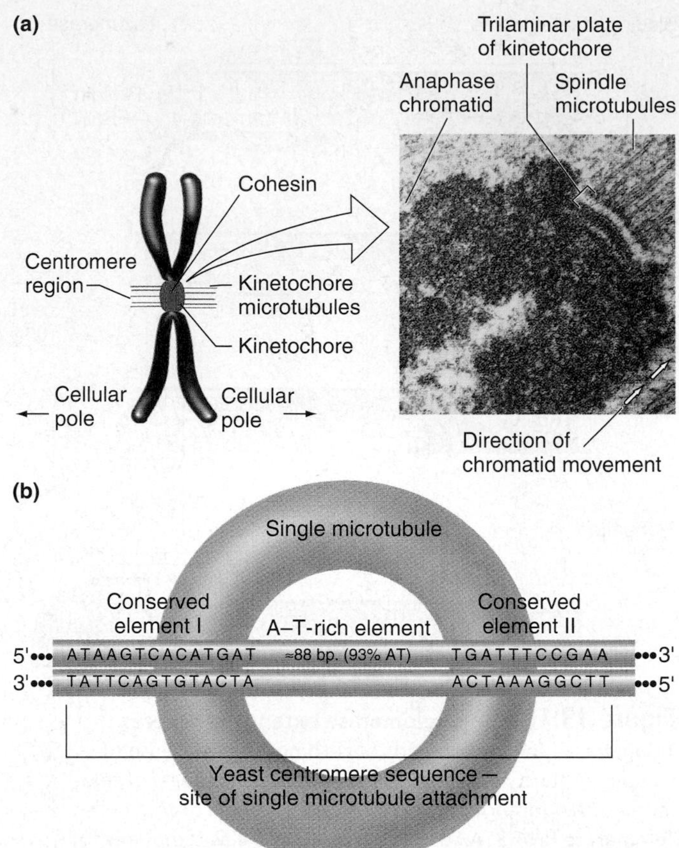

Figure 13.11 Centromere structure and function. **(a)** Structure of centromeres in higher organisms. Centromeres hold sister chromatids together and contain information for the construction of a kinetochore (*gold*), the structure that allows the chromosome to bind to spindle fibers. Cohesin (*yellow*) binds the sister chromatids together in the centromere region. In this micrograph, an anaphase chromatid is migrating toward the spindle pole via microtubules attached to the kinetochore. The kinetochore is organized into a complicated "trilaminar plate." **(b)** Structure and DNA sequence organization of yeast centromeres.

Using Centromere Function to Analyze Centromere Structure

Investigators can exploit the centromere's role in chromosome segregation to isolate and then analyze the specific chromosomal regions that make up a centromere. If removal of a DNA sequence disrupts chromosome segregation and reinsertion of that same sequence restores stable transmission, the sequence must be part of the centromere. In the yeast *S. cerevisiae,* centromeres consist of two highly conserved nucleotide sequences, each only 10–15 bp long, separated by approximately 90 bp of AT-rich DNA (Fig. 13.11b). This means that a short stretch of roughly 120 nucleotides is sufficient to specify a centromere in this organism. The centromere sequences of different yeast chromosomes are so closely related that the centromere of one chromosome can substitute for that of another. This indicates that while all centromeres play the same role in chromosome segregation, they do not help distinguish one chromosome from another.

The centromeres of higher eukaryotic organisms are much larger and more complex than those of yeast. In these multicellular organisms, the centromeres lie buried in a considerable amount of darkly staining, highly condensed chromatin, which makes it difficult to discover which specific DNA sequences are critical to centromere function. The kinetochores in higher eukaryotes attach to many spindle fibers instead of just one, as in yeast. Researchers think that these complex kinetochores are likely to consist of repeating structural subunits, with each subunit responsible for attachment to one fiber.

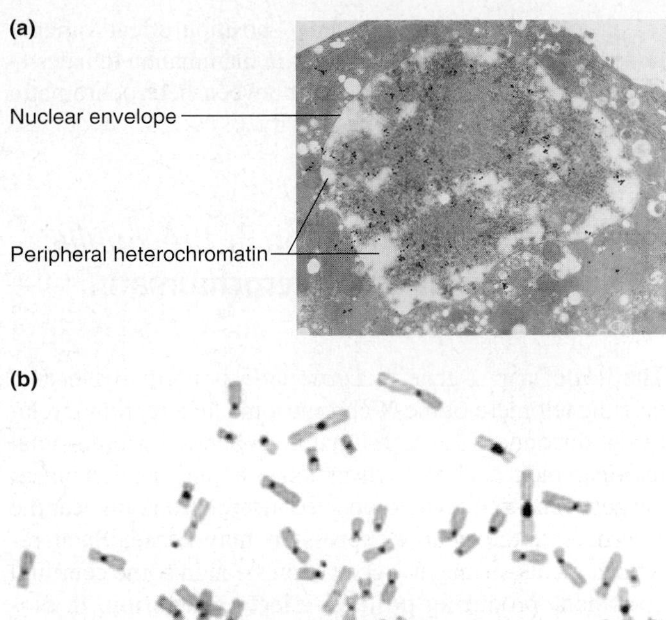

Figure 13.12 Heterochromatin versus euchromatin. (a) In this image of decondensed interphase chromatin, areas containing heterochromatin are bleached white and tend to be packed just inside the nuclear envelope, while ribonucleoprotein-containing nuclear domains remain darker and granular. The cell from which this thin section was taken was pulse-labeled with [^{3}H]uridine to mark sites of RNA synthesis (*dark spots*), most of which occur in euchromatin. **(b)** In this image, human metaphase chromosomes were stained by a special C-banding technique that darkens the constitutive heterochromatin, most of which localizes to regions surrounding the centromere.

13.4 How Chromosomal Packaging Influences Gene Activity

Origins of replication, telomeres, and centromeres are specialized chromosomal elements composed of specific noncoding DNA sequences packaged in specific ways for proper function. Chromosomal packaging also affects the function of coding sequences. We now describe examples of the different types of chromatin packaging along the chromosomes and how this packaging correlates with function.

Heterochromatin Contains More Highly Condensed DNA That Is Unavailable for Transcription

In cells stained with certain DNA-binding chemicals, a small proportion of chromosomal regions appear much darker than others when viewed under the light microscope. Geneticists call these darker regions **heterochromatin;** they refer to the contrasting lighter regions as **euchromatin.** The distinction between euchromatin and heterochromatin also shows up in the electron microscope, where the heterochromatin appears much more condensed than the euchromatin. Microscopists first identified dark-staining heterochromatin in the decondensed chromatin of interphase cells, where it tends to localize at the periphery of the nucleus **(Fig. 13.12a).** Even highly compacted metaphase chromosomes show the differential staining of heterochromatin versus euchromatin (Fig. 13.12b). Most of the heterochromatin in highly condensed chromosomes is found in regions flanking the centromere, but in some animals, heterochromatin forms in other regions of the chromosomes. In *Drosophila,* the entire Y chromosome, and in humans, most of the Y chromosome, is heterochromatic. Chromosomal regions that remain condensed in heterochromatin at most times in all cells are known as **constitutive heterochromatin.**

Autoradiography reveals that cells actively expressing their housekeeping and specialty genes incorporate radioactive RNA precursors into RNA almost exclusively in regions of euchromatin (see Fig. 13.12a). This observation indicates that euchromatin contains most of the sites of transcription and thus almost all of the genes. By contrast, heterochromatin appears to be transcriptionally inactive for the most part, probably because it is so tightly packaged that the enzymes required for transcription of the few genes it contains cannot access the correct DNA sequences.

Two specialized phenomena—position effect variegation in *Drosophila* and Barr bodies in mammalian females—clearly illustrate the correlation between heterochromatin formation and a loss of gene activity.

Position-Effect Variegation in *Drosophila:* Moving a Gene Near Heterochromatin Prevents Its Expression

The *white*$^+$ (w^+) gene in *Drosophila* is normally located near the telomere of the X chromosome, in a region of relatively decondensed euchromatin. When a chromosomal rearrangement such as an inversion or a translocation places the gene next to highly compacted heterochromatin near the centromere, the gene's expression may cease. Such rearrangements silence w^+ gene expression in some cells and not others, producing **position-effect variegation.** In flies carrying the wild-type w^+ allele, cells in the eye with an active w^+ gene are red, while cells with an inactive w^+ gene are white. Apparently, when normally euchromatic genes come into the vicinity of heterochromatin, the heterochromatin can spread into the euchromatic regions, shutting off gene expression in those cells where the heterochromatic "invasion" takes place. In such a situation, the DNA of the gene has not been altered, but the relocation has altered the gene's packaging in some cells. The phenomenon of position effect variegation thus reflects the existence of **facultative heterochromatin:** regions of chromosomes (or even whole chromosomes) that are heterochromatic in some cells and euchromatic in other cells of the same organism.

Position-effect variegation of red and white eye color in *Drosophila* produces eyes that are a mosaic of red and white patches of different sizes (**Fig. 13.13a**). The position and size of the patches vary from eye to eye. Such variation suggests that the "decision" determining whether heterochromatin spreads to the w^+ gene in a particular cell is the result of a random process. Because patches composed of many adjacent cells have the same color, the decision must be made early in the development of the eye. Once made, the decision determining whether the white gene will be on or off is transmitted to all the cell's mitotic descendants. These descendants occupy a particular region of the eye, forming patches of red or white cells respectively.

One interesting property revealed by position-effect variegation is that heterochromatin can spread over more than 1000 kb of previously euchromatic chromatin. For example, some rearrangements that bring the w^+ gene near heterochromatin also place the *roughest*$^+$ gene in the same vicinity, although a little farther away from the centromeric heterochromatin (Fig. 13.13b). The wild-type *roughest*$^+$ (rst^+) gene normally produces a smooth eye surface. In flies carrying the rearrangements, some white-colored patches have smooth surfaces while others have rough surfaces. In the latter patches, the heterochromatin inactivated both the w^+ and

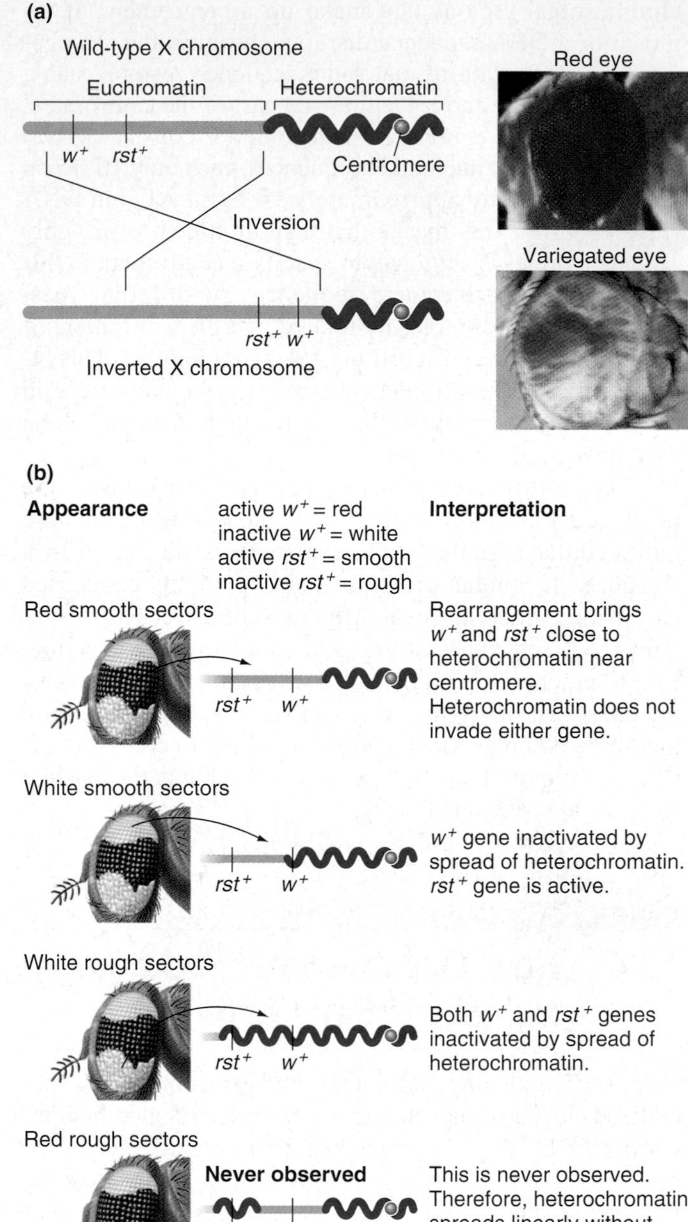

Figure 13.13 Position-effect variegation in *Drosophila* is a phenotypic effect of facultative heterochromatin. (a) When the w^+ eye color gene is brought near an area of heterochromatin through a chromosomal rearrangement such as an inversion, the eyes of the fly can become variegated, with some red cells and some white cells. **(b)** A model for position-effect variegation of the w^+ and *roughest* (rst^+) genes postulates that heterochromatin can spread from its normal location surrounding the centromere to nearby genes, causing their inactivation. Cells in which w^+ is active are red; those in which the gene is inactive are white. Cells in which rst^+ is active have a smooth surface; those in which this gene is inactive have a rough surface. Heterochromatin can spread different distances in different cells, but it usually does not skip genes.

then the rst^+ gene. Red-colored, rough-surfaced patches never form, which means that the heterochromatin does not skip over genes as it spreads along the chromosome.

Geneticists currently know very little about how heterochromatin invades euchromatin, what determines the extent of heterochromatin invasion in a particular cell, or how that cell's descendants inherit its specific heterochromatin pattern. They can, nevertheless, use the phenomenon of position-effect variegation to identify the molecules involved in heterochromatin formation. In one procedure, they obtained mutations that either enhance the amount of variegation produced by genes positioned near heterochromatin or diminish the amount of variegation. Enhancement reflects gene inactivation in more cells; diminishment reflects gene inactivation in fewer cells. The researchers later isolated by molecular cloning several of the genes that had mutated, and they raised antibodies against the mutant protein products of these genes. In this way, they discovered that at least some of the genes influencing heterochromatin formation encode proteins that localize selectively to the heterochromatin. The researchers hypothesize that if these proteins migrate from established heterochromatin to nearby DNA sequences, position-effect variegation results.

Inactivation of an X Chromosome by Heterochromatin Formation Is the Basis of Dosage Compensation in Female Mammals

In both fruit flies and mammals, normal males have one copy of the X chromosome, while females have two. The two sexes, however, require equal amounts of most proteins encoded by genes on the X chromosome. To compensate for the discrepancy in the dose of X-linked genes, male *Drosophila* double the rate at which they express the genes on their single X. Mammals have a different control mechanism for dosage compensation: the random inactivation of all but one X chromosome in each of the female's somatic cells. The inactive X chromosomes are observable in interphase cells as darkly stained heterochromatin masses. Geneticists call these densely staining X chromosomes **Barr bodies** after Murray Barr, the cytologist who discovered them.

The inactive X chromosome in mammalian females is another example of facultative heterochromatin. Here, a whole X chromosome becomes completely heterochromatic in some cells, while remaining euchromatic in others. Most genes on the X chromosome are available for transcription only in cells where the chromosome is euchromatic; very few genes are available for transcription when the X becomes heterochromatic.

An XX female has one Barr body in each somatic cell. The X chromosome that remains genetically active in these cells decondenses and stains as expected during interphase. XY male cells do not contain Barr bodies. As a result, normal male and female mammals have the same number of active X chromosomes. Females with XXX or XXXX karyotypes can survive because they have two or three Barr bodies and only a single active X. Barr bodies are not restricted to females, but their presence does require more than one X chromosome per cell. Cells from XXY Klinefelter males also contain a Barr body.

The "decision" determining which X chromosome in each cell becomes a Barr body occurs at random in the early stages of development and is inherited by the descendants of each cell. In humans, for example, two weeks after fertilization, when an XX female embryo consists of 500–1000 cells, one of the X chromosomes in each cell condenses to a Barr body. Each embryonic cell "decides" independently which X it will be. In some cells, it is the X inherited from the mother; in others, it is the X inherited from the father. Once the determination is made, it is clonally perpetuated so that all of the millions of cells descended by mitosis from a particular embryonic cell condense the same X chromosome to a Barr body and thereby inactivate it. Female mammals are thus a mosaic of cells containing either a maternally or a paternally derived inactivated X chromosome. In an individual female that is heterozygous for an X-linked gene, some cells will express one allele, while other cells will express the alternative. In females heterozygous for an X-linked mutation that would be lethal in a male, the relation between the two populations of activated X cells can make the difference between life and death. If the X chromosome carrying the wild-type allele is active in a high enough proportion of cells where expression of the gene is required, the individual will survive.

A Specific Gene Transcript Known as *Xist* Is Necessary for X Inactivation

About a dozen genes on the X chromosomes of both mice and humans escape Barr body inactivation, most likely because they are in small regions that do not become completely heterochromatic. Most of these genes remain active on both X chromosomes in a female. One, however, is inactive on the euchromatic X but active on the heterochromatic X, and in its active form, it contributes to the mechanism that causes X inactivation. This gene is referred to as *Xist* for X inactivation specific transcript. An X chromosome that does not contain *Xist* cannot be inactivated. Deletion of the *Xist* gene abolishes a chromosome's capacity for X inactivation. Cells carrying one X chromosome that lacks the *Xist* gene must inactivate the other X. The specific function of the *Xist* gene is to produce an unusually large *cis*-acting RNA transcript. Studies using fluorescent molecular probes show that this *cis*-acting *Xist* RNA, unlike most transcripts, never leaves the nucleus and is never translated into a protein. One model for how the *Xist* RNA produces inactivation proposes that the unusually large *cis*-acting RNA binds to the

X chromosome that produces it. Histones H3 and H4 are then modified by the addition of methyl groups and removal of some acetyl groups. The histone modifications in conjunction with other protein factors that bind to the *Xist*-coated DNA produce the inactive, condensed heterochromatic state.

Centromeric Heterochromatin Contains Altered Histones That Support Centromere Function

Centromeric DNA, we have seen, consists of repetitive sequences known as satellite DNAs. In higher eukaryotes, the central core of each centromere is composed of purely satellite DNA and is completely heterochromatic and not readily available for recombination and transcription. Surrounding this core are regions of heterochromatin interspersed with euchromatin. In all eukaryotes examined, the histone H3 protein has been replaced by a histone variant called CENP-A. This protein is very similar to histone H3 in its C-terminal region, but different from H3 in its N-terminal portion. The specialized chromatin in the centromere core marks this region for the attachment of the kinetochore protein complexes that are necessary for chromosome segregation.

The Flexible Packaging of Chromatin Plays a Role in Cellular Differentiation

Cells express their genes mainly during interphase when the chromosomes have decondensed, or decompacted, but even the relatively decompacted euchromatic interphase chromatin requires further unwinding to open up the higher orders of packaging and expose the DNA inside nucleosomes for transcription. Indeed, as we have seen, the location of nucleosomes along the chromosomes does not appear to be random; it is influenced by gene location and activity such that the DNA sequences required for transcription become exposed at the appropriate time. The decompaction of chromatin can be observed at the molecular level. We describe this phenomenon in greater detail in Chapter 18 on "Gene Regulation in Eukaryotes."

As cells differentiate to perform specific roles that require the synthesis of specific proteins, patterns of chromatin compaction and decompaction change to allow expression of the appropriate genes. Within each chromosome, inactive genes are compacted into heterochromatin, while highly active genes remain accessible in euchromatin. Once established, specific patterns of gene decompaction and exposure for expression persist in ensuing generations of cells. Thus, because of slight differences in packaging, different areas of chromatin unwind for expression during interphase in different types of cells.

Connections

Eukaryotic chromosomes package and manage the genetic information in DNA through a modular chromatin design whose flexibility allows back-and-forth shifts between different levels of organization. These reversible changes in chromatin structure reliably sustain a variety of chromosome functions, producing selective unwinding for gene expression, universal unwinding for replication, and coordinated compaction for segregation and transport. Histones and nonhistone proteins provide the framework for chromatin and help regulate changes in chromosome structure and function. Noncoding sequences that specify the origins of replication, centromeres, and telomeres are essential to chromosome duplication, segregation, and integrity.

Although the faithful function, replication, and transmission of chromosomes underlie the perpetuation of life within each species, chromosomal changes do occur. We have already described two mechanisms of change: mutation of individual nucleotides (Chapter 7) and homologous recombination, which exchanges bases between homologs (Chapters 4, 5, and 6). In Chapter 14, we examine broader chromosomal rearrangements that produce different numbers of chromosomes, reshuffle genes between nonhomologous chromosomes, and reorganize the genes of a single chromosome. These large-scale modifications, by altering the genetic content of a genome, provide some of the important variations that fuel evolution.

Essential Concepts

1. Each chromosome consists of one long molecule of DNA compacted by *histone* and *nonhistone* proteins. The five types of histones—H1, H2A, H2B, H3, and H4—are essential to the establishment of generalized chromosome structure. There are more than 1000 kinds of nonhistone proteins: Some are structural; others are active in replication and segregation; but most help determine the time and place of gene expression.

2. DNA-protein interactions create reversible levels of compaction.
   a. The naked DNA wraps around core histones to form *nucleosomes,* which are secured by H1.
   b. Models of higher-level compaction suggest that some sort of *supercoiling* condenses the nucleosomal fiber to a wider fiber. Nonhistone proteins then anchor this fiber to form *loops.* In metaphase chromosomes, higher levels of compaction (perhaps into rosettes and bunches of rosettes) compact DNA 10,000-fold.
   c. In fully compacted metaphase chromosomes, the *centromere* and *telomeres* become visible under the microscope. Giemsa staining of metaphase chromosomes reveals highly reproducible banding patterns that researchers can use to locate genes, analyze chromosomal differences between species, and diagnose some genetic diseases.
3. Specialized elements are required for normal chromosome function.
   a. *Origins of replication* are sites accessible for the binding of proteins that initiate replication. In eukaryotic chromosomes, many origins of replication ensure timely DNA replication.
   b. *Telomeres,* composed of repetitive base sequences, protect the ends of chromosomes, ensuring their integrity. The enzyme *telomerase* helps reconstruct the complete telomere with each cell division.
   c. *Centromeres,* which appear as constrictions in metaphase chromosomes, ensure proper segregation by holding sister chromatids together and by elaborating *kinetochores,* which properly attach sister chromatids to spindle fibers for mitosis and meiosis.
4. Chromosomal packaging influences function.
   a. Extreme condensation silences gene expression. Extremely condensed chromosomal areas appear as darkly staining *heterochromatin* under the microscope. *Position-effect variegation* in *Drosophila* and *Barr bodies* in mammals are examples of the correlation between heterochromatin formation and a loss of gene activity.
   b. Specific changes in the acetylation and methylation of histones H3 and H4 occur in regions of altered chromatin. The highly condensed heterochromatin of centromeres attracts specific protein complexes necessary for proper chromosome segregation.
   c. Chromatin is less compacted in regions of a chromosome that are being transcribed. As cells differentiate to perform specific functions, patterns of decompaction are established that allow expression of cell-type-specific genes.

On Our Website

www.mhhe.com/hartwell3
Chapter 13

Annotated Suggested Readings and Links to Other Websites

- Construction of artificial chromosomes
- Chromosome packaging and mechanics
- Recent findings on X-chromosome inactivation
- Telomeres, aging, and cancer

Specialized Topics

- Visualization of RNA transcripts associated with the chromosome

Social and Ethical Issues

1. Recent evidence has shown that the ends of chromosomes (telomeres) get shorter as somatic cells age because of a lack of the enzyme telomerase, which maintains the ends and length of chromosomes. In contrast, cancer cells that are "immortal" and divide rapidly have telomerase activity. Whether the lack of telomeres in older cells is a cause or a result of aging is still under investigation. A proposal that included experimentation on telomerase has been submitted to the National Institutes of Health. A stated goal of the proposal was exploration of ways to extend the human life span. Is this a valid way to spend federal research

monies given that many commercial companies are already exploring ways to use these findings to extend the human life span? What are the consequences for society if people live longer? What do you think of the goal of a longer life?

2. A city school board put forward a proposal to screen school children with learning disabilities for the fragile X mutation (see the Genetics and Society box "Amplified Trinucleotide Repeats May Have Medical Consequences" on pp. 216–217 in Chapter 7). Fragile X is the most common form of inherited mental retardation and is caused by an unusual expansion of a triplet nucleotide repeat in a gene on the X chromosome. (The triplet expansion is extreme enough to be seen as a cytological abnormality as well as by DNA sequence analysis.) The extent of mental impairment and other abnormal phenotypes varies with the size of the repeat. The proposal was well intended; the school board hoped to use the screening to identify students with potential disabilities and give them extra help and consideration. However, many parents and professionals opposed the testing. How could selection for testing harm a child? How could identification of fragile X harm a child? What are the benefits of such testing?

3. With available data on the DNA sequence of the human genome, what has changed about the information that a researcher might choose to share with other researchers? For example, a scientist who has worked for two years on identifying a particular human disease gene recently determined the 10 Mb region of a chromosome where the gene is located. The next steps are to use genomic sequence information to identify potential gene candidates and determine which of the candidates is altered by disease-causing mutations. The scientist hesitates to talk about his finding because he thinks many other researchers will immediately use the data and design PCR primers to investigate the potential genes and mutations in affected individuals. Does he have a right to withhold the information until he has completed his analysis of mutants? Does the answer to this question depend on whether his work is supported by public (government) funding agencies or by a private biotechnology company?

Solved Problems

I. One can construct YACs that range in size from 15 kb to 1 Mb. Based on DNA length, what level of chromosome compaction would you predict for a YAC of 50 kb compared with a YAC of 500 kb?

Answer

To answer this question, you need to apply information about the amount of DNA needed to get different types of chromosome condensation.

The 500 kb YAC would probably be more condensed than the 50 kb YAC based on its larger size. The DNA of both YACs would be wound around histones to form the nucleosome structure (160 bp around the core histones plus 40 bp in linker region). That DNA would be further compacted into 300 Å fibers that contain six nucleosomes per turn. *The 500 kb YAC would be compacted at a higher level of order, presumably in radial loops that occur every 60–100 kb in the chromosome, but the 50 kb YAC is not large enough to be packaged in this way.*

II. To clone and express genes in yeast, researchers constructed recombinant plasmids in the 1970s. Most of these early yeast plasmids did not contain an origin of replication and therefore were not maintained as autonomous plasmids separate from the chromosomes. If one of these plasmids integrated into the chromosome by recombination, it was replicated as part of the chromosome and was stably inherited from one division to the next. Later, when yeast origins of replication (ARSs) were characterized, researchers constructed autonomous yeast plasmids that could be maintained and transmitted independently of the chromosome. Describe how you would distinguish a plasmid integrated into a chromosome from an autonomous plasmid, using hybridization techniques.

Answer

This question requires an understanding of the use of hybridization (that is, the source of DNA and the probe used) and the interpretation of results. Think about differences between the two plasmid states (integrated and autonomous) and how the differences might be visualized using hybridization.

If plasmid DNA is used as a probe in a hybridization with a blot of restriction enzyme-digested DNA from cells containing integrated plasmids and from cells containing autonomous plasmids, the hybridizing bands will be different (see the diagram that follows). In the DNA sample containing the autonomous plasmid, the probe will hybridize to band(s) that together equal the total size of the original plasmid. If DNA from cells containing the integrated plasmid is digested with a restriction enzyme that cuts within the plasmid, two new fragments appear that contain both plasmid DNA and the chromosomal DNA on either side of the integration site in the chromosome. In the example in the figure, a 4 kb plasmid containing a *Eco*RI site recombined with the chromosome at a site within a 2.5 kb *Eco*RI chromosomal fragment. The 2.2 kb and 4.3 kb fragments are joint fragments containing both chromosomal DNA and plasmid DNA. Both bands will hybridize with the probe.

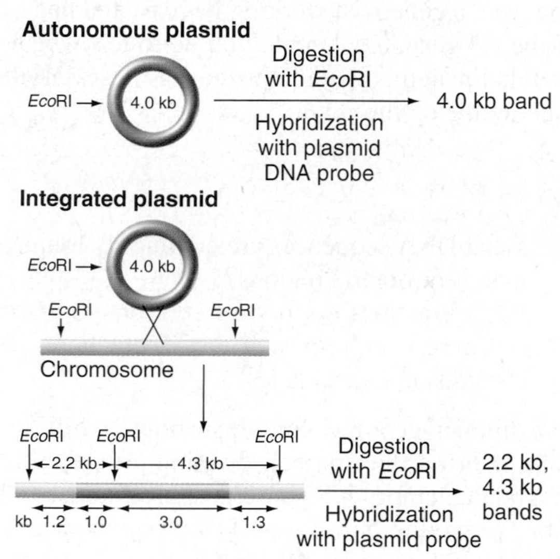

III. Mouse geneticist Mary Lyon proposed in 1961 that all but one copy of the X chromosome were inactivated in mammals.

a. What cytological finding supports the Lyon hypothesis?

b. What is a genetic result that supports the Lyon hypothesis?

Answer

This question requires an understanding of the experimental observations on X inactivation.

a. Microscopic examination of cells produces cytological evidence. *The number of Barr bodies seen in the cells of individuals with different numbers of X chromosomes supports the hypothesis.* For example, cells from XX females have one Barr body, whereas cells from an XXX female have two Barr bodies.

b. *Examination of the phenotype of cells from a female heterozygous for two different alleles of an X-linked gene produces genetic evidence that supports the Lyon hypothesis.* In these females, some cells have the phenotype associated with one allele, while other cells have the phenotype associated with the other allele.

Problems

Interactive Web Exercise

BLAST is a powerful tool developed by the National Center for Biotechnology Information (NCBI). BLAST compares a nucleotide or protein sequence with sequences in databases to identify similar nucleotide or protein sequences. Our website at www.mhhe.com/hartwell3 contains a brief exercise to introduce you to the use of this tool; once at the website, go to Chapter 13 and click on "Interactive Web Exercise".

Vocabulary

1. For each of the terms in the left column, choose the best matching phrase in the right column.

a. telomere	1. site at the base of chromatin loops that anchors loops to the scaffold
b. G bands	2. origin of replication in yeast
c. polytene chromosomes	3. repetitive DNA found near the centromere in higher eukaryotes
d. kinetochore	4. specialized structure at the end of a linear chromosome
e. nucleosome	5. giant chromosomes in *Drosophila* salivary glands
f. ARS	6. complexes of DNA and protein in the eukaryotic nucleus
g. satellite DNA	7. small basic proteins that bind to DNA and form the core of the nucleosome
h. chromatin	8. complex of DNA and proteins where spindle fibers attach to a chromosome
i. SAR	9. beadlike structure consisting of DNA wound around histone proteins
j. histones	10. regions of a chromosome that are distinguished by staining differences

Section 13.1

2. Many proteins other than histones are found associated with chromosomes. What roles do these nonhistone proteins play? Why are there more different types of nonhistone than histone proteins?

Section 13.2

3. What difference is there in the compaction of chromosomes during metaphase and interphase?

4. What is the role of the core histones in compaction compared to the role of histone H1?

5. a. About how many molecules of histone H2A would be required in a typical human cell just after the completion of S phase, assuming an average nucleosome spacing of 200 bp?
   b. During what stage of the cell cycle is it most crucial to synthesize new histone proteins?
   c. The human genome contains 60 histone genes, with 10–15 genes of each type (H1, H2A, H2B, H3, and H4). Why do you think the genome contains multiple copies of each histone gene?

6. a. What letters are used to represent the short and long arms of human chromosomes?
   b. Sketch a schematic diagram of a hypothetical chromosome 3 that has 3 regions with 2 bands each on the short arm and 5 regions with 3 bands each on the long arm. Label the arms, regions, and bands and indicate a gene at position 3p32.

7. There are ~2000 G bands visible in a high resolution karyotype of the 3 billion base pairs in the haploid human genome. If the genome contains ~30,000 genes, about how many genes would be removed by a deletion of DNA that could be detected by karyotype analysis?

8. The enzyme micrococcal nuclease can cleave phosphodiester bonds on single- or double-stranded DNAs, but DNA that is bound to proteins is protected from digestion by micrococcal nuclease. When chromatin from eukaryotic cells is treated for a short period of time with micrococcal nuclease and then the DNA is extracted and analyzed by electrophoresis and ethidium bromide staining, the pattern shown in lane A on the gel below is found. Treatment for a longer time results in the pattern shown in lane B, and treatment for yet more time yields that shown in lane C. Interpret these results.

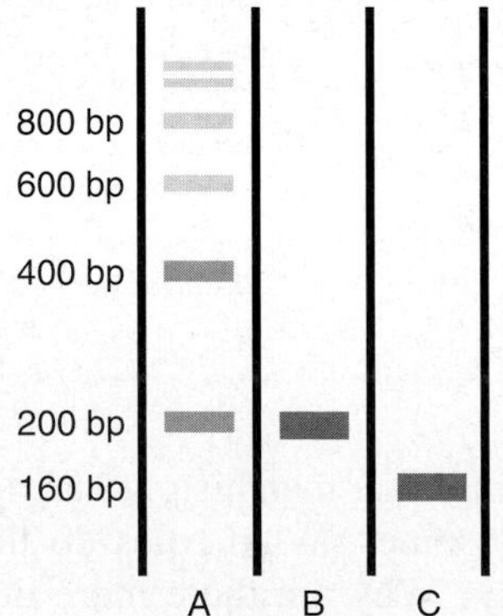

9. Histone H1 appears to play an important role in the formation of the 300Å fiber, while the other histone proteins do not appear to participate. Why do you think this is true?

10. Chromosome assembly factor (CAF-1) is a complex of proteins that was identified biochemically in extracts from human cells. The sequence of amino acids in the proteins was identified.
   a. How could you use this data to look for homologous genes in yeast?
   b. As a geneticist, why would you find it advantageous to identify the yeast genes to further your understanding of chromatin assembly?

11. The histone proteins H3 and H4 are modified in predictable and consistent ways that are conserved across species. One of the modifications is addition of an acetyl group to the twelfth lysine in the H4 protein. If you were a geneticist working in yeast and had a clone of the H4 gene, what could you do to test whether the acetylation at this specific lysine was necessary for the functioning of chromatin?

Section 13.3

12. a. What DNA sequences are commonly found at human centromeric regions?
   b. What functions do the two centromere-associated complexes, cohesin and the kinetochore, play in chromosome mechanics?

13. The human genome contains about 3 billion base pairs. During the first cell division after fertilization of a human embryo, S phase is approximately 3 hours long. Assuming an average DNA polymerase rate of 50 nucleotides/sec over the entire S phase, what is the minimum number of origins of replication you would expect to find in the human genome?

14. a. Give at least three examples of types of mutations that would disrupt the process of mitotic chromosome segregation. That is, explain in what DNA structures or in genes encoding what kinds of proteins would you find these segregation-disrupting mutations.
   b. How could you use yeast artificial chromosomes (YACs) to find such mutations in *S. cerevisiae*?

15. The mitotic cell divisions in the early embryo of *D. melanogaster* occur very rapidly (every 8 minutes).
   a. If there was one bidirectional origin in the middle of each chromosome, how many nucleotides would DNA polymerase have to add per second to replicate all the DNA in the longest chromosome (66 Mb) during the 8-minute early embryonic cell cycles? (Assume that replication occurs during the entire cell division cycle.)
   b. In fact, many origins of replication are active on each chromosome during the early embryonic divisions and are spaced approximately 7 kb apart. Calculate the average rate (per second) with which DNA polymerase adds complementary nucleotides to a growing chain in the early *Drosophila* embryo.

16. The extreme ends of chromosomes (telomeres) contain repeated DNA sequences (for example, 5′ TTAGGG 3′ in humans). If a blot of genomic DNA was hybridized using a probe of 3′ AATCCC 5′, all end fragments would hybridize with the probe.

a. If you wanted to examine an end of one specific chromosome by hybridization, what would you have to use as a probe?

b. When a probe specific for one chromosome end is used in Southern hybridization, a blurred band appears (see figure below). Why is the hybridizing band not a distinct sharp band as seen with probes that hybridize with internal sequences on the chromosome?

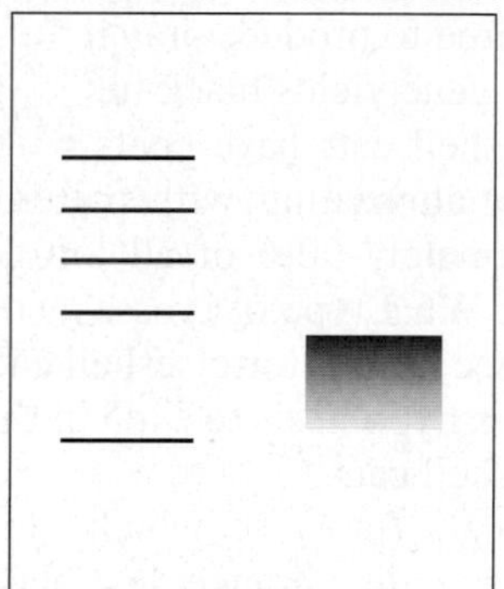

17. The enzyme telomerase consists of protein and an RNA containing a template sequence that directs the addition of an end sequence appropriate for the species. Telomere sequences (TTGGGG) from the ciliated protozoan *Tetrahymena* were cloned onto the ends of a linear YAC, which was then transformed into yeast. The YAC survived as a linear piece of DNA but the YAC now had TGGTGG sequences at the very ends in addition to TTGGGG. Why do you think these sequences were added?

18. The CENP-B and CENP-A proteins are both involved in centromere function. CENP-B is a nonessential protein but CENP-A is essential for viability.

a. In yeast, if you had a mutant containing a temperature-sensitive allele of CENP-B and another mutant containing a temperature-sensitive allele of CENP-A, what phenotype would you expect for viability and what phenotype for chromosome loss for each of the mutants when you raised the temperature?

b. Describe a test you could use to assay chromosome loss in these mutants.

19. A number of yeast-derived elements were added to the bacterial plasmid pBR322. Yeast that require uracil for growth (Ura^- cells) were transformed with these modified plasmids and Ura^+ colonies were selected by growth in media lacking uracil. For plasmids containing each of the elements listed, indicate whether you expect the plasmid to integrate into a chromosome or to be maintained separately as a plasmid. If the plasmid is maintained autonomously, is it stably inherited by all of the daughter cells of subsequent generations when you no longer select for Ura^+ cells (that is, grow the yeast in media containing uracil)?

a. URA^+ gene

b. URA^+ gene, ARS

c. URA^+ gene, ARS, CEN (centromere)

20. A protein called CBF1 was identified in yeast as a centromere binding protein. You want to know if similar proteins are present in human cells.

a. Starting with a human cDNA expression library from actively dividing cells, how could you isolate a clone containing a human gene similar to the yeast gene?

b. Imagine you obtained a clone containing a human gene similar to the yeast gene. How could you test to see if the protein encoded by the human gene is associated with a human centromere region?

21. A DNA fragment containing yeast centromere DNA was cloned into a TRP ARS plasmid, YRp7, causing the plasmid to become mitotically very stable (that is, the plasmid was transmitted during mitotic divisions to each daughter cell). The assay for mitotic stability consists of growing a transformed cell without selection for the plasmid and determining the number of Trp^+ colonies remaining after 20 generations of growth under conditions that are not selective for the plasmid. To identify the region of the cloned fragment that contained centromere DNA, you cut the initial fragment into smaller pieces, reclone those pieces into YRp7, and test for mitotic stability. Based on the map that follows and results of the mitotic stability assay, where is the centromeric DNA located?

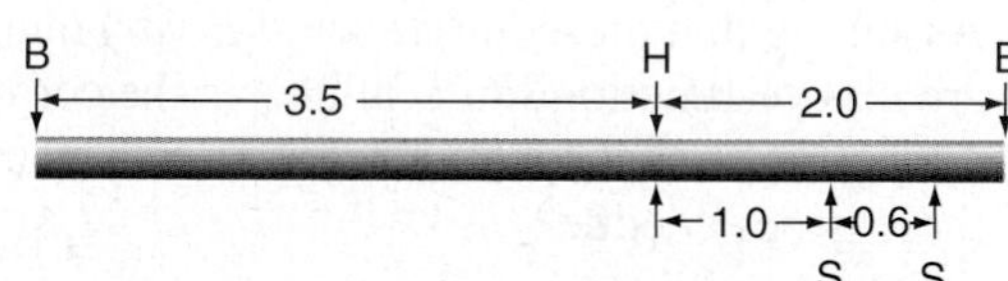

Results of Mitotic Stability Assay

Plasmid DNA	Percentage of Trp^+ colonies: after 20 generations
YRp7	0.9
YRp7 + 5.5 kb *Bam*HI (B)	68.1
YRp7 + 3.5 kb *Bam*HI-*Hin*dIII (H)	0.5
YRp7 + 2.0 kb *Bam*HI-*Hin*dIII	80.3
YRp7 + 0.6 kb *Sau*3A (S)	76.2
YRp7 + 1.0 kb *Hin*dIII-*Sau*3A	0.7

22. Another vector system for cloning large fragments of DNA that investigators developed for genome analysis uses a derivative of the baculovirus that infects arthropods. One of the advantages of these BAC vectors compared to YACs is that DNA cloned into these vectors is less likely to get rearranged. Clones of a human *Sfi*I fragment were isolated from BAC- and YAC-based libraries using the same probe but the *Hin*dIII digests of these clones were very different from each other. How could you determine which of these two clones, if either, has the same order and size as the *Hin*dIII fragments present in the human genome?

Section 13.4

23. For each of the following pairs of chromatin types, which is the most condensed?
   a. 100 Å fiber or 300 Å fiber
   b. 300 Å fiber or DNA loops attached to a scaffold
   c. Euchromatin or heterochromatin
   d. Interphase chromosomes or metaphase chromosomes

24. What element on the X chromosome has been identified as necessary for X inactivation to occur? What are the characteristics of this element?

25. Give examples of constitutive and facultative heterochromatin in
   a. *Drosophila*
   b. humans

26. *Drosophila* geneticists have isolated many mutations that modify position-effect variegation. Dominant *suppressors of variegation* [*Su(var)*s] cause less frequent inactivation of genes brought near heterochromatin by chromosome rearrangements, while dominant *enhancers of variegation* [*E(var)*s] cause more frequent inactivation of such genes.
   a. What effects would each of these two kinds of mutations have on position-effect variegation of the *white* gene in *Drosophila.*
   b. Assuming that these *Su(var)* and *E(var)* mutations are loss-of-function (null) alleles in the corresponding genes, what kinds of proteins do you think these genes encode?

27. How many Barr bodies are present in humans with the following karyotype?
   a. an XX female
   b. an XY male
   c. an XX male (known as an exceptional male*)
   d. an XXY male
   e. an XXXX female
   f. an XO female

28. A pair of twin sisters were believed to be identical until one was diagnosed with Duchenne muscular dystrophy, an X-linked trait. Her sister did not have the disease. Does this finding mean they are definitely not identical twins (derived from fertilization of one egg)? Why or why not?

29. Females with the genotype $X^{CB}\ X^{cb}$ are rarely colorblind although some have only partial color vision. Speculate on why this is true.

30. In cats, the dominant *O* allele of the X-linked *orange* gene is required to produce orange fur; the recessive *o* allele of this gene yields black fur.
   a. Tortoiseshell cats have coats with patches of orange fur alternating with patterns of black fur. Approximately 90% of all tortoiseshell cats are females. What type of crosses would be expected to produce female tortoiseshell cats?
   b. Suggest a hypothesis to explain the origin of male tortoiseshell cats.
   c. Calico cats (most of which are females) have patches of white, orange, and black fur. Suggest a hypothesis for the origin of calico cats.

31. In marsupials like the opposum or kangaroo, X inactivation selectively inactivates the paternal X chromosome.
   a. Predict the possible coat color phenotypes of the progeny of both sexes if a female marsupial homozygous for a mutant allele of an X-linked coat color gene was mated with a male hemizygous for the alternative wild-type allele of this gene.
   b. Predict the possible coat color phenotypes of the progeny of both sexes if a male marsupial hemizygous for a mutant allele of an X-linked coat color gene was mated with a female homozygous for the alternative wild-type allele of this gene.
   c. Why are the terms "recessive" and "dominant" not useful in describing the alleles of X-linked coat color genes in marsupials?
   d. Why would marsupials heterozygous for two alleles of an X-linked coat color gene not have patches of fur of two different colors as did the tortoiseshell cats described in the previous problem?

*In an exceptional XX male, one of the two X chromosomes has essentially all of the genes normally found on the X chromosome plus a small region from the Y chromosome that carries the testes-determining factor required for male development.

Chromosomal Rearrangements and Changes in Chromosome Number Reshape Eukaryotic Genomes

Chapter **14**

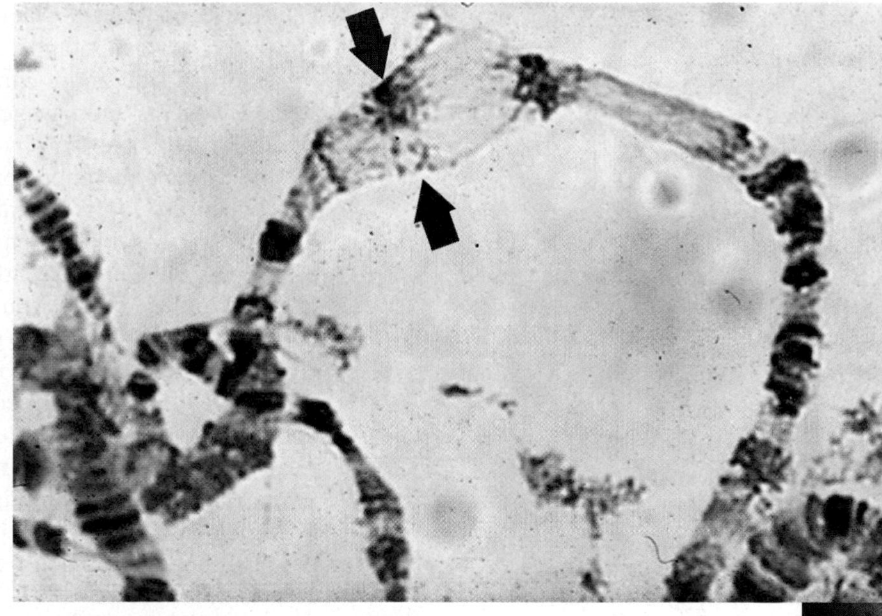

Chromosomal rearrangements can be mapped with high precision on Drosophila *polytene chromosomes. The* red arrow *points to a very large transposable element that has inserted into one chromosome but is not present in the paired homologous wild-type chromosome (*black arrow*).*

During the early days of genome sequencing in the 1990s, studies comparing the human genome with that of the laboratory mouse (*Mus musculus*) revealed a surprising evolutionary paradox: At the DNA level, there is a close similarity of nucleotide sequence across hundreds of thousands of base pairs; but at the chromosomal level, mouse and human karyotypes bear little resemblance to each other. These early genomic analyses focused considerable effort on the sequencing of regions encompassing more than 2000 kb of mouse and human DNA containing a complex of genes that encode proteins known as T-cell receptors. Key components of the mammalian immune system, these receptors become embedded in the membrane of the immune system's T cells, where they recognize foreign molecules and microorganisms and orchestrate the immune responses that dispose of them. Comparisons of the corresponding mouse and human regions show that the nucleotide sequences of the T-cell receptor genes are similar (though not identical) in the two species, as are the order of the genes and the relative positions of a variety of noncoding sequences (of unknown function) along the chromosome. Comparisons of mouse and human Giemsa-stained karyotypes, however, reveal no conservation of banding patterns between the 20 mouse and 23 human chromosomes.

Data for resolving this apparent paradox emerged with the 2002 publication of the nearly complete mouse genome sequence, which researchers could compare with the human genome sequence completed a year earlier. The data showed that each mouse chromosome consists of pieces of different human chromosomes, and vice versa. For example, mouse chromosome 1 contains large blocks of sequences found on human chromosomes 1, 2, 5, 6, 8, 13, and 18 (portrayed in different colors in **Fig. 14.1**). These blocks represent **syntenic segments** in which the identity, order, and transcriptional direction of the genes are almost exactly the same in the two genomes. In principle, scientists could "reconstruct" the mouse genome by breaking the human genome into 342 fragments, each an average length of about 16 Mb, and pasting these fragments together in a different order. Figure 14.1 illustrates this process in detail for mouse chromosome 1; Figure 10.12 on p. 367 shows the syntenic relationships between the entire mouse and human genomes at lower resolution. Since a 16 Mb fragment would occupy no more than one or two bands of a stained chromosome, this level of conservation is not visible in karyotypes. It does,

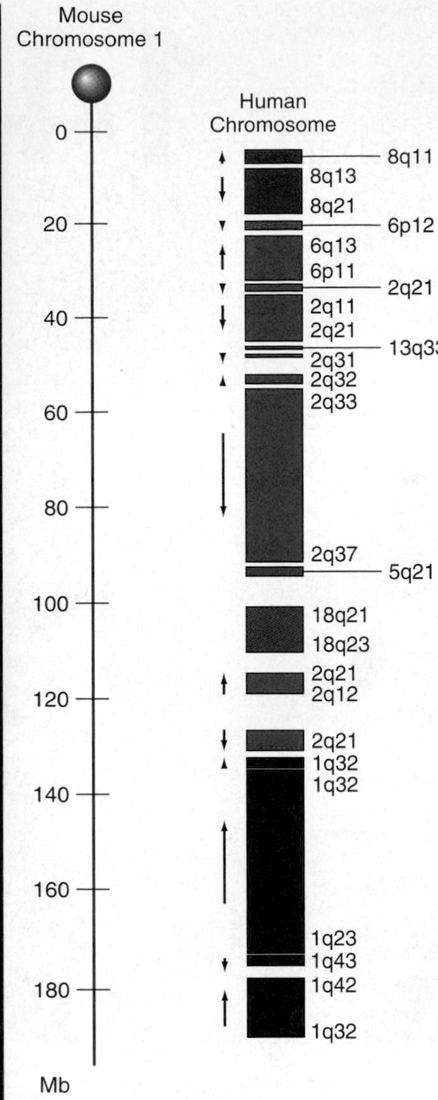

Figure 14.1 Comparing the mouse and human genomes. Mouse chromosome 1 contains large blocks of sequences found on human chromosomes 1, 2, 5, 6, 8, 13, and 18 (portrayed in different colors). *Arrows* indicate the relative orientations of multiple sequence blocks from the same human chromosome.

however, show up in the sequence of a smaller genomic region, such as that encoding the T-cell receptors.

These findings contribute to our understanding of how complex life-forms evolved. Although mice and humans diverged from a common ancestor about 65 million years ago, the DNA sequence in many regions of the two genomes is very similar. It is thus possible to hypothesize that the mouse and human genomes evolved through a series of approximately 300 reshaping events during which the chromosomes broke apart and the resulting fragments resealed end to end in novel ways. After each event, the newly rearranged chromosomes somehow became fixed in the genome of the emerging species. Both nucleotide sequence differences and differences in genome organization thus contribute to dissimilarities between the species.

In this chapter, we examine two types of events that reshape genomes: (1) **rearrangements,** which reorganize the DNA sequences within one or more chromosomes, and (2) **changes in chromosome number** involving losses or gains of entire chromosomes or sets of chromosomes (**Table 14.1**). Rearrangements and changes in chromosome number may affect gene activity or gene transmission by altering the position, order, or number of genes in a cell. Such alterations often, but not always, lead to a genetic imbalance that is harmful to the organism or its progeny.

Two general themes emerge from our discussion. First, karyotypes generally remain constant within a species, not because rearrangements and changes in chromosome number occur infrequently (they are, in fact, quite common), but because the genetic instabilities and imbalances produced by such changes usually place individual cells or organisms and their progeny at a selective disadvantage. Second, despite selection against chromosomal variations, related species almost always have different karyotypes, with closely related species (such as chimpanzees and humans) diverging by only a few rearrangements and more distantly related species (such as mice and humans) diverging by a larger number of rearrangements. These observations suggest there is some correlation between karyotypic rearrangements and the evolution of new species.

In our examination of events that reshape eukaryotic genomes, we discuss

- Rearrangements of DNA sequences within and between chromosomes: deletions, duplications, inversions, translocations, and movements of transposable elements.
- Changes in chromosome number: aneuploidy, including monosomy and trisomy; monoploidy; and polyploidy.
- Emergent technologies for analyzing chromosomal rearrangements and changes in chromosome number.

TABLE 14.1 Chromosomal Rearrangements and Changes in Chromosome Number

Chromosomal Rearrangements

	Before		After
Deletion: Removal of a segment of DNA	1 2 3 4 5 6 7 8	→	1 2 3 5 6 7 8
Duplication: Increase in the number of copies of a chromosomal region	1 2 3 4 5 6 7 8	→	1 2 3 2 3 4 5 6 7 8
Inversion: Half-circle rotation of a chromosomal region	1 2 3 4 5 6 7 8 (180° rotation)	→	1 4 3 2 5 6 7 8
Translocations:			
Nonreciprocal: Unequal exchanges between nonhomologous chromosomes	1 2 3 4 5 6 7 8 12 13 14 15 16 17 18	→	12 13 4 5 6 7 8 14 15 16 17 18
Reciprocal: Parts of two nonhomologous chromosomes trade places	1 2 3 4 5 6 7 8 12 13 14 15 16 17 18	→	12 13 14 15 5 6 7 8 1 2 3 4 16 17 18
Transposition: Movement of short DNA segments from one position in the genome to another	1 2 3 4 5 6 7 8	→	1 2 4 5 6 3 7 8

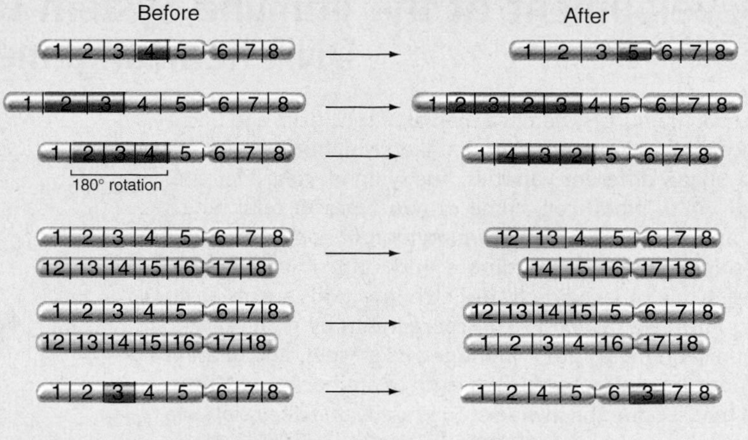

Changes in Chromosome Number

Chromosomes 1, 2, and 3 (represented in different colors) are non-homologous.

	Chromosome 1	Chromosome 2	Chromosome 3
Euploidy: Cells that contain only complete sets of chromosomes			
Diploidy (2*x*): Two copies of each homolog			
Monoploidy (*x*): One copy of each homolog			
Polyploidy: More than the normal diploid number of chromosome sets			
Triploidy (3*x*): Three copies of each homolog			
Tetraploidy (4*x*): Four copies of each homolog			
Aneuploidy: Loss or gain of one or more chromosomes producing a chromosome number that is not an exact multiple of the haploid number			
Monosomy ($2n - 1$)			
Trisomy ($2n + 1$)			
Tetrasomy ($2n + 2$)			

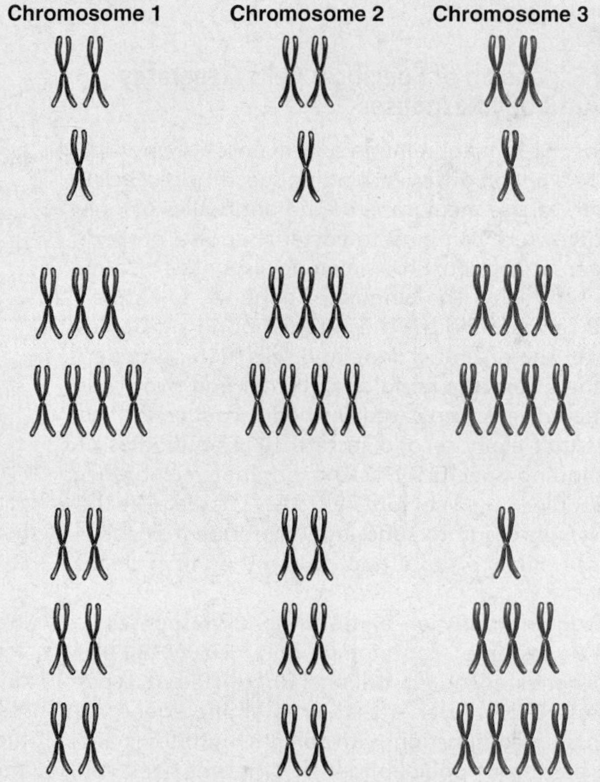

14.1 Rearrangements of DNA Sequences Within Chromosomes

All chromosomal rearrangements alter DNA sequence. Some do so by removing or adding base pairs. Others relocate chromosomal regions without changing the number of base pairs they contain. This chapter focuses on heritable rearrangements that can be transmitted through the germ line from one generation to the next, but it also explains that the genomes of somatic cells can undergo changes in nucleotide number or order. For example, the Fast Forward box "Development of the Immune System Depends on Programmed DNA Rearrangements" (pp. 492–494) describes how the normal development of the human immune system depends on noninherited, programmed rearrangements of the genome in somatic cells.

FAST FORWARD

Development of the Immune System Depends on Programmed DNA Rearrangements

The human immune system is a marvel of specificity and diversity, including as it does close to a trillion lymphocytes of more than a billion different varieties. The lymphocytes, one category of white blood cell, come in two types: B cells, which make antibodies, also called *immunoglobulins (Igs);* and T cells, some of which coordinate and control immune responses, while others directly dispatch infectious agents. Both B cells and T cells respond to microorganisms by proliferating, communicating with other immune-system cells, and gearing up to produce molecular missiles (such as antibodies and interferon) that subdue the invaders. To simplify our discussion, we focus on the maturation of B cells and the genetics of antibody formation, but variations of the mechanisms we describe also apply to the synthesis of receptor molecules critical for T cell function.

The Clonal Expansion of Specific B Cells Generates a Specific Antibody Response

The B cells of the human immune system produce antibodies of close to a billion different binding specificities. Each B cell, however, carries membrane-bound antibodies of only one specificity, which can bind to bacterial or viral proteins (called *antigens* in the context of immune responses) of complementary specificity. The binding of antibody to antigen causes the B cell carrying the membrane-bound antibodies to proliferate. The resulting clone of B cells matures generation by generation to a population containing two types of differentiated cells: dedicated antibody producers called *plasma cells* that every second secrete 1000 antibodies of the same binding specificity as the original membrane-bound antibodies, and *memory cells* that circulate in the blood and lymph ready to subdue in subsequent encounters antigens similar to those that caused the initial clonal expansion.

One intriguing question about antibody responses is, How can a genome containing only 20,000–30,000 ($2\text{–}3 \times 10^4$) genes encode a billion (10^9) different types of antibodies? The answer is that programmed gene rearrangements, in conjunction with somatic mutations and the diverse pairing of polypeptides of different sizes, can generate roughly a billion binding specificities from a much smaller number of genes. To understand the mechanism of this diversity, it is necessary to know how antibodies are constructed and how B cells come to express the antibody-encoding genes determining specific antigen-binding sites.

The Genetics of Antibody Formation Produce Specificity and Diversity

All antibody molecules consist of a single or multiple copies of the same basic molecular unit. That unit is composed of two pairs of polypeptides: two identical light and two identical heavy chains (**Fig. A**). Each light and each heavy chain has a constant (C) domain and a variable (V) domain. The C domain of the heavy chain determines whether the antibody falls into one of five major classes (designated IgM, IgG, IgE, IgD, and IgA), which influence where and how an antibody functions. For example, IgM antibodies form early in an immune response and are anchored in the B-cell membrane, where they help initiate an immune response; IgG antibodies emerge later in a response and are secreted into the blood serum, where they target antigens for destruction. The C domains of the light and heavy chains are not involved in determining the specificity of antibodies. That function falls to the V domains. The variable domains of light and heavy chains come together to form the antigen-binding site, which defines an antibody's specificity.

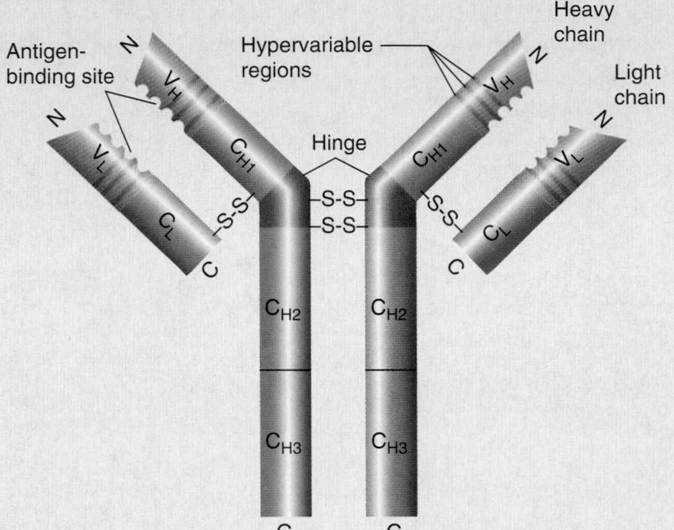

Figure A How antibody specificity emerges from molecular structure. Two heavy chains and two light chains held together by disulfide (–S–S–) bonds form the basic unit of an antibody molecule. Both heavy and light chains have variable (V) domains near their N termini, which associate to form the antigen-binding site. "Hypervariable" stretches of amino acids within the V domains vary extensively between antibody molecules. The remainder of each chain is composed of a C (constant) domain; that of the heavy chain has several subdomains (C_{H1}, hinge, C_{H2}, and C_{H3}).

The DNA for all domains of the heavy chain resides on chromosome 14 (**Fig. B**). This heavy-chain gene region consists of more than 100 V-encoding segments, each preceded by a promoter, several D (for diversity) segments, several J (for joining) segments, and nine C-encoding segments preceded by an enhancer (a short DNA segment that aids in the initiation of

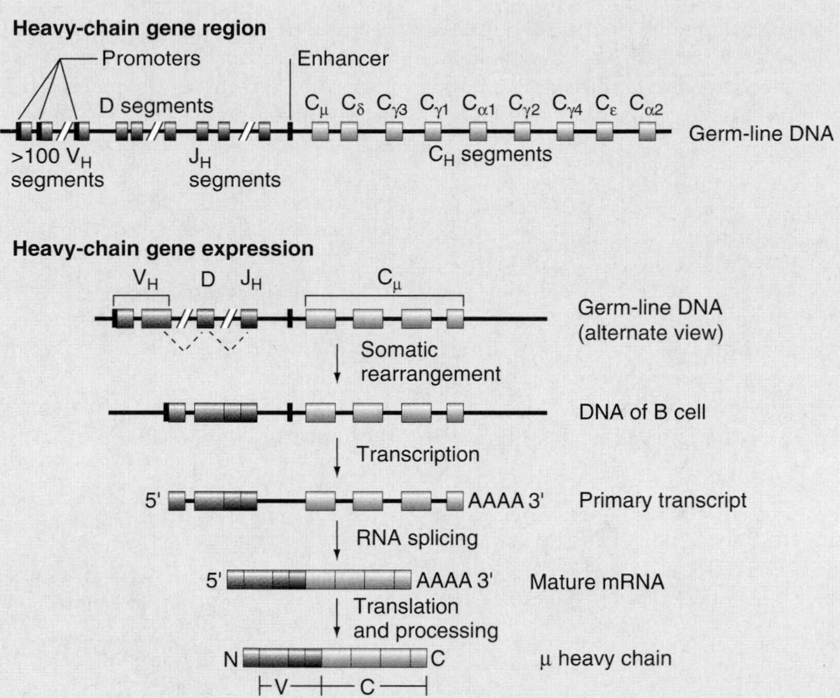

Figure B **The heavy-chain gene region on chromosome 14.** The DNA of germ-line cells (as well as all non-antibody-producing cells) contains more than 100 V_H segments, about 20 D segments, 6 J_H segments, and 9 C_H segments (*top*). Each V_H and C_H segment is composed of two or more exons, as seen in the alternate view of the same DNA on the next line. In B cells, somatic rearrangements bring together random, individual V_H, D, and J_H segments. The promoter of the selected V_H segment is now activated by the enhancer adjacent to $C\mu$,allowing transcription of the newly constructed heavy-chain gene into a primary transcript, which is subsequently spliced into a mature mRNA. The μ heavy chain translated from this mRNA is the type of heavy chain found in IgM antibodies. Later in B-cell development, other rearrangements (*not shown*) connect the same V-D-J variable region to other C_H segments such as C_δ, allowing the synthesis of other antibody classes.

transcription by interacting with the promoter; see Chapter 18 for details). In all germ-line cells and in most somatic cells, including the cells destined to become B lymphocytes, these various gene segments lie far apart on the chromosome. During B-cell development, however, somatic rearrangements juxtapose random, individual V, D, and J segments together to form the particular variable region that will be transcribed. These rearrangements also place the newly formed variable region next to a C segment and its enhancer, and they further bring the promoter and enhancer into proximity, allowing transcription of the heavy-chain gene. RNA splicing removes the introns from the primary transcript, making a mature mRNA encoding a complete heavy-chain polypeptide.

The somatic rearrangements that shuffle the same deck of V, D, J, and C segments at random in each B cell of each individual permit expression of one, and only one, specific heavy chain. Without the rearrangements, antibody gene expression cannot occur. Random somatic rearrangements also generate the actual genes that will be expressed as light chains. The somatic rearrangements allowing the expression of antibodies thus generate enormous diversity of binding sites through the random selection and recombination of gene elements.

Several other mechanisms add to this diversity. First, there is the imprecise joining of each gene's DNA elements, perpetrated by cutting and splicing enzymes that sometimes trim DNA from or add nucleotides to the junctions of the segments they join. Next, random somatic mutations in a rearranged gene's V region increase the variation of the antibody's V domain. Finally, in every B cell, two copies of a specific H chain that emerged from random DNA rearrangements combine with two copies of a specific L chain that emerged from random DNA rearrangements to create molecules with a specific, unique binding site. Each B cell will display antibodies of only this specificity in its membrane, and it will manufacture large quantities of these same antibodies when stimulated by antigen. The fact that any light chain can pair with any heavy chain exponentially increases the potential diversity of antibody types. For example, if there were 10^4 different light chains and 10^5 different heavy chains, there would be 10^9 possible combinations of the two.

Interestingly, the somatic rearrangements that lead to antibody expression may also be responsible for the phenomenon of "allelic exclusion" in which a cell expresses either the paternal or maternal allele of an antibody gene—but not both. One would expect each B cell to express several types of antibodies—some bearing a paternal H and L chain, some a maternal H and L chain, and some a mix of the two. But this rarely, if ever, happens. The "one cell, one antibody" phenomenon prevails; each B cell makes only one kind of H chain and one kind of L chain at a time. Immunogeneticists believe that allelic exclusion is one consequence of the rearrangement process. In B cells in which a successful rearrangement has occurred, the rearrangement apparatus shuts down, precluding the expression of more than one allele.

Mistakes by the Enzymes That Carry Out Antibody Gene Rearrangements Can Lead to Cancer

RagI and RagII are enzymes that interact with DNA sequences in antibody genes to help catalyze the rearrangements required for gene expression. In carrying out their rearrangement activities, however, the enzymes sometimes make a mistake that results in a reciprocal translocation between human chromosomes 8 and 14. After this translocation, the enhancer of the chromosome 14 heavy-chain gene lies in the vicinity of the unrelated *c-myc* gene from chromosome 8. Under normal circumstances, *c-myc* generates a transcription factor that turns on other genes active in cell division, at the appropriate time and rate in the cell cycle. The translocated antibody-gene enhancer accelerates expression of *c-myc,* causing B cells containing the translocation to divide out of control. This uncontrolled B-cell division leads to a cancer known as Burkitt's lymphoma (**Fig. C**).

Thus, although programmed gene rearrangements contribute to the normal development of a healthy immune system, misfiring of the rearrangement mechanism can promote disease.

Chapter 22 describes the evolution of the gene families that encode antibodies and other immune system proteins.

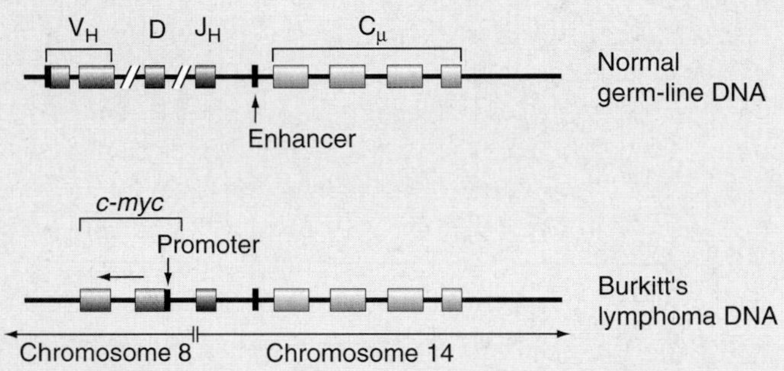

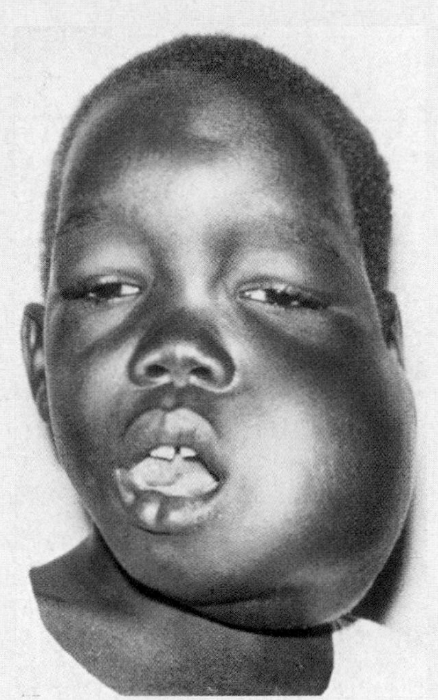

Figure C Misguided translocations can lead to Burkitt's lymphoma. In DNA from this Burkitt's lymphoma patient, translocations bring transcription of the *c-myc* gene under the control of the enhancer adjacent to $C\mu$. As a result, B cells produce abnormally high levels of the c-myc protein. Apparently, the RagI and RagII enzymes have mistakenly connected a J_H segment to the *c-myc* gene from chromosome 8, instead of to a heavy-chain D segment as would normally occur.

Deletions Remove Material from the Genome

We saw in Chapter 7 that **deletions** remove one or more contiguous base pairs of DNA from a chromosome. They may arise from errors in replication, from faulty meiotic or mitotic recombination, and from exposure to X-rays or other chromosome-damaging agents that break the DNA backbone (**Fig. 14.2a**). Here we use the symbol *Del* to designate a chromosome that has sustained a deletion. However, many geneticists, particularly those working on *Drosophila,* prefer the term *deficiency* (abbreviated as *Df*) to deletion.

Small deletions often affect only one gene, while large deletions can generate chromosomes lacking tens or even hundreds of genes. In higher organisms, geneticists usually find it difficult to distinguish small deletions affecting only one gene from point mutations; they can resolve such distinctions only through analysis of the DNA itself. For example, deletions can result in smaller restriction fragments or polymerase chain reaction (PCR) products, whereas most point mutations would not cause such changes (Fig. 14.2b). Larger deletions are sometimes identifiable because they affect the expression of two or more adjacent genes. Very large deletions are visible at the relatively low resolution of a karyotype, showing up as the loss of one or more bands from a chromosome.

Homozygosity for a Deletion Is Often, But Not Always, Lethal

Because many of the genes in a genome are essential to an individual's survival, homozygotes (*Del*/*Del*) or hemizygotes

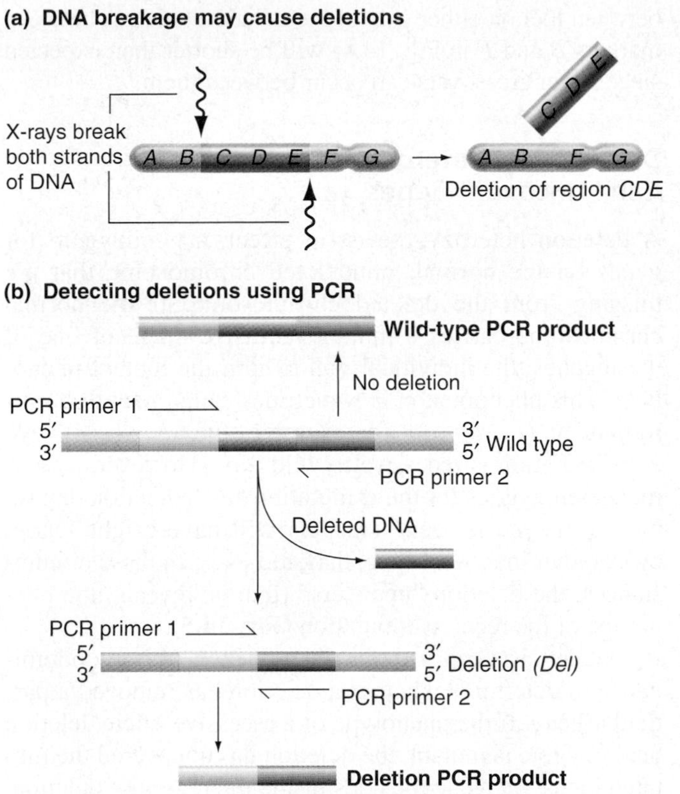

(a) DNA breakage may cause deletions

X-rays break both strands of DNA
A B C D E F G → A B F G
C D E
Deletion of region *CDE*

(b) Detecting deletions using PCR

Wild-type PCR product
No deletion
PCR primer 1
5′ 3′
3′ 5′ Wild type
PCR primer 2
Deleted DNA
PCR primer 1
5′ 3′
3′ 5′ Deletion (*Del*)
PCR primer 2
Deletion PCR product

Figure 14.2 Deletions: Origin and detection. (a) If a chromosome sustains two double-strand breaks, a deletion will result if the chromosomal fragments are not properly religated. **(b)** One way to detect deletions is by PCR. The two PCR primers shown will amplify a larger PCR product from wild-type DNA than from DNA with a deletion.

(*Del*/Y) for most deletion-bearing chromosomes do not survive. In rare cases where the deleted chromosomal region is devoid of genes essential for viability, however, a deletion hemi- or homozygote may survive. For example, *Drosophila* males hemizygous for an 80 kb deletion including the *white* (*w*) gene survive perfectly well in the laboratory; lacking the w^+ allele required for red eye pigmentation, they have white eyes.

Heterozygosity for a Deletion Is Often Detrimental

Usually, the only way an organism can survive a deletion of more than a few genes is if it carries a nondeleted wild-type homolog of the deleted chromosome. Such a *Del*/+ individual is known as a *deletion heterozygote.* Nonetheless, the missing segment cannot be too large, as heterozygosity for very large deletions is almost always lethal. Even small deletions can be harmful in heterozygotes. Newborn humans heterozygous for a relatively small deletion from the short arm of chromosome 5 have *cri du chat* syndrome (from the French for "cry of the cat"), so named because the symptoms include an abnormal cry reminiscent of a mewing kitten. The syndrome also leads to mental retardation.

Why should heterozygosity for a deletion have harmful consequences when the *Del*/+ individual has at least one wild-type copy of all of its genes? The answer is that changes in **gene dosage**—the number of times a given gene is present in the cell nucleus—can create a **genetic imbalance.** This imbalance in gene dosage alters the amount of a particular protein relative to all other proteins, and this alteration in the relative amounts of protein can have a variety of phenotypic effects, depending on how the proteins function and how critical the maintenance of a precise ratio of proteins is to the survival of the organism. For some rare genes, the normal diploid level of gene expression is essential to individual survival; fewer than two copies of such a gene results in lethality. In *Drosophila,* a single dose of the locus known as *Triplolethal* (Tpl^+) is lethal in an otherwise diploid individual. For certain other genes, the phenotypic consequences of a decrease in gene dosage are noticeable but not catastrophic. For example, *Drosophila* containing only one copy of the wild-type *Notch* gene have visible wing abnormalities but otherwise seem to function normally (**Fig. 14.3**). In contrast with these unusual examples, diminishing the dosage of most genes produces no obvious change in phenotype. There is a catch, however. Although a single dose of any one gene may not cause substantial harm to the individual, the genetic imbalance resulting from a single dose of many genes at the same time can be lethal. Humans, for example, cannot survive, even as heterozygotes, with deletions that remove more than about 3% of any part of their haploid genome.

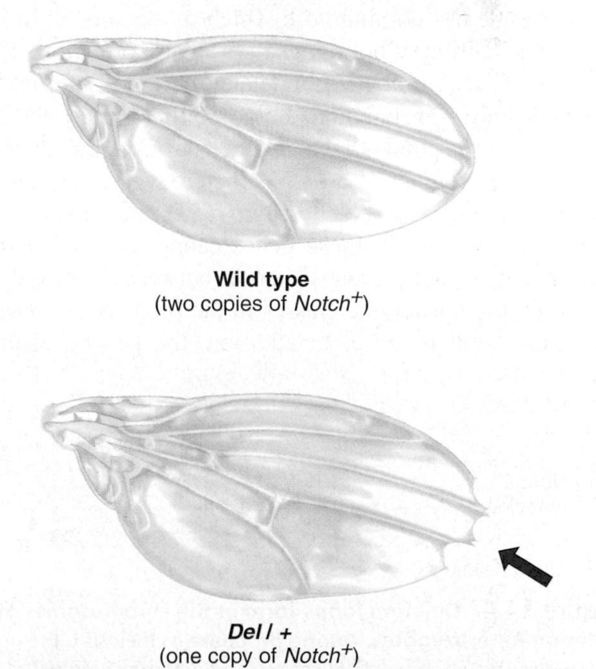

Figure 14.3 Heterozygosity for deletions may have phenotypic consequences. Flies carrying only one copy of the $Notch^+$ gene instead of the normal two copies have abnormal wings.

There is another answer to the question of why heterozygosity for a deletion can be harmful. If a somatic cell heterozygous for a deletion sustains a mutation in the remaining wild-type region of the homologous chromosome, the cell may become unable to make the protein encoded by a gene in the deleted region. If this protein helps control cell division, the cell may divide out of control and generate a tumor. Thus, individuals born heterozygous for certain deletions have a greatly increased risk of losing both copies of certain genes and developing cancer. One case in point is retinoblastoma (RB), the most malignant form of eye cancer, which was previously introduced in Chapter 5 (p. 154). Karyotypes of normal, noncancerous tissues from many people suffering from retinoblastoma reveal heterozygosity for deletions on chromosome 13. Cells from the retinal tumors of these same patients have a mutation in the remaining copy of the *RB* gene on the nondeleted chromosome 13. Chapter 19, "Cell-Cycle Regulation and the Genetics of Cancer," explains in detail how deletion of certain chromosomal regions greatly increases the risk of cancer and how researchers have used this knowledge to locate and clone some of the genes whose mutant forms cause cancer.

Heterozygosity for Deletions Affects Mapping Distances

Because recombination between maternal and paternal homologs can occur only at regions of similarity, map distances derived from genetic recombination frequencies in deletion heterozygotes will be aberrant. For example, no recombination is possible between genes *C, D,* and *E* in **Fig. 14.4** because the DNA in this region of the normal, nondeleted chromosome has nothing with which to recombine. In fact, during the pairing of homologs in prophase of meiosis I, the "orphaned" region of the nondeleted chromosome forms a **deletion loop**—an unpaired bulge of the normal chromosome that corresponds to the area deleted from the other homolog. The progeny of a *Del/+* heterozygote will always inherit the markers in a deletion loop (*C, D,* and *E* in Fig. 14.4) as a unit. As a result, these genes cannot be separated by recombination, and the map distances between them, as determined by the phenotypic classes in the progeny of a *Del/+* individual, will be zero. In addition, the genetic distance between loci on either side of the deletion (such as between markers *B* and *F* in Fig. 14.4) will be shorter than expected since fewer crossovers can occur between them.

Deletions in Heterozygotes Can "Uncover" Genes

A deletion heterozygote is, in effect, a hemizygote for genes on the normal, nondeleted chromosome that are missing from the deleted chromosome. If the normal chromosome carries a mutant recessive allele of one of these genes, the individual will exhibit the mutant phenotype. This phenomenon is sometimes called **pseudodominance.** In *Drosophila,* for example, the *scarlet* (*st*) eye color mutation is recessive to wild type. However, an animal heterozygous for the *st* mutation and a deletion that removes the *scarlet* gene (*st/Del*) will have bright scarlet eyes, rather than wild-type, dark red eyes. In these circumstances, the deletion "uncovers" (that is, reveals) the phenotype of the recessive mutation (**Fig. 14.5**).

Geneticists can use this phenomenon of pseudodominance to determine whether a deletion has removed a particular gene. If the phenotype of a recessive-allele/deletion heterozygote is mutant, the deletion has uncovered the mutated locus; the gene thus lies inside the region of deletion. If the trait determined by the gene is wild type in these heterozygotes, the deletion has not uncovered the recessive allele, and the gene must lie outside the deleted region. You can consider this experiment as a complementation test between the mutation and the deletion: the uncovering of a mutant recessive phenotype demonstrates a lack of complementation because neither chromosome can supply wild-type gene function.

Using Deletions to Locate Genes

Geneticists can use deletions that alter chromosomal banding patterns to map genes relative to specific regions of metaphase chromosomes. A deletion that results in the loss of one or more bands from a chromosome and also uncovers the recessive mutation of a particular gene places that gene within the missing chromosomal segment.

The greater the number of distinguishable bands in a chromosome, the greater the accuracy of gene localization

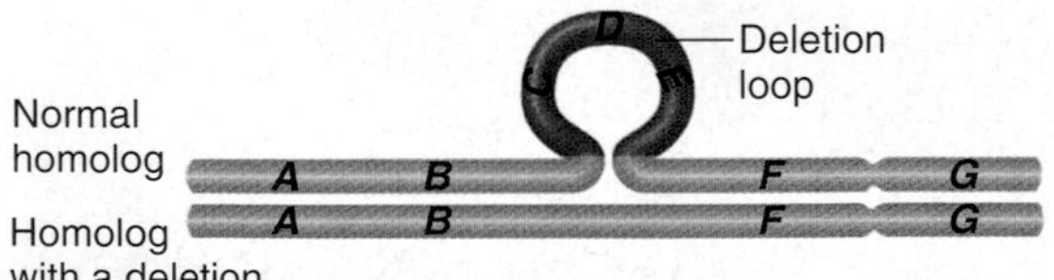

Figure 14.4 Deletion loops form in the chromosomes of deletion heterozygotes. During prophase of meiosis I, the undeleted region of the normal chromosome has nothing with which to pair and thus forms a deletion loop. No recombination can occur between genes within the deletion loop. In this simplified figure, each line represents two chromatids.

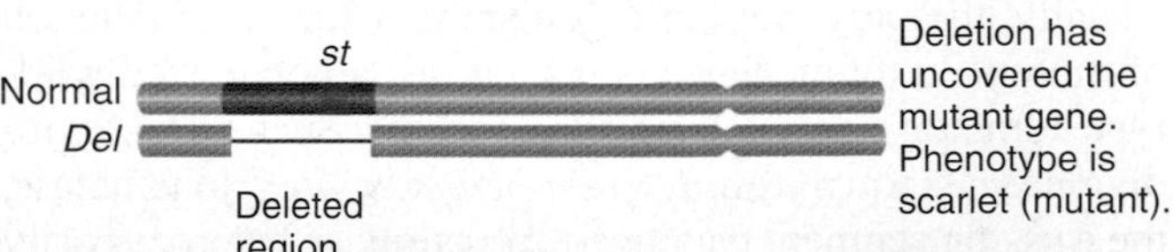

Figure 14.5 In deletion heterozygotes, pseudodominance shows that a deletion has removed a particular gene. A fly of genotype *st/Del* (where *Del* is a deletion of the region in which the *scarlet* [*st*] gene normally lies) displays the recessive scarlet eye color. The deletion has thus "uncovered" the scarlet mutation.

(a) Banding pattern of *Drosophila* polytene chromosomes

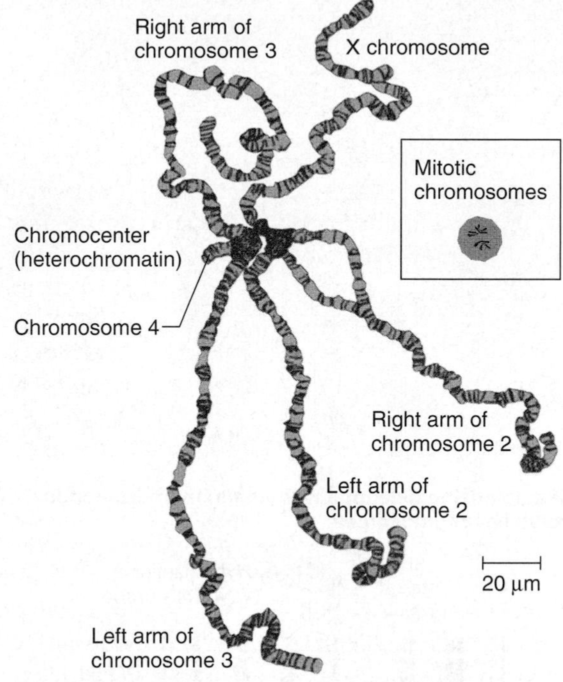

(b) Alignment of chromatids in polytene chromosomes

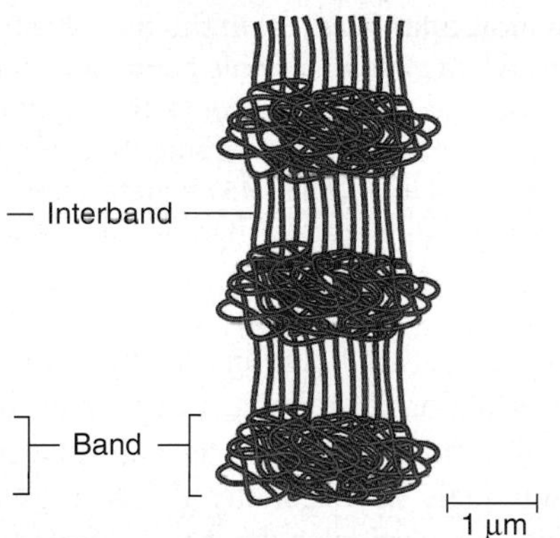

Figure 14.6 Polytene chromosomes in the salivary glands of *Drosophila* larvae. (a) A drawing of the banding pattern seen in polytene chromosomes. The inset shows the relative size of normal mitotic chromosomes. Note that the homologous polytene chromosomes are paired along their lengths. **(b)** A hypothetical model showing how the 1024 chromatids of each polytene chromosome are aligned in register, with the chromatin in the bands being more condensed than the chromatin of the interbands.

by this strategy. For this reason, specialized giant chromosomes found in the salivary gland cells of *Drosophila* larvae are a prized mapping resource. The interphase chromosomes in these cells go through 10 rounds of replication without ever entering mitosis. As a result, the sister chromatids never separate, and each chromosome consists of $2^{10}(=1024)$ double helices. In addition, because the homologous chromosomes in the somatic cells of *Drosophila* remain tightly paired throughout interphase, pairs of homologs form a cable of double thickness containing 2048 double helices of DNA (1024 from each homolog). These giant chromosomes consisting of many identical chromatids lying in parallel register are called **polytene chromosomes (Fig. 14.6a)**.

When stained and viewed in the light microscope, *Drosophila* polytene chromosomes have an irregular fine-grain banding pattern in which denser dark bands alternate with lighter interbands. The chromatin of each dark band is roughly 10 times more condensed than the chromatin of the lighter interbands (Fig. 14.6b). Scientists do not yet understand the functional significance of the division of polytene chromosomes into bands and interbands. One possibility is that the bands represent units of transcriptional regulation containing genes activated at the same time. In any event, the precisely reproducible banding patterns of polytene chromosomes provide a detailed physical guide to gene mapping. *Drosophila* polytene chromosomes collectively carry about 5000 bands that range in size from 3 kb to approximately 150 kb; investigators designate these bands by numbers and letters of the alphabet.

Since homologous polytene chromosomes pair with each other, deletion loops will form in the polytene chromosomes of deletion heterozygotes (**Fig. 14.7**). Scientists can pinpoint the region of the deletion by noting which bands are present in the wild-type homolog but missing in the deletion. If researchers find that a small deletion removing only a few polytene chromosome bands uncovers a gene or that several overlapping larger deletions affect the same gene, they can assign the gene to one or a small number of bands, often representing less than 100 kb of

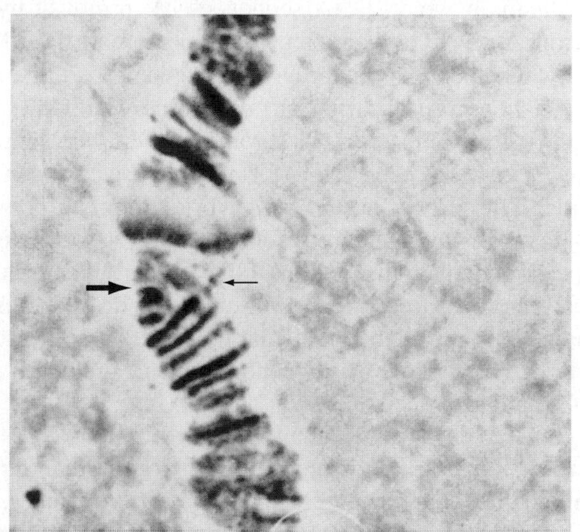

Figure 14.7 Deletion loops also form in the paired polytene chromosomes of *Drosophila* deletion heterozygotes. The *thick arrow* points to the wild-type chromosome; the corresponding region is missing from the *Del* homolog.

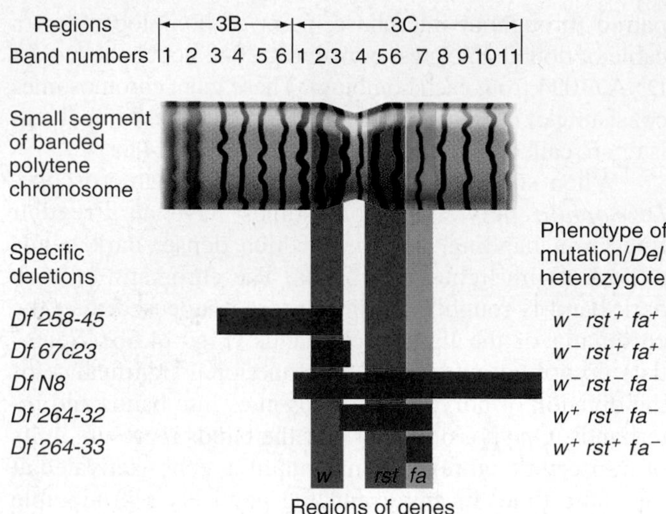

Figure 14.8 Using deletions to assign genes to bands on *Drosophila* polytene chromosomes. *Red bars* show the bands removed by various deletions; for example, *Df 258-45* eliminates bands 3B3–3C3. Complementation experiments determined whether these deletions uncovered the *white* (*w*), *roughest* (*rst*), or *facet* (*fa*) genes. For instance, *w*/*Df 258-45* females have white eyes, so the *w* gene is removed by this deletion. The *w* gene must lie within bands 3C2–3 (*green*) because that is the region common to the deletions that uncover *w*. Similarly, the *rst* gene must be in bands 3C5–6 (*yellow*) and the *fa* gene in band 3C7 (*purple*).

(a) *In situ* hybridization of the *white* gene to wild-type polytene chromosomes

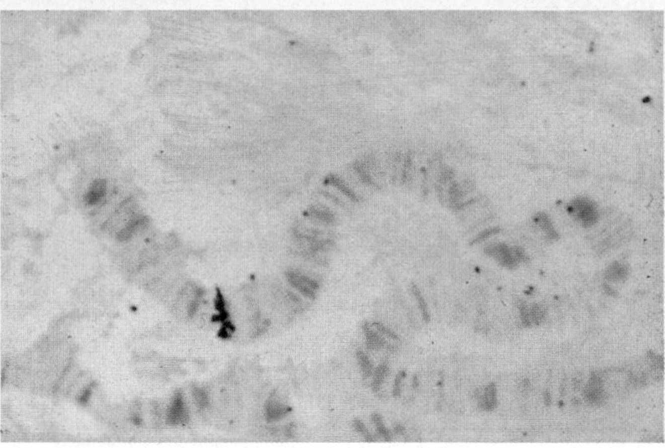

(b) Characterizing deletions with *in situ* hybridization to polytene chromosomes

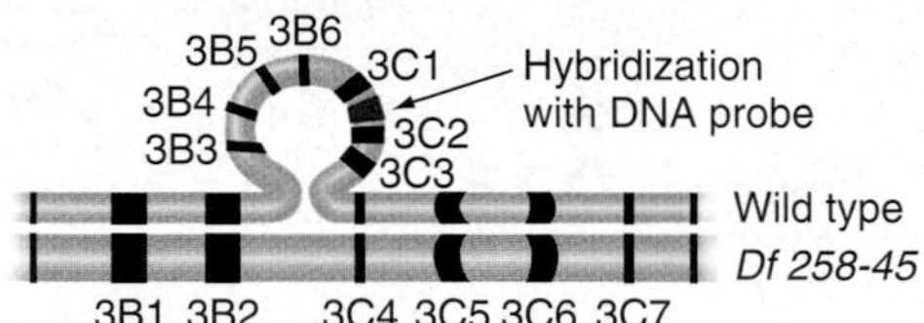

Figure 14.9 *In situ* hybridization as a tool for locating genes at the molecular level. (a) *In situ* hybridization of a probe containing the *white* gene to a single band (3C2) near the tip of the wild-type *Drosophila* X chromosome. **(b)** A particular labeled probe hybridizes to the wild-type chromosome but not to the deletion chromosome in a *Df 258-45* / + heterozygote. The *Df 258-45* deletion thus lacks DNA homologous to the probe.

DNA. **Figure 14.8** shows how geneticists used this strategy to assign three genes to regions containing only one or two polytene chromosome bands on the *Drosophila* X chromosome.

Geneticists can use deletions analyzed at even higher levels of resolution to help locate genes on cloned fragments of DNA. They must first determine whether a particular deletion uncovers a recessive allele of the gene of interest and then ascertain which DNA sequences are removed by the deletion. *In situ* hybridization provides a straightforward way to show whether a particular DNA sequence is part of a deletion. Suppose you are trying to determine whether a small segment of the *Drosophila* X chromosome in the vicinity of the *white* gene has been deleted. You could use purified DNA fragments as probes for *in situ* hybridization to polytene chromosomes prepared from female flies heterozygous for various deletions in this region of their X chromosomes. If a probe hybridizes to a *Del* chromosome, the deletion has not completely removed that particular fragment of DNA; lack of a hybridization signal on a *Del* chromosome, however, indicates that the fragment has been deleted (**Fig. 14.9**).

Geneticists can also localize deleted regions by asking whether particular bands are removed from human mitotic chromosomes, but since bands in these chromosomes containing less than 5 Mb of DNA cannot be detected, the resolution of this method is much lower than is possible with *Drosophila* polytene chromosomes. As the final section of this chapter on "A Glimpse of the Future" illustrates, new techniques nonetheless allow human geneticists to determine the molecular extent of deletions in human chromosomes. Once this information is available, *in situ* hybridization to human mitotic chromosomes serves as a useful tool to diagnose whether individuals have genetic diseases associated with heterozygosity for particular deletions. **Figure 14.10** shows an application of this strategy to the diagnosis of DiGeorge syndrome, which accounts for approximately 5% of all congenital heart malformations.

Duplications Add Material to the Genome

Duplications increase the number of copies of a particular chromosomal region. In **tandem duplications,** repeats of a region lie adjacent to each other, either in the same order or in reverse order (**Fig. 14.11a**). In **nontandem** (or *dispersed*) **duplications,** the two or more copies of a region are not adjacent to each other and may lie far apart on the same chromosome or on different chromosomes. Duplications arise by chromosomal breakage and faulty repair, unequal crossing-over, or errors in DNA replication (Fig. 14.11b). In this book, we use *Dp* as the symbol for a chromosome carrying a duplication.

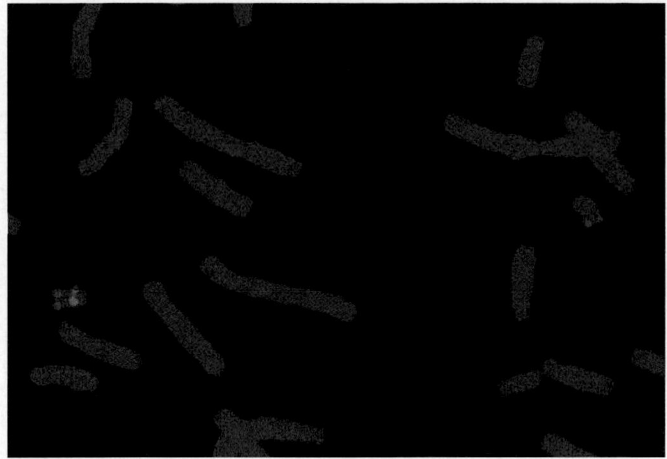

Figure 14.10 Diagnosing DiGeorge syndrome by fluorescence *in situ* hybridization (FISH) to human metaphase chromosomes. The *green* signal is a control probe that identifies both chromosome 22's. The *red* signal is a fluorescent probe from region 22q11, which is deleted in one of the chromosome 22's in DiGeorge syndrome patients. Note that the homologous metaphase chromosomes shown here do not pair with each other and thus do not form a deletion loop, as contrasted with the *Drosophila* polytene chromosomes shown in Fig.14.7.

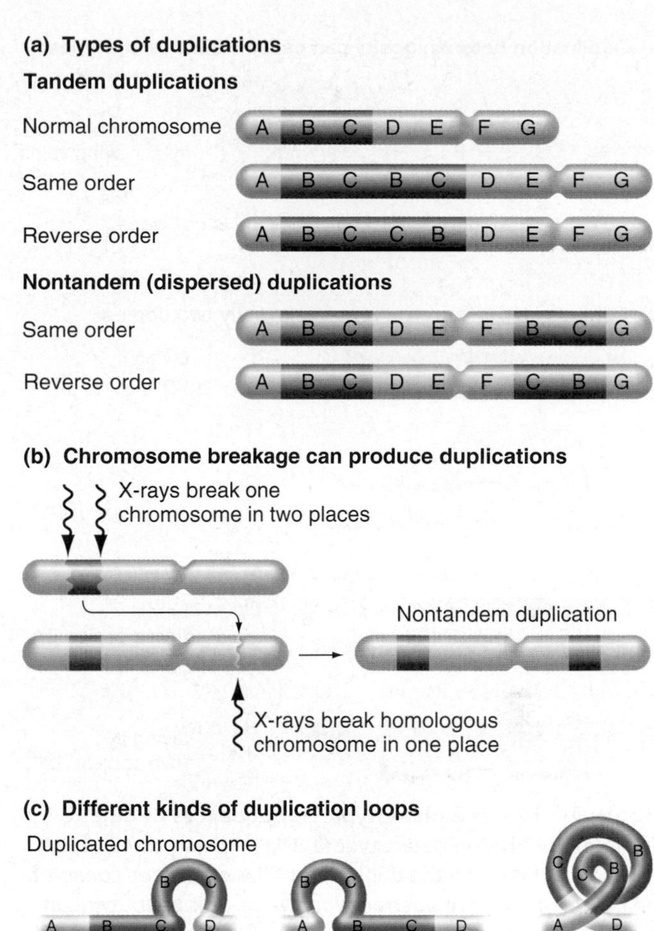

Figure 14.11 Duplications: Structure, origin, and detection. (a) In tandem duplications, the repeated regions lie adjacent to each other in the same or in reverse order. In nontandem duplications, the two copies of the same region are separated. **(b)** In one scenario for duplication formation, X-rays break one chromosome twice and its homolog once. A fragment of the first chromosome can then insert elsewhere on its homolog, producing a nontandem duplication. **(c)** Duplication loops form when chromosomes pair in duplication heterozygotes (*Dp*/+). During prophase I, the duplication loop can assume different configurations that maximize the pairing of related regions. A single line represents two chromatids in this simplified diagram.

Most duplications have no obvious phenotypic consequences and can be detected only by cytological or molecular means. Sufficiently large duplications, for example, show up as repeated bands in metaphase or polytene chromosomes. During the prophase of meiosis I in heterozygotes for such duplications (*Dp*/+), the repeated bands form a **duplication loop**—a bulge in the *Dp*-bearing chromosome that has no similar region with which to pair in the unduplicated normal homologous chromosome. Duplication loops can occur in several alternative configurations (Fig. 14.11c). Such loops also form in the polytene chromosomes of *Drosophila* duplication heterozygotes, where the pattern of the bands in the duplication loops is a repeat of that seen in the other copy of the same region elsewhere on the chromosome.

Duplications Can Affect Phenotype

Although duplications are much less likely to affect phenotype than are deletions of comparable size, some duplications do have phenotypic consequences for visible traits or for survival. Geneticists can use such phenotypes to identify individuals whose genomes contain the duplication. Duplications can produce a novel phenotype either by increasing the number of copies of a particular gene or set of genes or by placing the genes bordering the duplication in a new chromosomal environment that alters their expression. These phenotypic consequences often arise even in duplication heterozygotes (*Dp*/+). For example, *Drosophila* heterozygous for a duplication including the $Notch^+$ gene have abnormal wings that signal the three copies of $Notch^+$ (**Fig. 14.12a**); we have already seen that *Del*/+ flies with only one copy of the $Notch^+$ gene have a different kind of wing abnormality (review Fig. 14.3). In another example from *Drosophila,* the locus known as *Triplolethal* (Tpl^+) is lethal when present in one or three doses in an otherwise diploid individual

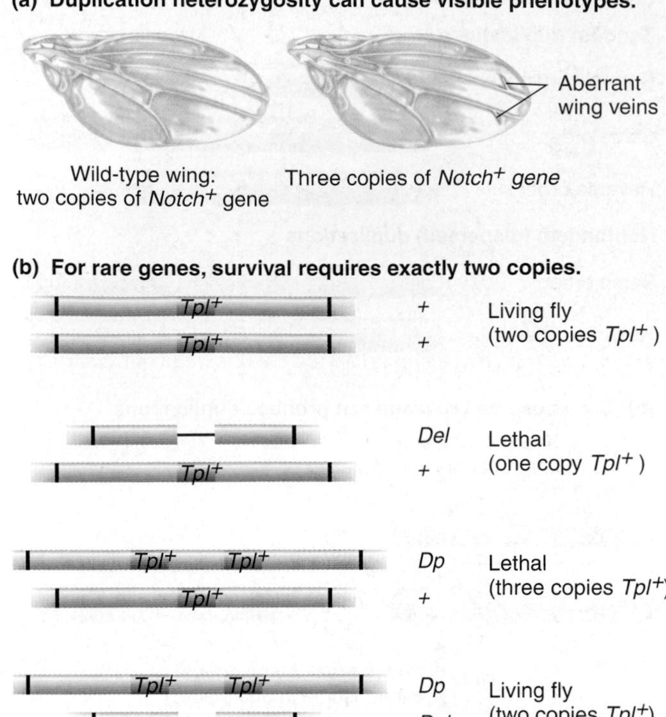

Figure 14.12 The phenotypic consequences of duplications. **(a)** Duplication heterozygotes (*Dp*/+), have three copies of genes contained in the duplication. Flies with three copies of the *Notch*$^+$ gene have aberrant wing veins. This phenotype differs from that caused by only one copy of *Notch*$^+$ (see Fig. 14.3). **(b)** In *Drosophila,* three copies or one copy of *Tpl*$^+$ are lethal.

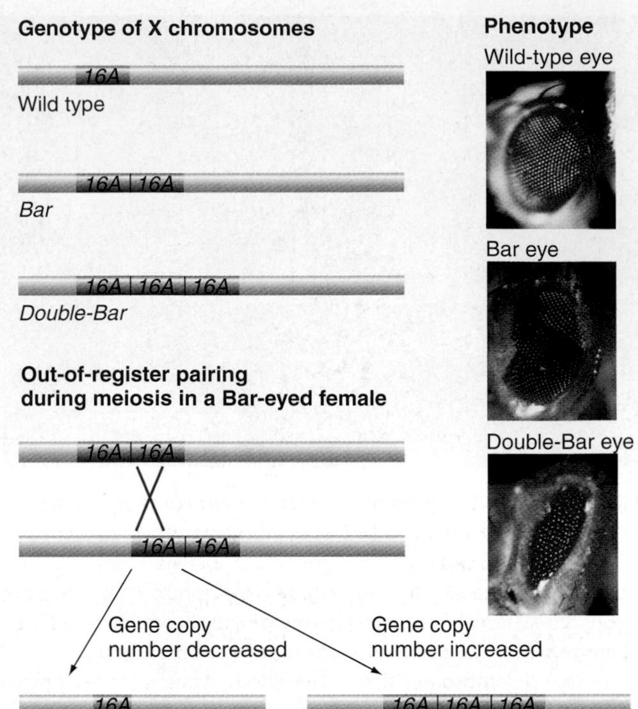

Figure 14.13 Unequal crossing-over can increase or decrease copy number. Duplication of the X chromosome polytene region 16A causes Bar eyes. Unequal pairing and crossing-over during meiosis in females homozygous for this duplication produce chromosomes that have either one copy of region 16A (conferring normal eyes) or three copies of 16A (causing the more abnormal double-Bar eyes).

(Fig. 14.12b). Thus, heterozygotes for a *Tpl* deletion (*Del*/+) or for a *Tpl*$^+$ duplication (*Dp*/+) do not survive. Heterozygotes carrying one homolog deleted for the locus and the other homolog duplicated for the locus (*Del*/*Dp*) are viable because they have two copies of *Tpl*$^+$.

Organisms are usually not so sensitive to additional copies of a gene; but just as for large deletions, imbalances for the many genes included in a very large duplication have additive deleterious effects that jeopardize survival. In humans, heterozygosity for duplications covering more than 5% of the haploid genome is most often lethal.

Unequal Crossing-Over Between Duplications Increases or Decreases Gene Copy Number

In individuals homozygous for a tandem duplication (*Dp*/*Dp*), homologs carrying the duplications occasionally pair out of register during meiosis. **Unequal crossing-over,** that is, recombination resulting from such out-of-register pairing, generates gametes containing increases to three and reciprocal decreases to one in the number of copies of the duplicated region. In *Drosophila,* tandem duplication of several polytene bands near the X chromosome centromere produces the Bar phenotype of kidney-shaped eyes. *Drosophila* females homozygous for the Bar eye duplication produce mostly Bar eye progeny. Some progeny, however, have wild-type eyes, while other progeny have double-Bar eyes that are even smaller than Bar eyes (**Fig. 14.13**). The genetic explanation is that flies with wild-type eyes carry X chromosomes containing only one copy of the region in question, flies with Bar eyes have X chromosomes containing two copies of the region, and flies with double-Bar eyes have X chromosomes carrying three copies. Unequal crossing-over in females homozygous for double-Bar chromosomes can yield progeny with even more extreme phenotypes associated with four or five copies of the duplicated region. Duplications in homozygotes thus allow for the expansion and contraction of the number of copies of a chromosomal region from one generation to the next.

How Duplications and Deletions Affect Phenotype and Evolution: A Summary

Both duplications and deletions alter the number of genes on a chromosome and, as a result, may affect the phenotypes of heterozygotes. Heterozygosity for deletions or duplications produces one or three copies of a gene in an otherwise diploid organism. These changes in gene dosage create an imbalance in gene products that can alter visible phenotypes; for a very few genes, the alterations in dosage are lethal to the organism. The deleterious effects of genetic imbalance are generally additive, such that heterozygotes for very large deletions or duplications of virtually any chromosomal region cannot survive. Both duplications and deletions can also alter phenotype by placing a gene in a new chromosomal location that modifies its expression. As a simple example, a deletion or duplication may relocate a gene close to a regulatory DNA sequence that will inappropriately suppress or enhance the transcription of the relocated gene.

Finally, deletions (by removing material from the genome) and duplications (by adding material to the genome) function as engines driving the evolution of the genome. Indeed, the duplication of chromosomal segments followed by the expansion or contraction of the number of copies of the duplicated regions has played a significant role in the evolution of present-day genomes. Families of tandemly repeated genes, such as those for hemoglobin and color perception, arose in this way (see the comprehensive examples in Chapters 7 and 9). We discuss the molecular basis of this evolutionary mechanism in more detail in Chapter 22.

Inversions Reorganize the DNA Sequence of a Chromosome

The half-circle rotation of a chromosomal region known as an **inversion (*In*)** can occur when radiation produces two double-strand breaks in a chromosome's DNA. The breaks release a middle fragment, which may turn 180° before religation to the flanking chromosomal regions, resulting in an inversion (**Fig. 14.14a**). Inversions may also result from rare crossovers between related DNA sequences present in two positions on the same chromosome in inverted orientation (Fig. 14.14b), or they may arise by the action of transposable genetic elements (discussed later). Inversions that include the centromere are **pericentric,** while inversions that exclude the centromere are **paracentric** (see Fig. 14.14a).

Most inversions do not result in an abnormal phenotype, because even though they alter the order of genes along the chromosome, they do not add or remove DNA and therefore do not change the identity or number of genes. Geneticists can detect some inversions that do not affect phenotype, especially those that cause cytologically visible changes in banding patterns or those that suppress recombination in heterozygotes (as described later) and thereby change the expected results of linkage analysis. In natural populations, however, many inversions that do not affect phenotype go undetected.

If one end of an inversion lies within the DNA of a gene (Fig. 14.14c), a novel phenotype can occur. Inversion following an intragenic break separates the two parts of the gene, relocating one part to a distant region of the chromosome, while leaving the other part at its original site. Such a split disrupts the gene's function. If that function is essential to viability, the inversion acts as a recessive lethal mutation, and homozygotes for the inversion will not survive.

(a) Chromosome breakage can produce inversions.

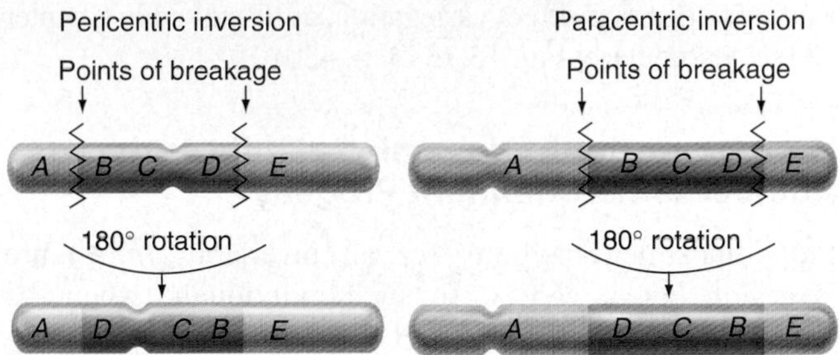

(b) Intrachromosomal recombination can also cause inversions.

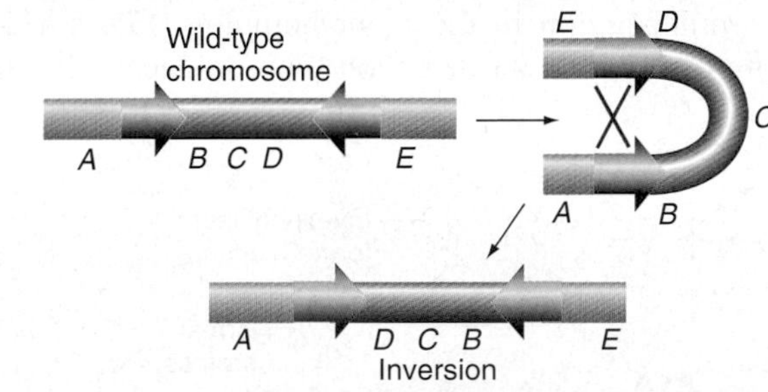

(c) Inversions can disrupt gene function.

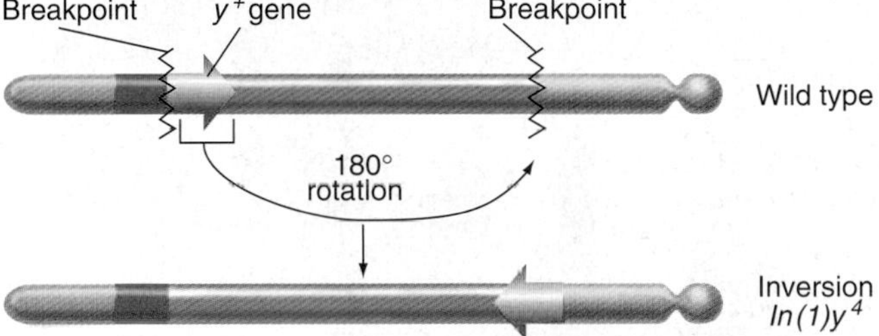

Figure 14.14 Inversions: Origins, types, and phenotypic effects. (a) Inversions can arise when chromosome breakage produces a DNA segment that rotates 180° before it reattaches to the chromosome. When the rotated segment includes the centromere, the inversion is *pericentric;* when the rotated segment does not include the centromere, the inversion is *paracentric.* **(b)** If a chromosome has two copies of a sequence in reverse orientation, rare intrachromosomal recombination between the two copies can give rise to an inversion. **(c)** An inversion can affect phenotype if it disrupts a gene. Here, the inversion $In(1)y^4$ inactivates the *y* (*yellow*) gene by dividing it in two, thereby altering the fly's body color.

Inversions can also produce unusual phenotypes by moving genes residing near the inversion breakpoints to chromosomal environments that alter their normal expression. For example, mutations in the *Antennapedia* gene of *Drosophila* that transform antennae into legs (review Fig. 7.•• on p. •••) are inversions that place the gene in a new regulatory environment, next to sequences that cause it to be transcribed in tissues where it would normally remain unexpressed. Inversions that reposition genes normally found in a chromosome's euchromatin to a position near a region of heterochromatin can also produce an unusual phenotype; spreading of the heterochromatin may inactivate the gene in some cells, leading to position-effect variegation, as discussed in Chapter 13 (see particularly Fig. 13.13 on p. 480).

Heterozygosity for Inversions Reduces the Number of Recombinant Progeny

Individuals heterozygous for an inversion (*In*/+) are *inversion heterozygotes.* In such individuals, when the chromosome carrying the inversion pairs with its homolog at meiosis, formation of an **inversion loop** allows the tightest possible alignment of homologous regions. In an inversion loop, one chromosomal region rotates to conform to the similar region in the other homolog (**Fig. 14.15**). Crossing-over within an inversion loop produces aberrant recombinant chromatids whether the inversion is pericentric or paracentric.

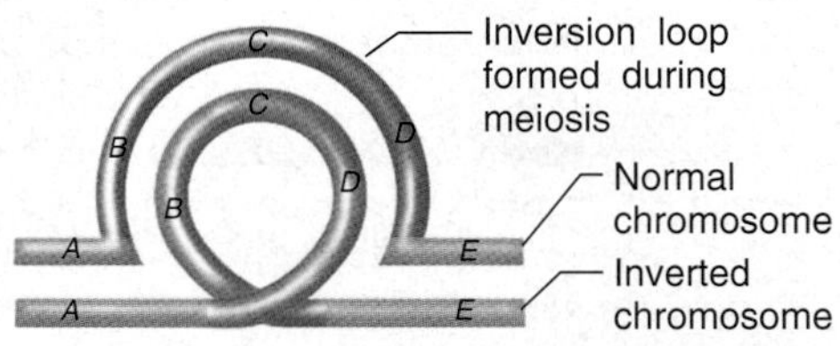

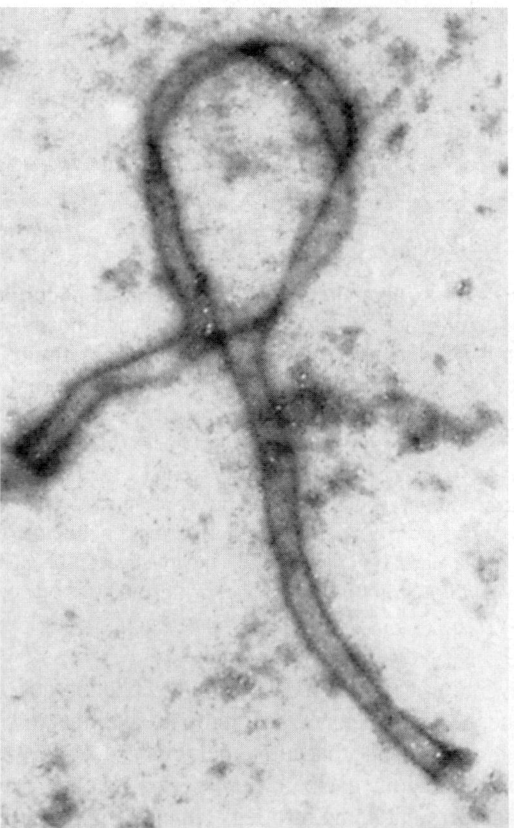

Figure 14.15 Inversion loops form in inversion heterozygotes. To maximize pairing during prophase of meiosis I in an inversion heterozygote (*In*/+), homologous regions form an inversion loop. (*Top*) Simplified diagram in which one line represents a pair of sister chromatids. (*Bottom*) Electron micrograph of an inversion loop during meiosis I in an *In*/+ mouse.

If the inversion is pericentric and a single crossover occurs within the inversion loop, each recombinant chromatid will have a single centromere—the normal number—but will carry a duplication of one region and a deletion of a different region (**Fig. 14.16a**). Gametes carrying these recombinant chromatids will have an abnormal dosage of some genes. After fertilization, zygotes created by the union of these abnormal gametes with normal gametes are likely to die because of genetic imbalance.

If the inversion is paracentric and a single crossover occurs within the inversion loop, the recombinant chromatids will be unbalanced not only in gene dosage but also in centromere number (Fig. 14.16b). One crossover product will be an **acentric fragment** lacking a centromere; while the reciprocal crossover product will be a **dicentric chromatid** with two centromeres. Because the acentric fragment without a centromere cannot attach to the spindle apparatus during the first meiotic division, the cell cannot package it into either of the daughter nuclei; as a result, this chromosome is lost and will not be included in a gamete. By contrast, at anaphase of meiosis I, opposing spindle forces pull the dicentric chromatid toward both spindle poles at the same time with such strength that the dicentric chromatid breaks at random positions along the chromosome. These broken chromosome fragments are deleted for many of their genes. This loss of the acentric fragment and breakage of the dicentric chromatid results in genetically unbalanced gametes, which at fertilization will produce lethally unbalanced zygotes that cannot develop beyond the earliest stages of embryonic development. Consequently, no recombinant progeny resulting from a crossover in a paracentric inversion loop survive. Any surviving progeny are nonrecombinants.

In summary, whether an inversion is pericentric or paracentric, crossing-over within the inversion loop of an inversion heterozygote has the same effect: formation of recombinant gametes that after fertilization prevent the zygote from developing. Since only gametes containing chromosomes that did not recombine within the inversion loop can yield viable progeny, inversions act as **crossover suppressors.** This does not mean that crossovers do not occur within inversion loops, but simply that there are no recombinants among the viable progeny of an inversion heterozygote.

Geneticists use crossover suppression to create **balancer chromosomes,** which contain multiple, overlapping inversions (both pericentric and paracentric), as well as a marker mutation that produces a visible dominant phenotype (**Fig. 14.17**). The viable progeny of a *Balancer*/+ heterozygote will receive either the balancer or the chromosome of normal order (+), but they cannot inherit a recombinant chromosome containing parts of both. Researchers can distinguish these two types of viable progeny by the presence or absence of the dominant marker phenotype. Geneticists often generate balancer heterozygotes to

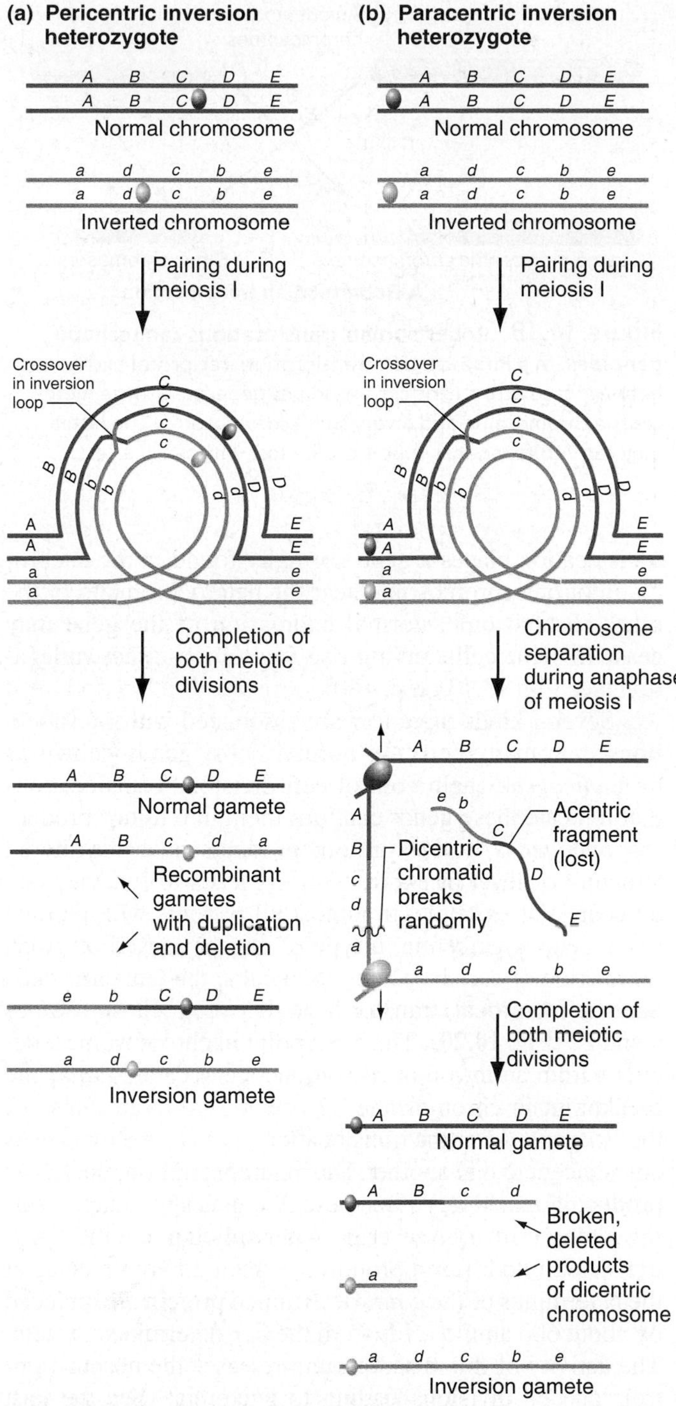

Figure 14.16 Why inversion heterozygotes produce few if any recombinant progeny. Throughout this figure, each line represents one chromatid, and different shades of *green* indicate the two homologous chromosomes. **(a)** The chromatids formed by recombination within the inversion loop of a pericentric inversion heterozygote are genetically unbalanced. Note that the duplicated or deleted genes in these recombinant chromatids lie outside the inversion loop. **(b)** The chromatids formed by recombination within the inversion loop of a paracentric inversion heterozygote are not only genetically unbalanced but also contain two or no centromeres, instead of the normal one.

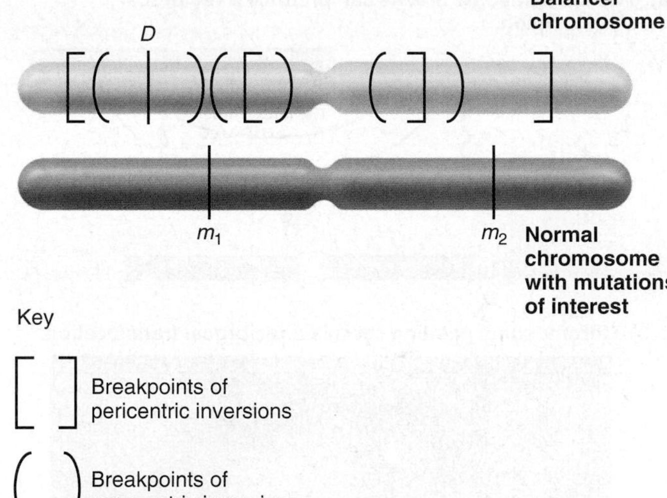

Figure 14.17 Balancer chromosomes are useful tools for genetic analysis. Balancer chromosomes carry both a dominant marker *D* as well as inversions (*brackets*) that prevent the balancer chromosome from recombining with an experimental chromosome carrying mutations of interest (m_1 and m_2). A parent heterozygous for the balancer and experimental chromosomes will transmit either the balancer or the experimental chromosome, but not a recombinant chromosome, to its surviving progeny.

ensure that a chromosome of normal order, along with any mutations of interest it may carry, is transmitted to the next generation unchanged by recombination. To help create genetic stocks, the marker in most balancer chromosomes not only causes a dominant visible phenotype, but it also acts as a recessive lethal mutation that prevents the survival of balancer chromosome homozygotes. The *Drosophila* portrait (on our website at www.mhhe.com/hartwell3) discusses this and other significant uses of balancer chromosomes in genetic analysis.

Translocations Attach Part of One Chromosome to Another Chromosome

Translocations are large-scale mutations in which part of one chromosome becomes attached to a nonhomologous chromosome or in which parts of two different chromosomes trade places. This second type of translocation is known as a **reciprocal translocation (Fig. 14.18a)**. It results when two breaks, one in each of two chromosomes, yield DNA fragments that do not religate to their chromosome of origin; rather, they switch places and become attached to the other chromosome. Depending on the positions of the breaks and the sizes of the exchanged fragments, the translocated chromosomes may be so different from the original chromosomes that the translocation is visible in a cytological examination (Fig. 14.18b).

(a) Two chromosome breaks can produce a reciprocal translocation.

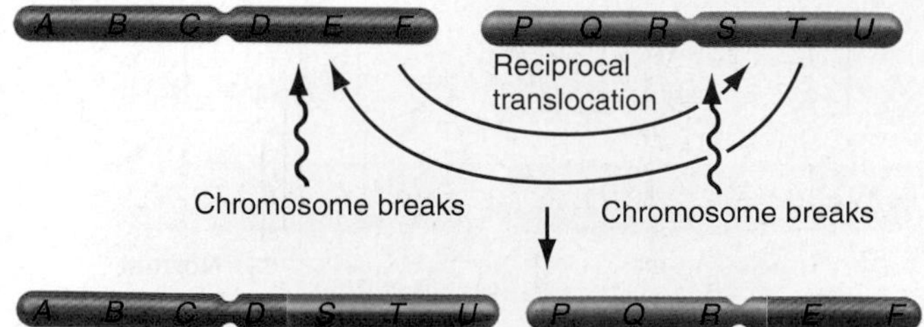

(b) Chromosome painting reveals a reciprocal translocation.

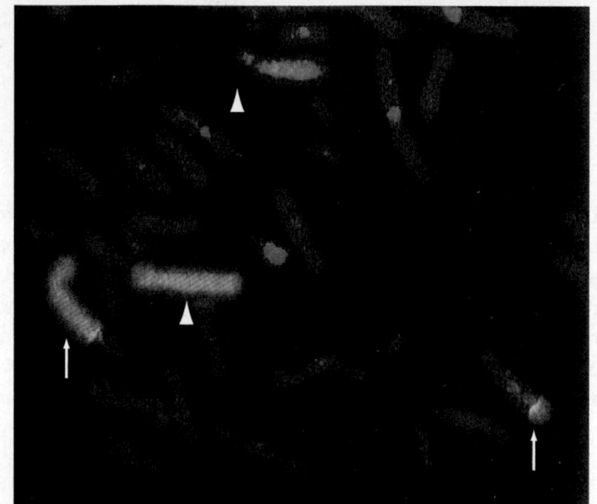

Figure 14.18 Reciprocal translocations are exchanges between nonhomologous chromosomes. (a) In a reciprocal translocation, the region gained by one chromosome is the region lost by the other chromosome. **(b)** Karyotype of a human genome containing a translocation. Researchers used a "chromosome-painting" hybridization technique that renders chromosomes in different colors. The two translocated chromosomes are stained both red *and* green (*arrows*). Two normal, nontranslocated chromosomes are stained entirely red *or* entirely green (*arrowheads*), indicating that this person is heterozygous for the translocation.

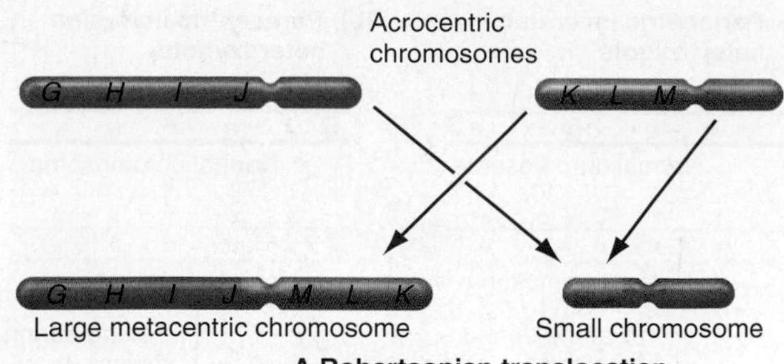

Figure 14.19 Robertsonian translocations can reshape genomes. In a Robertsonian translocation, reciprocal exchanges between two acrocentric chromosomes generate a large metacentric chromosome and a very small chromosome. The latter may carry so few genes that it can be lost without ill effect.

Robertsonian translocations are an important type of cytologically visible reciprocal translocations that arise from breaks at or near the centromeres of two acrocentric chromosomes (**Fig. 14.19**). The reciprocal exchange of broken parts generates one large metacentric chromosome and one very small chromosome containing few, if any, genes. This tiny chromosome may subsequently be lost from the organism. Robertsonian translocations are named after W. R. B. Robertson, who in 1911 was the first to suggest that during evolution, metacentric chromosomes may arise from the fusion of two acrocentrics.

Most individuals bearing reciprocal translocations are phenotypically normal because they have neither lost nor gained genetic material. As with inversions, however, if one of the translocation breakpoints occurs within a gene, that gene's function may change or be destroyed. Or if the translocation places a gene normally found in the euchromatin of one chromosome near the heterochromatin of the other chromosome, normal expression of the gene may cease in some cells, giving rise to position-effect variegation (see Fig. 13.13 on p. 480).

Several kinds of cancer are associated with translocations in somatic cells. In normal cells, genes known as *protooncogenes* help control cell division. Translocations that relocate these genes can turn them into tumor-producing *oncogenes* whose protein products have an altered structure or level of expression that leads to runaway cell division. For example, in almost all patients with chronic myelogenous leukemia, a type of cancer caused by overproduction of certain white blood cells, the leukemic cells have a reciprocal translocation between chromosomes 9 and 22 (**Fig. 14.20**). The breakpoint in chromosome 9 occurs within an intron of a protooncogene called *c-abl;* the breakpoint in chromosome 22 occurs within an intron of the *bcr* gene. After the translocation, parts of the two genes are adjacent to one another. During transcription, the RNA-producing machinery runs these two genes together, creating a long primary transcript. After splicing, the mRNA is translated into a fused protein in which 25 amino acids at the N terminus of the *c-abl*-determined protein are replaced by about 600 amino acids from the *bcr*-determined protein. The activity of this fused protein releases the normal controls on cell division, leading to leukemia. (See the Fast Forward box on pp. 492–494 of this chapter for another example of a translocation-induced cancer called Burkitt's lymphoma.)

Medical practitioners can exploit the rearrangement of DNA sequences that accompany cancer-related translocations for diagnostic and therapeutic purposes. To confirm a diagnosis of myelogenous leukemia, for example, they first obtain a blood sample from the patient, and they then use a pair of PCR primers derived from opposite sides of the breakpoint—one synthesized from the appropriate part of chromosome 22, the other from chromosome 9—to carry out a PCR on DNA from the blood cells. The PCR will

(a) Leukemia patients have too many white blood cells.

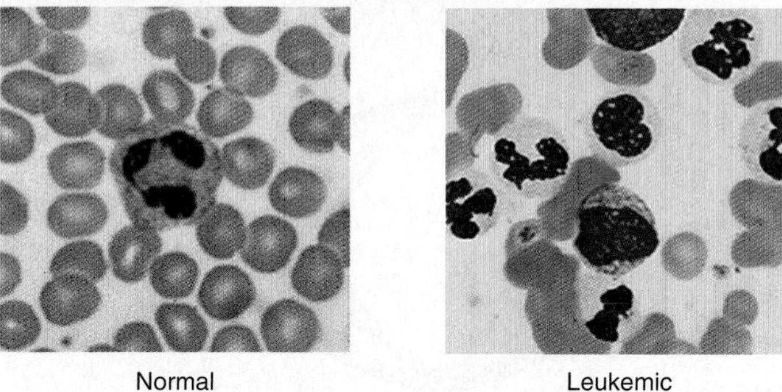

(b) The genetic basis for chronic myelogenous leukemia

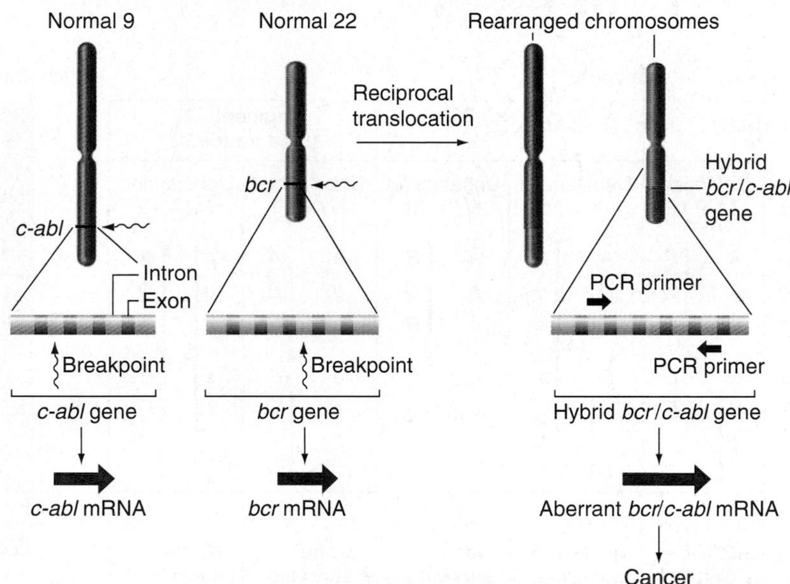

Figure 14.20 How a reciprocal translocation helps cause one kind of leukemia. (a) Uncontrolled divisions of large, dark-staining white blood cells in a leukemia patient (*right*) produce a higher ratio of white to red blood cells than that in a normal individual (*left*). **(b)** A reciprocal translocation between chromosomes 9 and 22 contributes to chronic myelogenous leukemia. This rearrangement makes an abnormal hybrid gene composed of part of the *c-abl* gene and part of the *bcr* gene. The hybrid gene encodes an abnormal fused protein that disrupts controls on cell division. *Black arrows* indicate PCR primers that will generate a PCR product only from DNA containing the hybrid gene.

amplify the region between the primers only if the DNA sample contains the translocation (Fig. 14.20b). To monitor the effects of chemotherapy, they again obtain a blood sample and extract genomic DNA from the white blood cells. If the sample contains even a few malignant cells, a PCR test with the same two primers will amplify the DNA translocation from those cells, indicating the need for more therapy. PCR thus becomes a sensitive assay for this type of leukemic cell.

Pharmaceutical researchers have recently exploited their understanding of the molecular nature of the translocation underlying chronic myelogenous leukemia to achieve a stunning breakthrough in the treatment of this cancer. The protein encoded by *c-abl* is a *protein tyrosine kinase,* an enzyme that adds phosphate groups to tyrosine amino acids on other proteins. This enzyme is an essential part of the set of signals that dictate cell growth and division. Normal cells closely regulate the activity of the c-abl protein, blocking its function most of the time but activating it in response to stimulation by growth factors in the environment. By contrast, the fused protein encoded by *bcr/c-abl* in cells carrying the translocation is not amenable to regulation. It is always active, even in the absence of growth factor, and this leads to runaway cell division. Pharmaceutical companies have developed a drug called Gleevec® that specifically inhibits the enzymatic activity of the protein tyrosine kinase encoded by *bcr/c-abl.* In clinical trials, 98% of participants experienced a complete disappearance of leukemic blood cells and the return of normal white cells. This drug is now the standard treatment for chronic myelogenous leukemia and is a model for new types of cancer treatments that home in on cancer cells without hurting healthy ones.

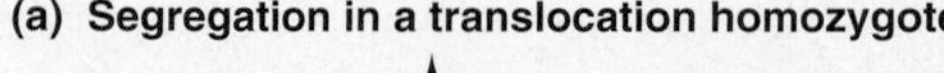

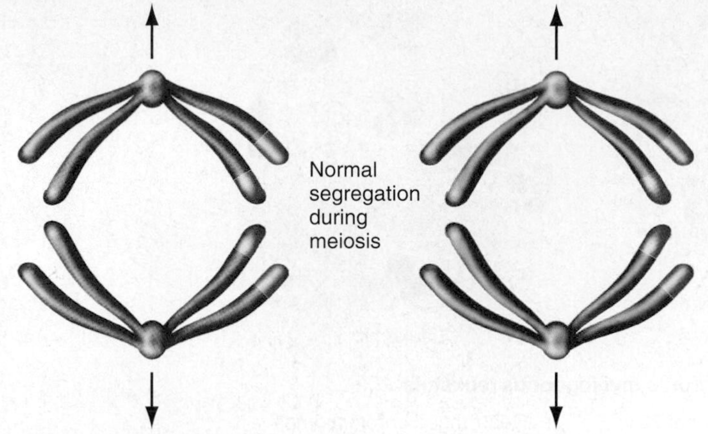

(c) Segregation in a translocation heterozygote

Segregation pattern	Alternate		Adjacent - 1		Adjacent - 2 (less frequent)	
	Balanced *N1* + *N2*	Balanced *T1* + *T2*	Unbalanced *T1* + *N2*	Unbalanced *N1* + *T2*	Unbalanced *N1* + *T1*	Unbalanced *N2* + *T2*
Gametes	abc•def; pqr•stu	ABC•DSTU; PQR•EF	ABC•DSTU; pqr•stu	abc•def; PQR•EF	abc•def; ABC•DSTU	pqr•stu; PQR•EF
Type of progeny when mated with normal *abcdefpqrstu* homozygote	*abcdef pqrstu*	*ABCDEF PQRSTU*	None surviving	None surviving	None surviving	None surviving

(d) Semisterility in corn

Figure 14.21 The meiotic segregation of chromosomes that have sustained reciprocal translocations. In all parts of this figure, each bar or line represents one chromatid. **(a)** In a translocation homozygote (*T*/*T*), chromosomes segregate normally during meiosis I. **(b)** In a translocation heterozygote (*T*/+), the four relevant chromosomes assume a cruciform (crosslike) configuration to maximize pairing. The alleles of genes on chromosomes in the original order (*N1* and *N2*) are shown in lowercase; the alleles of these genes on the translocated chromosomes (*T1* and *T2*) are in uppercase letters. **(c)** There are three possible segregation patterns in a translocation heterozygote from the cruciform configuration shown in part (b). Only the alternate segregation pattern gives rise to balanced gametes. As a result, genes that were originally on two different chromosomes will behave as if they are genetically linked. **(d)** This semisterile ear of corn comes from a plant heterozygous for a reciprocal translocation. It has fewer kernels than normal because unbalanced ovules are aborted.

Heterozygosity for Translocations Diminishes Fertility and Results in Pseudolinkage

Translocations, like inversions, produce no significant genetic consequences in homozygotes if the breakpoints do not interfere with gene function. During meiosis in a translocation homozygote, chromosomes segregate normally according to Mendelian principles (**Fig. 14.21a**). Even though the genes have been rearranged, both haploid sets of chromosomes in the individual have the same rearrangement. As a result, all chromosomes will find a single partner with which to pair at meiosis, and there will be no deleterious consequences for the progeny.

In translocation heterozygotes, however, certain patterns of chromosome segregation during meiosis produce genetically unbalanced gametes that at fertilization become deleterious to the zygote. In a translocation heterozygote, the two haploid sets of chromosomes do not carry the same arrangement of genetic information. As a result, during prophase of the first meiotic division, the translocated chromosomes and their normal homologs assume a crosslike configuration in which four chromosomes, rather than the normal two, pair to achieve a maximum of synapsis between similar regions (Fig. 14.21b). To keep track of the four chromosomes participating in this crosslike structure, we denote the chromosomes carrying translocated material with a *T* and the chromosomes with a normal order of genes with an *N*. Chromosomes *N1* and *T1* have homologous centromeres found in wild type on chromosome 1; *N2* and *T2* have centromeres found in wild type on chromosome 2.

During anaphase of meiosis I, the mechanisms that attach the spindle to the chromosomes in this crosslike configuration still usually ensure the disjunction of homologous centromeres, bringing homologous chromosomes to opposite spindle poles (that is, *T1* and *N1* go to opposite poles, as do *T2* and *N2*). Depending on the arrangement of the four chromosomes on the metaphase plate, this normal disjunction of homologs produces one of two equally likely patterns of segregation (Fig. 14.21c). In the **alternate segregation pattern,** the two translocation chromosomes (*T1* and *T2*) go to one pole, while the two normal chromosomes (*N1* and *N2*) move to the opposite pole. Both kinds of gametes resulting from this segregation (*T1, T2* and *N1, N2*) carry the correct haploid number of genes; and the zygotes formed by union of these gametes with a normal gamete will be viable. By contrast, in the **adjacent-1 segregation pattern,** homologous centromeres disjoin so that *T1* and *N2* go to one pole, while the *N1* and *T2* go to the opposite pole. As a result, each gamete contains a large duplication (of the region found in both the normal and the translocated chromosome in that gamete) and a correspondingly large deletion (of the region found in neither of the chromosomes in that gamete), which make them genetically unbalanced. Zygotes formed by union of these gametes with a normal gamete are usually not viable.

Because of the unusual cruciform pairing configuration in translocation heterozygotes, nondisjunction of homologous centromeres occurs at a measurable but low rate. This nondisjunction produces an **adjacent-2 segregation pattern** in which the homologous centromeres *N1* and *T1* go to the same spindle pole, while the homologous centromeres *N2* and *T2* go to the other spindle pole (Fig. 14.21c). The resulting genetic imbalances are lethal after fertilization to the zygotes containing them.

Thus, of all the gametes generated by translocation heterozygotes, only those arising from alternate segregation, which account for slightly less than half the total, can produce viable progeny when crossed with individuals who do not carry the translocation. As a result, the fertility of most translocation heterozygotes, that is, their capacity for generating viable offspring, is diminished by at least 50%. This condition is known as **semisterility.** Corn plants illustrate the correlation between translocation heterozygosity and semisterility. The demise of genetically unbalanced ovules produces gaps in the ear where kernels would normally appear (Fig. 14.21d); in addition, genetically unbalanced pollen grains are abnormally small (not shown).

The semisterility of translocation heterozygotes undermines the potential of genes on the two translocated chromosomes to assort independently. Mendel's second law requires that all gametes resulting from both possible metaphase alignments of two chromosomal pairs produce viable progeny. But as we have seen, in a translocation heterozygote, only the alternate segregation pattern yields viable progeny in outcrosses; the equally likely adjacent-1 pattern and the rare adjacent-2 pattern do not. Because of this, genes near the translocation breakpoints on the nonhomologous chromosomes participating in a reciprocal translocation exhibit **pseudolinkage:** They behave as if they are linked.

Figure 14.21c illustrates why pseudolinkage occurs in a translocation heterozygote. In the figure, lowercase *a b c d e f* represent the alleles of genes present on normal chromosome 1 (*N1*), and *p q r s t u* are the alleles of genes on a nonhomologous normal chromosome 2 (*N2*). The alleles of these genes on the translocated chromosomes *T1* and *T2* are in uppercase. In the absence of recombination, Mendel's law of independent assortment would predict that genes on two different chromosomes will appear in four types of gametes in equal frequencies; for example, *a p, A P, a P,* and *A p*. But alternate segregation, the only pattern that can give rise to viable progeny, produces only *a p* and *A P* gametes. Thus, in translocation heterozygotes such as these, the genes on the two nonhomologous chromosomes act as if they are linked to each other.

Translocations Can Sometimes Help Map Important Genes

In humans, approximately 1 of every 500 individuals is heterozygous for some kind of translocation. While most such people are phenotypically normal, their fertility is diminished because many of the zygotes they produce abort spontaneously. As we have seen, this semisterility results from genetic imbalances associated with gametes formed by adjacent-1 or adjacent-2 segregation patterns. But such genetic imbalances are not inevitably lethal to the zygotes. If the duplicated or deleted regions are very small, the imbalanced gametes generated by these modes of segregation may produce children.

An important example of this phenomenon is seen among individuals heterozygous for certain reciprocal translocations involving chromosome 21, such as the Robertsonian translocation shown in **Fig. 14.22.** These people are phenotypically normal but produce some gametes from the adjacent-1 segregation pattern that have two copies of a part of chromosome 21 near the tip of its long arm. At fertilization, if a gamete with the duplication unites with a normal gamete, the resulting child will have three copies of this region of chromosome 21. A few individuals affected by Down syndrome have, in this way, inherited a third copy of only a small part of chromosome 21. These individuals with **translocation Down syndrome**

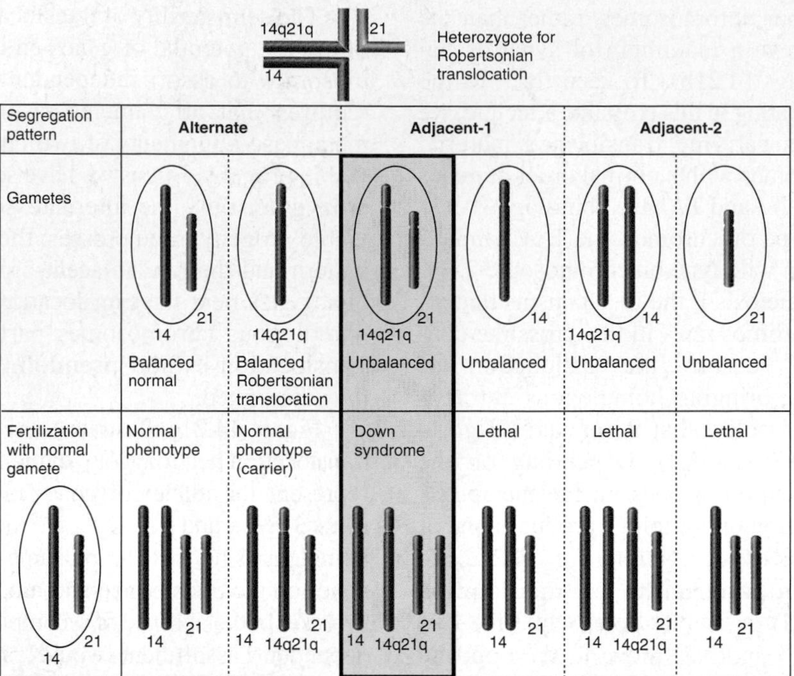

Figure 14.22 **How translocation Down syndrome arises.** In heterozygotes for a translocation involving chromosome 21, such as 14q21q (a Robertsonian translocation between chromosomes 21 and 14), adjacent-1 segregation can produce gametes with two copies of part of chromosome 21. If such a gamete unites with a normal gamete, the resulting zygote will have three copies of part of chromosome 21. Depending on which region of chromosome 21 is present in three copies, this tripling may cause Down syndrome. (In the original translocation heterozygote, the small, reciprocally translocated chromosome [14p21p] has been lost.)

provide evidence that the entirety of chromosome 21 need not be present in three copies to generate the phenotype.

Geneticists are now mapping the chromosome 21 regions duplicated in translocation Down syndrome patients to find the one or more genes responsible for the syndrome. Although chromosome 21 is the smallest human autosome, it nevertheless contains an estimated 350 genes, most of them in the 43 million base pairs of its long arm. The mapping of genes relative to the breakpoints of one or more such translocations considerably simplifies the task of identifying those genes that in triplicate produce the symptoms of Down syndrome. One way to locate which parts of chromosome 21 are responsible for Down syndrome is to obtain cloned chromosome 21 sequences from the Human Genome Project and then use these clones as FISH (fluorescence *in situ* hybridization, described (in Fig. 10.8 on p. 372) probes for the genome of the translocation Down syndrome patient. If the probe lights up the translocation chromosome as well as the two normal copies of chromosome 21, it identifies a region of the genome that is of potential importance to the syndrome.

Translocations and Inversions Have Several Effects in Common: A Summary

Reciprocal translocations, like inversions, bring together pieces of DNA that were not adjacent before the rearrangement, without altering the amount of DNA in the genome. Both reciprocal translocations and inversions may affect gene function and as a result, phenotype. This occurs either if one of the breakpoints of the rearrangement is within a gene or if the rearrangement places a gene in a chromosomal environment that modifies its expression. Homozygosity or heterozygosity for translocations and inversions that do not affect gene function is usually without obvious phenotypic effect. However, heterozygosity for both types of rearrangements can produce genetically imbalanced gametes that are deleterious to the zygotes they help create. Inversion heterozygotes produce genetically imbalanced gametes from crossing-over within an inversion loop; this results in crossover suppression. Translocation heterozygotes produce genetically imbalanced gametes from two of three possible meiotic segregation patterns; this leads to semisterility and pseudolinkage. Finally, translocations and inversions, by reducing production of viable progeny from heterozygotes, can be catalysts of speciation, as we explain later in this chapter's comprehensive example.

Transposable Genetic Elements Move from Place to Place in the Genome

Large deletions and duplications, as well as inversions and translocations, are major chromosomal reorganizations visible at the relatively low resolution of a karyotype. Small deletions and duplications are lesser chromosomal

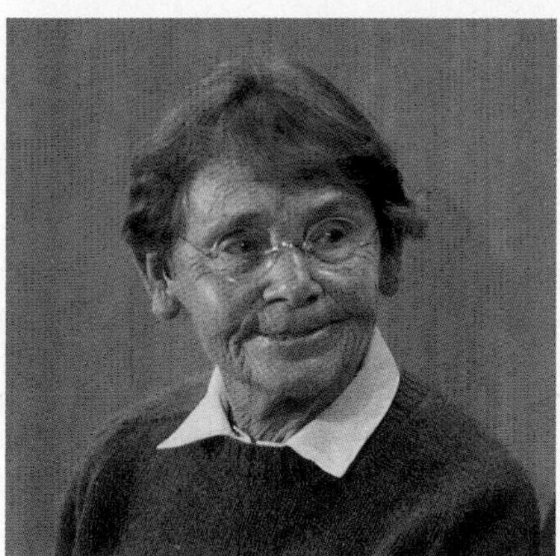

Figure 14.23 Barbara McClintock: Discoverer of transposable elements.

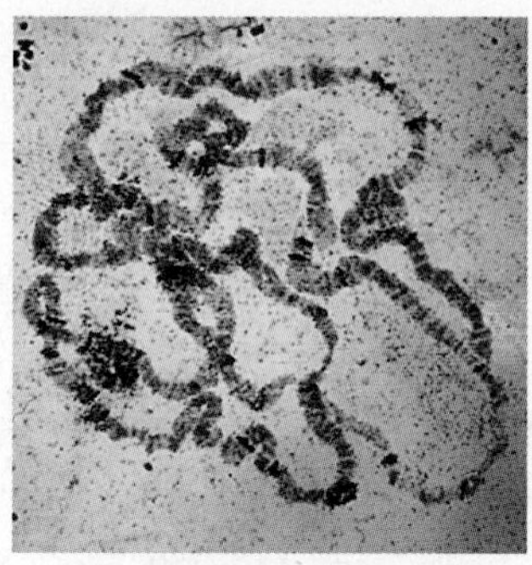

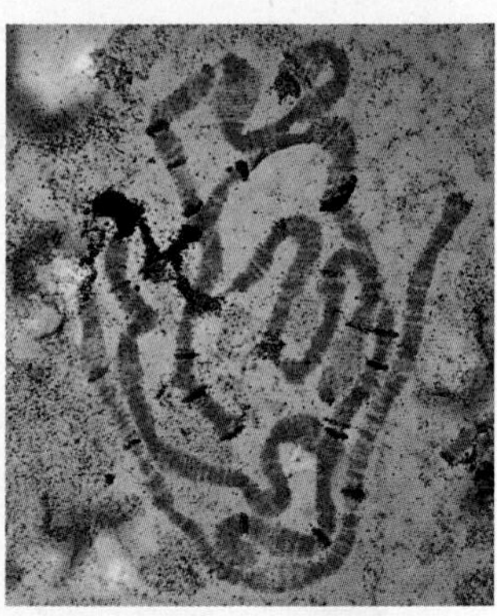

Figure 14.24 Transposable elements (TEs) can move to many locations in a genome. A probe for the *copia* TE hybridizes to multiple sites (*black bands* superimposed over the *blue* chromosomes) that differ in two different fly strains.

reorganizations that reshape genomes without any visible effect on karyotype. Another type of cytologically invisible sequence rearrangement with a significant genomic impact is **transposition:** the movement of small segments of DNA—entities known as **transposable elements (TEs)**—from one position in the genome to another.

Marcus Rhoades in the 1930s and Barbara McClintock in the 1950s inferred the existence of TEs from intricate genetic studies of corn. At first, the scientific community did not appreciate the importance of their work because their findings did not support the conclusion from classical recombination mapping that genes are located at fixed positions on chromosomes. Once the cloning of TEs made it possible to study them in detail, geneticists not only acknowledged their existence, but also discovered TEs in the genomes of virtually all organisms, from bacteria to humans. In 1983, Barbara McClintock received the Nobel Prize for her insightful studies on movable genetic elements (**Fig. 14.23**).

Copia is a transposable element in *Drosophila.* If you examined the polytene chromosomes from two strains of flies isolated from different geographic locations, you would find in general that the chromosomes appear identical. A probe derived from the *white* gene for eye color, for example, would hybridize to a single site near the tip of the X chromosome in both strains (review Fig. 14.9a). However, a probe including the *copia* TE would hybridize to 30–50 sites scattered throughout the genome, and the positions of *in situ* hybridization would not be the same in the two strains. Some sites would be identical in the two polytene sets, but others would be different (**Fig. 14.24**). These observations suggest that since the time the strains were separated geographically, the *copia* sequences have moved around (transposed) in different ways in the two genomes even though the genes have remained in fixed positions.

Any segment of DNA that evolves the ability to move from place to place within a genome is by definition a transposable element, regardless of its origin or function. TEs need not be sequences that do something for the organism; indeed, many scientists regard them primarily as "selfish" parasitic entities carrying only information that allows their self-perpetuation. Some TEs, however, appear to have evolved functions that help their host. In one interesting example, TEs maintain the length of *Drosophila* chromosomes. *Drosophila* telomeres, in contrast to those of most organisms, do not contain TTAGGG repeats that are extendable by the telomerase enzyme (see Fig. 13.10 on p. 477). Certain TEs, however, combat the shortening of chromosome ends that accompanies every cycle of replication by jumping with high frequency into DNA very near chromosome ends. As a result, chromosome size stays relatively constant.

Most transposable elements in nature range from 50 bp to approximately 10,000 bp (10 kb) in length. A particular TE can be present in a genome anywhere from one to hundreds of thousands of times. *Drosophila melanogaster,* for example, harbors approximately 80 different TEs, each an average of 5 kb in length, and each present an average of 50 times. These TEs constitute $80 \times 50 \times 5 = 20{,}000$ kb, or roughly 12.5% of the 160,000 kb *Drosophila* genome. Mammals carry two major classes of TEs: **LINEs,** or long interspersed elements; and **SINEs,** or short interspersed elements. The human genome contains approximately 20,000 copies of the main human LINE—*L1*—which is up to 6.4 kb in length. The human genome also carries 300,000 copies of the main human SINE—*Alu*—which is 0.28 kb in length (**Fig. 14.25a**). These two TEs alone thus constitute roughly 7% of the 3,000,000 kb human genome. Because some TEs exist in only one or a few closely related species, it is probable that some elements arise and then disappear rather frequently over evolutionary time. Chapter 22 describes the evolutionary origins of LINEs and SINEs.

Classification of TEs on the basis of how they move around the genome distinguishes two groups. **Retroposons** transpose via reverse transcription of an RNA intermediate.

(a) *Alu* SINEs in the human genome

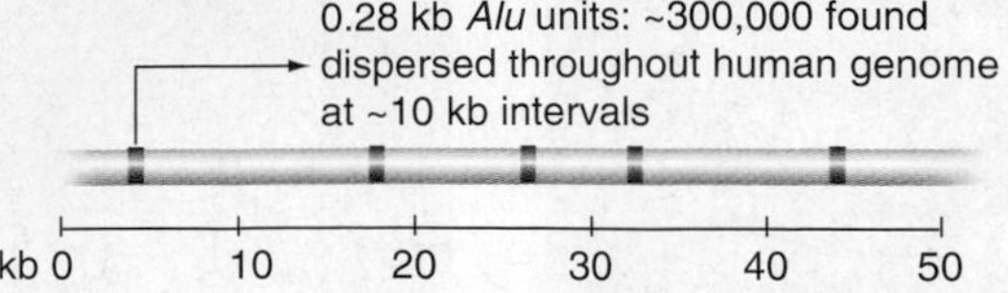

(b) TEs cause mottling in corn.

Figure 14.25 **TEs in human and corn genomes. (a)** The human genome carries about 300,000 copies of the 0.28 kb *Alu* retroposon, the major human SINE; these are spaced, on average, about 10 kb apart. The longer LINE elements are, on average, spaced further apart (*not shown*). **(b)** Movements of a transposon mottles corn kernels when the transposon jumps into or out of genes that influence pigmentation.

The *Drosophila copia* elements and the human SINEs and LINEs just described are retroposons. **Transposons** move their DNA directly without the requirement of an RNA intermediate. The genetic elements discovered by Barbara McClintock in corn responsible for mottling the kernels are transposons (Fig. 14.25b). Some biologists use the term "transposon" in the broader sense to refer to all TEs. In this book, we reserve it for the direct-movement class of genetic elements, and we use "transposable elements (TEs)" to indicate all DNA segments that move about in the genome, regardless of the mechanism.

Retroposons: Transcription Generates an RNA That Encodes a Reverse-Transcriptase-like Enzyme

This enzyme with reverse-transcriptase activity, like the reverse transcriptase made by the AIDS-causing HIV virus described in the Genetics and Society box on pp. 270–271 of Chapter 8, can copy RNA into a single strand of cDNA and then use that single DNA strand as a template for producing double-stranded cDNA. Many retroposons also encode polypeptides other than reverse transcriptase.

Some retroposons have a poly-A tail at the 3′ end of the RNA-like DNA strand, a configuration reminiscent of mRNA molecules (**Fig. 14.26a**). Other retroposons end in *long terminal repeats* (*LTR*s): nucleotide sequences repeated in the same orientation at both ends of the element (Fig. 14.26a). The structure of this second type of retroposon is similar to the integrated DNA copies of RNA tumor viruses (known as retroviruses), suggesting that retroviruses evolved from this kind of retroposon, or vice versa. In support of this notion, researchers sometimes find retroposon transcripts enclosed in viruslike particles.

The structural parallels between retroposons, mRNAs, and retroviruses, as well as the fact that retroposons encode a reverse-transcriptase-like enzyme, prompted investigators to ask whether retroposons move around the genome via an RNA intermediate. Experiments in yeast helped confirm that they do. In one study, a copy of the *Ty1* retroposon found on a yeast plasmid contained an intron in one of its genes; after transposition into the yeast chromosome, however, the intron was not there (Fig. 14.26b). Since removal of introns occurs only during mRNA processing, researchers concluded that the *Ty1* retroposon passes through an RNA intermediate during transposition.

The mechanisms by which various retroposons move around the genome resemble each other in general outline but differ in detail. Figure 14.26c outlines what is known of the process for the better understood LTR-containing retroposons. As the figure illustrates, one outcome of transposition via an RNA intermediate is that the original copy of the retroposon remains in place while the new copy inserts in another location. With this mode of transmission, the number of copies can increase rapidly with time. Human LINEs and SINEs, for example, occur in tens of thousands or even hundreds of thousands of copies within the genome. Other retroposons, however, such as the *copia* elements found in *Drosophila,* do not proliferate so profusely and exist in much more moderate copy numbers of 30–50. Currently unknown mechanisms may account for these differences by regulating the rate of retroposon transcription or by limiting the number of copies through selection at the level of the whole organism.

Transposons Encode Transposase Enzymes That Catalyze the Events of Transposition

A hallmark of transposons—TEs whose movement does not involve an RNA intermediate—is that their ends are inverted repeats of each other, that is, a sequence of base pairs at one end is present in mirror image at the other end (**Fig. 14.27a** on p. 512). The inverted repeat is usually 10–200 bp long.

DNA between the transposon's inverted repeats commonly contains a gene encoding a transposase, a protein that catalyzes transposition through its recognition of those repeats. As Fig. 14.27a illustrates, the steps resulting in transposition include excision of the transposon from its original genomic position and integration into a new location. The double-stranded breaks at the transposon's excision and integration sites are repaired in different ways in different cases. Figure 14.27b shows two of the possibilities. In *Drosophila,* after excision of a transposon known as a *P element,* DNA exonucleases first widen the resulting

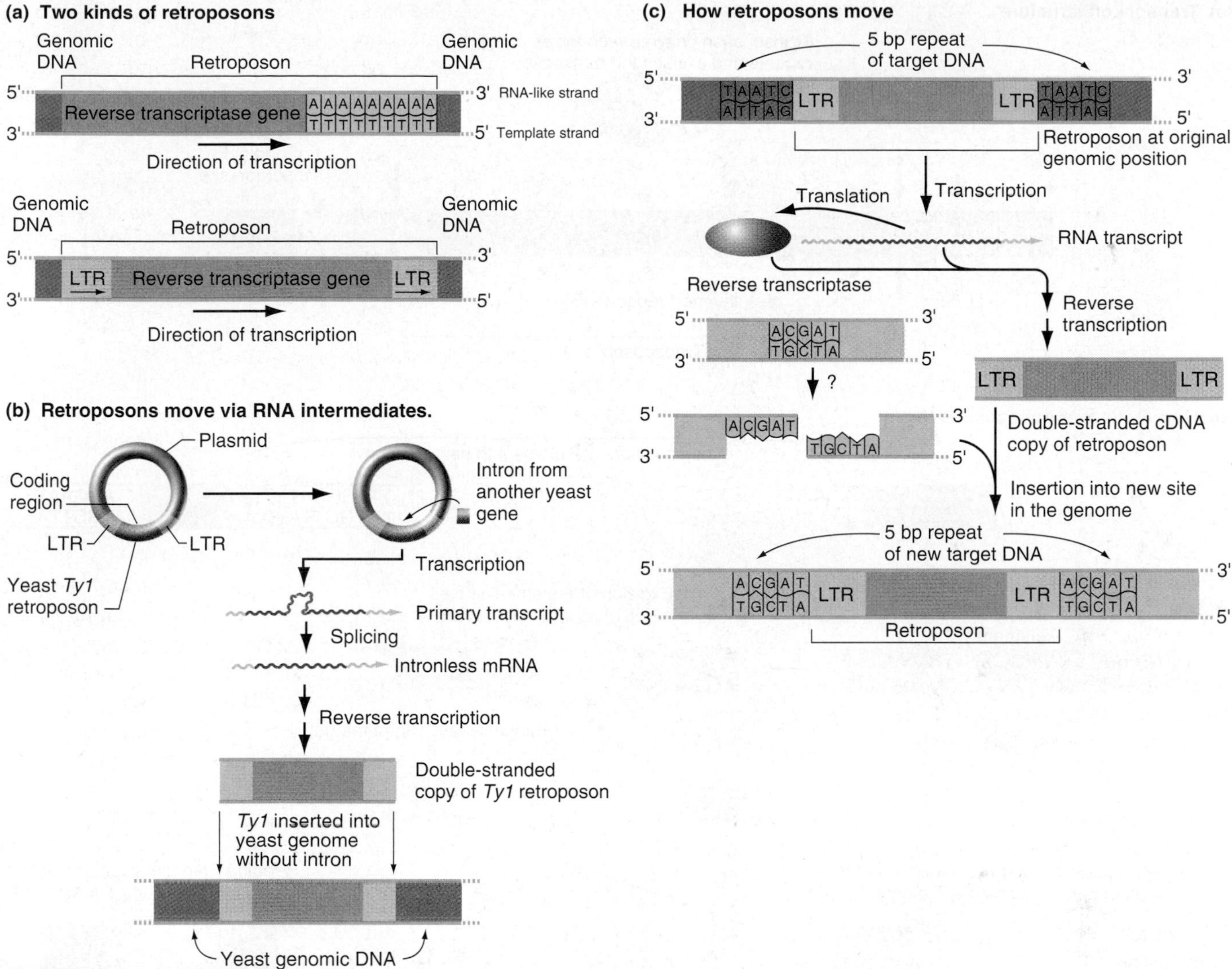

Figure 14.26 Retroposons: Structure and movement. (a) Some retroposons have a poly-A tail at the 3′ end of the RNA-like DNA strand (*top*); others are flanked on both sides by long terminal repeats (LTRs; *bottom*). **(b)** Researchers constructed a plasmid bearing a *Ty1* retroposon that contained an intron. When this plasmid was transformed into yeast cells, researchers could isolate new insertions of *Ty1* into yeast genomic DNA. The newly inserted *Ty1* did not have the intron, which implies that transposition involves splicing of a primary transcript to form an intronless mRNA. **(c)** The reverse-transcriptase-like enzyme synthesizes double-stranded retroposon cDNA in a series of steps. Insertion of this double-stranded cDNA into a new genomic location (*blue*) involves a staggered cleavage of the target site that leaves "sticky ends"; polymerization to fill in the sticky ends produces two copies of the 5 bp target site.

gap and then repair it using either a sister chromatid or a homologous chromosome as a template. If the template contains the *P* element and DNA replication is completely accurate, repair will restore a *P* element to the position from which it was excised; this will make it appear as if the *P* element remained at its original location during transposition (Fig. 14.27b, *left*). If the template does not contain a *P* element, the transposon will be lost from the original site after transposition (Fig. 14.27b, *right*).

Some strains of *D. melanogaster* are called "P strains" because they harbor many copies of the *P* element; "M strains" of the same species do not carry the *P* element at all. Virtually all commonly used laboratory flies are M strains, while many flies isolated from natural populations since 1950 are P strains. Because Thomas Hunt Morgan and coworkers in the early part of the twentieth century isolated the flies that have proliferated into most current laboratory strains, these observations suggest that *P* elements did not enter *D. melanogaster* genomes until around 1950. The prevalence of *P* elements in many contemporary natural populations attests to the rapidity with which transposable elements can spread once they enter a species' genome.

Interestingly, the mating of male flies from P strains with females from M strains causes a phenomenon called **hybrid dysgenesis,** which creates a series of defects including sterility of offspring, mutation, and chromosome

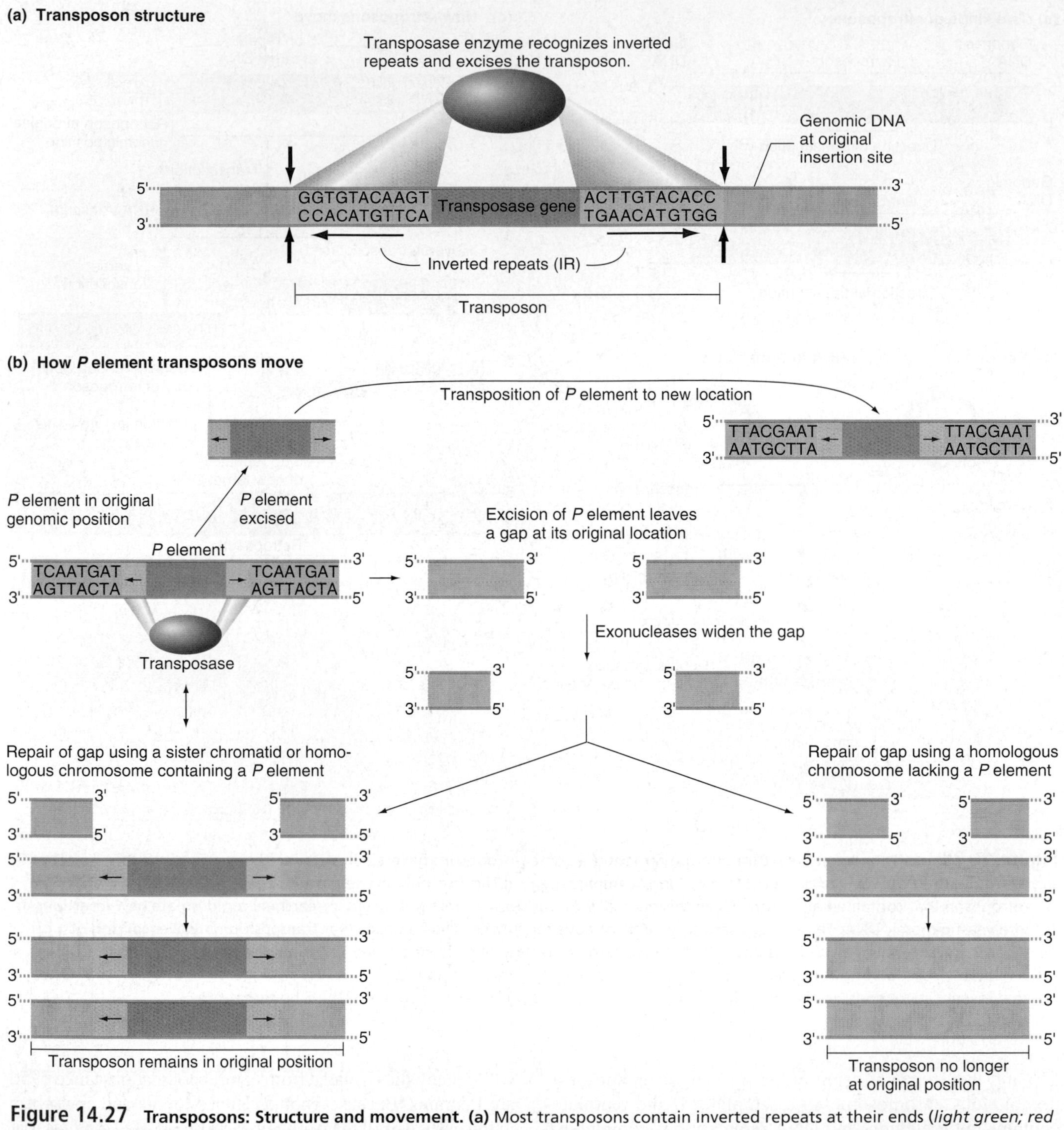

Figure 14.27 Transposons: Structure and movement. **(a)** Most transposons contain inverted repeats at their ends (*light green; red arrows*) and encode a transposase enzyme that recognizes these inverted repeats. The transposase cuts at the borders between the transposon and adjacent genomic DNA, and it also helps the excised transposon integrate at a new site. **(b)** Transposase-catalyzed integration of *P* elements creates a duplication of 8 bp present at the new target site. A gap remains when transposons are excised from their original position. After exonucleases widen the gap, cells repair the gap using related DNA sequences as templates. Depending on whether the template contains or lacks a *P* element, the transposon will appear to remain or to be excised from its original location during transposition.

breakage. One of the more interesting effects of hybrid dysgenesis is to promote the movement of *P* elements to new positions in the genome. Because elevated levels of transposition can foster many kinds of genetic changes (described in the following), some geneticists speculate that hybrid dysgenesis-like events involving various transposons in different species had a strong impact on evolution. The *Drosophila* portrait (on our website at www.mhhe.com/hartwell3) provides more information on the molecular mechanisms underlying hybrid dysgenesis and the ways in which fly geneticists use this phenomenon to introduce new genes into *Drosophila.*

Genomes Often Contain Defective Copies of Transposable Elements

Many copies of TEs sustain deletions either as a result of the transposition process itself (for example, incomplete reverse transcription of a retroposon RNA) or as a result of events following transposition (for example, faulty repair of a site from which a *P* element was earlier excised). If a deletion removes the promoter needed for transcription of a retroposon, that copy of the element cannot generate the RNA intermediate for future movements. If the deletion removes one of the inverted repeats at one end of a transposon, transposase will be unable to catalyze transposition of that element. Such deletions create defective TEs unable to transpose again. Most SINEs and LINEs in the human genome are defective in this way. Other types of deletions create defective elements that are unable to move on their own, but they can move if nondefective copies of the element elsewhere in the genome supply the deleted function. For example, a deletion inactivating the reverse transcriptase gene in a retroposon or the transposase gene in a transposon would "ground" that copy of the element at one genomic location if it is the only source of the essential enzyme in the genome. If reverse transcriptase or transposase were provided by other copies of the same element in the genome, however, the defective copy could move. Defective TEs that require the activity of nondeleted copies of the same TE for movement are called **nonautonomous elements;** the nondeleted copies that can move by themselves are **autonomous elements.**

Transposable Elements Can Generate Mutations in Adjacent Genes

Insertion of a TE near or within a gene can affect gene expression and alter phenotype. We now know that the wrinkled pea mutation first studied by Mendel resulted from insertion of a TE into the gene for starch-branching enzyme I. In *Drosophila,* a large percentage of spontaneous mutations, including the w^1 mutation discovered by T. H. Morgan in 1910, are caused by insertion of TEs (**Fig. 14.28**). Surprisingly, in light of the large numbers of LINEs and SINEs in human genomes, only a handful of mutant human phenotypes are known to result from insertion of TEs.

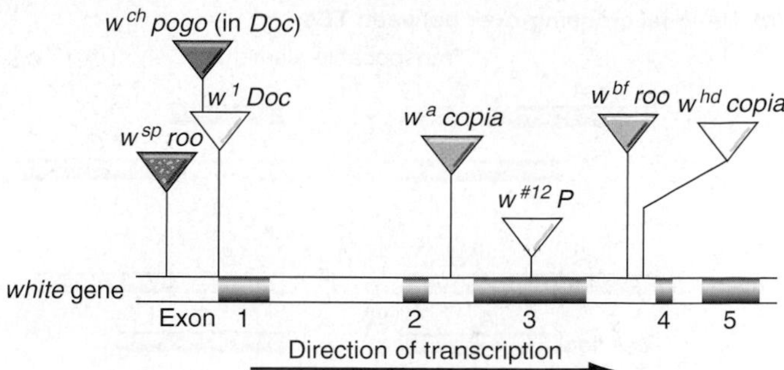

Figure 14.28 TEs can cause mutations on insertion into a gene. Many spontaneous mutations in the *white* gene of *Drosophila* arise from insertions of TEs such as *copia, roo, pogo,* or *Doc.* The consequences for the eye color phenotype (indicated by the color in the *triangles*) depend on the element involved and where in the *white* gene it inserts.

Among these is a B-type hemophilia caused by *Alu* insertion into a gene encoding clotting factor IX; recall that *Alu* is the main human SINE.

A TE's effect on a gene depends on what the element is and where within or near the gene it inserts (Fig. 14.28). If an element lands within a protein-coding exon, the additional DNA may shift the reading frame or supply an in-frame stop codon that truncates the polypeptide. If the element falls in an intron, it could diminish the efficiency of splicing. Some of these inefficient splicing events might completely remove the element from the gene's primary transcript; this would still allow some—but less than normal—synthesis of functional polypeptide. TEs that land within exons or introns may also provide a transcription stop signal that prevents transcription of gene sequences downstream of the insertion site. Finally, insertions into regions that regulate transcription, such as promoters, can influence the amount of gene product made in particular tissues at particular times during development. Some transposons insert preferentially into the upstream regulatory regions of genes, and some even prefer specific types of genes, such as tRNA genes.

Transposable Genetic Elements Can Also Generate Chromosomal Rearrangements

Retroposons and transposons can trigger spontaneous chromosomal rearrangements other than transpositions in several ways. Sometimes, deletion or duplication of chromosomal material adjacent to the transposon occurs as a mistake during the transposition event itself. In another mechanism, if two copies of the same TE occupy nearby but not identical sites in homologous chromosomes, the two copies of the TE in heterozygotes carrying both types of homolog may pair with each other and cross over (**Fig. 14.29a**). The recombination resulting from this unequal crossover would produce one chromosome deleted for the

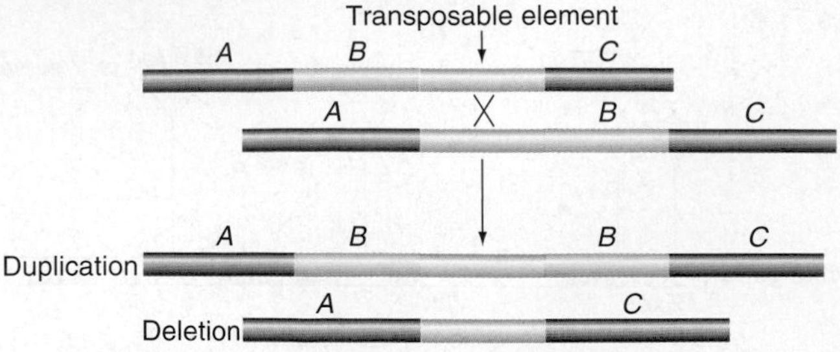

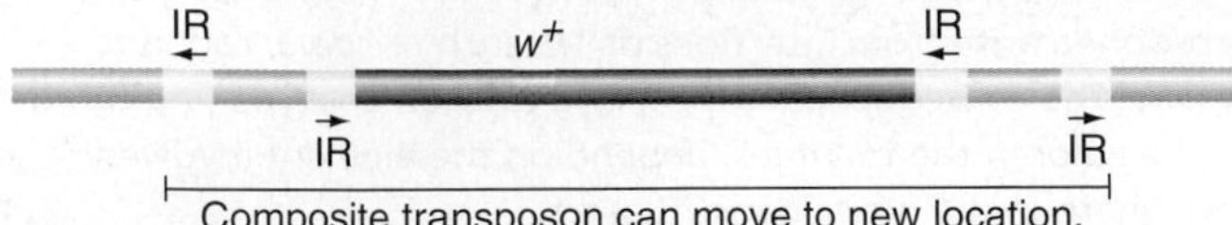

Figure 14.29 How TEs generate chromosomal rearrangements and relocate genes. (a) If a TE (*pink*) is found in slightly different locations on homologous chromosomes (here on opposite sides of segment *B*), unequal crossing-over between the TEs will produce reciprocal deletions and duplications of the intervening region. **(b)** If two copies of the same transposon are nearby on the same chromosome, transposase can recognize the outermost inverted repeats (IRs), creating a composite transposon that allows intervening genes such as w^+ (*red*) to jump to new locations.

region between the two TEs and a reciprocal homolog with a tandem duplication of the same region. The duplication associated with the *Bar* mutation in *Drosophila* (review Fig. 14.13) probably arose in this way.

Transposition Can Relocate Genes

When two copies of a transposon occur in nearby but not identical locations on the same chromosome, the inverted repeats of the transposons are positioned such that an inverted version of the sequence at the 5′ end of the copy on the left will exist at the 3′ end of the copy to its right (Fig. 14.29b). If transposase acts on this pair of inverted repeats during transposition, it allows the entire region between them to move as one giant transposon, mobilizing and relocating any genes the region contains. Some composite transposons, such as that pictured in the figure at the beginning of this chapter on p. 489, carry as much as 400 kb of DNA. In prokaryotes, the capacity of two TEs to relocate the intervening genes helps mediate the transfer of drug resistance between different strains or species of bacteria, as will be discussed in Chapter 15.

Summary of Transposition

Although pieces of DNA that can move from place to place in the genome might enhance or diminish a genome's ability to respond to a changing environment, it is usually helpful to think of TEs as segments of "selfish" DNA that exist for their own sake. The movement of TEs has three main genetic consequences.

1. The insertion of both retroposons and transposons can mutate a gene; precise removal of the element from the gene then causes a reversion to wild type.
2. Side effects of the transposition process itself or of unequal crossing-over between copies of the same TE can help generate chromosomal rearrangements.
3. Recognition by transposase of two nearby transposons on the same chromosome can relocate the intervening genes.

Because of these genetic consequences, TEs make a major contribution to the evolution of genomes.

Rearrangements and Evolution: A Speculative Comprehensive Example

We saw at the beginning of this chapter that roughly 300 chromosomal rearrangements could reshape the human genome to a form that resembles the mouse genome. Many of these rearrangements are transpositions that could construct a new chromosome from large blocks of sequences that were on different chromosomes in an ancestral organism. Figure 14.1 provides clear evidence that these reorganizations also include inversions. For example, mouse chromosome 1 contains two adjacent syntenic segments that are found in human chromosome 6, but in a reshuffled order, with one segment turned around 180° with respect to the other segment. Direct DNA sequence comparison of the mouse and human genomes further indicates that deletions, duplications, and transpositions have occurred in one or the other lineage since humans and mice began to diverge from a common ancestor 65 million years ago.

The occurrence of these various rearrangements over evolutionary time suggests two things. First, although most chromosomal variations, including single-base changes and chromosomal rearrangements, are deleterious to an organism or its progeny, a few changes are either neutral or provide an advantage for survival and manage to become fixed in a population. Second, some rearrangements almost certainly contribute to the processes underlying speciation. Although we still do not know enough to understand how any particular rearrangement that distinguishes the human from the mouse genome may have provided a survival advantage or otherwise helped guide speciation, it is nonetheless useful to consider in a general way how chromosomal rearrangements might contribute to evolution.

Deletions A small deletion that moves a coding sequence of one gene next to a promoter or other regulatory element of an adjacent gene may rarely

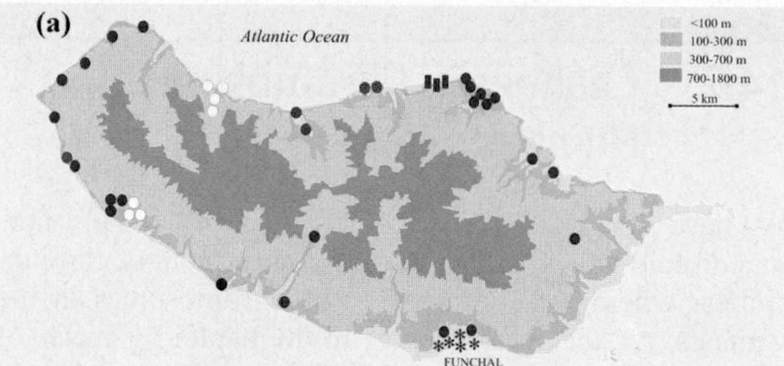

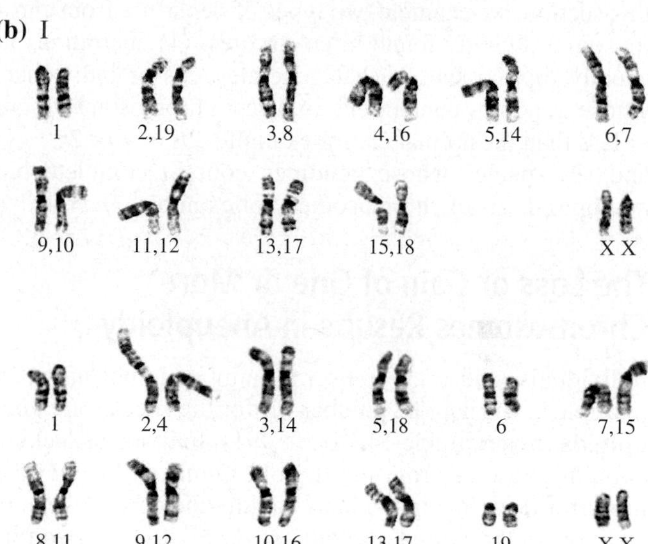

Figure 14.30 Rapid chromosomal evolution in house mice on the island of Madeira. (a) Distribution of mouse populations with different sets of Robertsonian translocations (indicated by circles of different colors). **(b)** Karyotypes of female mice from two different populations. The karyotype I at the *top* is from the population shown with *red dots* in part (a); the karyotype II at the *bottom* is from the population indicated by *green dots.* Robertsonian translocations are indicated by numbers separated by a comma (for example, *2,19* is a Robertsonian translocation between chromosomes 2 and 19 of the standard mouse karyotype.

allow expression of a protein at a novel time in development or in a novel tissue. If the new time or place of expression is advantageous to the organism, it might become established in the genome.

Duplications An organism cannot normally tolerate mutations in a gene essential to its survival, but duplication would provide two copies of the gene. If one copy remained intact to perform the essential function, the other would be free to evolve a new function. The genomes of most higher plants and animals, in fact, contain many **gene families**—sets of closely related genes with slightly different functions that most likely arose from a succession of gene duplication events. In vertebrates, some *multigene families* have hundreds of members.

Inversions Suppose one region of a chromosome has three mutations that together greatly enhance the reproductive fitness of the organism. In heterozygotes where one homolog carries the mutations and the other does not, recombination could undo the beneficial linkage. If, however, the three mutations are part of an inversion, crossover suppression will ensure that they remain together as they spread through the population.

Translocations On the tiny volcanic island of Madeira off the coast of Portugal in the Atlantic Ocean, two populations of the common house mouse (*Mus musculus*) are in the process of becoming separate species because of translocations that have led to reproductive isolation. The mice live in a few narrow valleys separated by steep mountains. Geneticists have found that populations of mice on the two sides of these mountain barriers have very different sets of chromosomes because they have accumulated different sets of Robertsonian translocations (**Fig. 14.30**). Mice in one Madeira population, for example, have a diploid number ($2n$) of 22 chromosomes, while mice in a different population on the island have 24; for most house mice throughout the world, $2n = 40$. (Recall from Fig. 14.19 that Robertsonian translocations can reduce chromosome number if the small chromosome that results from a translocation is lost.)

The hybrid offspring of matings between individuals of these two populations are completely sterile or infertile because chromosomal complements that are so different cannot properly segregate at meiosis. Thus, reproductive isolation has reinforced the already established geographical isolation, and the two populations are close to becoming two separate species. What is remarkable about this example of speciation is that mice were introduced into Madeira by Portuguese settlers only in the fifteenth century. This means that the varied and complicated sets of Robertsonian translocations that contributed to speciation became fixed in the different populations in less than 600 years.

Transpositions Movement of TEs may cause novel mutations, a small proportion of which might be selected for because they are advantageous to the organism. TEs can also help generate potentially useful duplications and inversions.

Because of these possibilities, many geneticists speculate that chromosomal rearrangements and transpositions are, like point mutations, important instruments of evolution.

14.2 Changes in Chromosome Number

We have seen that in peas, *Drosophila,* and humans, normal diploid individuals carry a $2n$ complement of chromosomes, where n is the number of chromosomes in the gametes. All the chromosomes in the haploid gametes of these diploid organisms are different from one another. In this section, we examine two types of departure from chromosomal diploidy found in eukaryotes: (1) aberrations in usually diploid species that generate cells or individuals whose genomes contain one to a few chromosomes more or less than the normal $2n$, for example, $2n + 1$ or $2n - 1$; and (2) species whose genomes contain complete but nondiploid sets of chromosomes, for example, $3n$ or $4n$.

The Loss or Gain of One or More Chromosomes Results in Aneuploidy

Individuals whose chromosome number is not an exact multiple of the haploid number (n) for the species are **aneuploids** (review Table 14.1 on p. 491). Individuals lacking one chromosome from the diploid number ($2n - 1$) are **monosomic,** while individuals having one chromosome in addition to the normal diploid set ($2n + 1$) are **trisomic.** Organisms with four copies of a particular chromosome ($2n + 2$) are **tetrasomic.**

Autosomal Aneuploidy Is Harmful to the Organism

Monosomy, trisomy, and other forms of aneuploidy create a genetic imbalance that is usually deleterious to the organism. In humans, monosomy for any autosome is generally lethal, but medical geneticists have reported a few cases of monosomy for chromosome 21, one of the smallest human chromosomes. Although born with severe multiple abnormalities, these monosomic individuals survived for a short time beyond birth. Similarly, trisomies involving a human autosome are also highly deleterious. Individuals with trisomies for larger chromosomes, such as 1 and 2, are almost always aborted spontaneously early in pregnancy. Trisomy 18 causes Edwards syndrome, and trisomy 13 causes Patau syndrome; both phenotypes include gross developmental abnormalities that result in early death.

The most frequently observed human autosomal trisomy, trisomy 21, results in Down syndrome. As one of the shortest human autosomes, chromosome 21 contains about 1.5% of the DNA in the human genome. Although there is considerable phenotypic variation among Down syndrome individuals, traits such as mental retardation and skeletal abnormalities are usually associated with the condition. Many Down syndrome babies die in their first year after birth from heart defects and increased susceptibility to infection. We saw earlier (in the discussion of translocations) that some people with Down syndrome have three copies of only part of, rather than the entire, chromosome 21. It is thus probable that genetic imbalance for only a few genes may be a sufficient cause of the condition. Unfortunately, as of late 2005, scientists had not yet been able to identify any of these genes unambiguously.

Humans Tolerate X Chromosome Aneuploidy Because X Inactivation Compensates for Dosage

Although the X chromosome is one of the longest human chromosomes and contains 5% of the DNA in the genome, individuals with X chromosome aneuploidy, such as XXY males, XO females, and XXX females, survive quite well compared with aneuploids for the larger autosomes. The explanation for this tolerance of X-chromosome aneuploidy is that X-chromosome inactivation equalizes the expression of most X-linked genes in individuals with different numbers of X chromosomes. As we saw in Chapter 13, X-chromosome inactivation represses expression of most genes on all but one X chromosome in a cell. As a result, even if the number of X chromosomes varies, the amount of protein generated by most X-linked genes remains constant. Human X-chromosome aneuploidies are nonetheless not without consequence. XXY men have Klinefelter syndrome, and XO women have Turner syndrome. The aneuploid individuals affected by these syndromes are usually infertile and display skeletal abnormalities, leading in the men to unusually long limbs and in the women to unusually short stature.

If X inactivation were 100% effective, we would not expect to see even the relatively minor abnormalities of Klinefelter syndrome, because the number of functional X chromosomes—one—would be the same as in normal individuals. One explanation is that during X inactivation, several genes near the telomere and centromere of the short arm of the human X chromosome escape inactivation and thus remain active. As a result, XXY males make twice the amount of protein encoded by these few genes as XY males (**Fig. 14.31**).

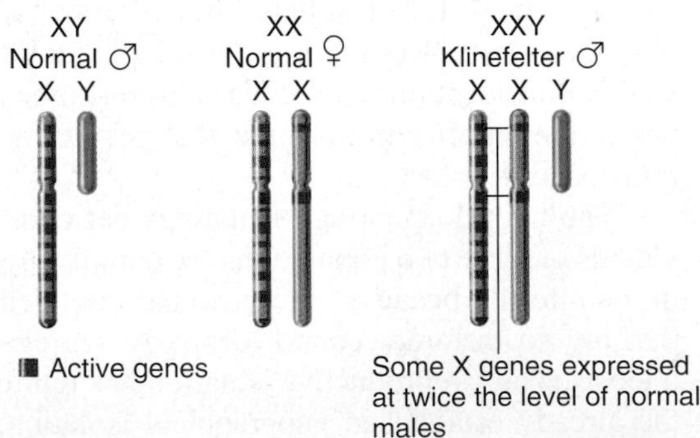

Figure 14.31 Why aneuploidy for the X chromosome can have phenotypic consequences. X-chromosome inactivation does not affect all genes on the X chromosome. As a result, in XXY Klinefelter males, a few X chromosome genes are expressed inappropriately at twice their normal level.

The reverse of X inactivation is X *reactivation;* it occurs in the oogonia, the female germ-line cells that develop into the oocytes that undergo meiosis (review Fig. 4.18 on p. 104). Reactivation of the previously inactivated X chromosomes in the oogonia ensures that every mature ovum (the gamete) receives an active X. If X reactivation did not occur, half of a woman's eggs (those with inactive X chromosomes) would be incapable of supporting development after fertilization. The phenomenon of X reactivation in the oogonia might help explain the infertility of women with Turner syndrome. With X reactivation, oogonia in normal XX females have two functional doses of X chromosome genes; but the corresponding cells in XO Turner women have only one dose of the same genes and may thus undergo defective oogenesis.

Aneuploidy Results from Meiotic Nondisjunction

How does aneuploidy arise? Mistakes in chromosome segregation during meiosis produce aneuploids of different types, depending on when the mistakes occur. If homologous chromosomes do not separate (that is, do not disjoin) during the first meiotic division, two of the resulting haploid gametes will carry both homologs, and two will carry neither. Union of these gametes with normal gametes will produce aneuploid zygotes, half monosomic, half trisomic (**Fig. 14.32a,** *left*). By contrast, if meiotic nondisjunction occurs during meiosis II, only two of the four resulting gametes will be aneuploid (Fig. 14.32a, *right*). Abnormal $n + 1$ gametes resulting from nondisjunction in a cell that

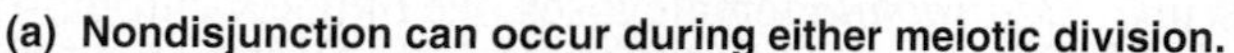

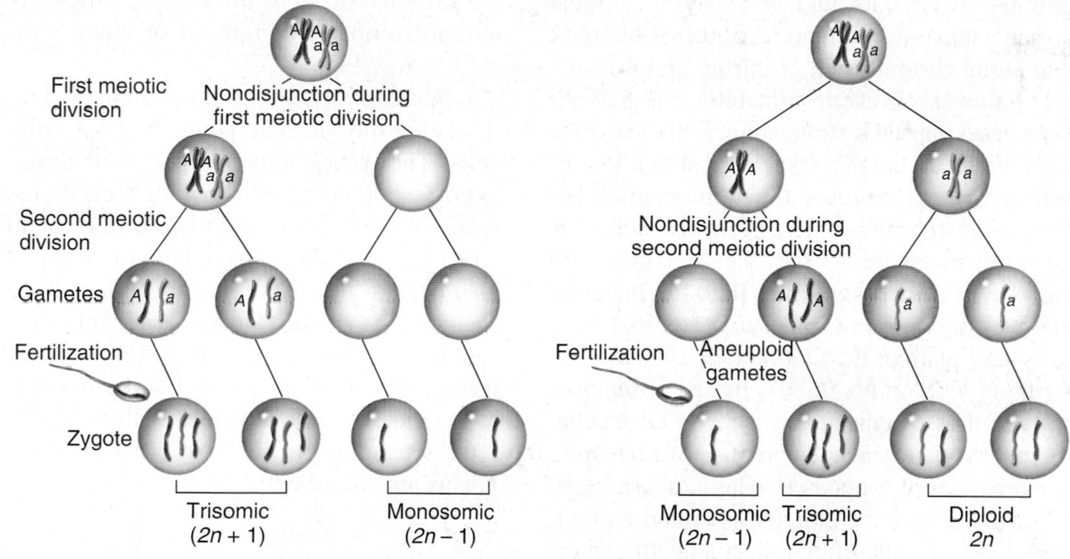

(b) Aneuploids beget aneuploid progeny.

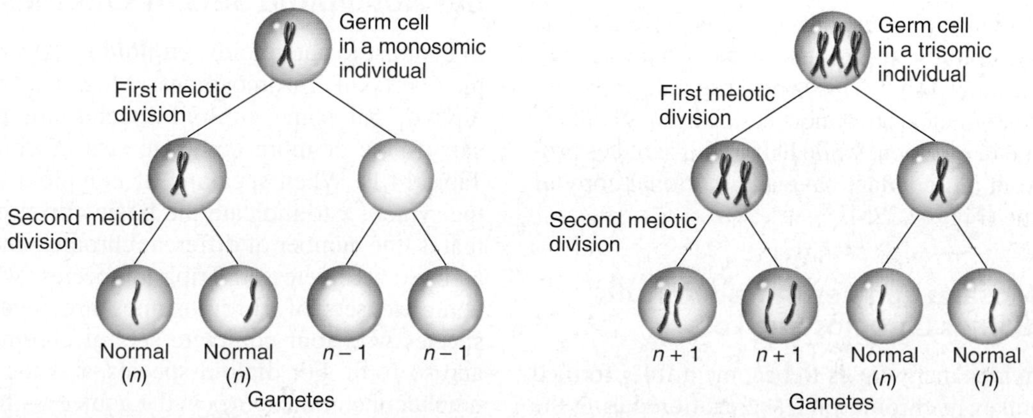

Figure 14.32 Aneuploidy is caused by problems in meiotic chromosome segregation. (a) If trisomic progeny inherit two different alleles (*A* and *a*) of a centromere-linked gene from one parent, the nondisjunction occurred in meiosis I (*left*). If the two alleles inherited from one parent are the same (*A* and *A;* or *a* and *a*), the nondisjunction occurred during meiosis II (*right*). **(b)** Because aneuploids carry chromosomes that have no homolog with which to pair, aneuploid individuals frequently produce aneuploid progeny.

is heterozygous for alleles on the nondisjoining chromosome will be heterozygous if the nondisjunction happens in the first meiotic division, but they will be homozygous if the nondisjunction takes place in the second meiotic division. (We assume here that no recombination has occurred between the heterozygous gene in question and the centromere, as would be the case for genes closely linked to the centromere.) It is possible to use this distinction to determine when a particular nondisjunction occurred (Fig. 14.32a). The nondisjunction events that give rise to Down syndrome, for example, occur much more frequently in mothers (90%) than in fathers (10%). Interestingly, in women, such nondisjunction events occur more often during the first meiotic division (about 75% of the time) than during the second. By contrast, when the nondisjunction event leading to Down syndrome takes place in men, the reverse is true.

Recently obtained data show that many meiotic nondisjunction events in humans result from problems in meiotic recombination. By tracking DNA markers, clinical investigators can establish whether recombination took place anywhere along chromosome 21 during meioses that created $n + 1$ gametes. In approximately one-half of Down syndrome cases caused by nondisjunction during the first meiotic division in the mother (that is, in about 35% of all Down syndrome cases), no recombination occurred between the homologous chromosome 21's in the defective meioses. This result makes sense because chiasmata, the structures associated with crossing-over, hold the maternal and paternal homologous chromosomes together in a bivalent at the metaphase plate of the first meiotic division (review Feature Figure 4.15 on pp. 96–97). In the absence of recombination and thus of chiasmata, there is no mechanism to ensure that the maternal and paternal chromosomes will go to opposite poles at anaphase I. The increase in the frequency of Down syndrome children associated with increasing maternal age may therefore reflect a decline in the effectiveness of the mother's machinery for meiotic recombination.

If an aneuploid individual survives and is fertile, the incidence of aneuploidy among his or her offspring will generally be extremely high. This is because half of the gametes produced by meiosis in a monosomic individual lack the chromosome in question, while half of the gametes produced in a trisomic individual have an additional copy of the chromosome (Fig. 14.32b).

Aneuploid Mosaics Can Result from Mitotic Nondisjunction or Chromosome Loss

As a zygote divides many times to become a fully formed organism, mistakes in chromosome segregation during the mitotic divisions accompanying this development may, in rare instances, augment or diminish the complement of chromosomes in certain cells. In **mitotic nondisjunction,** the failure of two sister chromatids to separate during mitotic anaphase generates reciprocal trisomic and monosomic daughter cells (**Fig. 14.33a**). Other types of mistakes, such as a lagging chromatid not pulled to either spindle pole at mitotic anaphase, result in a **chromosome loss** that produces one monosomic and one diploid daughter cell (Fig. 14.33b).

Aneuploid cells arising from either mitotic nondisjunction or chromosome loss may survive and undergo further rounds of cell division, producing clones of cells with an abnormal chromosome count. Nondisjunction or chromosome loss occurring early in development will generate larger aneuploid clones than the same events occurring later in development. The side-by-side existence of aneuploid and normal tissues results in a **mosaic** organism whose phenotype depends on what tissue bears the aneuploidy, the number of aneuploid cells, and the specific genes on the aneuploid chromosome. Many examples of mosaicism involve the sex chromosomes. If an XX *Drosophila* female loses one of the X chromosomes during the first mitotic division after fertilization, the result is a **gynandromorph** composed of equal parts male and female tissue (Fig. 14.33c).

Interestingly, in humans, many Turner syndrome females are mosaics carrying some XX cells and some XO cells. These individuals began their development as XX zygotes, but with the loss of an X chromosome during the embryo's early mitotic divisions, they acquired a clone of XO cells. Similar mosaicism involving the autosomes also occurs. For example, physicians have recorded several cases of mild Down syndrome arising from mosaicism for trisomy 21. In people with Turner or Down mosaicism, the existence of some normal tissue appears to ameliorate the condition, with the individual's phenotype depending on the particular distribution of diploid versus aneuploid cells.

Some Euploid Species Contain Complete but Nondiploid Sets of Chromosomes

In contrast to aneuploids, **euploid** cells contain only complete sets of chromosomes. Most euploid species are diploid, but some euploid species are **polyploids** that carry three or more complete sets of chromosomes (see Table 14.1). When speaking of polyploids, geneticists use the symbol x to indicate the **basic chromosome number,** that is, the number of different chromosomes that make up a single complete set. Triploid species, which have three complete sets of chromosomes are then $3x$; tetraploid species with four complete sets of chromosomes are $4x$; and so forth. For diploid species, x is identical to n—the number of chromosomes in the gametes—because each gamete contains a single complete set of chromosomes. This identity of $x = n$ does not, however, hold for polyploid species, as the following example illustrates. Commercially grown bread wheat has a total of 42 chromosomes:

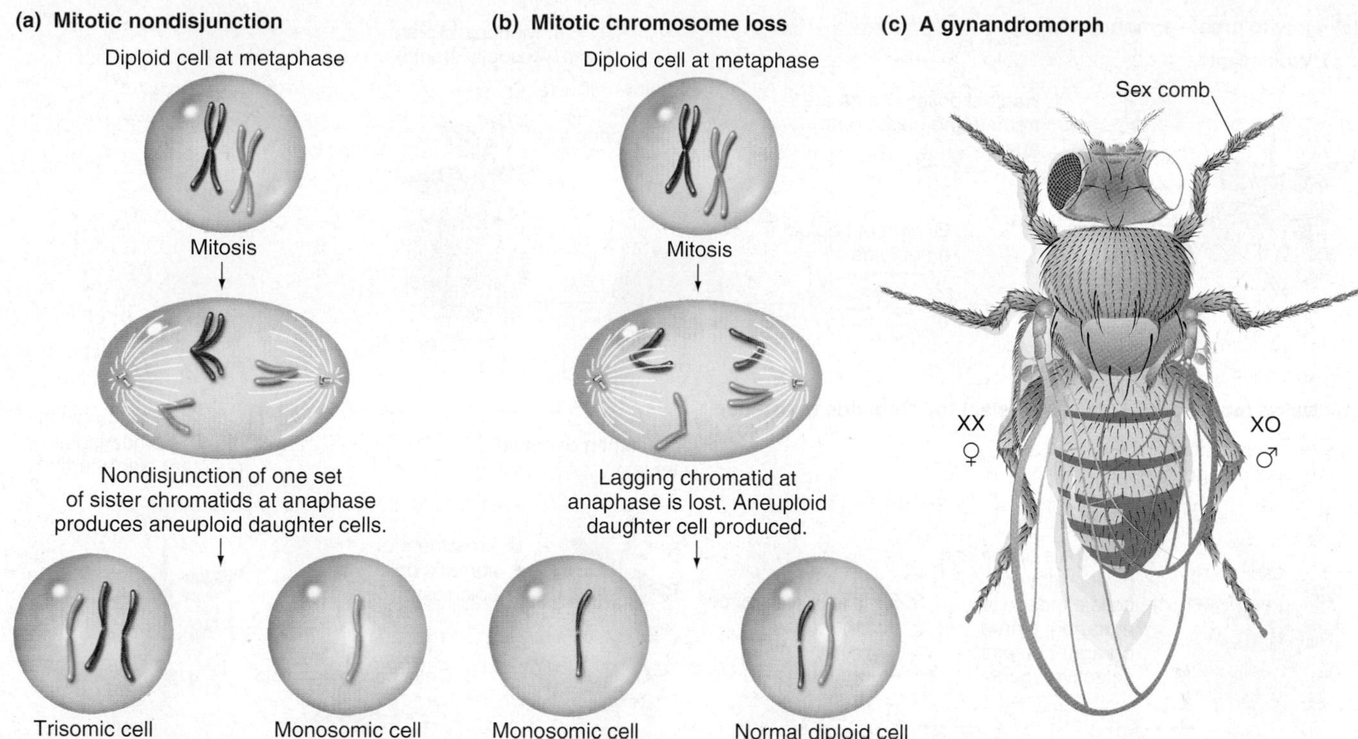

Figure 14.33 Mistakes during mitosis can generate clones of aneuploid cells. Mitotic nondisjunction **(a)** or chromosome loss during mitosis **(b)** can create monosomic or trisomic cells that can divide to produce aneuploid clones. **(c)** If an X chromosome is lost during the first mitotic division of an XX *Drosophila* zygote, one daughter cell will be XX (female), while the other will be XO (male). Such an embryo will grow into a gynandromorph. Here, the zygote was $w^+ \ m^+ \ / \ w \ m$, so the XX half of the fly (*left*) has red eyes and normal wings; loss of the $w^+ \ m^+$ X chromosome gives the XO half of the fly (*right*) white eyes (*w*), miniature wings (*m*), and a male-specific sex comb on the front leg.

6 nearly (but not wholly) identical sets each containing 7 different chromosomes. Bread wheat is thus a hexaploid with a basic number of $x = 7$ and $6x = 42$. But each triploid gamete has one-half the total number of chromosomes, so $n = 21$. Thus, for bread wheat, x and n are not the same. Another form of euploidy, in addition to polyploidy, exists in **monoploid** (x) organisms, which have only one set of chromosomes.

Monoploidy and polyploidy are rarely observed in animals. Among the few examples of monoploidy are some species of ants and bees in which the males are monoploid, while the females are diploid. Males of these species develop *parthenogenetically* from unfertilized eggs. These monoploid males produce gametes through a modified meiosis that in some unknown fashion ensures distribution of all the chromosomes to the same daughter cell during meiosis I; the sister chromatids then separate normally during meiosis II. Polyploidy in animals normally exists only in species with unusual reproductive cycles, such as hermaphroditic earthworms, which carry both male and female reproductive organs, and goldfish, which are parthenogenetically tetraploid species. In *Drosophila,* it is possible, under special circumstances, to produce triploid and tetraploid females, but never males. In humans, polyploidy is always lethal, usually resulting in spontaneous abortion during the first trimester of pregnancy.

Monoploid Organisms Contain a Single Copy of Each Chromosome and Are Usually Infertile

Botanists can produce monoploid plants experimentally by special treatment of germ cells from diploid species that have completed meiosis and would normally develop into pollen. (Note that monoploid plants obtained in this manner can also be considered haploids because $x = n$.) The treated cells divide into a mass of tissue known as an *embryoid.* Subsequent exposure to plant hormones enables the embryoid to develop into a plant (**Fig. 14.34a**). Monoploid plants may also arise from rare spontaneous events in a large natural population. Most monoploid plants, no matter how they originate, are infertile. Since the chromosomes have no homologs with which to pair during meiosis I, they are distributed at random to the two spindle poles during this division. Rarely do all chromosomes go to the same pole, and if they do not, the resulting gametes are defective as they lack one or more chromosomes. The greater the number of chromosomes

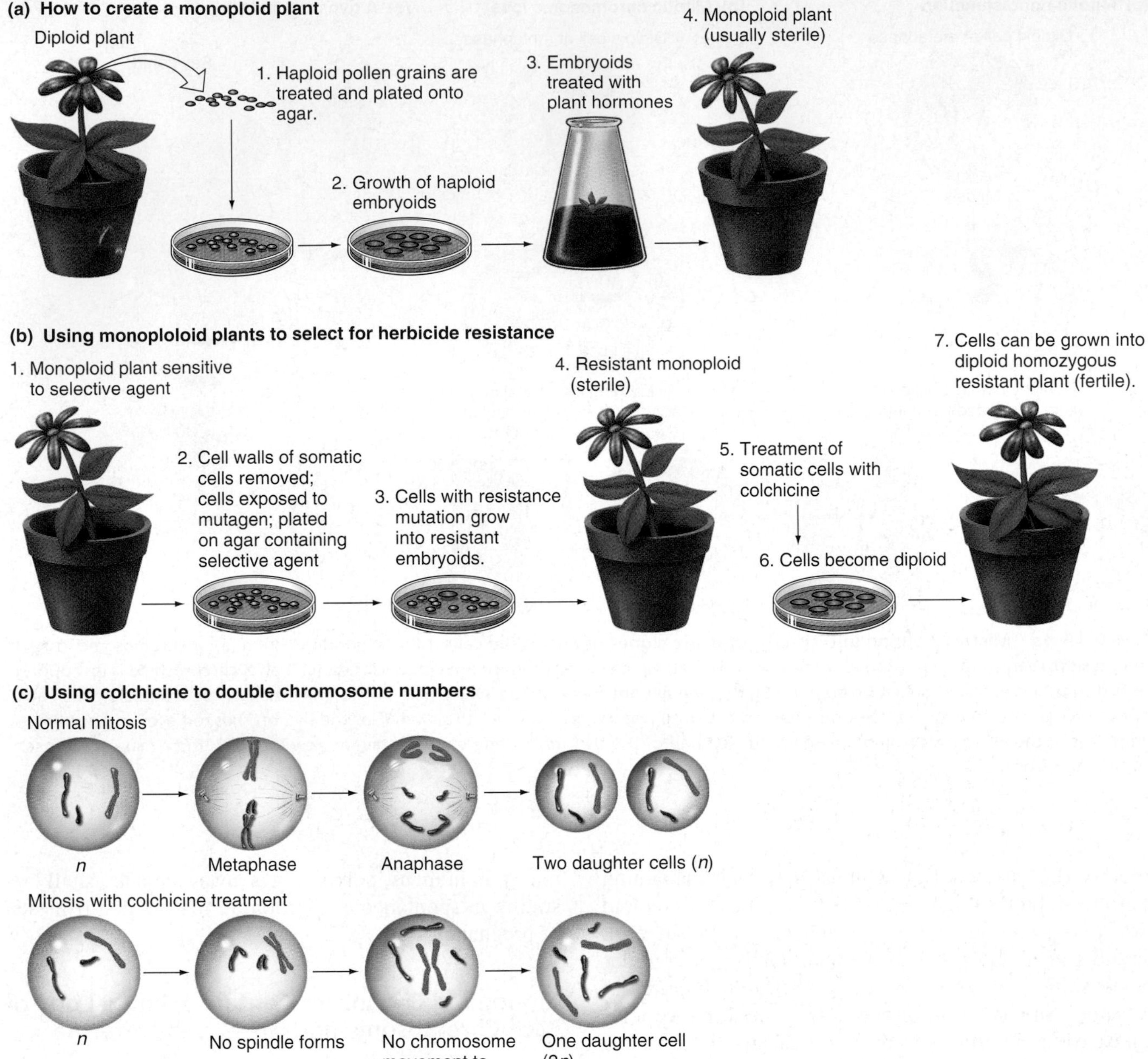

Figure 14.34 The creation and use of monoploid plants. (a) Under certain conditions, haploid pollen grains can grow into haploid embryoids. When treated with plant hormones, haploid embryoids grow into monoploid plants. **(b)** Researchers select monoploid cells for recessive traits such as herbicide resistance. They then grow the selected cells into a resistant embryoid, which (with hormone treatment) eventually becomes a mature, resistant monoploid plant. Treatment with colchicine doubles the chromosome number, creating diploid cells that can be grown in culture with hormones to make a homozygous resistant diploid plant. **(c)** Colchicine treatment prevents formation of the mitotic spindle and also blocks cytokinesis, generating cells with twice the number of chromosomes. *Blue, red,* and *green* colors denote nonhomologous chromosomes.

in the genome, the lower the likelihood of producing a gamete containing all of them.

Despite such gamete-generating problems, monoploid plants and tissues are of great value to plant breeders. They make it possible to visualize normally recessive traits directly, without crosses to achieve homozygosity. Plant researchers can also introduce mutations into individual monoploid cells; select for desirable phenotypes, such as resistance to herbicides; and use hormone treatments to grow the selected cells into monoploid plants (Fig. 14.34b). They can then convert monoploids of their choice into homozygous diploid plants by treating tissue with *colchicine,* an alkaloid drug obtained from the autumn crocus. By binding to tubulin—the major protein

Figure 14.35 Many polyploid plants are larger than their diploid counterparts. A comparison of octaploid (*left*) and diploid (*right*) strawberries.

component of the spindle—colchicine prevents formation of the spindle apparatus. In cells without a spindle, the sister chromatids cannot segregate after the centromere splits, so there is often a doubling of the chromosome set following treatment with colchicine (Fig. 14.34c). The resulting diploid cells can be grown into diploid plants that will express the desired phenotype and produce fertile gametes.

Polyploidy Has Accompanied the Evolution of Many Cultivated Plant Species

Roughly one out of every three known species of flowering plants is a polyploid, and because polyploidy often increases plant size and vigor, many polyploid plants with edible parts have been selected for agricultural cultivation. Most commercially grown alfalfa, coffee, and peanuts are tetraploids ($4x$). MacIntosh apple and Bartlett pear trees that produce giant fruits are also tetraploids. Commercially grown strawberries are octaploids ($8x$) (**Fig. 14.35**). The evolutionary success of polyploid plant species may stem from the fact that polyploidy, like gene duplication, provides additional copies of genes; while one copy continues to perform the original function, the others can evolve new functions. However, as we now show, the consequences of polyploidy depend in large part on whether the number of chromosome sets is even or odd.

Triploids Are Almost Always Sterile

Triploids ($3x$) result from the union of monoploid (x) and diploid ($2x$) gametes (**Fig. 14.36a**). The diploid gametes may be the products of meiosis in tetraploid ($4x$) germ cells, or they may be the products of rare spindle or cytokinesis failures during meiosis in a diploid.

Sexual reproduction in triploid organisms is extremely inefficient because meiosis produces mostly unbalanced gametes. During the first meiotic division in a triploid germ cell, three sets of chromosomes must segregate into two daughter cells; regardless of how the chromosomes align in pairs, there is no way to ensure that the resulting gametes obtain a complete, balanced x or $2x$ complement of chromosomes. In most cases, at the end of

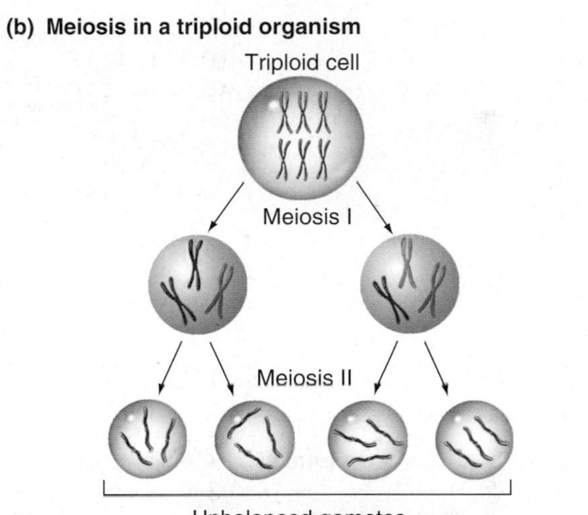

Figure 14.36 The genetics of triploidy. (a) Production of a triploid ($x = 3$) from fertilization of a monoploid gamete by a diploid gamete. Nonhomologous chromosomes are either *blue* or *red*. **(b)** Meiosis in a triploid produces unbalanced gametes because meiosis I produces two daughter cells with unequal numbers of any one type of chromosome. If x is large, balanced gametes with equal numbers of all the chromosomes are very rare.

anaphase I, two chromosomes of any one type move to one pole, while the remaining chromosome of the same type moves to the opposite pole. The products of such a meiosis have two copies of some chromosomes and one copy of others (Fig. 14.36b). If the number of chromosomes in the basic set is large, the chance of obtaining any balanced gametes at all is remote. Thus, fertilization with gametes from triploid individuals does not produce many viable offspring.

It is possible to propagate some triploid species, such as bananas and watermelons, through asexual reproduction. The fruits of triploid plants are seedless because the unbalanced gametes do not function properly in fertilization or, if fertilization occurs, the resultant zygote is so

(a) Generation of tetraploid (4*x*) cells

Diploid (2*x*) interphase cell (*x* = 2) → Mitotic metaphase in 2*x* cell → Defective mitosis, chromosomes remain in same cell

Tetraploid (4*x*) cell

(b) Pairing of chromosomes as bivalents

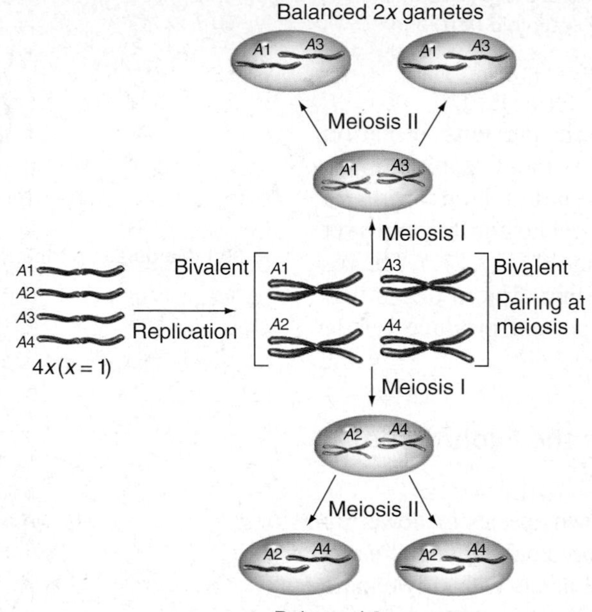

(c) Gametes formed by *A A a a* tetraploids

Chromosomes	Pairing	Gametes Produced by Random Spindle Attachment
1. *A* 2. *A* 3. *a* 4. *a*	1 *A* / 2 *A*; 3 *a* / 4 *a* or 1 *A* / 2 *A*; 4 *a* / 3 *a*	1 + 3 *A a* 2 + 4 *A a* or 1 + 4 *A a* 2 + 3 *A a*
	1 *A* / 3 *a*; 2 *A* / 4 *a* or 1 *A* / 3 *a*; 4 *a* / 2 *A*	1 + 2 *A A* 3 + 4 *a a* or 1 + 4 *A a* 2 + 3 *A a*
	1 *A* / 4 *a*; 2 *A* / 3 *a* or 1 *A* / 4 *a*; 3 *a* / 2 *A*	1 + 2 *A A* 3 + 4 *a a* or 1 + 3 *A a* 2 + 4 *A a*

Total:
8*A a*:2*A A*:2*a a* = 4*A a*:1*A A*:1*a a*

Figure 14.37 The genetics of tetraploidy. (a) Tetraploids arise from a failure of chromosomes to separate into two daughter cells during mitosis in a diploid. **(b)** In successful tetraploids, the pairing of chromosomes as bivalents generates genetically balanced gametes. **(c)** Gametes produced in an *A A a a* tetraploid heterozygous for two alleles of a centromere-linked gene, with orderly pairing of bivalents. The four chromosomes can pair to form two bivalents in three possible ways. For each pairing scheme, the chromosomes in the two pairs can assort in two different orientations. If all possibilities are equally likely, the expected genotype frequency in a population of gametes will be 1 *A A* : 4 *A a* : 1 *a a*.

genetically unbalanced that it cannot develop. Either way, no seeds form. Like triploids, all polyploids with odd numbers of chromosome sets (such as $5x$ or $7x$) are sterile because they cannot reliably produce balanced gametes.

Tetraploids Are Often the Source of New Species

During mitosis, if the chromosomes in a diploid ($2x$) tissue fail to separate after replication, the resulting daughter cells will be tetraploid ($4x$; **Fig. 14.37a**). If such tetraploid cells arise in reproductive tissue, subsequent meioses will produce diploid gametes. Rare unions between diploid gametes produce tetraploid organisms. Self-fertilization of a newly created tetraploid organism will produce an entirely new species, because crosses between the tetraploid and the original diploid organism will produce infertile triploids (review Fig. 14.36a). Tetraploids made in this fashion are **autopolyploids,** a kind of polyploid that derives all its chromosome sets from the same species.

Maintenance of a tetraploid species depends on the production of gametes with balanced sets of chromosomes. Most successful tetraploids have evolved mechanisms ensuring that the four copies of each group of homologs pair two by two to form two **bivalents**—pairs of synapsed homologous chromosomes (Fig. 14.37b). Because the chromosomes in each bivalent become attached to opposite spindle poles during meiosis I, meiosis regularly produces gametes carrying two complete sets of chromosomes. The mechanism requiring that each chromosome pair with only a single homolog suppresses other pairing possibilities, such as a 3:1, which cannot guarantee equivalent chromosome segregation.

Tetraploids, with four copies of every gene, generate unusual Mendelian ratios. For example, even if there are only two alleles of a gene (say, *A* and *a*), five different genotypes are possible: *A A A A, A A A a, A A a a, A a a a,* and *a a a a.* If the phenotype depends on the dosage of *A,* then five phenotypes, each corresponding to one of the genotypes, will appear. The segregation of alleles during meiosis in a tetraploid is similarly complex. Consider an *A A a a* heterozygote in which the *A* gene is closely linked to the centromere, and the *A* allele is completely dominant. What are the chances of obtaining progeny with the recessive phenotype, generated by only the *a a a a* genotype? As Fig. 14.37c illustrates, if during meiosis I, the four chromosomes carrying the gene align at random in bivalents along the metaphase plate, the expected ratio of gametes is $2\,A\,A : 8\,A\,a : 2\,a\,a = 1\,A\,A : 4\,A\,a : 1\,a\,a$. The chance of obtaining *a a a a* progeny during self-fertilization is thus $1/6 \times 1/6 = 1/36$. In other words, since *A* is completely dominant, the ratio of dominant to recessive phenotypes, determined by the ratio of *A - - -* to *a a a a* genotypes is 35:1. The ratios will be different if the gene is not closely linked to the centromere or if the dominance relationship between the alleles is not so simple.

New levels of polyploidy can arise from the doubling of a polyploid genome. Such doubling occurs on rare occasions in nature; it also results from controlled treatment with colchicine or other drugs that disrupt the mitotic spindle. The doubling of a tetraploid genome yields an octaploid ($8x$). These higher-level polyploids created by successive rounds of genome doubling are autopolyploids because all of their chromosomes derive from a single species.

Some Polyploids Have Agriculturally Desirable Traits Derived from Two Species

Polyploidy can also arise from crosses between members of two species, even if they have different numbers of chromosomes. Hybrids in which the chromosome sets come from two or more distinct, though related, species are known as **allopolyploids.** In crosses between octaploids and tetraploids, for example, fertilization unites tetraploid and diploid gametes to produce hexaploid progeny. Fertile allopolyploids arise only rarely, under special conditions, because chromosomes from the two species differ in shape, size, and number, so they cannot easily pair with each other. The resulting irregular segregation creates genetically unbalanced gametes such that the hybrid progeny will be sterile. Chromosomal doubling in germ cells, however, can restore fertility by creating a pairing partner for each chromosome. Organisms produced in this manner are termed **amphidiploids** if the two parental species were diploids; they contain two diploid genomes, each one derived from a different parent. As the following illustrations show, it is hard to predict the characteristics of an amphidiploid or other allopolyploids.

A cross between cabbages and radishes, for example, leads to the production of amphidiploids known as *Raphanobrassica.* The gametes of both parental species contain 9 chromosomes; the sterile F_1 hybrids have 18 chromosomes, none of which has a homolog. Chromosome doubling in the germ cells after treatment with colchicine, followed by union of two of the resulting gametes, produces a new species: a fertile *Raphanobrassica* amphidiploid carrying 36 chromosomes—a full complement of 18 (9 pairs) derived from cabbages and a full complement of 18 (9 pairs) derived from radishes. Unfortunately, this amphidiploid has the roots of a cabbage plant and leaves resembling those of a radish, so it is not agriculturally useful.

By contrast, crosses between tetraploid (or hexaploid) wheat and diploid rye have led to the creation of several allopolyploid hybrids with agriculturally desirable traits from both species (**Fig. 14.38**). Some of the hybrids combine the high yields of wheat with rye's ability to adapt to unfavorable environments. Others combine wheat's high level of protein with rye's high level of lysine; wheat protein does not contain very much of this amino acid, an essential ingredient in the human diet. The various hybrids

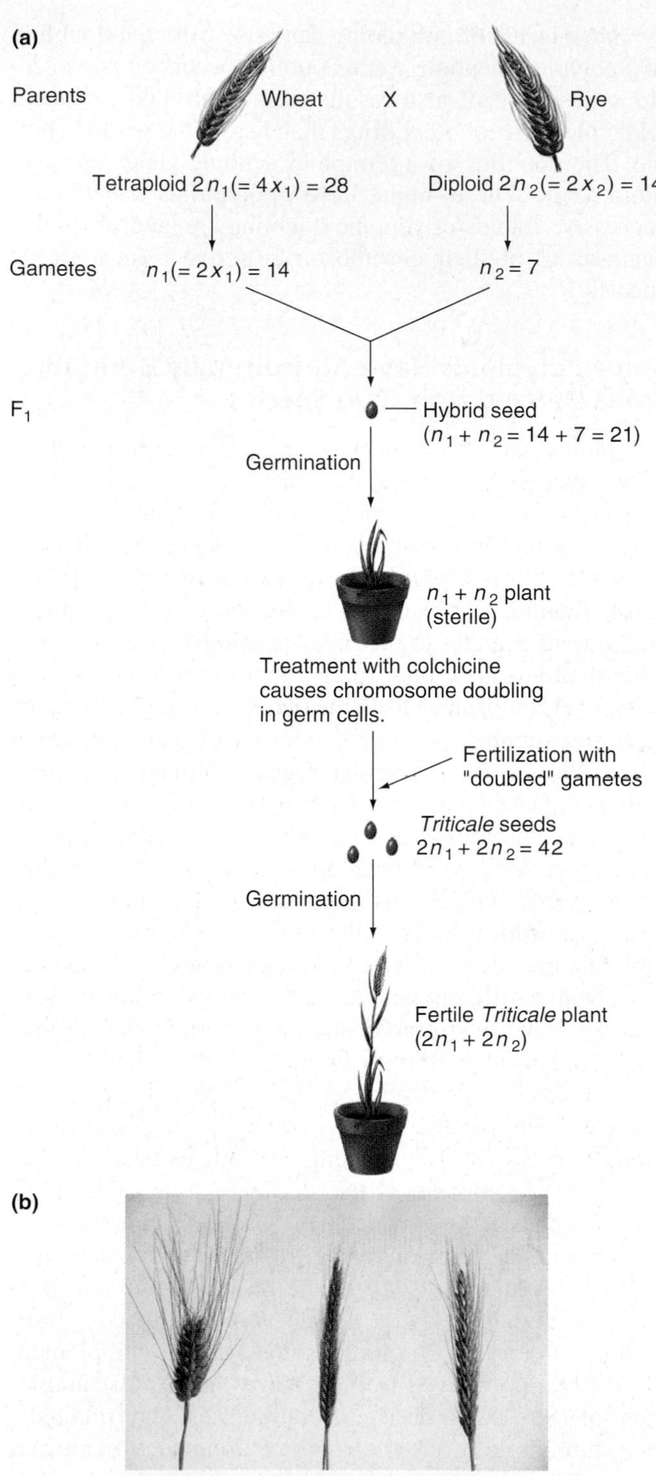

Figure 14.38 Amphidiploids in agriculture. (a) Plant breeders cross wheat with rye to create allopolyploid *Triticale*. Because this strain of wheat is tetraploid, x_1 (the number of chromosomes in the basic wheat set) is one-half n_1 (the number of chromosomes in a wheat gamete). For diploid rye, $n_2 = x_2$. Note that the F_1 hybrid between wheat and rye is sterile because the rye chromosomes have no pairing partners. Doubling of chromosome numbers by colchicine treatment of the F_1 hybrid corrects this problem, allowing regular pairing. **(b)** A comparison of wheat, rye, and *Triticale* grain stalks.

between wheat and rye form a new crop known as *Triticale*. Some triticale strains produce nutritious grains that already appear in breads sold in health food stores. Plant breeders are currently assessing the usefulness of various triticale strains for large-scale agriculture.

14.3 A Glimpse of the Future: Emergent Technologies in the Analysis of Chromosomal Rearrangements and Changes in Chromosome Number

Two main problems occur when searching for chromosomal rearrangements and changes in chromosome number by karyotype analysis. First, it is a tedious procedure that depends on highly trained technicians to identify chromosomal alterations under the microscope. Because of the subjective nature of the analysis, mistakes can reduce the accuracy of results. Second, even in the hands of the best technicians, there is a limit to the viewing resolution. Even under optimal circumstances, it is not possible to detect deletions or duplications of less than 5 Mb in human karyotypes. Human populations no doubt have many chromosomes with as yet undetected smaller deletions or duplications.

To overcome the limitations of karyotype analysis, researchers have developed a microarray-based hybridization protocol that can scan the genome for deletions, duplications, and aneuploidy with much greater resolution, very high accuracy, and much greater throughput and without the need for a subjective determination of the result. The technique is called *comparative genomic hybridization (CGH)*.

The protocol works as follows (**Fig. 14.39**). First, a series of 20,000 BAC clones with DNA inserts averaging 150 kb that collectively represent the entire human genome are spotted onto a microarray. These BAC clones were characterized in the course of the Human Genome Project. Next, genomic DNA from a control sample with a normal genome content is labeled with a yellow fluorescent dye, while the genomic DNA from the test sample is labeled with a red fluorescent dye. The two genomic DNA samples are mixed together in equal amounts, denatured, and applied to the microarray as a probe. After hybridization is complete and unhybridized material is washed away, the fluorescence emission from each microarray dot is analyzed automatically by a machine designed for this task.

If the genomic region probed with a particular BAC clone is present in two copies in the test sample, then the ratio of red to yellow dyes on that dot will be 1:1. However, if a particular genomic region is duplicated or deleted from

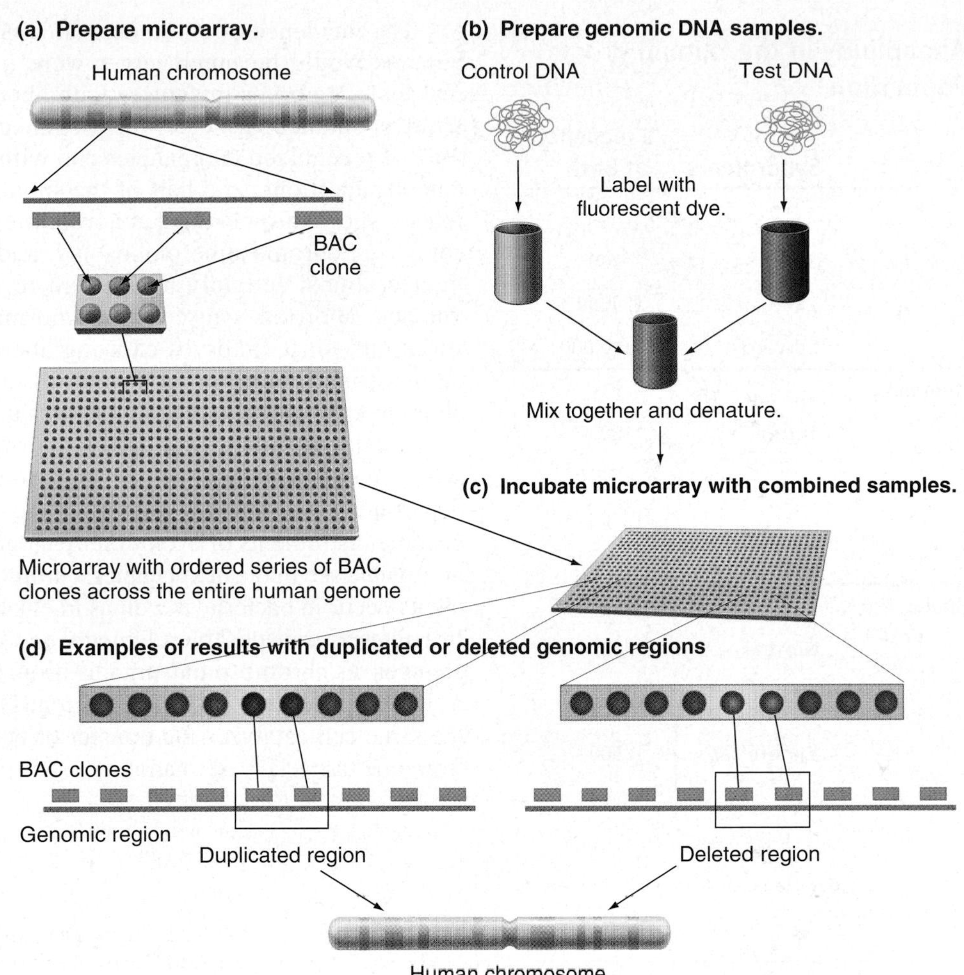

Figure 14.39 Comparative Genomic Hybridization detects duplications, deletions, and aneuploidy. **(a)** BAC clones representing the human genome are spotted in order onto a microarray. **(b)** The genomic sample to be tested is labeled with one color dye (here, *red*), and the control genome sample is labeled with a second color dye (*yellow*). **(c)** The two samples are mixed together, denatured, and then incubated on the microarray. **(d)** Automated analysis of each spot on the microarray detects the ratio of the two dyed probes that hybridize. *Orange* indicates a 1:1 ratio; other colors indicate deletion (0.5 : 1 ratio; *yellow*) or duplication (1.5 : 1 ratio; *red*) of BAC clone sequences in the test sample.

one homolog in the test sample, the ratio of red to yellow will be 1.5 : 1 or 0.5 : 1, respectively. An example of this analysis is shown in Fig. 14.39.

CGH provides a powerful clinical tool to detect any type of aneuploidy or any deletion or duplication of 50 kb or more anywhere in the genome. Clinicians can use it in conjunction with amniocentesis or preimplantation genetic analysis. They can also use CGH to screen tissue biopsies for cancerous cells that have deleted or duplicated regions containing oncogenes or tumor suppressor genes. The technique thus holds great promise for the detection of new genes that contribute to the genesis of cancer.

Connections

The detrimental consequences of most changes in chromosome organization and number cause considerable distress in humans (**Table 14.2**). Approximately 4 of every 1000 individuals has an abnormal phenotype associated with aberrant chromosome organization or number. Most of these abnormalities result from either aneuploidy for the X chromosome or trisomy 21. By comparison, about 10 people per 1000 suffer from an inherited disease caused by a single-gene mutation.

TABLE 14.2 Aneuploidy in the Human Population

Chromosomes	Syndrome	Frequency at Birth
Autosomes		
Trisomic 21	Down	1/700
Trisomic 13	Patau	1/5000
Trisomic 18	Edwards	1/10,000
Sex chromosomes, females		
XO, monosomic	Turner	1/5000
XXX, trisomic		1/700
XXXX, tetrasomic		
XXXXX, pentasomic		
Sex chromosomes, males		
XYY, trisomic	Normal	1/10,000
XXYY, tetrasomic	Klinefelter	1/500
XXXY, tetrasomic		
XXXXY, pentasomic		
XXXXXY, hexasomic		

About 0.4% of all babies born have a detectable chromosomal abnormality that generates a detrimental phenotype.

The incidence of chromosomal abnormalities among humans would be much larger were it not for the fact that many embryos or fetuses with abnormal karyotypes abort spontaneously early in pregnancy. Fully 15% to 20% of recognized pregnancies end with detectable spontaneous abortions; and half of the spontaneously aborted fetuses show chromosomal abnormalities, particularly trisomy, sex chromosome monosomy, and triploidy. These figures almost certainly underestimate the rate of spontaneous abortion caused by abnormal chromosomal variations, since embryos carrying aberrations for larger chromosomes, such as monosomy 2 or trisomy 5, may abort so early that the pregnancy goes unrecognized.

But despite all the negative effects of chromosomal rearrangements and changes in chromosome number, a few departures from normal genome organization survive to become instruments of evolution by natural selection.

As we see in the next chapter, chromosomal rearrangements occur in bacteria as well as in eukaryotic organisms. In bacteria, transposable elements catalyze many of the changes in chromosomal organization. Remarkably, the reshuffling of genes between different DNA molecules in the same cell catalyzes the transfer of genetic information from one bacterial cell to another.

Essential Concepts

1. *Rearrangements* reorganize the DNA sequences within genomes. Like point mutations, rearrangements are subject to natural selection and thus serve as instruments of evolution.
   a. *Deletions* remove DNA from a chromosome. Homozygosity for a large deletion is usually lethal, but even heterozygosity for a large deletion can create a deleterious *genetic imbalance.* Deletions may uncover recessive mutations on the homologous chromosome and are thus useful for gene mapping at the cytological and molecular levels.
   b. *Duplications* add DNA to a chromosome. The additional copies of genes contained in a duplication are a major source of new genetic functions. Duplications sometimes produce changes in phenotype that allow their detection. Homozygosity or heterozygosity for duplications causes departures from normal gene dosage that are often harmful to the organism. *Unequal crossing-over* between duplicated regions expands or contracts the copy number of genes and may lead to multigene families.
   c. *Inversions* alter the order, but not the number, of genes on a chromosome. They may produce novel phenotypes by modifying the activity of genes near the rearrangement breakpoints or by disrupting genes at the breakpoints. Inversion heterozygotes act as *crossover suppressors* because progeny formed from recombinant gametes are genetically imbalanced.
   d. In *reciprocal translocations,* parts of two chromosomes trade places without the loss or gain of chromosomal material. By relocating pieces of DNA, translocations may modify the function of genes at or near the translocation breakpoints. Heterozygosity for translocations in the germ line results in *semisterility* and *pseudolinkage.*
   e. *Transposable elements (TEs)* are short, mobile segments of DNA that reshape genomes by generating mutations, causing chromosomal rearrangements, and relocating genes.

2. Changes in chromosome number also reshape genomes.
   a. *Aneuploidy,* the loss or gain of one or more chromosomes, creates a genetic imbalance. Aneuploidy can result from mistakes in meiosis, which produce aneuploid gametes, or from mistakes in mitosis, which generate aneuploid clones of cells. Autosomal aneuploidy is usually lethal to the organism. Aneuploidy for sex chromosomes is better tolerated because of dosage compensation mechanisms; nevertheless, it can have significant consequences.
   b. *Euploid* organisms contain complete sets of chromosomes. Organisms with three or more sets of chromosomes are *polyploids. Autopolyploids* derive all their chromosome sets from the same species; *allopolyploids* are hybrids in which the chromosome sets come from two or more distinct, though related, species.
   c. Organisms containing odd numbers of complete chromosome sets are sterile because the chromosomes cannot pair properly during the first meiotic division.
   d. Polyploids with even numbers of chromosome sets can be fertile if proper chromosome segregation occurs sufficiently frequently. *Amphidiploids,* which are allopolyploids produced by chromosome doubling of two genomes derived from different diploid parental species, are often fertile and are sometimes useful in agriculture.

On Our Website

www.mhhe.com/hartwell3
Chapter 14

Annotated Suggested Readings and Links to Other Websites

- Historical articles describing early investigations on chromosomal rearrangements, transposable elements, and variations in chromosome number
- Recent reviews and research articles on these topics, with special emphasis on the use of transposable elements as tools for molecular genetic analysis
- A database cataloging human chromosomal abnormalities that have been characterized by fluorescence *in situ* hybridization (FISH) and comparative genomic hybridization (CGH)
- Online maps comparing the organization of human chromosomes with those from mice and rats

Social and Ethical Issues

1. Until recently, female athletes were gender tested to confirm they were female, using the Barr body test. In 1985, the Spanish hurdler Maria Patinez failed the Barr body test. She was encouraged by her coaches to fake an injury, quietly withdraw from the race, and retire from competition. Maria, indignant that someone suggested she was not a woman, spoke out against the testing policy. What are the problems scientifically with the Barr body test for gender identification? What do you think is a reasonable, accurate way to ascertain the sex of an individual? How should sex be defined? By karyotype? By examination of external genitalia? By internal gonadal structure?

2. Gary is a livestock farmer and has 2000 head of cattle. An outbreak of bacterial infection would be devastating to his livelihood, so he uses feed that contains antibiotics. Recently he heard a report indicating that the use of antibiotics in livestock food was contributing to the increase in antibiotic resistance in bacteria that infect humans. (The antibiotic resistance genes are contained on transposable elements that can be transferred from one bacterium to another.) Being a conscientious citizen, Gary is now considering eliminating the use of antibiotic feed. Will discontinuing the use of this feed make enough of a difference to justify the risk he takes? What should he do? What are the stakes for him and for the public in this issue?

3. Jason and Sylvia received some unsettling news from the amniocentesis of their fetus. Karyotype analysis showed a partial trisomy in which one chromosome 8 homolog carried a duplication of a particular region. However, this specific partial trisomy had not been reported before, and its observation led to considerable uncertainty because some partial trisomies lead to developmental abnormalities, while others do not. The parents decided to continue the pregnancy. Their

doctor alerted other medical colleagues who now would like to follow the progress of this child after birth in order to understand the consequences of this chromosomal rearrangement. Jason and Sylvia would prefer not to have their child subjected to any unnecessary medical scrutiny and are upset that their doctor gave information without their permission. Should the doctor have obtained permission before sharing the details with colleagues? Do Jason and Sylvia have a responsibility to provide information on the medical progress of their child so others in the future will know what to expect if the same duplication occurs?

Solved Problems

I. Male *Drosophila* from a true-breeding wild-type stock were irradiated with X-rays and then mated with females from a true-breeding stock carrying the following recessive mutations on the X chromosome: yellow body (*y*), crossveinless wings (*cv*), cut wings (*ct*), singed bristles (*sn*), and miniature wings (*m*). These markers are known to map in the order:

y - cv - ct - sn - m

Most of the female progeny of this cross were phenotypically wild type, but one female exhibited *ct* and *sn* phenotypes. When this exceptional *ct sn* female was mated with a male from the true-breeding wild-type stock, there were twice as many females as males among the progeny.

a. What is the nature of the X-ray-induced mutation present in the exceptional female?
b. Draw the X chromosomes present in the exceptional *ct sn* female as they would appear during pairing in meiosis.
c. What phenotypic classes would you expect to see among the progeny produced by mating the exceptional *ct sn* female with a normal male from a true-breeding wild-type stock? List males and females separately.

Answer

To answer this problem, you need to think first about the effects of different types of chromosomal mutations in order to deduce the nature of the mutation. Then you can evaluate the consequences of the mutation on inheritance.

a. Two observations indicate that *X-rays induced a deletion mutation.* The fact that two recessive mutations are phenotypically expressed in the exceptional female suggests that a deletion was present on one of her X chromosomes that uncovered the two mutant alleles (*ct* and *sn*) on the other X chromosome. Second, the finding that there were twice as many females as males among the progeny of the exceptional female is also consistent with a deletion mutation. Males who inherit the deletion-bearing X chromosome from their exceptional mother will be inviable (because other essential genes are located in the region that is now deleted), but sons who inherit a nondeleted X chromosome will survive. On the other hand, all of the exceptional female's daughters will be viable: Even if they inherit a deleted X chromosome from their mother, they also receive a normal X chromosome from their father. As a result, there are half as many male progeny as females from the cross of the exceptional female with a wild-type male.

b. During pairing, *the DNA in the normal (nondeleted) X chromosome will loop out* because there is no homologous region in the deletion chromosome. In the simplified drawing of meiosis I that follows, each line represents both chromatids comprising each homolog.

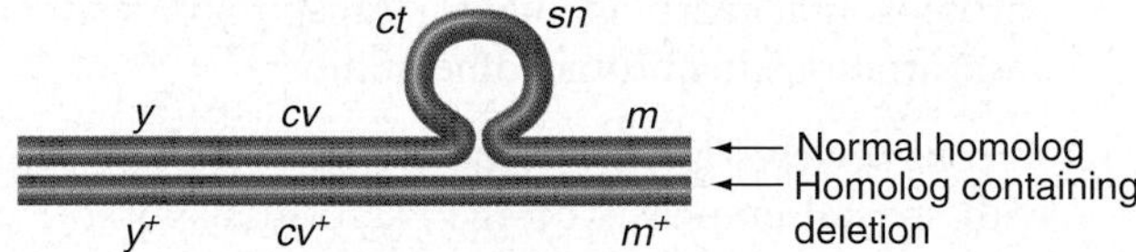

c. *All daughters of the exceptional female will be wild type* since the father contributes wild-type copies of all the genes. Each of the surviving sons must inherit a nondeleted X chromosome from the exceptional female. Some of these X chromosomes are produced from meioses in which no recombination occurred, but other X chromosomes are the products of recombination. *Males can have any of the genotypes listed here and therefore the corresponding phenotypes.* All contain the *ct sn* combination because no recombination between homologs is possible in this deleted region. Some of these genotypes require multiple crossovers during meiosis in the mother and will thus be relatively rare.

y	*cv*	*ct*	*sn*	*m*
+	+	*ct*	*sn*	+
+	*cv*	*ct*	*sn*	*m*
y	+	*ct*	*sn*	+
y	*cv*	*ct*	*sn*	+
+	+	*ct*	*sn*	*m*
+	*cv*	*ct*	*sn*	+
y	+	*ct*	*sn*	*m*

II. One of the X chromosomes in a particular *Drosophila* female had a normal order of genes but carried recessive alleles of the genes for yellow body color (y), vermilion eye color (v), and forked bristles (f), as well as the dominant X-linked Bar eye mutation (B). Her other X chromosome carried the wild-type alleles of all four genes, but the region including y^+, v^+, and f^+ (but not B^+) was inverted with respect to the normal order of genes. This female was crossed to a wild-type male in the cross diagrammed here.

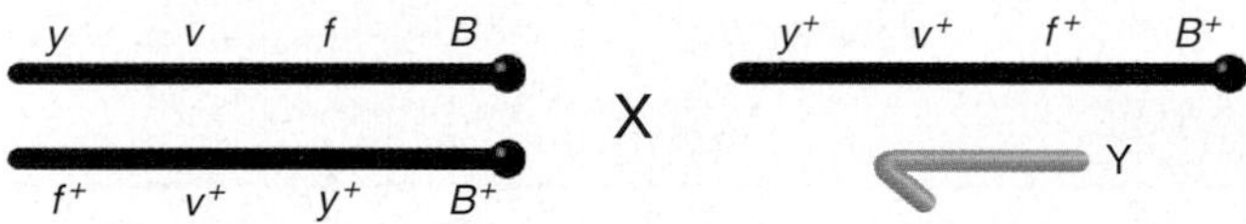

The cross produced the following male offspring:

y	v	f	B	48
y^+	v^+	f^+	B^+	45
y	v	f	B^+	11
y^+	v^+	f^+	B	8
y	v^+	f	B	1
y^+	v	f^+	B^+	1

a. Why are there no male offspring with the allele combinations $y\ v\ f^+$, $y^+\ v^+\ f$, $y\ v^+\ f^+$, or $y^+\ v\ f$, (regardless of the allele of the Bar eye gene)?

b. What kinds of crossovers produced the $y\ v\ f\ B^+$ and $v^+\ y^+\ f^+\ B$ offspring? Can you determine any genetic distances from these classes of progeny?

c. What kinds of crossovers produced the $y^+\ v\ f^+\ B^+$ and $y\ v^+\ f\ B$ offspring?

Answer

To answer this question, you need to be able to draw and interpret pairing in inversion heterozygotes. Note that this inversion is paracentric.

a. During meiosis in an inversion heterozygote, a loop of the inverted region is formed when the homologous genes align. In the following simplified drawing, each line represents both chromatids comprising each homolog.

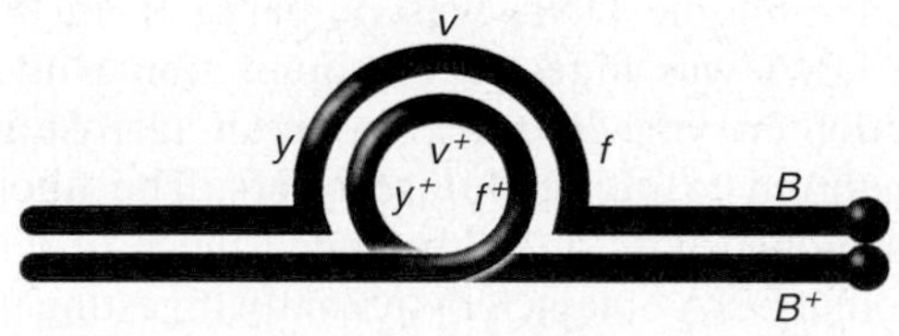

If a single crossover occurs within the inversion loop, a dicentric and an acentric chromosome are formed. Cells containing these types of chromosomes are not viable. The resulting allele combinations from such single crossovers are not recovered. The four phenotypic classes of missing male offspring would be formed by single crossovers between the y and v or between the v and f genes in the female inversion heterozygote and therefore are not recovered.

b. *The* y v f B$^+$ *and* y$^+$ v$^+$ f$^+$ B *offspring are the result of single crossover events outside of the inversion loop, between the end of the inversion (just to the right of* f *on the preceding diagram) and the* B *gene.* This region is approximately 16.7 m.u. in length (19 recombinants out of 114 total progeny).

c. *The* y$^+$ v f$^+$ B$^+$ *and* y v$^+$ f B *offspring would result from two crossover events within the inversion loop, one between the* y *and* v *genes and the other between the* v *and* f *genes.* You should note that these could be either two-strand or three-strand double crossovers, but they could not be four-strand double crossovers.

III. In maize trisomics, $n + 1$ pollen is not viable. If a dominant allele at the B locus produces purple color instead of the recessive phenotype bronze and a $B\ b\ b$ trisomic plant is pollinated by a $B\ B\ b$ plant, what proportion of the progeny produced will be trisomic and have a bronze phenotype?

Answer

To solve this problem, think about what is needed to produce trisomic bronze progeny: three b chromosomes in the zygote. The female parent would have to contribute two b alleles, since the $n + 1$ pollen from the male is not viable. What kinds of gametes could be generated by the trisomic $B\ b\ b$ purple female parent, and in what proportion? To track all the possibilities, rewrite this genotype as $B\ b_1\ b_2$, even though b_1 and b_2 have identical effects on phenotype. In the trisomic female, there are three possible ways the chromosomes carrying these alleles could pair as bivalents during the first meiotic division so that they would segregate to opposite poles: B with b_1, B with b_2, and b_1 with b_2. In all three cases, the remaining chromosome could move to either pole. To tabulate the possibilities,

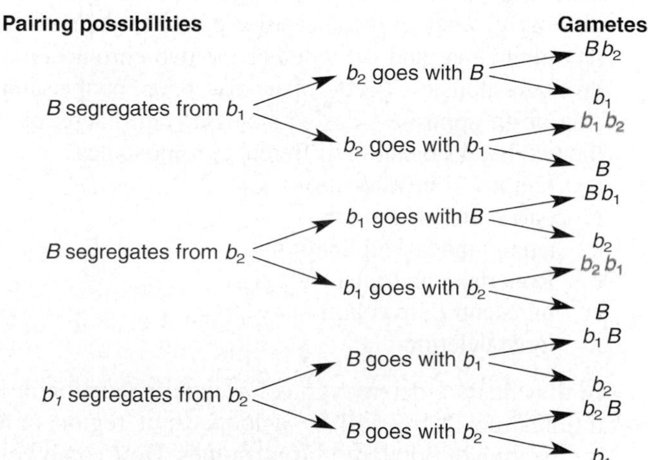

Of the 12 gamete classes produced by these different possible segregations, only the 2 classes written in red contain the two *b* alleles needed to generate the bronze (*b b b*) trisomic zygotes. There is thus a 2/12 = 1/6 chance of obtaining such gametes.

Although segregation in the *B B b* male parent is equally complicated, remember that males cannot produce viable $n + 1$ pollen. The only surviving gametes would thus be *B* and *b,* in a ratio (2/3 *B* and 1/3 *b*) that must reflect their relative prevalence in the male parent genome. *The probability of obtaining trisomic bronze progeny from this cross is therefore the product of the individual probabilities of the appropriate* b b *gametes from the female parent (1/6) and* b *pollen from the male parent (1/3): 1/6 × 1/3 = 1/18.*

Problems

Vocabulary

1. For each of the terms in the left column, choose the best matching phrase in the right column.

a. reciprocal translocation	1. lacking one or more chromosomes or having one or more extra chromosomes
b. gynandromorph	2. movement of short DNA elements
c. pericentric	3. having more than two complete sets of chromosomes
d. paracentric	4. exact exchange of parts of two nonhomologous chromosomes
e. euploids	5. excluding the centromere
f. polyploidy	6. including the centromere
g. transposition	7. having complete sets of chromosomes
h. aneuploids	8. mosaic combination of male and female tissue

Section 14.1

2. For each of the following types of chromosomal aberrations, tell: (i) whether an organism heterozygous for the aberration will form any type of loop in the chromosomes during prophase I of meiosis; (ii) whether a chromosomal bridge can be formed during anaphase I in a heterozygote, and if so, under what condition; (iii) whether an acentric fragment can be formed during anaphase I in a heterozygote, and if so, under what condition; (iv) whether the aberration can suppress meiotic recombination; and (v) whether the two chromosomal breaks responsible for the aberration occur on the same side or on opposite sides of a single centromere, or if the two breaks occur on different chromosomes.

a. reciprocal translocation
b. paracentric inversion
c. small tandem duplication
d. Robertsonian translocation
e. paracentric inversion
f. large deletion

3. In flies that are heterozygous for either a deletion or a duplication, there will be a looped-out region in a preparation of polytene chromosomes. How could you distinguish between a deletion or a duplication using polytene chromosome analysis?

4. For the following types of chromosomal rearrangements, would it theoretically ever be possible to obtain a perfect reversion of the rearrangement? If so, would such revertants be found only rarely, or would they be relatively common?

a. a deletion of a region including five genes
b. a tandem duplication of a region including five genes
c. a pericentric inversion
d. a Robertsonian translocation
e. a mutation caused by a transposable element jumping into a protein-coding exon of a gene

5. Four strains of *Drosophila* were constructed in which one autosome contained recessive mutant alleles of the four genes *rolled eyes, thick legs, straw bristles,* and *apterous wings,* and the homologous autosome contained one of four different deletions (deletions 1–4). The phenotypes of the flies were as follows:

Deletion	Phenotype
1	rolled eyes, straw bristles
2	apterous wings, rolled eyes
3	thick legs, straw bristles
4	apterous wings

Whole-genome DNA was prepared from the flies. The DNA was digested to completion with the restriction enzyme *Bam*HI, run on an agarose gel, and transferred to nitrocellulose filters. The filters were then probed with a 20 kb cloned piece of wild-type genomic DNA obtained by partially digesting the plasmid clone with *Bam*HI (so the ends of the probe were *Bam*HI ends, but the piece was not digested into all the possible *Bam*HI fragments). The results of this whole-genome Southern blot are shown at the top of the next page. Dark bands indicate fragments present twice in the diploid genome; light bands indicate fragments present once in the genome.

1	2	3	4
6.3	6.3	6.3	6.3
5.6	5.6	5.6	5.6
5.1			
			4.9
4.2	4.2	4.2	4.2
3.8			
	3.3		
3.0	3.0	3.0	3.0
		1.7	
0.9	0.9	0.9	0.9

Deletion 1 2 3 4

a. Make a map of the *Bam*HI restriction sites in this 20 kb part of the wild-type *Drosophila* genome, indicating distances in kilobases between adjacent *Bam*HI sites. (Hint: The genomic DNA fragments in wild type are 6.3, 5.6, 4.2, 3.0, and 0.9 kb long.)

b. On your map, indicate the locations of the genes.

6. A diploid strain of yeast was made by mating a haploid strain with a genotype w^-, x^-, y^-, and z^- with a haploid strain of opposite mating type that is wild type for these four genes. The diploid strain was phenotypically wild type. Four different X-ray-induced diploid mutants with the following phenotypes were produced from this diploid yeast strain. Assume there is a single new mutation in each strain.

Strain 1	w^-	x^+	y^-	z^+
Strain 2	w^+	x^-	y^-	z^-
Strain 3	w^-	x^+	y^-	z^-
Strain 4	w^-	x^+	y^+	y^+

When these mutant diploid strains of yeast go through meiosis, each ascus is found to contain only two viable haploid spores.

a. What kind of mutations were induced by X-rays to make the listed diploid strains?

b. Why did two spores in each ascus die?

c. Are any of the genes *w, x, y,* or *z* located on the same chromosome?

d. Give the order of the genes that are found on the same chromosome.

7. Human chromosome 1 is a large, metacentric chromosome. A map of a cloned region from near the telomere of chromosome 1 is shown at top of right column. Three probe DNAs (A, B, and C) from this region were used for *in situ* hybridization to human mitotic metaphase chromosome squashes made with cells obtained from individuals with various genotypes. The breakpoints of chromosomal rearrangements in this region are indicated on the map. The black bars for deletions (*Del*) 1 and 2 represent DNA that is deleted. The breakpoints of inversions (*Inv*) 1 and 2 not shown in the figure are near but not at the centromere. For each of the following genotypes, draw chromosome 1 as it would appear after *in situ* hybridization. An example is shown in the following figure for hybridization of probe A to the two copies of chromosome 1 in wild type(+/+).

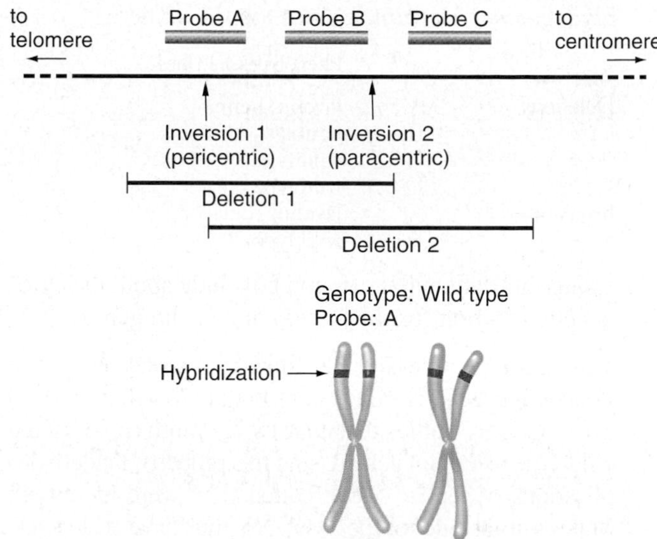

a. genotype: *Del1/Del2;* probe: B

b. genotype: *Del1/Del2;* probe: C

c. genotype: *Del1/+;* probe: A

d. genotype: *Inv1/+;* probe: A

e. genotype: *Inv2/+;* probe: B

f. genotype: *Inv2/Inv2;* probe: C

8. A series of chromosomal mutations in *Drosophila* were used to map the *javelin* gene, which affects bristle shape, and *henna,* which affects eye pigmentation. Both the *javelin* and *henna* mutations are recessive. A diagram of region 65 of the *Drosophila* polytene chromosomes is shown here.

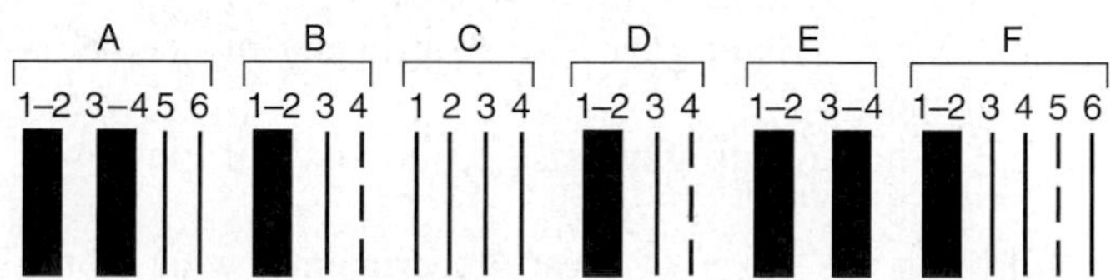

The chromosomal breakpoints for six chromosome rearrangements are indicated in the following table. (For example, deletion A has one breakpoint between bands A2 and A3 and the other between bands D2 and D3.)

		Breakpoints in region 65	
Deletion	A	A2–3;	D2–3
	B	C2–3;	E4–F1
	C	D2–3;	F4–5
	D	D4–E1;	F3–4
		Breakpoints	
Inversions	A	Band 65A6	Band 82A1
	B	Band 65B4	Band 98A3

Flies with a chromosome containing one of these six rearrangements (deletions or inversions) were mated to flies homozygous for both *javelin* and *henna.* The phenotypes of the heterozygous progeny (that is, *rearrangement / javelin, henna*) are shown here.

		Phenotypes of F_1 flies
Deletions	A	javelin, henna
	B	henna
	C	wild type
	D	wild type
Inversions	A	javelin
	B	wild type

Using this data, what can you conclude about the cytogenetic location for the *javelin* and *henna* genes?

9. The partially recessive, X-linked z^1 mutation of the *Drosophila* gene *zeste* (z) can produce a yellow (zeste) eye color only in flies that have two or more copies of the wild-type *white* (w) gene. Using this property, tandem duplications of the w^+ gene called w^{+R} were identified. Males with the genotype $z^1\ w^{+R}/\text{Y}$ thus have zeste eyes. These males were crossed to females with the genotype $y\ z^1\ w^{+R} spl / y^+ z^1\ w^{+R} spl^+$. (These four genes are closely linked on the X chromosome, in the order given in the genotype, with the centromere to the right of all these genes: y = yellow bodies; y^+ = tan bodies; spl = split bristles; spl^+ = normal bristles.) Out of 81,540 male progeny of these females, the following exceptions were found:

Class A	2430 yellow bodies, zeste eyes, wild-type bristles
Class B	2394 tan bodies, zeste eyes, split bristles
Class C	23 yellow bodies, wild-type eyes, wild-type bristles
Class D	22 tan bodies, wild-type eyes, split bristles

a. What were the phenotypes of the remainder of the 81,540 males from the first cross?
b. What events gave rise to progeny of classes A and B?
c. What events gave rise to progeny of classes C and D?
d. On the basis of these experiments, what is the genetic distance between *y* and *spl*?

10. Genes *a* and *b* are 21 m.u. apart when mapped in highly inbred strain 1 of corn and 21 m.u. apart when mapped in highly inbred strain 2. But when the distance is mapped by testcrossing the F_1 progeny of a cross between strains 1 and 2, the two genes are only 1.5 m.u. apart. What arrangement of genes *a* and *b* and any potential rearrangement breakpoints could explain these results?

11. In the following group of figures, the *pink* lines indicate an area of a chromosome that is inverted relative to the normal (*black* line) order of genes. The diploid chromosome constitution of individuals 1–4 is shown. Match the individuals with the appropriate statement(s) that follow. More than one diagram may correspond to the following statements, and a diagram may be a correct answer for more than one question.

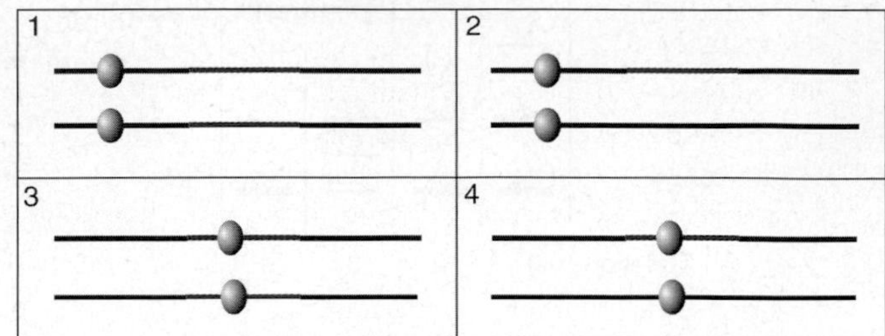

a. An inversion loop would form during meiosis I and in polytene chromosomes.
b. A single crossover involving the inverted region on one chromosome and the homologous region of the other chromosome would yield genetically imbalanced gametes.
c. A single crossover involving the inverted region on one chromosome and the homologous region of the other chromosome would yield an acentric fragment.
d. A single crossover involving the inverted region yields four viable gametes.

12. Three strains of *Drosophila* (Bravo, X-ray, and Zorro) are obtained that are homozygous for three variant forms of a particular chromosome. When examined in salivary gland polytene chromosome spreads, all chromosomes have the same number of bands in all three strains. When genetic mapping is performed in the Bravo strain, the following map is obtained (distances in map units).

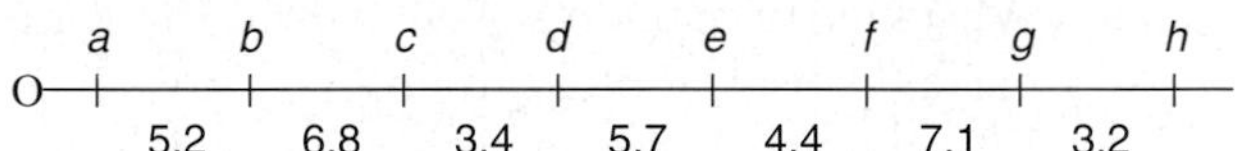

Bravo and X-ray flies are now mated to form Bravo/X-ray F_1 progeny, and Bravo flies are also mated with Zorro flies to form Bravo/Zorro F_1 progeny. In subsequent crosses, the following genetic distances were found to separate the various genes in the hybrids:

	Bravo / X-ray	Bravo / Zorro
a–b	5.2	5.2
b–c	6.8	0.7
c–d	0.2	<0.1
d–e	<0.1	<0.1
e–f	<0.1	<0.1
f–g	0.65	0.7
g–h	3.2	3.2

a. Make a map showing the relative order of genes *a* through *h* in the X-ray and Zorro strains. Do not show distances between genes.

b. In the original X-ray homozygotes, would the physical distance between genes *c* and *d* be greater than, less than, or approximately equal to the physical distance between these same genes in the original Bravo homozygotes?
c. In the original X-ray homozygotes, would the physical distance between genes *d* and *e* be greater than, less than, or approximately equal to the physical distance between these same genes in the original Bravo homozygotes?

13. Two yeast strains were mated and sporulated (allowed to carry out meiosis). One strain was a haploid with normal chromosomes and the linked genetic markers *ura3* (requires uracil for growth) and *arg9* (requires arginine for growth) surrounding their centromere. The other strain was wild type for the two markers (*URA3* and *ARG9*) but had an inversion in this region of the chromosome as shown here in *pink*:

During meiosis, several different kinds of crossover events could occur. For each of the following events, give the genotype and phenotype of the resulting four haploid spores. Assume that any chromosomal deficiencies are lethal in haploid yeast. Do not consider crossovers between sister chromatids.
a. a single crossover outside the inverted region
b. a single crossover between *URA3* and the centromere
c. a double crossover involving the same two chromatids each time, where one crossover occurs between *URA3* and the centromere and the other occurs between *ARG9* and the centromere

14. Suppose a haploid yeast strain carrying two recessive linked markers *his4* and *leu2* was crossed with a strain that was wild type for *HIS4* and *LEU2* but had an inversion of this region of the chromosome as shown here in *blue*.

Several different kinds of crossover events could occur during meiosis in the resulting diploid. For each of the following events, state the genotype and phenotype of the resulting four haploid spores. Do not consider crossover events between chromatids attached to the same centromere.
a. a single crossover between the markers *HIS4* and *LEU2*
b. a double crossover involving the same chromatids each time, where both crossovers occur between the markers *HIS4* and *LEU2*
c. a single crossover between the centromere and the beginning of the inverted region

15. In the mating between two haploid yeast strains depicted in Problem 14, describe a scenario that would result in a tetratype ascus in which all four spores are viable.

16. During ascus formation in *Neurospora*, any ascospore with a chromosomal deletion dies and appears white in color. How many ascospores of the eight spores in the ascus would be white if the octad came from a cross of a wild-type strain with a strain of the opposite mating type carrying
a. a paracentric inversion, and no crossovers occurred between normal and inverted chromosomes?
b. a pericentric inversion, and a single crossover occurred in the inversion loop?
c. a paracentric inversion, and a single crossover occurred outside the inversion loop?
d. a reciprocal translocation, and an adjacent-1 segregation occurred with no crossovers between translocated chromosomes?
e. a reciprocal translocation, and alternate segregation occurred with no crossovers between translocated chromosomes?
f. a reciprocal translocation, and alternate segregation occurred with one crossover between translocated chromosomes (but not between the translocation breakpoint and the centromere of any chromosome)?

17. In the following figure, *black* and *pink* lines represent nonhomologous chromosomes. Which of the figures matches the descriptions below? More than one diagram may correspond to the statements, and a diagram may be a correct answer for more than one question.

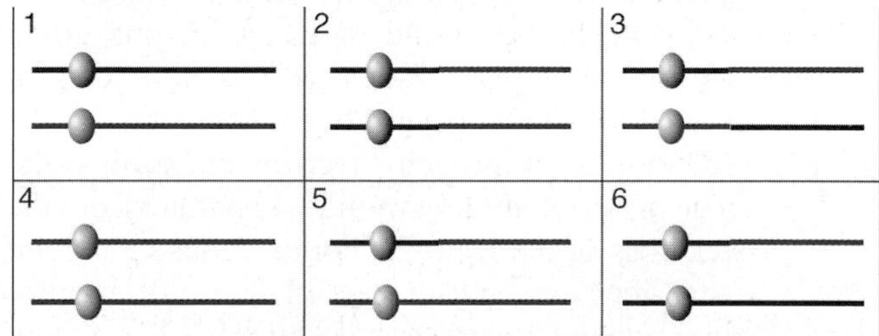

a. gametes produced by a translocation heterozygote
b. gametes that could not be produced by a translocation heterozygote
c. genetically balanced gametes produced by a translocation heterozygote
d. genetically imbalanced gametes that can be produced (at any frequency) by a translocation heterozygote

18. In *Drosophila*, the gene for cinnabar eye color is on chromosome 2, and the gene for scarlet eye color is on chromosome 3. A fly homozygous for both recessive *cinnabar* and *scarlet* alleles (*cn/cn; st/st*) is white-eyed.

a. If male flies (containing chromosomes with the normal gene order) heterozygous for *cn* and *st* alleles are crossed to white-eyed females homozygous for the *cn* and *st* alleles, what are the expected phenotypes and their frequencies in the progeny?

b. One unusual male heterozygous for *cn* and *st* alleles, when crossed to a white-eyed female, produced only wild-type and white-eyed progeny. Explain the likely chromosomal constitution of this male.

c. When the wild-type F_1 females from the cross with the unusual male were backcrossed to normal *cn/cn; st/st* males, the following results were obtained:

wild type	45%
cinnabar	5%
scarlet	5%
white	45%

Diagram a genetic event at metaphase I that could produce the rare cinnabar or scarlet flies among the progeny of the wild-type F_1 females.

19. Semisterility in corn, as seen by unfilled ears with gaps due to abortion of approximately half the ovules, is an indication that the strain is a translocation heterozygote. The chromosomes involved in the translocation can be identified by crossing the translocation heterozygote to a strain homozygous recessive for a gene on the chromosome being tested. The ratio of phenotypic classes produced from crossing semisterile F_1 progeny back to a homozygous recessive plant indicates whether the gene is on one of the chromosomes involved in the translocation. For example, a semisterile strain could be crossed to a strain homozygous for the *yg* mutation on chromosome 9. (The mutant has yellow-green leaves instead of the wild-type green leaves.) The semisterile F_1 progeny would then be backcrossed to the homozygous *yg* mutant.

a. What types of progeny (fertile or semisterile, green or yellow-green) would you predict from the backcross of the F_1 to the homozygous *yg* mutant if the gene was not on one of the two chromosomes involved in the translocation?

b. What types of progeny (fertile or semisterile, green or yellow-green) would you predict from the backcross of the F_1 to the homozygous mutant if the *yg* gene is on one of the two chromosomes involved in the translocation?

c. If the *yg* gene is located on one of the chromosomes involved in the translocation, a few fertile, green progeny and a few semisterile, yellow-green progeny are produced. How could these relatively rare progeny classes arise? What genetic distance could you determine from the frequency of these rare progeny?

20. A proposed biological method for insect control involves the release of insects that could interfere with the fertility of the normal resident insects. One approach is to introduce sterile males to compete with the resident fertile males for matings. A disadvantage of this strategy is that the irradiated sterile males are not very robust and can have problems competing with the fertile males. An alternate approach that is being tried is to release laboratory-reared insects that are homozygous for several translocations. Explain how this strategy will work. Be sure to mention which insects will be sterile.

21. A *Drosophila* male is heterozygous for a translocation between an autosome originally bearing the dominant mutation *Lyra* (shortened wings) and the Y chromosome; the other copy of the same autosome is $Lyra^+$, This male is now mated with a true-breeding, wild-type female. What kinds of progeny would be obtained, and in what proportions?

22. The picture at the beginning of this chapter on p. 489 shows a polytene chromosome preparation from a fruit fly heterozygous for a chromosome carrying a very large composite transposon and a wild-type homolog. Suppose you had a probe made from wild-type DNA sequences that span the site into which the transposon is inserted. Diagram the pattern of *in situ* hybridization you would expect on the polytene chromosome preparation shown on p. 489.

23. a. Among the selfed progeny of a semisterile corn plant that is heterozygous for a reciprocal translocation, what ratio do you expect for progeny plants with normal fertility versus those showing semisterility? In this problem, ignore the rare gametes produced by adjacent-2 segregation.

b. Among the selfed progeny of a particular semisterile corn plant heterozygous for a reciprocal translocation, the ratio of fertile to semisterile plants was 1:4. How can you explain this deviation from your answer to part a.?

24. Explain how transposable elements can cause movement of genes that are not part of the transposable element.

25. In the 1950s, Barbara McClintock found a transposable element in corn she called *Ds* (*Dissociator*). When inserted at a particular location, this element could often cause chromosomal breaks at that site, but these breaks occurred only in the presence of another unlinked genetic element she called *Ac* (*Activator*). She found further that in the presence of *Ac, Ds* could jump to other chromosomal locations. At some of these locations (and in the presence of *Ac*) *Ds* would now cause chromosomal breakage at the new position; at other positions, it appeared that *Ds* could cause new mutations that were unstable as shown by their patchy, variegated expression in kernels. Interestingly, the position of the *Ac* element seemed to be very different in various strains of corn. Explain these results in terms of our present-day understanding of transposons.

26. Gerasimova and colleagues in the former Soviet Union characterized a mutation in the *Drosophila* cut wing (*ct*) gene called ct^{MR2}, which is associated with the insertion of a transposable element called *gypsy*. This allele is very unstable: approximately 1 in 100 of the progeny of flies bearing ct^{MR2} show new *ct* variants. Some of these are ct^+ revertants, while others appear to be more severe alleles of *ct* with stronger effects on wing shape. When the ct^+ revertants themselves are mated, some of the ct^+ alleles appear to be stable (no new *ct* mutants appear), while others are highly unstable (many new mutations appear). What might explain the generation of stable and unstable ct^+ revertants as well as the stronger *ct* mutant alleles?

27. In sequencing a region of the human genome, you have come across a segment of about 200 A nucleotides. You suspect that the sequence preceding the A residues may have been moved here by a transposition event mediated by reverse transcriptase. If the adjacent sequence is in fact a retroposon, you might expect to find other copies in the genome. How could you determine if other copies of this DNA exist?

28. The *Eco*RI restriction map of the region in which a coat-color gene in mice is located is presented in the following. The left-most *Eco*RI site is arbitrarily labeled 0 and the other distances in kilobases are given relative to this coordinate. Genomic DNA was prepared from one wild-type mouse and 10 mice homozygous for various mutant alleles. This genomic DNA is digested with *Eco*RI, fractionated on agarose gels, and then transferred to nitrocellulose filters. The filters were probed with the radioactive DNA fragment indicated by the *purple* bar, extending from coordinate 2.6 kb to coordinate 14.5. The resultant autoradiogram is shown schematically.

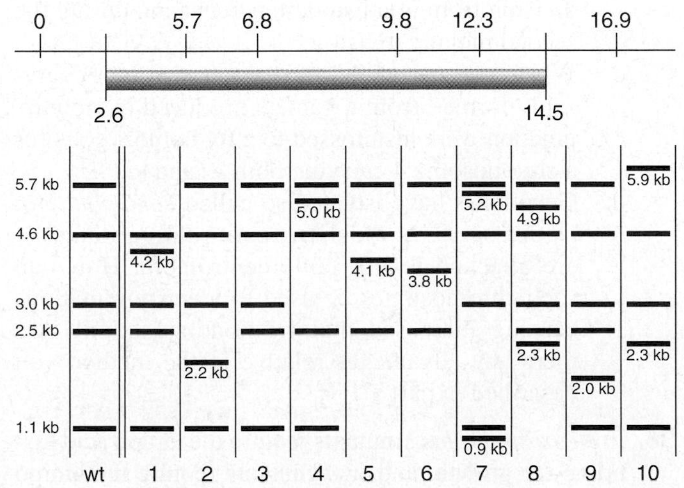

Assume that each of the mutations 1–10 is caused by one and only one of the events on the following list. Which event corresponds to which mutation?

a. a point mutation exactly at coordinate 6.8
b. a point mutation exactly at coordinate 6.9
c. a deletion between coordinates 10.1 and 10.4
d. a deletion between coordinates 6.7 and 7.0
e. insertion of a transposable element at coordinate 6.2
f. an inversion with breakpoints at coordinates 2.2 and 9.9
g. a reciprocal translocation with another chromosome with a breakpoint at coordinate 10.1
h. a reciprocal translocation with another chromosome with a breakpoint at coordinate 2.4
i. a tandem duplication of sequences between coordinates 7.2 and 9.2
j. a tandem duplication of sequences between coordinates 11.3 and 14.3

Section 14.2

29. The number of chromosomes in the somatic cells of several oat varieties (*Avena* species) are: sand oats (*Avena strigosa*)—14; slender wild oats (*Avena barata*)—28; and cultivated oats (*Avena sativa*)—42.

a. What is the basic chromosome number (x) in *Avena?*
b. What is the ploidy for each of the different species?
c. What is the number of chromosomes in the gametes produced by each of these oat varieties?
d. What is the n number of chromosomes in each species?

30. Common red clover, *Trifolium pratense,* is a diploid with 14 chromosomes per somatic cell. What would be the somatic chromosome number of

a. a trisomic variant of this species?
b. a monosomic variant of this species?
c. a triploid variant of this species?
d. an autotetraploid variant?

31. Somatic cells in organisms of a particular diploid plant species normally have 14 chromosomes. The chromosomes in the gametes are numbered from 1 through 7. Rarely, zygotes are formed that contain more or fewer than 14 chromosomes. For each of the zygotes below, (i) state whether the chromosome complement is euploid or aneuploid; (ii) provide terms that describe the individual's genetic makeup as accurately as possible; and (iii) state whether or not the individual will likely develop through the embryonic stages to make an adult plant, and if so, whether or not this plant will be fertile.

a. 11 22 33 44 5 66 77
b. 111 22 33 44 555 66 77
c. 111 222 333 444 555 666 777
d. 1111 2222 3333 4444 5555 6666 7777

32. Genomes A, B, and C all have basic chromosome numbers (x) of nine. These genomes were originally derived from plant species that had diverged from each other sufficiently far back in the evolutionary past that the chromosomes from one genome can no longer pair with the chromosomes from any other genome. For plants with the following kinds of euploid chromosome complements, (i) state the number of chromosomes in the organism; (ii) provide terms that describe the individual's genetic makeup as accurately as possible; (iii) state whether or not it is likely that this plant will be fertile, and if so, give the number of chromosomes (n) in the gametes.

a. AABBC
b. BBBB
c. CCC
d. BBCC
e. ABC
f. AABBCC

33. Fred and Mary have a child with Down syndrome. A probe derived from chromosome 21 was used to identify RFLPs in Fred, Mary, and the child (darker bands indicate signals of twice the intensity). Explain what kind of nondisjunction events must have occurred to produce the child if the child's RFLP pattern looked like that in lanes A, B, C, or D of the following figure:

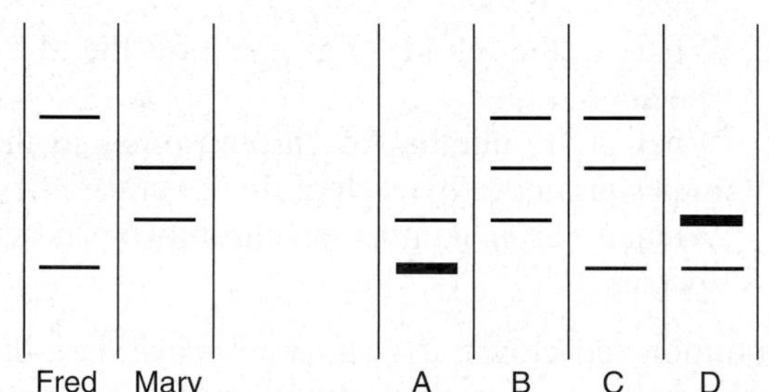

34. *Uniparental disomy* is a rare phenomenon in which only one of the parents of a child with a recessive disorder is a carrier for that trait; the other parent is homozygous normal. By analyzing DNA polymorphisms, it is clear that the child received both mutant alleles from the carrier parent but did not receive any copy of the gene from the other parent.

a. Diagram at least two ways in which uniparental disomy could arise. (Hint: These mechanisms all require more than one error in cell division, explaining why uniparental disomy is so rare.) Is there any way to distinguish between these mechanisms to explain any particular case of uniparental disomy?
b. How might the phenomenon of uniparental disomy explain rare cases in which girls are affected with rare X-linked recessive disorders but have unaffected fathers, or other cases in which an X-linked recessive disorder is transmitted from father to son?
c. If you were a human geneticist and believed one of your patients had a disease syndrome caused by uniparental disomy, how could you establish that the cause was not instead mitotic recombination early in the patient's development from a zygote?

35. Human geneticists interested in the effects of abnormalities in chromosome number often karyotype tissue obtained from spontaneous abortions. About 35% of these samples show autosomal trisomies, but only about 3% of the samples display autosomal monosomies. Based on the kinds of errors that can give rise to aneuploidy, would you expect that the frequencies of autosomal trisomy and autosomal monosomy should be more equal? Why or why not? If you think the frequencies should be more equal, how can you explain the large excess of trisomies as opposed to monosomies?

36. Among adults with Turner syndrome, it has been found that a very high proportion are genetic mosaics. These are of two types: In some individuals, the majority of cells are 45, XO, but a minority of cells are 46, XX. In other Turner individuals, the majority of cells are 45, XO, but a minority of cells are 46, XY. Explain how these somatic mosaics could arise.

37. The *Drosophila* chromosome 4 is extremely small; there is virtually no recombination between genes on this chromosome. You have available three differently marked chromosome 4's: one has a recessive allele of the gene *eyeless* (*ey*), causing very small eyes; one has a recessive allele of the *cubitus interruptus* (*ci*) gene, which causes disruptions in the veins on the wings; and the third carries the recessive alleles of both genes. *Drosophila* adults can survive with two or three, but not with one or four, copies of chromosome 4.

a. How could you use these three chromosomes to find *Drosophila* mutants with defective meioses causing an elevated rate of nondisjunction?
b. Would your technique allow you to discriminate nondisjunction occurring during the first meiotic division from nondisjunction occurring during the second meiotic division?
c. What progeny would you expect if a fly recognizably formed from a gamete produced by nondisjunction were testcrossed to a fly homozygous for a chromosome 4 carrying both *ey* and *ci?*
d. Geneticists have isolated so-called *compound 4th chromosomes* in which two entire chromosome 4's are attached to the same centromere. How can such chromosomes be used to identify mutations causing increased meiotic nondisjunction? Are there any advantages relative to the method you described in part a.?

38. In *Neurospora, his2* mutants require the amino acid histidine for growth, and *lys4* mutants require the amino acid lysine. The two genes are on the same arm of the same chromosome, in the order

centromere - *his2* - *lys4*.

A *his2* mutant is mated with a *lys4* mutant. Draw all of the possible ordered asci that could result from meioses in which the following events occurred, accounting for the nutritional requirements for each ascospore. Ascospores without any copy of a chromosome will abort and die, turning white in the process.

a. a single crossover between the centromere and *his2*
b. a single crossover between *his2* and *lys4*
c. nondisjunction during the first meiotic division
d. nondisjunction during the second meiotic division
e. a single crossover between the centromere and *his2*, followed by nondisjunction during the first meiotic division
f. a single crossover between *his2* and *lys4*, followed by nondisjunction during the first meiotic division

39. You have haploid tobacco cells in culture and have made transgenic cells that are resistant to herbicide. What would you do to obtain a diploid cell line that could be used to generate a new fertile herbicide-resistant plant?

40. An allotetraploid species has a genome composed of two ancestral genomes, A and B, each of which have a basic chromosome number (x) of seven. In this species, the two copies of each chromosome of each ancestral genome pair only with each other during meiosis. Resistance to a pathogen that attacks the foliage of the plant is controlled by a dominant allele at the F locus. The recessive alleles F^a and F^b confer sensitivity to the pathogen, but the dominant resistance alleles present in the two genomes have slightly different effects. Plants with at least one F^A allele are resistant to races 1 and 2 of the pathogen regardless of the genotype in the B genome, and plants with at least one F^B allele are resistant to races 1 and 3 of the pathogen regardless of the genotype in the A genome. What proportion of the self-progeny of an $F^A\ F^a\ F^B\ F^b$ plant will be resistant to all three races of the pathogen?

41. Using karyotype analysis, how could you distinguish between autopolyploids and allopolyploids?

42. Chromosomes normally associate during meiosis I as bivalents (a pair of synapsed homologous chromosomes) because chromosome pairing involves the synapsis of the corresponding regions of two homologous chromosomes. However, Fig. 14.21b on p. 506 shows that in a heterozygote for a reciprocal translocation, chromosomes pair as quadrivalents (that is, four chromosomes are associated with each other). Quadrivalents can form in other ways: For example, in some autotetraploid species, chromosomes can pair as quadrivalents rather than as bivalents.

a. How could quadrivalents actually form in these autotetraploids, given that chromosomal regions synapse in pairs? To answer this question, diagram such a quadrivalent.
b. How can these autotetraploid species generate euploid gametes if the chromosomes pair as quadrivalents rather than bivalents?
c. Could quadrivalents form in an amphidiploid species? Discuss.

The Prokaryotic Chromosome: Genetic Analysis in Bacteria

Chapter 15

Some species of bacteria can live in environments as hostile as hot springs (such as this beautiful pool in Yellowstone National Park, Wyoming). Comparative genome analyses of bacteria that live in unusual environments will increase our understanding of the adaptations that allow survival in different niches.

Gonorrhea, a sexually transmitted infection of the urogenital tract in men and women, is on the rise in many parts of the world. Caused by the bacterium *Neisseria gonorrhoeae,* gonorrhea is rarely fatal, but in men it can spread from an initial site, usually the urethra, to the prostate gland and the epididymis (two structures that play a role in sperm production), diminishing sperm count; and in women, it can move from the cervix to the uterine lining and fallopian tubes, leading to sterility. A common symptom is painful urination, although many infected people have no symptoms at all. Because infants passing through a gonorrhea-infected birth canal can contract severe eye infections, hospitals routinely treat the eyes of newborns with a few drops of silver nitrate solution or penicillin. If left untreated, gonorrhea can spread through the bloodstream to the joints, skin, bones, tendons, and heart. Until the late 1970s, a few shots of penicillin were a surefire cure for gonorrhea, but by 1995, more than 20% of *N. gonorrhoeae* bacteria isolated from patients worldwide were resistant to penicillin.

Geneticists now know that the agent of this alarming increase in antibiotic resistance was the transfer of DNA from one bacterium to another. According to epidemiologists, penicillin-resistant *N. gonorrhoeae* bacteria first appeared in Asia in the 1980s, in a patient receiving penicillin treatment for gonorrhea who was also fighting an infection caused by another species of bacteria—*Haemophilus influenzae.* Some of the patient's *H. influenzae* bacteria apparently carried a plasmid, a small, circular molecule of double-stranded DNA (see Chapter 9), that contained a gene encoding penicillinase, an enzyme that destroys penicillin. When the doubly infected patient mounted a specific immune response to *H. influenzae* that degraded these cells, the broken bacteria released their plasmids. Some of the freed circles of DNA entered *N. gonorrhoeae* cells, transforming them to penicillin-resistant bacteria. The transformed gonorrhea bacteria then multiplied, and successive exposures to penicillin selected for the resistant bacteria. As a result, the patient transmitted penicillin-resistant *N. gonorrhoeae* to subsequent sexual partners. Thus, while penicillin treatment does not create the genes for resistance, it accelerates the spread of those genes. After continuous, long-term exposure to the drug, only cells resistant to penicillin survive. Today in the United States, many *N. gonorrhoeae* are simultaneously resistant to penicillin and two other antibiotics—spectinomycin and tetracycline.

In this chapter, we survey the remarkable diversity of bacteria and discuss the structure of bacterial genomes. We also examine the mechanisms by which bacteria transfer genes between cells of the same species, between cells of distantly related species, and between bacterial cells and bacterial viruses. Geneticists can use their knowledge of bacterial genetics in general and of various forms of bacterial gene transfer in particular to identify, map, and characterize the genes that contribute to biological processes. Such genetic analyses have made it possible to dissect and reconstruct, for example, the biochemical pathways by which bacteria move toward food and away from toxic substances.

(a)

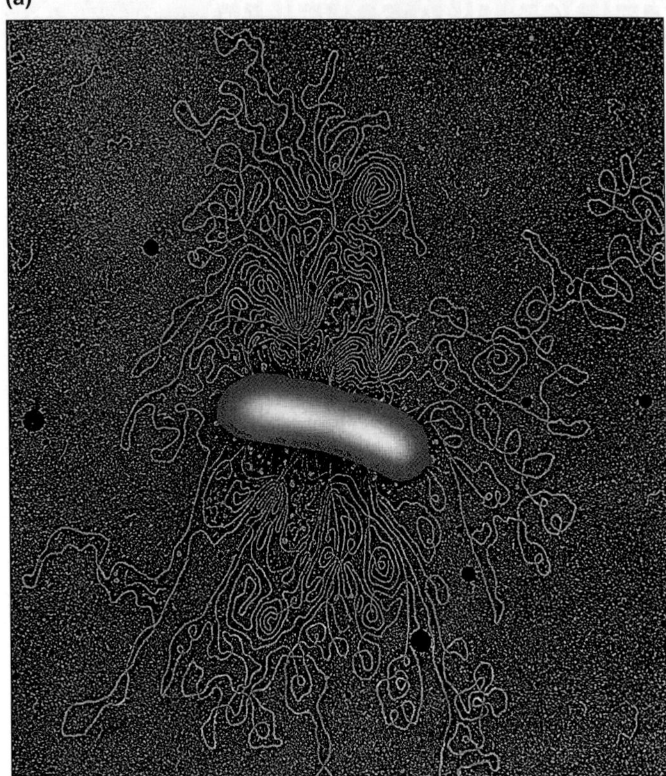

(b)

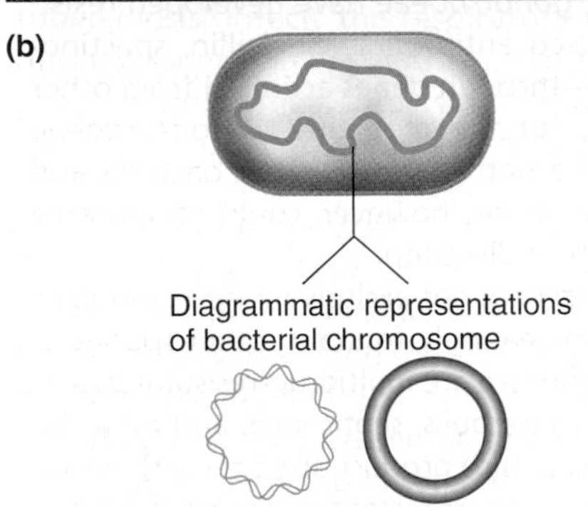

Figure 15.4 The bacterial chromosome: A circle of double-stranded DNA. **(a)** An electron micrograph of an *E. coli* cell that has been lysed, allowing its chromosome to escape. **(b)** Chromosomal DNA is shown either as a double helix or as a single ring in this chapter.

4. Mutations affecting the *ability of cells to break down and use complicated chemicals* in the environment; for example, the *lacZ* gene in *E. coli* encodes the enzyme β-galactosidase needed to break down the sugar lactose into glucose and galactose. Wild-type cells can grow if lactose, rather than glucose, is the sole source of carbon in the medium, but *lacZ*$^-$ mutants cannot.
5. Mutations in *essential genes* whose protein products are required for growth; because a null mutation in an essential gene would prevent a colony from growing in any environment, bacteriologists must work with conditional lethal mutations such as temperature-sensitive (*ts*) mutations that allow growth at one temperature but not at another.

Bacteriologists use different techniques to isolate rare mutations of the various classes described previously. With mutations conferring resistance to a particular agent, they can do a straightforward **selection,** that is, establish conditions in which only the desired mutant will grow. For example, if wild-type bacteria are streaked on a petri plate containing the antibiotic streptomycin, the only colonies to appear will be streptomycin resistant (Str^r). It is also possible to select for prototrophic revertants of strains carrying auxotrophic mutations by simply plating cells on minimal medium agar, which does not contain the compounds auxotrophs require for growth.

Because the key characteristic of most of the other types of mutants just described is their inability to grow under particular conditions, it is not possible to select for them directly. Instead, researchers must identify these mutations by a **genetic screen:** an examination of each colony in a population for its phenotype. They can, for example, use a toothpick to transfer cells from a colony growing on minimal medium supplemented with methionine to a petri plate containing minimal medium without methionine. Failure of those cells to grow on the unsupplemented medium would indicate that the corresponding colony on the original plate is auxotrophic for methionine.

Spontaneous mutations in specific bacterial genes occur very rarely, in 1 in 10^6 to 1 in 10^8 cells, depending on the gene. It would be virtually impossible to identify such rare mutations if you had to check the phenotype of a million to a hundred million colonies through the individual transfer of each one with a toothpick. A number of techniques simplify the process. We describe four.

1. *Replica plating* allows the simultaneous transfer of thousands of colonies from one plate to another (see Fig. 7.5 on p. 212).
2. *Treatments with mutagens* increase the frequency with which a mutation in a gene appears in the population (see Fig. 7.10 on pp. 218–219).
3. *Enrichment* increases the proportion of mutant cells in a population by exposing the population to agents that kill wild-type cells. Penicillin is one agent of enrichment; it acts by disrupting the formation of the cell wall in growing cells and thus kills cells that can grow but not cells that cannot grow. **Figure 15.5** shows how researchers use this property of penicillin to enrich the proportion of auxotrophic cells in a mixed population of auxotrophic and prototrophic cells.
4. *Testing for visible mutant phenotypes* on a petri plate. In an important example, *E. coli* producing functional

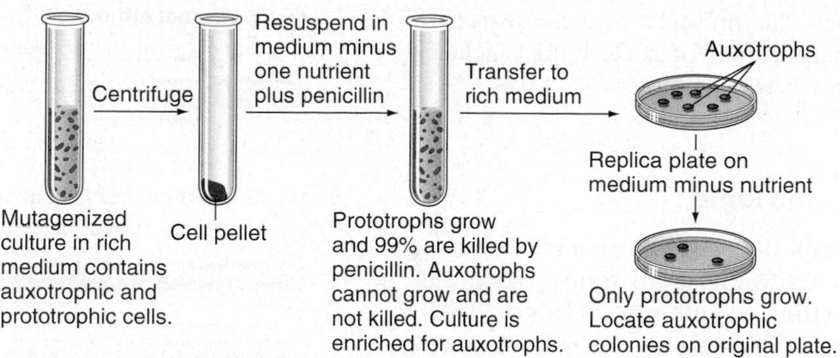

Figure 15.5 Penicillin enrichment for auxotrophic mutants. Penicillin selectively kills growing cells that are making new cell walls, but not bacteria whose growth is arrested. In the absence of nutrients, auxotrophs will not be killed by the penicillin. After enrichment, cells must be screened by replica plating to identify auxotrophs, because penicillin does not kill 100% of the prototrophs.

β-galactosidase (the product of the *lacZ* gene) cleaves the colorless artificial compound X-Gal, producing a blue product. Thus *lacZ*$^+$ colonies turn blue on medium containing X-Gal, while *lacZ*$^-$ colonies remain white (the usual color of the colonies; review Fig. 9.8 on p. 313).

Researchers designate the genes of bacteria by three lowercase, italicized letters that signify something about the function of the gene. For example, genes in which mutations result in the inability to synthesize the amino acid leucine are *leu* genes. In *E. coli,* there are four *leu* genes—*leuA, leuB, leuC,* and *leuD*—that correspond to the three enzymes (one constructed from two different polypeptides) needed for the synthesis of leucine from other compounds. A mutation in any one of the *leu* genes changes a bacterium into an auxotroph for leucine, that is, into a cell unable to synthesize leucine. Such a cell can grow only in media supplemented with leucine. Mutations in genes required for the breakdown of a sugar (for example, the *lacZ* gene) produce cells unable to grow in medium containing only that sugar (lactose) as a source of carbon. Other types of mutations give rise to antibiotic resistance; *str*^{r} is a mutation producing streptomycin resistance. To designate the alleles of genes present in wild-type bacteria, researchers use a superscript "+": *leu*$^+$, *str*$^+$, *lacZ*$^+$. To designate mutant alleles, they use a superscript "−", as in *leuA*$^-$ and *lacZ*$^-$, or a superscript description, as in *str*^{r}.

The phenotype of a bacterium that is wild type or mutant for a particular gene is indicated by the three letters that designate the gene, written, however, with an initial capital letter, no italics, and a superscript of minus, plus, or a one-letter abbreviation: Leu$^-$ (requires leucine for growth); Lac$^+$ (grows on lactose); Str^r (is resistant to streptomycin). A Leu$^-$ *E. coli* strain cannot multiply unless it grows in a medium containing leucine; a Lac$^+$ strain can grow if lactose replaces the usual glucose in the medium; a Str^r strain can grow in the presence of streptomycin.

Structure and Organization of the *E. coli* Chromosome

The circular *E. coli* chromosome consists of about 4.6 million base pairs. In 1997, molecular geneticists completed sequencing the genome of the *E. coli* strain known as K12. From previous genetic work, they knew many of the genes within this sequence; and they could identify others because the polypeptides they encode have amino acid sequences similar to those of sequenced proteins found in other bacteria or even eukaryotic species. Some of the sequences identified as genes, however, encode proteins with functions that have not yet been assigned. Such putative genes are known as **open reading frames,** or **ORFs;** they consist of long stretches of codons in the same reading frame uninterrupted by stop codons.

Close to 90% of *E. coli* DNA encodes proteins; on average, every kilobase (kb) of the chromosome contains one gene. This contrasts sharply with the human genome, in which less than 5% of the DNA encodes proteins and there is roughly one gene every 100 kb. One reason for this discrepancy is that *E. coli* genes have no introns. In addition, there is very little repeated DNA in bacteria, and intergenic regions tend to be very small.

The complete sequence of the *E. coli* genome identified 4288 genes; surprisingly, the function of 40% of the genes remains a mystery at this time. Given the small genome size and the tools developed over the years for genetic analysis in *E. coli,* however, researchers can easily mutate the genes and examine the resulting cells for phenotypic effects. So far, they have grouped genes whose function is known or has been deduced on the basis of sequence into broad functional classes. The 427 genes that are postulated to have a transport function make up the largest class. Other classes include the genes for translation, amino acid biosynthesis, DNA replication, and recombination.

Another interesting feature of the *E. coli* genome is the existence in eight different locations of remnants of bacteriophage genomes. The presence of these sequences suggests an evolutionary history of bacteria that includes invasion by viruses on several occasions.

Insertion Sequences Dot the *E. coli* Chromosome

DNA sequence analysis of bacterial genomes also revealed the position of several small transposable elements called **insertion sequences (IS's).** These elements, which dot the chromosomes of many types of bacteria, are transposons that do not contain selectable markers (such as genes conferring antibiotic resistance). Researchers have identified several distinct elements ranging in length from 700–5000 bp; they named the elements IS1, IS2, IS3, and so forth, with the numbers designating the order of discovery. Like the ends of transposons in eukaryotic cells (see Chapter 14), the ends of IS elements are inverted repeats of each other (**Fig. 15.6a**); and each IS includes a gene encoding a transposase that initiates transposition by recognizing these mirror-image ends. Because insertion sequences can move to other sites on the bacterial chromosome when they transpose, their distribution differs in different strains of the same bacterial species. For example, one strain of *E. coli* may have 15 insertion sequences of five different kinds, while a second strain isolated from a different population may have 25 insertion sequences, lack one of the types found in the first strain, and have a different distribution of IS's around the chromosome. Some bacterial species, such as *Bacillus subtilis,* carry no insertion sequences.

Insertion sequences were first identified in the 1970s as elements that caused inactivation of genes required for galactose metabolism (Gal$^-$ mutants). When an IS transposes and lands within the coding region of a gene, it disrupts the coding region and inactivates the gene (Fig. 15.6b). We now know that many of the spontaneous mutations isolated in *E. coli* are the result of IS transposition into a gene. Researchers have exploited this ability to cause mutation by using a more complex type of transposable element in bacteria: a Tn element. In addition to carrying a gene for transposase, Tn elements contain genes conferring resistance to antibiotics or toxic metals such as mercury. One Tn element known as Tn10 consists of two IS10 elements flanking a gene encoding resistance to tetracycline (Fig. 15.6c). After the introduction of Tn10 into a cell, its transposition into, for example, the *lacZ* gene, produces a *lacZ*$^-$ mutant that is phenotypically both Lac$^-$ and Tet^r (resistant to tetracycline). Because of these effects, the Tn element in the gene is an easily scored genetic marker for mapping experiments (as described later). In addition, because researchers know the sequence of Tn10, they can make a primer corresponding to the end of the Tn10 element to determine the base sequence of the adjacent DNA and thereby identify the gene that was mutated by the Tn10 insertion. This is only one of the many ingenious uses of transposons that researchers have devised.

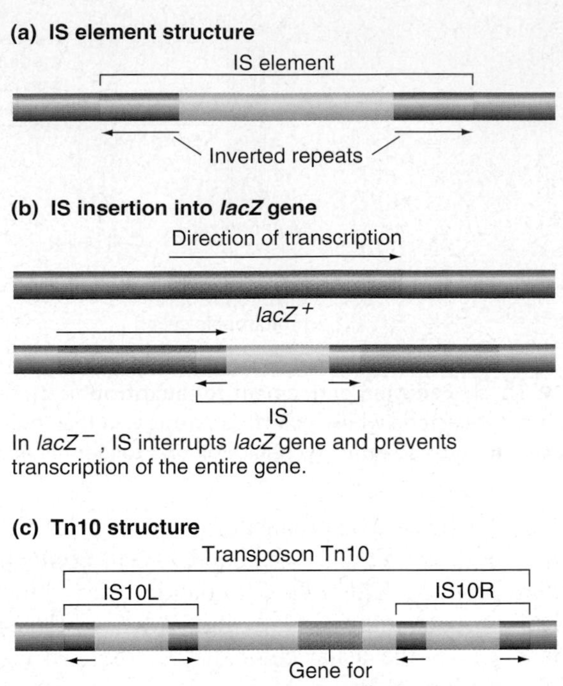

Figure 15.6 Transposable elements in bacteria. (a) An IS element showing the inverted repeats at each end. **(b)** Insertion of an IS into a gene. Here, insertion of an IS inactivates the *lacZ* gene because the IS contains a transcription termination signal. **(c)** The composite transposon Tn10, in which two slightly different IS10's (IS10L and IS10R) flank 7 kb of DNA including a gene for tetracycline resistance. Because it is flanked by IS10 inverted repeats, Tn10 can be mobilized by the IS10 transposase.

Transcription in *E. coli*

During the transcription of the *E. coli* chromosome, different strands of DNA serve as the template for different genes. The transcription machinery moves clockwise on one DNA strand for a few kilobases, transcribing several contiguous and often related genes. In an adjacent region, RNA polymerase transcribes the other strand of genomic DNA, moving counterclockwise for a few kilobases. The only rule seems to be that the most highly transcribed genes tend to be oriented in the direction in which the replication fork moves.

DNA Replication in *E. coli*

DNA replication begins at a site called *oriC,* for origin of chromosomal replication, and proceeds simultaneously in both directions (clockwise and counterclockwise) from that point to a termination region called *terC* (**Fig. 15.7a**).

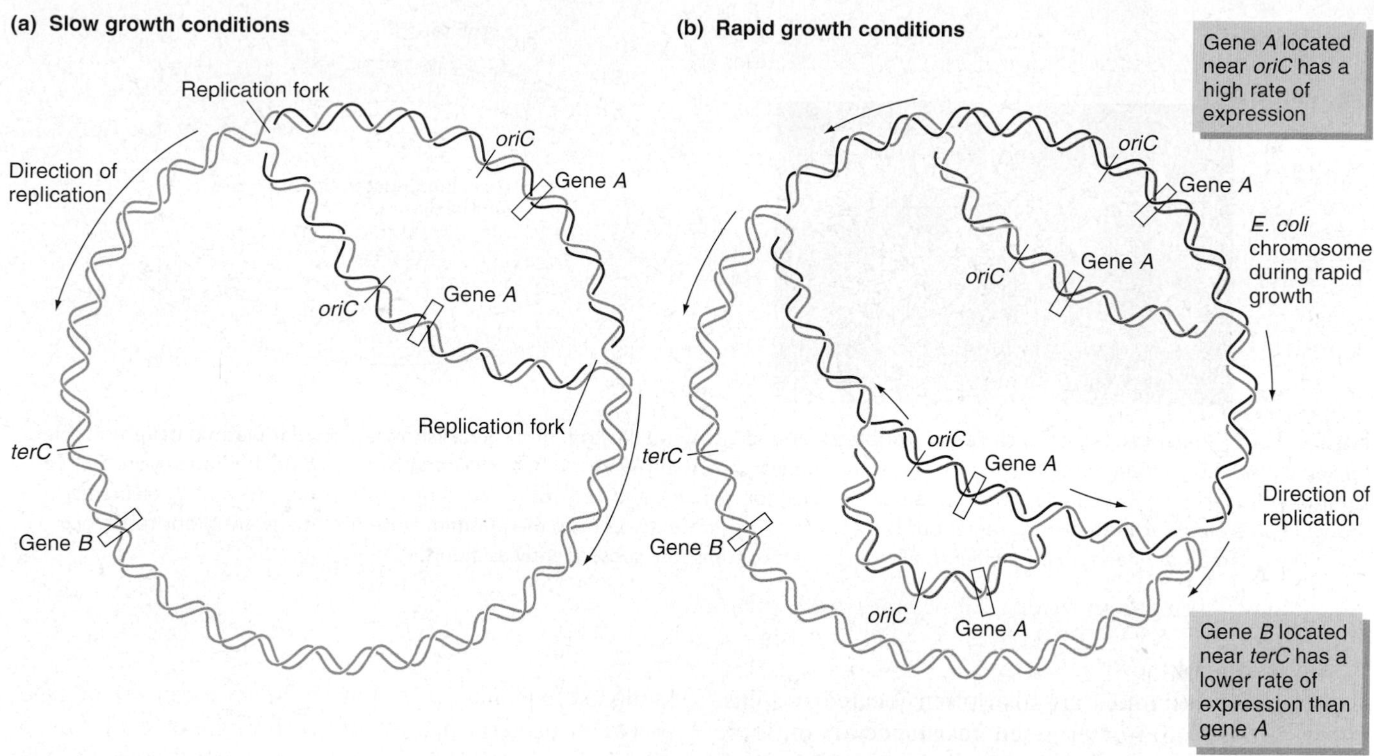

Figure 15.7 Replication of the *E. coli* chromosome. Replication begins at the origin (*oriC*) and proceeds to the terminator (*terC*). The *arrows* indicates the directions in which the replication forks move. **(a)** Under conditions of slow growth, the origin of replication "fires" once in each cell division cycle. **(b)** Under conditions of rapid growth, replication initiates more times than the cell divides. As a result, there will be more copies of genes near the origin than genes near the terminator.

Genes that must undergo a particularly high rate of expression, such as the genes for ribosomal RNA and ribosomal proteins, are often concentrated near the origin of replication. The reason is that during rapid growth, new rounds of replication are initiated before the previous one is completed. This means that there are several replication forks progressing along the chromosome from the single origin of replication at any one time (Fig. 15.7b). As a result, genes near the origin are present in four to eight copies, while genes near the terminator are present in only one copy. The increase in copy number of genes near the origin allows the cell to produce more of the corresponding proteins.

Plasmids Are Smaller Circles of DNA That Do Not Carry Essential Genes

Bacteria carry their essential genes—those necessary for growth and reproduction—in their large circular chromosome. In addition, some bacteria carry genes not needed for growth and reproduction under normal conditions in smaller circles of double-stranded DNA known as **plasmids (Fig. 15.8a)**. Plasmids come in a range of sizes. The smallest are 1000 bp long; the largest are several megabases (Mb) in length. Bacteria usually harbor no more than one extremely large plasmid, but they can house several or even hundreds of copies of smaller DNA circles.

Although plasmids carry genes not normally needed by their bacterial hosts for growth and reproduction, these same genes may benefit the host cell under certain conditions. For example, the plasmids in many bacterial species carry genes that protect their hosts against toxic metals such as mercury. The plasmids of various soil-inhabiting *Pseudomonas* species encode proteins that allow the bacteria to metabolize chemicals such as toluene, naphthalene, or petroleum products. Since the 1980s, natural and genetically engineered plasmids of this type have become part of the tool kit for cleaning up oil spills and other contaminated sites. Plasmids thus help expand the capabilities of bacteria in nature, and they also provide a rich source of unusual and useful proteins for commercial purposes.

Many of the genes that contribute to pathogenicity (that is, the capacity for causing disease in plants or animals) reside in plasmids. For example, the toxins produced by *Shigella dysenteriae,* the causative agent of dysentery, are encoded by plasmids. Genes encoding

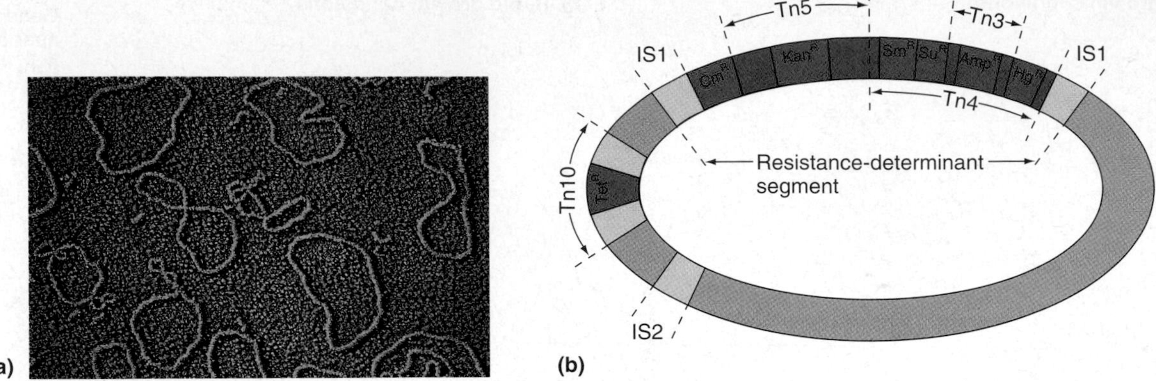

Figure 15.8 Plasmids are small circles of double-stranded DNA. **(a)** Electron micrograph showing circular plasmid DNA molecules. **(b)** Some plasmids contain multiple antibiotic resistance genes (shown in *yellow: Cm^R* for chloramphenicol, *Kan^R* for kanamycin, *Sm^R* for streptomycin, *Su^R* for sulfonamide, *Amp^R* for ampicillin, *Hg^R* for mercury, and *Tet^R* for tetracycline). Transposons (IS and Tn elements, shown in *pale pink*) facilitate the movement of the antibiotic resistance genes onto the plasmid. Note that many antibiotic resistance genes are located between two IS1 elements, allowing them to transpose subsequently as a unit.

resistance to antibiotics are also often located on plasmids. The plasmid-determined resistance to multiple drugs was first discovered in *Shigella* in the 1970s. Multiple antibiotic resistance is often due to composite IS/Tn elements on a plasmid (Fig. 15.8b). As we see later, plasmids can be transferred from one bacterium to another, sometimes even across species. Plasmids thus have terrifying implications for medicine. If transferred to new strains of pathogenic bacteria, the new hosts acquire resistance to many antibiotics in a single step. We encountered an example of this potential in the opening story on gonorrhea.

One important group of plasmids allows the bacterial cells that carry them to make contact with another bacterium and transfer genes—both plasmid and bacterial—to the second cell. We describe this cell-to-cell mating, known as conjugation, in the next section.

15.3 Gene Transfer in Bacteria

Gene transfer from one individual to another plays an important role in the evolution of new variants in nature. Many cases of such horizontal, or lateral, gene transfer have come to light through recent molecular and DNA sequencing analyses. The phrase **lateral gene transfer** means that the traits involved are not transferred by inheritance from one generation to the next; rather, they are introduced from unrelated individuals or from different species. Comparative genomic analysis of many different genes in various bacterial species has shown similarities of genes in species thought to be only distantly related. The simplest explanation is that significant transfer of DNA between bacteria has occurred throughout evolution. A close examination of the known mechanisms of DNA transfer will help illuminate this phenomenon. In addition, you will see that researchers can use the various methods of gene transfer to map genes and to construct bacterial strains with which to test the function and regulation of specific genes.

Bacteria can transfer genes from one strain to another by three different mechanisms: *transformation, conjugation,* and *transduction* (**Fig. 15.9**). In all three mechanisms, one cell—the **donor**—provides the genetic material for transfer, while a second cell—the **recipient**—receives the material. In **transformation,** DNA from a donor is added to the bacterial growth medium and is then taken up from that medium by the recipient. In **conjugation,** the donor carries a special type of plasmid that allows it to come in contact with the recipient and transfer DNA directly. In **transduction,** the donor DNA is packaged within the protein coat of a bacteriophage and transferred to the recipient when the phage particle infects it. The recipients of a gene transfer are known as **transformants, exconjugants,** or **transductants,** depending on the mechanism of DNA transfer that created them.

All bacterial gene transfer is asymmetrical in two ways. First, transfer goes in only one direction, from donor to recipient. Second, most recipients receive 3% or less of a donor's DNA; only some exconjugants contain a greater percentage of donor material. Thus, the amount of donor DNA entering the recipient is small relative to the size of the recipient's chromosome, and the recipient retains most of its own DNA. We now examine each type of gene transfer in detail.

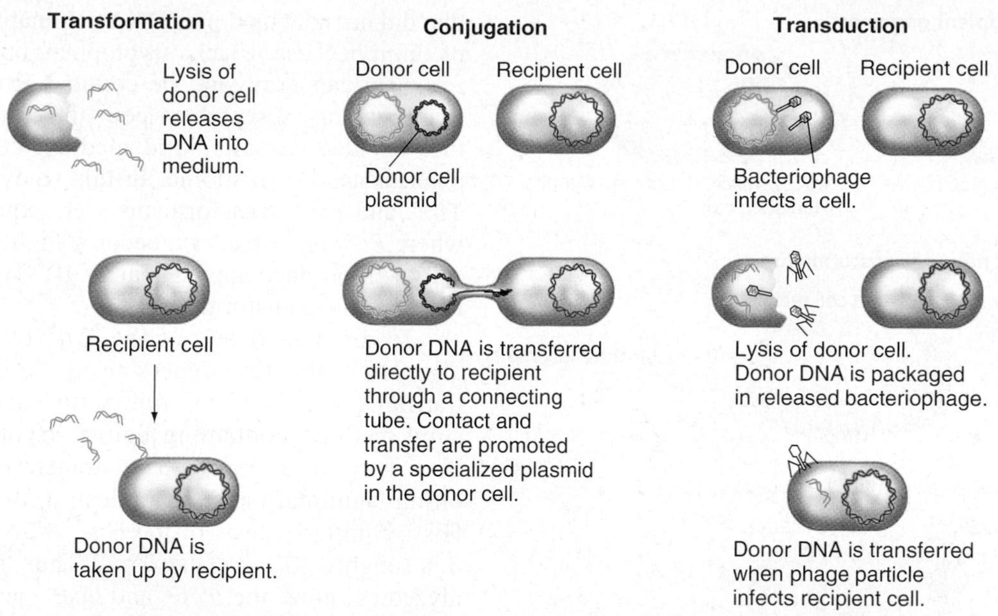

Figure 15.9 Three mechanisms of gene transfer in bacteria: An overview. In this figure, and throughout this chapter, the donor's chromosome is *blue,* and the recipient's chromosome is *orange.* In transformation, fragments of donor DNA released into the medium enters the recipient cell. In conjugation, a specialized plasmid (shown in *red*) in the donor cell promotes contact with the recipient and initiates the transfer of DNA. In transduction, DNA from the donor cell is packaged into bacteriophage particles that can infect a recipient cell, transferring the donor DNA into the recipient.

Transformation: Fragments of Donor DNA Enter the Recipient and Alter Its Genotype

A few species of bacteria spontaneously take up DNA fragments from their surroundings in a process known as **natural transformation.** By contrast, the large majority of bacterial species can take up DNA from the surrounding medium only after laboratory procedures make their cell walls and membranes permeable to DNA in a process known as **artificial transformation.**

In Natural Transformation, the Recipient Cell Has the Enzymatic Machinery for DNA Import

Researchers have studied several species of bacteria that undergo natural transformation, including *S. pneumoniae,* the pathogen in which transformation was discovered by Frederick Griffith (see Chapter 6) and that causes pneumonia in humans; *B. subtilis,* a harmless soil bacterium; *H. influenzae,* a pathogen causing various diseases in humans; and *N. gonorrhoeae,* the microbial agent of gonorrhea.

In one study of natural transformation, investigators isolated *B. subtilis* bacteria with two mutations—*trpC2* and *hisB2*—that made them Trp^-, His^- double auxotrophs. These double auxotrophs served as the recipients in the study; wild-type cells (Trp^+, His^+) were the donors (**Fig. 15.10a**). The experimenters extracted and purified donor DNA and grew the *trpC hisB* recipients in a suitable medium until the cells became **competent,** that is, able to take up DNA from the medium. Different bacterial species require different regimens to achieve competence. For *B. subtilis,* competence occurs only in nearly starving cells at very specific times in the growth of the culture. Investigators can starve the cells by growing them in a glucose-salts medium containing a limited amount of tryptophan and histidine. As growth of the culture slows toward the end of what is known as the stage of logarithmic growth, a fraction of the bacteria—1% to 5% in *B. subtilis*—become competent and will take up DNA added to the medium at this time.

The mechanism of DNA uptake in *B. subtilis* is more complicated than was originally anticipated (Fig. 15.10b). The donor DNA fed to the competent cells usually consists of linear double-stranded molecules with an average length of 20 kb. The competent cells have enzyme complexes at their surfaces that take up one strand of a DNA fragment and simultaneously break down the other strand to its constituent nucleotides. The wild-type DNA strand admitted to the doubly auxotrophic cell quickly finds a homologous region of the bacterial chromosome and recombines with it, replacing the almost identical portion of one of the two strands of the recipient DNA (this one strand is subsequently degraded). After the chromosome replicates and the cell divides, one of the two daughter cells becomes a transformant, carrying wild-type $trpC^+$ and/or $hisB^+$ genes from the donor.

(a) Donor and recipient genomes

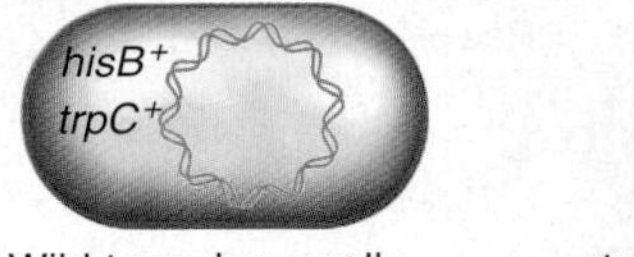
hisB+
trpC+
Wild-type donor cell

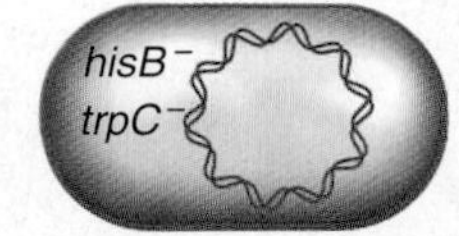
hisB−
trpC−
trpC− / hisB− double auxotrophs
Recipient cell

(b) Mechanism of natural transformation

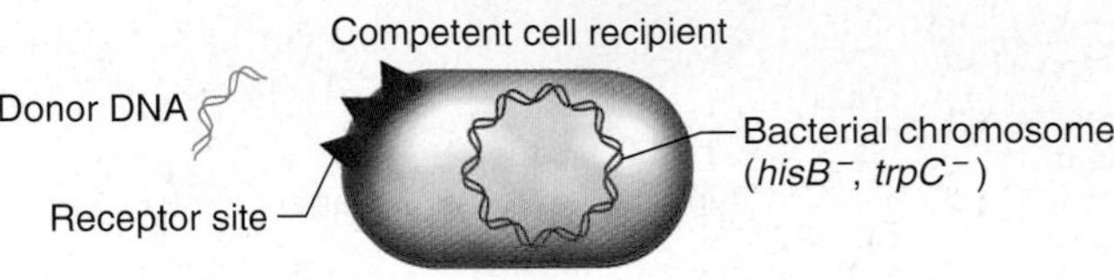
Competent cell recipient
Donor DNA
Bacterial chromosome
(hisB−, trpC−)
Receptor site
Donor DNA binds to recipient cell at receptor site.

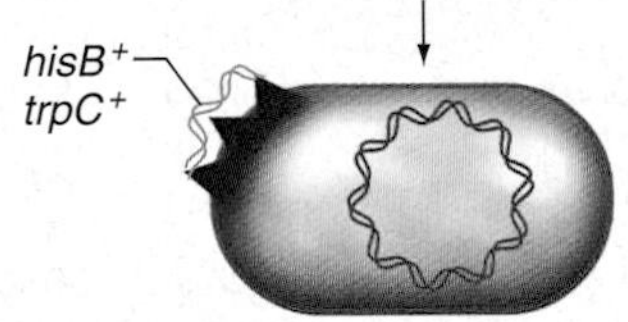
hisB+
trpC+
One donor strand is degraded. The admitted donor strand pairs with
homologous region of bacterial chromosome.
The replaced strand is degraded.

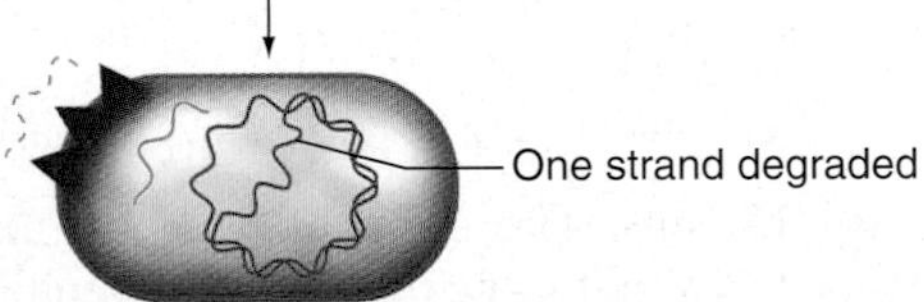
One strand degraded
Donor strand is integrated into bacterial chromosome.

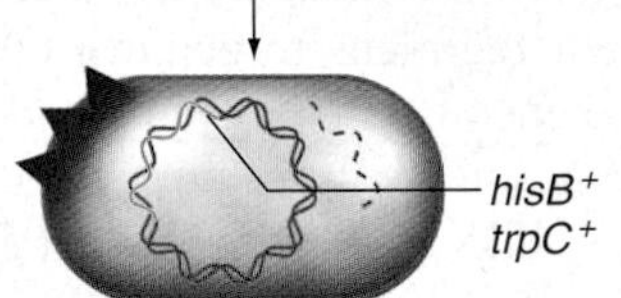
hisB+
trpC+
After cell replication, one cell is identical to original recipient;
the other carries the mutant genes.

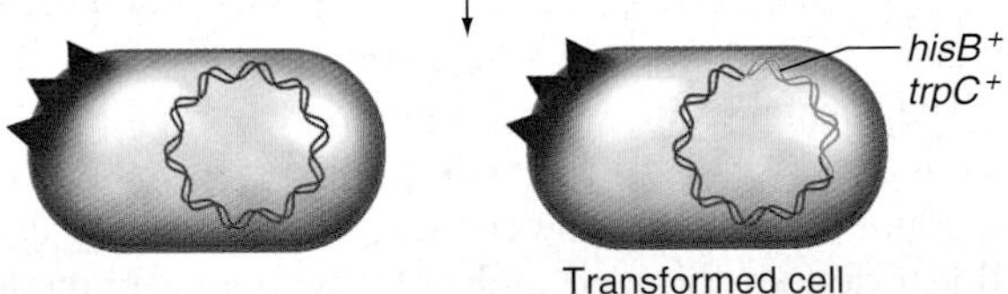
hisB+
trpC+
Transformed cell

Figure 15.10 **Natural transformation in *B. subtilis.*** **(a)** A wild-type donor and a *hisB*$^-$ *trpC*$^-$ double auxotroph recipient. Selection for His$^+$ and/or Trp$^+$ phenotypes identifies transformants. **(b)** Mechanism of natural transformation in *B. subtilis.* One strand of a fragment of donor DNA enters the recipient, while the other strand is degraded. The entering strand recombines with the recipient chromosome, producing a transformant when the recipient cell divides.

To observe and count Trp$^+$ transformants, researchers decant the liquid containing newly transformed recipient cells onto petri plates containing a simple glucose-salts solid medium with histidine. Recipient cells that did not take up donor DNA are unable to grow on this medium because it lacks tryptophan, but the Trp$^+$ transformants can grow and be counted. To select for His$^+$ transformants, researchers pour the transformation mixture on glucose-salts solid medium containing tryptophan, instead of histidine. In this study, the numbers of Trp$^+$ and His$^+$ transformants were equal. In conditions where *B. subtilis* bacteria become highly competent, 10^9 cells will produce approximately 10^5 Trp$^+$ transformants and 10^5 His$^+$transformants.

To discover if any of the Trp$^+$ transformants were also His$^+$, the researchers used sterile toothpicks to transfer colonies of Trp$^+$ transformants to a glucose-salts solid medium containing neither tryptophan nor histidine. Forty of every 100 Trp$^+$ transferred colonies grew on this minimal medium, indicating that they were also His$^+$. Similarly, tests of the His$^+$ transformants showed that roughly 40% are also Trp$^+$. Thus, in 40% of the analyzed colonies, the *trpC*$^+$and *hisB*$^+$ genes had been cotransformed. **Cotransformation** is the simultaneous transformation of two or more genes.

Since during transformation, donor DNA replaces only a small percentage of the recipient's chromosome, it might seem surprising that the two *B. subtilis* genes are cotransformed with such high frequency. The explanation is that the *trpC* and *hisB* genes lie very close together on the chromosome and are thus genetically linked. The entire *B. subtilis* chromosome is approximately 4700 kb long. Only genes in the same chromosomal vicinity can be cotransformed; the closer together the genes lie, the more frequently they will be cotransformed. Thus, although the donor chromosome is fragmented into small pieces of about 20 kb during its extraction for the transformation process, the wild-type *trpC*$^+$and *hisB*$^+$ alleles are so close that they often appear together in the same donor DNA molecule. Sequence analysis shows that the *trpC* and *hisB* genes are only about 7 kb apart. By contrast, genes sufficiently far apart that they cannot appear together on a fragment of donor DNA will almost never be cotransformed, because transformation is so inefficient that recipient cells usually take up only a single DNA fragment.

Transformation usually incorporates a single strand of a linear donor DNA fragment into the bacterial chromosome of the recipient through recombination. However, if the donor DNA includes plasmids, recipient cells may take up an entire plasmid and acquire the characteristics conferred by the plasmid genes. Bacteriologists suspect that penicillin-resistant *N. gonorrhoeae*, described in the introduction to this chapter, originated through transformation by plasmids. The donors of the plasmids were *H. influenzae* cells disrupted by the immune defenses of a doubly infected patient. The plasmids carried the gene for penicillinase; and the recipient *N. gonorrhoeae* bacteria, transformed by the plasmids, acquired resistance to penicillin.

In Artificial Transformation, Damage to Recipient Cell Walls Allows Donor DNA to Enter the Cells

Researchers have devised many methods to transform bacteria that do not undergo natural transformation. The existence of these techniques was critical for the development of the gene-cloning technology described in Chapter 9. All the methods include treatments that damage the cell walls and membranes of recipient bacteria so that donor DNA can diffuse into the cells. With *E. coli,* the most common treatment consists in suspending the cells in a high concentration of calcium at cold temperature. Under these conditions, the cells become permeable to single- and double-stranded DNA. Because *E. coli* cells have efficient DNA exonucleases that degrade free ends, circular DNA is much more efficient than linear DNA at transforming *E. coli.*

Another technique of artificial transformation is *electroporation,* in which researchers mix a suspension of recipient bacteria with donor DNA and then subject the mixture to a very brief high-voltage shock. The shock most likely causes holes to form in the cell membranes. With the proper shocking conditions, recipient cells take up the donor DNA very efficiently. Transformation by electroporation works with most bacteria.

Conjugation: Donor Cells Carrying Specialized Plasmids Establish Contact with and Transfer DNA to Recipients

In the late 1940s, Joshua Lederberg and Edward Tatum analyzed two *E. coli* strains that were each multiple auxotrophs and made the striking discovery that genes seemed to transfer from one type of *E. coli* cell to the other (**Fig. 15.11a**). Neither strain grew on a minimal glucose-salts medium. Strain A required supplementation with methionine and biotin; strain B required supplementation with threonine, leucine, and thiamine (vitamin B_1). Lederberg and Tatum grew the two strains together on supplemented medium. When they then transferred a mixture of the two strains to minimal medium, about 1 in every 10^7 transferred cells proliferated to a visible colony. What were these colonies, and how they did they arise?

To answer these questions, Lederberg and Tatum performed the necessary control for this experiment. If they plated either strain alone on a minimal medium, nothing grew. Because they observed colonies only when they plated a mixture of the two strains on a simple medium, they concluded that some sort of genetic exchange—rather than a highly unlikely set of simultaneous reversions—had taken place. *E. coli,* however, do not undergo natural transformation. Further experiments showed that instead, direct cell-to-cell contact was necessary for the exchange. In one experiment, Bernard Davis showed that the genetic exchange did not occur when cells of one strain were mixed with a filtrate containing DNA of the other strain or when a porous filter allowing the passage of liquid medium but not bacteria cells separated strain A from strain B in a U-shaped tube (Fig. 15.11b). More than a decade of further experiments confirmed that Lederberg and Tatum had observed what became known as bacterial conjugation: a one-way DNA transfer from donor to recipient initiated by **conjugative plasmids** in donor strains. Many different plasmids can

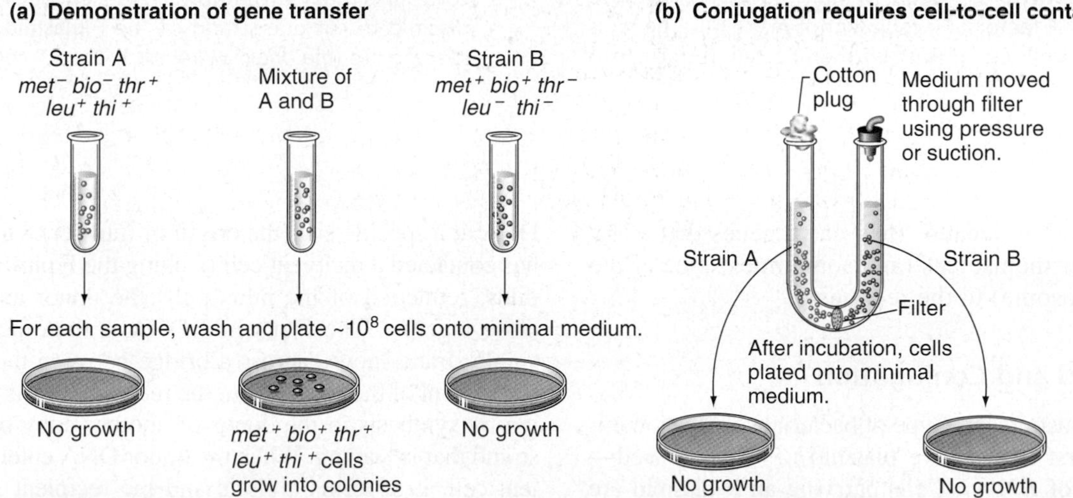

Figure 15.11 Conjugation: A type of gene transfer requiring cell-to-cell contact. (a) Neither of two multiple auxotrophic strains analyzed by Lederberg and Tatum formed colonies on minimal medium. When cells of the two strains were mixed, gene transfer produced some prototrophic cells that formed colonies on minimal medium. **(b)** The U-tube experiment demonstrating that direct contact between cells is needed for conjugation. Here, the two strains were separated by a filter that allowed the passage of all molecules in the medium, including DNA fragments, but prevented contact between the cells. The filter abolished gene transfer between the two strains.

Feature Figure 15.12

The F Plasmid and Conjugation

a. The F plasmid contains genes for synthesizing connections between donor and recipient cells. The F plasmid is a 100-kb-long circle of double-stranded DNA. Host cells that carry it generally have one copy of the plasmid. By analogy with sexual reproduction, researchers think of F^+ cells as *male bacteria* because the cells can transfer genes to other bacteria. About 35% of F plasmid DNA consists of genes that control the transfer of the plasmids. Most of these genes encode polypeptides involved in the construction of a structure called the F **pilus** (plural, **pili**): a stiff, thin strand of protein that protrudes from the bacterial cell. Other regions of the plasmid carry IS's and genes for proteins involved in DNA replication.

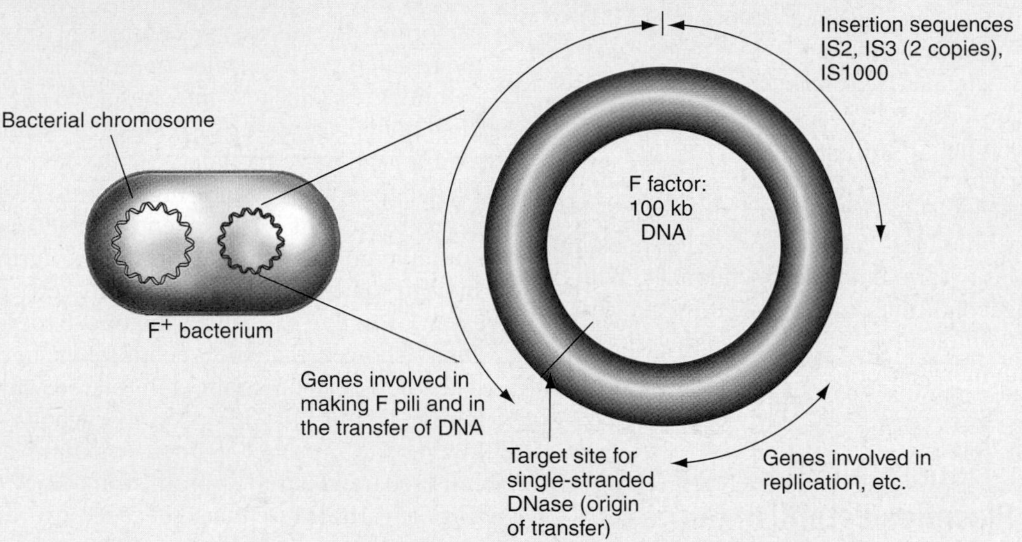

b. The process of conjugation.

1. The **pilus.** An average pilus is 1 μm in length, which is almost as long as the average *E. coli* cell. The distal tip of the pilus consists of a protein that binds specifically to the cell walls of F^- *E. coli* not carrying the F factor.
2. **Attachment to F^- cells (female bacteria).** Because they lack F factors, F^- cells cannot make pili. The pilus of an F^+ cell, on contact with an F^- cell, retracts into the F^+ cell, drawing the F^- cell closer. A narrow passageway forms through the now adjacent F^+ and F^- cell membranes.
3. **Gene transfer: A single strand of DNA travels from the male to the female cell.** Completion of the cell-to-cell corridor signals an endonuclease encoded by the F plasmid to cut one strand of the F plasmid DNA at a specific site (*the origin of transfer*). The F^+ cell extrudes

initiate conjugation because they carry genes that allow them to transfer themselves (and sometimes some of the donor's chromosome) to the recipient.

The F Plasmid and Conjugation

Figure 15.12 illustrates the type of bacterial conjugation initiated by the first conjugative plasmid to be discovered—the **F plasmid** of *E. coli.* Cells carrying an F plasmid are called F^+ cells; cells without the plasmid are F^-. The F plasmid carries many genes required for the transfer of DNA, including genes for formation of an appendage, known as the pilus, by which a donor cell contacts a recipient cell, and a gene encoding an endonuclease that nicks the F plasmid's DNA at a specific site (the origin of transfer). Once a donor has contacted a recipient cell (lacking the F plasmid) via the pilus, retraction of the pilus pulls the donor and recipient close together. The F plasmid DNA is then nicked, and a single strand moves across a bridge between the two cells. Movement of the F DNA into the recipient cell is accompanied by synthesis in the donor of another copy of the DNA strand that is leaving. When the donor DNA enters the recipient cell, it re-forms a circle and the recipient synthesizes the complementary DNA strand. In this $F^+ \times F^-$ mating, the recipient becomes F^+, and the donor remains F^+. By initiating and carrying out conjugation, the F plasmid acts in bacterial populations the way an agent of sexually transmitted disease acts in human populations. When introduced via

the cut strand through the passageway into the F^- cell at a rate of about 47,000 bases per minute. As it receives the single strand of F plasmid DNA, the F^- cell synthesizes a complementary strand so that after 2–3 minutes of contact, the formerly F^- cell contains a double-stranded F plasmid and is now an F^- cell.

4. **Meanwhile, back in the original F^+ cell, newly synthesized DNA replaces the single strand transferred to the previously F^- cell.** When the two bacteria separate at the completion of DNA transfer and synthesis, they are both F^+.

To summarize: Conjugation passes the F^+ plasmid and therefore maleness—the ability to transfer DNA to other cells—from F^+ to F^- bacteria.

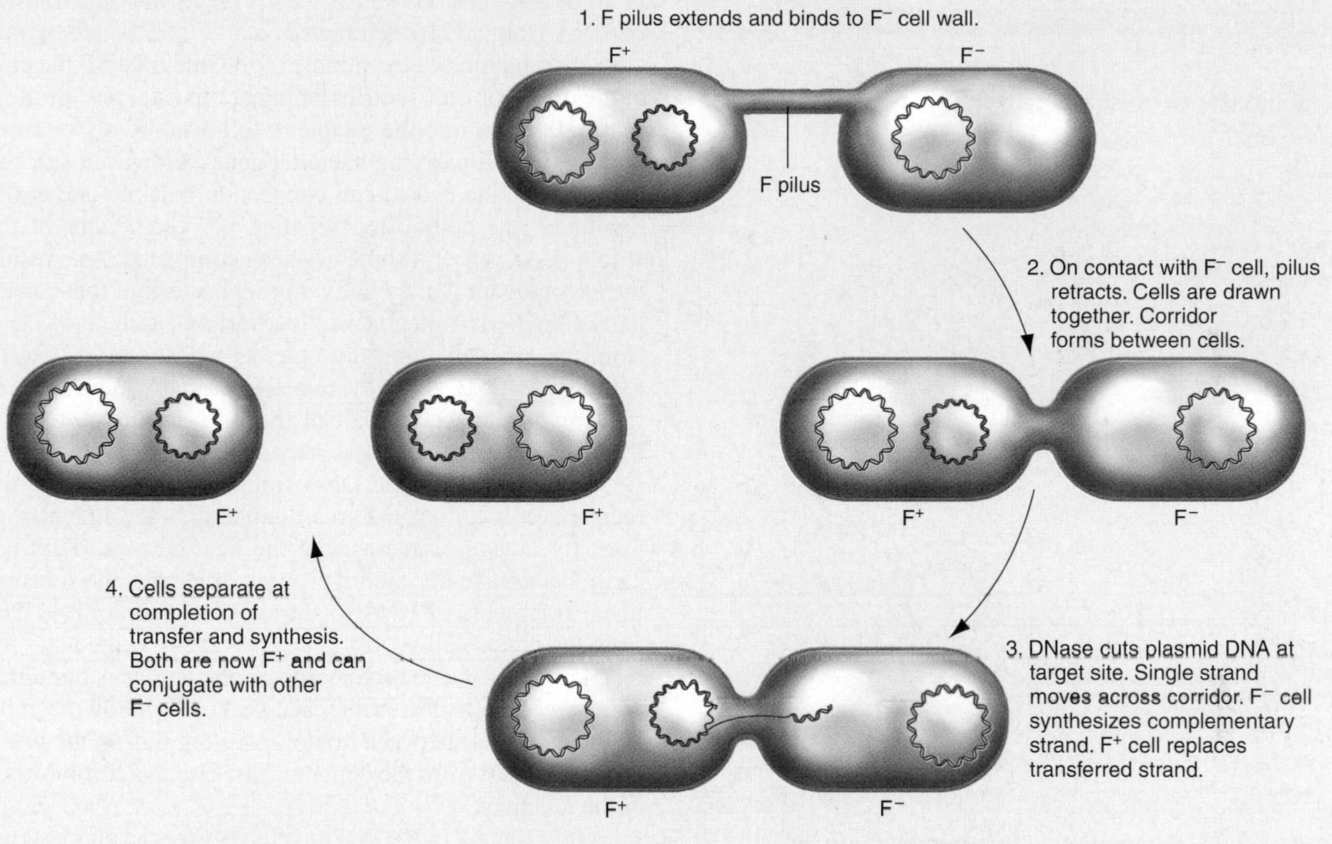
1. F pilus extends and binds to F⁻ cell wall.
F+
F−
F pilus
2. On contact with F⁻ cell, pilus retracts. Cells are drawn together. Corridor forms between cells.
F+
F−
3. DNase cuts plasmid DNA at target site. Single strand moves across corridor. F⁻ cell synthesizes complementary strand. F⁺ cell replaces transferred strand.
F+
F−
4. Cells separate at completion of transfer and synthesis. Both are now F⁺ and can conjugate with other F⁻ cells.
F+
F+

a few donor bacteria into a large culture of cells that do not carry the plasmid, the F plasmid soon spreads throughout the entire culture, and all the cells become F^+.

The F Plasmid Occasionally Integrates into the *E. coli* Chromosome Where It Promotes the Conjugational Transfer of Chromosomal Genes

The F plasmid contains three different IS transposable elements: one copy of IS2, two copies of IS3, and one copy of the particularly long IS1000. These IS sequences on the F plasmid are identical to copies of the same IS elements found at various positions along the bacterial chromosome. In roughly 1 of every 10^5 (100,000) F^+ cells, homologous recombination (that is, a crossover) between an IS on the plasmid and the same IS on the chromosome integrates the entire F plasmid into the *E. coli* chromosome (**Fig. 15.13a**). This insertion by recombination enlarges the 4700 kb chromosome by 100 kb of DNA. Cells whose chromosomes carry an integrated plasmid are called **Hfr** bacteria, because, as we will see, they produce a high frequency of recombinants for chromosomal genes in mating experiments with F^- strains. Since the recombination event that results in the F plasmid's insertion into the bacterial chromosome can occur between any of the IS elements on the F plasmid and any of the corresponding IS elements in the bacterial chromosome, geneticists can isolate 20–30 different strains

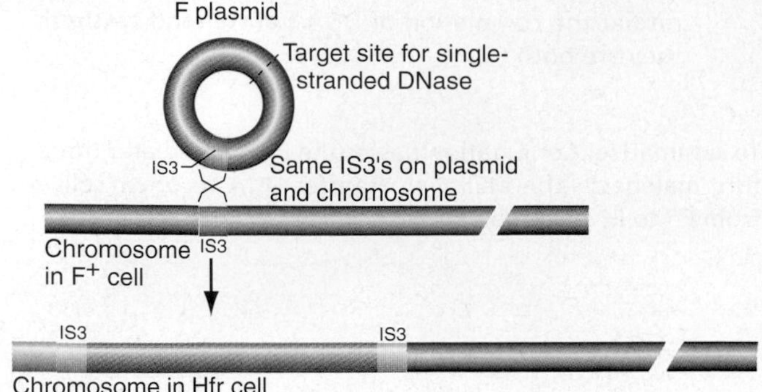

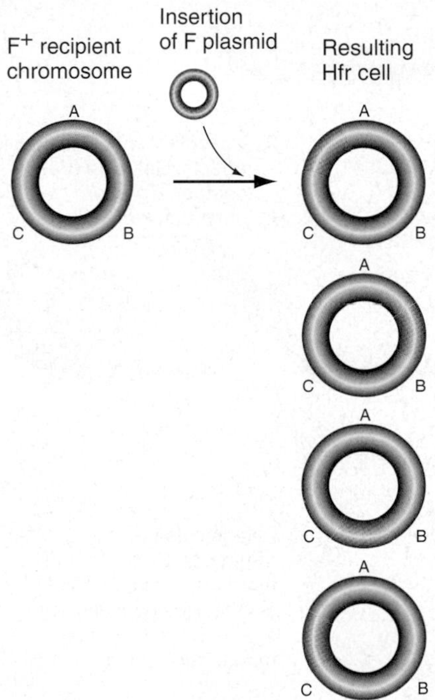

Figure 15.13 Integration of the F plasmid into the bacterial chromosome forms an Hfr bacterium. In this figure, the *filled bar* represents both strands of DNA. **(a)** Recombination between an IS on the F plasmid and the same kind of IS on the bacterial chromosome creates an Hfr chromosome. **(b)** Recombination can occur between any IS on the F plasmid and any corresponding IS on the bacterial chromosome to create many different Hfr strains.

of Hfr cells (Fig. 15.13b). Various Hfr strains are distinguished by the location and orientation (clockwise or counterclockwise) of the integrated F plasmid with respect to the bacterial chromosome. These different Hfr strains have different names; one, for example, is called HfrH after its discoverer William Hayes. Plasmids like the F plasmid that can integrate into the host chromosome are sometimes called **episomes.**

During bacterial reproduction, the integrated plasmid of an Hfr cell replicates with the rest of the bacterial chromosome. As a result, the chromosomes in daughter cells produced by cell division contain an intact F plasmid at exactly the same location that the plasmid originally integrated into the chromosome of the parental cell. All progeny of an Hfr cell are thus identical, with the F plasmid inserted into the same chromosomal location and in the same orientation.

The integrated F plasmid still has the capacity to initiate DNA transfer via conjugation, but now that it is part of a bacterial chromosome, it can promote the transfer of some or all of that chromosome as well (**Fig. 15.14**). The transfer of DNA from an Hfr cell mated to an F^- cell starts with a single-strand nick in the middle of the integrated F plasmid at the origin of transfer. Beginning at this site, half of the F plasmid moves into the recipient, followed by donor chromosomal DNA carrying bacterial genes. However, since in conjugation, the cell-to-cell connection is labile and easily disrupted, the cells often separate before transfer of the whole *E. coli* chromosome has been completed. As a result, the exconjugant is a partially diploid bacterium that carries half of the F plasmid attached to a variable amount of DNA from the Hfr cell's chromosome, as well as the F^- cell's entire genome. Because the transfer is usually interrupted in midstream, the second half of the F plasmid is not transferred and the recipient thus remains F^-.

The single-stranded DNA fragment transferred to the recipient cell is copied into a double-stranded linear fragment by DNA polymerases in the new host cell. But because bacterial cells cannot properly replicate and transmit linear fragments of DNA to subsequent generations, DNA from the donor will be lost unless it is incorporated into the recipient's chromosome. Incorporation can occur through a double recombination event (catalyzed by bacterial recombination enzymes) between homologous regions on the linear DNA fragment from the donor and the circular chromosome of the recipient.

Only about 1 in 10,000 Hfr cells transfers an entire strand of its chromosome to a recipient F cell. And only those rare F^- cells that receive the entire Hfr chromosome, including the second half of the F plasmid, have the full 100 kb of F plasmid DNA that make them F^+. Therefore, most of the progeny produced by an Hfr × F^- conjugation remain F^-. This contrasts sharply with the result of conjugation initiated by an F^+ cell, where almost all the exconjugants become F^+. Thus, Hfr cells are poor donors of the entire F plasmid but efficient donors of chromosomal genes. The situation is reversed for F^+ donors, which can transfer bacterial genes only in those rare cells in which the F plasmid integrates to form an Hfr.

Analyses of Hfr × F^- Matings Help Map Genes

Because genes in an Hfr chromosome are transferred into the recipient in a consistent order, researchers realized that they could use Hfr crosses to map genes. To gain useful information from a conjugational mating between an Hfr "male" and an F^- "female," they analyze strains that differ

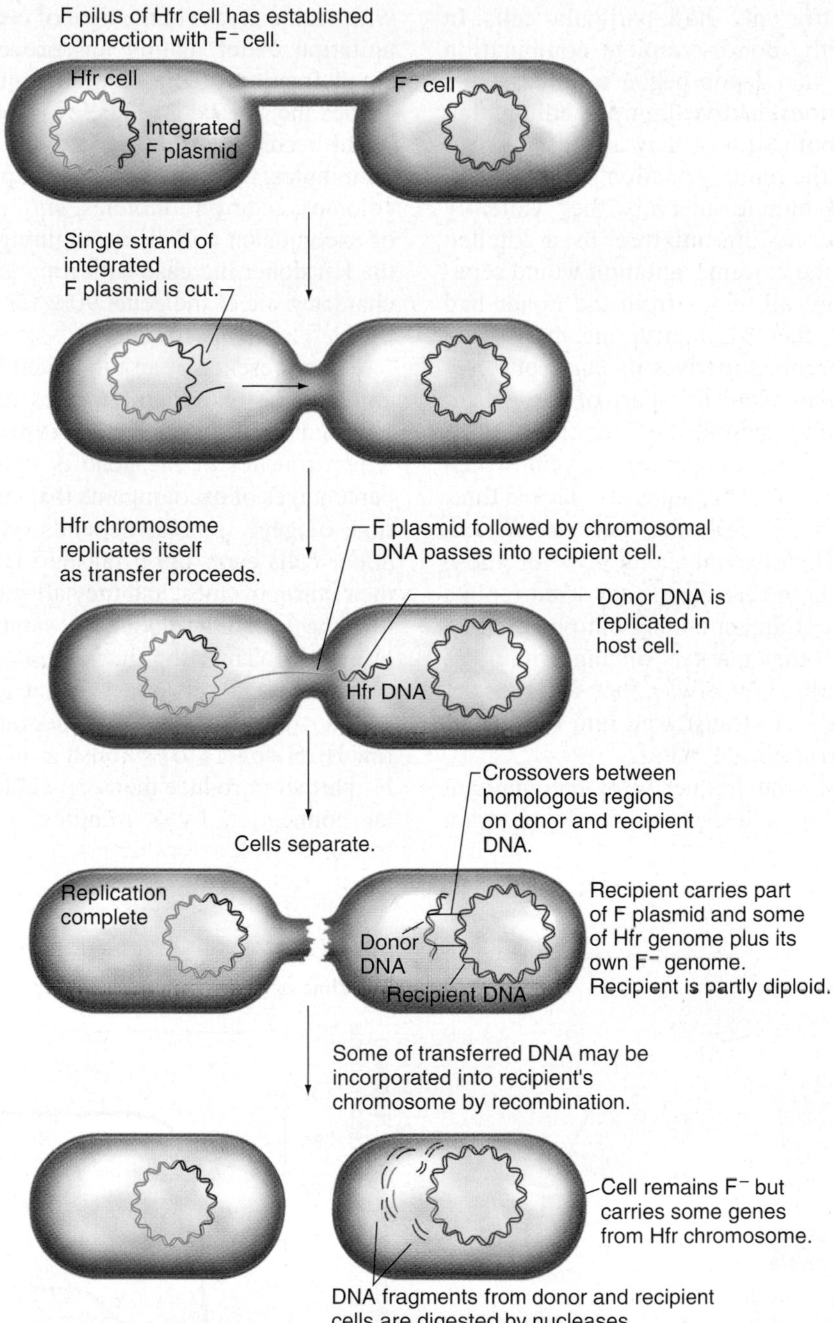
F pilus of Hfr cell has established connection with F⁻ cell.
Hfr cell
F⁻ cell
Integrated F plasmid
Single strand of integrated F plasmid is cut.
Hfr chromosome replicates itself as transfer proceeds.
F plasmid followed by chromosomal DNA passes into recipient cell.
Donor DNA is replicated in host cell.
Hfr DNA
Cells separate.
Crossovers between homologous regions on donor and recipient DNA.
Replication complete
Donor DNA
Recipient DNA
Recipient carries part of F plasmid and some of Hfr genome plus its own F⁻ genome. Recipient is partly diploid.
Some of transferred DNA may be incorporated into recipient's chromosome by recombination.
Cell remains F⁻ but carries some genes from Hfr chromosome.
DNA fragments from donor and recipient cells are digested by nucleases.

Figure 15.14 Gene transfer in a mating between Hfr donors and F^- recipients. In an Hfr × F^- mating, single-stranded DNA is transferred into the recipient, starting with the origin of transfer on the integrated F plasmid. Within the recipient cell, this single-stranded DNA is copied into double-stranded DNA. If mating is interrupted, the recipient cell will contain a double-stranded linear fragment of DNA plus its own chromosome. Genes from the donor are retained in the exconjugant only if they recombine into the recipient's chromosome.

from each other in the alleles of several genes. In early studies of the *E. coli* genome, for example, Elie Wollman and François Jacob used the following genotypes:

HfrH	F^-
str^s (sensitive to streptomycin)	str^r (resistant to streptomycin)
thr^+ (able to synthesize the amino acid threonine)	thr^- (threonine auxotroph)
azi^r (resistant to sodium azide)	azi^s (sensitive to sodium azide)
ton^r (resistant to bacteriophage T1)	ton^s (sensitive to phage T1)
lac^+ (able to grow with lactose as sole source of carbon)	lac^- (unable to grow on lactose)
gal^+ (able to grow with galactose as sole source of carbon)	gal^- (unable to grow on galactose)

To isolate and analyze the exconjugants in which transfer and recombination had occurred, the researchers had to be able to select for only those particular cells. In one experiment using the donor-recipient combination just outlined, Wollman and Jacob began by mixing the two strains in a rich nonselective liquid medium that allowed the growth of both strains; they also established conditions that favored the rapid formation of cell-to-cell connections. Next, at 1-minute intervals, they violently stirred samples of the mating mixture in a kitchen blender, reasoning that the extreme agitation would separate pairs of cells before all genes from the donor had passed to the recipient, thereby interrupting the mating (**Fig. 15.15a**). (The experiment derives its name of *interrupted-mating experiment* from this part of the procedure.) They then poured samples of the terminated matings onto petri plates containing streptomycin, which killed the original donor cells. The plates also lacked threonine, to select against F^- cells that had not mated. Finally, after the Str^r Thr^+ exconjugants grew on these streptomycin plates lacking threonine, they used replica plating to copy each exconjugant colony onto media that would test for the four other markers distinguishing the donor and recipient strains. In this way, they could determine which donor genes had transferred into the F^- recipients before the interruption of mating.

Figure 15.15b shows the frequency of recombinant colonies containing various alleles from the donor strain with each increase in the time of conjugation before violent agitation. After mating has proceeded for 8 minutes, a small fraction of the recombinants are *azi*^{r}, but not one carries the other donor alleles. At about 10 minutes, some of the recombinants also have the donor's *ton*^{r} allele. By 15 minutes, the first *lac*$^+$ alleles appear in the recombinant colonies; and by 17 minutes, *gal*$^+$ arrives. The percentages of exconjugant colonies containing a particular gene from the Hfr donor increase with time until they reach a plateau characteristic of the gene: 90% for *azi*^{r}, 85% for *ton*^{r}, 40% for *lac*$^+$ and 20% for *gal*$^+$.

The presence of an integrated F plasmid in Hfr strains explains the two characteristics of each transferred gene revealed by the interrupted mating experiment: the time at which transfer of the gene is first seen, and the plateau percentages of exconjugants that carry the donor gene. The time of gene transfer depends on the fact that all HfrH donor cells carry the F plasmid DNA at the same site on their chromosomes, and they all start to transfer their DNA into the F cells at about the same time and in the same orientation. Thus, the time a gene first enters the recipient cell reflects the distance of that gene from the origin of transfer located within the integrated F plasmid. The first few HfrH donors to establish a mating connection with an F^- partner introduce their *azi*^{r} allele through the intercellular connection by 8 minutes; more donors make the

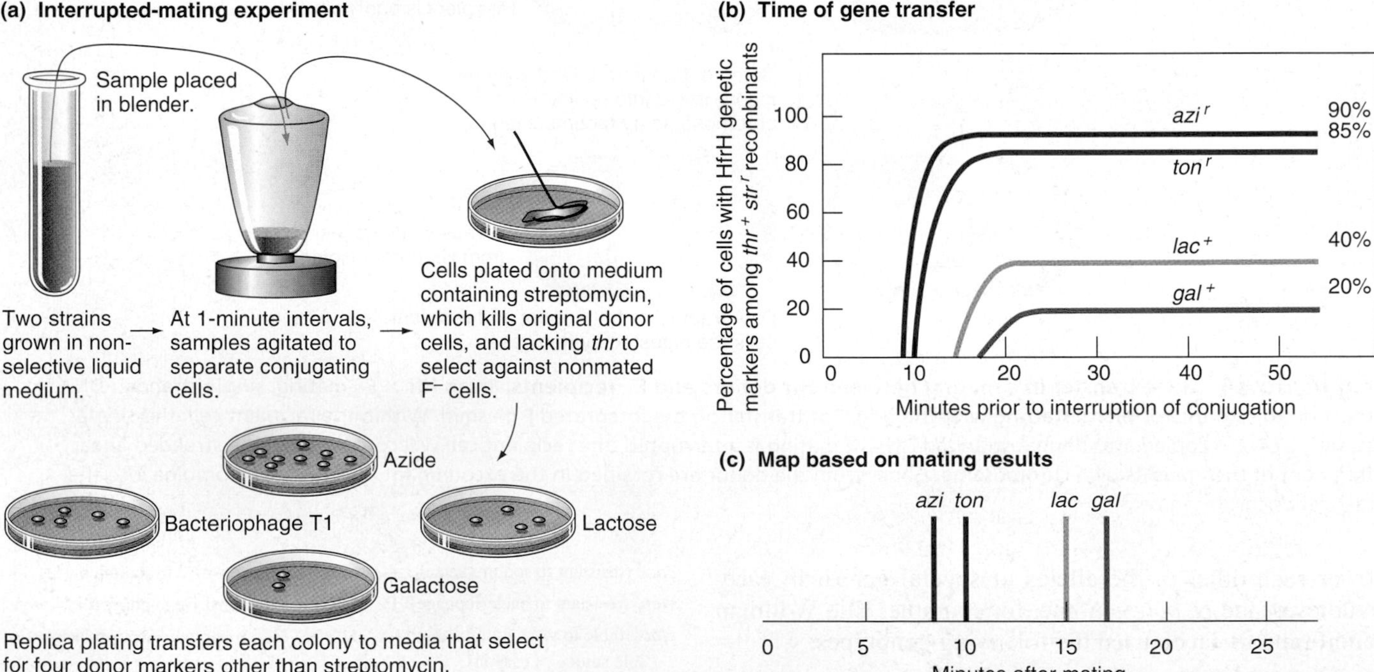

Figure 15.15 Mapping genes by interrupted-mating experiments. **(a)** Hfr and F^- cells were mixed to initiate mating. Samples were agitated at 1-minute intervals in a kitchen blender to disrupt gene transfer. Cells were plated onto a medium that contained streptomycin (to kill the Hfr donor cells) and that lacked threonine (to prevent growth of F^- cells that had not mated). The phenotypes of the exconjugants for other markers were established by replica plating. **(b)** Results of the Wollman-Jacob interrupted-mating experiment. **(c)** Gene order established from the data in part (b), with positions determined by the time a donor gene first enters the recipient.

connection and transfer *azi*^{r} thereafter. The first *ton*^{r} alleles pass into F$^-$ cells at about 10 minutes, the first *lac*$^+$ alleles at about 15 minutes, and the first *gal*$^+$ alleles at about 17 minutes after the mixing together of Hfr and F$^-$ cells. This interpretation not only predicts the order of genes on the *E. coli* chromosome, it also makes it possible to map the distances between the genes. The units of distance are defined as *minutes of chromosome transfer,* and each minute translates to approximately 47,000 bases transferred (Fig. 15.15c).

The plateaus for the percentage of colonies carrying each transferred gene derive from the fragility of cell-to-cell connections. Transfer between Hfr donors and F$^-$ recipients can spontaneously abort if the connected cells separate or if the transferring chromosome breaks. As a result, not all exconjugants receive even early arriving genes such as *azi*^{r}. In addition, the more time required to transfer a marker, the greater the chance that mating pairs will separate before the transfer occurs. As a result, the percentage of exconjugants receiving later-arriving genes becomes successively smaller.

Comparisons of Interrupted Mating Studies Using Different Hfr Strains Confirm That the Bacterial Chromosome Is a Circle

The HfrH strain transfers genes in the order *azi-ton-lac-gal.* A different Hfr strain (strain 1 in **Fig. 15.16a**) transfers these genes in the opposite order: *gal-lac-ton-azi.* The maps obtained using these and other Hfr strains as donors can be related on the basis of the fact that in various Hfr strains, the F plasmid integrates into the bacterial chromosome at different locations and in different orientations (Fig. 15.16b). The location of the origin of transfer within each type of F plasmid determines where transfer begins and thus the start of that particular map. The orientation of the plasmid with respect to the bacterial chromosome determines whether transfer occurs clockwise or counterclockwise.

Comparisons of maps drawn from many Hfr × F$^-$ crosses provided the first clues that the bacterial chromosome is circular (Fig.15.16b). The maps from different Hfr matings are all permutations of the same order of genes that would occur if the genes were located on a circular chromosome, but with transfer starting at a different position and moving in either the clockwise or counterclockwise direction. Many physical studies examining the DNA of the bacterial genome later confirmed the circularity of the bacterial chromosome.

Researchers now use well-studied collections of Hfr strains with different sites and orientations of F plasmid insertion to determine the approximate position on the *E. coli* genetic map of any gene they can follow on the basis of its phenotype. They designate the gene's position by the minute at which its transfer begins. The circumference of the entire circular map is 100 minutes, the time

(a) Gene transfer in different Hfr's

Hfr strain	Order of transfer ⟶
H	*thr* ***azi ton lac*** *pur* ***gal*** *his gly thi*
1	*thr thi gly his* ***gal*** *pur* ***lac ton azi***
2	***lac*** *pur* ***gal*** *his gly thi thr* ***azi ton***
3	***gal*** *pur* ***lac ton azi*** *thr thi gly his*

(b) Data interpretation

Strain H: thr, azi, ton, lac, pur, gal, his, gly, thi

Strain 1: thr, azi, ton, lac, pur, gal, his, gly, thi

Strain 2: thr, azi, ton, lac, pur, gal, his, gly, thi

Strain 3: thr, azi, ton, lac, pur, gal, his, gly, thi

■ F factor
▶ Direction of transfer

(c) The circular *E.coli* chromosome

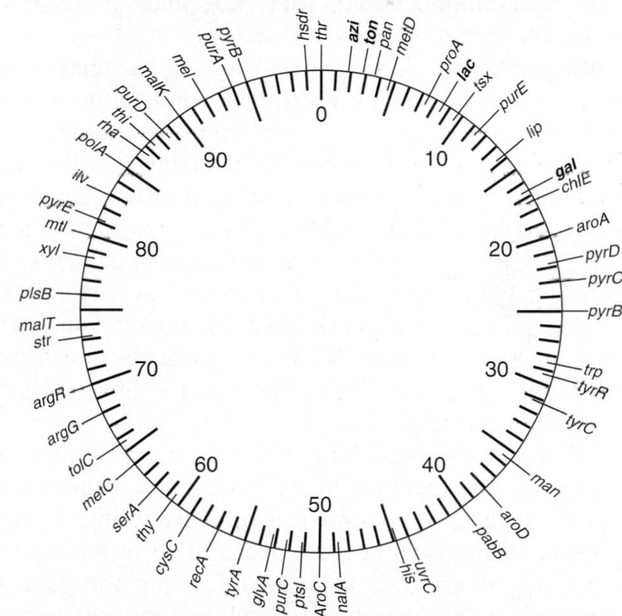

Figure 15.16 Order of gene transfers in different Hfr strains. **(a)** Comparison of the order of gene transfers in crosses between F$^-$ recipients and various Hfr strains. The genes from Fig. 15.15 are indicated in *bold;* genes mapped later by other investigators are in *bold face.* **(b)** Interpretation of the data shown in part (a). In different Hfr strains, the F plasmid integrates in different locations and in different orientations into the circular donor chromosome. **(c)** A partial genetic map of the *E. coli* chromosome.

required for transfer of the complete *E. coli* chromosome. The top of the circle, where 0 minutes meets 100 minutes, is arbitrarily set at the *thr* gene, the site of F plasmid insertion in HfrH. About 1400 of the estimated 4000–5000 *E. coli* genes have so far been located in this way. A selection of these mapped genes appears in Fig.15.16c.

Recombination Analyses of Hfr Crosses Improve Mapping Accuracy

Interrupted mating experiments cannot distinguish the relative positions of genes within about 2 minutes of each other and thus give only a crude idea of gene location. However, analyses of conjugation experiments can provide better resolution if they take into account what happens within an F^- exconjugant after it has received genes from the Hfr donor. Because bacteria cannot completely replicate and transmit the linear DNA from the Hfr donor remaining inside the F^- cell at the interruption of mating, the only way the progeny of the recipient F^- cell can inherit donor genes is if those genes recombine into the F^- cell's chromosome. Note that a single recombination event, or any odd number of recombination events, between the circular bacterial chromosome and the linear DNA fragment would be lethal: By making the chromosome linear, it would prevent multiplication of the cell. As a result, conjugation can produce recombinant cells only if an even number (two, four, six, and so forth) of crossovers occurs.

To map genes on the basis of the frequencies of recombinant classes emerging from a conjugation experiment, you must consider only exconjugants that received from the donor all of the genes to be mapped. Otherwise, the resulting progeny would reflect not only recombination events but also the likelihood that a particular gene from the Hfr donor was transferred to recipient cells. To ensure that all the recipient cells you are examining received all pertinent genes from the donor, you select exconjugants that contain the last of the genes transferred from the Hfr donor, as these cells must also have received the earlier arriving genes.

With this understanding of how to select suitable exconjugants and what type of recombination occurs inside those F^- cells, it is possible to map closely linked genes by a type of three-point cross. Recall that in three-point crosses in eukaryotes, you can infer that the gene in the middle is the gene that recombined with respect to the other two in the double-crossover class. You can do the same in prokaryotes, although here, the equivalent of the double-crossover class requires four crossovers. In the example shown in **Fig. 15.17,** the order of the closely linked genes *met* and *arg* is determined by a three-point cross in which you select for a nearby marker known to be transferred later—in this case, mal^+. After selecting for Mal^+ and then testing for the Arg and Met phenotypes, you tally the numbers for each type of exconjugant (Fig. 15.17b). The least frequently occurring class was produced by quadruple crossovers. Once the gene in the middle (*arg* in this experiment) has been defined, you can diagram the recombination events responsible for each class of recombinants (Fig. 15.17c). Comparisons of recombination frequencies between genes in a particular bacterial cross provide a measure of relative map distances.

(a) Hfr mapping: mal^+ transfers last

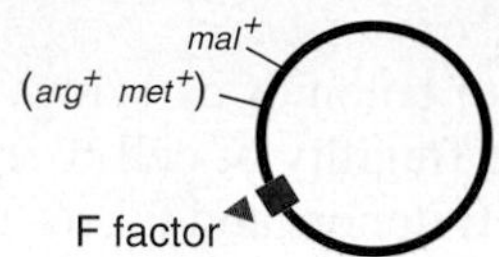

(b) Transfer of genes in mal^+ exconjugants

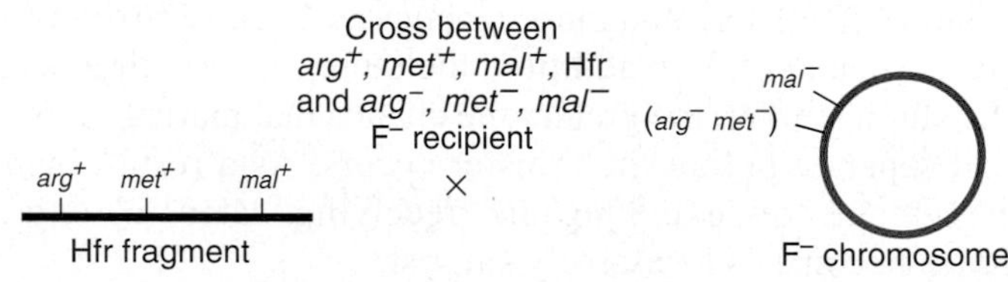

Exconjugant Phenotype	Number
Met^+ Arg^+ Mal^+	375
Met^- Arg^- Mal^+	149
Met^- Arg^+ Mal^+	32
Met^+ Arg^- Mal^+	2

(c) Types of crossover events

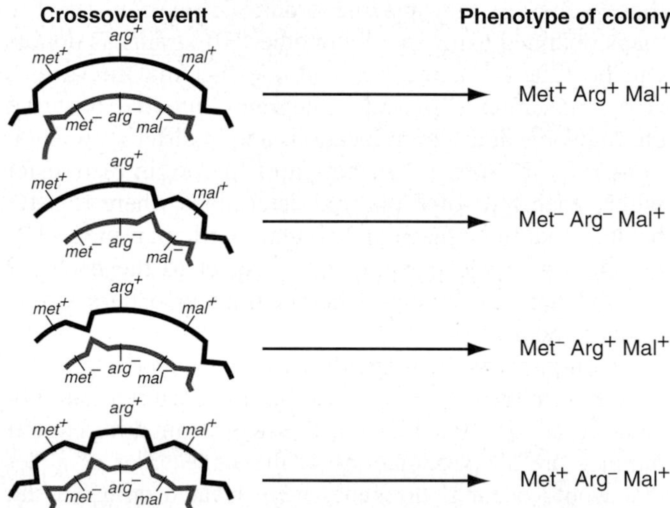

Figure 15.17 Mapping genes using a three-point cross. **(a)** The order of the closely linked genes *met* and *arg* can be determined by a three-point cross, selecting for a nearby marker known to be transferred later (in this case, mal^+). **(b)** The percentage of mal^+ exconjugants having all other combinations of the *arg* and *met* alleles are counted. **(c)** Crossover events that produce the various types of exconjugants. The met^+ arg^- mal^+ exconjugant requires quadruple crossovers, so it is the least frequently observed class of exconjugants, *arg* must be the gene in the middle since it recombines relative to both met and *mal* in this class.

Bacterial Geneticists Can Use a Special Type of F Plasmid Known As F′ for Genetic Complementation Studies

We saw earlier in this chapter that insertion of the F plasmid into the bacterial chromosome occurs in about 1 F^+ cell per 100,000 to produce an Hfr cell. In approximately that same proportion of Hfr cells, an excision event causes the Hfr cell to revert to an F^+ cell; and in a small fraction of these excision events, an error in recombination generates a plasmid containing most of the F plasmid genes plus a small region of the bacterial chromosome that had been adjacent to the integrated F plasmid. The newly formed plasmid carrying most of the genes of the F plasmid plus some bacterial DNA is known as an **F′ plasmid** (**Fig. 15.18a**). F′ plasmids replicate as discrete circles of DNA inside *E. coli* cells. They are transferred to recipient (F^-) cells in the same way that F plasmids are transferred. The difference is that a few chromosomal genes will always be transferred as part of the F′ plasmid. This ability to transfer chromosomal genes on a molecule that can replicate independently of the bacterial chromosome makes the F′ plasmid a perfect vehicle for creating partial diploids for complementation analysis.

Recall from Chapter 7 that complementation tests depend on cells that are diploid for the genes under analysis. Although bacteria are haploid, F′ plasmids carrying bacterial genes can create specific regions of partial diploidy. In fact, it is possible to find F′ plasmids carrying *E. coli* genes that lie close to any of the chromosomal positions where F plasmids integrate. For example, some F′ plasmids carry the *trp* genes that control the biosynthesis of tryptophan. An F′ plasmid carrying these genes is called an F′ *trp* plasmid.

To create partial diploids in bacterial cells using F′ plasmids, researchers must transfer the F′ plasmid into a strain whose chromosome is not deleted for the genes carried by the F′ plasmid (since the chromosome of the cell in which the aberrant plasmid excision occurred is now deleted for those genes). Because the F′ plasmids can still promote

(a) F' plasmid formation

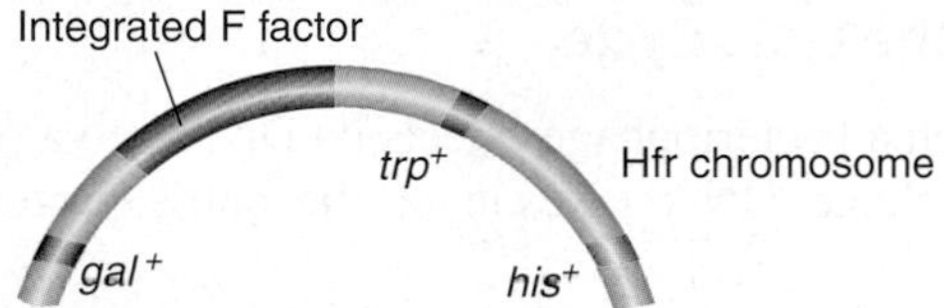

A rare recombination event between regions of limited sequence homology permits out-looping of F factor including *trp*+ locus.

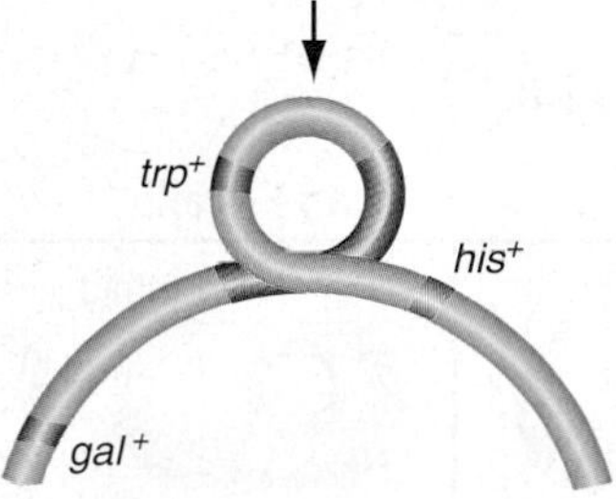

Separation of F creates F' *trp*+ plasmid and a chromosome deleted for the *trp* genes.

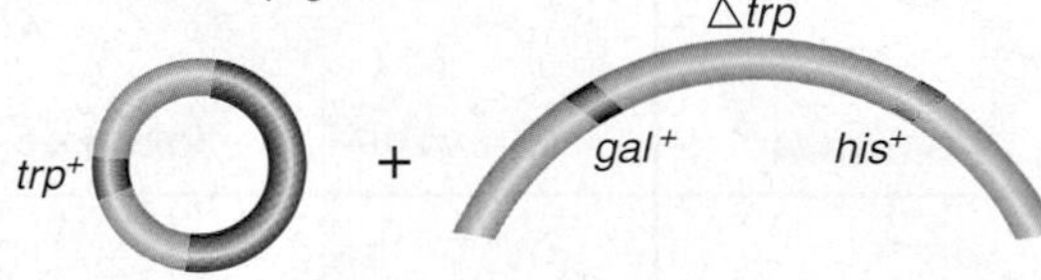

(b) F' plasmid transfer

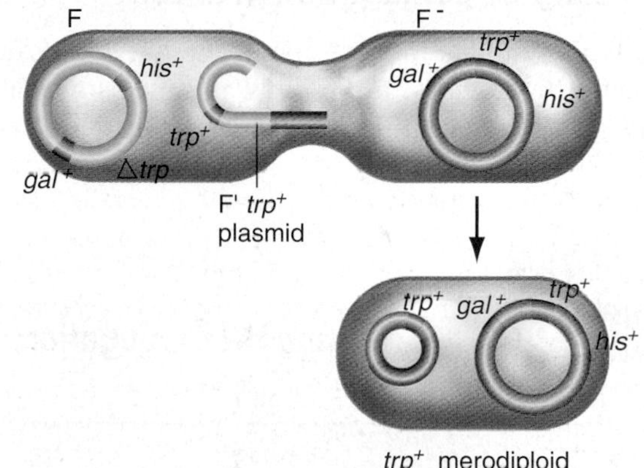

(c) Complementation testing using F' plasmids

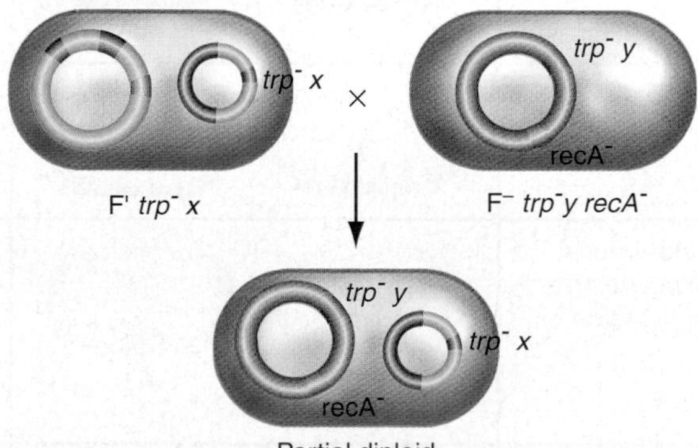

Figure 15.18 **F′ plasmids. (a)** Formation of an F′ plasmid by improper excision of an integrated F plasmid from an Hfr chromosome. **(b)** Transfer of an F′ *trp* plasmid into a trp^+ F^- recipient creates a merodiploid with two copies of trp^+. **(c)** Using F′ plasmids for complementation analysis. Different trp^- mutations are induced on the F′ *trp* plasmid (trp^-x) and in the F^- chromosome (trp^-y). Matings between these strains create partial diploids with both mutations; the phenotype of the partial diploid establishes whether the two mutations complement each other or not. The $recA^-$ mutation prevents recombination between the plasmid and the chromosome.

conjugation, experimenters easily accomplish this by mating F'-carrying cells with F^- cells (Fig. 15.18b). The exconjugants from these matings are partial diploids containing two copies of certain bacterial genes—one on the F' plasmid, the other on the bacterial chromosome. Partial diploids in which there are two copies of some genes are known as **merodiploids.**

As an example of complementation studies using F' plasmids, consider an analysis of mutations affecting tryptophan biosynthesis (Fig. 15.18c). All these mutations map very close to each other. You first construct a partial diploid by introducing an F' plasmid carrying the entire *trp* region with a particular trp^- mutation into a bacterial strain that carries a different trp^- mutation in the chromosome. Growth of a particular partial diploid on minimal medium without tryptophan would indicate that the mutations complement each other and are thus in different genes. If the cells do not grow, the mutations must be in the same gene. Complementation studies using F' *trp* merodiploids have shown that there are five different *trp* genes: *A, B, C, D,* and *E,* each one corresponding to one of the five polypeptides required for the biosynthesis of tryptophan.

Summary of Conjugation in *E. coli*

The F plasmid that initiates conjugation in *E. coli* functions in one of three ways. In F^+ cells, it remains a small, independent, self-replicating circle of DNA that during conjugation transfers only its own genes. In Hfr cells, it integrates via homologous recombination into the bacterial chromosome so that during conjugation, it transfers some or all of the chromosomal DNA. In F' cells, the F plasmid carries a small amount of chromosomal DNA, and during conjugation, only the chromosomal DNA on the plasmid is transferred, along with the plasmid itself. **Table 15.2** compares the conjugational cycles of the three F-plasmid incarnations.

Transduction: Gene Transfer via Bacteriophages

The bacteriophages, or phages, that infect, multiply in, and kill various species of bacteria are widely distributed in nature. Most bacteria are susceptible to one or more such viruses. During infection, a virus particle may incorporate a piece of the bacterial chromosome and introduce this piece of bacterial DNA into other host cells during subsequent rounds of infection. The process by which viral particles transfer bacterial DNA from one host cell to another is known as *transduction*.

Bacteriophage Particles Are Produced by the Lytic Cycle

When a bacteriophage injects its DNA into a bacterial cell, the phage DNA takes over the cell's protein synthesis

TABLE 15.2 Comparing the Conjugational Cycles of F^+, Hfr, and F' Cells

	In an F^+ Cell	In an Hfr Cell	In an F' Cell
Donor and Recipient Before Conjugation	F plasmid  F^+ F^-	 Hfr F^-	F' plasmid 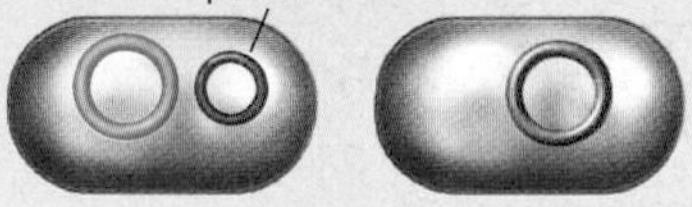 F' F^-
Events of Conjugation	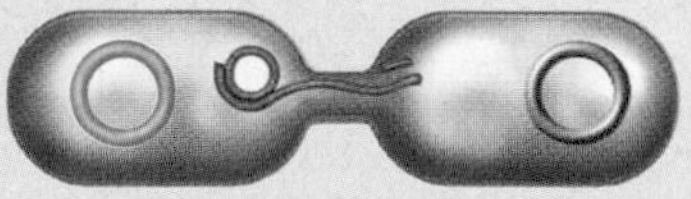	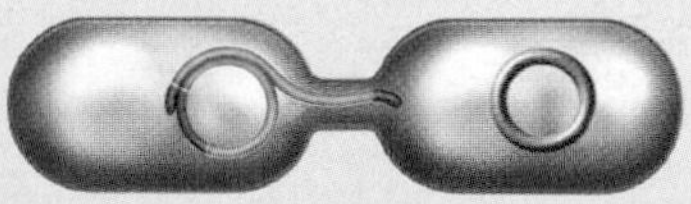	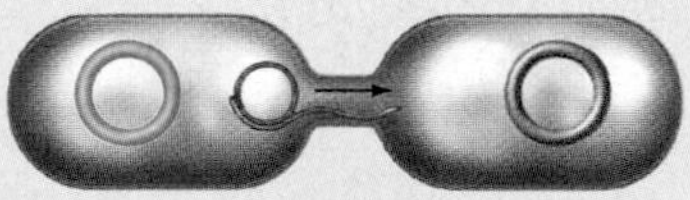
Donor and Recipient After Conjugation	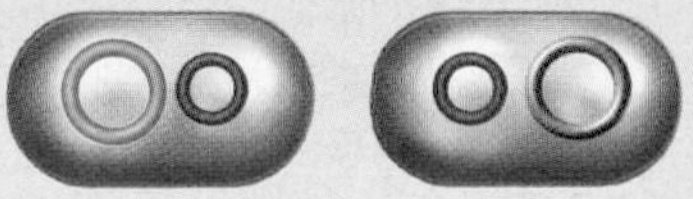 F^+ F^+	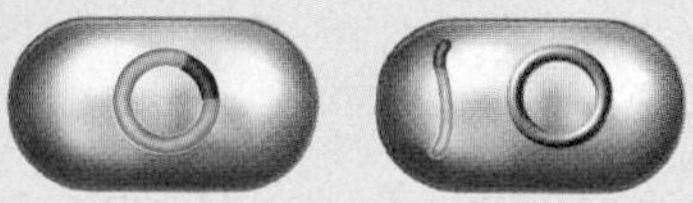 Hfr F^- partial diploid (transient)	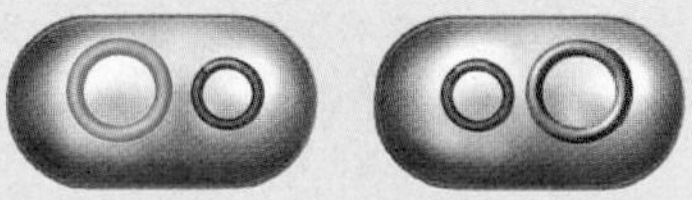 F' F' merodiploid
Comments	*Efficient transfer of F factor; poor transfer of bacterial genes.*	*Transfer of entire F factor only inefficiently. Efficient transfer of bacterial genes but requires an even number of crossovers in recipient (not shown).*	*Efficient transfer of F' element, which includes a few bacterial genes, generating stable partial diploids. Poor transfer of other chromosomal genes.*

and DNA replication machinery, forcing it to express the phage genes, produce phage proteins, and replicate the phage DNA (see Fig. 7.20 on pp. 228–229). The newly produced phage proteins and DNA assemble into phage particles, after which the infected cell bursts, or lyses, releasing 100–200 new viral particles ready to infect other cells. The cycle resulting in cell lysis and release of progeny phage is called the **lytic cycle** of phage multiplication. The population of phage particles released from the host bacteria at the end of the lytic cycle is known as a **lysate.**

Generalized Transduction

Many kinds of bacteriophages encode enzymes that destroy the chromosomes of the host cells. Digestion of the bacterial chromosome by these enzymes sometimes generates fragments of bacterial DNA about the same length as the phage genome, and these phage-length bacterial DNA fragments occasionally get incorporated into phage particles in place of the phage DNA (**Fig. 15.19**). After lysis of the host cell, the phage particles can attach to and inject the DNA they carry into other bacterial cells, thereby transferring genes from the first bacterial strain (the donor) to a second strain (the recipient). Recombination between the injected DNA and the chromosome of the new host completes the transfer. This process, which can result in the transfer of any bacterial gene between related strains of bacteria, is known as **generalized transduction.**

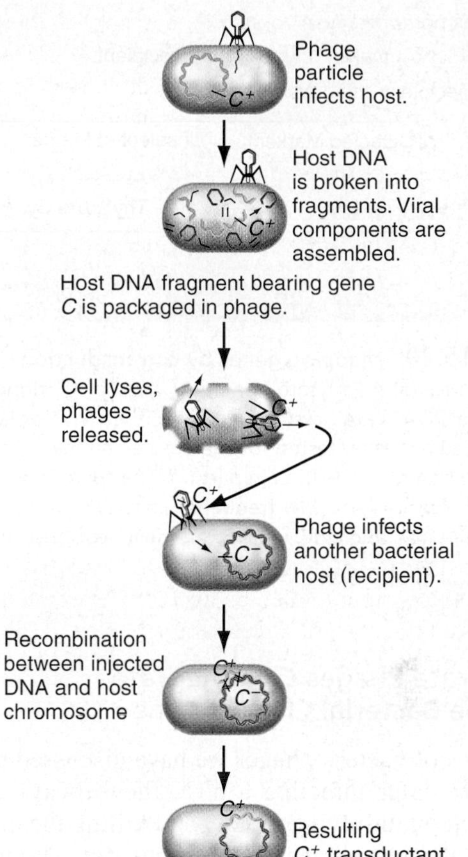

Figure 15.19 Generalized transduction. The incorporation of random fragments of bacterial DNA from a donor into bacteriophage particles yields generalized transducing phages. When these phage particles infect a recipient, donor DNA is injected into the recipient's cell. Recombination of donor DNA fragments with the recipient cell chromosome yields transductants, an even number of crossover events is required.

Mapping Genes by Generalized Transduction

As with cotransformation, two genes close together on the bacterial chromosome may be cotransduced. The frequency with which two bacterial genes are cotransduced depends directly on the distance between them: the closer two genes are, the more likely they are to appear on the same short DNA fragment and be packaged into the same transducing phage. Two genes that are farther apart than the amount of DNA that can be packaged into a single phage particle can never be cotransduced. For bacteriophage P1, a phage often used for generalized transduction experiments with *E. coli,* the maximum separation allowing cotransduction is about 90 kb of DNA, which corresponds to about 2% (or 2 minutes) of the bacterial chromosome. You will recall that researchers cannot use interrupted conjugation to define the relative location of genes injected less than 2 minutes apart. Thus, the two methods of genetic mapping can be used in concert to cover the entire spectrum of distances. In mapping a new *E. coli* mutation, investigators first find the approximate position of the mutation on the *E. coli* chromosome map by mating the mutant strain to a small collection of different Hfr strains. They then use P1 phage transduction to map the new mutation relative to other known genes in the region.

Consider, for example, the three genes—*thyA, lysA,* and *cysC*—that all map by interrupted mating experiments to the region between 53 and 56 minutes of the *E. coli* chromosome. Where do they lie in relation to one another? You can find out by using a P1 generalized transducing lysate from a wild-type strain to infect a $thyA^-$, $lysA^-$, $cysC^-$ strain and then selecting the transductants for either Thy^+ or Lys^+ phenotypes. After replica plating, you test each type of selected transductant for alleles of the two nonselected genes. As the phenotypic data in **Fig. 15.20a** indicate, *thyA* and *lysA* are close to each other but far from *cysC; lysA* and *cysC* are so far apart that they never appear in the same transducing phage particle; *thyA* and *cysC* are only rarely cotransduced. Thus, the order of the three genes must be *lysA, thyA, cysC* (Fig. 15.20b).

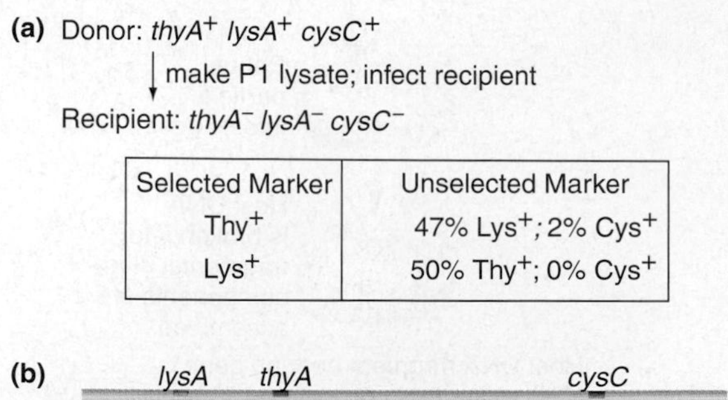

Selected Marker	Unselected Marker
Thy$^+$	47% Lys$^+$; 2% Cys$^+$
Lys$^+$	50% Thy$^+$; 0% Cys$^+$

Figure 15.20 Mapping genes by cotransduction frequencies. (a) A P1 lysate of a *thyA*$^+$ *lysA*$^+$*cysC*$^+$ donor is used to infect a *thyA*$^-$ *lysA*$^-$ *cysC*$^-$ recipient. Either Thy$^+$ or Lys$^+$ cells are selected and then tested for the unselected markers. **(b)** Genetic map based on the data in part (a). The *thyA* and *cysC* genes were cotransduced at a low frequency, so they must be closer together than *lysA* and *cysC*, which were never cotransduced.

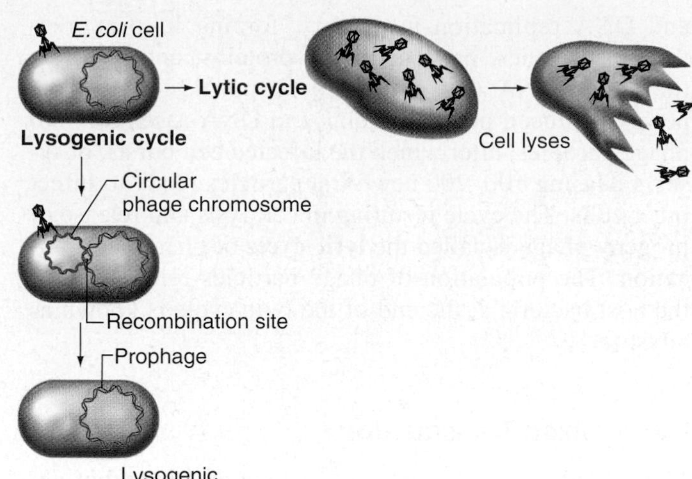

Figure 15.21 Temperate phages can choose between lytic and lysogenic modes of reproduction. Cells infected with temperate bacteriophages (whose chromosomes are shown in *green*) enter either the lytic or lysogenic cycles. In the lytic cycle, phages reproduce by forming new bacteriophage particles that lyse the host cell and can infect new hosts. In the lysogenic cycle, the phage chromosome becomes a prophage incorporated into the host chromosome.

Temperate Phages Can Integrate into the Bacterial Chromosome

The types of bacteriophages we have discussed so far are **virulent:** After infecting a host, they always enter the lytic cycle, multiplying rapidly and killing the cell. Other types of bacteriophages are **temperate:** Although they can enter the lytic cycle, they can also enter an alternative **lysogenic cycle,** during which their DNA integrates into the host genome and multiplies along with it, doing little or no harm to the host (**Fig. 15.21**). The integrated copy of the temperate bacteriophage is called a **prophage.** The choice of lifestyle—lytic or lysogenic—occurs when a temperate phage injects its DNA into a bacterial cell and depends on many factors, including environmental conditions. Normally when temperate phages inject their DNA into host cells, some of the cells undergo a lytic cycle, while others undergo a lysogenic cycle. One temperate phage commonly used in research is bacteriophage lambda (λ; **Fig. 15.22**).

Under certain conditions, it is possible to induce an integrated viral genome to excise from the chromosome, undergo replication, and form new viruses (Fig. 15.22c). In a small percentage of excision events, some of the bacterial genes adjacent to the site where the bacteriophage integrated may be cut out along with the viral genome and be packaged as part of that genome. Viruses produced by the faulty excision of a lysogenic virus from the bacterial genome are called **specialized transducing phage** (Fig. 15.22c). During the production of such phage, bacterial genes may become passengers along with the viral DNA. When the specialized transducing phage then infects other cells, these few bacterial genes may be transferred into the infected cells. The phage-mediated transfer of a few bacterial genes is known as **specialized transduction.** Temperate phages are thought to be a significant vehicle for the lateral transfer of genes from one bacterial strain to another or even from one species to another.

Comparing Generalized and Specialized Transduction

Phage particles that act as agents of generalized transduction differ in critical ways from particles that carry out specialized transduction.

1. Generalized transducing phages pick up donor bacterial DNA during the lytic cycle, at the point when DNA is packaged inside a phage protein coat; specialized transducing phages pick up the donor bacterial DNA during the transition from the lysogenic to the lytic cycle.
2. Generalized transducing phages can transfer any bacterial gene or set of genes contained in the right size DNA fragment into the bacterial chromosome; specialized transducing phages can transfer just those genes near the site where the bacteriophage inserted into the bacterial genome.

Bacterial Genetic Analysis Today

As geneticists learn more about genome structure and the mechanisms of gene transfer, including the transposition of DNA sequences, transduction, and conjugation, they are able to devise ever more clever ways of carrying out

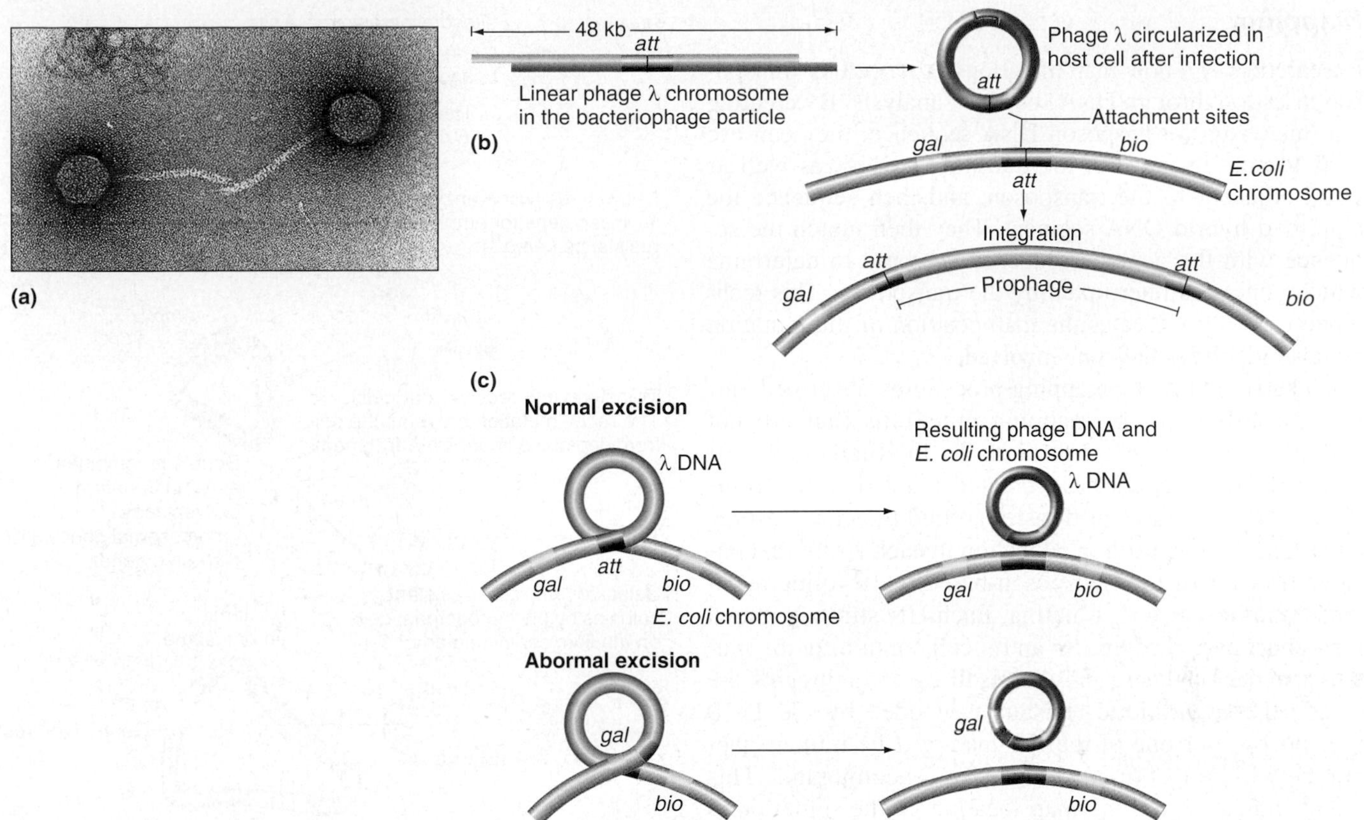

Figure 15.22 Bacteriophage lambda and lysogeny. **(a)** Electron micrograph of a temperate phage, bacteriophage lambda (λ). **(b)** Integration of the phage DNA initiates the lysogenic cycle. Recombination between *att* sites on the phage and bacterial chromosomes allows integration of the prophage. **(c)** Errors in prophage excision produce specialized transducing phages. Normal excision produces circles containing only lambda DNA. If excision is inaccurate, adjacent bacterial genes are included in the circles that form and in the resulting bacteriophages. Illegitimate recombination between the prophage and bacterial chromosome forms a circle that lacks some phage genes but has acquired the adjacent *gal* genes.

genetic analysis. Here we describe how an *E. coli* geneticist might approach the genetic dissection of a phenomenon. Many of the principles we present are applicable to other bacteria in which similar gene transfer mechanisms exist.

The Isolation of Mutants Is the Geneticist's Top Priority

Transposons have played the largest role in simplifying genetic analysis because they can create mutations. The insertion of a transposon into a gene, resulting in the gene's inactivation, is the basis of many mutant screens. Transposons are useful as mutagenic agents because they contain genes for easily selectable antibiotic resistance. To carry out transposon mutagenesis, geneticists introduce a transposon into a cell as part of a DNA molecule that is not able to replicate on its own inside the cell. The mechanism of gene transfer can be transformation, transduction, or conjugation. For the transposon to be passed on during cell division, it must transpose from the incoming DNA molecule to the bacterial chromosome. By growing cells on a medium containing antibiotics, it is possible to select for those cells in which transposition has occurred. One can then screen the resulting population of cells, which contain transposons at different locations around the chromosome, for the mutant phenotype of interest.

Transposon insertion usually inactivates the gene receiving the insertion, thereby creating a knockout or a null mutation. Such mutations can be useful, but if a gene is essential for a bacterium's survival, it will not be possible to isolate the knockout mutation. For genes encoding essential proteins, conditional mutants (for example, temperature-sensitive mutants) are isolated. Even for nonessential genes, conditional mutations may be the most informative, since cells grown under permissive conditions can be shifted to nonpermissive conditions and then observed for changes in phenotype.

Mapping

Researchers can now map mutations produced by transposon insertion through DNA sequence analysis. By creating a primer from a transposon DNA sequence, they can use PCR to amplify a bit of the transposon DNA, as well as DNA adjacent to the transposon, and then sequence the amplified hybrid DNA segment. They then match the sequence with the genome sequence database to determine which gene was interrupted by the transposon. This technique not only indicates the map position of the mutation but also identifies the gene involved.

The standard Hfr mapping procedures described earlier are still used for mapping mutations that are not caused by transposon insertion. To facilitate such Hfr mapping, researchers have created a set of Hfr strains with different origins of transfer around the chromosome and a Tn10 transposon in a location in each strain that ensures transfer of the transposon-based tetracycline resistance marker early in a mating. Each Hfr strain is mated for a short period of time to an F^- cell containing the mutation under analysis, and the resulting exconjugants are screened for antibiotic resistance encoded by the Tn10 transposon. For one of the Hfr matings, the mutant phenotype will be lost in most of the Tet^r exconjugants. This result indicates that the map location of the mutation is near the origin of transfer (the site at which the F plasmid integrated into the bacterial chromosome) in this particular Hfr strain. The use of 10–12 Hfrs can localize the mutation to a region of the chromosome. The use of P1 transductions can then refine the map position of the gene. This quick method for mapping a mutation eliminates the need for a series of interrupted matings and is thus much less labor intensive.

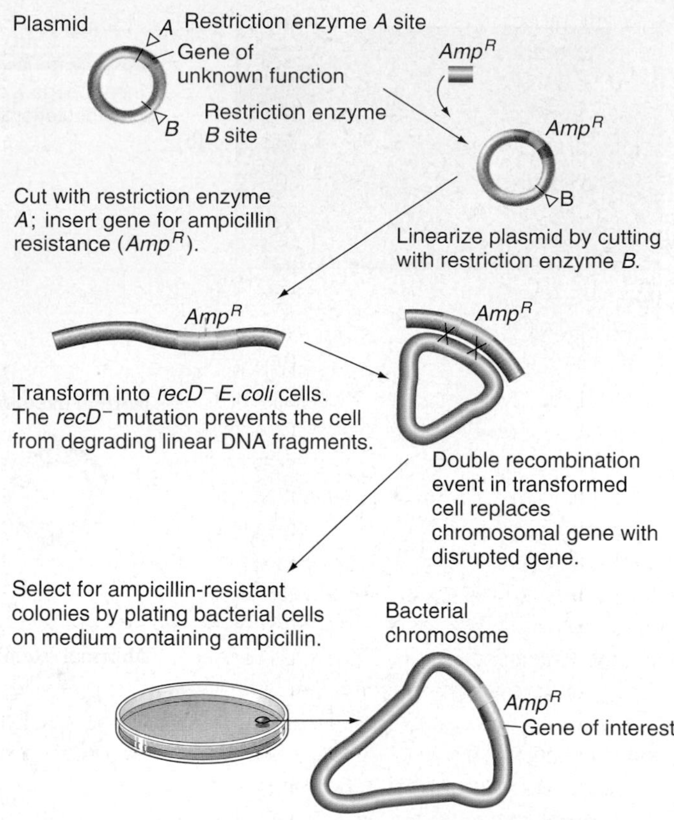

Figure 15.23 One way to characterize the roles of uncharted bacterial genes. First, an ampicillin resistance gene (Amp^R) is inserted into the gene of interest, located on a plasmid. The linearized plasmid is then transformed into a wild-type *E. coli* strain, where it recombines to leave a disrupted gene. The changed phenotype of the specific knockout can reveal the gene's function.

The Reverse Genetics Approach to the Study of Gene Function

The sequence analysis of bacterial genomes has led to the identification of genes whose functions are not yet known. One approach to determining the function of such genes is to make a knockout mutation in the chromosomal gene, using recombinant DNA techniques and the homologous recombination machinery of the cell (**Fig. 15.23**). This approach will work only if the gene is not essential. The gene to be mutated is first cloned in a plasmid. Next, a gene for antibiotic resistance is inserted into the middle of the cloned gene through use of a restriction enzyme that cuts only within the gene. The resulting circular molecule is then cut with a restriction enzyme that recognizes only a vector sequence; the cut creates a linear fragment of DNA that cannot replicate on its own in the cell and has to integrate into the chromosome to be passed on to further generations. This linear DNA fragment is used to transform a population of cells, after which selection for antibiotic resistance identifies those cells in which recombination between the mutated gene on the linear DNA fragment and the wild-type gene on the bacterial chromosome resulted in replacement of the wild-type gene with the mutated copy. The phenotype of the cell containing the knockout mutation provides clues about the function of the gene.

15.4 Comprehensive Example: Genetic Dissection Helps Explain How Bacteria Move

Bacteria can move up gradients of chemical attractants, such as food particles, and down gradients of chemical repellents, such as toxins. This sophisticated behavior of seeking out or avoiding chemicals diffusing through their growth medium is known as **bacterial chemotaxis.** Researchers have applied the powerful procedures for mutating, mapping, and cloning bacterial genes to identify the genes and proteins that function in bacterial chemotaxis and to discover how the proteins work together to accomplish their task.

How Bacteria Move to Achieve Chemotaxis

With specially constructed microscopes that make it possible to record the motion of individual bacteria, scientists have observed that cells swim rapidly in one direction for about 1 second and then abruptly reorient and swim in another direction. The 1-second, rapid, one-way movement is called a "straight run"; the abrupt reorienting movement is a "tumble" (**Fig. 15.24**). The change in direction is random; thus bacteria move through space in a **random walk.** With the addition of an attractant or repellent, the behavior of individual bacteria changes. They still move in a random walk, but the time spent in a straight run is longer immediately after the addition of an attractant and shorter immediately after the addition of a repellent. This reactive behavior is known as a **biased random walk:** Each bacterium still executes a random walk with frequent changes in direction, but its movement over time is biased toward or away from a chemical gradient. The biased behavior of individual bacteria determines the net movement of a population toward an attractant and away from a repellent.

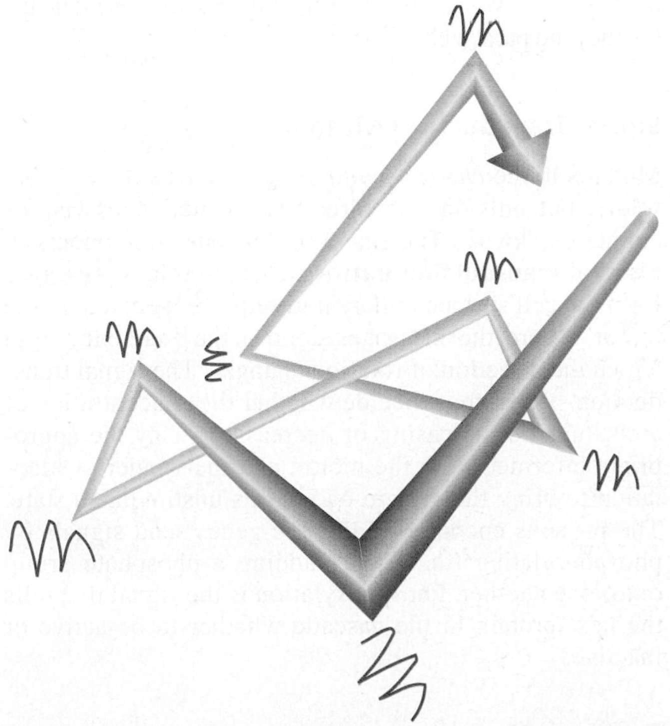

Figure 15.24 How bacteria move. The lack of a consistent direction is seen in the run and tumble style of movement. *Squiggles* indicate tumbles.

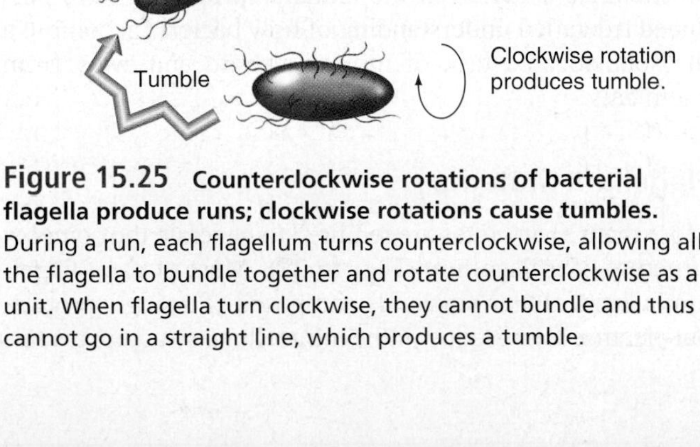

If the concentration of attractant or repellent does not change, within a short time, the bacteria resume the unbiased random walk they were doing before the addition of either type of chemical. The ability to stop responding when the stimulus is present in an unchanging concentration allows a bacterium to "sense" when it is moving up or down a concentration gradient. This aspect of chemotaxis is known as **adaptation.** It is analogous to the way you adjust to the bright light of day when you emerge from a dark movie theater—after a while in the new environment, your eyes adapt and you no longer blink.

Rotating Flagella Are the Mechanical Basis of the Random Walk

Bacteria have many long, rotating flagella distributed over their surface. When the flagella all rotate counterclockwise, they bundle together, propelling the cell in a forward run. When they rotate clockwise, the helical structure of each flagellum prevents bundling, and the cell undergoes a chaotic motion producing a tumble (**Fig. 15.25**).

You can observe the direction of rotation of the flagella on individual bacteria by attaching the cells to the surface of a petri plate with an antibody against the flagella and then looking down through a microscope at the rotating cells. As with runs and tumbles, the cells spend about 1 second rotating counterclockwise and then switch and rotate clockwise for a brief period. Immediately after the addition of an attractant, cells spend longer periods rotating counterclockwise; immediately after the addition of a repellent,

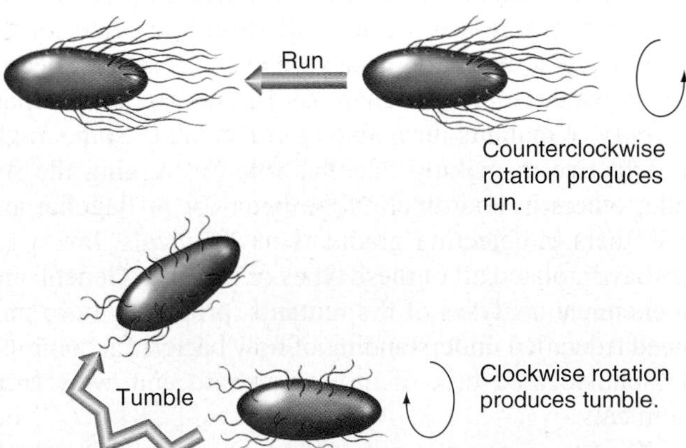

Figure 15.25 Counterclockwise rotations of bacterial flagella produce runs; clockwise rotations cause tumbles. During a run, each flagellum turns counterclockwise, allowing all the flagella to bundle together and rotate counterclockwise as a unit. When flagella turn clockwise, they cannot bundle and thus cannot go in a straight line, which produces a tumble.

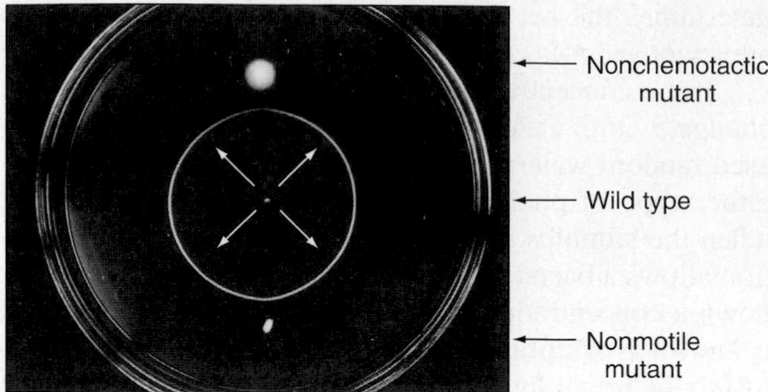

Figure 15.26 Isolating bacterial mutants that cannot move toward food. The wild-type bacteria move away from their original location at the center of the plate toward more food as nutrients are used up, and thus form a concentric ring. The nonchemotactic and nonmotile mutants are unable to sense or move toward food, respectively, and remain at their original locations.

they spend shorter periods rotating counterclockwise. From these observations, it is possible to conclude that gradients of attractants and repellents influence the direction of flagellar rotation.

Many Bacterial Mutants Cannot Carry out Chemotaxis

Bacteria inoculated into the center of a nutrient plate that is wet enough for them to swim in will multiply and use up the nutrients in the immediate vicinity. As they do, the population of cells swarms out in concentric circles, with the leading edge moving toward the unused nutrient supply. This behavior is the basis of a quick test for mutants that cannot move toward food; it also provides a way to isolate such mutants (**Fig. 15.26**). After the motile cells have swarmed, it is possible to lift from the center of the plate the cells that did not move. Given the known details of how bacteria move in chemotaxis, one would anticipate a variety of mutants incapable of chemotaxis. Some might be defective in making flagella; some in turning the flagella; others in coordinating the behavior of flagella; and still others in detecting gradients of chemicals. Investigators have isolated all of these types of mutants. Genetic and biochemical analyses of the mutants' properties have produced a detailed understanding of how bacteria accomplish the sophisticated task of moving toward and away from chemicals.

Flagellum Mutants

More than 20 *fla* genes are required to generate the complex structure of a flagellum (**Fig. 15.27**). Most of each flagellum's extracellular portion is composed of a single protein, but electron microscopy of the intracellular structure reveals

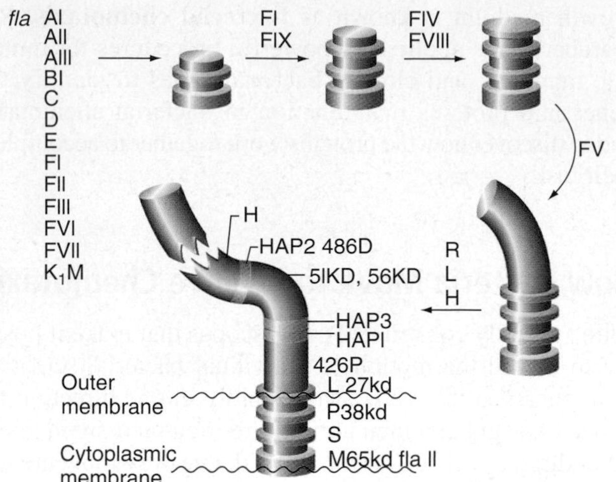

Figure 15.27 More than 20 genes are needed to generate a bacterial flagellum. Some proteins anchor the flagellum in the cell membrane; others implement the switch in rotational direction; still others serve as structural proteins for the part of the flagellum that is outside the cell.

a series of rings. Mutations in the *fla* genes usually prevent production of functional flagella.

Motor Mutants

The proteins encoded by the *mot* genes are required to turn the flagellum. Mutant *mot* cells have normal-looking flagella, but they are paralyzed.

Signal Transduction Mutants

Mutants in the *che* (*chemotaxis*) genes have flagella that rotate, but only in one direction—either clockwise or counterclockwise. The *che* genes encode components of a **signal transduction pathway,** which relays messages from the cell surface, where nutrients are detected, to the motor, where the messages control the frequency with which the direction of rotation changes. The signal transduction system must decide whether the concentration of a chemical is increasing or decreasing, relay the appropriate information to the motor, and then undergo adaptation to bring the system back to its unstimulated state. The proteins encoded by the *che* genes send signals by phosphorylating (that is, by adding a phosphate group onto) one another. Phosphorylation is the signal that tells the next protein in the cascade whether to be active or inactive.

Receptor Mutants

All the mutants described so far have a defective response to all attractants and repellents. But some mutants respond

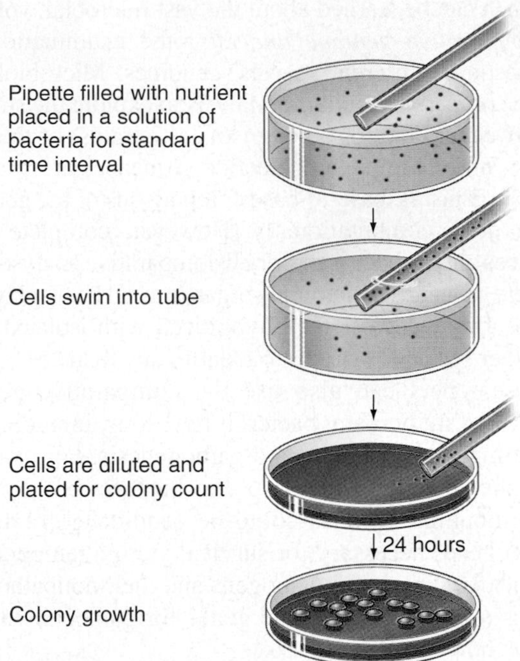

Figure 15.28 The capillary test for chemotaxis. The basis of this test is to see whether bacteria swim into a capillary tube containing attractants and/or repellents. The number of bacteria retained in the tube after a standard amount of time allows for quantitative measurements of chemotactic behavior.

in a defective way to a particular attractant or repellent, while responding normally to others. A simple "capillary test" reveals why. If a capillary pipette containing a nutrient is placed in a solution of bacteria, the cells swim into the tube (**Fig. 15.28**). Researchers leave the capillary tube in the solution for a standard amount of time and then remove it and determine the number of cells it contains by diluting and plating the bacteria for colony count. With this assay, they can discover the response of bacteria to specific nutrients and determine whether particular bacteria can respond to one nutrient in the capillary in the presence of a high concentration of another nutrient (both in the bacterial solution and in the capillary). Their observations confirm that bacteria normally respond to many different attractants and repellents independently. This suggests that the bacterial systems for detecting these chemicals are composed, in part, of different molecules.

Other studies have shown that mutants defective in specific responses have normal flagella that rotate normally and that switch from clockwise to counterclockwise with the expected frequency for an unstimulated cell. These mutants do not, however, respond when a particular stimulus is added to their growth medium, because the membrane proteins that act as receptors for the attractant or repellent in question are defective. These receptor proteins have an extracellular domain that interacts with one or a few attractants or repellents, a transmembrane domain, and an intracellular domain that interacts with the Che proteins to stimulate signal transduction (**Fig. 15.29**).

Not surprisingly, the cell surface receptors that bind particular chemicals are the molecules that are modified during adaptation to bring the cell back to an unstimulated state. The protein product of the *cheR* gene reversibly methylates positions on the intracellular domains of the receptors; the protein product of the *cheB* gene reversibly demethylates those same sites. The methylation status of the intracellular domain, in turn, determines the receptors' ability to respond to the binding of attractant or repellent by the extracellular receptor domain. The CheW protein phosphorylates CheB, CheY, and CheA, thereby activating them. Phosphorylated CheY sends a signal to the flagella for clockwise rotation. CheZ dephosphorylates CheY to restore the initial state (**Fig. 15.30**).

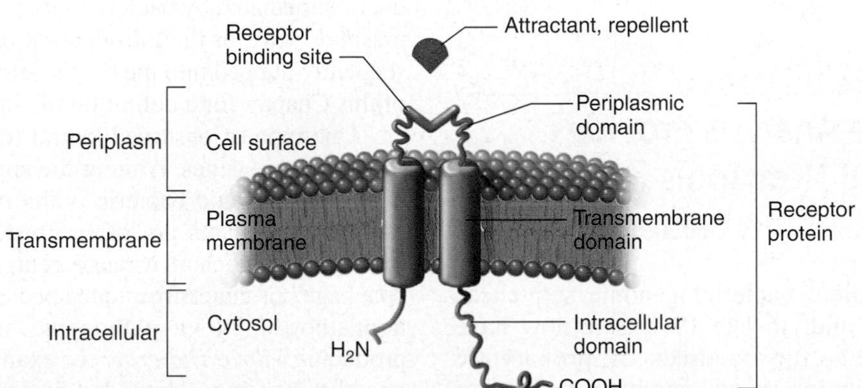

Figure 15.29 Bacteria have cell surface receptor proteins that recognize particular attractants or repellents. The binding of an attractant or repellent molecule to the membrane domain of a receptor specific for the compound causes changes in the conformation of the receptor's intracellular domain. The intracellular domain then interacts with the general components of the signaling pathway.

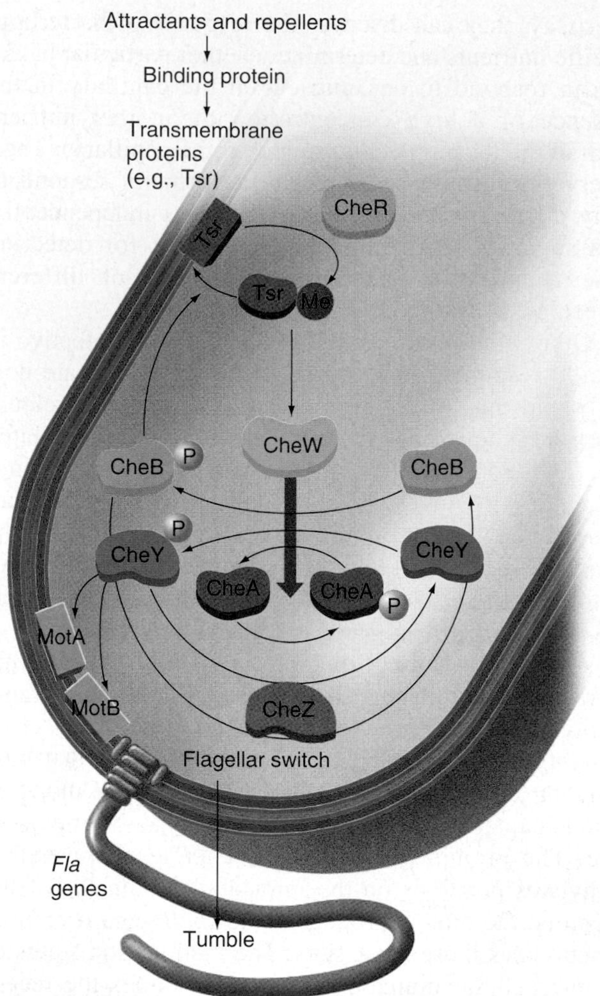

Figure 15.30 The genetic and molecular basis of bacterial chemotaxis. The *fla* genes are necessary for the structure of the flagella; *mot* genes are involved in the flagellar motor and determine the direction of rotation; *tsr* genes encode the transmembrane receptors; Che proteins process the signals from receptors and communicate with the Mot proteins to cause the appropriate response. The methylation (Me) of receptors prevents them from sustaining continual stimulation.

15.5 Genome Analysis Provides Powerful New Tools for Understanding Bacteria

While the first complete bacterial genome sequences were reported in the mid- to late 1990s, we now have complete genome data for hundreds of prokaryotic species. The recent explosion of genome data has not only provided intriguing information about pathogenesis, bacterial evolution, and unusual metabolic pathways and enzymes; it has also stimulated new avenues of inquiry and experimentation.

Much can be learned about the vast microbial world using *comparative genome analysis*—the examination and comparison of different species' genomes. Microbiologists can use comparative genome analysis to explore the similarities between species or among isolates (strains) of the same species. When complete nucleotide sequences are available for two organisms, base-to-base comparisons of the genomes can be done computationally. However, complete DNA sequences are not necessary for all comparative analyses. The complete sequence data for one organism can be displayed on a DNA array, which is then hybridized with isolated DNA from other species or strains to identify any matches.

Researchers can also use the comparative genome approach to study many bacterial functions, including survival at high temperatures and pathogenesis. For example, genes shared among thermophilic bacteria but not with nonthermophilic bacteria could be candidates for further study as genes necessary for survival at high temperatures. And one could compare pathogens and their nonpathogenic relatives to identify candidate genes for pathogenicity and examine how those genes arose.

Recent studies have suggested that several bacteria have streamlined their genomes and retained only genes essential for growth and pathogenesis. In these studies, comparative genome analysis of isolates of *Mycobacterium tuberculosis* showed deletions of genomic regions that must not be essential for growth and pathogenesis. Analyses of other obligate parasitic bacteria further narrowed the set of genes considered essential for growth.

Comparative Genome Analysis Reveals Strong Evidence for Lateral Gene Transfer

Comparative genome analyses of several pathogenic bacteria support the hypothesis that *lateral gene transfer* introduced many of the disease-causing genes found in these species. The transfers may have occurred by any of the mechanisms described earlier in this chapter, including mating between cells, the transduction of naked DNA or transduction mediated by bacterial viruses, the transformation of plasmid DNA, or the introduction of transposons that subsequently hopped into the host's genomic DNA. (See p. 550 of this Chapter for a definition of "lateral gene transfer.")

One type of observed lateral transfer involves groups of pathogenic genes. Among the significant genomic findings in pathogenic bacteria is the observation that pathogenic determinants are often clustered in **pathogenicity islands.** With such an arrangement, the lateral transfer of a "package" of genes from one species to another can turn a nonpathogen into a pathogenic strain. In the cholera-producing *Vibrio cholerae,* for example, a single region of one chromosome is bounded by inverted repeats, and the DNA of the region encodes a recombinase enzyme. The DNA organization and content suggest that the entire chromosomal region was transferred intact into *V. cholerae* from another source. Comparison of the genomes of *E. coli* and its close relative *S. enterica* var. Typhimurium confirm the presence of several large pathogenicity islands that had previously been identified by genetic mapping analyses.

In another visualization of lateral transfer, the genomes of several pathogenic strains show evidence that bacteriophages served as vectors in their evolution. For example, comparisons of the nonpathogenic laboratory *E. coli* K12 strain with two different strains of *E. coli* O157 (identified in the 1990s as the cause of food poisonings) showed that the O157 genomes contain several fragments of bacteriophage DNA that contribute to their pathogenicity. Apparently during the evolution of the pathogenic O157 strains, several DNA additions were required to produce virulence. Indeed, the O157 strains contain 1 Mb of DNA that is not present in the *E. coli* K12 genome; and approximately half the DNA present in both O157 isolates is bacteriophage DNA. The DNA introduced by bacteriophage encodes known toxins (similar to those produced by *Shigella*), an intestinal colonization protein, and an outer membrane protein that allows survival within macrophages (see p. 545 for definition). Other pathogens, including *Streptococcus* and *Staphylococcus* strains, contain bacteriophage DNA adjacent to their virulence genes (which consist of toxin-producing and host-evasion genes). The location of virulence genes next to bacteriophage DNA suggests that an imprecise excision event generated a specialized transducing phage able to transfer these genes to a new genome.

The glimpse into prokaryotic history made possible by comparative genome analysis has altered our view of evolution. Geneticists had thought that bacteria started out with a set of genes that slowly evolved through point mutation, deletion, and duplication within the species. But the data showing that some genes or sets of genes in one bacterial species are very similar to those in another species suggest that bacterial genomes have picked up DNA from several different sources during the course of their evolution. Biologists now recognize that lateral gene transfer between species is a significant evolutionary factor in pathogenicity and many other bacterial functions.

The Fruits of Genome Analysis May Help Protect Human Health

Many bacteriologists hope that genomic knowledge of pathogenic bacteria will lead to the identification of vaccine candidates. This is an ever more pressing concern in the current era of increasing bacterial resistance to antibiotics. Another hope in the fight against harmful pathogens is that genomic analysis will aid in the discovery of new drug targets. Identification of the genes and gene products essential for growth in a pathogenic species, for example, could allow rational drug design wherein pharmacologists synthesize compounds that target only those proteins not found in the host species.

Genomic technology and information may also help epidemiologists develop new strategies for the unambiguous identification of specific bacterial strains. They could then use this knowledge to trace the history of an infection. For example, during an outbreak of *V. cholerae,* DNA array analysis using characteristic, specific DNA from several isolates that have caused disease in the past could enable officials to determine if the outbreak is caused by a previously identified strain or a newly evolved one. If it turns out to be a new pathogen, investigators could identify key features of the new strain to understand how it evolved.

The Study of Microbial Ecology Benefits from Genomic Methodologies

Rapid DNA sequencing, large-scale PCR amplification, and DNA arrays have opened the door to investigations of various aspects of microbial ecology, including the composition of microbial communities in many settings. For example, bacteria that live in extreme environments have been difficult to culture in the laboratory. As a result, we know little about what organisms are present in these challenging environments, what their numbers are, what they do, and how they interact. In one recent study, researchers used PCR to amplify microbial DNA from various communities using primers for a conserved bacterial 16s rRNA (**Fig. 15.31**). Because different species show slight variations in the 16s rRNA sequence, the scientists could estimate the number of species present from comparisons of the amplified DNA. The data showed that there are many more species than had previously been identified by culturing and microscopic techniques.

Investigators also used a different large-scale approach to examine ocean microbes for metabolic as well as species diversity. They cloned and sequenced over 1 million random DNA fragments from the microbial genomes isolated from 200 liters of seawater and discovered many new proteins and metabolic activities (Fig. 15.31). Comparative analyses of these sequences revealed a vast metabolic diversity among oceanic microbes, including a very interesting rhodopsin-based photosynthesis capability in some bacteria.

A more directed approach to the study of microbial ecology is to determine the complete sequence of organisms that, though poorly understood, are recognized as major contributors to global nutrient cycles. Researchers have already sequenced the genomes of several marine cyanobacterial species critical for carbon fixation via photosynthesis. They can now analyze these genomes for their unique properties.

Combining Genetic and Genomic Approaches

Genomic experimentation adds an exciting new dimension to the impressive set of tools developed since the 1950s for the analysis of bacterial life. The marriage of genomic, genetic, and recombinant DNA approaches has led to elegant and innovative experiments. For example, a recent study of *Pseudomonas aeruginosa* used gene transfer techniques involving specially constructed transposons to produce a

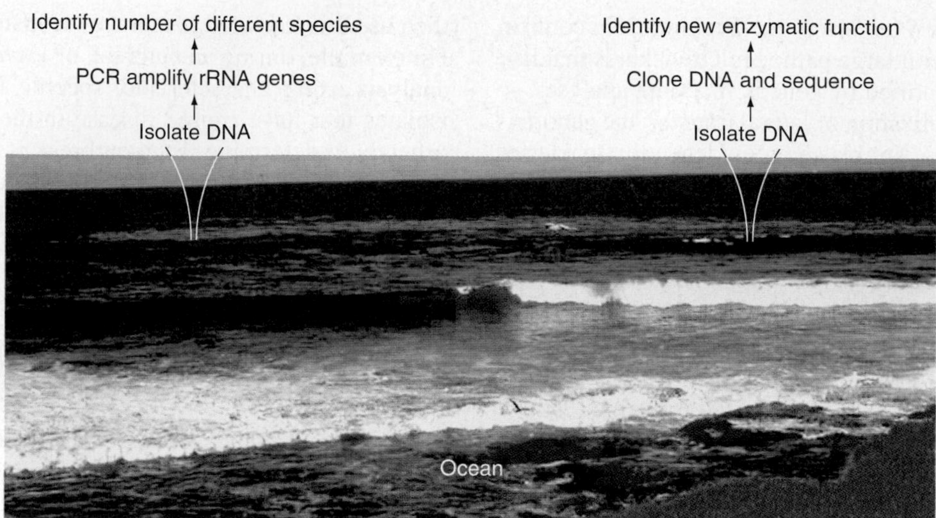

Figure 15.31 New analyses for assessing microbial diversity. Samples from the environment are analyzed to estimate either the number of different microbial species present or to identify new metabolic activities.

large-scale library of mutants for further genetic analysis. *P. aeruginosa* is an opportunistic pathogen that causes pulmonary infections in immune-compromised and cystic fibrosis patients. The sequence of its large 6.3 Mb genome was determined in 2000. However, since knowledge of DNA sequence does not immediately indicate what all the genes do, geneticists wanted to produce specific mutations in the genes to study their function. Taking a global approach, they generated a set of isolates, each containing a different gene mutated by insertion of a transposon carrying an antibiotic resistance marker into the open reading frame. To introduce the transposon-carrying DNA into the cell, the researchers mated *P. aeruginosa* with *E. coli*. The transposons in *E. coli* jumped into all regions of the *P. aeruginosa* chromosome. Mutants were selected on the basis of transposon-conferred drug resistance. The genes mutated by transposon insertion could be identified by PCR amplification followed by sequencing from the transposon into the adjacent DNA. The full genomic sequence was the reference material that allowed identification of the disrupted gene. Using this protocol, the investigators disrupted about 90% of the ORFs in the genome and characterized 36,000 mutations. The remaining 10% of the ORFs presumably included essential genes that could not be mutated if the bacteria were to survive. The mutant library now provides a resource for additional studies on the function of individual *P. aeruginosa* genes and demonstrates the effectiveness of a strategy that is broadly applicable to many other bacterial species. The genomic tools that allow us to approach questions on a large scale, combined with the tricks and tools developed through many years of genetic and recombinant experimentation, will speed our analysis of the microbial world.

Connections

The study of bacterial genetics underscores the unity of genetic phenomena in all types of living organisms. Double-stranded DNA serves as the genetic material in bacteria as it does in eukaryotes. The general mechanisms of gene expression, DNA replication, and recombination are also similar in prokaryotes and eukaryotes. The study of bacterial genetics shows, however, that within the unity of basic biological phenomena, there is a remarkable diversity of detail. Unlike most eukaryotic cells, bacteria do not produce gametes through meiosis or new generations of cells by the union of gametes. Nevertheless, bacteria can exchange genes between different strains through transformation, conjugation, and transduction. Together, these three modes of gene transfer increase the potential for the evolution of prokaryotic genetic material.

Knowledge of bacterial genetics raises the possibility of combatting bacteria harmful to humans, animals, and crops and using beneficial bacteria for positive purposes, such as degrading oil slicks and protecting crops from freezing. The genome sequences of hundreds of bacteria and the innovative use of genomic approaches assist in achieving these goals. Knowledge of bacterial genetics also increases our understanding of the organelles of eukaryotic cells. Biologists believe that mitochondria, the cell organelles that produce energy for metabolic processes, and chloroplasts, the photosynthetic organelles of plant cells, are descendants of bacteria that fused with the earliest nucleated cells. Mitochondria are similar in size and shape to today's aerobic bacteria and have their own DNA, which replicates independently of the cell's nuclear genetic material.

Chloroplasts are similar in shape and size to certain cyanobacteria and have self-replicating DNA carrying bacteria-like genes. Based in part on these observations, the **endosymbiont theory** proposes that chloroplasts and mitochondria originated when free-living bacteria were engulfed by primitive nucleated cells. Host and guest formed cellular communities in which each member adapted to the group arrangement and derived benefit from it. Chapter 16 examines the structure and function of the chromosomes of mitochondria and chloroplasts.

Essential Concepts

1. Bacteria are prokaryotic cells with no membrane-enclosed nucleus or other cell organelles. The bacterial genome consists of a single circular chromosome in which the genes are tightly packed, with about one gene per kilobase pair.
2. During transcription in *E. coli,* different strands of DNA serve as the template for different genes, with the most frequently transcribed genes oriented in the direction of replication fork movement. Replication begins at the origin of chromosomal replication (*oriC*) and proceeds bidirectionally around the circular genome to the termination region (*terC*).
3. In addition to their chromosome, most bacteria carry plasmids: small circles of double-stranded DNA. Plasmids may include genes that benefit the bacterial host under certain conditions. One important group of plasmids promotes conjugative gene transfer between two bacteria.
4. Bacterial genomes contain IS and Tn elements, transposons that can move between sites on any DNA molecule in the cell.
5. *Transformation* is a form of gene transfer in which donor DNA that is floating free in the growth medium enters a recipient cell. Some bacteria have cellular machinery that supports efficient natural transformation. Species that do not undergo natural transformation can be induced to take up DNA by treatments that disrupt their cell walls in a process known as artificial transformation.
6. *Conjugation* is a second form of gene transfer. It depends on direct cell-to-cell contact between a donor carrying a conjugative plasmid (the F plasmid is one example) and a recipient lacking such a plasmid. In crosses between F^+ and F^- strains, only the plasmid is transferred. In crosses between Hfr and F^- strains or between F' and F^- strains, the plasmid also transfers chromosomal genes.
7. *Transduction* is a third form of gene transfer in bacteria. It depends on the packaging of bacterial donor DNA in the protein coat of a bacteriophage. In generalized transduction, phages can package any part of the donor genome. Specialized transduction is a property of lysogenic bacteriophages, which can package only host genes adjacent to the integrated prophage genome.
8. It is possible to map bacterial genes using any of the three forms of gene transfer. Interrupted-mating experiments can map the approximate positions of genes. Measures of frequencies of cotransformation or cotransduction provide better resolution: The closer two genes are, the more likely they are to appear on the same short DNA fragment. Sequences of complete bacterial genomes are becoming increasingly important as mapping resources.
9. The tools of genomics, including comparative genome analyses and DNA arrays have provided new insights into pathogenesis, evolution, and microbial diversity.

On Our Website

www.mhhe.com/hartwell3
Chapter 15

Annotated Suggested Readings and Links to Other Websites

- Internet resources on bacterial genomics
- Recent papers on comparative genomics, lateral gene transfer, and the evolution of pathogens
- Recent papers on microbial communities using genomic analyses
- Recent papers on genetic analysis of a pathogen

Social and Ethical Issues

1. Bacteria have now been engineered to break down components of oil, potentially providing a biological way to clean up an oil spill. Many environmentalists have been upset by and are opposed to this action. Why would they be disturbed by this action? How widespread do you think the effects of seeding a spill site with oil-eating bacteria might be? Who would have to approve this type of action in cleaning up an oil spill? What characteristics might you want to engineer into these bacteria to make them safer for the environment?
2. Sarah has been using Southern blot analysis to determine whether genes homologous to specific ones found in *E. coli* are present in a rare thermophilic (heat-loving) bacterium. Since she had very little of the bacterial DNA, she was able to make only one filter, and she had to reprobe that filter several times. After many months of work, she is getting her answer on what genes are present. The final blot that she does gives a large smudge in an important lane, but she feels she can still tell where there are bands and where there are not bands. The blot is not of high enough quality for reproduction in a publication. Sarah knows that reviewers and editors would probably reject it, but she thinks she could refer to this piece of evidence in her paper without showing a photo of the blot (listed as data not shown in the paper). It would take several months to again grow enough of the bacteria to repeat the experiment, and she knows there are competitors with similar experiments to report. Should she go ahead and submit the paper for publication even though most people would say the data are messy?
3. *E. coli* strain O157 is a unique strain of this bacterial species that has caused serious outbreaks of human illness, in some cases leading to the death of small children. Yet *E. coli* is part of the normal flora of the gut of all individuals, and researchers work on harmless laboratory strains all the time. In a college microbiology course in a large university, one student got upset at the prospect of working with *E. coli* in the lab. Even with an explanation of the difference between the toxic and nontoxic bacteria, he was not convinced that his safety was ensured. He says he will not do the lab work himself but will do analysis of the lab data obtained by his partner. The laboratory is a required part of the course and accounts for 25% of the final grade. What should and/or could the professor do with this student? Should the student have to forfeit the laboratory points in the course? How far should the professor or university go to accommodate his needs?

Solved Problems

I. You have cloned the gene encoding the major protein in the flagella of a new bacterial strain. In screening for mutant bacteria that have a defective flagellar protein, you found mutants at an exceptionally high frequency (1 in 10^3 bacterial cells). You suspect these may have been caused by insertion of a transposable element into the gene. How could you determine if this had occurred?

Answer

One way to determine if the high-frequency mutants result from insertion into the gene is to perform a Southern hybridization. *The cloned gene would be used as a probe to hybridize with DNA from the wild-type and the mutant cells.* If the mutant arises from insertion of a transposable element, the size of fragments containing the interrupted gene will be different from fragments containing the normal gene.

II. You have an F^- strain of *E. coli* that is resistant to streptomycin but requires the following amino acids for growth on minimal medium: arginine, cysteine, methionine, phenylalanine, and proline. You do a series of interrupted-mating experiments with two different Hfr strains (Hfr1 and Hfr2), both of which are wild type for all of the amino acid markers but are Str^s.

a. The following data are the times in minutes at which you first observed recombinant progeny containing the indicated amino acid marker. Draw a chromosome map showing all the genes, the position of the origin of each Hfr, the direction of transfer of each Hfr, and the distances between each marker. Use a directional arrowhead to indicate each origin of transfer.

	Hfr1 × F^-	Hfr2 × F^-
arginine	5	40
cysteine	65	70
methionine	75	60
phenylalanine	35	10
proline	10	35

b. You do a new mating experiment with the same strains; you interrupt the mating after 60 minutes and plate on a medium that selects for progeny that are recombinant for the phenylalanine gene (phe^+) What medium would you use to select for these progeny?

Answer

a. The order of transfer of the five genes is indicated by the time at which cells showing the marker first appear. The origin of these two Hfr strains is different; that is, the location of the inserted F factor is different and the direction of transfer is different in the two Hfr strains.

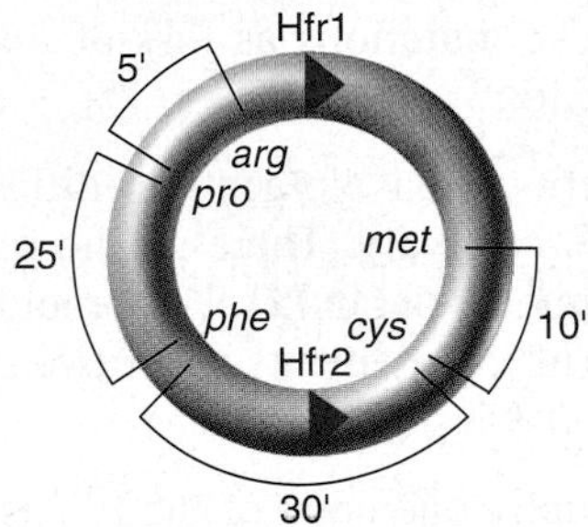

b. When you repeat the experiment selecting for progeny recombinant for phenylalanine, you plate *the cells on medium that contains arginine, cysteine, methionine, proline, and streptomycin.* The streptomycin selects against the Hfr strain, and only phenylalanine is left out of the media to select for F^- cells that have received genes from the donor.

III. Using bacteriophage P22 you performed a three-factor cross in *Salmonella typhimurium.* The cross was between an Arg^- Leu^- His^- recipient bacterium and bacteriophage P22, which was grown on an Arg^+ Leu^+ and His^+ strain. You selected for 1000 Arg^+ transductants and tested them on several selective media by replica plating. You obtained the following results:

Arg^+ Leu^- His^-	585
Arg^+ Leu^- His^+	300
Arg^+ Leu^+ His^+	114
Arg^+ Leu^+ His^-	1

a. What is the order of the three markers?
b. What are the cotransduction frequencies?

Answer

a. The order can be determined by looking at the relative frequencies of each phenotypic class. Arg^+ Leu^- His^- is the largest class, with only the *arg* gene transferred. The next largest class is Arg^+ and His^+. Therefore *arg* and *his* are closer to each other than *arg* and *leu. The order of the genes is* arg-his-leu.

b. Cotransduction frequency is the percentage of cells that received two markers. For *arg* and *his,* this includes the Arg^+ Leu^- His^+ cells (300) and Arg^+ Leu^+ His^+ cells (114). The cotransduction frequency of *arg* and *his* is 414/1000 = 41.4%. The cotransduction frequency of *arg* and *leu* is 114 + 1 or 115/1000 = 11.5%.

Problems

Interactive Web Exercise

The Institute for Genomic Research (TIGR) is a not-for-profit institute that hosts one of the most comprehensive sites for microbial genome research. The portion of the site devoted to comprehensive microbial research is known as CMR. Our website at www.mhhe.com/hartwell3 contains a brief exercise to introduce you to the use of TIGR-CMR; once at the web site, go to Chapter 15 and click on "Interactive Web Exercise".

Vocabulary

1. Choose the phrase from the right column that best fits the term in the left column.

a. transformation	1. requires supplements in medium for growth
b. conjugation	2. transfer of DNA between bacteria via virus particles
c. transduction	3. small circular DNA molecule that can integrate into the chromosome
d. lytic cycle	4. transfer of naked DNA
e. lysogeny	5. transfer of DNA requiring direct physical contact
f. episome	6. integration of phage DNA into the chromosome
g. auxotroph	7. infection by phages in which lysis of cells releases new virus particles

Section 15.1

2. The unicellular rod-shaped bacterium *E. coli* is ~2 μm long and 0.8 μm wide, and has a genome consisting of a single 5.6 Mb circular DNA molecule. The unicellular archaean *Methanosarcina acetivorans* is spherical (coccus-shaped) with a diameter of 3 μm and has a 5.7 Mb circular genome. The unicellular eukaryote *Saccharomyces cerevisiae* is roughly spherical, with a diameter of 5-10 μm. It has a haploid genome of 12 Mb divided among 16 linear chromosomes. Given these descriptions, how could you determine whether a new, uncharacterized microorganism was a bacterium, an archaean, or a eukaryote?

3. A liquid culture of *E. coli* at a concentration of 2×10^8 cells/ml was diluted serially, as shown in the following diagram, and 0.1 ml of cells from the last two test tubes were spread on agar plates containing rich media. How many colonies do you expect to grow on each of the two plates?

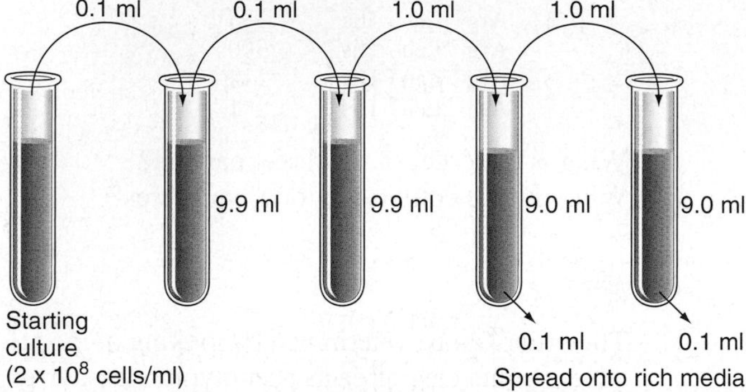

Section 15.2

4. Now that the sequence of the entire *E. coli* genome (about 5 Mb) is known, you can determine exactly where a cloned fragment of DNA came from in the genome by sequencing a few bases and matching that data with genomic information.

a. How many nucleotides of sequence information would you need to determine exactly where a fragment is from?

b. If you had purified a protein from *E. coli* cells, roughly how many amino acids of that protein would you need to know to establish which gene encoded the protein?

5. Pick out the media (i, ii, iii, or iv) onto which you would spread cells from a Lac^- Met^- *E. coli* culture to

a. select for Lac^+ cells
b. screen for Lac^+ cells
c. select for Met^+ cells

i. minimal media + glucose + methionine
ii. minimal media + glucose (no methionine)
iii. rich media + X-Gal
iv. minimal media + lactose + methionine

6. Linezolid is a new type of antibiotic that inhibits protein synthesis in several bacterial species by binding to the 50S subunit of the ribosome and inhibiting its ability to participate in the formation of translational initiation complexes. Physicians are particularly interested in this antibiotic for treating pneumonia caused by penicillin-resistant *Streptococcus pneumoniae* (also called ("pneumococci"). To explore the mechanisms by which pneumococci can develop resistance to linezolid, you want first to identify linezolid-resistant mutant strains. Next, using one of these strains as starting material, you now want to identify derivatives of these mutants that are no longer tolerant of linezolid.

a. Outline the techniques you would use to identify linezolid-resistant mutant pneumococci and linezolid-sensitive derivatives of these mutants. In each case, would your techniques involve direct selection, screening, replica plating, enrichment, treating with mutagens, or testing for a visible phenotype?

b. Suggest possible mutations that could be responsible for the two kinds of phenotypes you will identify. What types of events in the bacterial cells would be altered by the mutations? Can you classify these mutations as loss-of-function or gain-of-function?

7. The numbers of IS1 elements in different laboratory strains of *E. coli* vary. There are no recognition sites for the enzyme *Eco*RI in IS1. How could you determine the number of IS1 elements in the two strains *E. coli B* and in *E. coli K?*

8. There is usually one copy of the F plasmid per cell in an *E. coli* strain. You suspect you have isolated a cell in which a mutation increases the copy number of F to three to four per cell. (The copy number is determined by hybridization experiments.) How could you distinguish between the possibility that the copy number change was due to a mutation in the F plasmid versus a mutation in a chromosomal gene?

9. Genes encoding toxins are often located on plasmids. There has been a recent outbreak in which a bacterium that is usually nonpathogenic is producing a toxin. Plasmid DNA can be isolated from this newly pathogenic bacterial strain and separated from the chromosomal DNA. To determine if the plasmid DNA contains a gene encoding the toxin, you could determine the sequence of the entire plasmid and search for a sequence that looks like other toxin genes previously identified. There is an easier way to determine whether the plasmid DNA carries the gene(s) for the toxin that does not involve DNA sequence analysis. Describe an experiment using this easier method.

Section 15.3

10. a. In order to map the closely linked genes *ptsI* and *purC*, you select a strain in which the F factor integrated at about 45 minutes and the direction of transfer is toward increasing numerical "addresses" along the chromosome. Using Fig. 15.16 as a guide, identify a reasonable marker to use for selection of exconjugates for recombination analysis.

b. You perform a cross with the Hfr strain described above, which is also $TyrA^+$, $PtsI^+$, and $PurC^+$, and an F$^-$ strain negative for all three traits. You select exconjugates that are TyrA$^+$. Describe the relative frequencies with which you expect the following phenotypic categories of recombinant exconjugates.

i. TyrA$^+$, PtsI$^+$, PurC$^+$
ii. TyrA$^+$, PtsI$^+$, PurC$^-$
iii. TyrA$^+$, PtsI$^-$, PurC$^+$
iv. TyrA$^+$, PtsI$^-$, PurC$^-$

11. a. You want to perform an interrupted mating mapping with an Hfr strain that is Pyr$^+$, Met$^+$, Xyl$^+$, Tyr$^+$, Arg$^+$, His$^+$, Mal$^+$, and Str^s. Describe an appropriate bacterial strain to be used as the other partner in this mating.

b. In an Hfr × F$^-$ cross, the *pyrE* gene enters the recipient in 5 minutes, but at this time point there are no exconjugants that are Met$^+$, Xyl$^+$, Tyr$^+$, Arg$^+$, His$^+$, or Mal$^+$. The mating is now allowed to proceed for 30 minutes and Pyr$^+$ exconjugants are selected. Of the Pyr$^+$ cells, 32% are Met$^+$, 94% are Xyl$^+$, 7% are Tyr$^+$, 59% are Arg$^+$, 0% are His$^+$, 71% are Mal$^+$. What can you conclude about the order of the genes?

12. You have strains with the following genotypes:

Hfr *cys*$^-$ *trp*$^+$ *man*$^+$ *his*$^+$ *tyr*$^+$ *thr*$^+$ *str*^{s}
F$^-$ *cys*$^+$ *trp*$^-$ *man*$^-$ *his*$^-$ *tyr*$^-$ *thr*$^-$ *str*^{r}

These mutations have the following phenotype:

Cys$^-$	cannot synthesize cysteine
Str^s	sensitive to streptomycin
Trp$^-$	cannot synthesize tryptophan
Man$^-$	cannot utilize the sugar mannose
His$^-$	cannot synthesize histidine
Tyr$^-$	cannot synthesize tyrosine
Thr$^-$	cannot synthesize threonine
Str^r	resistant to streptomycin

a. What is the simplest medium you can use to grow the Hfr strain? the F$^-$ strain?

b. You mixed the two strains and did two experiments. In experiment 1, you took samples and then diluted and plated on minimal medium supplemented with histidine, tyrosine, threonine, and streptomycin and glucose as a carbon source. In experiment 2, you did essentially the same experiment, but before you diluted, you subjected the mating mixture to treatment in the food blender. In experiment 2, you never got any colonies until after 8 minutes. In experiment 1, you got some colonies at an earlier time. Why was there a difference in the results of these two experiments?

c. What medium would you use to select the exconjugants from experiment 2 that are Man$^+$?

d. The following table presents some of the data from experiment 2 in part *b*.

Marker	Time when first recombinants were found in minutes
Man$^+$	15.5
His$^+$	24
Thr$^+$	16
Tyr$^+$	16

Make the best map of the Hfr origin and genes.

e. To resolve ambiguities, you did the following crosses:

Cross 3 Hfr *man*$^+$ *thr*$^-$ *tyr*$^+$ × F$^-$ *man*$^-$ *thr*$^+$ *tyr*$^-$
Cross 4 Hfr *man*$^+$ *thr*$^+$ *tyr*$^-$ × F$^-$ *man*$^-$ *thr*$^-$ *tyr*$^+$

You note how many wild-type recombinants are produced and find many more in cross 3 than in cross 4. Show the order of these three genes.

13. Generalized and specialized transduction both involve bacteriophages. What are the differences between these two types of transduction?

14. In a cross between an Hfr that has the genotype *ilv*$^+$ *bgl*$^+$ *mtl*$^+$ and an F$^-$ that is *ilv*$^-$ *bgl*$^-$ *mtl*$^-$, the *ilv* gene is known to be transferred later than *bgl* and *mtl*. To determine the order of *bgl* and *mtl* with respect to *ilv*, Ilv$^+$ exconjugants were selected, and these colonies were screened for Bgl and Mtl phenotypes. Based on the following data, what is the order of the three genes?

Exconjugant type	Number of exconjugants
Ilv$^+$ Mtl$^+$ Bgl$^+$	220
Ilv$^+$ Mtl$^-$ Bgl$^-$	60
Ilv$^+$ Mtl$^+$ Bgl$^-$	0
Ilv$^+$ Mtl$^-$ Bgl$^+$	18

15. In two isolates (one is resistant to ampicillin and the other is sensitive to ampicillin) of a new bacterium, you found that genes encoding ampicillin resistance are being transferred into the sensitive strain. To determine if the gene transfer is transduction or transformation, you treat the mixed culture of cells with DNase. Why would this treatment distinguish between these two modes of gene transfer? Describe the results predicted if the gene transfer is transformation versus transduction.

16. Starting with an F$^+$ strain that was prototrophic (that is, had no auxotrophic mutations) and Str^s, several independent Hfr strains were isolated. These Hfr strains were mated to an F$^-$ strain that was Str^r Arg$^-$ Cys$^-$ His$^-$ Ilv$^-$ Lys$^-$ Met$^-$ Nic$^-$ Pab$^-$ Pyr$^-$ Trp$^-$. Interrupted mating experiments showed that the Hfr strains transferred the wild-type alleles in the order listed in the following table as a function of time. The time of entry for the markers within parentheses could not be distinguished from one another.

Hfr strain	Order of transfer →
HfrA	*pab ilv met arg nic (trp pyr cys) his lys*
HfrB	*(trp pyr cys) nic arg met ilv pab lys his*
HfrC	*his lys pab ilv met arg nic (trp pyr cys)*
HfrD	*arg met ilv pab lys his (trp pyr cys) nic*
HfrE	*his (trp pyr cys) nic arg met ilv pab lys*

a. From these data derive a map of the relative position of these markers. Indicate with labeled arrows the position and orientation of the integrated F plasmid for each Hfr strain.

b. To determine the relative order of the *trp, pyr,* and *cys* markers and the distances between them, HfrB was mated with the F^- strain long enough to allow transfer of the *nic* marker, after which Trp^+ recombinants were selected. The unselected markers *pyr* and *cys* were then scored in the Trp^+ recombinants, yielding the following results:

Number of recombinants	Trp	Pyr	Cys
790	+	+	+
145	+	+	–
60	+	–	+
5	+	–	–

Draw a map of the *trp, pyr,* and *cys* markers relative to each other. (Note that you cannot determine the order relative to the *nic* or *his* genes using this data.) Express map distances between adjacent genes as the frequency of crossing-over between them.

c. Detailed study of the genes in the *trp, pyr, cys* region would require an F′ plasmid carrying the three genes. Which of the preceding Hfr strain(s) would you use to isolate such an F′ plasmid?

17. You have recovered three independent mutations in one of the arginine biosynthesis genes. You now wish to map the site of the three mutations (*arg101, arg102,* and *arg103*) in the *arg* gene via the following Hfr × F^- crosses in which Lac^+ cells were selected and then screened for Arg phenotype.

Cross	Hfr (lac^+)	F^- (lac^-)	% of Lac^+ that are Arg^+
A	$arg101^-$, $arg102^+$	$arg101^+$, $arg102^-$	0.50
B	$arg101^+$, $arg102^-$	$arg101^-$, $arg102^+$	0.02
C	$arg101^+$, $arg103^-$	$arg101^-$, $arg103^+$	0.50
D	$arg101^-$, $arg103^+$	$arg101^+$, $arg103^-$	0.04
E	$arg102^+$, $arg103^-$	$arg102^-$, $arg103^+$	0.50
F	$arg102^-$, $arg103^+$	$arg102^+$, $arg103^-$	0.06

Draw a genetic map of the three *arg* mutations. Show the order of the three mutant sites relative to the *lac* gene.

18. You can carry out matings between an Hfr and F^- strain by mixing the two cell types in a small patch on a plate and then replica plating to selective media. This methodology was used to screen hundreds of different cells for a recombination-deficient $recA^-$ mutant. Why is this an assay for RecA function? Would you be screening for a $recA^-$ mutation in the F^- or Hfr strain using this protocol?

19. An F′ element containing the maltose (mal^+) gene is discovered in *E. coli.*

a. Starting with an F^+ strain of *E. coli,* diagram the chromosomal events that must have occurred to produce this element.

b. The F′ mal^+ element was introduced into a F^- mal^- strain. Most of the resulting cells transferred the element to other F^- cells. However, occasionally cells appeared that transferred the entire *E. coli* chromosome, beginning with the maltose gene. Draw the chromosomal events that must have occurred to produce such a strain.

20. Suppose you have two Hfr strains of *E. coli* (HfrA and HfrB), derived from a fully prototrophic streptomycin-sensitive (wild-type) F^+ strain. In separate experiments you allow these two Hfr strains to conjugate with an F^- recipient strain (Rcp) that is streptomycin resistant and auxotrophic for glycine (Gly^-), lysine (Lys^-), nicotinic acid (Nic^-), phenylalanine (Phe^-), tyrosine (Tyr^-). and uracil (Ura^-). By using an interrupted mating protocol you determined the earliest time after mating at which each of the markers can be detected in the streptomycin-resistant recipient strain, as shown here.

	Gly^+	Lys^+	Nic^+	Phe^+	Tyr^+	Ura^+
HfrA × Rcp	3	*	8	3	3	3
HfrB × Rcp	8	3	13	8	8	8

(The * indicates that no Lys^+ cells were recovered in the 60 minutes of the experiment.)

a. Draw the best map you can from these data, showing the relative locations of the markers and the origins of transfer in strains HfrA and HfrB. Show distances where possible.

b. To resolve ambiguities in the preceding map, you studied cotransduction of the markers by the generalized transducing phage P1. You grew phage P1 on strain HfrB and then used the lysate to infect strain Rcp. You selected 1000 Phe^+ clones and tested them for the presence of unselected markers, with the following results:

Number of transductants	Gly	Lys	Nic	Phe	Tyr	Ura
			Phenotype			
600	−	−	−	+	−	−
300	−	−	−	+	−	−
100	−	−	−	+	+	+

Draw the order of the genes as best you can based on the preceding cotransduction data.

c. Suppose you wanted to use generalized transduction to map the *gly* gene relative to at least some of the other markers. How would you modify the cotransduction experiment just described to increase your chances of success? Describe the composition of the medium you would use.

21. In *E. coli,* four streptomycin-sensitive, prototrophic Hfr strains were mated with a streptomycin-resistant F$^-$ strain auxotrophic for a number of nutritional requirements (indicated by letters of the alphabet). Matings were interrupted at various intervals, and cells were plated on streptomycincontaining minimal medium supplemented with particular nutrients to test for gene transfer. The results are plotted in the graphs shown.

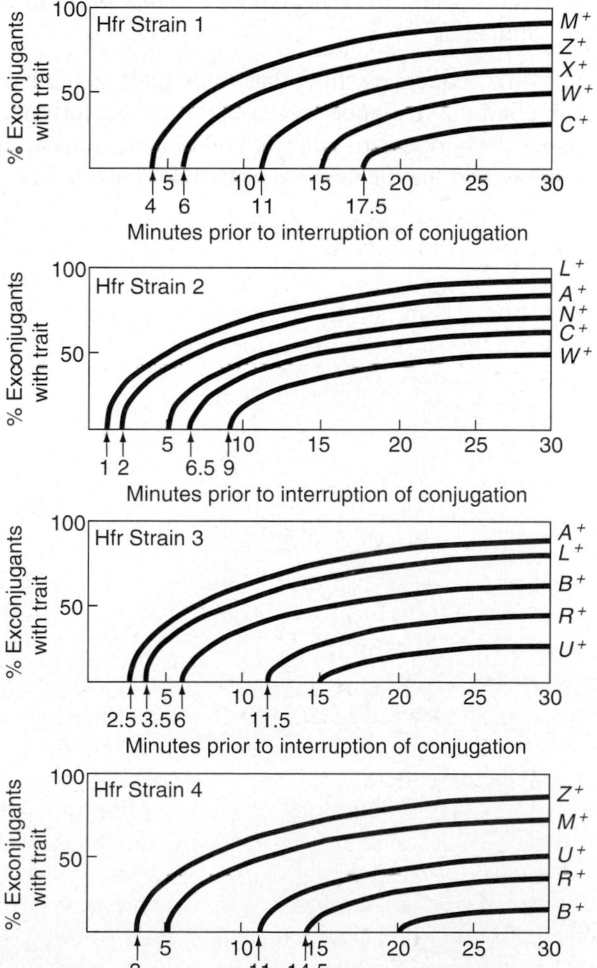

a. Draw the best map for the *E. coli* genome consistent with the data. Include distances between the genes, and label arrowheads to show the origin of transfer for the four Hfr strains.

b. Suppose you wanted to generate an F′ (N^+) plasmid for use in complementation experiments. Describe how you would obtain such a plasmid, indicating which of the four Hfr strains in the graphs you would use and briefly listing the steps involved in isolating the F′ (N^+) plasmid. (Hint: Think about how the time of transfer of gene *N* could differ in an Hfr and an F′ (N^+) strain.)

Section 15.4

22. DNA sequencing of the entire *H. influenzae* genome was completed in 1995. When DNA from the nonpathogenic strain *H. influenzae Rd* was compared to that of the pathogenic *b* strain, eight genes of the fimbrial gene cluster (located between the *purE* and *pepN* genes) involved in adhesion of bacteria to host cells were completely missing from the nonpathogenic strain. What effect would this have on cotransformation of *purE* and *pepN* genes using DNA isolated from the nonpathogenic versus the pathogenic strain?

23. a. Using the following pieces of technical information in the order given, explain how you would be able to identify the genes encoding proteins in *E.coli* cells that could bind directly to β-galactosidase: (1) β-galactosidase protein binds very tightly to a resin called APTG-agarose; (2) the 20 amino acids found in proteins vary widely in molecular weight; (3) the enzyme trypsin can cleave proteins into smaller peptides that are in the range of 3–40 amino acids long; the enzyme cleaves in a very predictable and reproducible way (after lysine and arginine amino acids); (4) modern techniques of mass spectrometry can measure the molecular weight of peptides to an accuracy of 0.01%; mass spectrometry machines measure the molecular weights of a large number of peptides in a complex mixture at the same time; (5) the entire *E. coli* genome has been sequenced.

b. Generalize the technique you described in part *a* to identify the genes encoding proteins that bind to any other particular protein in *E. coli.* (Hint: Use the fact that β-galactosidase binds to the APTG-agarose in your scheme.)

24. You have isolated a new temperature-sensitive mutation that affects the ability of *E. coli* to insert proteins into the membrane. At the restrictive temperature, the cells are unable to grow and thus do not form colonies.

a. Using a set of Hfr strains with origins every 5′ around the chromosome and a Tn10 insertion as a marker that is transferred in the first 5′, describe how you would rapidly narrow down the location of the new mutation. How many matings would you have to do?
b. If the Hfr set was not available, you could locate the mutation by transforming with a set of plasmids (average size: 20 kb). About how many plasmids would you have to transform into the mutant to just cover the genome? (Assume that your plasmids each overlap by 1 kb.)

25. List at least two examples in which bacterial strains have acquired new pathogenicity genes. State both the organism and mode of introduction of the gene.

26. *Streptococcus parasanguis* is a bacterial species that initiates dental plaque formation by adhering to teeth. To investigate ways to eliminate plaque, researchers constructed a plasmid, depicted in the figure shown, to mutagenize *S. parasanguis.* The key features of this plasmid include *repA-ts* (a temperature-sensitive origin of replication), *Kan*^{R} (a gene for resistance to the antibiotic kanamycin), and the transposon *IS256.* This transposon contains the *Erm*^{R} gene for resistance to the antibiotic erythryomycin and transposes in *S. parasanguis* thanks to a gene encoding a transposase enzyme that moves all DNA sequences located between the transposon's inverted repeats [IRs].

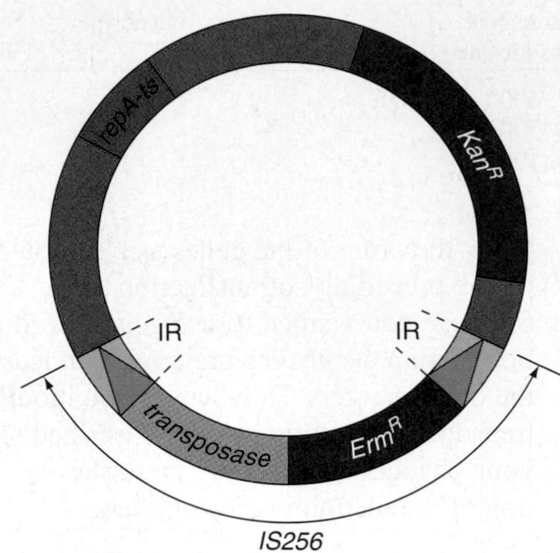

a. How could the researchers use this plasmid as a mutagen? Consider how they could get the transposon into the bacteria, and how they could identify strains that had new insertions of *IS256* into *S. parasanguis* genes. Your answer should explain why the plasmid has two different antibiotic resistance genes as well as a temperature-sensitive origin of replication.
b. Why would the researchers use this plasmid as a mutagen?

27. Genome sequences show that some pathogenic bacteria contain virulence genes next to bacteriophage genes. Why does this suggest lateral gene transfer, and what would the mechanism of transfer have been?

The Chromosomes of Organelles Outside the Nucleus Exhibit Non-Mendelian Patterns of Inheritance

Chapter 16

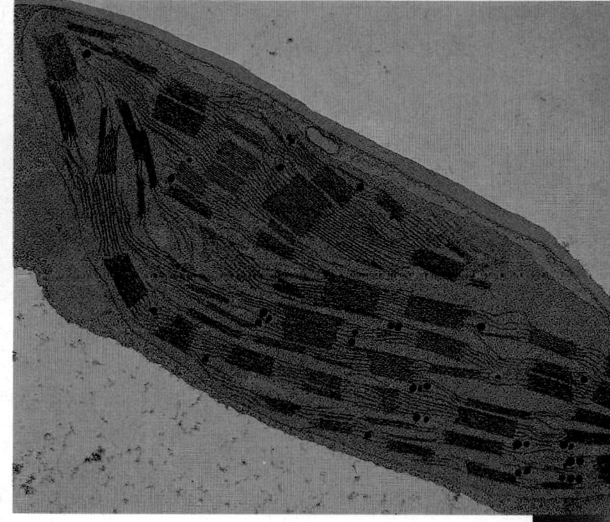

False color electron micrograph of an isolated chloroplast in a leaf cell of timothy grass (Phleum pratense) *(11,000×).*

Just nine years after the rediscovery of Mendel's laws, plant geneticists reported a perplexing phenomenon that challenged one of Mendel's basic assumptions. In a 1909 paper, they described the results of reciprocal crosses analyzing the transmission of green versus variegated leaves in flowering plants known as four-o'clocks (**Fig. 16.1**). Fertilization of eggs from a plant with variegated leaves by pollen from a green-leafed plant produced uniformly variegated offspring. Surprisingly, the reciprocal cross—in which the leaves of the mother plant were green and those of the father variegated—did not lead to the same outcome; instead all of the progeny from this cross displayed green foliage. From these results, it appeared that offspring inherit their form of the variegation trait from the mother only. This type of transmission, known as maternal inheritance, challenged Mendel's assumption that maternal and paternal gametes contribute equally to inheritance. Geneticists thus said that the trait in question exhibited **non-Mendelian inheritance.**

Another example of a non-Mendelian trait emerged 40 years later. In 1949, French researchers published studies on the size of yeast colonies in laboratory strains of the single-celled organism *Saccharomyces cerevisiae.* Mitotically dividing cultures of these cells, when grown on plates containing the fermentable sugar glucose as the source of carbon, produced colonies of two distinctly different sizes. Ninety-five percent of the colonies were large (in French, *grande*); the remaining 5% were small (*petite*). Cells from *grande* colonies, when separated and grown on fresh plates containing glucose, yielded some *petite* colonies, but cells from *petite* colonies never generated *grande* colonies. From these observations, the researchers deduced that the founder cells of *petite* colonies arose from frequent—1 in 20 cells—mutations in cells of the *grande* colonies.

The French researchers pursued their study of the genetic basis of this difference in colony size by analyzing various matings using haploid cells from *grande* and *petite* colonies. As described in Chapter 4, the diploid cells formed by mating haploid cells of opposite mating types may, under stressful conditions (not enough nutrients, for example), enter meiosis. When the French researchers mated *grande* cells of one mating type with *grande* cells of the opposite mating type, the resulting diploids were *grande;* and when these *grande* diploids sporulated via meiosis, each one yielded four *grande* spores (that is, spores that after germination produced *grande* colonies) and zero *petite* spores. A cross of two cells from *petite* colonies produced only *petite* diploids, which, however, could not sporulate because of their deficiency in respiration. Matings of *grande* with *petite* generated only *grande* diploids, and each sporulation of those diploids yielded four *grande* spores and zero *petites.* This 4:0 ratio consistently replaced the 2:2 ratio predicted by Mendelian genetics. From these observations, the researchers concluded that a genetic factor necessary for respiratory growth is present in *grande* cells but absent from *petite* cells. They named the factor "rho" (symbolized by "ρ"); and they designated grande

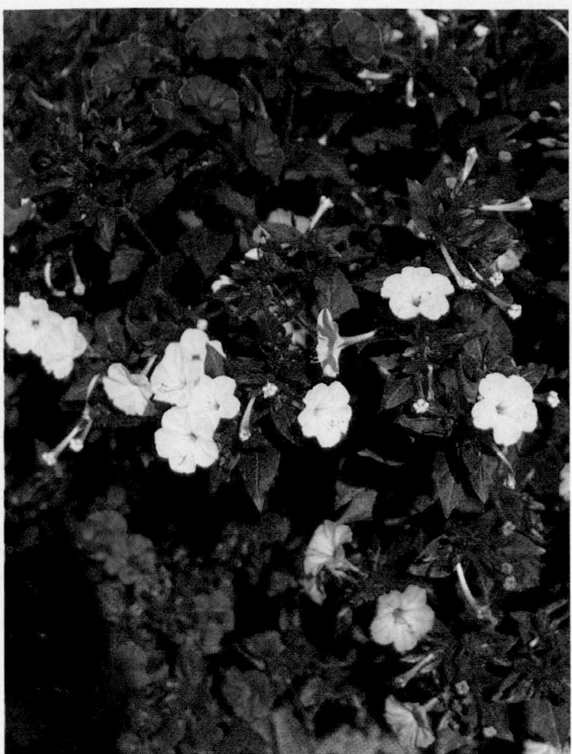

Figure 16.1 Non-Mendelian inheritance in four-o'clocks. The first example of non-Mendelian inheritance uncovered by geneticists was seen in the flowering plants known as four-o-clocks.

cells "ρ^+" and petite cells "ρ^-". They also noted that because of the non-Mendelian inheritance pattern, the rho factor did not segregate at meiosis.

Is there a connection between the maternal inheritance of leaf variegation in four-o'clocks and the unusual 4:0 inheritance pattern of *grande* and *petite* colony sizes in yeast? The answer is yes. Decades of experiments have shown that both traits are determined by genes that do not reside in the nucleus but instead lie in the genomes of nonnuclear organelles. Mutations resulting in leaf variegation occur in *chloroplast DNA* (*cpDNA*), in genes that encode proteins active in photosynthesis. Mutations that diminish yeast colony size occur in mitochondrial *DNA* (*mtDNA*), in regions of the genome that influence the efficiency of a cell's energy use. While chloroplasts are the sites of photosynthesis in plant cells, mitochondria are the sites of cellular respiration—the aerobic conversion of energy stored in nutrient molecules to energy stored in the high-energy bonds of ATP—in all eukaryotic cells. The *petite* mutants form smaller colonies because they are unable to carry out cellular respiration and must obtain the energy they need for survival from the less efficient, anaerobic energy conversion pathway of fermentation.

In this chapter, we see that non-Mendelian inheritance results from the fact that the genomes of chloroplasts and mitochondria are transmitted from generation to generation separately from the nuclear genome and in a very different fashion.

Three themes surface during our detailed discussion of the genes and genomes of mitochondria and chloroplasts. First, unlike the rules governing the transmission of nuclear genes, the formal rules for the transmission of organelle genomes can vary from organism to organism. Most diploid organisms, for example, inherit non-Mendelian traits from their mother, but in some species, the traits are transmitted by the father, and in other species, by both parents. Second, the maintenance of organelles requires cooperation between two genomes: that of the organelle itself and that in the nucleus of the cell in which the organelle functions. Some of the molecules in an organelle are encoded by organelle DNA and synthesized in the organelle; other molecules active in the organelle are encoded by nuclear genes and imported into the organelle. Finally, the genomes and biochemical processes of organelles are more similar in many ways to those of bacteria than to those in other parts of the eukaryotic cell. These observations formed the basis of the now accepted *endosymbiont theory* of the origin of extranuclear organelles. The theory proposes that organelles are the evolutionary remnants of bacteria that became symbionts in the ancient precursors of the earliest eukaryotes.

As we examine the structure, function, and transmission of the DNA carried by mitochondria and chloroplasts, we present

- The structure and function of mitochondrial and chloroplast genomes, including a description of the size, shape, replication, and expression of those genomes, an explanation of how the genomes can be modified, and a brief discussion of the evolutionary origin of mitochondria and chloroplasts.
- How genetic studies of transmission revealed and explained non-Mendelian patterns of inheritance, including descriptions of uniparental and biparental inheritance and mitotic segregation.
- A comprehensive example of how mutations in mitochondrial DNA affect human health.

16.1 The Structure and Function of Mitochondrial and Chloroplast Genomes

Although geneticists pieced together the molecular foundation of non-Mendelian inheritance—that is, the structure and function of organelle genomes—beginning only in the 1960s, we present a summary of that molecular context up front to provide a clear framework for understanding the earlier genetic evidence. We begin with a brief description of the structure and function of mitochondria and chloroplasts, and we then describe the genomes that encode the RNA and protein machinery the organelles use to sustain their activities.

Mitochondria and Chloroplasts Are Organelles of Energy Conversion That Carry Their Own DNA

Every time a bird takes flight, a person speaks, a worm turns, or a flower unfurls, the organism's cells use energy that a plant captured from sunlight and stored in nutrient molecules. The chloroplasts of plants and other photosynthesizers capture solar energy and store it in the chemical bonds of carbohydrates. The mitochondria found in most eukaryotic cells release energy from those nutrients and convert it to ATP.

Mitochondria Are Sites of the Krebs Cycle and an Electron Transport Chain That Carries Out the Oxidative Phosphorylation of ADP to ATP

Each eukaryotic cell houses many mitochondria, with the exact number depending on the energy requirements of the cell as well as the chance distribution of mitochondria during cell division. In humans, nerve, muscle, and liver cells each carry more than a thousand mitochondria.

Figure 16.2 reveals the organelle's detailed structure. Separating the mitochondrion from the cell's cytoplasm is a smooth *outer membrane* perforated by protein-lined pores; molecules up to the size of small proteins can readily pass through these pores to an *intermembrane space*—a gap between the outer membrane, and a second, *inner membrane.* The inner membrane has many infoldings called *cristae* and is studded with enzymes of the electron transport chain and ATP synthesis. This membrane is unusually rich in proteins (it is 75% protein by weight) and is impermeable to most substances, which can traverse it only by active transport. Tightly sequestered inside the inner membrane is a core compartment known as the *matrix,* which contains many enzyme systems, including one that carries out the tricarboxylic acid (TCA) cycle, commonly referred to as the *Krebs cycle* after its discoverer. The matrix also carries one or more circles of mitochondrial DNA.

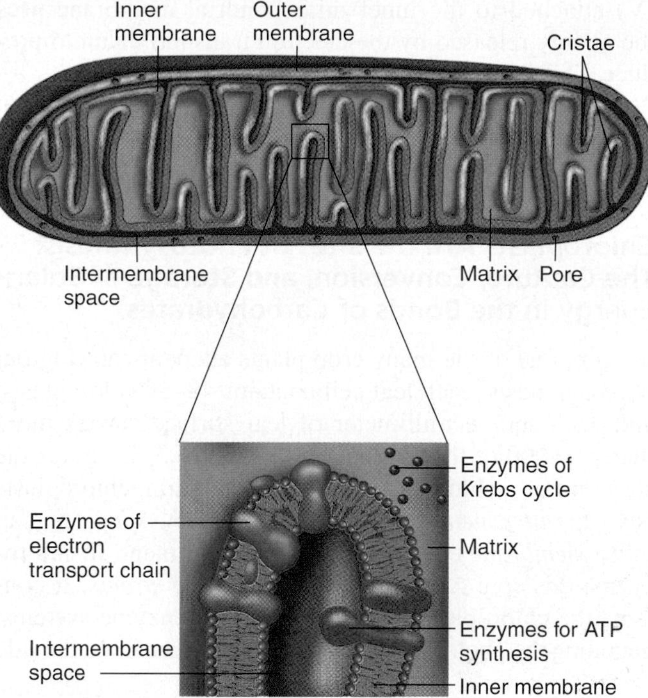

Figure 16.2 Anatomy of a mitochondrion. The organization and structure of a single mitochondrion is shown. Core regions that are entirely enclosed within an inner membrane are known as the matrix (*shown in blue*). The matrix contains the mitochondrial DNA and enzymes of the Krebs cycle. Individual inner-membrane foldings are called cristae. A single crista is magnified to show how enzymes of the electron transport chain carry out oxidative phosphorylation between the matrix and the inter-membrane space.

Biologists call mitochondria the powerhouses of the cell because these organelles produce most of the cell's usable energy in the form of ATP molecules. There are two stages by which mitochondria help convert food into energy. In the first, mitochondria employ the Krebs cycle to metabolize pyruvate and fatty acids (the breakdown products of carbohydrates and fats, respectively) and to produce the high-energy electron carriers NADH and $FADH_2$. In the second stage, four multisubunit enzyme complexes (referred to as I, II, III, and IV), which are embedded in the inner mitochondrial membrane, harness the energy in the high-energy electrons carried by the NADH and $FADH_2$. The process by which they harness this energy is **oxidative phosphorylation:** a set of reactions requiring oxygen that creates portable packets of energy in the form of ATP. In carrying out oxidative phosphorylation, molecular complexes I, II, III, and IV form a chain that transports the high-energy electrons from NADH and $FADH_2$ to their final electron acceptor, oxygen. A fifth protein complex

(V) attached to the inner mitochondrial membrane uses the energy released by the electron transport chain to produce ATP.

To summarize: Starting with O_2 and the breakdown products of fats and carbohydrates, mitochondria produce packets of energy in the form of ATP.

Chloroplasts Are the Sites of Photosynthesis: The Capture, Conversion, and Storage of Solar Energy in the Bonds of Carbohydrates

In corn, one of the many crop plants adept at carrying out photosynthesis, each leaf cell contains 40–50 chloroplasts, and each square millimeter of leaf surface carries more than 500,000 of the organelles. **Figure 16.3** illustrates the structure of a chloroplast. Like mitochondria, chloroplasts have an *outer membrane,* an *intermembrane space,* and an *inner membrane* enclosing a core compartment. In chloroplasts, this core compartment is called the *stroma.* It contains the chloroplast's DNA and various enzyme systems, including the photosynthetic enzymes that fix carbon dioxide into carbohydrates. Within the stroma lies a third set of membranes, the *thylakoid membranes,* which surround a space called the *thylakoid lumen.* In places, portions of the thylakoid membranes fuse into stacks of disklike sacs known as *grana* (singular, *granum*). Embedded in the thylakoid membranes are chlorophyll and other light-absorbing proteins as well as proteins of the photosynthetic electron transport system. The light-absorbing proteins capture light energy from the sun and transfer this trapped solar energy to a photosynthetic electron transport chain.

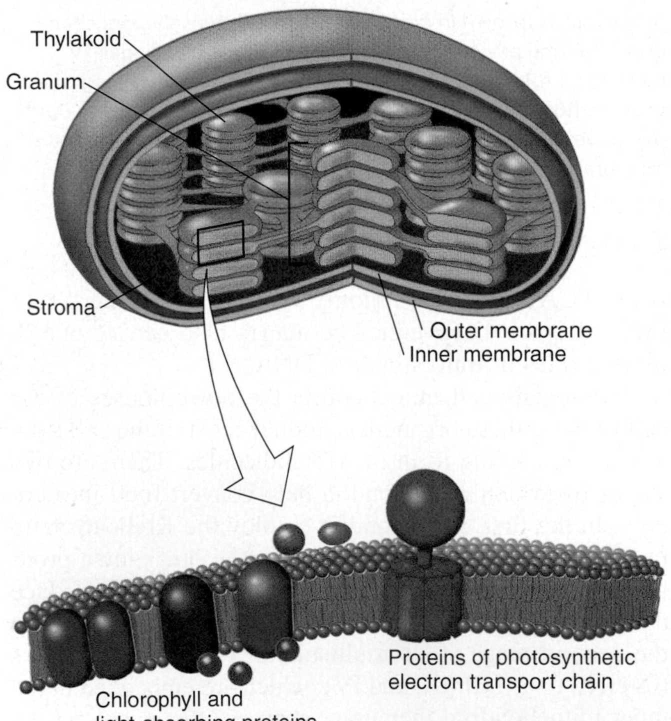

Figure 16.3 Anatomy of a chloroplast. Like a mitochondrion, a chloroplast also has an outer membrane and an inner membrane, but the inner membrane is not folded. The space found within the inner membrane—containing the chloroplast DNA and photosynthetic enzymes—is called a stroma. The stroma contains thylakoid membranes surrounding the thylakoid lumen. The magnified portion of a thylakoid membrane indicates the presence of photosynthetic enzymes.

Photosynthesis takes place in two parts: a light-trapping phase and a sugar-building phase. During the light-trapping phase, solar energy becomes trapped as it boosts electrons in chlorophyll or another pigment molecule to higher energy levels. The energized electrons are then conveyed to an electron transport system that uses the energy to convert water to oxygen and H^+. Photosynthetic electron transport forms NADPH and drives the synthesis of ATP via an ATP synthase similar to the one in mitochondria. During the sugar-building phase of photosynthesis, enzymes of the Calvin cycle, located in the chloroplast stroma, use the ATP and NADPH generated by the conversion of solar to chemical energy to fix atmospheric carbon dioxide into carbohydrates. The energy stored in the bonds of these nutrient molecules fuels the activities of both the plants that make the carbohydrates and the animals that ingest them.

Mitochondria and Chloroplasts Carry Their Own DNA

Cells stained with DNA-specific dyes, when viewed under the light microscope, reveal DNA molecules in the matrix of mitochondria and the stroma of chloroplasts, as well as in the nucleus. Using methods for purifying mitochondria and chloroplasts, researchers have extracted DNA directly from these organelles and shown by analyses of base composition and buoyant density that an organism's organelle DNA differs from its nuclear DNA.

With this background in mind, we examine the structure and function of mitochondrial and chloroplast genomes. As we do, we see that although both organelles replicate and express all the genes in their own DNA, their genomes encode only some of the proteins they require for their activities.

The Genomes of Mitochondria

Mitochondrial DNA lies within the matrix of the organelle, where it appears in highly condensed structures called *nucleoids.* Each haploid yeast cell, for example, has about 20 mtDNA molecules in a variable number of mitochondria and nucleoids. In cells with just one to a few mitochondria, each organelle contains a number of nucleoids. By contrast, in cells grown on lactate medium, which carry up to

TABLE 16.1 Mitochondrial DNA Sizes

Organism	Size (kb)
Plasmodium	6
Yeast	75
Drosophila	18
Pea	110
Human	16.5

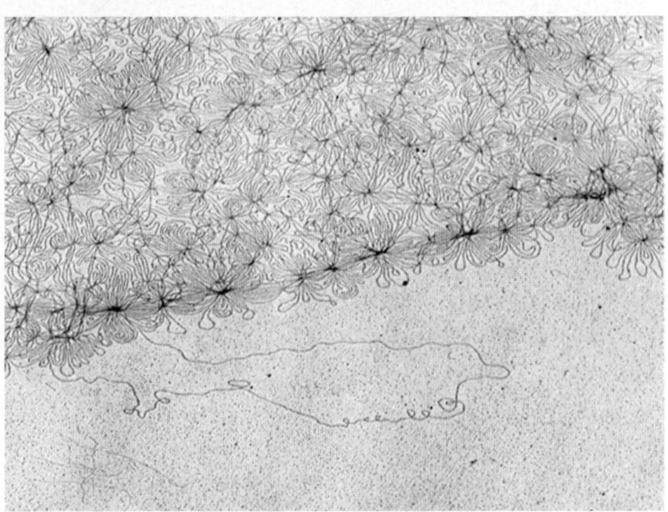

Figure 16.4 Kinetoplast DNA network. In certain protozoan parasites, there is a single mitochondrion, or kinetoplast, that contains a large interlocking network of DNA molecules present in mini- and maxicircles.

20 mitochondria, each organelle contains only one to a few nucleoids. Thus, the mtDNA of most cells does not reside in a single location.

Variations in the number of mitochondria, nucleoids, and mtDNA molecules are regulated by complicated means that researchers do not yet understand. Mitochondria can fuse with each other as well as divide. In general, however, mitochondria double in size and then divide in half in each cell generation. The replication of mtDNA molecules, as well as the division of the mitochondria, can occur throughout the cell cycle independent of the replication of genomic nuclear DNA (which occurs only during S phase) and of the cell division at the end of mitosis. Interestingly, which mtDNA molecules undergo replication seems to be determined at random; as a result, some molecules replicate many times in each cell cycle, while others do not replicate at all. This is one cause of the mitotic segregation of mitochondrial genomes discussed later in the chapter.

The size and gene content of mitochondrial DNA vary from organism to organism. The mtDNAs in the malaria parasite *Plasmodium falciparum* are only 6 kb in length; those in the free-living nematode *Ascaris suum* are 14.3 kb; those in the muskmelon *Cucumis melo* are a giant 2400 kb long. These mtDNA size differences do not necessarily reflect comparable differences in gene content. Although the large mtDNAs of higher plants do contain more genes than the smaller mtDNAs of other organisms, the 75 kb mtDNA of baker's yeast encodes fewer proteins of the respiratory chain than does the 16.5 kb mtDNA of humans. **Tables 16.1** and **16.2** summarize the size and gene content of mtDNAs from organisms representative of plants, animals, and fungi.

Like size and gene content, the shape of mtDNAs varies. Biochemical analyses and mapping studies have shown that the mtDNAs of most species are circular; but the mtDNAs of the ciliated protozoans *Tetrahymena* and *Paramecium,* the alga *Chlamydomonas,* and the yeast *Hansenula* are linear. The difficulty in isolating unbroken mtDNA molecules from some organisms makes it hard to tell the shape of their mtDNA *in vivo.*

To complete this snapshot of the mitochondrial genome's variation in size, gene content, and shape, it is necessary to mention the unusually organized mtDNAs of protozoan parasites of the genera *Trypanosoma, Leishmania,* and *Crithidia.* These single-celled eukaryotic organisms carry a single mitochondrion known as a kinetoplast. Within this structure, the mtDNA exists in one place (contrary to the mtDNA of most other cells) as a large network of 10–25,000 minicircles 0.5–2.5 kb in length interlocked with 50–100 maxicircles 21–31 kb long (**Fig. 16.4**). The maxicircles contain most of the genes usually found on mtDNA, while the minicircles play a role in RNA editing, as described later.

A comparison of mitochondrial genome organization and function in humans, yeast, and the liverwort (a species of moss) illustrates some of the details of mtDNA diversity. (**Fig. 16.5**) Molecular biologists have sequenced the mtDNAs of all three organisms.

TABLE 16.2 Comparison of Some Functions Encoded in mtDNA

Organism	Oxidative Phosphorylation Genes	tRNAs	Genome Size (kb)
Yeast	7	25	75
Marchantia (liverwort)	14	29	186.0
Human	13	22	16.5

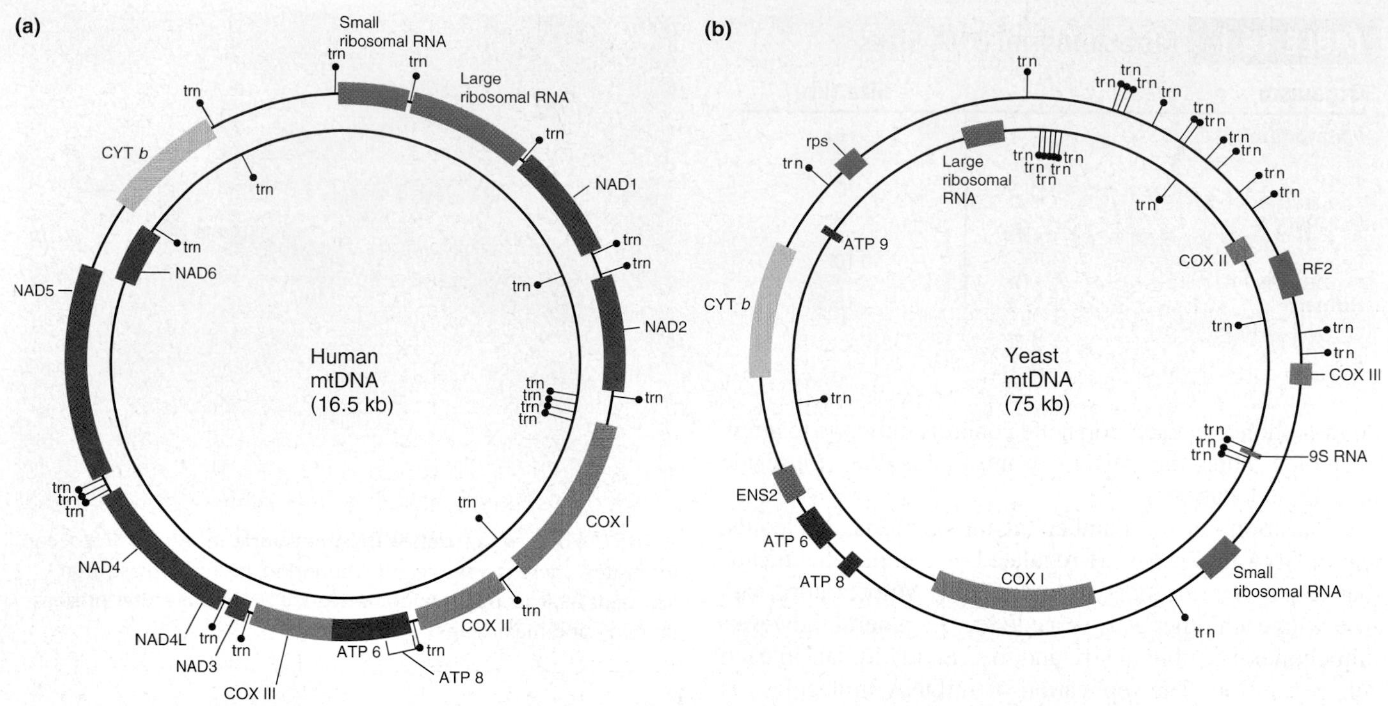

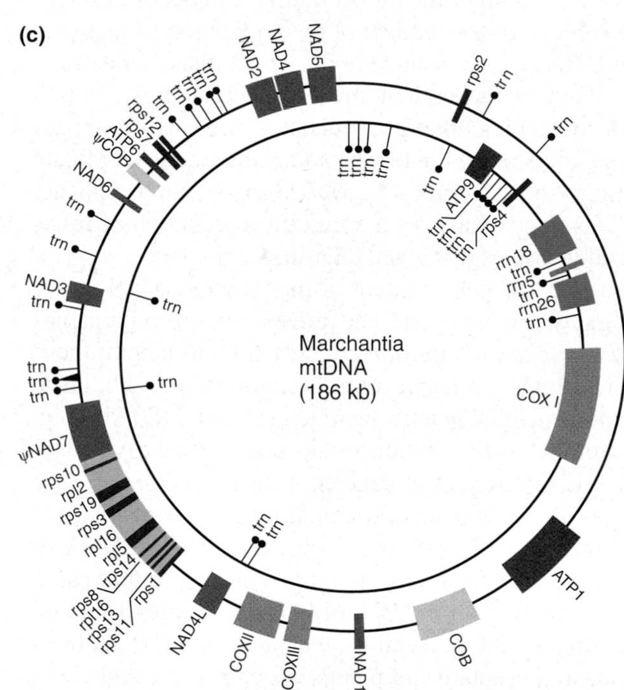

Figure 16.5 Mitochondrial genomes of three species. Comparison of mitochondrial genomes of humans, baker's yeast, and liverwort (circles not drawn to the same scale). Human mtDNA is 16.5 kb long, while that of baker's yeast is 75 kb long, and liverwort is 121 kb long. Important differences illustrated in the diagrams are the presence of long intergenic sequences in yeast mtDNA and their absence from human mtDNA and differences in the numbers of genes in the mt genomes from these three organisms. The color scheme allows you to quickly see differences in numbers of genes for various functions in these three species: *green* for genes coding for cytochrome oxidase proteins; *red* for genes coding for ATPase subunit proteins; *yellow* for genes coding for NADH complex proteins; *tan* for genes coding for cytochrome complex proteins; *purple* for genes coding for ribosomal proteins or ribosomal RNAs. Each tRNA gene is indicated by a *black ball and stick*. Genes shown on the outer and inner circles are transcribed in opposite directions.

Human mtDNA Carries Closely Packed Genes

The 16.5 kb human mitochondrial genome, which accounts for 0.3% of a cell's DNA, is a circular DNA molecule (like most mtDNAs) that carries 37 genes. Thirteen of these genes encode polypeptide subunits of the protein complexes that make up the oxidative phosphorylation apparatus embedded in the mitochondrial inner membrane. These same 13 genes are present in the mtDNAs of all metazoans analyzed to date. In addition to the 13 polypeptide-encoding genes, there are 22 tRNA genes. The remaining 2 genes are for the large and small rRNAs found in mitochondrial ribosomes (Fig. 16.5a).

A significant feature of the human mitochondrial genome is the compactness of its gene arrangement. As Fig. 16.5a shows, adjacent genes either abut each other or slightly overlap. With virtually no nucleotides between them and no introns within them, the genes are packaged very tightly. The reason for this compact arrangement is not yet known.

The Larger Yeast Mitochondrial Genome Contains Spacers and Introns

The mitochondrial genome of the yeast *S. cerevisiae* is more than four times longer than human and other animal mtDNAs (Fig. 16.5b). Two DNA elements account for the larger size of the yeast mitochondrial genome: long intergenic sequences and introns. Long AT-rich sequences called "spacers" separate the genes in yeast mtDNA and account for more than half of the additional DNA. Most of the spacer DNA is transcribed, and some of it is retained in the mRNAs as long untranslated 5′ and 3′ extensions. Introns, the second DNA-lengthening element, form about 25% of the yeast mitochondrial genome and account for most of the remaining size difference between yeast and human mtDNA.

The 186 kb mtDNA of the Liverwort *Marchantia polymorpha* Carries Many More Genes Than Its Animal and Fungal Counterparts

The mtDNA of *M. polymorpha* was the first plant mtDNA to be entirely sequenced. Although it is one of the smallest plant mitochondrial genomes, it is far larger and has many more genes than nonplant mtDNAs (Fig. 16.5c). The genes it carries encode 12 of the 13 electron-transport-chain proteins found in human mtDNA (lacking only the gene for subunit 8 of ATP synthase), as well as subunit 9 of ATP synthase (which is absent from humans) and subunit A of ATP synthase (which is in the nuclear genomes of all animals and fungi). In addition, the liverwort mitochondrial genome contains 16 genes for ribosomal proteins and 29 genes encoding proteins of unknown function.

Thus, although mitochondria in different eukaryotic organisms play similar roles in the conversion of food to energy, evolution has produced mtDNAs with an astonishing diversity in the content and organization of their genes. As we see next, mitochondrial evolution has also led to some remarkable variations on the basic mechanisms of gene expression.

Mitochondrial Transcripts Undergo RNA Editing, a Rare Variation on the Basic Theme of Gene Expression

Researchers discovered the unexpected phenomenon of RNA editing in the mitochondria of trypanosomes. As already noted, these protozoan parasites have a single, large mitochondrion—the kinetoplast—which contains much more DNA than the mitochondria of other organisms; this kDNA exists as a series of interlocking maxi- and minicircles. DNA sequencing shows that the minicircles carry no protein-encoding genes. The detection of transcripts from maxicircle DNA, however, confirms that these larger circles do carry and express genes.

Surprisingly, the sequencing of maxicircle DNA revealed only short, recognizable gene fragments, instead of whole mitochondrial genes. Furthermore, the sequencing of RNA molecules in the kinetoplast revealed both RNAs that looked like the strange fragments of kinetoplast genes and related RNAs that could encode recognizable mitochondrial proteins. From these observations, investigators concluded that kDNA encodes a precursor (the strange fragment observed) for each mRNA. After transcription, the cellular machinery turns these precursors into functional mRNAs through the insertion or deletion of nucleotides.

The process that converts pre-mRNAs to mature mRNAs is **RNA editing.** It is essential for the expression of these mitochondrial genes because without RNA editing, the pre-mRNAs do not encode polypeptides. Some pre-mRNAs lack a first codon suitable for translation initiation; others lack a stop codon for the termination of translation. RNA editing creates both types of sites, as well as many new codons within the genes.

In addition to the mitochondria of trypanosomes, the mitochondria of some plants and fungi carry out RNA editing. The extent of RNA editing varies from mRNA to mRNA and from organism to organism. In trypanosomes, the RNA editing machinery adds or deletes uracils. In plants, the editing adds or deletes cytosines. At present, researchers understand the general mechanism of uracil editing, but not that of cytosine editing. As **Fig. 16.6** shows, uracil editing occurs in stages in which enzymes use an RNA template as a guide for correcting the pre-mRNA. The guide RNAs are encoded by short stretches of kDNA on both maxi- and minicircles, and a structure known as an

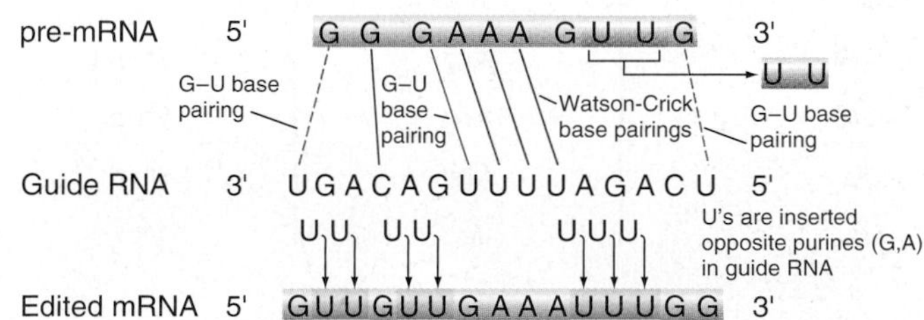

Figure 16.6 RNA editing in trypanosomes. Example of a portion of a pre-mRNA sequence is shown at the *top*. This pre-mRNA forms a double-stranded hybrid with a guide RNA through both standard Watson-Crick A–U and G–C base pairing, as well as atypical G–U base pairing. Unpaired G and A bases within the guide RNA initiate the insertion of U's within the pre-mRNA sequence, while unpaired U's in the pre-mRNA are deleted, bringing about the final edited mRNA.

"editosome" is the workbench where the RNA editing takes place.

Translation in Mitochondria Shows That the Genetic Code Is Not Universal

As mtDNA carrying its own rRNA and tRNA genes would suggest, mitochondria have their own distinct translational apparatus. Mitochondrial translation is quite unlike the cytoplasmic translation of mRNAs transcribed from nuclear genes in eukaryotes. Many aspects of the mitochondrial translational system resemble details of translation in prokaryotes. For example, as in bacteria, *N*-formyl methionine and $tRNA^{fMet}$ initiate translation in mitochondria. Moreover, inhibitors of bacterial translation, such as chloramphenicol and erythromycin, which have no effect on eukaryotic cytoplasmic protein synthesis, are potent inhibitors of mitochondrial protein synthesis.

We saw in Chapter 8 that the genetic code is almost, but not quite, universal. The mtDNA sequences of tRNAs and protein-encoding genes in several species cannot explain the sequences of the resulting proteins in terms of the "universal" code. For example, in human mtDNA, the codon UGA specifies tryptophan rather than stop (as in the standard genetic code); AGG and AGA specify stop instead of arginine; and AUA specifies methionine rather than isoleucine (**Table 16.3**). No single mitochondrial genetic code functions in all organisms, and the mitochondria of higher plants use the universal code. Moreover, while an f-Met-tRNA usually initiates translation in mitochondria by reading AUG or AUA, other triplets, which do not specify methionine, often mark the site of initiation. The genetic codes of mitochondria probably diverged from the universal code by a series of mutations occurring some time after the organelles became established components of eukaryotic cells.

As we see next, chloroplast DNA, although similar in many ways to mtDNA, has some remarkable features of its own.

TABLE 16.3 Variations in the Genetic Code of Human Mitochondria

Characteristic	Universal Code	mtDNA Code
Number of tRNAs	32	22
UGG	Trp	Trp
UGA	Stop	Trp
AGG	Arg	Stop
AGA	Arg	Stop
AUG	Met	Met
AUA	Ile	Met

Altered genetic code. The human mtDNA genetic code is simplified such that a modified U in the tRNA "wobble" position can read all four codons in a codon family (that is, UUU, UUC, UUA, and UUG). An unmodified U can read both purines, and G can read both pyrimidines. Tryptophan tRNA has a U in the wobble position, so it will read both the traditional UGG codon and the associated UGA stop codon as tryptophan. Similarly, the methionine codon reads both AUG and the associated AUA as methionine. Finally, human mtDNA has only a single arginine tRNA such that two of the six arginine codons (AGG and AGA) now function as stop codons.

The Genomes of Chloroplasts

Chloroplasts occur in plants and algae. The genomes they carry are much more uniform in size than the genomes of mitochondria. Although chloroplast DNAs range in size from 120–217 kb, most are between 120 and 160 kb long (**Table 16.4**). cpDNA contains many more genes than mtDNA. Like the genes of bacteria and human mtDNA, these genes are closely packed, with relatively few nucleotides between adjacent coding sequences. Like the genes of yeast mtDNA, they contain introns. The shape of most cpDNAs is a circle.

A close look at the chloroplast genome of the liverwort *M. polymorpha,* the first cpDNA to be sequenced completely, illustrates details of cpDNA structure and function (**Fig. 16.7**). Liverwort cpDNA contains 92 protein-encoding genes and 36 genes for tRNAs and rRNAs, which is more genes than are carried by the mitochondrial genomes of much larger plants. The rRNA genes of the chloroplast encode four ribosomal RNAs—16S, 23S, 4.5S, and 5S. These same four rRNAs are found in eubacteria. Another similarity with the bacterial genome is that some functionally related cpDNA genes are organized in clusters that resemble bacterial operons.

The 92 polypeptides encoded by *M. polymorpha* cpDNA account for only a small fraction of the proteins active in chloroplasts. However, cpDNA encodes proteins for a wider range of processes than does mtDNA. The cpDNA-encoded proteins include many of the molecules that carry out photosynthetic electron transport and other aspects of photosynthesis, as well as RNA polymerase, translation factors, ribosomal proteins, and other molecules

TABLE 16.4 Chloroplast DNA Sizes

Organism	Size (kb)
Chlamydomonas reinhardtii	196
Marchantia (liverwort)	121
Nicotiana tabacum (tobacco)	156
Oryza sativa (rice)	135

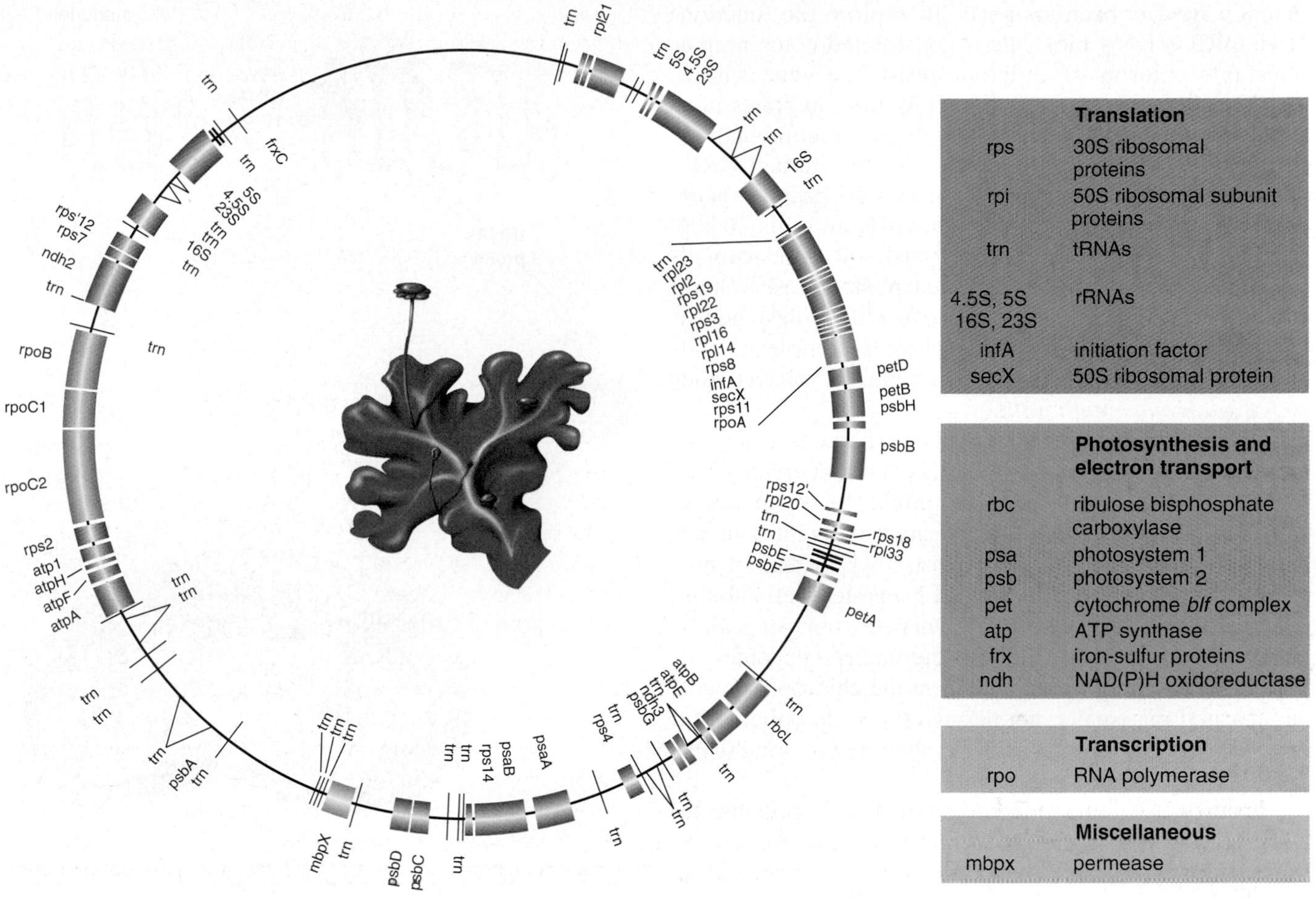

Figure 16.7 Chloroplast genome of the liverwort *M. polymorpha*. The relative locations and symbols of some of the 128 genes are indicated. Genes are color-coded according to general function.

active in chloroplast gene expression. The RNA polymerase of chloroplasts is similar to the multisubunit bacterial RNA polymerases. Inhibitors of bacterial translation, such as chloramphenicol and streptomycin, inhibit translation in chloroplasts, as they do in mitochondria.

The function of some chloroplast genes is not yet known. ORF's for which no function has been defined are called *URF*'s. Because most of these unidentified reading frames are well conserved in many cpDNAs and have counterparts in bacteria, it is likely that they are key elements in chloroplast gene expression and organelle function.

Manipulation of Chloroplast Genome Is Possible Through Genetic Engineering

In the early days of recombinant DNA technology, chloroplast researchers were frustrated by an inability to transfer cloned genes and mutated DNA fragments into organelle genomes. Development of the gene gun and a gene delivery method known as *biolistic transformation* in the late 1980s solved the problem. The basic idea is to coat small (1 μm) metal particles with DNA and then shoot these DNA-carrying "bullets" at cells. Biolistic transformation occurs when a particle lands within a cell, such as a plant protoplast, without killing it and releases the DNA from the metal; in rare instances, the DNA enters the nucleus or organelles where it may recombine into the genome. If the DNA shot into the cell contains a strong selectable marker, plant geneticists can isolate the rare transformed plant cells in which the released DNA has entered the organelle.

Introduction of DNA into nuclei, chloroplasts, or mitochondria has been successful for many organisms, but stable transformation, that is, the production of cells in which the injected DNA integrates into the genome, is highly successful only for chloroplasts. A variety of vectors containing different selectable markers and sequences that support the expression of introduced genes exists for many plant species. For plants whose chloroplast genome sequence has been determined, organelle transformation and the generation of mutants provide a way to determine the function of ORFs for which no function has yet been assigned. To explore the function of an ORF, a DNA molecule is constructed containing a selectable chloroplast antibiotic-resistance gene within the ORF. This DNA is shot into cells and integrates into the chloroplast genome via a homologous recombination that replaces the wild-type ORF with the mutant ORF. Selection for the marker gene increases the proportion of chloroplasts containing the mutant ORF, and this makes it possible to study the phenotype of the transformed cells in culture or in reconstituted plants. Researchers have used this protocol to identify chloroplast genes encoding novel subunits of cytochrome complexes and assembly factors for photosystem genes in tobacco and *Chlamydomonas reinhardtii*.

Transformation of the chloroplast genome is a suitable mechanism for altering the properties of commercially important crop plants. One goal might be to produce herbicide-resistant plants. The advantages to introducing herbicide resistance into chloroplast DNA instead of nuclear DNA are twofold. First, DNA integrates well into the chloroplast genome, whereas for reasons not yet understood, it does not integrate into the nuclear genomes of plants. Furthermore, foreign DNA in the chloroplasts will be inherited maternally, not through the male pollen. The risk that introduced genes will spread to neighboring plant populations is therefore low.

Chloroplast transformation also makes it possible to make plants into protein-production factories. One could, for example, produce a vaccine in the leaves of an edible plant by incorporating the genes encoding the vaccine into the chloroplast genome. A modified *E. coli* labile-toxin (LT) gene has already been introduced into the chloroplasts of tobacco. (The LT toxin protein causes diarrhea.) Transformation of the same gene into chloroplasts in edible leaves such as lettuce or spinach would generate an ingestible vaccine that could, in principle, stimulate the human immune system to respond to, and eliminate, any *E. coli* LT it encountered.

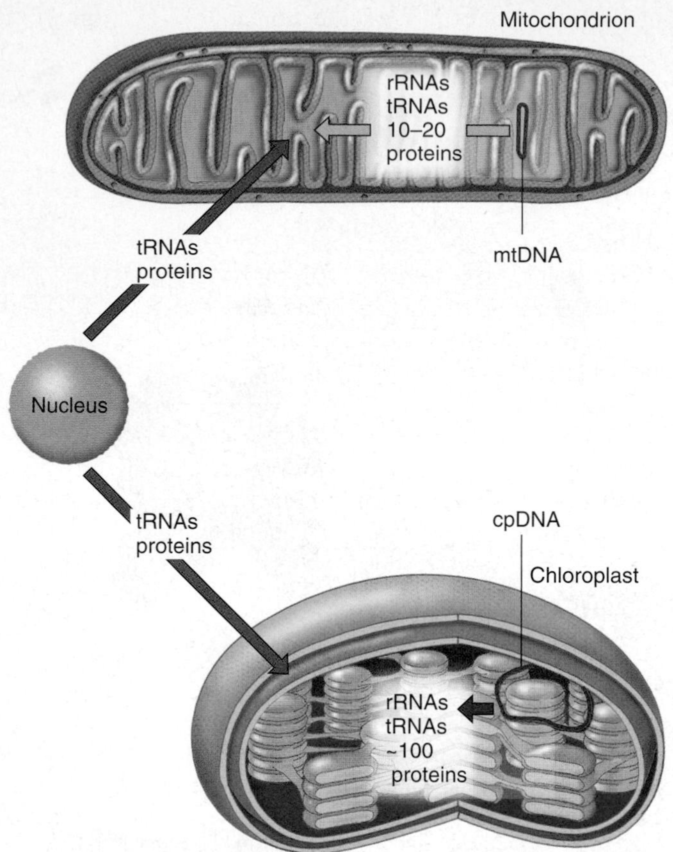

Number and genomic location of oxidative phosphorylation genes

	Number of polypeptides					
	Electron transport chain				ATP synthase	
Genomic location	I	II	III	IV	V	Total
Mitochondrion	7	0	1	3	2	13
Nucleus	≥33	4	10	10	10	≥67
Total	≥40	4	11	13	12	≥80

Figure 16.8 Mitochondria and chloroplasts depend on gene products from the nucleus. Although some organelles in some species have many more genes than others, all are dependent on RNA and protein products encoded by nuclear genes. The location of oxidative phosphorylation (OXPHOS) genes is shown.

Mitochondria and Chloroplast Functions Require Cooperation Between the Organelle and Nuclear Genomes

The maintenance and assembly of functional mitochondria and chloroplasts depend on gene products from both the organelles themselves and from the nuclear genome (**Fig. 16.8**). In most organisms, for example, cytochrome *c* oxidase, the terminal protein of the mitochondrial electron transport chain, is composed of seven subunits, three of which are encoded by mitochondrial genes, whose mRNAs are translated on mitochondrial ribosomes. The remaining four are encoded by nuclear genes whose messages are translated on ribosomes in the cytoplasm. In all organisms, nuclear genes encode the majority of the proteins active in mitochondria and chloroplasts. For example, although mitochondrial genomes carry the rRNA genes, nuclear genomes carry the genes for most (in yeast and plants) or all (in animals) of the proteins in the mitochondrial ribosome.

Because mitochondria and chloroplasts do not carry all the genes for the proteins (and in some organisms, the tRNAs) they need to function and reproduce, these organelles are *semiautonomous,* requiring the constant provision of proteins (and tRNAs) encoded by nuclear genes.

Origin and Evolution of Organelle Genomes: The Molecular Evidence

Mitochondria are remarkably similar in size and shape to today's aerobic bacteria, while chloroplasts in many ways resemble living photosynthetic bacteria. In addition, the genomes of both organelles bear many striking resemblances to the genomes of bacteria. Because of these morphological and molecular likenesses, it is probable that mitochondria and chloroplasts started out as free-living bacteria that merged with the ancestors of modern eukaryotic cells to form a cellular community in which host and guest benefited from the group arrangement.

The Endosymbiont Theory

In the 1970s, Lynn Margulis was one of the first biologists to propose that mitochondria and chloroplasts originated more than a billion years ago when ancient precursors of eukaryotic cells engulfed and established a symbiotic relationship with some bacteria. The primitive cells carrying a mitochondrion-like or chloroplast-like bacterial cell would have gained an edge in the fierce competition for energy production and eventually evolved into complex eukaryotes. So much evidence now supports this hypothesis that it is generally accepted as the **endosymbiont theory.** The molecular evidence for the theory includes the following facts: (1) Both mitochondria and chloroplasts have their own DNA, which replicates independently of the nuclear genome. (2) Like the DNA of bacteria, mtDNA and cpDNA are not organized into nucleosomes by histones. (3) Mitochondrial genomes, as we have seen, use *N*-formyl methionine and $tRNA^{fMet}$ in translation. And (4), inhibitors of bacterial translation, such as chloramphenicol and erythromycin, block mitochondrial and chloroplast translation but have no effect on eukaryotic cytoplasmic protein synthesis. Comparisons of DNA sequences from organelle and bacterial rRNA genes and comparisons of sequences from organelles of different organisms suggest that mitochondrial genomes derive from a common ancestor of present-day gram-negative nonsulfur purple bacteria, while chloroplast genomes derive from cyanobacteria (formerly referred to as blue-green algae).

It is likely that the precursors of eukaryotes and the ancient bacteria that evolved into organelles interacted many times and in many different ways en route to the present-day symbiotic relations. The first interactions may have been transient; indeed, many present-day eukaryotes containing mitochondria and chloroplasts also have bacteria living inside them. Once a stable symbiotic state formed in a given lineage, however, if the engulfed bacterial cell lost some genes critical for life on the outside, it would have also lost the ability to function and compete on its own. The loss of genes could have occurred through transfer to the nucleus or deletion of redundant sequences. We have already seen that some of the genes required for oxidative phosphorylation and photosynthesis reside in the nuclear genome; they may have been transferred there from the organelle genome. The deletion of redundant sequences might have occurred when organelle genes encoding a biosynthetic pathway critical for autonomous growth were discarded or lost from the organelle genome because the nuclear genome of the host encoded its own version of the pathway. In the evolution of the eukaryote's symbiotic community, the engulfed cell may have not only lost genes but also contributed coding sequences to the host cell; these sequences could have encoded enzymes that added to or replaced proteins encoded by the genes of the nuclear chromosomes.

Gene Transfer Occurs Through an RNA Intermediate or by Movement of Pieces of DNA

Researchers have some understanding of the mechanisms by which genes transfer between an organelle and the nucleus. In many plants, the mitochondrial genome encodes the *COXII* gene of the mitochondrial electron transport chain; in other plants, the nuclear DNA (nDNA) encodes that same gene; and in several plant species where the nuclear *COXII* gene is functional, the mtDNA still contains a recognizable, but nonfunctional, copy of the gene (that is, a *COXII* pseudogene). Remarkably, the mtDNA gene contains an intron, while the nuclear gene does not. Geneticists have interpreted this finding to mean that the *COXII* gene transferred from mtDNA to nDNA via an RNA intermediate; the RNA would have lacked the intron, and when the mRNA was copied into DNA and integrated into a chromosome in the nucleus, the resulting nuclear gene also had no intron.

Good evidence also exists for the transfer of many genes at the DNA level. The fact that some plant mtDNAs carry large fragments of cpDNA shows that pieces of cpDNA can move from one organelle to another. Similarly, nonfunctional, intact or partial copies of organelle genes litter the nuclear genomes of eukaryotes. DNA sequencing reveals strong similarities between the organelle and nuclear DNAs, which means that the nuclear copies are relatively young. This, in turn, suggests that the organelle-to-nucleus transfer of DNA is still going on.

In this evolutionary perspective, the properties of mitochondrial and chloroplast genomes that vary among the organelles of present-day species are probably relatively new. These recently established features include long stretches of cpDNA incorporated in the mtDNAs of many plants, as well as many of the introns in organelle genomes. Some of these introns may have originated in the earliest bacterial symbionts; or they may have been incorporated into the organelle genomes after horizontal transfers between organelle DNAs long after the organelles were established.

Mitochondrial DNA Has a High Rate of Mutation

In the 1980s, surveys of DNA sequence variations among individuals of a given species and between closely related species showed that the mtDNA of vertebrates evolves almost 10 times more rapidly than the nuclear DNA of those same organisms. The higher rate of DNA mutation in mitochondria probably reflects more errors in replication and less efficient repair mechanisms.

Because of mtDNA's high mutation rate, the variation among mitochondrial genomes provides a valuable tool for studying the evolutionary relationships of organisms whose nuclear DNAs are very similar. Conversely, mtDNA variation, because it accumulates so rapidly, is of little value in evaluating the relationships of distant evolutionary relations, but here sequence variation data for nuclear genomes are useful (see Chapter 21). Sequence analyses of mtDNA have shown that the maternal lineage of all present-day humans, no matter what ethnic group they belong to, traces back to a few women who lived in Africa some 200,000 years ago. The Fast Forward box on pp. 594–595 "Mitochondrial DNA Sequences Shed Light on Human Evolution" describes the evidence for this hypothesis of a "mitochondrial Eve." It also explains how the early representatives of modern humans may have interacted with archaic humans and other hominid species in the genus *Homo* during the evolution of modern human populations.

16.2 Genetic Studies of Organelle Genomes Clarify Key Elements of Non-Mendelian Inheritance

Mutations in organelle genes produce readily detectable whole-organism phenotypes because the altered proteins and RNAs they encode disrupt the production of cellular energy. The mtDNA mutations that cause *petite* colonies in yeast, for example, disrupt the oxidative-phosphorylation system; the cpDNA mutations that cause variegation in four-o'clocks incapacitate proteins essential for photosynthesis. Although most mutations in the genes for these energy-producing systems are lethal in both plants and animals, some organelle gene mutations yield detectable, nonlethal phenotypes that researchers can study genetically. Data from such genetic studies show that modes of organelle gene transmission vary among organisms. We describe the main modes of transmission from one generation to the next. We also examine how parental contributions to the zygote and mitotic segregation determine which organelles are passed on to progeny cells during cell divisions.

In Most Species, Progeny Inherit Organelle DNA from Only One—Usually the Maternal—Parent

Genetic and biochemical analyses of thousands of crosses have shown that most species transmit mtDNA and cpDNA via one parent, usually the mother; that is, they use a uniparental, mainly maternal, mode of transmission. Researchers have analyzed **uniparental inheritance** in several ways.

Maternal Inheritance of Preexisting Differences in Wild-Type mtDNAs

In a classic experiment documenting maternal inheritance in vertebrates, investigators purified mtDNA from frog eggs, which contain a large number of mitochondria, and used hybridization tests to distinguish the mtDNA of one frog species, *Xenopus laevis,* from the mtDNA of the closely related *Xenopus borealis.* In these tests, probes from *X. laevis* hybridized more efficiently with *X. laevis* mtDNA than with *X. borealis* DNA, and vice versa. Because interspecific crosses between the two species yield viable progeny, the analysis of F_1 mtDNA was one way to trace the inheritance of that DNA in frogs. **Figure 16.9** diagrams the reciprocal crosses and mtDNA typing that formed the basis of the study. The first-generation progeny of both crosses carried mtDNA like that of the maternal parent. Although the analysis might have missed small contributions from the paternal genome, these *Xenopus* crosses confirmed the predominantly maternal inheritance of mtDNA in these species. They also showed that it is possible to follow pre-existing differences in functionally wild-type mtDNAs in a cross. Since the 1980s, analysis of crosses using DNA polymorphisms confirmed maternal mtDNA inheritance among horses, donkeys, and many other vertebrates.

Maternal Inheritance of Specific Genes in cpDNA

Interspecific crosses tracing biochemically detectable, species-specific differences in several chloroplast proteins provided further evidence of maternal inheritance. In the mtDNA inheritance studies just cited, the identity of the organelle gene containing the markers was not known and did not matter. By contrast, in cpDNA studies,

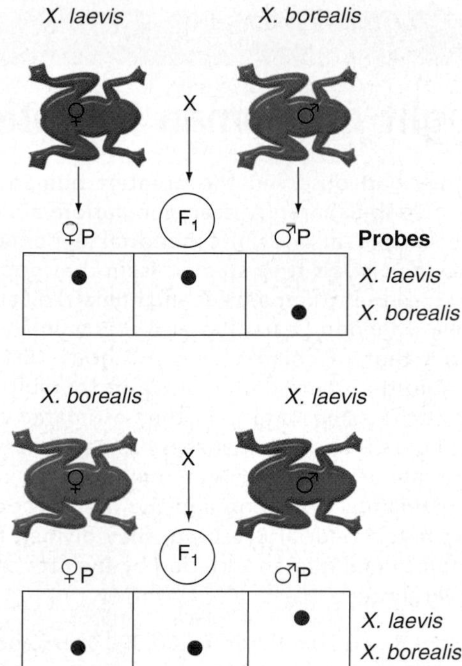

Figure 16.9 Maternal inheritance of *Xenopus* mtDNA. *X. laevis* and *X. borealis* mtDNA can be distinguished by strong hybridization only to probes made from the same species. Reciprocal crosses between two species produce F_1 hybrids. Each F_1 hybrid retains mtDNA only from its mother.

researchers identified specific organelle genes through the analysis of proteins. They began by isolating from tobacco plants (*Nicotiana* species) proteins in which interspecies differences could be distinguished by gel electrophoresis. To determine each protein's mode of inheritance, they evaluated the allele expressed in the progeny of a controlled cross, carefully noting the maternal (ovum) and paternal (pollen) contributions. In one set of experiments, ribulose bisphosphate carboxylase (Rubisco, for short), the first enzyme of photosynthetic carbon fixation in plants and the most abundant protein in tobacco leaves, was the protein under observation. Rubisco has a 55 kDa large subunit (called LSU) and a 12 kDa small subunit (called SSU). The researchers purified Rubisco from many strains of tobacco plants, digested the purified proteins with trypsin, and analyzed the digests for usable differences. When they followed the inheritance of these differences, they found that LSUs manifested patterns of maternal inheritance, while SSUs showed biparental inheritance. From these results, they hypothesized that a chloroplast gene encodes the LSU polypeptide, while a nuclear gene encodes the SSU.

These studies of the two Rubisco subunits reveal that organelle and nuclear genomes cooperate in specifying even a relatively simple enzyme with only two different subunits. Both genomes contribute essential information for most photosynthetic activities, including those whose elements are much more complex.

The inheritance studies just described followed preexisting differences in functionally wild-type organelle genomes. To verify and understand the details of uniparental inheritance, organelle geneticists followed the inheritance of mutations affecting phenotype at both the biochemical and the organismal levels, as seen in the following example.

A Mutation in a Human Mitochondrial Gene Generates a Maternally Inherited Neurodegenerative Disease

Leber's hereditary optic neuropathy, or LHON, is a disease in which flaws in the mitochondria's electron transport chain lead to optic nerve degeneration and blindness (**Fig. 16.10**). Family pedigrees show that LHON passes only from mother to offspring. In the late 1980s and early 1990s, a series of molecular studies showed that a G-to-A substitution at nucleotide 11,778 in the human mitochondrial genome is a main cause of the condition. The substitution alters an arginine-specifying codon in the NADH dehydrogenase subunit 4 gene to a histidine codon. The resulting protein product diminishes the efficiency of electron flow down the respiratory transport chain, reducing the cell's production of ATP sufficiently to cause a gradual decline in cell function and ultimately cell death. Since optic nerve cells have a relatively high requirement for energy, the genetic defect affects vision before it affects other physiological systems.

In large families, not all offspring show signs of the disease, and not all siblings manifesting the condition have symptoms of the same severity. The random allotment to daughter cells of a large number of mitochondria during mitosis (described in the following section) helps explain these observations.

Cells Can Contain One Type or a Mixture of Organelle Genomes

A diploid cell contains dozens to thousands of organelle DNAs. It is therefore not possible to use the terms "homozygous" and "heterozygous" to describe a cell's complement of mtDNA or cpDNA. Instead, geneticists use the

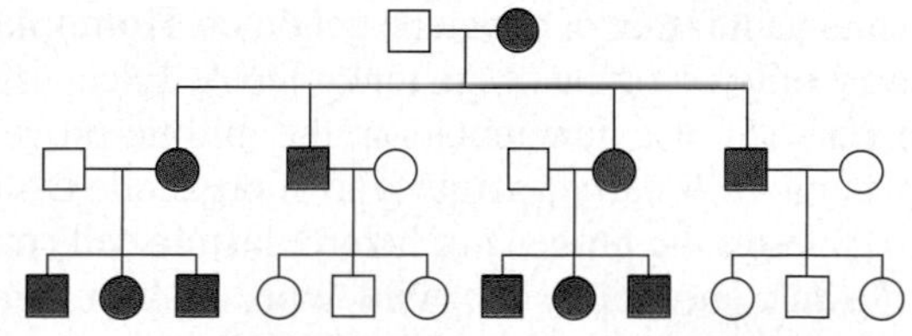

Figure 16.10 Hypothetical example of LHON pedigree. Characteristic pedigree of mitochondrial disease. All offspring of diseased mothers can express the disease phenotype, while none of the offspring of diseased fathers express the disease phenotype.

FAST FORWARD

Mitochondrial DNA Sequences Shed Light on Human Evolution

The mitochondrial DNA of all humans alive today traces back through maternal lineages to the mtDNA of a human population living in Africa some 200,000 years ago. Such is the startling conclusion of two papers published in 1987 and 1991 by Allan C. Wilson and colleagues. The carrier of this ancestral mtDNA, dubbed "mitochondrial Eve," probably lived in a population of 10,000–50,000 people. (In this context, a *population* is a group of interbreeding individuals of the same species who inhabit the same space.)

How mtDNA Variations Suggest the Region Where Modern Humans Emerged

In their studies supporting an African origin for modern humans, Wilson and coworkers first looked at restriction fragment length polymorphisms (RFLPs) in the mtDNA of 143 subjects, including Americans of African, Asian, European, and Middle Eastern origin, as well as Aboriginal women in New Guinea and Australia. Four years later, the group followed up their original 1987 study with sequence analyses of a rapidly evolving, highly polymorphic noncoding segment of mtDNA from 189 individuals, including 121 native Africans from various parts of the continent, Papua New Guineans, Europeans, Asians, African-Americans, and a native Australian. In both studies, the researchers found greater sequence differences among Africans, particularly sub-Saharan Africans, than among Asians or Europeans. Because mutations accumulate over time, they concluded that the African population has had the longest time to evolve variation and thus modern humans originated in Africa.

Statistical Calculations of the Number and Rate of Mutations Suggest When Modern Humans Appeared

Having proposed that modern humans first appeared in Africa, the researchers calculated the probable date of their origin by extrapolating the unknown from the known. They had observed the greatest human mtDNA variation in a sub-Saharan African population; about 2.8% of the base pairs in the mitochondrial genome varied among the individuals they studied from that population. They also knew that chimpanzees and humans diverged approximately 5 million years ago and that human mtDNA differs from that of chimpanzees in about 15% of the genome. Adjusting these data to account for multiple substitutions at the same base pair, they estimated that the mtDNA of humans and chimpanzees has been diverging at an average rate of about 13.8% per million years. To determine approximately how long ago the human population containing mitochondrial Eve lived, they divided the percentage of maximal human variation by the rate of chimp-human divergence:

$$2.8/13.8 = 0.20 \text{ million} = 200{,}000 \text{ years ago}$$

Although there has been some controversy over the statistical methods and assumptions that formed the basis of this analysis, most geneticists now agree with the conclusion that the women carrying our ancestral mtDNA lived roughly 200,000 years ago in sub-Saharan Africa. Recent studies of parts of the Y chromosome and other pieces of nuclear DNA support this conclusion. In one 1997 study, evolutionary geneticists examined two nuclear DNA segments on chromosome 12 in the genomes of 1600 individuals from 42 populations around the world. In populations outside of Africa, one combination of polymorphisms in the two elements showed up almost exclusively; by contrast, in sub-Saharan Africa, almost all the possible variants appeared. Recently, analyses of Y chromosome polymorphisms uncovered two ancient DNA markers shared by nonhuman primates and a small group of sub-Saharan African men. These markers sustained mutations about 100,000–200,000 years ago. Today, all men outside of Africa as well as most African men carry the mutated markers.

terms "heteroplasmic" and "homoplasmic" to describe the genomic makeup of a cell's organelles. **Heteroplasmic** cells contain a mixture of organelle genomes. **Homoplasmic** cells carry only one type of organelle DNA. Except for the rare appearance of a new mutation, the mitotic progeny of homoplasmic cells carry a single type of organelle DNA. By contrast, the mitotic progeny of heteroplasmic cells may be heteroplasmic, homoplasmic wild type, or homoplasmic mutant. In most people affected by LHON, for example, the optic nerve cells are homoplasmic for the disease mutation; but in some LHON patients, these optic nerve cells are heteroplasmic. Homoplasmy causes earlier appearance of the disease as well as more severe symptoms.

Mitotic Segregation Produces an Uneven Distribution of Organelle Genes in Heteroplasmic Cells

The random partitioning of organelles during cell divisions is the basis of the mitotic segregation of organelle genomes. This pattern of random segregation is in sharp contrast to the mitotic transmission pattern of nuclear genes: In diploid cells that are heterozygous for a nuclear gene, the alleles of the gene almost never segregate during mitotic divisions; the rare exception in which a heterozygous diploid produces homozygous wild-type and homozygous mutant progeny is the result of mitotic crossing-over.

Comparisons of Sequence Variations Point to How *Homo sapiens* Evolved

In addition to providing evidence for where and when modern humans first appeared, mtDNA may help evolutionary biologists determine what happened to the human populations that migrated out of Africa to other areas of the globe. At present, there are two models. One, the replacement theory, posits that as early modern humans (*Homo sapiens*) migrated to Europe, Asia, and elsewhere, they eventually replaced the archaic hominid groups living in those areas. In this view, although early modern *H. sapiens* spread to parts of the world already populated by archaic species in the genus *Homo* (such as *H. erectus*), the early moderns did not intermingle with these other hominid groups. Rather, they outcompeted and eventually replaced them such that anatomically modern *H. sapiens* became the only extant human (*Homo*) species (**Fig. A**). An opposing view, the regional continuity theory, proposes that humans evolved continuously in many parts of the world, with early modern humans interbreeding with archaic peoples in Europe and Asia. These archaic peoples would have included the Neanderthals, who coexisted with modern humans in the Middle East and parts of Europe and Asia for close to 70,000 years. The most recent DNA comparisons, which show significant differences between modern human and Neanderthal mtDNA, seem to support the replacement theory.

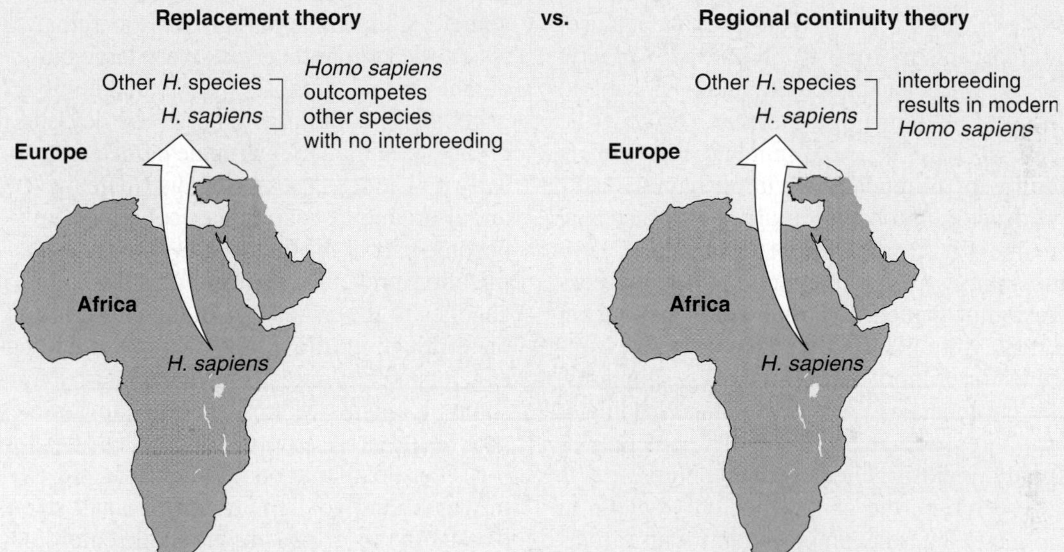

Figure A Human evolution. Two current theories pose that *Homo sapiens* evolved in Africa and migrated out of Africa to Europe and Asia, but in the regional continuity theory, the new migrants interbred with existing *Homo* species. In the replacement theory, the migrant *Homo sapiens* outcompeted the existing species. Mitochondrial evidence supports the replacement theory.

The mitotic segregation of organelle genomes has distinct phenotypic consequences. In a woman whose cells are heteroplasmic for the LHON mutation, some ova may carry a few mitochondria with the LHON mutation and a large number of mitochondria with the wild-type gene for subunit 4 of NADH dehydrogenase; other ova may carry mainly mitochondria with LHON mutations; still others may carry only wild-type organelles. The precise combination depends on the random partitioning of mitochondria during the mitotic divisions that gave rise to the germ line. After fertilization, as a result of the mitotic divisions of embryonic development, the random segregation of mutation-carrying mitochondria from heteroplasmic cells can produce tissues with completely normal ATP production and tissues of low energy production. If cells homoplasmic for low energy production happen to end up in the optic nerve, LHON will result.

In plants where cpDNA mutations leading to a defect in photosynthesis would be lethal, heteroplasmy for chloroplast genomes is prevalent. In fact, mitotic segregation of the chloroplasts of heteroplasmic cells explains the transmission of variegation in four-o'clocks, described at the beginning of this chapter (see Fig. 16.1). Most female gametes from a variegated plant are heteroplasmic for mutant and wild-type cpDNAs. Zygotes resulting from fertilization with pollen from a wild-type green plant will develop into variegated progeny. Segregation of wild-type and mutant chloroplasts during F_1 plant development may, however, generate some female gametes with only mutant cpDNA. Fertilization of these homoplasmic mutant gametes with wild-type pollen produces zygotes with only mutant cpDNA; the seedlings that develop from these zygotes cannot carry out photosynthesis and eventually die.

Experiments with Mutants of cpDNA in *Chlamydomonas reinhardtii* Reveal Uniparental Inheritance of Chloroplasts

Chlamydomonas, a unicellular alga, is easy to handle, grows abundantly in defined media, and has a tractable sexual reproduction cycle. It can live in different environmental conditions, functioning photosynthetically in the light on media where atmospheric CO_2 is the sole source of carbon, or without photosynthesis in the dark on media containing acetate as a source of carbon. These features make it a useful experimental organism. In the laboratory, *Chlamydomonas* usually grows as a haploid cell, and each haploid strain comes in one of two mating types: mt^+ or mt^- (the minus in this case does not denote a mutant). Researchers can induce cultures of mitotically dividing haploid cells (**Fig. 16.11a**) to develop into haploid gametes by a special physiological regime (Fig. 16.11b). When they then mix gametes of opposite mating types, an elaborate mating process results in the fusion of gametes of opposite types to form diploid zygotes. On suitable medium, these zygotes undergo meiosis and form unordered tetrads. For simplicity, we describe all experiments with this organism as leading to tetrads with four spores, even though in reality sometimes a mitotic division after meiosis generates eight spores.

Pioneering research on the genetics of chloroplasts in *Chlamydomonas* led to the isolation of mutants with various defects in photosynthetic growth. The first such mutant analyzed genetically carried the *sm-r* allele, which confers resistance to the antibiotic streptomycin, an inhibitor of prokaryotic (and also chloroplast) protein synthesis. These mutants could grow photosynthetically on medium containing streptomycin, while the drug-sensitive *sm-s* parental strain could not. The mutant strain also grew well in the absence of the drug. Researchers could thus carry out crosses in media lacking the drug and then grow the progeny on drug-containing media to determine their phenotype.

Consider a cross between mt^+ *sm-r* gametes and mt^- *sm-s* gametes in which researchers scored the resulting tetrads for the transmission of drug sensitivity (Fig. 16.11c). The investigators included a known nuclear marker affecting colony color (*yellow,* with alleles yl^+ and yl^-) in the parental strains and monitored the marker in the progeny. In this experiment, alleles of the mating-type gene and the *yellow* marker gene segregated 2:2 in the spores, but the ratio of *sm-r* to *sm-s* was 4:0. This lack of segregation was a striking departure from Mendelian principles. Subsequent reciprocal crosses showed that the *sm-r* allele did not itself determine the nonsegregating outcome. In a reciprocal cross of mt^+ yl^+ *sm-s* gametes with mt^- yl^- *sm-r* gametes, the tetrads were 4 *sm-s* : 0 *sm-r.* Because *Chlamydomonas* lacks distinct male and female gametes, it is an *isogamous* species in which the "sex" cells are similar in size and shape but come in two mating types. In speaking of isogamous species, geneticists say that a 4:0 ratio of spore genotypes reflects *uniparental,* rather than maternal, inheritance. Researchers have isolated many other *Chlamydomonas* mutants resistant to various antibiotics or deficient in photosynthesis, and all show the uniparental mode of transmission.

Many Mechanisms Contribute to Uniparental Inheritance of Mitochondrial and Chloroplast Genes

Differences in gamete size help explain maternal inheritance in some species. In most higher eukaryotes, the male gamete is much smaller than the female gamete. As a result, the zygote receives a very large number of maternal organelles and, at most, a very small number of paternal organelles. In some organisms, cells degrade the organelles or the organelle DNA of male gametes. In some plants, the early divisions of the zygote distribute most or all of the paternal organelle genomes to cells that are not destined to become part of the embryo. In some animals, details of fertilization prevent a paternal cell from contributing its organelles to the zygote. With tunicates, for example, events of fertilization allow only the sperm nucleus to enter the egg, physically excluding the paternal mitochondria. In some organisms where the complete gametes fuse, the zygote destroys the paternal organelles after fertilization.

These explanations do not apply to *Chlamydomonas,* in which the two gametes are of equal size and fuse completely. In this organism, the single chloroplast of one gamete fuses with the chloroplast of the other gamete shortly after cell fusion. However, the zygote, able to distinguish between cpDNAs from mt^+ and mt^- cells, degrades most of the cpDNAs from the mt^- parent. Interestingly, *Chlamydomonas* mtDNA is also inherited from one parent, but here, the mitochondrial genome from the mt^- cell type remains.

Summary of Uniparental Inheritance

The transmission of organelles and their genomes via one parent, usually the mother, in most plants and animals results in two significant departures from Mendelian patterns of inheritance. First, organelle genes show no meiotic segregation: All F_1 individuals carry the same organelle genotype—that of the transmitting parent. Second, when the parent transmitting the organelle genome is heteroplasmic and therefore passes along organelles of more than one genotype, mitotic segregation of those genotypes occurs in the offspring. This segregation of genotypes during mitosis is a consequence of the random partitioning of organelles during cell division.

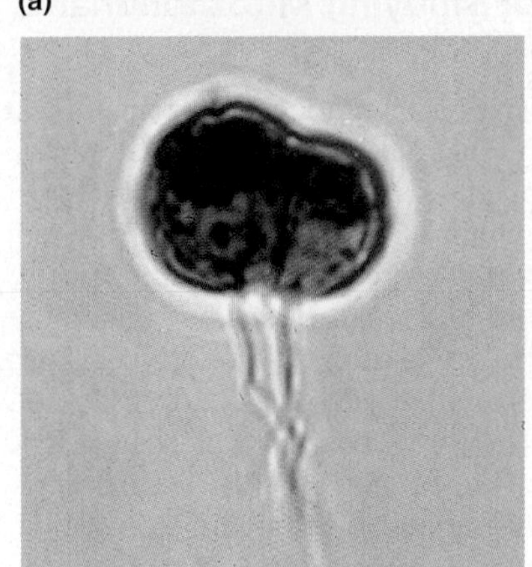

Figure 16.11 Genes in the cpDNA of *C. reinhardtii* do not segregate at meiosis. **(a)** When haploid cultures of gametes of opposite mating type are mixed, cells aggregate and form pairs, leading to another round of life cycles. **(b)** A cross of *C. reinhardtii* gametes is illustrated. Vegetative cells of this organism are usually haploid. Mating of those cells is controlled by the mating-type locus that exists in two alleles (mt^+ and mt^-), indicated by a nucleus colored *blue* or *yellow*, respectively. Each haploid cell has one large chloroplast (*light* or *dark green* for different alleles of cpDNA) and several mitochondria (*light* or *dark red* for different alleles of mtDNA). Mating occurs by pairwise fusion of gametes of opposite mating type. Shortly after cell fusion, the nuclei and chloroplasts fuse so that each zygote has one diploid nucleus and one large chloroplast. The zygote matures, during which time chloroplast markers from the mt^- parent and mitochondrial markers from the mt^+ parent are inactivated. Mature zygotes then undergo meiosis, forming four haploid meiotic products, each of which can germinate. If light or ammonium ions are present, the haploid cells will grow vegetatively. If ammonium ions are absent, the cells will undergo gametogenesis to prepare for another round of mating. **(c)** A nuclear gene (*yl*) segregates 2:2 in the gametes, but a chloroplast-encoded gene (*sm*) segregates 4:0.

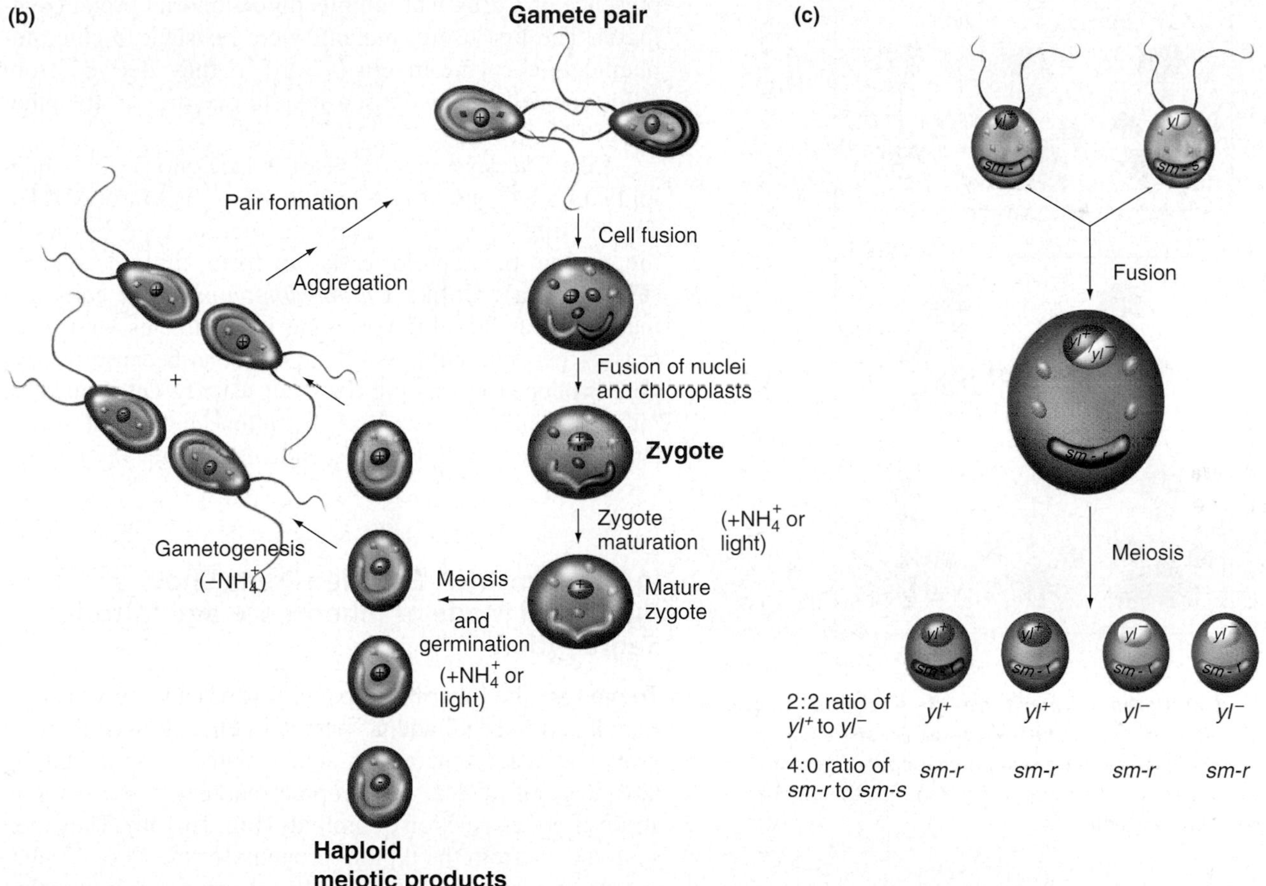

The Genetics and Society box "Mitochondrial DNA Tests Replace HLA Typing As Evidence of Kinship in Argentine Courts" on pp. 600–601 describes how a human rights organization in Argentina eventually used mtDNA sequences as the legal basis for reuniting kidnapped children with their biological families. The maternal inheritance of mitochondria makes it possible to compare and match the DNA of a grandmother and a grandchild.

Some Organisms Exhibit Biparental Inheritance

Although uniparental inheritance of organelles is the norm among most metazoans and plants, single-celled yeast and some plants inherit their organelle genomes from both parents in a **biparental** fashion. The earliest report of biparental organelle inheritance is a 1909 description of reciprocal

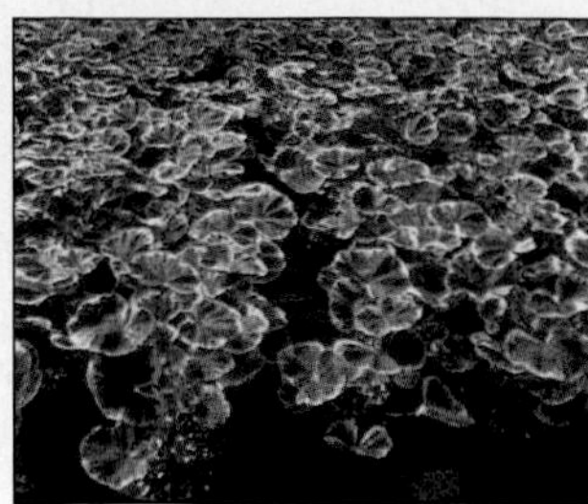

(a)

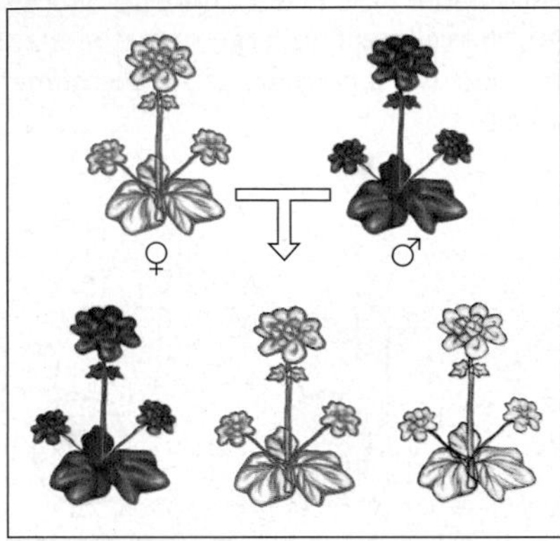

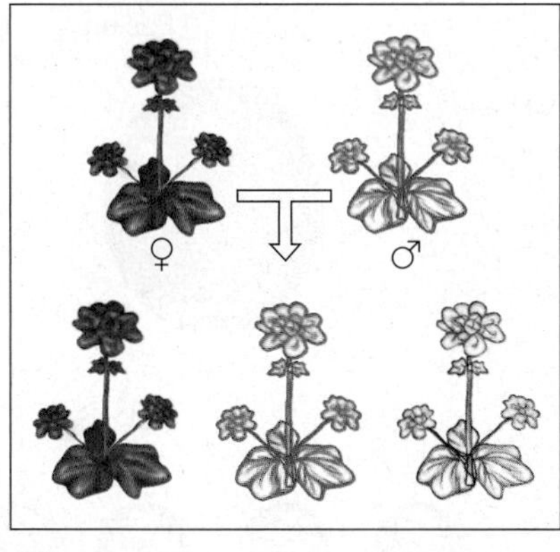

(b)

Figure 16.12 Biparental inheritance of variegation in geraniums. **(a)** Examples of green and variegated *P. zonale* plants. **(b)** Reciprocal crosses between green and variegated geraniums yield the same classes of offspring, indicating that the gene is inherited from both parents.

crosses between green and variegated geraniums (*Pelargonium zonale*). Unlike what happens in four-o'clocks (as described earlier), both reciprocal crosses yielded green, white, and variegated seedlings in varying proportions (**Fig. 16.12**). Thus, variegated leaves in geraniums are a chimeric condition that results from the chloroplast traits inherited from both parents. Many other plants, as well as yeast, similarly inherit their organelle genomes from both parents. Geneticists have devised experimental systems for analyzing biparental organelle inheritance in several organisms. We now look at some of the studies in yeast.

A Genetic System for Studying Mitochondrial Genes in Yeast

We have seen that translation in mitochondria resembles translation in bacteria and that inhibitors of bacterial translation, such as chloramphenicol and erythromycin, are potent inhibitors of mitochondrial protein synthesis. In the early 1960s, researchers discovered that chloramphenicol inhibits the growth of wild-type yeast on media containing a nonfermentable source of carbon (either glycerol or ethanol); but the same or higher levels of the drug do not inhibit yeast growth on medium containing glucose, a fermentable carbon source. They concluded that the inhibitor acts on the then recently discovered mitochondrial translation system and began to develop a system for studying the inheritance of yeast mtDNA. The initial step was to isolate mutants that could grow on medium containing glycerol in the presence of a drug that inhibits mitochondrial protein synthesis. The first useful mutants were resistant to chloramphenicol or erythromycin (C^r or E^r); they derived from wild-type cells that were sensitive to one drug or the other (C^s or E^s).

Like *Chlamydomonas,* yeast is isogamous. Each haploid strain has one of two mating types: MATa or MATα. The mixing of cells of opposite mating type results in the fusion of haploid cells to form diploid zygotes (**Fig. 16.13a**). Unlike *Chlamydomonas,* yeast cells can mate without first differentiating into gametes, and yeast crosses generate cultures of diploid cells because the zygote produced by mating does not usually enter meiosis; rather, it continues to multiply by mitosis until some sort of stress induces its diploid descendants to enter the meiotic cycle.

In Yeast, mtDNA-Encoded Traits Show a Biparental Mode of Inheritance and Mitotic Segregation

To analyze the inheritance pattern of mtDNA in yeast, researchers mixed C^r and C^s parental cells, allowed them to grow for many generations, and isolated diploids, taking advantage of nuclear auxotrophic markers in the parental strains for the selection of diploids (Fig. 16.13b). They then scored cells from the diploid progeny for the C^r or C^s phenotype by replica plating the cells to petri plates containing glycerol medium with or without an inhibitory level of chloramphenicol.

In most progeny resulting from the crosses, about half of the diploid cells were C^s, and half were C^r. This result shows that both parents transmitted organelles to the progeny; that is, inheritance was biparental. It also shows that the organelles from parental cells of different mating types segregated during the mitotic divisions of vegetative growth, so after several generations, colonies were either C^s or C^r, but not both. Importantly, all of the progeny scored, whether C^s or C^r, had the same nuclear-determined

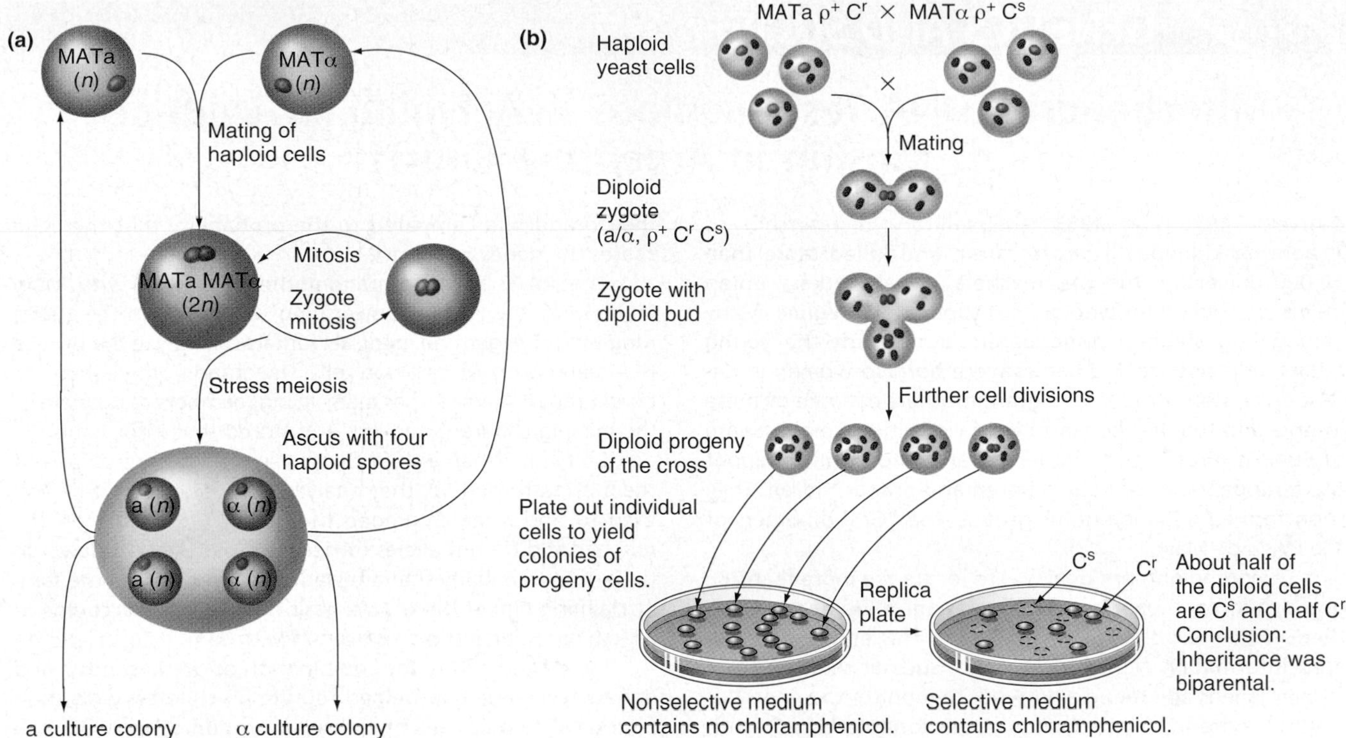

Figure 16.13 Mitochondrial transmission from the yeast *S. cerevisiae*. **(a)** General life cycle of *S. cerevisiae*. **(b)** One-factor crosses in *S. cerevisiae* show biparental inheritance of mtDNA.

phenotype (that is, they were prototrophic for two nutrients). This meant that the diploids carried nuclear genes from both parents in the cross, and those nuclear genes did not segregate during this period of vegetative growth.

Summary of the Genetic Principles of Non-Mendelian Inheritance

Three features distinguish the non-Mendelian traits encoded by organelle genomes from the Mendelian traits encoded by nuclear genomes.

1. In the inheritance of organelle genomes from one generation to the next, there is a 4:0 segregation of parental alleles, instead of the 2:2 pattern seen for the alleles of nuclear genes.
2. In most organisms, transmission of organelle-encoded traits is uniparental, mainly maternal, although in a few organisms, notably yeast, transmission is biparental.
3. With both uniparental and biparental inheritance, when the parents transmit organelles of more than one genotype, mitotic segregation of those genotypes occurs in the offspring. This segregation of genotypes during mitosis is a consequence of the random partitioning of organelles during cell division.

16.3 Comprehensive Example: How Mutations in mtDNA Affect Human Health

Some debilitating diseases of the human nervous system pass from mother to daughters and sons, from affected daughters to granddaughters and grandsons, and so on down through the maternal line. Unexpectedly, the symptoms of these diseases vary enormously among family members, even among very close relatives. Pedigrees tracing general transmission patterns, in combination with molecular studies, help clarify what is happening.

Individuals with Certain Rare Diseases of the Nervous System Are Heteroplasmic for Wild-Type and Mutant mtDNAs

People with a rare inherited condition known as myoclonic epilepsy and ragged red fiber disease (MERRF) have a range of symptoms: uncontrolled jerking (the myoclonic epilepsy part of the condition), muscle weakness, deafness, heart problems, kidney problems, and progressive dementia.

GENETICS AND SOCIETY

Mitochondrial DNA Tests Replace HLA Typing As Evidence of Kinship in Argentine Courts

Between 1976 and 1983, the military dictatorship of Argentina kidnapped, incarcerated, and killed more than 10,000 university students, teachers, social workers, union members, and others who did not support the regime. Many very young children disappeared along with the young adults, and close to 120 babies were born to women in detention centers. In 1977, the grandmothers of some of these infants and toddlers began to hold vigils in the main square of Buenos Aires to bear witness to and inform others about the disappearance of their children and grandchildren. They soon formed a human rights group—the "Grandmothers of the Plaza de Mayo."

The grandmothers' goal was to locate the more than 200 grandchildren they suspected were still alive and reunite them with their biological families. To this end, they gathered information from eyewitnesses, such as midwives and former jailers, and set up a network to monitor the papers of children entering kindergarten. Questionable birth certificates could reveal a possible missing grandchild. They also publicized their work inside Argentina and contacted organizations outside the country, including the United Nations Human Rights Commission and the American Association for the Advancement of Science (AAAS).

What the grandmothers asked of AAAS was help with genetic analyses that would stand up in court. By the time a democracy had replaced the military regime and the grandmothers could argue their legal cases before a relatively impartial court, children abducted at age 2 or 3 or born in 1976 were 7–10 years old. Although the grandmothers had compiled an enormous amount of circumstantial evidence, the Argentine courts did not accept such evidence as proof of a young person's identity and biological relatedness. The courts did acknowledge, however, that although the size and other external features of the children had changed, their genes—relating them unequivocally to their biological families—had not. The grandmothers, who had educated themselves about the potential of genetic tests, sought help with the details of obtaining and analyzing such tests. Starting in 1983, the courts agreed to accept their test results as proof of kinship.

In 1983, the best way to confirm or exclude the relatedness of two or more individuals (for instance, a child and her grandparents) was to compare the gene-encoded proteins called human lymphocyte antigens (HLAs). People carry a unique set of HLA markers on their white blood cells, or lymphocytes, and these markers are diverse enough to form a kind of molecular fingerprint. HLA analyses can be carried out even if a child's parents are no longer alive, because for each HLA marker, a child inherits one allele from the maternal grandparents and one from the paternal grandparents. Statistical analyses can establish the probability that a child shares genes with a set of grandparents because he or she is their grandchild compared to the probability that the child shares the genes by chance.

The AAAS put the grandmothers in touch with Mary Claire King, then at the University of California. In the 1980s, King taught Argentine medical workers to analyze the diverse HLA markers on white blood cells. The grandmothers then obtained the HLA types of as many living members as possible of the missing children's families and stored that information in an HLA bank. When a child whom they believed to be one of the missing turned up, they analyzed his or her HLA type and tried to find a match among their data. Depending on the number of different alleles carried and the rarity of the variations, the probability that a tested child belonged to the family claiming him or her on the basis of eyewitness accounts of a birth or abduction varied from 75% to 99%.

By the late 1980s, the combination of circumstantial and genetic evidence had helped reunite 49 children with their biological families. In one case, a grandmother who had been kidnapped September 5, 1976, along with her pregnant daughter and then released from prison contacted the Grandmothers of the Plaza de Mayo for help. Ten and a half years later, in April 1987, she was reunited with her granddaughter. The child had been born on November 5, 1976, and registered in the name of a detention center police officer. Her place of birth remains unknown. But, her grandmother recounted in an April 1987 interview, "It was very easy to prove she was my granddaughter because of the blood tests. They analyzed my blood, the blood of my husband and the other grandparents and they [the results] proved more than 99% positive."

As time passed and the "easier" cases had been settled, the limitations of the HLA approach became apparent. By the mid-1980s, for example, there were too few living relatives in some families to establish a reliable match through HLA typing. But the advent of new tools such as PCR and DNA sequencing made it possible to look at DNA directly.

King and two colleagues—C. Orrego and A. C. Wilson—used the new techniques to develop an mtDNA test based on the PCR amplification and direct sequencing of a highly variable noncoding region of the mitochondrial genome. Mitochondrial DNA has several characteristics that make it more powerful than HLA typing for confirming or excluding family relatedness: maternal inheritance, a lack of recombination, a highly variable noncoding region of 331 bp, and a large number of mtDNAs per cell. The maternal inheritance and lack of recombination mean that as long as a single maternal relative is available for matching, the approach can resolve cases of disputed relatedness. The extremely polymorphic noncoding region makes it possible to identify grandchildren through a direct match with the mtDNA of only one person—their maternal grandmother, or mother's sister or brother—rather than through statistical calculations assessing data from four people; the probability that any two unrelated individuals would be identical for all 331 bp by chance is almost nil. Finally, the large number of mtDNAs per cell increases the chances of isolating mtDNA from and identifying the bodies of presumed parents that are found after many years.

To validate their approach for matching maternal grandmothers to their grandchildren, King and colleagues amplified sequences from three children and their three maternal grandmothers without knowing who was related to whom. The mtDNA test unambiguously matched the children with their grandmothers. Thus, after 1989, the grandmothers included mtDNA data in their archives.

Today, the grandchildren—the children of "the disappeared"—have reached adulthood and attained legal independence. Although most of their grandmothers have died, the grandchildren may still discover their biological identity and what happened to their families through the HLA and mtDNA data the grandmothers left behind.

Affected individuals often have an unusual "ragged" staining pattern in regions of their skeletal muscles, which explains the ragged red fiber part of the condition's name (**Fig. 16.14a**). As the pedigree in Fig. 16.14b shows, family members inherit MERRF from their mothers; in the pedigree, none of the offspring of the affected male sibling (II-4) has symptoms of the disease. The family history also reveals individual variations in the number and severity of symptoms. From these two features of transmission, clinical researchers suspected that MERRF results from mutations in the mitochondrial genome.

Molecular analyses confirmed this hypothesis. The mtDNA from patients affected by MERRF carries a mutation in the gene for $tRNA^{Lys}$ or one of the other mitochondrial tRNAs. These tRNA mutations disrupt the synthesis of proteins in multiplexes I and IV of the mitochondrial electron transport chain, thereby decreasing the production of ATP.

The Proportion of Mutant mtDNAs and the Tissue in Which They Reside Influence Phenotype

In a second large family in which clinicians looked at the mitochondria of muscle cells, an individual carrying 73% mutant and 27% normal mtDNAs showed no symptoms of MERRF; a relative with 85% mutant mtDNA showed no external signs of the disease, but lab tests revealed some muscle tissue abnormalities; and two family members with 98% mutant mitochondria showed serious symptoms of MERRF. This suggests that a relatively small percentage of normal mitochondria can have a strong protective effect. The random partitioning of mitochondria during the mitotic cell divisions of development, both *in utero* and after birth, explains the variation in the percentage of mutant mitochondria in individuals of the second, third, and fourth generations of family pedigrees.

It is likely that many tissues in individuals affected by MERRF are heteroplasmic for the tRNA mutations causing the disease. Within each person, the ratio of mutant to wild-type mtDNA varies considerably from tissue to tissue, and because each tissue has its own energy requirements, even the same ratio can affect different tissues differently (**Fig. 16.15**). Muscle and nerve cells have the highest energy needs of all types of cells and are therefore the most dependent on oxidative phosphorylation. This explains why mitochondrial mutations that by chance segregate to these tissues generate the defining features of MERRF.

(a)

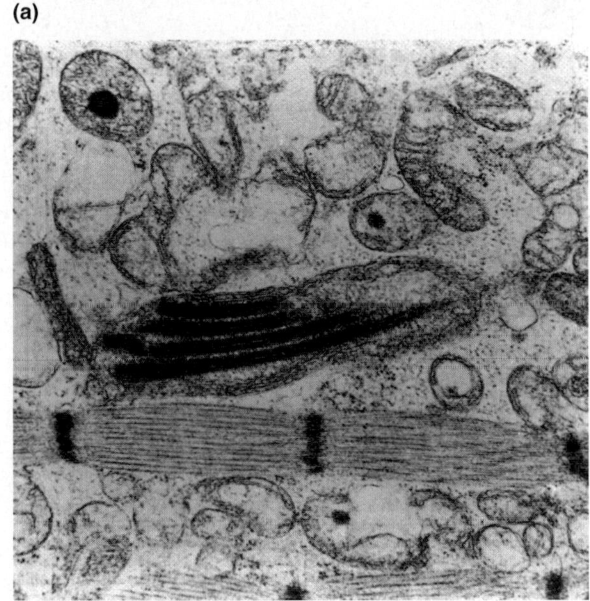

(b)

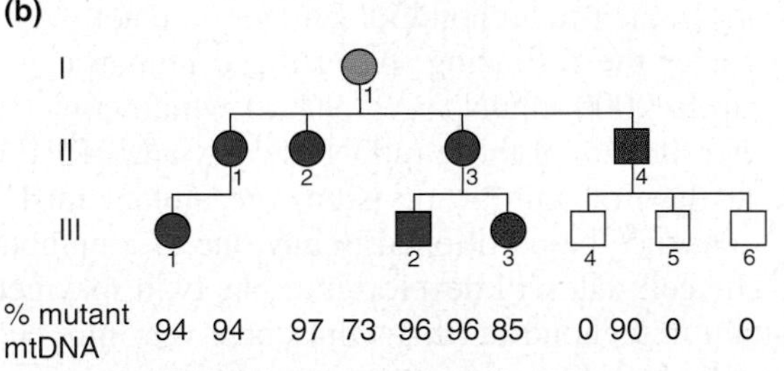

Figure 16.14 Maternal inheritance of the mitochondrial disease MERRF. (a) Transmission electron micrograph of muscle mitochondria from patients expressing MERRF. Mutant mitochondria are highly abnormal, showing paracrystalline arrays and crista degeneration. **(b)** Pedigree of family showing inheritance of MERRF. Pedigree shows typical pattern of maternal transmission observed with mitochondrial mutations. The percent of mutant mtDNA in the cells of individuals varies and corresponds with the severity of the condition (indicated by different color coding).

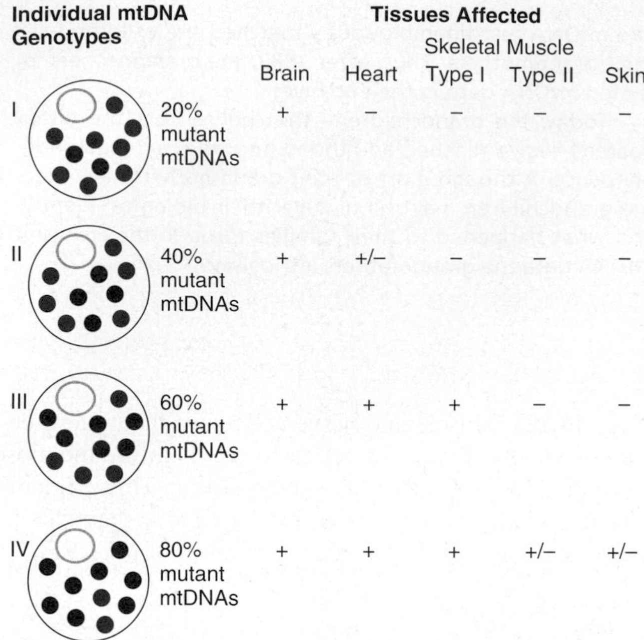

Individual mtDNA Genotypes
Tissues Affected
Skeletal Muscle
Brain
Heart
Type I
Type II
Skin
I
20% mutant mtDNAs
+ – – – –
II
40% mutant mtDNAs
+ +/– – – –
III
60% mutant mtDNAs
+ + + – –
IV
80% mutant mtDNAs
+ + + +/– +/–

Figure 16.15 Disease phenotypes reflect the ratio of mutant-to-wild-type mtDNAs and the reliance of cell type on mitochondrial function. The proportion of mutant mitochondria determines the severity of the MERRF phenotype and the tissues that are affected (+). Tissues with higher energy requirements (for example, brain) are least tolerant of mutant mitochondria. Tissues with low energy requirements (for example, skin) are affected only when the proportion of wild-type mitochondria is greatly reduced.

Mitochondrial Inheritance in Identical Twins

In some pairs of identical twins, one twin manifests symptoms of neurodegenerative disease, while the other twin does not. This is because even though their nuclear genomes are identical, their mitochondrial genomes are not.

Consider the following. A fertilized human egg cell carries up to 2000 mtDNAs, 99.9% of which come from the mother. If 10% of those mtDNAs carry a MERRF mutation, at the first mitotic division, the mutant mtDNAs could by chance be partitioned in any one of a number of ways: The cell that will develop into one twin may get all the mutant mitochondria (this would be a very rare occurrence); the mtDNAs with mutations may be equally divided between the two twin-producing cells; the mutated mtDNAs may be unevenly distributed to the two twin-generating cells. Even with an even distribution at the first cell division, during the subsequent divisions of development, the mutant mitochondria in one twin may end up in cells that eventually die during normal development, while the mutant mitochondria in the other twin may end up in cells that differentiate into heart and optic nerve tissue.

Mitochondrial Mutations and Aging

We have seen that like the nuclear DNA mutations that cause cancer, some mutations in mtDNA are inherited through the germ line, while others arise sporadically in somatic cells as a result of random events, such as radiation or chemical insults. We have also seen that the rate of somatic mutations is much higher in mitochondrial DNA than in nuclear DNA (nDNA). This is in part because the oxidative phosphorylation system in mitochondria generates a relatively high concentration of DNA-damaging free radicals in a membrane-confined space. In one study, mtDNA accumulated 16 times more oxidative damage than nDNA.

Some researchers focusing on the genetics of aging think that the accumulation of mutations (both base substitutions and deletions) in mtDNA over a lifetime, and the progressive enrichment of mtDNAs carrying deletions through their tendency to replicate more rapidly than longer, nondeleted mtDNAs, results in an age-related decline in oxidative phosphorylation. This decline, in turn, accounts for some of the symptoms of aging, such as decreases in heart and brain function.

Proponents of this hypothesis suggest that individuals born with deleterious mtDNA mutations start life with a diminished capacity for ATP production, and as a result, several of their tissues may cross the threshold from function to nonfunction early or in the middle of life; by comparison, people born with a normal mitochondrial genome start life with a high capacity for ATP production and may die before a large number of tissues dip below the required energy threshold.

Evidence in support of an association between mtDNA mutations and aging comes from a variety of studies. In one study, researchers looked at 140 hearts obtained from autopsies and found significant decreases in cytochrome *c* oxidase, a respiratory enzyme largely encoded by mtDNA. In another study, researchers analyzed a 7.4 kb and a 5 kb deletion in heart and brain mtDNAs in people of different ages. The percentage of hearts that had the 7.4 kb deletion increased with age, and the number of 5 kb deletions increased in normal heart tissue after age 40. Moreover, the 5 kb deletion was absent from the brain tissue of children but present in the brain tissue of adults. Finally, although biomedical researchers had known for decades that the brain cells of people showing symptoms of Alzheimer's disease (AD) have an abnormally low energy metabolism, they recently discovered that 20% to 35% of the mitochondria in the brain cells of most AD patients carry mutations in two of their three cytochrome *c* oxidase genes; the inefficiency of the resulting mutant cytochrome *c* oxidase enzyme impairs the brain's energy metabolism. To confirm an association between this enzymatic abnormality and AD, the researchers transferred mitochondria from AD patients into normal cultured cells from which they had removed the native mitochondria, and they found that the engineered cells had defective energy production.

These data suggest that if it were possible to assess all forms of mtDNA damage, it might turn out that a significant proportion of mtDNAs are defective in elderly people. On the basis of this hypothesis, clinicians have proposed that the restoration of enzymes encoded by wild-type mtDNA but absent from mitochondria with deleted and otherwise mutated genomes might ease some of the symptoms of aging. Further research will be necessary to discover whether mitochondrial damage makes a significant contribution to the aging process.

Connections

Mitochondria, the sites of oxidative phosphorylation, and chloroplasts, the sites of photosynthesis, exhibit non-Mendelian patterns of inheritance that are remarkably similar across a wide range of species.

Various regulatory mechanisms help correlate the day-to-day operations of a cell's organelles with the needs and circumstances of different tissues. One unusual example of the fine-tuning of these controls is seen in the voodoo lily (*Sauromatum guttatum*), a native of Southeast Asia that belongs to the same plant family as the skunk cabbage. The day the flower opens, a burst of mitochondrial activity in some of its cells increases the temperature inside the bloom by 22°C (72°F); the heat so generated causes odorous molecules to evaporate and attract insect pollinators, increasing the plant's chances for reproduction. Although researchers currently know very little about the regulation of mitochondrial genes, they believe that, given the endosymbiont theory of organelle evolution, the regulation of genes in prokaryotes may serve as a model for what happens in mitochondria.

In Chapter 17, we examine the molecular mechanisms that regulate gene expression in prokaryotes. Then, in Chapter 18, we explore the regulation of nuclear genes in eukaryotes. In both these chapters, we see that gene regulation controls the metabolic activities of all cells as well as the differentiation of cells in multicellular organisms.

Essential Concepts

1. Mitochondria and chloroplasts are semiautonomous organelles of energy conversion. They carry their own double-stranded DNA in circular or linear chromosomes whose size and gene content vary from species to species.
2. Translation in the mitochondria of many species depends on an alternative genetic code.
3. Chloroplast genomes are relatively uniform in size and carry many more genes than mitochondrial genomes. In some chloroplasts, the genes are organized in clusters that resemble the operons of bacteria.
4. According to the *endosymbiont theory,* mitochondria and chloroplasts evolved from bacteria engulfed by the precursors of eukaryotic cells. The genomes of these organelles have probably lost more than two-thirds of their original bacterial genes in the course of evolution.
5. In most species, organelle genomes show *uniparental inheritance,* mainly through the maternal line. Cells containing a mixture of organelle genomes are *heteroplasmic.* Cells carrying only one type of organelle DNA are *homoplasmic.* The genomes of heteroplasmic cells are not evenly partitioned at mitosis.
6. The organelles of yeast and several other species exhibit biparental inheritance. The genomes of organelles inherited in a biparental fashion show mitotic segregation.

On Our Website

www.mhhe.com/hartwell3
Chapter 16

Annotated Suggested Readings and Links to Other Websites

- Reviews of mitochondrial diseases in humans
- Recent papers on genetic engineering of chloroplasts
- Recent papers and reviews of RNA editing in mitochondria

Social and Ethical Issues

1. A researcher has promising results that could lead to a treatment for a neurodegenerative mitochondrial disease. When she applies for a grant from a federal agency, the reviewers respond that the research proposed is more applied than basic in nature and therefore should be funded by a commercial pharmaceutical company. When the researcher approaches drug companies with her proposal, she finds they are not interested in developing a drug that would have a rather limited market. Should the government create mechanisms to cover the expenses of development and clinical trials for drugs of relatively low use that do not have a potential for producing a profit but that will nonetheless greatly benefit affected people?
2. During repressive political regimes in many countries, children often are separated from their parents and/or become orphaned. Human rights groups have been in the forefront of assisting in the identification and reunion of relatives using genetic markers such as mitochondrial DNA. Should the government of the country in which the atrocities were committed be obligated to pay some or all of the costs of these searches, even if the current regime is far removed in political ideology and action from the repressive regime?
3. It has been suggested that a databank consisting of DNA fingerprints of previously convicted felons be established. Both nuclear and mitochondrial fingerprints would be included to accommodate situations in which the nuclear DNA was too degraded to be used. Who should have access to such a database? Law enforcers? Psychologists? Sociologists? Is this an invasion of privacy or a protection for the public?

Solved Problems

I. Differential hybridization of a probe to mitochondrial DNA from two *Xenopus* species was the methodology employed to demonstrate maternal inheritance in vertebrates (see Fig. 16.9 on p. 593). However, this hybridization technique was not sensitive enough to detect small amounts of paternal DNA. What technique that is more sensitive to small amounts of DNA could be used today? How could you use this technique to determine if paternal mitochondrial DNA was present in the progeny of the interspecies cross?

Answer

Polymerase chain reaction (PCR) is a sensitive technique that detects very small amounts of DNA. Oligonucleotide primers that are specific for each of the mitochondrial DNAs in each of the two different species could be used to determine if paternal DNA is present in the offspring from the interspecies cross.

II. A newly isolated *petite* (respiration deficient) mutant of yeast, when mated to a wild-type strain and sporulated, produced two *grande* and two *petite* spores. What does this meiotic segregation pattern tell you about the *petite* mutation?

Answer

The 2:2 segregation of mutant and wild-type phenotypes from meiosis indicates that the *mutation is in a nuclear gene.* The finding that a *petite* phenotype is caused by a nuclear mutation is not surprising, since functioning of the mitochondria requires both nuclear and mitochondrial genes. If the mutation were in mitochondrial DNA, a 4:0 segregation pattern would be seen instead of the 2:2 pattern.

III. a. Does the following pedigree suggest mitochondrial inheritance? Why or why not?

b. Is there another mode of inheritance that is consistent with this data?

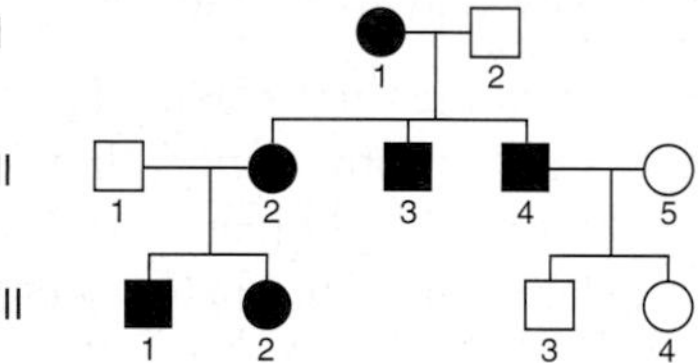

Answer

a. The data presented in this pedigree *are consistent with mitochondrial inheritance since the trait is transmitted by females; the affected males in this family did not transmit the trait; and all of the females' progeny have the trait.*

b. This inheritance pattern is *also consistent with transmission of an autosomal dominant trait.* According to this hypothesis, individuals I-1 and II-2 passed on the dominant allele to all children, but II-4 did not pass on the dominant allele to either child.

Problems

Vocabulary

1. Choose the phrase from the right column that best fits the term in the left column.

a. cristae	1. results in production of ATP
b. grana	2. yeast cells that can respire
c. *petite*	3. results in production of sugar
d. *grande*	4. cell containing organelles that are genetically identical
e. oxidative phosphorylation	5. contains multiple linked mini- and maxicircles of DNA
f. Calvin cycle	6. folded inner membrane in mitochondria
g. kinetoplast	7. yeast cells that cannot respire
h. heteroplasmic	8. stacked thylakoid membranes in chloroplasts
i. homoplasmic	9. cell containing two genetically different organelles

Section 16.1

2. Does each of the following processes occur in the chloroplast, mitochondria, both, or neither?
- a. electron transport
- b. pyruvate and fatty acid breakdown
- c. sugar synthesis
- d. O_2 generation
- e. O_2 consumption

3. Is each of these statements true of chloroplast or mitochondrial genomes, both, or neither?
- a. contain tRNA genes
- b. exist as condensed structures called nucleoids
- c. all genes necessary for function of the organelle are present
- d. vary in size from organism to organism

4. Some genes required for chloroplast function are encoded in the nuclear genome; others are encoded in the chloroplast genome. Nuclear and chloroplast DNA have different buoyant densities and can therefore be separated from each other by centrifugation based on these differences. There is a small amount of cross-contamination in the separation of nuclear and chloroplast DNAs using this technique. You have just found that a probe for a photosynthetic gene that is present in the chloroplast genome of plants hybridizes to nuclear DNA of a red alga.
- a. Do these results clearly show that the gene of interest is nuclear in the red alga? Why or why not?
- b. What additional DNA hybridization information would allow you to clarify your answer to *a*?
- c. Assuming this red alga shows uniparental inheritance of chloroplast genes and can be used in reciprocal crosses, design an experiment to confirm the genomic location of the gene discussed in *a*.

5. "Reverse translation" is a term given to the process of deducing the DNA sequence that could encode a particular protein. If you had the amino acid sequence Trp His Ile Met,
- a. What mammalian nuclear DNA sequence could have encoded these amino acids? (Include all possible variations.)
- b. What mammalian mitochondrial DNA sequence could have encoded these amino acids? (Include all possible variations.)

6. a. Results from hybridization using a probe for the small subunit gene of the Rubisco protein and a probe for the large subunit gene of the Rubisco protein to chloroplast (cp) and nuclear DNAs from a green, a red, and a brown alga are shown here. What conclusions would you reach about the location of the small and large subunit genes in each of the three types of algae?

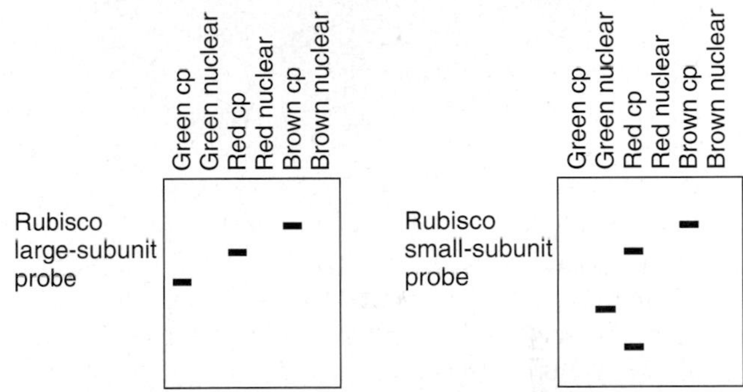

b. When RNA was extracted from the same three algal species and hybridized with a large subunit Rubisco probe and also with a small subunit probe, the following results were obtained. What conclusion would you reach about large and small subunit gene transcription in red and golden brown algae? Is this consistent with your answer in part *a?*

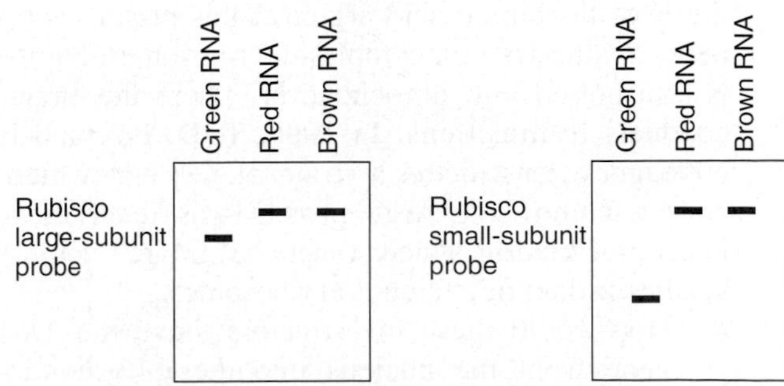

7. Which of the characteristics of chloroplasts and/or mitochondria listed at the top of p. 606 make them seem more similar to bacterial cells than to eukaryotic cells?

a. Translation is sensitive to chloramphenicol and erythromycin.
b. Alternate codons are used in mitochondria genes.
c. Chloroplasts and mitochondria have inner and outer membranes.
d. Introns are present in organelle genes.
e. DNA in organelles is not arranged in nucleosomes.

8. A chloroplast translation elongation factor gene (*tufA*) hybridized to a fragment of nuclear DNA in the green alga *Coleochaete.* How could you determine if the entire coding sequence of the gene was intact in the nuclear fragment that hybridized with the probe?

9. An example of a cloning vector used for biolistic transformation of chloroplasts is shown in the following diagram. The vector DNA can be prepared in large quantities in *Escherichia coli.* Once "shot" into a chloroplast, the vector DNA integrates into the genome. Match the component of the vector with its function.

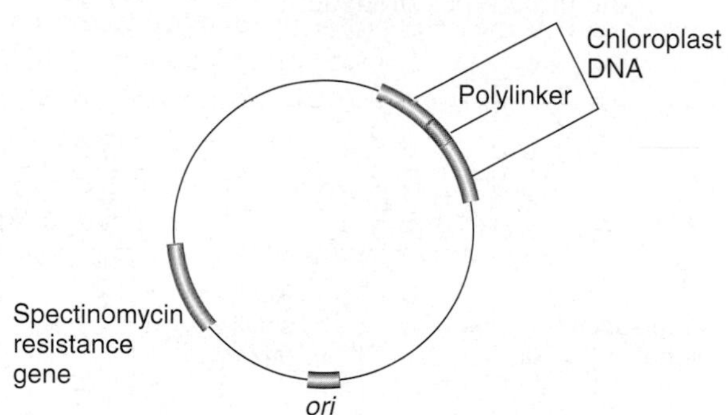

a. spectinomycin resistance gene	1. homologous DNA that mediates integration
b. chloroplast DNA	2. gene used to select chloroplast transformants
c. polylinker (multiple restriction sites)	3. sequence for replication in *E. coli*
d. ori	4. site at which DNA can be inserted

10. The *Saccharomyces cervevisiae* nuclear gene *ARG8* encodes an enzyme that catalyzes a key step in biosynthesis of the amino acid arginine; this protein is normally synthesized on cytoplasmic ribosomes, but then is transported into mitochondria, where the enzyme conducts its functions. In 1996, T. D. Fox and his colleagues constructed a strain of yeast in which a gene encoding the Arg8 protein was itself moved into mitochondria, where functional protein could be synthesized on mitochondrial ribosomes.

a. How could these investigators move the *ARG8* gene from the nucleus into the mitochondria, while permitting the synthesis of active enzyme? In what ways would the investigators need to alter the *ARG8* gene to allow it to function in the mitochondria instead of in the nucleus?
b. Why might these researchers have wished to move the *ARG8* gene into mitochondria in the first place?

Section 16.2

11. Studies distinguishing between uniparental and biparental inheritance of organelles employed a variety of detection methods. Match the system studied with the method used.

a. *Xenopus laevis* and *X. borealis* mitochondrial DNA	1. protein analysis
b. *Nicotiana* large and small subunits of Rubisco	2. pedigrees
c. LHON phenotype in humans	3. differential hybridization

12. Describe two ways in which the contribution of mitochondrial genomes from male parents is prevented in different species.

13. If a human trait is determined by a factor in the cytoplasm, would an offspring more resemble its mother or its father? Why?

14. Why are very severe mitochondrial or chloroplast mutations usually found in heteroplasmic cells instead of homoplasmic cells?

15. Which of the two methods listed would you choose to determine if organelles in an organism are heteroplasmic or homoplasmic and why?

a. hybridize probes to cells immobilized on a slide
b. PCR amplify DNA isolated from a population of cells

16. You have isolated a new C^r mutant in yeast, starting with a MATa strain. How could you test the new C^r mutant to determine if the mutation is in nuclear DNA or mitochondrial DNA?

17. A haploid yeast strain of mating type a that was chloramphenicol resistant (C^r) and erythromycin sensitive (E^s) was crossed with a haploid strain of mating type α that was chloramphenicol sensitive (C^s) and erythromycin resistant (E^r). The diploid cells were allowed to divide for several generations and then induced to enter meiosis. Of 100 resultant tetrads, 66 had four $C^r E^s$ spores, 29 had four C^sE^r spores, 3 had four C^rE^r spores, and 2 had four C^sE^s spores.

a. Explain these results. Why is the segregation always 4:0? How were the four different types of tetrads generated?
b. In the class of tetrads with four C^sE^s spores, how many spores would be of mating type a or α?

18. What characteristics in a human pedigree suggest a mitochondrial location for a mutation affecting the trait?

19. You have isolated a new erythromycin-resistant mutant of *Chlamydomonas* in an mt^+ strain.

a. If the mutation was in a chloroplast gene, what predictions would you make about genotype and phenotype of the diploid after mating the mutant with a mt^- er^s strain?

b. What would be the genotype and phenotype of the diploid if the mutation was in a nuclear gene?

c. What additional step would enable you to determine if the mutation was in chloroplast or nuclear DNA?

20. In the early 1900s, Carl Correns reported the results of observations he made on the inheritance of leaf color in the four-o-clock plant *Mirabilus jalapa.* He noticed that on the same plant, some branches contained all green leaves, some branches contained all white leaves, and some branches contained variegated leaves that had patches of green and white tissue.

a. Explain why some branches have green leaves, some have white leaves, and some have variegated leaves. Explain why variegated leaves have some patches of white and some patches of green tissue.

b. When Correns fertilized ovules from a green-leafed branch with pollen from flowers on any type of branch, he found that all the leaves in all of the progeny were green. When he fertilized ovules from a variegated branch with pollen from flowers on any type of branch, 90% of the progeny had some branches with green, some with white, and some with variegated leaves. 5% of the progeny had only green leaves, and the remaining 5% of the progeny had white leaves but were severely stunted and died soon after germination. Explain these results. How could ovules from a variegated branch produce progeny with all green or all white leaves? Why did the completely white-leaved plants die early?

c. Given your answer to part *b*, how could variegated plants have branches with apparently healthy white leaves?

21. A form of male sterility in corn is inherited maternally. Marcus Rhoades first described this cytoplasmic male sterility by crossing female gametes from a male sterile plant with pollen from a male fertile plant. The resulting progeny plants were male sterile.

a. Diagram the cross, using different colors on lines to distinguish between nuclear and cytoplasmic genomes from the male sterile and male fertile strains.

b. Female gametes from the male sterile progeny were backcrossed with pollen from the same male fertile parent of the first cross. The process was repeated many times. Diagram the next two generations including possible crossover events.

c. What was the purpose of the series of backcrosses? (Hint: Look at your answer to part *b* and think about what is happening to the nuclear genome.)

22. Plant breeders have long appreciated the phenomenon called *hybrid vigor* or *heterosis,* in which hybrids formed between two inbred strains have increased vigor and crop yield relative to the two parental strains. Starting in the 1930s, seed companies exploited cytoplasmic male sterility (CMS) in corn so that they could cheaply produce hybrid corn seed to sell to farmers. This type of CMS is caused by mutant mitochondrial genomes that prevent pollen formation.

a. How would CMS aid seed companies in producing hybrid corn seed?

Dominant *Rf* alleles of a nuclear gene called *Restorer* suppress the CMS phenotype, so that *Rf*-containing plants with mutant mitochondrial genomes are male fertile.

b. Describe a cross generating hybrid corn seed that would grow into fertile (self-fertilizing) plants. (Farmers planting hybrid seed want fertile plants since corn kernels result from fertilized ovules.)

c. One of the historical challenges in the commercialization of hybrid corn produced through CMS was the maintenance of strains with CMS mitochondria: How could the seed companies keep producing male sterile corn plants if they never themselves produced pollen? Suggest a strategy by which they could continue to obtain male sterile plants every breeding season.

d. Are there any potential disadvantages to the use of hybrid corn?

23. A mutant haploid strain of yeast called *cox2-1* was found that was unable to grow on media containing glycerol as the sole source of carbon and energy (glycerol is a nonfermentable substrate for yeast). This strain could however grow on the fermentable substrate glucose. Researchers discovered that *cox2-1* cells lack a mitochondrial protein called cytochrome *c* oxidase.

a. Explain why *cox2-1* cells can grow on medium containing glucose but not on glycerol medium.

When *cox2-1* was crossed with a wild type yeast strain and the resultant diploid cells were allowed to grow mitotically, it was found that about half the diploid clones were able to grow on glycerol, while the other half could not. The diploid clones that could grow on glycerol were induced to sporulate, and yielded tetrads with 4 spores that were all able to grow on glycerol medium. In all of these tetrads 2 of the haploid progeny were of mating type a and two of mating type α. The diploids that could not grow on glycerol would not sporulate.

b. What do the results of the mating say about the location of the *cox2-1* mutation? Explain your answer.

A different mutant haploid strain of yeast called *pet111-1* is also unable to grow on glycerol medium but can grow on glucose medium. These mutant cells similarly lacked the cytochrome *c* oxidase. When *pet111-1* was crossed with a wild-type haploid strain of the opposite mating type, the resultant diploids were able to grow on gycerol and yielded asci that all showed a 2:2 segregation of haploid cells that could or could not grow on glycerol.

c. Explain these last results in light of your answer to part *b*.

24. Deletions of various sizes in the mitochondrial genome have been found to increase with age in humans. Which technique would you use to analyze human mtDNA samples for deletions—PCR or gel electrophoresis? Why?

Section 16.3

25. The first person in the family represented by the pedigree shown here who exhibited symptoms of the mitochondrial disease MERFF was II-2.

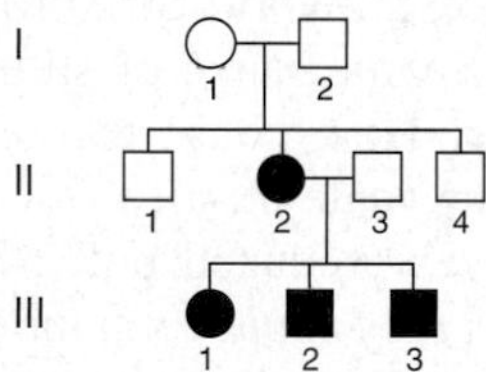

a. What are two possible explanations of why the mother I-1 was unaffected but daughter II-2 was affected?

b. How could you differentiate between the two possible explanations?

26. In 1988, neurologists in Australia reported the existence of identical twins who had developed myoclonic epilepsy in their teens. One twin remained only mildly affected by this condition, but the other twin later developed other symptoms of full-blown MERRF, including deafness, ragged red fibers, and ataxia (loss of the ability to control muscles). Explain the phenotypic dissimilarity in these identical twins.

27. Kearns-Sayre is a disease in which mitochrondrial DNA carries deletions of up to 7.6 kb of the mitochrondrial genome. Although Kearns-Sayre is due to a mitochondrial DNA defect, it does not show maternal inheritance but arises as a new mutation in an individual. The severity of symptoms range from mild to severe and affected people can have defects in some tissues but not in others. How can you explain the variation in tissues affected and severity of symptoms? (Assume that the size of the deletion does not contribute to phenotypic differences.)

28. If you were a genetics counselor and had a patient with MERRF who wanted to have a child, what kind of advice could you give about the chances the child would also have the disease? Are there any tests you could suggest that could be performed prenatally to determine if a fetus would be affected by MERRF?

Gene Regulation in Prokaryotes

Chapter 17

Among the many types of bacteria that thrive in sewage water is the species *Vibrio cholerae,* the cause of the life-threatening diarrheal disease cholera. The last worldwide cholera pandemic began in 1961. Today, cholera is nearly absent from areas of the globe where secure sanitation systems are in place, but epidemics of the disease still devastate human populations in regions where sewage treatment and water purification programs are inadequate or nonexistent.

When a person drinks water contaminated by a disease-causing *V. cholerae,* the bacteria enter the digestive tract. Soon after, the bacteria encounter the "perilous" environment of the stomach, whose acidity kills the majority of them. The bacteria respond to this hostile environment by curtailing production of several proteins they will use later but do not need for passage through the stomach. Only a large initial *V. cholerae* population ensures that at least a small group of cells will survive to exit the stomach and enter the small intestine.

Upon arrival in the small intestine, these survivors come face to face with a thick mucus that coats their ultimate target—the intestinal epithelial cells. To penetrate this protective mucous layer, the *V. cholerae* cells navigate by chemotaxis (see the comprehensive example in Chapter 15) and by using flagella (**Fig. 17.1**). They also make and secrete proteases that ease their passage by degrading the protein component of mucus.

When the bacteria at last reach their destination, they stop fabricating flagellin (the chief protein component of flagella) and begin production of several virulence proteins, including a pilus (by which they attach to epithelial cells of the small intestine) and a potent toxin that is the actual agent of cholera. Mutant bacteria that produce no toxin do not generate symptoms of disease.

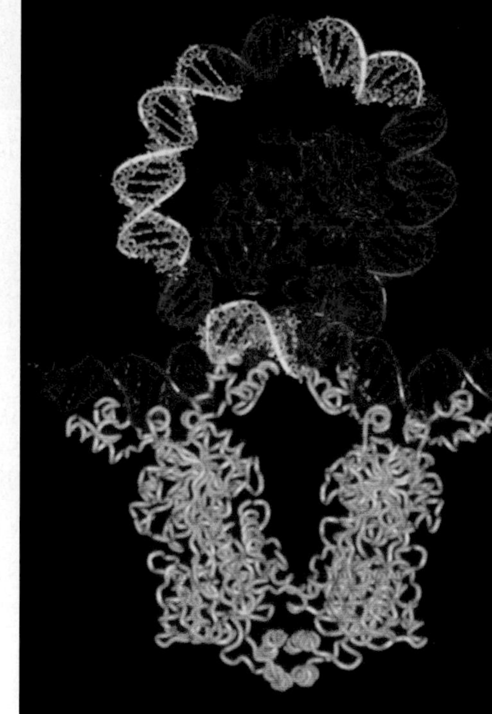

Lac repressor protein (violet) *binds to specific sites in the DNA to turn off expression of the lac operon in E. coli. Lac repressor is a tetramer with two subunits binding to each of two operator sites 93 bases apart, causing a loop* (blue *and* green) *to form in the DNA. This model also shows where the CRP protein* (dark blue) *binds to lac DNA.*

The toxin secreted by cholera bacteria causes chloride ions (Cl^-) to leak from the intestinal cells. To reestablish the osmotic balance, these same cells secrete water. Symptoms of this ionic disruption and fluid flow are watery diarrhea and severe dehydration, which can lead to death within a few hours. The most effective life-saving therapy is oral rehydration: administration of an electrolyte solution consisting of glucose, table salt (NaCl), sodium bicarbonate ($NaHCO_3$), and potassium chloride (KCl) dissolved in purified water. Once toxin production is in full swing, antibiotics without oral rehydration are of little benefit.

The story of *V. cholerae* infection illustrates two key aspects of the life of a unicellular prokaryote: direct contact with the external environment and the ability to respond to changes in that environment by changes in gene expression. *V. cholerae* bacteria that have entered a human host respond to rapid changes in external conditions in part by diminishing or abolishing the production of proteins not required for survival (thereby conserving energy and nutrients) and in part by initiating or increasing synthesis of proteins required in new environments (such as proteases when they contact mucus, and toxin and other virulence proteins in the vicinity of intestinal epithelial cells). Such coordinated control of gene expression in a bacterial cell is an example of **prokaryotic gene regulation,** the subject of this chapter.

Prokaryotes regulate gene expression by activating, increasing, diminishing, or preventing the transcription and translation of specific genes or groups of genes.

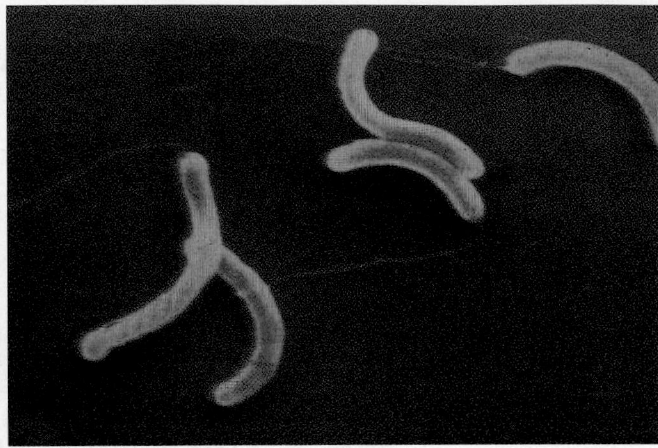

Figure 17.1 ***V. cholerae*** **bacteria.** *V. cholerae* invade cells in the intestine.

Most metabolic pathways in bacteria require several gene products, and bacterial cells have evolved coordinate controls that enable them to turn the genes encoding those products on and off in concert. Bacteria attune their controls to gene function. With the catabolic pathways that break down complex substances to usable units—for example, lactose to glucose and galactose—the appearance of the substrate (lactose) induces, or turns on, coordinate expression of genes whose protein products will accomplish the substrate's breakdown. With anabolic (biosynthetic) pathways that build needed compounds—for example, the amino acid tryptophan—the control mechanisms coordinately repress expression of the pathway's multiple genes when the product of the pathway (tryptophan) is present. This means that bacterial cells do not waste energy making unneeded proteins, such as the enzymes in a catabolic pathway whose substrate is absent or the enzymes in an anabolic pathway whose end product is already present in the environment. In fact, many aspects of prokaryotic gene regulation enable bacterial cells to conserve energy by distinguishing housekeeping proteins that are synthesized continuously, irrespective of environmental conditions, from proteins required only in situations signaled by specific cues.

One overarching theme emerges from our discussion. In unicellular organisms like bacteria, the regulatory mechanisms that turn genes on and off in response to environmental conditions enable the organisms to adapt and survive in a constantly changing world.

As we examine the regulation of gene expression in bacteria we describe

- An overview of prokaryotic gene regulation: There are many steps in gene expression, and regulation can occur at any one of them.
- The regulation of gene transcription: Genetic and molecular studies, particularly of the genes for lactose utilization, show that most regulation affects the initiation of RNA transcripts; "negative regulation" blocks transcription, while "positive regulation" increases it. DNA-binding proteins able to inhibit or enhance the activity of RNA polymerase at the promoter are the main agents of both positive and negative regulation.
- The attenuation of gene expression: Controls regulating the genes of the tryptophan pathway include mechanisms that fine-tune gene expression by prematurely terminating transcription.
- Global regulatory mechanisms: *Escherichia coli*'s response to heat shock is an example of the bacterial ability to coordinate the expression of different sets of genes dispersed around the chromosome. Microarrays and comparative genomic analyses are important new tools for increasing our understanding of the global regulatory systems in bacteria.
- A comprehensive example of how *Vibrio cholerae* regulate their virulence genes.

17.1 An Overview of Prokaryotic Gene Regulation

We saw in Chapter 8 that *gene expression* is the production of proteins according to instructions encoded in DNA. During gene expression, the information in DNA is transcribed into RNA, and the RNA message is translated into a string of amino acids.

RNA Polymerase Is the Key Enzyme for Transcription

To begin the process of gene expression in prokaryotes, RNA polymerase transcribes a gene's DNA into RNA. RNA polymerase participates in all three phases of transcription: initiation, elongation, and termination. Initiation requires a special subunit of RNA polymerase—the sigma (σ) subunit—in addition to the two alpha (α), one beta (β), and one β' subunits that make up the core enzyme (**Fig. 17.2**). When bound to the core enzyme, the σ subunit recognizes and binds specific DNA sequences at the promoter; in its free form, σ does not bind DNA because the σ DNA-binding site is obscured by its own C-terminal tail. The full RNA polymerase—core enzyme plus σ—when bound to the promoter, functions as a complex that both initiates transcription by unwinding the DNA and begins polymerization of bases complementary to the DNA template strand.

The switch from initiation to elongation requires the movement of RNA polymerase away from the promoter and the release of σ. Interaction between the short emerging RNA chain and a site on the polymerase surface probably triggers this release. During elongation, the core RNA polymerase catalyzes the polymerization of ribonucleotides onto the 3′ end of the growing chain, eventually forming a complete mRNA. Elongation continues until the RNA polymerase encounters a signal in the RNA sequence that triggers termination (Fig. 17.2). Two types of termination signals are found in prokaryotes: Rho dependent and Rho independent. In Rho-dependent termination, a protein factor called Rho recognizes a sequence in the newly transcribed mRNA and terminates transcription by binding to the RNA and pulling it away from the RNA polymerase enzyme. In Rho-independent termination, a sequence of about 20 bases in the RNA, with a run of 6 or more U's at the end, forms a secondary structure, known as a *stem loop,* that serves as a signal for the release of RNA polymerase from the completed RNA.

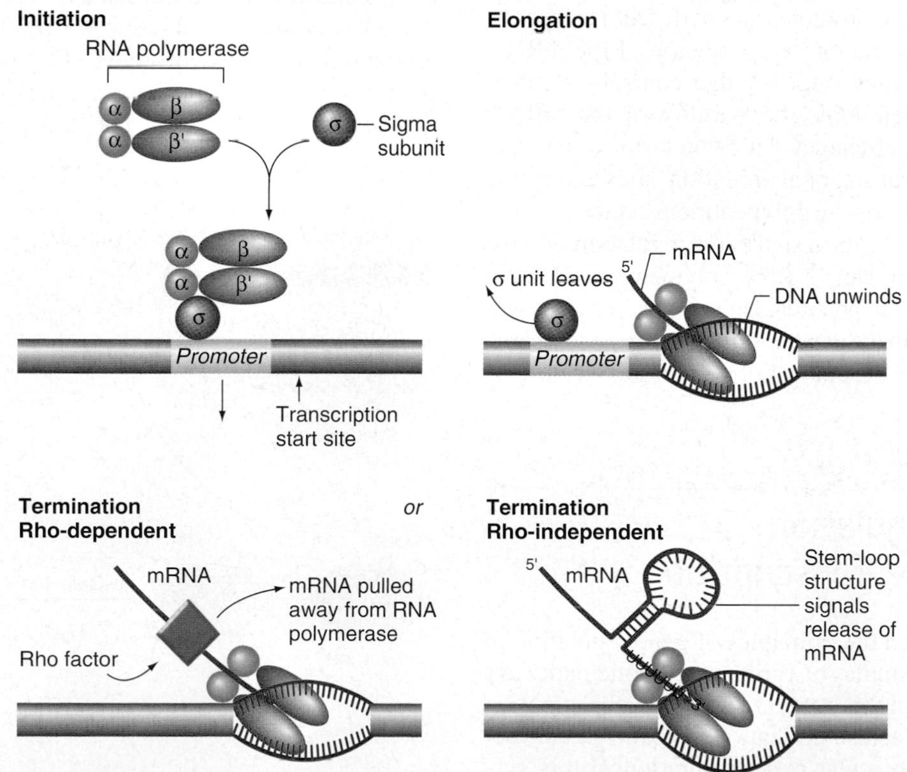

Figure 17.2 Role of RNA polymerase in transcription. The core RNA polymerase enzyme plus sigma factor bind to a promoter sequence to initiate transcription. Once productive copying of the DNA into RNA begins, RNA polymerase moves along the DNA to elongate the transcript, leaving sigma factor behind. Transcription is terminated when rho factor recognizes a sequence on the mRNA and pulls the message away from the enzyme or a stem loop (Rho-independent signal) forms in the mRNA, causing release of the enzyme and message.

Translation in Prokaryotes Starts Before Transcription Ends

Because there is no membrane enclosing the bacterial chromosome, translation of the RNA message into a polypeptide can begin while mRNA is still being transcribed. Ribosomes bind to special initiation sites at the 5′ end of the reading RNA frame while transcription of downstream regions of the RNA is still in progress. Signals for the initiation and termination of translation are distinct from signals for the initiation and termination of transcription. Because prokaryotic mRNAs are often polycistronic, that is, contain the information of several genes, ribosomes can initiate translation at several positions along a single mRNA molecule. See Fig. 8.25 on pp. 279–281 for a review of how ribosomes, tRNAs, and translation factors mediate the initiation, elongation, and termination phases of mRNA translation to produce a polypeptide that grows from its N terminus to its C terminus, according to instructions embodied in the sequence of mRNA codons.

The Regulation of Gene Expression Can Occur at Many Steps

Many levels of control determine the amount of a particular polypeptide in a bacterial cell at any one time. Some controls affect an aspect of transcription: the binding of RNA polymerase to the promoter, the shift from transcriptional initiation to elongation, or the release of the mRNA at the termination of transcription. Other controls are post-transcriptional and determine the stability of the mRNA after its synthesis, the efficiency with which ribosomes recognize the various translational initiation sites along the mRNA, or the stability of the polypeptide product.

As we see next, the critical step in the regulation of most bacterial genes is the binding of RNA polymerase to DNA at the promoter. The other potential points of control, while sometimes important in the expression of certain genes, serve more often to fine-tune the amount of protein produced.

17.2 The Regulation of Gene Transcription

Researchers delineated the principles of gene regulation in prokaryotes through studies of various metabolic pathways in *Escherichia coli.* In this section, we focus our attention on regulation of the lactose utilization genes in *E. coli* because genetic and molecular experimentation in this system established a fundamental principle of gene regulation: The binding of regulatory proteins to DNA targets controls transcription. The DNA binding of these regulatory proteins either inhibits or enhances the effectiveness of RNA polymerase in initiating transcription. In our discussion, we consider the inhibition of RNA polymerase activity as "negative regulation" and the enhancement of RNA polymerase activity as "positive regulation."

The Utilization of Lactose by *E. coli:* A Model System for Studying Gene Regulation

Proliferating *E. coli* can use any one of several sugars as a source of carbon and energy. If given a choice, however, they prefer glucose. *E. coli* grown in medium containing both glucose and lactose, for example, will deplete the glucose before gearing up to utilize lactose.

Lactose is a complex sugar composed of two monosaccharides: glucose and galactose. A membrane protein, lac permease, transports lactose in the medium into the *E. coli* cell. There, the enzyme β-galactosidase splits the lactose into galactose and glucose (**Fig. 17.3**).

The Presence of Lactose Induces Expression of the Genes Required for Lactose Utilization

The two proteins lac permease and β-galactosidase, both required for lactose utilization, are present at very low levels in cells grown without lactose. The addition of lactose to the bacterial medium induces a 1000-fold increase in the production of these proteins. The process by which a specific molecule stimulates synthesis of a given protein is known as **induction.** The molecule responsible for

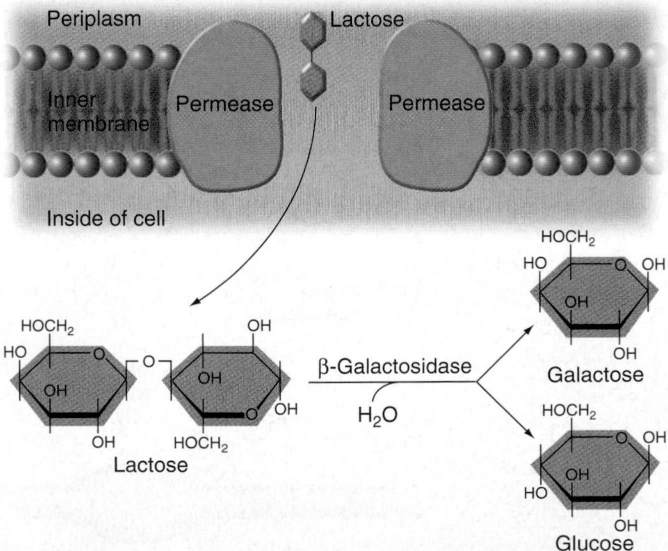

Figure 17.3 Lactose utilization in an *E. coli* cell. Lactose passes through the membranes of the cell via an opening formed by the lactose permease protein. Inside the cell, β-galactosidase splits lactose into galactose and glucose.

stimulating production of the protein is called the **inducer.** In the regulatory system under consideration, lactose modified to a derivative known as allolactose is the inducer of the genes for lactose utilization.

How lactose in the medium induces the simultaneous expression of the proteins required for its utilization was the subject of a major research effort in the 1950s and 1960s—a period some refer to as the golden era of bacterial genetics.

Lactose Utilization in *E. coli:* A Wise Choice for the Study of Gene Regulation

Like Mendel's decision to study seven easily identifiable characteristics in peas, the decision to focus on the induction of genes for lactose utilization in *E. coli* was well considered. The possibility of culturing large numbers of the bacteria made it easy to isolate rare mutants. Once isolated, the mutations responsible for the altered phenotypes could be located by mapping techniques developed in the 1950s (see Fig. 15.15 on p. 558). Another advantage was that the lactose utilization genes are not essential for survival because the bacteria can grow using glucose as a carbon source. In addition, there is a striking 1000-fold difference between lactose utilization protein levels in induced and uninduced cells. This makes it easy to see the difference between the mutant and wild-type states, and it also allows the identification of mutants that have partial—not just all-or-none—effects.

The ability to measure levels of expression was critical for many of these experiments. To this end, chemists synthesized compounds other than lactose, such as *o*-nitrophenyl-galactoside (ONPG), that could be split by β-galactosidase into products that were easy to assay. One product of ONPG splitting has a yellow color, whose intensity is proportional to the amount of product made and thus reflects the level of activity of the β-galactosidase enzyme. A spectrophotometer can easily measure the amount of cleaved yellow product in a sample. Another substrate of the β-galactosidase enzyme that produces a color change upon cleavage is X-Gal, whose cleavage produces a blue substance; as we have seen, X-Gal is often used to indicate whether a piece of DNA has been cloned into plasmid vectors containing parts of the β-galactosidase gene (see Fig. 9.8 on p. 313).

The Operon Theory

Jacques Monod (**Fig 17.4**), a man of diverse interests, was a catalyst for research on the regulation of lactose utilization. A political activist and a chief of French Resistance operations during World War II, he was also a fine musician and esteemed writer on the philosophy of science. Monod led a research effort centered at the Pasteur Institute in Paris, where scientists from around the world came to study enzyme induction. Results from many genetic studies led Monod and his close collaborator François Jacob to propose a model of gene regulation known as the **operon theory,** which suggested that a single signal can simultaneously regulate the expression of several genes that are clustered together on a chromosome and involved in the same process. They reasoned that because these genes form a cluster, they can be transcribed together into a single mRNA, and thus anything that regulates the transcription of this mRNA will affect all the genes in the cluster. Clusters of genes regulated in this way are called **operons.** We first summarize the theory itself and then describe key experiments that influenced Jacob and Monod's thinking, as well as data that supported components of their theory.

Figure 17.4 **Jacques Monod.** A key scientist in discovering principles of gene regulation, Jacques Monod was also a talented musician, philosopher, and political activist.

Based on genetic evidence for three lactose-utilization genes that function as an ensemble, as well as genetic evidence for a repressor that binds both an operator site in the DNA and an inducer (described in the following), Monod and Jacob proposed the operon theory of gene regulation. **Figure 17.5** presents the players in the theory and how they interact to achieve the coordinate regulation of the genes for lactose utilization. As shown, the structural genes (*lacZ, lacY,* and *lacA*) encoding proteins needed for lactose utilization, together with two regulatory elements—the promoter (*P*) and the operator (*O*)—make up the ***lac*** **operon:** a single DNA unit enabling the simultaneous regulation of the three structural genes in response to environmental changes. Molecules that interact with the operon include the repressor, which binds to the operon's operator, and the inducer, which when present, binds to the repressor and prevents it from binding to the operator.

The 1961 Paper Describing the Operon Theory Was a Landmark

In proposing the operon theory, Jacob and Monod took enzyme induction, which most of their contemporaries considered a biochemical problem, and used genetic analysis to develop a molecular model explaining how environmental changes could provoke changes in gene activity. Their theory

Feature Figure 17.5

The Lactose Operon in *E. coli*

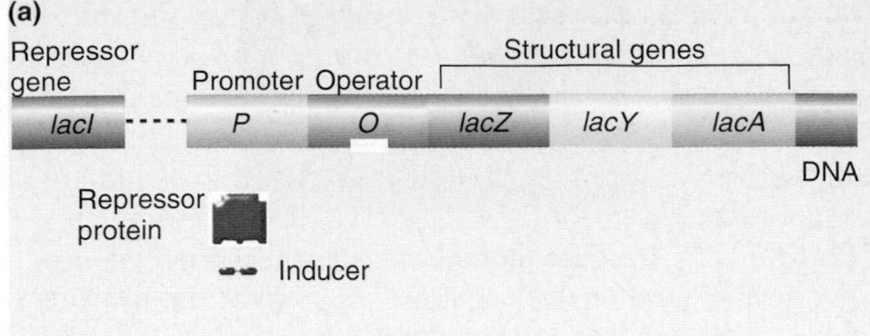

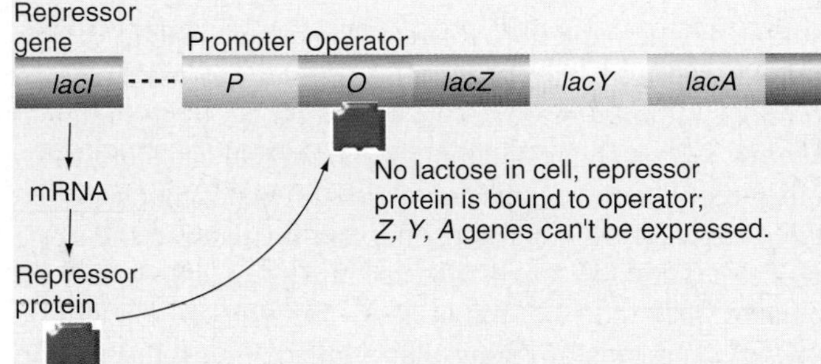

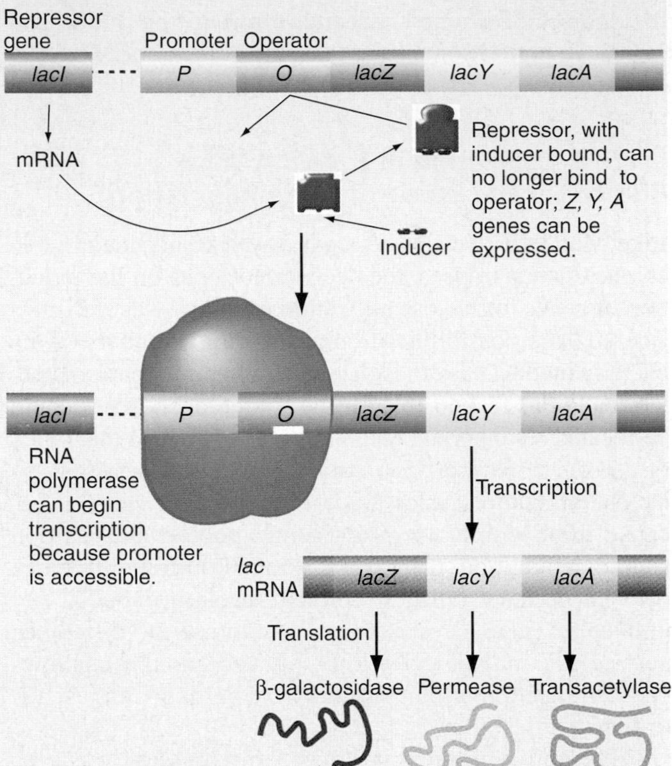

a. The players

The coordination of various elements enables bacteria to utilize lactose in an energy-efficient way. These elements include

1. A closely linked cluster of three structural genes—*lacZ, lacY,* and *lacA*—that encode the enzymes active in splitting lactose into glucose and galactose.
2. A promoter site, from which RNA polymerase initiates transcription of a polycistronic mRNA. The promoter acts in *cis,* affecting the expression of only downstream structural *lac* genes on the same DNA molecule.
3. A *cis*-acting DNA operator site lying very near the *lac* operon promoter on the same DNA molecule.
   The three structural genes together with the promoter and the operator constitute the *lac operon:* a regulatory unit that allows the *lacZ, lacY,* and *lacA* genes to respond to changes in the environment in the same way.
4. A *trans*-acting repressor that can bind to the operator; the repressor is encoded by the *lacI* gene, which is separate from the operon and is unregulated. After synthesis, the repressor diffuses through the cytoplasm and binds with its target.
5. An inducer that prevents the repressor's binding to the operator. Although early experimenters thought lactose was the inducer, we now know that the inducer is actually allolactose, a molecule derived from and thus related to lactose.

How the Players Interact to Regulate the Lactose-Utilization Genes

b. Repression

In the absence of lactose, the repressor binds to the DNA of the operator, and this binding prevents transcription. The repressor thus serves as a negative regulatory element.

c. Induction

1. When lactose is present, allolactose, an inducer derived from the sugar, binds to the repressor. This binding changes the shape of the repressor, making it unable to bind to the operator.
2. With the release of the repressor from the operator, RNA polymerase gains access to the *lac* operon promoter and initiates transcription of the three lactose-utilization genes into a single polycistronic mRNA.

is remarkable because the authors were working with a very abstract sense of the molecules in the bacterial cell: The Watson-Crick model of DNA structure was only eight years old, mRNA had only recently been identified, and the details of transcription had not yet been described. In 1961, the details of information flow from DNA to RNA to protein were still being established and knowledge of protein roles in the cell was limited. For example, although Monod was a biochemist with a special interest in allostery and its effects, the repressor itself was a purely conceptual construct; at the time of publication, it had not yet been isolated, and it was unknown whether it was RNA or protein. Jacob and Monod thus made a major leap in understanding to propose the theory. In their endeavor, they applied the philosophic rule known as Occam's razor: The simplest, most economical explanation is preferable to a more complex one. As is often true in nature, there is a beauty, or elegance, in the simplicity of the solution.

We now know that a key concept of the theory—that proteins bind to DNA to regulate gene expression—holds true for the positive as well as the negative regulation of the *lac* operon. It also applies to many prokaryotic genes outside the *lac* operon and to eukaryotic genes as well. In the next section, we look at some of the experiments that suggested how the presence of lactose induces expression of the genes required for its own utilization.

Experiments Analyzing the Behavior of Lactose-Utilization Mutants Reveal the Coordinate Repression and Induction of Three Genes

Complementation Analysis of Dozens of Mutants Identifies a Set of Lactose-Utilization Genes

On the way to developing the operon theory of gene regulation, Monod and his collaborators isolated many different Lac$^-$ mutants, that is, bacterial cells unable to utilize lactose. Then, using complementation analysis, they showed that the cells' inability to break down lactose resulted from mutations in two genes: *lacZ,* which encodes β-galactosidase, and *lacY,* which encodes lac permease. They also discovered a third *lac* gene, *lacA,* which encodes a transacetylase enzyme that adds an acetyl (CH_3CO) group to lactose and other β-galactoside sugars. Genetic mapping showed that the three genes appear on the bacterial chromosome in a tightly linked cluster, in the order *lacZ-lacY-lacA* (**Fig. 17.6**). Because the *lacA* gene product is not required for the breakdown of lactose, most studies of lactose utilization have focused on *lacZ* and *lacY.*

Figure 17.6 Lactose utilization genes in *E. coli.* Three genes, *lacZ, lacY,* and *lacA,* are involved in lactose metabolism in *E. coli.*

Experimental Evidence for a Repressor Protein

Mutations in another gene named *lacI* produce **constitutive mutants** that synthesize β-galactosidase and lac permease even in the absence of lactose. *Constitutive mutants* synthesize certain enzymes irrespective of environmental conditions. The existence of these constitutive mutants suggested that *lacI* encodes a negative regulator, or **repressor.** Cells would need such a repressor to prevent expression of *lacY* and *lacZ* in the absence of inducer. In constitutive mutants, however, a mutation in the *lacI* gene generates a defect in the repressor protein that prevents it from carrying out this negative regulatory function.

The historic PaJaMo experiment—named after Arthur Pardee (a third collaborator), Jacob, and Monod—provided further evidence that *lacI* indeed encodes this hypothetical negative regulator of the *lac* genes. Matings in which the chromosomal DNA of a donor cell is transferred into a recipient cell served as the basis of the PaJaMo study. The researchers transferred the *lacI*$^+$ and *lacZ*$^+$ alleles into a cytoplasm devoid of LacI and LacZ proteins in a medium containing no lactose (**Fig. 17.7**). Shortly after the transfer of the *lacI*$^+$ and *lacZ*$^+$ genes, the researchers detected synthesis of β-galactosidase. Within about an hour, this synthesis stopped.

Pardee, Jacob, and Monod interpreted these results as follows. When the donor DNA is first transferred to the recipient, there is no repressor (LacI protein) in the

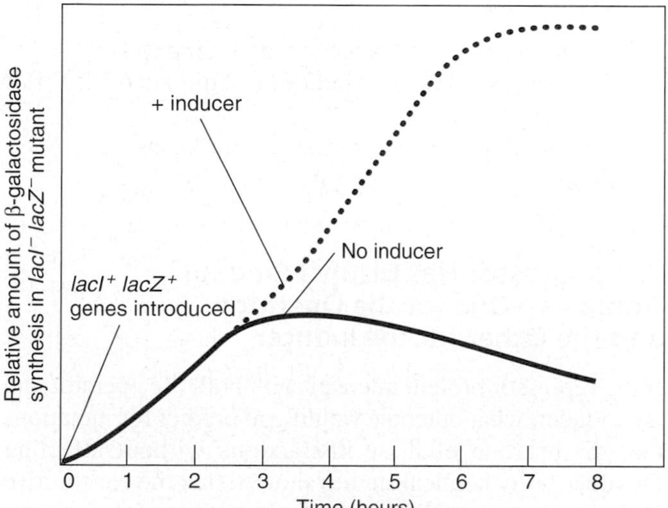

Figure 17.7 The PaJaMo experiment. When DNA carrying *lacI*$^-$ *lacZ*$^+$ genes was introduced into a *lacI*$^-$ *lacZ*$^-$ cell, β-galactosidase was synthesized from the introduced *lacZ*$^+$ gene initially, but as repressor (made from the introduced *lacI*$^+$ copy of the gene) accumulates, the synthesis of β-galactosidase stops and only residual β-galactosidase is seen. If inducer is added (*dotted line*), the synthesis of β-galactosidase resumes.

cytoplasm because the recipient cell's chromosome is *lacI*$^-$. In the absence of repressor, the *lacY* and *lacZ* genes are expressed. Over time, however, the host cell begins to make the LacI repressor protein from the *lacI*$^+$ gene introduced by the mating.

On the basis of the described experiments, Monod and company proposed that the repressor protein prevents further transcription of *lacY* and *lacZ* by binding to a hypothetical **operator site:** a DNA sequence near the promoter of the lactose-utilization genes. They suggested that the binding of repressor to this operator site blocks the promoter and occurs only when lactose is not present in the medium.

How the Inducer Releases Repression to Trigger Enzyme Synthesis

In the final step of the PaJaMo experiment, the researchers added the lactose inducer to the culture medium. With this addition, the synthesis of β-galactosidase resumed.

Their interpretation of this result was that the inducer binds to the repressor. This binding changes the shape of the repressor so that it can no longer bind to DNA. When the inducer is removed from the environment, the repressor, free of inducer, reverts to its DNA-bindable shape. Proteins that undergo reversible changes in conformation when bound to another molecule are called **allosteric proteins.** The binding of inducer to repressor causes an allosteric effect that abolishes the repressor's ability to bind the operator. In this sequence of events, the inducer is an effector molecule that releases repression without itself binding to the DNA.

In summary,

- Repressor bound to inducer cannot bind to DNA.
- Repressor not bound to inducer can bind to the DNA of the operator.
- The binding of repressor to operator keeps RNA polymerase from recognizing the promoter.

The Repressor Has Distinct Binding Domains—One for the Operator and the Other for the Inducer

If the repressor protein interacts with both the operator and the inducer, what outcome would you predict for mutations that disrupt one of these interactions without affecting the other? Biochemical studies showed that the constitutive *lacI*$^-$ mutations we discussed earlier produce defects in the repressor's ability to bind DNA (**Fig. 17.8a**). A different type of *lacI* mutant cannot undergo induction. Researchers designate the noninducible mutations as *lacI*^{s} or superrepressor (Fig. 17.8b). The *lacI*^{s} mutants, although they cannot bind inducer, can still bind to DNA and repress transcription of the operon. The mapping of large numbers of these two types of mutations by DNA sequencing has shown that *lacI*$^-$ missense mutations, which generate proteins incapable of binding DNA, are clustered in the codons that determine the amino (N) terminus of the repressor, while the *lacI*^{s} mutations, which generate proteins incapable of binding inducer, cause amino acid alterations throughout much of the rest of the repressor. Subsequent structural analyses of the repressor protein confirmed what these

(a) Repression

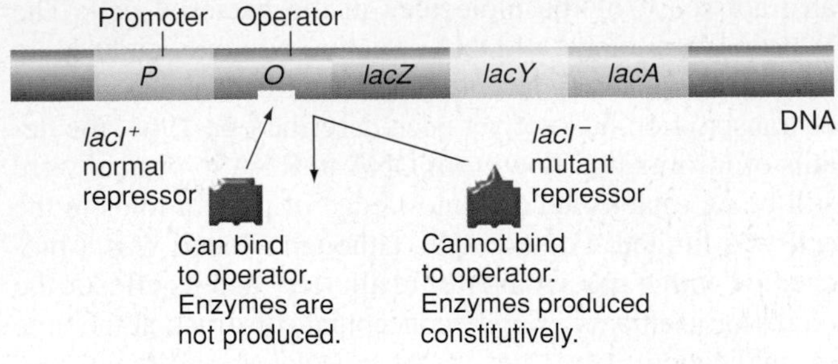

(b) Induction

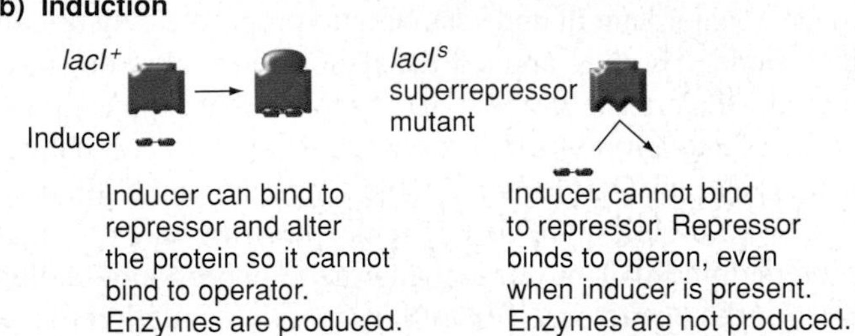

(c)

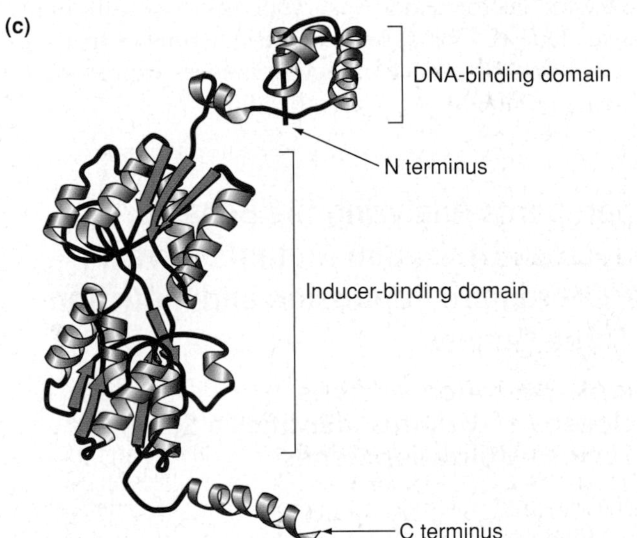

Figure 17.8 Mutational and structural analysis of the repressor. (a) In the *lacI*$^-$ mutants, the repressor cannot bind to the operator site and therefore cannot repress the operon. **(b)** Superrepressor mutants *lacI*^{s} bind to operator but cannot bind inducer, so the repressor cannot be removed from the operator and genes are continually repressed. **(c)** X-ray crystallographic data enable the construction of a model of repressor structure that shows a region to which operator DNA binds and another region to which inducer binds.

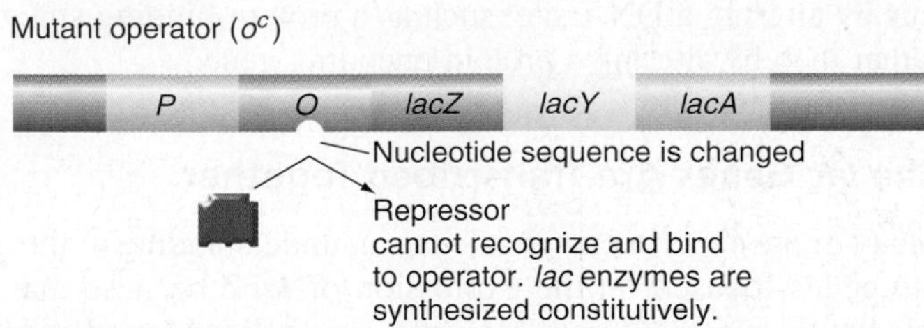

Figure 17.9 Operator mutants. The repressor cannot recognize the altered DNA sequence in the *lacO*^{c} mutant site, so it cannot bind and repress the operon. The result: Lactose genes are constitutively expressed.

mutational analyses suggest: The repressor protein has at least two separate domains, one that binds to DNA, another that binds inducer (Fig. 17.8c).

To summarize,

- The repressor has separate regions, or domains, correlating with two functions defined by mutations.
- Defects in either domain as well as the presence or absence of the inducer can affect the function of the repressor.

Changes in the Operator DNA to Which the Repressor Binds Can Affect Repressor Activity

While mutations in the DNA-binding domain of the repressor can erase repressor activity, mutations that alter the specific nucleotide sequence of the operator recognized by the repressor can have the same effect (**Fig. 17.9**). When mutations change the nucleotide sequence of the operator, the repressor is unable to recognize and bind to the site; the resulting phenotype is the constitutive synthesis of the lactose-utilization proteins. Researchers have isolated constitutive mutants whose genetic defects map to the *lac* operator site, which is adjacent to the *lacZ* and *lacY* genes. They call the constitutive operator DNA alterations *lacO*^{c} mutations.

Proteins Act in *Trans,* but DNA Sites Act Only in *Cis*

How can one distinguish the constitutive operator (*lacO*^{c}) mutants from the previously described constitutive *lacI*$^-$ mutants, considering that both prevent repression? The answer is found in a *cis/trans* test.

Elements that act in ***trans*** can diffuse through the cytoplasm and act at target DNA sites on any DNA molecule in the cell. Elements that act in ***cis*** can influence only the expression of adjacent genes on the same DNA molecule. Studies of partial diploids constructed using F′ plasmids (see Fig. 15.18 on p. 561) helped distinguish mutations in the operator site (*lacO*^{c}), which act in *cis,* from mutations in *lacI,* which encodes a protein that acts in *trans.*

F′ factors are plasmids that carry a few chromosomal bacterial genes (see pp. 561–562 of Chapter 15 for details). When F′ *lac* plasmids are present in a bacterium, the cell has two copies of the lactose-utilization genes—one on the plasmid and one on the bacterial chromosome. Using F′ *lac* plasmids, Monod's group could create bacterial strains with diverse combinations of regulatory (*lacO*^{c} and *lacI*) and structural-gene (*lacZ* and *lacY*) mutations. The phenotype of these partially diploid cells allowed Monod and his collaborators to determine whether particular constitutive mutations were in the genes that produce diffusible, *trans*-acting proteins or at *cis*-acting DNA sites that affect only genes on the same molecule.

In one experiment, Monod and colleagues used a *lacI*$^-$ Z^+ bacterial strain that was constitutive for β-galactosidase production because it could not synthesize repressor (**Fig. 17.10a**). The introduction of an F′ *lacI*$^+$ Z^- plasmid into this strain created a partial diploid that was phenotypically wild type with respect to β-galactosidase expression: both repressible in the absence of lactose and inducible in its presence. Its capacity for repression and subsequent induction indicated that *lacI*$^+$ is dominant to *lacI*$^-$ and that the LacI protein produced from the *lacI*$^+$ gene on the plasmid can bind to the operator on the bacterial chromosome. Thus, the product of the *lacI* gene is a *trans*-acting protein able to diffuse inside the cell and bind to any operator site it encounters, regardless of the chromosomal location of the operator.

In a second experiment, introduction of a *lacI*^{s} plasmid into a *lacI*$^+$ strain of bacteria that was both repressible and inducible created bacteria that were still repressible but were no longer inducible (Fig. 17.10b). This was because the mutant LacI^s repressor, while still able to bind to the operator, could no longer bind inducer. The noninducible repressor was dominant to the wild-type repressor because after a while, the mutant repressor, unable to bind inducer, occupied all the operator sites and blocked all *lac* gene transcription in the cell.

In a third set of experiments, the researchers used *lacI*$^+$ *lacO*^{c} *lacZ*$^+$ bacteria that were constitutive for lactose utilization because the wild-type repressor they produced could not bind to the altered operator (Fig. 17.10c). Introduction of an F′ *lacI*$^+$ *lacO*$^+$ *lacZ*$^-$ plasmid did not change this state of affairs—the cells remained constitutive for β-galactosidase production. The explanation is that the *lacO*$^+$ operator on the plasmid had no effect on the *lacZ*$^+$ gene on the chromosome DNA because the operator DNA acts only in *cis.* Since it was able to influence gene expression only of the *lacZ*$^-$ gene on its own DNA molecule, the wild-type operator on the plasmid could not override the mutant chromosomal operator to allow repression of genes on the bacterial chromosome.

A general rule derived from these experiments is that if a gene encodes a diffusible element—usually a protein—that can bind to target sites on any DNA molecule in the cell, whichever allele of the gene is dominant will override any other allele of that gene in the cell (and therefore act in *trans*). If a mutation is *cis*-acting, it affects only the expression of adjacent genes on the same DNA molecule; it does

(a) F' *lacI*$^+$ *o*$^+$ *Z*$^-$ plasmid in *lacI*$^-$ *o*$^+$ *Z*$^+$ bacteria
lacI$^+$ gene encodes a diffusible element that acts in *trans.*
Inducible synthesis

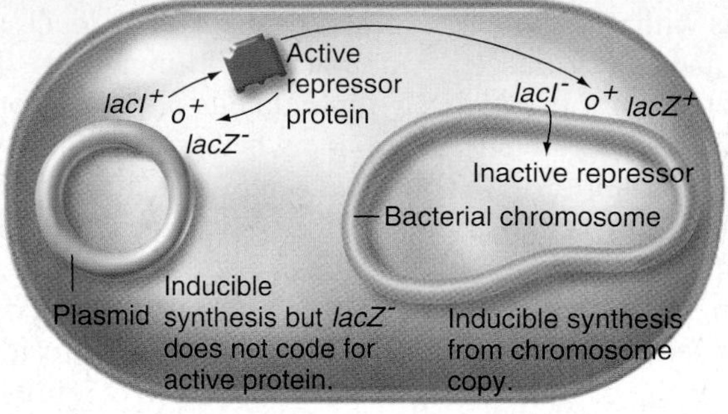

(b) F' *lacI*^{s} *Z*$^-$ plasmid in *lacI*$^+$ *Z*$^+$ bacteria
Noninducible—all *o*$^+$ sites eventually occupied by superrepressor.

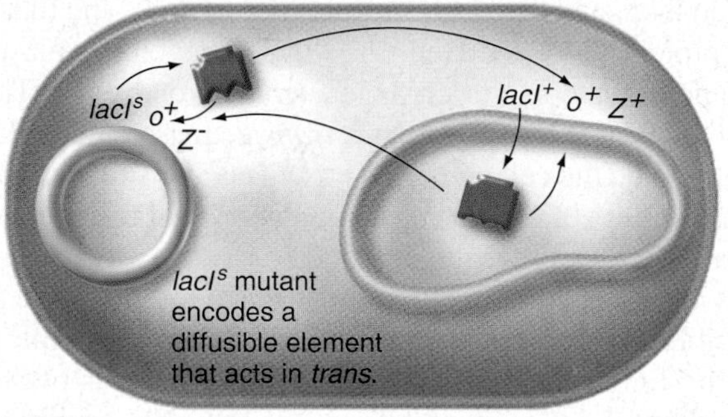

(c) F' *lacI*$^+$ *o*$^+$ *lacZ*$^-$ plasmid in *lacI*$^+$ *o*^{c} *lacZ*$^+$ bacteria
Constitutive—presence of *o*$^+$ in plasmid has no effect on expression of *lacZ*$^+$ gene in bacterial chromosome.

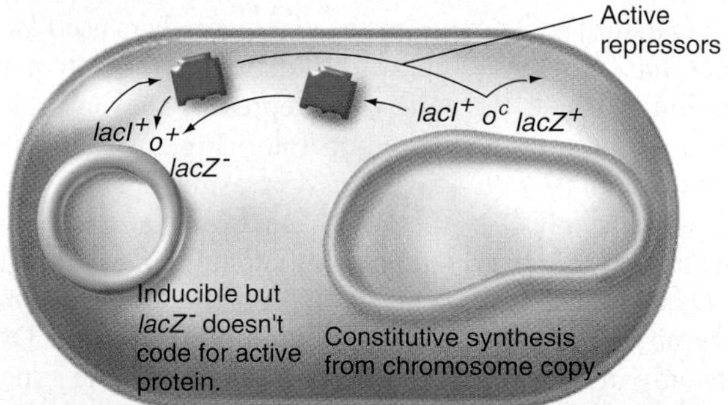

Figure 17.10 ***Trans*-acting proteins and *cis*-acting sites.** **(a)** LacI$^+$ protein acts in *trans.* Repressor protein, made from the *lacI*$^+$ gene on the plasmid, can diffuse in the cytoplasm and bind to the operator on the chromosome as well as to the operator on the plasmid. **(b)** LacI^s protein acts in *trans.* The superrepressor encoded by lacI^s on the plasmid diffuses and binds to operators on both the plasmid and chromosome to repress the *lac* operon even if the inducer is present. **(c)** *lacO*^{c} acts in *cis.* The *lacO*^{c} constitutive mutation affects only the operon of which it is a part. In this cell, only the chromosomal copy will be expressed constitutively.

this by altering a DNA site, such as a protein-binding site, rather than by altering a protein-encoding gene.

The *lac* Genes Are Transcribed Together

Many of the experiments that led to an understanding of the *lac* genes focused on the expression of *lacZ* because the level of β-galactosidase is easy to measure. But Monod and coworkers also showed that the repression and induction of *lacY* and *lacA* occur in tandem with the repression and induction of *lacZ.*

Observation of the coordinate expression of the genes for lactose utilization led to the proposal that the three genes are transcribed as part of the same polycistronic mRNA. Although researchers in the 1960s hypothesized that RNA was the intermediate between DNA and protein, they had not yet demonstrated the existence of such an intermediary for any gene. In the 1970s, however, biochemical studies showed that RNA polymerase initiates transcription of the tightly linked *lac* gene cluster from a single promoter. During transcription, the polymerase produces a single polycistronic mRNA containing the *lac* gene information in the order 5′-*lacZ-lacY-lacA*-3′. As a result, mutations in the promoter (which must be located just upstream of *lacZ*) affect the transcription of all three genes.

Operons Can Also Be Regulated by Positive Controls

In focusing on the repression and subsequent induction of the genes for lactose utilization, the Jacob and Monod model did not address a key question: Why do *E. coli* grown in a medium containing both glucose and lactose not initiate *lac* gene transcription? If the lactose is present, why doesn't it act as an inducer? Answers to these questions emerged from molecular studies carried out long after publication of the theory. These studies showed that transcriptional initiation at the *lac* operon is a complex event. In addition to the release of repression, initiation depends on a positive regulator protein that assists RNA polymerase in the start-up of transcription. Without this assist, the polymerase does not open up the double helix very efficiently. As we see next, the presence of glucose indirectly blocks the function of this positive regulator.

The CRP Protein in a Complex with cAMP Is a Positive Regulator of Catabolic Operons

Inside bacterial cells, the small nucleotide known as cAMP (cyclic adenosine monophosphate) binds to a protein called cAMP receptor protein, or CRP. The binding of cAMP to CRP enables CRP to bind to DNA in the regulatory region of the *lac* operon, and this DNA binding of CRP increases the ability of RNA polymerase to transcribe the *lac* genes

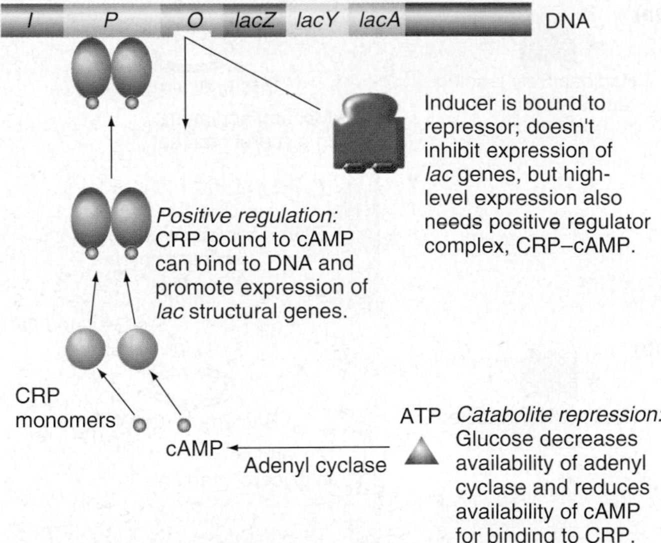

Figure 17.11 Positive regulation of the *lac* operon by CRP–cAMP and catabolite repression by glucose. High-level expression of the *lac* operon requires that a positive regulator, the CRP–cAMP complex, be bound to the promoter region. When little or no glucose is present in the cell, adenyl cyclase activity is high and causes synthesis of cAMP, which binds to CRP to make up the positive regulator. When the level of glucose is high, cAMP remains at a low level, and catabolite repression occurs.

(**Fig. 17.11**). Thus, CRP functions as a positive regulator that enhances the transcriptional activity of RNA polymerase at the *lac* promoter, while cAMP is an effector whose binding to CRP enables CRP to bind to DNA near the promoter and carry out its regulatory function.

Glucose indirectly controls the amount of cAMP in the cell by decreasing the activity of adenyl cyclase, the enzyme that converts ATP into cAMP. Thus, when glucose is present, the level of cAMP remains low; when glucose is absent, cAMP synthesis increases. As a result, when glucose is present in the culture medium, there is little cAMP available to bind to CRP and therefore little induction of the *lac* operon, even if lactose is present in the culture medium. The overall effect of glucose in preventing *lac* gene transcription is known as **catabolite repression,** because the presence of a preferred catabolite (glucose) represses transcription of the operon.

In addition to functioning as a positive regulator of the *lac* operon, the CRP–cAMP complex increases transcription in several other catabolic gene systems, including the *gal* operon (whose protein products help break down the sugar galactose) and the *ara* operon (contributing to the breakdown of the sugar arabinose). As you would expect, these other catabolic operons are also sensitive to the presence of glucose, exhibiting a low level of expression when glucose is present and cAMP is in short supply. Mutations in the gene encoding CRP that alter the DNA-binding domain of the protein reduce transcription of the *lac* and other catabolic operons. The binding of the CRP–cAMP complex is an example of a global regulatory strategy in response to limited glucose in the environment.

Some Positive Regulators Increase Transcription of the Genes in Only One Pathway

AraC, for example, is a positive regulatory protein specific for all the arabinose genes responsible for the breakdown of the sugar arabinose. Three arabinose structural genes, *araB, araA,* and *araD,* appearing on the chromosome in that order, constitute an operon (*araBAD*) that is regulated as a single transcription unit. Like the genes for lactose utilization, the arabinose genes are induced when their substrate (arabinose) is present. Evidence that the AraC protein is a positive regulator of the *araBAD* operon came from studies in which $araC^-$ mutants did not express high levels of these three arabinose genes in either the presence or absence of arabinose (**Fig. 17.12**). (We describe later the mechanisms by which AraC acts.) The mutations were recessive, loss-of-function mutations. When the loss of function of a regulatory protein results in little or no expression of the regulated genes, the protein must be a positive regulator. (By contrast, loss of function of a negative regulator causes constitutive production of the operon's gene products.)

Summary of How DNA-Binding Proteins Control the Initiation of Transcription at the Lactose and Other Operons

In bacteria, the initiation of transcription by RNA polymerase is under the control of regulatory genes whose products bind to specific DNA sequences in the vicinity of the promoter. The binding of negative regulatory proteins prevents the initiation of transcription; the binding of positive regulators assists the initiation of transcription.

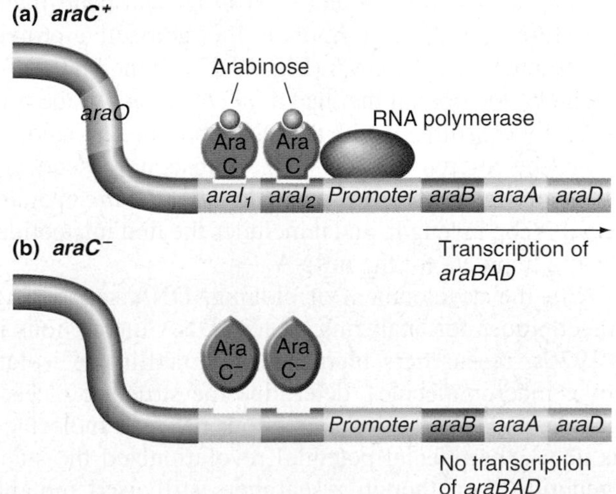

Figure 17.12 Loss-of-function mutations show that AraC is a positive regulator. Expression of the arabinose genes in *E. coli* requires the AraC protein to be bound next to the promoter. In an $araC^-$ mutant, the defective protein cannot bind, and RNA polymerase will not transcribe the genes.

Regulation of the *lac* operon depends on at least two proteins: the repressor (a negative regulator) and CRP (a positive regulator). Maximum induction of the *lac* operon occurs in media containing lactose but lacking glucose. Under these conditions, the repressor binds inducer and becomes unable to bind to the operator, while CRP complexed with cAMP binds to a site near the promoter to assist RNA polymerase in the initiation of transcription.

Operons that function in the breakdown of other sugars are also under the control of negative and positive regulators. Transcription of the arabinose operon, for example, which is induced in the presence of arabinose, receives a boost from two positive regulators: the CRP–cAMP complex and AraC.

Thus, proteins that bind to DNA affect RNA polymerase's ability to transcribe a gene. The activity of multiple regulators that respond to different cues increases the range of gene regulation.

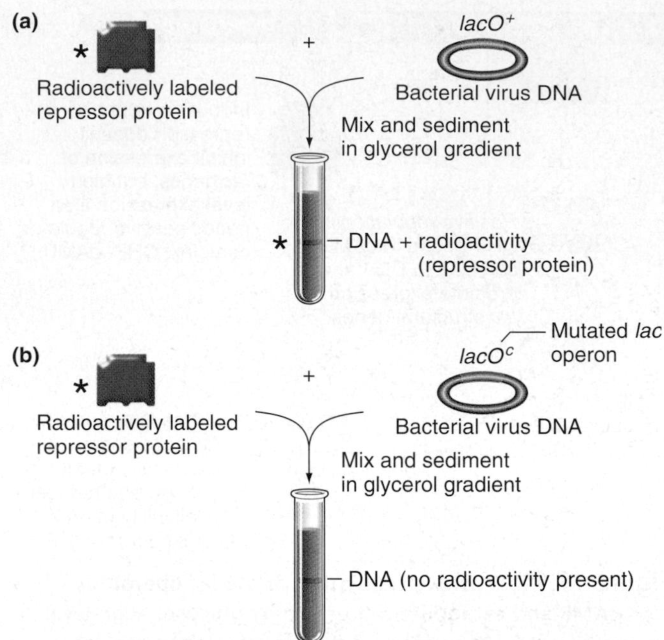

Figure 17.13 The *lac* repressor binds to operator DNA. A radioactive tag is attached to the *lac* repressor protein so it can be followed in the experiment. **(a)** When repressor protein from $lacI^+$ cells was purified and mixed with DNA containing the *lac* operator (on bacterial virus DNA), the protein cosedimented with the DNA. **(b)** When wild-type repressor was mixed with DNA containing a mutant operator site, no radioactivity sedimented with the DNA.

Molecular Studies Help Fill in the Details of Control Mechanisms

In 1966, Walter Gilbert and coworkers purified the *lac* repressor protein and determined that it is a tetramer of four identical *lacI*-encoded subunits, with each subunit containing an inducer-binding domain as well as a domain that recognizes and binds to DNA. (Note that we use the term "domains" for the functional parts of proteins but the term "sites" for the DNA sequences with which a protein's DNA-binding domain interacts.) Gilbert and colleagues then used radioactively labeled repressor protein and a bacterial virus DNA that contained the *lac* operon to show that the repressor binds to operator DNA. When they combined the labeled protein and viral DNA and centrifuged the mixture in a glycerol gradient, the radioactive protein cosedimented with the DNA (**Fig. 17.13**). If the viral DNA contained a *lac* operon that had a $lacO^c$ mutation, the protein did not cosediment with the DNA, because it could not bind to the altered operator site. Subsequent sequence analysis of the isolated DNA revealed that the *lac* operator is about 26 bp in length, and it includes the first nucleotides used as a template for the mRNA.

With the development of cloning, DNA sequencing, and techniques for analyzing protein-DNA interactions in the 1970s, researchers increased their ability to isolate specific macromolecules, determine the structure of each molecule, and analyze the interactions between molecules. This new experimental potential revolutionized the study of genetics even though researchers still used mutants isolated in earlier genetic analyses as the basis of many of their studies. Such mutants continue to be valuable research tools today. The new generation of gene-regulation analyses confirmed the basic principles of the operon theory and elucidated compelling details of their components.

Many DNA-Binding Proteins Contain a Helix-Turn-Helix Motif

It is now possible to predict a protein's secondary structures—such as α helices and β sheets—by comparing the amino acid sequence of a newly isolated protein with sequences of proteins whose secondary structures have already been determined by X-ray crystallography. Several of the polypeptides that make up repressor proteins, including the subunits of the *lac* repressor, have the identifiable feature of two α-helical regions separated by a turn in the protein structure. This helix-turn-helix (HTH) motif in the protein fits well into the major groove of the DNA. One of the α helices in an HTH carries amino acids that recognize and interact with a specific DNA sequence of nucleotides; thus, each HTH has a specificity for DNA binding based on its sequence of amino acids (**Fig. 17.14a**).

Mark Ptashne and colleagues used cloned DNA to construct made-to-order mutations in the gene encoding the repressor of a bacterial virus known as 434. (This bacterial virus is similar to the temperate bacteriophage λ, which infects *E. coli,* and bacteriophage P22, which infects *Salmonella enterica.*) The 434 repressor binds to DNA of the 434 viral DNA that has integrated into the bacterial genome and prevents transcription and production of viral particles. After predicting that a region of the α helix of the 434 repressor

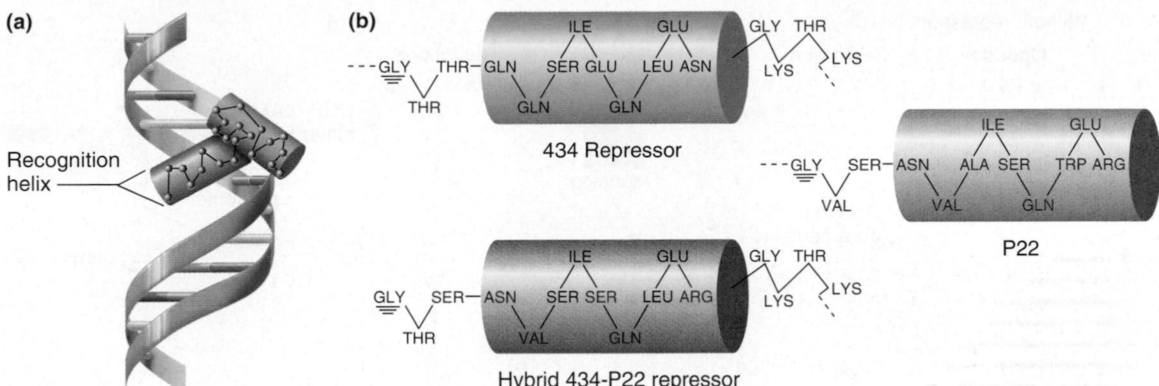

Figure 17.14 **DNA recognition sequences in a helix-turn-helix section of a protein. (a)** A protein motif that has the shape of a helix-turn-helix (*helices shown here inside a cylindrical shape*) fits into the major groove of the DNA helix. Specific amino acids within the helical region of the protein recognize a particular base sequence in the DNA. **(b)** The amino acids inside the recognition helix for phages 434 and P22 and for the hybrid 434-P22 repressor. The amino acids shown in *red* in the hybrid repressor helix section are ones that were modified to be like those of the P22 repressor.

recognizes the DNA of its specific operator site, Ptashne and coworkers altered the DNA sequence of the gene region encoding this α helix so that it now encoded most of the amino acids in the corresponding α helix of the repressor for another bacterial virus P22 (Fig. 17.14b). The resulting hybrid 434-P22 repressor protein, encoded by the altered gene, contained a P22 α helix that recognized the P22 operator *in vivo*. Binding of the hybrid repressor to the P22 prophage DNA operator region that had integrated into the bacterial host genome shut down transcription of most P22 viral proteins and prevented subsequent infection by the P22 virus. This experiment showed that specific amino acids in this α helix determine the binding specificity of the repressor protein. The P22-like *α* helix in the hybrid protein is sufficient to convert the binding specificity of the 434 repressor to that of the P22 repressor.

The HTH motif is found in hundreds of DNA-binding proteins. Surprisingly, more than 20 different DNA-binding proteins in bacteria are very similar to the LacI repressor, not only in the HTH DNA-binding domain but throughout much of the protein. This group of repressors is known as the LacI repressor family of proteins. Their structural similarity suggests that they evolved from a common ancestral gene whose duplication and divergence produced a family of transcriptional repressor proteins with similar overall structures but unique recognition regions. The uniqueness of their DNA-recognition regions means that they interact with different operators to regulate different groups of genes.

Most Regulatory Proteins Are Oligomeric and Contain More Than One Binding Domain

In the 1970s, geneticists studying gene regulation developed new *in vitro* techniques to determine where regulatory proteins bind to the DNA. Purified proteins that bind to fragments of DNA protect the region of the double helix to which they bind from digestion by enzymes such as DNase I that break the phosphodiester bonds between nucleotides. If a sample of DNA, labeled at one end of one strand and bound by a purified protein, is partially digested with DNase I, the enzyme will cleave phosphodiester bonds in at least some DNA molecules in the sample, except for those phosphodiester bonds that are in regions protected by the bound protein. After the DNase I and binding proteins are removed and the DNA is denatured into single strands, gel electrophoresis of the DNA followed by autoradiography of the gel will reveal bands at positions corresponding to the cleavage at each phosphodiester bond, except in the region where bound protein protected the DNA. Portions of the gel without bands are thus "footprints," indicating the nucleotides of the DNA fragment that were protected by the DNA-binding protein (**Fig. 17.15a**).

When analyzed in this way, many of the DNA sequences to which a negative or positive regulator protein binds exhibit rotational symmetry; that is, their two DNA strands have a similarity of sequence when read in the 5′-to-3′ direction (although these sequences are usually not perfect palindromes). An example of such symmetry is in the *lac* operon's CRP-binding site whose sequence is

5′ TGTGAGTTAGCTCACA 3′

3′ ACACTCAATCGAGTGT 5′

Most regulatory proteins that bind to DNA exist as oligomers composed of two to four polypeptide subunits. The regulatory proteins, which are present in very low numbers in the cell, gain an advantage from this multimeric form: Since each polypeptide subunit of an oligomer has a DNA-binding domain, an assembled oligomer has multiple DNA-binding domains. If the sites to which an oligomeric protein can bind are clustered in a gene's regulatory region, many contacts can be established between the protein and the regulatory region. By increasing the stability of

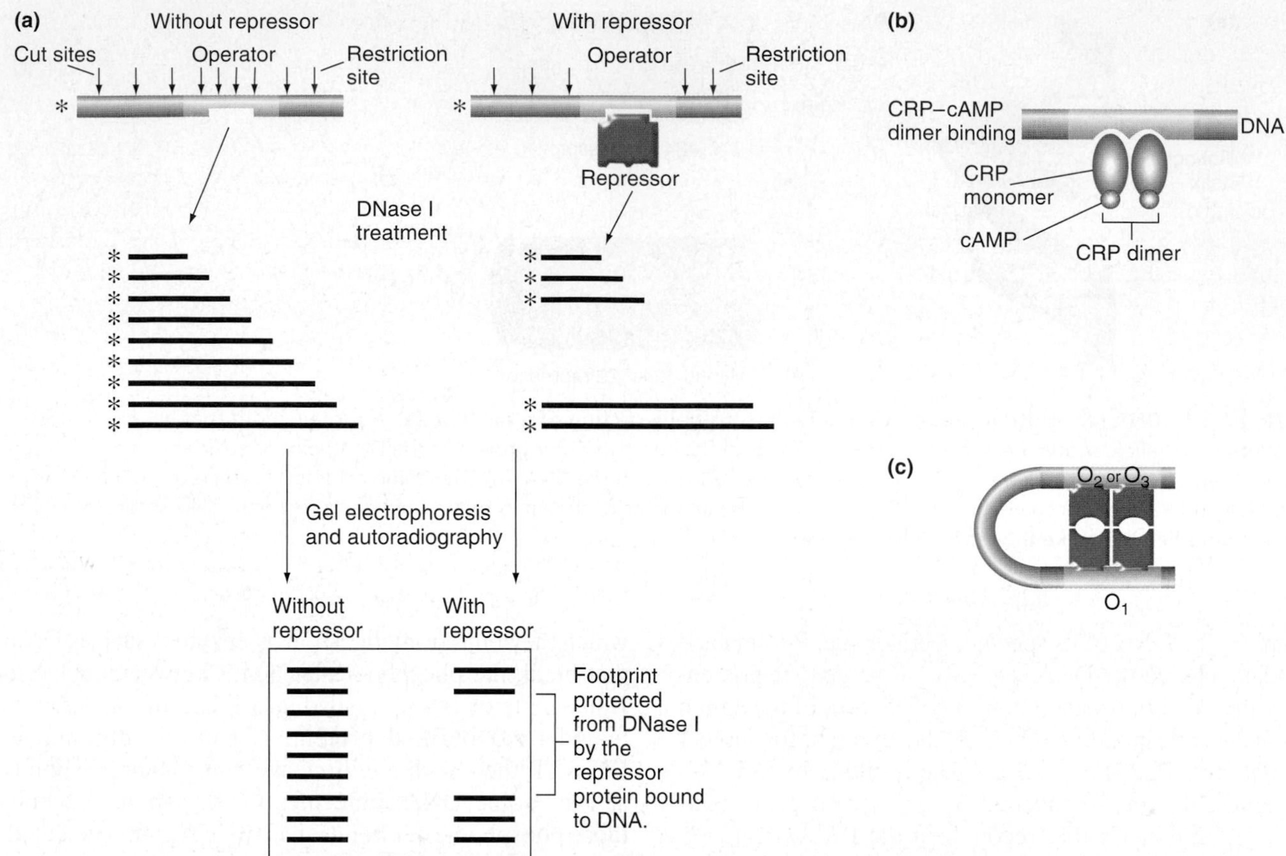

Figure 17.15 Proteins binding to DNA. (a) DNase footprint establishes the region to which a protein binds. A partial digestion with DNase I produces a series of fragments. If a protein is bound to DNA, DNase cannot digest at sites covered by the protein. Gel electrophoresis of digested products shows which products were not generated and indicates where the protein binds. **(b)** CRP–cAMP binds as a dimer to a regulatory region. **(c)** The Lac repressor is a tetrameric protein. For simplicity we previously showed a single repressor object binding to one operator site, but in reality, there are two identical LacI subunits that bind to each operator site. Two of the subunits bind to the sequence in one operator site (O_1), and the other two subunits bind to a second operator (either O_2 or O_3).

protein-DNA interactions, these multiple binding domains collectively produce the strength of binding necessary to maintain repression or activate transcription.

The CRP protein binds to DNA as a dimer at a sequence with rotational symmetry, with one monomer of CRP binding to each side of the sequence (Fig. 17.15b). Thus, CRP-binding sites (such as the site in the *lac* operon whose sequence was just shown) actually consist of two recognition sequences, each able to bind one subunit of the CRP dimer.

The *lac* repressor exists as a tetramer with each of its four subunits containing a DNA-binding HTH motif. This tetramer binds to two operators located far apart on the DNA, with each operator containing two recognition sequences. The binding of the tetrameric repressor to the two operators causes a loop of DNA to form between the two operator sites (Fig. 17.15c); formation of the loop, in turn, facilitates the two-position binding. There are actually three operator sites in the *lac* operon to which the repressor can bind: O_1 (the site originally identified by *lacO*^{c} mutations), O_2, and O_3. Site O_1 has the strongest binding affinity for the repressor, and two subunits of the tetramer always bind at this site. The other two subunits bind at either O_2 or O_3. The distance between operator sites—multiples of 10 bases—allows repressor binding to the same side of the helix and thus formation of the loop. Mutations in *either* O_2 *or* O_3 have very little effect on repression. By contrast, mutations in *both* O_2 *and* O_3 make repression 50 times less effective. The conclusion is that for maximal repression, all four of the repressor's subunits must bind DNA simultaneously. Binding at four recognition sequences (in two operator sites) increases the stability of the protein-DNA interactions. In fact, the DNA binding of the *lac* repressor is so efficient that only 10 repressor tetramers per cell are sufficient to maintain repression.

The Looping of DNA Is a Common Feature of Regulatory Systems

We have seen that the DNA binding of the *lac* repressor tetramer to two distant sites on the DNA results in formation of a loop. This ability to form a loop shows that DNA is not a static, stiff molecule. It has a flexibility that enables it to bend, and this potential for bending enables all four DNA-binding domains of the *lac* repressor tetramer to work in tandem for maximal repression.

Looping first came to light in work on AraC, the regulatory protein that helps control the arabinose operon described previously. AraC functions as a dimer. As we have seen, in the presence of the inducer arabinose, AraC is a positive regulator that helps initiate transcription of the *araBAD* operon. Unexpectedly, in the absence of arabinose, AraC acts as a repressor. In this capacity, the AraC dimer binds to two sites—*araO* and *araI*—that are 194 nucleotide pairs apart (**Fig. 17.16a**). In one set of experiments analyzing the binding to two sites with concomitant looping of DNA, researchers altered the distance between *araO* and *araI* by inserting several base pairs. The introduction of 11 or 31 bp—alterations that are close to integral changes in the number of turns of the double helix (a full turn of the helix = 10.5 bp)—had little effect on repression. The introduction of 5, 15, or 24 bp, however, noticeably reduced repression. These results suggest that the orientation of the binding sites is a significant variable; only when two sites have an orientation that puts them on the same side of the helix can a dimer bind simultaneously to both and cause formation of a DNA loop.

What enables the AraC dimer to function as both an activator and a repressor? The answer most likely involves allostery and AraC's different binding affinities for recognition sequences at *araI* and *araO*. The *araI* site contains two recognition sequences (*araI*$_1$ and *araI*$_2$) to which AraC can bind; the *araO* site contains one recognition sequence. The AraC dimer is an allosteric protein whose structure changes with the binding of inducer (arabinose). One hypothesis of how this allosteric change alters function is that in the absence of arabinose, the size and shape of the protein unbound to inducer allow the AraC dimer to bind to two recognition sequences—*araO* and one of the two sites within *araI*—at the same time; this double binding prevents AraC from binding in a way that would enable it to assist RNA polymerase in the initiation of transcription. When arabinose is present, binding to the inducer changes the shape of the AraC dimer. In this inducer-bound conformation, the regulatory molecule does not bind to *araI* and *araO* at the same time; instead, the inducer–dimer complex binds exclusively to recognition sequences in *araI* (Fig. 17.16b). When bound to DNA at only this site, AraC's positive regulatory domain is free to interact with RNA polymerase and increase transcription.

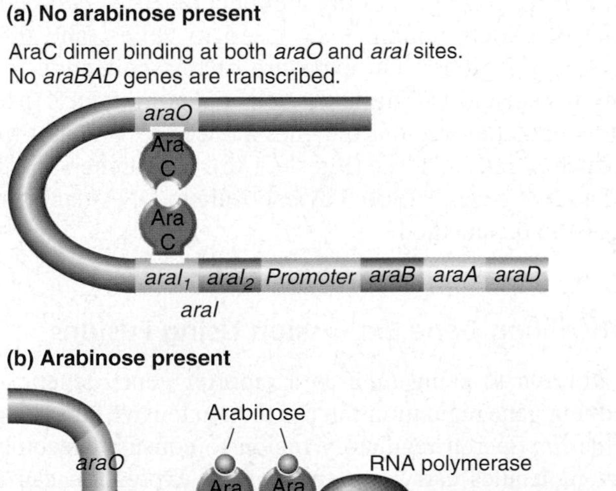

(a) No arabinose present

AraC dimer binding at both *araO* and *araI* sites. No *araBAD* genes are transcribed.

(b) Arabinose present

Figure 17.16 How AraC acts as both a repressor and an activator. The AraC protein can bind to sites *araI*$_1$, *araI*$_2$, and *araO*. **(a)** When no arabinose is present, the binding of AraC to *araO* and the *araI*$_1$ sites causes looping of the DNA and prevents RNA polymerase from transcribing the genes. **(b)** When arabinose (inducer) is present, AraC binds to *araI*$_1$ and *araI*$_2$ but not to *araO*. RNA polymerase interacts with AraC at the *araI* sites and transcribes the genes.

How Regulatory Proteins Interact with RNA Polymerase

Many negative regulators, such as the *lac* repressor, prevent initiation by blocking the functional binding of RNA polymerase. For example, the O_1 operator site to which the *lac* repressor tightly binds consists of 27 nucleotides centered 11 bp downstream from the transcriptional start site. The operator thus includes part of the region where RNA polymerase has to bind to initiate transcription (**Fig. 17.17a**). When repressor is bound to the operator, its presence on the DNA prevents RNA polymerase from binding in the way needed to initiate transcription.

Positive regulators, by contrast, usually establish a physical contact with RNA polymerase that enhances the enzyme's ability to initiate transcription (Fig. 17.17b). For several positive regulators, researchers have identified points of contact between the regulator and the α, β, or σ subunits of RNA polymerase. Although RNA polymerase will bind to a promoter in the absence of a positive regulator, it is less likely to unwind DNA and begin polymerization than when it receives assistance from a positive regulator.

Using the *lacZ* Gene As a Reporter of Gene Expression

Extensive molecular knowledge of the *lacZ* gene and assays to measure its expression have enabled its use as a "reporter" gene in the study of a large variety of regulatory regions in both prokaryotes and eukaryotes. A **reporter gene** is a protein-encoding gene whose expression in the cell is quantifiable by sensitive and reliable techniques of protein detection.

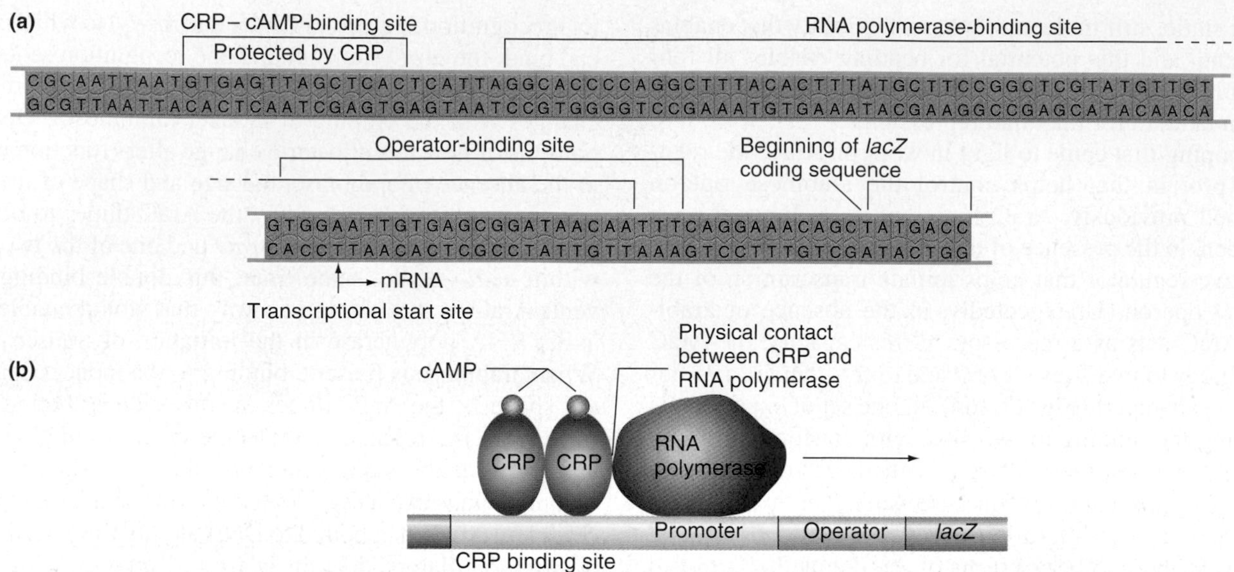

Figure 17.17 **How regulatory proteins interact with RNA polymerase.** **(a)** The *lac* repressor bound to the operator prevents RNA polymerase from binding. The binding sites for RNA polymerase and repressor (determined by DNase digestion experiments) show that there is overlap between the two sites. **(b)** The CRP–cAMP complex contacts RNA polymerase directly to help in transcription initiation.

Measuring Gene Expression Using Fusions

Fusion of the coding region of the reporter gene to *cis*-acting regulatory regions (including promoters and operators) of other genes creates a DNA molecule that enables researchers to assess the activity of the regulatory elements by monitoring the amount of reporter gene product appearing in the cell. For example, with the fusion of gene *X*'s regulatory region to the *lacZ* gene, one can assess the activity of the regulatory elements of gene *X* by monitoring the level of β-galactosidase expression. With this fusion molecule, conditions that induce expression of gene *X* will generate β-galactosidase (**Fig. 17.18a**).

Identifying Regulatory Sites Using Fusions

The use of reporter genes makes it possible to identify the DNA sites necessary for regulation as well as the genes and signals involved in that regulation. For example, you could mutate gene *X*'s control region *in vitro* and then transform the gene *X-lacZ* fusion molecule back into bacterial cells and look for mutations that disrupt a particular aspect of control (as measured by levels of *lacZ*); this protocol would identify *cis*-acting sites important for the regulation. Or, you could mutate the bacteria themselves and then introduce a reporter fusion molecule into the mutated cells and look for changes in level of *lacZ* expression as measured by blue or white colony color; this protocol would identify *trans*-acting genes.

Identifying Sets of Genes Regulated by the Same Stimulus

Reporter fusion molecules not only provide the basis for analyzing the regulation of one specific gene, they also make it possible to identify many genes regulated by the same stimulus. To this end, researchers can use transposition to insert the *lacZ* gene without its regulatory region at various sites around the bacterial chromosome. For example, they can engineer the bacteriophage Mu to contain *lacZ* inside its genome and then infect a population of *E. coli* cells with the specially engineered phage. The phage will insert at sites around the bacterial chromosome at random, generating a collection of *E. coli* cells, some of which contain *lacZ* fused to genes (and their regulatory regions). On exposure of this collection to a stimulus such as UV light, the cells containing *lacZ* fused to the regulatory regions of genes induced by UV light will produce β-galactosidase (Fig. 17.18b). Researchers identified a set of genes activated by exposure to DNA-damaging agents by this method.

Controlling Gene Expression Using Fusions

In addition to using *lacZ* as a reporter gene, geneticists studying gene regulation can use their extensive knowledge of the *lac* operon regulatory region to construct recombinant molecules carrying genes whose expression can be controlled. For example, by fusing the *lac* operon control DNA to a human gene expressed in *E. coli,* they could cause overproduction of the human protein in response to an induction cue. The ability to control expression of a foreign gene is important because it provides a way to ensure that protein production is not turned on until cells have proliferated to a high density. The culture will thus contain many cells making the desired foreign protein; and even if the protein has deleterious effects on the growth of *E. coli,* the culture can still grow to high density before addition of

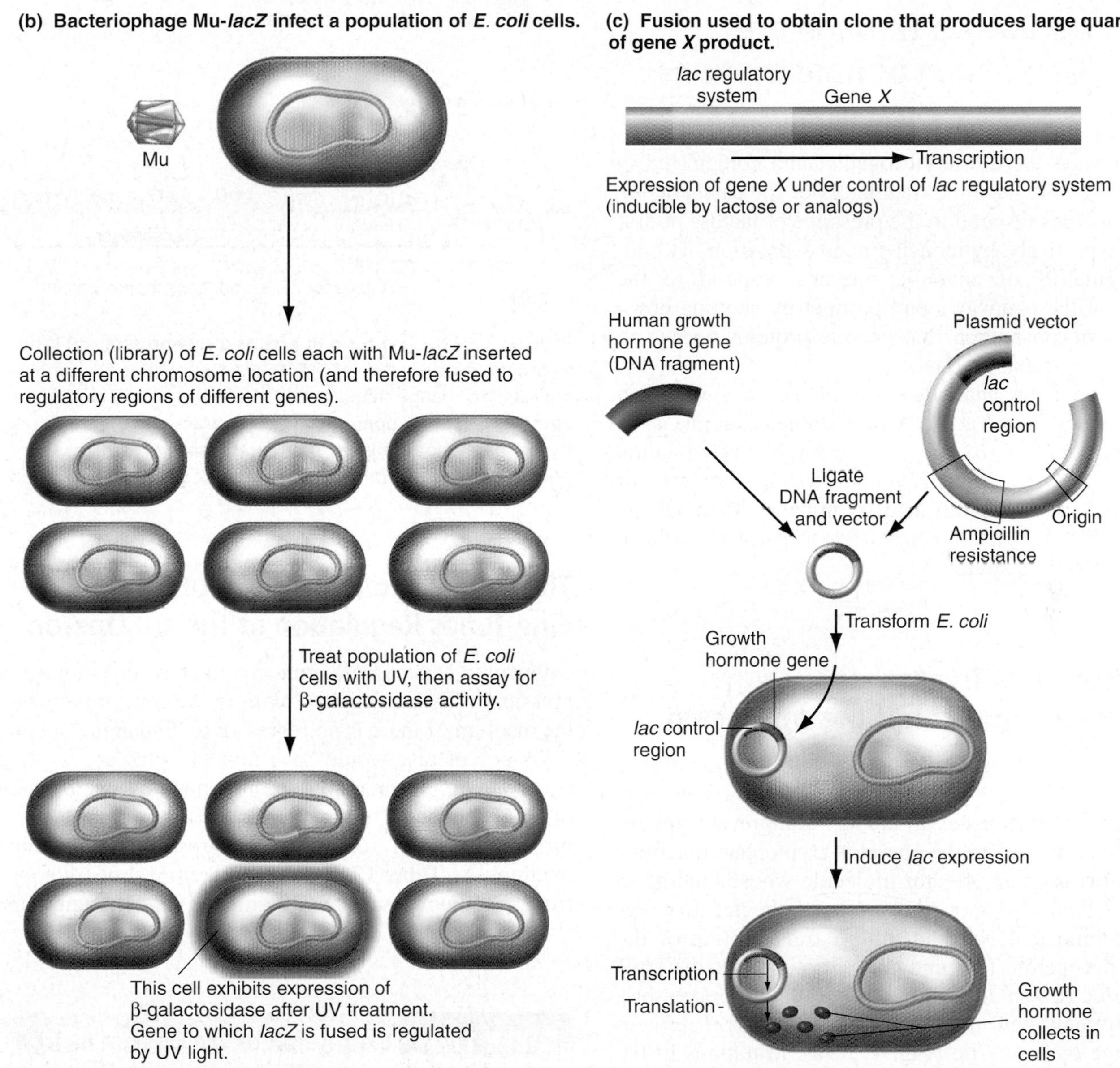
(a) Fusion used to perform genetic studies of the regulatory region of gene X.
cis-acting gene X regulatory region
lacZ (Reporter gene)
Transcription
Translation
lacZ expression is under control of gene X regulatory region.
β-galactosidase
(b) Bacteriophage Mu-lacZ infect a population of E. coli cells.
Mu
Collection (library) of E. coli cells each with Mu-lacZ inserted at a different chromosome location (and therefore fused to regulatory regions of different genes).
Treat population of E. coli cells with UV, then assay for β-galactosidase activity.
This cell exhibits expression of β-galactosidase after UV treatment. Gene to which lacZ is fused is regulated by UV light.
(c) Fusion used to obtain clone that produces large quantities of gene X product.
lac regulatory system
Gene X
Transcription
Expression of gene X under control of lac regulatory system (inducible by lactose or analogs)
Human growth hormone gene (DNA fragment)
Plasmid vector
lac control region
Ligate DNA fragment and vector
Origin
Ampicillin resistance
Transform E. coli
Growth hormone gene
lac control region
Induce lac expression
Transcription
Translation
Growth hormone collects in cells

Figure 17.18 **Use of *lac* gene fusions to study gene regulation.** **(a)** The *lacZ* structural gene can be fused to a regulatory region of gene *X*. Expression of β-galactosidase will be dependent on signals in the regulatory region to which *lacZ* is fused. **(b)** Creating a collection of *lacZ* insertions in the chromosome. Bacteriophage Mu containing the *lacZ* gene without its promoter integrates randomly around the chromosome. If the Mu-*lacZ* integrates within a gene in the orientation of transcription, *lacZ* expression will be controlled by that gene's regulatory region. The library of clones created can be screened to identify insertions in genes regulated by a common signal. **(c)** Use of fusions to overproduce a gene product. The *lac* regulatory region can be fused to gene *X* to control expression of genes. The gene encoding human growth hormone is cloned next to the *lac* control region and transformed into *E. coli*. Conditions that induce *lac* expression will cause expression of growth hormone that can be purified from the cells.

the inducer. The production in *E. coli* of human proteins, such as human growth hormone, human insulin, and other pharmacologically useful proteins, is based on this strategy (Fig. 17.18c).

17.3 The Attenuation of Gene Expression: Fine-Tuning the *trp* Operon Through the Termination of Transcription

In bacteria, the multiple genes of both catabolic and anabolic pathways are clustered together and coregulated in operons. We have seen that regulators of the catabolic *lac* and *ara* operons respond to the presence of lactose or arabinose, respectively, by inducing gene expression. By contrast, regulators of anabolic operons respond to the presence of the pathway's end product by shutting down the genes of the operon that encode proteins needed to manufacture the end product.

There are many anabolic bacterial operons involved in the synthesis of amino acids. A well-studied example is the *E. coli* tryptophan (*trp*) operon, a group of five structural genes—*trpE, trpD, trpC, trpB,* and *trpA*—required for construction of the amino acid tryptophan. Maximal expression of the *trp* genes occurs when tryptophan is absent from the growth medium.

The Presence of Tryptophan Activates a Repressor of the *trp* Operon

The *trp* operon is regulated by a protein repressor that is the product of the *trpR* gene. In contrast to the *lac* operon, where lactose functions as an inducer that prevents the repressor from binding to the operator, tryptophan functions as a **corepressor:** an effector molecule whose binding to the actual TrpR repressor protein allows the negative regulator to bind to DNA and inhibit transcription of the genes in the operon. The binding of tryptophan to the TrpR repressor causes an allosteric alteration in the repressor's shape, and only with this alteration can the TrpR protein bind to the operator site (**Fig. 17.19a**). Mutations in the *trpR* gene that change either the protein's tryptophan-binding domain or its DNA-binding domain destroy the TrpR repressor's ability to bind DNA, and they result in the constitutive expression of the *trp* genes even when tryptophan is present in the growth medium. The TrpR-mediated repression of the *trp* operon is critical, but it is only one of the regulatory components controlling expression of the *trp* genes in *E. coli.*

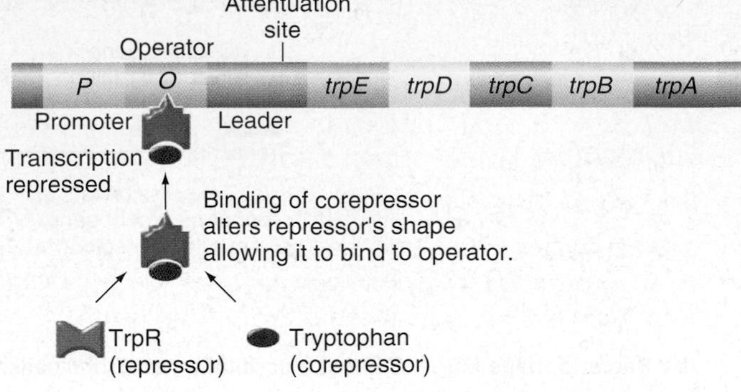

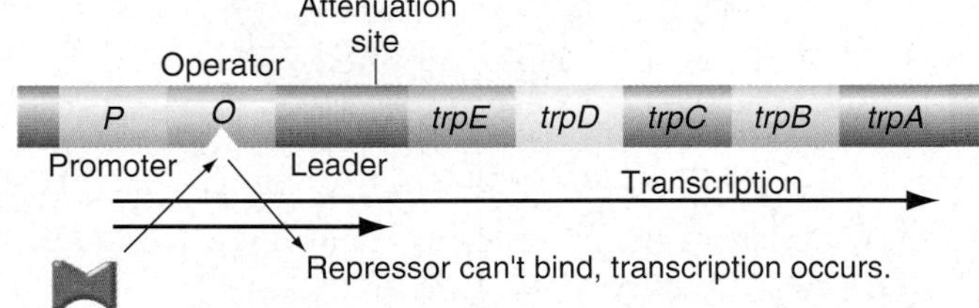

Figure 17.19 Tryptophan acts as a corepressor. (a) When tryptophan is available, it binds to the *trp* repressor, causing the molecule to change shape so that the repressor can bind to the operator of the *trp* operon and repress transcription. **(b)** When tryptophan is not available, the repressor cannot bind to the operator, and the tryptophan biosynthetic genes are expressed.

The Termination of Transcription Fine-Tunes Regulation of the *trp* Operon

One would expect $trpR^-$ mutants to show constitutive expression of their *trp* genes. With or without tryptophan in the medium, if there is no repressor to bind at the operator, RNA polymerase would have uninterrupted access to the *trp* promoter. Surprisingly, studies show that the *trp* genes of $trpR^-$ mutants are not completely de-repressed (that is, turned on) when tryptophan is present in the growth medium. As **Table 17.1** shows, the removal of tryptophan from a medium in which $trpR^-$ mutants are growing causes

TABLE 17.1 Expression of *trp* Operon in $trpR^+$ and $trpR^-$ Strains

	With Tryptophan* (%)	Without Tryptophan* (%)
$trpR^+$	8	100
$trpR^-$	33	100

*In the growth medium

expression of the *trp* genes to increase threefold. What control mechanism is responsible for this repressor-independent change in *trp* operon expression?

Two Alternative Transcripts Lead to Different Transcriptional Outcomes

In a series of elegant experiments analyzing transcription of the *trp* operon, Charles Yanofsky and coworkers found that initiation at the *trp* promoter can produce two alternative transcripts (Fig. 17.19 and **Fig. 17.20a**). Sometimes initiation at the promoter leads to transcription of a truncated mRNA about 140 bases long from a short DNA region immediately preceding the first *trp* structural gene (*trpE*); this pregene DNA region is called a **leader sequence,** and the RNA transcribed from it is the *RNA leader*. At other times, transcription continues beyond the end of the leader sequence to produce a full operon-length transcript. In analyzing why some mRNAs terminate before they can transcribe the structural *trp* genes, while others do not, the researchers discovered **attenuation:** control of gene expression by premature termination of transcription. Whether or not transcription terminates prematurely depends on how the translation machinery reads the secondary structure of the RNA leader.

The RNA leader can fold into two different stable conformations, each one based on the complementarity of bases in the same molecule of RNA. The first structure contains two stem-loop structures with regions 1 and 2 associated by base pairing and regions 3 and 4 similarly associated by base pairing. When the transcriptional machinery "sees" the 3–4 stem-loop configuration, which has seven U's at the end, it stops transcription, producing a short, "attenuated" RNA. The alternative RNA structure forms by base pairing between regions 2 and 3. In this conformation, the leader RNA does not display the 3–4 stem-loop termination signal, and as a result, the transcription machinery reads right through it to produce a full-length transcript that includes the structural-gene sequences.

The Simultaneous Translation of a Small Peptide in the Leader Determines Which Outcome Occurs

The early translation of a short portion of the RNA leader (while transcription of the rest of the leader is still taking

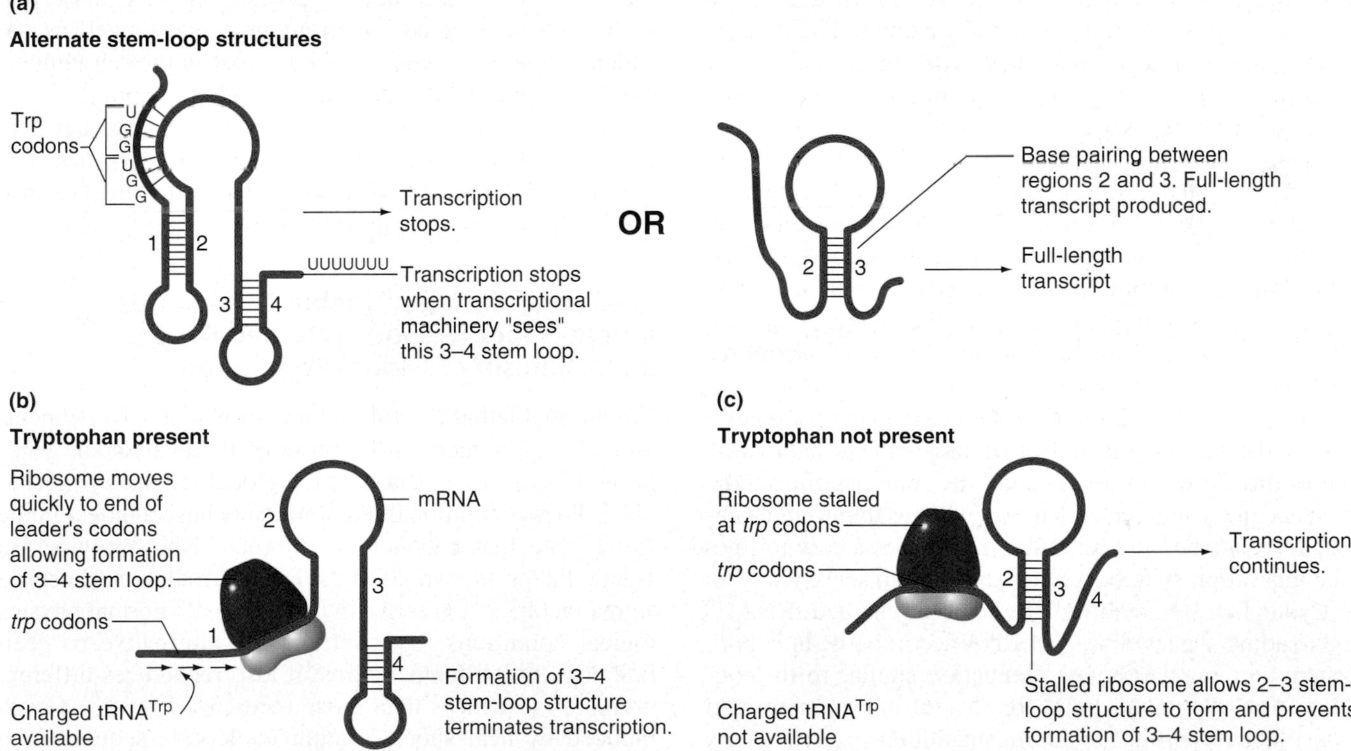
(a)
Alternate stem-loop structures
Trp codons
U
G
G
U
G
G
1
2
3
4
UUUUUUU
Transcription stops.
Transcription stops when transcriptional machinery "sees" this 3–4 stem loop.
OR
2
3
Base pairing between regions 2 and 3. Full-length transcript produced.
Full-length transcript
(b)
Tryptophan present
Ribosome moves quickly to end of leader's codons allowing formation of 3–4 stem loop.
trp codons
Charged tRNATrp available
mRNA
Formation of 3–4 stem-loop structure terminates transcription.
(c)
Tryptophan not present
Ribosome stalled at trp codons
trp codons
Charged tRNATrp not available
Transcription continues.
Stalled ribosome allows 2–3 stem-loop structure to form and prevents formation of 3–4 stem loop.

Figure 17.20 Attenuation in the tryptophan operon of *E. coli.* **(a)** Stem loops form by complementary base pairing in the *trp* leader RNA. Two different secondary structures are possible in the mRNA from the *trp* operon. Stem loops using complementary base pairs between regions 1 and 2 will enable the formation also of the stem-loop 3–4, which is a termination signal for RNA polymerase. Base pairing between regions 2 and 3 leads to a stem loop that prevents formation of stem-loop 3–4. In the early portion of the transcript, there are two codons for tryptophan. **(b)** When tryptophan is present, the ribosome follows quickly along the transcript, preventing stem-loop 2–3 from forming. Stem-loop 3–4 can form and transcription is terminated. **(c)** If tryptophan is absent, the ribosome stalls at the *trp* codons, allowing formation of stem-loop 2–3, preventing stem-loop 3–4 formation. Transcription continues.

place) determines which of the two alternative RNA structures forms. That key portion of the RNA leader includes a short open reading frame containing 14 codons, 2 of which are *trp* codons (Fig. 17.20b). When tryptophan is present, the ribosome moves quickly past the *trp* codons in the RNA leader and proceeds to the end of the leader's codons, allowing formation of the stem-loop 3–4 structure and preventing formation of the alternative RNA structure. As we have seen, this 3–4 RNA structure causes the termination of transcription and hence the attenuation (that is, lessening) of *trp* gene expression. In the absence of tryptophan, the ribosome stalls at the two *trp* codons in the RNA leader because of the lack of charged $tRNA^{Trp}$ in the cell. The 2–3 stem-loop structure depicted in Fig. 17.20c is then able to form, and its formation prevents formation of the 3–4 stem-loop RNA structure recognized by the transcriptional terminator. As a result, transcription proceeds through the leader into the structural genes.

How do we know these secondary RNA structures exist *in vivo* and that the translation of the leader RNA plays a significant role in their formation? Several experiments support this model of attenuation. First, deletion of the complete leader sequence results in the loss of control by attenuation; in $trpR^-$ mutants that also do not contain the leader sequence, there is no difference in *trp* expression with or without tryptophan in the medium. This double mutant makes the *trp* biosynthetic enzymes constitutively at maximal levels. Second, mutations that weaken the stems of the RNA stem-loop secondary structures alter regulation, but they can be compensated for by a second-site mutation that restores base pairing. Third, mutations that change the RNA terminator structure increase the readthrough of transcription and thus enhance gene expression. Finally, mutations of the translation-initiating AUG codon that prevent translation of the leader sequence produce an increase in the amount of short transcript, as predicted from the model.

Why has such a complex system evolved in the regulation of the *trp* operon and other biosynthetic pathways? While the TrpR repressor shuts off transcription in the presence of tryptophan and allows it in the amino acid's absence, the attenuation mechanism provides a way to fine-tune this off/on switch. It allows the cell to sense the level of tryptophan by "reading" the level of charged $tRNA^{Trp}$ and to adjust the level of *trp* mRNA accordingly. In *E. coli,* systems for regulation by attenuation similar to that observed for tryptophan exist for several other amino acid biosynthetic operons, including histidine, phenylalanine, threonine, and leucine.

The attenuation mechanism is unique to prokaryotes because only in cells without a membrane-enclosed nucleus can the expression machinery couple transcription and translation. The opportunity for some aspect of the translational apparatus to directly affect the outcome of transcription does not exist in eukaryotes.

17.4 Global Regulatory Mechanisms Coordinate the Expression of Many Sets of Genes

Dramatic shifts in environmental conditions can trigger the expression of sets of genes or operons dispersed around the chromosome. The absence of glucose, we have already seen, increases the expression of several catabolic operons that are at least partially controlled by a common factor, the CRP—cAMP complex. A group of genes whose expression is regulated by the same regulatory proteins is called a **regulon.** Another example of such global regulation is *E. coli*'s response to heat shock, which results from exposure to extremely high temperatures (up to 45°C).

An Alternative Sigma (σ) Factor Mediates *E. coli's* Global Response to Heat Shock

At high temperatures, most proteins denature or aggregate, or both. In *E. coli,* exposure to high temperature induces the expression of several proteins that alleviate heat-shock-related damage. The induced proteins include those that recognize and degrade aberrant proteins as well as so-called *chaperone proteins,* which assist in the refolding of other proteins and also prevent their aggregation.

E. coli's induction of the proteins that combat heat shock is a highly conserved stress response. Organisms as different as bacteria, flies, and plants induce similar proteins, notably the chaperones, in response to high temperatures.

Studies of Mutants Unable to Generate a Heat-Shock Response Helped Reveal a Mechanism of Global Regulation

Conditional lethal *E. coli* mutants in which high temperatures do not induce transcription of the heat-shock genes provided critical evidence for the global regulatory mechanism. These conditional lethal mutants have a defect in the *rpoH* gene that encodes an alternate RNA polymerase sigma factor known as σ^{32}. The normal housekeeping sigma factor, σ^{70}, is active in the cell under normal physiological conditions. By contrast, the alternative σ^{32} can function at high temperatures; it also recognizes different promoter sequences than those recognized by σ^{70}. Genes induced by heat shock contain nucleotide sequences in their promoters that are recognized by σ^{32} (**Fig. 17.21a**). σ^{32} mediates the heat-shock response by binding to the core RNA polymerase, thereby allowing the polymerase to initiate transcription of the genes encoding the heat-shock proteins. The RNA polymerase σ^{32} holoenzyme is relatively resistant to heat inactivation compared to the heat-sensitive σ^{70}-dependent RNA polymerase. As a result,

(a)

σ^{70} recognizes this promoter sequence. TTGACA 16–18 bp TATAAT

σ^{32} recognizes this promoter sequence. CTTGAA 13–15 bp CCCCATNT

(b) At high temperatures, σ^{24} recognizes different promoter sequence on *rpoH* gene, and σ^{32} is transcribed.

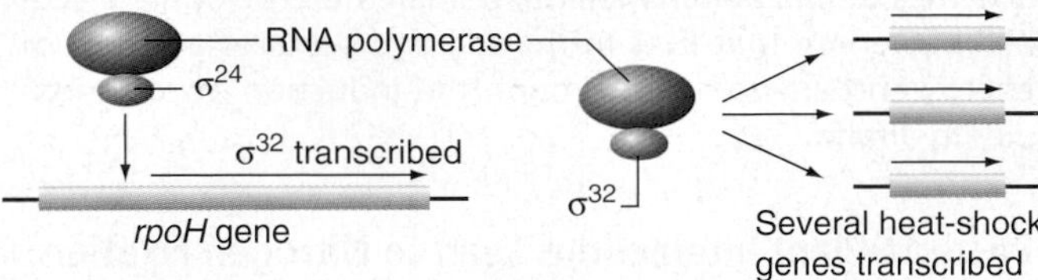

Figure 17.21 Alternate sigma factors. **(a)** Base sequences recognized by σ^{32} and σ^{70}. (The *N* indicates that any base can be found at this position.) **(b)** At high temperature, the *rpoH* gene (encoding σ^{32}) is transcribed. The σ^{32} interacts with RNA polymerase and transcribes the heat-shock genes.

when temperatures rise, genes with a σ^{32} promoter undergo transcription, while genes with a σ^{70} promoter do not.

Levels of the σ^{32} protein and the σ^{32} RNA polymerase holoenzyme increase immediately after heat shock. Several factors cause this increase in σ^{32} activity, including

- An increase in the transcription of the *rpoH* gene.
- An increase in the translation of σ^{32} mRNA stemming from greater stability of the *rpoH* mRNA.
- An increase in the stability and activity of the σ^{32} protein. Chaperones DnaJ/K bind to and inhibit σ^{32} under normal physiological conditions. When the temperature rises, these proteins bind to the large number of cellular proteins that become denatured, leaving σ^{32} free to associate with RNA polymerase.
- The inactivity of σ^{70} at high temperatures. Because of this inactivity, σ^{70} does not compete with σ^{32} in forming the RNA polymerase holoenzyme.

Given that high temperatures render σ^{70} inactive, what enables the transcription of σ^{32} during heat shock? The *rpoH* gene, which encodes σ^{32}, has a promoter sequence that is recognized by σ^{70} and used for transcription at lower temperatures. However, at high temperatures, another sigma factor, σ^{24}, recognizes a different promoter sequence at *rpoH* and transcribes the *rpoH* gene from that promoter (Fig. 17.21b). Although σ^{24} is always present in the cell, its own transcription (mediated by the σ^{24} holoenzyme) increases with heat shock and the appearance of denatured proteins.

In summary, *E. coli*'s heat-shock response depends on the regulation of transcription by alternative sigma factors. These alternative factors, which recognize different promoter sequences, complex with the core RNA polymerase as the temperature rises, leading to the transcription of heat-shock proteins.

The Induction of Alternative σ Factors That Recognize Different Promoter Sequences Serves As a Global Regulatory Mechanism in Many Bacteria

By coordinating the expression of different sets of bacterial genes in response to specific cues from the environment, this global regulatory mechanism contributes to the control of such complex processes as sporulation, the synthesis of flagella, and nitrogen fixation (see the Genetics and Society box "Nitrogen Fixation Depends on Many Levels of Gene Regulation" on pp. 630–631). For example, the bacterium *Bacillus subtilis,* under the adverse conditions of nutrient deprivation (such as nitrogen or carbon starvation), uses a cascade of sigma factors, induced in a temporal order, to turn on successive sets of genes needed to form spores. With the proper expression of these genes, the bacterial cell becomes a metabolically inert spore able to withstand heat, aridity, extreme cold, toxic chemicals, and radiation.

Microarrays Provide a New Tool for Studying Genes Regulated As Part of a Global Response

Microarrays (review Fig. 1.13, p. 10 and pp. 376–377 of Chapter 10) are an important new tool for microbial geneticists studying cellular responses to changing environmental conditions. These environmental variations include such phenomena as the rising temperatures that produce heat shock, changes in osmotic pressure, and the presence or absence of glucose. The cellular responses to these conditions often involve a global change in gene expression that is measurable by microarray analysis of mRNA isolated from cultures of cells grown under different conditions.

For example, to study changes in gene expression when lactose is substituted for glucose in the extrinsic environment (growth medium), scientists grow one culture of *E. coli* in medium containing glucose as a carbon source and another culture in medium containing lactose as a carbon source. They then isolate RNA from each culture and synthesize labeled DNA complementary to the collection of RNAs from the two different cultures. By comparing the hybridization of the two sets of cDNAs to microarrays containing oligonucleotide spots for each *E. coli* gene, they can see which genes are turned on and which are turned off in cells grown in lactose versus glucose. In the lactose-treated cells, one would expect to see an increase in the mRNA of the lactose operon genes and therefore an increase in hybridization to the *lacZ, Y,* and *A* gene spots on the microarray.

Experiments using DNA arrays to compare the gene expression patterns of cells grown in media containing as their carbon source glucose, glycerol, succinate, or alanine provide an interesting glimpse into the cellular response to poorer energy sources. The cells grown on glycerol,

GENETICS AND SOCIETY

Nitrogen Fixation Depends on Many Levels of Gene Regulation

Nitrogen, an essential component of amino acids, chlorophylls, and nucleic acids, is a growth-limiting plant nutrient—the more nitrogen available, the faster most plants grow. However, although gaseous nitrogen (N_2) makes up 78% of Earth's atmosphere, plants cannot use nitrogen in this form. They can use only nitrogen that has been *fixed,* that is, converted to ammonia (NH_3) or another nitrogen-containing compound.

Plants obtain fixed nitrogen from three main sources: (1) the decayed organic matter in soils, which releases nitrate and ammonium; (2) the activity of nitrogen-fixing bacteria, which fix atmospheric N_2 into ammonium and other biologically available forms of nitrogen; and (3) inorganic nitrogen fertilizers. Because the amount of nitrogen available often limits plant growth in agricultural settings, the last 50 years has seen a 10-fold increase in the application of inorganic fertilizer. This excessive use of fertilizer has produced runoff that increases the mineral content of rivers and coastal waters; and the increase in mineral content has led to algal blooms and a depletion of oxygen that seriously disrupt aquatic ecosystems.

In an attempt to reduce the amount of inorganic fertilizers used in agriculture, many scientists are studying how bacteria fix nitrogen. The goal is to engineer the bacteria to fix nitrogen in a broader range of plant hosts and to engineer crop plants that fix their own nitrogen. Several types of bacteria are agents of nitrogen fixation. These bacteria may be free-living cells (such as those in the genus *Klebsiella*) or plant symbionts (such as those in the genus *Rhizobium*). Of the many symbiotic rhizobial species, each one is able to form a working relationship with only one or a few plants, mainly legumes like peas, beans, and alfalfa. For hundreds of centuries, without understanding the microbiological and molecular bases of nitrogen fixation, farmers made use of the nitrogen-fixing abilities of rhizobial bacteria via the rotation of crops, observing that if they planted alfalfa one year, the next year's wheat crop, for example, would be much more abundant. Here we describe a few aspects of the multilayered regulatory mechanisms controlling the nitrogen-fixation activities of *Rhizobium meliloti,* a small, heterotrophic, rhizobial bacterium that lives in the soil and acquires a source of energy and amino acids through the induction of root nodules in alfalfa.

Bacteria-Plant Interactions Lead to Nitrogen Fixation

In the symbiotic relationship that develops between *R. meliloti* and alfalfa, bacterial genes produce the enzymatic machinery for nitrogen fixation, while the plants provide a low-oxygen environment that allows the nitrogen-fixation enzymes to function. Achievement of the symbiotic state depends on a series of communications between plant and bacteria that lead to dramatic changes in both the anatomy of the plant and the structure of the bacteria.

Alfalfa's secretion of flavonoids triggers the events leading to nitrogen fixation. *R. meliloti* responds to this environmental signal by expressing *nodulation* (*nod*) genes, whose protein products are enzymes active in the synthesis of liposaccharides known as Nod factors. Different Nod factors determine the specificity of interaction between different bacterial species and the plants with which they establish symbiosis. Release of the Nod factors from the rhizobial cells elicits a curling of root hairs and cell division in the meristem, which lead to the formation of root nodules in the alfalfa plant (**Fig. A**). The bacteria now navigate by chemotaxis to the alfalfa host nodules and penetrate the host's root cortex with the help of a gelatinous filament secreted by the plant itself. Once inside the plant, the *R. meliloti* enter root cells, where they divide and differentiate into *bacteroids,* cells that produce nitrogenase, an enzyme complex that catalyzes the conversion of N_2 to NH_3. The host plant monitors the concentration of oxygen in the area of the nodule where the bacteroids thrive to ensure that it is much lower than in the surrounding plant cells or soil. A nearly anaerobic environment

succinate, or alanine not only turned on the few genes specifically required to use the poorer carbon source, which is associated with a slower growth rate, they also turned on a hierarchy of large sets of additional genes. Cells grown on glycerol or succinate showed increased expression of 40 genes; cells grown on alanine turned on 188 genes, including the set of 40 that were expressed in the glycerol- and succinate-grown cultures. These 40 genes included those of the stress response as well as those of the CRP regulon. Recall that when glucose is not present, cAMP levels rise, and the CRP–cAMP complex binds and increases expression of other genes, including several catabolic genes. Unexpectedly, the cells grown on poorer carbon sources also turned on genes encoding proteins for motility and for the transport of many compounds in addition to the compound in their medium. The cellular response to carbon sources other than glucose seems to be the expression of genes that allow the cell to search out and use any alternative energy source. Motility genes, for example, might be turned on to allow the cell to move about in search of food. Another global transcription change is that RNA polymerase transcribes rRNA at a lower rate in the cells with poorer carbon sources, because these cells channel their energy into the search for and use of an alternative carbon source.

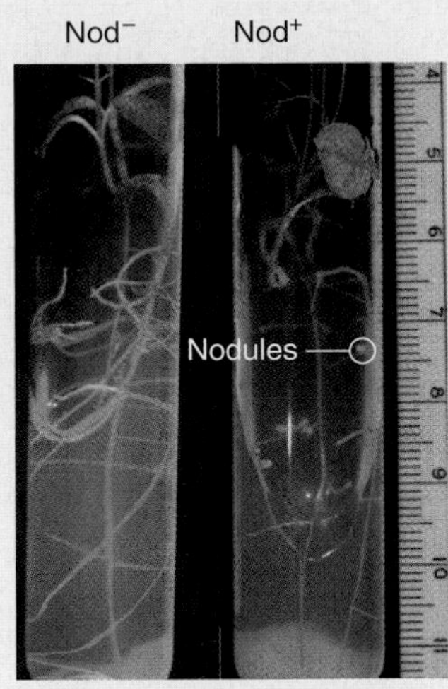

Figure A ***R. meliloti's* release of Nod factors induces the formation of root nodules in alfalfa.** Nod⁻ mutants cannot form nodules.

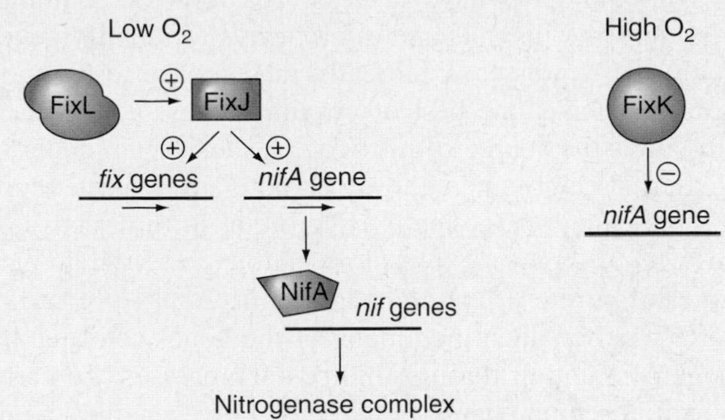

Figure B How environmental signals influence the expression of *nif* genes in *R. meliloti*. Low oxygen activates the *nif* genes via RNA polymerase associated with a σ^{54} factor. A regulatory cascade promotes expression of the *nifA* gene only under appropriate conditions. Pluses and minuses indicate the turning on or off of genes.

in the nodule is crucial to the survival and function of nitrogenase, which is extremely sensitive to oxygen and becomes irreversibly inactivated in air. The alfalfa plant uses the nitrogen fixed by *R. meliloti* as its source of nitrogen and, in return, provides the bacteria with photosynthetic products and amino acids.

The Genetic Components and Mechanisms That Mediate Nitrogen Fixation in Rhizobial Bacteria

The steps of nodule formation and nitrogen fixation just outlined require the coordinated expression of at least three types of *R. meliloti* genes: *nod* genes, which elicit the early steps of nodule formation; *fix* genes, which contribute to the development and metabolism of bacteroids and are essential to nitrogen fixation (a mutation in a *fix* gene destroys the ability to fix nitrogen); and *nif* genes, which encode the polypeptide subunits of the nitrogenase complex. The FixL protein in the membrane senses the O_2 concentration and activates the transcription factor FixJ, which then turns on expression of other *fix* genes as well as *nifA*. The *nif* genes carry a special type of promoter that RNA polymerase recognizes only when the polymerase is associated with a specific σ factor called σ^{54} factor. Initiation of transcription at the σ^{54}-dependent promoters of the *nif* genes depends on NifA, an activator protein responsive (via various intermediaries) to specific environmental signals that reflect the concentration of oxygen (**Fig. B**).

This brief description of the process of nitrogen fixation by *R. meliloti* gives some idea of the complex layering of gene regulation that ensures nitrogen fixation moves forward under favorable conditions but comes to a halt under conditions of too much oxygen. Other symbiotic, nitrogen-fixing bacteria also have intricate controls over the expression of their *fix* and *nif* genes. The coordinate expression of these bacterial genes contributes to the bacteria's ability to respond to environmental signals and become nitrogen-generating symbionts in their host plants.

Mutants Make It Possible to Study More Specific Response Mechanisms

The experiments previously described enhanced our understanding of the genetic basis of the physiological reaction to a change in carbon source. However, they were not specific enough to reveal changes in gene expression related only to the change from glucose to an alternative carbon source. To circumvent the experimental problem arising from the fact that environmental changes often produce a general physiological reaction, investigators use bacterial strains with mutations in specific genes: the regulatory genes that serve as the main on/off switch for numerous genes in the pathway they are analyzing. Instead of treating cells to two different growth conditions and measuring RNA levels under the different conditions, they use microarrays to compare RNA levels in a wild-type culture to RNA levels in a culture of cells containing a mutation in the key regulatory gene. They can grow the two cultures under the same environmental conditions because the mutation itself simulates a different environmental condition. For example, a *lacI*⁻ cell behaves as if lactose is in the medium, even if the mutant is grown in a glucose medium.

Microbiologists have successfully used the combination of specific mutants and microarrays to identify genes that *E. coli* expresses specifically in response to nitrogen limitation. Under the best of circumstances, *E. coli* uses ammonia as its source of nitrogen. A lack of ammonia in the external environment, however, activates a master control gene called *ntrC*. The NtrC protein, in turn, activates many genes whose expression enables *E. coli* to use sources of nitrogen other than ammonia.

Researchers identified many of the genes activated by nitrogen limitation through molecular analyses. To carry out a more comprehensive analysis, however, they compared RNA levels of all *E. coli* genes in a cell containing a null mutation in *ntrC* to RNA levels in a mutant strain that produced a greater than normal amount of the NtrC protein. The resulting microarray data confirmed a finding from earlier studies that the *glnA* gene, which encodes glutamine synthetase, is regulated by NtrC. A second, very striking finding from the microarray analysis was that about 2% of the *E. coli* genome is under NtrC control.

Arranging the spots on microarrays in the order in which the genes occur in the genome makes it easier to identify coregulated adjacent genes that might form an operon. This type of microarray analysis has revealed several additional genes that are regulated by NtrC. The additional genes are involved in the scavenging of proteins and nitrogen through the transport of nitrogen-containing compounds from the cell wall into the cell and the breakdown of amino acids. NtrC's regulation of these genes has been confirmed by other types of analysis. Microarrays have thus provided an avenue for uncovering changes in gene expression that investigators can confirm and expand on through other methodologies.

Computer Analyses of Genomic Data Can Identify Regulatory Proteins and Their DNA Binding Sites

A goal in the postgenomic era is to identify the complete set of proteins that regulate transcription in an organism, as well as their DNA binding sites and the genes they regulate. This information will help researchers discover molecular targets for controlling cell proliferation and the production of harmful or helpful metabolites; it will also make it possible to model how a cell works. A first step in uncovering the regulatory machinery in bacteria is to identify operons. It is easy to correlate genomic DNA sequence with potential open reading frames, but how can you find genes that are cotranscribed? Since operons have a single promoter for several genes, you can look at clusters of genes that have promoter sequences before the first but not the subsequent genes. You can also look for genes with almost no separation between them. Cotranscribed genes have little space between them in the genome because no nucleotides are needed to regulate the expression of each gene separately. Computer experts have developed algorithms that search for one promoter for several closely spaced genes, as well as for transcription termination signals. While not perfect, these algorithms appear to be good operon predictors because known operons are among the results. The predictions can be further assessed by comparative species analysis. With genes that are cotranscribed and regulated as an operon in one species, a homologous set of adjacent genes in another species is also likely to be cotranscribed as an operon. As with all *in silico* analyses, predictions made with computational tools should be tested experimentally.

Genes encoding regulatory proteins can be identified in the genome by searching for sequences encoding DNA binding motifs, such as HTH. Of 314 putative regulator proteins identified by the presence of transcription factor domains in *E. coli,* 248 contained the HTH motif. Comparative genomic analyses by computer can help identify the genes that encode these regulatory proteins. When a set of coregulated genes is present in different species, the proteins that regulate these genes are often conserved, as are their DNA binding sites. Researchers can use information from organisms such as *E. coli,* in which extensive genetic and biochemical analyses have defined regulatory pathways, to discover regulatory components in less well understood bacteria.

17.5 A Comprehensive Example: The Regulation of Virulence Genes in *V. cholerae*

The principles and mechanisms of gene regulation in *E. coli* apply to gene regulation in other prokaryotics as well, including the bacteria *V. cholerae,* which we described at the beginning of this chapter. As we saw, these bacterial agents of cholera are able to sense changes in their environment and transmit signals about those changes to regulators. These regulators then initiate, enhance, diminish, or repress the expression of various genes as the bacteria pass through the stomach, colonize the intestine, and finally produce a toxin. Of particular interest to epidemiologists and medical practitioners seeking to prevent or treat the symptoms of cholera are the genes bestowing virulence.

Identifying the Regulators of Toxin Production

To understand the regulation of the genes for virulence, researchers first cloned the two genes that encode the polypeptide subunits of cholera toxin: *ctxA* and *ctxB,* which are transcribed and regulated together as an operon. They next made a *ctxA-lacZ* reporter gene fusion molecule that could detect changes in regulation of the operon through changes in levels of β-galactosidase expression. *LacZ* would be expressed and β-galactosidase produced when the cholera toxin promoter was being used, and no β-galactosidase would be produced when the promoter was shut off. They then cut *V. cholerae* genomic DNA into pieces and cloned these into a vector that would replicate in *E. coli.* With the construction of these tools, they were able to perform experiments in *E. coli* cells, which are more amenable than *V. cholerae* to some types of genetic manipulation.

To isolate a gene that regulates expression of the *ctx* operon, they transformed *E. coli* cells already containing the *ctxA-lacZ* fusion molecule with clones containing *V. cholerae* DNA. A clone that contains a gene encoding a positive cholera toxin regulatory protein should turn on expression of the *lacZ* fusion molecule in *E. coli.* Clones that turned on expression contained the regulatory *toxR* gene, which encodes a membrane protein (ToxR) with an N-terminal end in the cytoplasm and a C-terminal end in the periplasm (the space between the inner and outer membranes of the bacterium). In *V. cholerae, toxR*$^-$ mutants do not induce virulence, and the *toxR*$^-$ mutation is recessive as you would predict of a positive regulator.

Identifying Genes Regulated by ToxR and ToxT

To determine what genes ToxR regulates other than those in the *ctx* operon, researchers fused the *toxR* gene to a constitutive promoter, and they introduced this fusion molecule into a collection of *V. cholerae* strains in which copies of the *lacZ* gene had randomly inserted around the chromosome. Those colonies expressing β-galactosidase (as shown by the blue color resulting from the splitting of the X-Gal substrate) contained *lacZ* genes adjacent to a promoter region regulated by *toxR.*

In *V. cholerae,* these *lacZ* fusion genes must have been regulated by ToxR (at least indirectly) because all bacteria in the study contained the *toxR* gene fused to a constitutive promoter and thus were constitutive synthesizers of ToxR. However, when transferred into *E. coli,* this collection of genes was not regulated by ToxR. Something required to make these genes respond to ToxR was present in the *V. cholerae* genome but was missing in *E. coli.* The lack of direct regulation by ToxR in *E. coli* triggered a search that culminated in the identification of an intermediate regulatory gene named *toxT.* The ToxT protein is a transcriptional activator that carries out its function by binding to the promoters of many genes, including *ctx* and the other virulence genes. While either ToxR or ToxT can activate the *ctx* genes that produce toxin, ToxT alone activates the additional virulence genes, which encode pili and other proteins that enable the bacteria to colonize the small intestine.

ToxT is a major regulator of several virulence genes, but how is it regulated? Mutations in the *tcpP* gene lead to loss of ToxT transcription, as do mutations in *toxR.* Analyses of the promoter region of *toxT* showed that TcpP binds to the *toxT* promoter close to the transcription start point, while ToxR binds further upstream. This upstream binding suggests that ToxR helps recruit the TcpP protein to the promoter, where TcpP acts as the positive regulator of *toxT.* Both ToxR and TcpP are membrane bound, with the N terminal region in the cytoplasm available to bind to DNA and the C terminus in the periplasmic space able to receive environmental signals about the location of the bacterium in the body. Expression of cholera toxin and the pilus is induced when the bacteria have reached the intestine after passing through the bile-laden stomach. Information about the environment is probably transmitted through the activation of *tcpP,* as the transcription of this gene is temperature- and pH-dependent. By comparison, *toxR* is transcribed independently of both temperature and pH.

A Model of Virulence Regulation Includes a Cascade of Regulators

On the basis of these studies, researchers proposed the following model (**Fig. 17.22**):

- ToxR is a positive regulator of the *ctx* genes and acts as an auxiliary factor in the regulation of ToxT.
- ToxT is a positive regulator of the many virulence genes that make up the virulence regulon in *V. cholerae.*

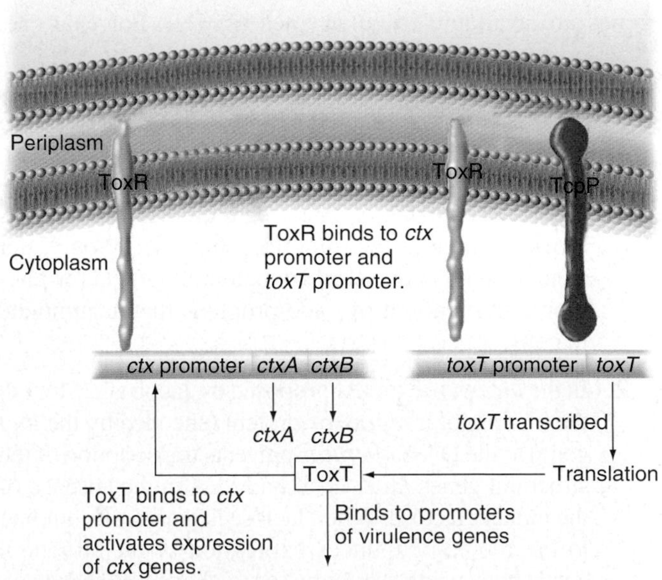

Figure 17.22 Model for how *V. cholerae* regulates genes for virulence. In the cytoplasm, ToxR interacts with promoter of the *ctxA* and *ctxB* genes. ToxR and TcpP both bind to the *toxT* gene. ToxT in turn regulates the expression of many other virulence genes.

- Maximal transcription of *toxT* requires the TcpP regulator, with assistance from ToxR.
- The sensing of environmental change is a part of the regulator gene activation process that is mediated by TcpP.

Experiments in which investigators monitored gene expression in the animal (mouse) disease model for *V. cholerae* confirmed the requirement of ToxR as a positive regulator of *ctxA* expression during pathogenesis in the animal. In these animal studies, when *V. cholerae* strains containing the *ctx* reporter gene fusion were injected into mice, expression of *ctxA-lacZ* occurred in the *toxR*$^+$ strain but not in the *toxR*$^-$ mutant. Interestingly, in the mouse model, the specifics of some of the other regulatory pathways for the *V. cholerae* pathogenesis genes did not coincide with the results obtained from the isolated bacterial cultures. From these studies, we can conclude that studying gene expression in pathogenic strains in culture, where it is easier to manipulate genes and measure their expression, provides a valuable first analysis. Once the potential regulators and pathways have been identified, experiments to detect gene expression during pathogenesis in model animal systems provide critical tests of the models.

Several intriguing questions remain about the regulatory system that controls the expression of the virulence genes in *V. cholerae.* What is the signal that makes the cholera bacteria stop swimming and start to colonize (that is, adhere to the cells of) the small intestine? What molecular events differentiate swimming and colonization? Why is there a cascade (ToxR, ToxT) of regulatory factors?

Answers to these questions will help scientists complete the picture of how *V. cholerae* generate disease. With a better understanding of pathogenesis, they will be able to devise more effective treatments for cholera as well as measures to prevent it.

Connections

Regulation in prokaryotes depends on the binding of regulatory proteins to specific DNA segments in the vicinity of a gene or group of genes. The existence of these regulatory elements adds another notch to the concept of the gene. Most geneticists would say that a gene consists of the nucleotides that specify amino acids in the gene's protein product or the ribonucleotides in the gene's RNA product, as well as the regulatory elements that influence the gene's transcription.

Some of the ways in which bacteria regulate their genes are available to eukaryotes as well. For example, both types of organisms can use diffusible regulatory proteins to start or stop transcription. By contrast, eukaryotes cannot regulate transcription by the attenuation mechanism described for the *trp* operon because their nuclear membrane prevents access to the growing transcript by the translational machinery. However, eukaryotic cells, with their larger genomes, have evolved many other mechanisms of gene regulation that go beyond those found in prokaryotic systems. In Chapter 18, we examine the special regulatory needs of eukaryotes and some of the solutions they have evolved.

Essential Concepts

1. Most mechanisms of gene regulation in prokaryotes block or enhance the initiation of transcription. Later steps in gene expression are potential targets for fine-tuning the amount of gene products that accumulate in cells.
2. In the *lac operon* model proposed by Jacob and Monod, the binding of a *repressor* protein (encoded by the *lacI* gene) to the DNA *operator* prevents transcription of the structural genes *lacZ, lacY,* and *lacA* in the absence of the inducer lactose. When lactose is present, its binding to the repressor induces expression of the structural genes by causing the repressor to change its shape and lose its ability to bind to the operator.
3. A critical, general principle emerges from the *lac* operon studies: Regulatory genes usually encode *trans*-acting regulatory proteins that interact with *cis*-acting regulatory DNA elements located near the promoter (such as the *operator*). Negative regulatory proteins prevent or diminish the rate of transcription, while positive regulatory proteins enhance transcription.
4. Many types of coordinate gene regulation result from the clustering of genes into *operons* that are transcribed into a single polycistronic mRNA from a single promoter.
5. The binding of *repressor* proteins to *operators* can be influenced by either inducers (as for the *lac* repressor) or corepressors (as for the *trp* operon).
6. *Catabolite repression* regulates certain catabolic operons by preventing the CRP protein, a positive regulator,

from binding to the operons' promoter region in the presence of high concentrations of glucose.

7. Many regulatory proteins, both positive and negative, contain a helix-turn-helix motif, function as oligomers that bind to more than one DNA site, and interact with RNA polymerase to prevent or assist its function.

8. *Attenuation,* a form of fine-tuning for operons involved in the biosynthesis of amino acids, is based on premature termination of mRNA transcription. The termination, in turn, is determined by the intracellular concentration of tRNAs charged with the amino acid produced by the enzyme products of the structural genes in the operon.

9. Cells can express different sets of genes at different times or under different conditions by using alternate sigma factors or by producing novel RNA polymerases that recognize different classes of promoters.

10. DNA sequences from many bacterial species are raw data that can be analyzed computationally to identify regulatory features in the bacterial genome.

On Our Website

www.mhhe.com/hartwell3
Chapter 17

Annotated Suggested Readings and Links to Other Websites

- Landmark papers on the lactose operon in *E. coli*
- Recent papers on the use of microarrays to understand gene regulation
- Recent papers on regulation of *V. cholerae* pathogenesis genes

Social and Ethical Issues

1. The bacterium *Bacillus thuringiensis* (Bt) produces a toxin capable of killing many insects that attack crop plants. For more than 20 years, Bt has served as a biological pesticide that is sprayed on crops. The constant spraying, however, has led to the development of resistant insect strains in the field. A cloned version of Bt toxin consists of the toxin gene fused to a regulatory region that allows expression of the toxin only when crops carrying the gene are under stress—as in an insect infestation. Should farmers be forced to plant the transgenic crops rather than spray their fields so that this biological pesticide will remain effective? Are there ways to enforce this policy in the United States or worldwide?

2. Many scientists who are employed as professors at universities have become involved in biotechnology ventures outside the university. Companies often set up their laboratories near universities to take advantage of the intellectual resources there. Industries that develop near universities employ nonscientists in the community as well as scientists and are generally looked upon as an asset to the community. The president of a major state university is encouraging faculty members to become active in the development of associated industries, viewing their mission as academics to include community-related activities. But legislators and the public, whose tax dollars fund the university, believe that the professors should concentrate on teaching. Should faculty be encouraged to participate in outside ventures because it is a service to the community and potentially to humankind through development of new biotechnology, or should they be held more strictly to a solely academic, teaching mission?

3. In studying virulence in bacteria, a group of researchers stumbled on ways of modifying a particular bacterial species in a way that increased the virulence. While their finding helped them understand the mechanism of virulence better, the group of researchers were split on whether to publish their findings or not. Some fear that this new information could be misused in developing biological warfare. Do scientists have a responsibility to withhold information if it could be used in a harmful way?

Solved Problems

I. In the galactose operon in *E. coli,* a repressor, encoded by the *galR* gene, binds to an operator site, *O,* to regulate expression of three structural genes, *galE, galT,* and *galK.* Expression is induced by the presence of galactose in the media. For each of the merodiploids listed, would the cell show constitutive, inducible, or no expression of each of the structural genes? (Assume that $galR^-$ is a loss of function mutation.)

a. $galR^-\ galO^+\ galE^+\ galT^+\ galK^+$

b. $galR^+\ galO^c\ galE^+\ galT^+\ galK^+$

c. $galR^-\ galO^+\ galE^+\ galT^+\ galK^-/$
$galR^+\ galO^+\ galE^-\ galT^+\ galK^+$

d. $galR^-\ galO^c\ galE^+\ galT^+\ galK^-/$
$galR^+\ galO^+\ galE^-\ galT^+\ galK^+$

Answer

This problem requires an understanding of how regulatory sites and proteins that bind to regulatory sites behave. To predict expression in merodiploid strains, look at each copy of the operon individually, and then assess what effect alleles present in the other copy of the operon could have on the expression. After doing that for each copy of the operon, combine the results.

a. The *galR* gene encodes a repressor, so the lack of a *GalR* gene product would lead to constitutive expression of the *galE, T,* and *K.*

b. The $galO^c$ mutation is an operator site mutation. By analogy with the *lac* operon, the designation $galO^c$ indicates that repressor cannot bind and there is constitutive expression of *galE, T,* and *K.*

c. The first copy of the operon listed has a $galR^-$ mutation. Alone, this would lead to constitutive synthesis *galE* and *galT.* (*galK* is mutant, so there will not be constitutive expression of this gene.) The other copy is wild type for the *galR* gene, so it produces a repressor that can act in *trans* on both copies of the operon, overriding the effect of the $galR^-$ mutation. Overall, there will be *inducible expression of the three* gal *genes.*

d. The first copy of the operon contains an $galO^c$ mutation, leading to constitutive synthesis of *galE* and *galT.* The other copy has a wild-type operator, so it is inducible, but neither operator has effects on the other copy of the operon. The net result is constitutive *galE* and *galT* and inducible *galK* expression.

II. The *araI* site is required for induction of *araBAD.* I^- mutants do not express *araBAD.* In an I^- mutant, a second mutation arose that resulted in constitutive arabinose synthesis. A Southern blot using a probe from the regulatory region and early part of the *araB* gene showed a very different set of restriction fragments than were seen in the starting strain. Based on the altered restriction pattern and constitutive expression, propose a hypothesis about the nature of the second mutation.

Answer

To answer this question, you need to consider how changes in restriction patterns could arise, what effects they could have, and what is necessary to get expression. The fact that the experiment began with a strain that lacked the inducing site, *I,* and that there is constitutive synthesis mean that the normal regulation is lacking. Constitutive synthesis could result from a *deletion that fused the* araBAD *genes to another promoter* (one that is on under the growth conditions used). A deletion would lead to a different pattern of restriction fragments that could be observed in the Southern hybridization analysis.

III. Bacteriophage λ, after infecting *E. coli,* can take one of two routes. It will either produce many progeny that are released by lysis of the cell (lytic growth), or the phage DNA will integrate into the chromosome because transcription from the major phage promoters of the phage will have been shut down. The repressor protein cI, encoded by phage λ, binds to two operator regions to shut down expression, and, therefore, no phages are produced. Mutations in the *cI* gene that destroy the binding ability of the repressor lead to the lytic type of life cycle exclusively; that is, all cells infected by the phage will burst and release progeny phages. Another type of mutation gives the same phenotype-lytic growth only. Such mutations, called λ*vir,* arise at a much lower frequency than the *cI* mutations (about 1 in 10^{12} compared with 1 in 10^6 for *cI* mutants). What do you think these mutations are, and why are they less frequent than *cI* mutations?

Answer

This problem requires an understanding of the types of regulatory mutations that can affect negative regulation. The lack of negative regulation (by *cI*) in the life cycle leads to the lytic cycle of growth only. Such mutations could be either in the gene encoding the negative regulator or in the site to which the repressor binds. You were told that the *cI* mutations are defects in the gene encoding the repressor. The λ*vir* mutations could be mutations in the site to which the repressor binds, but since the repressor has to bind to two sites, *there must be two mutations in a* λvir *mutant. Therefore, these would arise less frequently* (at a frequency predicted for two independent mutational events combined: 1 in $10^6 \times$ 1 in 10^6, or 1 in 10^{12}).

Problems

Interactive Web Exercise

SWISS-PROT is a central resource of annotated protein sequences. Our website at www.mhhe.com/hartwell3 contains a brief exercise in which you use this resource to explore families of bacterial transcription factors; once at the website, go to Chapter 17 and click on "Interactive Web Exercise".

Vocabulary

1. For each of the terms in the left column, choose the best matching phrase in the right column.

a. induction	1. glucose prevents expression of catabolic operons
b. repressor	2. protein undergoes a reversible conformational change
c. operator	3. often fused to regulatory regions of genes whose expression is being monitored
d. allostery	4. stimulation of protein synthesis by a specific molecule
e. operon	5. site to which repressor binds
f. catabolite repression	6. gene regulation involving premature termination of transcription
g. reporter gene	7. group of genes transcribed into one mRNA
h. attenuation	8. negative regulator

Section 17.1

2. The following statement occurs early in this chapter: ". . . the critical step in the regulation of most bacterial genes is the binding of RNA polymerase to DNA at the promoter." Why might it be advantageous for bacteria to regulate the expression of their genes at this step?

3. One of the main lessons of this chapter is that several bacterial genes are often transcribed from a single promoter into a large multigene transcript. The region of DNA containing the set of genes that are cotranscribed, along with all of the regulatory elements that control the expression of these genes, is called an *operon*.

a. Which of the mechanisms in the list below could explain differences in the levels of the mRNAs for different operons?

b. Which of the mechanisms in the list below could explain differences in the levels of the protein products of different genes in the same operon?

i. Different promoters might have different DNA sequences.

ii. Different promoters might be recognized by different types of RNA polymerase.

iii. The secondary structures of mRNAs might differ so as to influence the rate at which they are degraded by ribonucleases.

iv. In an operon, some genes are further away from the promoter than other genes.

v. The translational initiation sequences at the beginning of different open reading frames in an operon might result in different efficiencies of translation.

vi. Proteins encoded by different genes in an operon might have different stabilities.

4. All mutations that abolish function of the Rho termination protein in *E. coli* are conditional mutations. What does this tell you about the *rho* gene?

Section 17.2

5. The promoter of an operon is the site to which RNA polymerase binds to begin transcription. Some base changes in the promoter result in a mutant site to which RNA polymerase cannot bind. Would you expect mutations in the promoter that prevent binding of RNA polymerase to act in *trans* on another copy of the operon on a plasmid in the cell, or only in *cis* on the copy immediately adjacent to the mutated site?

6. You are studying an operon containing three genes that are cotranscribed in the order *hupF, hupH,* and *hupG.* Diagram the mRNA for this operon, showing the location of the 5′- and 3′-ends, all open reading frames, translational start sites, stop codons, transcription termination signals, and any regions that might be in the mRNA but not serve any of these functions.

7. You have isolated a protein that binds to DNA in the region upstream of the promoter sequence of the *sys* gene. If this protein is a positive regulator, which of the following would be true?

a. Loss-of-function mutations in the gene encoding the DNA-binding protein would cause constitutive expression.

b. Loss-of-function mutations in the gene encoding the DNA-binding protein would result in little or no expression.

8. You have isolated two different mutants (*reg1* and *reg2*) causing constitutive expression of the *emu* operon (*emu1 emu2*). One mutant contains a defect in a DNA-binding site, and the other has a loss-of-function defect in the gene encoding a protein that binds to the site.

a. Is the DNA-binding protein a positive or negative regulator of gene expression?

b. To determine which mutant has a defect in the site and which one has a mutation in the binding protein, you decide to do a merodiploid analysis. Assuming you can assay levels of the Emu1 and Emu2 proteins, what results do you predict for the two strains (*i* and *ii*) (see descriptions below) if *reg2* encodes the regulatory protein and *reg1* is the regulatory site?

i. F' $reg1^- reg2^+ emu1^- emu2^+ / reg1^+ reg2^+ emu1^+ emu2^-$

ii. F' $reg1^+ reg2^- emu1^- emu2^+ / reg1^+ reg2^+ emu1^+ emu2^-$

c. What results do you predict for the two strains (*i* and *ii*) if *reg1* encodes the regulatory protein and *reg2* is the regulatory site?

9. Bacteriophage λ, after infecting a cell, can integrate into the chromosome of the cell if the repressor protein, cI, binds to and shuts down phage transcription immediately. (A strain containing a bacteriophage integrated in the chromosome is called a lysogen.) The alternative fate is the production of many more viruses and lysis of the cell. In a mating, a donor strain that is a lysogen was crossed with a lysogenic recipient cell and no phages were produced. However, when the lysogen donor strain transferred its DNA to a nonlysogenic recipient cell, the recipient cell burst, releasing a new generation of phages. Why did mating with a nonlysogenic cell result in phage growth and release but infection of a lysogenic recipient did not?

10. Mutants were isolated in which the constitutive phenotype of a missense *lacI* mutation was suppressed. That is, the operon was now inducible. These mapped to the operon but were not in the *lacI* gene. What could these mutations be?

11. For each of the *E. coli* merodiploid strains containing the *lac* operon alleles listed, indicate whether the strain is inducible, constitutive, or unable to express β-galactosidase and permease.

a. $I^+ O^+ Z^- Y^+ / I^+ O^c Z^+ Y^+$

b. $I^+ O^+ Z^+ Y^+ / I^- O^c Z^+ Y^-$

c. $I^+ O^+ Z^- Y^+ / I^- O^+ Z^+ Y^-$

d. $I^- P^- O^+ Z^+ Y^- / I^+ P^+ O^c Z^- Y^+$

e. $I^s O^+ Z^+ Y^+ / I^- O^+ Z^+ Y^-$

12. For each of the growth conditions listed, what proteins would be bound to *lac* operon DNA? (Do not include RNA polymerase.)

a. glucose

b. glucose + lactose

c. lactose

13. For each of the following mutant *E. coli* strains, plot a 30-minute time course of concentration of β-galactosidase, permease, and acetylase enzymes grown under the following conditions. For the first 10 minutes, no lactose is present; at 10′ lactose becomes the sole carbon source. Plot concentration on the y-axis, time on the x-axis. (Don't worry about the exact units for each protein on the y-axis.)

a. $I^- P^+ O^+ Z^+ Y^+ A^+ / I^+ P^+ O^+ Z^- Y^+ A^+$

b. $I^- P^+ O^c Z^+ Y^+ A^- / I^+ P^+ O^+ Z^- Y^+ A^+$

c. $I^s P^+ O^+ Z^+ Y^+ A^+ / I^- P^+ O^+ Z^- Y^+ A^+$

d. $I^- P^- O^+ Z^+ Y^+ A^+ / I^- P^+ O^c Z^+ Y^- A^+$

e. $I^- P^+ O^+ Z^- Y^+ A^+ / I^- P^- O^c Z^+ Y^- A^+$

14. Maltose utilization in *E. coli* requires the proteins encoded by genes in three different operons. One operon includes the genes *malE, malF,* and *malG;* the second includes *malK* and *lamB;* and the genes in the third operon are *malP* and *malQ*. The MalT protein is a positive regulator that regulates the expression of all three operons; expression of the *malT* gene itself is catabolite sensitive.

a. What phenotype would you expect to result from a loss-of-function mutation in the *malT* gene?

b. Do you expect the three maltose operons to contain binding sites for CRP (oAMP receptor protein)? Why or why not?

In order to infect *E. coli,* bacteriophage λ binds to the maltose transport protein LamB (also known as the λ receptor protein) that is found in the outer membrane of the bacterial cell. The synthesis of LamB is induced by maltose in the medium via expression of the MalT protein, as described above.

c. List the culture conditions under which wild-type *E. coli* cells would be sensitive to infection by bacteriophage λ.

d. *E. coli* cells that are resistant to infection by bacteriophage λ have been isolated. List the types of mutations in the maltose regulon that λ-resistant mutants could contain.

15. Clones of three adjacent genes involved in arginine biosynthesis have been isolated from a bacterium. If these three genes together make up an operon, what result do you expect when you use the DNA from each of these genes as probes in a Northern analysis? What result do you expect if the three genes do not make up an operon?

16. Given the following data, explain which strains and growth conditions are important for reaching the following conclusions.

a. Arabinose induces coordinate expression of the *araBAD* genes (encoding kinase, isomerase, and epimerase).

b. The *araC* gene encodes a positive regulator of *araBAD* expression.

Genotype	Arabinose in medium	Kinase	Isomerase	Epimerase
1. $C^+ B^+ A^+ D^+$	no	–	–	–
2. $C^+ B^+ A^+ D^+$	yes	+	+	+
3. $C^- B^+ A^+ D^+$	no	–	–	–
4. $C^- B^+ A^+ D^+$	yes	–	–	–

17. Seven *E. coli* mutants were isolated. The activity of the enzyme β-galactosidase produced by cells containing each mutation alone or in combination with other mutations was measured when the cells were grown in medium supplemented with different carbon sources.

	Glycerol	Lactose	Lactose + Glucose
Wild type	0	1000	10
Mutant 1	0	10	10
Mutant 2	0	10	10
Mutant 3	0	0	0
Mutant 4	0	0	0
Mutant 5	1000	1000	10
Mutant 6	1000	1000	10
Mutant 7	0	1000	10
F′ *lac* from mutant 1/ mutant 3	0	1000	10
F′ *lac* from mutant 2/ mutant 3	0	10	10
Mutants 3 + 7	0	1000	10
Mutants 4 + 7	0	0	0
Mutants 5 + 7	0	1000	10
Mutants 6 + 7	1000	1000	10

Assume that each of the seven mutations is one and only one of the genetic lesions in the following list. Identify the type of alteration each mutation represents.

a. superrepressor
b. operator deletion
c. nonsense (amber) suppressor tRNA gene (assume that the suppressor tRNA is 100% efficient in suppressing amber mutations)
d. defective CRP–cAMP binding site
e. nonsense (amber) mutation in the β-galactosidase gene
f. nonsense (amber) mutation in the repressor gene
g. defective *crp* gene (encoding the CRP protein)

18. Cells containing mutations in the *crp* gene (encoding the positive regulator CRP) are Lac^-, Mal^-, Gal^-, etc. To find suppressors of the *crp* mutation, cells were screened to find those that were both Lac^+ and Mal^+.

a. What types of suppressors would you expect to get using this screen compared with a screen for Lac^+ only?
b. All suppressors isolated were mutant in the gene for the α-subunit of RNA polymerase. What hypothesis could you propose based on this analysis?

19. Six strains of *E. coli* (mutants 1–6) that had one of the following mutations affecting the *lac* operon were isolated.

i. deletion of *lacY*
ii. *lacO*^{c} mutation
iii. missense mutation in *lacZ*
iv. inversion of the *lac* operon (but not an inversion of the *lacI* gene)
v. superrepressor mutation
vi. inversion of *lacZ, Y,* and *A* but not *lacI, P, O*

a. Which of these mutations would prevent the strain from utilizing lactose?
b. The entire *lac* operon (including the *lacI* gene and its promoter) from each of the six *E. coli* strains was cloned into a plasmid vector containing an ampicillin resistance gene. Each recombinant plasmid was transformed into each of the six strains to create merodiploids. In the merodiploid analysis, mutant 1 was found to carry a deletion of *lacY*, so this strain corresponds to mutation i in the list above. Which of the other types of mutations would be expected to complement mutant 1 in these merodiploids so as to allow lactose utilization?
c. In the merodiploid analysis described in part b, each merodiploid strain was plated on ampicillin media in which lactose is the only carbon source. (Ampicillin was included to ensure maintenance of the plasmid.) Growth of the transformants is scored below (a + sign = growth, a − sign = no growth). Synthesis of β-galactosidase and permease are required for growth on this medium. Results of this merodiploid analysis are shown here. Which mutant bacterial strain (1–6) contained each of the alterations (i–vi) listed previously?

	1	2	3	4	5	6
1	−	+	−	+	−	+
2	+	−	−	+	−	+
3	−	−	−	+	−	+
4	+	+	+	+	−	+
5	−	−	−	−	−	+
6	+	+	+	+	+	+

20. The following data (top of p. 640) are from a DNaseI footprinting experiment in which either RNA polymerase or a repressor protein was added to a labeled DNA fragment, and then the complex was digested with DNaseI. DNA sequencing reactions were also performed on the same DNA so the bases that were protected by proteins binding could be identified. (Notice that DNaseI does not cut after each base in the DNA fragment.)

a. What is the sequence of the DNA in this fragment?
b. Mark on the sequence the region where the repressor binds.
c. Mark on the sequence the region where RNA polymerase binds.

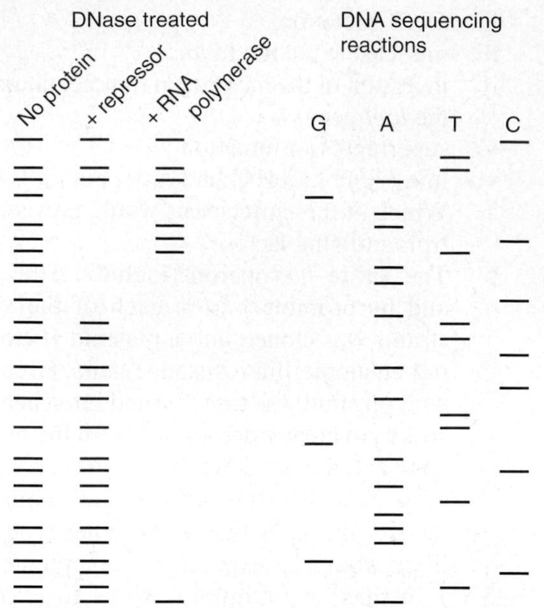

21. a. The original constitutive operator mutations in the *lac* operon were all base changes in O_1. Why do you think mutations in O_2 or O_3 were not isolated in these screens?

b. Explain how a mutagen that causes small insertions could produce an O^c mutation.

c. Would the O^c mutation described above in part *b* be sensitive to lacI^s? Why or why not?

22. In an effort to determine the location of an operator site for a negatively regulated gene, you have made a series of deletions within the regulatory region. The extent of each deletion is shown by the line underneath the sequence, and the resulting expression from the operon (i = inducible; c = constitutive; – = no expression is also indicated).

```
      . . . GGATCTTAGCCGGCTAACATGATAAATATAA...
      . . . CCTAGAATCGGCCGATTGTACTATTTATATT...
1 i   ──────────
2 –   ──────────────────────────────────────────
3 c   ────────────────────────────────────
4 –   ────────────────────────────────────────
5 c   ────────────────
```

a. What can you conclude from this data about the location of the operator site?

b. Why do you think deletions 2 and 4 show no expression?

23. An operon fusion consists of a regulatory region cloned next to the coding region of the genes of an operon. A gene fusion consists of a regulatory region of a gene such as *lacZ* and the DNA encoding the first amino acids of the β-galactosidase protein cloned next to the coding region of another gene. What additional feature do you have to consider to create a functional gene fusion that is not necessary for an operon fusion?

Section 17.3

24. a. How many ribosomes are required at a minimum for the translation of *trpE* and *trpC* from a single transcript of the *trp* operon?

b. How would you expect deletion of the two tryptophan codons in the RNA leader to affect expression of the *trpE* and *trpC* genes?

25. The following is a sequence of the leader region of the *his* operon mRNA in *Salmonella typhimurium.* What bases in this sequence could cause a ribosome to pause when histidine is limiting (that is, when there is very little of it) in the medium?

5′ AUGACACGCGUUCAAUUUAAACACCACCAUCAUCACCAUCA UCCUGACUAGUCUUUCAGGC 3′

26. For each of the *E. coli* strains that follow, indicate the effect of the genotype on expression of the *trpE* and *trpC* genes in the presence and absence of tryptophan. (In the wild type $[R^+\ P^+\ O^+\ att^+\ trpE^+\ trpC^+]$, *trpC* and *trpE* are fully repressed in the presence of tryptophan and are fully induced in the absence of tryptophan.)

R = repressor gene; R^n product cannot bind tryptophan; R^- product cannot bind operator
O = operator; O^- cannot bind repressor
att = attenuator; att^- is a deletion of the attenuator
P = promoter; P^- is a deletion of the *trp* operon promoter
$trpE^-$ and $trpC^-$ are null (loss-of-function) mutations

a. $R^+\ P^-\ O^+ att^+ trpE^+\ trpC^+$
b. $R^-\ P^+\ O^+ att^+ trpE^+\ trpC^+$
c. $R^n\ P^+\ O^+ att^+ trpE^+\ trpC^+$
d. $R^-\ P^+\ O^+ att^-\ trpE^+\ trpC^+$
e. $R^+P^+O^-\ att^+ trpE^+ trpC^-\ /\ R^-\ P^+O^+ att^+ trpE^-\ trpC^+$
f. $R^+P^-O^+ att^+ trpE^+ trpC^-\ /\ R^-\ P^+O^+ att^+ trpE^-\ trpC^+$
g. $R^+P^+O^-\ att^-\ trpE^+ trpC^-\ /\ R^-\ P^+O^-\ att^+ trpE^-\ trpC^+$

27. A molecular geneticist is investigating an operon by measuring the amount of expression of the four structural genes (*A, B, C,* and *D*) produced in wild-type and mutant bacterial cells after the addition of compound Z to a minimal medium. An additional protein (E) is of very small size (less than 20 amino acids) and cannot be measured by the same analytical system employed for the other proteins. Several of the mutations are nonsense mutations that have an effect on the genes transcribed after them in the operon. In addition to stopping translation of the gene in which the mutations lie, these so-called *nonsense polar mutations* prevent the expression of genes downstream of the mutation. (For example, in the *lac* operon, some *lacZ* nonsense mutations can result in no expression of *lacY* and *lacA.*) The investigator has also obtained mutations in two other sites, *F* and *G*, closely linked to *A–D*. The graphs shown are all semilogarithmic. The percentage of maximal possible expression for a particular protein is plotted on the y-axis, while the x-axis coordinate is time. Compound Z is added at the point specified by the arrow.

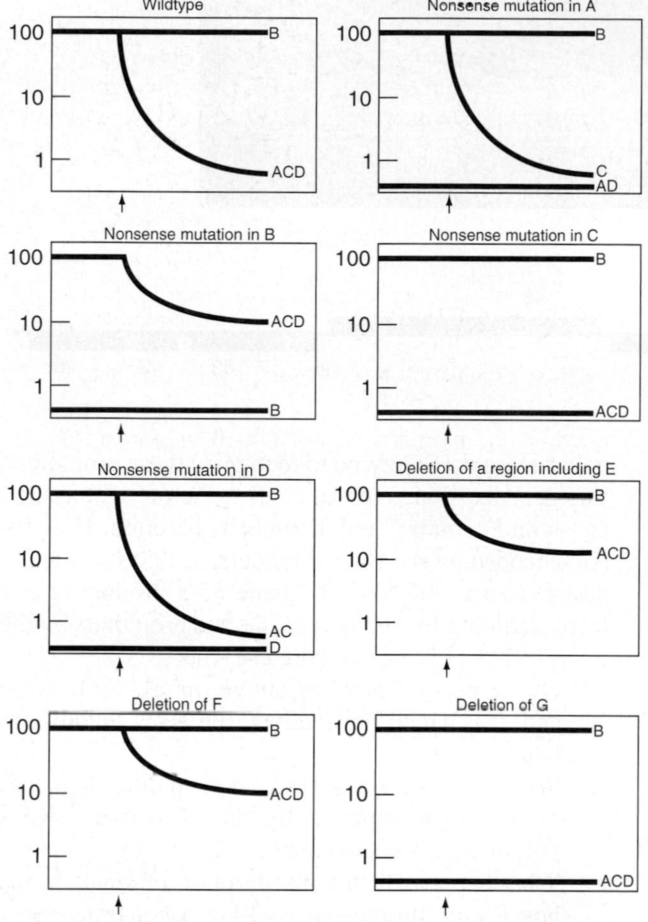

a. Is this operon likely to be involved in a pathway of biosynthesis or a pathway of degradation? Is the operon inducible or repressible?
b. For each of the conditions graphed, state if genes A–D are constitutive, completely repressible, partly repressible, or not expressed.
c. Construct a map of this operon. Indicate the relative positions of genes *A, B, C, D,* and *E* as well as the sites *F* and *G.* List possible functions for all the genes and sites. (Possible functions for the genes and sites *A–G:* promoter, operator, enzyme structural gene, CRP-binding site, *crp* gene, attenuator region. This list is not necessarily all inclusive.)

28. The previous problem (#27) introduced the concept of polar mutations in bacterial operons: nonsense mutations in a "proximal" gene nearer the promoter of the operon can abolish the expression of a "distal" gene in the same operon that is further from the promoter. Essentially all polar mutations are nonsense mutations; missense mutations do not have this property.

a. Suggest a model to explain why nonsense but not missense mutations might exhibit polarity.
b. Interestingly, in strains that simultaneously carry a polar mutation and a nonsense suppressor mutation in a tRNA gene, the expression both of the gene with the nonsense mutation and of the distal genes in the operon can be restored. However, in strains with both a polar mutation and a loss-of-function mutation in the gene encoding the Rho transcription termination factor, expression of the distal genes can be restored but that of the gene with the nonsense mutation cannot. How might these results influence your model for the underlying cause of polarity?

Section 17.4

29. Many genes whose expression is turned on by DNA damage have been isolated. Loss of function mutations in the *lexA* gene leads to expression of many of these genes, even when there has been no DNA damage. Would you hypothesize that LexA protein is a positive or negative regulator? Why?

30. In 2005, Frederick Blattner and his colleagues found that *E. coli* have a global transcriptional program that helps them "forage" for better sources of carbon. Many genes, including genes needed for bacterial motility, are turned on in response to poorer carbon sources so that the bacteria can search for better nutrition. You now want to search for genes that regulate this response. How could you use *lacZ* fusions to try to identify such regulatory genes?

31. To find genes that are turned on or off in response to changes in osmolarity (the total concentration of solutes in solution), you grow a culture of *E. coli* in a medium with high osmolarity and another culture in a medium with low osmolarity. You now want to perform a DNA microarray analysis.

a. What would you use as your probe(s) for the DNA array analysis?
b. What nucleic acids would you spot on the DNA array? How many spots should the DNA array contain?
c. It is possible that osmotic changes may induce a general stress response that may be seen with other stresses as well (for example, heat-shock)? How could you distinguish the genes that might be involved in a general stress response from those that are specific for the osmolarity change?

32. Figure A on p. 642 shows the results of a recent microarray analysis measuring the relative abundance of mRNAs for all of the genes of bacteriophage T4 as a function of time after the infection of *E. coli* cells. The genes can be subdivided into three main classes: early genes (*blue*) transcribed almost immediately after

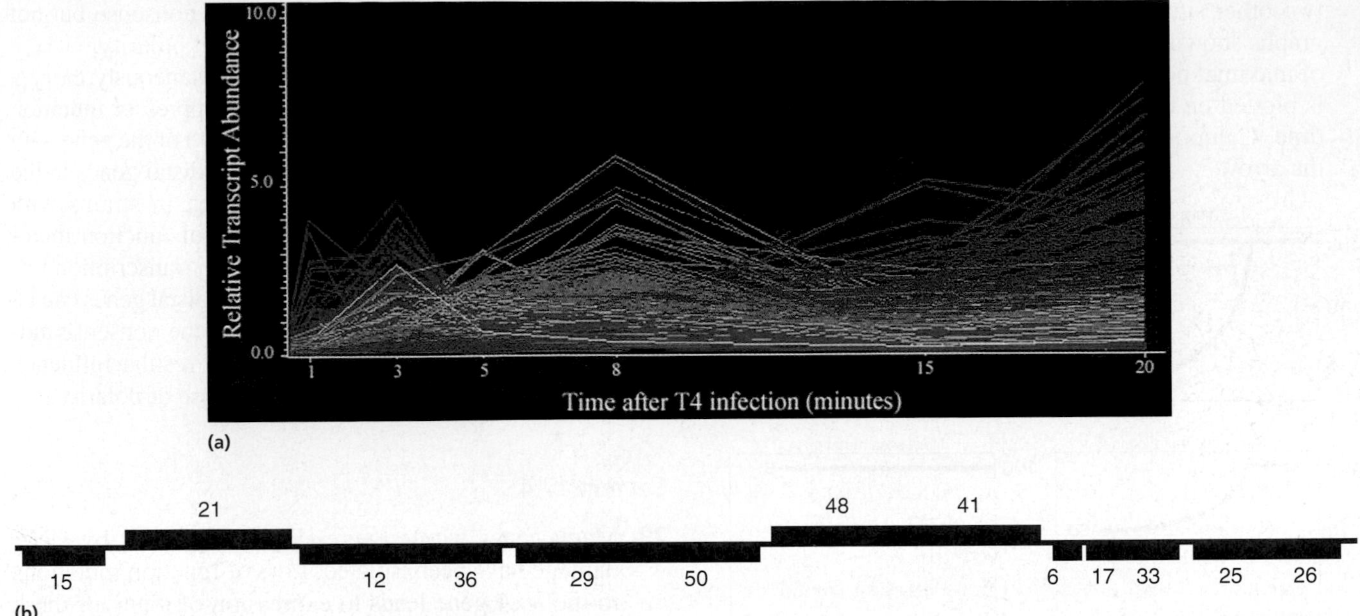

infection, middle genes (*green*) transcribed somewhat later, and late genes (*red*) transcribed later still. Figure B depicts a 10 kb region of the 170 kb T4 genome showing the extent of several genes, each indicated by a number and classified by the same color scheme. Colored boxes above the black line indicate genes transcribed from left to right, while those below the line indicate genes transcribed in the opposite direction.

a. What is the minimal number of promoters in the 10 kb region depicted in Figure B? Which genes could be transcribed as part of the same operon(s)?
b. The *e* gene of bacteriophage T4 encodes an enzyme called endolysin, which helps lyse the *E. coli* host cell to release progeny bacteriophage particles. Would you expect *e* to be an early, middle or late class gene? Explain your reasoning.

33. Several T4 genes participate in regulating the bacteriophage T4 life cycle described in the preceding problem (#32) and in Figures A and B below. The product of the *motA* gene is a protein that binds to DNA near the promoters for middle genes, enabling the *E. coli* RNA polymerase core enzyme to recognize these promoters. The gene *asiA* encodes an "anti-σ factor" that associates with *E. coli* σ^{70} and disrupts its function. The protein encoded by *regA* is a ribonuclease that specifically destroys early mRNAs. T4 gene *55*'s product is a σ factor required for recognition of late promoters by the *E. coli* RNA polymerase core enzyme.

a. Of the genes described above (*motA, asiA, regA,* and *55*), which are likely to be early, middle, or late?
b. What class of T4 genes (early, middle, late) has promoters recognized by the *E. coli* σ^{70} RNA polymerase holoenzyme?
c. What happens to the transcription of genes in the host *E. coli* chromosome as T4 infection progresses?
d. Predict the results of loss-of-function mutations in the *motA, asiA,* and *55* genes on the transcription of early, middle, and late mRNAs as well as the mRNAs for host *E. coli* genes.
e. What aspect of Figure A is explained by the function of the RegA ribonuclease?

Gene Regulation in Eukaryotes

Chapter 18

When a *Drosophila* male courts a *Drosophila* female, he sings a species-specific song and dances an ancient dance. If successful, his instinctive behaviors culminate in mating. The male senses a female's presence by visual and tactile cues as well as by the pheromones she produces. After orienting himself at a precise angle with respect to his prospective mate, he taps his partner's abdomen with his foreleg and then performs his song by stretching out his wings and vibrating them at a set frequency; when the song is over, he begins to follow the female. If she is unreceptive (perhaps because she has recently mated with another male), she will run away, but if she is receptive, she will let him overtake her. When he does, he licks her genitals with his proboscis, curls his abdomen, mounts the female, and copulates with her for about 20 minutes.

A Drosophila *male (right center) courting a female (on the left).*

Various mutations in a gene called *fruitless* produce behavioral changes that prevent the male from mating properly. Some mutant alleles alter the song to an unfamiliar air. Others diminish the male's ability to distinguish females from males; male flies with this mutation court each other. Still others reduce the male's ability to court either sex. Finally, some mutations create lethal null alleles that cause male flies to die just before they emerge from the pupal case, showing that the *fruitless* gene has other functions in addition to its effects on courtship and mating.

Cloned in 1996, *fruitless* is a large gene—roughly 150 kb in length—encoding a product that can regulate transcription at multiple promoters. Alternative RNA splicing of the gene's several different transcripts can produce many variations of the fruitless protein. All of these gene products contain zinc-finger peptide motifs that facilitate binding to DNA, suggesting a role for the protein in the regulation of transcription. Some forms of the fruitless protein are sex-specific, appearing only in males or only in females. By contrast, some fruitless proteins are expressed at low levels in many kinds of cells in both sexes.

The male version of the fruitless protein, although very similar to the female protein in amino acid sequence, has an extra 101 amino acids at its N terminus, and this addition almost certainly determines the observed differences in male and female behavior. Remarkably, the male-specific fruitless mRNA is synthesized in only a few hundred of the tens of thousands of neurons that make up the male *Drosophila*'s nervous system. Most of these *fruitless*-expressing cells are located near motor neurons that control either wing movements (and thus possibly the song) or abdominal movements (and thus possibly the abdominal curling that immediately precedes mating). Work on the *Drosophila fruitless* gene has provided strong evidence that differences in gene expression can directly influence complex behaviors.

In this chapter, we see that **eukaryotic gene regulation**—the control of gene expression in the cells of eukaryotes—depends on an array of interacting regulatory elements that turn genes on and off in the right places at the right times. Some of these elements are specific DNA sequences in the vicinity of the gene to be regulated; others are DNA-binding proteins encoded by genes located elsewhere in the genome; and others, scientists have recently discovered, are micro-RNA molecules (miRNAs) that use the specificity of base paring to down-regulate, or curtail, gene expression posttranscriptionally.

During the embryonic development of multicellular eukaryotic organisms such as *Drosophila* or humans, gene regulation controls not only the elaboration of sex-related characteristics and behaviors but also the differentiation of tissues and organs, as well as the precise positioning of these tissues and organs within the multicellular animal. For example, because of the tightly controlled differential regulation of genes in different types of cells, red blood cell precursors selectively turn on the genes for making hemoglobin, while pancreatic cells turn on the genes for making insulin, even though both types of cells, like nearly all somatic cells, carry the same number of chromosomes with the same amount of DNA and the same complete set of genes.

In exploring the intricacies of eukaryotic gene regulation, it is helpful to bear in mind key similarities and differences between eukaryotes and prokaryotes (**Table 18.1**). In both types of cells, transcription is regulated through the attachment of DNA-binding proteins to specific DNA sequences that are adjacent to or near the transcription unit itself; and in both eukaryotes and prokaryotes, the same polypeptide motifs appear in many DNA-binding proteins. Indeed many, though not all, of the basic principles of gene regulation in prokaryotes apply to eukaryotes, but additional levels of complexity tailor regulation in eukaryotes to their requirements. First, because eukaryotes have up to 700 times more DNA and 10 times more genes than prokaryotes, they require additional biochemical mechanisms to achieve specificity; the mechanisms they have evolved include the higher-affinity binding of proteins to DNA. Second, because eukaryotic DNA is packaged in chromatin, the transcriptional machinery must be able to control chromatin compaction and decompaction. The control of chromatin structure over defined distances along the chromosomes provides eukaryotic cells with a level of gene regulation that is unavailable to prokaryotic cells. Third, eukaryotes accomplish a type of regulation, through the differential splicing of exons, that is also unavailable to prokaryotes. Moreover, although the separation of transcription and translation into different cellular compartments in eukaryotes precludes certain forms of prokaryotic regulation, such as attenuation, it opens up the possibility of regulating gene function at different points in the expression process, from the initiation of transcription, through mRNA processing before transport from the nucleus, to protein synthesis and modification in the cytoplasm. Along this entire pathway, short miRNAs modulate gene expression through a process called *RNA interference*. In addition to these basic differences between

TABLE 18.1 Key Regulatory Differences Between Eukaryotes and Prokaryotes

Characteristic	Prokaryote	Eukaryote
Control of transcription through specific DNA-binding proteins	Yes	Yes
Reutilization of same DNA-binding motifs by different DNA-binding proteins	Yes	Yes
Activator proteins	Yes	Yes
Repressor proteins	Yes	Yes
Specificity of binding to DNA by regulatory protein	Specific	Highly specific
Affinity of binding	Strong	Very strong
Role played by chromatin structure	No	Yes
Coordinate control achieved with operons	Yes	Rare
Differential splicing	No	Yes
Attenuation	Yes	No
mRNA processing	No	Yes
Differential polyadenylation	No	Yes
Differential transport of RNA from nucleus to cytoplasm	No	Yes
RNA interference carried out by micro-RNAs	No	Yes

eukaryotes and prokaryotes, multicellular eukaryotes must be able to use gene regulation to control cellular differentiation and the complex interactions of various types of differentiated cells within tissues and organs.

One multifaceted overarching theme emerges from our discussion of eukaryotic gene regulation. Like complex behaviors, most biological functions arise from the regulated interactions of large networks of genes. And because each gene in a network has multiple potential points of regulation, the possibilities for regulatory refinement are enormous. Thus, eukaryotic gene regulation is a modular system in which different combinations of elements interact with each other at specified times and places, as well as in response to changes in the cellular environment, to allow the synthesis of precisely modulated amounts of gene products.

As we examine the components and mechanisms of eukaryotic gene regulation, we describe

- The use of genetics to study gene regulation, focusing on mutations that identify *cis*-acting control elements and *trans*-acting proteins.
- Gene regulation at the initiation of transcription, including a discussion of how three different RNA polymerases recognize three different classes of promoters; how *trans*-acting proteins help control class II promoters, causing the activation or repression of gene expression; how chromatin structure affects expression; how signal transduction systems work; and how DNA methylation, an epigenetic process, can regulate gene expression through silencing and genomic imprinting.
- Regulation after transcription, including a discussion of the control of RNA splicing, RNA interference, and translation, as well as the control of protein activity by posttranslational modification.
- A comprehensive example of gene regulation during sex determination in *Drosophila.*

18.1 The Use of Genetics to Study Gene Regulation

Until recently, molecular biologists assumed that eukaryotic gene expression was regulated predominantly at the point of transcriptional initiation. With the discovery and elucidation of RNA interference (discussed later in this chapter), scientists now understand that this view is incorrect. Nevertheless, it is still true that the critical decisions controlling the amount of gene product synthesized are made during the initiation of transcription, when RNA polymerase starts to make a primary transcript, or RNA copy, of a gene's template strand (review Fig. 8.11 on pp. 266–267).

The Analysis of Regulatory Components Focuses on Mutations That Affect a Gene's Function but Do Not Affect the Amino Acids in the Gene's Product

Geneticists can study regulation by choosing a gene whose control they want to study (the "target gene") and then looking for mutations that affect the expression of that gene. To find such mutations, they observe changes in a phenotype that is extremely sensitive to the amount of gene product synthesized.

Regulatory Mutations That Map at or Near the Target Gene Help Define *cis*-Acting DNA Sequences That Influence Transcription

***cis*-acting elements** at a gene are DNA sequences that serve as attachment sites for the DNA-binding proteins (and perhaps micro-RNAs) that regulate the initiation of transcription. The *promoter* is a *cis*-acting element that is very close to a gene's initiation site; its binding of RNA polymerase allows a basal level of transcription (**Fig. 18.1a**). *Enhancers* are another class of *cis*-acting elements; their binding of proteins augments or represses basal levels of transcription. Enhancers are defined by their ability to retain original function even when moved far from the gene whose transcription they influence or when reversed in orientation.

Reporter constructs are one tool for identifying and characterizing *cis*-acting regulatory elements. They are DNA molecules synthesized in the laboratory to contain a gene's postulated regulatory region but with a "reporter"

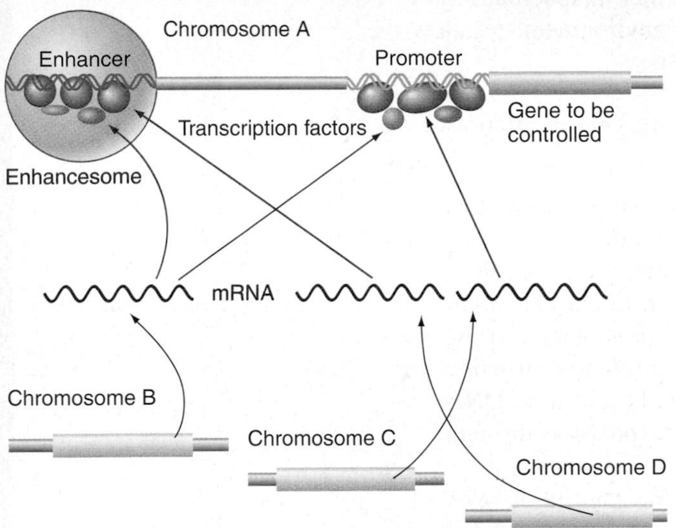

Figure 18.1 How *cis*-acting and *trans*-acting elements influence transcription. (a) *cis*-acting regulatory elements are regions of DNA sequence that lie nearby on the same DNA molecule as the gene they control. Promoter elements typically lie directly adjacent to the gene that they control. Enhancers that regulate expression can sometimes lie thousands of base pairs away from a gene. **(b)** *trans*-acting genetic elements encode products called transcription factors that interact with *cis*-acting elements, either directly through DNA binding or indirectly through protein-protein interactions. A stable multimeric transcription factor–enhancer complex is called an enhancesome.

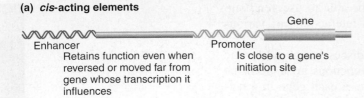

coding region inserted in place of the gene's own coding region. Investigators can systematically identify promoters and enhancers by altering reporter constructs through *in vitro* mutagenesis across a presumed regulatory region and then reintroducing the reporter constructs into the genome by transformation. Cells transformed with the reporter construct "report" the presence or absence of regulatory elements. In assembling a reporter construct for this purpose, scientists replace the coding region of the gene whose regulation they are studying with the coding region of an easily identifiable product (the "reporter"), such as β-galactosidase or green fluorescent protein (GFP). (Recall that β-galactosidase turns blue in the presence of a substrate known as X-Gal [review Fig. 9.8 on p. 313]; similarly, GFP luminesces green when exposed to light of a particular wavelength.) Reporter constructs are particularly valuable for looking at mutations that affect gene expression rather than the amino acid composition of the gene's polypeptide product. Mutations that alter the amount of reporter synthesized help define the elements necessary for a gene's regulation.

Regulatory Mutations That Map Far from a Target Gene Help Define *trans*-Acting Genetic Elements

***trans*-acting elements** are genes located elsewhere in the genome, that is, somewhere other than at or near the target locus; they encode proteins or micro-RNAs that interact directly or indirectly with the target gene's *cis*-acting elements to activate or repress expression of the target gene (Fig. 18.1b). *trans*-acting genes operate through their gene products, known as *trans-acting factors.* The *trans*-acting proteins and RNAs that influence transcription are known generically as **transcription factors.**

Researchers can identify *trans*-acting elements in many ways. Mutations that alter phenotypes dependent on either the level of target gene expression or the level of expression of a reporter construct previously introduced into the genome and that map far from the target gene or reporter construct, are likely to reside in *trans*-acting elements. Biochemical procedures are useful for isolating proteins that bind *in vitro* to *cis*-acting DNA sequences. Once researchers identify a *trans*-acting element, they can clone it for further study.

We now examine how *trans*-acting proteins interact with *cis*-acting regulatory elements to initiate, increase, diminish, or prevent gene expression.

18.2 Gene Regulation Begins with Control Over the Initiation of Transcription

During transcription in eukaryotes, RNA polymerase catalyzes the synthesis of a single-stranded RNA molecule—known as the primary transcript—that is complementary in base sequence to a gene's DNA template strand (see Fig. 8.11a on pp. 266–267). Many *cis*-acting DNA elements and *trans*-acting regulatory proteins help initiate or prevent this first step of gene expression.

In Eukaryotes, Three RNA Polymerases Transcribe Different Sets of Genes

One of the first indications that events at the initiation of transcription play a primary role in gene regulation was the finding that different classes of genes in eukaryotic cells

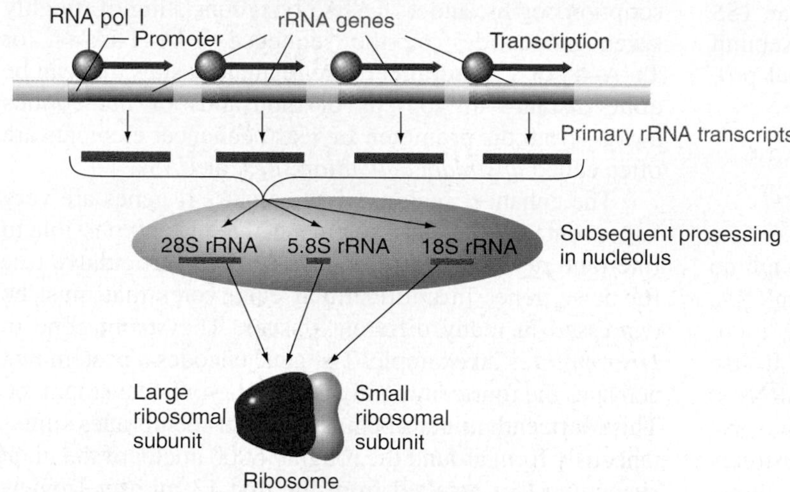

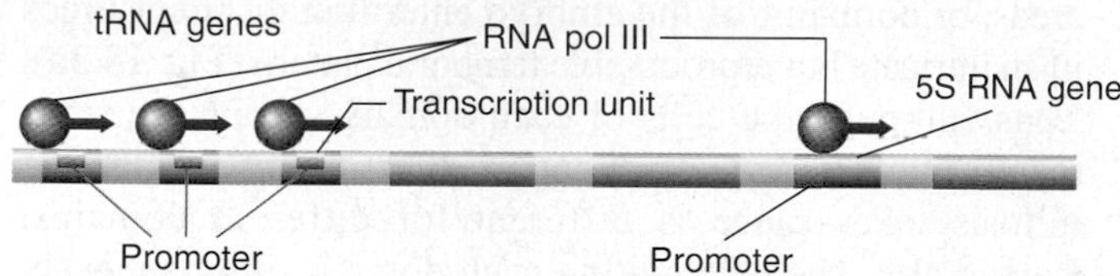

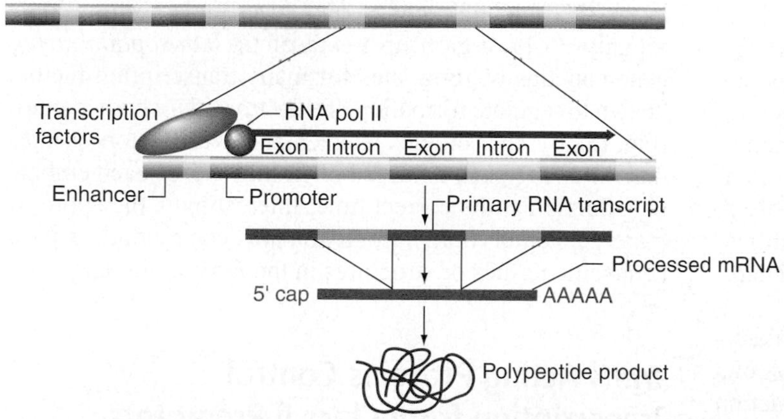

Figure 18.2 The three RNA polymerases of eukaryotic cells have different functions and recognize different promoters. (a) RNA polymerase I recognizes promoters associated with ribosomal RNA (rRNA) genes, which always exist in tandem arrays. The primary rRNA transcripts are processed into smaller RNA molecules that form components of the ribosome. **(b)** RNA polymerase III recognizes promoters associated with a limited number of genes that encode small RNA molecules involved in the translation process. **(c)** RNA polymerase II recognizes promoters associated with all of the diverse protein- and micro-RNA-encoding genes in the genome.

are transcribed by different RNA polymerases. The three types of RNA polymerases have some subunits in common, but they contain other subunits that are distinct to each type. Each of the three RNA polymerases transcribes a nonoverlapping class of genes (**Fig. 18.2**). RNA polymerase I (pol I) transcribes class I genes, which encode the major RNA components of ribosomes (rRNAs). RNA polymerase II (pol II) transcribes class II genes, by far the largest class; they encode all proteins and micro-RNAs. RNA polymerase III (pol III) transcribes class III genes, which encode the tRNAs as well as certain other, small RNA molecules.

RNA Polymerase I Is Responsible for the Transcription of rRNA

In most, if not all, eukaryotic genomes, the genes for rRNA are arranged in tandem repeats at sites on one or more chromosomes. These sites loop out from their respective chromosomes and come together in the nucleolus (as described under Chapter 13 on our website: www.mhhe.com/hartwell3). Pol I transcribes rRNA from these clusters of class I genes in the nucleolus. Each tandem repeat has a promoter region just upstream of the transcription unit (Fig. 18.2a). The transcription of rRNA by pol I produces a large rRNA primary transcript.

Subsequent RNA processing in the nucleolus breaks that transcript into a 28S rRNA and a 5.8S rRNA that become incorporated into the large ribosomal subunit and an 18S rRNA that becomes incorporated into the small subunit (Fig. 18.2a). These rRNAs associate with ribosomal proteins during assembly of the ribosome.

RNA Polymerase III Transcribes the tRNAs and Other Small RNAs

In addition to transcribing the tRNAs, which bring amino acids to the ribosome in response to codons in the mRNA, pol III transcribes a 5S rRNA that is yet another component of the large ribosomal subunit (Fig. 18.2b). It also transcribes some of the small nuclear RNAs (snRNAs) found in spliceosomes. As with class I genes, the promoter elements in some class III genes are just upstream of the gene's transcribed sequence. Surprisingly, however, studies of the transcriptional effects of small deletions in tRNA genes showed that the pol III promoter for these genes actually lies within the transcription units themselves. These sequences thus serve two functions: They are promoter elements recognized by specific transcription factors; and, after the initiation of transcription, they serve as part of the tRNA gene that is transcribed into the primary transcript.

RNA Polymerase II Transcribes All Protein-Encoding Genes

Most of the primary transcripts produced by RNA polymerase II (pol II) undergo further processing to generate mRNAs (Fig. 18.2c and review pp. 269–275 of Chapter 8). During mRNA formation, introns are spliced out. In addition, ribonuclease cleaves pol II–transcribed primary transcripts to form a new 3′ end, to which the enzyme poly-A polymerase adds a poly-A tail; and the chemical modification of the 5′ end of the transcript produces a "5′GTP cap," which protects the molecule from degradation.

Measurements of mRNA and protein levels in eukaryotic cells have revealed that, usually, the more mRNA of a gene that accumulates in the cell, the greater the production of that gene's protein product. Further biochemical experiments showed that the control of mRNA production almost always reflects the rate at which transcription is initiated, rather than the rate of elongation or termination of transcription, or the stability of the mRNA.

cis-Acting Regulatory Regions Recognized by Pol II Consist of a Promoter and One or More Enhancers

Although each of the regulatory regions of the thousands of class II genes in a eukaryotic genome is unique, they all contain two kinds of essential DNA sequences (Fig. 18.2c). The **promoter** is always very close to the gene's protein-coding region. It includes an initiation site, where transcription begins, and a "TATA" box, consisting of roughly seven nucleotides of the sequence T–A–T–A–(A or T)–A–(A or T). **Enhancers** are regulatory sites that can be quite distant—up to tens of thousands of nucleotides away—from the promoter. In yeast, enhancer elements are often called *upstream activation sites,* or *UAS*s.

The enhancer regions of some class II genes are very large, containing multiple elements that make it possible to fine-tune regulation of the gene. This is particularly true for those genes in multicellular eukaryotes that must be expressed in many different tissues. The *string* gene in *Drosophila* is an example. The gene encodes a protein that activates the fourteenth mitosis of embryonic development. This fourteenth mitosis begins just after membranes simultaneously form around the roughly 6000 nuclei of the giant syncytium that resulted from the first 13 mitoses (review the Fast Forward box in Chapter 4 on p. 95). What is interesting about the fourteenth mitosis is that cells in different areas, or domains, of the embryo enter it at different times in an intricate but reproducible temporal pattern (**Fig. 18.3a**). Thus, although the cells of each domain sustain the fourteenth nuclear division together, the time at which that mitosis takes place is different for different domains. Remarkably, the cells within each domain simultaneously express the *string* gene just before they enter mitosis; in fact, expression of the *string* gene induces their entry into the mitotic cycle. In *string* mutants, all embryonic cells arrest in the G2 stage of cycle 14 and never undergo mitosis. A roughly 35 kb region upstream of the *Drosophila string* gene contains binding sites for many transcription factors known to regulate formation of the *Drosophila* body pattern (Fig. 18.3b). The complex interaction of these factors ensures that the *string* gene is turned on in the cells of each embryonic domain at the correct time. Interestingly, the approximately 25 embryonic mitotic domains correspond, in part, to specific tissues or structures in the *Drosophila* larva.

trans-Acting Proteins Control Transcription from Class II Promoters

The binding of *trans*-acting proteins (transcription factors) to a gene's promoter and enhancer (or enhancers) controls the initiation of transcription from each class II gene. Different types of proteins bind to each of the *cis*-acting regulatory regions: *Basal factors* bind to the promoter; *activators* and *repressors* bind to the enhancers.

Basal Factors Are Required to Maintain a Basal Level of Transcription

Basal factors assist the binding of RNA polymerase II to the promoter and the initiation of a low level of transcription

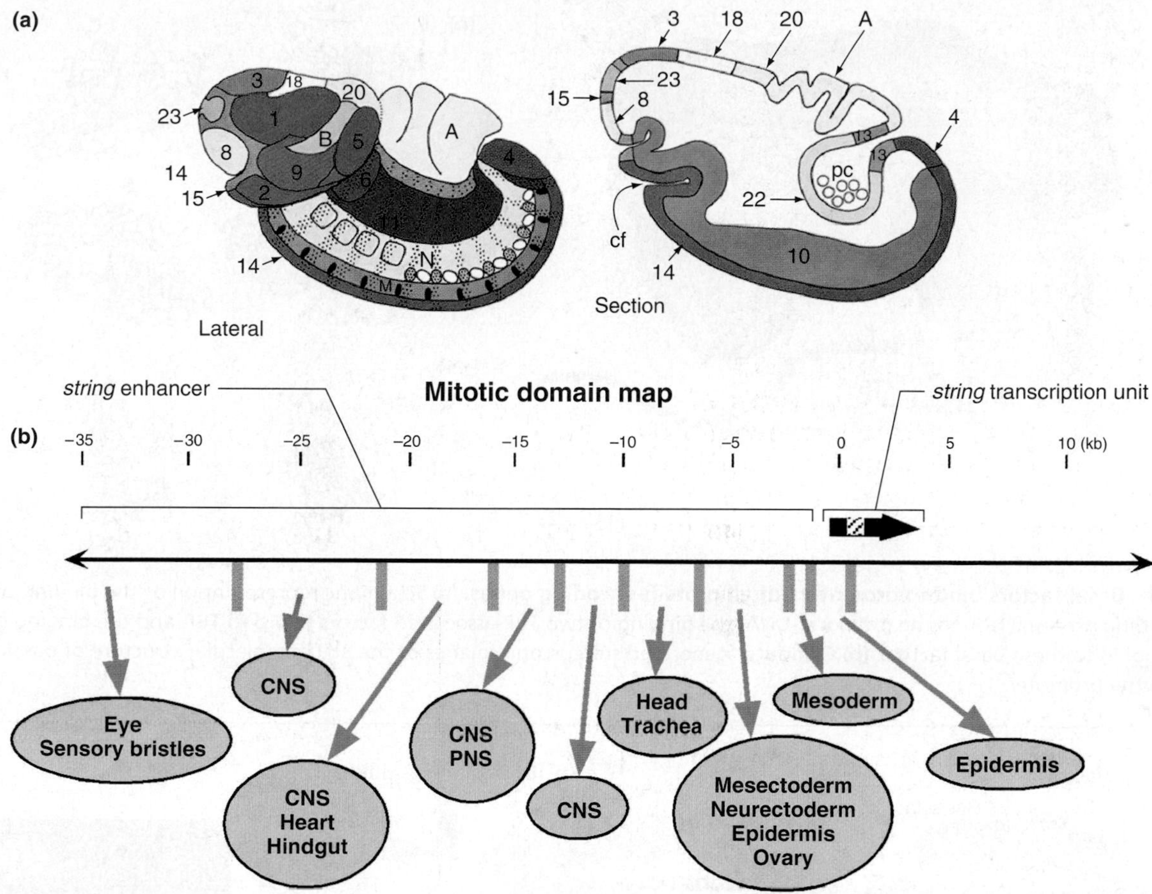

Figure 18.3 The large enhancer region of the *Drosophila string* gene helps create discrete mitotic domains during embryogenesis. (a) *Colors* indicate individual mitotic domains during the fourteenth cell cycle of the fruit fly embryo. The cells within each domain divide synchronously, but different domains initiate their divisions at different times. Both lateral and cross-sectional views of the embryo are shown. **(b)** A variety of enhancer regions ensure that *string* is turned on at the right time in each mitotic domain and tissue type (CNS: central nervous system; PNS: peripheral nervous system).

called basal transcription (from which the basal factors get their name). The key component of the basal factor complex that forms on most promoters is the TATA box–binding protein, or **TBP** (so named because it binds to the TATA box described previously). The TBP is essential to the initiation of transcription from all class II genes that have a TATA box in their promoter. TBP associates with several other basal factors called TBP-associated factors, or **TAFs** (**Fig. 18.4**). The complex of basal factors bind to the proximal promoter in an ordered pathway of assembly. Once the complex has formed, basal transcription is initiated. Researchers have determined the structure of the TBP-TAF complex on the DNA at the TATA box (Fig. 18.4). The most striking feature of the structure is a sharp bend in the DNA at the TATA box, induced by TBP.

The primary sequence and three-dimensional structure of the basal factors are highly conserved in all eukaryotes, from yeast to humans, and this sequence conservation has facilitated identification of the factors. For example, researchers isolated yeast TBP—the first basal factor to be purified—through its ability to substitute for mammalian TBP *in vitro*.

Activators Bind to Enhancer Sequences and Can Increase Transcription 100-Fold Above the Basal Level

Although similar sets of basal factors bind to all the promoters of the tens of thousands of genes in the eukaryotic genome, a cell can transcribe different genes into widely varying amounts of mRNA. This enormous range of transcriptional regulation occurs through the binding of different transcription factors to different enhancer elements associated with different genes. When the transcription factors bind to an enhancer element, they can interact directly or indirectly with basal factors at the promoter to cause an increase in transcriptional initiation (**Fig. 18.5a**). Because of this transcription-increasing activity, the factors are called **activators.** Researchers have already identified

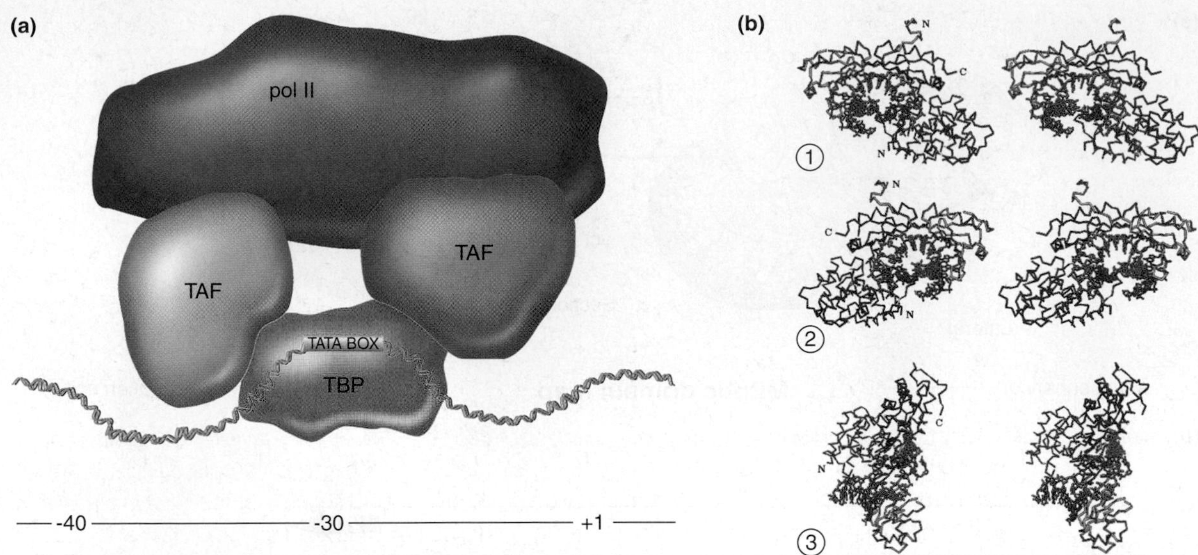

Figure 18.4 **Basal factors bind to promoters of all protein-encoding genes. (a)** Schematic representation of the binding of the TATA box–binding protein (TBP) to the promoter DNA, the binding of two TBP-associated factors (TAFs) to TBP, and the binding of RNA polymerase (pol II) to these basal factors. **(b)** Computer-generated stereoscopic images of the actual molecular structure of basal factors complexed to the promoter.

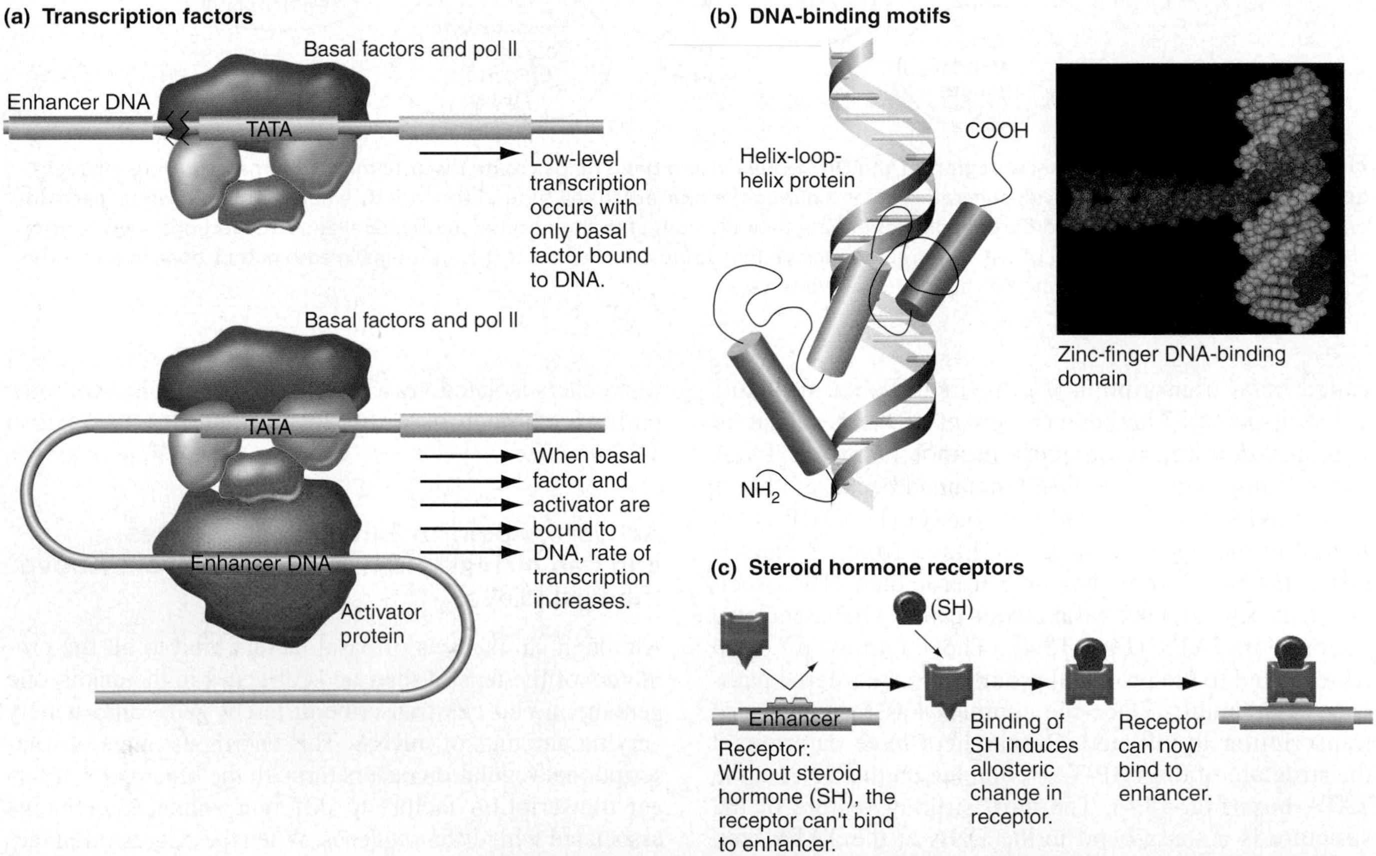

(a) Transcription factors

Basal factors and pol II

Enhancer DNA

TATA

Low-level transcription occurs with only basal factor bound to DNA.

Basal factors and pol II

TATA

Enhancer DNA

Activator protein

When basal factor and activator are bound to DNA, rate of transcription increases.

(b) DNA-binding motifs

Helix-loop-helix protein

COOH

NH_2

Zinc-finger DNA-binding domain

(c) Steroid hormone receptors

(SH)

Enhancer

Receptor: Without steroid hormone (SH), the receptor can't bind to enhancer.

Binding of SH induces allosteric change in receptor.

Receptor can now bind to enhancer.

Figure 18.5 **Transcriptional activators bind to specific enhancers at specific times to increase transcriptional levels. (a)** In the presence of basal factors alone bound to the promoter, low levels of transcription occur. The binding of activator proteins to an enhancer element leads to an increase in transcription beyond the basal level. **(b)** Examples of peptide motifs that appear in the DNA-binding domains of activator proteins and their interaction with enhancer elements. A zinc-finger domain extends into the groove of the double helix. A helix-loop-helix protein wraps around the DNA molecule. **(c)** Some activator domains are themselves activated into a DNA-binding conformation through allosteric changes caused by the binding of a steroid hormone molecule to another domain within the activator protein.

hundreds of eukaryotic activators, and it is likely that each eukaryotic genome encodes several thousand of them.

To carry out their function, activator proteins (the transcription factors) must bind to enhancer DNA in a sequence-specific way; and after binding, they must be able to interact with other proteins to activate transcription. Two structural domains within the activator protein—the DNA-binding domain and the transcription-activator domain—mediate these two biochemical functions.

A small number of peptide motifs appear over and over again in the DNA-binding domains of many different transcription factors (Fig. 18.5b). The best characterized of these motifs are the zinc-finger, the helix-loop-helix, and the helix-turn-helix conformations. The general function of each of these motifs is to promote binding to the DNA double helix. Subtle differences in amino acid sequence among activators can specify high-affinity binding to different DNA sequences associated with different enhancer elements. Extensive repetition of coding regions for the same peptide motif increases the genomic potential for evolving new activator proteins with new DNA-binding specificity. The duplication of an existing gene, followed by a slight divergence that by chance increases the binding affinity for a different DNA sequence, could generate a new protein that acts at a different target gene.

Many of the transcription-activation domains associated with different activators also have features in common. One subset is relatively acidic and rich in the amino acid asparagine; another is rich in prolines. Other transcription-activation domains are unique to the protein in which they occur.

Some activators have a third domain that is responsive to specific signals from the environment. An example of activators with this type of domain are the steroid hormone receptors (Fig. 18.5c). Each receptor has a domain that is unique for a particular steroid. The binding of this steroid causes an allosteric change that greatly increases the affinity of the DNA-binding domain of the protein for its target enhancer sequence. Once bound, the hormone-receptor complex activates transcription of its target genes. In the absence of hormone, DNA binding does not occur, and target genes remain unactivated, that is, transcribed only at basal rates. A steroid hormone gene regulation system allows one organ in the human body (a hormone-producing gland) to control gene activity in other organs. There are no universal features of signal response domains.

Polypeptides and other molecules that play a role in transcriptional activation without binding directly to DNA are called **coactivators.** The hormone component of a DNA-bound hormone-receptor activation complex is one example of a coactivator. The term **enhancesome** is used to describe a multimeric complex of proteins and other small molecules associated with an enhancer element (see Fig. 18.1b); the multimeric complex of proteins can include activators, coactivators, and other types of transcription factors known as *repressors* and *corepressors* (which we define shortly).

Molecular biologists have identified and characterized the different domains associated with different activators through the use of recombinant DNA constructs. These constructs encode proteins with a well-defined domain from a well-characterized transcription factor fused to a series of polypeptide regions from the activator under analysis (**Fig. 18.6**). For example, the bacterial repressor lexA, which contains only a DNA-binding domain and no activator domain, binds to the same well-defined DNA sequence whether it is in its native prokaryotic cell or in an engineered eukaryotic cell. One can fuse different segments of an activator protein gene to *lexA* and produce this fusion protein in an engineered yeast cell in which the lexA-binding site has been placed upstream of a special reporter gene. Activation (or lack of activation) of the reporter gene can then signal that the segment from the activator under analysis does (or does not) have a complete activation domain.

Most Eukaryotic Activators Must Form Dimers to Function

Molecular analyses have shown that many eukaryotic transcription factors are **homomers** (that is, multimeric proteins composed of identical subunits) or **heteromers** (multimeric proteins composed of nonidentical subunits; review Fig. 7.27 on p. 239). Among the best-characterized transcription factors of this type is Jun, which can form dimers (multimers composed of two subunits) with either itself or with another protein called Fos (**Fig. 18.7a**). The Jun-Jun dimers are *homodimers;* the Jun-Fos dimers are *heterodimers.*

Dimerization occurs through yet another transcription factor domain, the **dimerization domain,** which is specialized for specific polypeptide-to-polypeptide interactions. As with other transcription factor domains, certain motifs recur in dimerization domains. One of the most common is the leucine zipper motif (Fig. 18.7b), an amino acid sequence that twirls into an α helix with leucine residues protruding at regular intervals. The motif received its name from the propensity of one leucine zipper motif to interlock like a zipper with a leucine zipper motif on another polypeptide. The ability of two leucine zippers to interlock depends on the specific amino acids that lie between the leucines.

The Jun and Fos polypeptides both contain leucine zippers in their dimerization domains. A Jun leucine zipper can interact with another Jun leucine zipper or with a Fos leucine zipper. But the Fos leucine zipper *cannot* interact with its own kind to form a homodimer. Neither Jun nor Fos alone can bind DNA, so neither can act as a transcription factor as a monomer. Thus the Jun-Fos transcription factor system can produce only two types of transcription factors: Jun-Jun proteins or Jun-Fos proteins. Both bind to the same enhancer elements, but with different affinities.

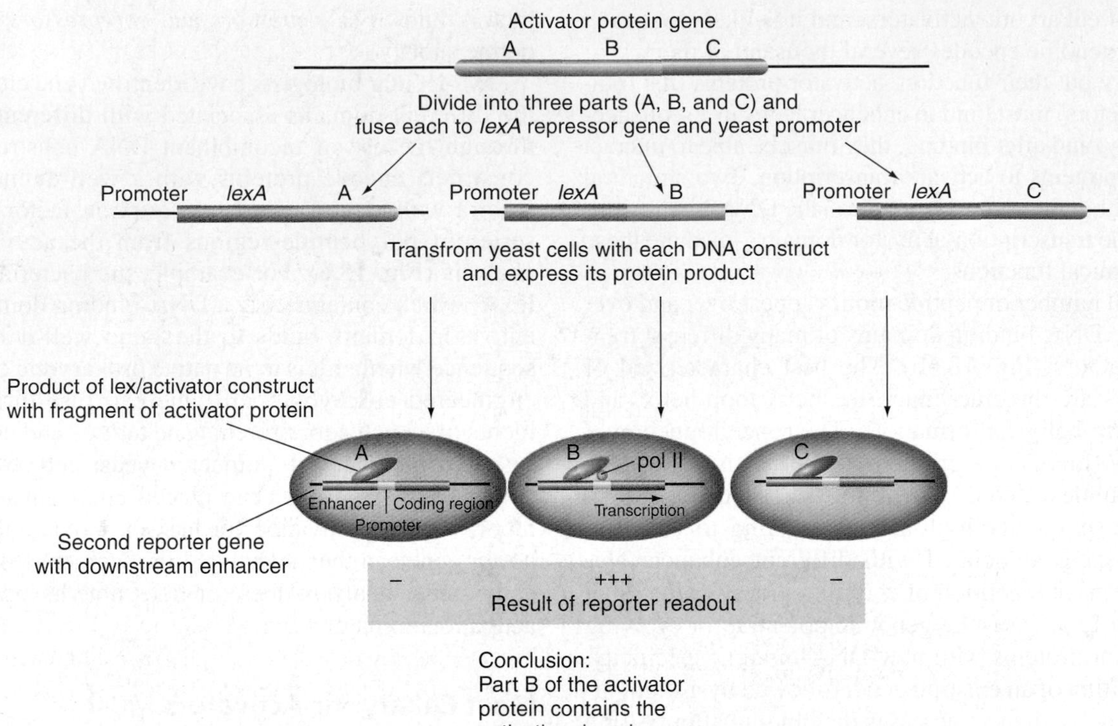

Figure 18.6 Localization of activator domains within activator proteins can be achieved with recombinant DNA constructs. The figure shows a strategy to determine which portion of an activator protein contains the actual domain that interacts with basal proteins to bring about activation. In this simplified example, the gene encoding the activator protein is divided into three parts, A, B, and C, and each part is fused to a portion of the bacterial *lexA* repressor gene that encodes a well-defined DNA-binding domain, and to a yeast cell promoter. Each three-part fusion construct is placed into different yeast cells, where the production of a fusion protein takes place. The yeast cell also contains a reporter gene with a downstream enhancer element that will bind to all three fusion proteins through their lexA-encoded binding domain. The reporter gene can only be transcribed if an activator domain is also present in the fusion construct. In this example, the results of the reporter gene readout demonstrate the presence of a complete activation domain in part B of the activator protein, but not in parts A or C.

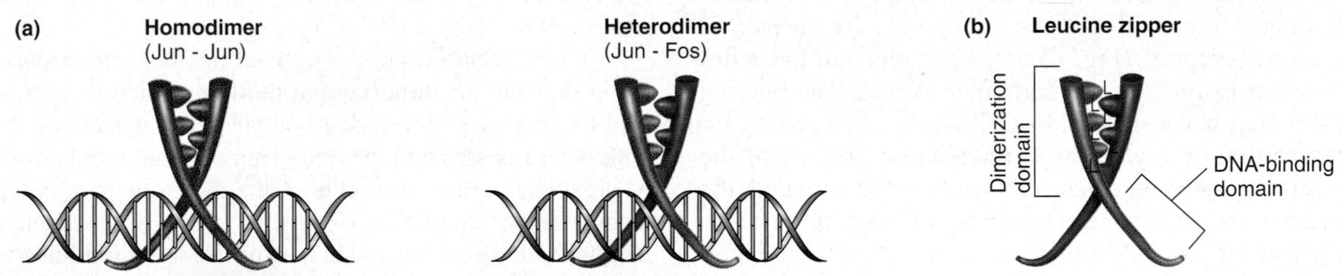

Figure 18.7 Most activator proteins function in the cell as dimers. (a) Homodimers contain two identical polypeptides, while heterodimers contain two different polypeptides. **(b)** The polypeptides within a dimer bind to each other through a dimerization domain. A common peptide motif present within dimerization domains is the leucine zipper.

The ability to form heterodimers greatly increases the number of potential transcription factors a cell can assemble from a set number of gene products. In theory, 100 polypeptides could combine in different ways to form 5000 different transcription factors; with 500 polypeptides, the number jumps to 125,000.

Repressors Diminish Transcriptional Activity

Some transcription factors suppress the activation of transcription caused by activator proteins. Any transcription factor that has this effect is considered a **repressor.** Different repressors act in different ways. Some compete with activator proteins for binding to the same enhancer (**Fig. 18.8a**). When a repressor binds to an enhancer, it blocks the activator's access to the same sequence. The Myc-Max system described in the following section provides an example of this type of activator-repressor competition.

Some repressors operate without binding DNA at all. Instead, in a mechanism called *quenching,* they bind directly to a specific activator (Fig. 18.8b). In one type of quenching, a repressor binds to and blocks the DNA-binding region of an

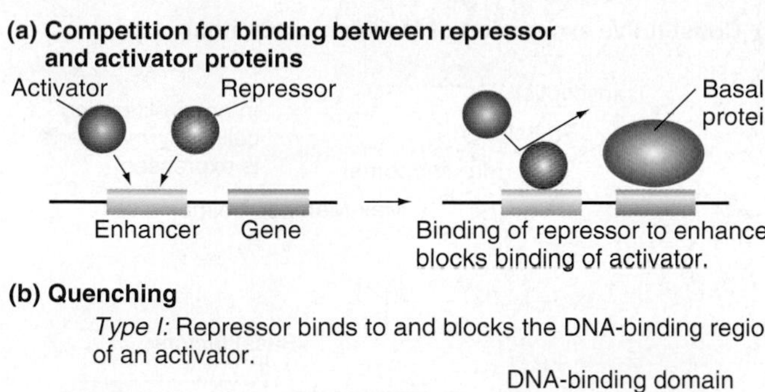
(a) Competition for binding between repressor and activator proteins
Activator
Repressor
Basal protein
Enhancer
Gene
Binding of repressor to enhancer blocks binding of activator.
(b) Quenching
Type I: Repressor binds to and blocks the DNA-binding region of an activator.
DNA-binding domain is blocked. Activator cannot bind to enhancer.
Activator
Repressor
Basal protein
DNA-binding domain

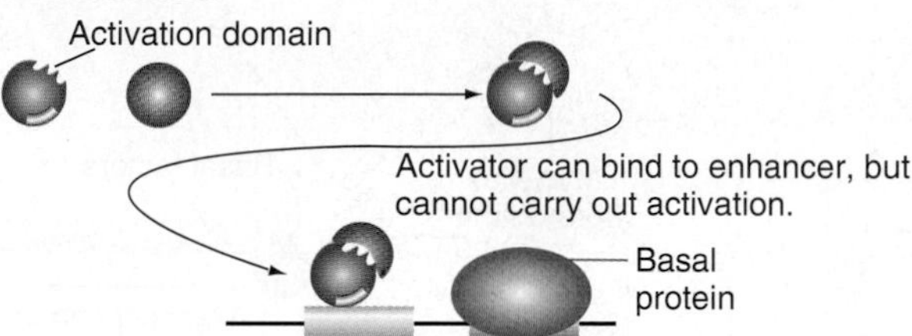
Type II: Repressor binds to and blocks the activation domain of an activator.
Activation domain
Activator can bind to enhancer, but cannot carry out activation.
Basal protein

Figure 18.8 Repressor proteins reduce transcriptional levels through competition or quenching. **(a)** Some repressor proteins act by *competing* for the same enhancer elements as activator proteins. But repressor proteins have no activation domain, so when they bind to enhancers, no activation of transcription can occur. **(b)** A second class of repressors act by binding directly to the activator proteins themselves to *quench* activation in one of two ways. Type I quenching is achieved when the repressor prevents the activator from reaching the enhancer. Type II quenching is achieved when the activator can bind to the enhancer, but the repressor prevents the activation domain from binding to basal proteins.

activator, thereby preventing the activator from attaching to its enhancer. In another type of quenching, a repressor binds to and blocks the activation domain of an activator. These blocked activators still bind to their enhancers, but once bound, they are unable to carry out activation. Quenching polypeptides that operate in this manner are **corepressors.** Just like coactivators, corepressors associate indirectly with enhancers through their interaction with DNA-binding proteins. An example of this type of repression is seen in the yeast GAL system, which we discuss later.

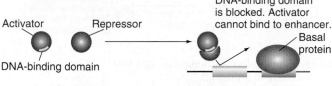

The repression resulting from both activator-repressor competition and quenching reduces activation, but it has no effect on basal transcription. As in prokaryotes, however, some eukaryotic repressors act directly on the promoter to eliminate almost all transcriptional activity. They can do this by binding to DNA sequences very close to the promoter and thereby blocking RNA polymerase's access to the promoter. Or they can bind to DNA sequences farther from the promoter and then reach over and contact the basal factor complex at the promoter, causing the DNA between the enhancer and promoter to loop out and allow contact between the repressor and the basal factor complex. This second mechanism also denies RNA polymerase access to the promoter and reduces transcription below the basal level.

Whether a transcription factor acts as an activator or a repressor, or has no effect at all, depends not only on the cell type in which it is expressed but also on the gene it is regulating. This is one reason why all *cis*-acting elements that bind either activators or repressors are referred to as enhancers, even though some may actually repress transcription when associated with the appropriate protein.

The specificity of transcription factors can be altered by other molecules in the cell. One example of this phenomenon is observed with the yeast α2 repressor, which helps determine the mating type of a cell. Yeast cells can be either haploid or diploid, and haploid cells come in two mating types: α and a. In α cells, the α2 repressor binds to enhancers that control the activity of a set of a-determining genes, whose expression would make the cell type a. The binding of the α2 repressor to these a-determining genes is one step in the generation of α cells (**Fig. 18.9a**). In diploid yeast cells, however, the same α2 repressor plays a different role. In such cells, expression of the polypeptide known as a1 occurs; the binding of a1 to the α2 repressor alters the repressor's DNA-binding specificity such that α2 now binds to enhancers associated with a set of haploid-specific genes, repressing the expression of those genes (Fig. 18.9b). To summarize, in diploid cells, the α2 repressor maintains the diploid state by repressing haploid-specific genes.

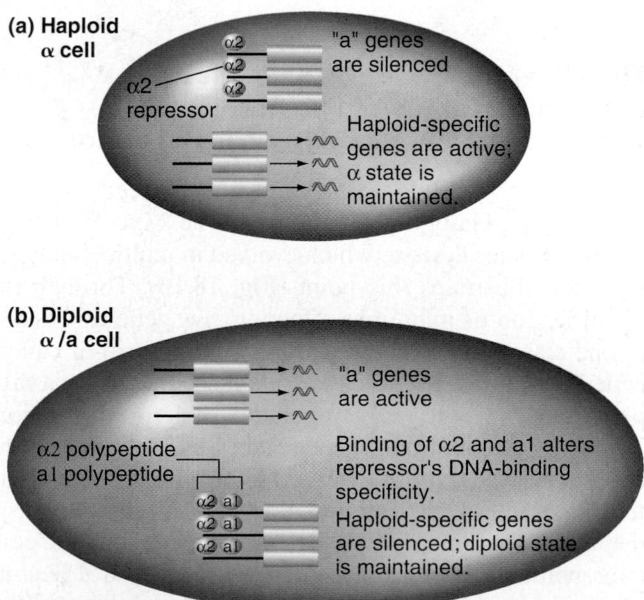
(a) Haploid α cell
α2
α2 repressor
"a" genes are silenced
Haploid-specific genes are active; α state is maintained.
(b) Diploid α /a cell
"a" genes are active
α2 polypeptide
a1 polypeptide
α2 a1
Binding of α2 and a1 alters repressor's DNA-binding specificity.
Haploid-specific genes are silenced; diploid state is maintained.

Figure 18.9 The same transcription factors can play different roles in different cells. **(a)** In haploid α yeast cells, the α2 factor acts to silence the set of "a" genes. **(b)** In α/a diploid yeast cells, the α2 factor dimerizes with the a1 factor and acts to silence the set of haploid-specific genes.

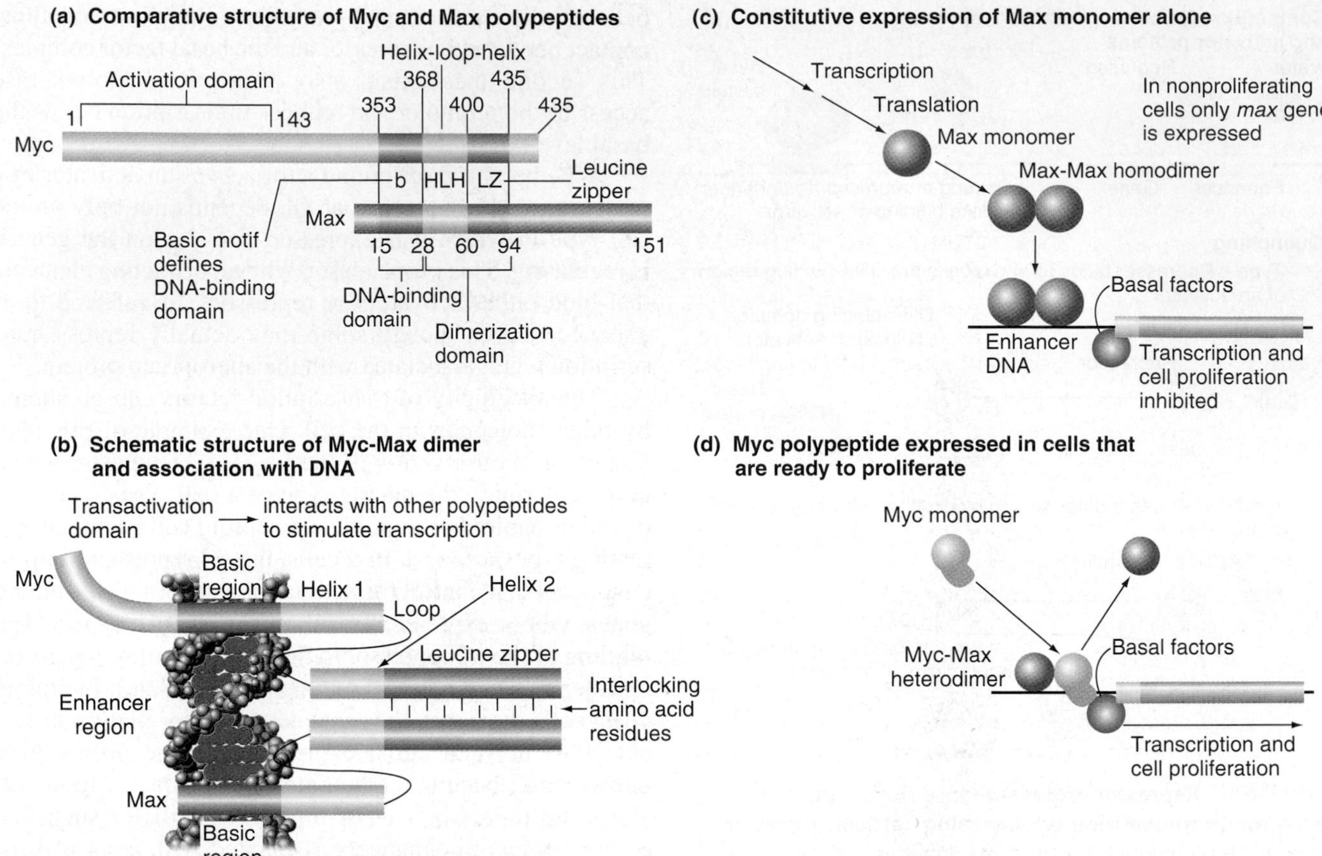

Figure 18.10 **The Myc-Max system of activation and repression.** **(a)** Linear illustrations of the Myc and Max polypeptides and the locations of different domains. The Myc polypeptide has an activation domain, while the Max polypeptide does not. Both polypeptides have a DNA-binding domain with a basic amino acid motif and a dimerization domain with adjacent helix-loop-helix and leucine-zipper motifs. **(b)** Schematic diagram of the Myc-Max heterodimer binding to DNA. **(c)** Gene repression results when a cell makes only the Max polypeptide. **(d)** Gene activation occurs when a cell makes both Myc and Max.

The Myc-Max System Is a Regulatory Mechanism for Switching Between the Activation and Repression of Transcription

The ratio of heterodimer partners within a cell can have profound regulatory consequences. The Myc-Max transcription factor system, which evolved in multicellular eukaryotes, illustrates this point (**Fig. 18.10**). Through the identification of mutations affecting *myc* gene expression in one class of lymphocytes (found in the form of cancer called Burkitt's lymphoma), researchers showed that *myc* plays a critical role in the regulation of cell proliferation. However, although the genetic data suggested that the Myc protein is a transcription factor, biochemists could find no evidence for this function *in vitro*. Their experiments revealed that even though the Myc polypeptide contains both a helix-loop-helix (HLH) motif and a leucine zipper, it cannot bind to DNA or form homodimers. The apparent contradiction between genetic and biochemical results stymied the scientists who first associated mutations in the *myc* gene with Burkitt's lymphoma and other forms of cancer.

The discovery of the *max* gene product helped resolve this dilemma. Like Myc, Max contains a HLH motif and a leucine zipper (Fig. 18.10a). Moreover, both Myc and Max contain another, more recently defined DNA-binding motif called a "basic motif" (because it contains mostly basic amino acids). Unlike Myc, however, the Max polypeptide can form homodimers. When one mixes Max with Myc, heterodimers of the two polypeptides form.

In the Myc-Max system, the HLH and leucine zipper are both components of the dimerization domain (Fig. 18.10a). The basic motif, by comparison, defines the DNA-binding domain, which functions only in dimeric proteins, never in monomers. This restriction of binding ability is a common feature of all transcription factor systems based on dimerization.

The Myc polypeptide contains an activation domain, but when the molecule is on its own, it cannot bind DNA and thus cannot serve as an activator. The Max polypeptide, on the other hand, can form homodimers and can bind DNA when present without Myc; but Max has no activation domain, so it cannot function as an activator even when it does bind to DNA. It is only when Myc and Max come together in a heterodimer that both DNA binding and Myc-directed activation become possible, and a transcriptional activator is born (Fig. 18.10b).

Myc-Max heterodimers and Max-Max homodimers both bind to the same enhancer sequences associated with multiple genes that contribute to cell proliferation. The binding of a heterodimer results in transcriptional activation, whereas the binding of a homodimer results in transcriptional repression.

One final characteristic of this system is that Myc polypeptides have a much higher affinity for Max polypeptides than Max has for its own polypeptides. Thus, when Myc and Max are in solution together, the predominant dimer is a heterodimer. With this extensive background, we are ready to see how the cell uses the Myc-Max system to respond rapidly to signals that tell it to proliferate or stop proliferating.

The *max* gene is expressed in all cells at all times, but since its protein product does not carry an activation domain, Max-Max homodimers, when bound to enhancer DNA, inhibit transcription and therefore inhibit cell proliferation (Fig. 18.10c). By contrast, the *myc* gene is not universally expressed; rather the Myc polypeptide is normally synthesized in cells undergoing proliferation but not in cells at rest.

As soon as a cell expresses its *myc* gene, all its Max-Max homodimers convert to Myc-Max heterodimers that bind to the enhancers previously bound by Max-Max homodimers. Since the heterodimers include the Myc activation domain, the binding of Myc-Max complexes induces the expression of genes required for cell proliferation. Although researchers have not yet characterized all the genes activated by the Myc-Max dimer, they know that the genes guide the cell through its mitotic cycle. Thus, each cell in which *myc* is active divides to produce two daughter cells.

Interestingly, when cells reach a state of terminal differentiation, they stop expressing *myc,* and since the Myc protein is unstable and easily degraded, any Myc remaining from earlier *myc* expression quickly disappears. The loss of Myc allows the remaining Max polypeptides to form homodimers, whose binding to the enhancer once again represses the expression of genes required for cell proliferation. When this happens, terminally differentiated cells stop dividing.

In short, the Myc-Max system provides a rapid genetic switch for regulating cell division during the cell proliferation and terminal differentiation phases of development. The system functions by keeping the *max* gene turned on in all cells, carefully regulating the expression of *myc,* allowing homodimerization of Max but not Myc, allowing Myc-Max heterodimerization to outcompete Max-Max homodimerization, and allowing the rapid degradation of Myc. As a result, the expression of *myc* quickly turns on genes required for cellular proliferation, and the cessation of *myc* expression just as quickly turns those same genes off. Mutations in the *myc* regulatory region can lead to constitutive expression of this critical gene, causing abnormal cellular proliferation and the potential for cancer.

The Yeast GAL System Is Another Complex Regulatory Mechanism

Like the Myc-Max system in multicellular eukaryotic organisms, the GAL system in unicellular yeast has been well characterized genetically and biochemically. The biochemical function of the GAL system is to modulate the expression of three linked genes—*GAL1, GAL7,* and *GAL10*—that encode the enzymes galactokinase, transferase, and epimerase (**Fig. 18.11a**). A cell must produce all three enzymes to use galactose as a source of energy. When galactose is present, the cell activates the genes encoding these enzymes. When galactose is absent, repression of the three genes allows the cell to avoid the unnecessary expenditure of energy and materials.

Since yeast cells always need the three enzymes at the same time, they economize by controlling the expression of all three structural genes from a single enhancer. Mutations in this enhancer can be detected by their effect on the regulation of all three gene products rather than just one.

Researchers identified the *GAL4* locus as a *trans*-acting genetic regulator of *GAL1, GAL7,* and *GAL10* through mutations that prevent galactose from activating expression of the three structural genes. Biochemical studies showed that the *GAL4* product has a DNA-binding domain containing multiple zinc-finger motifs and an acidic activation domain; it binds to an enhancer several hundreds of base pairs upstream of the *GAL1, GAL7, GAL10* genes and activates their transcription. Mutations in the *GAL4* coding region prevent its product from functioning as an activator.

Researchers identified another *trans*-acting genetic element—*GAL80*—through mutations that caused continuous expression of the three GAL enzymes, even in the absence of galactose. This genetic result suggested that the product of the *GAL80* gene is a repressor of *GAL1, GAL7,* and *GAL10.* When this repressor is absent, constitutive expression can occur. (Recall that a *constitutive mutation* is one that causes a specified gene, or group of genes, to be expressed at high levels irrespective of environmental changes that can affect transcription in the wild-type cell. *Constitutive expression* refers to an active state of gene expression that is unaffected by environmental conditions.)

On the basis of further genetic and biochemical studies, the researchers proposed that the GAL80 product binds to the activation domain of *GAL4* (Fig. 18.11b). GAL80 would thus be a *corepressor,* which does not bind directly to DNA; instead it quenches the activating ability of the GAL4 product by sequestering its activation domain. To test this model, they constructed double mutants in which they had knocked out both *GAL4* and *GAL80.* In these $GAL4^-/GAL80^-$ cells, no induction of GAL1, GAL7, and GAL10 occurred. This means that, in genetic

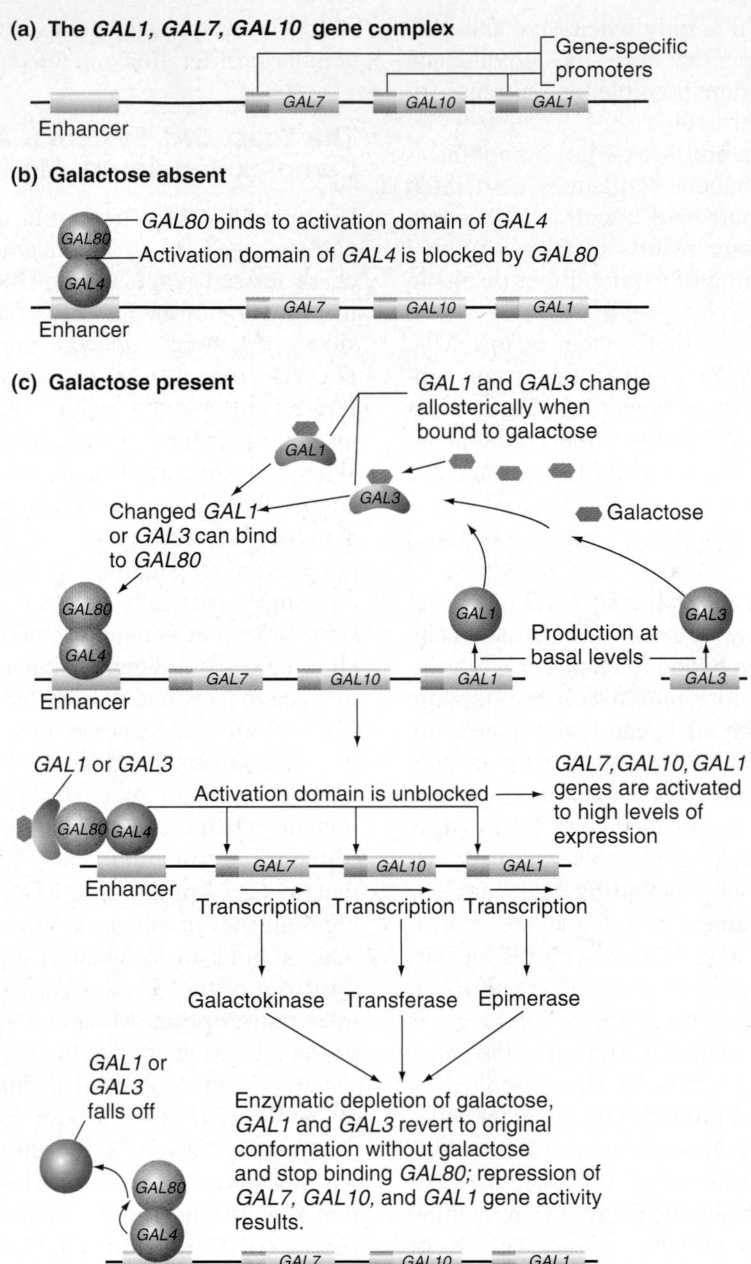

(a) The GAL1, GAL7, GAL10 gene complex
Gene-specific promoters
Enhancer
GAL7
GAL10
GAL1
(b) Galactose absent
GAL80 binds to activation domain of GAL4
Activation domain of GAL4 is blocked by GAL80
GAL80
GAL4
Enhancer
(c) Galactose present
GAL1 and GAL3 change allosterically when bound to galactose
GAL3
Galactose
Changed GAL1 or GAL3 can bind to GAL80
Production at basal levels
GAL1 or GAL3
Activation domain is unblocked
GAL7, GAL10, GAL1 genes are activated to high levels of expression
Transcription
Transcription
Transcription
Galactokinase
Transferase
Epimerase
GAL1 or GAL3 falls off
Enzymatic depletion of galactose, GAL1 and GAL3 revert to original conformation without galactose and stop binding GAL80; repression of GAL7, GAL10, and GAL1 gene activity results.

Figure 18.11 The *GAL1, GAL7, GAL10* gene system is a model of complex transcriptional regulation. (a) The three genes have independent promoters but are all controlled by the same enhancer element. **(b)** The activator protein GAL4 binds to this enhancer, but its activity is quenched by the repressor protein GAL80. **(c)** Galactose molecules in the cell bind to GAL1 and GAL3 polypeptides, which undergo allosteric changes that allow them to bind to GAL80, causing GAL80 to move into a position that is no longer quenching the activator GAL4. This allows activation of the *GAL1, GAL7,* and *GAL10* genes, whose products bring about the enzymatic digestion of galactose. Once galactose is depleted, GAL1 and GAL3 can no longer bind to GAL80, and it moves back into position to quench GAL4.

terms, the GAL4 phenotype is epistatic to the GAL80 phenotype. Such a result is consistent with the biochemical model, because to function as a repressor, the GAL80 product needs to operate on the GAL4 product. In the absence of *GAL4,* mutations in *GAL80* will have no effect on gene regulation.

How does a normal yeast cell use this genetic regulatory system to control the expression of the *GAL1, GAL7,* and *GAL10* structural genes in response to the presence or absence of galactose? The answer is by controlling the physical placement of the GAL80 gene product. When galactose is present in the cell, it attaches to both the GAL1 product (a very small amount of which is produced through basal transcription), as well as to another constitutively expressed polypeptide called GAL3. Association with galactose causes an allosteric change in GAL1 and GAL3 that gives both polypeptides a high-affinity binding site for the GAL80 product. When either polypeptide binds to the GAL80 product, this binding somehow dislodges GAL80 from the GAL4 activation domain. With its activation domain now available, GAL4 is able to activate the *GAL1, GAL7, GAL10* gene complex (Fig. 18.11c). Later, when the cell has depleted its supply of galactose, the GAL1 and GAL3 products again change shape, reverting to the conformation in which they do not have a high affinity for GAL80 and falling off. GAL80, when no longer bound to GAL1 or GAL3, moves back into position to block the GAL4 activation domain. The binding of GAL80 to the activation domain of GAL4 once again reduces transcription of the three *GAL* structural genes to basal levels.

A Locus Control Region Is a *cis*-Acting Regulatory Sequence That Operates Sequentially on a Cluster of Related Genes

Promoters and enhancers are not the only kinds of *cis*-acting regulatory regions active in eukaryotic genomes. Clues that another kind of *cis*-acting element exists came from experiments in which the promoter and adjacent enhancer region of the β-globin gene were not sufficient to mediate its full transcriptional activation.

As we saw in Chapter 9 (introduction and comprehensive example), the β-globin gene is part of a cluster of related genes that function at different times during mammalian development to produce polypeptides that become incorporated into hemoglobin molecules with different oxygen-binding kinetics (**Fig. 18.12a**). The β-globin gene was among the first mammalian genes to be cloned, characterized, and used in the production of transgenic mice. Results from the earliest transgenic studies were disappointing. Even when researchers introduced into transgenic mice the entire region of noncoding genomic DNA 5′ of the β-globin gene (which separates that gene from the flanking gene in the cluster), the level of β-globin transgene transcription in the mice was 10 times less than that observed with the endogenous gene.

Subsequent experiments showing that other transgenes with just the 5′-flanking sequences (containing the promoter and all of the enhancer elements) were fully expressed suggested that the β-globin gene is different. For full transcription, it requires a *cis*-acting element that was missing from the DNA containing both the gene and the flanking sequence that separates it from its nearest neighbor (Fig. 18.12b).

Surprisingly, investigators found that the missing element, located 50 kb upstream of the β-globin gene, is separated from the β-globin gene by four other genes in the cluster. This regulatory element is called the **locus control region (LCR)** because it exerts a higher level of control,

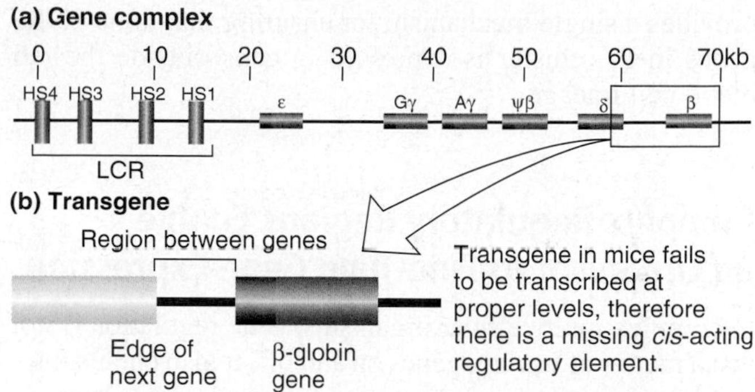

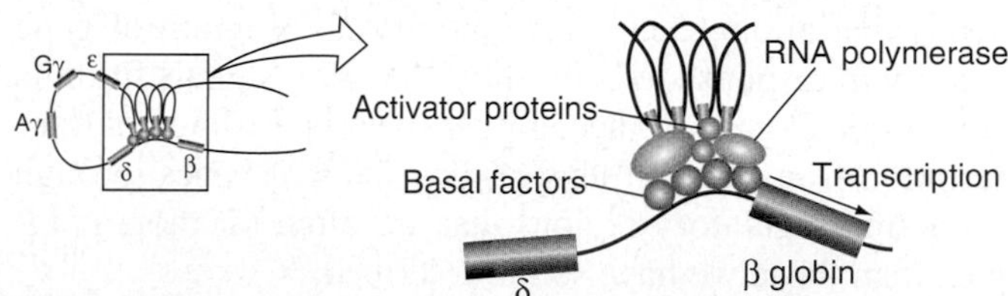

Figure 18.12 A locus control region (LCR) operates over a cluster of genes to activate transcription. (a) The human β-globin gene cluster contains five functional genes that can all undergo *cis*-regulation by a distant LCR. The LCR associated with this particular gene cluster contains four separate *cis*-acting DNA elements designated HS1 through HS4. **(b)** A transgene composed of an isolated β-globin gene together with the entire 5′ flanking region between it and the δ gene fails to undergo proper expression when it is inserted into the mouse genome. **(c)** The LCR at the 5′ end of the β-globin gene cluster is required for transcriptional activation of each of the genes in the cluster in hemoglobin-producing cells at different points in development. The four DNA elements of the LCR (HS1 through HS4) could effect activation of the distant δ-globin gene through DNA looping, which brings each of these elements and their associated transcription factors into contact with the β-globin promoter and its associated basal factors and RNA polymerase.

influencing the transcription of all genes in the β-globin cluster. Like an enhancer, the LCR functions by binding to transcription factors with activation domains. Unlike most enhancers, however, the LCR–transcription factor complex interacts sequentially during development with other transcription factors at *cis*-regulatory regions that are directly adjacent to each gene within the β-globin gene cluster (Fig. 18.12c). An assembled LCR–transcription factor complex—like an assembled enhancer-transcription factor complex—is known as an *enhancesome.* If any component of the enhancesome is missing, the program does not work, and no activation of any genes in the cluster takes place. Cells other than the precursors of red blood cells do not express the genes for β-globin-LCR-specific enhancesome components; as a result, there is no functional enhancesome and no expression of any of the β-globin genes in these cells. The LCR for the β-globin cluster thus provides a single mechanism for ensuring that none of the genes in the cluster is expressed in cells outside the red blood cell lineage.

Complex Regulatory Regions Enable an Organism to Fine-Tune Gene Expression

In complex multicellular organisms, gene regulation is not just a matter of turning genes on and off. It also entails fine-tuning the precise level of transcription—higher or lower in different cells, higher or lower in cells of the same tissue at different stages of development. It also includes mechanisms that allow each cell to modify its program of gene activity in response to constantly changing signals from its neighbors. Organisms accomplish the orchestration of transcription from each of tens of thousands of genes through *cis*-acting regulatory regions that are often far more complex than those we have so far described.

A regulatory region may contain a dozen or more enhancer elements, each with the ability to bind different activators and repressors, with varying affinities. At any moment, there may be dozens of transcription factors in the cell whose affinities for DNA or other polypeptides are being modulated by binding to hormones or other molecules. Different sets of these transcription factors compete for different enhancers within the regulatory region. And different sets of coactivators and corepressors compete with each other for binding to different activators or repressors. The biochemical integration of all this information yields a precise level of transcriptional activation or repression.

As we saw earlier, a complex of regulatory factors together with the enhancer to which they are bound is an enhancesome. Slight changes in a cell's environment can dispatch signal molecules that cause changes in the balance of transcription factors or in their relative affinities for DNA or for each other. These changes, in turn, lead to the assembly of an altered enhancesome, which recalibrates gene activity.

Molecular biologists and biochemists are not yet able to unravel the details of such complex regulatory networks, but their knowledge of them increases every day.

Chromatin Structure Plays a Role in Eukaryotic Gene Regulation

We know from Chapter 13 that the DNA of eukaryotic genomes does not float freely in the nucleus but is normally packaged into chromatin. The basic repeating structural unit of chromatin is the nucleosome, which consists of a ball of histone proteins (two each of H2A, H2B, H3, and H4) around which is wrapped approximately 200 bp of DNA. Histone H1, which binds to short stretches of DNA that lie between nucleosomes, helps maintain chromatin structure. The binding of nonhistone proteins is the basis of higher-order structures that further compact DNA (review Table 13.2 on p. 469 and Fig. 13.4 on p. 471).

The Normal Structure of Chromatin Provides a Brake on Runaway Basal Transcription

In vitro experiments show that basal factors and RNA polymerase readily bind to promoters on naked DNA and initiate high levels of transcription in the absence of activator proteins (**Fig. 18.13a**). One significant function of chromatin, then, is the reduction of basal transcription from all genes to a very low level. In contrast to transcriptional modulation in prokaryotes, which requires active repression through the binding of repressors to *cis*-acting elements, in eukaryotes, the normal structure of chromatin is sufficient in and of itself to maintain transcriptional activity at the minimal, basal level (Fig. 18.13b). Thus, eukaryotic repressors play a different role than their prokaryotic counterparts. The main role of eukaryotic repressors is to modulate the activation of transcription caused by transcriptional activators. They do this through competition for DNA-binding sites, as described earlier for the Myc-Max system, or through quenching, as described for the yeast GAL system.

The Remodeling of Chromatin Mediates the Activation of Transcription

An initial component of gene activation in eukaryotes is the remodeling of chromatin structure in the promoter region. Specialized proteins that unravel promoter DNA sequences away from the histone core are the agents of this remodeling. The freed DNA becomes much more accessible to transcription factors that then control transcription (Fig. 18.13c).

Chromosomal regions from which the nucleosomes have been eliminated are experimentally recognizable through their hypersensitivity to the enzyme DNase. When one scans a chromosome with the enzyme for the presence of DNase hypersensitive (DH) sites, the sites show up at the 5′ ends of genes that are either undergoing transcription or are being prepared for transcription in a later step of cellular differentiation (Fig. 18.13d). For example, DH sites appear at the 5′ end of the β-globin gene in human stem cells that are precursors to the hematopoietic cells in which the gene will be activated, but not in cells from other differentiative pathways.

Studies of chromatin structure show that nucleosomes sit atop the promoters of most inactive genes. Protein complexes that remove these promoter-blocking nucleosomes, or reposition them in relation to the gene, help prepare a gene for activation. For example, the SWI and SNF proteins in yeast form a multisubunit complex that disrupts chromatin structure by removing or repositioning nucleosomes. The resulting chromatin decompaction gives basal factors much greater access to promoter regions, and consequently,

(a) Naked promoter binds RNA polymerase and basal factors.

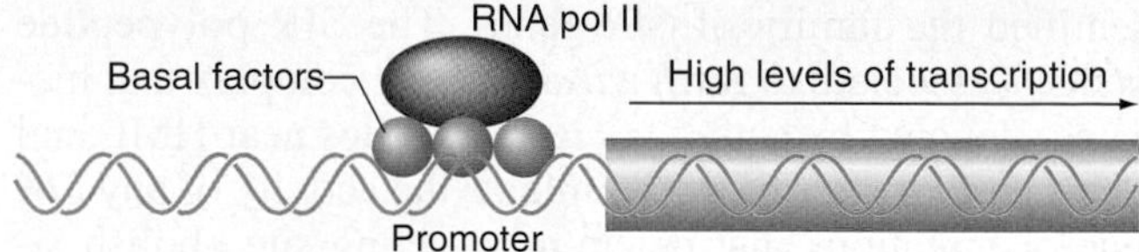

(b) Chromatin reduces binding to basal factors and RNA pol II to very low levels.

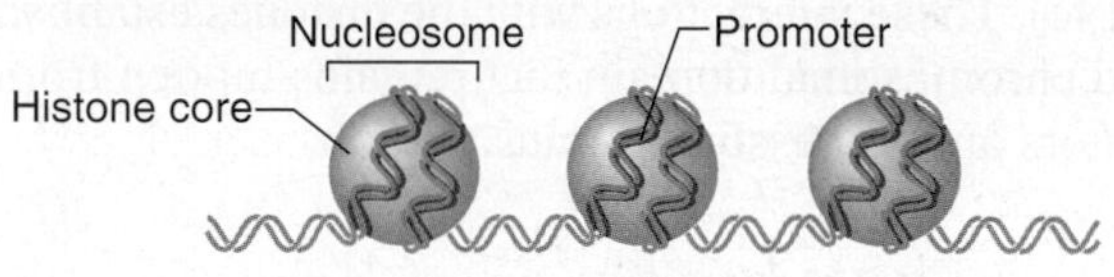

(c) Chromatin remodeling can expose promoter.

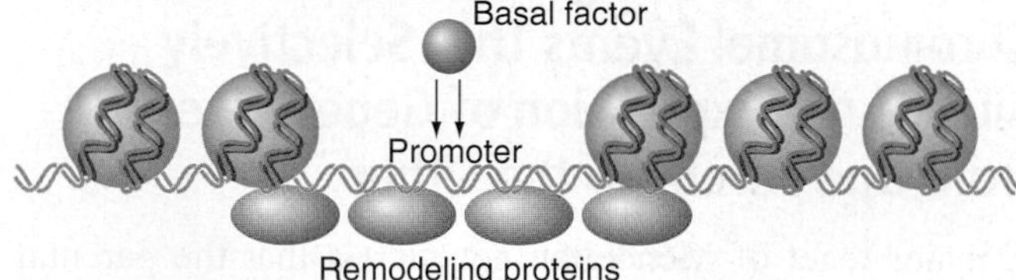

(d) DNase hypersensitive sites are at 5' ends of genes.

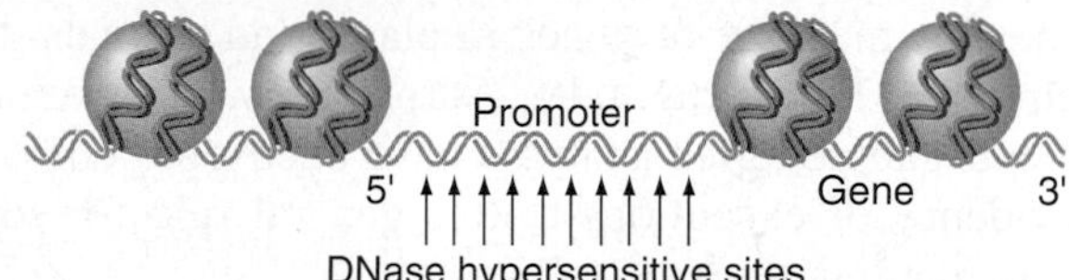

(e) The SWI-SNF multisubunit complex is an example of a remodeling complex.

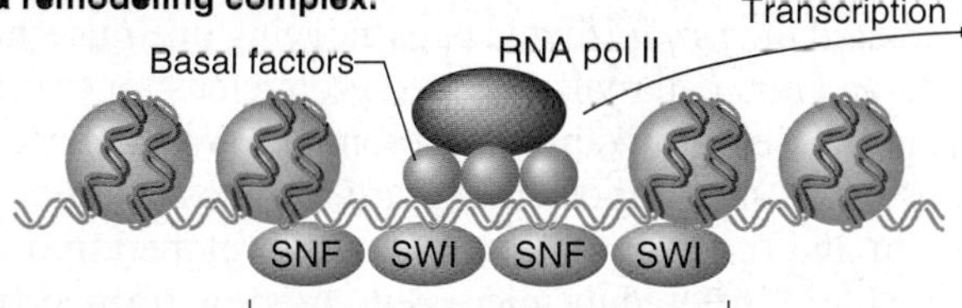

Figure 18.13 Chromatin structure plays a critical role in eukaryotic gene regulation. (a) DNA molecules containing a promoter and an associated gene can be purified away from chromatin proteins *in vitro.* The addition of basal factors and RNA polymerase to this purified DNA induces high levels of transcription. **(b)** Within the eukaryotic nucleus, DNA is present within chromatin. The underlying structure of chromatin is based on repeating nucleosomes that each contain two loops of DNA wrapped around a histone core. Promoter regions are generally sequestered within the nucleosome and only rarely bind to basal factors and RNA polymerase. Thus, the chromatin structure maintains basal transcription at very low levels. **(c)** Chromatin remodeling can expose the promoter region. Remodeling proteins cause specific nucleosomes to unravel in specific cells at specific times during differentiation or development. Exposed promoter regions more readily bind basal factors. **(d)** Exposed regions of DNA can be recognized by their hypersensitivity to DNase digestion. **(e)** The SWI-SNF protein complex is a well-characterized remodeling apparatus that functions within yeast cells to expose promoter regions to basal factors, RNA polymerase, and transcriptional activation.

transcription rapidly accelerates (Fig. 18.13e). Human cells contain related multisubunit protein complexes that also influence nucleosome position or structure, suggesting that this particular nucleosome-disrupting machinery has been conserved throughout evolution. The SWI-SNF protein complex represents just one of the many that help remodel chromatin at specific chromosomal locations in specific cells at particular points of development.

Hypercondensation over Chromatin Domains Causes Transcriptional Silencing

The heterochromatic regions of eukaryotic chromosomes, including parts of centromeres and telomeres and the whole of Barr bodies, are highly condensed. As a result, the genes contained in heterochromatin are completely silent without even basal transcription (review Fig. 13.13).

The **transcriptional silencing** of heterochromatin makes it impossible to activate genes within a heterochromatic region, no matter what transcription factors are active in the cell, as long as the silencing structure is in place. By contrast, as we have seen, substantial expression of genes in unsilenced regions depends on the outcome of competition between activators and repressors. The appropriate combination of signals can always activate these genes.

In mammals, the chromosomal regions that are silenced by heterochromatin vary with the cell type, but they are often associated with DNA methylation (**Fig. 18.14a**). Methylation—the addition of a methyl (CH_3) group—is specific for CG dinucleotides in the DNA and occurs at the fifth carbon of the cytosine base (see Fig. 6.7a on p. 174). It is possible to determine the state of methylation of a DNA region by using two restriction enzymes that both cleave at a sequence containing a CG dinucleotide but have different sensitivities to the methylation of the DNA substrate. For example, *Hpa*II and *Msp*I both cleave at CCGG, but *Hpa*II will not cleave if the middle C of this site is methylated, while *Msp*I will (Fig. 18.14b). Thus, by digesting genomic DNA with *Hpa*II and *Msp*I and using a specific DNA probe on a Southern blot, you can determine whether a given CCGG sequence is methylated. Although methylation is associated with silencing, we know it cannot be the only mechanism, because some organisms that show silencing, such as yeast, do not contain methylated DNA.

Deeper insight into the mechanism of silencing comes from studies of mutations that give rise to sterility in yeast. Recall that yeast cells come in two mating types—α and a—and the α2 gene product represses certain a-determining genes. The chromosomal locus of the α2 gene is known as MAT (for mating type). In normally mating yeast cells, there are two additional copies of the MAT locus called HML and HMR; located near the telomeres on each arm of chromosome III, these loci are transcriptionally silent. Mutations that reduce or destroy silencing at these loci cause sterility because they allow the simultaneous expression of α and a information, and the resulting cells, which

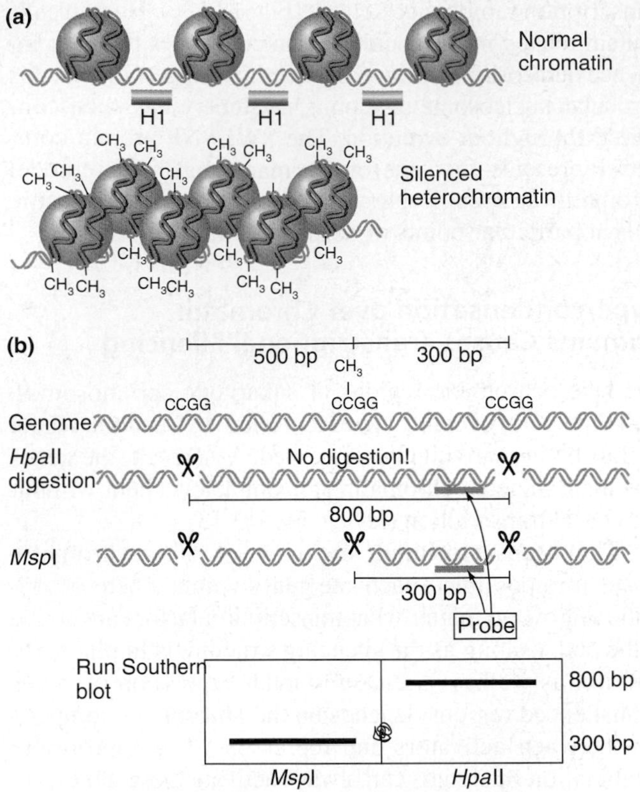

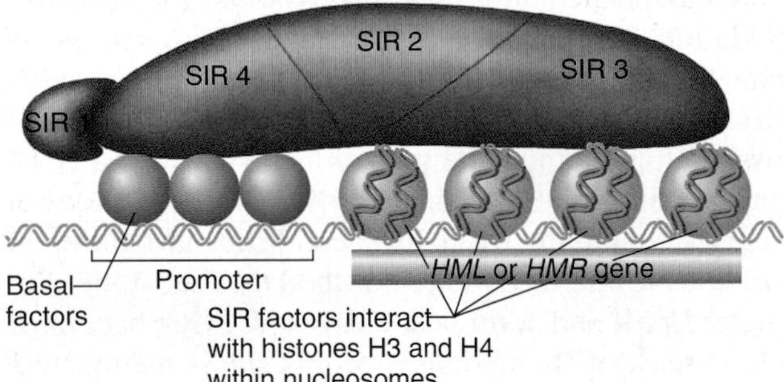

Figure 18.14 Heterochromatin formation can lead to transcriptional silencing. (a) Normal chromatin in standard nucleosome conformation can be converted into tightly packed heterochromatin with the addition of methyl groups to a series of cytosine bases within a local DNA region. **(b)** A determination of the methylation status of a DNA region can be made using a pair of restriction enzymes that both recognize the same base sequence, with one being able to digest methylated DNA, while the other can't. In this example, the restriction site is CCGG, and the enzymes are *Msp*I, which can digest both methylated and unmethylated sites, and *Hpa*II, which cannot digest methylated sites. If a methylated site is present between two unmethylated sites, *Hpa*II digestion will leave a larger fragment than *Msp*I. After electrophoresis and Southern blot analysis with a probe that hybridizes to a sequence on one side of the methylated site, there will be a clearly observable difference in band size. **(c)** The SIR complex of polypeptides can bind to basal factors associated with the promoters of the *HML* and *HMR* genes. This binding, in turn, causes the SIR complex to interact with the histones H3 and H4 present in downstream nucleosomes associated with the gene itself. The result is the complete silencing of transcription.

behave as diploids, do not mate. Analysis of these mutations identified the family of *SIR* genes. The SIR polypeptide products associate to form a *trans*-acting complex that mediates silencing by acting at *cis*-acting sites near HML and HMR. Null mutations that eliminate the activity of any *SIR* gene or mutations that delete a *cis*-acting site abolish silencing. The SIR complex binds to other polypeptides, and these larger complexes interact with histones H3 and H4 (Fig. 18.14c). These interactions with the histones establish a silenced chromosomal domain that remains hidden from the activators and repressors of transcription.

Genomic Imprinting Results from Chromosomal Events That Selectively Silence the Expression of Genes Inherited from One Parent or the Other

A major tenet of Mendelian genetics is that the parental origin of an allele—whether it comes from the mother or the father—does not affect its function in the F_1 generation. For the vast majority of genes in plants and animals, this principle still holds true today. Surprisingly, however, experiments and pedigree analyses have uncovered convincing evidence of exceptions to this general rule for some genes in mammals.

In the 1980s, when embryologists learned to manipulate the haploid nuclei, or pronuclei, of eggs and sperm, they created *in vitro* fertilized eggs carrying one maternally derived and one paternally derived pronucleus; or two maternal pronuclei; or two paternal pronuclei. When placed in a mouse uterus, embryos emerging from the fusion of one maternal and one paternal pronucleus developed into normal, viable, fertile adult mice. But embryos formed from two maternal or two paternal haploid nuclei did not develop normally. Since all of the fertilized eggs providing the pronuclei were produced by matings between inbred animals of the same strain, the only possible genetic differences between the embryos lay in the number and type of sex chromosomes. Yet, even XX embryos with two maternal or two paternal pronuclei failed to thrive. The experimenters concluded that something other than the DNA itself differentiated maternal and paternal pronuclei.

An understanding of the mechanism behind these failures to develop normally came from studies of the transmission of a deletion in the chromosome 7 insulin-like growth factor gene (*Igf2*) in mice. Mice inheriting the deletion from the paternal side were small, while mice inheriting the same deletion from the maternal side were normal size. The simplest explanation of these results is based on a model in which the *Igf2* gene copy inherited from the mother is normally silenced (**Fig. 18.15a** on pp. 662–663). Thus, a deletion inherited from the mother produces no phenotypic effect because the maternal allele is not expressed anyway. If the deletion comes from the father, however, it produces a phenotypic effect because the animal is now unable to make any IGF2 products.

With the advent of molecular tools able to distinguish the transcription of a gene on a maternally derived chromosome from transcription of the same gene on the paternally derived homolog, geneticists observed that expression of a small number of genes—scattered around the genome, but often found in clusters—depends on whether the copy of the gene comes from the mother or the father. The phenomenon in which a gene's expression depends on the parent that transmits it is known as **genomic imprinting.** In most cases of genomic imprinting, the copy of a gene inherited from one parent is transcriptionally inactive in all or most of the tissues in which the copy from the other parent is active. The term "imprinting" signifies that whatever silences the maternal or paternal copy of an imprinted gene is not encoded in DNA sequence; rather it exercises its effect through some **epigenetic** (that is, outside-the-genes, but heritable) alteration of the DNA or chromatin during gametogenesis. This epigenetic imprint remains throughout the life of the mammal, but it is erased and regenerated during each passage of the gene through the germ line into the next generation (Fig. 18.15b). Some genes receive an imprint in the maternal germ line; others receive it in the paternal germ line. For each gene, imprinting occurs in either the maternal or paternal line, never in both.

Although the biochemical mechanism of genomic imprinting is not yet completely understood, one important component is the methylation of cytosines in CG dinucleotides within the imprinted region (Fig. 18.15c). The methylated C's silence the gene or genes in the region by preventing RNA polymerase and other transcription factors from gaining access to the DNA. A particular pattern of methylation can be transmitted during DNA replication, with the presence of a methyl group on one strand of a newly synthesized double helix signaling methylase enzymes to add a methyl group to the other strand.

While this simple model accounts for many instances of imprinting, some cases are more complex. One example involves *Igf2* and a closely linked gene called *H19.* Insight into the mechanism of *Igf2* imprinting came from the surprising finding that *H19,* found just 70 kb downstream of *Igf2,* is also imprinted, but in the opposite way. With *H19,* it is the copy inherited from the father that is silenced and the copy inherited from the mother that is active in normal mice.

A model of how imprinting works at both *H19* and *Igf2* is based on detailed biochemical and genetic studies of a 100 kb region encompassing both genes. Researchers identified an enhancer region downstream of the *H19* gene that can interact with promoters for both genes (Fig. 18.15d). In the region between the two genes lies another type of transcriptional regulation element called an **insulator.** When an insulator becomes functional, it stops communication between enhancers on one side of it with promoters on the other side. Insulators exist throughout the genome, where they play an important role in limiting the chromatin region over which an enhancer can operate. Without insulators, enhancers could wreak havoc in a cell by turning on genes at DNA distances of hundreds of kilobases.

In the *Igf2-H19* region, the insulator DNA becomes functional by binding a protein called CTCF. The binding normally occurs on the maternal chromosome. As a result, the enhancer element on the maternal chromosome can only interact with the promoter of *H19;* this interaction, of course, turns on the *H19* gene. In such a situation, the *Igf2* gene remains unexpressed. On the paternal chromosome, by comparison, both the insulator and the *H19* promoter are methylated. Since methylation of the insulator prevents the binding of CTFC, the insulator is not functional; and without a functional insulator, the enhancer downstream of *H19* can reach over a great distance to activate transcription from the *Igf2* promoter. In addition, methylation of the *H19* promoter suppresses transcription of the paternal *H19* gene. Imprinting of the paternal chromosome by methylation thus turns on transcription of *Igf2* and prevents transcription of *H19.*

Although biochemical analyses suggested the preceding model, results of a genetic experiment greatly strengthened the evidence for the model. In the experiment, researchers knocked out the gene for methylase—the enzyme that adds methyl groups to DNA—in mouse embryos. The experimental embryos lacking the methylase gene died before birth. The reason is as follows: Since methylation of the paternally inherited *H19* gene and insulator region could not occur, the CTCF protein could bind to the insulator on both maternal and paternal chromosomes; this binding induced transcription of the *H19* gene from both chromosomes, while preventing transcription of both copies of *Igf2.*

Before the late 1980s, clinical geneticists were used to seeing sex-linked differences in inherited phenotypes related to the sex of the affected individual. With imprinting, however, it is the sex of the parent carrying a mutant allele that counts, and not the sex of the individual inheriting the mutation. After the discovery of imprinting in mice, medical geneticists reanalyzed human pedigrees and determined retroactively that what appeared to be instances of incomplete penetrance were actually manifestations of imprinting (Fig. 18.15e). An inactivating mutation in a maternally imprinted gene could pass unnoticed from mother to daughter for many generations (since the maternally derived gene copy is inactive in any case because of imprinting). If, however, the mutation passed from mother to son, the son would have a normal phenotype (because he received an active wild-type allele from his father), but the son's children, both boys and girls, would each have a 50% chance of receiving a mutant paternal allele and of expressing the mutant phenotype resulting from the absence of any gene activity.

Evidence for imprinting as a contributing factor now exists for a variety of human developmental disorders, including the related pair of syndromes known as Prader-Willi syndrome and Angelman syndrome. Children with Prader-Willi syndrome have small hands and feet, underdeveloped gonads and genitalia, a short stature, and mental retardation; they are also obese and are compulsive overeaters. Children affected by Angelman syndrome have red cheeks,

Feature Figure 18.15

Genomic Imprinting

(a) Deletion of *Igf2* causes mutant phenotype only when transmitted by father.

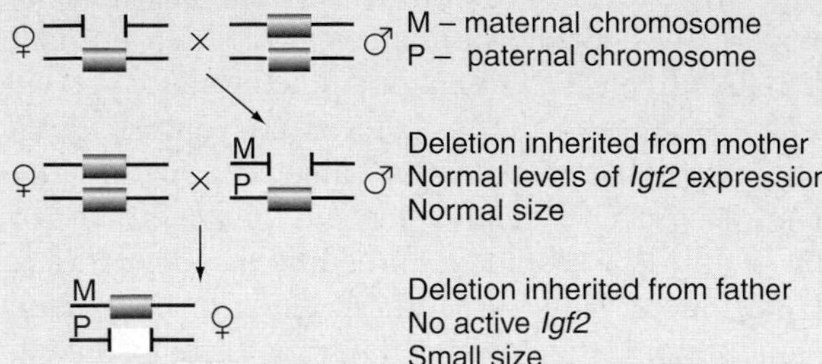

The phenotypic effect of an *Igf2* deletion is determined by the parent transmitting the mutant locus. This parent-of-origin effect can be demonstrated in the two-generation cross illustrated here.

(b) The resetting of genomic imprints during meiosis

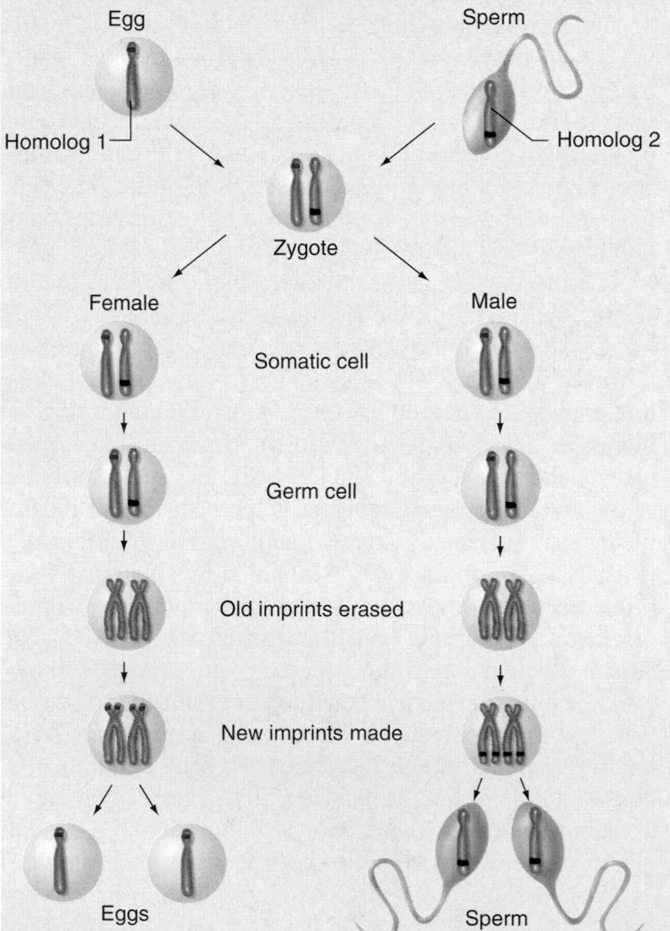

Follow the transmission of a pair of homologous chromosomes (homolog 1 from the mother and homolog 2 from the father) from gametes through fertilization and the development of female and male progeny, to meiosis and the creation of a new set of gametes. Maternally imprinted genes are shown in red, paternally imprinted genes in black. The cellular machinery erases the old imprints and establishes new ones in germs cells during meiosis. Note that in the second generation, one of the chromosomes in both egg (homolog 2) and sperm (homolog 1) will be differently imprinted than the way it was in the first generation.

(c) Methylation of complementary strands of DNA

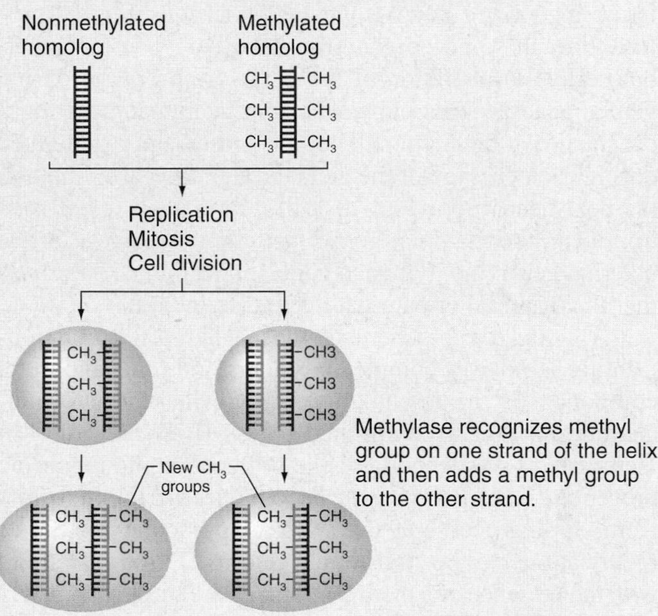

An epigenetic state of DNA methylation can be maintained across cell generations. This is accomplished by the activity of DNA methylases that recognize methyl groups on one strand of a double helix and respond by methylating the opposite strand.

(d) Methylation of paternally inherited *H19* promoter

CTCF
H19
Igf2
Maternal
Enhancers
Insulator elements
Paternal
Methylation
Enhancers

Mouse embryo lacking methylase

CTCF
H19
Igf2
Maternal
Enhancers
CTCF
H19
Paternal
Igf2
Insulator elements
Enhancers

Reciprocal parent-of-origin expression occurs with the *Igf2–H19* gene pair. Only the unmethylated insulator between the two genes can bind to the protein CTCF. On the maternal chromosome, the enhancer only has access to the *H19* promoter. On the paternal chromosome, methylation occurs only at the insulator and the *H19* promoter (indicated with darkened circles). This serves the double purpose of blocking transcription of *H19* and allowing access of the enhancer to *Igf2*. In mouse embryos lacking methylase, the paternal chromosome behaves biochemically like the maternal chromosome.

(e) Genomic imprinting and human disease

Paternal imprinting

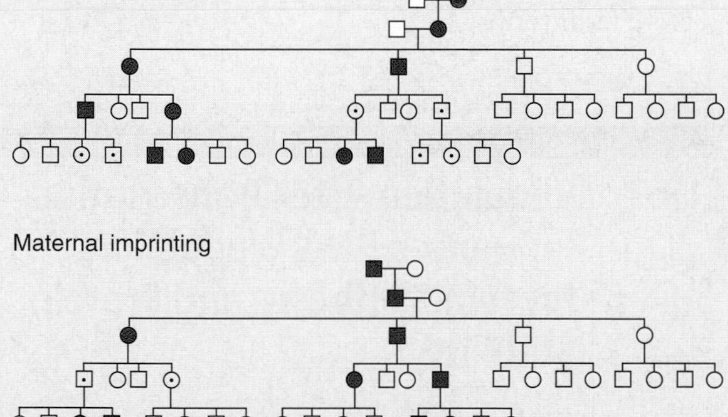

In each pedigree, affected individuals (represented by filled-in, yellow circles and squares) are heterozygotes for a deletion removing a gene that has either a paternal or a maternal imprint. In these pedigrees, a dotted symbol indicates individuals carrying a deleted chromosome but not displaying the mutant phenotype.

a large jaw, a large mouth with a prominent tongue, and a happy disposition accompanied by excessive laughing; they also show severe mental and motor retardation. Both syndromes are often associated with small deletions in the q11–13 region of chromosome 15. When the deletions are inherited from the father, the child gets Prader-Willi syndrome; when the deletions come from the mother, the child has Angelman syndrome. The explanation is that at least two genes in the region of these deletions are differently imprinted. One gene is maternally imprinted: Children receiving a deleted chromosome from their father and a wild-type (nondeleted) chromosome with an imprinted (inactivated) copy of this gene from their mother will have Prader-Willi syndrome. The other gene is paternally imprinted: Children receiving a deleted chromosome from their mother and an imprinted (inactivated) gene from their father will have Angelman syndrome.

Among vertebrate animals, only mammals imprint their genomes. Thus, imprinting is a rather late evolutionary development. Researchers estimate that about 1% of the mammalian genome—roughly 500 genes—will turn out to be imprinted. Most genes imprinted in one mammalian species are also imprinted in other species of mammals, but there are exceptions.

The biological role of genomic imprinting is still under debate. Some scientists believe it is one of the many specialized tools that cells use for gene regulation. Other scientists believe that genomic imprinting is not fundamentally a mechanism of gene regulation but, instead, an evolutionary response to competition between the sexes. In support of this evolutionary view, they point out that genomic imprinting is not universal, since as mentioned, it is found only among mammals.

From an evolutionary standpoint, genomic imprinting seems to place individuals at a disadvantage for survival. Individuals with two potentially active copies of a gene have one copy to spare to random mutations. Most null mutations, for example, are recessive because with only one copy mutated, the remaining functional copy of the gene provides enough gene product for viability. By comparison, with imprinted genes, the single functional gene copy makes an individual susceptible to all mutations.

Why then does imprinting exist? Does it confer some advantage on a population of organisms that counterbalances the obvious evolutionary disadvantages? One proposed answer is that imprinting represents the endgame in a struggle between the sexes. The argument is as follows: In most mammalian species, a female in estrus can mate with multiple males, generating multiple embryos fathered by different males. If a male could cause the embryos he sires to underexpress genes that normally retard embryonic growth, those embryos would use more maternal resources and grow more rapidly *in utero* than the sibling embryos; at birth, the larger, more robust offspring would have a fitness advantage. But the overgrowth of embryos and fetuses, while potentially beneficial to some offspring, would be severely draining to the mother. She would compensate by down-regulating her copies of genes that normally enhance growth. This down-regulation would give all the embryos she is carrying an equal chance of survival and prevent a multiple pregnancy from sapping all her strength. At the evolutionary endpoint of this male-versus-female tug-of-war, imprinting will have silenced growth-enhancing genes in all females and growth-retarding genes in all males.

Known as the Haig hypothesis for David Haig, its originator, this line of reasoning explains why imprinting is found only in mammals: It is only in mammals that embryos can compete for maternal resources *in utero*. Interestingly, many of the imprinted genes uncovered to date are involved in embryonic growth, as predicted by the hypothesis. However, other imprinted genes are not. Only many more hours of experiment and analysis will confirm or reject the Haig hypothesis.

Irrespective of the reason for its original emergence, genomic imprinting now plays a role in gene regulation for a tiny number of genes in a small number of eukaryotic species.

As we saw earlier, genomic imprinting is an example of an epigenetic phenomenon: a variant condition that does not involve a change in DNA sequence but is nevertheless inherited from one somatic cell generation to the next during the growth of the organism. Other examples of epigenetic phenomena described in earlier chapters include X-chromosome inactivation and position-effect variegation. Methylation seems to play an important role in maintaining all three of these epigenetic conditions, which arise from gene or chromosomal inactivation events that occur early in development or in each gamete before fertilization.

18.3 Regulation After Transcription Influences RNA Production, Protein Synthesis, and Protein Stability

Gene regulation can take place at any point in the process of gene expression. While most regulatory mechanisms influence the initiation of transcription, some affect post-transcriptional events such as RNA splicing, micro-RNA production, and protein synthesis and stability.

RNA Splicing Helps Regulate Gene Expression

One example of how RNA splicing contributes to the regulation of gene expression is the role it plays in maintaining sexual identity throughout the life of *Drosophila.* As we see in the comprehensive example at the end of this chapter, transcription factors in very early female (XX) embryos activate the expression of a key gene called *Sxl* through a promoter called the early promoter (P_e). The cellular machinery splices the resulting transcript to create an mRNA that is translated into the Sxl protein, which is essential to the female-specific developmental program. The *Sxl* gene is not transcribed in early male (XY) embryos, so these embryos do not make the Sxl protein (**Fig. 18.16a**).

Later in development, the transcription factors activating the *Sxl* early promoter in females disappear; yet, to develop as females, these animals still need the Sxl protein. How can they still make the Sxl protein they need? The answer is that later in embryogenesis, the *Sxl* gene in both males and females is transcribed from another promoter—the late promoter (P_L) (Fig. 18.16b). In males, splicing of the primary Sxl transcript generates an RNA including an exon (exon 3) that contains a stop codon in the reading frame of the protein. As a result, this RNA in males is not productive; that is, it does not generate any Sxl protein. In females, however, the Sxl protein previously produced by transcription from the early promoter influences the splicing of the primary transcript initiated at the late promoter. When the earlier-made Sxl protein binds to the later-transcribed RNA, this binding alters the splicing pathway such that exon 3 is no longer part of the final mRNA. Without exon 3, the mRNA is productive; that is, it can be translated to make more Sxl protein. Thus, a small amount of Sxl protein synthesized very early in development establishes a positive feedback loop that ensures more fabrication of Sxl protein later in development.

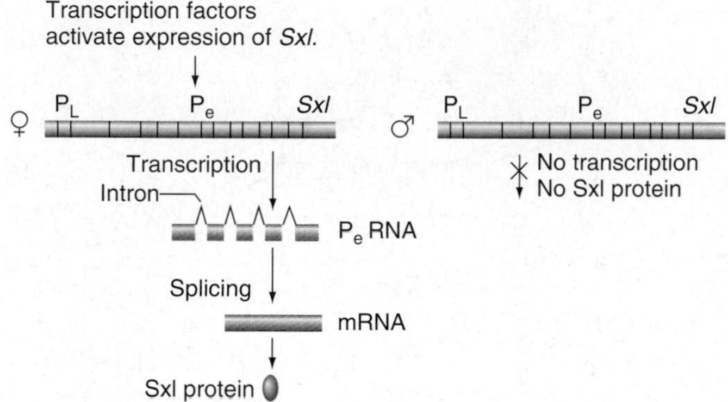

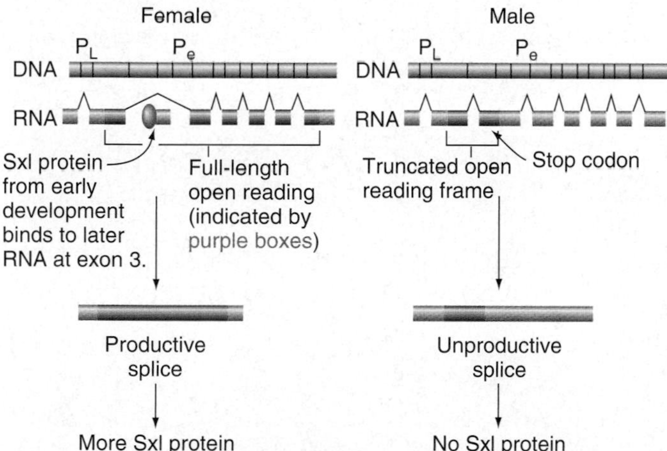

Figure 18.16 Differential RNA splicing can regulate gene expression at the polypeptide level. (a) In the early female—but not the male—*Drosophila* embryo, transcriptional activators initiate transcription from the P_e promoter of *Sxl* to produce an mRNA that encodes the Sxl protein. **(b)** Later in development, transcriptional activators that bind the P_L promoter are produced in both male and female animals. When the Sxl protein is present, as it is in females, it causes the splicing apparatus to skip over this exon and splice exon 2 directly to exon 4. The resulting RNA molecule has an intact coding sequence and can be translated into more Sxl protein. This results in a feedback loop that maintains the presence of Sxl protein in females but not in males.

Micro-RNAs Mediate RNA Interference

In the first five years of the twenty-first century, molecular biologists elucidated a previously unrecognized, yet enormously important and widespread aspect of eukaryotic gene regulation called **RNA interference (RNAi).** In this regulatory mechanism, small specialized **RNAs** prevent the expression of specific genes through complementary base pairing. The RNAs that mediate RNA interference are *trans*-acting single-stranded **micro-RNAs (miRNAs),** 21 to 24 nucleotides in length.

Micro-RNAs are products processed from much longer primary transcripts (**Fig. 18.17a**). About a quarter of the miRNAs characterized to date are generated from the introns of protein-coding transcripts; the remaining 75% represent the only functional products of primary transcripts that are completely devoid of an open reading frame (ORF). Specialized ORF-less micro-RNA genes are transcribed from their own pol II promoters, and as with most protein-coding genes, their expression is regulated by *cis*-acting enhancers and are tissue-specific. Primary miRNA transcripts (pri-miRNAs) receive 5′ GTP caps and 3′ poly-A tails, the same as mRNAs.

It now appears likely that all fungal, plant, and animal species encode and express thousands or tens of thousands of distinct miRNAs, which play some role in the regulation of at least 70% of each organism's protein-coding genes. The actual number of transcribed, ORF-less, miRNA-containing loci in plant or animal genomes is not yet know, but recently accumulated data suggest it could exceed the number of protein-coding genes. The critical nature of miRNA function is demonstrated by the conservation of miRNAs throughout evolution. All miRNAs uncovered to date in mice have homologs in humans and other mammals. In addition, at least 80% of mammalian miRNAs are conserved in the genomes of fish, and at least one-third are

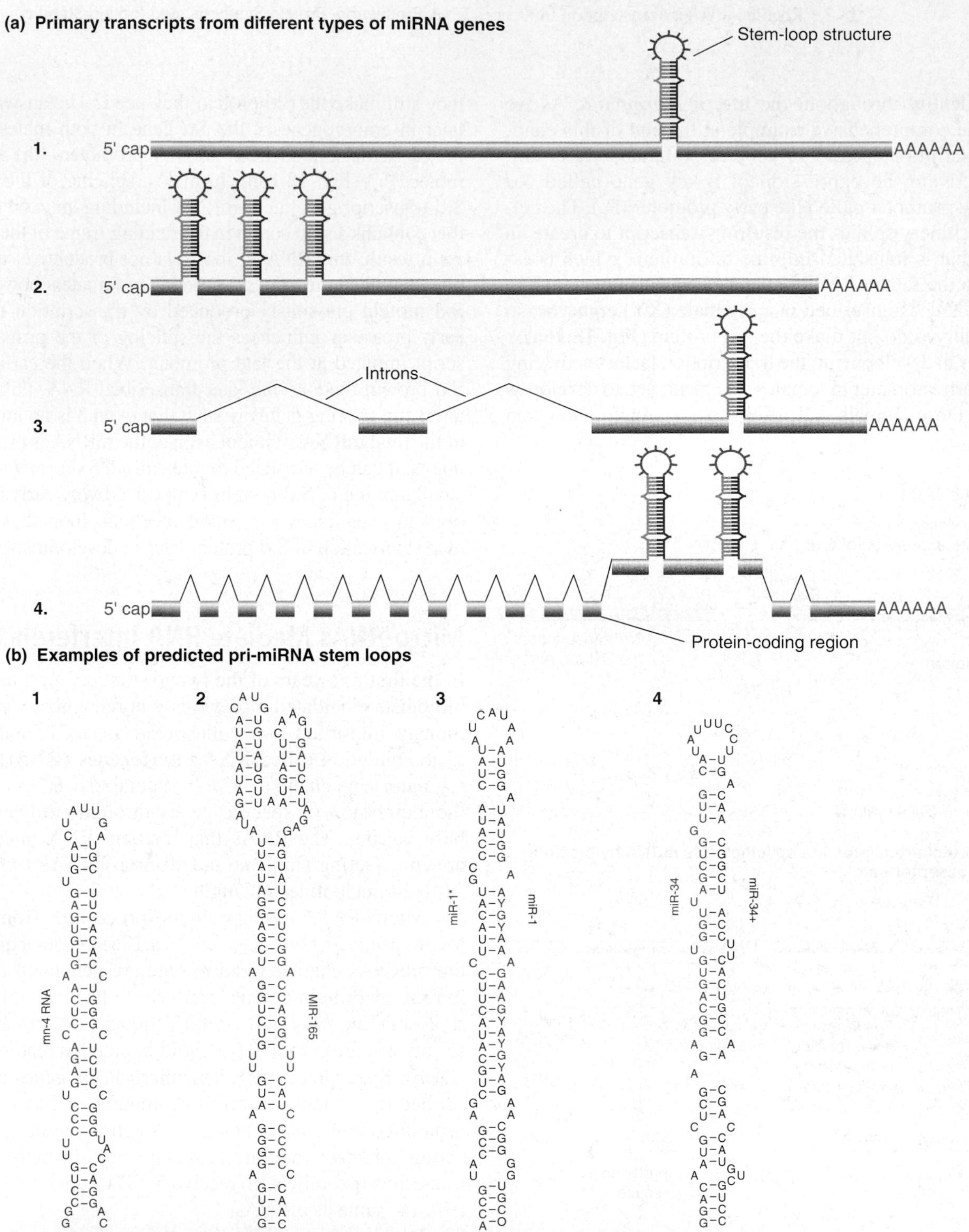

Figure 18.17 **Examples of primary transcripts from micro-RNA-containing genes.** **(a)** Most primary (pri-) miRNA transcripts do not contain an open reading frame (examples 1–3), but some miRNAs are present within the introns of protein-coding mRNAs, as shown in the fourth example. Some pri-miRNA transcripts contain just one miRNA (examples 1 and 3); others contain multiple miRNAs (examples 2 and 4). Some undergo splicing (examples 3 and 4); others do not (examples 1 and 2). If a pri-miRNA transcript undergoes splicing, the miRNA may be in an exon (example 3) or in an intron (example 4). Many variations upon these basic types of primary miRNA transcripts have been observed. **(b)** Ribonucleotide sequences and predicted duplex structures of stem loops in different pri-miRNA transcripts. Examples 1, 3, and 4 are from *C. elegans*; nearly identical homologs of these stem-loop structures have been found in other animals, including flies and mammals. Example 2 is from the plant *Arabidopsis;* a nearly identical homolog has been uncovered in rice and other plants. The final processed miRNA obtained from each stem loop is indicated in *red*. The opposite strand sequence that remains after dicer activity (called an miRNA*) is shown in *purple;* it is eventually degraded (see Fig.18.18).

Source: Part (a) is created by Lee Silver; Part (b) is extracted directly from figure 1 of {MicroRNAs: Genomics, Biogenesis, Mechanism, and Function; David P. Bartel; CELL, 23 January 2004, Pages 281–297

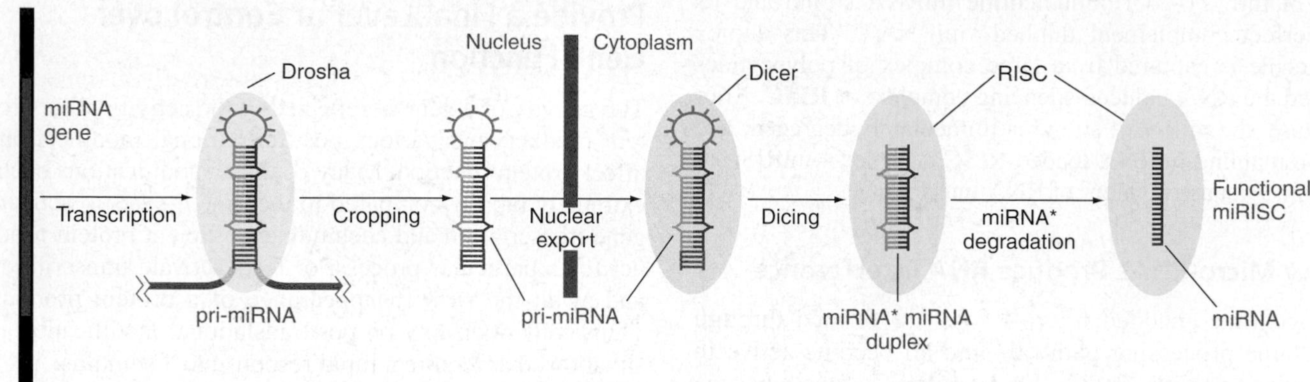

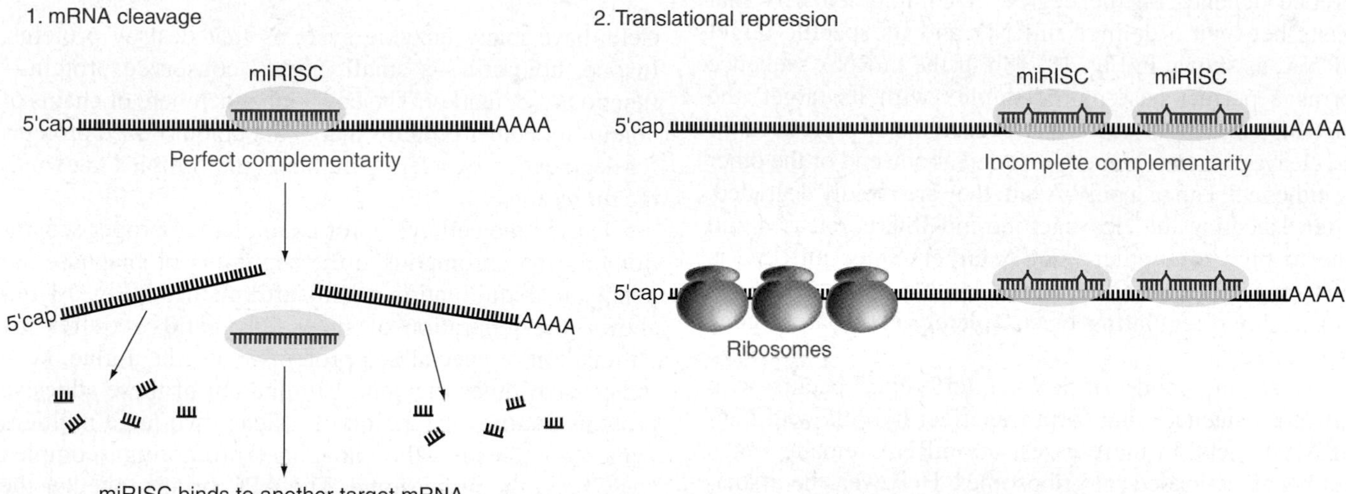

Figure 18.18 **Micro-RNA processing and modes of action. (a)** Immediately after transcription, micro-RNA-containing primary transcripts (pri-miRNAs) are recognized by the nuclear enzyme Drosha, which crops out pre-miRNA stem-loop structures from the larger RNA. The pre-miRNAs undergo active transport from the nucleus into the cytoplasm where they are recognized by the enzyme Dicer. Dicer reduces the pre-miRNA into a short-lived miRNA:miRNA* duplex, which is released and picked up by an RNA-induced silencing complex (RISC). RISC eliminates the miRNA* strand from the duplex and becomes a functional and highly specific miRISC. **(b)** The miRISC can down-regulate gene expression through two different modes of action that are both based on specific binding to a target mRNA. 1. If the miRNA and its target mRNA contain perfectly complementary sequences, miRISC cleaves the mRNA. The two cleavage products are no longer protected from RNase and are repidly degraded. 2. If the miRNA and its target mRNA have only partial complementarity, cleavage does not occur. However, the miRISC remains bound to its target and represses its movement across ribosomes. This mode of down-regulation is less efficient than cleavage.

conserved in the distantly related nematode *C. elegans.* The same broad degree of conservation holds in the plant kingdom as well: Most miRNAs uncovered in *Arabidopsis* have homologs in rice and maize.

How Cells Generate Micro-RNAs

The fundamental distinguishing attribute of miRNA-containing transcripts is the presence of one or more 60–120 ribonucleotide-long segments exhibiting a sufficient degree of reverse sequence complementarity to snap back spontaneously into hairpin stem-loop structures of the kind shown in Fig. 18.17b. These stem-loop structures trigger a cascade of biochemical processing events that lead ultimately to a functional miRNA complex.

Figure 18.18a illustrates the biogenesis of microRNAs. The first step, called *cropping,* is performed in the nucleus by *drosha,* a specialized RNase that recognizes and binds to the stem-loop structure. Drosha crops the stem loop by cleaving the RNA sequences at the base of the structure. The released pre-miRNA molecule is actively transported from the nucleus to the cytoplasm where it is recognized by *dicer,* another type of RNase. Dicer cuts the loop off the top of the pre-miRNA and trims back ribonucleotides from both ends of the stem to produce an RNA duplex consisting of the mature 21–24 ribonucleotide miRNA strand and its imperfect complement dubbed "miRNA*." This duplex molecule is captured by a large complex of polypeptides called the RNA-induced silencing complex, or **RISC.** After capture, the miRNA* strand is immediately degraded, and the remaining miRNA-loaded RISC complex—miRISC—becomes an active agent of RNA interference.

How Micro-RNAs Produce RNA Interference

All genome-encoded micro-RNAs are formed through the same processing pathway, and all become active in association with RISC. Each miRISC functions via miRNA base pairing to a specific target sequence in a cytoplasmic mRNA molecule. The precise mode of RNA interference depends on the degree of complementarity that exists between a defined miRNA and its specific target mRNA, as shown in Fig. 18. 18b. If the miRNA sequence forms a perfect base-paired duplex with its target, the miRISC cleaves the mRNA. Because the two products of the cleavage are no longer protected at one end or the other by either a 5′ cap or a poly-A tail, they are rapidly degraded. After executing the cleavage, the miRISC is released and free to bind to another mRNA target. Since miRISC is thus recyclable, small amounts can have a dramatic effect in down-regulating or completely eliminating gene expression.

A second mode of RNA interference occurs with miRNA sequences that form imperfect hybrids with their mRNA targets. In these cases, the mRNA remains intact and becomes loaded into ribosomes. However, the associated miRISC represses the movement of the ribosomes, and, as a consequence, translation is down-regulated. This more subtle mode of RNA interference is fine-tuned by the amount of the specific miRNA produced by the cell; the degree of miRNA:mRNA complementarity; and the number of independent miRISC binding sites that exist on the target mRNA.

Scientific understanding of micro-RNAs and RNA interference is still in its infancy, with many unanswered questions and intriguing experimental results that suggest further modes of RNA interference. In addition to their role in posttranscriptional regulation, miRNAs have been implicated as *trans*-acting transcription factors. miRNAs that function in this way remain sequestered in the nucleus and do not bind to RISC. They cause gene silencing by displacing and replacing one DNA strand at a target locus that contains a complementary sequence. Finally, as we explain in the Tools of Genetics box on pp. 670–672, the biochemical machinery of posttranscriptional RNA interference—which exists naturally in all eukaryotic cells—provides a convenient handle for the development of powerful new RNA-based therapies. In the future, these therapies may make it possible to treat many types of disease that result from aberrant gene expression.

Protein Modifications After Translation Provide a Final Level of Control over Gene Function

The activity of a gene is reflected in the activity of its protein product, and various posttranslational modifications affect protein function. Many of these modifications occur extremely rapidly compared to the time it takes to activate gene transcription and accumulate sufficient protein product for a particular process, or to deactivate transcription and await the slow disappearance of a protein product. Thus, cells often rely on posttranslational modification in situations that require a rapid response to a stimulus.

Ubiquitination Targets Proteins for Degradation

Cells have many enzyme systems that destroy proteins. In one, ubiquitin—a small, highly conserved protein—functions as a marker. The covalent attachment of chains of ubiquitin to other proteins marks the ubiquitinized proteins for degradation by a large multienzyme complex known as the *proteosome.*

During the cell cycle, for example, the proper separation of sister chromatids at the beginning of anaphase depends on ubiquitination and its aftermath. As **Fig. 18.19a** shows, the separation of sister chromatids requires the elimination of specialized proteins that, like a glue, keep the chromosomes together. Elimination of these adhesive proteins requires the action of at least two large multienzyme complexes—the anaphase-promoting complex (APC) and the proteosome. The APC brings together the gluelike proteins that must undergo ubiquitination with enzymes that catalyze the attachment of chains of ubiquitin to these targets. Inhibitors of sister chromatid separation are among the various targets of the APC. Ubiquitination of these inhibitors at the onset of anaphase marks them for degradation by proteosomes, which, in turn, allows the sister chromatids to separate. Researchers do not yet know exactly how the inhibitors carry out their function. One hypothesis is that some act as part of a proteinaceous glue that attaches sister chromatids at their centromeres; degradation of this glue as a result of APC and proteosome activity allows the sister chromatids to separate. Whatever the means, ubiquitination swiftly lifts the inhibition on sister chromatid separation, allowing the cell cycle to proceed.

Phosphorylation and Dephosphorylation Are Other Regulators of Protein Activity

In a process known as *sensitization,* many tissues exposed to hormones for a long time lose their ability to respond to the hormone. An example is the exposure of heart muscle to the stress hormone epinephrine. The binding of epinephrine to proteins called β-adrenergic receptors, which

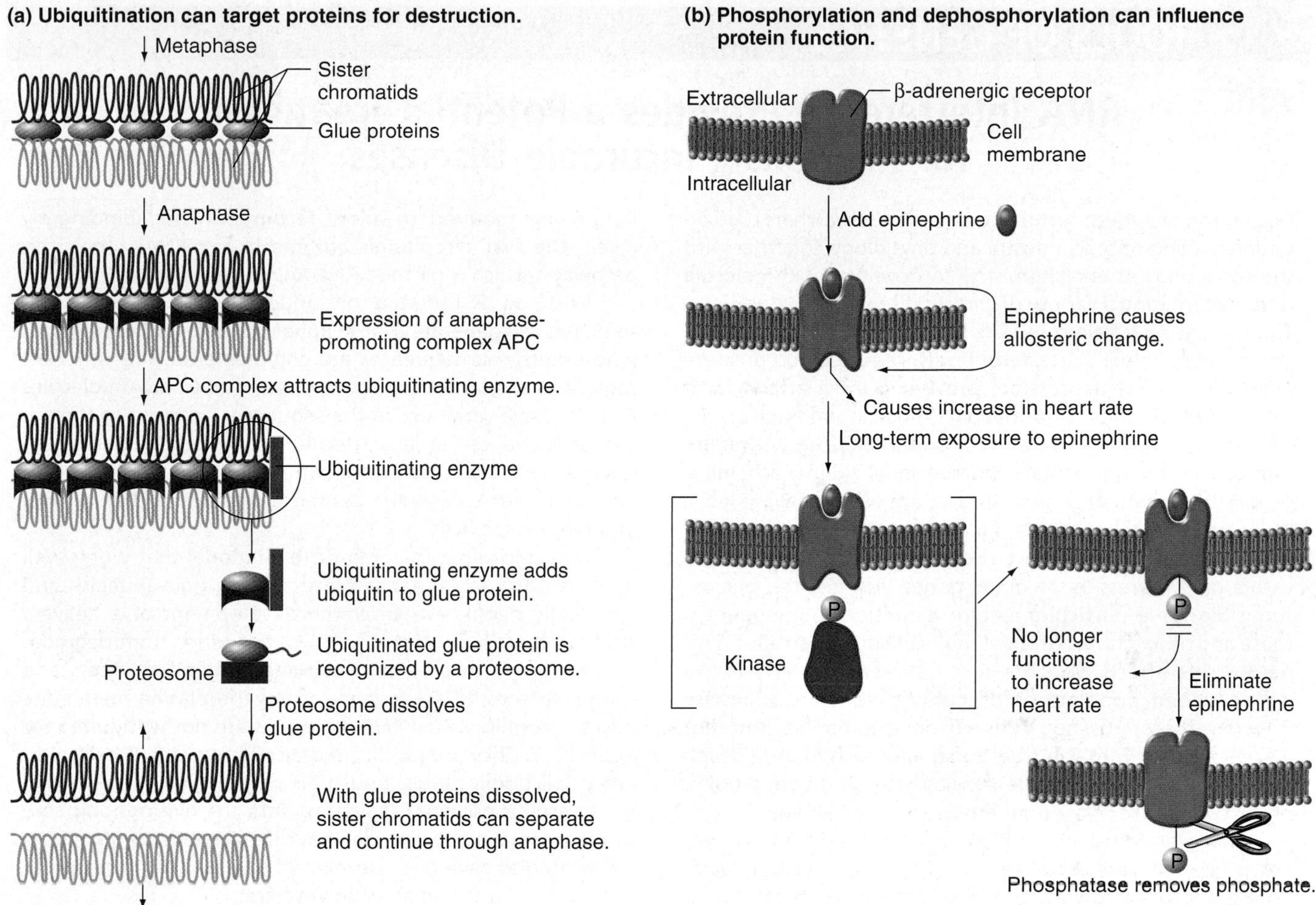

Figure 18.19 Protein modifications can affect the final level of gene function. (a) The covalent addition of ubiquitin peptide chains to the gluelike proteins that hold together sister chromatids during the metaphase stage of mitosis marks them for digestion by proteosomes. **(b)** Covalent phosphorylation of the β-adrenergic receptor has no effect on its binding to epinephrine, but it blocks its downstream function of modulating heart rate.

are located in the plasma membrane of heart muscle cells, normally increases the rate at which the heart contracts. But after several hours of continuous exposure to epinephrine, the heart muscle cells no longer respond in this way. Their sensitization is due to phosphorylation of the β-adrenergic receptors. The phosphorylation (addition of a phosphate) does not affect a receptor's ability to bind epinephrine, but it does prevent the receptor from transmitting the hormone signal into the heart muscle cells (see Chapter 19 for a detailed discussion of signal transduction). The phosphorylation itself depends in large part on the activity of kinase (phosphate-adding) enzymes that phosphorylate the β-adrenergic receptor only when the receptor is bound to epinephrine (Fig. 18.19b). With the removal of epinephrine from the heart tissue, the kinases no longer act on the receptors, and phosphatase enzymes remove any phosphates already on them. The removal of phosphate from the β-adrenergic receptors eventually restores the heart muscle's ability to respond to new doses of epinephrine.

18.4 Sex Determination in *Drosophila:* A Comprehensive Example of Gene Regulation

Male and female *Drosophila* exhibit many sex-specific differences in morphology, biochemistry, behavior, and function of the germ line (**Fig. 18.20** on p. 673). By examining the phenotypes of flies with different chromosomal constitutions, researchers confirmed that the ratio of X to autosomal chromosomes (X:A) helps determine sex, fertility, and viability (**Table 18.2** on p. 673). They then carried out genetic experiments that showed that the X:A ratio influences sex through three independent pathways: One determines whether the flies look and act like males or females; another determines whether germ cells develop as eggs or sperm; and a third produces dosage compensation through doubling the rate of transcription of X-linked genes in

TOOLS OF GENETICS

RNA Interference Provides a Potential Treatment for Previously Incurable Diseases

During the twentieth century, biomedical researchers used an understanding of biochemistry and physiology, together with the tools of synthetic chemistry, to develop small molecule drugs for treating a variety of chronic diseases and conditions. These drugs now help hundreds of millions of people reduce chronic pain, lower cholesterol levels, reduce blood pressure overcome clinical depression, prevent asthma attacks, and treat a host of other abnormal physiological conditions.

However, although these traditional drugs have improved the quality of many human lives, their biochemical properties and physiological impact are almost always more broad-ranging than is ideal for the treatment of a specific disease. Steroidal hormones, for example, operate as coactivators or corepressors of many genes, not just the one responsible for a particular health condition. Consequently, these and other classes of small molecule drugs can elicit dangerous side effects, which are sometimes so severe or frequent among populations of patients that an otherwise effective therapy is not allowed on the market. Another global problem with the traditional approach to drug development is that many well-defined disease conditions are simply not amenable to modulation by small molecules.

The recent discovery of RNA interference (RNAi) as a natural process of gene regulation in all eukaryotic cells suggests a new approach toward the development of therapies to combat disease. The general idea is to co-opt the existing cellular RNAi machinery into working with laboratory-designed **short interfering RNA (siRNA)** molecules that target specific mRNAs from the disease-causing gene for destruction through the mechanism shown in **Fig. A,** parts 1 and 2. The power of RNAi therapy—and the cause of excitement about it among biomedical scientists in many different fields—lies both in its specificity and its broad range of applications. In theory, RNAi could be exploited in the development of therapies for any disease or human condition that is caused by over expressed or aberrant RNA transcripts from a specific gene in either the patient's own DNA or the genome of an infectious virus or microbe. Among the diseases researchers are currently targeting for RNAi therapy are incurable conditions such Huntington disease, Lou Gehrig's syndrome, AIDS, and a variety of cancers. They are also targeting conditions such as hypertension and hypercholesterolemia, for which current treatments are not specific enough.

The first step in the development of an RNAi therapy involves the design, construction, and experimental validation of an siRNA that can function inside living cells to eliminate disease-causing *target transcripts,* while not affecting the transcripts of any other gene. A well-designed siRNA will contain a 21–23 base-long **antisense** sequence that is perfectly complementary to a unique sequence within the target transcript. This antisense sequence must be contained within a longer RNA strand that is itself part of a duplex with a complementary **sense** strand. The duplex structure and extended RNA length are required to allow recognition and binding by *dicer*—the first cytoplasmic enzyme in the RNAi processing pathway—which trims the siRNA and passes it on to RISC.

With an automated oligonucleotide synthesizer, researchers can generate a large panel of duplex RNA molecules whose antisense sequences are complementary to different regions of the target transcript. These duplex RNA molecules will also have variations in the sequences adjacent to the antisense sequences, as illustrated in Fig. A.1. Experiments conducted on tissue culture cells can be used to identify which particular siRNAs have the desired properties of high activity and target specificity.

Moving from a therapeutic RNAi model that works well in tissue culture to one that is effective in whole animals—and eventually people—requires the development of a delivery strategy that (1) protects the siRNA sequence from degradation before it reaches cells carrying target transcripts and (2) guides the siRNA sequence across the plasma membrane into those cells. Naked RNA molecules are not well-suited for either task: They are rapidly degraded by sentry RNases present in all bodily fluids, and their negatively charged phosphate groups prevent ready entry into the hydrophobic core of the plasma membrane. Several clever chemical and biological strategies have been developed to overcome these problems. We present two alternatives here.

Hans-Peter Vornlocher and his group from Kulmbach, Germany developed a chemical strategy for overcoming siRNA delivery problems in a mouse model for hypercholesterolemia, a condition caused by excess low-density lipoproteins (LDL) in blood serum. The liver protein apolipoprotein B (apoB) functions only in LDL biogenesis and thus presents an excellent target for the development of an LDL-specific RNAi therapy. In tissue culture experiments, Vornlocher et al. identified an siRNA with high activity and specificity for mouse apoB transcripts. He then synthesized the two strands of the siRNA, but rather than using them directly, he created chemical modifications at the 5′ and 3′ ends of both strands, as illustrated in Fig. A.2. The modified siRNA product could not be recognized by serum RNases, which do their work by digesting naked RNA from one end or the other.

Vornlocher's second chemical trick was to employ the lipid cholesterol as the chemical entity that was attached to the 3′ end of the sense strand in the siRNA duplex. Cholesterol not only protects one end of the RNA from degradation but also tends to incorporate itself into plasma membranes (as a result of its hydrophobicity), which facilitates the passage of its siRNA cargo into the cell proper. Evidence that this strategy can actually work in a living animal was obtained after injection of the specially modified anti-apoB siRNA into the tail veins of normal mice. Within 24 hours, their serum LDL levels had fallen by over 50%. This remarkable result serves as a proof-of-principle for the use of chemical-based siRNA delivery systems in the development of RNAi therapies for treating chronic human diseases.

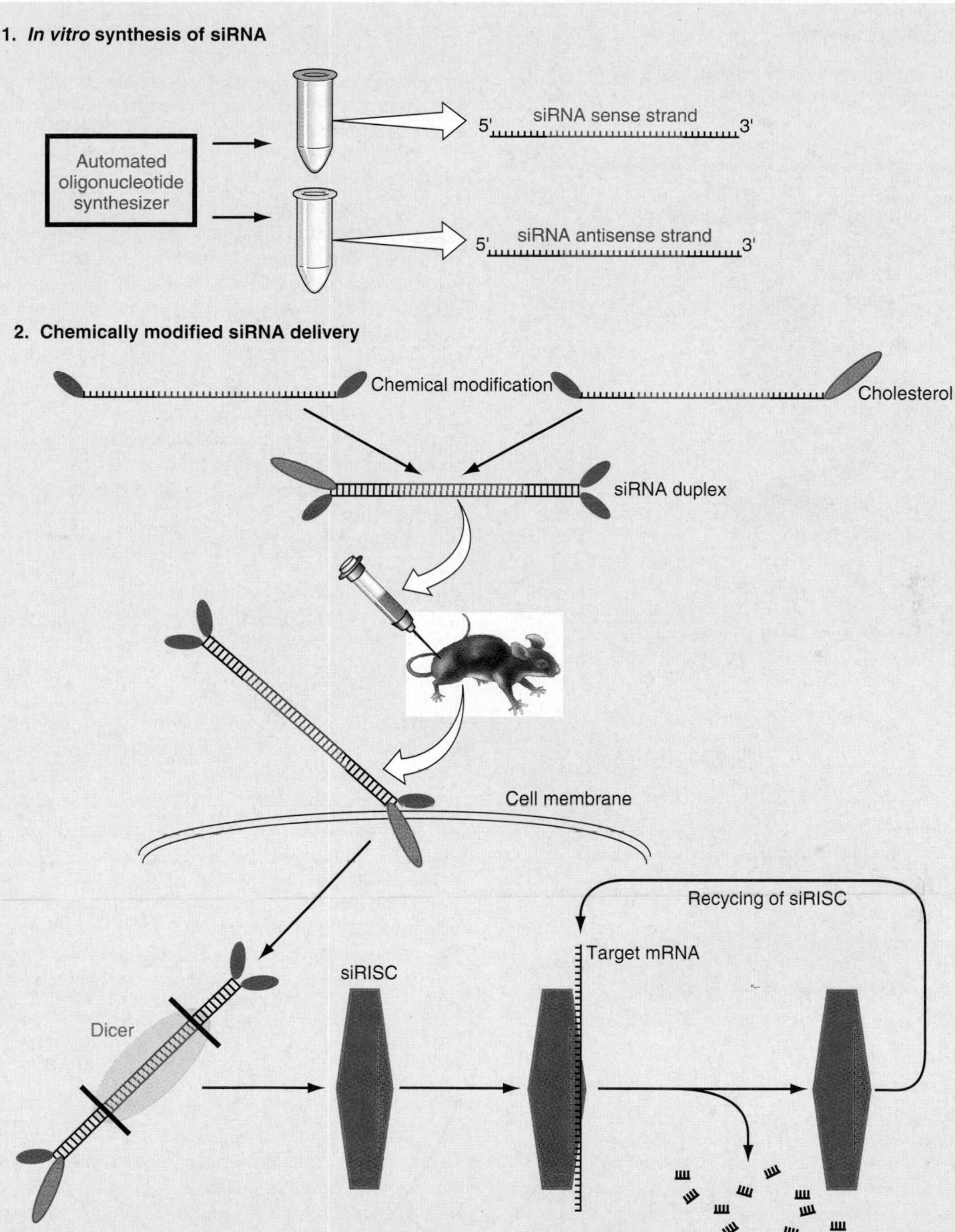

Figure A The development of RNAi therapy (1) An automated oligonucleotide synthesizer is used to create the antisense and sense strands of a potential siRNA molecule. (2) The two strands are chemically modified at their 5′ and 3′ ends and then brought together to form an siRNA duplex, which is injected into experimental animals. This chemical modification includes attachment of a cholesterol molecule to the 3′ end of the sense strand. The cholesterol moiety incorporates itself into the plasma membrane of cells and facilities entry of the whole siRNA duplex. In the cytoplasm, the RNAi enzyme dicer recognizes the siRNA, cleaves off its ends, and passes it to RISC. The siRNA-loaded RISC attaches to target mRNA transcripts and destroys them. The RISC complex is then recycled to attack further target mRNAs. (3) In the biological delivery system depicted on p. 30, the sense and antisense strands are ligated to each other and then integrated into a retroviral expression vector, which is packaged in a retroviral coat. The recombinant virus particles are injected into experimental animals. The particles fuse with the cell's plasma membrane, and after reverse transcription, the recombinant genomes integrate into the host chromosome. Transcripts from the recombinant retroviral genomes are processed by the cell's natural RNAi machinery.

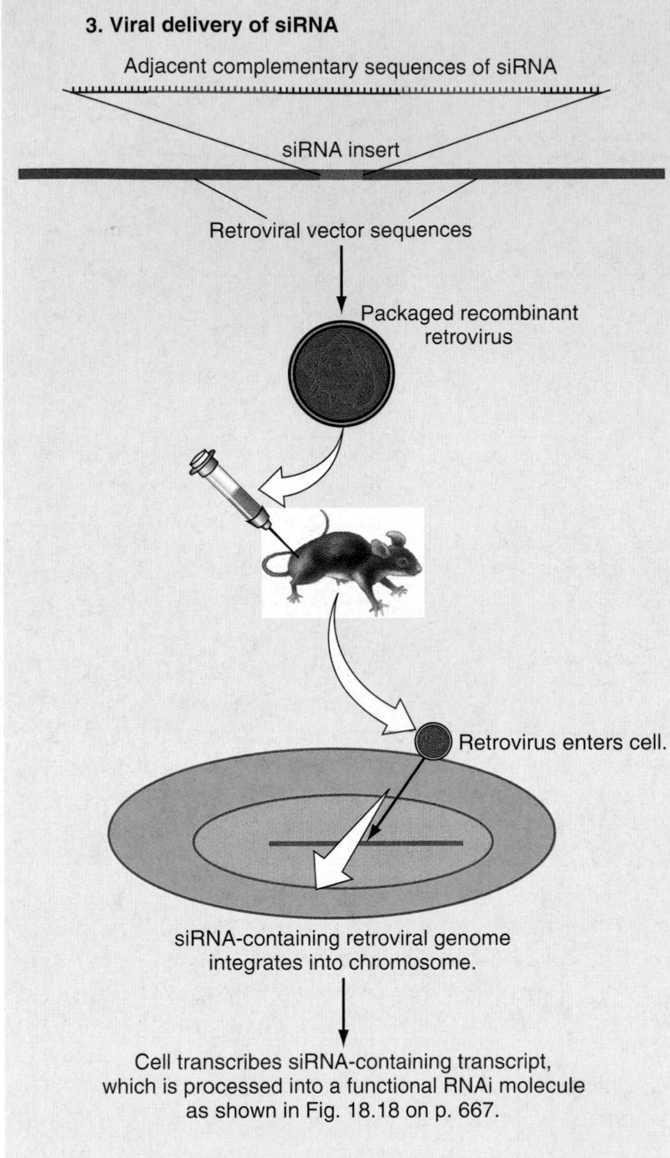
3. Viral delivery of siRNA
Adjacent complementary sequences of siRNA
siRNA insert
Retroviral vector sequences
Packaged recombinant retrovirus
Retrovirus enters cell.
siRNA-containing retroviral genome integrates into chromosome.
Cell transcribes siRNA-containing transcript, which is processed into a functional RNAi molecule as shown in Fig. 18.18 on p. 667.

An alternative strategy for delivering siRNA sequences into an animal is based on biological rather than chemical manipulation, as illustrated in Fig. A.3. The biological strategy takes advantage of certain natural attributes of the retroviral class of viruses (which includes the HIV virus that causes AIDS). During their journeys through bodily fluids, retroviruses protect their genetic cargo inside a plasma membrane coat. The tendency of the retroviral membrane to fuse with cellular plasma membranes provides a mechanism for delivering the viral genetic material into cells. Finally, the most significant advantage of incorporating an siRNA sequence into a retroviral vector within a viral coat is that the recombinant genome can integrate into cellular chromosomes and continue to be expressed for the remainder of the animal or person's life.

The biological strategy is being pursued by a number of research groups including ones headed by Patrick Aebischer in Lausanne, Switzerland, and G. Scott Ralph in Oxford, England. These two teams are working with a mouse model of the late-onset neurodegenerative disease *amyotrophic lateral sclerosis (ALS),* commonly known as Lou Gehrig's disease because of the role played by this early twentieth-century major league baseball star in eliciting public awareness of the condition. One form of the disease is caused by dominant mutations in the coding region of the gene *SOD1.* Mutant SOD1 protein is toxic to motor neurons. As the neurons die, progressive paralysis sets in. Within a few years, death comes to the patient. (Lou Gehrig died just 2 years after he was diagnosed.) No treatment currently exists for curing or delaying the onset of ALS.

The groups led by Aebischer and Ralph first determined optimal sequences for functional anti-SOD1 siRNA through experiments on tissue culture cells. Next, instead of administering the siRNA directly, they incorporated the sequence into a retroviral expression vector and encapsulated the recombinant RNA in a membrane coat. They then injected siRNA-encoding virus particles into the spinal cords of mutant mice. The viruses fused with the plasma membranes of motor neurons and released their recombinant RNA genomes into the neuronal cytoplasm, where the RNA was reverse transcribed into cDNA, which then integrated into the cell's genome. The cell's own transcriptional machinery initiated the production of siRNA-containing transcripts, which were recognized and processed like natural miRNA transcripts. RISCs loaded with the mature siRNA degraded mutant SOD1 transcripts, leading to a retardation in the onset and progression of neurodegenerative disease. Again, this experimental result—which has been reproduced in three independent labs—provides a strong proof-of-principle for the use of RNAi therapy as treatment for a multitude of now incurable human diseases caused by aberrant or overexpressed gene products.

So far, the effectiveness of RNAi therapies has been demonstrated only in experimental animals. Although the results are very encouraging, such therapies are not yet ready for use in people. To develop human therapies, researchers must design siRNAs that work in human cells and then conduct full-scale clinical trials to assure the effectiveness and safety of each RNAi protocol.

males. (Note that this strategy of dosage compensation is just the opposite of that seen in mammals, where the inactivation of one X chromosome in females equalizes the expression of X-linked genes with that in males.)

To simplify our discussion of sex determination in *Drosophila,* we focus on the first-mentioned pathway: the determination of somatic sexual characteristics. An understanding of this pathway emerged from analyses of mutations affecting particular sexual characteristics in one sex or the other. For example, as we saw at the beginning of the chapter, XY flies carrying mutations in the *fruitless* gene exhibit aberrant male courtship behavior,

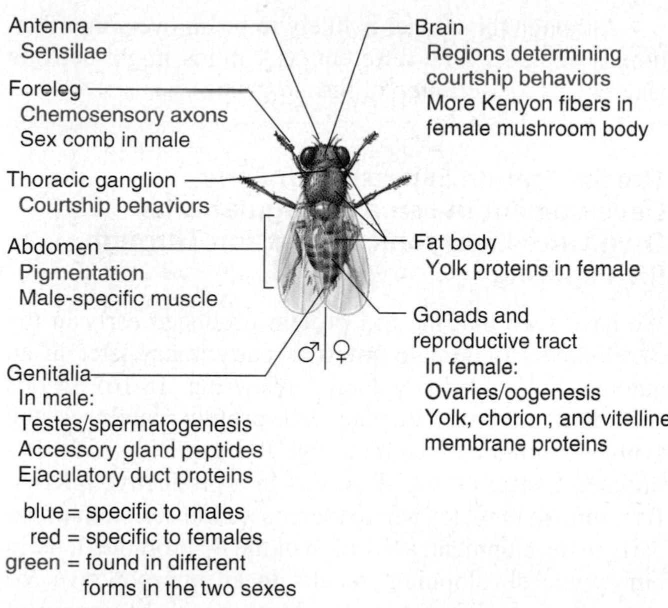

Figure 18.20 Sex-specific traits in *Drosophila*. Objects or traits shown in *blue* are specific to males. Objects or traits shown in *red* are specific to females. Objects or traits shown in *green* are found in different forms in the two sexes.

while XX flies with the same *fruitless* mutations appear to behave as normal females. **Table 18.3** shows that mutations in other genes also affect the two sexes differently. Clarification of how these mutations influence somatic sex determination came from a combination of genetic experiments (studying, for example, whether one mutation in a double mutant is epistatic to the other) and molecular biology experiments (in which investigators cloned mutant and normal gene products for analysis). Through such experiments, *Drosophila* geneticists dissected various stages of sex determination to delineate the following complex regulatory network.

TABLE 18.2 How Chromosomal Constitution Affects Phenotype in *Drosophila*

Sex Chromosomes	X:A	Sex Phenotype
Autosomal Diploids		
XO	0.5	Male (sterile)
XY	0.5	Male
XX	1.0	Female
XXY	1.0	Female
Autosomal Triploids		
XXX	1.0	Female
XYY	0.33	Male
XXY	0.66	Intersex

TABLE 18.3 *Drosophila* Mutations That Affect the Two Sexes Differently

Mutation	Phenotype of XY	Phenotype of XX
Sx^{fl}*	Male	Dead
Sxl^{ML}**	Dead	Female
transformer (tra)	Male	Male (sterile)
doublesex (dsx)	Intersex	Intersex
fruitless	Male with aberrant courtship behavior	Female

*Sx^{fl} is a recessive mutation of *Sex lethal.*
**Sxl^{ML} is a dominant mutation of *Sex lethal.*

The X:A Ratio Regulates Expression of the *Sex Lethal* (*Sxl*) Gene

Recall from Chapter 4 that it is the ratio of X chromosomes to autosomes (A) that determines sex in *Drosophila.* Since in normal diploids, there are two copies of each autosome, the X:A ratio is 2/2 = 1.0 in a normal XX female and 1/2 = 0.5 in a normal XY male. In short, when the X:A ratio is 1.0, females develop; when the ratio is 0.5, males develop.

Key factors of sex determination are helix-loop-helix proteins encoded by genes on the X chromosome. Sisterless-A (Sis-A) and sisterless-B (Sis-B) are two such proteins. Referred to as *numerator elements,* these two proteins monitor the X:A ratio through the formation of homodimers containing two of the same kind of subunit or heterodimers containing two different subunits. The homodimers consists of two numerator elements, while the heterodimers are composed of one numerator element and one denominator element. *Denominator elements* are helix-loop-helix proteins that are encoded by genes on autosomes. Since the number of X chromosomes determines the ratio of numerator homodimers to numerator/denominator heterodimers, the homodimers of numerator elements provide a measure of the X:A ratio (**Fig. 18.21**).

The observation that in flies with a greater number of numerator homodimers, transcription of the *Sxl* gene occurs early in development suggests the following hypothesis for operation of the X:A ratio.

Numerator Subunit Homodimers May Function As Transcription Factors That Turn On *Sxl*

In this hypothesis, the association of denominator subunits with numerator subunits sequesters the numerator elements in inactive heterodimers that cannot activate transcription. Females produce enough numerator subunits, however, that some remain unbound by denominator

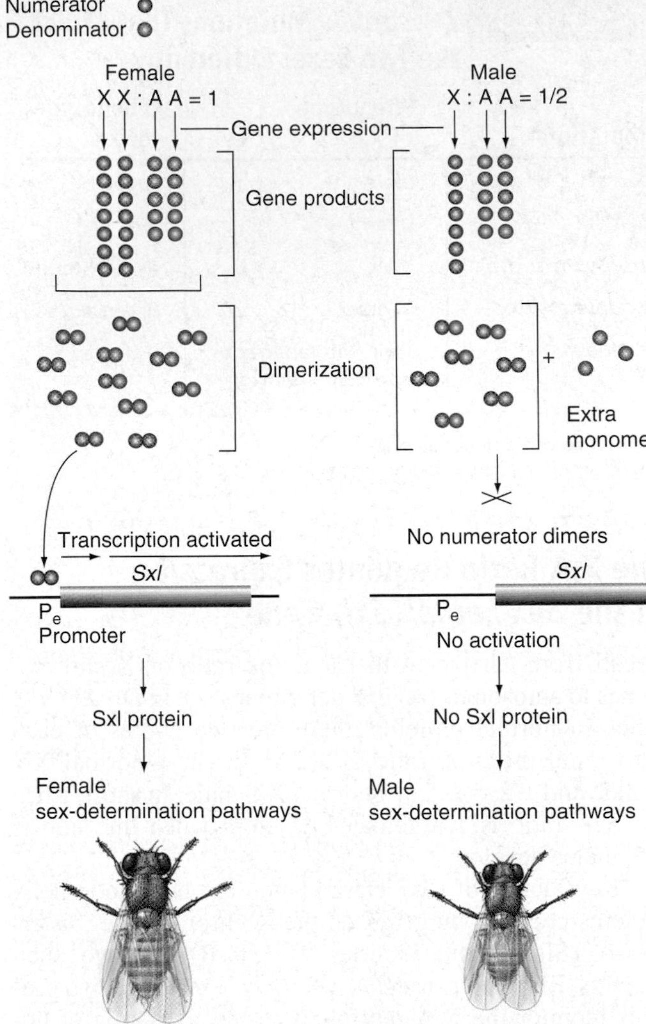

Figure 18.21 **The X:A ratio determines the expression of *Sxl*, the first gene in the sex-determination pathway.** Numerator elements are produced by the X chromosome at a slightly higher level than denominator elements are produced by autosomes. Denominator elements cannot form homodimers, but they can bind tightly to, and sequester, numerator elements. When the X:A ratio is 1 (in females), there are too many numerator elements to be occupied by denominators, and those not sequestered can form homodimers, which act as activators of the *Sxl* gene. When the X:A ratio is 1/2 (in males), there are fewer numerators than denominators, and all the numerators become sequestered. Thus males are unable to produce numerator homodimers, and *Sxl* remains turned off.

elements. Homodimers formed from these free numerator elements act as transcriptional activators of *Sxl* at the P_e promoter early in development. Males, by contrast, carry only half as many X-encoded numerator subunits; thus, the abundant denominator proteins tie up all the numerator elements, and as a result, there are no free numerator elements in males to turn on the P_e promoter of the *Sxl* gene.

Although this model is likely to be an oversimplification, it suggests how different X:A ratios might activate and repress transcription of the *Sxl* gene.

The Sxl Protein Expressed in Early Development in Females Regulates Its Own Later Embryonic Expression Through RNA Splicing

We have seen that the Sxl protein produced early in the development of female embryos participates later in an autoregulatory feedback loop (review Fig. 18.16). In this self-regulating system, the Sxl protein catalyzes the synthesis of more of itself through RNA splicing of the P_L-initiated transcript, which results in a productive mRNA. By contrast, in males where there is no transcription of *Sxl* early in development, activation of the P_L promoter later in embryonic development results in an unproductive *Sxl* transcript containing a stop codon near the beginning of the message. Since no Sxl protein is present to splice out the problem stop codon, this unproductive transcript is not translated to protein; and males thus have no Sxl protein at any point in development.

The Effects of *Sxl* Mutations

Recessive *Sxl* mutations that produce nonfunctional gene products have no effect in XY males, but they are lethal in XX females (see Table 18.3). This is because males, which do not normally express the *Sxl* gene, do not miss its functional product, but females, which depend on the Sxl protein for sex determination, do. The absence of the Sxl protein in females allows the aberrant expression of certain male-specific dosage-compensation genes that increase transcription of genes on the X chromosome; and the hypertranscription of these X-linked genes on two X chromosomes in mutant females proves lethal.

By comparison, rare dominant *Sxl* mutations that allow production of Sxl protein even in XY embryos are without effect in females, which normally produce the product, but lethal to males, which normally do not. In these mutants, the *Sxl* gene product indirectly represses transcription of genes that males need to express for dosage compensation. Without the products of these male-specific dosage-compensation genes, males cannot hypertranscribe X-linked genes and thus do not have enough X-linked gene products to survive.

Summary: The *Sxl* Gene Acts As an On/Off Switch That Is Sensitive to the X:A Ratio

Different proportions of numerator and denominator proteins reflecting the X:A ratio determine whether or not the *Sxl* P_e promoter is active in the early embryo. Later, when other transcription factors activate the upstream P_L

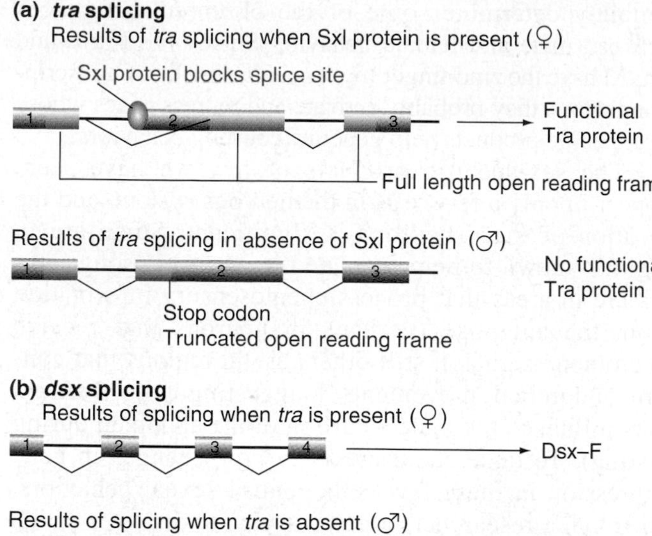

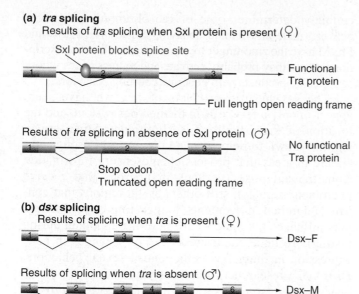

Figure 18.22 The expression or nonexpression of *Sxl* leads to downstream differences in expression of products that also play a role in the sex-determination pathway. (a) The presence of *Sxl* alters the splicing of *tra* mRNA. Female transcripts produce functional Tra protein, while male transcripts have a truncated open reading frame and are unable to produce Tra. **(b)** Tra protein, in turn, plays a role in altering the splicing pattern of the *dsx* mRNA. A different Dsx product results in males (Dsx-M) rather than in females (Dsx-F).

promoter in both males and females, alternative RNA splicing after transcription generates more Sxl protein in females and none in males. Female development requires Sxl protein; male development requires its absence.

Sxl Triggers a Cascade of Splicing

In addition to splicing its own transcript, the Sxl protein influences the splicing of RNAs transcribed from other genes. Among these is the *transformer* (*tra*) gene. In the presence of the Sxl protein, the *tra* primary transcript undergoes productive splicing that produces an mRNA translatable to a functional protein. In the absence of Sxl protein, the splicing of the *tra* transcript results in a nonfunctional protein (**Fig. 18.22a**).

The cascade continues. The functional Tra protein synthesized only in females, along with another protein encoded by the *tra2* gene (which is transcribed in both males and females), influences the splicing of the *doublesex* (*dsx*) gene's primary transcript. This splicing pathway results in the production of a female-specific Dsx protein called Dsx-F. In males, where there is no Tra protein, the splicing of the *dsx* primary transcript produces the related, but different, Dsx-M protein (Fig. 18.22b). The N-terminal parts of the Dsx-F and Dsx-M proteins are the same, but the C-terminal parts of the proteins are different.

Effects of Mutations in *tra*

Like mutations in *Sxl,* mutations in *tra* affect the two sexes differently. Since males do not normally express *tra,* null mutations of the gene have no effect on them. In females, however, the Tra protein helps implement female-specific patterns of RNA splicing. In the absence of Tra protein, *dsx* primary transcripts can only be spliced to make Dsx-M. XX *Tra2* flies expressing Dsx-M are converted into flies having the phenotype of a male. The resulting XX males exhibit proper male sexual behavior, but without the Y chromosome, they are sterile because certain Y-linked genes are prerequisites of male fertility.

The Dsx-F and Dsx-M Proteins Are Transcription Factors That Determine Somatic Sexual Characteristics

Although both Dsx-F and Dsx-M function as transcription factors, they have opposite effects. In conjunction with the protein encoded by the *intersex* (*ix*) gene, Dsx-F functions mainly as a repressor that prevents the transcription of genes whose expression would generate the somatic sexual characteristics of males. Dsx-M, which works independently of the intersex protein, accomplishes the opposite: the activation of genes for the somatic sexual characteristics of males and the repression of genes that determine female somatic sexual characteristics. Interestingly, the two Dsx proteins can bind to the same enhancer elements, but their binding produces opposite outcomes (**Fig. 18.23**). For example, both bind to an enhancer upstream of the promoter for the *YP1* gene, which encodes a yolk protein; females make this protein in their fat body organs and then transfer it to developing eggs. The binding of Dsx-F

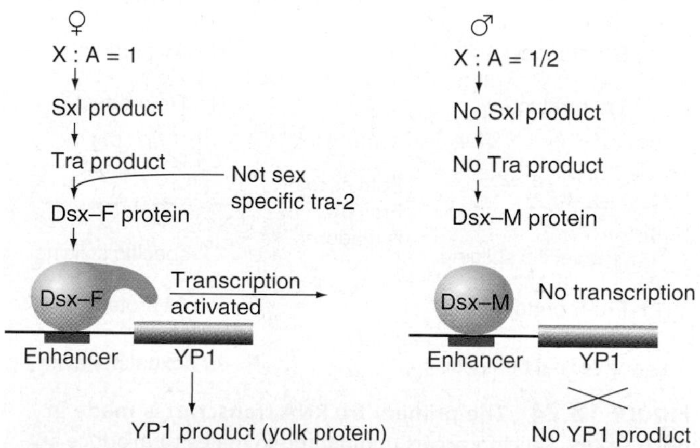

Figure 18.23 Alternative forms of the *dxs* gene product both bind to the same *YP1* enhancer, but they have opposite effects on the expression of the *YP1* gene. Dsx-F acts as a transcriptional activator, while Dsx-M acts as a transcriptional repressor.

stimulates transcription of the *YP1* gene in females; the binding of Dsx-M to the same enhancer region inactivates (in conjunction with other transcription factors) transcription of *YP1* in males.

The Effect of *dsx* Mutations

Mutations in *dsx* affect both sexes because in both males and females, the production of Dsx proteins represses certain genes specific to development of the opposite sex. Null mutations in *dsx* that make it impossible to produce either functional Dsx-F or Dsx-M result in intersexes that cannot repress either certain male-specific or certain female-specific genes.

The Tra and Tra-2 Proteins Also Help Regulate Expression of the *Fruitless* Gene

We saw at the beginning of this chapter that the courting song and dance of male *Drosophila* are among the sexual behaviors under the control of the *fruitless* (*fru*) gene. As it turns out, the *fru* primary transcript is another regulatory target of the Tra and Tra2 splicing factors (**Fig. 18.24**). In females, whose cells make both Tra and Tra2 proteins, splicing of the *fru* transcript produces an mRNA that encodes a protein we refer to as Fru-F. In males, whose cells carry no Tra protein, alternative splicing of the *fru* transcript generates a related Fru-M protein with 101 additional amino acids at its N terminus. As we mentioned at the beginning of this chapter, these additions almost certainly determine some of the observed differences between male and female behavior. Since both Fru-F and Fru-M have the zinc-finger motifs characteristic of transcription factors, they probably activate and repress genes whose sex-specific products help generate courting behaviors.

The sex-specific products of *fru,* we have seen, appear in only a few cells in the nervous system, and the location of these neurons is significant. Some are in regions known to help regulate the courtship song; others are in areas that process chemosensory information from the antennae (perhaps in neurons that receive pheromone signals); still others are in regions that control abdominal movements (suggesting how *fruitless* may influence the male's curling of the abdomen during mating). To understand precisely how changes in gene expression in these few cells control sexual behaviors, *Drosophila* researchers are now trying to discover which genes are the targets of transcriptional regulation by the fruitless protein.

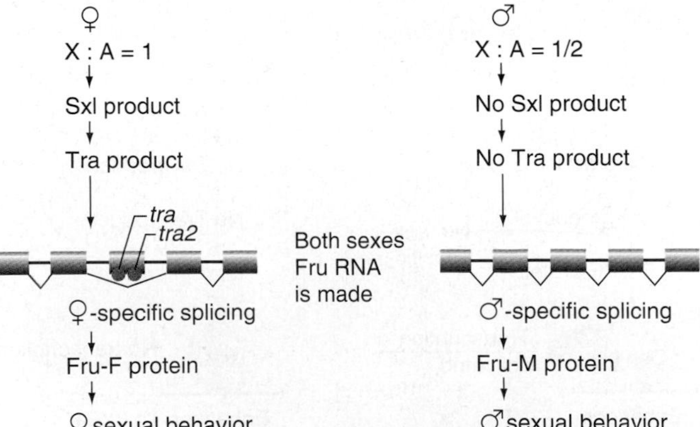

Figure 18.24 The primary *fru* RNA transcript is made in both sexes. Splicing occurs unhindered in males to produce an mRNA, which is translated into the Fru-M protein product. But tra protein (present only in females) causes alternative splicing of the *fru* transcript to produce an alternative mRNA, which encodes an alternative protein product Fru-F.

Summary: A Complex Network of Molecular Interactions Regulates the Determination of Somatic Sexual Characteristics in *Drosophila*

Several kinds of gene regulation make the embryonic "decisions" that determine the sex of the fly. Controls of both transcriptional initiation and RNA splicing play critical roles in this biological process. First, X-encoded numerator and autosome-encoded denominator elements allow the early embryo to measure the X:A ratio. Presumably, numerator homodimers, present in sufficient quantity only in females, activate *Sxl* transcription in early embryos. Next, the Sxl protein acts as a splicing factor that perpetuates its own synthesis in females. There it initiates a cascade of splicing factors, each of which determines the splicing patterns of the primary transcript of the next gene in the pathway. These splicing cascades eventually produce female-specific versions of at least two transcription factors: Dsx-F and Fru-F. In males, default splicing of these genes produces the Dsx-M and Fru-M transcription factors. The female- and male-specific versions of the Dsx and Fru proteins help turn on and off genes whose products elaborate the sexual appearance and behavior of male and female *Drosophila*.

Genes other than those just described also contribute to sex-specific behaviors. For example, a gene called *dissatisfaction* affects courtship behavior in males and females. Mutant females resist copulating with males; mutant males are maladroit in copulation because they don't properly curl their abdomen. Genetic experiments suggest that the *dissatisfaction* gene is part of the same sex-determination pathway as *fru*.

Connections

Multiple controls regulate gene activity and function from imprinting, to chromatin remodeling, to the initiation of transcription, to the processing of RNA transcripts to RNA interference, to the chemical modification of final gene products. At the outset, the regulation of transcription occurs through the interaction of *cis*-acting DNA regions and a variety of transcription factors.

Accurate regulation of gene function is crucial for proper control of the cell cycle. Indeed, a critical network of *cis*-acting control regions, *trans*-acting factors, and protein modifications promotes cell growth, DNA replication, and cell division in response to certain environmental signals and delays these events in response to other signals, such as DNA damage. In Chapter 19, we describe the regulatory network controlling the cell cycle in normal cells and explain how mutations that disrupt one or more aspects of that network can result in cancer.

Essential Concepts

1. Transcriptional initiation is a critical point in the regulation of gene activity. Analyses of mutations that affect a gene's function without changing the sequence of its product provided insight into this level of regulation. Through these mutations, researchers defined *cis*-acting DNA regulatory elements and *trans*-acting *transcription factors.*
2. The eukaryotic genome contains three classes of genes, each associated with a different type of promoter and transcribed by a different RNA polymerase. Class I genes encode the rRNAs. Class III genes encode the tRNAs as well as some other small RNA molecules. Class II genes, by far the largest class, encode all proteins and micro-RNAs.
3. Two types of *cis*-acting regulatory regions—*promoters* and *enhancers*—are associated with class II genes. All class II promoters are located at the 5′ end of the gene they influence. The enhancers have a more variable location in relation to the genes they control.
4. The binding of nonspecific *basal factors* to promoters is a prerequisite for the transcription of a gene. Basal factors alone allow a low, nonspecific basal level of gene transcription.
5. The association of transcription factors with enhancer elements can modulate levels of transcriptional initiation. Activation is mediated by transcription factors called *activators,* which bind to enhancers, and *coactivators,* which bind to activators. Activators and coactivators can interact with basal factors at the promoter to increase transcription above the basal level. *Repressors* can compete with activators for enhancer binding or quench the ability of activators to carry out their function. *Corepressors* operate by binding to DNA-bound transcription factors. Many transcription factors can function as either activators or repressors, depending on the situation. Activators and repressors that bind directly to DNA typically form homodimers and/or heterodimers. Formation of these dimers is a prerequisite for them to function as transcription factors.
6. Although promoters and enhancers are the most common and best characterized types of *cis*-acting regulatory elements, there are other elements that do not fit into either of these classes. One such element is the *locus control region* associated with the β-globin gene cluster. Another is the *insulator,* which restricts the genomic region over which an enhancer can operate.
7. The regulation of transcription is a complex biochemical process dependent on the interactions of multiple enhancer elements with many competing transcription factors, and responsive to signals originating within and outside the cell. The integration of signals and regulatory components determines the precise level of transcription of a particular gene in a particular cell at a particular time.
8. The structure of chromatin around a gene contributes to the regulation of transcription. Normal chromatin structure prevents runaway transcription from genes that are not under activation. The unraveling of the DNA in chromatin is an initial step in activation. Hypercompaction of chromatin domains causes *transcriptional silencing* by blocking access to the promoter and enhancers of a gene and thereby preventing its activation even in the presence of activator proteins.

9. *Genomic imprinting* is an example of epigenetic control over gene expression. Imprinting operates on the copy of a gene received from one parent but not the other. *DNA methylation* plays a role in the maintenance of imprinting from one mammalian somatic cell generation to the next.

10. Although the regulation of most genes depends primarily on controls over transcription, in some cases, further regulation down the path to protein production also plays a role. Modulation of gene function can occur through changes in RNA splicing, RNA interference, changes in the efficiency of translation, and chemical modification of the gene product.

Social and Ethical Issues

1. Since the cloning of the human growth hormone gene (*HGH*), the hormone has become available as a drug and is prescribed for children with growth hormone defects. Fred and Susannah's 12-year-old son Sam is below average in height, and his parents want the family doctor to prescribe growth hormone. They feel that Sam is suffering socially because of his small size and that his height will have a lasting psychological impact on him. The doctor believes that medicine with a lifelong effect is only appropriate when actual disease is present, and he does not consider Sam to have a disease. Who should set the limits for the use of HGH to treat potential growth problems in children—a governmental agency, the child's physician, or the parents? If a governmental agency or an individual physician regulates the use of HGH, how should they decide on a cutoff height at which to begin treatment, and how should they determine a final height at which treatment should end? In thinking about these questions, take into consideration the fact that different human populations have different distributions of height, with mean values that can differ by as much as 7 inches.

2. Statins are a class of drugs that lower the level of a harmful form of cholesterol called LDL, which plays a significant role in atherosclerosis, a form of heart disease. The drug acts by inhibiting the liver enzyme HMG-CoA reductase, which converts HMG-CoA to mevalonate, an early step in the biosynthetic pathway for cholesterol. Unfortunately, in some people, some forms of the drug can cause grossly elevated levels of the muscle enzyme creatine kinase, leading to muscle failure and even, rarely, death. Sean has borderline high levels of LDL and wants to begin taking statins. His doctor thinks that the side effects of the drug are not well enough known, and she doesn't want to prescribe the drug to Sean. Should Sean listen to his doctor, or should he try to find another doctor who is more willing to prescribe the drug?

3. Experimentation has shown that fetal tissue can be used to treat conditions such as the neurological disorder of Parkinson disease. However, because the fetal tissue is at a relatively early stage in development, it could contain many proteins that will affect the expression of genes in the tissue into which it is transplanted. If this is an experimental technology, should the individuals taking part in the trials be obliged to undergo testing that monitors changes in their physiology, or should they be able to choose which follow-up studies they want to participate in?

Solved Problems

I. You are studying expression of a gene whose protein product is made after UV irradiation. You cloned the gene and made antibody to the protein.
   a. If expression is regulated by turning on transcription after UV exposure, what results would you predict from hybridizing a DNA probe to RNA isolated from cells before and after UV irradiation (Northern analysis) and from incubating the antibody to proteins isolated from cells before and after UV treatment (Western analysis)?
   b. If expression is regulated by preventing translation, what results would you predict from doing similar Northern and Western analyses?

Answer

To answer this question, you need to consider the consequences of transcriptional and translational regulation on expression and think through what happens experimentally in Northern and Western analyses.

a. If a gene is transcriptionally regulated, the mRNA will not be present in cells that were not exposed to UV. *There will be no hybridizing band in the Northern analysis of mRNA from unexposed cells. The mRNA will be present in cells that have been treated with UV, and there will be a hybridizing band.* Similarly, the protein will only be found in cells that were exposed to UV, and *the antibody will bind to its protein target only in the protein preparation from exposed cells.*

b. If expression is regulated at the translation step, mRNA will be present in the cells whether they have been exposed to UV or not. *Hybridizing bands will be found in both RNA samples. The protein will be present only in those cells that were exposed to UV, so signal will be seen only in the exposed preparation.*

II. The retinoic acid receptor (RAR) is a transcription factor that is similar to steroid hormone receptors. The substance (ligand) that binds to this receptor is retinoic acid. One of the genes whose transcription is activated by retinoic acid binding to the receptor is *myoD*. The diagram at the end of this problem shows a schematic of the RAR protein produced by a gene into which two different 12-base oligonucleotides had been inserted in the sequences encoding sites indicated by a—m. For constructs encoding a–e, oligonucleotide 1 (TTAATTAATTAA) was inserted into the *RAR* gene. For constructs encoding f–m, oligonucleotide 2 (CCGGCCGGCCGG) was inserted into the gene. Each mutant protein was tested for its ability to bind retinoic acid, bind to DNA, and activate transcription of *myoD* gene. Results are tabulated as follows; the insertion site associated with each mutant protein is indicated with the appropriate letter on the polypeptide map.

NH_2 f g h i j k l m COOH
a b c d e

Mutant	Retinoic acid binding	DNA binding	Transcriptional activation
a	–	–	–
b	–	–	–
c	–	–	–
d	–	+	+
e	+	+	+
f	+	+	+
g	+	+	–
h	+	+	–
i	+	–	–
j	+	–	–
k	–	+	+
l	–	+	+
m	+	+	+

a. What is the effect of inserting oligonucleotide 1 anywhere in the protein?

b. What is the effect of inserting oligonucleotide 2 anywhere in the protein?

c. Indicate the three protein domains on a copy of the preceding drawing.

Answer

This question involves the concepts of domains within proteins and use of the genetic code to understand effects of oligonucleotide insertions.

a. Oligonucleotide 1 contains a stop codon in any of its three reading frames. This means it will *cause termination of translation of the protein wherever it is inserted.*

b. Oligonucleotide 2 does not contain any stop codons and so will *just add amino acids to the protein.* Because there are 12 bases in the oligonucleotide, it will not change the reading frame of the protein. *Insertion of the oligonucleotide can disrupt the function of a site in which it inserts.*

c. Looking at the data overall, notice that all mutants that are defective in DNA binding are also defective in transcriptional activation, as would be expected for a transcription factor that binds to DNA. The mutants that will be informative about the transcriptional activation domain are those that do not have a DNA-binding defect. Inserts a, b, and c using oligonucleotide 1, which truncates the protein at the site of insertion, are defective in all three activities. The protein must be made at least as far as point d before DNA binding or transcription activation are seen. These two activities must lie before d. Truncation at d is negative for retinoic acid binding, but the truncation at e does bind to retinoic acid. The retinoic acid-binding activity must lie before e. Using the oligonucleotide 2 set of insertions, transcriptional activation was disrupted by insertions at sites g and h, indicating that this region is part of the transcriptional domain; i and j insertions disrupted the DNA binding; and k and l insertions disrupted the retinoic acid binding. The minimal endpoints of domains as determined from these data are summarized in the following schematic.

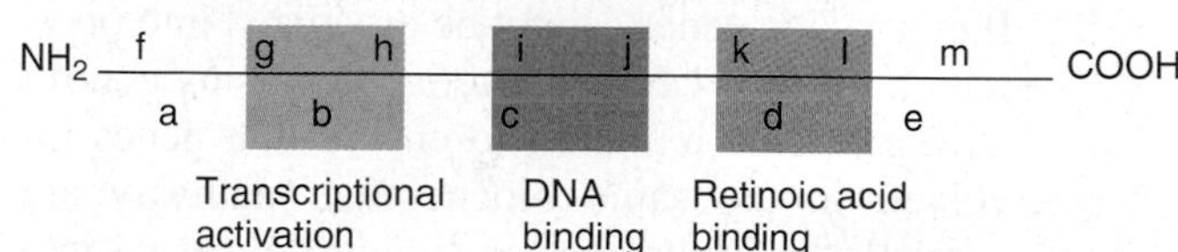

III. A cDNA clone that you isolated using pituitary gland mRNA from mice was used as a probe against a blot containing RNAs from embryonic heart (EH), adult heart (AH), embryonic pituitary (EP), adult pituitary (AP), and testis (T). The results of the hybridization are shown here.

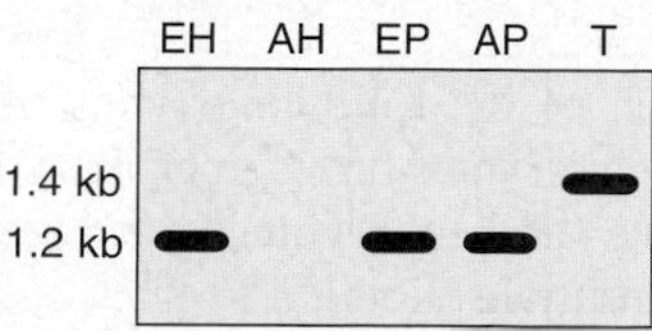

a. What would you conclude about this gene based on the result with AH RNA?

b. How would you explain the result with testis RNA?

Answer

This problem requires an understanding of RNA and transcription.

a. No RNA from adult heart (AH) hybridized with the probe, indicating that the *gene is not transcribed in this tissue.*

b. *A different-sized RNA is seen in the testis sample.* This could be due to *alternate splicing of the transcript or a different start site* in testis compared to other tissues.

Problems

Vocabulary

1. For each of the terms in the left column, choose the best matching phrase in the right column.

a. basal factors	1. influences transcription from multiple genes over a large region of DNA
b. transcriptional silencing	2. marks a protein for degradation
c. locus control region	3. pattern of expression is dependent on which parent transmitted the allele
d. activators	4. multimers of nonidentical subunits
e. imprinting	5. heterochromatin
f. RNAi	6. multimers of identical subunits
g. coactivators	7. bind to enhancers
h. homomers	8. bind to promoters
i. heteromers	9. bind to activators
j. ubiquitination	10. prevents or reduces gene expression post transcriptionally

Section 18.1

2. a. Is there a fundamental reason that would prevent eukaryotic genes from being organized into operons as they are in prokaryotes? If so, what is the reason?

b. If eukaryotic genes cannot be organized into operons, how then can eukaryotic organisms control transcriptional initiation to ensure that genes involved in the same biochemical pathway are coregulated? Your answer should consider separately coregulated genes scattered in different regions of the genome and coregulated genes that are clustered in the same region.

c. Several different proteins can be expressed from a single operon mRNA molecule in prokaryotes. Is there any way that a single mRNA molecule in a eukaryotic cell could produce several different proteins?

3. Does each of the following types of gene regulation occur in eukaryotes only? in prokaryotes only? in both prokaryotes and eukaryotes?

a. differential splicing
b. positive regulation
c. chromatin compaction
d. RNA interference
e. attenuation
f. negative regulation

4. Which of the following types of fusion gene would you use for which purpose? (The slash indicates the fusion, and the parts of each type of fusion are given in the order in which they would occur.)

Types of fusions:

a. DNA-binding domain of *lex*A gene/random mouse genome fragments
b. random mouse sequences/*lacZ* gene
c. mouse metallothionein promoter/a mouse gene

Uses:

i. to identify genes turned on in specific cell types
ii. to turn on expression of a gene by including Zn in the diet
iii. to search for transcription-activation domains

Section 18.2

5. Which eukaryotic RNA polymerase (RNA pol I, pol II, or pol III) transcribes which genes?

a. tRNAs
b. mRNAs
c. rRNAs
d. miRNAs

6. You isolated a gene expressed in differentiated neurons in mice. You then fused the upstream DNA and beginning of the gene to *lacZ* (reporter gene) so that you could monitor expression. Different fragments (shown as *dark lines* in the following figure) were cloned next to the *lacZ* gene that lacked a promoter. The clones were introduced into neurons in tissue culture to monitor expression. From the results that follow, which region contains the promoter and which contains an enhancer?

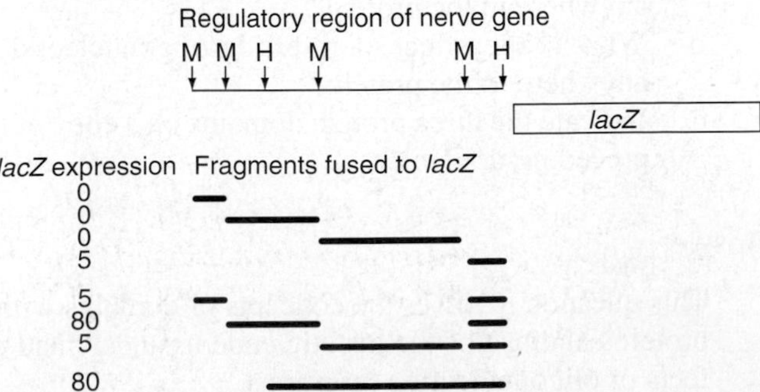

7. Which of the listed motifs is associated with DNA binding, transcription activation, or dimer formation?
   a. zinc finger
   b. helix-loop-helix
   c. leucine zipper
   d. acidic region
   e. helix-turn-helix

8. In yeast, the GAL4 protein binds to DNA to activate transcription of GAL7 or GAL10. GAL80 represses expression by binding to GAL4 protein and preventing it from binding to DNA. In which gene(s) should you be able to isolate galactose constitutive mutations, and in each case, what characteristics of the protein would the mutation disrupt?

9. A single enhancer site regulates expression of three adjacent genes *GAL1, GAL7,* and *GAL10,* but the genes are not cotranscribed as one mRNA. How could you show experimentally that each gene is transcribed separately?

10. How could you make a library of genes expressed during sporulation in yeast?

11. *MyoD* is a transcriptional activator that turns on the expression of several muscle-specific genes in human cells. The *Id* gene product inhibits *MyoD* action. How could you determine if *Id* acts by quenching *MyoD* (as *GAL80* does to *GAL4* in yeast) or by blocking access to the enhancer? What differences would you expect to see experimentally?

12. You isolated nuclei from liver cells, treated them with increasing amounts of DNase-I, stopped the reactions, and isolated the DNA from each sample. You next treated the DNAs with the restriction enzyme *Eag*I, electrophoresed the DNAs, transferred them to a blot, and hybridized the blot with a probe from the gene you are studying. With no DNase-I treatment, there was a 20 kb *Eag*I fragment that hybridized with your probe. With trace amounts of DNase-I, two bands of 16 and 4 kb were present. The same DNase-I treatment of DNA from muscle cells produced only a 20 kb fragment. What does this result tell you about the region of DNA?

13. What experimental evidence indicates that chromatin structure acts to reduce basal levels of transcription?

14. Match the gene expression phenomenon with molecular components that modulate each.

a.	transcriptional silencing	1.	insulator
b.	imprinting	2.	heterochromatin
c.	restricting the range of enhancer activity	3.	TBF, TAFs
d.	basal transcription	4.	methylation

15. Which of the following would be suggested by a DNase hypersensitive site?
   a. No transcription occurs in this region of the chromosome.
   b. The chromatin is in a more open state than a region without the hypersensitive site.
   c. Transcription terminates at this site.

16. From Northern analysis, you find that the *ADAG* gene is expressed only in the brain. You examine expression in glial and neuronal cell lines (two types of cells in the brain) and find that only glial cells make *ADAG* mRNA. No one has characterized the *cis*- or *trans*-acting elements required for glial specific expression, so you decide to do so. You make a set of deletions in the regulatory region and fuse these to the *lacZ* gene so you can easily monitor the expression after introducing the clones into tissue culture cells derived from a glial tumor. Deletions beginning at a site upstream of the gene and extending to base −85 (with the transcription start site considered −1 and bases prior to the start having negative designations) still retain full activity, but a deletion to −75 leaves only 1% of the original activity.
   a. What do these findings tell you?

You now mix a DNA fragment from this region that is labeled with ^{32}P at the 5′-end of one strand with a purified glial-specific transcription factor. You perform a DNase-I footprinting experiment (as described in Figure 17.15 on p. 622), and obtain the results tabulated below. Lane 1 shows the DNase-I reaction of the labeled DNA alone, Lane 2 is the reaction involving the mixture of DNA and protein. You also analyze the DNA sequence of the same DNA fragment on the same gel.

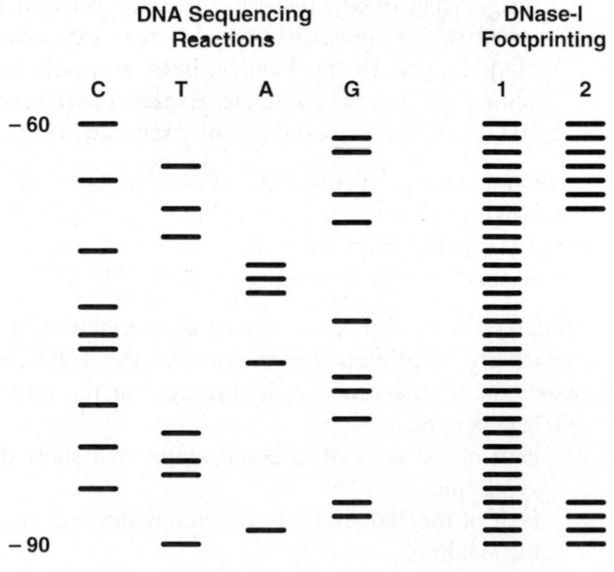

DNA Sequencing Reactions — C T A G
DNase-I Footprinting — 1 2
−60
−90

   b. What is the sequence of the segment containing the binding site(s) for the glial-specific transcription factor?
   c. Additional evidence indicates that the glial-specific factor binds to DNA as a dimer, and each monomer binds the same target sequence. Identify the likely binding sites for the two monomers.

17. a. Assume that two transcription factors are required for expression of the blue pigmentation genes in pansies. (Without the pigment, the flowers are white.) What phenotypic ratios would you expect from crossing strains heterozygous for each of the genes encoding these transcription factors?

b. Now assume that either transcription factor is sufficient to get blue color. What phenotypic ratios would you expect from crossing strains heterozygous for each of the genes encoding these transcription factors?

18. a. You want to create a genetic construct that will express the β-galactosidase enzyme encoded by *E. coli*'s *lacZ* gene in *Drosophila.* In addition to the *lacZ* coding sequence, what DNA element(s) must you include in order to express this protein in flies if the construct could somehow become intergrated into the *Drosophila* genome? Where should such DNA element(s) be located?

b. In making your construct, you place inverted repeats found at the ends of a particular type of transposable element on either side of the *lacZ* coding region and all of the DNA elements required by the answer to part *a.* Since all the DNA sequences located between these inverted repeats can move from place to place in the *Drosophila* genome, it is possible to generate many different fly strains, each with the construct integrated at a different location in the genome. You treat animals from each strain with a chemical that turns blue in the presence of β-galactosidase. Animals from different strains show different patterns: some show blue staining in the head, others in the thorax, some show no blue color, etc. Explain these results and describe a potential use of your construct.

19. In the previous problem (#18), you identified a region that is likely to behave as an enhancer. What experiments could you perform to verify that these DNA sequences indeed share all the characteristics of an enhancer?

20. Prader-Willi syndrome is caused by a mutation in a maternally imprinted gene. Answer the following questions as true or false, assuming that the trait is 100% penetrant.

a. Half of the sons of affected males will show the syndrome.

b. Half of the daughters of affected males will show the syndrome.

c. Half of the sons of affected females will show the syndrome.

d. Half of the daughters of affected females will show the syndrome.

21. A boy expresses a mutant phenotype because he has received a mutation in a paternally imprinted gene. From which parent did the boy inherit the mutant allele?

22. The *IGF-2R* gene is autosomal and maternally imprinted. Copies of the gene received from the mother are not expressed, but copies received from the father are expressed. You have found two alleles of this gene that encode two different forms of the IGF-2R protein distinguishable by gel electrophoresis. One allele encodes a 60K blood protein; the other allele encodes a 50K blood protein. In an analysis of blood proteins from a couple named Bill and Joan, you find only the 60K protein in Joan's blood and only the 50K protein in Bill's blood. You then look at their children, Jill and Bill Jr. Jill is producing only the 50K protein, while Bill Jr. is producing only the 60K protein.

a. With these data alone, what can you say about the *IGF-2R* genotype of Bill Sr. and Joan?

b. Bill Jr. and a woman named Sara have two children, Pat and Tim. Pat produces only the 60K protein and Tim produces only the 50K protein. With the accumulated data, what can you now say about the genotypes of Joan and Bill Sr.?

23. Assume that the disease illustrated with the pedigree below is due to expression of a rare allele of an autosomal gene that is a paternally imprinted. What would you predict is the genotype of individuals (a) I-1, (b) II-1, and (c) III-2?

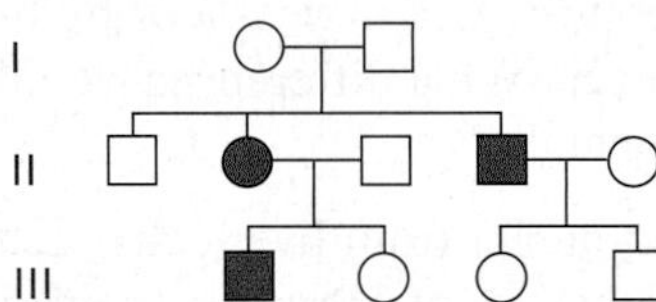

24. Follow the expression of a paternally imprinted gene through three generations. Indicate whether the copy of the gene from the male in generation I is expressed in the germ cells and somatic cells of the individuals listed.

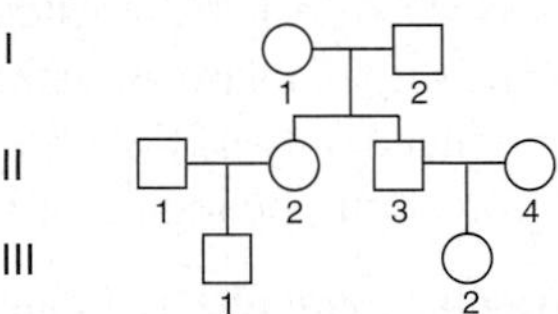

a. generation I male (I-2) germ cells
b. generation II daughter (II-2): somatic cells
c. generation II daughter (II-2): germ cells
d. generation II son (II-3): somatic cells
e. generation II son (II-3): germ cells
f. generation III grandson (III-1): somatic cells
g. generation III grandson (III-1): germ cells

Section 18.3

25. What events occur during processing of a primary transcript?

26. You are studying muscle cells and have found a protein that is only made in this tissue. The data here are from analysis of RNA (Northern blot) using a DNA probe from the gene and analysis of protein (Western blot) using antibody directed against the protein. Is this gene transcriptionally or translationally regulated?

RNA analysis
Muscle cell RNA | Nerve cell RNA | Skin cell RNA

Protein analysis
Muscle cell protein | Nerve cell protein | Skin cell protein

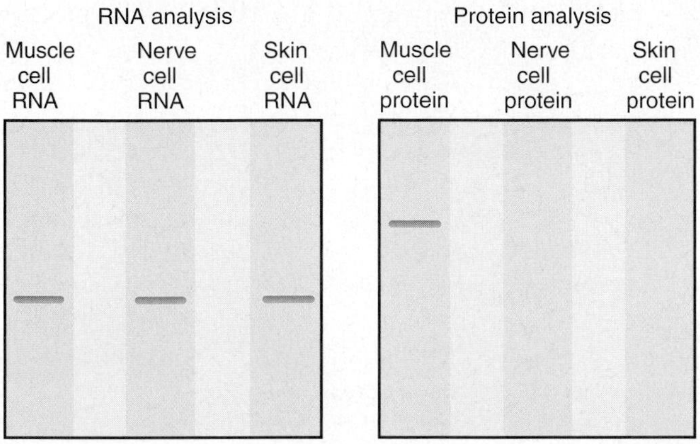

27. The *hunchback* gene, one of the genes necessary for setting up the dorsal-ventral axis of the *Drosophila* embryo, is translationally regulated. The position of the coding region within the transcript is known, and there is additional sequence beyond the coding region at the 5′ and 3′ ends of the mRNA. How could you determine if the sequences at the 5′ or 3′ or both ends are necessary for proper regulation of translation?

28. You isolated a cDNA from skin cells, and when you hybridized that cDNA as a probe with a blot containing mRNAs from skin cells and nerve cells you saw a 1.2 and a 1.3 kb fragment, respectively. How could you explain the different-sized cDNAs?

29. From Northern and Western hybridization, you know that the mRNA and protein produced by a tissue-specific gene are present in brain, liver, and fat cells, but you detect an enzymatic activity associated with this protein only in fat cells. Provide an explanation for this phenomenon.

30. Modern-day geneticists are very excited by the prospect of using RNA interference as a way to find genes involved in various biological processes, such as mitosis or the development of specific body parts like the pancreatic cells that make insulin.

a. How would you perform an RNAi-based screen to find genes involved in these processes?

b. What is the advantage of doing an RNAi-based screen as opposed to a classical genetic screen involving mutagens?

c. What are the disadvantages of performing an RNAi-based screen?

31. You are studying a strain of transgenic mice that express an *E. coli lacZ* reporter gene under the control of *cis*-acting regulatory elements that normally control an interesting gene needed for the early development of mice. Previous evidence from Northern blots indicates that mRNA for the gene of interest can be identified between days 8.5 and 10.5 of gestation. In your strain, staining for β-galactosidase (the protein product of *lacZ*) can be seen from about day 8.75 until at least day 12.

a. Explain the discrepancy between mRNA and protein expression.

b. Would you expect β-galactosidase protein expression to indicate more accurately the normal onset of activity for this gene, or the normal cessation of this gene's activity? Explain.

Section 18.4

32. The *Drosophila* gene *Sex lethal (Sxl)* is very deserving of its name. Certain alleles have no effect on XY animals, but cause XX animals to die early in development. Other alleles have no effect on XX animals, but XY animals with these alleles die early in development. Thus, some *Sxl* alleles are lethal to females, while others are lethal to males.

a. Would you expect a null mutation in *Sxl* to cause lethality in males or in females? What about a constitutively active *Sxl* mutation?

b. Why do *Sxl* alleles of either type cause lethality in a specific sex?

The gene *transformer (tra)* gets its name from "sexual transformation," since some *tra* alleles can change XX animals into sterile males, while other *tra* alleles can change XY animals into normal-appearing females.

c. Which of these sex transformations would be caused by null alleles of *tra* and which would be caused by consitutively active alleles of *tra*?

d. XX animals carrying particular alleles of *tra* develop as males, but they are sterile. Why?

e. In contrast with *Sxl,* null *tra* mutations do not cause lethality either in XX or in XY animals. However, the Sxl protein regulates the production of the Tra protein. Why then do all *tra* mutant animals survive?

f. Predict the consequences of null mutations in *tra-2* on XX and XY animals. (See Fig. 18.23 on p. 675.)

g. XY males carrying loss-of-function mutations in the *fruitless* (*fru*) gene display aberrant courtship behavior. Would you predict that either XX or XY animals with wildtype alleles of *fru* but loss-of-function mutations of *tra* would also court abnormally?

Cell-Cycle Regulation and the Genetics of Cancer

Chapter 19

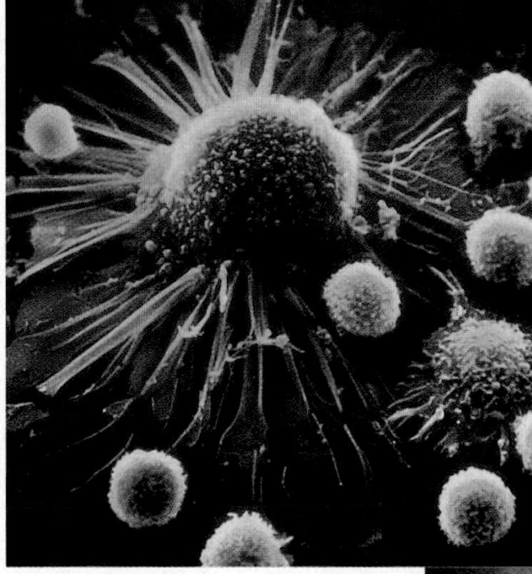

Killer cells surround a large cancer cell. Note the extended cellular processes of the cancer cell.

The roughly 300 different types of cells that compose the brain, heart, lungs, liver, immune system, and other tissues of a healthy human adult proliferate only when necessary—to replace cells lost through normal attrition or injury and, in the case of the immune system, to attack infectious agents. Many cells, such as those that constitute the inner skin, blood, and intestinal lining, multiply daily to compensate for the senescence and death of their counterparts. Other cells, such as those in the liver, rarely divide under normal circumstances but proliferate rapidly when toxins destroy some among them. Still others, such as specific cells of the immune system, reproduce in response to particular bacteria and viruses. Finally, some cells, such as the terminally differentiated cells of nerves and surface skin, never divide.

Most of us give little thought to this life-sustaining cycle of cell death and renewal, because despite the continuous comings and goings of our body's cells, we generally remain the same shape and size. But the appearance of an abnormal growth or a set of symptoms diagnosed as cancer startles us into the realization that we take for granted the intricate checks and balances that control cell division and behavior. These controls enable all cells of the body to function as part of a tightly organized, cooperative society. Cancer results when some cells divide out of control and eventually acquire the ability to spread beyond their prescribed boundaries. Cancer rarely occurs in children, but it is a common disease of older adults, in whom it accounts for about one-fourth of all deaths. Although there are many types of cancer, they all result from excessive and inaccurate cellular proliferation (**Fig. 19.1**). Thus, the body's ability to regulate cell division is a foundation of health.

Scientists who study cancer seek information about the normal controls that govern cell proliferation as a basis for understanding what goes wrong in cancer. But they often work in reverse, using the abnormalities of cancer to retrieve information about the normal operation of the cell cycle. Observations of the differences in behavior between normal and cancerous cells, obtained from diseased individuals or from cultured cell lines, have made it clear that cancer cells can divide when normal cells cannot, and cancer cells can migrate from one tissue to another, whereas normal cells cannot. Unfortunately, it is hard to study cellular behavior over time *in vivo* and *in vitro.* To study cells *in vivo,* it is necessary to see and manipulate them; but this requires cutting open an animal and carrying out extended observations, an approach that is rarely possible in humans and not easy even in model organisms like mice. It is also difficult to compare most types of tumor cells with normal laboratory-grown cells because most normal cell types do not grow in culture, and even those that do have limited life spans. Genetic studies on the genomes of various eukaryotic organisms, from yeast to humans, have helped overcome these experimental impasses by identifying genes in which mutations lead to cancer; such studies continue to provide insights that are unavailable from other types of analyses.

In this chapter, we examine the genes and gene products that control cell proliferation, including molecules that control the machinery of cell division, molecules that integrate the repair of DNA damage with progression through the cell cycle, and molecules that relay messages about whether conditions are right for cell division. We then describe how mutations in these genes and gene products lead to cancer.

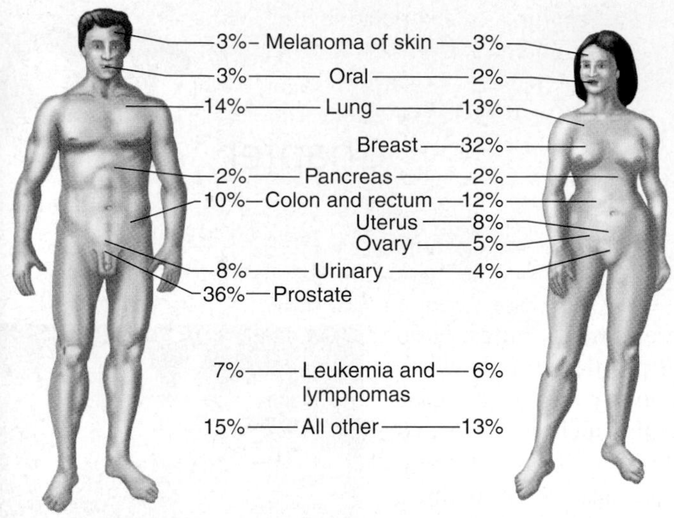

Figure 19.1 The relative percentages of new cancers in the United States that occur at different sites in the bodies of men and women.

Two unifying themes emerge from our discussion. First, cancer is ultimately a disease of the genes: The multiple phenotypes collectively referred to as cancer all result from mutations in genes that regulate a cell's passage through the cycle of growth and division. Chemicals in the environment that raise the rate of gene mutation increase the probability of cancer incidence. Second, cancer differs in two ways from cystic fibrosis, Huntington disease, and other genetic conditions caused by the inheritance of one or two copies of a single defective gene: (1) Although some people inherit mutations that predispose them to cancer, most mutations that lead to cancer occur in the somatic cells of one tissue, and (2) multiple mutations in an array of genes must accumulate over time in the clonal descendants of a single cell before the cancer phenotype appears. By contrast, the mutations that cause cystic fibrosis and Huntington disease are transmitted through the germ line; thus, the mutant alleles of one particular gene appearing in all cells of all somatic tissues cause the disease in an affected individual.

Our presentation of the correlation between cell-cycle regulation and cancer covers

- The normal control of cell division, including a description of the normal cell cycle and the molecular signals, machinery, and checkpoints that regulate passage through the cell cycle.
- How cancer arises from malfunctions in controls over cell division, including a description of cancer phenotypes, an analysis of the clonal nature of tumors, and an examination of the mutations in protooncogenes and tumor-suppressor genes that underlie all cancers.

19.1 The Normal Control of Cell Division

A variety of genes and proteins control the events of the cell cycle. These genes and proteins allow progression to the next stage of the cycle when all is well, but they cause the cellular machinery to slow down when damage to the genome or to the machinery itself requires repair. We now describe the molecules that control cell division.

Cyclin-Dependent Kinases Collaborate with Cyclins to Ensure the Proper Timing and Sequence of Cell-Cycle Events

Cell division, we saw in Chapter 4, requires the duplication of chromosomes and other cellular components as well as the precise partitioning of the duplicated elements to two daughter cells. During this complicated process, the cell coordinates the function of hundreds of different proteins. To see how the cell orchestrates the events of cell division, we first review the stages of the cell cycle and then look at some of the proteins that control progression through that cycle.

The Cell Cycle Has Four Phases: G_1, S, G_2, and M

G_1 is the period, or gap, between the end of a mitosis and the DNA synthesis that precedes the next mitosis (**Fig. 19.2**). During G_1, the cell grows in size, imports materials to the nucleus, and prepares in other ways for DNA replication. S is the period of DNA synthesis, or replication. It requires the timely, sequential activation of proteins that function at the replication fork, as well as the timely inactivation of these same proteins. G_2 is the gap between DNA synthesis and mitosis. During G_2, the cell prepares for division. M, the phase of mitosis, includes the breakdown of the nuclear membrane, the condensation of the chromosomes, their attachment to the mitotic spindle, and the segregation of chromosomes to the two poles; at the completion of

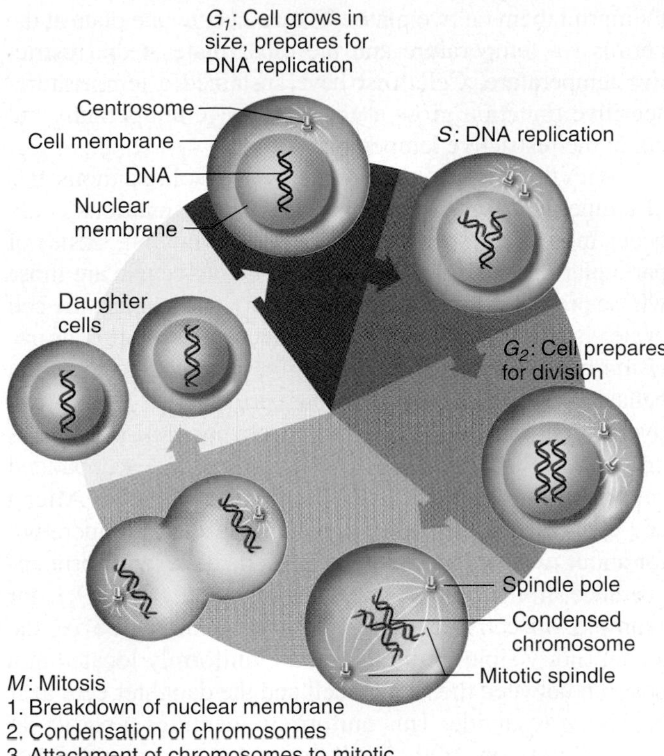

Figure 19.2 The cell cycle is the series of events that transpire between one cell division and the next. After division, a cell begins in the G_1 phase, progresses into *S* phase, where the chromosomes replicate, to the G_2 phase, and to the *M* phase, where replicated sister chromosomes segregate to daughter cells. In M phase, the nuclear membrane breaks down, centrosomes form the poles of the spindle, and microtubules construct a scaffold on which chromosomes migrate.

mitosis, the cell divides by cytokinesis. During M, the cell must coordinate the activities of a variety of proteins: those that cause chromosome condensation (see Chapters 4 and 13), tubulins that polymerize to form the mitotic spindle on which the chromosomes move, motor proteins in the kinetochoros that power chromosome movement, proteins that dissolve and re-form the nuclear membrane at the beginning and end of mitosis, and others.

Experiments with Yeast Helped Identify Genes That Control Cell Division

The budding yeast *Saccharomyces cerevisiae* and the fission yeast *Schizosaccharomyces pombe* have been instrumental in identifying the genes that control cell division. Several properties of both yeast species make them particularly useful for the genetic dissection of cellular processes. Both can grow as haploid or diploid organisms. As a result, it is possible to identify recessive mutations in the haploid cells, which carry no second, wild-type allele to obscure the mutant phenotype; and it is possible to construct diploid cells containing two mutations, allow them to proliferate, and test the resulting cell populations to determine the number of complementation groups defined by the mutations. The budding yeast *S. cerevisiae* has yet another property that facilitates cell-cycle analysis. At the beginning of the cell cycle, toward the end of G_1, a new daughter cell arises as a bud on the surface of the mother cell. As the mother cell progresses through the division cycle, the bud grows in size; it is small during S phase and large during mitosis. Bud size thus serves as a marker of progress through the cell cycle (**Fig. 19.3a**). One can order cells in an asynchronously cycling population according to position in the cell cycle by

(a)

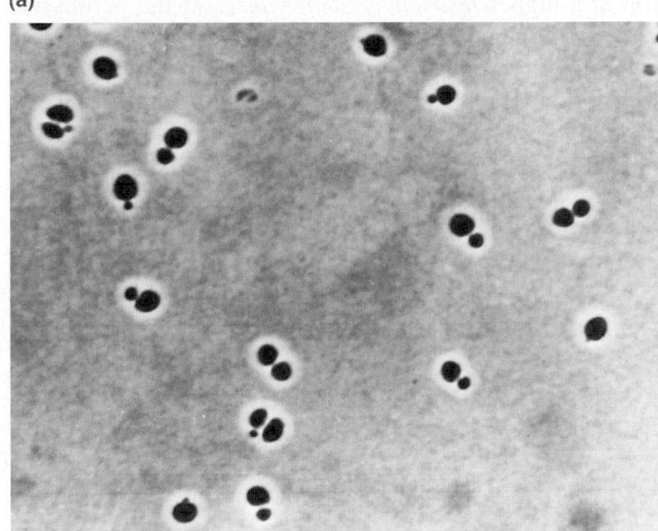

(b)

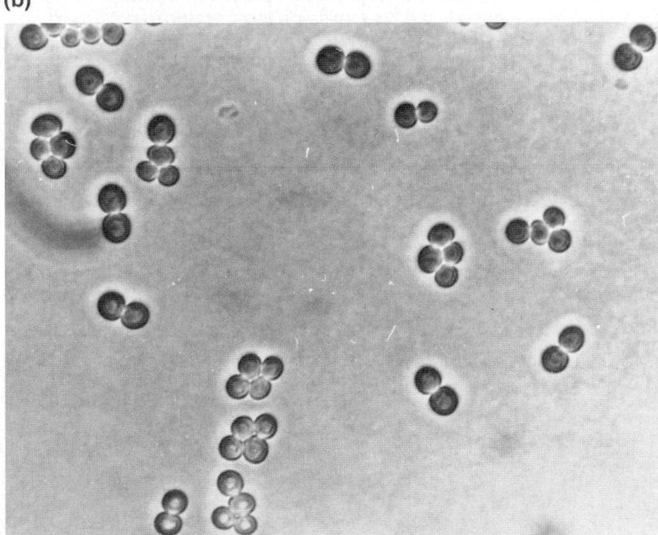

Figure 19.3 A cell-cycle mutant of yeast. Cells of a temperature-sensitive mutant growing at the permissive temperature **(a)** display buds of all sizes. After incubation at the restrictive temperature **(b)**, the same cells have arrested—all with a large bud. Cells that are early in the cell cycle at the time of the temperature shift (small buds in [a]) arrest in the first cell cycle. Cells that are later in the cell cycle (large buds in [a]) finish the first cell cycle and arrest in the second, producing clumps with two large-budded cells.

observing the relative sizes of their buds. A normal population of growing yeast cells contains nonbudding cells as well as cells with buds of all sizes.

Mutations that interfere with the cell cycle are lethal; and because cell proliferation depends on successive repeats of the cell cycle, a mutant unable to complete the cell cycle cannot grow into a population of cells (Fig. 19.3b). Researchers have obtained cell-cycle-defective mutants by isolating cells with temperature-sensitive mutations (see the Fast Forward box in Chapter 7 on pp. 240–241). In these mutants, a protein needed for cell division functions normally at a low *permissive temperature* but loses function at a higher *restrictive temperature.* At the permissive temperature, the mutants grow nearly normally, producing a population of cells for study. A shift to the restrictive temperature causes the temperature-sensitive gene in the mutant population to become nonfunctional; researchers can then study the consequences of its loss.

To isolate temperature-sensitive mutations, investigators expose haploid cells to a mutagen and then plate them at the permissive temperature, allowing them to form colonies (**Fig. 19.4**). After the colonies grow up, each from a single mutant cell, the experimenters use replica plating to imprint them on two plates. They incubate one plate at the permissive temperature and the other plate at the restrictive temperature. Cells that have sustained a temperature-sensitive mutation grow at the permissive temperature but not at the restrictive temperature.

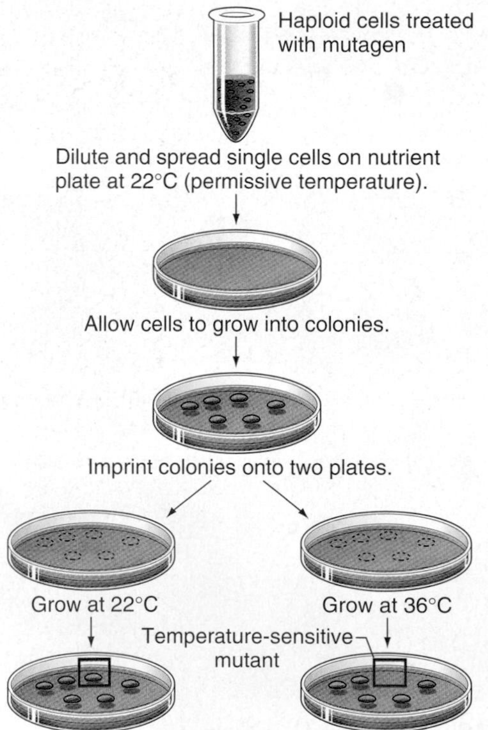

Figure 19.4 The isolation of temperature-sensitive mutants of yeast. Mutations are induced in a culture of haploid cells by exposure to a chemical mutagen. The treated cells are distributed onto solid medium. Each cell proliferates to a colony (clone) of cells, passing on the mutation. Replicas of the colonies are imprinted onto solid medium. One is grown at the permissive temperature (22°C), one at the restrictive temperature (36°C). Colonies that grow on the former, but not the latter, carry a temperature-sensitive mutation.

With this protocol, researchers have isolated thousands of temperature-sensitive mutations. These mutations could occur in any gene required for cell reproduction. Genes of particular interest for understanding the cell cycle are those whose protein product functions at only one stage in the cell cycle. It was possible to identify such mutants in *S. cerevisiae* by observing in the light microscope the shape and behavior of cells shifted from the permissive to the restrictive temperatures (see Fig. 19.3). A population of cells growing at the permissive temperature includes unbudded cells as well as cells with the full range of bud sizes. After a cell-cycle mutant has grown at the restrictive temperature for about two cycles, however, the cells have a uniform appearance. In the mutant population shown in Fig. 19.3, for example, all cells have a single large bud. Moreover, the nuclei (not visible in the figure) are uniformly located at a position between the mother cell and the daughter cell, as if beginning to divide. This uniformity identifies a particular cell-cycle mutant. Other cell-cycle mutants would arrest with different but also uniform morphologies, for example, with all unbudded cells. Thus, mutants that acquire a uniform bud-related morphology at the restrictive temperature are each defective at one stage of the cell cycle.

Further examination of cells transferred from permissive to restrictive temperatures illustrates another property of cell-cycle mutants—a requirement for the normal gene product at a particular stage of the cell cycle. Note that some of the cells in Fig. 19.3 formed one cell with a large bud at the restrictive temperature, while others formed two cells, each with a large bud. Note also that the former all had smaller buds than the latter at the time of the shift to the restrictive temperature. This observation indicates that cells early in the cell cycle at the moment of temperature shift arrested division in the first cell cycle, while those later in the cell cycle at the time of temperature shift finished the first cycle and became arrested only in the second cell cycle. The point at which a cell acquires the ability to complete a cell cycle is the moment at which the temperature-sensitive protein has fulfilled its function in that cycle.

By analyzing the morphology of buds on cells shifted from permissive to restrictive temperatures and using other methods, yeast geneticists have identified over 100 cell-cycle genes (**Table 19.1**). Epistasis experiments comparing the phenotype of a double mutant to that of each of the two single mutants have enabled them to order some of these mutations with respect to one another (**Fig. 19.5**). One of the most important mutations, in the *CDC28* gene, identified the first step in the cell cycle, a step occurring prior to that catalyzed by the *CDC7* gene: After the *CDC28* step, the cell is committed to finishing the cycle it has begun before pursuing an alternative fate. Alternative fates include the arrest of cell division in response to nutrient starvation

TABLE 19.1 Some Important Cell-Cycle and DNA Repair Genes

Genes	Gene Products and Their Function
CDKs	Enzymes known as cyclin-dependent protein kinases that control the activity of other proteins by phosphorylating them
CDC28	A CDK discovered in the yeast *Saccharomyces cerevisiae* that controls several steps in the *S. cerevisiae* cell cycle
CDC2	A CDK discovered in the yeast *Schizosaccharomyces pombe* that controls several steps in the *S. pombe* cell cycle; also the designation for a particular CDK in mammalian cells
CDK4	A CDK of mammalian cells important for the G_1-to-S transition
CDK2	A CDK of mammalian cells important for the G_1-to-S transition
cyclins	Proteins that are necessary for and influence the activity of CDKs
cyclinD	A cyclin of mammalian cells important for the G_1-to-S transition
cyclinE	A cyclin of mammalian cells important for the G_1-to-S transition
cyclinA	A cyclin of mammalian cells important for S phase
cyclinB	A cyclin of mammalian cells important for the G_2-to-M transition
E2F	A transcription factor of mammalian cells important for the G_1-to-S transition
RB	A mammalian protein that inhibits E2F
p21	A protein of mammalian cells that inhibits CDK activity
p16	A protein of mammalian cells that inhibits CDK activity
p53	A transcription factor of mammalian cells that activates transcription of DNA repair genes as well as transcription of *p21*
RAD9	A protein that inhibits the G_2-to-M transition of *S. cerevisiae* in response to DNA damage
E6	A protein of the HPV virus that inhibits p53
E7	A protein of the HPV virus that inhibits Rb

or fusion with a cell of opposite mating type to form a diploid cell (**Fig. 19.6**). The decision to pursue a particular fate occurs in the G_1 phase of the cell cycle at a step called "start." If cells have completed this *CDC28*-determined step, they cannot consider the alternatives until they have finished the cell cycle and arrived at the next G_1. The *CDC28* gene thus controls the commitment to proceed through the cell cycle.

The significance of the *CDC28* gene became apparent when geneticists identified related genes in other organisms. They found, for example, that the *CDC2* gene in fission yeast controls a step of commitment in that cell. They also

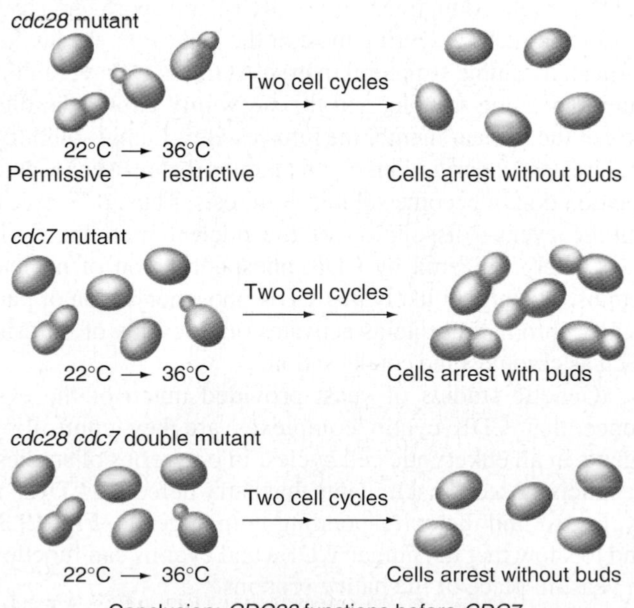

Figure 19.5 A double mutant reveals which gene is needed first. After a shift from permissive to restrictive temperatures, a temperature-sensitive *cdc28* mutant arrests as unbudded cells and a *cdc7* mutant as budded cells. The double mutant arrests with the phenotype of the *cdc28* mutant, revealing that this step occurs before the *cdc7* step.

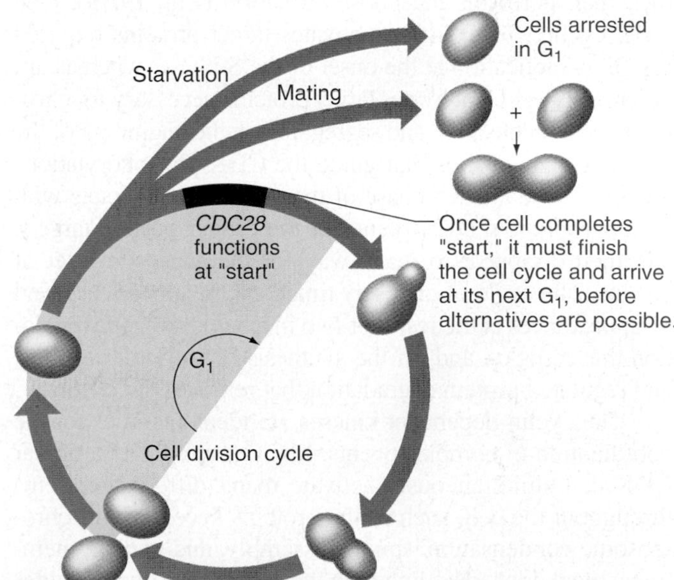

Figure 19.6 Yeast cells become committed to the cell cycle in G_1. A haploid yeast cell prior to "start" in the G_1 phase of the cell cycle is capable of arresting growth due to starvation, mating with a cell of opposite mating type, or dividing. Once the cell has passed "start," it is committed to completing the cell cycle.

learned that in extracts of *Xenopus laevis* (African clawed frog) embryos, the activity of a protein known as MPF (for maturation-promoting factor) controls the rapid early divisions. Sequences of the cloned budding and fission yeast genes revealed that they encode *protein kinases:* enzymes that add phosphate groups to their substrates. The *Xenopus* MPF also turned out to be a protein kinase. Moreover, genetic swapping experiments showed that the budding yeast *CDC28* gene and the fission yeast *CDC2* gene can replace one another in either organism, demonstrating that they encode proteins that carry out the same activity; the same is true of the *Xenopus* MPF-encoding gene.

Thus, in three different organisms, genes that seem to be the central controlling element of the cell cycle encode functionally homologous protein kinases. Further work has shown that these kinases are **cyclin-dependent kinases (CDKs);** that is, they require another protein known as a **cyclin** for their activity.

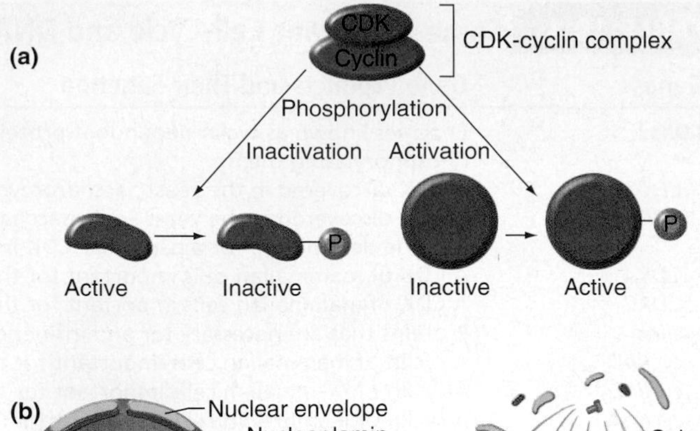

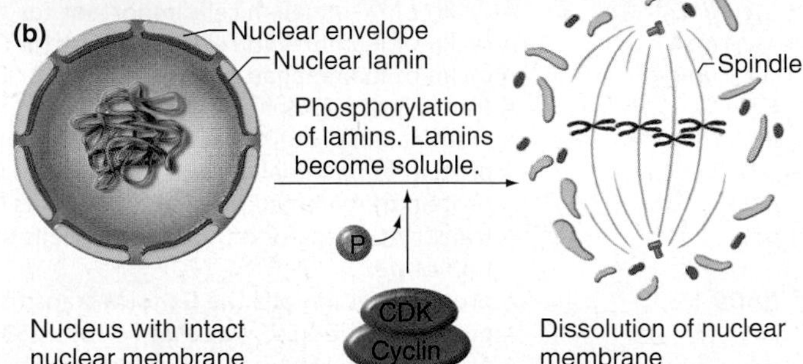

Figure 19.7 The cyclin-dependent kinases (CDK) control the cell cycle by phosphorylating other proteins. (a) A CDK combines with a cyclin and acquires the capacity to phosphorylate other proteins. Phosphorylation of a protein can either inactivate or activate it. **(b)** CDK phosphorylation of the nuclear structure proteins, lamins, is responsible for the dissolution of the nuclear membrane at mitosis.

Cyclin-Dependent Kinases and Cyclins Set the Times of Cell-Cycle Events

A protein kinase is an enzyme that adds a phosphate group to a target protein molecule; the addition either activates or inactivates the target protein. The CDKs are a family of kinases that regulate the transition from G_1 to S and from G_2 to M through phosphorylations that activate or inactivate target proteins. As mentioned and as their name implies, cyclin-dependent kinases function only after associating with another type of protein known as a cyclin. The cyclin portion of a CDK-cyclin complex specifies which set of proteins a particular CDK will phosphorylate; the CDK portion of the complex then performs the phosphorylation (**Fig. 19.7a**). One CDK-cyclin, for example, activates target proteins required for DNA replication at the onset of the S phase, whereas another CDK-cyclin activates target proteins necessary for chromosome condensation and segregation at the beginning of the M phase. The cyclins that guide the CDK phosphorylations appear on cue at each phase of the cell cycle, associate with the appropriate CDKs, point out the proper protein targets, and then disappear to make way for the succeeding set of cyclins. The cycle of precisely timed cyclin appearances and disappearances is the result of two mechanisms: gene regulation that turns on and off the synthesis of particular cyclins, and regulated protein degradation that removes the cyclins.

The cyclin-dependent kinases are ideal enzymes for the coordination of complex events. Many copies of a particular CDK can simultaneously activate many different proteins throughout the cell, such as the proteins necessary for chromosome condensation, spindle assembly, and nuclear membrane breakdown. Binding with the appropriate cyclin guides the CDK to these targets as the cell prepares for mitosis.

The target of an enzyme is its **substrate.** For example, the **nuclear lamins,** a group of proteins that underlie the inner surface of the nuclear membrane, are substrates for CDK phosphorylation (Fig. 19.7b). The nuclear lamins probably provide structural support for the nucleus and possibly provide sites for the assembly of proteins that function in DNA replication, transcription, RNA transport, and chromosome structure. During most of the cell cycle, the lamins form an insoluble structural matrix. At mitosis, however, the lamins become soluble, and this solubility allows dissolution of the nuclear membrane into vesicles. Lamin solubility requires phosphorylation; mutant lamins that resist phosphorylation do not become soluble at mitosis. Thus, one critical mitotic event—dissolution of the nuclear membrane—is most likely triggered by CDK phosphorylation of nuclear lamins. It is highly likely that CDK phosphorylation of particular chromatin proteins activates other events of mitosis, such as chromosome condensation.

Genetic studies of yeast provided much of the evidence that CDK-cyclin complexes are key controlling agents in all eukaryotic cell cycles. In one series of studies, geneticists used yeast mutants that carry defective CDKs or cyclins to find the corresponding human genes (**Fig. 19.8**) and to show that the human CDKs and cyclins can function in yeast in place of the native proteins.

How Molecular Interactions Control the G_1-to-S Transition

Cell-cycle investigators have identified many of the molecular events controlling the transition from G_1 to S in human

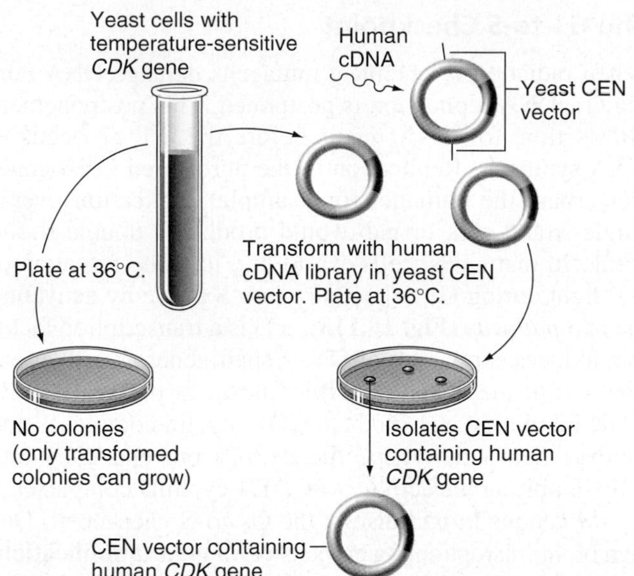

Figure 19.8 Mutant yeast permit the cloning of a human CDK gene. A culture of yeast cells containing a temperature-sensitive mutation in the yeast CDK gene was transformed with a library composed of human cDNA cloned into a yeast centromere-containing (CEN) vector. The transformed yeast cells were spread on solid medium at the restrictive temperature. Only the rare transformants with a functional copy of the human CDK gene were able to grow.

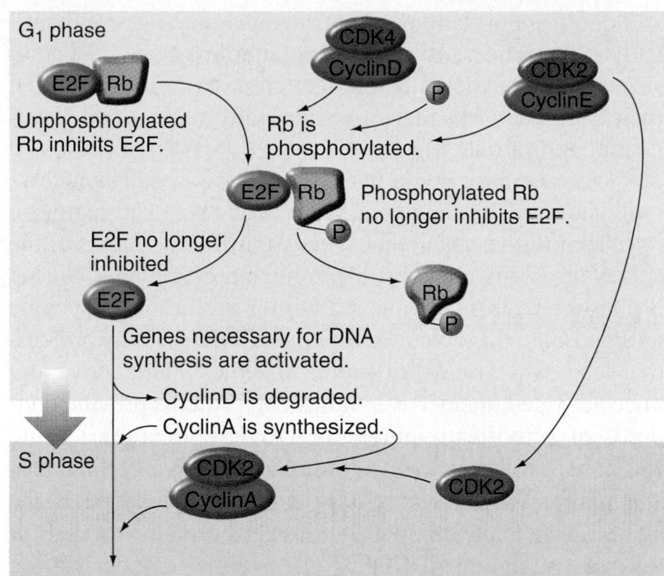

Figure 19.9 CDKs mediate the transition from the G_1 to the S phase of the cell cycle. In human cells, CDK4 complexed to cyclinD, and CDK2 complexed to cyclinE phosphorylate the Rb protein, causing it to dissociate from, and activate, the E2F transcription factor. E2F stimulates transcription of many genes needed for DNA replication. At the transition into S phase, cyclinD is destroyed, cyclinA is synthesized, and the CDK2–cyclinA complex activates DNA replication.

cells by analogy with similar events in the cell cycle of yeast. From their analyses, they have pieced together the following scenario.

The first CDK-cyclin complexes to appear during G_1 in humans are CDK4-cyclinD and CDK2-cyclinE (**Fig. 19.9**). These complexes initiate the transition to S by a programmed succession of specific phosphorylations, among which are phosphorylations of the protein product of the retinoblastoma (*RB*) gene. Unphosphorylated Rb protein inhibits a transcription factor, E2F. Phosphorylated Rb no longer inhibits E2F. Rb phosphorylation thus indirectly activates DNA synthesis by releasing the brakes on E2F and thereby allowing it to activate the transcription of genes necessary for DNA synthesis.

How Molecular Interactions Control the G_2-to-M Transition

Human cells appear to make the transition from G_2 to mitosis much as the well-studied cells of the yeast *S. pombe* accomplish the same transition. In the yeast, a CDK known as CDC2 (the second C replaces the K for historical reasons) forms a complex with cyclinB. Both the CDC2 kinase and cyclinB are present throughout G_2, but phosphorylation of a specific tyrosine residue on the cyclin-dependent kinase (by another protein kinase) keeps it inactive. When the time comes to initiate mitosis, a phosphatase enzyme removes the phosphate group from the CDC2 tyrosine; this removal activates the CDK, and the cell enters mitosis (**Fig. 19.10**).

Summary of Controls Intrinsic to the Cell Cycle

The activation and inactivation of CDK-cyclin complexes trigger transitions of the cell cycle. A variety of controls set the timing of these events. For example, phosphorylation

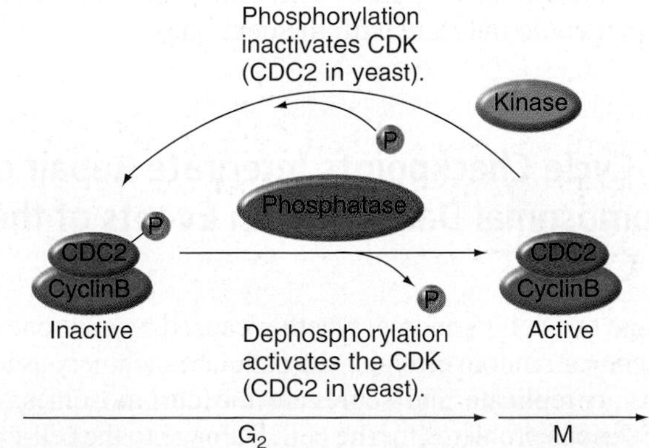

Figure 19.10 CDK activity in yeast is controlled by phosphorylation and dephosphorylation. The CDC2 protein complexed with cyclinB is inactivated prior to mitosis through phosphorylation by a specific kinase and then activated at the onset of mitosis through dephosphorylation by a specific phosphatase.

and dephosphorylation at specific sites inactivate and activate CDKs. The transcriptional regulation of cyclin synthesis and the rapid degradation of cyclins at the completion of their specified tasks make these proteins appear and disappear at appropriate times.

We now have some insight into how a cell is able to replicate its DNA at one time in the cell cycle and segregate its chromosomes at another. The two different phases of the cell cycle are governed by different kinase activities. During S phase, a CDK is complexed with a cyclin that is specific to S phase. In this complex, it phosphorylates many proteins that lead to a cascade of protein synthesis and activation; once activated, the newly synthesized proteins provide hundreds of activities required for DNA replication. During M phase, a different CDK is complexed with a cyclin that is specific for M phase. Its activity in this complex leads to the synthesis and activation of hundreds of proteins needed for mitosis. In summary, CDKs and cyclins together set the "state" of the cell: S phase or M phase.

How does the cell change from one state to the other? Among the cellular processes activated by CDKs are those that irreversibly destroy key regulatory proteins, including the cyclins. Thus, as the cell enters either S phase or M phase, it sets in place the end of each phase by removing cyclins and many other proteins whose activities must be limited to either S or M phase.

Just as attachment of a phosphate group activates or deactivates proteins, the covalent attachment of a *ubiquitin* tag marks proteins for degradation. Proteins tagged with ubiquitin are rapidly degraded by the cell in the large multiprotein complex called the *proteosome*. During S phase, activation of a group of proteins called *SCF* occurs. The activated SCF adds ubiquitin to proteins such as the S phase cyclins. During M phase, activation of a group of proteins called *APC* takes place. The activated APC adds ubiquitin to proteins such as the M phase cyclins. Thus, the cell cycle has an intrinsic ratchetlike mechanism, ensuring that activation of one phase (S or M) leads inevitably to the irreversible end of that phase and elimination of any proteins that could interfere with the next phase.

Cell-Cycle Checkpoints Integrate Repair of Chromosomal Damage with Events of the Cell Cycle

Damage to a cell's genome, whether caused by environmental agents or random errors of the cellular machinery as it attempts to replicate and segregate the chromosomes, can cause serious problems for the cell. Damage to the cell-cycle machinery can also cause problems. It is therefore not surprising that elaborate mechanisms have evolved to arrest the cell cycle while repair takes place. These additional controls are called **checkpoints** because they check the integrity of the genome and cell-cycle machinery before allowing the cell to continue to the next phase of the cell cycle.

The G1-to-S Checkpoint

When radiation or chemical mutagens damage DNA during G_1, DNA replication is postponed. This postponement allows time for DNA repair before the cell proceeds to DNA synthesis. Replication of the unrepaired DNA could exacerbate the damage; for example, replication over a single-strand nick or gap would produce a double-strand break. In mammals, cells exposed to ionizing radiation or UV light during G_1 delay entry into S phase by activating the *p53 pathway* (**Fig. 19.11a**). p53 is a transcription factor that induces expression of DNA repair genes as well as expression of the **CDK inhibitor** known as p21. Like other CDK inhibitors, p21 binds to CDK–cyclin complexes and inhibits their activity; specifically, p21 prevents entry into S by inhibiting the activity of CDK4-cyclinD complexes.

Mutations in *p53* disrupt the G_1-to-S checkpoint. One sign of this disruption is a propensity for **gene amplification:** an increase from the normal two copies to hundreds of copies of a gene. This amplification is visible under the microscope, appearing as an enlarged area within a chromosome known as a *homogeneously staining region* (*HSR*) or as small chromosome-like bodies (called double *minutes*) that lack centromeres and telomeres (Fig. 19.11b). Normal human cells do not generate gene amplification in culture, but *p53* mutants exhibit high rates of such amplification; *p53* mutants also exhibit many types of chromosome rearrangements. The explanation is as follows. Cells carrying mutations in the *p53* gene most likely have a defective G_1-to-S checkpoint that allows the replication of single-strand nicks. This replication produces double-strand breaks, which, in turn, lead to chromosome rearrangements. Some of the rearrangements generate gene amplification (Fig. 19.11c).

Wild-type cells able to produce functional p53 not only arrest in G_1 in the presence of DNA damage; if the damage is great enough, they also commit suicide in a process known as **programmed cell death (PCD),** or **apoptosis.** During apoptosis, the cellular DNA is degraded, and the nucleus condenses. The cell may then be devoured by neighboring cells or by phagocytes (Fig. 19.11d). Programmed cell death and the proteins that regulate it—including those that are part of the p53 pathway—appear in multicellular animals from roundworms to humans. It makes sense for multicellular organisms to have a mechanism for eliminating cells that have sustained chromosomal damage. The survival and reproduction of such cells could generate cancers.

The G_2-to-M Checkpoint

Damage to DNA during G_2 delays mitosis, allowing time for repair before chromosome segregation (**Fig. 19.12a**). Researchers have identified many genes in mammalian and yeast cells that mediate this control. One of these genes is *RAD9*. Whereas wild-type yeast cells can pause to repair as many as 100 double-strand breaks before entering mitosis, *RAD9* mutants fail to arrest in G_2 and die as a result of any double-strand breaks that were not repaired before mitosis.

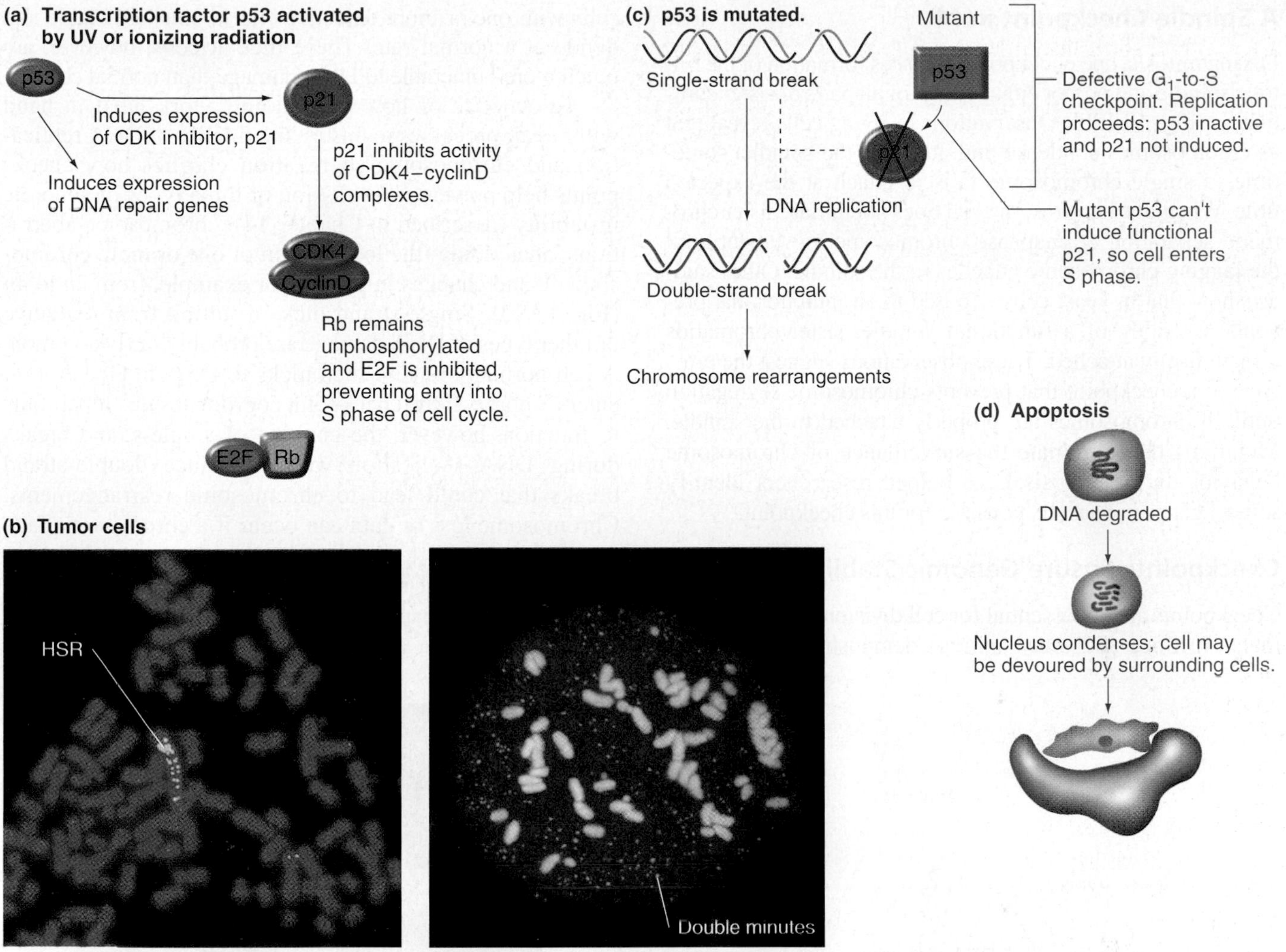

(a) Transcription factor p53 activated by UV or ionizing radiation

p53
Induces expression of CDK inhibitor, p21
p21
Induces expression of DNA repair genes
p21 inhibits activity of CDK4–cyclinD complexes.
CDK4
CyclinD
Rb remains unphosphorylated and E2F is inhibited, preventing entry into S phase of cell cycle.
E2F
Rb

(b) Tumor cells

HSR
Double minutes

(c) p53 is mutated.

Single-strand break
Mutant
p53
Defective G_1-to-S checkpoint. Replication proceeds: p53 inactive and p21 not induced.
p21
DNA replication
Mutant p53 can't induce functional p21, so cell enters S phase.
Double-strand break
Chromosome rearrangements

(d) Apoptosis

DNA degraded
Nucleus condenses; cell may be devoured by surrounding cells.

Figure 19.11 **Cellular responses to DNA damage.** **(a)** DNA damage activates the p53 transcription factor, which, in turn, induces expression of the *p21* gene. The p21 protein inhibits CDK activity, producing an arrest of the cell cycle in the G_1 phase. **(b)** Tumor cells exhibit amplified regions of DNA, unlike normal cells, that can appear as homogeneously staining regions (HSR) within a chromosome or as double minutes, small pieces of extrachromosomal DNA. **(c)** When the *p53* gene is mutated in cancers, *p21* expression is not induced, cell-cycle progress is not arrested, and cells replicate damaged DNA, producing DNA double-strand breaks from single-strand nicks or gaps. **(d)** DNA damage in normal cells often leads to programmed cell death, or apoptosis, in which cellular DNA is degraded and the cell is engulfed and digested by neighboring cells.

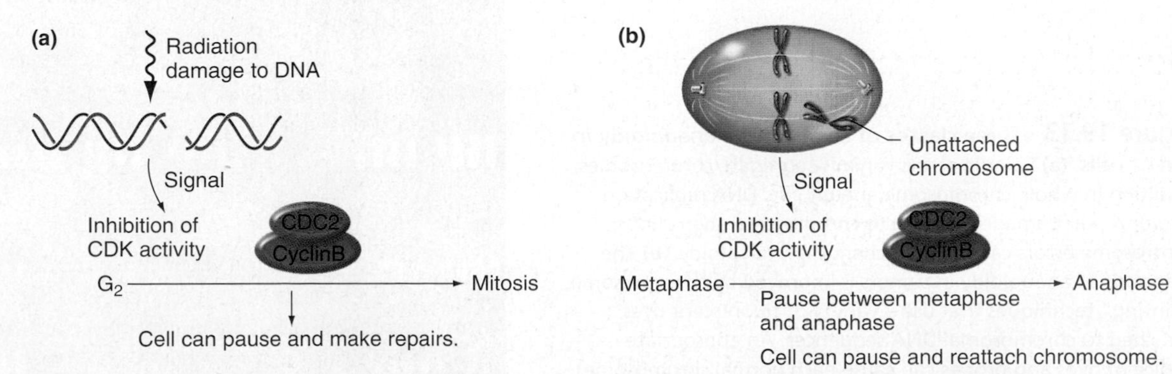

Figure 19.12 **Checkpoints acting at the G_2-to-M cell-cycle transition or during M phase.** **(a)** DNA damage, particularly double-strand breaks, induce a signal that inhibits CDK activity, preventing entry into mitosis. **(b)** Spindle damage resulting from the failure of a chromosome to attach to the mitotic spindle generates a signal that inhibits CDK activity and thereby prevents the metaphase-to-anaphase transition.

A Spindle Checkpoint in M

During mitosis, one checkpoint oversees formation of the mitotic spindle and proper engagement of all pairs of sister chromatids (Fig. 19.12b). Observations of living cells reveal that as chromosomes condense and attach to the spindle, sometimes a single chromosome fails to attach at the expected time. When this happens, the cell does not initiate sister chromatid separation or anaphase chromosome movement until the lagging chromosome attaches to the spindle. Other studies show that in yeast cells exposed to an inhibitor that prevents assembly of a functional spindle, sister chromatids remain firmly attached. These observations suggest the presence of a checkpoint that prevents chromosome segregation until all chromosomes are properly attached to the spindle. Mutations that eliminate the surveillance of chromosome behavior during mitosis have helped researchers identify several genes in yeast responsible for this checkpoint.

Checkpoints Ensure Genomic Stability

Checkpoints are not essential for cell division. In fact, experiments in mice and other animals demonstrate that mutant cells with one or more defective checkpoints are viable and divide at a normal rate. These mutant cells, however, are much more vulnerable to DNA damage than normal cells.

Knowledge of how checkpoints work hand in hand with repair processes to ensure the fidelity of DNA replication and chromosome segregation clarifies how checkpoints help prevent transmission of three types of genomic instability (described in Chapter 14): chromosome aberrations; aneuploidy (the loss or gain of one or more chromosomes); and changes in ploidy, for example, from $2n$ to $4n$ (**Fig. 19.13**). Single-strand nicks resulting from oxidative or other types of DNA damage are probably fairly common. A cell normally repairs such nicks to DNA in G_1 before it enters S phase. If the checkpoint coordinates this repair fails to function, however, the copying of single-strand breaks during DNA replication would produce double-strand breaks that could lead to chromosome rearrangements. Chromosome loss or gain can occur if a chromosome fails to attach properly to the spindle. Normally, the M phase spindle checkpoint recognizes such failures and prevents the initiation of anaphase until the cell has fixed the problem. Cells without a functional checkpoint will produce

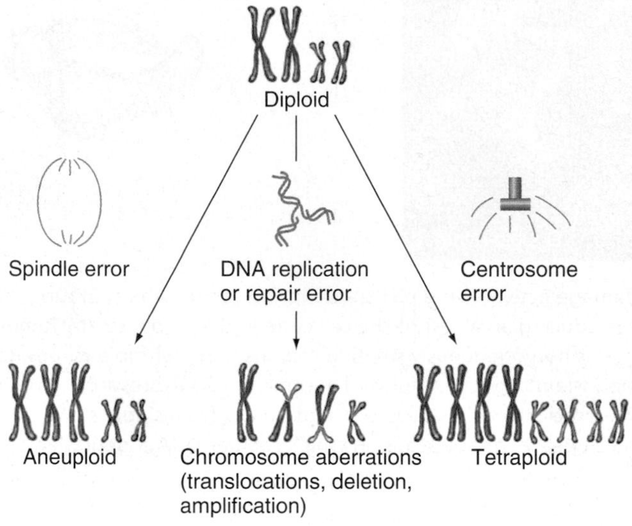

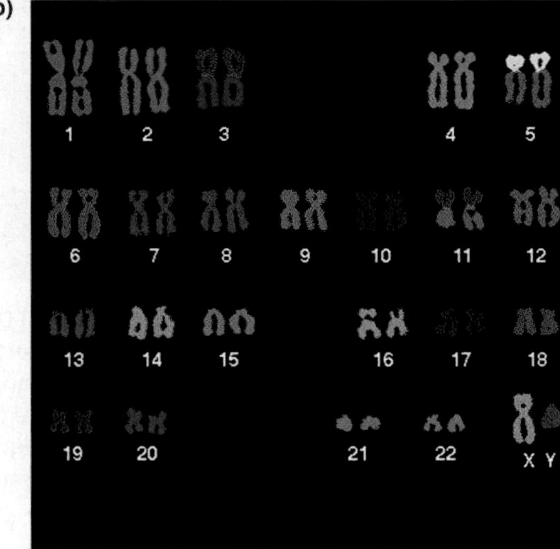

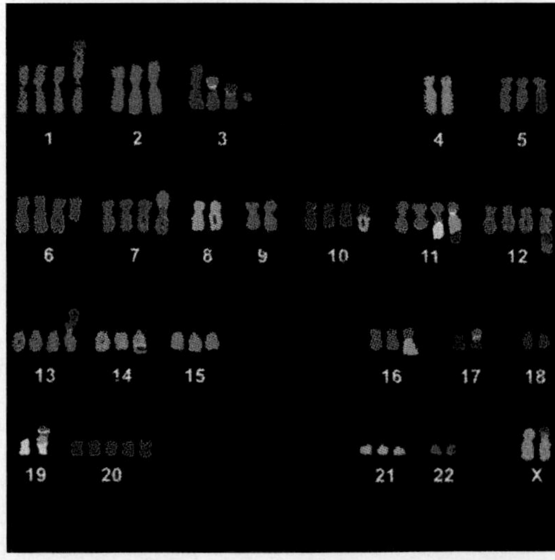

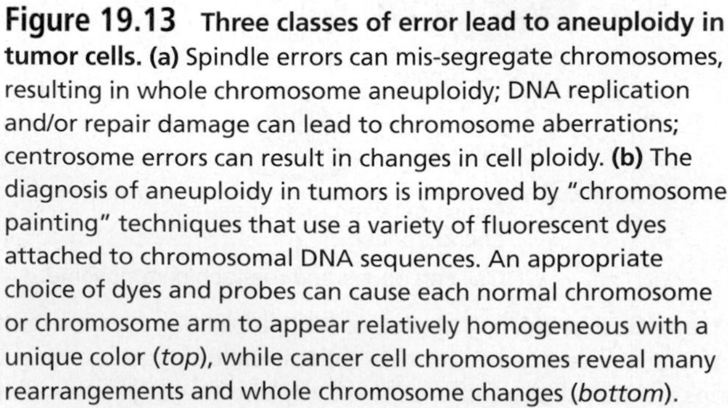

Figure 19.13 Three classes of error lead to aneuploidy in tumor cells. (a) Spindle errors can mis-segregate chromosomes, resulting in whole chromosome aneuploidy; DNA replication and/or repair damage can lead to chromosome aberrations; centrosome errors can result in changes in cell ploidy. **(b)** The diagnosis of aneuploidy in tumors is improved by "chromosome painting" techniques that use a variety of fluorescent dyes attached to chromosomal DNA sequences. An appropriate choice of dyes and probes can cause each normal chromosome or chromosome arm to appear relatively homogeneous with a unique color (*top*), while cancer cell chromosomes reveal many rearrangements and whole chromosome changes (*bottom*).

daughter cells carrying one too few or one too many chromosomes. Finally changes in ploidy can occur if a cell begins S phase before completing mitosis or if a cell fails to replicate or to properly segregate its microtubule-organizing centers, or centrosomes. Checkpoints also recognize these errors, ensuring integration of the centrosome cycle with DNA replication and the formation and function of the mitotic spindle.

A Cascade of External and Internal Molecules Tells Cells Whether or Not to Initiate Division

How do cells know when to divide? To function according to the needs of the body as a whole, cells depend on signals sent from one tissue to another. These signals tell them whether to divide or metabolize (that is, make the product[s] they are programmed to make) or die. There are two basic types of signals: extracellular and cell-bound. *Extracellular signals* in the form of steroids, peptides, and proteins act over long or short distances and are collectively known as *hormones* (**Fig. 19.14a**). The thyroid-stimulating hormone (TSH) produced by the brain's pituitary gland, for example, travels through the bloodstream to the thyroid gland, where it stimulates cells to produce another hormone, thyroxine, which increases metabolic rate. *Cell-bound signals,* such as the histocompatibility proteins that, like fingerprints, distinguish an individual's cells from all foreign cells and molecules, require direct contact between cells for transmission (Fig.19.14b). The macrophages, helper T cells, and antibody-producing B cells of the immune system communicate via cell-bound signals about the presence of viral particles, bacteria, and toxins.

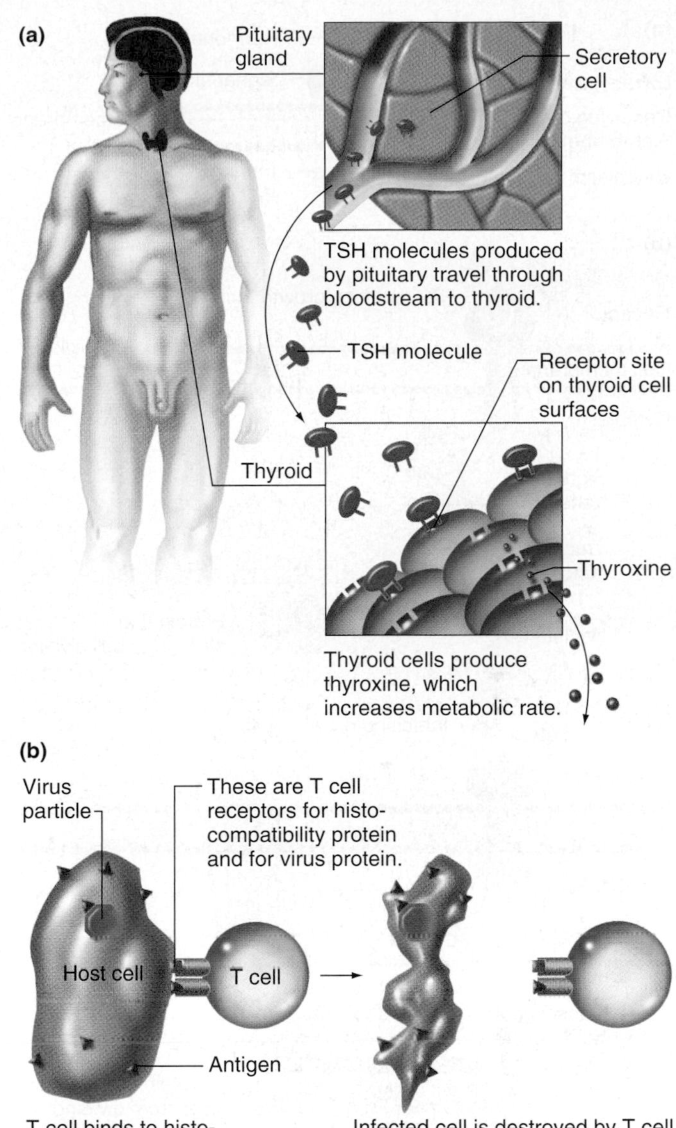

Figure 19.14 Extracellular signals can diffuse from one cell to another or be delivered by cell-to-cell contact. **(a)** The pituitary gland produces thyroid-stimulating hormone (TSH) that diffuses through the circulation to the thyroid gland, which produces another hormone, thyroxine, that acts on many cells throughout the body. **(b)** A killer T cell recognizes its target cell by direct cell-to-cell contact.

The Molecular Components of Each Signaling System: Growth Factors, Receptors, Intracellular Transducers, and Transcription Factors

Extracellular hormones and cell-bound signals that stimulate or inhibit cell proliferation are known as **growth factors.** Most growth factors deliver their message to specific **receptors** embedded in the membrane of the receiving cell (**Fig. 19.15a**). The receptors are proteins that have three parts: a signal-binding site outside the cell, a transmembrane segment that passes through the semipermeable cell membrane, and an intracellular domain that relays the signal (that is, the binding of growth factor) to proteins inside the cell's cytoplasm. These cytoplasmic proteins, which relay signals inside the cell, are known as **signal transducers.** The final link in the relay system is usually a **transcription factor** that activates the expression of specific genes in the nucleus, either to promote or to inhibit cell proliferation (Fig. 19.15b and c).

How the Molecules Interact to Relay a Signal

Binding of a growth factor to its specific receptor elicits a cascade of biochemical reactions inside the cell, often involving a large number of molecules. Each molecule in the cascade transmits the receptor's binding-of-messenger signal by activating or inhibiting another molecule. The activation and inhibition of intracellular targets after growth-factor binding is called **signal transduction.**

One example of a signal transduction system includes the product of the *RAS* gene (Fig. 19.15d). The RAS protein is a molecular switch that exists in two forms: an inactive

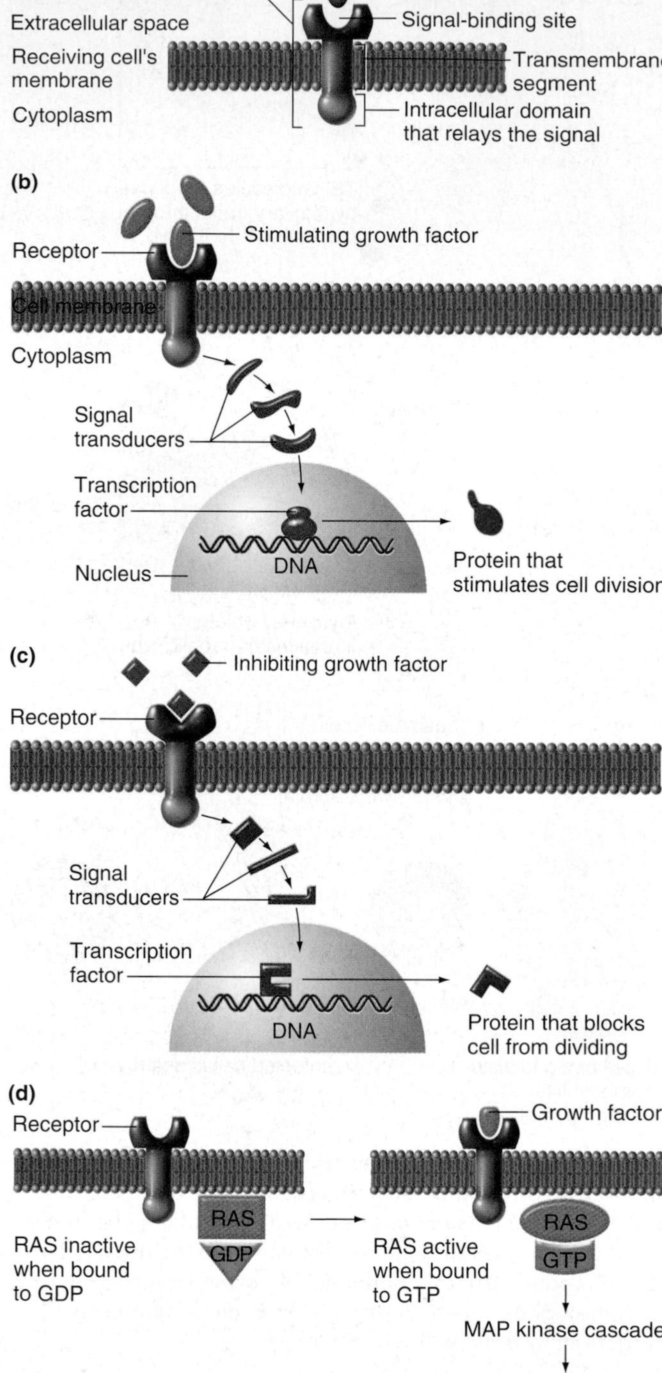

Figure 19.15 Many hormones transmit signals into cells through receptors that span the cellular membrane. (a) Hormones bind to a cell surface receptor that specifically recognizes their structure. The extracellular surface of the receptor transmits a signal to the intracellular domain of the receptor, which, in turn, interacts with other signaling molecules in the cell either to stimulate growth **(b)** or to inhibit growth **(c)**. The RAS protein is an intracellular signaling molecule that is induced to exchange a bound GDP (inactive) for a bound GTP (active) when a growth factor binds to the cellular receptor with which RAS interacts **(d)**.

form in which it is bound to guanosine diphosphate (RAS-GDP) and an active form in which it is bound to guanosine triphosphate (RAS-GTP). (GTP is also the nucleotide used to insert a G residue into an RNA molecule during transcription.) Once a growth factor activates a receptor, the receptor flicks the RAS switch to active by exchanging GDP for GTP. Next, RAS-GTP activates a series of three protein kinases, and this trio, known as a MAP kinase cascade, activates a transcription factor.

The proteins in a signal transduction system are like the neurons in a nerve fiber: Each one serves as a link in a message-relay chain. In deciding whether or not to divide, the cell, like the brain, combines messages from many signal transduction systems and adjusts its behavior in response to the integrated information.

As we see in the next section, cells that have lost the ability to reproduce their genomes faithfully, because of defects in the DNA repair systems described in Chapter 7 or defects in checkpoints, are prone to mutations that cause cancer.

19.2 Cancer Arises When Controls Over Cell Division No Longer Function Properly

An understanding of the molecular basis of cell-cycle regulation sheds light on the life-threatening proliferative disease of cancer. The many genes contributing to the normal control of cell proliferation are all subject to mutations. Such mutations may produce inappropriate signals about the need for cell division, malfunctions in the CDK-cyclin complexes that control cell-cycle transitions, or checkpoint breakdowns that lead to genomic instability.

The Cancer Phenotype Results from the Accumulation of Multiple Mutations in the Clonal Progeny of a Cell

Cancer biologists now believe that most cancers result from the accumulation of many mutations during the proliferation of somatic cells. When enough mutations accumulate in genes controlling proliferation and other processes within a single clone of cells, that clone overgrows the normal cells that surround it, disseminates through the bloodstream to other parts of the body, and forms a life-threatening tumor, or cancer. (In this chapter, we use the term "tumor" to designate cancerous tissue and the term "growth" to designate a benign mass.)

Epidemiological data, clinical studies, and experimental analyses of a range of cell types in a variety of species provide evidence for this gene-based view of cancer. Here we present some of that evidence as well as an interpretation of the evidence that explains how a cell progresses from normal to cancerous.

The General Cancer Phenotype Includes Many Types of Cellular Abnormalities

Theodor Boveri, one of the architects of the chromosome theory of inheritance, observed as early as 1914 that cells excised from malignant tumors have abnormal chromosomes. By the 1970s, when new staining techniques and improved equipment made it possible to distinguish each of the 23 different chromosome pairs in the human genome by their specific banding patterns, investigators noted that many different chromosomal abnormalities appear in tumor cells. Using tools developed in the 1980s, geneticists confirmed that most tumor cells exhibit karyotypic instability.

Figure 19.16 shows the main characteristics that distinguish tumor cells from normal cells. The cancer phenotype includes uncontrolled cell growth, genomic and karyotypic instability, the potential for immortality, and the ability to invade and disrupt local and distant tissues. Although no one cancer cell necessarily manifests all the phenotypic (Text continues on p. 700)

Feature Figure 19.16

Phenotypic Changes That Distinguish Tumor Cells From Normal Cells

MOST NORMAL CELLS | MANY CANCER CELLS

a.1 Autocrine stimulation

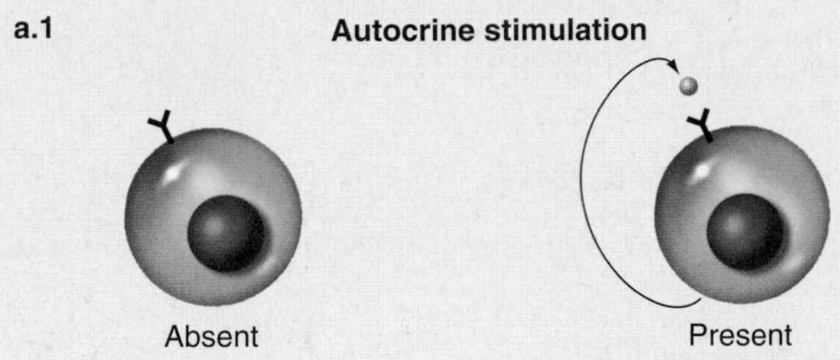

a.2 Contact inhibition

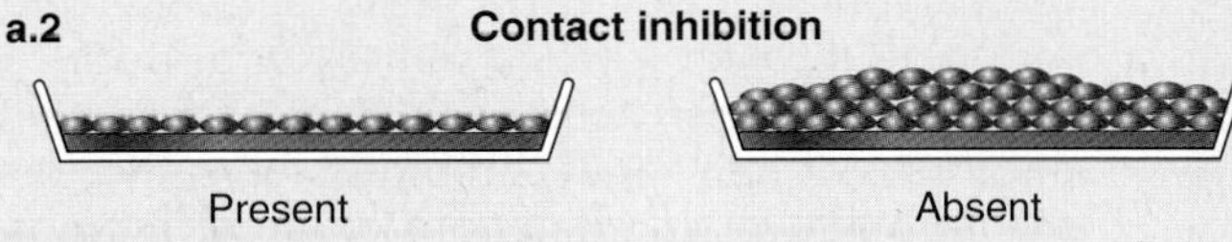

a.3 Cell death

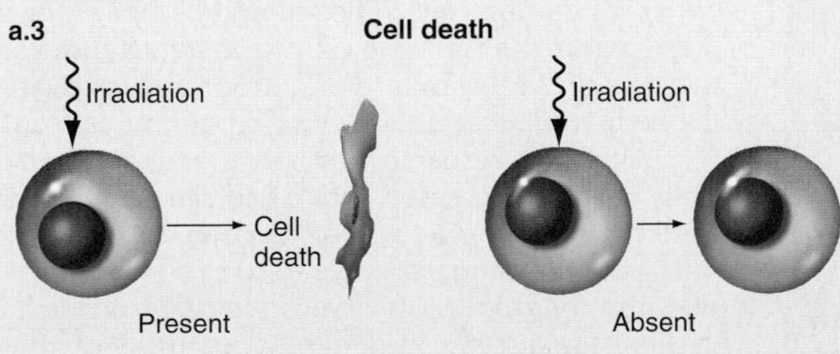

a.4 Gap junctions

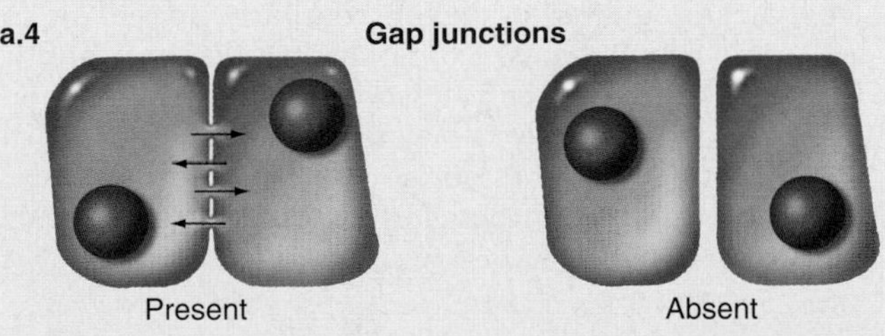

a. Changes that produce uncontrolled cell growth

1. *Autocrine stimulation.* Most cells "decide" whether or not to divide only after receiving signals from neighboring cells, either positive signals that stimulate division or negative signals that prevent proliferation. Many tumor cells, by contrast, make their own stimulatory signals, in a process known as **autocrine stimulation**, or are insensitive to negative signals.
2. *Loss of contact inhibition.* Normal cells stop dividing when they come in contact with one another, as evidenced by the fact that the few normal cell types that grow in culture form sheets one cell thick. Tumor cells, which have lost the property of contact inhibition, climb all over each other to produce piles that are many cells thick. This change in behavior contributes to the disordered array of cells seen in tumors, which is a significant departure from the highly ordered patterns seen in normal tissues.
3. *Loss of cell death.* Normal cells die when starved of growth factors or when exposed to agents such as toxins or X-rays that damage them. Their programmed cell death is activated by the expression of certain genes in the cell. Because it keeps cells from proliferating when they should not and eliminates badly damaged cells, apoptosis is probably a safeguard against the early stages of cancer. Most cancer cells are much more resistant than normal cells to pro-grammed cell death.
4. *Loss of gap junctions.* Normal cells connect to their neighbors by small pores, or gap junctions, in their membranes. The gap junctions permit the transfer of small molecules that may be important in controlling cell growth. Most tumor cells have lost these channels of communication.

(*Continued*)

Feature Figure 19.16 (*Continued*)

b.1

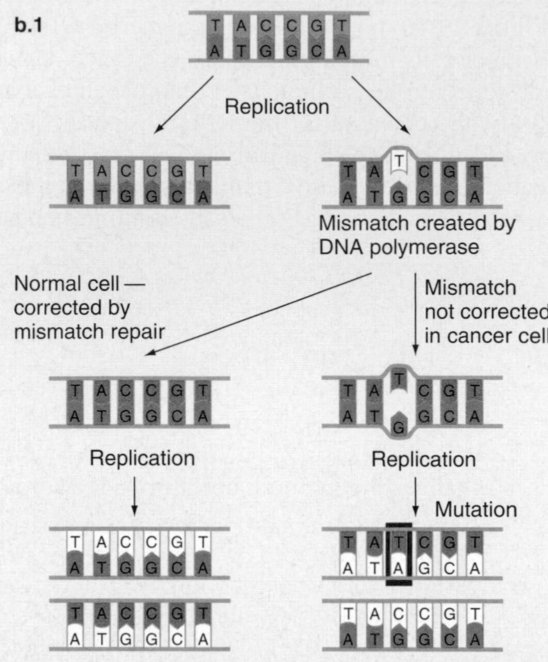

b. Changes that produce genomic and karyotypic instability

1. *Defects in the DNA replication machinery.* Cancer arises most often in cells that have lost the ability to reproduce their genomes faithfully. We saw in Chapters 6 and 7 that cells have elaborate systems for repairing DNA damage; these systems include the enzymatic machinery for mismatch repair and the repair of damage caused by radiation or ultraviolet light. Work on yeast and bacteria has shown that mutant organisms defective in DNA repair have enormously increased rates of mutation. And these increased mutation rates often lead to cancer. An example is the hereditary syndrome of xeroderma pigmentosum, which predisposes to skin cancer.

b.2

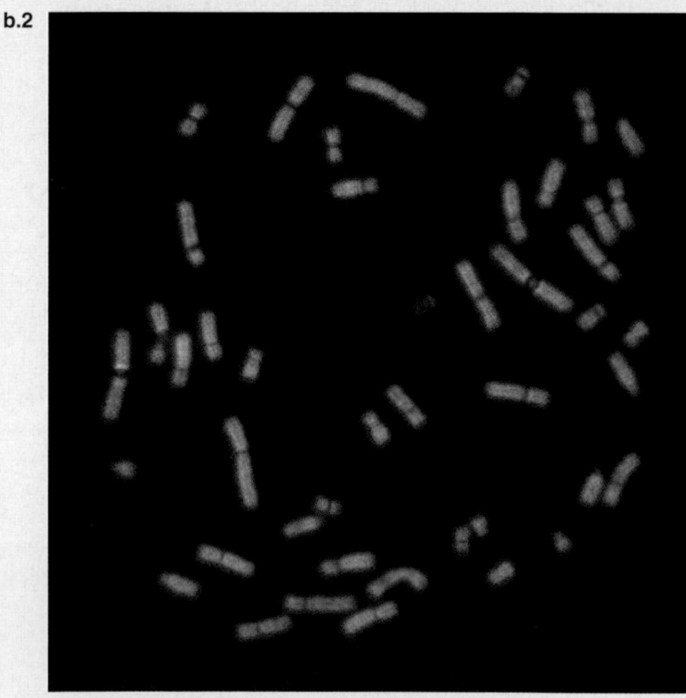

2. *Increased rate of chromosomal aberrations.* Tumor-cell karyotypes often carry gross rearrangements, including broken chromosomes, with some of the pieces rejoined to other chromosomes; multiple copies of individual chromosomes, rather than the normal two; and deletions of large chromosomal segments and of whole chromosomes. The changes can be revealed by chromosome painting. Studies have confirmed that the fidelity of chromosome reproduction is greatly diminished in tumor cells. Normal fibroblast cells, for example, have an undetectable rate of gene amplification (an increase in the number of copies of a gene), whereas tumor cells have amplification rates as high as 1 in 100 cells. Some of the genes that control cell proliferation are inappropriately activated by such amplification.

   Although research has confirmed that tumor cells have an increased rate of chromosomal aberration, probably only a small fraction of these chromosomal rearrangements lead to cancer; for example, tumors from solid tissues typically carry many chromosomal rearrangements, but most of these aberrations do not recur in all tumors. A few rearrangements, however, regularly appear in specific tumor types. Examples include the translocation between chromosomes 8 and 14 found in patients with certain kinds of lymphoma and the translocation between chromosomes 9 and 22 found in certain types of leukemias (see Fig. 14.20 on p. 505).

c. Changes that produce a potential for immortality

1. *Loss of limitations on the number of cell divisions.* Most normal cells (except for the rare stem cells) die spontaneously after a specifiable number of cell divisions. The senescence and natural death of normal cells is evident both in culture and in the body. Tumor cells, by contrast, can divide indefinitely.
2. *Ability to grow in culture.* Cells derived from tumor cells usually grow readily in culture, making cancerous cell lines available for study. Normal cells do not grow well in culture. This difference between abnormal and normal is greatest in cells grown as isolated clones in soft agar, most likely because the normal cells must undergo more genetic changes to become competent to divide under the artificial conditions in culture than tumor cells.
3. *Restoration of telomerase activity* (not shown). Most normal human somatic cells do not express the enzyme telomerase, and this lack of telomerase expression prevents them from replicating the repeated sequences in the telomeres at the ends of their chromosomes (see Chapter 13, pp. 475–477). As a result, after a certain number of cell divisions, the telomeres shorten to the point where they contribute to cell senescence and death. Tumor cells have the ability to express telomerase, a feature that most likely contributes to their immortality.

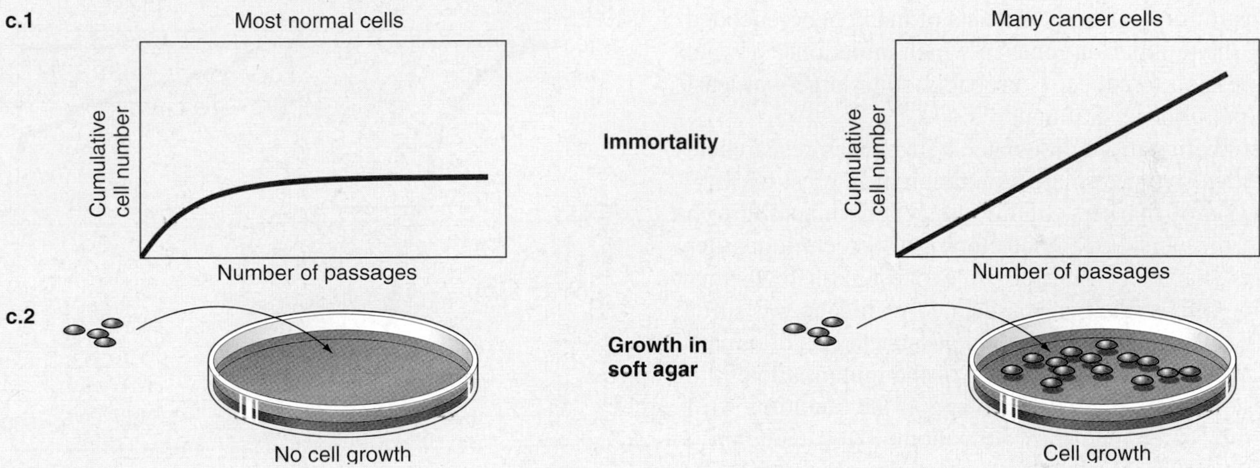

d. Changes that enable a tumor to disrupt local tissue and invade distant tissues

1. *The ability to metastasize.* Normal cells stay within rigidly defined boundaries. Tumor cells, by comparison, often acquire the capacity to invade surrounding tissues and eventually to travel through the bloodstream to colonize distant tissues. Metastasis—the invasion of other tissues—is a complicated behavior requiring many genetic changes.
2. *Angiogenesis.* Once the adult human body has developed, new blood vessels do not normally form except to heal a wound. Tumor cells, however, secrete substances that cause blood vessels to grow toward them. The new vessels serve as supply lines through which the tumor can tap new sources of nutrients; they also serve as escape routes through which tumor cells can metastasize.
3. *Evasion of immune surveillance* (not shown). The human immune system may recognize cancer cells as foreign and attack them, thereby helping to eliminate tumors before they become large enough for clinical detection. As evidence, cancer patients often have antibodies and/or killer T cells directed against their cancer cells. Successful tumor cells, however, somehow develop the ability to evade detection by the immune system.

d.1 Metastasis

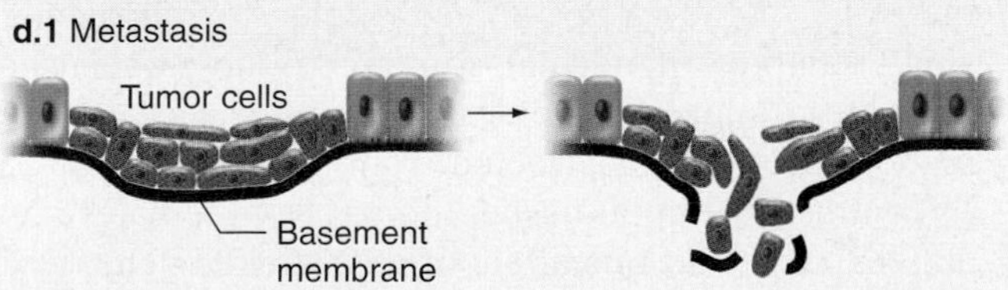

d.2 Angiogenesis

changes illustrated in Fig. 19.16, each cancer cell displays a number of them.

More Than One Mutation Must Fuel the Progression from Normal to Cancerous

The large catalog of phenotypic changes seen in tumor cells suggests that many mutations—at least one for each phenotypic change—are necessary to convert a normal cell into a cancerous cell. In other words, a normal cell must change many aspects of its phenotype to become a life-threatening cancer cell, and this set of changes probably requires the alteration of many genes. Clinical analyses of human cells from colon tumors show that these cancer cells contain mutations in at least 5–10 genes. Since the intricate web of controls that organize and maintain the society of cells constituting our body consists of many gene-encoded proteins, the number of genes in which mutations can fuel the progression to cancer is probably quite large—at least 100 and possibly several hundred.

To study mutations associated with cancer, researchers initially identify and isolate a mutation of interest by linkage analysis of markers, traditional genetic mapping to a chromosome, and positional cloning (all techniques described in Chapter 11). It is possible to test in mice whether a mutation in a single gene associated with cancer is sufficient to induce a tumor. If the mutation acts in a dominant fashion, researchers insert a copy of the mutant allele into the mouse genome of a fertilized egg; if the mutation is recessive, they delete one copy of the homologous gene from the early embryonic mouse genome and then breed animals homozygous for the deletion. In gene transfer experiments where a dominant cancer-causing mutation was inserted into a mouse genome under the control of a breast-cell-specific promoter, the transgenic mice produced a few breast tumors (**Fig. 19.17a**). Doubly transgenic mice made by breeding these transgenic mice carrying one mutated gene with transgenic mice carrying a different mutated gene implicated in cancer generated more tumors earlier. Even in these mice, however, only a small percentage of the transformed cells proliferated abnormally to form a tumor. These results support the idea that it takes several mutations in different genes to produce a cancer.

Studies of recessive mutations in the *p53* gene point to the same conclusion. Mice with both copies of the *p53* gene deleted from their genome develop relatively normally. The *p53* mutant mice, however, have shortened life spans and get a variety of tumors more frequently than wild-type mice (Fig. 19.17b). This experiment shows that the *p53* gene is not essential for development or for normal cell function, but it does play a role in preventing tumor formation. Consequently, deleting both copies of the wild-type gene from all of a mouse's cells does not convert every cell to a tumor cell, but it does increase the probability that at least one cell will become cancerous. The conclusion is that mutations in *p53* are just one of the many genetic changes that may occur in a cell to produce cancer.

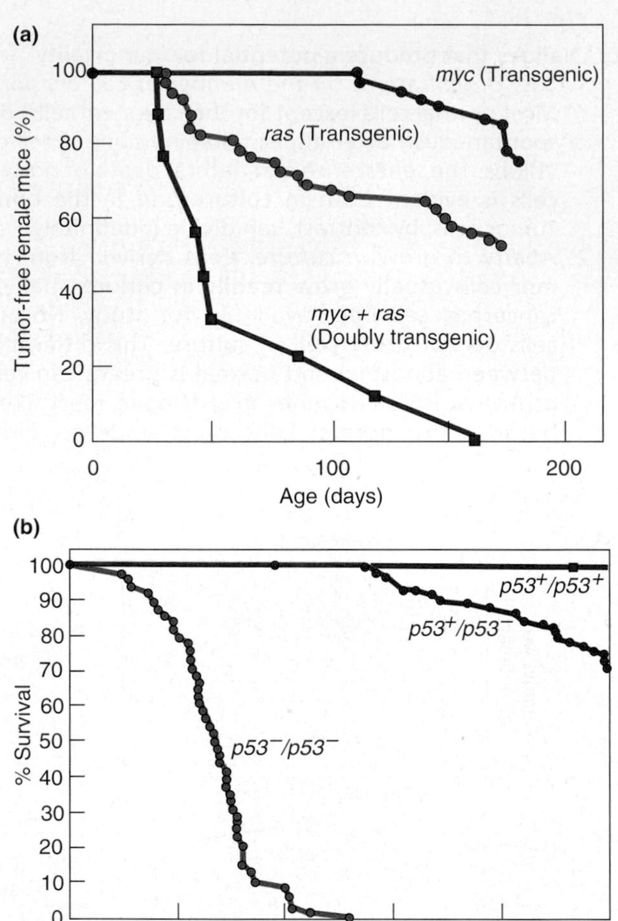

Figure 19.17 The percent of mice still alive as a function of age. (a) The activated *myc* oncogene produces tumors more slowly than the *ras* oncogene. Mice containing both oncogenes develop tumors even faster than mice with *ras*. **(b)** Homozygous $p53^+$ mice rarely get life-threatening tumors, while those heterozygous for a $p53^-$ mutation develop tumors late in life. Mice homozygous for the $p53^-$ mutation develop tumors early in life.

Cancers Are Clonal Descendants of One Cell

Examination of cells from women heterozygous for X-linked alleles provide evidence that cancer originates in a single somatic cell (**Fig. 19.18**). Although the random inactivation of one of the two X chromosomes in each cell of a female means that individual cells express only one of the two X-linked alleles for any gene, in small samples of normal somatic tissues, one usually finds both alleles expressed. This is because most somatic tissues are constructed from many clones of cells. In contrast to normal tissue, tumors from females invariably express only one allele of an X-linked gene (review discussion of X inactivation on pp. 481–482 of Chapter 13). This finding suggests that the cells of each tumor are the clonal descendants of a single somatic cell that sustained a rare mutation.

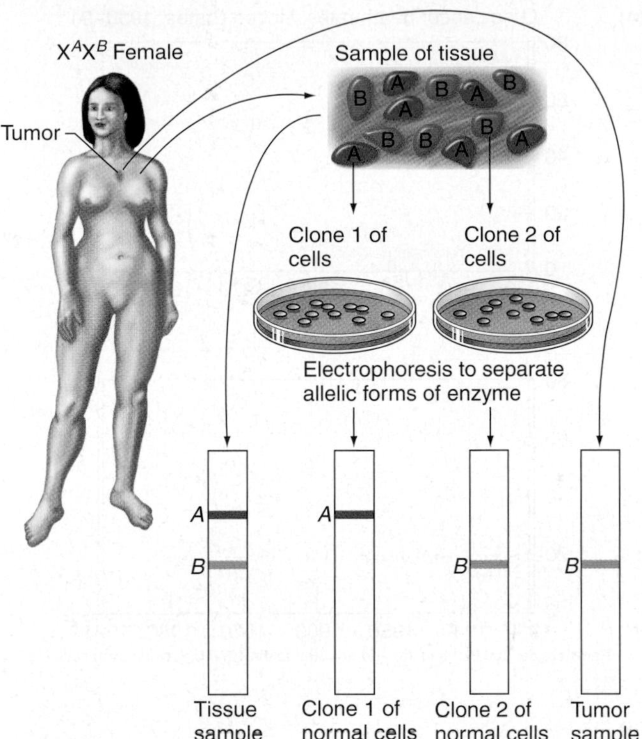

Figure 19.18 Polymorphic enzymes encoded by the X chromosome reveal the clonal origin of tumors. Each individual cell in a female expresses only one form of a polymorphic X-linked gene because of X-chromosome inactivation. A patch of tissue will usually contain both types of cells. If single cells are grown into a clone, they will exhibit one or the other enzyme form. A tumor also exhibits only one form, demonstrating that it arose from a single cell. The two allelic forms of the gene's protein product are distinguished by electrophoresis.

Most Cancers Result from Exposure to Environmental Mutagens

Several epidemiological surveys support the hypothesis that most cancers arise by chance in somatic cells during their division and differentiation from fertilized egg to adult. The mutations that produce these cancers are not inherited through the germ line in a dominant or recessive pattern; rather, they arise sporadically in a population as a result of chemicals or viruses in the environment. The evidence is as follows. First, the degree of concordance for cancers of the same type among first-degree relatives, such as sisters and brothers or even identical twins, is low for most forms of cancer in the population as a whole (we discuss specific exceptions later). If one sibling or twin gets a cancer, the other usually does not. Second, although rates for the incidence of specific cancers vary worldwide (**Table 19.2**), when populations migrate from one place to another, their profile of cancer incidence becomes more like that of the people indigenous to the new location. The change in cancer profile often takes decades, suggesting that the environment acts over a long period of time to induce the cancer. Third, epidemiological studies have established that numerous environmental agents increase the likelihood of cancer, and many of these agents are mutagens. These mutagens include cancer-causing viruses, some of which carry mutant forms of normal genes that control the cell cycle, as well as cigarette smoke. People who smoke for many years have a higher risk of lung cancer than people who do not smoke, and their risk increases with the number of cigarettes and the length of time they smoke.

TABLE 19.2 The Incidence of Some Common Cancers Varies Between Countries

Site of Origin of Cancer	High-Incidence Population		Low-Incidence Population	
	Location	*Incidence**	*Location*	*Incidence**
Lung	USA (New Orleans, blacks)	110	India (Chennai)	5.8
Breast	Hawaii (Hawaiians)	94	Israel (non-Jews)	14.0
Prostate	USA (Atlanta, blacks)	91	China (Tianjin)	1.3
Cervix	Brazil (Recife)	83	Israel (non-Jews)	3.0
Stomach	Japan (Nagasaki)	82	Kuwait (Kuwaitis)	3.7
Liver	China (Shanghai)	34	Canada (Nova Scotia)	0.7
Colon	USA (Connecticut, whites)	34	India (Chennai)	1.8
Melanoma	Australia (Queensland)	31	Japan (Osaka)	0.2
Nasopharynx	Hong Kong	30	UK (southwestern)	0.3
Esophagus	France (Calvados)	30	Romania (urban Cluj)	1.1
Bladder	Switzerland (Basel)	28	India (Nagpur)	1.7
Ovary	New Zealand (Polynesian Islanders)	26	Kuwait (Kuwaitis)	3.3
Pancreas	USA (Los Angeles, Koreans)	16	India (Pune)	1.5
Lip	Canada (Newfoundland)	15	Japan (Osaka)	0.1

*Incidence indicates number of new cases per year per 100,000 population, adjusted for a standardized population age distribution (so as to eliminate effects due merely to differences of population age distribution). Figures for cancers of breast, cervix, and ovary are for women; other figures are for men. Adapted from V. T. DeVita, S. Hellman, and S. A. Rosenberg (eds.), *Cancer: Principles and Practice of Oncology,* 4th ed. Philadelphia: Lippincott, 1993; based on data from C. Muir et al., *Cancer Incidence in Five Continents,* Vol. 5. Lyon: International Agency for Research on Cancer, 1987.

Cancer Develops Over Time

The data on lung cancer show that decades elapse between the time a population begins smoking and the time that lung cancer begins increasing. In the United States, cancer incidence in men rose dramatically after 1940, roughly two decades after men began frequent smoking; women did not begin frequent smoking until several decades after men, and lung cancer incidence in women did not begin its dramatic increase until after 1960 (**Fig. 19.19a**).

Epidemiological data also show that the incidence of cancer rises with age. The prevalence of cancer in older people supports the idea that cancer develops over time as well as the idea that the accumulation of many mutations in the clonal descendants of a somatic cell fuels the progression from normal to cancerous. If you assume that the rate of accumulation of cancer-causing mutations is constant over a lifetime, the slope of a logarithmic curve plotting cancer incidence against age is a measure of the number of mutations required for cancer (Fig. 19.19b). Interestingly, the data for many types of tumors generate a similar curve in which the evolution of cancer requires 6–10 mutations. Thus, the correlation between cancer incidence and aging, as well as the time lag between exposure to carcinogens and the appearance of tumors, suggests that the mutations that produce cancer accumulate over time. However, this simple interpretation is only part of the picture because cells increase their mutation rate at some point during their progression to cancer.

(a)

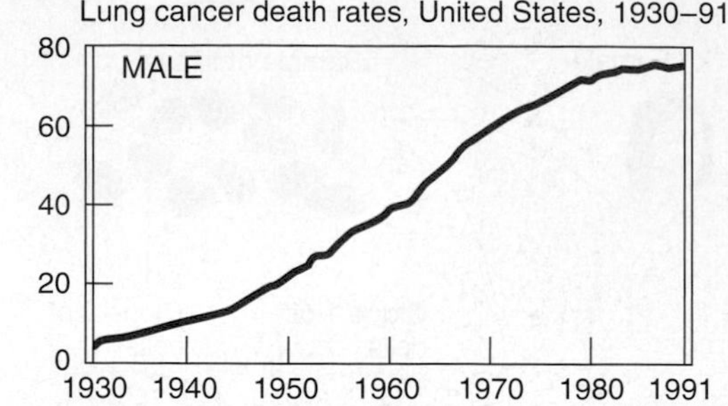

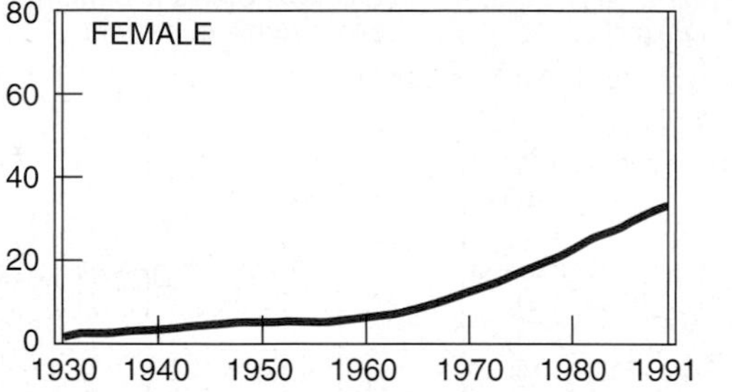

Rates are per 100,000 and are age-adjusted to the 1970 U.S. census population.

(b)

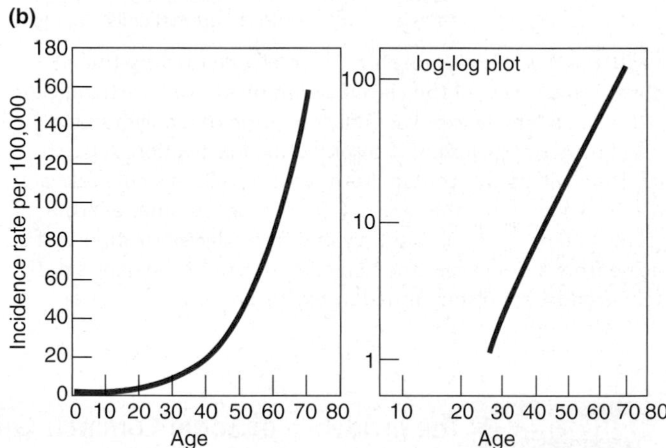

Figure 19.19 Lung cancer death rates and incidence of cancer with age. (a) Lung cancer death rates in the United States during the twentieth century began increasing rapidly for men in the 1940s and for women in the 1960s. This reflects the fact that smoking became prevalent among men about 20 years before it did among women. **(b)** The incidence of most cancers shows a dramatic increase with age, a result thought to reflect the accumulation of mutations in somatic cells.

Although Most Cancers Are Sporadic, Some Cancers Run in Families

In some families, a specific type of cancer recurs in many members, indicating the inheritance of a predisposition through the germ line. Retinoblastoma is an example of this type of cancer (see the Genetics and Society box on p. 154 of Chapter 5). Half the individuals in families affected by retinoblastoma inherit a mutation in the *RB* gene from one parent. Since all their somatic cells carry one defective copy of the gene, a mutation in the single remaining wild-type copy of the *RB* gene in the cells that proliferate to produce the retina predisposes these cells to develop retinal cancer (**Fig. 19.20**). People who do not inherit a mutation in the *RB* gene need to experience a mutation in both copies of the gene in the same cell to develop cancer; this type of double hit is very rare. Interestingly, for nearly all common types of cancer that occur sporadically in a population, rare families can be found that exhibit an inherited predisposition to that cancer.

Interpretation of the Evidence: Cancer Develops When Cells in a Single Lineage Accumulate Enough Mutations to Overgrow Normal Cells

The epidemiological, clinical, and experimental evidence demonstrates that cancer cells have many phenotypic abnormalities. Underlying these phenotypic abnormalities are mutations. More than one mutation in the clonal descendants of a single cell is necessary to produce cancer. These mutations occur primarily in somatic cells and accumulate over time. Most mutations result from exposure to environmental mutagens and appear sporadically in a population, but the inheritance of certain mutations predisposes some families to cancers associated with those mutations.

A multistep model of cancer formation accounts for the evidence and provides a coherent interpretation of the disparate observations. According to this model,

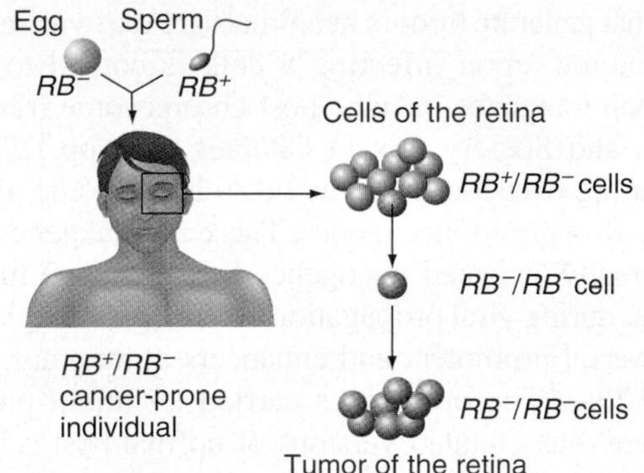

Figure 19.20 Individuals who inherit one copy of the RB^- allele are prone to cancer of the retina. During the proliferation of retinal cells, the RB^+ allele is lost or mutated, and cancers grow out of the RB^-/RB^- clone of cells.

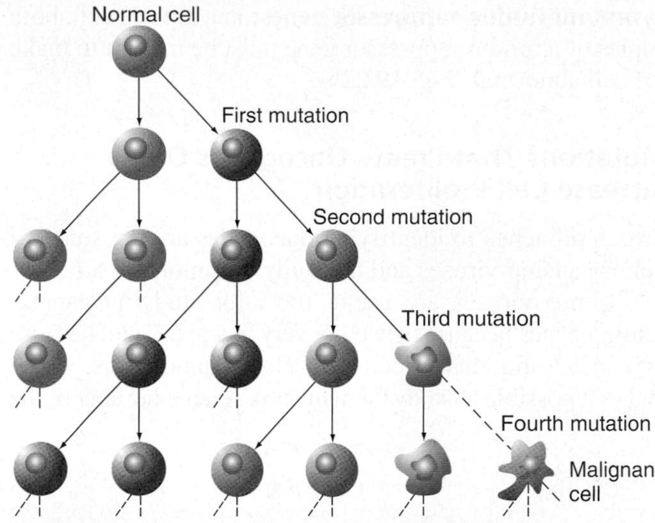

Figure 19.21 Cancer is thought to arise by successive mutations in a clone of proliferating cells.

mutations accumulate over time in the cells of a clone, and this accumulation fuels the progression from normal to cancerous (**Fig. 19.21**).

The Mutations That Lead to Cancer Create Dominant Oncogenic Alleles or Recessive Tumor-Suppressor Alleles

Research has not only revealed that cancer results from multiple genetic changes in the clonal descendants of one cell, it has also established that the mutations found in tumors are of two general types: those that improperly activate genes (for example, the genes responsible for promoting cell proliferation) and those that improperly inactivate genes (for example, the genes responsible for preventing excessive cell proliferation).

The mutant alleles that lead to cancer are referred to as **cancer genes;** but the term "genes" is a misnomer. *All cancer genes are, in fact, mutant alleles of normal genes.* When present in all or a subset of cells within an organism, these mutant alleles predispose the individual to develop cancer over a lifetime. Mutant alleles that act dominantly are known as **oncogenes;** in a diploid cell, one mutant oncogenic allele is sufficient to alter the cell phenotype (**Fig. 19.22a**). Mutant alleles that are recessive are known

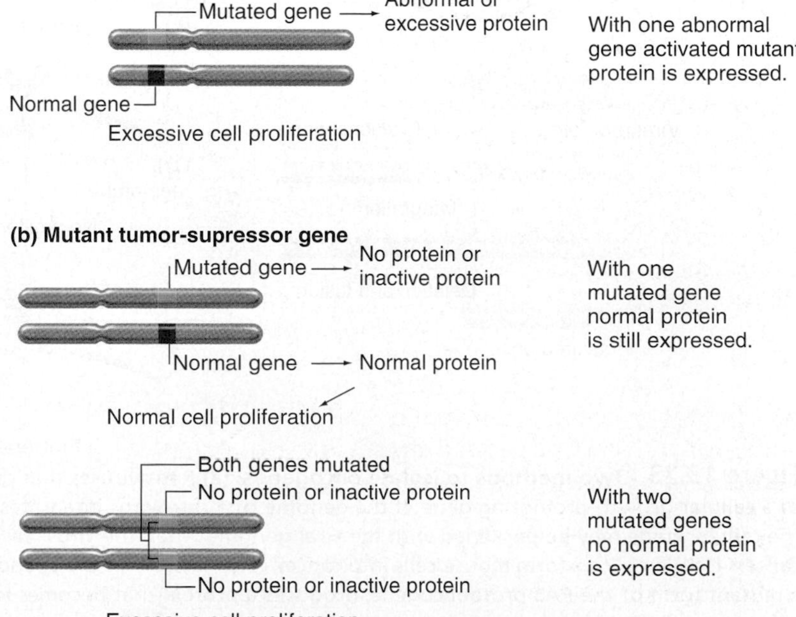

Figure 19.22 Cancer-producing mutations occur in two forms. (a) Dominant mutations generate oncogenes that exhibit abnormal activity or produce an excessive amount of protein. **(b)** Recessive mutations produce altered tumor suppressors that usually generate little or no phenotype when they are heterozygous with a wild-type allele, but they affect cell proliferation when a second mutation inactivates the wild-type allele.

as **mutant tumor-suppressor genes;** in a diploid cell, both copies of a tumor-suppressor gene must be mutant to make the cell abnormal (Fig. 19.22b).

Mutations That Create Oncogenes Often Increase Cell Proliferation

Two approaches to identifying oncogenes are the study of tumor-causing viruses and the study of tumor DNA itself.

Tumor viruses are useful tools for studying cancer-causing genes because they carry very few genes and because they infect and change cultured cells to tumor cells, which makes it possible to study them *in vitro.* A large number of the viruses that generate tumors in animals are retroviruses whose RNA genome, upon infecting a cell, is copied to cDNA, which then integrates into the host chromosome (review the Genetics and Society box in Chapter 8 on pp. 270–271). Later, during excision from the host chromosome, the virus can pick up copies of host genes. These normal genes change to abnormally activated oncogenes either through mutations that occur during viral propagation or through their placement near powerful promoters and enhancers in the viral genome (**Fig. 19.23a**). The oncogenes carried by tumor-producing viruses are thus mutated versions of normal host cell genes. The wild-type, or normal, genes that become oncogenes upon mutation are known as **protooncogenes.** When a virus

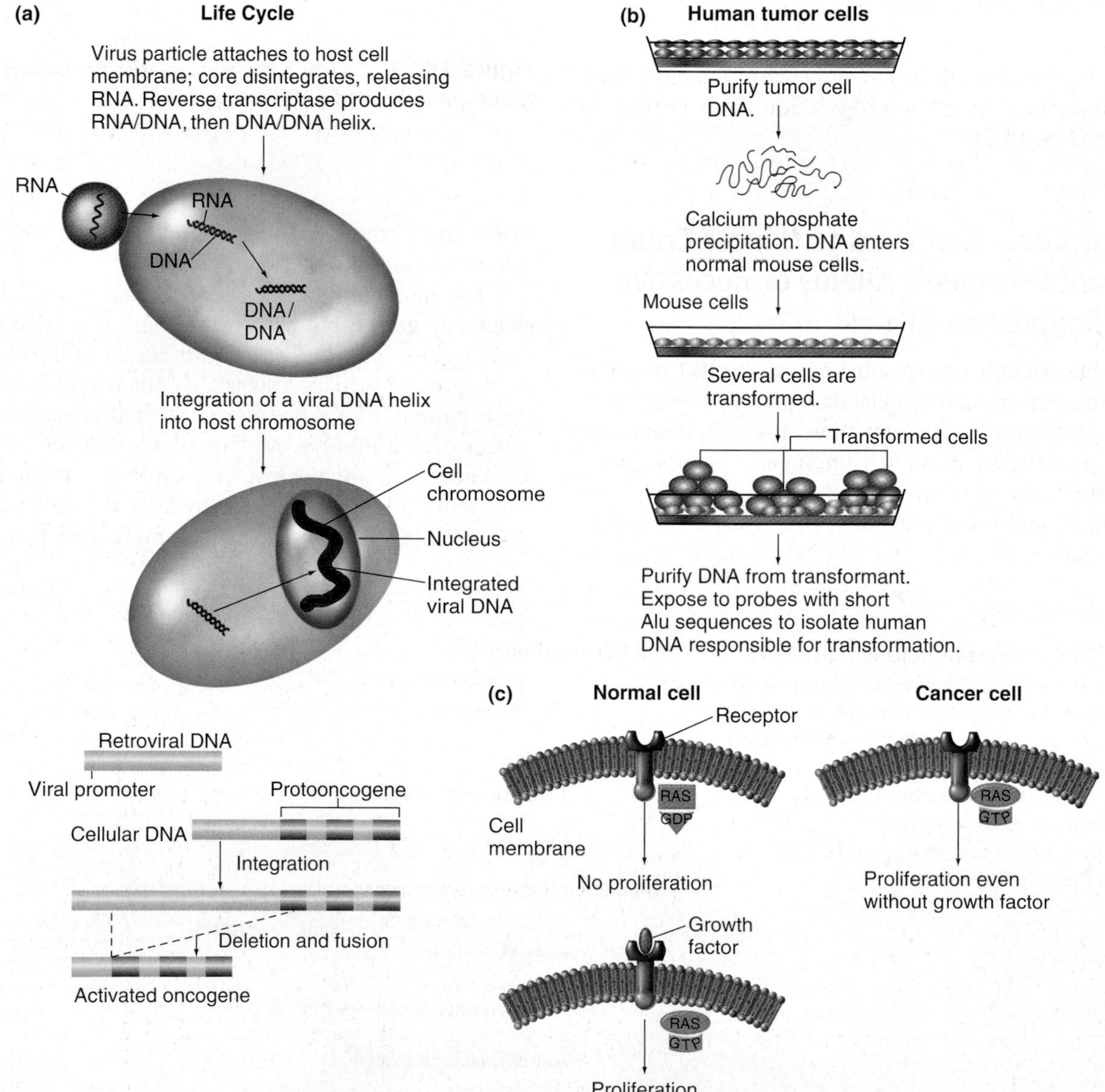

Figure 19.23 **Two methods to isolate oncogenes. (a)** Retroviruses that cause cancer carry a mutant or overexpressed copy of a cellular growth-promoting gene. If the genome of a retrovirus integrates into the host chromosome near a protooncogene, the cellular gene may be packaged with the viral genome when the virus leaves the cell. **(b)** DNA isolated from some human cancers is able to transform mouse cells into cancer cells. These cells are found to contain a human oncogene. **(c)** The *RAS* oncogene, a mutant form of the *RAS* protooncogene, produces a protein that becomes locked into the GTP-activated form.

TABLE 19.3 Retroviruses and Their Associated Oncogenes*

Virus	Species	Tumor	Oncogene
Rous sarcoma	Chicken	Sarcoma	*src*
Harvey murine sarcoma	Rat	Sarcoma and erthyroleukemia	*H-ras*
Kristen murine sarcoma	Rat	Sarcoma and erthyroleukemia	*K-ras*
Moloney murine sarcoma	Mouse	Sarcoma	*mos*
FBJ murine osteosarcoma	Mouse	Chondrosarcoma	*fos*
Simian sarcoma	Monkey	Sarcoma	*sis*
Feline sarcoma	Cat	Sarcoma	*sis*
Avian sarcoma	Chicken	Fibrosarcoma	*jun*
Avian myelocytomatosis	Chicken	Carcinoma, sarcoma, and myleocytoma	*myc*
Ableson leukemia	Mouse	B cell lymphoma	*abl*

*Retroviruses identified as causative agents of tumors in animals contain oncogenes that were derived from a cellular gene. Adapted from Lewin, *Genetics,* 1e, Oxford University Press, Inc. by permission.

carrying one or more oncogenes infects a cell, the oncogenes cause abnormal proliferation that can lead to the accumulation of more mutations and eventually to cancer. The analysis of tumor-causing retroviruses led to the discovery of oncogenes in a variety of species (**Table 19.3**).

Some DNA viruses also carry oncogenes. An example is the human papilloma virus (HPV). HPV infection of a woman's cervical cells is probably the first step in the development of cervical cancer. The papilloma virus carries at least two oncogenes capable of transforming appropriate recipient cells in culture: *E6* and *E7.* The E6 and E7 proteins bind to and inactivate the normal products of the *p53* and *RB* genes. Only those HPV subtypes whose E6 and E7 proteins bind p53 and RB proteins are associated with cervical cancer in women. Progression of HPV-initiated cells to cancer requires additional mutations in genes not yet identified.

Scientists also identify oncogenes by isolating DNA from tumor cells and exposing noncancerous cells in culture to this tumor DNA. Some tumor DNA transforms cultured cells into cells capable of producing tumors (Fig. 19.23b). If one uses human tumor DNA in the transformation of mouse cells, for example, it is possible to identify the DNA responsible for the transformation by reisolating the human DNA from the transformed mouse cells with probes for the short interspersed elements known as Alu sequences that appear only in the human genome (see Fig. 14.25a on p. 510). The oncogenes identified in this way, like those discovered in studies of tumor viruses, are oncogenic alleles of normal cellular genes that have mutated to abnormally active forms. Sometimes the two approaches have identified the same oncogene, for example, *RAS.* The oncogenic forms of the *RAS* gene generate proteins that are always (or constitutively) in the GTP-activated form; hence, whether or not growth factor is present, a cell carrying a *RAS* oncogene receives signals to divide (Fig. 19.23c). Like mutated *RAS,* many oncogenes continuously turn on one or more of a cell's many signal transduction systems. They do this by encoding receptors, signal transmitters, and transcription factors that are active with or without growth factor (**Table 19.4**).

TABLE 19.4 Oncogenes Are Members of Signal Transduction Systems*

Name of Oncogene	Tumor Associations	Mechanism of Activation	Properties of Gene Product
hst	Stomach carcinoma	Rearrangement	Growth factor
erb-B	Mammary carcinoma, glioblastoma	Amplification	Growth factor receptor
trk	Papillary thyroid carcinomas	Rearrangement	Growth factor receptor
Ha-ras	Bladder carcinoma	Point mutation	GDP/GTP binding signaling protein
raf	Stomach carcinoma	Rearrangement	Cytoplasmic serine/threonine kinase
myc	Lymphomas, carcinomas	Amplification, chromosomal translocation	Nuclear transcription factor

*The roles of several oncogene products that are members of the signal transduction pathway and the ways in which they get activated in human cells are shown.

Excessive Proliferation Enhances the Potential for Mutation

Like the oncogenic *RAS* gene, many of the oncogenes so far identified affect cell-signaling pathways that tell a cell whether or not to proliferate. The importance of these genes in generating cancer is not just that they cause cells to proliferate, because an increase in proliferation alone, without other changes, generates benign growths that are not life threatening and can be removed by surgery. Rather, increased proliferation provides a large clone of cells within which further mutations can occur, and these further mutations may eventually lead to malignancy. The more cells that exist in a clone, the more likely that rare mutations will occur in the clone; and these additional mutations appear in a clone that already has the potential for rapidly propagating them. Although not all cancer-causing genes are dominant oncogenes, oncogenic mutations have been the easiest to identify for technical reasons.

Mutations That Create Mutant Tumor-Suppressor Alleles Release a Brake on Cell Division and Often Decrease the Accuracy of Cell Reproduction

Mutant tumor-suppressor genes are recessive alleles of genes whose normal alleles help put cell division on hold, whether in terminally differentiated cells or in cells with DNA damage. Targets for tumor-suppressor mutations include *RB, p53,* and *p16.* One wild-type copy of these genes apparently produces enough protein to regulate cell division; the loss of both wild-type copies releases a brake on proliferation (see Fig. 19.22b). Researchers have identified dozens of tumor-suppressor genes through the genomic analysis of families with an inherited predisposition to specific types of cancer or through the analysis of specific chromosomal regions that are reproducibly deleted in certain tumor types.

Retinoblastoma provides an example of this identification process. A cancer of the color-perceiving cone cells in the retina, retinoblastoma is one of several cancers inherited in a dominant fashion in human families (**Fig. 19.24a**). Roughly half the children of a parent with retinoblastoma get the disease. Retinoblastoma tumors are easy to diagnose and remove before they become invasive. As we saw in Chapter 14, karyotypes of normal, noncancerous tissues from many people suffering from retinoblastoma reveal heterozygosity for deletions in the long arm of chromosome 13; that is, the patients carry one normal and one partially deleted copy of 13q. Karyotypes of the cancerous retinal cells from some of these same patients show homozygosity for the same chromosome 13 deletions that are heterozygous in the noncancerous cells (Fig. 19.24b). Although the deletions vary in size and position from patient to patient, they all remove band 13q14.

These observations indicate that band 13q14 includes a gene whose removal contributes to the development of retinoblastoma. *RB* is the symbol for this gene. The

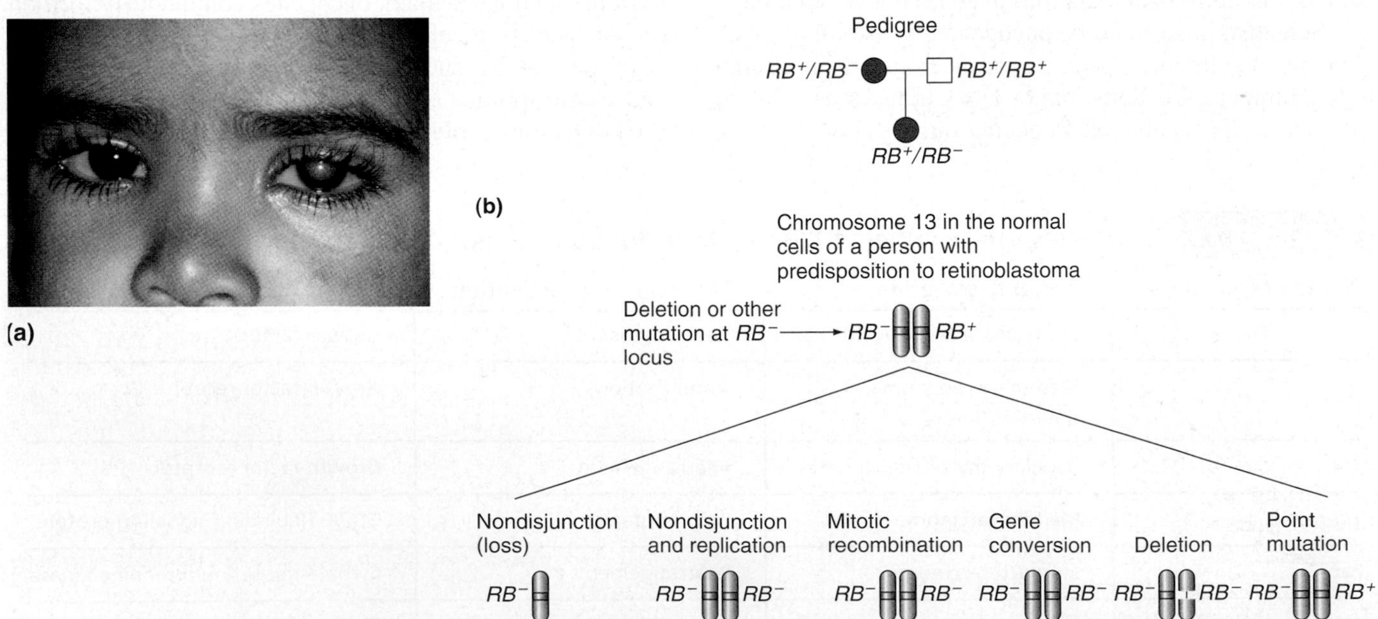

Figure 19.24 The retinoblastoma tumor-suppressor gene. (a) A child with a retinoblastoma tumor in the left eye. **(b)** The RB^- gene is inherited through the germ line as an autosomal recessive mutation. Subsequent changes to the RB^+ allele during somatic divisions generate a clone of cells homozygous or hemizygous for the RB^- allele.

heterozygous cells in a patient's normal tissues carry one copy of the gene's wild-type allele (RB^+), and this one copy prevents the cells from becoming cancerous. Tumor cells homozygous for the deletion, however, do not carry any copies of RB^+, and without it, they begin to divide out of control.

Geneticists used their understanding of retinoblastoma inheritance to find the *RB* gene. They cloned DNA carrying the gene by looking for DNA sequences in band 13q14 that were lost in all of the deletions associated with the hereditary condition. They then identified the gene by characterizing a very small deletion that affected only one transcriptional unit—the *RB* gene itself. Analysis of the gene's function showed that it encodes a protein involved (along with many other proteins) in regulating the cell cycle (Review Fig. 19.9). *RB* thus fits our definition of a tumor-suppressor gene: The protein it determines helps prevent cells from becoming cancerous. Cancer can arise when cells heterozygous for an *RB* deletion lose the remaining copy of the gene.

This picture of the genetics of retinoblastoma raises a perplexing question: How can the retinoblastoma trait be inherited in a dominant fashion if a deletion of the *RB* gene is recessive to the wild-type RB^+ allele? The answer depends on the specific phenotype under consideration, in this case, susceptibility to cancer versus the progression from a normal to a cancerous cell. At the level of the organism, *RB* deletions are dominant because there is a strong likelihood that in at least one of the hundreds of thousands of retinal cells heterozygous for the deletion, a subsequent genetic event, such as mitotic nondisjunction, mitotic recombination, or a new mutation, will disable the single remaining RB^+ allele, resulting in a mutant cell with no functional tumor-suppressor gene. This one cell lacking any RB^+ gene will then multiply out of control, eventually generating a clone of cancerous cells. Although the aberrant events that create mutant cells with no functional *RB* gene happen very rarely, there are enough cells in the retina that most patients who inherit one 13q14-deleted chromosome 13 develop multiple retinoblastoma tumors in both eyes. At the level of a single cell, however, *RB* deletions are recessive to RB^+ because a cell with only one copy of RB^+ produces enough functional protein to remain normal. Hence, mutations in tumor-suppressor genes appear dominant when the phenotype under consideration is susceptibility of individuals to cancer, but they act in a recessive manner when the phenotype under examination is the progression from normal to cancerous of individual cells in a clone.

Geneticists first recognized the recessive *RB* mutation that leads to retinoblastoma through the genomic analysis of families inheriting a predisposition to the cancer. More recently, they noted that both copies of the *p16* gene on chromosome 9 are deleted in roughly 75% of all melanomas (a malignant skin cancer) and in approximately 85% of all gliomas (the most common form of brain cancer). The *p16* gene encodes a protein that binds to and inactivates CDK4. In another example, observations of deletions of both copies of a specific region of chromosome 18 in all colon cancers led to identification of the *DCC* (deleted in colorectal cancer) gene.

Many tumor-suppressor mutations occur in genes that control the cell cycle and, with it, the accuracy of genomic replication. It is important to distinguish those mutations that determine how the cell cycle is completed from mutations in genes that control proliferation. Alterations in genes that control proliferation result in an enlarged clone of cells, but aside from their increase in number, these cells—if they sustain no further mutations—are normal and thus form a benign growth. By contrast, mutations in genes that control the cell cycle can alter the accuracy with which a cell reproduces its genome. The resulting mutant cells can produce offspring with many more mutations than occur in normal cells, and this increase in the frequency of mutation vastly increases the probability that the cascade of mutations necessary to produce the phenotypic changes of tumor cells will occur.

Because cancer arises in cells that have lost the ability to reproduce their genomes faithfully, it seems reasonable to conclude that a cell's primary safeguard against cancer lies in maintaining the integrity of its genome. Cells have extensive, elaborate systems for repairing damage to their DNA. Studies of yeast and bacteria have shown that mutant organisms defective in DNA repair often have enormously increased rates of mutation. One important system for repairing DNA damage is the mismatch repair system, described in Chapter 7. This system recognizes mismatches that result from errors in DNA replication and removes the mismatched bases, allowing substitution of the correct bases. Proper functioning of the mismatch repair system increases the accuracy of DNA replication 100- to 1000-fold over the accuracy obtained by DNA polymerase alone. Mutants with a defective system have mutation rates several orders of magnitude greater than wild-type cells. In the 1990s, cancer researchers discovered that some people with a hereditary predisposition to colon cancer are heterozygous for a mutation that inactivates a gene required for the normal functioning of the mismatch repair system; the cancers that develop in these individuals consist of cells that have lost the single remaining wild-type allele. This mismatch repair gene thus behaves like a classical tumor-suppressor gene. Presumably, the greatly increased mutation rate in a homozygous cell that has lost both wild-type alleles makes it easier for progeny cells to accumulate the large number of mutations necessary to produce a cancer cell. Why these cancers develop mainly in the colon rather than in other tissues is not clear. **Table 19.5** describes several other tumor-suppressor genes that affect the accuracy of cell division.

GENETICS AND SOCIETY

The Uses of Genetic Testing in Predicting and Treating Cancer

Genetic tests for mutations in protooncogenes and tumor-suppressor genes can reveal whether a person has a higher probability of getting cancer at some point in his or her lifetime than a person without the mutations. But of those with an increased risk, some will and some will not get cancer. While a person who acquires one of these mutations through inheritance is pushed one step along the road to cancer, other mutations must occur in one clone of cells by chance; and nongenetic factors, such as exposure to radiation, influence if, where, and when the mutations occur. Given this situation, what good is it to learn from a genetic test that you have an increased probability of getting cancer sometime in your life?

Predictive testing is useful if the means of medical surveillance make it possible to detect the cancer to which a mutation predisposes at an early stage. Thus, testing for a genetic predisposition to skin, breast, or colon cancers is useful because careful examination of these tissues with existing technologies (microscopes, mammograms, or colonoscopies) can often detect cancers in their earliest stages. A person whose genetic test shows a predisposition to colon cancer, for example, could undergo a colonoscopy each year, and if one of these colon exams disclosed a small cancer, doctors could remove it by surgery or treat it by other means. Predictive testing is not yet useful for pancreatic cancer because there is as yet no way to detect tumors of the pancreas when they are small. By the time this cancer is identified it has almost always reached the aggressive state and metastasized to other tissues.

Once a cancer has been diagnosed, genetic testing of tumor cells can provide information that is useful for making a prognosis and determining a course of therapy. Completion of the Human Genome Project has opened up new possibilities for identifying and tracking the effects on survival of specific

TABLE 19.5 Mutant Alleles of These Tumor-Suppressor Genes Decrease the Accuracy of Cell Reproduction*

Gene	Normal Function of Gene (if known), or Disease Syndrome Resulting from Mutation	Function of Normal Protein Product
p53	Controls G_1-to-S checkpoint	Transcription factor
RB	Controls G_1-to-S transition	Inhibits a transcription factor
ATM	Controls G_1-to-S phase, and G_2-to-M checkpoint	DNA-dependent protein kinase
BS	Recombinational repair of DNA damage	DNA/RNA ligase
XP	Excision of DNA damage	Several enzymes
hMSH2, hmLH1	Correction of base-pair matches	Several enzymes
FA	Fanconi anemia	Unknown
BRCA1	Repair of DNA breaks	Unknown
BRCA2	Repair of DNA breaks	Unknown

*Many tumor-suppressor genes have been associated with a specific function in the cell cycle necessary for accuracy of cell division.

Summary: The Accumulation of Oncogenic and Tumor-Suppressor Mutations Produces Cancer Cells with Grossly Altered Genomes

Cancer-causing mutations disrupt the normal controls that create a balance between activation and inhibition of cell division. Dominant mutations that change protooncogenes to oncogenes may overactivate expression of proteins that promote proliferation. Recessive mutations in tumor-suppressor genes may release the brakes that keep cells from proliferating. Both types of mutations may tip the balance toward excessive and inaccurate cell proliferation.

Mutations that disable one part of a cell's elaborate DNA repair system increase its mutation rate and thus its likelihood of becoming cancerous. Although no single mutation converts a normal cell to a cancer cell, if a cell has a mutation in one gene that predisposes to cancer, that cell has a higher than normal probability of becoming cancerous because it is already one step along the way. The early mutations in a cell's progression from normal to cancerous may lead to increased proliferation and affect the accuracy of cellular reproduction, allowing the accumulation of several mutations. Other subsequent or simultaneous mutations may enable the abnormally and inaccurately proliferating cells of a single lineage to avoid programmed cell death, evade the immune system, increase formation of blood vessels supplying the abnormal clone, alter the proteins that control tissue architecture, and invade nearby or distant tissues (that is, metastasize). Environmental factors such as radiation and mutagenic chemicals cause most of the mutations that result in cancer, but rare inherited defects can contribute the first step (see the Genetics and Society box "The Uses of Genetic Testing in Predicting and Treating Cancer").

cancer-cell mutations. Indeed, comparative microarray analyses (see Chapter 10) of cancerous and normal tissues have revealed molecular markers (RNA and DNA changes) for leukemia, prostate cancer, breast cancer, and melanoma (an aggressive form of skin cancer). These markers make it possible to separate phenotypically similar cancers into distinct groups that probably arose in different ways, have different prognoses, and require different treatments. Mutations in these genes are not inherited like those in *BRCA1* and *BRCA2;* rather, they arise during a person's lifetime.

With breast cancer, for example, if a person has a small tumor (less than 2 cm in diameter) that has not spread to the lymph nodes, surgical removal and follow-up radiation (sometimes in conjunction with chemotherapy) usually give the patient a good chance of overcoming the cancer. However, if the tumor cells carry mutations in the *p53* gene, the prognosis is poorer, because breast tumors with absent or mutated p53 proteins tend to resist treatment with radiation and many anticancer drugs. On the other hand, large, fast-growing tumors that have already metastasized—if (and only if) their cells carry a mutation in the *HER2/neu* protooncogene—respond well to a recently developed drug named Herceptin. The *HER2/neu* gene encodes a human epidermal growth factor receptor that helps control how cells grow, divide, and repair themselves. Breast cancers with a mutated *HER2/neu* are very aggressive and more likely to recur than some other types of breast cancer. Herceptin, an antibody-based drug, shrinks and even gets rid of HER2/neu-positive breast cancers that have spread; it also shrinks medium to large tumors in the breast tissue itself and reduces the risk of recurrence.

Genetic tests for oncogenic and mutated tumor-suppressor alleles thus have specific predictive and therapeutic uses; but understanding the limitations of such tests is important. The predictive tests can indicate only increased risk. If knowledge of such risk makes it possible to take preventive measures, the tests make sense. Tests to help determine the course of therapy are currently possible for only a few cancers in which specific mutations have been linked to specific prognoses. However, with the application of new genomic and proteomic tools, the number and scope of such tests will increase, which in conjunction with more precisely targeted drugs, will enable doctors to hone their diagnoses and tailor their treatments for individual cancer patients.

Connections

The existence of numerous controls in each of several cell-cycle pathways suggests that evolution has erected many barriers in multicellular animals to the uncontrolled reproduction of "selfish" cells. At the same time, the hundreds of genes contributing to normal cell-cycle regulation provide hundreds of targets for cancer-producing mutations.

Variations on the theme of cell-cycle regulation play a key role in the development of eukaryotic organisms. During the development of multicellular organisms, cells must not only control their cell cycles, they must also adopt different fates and differentiate into different tissues. In *Drosophila,* for example, after fertilization, nuclear division occurs without cell division for the first 13 cycles; during these cycles, the nuclei go through many rapid S and M phases without any intervening G_1 or G_2 (**Fig. 19.25**). In cycles 10–13, the synthesis and degradation of cyclinB regulates mitosis. Sometime during cycles 14–16, a G_2 phase appears, and distinct patches of cells with different-length cycles become evident within the embryo. The differences in cycle time between the different cell types are the result of variable G_2 phases. Late in G_2, *CDC25* activates cyclin-dependent kinases to control the timing of mitosis. Many tissues stop dividing at cycle 16, but a few continue. In the still-dividing cells, a G_1 phase appears. Some of these cells will arrest in G_1 during larval growth, only to start dividing again in response to signals relayed during metamorphosis.

In Chapter 20, we present the basic principles of development and describe how biologists have used genetic analysis in various model organisms to examine development at the cellular and molecular levels.

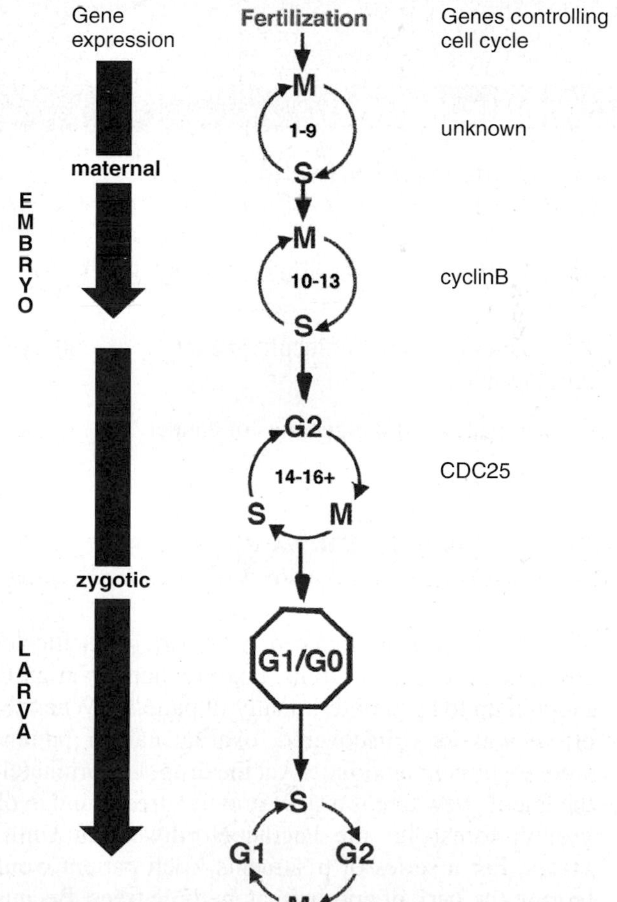

Figure 19.25 Regulation of the cell-cycle changes during *Drosophila* development. Each step of development has built-in regulators that act as barriers to uncontrolled reproduction of "selfish" cells. Some of these regulators, such as cyclinB and *CDC25,* are known; others are not.

Essential Concepts

1. Several genetic pathways help control cell division.
   a. The inhibition or activation of CDKs inhibits or activates G_1-to-S and G_2-to-M transitions.
   b. The measured synthesis and degradation of different cyclins guides CDKs to the appropriate targets at the appropriate times.
   c. Checkpoints that integrate repair of chromosomal damage with events of the cell cycle minimize the replication of damaged DNA.
   d. The genes and proteins of various signal transduction systems relay signals about whether or not to enter the cell cycle.
2. Cancer is a genetic disease resulting from the growth of a clone of mutant cells.
3. A cell requires many mutations to become cancerous. Exposure to environmental mutagens probably generates most of these mutations.
4. Many mutations that lead to cancer jeopardize cell-cycle regulation.
   a. Mutations in growth factors, receptors, and other elements of signal transduction pathways can release cells from the signals normally required for proliferation.
   b. Mutations in CDKs and the proteins that control them may also lead to inappropriate proliferation or genomic instability.
   c. Mutations in DNA repair and checkpoint controls lead to genomic instability and, often, to loss of the surveillance system that kills aberrant cells by apoptosis.
   d. Mutations that lead to genomic instability permit the rapid evolution of abnormal tumor cells.

On Our Website

www.mhhe.com/hartwell3
Chapter 19

Annotated Suggested Readings and Links to Other Websites

- Key papers on the molecular genetics of cell-cycle regulation
- Classic papers on the genetics of cancer

Specialized Topics

- Comprehensive example of the genetics of brain cancer

Social and Ethical Issues

1. Taxol is a chemical produced in the bark of Pacific yew trees. In experimental trials, taxol reduced ovarian tumors in up to 50% of terminally ill patients. When this effect was first discovered, ovarian cancer patients were extremely anxious to get the drug. Unfortunately, the Pacific yew tree is a slow-growing tree found in old growth forests of the Pacific Northwestern United States. For a series of treatments, each patient would require the bark of approximately three trees. Because of the scarcity of trees and the difficulty in obtaining taxol from the bark, the expense was high. In 1993, the cost of production by the National Cancer Institute was $1200 for the 2 grams needed for each patient. When a drug is scarce, who should get the drug? The clinical trials involved patients for whom no other treatment had worked. Is this the best use of limited resources of a potentially powerful anticancer drug? Who should decide who gets the drug? On what basis should they make their decision?

2. In 1976, cancer patient John Moore had his spleen removed at a university teaching hospital as a treatment for the hairy cell leukemia from which he suffered. Some of his spleen cells were cultured at that time. During subsequent treatment other cell types were removed from the same patient and preserved in culture. Researchers eventually used some of these cells to establish a very useful T-cell line. The cell line produced

the lymphokine GM-CSF, used in treating cancers and AIDS. A patent, granted to the university and the doctors for the cell line, was sold to two biotechnology companies. Moore filed a claim against the doctors, companies, and the university stating that he was entitled to some of the profits and that there had been a lack of informed consent. He had signed a consent form but was never informed of the potential for commercial exploitation. Several questions arise from the landmark legal case, *Moore v. Regents of California.* To what extent must a doctor disclose interests in a commercial venture to a patient? Does a person own the rights to cells that have been removed from his or her body?

3. Suitable animal models for the effects of exposure to radiation on humans are not available. Only tests on humans can make it possible to evaluate the dangers of chromosome damage and cancer resulting from various types of exposure. Under these circumstances, is human experimentation acceptable? Who should participate in human trials? Volunteers who are well apprised of the risks? Prisoners on death row? Are there valid alternatives?

Solved Problems

I. The addition of growth factors to tissue culture cells stimulates cell division. A number of candidate drugs can be tested for their ability to stop this stimulation of cell division. What do you think the target of these drugs could be?

Answer

This question concerns the regulation of cell division. Growth factors are made by one cell and bind to receptors of another cell to stimulate the cell division cycle. *A drug that binds to receptors* would block access and prevent growth factors from binding. Alternatively, *the drug could bind to the growth factor,* thereby preventing its interaction with the receptor. (These are the most obvious targets. If you are familiar with the signal transduction pathway inside the cell, you might also propose that proteins in this pathway could be targets for drug development.)

II. The *p53* gene has been cloned, and you are using it to analyze DNA in patients in which *p53* defects are involved in the development of their tumors. DNA samples were obtained from normal and tumor tissue of three different cancer patients, digested with *Bam*HI, electrophoresed on an agarose gel, and transferred to filter paper that was probed with a labeled *p53* fragment. Each of the patients inherited a *p53* mutation.

Wild-type | Individual 1: Normal tissue, Tumor tissue | Individual 2: Normal tissue, Tumor tissue | Individual 3: Normal tissue, Tumor tissue

Thin bands indicate half the DNA content of thick bands. DNA from an individual who did not inherit a *p53* mutation is shown in the lane labeled wild type. All wild-type alleles in this study produce the same three fragments. Assuming the model of *p53* acting as a tumor-suppressor gene is correct and that *p53* defects are involved in each of these cancer patients, how would you describe the genetic makeup of the *p53* gene in the normal and tumor tissue of each of the three patients?

Answer

This question requires knowledge of tumor-suppressor genes. The wild-type *p53* region (as seen in the "wild-type" individual) has three hybridizing bands. Because *p53* is a tumor suppressor, it is recessive at the cellular level, and both copies must be defective in the tumor cells. No observable changes are apparent in patient 1, so this patient must have inherited a point mutation in *p53,* and in the tumor cells, the second copy would also contain a small mutation, thereby inactivating both copies of *p53* in the tumor. In patient 2, a point mutation must have been inherited. In the tumor, the whole region containing *p53* was deleted from the wild-type copy (thereby removing the second copy of the gene), as seen by the loss of restriction fragments. In patient 3, a mutation is evident in one copy of gene from the altered restriction pattern, and in the tumor, the wild-type copy of the gene was deleted (probably by gene conversion since the tumor has two mutant copies of the gene).

III. The CDC28 protein of budding yeast *S. cerevisiae* and the CDC2 protein of fission yeast *S. pombe* are protein kinases required at the "start" of the cell cycle. The genes for both proteins were identified by mutational analysis (temperature-sensitive mutations in each gene cause cell cycle arrest), and both genes have been cloned. How could you determine if one could substitute for the other functionally? (Be sure to mention sources of DNA and genotypes involved.)

Answer

The *CDC28* gene of *S. cerevisiae* could be cloned into a vector and transformed into a temperature-sensitive *cdc2* mutant of *S. pombe.* If *CDC28* has the same role (function) as *CDC2,* the transformed cell will now grow and divide at nonpermissive temperatures. Conversely, a clone of the *CDC2* gene of *S. pombe* could be cloned into a vector able to transform a temperature-sensitive *cdc28* mutant of *S. cerevisiae.* If *CDC2* of *S. pombe* can substitute for *CDC28* of *S. cerevisiae,* the transformed cells would grow and divide at nonpermissive temperatures.

Problems

Interactive Web Exercise

The National Cancer Institute has gathered information and data about normal, precancerous, and cancerous cells at a site called the Cancer Genome Anatomy Project (CGAP). Our website at www.mhhe.com/hartwell3 contains a brief exercise to introduce you to the use of CGAP. Once at our site, go to Chapter 19 and click on "Interactive Web Exercise".

Vocabulary

1. For each of the terms in the left column, choose the best matching phrase in the right column.

a. growth factor	1. mutations in these genes are dominant for cancer formation
b. tumor-suppressor genes	2. programmed cell death
c. cyclin-dependent protein kinases	3. series of steps by which a message is transmitted
d. apoptosis	4. proteins that are active cyclically during the cell cycle
e. oncogenes	5. control progress in the cell cycle in response to DNA damage
f. receptor	6. mutations in these genes are recessive at the cellular level for cancer formation
g. signal transduction	7. signals a cell to leave G_0 and enter G_1
h. checkpoints	8. cell-cycle enzymes that phosphorylate proteins
i. cyclins	9. protein that binds a hormone

Section 19.1

2. During which phase(s) of the cell cycle would the following enzymes or proteins be most active?

a. tubulins in the spindle fibers
b. centromere motor
c. DNA polymerase
d. CDC28 of *S. cerevisiae* or CDC2 of *S. pombe*

3. Conditional mutations are useful for genetic analysis of essential processes. For example, temperature-sensitive cell-cycle mutations in yeast do not divide at 37°C (nonpermissive temperature) but will divide at 30°C. An alternative type of conditional mutation is a cold-sensitive mutation in which the nonpermissive temperature is low (23°C). List the steps you would go through to isolate cold-sensitive cell-cycle mutants of yeast.

4. Many temperature-sensitive yeast mutants that showed defects in the cell cycle were isolated in the 1970s. The mutants that arrested at the unbudded stage were mated with each other to do a complementation analysis. A + sign on the chart indicates that the resulting diploids grew at the high (nonpermissive) temperature. How many complementation groups (that is, how many genes) are represented by these mutants?

	1	2	3	4	5	6	7	8	9
1	−	+	+	−	−	+	+	+	+
2	+	−	+	+	+	+	+	−	+
3	+	+	−	+	+	−	−	+	−
4	−	+	+	−	−	+	+	+	+
5	−	+	+	−	−	+	+	+	+
6	+	+	−	+	+	−	−	+	−
7	+	+	−	+	+	−	−	+	−
8	+	−	+	+	+	+	+	−	+
9	+	+	−	+	+	−	−	+	−

5. In 1951 a woman named Henrietta Lacks died of cervical cancer. Just before she died a piece of her tumor was taken and put into culture in a laboratory in an attempt to induce the cells to grow *in vitro.* The attempt succeeded, and the resulting cell line (known as HeLa cells) is still used today in laboratories around the world for studies of various aspects of cell biology. In the cell cycle of typical HeLa cells, G_1 lasts about 11 hours, S lasts about 8 hours, G_2 lasts 4 hours, and mitosis (M) takes about 1 hour.

a. Cultured cells do not typically grow synchronously. That is, the individual cells in a culture are randomly distributed throughout the cell cycle. If you looked through the microscope at a sample of HeLa cells, in approximately what proportion of them would you expect the chromosomes to be visible? (The cells do not split apart completely after cytokinesis, and each joined double cell should be counted as one.)

b. Approximately what proportion would be in interphase?

6. The activity of key cell-cycle regulatory proteins is cyclical, appearing only when needed. What are three ways by which a cell can achieve this cyclical nature of protein activity?

7. True or false?
   a. CDKs phosphorylate proteins in the absence of cyclins.
   b. Degradation of cyclins is required for the cell cycle to proceed.
   c. CDKs are involved in checking for aberrant cell-cycle events.

8. Checkpoints occur at several different times during the cell cycle to check that the DNA content of the cell has not been damaged or altered. Match the defect in a checkpoint with the consequences of that defective checkpoint.

Defective checkpoint	Consequences of defect in checkpoint
a. G_1 to S	1. aneuploidy
b. G_2 to M	2. single-strand nicks get replicated and amplification occurs
c. M	3. unrepaired double-strand breaks result in broken chromosomes

9. Molecules outside and inside the cell regulate the cell cycle, making it start or stop.
   a. What is an example of an external molecule?
   b. What is an example of a molecule inside the cell that is involved in cell cycle regulation?

10. a. Would you expect a cell to continue or stop dividing at a nonpermissive high temperature if you had isolated a temperature-sensitive *RAS* mutant that remained fixed in the GTP-bound form at nonpermissive temperature?
    b. What would you expect if you had a temperature-sensitive mutant in which *RAS* stayed in the GDP-bound form at high temperature?

11. Put the following steps in the correct ordered sequence.
    a. kinase cascade
    b. activation of a transcription factor
    c. hormone binds receptor
    d. expression of target genes in the nucleus
    e. RAS molecular switch

12. Draw a diagram illustrating the accumulation of S phase and M phase cyclins during the cell cycle. When are SCF and APC (the protein complexes that add ubiquitin to S phase and M phase cyclins, respectively) activated?

13. One of the hallmarks of mitotic anaphase is the separation of sister chromatids. Sister chromatids are held together by a protein complex called "cohesin." Based on your answer to the previous problem (# 12), propose a mechanism that would allow sister chromatids to separate during anaphase. How might your proposed mechanism also explain the M phase checkpoint that prevents sister chromatid separation until all the chromosomes have connected properly to the mitotic spindle?

Section 19.2

14. Mouse tissue culture cells infected with the SV40 virus lose normal growth control and become transformed. If transformed cells are transferred into mice, they grow into tumors. The SV40 protein responsible for this transformation is called T antigen. T antigen has been found to associate with the cellular protein p53. If the *p53* gene fused to a high-level expression promoter is transfected into tissue culture cells, the cells are no longer transformed by infection by SV40.
    a. Propose a hypothesis to explain how the high expression of *p53* saves the cells from transformation by T antigen.
    b. You have decided to examine the functional domains of the p53 protein by mutagenizing the cDNA, fusing it to the high-level promoter, and transfecting into cells. Results are shown in the following table. How would you explain the effects of mutations 1 and 2 on p53 function?
    c. What is the effect of mutation 3 on p53 function?

	Morphology	
p53 construct	Noninfected cells	SV40-infected cells
None	Normal	Transformed
Wild type	Normal	Normal
Mutation 1	Normal	Transformed
Mutation 2	Normal	Transformed
Mutation 3	Normal	Normal

15. What are four characteristics of the cancer phenotype?

16. Amplification of DNA sequences in *p53* mutants can be visualized using electron microscopy.
    a. Using a different technique, how could you detect amplification of a specific sequence?
    b. How could you detect gross rearrangements (>10 Mb) of chromosomal DNA?

17. Some germ-line mutations predispose to cancer, yet often environmental factors (chemicals, exposure to radiation) are considered major risks for developing cancer. Are these conflicting views of the cause of cancer or can they be reconciled?

18. The incidence of colon cancer in the United States is 30 times higher than it is in India. Differences in diet and/or genetic differences between the two populations may contribute to these statistics. How would you assess the role of each of these factors?

19. Put the following steps in the order appropriate for the positional cloning of *BRCA1*, a gene involved in predisposition to breast cancer.
    a. Locate transcripts corresponding to the DNA.
    b. Use the physical map to get clones.
    c. Determine the tissues in which the transcripts are present.

d. Look for homologous DNA in other organisms.
e. Determine linkage to RFLPs and other molecular markers.
f. Sequence the DNA from affected individuals.

20. Because mutations occur in the development of cancer, researchers suspected that defects in DNA repair machinery might lead to a predisposition to cancer. Place the following steps in appropriate order for following a candidate gene approach to determine if defects in mismatch repair genes lead to cancer.
a. Use molecular markers near the homologous gene to determine if the candidate gene is linked to a predisposition phenotype.
b. Isolate a human homolog of a yeast mismatch repair gene.
c. Compare the DNA sequence of the mismatch repair gene of affected and unaffected persons in a family with predispositions.
d. Determine the map location of the human homolog of the yeast mismatch gene.

21. Which of the following events is unlikely to be associated with cancer?
a. mutations of a cellular protooncogene in a normal diploid cell
b. chromosomal translocations with breakpoints near a cellular protooncogene
c. deletion of a cellular protooncogene
d. mitotic nondisjunction in a cell carrying a deletion of a tumor-suppressor gene
e. incorporation of a cellular oncogene into a retrovirus chromosome

22. You have decided to study genetic factors associated with colon cancer. An extended family from Morocco in which the disease presents itself in a large percentage of family members at a very early age has come to your attention. (The pedigree is shown below.) In this family, individuals either get colon cancer before the age of 16, or they don't get it at all.

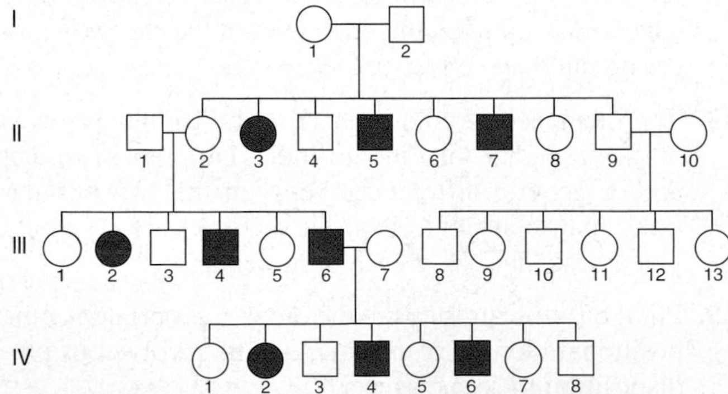

a. Based on the information you have been given, what evidence, if any, suggests an inherited contribution to the development of this disease?
b. You decide to take a medical history of all of the 36 people indicated in the pedigree and discover that a very large percentage drink a special coffee on a daily basis, while the others do not. The only ones who don't drink coffee are individuals numbered I-1, II-2, II-4, II-9, III-7, III-13, IV-1, and IV-3. Could the drinking of this special coffee possibly play a role in colon cancer? Explain your answer.

23. To further understand the basis for colon cancer, you find a family from the United States in which two members also get the disease before the age of 16. If there were a dominant inherited mutation segregating in this family, which of the individual(s) would you predict had the mutation in their colon cells but did not develop the disease?

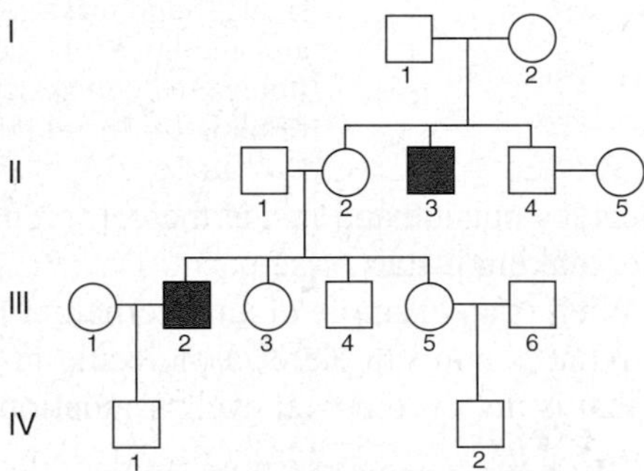

24. You suspect that a very specific point mutation in the *p53* gene is responsible for the majority of *p53* mutations found associated with tumors. Which combination of these techniques would you be most likely to use in developing a simple assay for predictive testing?
a. polymerase chain reaction with oligonucleotide primers flanking the mutation
b. restriction enzyme digestion followed by Southern blot
c. RNA isolation followed by Northern blot
d. hybridization with allele-specific oligonucleotides

25. A female patient 19 years old is diagnosed with chronic myelogenous leukemia (CML), whose symptoms are anemia and internal bleeding due to a massive buildup of leukemic white blood cells. Karyotype analysis shows that the leukemic cells of this patient are heterozygous for a reciprocal translocation involving chromosomes 9 and 22. However, none of the normal, non-leukemic cells of this patient contain the translocation. Which of the following statements is true and which is false?
a. The translocation results in the inactivation (loss-of-function) of a tumor suppressor gene.
b. The translocation results in the inactivation (loss-of-function) of an oncogene.
c. There is a 50% chance that any child of this patient will have CML.
d. This patient is a somatic mosaic in terms of the karyotype.

e. DNA extracted from leukemic cells of this patient, if taken up by normal mouse tissue culture cells, could potentially transform the mouse cells into cells capable of causing tumors.
f. The normal function of the affected tumor suppressor gene or proto-oncogene at the translocation breakpoint could potentially block the function of the cyclin proteins that drive the cell cycle forward.
g. This woman is heterozygous for an X-linked gene; the two alleles encode two distinguishable variant forms of the protein product of the gene. If you looked at different normal cells from different parts of her body, some would express exclusively one variant form of the protein, and other normal cells would express exclusively the other variant form.
h. If you examined different leukemic cells from this patient for the protein described in part *g*, all would express the same variant form of the protein.
i. Two rare events must have occurred to disrupt both copies of the tumor suppressor gene or proto-oncogene at the translocation breakpoint in the leukemic cells.
j. A possible treatment of the leukemia would involve a drug that would turn on the expression of the tumor suppressor gene or proto-oncogene at the translocation breakpoint in the leukemic cells.

26. Describe a molecular test to determine if chemotherapy given to the patient described in the previous problem (#25) was completely successful. That is, devise a method to make sure that the patient's blood is now free of leukemic cells.

27. A generic signaling cascade is shown in the following figure. A Growth Factor (GF) binds to a Growth Factor Receptor, activating the kinase function of an intracellular domain of the Growth Factor Receptor. One substrate of the Growth Factor Receptor kinase is another kinase, Kinase A, that has enzymatic activity only when it is itself phosphorylated by the Growth Factor Receptor kinase. Activated Kinase A adds phosphate to a Transcription Factor. When it is unphosphorylated, the Transcription Factor is inactive and stays in the cytoplasm. When it is phosphorylated by Kinase A, the Transcription Factor moves into the nucleus and helps turn on the transcription of a *Mitosis factor* gene whose product stimulates cells to divide.

a. The following list contains the names of the genes encoding the corresponding proteins. Which of these could potentially act as a proto-oncogene? Which might be a tumor suppressor gene?
   i. *Growth Factor*
   ii. *Growth Factor Receptor*
   iii. *Kinase A*
   iv. *Transcription Factor*
   v. *Mitosis factor*

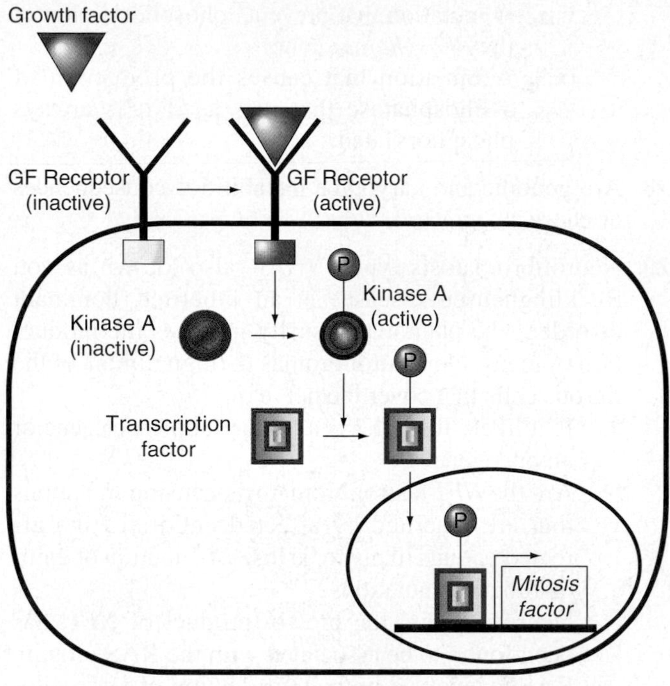

Though it is not pictured, the cell in the preceding figure also has a Phosphatase, an enzyme that removes phosphates from proteins—in this case, from the Transcription Factor. This Phosphatase is itself regulated by Kinase A.

b. What would you expect to be the effect when KinaseA adds a phosphate group to the Phosphatase? Would this activate the Phosphatase enzyme or inhibit it? Explain.
c. Is the *Phosphatase* gene likely to be a proto-oncogene or a tumor suppressor gene?
d. Several mutations are listed below. For each, indicate whether the mutation would lead to excessive cell growth or decreased cell growth if the cell were either homozygous for the mutation or heterozygous for the mutation and a wild-type allele. Assume that 50% of the normal activity of all these genes is sufficient for normal cell growth.
   i. A null mutation in the *Phosphatase* gene
   ii. A null mutation in the *Transcription Factor* gene
   iii. A null mutation in the *Kinase A* gene
   iv. A null mutation in the *Growth Factor Receptor* gene
   v. A mutation that causes production of a constitutively active Growth Factor Receptor whose kinase function is active even in the absence of the Growth Factor
   vi. A mutation that causes production of a constitutively active Kinase A
   vii. A reciprocal translocation that places the *Transcription Factor* gene downstream of a very strong promoter

viii. A mutation that prevents phosphorylation of the *Phosphatase* gene
ix. A mutation that causes the production of a Phosphatase that acts as if it is always phosphorylated.

28. Are genome and karyotype instabilities consequences or causes of cancer?

29. Neurofibromatosis type 1 (NF1; also known as von Recklinghausen disease) is an inherited dominant disorder. The phenotype usually involves the production of many skin neurofibromas (benign tumors of the fibrous cells that cover the nerves).

a. Is it likely that *NF1* is a tumor suppressor gene or an oncogene?

b. Are the *NF1* neurofibromatosis-causing mutations that are inherited by affected children from affected parents likely to be loss-of-function or gain-of-function mutations?

c. Neurofibromin, the protein product of *NF1*, has been found to be associated with the RAS protein. RAS is involved in the transduction of extracellular signals from growth factors. The active form of RAS (the form initiating the signal transduction cascade causing proliferation) is complexed with GTP; the inactive form of RAS is complexed with GDP. Would the wild-type neurofibromin protein favor the formation of RAS-GTP or RAS-GDP?

d. Which of the following events in a normal cell from an individual inheriting a neurofibromatosis-causing allele could cause the descendents of that cell to grow into a neurofibroma?

i. A second point mutation in the allele of *NF1* inherited from the afflicted parent.
ii. A point mutation in the allele of *NF1* inherited from the normal parent.
iii. A large deletion that removes the *NF1* gene from the chromosome inherited from the afflicted parent.
iv. A large deletion that removes the *NF1* gene from the chromosome inherited from the normal parent.
v. Mitotic chromosomal non-disjunction or chromosome loss.
vi. Mitotic recombination in the region between the *NF1* gene and the centromere of the chromosome carrying *NF1*
vii. Mitotic recombination in the region between the *NF1* gene and the telomere of the chromosome carrying *NF1*

e. The *American Journal of Medical Genetics* published a report in 1999 that certain patients with neurofibromatosis type I who had an affected parent also inherited specific facial anomalies from that parent. Formulate a succinct hypothesis to explain why these patients inherit this additional phenotype, but most other patients with inherited neurofibromatosis I do not.

f. There is a much rarer form of NF1 called segmental NF1. In this form of the disease, neither parent of the patient has any clinical sign of the disease. The tumors in the patient are restricted to one part of the body, like the right leg. Suggest an explanation for the genesis of segmental NF1 and why it is restricted to one part of the body.

Using Genetics to Study Development

Chapter 20

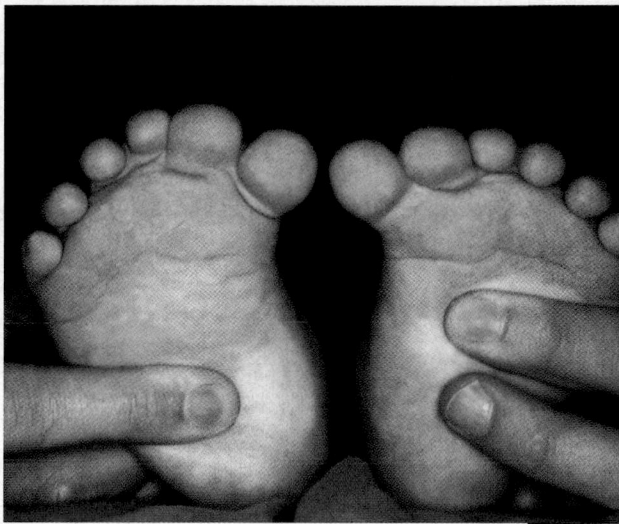
A young patient with polydactyly, a genetically caused developmental defect of humans that produces extra toes and/or fingers.

The union of a human sperm and egg (**Fig. 20.1a**) initiates the amazing process of development in which a single cell—the fertilized egg—divides by mitosis into trillions of genetically identical cells. These cells differentiate from each other during embryonic development to form hundreds of different cell types. Cells of various types assemble into wondrously complex yet carefully structured systems of organs, including two eyes, a heart, two lungs, and an intricate nervous system. Within a period of three months, the human embryo develops into a fetus whose form anticipates that of the baby who will be born six months later (Fig. 20.1b). At birth, the baby is already capable of crying, breathing, and eating; and the infant's development does not stop there. New cells form and differentiate throughout a person's growth, maturation, and even senescence.

The development of a child from a fertilized egg is so miraculous that it is difficult to look at the process from a reductionist point of view, seeking to explain it on the basis of physical, chemical, and biological principles. Yet the more biologists study development at the microscopic level, the more clearly the miracle emerges as a precisely patterned set of cellular behaviors controlled by the expression of genes. Cells divide, change shape, specialize, and interact in highly reproducible ways.

Biologists now accept that genes direct the cellular behaviors underlying development, but as recently as the 1940s, this idea was controversial. Many embryologists could not understand how cells with identical chromosome sets, and thus the same genes, could form so many different types of cells if genes were the major determinants of development. As we now know, the answer to this riddle is very simple: Not all genes are "turned on" in all tissues. Cells regulate the expression of their genes so that each gene's protein product appears only when and where it is needed. Two central challenges for scientists studying development are to identify which genes are critical for the development of particular cell types or organs and to figure out how these genes work together to ensure that each is expressed at the right time, in the right place, and in the right amount.

Biologists who use genetics to study how the fertilized egg of a multicellular organism becomes an adult are called **developmental geneticists.** Like other geneticists, they analyze mutations; in this case, mutations that produce developmental abnormalities. An understanding of such mutations helps clarify how normal genes control cell growth, cell communication, and the emergence of specialized cells, tissues, and organs.

Given the great diversity of the 6 billion humans on our planet, it is not hard to find rare individuals carrying mutations that alter their development. For example, pedigree analysis shows that one form of *polydactyly* (hands or feet with more than five fingers or toes) is inherited in families as a simple Mendelian autosomal dominant trait (see photograph on this page). There are, however, significant ethical and practical limitations on the study of developmental genetics in humans that prevent scientists from testing hypotheses or conducting systematic investigations. These include taboos on the deliberate production of mutants, on the experimental manipulation of affected individuals, and on forced matings between specific individuals with various

(a) Fertilization

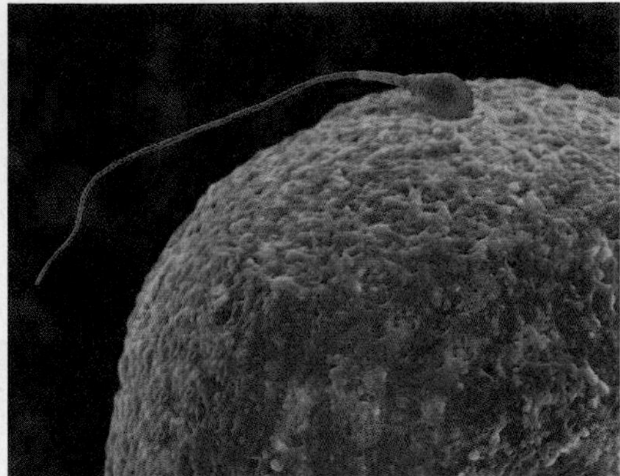

(b) A human fetus three months after fertilzation

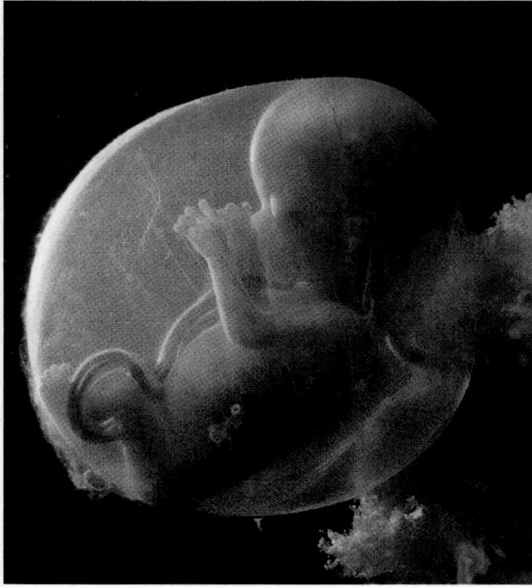

Figure 20.1 Human development. Fertilization of an egg by a sperm **(a)** creates a single cell (the zygote), which undergoes many rounds of division and cell differentiation to produce a recognizable fetus **(b)** by the end of the first trimester of pregnancy.

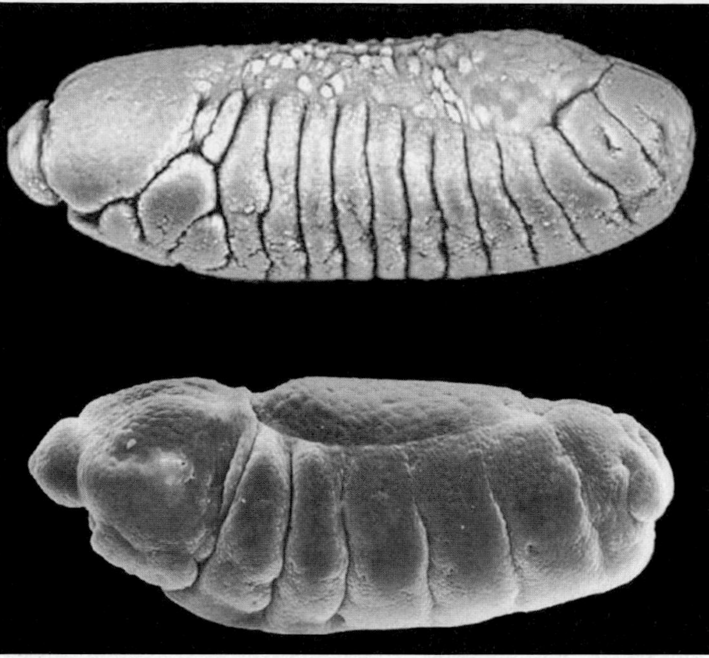

Figure 20.2 Mutations in specific *Drosophila* genes can affect early development. Wild-type embryo (*top*); embryo homozygous for a mutation in a gene called *ftz* (*bottom*). The mutant embryo has fewer body segments than normal.

abnormalities. But perhaps the most important limitation is not so obvious: Mutations that disrupt the earliest (and to some, the most interesting) stages of development almost always cause the spontaneous abortion of the affected embryo or fetus, often before the mother knows she is pregnant.

As a result, most modern developmental geneticists, even those whose primary interest is in human development, study mutations affecting the development of model organisms, which are more amenable to experimentation. In *Drosophila,* for example, only a few dozen genes guide the formation of the early embryo's segmented body plan. Mutations in some of these genes eliminate specific body segments (**Fig. 20.2**). Once the embryo has divided into segments that will become parts of the head, thorax, and abdomen, the activation and inactivation of different sets of genes direct the development of specialized organs, such as wings and legs, in each segment. Surprisingly, many of the genes controlling segmentation and body structure in *Drosophila* have counterparts in humans, and some of the human genes can substitute for the corresponding insect genes during fruit fly development. From this observation, we can infer that evolution has conserved many of the genes and genetic pathways underlying development. Insights from model organisms can thus illuminate general principles applicable to many species, including humans.

We examine in this chapter how the single cell of the fertilized egg, or zygote, differentiates into hundreds of cell types. It is impossible to do justice to this complicated issue in a short section of a textbook. What we offer here is an overview of some of the experimental strategies scientists use to study this question and a synopsis of key results illustrating some of the genetically controlled mechanisms that direct development. The genetic portraits on our website (www.mhhe.com/hartwell3) provide more details; each portrait focuses on one model organism's contribution to our understanding of development.

In this chapter, we describe

- How model organisms serve as prototypes for research in developmental genetics, and how investigators extrapolate the insights from this research to all living forms.
- Important experimental strategies developmental geneticists use to dissect the development of complex multicellular organisms.
- A comprehensive example describing the genetic analysis of body plan development in *Drosophila* and the relevance of results from this analysis to the study of body plan specification in mammals, including humans.
- Some general conclusions that provide a framework for understanding how genes help control cellular differentiation during development, that is, how they direct most of the changes that turn a fertilized egg into a functioning adult.

20.1 Model Organisms: Prototypes for Developmental Genetics

Throughout the twentieth century, developmental geneticists concentrated their research efforts on a small number of organisms that sampled a range of species from different phyla. The organisms that have contributed most to our understanding of development include the yeast *Saccharomyces cerevisiae;* the plant *Arabidopsis thaliana;* two invertebrate animals (the fruit fly *Drosophila melanogaster* and the nematode [or round worm] *Caenorhabditis elegans*); and one vertebrate (the house mouse *Mus musculus*). Although we focus here on these five eukaryotic organisms, some researchers have made major findings in other model systems, such as corn and the zebrafish. Even prokaryotic organisms and viruses have provided paradigms for tackling certain developmental problems in eukaryotes.

Why These Model Organisms?

The five model organisms we discuss in this chapter and in the genetic portraits on our website (www.mhhe.com/hartwell3) are easy to cultivate and rapidly produce large numbers of progeny. Geneticists can thus find rare mutations and study their segregation and behavior through successive generations. Each organism has attracted a dedicated cadre of researchers who share information, mutants, and other reagents. Stock centers maintain these mutants and make them available to the whole community of geneticists even decades after their original isolation. Recently, each model organism's genome has been completely sequenced, and the results have been collated and annotated on computer databases. The completion of these genome projects makes it much easier for geneticists to identify genes whose alteration by mutation produces a phenotypic effect on the organism's development.

In addition to these shared advantages, each model organism also possesses idosyncratic features that make it valuable for particular types of genetic or developmental analyses. Take yeast, for example. Although *S. cerevisiae* is a single-celled eukaryote, yeast cells signal to each other and differentiate into two mating types using variations of processes involved in the development of multicellular eukaryotes. Because *S. cerevisiae* cells can grow as haploids or diploids, researchers can identify extremely rare mutations in very large populations of haploid cells and then combine mutations in diploid cells for complementation analysis. In another example, the roundworm *C. elegans* is transparent (**Fig. 20.3**) and contains an invariant number of somatic cells as an adult—959 in the female/hermaphrodite and 1031 in the male. Because of these unusual properties, researchers can discern the lineage of every cell as the fertilized egg develops into the multicellular adult.

All Living Forms Are Related. . .

In the last 150 years, biologists have come to realize that life-forms are related on many levels. For example, the cells of all eukaryotic organisms have many features in

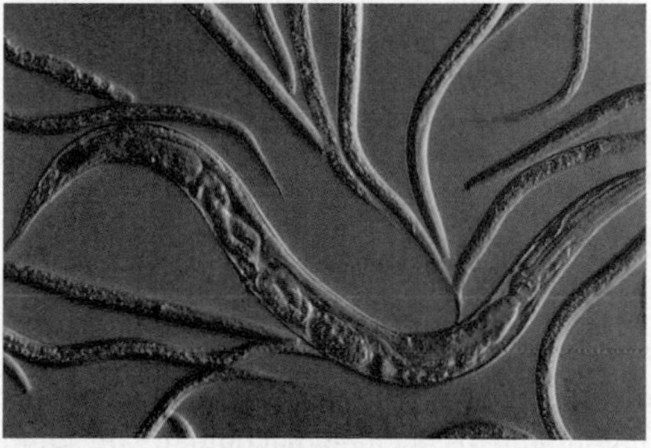

Figure 20.3 **The transparency of *C. elegans* facilitates study of the worm's development.**

common that are recognizable in the light or electron microscope. These include a membrane-bounded nucleus containing linear chromosomes; mitochondria containing their own chromosomes; and a network of membranous structures such as the endoplasmic reticulum, which facilitates the intracellular transport of molecules. Moreover, the metabolic pathways by which cells make or degrade organic molecules are virtually identical in all living organisms, and almost all cells use the same genetic code to synthesize proteins. The relatedness of organisms is even visible at the level of the amino acid sequence of individual proteins. For example, evolution has conserved the sequence of the histone protein H4, so the H4 proteins of widely divergent species are identical at all but a few amino acids. Natural selection has thus closely maintained the sequence of the H4 gene throughout the roughly 2 billion years since eukaryotic organisms first emerged. Most other proteins are not as invariant as H4, but nonetheless, scientists can often trace the evolutionary descent of a protein through the amino acid similarities of its homologs in various species.

Of particular importance to this chapter is the fact that evolution has conserved many basic strategies of development in all multicellular eukaryotes, even in organisms with body plans that look quite different. A graphic example is seen in studies of the genetic control of eye development in fruit flies, mice, and humans. *Drosophila* homozygous for mutations of the *eyeless* (*ey*) gene have either no eyes at all or, at best, very small eyes (**Fig. 20.4a**). Mutations in the *Pax-6* gene in mice (Fig. 20.4b) and the *Aniridia* gene in humans also reduce or totally abolish eye formation. When researchers cloned the *ey, Pax-6,* and *Aniridia* genes, they found that the amino acid sequences of all three encoded proteins were closely related. This is surprising because the eyes of vertebrates and insects are so dissimilar: Insect eyes are composed of many facets called ommatidia, while the vertebrate eye is a single camera-like organ (see Fig. 1.9 on p. 7). Biologists had thus long assumed that the two types of eyes evolved independently. However, the homology of *ey, Pax-6,* and *Aniridia* suggests instead that the eyes of insects and vertebrates evolved from a single prototypical light-sensing organ whose development required a gene ancestral to *ey* and its mouse and human homologs. Providing an even more emphatic demonstration of the conservation of this genetic program through more than 300 million years of evolution, the wild-type mouse and human genes can function in flies to direct the formation of *Drosophila* eyes. We will describe this experiment in more detail later in this chapter.

(a)

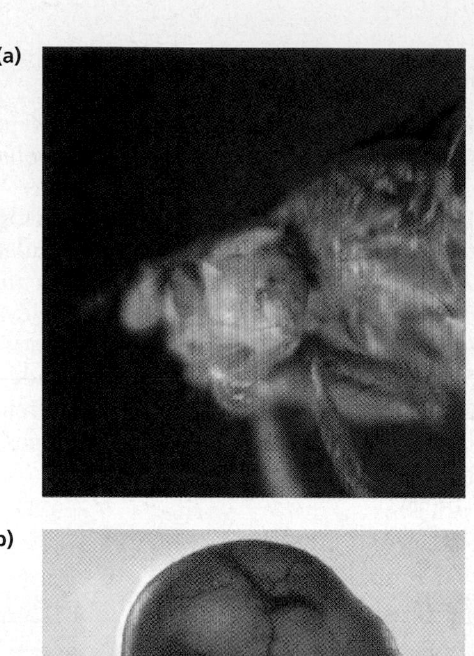

(b)

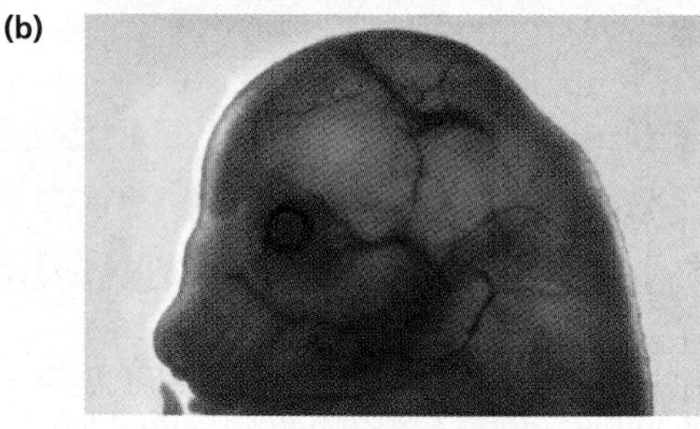

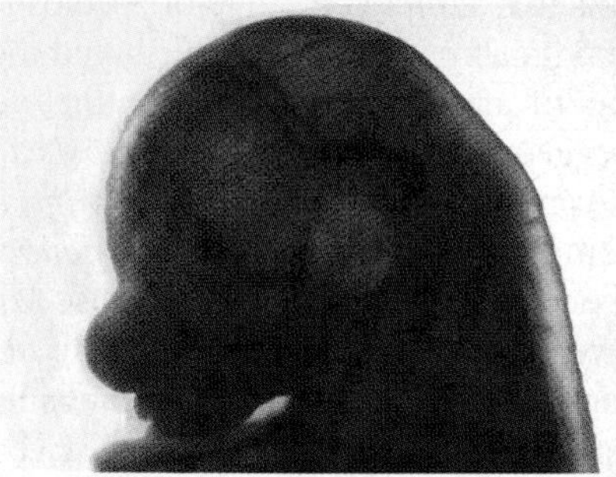

Figure 20.4 The *eyeless/Pax-6* gene is critical for eye development. (a) Hypomorphic or null mutations of the *eyeless* gene reduce the size of eyes or completely abolish them in adult flies. **(b)** Hypomorphic or null mutations in the homologous mouse *Pax-6* gene also reduce or abolish eye development. *Top:* wild-type mouse fetus. *Bottom: Pax-6* mutant fetus.

. . .Yet All Species Are Unique

Although the conservation of developmental pathways makes it tempting to conclude that humans are simply large fruit flies, this is obviously not true. Evolution is not only conservative, but it is also innovative. Organisms sometimes use disparate strategies to accomplish the same developmental goal. One example is the difference between the two-cell embryos that form in *C. elegans* and humans upon completion of the first mitotic division in the zygote. If one of the two cells is removed or destroyed in a *C. elegans* embryo at this stage, a complete nematode cannot develop. Because each of the two cells has already received a different set of molecular instructions to guide development, the descendents of one of the cells can differentiate into only certain cell types and the descendent of the other cell into other types. The situation is very different in humans: If the two embryonic cells are separated from each other, two complete individuals (identical twins) will develop. In fact, as we saw in Fig. 11.1 on p. 391, removal of a cell from a

6–10-cell human embryo has no effect on the development of the remainder of the embryo. There is thus an intrinsic difference in the way worm and human embryos develop at these early stages. As soon as it has been formed by mitosis, each cell in the early *C. elegans* embryo has already been assigned a specific fate; this pattern of development is often called **mosaic determination.** In contrast, the cells of a human embryo can alter or "regulate" their fates according to the environment, for example to make up for missing cells; this is called **regulative determination.**

To summarize, genetic studies of development in model organisms often provide important lessons that can be generalized to all eukaryotes. The differences observed in the general strategies directing cell fates and in the molecular underpinnings of developmental processes highlight the distinctive ways in which evolution has molded the action of conserved genes to produce diverse development programs.

20.2 Genetics Simplifies the Study of Development

Because proteins are the basic elements of cellular function, biologists can try to understand development by defining the roles played by individual proteins. To do this, they eliminate all copies of a single type of protein from a cell or organism and determine the consequences. From these consequences, they can often infer the function of the normal protein in normal development.

Genetics makes this experimental strategy possible. All an investigator has to do is isolate a mutant cell or organism with a specific, inactivated gene. Such mutants are usually found in the course of genetic screens to look for animals whose development is aberrant in interesting ways. A mutant with an altered gene will lack the wild-type protein encoded by that gene. Careful analysis of the mutant phenotype can pinpoint what the protein does in development. As we will see, this basic strategy, while not the only way researchers can harness genetics to study development, is almost always the first and most important step.

The Genetic Dissection of Development Depends on a Comprehensive Set of Mutants

Once a gene affecting a developmental process has been identified by a mutation, geneticists try to isolate many additional mutant alleles of that gene. If the mutations affect the function of the corresponding protein in a different way, studies of the phenotypes associated with the various mutations may shed light on the diverse roles the protein plays in the organism. The appendix *Genetic Tools for the Analysis of Development* (on our website) discusses in detail the types of genetic screens that researchers can perform to identify the genes important to development, as well as various useful alleles of these genes.

Loss-of-Function Mutations

Most mutations are loss-of-function mutations: They disrupt gene function by altering the amino acid sequence (and thus the three-dimensional structure) of the protein product or by interfering with any step of gene expression (transcription, translation, or RNA processing). As a result, such mutations give rise to proteins with diminished (or no) biochemical activity; or they decrease (or stop) production of an otherwise normal protein. We describe four kinds of loss-of-function mutations.

Null Mutations. The best way to draw legitimate conclusions about the importance of a protein in development is to study an organism that completely lacks the function provided by that protein. In an analogy, if you tried to ride a bicycle with no chain, you might conclude that the chain is required to move the wheels. If, however, you tried to ride a bicycle whose chain had a damaged link (the equivalent of a partially defective protein) such that the bicycle would move but respond only erratically to your peddling, you might conclude that the chain is not critical to wheel movement and instead affords the cyclist some control over wheel movement. In Chapter 7, we saw that mutations that remove all function act as *null alleles* and that such null alleles are usually (but not always) recessive to wild-type alleles.

It is unfortunately not always easy or even possible to find null mutations in a genetic screen, even if mutagens are employed to increase the mutation rate. However, at least in some model organisms, scientists can use targeted mutagenesis to construct animals bearing null mutations in genes suspected of playing a critical role in development. The idea is to take a cloned gene, use recombinant DNA techniques to destroy its function, and then replace the wild-type gene in the genome with the inactivated cloned copy.

Knockout mice provide key examples of this kind of targeted mutagenesis (**Fig. 20.5**). The formation of knockout mice depends on the existence of **embryonic stem (ES) cells,** which are undifferentiated cells, originally derived from early embryos, that can grow in cell culture and remain undifferentiated. If these cultured cells are injected into a different early embryo, they can contribute to any and all of the tissues in the mouse that develops from that embryo. (See the Genetics and Society box on pp. 724–725 for a discussion of "Stem Cells and *Human* Cloning.")

To generate knockout mice, scientists first disrupt the cloned gene of interest by inserting foreign DNA into the middle of the gene. They then treat a culture of ES cells with the altered gene. Some of the cells will "take up" the altered cloned gene, and in a small fraction of these cells,

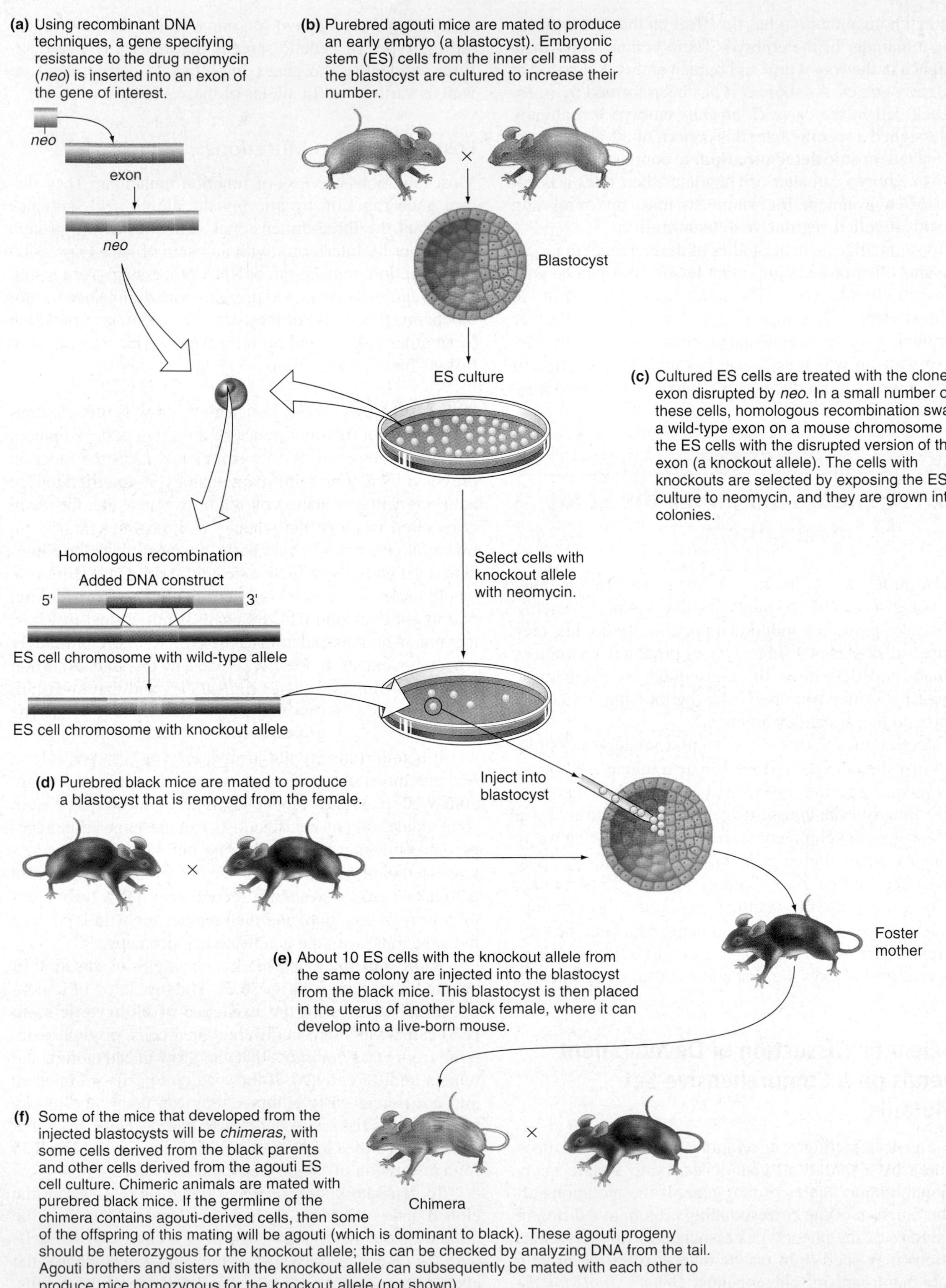

(a) Using recombinant DNA techniques, a gene specifying resistance to the drug neomycin (*neo*) is inserted into an exon of the gene of interest.
neo
exon
neo
(b) Purebred agouti mice are mated to produce an early embryo (a blastocyst). Embryonic stem (ES) cells from the inner cell mass of the blastocyst are cultured to increase their number.
×
Blastocyst
ES culture
(c) Cultured ES cells are treated with the cloned exon disrupted by *neo.* In a small number of these cells, homologous recombination swaps a wild-type exon on a mouse chromosome in the ES cells with the disrupted version of the exon (a knockout allele). The cells with knockouts are selected by exposing the ES culture to neomycin, and they are grown into colonies.
Homologous recombination
Added DNA construct
5'
3'
ES cell chromosome with wild-type allele
ES cell chromosome with knockout allele
Select cells with knockout allele with neomycin.
Inject into blastocyst
(d) Purebred black mice are mated to produce a blastocyst that is removed from the female.
×
Foster mother
(e) About 10 ES cells with the knockout allele from the same colony are injected into the blastocyst from the black mice. This blastocyst is then placed in the uterus of another black female, where it can develop into a live-born mouse.
(f) Some of the mice that developed from the injected blastocysts will be *chimeras,* with some cells derived from the black parents and other cells derived from the agouti ES cell culture. Chimeric animals are mated with purebred black mice. If the germline of the chimera contains agouti-derived cells, then some of the offspring of this mating will be agouti (which is dominant to black). These agouti progeny should be heterozygous for the knockout allele; this can be checked by analyzing DNA from the tail. Agouti brothers and sisters with the knockout allele can subsequently be mated with each other to produce mice homozygous for the knockout allele (not shown).
×
Chimera

Figure 20.5 Constructing knockout mice.

homologous recombination will allow the altered gene to replace the original gene (Fig. 20.5). Researchers can use one of several strategies to select those rare ES cells in which homologous recombination has occurred. These cells, now containing a null allele in place of a wild-type allele, are allowed to multiply in cell culture, and some of the resulting cells are injected into mouse embryos. Mice with tissues carrying the mutation are then used to begin a series of matings that culminate in the generation of animals homozygous for the null knockout mutation. The article *Mus Musculus: Genetic Portrait of the House Mouse* on our website discusses the protocols for creating knockout mice in more detail.

Hypomorphic Mutations. Although the use of null mutations allows investigators to infer the most straightforward explanations for the function of the wild-type protein in development, there are situations in which it is actually more desirable to have a partial loss-of-function (hypomorphic) mutant allele. This is because many molecules function at multiple times in development. For example, the *wingless* (*wg*) gene in *Drosophila* is needed both for the formation of a proper embryo early in development and for the formation of an adult wing much later in development. An animal homozygous for a null allele of *wg* will die during embryogenesis. This allele thus points to the importance of *wg* for early development. But because the animal dies before the wings form, you could not infer from this homozygote that the gene also functions to generate wing structures. On the other hand, flies homozygous for a certain hypomorphic allele of the *wg* gene survive to adulthood, but they have no wings. Observing the effects of this allele alone, you would conclude that the gene is involved in wing formation, but you could not infer its role in early development. This example illustrates the importance of obtaining several different mutant alleles of a gene whose function you wish to study.

Conditional Mutations. Another way to study genes with effects on diverse developmental processes is to isolate *conditional mutations* that cause a loss of function only under special circumstances. The most commonly studied type of conditional mutation, the temperature-sensitive mutation, produces a protein that is functional at a lower, *permissive temperature* but defective at a higher, *restrictive temperature.* In contrast, the protein product of the wild-type allele functions at both temperatures. It is best if the conditional mutation produces completely nonfunctional proteins at the restrictive temperature, but it is sometimes difficult to determine whether the resulting protein is completely nonfunctional or remains partly functional.

Temperature-sensitive mutations have one main experimental advantage. They make it possible to raise an animal at the permissive temperature (which allows the early stages of development to proceed normally) and then to increase the temperature to assess the importance of the gene

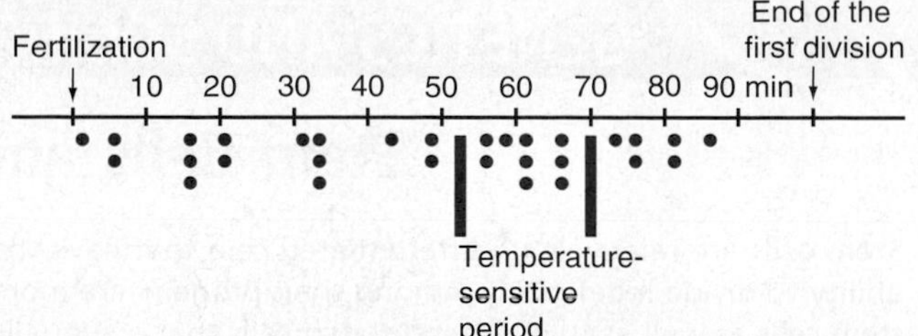

Figure 20.6 Time-of-function analysis. *C. elegans* embryos from mothers homozygous for a temperature-sensitive allele of the *zyg-9* gene develop properly if they are subjected to a 15-minute pulse of high temperature starting at any of the times indicated by the *green circles.* They develop incorrectly and subsequently die only if the high temperature begins at one of the times indicated by the *red circles.* (Each circle represents an experiment with one embryo.) These data show that the ZYG-9 protein is needed only during 15 of the first 100 minutes of development.

product at later developmental stages. This temperature-shift procedure can help define the time at which the protein product is needed for any developmental process in which it participates. **Figure 20.6** shows a temperature-shift analysis of a mutant strain of *C. elegans* carrying a temperature-sensitive lethal allele of the *zyg-9* gene, which helps determine the basic polarity of the early embryo. This temperature-shift study established that the ZYG-9 protein is required only in a very narrow window of about 15 minutes during the period between fertilization and the completion of the first mitotic division. If the protein is inactivated prior to the end of the first mitosis at any time outside the 15-minute window, development is normal.

Dominant-Negative Mutations. Most loss-of-function mutations have recessive effects because heterozygotes have about 50% of the gene function of wild-type homozygotes, and this level of gene function is sufficient for a normal phenotype. There are two exceptions to this rule. First, as discussed on pp. 288–289 in Chapter 8, for a small number of developmentally important genes, one wild-type copy is insufficient for normal development. The mutant allele will thus be dominant to wild type; this type of dominance is called **haploinsufficiency.** A second situation in which a loss-of-function mutation can have a dominant effect occurs with so-called **dominant-negative mutants.** Here, the inactive protein encoded by a mutant allele "poisons" or otherwise counteracts the function of the protein encoded by the wild-type allele. Figure 8.31b on p. 288 illustrates one of several ways in which this can occur: In a multimeric protein, the presence of one abnormal subunit might block the function of the protein even if the protein's other subunits are wild type.

Dominant-negative mutations can be particularly valuable when researchers suspect that a gene has an impact on development but they have not yet found a loss-of-function

Stem Cells and Human Cloning

Stem cells are relatively undifferentiated cells that have the ability to divide indefinitely. Among their progeny are more stem cells as well as fully differentiated cells that eventually cease dividing. *Embryonic stem (ES) cells,* which are obtained from the undifferentiated inner mass cells of a blastocyst (an early-stage embryo), are *pluripotent.* Their progeny can develop into many different cell types in the body. *Adult stem cells,* which are involved in tissue renewal and repair and are found in specific locations in the body, are *multipotent:* They can give rise only to specific types of cells. For example, hematopoietic stem cells in the bone marrow give rise to an array of red and white blood cells. While many investigators value embryonic stem cells because of their pluripotency, research with human embryonic stem cells is controversial because in order to start a stem cell culture, a blastocyst must be destroyed. Medical research with adult stem cells is relatively noncontroversial because these cells can be harvested from a patient's own tissues. However, adult stem cells have significant limitations. They are present in only minute quantities and are thus difficult to isolate, and they can give rise to only certain kinds of differentiated cells.

For medical researchers, the greatest excitement surrounding the use of embryonic stem cells is the potential for human **therapeutic cloning** to replace lost or damaged tissues. In a protocol known as *somatic cell nuclear transfer,* researchers create a cloned embryo by taking the nucleus of a somatic cell from one individual and inserting it into an egg cell whose own nucleus has been removed (**Fig. A**). This hybrid egg is then stimulated to begin embryonic divisions by treatment with electricity or certain ions. The embryo is not allowed to develop to term; instead, it is cultured for about five days in a petri plate to the blastocyst stage, at which point the ES cells in the inner cell mass are collected and placed in culture. The cultured ES cells can be induced to differentiate into many kinds of cells that might be of therapeutic value, such as nerve cells to treat Parkinson disease (Fig. A).

One of the major advantages of therapeutic cloning is that the ES cells and the differentiated cells derived from them are genetically identical to the patient's own cells. Thus, there should be no chance of tissue rejection when these cells are transplanted into the patient's body. Some studies with animal models suggest that therapeutic cloning may be effective, at least in certain situations, but therapeutic cloning in humans may be more technically challenging. This field of research was badly shaken late in 2005 when claims that a laboratory in South Korea had cloned human embryos and extracted ES cells from them were proven fraudulent. Other recent studies have raised worries that even if therapeutic cloning in humans is possible, stem cells may accumulate cancer-causing mutations. The potential of stem cells to treat diseases is thus, at present, extremely promising but far from demonstrated.

Therapeutic cloning, which is specifically intended to produce stem cells for the treatment of ailing patients, must not be confused with **reproductive cloning,** a type of cloning designed to make genetically identical complete organisms. The idea here is to create a cloned embryo by the same method just described for therapeutic cloning. In this case, however, the embryo is implanted into the uterus of a foster mother and allowed to develop to term (Fig. A). Reproductive cloning has been successfully performed in several mammalian species, such as sheep and cats. Many cloned animals exhibit puzzling developmental defects such as obesity that may be due to the inability of the cloning procedure to compensate for epigenetic phenomena including gene imprinting (described in Chapter 18). No country or group of scientists has yet condoned the reproductive cloning of humans.

The ethical controversy over embryonic stem cell research arises from the destruction of the blastocyst when the ES cells are harvested for therapeutic cloning. To most opponents of abortion, this is the destruction of a human life, whether the embryo was cloned or whether it was left over from *in vitro* fertilization attempts and would eventually be discarded. President George W. Bush shared these concerns and announced on August 11, 2001, that federal funding for ES cell research would be limited to ES cell lines that had already been established by that date. However, most researchers believe that this restriction is too limiting, particularly in light of a report published in 2005 that all of the ES cell lines available for federally funded research were contaminated with nonhuman molecules from the bovine serum used to feed the cells. Many humans naturally carry antibodies against these molecules, so cells from these lines might be subject to hyperacute immune attack if ever transplanted into the body for therapy. In response to these restrictions, Californian voters in 2004 approved Proposition 71, which bypasses the federal funding ban. This Proposition established the California Institute on Regenerative Medicine, which will provide up to $350 million a year from state funds for stem cell research. Several other states currently have similar legislation pending.

Ironically, decisions in the United States may have only a limited impact on the progress of human stem cell research worldwide, given that other countries have a more permissive environment for this kind of work. Future scientific developments may eventually also make the ES cell controversy in the United States moot. For example, some progress has been reported on techniques to "reprogram" adult stem cells or even adult somatic cells to behave more like pluripotent ES cells. The next few years should help clarify the scientific and political issues surrounding the potential for stem-cell-based therapies.

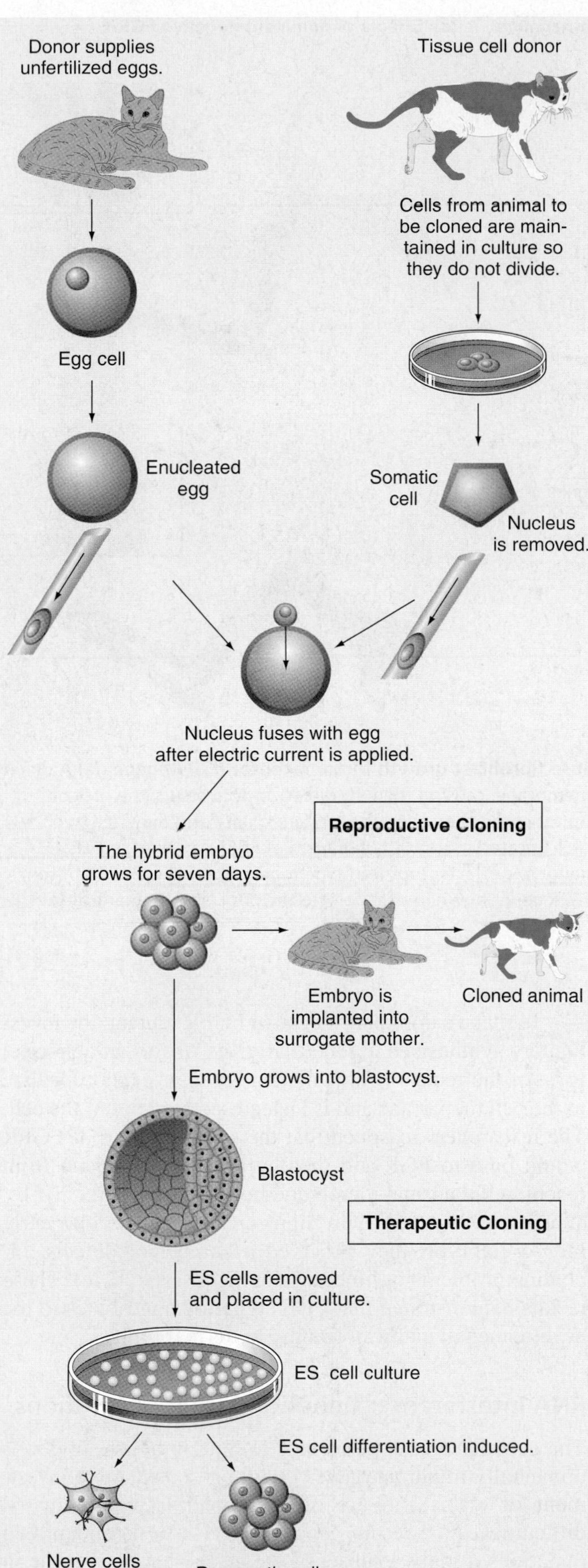
Donor supplies unfertilized eggs.
Tissue cell donor
Cells from animal to be cloned are maintained in culture so they do not divide.
Egg cell
Enucleated egg
Somatic cell
Nucleus is removed.
Nucleus fuses with egg after electric current is applied.
Reproductive Cloning
The hybrid embryo grows for seven days.
Embryo is implanted into surrogate mother.
Cloned animal
Embryo grows into blastocyst.
Blastocyst
Therapeutic Cloning
ES cells removed and placed in culture.
ES cell culture
ES cell differentiation induced.
Nerve cells
Pancreatic cells

Figure A **Reproductive cloning and therapeutic cloning.** Both procedures begin with the fusion of a somatic cell nucleus and an enucleated egg, producing a hybrid egg that is allowed to divide in culture into an early embryo. In reproductive cloning, this embryo is implanted into a surrogate mother and allowed to develop until birth. In therapeutic cloning, the early embryo continues to develop in culture to the blastocyst stage, when the embryonic stem (ES) cells of the inner cell mass are harvested. These ES cells are grown in culture and induced to differentiate into various cell types.

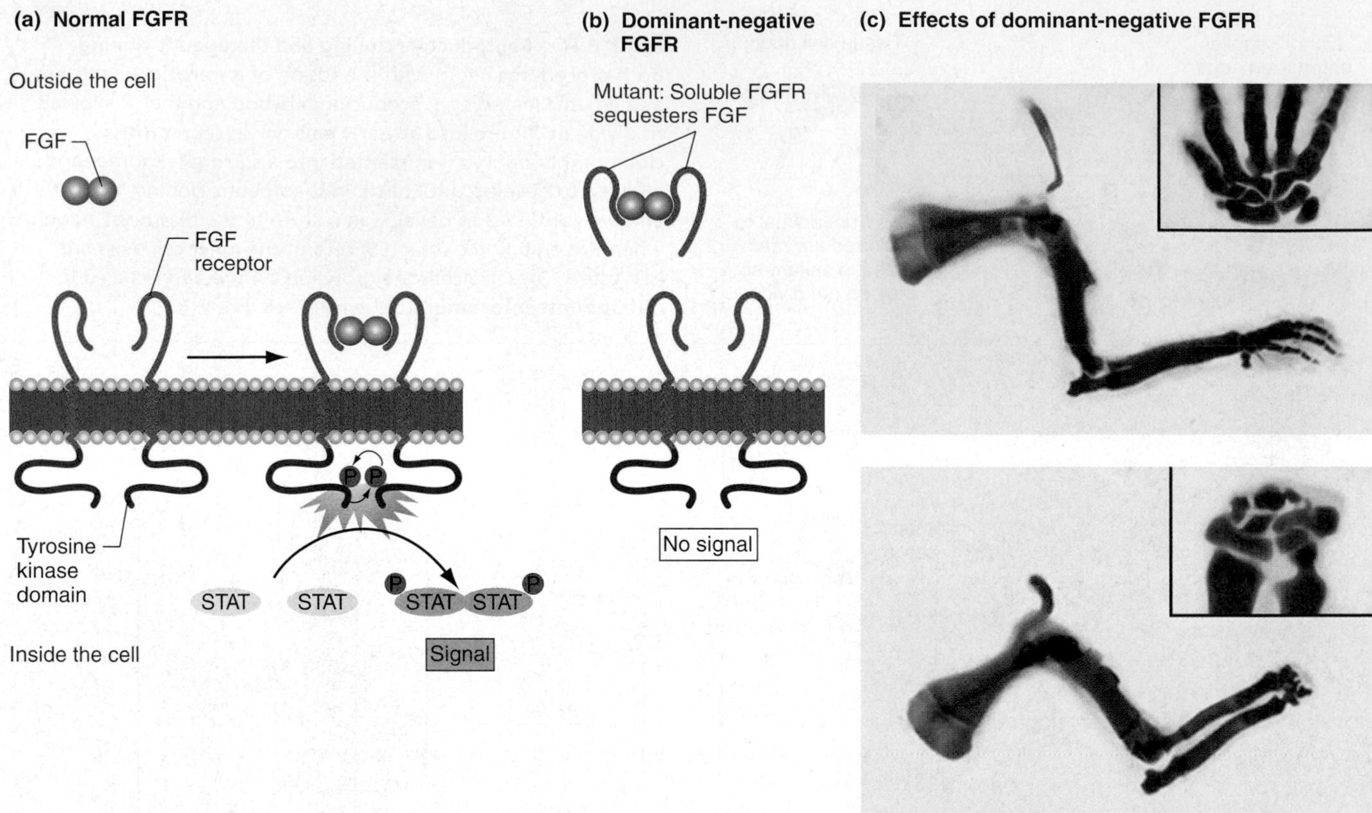

Figure 20.7 Engineering a dominant-negative mutation in a mouse fibroblast growth factor receptor (FGFR) gene. (a) A dimer of fibroblast growth factor (FGF) binds two FGFR molecules in the cell membrane, causing them to dimerize. As a result, the protein kinase domains of the two FGFRs phosphorylate (add a phosphate group to) each other; they also phosphorylate and dimerize two STAT molecules, which initiates a signal necessary for development. **(b)** A truncated mutant soluble form of FGFR can bind to FGF, preventing it from binding to normal FGFR in the cell membrane. **(c)** Phenotypic effects. *Top:* wild-type mouse limb. *Bottom:* limb from a mouse engineered to contain the dominant-negative version of the *FGFR* gene shown in part *b*. Note the poor development of the digits in the mutant mouse (*insets*).

mutation in the gene to test their hypotheses. In such a situation, it is sometimes possible to engineer a dominant-negative mutant transgene *in vitro,* and then introduce this transgene back into a wild-type organism. In an interesting example of this technology, one research group made a dominant-negative fibroblast growth factor receptor (FGFR) in mice. These receptors are normally found on the surface membranes of many cell types (**Fig. 20.7a**). One part of the receptor molecule faces the outside of the cell, where it can bind to a molecule called fibroblast growth factor (FGF). FGF is a **ligand:** a molecule involved in cell-to-cell communication that is produced by the cell sending the signal. Binding of the ligand to a receptor in the membrane of the cell receiving the signal alters the behavior of that cell. The binding of FGF to the extracellular part of FGFR causes several changes in the receptor, including the dimerization of two FGFR subunits and the activation of a kinase in each subunit that adds a phosphate group to the other subunit. This reciprocal phosphorylation, in turn, initiates a complicated intracellular signaling mechanism that changes the receiving cell's developmental fate.

To make a dominant-negative FGFR mutant, the investigators synthesized a gene that gives rise to an abnormal form of the receptor. This mutant receptor cannot localize to the cell membrane and is instead secreted out of the cell. The researchers reasoned that the secreted form of FGFR would bind to FGF and thereby prevent the ligand from reaching the normal membrane-bound FGFR (Fig. 20.7b). When they injected their engineered transgene into early mouse embryos, they observed a number of defects, including problems in limb development (Fig. 20.7c). These results demonstrated that FGF signaling contributes to the developmental pathway leading to normal limbs.

RNA Interference: Genetics Without Mutations

The genetic screens performed to find mutations in developmentally important genes require a considerable investment of effort and are often subject to unanticipated difficulties. As a result, geneticists have to date identified mutations in only a subset of the genes that play a role in development. But within the last several years, researchers have been able to employ a new strategy to deplete the protein products of specific genes from developing organisms. This strategy, which makes use of *RNA interference (RNAi),* is based on the following discovery. When cells ingest or are injected with double-stranded RNA (dsRNA) corresponding to the sequence of a gene's mRNA, the intracellular presence of the dsRNA triggers the degradation of the corresponding mRNA into short fragments. In the absence of intact mRNA, the cell cannot synthesize the protein. The details of RNAi are just being worked out, but it appears that many kinds of cells have enzymes that degrade long dsRNAs into shorter dsRNAs roughly 21 nucleotides in length. These shorter dsRNAs then serve as templates for the degradation of homologous mRNAs into similar 21 bp fragments (see Chapter 18 as well as the *C. elegans* portrait on our website).

To employ this RNAi strategy, researchers first synthesize a dsRNA and then deliver it into the cells of a developing organism. They usually carry out the dsRNA synthesis *in vitro.* For example, they clone a cDNA corresponding to a gene's mRNA into a plasmid vector such that the cDNA is located between strong promoters (**Fig. 20.8a**). They next use purified DNA from the recombinant clone as a template for transcribing the cDNA. The addition of RNA polymerase and the four nucleotide triphosphates (ATP, CTP, GTP, and UTP) initiates transcription, which then proceeds in both directions and produces RNAs from both strands of the cDNA. These complementary RNA strands can anneal together to form dsRNA. There are several methods for getting the dsRNA into developing animals, including injection of dsRNA into the body cavity or soaking the animal in a dsRNA-containing solution. Investigators working with *C. elegans* can simply feed larvae with *E. coli* cells that contain a plasmid like the one shown in Fig. 20.8a. RNA polymerase within the *E. coli* cells containing such a plasmid will synthesize the desired dsRNA, which is then taken up by *C. elegans* larval cells as the bacteria are digested in the worm's gut.

Figure 20.8b shows an RNAi experiment in which the dsRNA corresponded to the mRNA for a *C. elegans* gene called *par-1.* The result of this dsRNA treatment was an abnormal vulva (the structure through which fertilized eggs are released) that protruded outside of the animal. Though worm researchers already knew that *par-1* functions very early in development to help establish the anterior-posterior axis of the animal, the results shown in Fig. 20.8b showed that the gene also functions later in development in the patterning of the vulva.

RNA interference is an extremely useful technique for creating a phenocopy that mimics a loss-of-function mutation. However, a phenocopy is not a true, heritable mutation; thus, RNAi-based protein depletion is not passed from generation to generation. Another slight drawback of the RNAi method is that results may vary from experiment to experiment because they depend on the relative level of dsRNA uptake. Consequently, even if

(a) Synthesis of dsRNA

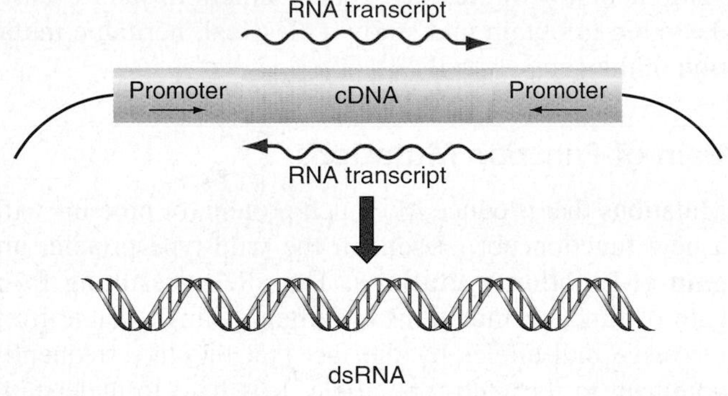

(b) A result of *par-1* dsRNA treatment

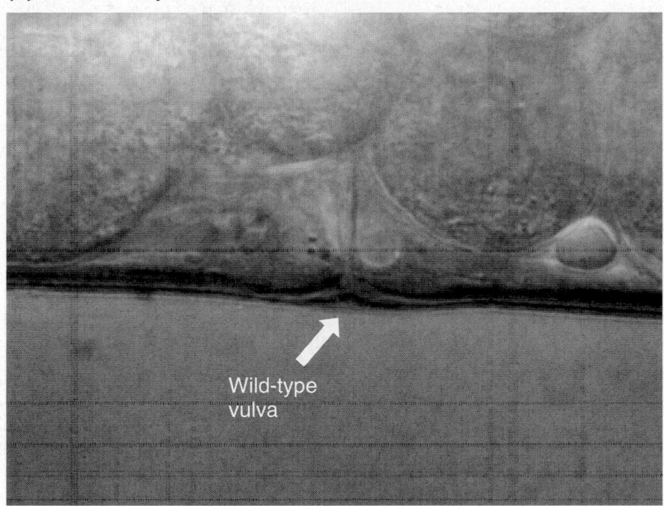

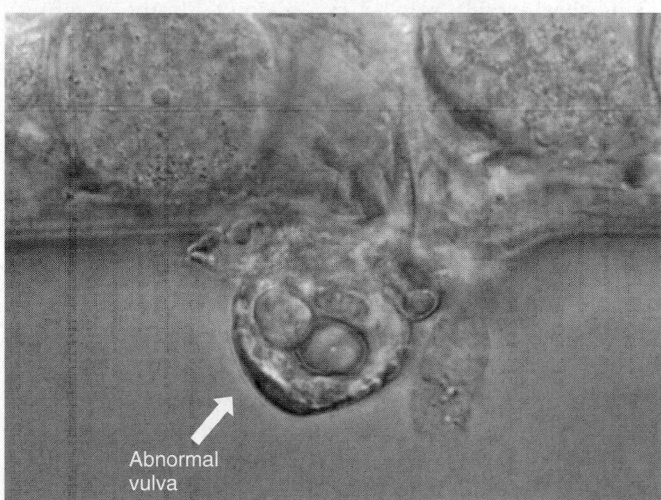

Figure 20.8 RNA interference (RNAi): A new tool for studying development. (a) How to make double-stranded RNA (dsRNA). A cDNA is cloned between two promoters, allowing transcription from both cDNA template strands. Complementary RNA transcripts will anneal with each other to make dsRNA. **(b)** Abnormal structure of the vulva in *C. elegans* treated with dsRNA for the *par-1* gene. *Top:* wild-type vulva. *Bottom:* protruding vulva in an animal treated with *par-1* dsRNA.

the results of an RNAi experiment provide clues to a developmentally interesting gene's function, it is usually desirable to obtain and study a classical, heritable mutation in the gene as well.

Gain-of-Function Mutations

Mutations that produce too much protein, or proteins with a new function not present in the wild-type protein, are **gain-of-function mutations.** The alleles resulting from gain-of-function mutations are often distinguishable from recessive null alleles by the fact that they are frequently dominant to the wild-type allele. It is hard to understand unambiguously the role of a protein in development from a gain-of-function allele. This is because the mutation, rather than taking something away, adds something unusual to the organism, which might behave in an unpredictable way. Nevertheless, gain-of-function mutants can help identify developmentally important genes and clarify the roles they play in development. We now look at two kinds of gain-of-function mutations.

Mutations Causing Excessive Gene Activity. Such mutations are rare because they result only from highly specific changes in a gene, in contrast with loss-of-function mutations, which can disrupt gene function in many ways. Nonetheless, there are several ways in which mutations can lead to increased gene activity. One mechanism involves changes to promoters that make the promoters more accessible to transcription factors and RNA polymerase. Other possible scenarios are illustrated by various dominant mutations in *FGFR3,* one of the four genes in mice and humans that encode related yet distinct fibroblast growth factor receptors (review Fig. 20.7a). Some of these mutations increase the affinity of the FGFR3 receptor protein for its ligand FGF, inappropriately turning on the developmental signal when the concentration of the FGF ligand would normally be too low to accomplish this. Other *FGFR3* mutations allow the developmental signal to be turned on in the absence of FGF. These mutations cause the constitutive activation of the phosphate-group-adding kinase domain of FGFR3; they accomplish this by altering a part of the protein that normally blocks the kinase function in the absence of the ligand.

Interestingly, a single amino acid substitution in the FGFR3 protein is sufficient to cause a gain-of-function dominant phenotype through the constitutive activation of the kinase domain. This substitution causes achondroplasia, the most common form of short-limb dwarfism in humans. Researchers have recently engineered mice with exactly the same amino acid substitution in their homologous *FGFR3* gene. Remarkably, this mutant gene produces what appears to be the same dominant dwarf phenotype seen in human achondroplasia (**Fig. 20.9**). You might wonder how a gain-of-function mutation in a growth factor receptor gene results in dwarfism, since this is not what you

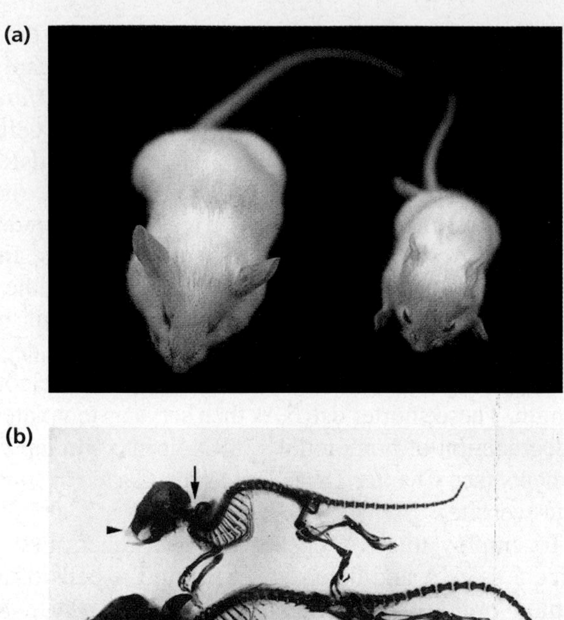

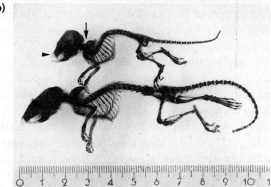

Figure 20.9 Achondroplastic dwarfism in the mouse. **(a)** The dwarf mouse at the *right* is heterozygous for an *FGFR3* allele with the same amino acid change as that causing achondroplasia in humans. A control littermate is at the *left.* **(b)** Skeletal abnormalities in the dwarf mouse at the *top* include a shorter face, overgrowth of the incisor teeth (*arrowhead*), and improper connection of the head to the spine (*arrow*) as compared to a control littermate at the *bottom.*

would expect at first glance. The answer is that the FGFR3 protein is made in cartilage precusor cells called chondrocytes. Activated FGFR3 signals the chondrocytes to stop dividing and begin differentiating first into cartilage and then into bone. A gain-of-function FGFR mutation thus causes the chondrocytes to stop proliferating sooner than normal, which leads to fewer bone precursor cells and thus less bone.

Mutations Causing Ectopic Gene Expression. Suppose you suspect that a particular protein plays an important role in initiating the development of some structure like the legs or eyes. If that were true, it might be possible that expression of this protein in tissues in which it is not normally made could lead to the development of legs or eyes in unusual locations in the animal. The expression of a gene at an abnormal place or time is called **ectopic gene expression.** Rarely, spontaneous mutations cause ectopic expression of genes important to development. An interesting example of such a mutation occurred in *Drosophila* when a chromosomal inversion moved the *Antennapedia* gene (normally transcribed in tissues destined to become legs) next to a specific kind of enhancer, which turned the gene on in tissues normally destined to develop into antennae. Animals carrying this *Antennapedia* mutation have legs growing out of their heads in place of antennae (review Fig. 8.31d on p. 288). The phenotype of this ectopic mutant shows that the wild-type protein encoded by *Antennapedia* plays a critical role in leg development.

Instead of relying on rare and unpredictable mutations that might cause ectopic gene expression, researchers can now use recombinant DNA technology to make such mutations in a systematic way. They can change the promoter of a cloned gene by adding enhancers or other elements that might cause it to be transcribed at inappropriate places or times and then introduce this altered gene back into the organism's genome by transformation. One research group placed the *eyeless* gene of *Drosophila* (review Fig. 20.4a) under the control of a promoter for a "heat-shock" gene whose transcription in any tissue is turned on by higher than normal temperatures. Flies bearing this recombinant gene that were grown at high temperature made the *eyeless*-encoded protein throughout their bodies. These animals had eye tissue growing at many different locations, even on their wings and antennae (**Fig. 20.10**). This result demonstrates that the Eyeless protein is a master developmental switch that can activate a cellular program causing eye development. Ectopic eyes also arise when the mouse *Pax-6* or the human *Aniridia* gene is expressed in *Drosophila* under the control of the same heat-shock gene promoter. This means that both elements of the amino acid sequence and the actual function of this master switch have been conserved throughout animal evolution.

Figure 20.10 Ectopic expression of the *eyeless* gene produces ectopic eye tissue. This fly carries a synthetic *eyeless* gene that is turned on inappropriately. As a result, eye tissue grows in unexpected places, such as at the end of the antennae and on the thorax above the wings.

Analyzing How Genes Work Together in Developmental Pathways

Once you have isolated a comprehensive set of mutations and identified as many as possible of the genes involved in the biological process of interest, the next step is to establish the functions performed by these genes. The ultimate aim of such studies is to discern a *developmental pathway:* a detailed description of how the products of these many genes interact and cooperate with each other to produce a particular outcome in development.

Characterizing Genes Important for Development

Before looking at complicated pathways, however, it is crucial to learn as much as possible about each of the genes identified in the screen. Answers to the following five questions can help fill in significant details.

What Is the Developmental Phenotype Caused by Mutations in the Gene? There are many different ways to look at phenotypes in order to understand which tissues a mutation affects and how it affects them. It is particularly important to define the earliest stage of development disrupted by a mutation, as this may provide clues about the order in which the gene functions in a developmental pathway.

What Is the Nature of the Protein Encoded by the Mutant Gene? With the completion of genome projects for key model organisms, it is now often possible to identify the mutant gene within a few months of finding the mutation. Once you know the nucleotide sequence of a gene, you automatically know the amino acid sequence of the protein it encodes. You can then use computer programs to search the amino acid sequence for motifs that offer clues to the protein's function. For example, computer programs can often predict whether a protein resembles known membrane-bound receptors, or whether a protein acts as a kinase that phosphorylates other proteins. One motif seen in many proteins with developmental significance is the **homeodomain** (**Fig. 20.11**). It is found in the proteins encoded by the *eyeless/Pax6* and *Antennapedia* genes discussed earlier in this chapter. The homeodomain is a region of about 60 amino acids that is structurally related to the helix-turn-helix motif of many bacterial regulatory proteins. The homeodomain binds to specific DNA sequences, so its presence suggests that a protein might be a transcription factor.

When and in What Tissues Is the Gene Expressed? One way to answer this question is to perform an RNA *in situ* hybridization experiment. To do this, you label cDNA sequences corresponding to the gene's mRNA and then use

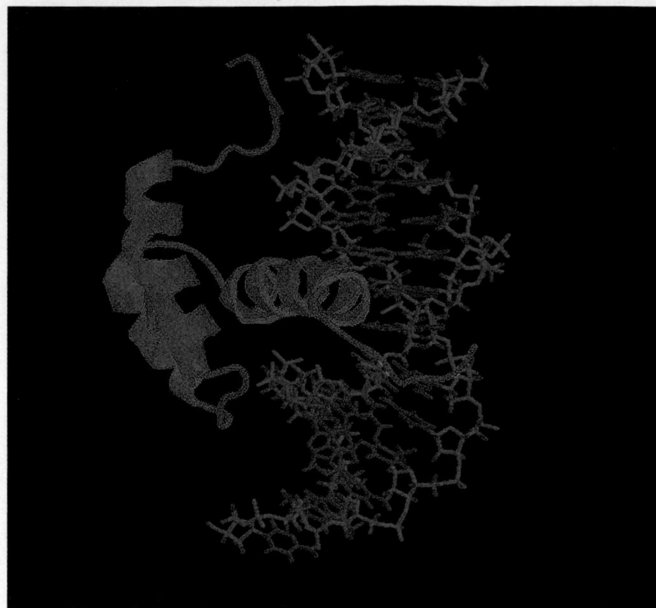

Figure 20.11 The homeodomain: A DNA-binding motif found in many transcription factors that regulate development. The amino acid backbone of a homeodomain (*yellow*) interacts with specific sequences in a DNA double helix (*red* and *blue*).

the labeled cDNA as a probe for the mRNA on preparations of thinly sectioned tissues. Where the probe is retained, the signal indicates cells containing the gene's mRNA (**Fig. 20.12**). Defining the tissues in which the gene is expressed can help formulate hypotheses concerning the gene's role in development. For example, if a mutation in the gene affects the development of a tissue other than that in which the gene is transcribed, you might hypothesize that the gene encodes a signaling molecule like a hormone. Such molecules of cellular communication are made in one tissue but influence the fate of cells in other tissues that contain receptors for the hormone.

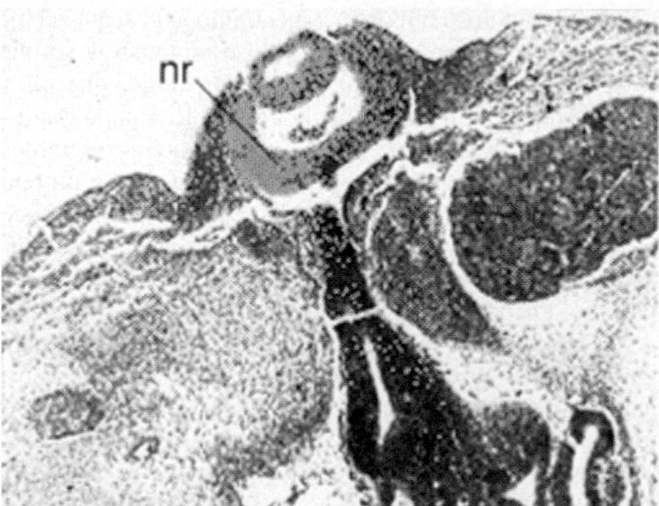

Figure 20.12 ***In situ*** **hybridization locates cells expressing a gene of interest.** This example shows that mRNA for the *Pax-6* gene (*yellow signals*) accumulates in the eye of a human fetus in the seventh week of gestation. Hybridization is specific to the developing neural retina (nr) and the developing eye lens above it.

Where Is the Protein Product of the Gene Within the Animal and Within the Cell? It is often technically easier to find and evaluate the tissues in which a gene is expressed by following the gene's protein product rather than by using RNA *in situ* hybridization to look for the gene's mRNA. In addition, an mRNA may be found in a tissue that doesn't contain the protein. This would point to the existence of regulatory controls that prevent translation of the mRNA. Finally, the intracellular localization of a protein often provides clues to its function. For example, concentration of the protein in the nucleus would be consistent with a role as a transcription factor.

Methods to follow a protein usually involve the generation of antibodies against parts of the protein. One way to do this is to use recombinant DNA techniques to construct a fused gene (**Fig. 20.13a**). In this construct, part of a cDNA for the gene of interest is cloned downstream of, and in the same reading frame as, part of a gene encoding a protein that can be made at high levels in bacteria. If you transform a plasmid containing this fused gene into *E. coli,* the bacterial cells will make large amounts of a **fusion protein** whose N-terminal amino acids are from the bacterial protein and whose C-terminal amino acids are from the eukaryotic developmental protein. If you inject this fusion protein into rabbits or other animals, they will synthesize antibodies against it. And once you label these antibodies with a fluorescent tag, the tagged antibodies will react with the corresponding protein of developmental interest in preparations of tissues and cells (Fig. 20.13b).

A new way to track a protein is to use recombinant DNA technology to construct a gene encoding a tagged protein that will itself fluoresce. The idea, illustrated in Fig. 20.13c, is to synthesize an open reading frame that encodes not only the entire protein of interest, but also (at the protein's N or C terminus) the amino acids composing a naturally fluorescent protein from jellyfish called *green fluorescent protein* (*GFP*). When this recombinant gene is reintroduced into the genome by transformation, the organism will make the GFP fusion protein in the same places and times it makes the normal untagged protein. Investigators can keep track of the fusion protein by following GFP fluorescence (Fig. 20.13d). A major advantage of this approach is that researchers can use it to follow a GFP-tagged protein in living cells or animals, which is generally not possible with tagged antibodies for technical reasons. With the GFP fusion protein, researchers can even make movies that reveal subtle changes in the location of the protein over time.

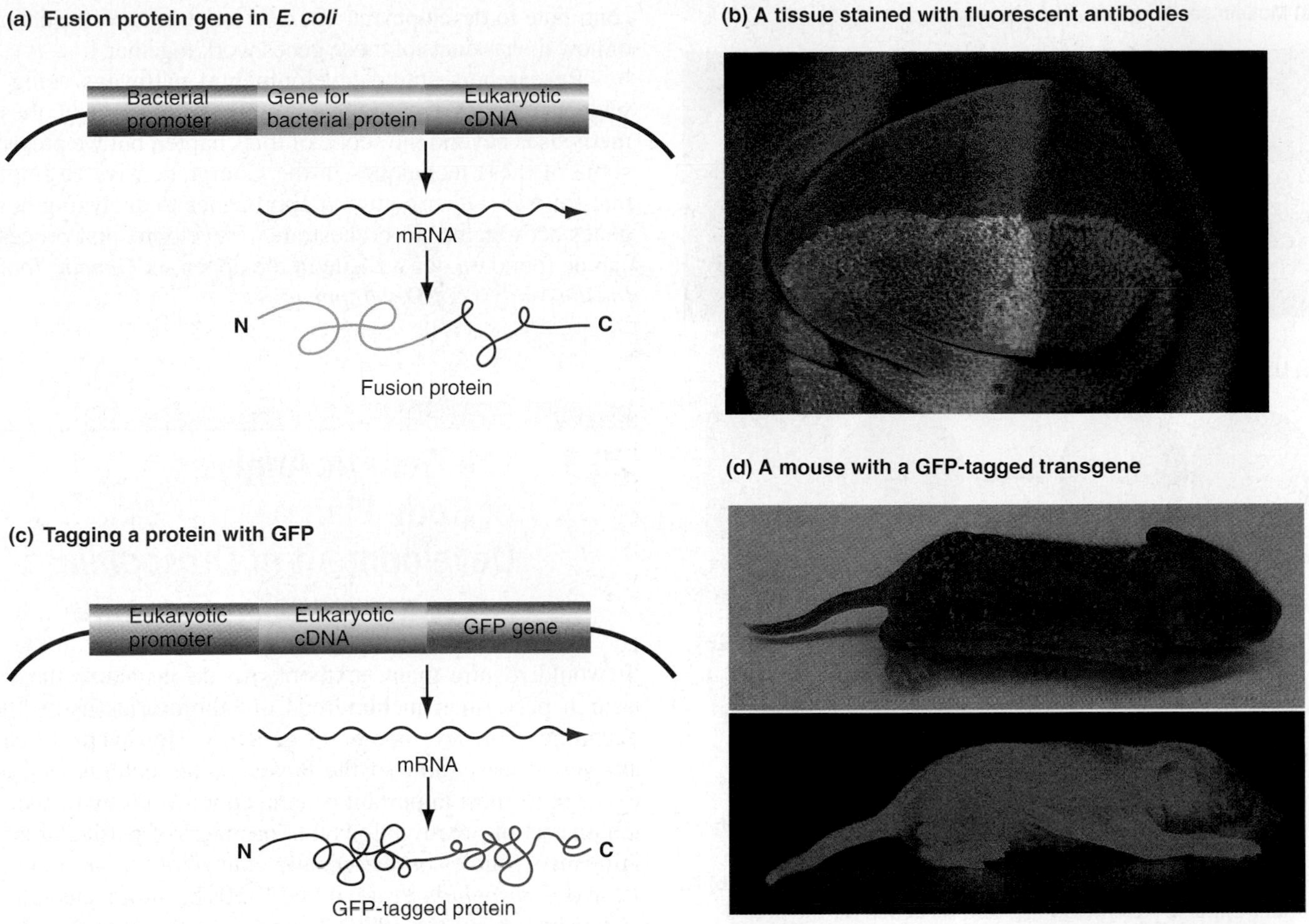

(a) Fusion protein gene in *E. coli*

Bacterial promoter | Gene for bacterial protein | Eukaryotic cDNA

mRNA

N — C

Fusion protein

(b) A tissue stained with fluorescent antibodies

(c) Tagging a protein with GFP

Eukaryotic promoter | Eukaryotic cDNA | GFP gene

mRNA

N — C

GFP-tagged protein

(d) A mouse with a GFP-tagged transgene

Figure 20.13 Using antibodies and GFP tagging to follow the localization of proteins. **(a)** This synthetic gene encodes a fusion protein that will be made at high levels when transformed into *E. coli* cells. Animals injected with purified fusion protein will make antibodies against the protein of interest. **(b)** A *Drosophila* "imaginal disc" is stained with antibodies against several proteins. Each antibody is tagged with a dye that fluoresces in a particular color. **(c)** Making a GFP-tagged protein. The recombinant gene encodes a fusion protein that contains GFP at its C terminus. **(d)** This mouse contains a GFP-labeled transgene expressed thoughout the body; the entire mouse becomes fluorescent when illuminated with UV light as at the *bottom*. The same mouse is shown in normal light at the *top*.

What Cells or Tissues Are Affected by the Loss of Gene Function? The answer to this question might seem obvious: Only cells that make a particular protein would show the phenotypic effects of mutations that prevent synthesis of that protein. But this "obvious" answer is misleading because cells often communicate with each other to influence developmental decisions. In one simple example, if a gland synthesizes a hormone that circulates through the blood and the gland can no longer make the hormone, the phenotypic effects might not show up in the gland itself but rather in target cells elsewhere in the body that contain receptors for the hormone.

To address a variety of issues involving communication between cells, developmental geneticists construct **genetic mosaics:** organisms in which some cells (like those in the gland just described) have one genotype, while other cells (such as those in the hormone's target tissues) have a different genotype. Researchers can use several techniques to make such genetic mosaics. The technique chosen often depends on the species. *Drosophila* geneticists usually employ mitotic recombination, those working with *C. elegans* use methods based on the loss of small extra chromosomes during mitosis, and investigators studying mice mix embryonic cells from mutant and wild-type strains to make *chimeric mice* with two different cell types. (For more species-specific details, see the genetic portraits of flies, worms, and mice on our website [www.mhhe.com/hartwell3].) **Chimeras** are genetic mosaics in which cells of different genotype originate from two different individuals.

In the genetic mosaics most useful for studying development, some cells have a mutant genotype, while other cells have a wild-type genotype. Most mosaics are constructed with markers, different from the mutant developmental gene, that allow investigators to differentiate between the two types of tissues. **Figure 20.14a** shows mosaic seedlings of the plant *Arabidopsis* in which blue tissue contains both a marker gene resulting in blue color and a wild-type gene called *AGAMOUS*$^+$, while white tissue

(a) Mosaic seedlings of *Arabidopsis*

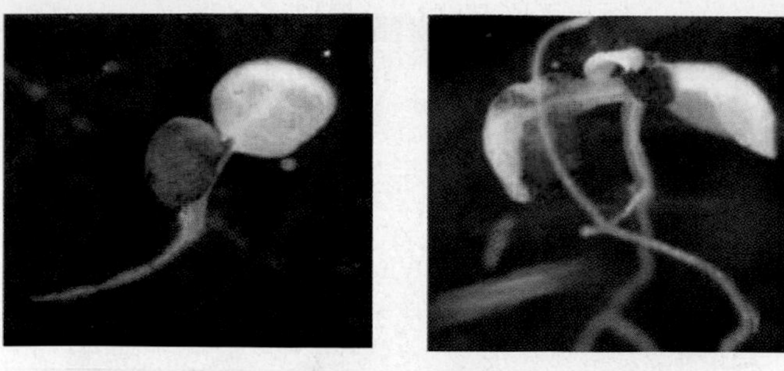

(b) Using mosaics to study cell signaling

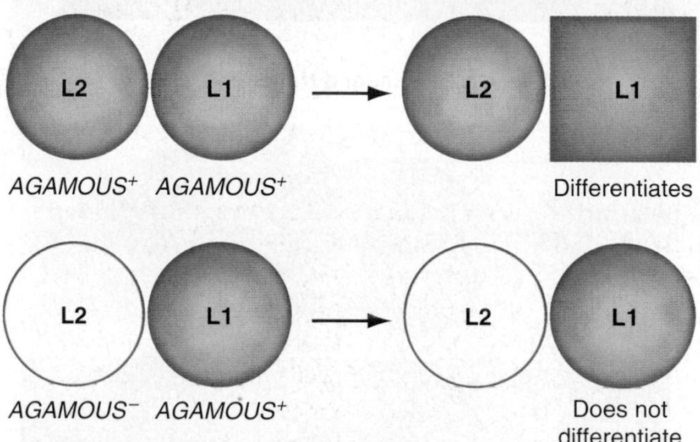

Figure 20.14 Mosaic analysis. (a) In these mosaic seedlings, *blue* tissue contains both a marker gene and the $AGAMOUS^+$ gene, while *white* tissue contains neither (it is $AGAMOUS^-$). **(b)** A signal from blue $AGAMOUS^+$ L2 cells is needed for the proper differentiation of nearby L1 cells (symbolized by a change from a *circle* to a *square*). If L2 cells lack the $AGAMOUS^+$ gene (*white*), nearby L1 cells do not differentiate properly, even if they are themselves $AGAMOUS^+$.

lacks the marker gene and is simultaneously mutant for *AGAMOUS*. Figure 20.14b diagrams how researchers used such marked mosaics to show that cells from a particular layer of undifferentiated cells (called L2) in the apical meristem send a signal needed for the proper differentiation of cells in a different layer (L1). This signal depends on the presence of a wild-type $AGAMOUS^+$ allele in L2 cells. In other words, even $AGAMOUS^+$ genotypically wild-type L1 cells develop abnormally if the adjacent L2 cells are mutant for this gene.

Discovering How Genes Interact to Generate Developmental Pathways

As you already know, genes do not work in isolation. Instead, complicated biological events demand the coordinated action of many genes. A process as complex as the development of an eye or an embryo must involve arrays of genes interacting in developmental pathways. A full description of development from a genetic perspective will thus require not only the identification and analysis of the individual genes that contribute to development, but also the eventual elucidation of how the products of those genes work together.

Researchers study developmental pathways using a wide variety of techniques. A full description of these methods is beyond the scope of this chapter, but we present some of these techniques in the Comprehensive Example that follows. Other common approaches to analyzing how genes act together to orchestrate a developmental process can be found on our website in the appendix *Genetic Tools for the Analysis of Development.*

20.3 The Genetic Analysis of Body-Plan Development in *Drosophila:* A Comprehensive Example

It would require many textbooks to do justice to the research performed in hundreds of laboratories using the techniques of developmental genetics. However, we can convey at least some of the flavor of this field as well as some of its most important general conclusions by focusing on one set of investigations that has been of particular significance. Studies on the genetic control of the basic body plan of *Drosophila,* begun in the 1950s but conducted most extensively since the 1970s, have revolutionized our understanding of development. Here, we focus on the aspect of this work that explains how the fly's body becomes differentiated and specialized along the *anterior-posterior (AP) axis,* the line running from the animal's head to its tail.

The research we describe was based on the observation that a fertilized *Drosophila* egg becomes subdivided into several clearly defined segments (review Fig. 20.2), each of which eventually has a specific appearance and function: some segments become parts of the head; others parts of the thorax; and still others, parts of the abdomen. Scientists designed experiments to answer two fundamental questions about this segmentation. First, how does the developing animal establish the proper number of body segments? And second, how does each body segment "know" what kinds of structures it should form and what role it should play in the animal's biology? Their results showed that very early in development, the action of a large group of genes called the **segmentation genes** subdivides the body into an array of essentially identical body segments. Later in development, the expression of a different set of genes called **homeotic genes** assigns a unique identity to each body segment.

Early Development of the Basic Body Plan

To understand how the segmentation and homeotic genes function, it is helpful to consider some of the basic events

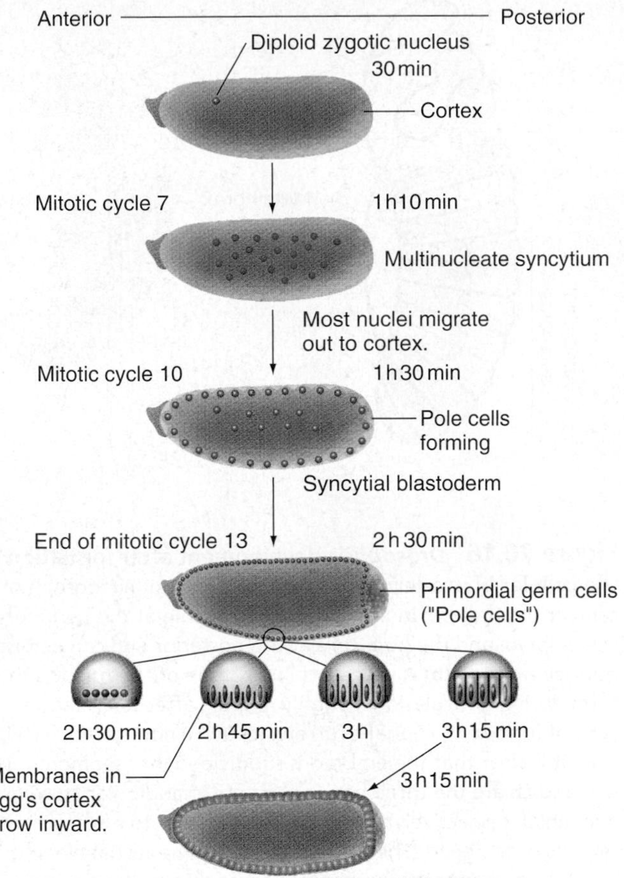

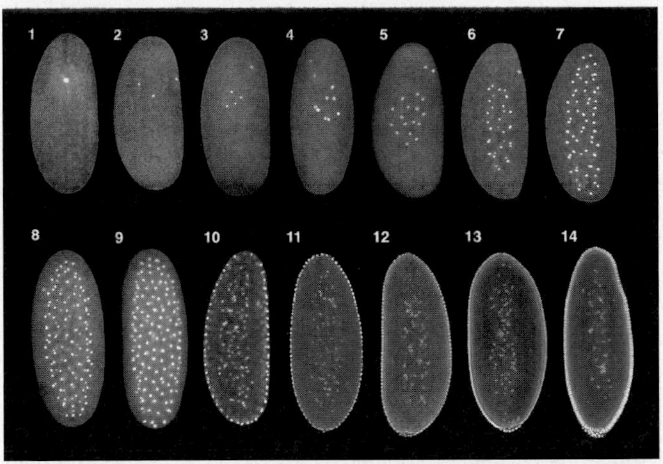

Figure 20.15 Early *Drosophila* development: From fertilization to formation of the cellular blastoderm. (a) The original zygotic nucleus undergoes 13 very rapid mitotic divisions in a single syncytium. A few nuclei at the posterior end of the embryo become the germ-line *pole cells.* At the *syncytial blastoderm* stage, the egg surface is covered by a monolayer of nuclei (except for the pole cells at the posterior end). At the end of the thirteenth division cycle, cell membranes enclose the nuclei at the cortex into separate cells to produce a *cellular blastoderm.* **(b)** Photomicrographs of early embryonic stages stained with a fluorescent dye for DNA. (For another view of the syncytial blastoderm, see Fig. 4.10 on p. 93.)

that take place in the first few hours of *Drosophila* development (**Fig. 20.15**). The egg is fertilized in the uterus as it is being laid, and the meiotic divisions of the oocyte nucleus, which had previously arrested in the metaphase of meiosis I, resume at this time. After fusion of the haploid male and female pronuclei, the diploid zygotic nucleus of the embryo undergoes 13 rounds of nuclear division at an extraordinarily rapid rate, with the average time of mitotic cycles 2 through 9 being only 8.5 minutes. Nuclear division in early *Drosophila* embryos, unlike most mitoses, is not accompanied by cell division, so the early embryo becomes a multinucleate syncytium. During the first eight division cycles, the multiple nuclei are centrally located in the egg; during the ninth division, most of the nuclei migrate out to the cortex—just under the surface of the embryo—to produce the **syncytial blastoderm.** During the tenth division, nuclei at the posterior pole of the egg are enclosed in membranes that invaginate from the egg cell membrane to form the first embryonic cells; these "pole cells" are the primordial germ cells. At the end of the thirteenth division cycle, about 6000 nuclei are present at the egg cortex.

During the interphase of the fourteenth cycle, membranes in the egg's cortex grow inward between these nuclei, creating an epithelial layer called the **cellular blastoderm** that is one cell deep (**Figs.** 20.15 and **20.16a**). The embryo completes formation of the cellular blastoderm about 3 hours after fertilization. At the cellular blastoderm stage, no regional differences in cell shape or size are apparent (with the exception of the pole cells at the posterior end). Experiments in which blastoderm cells have been transplanted from one location to another, however, show that despite this morphological uniformity, the segmental identity of the cells has already been determined. Consistent with this finding, molecular studies reveal that most segmentation and homeotic genes function during or even before the cellular blastoderm stage.

Immediately after cellularization, **gastrulation** and establishment of the embryonic germ layers begin. The *mesoderm* forms by invagination of a band of midventral cells that extends most of the length of the embryo. This infolding (the ventral furrow; Fig. 20.16b) produces an internal tube whose cells soon divide and migrate to produce a mesodermal layer. The *endoderm* forms by distinct invaginations anterior and posterior to the ventral furrow; one of these invaginations is the cephalic furrow seen in Fig. 20.16b. The cells of the endodermal infoldings migrate over the yolk to produce the gut. Finally, the nervous system arises from neuroblasts that segregate from bilateral zones of the ventral *ectoderm.* The first visible signs of segmentation are periodic bulges in the mesoderm, which appear about 40 minutes after gastrulation begins. Within a few hours of gastrulation, the embryo is divided into clear-cut body segments that will become the three head

(a) Cellular blastoderm

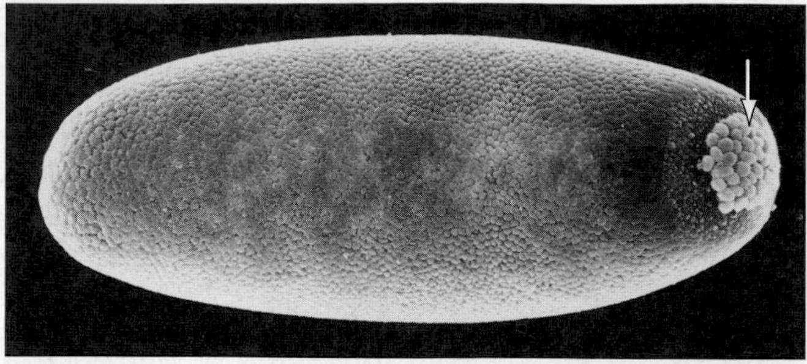

(b) Gastrulation

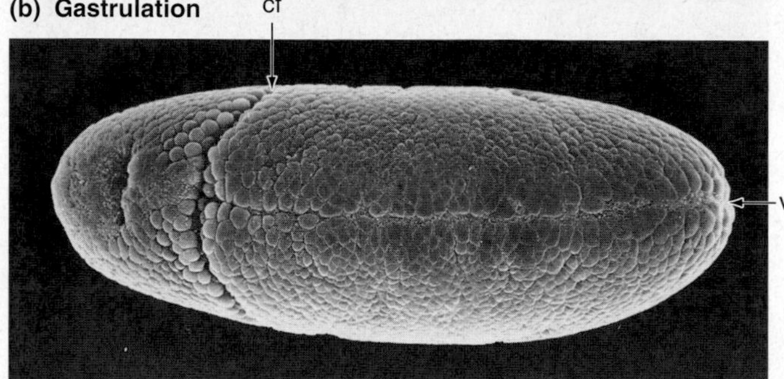

(c) Segmentation

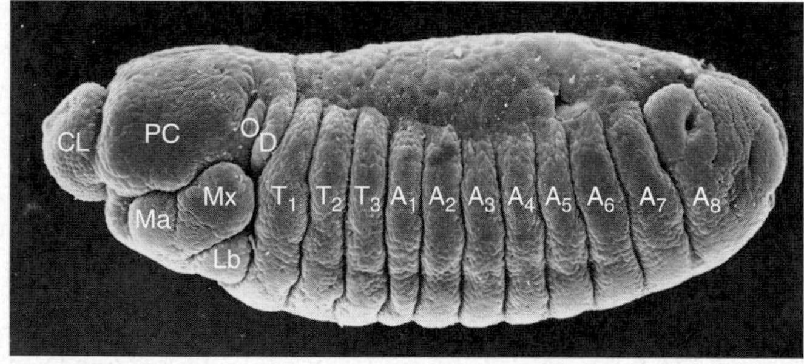

(d) Segment identity is preserved throughout development.

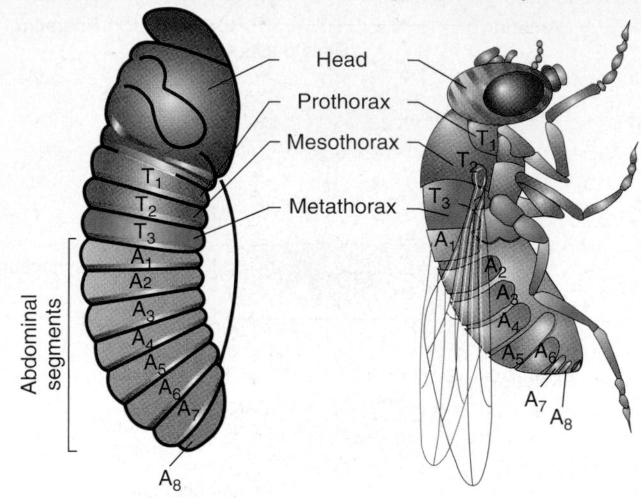

Figure 20.16 ***Drosophila*** **development after formation of the cellular blastoderm. (a)** Scanning electron micrograph of a cellular blastoderm; individual cells are visible at the periphery of the embryo, and the pole cells at the posterior end can be distinguished (*arrow*). **(b)** A ventral view of some of the furrows that form during gastrulation, roughly 4 hours after fertilization: vf, ventral furrow; cf, cephalic furrow. **(c)** By 10 hours after fertilization, it is clear that the embryo is subdivided into segments. Ma, Mx, and Lb are the three head segments (mandibular, maxillary, and labial, respectively). CL, PC, O, and D refer to nonsegmented regions of the head. The three thoracic segments (labeled T_1, T_2, and T_3) are the prothorax, the mesothorax, and the metathorax, respectively, while the abdominal segments are labeled A_1-A_8. **(d)** The identity of embryonic segments (*left*) is preserved through the larval stages and is also retained through metamorphosis into the adult (*right*). For simplicity, head segments are not distinguished here.

segments, three thoracic segments, and eight major abdominal segments of the larva that hatches from the eggshell (Fig. 20.16c). Even though the animal eventually undergoes metamorphosis to become an adult fly, the same basic body plan is conserved in the adult stage (Fig. 20.16d).

The genes responsible for the formation of segments fall into four classes and function in a regulatory hierarchy that progressively subdivides the embryo into successively smaller units. In the order of their expression, these classes are (1) maternal genes, (2) gap genes, (3) pair-rule genes, and (4) segment-polarity genes.

Specification of Segment Number: Maternal Genes Interact to Produce Morphogen Gradients

There is very little transcription of genes in the embryonic nuclei between fertilization and the end of the 13 rapid syncytial divisions that immediately follow fertilization. Because of this near (but not total) lack of transcription, developmental biologists suspected that formation of the basic body plan initially requires **maternally supplied components** deposited by the mother into the egg during oogenesis. How could they identify the genes encoding these maternally supplied components? Christiane Nüsslein-Volhard and Eric Wieschaus realized that the embryonic phenotype determined by such genes does not depend on the embryo's own genotype; rather it is determined by the genotype of the mother. By asking whether mothers homozygous for candidate mutations would produce defective embryos, they devised genetic screens to identify recessive mutations in maternal genes that influence embryonic development; these recessive mutations are often called **maternal-effect mutations.**

To carry out their screens, Nüsslein-Volhard and Wieschaus established individual balanced stocks for thousands of mutagen-treated chromosomes, and they then examined the phenotypes of embryos obtained from homozygous mutant mothers. They focused their attention on stocks in which homozygous mutant females were sterile, because they anticipated that the absence of maternally supplied components needed for the earliest stages of development would result in embryos so defective that they could never grow into adults. Through these large-scale screens, Nüsslein-Volhard and Wieschaus identified a large number of maternal genes that are required for the normal patterning of the body. For this and other contributions, they shared the Nobel Prize for physiology or medicine with Edward B. Lewis—whose work we describe later. We focus here on two groups of the genes they found. One group is required for normal patterning of the embryo's anterior; the other is required mainly for normal posterior patterning. The genes in these two groups are the first genes activated in the process that determines segment number.

The finding that separate groups of maternal genes control anterior and posterior patterning in the embryo is consistent with the conclusions of classical embryological experiments. Studies in which polar cytoplasm from the embryo's ends was transplanted, or in which preblastoderm embryos were separated into two halves by constriction of the embryo with a fine thread, suggested that the insect body axis is patterned during cleavage by the interaction of two signaling centers located at the anterior and posterior poles of the egg. In a specific model, Klaus Sander proposed that each pole of the egg produces a different substance and that these substances form opposing gradients by diffusion. He suggested that the concentrations of these substances then determine the types of structures produced at each position along the body axis. Molecular characterization of the maternal genes of the anterior and posterior groups indicates that the Sander model for body axis patterning is correct. Substances that define different cell fates in a concentration-dependent manner are known as **morphogens.**

In *Drosophila,* the *bicoid* (*bcd*) Gene Encodes the Anterior Morphogen

Embryos from mothers homozygous for null alleles of *bcd* lack all head and thoracic structures. The protein product of *bcd* is a DNA-binding transcription factor whose transcript is localized near the anterior pole of the egg cytoplasm (**Fig. 20.17a**). Translation of the *bcd* transcripts takes place

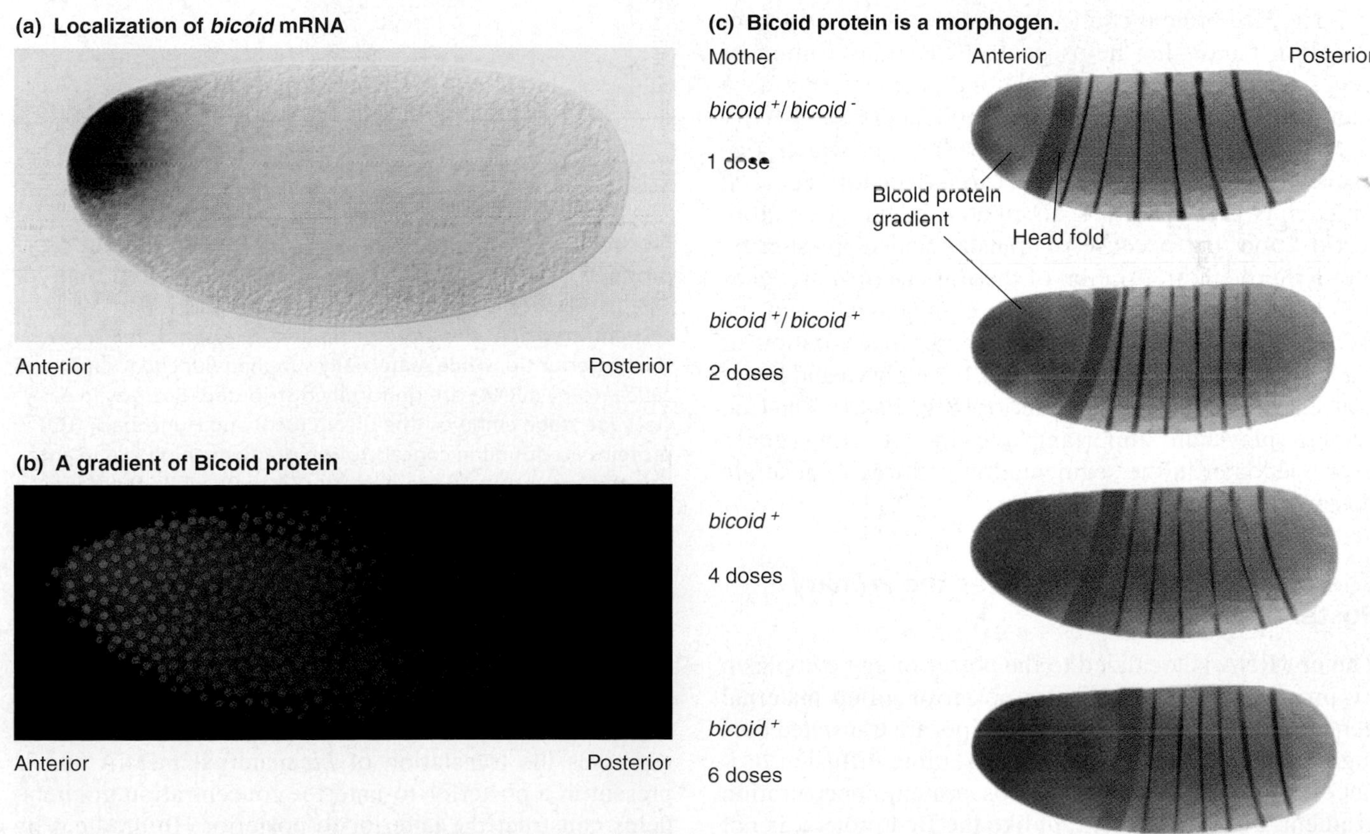

Figure 20.17 The Bicoid protein is the anterior morphogen. (a) The *bicoid* (*bcd*) mRNA (visualized by *in situ* hybridization in *purple*) concentrates at the anterior tip of the embryo. **(b)** The Bicoid (Bcd) protein (seen by *green* antibody staining) is distributed in a gradient: high at the anterior end and trailing off toward the posterior end. It can be seen that the Bcd protein (a transcription factor) accumulates in the nuclei of this syncytial blastoderm embryo. **(c)** The greater the maternal dosage of bcd^+, the higher the concentration of Bcd protein in the embryo, and the more of the embryo that is devoted to anterior structures. Head structures will develop anterior of the head fold invagination; thoracic and abdominal structures posterior to it.

after fertilization. The newly made Bcd protein diffuses from its source at the pole to produce a high-to-low, anterior-to-posterior concentration gradient that extends over the anterior two-thirds of the embryo by the ninth division cycle (Fig. 20.17b). This gradient determines most aspects of head and thorax development.

One of the first lines of evidence that the Bcd protein functions as a morphogen came from experiments in which the maternal dosage of the *bcd* gene varied (Fig. 20.17c). Mothers that carried only one dose of the *bcd* gene, instead of the normal diploid dose, incorporated about half the normal amount of *bcd* RNA into their eggs. As a result, translation yielded less Bcd protein, and the Bcd gradient was shallower and shifted to the anterior. In these Bcd-deficient embryos, the thoracic segments developed from more anterior regions than normal, and less of the body was devoted to the head. The opposite effect occurred in mothers carrying extra doses of the *bcd* gene. These and other observations suggested that the level of Bcd protein is a key to the determination of head and thoracic fates in the embryo. Three other genes work with *bcd* in the anterior group of maternal genes; the function of the protein products of these three genes is to localize *bcd* transcripts to the egg's anterior pole.

The Bcd protein itself works in two ways: as a transcription factor that helps control the transcription of genes farther down the regulatory pathway, and as a translational repressor. The target of its repressor activity is the transcript of the *caudal* (*cad*) gene, which also encodes a DNA-binding transcription factor. The *cad* transcripts are uniformly distributed in the egg before fertilization, but because of translational repression by the Bcd protein, translation of these transcripts produces a gradient of Cad protein that is complementary to the Bcd gradient. That is, there is a high concentration of Cad protein at the posterior end of the embryo and lower concentrations toward the anterior (**Fig. 20.18**). The Cad protein plays an important role in activating genes expressed later in the segmentation pathway to generate posterior structures.

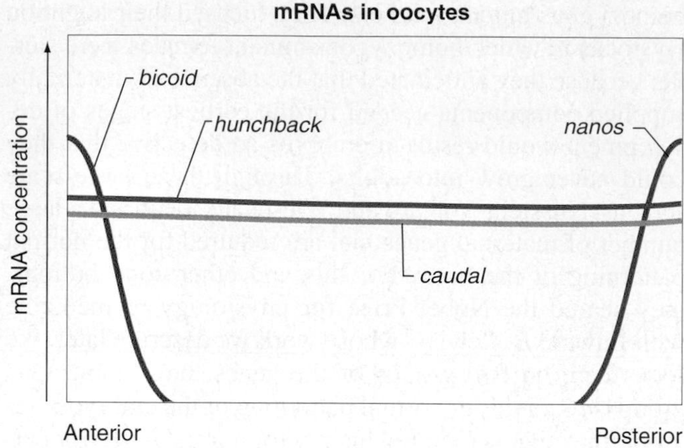

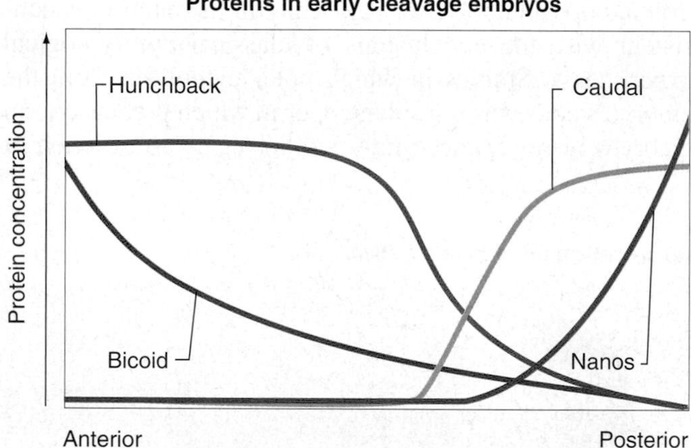

Figure 20.18 Distribution of the mRNA and protein products of maternal-effect genes within the early embryo. *Top:* In the oocyte prior to fertilization, *bicoid* (*bcd*) mRNA is concentrated near the anterior tip and *nanos* (*nos*) mRNA at the posterior tip, while maternally supplied *hunchback* (*hb*) and *caudal* (*cad*) mRNAs are uniformly distributed. *Bottom:* In early cleavage stage embryos, the Bicoid (Bcd) and Hunchback (Hb) proteins are found in concentration gradients high at the anterior and lower toward the posterior (A to P), while the Nanos (Nos) and Caudal (Cad) proteins are distributed in opposite P-to-A gradients.

The *nanos* (*nos*) Gene Encodes the Primary Posterior Morphogen

The *nos* RNA is localized to the posterior egg cytoplasm by proteins encoded by other posterior group maternal genes. Like *bcd* RNAs, *nos* transcripts are translated during the cleavage stages. After translation, diffusion produces a posterior-to-anterior Nos protein concentration gradient. The Nos protein, unlike the Bcd protein, is not a transcription factor; rather, the Nos protein functions only as a translational repressor. Its major target is the maternally supplied transcript of the *hunchback* (*hb*) gene, which is deposited in the egg during oogenesis and is uniformly distributed before fertilization. For development to occur properly, the Hb protein (which is another transcription factor) must be present in a gradient with high concentrations at the embryo's anterior and low concentrations at the posterior. The Nos protein, which represses the translation of *hb* maternal mRNA and is present in a posterior-to-anterior concentration gradient, helps construct the anterior-to-posterior Hb gradient by lowering the concentration of the Hb protein toward the embryo's posterior pole (Fig. 20.18). The embryo also has a second mechanism for establishing the Hb protein gradient that functions somewhat later: It only transcribes the *hb* gene from zygotic nuclei in the anterior region (see following).

Figure 20.18 summarizes how the coordinated activity of the maternal genes establishes polarities in the embryo that eventually result in proper embryonic segmentation. In addition to distributing the Bcd and Cad transcription factors in gradients within the embryo, the maternal genes ensure that the maternally supplied *hb* RNA is not translated in the posterior part of the embryo.

Specification of Segment Number Through the Activation of Zygotic Genes in Successively More Sharply Defined Regions

The maternally determined Bcd, Hb, and Cad protein gradients control the spatial expression of zygotic segmentation genes. Unlike the products of maternal genes, whose mRNAs are placed in the egg during oogenesis, the products of zygotic genes are transcribed and translated from DNA in the nuclei of embryonic cells descended from the original zygotic nucleus. The expression of zygotic segmentation genes begins in the syncytial blastoderm stage, a few division cycles before cellularization (roughly cycle 10).

Most of the zygotic segmentation genes were identified in a second mutant screen also carried out in the late 1970s by Christiane Nüsslein-Volhard and Eric Wieschaus. In this screen, the two *Drosophila* geneticists placed individual ethyl methane sulfonate (EMS)-mutagenized chromosomes into balanced stocks and then examined homozygous mutant embryos from these stocks for defects in the segmentation pattern of the embryo. Such homozygous mutant embryos were so aberrant that they were unable to grow into adults; thus, the mutations causing these defects would be classified as recessive lethals. After screening several thousand such stocks for each of the *Drosophila* chromosomes, they identified three classes of zygotic segmentation genes: gap genes (9 different genes); pair-rule genes (8 genes); and segment-polarity genes (about 17 genes). These three classes of zygotic genes fit into a hierarchy of gene expression.

Gap Genes Are Expressed in Broad Zones Along the Anterior-Posterior Axis

The gap genes are the first zygotic segmentation genes to be transcribed. Embryos homozygous for mutations in the gap genes show a gap in the segmentation pattern caused by an absence of particular segments that correspond to the position at which each gene is transcribed (**Fig. 20.19**). How do the maternal transcription factor gradients ensure that the various gap genes are expressed in their broad zones at the proper position in the embryo? Part of the answer is that the binding sites in the promoter regions of the gap genes have different affinities for the maternal transcription factors. For example, some gap genes are activated by the Bcd protein (the anterior morphogen). Gap genes such as *hb* with low-affinity Bcd protein-binding sites will be activated only in the most anterior regions, where the concentration of Bcd is at its highest; by contrast, genes with high-affinity sites will also be activated farther toward the posterior pole.

Another part of the answer is that the gap genes themselves encode transcription factors that can influence the expression of other gap genes. The *Krüppel* (*Kr*) gap gene, for example, appears to be turned off by high amounts of Hb protein at the anterior end of its band of expression; activated within its expression band by Bcd protein in conjunction with lower levels of Hb protein; and turned off at the posterior end of its expression zone by the products of the *knirps* (*kni*) gap gene (Fig. 20.19c). (Note that the *hb* gene is usually classified as a gap gene, despite the maternal supply of some *hb* RNA, because the protein translated from the transcripts of zygotic nuclei actually plays the more important role.)

Pair-Rule Genes Subdivide the Body into Units That Are Two Segments in Length

After the gap genes have divided the body axis into rough, generalized regions, activation of the pair-rule genes generates more sharply defined sections. These genes encode transcription factors that are expressed in seven stripes in preblastoderm and blastoderm embryos (**Fig. 20.20a**). The stripes have a two-segment periodicity; that is, there is one stripe for every two segments. Mutations in pair-rule genes cause the deletion of similar pattern elements from every alternate segment. For example, larvae mutant for *fushi tarazu* ("segment deficient" in Japanese) lack parts of abdominal segments A1, A3, A5, and A7 (see Fig. 20.2). Mutations in *even-skipped* cause loss of even-numbered abdominal segments.

There are two classes of pair-rule genes: primary and secondary. The striped expression pattern of the three primary pair-rule genes depends on the transcription factors encoded by the maternal genes and the zygotic gap genes. Specific elements within the upstream regulatory region of each pair-rule gene drive the expression of that pair-rule gene within a particular stripe. For example, as Fig. 20.20b and c shows, the DNA region responsible for driving the expression of *even-skipped* (*eve*) in the second stripe contains multiple binding sites for the Bcd protein and the proteins encoded by the gap genes *Krüppel, giant* (*gt*), and *hb.* The transcription of *eve* in this stripe of the embryo is activated by Bcd and Hb, while it is repressed by Gt and Kr; only in the stripe 2 region are Gt and Kr levels low enough and Bcd and Hb levels high enough to allow activation of the element driving *eve* expression. In contrast with the primary pair-rule genes, the five pair-rule genes of the secondary class are controlled by interactions with transcription factors encoded by other pair-rule genes.

(a) Zones of gap gene expression

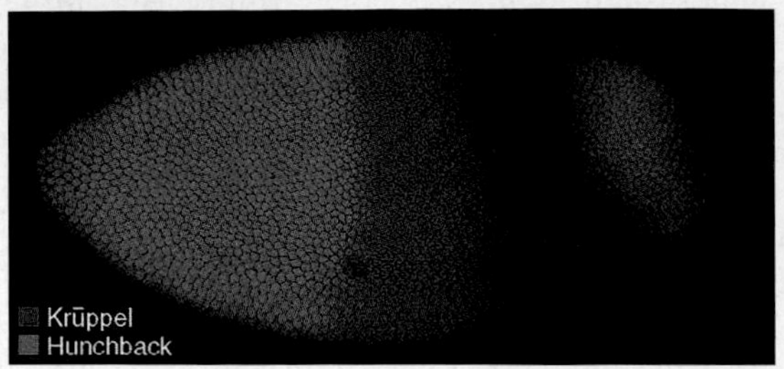

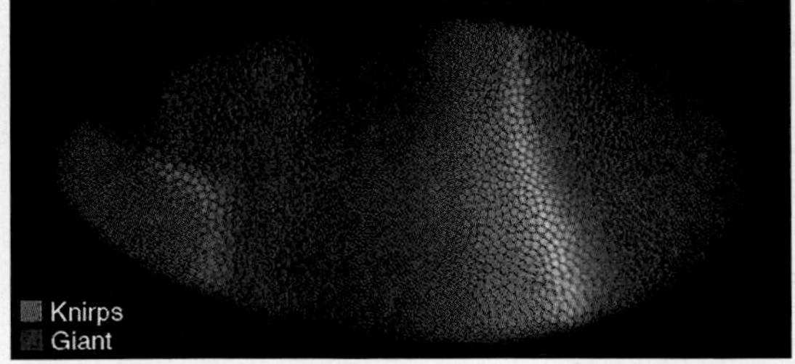

(b) Phenotypes caused by gap gene mutations

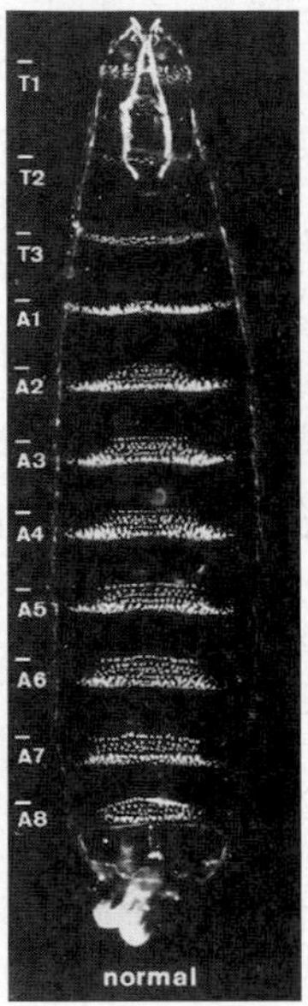

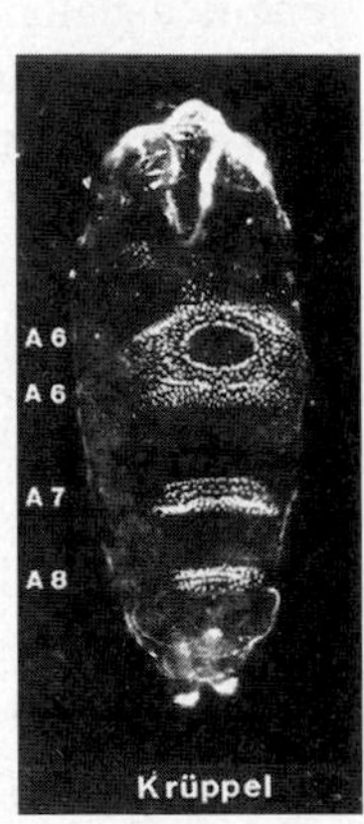

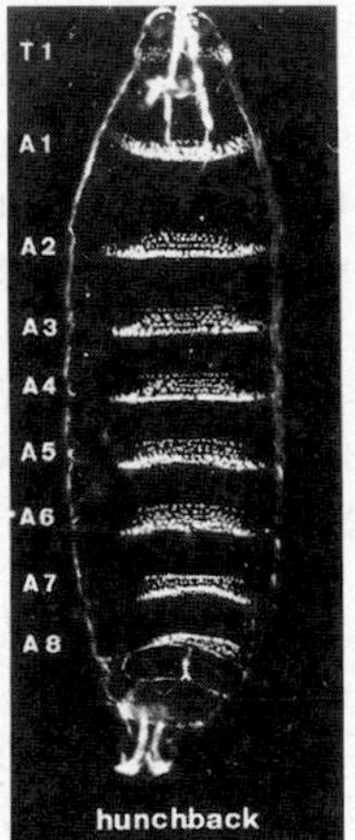

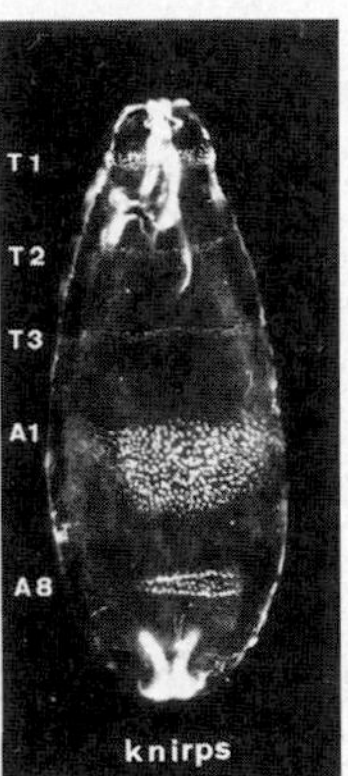

(c) Gap genes: a summary

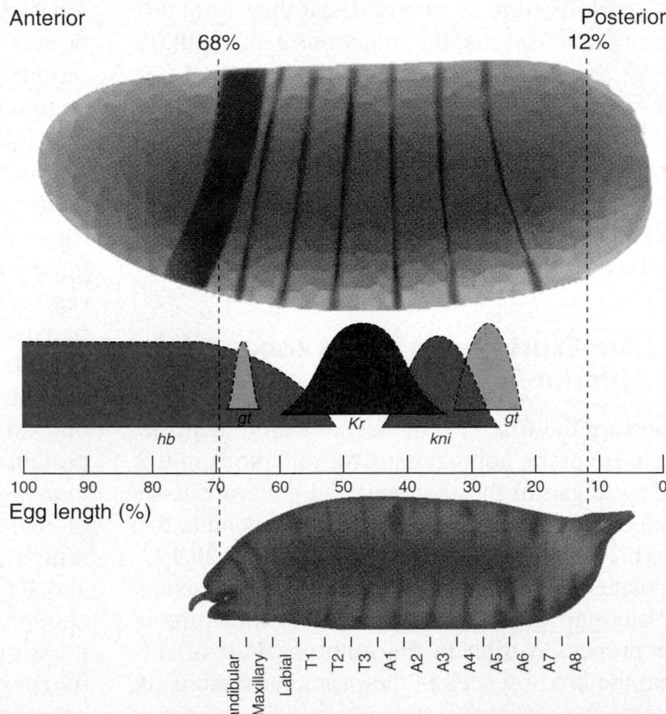

Figure 20.19 Gap genes. (a) Zones of expression of four gap genes (*hunchback* [*hb*], *Krüppel* [*Kr*], *knirps* [*kni*], and *giant* [*gt*]) in late syncytial blastoderm embryos, as visualized with fluorescently labeled antibodies. **(b)** Defects in segmentation caused by mutations in selected gap genes, as seen in late embryos. Only the remaining thoracic and abdominal segments are labeled; the head segments at the anterior end are highly compressed and not labeled. **(c)** Mutation of a particular gap gene results in the loss of segments corresponding to the zone of expression of that gap gene in the embryo.

(a) Distribution of pair-rule gene products

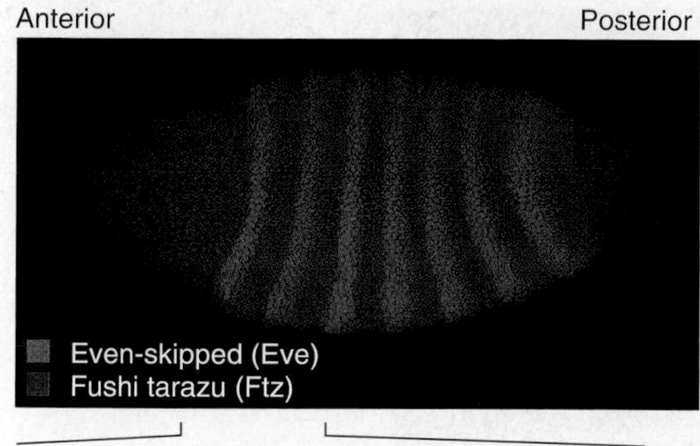

(b) Proteins regulating *eve* transcription

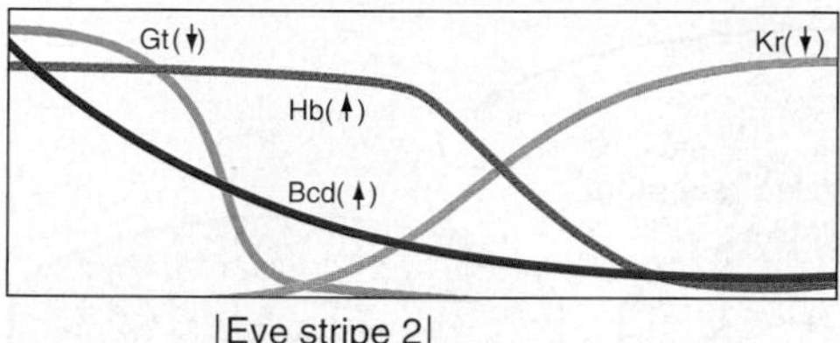

(c) Upstream regulatory region of *eve*

-1500 base pairs -800

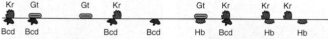

Figure 20.20 **Pair-rule genes. (a)** Zones of expression of the proteins encoded by the pair-rule genes *fushi tarazu* (*ftz*) and *even-skipped* (*eve*) at the beginning of the cellular blastoderm stage. Each gene is expressed in seven stripes. Eve stripe 2 is the second green stripe from the *left*. **(b)** The formation of Eve stripe 2 requires activation of *eve* transcription by the Bcd and Hb proteins and repression at its left and right ends by Gt and Kr proteins, respectively. **(c)** The 700 bp upstream regulatory region of the *eve* gene that directs the Eve second stripe contains multiple binding sites for the four proteins shown in part (b).

Segment Polarity Genes Occupy the Lowest Level of the Segmentation Hierarchy

Many segment polarity genes are expressed in stripes that are repeated with a single segment periodicity; that is, there is one stripe per segment (**Fig. 20.21a**). Mutations in segment polarity genes cause deletion of part of each segment, often accompanied by mirror-image duplication of the remaining parts. The segment polarity genes thus function to determine certain patterns that are repeated in each segment.

The regulatory system that directs the expression of segment polarity genes in a single stripe per segment is quite complex. In general, the transcription factors encoded by pair-rule genes initiate the pattern by directly regulating certain segment polarity genes. Interactions between various cell polarity genes then maintain this periodicity later in development. Significantly, activation of segment polarity genes occurs after cellularization of the embryo is complete. Hence, the diffusion of transcription factors within the syncytium, of central importance in segmentation up to this point, ceases to play a role. Instead, intrasegmental patterning is determined mostly by the diffusion of secreted proteins between cells.

Two of the segment polarity genes, *hedgehog* (*hh*) and *wingless* (*wg*), encode secreted proteins. These proteins, together with the transcription factor encoded by the *engrailed* (*en*) segment polarity gene, are responsible for many aspects of segmental patterning (Fig. 20.21b). A key component of this control is that a one-cell-wide stripe of cells secreting the Wg protein is adjacent to a stripe of cells expressing the En protein and secreting the Hh protein. The interface of these two types of cells is a self-reinforcing, reciprocal loop. The Wg protein secreted by the more anterior of the two adjacent stripes of cells is required for the continued expression of *hh* and *en* in the adjacent posterior stripe. The Hh protein secreted by the more posterior stripe of cells maintains expression of *wg* in the anterior stripe. Gradients of Wg and Hh proteins made from these adjacent stripes of cells control many aspects of patterning in the remainder of the segment. This is because the products of both *wg* and *hh* appear to function as morphogens; that is, responding cells appear to adopt different fates depending on the concentration of Wg or Hh protein to which they are exposed.

Other segment polarity genes encode proteins involved in **signal transduction pathways** initiated by the binding of Wg and Hh proteins to receptors on cell surfaces. Signal transduction pathways enable a signal received from a receptor on the cell's surface to be converted through a series of intermediate steps to a final intracellular regulatory response, usually the activation or repression of particular target genes. The signal transduction pathways initiated by the Wg and Hh proteins determine the ability of cells in portions of each segment to differentiate into the particular cell types characteristic of those locations.

Homologs of the segment polarity genes are key players in many important patterning events in vertebrates. For example, the chicken *sonic hedgehog* gene (related to the fly *hh*) is critical for the initiation of the left-right asymmetry in the early chicken embryo as well as for the processes that determine the number and polarity of digits produced by the limb buds. The mammalian homolog of this gene has the same conserved functions.

Summary of Segment Number Specification

The pattern of expression for members of each class of segmentation genes is controlled either by genes higher in the hierarchy or by members of the same class, never by genes

(a) Distribution of Engrailed protein

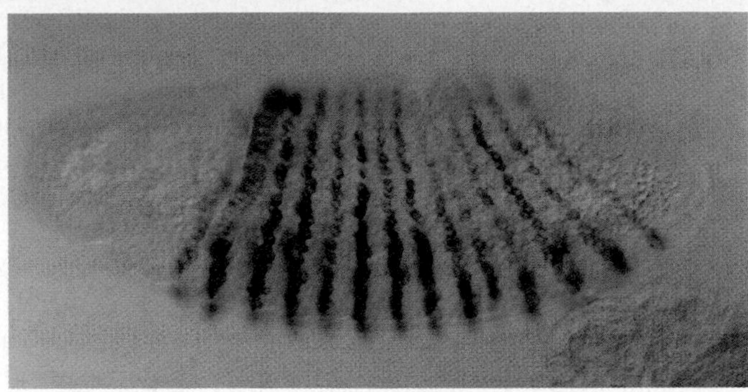

(b) Segment polarity genes establish compartment borders.

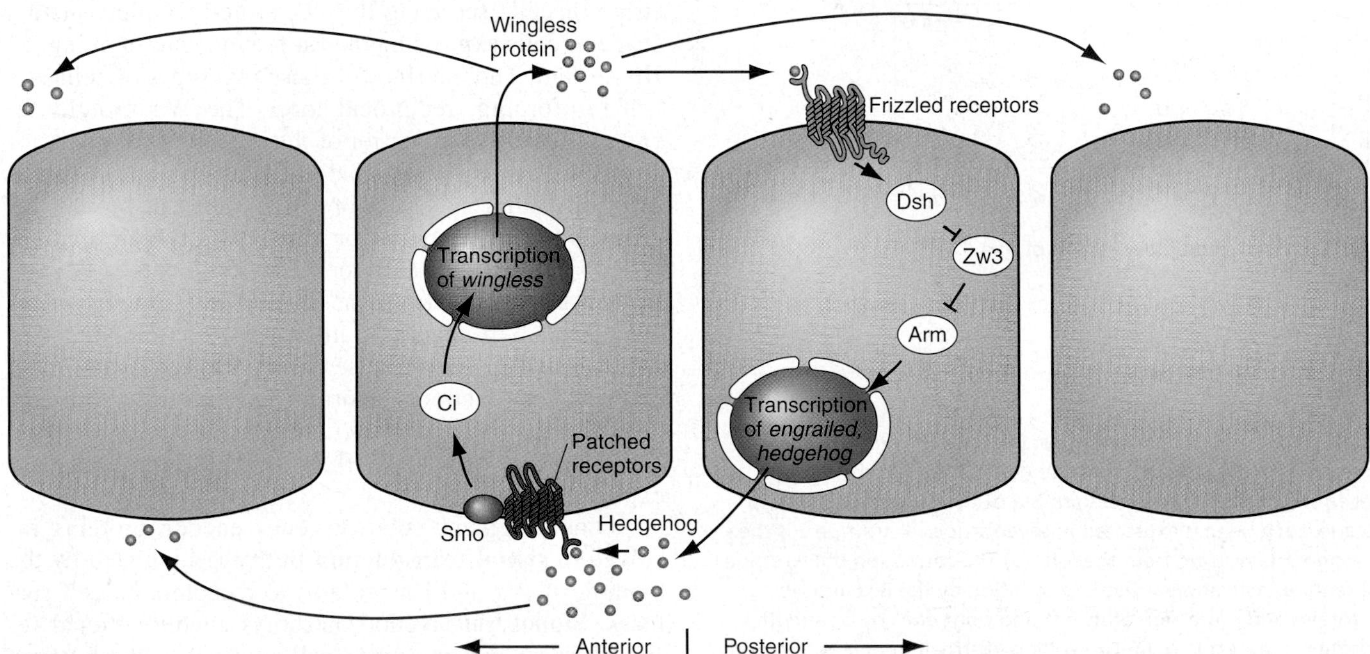

Figure 20.21 Segment polarity genes. (a) Wild-type embryos express the segment polarity gene *engrailed* in 14 stripes. **(b)** The border between a segment's posterior and anterior compartments is governed by the *engrailed* (*en*), *wingless* (*wg*), and *hedgehog* (*hh*) segment polarity genes. Regulation by pair-rule genes ensures that cells in posterior compartments express *en.* The En protein activates the transcription of the *hh* gene, which encodes a secreted protein ligand. Binding of this Hh protein to the Patched receptor in the adjacent anterior cell initiates a signal transduction pathway (through the Smo and Ci proteins) leading to the transcription of the *wg* gene. Wg is also a secreted protein that binds to a different receptor in the posterior cell, which is encoded by *frizzled.* Binding of the Wg protein to this receptor initiates a different signal transduction pathway (including the Dsh, Zw3, and Arm proteins) that stimulates the transcription of *en* and of *hh.* The result is a reciprocal loop stabilizing the alternate fates of adjacent cells at the border.

of a lower class (**Fig. 20.22**). In this regulatory cascade, the maternal genes control the gap and pair-rule genes, the gap genes control themselves and the pair-rule genes, and the pair-rule genes control themselves and the segment polarity genes.

The expression of genes in successively lower parts of the hierarchy is restricted to more sharply defined embryonic regions. During the early embryonic cycles of nuclear divisions, the spatial restriction of gene expression is a response to maternally established transcription factor gradients. In the syncytial blastoderm embryo, refinement of the spatial restriction of gene expression depends on the hierarchical action of the transcription factors encoded by the zygotic genes. Once cellularization is complete, intercellular communication mediated by secreted proteins encoded by certain segment polarity genes becomes the major mode of pattern formation. Although the cellular blastoderm, when viewed from the outside, looks like a uniform layer of cells (as seen in Fig. 20.16a), the coordinated action of the segmentation genes has already divided the embryo into segment primordia. A few hours after gastrulation, these primordia become distinguishable morphologically as clear-cut segments (as in Fig. 20.16c).

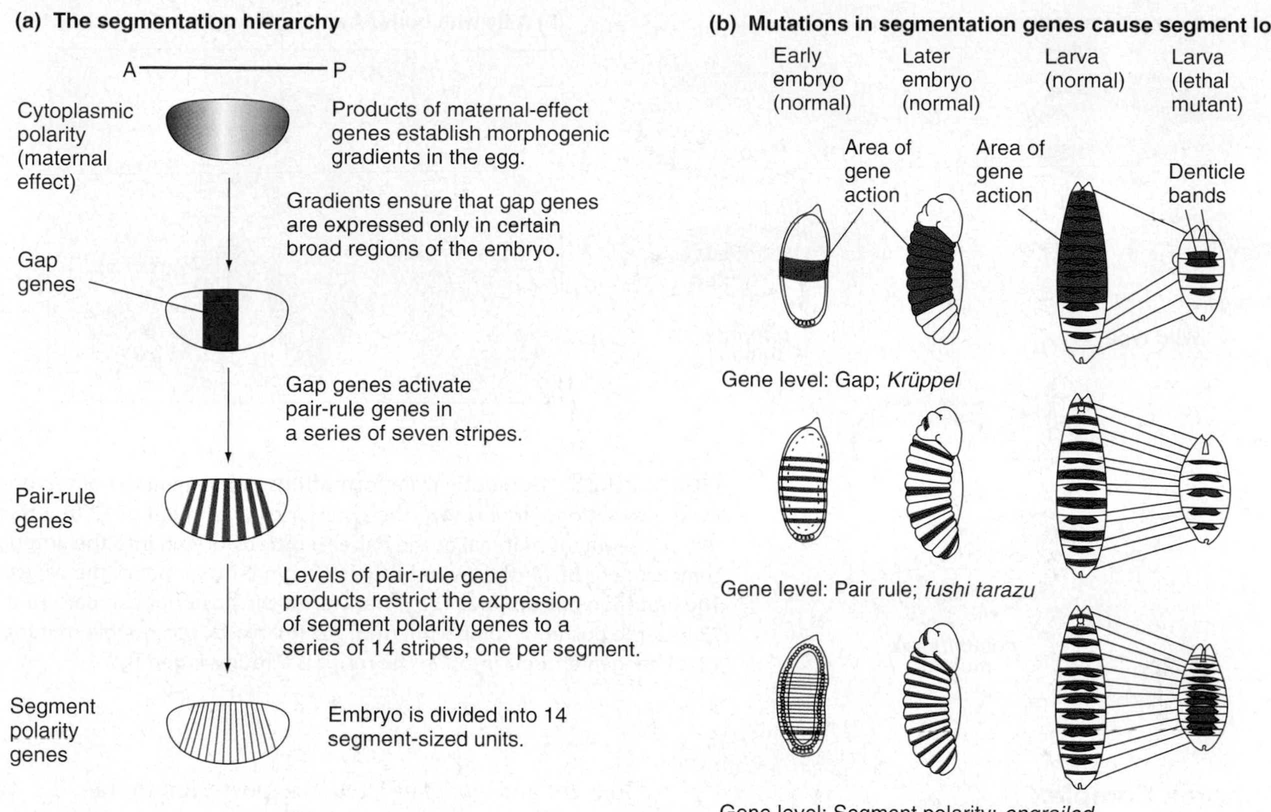
(a) The segmentation hierarchy
A
P
Cytoplasmic polarity (maternal effect)
Products of maternal-effect genes establish morphogenic gradients in the egg.
Gradients ensure that gap genes are expressed only in certain broad regions of the embryo.
Gap genes
Gap genes activate pair-rule genes in a series of seven stripes.
Pair-rule genes
Levels of pair-rule gene products restrict the expression of segment polarity genes to a series of 14 stripes, one per segment.
Segment polarity genes
Embryo is divided into 14 segment-sized units.
(b) Mutations in segmentation genes cause segment loss.
Early embryo (normal)
Later embryo (normal)
Larva (normal)
Larva (lethal mutant)
Area of gene action
Area of gene action
Denticle bands
Gene level: Gap; Krüppel
Gene level: Pair rule; fushi tarazu
Gene level: Segment polarity; engrailed

Figure 20.22 **A summary of the genetic hierarchy leading to segmentation in *Drosophila*.** **(a)** Genes in successively lower parts of the hierarchy are expressed in narrower bands within the embryos. **(b)** Mutations in segmentation genes cause the loss of segments that correspond to regions where the gene is expressed (*shown in yellow*). The denticle bands (*dark brown*) are features that help researchers identify the segments remaining in the mutant animals.

Each Segment Establishes Its Own Unique Identity Through the Activation of Homeotic Genes

After the segmentation genes have subdivided the body into a precise number of segments, the homeotic genes help assign a unique identity to each segment. They do this by functioning as master regulators that control the transcription of batteries of genes responsible for the development of segment-specific structures. The homeotic genes themselves are regulated by the gap, pair-rule, and segment polarity genes so that at the cellular blastoderm stage, or shortly thereafter, each homeotic gene becomes expressed within a specific subset of body segments. Most homeotic genes then remain active through the rest of development, functioning continuously to direct proper segmental specialization.

Mutations in homeotic genes, referred to as **homeotic mutations,** cause particular segments, or parts of them, to develop as if they were located elsewhere in the body. Because some of the mutant homeotic phenotypes are quite spectacular, researchers noticed them very early in *Drosophila* research. In 1915, for example, Calvin Bridges found a mutant he called *bithorax* (*bx*). In homozygotes for this mutation, the anterior portion of the third thoracic segment (T3) develops like the anterior second thoracic segment (T2); in other words, this mutation transforms part of T3 into the corresponding part of T2, as illustrated in **Fig. 20.23a.** This mutant phenotype is very dramatic, as T3 normally produces only small club-shaped balancer organs called *halteres,* whereas T2 produces the wings. Another homeotic mutation is *postbithorax* (*pbx*), which affects only posterior T3, causing its transformation into posterior T2. (Note that in this context, *Drosophila* geneticists use the term "transformation" to mean a change of body form.) In the *bx pbx* double mutant, all of T3 develops as T2 to produce the now famous four-winged fly (Fig. 20.23b).

In the last half of the twentieth century, researchers isolated many other homeotic mutations, most of which map within either of two gene clusters. Mutations affecting segments in the abdomen and posterior thorax lie within a cluster known as the **bithorax complex (BX-C);** mutations affecting segments in the head and anterior thorax lie within the **Antennapedia complex (ANT-C)** (**Fig. 20.24**).

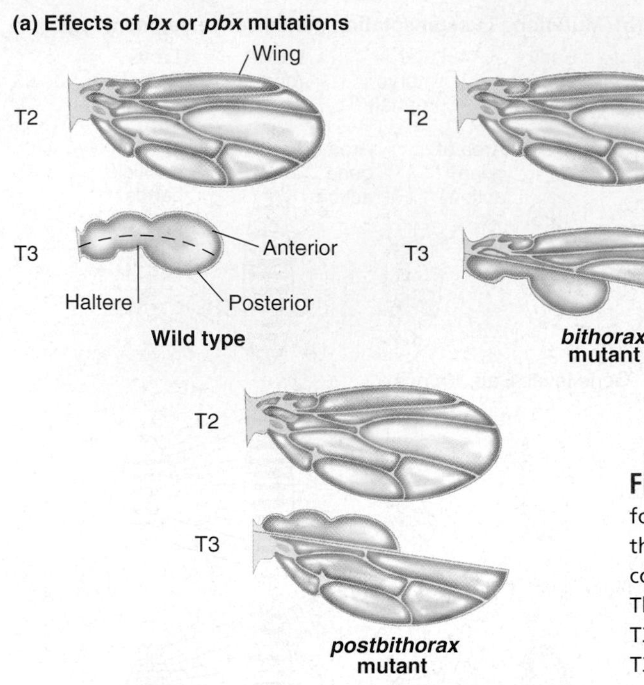

(b) A fly with both bx and pbx mutations

Figure 20.23 Homeotic transformations. (a) In animals homozygous for the mutation *bithorax* (*bx*), the anterior compartment of T3 (the third thoracic segment that makes the haltere) is transformed into the anterior compartment of T2 (the second thoracic segment that makes the wing). The mutation *postbithorax* (*pbx*) transforms the posterior compartment of T3 into the posterior compartment of T2. **(b)** In a *bx pbx* double mutant, T3 is changed entirely into T2. The result is a four-winged fly.

The Bithorax Complex

Edward B. Lewis shared the 1995 Nobel Prize for physiology or medicine with Christiane Nüsslein-Volhard and Eric Wieschaus for his extensive genetic studies of the BX-C. In his work, Lewis isolated BX-C mutations that, like *bx* and *pbx,* affected the posterior thorax; he also found novel BX-C mutations that caused anteriorly directed transformations of each of the eight abdominal segments. Lewis named mutations affecting abdominal segments *infra-abdominal* (*iab*) mutations, and he numbered these according to the primary segment they affect. Thus, *iab-2* mutations cause transformations of A2 toward A1, *iab-3* mutations cause transformations of A3 toward A2, and so forth.

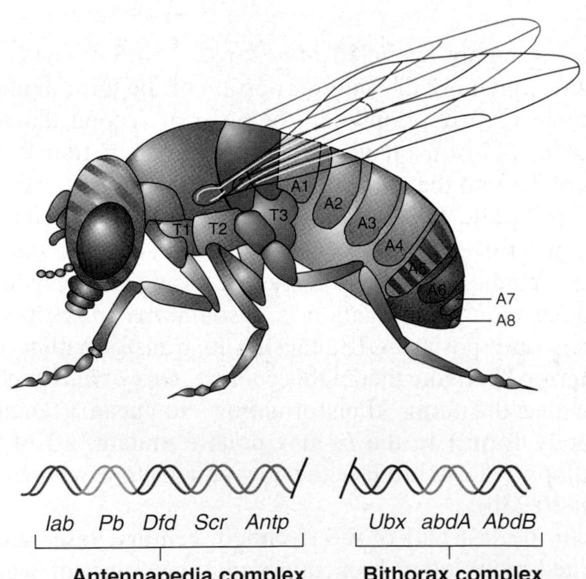

Figure 20.24 Homeotic selector genes. Two clusters of genes on *Drosophila's* chromosome 3—the Antennapedia complex and the bithorax complex—are responsible for determining most aspects of segment identity. Interestingly, the order of genes in these complexes is the same as the order of the segments each gene controls.

Researchers initiated molecular studies of the BX-C in the early 1980s, and in 15 years, they not only extensively characterized all of the genes and mutations in the BX-C at the molecular level, but they also completed the sequencing of the entire 315 kb region. **Figure 20.25** summarizes the structure of the complex. A remarkable feature of the BX-C is that mutations map in the same order on the chromosome as the anterior-posterior order of the segments each mutation affects. Thus, *bx* mutations, which affect anterior T3, lie near the left end of the complex, while *pbx* mutations, which affect posterior T3, lie immediately to their right. In turn, *iab-2,* which affects A2, is to the right of *pbx* but to the left of the A3-determining *iab-3*.

Because the *bx, pbx,* and *iab* elements are independently mutable, Lewis thought that each was a separate gene. However, the molecular characterization of the region revealed that the BX-C actually contains only three protein-coding genes: *Ultrabithorax* (*Ubx*), which controls the identity of T3; *abdominal-A* (*abd-A*), which controls the identities of A1–A5; and *Abdominal-B* (*Abd-B*), which controls the identities of A5–A8 (Fig. 20.25). The expression patterns of these genes are consistent with their roles. *Ubx* is

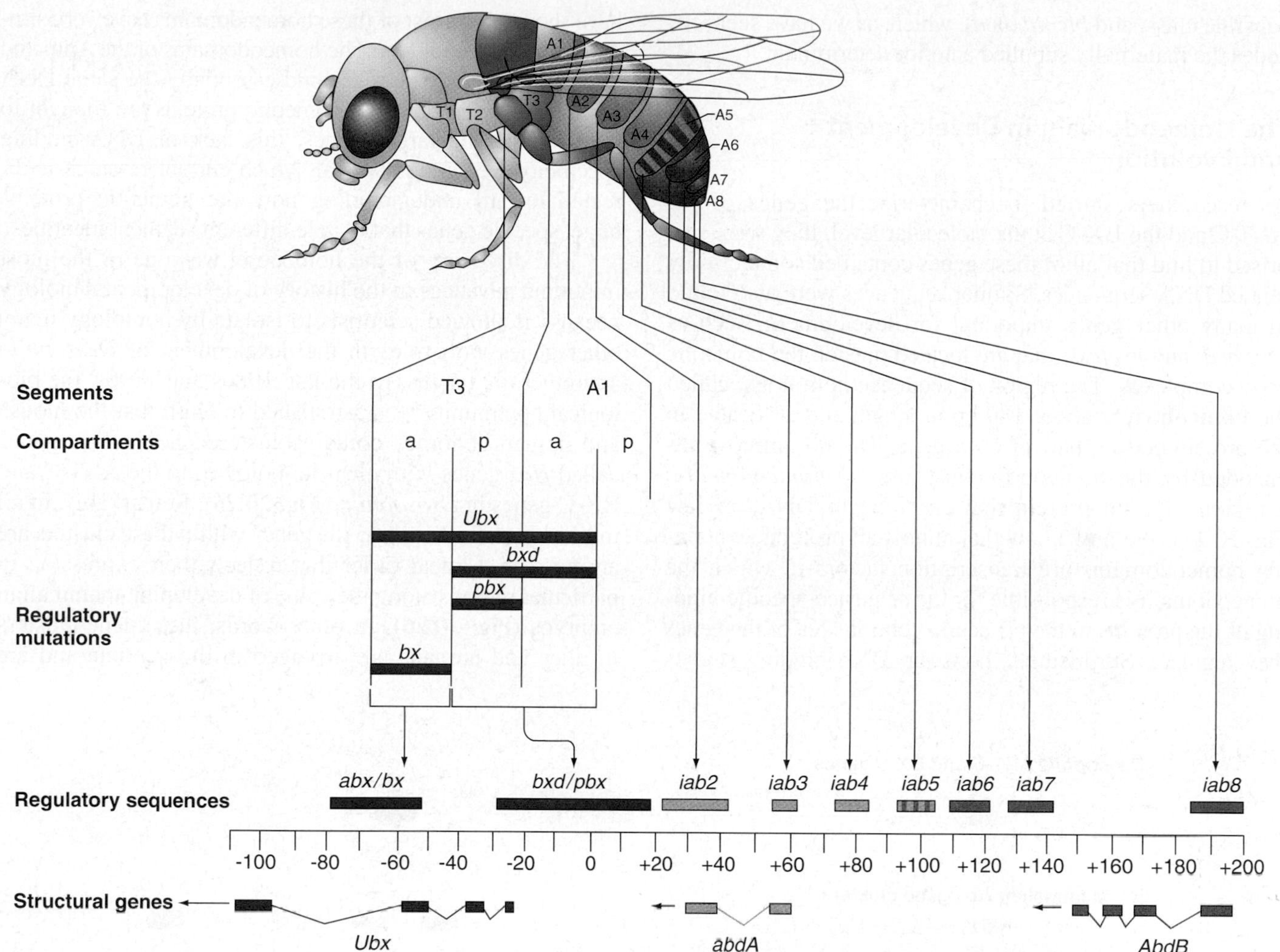

Figure 20.25 A close-up view of the 300 kb of the bithorax complex. There are only three homeotic genes in this complex: *Ubx, abd-A,* and *Abd-B.* Many homeotic mutations such as *bx* and *pbx* affect regulatory regions that influence the transcription of one of these three genes in particular segments. For example, *bx* mutations (*at the left end of the complex*) prevent the transcription of *Ubx* in the anterior compartment of the third thoracic segment, while *iab-8* mutations (*far right*) affect the transcription of *Abd-B* in segment A8. Note that the order of these regulatory regions corresponds to the anterior-to-posterior order of segments in the animal. *iab-5* mutations disrupt a regulatory sequence needed for the expression of both *abd-A* and *Abd-B* in segment A5.

expressed in segments T3–A8 (but most strongly in T3); *abd-A* is expressed in A1–A8 (most strongly in A1–A4); and *Abd-B* is expressed in A5–A8. The *bx, pbx,* and *iab* mutations studied by Lewis affect large *cis*-regulatory regions that control the intricate spatial and temporal expression of these genes within specific segments. The average size of each of these regulatory regions is about 15 kb. Researchers have not yet completely worked out the mechanisms by which single BX-C genes control multiple segment identities. However, it is likely that segment identity is determined primarily by the level or timing of BX-C gene expression.

The Antennapedia Complex

Genetic studies in the early 1980s showed that a second homeotic gene cluster, the Antennapedia complex (ANT-C), specifies the identities of segments in the head and anterior thorax of *Drosophila.* The five homeotic genes of the ANT-C are *labial* (*lab*), which is expressed in the intercalary region; *proboscipedia* (*Pb*), expressed in the maxillary and labial segments; *Deformed* (*Dfd*), expressed in the mandibular and maxillary segments; *Sex combs reduced* (*Scr*), expressed in the labial and T1 segments; and *Antennapedia* (*Antp*), expressed mainly in T2, although it is also active at lower levels in all three thoracic and most abdominal segments. (Figure 20.16c shows these head and thoracic segments, while Fig. 20.24 illustrates the order of the homeotic genes in the ANT-C.) As with the BX-C, the order of genes in the ANT-C is the same (with the exception of *pb*) as the order of segments each controls. In addition to homeotic genes, the ANT-C contains several other important genes, including *zerknullt* (*zen*), which specifies dorsal embryonic structures; *fushi tarazu* (*ftz*), a segmentation gene of the pair-rule class; and *bicoid* (*bcd*), which, as we have seen, encodes the maternally supplied anterior determinant.

The Homeodomain in Development and Evolution

As researchers started to characterize the genes of the ANT-C and the BX-C at the molecular level, they were surprised to find that all of these genes contained some closely related DNA sequences. Similar sequences were also found in many other genes important for development, such as *zen, bcd,* and *eyeless,* that are located outside the homeotic gene complexes. The region of sequence homology, called the **homeobox,** is about 180 bp in length and is located in the protein-coding part of each gene. The 60 amino acids encoded by the homeobox constitute the *homeodomain,* a region of each protein that can bind to DNA (review Fig. 20.11). We now know that almost all proteins containing homeodomains are transcription factors in which the homeodomain is responsible for the sequence-specific binding of the proteins to the *cis*-acting control sites of the genes they regulate. Surprisingly, however, DNA-binding studies have shown that most of these homeodomains have very similar binding specificities. The homeodomains of the Antp and Ubx proteins, for example, bind essentially the same DNA sequences. Since different homeotic proteins are thought to regulate specific target genes, this lack of DNA-binding specificity seems paradoxical. Much current research is directed toward understanding how the homeotic proteins target specific genes that dictate different segment identities.

The discovery of the homeobox was one of the most important advances in the history of developmental biology because it allowed scientists to isolate by homology many other genes with roles in the development of *Drosophila* and other organisms. In the late 1980s and 1990s, the biological community was astonished to learn that the mouse and human genomes contain clustered homeobox genes called *Hox* genes with clear homologies to the ANT-C and BX-C genes in *Drosophila* (**Fig. 20.26**). Remarkably, in all mammals studied to date, the genes within these clusters are arranged in a linear order that reflects their expression in particular regions along the spine of developing mammalian embryos (Fig. 20.26). In other words, these gene clusters in mice and humans are arranged in the genome and are

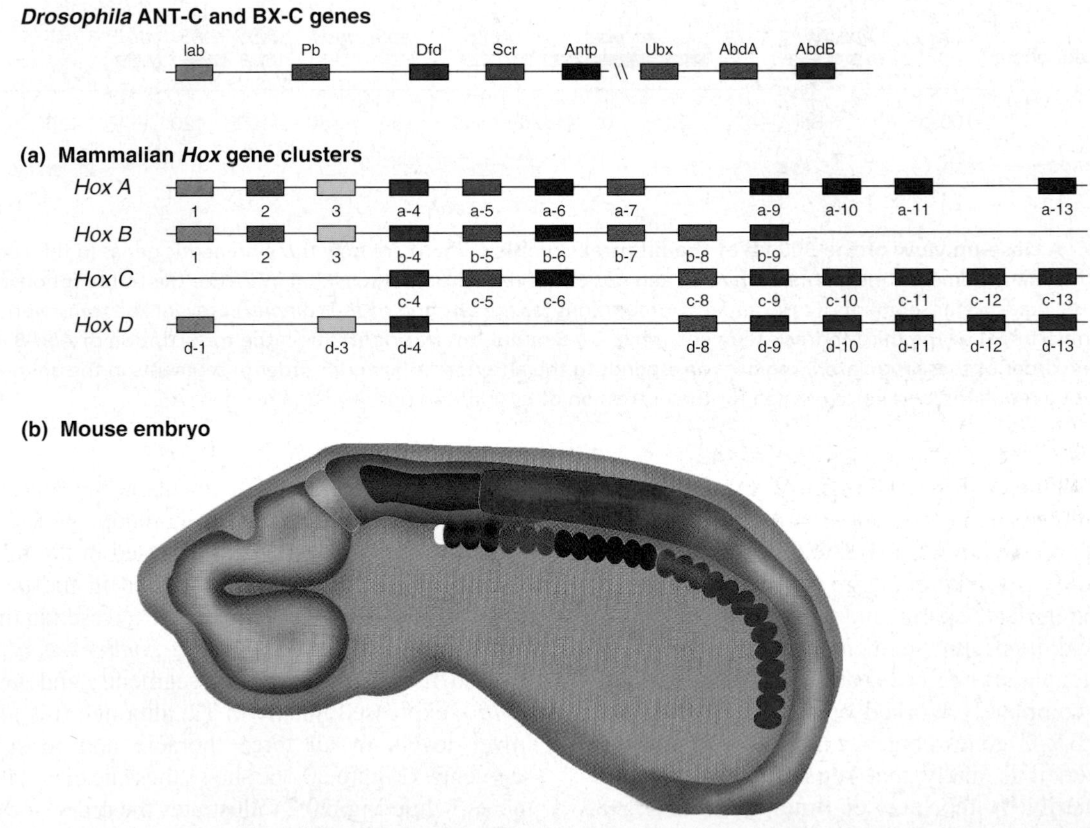

Figure 20.26 The mammalian *Hox* gene superfamily is organized into four clusters. (a) Mammalian genomes contain multiple homologs of each of the ANT-C and BX-C homeobox genes in *Drosophila.* Not all of the four clusters have homologs of every *Drosophila* gene. Some of the clusters have multiple genes related to fly *Pb* and *AbdB.* **(b)** Just as in *Drosophila,* the mammalian (mouse) *Hox* genes in each cluster are arranged in the order they are expressed along the anterior-posterior axis of the embryo. The *colored disks* represent somites—precursors of the vertebrae and other structures. The other colored areas are regions of the central nervous system. The colors represent the *Hox* gene expressed in that tissue.

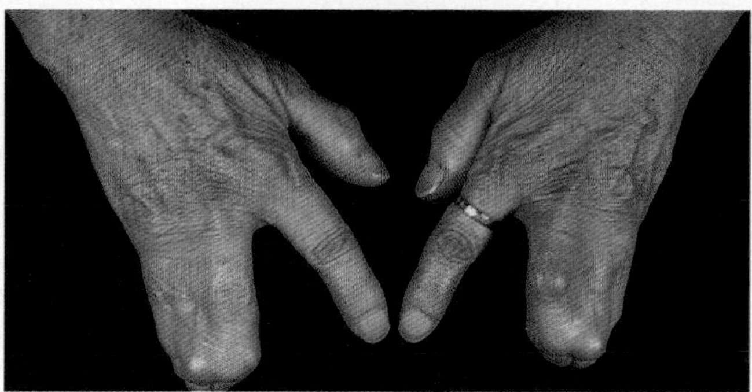

Figure 20.27 Synpolydactyly caused by mutations in the human *HoxD13* gene. The milder phenotype to the *left* is seen in a heterozygous individual.

regulated along the anterior-posterior axis in almost exactly the same way as the fly ANT-C and BX-C complexes.

As it turns out, all animal genomes, even those of sponges, the most primitive animals, contain *Hox* genes, so these genes are ancient and have played important (though not necessarily identical) roles in the developmental patterns of all animals. Sponges are radially symmetric and thus have no anterior-posterior axis; they are also unsegmented. The single *Hox* gene found in most sponge genomes helps regulate the formation of channels through the sponge. All other animal phyla have multiple *Hox* genes arranged in clusters. Generally, the more complex the body plan, the more *Hox* genes: Humans and other mammals have four *Hox* clusters that together contain 38 *Hox* genes (see Fig. 20.26). In just one demonstration that *Hox* genes mediate the developmental fate of specific regions in the body of animals other than *Drosophila,* it has recently been shown that the malformation of the digits in humans, in a condition called *synpolydactyly,* is caused by mutations in *HoxD13,* one of these 38 *Hox* genes (**Fig. 20.27**).

20.4 How Genes Help Control Development: A Mechanistic Framework

The previously described analysis of *Drosophila* body plan development revealed some of the strategies by which genes control the development of multicellular organisms. Although discovered in studies of anterior-posterior axis determination in flies, these strategies form the basic underpinnings for many diverse developmental pathways in many organisms. Here are highlights of the lessons learned from *Drosophila.*

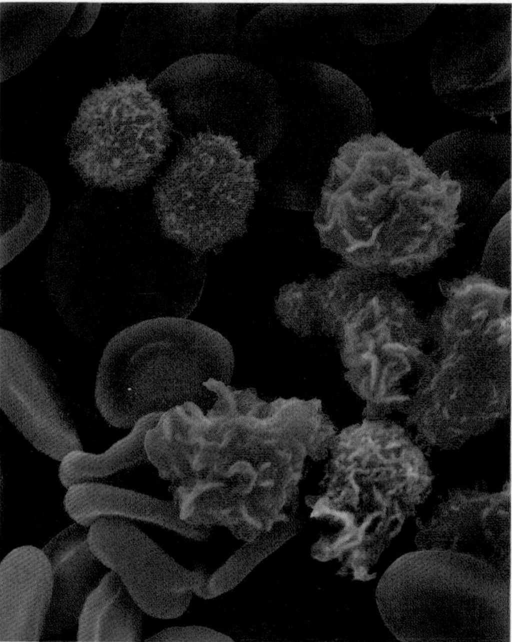

Figure 20.28 Different types of blood cells. The *red* cells are erythrocytes, the oxygen-carrying red blood cells. The cells colored in *green* are macrophages that ingest and destroy invading microbes. The *yellow* cells are T lymphocytes involved in the immune system. The single *blue* cell is a monocyte, an immature cell that can develop into a macrophage.

Development Requires Sequential Changes in Gene Expression

The enormous diversity of cells within the body of a multicellular organism results in a remarkable variety of cell shapes and functions. Even a single tissue, such as the blood, harbors many different kinds of cells (**Fig. 20.28**).

How do cells that contain the same genes make such varied developmental decisions and become so different? As we saw in the introduction to this chapter, the reason is simple: Different cell types express different, characteristic subsets of genes. It is easy to understand this point when comparing cells whose function depends on the production of a large amounts of a particular gene product. Red blood cells produce copious amounts of hemoglobin, the cone and rod cells in our eyes synthesize vast numbers of photoreceptor molecules, and certain pancreatic cells produce insulin and secrete it into the bloodstream. But the biochemical differences between cell types are not restricted to the expression of a single key gene. Instead, the differentiation of these various cells requires changes in the expression of many genes. **Figure 20.29** illustrates how complex these developmental patterns of expression can be. It shows that in *Drosophila,* for example, the many different proteins necessary for generating the structure of an adult wing are expressed in very precise, partially overlapping subsets of cells in larval structures called *imaginal discs* that give rise to the wing.

Figure 20.29 Development requires precise control of the expression of many genes. Each imaginal disc was stained with a fluorescently tagged antibody against a different specific protein important for patterning of the wing. Each protein is expressed in a unique set of cells in these imaginal discs.

Cell Fate Is Progressively Refined During Embryonic Development

Differentiation into many types of cells and tissues requires that cells undergo a successive restriction in developmental potential that affects both themselves and their descendents. For example, the two daughter cells of a human zygote can each generate descendents able to fulfill any fate in the adult. But later in development, cells must "decide" whether they and their descendents will adopt one kind of fate (say that of neurons) or a different kind of fate (say that of epidermal cells). Once a developmental decision is made, a cell and its descendents embark on a particular pathway of differentiation that excludes them from an alternative fate.

The hierarchic developmental system that determines the number of segments in *Drosophila* embryos provides a clear example of how cells (or nuclei in the syncytial fly embryo) successively obtain more refined information for the specification of cell fates (review Fig. 20.22). We have seen that the gap genes such as *Krüppel* are expressed in broad regions covering roughly one-quarter to one-third of the embryo, but later, pair-rule genes are expressed in a fashion that subdivides the regions in which each gap gene was expressed. And later still, the segment polarity genes are expressed in even more sharply defined areas.

Most Developmental Decisions Ultimately Affect Transcription

To decide which proteins a particular cell type will or will not produce, it makes the most sense for those cells to regulate the first step in gene expression: the initiation of transcription. Indeed, most of the processes that influence cellular fates culminate in decisions to turn on or off the transcription of "target" genes (such as those for hemoglobin) whose expression is important to that cell type (the red blood cell precursor).

We can make this generalization based on three kinds of observations. First, RNA *in situ* hybridization experiments, such as the one shown in Fig. 20.12, demonstrate that the mRNA for many developmentally important genes appears only in certain cells at certain times in development. Second, measurements of mRNA levels in many kinds of differentiated cells using techniques such as microarrays or quantitative PCR show that the levels of almost all proteins in those cells reflect the abundance of the mRNA encoding that protein. Third, many of the genes that play key roles in developmental decision-making encode proteins that function as transcription factors. For example, *bicoid* and most gap and pair-rule genes encode transcription factors. We have also seen that the homeodomain characteristic of proteins like the products of genes in the BX-C and ANT-C, as well as other genes such as *eyeless/Pax-6,* allows these proteins to bind to DNA and thus act as regulatory transcription factors. Hierarchies of transcription factors allow an organism to provide its cells with increasingly specific information that guides them to specific fates.

Some Developmental Processes Require Other Types of Gene Regulation

Although regulation of transcriptional initiation is the most general strategy by which cells control gene expression during development, it is by no means the only one. The progression from gene to protein involves many subsequent steps, each of which is amenable to regulation. A gene's primary transcript has to be spliced into a mature mRNA. This mRNA must be translocated from the nucleus into the cytoplasm, and then it must be translated into a protein. The relative stability of an mRNA or protein can affect its concentration in the cell. And finally, once made, a protein can be altered after translation in ways that affect its activity. Researchers have learned of various molecular mechanisms underlying development that exploit each of these steps of gene regulation. You saw one case of developmentally important posttranscriptional regulation in the comprehensive example: Bicoid and Nanos, two proteins encoded by maternal-effect genes functioning at the beginning of the segmentation hierarchy in *Drosophila,* act as repressors of translation.

It's the Chicken *and* the Egg: Both Maternal and Zygotic Genes Impact Early Development

The earliest stages of development require not only the regulation of the expression of genes in the developing individual's genome but also the regulation of gene expression in the mother's genome. Before fertilization, the egg in most organisms already contains many of the mRNAs and proteins needed for the earliest stages of development. The egg must load up on these molecules because transcription of the zygotic genome (that is, of the diploid genome containing an equal number of maternal and paternal chromosomes) does not begin immediately after fertilization. In *Drosophila,* for example, transcription of zygotic genes does not usually begin in earnest until the embryo contains roughly 6000 cells and has completed some of the earliest steps determining cell fates. Nüsslein-Volhard and Wieschaus recognized the importance of components placed into the egg by the mother during oogenesis when they screened for maternal-effect mutations and found the anterior and posterior morphogens Bicoid and Nanos.

Development Exploits Asymmetries

For cells to differentiate into different types, they must either be exposed to different signals from their environment or they must be intrinsically biochemically distinct. Nature has used both strategies for the differentiation of cell types during the development of multicellular organisms.

In some species, the egg is inherently asymmetric, providing a way for cells in the early embryo to receive information about their relative position. The *Drosophila* egg cell, for example, is part of a more complicated structure called an *egg chamber* (**Fig. 20.30**). (You should note that we use the term "egg cell" somewhat incorrectly for simplicity's sake: it is more properly called the oocyte.)

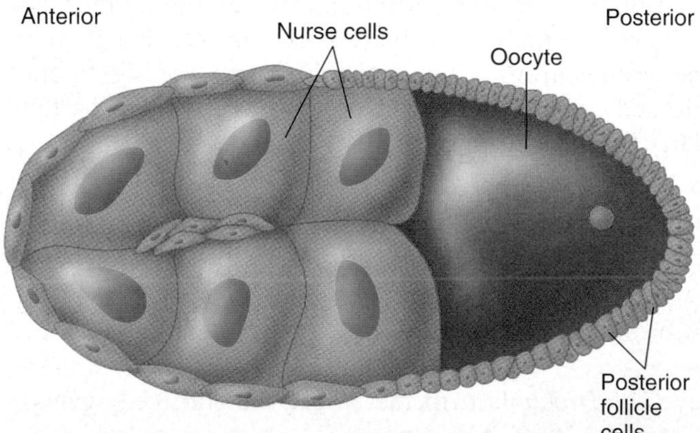

Figure 20.30 A *Drosophila* egg chamber. Large nurse cells at the anterior of the egg chamber synthesize mRNA and proteins and transport them to the oocyte. The nurse cells and the oocyte are surrounded by a layer of *follicle cells.*

Certain cells known as *nurse cells* act as factories that synthesize large amounts of mRNAs and proteins; the nurse cells then deposit these molecules into the egg. The egg has an anterior-to-posterior (that is, head-to-tail) sense of direction in large part because it is connected to the nurse cells only at its anterior end. The mRNA of the *bicoid* gene is transcribed in the nuclei of the nurse cells and then transported into the egg. The *bicoid* mRNA, in association with certain proteins that bind to its 3′-UTR, appears to become ensnared by microtubules within the egg cell. These microtubules act as tracks along which the mRNA and its associated proteins are transported to the cortex (the cytoplasm just beneath the cell membrane) at the egg cell's anterior end.

In other species, the first asymmetries important for development occur after fertilization. For example, in *C. elegans,* the site at which the sperm enters the egg to effect fertilization defines the posterior end of the embryo. Before fertilization, the egg has no polarity; sperm entry initiates rearrangements of the cytoplasm that establish the anterior-to-posterior axis of the embryo. The asymmetries affecting early mammalian development emerge even later, after four rounds of mitosis have produced a 16-cell embryo.

Cell-to-Cell Communication Is Essential for Proper Development

Construction of a large, complicated multicellular organism depends on more than broad, asymmetric cues such as the morphogen gradients laid down by the mother and the asymmetric distribution of molecules during cell division. Cells must "talk" to each other to obtain enough information about their relative positions in the organism. The information obtained from cell-to-cell communication enables cells to refine the decisions that guide their subsequent development.

Cells can communicate with each other either by direct contact or by diffusible factors (usually proteins) released from one cell and received by a second cell. Cell-to-cell communication usually takes place at the surface of the second cell when a ligand made by the signaling cell binds to a receptor embedded in the membrane of the receiving cell. One type of cell-to-cell communication, called *juxtacrine signaling,* takes the form of direct contact. In such signaling, the ligand is a cell surface molecule anchored in the membrane and extending outside of the signaling cell.

Other cellular interactions are mediated by *paracrine factors:* ligands secreted by the signaling cell. Ligands called *hormones,* or *endocrine factors,* circulate throughout the body in the blood and can affect tissues far removed from the gland that produces them. By contrast, some ligands diffuse only over short distances. The reciprocal interactions of *Drosophila* embryonic cells making the Wingless and Hedgehog segment polarity proteins (review Fig. 20.21b) illustrate this kind of short-range paracrine signaling. Both Wingless and Hedgehog are secreted by certain cells, and only nearby cells with appropriate receptors can respond to these ligands. Fig. 20.21b emphasizes another feature common to most kinds of cell-to-cell communication: The binding of the ligand to a cell-surface receptor initiates a signal transduction pathway that culminates in changes to the transcriptional regulation of suites of genes in the receiving cell's nucleus. Different ligand/receptor combinations activate different signal transduction pathways. For example, in Fig. 20.21b, the Smo and Ci proteins participate in the pathway activated by the binding of the Hedgehog ligand to its receptor (Patched), while the Dsh, Zw3, and Arm proteins are part of the pathway initiated by the binding of the Wingless ligand to its receptor (Frizzled).

Genes Explain Much, but Not Everything, About Development

Throughout this chapter, we have considered cells in developing multicellular organisms as complex computers. These cellular computers integrate a variety of inputs: the cell's history, its location within the organism, signals from neighboring cells, and signals from more distant cells. The outputs of the cellular computer are alterations to the transcription of a large suite of target genes, which determine the developmental fate of the cell. The central processors that convert the inputs into the outputs are located near the promoters of the target genes, where assessment of the combinatorial effects of many transcription factors determines the time and rate of target gene transcription.

This reductionist point of view has been remarkably successful in building our understanding of development. We now have lists of many genes that play important roles in development, and we are beginning to fathom how each of these genes work and how they interact with each other. Particularly remarkable in the recent past has been our growing appreciation for the way in which evolution has conserved critical genes and pathways, while at the same time creating new twists that underlie the enormous complexity and diversity of life-forms on earth.

Although genes clearly set the ground rules for an individual's development, the same set of genes does not inevitably lead to precisely the same result. Many events in development reflect the strong influence of environmental factors or chance on the execution of the genetic blueprint. For example, the name of the first kitten cloned from an adult cell (**Fig. 20.31**) is "cc" for "carbon copy," but this is somewhat of a misnomer. Though the kitten has exactly the same alleles of all genes as the cat that donated the adult cell, the coats of the two animals are dissimilar due to different prenatal environments. Chance occurrences often influence expression of the genome as well as cellular behaviors. In mammalian females, for example, the decision of which X chromosome is inactivated in which cell is determined by stochastic (chance) events. Similarly, the choice of which cells adopt particular fates depends on small chance fluctuations in the concentrations of certain ligands and receptors. Finally, the incredibly complex connections between neurons in the developing brain are highly plastic and can be influenced by the environment, particularly through learning.

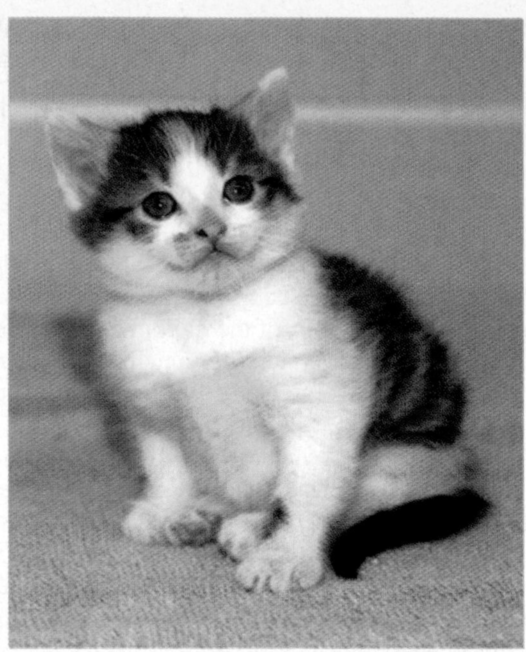

Figure 20.31 "cc," the cloned kitten.

Connections

This chapter has presented ample evidence for the conservation of genes that play important roles in development. The *eyeless/Pax-6/Aniridia* gene, for example, acts as a master switch to initiate the development of eyes in many types of organisms. Yet the eyes of various species show tremendous differences, from the compound eyes of *Drosophila* to the single camera-like organs of humans.

The themes of conservation and change have been central to our understanding of evolution since Darwin. Evolution creates and then preserves genetic solutions to problems organisms encounter in their development, biochemistry, physiology, and behavior; but evolution also tinkers with these solutions to produce novel outcomes.

In the next two chapters, we shift our focus from the analysis of gene activity in individuals to an analysis of gene transmission in whole populations and an examination, at the molecular level, of how genes and genomes evolve over time. Chapter 21 describes why an understanding of evolution requires knowledge of gene transmission in populations. Chapter 22 then builds on ideas presented throughout the book to reconstruct the molecular strategies by which genes and genomes have evolved throughout the roughly 4 billion years of life on earth.

Essential Concepts

1. Developmental geneticists use model organisms as the basis for studying how a fertilized egg becomes a multicellular adult. The evolutionary relatedness of all organisms, revealed by the conservation of genes, structures, and functions often makes it possible to extrapolate from model organisms to all living forms.
2. A key to the genetic dissection of development is the isolation of a comprehensive set of mutations in a gene. To understand the gene's role in development, researchers analyze mutant phenotypes. They also determine where the gene is expressed through RNA *in situ* hybridization experiments or by following the protein product with fluorescently tagged antibodies or with GFP labels. The construction of *genetic mosaics* can help determine which cells need to express the gene so that the organism can develop normally.
3. Genetic analysis of the *Drosophila* body plan revealed several basic mechanisms by which genes help control development. A hierarchy of *segmentation genes* subdivides the body into an array of body segments; the expression of *homeotic genes* assigns a unique identity to each segment.
4. Cellular differentiation requires progressive changes in gene expression. These changes usually result from decisions concerning the transcription of batteries of genes. Some developmental decisions, however, involve posttranscriptional gene regulation. The earliest stages of development require control of gene expression in both the maternal and zygotic genomes.
5. Differentiation requires either that cells have intrinsic differences at the biochemical level or that they are exposed to different information in their environment. Asymmetries in early embryonic development, or the asymmetric distribution of molecules during cell division, can generate intrinsic differences. Cell-to-cell communication, effected by the binding of ligands to receptors and mediated by signal transduction pathways, supplies cells with complex information about their relative position in the organism.

On Our Website

www.mhhe.com/hartwell3
Chapter 20

Annotated Suggested Readings and Links to Other Websites

- Historical articles by Nüsslein-Volhard, Wieschaus, Lewis, and others, reporting their pathbreaking work on the development of the *Drosophila* embryo
- Links to images, databases, and interactive sites depicting details of *Drosophila* development

Specialized Topics

- Genetic portraits: Many of the fundamental findings in the genetics of development have depended on genetic manipulations specific to particular organisms. Our website contains chapter-length "genetic portraits" that discuss experimental techniques and key findings for the five most intensively investigated model organisms: the yeast *Saccharomyces cerevisiae,* the plant *Arabidopsis thaliana,* the worm *Caenorhabditis elegans,* the fruit fly *Drosophila melanogaster,* and the house mouse *Mus musculus.*
- More on the genetic analysis of development: a description of advanced techniques for the discovery and analysis of genes important to development

Social and Ethical Issues

1. Most people have an instinctive aversion to the idea of human reproductive cloning. As the Genetics and Society box on p. 724–725 states, "No country or group of scientists has yet condoned the reproductive cloning of humans." However, there has been little public resistance to research on reproductive cloning in other mammalian species. As of 2005, such research had encountered technical problems that hampered the progress of reproductive cloning in these other species. For example, the technology is very inefficient; usually more than 100 cloned embryos must be implanted in surrogate mothers to obtain one live-born animal. It also appears that cloned animals are not completely normal; Dolly, the first cloned animal (a sheep), was considerably overweight, and many cloned cows have grossly enlarged hearts and lungs. Few people think it is ethical to proceed with human cloning given these problems. But if these issues could ever be surmounted, is human reproductive cloning inherently unethical or injurious to society? Why or why not?

2. The embryonic stem cell controversy in the United States has focused attention on the fate of embryos left over from *in vitro* fertilization attempts. As presently performed, *in vitro* fertilization usually involves the harvesting of multiple eggs from the prospective mother and their fertilization with the father's sperm. To avoid dangerous multiple pregnancies (quadruplets or more), most clinics restrict the number of embryos that can be implanted into the mother's womb to three or less. Additional embryos are usually frozen, as they can be implanted into the mother at later times. However, many of these embryos are never used. What should be the fate of these embryos?

3. To overcome the objection of at least some abortion opponents to embryonic stem cell research, one member of the President's Council on Bioethics proposed in 2004 an unconventional solution. His idea was to create a cloned embryo by transplanting a somatic cell nucleus into an enucleated egg, while disabling a gene in the transplanted nucleus that is needed for proper development of the embryo soon after the blastocyst stage. Such a cloned embryo could grow into a blastocyst from which embryonic stem cells could be harvested. The originator of this idea maintained that because of the disabled gene, the embryo could not be considered a human being: It could never develop into an adult because it was, in effect, engineered to die. Is this proposition reasonable, or does this idea place us on a "slippery slope" where the definition of what is human becomes too problematic?

Problems

Vocabulary

1. Match each of the terms in the left column to the best-fitting phrase from the right column.

a. mosaic determination	1. divide the body into identical units (segments)
b. regulative determination	2. initiated by the binding of ligand to receptor
c. haploinsufficiency	3. individuals with cells of more than one genotype
d. RNAi	4. the fate of early embryonic cells can be altered by the environment
e. ectopic expression	5. assign identity to body segments
f. homeodomain	6. substance whose concentration determines cell fates
g. green fluorescent protein	7. suppression of gene expression by double-stranded RNA
h. genetic mosaics	8. when a null allele is dominant to a wild-type allele
i. segmentation genes	9. a DNA-binding motif found in certain transcription factors
j. homeotic genes	10. encode proteins that accumulate in unfertilized eggs and are needed for embryo development
k. morphogen	11. early embryonic cells are assigned specific fates
l. maternal effect genes	12. a gene is turned on in an inappropriate tissue or at the wrong time
m. signal transduction pathways	13. a tag used to follow proteins in living cells

Section 20.1

2. a. If you were interested in the role of a particular gene in the embryonic development of the human heart, why would you probably study this role in a model organism, and which model organism(s) would you choose?
   b. If you were interested in finding new genes that might be required for human heart development, why would you try to find these genes in a model organism, and which model organism(s) would you choose?

3. Early *C. elegans* embryos display mosaic determination, while early mouse embryos exhibit regulative determination. Predict the results you would expect if

the following treatments were performed on four-cell embryos of each of these species (assuming these manipulations could actually be performed):

a. A laser is used to destroy one of the four cells (this technique is called *laser ablation*).
b. The four cells of the embryo are separated from each other and allowed to develop.
c. The cells from two different four-celled embryos are fused together to make an eight-celled embryo.

Section 20.2

Problems #4–7 concern a *Drosophila* gene called *rugose* (*rg*). Adult flies homozygous for recessive mutations in this gene have rough eyes in which the regular pattern of the eye segments called *ommatidia* are disrupted. The scanning electron micrographs below contrast the smooth eyes of wild type flies on the *left* with the rough eyes of *rugose* mutants on the *right.* The disruption of the eye segment pattern is caused by the absence of one or more so-called *cone cells* from ommatidia; in the wild type, each ommatidium has four cone cells.

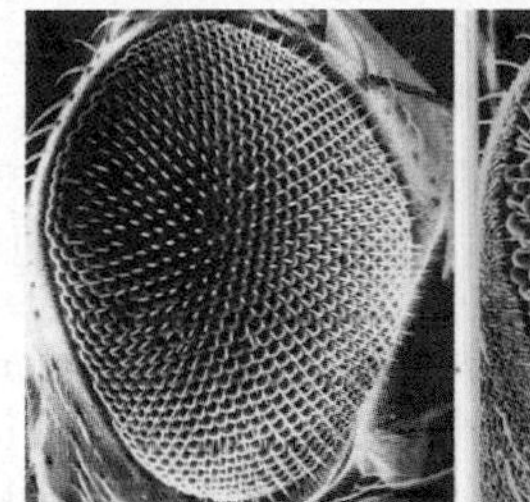
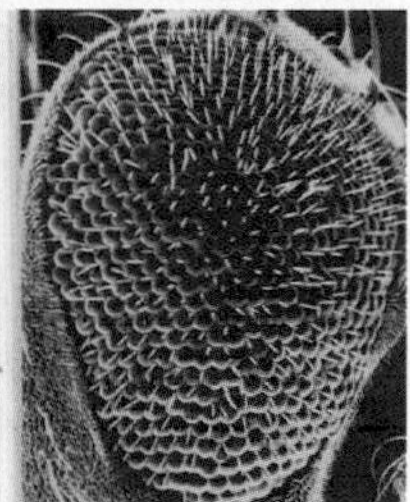

4. In 1932, H. J. Muller suggested a genetic test to determine whether a particular mutation whose phenotypic effects are recessive to wild type is a null (Muller called this an *amorphic*) or hypomorphic allele of a gene. Muller's test was to compare the phenotype of homozygotes for the recessive mutant alleles to the phenotype of a heterozygote in which one chromosome carries the recessive mutation in question and the homologous chromosome carries a deletion for a large region including the gene. In a recent study utilizing Muller's test, investigators examined two mutant alleles of *rugose* named rg^{41} and $rg^{\gamma 3}$. The eye phenotypes displayed by flies of several genotypes is indicated in the following table. *Df(1)JC70* is a large deletion that removes *rugose* and several genes to either side of it.

Genotype	Eye surface	Cone cells per ommatidium
wild type	smooth	4
rg^{41} / rg^{41}	mildly rough	2–3
rg^{41} / *Df(1)JC70*	moderately rough	1–2
$rg^{\gamma 3}$ / $rg^{\gamma 3}$	very rough	0–1
$rg^{\gamma 3}$ / *Df(1)JC70*	very rough	0–1

a. Which allele (rg^{41} or $rg^{\gamma 3}$) is "stronger" (that is, which causes the more severe phenotype)?
b. Which allele directs the production of higher levels of functional Rugose protein?
c. How would Muller's test discriminate between a null allele and a hypomorphic allele? Suggest a theoretical explanation for Muller's test. Based on the results shown in the table, is either of these two mutations likely to be a null allele of *rugose?* If so, which one?

5. The molecular identity of the fruit fly *rugose* gene is now known. cDNA clones corresponding to the *rugose* gene mRNA and antibodies that recognize the Rugose protein are also available. Outline several alternatives to the approach described in Problem #4 that might help you decide whether a newly discovered recessive allele of *rugose* is a null or a hypomorphic mutation.

6. In a *Drosophila* population of genotype $rg^{\gamma 3}/rg^{\gamma 3}$, it was noticed that about 35% of fertilized eggs develop into defective embryos that are unable to hatch into larvae. In contrast, only about 3% of fertilized wild-type eggs fail to develop into larvae.

a. In the light of this information as well as the data presented in Problem #4, predict which fly tissues at which stages of development require the function of the Rugose protein. (Note: The eyes of adult *Drosophila* are not pre-formed in embryos or larvae; instead, they develop from sacs of tissue in larvae called *imaginal discs.*)
b. How could you determine whether the *rugose* gene was expressed in the tissues you predicted in part *a*? Does the expression of the gene in those tissues establish that the Rugose protein plays an essential function there?

7. The *rugose* gene (*rg*) is located about midway between the centromere and telomere of the acrocentric *Drosophila* X chromosome. The *white* gene for eye color is located near the X chromosome telomere; the dominant w^+ allele specifies red color in eye cells, while w^- causes eye cells to be white. Mitotic recombination like that shown in Fig. 5.24 on p. 152 can be induced by exposing *Drosophila* larvae to X-rays.

a. Scientists can use mitotic recombination to create adult flies with mosaic eyes in which some eye cells would be simultaneously homozygous for mutant alleles of *rugose* and *white,* while the other cells in the eye would be heterozygous for the mutant and wild type alleles of both genes. Diagram an arrangement of mutant and wild type alleles of these two genes that would create such mosaic eyes upon X-ray-induced mitotic recombination.

b. How could you use this system of mitotic recombination to determine whether the lack of the Rugose protein in one ommatidium might affect the proper development of an adjacent ommatidium?
c. Suppose for the sake of argument that all animals homozygous for a true null mutation of *rugose* would die as embryos. How could you use this system of mitotic recombination to determine the effect of a complete lack of the Rugose protein on development of the adult eye?

Problems #8–11 concern a recombinant DNA construct called *myo-2::GFP* that *C. elegans* developmental geneticists have transformed into worms. Worms containing this construct express green fluorescent protein (GFP) in their pharynx, as shown in the following picture. The pharynx is an organ located between the mouth and the gut that grinds up the bacteria *C. elegans* eats so that these bacteria can be used as a food source. The *myo-2::GFP* construct was made by cloning the open reading frame for jellyfish GFP downstream of the promoter for *myo-2,* a gene that is specifically expressed in the muscle cells of the pharynx.

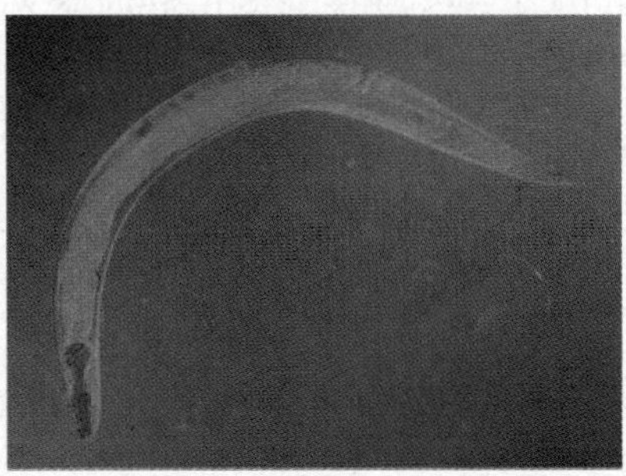

8. a. Explain how you could use worms transformed with *myo-2::GFP* to find mutations that disrupt the structure of the pharynx.
b. Nematodes homozygous for loss-of-function mutations in a gene called *pha-4* have no detectable pharyngeal structures. What do you think will be the fate of these worms?
c. How could you use *myo-2::GFP* to determine if *pha-4* is a master regulatory gene that directs development of the pharynx in a manner similar to the way *Pax-6/eyeless* controls eye development?

9. How could you use the pictured *myo-2::GFP* construct to find out what DNA sequence elements in the *myo-2* gene promoter are required for the pharynx-specific expression of the *myo-2* gene?

10. Suppose you wanted to determine whether a particular gene *X* was important for specification of the pharynx, but mutations in this same gene disrupt embryonic development well before pharyngeal structures appear. How could you use *myo-2::GFP*, the *myo-2* promoter, the DNA sequence of gene *X,* and your knowledge of RNA interference (RNAi) to generate worms that lack gene *X* expression in the pharynx but express gene *X* in all other tissues in which it is expressed in wild-type *C. elegans*?

11. The procedure normally used to transform *C. elegans* involves injection of DNA into the gonads of hermaphrodites. The DNA is incorporated into oocytes, but the injected DNA molecules usually recombine with each other, forming extrachromosomal arrays. These extrachromosomal arrays can be lost during mitosis at a low frequency, producing cells that lack the arrays. How could you use *myo-2::GFP* to create nematodes with mosaic pharynxes, such that some cells are homozygous for null mutations of gene *X* while other cells in the same pharynx have gene *X* activity? (Assume that null mutations and the genomic DNA of gene *X* are both available.)

12. Figure 20.5 on p. 722 shows how scientists can knock out any gene in mice using homologous recombination. An alternative and technically much simpler methodology to manipulate the mouse genome is an "add-on" strategy in which DNA is injected into a pronucleus of a fertilized egg, and the injected one-cell embryo is placed into an oviduct of a receptive female. In this add-on strategy, the injected DNA will integrate into various locations in the genome at random. For each of the following situations, indicate whether it would be preferable to use a knockout or add-on strategy, and explain both your decision and how you would employ the technology of your choice.
a. You want to create a mouse model of a human genetic disease in which a particular missense mutation has a recessive deleterious effect on development.
b. You want to create a mouse model of a human genetic disease in which a particular missense mutation has a dominant deleterious effect on development.
c. You want to explore the potential effects of the ectopic expression of a gene in a tissue in which it is normally not expressed.
d. You want to explore the potential deleterious haploinsufficient effects of the deletion of a gene.
e. You want to explore the potential deleterious effects of homozygosity for the deletion of a particular gene.
f. You want to explore the potential effects of the absence of gene function associated with the expression of a dominant-negative allele of a gene.
g. You want to suppress the function of a particular gene by RNA interference.
h. You want to find *cis*-acting regulatory sequences that cause a certain gene to be expressed only in particular tissues.

i. You want to prove that a polymorphism you have detected in the DNA of a particular candidate gene is responsible for a specific phenotype of abnormal development seen in mutant animals. (Assume that the mutation actually causing the phenotype is associated with a loss of function, but consider mutations that are recessive or dominant to wild type separately.)

13. As explained in Problem #12 above, when the "add-on" strategy is used to create transgenic mice, the injected DNA can insert at random into any chromosome. Subsequent matings produce animals homozygous for the transgene insertion, and sometimes an interesting developmental phenotype is generated by the insertion event itself. In one case, after injection of DNA containing the mouse mammary tumor virus (MMTV) promoter fused to the *c-myc* gene, investigators identified a recessive mutation that causes limb deformity. In this mouse, the distal bones were reduced and fused together; the mutation also caused kidney malfunction.

a. The mutant phenotype could be due to insertion of the MMTV/*c-myc* transgene in a particular region of the chromosome or a chance point mutation that arose in the mouse. How could you distinguish between these two possibilities?

b. The mutation in this example was in fact caused by insertion of the transgene. How could you use this transgene insertion as a tag for cloning?

c. The insertion mutation was mapped to chromosome 2 of mice in a region where a mutation called limb deformity (*ld*) had previously been identified. Mice carrying this mutation are available from a major mouse research laboratory. How could you tell if the *ld* mutation was in the same gene as the transgenic insertion mutation?

Section 20.3

14. Which of the following is not a property of the *hunchback* gene in *Drosophila*?

a. The *hunchback* mRNA is uniformly distributed in the egg by the mother.

b. Transcription of *hunchback* is enhanced by Bicoid (the anterior morphogen).

c. Translation of the *hunchback* mRNA is inhibited by Nanos (the posterior morphogen).

d. The Hunchback protein eventually is distributed in a gradient (anterior high; posterior low).

e. Hunchback protein directs the distribution of *bicoid* mRNA.

15. The *hunchback* gene contains a promoter region, the structural region (the amino acid coding sequence), and a 3′ untranslated region (DNA that will be transcribed into sequences appearing at the 3′ end of the mRNA that are not translated into amino acids).

a. What important sequences required to control *hunchback* gene expression are found in the promoter region of *hunchback?*

b. What sequence elements that encode specific protein domains are found in the structural region of *hunchback?*

c. There is another important kind of sequence that turns out to be located in the part of the gene transcribed as the 3′ UTR (untranslated region) of the *hunchback* mRNA. What might this sequence do?

16. How do the segment polarity genes differ in their mode of action from the gap and pair-rule genes?

17. One important demonstration that Bicoid is an anterior determinant came from injection experiments analogous to those done by early embryologists. Injection experiments involve introduction of components such as cytoplasm from an egg or mRNA that is synthesized *in vitro* into the egg by direct injection. Describe injection experiments that would demonstrate that Bicoid is the anterior determinant.

18. In flies developing from eggs laid by a *nanos*⁻ mother, development of the abdomen is inhibited. Flies developing from eggs that have no maternally supplied *hunchback* mRNA are normal. Flies developing from eggs laid by a *nanos*⁻ mother that also have no maternally supplied *hunchback* mRNA are normal. If there is too much Hunchback protein in the posterior of the egg, abdominal development is prevented.

a. What do these findings say about the function of the Nanos protein and of the *hunchback* maternally supplied mRNA?

b. What do these findings say about the efficiency of evolution?

19. Mutant embryos lacking the gap gene *knirps* (*kni*) are stained at the syncytial blastoderm stage to examine the distributions of the Hunchback and Krüppel proteins. The results of the *knirps*⁻ and wild-type embryos stained for the Hunchback and Krüppel proteins is shown schematically here.

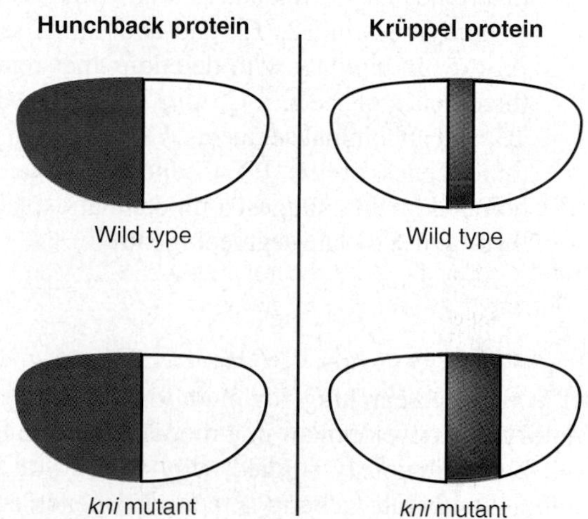

a. Based on these results, what can you conclude about the relationships among these three genes?
b. Would the pattern of Hunchback protein in embryos from a *nanos*⁻ mutant mother differ from that shown? If yes, describe the difference and explain why. If not, explain why not.

20. In *Drosophila* with loss-of-function mutations affecting the *Ubx* gene, transformations of body segments are always in the anterior direction. That is, in *bx* mutants, the anterior compartment of T3 is transformed into the anterior compartment of T2, while in *pbx* mutants, the posterior part of T3 is transformed into the posterior compartment of T2. In wild type, the *Ubx* gene itself is expressed in T3-A8, but most strongly in T3.
a. The *Abd-B* gene is transcribed in segments A5-A8. Assuming the mode of function of *Abd-B* is the same as that of *Ubx*, what is the likely consequence of homozygosity for a null allele of *Abd-B* (that is, what segment transformations would you expect to see)?
b. Since *Abd-A* is expressed in segments A1-A8, there is some transcription of all three genes of the BX-C (*Ubx, Abd-A*, and *Abd-B*) in segments A5-A8. Why then are segments A5, A6, A7, and A8 morphologically distinguishable?
c. What segment transformations would you expect to see in an animal deleted for all three genes of the BX-C (*Ubx, Abd-A*, and *Abd-B*)?
d. Certain *contrabithorax* mutations in the BX-C cause transformations of wing to haltere. Propose an explanation for this phenotype based on the transcription of the *Ubx* gene in particular segments. Do you anticipate that *contrabithorax* mutations would be dominant or recessive to wild type? Explain.
e. During wild-type development, *Antp* is expressed in T1, T2, and T3, but most strongly in T2 and only weakly in T3. In animals with *Ubx* null mutations, *Antp* is expressed at much higher levels in T3 as compared with wild type. In animals with deletions that remove both *Ubx* and *Abd-A*, *Antp* is expressed at high levels in T2, T3, and abdominal segments A1-A5. In animals with deletions that remove all three genes of the BX-C, *Antp* is expressed in T2, T3, and abdominal segments A1-A8. Given that the three genes of the BX-C encode proteins with homeodomains, suggest a model that explains how these genes dictate segment identity.

Section 20.4

21. If you were searching for mutations that affect early embryonic development in a model organism that had not been previously studied, why would you need to conduct separate genetic screens for genes encoding maternally supplied components and for genes whose transcription begins only after fertilization? What kinds of screens would you employ in both cases?

22. The unfertilized eggs of *C. elegans* have no predetermined anterior or posterior end. The polarity of the embryo instead depends on the site of sperm entry, which becomes the posterior end. Very soon after fertilization, so-called PAR (for "partitioning") proteins, which are uniformly distributed in unfertilized eggs, become localized to the embryonic cortex (the layer of cytoplasm just under the cell membrane) at one or the other end of the embryo. The following figure shows the distribution of two of these proteins, PAR-2 (which becomes localized to the posterior cortex) and PAR-3 (which goes to the anterior cortex). Once the redistribution of these proteins has been achieved, the zygote divides so as to produce a two-cell embryo with an anterior cell and a posterior cell.

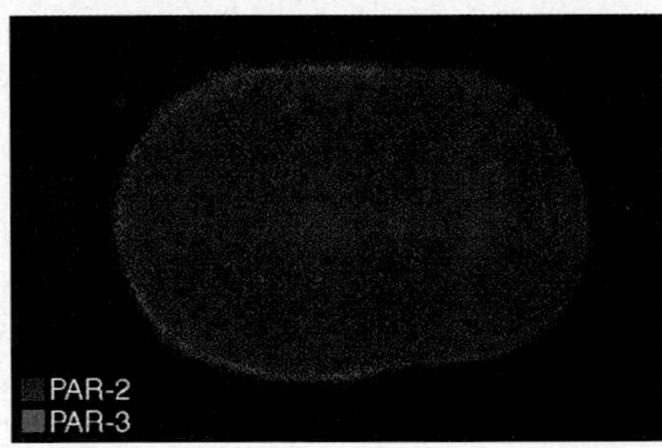

a. How do these findings help explain why the early development of *C. elegans* embryos display mosaic, rather than regulative, determination?
b. Mutations in the *par-2* or *par-3* genes cause the arrest of development in early embryonic stages. Would you be more likely to find mutations in these genes in screens looking for maternal-effect genes or in screens for zygotic genes?
c. In zygotes produced by hermaphrodite mothers homozygous for loss-of-function *par-2* alleles, the PAR-3 protein is distributed uniformly around the cortex; the same is true of the PAR-2 protein in zygotes made by *par-3* mutant hermaphrodites. What does this information say about the establishment of early polarity in *C. elegans*?

23. At the end of two rounds of mitosis, the *C. elegans* embryo has four cells named ABa, ABp, P_2, and EMS (see the following figure). The ABa and ABp cells are originally developmentally equivalent, but they become different as a result of interactions between ABp and P_2 that involve two proteins called GLP-1 and APX-1, as shown in the figure. Both GLP-1 and APX-1 are membrane bound, with domains that lie outside of the cell. GLP-1 is expressed around the entire surface of ABa and ABp, while APX-1 is found at the membrane junction between ABp and P_2 as shown in the figure.

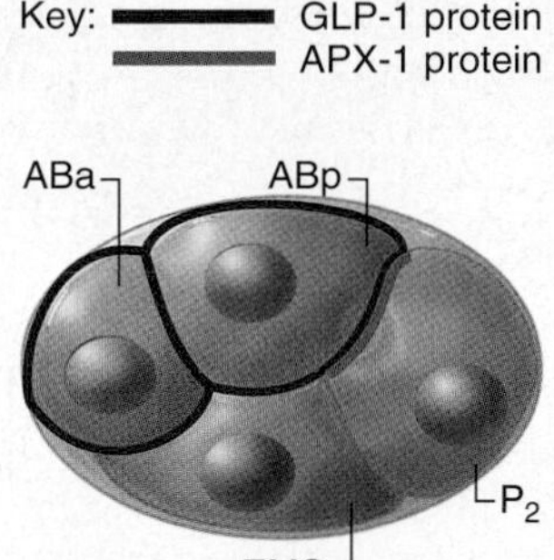

a. The mRNA for *glp-1* is found in all four cells, but the GLP-1 protein is found only in ABa and ABp. In light of your answer to Problem #22, provide an explanation for this observation.

b. Based on the information in the figure, suggest an hypothesis to explain the localization of APX-1 in only one region of the membrane of the P_2 cell.

c. Assuming that the effect of these proteins on the fate of ABp is caused by a signal transduction pathway, which of the two proteins GLP-1 and APX-1 is likely to be a ligand and which a receptor for this ligand?

d. Describe the effects on the fate of the ABp cell of the following: (i) Laser ablation of the P_2 cell; (ii) a null mutation in the *apx-1* gene; (iii) a null mutation in the *glp-1* gene; (iv) a null mutation in a gene encoding a component of the signal transduction pathway initiated by binding of the ligand to its receptor. (Assume here that the mutations in part ii-iv only affect the fate of the ABp cell and not other processes in the early nematode embryo.)

The Genetic Analysis of Populations and How They Evolve

Chapter 21

The enormous range of genetic diversity within our own species is easy to see.

Tuberculosis (TB) is an ancient and persistent human disease. Bone deformities typical of those produced by the infection are found in Egyptian mummies dating to 2000–4000 B.C.; and as recently as the mid–nineteenth century, TB was the leading cause of death in Europe and the urban United States. The microbe that causes TB is the bacterium *Mycobacterium tuberculosis.* In humans, populations of *M. tuberculosis* most often infect the lungs and lymph nodes, but sometimes they colonize the bones and skin of a patient (**Fig. 21.1a**). Severe infection by the tuberculosis bacterium damages the lungs, resulting in weakness, a persistent cough, bloody sputum, and ultimately death. *M. tuberculosis* bacteria can spread from person to person through the air when an infected individual exhales bacteria from his or her lungs during coughing.

Beginning in the late nineteenth century, improved sanitation and the quarantine of TB patients in sanitaria led to a steady decline in the death rate from TB in Europe and the United States (Fig. 21.1b). The introduction of antibiotics during the 1940s and 1950s further reduced TB mortality in those areas, and in the 1960s, when the death rate from TB had decreased to less than 2 deaths per 100,000, many people believed that the disease had been eradicated as a threat to public health, at least in the United States, where the drugs were available. But 25 years later, the incidence of TB began to rise in urban areas around the globe. By 1994, in the Harlem area of New York City, there were 700 cases per 100,000 people. By 2000, TB accounted for more deaths worldwide than any other identifiable infectious disease, claiming close to 3 million lives annually; and epidemiologists estimated that *M. tuberculosis* had infected an astonishing 2 billion people, roughly one-third of the global population. Three factors contributed to this rapid increase in TB incidence: the emergence of AIDS, which among other effects, weakens the human immune system; protein deficiencies among the malnourished in many areas of the globe, which also weaken the immune system; and the widespread occurrence of *M. tuberculosis* strains that are resistant to one or more antibiotics.

Despite the cultural, technological, and medical advances of the twentieth century, infectious microorganisms such as *M. tuberculosis* are still among the leading causes of death in many human populations. And in a related arena, populations of plant pests (such as the mites that prey on strawberry plants and almond trees) destroy a substantial fraction of human food supplies.

How do new diseases emerge in human populations? Why do diseases persist in all living organisms? What causes diseases and pests long under control to resurge in frequency and intensity?

(a)

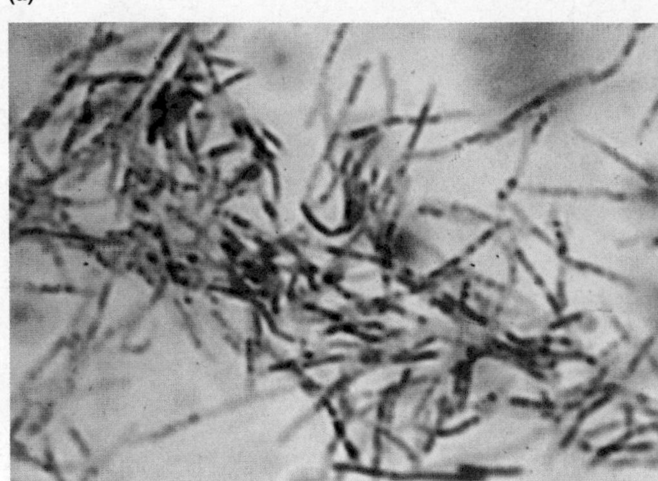

(b)

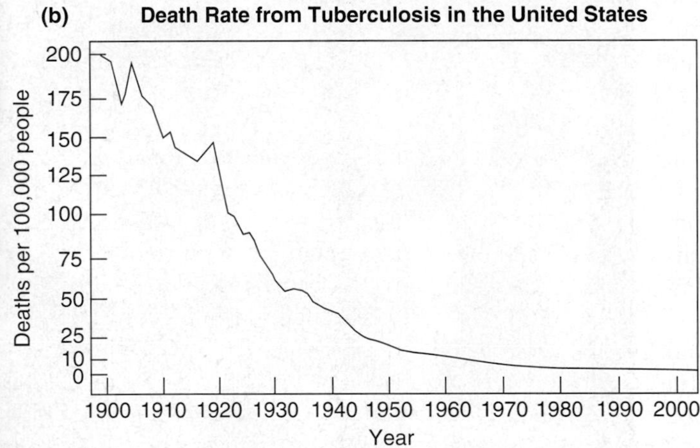

Figure 21.1 **Tuberculosis in human populations. (a)** Photograph of *M. tuberculosis* bacteria colonizing the lungs and bones. **(b)** Death rate from tuberculosis in the United States from 1900–2000.

One way to answer these questions is to examine genetic variation and its expression as phenotypes within populations of organisms. The scientific discipline that studies what happens in whole populations at the genetic level is known as **population genetics.** It encompasses the evolutionary ideas of Darwin, the laws of Mendel, and the insights of molecular biology. To population geneticists, a **population** is a group of interbreeding individuals of the same species that inhabit the same space at the same time. An example would be all the white-tailed deer on Angel Island in San Francisco Bay in 1990 or all the rock cod at the mouth of the bay. The sum total of all alleles carried in all members of a population is that population's **gene pool.** In nature, the genetic makeup of a population changes over time as new alleles arise by mutation or are introduced by immigration and as rare, preexisting alleles disappear when all individuals carrying them leave the population or die. Changes in the frequency of alleles within a population are the basis of **microevolution:** alterations of a population's gene pool.

In this chapter, we explore the nature of genotypic and phenotypic variation within populations and the role of this variation in evolution. We know from Chapter 2 that variation exists in all populations. To begin our examination of this variation, we analyze the incidence of genetic diseases, such as cystic fibrosis and retinoblastoma, that are determined by a single gene. Variations in the nature and number of wild-type and disease alleles determine the frequency of disease and how this frequency may change. Through our relatively simple analysis of single-gene disorders, we develop a framework for understanding how the frequency of a disease-causing allele determines the frequency of diseased individuals in a population.

We next examine how genetic variation within populations of pathogens and pests arises and how human efforts to control these destructive organisms force changes in allele frequencies that determine the rate at which the pathogenic and pest populations evolve.

Finally, we consider variation in multifactorial traits, that is, in traits determined by two or more genes and their interaction with the environment (see Chapter 3). In fact, most aspects of disease susceptibility are multifactorial. To analyze the inheritance of multifactorial traits, population geneticists use quantitative tools that enable them to assess the relative contributions of genetic and environmental factors to individual differences in phenotype.

One general theme emerges from our discussion: Population geneticists rely on mathematical models in predicting a population's potential for stasis or change because most of the scientifically useful questions they ask are statistical. These questions include, What is the frequency of genetic disease in a population? What fraction of the phenotypic variation in a trait is the result of genetic variation? Given certain quantifiable variables, how rapidly can a disease gain a foothold in a population, and how long is it likely to persist? Simple mathematical models not only clarify the questions, they also serve as tools for analyzing data obtained from the field and for making predictions about future populations.

As we present ways to examine the gene pools of stable and evolving populations over time, we describe

- The Hardy-Weinberg law: a model for understanding allele, genotype, and phenotype frequencies for single-gene traits in a genetically stable population.
- Calculations beyond Hardy-Weinberg: measuring how selection and mutations cause changes in the allele frequencies of a gene pool over time, including a comprehensive example of the evolution of drug and pesticide resistance.
- How to use the tools of quantitative genetics to analyze the inheritance of multifactorial traits.

21.1 The Hardy-Weinberg Law: A Model for Understanding Allele, Genotype, and Phenotype Frequencies for a Single-Gene Trait in a Genetically Stable Population

Suppose you were to look at a human population of 16 in which 6 people suffer from the recessive disease of cystic fibrosis caused by the *r* variant of the wild-type *R* allele. To predict how the number of diseased individuals in the population will change over time, you need to determine the frequencies of each genotype (homozygous *RR,* heterozygous *Rr,* and homozygous *rr*), each phenotype (normal and diseased), and each allele (*R* and *r*). Population geneticists define **phenotype frequency** as the proportion of individuals in a population that express a particular phenotype. For our hypothetical population, the phenotype frequencies are $6/16 = 3/8$ diseased (the number of homozygous *rr* individuals affected by the recessive condition) and $10/16 = 5/8$ normal (the remaining fraction with either *RR* or *Rr* genotypes).

Genotype frequency is the proportion of total individuals in a population that carry a particular genotype. To determine genotype frequencies, you simply count the number of individuals of each genotype and divide by the total number of individuals in the population. For recessive traits such as cystic fibrosis, it is not possible to distinguish between wild-type and heterozygous genotypes: Both give rise to nondiseased individuals. Thus, the only way to determine genotype frequencies directly at loci like the cystic fibrosis locus is to use molecular probes that distinguish between different alleles. For our hypothetical population, molecular analyses showed that 8 individuals (of 16) are of genotype *RR*, 2 are of genotype *Rr*, and 6 are *rr*. This means that the *RR* genotype frequency is $8/16 = 4/8 = 1/2$; the *Rr* genotype frequency is $2/16 = 1/8$; and the *rr* genotype frequency is $6/16 = 3/8$. Note that these three frequencies $(4/8 + 1/8 + 3/8)$ sum to 1, the totality of genotypes in the whole population.

Finally, the definition of **allele frequency** is the proportion of all copies of a gene in a population that are of a given allele type. Because each individual in a population has two copies of each gene, the total number of gene copies is two times the number of individuals in the population. Thus, for our hypothetical population of 16 people, there would be 32 copies of the cystic fibrosis gene. Of course, both homozygotes and heterozygotes contribute to the frequency of an allele. But homozygotes contribute twice to the frequency of a particular allele, while heterozygotes contribute only once. To find the frequencies of *R* and *r*, you first use the number of people with each genotype to compute the number of *R* and *r* alleles.

$$8\ RR \rightarrow 16 \text{ copies of } R$$
$$2\ Rr \rightarrow 2 \text{ copies of } R$$
$$6\ rr \rightarrow 0 \text{ copies of } R$$

Together, $16 + 2 + 0 = 18$ copies of the *R* allele.

$$8\ RR \rightarrow 0 \text{ copies of } r$$
$$2\ Rr \rightarrow 2 \text{ copies of } r$$
$$6\ rr \rightarrow 12 \text{ copies of } r$$

Together, $0 + 2 + 12 = 14$ copies of the *r* allele.

Next, you add the 18 *R* alleles to the 14 *r* alleles to find the total number of gene copies.

$18 + 14 = 32$ copies of the gene, which is twice the number of people in the population

Finally, you divide the number of each allele by the total number of gene copies to find the proportion, or frequency, for each allele.

For the *R* allele, it is $18/32 = 9/16 = 0.56$

For the *r* allele, it is $14/32 = 7/16 = 0.44$

Note that here again, the frequencies sum to a 1, representing all the alleles in the gene pool.

We have spent time defining and deriving the genotype, phenotype, and allele frequencies of a hypothetical population because population geneticists use these frequencies as the basis for calculating the genetic and phenotypic variations of a gene in a population and how they may change over time.

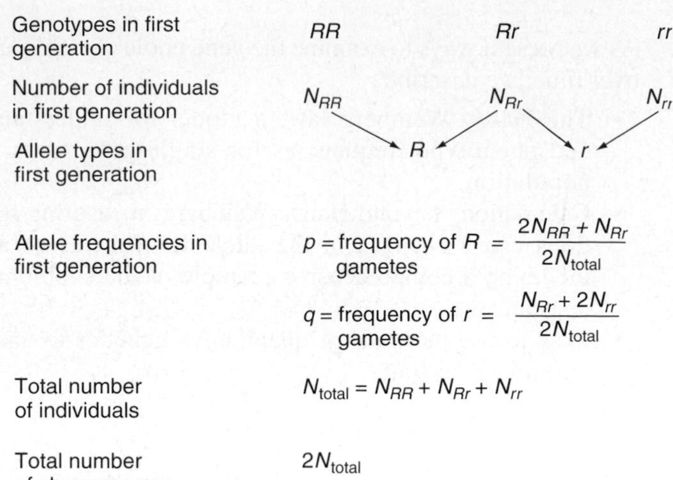

Figure 21.2 Computing allele frequencies from genotype frequencies.

The Hardy-Weinberg Law Correlates Allele and Genotype Frequencies Through a Binomial Equation

Many inherited diseases result from disease-causing alleles that are recessive to wild-type alleles. For scientists seeking to predict the potential incidence of such recessive conditions, an important question is, How common are the heterozygous carriers of the disease-causing allele in a population? At the start, the scientists know only the phenotypic frequencies of healthy and diseased individuals. Can they use this information to predict the frequency of heterozygous carriers?

The key to answering this question lies in establishing a quantitative relationship between phenotype, genotype, and allele frequencies within and between generations. The **Hardy-Weinberg law,** named for the two men—G.H. Hardy and W. Weinberg—who independently developed it in 1908, clarifies the relationships between genotype and allele frequencies within a generation and from one generation to the next. To derive a general law describing the inheritance of alleles in a population, it is necessary to make certain simplifying assumptions about the nature of the population, the individuals within it, and the genes these individuals carry. Derivation of the Hardy-Weinberg law depends on five such assumptions:

1. The population includes a very large, virtually infinite number of individuals.
2. The individuals mate at random in the sense that each individual's genotype at the locus of interest does not influence his or her choice of a mate.
3. No new mutations appear in the gene pool.
4. There is no migration into or out of the population.
5. There are no genotype-dependent differences in the ability to survive to reproductive age and transmit genes to the next generation.

Populations that satisfy all five of these assumptions are said to be at Hardy-Weinberg equilibrium.

The assumptions behind the Hardy-Weinberg equilibrium enable the mathematical derivation of an equation for calculating genotype (and thence phenotype) frequencies, even though no actual population is at Hardy-Weinberg equilibrium. All populations are finite; mating is not always at random; mutations occur constantly; migration into and out of a population is common; and many genotypes of interest, such as those that cause lethal childhood diseases, affect the ability to survive and reproduce. Nevertheless, even when many of the assumptions of the Hardy-Weinberg equilibrium do not apply, the equation derived on the basis of these assumptions is remarkably robust at providing estimates of genotype and phenotype frequencies in real populations. Indeed, the equation has always been the most powerful mathematical tool available to population geneticists.

For a population of sexually reproducing diploid organisms, there are two steps in translating the genotype frequencies of one generation into the genotype frequencies of the next generation (**Fig. 21.2**).

First, if the likelihood that an individual will grow into an adult does not depend on the genotype (that is, if there is no difference in fitness among individuals), then the allele frequencies in the adults should be the same as in their gametes. For example, if p is the frequency of allele R, and q is the frequency of allele r in the adults, p and q will also be the frequencies of the two alleles in the gametes produced by the whole population of those adults.

Next, you can use the allele frequencies in the gametes to calculate genotype frequencies in the zygotes of the next generation. An enhanced version of the Punnett square, which provides a systematic means of considering all possible combinations of uniting gametes, is the tool of choice (**Fig. 21.3**). For example, if the probability that an egg and a sperm unite does not depend on the genotype of the egg or sperm (that is, if fertilization is random among individuals with any genotype) and if the population of gametes is very large, then the following pattern emerges. Recall that *RR* zygotes result from fertilization of *R*-carrying eggs by *R*-carrying sperm. If p is the frequency of R gametes (eggs and sperm), then, applying the product rule, the frequency of *RR* zygotes is p (frequency of R eggs) $\times$ p (frequency of R sperm) $= p^2$. Similarly, *rr* zygotes result from fertilization of *r*-carrying eggs by *r*-carrying sperm. If q is the frequency of r gametes (eggs and sperm), the frequency

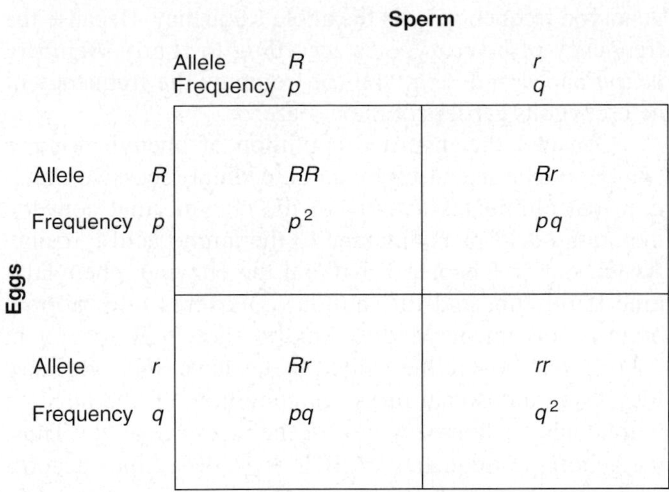

Figure 21.3 Derivation of the Hardy-Weinberg equation. After mating, the genotype frequencies of the next generation will be p^2 for *RR*, $2pq$ for *Rr*, q^2 for *rr*.

of *rr* zygotes will be q (the frequency of *r* eggs) $\times$ q (the frequency of *r* sperm) $= q^2$. Finally, *Rr* zygotes result either from fertilization of *R* eggs by *r* sperm, with a frequency of $p \times q = pq$, or from the fertilization of *r* eggs by *R* sperm, occurring at a frequency of $q \times p = pq$. The total frequency of *Rr* zygotes is thus $pq + pq = 2pq$.

To summarize: The genotype frequencies of zygotes arising in a large population of sexually reproducing diploid organisms are p^2 for *RR*, $2pq$ for *Rr*, and q^2 for *rr* (see Fig. 21.3). These genotype frequencies are known as the Hardy-Weinberg proportions; they exist in populations that satisfy the Hardy-Weinberg assumptions of a large number of individuals, mating at random, with no new mutations, no migration, and no genotype-dependent differences in fitness. Since these genotype frequencies represent the totality of genotypes in the population, they must sum to 1. Thus the binomial equation representing the Hardy-Weinberg proportions is

$$p^2 + 2pq + q^2 = 1 \qquad (21.1)$$

Because we have assumed no differences in fitness, the genotype frequencies of the zygotes will be the genotype frequencies of the adult generation that develops from those zygotes.

This equation thus enables us to use information on genotype and allele frequencies to predict the genotype frequencies of the next generation. Suppose, for example, that in a population of 100,000 people carrying the recessive allele *a* for albinism, there are 100 *aa* albinos and 1800 *Aa* heterozygous carriers. To find what the frequency of heterozygous carriers will be in the next generation, you compute the allele frequencies in the parent population.

98,100 *AA* individuals; 1800 *Aa* individuals, and 100 *aa* individuals $\rightarrow$ 196,200 *A* alleles + 1800 *A* alleles; 1800 *a* alleles + 200 *a* alleles

Out of 200,000 total alleles the frequency of the *A* allele is

$$198{,}000/200{,}000 = 99/100 = 0.99; \text{ thus } p = 0.99$$

and the frequency of the *a* allele is

$$2{,}000/200{,}000 = 1/100 = 0.01; \text{ thus, } q = 0.01$$

The Hardy-Weinberg equation for the albino gene in this population is

$$p^2 + 2pq + q^2 = 1$$
$$(0.99)^2 + 2(0.99 \times 0.01) + (0.01)^2 = 1$$
$$0.9801 + 0.0198 + 0.0001 = 1$$

It thus predicts that in the next generation of 100,000 individuals, there will be

$$100{,}000 \times 0.9801 = 98{,}010\ AA \text{ individuals}$$
$$100{,}000 \times 0.0198 = 1980\ Aa \text{ individuals}$$
$$100{,}000 \times 0.0001 = 10\ aa \text{ individuals}$$

This example shows that in one generation, the genotype frequencies have changed slightly. A natural question is, Have the allele frequencies also changed? Recall that the initial frequencies of the *R* and *r* alleles are p and q, respectively, and that $p + q = 1$. You can use the rules for computing allele frequencies from genotype frequencies (see Figs. 21.2 and 21.3) to compute the allele frequencies of the next generation. From the Hardy-Weinberg equation, you know that p^2 of the individuals are *RR*, whose alleles are all *R*, and $2pq$ of the individuals are *Rr*, 1/2 of whose alleles are *R*. Similarly, q^2 of the individuals are *rr*, whose alleles are all *r*, and $2pq$ of the individuals are *Rr*, 1/2 of whose alleles are *r*. If $p + q = 1$, then $q = 1 - p$, and the frequency of the *R* allele in the next-generation population is

$$\begin{aligned} p^2 + 1/2[2p(1-p)] &= p^2 + p(1-p) \\ &= p^2 + p - p^2 = p \end{aligned} \qquad (21.2)$$

Similarly, $p = 1 - q$, and the frequency of the *r* allele in the next-generation population is

$$\begin{aligned} q^2 + 1/2[2q(1-q)] &= q^2 + q(1-q) \\ &= q^2 + q - q^2 = q \end{aligned} \qquad (21.3)$$

Using these equations to calculate the allele frequencies of *A* and *a* in the second generation of 100,000 individuals, some of whom are albinos, we find

for p $0.9801 + 0.99 - 0.9801 = 0.99$ = the frequency of the *A* allele

for q $0.0001 + 0.01 - 0.0001 = 0.01$ = the frequency of the *a* allele

These frequencies are the same as those in the previous generation. Thus, even though the genotype frequencies have changed slightly from the first generation to the next, the allele frequencies have not. Note that this is true of both the dominant and the recessive alleles.

Consider, for example, the dominant human condition of *brachydactyly,* which causes people to have short, stubby fingers. If a group of *BB* persons homozygous for the allele causing short, stubby digits were all to conceive children with *bb* mates homozygous for the allele determining normal finger length, all the progeny will be *Bb* heterozygotes showing the brachydactyly phenotype. But even though the phenotype and genotype frequencies have changed drastically, the allele frequencies have not. In both generations, one-half of the alleles are *B*, and one-half are *b*. If no new alleles appear in the population by mutation or immigration and the heterozygotes of the second generation mate with each other at random (with respect to the presence or absence of brachydactyly) the third generation produced by $Bb \times Bb$ matings will contain the 3:1 ratio of brachydactyly to normal phenotypes, determined by the Mendelian genotype ratios of 1*BB* : 2*Bb* : 1*bb*. Within this third generation, the allele frequency is 4/8*B* and 4/8*b*, the same as in the first generation in which all *BB* individuals mated with all *bb* individuals only.

The Hardy-Weinberg law generates two significant conclusions.

Allele Frequencies Do Not Change from Generation to Generation in a Population at Hardy-Weinberg Equilibrium

This is one of the most important insights of the Hardy-Weinberg law: As long as a population is free of all influences that introduce or remove alleles, that is, as long as it is at Hardy-Weinberg equilibrium, its allele frequencies remain unchanged. The Hardy-Weinberg law thus provides a model for looking at genotype, allele, and phenotype frequencies in a genetically stable population.

A Hardy-Weinberg Population Achieves the Genotype Frequencies of p^2, $2pq$, and q^2 in Just One Generation and Maintains These Frequencies in Subsequent Generations

This second significant implication of the Hardy-Weinberg law, which applies regardless of the genotype frequencies in the initial generation, enables us to answer the question posed at the beginning of this discussion: If you know only the phenotypic frequencies of healthy and diseased individuals in a population, can you use this information to predict the frequency of heterozygous carriers? For an autosomal recessive condition, all affected individuals are homozygotes for the recessive disease-causing allele. If you assume that the genotype and allele frequencies for the condition follow the Hardy-Weinberg law and that the frequencies of the normal and disease-causing alleles are p and q, respectively, then the genotype frequency of affected homozygotes is simply q^2, and the square root of this genotype frequency, q, is the allele frequency. Because the frequency of heterozygotes according to Hardy-Weinberg is $2pq$ and $p = 1 - q$, you can compute the frequency of heterozygous carriers of the disease.

Consider the inherited condition of phenylketonuria (PKU), which is caused by an autosomal recessive mutation that eliminates activity of the enzyme that converts the amino acid phenylalanine to the amino acid tyrosine. As we saw in Chapter 3, without the enzyme, phenylalanine builds up, and this buildup interferes with normal brain development. Medical studies show that about 1 in 3600 Caucasians in the United States have PKU. Because PKU is an autosomal recessive condition, all the affected individuals are homozygotes for the recessive allele. Thus, the genotypic frequency of PKU is 1/3600. Since according to the Hardy-Weinberg equation, the genotypic frequency of homozygotes for the recessive allele is q^2, you can say that $q^2 = 1/3600$ and then compute the frequency of the recessive PKU allele in the whole population by finding the square root of this number.

$$q = \sqrt{q^2} = \sqrt{1/3600} = 1/60 = 0.0167$$

Next, because $p = 1 - q$, you can compute p.

$$p = 1 - 1/60 = 59/60 = 0.9833$$

Knowing the frequencies of both p and q, we can compute the frequency of heterozygotes according to Hardy-Weinberg.

$$2pq = 2 \times (0.0167) \times (0.9833) = 0.0328$$

Thus, if one assumes that the U.S. Caucasian population can be approximated by the conditions that define a Hardy-Weinberg equilibrium, it becomes possible to determine simply from observations of phenotypes that within this population, the frequency of heterozygous carriers is 0.0328, or roughly 1 in 30 people.

21.2 Beyond Hardy-Weinberg: Measuring How Mutation and Selection Cause Changes in Allele Frequencies

One definition of **evolution** is changes in allele frequency in a population over multiple generations. Evolutionary scientists refer to this level of evolution as *microevolution* in comparison with **macroevolution,** which occurs over geologic time. Macroevolution refers to the process by which new species emerge from existing species, while microevolution refers to genetic changes that occur within a species. Many evolutionary biologists believe that much macroevolution is simply a consequence of extended periods of microevolution.

Population geneticists can use the Hardy-Weinberg law to look at microevolution. Because the law models populations in which genetic changes do not occur, it provides a theoretical standard against which to measure changes that do occur. In natural populations, conditions almost always deviate at least slightly from the Hardy-Weinberg assumptions: Some populations are not very large, individuals do not always mate at random, new mutations do arise, there is migration into and out of the population, and different genotypes do generate differences in survival rates. We now examine how simple modifications of the Hardy-Weinberg law make it possible to analyze the effects of these deviations, particularly those resulting from genotype-dependent differences in fitness, and new mutations.

Figure 21.4 Giraffes on the savanna. A visible example of one outcome of natural selection for increased fitness.

Natural Selection Acts on Differences in Fitness to Alter Allele Frequencies

Contrary to the Hardy-Weinberg assumption that the probability of survival and reproduction does not vary with the genotype, for many phenotypic traits, including inherited diseases, genotype does influence survival and the ability to reproduce. Thus, in real populations, as opposed to model populations constructed according to Hardy-Weinberg assumptions, not all individuals survive to adulthood, and there is always some probability that an individual will not live long enough to reproduce. As a result, the genotype frequencies of real populations change as their individual members mature from zygotes to adults.

To population geneticists, an individual's relative ability to survive and transmit its genes to the next generation is its **fitness.** But although fitness is an attribute associated with each genotype, it cannot be measured within the individuals of a population because each animal with a particular genotype survives and reproduces in a manner that is greatly affected by chance circumstances. However, by considering all the individuals of a particular genotype together as a group, it becomes possible to measure the relative fitness for that genotype. Thus, for population geneticists, fitness is a statistical measurement only. Nevertheless, differences in fitness can have a profound effect on the allele frequencies of a population.

Fitness has two basic components: viability and reproductive success. The fitness of individuals possessing variations that help them survive and reproduce in a changing environment is relatively high; the fitness of individuals without those adaptive variations is relatively low. In nature, the process that progressively eliminates individuals whose fitness is lower and chooses individuals of higher fitness to survive and become the parents of the next generation is known as **natural selection.** The mechanisms of selection act independently of any individual; often, interactions between genetically determined phenotypes and environmental conditions are the agents of natural selection.

For example, in a hypothetical population of giraffes browsing on the leaves of an ancient savanna, suppose that some of the animals had long necks and some short necks, and that each of these phenotypes resulted from variations in the genes contributing to neck length. If during a long drought, there were fewer low-hanging leaves than usual on the flora of the savanna, those long-necked individuals able to reach the higher morsels would have been able to harvest more food and survive to produce more offspring than individuals with shorter necks. Natural conditions would thus have selected the better adapted, long-necked giraffes—those with the higher fitness—to survive and become the parents of the next generation (**Fig. 21.4**). Similarly, in the laboratory, scientists can establish experimental conditions, such as the absence of a nutrient or the presence of an antibiotic, that become the agents of **artificial selection.** In both natural and artificial selection, environmental and other external conditions interact with preexisting variations to "select" the parents of the next generation. The more extreme a phenotype becomes, the more likely it will have negative effects on viability, which counteracts any advantage it might confer. As a result, giraffe necks, for example, do not just keep getting longer.

Field studies show that natural selection occurs for many phenotypic traits in a variety of natural populations. Since in human populations, genetic diseases frequently affect a person's survival and reproduction, the potential for natural selection in people affected by such diseases is considerable. How does selection alter our conclusions from Hardy-Weinberg?

We begin to answer this question with an analysis of the *R* gene in a population of zygotes in Hardy-Weinberg equilibrium. In this population, the genotype frequencies *RR*, *Rr*, and *rr* are p^2, $2pq$, and q^2, respectively. Now suppose that the viability, that is, the probability of surviving from zygote to adult, depends on genotype, while the second component of fitness—success at productive mating—is

Parental Gametes

Allele	R	r
Frequency	p	q

First Generation Offspring

Genotypes	RR	Rr	rr
Zygote frequency	p^2	$2pq$	q^2
Relative fitness	W_{RR}	W_{Rr}	W_{rr}
Frequency after selection	p^2W_{RR}	$2pqW_{Rr}$	q^2W_{rr}
Calibrated adult frequency	$p^2W_{RR}/\overline{W}$	$2pqW_{Rr}/\overline{W}$	$q^2W_{rr}/\overline{W}$

Gametes from First Generation

Allele	R	r
Frequency	$p' = (p^2W_{RR} + pqW_{Rr})/\overline{W}$	$q' = (q^2W_{rr} + pqW_{Rr})/\overline{W}$

$\overline{W} = p^2W_{RR} + 2pqW_{Rr} + q^2W_{rr}$

Figure 21.5 Changes in allele frequencies caused by selection. The uncalibrated frequency after selection is calculated by multiplying the zygote frequency by the relative fitness value. Adult frequencies are calibrated through division by the sum of the relative fitness values ($\overline{w}$).

independent of genotype. If we define the relative fitness of the three genotypes as w_{RR}, w_{Rr}, and w_{rr}, respectively, the relative frequencies of the three genotypes at adulthood is p^2w_{RR}, $2pqw_{Rr}$, and q^2w_{rr} (**Fig. 21.5**). These frequencies do not sum to 1 because they are relative frequencies for the three genotypes. To make the frequencies sum to 1, even with the inclusion of modifying terms that change the value of each genotypic frequency, it is necessary to "recalibrate" the Hardy-Weinberg equation.

This recalibration is accomplished by calculating the value of a numerical factor that each Hardy-Weinberg term can be divided into so that the terms all add up to 1. The calculation consists simply in setting the sum of the terms in the modified equation to a new variable, designated $\overline{w}$.

$$p^2w_{RR} + 2pqw_{Rr} + q^2w_{rr} = \overline{w} \qquad (21.4a)$$

Since $\overline{w}$ represents the sum of the individual fitness values, each multiplied by their relative occurrence in the population, it in fact represents the average fitness of the population.

In populations that satisfy the conditions of the original Hardy-Weinberg equilibrium, when the fitness for each genotype is 1, the value of $\overline{w}$ is also 1. However, when fitness varies from one genotype to another, $\overline{w}$ can also vary but in a way that can be calculated as long as the value of each of the variables in the equation is known. The modified Hardy-Weinberg equation is thus recalibrated by dividing each term by $\overline{w}$ such that the new equation becomes

$$\frac{p^2w_{RR}}{\overline{w}} + \frac{2pqw_{Rr}}{\overline{w}} + \frac{q^2w_{rr}}{\overline{w}} = 1 \qquad (21.4b)$$

Each term in this recalibrated equation represents the actual frequency that each genotype will assume in the generation following the one used for the original calculation. Figure 21.5 summarizes this process of calculation.

As an example, let us use the variables p' and q' to represent the frequencies of the R and r alleles in this next generation. Among the gametes produced by the original population, the frequency of allele r will be the result of contributions of r alleles from both Rr and rr adults relative to the number of individuals in the entire adult population. If q' represents the frequency of the r allele in the next generation adults, then

$$q' = \frac{q^2w_{rr} + \frac{1}{2}(2pqw_{Rr})}{\overline{w}} = \frac{q(qw_{rr} + pw_{Rr})}{\overline{w}} \qquad (21.5)$$

Keep in mind,

1. Δq is defined as $q' - q$.
2. A useful identity in moving from Equation 21.5 to Equation 21.6 is $1 - 2q = 1 - q - q = (p - q)$.

Thus, in one generation of selection, the allele frequency of r has changed from q to q'. It is often useful to know the change in allele frequency over one generation of selection. We can represent this change as

$$\Delta q = \frac{pq[q(w_{rr} - w_{Rr}) + p(w_{Rr} - w_{RR})]}{\overline{w}} \qquad (21.6)$$

As these equations (21.5 and 21.6) show, selection can cause the frequency of an allele to change from one generation to the next. Equation 21.6 shows that the change in allele frequency resulting from one generation of selection depends on both the allele frequencies and the relative fitnesses (in this case, the viability component of fitness) of the three genotypes. Note that if the fitnesses of all genotypes are the same, as in populations at Hardy-Weinberg equilibrium, then the change in q $(\Delta q) = 0$; in other words, if there are no genotype-related differences in fitness, there is no possibility of selection, and if there is no possibility of selection and a population is large enough to be considered essentially "infinite," there can be no evolution (defined as changes in allele frequency).

As an example, let us use Equation 21.6 to look at how a recessive genetic condition, such as thalassemia, influences the allele frequencies of a population. If the disease, which results from an *rr* genotype for the *R* gene, decreases fitness by decreasing the probability of surviving to adulthood, then the fitness of *RR* and *Rr* individuals is the same, while the fitness of *rr* individuals is reduced. Because only the relative values of the fitnesses are important, it is useful to set the values of $w_{RR} = 1$, $w_{Rr} = 1$, and $w_{rr} = 1 - s$, where s is the *selection coefficient* against the *rr* genotype. This selection coefficient can vary from 0 (no selection against *rr*) to 1 (*rr* is lethal, and no *rr* individuals survive to adulthood). For this example, we can rewrite Equation 21.6 as

$$\Delta q = \frac{pq[q(1 - s - 1) + q(1 - 1)]}{\overline{w}} = \frac{-spq^2}{\overline{w}} \qquad (21.7)$$

Equation 21.7 has three interesting features. First, unless there is no selection and $s = 0$, Δq is always negative, and the frequency of the *r* allele decreases with time.

Second, the rate at which q decreases over time depends on the allele frequencies; in particular, because Δq varies with q^2, the rate at which q decreases diminishes as q becomes smaller. (Recall that because q is always less than 1, $q^2 < q$.) To understand the effect of this correlation between the allele frequency and the rate at which q decreases over time, consider the special case of a lethal recessive disease for which $s = 1$. The dotted line in **Fig. 21.6** shows the decrease in allele frequency predicted by Equation 21.7, starting from an initial allele frequency of 0.5. The decrease in allele frequency is rapid at first, and then slows. After 10 generations, the predicted frequency of the recessive disease allele is nearly 10%, even though the homozygous recessive genotype is lethal. The solid line in Fig. 21.6 plots actual data for the decrease in frequency of an autosomal lethal allele in an experimental population of *Drosophila melanogaster.* In

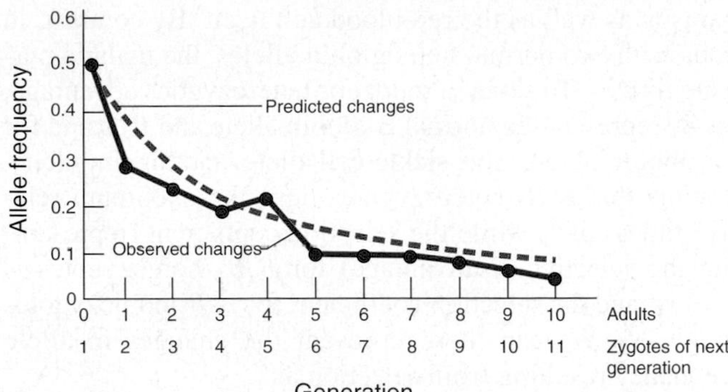

Figure 21.6 Decrease in the frequency of a recessive allele over time. The *dotted line* represents the mathematical prediction. The *blue line* represents the actual data obtained with an autosomal recessive allele.

Fig. 21.6, the predicted and observed changes in allele frequency match quite closely.

Why is selection unable to reduce the frequency of a recessive lethal allele to zero? The answer goes back to our consideration of the frequency of heterozygous carriers of a recessive disease allele. When q is small, individuals homozygous for the disease allele (at a frequency of q^2) are very rare because most copies of the *r* allele occur in *Rr* heterozygotes (at a frequency of $2pq$) who do not experience negative selection. (In mathematical terms, the ratio of q^2 to q decreases exponentially for all values of q less than 1.) By contrast, a lethal dominant allele will disappear from a population in a single generation of selection.

The third feature of Equation 21.7 is that it predicts that the allele frequency q should continue to decline, albeit more and more slowly over time as q moves closer and closer to a value of zero. However, as we noted earlier, the reality is that many recessive genetic diseases persist in human populations at low but stable frequencies. What maintains these diseases despite continuing selection against them? One answer is that sometimes heterozygotes have a higher fitness than either homozygote, a situation referred to as **heterozygous advantage.**

It Is Possible to Calculate the Effect of Heterozygous Advantage on Allele Frequencies

We have seen that sickle-cell anemia, which includes episodes of severe pain, serious anemia, and a probability of early death, is a recessive condition resulting from two copies of the sickle-cell allele at the β-globin locus. The disease allele has not disappeared from several African populations, where it seems to have existed for a very long time. One clue to the maintenance of the sickle-cell allele in human populations lies in the observation that heterozygotes for the normal and sickle-cell alleles are resistant to malaria. This resistance is due, in part, to the fact that red blood cells infected by the malaria parasite, if they also contain a sickle-cell allele, break open, destroying the parasite as well as the red blood cell itself. By contrast, in cells with two normal hemoglobin alleles, the malaria parasite thrives. To set up a model of heterozygous advantage, let B_1 represent the normal β-globin allele and B_2 stand for the abnormal recessive sickle-cell allele; and for simplicity, assume that B_1B_2 heterozygotes have the maximum relative fitness of 1, while the selection coefficient (representing the selective disadvantage) for B_1B_1 homozygotes is $1 - s_1$, and the selection coefficient for B_2B_2 homozygotes is $1 - s_2$. We can then represent the changes in allele frequency resulting from selection as

$$\Delta q = \frac{pq(s_1p - s_2q)}{\overline{w}} \quad (21.8)$$

To maintain both alleles in the population using this equation, Δq must be 0 for some value of q between 0 and 1. The q value at which $\Delta q = 0$ is known as the **equilibrium frequency** of allele B_2. The value of q when $\Delta q = 0$ and both alleles are present occurs when the term inside the parentheses of Equation 21.8 is 0, that is, when

$$s_1p - s_2q = 0$$

Substituting $1 - q$ for p $(p = 1 - q)$ and solving this equation for q reveals that the equilibrium frequency of B_2 (represented by q_e) is reached when

$$q_e = \frac{s_1}{s_1 + s_2} \quad (21.9)$$

Note that to find the equilibrium frequency, that is, the value of q at which $\Delta q = 0$ such that both alleles B_1 and B_2 persist in the population, you need know only the selection coefficients for the two homozygotes.

To understand the relationship between q, the change in q, and the equilibrium frequency q_e, you can formulate Δq using q_e.

$$\Delta q = \frac{-pq(s_1 + s_2)(q - q_e)}{\overline{w}} \quad (21.10)$$

From this formulation, you can see that when q is greater than q_e, Δq is negative. Under these circumstances, q, or the frequency of allele B_2, will decrease toward the equilibrium frequency. By contrast, when q is less than q_e, Δq is positive and the frequency of B_2 will increase toward the equilibrium. Thus, the equilibrium frequency stabilizes, because a change away from it is always followed by a change toward it.

Now, if you assume that the African populations in which sickle-cell anemia is prevalent are currently at equilibrium relative to their alleles at the β-globin locus, you can use the observed frequency of the sickle-cell allele in these populations to calculate the relative values of the selection coefficients. Field studies show that the actual value of q_e lies between 0.15 and 0.2 for an average value of 0.17. If you plug this number into Equation 21.9, you get

$$0.17 = \frac{s_1}{(s_1 + s_2)}$$

This equation makes it possible to express either selection coefficient in terms of the other. For example, $s_1 = 0.2s_2$.

If you assume that $s_2 = 1$ (that is, those with sickle-cell trait never reproduce), as was essentially true before medical advances enabled the survival of children expressing the sickle-cell trait, you will find that $s_1 = 0.2$, which, in turn, means that the relative fitness of the wild-type genotype is 0.8. Recall, however, that we set the fitness of the heterozygote at 1.0. By dividing 1.0 by 0.8, you get 1.25, which represents the relative advantage in fitness that heterozygotes for the sickle-cell allele have over people who do not carry this allele in African populations exposed to malaria. The use of simple statistical methods to calculate this heterozygous advantage demonstrates how medical geneticists can use the tools of population genetics (**Fig. 21.7**).

Heterozygous advantage is one phenomenon that maintains recessive disease alleles in human populations. We now look at other phenomena that can counteract the effects of selection and maintain recessive alleles within a population.

Evolutionary Equilibrium: A Balance Between Mutation to a New Allele and Selection Against That Allele

Mutations, we have seen, are the ultimate source of genetic variation in a population. They arise from DNA damage due to environmental agents and from errors during DNA replication and the transmission of genetic information during cell division. Various studies show that in most animals, the mutation rate is roughly 10^{-4} to 10^{-6} mutations per gene per generation. Since we know that the human genome carries about 20,000–30,000 genes, if we take the average of 25,000, we can infer that 0.025–2.5 new mutations arise somewhere in the genes of each human gamete per generation.

Many mutations that alter the amino acid sequence of a gene product have deleterious consequences on survival and reproduction, that is, on fitness. Indeed, many genetic diseases result from mutations that give rise to a nonfunctional gene product. Individuals homozygous for such disease alleles produce no functional copies of an enzyme or other polypeptide required for normal development or metabolism.

How Selection Against a Disease Interacts with Mutation to the Disease Allele to Influence the Frequency of the Disease Allele in a Population

Let us again consider the example of a recessive genetic disease that occurs at low frequency in a population. Suppose that mutations from allele R (the dominant, normal allele) to allele r (the recessive, disease allele) occur at a rate of μ per generation. As a first approximation, we can ignore the very

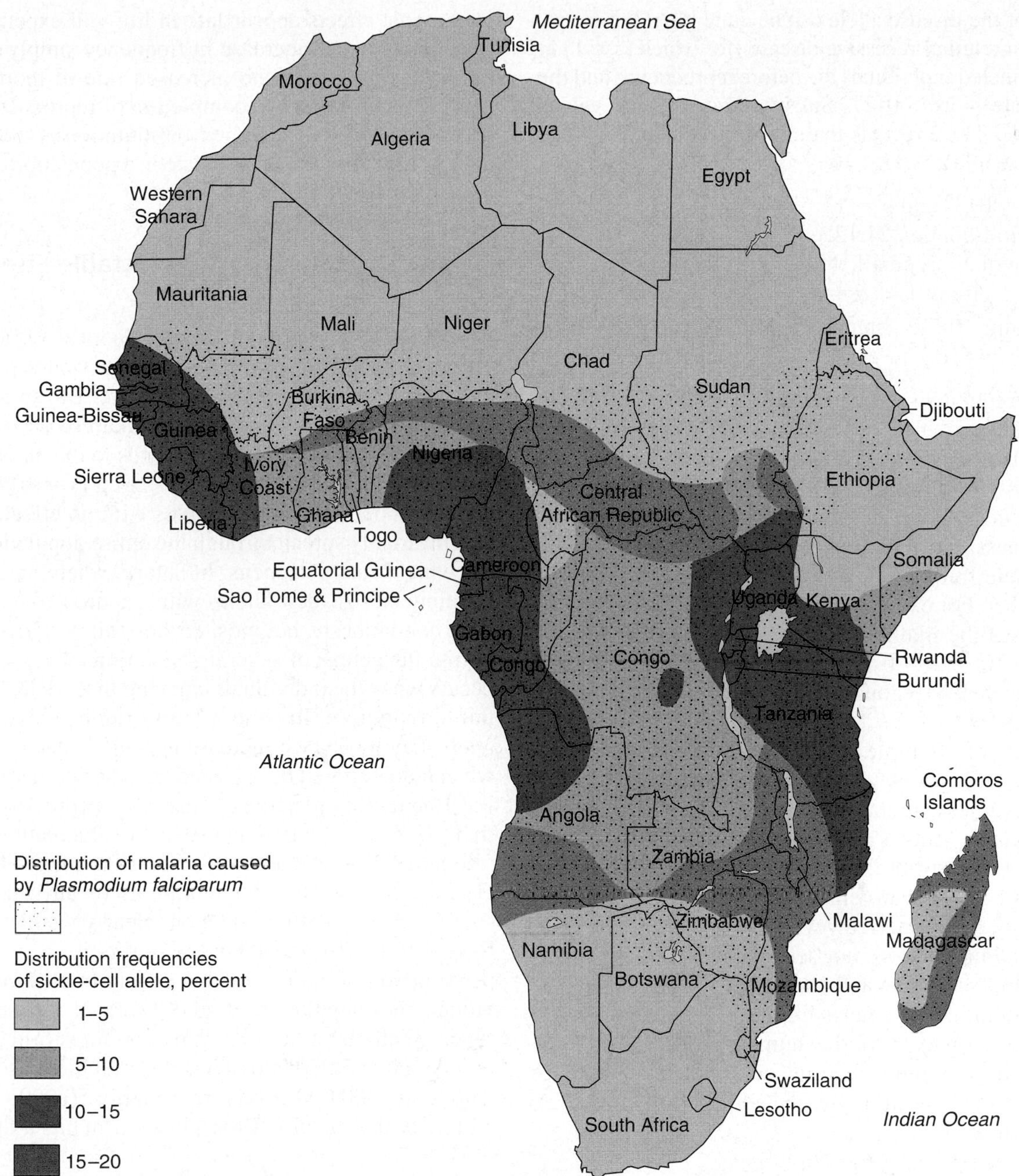

Figure 21.7 **Frequency of the sickle-cell allele across Africa where malaria is prevalent.**

low level of reverse mutations from r to R because the frequency of r is so small. With this background, we can recalculate Δq as the sum of two terms representing effects due to selection and mutation. The first term is derived from Equation 21.7. The second term is simply the product of the mutation rate (μ) with the frequency of the wild-type allele $\mu(1 - q)$ divided by the average fitness ($\overline{w}$) of the population. In this equation, we multiply μ by $(1 - q)$ because only R alleles can mutate to r.

$$\Delta q = \frac{-spq^2 + \mu(1 - q)}{\overline{w}} \qquad (21.11)$$

Note that the effects of selection (s) and mutation (μ) on allele frequency are in opposite directions. Allele frequencies will thus evolve until $\Delta q = 0$. The allele frequency at which evolution no longer continues, labeled $\hat{q}$, is called the **evolutionary equilibrium.** By rearranging Equation 21.11, you can see that this equilibrium occurs when

$$\hat{q} \,/\, \sqrt{(1 - \hat{q})} = \sqrt{(\mu/sp)} \qquad (21.12)$$

In this equation, $\hat{q}$ represents the evolutionary equilibrium.

This simple equation shows how a balance between selection and mutation can maintain recessive genetic diseases at a low level in a population. Note that the equilibrium frequency of the disease allele can be quite high. For example, consider a lethal recessive disease (for which $s = 1$) in which all affected individuals die before reproducing and the mutation rate $= 1 \times 10^{-6}$. Substitution of these values into Equation 21.12 reveals that the equilibrium frequency of the disease allele is 1×10^{-3}, or 1 in 1000.

1. Let $\mu = 10^{-6}$ and $s = 1$.
2. Plug into Equation 21.12.
3. Assume that $\hat{q}$ is small, which enables you to approximate $p = 1$ and $1 - \hat{q} = 1$.
4. Following these steps, you can conclude that $\hat{q} = \sqrt{\mu} = 10^{-3}$.

Since the frequency of heterozygotes among zygotes is $2pq$, as predicted by the Hardy-Weinberg law, the frequency of heterozygous carriers of this deadly recessive condition will be approximately 2 per 1000.

For the many genetic diseases that have less severe effects on fitness and therefore sustain less negative selection, the allele frequency at evolutionary equilibrium will be even higher. For example, an inherited recessive disease that decreases the likelihood of surviving to reproductive age by only 10% ($s = 0.1$) would result in an equilibrium frequency of heterozygous carriers of about 63 per 1000 individuals.

In short, this simple mathematical model suggests that, like heterozygous advantage (that is, selection for the heterozygotes), mutation is quite effective in maintaining recessive alleles in a population. As a result, at evolutionary equilibrium, substantial numbers of deleterious recessive alleles can remain in the heterozygous carriers of a population. However, as we see later, the mathematical model does not accurately predict what actually happens in human and other populations, because a critical assumption—namely, infinite size—does not reflect reality. However, if the human population keeps growing and free interbreeding continues, the human population might essentially reach equilibrium in several thousand years.

The Time of Onset of a Disease Can Also Influence the Frequency of Disease Alleles

The effects of many inherited human diseases become manifest only during middle age or later in life; the effects of others may appear earlier than middle age but not cause death until much later. From an evolutionary viewpoint, the important issue is how a disease reduces survival to reproductive age, for it is its impact on reproduction that determines a disease's potential effect on the strength of selection. Diseases that cause death after the completion of reproduction will sustain little or no negative selection; and with lower selection against them, their equilibrium frequency will be higher. One significant consequence of the relationship between the time of onset of disease and the strength of selection is that genetic diseases whose phenotypic effects appear late in life will experience little selection and may increase in frequency simply as a result of mutation. Aging—the increased rate of mortality with age that occurs after the completion of reproduction—may be the natural consequence of mutations whose effects appear late in life and accumulate in a population unopposed by natural selection.

Genetic Drift Has an Unpredictable Effect on the Evolutionary Equilibrium

Equation 21.12 models the balance between selection and mutation in a Hardy-Weinberg world where populations are very, very large—virtually infinite. But in addition to clarifying how mutations help maintain recessive diseases in a population, Equation 21.12 leads to the untenable conclusion that in the absence of selection ($s = 0$), $\hat{q}$ will become infinite; that is, new alleles with no effect on fitness will invariably spread through the entire population. This is not what actually happens. In nature, where no population is infinite in size, new alleles with a neutral effect on fitness arise continuously, but most of these alleles drift to extinction in the course of several generations. Loss of an allele occurs when the individuals carrying that allele, by chance, fail to reproduce. In contrast, very rarely, a new recessive allele may by chance increase to high frequency, but even when it does, it will be replaced later by another new allele.

Population geneticists use the expression **genetic drift** to refer to unpredictable, chance fluctuations in allele frequency that have a neutral effect on fitness (**Fig. 21.8**). Genetic drift occurs in populations of any size, but the smaller the population, the more clearly visible are the effects of drift. To understand why this is so, consider the perpetuation of allele *r*, which occurs in 10% of the individuals in a population of sunflowers. If a chance occurrence such as a severe summer hailstorm strikes a population of 1,000,000 plants and half die, it is likely that among the 500,000 survivors, roughly 50,000 (10%) will still carry the *r* allele. If the same storm hits a field of just

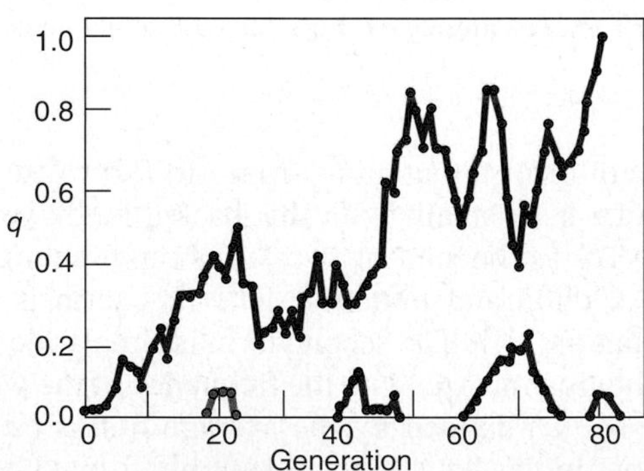

Figure 21.8 The effect of genetic drift. New alleles with no effect on fitness are usually lost rapidly from a population (*red* and *green lines*). Rarely, a new allele will increase in frequency (*blue line*).

10 plants, only 1 of which carries the *r* allele, there is a 50% chance that the single plant bearing the *r* allele will not be among the survivors. If it isn't, the total loss of this one allele from the population will reduce the allele frequency in an unpredictable way to 0. If, however, the lone individual carrying *r* does by chance survive, the frequency of allele *r* will increase to 20%.

One example of genetic drift in human populations is the **founder effect,** which occurs when a few individuals separate from a larger population and establish a new one that is isolated from the original population. The small number of founders in the new population carry only a fraction of the alleles from the original population, and this discrepancy may change the allele frequencies. The 14,000 people living in the Amish community in eastern Pennsylvania are the descendants of about 200 individuals who emigrated from Germany in the early eighteenth century. Among these Amish, there is a much higher incidence of manic-depressive illness than among the larger original European population, most likely because several founders carried alleles producing this disease.

Summary of the Balance Between Mutation and Selection

New alleles (some of which cause disease) arise in a population by mutation. When a new allele has an effect on fitness that differs from the effect of the wild-type allele, selection will drive the frequency of the new allele toward an equilibrium with the wild-type allele. The equilibrium value will be determined by the relative selection coefficients of individuals heterozygous and homozygous for the new allele compared to individuals who are wild type. In contrast, when an allele has a neutral effect on fitness, its frequency will depend on unpredictable genetic drift. Most neutral alleles become extinct, but occasionally one "drifts up" to become the new wild-type allele in the population.

Comprehensive Example: How Human Activity Affects the Evolution of Human Pathogens and Crop Pests

Infectious diseases have been a major killer throughout human history, and as the AIDS epidemic illustrates, previously unknown diseases continue to emerge. In the twentieth century, discovery of a variety of vaccines, antibiotics, and other drugs made it possible to combat infectious diseases such as smallpox (vaccines) and tuberculosis (antibiotics) with great success. In the last 25 years, however, many formerly surefire drugs have lost their effectiveness because populations of pathogens have evolved resistance to them. Similarly, populations of agricultural pests have evolved resistance to the pesticides used to control or eradicate them.

At the beginning of this chapter, we posed three questions: How do new diseases emerge in human populations? Why do diseases persist in all living organisms? What causes diseases and pests long under control to resurge in frequency and intensity? We have answered the first two. New diseases emerge in human populations as a consequence of new mutations. Diseases persist because changes in allele frequency tend toward an evolutionary equilibrium in which mutation balances selection. To answer the third question, we turn to an examination of how pathogens and pests interact with their hosts.

The Evolution of Drug Resistance in Pathogens

We have seen that many of the bacterial agents of tuberculosis are resistant to several antibiotics. We now know that a major factor contributing to the evolution of multidrug-resistant TB strains is the failure of patients to complete the lengthy drug regimens required for a cure. The two most widely used drugs for TB, isoniazid and rifampicin, require ingestion for six months and have side effects that include nausea and loss of appetite. However, the symptoms of TB can begin to disappear after only two to four weeks of treatment.

Imagine a TB patient with a persistent cough, shortness of breath, and general weakness. This individual harbors a large, actively growing and dividing population of TB bacteria in his lungs. At first, these bacteria are susceptible to antibiotics, but occasional mutations conferring partial resistance appear at random (**Fig. 21.9**). The patient's physician prescribes a six-month course of treatment with the antibiotic isoniazid. After a few weeks of treatment, the bacterial population in the patient's lungs has decreased considerably, and the patient's symptoms have abated, although the negative side effects of the drug continue to cause discomfort. However, the composition of the bacterial population has now changed so that the remaining bacteria are likely to include a high proportion of mutant bacteria possessing partial resistance to the antibiotic. If the patient continues his course of treatment, the persistent dose of antibiotic will eventually kill all of the bacteria, even those with partial resistance, eliminating the infection. By contrast, if the patient stops treatment prematurely, the remaining (partially resistant) bacteria will proliferate and within three to four weeks reestablish a large population. Subsequent treatment of the same individual upon relapse or of a new patient to whom the partially resistant bacteria have spread would permit a second cycle of selection. New mutations could then convert the partially resistant bacteria to fully resistant microbes.

It is easy to see how repeated cycles of antibiotic treatment with multiple drugs, coupled with premature cessation of treatment, can promote the evolution of fully resistant bacterial populations and even bacterial strains resistant to more than one drug within a single patient. Initially individual mutant cells that express genes conferring partial resistance

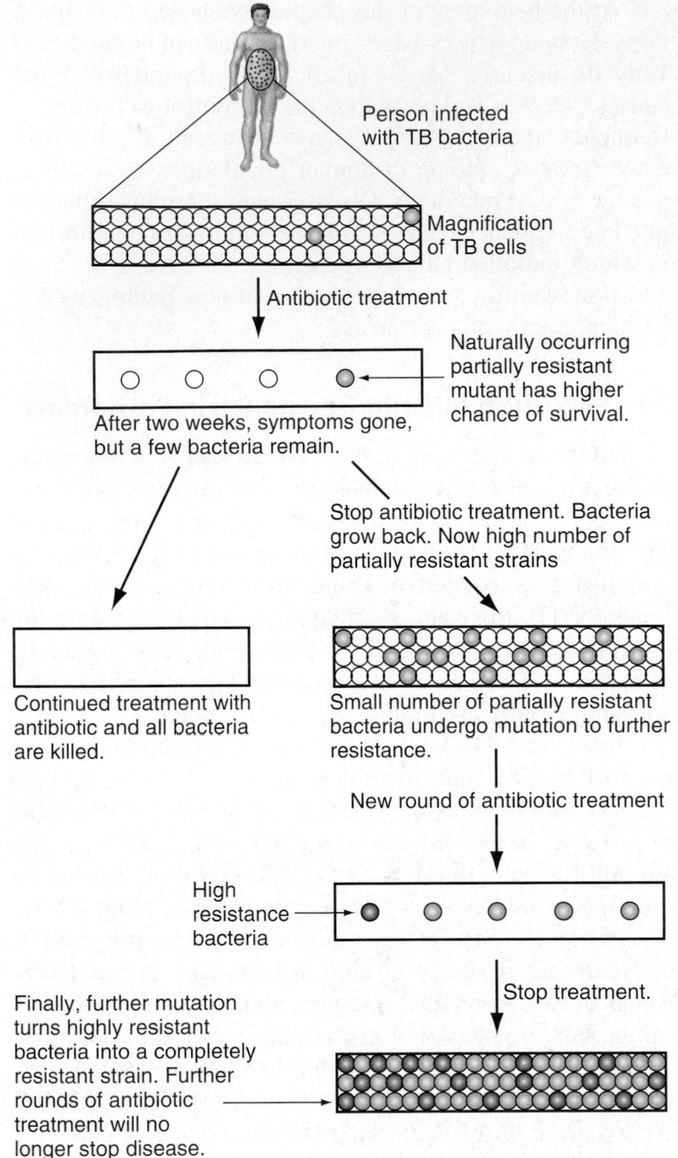
Person infected with TB bacteria
Magnification of TB cells
Antibiotic treatment
Naturally occurring partially resistant mutant has higher chance of survival.
After two weeks, symptoms gone, but a few bacteria remain.
Stop antibiotic treatment. Bacteria grow back. Now high number of partially resistant strains
Continued treatment with antibiotic and all bacteria are killed.
Small number of partially resistant bacteria undergo mutation to further resistance.
New round of antibiotic treatment
High resistance bacteria
Stop treatment.
Finally, further mutation turns highly resistant bacteria into a completely resistant strain. Further rounds of antibiotic treatment will no longer stop disease.

Figure 21.9 **The evolution of resistance in TB bacteria.**

increase in frequency; subsequent mutations in some of the mutant bacteria will increase resistance, and incomplete drug dosages will select for the resistant strains. Surveys of resistance illustrate the effects of this process on the frequency of resistance. A 1982–1986 U.S. Centers for Disease Control (CDC) survey showed that 9% of TB strains isolated from patients who had not previously received treatment were resistant to antibiotics, compared with 25% of strains isolated from previously treated, relapsed patients. More recently, a 1991 survey in New York City showed that 23% of strains isolated from previously untreated patients were resistant to one or more antibiotic drugs and 7% were resistant to both isoniazid and rifampicin; moreover, 44% of isolates from relapsed patients were resistant to one or more drugs and 30% were resistant to both isoniazid and rifampicin. These data support the idea that patient noncompliance with drug treatments is a major factor in the evolution of antibiotic resistance.

Several factors contribute to the rapid evolution of resistance in bacterial pathogens. The short generation times—often only a few hours—and rapid rate of reproduction under optimal conditions allow evolution to proceed quickly relative to a human life span. The large population densities typical of bacteria, which may exceed $10^9/cm^3$, ensure that rare resistance-conferring mutations will appear by chance in the population. The strong selection imposed by antibiotics increases the rate of evolution in each generation, unless the bacterial population is entirely eliminated.

The large variety of ways by which bacteria can acquire genes also contributes to the rapid evolution of resistance. Many genes for resistance are found on plasmids, and, as we saw in Chapters 9 and 15, the capacity of plasmids to replicate and be transmitted among bacteria allows the amplified expression of resistance genes in bacterial populations. Plasmids also provide a means for the genetic exchange of resistance genes among bacterial populations and species through transformation, conjugation, and transduction. Laboratory studies have demonstrated the ready transfer of plasmid-borne resistance genes to new bacterial species.

The Evolution of Pesticide Resistance

Like infectious bacteria, many agricultural pests spawn large populations because of their short generation times and rapid rates of reproduction. These large, rapidly reproducing populations evolve resistance to the chemical pesticides used to control them via selection for resistance-conferring mutations. Our understanding of this familiar pattern of selection and rapid evolution is most complete for certain insecticides, that is, pesticides that target insects.

The large-scale, commercial use of DDT and other synthetic organic insecticides, begun in the 1940s, was initially highly successful at reducing crop destruction by agricultural pests, such as the boll weevil, and medical pests, such as the mosquitoes that transmit malaria and yellow fever. Within a few years, however, resistance to these insecticides was detectable in the targeted insect populations. Since the 1950s, resistance to every known insecticide has evolved within 10 years of its commercial introduction. By 1984 there were reports of more than 450 resistant species of insects and mites (**Fig. 21.10**). Because different populations within a species can become resistant independently of other populations, the number of times insecticide resistance has evolved probably exceeds 1000.

Genetic studies show that insecticide resistance often results from changes in a single gene, and that several significant mechanisms of resistance are similar to those seen in infectious bacteria. DDT, for example, is a nerve toxin in insects. House flies and some mosquitoes develop resistance to DDT from dominant mutations in a single,

(a)

(b)

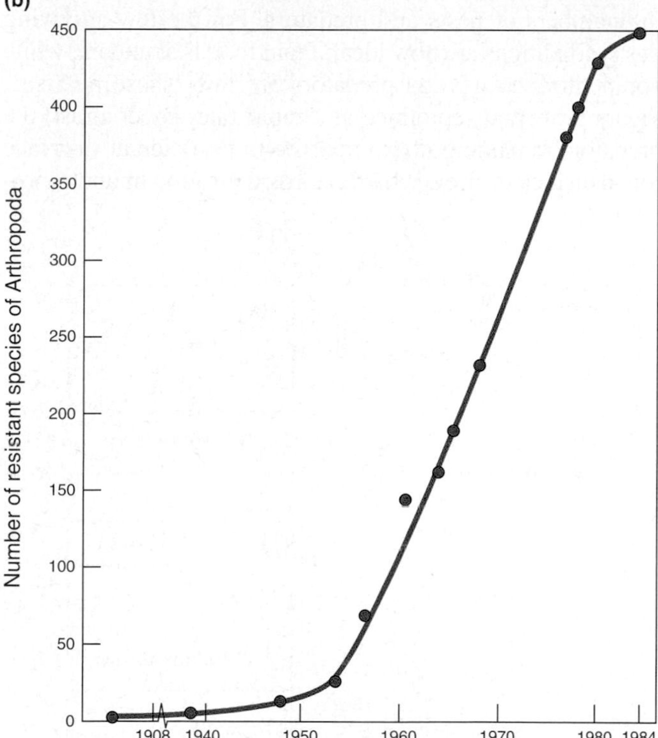

Figure 21.10 Increase of insecticide resistance from 1908–1984. (a) Insecticide resistance evolved with the aerial spraying of DDT, which began in the 1940s. Within 10 years, resistance to an insecticide becomes widespread as the insects evolve defenses to the insecticides just as they would evolve defenses against infectious bacteria. **(b)** The evolution of resistance among Arthropoda.

(a)

(b)

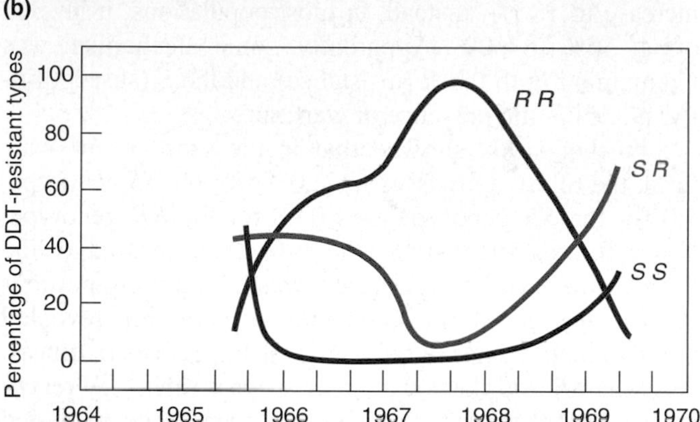

Figure 21.11 How genotype frequencies among populations of *A. aegypti* mosquito larvae change in response to insecticide. (a) Mosquitoes and larvae. **(b)** Changing proportions of resistance genotypes of *A. aegypti* (larvae) under selection with DDT (1964–1967), and after selection was relaxed (1968), in a suburb of Bangkok, Thailand.

enzyme-encoding allele. The mutant enzyme detoxifies DDT, rendering it harmless to the insect. As we saw earlier, even at low frequencies, dominant alleles can experience strong selection, because of heterozygous advantage. Consider, for example, dominant mutation *R* (for insecticide resistance), which occurs initially at low frequency in a population. Soon after the mutation appears, most of the *R* alleles are in *SR* heterozygotes (in which *S* is the wild-type susceptibility allele). With the application of insecticide, strong selection favoring *SR* heterozygotes rapidly increases the frequency of the resistance allele in the population. A field study of the use of DDT in Bangkok, Thailand, to control *Aedes aegypti* mosquitoes, the carriers of yellow fever, illustrates the rapid evolution of resistance. Spraying of the insecticide began in 1964 (**Fig. 21.11**). Within a year, DDT-resistant genotypes emerged and rapidly increased in frequency. By mid-1967, the frequency of resistant *RR* homozygotes was nearly 100%.

The Biological Costs of Resistance

Since DDT no longer controlled mosquito populations in the region, the insecticide program was stopped. The response of the mosquito population to the cessation of spraying was intriguing: The frequency of the *R* allele decreased rapidly, and by 1969, *RR* genotypes had virtually disappeared. The precipitous decline of the *R* allele suggests that in the absence of DDT, the *RR* genotype produces a lower fitness than the *SS* genotype. In other words, the homozygous resistance genotype imposes a **fitness cost** on individuals such that in the absence of insecticide, resistance is subject to a negative selection that decreases the frequency of *R* in the population.

To understand the biological basis of fitness costs, consider how rats evolve resistance to warfarin, a pesticide introduced in the 1940s and 1950s to control small mammals, among other pests. Warfarin interferes with blood clotting by blocking the recycling of vitamin K (a cofactor in the clotting cascade). When a rat ingests warfarin, the inability to form a clot leads to a fatal loss of blood following any internal or external injury. In Europe in the 1960s, the extensive use of warfarin for rat control fueled the evolution of a single-gene resistance allele in many rat populations. The frequency of resistance, however, did not increase to 100%; instead, in most populations, it leveled off at 30% to 60%. Apparently, some mechanism was maintaining both the *R* (resistance) and the *S* (susceptibility) alleles in the presence of warfarin.

Further study showed that in the presence of warfarin, the relative fitnesses were 0.37 for the *SS* genotype, 1.0 for the *SR* genotype, and 0.68 for the *RR* genotype. These figures suggest that heterozygous advantage, the higher fitness of heterozygotes, was maintaining both alleles in the populations. Further investigation revealed that two factors caused the observed differences in fitness. Both the *SR* and the *RR* genotypes were relatively resistant to the effects of warfarin, and thus, they provided higher fitnesses than the susceptible *SS* genotypes. However, *RR* homozygotes suffered from a vitamin K deficiency because of the less efficient vitamin K recycling during blood clotting, and this deficiency reduced the rate of survival when the diet did not contain a large amount of vitamin K. In the absence of warfarin, therefore, *RR* homozygotes had a lower fitness than *SS* homozygotes because of a nutritional deficiency. The biological costs of fitness, which are widespread and occur by various mechanisms, are very likely a major reason why resistance alleles occurred in very low or undetectable frequencies before the routine use of pesticides.

Ecological Considerations

Selection for resistance in the presence of pesticides is not the only factor influencing the rate of pesticide resistance evolution among pest populations. The ecological context also plays a significant role. One illustration of this phenomenon comes from studies of insect predators and parasites, and insect pests. Most predators and parasites on agricultural insect pests are also insects. It would be logical to expect that applications of insecticides to agricultural fields would expose pests and predators alike, fostering the evolution of resistance in both classes of insects. Observations, however, do not support this logic. Of the more than 450 species of insects, mites, and ticks known to be resistant to insecticides, less than 10% are predators or parasites. What causes this pattern of resistance, and what are its consequences?

Simple ecological and evolutionary models provide an explanation. Consider a pest species and a predator species that coexist in an agricultural field. In the absence of insecticide, the population densities of pest and predator change over time, but both persist (**Fig. 21.12a**). Suppose that in each species, a single allele *R* at one locus determines resistance to a particular insecticide and that initially, the frequency of the *S* susceptibility allele is nearly 100% in both populations. The grower then applies heavy doses of insecticide to the field every two weeks (Fig. 21.12b). Because most individuals in both the pest and the predator populations are *SS*, high mortality from the insecticide decreases the numbers of pests and predators. For the few surviving pests, conditions are now ideal: Plant food is abundant, while competitors as well as predators are few. These pest survivors grow and reproduce at a rapid rate. By contrast, the predators remaining after exposure to pesticide at first face low densities of prey, which retards their growth and repro-

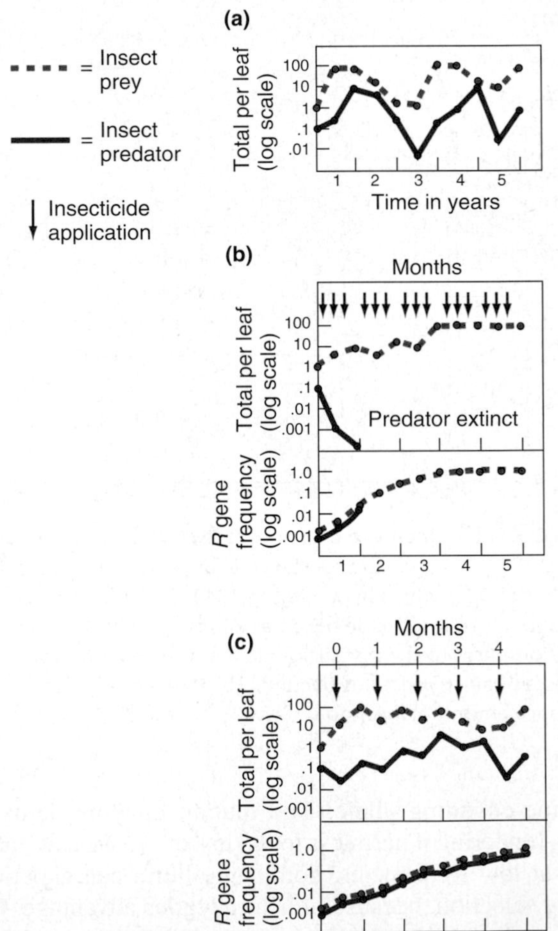

Figure 21.12 Changes in population density (insects per leaf) of prey and pest over time, with and without insecticide spraying. Fluctuations **(a)** in the absence of insecticide, **(b)** with biweekly applications of insecticide, and **(c)** with monthly applications of insecticide.

duction. This initial impairment may be so great that the predator population dies out within a few generations, before it has had a chance to evolve pesticide resistance. Now, the pest population, which has no predators to contend with, goes on to evolve a high frequency of resistance as a result of strong selection promoted by the insecticide. In these circumstances, the application of large doses of insecticide yields a worst-case scenario of a resistant pest population and a suppressed or extinct predator population. In sum, the different ecological consequences of insecticide application for prey and predators have created differences in the rate and likelihood that each population will evolve resistance. As Fig. 21.12c shows, changes in the intensity and timing of insecticide application may diminish these differences by allowing predators more time to recover before the next exposure. These models suggest that both genetics and ecology have an impact on the use of predators to control pests.

21.3 Analyzing the Quantitative Variation of Multifactorial Traits

Unlike cystic fibrosis, albinism, PKU, and thalassemia, the majority of inherited human disorders are influenced by more than one gene. For example, at least 12 separate regions of the human genome are associated with the development of insulin-dependent diabetes. Indeed, in addition to the many inherited abnormalities found in a population, most aspects of normal size, shape, physiology, and behavior in humans and other organisms are complex traits determined by multiple genes that interact with each other as well as with the environment.

Complex traits of this type are called **multifactorial** because more than one factor, either genetic or environmental, contributes to the phenotype. We saw in Chapter 3 that many multifactorial traits are continuous, or **quantitative;** that is, they vary over a continuous range of measurement—such as the length of a tobacco flower in millimeters, the amount of milk produced by a cow per day in liters, or the height of a person in meters. Figure 3.22 on p. 71 shows how quantitative variation can arise when the alleles of two or more genes affect the same trait. From a re view of the Chapter 3 figure, you can see that two alleles—one dominant and one recessive—at one locus produce two discrete (discontinuous) phenotypes (*AA* and *Aa* versus *aa*); two alleles at two genes generate five discrete phenotypes; two alleles at six genes can produce 13 phenotypes that will often appear to vary over a continuous range. As the number of phenotypic categories for a trait increases, the trait's variation appears more and more continuous.

Quantitative traits controlled solely by the alleles of two or more genes are **polygenic traits** that are, by definition, multifactorial. But not all multifactorial traits are polygenic (some may be determined by interactions of the alleles of one gene with each other and with the environment). Moreover, not all quantitative traits are polygenic (continuous variation may result from environmental effects on the expression of one gene), and not all polygenic traits are quantitative (some traits determined by the interaction of several genes may have discrete, discontinuous phenotypes).

With these subtle distinctions in mind, we now examine how population geneticists study quantitative traits. The continuous variation of such traits depends on the number of genes that generate the trait, as well as the genetic and environmental factors that affect the penetrance and expressivity of these genes (see Chapters 3 and 11). One of the goals of quantitative analysis is to discover how much of the variation in a particular trait is the result of genotypic differences among individuals in a population and how much arises from differences in the environment.

Genes Versus the Environment

To sort out the genetic and environmental determinants of phenotypic variation in a population, consider a series of experiments on a population of dandelions, a common weedy plant in lawns and other disturbed areas throughout North America (**Fig. 21.13a**). Dandelions have a long tubular stem and a large, yellow composite flowering structure composed of many small individual flowers; each of these flowers can produce a single, tufted, diploid seed, dispersible by the wind. Most dandelion seeds arise from mitotic, rather than meiotic, divisions such that all the seeds from a single plant are genetically identical. Your goal is to compare the influence of genes and environment on the length of the stem at flowering.

To distinguish environmental from genetic effects on phenotypic variation, you need to quantify one variable, say the environment, while controlling for the other one; that is, while holding the genetic contribution steady. You could begin by planting half of the genetically identical seeds from a single inbred plant on a grassy hillside and allowing them to grow undisturbed until they flower. You then measure the length of the stem of each flowering plant and determine the mean and variance of the distribution of values for this trait in this dandelion population. As Fig. 21.13b shows, you find the **mean** by summing the values of all stem lengths and dividing by the number of stems. You then find the **variance** by expressing the stem lengths as plus or minus deviations from the mean, squaring those deviations, and again dividing by the number of stems. Because all members of this population are genetically identical, any observed variation in stem length among individuals should be a consequence of environmental variations, such as different amounts of water and sunlight at different locations on the hillside (if we ignore rare mutations). When represented as a variance from the mean, these observed environmentally determined differences in stem length are called the **environmental variance,** or V_E.

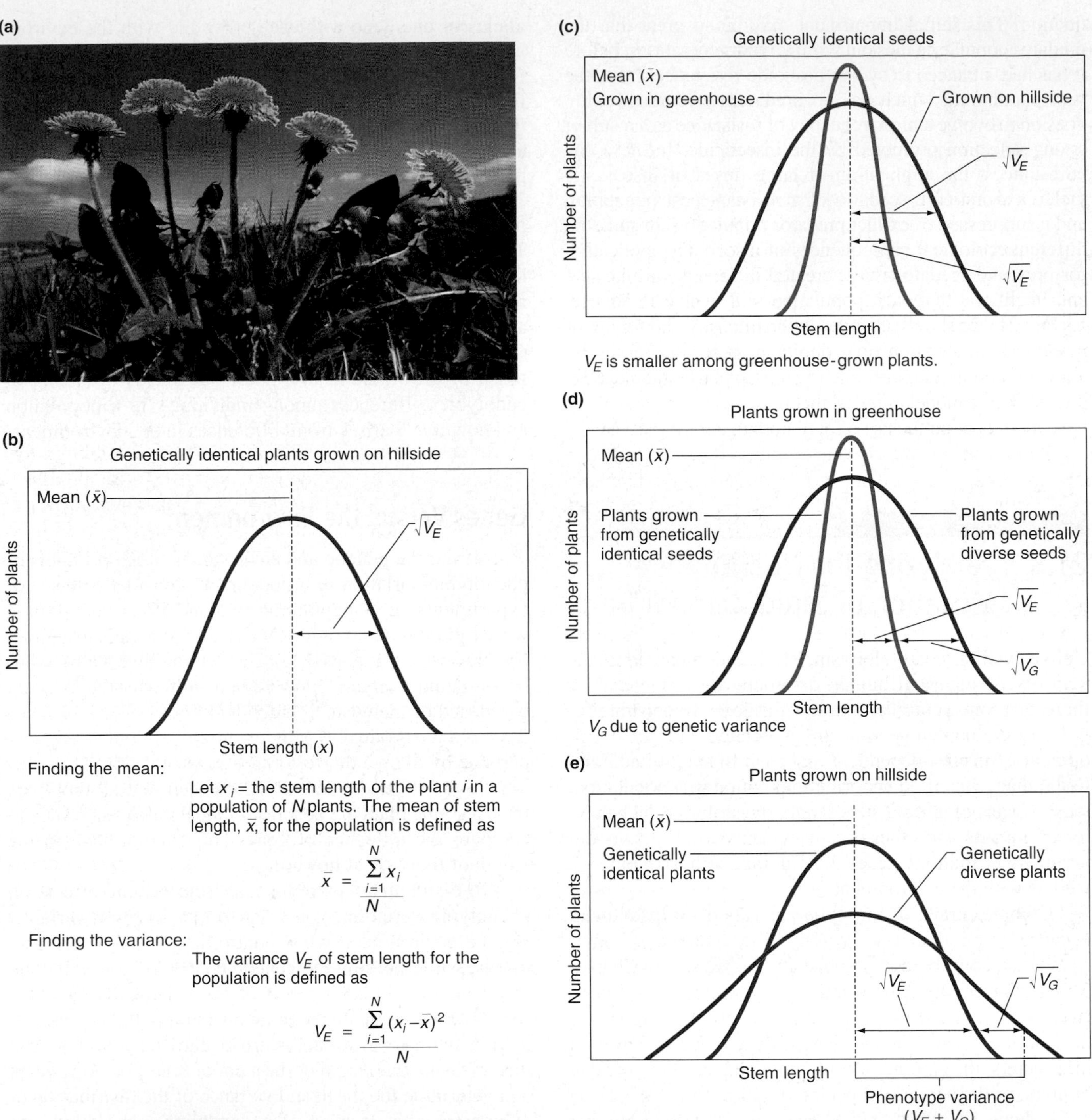

Figure 21.13 **Studies of dandelions can help sort out the effects of genes versus the environment.** **(a)** The familiar dandelion (*Taxaracum* sp.) is a useful model when studying population variations. **(b)** Finding the mean and variance of stem length. **(c)** Variance of genetically identical plants grown in a greenhouse and grown on a hillside. **(d)** Genetic variance of plants grown in a greenhouse. **(e)** Phenotype variance of plants grown on a hillside.

To refine your estimate of environmental variance, you plant the second half of the genetically identical seeds from the single test plant in a controlled greenhouse in which growth conditions are everywhere the same (Fig. 21.13c). Because environmental conditions are much more similar for all these genetically identical plants, the amount of environmental variance (V_E) among greenhouse plants is much smaller than among hillside plants. In theory, in a perfectly controlled greenhouse, growth conditions would be the same for all plants, the V_E would be zero, and all plants would have identical stem lengths (within measurement error). In reality, there is no such thing as perfect control or a homogeneous environment, and the greenhouse V_E will have some value greater than zero. Nonetheless, the difference between the V_E of the dandelions grown on the hillside and the V_E of dandelions grown in the greenhouse is a measure of the impact of the more diversified hillside environment on the phenotypic variation of stem length.

Even though the greenhouse V_E will have some value greater than zero, for the sake of simplicity in the following discussion, we assume that it is small.

To examine the impact of genetic differences on stem length, you take seeds from many different dandelion plants produced in many different locations, and you plant them in a controlled greenhouse (Fig. 21.13d). Because you are raising genetically diverse plants in a relatively uniform environment, observed variation in stem length—beyond that found in the genetically identical population—is the result of genetic differences promoting **genetic variance,** or V_G.

Now, to determine the total impact on phenotype of genes and environment, you take the seeds of many different plants from many different locations and grow them on the same hillside (Fig. 21.13e). For the population of dandelions that grow up from these seeds, the **total phenotype variance** (V_P) in stem length will be the sum of the genetic variance (V_G) and the environmental variance (V_E). The environmental variance is determined directly from the phenotypic variance found in the initial population of genetically identical plants grown on the hillside. It thus becomes possible for you to calculate the genetic variance in the second, mixed population as the difference between the phenotypic variance found in this genetically mixed population and the phenotypic variance found in the genetically identical population. For natural populations of dandelions, both genetic variation among individuals and variation in the environmental conditions experienced by each plant contribute to the total phenotypic variation.

Heritability Is the Proportion of Total Phenotype Variance Attributable to Genetic Variance

With the ability to determine the relative contributions to phenotypic variation of genes and environment, geneticists have developed a mathematical definition of the **heritability** (***h***^{2}) of a trait: It is the proportion of total phenotypic variance ascribable to genetic variance.

$$h^2 = \frac{V_G}{V_G + V_E} = \frac{V_G}{V_P} \qquad (21.13)$$

Because the amounts of genetic, environmental, and phenotypic variation may differ among traits, among populations, and among different environments, the heritability of a trait is always defined for a specific population and specific set of environmental conditions. If you know any two of the three variables of total phenotypic variance, genetic variance, and environmental variance, you can find the remaining unknown variance.

How to Measure Heritability

In analyzing the contributions of genes and environment to dandelion stem length, you measured the phenotypic variation among genetically identical individuals in a range of specified environments and compared it to the phenotypic variation among all individuals in the population. Of course, most organisms are not as easy to clone as dandelions. The key to generalizing from the dandelion example is to recognize that genetic clones are simply a special case of the broader notion of genetically related individuals, or genetic relatives. Various kinds of genetic relatives share certain alleles because they have one or more common ancestors. To quantify this idea, we can define the **genetic relatedness** of two individuals as the average fraction of common alleles at all gene loci that the individuals share because they inherited them from a common ancestor.

To determine the genetic relatedness of two siblings, for example, all you have to do is calculate the probability they received the same allele at any locus from the same parent. If you assume that one sibling received allele *A1* from an *A1* / *A2* heterozygous parent, the probability that the second sibling received the same allele is 0.5. Because this simple calculation holds for every locus transmitted by both parents, the total genetic relatedness of two siblings is 0.5. With an extension of this probabilistic analysis, we can see that an aunt and niece have 0.25 genetic relatedness, and first cousins 0.125.

If genetic similarity contributes to phenotypic similarity for some trait, it is logical to expect that a pair of close genetic relatives will be more phenotypically similar than a pair of individuals chosen at random from the population at large. Thus, by comparing the phenotypic variation among a well-defined set of genetic relatives with the phenotypic variation of the entire population over some range of environments, it is possible to estimate the heritability of a trait.

The finches observed by Darwin in the Galápagos Islands (often referred to as "Darwin's finches") provide an example of a population for which geneticists have measured the heritability of a trait under natural conditions in the field. Scientists studied the medium ground finch, *Geospiza fortis,* on the island of Daphne Major by banding many of the individual birds in the population (**Fig. 21.14a**). They then measured the depth of the bill for the mother, father, and offspring in each nest on the island and calculated how the bill depth of the offspring correlated with the average bill depth of the mother and father (called the *midparent value*). The results, depicted in Fig. 21.14b, show a clear correlation between parents and offspring; parents with deeper bills had offspring with deeper bills, while parents with smaller bill depth had offspring with smaller bill depth. In the figure, the heritability of bill depth, as represented by the slope of the line correlating midparent bill depth to offspring bill depth, is 0.82. This means that roughly 82% of the variation in bill depth in this population of Darwin's finches is attributable to genetic variation among individuals in the population; the other 18% results from variation in

(a)

(b) Correlation between parents and offspring

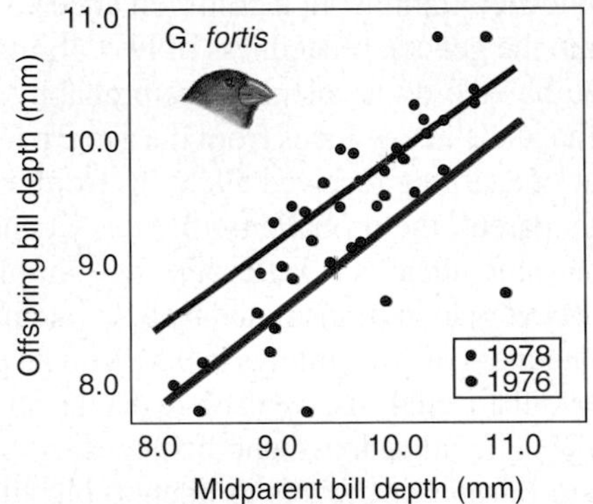

(c) If the heritability were 1.0

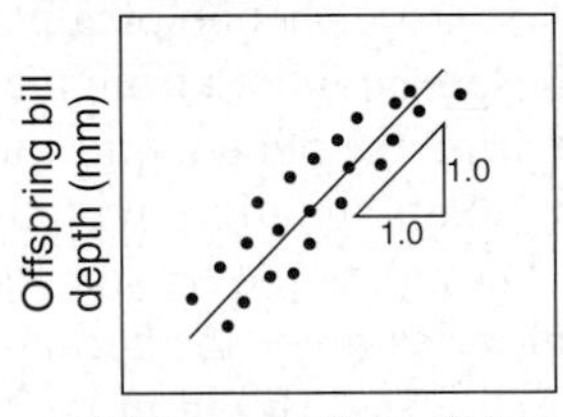

(d) If the heritability were 0.0

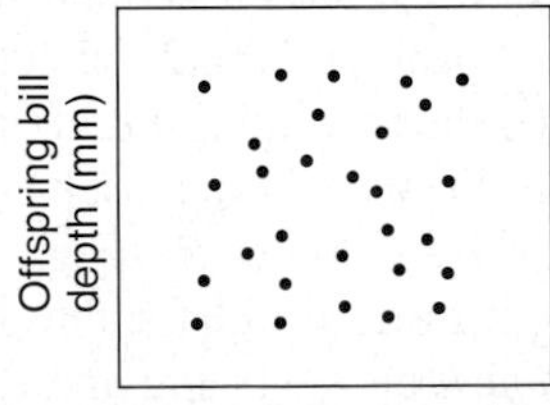

Figure 21.14 Measuring the heritability of bill depth in populations of Darwin's finches. (a) *G. fortis* with bands placed by scientists. **(b)** The correlation between beak size of offspring and their midparent value (the average of the parents' beak size) is 0.90 both in 1976 (*red circles*) and 1978 (*blue circles*), even though the mean beak size increased due to a drought in 1978. This correlation shows constant high heritability independent of environmental change. Note that high heritability does *not* mean that a trait is constant: beak size is highly variable (note range of axes) and varies over time (displacement of slopes). **(c)** Results if heritability were 1.0. **(d)** Results if heritability were 0.0.

the environment. If the environment had no influence at all on the trait, the slope of the line representing the heritability of bill depth, that is, correlating bill depth in parents with bill depth in offspring, would be 1.0 (Fig. 21.14c).

Now consider a population in which the bill depth for parents and their offspring is, on average, no more or less similar than the bill depths for any pair of individuals chosen from the population at random. In such a population, there is no correlation between the bill depth trait in parents and in offspring, and a plot of midparent and offspring bill depths produces a circular "cloud" of points (Fig. 21.14d).

From these examples, you can conclude that phenotypic similarity among genetically related individuals may provide evidence for the heritability of a trait. However, conversion of the phenotypic similarity among genetic relatives to a measure of heritability depends on a crucial assumption: that there is a random distribution of genetic relatives with respect to environmental conditions experienced by the population. In the finch example, we assumed that parents and their offspring do not experience environments that are any more similar than the environments of unrelated individuals. In nature, however, there may be reasons why genetic relatives violate this assumption by inhabiting similar environments. With finches, for example, all offspring produced by a mother and father during a breeding season normally hatch and grow in a single nest where they receive food from their parents. Because bill depth affects a finch's capacity to forage for food, the amount of feeding in a nest correlates with parental bill depth, for reasons quite distinct from genetic similarities. One way to circumvent the confounding variable of environmental similarity is to remove eggs from the nest of one pair of parents and randomly place them in nests built by other parents in the population; this random relocation of eggs is called **cross-fostering.** In heritability studies of animals that receive parental care, cross-fostering helps randomize environmental conditions. Controlling for both environmental conditions and breeding crosses is a fundamental part of the experimental design of heritability studies carried out on wild and domesticated organisms.

Measuring the Heritability of Polygenic Traits in Humans

Mating does not occur at random with respect to phenotypes in human populations, yet researchers cannot apply techniques for controlling environmental conditions and breeding crosses to studies of such populations. Nonetheless, in most human societies, family members share similar family and cultural environments. Thus, phenotypic similarity between genetic relatives may result either from genetic similarities or similar environments or, most often, both. How can you distinguish the effects of genetic similarity from the effects of a shared environment?

One way is to study monozygotic, "identical" twins given up for adoption shortly after birth and raised in different families. In such a pair of identical twins, any phenotypic similarity should be the result of genetic similarity.

At first glance, then, the study of adopted identical twins eliminates the confounding effects of a similar family environment. Further scrutiny shows that this is often not true. Many pairs of twins are adopted by different genetic relatives; the adoptions often occur in the same geographic region (usually in the same state and even the same city); and families wishing to adopt must satisfy many criteria, including job and financial stability and a certain family size. As a result, the two families adopting a pair of twins are likely to be more similar than a pair of families chosen at random, and this similarity can reduce the phenotypic differences between the twins. A valid scientific study of separated twins will take these factors into consideration.

A related approach is to compare the phenotypic differences between different sets of genetic relatives, particularly different types of twins (**Fig. 21.15a**). For example, monozygotic (MZ) twins, which are the result of a split in the zygote after fertilization, are genetically identical because they come from a single sperm and a single egg; they share all alleles at all loci and thus have a genetic relatedness of 1.0. By contrast, dizygotic (DZ) twins, which are the result of sperm from a single father fertilizing two different maternal eggs, are like any pair of siblings born at separate times; they have a genetic relatedness of 0.5 (which actually means that their dissimilarity is only 50% of the average dissimilarity between two unrelated individuals). Comparing the phenotypic differences between a pair of MZ twins with the phenotypic differences between a pair of DZ twins can help distinguish between the effects of genes and family environment.

Consider a trait in which the differences in phenotype among individuals in the population arise entirely from differences in the environment experienced by each individual, that is, a trait for which the heritability is 0.0 (Fig. 21.15b1). For this trait, you would expect the phenotypic differences among many pairs of MZ twins to be as great as the differences among many pairs of DZ twins. The fact that the MZ twins are more closely related genetically has no effect.

Now consider a trait for which differences in phenotypes among individuals in a population arise entirely from genetic differences, that is, a trait for which the heritability is 1.0 (Fig. 21.15b2). Since MZ twins have a genetic relatedness of 1.0, they will always show 100% concordance in expression: If one expresses the trait, the other will as well. The concordance of trait expression between unrelated individuals will vary based on the commonality of the trait (as shown in Fig. 21.15b2). Dizygotic twins will display greater concordance than genetically unrelated individuals, but less than monozygotic twins. In the highly simplified case of a dominant trait caused by an allele at a single autosomal gene, dizygotic twins will show a level of concordance that is halfway between the unrelated value and 100%. In reality, nearly all traits are affected by multiple genes that may have dominant, recessive, semidominant, and interacting effects, and the heritabilities of nearly all traits lie between 0.0 and 1.0. By comparing measures of differences among MZ and DZ twins, it is

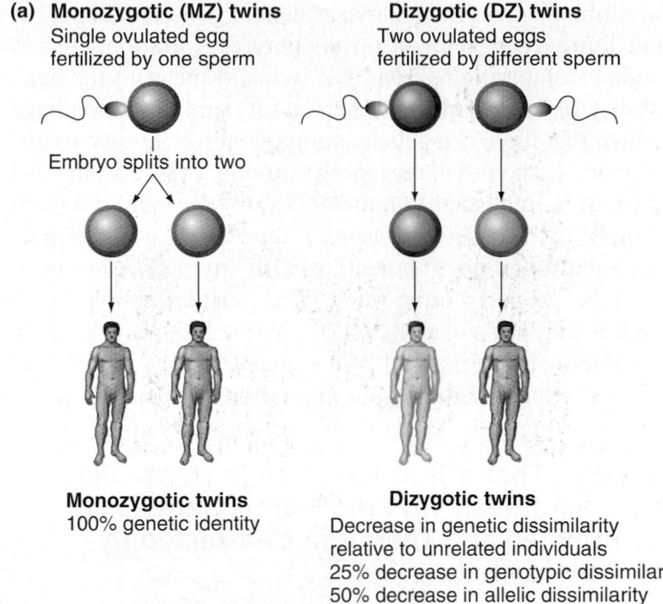

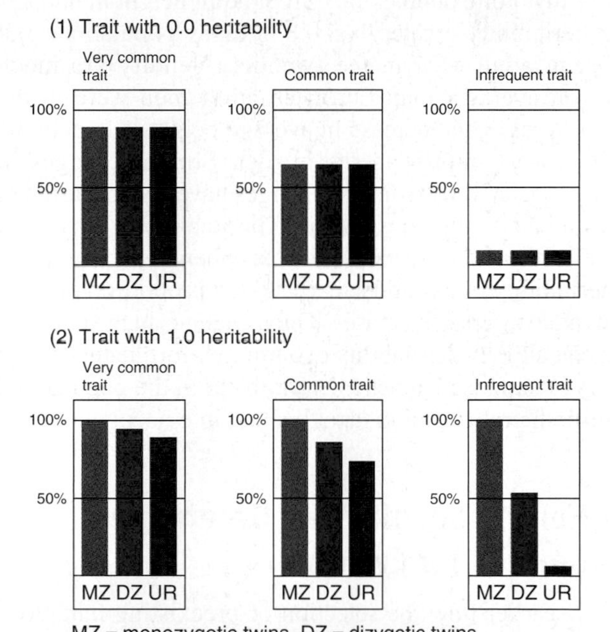

Figure 21.15 The impact of heritability on the concordance of dominant trait expression in two children raised in the same family environment. (a) Monozygotic and dizygotic twins have different genetic origins. Dizygotic twins have the same genetic relationship as any two full siblings. **(b)** The frequency with which a second child will share a trait expressed by a first child. *(1)* At one extreme are hypothetical traits associated with a heritability of 0.0. Irrespective of the frequency of trait expression, no differences will be observed in a comparison of monozygotic twins, dizygotic twins, or situations where one or both children are adopted. *(2)* At the opposite extreme are hypothetical traits associated with a heritability of 1.0. In all such cases, monozygotic pairs will be concordant, whereas dizygotic pairs of twins will show a concordance halfway between 100% and the concordance found between genetically unrelated children (because of their 50% decrease in genetic dissimilarity relative to the population at large).

possible to assess the relative effects of genetic relatedness and family environment on phenotypic variation. Recent studies comparing MZ and DZ twins to measure the heritability of some aspects of personality and cognition have shown that there is a genetic component to memory, extroversion, and verbal reasoning, among other behavioral traits. It is important to remember, however, that although heritability is a useful measure of the relative contributions of genetics and environment, as with any measurement, it must be properly interpreted. The heritability of a trait applies only for a specific set of environments and specific population; the heritability of the same trait may be different in a different environment or in a different population.

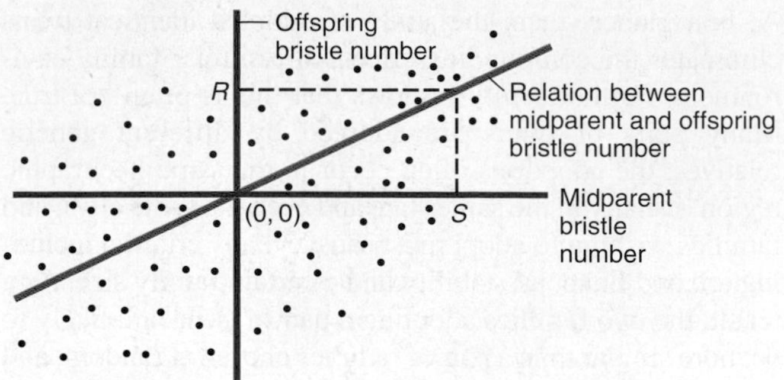

Figure 21.16 **Relationship between midparent number of abdominal bristles and bristle number in offspring for a hypothetical laboratory population of *Drosophila*.**

The Fact That a Trait Has a Large Heritability Does Not Mean That the Trait Is Genetically Determined and Therefore Unaffected by the Environment

Changes in the environment can, in fact, cause large changes in a phenotypic trait, even if that trait has a high heritability. For example, the trait of adult height in humans has a heritability greater than 0.8 in many populations; but European adult men in the twentieth century are much taller, on average, than European adult men were in the Middle Ages. This increase in average height is largely the result of vast improvements in diet. Similar changes in height associated with dietary changes have also occurred in other human populations. Because heritability is defined as the ratio of genetic variation to total phenotypic variation, the heritability of a trait decreases as the proportion of environmental variation increases. Thus, changes in the environment can alter the heritability of a trait. Unfortunately, many reports of supposed genetic determinism in the popular and scientific literature fail to take this fact into account.

The Heritability of a Trait Determines Its Potential for Evolution

We saw earlier how the selection of preexisting mutations generates evolutionary change. Since the heritability of a multifactorial trait is a measure of the genetic component of its variation, heritability quantifies the potential for selection and thus the potential for evolution from one generation to the next. A trait with high heritability has a large potential for evolution via selection, because the amount of evolution (change in allele frequency) generated by selection depends on the degree of heritability. To grasp the relationship between heritability, selection, and evolution, consider the number of bristles on the abdomens of fruit flies in a laboratory population of *D. melanogaster.* This fruit fly population exhibits some phenotypic variation in the trait of bristle number. If the trait has a high heritability in the population, the offspring of this original population will closely resemble their parents in bristle number (**Fig. 21.16**). If, however, you select as parents of the next generation only those flies among the top 15% in bristle number, the average bristle number among these breeders of the next generation will be greater than the average bristle number in the population as a whole. This artificial selection in conjunction with the high heritability of the trait will produce an F_1 generation in which the average bristle number is greater than the average bristle number in the previous generation. In other words, the artificial selection imposed by the experimenter will induce an evolutionary change whose magnitude is related to the heritability of the trait. If the heritability of bristle number were zero, there would be no evolutionary change. (Figure 21.14b also shows the impact of natural selection on a trait with high heritability.)

A Mathematical Model of the Relation Between Heritability and Evolution

If you let S represent the average trait value (in this case, the average bristle number) of breeding individuals in the parental population, measured as the difference between the value of this trait for *parents* and the value of the trait in the entire parental population of both breeding and nonbreeding individuals, then $\boldsymbol{S}$ is a measure of the strength of selection on the trait; as such, it is called the **selection differential.** Now let R represent the average trait value in the offspring of these breeding parents, this time measured as the difference between the trait's value for *offspring* and its value in the entire parental population of breeding and nonbreeding individuals. Used in this way, $\boldsymbol{R}$ signifies the **response to selection,** that is, the amount of evolution, or change in mean trait value, resulting from selection. The heritability of the trait (arbitrarily designated as h^2), as seen in the slope of the line relating parental to offspring trait values in Fig. 12.16, determines the relationship between S and R.

$$R = h^2S \qquad (21.14)$$

In other words, the strength of selection (represented by S) and the heritability of a trait (h^2) directly determine the trait's amount or rate of evolution in each generation. This is the primary reason population and evolutionary geneticists consider the ability to measure heritability so important.

Variations in Polygenic Traits Arise at a Rapid Rate Because a Change at Any One of Many Loci Can Cause a Change in Phenotype

Geneticists have long used bristle number in *Drosophila* as a model for understanding the variation, selection, and evolution of quantitative traits. Early laboratory studies of selection acting on bristle number showed that the trait has substantial heritability in *Drosophila,* and it evolves rapidly in response to selection for either high or low bristle number (**Fig. 21.17**). Two results from these early studies were particularly striking. First, selection can rapidly lead to phenotypes not seen in the original population. After 35 generations of artificial selection for high or low bristle number, there was no overlap in the unselected and the selected populations. Some of the change in bristle number probably arose from reassortment and changes in frequency of existing alleles at multiple genes affecting the trait, without the appearance of new alleles. However, traits such as bristle number continue to evolve in response to selection for many generations. This observation suggests that new mutation is an additional source of variation in the population.

Experimenters have examined the contribution of mutation to genetic variation in bristle number (and by extension, other quantitative traits) through studies of highly inbred lines of *Drosophila* that at first had low or no genetic variation in bristle number. With these inbred lines, selection could occur only in the presence of new mutations affecting bristle number. Quantitative analyses revealed a significant selection-driven evolution of bristle number, which means that new mutations affecting bristle number arise in a population at a substantial rate (**Fig. 21.18**).

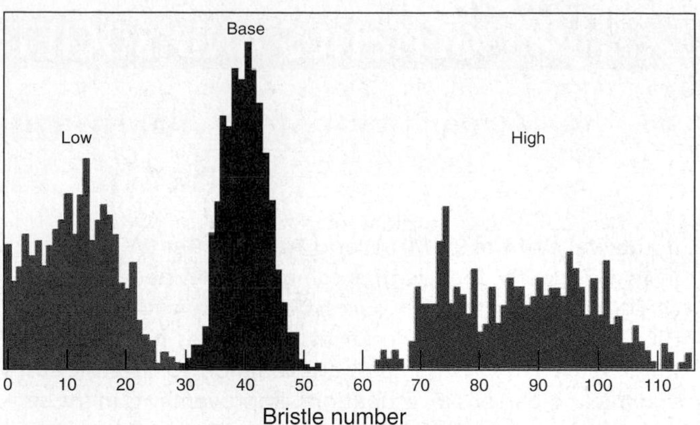

Figure 21.17 Evolution of abdominal bristle number in response to artificial selection in *Drosophila*. Bristle number distributions in different populations under selection for low (*green*) or high (*orange*) values, compared to distributions in a population not subjected to selection (base in *blue*).

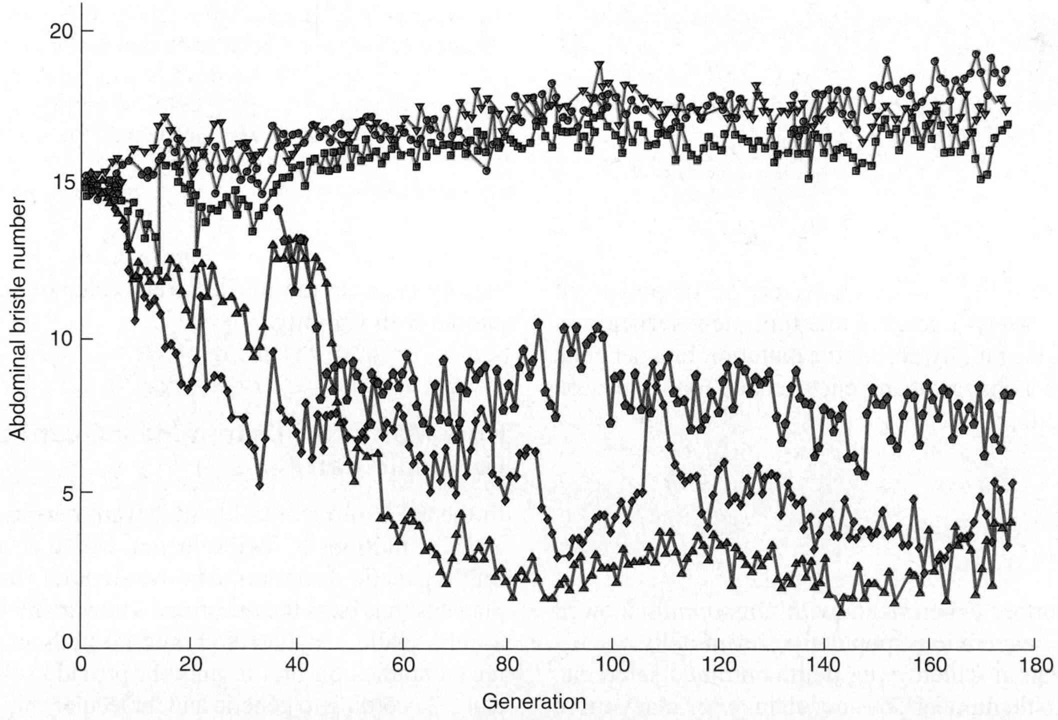

Figure 21.18 The effect of new mutations on mean bristle number in Drosophila. The average bristle number in a population under artificial selection for a reduced number over many generations is indicated with the *diamond, pentagon,* and *up-triangle* data points (*green*): Populations not under selection are indicated with the *down-triangle, circle,* and *square* data points (*red*).

How Improved Techniques of DNA Analysis Helped Identify Victims of 9/11

In the wake of the 9/11/01 World Trade Center (WTC) attacks in New York City, the scientific community refined techniques of DNA analysis to work successfully with badly damaged DNA and the lack of any prior tissue samples from some victims. Degraded DNA samples are common to forensic cases and missing persons investigations. Improvements in the sensitivity and scope of the methods for analyzing DNA extend the range of information recoverable from highly degraded specimens.

The crime scene following the WTC disaster covered 17 acres (**Fig. A**), from which searchers retrieved 19,893 separate body parts, including a single tooth. Fewer than 300 bodies were intact. Only 12 could be identified by sight. In addition, the DNA fragments in many of the recovered samples were very scarce and very small. Thus, shortly after the recovery efforts began, it became apparent that traditional means of DNA fingerprinting, which require long, intact pieces of DNA, would not be sufficient to identify the victims of 9/11. To characterize the small pieces with great enough accuracy, it would be necessary to develop new strategies and optimize existing protocols for use with small DNA fragments.

To this end, in October of 2001, the National Institute of Justice established the WTC Kinship and Data Analysis Panel (KADAP), a working group composed of 25 scientific experts. Among its members were analysts from the New York State Police, New York Department of Health, U.S. Department of Defense, National Institutes of Health, National Institute of Standards and Technology, the U.S. Department of Justice, and academia. Over the next three and a half years, KADAP met monthly to develop and evaluate new technologies for the analysis of DNA remains.

The DNA analysis and victim identification process had three main phases. First, the researchers gathered as much information as possible about the missing individuals. Personal effects, such as the victims' hairbrushes, razors, combs, dirty clothes, and toothbrushes, were one source of information; DNA reference samples from family members—including children, spouses, siblings, and parents—were another.

In phase two, these personal effects and family reference samples were subjected to DNA typing tests, including PCR analysis of various microsatellite and SNP alleles, described previously in Chapter 11. In some cases, the allele constitution of the missing person could be reconstructed directly from left-behind biological material. In other cases, a victim's microsatellite and SNP alleles could be inferred only through genetic analysis of surviving children and spouses: Any allele that was present in a child and not in the surviving parent must have come from the victim's genome.

In phase three, the resultant DNA profiles were loaded into computer databases for matching with DNA extracted from the human remains found at the WTC site. Several companies contributed to this third phase of the WTC identification efforts. A standardized, highly sensitive, automated microsatellite analysis was developed by the Bode Technology Group (Springfield, VA) for use on DNA extracted from the bones of WTC victims, as well as family-provided personal effects (see Fig. 11.12 on p. 405). This analysis retrieved genotypic information from 13 specific microsatellite loci that have been designated in the FBI's CODIS (Combined DNA Index System) database as representative of an individual's DNA fingerprint. Mitochondrial DNA sequencing was performed by Celera Genomics

These results highlight a key characteristic of polygenic traits: If many genes contribute to a trait, new variation in the trait may arise rapidly even if the mutation rate per gene is low, because a change at any one of many loci can cause a phenotypic difference.

Populations Eventually Reach a Selective Plateau

The bristle-number experiments with *Drosophila* showed that after many generations, populations eventually reach a selective plateau at which, even with continued selection, the average bristle number does not change for many more generations. The existence of such evolutionary plateaus suggests that selection can, for a time, eliminate all genetic variation in a trait and that the potential for new mutations allowing further extremes in phenotypes has been exhausted, usually because the most extreme phenotypes are incompatible with viability.

How Many Loci Determine a Quantitative, Polygenic Trait?

In theory, if a quantitative trait were determined by thousands or millions of loci, selection would never exhaust the trait's genetic variation. The observation of evolutionary plateaus thus begs the question of how many loci determine a trait. While it is hard to come up with an exact answer, recent studies on bristle number provide some tantalizing clues. According to genetic and molecular characterization of QTLs (quantitative trait loci, see Chapter 11, p. 421) affecting bristle number in *Drosophila,* 10–20 loci contribute the majority of genetic variation in bristle number, and substantial variation is attributable to alleles with large phenotypic

(Rockville, MD) (see Chapter 16), and Orchid GeneScreen (Dallas, TX) developed more sensitive methods to detect single nucleotide polymorphisms (SNPs) (see Fig. 11.11 on p. 403). Finally, to analyze and match genetic data obtained from the WTC victims, their personal effects, and their family reference samples, Gene Codes Forensics (Ann Arbor, MI) developed the Mass Fatality Identification System (M-FISys) software.

The microsatellite loci chosen for analysis are highly polymorphic, with four or more common and distinguishable alleles. SNP loci have only two alleles. To be reliable enough for use as forensic evidence, genotypic analysis usually requires the analysis of 13 microsatellite loci *or* the evaluation of 70 SNP. With the degraded DNA from the WTC, it was often impossible to meet either of these standards. To resolve this impasse, investigators often combined information obtained from microsatellite and SNP analysis. In tandem, these data provided a DNA profile that was useful in identifying more remains than either method alone.

In February 2005, the New York medical examiner's office announced that it had "exhausted all current technologies" and ended efforts to identify the remains of those killed three and a half years earlier at the World Trade Center. Of the 2749 people who died, 1100 remained unidentified. In addition, about 10,000 unidentified bone and tissue fragments had not yet been matched with the list of the missing. These will be held in the New York medical examiner's laboratory until identification is possible.

Lessons learned from the World Trade Center recovery and identification process were applicable in the aftermath of the tsunami that struck Thailand on December 26, 2004. The result of an undersea earthquake in the Indian Ocean, it killed an estimated 174,000–275,000 people. Thai forensic experts and disaster teams from more than 25 other nations have made positive matches on 2156 bodies, most of them tourists killed while on vacation. These identification efforts, based on continuously improving strategies of DNA analysis, will continue well into the future.

Figure A Wreckage of the World Trade Center in New York City, October 4, 2001.

effects at a small number of loci. It is thus likely that many polygenic traits are determined by tens, rather than hundreds or thousands, of loci, and that alleles of large effect at relatively few loci have considerable influence on such quantitative traits. If this turns out to be a general pattern, it will greatly simplify genetic analysis of normal and disease-causing variation in human and other populations.

Connections

We have seen that populations at Hardy-Weinberg equilibrium have unchanging allele frequencies, and in one generation, they achieve genotype frequencies of p^2, $2pq$, and q^2, which are subsequently maintained. In nature, where populations are rarely at complete Hardy-Weinberg equilibrium, however, natural selection acts on differences in fitness to alter allele frequencies. New mutations, genetic drift, and heterozygous advantage can also alter the allele frequencies of a population. For quantitative traits influenced by the alleles of two or more genes as well as by the interaction of those alleles with the environment, the heritability of a trait, that is, the proportion of its phenotypic variation attributable to genetic variation, determines its potential for evolution by mutation, selection, and genetic drift.

In Chapter 22, we look at evolution from the point of view of the molecular mechanisms that propel it. In that chapter, we examine the various ways in which changes at the genomic level continually reshuffle the genetic deck to create the ever-changing abundance of life-forms that inhabit the earth.

Essential Concepts

1. With the simplifying assumptions of an ideal population of very large size, where individuals mate at random, no new mutations appear, no individuals enter or leave, and there are no genotype-dependent differences in fitness, it becomes possible to derive a simple binomial equation that describes the precise relationships existing between allele, genotype, and phenotype frequencies. This equation, $p^2 + 2pq + q^2 = 1$, is called the Hardy-Weinberg law.
2. A population satisfying the Hardy-Weinberg assumptions is said to be at *Hardy-Weinberg equilibrium.* In such a population, allele frequencies remain constant from one generation to the next, and the genotype frequencies of p^2, $2pq$, and q^2 appear in one generation, after which they are maintained.
3. *Evolution* consists of changes in allele frequency over time. *Selection* acting on genotype-dependent differences in fitness can drive evolution. Selection does not entirely eliminate deleterious recessive alleles from a population. One reason for this is *heterozygous advantage.*
4. The existence of an *evolutionary equilibrium* is another reason deleterious recessive alleles persist in populations. The evolutionary equilibrium is a balance between mutation to a new allele and selection against that allele. Late onset of a disease undermines selection against the disease allele, while *genetic drift* has an unpredictable effect on the evolutionary equilibrium.
5. For quantitative traits, the *environmental variance* is a measure of the influence of environment on phenotypic variation. Similarly, *genetic variance* measures the contribution of genes to phenotypic variation. *Total phenotype variance* is the sum of genetic variance and environmental variance.
6. Measures of environmental, genetic, and total phenotype variance make it possible to define the *heritability* of a trait as the proportion of total phenotype variance attributable to genetic variance. With traits for which the number and identity of contributing genes remain unknown and there is no way to obtain genetic clones, it is possible to correlate phenotypic variation with the genetic relatedness of individuals—that is, the average fraction of common alleles at all gene loci that the individuals share because they inherited them from a common ancestor—to measure the heritability of a trait.
7. To ascertain the heritability of a human trait, population geneticists often turn to studies of twins. The most useful approach is to compare the phenotypic differences between pairs of monozygotic and dizygotic twins. Environmental changes can always influence the degree of heritability.
8. The heritability of a trait quantifies its potential for sustaining selection and thus its potential for evolution from one generation to the next.
9. Variations in polygenic traits arise at a rapid rate through selection because changes at any one of many loci can contribute to changes in phenotype. Nevertheless, it is possible that many polygenic traits are influenced primarily by tens, rather than thousands, of loci.

On Our Website

www.mhhe.com/hartwell3
Chapter 21

Annotated Suggested Readings and Links to Other Websites

- More on correlations between mutant hemoglobin alleles and malaria
- Monte Carlo simulations of effects of population size, drift, and unequal allelic fitness on changes in allele frequencies over time

Specialized Topics

- Modern changes in allele frequencies over time in Darwin's finches
- The use of data obtained from identical twins raised apart to determine genetic and nongenetic contributions to behavioral traits in human beings

Social and Ethical Issues

1. Zoos often house endangered species and breed them to increase the population size and reintroduce individuals into the wild. A problem with some endangered species is that, in the wild as well as in the zoos, there is very little genetic diversity remaining. Should it be a policy to try to save a species by mating individuals from different geographic locales in the hope of generating more genetic diversity and thereby increasing the possibility of survival? Or should the focus instead be to breed existing animals to try to increase their numbers even though the inbreeding may lead to the expression of additional recessive diseases? Are these activities that you would want your city zoo to spend money on? What is the human responsibility in preserving other species?

2. Periodically in the last 30 years, there have been efforts to determine the heritability of IQ. Unfortunately most people without any knowledge of genetics do not understand that the concept of heritability is applicable only to populations, not individuals, and that the actual value is dependent on environmental conditions. Is it ethical to even do studies on the heritability of IQ or other politically or socially charged topics when the data are easily misinterpreted or misused? Is there a way to prevent these studies from being done or to prevent their inappropriate application?

Solved Problems

I. A population called the "founder generation," consisting of 2000 *AA* individuals, 2000 *Aa* individuals, and 6000 *aa* individuals is established on a remote island. Mating within this population occurs at random, the three genotypes are selectively neutral, and mutations occur at a negligible rate.

a. What are the frequencies of alleles *A* and *a* in the founder generation?

b. Is the founder generation at Hardy-Weinberg equilibrium?

c. What is the frequency of the *A* allele in the second generation (that is, the generation subsequent to the founder generation)?

d. What are the frequencies for the *AA, Aa,* and *aa* genotypes in the second generation?

e. Is the second generation at Hardy-Weinberg equilibrium?

f. What are the frequencies for the *AA, Aa,* and *aa* genotypes in the third generation?

Answer

This question requires calculation of allele and genotype frequencies and an understanding of the Hardy-Weinberg equilibrium principle.

a. To calculate allele frequencies, count the total alleles represented in individuals with each genotype and divide by the total number of alleles.

Number of individuals	Number of A alleles	Number of *a* alleles
2000 AA	4000	0
2000 A*a*	2000	2000
6000 *aa*	0	12,000
Total	6000	14,000

The frequency of the A *allele* (p) = *6000/20,000* = *0.3.*
The frequency of the a *allele* (q) = *14,000/20,000* = *0.7.*

b. If a population is at Hardy-Weinberg equilibrium, the genotype frequencies are p^2, $2pq$, and q^2. We calculated in part *a* that $p = 0.3$ and $q = 0.7$ in this population. Therefore,

$$p^2 = 0.09$$

$$2pq = 2\,(0.3)(0.7) = 0.42$$

$$q^2 = 0.49$$

For a population of 10,000 individuals, the number of individuals with each genotype, if the population were at equilibrium and the allele frequencies were $p = 0.3$ and $p = 0.7$, would be *AA,* 900; *Aa,* 4200; and *aa,* 4900. *The founder population described therefore is not at equilibrium.*

c. Given the conditions of random mating, selectively neutral alleles, and no new mutations, allele frequencies do not change from one generation to the next; $p = 0.3$, *and* $q = 0.7$.

d. *The genotype frequencies for the second generation would be those calculated for part* b because in one generation the population will go to equilibrium. $AA = p^2 = 0.09$; $Aa = 2\,pq = 0.42$; and $aa = q^2 = 0.49$.

e. *Yes,* in one generation a population not at equilibrium will go to equilibrium if mating is random and there is no selection or significant mutation.

f. *The genotype frequencies will be the same in the third generation as in the second generation.*

II. Two alleles have been found at the X-linked phosphoglucomutase gene (*Pgm*) in *Drosophila persimilis* populations in California. The frequency of the Pgm^A allele is 0.25, while the frequency of the Pgm^B allele is 0.75. Assuming the population is at Hardy-Weinberg equilibrium, what are the expected genotype frequencies in males and females?

Answer

This problem requires application of the concept of allele and genotype frequencies to X-linked genes. For X-linked genes, males (XY) have only one copy of the X chromosome, so the genotype frequency is equal to the allele frequency. Therefore, $p = 0.25$ and $q = 0.75$. *The frequency of male flies with genotype $X^{PgmA}Y$ is 0.25; the frequency of males with genotype $X^{PgmB}Y$ is 0.75.* Three genotypes exist for females: $X^{PgmA}X^{PgmA}$, $X^{PgmA}X^{PgmB}$, X^{PgmB} and X^{PgmB} corresponding to p^2, $2pq$, and q^2. *The frequencies of female flies with these three genotypes are* $(0.25)^2$, $2(0.25)(0.75)$, and $(0.75)^2$; or 0.0625, 0.375, and 0.5625 respectively.

III. Two hypothetical lizard populations found on opposite sides of a mountain in the Arizonan desert have two alleles (A^F, A^S) of a single gene A with the following three genotype frequencies:

	$A^F A^F$	$A^F A^S$	$A^S A^S$
Population 1	38	44	18
Population 2	0	80	20

a. What is the allele frequency of A^F in the two populations?
b. Do either of the two populations appear to be at Hardy-Weinberg equilibrium?
c. A huge flood opened a canyon in the mountain range separating populations 1 and 2. They were then able to migrate such that the two populations, which were of equal size, mixed completely and mated at random. What are the frequencies of the three genotypes (A^FA^F, A^FA^S, and A^SA^S) in the next generation of the single new population of lizards?

Answer

This question requires calculation of allele frequencies and genotype frequencies in existing and in newly created populations.

a. The frequency of allele A is calculated in the following way:

38 A^FA^F	× 2	76 A^F *alleles*
44 A^FA^S	× 2	44 A^F *alleles*
		120 A^F *alleles*

120 A^F alleles/200 total alleles = 0.6.

b. For population 1, the allele frequencies are $p = 0.6$ and $q = 0.4$. Genotype frequencies when the population is in equilibrium are

$$p^2 = (0.6)^2 = 0.36$$
$$2pq = 2(0.6)(0.4) = 0.48$$
$$q^2 = (0.4)^2 = 0.16$$

For population 1, which consists of 100 individuals, the equilibrium would be 36 A^FA^F, 48 A^FA^S, and 16 A^SA^S lizards. *Population 1 does seem to be at equilibrium.* (Sampling error and small population size could lead to slight variations from the expected frequencies.) For population 2, the allele frequency (p) is based solely on the number of A^F alleles from the 80 A^FA^S individuals. The total number of alleles = 200, so the frequency of A^F alleles is 80/200 or 0.4. The genotype frequencies for a population at equilibrium would be

$$p^2 = (0.4)^2 = 0.16$$
$$2pq = 2(0.4)(0.6) = 0.48$$
$$q^2 = (0.6)^2 = 0.36$$

Population 2 does not seem to be at equilibrium.

c. The combination of the two populations of lizards results in one population with the following allele frequencies:

A^F *alleles*		
$A^F A^F$	38 × 2	76
$A^F A^S$	44	44
$A^F A^S$	80	80
Total:		200

A^S *alleles*		
$A^F A^S$	44	44
$A^F A^S$	80	80
$A^S A^S$	18 × 2	36
$A^S A^S$	20 × 2	40
Total:		200

The allele frequencies are 200/400, or 0.5, for both p and q. *The genotype frequencies in the next generation will therefore be*

$$p^2 = (0.5)^2 = 0.25$$
$$2pq = 2(0.5)(0.5) = 0.5$$
$$q^2 = (0.5)^2 = 0.25$$

Problems

Vocabulary

1. Choose the best matching phrase in the right column for each of the terms in the left column.

a. fitness	1. the genotype with the highest fitness is the heterozygote
b. gene pool	2. chance fluctuations in allele frequency
c. fitness cost	3. ability to survive and reproduce
d. allele frequency	4. proportion of total phenotypic variance representation attributed to genetic variance
e. genotype frequency representation	5. collection of alleles carried by all members of a population
f. heterozygote advantage	6. p^2 and q^2
g. equilibrium frequency	7. p and q
h. genetic drift	8. the advantage of a particular genotype in one situation is a disadvantage in another situation
i. heritability	9. frequency of an allele at which $\Delta q = 0$

Section 21.1

2. In a certain population of frogs, 120 are green, 60 are brownish green, and 20 are brown. The allele for brown is denoted G^B, while that for green is G^G, and these two alleles show incomplete dominance relative to each other.

a. What are the genotype frequencies in the population?

b. What are the allele frequencies of G^B and G^G in this population?

c. What are the expected frequencies of the genotypes if the population is at Hardy-Weinberg equilibrium?

3. Which of the following populations are at Hardy-Weinberg equilibrium?

Population	*AA*	*Aa*	*aa*
a	0.25	0.50	0.25
b	0.10	0.74	0.16
c	0.64	0.27	0.09
d	0.46	0.50	0.04
e	0.81	0.18	0.01

4. A dominant mutation in *Drosophila* called *Delta* causes changes in wing morphology in *Delta* / + heterozygotes. Homozygosity for this mutation (*Delta* / *Delta*) is lethal. In a population of 150 flies, it was determined that 60 had normal wings and 90 had abnormal wings.

a. What are the allele frequencies in this population?

b. Using the allele frequencies calculated in part *a*, how many total zygotes must be produced by this population in order for you to count 160 viable adults in the next generation?

c. Given that there is random mating, no migration, and no mutation, and ignoring the effects of genetic drift, what are the expected numbers of the different genotypes in the next generation if 160 viable offspring of the population in part *a* are counted?

d. Is this next generation at Hardy-Weinberg equilibrium? Why or why not?

5. A large, random mating population is started with the following proportion of individuals for the indicated blood types:

0.5 MM
0.2 MN
0.3 NN

This blood type gene is autosomal and the *M* and *N* alleles are codominant.

a. Is this population at Hardy-Weinberg equilibrium?

b. What will be the allele and genotype frequencies after one generation under the conditions assumed for the Hardy-Weinberg equilibrium?

c. What will be the allele and genotype frequencies after two generations under the conditions assumed for the Hardy-Weinberg equilibrium?

6. A gene called *Q* has two alleles, Q^F and Q^G, that encode alternative forms of a red blood cell protein that allows blood group typing. A different, independently segregating gene called *R* has two alleles, R^C and R^D, permitting a different kind of blood group typing. A random, representative population of football fans was examined, and on the basis of their blood typing, the following distribution of genotypes was inferred (all genotypes were equally distributed between males and females):

$Q^FQ^F\ R^CR^C$	202
$Q^FQ^G\ R^CR^C$	101
$Q^GQ^G\ R^CR^C$	101
$Q^FQ^F\ R^CR^D$	372
$Q^FQ^G\ R^CR^D$	186
$Q^GQ^G\ R^CR^D$	186
$Q^FQ^F\ R^DR^D$	166
$Q^FQ^G\ R^DR^D$	83
$Q^GQ^G\ R^DR^D$	83

This sample contains 1480 fans.

a. Is the population at Hardy-Weinberg equilibrium with respect to either or both of the *Q* and *R* genes?

b. After one generation of random mating within this group, what fraction of the *next* generation of football fans will be Q^FQ^F (independent of their *R* genotype)?

c. After one generation of random mating, what fraction of the *next* generation of football fans will be $R^C R^C$ (independent of their Q genotype)?
d. What is the chance that the first child of a $Q^F Q^{G-}$ $R^C R^D$ female and a $Q^F Q^F R^C R^D$ male will be a Q^F Q^G $R^D R^D$ male?

7. A population with an allele frequency (p) of 0.5 and a genotype frequency (p^2) of 0.25 is at equilibrium. How can you explain the fact that a population with an allele frequency (p) of 0.1 and a genotype frequency (p^2) of 0.01 is also at equilibrium?

8. When an allele is dominant, why does it not always increase to produce the phenotype proportion of 3:1 (3/4 dominant : 1/4 recessive individuals) in a population?

9. It is the year 1998, and the men and women sailors (in equal numbers) on the American ship the *Medischol Bounty* have mutinied in the South Pacific and settled on the island of Bali Hai, where they have come into contact with the local Polynesian population. Of the 400 sailors that come ashore on the island, 324 have MM blood type, 4 have the NN blood type, and 72 have the MN blood type. Already on the island are 600 Polynesians between the ages of 19 and 23. In the Polynesian population, the allele frequency of the M allele is 0.06, and the allele frequency of the N allele is 0.94. No other people come to the island over the next 10 years.
a. What is the allele frequency of the N allele in the sailor population that mutinied?
b. It is the year 2008, and 1000 children have been born on the island of Bali Hai. If the mixed population of 1000 young people on the island in 1998 mated randomly and the different blood group phenotypes had no effect on viability, how many of the 1000 children would you expect to have MN blood type?
c. In fact, 50 children have MM blood type, 850 have MN blood type, and 100 have NN blood type. What is the observed frequency of the N allele among the children?

10. Alkaptonuria is a recessive autosomal genetic disorder associated with darkening of the urine. In the United States, approximately one out of every 250,000 people have alkaptonuria.
a. Assuming Hardy-Weinberg equilibrium, estimate the frequency of the allele responsible for this trait.
b. What proportion of people in the U.S. population are carriers for this trait? In this population, what is the ratio of carriers to individuals affected by alkaptonuria?
c. If a woman without alkaptonuria who had a child with this trait with one husband then remarried, what is the chance that a child produced by her second marriage would have alkaptonuria?
d. Alkaptonuria is a relatively benign condition, so there is little selective advantage to individuals with any genotype; as a result, your assumption of Hardy-Weinberg equilibrium in part *a* is reasonable. Could you also use the assumption of Hardy-Weinberg equilibrium to estimate the allele frequencies and carrier frequencies of more severe recessive autosomal conditions such as cystic fibrosis? Explain.

11. The equation $p^2 + 2pq + q^2 = 1$ representing the Hardy-Weinberg proportions examines genes with only two alleles in a population.
a. Derive a similar equation describing the equilibrium proportions of genotypes for a gene with three alleles. [*Hint*: Remember that the Hardy-Weinberg equation can be written as the binomial expansion $(p + q)^2$.]
b. A single gene with three alleles (I^A, I^B, and i) is responsible for the ABO blood groups. Individuals with blood type A can be either I^A I^A or I^A i; those with blood type B can be either I^B I^B or I^B i; people with AB blood are I^A I^B, and type O individuals are ii. Among Armenians, the frequency of I^A is 0.360, the frequency of I^B is 0.104, and the frequency of i is 0.536. Calculate the frequencies of individuals in this population with the four possible blood types, assuming Hardy-Weinberg equilibrium.

12. a. Alleles of genes on the X chromosome can also be at equilibrium, but the equilibrium frequencies under the Hardy-Weinberg assumptions must be calculated separately for the two sexes. For a gene with two alleles A and a at frequencies of p and q, respectively, write expressions that describe the equilibrium frequencies for all the genotypes in men and women.
b. Approximately 1 in 10,000 males in the United States is afflicted with hemophilia, an X-linked recessive condition. If you assume that the population is at Hardy-Weinberg equilibrium, what proportion of American females would be hemophiliacs? About how many female hemophiliacs would you expect to find among the 100 million women living in the United States?

13. In 1927, the opthamologist George Waaler tested 9049 schoolboys in Oslo, Norway for red-green colorblindness and found 8324 of them to be normal and 725 to be colorblind. He also tested 9072 schoolgirls and found 9032 that had normal color vision while 40 were colorblind.
a. Assuming that the same sex-linked recessive allele c causes all forms of red-green colorblindness, calculate the allele frequencies of c and C (the allele for normal vision) from the data for the schoolboys. (*Note*: Refer to your answer to problem #12a above.)
b. Does Waaler's sample demonstrate Hardy-Weinberg equilibrium for this gene? Explain your answer

by describing observations that are either consistent or inconsistent with this hypothesis.

On closer analysis of these schoolchildren, Waaler found that there was actually more than one *c* allele causing colorblindness in his sample: one kind for the "prot" type (c^p) and one for the "deuter" type (c^d) (protanopia and deuteranopia are slightly different forms of red-green colorblindness). Importantly, some of the "normal" females in Waaler's studies were probably of genotype c^p/c^d. Through further analysis of the 40 colorblind females, he found that 3 were prot (c^p/c^p), and 37 were deuter (c^d/c^d).

c. Based on this new information, what is the frequency of the c^p, c^d, and *C* alleles in the population examined by Waaler? Calculate these values as if the frequencies obey the Hardy-Weinberg equilibrium. (*Note:* Refer to your answer to problem 11.a above.)
d. Calculate the frequencies of all genotypes among men and women expected if the population is at equilibrium.
e. Do these results make it more likely or less likely that the population in Oslo is indeed at equilibrium for red-green colorblindness? Explain your reasoning.

14. A new university on a Caribbean island has recruited its 700 faculty members from colleges in France and Kenya. Five hundred came from France and 200 came from Kenya, with equal numbers of men and women in both groups. Upon arrival, you notice that 90 of the French and 75 of the Kenyans express a peculiar trait of rolling their eyes up into their sockets when asked a stupid question. Upon studying this trait, you discover that it is always due to the expression of a dominant allele at a single gene called *Ugh.* Field trips taken to both Kenya and France indicate that the two alleles at the *Ugh* locus are at Hardy-Weinberg equilibrium in both of these separate populations. All of the faculty members arrived on the island single, but after teaching for a few years, they all married other faculty members in a random manner. Among 1000 progeny from these marriages, how many children do you expect will express the eye-rolling phenotype?

15. In *Drosophila,* the vestigial wings recessive allele, *vg,* causes the wings to be very small. A geneticist crossed some true-breeding wild-type males to some vestigial virgin females. The male and female F_1 flies were wild type. He then allowed the F_1 flies to mate and found that 1/4 of the male and female F_2 flies had vestigial wings. He dumped the vestigial F_2 flies into a morgue and allowed the wild-type F_2 flies to mate and produce an F_3 generation.

a. Give the genotype and allele frequencies among the wild-type F_2 flies.
b. What will be the frequencies of wild-type and vestigial flies in the F_3?
c. Assuming the geneticist repeated the selection against the vestigial F_3 flies (that is, he dumped them in a morgue and allowed the wild-type F_3 flies to mate at random), what will be the frequency of the wild-type allele and mutant alleles in the F_4 generation?
d. Now the geneticist lets all of the F_4 flies mate at random (that is, both wild-type and vestigial flies mate). What will be the frequencies of wild-type and vestigial F_5 flies?

16. A mouse mutation with incomplete dominance (*t* = *tailless*) causes short tails in heterozygotes (t^+/t). The same mutation acts as a recessive lethal that causes homozygotes (t/t) to die in utero. In a population consisting of 150 mice, 60 are t^+/t^+ and 90 are heterozygotes.

a. What are the allele frequencies in this population?
b. Given that there is random mating among mice, no migration, and no mutation, and ignoring the effects of random genetic drift, what are the expected numbers of the different genotypes in this next generation if 200 offspring are born?
c. Two populations (called Dom 1 and Dom 2) of mice come into contact and interbreed randomly. These populations initially are composed of the following numbers of wildtype (t^+/t^+) homozygotes and tailless (t^+/t) heterozygotes.

	Dom 1	*Dom 2*
Wild type	16	48
Tailless	48	36

What are the frequencies of the two genotypes in the next generation?

Section 21.2

17. Why is the elimination of a fully recessive deleterious allele by natural selection difficult?

18. Would you expect to see a greater Δq from one generation to the next in a population with an allele frequency (q) of 0.2 or in a population with an allele frequency of 0.02? Assume relative fitness is the same in both populations and that the equilibrium frequency for q is 0.01.

19. You have identified an autosomal gene that contributes to tail size in male guppies, with a dominant allele *B* for large tails and a recessive allele *b* for small tails. Female guppies of all genotypes have similar tail sizes. You know that female guppies usually mate with males with the largest tails, but the effects of population density and the ratio of the sexes on this preference have not been studied. You therefore place an equal number of males in three tanks. In tank 1, the number of females is twice the number of males. In tank two, the numbers of males and females are equal. In tank 3, there are half as many females as males. After mating, you find the following proportions of small-tailed males among the progeny: tank 1, 16%; tank 2, 25%; tank 3, 30%.

a. In your original population, 25% of the males have small tails. Assuming that the allele frequencies in males and females are the same, calculate the frequencies of *B* and *b* in your original population.
b. Calculate Δq for each tank.
c. If $w_{BB} = 1.0$, what is w_{Bb} for each tank?
d. If $w_{BB} = 1.0$, is w_{bb} less than, equal to, or greater than 1.0 for each tank?

20. An allele of the *G6PD* gene acts in a recessive manner to cause sensitivity to fava beans, resulting in a hemolytic reaction (lysis of red blood cells) after ingestion of the beans. The same allele also confers dominant resistance to malaria. The heterozygote has an advantage in a region where malaria is prevalent. Will the equilibrium frequency (q_e) be the same for an African and a North American country? What factors affect q_e?

21. In Europe, the frequency of the CF^- allele causing the recessive autosomal disease cystic fibrosis is about 0.04. Cystic fibrosis causes death before reproduction in virtually all cases.

a. Determine values of fitness (w) and of the selection coefficient (s) for the unaffected, carrier, and affected genotypes.
b. Determine the average fitness at birth of the population as a whole with respect to the cystic fibrosis trait ($\overline{w}$) and the expected change in allele frequency over one generation (Δq) when measured at the birth of the next generation.

Now suppose that the mutation rate from CF^+ to CF^- alleles is 1×10^{-6}.

c. What is the expected evolutionary equilibrium frequency ($\hat{q}$) of the CF^- allele? Is this larger or smaller than the observed frequency?
d. Without changing the value of s for the CF^-/CF^- genotype you calculated in part *a*, propose an explanation that might resolve the discrepancy between the observed and expected frequencies of the CF^- allele you noted in part *c*.

Section 21.3

22. How can each of the following be used in determining the role of genetic and/or environmental factors in phenotypic variation in different organisms?

a. genetic clones
b. human monozygotic versus dizygotic twins
c. cross-fostering

23. Which of the following statements would be true of a human trait that has high heritability in a population of one country?

a. The phenotypic difference within monozygotic twin pairs would be about the same as the phenotypic differences among members of dizygotic twin pairs.
b. There is very little phenotypic variation between monozygotic twins but high variability between dizygotic twins.
c. The trait would have the same heritability in a population of another country.

24. a. Studies have indicated that for pairs of twins raised in the same family, the environmental similarity for monozygotic (MZ) twins is not significantly different from the environmental similarity for fraternal (dizygotic or DZ) twins. Why is this an important fact for calculations of heritability?
b. If you wished to determine the heritability of a particular trait in humans, would it be more useful to study MZ or DZ twins? Explain.

25. A study published in 1937 examined the average differences between pairs of twins [either monozygotic (MZ) or dizygotic (DZ)] and pairs of siblings for three different traits: height, weight, and intelligence quotient (IQ) as measured by the Stanford-Binet test. (The concept of "IQ" is extremely controversial as it is unclear to what extent IQ tests meaure native intelligence, but for this problem, consider IQ as a measurable phenotype even if its significance is unknown.) Some of the MZ twins were raised together in the same household (RT), while other MZ twins were raised apart in different families (RA). The results of this study are as follows:

	MZ/RT	MZ/RA	DZ	Siblings
Height	1.7 cm	1.8 cm	4.4 cm	4.5 cm
Weight	1.86 kg	4.49 kg	4.54 kg	4.72 kg
IQ	5.9	8.2	9.9	9.8

a. Which of these three traits appears to have the highest heritability? The lowest heritability?
b. The Centers for Disease Control and Prevention (CDC) of the National Institutes of Health recently reported that in the United States during the period 1960–2002, the average weight of a 15-year-old boy increased from 135.5 pounds (61.46 kg) to 150.3 pounds (68.17 kg). During the same period, the average height of a 15-year-old boy increased from 67.5 inches (171.5 cm) to 68.4 inches (173.7 cm). How do these statistics match your estimates of heritability from part *a*?

26. Two different groups of scientists studying a rare trait in ground squirrels report very different heritabilities. What factors influencing heritability values make it possible for both conclusions to be correct?

27. Human geneticists have found the Finnish population to be very useful for studies of a variety of conditions. The Finnish population is small; Finns have extensive church records documenting lineages; and few people have migrated into the population. The frequency of some recessive disorders is higher in the Finnish population than elsewhere in the world; and diseases such as PKU and cystic fibrosis that are common elsewhere do not occur in the Finnish population.

a. How would a population geneticist explain these variations in disease occurrence?

b. The Finnish population is also a source of information for the study of quantitative traits. The genetic basis of schizophrenia is one question that can be explored in this population. What advantage(s) and disadvantage(s) can you imagine for studying complex traits based on the Finnish population structure?

28. Two traits with similar phenotype variance exist in a population. If one trait has two major genes and six minor loci that influence the phenotype, and the second trait has 12 minor loci and no major genes affecting the phenotype, which trait would you expect to respond most consistently to selection? Explain.

29. Two alleles at one locus produce 3 distinct phenotypes. Two alleles of two genes lead to 5 distinct phenotypes. Two alleles of six genes lead to 13 distinct phenotypes. (These statements assume that the alleles at any one locus are codominant and that each gene makes an equal contribution to the phenotype.)

a. Derive a formula to express this relationship. (Let n equal the number of genes.)

b. Each of the most extreme phenotypes for a trait determined by two alleles at one locus are found in a proportion of 1/4 in the F_2 generation. If there are two alleles of two genes that determine a trait, each extreme phenotype will be present in the F_2 as 1/16 of the population. In common wheat (*Triticum aestivum*), kernel color varies from red to white and the genes controlling the color act additively, that is, alleles for each gene are semidominant and each gene contributes equally to the color. A true-breeding red variety is crossed to a true-breeding white variety, and 1/256 of the F_2 have red kernels and 1/256 have white kernels. How many genes control kernel color in this cross?

30. In a certain plant, leaf size is determined by four independently assorting genes acting additively. Thus, alleles *A*, *B*, *C*, and *D* each adds 4 cm to leaf length and alleles *A′*, *B′*, *C′*, and *D′* each adds 2 cm to leaf length. Therefore, an *A/A, B/B, C/C, D/D* plant has leaves 32 cm long and an *A′/A′, B′/B′, C′/C′, D′/D′* plant has leaves 16 cm long.

a. If true-breeding plants with leaves 32 cm long are crossed to true-breeding plants with leaves 16 cm long, the F_1 will have leaves 24 cm long, that is, *A/A′, B/B′, C/C′, D/D′*. List all possible leaf lengths and their expected frequencies in the F_2 generation produced from these F_1 plants.

b. Now assume that in a randomly mating population the following allele frequencies occur:

frequency of A = 0.9
frequency of A′ = 0.1
frequency of B = 0.9
frequency of B′ = 0.1
frequency of C = 0.1
frequency of C′ = 0.9
frequency of D = 0.5
frequency of D′ = 0.5

Calculate separately the expected frequency in this population of the three possible genotypes for each of the four genes.

c. What proportion of the plants in the population described in part *b* will have leaves that are 32 cm long?

Evolution at the Molecular Level

Chapter 22

From December 1831 to October 1836, Charles Darwin circled the globe as naturalist for the HMS *Beagle* (**Fig. 22.1**). He was 22 years of age when he set sail. With indefatigable energy and insatiable curiosity, he dredged the oceans of the world for samples of the myriad organisms they concealed. He scoured the pampas of Uruguay for fossils and unusual contemporary species of lizards, birds, and mammals; scouted the no-man's-land of Tierra del Fuego for signs of unexpected life-forms; and climbed the highest peaks of the Andes, where he found deposits of seashells thousands of feet above sea level. Darwin also collected sundry specimens of the various species inhabiting different islands of the Galápagos. Whenever possible, he preserved and shipped his discoveries back to England. Some 23 years later, in 1859, he published *On the Origin of Species by Means of Natural Selection,* a comprehensive distillation of his thinking on what he had observed. The first printing of 1250 copies sold out in one day.

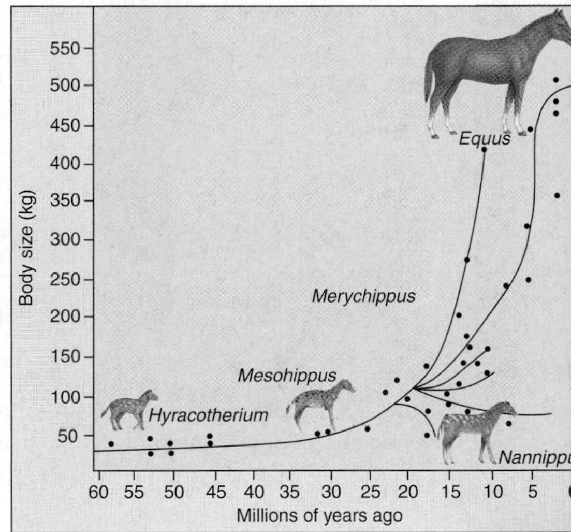

Lines show the broad outline of evolutionary relationships in the body size of horses. Most changes involved an increase in body size, but a few actually reflect a decrease in body size.

In *The Origin of Species,* Darwin uses an extensive comparative analysis of thousands of specimens and fossils as the basis for proposing that "the similar framework of bones in the hand of a man, wing of a bat, fin of the porpoise, and leg of the horse—the same number of vertebrae forming the neck of the giraffe and of the elephant—and innumerable other such facts, at once explain themselves on the theory of descent with slow and slight successive modifications." This observation leads him to the stunning conclusion that "all organic beings which have ever lived on this earth may be descended from some one primordial form."

While Darwin was not the first to suggest that species could undergo evolution, he was the first to suggest a mechanism by which evolution could occur. He based his theory of evolution on three principles—each obvious in and of itself—whose combination had revolutionary implications. First, within any species, there is variation among the individuals of a population in the expression of numerous traits. Second, variant forms of traits can be passed down through inheritance from one generation to the next. Third, some variant traits give the individuals that express them a greater chance of surviving and reproducing. (This is the so-called "survival of the fittest" principle.)

Darwin recognized that an advantageous trait that at first appears in only one or a few individuals could allow those individuals and their descendants to outcompete individuals that do not express the trait. This process of "natural selection," that is, of natural conditions selecting for an advantageous trait, would result in the transmission of the trait to a greater proportion of the population in each successive generation. Ultimately, many generations of natural selection would produce a population whose members all expressed a particular advantageous trait. Meanwhile, variant forms of other traits would arise in different individuals in each generation, and natural selection would, for each trait, determine which variants survive over time. Darwin's understanding of the process of continuous evolution by natural selection became the cornerstone of his theory of evolution. In *The Origin of Species,* he nevertheless cautions that "the natural selection of numerous successive, slight, favorable variations . . . has been the most important, but not the exclusive, means of modification."

Despite the enormous range of his revolutionary insight, Darwin was at a loss to explain the source of the visible variation on which natural selection acts. He knew inheritance plays a role: "Everyone must have heard of cases of albinism, prickly skin, hairy bodies, etc., appearing in several members of the same family. If strange

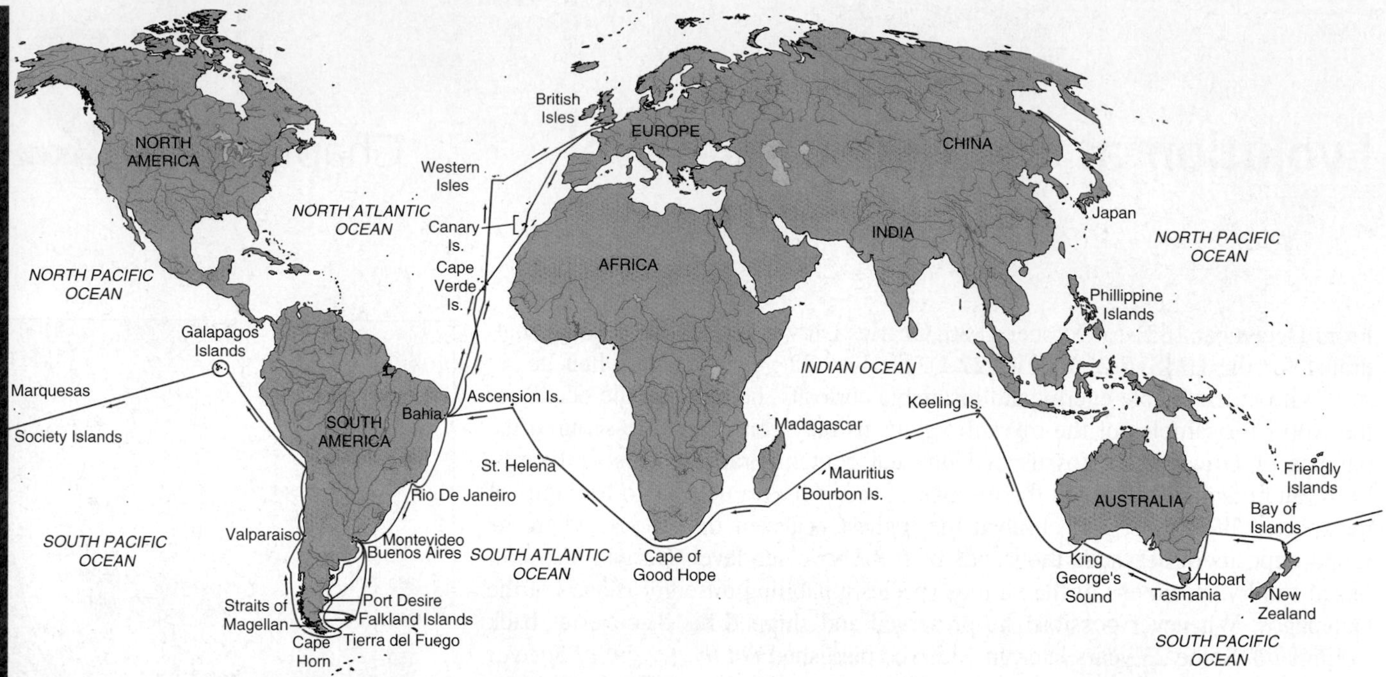

Figure 22.1 The voyage of the HMS *Beagle*. Charles Darwin spent 5 years circling the globe collecting specimens. The material acquired on this grand tour provided the basis for his comparative analysis in *The Origin of Species*, published some 23 years after he returned to England.

and rare deviations of structure are really inherited, less strange and commoner deviations may be freely admitted to be inheritable." But he did not understand how: "The laws governing inheritance are for the most part unknown." Mendel published "Experiments on Plant Hybrids" in 1866 (just seven years after Darwin's *The Origin of Species* appeared), but as we saw in Chapter 2, although Darwin received a copy of Mendel's paper, he most likely never read it.

Today, over 140 years after the publication of *The Origin of Species,* biologists accept Darwin's theory as a foundation of modern biology. Moreover, thanks to Mendel (and many other geneticists), they now understand the basic principles of heredity. And thanks to Watson and Crick (and many other molecular biologists), they know that ultimately, evolution is a process that begins at the molecular level, inside the double helix of DNA.

An example of short-term competitive evolution at the molecular level is seen in the battle between cells of the human immune system and the AIDS virus. Both evolve by the diversification of progenitor populations followed by the selective amplification of some divergent forms. When an invading virus activates the immune system, a virus-specific immune response ensues. As we saw in the Fast Forward box on pp. 492–494 of Chapter 14, specific human immune responses depend on the cellular diversity of the immune system, which includes a trillion (10^{12}) circulating white blood cells, or lymphocytes, divided into two types—T cells and B cells. Populations of lymphocytes have the genetic capacity to synthesize a large and varied group of cell surface immune receptors collectively capable of recognizing foreign (that is, nonself) macromolecular structures on virtually any virus, bacterium, or other pathogenic microorganism (**Fig. 22.2**). Somatic mutations and gene rearrangements in the cells of the developing immune system enable this enormous diversification of receptors from a relatively small repertoire of genes.

However, although populations of lymphocytes carry a great diversity of immune receptors, each individual lymphocyte synthesizes only one type of immune receptor. This attribute of one type of immune receptor per cell is the key to the specificity of an immune response. When the multiple copies of that one type of receptor in a cell's membrane encounter complementary foreign structures (called *antigenic determinants*), interaction between the immune receptors and the antigenic determinants triggers the lymphocyte to divide and differentiate.

The rapid proliferation of just a few selected lymphocytes to a few clones of genetically identical cells is the first step of an immune response. Among the cells in each expanded clone are *memory cells,* which may live as long as 40 years, and *effector cells,* which actually carry out (or effect) the immune reactions that help dispose of the microorganisms. These effector cells include T effector cells that bind to the antigenic determinants of a bacterium or virus and B effector cells that secrete antibodies, which in turn bind to antigenic determinants. As the immune response progresses, some of the B effector cells mutate their membrane receptors (generating further diversification). The altered receptors with the closest fit to the antigenic determinants bind the determinants more tightly, and this high-affinity binding drives the amplification of the lymphocytes that carry the altered receptors (selective amplification). The differentiated effector cells become involved in a multitude of effector mechanisms that ultimately destroy the pathogenic organism carrying the antigenic determinants targeted by the immune response. Thus, the immune system's generation of specific immune responses is a marvelous example of molecular evolution: the diversification of lymphocytes into many variants and the selection by antigenic determinants of just one or a few of these variants for amplification over a period of weeks.

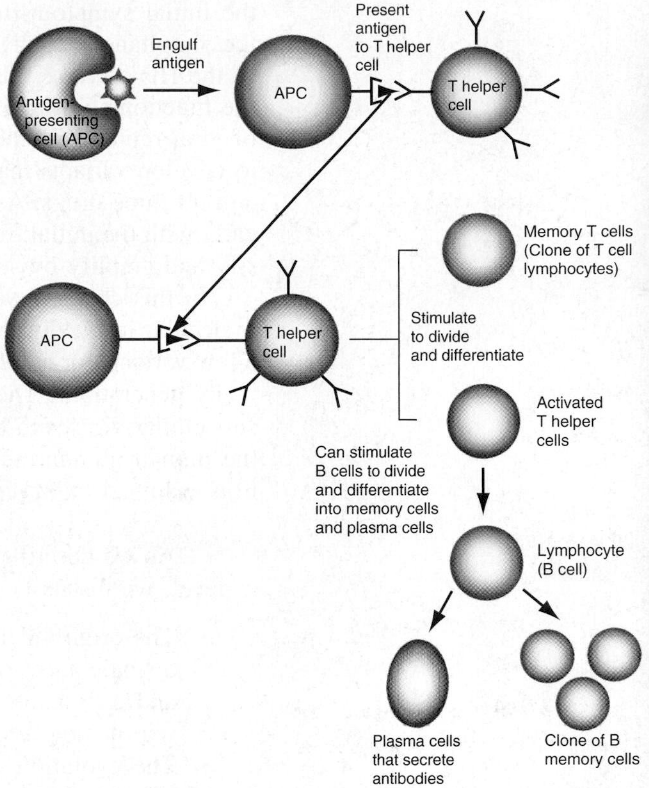

Figure 22.2 The immune response. Diverse receptors on lymphocytes recognize different molecular structures from invading pathogens. T cells employ T-cell receptors; B cells employ antibodies. Differentiated B cells, or plasma cells, secrete antibodies that destroy or neutralize antigen by a variety of mechanisms. Expanded numbers of memory T cells and B cells form the basis of a memory response that can rapidly respond to an antigen encountered a second time.

The AIDS virus (HIV) is a worthy adversary for the human immune system because it is able to diversify and amplify via selection far faster than the immune response itself. HIV, we have seen, is a retrovirus with an RNA genome (review Fig. A in the Genetics and Society box on p. 270 of Chapter 8). The virus makes its own reverse transcriptase available to the cells it infects (which include the T lymphocytes), forcing the cellular machinery first to copy the viral RNA genome into a DNA copy and then to generate a complementary DNA strand to form a proviral double helix. The infected cells integrate the double-stranded viral DNA into their own genomes (review Fig. B in the Genetics and Society box on p. 270 of Chapter 8). We have seen that the HIV reverse transcriptase has a very high mutation rate of roughly 1 in 5000 nucleotides incorporated into the viral cDNA. Since the HIV genome is about 10 kb in length, each replicated virus carries an average of two mutations (the basis of rapid diversification).

When HIV replicates explosively on infection, it generates billions of variant viruses before the host's immune response has a chance to take hold. Many of the variant viral particles are beyond the reach of the immune response. This is because the antigenic determinants targeted by the initial response have mutated beyond recognition by the time the originally activated T cells and B cells have proliferated (selection by lack of immune-cell recognition). Although some of the new antigenic determinants may be recognized by other cells among the diverse population of lymphocytes, subsequent viral mutations may alter these viral targets in good time. Eventually, the speed of viral evolution outstrips the ability of the immune response to keep pace not only because the viral genome mutates at a high frequency but also because a viral generation is so short.

A new therapeutic approach to AIDS, which tries to tip the evolutionary balance in favor of the human immune system, has been remarkably effective in prolonging the initial symptom-free phase of the disease. Called triple drug therapy, it entails the simultaneous delivery of three different anti-HIV drugs; two block the function of the HIV reverse transcriptase through different mechanisms, and the third blocks the functioning of a viral protease critical to HIV reproduction. By reducing the rate of viral replication, the triple drug therapy significantly decreases the virus's ability to develop variants that are simultaneously resistant to the human immune response and all three drugs. As a result, the HIV population in patients diminishes dramatically with the initial treatment. The triple drug therapy thus shifts the ability to diversify and amplify by selection in favor of the human immune response.

In this chapter, we examine in detail the basic components of evolution at the molecular level: **diversification** into many variants, followed by **selection** of one or a few variants for amplification (that is, differential reproduction) in a population over many generations. The same process is at work in nearly all organisms, from tiny, subcellular viruses to large, multicellular plants and animals. Unlike the evolution of the human immune response and the AIDS virus, however, most molecular evolution in higher, more complex organisms occurs over millions of years.

As we describe how diversification and selection operate at the molecular level, we discuss

- The origin of life on earth, including a possible scenario for descent from a single ancestor; the idea that RNA may have been the original replicator; and a probable timetable for the evolution of complex life-forms from a sea of molecules.
- The evolution of genomes, including a review of the various types and effects of mutations; an exploration of how evolutionary changes in gene regulatory networks affect development; a discussion of how an increase in genome size correlates with the evolution of complexity; a description of how larger genomes evolve through duplication and divergence; and an overview of molecular archaeology based on an understanding of gene duplication, diversification, and selection.
- The organization of genomes, including a description of genes, gene families, and gene superfamilies; repetitive nonfunctional DNA families; simple sequence repeats that dot the genome at random; and repeat sequences in centromeres and telomeres.
- The immunoglobulin gene superfamily: a comprehensive example of molecular evolution.

22.1 The Origin of Life on Earth

To many biologists, the similarity of all living things on earth is more striking than the differences among them. Not only are all organisms composed of cells, but these cells work in essentially the same way, using the same complex molecules and the same type of genetic material read according to virtually the same genetic code. As we have seen, the flow of information in most living organisms from DNA via RNA to protein follows a well-defined pathway. DNA contains a digital code with a four-letter alphabet encoding the instructions for the construction and development of organisms. The units of information in the DNA code are genes that can be differentially expressed as RNA. Some RNA molecules (for example, mRNA) contain codes for proteins; other RNA molecules can fold into three-dimensional molecular machines and function as enzymes. The proteins synthesized from the information carried in RNA are three-dimensional objects that catalyze the chemistry of life and give organisms their shape and form. In complex organisms, the proteins, together with other macromolecules, assemble to form organelles, cells, tissues, and organ systems.

Descent from a Single Ancestor: Speculations on How the First Cell Arose

No fundamental law of biochemistry says that all living cells have to be constructed in the way that they are; indeed, an imaginative biochemist could think of an almost infinite number of ways to build a functioning cell that, like the organic life-forms around us, is based on the laws of chemistry. The observation that the cells of all plants, animals, fungi, and microorganisms analyzed so far are extremely similar in subcellular organelles, biochemistry, and genetic processes suggests that the abundant variety of life-forms alive today descend from the same original cell that happened to begin life with the particular genetic code scientists now consider universal (even with the minor differences seen in some organelles and microorganisms).

The First Step on the Pathway to Life Must Have Been a Replicator Molecule That Could Copy Itself

While the molecular evidence for the evolution of viral genomes and immune responses is accessible and available for examination, no evidence remains of the earliest steps of molecular evolution, during which self-replicating macromolecules became enclosed in membranes and ultimately formed cells, the basic units of life as we know it today. Scientists have therefore inferred and imagined these earliest steps, building their inferences and imaginings on a foundation of macro- and microbiological observations and analyses.

A key step in molecular evolution was the emergence of a molecule that could replicate itself. Although one can only speculate on the nature of the original replicator, it must have been simple enough to form spontaneously after a half billion years of atoms bumping into each other on the surface of the primordial earth. To give rise to the living cells we see today, it would have had to fulfill three requirements: (1) encode information by the variation of letters in strings of a simple digital alphabet; (2) fold in three dimensions to create molecules capable of self-replication and ultimately other functions; and (3) expand the population of successful molecules through selective self-replication. What was the original self-replicating molecule and where did it come into existence?

The Original Replicator May Have Been RNA

In the 1980s, Thomas Cech and colleagues discovered that RNA, in addition to its ability to carry genetic information like DNA, can catalyze chemical reactions. Before this discovery, molecular biologists had assumed that only proteins have the chemical flexibility to fold into enzymes. RNA molecules that can act as enzymes to catalyze specific chemical reactions are called **ribozymes.** Although the substrates of most naturally occurring ribozymes are other RNA molecules, their potential range of enzymatic activity may be much broader.

The discovery of ribozymes gave molecular biologists an ideal candidate for the original replicator: a hypothetical RNA molecule that on its own could have (1) encoded information as linear strings in a four-letter alphabet; (2) folded into three-dimensional molecular machines able to execute critical functions of life (such as those currently carried out by polymerases, nucleases, and ligases); and (3) reproduced itself. The hypothetical primordial world in which this RNA became the first replicator has been termed the **RNA world.** The fact that all living organisms share the four letters of the RNA alphabet (A, U, G, and C) is testimony to the idea that successful, self-replicating RNA strings probably had a single origin, which, in turn, could have provided the common evolutionary origin of the information in all living organisms.

The earliest RNA may have had informational motifs, or coding regions (for example, encoding a simple polymerase), separated by noncoding regions of "background noise." The ribozymes of contemporary organisms that can cleave, join, copy, and even modify informational strings may reflect these early RNA activities. The pattern of coding regions separated by noncoding introns could have been established by the early random assembly of RNA strings. The evolutionary challenge was how to use the information and separate it from the noninformational regions. Accordingly, there were evolutionary pressures to evolve RNA coding regions that could cut (like an endonuclease) and splice (like a ligase), as well as synthesize (like a polymerase) RNA molecules.

However, multipurpose RNA strings that both stored information and folded into molecular machines would have had several intrinsic disadvantages. First, RNA is a relatively unstable molecule, readily susceptible to chemical and enzymatic hydrolysis. Second, RNA has only four letters in its alphabet, which are chemically similar, and is thus less capable of folding into complex three-dimensional structures than other polymers with more complex alphabets. This is so because each letter in a molecular alphabet represents a particular shape and charge; proteins with their 20 chemically divergent amino acid letters can fold into more complex structures than RNA with only four. Accordingly, there were evolutionary pressures to evolve more stable storage molecules (DNAs) and more flexible molecular machines (proteins).

No Record Exists of the Intermediates Between the RNA World and the Organized Complexity of the Cell

This lack of evidence makes it difficult to provide solid answers to questions about evolutionary events at the precellular stage. If RNA was indeed the original replicator, was it available for the first cells to use as an informational molecule, with the DNA and protein alphabets evolving subsequently? How did membrane compartments that concentrate the molecules of life and facilitate their sorting arise? What metabolic pathways gave some of the early cells an evolutionary edge?

Many biologists speculate that life began around the volcanic thermal vents of the ancient seas more than 3.5 billion years ago. In this high-temperature environment, informational subunits could have emerged and then polymerized into strings of informational molecules. Each of these informational intermediates was a proto-life-form that reproduced itself into a large number of slightly different copies (diversification) until by chance one gained a better mode of survival and outcompeted all its cousins in the next round of reproduction (selection). This two-step process of diversification and selection repeated itself over and over again until the first cell appeared. From that point on, the history of life's evolution is easier to discern.

The Evolution of Living Organisms: Inferences from the Fossil Record

It is an enormous jump from a microscopically small single-celled organism to, for example, a human being composed of 100 trillion cells. However, like Darwin, no current biologist doubts that the one evolved into the other. This confidence arises not just from what Darwin's theory says *should* happen but from the possibility of seeing the critical intermediate stages of biological complexity, all the way up the ladder from single cells to sponges, to worms, to fish, to reptiles, to mammals, to primates and humans.

The Intermediate Stages of Evolution from a Single Cell to Complex Multicellular Organisms Exist in the Fossil Record and in the Living Fossils That Surround Us

While the fossil record enables us to date the initial appearance of each intermediate stage, the living representatives of each stage, in conjunction with our ability to study them with the tools of molecular biology, give us a glimpse into how evolution occurs step by step. It is fortunate for the scientists who study evolution that so many critical intermediate forms of life kept on reproducing in their less complex state, even as their cousins went on to evolve to the next stage of complexity.

Single-Celled Organisms Without a Membrane-Bounded Nucleus Emerged First

Scientists agree that planet earth coalesced some 4.5 billion years ago. By 4.2 billion years ago, enormous oceans covered the planet, and the first informational RNA molecules may have emerged around the high-temperature volcanic ocean vents. The first living organisms, consisting of a membrane surrounding information-replicating and information-executing machinery, evolved about 3.7 billion years ago; these were the precursors of present-day cells.

Fossil cells laid down 3.5 billion years ago near North Pole, Australia (a small town in an arid, rocky region) are the earliest evidence so far uncovered of distinct cells (**Fig. 22.3a**). Once life in cellular form emerged, living organisms evolved into three distinct domains: archaea, bacteria, and eukarya (Fig. 22.3b). Contemporary representatives of archaea and bacteria include only single-celled organisms whose genomes carry tightly packed genes. Because their genes do not contain introns, some biologists have concluded that introns were a late evolutionary elaboration, appearing with the emergence of eukaryotes. An

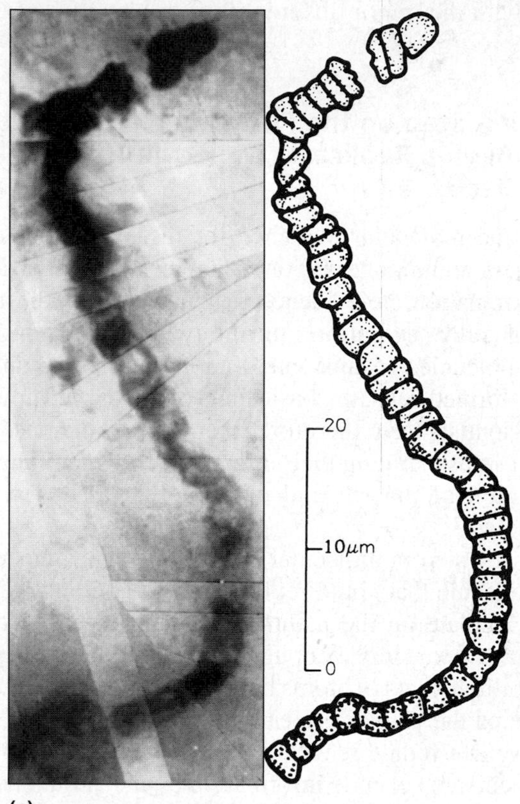

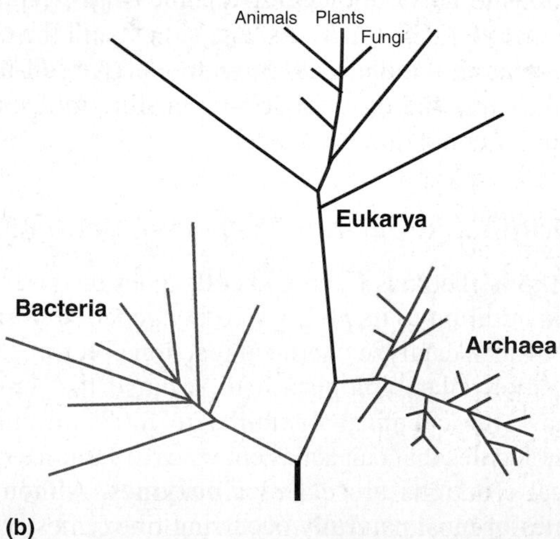

Figure 22.3 The earliest cells evolved into three kingdoms of living organisms. (a) The oldest fossilized cells. **(b)** The distinct branches represent different organisms in each kingdom. The length of the branches is proportional to the times of species divergence.

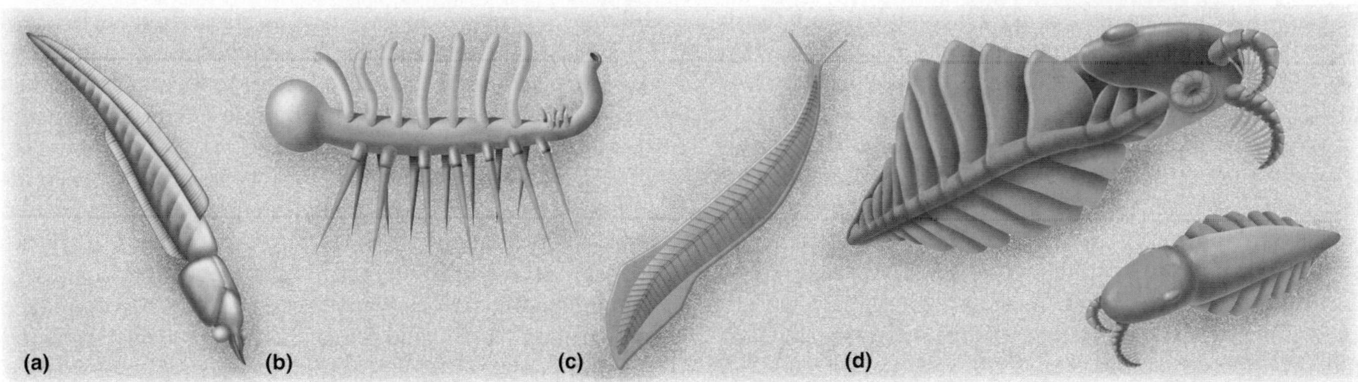

Figure 22.4 **Some of the body plans found in Burgess shale organisms.** Although all these life-forms are now extinct, fossils have revealed the enormous diversity of their body plans: **(a)** the *Nectocaris,* **(b)** the *Hallucigenia,* **(c)** the *Pikaia,* and **(d)** the *Anomalocaris.* The *Pikaia* is a vertebrate with a notochord, which makes it an ancestor to modern-day vertebrates.

alternative explanation is that archaea and bacteria, requiring great efficiency in their use of energy, relinquished, in the course of evolution, the flexibility of discontinuous coding segments for the more energy-efficient synthesis of genes without introns. (See Chapter 8, pp. 271–273 for a definition and discussion of introns.)

More Complex Cells and Multicellular Organisms Appeared After More Than 2 Billion Years of Cellular Evolution

Eukaryotes emerged about 1.4 billion years ago with the symbiotic incorporation of certain single-celled organisms into other single-celled organisms and the complex compartmentalization of the cell's interior, including the segregation of DNA molecules into the nucleus. The incorporated single-celled organisms evolved to become intracellular organelles. The evolution of these relatively complex eukaryotes from the earliest cells thus took almost 2.3 billion years. About 1 billion years ago, the single-celled ancestors of contemporary plants and animals diverged. The first primitive multicellular organisms appeared 600–900 million years ago. Then 570 million years ago, one of the most remarkable events in the evolution of life occurred: the explosive appearance of a multitude of multicellular organisms, both plants and animals. The multicellular animals are often referred to as **metazoans.**

The Burgess Shale of Southeastern British Columbia in Canada Captured the Drama of Metazoan Evolution

The shale arose from a mud slide that trapped a wide variety of different organisms in a shallow Cambrian Sea. Amazingly, events and conditions during and after the slide conspired to achieve the nearly perfect preservation of the three-dimensional structure of the entrapped specimens' soft body parts. Three aspects of the Burgess organisms are remarkable. First they represent a myriad of very different body plans (**Fig. 22.4**). For example, paleobiologists have distinguished 20–30 classes of arthropods in the Burgess sea shale, a striking contrast to the three contemporary classes of arthropods. Second, this emergence of metazoan organisms occurred over a remarkably short (in evolutionary terms) period, perhaps just 20–50 million years. This rapid evolution is an example of **punctuated equilibrium:** the tendency of evolution to proceed through long periods of stasis (lack of change) followed by short periods of explosive change. It was as if the evolution of life represented a supersaturated salt solution where the addition of a final salt crystal catalyzed a remarkable solutionwide cooperative crystallization of many different body plans. As we will see later, this rapid change in body plans reflects an equally rapid change in the regulatory networks that control the development of organisms. Third, it seems that all the basic body plans of contemporary organisms initially established themselves in the metazoan explosion. For example, the ancestor of contemporary vertebrates depicted in Fig. 22.4c probably emerged at about the same time as the ancestor of all contemporary invertebrates.

Many Species Resulting from the Metazoan Explosion Have Disappeared

The enormous diversity of metazoan body plans that materialized about 500 million years ago has by now become tremendously reduced, in part through four to six abrupt extinction events that each destroyed 70% to 95% of the existing organisms. The most recent example was the global decimation 65 million years ago that led to the extinction of the dinosaurs. Many scientists believe this extinction was a consequence of a large meteorite impact in the Yucatan region of present-day Mexico that dramatically changed earth's climate by propelling enormous amounts of dust into the higher atmosphere. Scientists hypothesize that this thick cloud of dust dispersed and shrouded the globe for several years, preventing solar rays from reaching earth's surface. The lack of solar energy led to a nuclear-winter-like scenario in which the demise of all green life caused the demise of all large animals, such as dinosaurs, that depended for survival either directly on plants or on animals that ate plants.

Some smaller animals (like our mouse-sized ancestors) presumably survived this long sunless winter because their lesser size allowed them to get by on seeds alone. When the sun returned, the seeds lying dormant on the ground sprang to life and the world again became an abundantly fertile environment. In the absence of competition from dinosaurs, mammals became the dominant large animal group, diverging into numerous species that could take advantage of all the newly unoccupied ecological niches. Some eventually evolved into our own species.

The Evolution of Humans

Humans arose from an ancestor common to most contemporary primates that existed 35 million years ago. They diverged from the ancestors of their closest primate relatives, the chimpanzees, about 6 million years ago (**Fig. 22.5**). While paleobiologists have not yet sorted out the immediate evolutionary ancestors of *Homo sapiens,* the recent typing of fossil DNA suggests that one previous candidate, the Neanderthal lineage, is not on the direct human evolutionary line (see the Fast Forward box on pp. 594–595 of Chapter 16). Perhaps the single most fascinating aspect of human evolution is the striking changes in the evolution of the brain since the human lineage diverged from the chimpanzee line. These changes include an increase in size, folding complexity, and nerve cell density, which together provide people with the capacity for a level of intelligence sufficient to appreciate the universe and the history of life.

Remarkably, on average, the chimpanzee and human genomes are approximately 99% similar. Moreover, as we saw in Chapter 13, the chimpanzee and human karyotypes are nearly the same. And in every comparison to date of chimpanzee and human DNA sequences, the observed differences between the two have been insignificant in terms of gene function. What these data suggest is that the evolution from a common primate ancestor to the modern human species might be accounted for by a few thousand isolated genetic changes yet to be uncovered.

While these changes may have occurred in protein-coding sequences, many evolutionary biologists think it more likely that the changes occurred in regulatory sequences. Such changes would alter the regulatory sequences controlling when and how master regulatory genes produce transcription factors, and when and how ordinary structural genes or batteries of genes respond to these regulatory molecules. For example, the brains of humans and chimps are quite different.

The human brain is larger, is far more convoluted (folded), and contains a significantly greater density of neurons (brain cells). Hence, the regulatory networks guiding chimp and human brain development have diverged strikingly in the 6 million years since the two species split from a common ancestor. These regulatory changes are presumably reflected in modified patterns of transcription factor binding sites in the promoter regions of genes that specify brain development. (See the discussion of gene regulatory networks on pp. 441–443 of Chapter 12.) The rewiring of the regulatory networks in less than 6 million years is a very rapid change in terms of evolutionary time. The same is true of the diversity of body plans generated during the Cambrian explosion of metazoa 570 million years ago. The idea that evolution occurs primarily because of changes in regulatory networks and not structural genes is supported by the amazing ability of genes from one species to substitute for the absence of homologous genes in other species, in some cases even when the species are as different as yeast and humans. If homologous coding sequences from very different species are functionally indistinguishable, it is reasonable to speculate that species-specific differences in phenotype may arise, to a large degree, from species-specific differences in gene expression.

(a)

(b)

(c)

(d)

Figure 22.5 Humans diverged from an ancestor shared with chimpanzees about 6 million years ago. Representatives of primates alive today: **(a)** orangutan, **(b)** gorilla, **(c)** chimpanzee, and **(d)** human.

Thus, what makes humans and chimpanzees different may not be the proteins that their genes encode but rather when, where, and at what level those proteins are expressed during development. It seems likely, for example, that the difference between chimpanzee and human consciousness can be explained simply by changes in the regulatory sequences of master regulatory genes and other genes encoding proteins that play a role in the development of the cerebral cortex.

22.2 The Evolution of Genomes

Although Darwin developed his theory of evolution—based on the selection of preexisting variations—without any knowledge of the molecules that make up living systems, evolution is very much a molecular process that operates on genetic information. In particular, the variation that initiates each step in the evolutionary process occurs within the genetic material itself in the form of new mutations. Indeed, new mutations provide a continuous source of variation.

DNA Alterations Form the Basis of Genomic Evolution

We have seen that mutations arise in several different ways (review Figs. 7.6–7.11 on pp. 213–220). One is the replacement of individual nucleotides by other nucleotides. Substitutions occurring in a coding region are silent, or *synonymous,* if because of the degeneracy of the genetic code, they have no effect on the amino acid encoded; by contrast, they are *nonsynonymous* if the change in nucleotide determines a change in amino acid or a premature termination codon, leading to a truncated gene product. Molecular biologists further distinguish between nonsynonymous changes that cause conservative amino acid changes (for example, from one acidic amino acid to another) and those that cause nonconservative changes (for example, from a charged amino acid to a noncharged amino acid). Other gene mutations, arising from errors in replication or recombination, consist in the deletion from or insertion into genes of a DNA sequence of any length.

Changes can also occur in the order and types of transcription factor binding sites in the promoters of genes; such changes alter the patterns of gene expression. These expression-altering changes may occur very rapidly, which raises the question of whether such regulatory evolution can be explained merely by single-base mutations followed by selection.

Different mutations can be *deleterious, neutral,* or *favorable* to the organisms that inherit them. In multicellular organisms with large genomes, such as corn or humans, genes and their regulatory sequences make up only a small fraction (<3%) of the total genetic material. As a result, random mutations occur most often in DNA that plays no role in the development or function of an organism. Such mutations are presumably neutral. It might seem that synonymous mutations within coding regions would also be neutral, but there is some evidence that even changes in the codon used to produce a particular amino acid can provide a minute advantage or disadvantage to the organism possibly based on the availability of different tRNA molecules and their associated synthetases. While conservative amino-acid replacements were once considered neutral, current evidence suggests that they can have an impact on the growth and survival of an organism. Nonconservative amino-acid changes and changes such as deletions and insertions involving larger portions of a gene almost certainly have an impact on gene function.

Genetic Changes That Are Truly Neutral Are Unaffected by the Agents of Selection

Such mutations have a neutral effect on fitness and thus are not subject to selection. They survive or disappear from a population through *genetic drift,* which is the result of chance reproductive events. The smaller the population, the more rapidly genetic drift exerts its effect.

Mutations That Have Only Deleterious Effects Will Disappear from a Population by Negative Selection, That Is, Selection Against an Allele

As we saw in Chapter 21, some mutations (such as those that cause sickle-cell anemia) are deleterious to homozygotes for the mutation but advantageous to heterozygotes. Because of this *heterozygous advantage,* selection retains these mutations in a population at a low equilibrium level (review Fig. 21.7 on p. 767). However, even deleterious mutations for which there is no heterozygous advantage are eliminated very slowly if they are fully recessive.

Some Extremely Rare Mutations Give an Organism a Significant Advantage

Because of this advantage, individuals carrying the mutation are more likely to reproduce, and in each succeeding generation, the frequency of the mutation increases (review Fig. 21.4 on p. 763). This is positive selection (selection for an allele). Ultimately, the allele that began as a mutation is present in nearly every member of the population on both chromosomes. At this stage, the allele has become *fixed* in the population.

Evidence Suggests That Changes in Gene Regulatory Networks May Dominate Developmental (and Therefore Phenotypic) Evolution

In Chapter 12 we learned that the gene regulatory networks active in development receive informational signals from signal transduction pathways, integrate and modulate those

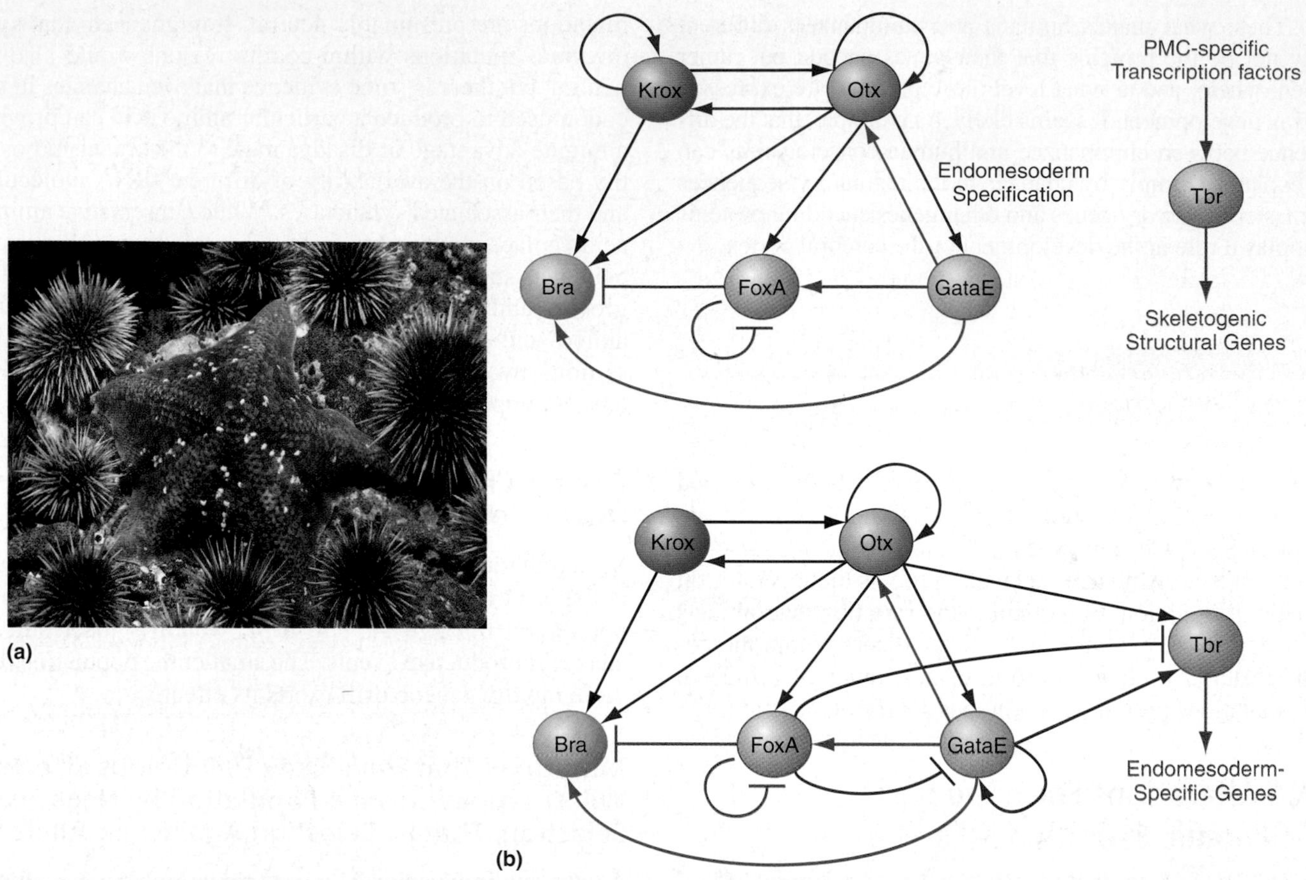

Figure 22.6 Comparison of sea urchin and starfish phenotypes and of subnetworks of the gene regulatory network controlling endomesodermal development in the two organisms. (a) Pictures of sea urchins and a starfish, organisms that diverged about 500 million years ago. **(b)** Subnetwork of the gene regulatory networks governing mesoendodermal development in the sea urchin (top) and starfish (bottom). The transcription factors controlling endomesodermal development in this subnetwork—there are five in the sea urchin and six in the starfish—are in different colors; the *arrows* (positive influence) and *bars* (negative influence) indicate the effect each transcription factor has on the others with which it interacts. The influence lines are in the same color as the transcription factors from which they arise. (Although Bra receives inputs from four transcription factors in this subnetwork, its output influences transcription factors not included in this figure.) PMC-specific transcription factors play a role in the expression of skeletal proteins. Note that the lines outlined in *red* represent the only interactions that have not been highly conserved. Note also that whereas the Tbr transcription factor in the sea urchin controls the expression of skeletogenic structural proteins, in the starfish it has been coopted into the endomesodermal network. Tbr's influences are represented by the *brown* arrows.

signals, and then transmit them to protein networks that mediate various aspects of development. We also learned that one of the best-studied gene regulatory networks specifies development of the endomesoderm in the sea urchin (see Figures 12.9 and 12. 10 on pp. 443–444). Eric Davidson, the scientist who pioneered the sea urchin studies, has begun to study the evolution of a portion of the endomesodermal gene regulatory network in two invertebrate organisms with markedly different phenotypes: the sea urchin and the starfish (**Fig. 22.6a**) The lineages leading to these two invertebrates diverged about 500 million years ago (about the same time vertebrates diverged from invertebrates). As Figure 22.6b shows, the subnetwork analyzed by Davidson contains five transcription factors in the sea urchin and six in the starfish. Two observations are striking. First, except for the changes indicated by the lines outlined in red, the interactions among these transcription factors are highly conserved. Second, the Tbr transcription factor, which in the see urchin specifies the expression of skeletogenic structural proteins, functions in the starfish as part of the endomesoderm network (as indicated by the brown arrows in Fig. 22.6). Changes in transcription factor specificity and function can arise from changes to the DNA of *cis*-control elements or from changes to the amino acid sequences of the transcription factors themselves. Changes to the DNA include the gain or loss of cognate *cis*-control sequences to which the factors bind, repositioning of *cis*-control elements on the regulatory DNA, increases or decreases in a *cis*-control element's binding affinity for a particular transcription factor, and mutations that make a *cis*-control element serve as the binding site for a different transcription factor. The relevant amino acid modifications change the DNA binding specificity of the transcription factor.

Although these comparative evolutionary studies are preliminary and incomplete, they suggest that a subtle rewiring of a gene regulatory network and the joining together of elements from previously separate subnetworks can encode enormous phenotypic changes, including those that give rise to sea urchin versus starfish body plans. A fascinating possibility is that gene regulatory networks in humans may share much of their basic wiring with invertebrates. If this turns out to be true, a detailed study of invertebrate gene regulatory networks could provide powerful insights into the basic gene regulatory networks controlling development in higher organisms. Moreover, if the subtle rewiring of gene regulatory networks can lead to enormous phenotypic changes, it can also explain the evolution of considerable biological complexity. (See the Tools of Genetics box on pp. 802–803 for a discussion of Darwinian evolution versus creationism and the key role of network evolution in the evolution of biological complexity.)

An Increase in Genome Size Generally Correlates with the Evolution of Complexity

Consider that although both bacteria and mammals evolved from a common cellular ancestor, the contemporary *Escherichia coli* genome is about 5 Mb in length, while in humans, the genome is about 3000 Mb long. How has evolution fashioned such different genomes from the same original material? The answer lies in the evolutionary potential for increasing the size of the genome through the duplication and diversification of genomic regions and, even more strikingly, through the acquisition of repetitive sequence elements that may represent more than 50% of the genome. In these processes, new DNA is born. Note, however, that although the increase in genome sizes from yeast to flies to vertebrates does reflect the increasing complexity of the organisms, there are examples of amoeba, plants, and amphibians with considerably more genomic DNA than humans.

Genomes Grow in Size Through Repeated Duplications

Duplications can occur at random throughout the genome, and the size of the duplication unit can vary from a few nucleotides to the entire genome. When a duplication segment contains one or more genes, either the original or the duplicated copy of each gene is free to accumulate function-destroying mutations (diversify) without harm to the organism, since the other "good" copy with original function is still present. With duplications acting as such an important force in evolution, it is critical to understand the two main ways in which they arise.

Some Duplications Result from Transpositions As we saw on pp. 509–511 of Chapter 14, transposition, the transfer of one copy of a chromosomal sequence from one chromosomal site to another, can occur in various ways: through the direct movement of a DNA sequence, through an RNA intermediate that is copied into a DNA intermediate, leaving the original DNA site intact; or through a DNA intermediate (**Fig. 22.7;** review

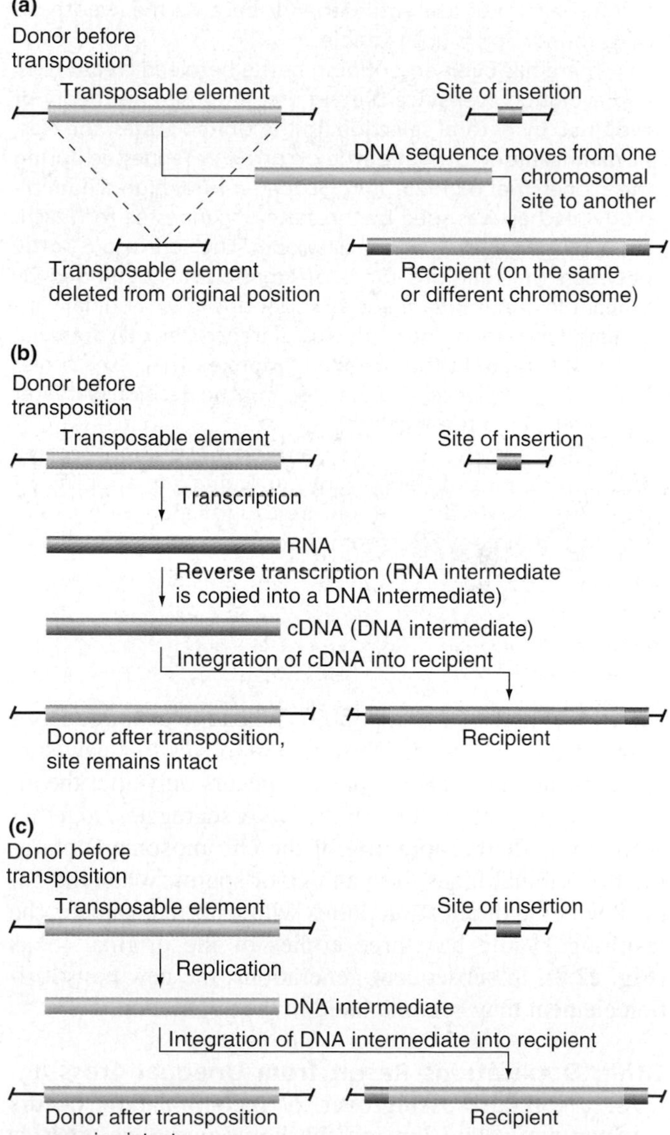

Figure 22.7 Duplication by transposition: Three possibilities. Transposition may occur by **(a)** excising and reinserting the DNA segment; **(b)** by making an RNA copy, which is then converted to a DNA copy for insertion; or **(c)** by making a DNA copy for integration. In the second and third examples, the transposon is duplicated.

GENETICS AND SOCIETY

The Scientific Theory of Darwinian Evolution Versus the Religious Idea of Intelligent Design

Darwin's evolutionary theory maintains that the diversity and complexity of living organisms are the result of evolution by the natural selection of pre-existing variations encoded by the genome in combination with random drift. Creationism is the belief that certain aspects of life are too complex to have evolved in this way; instead, it maintains that a divine being created the world, including the diversity and complexity of all life from microbes to people. Creationism is a religious belief, not a scientific theory, because the hypothesis at its foundation is not testable.

There has been an ongoing battle between creationism and evolution ever since Darwin first proposed his theory of evolution by natural selection. In the United States, the classic manifestation of that battle occurred in Tennessee during the Scopes Trial of 1925. John Scopes, a rural high school biology teacher, was sued by the state of Tennessee for teaching evolution in the science classroom. The courtroom battle between the religious ideas of creationism (defended by William Jennings Bryan) and the scientific ideas originally articulated by Darwin (defended by Clarence Darrow) drew national attention. In the so-called "Monkey Trial," the forces for evolution (science) actually lost, but the decision was later overturned on a technicality.

In the ensuing years, as Darwinian evolution became an accepted part of the biology curriculum, creationists argued that creationism should be taught alongside evolution in the biology classroom, because according to them, both were unproven theories developed to explain the complexity of living things. In 1987 the Supreme Court finally decided this issue when it struck down a Louisiana law that required schools teaching evolution to teach creationism at the same time. In that ruling, the Court said that (1) the Louisiana law was an unconstitutional establishment of religion and (2) schools couldn't teach creationism as science.

In the next stage of this persistent controversy, the proponents of creationism cast their argument in a new form known as Intelligent Design (ID). Life, they said, is too complex to have been created by Darwinian evolution, instead, it must have been created by a higher agent, or intelligence. That intelligence is generally taken to be God. This new brand of creationism is clearly in evidence in the supplemental biology textbook entitled *Of Pandas and People.* When the book was originally written, the term "creationism" was used throughout. In subsequent versions (published after the Supreme Court decision of 1987), the word creationism has been replaced with "intelligent design", with virtually no change in any of the arguments. This book is misleading in suggesting that the "theory of evolution" is nothing more than an educated guess, when many of the theory's assumptions have been tested by experiment and borne out by observation. It is also wrong in many of its facts, including descriptions of the fossil and molecular evidence of Darwinian evolution.

Figs. 14.23 and 14.24 on p. 509). When the genomic DNA (rather than its RNA or DNA proxy) moves to a new site, the duplication of genetic material occurs only after the altered chromosome receiving the DNA segregates, together with the unaltered homolog of the chromosome containing the original locus, into an egg or sperm. When the gamete with the duplication unites with a normal gamete, the resulting zygote has three copies of the original locus (**Fig. 22.8**). In subsequent generations, the new transposition element may become fixed in the population.

Other Duplications Result from Unequal Crossing-Over Normal crossing-over, or recombination, occurs between equivalent loci on the homologous chromatids present in a synaptonemal complex that forms during the pachytene stage of meiosis. *Unequal crossing-over,* also referred to as *illegitimate recombination,* occurs between nonequivalent loci (review Fig. 7.10a on p. 218). Unequal crossing-over is most often initiated by related sequences located close to each other in the genome. Although the event is unequal in the exchange of nonequivalent segments of DNA, it is still mediated by the sequence similarities at the two separate loci.

So-called **nonhomologous unequal crossing-over** also occurs, although much less often than homologous crossing-over. A nonhomologous crossover may be mediated by at least a short stretch of sequence homology, coding or noncoding, at the crossover's two sites of initiation.

An initial duplication by unequal crossing-over that produces a two-unit cluster may be either homologous or nonhomologous, but as **Fig. 22.9** illustrates, once two units of related sequence are present in tandem, further rounds of homologous unequal crossing-over between nonequivalent members of the pair readily occur. Thus small clusters can easily expand to contain three, four, and many more copies of an original DNA sequence.

The result of unequal crossing-over between homologous chromosomes is always two reciprocal chromosomal products: one carrying a duplication of the region located between the two crossover sites and the other

In 2005 the parents of eleven students sued the Dover, Pennsylvania school board over the board's mandate that all science teachers must tell their students that evolution is "just a theory" and that the book *Of Pandas and People* is available to provide an alternative hypothesis about the creation of life. When many teachers in the district refused to follow the board's order, their principals read the dictum. The parental lawsuit argued that the mandate to give evolution and intelligent design equal time in the science classroom was inappropriate because intelligent design is not science, but religious belief. On December 20, 2005 Judge John E. Jones of the federal district court hearing the case issued his decision: a sweeping repudiation of the teaching of intelligent design in the science classroom as an alternative to Darwinian evolution. First and foremost, he argued that intelligent design is not a scientific theory because its supernatural premises cannot be tested; rather it is a religious belief that has no place in a science course. In his words: "ID violates the centuries-old ground rules of science by invoking and permitting supernatural causation; the argument of irreducible complexity, central to ID, employs the same flawed and illogical contrived dualism that doomed creation science in the 1980s; and ID's negative attacks on evolution have been refuted by the scientific community." In addition, the proponents of intelligent design have "not generated peer-reviewed publications, nor has [the concept] been the subject of testing and research." Judge Jones also specifically debunked *Of Pandas and People,* describing it as thinly disguised creationism, and chastised the school board members supporting the teaching of intelligent design in the science classroom as hypocritical in their claim that they only wanted to teach science properly: "We find that ID . . . cannot be adjudged a valid, accepted scientific theory. . . . [It] is grounded in theology, not science . . . . It has no place in a science curriculum . . . . The goal of the ID movement is not to encourage critical thought, but to foment a revolution that would supplant evolutionary theory with ID." This decision stands as a landmark in the evolution versus creationism battle—a controversy that is likely to continue, at least in the US, as the creationists continue to seek a formulation of creationism that is acceptable to the courts.

Interestingly, in a November school board election held about a month before Judge Jones announced his decision, eight of the nine conservative members of the Dover school board were defeated by candidates opposed to the teaching of intelligent design in the science classroom. The new board will not appeal the Jones decision. Instead, they plan to revise the curriculum so that intelligent design will be taught in a class on religion rather than as a scientific theory in the biology classroom.

Even as the proponents of Intelligent Design base their anti-Darwinian evolution arguments on a supernatural explanation of how biological complexity emerged, biologists are making great strides toward understanding that complexity scientifically through the systems approach to biology. Perhaps the best example to date is Davidson's study of a portion of a gene regulatory network in the sea urchin and the starfish (see Fig. 22.6 on p. 800). Apparently rather simple evolutionary changes in this network gave rise to profound changes in the endomesodermal tissue in these two organisms. Indeed, it is probable that the comparative study of life's biological circuits (the gene regulatory and protein networks) through systems analysis will lead to scientific explanations of the evolution of life's complexity.

carrying a deletion that covers the exact same region (review Fig. 7.10 on pp. 218–219). Unlike retrotransposition, unequal crossing-over operates on genomic regions without regard to functional boundaries. Regions duplicated by unequal crossing-over can vary from a few base pairs to hundreds of kilobases in length, and they may contain no genes, a portion of a gene, a few genes, or many genes.

Genetic Drift and Mutations Can Turn Duplications into Pseudogenes That Eventually Dissolve into Random DNA Sequences

Duplicated regions, like all other genetic novelties, must originate in the genome of a single individual. At first, the survival of a duplication in at least some animals in each subsequent generation of a population is, most often, a matter of chance. This is because the addition of one extra copy of a chromosomal region, including most genes, to the two already present in the diploid genome usually causes no significant harm to the individual animal. In the terminology of population genetics, the duplicated units are neutral with regard to genetic selection. They are thus subject to genetic drift, inherited at random by some offspring but not others. By chance again, most neutral genetic elements disappear from a population in several generations.

When a duplicated region that includes a functional gene survives for hundreds or thousands of generations, random mutations in the gene may turn it into a related gene with a different function, or into a nonfunctional *pseudogene.* Some of the mutations generating a pseudogene lead to a loss of regulatory function; others change one or more critical amino acids in the gene product; still others cause premature termination of the growing polypeptide chain, or change the translational reading frame of the gene, or alter the RNA splicing patterns. Pseudogenes, because they serve no function, are subject to mutation without selection and thus accumulate mutations at a far faster pace than the

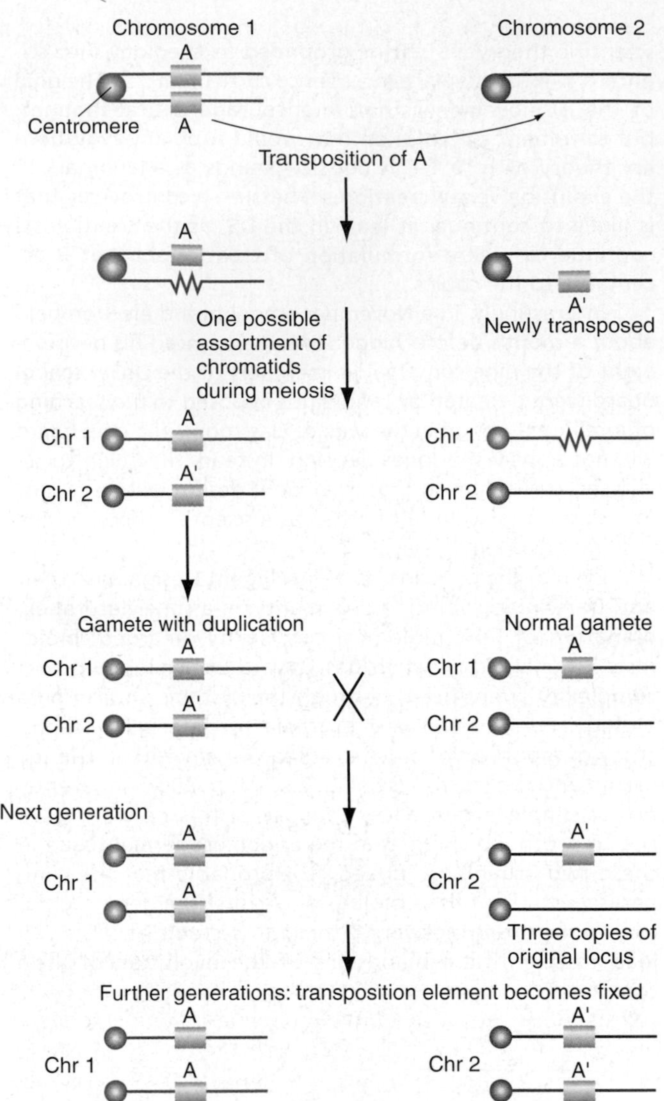

Figure 22.8 Transposition through direct movement of a DNA sequence. A DNA sequence may transpose from one chromosome to a second chromosome in a sex cell. In subsequent generations, this transposed element may become fixed.

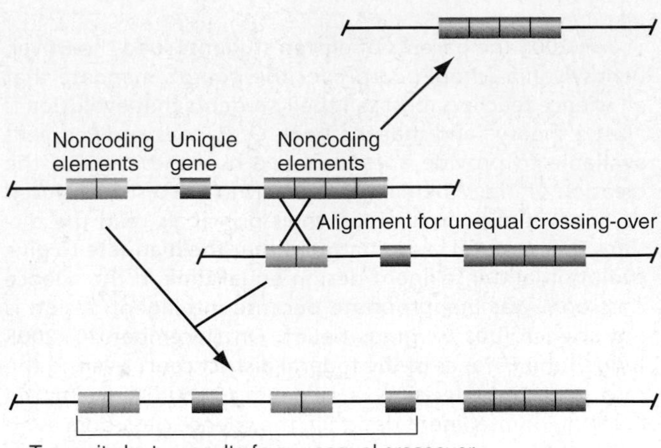

Figure 22.9 Duplications arise from unequal crossing-over. The crossover shown involved dispersed repeat elements (*blue boxes*). The *pink boxes* denote a unique gene. In the crossover event, this originally unique gene is deleted in one chromosome and duplicated in the other.

coding or regulatory regions of a functional gene. Eventually, nearly all pseudogene sequences mutate past a boundary beyond which it is no longer possible to identify the functional genes from which they derived. Thus, continuous mutation can turn a once functional sequence into an essentially random sequence of DNA.

Diversification of a Duplicated Gene Followed by Selection Can Produce a New Gene

Every so often, the accumulation of a set of random mutations in a spare copy of a gene leads to the emergence of a new functional gene that provides benefit and, consequently, selective advantage to the organism in which it resides. Because it provides a selective advantage, the new gene persists in the population. Although its function is usually related to that of the original gene, it almost always has a novel pattern of expression—in time, in space, or both—which most likely results from alterations in *cis*-regulatory sequences that occur along with codon changes. For example, a duplicated copy of the original human β-globin gene evolved into the myoglobin gene, whose protein product has a higher affinity for oxygen than the hemoglobin molecules composed in part of β-globin polypeptides. The myoglobin gene is active only in muscle cells, while the β-globin gene is expressed only in red blood cell precursors. Thus, duplication, divergence, and selection generated a new gene function from a previously functional gene.

Molecular Archaeology Based on an Understanding of Gene Diversification and Selection

Before the advent of molecular biology, researchers determined the genealogies of organisms by calculating the rates of evolution in phenotypic traits such as teeth and vertebrae. Then, in the 1950s, Linus Pauling and Emile Zuckerkandl analyzed hemoglobin and cytochrome C protein sequences from different species and noted that the rates of amino acid substitution in each type of protein are similar for various mammalian lineages. On the basis of this observation, they postulated that for a given protein, the rate of evolution is constant across all lineages. They called this idea of a constant rate of change for each type of molecule a **molecular clock.** Its existence enables biologists to determine from molecular data the approximate times when species diverged and then use these dates to reconstruct genealogical phylogenetic trees. Although the molecular clock hypothesis does not keep perfect time, in many instances where it has been tested, it has produced a reasonably good estimate of the time of divergence between two types of organisms. Today molecular biologists compare the nucleotide sequences of genes as well as the amino acid sequences of proteins to determine phylogenetic divergences.

Scientists use molecular data in various ways to construct **phylogenetic trees** that illustrate the relatedness of homologous genes or proteins. A phylogenetic tree consists of **nodes** and **branches** (**Fig. 22.10**). The nodes represent the taxonomic units, which may be species, populations, individuals, or genes, while the branches define the relationship of these units. The branch length suggests the amount of time that has elapsed based on the number of molecular changes that have occurred. Because different genes accumulate changes at different rates, different types of genes are best suited to the construction of different types of phylogenetic trees. For example, in fibrinopeptides, the major components of wound-responsive blood clots, the exact amino acid sequences are not critical to function. As a result, the fibrinopeptides are not under strong selective pressure, and their genes evolve quickly. In contrast, the ribosomal genes, which are highly conserved, evolve slowly. Thus, the genes encoding fibrinopeptides are useful for looking at recent evolutionary events among very closely related species, while the ribosomal genes are useful for looking at ancient evolutionary events such as the relationships of phyla to each other. Indeed, phylogenetic trees based on ribosomal genes helped rewrite the most fundamental domain classifications of multicellular organisms (see Fig. 22.3).

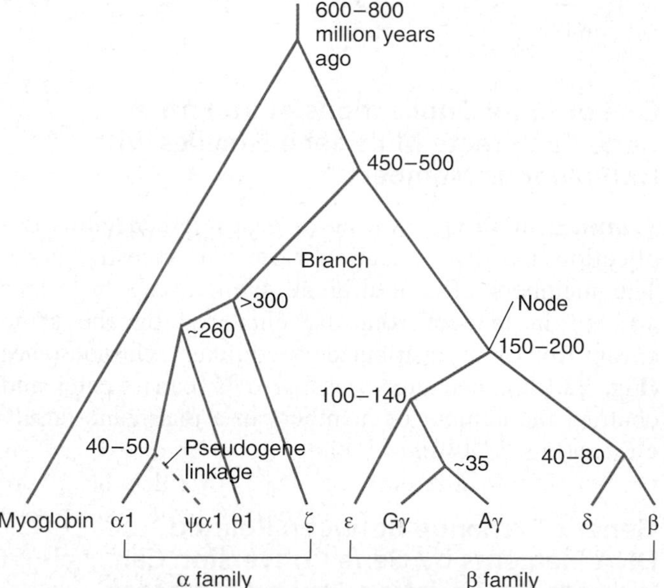

Figure 22.10 **A phylogenetic tree.** This phylogenetic tree diagrams the evolutionary history of human hemoglobin genes. The *broken line* denotes a pseudogene linkage. Only one of the two α-hemoglobin genes is shown because their date of divergence is so recent.

22.3 The Organization of Genomes

With this understanding of the basic mechanisms by which genomes evolve, we turn our attention to the results of genome evolution as seen in the organization of contemporary genomes. Our focus is on the various organizational features of the enormous mammalian genome, which evolved from the much simpler bacteria-like genomes through eons of duplication, diversification, and selection.

Molecular geneticists have accumulated data on genome organization by analyzing the genomic sequences of chromosomal DNAs from completed sequences of more than 100 genomes, including those of humans, mice, puffer fish, rice, *Drosophila, C. elegans* (the first animal to have its genome completely sequenced), Arabidopsis (the first plant to yield a complete genome sequence), yeast, *E. coli,* and approximately 300 other organisms (mostly microbes). The single-celled organisms exhibit genomes with densely packed genes and few, if any, introns. The mammalian genomes, with their far less densely packed genes, have several distinct features dominating their landscape: genes and families of genes; dispersed repetitive elements constituting more than a third of the genome; simple sequence repeats composed of single nucleotides or di-, tri-, tetra-, pentamers, and so forth; simple repetitive elements serving as a core for centromeres and telomeres; and unique non-gene sequences. We now describe how these genomic features could have evolved.

Mechanisms Behind the Expansion from Genes to Multigene Families to Gene Superfamilies

Four levels of duplication (followed by diversification and selection) have fueled the evolution of complex genomes. At the lowest level, exons duplicate or shuffle to change the size or function of genes. At the next three increasingly complex levels, entire genes duplicate to create multigene families; multigene families duplicate to produce gene superfamilies; and the entire genome duplicates to double the number of copies of every gene and gene family (**Fig. 22.11**). At each of these successively higher levels of organization, the duplication of larger and larger units leads to the hierarchical generation of greater and greater amounts of new information.

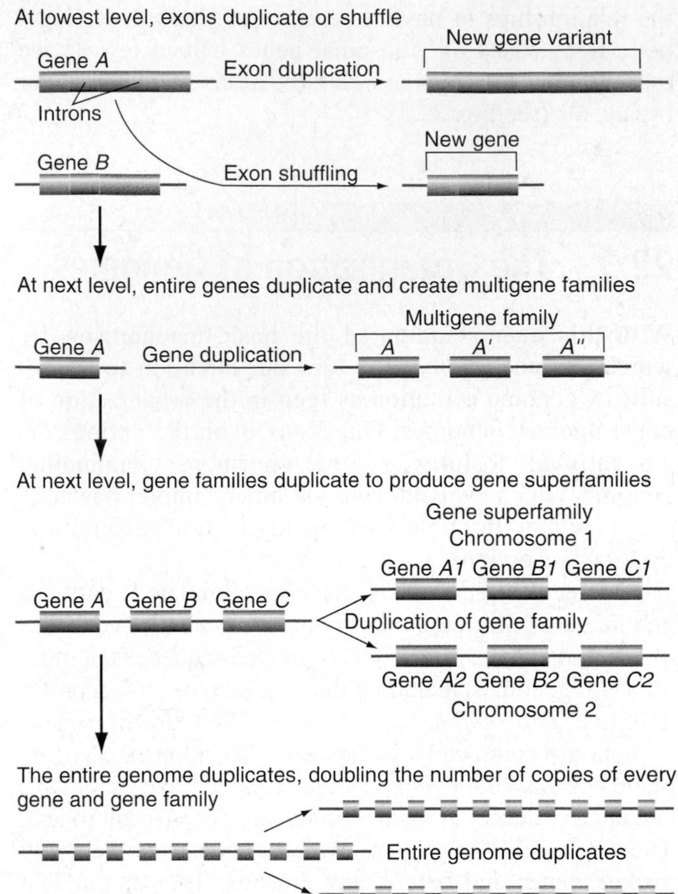

Figure 22.11 Four hierarchic levels of duplications increase genome size. Genome size increases through duplication of exons, genes, gene families, and finally of entire genomes.

Basic Gene Structure: All Genes Have Several Components

We have seen that genes consist of many different components: exons containing coding regions and 5′ and 3′ untranslated regions; introns that will be spliced out; and associated control regions (**Fig. 22.12**). Many of the control regions lie just 5′ to the transcribed region; but some, such as the locus control region of the β-globin gene family

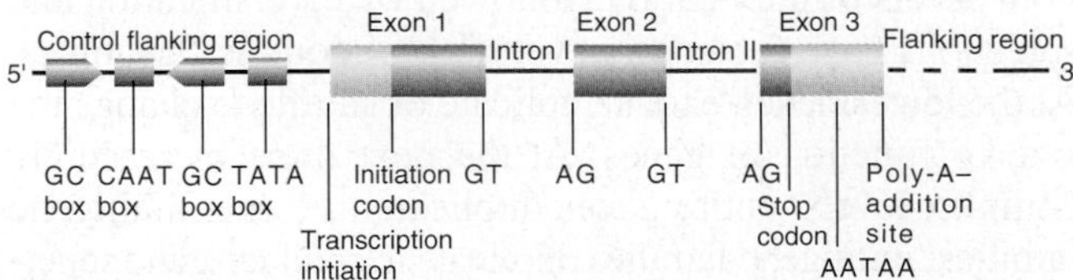

Figure 22.12 The basic structure of a gene. A typical eukaryotic protein-coding gene. The *light green boxes* represent the 5′ and 3′ untranslated mRNA sequences. The small *pink boxes* represent regulatory sequences where gene-controlling transcription factors bind. A eukaryotic gene may range in size from a few hundred base pairs to more than 2 million base pairs.

(described on p. 657 of Chapter 18), lie far outside the gene and appear to play a role in opening up the chromatin of the gene family locus so that gene expression can proceed at the appropriate time and level. Until all associated regulatory elements have been defined, the boundaries of a genetic locus remain uncertain.

The Duplication and Shuffling of Exons Can Elongate Genes and Enable Them to Encode Proteins with More Than One Functional Domain

Many proteins carry discrete compact domains, some of which perform a specific function, while others sustain molecular structure. In many genes, the discrete exons encode the structural and functional domains of a protein. Genes may elongate by the duplication of these exons to generate tandem exons that determine tandem functional domains such as those seen in antibody molecules (**Fig. 22.13a**). The functions of tandem domains may eventually diverge.

Entirely new genes may arise from **exon shuffling:** the exchange of exons among different genes. Exon shuffling produces mosaic proteins such as tissue plasminogen activator (TPA), a molecule with four domains of three distinct types: kringle (K), growth factor (G), and finger (F). The gene for TPA captured exons governing the synthesis of four domains from the genes for three other proteins: K from the gene for plasminogen, G from the gene for epidermal growth factor, and F from the gene for fibronecten (Fig. 22.13b). The mechanism by which exon shuffling occurs is unclear: Since each distinct domain has a different function, one protein may have two or more different functions. Exon shuffling may create new proteins with different combinations of functions.

One or More Duplications of an Entire Gene Can Create Multigene Families with Homologous Members

A **multigene family** is a set of genes descended by duplication and diversification from one ancestral gene. The members of a multigene family may be either arrayed in tandem (that is, clustered on the same chromosome) or distributed on different chromosomes (**Fig. 22.14a**). Unequal crossing-over can expand and contract the number of members in a multigene family cluster (Fig. 22.14b).

Genetic Exchange Between Related DNA Elements by Gene Conversion Can Increase the Variation Among Members of a Multigene Family

There are many places in the genome where there appears to have been a flow of genetic information from one DNA element to other related, but nonallelic, elements located

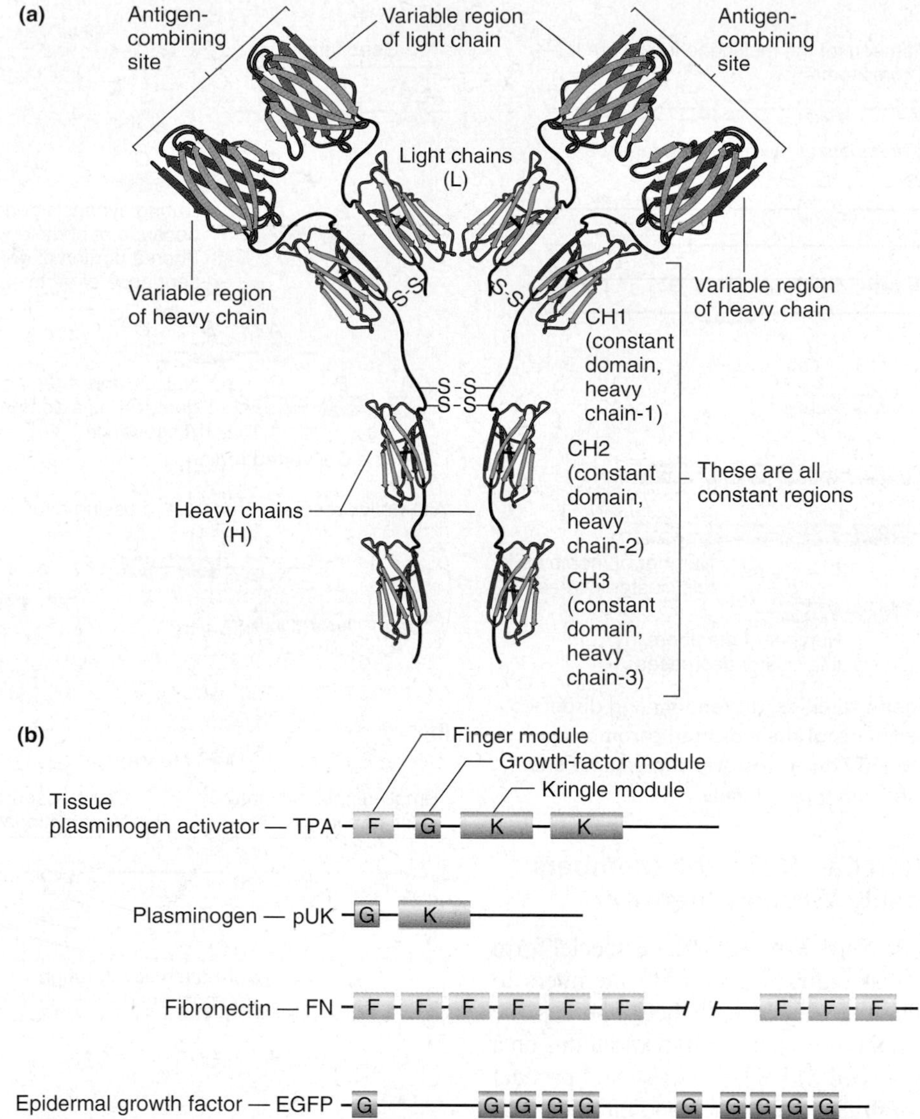

Figure 22.13 Genes can change by the duplication and shuffling of exons. (a) The antibody molecule is composed of two identical pairs of light (*inner*) and heavy (*outer*) chains. The variable regions of each chain, which are modified by somatic mutation, are shown in *tan;* the constant regions in *lavender.* The variable regions of a light and a heavy chain come together to form a V domain. Likewise, the pairs of the CH1, CH2, and CH3 regions come together to generate the CH1, CH2, and CH3 domains, respectively. Each domain carries out separate functions. **(b)** The tissue plasminogen activator (TPA) gene has evolved by the shuffling of exons from the genes for plasminogen, fibronectin, and epidermal growth factor.

nearby or on different chromosomes. Such information flow between related DNA sequences occurs through an alternative outcome of the process responsible for unequal crossing-over. This alternative is known as **intergenic gene conversion (Fig. 22.15a)**. Recall from Fig 6.23, steps 7 and 8 on p. 199 that a recombination intermediate can resolve itself in two ways: one leads to crossing-over and the other allows gene conversion to occur without crossing-over. The same alternative outcomes can occur with unequal recombination intermediates. The gene conversion outcome of unequal crossing-over allows the transfer of information from one gene to another. In special cases (such as the ribosomal gene family that exists in every eukaryotic species), the flow of information from such intergenic gene conversion has been so extreme that it has caused all members of a gene family to coevolve with near identity. And in at least one case—that of the class I genes of the major histocompatibility complex (MHC)—selection has acted on information flow in only one direction, causing information transfer from a series of nonfunctional pseudogenes to a small subset of just two to three functional genes (Fig. 22.15b). In this unusual case, the pseudogene family members served as a reservoir of genetic information that produced a dramatic increase in the amount of polymorphism (that is, the number of alleles) in the small number of functional gene members.

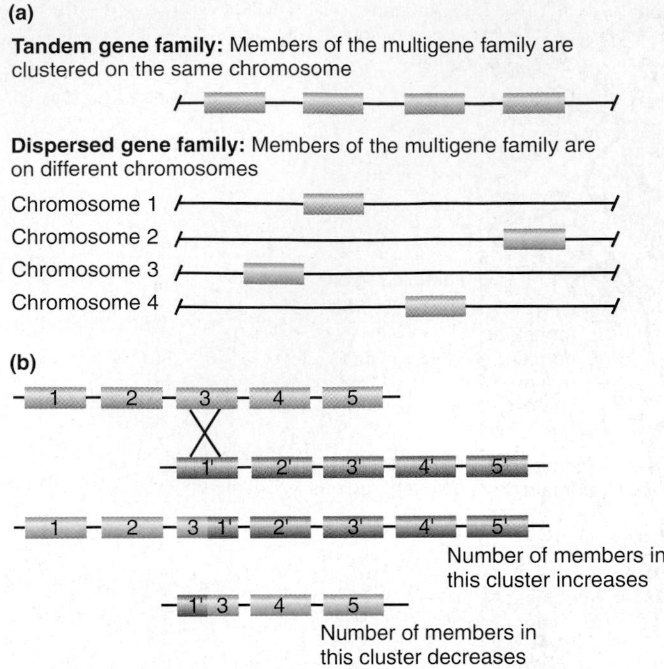

Figure 22.14 Multigene families. (a) Tandem and dispersed multigene families on segments of the indicated chromosomes. **(b)** A schematic illustration of how a crossover can expand and contract gene numbers in a multigene family.

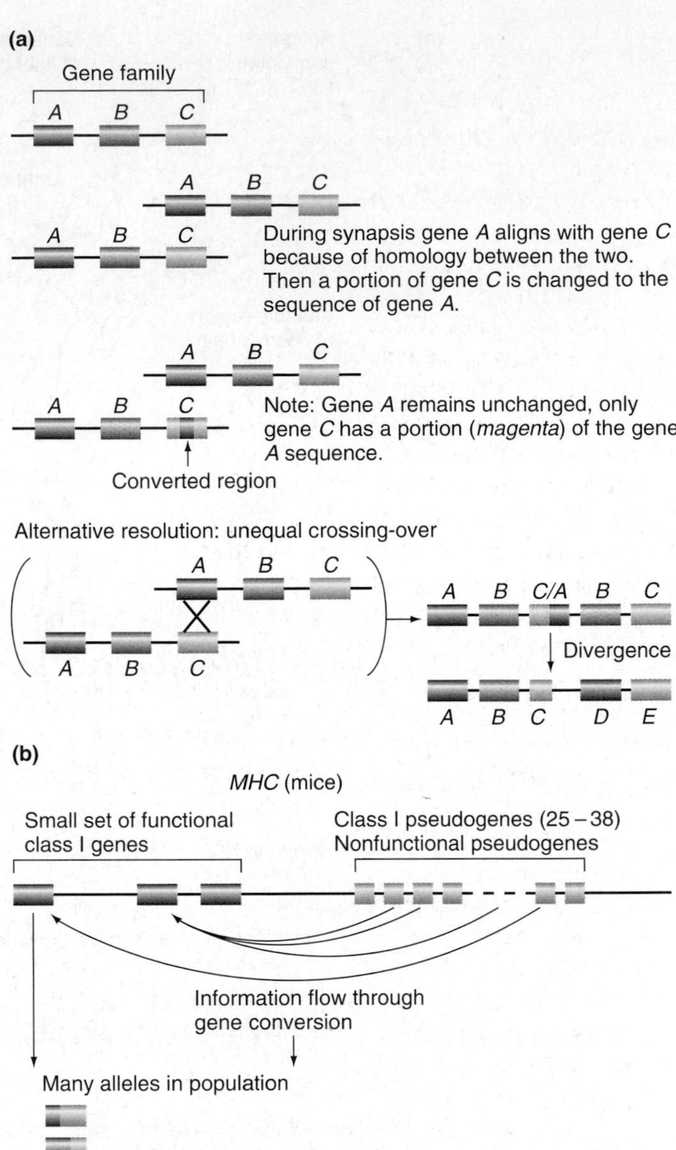

Figure 22.15 Intergenic gene conversion. (a) In intergenic gene conversion, one gene is changed, the other is not. **(b)** How gene conversion events from pseudo-class I genes could increase the polymorphism in functional MHC class 1 genes in mice.

Concerted Evolution Can Make the Members of a Multigene Family Virtually Identical

A few multigene families have evolved under a special form of selective pressure that requires all family members to maintain essentially the same sequence. In these families, the high number of gene copies does not result in variations on a theme; rather it supplies a cell with a large amount of product within a short period of time. Among the gene families with identical elements is the one that produces RNA components of the cell's ribosomes, the one that produces tRNAs, and the one that produces the histones (which must rapidly generate enough protein to coat the new copy of the whole genome replicated during the S phase of every cell cycle).

Each of these gene families consists of one or more clusters of tandem repeats of identical elements. There is strong selective pressure to maintain the same sequence across all members of each family because all must produce the same product. Optimal functioning of the cell requires that the products of any individual gene be interchangeable in structure and function with the products of all other members of the same family. The problem is that the natural tendency of duplicated sequences is to drift apart over time. How does the genome counteract this natural tendency?

When researchers first compared ribosomal RNA and other gene families in this class, both between and within species, a remarkable picture emerged: Between species, there was clear evidence of genetic drift, but within species, all sequences were essentially equivalent. Thus it is not simply that some mechanism suppresses mutational changes in these gene families. Rather, there appears to be an ongoing process of **concerted evolution,** which allows changes in single genetic elements to spread across a complete set of genes in a particular family. With this knowledge, we can narrow the previously posed question about counteracting the natural tendency toward drifting apart to: How does concerted evolution occur?

Concerted evolution appears to occur through two different processes. The first is based on the expansion and contraction of gene family size through sequential rounds of unequal crossing-over between homologous sequences (**Fig. 22.16a**). Selection acts to maintain the absolute size of the gene family within a small range around an optimal mean. As the gene family becomes too large, the shorter of the unequal crossover products becomes selected; as the family becomes too small, the longer of the products

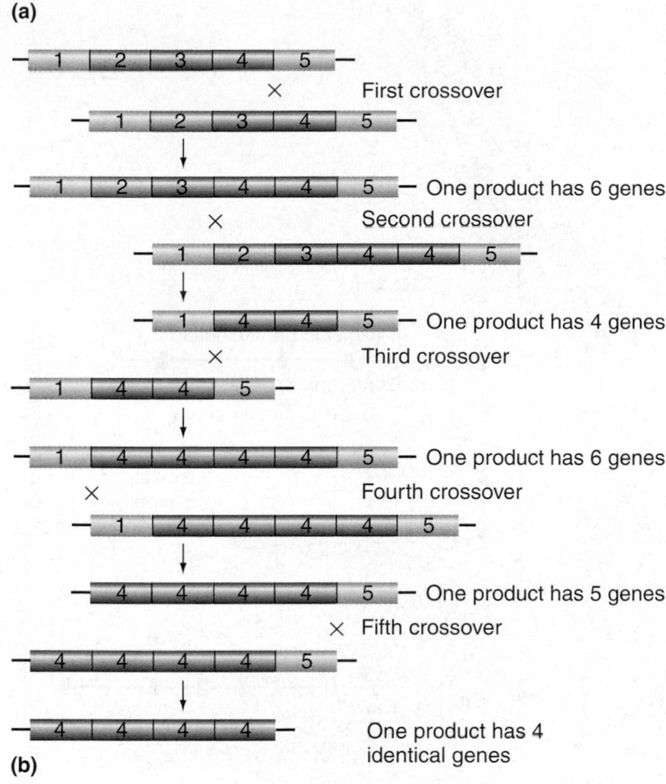

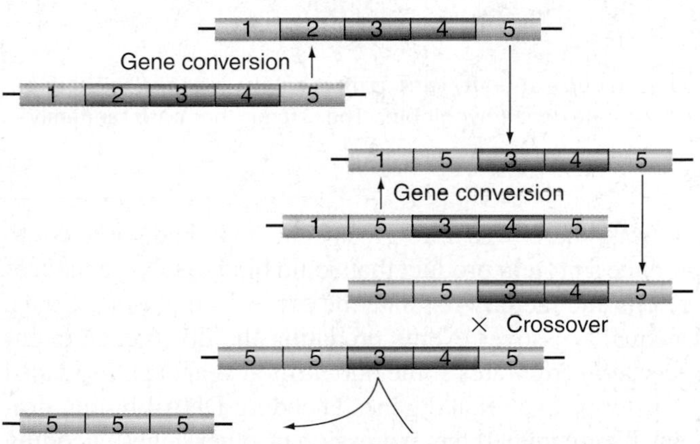

Figure 22.16 **Concerted evolution can lead to gene homogeneity.** Boxes with different colors and numbers represent gene family members with variant DNA sequences. Repeat cycles of unequal crossover events **(a)** or gene conversion **(b)** cause the duplicated genes on each chromosome to become progressively more homogenized.

becomes selected. This cyclic process causes a continuous oscillation around a size mean. However, each contraction results in the loss of divergent genes, whereas each expansion results in the indirect "replacement" of those lost genes with identical copies of other genes in the family. With unequal crossovers occurring at random positions throughout the cluster and with selection acting in favor of the least divergence among family members, this process can act to slow down dramatically the natural tendency of genetic drift between family members.

The second process responsible for concerted evolution is intergenic gene conversion between nonallelic family members. Although in each case, the direction of information transfer from one gene copy to the next is random, selection will act on this molecular process to ensure an increase in homogeneity among different gene family members (Fig. 22.16b). Information transfer (presumably by means of intergenic gene conversion) can also occur across gene clusters that belong to the same family but are distributed to different chromosomes.

Thus, with unequal crossing-over and intergenic gene conversion (which are actually two alternative outcomes of the same initial process) as well as selection for homogeneity, all members of a gene family can retain nearly the same DNA sequence. However, although concerted evolution maintains the close similarity of the members of multigene families within a species, divergences between the whole gene families present in different species nevertheless occur. This is because different sequences serve as the prototype for gene family homogenization in different species.

The Evolution of Gene Superfamilies

Molecular geneticists use the phrase **gene superfamily** to describe a large set of related genes that is divisible into smaller sets, or families, with the genes in each family being more closely related to each other than to other members of the larger superfamily. The multigene (or single-gene) families that compose a gene superfamily reside at different chromosomal locations. A prototypical small-size gene superfamily is the very well-studied globin genes illustrated in **Fig. 22.17.** The superfamily has three branches: the multigene family of β-like genes, the multigene family of α-like genes, and the single myoglobin gene. The duplications and divergences that produced the three superfamily branches occurred early in the evolution of vertebrates; as a result, the three branches of the superfamily are found in all mammals. All functional members of this superfamily play a role in oxygen transport (see Chapter 9 introduction and comprehensive example). As we have seen, the products of the α- and β-globin genes are active in red blood cells, while the product of the myoglobin gene transports oxygen in muscle tissue.

The primordial globin gene gave rise to the myoglobin and α-/β-globin precursor genes by gene duplication and transposition. The primordial α- and β-globin genes probably arose by a large-scale genome duplication (tetraploidization).

The β-like branch of this gene superfamily arose by duplication via multiple unequal crossovers. In the mouse, all the β-like genes are present in a single cluster on chromosome 7, which contains four functional genes and three pseudogenes (Fig. 22.17). The β-like chains encode similar polypeptides, each of which has been selected for optimal function at a specific stage of mouse development: one functions during early embryogenesis, one during a later stage of embryogenesis, and two function in the adult.

The α-like branch also evolved by unequal crossovers and divergences that generated a cluster of three genes on

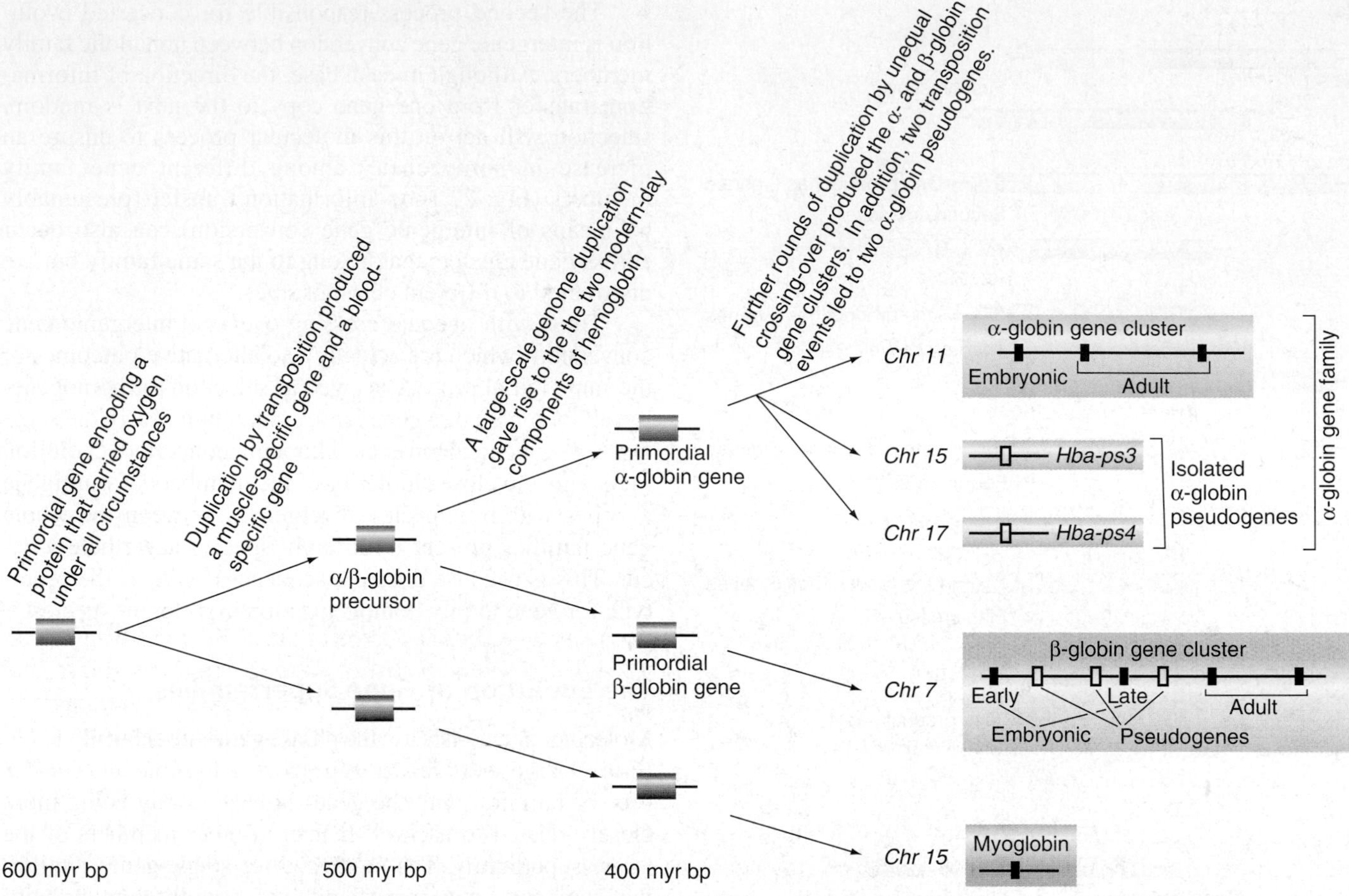

Figure 22.17 Evolution of the mouse globin superfamily. Repeated gene duplication by various mechanisms gave rise to the globin supergene family in mice, with two multigene families (α and β) and one single gene (myoglobin). The α family has both tandemly arrayed and dispersed gene members.

mouse chromosome 11: one functions during embryogenesis, and two function in the adult (Fig. 22.17). The two adult α-globin genes are virtually identical at the level of DNA sequence, which suggests that the duplication producing them occurred very recently (on the evolutionary time scale). In addition to the primary α-like cluster, there are two nonfunctional α-like genes—pseudogenes—that have dispersed via transposition to locations on chromosomes 15 and 17. Pseudogenes existing in isolation from their parental families are called *orphons*. Interestingly, the α-globin orphon on mouse chromosome 15 (named *Hba-ps3*) has no introns and thus appears to have arisen through a retrotransposition event involving mRNA copied back to DNA; whereas the α-globin orphon on chromosome 17 (*Hba-ps4*) contains introns and may have arisen by a direct DNA-mediated transposition. The single mouse myoglobin gene on chromosome 15 has no close relatives either nearby or far away. The globin gene superfamily provides a view of the many different mechanisms that the genome can employ to evolve structural and functional complexity.

The mouse *Hox* gene superfamily (discussed in the genetic portrait of the house mouse on our website: www.mhhe.com/hartwell3) provides an alternative prototype for the evolution of a gene superfamily (**Fig. 22.18**). At first, there was one *Hox* gene that had evolved to produce a protein product that could bind to DNA enhancer regions and thereby regulate the expression of other genes. Unequal crossover events predating the divergence of insects and vertebrates some 600 million years ago produced a cluster of five related genes encoding DNA-binding proteins that regulated the expression of other genes encoding spatial information (that is, instructions for the spatial positioning of tissues and organs) in the developing embryo. The original *Hox* gene family then duplicated en masse and dispersed to four locations—on chromosomes in a common ancestor to all vertebrates. Because of the order of duplication events leading to the superfamily—unequal crossing-over to expand the original cluster size, followed by transposition en masse of the expanded cluster to create the superfamily—an evolutionary tree would show that a single gene family within the superfamily has actually splayed out physically across the four gene clusters, as shown in Fig. 22.18. After the en masse duplication that generated the gene superfamily, smaller duplications by unequal crossing-over added genes to some of the dispersed clusters and subtracted genes from others, thereby generating differences in gene number and type within a basic framework of homology among the different clusters.

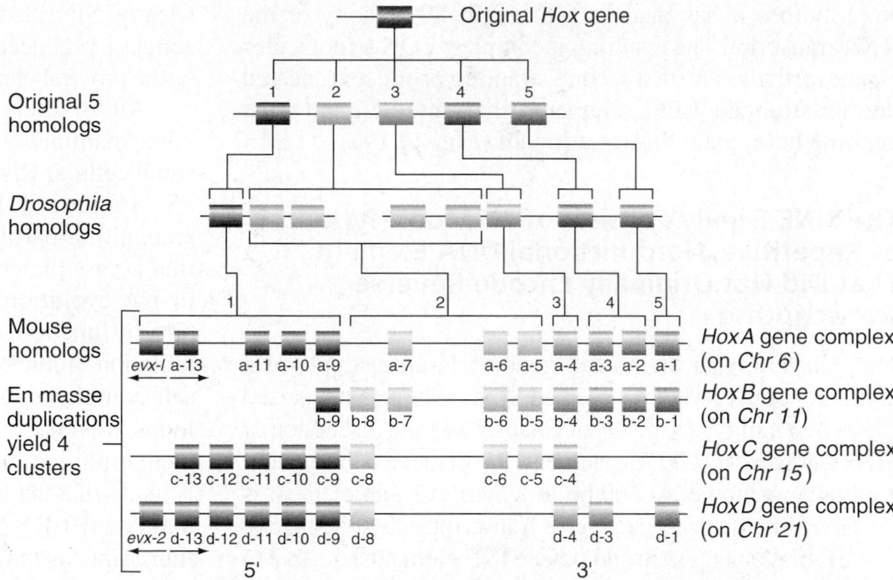

Figure 22.18 Evolution of the *Hox* gene superfamily of mouse and *Drosophila*. This gene superfamily arose by a series of gene duplications. Four multigene families are present in the mouse and one in *Drosophila.*

Each of the four *Hox* clusters in the mouse superfamily currently contains 9–12 homologous genes.

Repetitive "Nonfunctional" DNA Families Constitute Nearly One-Half of the Genome

Many repetitive nonfunctional DNA families consist of retroviral elements. Retroviruses, we have seen, are RNA-containing viruses that can convert their RNA genome into circular DNA molecules through the viral-associated RNA-dependent DNA polymerase known as reverse transcriptase, which becomes activated upon cell infection (see Figs. A and B in the Genetics and Society box in Chapter 8 on p. 270). The resulting DNA can integrate itself at random into the host genome, where it becomes a provirus that retains the genetic information of the retroviral genome. Under certain conditions, the provirus can become activated to produce new viral RNA genomes and associated proteins, including reverse transcriptase, which can come together to form new virus particles that are ultimately released from the cell surface by exocytosis. By contrast, many stably integrated retroviral elements appear to be inactive.

Once integrated into a host chromosome, the provirus replicates with every round of host DNA replication, irrespective of whether the provirus itself is expressed or silent. Moreover, proviruses that integrate into the germ line, through the sperm or egg genome, segregate along with their host chromosome into the progeny of the host animal and into subsequent generations of animals as well. The genomes of all species of mammals contain inactive integrated proviral elements.

It is of evolutionary interest to ask where retroviruses originated. Since retroviruses cannot propagate in the absence of cells, but cells can propagate in the absence of retroviruses, it seems likely that retroviruses evolved from nucleotide sequences that were originally present in the cell genome. The first retrovirus might have been able to free itself from the confines of the cell nucleus through an association with a small number of proteins that allowed it to coat and thus protect itself from the harsh extracellular environment. Of course the protein most critical to the propagation of the retrovirus would have been the reverse transcriptase that allows the retrovirus to reproduce. But where did this enzyme come from? Reverse transcriptase catalyzes the production of single-stranded complementary DNA molecules from an RNA template, but this enzymatic activity is not required for any normal cellular process known in mammals. So, how could an activity without any apparent benefit to the host organism arise *de novo* in a normal cell? One possible answer is that reverse transcriptase did not evolve because it benefited the organism itself; rather it evolved because it benefited selfish DNA elements within the genome that use it to propagate themselves within the confines of the genome. We now describe one such selfish element.

The LINE Family of DNA Elements Are Retroposons That Appear to Derive from a Selfish DNA Sequence Encoding Reverse Transcriptase

As we saw on p. 509 of Chapter 14, LINE is an acronym for Long INterspersed Elements. The LINE family of DNA elements is very old. Homologous families of repetitive elements exist in a wide variety of organisms, including protists and plants. Thus, LINE-related elements, or other DNA elements of a similar nature, are likely to have been the source material that gave rise to retroviruses.

Dispersion to new positions in the germ-line genome presumably begins with the transcription of LINE elements in spermatogenic or oogenic cells. The reverse-transcriptase-encoding region on the transcript is translated into an enzyme that preferentially associates with and uses the transcript it came from as a template to produce LINE cDNA sequences; however, the reverse transcriptase often stops before it has made a full-length DNA copy of the RNA transcript. The resulting incomplete cDNA molecules can nevertheless form a second strand to produce truncated double-stranded LINE elements that integrate into the genome but remain forever dormant (**Fig. 22.19a**).

The SINE Family Consists of a Second Type of Repetitive, Nonfunctional DNA Element That Did Not Originally Encode Reverse Transcriptase

The Alu family in the human genome is an example of a highly repetitive, widely dispersed SINE (Short INterspersed Elements) family (see p. 509 of Chapter 14 for a discussion of SINEs). Over 500,000 Alu elements are dispersed throughout the human genome. At 300 bp in length, the Alu element is far too short to encode a reverse transcriptase. Nonetheless, like LINE elements, Alu and other SINE elements are able to disperse themselves throughout the genome by means of an RNA intermediate that undergoes reverse transcription.

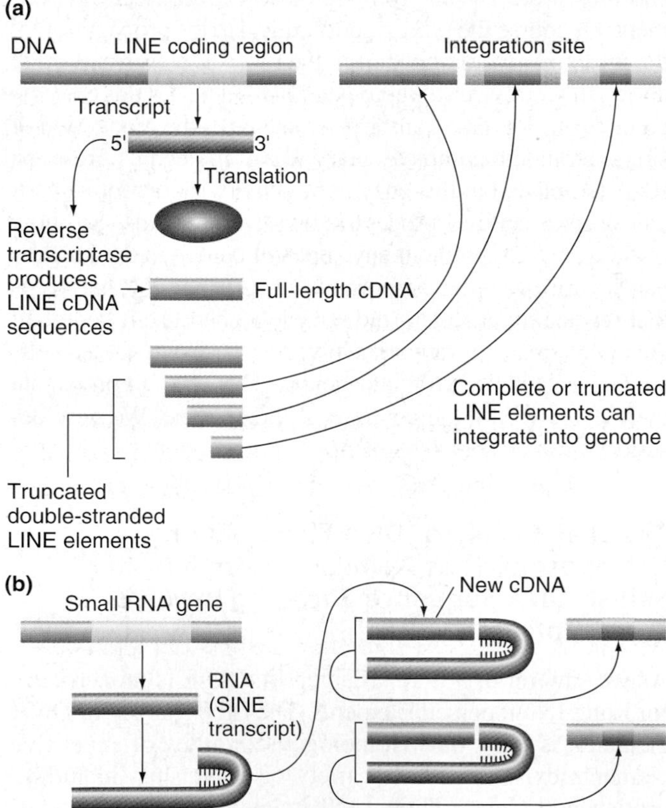

Figure 22.19 The creation of LINE and SINE gene families. **(a)** A complete LINE sequence can be copied into RNA. It encodes a reverse transcriptase that can make cDNA copies from the RNA. These copies may be complete or truncated. Either kind may integrate into other sites on any chromosome. **(b)** SINE elements can be transcribed and, because they form 3′ hairpin loops, can be copied into cDNAs by LINE-encoded reverse transcriptase. These cDNA copies then integrate into the genome.

Clearly, SINES depend on the availability of reverse transcriptase produced elsewhere, perhaps from LINE transcripts or the proviral elements of retroviruses.

All SINE elements in the human genome, as well as in other mammalian genomes, appear to have evolved from small cellular RNA species, most often tRNAs, but also the 7S cytoplasmic RNA that is a component of the signal recognition particle (SRP) essential for protein translocation across the endoplasmic reticulum. The defining event in the evolution of a functional cellular RNA into an altered-function, self-replicating SINE element is the accumulation of nucleotide changes in the 3′ region that lead to self-complementarity with the propensity to form hairpin loops. Reverse transcriptase can recognize the open end of the hairpin loop as a primer for strand elongation. Since it is likely that the hairpin loops form only rarely among normal cellular RNAs, a cell will preferentially use its SINE transcripts as templates for the production of cDNA molecules that are somehow able to integrate into the genome at random sites (Fig. 22.19b).

Comparing LINE and SINE Families

Like LINE families, the Alu SINE family appears to be evolving by episodic amplification followed by sequence degradation. Unlike LINE families, however, SINE families present in the genomes of different organisms appear, for the most part, to have independent origins.

Some Selfish Elements May Confer a Selective Advantage

While SINE and LINE elements may have amplified themselves for selfish purposes, they have had a profound impact on whole-genome evolution. In particular, homologous SINE or LINE elements located near each other can, and will, catalyze unequal but homologous crossovers that result in the duplication of single-copy genes located between the homologous elements. Such duplications initiate the formation of multigene families. In addition, some selfish elements appear to have evolved a regulatory role—acting as enhancers or promoters—through chance insertions next to open reading frames. Thus, selfish elements may confer a selective advantage by facilitating duplication through unequal crossing-over or by becoming regulatory elements.

Genomic Stutters—the Simple Sequence Repeats Known As Microsatellites, Minisatellites, and Macrosatellites—Dot the Mammalian Genome

Through large-scale sequencing and hybridization analyses of mammalian genomes, researchers have found tandem repeats of DNA sequences with no apparent function scattered throughout the genome. The size of the repeating units ranges from two nucleotides (CACACACA . . .) to

20 kb or more, and the number of tandem repeats varies from two to several hundred.

The mechanism by which these tandem repeats originate may be different for loci having very short repeat units as compared with loci having longer repeat units. Tandem repeats of short di- or trinucleotides can originate through random changes in nonfunctional sequences. By contrast, the initial duplication of larger repeat units is likely to be a consequence of unequal crossing-over. Once two or more copies of a repeat unit exist in tandem, an increase in the number of repeat units in subsequent generations can occur through unequal crossovers or errors in replication (see Figs. 22.9 and 22.14b). It is not yet clear whether random mechanisms alone can account for the rich variety of tandem repeat loci in mammalian genomes or whether other selective forces are at play. Either way, tandem repeat loci continue to be highly susceptible to unequal crossing-over and, as a result, tend to be highly polymorphic in overall size.

Repeat Sequences in Centromeres and Telomeres

Highly repetitive, noncoding sequences shorter than 200 bp are found in and around centromeres. For example, in the human genome, *alphoid,* a noncoding sequence 171 bp in length, is present in tandem arrays extending over a megabase in the centromeric region of each chromosome. In addition, several similar-sized repetitive sequences unrelated to alphoid are found in some centromeres (review Fig. 13.11 on p. 478). These regions are sites of interaction with the spindle fibers that segregate chromosomes during meiosis and mitosis. Selection may have acted to retain the thousands of copies of centromeric repeat elements in each centromeric region because they increase the efficiency and/or accuracy of chromosome segregation.

A second type of repeated genomic element with a special location is the hexamer TTAGGG found in the telomeres of all human chromosomes. The six-base unit is repeated in tandem arrays 5–10 kb in length at the ends of human and all other mammalian chromosomes. Selection may have conserved this repeat element because it plays an essential role in maintaining chromosomal length (review Figs. 13.9 and 13.10 on pp. 476 and 477).

Unique Nongene Sequences

The mutation and informational degradation of extra copies of duplicated genes, which produces pseudogenes, can ultimately result in unique nongene sequences. The mutation and divergence of genomic repeat elements can also lead to unique nongene sequences.

In summary, the organization of genomes in higher organisms exhibits several cardinal features. Genes, constituting less than 3% of the genome, are often organized into multigene families or even gene superfamilies. A variety of mechanisms, such as homologous but unequal crossing-over and gene conversion, permit the expansion and contraction of gene families, as well as their concerted evolution. In some organisms, such as human and mouse, approximately 50% of the genome consists of repeat elements. These elements may facilitate chromosomal rearrangements or duplications and, on occasion, evolve to become genes or regulatory sequences. The telomeres and centromeres are largely composed of repeat sequence elements. Close to half of the genome may represent genes or repeat sequences that have degraded to unrecognizable unique sequences.

22.4 The Immunoglobulin Gene Superfamily: A Comprehensive Example of Molecular Evolution

More than 200 different types of genes belong to the immunoglobulin gene superfamily (**Table 22.1**). Most encode cell-membrane proteins that recognize either soluble molecules or molecules on the surface of other cells. All members of the immunoglobulin gene superfamily share a common feature: the **immunoglobulin homology unit,** so-called because it was first identified in the genes for immunoglobulin (antibody) molecules (**Fig. 22.20**). The homology unit encodes about 100 amino acids, which fold into a characteristic three-dimensional structure termed the *immunoglobulin fold.* The unique design of the immunoglobulin fold enables it to (1) interact with other immunoglobulin folds to form dimeric molecules, (2) display an enormous diversity of recognition sites while still maintaining a highly conserved three-dimensional framework, and (3) resist proteolysis in the blood. The folds encoded by different homology units may vary by as much as 90% of their sequences while still preserving their folded framework because of a small but critical number of highly conserved structural features. The evolutionary success of the immunoglobulin homology unit is reflected in its deployment in more than 200 types of genes.

Evolution of the Immunoglobulin Homology Unit

The immunoglobulin homology unit probably emerged early in the evolution of metazoans. Scientists believe this to be true for three reasons. First, both vertebrates and invertebrates employ cell surface recognition molecules containing the immunoglobulin homology unit, but such

TABLE 22.1 Examples of Proteins Encoded by Immunoglobulin Gene Superfamily Members

Protein	Molecular wt. (kDa)	Ig domains C1	Ig domains C2	Ig domains V	Gene on Human Chromosome	Function
Ig heavy chain	55	3–4	—	1	14q32.33	Recognition of Ag
Ig κ light chain	24	1	—	1	2p12	Recognition of Ag
Ig λ light chain	24	1	—	1	32q11.12	Recognition of Ag
B29 (Ig-β/Ig-γ)	39	(1 Ig domain)			(Mouse)	Component of surface Ig complex
MB-1 (IgM-α)	34	(1 Ig domain)			(Mouse)	Component of surface Ig complex
TCR α chain	45–60	1	—	1	14q11.12	Recognition of peptide-MHC
TCR β chain	40–50	1	—	1	7q32	Recognition of peptide-MHC
TCR γ chain	45–60	1	—	1	7p15	Recognition of peptide-MHC
TCR δ chain	40–60	1	—	1	14q11.2	Recognition of peptide-MHC
CD3 ϵ chain	25–28	—	1	—	11q23	TCR signal transduction
CD3 δ chain	20	—	1	—	11q23	TCR signal transduction
CD3 ϵ chain	20	—	1	—	11q23	TCR signal transduction
MHC class I protein	44	1	—	—	6p21.3	Peptide binding
β2 microglobulin	12	1	—	—	15q22-q22	Stabilize MHC class I α
MHC class II protein	32–34	1	—	—	6p21.3	Peptide binding
MHC class II β chain	29–32	1	—	—	6p21.3	Peptide binding
CD2	50	—	2	—	1p13	Ligand for LFA-S (CD58)
CD4	55	—	1	2	12pter-q12	Binds MHC class II
CD8 α chain	34	—	—	1	2p12	Binds MHC class I
CD8 β chain	34	—	—	1	2p12	Binds MHC class I
CD7	40	—	—	1	17q25	T-cell activation
CD22	130/140	—	5	—	Unassigned	B-cell adhesion
CD28	44	—	—	1	2q33-q34	T-cell activation
CD48 (blast-2)	41–45	—	1	1	1q22-q23	B-cell adhesion
Thy-1	17.5–18.7	—	—	1	11q23	T-cell activation and adhesion
MRC OX2	41–47	—	1	1	3	Unknown
MAG	67/72	—	5	—	(Rat)	Axon-glial interactions of myelination
P_o myelin protein	28–30	—	—	1	(Rat)	Myelin compaction
Falciculum II	96	—	5	—	(Grasshopper)	Axonal fasciculation
Almagam	38–50	—	2	1	(*Drosophila*)	Neural cell adhesion
Contactin	180	—	6	—	(Aves)	Neurite fasciculation
L1	200	—	6	—	(Mouse)	Neurite fasciculation and elongation, neuronel migration
Carcinoembryonic Ag	175–200	—	2–6	1	19q13.1-q13.2	Family of molecules; adhesion?
B, Memb. link, protein	44.5–48.8	—	—	1	Unassigned	Locks proteoglycan to haluronate chain

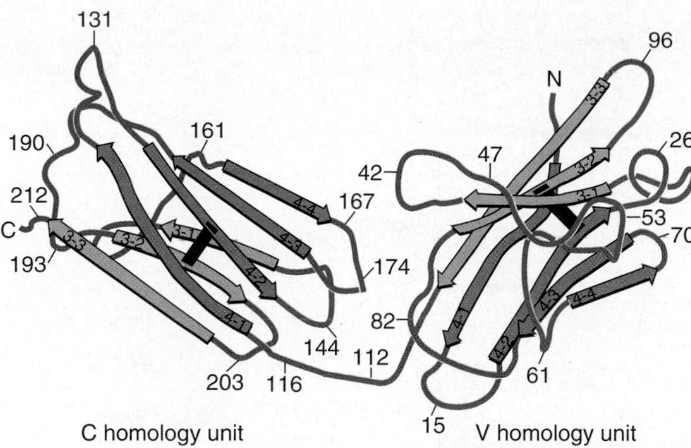

Figure 22.20 Three-dimensional structures encoded by immunoglobulin homology units. Tertiary organization of V and C homology units in an immunoglobulin light chain. The β strands and their orientation are shown as *flat arrows.* Opposite faces of the β strands are *light lavender.* The disulfide bond is indicated by a *solid* black *bar.* The V and C homology units are remarkably similar in structure. The variable loops indicated by *96, 26,* and *53* fold to form the walls of the antigen-binding site.

molecules are not found in prokaryotes or single-cell eukaryotes. Since the common ancestor of present-day vertebrates and invertebrates was present when metazoan organisms first emerged, it is likely that the immunoglobulin gene superfamily also came into existence at that time. Second, the need for molecular recognition at the cell surface increased exponentially with the evolution of multicellular organisms, and it appears that the immunoglobulin homology unit evolved coincidentally, allowing metazoans to fill their need for diverse cell surface recognition units. Third, the early primordial homology unit quickly duplicated to make three distinct types of homology units: variable (V), constant (C), and homology (H) (**Fig. 22.21a**). The V, C, and H units are equally related to one another and are deployed in the diverse cell surface receptors used by all metazoans (Fig. 22.21b). The inescapable conclusion from these observations is that the highly diverse immunoglobulin gene superfamily evolved relatively rapidly, and its rapid creation may have been a driving force in generating the cell surface recognition cababilities required by emerging metazoan organisms.

Evolution of the Immunoglobulin Gene Superfamily

Molecular evolution, we have seen, can occur very rapidly because of the hierarchical nature of changes in structural information: Nucleotide substitutions can change exons; the duplication, divergence, and shuffling of exons can create more complex genes; the duplication of genes can create multigene families; and the duplication of multigene families (and single-gene members) can create gene superfamilies.

The immunoglobulin gene superfamily evolved by exon shuffling to create mosaic proteins (Fig. 22.21c), by exon duplication to create very long genes, by gene duplication to create multigene families, and by the duplication of whole gene families to create the gene superfamily depicted in Fig. 22.21 and Table 22.1.

The First Members of the Superfamily to Evolve Encoded Specific Antigen Receptors

These receptors included antibodies, T-cell receptors, and MHC molecules. Antibodies recognize the three-dimensional shapes of their cognate antigens. T-cell receptors recognize peptide fragments of antigen presented by MHC molecules; the T-cell receptors also react with coreceptors, including the CD4 and CD8 molecules (which are also members of the immunoglobulin gene superfamily), to orchestrate immune responses. Elaborate recognition mechanisms operating at the cell surface are the most common functions of the immunoglobulin superfamily members. Other members of the immunoglobulin gene superfamily evolved to recognize immunoglobulin molecules at the cell surface and/or to transport immunoglobulins across membrane compartments.

Surprisingly, Many Members of the Immunoglobulin Superfamily Carry out Functions Unrelated to Vertebrate Immune Recognition

Some components, such as link protein, help recognize the cellular matrix. Other components make up the carcinoembryonic antigen (CEA) gene family, which has 10 or more members all involved in specific cell adhesions; the recognition of specific cells by other cells is critical for the development and structural integrity of tissues and organs. Several cell-adhesion molecules guide the growth of neurons in vertebrates (L1 and contactin) and invertebrates (falciculum II and almagam). Two other members of the superfamily (MAG and P_0) appear to play a role in the wrapping and compaction of the myelin sheath. Interestingly, several other superfamily molecules, including Thy I and MRC OX2, are part of the vertebrate immune and nervous systems, raising the possibility that these molecules play a key role in communication between these two complex systems. Other members can encode growth factors or their receptors (PDGF receptor). The existence of so many immunoglobulin superfamily members that do not play a role in the immune response demonstrates how powerful the forces of duplication, divergence, and selection are at making new functional units from older units with a different function.

Immune-Cell Receptors Provide the Backbone of the Immune Response

We saw at the beginning of this chapter that the vertebrate immune response develops the capacity to recognize

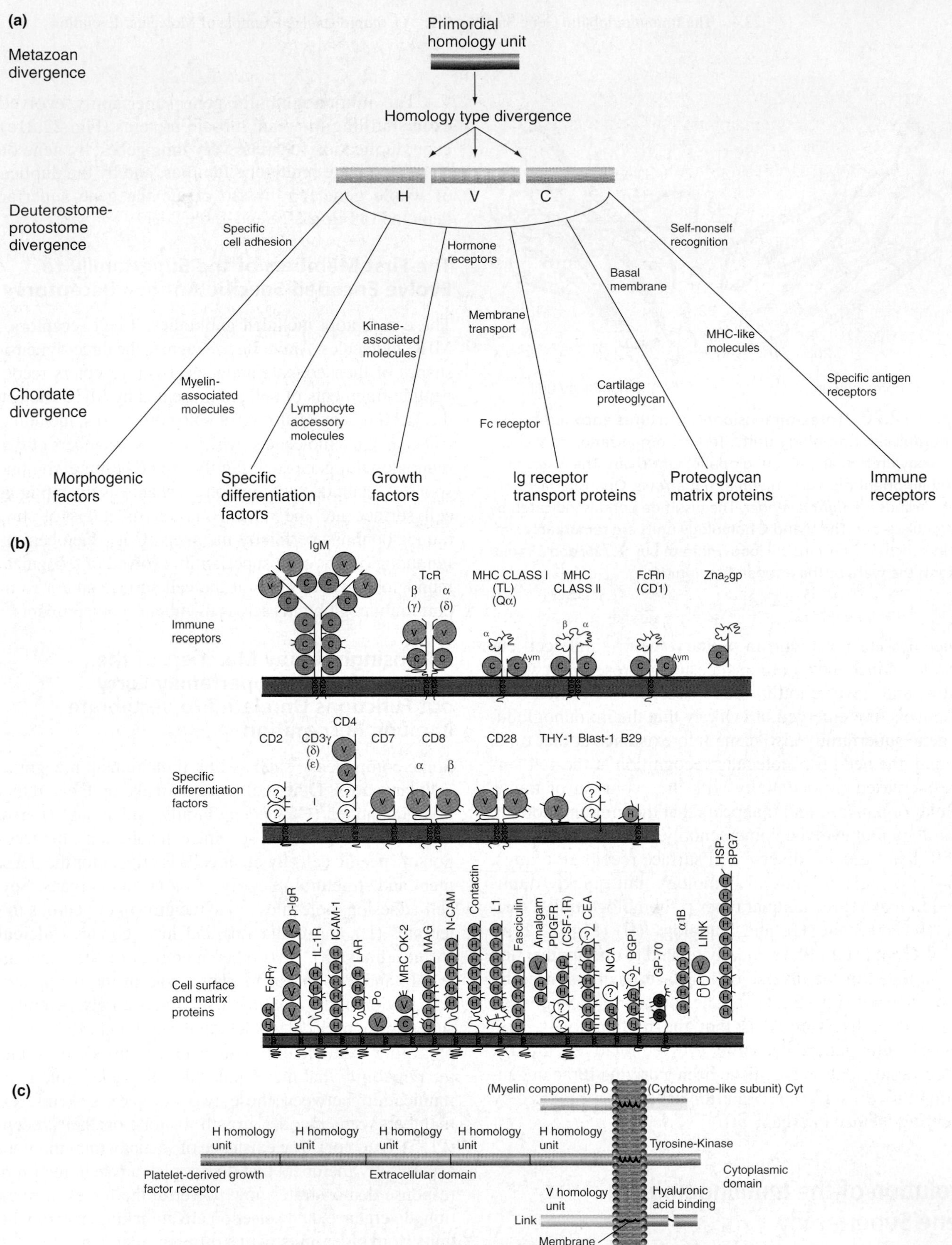

Figure 22.21 Evolution of the immunoglobulin homology unit and immunoglobulin gene superfamily. (a) Model of the evolution of different functional branches of the immunoglobulin gene superfamily showing the emergence of special functional properties. (b) Examples of the structures of members of the immunoglobin gene superfamily. Homology units are indicated as loops labeled *V, C,* or *H.* Loops of uncertain relationship to homology units are labeled with a question mark. Different-sized loops illustrate the relative differences in length between the conserved disulfide bond of the labeled types. Membrane-spanning peptides are shown as simple helices. Functionally related members are illustrated with a single structure, as indicated by the name labels above each structure. (c) Examples of exon shuffling using immunoglobulin homology units.

virtually any foreign macromolecular pattern (antigenic determinant) through the maturation of its two major functional branches. The B cell, or humoral, system manufactures antibodies that bind with and clear unprocessed antigen in viral particles and microbial macromolecules. The T cell, or cellular, immune system uses T-cell receptors and two classes of major histocompatibility complex (MHC) receptors to recognize the processed peptide products of protein antigens. There are three general types of T cells: T cytotoxic cells, which destroy foreign cells as well as intracellular infections and neoplastically transformed cancer cells; T helper cells, which facilitate the differentiation of B-cell and T-effector-cell responses; and T regulatory cells, which help control immune responses.

A Close Look at Immune-Cell Receptors

The immune receptors on both B cells and T cells are heterodimers. The antibodies on B cells consist of light and heavy chains; T-cell receptors are composed of α and β chains; class II MHC molecules carry α and β chains, while class I MHC molecules consist of a heavy chain plus a β-microglobulin chain (which is an immunoglobulin superfamily member containing only a single immunoglobulin fold) (**Fig. 22.22**).

The T-cell and B-cell systems employ very different strategies for the recognition of antigen (see Figs. 22.13a and 22.22). The antibodies of B cells view the native three-dimensional surface of their corresponding antigen by a direct molecular complementarity. T-cell receptors recognize the peptide fragment of their corresponding antigens as presented in a molecular groove on the surface of the MHC molecules. Class I MHC molecules on antigen-presenting cells generally present peptide fragments to T cytotoxic cells, whereas class II MHC molecules generally present peptides to T helper cells. Antibodies and T-cell receptors both employ two types of homology units: variable (V) and constant (C).

The primary structure of the antibody molecule provided the first evidence of the homologous repeat units that are the functional building blocks of the immunoglobulin gene superfamily. In three dimensions, the homology units form a sandwich of β-pleated sheets connected by loops that can be highly variable. It is in the variation of these loops that much of the immune cells' ability to recognize specific antigen resides (review Fig. 22.20).

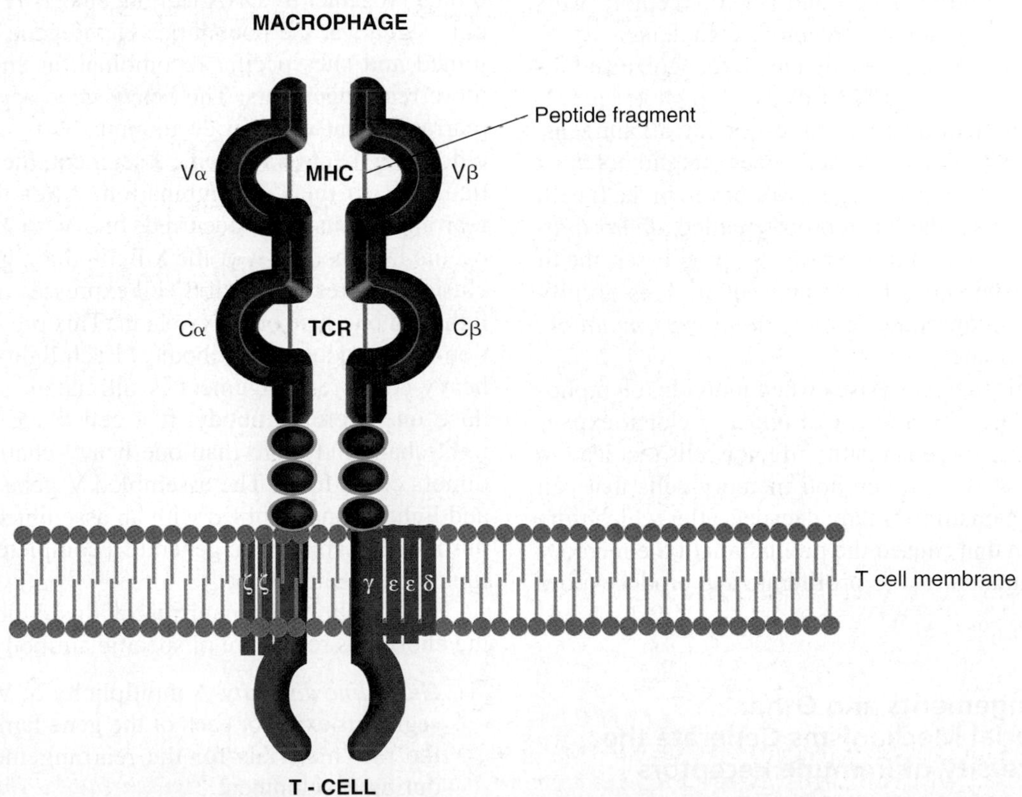

Figure 22.22 Schematic diagram of a T-cell receptor (TCR) interacting with a peptide fragment enclosed within an MHC (major histocompatibility) receptor on a macrophage (a type of white blood cell that presents antigen to lymphocytes). The association of the γ, ε, δ, and ζ chains of the T-cell receptor with its α and β chains facilitates the process of signal transduction into the T cell by the three-part (MHC-peptide-TCR) interaction complex.

Immune-Cell Receptors Define the Basic Features of an Immune Response

An immune response has three characteristics: it is highly **specific,** it is **adaptive,** and it has a **memory.** Immunologic *specificity* arises because each type of T cell or B cell expresses a single type of receptor. The binding of receptor to complementary, or cognate, antigen drives the clonal expansion of B cells and T cells carrying suitable receptors and thus provides the basis for the specificity of immune responses (see Fig. 22.2).

The fact that each lymphocyte expresses only a single receptor is interesting in two regards. It signifies that in a large multigene family, only one gene undergoes expression; that is, of the two copies of this gene—on the paternal and maternal chromosomes—only one copy is expressed. (This second phenomenon—the expression of only one allele of the gene—is known as *allelic exclusion.*) The DNA rearrangement process for creating the highly variable and thus highly specific antigen-recognition regions is probably responsible for the expression of just one immune receptor gene and just one allele of this gene (review the Fast Forward box on pp. 492–494 of Chapter 14).

The immune response is *adaptive* because it has a large repertoire of different lymphocytes (approximately 10^{12}) that circulate in the blood and lymph. With this constant circulation, foreign organisms receive continuous exposure to many different T-cell and B-cell receptors. This enables specific lymphocyte receptors complementary to the pathogen's antigenic determinants to recognize and interact with them. The large repertoire of lymphocytes collectively has sufficient diversity to respond to all antigens, even the many natural and synthetic ones that did not exist when the immune system emerged. Moreover, in the B cell, or antibody response, there is a process called *affinity maturation* in which as the immune response progresses, the fit of the antibody becomes better and better. This affinity maturation is the consequence of *somatic hypermutation,* which we discuss later.

Immunologic *memory* arises when individual lymphocytes are stimulated by antigen to undergo clonal expansion and two types of cells result: effector cells that lead to the destruction of the antigen and memory cells that can remain in the organism for many decades, after which time the same antigen that caused the production of the memory cells can restimulate them to proliferate and produce more effector cells.

Gene Rearrangements and Other Recombinatorial Mechanisms Generate the Immense Diversity of Immune Receptors

Immunoglobulin (or antibody) molecules are encoded by three families of genes: one heavy-chain gene and two types of light-chain genes, kappa (κ) and lambda (λ) (**Fig. 22.23**). The genes for the variable regions of each

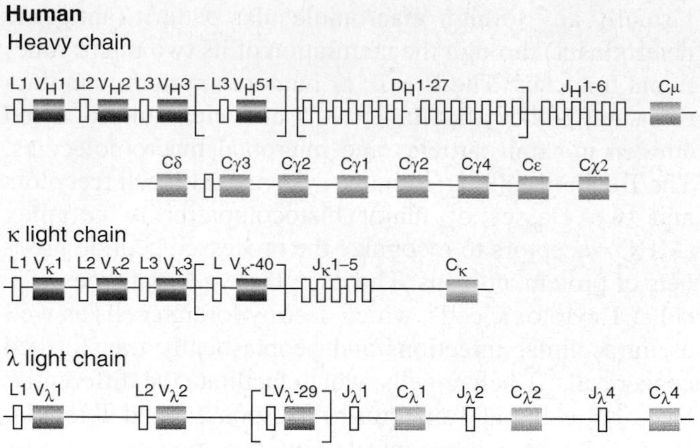

Figure 22.23 Three families of antibody genes encode three types of antibody chains: One heavy and two light—λ and κ, The letters *L, V, D, J,* and *C* indicate leader, variable, diversity, joining, and constant regions of the genes, respectively.

heavy chain contain three distinct types of segments: variable (V), diversity (D), and joining (J). The genes for the light-chain variable regions encode V and J gene segments. During the differentiation of lymphocytes that occurs as an individual's immune system develops, the gene segments that will make up the gene to be expressed in each lymphocyte are brought together by DNA rearrangements (**Fig. 22.24**). Specific signals at the boundaries of the gene segments to be joined and site-specific recombination enzymes mediate these rearrangements. The heavy gene segments undergo rearrangement early in development—first joining a specific D segment to a specific J segment, then joining a particular V to the DJ combination. After the heavy gene rearrangements have occurred, one V to J rearrangement occurs in either the κ or the λ light-chain gene. Allelic exclusion ensures that each B cell expresses only one type of light and one type of heavy chain. This prevents the formation of unproductive antibodies. Each light-chain and each heavy-chain rearrangement is ultimately selected to produce one useful antibody; if a cell made more than one light chain and more than one heavy chain, unproductive dimers could form. The assembled V gene for both heavy and light chains is joined with an assembled constant gene during RNA splicing to generate a complete mRNA for the immunoglobulin chain.

Five distinct mechanisms of diversification generate the enormous repertoire of specific antibody receptors:

1. *Germ-line diversity.* A multiplicity of V, D, and J gene segments exist for each of the gene families. These are the raw materials for the rearrangements that occur during development.
2. *Combinatorial joining.* With heavy chains, any D can join to any J and any V can join to any DJ; with light chains, any V can join to any J. Thus, there are a large number of possible V, D, and J combinations.

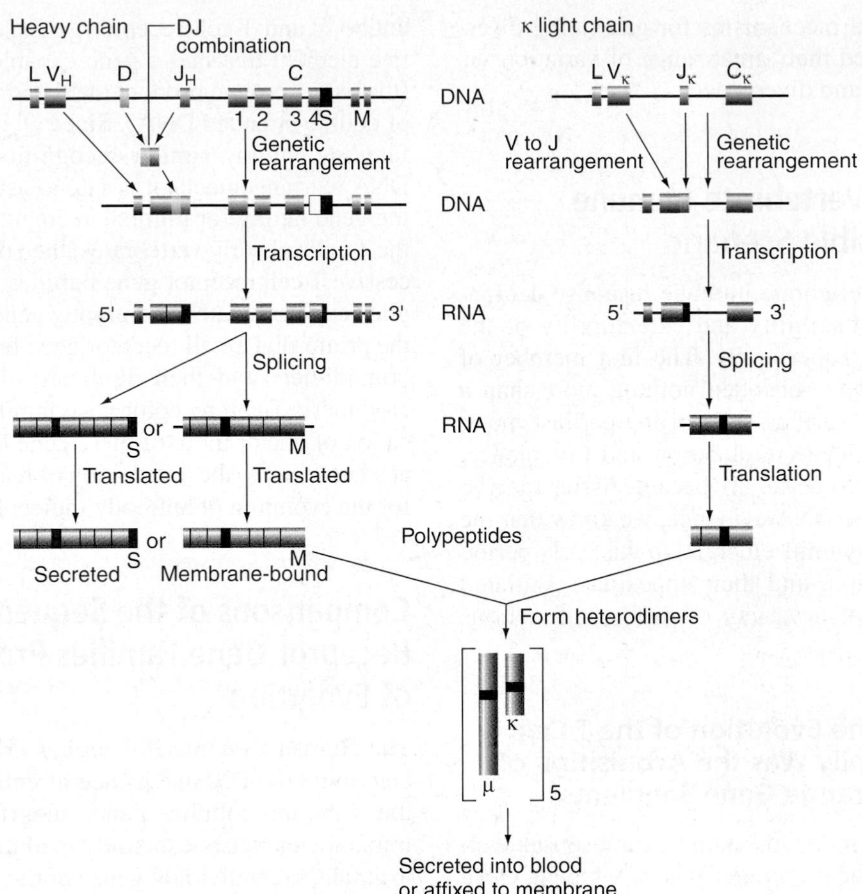

Figure 22.24 DNA rearrangements bring together segments of a gene for expression. Expression of rearranged heavy- and light-chain genes in Bμ cells. The genetic rearrangements occur during development of the Bμ. For simplicity, only a single-gene segment of each kind is represented. The letters *L, V, D, J,* and *C* are leader, variable, diversity, joining, and constant regions, respectively; *S* and *M* indicate the secreted and membrane-bound tails of the μ chain, which are generated by alternative patterns of RNA splicing. Thus the IgM molecule (made in Bμ cells) may be attached to the membrane (M), where it will serve as a receptor whose binding with antigen triggers B-cell differentiation, or secreted from the cell to the circulation, where it will serve as a soluble antibody molecule.

3. *Junctional diversity.* Two mechanisms operate to create diversity at the junctions of joined gene segments. Exonucleases can trim back one or both of the gene segments to be joined; and the enzyme terminal deoxynucleotide transferase can add random nucleotides not encoded in the chromosome to the junctional region. These two mechanisms together generate enormous diversity in the joined regions, which in the resulting folded protein happen to represent a major portion of the antigen-binding site.
4. *Somatic hypermutation.* As the immune response proceeds with the proliferation of specific cells into large clones of millions of identical cells, the joined V genes undergo random *somatic hypermutation,* that is, somatic mutation at a rate that is orders of magnitude higher than the typical mutation rate. A single V gene may have 10–15 changes. These somatic hypermutations are often rather extensive. Hypermutation produces many different forms of the gene; those mutations encoding V regions with a better fit for the antigen will be selectively amplified through tighter binding to create "affinity maturation" during the immune response.
5. *Joining light and heavy chains.* Any light chain can associate with any heavy chain, so once again, there is a combinatorial amplification of diversity through the manipulation of information at the protein level. For example, if two cells happen to produce the same light chain, it is highly unlikely they will also produce the same heavy chain; and the joining of a different heavy chain with the same light chain produces two distinct immunoglobulin molecules with distinct antigen-binding sites.

Most of these mechanisms of diversification also apply to T-cell receptors. The two types of T-cell receptors, α/β and γ/δ, are encoded by three distinct gene families: α/δ, β, and γ. The T-cell receptors employ all the diversification mechanisms discussed for the immunoglobulin receptors, except for somatic hypermutation. The MHC receptors do not employ any special mechanisms for generating diversity; they have achieved their great range of variation via evolutionary mutation and divergence.

Evolution of the Vertebrate Immune Response: A Possible Scenario

The evolution of the vertebrate immune response demonstrates the enormous flexibility and extendibility of the immunoglobulin gene superfamily. The first member of the superfamily may have encoded nothing more than a cellular glue to adjoin various cells of the earliest metazoan organisms. With gene duplication and divergence, various members of the superfamily became tissue-specific cell adhesives. From the DNA evidence, we know that the three types of homology units emerged in this early period of superfamily evolution, and their appearance initiated the rapid acquisition of a variety of different functions (review Figure 22.21).

The Key Event in the Evolution of the T-Cell Receptor Gene Family Was the Acquisition of the Ability to Rearrange Gene Segments

This probably arose from the insertion of a transposable element in the 3′ end of the primordial V gene segment. Once the DNA rearrangement machinery of this transposon came under precise developmental control, the first T-cell receptor gene family acquired the ability to rearrange gene segments. We know now that Rag1 and Rag2, the two enzymes that mediate the rearrangement and joining events of antibody and T-cell receptor genes, evolved from a repetitive element that had a gene capable of transposing DNA (that is, its protein product mediated the cutting and joining of double-stranded DNA). Since all of the rearranging gene families employ similar recognition sequences for their DNA rearrangements, it is safe to assume that the rearranging gene family for immune recognition emerged once—in the T cells of early vertebrates; then duplicated to make successive T-cell receptor gene families; and subsequently, duplicated to generate the antibody gene families. Presumably the primordial T-cell receptor gene family was able to make homodimers and then duplicate to make the early heterodimeric T-cell receptor gene families. Subsequent duplication of one or more of these gene families could have led at a later time to the acquisition of rearranging gene families for the evolution of antibody molecules.

Comparisons of the Sequences of T-Cell Receptor Gene Families Provide a Snapshot of Evolution

The Human Genome Project has created the resources to determine the DNA sequence of entire family branches of the immunoglobulin gene superfamily. As a result, immunogeneticists can study evolution by comparing, for example, the individual genes of the β T-cell receptor family in one species to one another and then comparing β T-cell receptor families of different species (**Fig. 22.25**). The availability of large-scale DNA sequence analyses that enable one to compare gene families is transforming our understanding of molecular evolution.

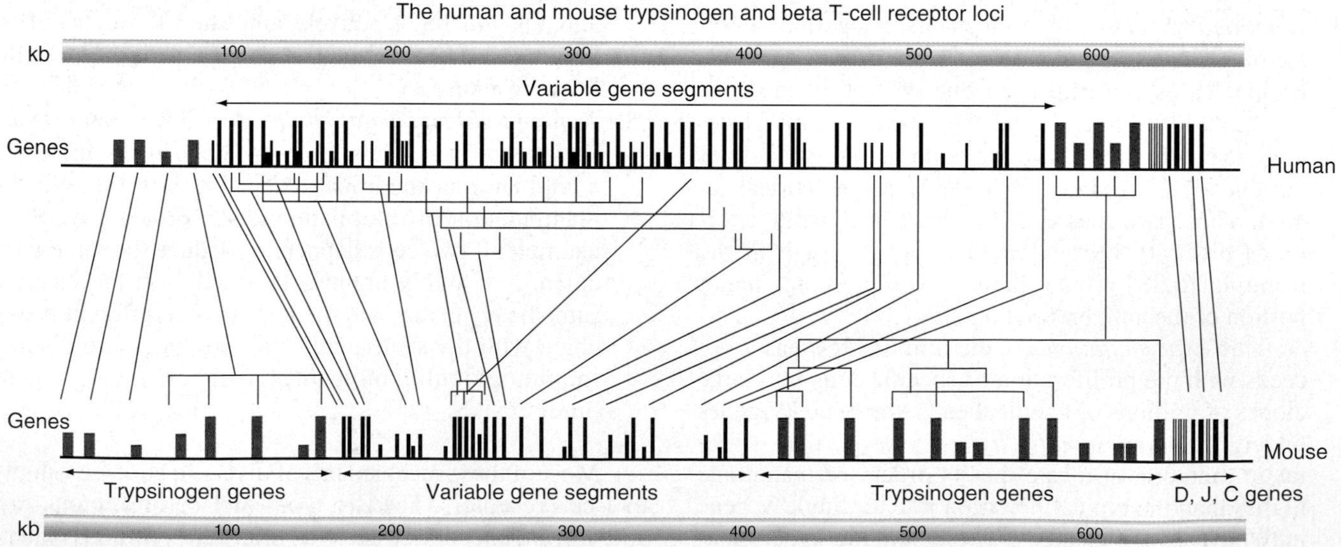

Figure 22.25 **Diagrams of the human and mouse β T-cell receptor loci.** The *black bars* indicate T-cell receptor elements and the *orange bars* trypsinogen genes. The *full-length bars* represent functional gene elements; the *half bars,* pseudogenes; and the *quarter bars,* badly decayed genes. The *vertical lines* between the mouse and human gene families indicate orthologs.

The Human β T-Cell Receptor Family

The β T-cell receptor family of humans spans 700,000 bp of DNA on chromosome 7. It contains 45 functional V β gene segments and 19 pseudogene segments, 2 clusters consisting of 1 D segment and 5 or 6 functional J gene segments, with each DJ region associated with a different C β gene (Fig. 22.25). The β locus also has trypsinogen genes or pseudogenes on either side of the V β cluster of genes. This gene family has several interesting features.

Approximately 36% of the locus consists of repeat units such as LINE and Alu sequences. Since the age of many of these repeat sequences can be approximated by noting the extent of their divergence from the ancestral repeat sequence (which itself can be determined because so many repeat element sequences are available), the repeat sequences have been very useful in reconstructing the complicated molecular archaeology of this family.

Within the human β T-cell receptor family, there are seven different genomic modules (duplicated chromosomal segments) ranging in size from 0.7–32 kb. One contains the two D, J, C clusters; a second the five trypsinogen genes that are intercalated between the V and D gene segments. The remaining five modules are associated with the amplification of two or more V gene segments. Indeed, more than 47% of the locus is composed of these modular units.

A portion of the 3′ end of the β T-cell receptor locus has been duplicated and translocated from human chromosome 7 to human chromosome 9. The translocated segment on chromosome 9 contains seven V gene segments and one functional trypsinogen gene. The V gene segments are not functional because the corresponding D, J, and C elements were not translocated. Thus, translocation is one mechanism for the birth of a new multigene family.

Analysis of the promoter regions of the 45 functional V β gene segments reveals that a conserved motif is present adjacent to most of the genes. Presumably this motif represents a site at which transcriptional factors bind to regulate the expression of these genes.

Rearrangement signals from the V, D, and J gene segments are strikingly similar to their counterparts in antibody genes, which reaffirms the idea that these rearranging gene families had a common origin.

Comparing Human and Mouse β T-Cell Receptor Gene Families Enables One to Ask and Tentatively Answer Questions About Evolution

One can search the sequences of the two species for conserved regions where recognizable sequence homology has been retained through 65 million years of mammalian species divergence. The conserved regions might represent coding regions, regulatory regions, or regions that are associated with various chromosomal functions, such as DNA replication or compaction.

Detailed comparisons of the V, D, and J genes in the two species can help reveal *orthologs*—the most closely related copies in the V gene families of different species—and make it possible to determine how these genes and their organization have changed in 65 million years of divergence. Preliminary comparisons have shown that very different repetitive sequences have been sprinkled through mouse and human loci (about 45% of the mouse locus consists of repetitive sequences), and the nature of the homology units that have expanded in each species is quite different. Nevertheless, identification of a number of different orthologs suggests that the general organization of these genes has been conserved in the two species.

In both mouse and human genomes, trypsinogen genes are associated with the V β genes. However, there has been an enormous expansion of trypsinogen genes in the mouse as compared to humans. Conversely, there has been an enormous expansion of V genes in the human genome as compared to that in the mouse. The net result of these two expansions is that the loci in both species are roughly the same 700,000 bp in length (Fig. 21.25). The trypsinogen genes are also associated with the T-cell receptor β genes in the chicken, a species that diverged from mammals more than 350 million years ago.

One as yet unanswered question raised by the conserved association of T-cell receptor genes and trypsinogen genes is whether genes that live together work together; that is, do the trypsinogen and T-cell receptor genes share common functional and/or regulatory strategies? This question must be answered separately for each genomic region. Its answer is crucial for understanding the regulation of the genes in a particular region.

Connections

A retrospective bird's-eye view of key events that led to an understanding of evolution at the molecular level goes something like this. In 1859, Charles Darwin published *The Origin of Species* in which he inferred from the visible evidence of descent through modification that the diverse organisms alive today evolved from a single primordial form, in large part, by a process of natural selection. Several years later, Mendel published "Experiments on Plant Hybrids" in which he applied the laws of probability to the visible evidence of heredity, inferring the existence of hereditary units that segregate during gamete formation and assort independently of each other. In the early twentieth century, Thomas Hunt Morgan and coworkers gave Mendel's units of heredity a physical location in the cell, establishing the chromosomal basis of heredity and showing not only that genes have chromosomal addresses but also that recombination can separate otherwise linked genes. In the 1940s several people, including Oswald Avery, Martha Chase, and Alfred Hershey, showed that the molecule of heredity is DNA. Then in 1953, James Watson and Francis Crick deciphered the structure of DNA and proposed a mechanism by which the molecule replicates. By the end of the twentieth century, extensive genomic analyses had made it possible to explain how DNA mutates, duplicates, diverges, and is acted on by selection to generate the diversity of life we see around us.

Essential Concepts

1. All forms of life on earth are descendants of a single cell—a common ancestor that existed approximately 3.7 billion years ago.
2. Charles Darwin explained how biological evolution occurs through a process of *natural selection.*
3. Natural selection operates on variant forms of inherited traits. The particular variant of a trait that provides the highest degree of reproductive fitness is selected over many generations to become the predominant form in the entire population.
4. New mutations provide a continuous source of variation.
5. Mutations with no effect on fitness are considered neutral. *Neutral mutations* are not acted on by selection and are subject instead to *genetic drift.* Mutations with a deleterious effect on fitness are selected against, while extremely rare mutations with a positive effect on fitness are selected for. Selection can operate simultaneously at hundreds or thousands of variant loci within a population.
6. RNA has a unique combination of properties: It can carry genetic information as well as catalyze chemical reactions. These two properties led scientists to speculate that RNA may have predated the cell as the original independent replicator, or proto-life-form.
7. The fossil record as well as living organisms of all levels of complexity provide scientists with a detailed picture of the evolution of complex life from the first cell to human beings.
8. Preliminary studies on the evolution of gene regulatory networks suggest that it can account for the evolution of biological complexity.
9. The evolution of organismal complexity depends on an increase in genome size, which occurs through repeated duplications. Some duplications result from transpositions, while others arise from unequal crossing-over.
10. Mutations rendering genes nonfunctional turn many duplicated genes into *pseudogenes* that over time diverge into random DNA sequences. However, rare advantageous mutations can turn a second copy of a gene into a new functional unit able to survive and spread through positive selection.
11. Sequence comparisons make it possible to construct *phylogenetic trees* illustrating the relatedness of species, populations, individuals, or molecules.
12. The mammalian genome contains genes, *multigene families, gene superfamilies,* genomewide repetitive elements, and simple sequence repeats; repetitive elements in centromeres and telomeres; and unique nongene sequences.
13. Complex genomes arose from four levels of duplication followed by diversification and selection: exon duplication to create larger, more complex genes; gene duplication to create multigene families; multigene family duplication to create gene superfamilies; and the duplication of entire genomes.
14. Genetic exchange between related DNA elements by intergenic gene conversion most often increases the variation among members of a multigene family. Sometimes, however, it can contribute to *concerted evolution,* which creates a family of nearly identical genes.
15. The enormously diverse immunoglobulin gene superfamily encodes cell receptors that carry out different recognition functions at the cell surface in different types of cells. In the cells of the vertebrate immune system, these receptors enable both the specificity and diversity of immune responses. Gene rearrangements and other recombinatorial mechanisms generate the immense diversity of immune receptors during the development of each individual.

On Our Website

www.mhhe.com/hartwell3
Chapter 22

Annotated Suggested Readings and Links to Other Websites

- Interesting historical papers on evolution
- Contemporary papers on evolution
- Papers on the ethical and social implications of evolution
- Interesting books on evolution

Social and Ethical Issues

1. Jill's husband's boss Mary always seems to be going to the doctor for some ailment and has built up quite a store of antibiotics. Whenever Mary gets a cold, she starts taking antibiotics. She continues them only until she begins to feel a little better. She is now starting to give her children medications in a similar way even though many of the illnesses may be caused by viruses, which do not respond to antibiotics. Jill knows from her training as a physician's assistant that this practice can lead to an increase in bacteria resistant to antibiotics, since the bacterial population evolves by selection. However, she hesitates to interfere with her husband's boss. Should she intervene? What should she do?

2. In the novel *Jurassic Park,* extinct dinosaurs were re-created from bits of DNA found in dinosaur blood from fossilized parasitic insects. While there were several unrealistic steps in the re-creation of the dinosaurs, it raised the issue of whether there should be controls or laws that would prevent individuals from pursuing this type of re-creation. Who could or would enforce them? What are the problems associated with introducing an extinct organism into today's environment?

3. Evolution is a challenging area of study because many of its hypothesized events are not directly observable or testable; instead they require us to draw inferences about what happened in the past from what is present today. But even so, some evolutionary hypotheses are testable with today's tools and technology. Indeed, there are several exciting avenues of evolutionary research today, including the comparative analysis of complete genomes and studies of the evolution the of protein and gene regulatory networks controlling development and physiological responses. In light of these limitations and possibilities, consider the following: When new data or new analyses seem to contradict or refute an existing major hypothesis, who decides when the new data are sufficient to invalidate the particular hypothesis?

Solved Problems

I. The sequence of two different forms of a gene starting with the ATG is shown here. Which of the base differences in the second sequence are synonymous changes and which are nonsynonymous changes?

```
form 1    ATGTCTCATGGACCCCTTCGTTTG
form 2    ATGTCTCAAAGACCACATCGTCTG
```

Answer

The key to answering this question is understanding the difference between synonymous and nonsynonymous changes in DNA sequence.

Synonymous changes are nucleotide substitutions that do not change the amino acid specified by the DNA sequence. Nonsynonymous changes are nucleotide changes that result in a different amino acid in the protein.

Looking at the amino acids specified by the base sequence,

	Met	Ser	His	Gly	Pro	Leu	Arg	Leu
form 1	ATG	TCT	CAT	GGA	CCC	CTT	CGT	TTG
	Met	Ser	Gln	Arg	Pro	His	Arg	Leu
form 2	ATG	TCT	CA<u>A</u>	<u>A</u>GA	CC<u>A</u>	C<u>A</u>T	CGT	<u>C</u>TG

(The base substitutions in form 2 are underlined.) *The first, second, and fourth A substitutions are nonsynonymous changes; the third A substitution and the C substitution are synonymous changes.*

II. What difficulties would arise if you tried to derive a molecular clock rate using a noncoding sequence in some species and a coding sequence in other species?

Answer

To answer this question, you need to think about how molecular clocks are derived and the constraints on base changes in coding and noncoding sequences.

The evolution of coding sequences is restricted by the fact that the gene sequence needs to be maintained for the gene product to function. The sequence of noncoding regions generally can tolerate many base substitutions without selection acting on these sequences. *Therefore, you would expect more substitutions in the noncoding sequence. The result would be an inconsistency in your clock rate* if you are using coding region in some species and noncoding DNA in other species.

II. If the chromosomes diagrammed below misalign by the pairing of repeated sequences (shown as solid blocks) and crossing-over occurs, what will be the products?

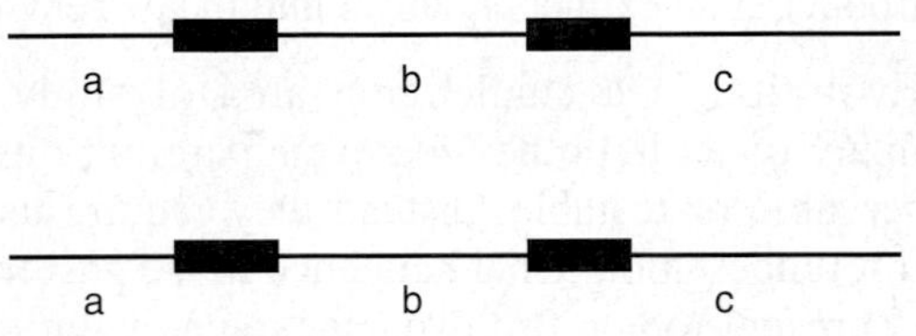

Answer

This question requires an understanding of crossing-over via homology. When you align the homologous repeated sequences out of register, one of the resulting products will have a duplication of the region between the repeats and three copies of the repeated sequence and the other product will be deleted for the DNA between the repeats and contain only one repeated sequence.

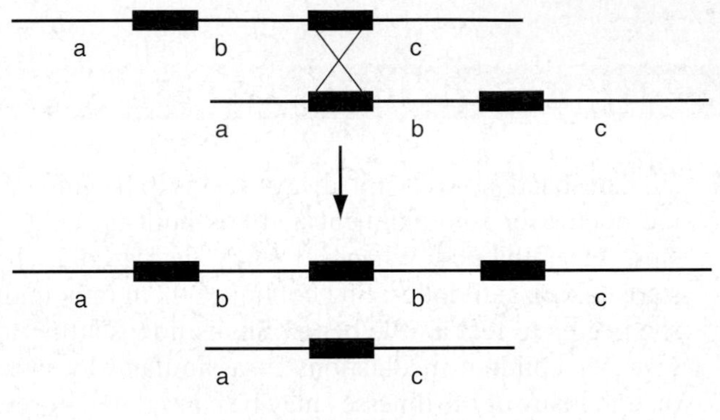

Problems

Vocabulary

1. For each of the terms in the left column, choose the best matching phrase in the right column.

a. ribozymes	1. constant rate of change in amino acid sequence
b. retrotransposition	2. sudden explosive evolutionary changes
c. SINES	3. exchange of pieces of genes
d. punctuated equilibrium	4. RNA molecules that catalyze specific chemical reactions
e. molecular clock	5. short repeated sequences in human genomes
f. phylogenetic trees	6. RNA intermediate in duplication of genetic material
g. exon shuffling	7. representation of evolutionary relationships

Section 22.1

2. What observations support the unity of life concept that all life-forms evolved from a common ancestor?

3. Which of the following statements is support for RNA being the first replicator molecule?
a. RNA molecules can function as enzymes.
b. DNA is more stable than RNA.
c. Information can be encoded in RNA.

4. a. In what ways is RNA not a good information storage molecule?
b. In what ways is RNA not as good as protein as a molecular machine?

5. a. A particular chemical reaction that occurs between two proteins is sensitive to RNase treatment *in vitro* (that is, it will not occur if the reaction mix is pretreated with RNase), but it is not sensitive to protease treatment. What would you propose about the nature of the enzyme that carries out the reaction?
b. Another chemical reaction is sensitive to both RNase and protease treatment. What would you propose about the nature of the enzyme that carries out this reaction?

6. Humans and chimps have a 1% difference in their genomic sequence, while two humans have a 0.1% sequence difference. How could something as small as a 1% difference in DNA sequence lead to dramatic differences seen in chimps versus humans? Speculate on the types of differences the 1% variation may represent.

Section 22.1

7. Rates of nonsynonymous and synonymous amino acid substitutions for three genes that were compared in humans and mice or rats are shown here. (The rates are expressed as the average number of substitutions per base per year, with the standard deviation given. In each case, the number shown is the rate at which the human and rat sequences have diverged from each other, by either a nonsynonymous or a synonymous substitution.)

a. Why do the rates of nonsynonymous substitutions vary among these genes?
b. Why are the rates of synonymous substitutions similar?

Protein	Nonsynonymous substitutions	Synonymous substitutions
Histone 3	$0.0 \pm 0.0 \times 10^{-9}$	$6.38 \pm 1.10 \times 10^{-9}$
Growth hormone	$1.23 \pm 0.15 \times 10^{-9}$	$4.95 \pm 0.77 \times 10^{-9}$
β globin	$0.8 \pm 0.13\ 3 \times 10^{-9}$	$3.05 \pm 0.56 \times 10^{-9}$

8. Synonymous mutations are more prevalent than nonsynonymous mutations in most genes. The immunoglobulin (Ig) genes encoding antibody subunits are an exception where nonsynonymous base changes outnumber synonymous changes. Based on the function of the Ig genes, why do you think this might be true?

9. Mutations in the *CF* gene that cause cystic fibrosis are carried by 1 in 20 individuals of Caucasian ancestry. The disease is clearly detrimental, yet the allele is maintained at a relatively high level in the population. What does this paradox suggest about the effect of the mutation?

10. Unequal crossing-over between two copies of a gene can lead to duplication and deletion on the two homologs involved. How could a single gene become duplicated?

11. Human beings have three-color vision, while most other species of animals have two-color vision or one-color vision (that is, black and white). Three-color vision is produced by the products of a three-member, cross-hybridizing gene family that encodes light-sensitive pigments active in different ranges of the color spectrum (red, green, and blue). What is the most likely molecular explanation for the evolution of three-color vision in the ancestor to human beings?

12. How do transposition and unequal crossing-over differ based on the location of final copies of the duplicated sequence?

13. You have identified an interesting new gene that appears to be involved in human brain development. You have discovered three cross-hybridizing copies of this gene within the human genome (*A, B,* and *C*). In the mouse genome, there is only a single copy of this gene (named *M*) and in the frog *Xenopus,* there is also only a single copy (named *X*). You have sequenced the same 10,000 bp of open reading frame from each of these genes and calculated the number of base-pair differences that exist between different pairs with the following results:

Comparison	Number of base-pair differences
A versus *B*	300
B versus *C*	10
A versus *C*	300
A versus *M*	600
A versus *X*	3000

If you assume that a constant rate of evolution has occurred with all members of this gene family, and you assume that mice and humans evolved apart 60 million years ago:

a. How long ago did frogs split from the line leading to mice?
b. How many gene duplication events are observable within these data? At what time in the past did each of these duplication events occur?
c. Mapping of the *A, B,* and *C* genes shows that *A* and *C* are very closely linked, while *B* assorts independently from either of these genes. With this linkage information, what can you say about the molecular nature of the duplication events that occurred along the evolutionary line leading to human beings?

14. Phylogenetic trees of primates constructed using chromosome mutations (deletions, insertions, inversions, etc.) show the same relationships of species as those constructed using base substitutions within a gene. Which type of genetic alteration would you expect to have the greater impact on the evolution of chimp and human from a common ancestor. Why?

15. When a phylogenetic tree was constructed using comparisons of glucose-6-phosphate isomerase amino acid sequences from a wide variety of species including animals, plants, and bacteria, the bacterium *E. coli* was placed on a branch of the phylogenetic tree with a flower. What explanation could there be for this unusual association based on this one protein-coding sequence? (Recall that bacteria and plants are in very different locations on the evolutionary trees derived from analyses of several gene sequences or morphology and physiology.)

Section 22.3

16. Organisms have a characteristic percentage of GC bases in their genomes. In a species of halophilic (salt-loving) bacteria containing two essential plasmids and one chromosome, the plasmids have a different GC content than the chromosome. What could you hypothesize about the origin of the plasmids?

17. What is the unit that can be duplicated and modified to form

a. a new gene?
b. a multigene family?

18. It has been hypothesized that in the evolution of vertebrates, there were two successive doublings of the genome (tetraploidization) to produce the vertebrate genome.

a. How does the *Hox* superfamily fit with this hypothesis?

b. Among vertebrates, there is variation in the numbers of genes present within the *Hox* gene family clusters. How could this variation arise?

19. LINEs and SINEs are considered to be selfish DNA, yet they can, in some instances, confer a selective advantage on the organism. What are two ways in which a LINE or SINE can change the genome?

20. How is the size polymorphism of dinucleotide repeated sequences thought to occur?

21. a. What is an example of repetitive noncoding DNA for which we know no physiological function?

b. What is an example of repetitive noncoding DNA for which we know a cellular function?

22. Match the observation on the right with the event that it suggests occurred during evolution.

Event	Evidence in the genome
a. concerted evolution	1. genes in different individuals of the same species have 6, 8, or 10 homology units in an immunoglobulin gene family member
b. exons and introns of a gene	2. several copies of the same gene that are identical in the genome
c. unequal crossing-over	3. poly-A sequences at the end of coding regions in genomic DNA
d. retrotransposition	4. blocks of DNA sequence conserved between species separated by nonconserved blocks

Section 22.4

23. The figure at the end of this problem diagrams an IgG antibody molecule. (See Fig. 22.13 for details.)

a. How many genes are required to encode this IgG molecule? What genes are required?

b. All of the genes encoding immunoglobulins are autosomal, so any individual will contain two alleles of each immunoglobulin gene. How many different kinds of IgG could be expressed by a cell making the molecule shown in the figure? Explain.

c. What kind of cell produces this IgG molecule? (B lymphocyte, T helper cell, T cytotoxic cell, T regulatory cell)

d. On the figure, show the locations of all of the following if present: (i) light chains, (ii) heavy chains, (iii) antigen bindings sites, (iv) V, C, and H immunoglobulin homology units; (v) disulfide (-S-S-) bonds; (vi) light chain gene promoter

e. A protease (an enzyme that cleaves peptide bonds) called papain divides the IgG molecule into exactly three pieces. Two of these pieces are identical to each other and bind antigen well; the third piece does not bind antigen at all. Papain cleaves polypeptides only at a very specific sequence of amino acids. On your diagram, show the position(s) on the IgG molecule cleaved by papain.

f. IgG molecules that bind to certain sugar antigens on the surface of red blood cells from individuals with type A blood can precipitate (or "agglutinate") these red blood cells; this precipitation test allows blood to be "typed." Why will this agglutination test not work with the same IgGs after they have been treated with papain?

g. If the diagrammed molecule was an IgM (see Fig. 22.21b) rather than an IgG, what would be different about your answer to part d?

h. Some B cells switch from making an IgM molecule to an IgG molecule with exactly the same antigen binding site. Propose an hypothesis to explain this switch.

i. After initial exposure to an antigen, the average affinity of antibodies in the blood for the antigen is low, but with time, the average affinity of antibodies in the blood for the same antigen increases. Propose an hypothesis for this phenomenon.

j. Which of the following is *not* a molecular mechanism leading to antibody diversity (that is, the generation of many antibodies with different antigen specificities): (i) combinatorial joining of different V, D, and J regions; (ii) junctional diversity (the imprecise joining of V with D or D with J in the heavy chain gene); (iii) the availability of different C regions (Cμ, Cδ etc.); (iv) somatic hypermutation in the V region; (v) the possibility of association between particular heavy and light chains?

k. Why do you think B cells recombine their DNA? Why don't they just use alternative splicing to express immunoglobulins with different specificities?

l. In 1972, Rodney Porter and Gerald Edelman won the Noble Prize in Physiology and Medicine for working out the structure of antibody molecules as depicted in the diagram. One problem they faced is that blood contains millions of different IgG molecules. Without a preparation of a single kind of IgG molecule whose purity is nearly 99%, they could not determine much about antibody structure. Suggest a strategy that Porter and Edelman might have used to overcome this problem.

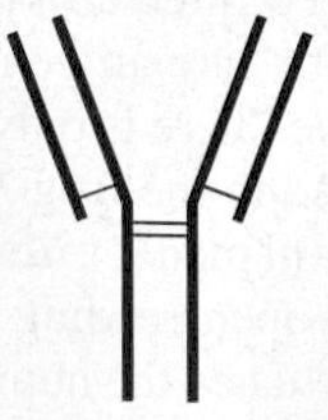

Guidelines for Gene Nomenclature

There are inconsistencies within the various branches of genetics on some nomenclature—because it is a relatively new area of scientific investigation, the consistency present in more basic sciences has not been established. The authors debated whether they should try to impose a consistency on the entire topic area and decided against that path. As the study of genetics matures, the process itself will create a more consistent nomenclature. The following guidelines can be applied to all chapters in this book.

General Rules

- Names of genes are in italics (*lacZ, CDC28*)
- Names of proteins are in regular (Roman) type with an initial cap (LacZ, Cdc28)
- Chromosomes: sex chromosomes are represented by a capital letter in Roman type (X, Y); autosomes are designated by a cardinal number (1, 2, 21, 22)
- Names of transposons are in roman type (Tn10)

Specific Rules for Different Organisms

- **Bacteria:** lowercase italics for genes (*lac, ara*), with the addition of a capital letter to designate a specific gene in a pathway or operon (*lacZ, lacA, araB*); numbers (not superscript) for alleles (*trpC2, hisB2*); superscript "+" (plus) for wild-type alleles; superscript "−" (minus) for mutant alleles ($lacZ^{+}$, lacZ^{-})
- **Yeast:** all caps for wild-type alleles (*CDC28*); all lowercase for mutant alleles, with a cardinal number indicating the specific mutation (*cdc28-1*)
- **Arabidopsis:** all caps for wild-type alleles (*LEAFY,* abbreviated *LFY*); all lowercase for mutant alleles (*lfy*)
- ***C. elegans:*** lowercase italics for genes and wild-type alleles (*dpy-10*); mutant alleles in parentheses following the gene [*dpy-10(e128)*]
- ***Drosophila:*** Many genes are named for the mutant phenotypes that revealed them. If the mutation causing the phenotype is dominant, the allele designation has an initial cap (*Deformed,* abbreviated *Dfd*); if the mutation that revealed the gene is recessive, the allele designation is all lowercase (*white,* abbreviated *w; wingless,* or *wg*)
- **Mice:** Genes are now named for the protein they encode and designated by an initial cap followed by any mix of letters or numbers (for example, *Tcp1* designates the gene for T complex protein 1); alleles are designated by superscripts. For multiple wild-type alleles, the superscripts may be *a, b, c,* or *1, 2, 3* ($Tcp1^{a}$, Tcp1^{b}); some wild-type alleles are indicated by a superscript plus "+" (Kit^{+}). For mutant alleles, the superscripts often describe the resulting phenotype (the Kit^{W} mutant allele causes white spotting).
- **Humans:** All caps for genes (*HD, CF*); wild-type alleles designated by a superscript plus, "+" (HD^{+}, CF^{+}); mutant alleles designated by a superscript minus "−" (HD^{-}, CF^{-}).

Brief Answer Section

Chapter 2

1. a. 4; b. 3; c. 6; d. 7; e. 11; f. 13; g. 10; h. 2; i. 14; j. 9; k. 12; l. 8; m. 5; n. 1.
3. For peas: (1) rapid generation time; (2) can either self-fertilize or be artificially crossed; (3) large numbers of offspring; (4) can be maintained as pure-breeding lines; (5) maintained as inbred stocks and two discrete forms of many phenotypic traits are known; (6) easy and inexpensive to grow. In contrast, for humans (1) generation time is long; (2) no self-fertilization, it is not ethical to manipulate crosses; (3) produce only a small number of offspring per mating; (4) although people that are homozygous for a trait exist, homozygosity cannot be maintained; (5) populations are not inbred so most traits show a continuum of phenotypes; (6) require a lot of expensive care to "grow" One advantage to the study of genetics in humans-a very large number of individuals with variant phenotypes can be recognized. Thus, the number of genes identified in this way is rapidly increasing.
5. Short hair is dominant to long hair.
7. The genotype can be determined by performing a testcross; that is, crossing your fly with the dominant phenotype (but unknown genotype) to a fly with the recessive (short wing) phenotype. If your fly has the homozygous dominant genotype the progeny in this case would be *Ww* and would have the dominant phenotype. If your fly had a heterozygous genotype, 1/2 of the progeny would be normal (Ww*)* and 1/2 of the progeny would be short (ww*)*.
9. The dominant trait (short tail) is easier to eliminate from the population by selective breeding. You can recognize every animal that has inherited the allele, because only one dominant allele is needed to see the phenotype. Those mice that have inherited the dominant allele can be prevented from mating.
11. a. Dry is recessive, sticky is dominant; b. The 3:1 and 1:1 ratios are obscured because the offspring are then combined results of different crosses.
13. a. 1/6; b. 1/2; c. 1/3; d. 1/36; e. 1/2; f. 1/6; g. 1/9.
15. a. 2; b. 4; c. 8; d. 16.
17. a. *aa Bb Cc DD Ee;* b. *a B C D E* or *a B c D E* or *a B C D e* or *a B c D e* or *a b C D E* or *a b C D e* or *a b c D E* or *a b c D e.*
19. They must both be carriers (*Pp*); the probability that their next child will have the *pp* genotype is 1/4.
21. a. Rough and black are the dominant alleles (R = rough, r = smooth; B = black, b = white); b. a ratio of 1/4 rough black: 1/4 rough white: 1/4 smooth black: 1/4 smooth white.
23. a. 3/16; b. 1/16.
25. P = purple, p = white; S = spiny, s = smooth. a. *Pp Ss* × *Pp Ss;* b. *PP Ss* × *P– ss* or *P– Ss* × *PP ss;* c. *Pp S–* × *pp SS* or *Pp SS* × *pp S–;* d. *pp Ss* × *Pp Ss;* e. *Pp ss* × *Pp ss;* f. *pp Ss* × *pp Ss.*
27. Cross 1: male: *tt Nn,* female: *tt Nn;* Cross 2: male: *Tt nn,* female: *tt Nn;* Cross 3: male: *Tt nn,* female: *Tt Nn;* Cross 4: male: *Tt nn,* female: *Tt NN*
29. a. Recessive. Two unaffected individuals have an affected child. It was a consanguineous marriage that produced the affected child. II-1 and V-2 are affected (*aa*); all unaffected individuals except II-2, II-4, III-4, III-5, and possibly V-1 are carriers (*Aa*). b. Dominant. The trait is seen in each generation and each affected child has an affected parent; if the trait were recessive it would not be possible for III-3 to be unaffected even though both his parents are affected. All affected individuals are *Aa,* though III-4, III-5, and III-6 could be *AA*; carrier, is not applicable when the mutation is dominant; c. Recessive. Unaffected parents have an affected child. I-2 and III-4 are affected (*aa*); II-4 and II-5 are carriers (*Aa*); all others could be *AA* or *Aa,* but I-1 is almost certainly *AA* if the disease is rare.
31. a. 2/3; b. 1/9; c. 4/9.
33. Recessive; common.
35. a. 1/16 = 0.0625; b. 0.067.
37. In about 40% of the families, both parents were *Mm* heterozygotes. In the remaining 60% of the families, at least one parent was *MM.*

Chapter 3

1. a. 2; b. 6; c. 11; d. 8; e. 7; f. 9; g. 12; h. 3; i. 5; j. 4; k. 1; l. 10.
3. One gene, 2 alleles, incomplete dominance; 1/2 c^rc^w (yellow): 1/4 c^rc^r (red): 1/4 c^wc^w (white).
5. Long is completely dominant to short. Flower color trait shows incomplete dominance of two alleles.
7. a. Single-gene inheritance with incomplete dominance. Heterozygotes have intermediate serum cholesterol levels; homozygotes have elevated levels. The following people must have the mutant allele but do not express it (incomplete penetrance): family 2 I-3 or I-4; family 4 I-1 or I-2. b. Other factors are involved, including environment (particularly diet) and other genes.
9. a. ii (phenotype O) or iI^A (phenotype A) or iI^B (phenotype B); b. I^BI^B, I^Bi or I^BI^A; c. ii (phenotype O).
11. a. 1/4 spotted dotted: 1/2 marbled: 1/4 spotted; b. marbled and dotted.
13. a. Coat color is determined by three alleles of a single gene arranged in a dominance series with C (for chinchilla) > c^h (for himalaya) > c^a (for albino). b. Cross 1: $c^hc^a \times c^hc^a$; Cross 2: $c^hc^a \times c^ac^a$; Cross 3: $Cc^h \times Cc^h$; Cross 4: $CC \times c^hc^h$ or $CC \times c^hc^a$; Cross 5: $Cc^a \times Cc^a$; Cross 6: $c^hc^h \times c^ac^a$; Cross 7: $Cc^a \times c^ac^a$; Cross 8: $c^ac^a \times c^ac^a$; Cross 9: $Cc^h \times c^hc^h$; Cross 10: $Cc^a \times c^hc^a$. c. 3/4 chinchilla (CC, Cc^h, and Cc^a) and 1/4 himalaya (c^hc^a).
15. a. 2/3 Curly: 1/3 normal; b. *Cy/Cy* is lethal; c. 90 Curly winged and 90 normal winged flies.
17. a. The 2:1 phenotypic ratio shows that the montezuma parents were heterozygous, *Mm* and homozygosity for *M* is lethal; b. 1/2 montezuma: 1/2 greenish, normal fin; c. 6/12 montezuma normal fin: 2/12 montezuma ruffled fin: 3/12 green normal fin: 1/12 green ruffled fin.
19. Incomplete penetrance or a spontaneous mutation during gamete or the father of the child is not the male parent of the couple.
21. Two genes are involved. The black mare was *AAbb* and the chestnut stallion was *aaBB,* the liver horses were *aabb* and the bay horses were *AaBb.*
23. a. There are two genes involved; homozygosity for the recessive allele of either or both genes causes yellow color. Green parent is *AABB;* yellow parent is *aabb.* b. *AaBb, aaBb, Aabb* and *aabb* in equal proportions: 1/4 green:3/4 yellow fruit.

25. Dominance relationships are between alleles of the same gene. Only one gene is involved. Epistasis involves two genes. The alleles at one gene affect the expression of a second gene.
27. 1/4 would appear to have O type blood, 3/8 have A, 3/8 have AB.
29.

	I-1	I-2	I-3	I-4	II-1	II-2	II-3	III-1	III-2
Phenotypes	AB	A	B	AB	O	O	AB	A	O
Genotypes	I^AI^B	I^A *or* I^AI^A	I^B *or* I^BI^B	I^AI^B	*ii*	I^B *or* I^BI^B	I^AI^B	I^Ai	I^AI^A *or* I^Ai I^AI^B *or* I^Bi
	Hh	*Hh*	*H–*	*H–*	*H–*	*hh*	*Hh*	*Hh*	*hh*

One or both of I-3 and I-4 must carry *h.*

31. 2/6 yellow: 3/6 albino: and 1/6 agouti progeny.
33. a. $A^yaBbCc \times AabbCc$; b. six phenotypes: albino, yellow, brown agouti, black agouti, brown, black.
35. a. 27/64 wild type, 37/64 mutant; b. *AA Bb Cc.*
37. a. Two genes are involved. *A–B–* and *aa B–* are WR, *A–bb* is DR, and *aa bb* is LR. b. For these true-breeding strains, WR-1 is *AA BB;* WR-2 is *aa BB;* DR is *AA bb,* and LR is *aa bb.* c. The cross was *Aa Bb* (WR) × *aa bb* (LR).
39. 44/56.

Chapter 4

1. a. 13; b. 7; c. 11; d. 10; e. 12; f. 8; g. 9; h. 1; i. 6; j. 15; k. 3; l. 2; m. 16; n. 4; o. 14; p. 5.
3. a. 7 centromeres; b. 7 chromosome; c. 14 chromatids; d. 3 pairs; e. 4 metacentric and 3 acrocentric; f. 4 colors; g. females are XX.
5. a. iii; b. i; c. iv; d. ii; e. v.
7. a. 1, 1 → 2, 2, 2, 2, 1, 1, b. yes, yes, yes, yes → no, no, no, no → yes; c. no, no, no, no → yes, yes, yes, yes → no; d. yes, yes, yes, yes → no, no, no, no → yes.
9. Meiosis produces 4 cells (*n,* haploid), each with 7 chromosomes.
11. a. Mitosis, meiosis I, II; b. mitosis, meiosis I; c. mitosis; d. meiosis II and meiosis I; e. meiosis I; f. none; g. meiosis I; h. meiosis II, mitosis; i. mitosis, meiosis I.
13. a. metaphase or early anaphase of meiosis I in a male (assuming X-Y sex determination in *Tenebrio molitor*); b. sister chromatids, centromeres, and telomeres (among others); c. five.
15. It is very realistic to assume that homologous chromosomes carry different alleles of some genes. In contrast, recombination almost always occurs between homologous chromosomes in any meiosis; thus the second assumption is much less realistic. The couple could potentially produce $2^{23} \times 2^{23} = 2^{46}$ or 70,368,744,177,664 different zygotic combinations.
17. Meiosis requires the pairing of homologous chromosomes during meiosis I.
19. a. 400 spermatozoa; b. 200; c. 100; d. 100; e. 100; f. none.
21. a. Only females; b. males; c. males; d. 1/5 ZZ males and 4/5 ZW females.
23. a. brown females and ivory-eyed males; b. females with brown eyes and males with ivory or brown eyes in a 1/2 to 1/2 ratio.
25. a. Nonbarred females and barred males; b. barred and nonbarred females and barred and nonbarred males.
27. The bag-winged females have one mutation on the X chromosome that has a dominant effect on wing structure and that also causes lethality in homozygous females or hemizygous males.

29.

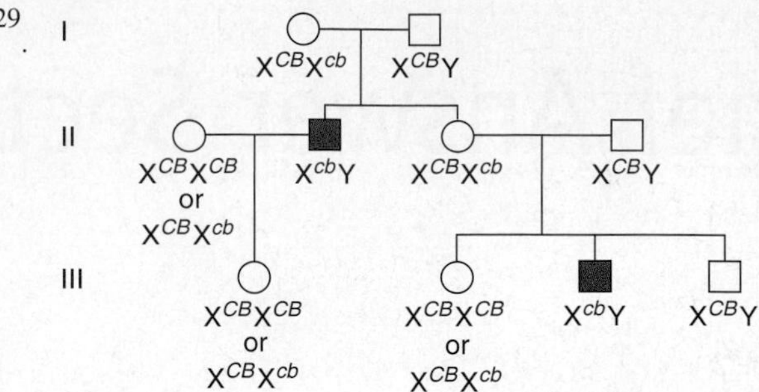

31. a. Recessive; b. autosomal; c. *aa;* d. *Aa;* e. *Aa;* f. *Aa;* g. *Aa;* h. *Aa.*
33. Vestigial wings is autosomal; body color is X-linked recessive.
35. a. X-linked dominant inheritance. b. Can exclude sex-linked recessive inheritance because affected females have unaffected sons. Can exclude autosomal recessive inheritance because the trait is rare and affected females have affected children with multiple husbands. Can exclude autosomal dominant inheritance because all the daughters but none of the sons of an affected male are affected. c. III-2 had four husbands and III-9 had six husbands.
37. a. 3; b. 1 or 3.
39. a. Purple is caused by homozygosity for a recessive allele of an autosomal gene (*p*), but the X-linked recessive white mutation is epistatic to *p* and to p^+. b. F_1 progeny: 1/2 white-eyed males and 1/2 wild-type (red) females; F_2 progeny: 1/4 white males, 1/4 white females, 3/16 red males, 3/16 red females, 1/16 purple males, 1/16 purple females.

Chapter 5

1. a. 8; b. 4; c. 1; d. 11; e. 2; f. 5; g. 6; h. 3; i. 10; j. 12; k. 9; l. 7.
3. a. *Oo Bb;* b. 9:3:3:1; c. not significant; d. between 0.5 and 0.1.
5. Both raspberry and sable are caused by recessive alleles of two X-linked genes that are separated by about 11.7 m.u. The parental generation was r^+ s / r^+ s females and *r* s^+ / Y males.
7. a. *Gs* Bhd^+ / Gs^+ *Bhd* ♀ × Gs^+ Bhd^+ / Y ♂ → 51 *Gs* Bhd^+ ♂: 48 Gs^+ *Bhd* ♂: 2 *Gs Bhd* ♂: 1 Gs^+ Bhd^+ ♂. The rf = 3mu; b. genotypes, phenotypes and frequencies of the female progeny would be the same as their brothers.
9. 10%
11. a. *A* = normal pigmentation, *a* = albino allele, $Hb\beta^A$ = normal globin, $Hb\beta^S$ = sickle allele 49.5% $aHb\beta^A$, 49.5% $AHb\beta^A$, 0.5% $aHb\beta^A$, 0.5% $AHb\beta^A$; b. 49.5% $aHb\beta^A$, 49.5% $AHb\beta^A$, 0.5% $aHb\beta^A$, 0.5% $AHb\beta^A$; c. 0.0025.
13. a. 1/4 black, 1/2 albino, 1/4 brown; b. 34 m.u. apart.
15. a. Gametes: 20% *Ab* and *aB,* 30% *AB* and *ab.* F_2 generation: 59% *A–B–,* 16% *A–bb* and *aa B–,* 9% *aa bb.* b. Gametes: 30% *Ab* and *aB,* 20% *AB* and *ab.* F_2 generation: 54% *A–B–,* 21% *A–bb* and *aa B–,* 4% *aa bb.*
17. a. two genes are assorting independently; b. the two genes are on the same chromosome; yes; c. recombination occurs at the four strand stage of meiosis, and so many crossovers occur between genes when they are far apart on the same chromosome that the linkage between alleles of these genes will be randomized; d. by summing up the values obtained for smaller distances separating other genes in between those at the ends.
19.

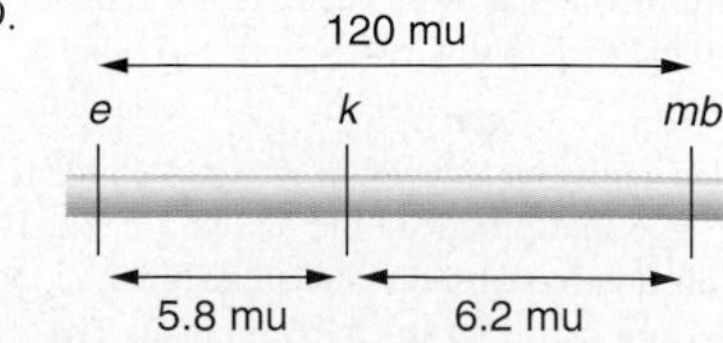

21. a. 360 $a^+b^+c^+$; 360 *abc;* 90 a^+bc; 90 ab^+c^+; 40 a^+b^+c; 40 abc^+; 10 a^+bc^+; 10 ab^+c; b. 500 $a^+b^+c^+$; 500 *abc.*

23. a. *sceccv* / + + + and *b* / +; b. *sceccv* / + + +, *sc–cv* = 9 m.u., *ec–cv* = 10.5 m.u.; c. predicted DCO = 0.009, observed DCO = 0.001, interference = 0.89.

25. a. 39%; b. 39%; c. 0.5%; d. 8%.

27. $\dfrac{dwp/pld^+\ rv^+\ rmp}{dwp^+/pld\ rv\ rmp^+}$ *rv–rmp* 10 m.u.; *pld/dwp–rv* 5 m.u.

29. a. 334;

b.

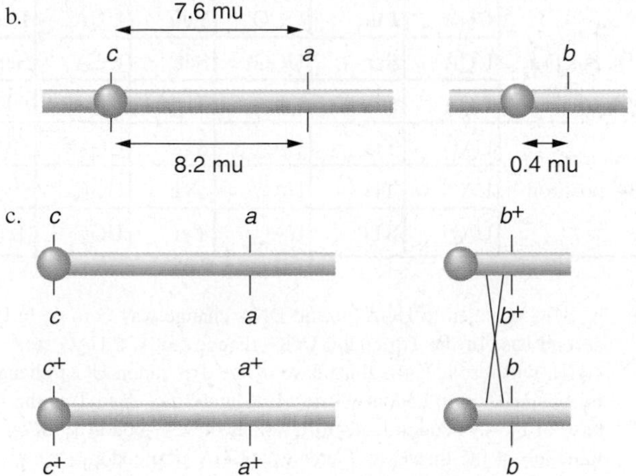

31. a. First group: met^+lys^+ and met^-lys^-; second group: $met^+\ lys^+$, $met^+\ lys^-$, met^-lys^+, met^-lys^-. b. 5.5 m.u. c. met^-lys^+ and met^+lys^-.

33. a. 100% 4; b. 100% 2; c. 50% 0 and 50% 2; d. 40% 0, 20% 1, and 40% 2; e. 100% 0; f. 80% 0 and 20% 1; g. 2.

35. a. (Unordered tetrads): PD = 2 $lys^-\ his^+$ and 2 $lys^+\ his^-$; NPD = 2 $lys^-\ his^-$ and 2 $lys^+\ his^+$; T = 1 $lys^-\ his^+$, 1 $lys^+\ his^-$, 1 $lys^-\ his^-$ and 1 $lys^+\ his^+$. b. 22.3 m.u. c. NCO = 222, SCO = 134; DCO = 44. d. 0.555 crossovers/meiosis. e. Two strand and three strand DCOs are missed. Map distance in map units = 1/2(T) + 3(NPD)/total asci; f. 27.8 m.u.

37. a. The sectors consist of $ade2^-/ade2^-$ cells generated by mitotic recombination. b. The sector size depends on when the mitotic recombination occurred during the growth of the colony. There should be many more small sectors because the mitotic recombinations creating them occur later in colony growth when there are many more cells.

39. a. Two mitotic crossovers occurred in succession in the same cell lineage. The first was between the *sn* and *y* genes, creating a patch of yellow tissue. The second was between the centromere and *sn,* creating a "clone within a clone" of yellow, singed cells. b. Yes.

Chapter 6

1. a. 6; b. 11; c. 9; d. 2; e. 4; f. 8; g. 10; h. 12; i. 3; j. 13; k. 5; l. 1; m. 7.

3. c.

5. Tube 1, nucleotides; tube 2, base pairs (without the sugar and phosphate) and sugar phosphate chains without the bases; tube 3, single strands of DNA.

7. a. 20% C; b. 30% T; c. 20% G.

9. Single stranded.

11. 5′.CAGAATGGTCTCTGCTAT.3′.

13. 3′ GGGAACCTTGATGTTTCGGCTCTAATT 5′.

15. a. once every 4,096 nucleotides; b. once every 4,096 nucleotides; c. 256 nucleotides apart.

17. After one additional generation, 1/4 intermediate; after two additional generations, 1/8 intermediate.

19.

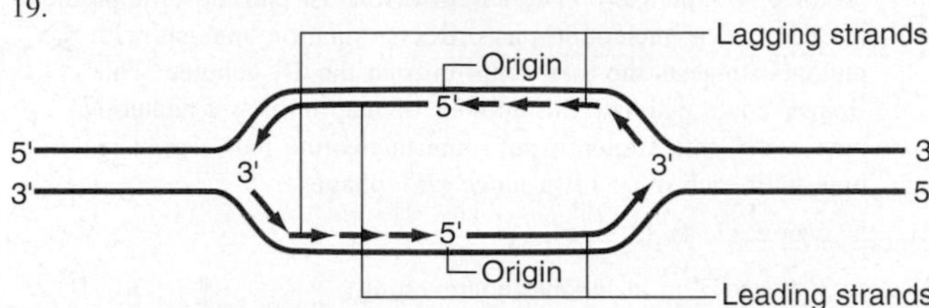

21. 5′ UAUACGAAUU 3′.

23. a. Relieves the stress of the overwound DNA ahead of the replication fork; b. unwinds the DNA; c. synthesizes a short RNA oligonucleotide; d. joins the sugar phosphate backbones.

25. The figure shows both strands of DNA are being replicated in the same direction relative to the replication fork.

27. a. no new DNAs will be formed; b. no new DNAs will be formed; c. the two DNA strands can pair with each other so two new DNA molecules will be formed; d. this single strand of DNA has two regions that have complementary base sequence so the DNA can form a so-called *hairpin loop* and the product will therefore be 5 nucleotides longer than the original:

29. Would not undergo recombination.

31. Regardless of which strands are cut during resolution (to result in crossing-over or no crossing-over) mismatches within the heteroduplex region can be corrected to the same allele, resulting in gene conversion.

33. If many short repeats are present in the double helix at the point where the invading strand is pairing, it is likely that the invading strand will not line up perfectly.

Chapter 7

1. 1. a, b, h; 2. b, c; 3. f, j; 4. g, j; 5. a, b, h; 6. f, i, k; 7. g, k; 8. d, i; 9. e, f, i.

3. 9.5×10^{-5}; higher than normal rate.

5. If phages induce resistance, several appear in random positions on each of the replica plates. If the mutations preexist, the resistant colonies would appear at the same locations on each of the three replica plates.

7. Female A has a white-eyed mutation on the X. Female B has a recessive lethal mutation on the X. Female C is mosaic with a lethal mutation on one strand and wild-type sequence on the other strand of one X chromosome, or she on the other strand of one X chromosome, or she is heterozygenes for an incompletely penetrant lethal mutation.

9. a. 857 essential X-linked genes; b. 37.6% of the genes on the X chromosome are essential; c. the X-ray induced mutation rate = 1.4×10^{-4} a 40-fold increase.

11. a. two-way mutagen; b. one-way mutagen; c. two-way mutagen; d. two-way mutagen; e. two-way mutagen.

13. a. nucleotide excision repair and the SOS-type error-prone repair; b. AP endonuclease and other enzymes in the base excision repair system could remove the damage and the SOS repair systems can work at AP sites, adding any of the 4 bases at random.

15. Yes; Liver converts substance X into a mutagen.

17. a. complementation test; − is a lack of complementation; + means that the two mutations complemented each other; b. 1 × 4 =−, 1 × 6 =+, 2 × 3 =+, 2 × 4 =+, 2 × 5 =+, 3 × 5 =+, 3 × 6 =+, 4 × 5 =+, 4 × 6 =+; c. 3 genes (1, 3, and 4), (2 and 6), (5).

19. a. Deletions do not revert. Also, deletions will fail to recombine with either of two rII^- mutations that can recombine with each other to produce rII^+ phage. b. The length of the T4 chromosome predicts the number of nucleotide pairs. Recombination analysis with the mutants suggests the total map units in the T4 genome. Thus, Benzer could estimate the number of map units per nucleotide pair. c. rII^- mutations in the same nucleotide pair cannot recombine with each other to produce rII^+ phage.

21. a. two; b. (1, 4), (2, 3, 5).

23. a. 3, 6, and 7 are deletions (nonreverting);

b.

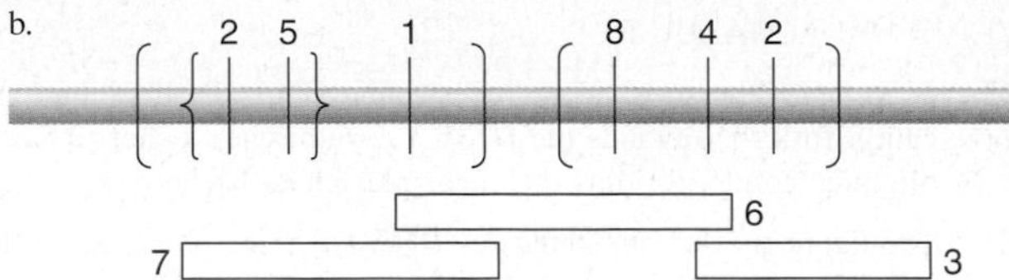

c. Use other deletions in crosses with mutants 2 and 5.

25. a. Parental ditype: All spores Arg^-. Nonparental ditypes: 2 Arg^-: 2 Arg^+; b. Two of the PD spores grow on either ornithine, citrulline, arginosuccinate, or arginine; the other two grow with arginine only; NPD Arg^- spores grow with arginine only.

27. 45 purple: 16 green: 3 blue.

29. a. In all four crosses, there are two unlinked genes involved with complete dominance at both loci. b. (Each arrow represents a biochemical reaction catalyzed by one of the two gene products.) (Cross 1) colorless → blue → purple; (Cross 2) colorless1 → colorless2 → purple; (Cross 3) colorless1 → red and colorless2 → blue, with red + blue = purple; (Cross 4) colorless1 → purple and colorless2 → purple. c. Cross 2. d. F_2 only. (Cross 1) 2 purple: 1 blue: 1 white; (Cross 2) 1 purple: 1 white; (Cross 3) 2 purple: 1 red: 1 blue; (Cross 4) all purple.

31. a. $X \xrightarrow{18} D \xrightarrow{14} B \xrightarrow{9} A \xrightarrow{10} C \xrightarrow{21}$ thymidine

b. 9 and 10 accumulates B; 10 and 14 accumulates D.

33 a. successful, immediate, prolonged; b. unsuccessful; c. successful, delayed, prolonged; d. successful, immediate, prolonged; e. unsuccessful; f. unsuccessful; g. successful, delayed, prolonged; h. unsuccessful; i. successful, immediate, short term; j. successful, immediate, prolonged.

35. a. two; b. 1/16 α1α1 β1β1: 1/8 α1α2 β1β1: 1/16 α2α2 β1β1: 1/8 α1α1 β1β2: 1/4 α1α2 β1β2: 1/8 α1α2 β2β2: 1/16 α1α1 β2β2: 1/8 α1α2 β2β2: 1/16 α2α2 β2 β2.

37. One chromosome with β β/δ δ; another with β/δ only (where / signifies a protein part of which, for example, the N-terminal part, is one type of globin and the other part the other type of globin.)

Chapter 8

1. a. 5; b. 10; c. 8; d. 12; e. 6; f. 2; g. 9; h. 14; i. 3; j. 13; k. 1; l. 7; m. 15; n. 11; o. 4; p. 16.

3. a. GU GU GU GU GU or UG UG UG UG;

b. GU UG GU UG GU UG GU UG GU;

c. GUG UGU GUG U etc.;

d. GUG UGU GUG UGU GUG UGU GUG UGU GU (depends on where you start);

e. GUG UGU GU or UGU GUG UG (depends on where you start).

5. Hb^C therefore precedes Hb^S in the map of β-*globin* gene.

7. 5′ GGN GCA CCA AGG AAA 3′

9.

Change \ Stop Codon	UAA		UAG		UGA	
1st position	AAA	**Lys**	AAG	**Lys**	AGA	**Arg**
	CAA	**Gln**	CAG	**Gln**	CGA	**Arg**
	GAA	**Glu**	GAG	Glu	GGA	**Gly**
2nd position	UUA	**Leu**	UUG	**Leu**	UUA	**Leu**
	UCA	**Ser**	UCG	**Ser**	UCA	**Ser**
	UGA	STP	UGG	**Trp**	UAA	STP
3rd position	UAU	**Tyr**	UAA	STP	UGU	**Cys**
	UAC	**Tyr**	UAC	**Tyr**	UGC	**Cys**
	UAG	STP	UAU	**Tyr**	UGG	**Trp**

11. a. UGG changed to UGA, so the DNA change was G to A. b. If the second base of the Trp codon UGG changes to A, a UAG stop codon will result. If the third base of the Trp codon UGG changes to A, a UGA stop codon will result. Mutation of A to T in the first base of the Lys codon leads to UAA. If the Gly codon is GGA, mutation of the first G to T creates a UGA stop codon.

13. Three.

15. a. Mutant 1: transversion changes Arg to Pro; mutant 2: single-base-pair deletion changes Val to Trp and then stop; mutant 3: transition Thr (silent); mutant 4: single-base-pair insertion changes several amino acids then stop; mutant 5: transition changes Arg to stop; mutant 6: inversion changes identity of 6 amino acids. b. EMS: 1, 3, 5; Proflavin: 2, 4.

17. Required to add the appropriate ribonucleotide to a growing RNA chain.

19. Gene *F:* bottom strand; gene *G:* top strand.

21. Base pairing between the codon in the mRNA and the anticodon in the tRNA is responsible for aligning the tRNA that carries the appropriate amino acid to be added to the polypeptide chain.

23. a. Translation. b. Tyrosine (Tyr) is the next amino acid to be added to the C terminus of the growing polypeptide, which will be nine amino acids long when completed. c. The carboxy-terminus of the growing polypeptide chain is tryptophan. d. The first amino acid at the N terminus would be f-met in a prokaryotic cell and met in a eukaryotic cell. The mRNA would have a cap at its 5′ end and a poly-A tail at its 3′ end in a eukaryotic cell but not in a prokaryotic cell. If the mRNA were sufficiently long, it might encode several proteins in a prokaryote but not in a eukaryote.

25 a. 1431 base pairs; b. 5′ ACCCUGGACUAGUCGAAAGUUAACU-UAC 3′; c. N Pro Trp Thr Ser Gly Lys Leu Thr Tyr.C.

27. Mitochondria do not use the same genetic code; mutate the 5′ CUA 3′ codons in the mitochondrial gene to 5′ ACN 3′.

29. Order: c e i f a k h d b j g.

31. a. Very severe; b. mild; c. very severe; d. mild; e. no effect; f. mild to no effect; g. severe; h. severe or mild.

33. Mutations possibly causing a detectable change in protein size: d, e, g, and i. In protein amount (assumes all mutant proteins are equally stable): e, f, j, and k. In mRNA size: i and j. In mRNA amount (assumes all mutant mRNAs with poly-A tails are equally stable): f and j.

35. If the met$^+$ phenotype is due to a true reversion, then: met^- x $met^+ \rightarrow met^+ / met^- \rightarrow 2\ met^+;\ 2\ met^-$. If there is an unlinked suppressor mutation: $met^-\ su^-$ (phenotypically met$^+$) $\times$ $met^+\ su^+$ (wild type) $\rightarrow met / met^+;\ su^- / su^+ \rightarrow$ 3/4 met$^+$; 1/4 met$^-$.

37. a. 3′ AUC 5′; b. 5′ CAG 3′, c. minimum two genes.

39. a. Missense mutations change identity of a particular amino acid inserted many times in many normal proteins but nonsense suppressors only make proteins longer b. (i) a mutation in a tRNA gene in a region other than that encoding the anticodon itself, so that the wrong aminoacyl-tRNA synthetase would sometimes recognize the tRNA and charge it with the wrong amino acid; (ii) a mutation in an aminoacyl-tRNA synthetase gene, making an enzyme that would sometimes put the wrong amino acid on a tRNA; (iii) a mutation in a gene encoding either a ribosomal protein, a ribosomal RNA or a translation factor that would make the ribosome more error-prone, inserting the wrong amino acid in the polypeptide; (iv) a mutation in a gene encoding a subunit of RNA polymerase that would sometimes cause the enzyme to transcribe the sequence incorrectly.

41. a. 5′ UUA 3′; b. 5′ UAG 3′ and 5′ UAA 3′ (due to wobble at the codon's 3′-most nucleotide); c. Gln, Lys, Glu, Ser, Leu, and Tyr.

Chapter 9

1. a. 10; b. 1; c. 9; d. 7; e. 6; f. 2; g. 8; h. 3; i. 5; j. 4.

3. Shorter molecules slip through pores more easily; large molecules get caught.

5. a. A; b. 10 kb; c. 10 kb.
d.

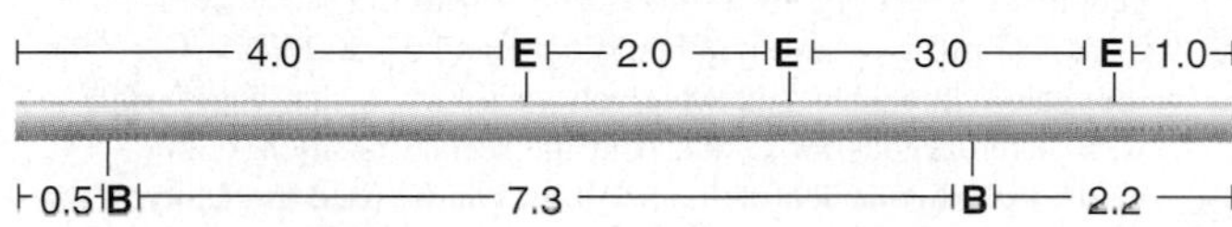

7. a. 1.83 kb.
b.

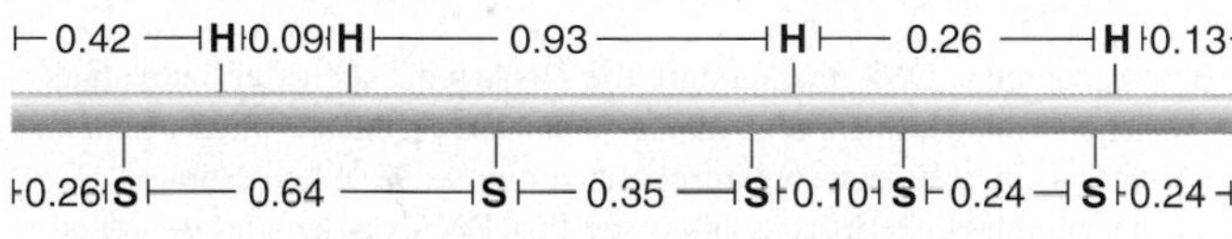

c.

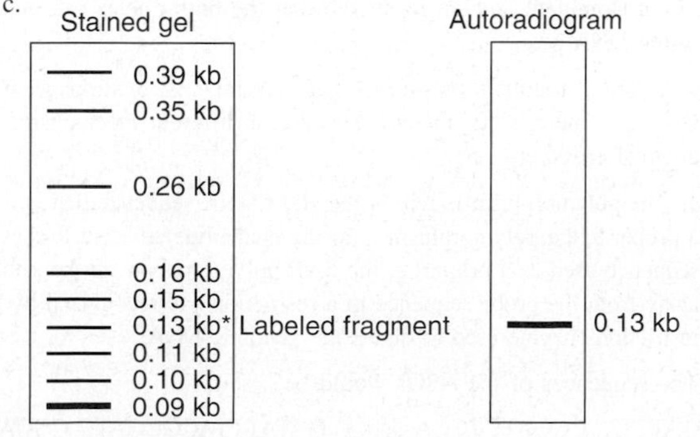

9. Selectable markers are genes that allow a vector to impart protection from an antibiotic on a host cell. When cells are transformed by a vector with a selectable marker, and then exposed to the appropriate antibiotic, only cells that have the vector will survive.

11. a. all; b. 1/4; c. none; d. 1/2 chance; e. 3/4.

13. a. Five; b. divide the number of base pairs in the genome by the average insert size, then multiply by 5.

15. After cloning: EcoRI: 42 and 2400 bp fragments; MboI: 705, 944, 500, and 300 bp or 905, 744, 500, and 300 bp fragments.

17. a. Such a vector is able to be transformed, selected, and maintained in animal cells and *E. coli* must contain a selectable marker in *E. coli,* a selectable marker in animal cells and replication origins for each type of cell; b. human gene must be a cDNA next to regulatory sequences for bacterial transcription and translation.

19. Where order doesn't matter, letters are bracketed: c, [j, f], a, k, i, g, b, e, h, d.

21. Use a probe of the gene from another organism, or the amino acid sequence could be 'reverse translated' into potential DNA or a mixture of oligonucleotides that could encode that peptide sequence could be used as a probe to hybridize to a library of genomic or cDNA clones, or the cDNA clones can be cloned into an expression vector and the clones can then be screened for expression of the ozonase enzyme be screening with the antibody.

23. a. DNA fragments that are 4096 bp long on average; fragments of this size are best resolved by electrophoresis in agarose gels; b. a broadly distributed smear centered around 4 kb; c. the DNA that is digested is very long – a human chromosome. Some fragments of DNA produced by digestion with restriction enzymes other than *Eco*RI will have either one or both ends at a position beyond the *Eco*RI sites defining the probe; d.

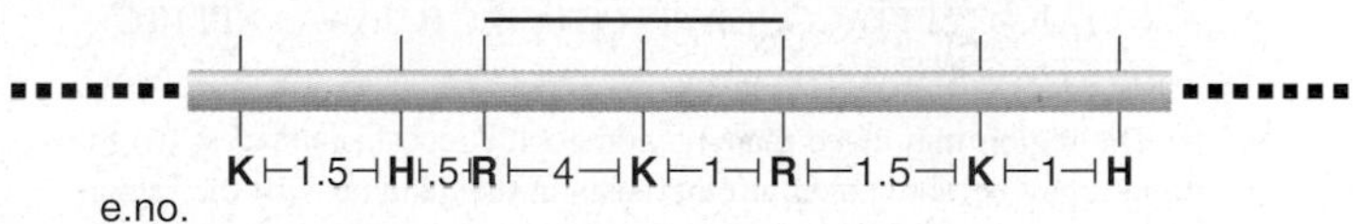

e. no.

25. b.

27. a. the N-terminus of the fusion protein contains most of the MBP protein sequence and the C-terminal end of the fusion protein will contain the C-terminal end of CFTR; b. the CFTR gene can only be inserted into the vector with the desired orientation; c. 5′CCCCCGAATTCGGGCTAAGATCTGAATTTTC3′and 3′ACCCGTTATTACATCGCGCAGCTGGGGGG5′; d. Make extracts of bacterial cells expressing the fusion protein and add these extracts to amylose resin. To get the fusion protein off the resin you can add the sugar maltose.

29. a. Enzyme-based; DNA ligase; b. Enzyme-based; restriction enzymes; c. Not necessarily enzyme-based; cut up DNA into smaller random fragments by mechanical shearing, by chemical treatments, or certain DNases; d. Non-enzymatic, hybridization relies on complementary base pairing; e. Enzyme-based; DNA polymerase; f. Enzymatic; reverse transcriptase for the first strand of cDNA and DNA polymerase for the complementary strand; g. Enzyme-based; DNA polymerases from thermophilic bacteria.

31. a.

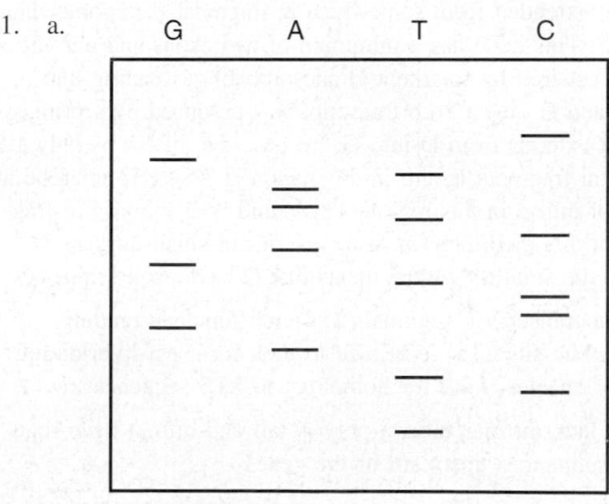

b. The newly synthesized strand has stop codons in all three frames (underlined) and therefore would not be the coding (exon) sequence. On the DNA sequencing template strand the reading frame that starts with the first nucleotide does not contain a stop codon and therefore is the ORF in this RNA-like strand; c. N Ala-Leu-Val-Gln-Ala-Arg.

Chapter 10

1. a. 8; b. 4; c. 2; d. 6; e. 3; f. 1; g. 5; h. 7.
3. a. Cut the clone into subfragments to use as probes and look for single bands in hybridization; b. oligonucleotides used to detect STSs surrounding a microsatellite could be used to amplify DNA from a large number of individuals in the population.
5. The other 2 cosmids probably contain repeated DNA also found in the BAC.
7. Two non-overlapping contigs:

   5′ CAAATAGCAGCAAATTACAGCAATATGAAGAGATCATAC AGTCCACTGAA 3′

   3′ GTTTATCGTCGTTTAATGTCGTTATACTTCTCTAGTATGTC AGGTGACTT 5′

   and:

   5′ GAATTTGAAAATGCCCTAAAGGAAATGAGATTTTTAAA AGGAGATACTAC 3′

   3′ CTTAAACTTTTACGGGATTTCCTTTACTCTAAAAATTTTC CTCTATGATG 5′
9. (i) The region may have many genome-wide repeat sequences; (ii) the region may be duplicated at other sites in the genome; (iii) the region may have long stretches of repeated DNA (iv) centromeric and telomeric regions of the genome can not be cloned.
11. a. the two sequences would have to be approximately 7.5 Mb apart; b. you use a computer to search the genome to find the location of the sequence. FISH is still useful to help characterize the genomes of organisms whose genome projects are not yet completed. In addition, FISH can still be useful in analyzing the genomes of individual people.
13. a. Beta; b. 16 mu between alpha and delta.
15. design PCR primers that would amplify small regions randomly scattered around the sequenced human genome; any two randomly selected individuals would have on average one SNP variation roughly every 1000 nucleotides.
17. Alternate splicing of the same primary transcript, or duplicated genes where the sequence in the middle has diverged.
19. (i) 4.1 kb mRNA: the primary transcript includes all the DNA between the B, C, D, E and F bands and part or all of A and G. Bands A, D, F and G must contain exon sequences. There are a minimum of 2 introns − one that includes bands B + C and a second that includes band E. (ii) 3.4 kb mRNA was produced by a primary transcript that extended from somewhere in fragment B to somewhere in fragment E. This gene has a minimum of two exons and one intron in the region covered by fragment D and probably extending into fragments C and E. (iii) 1.8 kb transcript was produced by a primary transcript that extends from D into G. Because the mRNA is only 1.8 kb but the total fragment length in this region is 3.5, there must be at least 1.7 kb of intron in this region. This could be a 1 intron in fragment E. If not this then there must be a series of small introns because even the small DNA fragments like G have exon sequences.
21. (1) Search for transcribed regions; (2) search for open reading frames and splice sites; (3) hybridize to look for cross-hybridizing DNA in other species; look for homology to STS sequences.
23. Pseudogenes lack introns, have a poly-A tail and do not have regulatory DNA sequences upstream of the gene.
25. a. Protein motifs; b. tissue- and development-specific expression; knocking out the gene.
27. a. the best consensus: 5′ XXATATAAAAXXXXXX 3′

    G T

    b. shows the nucleotides that are likely to be the most important to gene or chromosome function; you must choose DNA sequence that are in the same evolutionarily conserved orthologous gene and in the same relative position (like the same exon) in the four species.
    c. compare the amino acids of the proteins encoded by homologous genes, you might be able to find particular conserved amino acids that you can use to define a consensus.
29. a. All of these genes, in both humans and chimpanzees would be considered homologous; the α_1, α_2, β, G_γ, A_γ, δ, ϵ and ζ genes in humans would comprise one set of paralogous genes; the α_1 gene in humans would be orthologous to the α_1 gene in chimpanzees.
    b. orthologous genes are more likely to have closely related functions; c. orthologous genes (the human β gene and the chimpanzee β gene) should have a greater degree of nucleotide homology, since they were more recently derived from a common ancestor. d. Most of the gene duplication events must have occurred before the human and chimpanzee lineages diverged.
31. mRNAs present in all tissues could represent alternately spliced messages leading to different proteins that are not recognized by the antibody. Or, the two other messages are not translated.
33. a. Cloning a cDNA; b. clone into an expression vector.
35. a. The person with the red label is a heterozygote with the M1 sequence on one homologue and the M2 sequence on the other homologue; the other person is heterozygous for M3 and M6. b. The sequencing would show that both people have the sequence: 5′ ACT(TG)ACCGAGAGA(GA)CCTGCG 3′; you could tell that both people are heterozygous at this position with the sequence 5′ ACT(T)ACC . . . on one homologue and 5′ ACT(G)ACC . . . on the other homologue. Likewise you would know that both people were heterozygous for G and A at the second position. c. this method of oligonucleotide microarrays can be used to genotype single nucleotide polymorphisms.

Chapter 11

1. a. 5; b. 3; c. 8; d. 6; e. 2; f. 7; g. 1; h. 4.
3. Anonymous DNA markers are the DNA sequence of an individual. The terms dominant and recessive can only be used when discussing the phenotype of an organism, so in one sense this question is meaningless. Geneticists often say that DNA markers are inherited in a codominant fashion to denote that the both alleles can be seen in the DNA sequence.
5. a. Different numbers of simple sequence repeats; b. slippage of DNA polymerase during replication; c. a different mechanism: unequal crossing-over.
7. a. The polymorphism is within the short DNA sequence that is used as a probe; b. the polymorphism is in the nucleotide adjacent to the sequence used as a primer; c. the SNP polymorphism can be kilobases away from the probe sequence in a restriction site recognized by the restriction enzyme used to digest the genomic DNA.
9. The sequences of the ASOs would be

3′ GATATTTACCCGATCCGCA and 3′ GATATTTACGCGATC CGCA.

11. Sperm collected from man, eggs are collected from woman. After *in vetro* fertilization, embryos are allowed to develop to the eight-cell stage. A single cell from each eight-cell embryo is removed. DNA is prepared and genotype is analyzed using PCR and *Mst* II digestion. Embryos with the desired genotype are implanted into the woman's uterus.
13. Coworker 3 has the same DNA fingerprint as the crime sample and must be the perpetrator of the crime. The probability is essentially 100%.

15. a. Individuals A, B, C and E; b. Individuals D and F; c. 48 bp.
17. a. 10 kb; b. 10 kb; c. 0%; d. 50%.
19. Members of the disease family must be segregating two or more alleles at each DNA marker that is chosen.
21. a. 0% chance; b. 0.0075 probability of an affected child.
23. Mating W is not informative; mating X is informative – both parents are doubly heterozygous; mating Y is non-informative; mating Z is non-informative.
25. Identify sequences that are transcribed into RNA; use computational analysis to identify sequences that are conserved between distantly related species; use computational analysis to identify sequences that are open reading frames with appropriate codon usage and splice sites.
27. a. A, C and E, b. three different genes have been identified; c. Yes; d. fragments C and E; e. gene recognized by fragment E; f. If there is a mouse model of this disease you would transform the mice with the cDNA clone of the candidate gene and look for the normal human gene to rescue the mutant phenotype in the mice.
29. a. The disease is autosomal dominant; b. Yes, II-2 and III-1.
31. a. 12,500 different haplotypes; b. 156,250,000 possible diplotypes; c. The father's genotype is A25 C4 B7 / A23 C2 B35; the mother's genotype is A24 C5 B8 / A3 C9 B44; d. 1/4.

Chapter 12

1. a. 4; b. 9; c. 7; d. 10; e. 11; f. 13; g. 15; h. 14; i. 18; j. 12; k. 2; l. 16; m. 1; n. 17; o. 6; p. 3; q. 5; r. 8.
3. The elements of the system; the physical associations among the elements; the biological context of the system; how the association of the system's elements and their relation to changes in the biological context explain its emergent property.
5. A gene regulatory network; to gain a further understanding of the network you need to understand what environmental information affects the network via signal transduction pathways and other inputs, how the network integrates and modifies these inputs and the output of the transformed information to various protein networks.
7. a. The first half of the system involves fusing the DNA binding domain (DBD) portion of Gal 4 to the DNA for the ZW 10 protein. The fused Gal 4 DBD/ZW 10 gene (the bait) is under the control of a yeast promoter that is constitutively expressed in yeast cells. The bait construct is transformed into yeast cells. The other half of the system involves fusing cDNAs from each of the ~18,000 genes of *Drosophila* (ORFs) to the activator domain (AD) of Gal 4. These AD/ORF fusions are also cloned into a yeast expression vector and used to individually transform yeast cells forming an AD/ORF library. The cells containing the bait construct are mated to the cells containing the AD/ORF library and the diploids are plated on media lacking histidine. The mated yeast cells will express two different fusion proteins – the bait and a *Drosophila* protein fused to the Gal 4 activator. If the fused yeast protein can bind to ZW10 then the two parts of the Gal 4 protein will be brought together forming a functional Gal 4 protein. This functional Gal 4 protein can bind the Gal 4 enhancer sequence and activate expression of the histidine complex, so these diploid yeast cells will be able to grow in the absence of histidine; b. advantages: you can detect proteins that interact with ZW10 but don't bind tightly enough to be retained on a column. You can also detect interactions with proteins that are found in very small quantities which will not be able to be visualized on a gel; disadvantages: many different reasons why the correct protein-protein interactions may not occur between the bait and the AD/ORF in yeast cells. Also, the yeast two hybrid system gives many false positives; c. fuse each of these to the DBD of Gal 4, giving you ~18,000 different baits. Then these same cDNA clones must be independently fused to the AD region of Gal 4. The next step involves creating a matrix of all possible pairwise combinations of the bait and the AD/ORFs.
9. For these 80 proteins the regulation of protein levels must occur at the post-transcriptional level. This can involve differences in the rates of translation of different mRNAs or the differences in the rates at which different proteins are degraded.
11. You could increase the resolution by fragmenting the DNA more, so the fragments are only 50–100 bp in.
13. a. True; b. True; c. False; d. False; e. True; f. True; g. True; h. False; i. False; j. True; k. True.
15. The regulation of concentration is at the post-transcriptional level.

Chapter 13

1. a. 4; b. 10; c. 5; d. 8; e. 9; f. 2; g. 3; h. 6; i. 1; j. 7.
3. Interphase: 40-fold compaction; metaphase: 10,000-fold compaction.
5. a. 1.2×10^8 molecules of H2A protein; b. during or just after S phase; c. more templates that the cells can transcribe simultaneously, allowing the more rapid production of histone proteins.
7. A deletion of one G band removes about 15 genes.
9. H1 is one the outside of the complex and locks the DNA to the core and interacts with H1 proteins from other nucleosomes to forming the center of the coil that is thought to form the 300A fiber. The other histone proteins are coated with DNA and can not form the 300A fiber.
11. Mutate the DNA sequence so that the twelfth amino acid encoded is not lysine but another similar amino acid.
13. 5500 origins of replication.
15. a. 68.75 kb per second; b. 7.3 bases per second.
17. The sequences added were specific for the species adding them (in this case, yeast)
19. a. Integrates; b. maintained separately but mitotically unstable; c. maintained separately but mitotically unstable.
21. 0.6 kb *Sau*3A.
23. a. 300 Å fiber; b. DNA loops attached to a scaffold; c. heterochromatin; d. metaphase chromosomes.
25. a. Drosophila: Constitutive heterochromatin: centromeric DNA, Y-chromosome facultative heterochromatin: PEV with the *white* gene; b. Human: constitutive heterochromatin: centromeric DNA, Y chromosome facultative heterochromatin: inactived X (Barr body).
27. a. 1; b. 0; c. 1; d. 1; e. 3; f. 0.
29. Some patches of cone cells in the eye in which the X chromosome carrying the *CB* allele was inactivated would be defective in color vision. If several patches occurred next to each other, there might be partial color vision.
31. a. mutant coat color; b. wild-type coat color; c. the heterozygotes are all females whose phenotype was determined by the allele received by the mother; d. in marsupials, the paternal X chromosome is always inactivated.

Chapter 14

1. a. 4; b. 8; c. 6; d. 5; e. 7; f. 3; g. 2; h. 1.
3. In a duplication, there would be a repeated set of bands; in a deletion, bands normally found would be missing.
5.

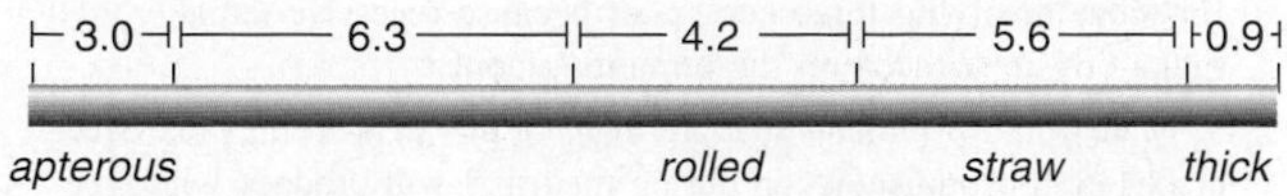

7.

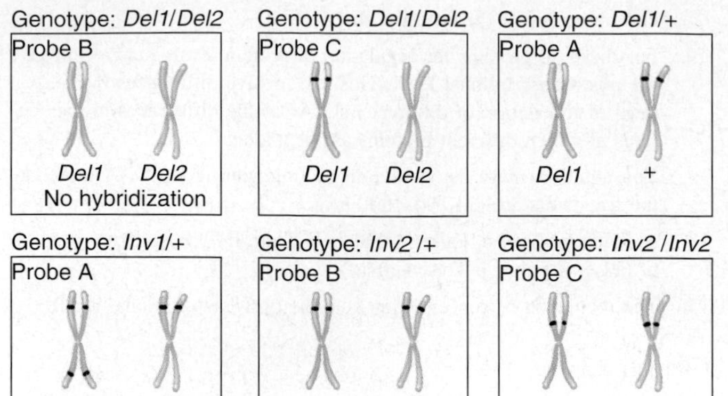

9. a. the parental types $y^+ z^l w^{+R} spl^+$/ Y (zeste) and $y z^l w^{+R} spl$ / Y (yellow zeste split); b. crossing over anywhere between the *y* and *spl* genes; c. mispairing and unequal crossing over between the two copies of the w^+ gene; d. 5.9 mu.

11. a. 2, 4; b. 2, 4; c. 2; d. 1, 3.

13. a. 2 *URA3 ARG9* spores and 2 *ura3 arg9* spores; b. 2 spores die, 1 *URA3 ARG9* and one *ura3 arg9*; c. 4 viable spores, 2 *URA3 ARG9* spores and 2 *ura3 arg9* spores.

15. A two-strand double crossover with both crossovers in the inversion loop. One crossover must occur between *LEU2* and *HIS4*. The other crossover must occur on the other side of either of the two genes but still within the inversion loop.

17. a. 1, 3, 5 and 6; b. 2 and 4; c. 1 and 3; d. 5 and 6, 2 and 4.

19. a. 1/4 fertile green, 1/4 fertile yellow-green, 1/4 semisterile green, 1/4 semisterile yellow-green; b. 1/2 fertile yellow-green, 1/2 semisterile green; c. from crossing-over events between the translocation chromosome and homologous region on the normal chromosome.

21. 1/2 *Lyra* males: 1/2 $Lyra^+$ (wild type) females.

23. a. 1/3 fertile: 2/3 semisterile. b. Most likely, translocation homozygotes do not survive because the translocation breakpoint disrupted an essential gene.

25. *Ds* is a defective transposable element and *Ac* is a complete, autonomous copy.

27. Use a probe made of DNA from the sequence preceding the 200 A residues to hybridize to genomic DNA on Southern blots or to chromosomes by *in situ* hybridization.

29. a. 7; b. sand oats: diploid, slender wild oats: tetraploid, cultivated oats: hexaploid; c. sand oats: 7, slender wild oats: 14, cultivated wild oats: 21; d. same answer as c.

31. a. (i) aneuploid, (ii) monosomic for chromosome 5, (iii) embryonic lethal; b. (i) aneuploid, (ii) trisomic for chromosomes 1 and 5, (iii) embryonic lethal; c. (i) euploid, (ii) autotriploid, (iii) viable but infertile; d. (i) euploid, (ii) autotetraploid, (iii) viable and fertile.

33. A: meiosis II in father; B: meiosis I in mother; C: meiosis I in father; D: meiosis II in mother.

35. You would actually expect more monosomies than trisomies, because meiotic nondisjunction would produce equal frequencies of monosomies and trisomies, but chromosome loss would produce only monosomies. The low frequency of monosomies observed is because monosomic zygotes usually arrest development so early that a pregnancy is not recognized. This may be due to a lower tolerance for imbalances involving only a single copy of a chromosome than for those involving three copies, or because recessive lethal mutations are carried on the remaining copy.

37. a. Mate putative mutants that are $ey\ ci^+$ / $ey^+\ ci$ with flies that are *ey ci* / *ey ci*. b. Nondisjunction during meiosis I will produce wild-type progeny; nondisjunction during meiosis II cannot be recognized. c. 2 eyeless: 2 cubitus interruptus: 1 eyeless, cubitus interruptus: 1 wild type. d. Mate putative mutants that are $ey\ ci^+$/ $ey^+\ ci$ with flies that have an unmarked attached chromosome 4. Nondisjunction during meiosis II would yield eyeless or cubitus interruptus progeny, but you could not recognize progeny resulting from nondisjunction during meiosis I. If the attached chromosome 4 carried two copies of *ey* and two copies of *ci*, you could recognize and discriminate some of the products of nondisjunction during the two meiotic divisions.

39. Treat with colchicine.

41. In autopolyploids, the banding patterns of homologs should be the same; in allopolyploids, different banding patterns will be seen for chromosomes from different species.

Chapter 15

1. a. 4; b. 5; c. 2; d. 7; e. 6; f. 3; g. 1.

3. 200 colonies on the first plate and 20 colonies on the second plate.

5. a. iv; b. iii; c. ii.

7. Southern hybridization of *Eco*RI digested DNA using an IS1 DNA as a probe.

9. Transform the plasmid into a nontoxin-producing recipient strain, assay for toxin production.

11. a. F^-, Str^R, pyr^-, met^-, xyl^-, tyr^-, arg^-, his^- and mal^-; b. *pyrE xyl mal arg met tyr his.*

13. Generalized transduction: transducing phage particles contain chromosomal DNA only; specialized transduction particles contain phage and chromosomal DNA.

15. DNA being transferred by transformation is degraded; DNase has no effect on DNA being transduced.

17. *lac arg102 arg101 arg103.*

19. a. F plasmid must have integrated into the chromosome near the *mal* genes and then exased incorrectly.

a.

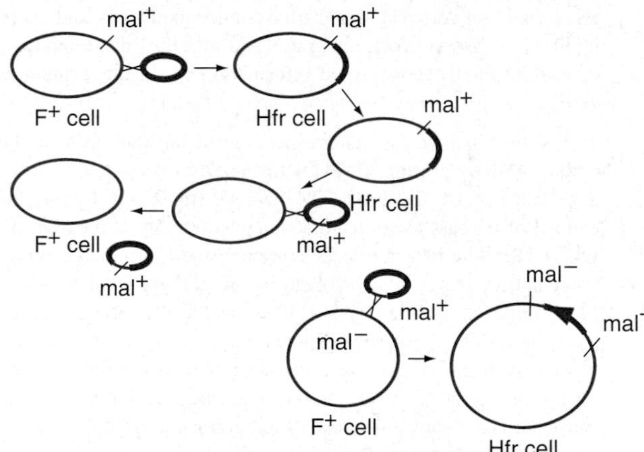

21. a.

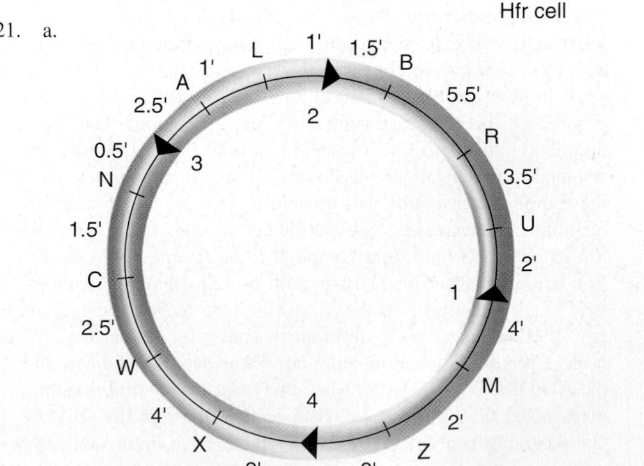

b. Use an Hfr strain that transfers the gene late (Hfr 3) and screen for a derivative that transfers the gene early.

23. a. (1) Pass extract of cells through APTG-agarose resin. (2) Digest with trypsin. (3) Use mass spectrometry to get molecular weights. (4) Compare profile of molecular weights with predicted MW from genome data. b. Make a fusion between protein coding region and lacZ and use APTG-agarose resin again.

25. plasmid transformation into *Shigella dysenteriae*, bacteriophage infection of *Staphylococcus, Streptococcus* or *E. coli* species and transposition of DNA (pathogenicity island) into *Vibrio cholerae.*

27. imprecise excision of a bacteriophage from a bacterial strain with virulence genes produced a specialized transducing bacteriophage containing both bacterial DNA (virulence genes in this case) and bacteriophage genes.

Chapter 16

1. a. 6; b. 8; c. 7; d. 2; e. 1; f. 3; g. 5; h. 9; i. 4.

3. a. both; b. both; c. neither; d. both.

5. a.

	Trp	His	Ile	Met
mRNA	5′ UGG	CAU/C	AUU/C/A	AUG
nDNA mRNA-like	5′ TGG	CAT/C	ATT/C/A	ATG
nDNA template	3′ ACC	GTA/G	TAA/G/T	TAC

b.

	Trp	His	Ile	Met
mRNA	5′ UGA/G	CAU/C	AUC/A	AUG/A
mtDNA mRNA-like	5′ TGA/G	CAT/C	ATC/T	ATG/A
mtDNA template	3′ ACT/C	GTA/G	TAG/A	TAC/T

7. a, c, e.

9. a. 2; b. 1; c. 4; d. 3.

11. a. 3; b. 1; c. 2.

13. Resemble mother because sperm does not contribute cytoplasmic material to the zygote.

15. method a.

17. a. segregation of mitochondrial, most asci have parental mitochondrial genomes and some have recombinant chloroplast genomes; b. 2:2.

19. a. Resistant to erythromycin (genotype ery^r); b. genotype ery^r / ery^s and the phenotype reflects the dominant allele; c. Sporulate the diploid. If the gene is nuclear, two spores would be ery^s and two would be ery^r. If the gene were mitochondrial, all four spores would be ery^r.

21. a. the female parent, which is the male sterile strain, gives the organelle genomes and a haploid set of chromosomes while the male parent, which is the male fertile strain, only gives a haploid set of chromosomes; b. the F_1 undergoes meiosis to make the female gametes recombination will occur between the homologous nuclear chromosomes; c. each generation of backcrossing increases the percentage of the nuclear genome from the male fertile strain. These backcrosses allow the researchers to study interactions of chloroplast and nuclear gene(s) that cause male sterility.

23. a. respiration requires oxygen and functional mitochondria. Mitochondrial function is deficient in *cox2-1* mutant cells; b. After mitotic growth about half of the diploid cells could grow on a nonfermentable substrate and the rest of the cells could not grow on this substrate. The cells that grew showed 4:0 segregation of the phenotype upon meiosis, so the *cox2-1* gene is a mitochondrial gene; c. The *pet111-1* mutation is recessive to the wild type allele, does not show mitotic segregation and does show 2:2 segregation after meiosis. Thus the *pet111-1* mutation behaves as expected for a trait controlled by a single nuclear gene.

25. a. Mother could have a small proportion of mutant genomes and the proportion in her daughter is higher or the mutation occurred in the germ line of I-2. b. Examine mother's somatic cells.

27. Differences in the time during development when mutation occurred or the proportion of mutant genomes in the cells in each tissue.

Chapter 17

1. a. 4; b. 8; c. 5; d. 2; e. 7; f. 1; g. 3; h. 6.

3. a. i, ii, iii; b. iv, v, vi.

5. Mutations in the promoter region can only act in cis to the structural genes immediately adjacent to this regulatory sequence. This promoter mutation will not affect the expression of a second, normal operon.

7. b.

9. Nonlysogenic recipient cell did not have the cI (repressor) protein, so incoming infecting phage could go into the lytic cycle.

11.

	β-galactosidase	*Permease*
a.	constitutive	constitutive
b.	constitutive	inducible
c.	inducible	inducible
d.	no expression	constitutive
e.	no expression	no expression

13.

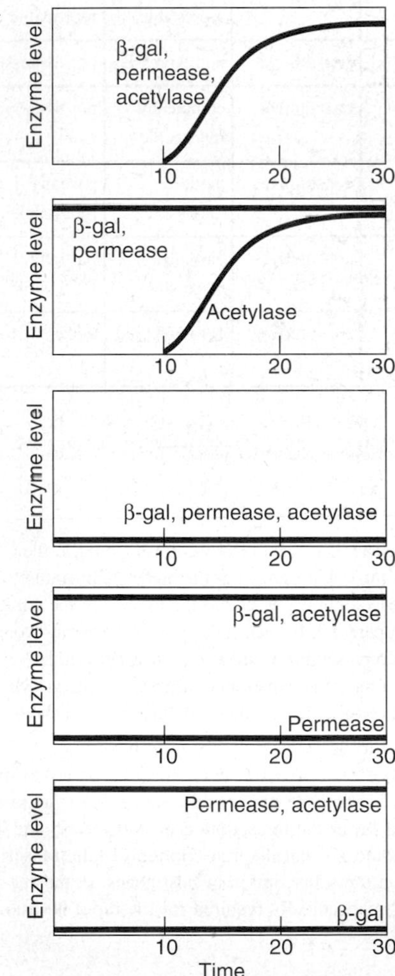

15. If the three genes make up an operon, they are cotranscribed to one mRNA and only one band should appear on a hybridization analysis using any of the three genes as a probe versus mRNA. If the genes are not part of an operon, there would be three differently sized hybridizing bands.

17. a. 4; b. 6; c. 7; d. 2; e. 3; f. 5; g. 1.

19. a. i, iii, v and vi, b. mutations ii, iii and iv; c. mutation 1 is i; mutation 6 is ii; mutation 2 is iii; mutation 4 is iv; mutation 5 is v; and mutation 3 is vi.

21. a. Mutations in O_2 or O_3 alone have only small effects on synthesis levels; b. Small DNA insertions between O_1 and O_2 may change the

face and either change the ability of the repressor to bind one of the sites or change the ability of the bound repressor to bend the DNA leading to an O^c mutant phenotype; c. insensitive to a I^s repressor protein.

23. The protein-coding region of your gene must be in the same frame as the *lacZ* gene.

25. Seven His codons (CAC or CAU), in a row.

27. a. This seems to be a biosynthetic operon, the operon is repressible. b.

Condition	Gene A	Gene B	Gene C	Gene D
Wildtype	completely repressible	constitutive	completely repressible	completely repressible
Nonsense in A	not expressed	constitutive	completely repressible	not expressed
Nonsense in B	partially repressible	not expressed	partially repressible	partially repressible
Nonsense in C	not expressed	constitutive	not expressed	not expressed
Nonsense in D	completely repressible	constitutive	completely repressible	not expressed
Deletion of region incl. E	partially repressible	constitutive	partially repressible	partially repressible
Deletion of F	partially repressible	constitutive	partially repressible	partially repressible
Deletion of G	not expressed	constitutive	not expressed	not expressed

c.

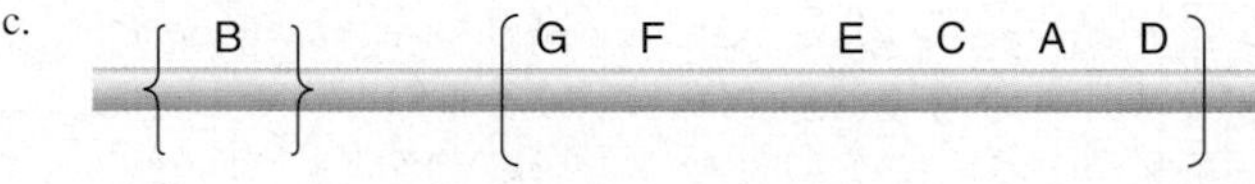

29. negative regulation.

31. a. two probes, one consists of labeled cDNA corresponding to the mRNA extracted from the culture grown at the higher temperature, and the other consists of cDNA corresponding to the mRNA in the culture grown at the lower temperature; b. each spot on the microarray would have a DNA sequence representing a single *E. coli* gene; c. use microarrays to compare the gene expression changes in cells grown under different osmotic conditions and those that are heat-shocked.

33. a. All of these turn out to be early genes; b. early genes; c. transcription of the large majority of *E. coli* genes would be drastically decreased; d. the *motA* gene prevents transcription of the middle genes, *asiA* should lower the transcription of middle and late T4 genes, the 55 gene should prevent the transcription of late transcripts but have little effect on the transcription of host genes; e. the *reg-A*-encoded ribonuclease is specifically required for the rapid destruction of T4 early mRNAs.

Chapter 18

1. a. 8; b. 5; c. 1; d. 7; e. 3; f. 10; g. 9; h. 6; i. 4; j. 2.

3. a. eukaryotes; b. both prokaryotes and eukaryotes; c. eukaryotes; d. eukaryotes; e. prokaryotes; f. both prokaryotes and eukaryotes.

5. a. pol III, b. pol II; c. pol I; d. pol II.

7. a. DNA binding; b. DNA binding; c. protein dimerization; d. transcription activation; e. DNA binding.

9. Using probes from each of the three genes you would expect to see different signals corresponding to three different transcripts on a Northern blot.

11. If *Id* acts by quenching it interacts with *MyoD,* whereas if it blocks access to an enhancer it binds to DNA. Experimentally, look for binding to the regulatory DNA of a gene regulated by *MyoD.*

13. RNA polymerase transcribes naked eukaryotic DNA at high levels *in vitro;* PEV and heterochromatization shut off gene expression.

15. b.

17. a. 9:7 ratio of blue to white flowered plants; b. 15:1 ratio of blue to white.

19. An enhancer is an expression-promoting DNA element that helps activate transcription from a nearby weak promoter. Function of the enhancer is usually independent of its orientation and its distance from the promoter. Create a construct in which this region is inverted with respect to the *lacZ* coding sequences–this should not affect expression of lacZ. You could add some extra DNA sequences between the minimal promoter and the presumptive enhancer - *lacZ* expression should remain high.

21. The mother.

23. The mutant allele is *a.* a. *Aa;* b. *AA;* c. *AA.*

25. Introns are spliced out, ribonuclease cleaves the primary transcript near the 3′ end and a poly-A tail is added, 5′ methl CAP is added.

27. The 5′ and 3′ untranslated regions could be cloned at the 5′ or 3′ ends of a reporter gene that is transformed back into *Drosophila* early embryos to see if either of the sequences affect the translatability of the reporter protein.

29. The protein in the fat cells may be post-translationally modified (for example, phosphorylated or de-phosphorylated) so that it is only active in fat cells. Alternatively, the protein may need a cofactor to be activated, and this cofactor is only transcribed in fat cells.

31. a. the difference in first detection of the mRNAs probably results from the different sensitivity in detecting mRNA versus protein. The difference in duration of the mRNA vs protein: the proteins are more stable than the mRNAs so they remain in the cells for several days longer; if the normal protein disappears at day 10.5 then the *lacZ* mRNA is more stable; or the β-galactosidase protein is more stable; or the transgene is transcribed until day 12; b. onset.

Chapter 19

1. a. 7; b. 6; c. 8; d. 2; e. 1; f. 9; g. 3; h. 5; i. 4.

3. Replica plate colonies onto two sets of plates—one incubated at 30°C and the other at 23°C. Cold-sensitive mutants defective in cell-cycle genes should die at the nonpermissive temperature (23°C).

5. a. 4%; b. 96%.

7. a. F; b. T; c. F.

9. a. growth factors; b. cyclin, cyclin-dependent protein kinase, any molecule in the signal transduction pathway, receptor or molecule that transmits a second signal.

11. c; e; a; b; d.

13. In the M phase checkpoint molecules made by unattached kinetochores prevent the anaphase promoting complex (APC) from being activated. The APC becomes activated at the beginning of anaphase to destroy M phase cyclin, allowing cells to leave M phase. The activated APC adds ubiquitin to protein substrates. When this happens the ubiquinylated proteins are rapidly destroyed by the proteosome. One simple hypothesis is that cohesin is also targeted by the APC since it must be destroyed at the beginning of anaphase.

15. Uncontrolled growth, genomic instability, potential for immortality, ability to invade and disrupt local and distant tissues.

17. The commonality is an accumulation of mutations necessary to cause cancer.

19. e; b; a or d or c; f.

21. Choice c. will not be associated with cancer.

23. individual II-2 carries the mutant allele; either I-1 or I-2 also carries the mutation.

25. a. F; b. F; c. F; d. T; e. T; f. F; g. T; h. T; i. F; j. F.

27. a. All of these are potential oncogenes; b. If adding a phosphate inhibited the phosphatase then all the actions of KinaseA would ensure that the transcription factor was activated; c. a tumor suppressor; d. homozygous means homozygous for the mutation, heterozygous means one mutant allele and one wild type allele, E = excessive cell growth and D = decreased cell growth.

Mutation	homozygous	heterozygous
i.	E	
ii.	D	
iii.	D	
iv.	D	
v.	E	E
vi.	E	E
vii.	E	E
viii.	D	D
ix.	E	

29. a. tumor suppressor; b. loss-of-function mutation; c. RAS-GDP; d. ii, iv, v and vi; e. inherited a loss-of-function NF1 allele *and* another mutant gene on the same chromosome; f. there must be a clone of cells in which one of the copies of NF1 is inactivated by some rare event then the other allele of NF1 is inactivated or lost.

Chapter 20

1. a. 11; b. 4; c. 8; d. 7; e. 12; f. 9; g. 13; h. 3; i. 1; j. 5; k. 6; l. 10; m. 2.

3. a. In *C. elegans,* laser ablation at this early stage of development would almost certainly be lethal, while in mice the loss of one out of four early embryonic cells would have no effect; b. lethal to *C. elegans* and it is possible that the separated cells could develop into a mouse; c. in C. *elegans* would likely be lethal, in mice, such a fusion would be tolerated giving rise to a chimeric animal.

5. Make RNA preparations from homozygotes for the new null allele and then analyzing these preparations on Northern blots; RT-PCR or the mutation could be null but the gene would still be transcribed. Analyze protein extracts from the homozygous mutant animals by Western blot using the antibody against the rugose protein as a probe.

7. a.

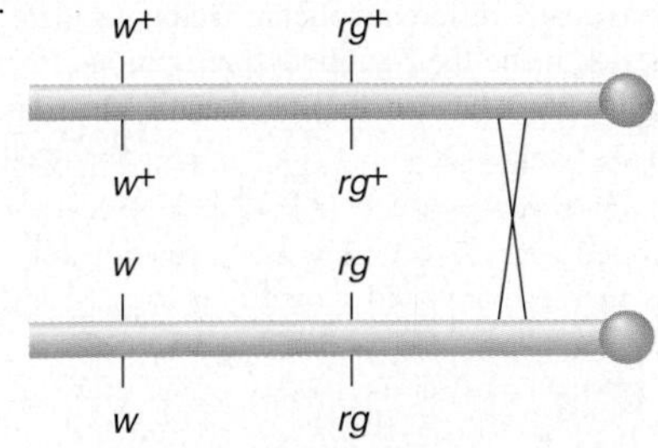

b. As a result of the mitotic crossover developing ommatidia in the eye would be simultaneously homozygous for the mutations in *rugose* and *white,* while adjacent ommatidia would be heterozygous for the wild type and mutant alleles of both genes. If the red ommatidia are abnormal even though their genotype predicts a normal structure, then the lack of rugose in the adjacent white ommatidia affects the red ommatidia; c. If these patches were normal in appearance, then rugose does not have an important role in eye development. If the white patch is abnormal, then rugose is important for eye development.

9. Mutate possible regulatory DNA elements.

11. Make DNA constructs that place a wild type genomic copy of gene *X* adjacent to *myo-2::GFP.* You then transform these constructs into worms that are homozygous for a null allele of gene *X* (and that did not contain any *GFP* source). The constructs form extrachromosomal arrays as described. Pharyngeal cells containing the arrays would be wild type for gene *X* and express GFP. Pharygeal cells that had lost the arrays would be homozygous mutant for gene *X* and would not express GFP.

13. a. If the mutation was due to an insertion of the transgene the MMTV *c-myc* gene should segregate with the phenotype; b. Clones containing the *c-myc* fusion could be identified by hybridization of MMTV sequences versus a library of genomic clones produced from the cells of the mutant mouse; c. The sequence of the gene into which the MMTV *c-myc* fusion inserted could be analyzed in the *ld* mutant to determine if there were mutations in the gene.

15. a. promoter, binding sites for transcription factors such as Bicoid and binding sites for other transcription factors that ensure the *hb* gene is transcribed in the proper cells in the mother; b. the amino acids in Hunchback that comprise DNA binding domains and domains involved in the transcriptional regulation of gap and pair rule genes; c. translational repression carried out by Nanos protein.

17. The cytoplasm from the anterior of a wild type embryo could be injected into the anterior end of a *bicoid* mutant embryo to see if there was rescue of the mutant phenotype. Alternately, purified *bicoid* mRNA injected into the anterior end of a *bicoid* mutant embryo would be a more definitive experiment. Finally, purified *bicoid* mRNA could be injected into the posterior end of a wild-type embryo.

19. a wild-type Knirps protein is needed to restrict the posterior limit of the zone of Kruppel expression; b. Hunchback protein would be seen throughout the embryo.

21. A mutation in the genes encoding a maternally supplied component which affects early development must be in the mother's genome. If the mutation affecting early development is in a gene whose transcription begins after fertilization then the mutation must be in the genome of the zygote (these are thus sometimes called "zygotic genes"). You would need two different kinds of genetic screens to make mutations either in the mothers' genome or the zygotes' genome.

23. a. the presence of PAR-2 and absence of PAR-3 from these cells indirectly dictates their ability to translate *glp-l* mRNA into GLP-1 protein; b. Such an interaction could occur through the extracellular domains of both proteins; c. receptor is the GLP-1 protein. Thus APX-1 would be the ligand; d. (i) the ablation of P_2 would make ABp and its descendants would have the same fate shown by ABa and its descendants, (ii) a null mutation of *apx-1* would have the same effect (iii) same, (iv) same.

Chapter 21

1. a. 3; b. 5; c. 8; d. 7; e. 6; f. 1; g. 9; h. 2; i. 4.

3. a, e.

5. a. the initial population is not in equilibrium; b. genotype frequencies in the F_1 will be 0.36 *MM* + 0.48 *MN* + 0.16 *NN* = 1, allele frequencies in the F_1 generation *M* = 0.6 and *N* = 0.4; c. the same as in part b.

7. Each allele frequency has a different set of genotype frequencies at equilibrium.

9. a. N = 0.1; b. 478 MN children on the island; c. N = 0.525.

11. a. $p^2 + 2pq + q^2 + 2pr + r^2 + 2qr = 1$; b. 0.516 A, 0.122 B, 0.075 AB and 0.287 O.

13. a. $C = 8324/9049 = 0.92$, $c = 725/9049 = 0.08$; b. this sample does not demonstrate Hardy-Weinberg equilibrium; c. the frequency of cP = 0.018, the frequency of c^d = 0.064, frequency of the C allele = 0.918; d. in boys C = 0.918 (normal vision), c^d = 0.064 (colorblind) and cP = 0.018 (colorblind). In the girls the genotype frequencies are: CC = 0.843 (normal vision), Cc^d = 0.118 (normal vision), CcP = 0.033 (normal vision), $cPcP = 3.3 \times 10\text{-}4$ (colorblind), c^dc^d = 0.004 (colorblind) and c^dcP = 0.002 (normal vision); e. the population is in equilibrium. As seen in part c, the allele frequency of C is the same in boys and girls and the allele frequency of c in the boys is the same as the total frequencies of $c^d + cP$ in girls.

15. a. the genotype frequencies in the F_2 are 0.33 vg^+ vg^+ and 0.67 vg^+ vg; the allele frequencies in the F_2 for vg^+ = 0.33 + 1/2 (0.67) = 0.67 and for vg = 1/2 (0.67) = 0.33; b. genotype frequencies in the F_3 progeny are 0.449 vg^+ vg^+ + 0.442 vg^+ vg + 0.109 vg vg = 1, or 0.891 wild type and 0.109 vestigial; c. F_4 allele frequencies are vg^+ = 0.743 and vg = 0.249; d. If all of the F_4 flies are allowed to mate at random then there is no selection and the population will be in Hardy-Weinberg equilibrium −0.566 vg^+ vg^+ + 0.373 vg^+ vg + 0.062 vg vg = 1; vg^+ = 0.743 and vg = 0.249.

17. Selection against the homozygous recessive genotype will decrease the frequency of the recessive allele in the population, but it will never totally remove it, as the recessive allele is hidden in the heterozygote, recessive allele sometimes confers an advantage when present in the heterozygote, mutation can produce new recessive alleles in the population.

19. a. b = √0.25 = 0.5, B = 0.5; b. Δq for tank 1 = −0.1, q for all tanks = 0.5; Δq for tank 2 = 0; Δq for tank 3 = 0.05.

	Tank 1	Tank 2	Tank 3
b. Δq	−0.1	0.0	0.05
c. w_{Bb}	1.0	1.0	1.0
d. w_{bb}	<1.0	1.0	>1.0

21. a. fitness value (w) = 0 and the selection coefficient (s) = 1 for the affected genotype. There is no selection pressure against the carrier or the homozygous normal genotypes, so for both of these w = 1 and s = 0; b. $\Delta q = -1.54 \times 10^{-3}$; c. 1.02×10^{-3}. This number (1.02×10^{-3}) is smaller than the observed q which is 0.04; d. CF^+/CF^- heterozygotes may be better able to survive outbreaks of cholera.

23. b.

25. a. Height has the highest heritability and weight has the lowest heritability; b. The data from the CDC is roughly in line with the conclusions from part a.

27. a. founder effect; b. advantages: genetic homogeneity and fewer genes that may affect a polygenic trait; disadvantages: some mutations are not found in the population that are in the general population.

29. a. 2n + 1 where n = number of genes; b. $(1/4)^n$ = 1/256, so n = 4.

Chapter 22

1. a. 4; b. 6; c. 5; d. 2; e. 1; f. 7; g. 3.

3. a, c.

5. a. The enzyme consists of an RNA molecule; b. the enzyme has both an RNA and a protein component.

7. a. Different constraints on the functions of each of the proteins; b. rates are more constant because these base changes do not affect function of the gene product.

9. Suggests there is some benefit to the CF allele in the heterozygous state.

11. Duplication followed by evolutionary divergence.

13. a. 240 million years; b. two; C allele arose 30 million years ago; B allele arose 1 million years ago; c. duplication of B: transposition; duplication of C: misalignment and crossing-over.

15. This gene was introduced from a different species.

17. a. Exons; b. genes.

19. They mediate genome rearrangements or contribute regulatory elements adjacent to a gene.

21. a. SINEs or LINEs; b. centromere satellite DNA.

23. a. Two genes– one for the heavy chain and one for the light chain; b. one kind due to allelic exclusion; c. B lymphocyte;

d.

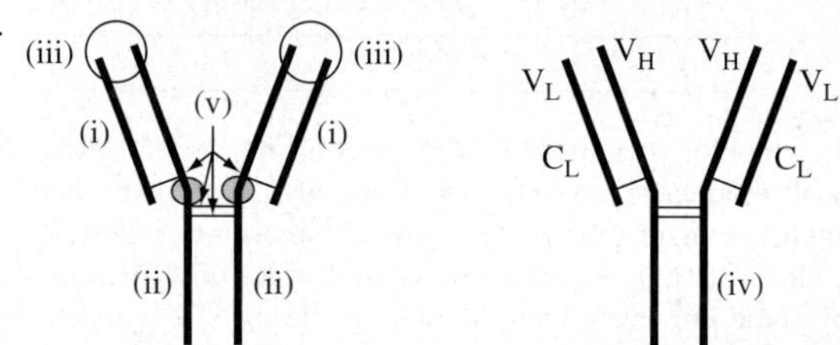

e. cleaves at the hinge, the hatched circles in the figure above; f. digestion with papain gives you single antigen binding sites; g. the constant regions (CH) will change;

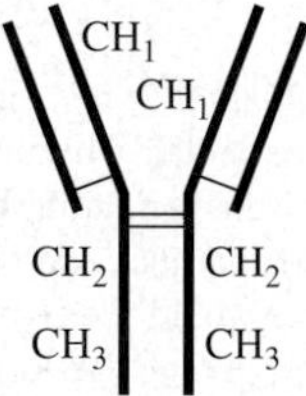

h. new rearrangement in the heavy chain gene such that the VDJ that are joined together switch from in front of a Cμ chain (IgM) to a Cγ (IgG); i. B cells with higher affinity antigen receptors are selected to produce larger clones of antibody secreting plasma cells; j. (iii); k. DNA recombination results in a B cell that only makes one kind of antibody because the DNA rearrangement is permanent or you would need as many different splicing factors as there are antibody specificities; l. made their antibody preparations from the blood of a patient with a cancer of the B cells, which is called a myeloma.

Glossary

Note: An italicized word in a definition indicates that word is defined elsewhere in the glossary.

A

acentric fragment a *chromatid* fragment lacking a *centromere;* usually the result of a crossover in an *inversion loop.*

acrocentric a *chromosome* in which the *centromere* is close to one end.

activator a type of *transcription factor* that can bind to specific cis-*acting* enhancer elements and increase the level of transcription from a nearby gene.

adaptation the ability to stop responding when the stimulus is present.

adenine (A) a nitrogenous base; one member of the *base pair* A–T (adenine–thymine).

adjacent-1 segregation pattern one of two patterns of *segregation* resulting from the normal disjunction of homologs during *meiosis I.* Homologous *centromeres* disjoin so that one *translocation* chromosome and one normal chromosome go to each pole. Contrast with *alternate segregation pattern.*

adjacent-2 segregation pattern a pattern of *segregation* resulting from a *nondisjunction* in which homologous centromeres go to the same pole during meiosis I.

allele frequency the proportion of all copies of a *gene* in a *population* that are of a given allele type.

alleles alternative forms of a single *gene.*

allele-specific oligonucleotides (ASOs) short *oligonucleotides* that hybridize with alleles distinguished by a single base difference.

allopolyploids *polyploid* hybrids in which the chromosome sets come from two or more distinct, though related, species.

allosteric proteins proteins that undergo reversible changes in conformation when bound to another molecule.

α-globin locus chromosomal region carrying all of the α-globin-like genes.

alternate segregation pattern one of two patterns of *segregation* resulting from the normal disjunction of homologs during *meiosis I.* Two *translocation* chromosomes go to one pole, while two normal chromosomes move to the opposite pole, resulting in *gametes* with the correct haploid number of genes.

alternative splicing production of different mature *mRNAs* from the same *primary transcript* by joining different combinations of *exons.*

amino acids the building blocks of proteins.

aminoacyl (A) site site on a *ribosome* to which a *charged tRNA* first binds.

aminoacyl-tRNA synthetases enzymes that catalyze the attachment of tRNAs to their corresponding amino acids, forming *charged tRNAs.*

amniocentesis medical procedure in which a sample of amniotic fluid is taken from a pregnant woman to determine the condition of an unborn baby. A hollow needle is inserted through the woman's abdomen and uterine wall, and the fluid, which contains cells shed by the embryo, is drawn off.

amorphic mutations changes that completely block the function of a *gene* (synonym for *null mutations*).

amphidiploids organisms produced by two *diploid* parental species; they contain two diploid genomes, each one derived from a different parent.

amplification spread of a molecular variant throughout a population over many generations.

anaphase the stage of *mitosis* in which the connection of *sister chromatids* is severed, allowing the chromatids to be pulled to opposite spindle poles.

anaphase I phase of *meiosis I* during which the *chiasmata* joining *homologous chromosomes* dissolve, allowing maternal and paternal homologs to move toward opposite spindle poles; the *centromeres* do not divide so that the *chromosomes* moving toward the poles each consist of two *chromatids.*

anaphase II phase of *meiosis II* when the dismantling of *cohesin* complexes allows *sister chromatids* to move to opposite spindle poles.

aneuploid an individual whose chromosome number is not an exact multiple of the *haploid* number for the species.

angiosperm plant descriptive meaning the seeds of the plant are enclosed in an ovary within the flower.

angstrom (Å) a unit of length equal to one ten-billionth of a meter.

anonymous locus a designated position on a chromosome with no known function; see *locus.*

antennapedia complex (ANT-C) in *Drosophila,* a region containing several *homeotic genes* specifying the identity of segments in the head and anterior thorax.

anticodons groups of three *nucleotides* on *transfer RNA (tRNA)* molecules that recognize codons on the mRNA by *complementary base pairing* and *wobble.*

antigen foreign substance (or a particular part of a foreign substance) that induces an *immune response* when introduced into the body.

antisense the sequence of a single-stranded RNA or DNA molecule that is complementary to—and can base-pair with—a portion of a transcribed RNA molecule.

apical meristem a group of undifferentiated plant cells that divide continuously, located at the growing points of both shoots and roots.

apoptosis programmed cell death; a process in which DNA is degraded and the nucleus condenses; the cell may then be devoured by neighboring cells or phagocytes.

archaea one of the three major evolutionary lineages (domains) of living organisms.

artificial selection the purposeful control of mating by choice of parents for the next generation. Contrast with *natural selection.*

artificial transformation a process to transfer genes from one bacterial strain to another, using laboratory procedures to weaken cell walls and make membranes permeable to DNA. Contrast with *natural transformation.*

ascospores in some fungi, the *haploid* cells that result from *meiosis*. Also known as *haplospores.*

ascus saclike structure in some varieties of fungi that houses all four *haploid* products of *meiosis*.

astral microtubules short, unstable microtubules that extend out from a *centrosome* toward the cell's periphery to stabilize the *mitotic spindle*.

attenuation a type of gene regulation in which *transcription* of a gene terminates in the regulatory region before a complete mRNA *transcript* is made.

autocrine stimulation process by which many tumor cells make their own signals to divide, rather than waiting for signals from neighboring cells.

autonomous elements intact transposable elements that can move from place to place in the genome by themselves. Contrast with *nonautonomous elements.*

autopolyploids a kind of *polypoid* that derives all of its chromosome sets from the same species.

autoradiograph an X-ray film image that records the location of radioactively labeled DNA.

autoradiography process of creating an *autoradiograph*.

autosome a *chromosome* not involved in sex determination. The *diploid* human genome consists of 46 chromosomes, 22 pairs of autosomes, and 1 pair of sex chromosomes (the X and Y chromosomes).

autotroph a plant that is nutritionally self-sufficient; it produces its own food via photosynthesis.

auxotroph a mutant microorganism that can grow on minimal medium only if it has been supplemented with one or more growth factors not required by wild-type strains.

B

bacterial chemotaxis bacterial movement up and down gradients of chemical attractants or repellents.

bacterial chromosome essential component of a bacterial *genome;* usually a single circular molecule of double-helical DNA.

bacteriophage See *phage.*

balancer chromosome special *chromosome* created for use in genetic manipulations; helps maintain *recessive lethal* mutations in stocks.

Barr bodies inactive X chromosomes observable at interphase as darkly stained *heterochromatin* masses.

basal factor a type of *transcription factor* that can associate nonspecifically with all *promoters* in a *genome*. A complex of basal factors associated with a promoter is required for the initiation of *transcription* by *RNA polymerase*. See *basal transcription apparatus.*

basal transcription apparatus complex protein machine that interacts with complexes of *transcription factors* and cis-*control elements* to mediate the synthesis of an RNA *transcript*. This RNA transcript is spliced and edited to produce a *messenger RNA* (*mRNA*).

base analogs *mutagens* that are so similar in chemical structure to the normal nitrogenous bases in DNA that the replication mechanism can incorporate them into DNA in place of the normal bases. This can cause base *substitutions* on the complementary strand synthesized in the next round of DNA replication.

base excision repair homology-dependent mechanism in which specific enzymes cleave an altered base from the sugar of its *nucleotide* to create an apurinic or apyrimidinic (AP) site in the DNA chain; nick the DNA backbone at the AP site; and remove nucleotides from the vicinity of the nick. DNA polymerase fills in this gap by copying the undamaged strand, restoring the original DNA sequence. See *nucleotide excision repair*.

base pair (bp) two nitrogenous bases on complementary DNA strands held together by *hydrogen bonds. Adenine* pairs with *thymine,* and *guanine* pairs with *cytosine*.

base sequence the order of nucleotide bases in a DNA or RNA strand.

base sequence analysis a method for determining a *base sequence.*

basic chromosome number (*x*) the number of different chromosomes that make up a single complete set. See *x*.

β-globin locus chromosomal region carrying all the β-globin-like genes.

B-form DNA the most common form of DNA in which molecular configuration spirals to the right. Compare with *Z-form DNA.*

biased random walk bacterial movement resulting from the addition of an attractant or repellant. The time spent in a straight run is longer immediately after the addition, so the movement over time is biased toward or away from the chemical gradient, though each direction change continues to be random.

bidirectional describes a mechanism of *DNA replication* in which two *replication forks* move in opposite directions away from the same *origin of replication.*

biochemical pathway an orderly series of reactions that allows an organism to obtain simple molecules from the environment and convert them step by step into successively more complicated molecules.

biological context the context in which a *biological system* operates within a cell or organism.

biological information consists of the digital information of the *genome* and environmental signals that activate or modulate the output of genomic information. These two types of information interact to mediate biological activity across the three time scales of *evolution,* development, and physiological responses.

biological system collection of interacting elements that carry out a specific biological task within a cell or among groups of cells within an organism.

biotechnology a set of biological techniques developed through basic research and now applied to research and product development. In particular, the use by industry of recombinant DNA, cell fusion, and new bioprocessing techniques.

biparental inheritance inheritance of organelles from both parents. Occurs in single-celled yeast and some plants.

bithorax complex (BX-C) in *Drosophila,* a cluster of *homeotic genes* that control the identity of segments in the abdomen and posterior thorax.

bivalent a pair of synapsed homologous chromosomes during *prophase* of *meiosis I*.

blastocysts describes the embryo at the 16-cell stage of development through the 64-cell stage, when the embryo implants.

blastomeres early embryonic cells.

blunt end the 5′ or 3′ end of a double-stranded DNA molecule without *sticky ends.*

branched-line diagram a method for systematically listing the expected results of multigene crosses.

branches (of a *phylogenetic tree*) lines that define the relationship between the taxonomic units represented by the *nodes* of the tree; their length represents an estimate of the time that has elapsed based on the number of molecular changes that have occurred.

branch sites special sequence of RNA nucleotides within an *intron* that helps form the "lariat" intermediate required for *RNA splicing.*

C

cancer genes *genes* having mutant *alleles* that lead to cancer.

carpels structures in a plant comprising the fourth whorl; usually two are fused together; house the female gametes in the form of ovules. The fused carpels are part of a cylinder known as the *pistil.*

carriers *heterozygous* individuals of normal *phenotype* that have a *recessive allele* for a trait.

catabolite repression repression of expression in sugar-metabolizing *operons* like the *lac* operon when glucose or another preferred catabolite is present.

CDK inhibitors inhibit the activity of *cyclin-dependent kinases.*

cDNA complementary DNA has a base sequence that is complementary to that of the mRNA template and contains no *introns.*

cDNA library a large collection of cDNA clones that are representative of the *mRNAs* expressed by a particular cell type, tissue, organ, or organism.

cell autonomous trait a trait for which the *phenotype* expressed by a cell depends solely on the *genotype* of that cell and not that of any other nearby cell. Compare with *nonautonomous trait.*

cell cycle repeating pattern of cell growth, replication of genetic material, and *mitosis.*

cell plate membrane-enclosed disk that forms inside a plant cell near the equator of the *mitotic spindle* and grows rapidly outward to divide the cell in two during *cytokinesis.*

cellular blastoderm in *Drosophila* embryos, the one-cell-deep epithelial layer resulting from cellularization of the *syncytial blastoderm.*

cellular clone an isolated colony of cells that are all descendants from a single progenitor cell.

centimorgan (cM) a unit of measure of *recombination frequency*. One cM is equal to a 1% chance that a marker at one genetic *locus* will be separated from a marker at a second locus due to *crossing-over* in a single generation.

centrioles short cylindrical structures that help organize microtubules. Two centrioles at right angles to each other form the core of a *centrosome*. Each centrosome serves as a pole of the *mitotic spindle*.

centromere a specialized chromosome region at which *sister chromatids* are connected and to which spindle fibers attach during cell division.

centrosomes microtubule organizing centers at the poles of the *spindle apparatus.*

charged tRNA a tRNA molecule to which the corresponding *amino acid* has been attached by an *aminoacyl-tRNA synthetase.*

checkpoints mechanisms that prevent cells from continuing to the next phase of the *cell cycle* until a previous stop has been successfully completed, thus safeguarding genomic integrity.

chiasmata observable regions in which nonsister chromatids of homologous chromosomes cross over.

chimera an embryo or animal composed of cells from two or more different organisms.

chi-square (χ^2) test a statistical test to determine the probability that an observed deviation from an expected outcome occurs solely by chance.

chromatid one of two copies of a *chromosome* that exist immediately after *DNA replication*. See *sister chromatids.*

chromatin the generic term for any complex of *DNA* and *protein* found in a cell's nucleus.

chromocenter the dense heterochromatic mass formed by the fusing of the *centromeres* of *polytene chromosomes.*

chromosomal interference the phenomenon of crossovers not occurring independently. Refer to *crossing-over.*

chromosomal puff the region of a *polytene chromosome* that swells to form a large, diffuse structure when high rates of gene *transcription* cause bands to decondense.

chromosomal rearrangement change in the order of *DNA sequence* along one or more *chromosomes.*

chromosome loss a mechanism causing *aneuploidy* in which a particular *chromatid* or *chromosome* fails to become incorporated into either daughter cell during cell division.

chromosome theory of inheritance the idea that *chromosomes* are the carriers of *genes.*

chromosomes the self-replicating genetic structures of cells containing the *DNA* that carries in its *nucleotide* sequence the linear array of *genes.*

chromosome walking using the ends of unconnected *contigs* as probes to retrieve *clones* that extend into an unmapped region.

***cis*-control elements** (sometimes referred to as *cis-acting elements*) short DNA sequences (6-15 base pairs long) that constitute the control elements adjacent to *genes*. Through their binding to *transcription factors, cis*-control elements control or modulate transcription initiation at one or more nearby *genes. Promoters, enhancers,* and *locus control regions* are three types of *cis*-control elements.

cistron a term sometimes used as a synonym for *complementation group* or *gene.*

cleavage stage in early embryonic development, the stage of the first four equal cell divisions.

clone a group of biological entities—cells or DNA molecules—that are genetically identical to each other.

cloning the process by which cellular clones or DNA clones are formed.

cloning vector DNA molecule into which another DNA fragment of appropriate size can be integrated without loss of the vector's capacity for replication. Vectors introduce foreign DNA into host cells, where it can be reproduced in large quantities. See *vector.*

coactivator a type of *transcription factor* or other molecule that binds to *transcription factors* rather than to *DNA* and plays a role in increasing levels of *transcription.*

codominant expression of *heterozygous* genotype resulting in hybrid offspring that resemble both parents equally for a particular trait.

codon nucleotide triplet that represents a particular *amino acid* to be inserted in a specific position in the growing amino acid chain during *translation*. Codons can be either in the *mRNA* or in the *DNA* from which the RNA is transcribed.

coefficient of coincidence the ratio between the actual frequency of double crossovers observed in an experiment and the number of double crossovers expected on the basis of independent probabilities.

coenocyte a plant cell containing multiple nuclei.

cohesin a multisubunit protein complex that associates with *sister chromatids* in eukaryotic cells and holds the chromatids together until *anaphase;* can be found at both the *centromere* and along the *chromosome* arms.

cohesive ends short, single-stranded unpaired flaps protruding from the ends of a cut DNA molecule. Each flap can reform *hydrogen bonds* with a complementary sequence protruding from the end of another piece of DNA.

colinearity the parallel between the sequence of *nucleotides* in a *gene* and the order of *amino acids* in a *polypeptide*.

colony a mound of genetically identical cells.

competent description of state of cells able to take up DNA from the medium.

complement (used as a verb) when a *heterozygote* for two *recessive mutations* displays a normal *phenotype* because the dominant *wild-type alleles* on each of the two homologs make up for a defect in the other *homologous chromosome*. Mutations that complement are usually in different genes, whereas mutations that fail to complement are usually in the same gene.

complementation process in which heterozygosity for *chromosomes* bearing mutant *recessive alleles* for two different *genes* produces a normal *phenotype*.

complementary base pairing during *DNA replication,* base pairing in which a complementary strand aligns opposite the exposed bases on the parent strand to create the nucleotide sequence of the new strand of DNA. Refer to *base pair, complementary sequences.*

complementary gene action genes working in tandem to produce a particular trait.

complementary sequences nucleic acid base sequences that can form a double-stranded structure by matching base pairs; the complementary sequence to 5′ GTAC 3′ is 3′ CATG 5′.

complementation group a collection of *mutations* that do not complement each other. Often used synonymously for *gene*.

complementation table a method of collating data that helps visualize the relationship among a large group of mutants. See Fig. 7.18b.

complementation test method of discovering whether two mutations are in the same or separate genes.

complete coverage in mapping of DNA, when the number of *markers* on a *linkage map* is sufficient so that the *locus* controlling any *phenotype* can be linked to at least one of those markers.

complete digest digesting a sample of DNA molecules with a *restriction enzyme* such that cleavage has occurred at every DNA site recognizable by the enzyme.

complete genomic library a hypothetical collection of *DNA clones* that includes one copy of every sequence in the entire *genome*.

complex refers to the multiple types of variation that can exist at alternative *alleles,* including more than one nucleotide *substitution,* a substitution in combination with a small *deletion,* a *duplication,* or another *insertion*.

complex haplotype a set of linked DNA variations along a *chromosome,* with the possibility of many differences in alternative *alleles*. See *haplotype*.

concerted evolution process that allows changes in single genetic elements to spread across a complete set of *genes* in a particular *gene family*.

condensation cellular process of *chromatin* compaction that results in the visible emergence of individual *chromosomes*.

condensin a multisubunit complex of proteins in eukaryotic cells that compacts *chromosomes* during *mitosis*.

conditional lethal an *allele* that is lethal only under certain conditions.

conjugation one of the mechanisms by which bacteria transfer genes from one strain to another; in this case, the *donor* carries a special type of *plasmid* that allows it to transfer DNA directly when it comes in contact with the recipient. The recipient is known as an exconjugant. Contrast *transformation*.

conjugative plasmids *plasmids* that initiate *conjugation* because they carry the genes that allow the *donor* to transfer genes to the *recipient*.

consanguine related by a common ancestor.

consanguineous mating mating between blood relatives sharing a recent common ancestor.

conservative substitutions mutational changes that substitute an amino acid in a protein with a different amino acid with similar chemical properties. Compare with *nonconservative substitutions.*

conserved synteny state in which the same two or more *loci* are found to be linked in several species. Compare *syntenic, syntenic segments.*

constitutive expression refers to a state of gene activation that remains at a constant high level and is not subject to modulation.

constitutive heterochromatin chromosomal regions that remain condensed in *heterochromatin* at most times in all cells.

constitutive mutants synthesize certain enzymes all the time, irrespective of environmental conditions.

constitutive mutation a mutation in a *cis*-acting or *trans*-acting element that causes an associated gene to remain in a state of activation irrespective of environmental or cellular conditions that modulate gene activity in nonmutant cells.

contig a set of two or more partially overlapping cloned DNA fragments that together cover an uninterrupted stretch of the *genome.*

continuous trait inherited trait that exhibits many intermediate forms; determined by segregating alleles of many different *genes* whose interaction with each other and the environment produces the *phenotype*. Also called a quantitative trait. Compare with *discontinuous trait.*

contractile ring transitory organelle composed of actin microfilaments aligned around the circumference of a dividing animal cell's equator; contraction of the filaments pinches the cell in two.

core histones proteins that form the core of the *nucleosome:* H2A, H2B, H3, and H4.

corepressor a type of *transcription factor* or other molecule that binds to transcription factors rather than DNA and prevents transcription above basal levels.

cosmids hybrid *plasmid-phage vectors* that make use of a virus capsule to infect bacteria; constructed with plasmid-derived selectable markers and two specialized segments of phage λ DNA known as *cos* (for *co*he*s*ive end) sites.

cotransformation simultaneous transformation of two or more genes. See *transformation.*

crisscross inheritance inheritance pattern in which males inherit a trait from their mothers, while daughters inherit the trait from their fathers.

cross the deliberate mating of selected parents based on particular genetic traits desired in the offspring.

cross-disciplinary biology a type of biology in which teams of biologists, computer scientists, chemists, engineers, mathematicians, physicists, and others work together on the problems of *systems biology*.

cross-fertilization brushing the pollen from one plant onto the female organ of another plant.

cross-fostering random relocation of offspring to the care of other parents, typically done with animal studies to randomize the effects of environment on outcome.

crossing-over during *meiosis,* the breaking of one maternal and one paternal *chromosome,* resulting in the exchange of corresponding sections of DNA and the rejoining of the chromosome. This process can result in the exchange of *alleles* between chromosomes. Compare *recombination.*

crossover suppression result of heterozygosity for *inversions,* in which no viable recombinant progeny are possible.

C terminus the end of the *polypeptide* chain that contains a free carboxylic acid group.

cyclin-dependent kinases (CDKs) a protein kinase is an enzyme that adds a phosphate group to a target protein molecule; in this case, the kinases are dependent on proteins known as cyclins for the targeting of their activity to a specific substrate. CDKs regulate the transition from G_1 to S and from G_2 to M through phosphorylations that activate or inactivate target proteins.

cyclins family of proteins that combine with *cyclin-dependent kinases* and thereby determine the substrate specificity of the kinases. By directing kinases to the right substrates, the cyclins help regulate passage of the cell through the *cell cycle.* Concentrations of the various cyclins rise and fall throughout the cell cycle.

cytokinesis the final stage of cell division, which begins during *anaphase* but is not completed until after *telophase*. In this stage, the daughter nuclei emerging at the end of telophase are packaged into two separate daughter cells.

cytosine(C) a nitrogenous base; one member of the base pair G–C (guanine–cytosine).

D

dauer larva in nematodes, an alternate L3 form that does not feed and has a specialized cuticle that resists desiccation; dauer larvae can survive more than six months when food is scarce.

deamination the removal of an amino ($-NH_2$) group from normal DNA.

degeneracy property of the *genetic code* in which several different *codons* can specify the same amino acid.

degrees of freedom (df) the measure of the number of independently varying parameters in an experiment.

deletion occurs when a block of one or more nucleotide pairs is lost from a DNA molecule. Compare *insertion.*

deletion loop an unpaired bulge of the normal *chromosome* that corresponds to an area deleted from a paired homolog. Contrast *duplication loop.*

denaturation (denature, denatured) the disruption of hydrogen bonds within a macromolecule that normally uses hydrogen bonds to maintain its structure and function. Hydrogen bonds can be disrupted by heat, extreme conditions of pH, or exposure to chemicals such as urea. When normally soluble proteins are denatured, they unfold and expose their nonpolar amino acids, which can cause them to become insoluble. When DNA is denatured, double-stranded molecules break apart into two separate strands.

deoxyribonucleic acid (DNA) See *DNA.*

deoxyribonucleotide See *nucleotide.*

depurination DNA alteration in which the hydrolysis of a purine base, either A or G, from the deoxyribose-phosphate backbone occurs.

developmental geneticists biologists who use *genetics* to study how the fertilized egg of a multicelular organism becomes an adult.

developmental genetics the use of *genetics* to study how the fertilized egg of a multicellular organism becomes an adult.

developmental hierarchy a *developmental pathway* in which the product of one *gene* regulates the expression of another gene.

developmental pathway a detailed description of how many *genes* interact biochemically to produce a particular outcome in development.

diakinesis substage of *prophase I* during which *chromosomes* condense to the point where each *tetrad* consists of four separate *chromatids;* at the end of this substage, the nuclear envelope breaks down and the microtubules of the *spindle apparatus* begin to form.

dicentric chromatid a *chromatid* with two *centromeres.*

Dicer a cytoplasmic enzyme component of the *RNAi* machinery present in all eukaryotic cells. Dicer recognizes double-stranded RNA duplexes and trims off both ends to create duplex products 21–24 base pairs long that contain a *miRNA* strand and a complementary miRNA* strand. Dicer passes the miRNA:miRNA* to *RISC.*

dicotyledonous plant descriptive meaning that the mature embryo carries two leaves.

dideoxynucleotide nucleotide analogue lacking the 3′-hydroxyl group that is critical for the formation of *phosphodiester bonds*. The four types of dideoxynucleotides are abbreviated ddTTP, ddATP, ddGTP, and ddCTP. Dideoxynucleotides are key components of the most common methods of DNA sequencing.

digestion the enzymatic process by which a complex biological molecule (DNA, RNA, protein, or complex carbohydrate) is broken down into smaller components.

dihybrid an individual that is *heterozygous* for two *genes* at the same time.

dimerization domain region of a *polypeptide* that facilitates interactions with other molecules of the same polypeptide or with other polypeptides. Certain motifs such as the leucine zipper often serve as dimerization domains.

dipeptide two *amino acids* connected by a *peptide bond.*

diploid *zygotes* and other cells carrying two matching sets of chromosomes are described as diploid (compare *haploid*).

diplotene substage of *prophase I* during which there is a slight separation of regions of *homologous chromosomes* but the aligned homologous chromosomes of each *bivalent* remain tightly merged at *chiasmata.*

discontinuous traits *phenotypes* that are expressed in clear-cut variations. Compare with *continuous trait.*

discovery science an approach to biology in which one seeks to identify all the elements of a *biological system* and place them in a database to enrich the infrastructure of biology.

discrete trait inherited trait that clearly exhibits an either/or status (that is, purple versus white flowers).

disease stratification the process of dividing a single general condition into different diseases on the basis of the underlying molecular defects.

diversification evolution of many variants from one progenitor molecule.

DNA *d*eoxyribo*n*ucleic *a*cid; the molecule of heredity that encodes genetic information.

DNA clone a purified sample containing a large number of identical *DNA* molecules.

DNA fingerprint the multilocus pattern produced by the detection of *genotype* at a group of unlinked, highly *polymorphic* loci.

DNA marker See *marker.*

DNA polymerase III complex enzyme that forms a new DNA strand during *replication* by adding *nucleotides,* one after the other, to the 3′ end of a growing strand.

DNA polymorphisms two or more *alleles* at a *locus* detected with any method that directly distinguishes differences in DNA sequence. The sequence variations of a DNA polymorphism can occur at any position on a chromosome and may, or may not, have an effect on *phenotype*. See *polymorphisms.*

DNA probe a purified fragment of DNA labeled with a radioactive isotope or fluorescent dye and used to identify complementary sequences by means of *hybridization.*

DNase hypersensitive (DH) sites sites on DNA that contain few, if any, *nucleosomes;* these sites are susceptible to cleavage by DNase enzymes.

DNA replication: process by which a double helical DNA molecule is duplicated into two identical double helical DNA molecules.

DNA sequence the relative order of *base pairs,* whether in a fragment of DNA, a *gene,* a *chromosome,* or an entire *genome.*

DNA topoisomerases a group of enzymes that help relax *supercoiling* of the DNA helix by nicking one or both strands to allow the strands to rotate relative to each other.

domain a discrete region of a *protein* with its own function. The combination of domains in a single protein determines its overall function.

domain architecture the number and order of a protein's functional domains.

dominance series dominance relations of all possible pairs of *alleles* are arranged in order from most dominant to most recessive.

dominant allele an *allele* whose *phenotype* is expressed in a *heterozygote.* See *recessive allele.*

dominant epistasis the effects of a *dominant allele* at one *gene* hide the effects of alleles at another gene. Compare with *recessive epistasis.*

dominant negative mechanism of dominance in which some alleles of genes encode subunits of *multimeric proteins* that block the activity of the subunits produced by *wild-type alleles.*

dominant-negative (or antimorphic) alleles (or mutations) *alleles* that block the activity of *wild-type alleles* of the same gene, causing a loss of function even in *heterozygotes.*

dominant trait the trait that appears in the F_1 hybrids (*heterozygotes*) resulting from a mating between pure-breeding parental strains showing antagonistic *phenotypes.*

donor in gene transfer in bacteria, the cell that provides the genetic material. See *recipient, transformation, conjugation, transduction.*

dosage compensation mechanism that equalizes levels of *X-linked* gene expression independent of the number of copies of the X chromosome; in mammals, the dosage compensation mechanism is *X chromosome inactivation.*

double helix the shape that two linear strands of DNA assume when bonded together.

downstream the direction traveled by RNA polymerase as it moves from the promoter to the terminator. Compare with *upstream.*

Drosha a nuclear enzyme component of the *RNAi* machinery present in all eukaryotic cells. Drosha recognizes and processes stem-loop structures associated with primary *miRNA*-containing transcripts. Drosha products are transported into the cytoplasm where they are further processed by *Dicer* into mature miRNAs.

duplication events that result in an increase in the number of copies of a particular chromosomal region. See *tandem duplications, nontandem duplications.*

duplication loop a bulge in the *duplication*-bearing chromosome that has no similar region with which to pair in the unduplicated normal homologous chromosome. Contrast with *deletion loop.*

E

ecotypes plant varieties analogous to animal strains.

ectopic expression gene expression that occurs outside the cell or tissue where the gene is normally expressed.

electrophoresis a method of separating large molecules (such as DNA fragments or proteins) from a mixture of similar molecules. An electric current is passed through a medium containing the mixture, and each kind of molecule travels through the medium at a different rate, depending on its size and electrical charge. See *gel electrophoresis.*

elements components of a *biological system.*

elongation phase of *DNA replication, transcription,* or *translation* that successively adds nucleotides or amino acids to a growing macromolecule. Compare *initiation.*

elongation factors proteins that aid in the *elongation* phase of *translation.*

embryonic stem cells (ES cells) cultured embryonic cells that continue to divide without differentiating.

emergent properties traits and behaviors that arise from the operation of a *biological system* as a whole; *immunity* and *tolerance* are two emergent properties of the immune system.

endosymbiont theory proposes that chloroplasts and mitochondria originated when free-living bacteria were engulfed by primitive nucleated cells. Host- and guest-formed cellular communities in which each member adapted to the group arrangement and derived benefit.

enhancer mutations mutations in a *modifier gene* that worsen the phenotypic effects of a mutation in another gene.

enhancer trapping in *Drosophila,* the identification of P-element insertion lines with particular β-galactosidase expression patterns.

enhancers *cis*-acting elements that can regulate *transcription* from nearby genes. In yeast, enhancers are called upstream activation sites, or UASs. Enhancers function by acting as binding sites for *transcription factors*.

enhancesome a completely assembled set of *transcription factors* (which may include *activators, repressors, coactivators,* and *corepressors*) associated with an *enhancer* or *locus control region* in a structure that is able to modulate transcription activity.

environmental variance (V_E) deviation from the mean attributed to the influence of external, noninheritable factors. Compare with *genetic variance.*

epigenetic a state of gene functionality that is not encoded within the DNA sequence but that is still heritable from one generation to the next. It can be accomplished and maintained through a chemical modification of DNA such as methylation.

episomes *plasmids,* like the F plasmid, that can integrate into the host chromosomes.

epistasis a gene interaction in which the effects of an *allele* at one gene hide the effects of alleles at another gene. See *dominant epistasis* and *recessive epistasis.*

epistatic describes an *allele* of one gene that masks the effects of one or more alleles of another gene.

equational division cell division that does *not* reduce the number of *chromosomes,* but instead distributes *sister chromatids* to the two daughter cells. *Mitosis* and *meiosis II* are both equational divisions.

equilibrium frequency the q value at which $\Delta q = 0$; the *allele* frequency required to maintain an allele in the *population.*

euchromatin chromosomal region of cells that appears much lighter and less condensed when viewed under a light microscope. Contrast with *heterochromatin.*

eukaryotes one of the three major evolutionary lineages of living organisms known as domains; organisms whose cells have a membrane-bounded nucleus. Contrast with *prokaryotes.*

eukaryotic gene regulation the control of gene expression in the cells of *eukaryotes.*

euploid describes cells containing only complete sets of *chromosomes.*

evolution in the study of genetics, changes in allele frequency in a population over multiple generations; the basic components of evolution at the molecular level include *diversification, selection,* and *amplification.*

evolutionary equilibrium the *allele* frequency at which evolution no longer continues; represented by $\hat{p}$.

excision repair DNA repair mechanism in which specialized proteins recognize damaged base pairing and remove the damaged section so that DNA *polymerase* can fill in the gap by complementary base pairing with the information from the undamaged strand of DNA. See *base excision repair* and *nucleotide excision repair.*

exconjugants *recipient* cells resulting from gene transfer in which *donor* cells carrying specialized *plasmids* establish contact with and transfer DNA to the recipients.

exit (E) site one of three *transfer RNA* binding sites in ribosomes. The E site is occupied by tRNAs during the period just after their disconnection from *amino acids* by the action of *peptidyl transferase* and just before the release of the tRNAs from the *ribosome.*

exons sequences that are found both in a gene's DNA and in the corresponding mature *messenger RNA (mRNA).* See *introns.*

exon shuffling the exchange of *exons* among different genes during evolution, producing mosaic proteins with two or more distinct functions.

expression vectors *cloning* vehicles that promote the expression of a *polypeptide* product from the gene inserts they carry.

expressivity the degree or intensity with which a particular *genotype* is expressed in a *phenotype.*

F

facultative heterochromatin regions of *chromosomes* (or even whole chromosomes) that are heterochromatic in some cells and euchromatic in other cells of the same organisms.

filial generations the successive offspring in a controlled sequence of breeding, starting with two parents (the P generation) and selfing or intercrossing the offspring of each subsequent generation.

fine structure mapping recombination mapping of *mutations* in the same gene; in some *bacteriophage* experiments, fine structure mapping can resolve mutations in adjacent nucleotide pairs.

first-division segregation pattern a *tetrad* in which the arrangement of *ascospores* indicates that the two *alleles* of a *gene* segregated from each other in the first meiotic division.

first filial (F_1) generation progeny of the *parental generation* in a controlled series of crosses.

FISH (fluorescence *in situ* hybridization) a *physical mapping* approach that uses fluorescent tags to detect *hybridization* of nucleic acid probes with *chromosomes.*

fitness the relative advantage or disadvantage in reproduction that a particular *genotype* provides to members of a *population* in comparison to alternative genotypes at the same *locus.*

fitness cost negative impact of the development of a homozygous resistance *genotype.*

5′-untranslated leader in *eukaryotes,* mRNA region between the 5′ *methylated cap* and the *initiation codon.*

fluctuation test the Luria-Delbrück experiment to determine the origin of bacterial resistance. Fluctuations in the numbers of resistant colonies growing in different petri plates showed that resistance is not caused by exposure to bactericides.

focus of action the cells in which a gene of interest must be active to allow the animal to develop and function normally.

forward mutation a mutation that changes a *wild-type allele* of a gene to a different *allele.*

founder cells in nematode development, progenitors of the major embryonic lineages.

founder effect variation of *genetic drift,* occurring when a few individuals separate from a larger *population* and establish a new one that is isolated from the original population, resulting in altered *allele frequencies* in the new population.

F plasmid a *conjugative plasmid* that carries many genes required for the transfer of DNA. Cells carrying F plasmid are called F^+ cells. Cells without the plasmid are called F^- cells.

F′ plasmids *F plasmid* variants that carry most F plasmid genes plus some bacterial DNA; particularly useful in genetic complementation studies.

frameshift mutations *insertions* or *deletions* of base pairs that alter the grouping of *nucleotides* into *codons.* Refer to *reading frame.*

free duplications small DNA fragments maintained extrachromosomally in a genetic stock.

fusion proteins proteins encoded by parts of more than one gene.

G

G_0 in the cell cycle, a resting form of G_1. Cells in G_0 normally do not divide.

gain-of-function alleles (or mutations) rare *mutations* that enhance a gene's function or confer a new activity on the gene's product.

gametes specialized *haploid* cells (eggs and sperm or pollen) that carry genes between generations.

gametogenesis the formation of *gametes.*

gastrulation folding of the cell sheet early in embryo formation; usually occurs immediately after the blastula stage of development.

gel electrophoresis a process used to separate DNA fragments, RNA molecules, or polypeptides according to their size. Electrophoresis is accomplished by passing an electrical current through agarose or polyacrylamide gels. The electrical current forces molecules to migrate into the gel at different rates dependent on their sizes.

gene basic unit of biological information; specific segment of DNA in a discrete region of a chromosome that serves as a unit of function by encoding a particular RNA or protein.

gene amplification an increase from the normal two copies to hundreds of copies of a *gene;* often due to *mutations* in *p53*, which disrupt the G_1-to-S *checkpoint.*

gene conversion any deviation from the expected 2:2 *segregation* of parental *alleles.*

gene dosage the number of times a given *gene* is present in the cell nucleus.

gene expression the process by which a gene's information is converted into *RNA* and then (for protein-coding genes) into a *polypeptide.*

gene family set of closely related genes with slightly different functions that most likely arose from a succession of gene *duplication* events.

gene function generally, to govern the synthesis of a *polypeptide;* in Mendelian terms, a gene's specific contribution to *phenotype.*

generalized transduction a type of *transduction* (gene transfer mediated by bacteriophages) that can result in the transfer of any bacterial gene between related strains of bacteria.

gene pool the sum total of all *alleles* carried in all members of a *population.*

gene regulatory network a set of interacting *transcription factors* and their cognate cis-*control elements* that receive diverse inputs of biological information; integrate and modify those inputs; and transmit the transformed information to various *protein networks*. The fundamental link in a gene regulatory network is the interaction of a transcription factor with its cognate *cis*-control element.

gene superfamily a large set of related genes that is divisible into smaller sets, or families, with the genes in each family being more closely related to each other than to other members of the larger superfamily. The single-gene or *multigene families* that compose a superfamily reside at different chromosomal locations. The families of genes encoding the globins and the *Hox* transcription factors are examples of gene superfamilies.

genetic code the sequence of *nucleotides,* coded in triplets (*codons*) along the *mRNA,* that determines the sequence of *amino acids* in *protein* synthesis.

genetic drift unpredictable, chance fluctuations in *allele* frequency that have a neutral effect on *fitness* of a *population.*

genetic imbalance situation when the *genome* of a cell or organism has more copies of some genes than other genes due to *chromosomal rearrangements* or *aneuploidy.*

genetic linkage See *linkage.*

genetic markers *genes* identifiable through phenotypic variants that can serve as points of reference in determining whether particular progeny are the result of *recombination*. Compare with *physical markers.*

genetic mosaic See *mosaic.*

genetic relatedness the average fraction of common *alleles* at all gene loci that individuals share because they inherited them from a common ancestor.

genetics the science of heredity.

genetic screen an examination of each individual in a *population* for its *phenotype.*

genetic variance (V_G) deviation from the mean attributable to inheritable factors. Compare with *environmental variance.*

genome the sum total of genetic information in a particular cell or organism.

genomic equivalent the number of clones—with inserts of a particular size—that would be required to carry a single copy of every sequence in a particular genome.

genomic imprinting the phenomenon in which a gene's expression depends on the parent that transmits it.

genomic library a collection of *DNA clones* that together carry a representative copy of every DNA sequence in the *genome* of a particular organism.

genomics the study of whole *genomes.*

genotype the actual *alleles* present in an individual.

genotype frequency proportion of total individuals in a *population* that are of a particular *genotype*.

genotypic class a grouping defined by a set of related *genotypes* that will produce a particular *phenotype*. The term is most useful in describing progeny of *dihybrid* or *multihybrid* crosses involving complete dominance; for example, in a cross between *Aa Bb* individuals, the genotypic classes are *A- B-*, *A- bb*, *aa B-*, and *aa bb*.

germ cells specialized cells that incorporate into the reproductive organs, where they ultimately undergo *meiosis,* thereby producing *haploid gametes* that transmit genes to the next generation Compare *somatic cells.*

germ line all the *germ cells* in a sexually reproducing organism. In animals, the germ line is set aside from the *somatic cells* during embryonic development. The germ cells in the germ line divide by *mitosis* to produce a collection of specialized *diploid* cells that then divide by *meiosis* to produce *haploid* cells, or *gametes*. The germ line thus includes the precursors of the gametes such as *oogonia, spermatagonia, primary* and *secondary oocytes,* and *primary* and *secondary spermatocytes* as well as the *gametes*.

germ-line gene therapy a genetic engineering technique that modifies the DNA of *germ cells* that are passed on to progeny.

globular stage embryo stage of development in a plant in which cell divisions give rise to the first evidence of differentiation found in the mature embryo. The innermost group of cells acquire an elongated shape, and a discrete outer cell layer is present.

G_1 phase stage of the cell cycle from the birth of a new cell until the onset of chromosome replication at *S phase.*

G_2 phase stage of the cell cycle from the completion of *chromosome replication* until the onset of cell division.

growth factors extracellular hormones and cell-bound *ligands* that stimulate or inhibit cell proliferation.

guanine (G) a nitrogenous base; one member of the base pair G–C (guanine–cytosine).

gynandromorph a rare genetic *mosaic* with some male tissue and some female tissue, usually in equal amounts.

H

hairpin loops structures formed when a single strand of DNA or RNA can fold back on itself because of *complementary base pairing* between different regions in the same molecule.

haploid a single set of *chromosomes* present in the egg and sperm cells of animals and in the egg and pollen cells of plants (compare with *diploid*).

haploinsufficiency a rare form of dominance in which an individual *heterozygous* for a *wild-type allele* and a *null* allele shows an abnormal *phenotype* because the level of gene activity is not enough to produce a normal phenotype.

haplospores See *ascospores.*

haplotype specific combination of *linked alleles* in a cluster of related *genes.* A contraction of the phrase "haploid genotype."

Hardy-Weinberg law defines the relationships between *genotype* and *allele frequencies* within a generation and from one generation to the next.

hemizygous describes the *genotype* for genes present in only one copy in an otherwise *diploid* organism, such as *X-linked* genes in a male.

heredity the way *genes* transmit biochemical, physical, and behavioral traits from parents to offspring.

heritability (h^2) the proportion of *total phenotype variance* ascribable to *genetic variance.*

hermaphrodite an organism that has both male and female organs and produces both male *gametes* (sperm) and female gametes (eggs). The organism can have both types of organs at the same time (simultaneous hermaphrodite) or have one type early in life and the other type later in life (sequential hermaphrodite).

heterochromatin highly condensed chromosomal regions within which genes are usually transcriptionally inactive.

heterochromatic DNA genomic DNA from heterochromatic regions; this DNA is often difficult to clone.

heterochronic mutations mutations resulting in the inappropriate timing of cell division and cell-fate decisions during development.

heteroduplex region a region of double-stranded DNA in which the two strands have nonidentical (though similar) sequences. Heteroduplex regions are often formed as intermediates during *crossing-over.*

heterogametic sex the gender of a species in which the two *sex chromosomes* are dissimilar; e.g., males are the heterogametic sex in humans because they have an X and a Y chromosome. Compare with *homogametic sex.*

heterogeneous trait occurs when a *mutation* at any one of a number of genes can give rise to the same *phenotype.*

heteromers multimeric proteins composed of nonidentical *subunits.* Compare *homomers.*

heteroplasmic genomic makeup of a cell's organelles characterized by a mixture of organelle *genomes.* Contrast *homoplasmic.*

heterothallic strains of organisms with stable *haploid* mating types; progeny of successive mitotic divisions always have the same mating type as their parents. Contrast *homothallic.*

heterozygote individual with two different *alleles* for a given *gene* or *locus.*

heterozygous a *genotype* in which the two copies of the gene that determine a particular trait are different *alleles.* See *hybrid.*

heterozygous advantage the situation where *heterozygotes* have a higher *fitness* than either *homozygote;* a possible explanation for the survival of recessive genetic diseases in a *population.*

heterozygous carrier unaffected parents who bear a *dominant* normal *allele* that masks the effects of an abnormal *recessive* one.

Hfr bacteria that produce a *h*igh *f*requency of *r*ecombinants for chromosomal genes in mating experiments because their *chromosomes* contain an integrated *F plasmid.*

high-density linkage map a *linkage map* that shows one *gene* or *marker* for each centimorgan of a *genome.*

histocompatibility antigens cell surface molecules that play a critical role in stimulating a proper *immune response.*

histones small DNA-binding proteins with a preponderance of the basic, positively charged amino acids lysine and arginine. Histones are the fundamental protein components of *nucleosomes.*

homeobox in *homeotic genes,* the region of homology, usually 180 bp in length, that encodes the *homeodomain.*

homeodomain a conserved DNA-binding region of *transcription factors* encoded by the *homeobox* of *homeotic genes.*

homeotic gene a gene that plays a role in determining a tissue's identity during development.

homeotic mutation mutation that causes cells to misinterpret their position in the blueprint and become normal organs in inappropriate positions; such a mutation can alter the overall body plan.

homeotic selector genes genes that control the identity of body segments.

homogametic sex the gender of a species in which the two *sex chromosomes* are identical; in humans, females are the homogametic sex because the have two X chromosomes. Compare with *heterogametic sex.*

homologous chromosomes (homologs) chromosomes that match in size, shape, and banding. A pair of chromosomes containing the same linear gene sequence, each derived from one parent.

homologs genes or regulatory DNA sequences that are similar in different species because of descent from a common ancestral sequence.

homologous genes genes in different species with enough sequence similarity to be evolutionarily related.

homomers *multimeric proteins* composed of identical *subunits*. Compare with *heteromers.*

homoplasmic genomic makeup of a cell's organelles characterized by a single type of organelle DNA. Contrast with *heteroplasmic.*

homothallic strains of organisms that can switch mating types; progeny of successive mitotic divisions may not have the same mating type as their parents. Compare with *heterothallic.*

homozygote individual with identical *alleles* for a given *gene* or *locus*.

homozygous a *genotype* in which the two copies of the gene that determine a particular trait are the same *allele.*

hormone a small molecule or polypeptide that is made and released by certain secretory cells in multicellular organisms. Secreted hormones diffuse or move via bodily fluids to other cell types that contain specific receptors. Hormone-receptor binding can elicit changes in gene function and differentiation in the target cells.

hot spots sites within a gene that mutate more frequently than others, either spontaneously or after treatment with a particular *mutagen.*

Human Genome Project initiative to determine the complete sequence of the human *genome* and to analyze this information.

hybrid dysgenesis phenomenon in which high *transposon* mobility causes reduced fertility in hybrid progeny; the result of crossing *Drosophila* males carrying the P element with females that lack the P element.

hybridization the propensity of complementary single strands of nucleic acid—either DNA or RNA—to form stable double helices. A powerful tool, hybridization can be used to identify sequences that are closely related to a DNA probe.

hybrids offspring of genetically dissimilar parents; often used as synonym for *heterozygotes.*

hydrogen bonds weak electrostatic bonds that result in a partial sharing of hydrogen atoms between reacting groups.

hypermorphic mutation produces an allele generating either more protein than the *wild-type allele* or the same amount of a more efficient protein. If excess protein activity alters *phenotype,* the hypermorphic allele is dominant. Compare with *hypomorphic mutation.*

hypomorphic mutation produces either less of a protein or a protein with a weak but detectable function. Compare with *hypermorphic mutation.*

I

imaginal discs flattened epithelial sacs that develop from small groups of cells set aside in the early embryo from which adult-specific structures like wings, legs, eyes, and genitalia develop; in *Drosophila,* they undergo extensive growth and development during the larval and pupal stages.

immune response physiological response to the immune system's activation by *antigen.* Immune responses include the production of antibodies by B lymphocytes and the ability of killer T cells to destroy foreign or cancerous cells by direct cell-to-cell contact.

immunity ability to generate immune responses to infectious agents or vaccines. Such a response protects against serious disease.

immunoglobulin homology unit common feature of the immunoglobulin *gene superfamily;* encodes about 100 amino acids, which fold into a characteristic three-dimensional structure termed the immunoglobulin fold.

incomplete dominance expression of *heterozygous phenotype* resulting in offspring whose phenotype is intermediate between those of the parents.

indels *deletions, duplications,* and *insertions* at non-repeat loci.

independent assortment the random distribution of different genes during gamete formation. See *Mendel's second law.*

inducer small molecule that causes *transcription* from a gene or set of genes.

induction process by which a signal induces expression of a gene or set of genes.

inflorescence in a plant, the flower or group of flowers at the tip of a branch.

initiation first phase of *DNA replication, transcription,* or *translation* needed to set the stage for the addition of nucleotide or amino acid building blocks during *elongation.*

initiation codon nucleotide triplet that marks the precise spot in the nucleotide sequence of an *mRNA* where the code for a particular *polypeptide* begins. Compare *nonsense codon.*

initiation factors a term usually applied to proteins that help promote the association of *ribosomes, mRNA,* and initiating *tRNA* during the first phase of *translation.*

insertion the addition to a DNA molecule of one or more nucleotide pairs.

insertional mutagenesis method of mutagenesis in which a foreign DNA sequence (viral, *plasmid, transposon,* or cloned fragment) is used as the mutagenic agent. Mutations result when the foreign sequence integrates into a gene. The disrupted gene can be easily identified and cloned based on its association with the foreign DNA.

insertion sequences (ISs) small *transposable elements* that dot the chromosomes of many types of bacteria; they are *transposons* that do not contain selectable markers.

insulator a transcriptional regulation element in *eukaryotes* that stops communication between *enhancers* on one side of it with *promoters* on the other side. Insulators play an important role in limiting the *chromatin* region over which an enhancer can operate.

interactome the network of all or many of the protein interactions in a cell or organism.

intercalators class of chemical *mutagens* composed of flat, planar molecules that can sandwich themselves between successive *base pairs* and disrupt the machinery of *replication, recombination,* or repair.

intergenic gene conversion information flow between related DNA sequences that occurs through an alternative outcome of the process responsible for *unequal crossing-over.*

Interkinesis brief *interphase* between *meiosis I* and *meiosis II.*

interphase the period in the cell cycle between divisions.

intragenic suppression the restoration of gene function by one *mutation* canceling the effects of another mutation in the same *gene.*

introns the DNA base sequences of a *gene* that are spliced out of the *primary transcript* and are therefore not found in the mature mRNA. See *exons.*

inversion a 180-degree rotation of a segment of a *chromosome* relative to the remainder of the chromosome.

inversion heterozygotes cells or organisms in which one *chromosome* is of a normal gene order, while the *homologous chromosome* carries an *inversion.*

inversion loop formed in the cells of an *inversion heterozygote* when the inverted region rotates to pair with the similar region in the normal homolog.

K

karyotype the visual description of the complete set of *chromosomes* in one cell of an organism; usually presented as a photomicrograph with the chromosomes arranged in a standard format showing the number, size, and shape of each chromosome type.

kinetochore a specialized chromosomal structure composed of DNA and proteins that is the site at which *chromosomes* attach to the spindle fibers. See *kinetochore microtubules.*

kinetochore microtubules microtubules of the *mitotic spindle* that extend between a *centrosome* and the *kinetochore* of a *chromatid. Chromosomes* move along the kinetochore microtubules during cell division.

knockout constructs cloned genes modified so that they no longer function. Used in *targeted mutagenesis.*

knockout mice mice homozygous for an induced mutation in a targeted gene; the mutation destroys (knocks out) the function of the gene. See *targeted mutagenesis.*

L

***lac* operon** a single DNA unit in *E. coli,* composed of the *lacZ, lacY,* and *lacA* genes together with the *promoter* (*p*) and *operator site* (*o*), that enables the simultaneous regulation of the three structural genes in response to environmental changes.

lagging strand during *replication,* DNA strand whose polarity is opposite to that of the *leading strand.* The lagging strand must be synthesized discontinuously as small *Okazaki fragments* that are ultimately joined into a continuous strand.

lambda (λ) see *phage lambda.*

λ repressor protein protein that binds to *operator sites* on phage λ DNA, preventing the transcription of genes needed for the lytic cycle; this protein makes λ lysogens immune to infection with incoming λ phage.

late-onset a genetic condition in which symptoms are not present at birth, but manifest themselves later in life.

lateral gene transfer the introduction and incorporation of DNA from an unrelated individual or from a different species.

law of independent assortment See *Mendel's second law.*

law of segregation See *Mendel's first law.*

lawn bacteria immobilized in a nutrient agar, used as a field on which to test for the presence of viral particles.

leader sequence DNA sequence that precedes the coding sequence and contains signals that regulate *transcription termination* in an *operon* controlled by *attenuation.*

leading strand during *replication,* DNA strand synthesized continuously 5′ to 3′ toward the unwinding Y-shaped *replication fork.* Compare with *lagging strand.*

leptotene first definable substage of *prophase I* during which the long, thin, already duplicated *chromosomes* begin to thicken.

library see *genomic library* or *cDNA library.*

ligands molecules such as *growth factors* or *hormones* that are produced by one cell and bind to receptors on a different cell, initiating a *signal transduction pathway.*

LINE *l*ong *in*terspersed *e*lements; one of the two major classes of *transposable elements* in mammals. Contrast with *SINE.*

lineage compartments regions of an organism in which cells have a restricted developmental potential.

linkage the proximity of two or more *markers* on a *chromosome;* the closer together the markers are, the lower the probability that they will be separated by *recombination.* Genes are *linked* when the frequency of parental type progeny exceeds that of recombinant progeny.

linkage disequilibrium when *alleles* at separate *loci* (such as *marker* alleles and disease alleles) are associated with each other at a significantly higher frequency than would be expected by chance. Linkage disequilibrium at particular loci in a *population* can be evidence of common ancestry.

linkage group a group of genes chained together by linkage relationships. See *linkage.*

linkage map depicts the distances between *loci* as well as the order in which they occur on the organism.

linked describes *genes* whose *alleles* are inherited together more often than not; linked genes are usually located close together on the same *chromosome.*

linker DNA a stretch of ≈40 base pairs of DNA that connect one *nucleosome* with the next.

locus a designated location on a *chromosome.* See *α-globin locus* and *β-globin locus.*

locus control region (LCR) a *cis*-acting regulatory element that operates to enhance *transcription* from individual genes within a gene complex such as the β-globin complex.

LOD score analysis *l*ogarithm of the *od*ds analysis, a statistical method of formal genetics for determining the likelihood that two or more *loci* are *linked* together on the same *chromosome.* LOD score analysis is most often used to determine linkage from natural mating data of the kind available for human populations.

loss-of-function mutation (or allele) DNA *mutation* that reduces or abolishes the activity of a gene; most (but not all) loss-of-function alleles are *recessive.*

lysate population of *phage* particles released from the host bacteria at the end of the *lytic cycle.*

lysogenic bacterium bacterial cell that carries a *prophage;* lysogen.

lysogenic cycle occurs when *bacteriophage* integrate their DNA into the host *genome* such that it multiplies along with that genome, but does little harm to the host.

lysogeny the integration of *bacteriophage* DNA into the host *chromosome.*

lytic cycle bacterial cycle of *phage*-infected cells resulting in cell lysis and release of progeny phage. Compare with *lysogenic cycle.*

M

macroarray microtiter plate-based DNA array.

macroevolution in the study of genetics, the process by which new species emerge from existing species. Contrast with *microevolution.*

major groove in a space-filling representation of the DNA *double helix* model, the wider of the two grooves resulting from the vertical displacement of the two backbone threads. See *minor groove*.

mapping the process of determining the *locus* of a *gene* on a particular *chromosome.*

mapping function mathematical equation that compensates for the inaccuracies inherent in relating *recombination frequencies* to physical distance.

mapping panels a set of DNA samples used in multiple *linkage* analysis tests.

map unit synonymous with *centimorgan.*

marker an identifiable physical location on a *chromosome,* whose inheritance can be monitored. Markers can be expressed regions of DNA (genes) or any segment of DNA with variant forms that can be followed.

maternal-effect mutations mutations in genes encoding maternal components (those supplied by the egg to the mother) that disrupt the development of her progeny.

maternally supplied components molecules synthesized by the mother that are supplied to the egg and that are needed for early development of the progeny.

mean statistical average; the middle point.

meiosis the process of two consecutive cell divisions starting in the *diploid* progenitors of sex cells. Meiosis results in four daughter cells, each with a *haploid* set of *chromosomes*.

meiosis I (or division I of meiosis) the parent nucleus divides to form two daughter nuclei; during meiosis I, the previously replicated *homologous chromosomes* segregate to different daughter cells.

meiosis II (or division II of meisosi) each of the two daughter nuclei resulting from *meiosis I* divide to produce four nuclei; because the *chromosomes* do not duplicate at the start of meiosis II, these four daughter nuclei are *haploid*.

Mendel's first law the law of segregation states that the two alleles for each trait separate (segregate) during gamete formation and then unite at random, one from each parent, at fertilization.

Mendel's second law the law of independent assortment states that during gamete formation, different pairs of *alleles* (genes) segregate independently of each other.

merodiploids partial diploids in which there are two copies of some genes.

messenger RNA (mRNA) RNA that serves as a template for protein synthesis. See *genetic code.*

metabolism the chemical and physical reactions that convert sources of energy and matter to fuel growth and repair within a cell.

metacentric a *chromosome* in which the *centromere* is at or near the middle.

metamorphosis a dramatic reorganization of an organism's body plan; for example, transition from larval stage to adult insect stage.

metaphase a stage in *mitosis* or *meiosis* during which the *chromosomes* are aligned along the equatorial plane of the cell.

metaphase I phase of *meiosis I* when the *kinetochores* of *homologous chromosomes* attach to microtubules from opposite spindle poles, positioning the *bivalents* at the equator of the *spindle apparatus*.

metaphase II second phase of *meiosis II* during which *kinetochores* of *sister chromatids* attach to microtubule fibers emanating from opposite poles of the *spindle apparatus*. Two characteristics distinguish metaphase II from its counterpart in *mitosis:* (1) the number of chromosomes is one-half that in mitotic metaphase of the same species, and (2) in most chromosomes, the two sister chromatids are no longer identical because of the recombination through crossing-over that occurred in *meiosis I*.

metaphase plate imaginary equator of the cell toward which *chromosomes* move during *metaphase.*

metazoans multicellular animals that first appeared about 0.57 billion years ago.

methylated cap formed by the action of capping enzyme and methyl transferases at the 5′ end of eukaryotic *mRNA,* critical for efficient *translation* of the mRNA into protein.

methyl-directed mismatch repair DNA repair mechanism that corrects mistakes in *replication,* discriminating between newly synthesized and parental DNA by the methyl groups on the parental strand.

microarray small glass-slide DNA array.

microevolution alterations of a population's *gene pool.* Contrast *macroevolution.*

micro-RNA (miRNA) an *RNA* molecule 21–24 bases in length that is encoded in the genome of an organism and used by a cell to modulate gene expression through the process of *RNA interference.*

microsatellite DNA element composed of 15–100 tandem repeats of one-, two-, or three-base-pair sequences.

minimal tiling path the result of the final step in the *shotgun* sequencing strategy, a minimally overlapping set of BAC clones.

minisatellite DNA element composed of 10–40 bp tandem repeating units of identical sequence.

minor groove in a space-filling representation of the DNA *double helix* model, the narrower of the two grooves resulting from the vertical displacement of the two backbone threads. See *major groove*.

miRISC a functional *RNA-induced silencing complex (RISC)* loaded with a *micro-RNA*.

missense mutations changes in the *nucleotide* sequence of a *gene* that change the identity of an *amino acid* in the *polypeptide* encoded by that gene.

mitosis the process of division that produces daughter cells that are genetically identical to each other and to the parent cell.

mitotic nondisjunction the failure of two *sister chromatids* to separate during mitotic *anaphase* generates reciprocal trisomic and monosomic daughter cells.

mitotic spindle structure composed of three types of microtubules (*kinetochore microtubules, polar microtubules,* and *astral microtubules*). The mitotic spindle provides a framework for the movement of *chromosomes* during cell division.

model organisms used in genomic analysis because they have many genetic mechanisms and cellular pathways in common with each other and with humans. These organisms lend themselves well to classical breeding experiments and direct manipulation of the genome.

modification the phenomenon in which growth on a restricting host changes a *phage* so that succeeding generations grow more efficiently on that same host.

modification enzymes enzymes that add methyl groups to specific DNA sequences, preventing the action of specific *restriction enzymes* on that DNA.

modifier genes genes that produce a subtle, secondary effect on *phenotype*. There is no formal distinction between major and modifier genes, rather it is a continuum of degrees of influence.

molecular clock hypothesis stating that one can assume a constant rate of *amino acid* or *nucleotide* substitution as a means of determining the genealogies of organisms.

molecular cloning the process by which a single DNA fragment is purified from a complex mixture of DNA molecules and then amplified into a large number of identical copies.

molecular machine complexe of different *proteins* or of proteins and *RNAs* that carry out a particular process; examples are the *ribosome* and the *spliceosome*.

molecular marker a segment of DNA found at a specific site in a *genome* that has variants which can be recognized and followed. See *physical marker, genetic marker.*

monohybrid crosses crosses between parents that differ in only one trait.

monohybrids individuals having two different alleles for a single trait.

monomorphic a gene with only one *wild-type allele*.

monoploid describing cells, nuclei, or organisms that have a single set of unpaired *chromosomes*. For *diploid* organisms, monoploid and *haploid* are synonymous.

monosomic individual lacking one *chromosome* from the *diploid* number for the species.

morphogens substances that define different cell fates in a concentration-dependent manner.

mosaic an organism containing tissues of different *genotypes*.

mosaic determination where the embryo is a collection of self-differentiating cells that at their formation receive a specific set of molecular instructions governing their unique fates. Contrast with *regulative determination.*

mRNA See *messenger RNA*.

multifactorial traits determined by two or more factors, including multiple genes interacting with each other or one or more genes interacting with the environment.

multigene family set of *genes* descended by *duplication* and *diversification* from one ancestral gene. The members of a multigene family may be either clustered on the same chromosome or distributed on different chromosomes.

multihybrid crosses crosses between parents that differ in three or more traits.

multimeric protein a protein made from more than one *polypeptide;* each polypeptide in the multimeric protein is called a *subunit.*

multiple alleles set of *alleles* of a gene with more than two variant forms.

mutagen any physical or chemical agent that raises the frequency of *mutations* above the spontaneous rate.

mutant allele (1) an *allele,* or DNA variant, whose frequency is less than 1% in a *population;* (2) an allele that dictates a phenotype seen only rarely in a population. See *allele frequency*.

mutant tumor-suppressor genes recessive *mutant alleles* that contribute to the formation of cancer.

mutations heritable alterations in *DNA sequence.*

N

n number of *chromosomes* in a normal *gamete;* for organisms that are not *polyploids, n* is the number of chromosomes in any *haploid* cell and is also equal to *x* (the number of chromosomes in a single complete set of hon-homologous chromosomes). *n* and *x* are *not* identical in polyploid organisms.

natural selection in nature, the process that progressively eliminates individuals whose fitness is low and chooses individuals of high fitness to survive and become the parents of the next generation. Contrast with *artificial selection.*

natural transformation a process by which a few species of bacteria transfer genes from one strain to another by spontaneously accepting DNA from their surroundings. Contrast with *artificial transformation.*

neomorphic mutations rare mutations that produce a novel *phenotype* due to production of a protein with a new function or due to *ectopic expression* of the protein.

N-formylmethionine (fMet) a modified methionine whose amino end is blocked by a formyl group; fMet is carried by a specialized *tRNA* that functions only at a *ribosome*'s *translation* initiation site.

nodes (of a *phylogenetic tree*) representations of taxonomic units such as species, *populations,* individuals, or *genes.*

nonallelic noncomplementation (also called second-site noncomplementation) failure of *mutations* in two different genes to complement each other, causing a mutant *phenotype* in *heterozygotes*. Usually indicates that the two polypeptide products cooperate with each other, for example as *subunits* of the same *multimeric protein.*

nonautonomous elements defective *transposable elements* that cannot move unless the genome contains non-defective *autonomous elements* that can supply necessary functions like tranposase enzymes.

nonautonomous trait when the *phenotype* expressed by a particular cell depends upon the genotypes of neighboring cells. Compare with *cell autonomous trait.*

nonconservative substitutions mutational changes that substitute an *amino acid* in a *protein* with a different amino acid with different chemical properties. Compare with *conservative substitutions.*

nondisjunction failures in *chromosome segregation* during *meiosis*; responsible for defects such as trisomy. (See trisomic).

nonhomologous end-joining mechanism for stitching back together ends formed by double-strand breaks. It relies on proteins that bind to the ends of the broken DNA strands and bring them close together. Overhanging ends are often "trimmed during nonhomologous end-joining, resulting in DNA loss.

nonhomologous unequal crossing-over exchange between nonequivalent chromosomal segments with little sequence homology. A nonhomologous crossover may be mediated by at least a short stretch of sequence homology—coding or noncoding—at the crossover's two sites of initiation. For example, related repeat sequences in the *genome* may mediate nonhomologous unequal crossing-over.

non-Mendelian inheritance pattern of inheritance that does not follow *Mendel's laws* and does not produce Mendelian ratios among the progeny of various crosses.

nonparental ditype (NPD) a *tetrad* containing four recombinant spores.

nonsense codons the three stop codons that terminate *translation.* Compare with *initiation codon.*

nonsense mutations mutational changes in which a codon for an amino acid is altered to a *stop codon,* resulting in the formation of a *truncated protein.*

nonsense suppressor tRNAs *tRNAs* encoded by mutant tRNA genes; these tRNAs contain *anticodons* that can recognize *stop codons,* thus suppressing the effects of *nonsense mutations* by inserting an *amino acid* into a *polypeptide* in spite of the stop codon.

nontandem duplications two or more copies of a region that are not adjacent to each other and may lie far apart on the same chromosome or on different chromosomes. Contrast with *tandem duplications.*

Northern blot analysis protocol for determining whether a fragment of DNA is transcribed in a particular tissue. In this protocol, RNA *transcripts* in the cells of a particular tissue are separated by *gel electrophoresis*. Compare *Southern blot.*

N terminus the end of a *polypeptide* chain that contains a free amino group that is not connected to any other amino acid.

nuclear envelope envelope composed of two membranes that surrounds the nucleus of a eukaryotic cell.

nuclear lamins proteins that underlie the inner surface of the nuclear membrane and serve as CDK substrates.

nucleoid body a folded bacterial *chromosome.*

nucleolar organizer clusters of *rRNA* genes on long loops of DNA within a *nucleolus.*

nucleolus large sphere-shaped organelle visible in the nucleus of *interphase* eukaryotic cells with a light microscope; formed by the *nucleolar organizer.*

nucleosome rudimentary DNA packaging unit; composed of DNA wrapped around a *histone protein* core.

nucleotide a subunit of DNA or RNA consisting of a nitrogenous base (adenine, guanine, thymine, or cytosine in DNA; adenine, guanine, uracil, or cytosine in RNA), a phosphate group, and a sugar (deoxyribose in DNA; ribose in RNA).

nucleotide excision repair homology-dependent mechanism that removes DNA alterations/errors, such as thymine-thymine dimers that *base excision repair* cannot take care of. Depends on process-specific enzyme complexes that patrol the DNA for irregularities and cut the damaged strand in two places that flank the error, releasing s short single-stranded region containing the alteration. DNA polymerase fills in the resultant gap.

null hypothesis a statistical hypothesis to be tested and either accepted or rejected in favor of an alternative.

null mutations (or **alleles**) mutations that abolish the function of a protein encoded by the *wild-type allele.* Such mutations either prevent synthesis of the protein or promote synthesis of a protein incapable of carrying out any function.

O

octads asci produced in *Neurospora* containing eight *ascospores* because of an extra round of *mitosis* that occurs after *meiosis* is completed.

octant stage embryo early stage of embryo formation during cell divisions of a plant.

Okazaki fragments during *DNA replication,* small fragments of about 1000 bases that are joined after synthesis to form the *lagging strand.*

oligonucleotide a short single-stranded DNA molecule (containing less than 50 bases); can be synthesized by an automated DNA synthesizer. Oligonucleotides are used as *DNA probes* and as *primers* for DNA sequencing or PCR.

oligopeptide several *amino acids* linked by *peptide bonds.*

oncogene a gene, one or more forms of which is associated with cancer. Many oncogenes are involved, directly or indirectly, in controlling the rate of cell growth.

oogenesis formation of the female gametes (eggs).

oogonia *diploid germ cells* in the ovary.

open reading frames (ORFs) DNA sequences with long stretches of codons in the same reading frame uninterrupted by *stop codons;* suggest the presence of genes.

operator site a short DNA sequence near a *promoter* that can be recognized by a *repressor* protein; binding of repressor to the operator blocks *transcription* of the gene.

operon a unit of DNA composed of specific genes, plus a *promoter* and/or *operator,* that acts in unison to regulate the response of the structural genes to environmental changes.

Operon theory theory explaining the repression and induction of genes in *E. coli.*

ordered tetrads *tetrads* in fungi such as *Neurospora* in which the order of *ascospores* in the *ascus* reflects the geometry of the meiotic divisions.

origin of replication short sequence of *nucleotides* at which the *initiation* of DNA *replication* begins.

orthologous genes genes with sequence similarities in two different species that arose from the same gene in the two species' common ancestor.

ovum *haploid* female *germ cell* (the egg).

oxidative phosphorylation a set of reactions requiring oxygen that creates portable packets of energy in the form of ATP.

P

pachytene substage of *prophase I* that begins at the completion of *synapsis* and includes the *crossing-over* of genetic material that results in *recombination.*

paracentric inversions *inversions* that exclude the *centromere.* Compare with *pericentric inversions.*

paralogous genes genes that arise by *duplication* within the same species, often within the same *chromosome;* paralogous genes often constitute a *multigene family.*

parental classes combinations of *alleles* present in the original *parental generation.*

parental ditype (PD) a *tetrad* that contains four parental class *haploid* cells.

parental (P) generation pure-breeding individuals whose progeny in subsequent generations will be studied for specific traits. Refer to *filial generations.*

parental types *phenotypes* that reflect a previously existing parental combination of genes that is retained during gamete formation.

parthenogenesis reproduction in which offspring are produced by an unfertilized female. Parthenogenesis is common in ants, bees, wasps, and certain species of fish and lizards.

partial digest the result of an enzymatic *digestion* of a DNA sample in which only some of the available restriction sites have been cut. Compare with *complete digest.*

pathogenicity islands segments of DNA in disease-causing bacteria that encode several genes involved in pathogenesis. Pathogenicity islands appear to have been transferred into the bacteria by lateral gene transfer from a different species.

pedigree an orderly diagram of a family's relevant genetic features, extending through as many generations as possible.

penetrance indicates how many members of a *population* with a particular *genotype* show the expected *phenotype.*

peptide bond a covalent bond that joins *amino acids* during protein synthesis.

peptide motifs stretches of *amino acids* conserved in many otherwise unrelated *polypeptides.*

peptidyl (P) site site on a *ribosome* to which the initiating *tRNA* first binds and at which the tRNAs carrying the growing *polypeptide* are located during *elongation.*

peptidyl transferase the enzymatic activity of the *ribosome* responsible for forming *peptide bonds* between successive *amino acids.*

pericentric inversions *inversions* that include the *centromere*. Compare with *paracentric inversions.*

permissive condition an environmental condition that allows the survival of an individual with a *conditional lethal* allele. Contrast with *restrictive condition.*

petals the structure composing the second *whorl* of a flower, leaf-shaped but containing no photosynthetic cells, therefore not green.

phage short for *bacteriophage;* a *virus* for which the natural host is a bacterial cell; literally "bacteria eaters."

phage induction process of phage DNA excision from the bacterial chromosome in a *lysogen* and entry into the *lytic cycle.*

phage lambda (λ) a naturally occurring double-stranded DNA virus that infects *E. coli.* A common *plasmid vector* used to clone DNA from other organisms.

phase of linkage when two or more linked loci are *heterozygous* in the *genotype* of a particular individual—for example, Aa Bb Cc—the phase of linkage refers to the combination of *alleles* present on each homolog. In the preceding example, one possible phase of linkage is [a-B-c]/[A-b-C].

phenocopy a change in *phenotype* arising from environmental agents that mimic the effects of a mutation in a gene. Phenocopies are not heritable because they do not result from a change in a gene.

phenotype an observable characteristic.

phenotype frequency the proportion of individuals in a population that are of a particular phenotype.

phenotype variance (V_P) see *total phenotype variance.*

pheromones molecules produced by one sex that serve as agents to elicit mating behaviors in individuals of the opposite sex.

phosphodiester bonds covalent bonds joining one *nucleotide* to another. Phosphodiester bonds between nucleotides form the backbone of DNA.

photomorphogenesis light-regulated developmental program of a plant.

photoperiod day length; period of light exposure for plants.

phylogenetic tree diagram composed of *nodes,* which represent the taxonomic units, and *branches,* which represent the relationship of these units.

physical associations protein or regulatory interactions in a *biological system.*

physical map a map of locations of identifiable landmarks on DNA (for example, restriction enzyme cutting sites, genes). For the human genome, the lowest-resolution physical map is the banding patterns on the 24 different chromosomes; the highest resolution map is the complete *nucleotide* sequence of the *chromosomes.* See *karyotype.*

physical markers cytologically visible abnormalities that make it possible to keep track of specific *chromosome* parts from one generation to the next. Compare *genetic markers.*

pilus hollow protein tube that protrudes from an F^+, *Hfr*, or F' bacterial cell and binds to the cell wall of an F^- cell. Retraction of the pilus into the F^+ cell draws the two cells close together in preparation for gene transfer.

pistil floral structure that consists of fused *carpels,* with pollen receptive *stigma* at the top, and a short neck or style, leading to the ovary.

plaque a clear area on a bacterial *lawn,* devoid of living bacterial cells, containing the genetically identical descendants of a single *bacteriophage.*

plasmids small circles of double-stranded DNA that can replicate in bacterial cells independently of the *bacterial chromosome;* commonly used as *cloning vectors.*

pleiotropy phenomenon in which a single gene determines a number of distinct and seemingly unrelated characteristics.

point mutation a mutation of one base pair.

polar body a cell produced by *meiosis I* or *meiosis II* during oogenesis that does not become the *primary* or *secondary oocyte.*

polarity an overall direction.

polar microtubules microtubules that originate in *centrosomes* and are directed toward the middle of the cell; polar microtubules that arise from opposite centrosomes interdigitate near the cell's equator and push the spindle poles apart during *anaphase.*

poly-A tail the 3′ end of eukaryotic *mRNA* consisting of 100–200 As, believed to stabilize the mRNA and increase the efficiency of the initial steps of *translation.*

polycistronic mRNA an *mRNA* that contains more than one protein coding regions; often the transcriptional product of an *operon* in bacterial cells.

polygenic trait trait controlled by multiple genes. See *continuous trait.*

polymer a linked chain of repeating subunits that form a larger molecule; DNA is a type of polymer.

polymerase chain reaction (PCR) a fast inexpensive method of replicating a DNA sequence once the sequence has been identified; based on a reiterative loop that amplifies the products of each previous round of *replication.*

polymerases, DNA or RNA enzymes that catalyze the synthesis of nucleic acids on preexisting nucleic acid templates, assembling *RNA* from ribonucleotides or *DNA* from deoxyribonucleotides.

polymerization the linkage of subunits to form a multi-unit chain. In DNA *replication,* the polymerization of nucleotides occurs through the formation of *phosphodiester bonds* by DNA polymerase III.

polymorphic a *locus* with two or more distinct *alleles* in a *population.*

polymorphism variant of a gene or noncoding region that has two or more *alleles.* Molecular geneticists use this term to describe a variant of a *locus* within a *population* of organisms that has two or more alleles. Population geneticists reserve the term for variants at a locus where two or more alleles are present at a frequency of 1% or greater; for example, to describe the alternative forms a gene that has more than one *wild-type allele.*

polypeptides amino-acid chains containing hundreds to thousands of *amino acids* joined by *peptide bonds.*

polyploids *euploid* species that carry three or more complete sets of *chromosomes.*

polyproteins *polypeptides* produced by *translation* that can subsequently be cleaved by protease enzymes into two or more separate proteins.

polyribosomes structures formed by the simultaneous *translation* of a single *mRNA* molecule by multiple *ribosomes.*

polytene chromosome giant *chromosome* consisting of many identical *chromatids* lying in parallel register.

population a group of interbreeding individuals of the same species that inhabit the same space at the same time.

population genetics scientific discipline that studies what happens in whole *populations* at the genetic level.

positional cloning the process that enables researchers to obtain the clone of a gene without any prior knowledge of its protein product or function. It uses genetic and physical maps to locate *mutations* responsible for particular *phenotypes.*

position-effect variegation variable expression of a gene in a population of cells, caused by the gene's location near highly compacted *heterochromatin.*

posttranslational modifications changes such as phosphorylation that occur to a *polypeptide* after *translation* has been completed.

primary oocytes/spermatocytes *germ-line* cells in which *meiosis I* occurs.

primary structure the linear sequence of *amino acids* within a *polypeptide.*

primary transcript the single strand of *RNA* resulting from *transcription.*

primer short, preexisting *oligonucleotide* chain to which new DNA can be added by DNA polymerase.

primer walking a common approach to directed sequencing of DNA in which *primers* are synthesized using the information from each previous round of DNA sequencing.

product rule states that the probability of two or more independent events occurring together is the product of the probabilities that each event will occur by itself.

programmed cell death (PCD) see *apoptosis.*

prokaryotes one of the three major evolutionary lineages of living organisms known as domains; characterized by the lack of a nuclear membrane. Contrast with *eukaryotes.*

prokaryotic gene regulation control of gene expression in a bacterial cell via mechanisms to increase or decrease the *transcription* and *translation* of specific *genes* or groups of genes.

prometaphase the stage of *mitosis* or *meiosis* just after the breakdown of the *nuclear envelope,* when the chromosomes connect to the *spindle apparatus* and begin to move toward the *metaphase plate.*

promoters DNA sequences near the beginning of *genes* that signal *RNA polymerase* where to begin transcription. Compare with *terminators.*

prophage the integrated *phage genome.*

prophase the phase of the *cell cycle* marked by the emergence of the individual *chromosomes* from the undifferentiated mass of *chromatin,* indicating the beginning of *mitosis.*

prophase I the longest, most complex phase of *meiosis* consisting of several substages.

prophase II first phase of *meiosis II;* if the *chromosomes* decondensed during interkinesis, they recondense. At the end of prophase II, the *nuclear envelope* breaks down and the *spindle apparatus* re-forms.

protein domains discrete functional units of a protein, encoded by discrete regions of DNA.

protein network set of proteins (and often other molecules such as metabolites) that interact in executing a particular biological function.

proteins large polymers composed of hundreds to thousands of *amino acid* subunits strung together in a specific order into long chains. Proteins are required for the structure, function, and regulation of the body's cells, tissues, and organs.

proteome the complete set of *proteins* encoded by a *genome.*

proteomics global analysis of most (or ideally, all) the *proteins* in a particular cell type or organism.

proteosome large multiprotein complex in the cytoplasm of eukaryotic cells that contains proteolytic enzymes that degrade proteins tagged with *ubiquitin.*

protooncogene a *gene* that can mutate into an *oncogene*—an *allele* that causes a cell to become cancerous.

prototroph a microorganism (usually wild type) that can grow on minimal medium in the absence of one or more growth factors. Compare with *auxotroph.*

pseudodominance expression of a *phenotype* caused by a *recessive allele* in a deletion heterozygote because the other homolog has no copy of that gene.

pseudogene a nonfunctioning gene. The result of *duplication* and divergence events in which one copy of an originally functioning *gene* has undergone *mutations* such that it no longer has an intact *polypeptide* coding sequence.

pseudolinkage characteristic of a heterozygote for a *reciprocal translocation,* in which genes located near the translocation breakpoint behave as if they are linked even though they originated on nonhomologous chromosomes.

pulse-field gel electrophoresis a special type of electrophoretic protocol that allows an extreme extension of the normal separating capacity of the gels. In this protocol, the DNA sample is subjected to pulses of electrical current that alternate between two directions. The range of sizes separated in this manner is a function of pulse length. Compare *electrophoresis.*

punctuated equilibrium the tendency of evolution to proceed through long periods of stasis (lack of change) followed by short periods of explosive change.

pure-breeding lines families of organisms that produce offspring with specific parental traits that remain constant from generation to generation.

***p* value** numerical probability that a particular set of observed experimental results represents a chance deviation from the values predicted by a particular hypothesis.

Q

quantitative trait See *continuous trait.*

quantitative trait loci (QTLs) loci that control the expression of *continuous traits.*

quaternary structure structure made up of the three-dimensional configuration of subunits in a *multimeric protein.*

R

radial loop-scaffold model the model of looping and gathering of DNA by nonhistone proteins that results in high compaction of *chromosomes* at *mitosis.*

random amplification of polymorphic DNA (RAPD) a protocol designed to detect single-base changes at *polymorphic* loci throughout a *genome.*

random walk description of movement of bacteria to achieve chemotaxis reflecting nonpredictable changes in the direction of movement. Contrast *biased random walk.*

reading frame the partitioning of groups of three *nucleotides* from a fixed starting point such that the sequential interpretation of each succeeding triplet generates the correct order of *amino acids* in the resulting *polypeptide chain.*

rearrangements events in which the *genome* is reshaped by the reorganization of *DNA sequences* within one or more *chromosomes.*

receptors proteins embedded in the membrane of a cell that bind to *growth factors* and other *ligands,* initiating a *signal transduction pathway* in the cell.

recessive allele an *allele* whose *phenotype* is not expressed in a heterozygote. See *dominant allele.*

recessive epistasis special case of *epistasis,* in which the *allele* causing the epistasis is *recessive*. Compare with *dominant epistasis.*

recessive lethal allele an *allele* that prevents the birth or survival of *homozygotes,* though *heterozygotes* carrying the allele survive.

recessive trait the trait that remains hidden in the F_1 hybrids (*heterozygotes*) resulting from a mating between pure-breeding parental strains showing antagonistic *phenotypes;* the recessive trait usually reappears in the *second filial* (F_2) generation.

recipient during gene transfer in bacteria, the cell that receives the genetic material. See *donor, transformation, conjugation,* and *transduction.*

reciprocal crosses *crosses* in which the traits in the males and females are reversed, thereby controlling whether a particular trait is transmitted by the egg or the pollen.

reciprocal translocation results when two breaks, one in each of two *chromosomes,* yield DNA fragments that do not re-ligate to their chromosome of origin; rather, they switch places and become attached to the other chromosome. Compare with *translocation.*

recombinant classes reshuffled combinations of *alleles* that were not present in the *parental generation.*

recombinant DNA molecules a combination of DNA molecules of different origin that are joined using *recombinant DNA technologies.*

recombinant DNA technology modern tools of genetic analysis based on the isolation, analysis, and use of DNA fragments purified from complex *genomes*. Recombinant DNA technology is used in basic research to characterize genes and other genomic elements and also as a tool of biotechnology to produce important medical products.

recombinants *chromosomes* that carry a mix of *alleles* derived from different *homologous chromosomes.*

recombinant types *phenotypes* reflecting a new combination of genes that occurs during gamete formation.

recombination the process by which offspring derive a combination of genes different from that of either parent; the generation of new allelic combinations. In higher organisms, this can occur by *crossing-over.*

recombination frequency (RF) the percentage of *recombinant* progeny; can be used as an indication of the physical distance separating any two genes on a chromosome. See *centimorgan.*

recombination nodules structures that appear during *pachytene* of *prophase I.* An exchange of parts between nonsister *chromatids* occurs at recombination nodules.

reduction(al) division cell division that reduces the number of *chromosomes,* usually by segregating *homologous chromosomes* to two daughter cells. *Meiosis I* is a reductional division.

regulative determination where the embryo is a collection of cells that can alter, or "regulate", their fates according to the environment, for example, to make up for missing cells. Contrast with *mosaic determination.*

regulon a group of *genes* that are regulated by a common signal and regulator.

release factors *proteins* that recognize *stop codons* and help end translation.

replication see *DNA replication.*

replication bubble unwound area of the original DNA *double helix* during *replication.*

replication fork Y-shaped area consisting of the two unwound DNA strands branching out into unpaired (but complementary) single strands during *replication.*

replicon (replication unit) the DNA running both ways from one *origin of replication* to the endpoints where it merges with DNA from adjoining *replication forks.*

reporter gene a protein-coding gene incorporated into a recombinant DNA molecule along with putative DNA regulatory elements. After transformation of cells, the reporter gene "reports" the activity of the putative regulatory elements to molecular biologists.

repressor a type of *transcription factor* that can bind to specific *cis*-acting elements such as *operator sites* and thereby diminish or prevent *transcription.*

reproductive cloning creation of a cloned embryo by insertion of the nucleus of a *somatic cell* from one individual into an egg cell whose nucleus has been removed. The hybrid egg is stimulated to begin embryonic cell divisions, and the resulting cloned embryo is transplanted into the uterus of a foster mother and allowed to develop to term. Contrast with *therapeutic cloning.*

reproductive development period in plant development that produces flower and seed. Contrast *vegetative development.*

response to selection (R) the amount of evolution, or change in mean trait value, resulting from *selection.* Compare with *selection differential.*

restriction the bacterial capacity for limiting viral growth.

restriction enzymes proteins made by bacteria that recognize specific, short *nucleotide* sequences and cut DNA at those sites.

restriction fragment length polymorphism (RFLP) variation between individuals in DNA fragment size cut by specific *restriction enzymes; polymorphic* sequences that result in RFLPs are used as markers on both *physical maps* and *genetic linkage* maps.

restriction fragments DNA fragments generated by the action of *restriction enzymes.*

restriction map a linear diagram that illustrates the positions of *restriction enzyme* recognition sites along a DNA molecule.

restrictive condition an environmental condition that prevents the survival of an individual with a *conditional lethal* allele. Contrast with *permissive condition.*

retroposons genetic elements that transpose via *reverse transcription* of an RNA intermediate. One class of *transposable elements.* Contrast with *transposons.*

retroviruses viruses that hold their genetic information in a single strand of *RNA* and carry the enzyme *reverse transcriptase* to convert that RNA into DNA within a host cell.

reverse mutation a mutation that causes a novel mutant to revert back to wild type.

reverse transcriptase an RNA-dependent DNA *polymerase* that synthesizes DNA strands complementary to an RNA *template.* The product of reverse transcriptase is a *cDNA* molecule.

reverse transcription the process by which *reverse transcriptase* synthesizes DNA strands complementary to an RNA *template.* The product of reverse transcription is a cDNA molecule.

reversion See *reverse mutation.*

ribonucleic acid (RNA) a *polymer* of ribonucleotides found in the nucleus and cytoplasm of cells; it plays an important role in protein synthesis. There are several classes of RNA molecules, including *messenger RNA (mRNA), transfer RNA (tRNA),* ribosomal RNA (rRNA), and other small RNAs, each serving a different purpose.

ribosomal RNAs (rRNAs) *RNA* components of *ribosomes,* which are composed of both rRNAs and proteins.

ribosome binding sites regions on *prokaryotic mRNAs* containing both an *initiation codon* and a *Shine-Dalgarno box; ribosomes* bind to these sites to start *translation.*

ribosomes cytoplasmic structures composed of *ribosomal RNA* (rRNA) and protein; the sites of protein synthesis.

ribozymes RNA molecules that can act as enzymes to catalyze specific chemical reactions.

RISC See *RNA-induced silencing complex.*

RNA See *ribonucleic acid.*

RNA-dependent DNA polymerase See *reverse transcriptase.*

RNA editing specific alteration of the genetic sequence carried within an *RNA* molecule after *transcription* is completed.

RNA-induced silencing complex (RISC) a large enzymatic complex in the cytoplasm of all eukaryotic cells that binds to an *miRNA* and performs sequence-specific *RNA interference.*

RNA interference (RNAi) the sequence-specific modulation of eukaryotic gene expression by a 21–24 nucleotide-long RNA molecule referred to as *micro-RNA* if it is encoded within the genome and *short-interfering RNA* if it is introduced into the cell by scientists or infectious agents. These small, specialized RNAs prevent the expression of specific genes through complementary base pairing. In the most common natural form of RNAi, primary miRNA-containing transcripts are processed sequentially with the enzymes *Drosha* and *Dicer* to produce a mature miRNA that is loaded onto an *RNA-induced silencing complex (RISC).* The RISC complex binds to complementary mRNA targets, causing mRNA degradation or reduced translational activity.

RNA-like strand strand of a double-helical DNA molecule that has the same nucleotide sequence as an *mRNA* (except for the substitution of T for U) and that is complementary to the *template strand.*

RNA polymerases enzymes that transcribe a DNA sequence into an RNA *transcript. Eukaryotes* have three types of RNA polymerases called pol I, pol II, and pol III that are responsible for transcribing different classes of genes.

RNA splicing a process that deletes *introns* and joins together adjacent *exons* to form a mature *mRNA* consisting of only exons.

RNA world hypothetical primordial world in which RNA became the first replicator.

Robertsonian translocation *translocation* arising from breaks at or near the *centromeres* of two *acrocentric chromosomes.* The reciprocal exchange of broken parts generates one large *metacentric* chromosome and one very small chromosome.

S

satellite DNAs blocks of repetitive, simple noncoding sequences, usually around *centromeres;* these blocks have a different *chromatin* structure and different higher-order packaging than other chromosomal regions.

saturation mutagenesis an attempt to isolate *mutations* in all of the *genes* that direct a particular biological process.

scaffold-associated regions (SARs) special, irregularly spaced repetitive base sequences of DNA that associate with nonhistone proteins to define chromatin loops. SARs are most likely the sites at which DNA is anchored to the condensation scaffold.

secondary oocytes/spermatocytes *germ-line* cells in which meiosis II occurs.

secondary structure localized region of a *polypeptide* chain with a characteristic geometry, such as an α-helix or β-pleated sheet.

second-division segregation pattern a *tetrad* in which the arrangement of *ascospores* indicates that the two *alleles* of a gene segregated from each other in the second meiotic division.

second filial (F_2) generation progeny resulting from self-crosses or intercrosses between individuals of the F_1 generation in a series of controlled matings.

sectors portions of a growing *colony* of microorganisms that have a different *genotype* than the remainder of the colony.

segmentation gene in *Drosophila,* a large group of genes responsible for subdividing the body into an array of body segments.

segregation equal separation of alleles for each trait during *gamete* formation, in which one *allele* of each *gene* goes to each gamete.

selectable markers *vector* genes that make it possible to pick out cells harboring a *recombinant DNA molecule* constructed with that vector.

selection a process that establishes conditions in which only the desired mutant will grow.

selection differential (S) measure of the strength of selection on a trait. Compare *response to selection.*

self-fertilization (selfing) fertilization in which both egg and pollen come from the same plant.

semiconservative replication a pattern of double helix duplication in which *complementary base pairing* followed by the linkage of successive nucleotides yields two daughter double helices that each contain one of the original DNA strands intact (conserved) and one completely new strand.

semisterility a condition in which the capacity of generating viable offspring is diminished by at least 50%.

sense describes a laboratory-designed single-strand RNA molecule, or portion of an RNA molecule, with a sequence equivalent to that present in a cellular mRNA. A laboratory-designed *short interfering RNA* (*siRNA*) mediator of RNA interference will contain both sense and *antisense* components.

sepals green leaflike structure composing the first *whorl* of a flower.

sequence tagged sites (STSs) one-of a-kind markers that tag positions along the DNA molecule.

sequencing determining the order of *nucleotides* (base sequences) in a *DNA* or *RNA* molecule or the order of *amino acids* in a *protein.*

serial analysis of gene expression (SAGE) sequencing technique for determining the quantities of different RNAs in a mixture.

sex chromosomes the X and Y *chromosomes* in human beings, which determine the sex of an individual. Compare with *autosome.*

sex-influenced traits traits that can show up in both sexes but are expressed differently in each sex due to hormonal differences.

sex-limited traits traits that affect a structure or process that is found in one sex but not the other.

Shine-Dalgarno box a sequence of six nucleotides in *mRNA* that is one of two elements comprising a *ribosome binding site* (the other element is the *initiation codon*).

short-interfering RNA (siRNA) an RNA molecule 21–24 bases in length that originates outside an organism and can co-opt the natural *RNAi* machinery to effect *RNA interference.*

shotgun approach sequencing approach in which the overlapping insert fragments to be sequenced have been randomly generated in one of three ways: from large insert clones known as BACs; from the shearing with sound of the whole *genome;* or from partial digestion of the whole genome with *restriction enzymes.*

signal transducers cytoplasmic proteins that relay signals inside the cell.

signal transduction the activation and inhibition of intracellular targets after binding of *growth factors* or other *ligands* to *receptors.*

signal transduction pathway form of molecular communication in which the binding of proteins to receptors on cell surfaces constitutes a signal that is converted through a series of intermediate steps to a final intracellular regulatory response, usually the activation or repression of *transcription* in target genes.

silent mutations *mutations* without effects on *phenotype;* usually denotes *point mutations* that change one of the three bases in a *codon* but that do not change the identity of the specified *amino acid* because of the *degeneracy* of the *genetic code.*

SINE *s*hort *in*terspersed *e*lements; one of the two major classes of *transposable elements* in mammals. Contrast with *LINE.*

single nucleotide polymorphism (SNP) a single *nucleotide* locus with two naturally existing *alleles* defined by a single base pair *substitution*. SNP loci are useful as DNA-based *markers* for formal genetic analysis.

sister chromatids the two identical copies of a chromosome that exist immediately after DNA replication. Sister chramatids are held together by protein complexes called *cohesins.*

SNP see *single nucleotide polymorphism.*

somatic cells any cell in an organism except *gametes* and their precursors. Compare with *germ cells.*

somatic gene therapy remedial measures in which a replacement *gene* is inserted into affected tissue to compensate for a faulty gene.

SOS system an emergency repair system in bacteria that relies on error-prone ("sloppy") DNA *polymerases;* these special SOS polymerases allow cells with damaged DNA to divide, but the daughter cells carry many new mutations.

Southern blot protocol for transferring DNA sequences separated by *gel electrophoresis* onto a nitrocellulose filter paper for analysis by *hybridization* with a *DNA probe.* Compare *Northern blot analysis.*

specialized transduction *bacteriophage*-mediated transfer of a few bacterial genes located next to the bacteriophage DNA in the bacterial chromosome.

specialized transducing phages *bacteriophage* carrying mainly phage DNA but also one or a few of the bacterial genes that lie near the site of *prophage* insertion. They can transfer these genes to another bacterium in the process known as *specialized transduction.*

sperm a *haploid* male *gamete* produced by *meiosis.*

spermatids *haploid* cells produced at the end of *meiosis* that will mature into *sperm.*

spermatogenesis the production of *sperm.*

spermatogonia *diploid* germ cells in the testes.

S phase stage of the *cell cycle* during which *chromosome replication* occurs.

spindle apparatus microtubule-based structure responsible for *chromosome* movements and *segregation* during cell division.

splice acceptors *nucleotide* sequences in a *primary transcript* at the border between an *intron* and the downstream *exon* that follows it; required for proper *RNA splicing.*

splice donors *nucleotide* sequences in a *primary transcript* at the border between an *intron* and the upstream *exon* that precedes it; required for proper *RNA splicing.*

spliceosome a complicated intranuclear machine that ensures that all of the splicing reactions take place in concert.

stamen structure comprising the third *whorl* of a flower that bears the male *gametes* in the form of pollen.

stem cells relatively undifferentiated cells that undergo asymmetric mitotic divisions. One of the daughter cells produced by such a division is another stem cell, while the other daughter cell can differentiate. In this way, stem cells are self-renewing but can also give rise to differentiated cells.

sticky end the result achieved after digestion by many *restriction enzymes* that break the *phosphodiester bonds* on the two strands of a double helix DNA molecule at slightly different locations. The resulting double-stranded DNA molecule has a single protruding strand at each end that is usually one to four bases in length.

stigma pollen-receptive structure at the top of the *pistil* of a flower.

stop codons see *nonsense codons.*

subclone the product of *subcloning*.

subcloning process by which a portion of a cloned DNA sequence is extracted and isolated within a new *DNA clone* containing a smaller insert. Subcloning allows an investigator to focus experimental analysis on a smaller DNA region of interest that was originally located in a larger clone.

substitution occurs when a base at a certain position in one strand of the DNA molecule is replaced by one of the other three bases.

substrate the target of an enzyme, usually cellular proteins.

subunit a single *polypeptide* that is a constituent of a *multimeric protein.*

sum rule the probability that any of two or more mutually exclusive events will occur is the sum of their individual probabilities.

supercoiling additional twisting of the DNA molecule caused by movement of the *replication fork* during unwinding. Refer to *DNA topoisomerases.*

suppressor mutations *mutations* that alleviate the phenotypic abnormality caused by another mutation. The two mutations can be in the same gene (see *intragenic suppression*), or the suppressor mutation can be in a second, *modifier gene.*

suspensor in plants, a structure analogous to the umbilical cord in mammals.

synapsis process during which *homologous chromosomes* become aligned and zipped together; occurs during *zygotene* of *prophase I.*

synaptonemal complex structure that helps align *homologous chromosomes* during *prophase* of *meiosis I.*

syncytial blastoderm in *Drosophila* embryos, formed when most of the nuclei migrate out to the cortex just under the surface of the embryo.

syncytium an animal cell with two or more nuclei.

syntenic relationship of two or more *loci* found to be located on the same chromosome. Compare with *conserved synteny* and *syntenic segments.*

syntenic blocks blocks of *linked loci* within a *genome.*

syntenic segments in the comparison of two *genomes,* large blocks of DNA sequences in which the identity, order, and transcriptional direction of the *genes* are almost exactly the same.

synthetic lethality when an individual carrying mutations in two different genes dies, even though individuals carrying either mutation alone survive.

systems biology definition of all (or as many as possible) of the components of a *biological system* followed by an attempt to understand how they interact among themselves and with the components of other systems.

T

TAFs *T*BP-*a*ssociated *f*actors; one type of basal factors.

tandem duplications repeats of a chromosomal region that lie adjacent to each other, either in the same order or in reverse order. Contrast with *nontandem duplications.*

targeted mutagenesis a form of transgenic technology that investigators use to alter the *DNA sequence* of a particular *gene* or regulatory region (animal or plant) in a predetermined way. It can include *mutations* introduced into a cloned gene or into a specific site in the *genome*. In mice, targeted mutagenesis often consists of *cloning* and altering a gene of interest, inserting the cloned mutated gene into the genome, and breeding the mice carrying the mutated gene to produce animals *homozygous* for the mutation.

TBP *T*ATA *b*ox-binding *p*rotein; key basal factor that assists the binding of RNA polymerase II to the promoter and the initiation of basal levels of *transcription.*

telomerase an enzyme critical to the successful replication of *telomeres* at *chromosome* ends.

telomeres specialized terminal structures on eukaryotic *chromosomes* that ensure the maintenance and accurate replication of the two ends of each linear chromosome.

telophase the final stage of *mitosis* in which the daughter *chromosomes* reach the opposite poles of the cell and reform nuclei.

telophase I phase of *meiosis I* when nuclear membranes form around the *chromosomes* that have moved to the poles; each incipient daughter nucleus contains one-half the number of chromosomes in the original parent cell nucleus, but each of these chromosomes consists of two *sister chromatids* held together by *cohesin* protein complexes.

telophase II final phase of *meiosis II* during which membranes form around each of the four daughter nuclei, and *cytokinesis* places each nucleus in a separate cell.

temperate bacteriophages after infecting the host, these *phages* can enter either the *lytic cycle* or the alternative *lysogenic cycle,* during which their DNA integrates into the host *genome.* Compare with *virulent bacteriophages.*

template a strand of *DNA* or *RNA* that is used as a model by DNA or RNA *polymerase* or by *reverse transcriptase* for the creation of a new complementary strand of DNA or RNA.

template strand the strand of the double helix that is complementary to both the RNA-like *DNA* strand and the *mRNA*.

10-fold sequence coverage the sequencing of every chromosomal region from 10 independently cloned inserts results in 10-fold sequence coverage.

teratype (T) a *tetrad* carrying four kinds of *haploid* cells; two different *parental class* spores and two different recombinants.

terminalization shifting of the *chiasmata* from their original position at the *centromere* toward the *chromosome* end or *telomere.*

termination phase of *translation* that brings *polypeptide* synthesis to a halt.

terminators sequences in the *RNA* products that tell RNA *polymerase* where to stop *transcription.* Compare with *promoters.*

tertiary structure ultimate three-dimensional shape of a *polypeptide.*

testcross a *cross* used to determine the *genotype* of an individual showing a *dominant phenotype* by mating with an individual showing the *recessive phenotype.*

tetrad (1) in some fungi, the assemblage of four *ascospores* (resulting from *meiosis*) in a single *ascus.* (2) a pair of synapsed homologous *chromosomes* during *prophase* of *meiosis I,* also known as a *bivalent.*

tetrad analysis the use of *tetrads* to study gene *linkage* and *recombination* during *meiosis* in fungi.

tetraploid describes cells or organisms with four complete sets of chromosomes. See *haploid, diploid,* and *triploid.*

tetrasomic otherwise *diploid* organism with four copies of a particular *chromosome.*

tetratype (T) a fungal *ascus* that carries four kinds of spores, or haploid cells: two different *parental types* and two different *recombinant types.*

therapeutic cloning creation of an embryo by the method described for *reproductive cloning,* with the exception that the embryo is not allowed to develop to term. Instead it is cultured in a Petri plate to the *blastocyst* stage, at which point the *embryonic stem (ES) cells* are collected and placed in culture. The cultured ES cells can be induced to differentiate into many kinds of cells that might be of therapeutic value.

thymine a nitrogenous base; one member of the base pair A–T (adenine–thymine).

tolerance ability to prevent the body from making *immune responses* to its own proteins.

torpedo stage embryo stage of embryonic development in plants where two protuberances expand and differentiate into two well-defined, discrete cotyledons.

total phenotype variance (V_P) population deviation calculated as the sum of the *genetic variance* and the *environmental variance.*

totipotent description of cell state during early embryonic development in which the cells have not yet differentiated and retain the ability to produce every type of cell found in the developing embryo and adult animal.

***trans*-acting element** a gene that codes for a *transcription factor.*

transcript the product of *transcription.*

transcription the conversion of *DNA*-encoded information to its *RNA*-encoded equivalent.

transcriptional silencing hypercondensation of *chromatin* domains makes it impossible to activate genes within those domains, no matter what *transcription factors* are active in the cell.

transcription bubble the region of DNA unwound by RNA *polymerase.*

transcription factor protein (or RNA) whose binding to or indirect association with a cis-*control element* helps regulate the timing, location, and level of a particular gene's *transcription.* Functional categories include *activators, repressors, coactivators,* and *corepressors.*

transcriptome the population of *mRNAs* expressed in a single cell or cell type.

transductants cells resulting from gene transfer mediated by *bacteriophages.*

transduction one of the mechanisms by which bacteria transfer genes from one strain to another; *donor* DNA is packaged within the protein coat of a *bacteriophage* and transferred to the recipient when the phage particle infects it. Recipient cells are known as *transductants.*

transfection transformation of mammalian cells via the uptake of DNA from the medium.

transfer RNA (tRNA) small RNA adaptor molecules that place specific *amino acids* at the correct position in a growing *polypeptide* chain.

transformants cells that have received naked *donor* DNA.

transformation one of the mechanisms by which bacteria transfer genes from one strain to another; occurs when DNA from a *donor* is added to the bacterial growth medium and is then taken up from the medium by the recipient. The recipient cell is known as a *transformant.* Contrast with *conjugation.*

transgene any piece of foreign *DNA* that researchers have inserted into the *genome* of an organism.

transgenic any individual carrying a *transgene.*

transgenic technology the tools for inducing a specific change in a gene and confirming that this change causes the predicted *phenotype.* Transgenic technology includes experimental methods that allow scientists to add laboratory-constructed *DNA sequences* to the *genomes* of animals or plants.

transient- (or triangular-) stage embryo stage following *globular stage* in plant development; embryo becomes self-sufficient for growth, having used up most of its maternally deposited reserves.

transition a type of *substitution* mutation that occurs when one purine (A or G) replaces the other purine, or one pyrimidine (C or T) replaces the other pyrimidine. Contrast with *transversions.*

translation the process in which the *codons* carried by *mRNA* direct the synthesis of *polypeptides* from *amino acids* according to the *genetic code.* Compare with *transcription.*

translocation a rearrangement that occurs when parts of two nonhomologous *chromosomes* change places. Compare with *reciprocal translocation.*

translocation Down syndrome occurs in individuals affected by Down syndrome who have inherited three copies of a part (rather than all) of chromosome 21 because one of their parents was *heterozygous* for a *translocation* involving chromosome 21.

***trans*-splicing** a rare type of *RNA splicing* that joins together *exons* of the *primary transcripts* of two different genes.

transposable elements (TEs) all DNA segments that move about in the *genome,* regardless of mechanism.

transposition the movement of *transposable elements* from one position in the *genome* to another.

transposons units of DNA that move from place to place within the *genome* without an RNA intermediate, sometimes causing a change in gene function when they insert themselves in a new chromosomal location.

transversions a type of *substitution* mutation that occurs when a purine (A or G) replaces a pyrimidine (C or T) or when a pyrimidine replaces a purine.

triploid describes cells or organisms with three complete sets of *chromosomes.* See *haploid, diploid,* and *tetraploid.*

trisomic individual having one extra *chromosome* in addition to the normal *diploid* set of the species.

truncated proteins *polypeptides* with fewer amino acids than normal encoded by genes containing *nonsense mutations.*

twin spots adjacent patches of tissue that are phenotypically distinct from each other and from the surrounding tissue; can be produced as a result of mitotic recombination.

2*n* number of *chromosomes* in a normal *diploid* cell.

U

ubiquitin highly conserved protein whose covalent attachment to other proteins marks them for degradation by the *proteosome.*

unequal crossing-over change in DNA caused by erroneous *recombination* in which one *homologous chromosome* ends up with a *duplication,* while the other homolog sustains a *deletion.*

uniparental inheritance transmission of organelle genes via one parent. Most species transmit mitochondrial DNA and chloroplast DNA through the mother. (See *biparental inheritance.*)

unordered tetrads *tetrads* in yeast in which the four *ascospores* are randomly arranged in the *ascus.*

5′-untranslated region (5′ UTR) portion of an *mRNA* between its 5′ end and the *initiation codon;* the 5′ UTR is thus by definition contained within one or more *exons* but does not encode any amino acids in the mRNA's protein product.

upstream movement opposite the direction *RNA* follows when moving along a gene. Compare with *downstream.*

uracil a nitrogenous base normally found in *RNA* but not in *DNA;* uracil is capable of forming a *base pair* with adenine.

V

vaccination inoculation with a nonvirulent or virulent but attenuated infectious agent that stimulates a protective *immune response.*

variance statistical measurement of deviation from the mean (middle); typically expressed in plus or minus terms referring to the relationship to the mean.

variolation injection with live smallpox virus obtained from a patient with a mild case of smallpox.

vector a specialized DNA sequence that can enter a living cell, signal its presence to an investigator by conferring a detectable property on the host cell, and provide a means of replication for itself and the foreign DNA inserted into it. A vector must also possess distinguishing physical traits by which it can be purified away from the host cell's *genome.* See *cloning vector.*

vegetative development in plants, the period of growth before flowering. Contrast with *reproductive development.*

vernalization the process of exposing plants to cold early in vegetative development to promote earlier flowering.

virulent bacteriophages after infecting the host, these phages always enter the *lytic cycle,* multiply rapidly, and kill the host. Compare with *temperate bacteriophages.*

virus a noncellular biological entity that can reproduce only within a host cell. Viruses consist of nucleic acid covered by protein. Inside an infected cell, the virus uses the synthetic capability of the host to produce progeny virus.

W

whorls in the structure of a flower, the concentric regions of modified leaves including *sepals, petals, stamens,* and *carpels.*

wild-type allele (1) an *allele,* or DNA variant, whose frequency is more than 1% in a *population;* (2) an allele that dictates the most frequently observed phenotype in a population. Wild-type alleles are often designated by a superscript "plus" sign ($^+$).

wobble ability of the 5′-most nucleotide of an anticodon to interact with more than one nucleotide at the 3′-end of codons; helps explain the *degeneracy* of the *genetic code.*

X

x indicates the number of chromosomes in a complete set of non-homologous chromosomes.

X chromosome inactivation in mammals, a mechanism of *dosage compensation* in which all X *chromosomes* in a cellular genome but one are inactivated at an early stage of development through the formation of heterochromatic *Barr bodies.*

X chromosome reactivation in mammals, a mechanism by which X *chromosomes* that were inactivated become reactivated in *oogonia* so that the *haploid* cells in the *germ line* all have an active X chromosome.

X-linked carried by the X *chromosome.*

Y

YAC See *yeast artificial chromosome.*

yeast artificial chromosome (YAC) a vector used to clone DNA fragments up to 400 kb in length; it is constructed from *telomeric, centromeric,* and *origin-of-replication* sequences needed for replication in yeast cells. Compare with *cloning vector, cosmid.*

Z

Z-form DNA DNA in which the *nucleotide sequences* cause the structure to assume a zigzag shape due to the helices spiraling to the left. Compare with *B-form DNA.* The significance of this variation on DNA structure is unknown at this time.

zygote the *diploid* cell formed by the fertilization of the egg by the *sperm* during sexual reproduction.

zygotene substage of *prophase I* when *homologous chromosomes* become zipped together in *synapsis.*

Credits

Text and Line Art Permissions

Chapter 1

Fig. 1.13 / Stephen H. Friedn, The Magic of Microarrays. Feb 2002 Scientific American, courtesy Jared Schneidman Designs.

Chapter 3

Genetics & Society / Provided by the Institut National d'Etudes Demographiques. Reprinted with permission from SCIENCE: "Fragment of a Family Tree Showing the Transmission of Juvenile Glaucoma" April 19, 1991, p.369 © American Association for the Advancement of Science.

Chapter 4

Problem 4-35 / From "Pedigree of Family with Congenital Hypertrichos," by E. J. Mange and A. P. Mange, Basic Human Genetics, 2e. Copyright © 1999 Sinauer Associates, Inc. Sunderland, MA

Chapter 6

Fig. 6.10 / From *Molecular Cell Biology,* 2e, by Lodish, et al. © 1986, 1990, 1995 by Scientific American Books. Used with permission by W.H. Freeman and Co.

Chapter 7

Fig. 7.10c / From Principles of Genetics, 2e by Snustad, Simmons, Copyright © 2000 John Wiley & Sons. Reprinted with permission of John Wiley & Sons, Inc. New York.

Chapter 8

Fig. 8.11c / From *An Introduction to Genetic Analysis,* by Griffiths, et al. © 2000 by W.H. Freeman and Co. Reprinted with permission. **Fig. 8.12** / From Molecular Cell Biology, 3e, by Lodish, et al. © 1999 by W.H. Freeman and Co. Reprinted with permission.

Chapter 9

Fig. 9.1b / Weatherall, D. G., and Clegg, J. B., The Thalassemia Syndrome, 3e, Copyright © 1981, with permission of Blackwell Scientific Publications. **Fig. 9.1c** / Weatherall, D. G., and Clegg, J. B., The Thalassemia Syndrome, 3e, Copyright © 1981, with permission of Blackwell Scientific Publications.

Chapter 10

Fig. 10.9, 10.10, 10.11, 10.12, 10.13, 10.14, 10.15 / "Initial Sequencing and Analysis of the Human Genome", Nature: Vol. 409, February 15th, 2001, with permission of the International Human Genome Sequencing Consortium. **Fig. 10.16** / Glusman, Lancet: "The Complete Human Olfactory Subgenome," Genome Research, May 2001, Vol 11 (3): 685-702. Courtesy Cold Spring Harbor Laboratory Press. **Fig. 10.20** / Reprinted from Genomics, vol. 1, Lee Hood, M. Hunkapillar and L.M. Smith, Automated DNA Sequencing and Analysis of the Human Genome, pp. 201-212, copyright 1987, with permission of Elsevier. **Fig. 10.22** / Reprinted with permission from Joseph L. DeRisi, et al: Exploring the Metabolic and Genetic Control of Gene Expression on a Genomic Scale, *Science,* 278:680:686. Copyright 1997 AAAS.

Chapter 11

Fig. 11.8a / Reprinted with permission from Gusella, J. F., et al: DNA markers for nervous system disease, from *Science,* 225:1320. Copyright 1984 AAAS. **Fig. 11.8b** / Human Genetics, 39: 382-391 © 1980 by Magenis et al (the James Gusella Lab). Published by the University of Chicago Press. **Fig. 11.18a** / Gusella, JF et al: DNA Markers for Nervous System Disease, 1984, Science 225: 1320; with permission AAAS.

Chapter 12

Fig. 12.16 / Courtesy Stan Fields, University of Washington Yeast Resource Center. **Art, page 462** / A Global Map of p53 Transcription-Factor Binding Sites in the Human Genome from Cell 124 pp.207–219 (2006).

Chapter 13

Fig. 13.11b / Courtesy of Alberts et al, Molecular Biology of the Cell, 3rd edition. New York: Garland Publishing, 1994. Courtesy of Alberts et al, Essential Cell Biology. New York: Garland Publishing, 1998.

Chapter 14

Fig. 14.39 / Snijders et al; Assembly of microarrays for genome wide measurement of DNA copy number, with permission Nature Genetics Vol. 29 No. 3: 263-269 (2001).

Chapter 15

Fig. 15.24 / H.C. Berg; "The helical filaments of the thin flagella that propel bacteria do not wave or beat but instead rotate rigidly like propellers" With permission Sci Am. 1975 Aug; 233(2):36-44.

Chapter 16

Fig. 16.14b / Reprinted from CELL, vol. 67, Shoffner and Wallace, et al., "Myclonic Epilepsy and Ragged-Red Fiber Disease (MERRF) Is Associated with a Mitochondrial DNA tRNA (LYS) Mutation, pp. 931-937, copyright 1990, with permission from Elsevier.

Chapter 17

Fig. 17.8 / Reprinted with permission from Lewis, et al., "View of the Lac-Repressor-DNA Complex Monomer," *Science,* vol. 271, pp. 1247-1254, March 1996. Copyright 1996 AAAS.

Chapter 19

Table 19.2 / Adapted from V.T. DeVita et al., Cancer: Principles and Practices of Oncology, 4e © 1993 with permission of Lippincott, Williams & Wilkins. **Table 19.3** / Retroviruses and Their Associated Oncogenes, adapted from Lewin, Genetics, 1e, Oxford University Press, Inc. Reprinted by permission. **Fig. 19.17** / Reprinted from Trends in Genetics, vol. 11, Copyright 1995, with permission from Elsevier. **Fig. 19.19** / Reprinted by permission of the American Cancer Society, Inc. **Fig. 19.21** / Scientific American, March 1995, p. 77; Jared Schneidman Design, with permission of Scientific American. **Fig. 19.24** / From *Introduction to Genetic Analysis,* by Griffiths, et al. © 1996 by W.H. Freeman and Co. Reprinted with permission. **Fig. 19.25** / The Development of Drosophila Nekanigaster / "Mitosis and Morphogenesis in Drosophila Embryo: Point and Counterpoint," by V. E. Foe, G. M. Odell, B. A. Edgar. Copyright © 1993 Cold Spring Harbor Laboratory Press. Used with permission.

Chapter 20

Fig. 20.7 / From "Engineering a Dominant-Negative Mutation in the Mouse Fibroblas Growth Factor Receptor (FGFR) Gene," by S. Gilbert, Development Biology, 5e, p. 10. Copyright © 1997 Sinauer Associates, Inc. Sunderland, MA. **Fig. 20.30** / From "Bicoid mRNA Is Localized to the Anterior Cortex of the Oocyte," by S. Gilber, Development Biology, 5e, p. 867. Copyright © 1997 Sinauer Associates, Inc. Sunderland, MA.

Photo Credits
About the Authors

Hartwell: Photo courtesy Fred Hutchinson Cancer Research Center; **Hood:** Photo by Greg Nystrom; **Silver:** © Jon Roemer.

Table of Contents

Part I: © Brian Stablyk/Stone/Getty Images; **Part II:** © Stone Imaging/Stone/Getty Images; **Part III:** © Cytographics/Visuals Unlimited; **Part IV:** © David M. Phillips/Visuals Unlimited; **Part V:** © & Courtesy of Mitchell Lewis, University of Pennsylvania. Reprinted with permission from *Science* 271:1247, from M. Lewis, et al, © 1996 American Association for the Advancement of Science; **Part VI:** © Jim Pikerell/Stock Connection.

Chapter 1

Opener: © James Strachan/Stone/Getty Images; **1.1(a):** © David M. Phillips/Visuals Unlimited; **1.1(b):** © T.E. Adams, Visuals Unlimited; **1.1(c):** © Vol. 44/Photo Disc; **1.1(d):** © Doug Sokell/Visuals Unlimited; **1.1(e):** GenomeSystems Inc./Photo provided by Kearns Communication Group; **1.1(f):** © Vol. 24/PhotoDisc; **1.3:** Applied Biosystems/Peter Arnold, Inc.; **1.4:** © Biophoto Associates/Photo Researchers, Inc.; **1.9a:** © Carolina Biological/Photo Researchers, Inc.; **1.9b:** © Vol. OS02/PhotoDisc; **Table 1.1a:** © J. William Schopf, UCLA. Reprinted with permission from *Science* 260: 640-646, 1993. "Microfossils of the Early Archean Apex Chert: New Evidence of the Antiquity of Life." © 1993 American Association for the Advancement of Science; **Table 1.1c&d:** Prof. Andrew Knoll; **Table 1.1e:** © Brand X Pictures/PunchStock; **Table 1.1f:** © Alan Sirulnikoff/Photo Researchers, Inc.; **1.11a&b:** Edward B. Lewis, CIT; **1.12(1):** © David M. Phillips/Visuals Unlimited; **1.12(2):** Lee Hartwell; **1.12(3):** © Sinclair Stammers/ SPL/Photo Researchers, Inc.; **1.12(4):** Courtesy Debra Nero/Cornell University; **1.12(5):** © Myung Shin/Bergman Collection.

Chapter 2

Opener: © Brian Stablyk/Stone/Getty Images; **2.1:** © Bruce Aryes/Stone/Getty Images; **2.2:** © Science Photo Library/Photo Researchers, Inc.; **2.3:** © Saudjie Cross Siino/Weathertop Labradors; **2.4:** © The Metropolitan Museum of Art, Gift of John D. Rockefeller, Jr., 1932 (32.143.2), Photograph ©1996 The Metropolitan Museum of Art; **2.5a:** © Malcolm Gutter/Visuals Unlimited; **2.5b:** © James King-Holmes/Photo Researchers, Inc.; **2.6:** © Klaus Gulbrandsen/SPL/Photo Researchers, Inc.; **2.7a:** © Dwight Kuhn Photography; **Page 22:** Reprinted from Bhattacharyya MK, et al. *Cell.* 1990 Jan 12: 60(1):115-22, with permission from Elsevier; **2.19a:** © Science Photo Library/Photo Researchers, Inc.; **2.19b-d:** © Mendelianum Institute, Moravian Museum.

Chapter 3

Opener: © Vol. 8/PhotoDisc/Getty Images; **3.1:** © Jerry Marshall; **3.3a:** © John D. Cunningham/Visuals Unlimited; **3.7a(left):** © McGraw-Hill Higher Education Group, Inc./Jill Birschbach, Photographer; Arranged by Alexandra Dove, McArdle Laboratory, University of Wisconsin-Madison; **3.7a(right), 3.7a(bottom):** © Charles River Laboratories; **3.10a:** © Stanley Flegler/Visuals Unlimited; **3.12a:** © William H. Allen, Jr./Allen Stock Photography; **3.19a:** © RADU SIGHETI/Reuters/Landov; **3.20a:** © Renee Lynn/Photo Researchers, Inc.; **3.21a:** © Rudi Von Briel/Photo Edit.

Chapter 4

Opener: © Adrian T. Sumner/Stone/Getty Images; **4.1:** © Richard Hutchings/Photo Researchers, Inc.; **4.4:** © Scott Camazine/Photo Researchers, Inc.; **4.5:** © L. West/Photo Researchers, Inc.; **4.6:** © Biophoto Association/Photo Researchers, Inc.; **4.8a-f:** Photographs by Dr. Conly L. Rieder, Division of Molecular Medicine, Wadsworth Center, NYS Dept. Of Health, Albany, NY; **4.9a:** © David M. Phillip/Visuals Unlimited; **4.9b:** © R. Calentine/Visuals Unlimited; **4.10:** © Dr. Byron Williams/Cornell University; **Figure A & B:** © Dr. Michael Goldberg/Cornell University; **4.16:** © Dr. Leona Chemnick, Dr. Oliver Ryder/San Diego Zoo, Center for Reproduction of Endangered Species; **4.22 (both):** Color deficit simulation courtesy of Vischeck (www.vischeck.com). Source image courtesy of NASA; **4.24:** © Gene Trindl/MPTV Photo Archive.

Chapter 5

Opener: © Rudy Von Briel/Photo Edit; **5.14a:** © J. Forsdyke, Gene Cox/SPL/Photo Researchers, Inc.; **5.14b:** © James W. Richardson/Visuals Unlimited; **5.19:** © Dr. Eric Alani/Cornell University; **5.25:** Image courtesy of B.A. Montelone, Ph.D. and T.R. Manney, Ph.D.

Chapter 6

Opener: © Stone Imaging/Stone/Getty Images; **6.1a:** © George Bernard/Animals Animals; **6.1b:** © William Hauswirth; **6.1c:** ©Archivo Iconografico, S.A./CORBIS; **6.3a(both):** © Evanston Northwestern Healthcare, Evanston, IL; **6.4b:** © The Bergman Collection; **6.6:** © Science Source/Photo Researchers, Inc.; **6.9a:** © A. Barrington Brown/Science Source/Photo Researchers, Inc.; **6.11a:** © Biophoto Associates/Science Source/Photo Researchers,

Inc.; **6.11b:** © Microworks/Dan/Phototake; **6.11c:** © Ross Inman & Maria Schnös, University of Wisconsin, Madison, WI; **6.11d:** © Jack D. Griffith/University of North Carolina Lineberger Comprehensive Cancer Center.

Chapter 7

Opener: © Milkie Studio, Inc.; **7.1:** © Dr. Don Fawcett/J.R. Paulson & U.K. Laemmli/Photo Researchers, Inc.; **7.3(both):** © Charles River Laboratories; **Figure A:** © Science VU/Visuals Unlimited; **7.15:** © Dr. Ken Greer/Visuals Unlimited; **7.17(all):** © Carolina Biological/ Photo Researchers, Inc.; **7.20a.1:** © The Bergman Collection; **7.20b.1:** © Seymour Benzer; **7.30:** Color deficit simulation courtesy of Vischeck (www.vischeck.com). Source image courtesy of NASA.

Chapter 8

8.1: © Sinclair Stammers, SPL, Photo Researchers, Inc.; **8.11b-2:** © Professor Oscar Miller/ SPL/Photo Researchers, Inc.; **8.17b:** © Dr. Thomas Maniatis, Thomas H. Lee Professor of Molecular and Cellular Biology, Harvard University; **8.24:** Cech TR. *Science* 11 August 2000:Vol. 289. no. 5481, pp. 878–879. © 2000 American Association for the Advancement of Science; **8.31a:** Courtesy of & © Dr. Karen Artzt/Artzt Lab/University of Texas at Austin; **8.31c:** © Tom Vasicek; **8.31d1:** © Science VU/Dr. F. R. Turner/Visuals Unlimited; **8.31d2:** © Eye of Science/Photo Researchers, Inc.

Chapter 9

Opener: © Cytographics/Visuals Unlimited; **9.1a:** © Stanley Flegler/ Visuals Unlimited; **9.4b, 9.5b(both):** © Lee Silver, Princeton University; **9.14a:** © Hank Morgan/Photo Researchers, Inc.; **9.15(all):** © Lee Silver, Princeton University; **9.18b:** © Jean Claude Revy/Phototake; **9.21a.3:** © Omikron/Science Source/Photo Researchers, Inc; **9.21b.3:** © Professor Sir David Weatherall, University of Oxford.

Chapter 10

10.1(top): Reprinted by permission from *Nature* 409:15th February 2001 © 2001 Macmillan Publishers Ltd.; **10.1(bottom):** Reprinted Full cover of *Science* 2/16/01. © 2001 American Association for the Advancement of Science; **10.6a:** © Lee Silver, Princeton University; **10.7:** Section of Cancer Genomic Genetics Branch/CCR/NCI/NIH; **10.8b:** © Dr. Gopal Murti/SPL/Photo Researchers, Inc.; **10.22b:** Image courtesy Patrick O. Brown and Joe Derisi;

Chapter 11

Opener: © Lee Silver, Princeton University; **11.1b:** © M. I. Walker/Photo Researchers, Inc.; **11.1c:** © Dr. David M. Phillips/Visuals Unlimited; **11.1d:** Permission to use this image has been granted by Bio-Rad Laboratories, Inc.; **11.1e:** Courtesy of Ronald Carson, The Reproductive Science Center of Boston/ IntegraMed America, Inc.; **11.1f-h:** Permission to use this image has been granted by Bio-Rad Laboratories, Inc. **11.6b:** © Lee Silver, Princeton University; **11.10:** Reprinted by permission of Oxford University Press. *Nucleic Acids Research,* Vol 26(16), 1998, fig 1 "Two Color Hybridization Analysis using High Density Oligonucleotide Arrays and Energy Transfer Dyes," Hacia et al. Image courtesy Joseph G. Hacia, University of Southern California; **11.14b:** © Lee Silver, Princeton University; **11.15:** Reprinted by permission from *Nature,* 394, figure 1, (1998), © 1998 Macmillan Publishers Ltd. Image courtesy Esther N. Signer; **11.20(4):** Reprinted with permission from *Nature* 1990 Jul 19; 346(6281):216-7, Sinclair et al. © 1990 Macmillian Magazines Limited; **11.21(top):** © Brigid Hogan, Vanderbilt University; **11.21(bottom):** Reproduced with permission from Dr. Robin Lovell-Badge/MRC National Institute for Medical Research; **11.22b:** © Johanna Rommens/Hospital for Sick Children, Toronto Reprinted from *Science* 245:1066, 1989. © 2001 American Association for the Advancement of Science; **11.24a:** © Calgene, Inc.; **Page 431:** Lee Silver from http://www.geninfo.no/undervisning/ DNA-fingerprinting.asp; **Page 432:** © Lee Silver, Princeton University.

Chapter 12

Opener: Patent/s Pending & Copyright © Lumeta Corporation 2006. All Rights Reserved.; **12.1a:** The Granger Collection, New York; **12.1b:** © Paul Almasy/Corbis; **12.4, 12.5:** John Aitchison; **12.6a&b:** McGraw-Hill Higher Education; **12.11b(top):** shutterstock.com; **12.11c(bottom):** shutterstock.com; **12.13a:** © Phototake; **12.17a&b:** Courtesy of & © Dr. Mike Snyder, Molecular Biophysics and Biochemistry Department, Yale University; **12.20b:** © Scimat/Photo Researchers, Inc.

Chapter 13

Opener: © David M. Phillips/Visuals Unlimited; **13.1:** © Biophoto Associates/Photo Researchers, Inc.; **13.2a:** © Dr. Don Fawcett, U.K. Laemmli/Photo Researchers, Inc.; **13.2b:** © Daniel A. Starr/University of Colorado; **13.3a:** © Ada L. Olins, Univ of Tenn /Biological Photo Services; **13.3c** © Dr. Gerard J. Bunick, Oak Ridge National Laboratory; **13.4a1:** © Dr. Barbara Hamkalo/University of California-Irvine, Department of Biochemistry; **13.4a2:** © Dr. Don Fawcett/H. Ris and A. Olins/Photo Researchers, Inc.; **13.5:** © Dr. Don Fawcett/J.R. Paulson, U.K. Laemmli/Photo Researchers, Inc.; **13.6a:** © Jacques Giltay, Center for Medical Genetics, Utrecht; **13.7a:** © H. Kreigstein and D.S. Hogness "Mechanism of DNA Repication in Drosophila Chromosomes: Structure of Replication Forks and Evidence of Bidirectionality", *Proceedings of the National Academy of Sciences* USA, 71(1974): 135-139; **13.8:** © Dr. Robert Moyzis/ University of California-Irvine, Department of Biochemistry; **13.11a-2:** © Dr. Jeremy David Pickett Heaps, School of Botany, University of Melbourne; **13.12a:** © Stanislav Fakan, University of Lausanne; **13.12b:** © Doug Chapman, University of Washington Medical Center Cytogenetics Laboratory; **13.13a(both):** © Dr. Clinton Bishop, Department of Biology, West Virginia University.

Chapter 14

Opener: © Dr. Michael Goldberg/Cornell University; **Figure C:** © Courtesy of The Centers for Disease Control; **14.7:** © Dr. Ross MacIntyre, Cornell University; **14.9a:** © Dr. Michael Goldberg, Cornell University; **14.10:** Courtesy of Dr. Philip Cotter; **14.13a:** © Cabisco/Visuals Unlimited; **14.13b:** Courtesy of Dr. Brian R. Calvi; **14.13c:** © Cabisco/Visuals Unlimited; **14.15:** © Provided by M.J. Moses, Duke University, from Poorman, Moses, Davisson and Roderick: *Chromosoma (Berl.)* 83:419-429 (1981); **14.18b:** © Lisa G. Shaffer, PhD/Baylor College of Medicine; **14.20(left):** © Dr. E. Walker/SPL/Photo Researchers, Inc.; **14.20(right):** © J. Carrillo-Farg/ Photo Researchers, Inc.; **14.21d:** © M.G. Neuffer, University of Missouri; **14.23:** © CORBIS; **14.24(both):** © Dr. Michael Goldberg, Cornell University; **14.25b:** © Dr. Nina Fedoroff; **14.30a&b:** Britton-Davidian et al, "Environmental genetics: Rapid chromosomal evolution in island mice," *Nature* 403, 158 (13 January 2000). Reproduced with permission; **14.35:**

© Leonard Lessin/Peter Arnold, Inc.; **14.38b:** © Davis Barber/PhotoEdit.

Chapter 15

Opener: © Pat O'Hara/Stone/Getty Images; **15.2a:** © Dr. Jeremy Burgess/SPL/Photo Researchers, Inc.; **15.2b:** © Stephen Frish Photography; **15.3:** © David M. Phillips/Visuals Unlimited; **Figure A:** © Veronika Burmeister/Visuals Unlimited: **Figure B:** © Oliver Meckes/Photo Researchers, Inc.; **15.4a:** © Dr. Gopal Murti/SPL/Photo Researchers, Inc.; **15.8a:** © SPL/Science Source/ Photo Researchers, Inc.; **15.22a:** © Jack D. Griffith/University of North Carolina Lineberger Comprehensive Cancer Center; **15.26:** © Julius Adler, University of Wisconsin.

Chapter 16

Opener: © Newcomb & Wergin; **16.1:** © Eric L. Heyer/Grant Heilman Photography; **16.4:** Electron micrograph by Dr. Stephen Hajduk/University of Alabama at Birmingham; **16.11a:** © Carolina Biological Supply/Phototake; **16.12a (both):** © Jim Strawser/Grant Heilman Photography; **16.14a:** © Reprinted from *Trends in Genetics,* January 1989, Vol. 5, No.1, Dr. Douglas C. Wallace, p.11. © 1989 with permission from Elsevier Science.

Chapter 17

Opener: © & Courtesy of Mitchell Lewis, University of Pennsylvania. Reprinted with permission from *Science* 271:1247, from M. Lewis, et al, © 1996 American Association for the Advancement of Science; **17.1:** © London School of Hygiene & Topical Medicine/SPL/Photo Researchers, Inc.; **17.4:** © Bettmann/ UPI/CORBIS; **Figure A:** © Dr. Ann Hirsch, UCLA; **Page 642:** From Luke K., Kogan Y. et al, Microarray analysis of gene expression during bacteriophage T4 infection. *Virology.* 2002 Aug1;299(2): 181-91. Copyright © 2002, with permission from Elsevier Science.

Chapter 18

Opener: © Dr. Raymond Mendez/Animals Animals - Earth Scenes **18.4b:** © Reprinted with permission from *Nature,* Vol. 377, 14 Sept. 1995, fig. 2, p. 124 © 1995 Macmillan Magazines Limited; **18.5b:** © S.L. McKnight.

Chapter 19

Opener: © Lennart Nilsson/Albert Bonniers Forlag/Boehringer Ingelheim International GmbH; **19.3(both):** © Lee Hartwell, Fred Hutchinson Cancer Research Center; **19.11b(both):** © Thea Tlsty, University of California-San Francisco Medical Center-Pathology; **19.13b** © Michael R. Speicher and David C. Ward,, "The Coloring of Cytogenetics." *Nature Genetics,* 2:1046-1048, 1996, figs 2 and 3. Photos courtesy of David C. Ward; **19.16b.2:** © Dr. Joe Gray, University of California-San Francisco; **19.24a:** © Custom Medical Stock Photo.

Chapter 20

Opener: Courtesy of The Indiana Hand Center and Gary Schnitz; **20.1a:** © Dennis Kunkel Microscopy, Inc.; **20.1b:** © Lennart Nilsson/Albert Bonniers Forlag, From *A Child is Born,* Dell Publishing Company; **20.2:** © Photomicrographs by F. Rudolph Turner/The FlyBase Consortium, 1999; **20.3:** © Sinclair Stammers/SPL/Photo Researchers, Inc; **20.4a:** Courtesy of Dr. Walter Gehring; **20.4b(both):** © Helen Pearson, Western General Hospital/MRC Human Genetics Unit; **20.7c(both):** © EMBO J 1998 Mar 17:17 (6):1642-55, Fig 3 parts B&C. By permission of Oxford University Press. Image courtesy of Glen Merlino, National Institute of Health; **20.8b(both):** Courtesy of Daryl Hurd and Ken Kemphues, Cornell University; **20.9a&b:** Reprinted from *Developmental Biology,* Vol. 96, No 8, Wang et al: " A mouse model for Achondroplasia produced by targeting fibroblast growth factor receptor 3", pp 4455-4460. © 1999, with permission from Elsevier Science; **20.10:** Courtesy Dr. Walter Gehring; **20.11:** Modeled by Thomas R. Bürglin, based on the data of Otting et al. EMBO J 1990 Oct. 9 (10):3085-92; **20.12:** Photo courtesy of Robert Hill. Reprinted from *Development* 1995 121:(5) 1433-1442, © 1995 with permission from Company of Biologists Ltd.; **20.13b&d:** Courtesy Andras Nagy PhD; **20.14a:** Reprinted from *Development* 1998 125: (23) 4303-4312, © 1998 reprinted with permission from Company of Biologists Ltd.; **20.15b:** © Bill Sullivan, University of California-Santa Cruz; **20.16a-c:** © Dr. Rudi Turner and Dr. Tom Kaufman, Indiana University; **20.17a:** © Steve Small, New York University; **20.17b:** © David Kosman, John Reinitz; **20.19a(both):** © David Kosman, Hohn Reinitz; **20.19b(all):** © Dr. Eric Wieschaus, Princeton University. Reprinted with permission from *Nature* 287: 795-801 "Mutations affecting Segment Number and Polarity in Drosophila", Nusslein- Volhard, C. and E. Wieschaus, 1980. © 1980 Macmillan Magazines Limited; **20.20a:** © David Kosman and John Reinitz; **20.21a:** © Steve Small, New York University; **20.23b:** © Edward Lewis, California Institute of Technology; **20.27:** © St. Bartholomew Hospital/Photo Researchers, Inc.; **20.28:** © Dennis Kunkel Microscopy, Inc.; **20.29:** © Scott Weatherbee; **20.31:** © AP/Wide World Photo. **Page 751:** From: Hoda K. Shamloula, Mkajuma P. Mbogho, Angel C. Pimentela, Zosia M. A. Chrzanowska-Lightowlersb, Vanneta Hyatta, Hideyuki Okanoc, and Tadmiri R. Venkatesh. "Rugose (rg), a Drosophila A kinase Anchor Protein, Is Required for Retinal Pattern Formation and Interacts Genetically with Multiple Signaling Pathways." *Genetics,* Vol. 161, 693-710, June 2002. © Genetics Society of America. **Page 752:** Image courtesy of John M. Kemner; **Page 754:** © Dr. Ken Kemphues and Bijan Etemad-Moghadam, Cornell University.

Chapter 21

Opener: © Jim Pikerell/Stock Connection; **21.1a:** © A.M. Siegelman/Visuals Unlimited; **21.4:** © Schafer & Hill/Peter Arnold, Inc.; **21.10a:** © PhotoDisc; **21.11a:** © Robert Noonan/Photo Researchers, Inc.; **21.13a:** © Dr. Eckart Pott/OKAPIA/Photo Researchers, Inc.; **21.14a:** © Frans Lanting/Photo Researchers, Inc.; **Figure A:** © AP/Wide World Photos.

Chapter 22

Figure 22.3a: © J. William Schopf, UCLA. Reprinted with permission from *Science* 260: 640-646, 1993. "Microfossils of the Early Archean Apex Chert: New Evidence of the Antiquity of Life." © 1993 American Association for the Advancement of Science; **22.5a:** © Daniel J. Cox/Stone/Getty Images; **22.5b:** © Joe McDonald/Animals Animals; **22.5c:** © Renee Lynn/Stone/Getty Images; **22.5d:** © Roger De La Harpe/Animals Animals; **22.6a:** © Marc Chamberlain/SeaPics.com.

Index

A

B

D

E

F

G

H

I

J

K

L

N

P

Q

R

S

T

U

V

W

X

Y

Z